ENCYCLOPEDIA OF SURFACTANTS

VOLUME I

Compiled by

Michael and Irene Ash

Chemical Publishing Co.
New York, N.Y.

ISBN 0-8206-0249-3

Printed in the United States of America

PREFACE

This comprehensive, modern encyclopedia of surfactant products has been compiled with the cooperation of the major national and international surfactant manufacturers in order to give the user quick and thorough information on each tradename product that is included. Information from thousands of brochures, technical bulletins, and data sheets has been gleaned in order to make a complete profile on each surfactant product available from one source—this three volume compendium. Though sometimes limited by the information provided by the manufacturers, in most instances, the chemical ingredients, properties, form, applications, etc. of each product are discussed in detail.

The use of surfactants pervades almost every chemical industrial concern—including textiles, paints, pesticides, foods, cosmetics, pharamceuticals, etc. We hope that because of the universality of surfactants, this reference work will be able to serve as a useful tool to the many chemical engineers and salespeople here and abroad.

We want to extend our thanks to Roberta Dakan for her dedication and tireless efforts in helping make this encyclopedia as accurate and consistent as possible.

When the temperature at which properties such as viscosity, density, surface tension, solubilities, etc. is not included, a standard temperature of 25 C is to be assumed. The information in this publication is reliable to the best of our knowledge. We would be glad to be informed of any errors or omissions so that these can be integrated into subsequent editions of this encyclopedia.

M. and I. Ash

CONTRIBUTORS

Aceto Chem. Co., Inc.
126-02 Northern Blvd.
Flushing, N.Y. 11368

Ajinomoto USA, Inc.
9 West 57th St.
New York, N.Y. 10019

Akzo Chemie bv
Stationsstraat 48
P.O. Box 247
3800 AE-Amersfoort
The Netherlands

Akzo Chemie Italia SpA.
Via Vismara
20020 Arese
Milano, Italy

Albright & Wilson Ltd.
Detergent Sector
P.O. Box 15
White Haven, Cumberland
CA 28 9QQ, England

Alcolac Inc.
3440 Fairfield Rd.
Baltimore, Md. 21226

Alconox Inc.
215 Park Ave. So.
New York, N.Y. 10003

Alkaril Chemicals Ltd.
3265 Wolfedale Road
Mississauga, Ont., Canada

Allied Colloids Inc.
One Robinson Lane
Ridgewood, N.J. 07450

Alox Corp.
3934 Buffalo Ave.
P.O. Box 517
Niagara Falls, N.Y. 14302

Amerchol Corp.
(Unit of CPC International Inc.)
P.O. Box 351
Talmadge Road
Edison, N.J. 08902

American Color & Chem. Corp.
11400 Westinghouse Blvd.
Charlotte, N.C. 28232

American Cyanamid Co.
Fine Chemicals Dept.
Pearl River, N.Y. 10965

American Cyanamid Co.
Industrial Chemicals Div.
Berdan Ave.
Wayne, N.J. 07470

American Cyanamid Co.
Organic Chemicals Div.
Bound Brook, N.J. 08805

American Hoechst Corp.
Industrial Chemicals Div.
Rt. 202–206 North
Somerville, N.J. 08876

American Lecithin Co., Inc.
32-34 61st St.
Woodside, L.I., N.Y. 11377

Apex Chem. Co., Inc.
200 South First St.
Elizabethport, N.J. 07206

Arizona Chemical Co.
840 Valley Brook Ave.
Lyndhurst, N.J. 07071

Arjay Inc.
P.O. Box 45045
Houston, Texas 77045

Arkansas Co., Inc.
185 Foundry St.
P.O. Box 210
Newark, N.J. 07101

Armak—Industrial Chemicals
P.O. Box 1805
Chicago, Ill. 60690

Armstrong Chem. Co., Inc.
1530 South Jackson St.
Janesville, Wis. 53545

Arol Chem. Prod. Co.
649 Ferry St.
Newark, N.J. 07105

Atlas Chem.
See ICI United States Inc.

Atlas Refinery, Inc.
142 Lockwood St.
Newark, N.J. 07105

BASF Wyandotte Corp.
Industrial Chemicals Group
1609 Biddle Ave.
Wyandotte, Mich. 48192

BASF Wyandotte Corp.
Textile Colors and Chem. Dept.
100 Cherry Hill Road
Parsippany, N.J. 07054

BASF Aktiengesellschaft
Carl-Bosch-Str. 36
6700 Ludwigshafen
Germany

Berol Kemi AB
S-444 01 Stenungsund 1
Sweden

Bostik South
Div. of USM Corp.
P.O. Box 5695
Greenville, S.C. 29606

Carson Chemicals, Inc.
2779 East El Presidio
Long Beach, Calif. 90810

Catawba-Charlab, Inc.
P.O. Box 948
Charlotte, N.C. 28231

Central Soya Co., Inc.
Chemurgy Div.
1300 Fort Wayne National Bank Bldg.
Fort Wayne, Ind. 46802

Chemform Corp.
Vertac Bridgeport Ind. Park
P.O. Box 1745
West Memphis, AR 72301

Ciba-Geigy Corp.
Dyestuff & Chemicals Div.
P.O. Box 11422
Greensboro, N.C. 27409

Cincinnati Milacron Chem. Inc.
West St.
Cincinnati, Ohio 45215

Cities Service Co.
ICD, Cities Service Bldg.
3445 Peachtree Rd. N.E.
Atlanta, Georgia 30326

Clintwood Chemical Co.
4342 S. Wolcott Ave.
Chicago, Ill. 60609

Conoco Chemicals
Continental Oil Co.
5 Greenway Plaza East
P.O. Box 2197
Houston, TX 77001

Croda Chemicals Ltd.
Cowick Hall
Snaith Goole
North Humberside DN14 9AA
England

Croda Inc.
51 Madison Ave.
New York, N.Y. 10010

Cyclo Chemicals Corp.
7500 N.W. 66th St.
Miami, Fla. 33166

Diamond Shamrock Corp.
Process Chemicals Div.
350 Mt. Kemble Ave.
Morristown, N.J. 07960

Dixo Co., Inc.
158 Central Ave.
Rochelle Park, N.J. 07662

Dow Chemical Europe S.A.
Leland I. Doan Str. 3 CH-8810
Horgen
Switzerland

Dow Corning Corp.
Midland, Mich. 48640

Du Pont de Nemours, E.I. & Co.
Specialty Chemicals & Products
Nemours Bldg.
Wilmington, DE 19898

Durkee Industrial Foods Group/
SCM Corp.
900 Union Commerce Bldg.
Cleveland, Ohio 44115

Dutton & Reinisch Ltd.
Crown House, London Road
Morden, Surrey
SM45DU, England

Emery Industries, Inc.
1501 W. Elizabeth Ave.
Linden, N.J. 07036

Emkay Chemical Co.
319-325 Second St.
Elizabeth, N.J. 07206

Emulsion Systems Inc.
58 Metropolitan Ave.
Brooklyn, N.Y. 11211

Essential Chemicals Corp.
28391 Essential Rd.
Merton, Wis. 53056

Exxon Company, USA
P.O. Box 2180
Houston, Texas 77001

GAF Corp.
Chemical Products
140 W. 51st St.
New York, N.Y. 10020

General Electric Co.
Silicone Prod. Dept.
Waterford, N.Y. 12188

General Mills Chemicals, Inc.
See Henkel Inc.
Chemical Specialties Div.

Georgia-Pacific Corp.
P.O. Box 1236
Bellingham, Wash. 98225

Glyco Chemicals, Inc.
51 Weaver St.
Greenwich, Conn. 06830

W.R. Grace & Co.
Organic Chemicals Div.
55 Hayden Ave.
Lexington, MA 02173

Graden Chemical Co., Inc.
426 Bryan St.
Havertown, Pa. 19083

Grindsted Products
Grindstedvaerket A/S
38 Edwin Rahrs Vej
DK-8220 Brabrand
Denmark

C.P. Hall Co.
7300 S. Central Ave.
Chicago, Ill. 60638

Hart Chemical Ltd.
Guelph, Canada N1H 6K8

Henkel Chemicals (Canada) Ltd.
9550 Blvd. Ray Lawson
Ville d'Anjou (Montreal)
P.Q. Canada

Henkel Inc.
Chemical Specialties Div.
1301 Jefferson St.
Hoboken, N.J. 07030

Hercules Inc.
Organics Dept.
Hercules Tower
910 Market St.
Wilmington, Del. 19894

Hilton-Davis Chemical Co.
2235 Langdon Farm Rd.
Cincinnati, Ohio 45237

ICI United States Inc.
New Murphy Rd. & Concord Pike
Wilmington, Del. 19897

IMC Chem. Group—NP Div.
4415 W. Harrison St.
Hillside, Ill. 60162

Lankro Chemicals Ltd.
Eccles
Manchester M30 OBH
England

Monsanto Co.
800 N. Lindbergh Blvd.
St. Louis, Mo. 63166

NL Industries
NL Baroid
P.O. Box 1675
2404 Southwest Freeway
Houston, TX 77001

NL Industries
Industrial Chemicals Div.
P.O. Box 700
Hightstown, N.J. 08520

Onyx Chemical Co.
Millmaster Onyx Group
190 Warren St.
Jersey City, N.J. 07302

Proctor & Gamble Co.
Sharonwood Technical Center
Cincinnati, Ohio 45217

PVO International Inc.
416 Division St.
Boonton, N.J. 07005

Reilly-Whiteman Inc.
Washington, & Righter Sts.
Conshohocken, Pa. 19428

Rewo Chemicals, Inc.
125 Wilbur Pl.
Bohemia, N.Y. 11716

Sandoz Colors & Chemicals
Hanover, N.J. 07936

Sandoz Ltd.
CH-4002
Basle, Switzerland

Shamton Lubricants & Chem. Ltd.
Tel Aviv, Israel

Shell Chemical Co.
One Shell Plaza
Houston, TX 77002

Sherex Chemical Co., Inc.
P.O. Box 646
Dublin, Ohio 43017

Standard Chemical Co.
Mill Lane
Cheadle, Chesire SK8 2NX
England

Stepan Chemical Co.
Edens & Winnetka Rds.
Northfield, Ill. 60093

Swastik Household & Industrial Products Ltd.
Shaghibag House
13 Walchand Hirachand Marg
Bellard Estate
Bombay 400 038 India

Swift Chemical Co.
383 Orenda Road
Bramaleg, Ontario, Canada

Thompson-Hayward Chem. Co.
5200 Speaker Rd.
Kansas City, Kansas 66106

3M Co.
3M Center
Commercial Chemical Div.
St. Paul, MN 55105

Toho Chemical Industry Co., Ltd.
1-14-9 Nihonbashi
Kakigara-cho, Chuo-ku
Tokyo, Japan

Van Schuppen Chemie
Nieuweweg, Veenendaal
Netherlands

R.T. Vanderbilt Co., Inc.
30 Winfield St.
Norwalk, Conn. 06855

Witco Chemical Corp.
Organics Div. & Sonneborn Div.
277 Park Ave.
New York, N.Y. 10017

Witco Chem. Corp.
Ultra Div.
2 Wood St.
Paterson, N.J. 07524

ABBREVIATIONS

aq.	aqueous
BHA	butylated hydroxyanisole
B.P.	boiling point
C	degrees Centigrade
cc	cubic centimeter(s)
CC	Closed cup
cm	centimeter(s)
COC	Cleveland open cup
conc.	concentration, concentrated
cps	centipoise(s)
cs or cst	centistoke(s)
disp.	dispersion, dispersible
dist.	distilled
EO	ethylene oxide
esp.	especially
ETO	ethylene oxide
F	degrees Fahrenheit
FDA	Food & Drug Administration
F.P.	freezing point
g	gram(s)
gal	gallon(s)
HLB	hydrophilic lipophilic balance
ingred.	ingredient(s)
IPA	isopropyl alcohol
IPM	isopropyl myristate
kg	kilogram(s)
l	liter(s)
lb	pound(s)
max.	maximum
MEK	methyl ethyl ketone
min.	minimum
ml	milliliter(s)
mm	millimeter(s)
M.P.	melting point
m.w.	molecular weight
NF	National Formulary

no.	number
O/W	oil in water
PEG	polyethylene glycol
pH	hydrogen ion concentration
PMCC	Pensky-Martens closed cup
POE	polyoxyethylated
pt.	point
PVC	polyvinyl chloride
quat.	quaternary
ref.	refractive
sapon.	saponification
sol.	soluble
solid.	solidification
sol'n.	solution
sp. gr.	specific gravity
SSU	Saybolt Seconds Universal
Stod.	Stoddard
SUS	Saybolt Universal Seconds
TCC	Taggart closed cup
tech.	technical
T/L	tank load
TOC	Taggart open cup
T/T	tank truck
USP	United States Pharmacopeia
uv	ultraviolet
visc.	viscosity
W/O	water in oil
wt	weight
%	percent
$\approx$	approximately
$<$	less than
$>$	greater than
$\leqslant$	less than or equal to
$\geqslant$	greater than or equal to
$\gg$	much larger than
$\ll$	much smaller than
#	number

The following products, which are included in this volume, are registered with the U.S. Patent Office.

Abex®
Actiflo®
Acto®
Adsee®
Aerosol®
Agrimul®
Ahco®
Alacsan®
Alcojet®
Alconox®
Alfol®
Alkaterge®
Alipal®
Amacarrier®
Amasoft®
Amawet®
Amerchol®
Ameroxol®
Aminogen®
Antara®
Antaron®
Antarox®
Arlacel®
Arlasolve®
Arlatone®
Arlex®
Armac®
Armeen®
Armol®
Armohib®
Arquad®
Atlas®
Atlox®
Atmer®
Atmos®
Atmul®
Atpet®
Atplus®
Atsurf®
Bio Soft®
Blancol®
Brij®
Casul®
Carbonox®
CC-16®
Centrol®
Centrolene®
Centrolex®
Centromix®
Centrophil®
Chemcol®
Cremophor®
Cutina®
Cyncal®
Darvan®
Daxad®
Deceresol®
Det-O-Jet®
Dow Corning®
Dresinate®
Drewmulse®
Duomac®
Duomeen®
Duoquad®
Duponol®
Emerest®
Emery®
Emid®
Emulphogene®
Emulphor®
Ethoduomeen®
Ethofat®
Ethomeen®
Ethoquad®
Flexricin®
Fluorad®

A

A-641. ICI Unites States Inc.

Chemical Description: Noncrystallizing sorbitol sol'n.
Category: Polyhydric alcohol (Polyol)
Form: Clear viscous liquid; Color: Water white to pale yellow; Odor: None; Taste: Pleasantly sweet
Properties: Ref. Index: 1.455–1.465; Sp. gr.: ≈ 1.31; Visc.: ≈ 140 cps
Composition: 53% sorbitol content

AA-10. Cities Service Co.

Category: Detergent, wetting agent, emulsifier
Applications: Emulsifier
Form: Liquid, flake; Color: White
Properties: Odor: None; Solubility: Highly soluble in water; Density: 2 lb/gal; Sp. gr.: 1/5 H_2O; Stability: Solid
Composition: 98% active ingredient
Reaction: Neutral
Toxicity: Nontoxic
Containers: Fiber drums

Abex 18-S. Alcolac

Category: Emulsifier, anionic
Applications: Polymer emulsion as binders for coated paper and paperboard; in paint and floor coatings; in polymerization of vinyl acetate and acrylics
Form: Liquid; Color: Pale yellow
Properties: Density: 8.5 lb/gal; Visc.: @ 15 C, 200 cps;Stability: Stable for broad range of pH and temperature; Surface Tension: 1.0% sol'n.–48 dynes/cm; 0.1% sol'n.–45–47 dynes/cm
Composition: 35% active ingredient

Abex 22-S. Alcolac

Category: Emulsifier, anionic
Applications: In emulsion polymer latices for floor polish, latex paints, paper coatings, pigment binders and adhesives
Form: Liquid; Color: Pale yellow
Properties: Density: 8.7 lb/gal; Visc.: @ 25 C, 1000 cps; Stability: Good; Surface Tension: 40 dynes/cm at critical micelle concentration

Composition: 25% active ingredient

Abex 26-S. Alcolac
Category: Emulsifier, anionic surfactant
Applications: Emulsifier for acrylics, vinyl acetates and other vinyl monomers
Form: Liquid; Color: Pale yellow
Properties: Density: 8.8 lb/gal; Visc.: @ 25 C, 100 cps; Stability: Good in presence of multivalent salts; Surface Tension: 55 dynes/cm at critical micelle concentration
Composition: 33% active ingredient

Abex JKB. Alcolac
Category: Emulsifier, anionic surfactant
Applications: Emulsifier for highly carboxylated acrylic monomers; latex thickeners, alkali-soluble resins; metalized floor finishes; textile sizing agents and handmodifiers
Form: Liquid; Color: Clear
Composition: 30% active ingredient

Abex VA-40. Alcolac
Category: Emulsifier, anionic surfactant
Applications: Emulsifier for high solids vinyl acetates
Form: Liquid; Color: Pale yellow
Properties: Visc.: @ 25 C, 600 cps
Composition: 40% active ingredient

Abex VA-50. Alcolac
Category: Emulsifier, anionic surfactant
Applications: Emulsifier for high solids vinyl acetates
Form: Liquid; Color: Pale yellow
Properties: Visc.: @ 25 C, 400 cps; Surface Tension: 47.4 dynes/cm (0.01% sol'n.)
Composition: 46% active ingredient

Absorption bases. Croda Chem.
Chemical Description: Cholesterol rich extracts of lanolin compounded into inert hydrocarbon bases
Category: Emollient
Applications: Cosmetics, especially emollient skin creams, lotions; baby preparations; make up
Form: Solid, liquid

Accobetaine CL. Armstrong Chem.
Chemical Description: Complex coco betaine
Category: Detergent, wetting agent, emulsifier, high foaming agent, solubilizer
Applications: Household and cosmetics uses (hair shampoos, liquid hand cleaners, dishwashing compounds, rug shampoo, etc.)
Form: Liquid; Color: Yellow; Odor: Pleasant
Properties: Solubility: Good in water; F.P.: 0 C
Composition: 35% active ingredient
Toxicity: Nontoxic

Accomeen C2. Armstrong Chem.
Chemical Description: Ethoxylated coco amine
Category: Emulsifier, antistat, nonionic surfactant

Form: Liquid; Color: 4 (Gardner); Odor: Amine
Properties: Solubility: Organic solvent; insoluble water; Density: 7.25 lb/gal; Sp. gr.: 0.87
Composition: 99% active ingredient; 2 moles ethylene oxide on primary coco amine
Reaction: Acid
Warnings and Caution: Corrosive

Accomeen C5. Armstrong Chem.
Chemical Description: Ethoxylated coco amine
Category: Emulsifier, nonionic surfactant
Applications: Emulsifier, antistat
Form: Liquid; Color: 4 (Gardner); Odor: Amine
Properties: Solubility: Organic solvents and water; Density: 8.15 lb/gal; Sp. gr.: 0.98
Composition: 99% active ingredient; 5 moles ethylene oxide on primary coco amine
Reaction: Alkaline
Warnings and Caution: Corrosive

Accomeen C10. Armstrong Chem.
Chemical Description. Ethoxylated coco amine
Category: Emulsifier, nonionic surfactant
Applications: Emulsifier, antistat
Form: Liquid; Color: 6 (Gardner); Odor: Amine
Properties: Solubility: Organic solvents and water; Density: 8.3 lb/gal; Sp. gr.: 1.0
Composition: 99% active ingredient; 10 moles ethylene oxide on primary coco amine
Reaction: Acid
Warnings and Caution: Corrosive

Accomeen C15. Armstrong Chem.
Chemical Description: Ethoxylated coco amine
Category: Emulsifier, nonionic surfactant
Applications: Emulsifier, antistat, textile spinning aid
Form: Liquid; Color: 6 max (Gardner); Odor: Mild amine
Properties: Density: 8.7 lb/gal; Sp. gr.: 1.04
Composition: 99% active ingredient; 15 moles ethylene oxide on primary coco amine
Reaction: Acid
Toxicity: Corrosive
Warnings and Caution: Corrosive

Accomeen S2. Armstrong Chem.
Chemical Description: Ethoxylated soya amine
Category: Emulsifier, nonionic surfactant
Applications: Emulsifier, antistat
Form: Liquid; Color: 5 (Gardner); Odor: Amine-type
Properties: Solubility: Organic solvents, insoluble water; Density: 7.6 lb/gal; Sp. gr.: 0.91; Surface Tension: 31.3 @ 0.1% sol'n.
Composition: 99% active ingredient; 2 moles ethylene oxide on primary soya amine
Reaction: Acid
Warnings and Caution: Corrosive

Accomeen S5. Armstrong Chem.
Chemical Description: Ethoxylated soya amine
Category: Emulsifier, nonionic surfactant

Applications: Emulsifier, antistat
Form: Liquid; Color: 10 (Gardner); Odor: Amine-type
Properties: Solubility: Organic solvents; disp. in water; Density: 7.9 lb/gal; Sp. gr.: 0.95
Composition: 99% active ingredient; 5 moles ethylene oxide on primary soya amine
Reaction: Acid
Warnings and Caution: Corrosive

Accomeen S10. Armstrong Chem.
Chemical Description: Ethoxylated soya amine
Category: Emulsifier, antistat, nonionic surfactant
Form: Liquid; Color: 10 (Gardner); Odor: Amine
Properties: Solubility: Organic solvents and water; Density: 8.5 lb/gal; Sp. gr.: 1.02; Surface Tension: 43 @ 0.1%
Composition: 99% active ingredient; 10 moles ethylene oxide on primary soya amine
Reaction: Acid
Warnings and Caution: Corrosive

Accomeen T2. Armstrong Chem.
Chemical Description: Ethoxylated tallow amine
Category: Emulsifier, antistat, nonionic surfactant
Form: Paste; Color: 3 (Gardner); Odor: Amine type
Properties: Organic solvents, insoluble in water; Density: 7.7 lb/gal; Sp. gr.: 0.92
Composition: 99% active ingredient; 2 moles ethylene oxide on primary tallow amine
Reaction: Acid
Warnings and Caution: Corrosive

Accomeen T15. Armstrong Chem.
Chemical Description: Ethoxylated tallow amine
Category: Emulsifier, nonionic surfactant
Applications: Emulsifier dispersant
Form: Liquid; Color: 5 (Gardner); Odor: Amine type
Properties: Refractive Index: 1; Solubility: Organic solvents, insoluble in water; Density: 8.1 lb/gal; Sp. gr.: 0.97
Composition: 99% active ingredient; 5 moles ethylene oxide on primary tallow amine
Reaction: Acid
Warnings and Caution: Corrosive

Accomid C. Armstrong Chem.
Chemical Description: Coconut oil diethanolamide
Category: Detergent, stabilizer, nonionic surfactant
Applications: Shampoo, bubble baths, diswashing detergent
Form: Liquid; Color: 4 (Gardner); Odor: Mild
Properties: Solubility: Completely soluble in water at low concentration; Density: 8.3 lb/gal; Sp. gr.: 0.99
Composition: 98% active ingredient

Acconon 300 MO. Armstrong Chem.
Chemical Description: Polyethylene glycol 300 monooleate
Category: Emulsifier, nonionic surfactant
Applications: Surface activity; cosmetics, food, agriculture, plastics

Form: Liquid; Color: 6 (Gardner)
Properties: Solubility: Organic solvents; Density: 8.3 lb/gal; Sp. gr.: 0.99; M.P.: <–5 C
Composition: 99% active ingredient; 6.8 moles ethylene oxide added to oleic acid
Containers: Drums

Acconon 400 MO. Armstrong Chem.
Chemical Description: Polyethylene glycol 400 monooleate
Category: Emulsifier, nonionic surfactant
Applications: Surface activity, lubricant, chemical intermediary; cosmetics, food, agriculture, plastics
Form: Liquid; Color: 6 (Gardner)
Properties: Solubility: Organic solvents; Density: 8.4 lb/gal; Sp. gr.: 1.01; M.P.: < 10 C; Biodegradable
Composition: 99% active ingredient; 9.1 moles ethylene oxide added to oleic acid

Acconon 1300. Armstrong Chem.
Chemical Description: Polyethoxylated-polypropoxylated C_{12-20} alcohols
Form: Liquid; Color: Light yellow, 3 max (Gardner)
Properties: Density: 8.35–8.45 lb/gal; Sp. gr.: 1.016–1.019; Cloud Pt.: °F, 135–145

Acconon TGH. Armstrong Chem.
Chemical Description: Complex tallow glyceride ethoxylate
Category: Wetting agent, foaming agent, nonionic surfactant
Applications: personal care and household products
Form: Liquid; Color: Yellow; Odor: Mild

Accoquat 2C-75. Armstrong Chem.
Chemical Description: Dicoco dimethyl ammonium chloride
Category: Emulsifier, coupling agent, cationic
Applications: Car spray waxes, dust control oil, removal of spots
Form: Liquid; Color: Amber; Odor: Mild
Properties: Solubility: Solvent is isopropanol; Density: 7.2 lb/gal; Sp. gr.: 0.86
Composition: 75% active ingred.
Warnings and Caution: Flammable liquid

Accoquat 2C-75–H. Armstrong Chem. Co.
Chemical Description: Dicoco dimethyl ammonium chloride–high flash
Category: Emulsifier, cationic
Applications: Car spray waxes, dust control oils, removal of spots
Form: Liquid; Color: Amber; Odor: Mild
Properties: Solubility: Solvent is hexylene glycol; Flash Pt.: High
Composition: 75% active ingred.

Accosoft 550. Armstrong Chem.
Chemical Description. Tallow amido amine quaternary compound
Category: Cationic surfactant
Applications: Household and industrial fabric softener, dispersion for household fabric softener, fabric antistat agent
Form: Liquid, gel; Color: 6 (Gardner)

Properties: Solubility: Solvent–Isopropanol; Density: 8.0; Sp. gr.: 0.96; Biodegradable
Composition: 75 or 90% active ingredient
Toxicity: Nontoxic
Warnings and Caution: Flammable, supports bacterial growth

Accosoft 620. Armstrong Chem.
Chemical Description: Tallow amido amine quaternary compound
Category: Cationic surfactant
Applications: Household or industrial fabric softener, fabric antistat agent
Form: Liquid; Color: 6 (Gardner)
Properties: Solubility: Solvent–IPA; Density: 8.0; Sp. gr.: 0.96; Biodegradable
Composition: 75 or 90% active ingredient
Toxicity: Nontoxic
Warnings and Caution: Flammable, supports bacterial growth

Accosoft 707. Armstrong Chem.
Chemical Description. Dimethyl dihydrogenated tallow ammonium chloride
Category: Cationic surfactant
Applications: Household or industrial fabric softener, fabric antistat agent
Form: Paste; Color: 3 (Gardner)
Properties: Solubility: Solvent–IPA, water; Density: 7.3; Sp. gr.: 8.7
Composition: 75% active ingredient
Toxicity: Nontoxic
Warnings and Caution: Flammable, supports bacterial growth

Accosoft 748. Armstrong Chem.
Chemical Description. Dimethyl dihydrogenated tallow ammonium metho sulfate
Category: Cationic surfactant
Applications: Household or industrial fabric softener or antistat particularly suited to drier use
Form: Paste, solid; Color: 4 (Gardner)
Properties: Solubility: Solvent–IPA, water, M.P.: 125–135 C; Biodegradable
Composition: 90% active ingredient
Toxicity: Nontoxic
Containers: Tankwagons, RR tank cars and drums
Warnings and Caution: Flammable, supports bacterial growth

Accosoft A-155. Armstrong Chem.
Chemical Description. Tallow amido amine quaternary compound
Category: Cationic surfactant
Applications: Household fabric softener base; dispersions for fabric softener; fabric antistat agent
Form: Gel; Color: 7 (Gardner)
Properties: Biodegradable
Composition: 75 or 90% active ingredient
Toxicity: Nontoxic
Containers: 4000–6000 gal stainless steel tankwagons
Category: Emulsifier, solubilizer
Form: Liquid; Color: Yellow
Properties: Sp. gr.: 1.1; Visc.: 360–400 cps; HLB: 16.7; Sapon. Value: 40–52; Acid No.: 2.0 max.; Hydroxyl No.: 97–113
Composition: 98% active ingred.
Containers: 55 gal lined bung

drums

Accosperse 20. Armstrong Chem.
Chemical Description: POE-20 sorbitan monolaurate

Accosperse 60. Armstrong Chem.
Chemical Description: POE-20 sorbitan monostearate
Category: Emulsifier
Form: Liquid, gel; Color: Amber
Properties: Sp. gr.: 1.1; Visc.: 550–600 cps; HLB: 14.9; Sapon. Value: 45–55; Acid No.: 2.0 max.; Hydroxyl No.: 80–96
Composition: 98% active ingred.

Accosperse 80. Armstrong Chem.
Chemical Description: POE-20 sorbitan monooleate
Category: Emulsifier
Form: Liquid; Color: Amber
Properties: Sp. gr.: 1.1; Visc.: 400–450; HLB: 15.0; Sapon. Value: 45–55; Acid No.: 2.0 max.; Hydroxyl No.: 65–81
Composition: 98% active ingred.
Containers: 55 gal lined bung drums

Accosperse TGH. Armstrong Chem.
Chemical Description: Alkoxylated mono-, di-glyceride
Category: Emulsifier, foaming and wetting agent, nonionic
Applications: Personal care and cosmetic products, conditioners, cleansing creams; industrial emulsions
Form: Liquid; Color: Yellow; Odor: Mild
Properties: F.P.: 30 F; Cloud Pt.: 140–146 F (1% in dist. water); Biodegradable
Composition: 98–100% active
Reaction: Neutral

Acetoquat CPB. Aceto Chem.
Chemical Description: Cetyl pyridinium bromide
Category: Germicide, cationic
Applications: Germicide, sanitizing agent
Form: Powder
Composition: 95% active ingredient

Acetoquat CPC. Aceto Chem.
Chemical Description: Cetyl pyridinium chloride
Category: Germicide, cationic
Applications: Germicides, sanitizing agent
Form: Powder
Composition: 100% active ingredient

Acetoquat CTAB. Aceto Chem.
Chemical Description: Cetyl trimethyl ammonium bromide
Category: Germicide, cationic
Applications: Germicide, sanitiz-agent
Form: Powder
Composition: 95% active ingredient

Acidan. Grindsted Products
Chemical Description: Citric acid esters of monoglycerides
Category: Emulsifier, anionic surfactant
Applications: Food industry in coffee whiteners, frying margarine, meat emulsions to improve dispersion, whitening and emulsification

Form: Powder, flakes
Composition: 100% active ingredient

ACL 56. Monsanto
Chemical Description: Sodium dichloroisocyanurate dihydrate
Category: Bleaching compound, sanitizer
Applications: Detergents, bleaches, pool disinfection
Composition: 56% available chlorine

ACL 59. Monsanto
Chemical Description: Potassium dichloroisocyanurate
Category: Bleaching compound, sanitizer
Applications: Pool disinfectant and sanitizer; oxidizer in dishwashing compounds
Composition: 59% available chlorine

ACL 60. Monsanto
Chemical Description: Sodium dichloroisocyanurate
Category: Bleaching compound, sanitizer
Applications: Detergent, bleach, pool disinfectant
Composition: 63% available chlorine

ACL 66. Monsanto
Chemical Description: (monotrichloro) tetra-(monopotassium dichloro)-penta–isocyanurate
Category: Bleaching compound, sanitizer
Applications: sanitizers, detergent
Composition: 66.5% available chlorine

ACL 85. Monsanto
Chemical Description: trichloroisocyanuric acid
Category: Bleaching compound, sanitizer
Applications: Pool disinfectant in compacted forms; detergent
Composition: 90% available chlorine

Aconol X10. Hart Chem.
Chemical Description: Tall oil ethoxylate
Category: Emulsifier, nonionic surfactant
Applications: Low-foaming nonionic for Built detergent systems
Form: Liquid; Color: Clear
Properties: Refractive Index: 1.4655; Solubility: Clear soluble; Density: 1.039; Visc.: 140 cps
Composition: 95% active ingredient
Containers: Tankcars, drums
Warnings and Caution: Avoid contact with skin and eyes

Acrawax. Glyco Chem.
Applications: Synthetic waxes and plastic lubricants
Form: Hard brittle solid; Color: Tan
Properties: Sp. gr.: 1.04 @ 20 C; M.P.: 90–96 C; Flash Pt.: ≈ 205 C

Acrawax B. Glyco Chem.
Form: Hard solid; Color: Light brown
Properties: Sp. gr.: 0.97 @ 20 C;

M.P.: 82–90 C; Flash Pt.: ≈ 200 C

Acrawax C. Glyco Chem.
Form: Hard solid; Color: Light tan
Properties: Sp. gr.: 0.97 @ 20 C; M.P.: 142–145 C; Flash Pt.: ≈ 285 C

Acrawax HM-S. Glyco Chem.
Form: Coarse powder; Color: Tan
Properties: Sp. gr.: 0.97 @ 20 C; M.P.: 243–246 C; Flash Pt.: ≈ 298 C

Actiflo 68, 70. Central Soya Co.
Chemical Description: Natural lecithin
Category: Emulsifier, wetting agent, amphoteric
Applications: Emulsifier, dispersant, wetting agent; food, cosmetics, textiles, plastics and rubber
Form: Liquid
Composition: 66–72% active ingredient

Acto 450. Exxon Corp.
Chemical Description: Alkylaryl sodium sulfonate
Category: Detergent, wetting agent, emulsifier, anionic
Applications: Detergents for textiles, skins, leather, dyeing, mines and minerals, paints; Wetting agents for dyeing, rewetting dried skins, tanning, insecticides, herbicides, fungicides; Emulsifiers in fat liquoring, cutting and soluble oils, dispersing waxes, pigments, resins; printing, preparation of emulsified paints; making of bitumen emulsions; rust preventative
Form: Liquid; Color: Reddish-amber; Odor: Alcohol
Properties: m.w.: 470; Solubility: oil-soluble, slight sol. in water but readily dispersible; Density: 8.24 lb/gal; Sp. gr.: 0.99; Stability: Good < 180 F
Composition: 44% active ingredient
Toxicity: Causes eye burn on contact; may cause mild skin irritation
Warnings and Cautions: Flammable
References: Data sheet DG-4F

Acto 500. Exxon Corp.
Chemical Description: Alkylaryl sodium sulfonate
Category: See Acto 450
Applications: See Acto 450
Form: Viscous liquid; Color: Amber; Odor: Slight bland petroleum oil
Properties: m.w.: 460; Solubility: oil-soluble; slight sol. in water but readily dispersible; Density: 8.33 lb/gal; Sp. gr.: 1.00; Stability: Good < 180 F
Composition: 50% active ingredient
Toxicity: See Acto 450
References: Data sheet DG-4F

Acto 630. Exxon Corp.
Chemical Description: Alkylaryl sodium sulfonate
Category: See Acto 450
Applications: See Acto 450
Form: Viscous liquid; Color: Amber; Odor: Slight bland petroleum oil odor
Properties: m.w.: 470; Solubility:

oil-soluble, slight sol. in water but readily dispersible; Density: 8.41 lb/gal; Sp. gr.: 1.01; Stability: Good < 180 F
Composition: 63% active ingredient
Toxicity: See Acto 450
References: Data sheet DG-4F

Acto 632. Exxon Corp.
Chemical Description: Alkylaryl sodium sulfonate
Category: See Acto 450
Applications: See Acto 450
Form: See Acto 450
Properties: m.w.: 430; Solubility: oil-soluble, slight sol. in water but readily dispersible; Density: 8.41 lb/gal; Sp. gr.: 1.01; Stability: Good < 180 F
Composition: 63% active ingredient
Toxicity: See Acto 450
References: Data sheet DG-4F

Acto 636. Exxon Corp.
Chemical Description: Alkylaryl sodium sulfonate
Category: See Acto 450
Applications: See Acto 450
Form: Viscous liquid; Color: Orange; Odor: Slight bland petroleum oil odor
Properties: m.w.: 470; Solubility: oil-soluble, slight sol. in water but readily dispersible; Density: 8.5 lb/gal; Sp. gr.: 1.02; Stability: Good < 180 F
Composition: 63% active ingredient
Toxicity: See Acto 450
References: Data sheet DG-4F

Acto 639. Exxon Corp.
Chemical Description: Alkylaryl sodium sulfonate
Category: See Acto 450
Applications: See Acto 450
Form: See Acto 500
Properties: m.w.: 520; Solubility: oil-soluble, slight sol. in water but readily dispersible; Density: 8.5 lb/gal; Sp. gr.: 1.02; Stability: Good < 180 F
Composition: 63% active ingredient
Toxicity: See Acto 450
References: Data sheet DG-4F

Acylan. Croda Chem.
Chemical Description: Acetylated lanolin
Category: Emollient, conditioner
Applications: Cosmetics, especially hypoallergenic and baby cosmetics
Form: Paste; Color: Yellow
Composition: 100% active ingred.
Reaction: Nonionic

Acylglutamate CS-11. Ajinomoto
Chemical Description: Monosodium N-cocoyl-L-glutamate
Category: Anionic detergent
Applications: Personal care; high detergency and foaming power; emollient for skin; bacteriostatic effect
Form: powder
Properties: Stability: Good in hard water; Biodegradable
Composition: 100% active ingredient
Toxicity: Nonskin irritation even for infants and eczema cases

Acylglutamate CT-12. Ajinomoto

Chemical Description: Monotriethanolamine N-cocoyl-L-glutamate
Category: See Acylg. CS-11.
Applications: See Acylg. CS-11.
Form: Liquid
Properties: See Acylg. CS-11.
Compositon: 20% active ingred.
Toxicity: See Acylg. CS-11.

Acylglutamate GS-11. Ajinomoto
Chemical Description: Monosodium N-mixed fatty acid acyl glutamate
Category: See Acylg. CS-11.
Applications: See Acylg. CS-11.
Properties: See Acylg. CS-11.
Toxicity: See Acylg. CS-11.

Acylglutamate HS-11. Ajinomoto
Chemical Description: Monosodium N-hydrogenated tallowyl-L-glutamate
Category: See Acylg. CS-11.
Applications: See Acylg. CS-11.
Form: Powder
Properties: See Acylg. CS-11.
Composition: 100% active ingred.
Toxicity: See Acylg. CS-11.

Acylglutamate LS-11. Ajinomoto
Chemical Description: Monosodium N-lauroyl-L-glutamate
Category: See Acylg. CS-11.
Applications: See Acylg. CS-11.
Form: Powder
Properties: See Acylg. CS-11.
Composition: 100% active ingred.
Toxicity: See Acylg. CS-11.

Adinol T. Croda Chem.
Chemical Description: Sodium N methyl N oleoyl taurate
Category: Detergent, wetting agent, emulsifier, dispersant, foaming agent
Applications: Shampoo, foam bath and skin cleansing compositions
Form: Liquid, paste, gel, powder;
Properties: Solubility: Good in hard water

Admex 433. Sherex
Chemical Description: Low-viscosity polyester
Category: Plasticizer
Applications: Foam, extrusion, calendered films, coated fabric; solvent resistant, compatible in high humidity

Admex 515. Sherex
Chemical Description: Low-viscosity polyester
Category: Plasticizer
Applications: Calendered and coated fabric, vinyl foam, plastisols, organosols, pigment grinding, wire coating

Admex 522. Sherex
Chemical Description: Low-viscosity polyester
Category: Plasticizer
Applications: Extrusion, calendered film and sheet, wire coating, organosols, plastisols, flooring
Properties: Stability: Stable to light and heat

Admex 523. Sherex
Chemical Description: Medium-viscosity polyester
Category: Plasticizer
Applications: Transfer hose and tubing, roof flashing and edging,

vinyl foams and gaskets requiring gas and oil resistance

Admex 525. Sherex
Chemical Description: Primary polyester
Category: Plasticizer
Applications: Coating fabrics, paper coatings, pigment dispersions, plastisols, organosols, tapes, vinyl foams, extruded profiles
Properties: Visc.: Low; Good low temperature flexibility and pigment wetting

Admex 529. Sherex
Chemical Description: General purpose polyester
Category: Plasticizer
Applications: Calendered and coated fabric, wire coatings, flooring

Admex 710. Sherex
Chemical Description: Epoxidized soybean oil
Category: Plasticizer
Applications: Plasticizer/stabilizer for PVC, food wrappings, beverage transfer tubing, floor tile, baby and nursery goods, nitro-cellulose and butyrate coatings; reactive intermediate

Admex 711. Sherex
Chemical Description: Epoxidized soybean oil
Category and Applications: See Admex 710

Admex 746. Sherex
Chemical Description: Epoxidized tallate ester
Category: Plasticizer, stabilizer
Applications: Calendered film and sheeting; PVC compounds; plastisol viscosity depressant

Admex 752. Sherex
Chemical Description: High molecular weight ester
Category: Plasticizer
Applications: Upholstery sheeting, coated fabric, headliner stock, side panel trim, armrest fabric, crash-pad molding
Properties: Excellent fog resistance, low temperature flexibility, low viscosity

Admex 760. Sherex
Chemical Description: High molecular weight polyester
Category: Plasticizer
Applications: Pressure-sensitive tape, electrical sleeving, organosols, vinyl solution coatings, drycleanable film

Admex 761. Sherex
Chemical Description: Medium molecular weight polyester
Category: Plasticizer
Applications: 105 C wire coating, electrical tape and sleeving, tape and mattress covers

Admex 770. Sherex
Chemical Description: Medium molecular weight polyester
Category: Plasticizer
Applications: Refrigerator gaskets, baby pants, nursery goods, highclarity film and sheeting

Admex 775. Sherex
Chemical Description: Medium molecular weight polyester

Category: Plasticizer
Applications: Refrigerator and freezer gaskets, coated fabrics, saturated paper coatings, infant wear, tapes

Admex 890. Sherex
Chemical Description: Medium molecular weight polyester
Category: Plasticizer
Applications: Refrigerator and freezer gaskets, coated fabric, appliance wire coating, automotive gasketing, tape applications

Admex ELO. Sherex
Chemical Description: Epoxidized linseed oil
Category: Stabilizer, plasticizer
Applications: Food wrap, beverage transfer tubing, PVC, film

Adogen 415. Sherex
Chemical Description: Trimethyl soya quaternary ammonium chloride
Category: Antistat, emulsifier, dispersant, cationic
Form: Liquid
Composition: 50% active ingred.

Adogen 432. Sherex
Chemical Description: Dialkyl dimethyl ammonium chloride
Category: Fabric softener, cationic
Applications: Retail, commercial laundry and textile mill formulations and paper
Form: Liquid
Composition: 75% solids content

Adogen 432CG. Sherex
Chemical Description: Dialkyl dimethyl ammonium chloride
Category: Creme rinse concentrate, cationic
Applications: Hair creme rinse
Form: Liquid
Composition: 75% solids content

Adogen 442. Sherex
Chemical Description: Dimethyl dihydrogenated tallow ammonium chloride
Category: Fabric softener, cationic
Applications: Home and commercial launderies, textile processing
Form: Paste
Composition: 75% solids content

Adogen 462. Sherex
Chemical Description: Dimethyl dicoco ammonium chloride
Category: Antistat, emulsifier, flocculating agent, cationic
Applications: Dispersant for oilfield chemicals
Form: Liquid
Composition: 75% quat. content

Adogen 464. Sherex
Chemical Description: Methyl tri (C_8-C_{10}) ammonium chloride
Category: Flotation agent, cationic
Applications: Solvent extraction of metal
Form: Liquid
Composition: 90% quat. content

Adogen 471. Sherex
Chemical Description: Trimethyl tallow ammonium chloride
Category: Emulsifier, dispersant, antistat, cationic
Form: Liquid
Composition: 50% quat. content

Adogen 477. Sherex
Chemical Description: N-tallow pentamethyl propane diammonium dichloride
Category: Emulsifier, dispersant, cationic
Form: Liquid
Composition: 50% quat. content

Adsee 775. Witco Chem. Corp., Organics Div.
Chemical Description: Blend of polyoxyethylene ethers and special resins
Category: Agricultural surfactant, nonionic
Applications: Spreader sticker and penetrant; adjuvant for spray-tank mixtures
Form: Liquid
Composition: 100% conc.

Adsee 799. Witco Chem. Corp., Organics Div.
Chemical Description: Alkylpolyoxyethylene ether
Category: Agricultural surfactant, nonionic
Applications: Soil penetrant
Form: Liquid
Composition: 100% conc.

Adsee 801. Witco Chem. Corp., Organics Div.
Chemical Description: Surfactant blend
Category and Applications: See Adsee 799

Adsee 2096. Witco Chem. Corp., Organics Div.
Chemical Description: Surfactant blend
Category: Agricultural surfactant, nonionic
Applications: Spreader penetrant for wetting of pesticide formulations; adjuvant for spray-tank mixtures

Adsee 2141. Witco Chem. Corp., Organics Div.
Chemical Description: Surfactant blend
Category: Agricultural surfactant, anionic/nonionic
Applications: Premix surfactant for high- and low-content monosodium methylarsonate (MSMA) formulations
Form: Liquid
Composition: 100% conc.

Adsee AC62-38. Witco Chem. Corp., Organics Div.
Chemical Description: Surfactant blend
Category: Agricultural surfactant, anionic/nonionic
Applications: Premix surfactant for high- and low-content monosodium methylarsonate (MSMA) formulations

AE-1214/3. Procter & Gamble Co.
Chemical Description: Ethoxylated linear alcohols, nonionic
Category: Detergent, emulsifier
Form: Liquid
Composition: 100% active ingred.

AE-1214/6. Procter & Gamble Co.
Chemical Description: Ethoxylated linear alcohols, nonionic
Category: Lubricant, wetting agent
Applications: Textiles

Form: Solid
Composition: 100% active ingred.

AEPD. IMC Chem.–NP Div.
Chemical Description: 2-amino-2-ethyl-1,3-propanediol
Category: Emulsifier, catalyst, anionic
Applications: As raw materials in polymer applications, e.g. oxazoline structures (as corrosion inhibitors, antifoam compounds, synthetic waxes); in coatings as pigment dispersants; in emulsions for cosmetics, cleaners, industrial insecticides, asphalt emulsions; in boiler-water treatment; as catalysts; in formaldehyde scavenging
Form: Color: 2 (Gardner)
Properties: m.w.: 119.17; Density: 9.15 lb/gal; Sp. gr.: 1.101 @ 40/40 C; M.P.: 38 C; B.P.: 152 C
Toxicity: Causes severe eye burns
Containers: Bulk or 450 lb drums
References: TDS No. 10

Aerosol 18. American Cyanamid
Chemical Description: Disodium N-octadecyl sulfosuccinamate
Category: Emulsifier, dispersant, foaming agent; anionic surfactant
Applications: Cosmetics, foaming latexes; emulsion polymerization, lubrication; scouring powders and soaps; industrial cleaners; agricultural foams
Form: Creamy paste; Color: 10 (Gardner)
Properties: Solubility: Good in water; Density: ≈ 8.9 lb/gal; Sp. gr.: ≈ 1.07; Visc.: Variable, semi-liquid; F.P.: Solid paste < 21 C; Stability: Good in acid and alkaline media; Surface Tension: 41 dynes/cm; Biodegradable
Composition: 35% active ingred.
References: Circular PRC-32

Aerosol 19. American Cyanamid
Chemical Description: Disodium N-alkylsulfosuccinamate
Category: Emulsifier, foaming and wetting agent
Applications: Foaming agent in foaming latexes, wallboard, cement, plastics, carpet & upholstery shampoos, hand soaps; Emulsifier in emulsion polymerization
Form: Clear liquid; Color: 10 (Gardner)
Properties: Solubility: Good in water; Density: ≈ 8.9 lb/gal; Sp. gr.: ≈ 1.07; Stability: Good in acid and alkaline media; Surface Tension: 41 dynes/cm; Biodegradable
Composition: 35% active ingred.
References: Circular PRC-145

Aerosol 22. American Cyanamid
Chemical Description: Tetrasodium N-(1,2-dicarboxyethyl)-N-octadecyl sulfosuccinamate
Category: Emulsifier, dispersant, anionic surfactant
Applications: Agricultural emulsions; emulsion polymerization; industrial, household and cosmetic cleaners; alkaline metal cleaners; cosmetics; anti-gelling agent; latex stabilizer; textile nonwoven applications
Form: Clear to cloudy liquid; Color: 8 (Gardner)
Properties: Solubility: Excellent in water; Density: 9.4 lb/gal; Sp. gr.: 1.12; Visc.: at 25 C, 53 cps; F.P.: Separates below 10 C; M.P.: > 200 C; Surface Tension: 41 dynes/

cm; Biodegradable
Composition: 35% active ingred.
References: Circular PRC-90

Aerosol 30. American Cyanamid
Chemical Description: Cocoamidopropyl betaine
Category: Amphoteric surfactant; foam booster; viscosity builder
Applications: hair and bath products, skin cleansers, disinfectants
Form: Clear, yellow liquid; Color: 4 max (Gardner)
Properties: Solubility: Excellent in water; Sp. gr.: 1.04; F.P.: 0 C; Surface Tension: 35 dynes/cm; Partially biodegradable
Composition: 35% active ingred.
Toxicity: Very mild on skin, eyes
References: Circular PRT-160
Patent: 3,225,074 (U.S.)

Aerosol 200. American Cyanamid
Chemical Description: Disodium (ethoxylated alkylamide) sulfosuccinate
Category: Emulsifier, foaming agent, dispersant
Applications: Cosmetics: shampoos, shaving creams, liquid hand soaps, bubble baths; carpet and upholstery cleaners; vinyl acetate and acrylic emulsions; foaming applications
Form: Clear, yellow liquid; Color: 7 max (Gardner); Odor: Mild
Properties: Solubility: Excellent in water; Density: 8.93 lb/gal; Sp. gr.: 1.07; Visc.: 200 cps max (Brookfield, #1 spindle, 60 rpm); Cloud Pt.: $<$0 C; Surface Tension: 35 dynes/cm; Biodegradable
Composition: 30% active ingred.
Toxicity: Very mild
References: Circular PRT-153

Aerosol 413. American Cyanamid
Chemical Description: Disodium ethanol alkyl (C_{18}) amide sulfosuccinate
Category: Emulsifier, foaming agent, dispersant, wetting agent
Applications: Vinyl acetate and acrylate emulsions; shampoos; foaming applications
Form: Clear, yellow liquid; Color: 5–7 (Gardner)
Properties: Solubility: Excellent in water; Density:8.95 lb/gal; Visc.: at 25 C, 70–90 cps; Stability: Good; Surface Tension: 38 dynes/cm; Biodegradable
Composition: 35% active ingred.
References: Circular PRT-154

Aerosol 501. American Cyanamid
Chemical Description: Disodium alkyl sulfosuccinate
Category: Emulsifier, wetting agent, dispersant, foaming agent, anionic-nonionic surfactant
Applications: Acrylic and vinyl acetate emulsions; self-crosslinking latexes; textile wetting and foaming applications
Form: Clear liquid; Color: 60 max (APHA)
Properties: Solubility: Excellent in water; Density: 9.66 lb/gal; Sp. gr.: 1.16; Visc.: 25 C, $\approx$ 260 cps; F.P.: -3 C; Stability: Good; Surface Tension: 28 dynes/cm
Composition: 50% active ingred.
References: Circular PRC-37
Patent: 3,947,400 (U.S.)

Aerosol A-102. American Cyanamid
Chemical Description: Disodium ethoxylated alochol half ester of sulfosuccinic acid
Category: Emulsifier, solubilizer, foaming agent, dispersant, anionic-nonionic surfactant
Applications: Emulsion polymerization of polyvinyl acetate/acrylics; textile industry–emulsions, wetting agent, pad-bath additive; solubilizing agent
Form: Clear liquid; Color: Colorless-to-light-yellow
Properties: Solubility: Excellent in water; Density: 9.01 lb/gal; Sp. gr.: 1.08; Visc.: @ 25 C, ≈ 40 cps; F.P.: –4 C; Stability: Excellent in acid, fair in base; Surface Tension: 29 dynes/cm
Composition: 30% active ingred.
References: Circular PRC-66

Aerosol A-103. American Cyanamid
Chemical Description: Disodium salt of half ester of sulfosuccinate acid derived from an ethoxylated nonyl phenol
Category: Emulsifier, wetting agent, anionic-nonionic surfactant
Applications: Polyvinyl acetate acrylic emulsions; Textiles–emulsions, pad-bath additive, wetting agent; solubilizer, surface tension depressant
Form: Liquid; Color: Colorless to light yellow
Properties: Solubility: Excellent in water; Density: ≈ 9.1 lb/gal; Sp. gr.: 1.09; Visc.: 25 C, 170–190 cps; F.P.: –9 C; Surface Tension: 34 dynes/cm
Composition: 35% active ingred.
References: Circular PRC-88

Aerosol A-196. American Cyanamid
Chemical Description: Sodium dicyclohexyl sulfosuccinate
Category: Emulsifier, dispersant, anionic surfactant
Applications: Sole emulsifier for modified styrene butadiene; post additive to stabilize latex and promote adhesion; potential low foaming surfactant
Form: White pellets, clear to cloudy liquid
Properties: Stability: Good mechanical stability; Surface Tension: 39 dynes/cm, high critical micelle concentration; Biodegradable
Composition: 85% active pellets, 15% water; 40% active solution in water (110 F)
References: Circular PRC-49

Aerosol A-268. American Cyanamid
Chemical Description: Disodium isodecyl sulfosuccinate
Category: Emulsifier, anionic surfactant
Applications: Sole emulsifier for PVC latexes–vinyl, vinylidene chloride, acrylics; surface tension depressant; solubilizer
Form: Clear liquid
Properties: Solubility: Excellent in water; Density: 9.96 lb/gal; Sp. gr.: 1.19; Visc.: 25 C, ≈ 150 cps; F.P.: –5 C; Stability: Good; Surface Tension: 28 dynes/cm
Composition: 50% active ingred.
References: Circular PRC-50

Aerosol AY-65. American Cyanamid
Chemical Description: Sodium diamyl sulfosuccinate
Category: Wetting agent, dispersant, anionic surfactant
Applications: Wetting agent esp. effective in conc. salt sol'ns.; agriculture; emulsion polymerization; electroplating; ore leaching; cleaning of porcelain, tile, brick, cement
Form: Clear liquid
Properties: Solubility: Solvent–water/ethanol; Density: 9.0 lb/gal; Sp. gr.: 1.081; F.P.: < 18 C; Surface Tension: 29 dynes/cm; Biodegradable
Composition: 65% active ingred.
References: Circular PRC-44

Aerosol AY-100. American Cyanamid
Chemical Description: See Aerosol AY–65
Category: See Aerosol AY–65
Applications: See Aerosol AY–65
Form: Hard waxy solid
Properties: Solubility: Sol. in water and organic solvents; Density: 10.0 lb/gal; Sp. gr.: 1.2; M.P.: 105–108 C; Stability: Good; Surface Tension: 29 dynes/cm; Biodegradable; good salt tolerance
Composition: 100% active ingred.
References: Circular PRC-44

Aerosol C-61. American Cyanamid
Chemical Description: Ethoxylated alkyl guanidine-amine complex
Category: Wetting agent, dispersant, settling agent, cationic surfactant
Applications: Textile softener, pigment dispersant, de-emulsifying agent, settling agent, antistatic agent for plastics, dye penetrant, foaming agent, acid cleaner formulations
Form: Creamy paste or semiliquid; Color: Light-tan
Properties: Density: ≈ 8 lb/gal; Sp. gr.: ≈ 1.0; F.P.: loses fluidity < 16 C; Stability: Good in acid and alkaline media; Surface Tension: 34 dynes/cm, Partially biodegradable
Composition: 70% active ingred.
Toxicity: Irritating to eyes, skin
References: Circular PRC-67

Aerosol GPG. American Cyanamid
Chemical Description: See Aerosol OT-70-PG
Category: General purpose grade OT surfactant; see Aerosol OT-70-PG
Applications: See Aerosol OT-70-PG
Form: Clear slightly viscous liquid
Properties: Solubility: Limited sol. in water, good organic solvents; Visc.: ≈ 200 cps; F.P.: –40 C; Surface Tension: 26 dynes/cm; Biodegradable
Composition: 70% active in water and alcohol
References: Circular PRT-73

Aerosol IB-45. American Cyanamid
Chemical Description: Sodium diisobutyl sulfosuccinate
Category: Emulsifier, wetting and dispersing agent, anionic surfactant
Applications: Emulsion polymerization of styrene butadiene; wet-

ting in conc. salt sol'ns.; dispersing dyes and pigments
Form: Clear liquid
Properties: Solubility: Sol. in water; F.P.: 20–21 C; Stability: Excellent in conc. salt sol'ns.; Surface Tension: 49 dynes/cm; Biodegradable; extremely hydrophilic
Composition: 45% active ingred.
Toxicity: May be irritating to eyes
References: Circular PRC-74

Aerosol MA-80. American Cyanamid
Chemical Description: Sodium dihexyl sulfosuccinate
Category: Wetting agent, emulsifier, dispersant, solubilizing agent, anionic surfactant
Applications: Emulsion polymerization; textile wetting applications; battery separators; electroplating; ore leaching; enhances germicidal activity
Form: Clear, slightly viscous liquid; Color: 50 max (APHA)
Properties: Solubility: Good in water, alcohol, and organic solvents; Density: ≈ 9.4 lb/gal; Sp. gr.: ≈ 1.13; F.P.: –28 C; M.P.: 199–292 C; Surface Tension: 28 dynes/cm; Slowly biodegradable
Composition: 80% active ingred.
Toxicity: Irritating to eyes, skin
References: Circular PRC-38

Aerosol OS. American Cyanamid
Chemical Description: Sodium isopropyl naphthalene sulfonate
Category: Emulsifier, wetting and dispersing agent, anionic surfactant
Applications: Pigment dispersant, wettable powders, agricultural herbicides, alkaline metal cleaners, industrial and consumer cleaners, etching solutions, emulsion polymerization
Form: White to light cream, free-flowing powder
Properties: Solubility: 50% in water; Density: 11.93 lb/gal; Sp. gr.: 1.43; M.P.: 205 C; Stability: excellent in concentrated acids and alkalis; Surface Tension: 37 dynes/cm; Slowly biodegradable
Composition: 75% active powder, 25% sodium sulfate
Toxicity: Irritating to eyes, skin; avoid dust inhalation
References: Circular PRC-122

Aerosol OT-70 PG. American Cyanamid
Chemical Description: Sodium dioctyl sulfosuccinate
Category: Wetting agent, surface tension depressant, emulsifier, anionic surfactant
Applications: Wetting agent in textiles, rubber, paper, petroleum, metal, paint, plastics and agriculture industries; antifog compositions; battery separators; coal dust control; cosmetics; dry cleaning detergents; emulsifying waxes; emulsion & suspension polymerization; industrial cleaners; paint production; pharmaceutical–topical; wettable powders
Form: Liquid; Color: Clear to slightly yellow
Properties: Solubility: limited water sol., good organic sol.; Solvent: Water, alcohol; Sp. gr.: 1.09; Visc.: 200–400 cps; Surface Tension: 26 dynes/cm; Biodegradable; Flash

Pt.: High
Composition: 70% active in water and propylene glycol
References: Circular PRT-73B

Aerosol OT-75. American Cyanamid
Chemical Description: See Aerosol OT-70-PG
Category: See Aerosol OT-70-PG
Applications: See Aerosol OT-70-PG
Form: Clear slightly viscous liquid
Properties: Solubility: Limited water sol., good organic solvents; Solvent–water, alcohol; Sp. gr.: 1.09; Visc.: ≈ 200 cps; F.P.: –40 C; Surface Tension: 26 dynes/cm; Biodegradable
Composition: 75% active ingred.
References: Circular PRT-73B

Aerosol OT-100. American Cyanamid
Chemical Description: See Aerosol OT-70-PG
Category: See Aerosol OT-70-PG
Applications: See Aerosol OT-70-PG
Form: Waxy solid
Properties: Solubility: Limited sol. in water, excellent for organic solvents; Sp. gr.: 1.1; M.P.: 153–157 C; Biodegradable
Composition: 100% active ingred.
References: Circular PRT-73B

Aerosol OT-B. American Cyanamid
Chemical Description: See Aerosol OT-70-PG
Category: See Aerosol OT-70-PG
Applications: See Aerosol OT-70-PG
Form: Granular solid; Color: White
Properties: Solubility: Limited sol. in water, insol. organic solvents; Solvent–Sodium benzoate; Sp. gr.: 1.1; M.P.: 300 C; Surface Tension: 26 dynes/cm; Biodegradable
Composition: 85% active in 15% sodium benzoate
References: Circular PRT 73-B

Aerosol OT-S. American Cyanamid
Chemical Description: See Aerosol OT-70–PG
Category: See Aerosol OT-70-PG
Applications: See Aerosol OT-70-PG
Form: Transparent liquid; Color: Light amber
Properties: Solubility: Limited sol. in water, good for organic solvents; Sp. gr.: 1.0; Visc.: 200–300 cps; Surface Tension: 26 dynes/cm; Biodegradable
Composition: 70% active in odorless mineral spirits
References: Circular PRT 73-B

Aerosol TR-70. American Cyanamid
Chemical Description: Sodium bistridecyl sulfosuccinate
Category: Emulsifier, Dispersant, anionic surfactant
Applications: Emulsion and suspension polymerization of vinyl chloride; emulsion polymerization of vinyl acetate; dispersing agent for resins, pigments, polymers and dyes in organic systems; pigment dispersing–printing inks; rust preventative
Form: Clear solution
Properties: Solubility: High sol. in organic media, limited water sol.; *Density:* ≈ 8.3; Sp. gr.: 0.995; Visc.: ≈ 110 cps; F.P. –40 C; Sta-

bility: Good; Surface Tension: 26 dynes/cm, low critical micelle concentration; Biodegradable
Composition: 70% active in water and alcohol
References: Circular PRC-66

AF 60. General Electric
Chemical Description: Dimethyl polysiloxane antifoam emulsion in water
Category: Defoaming agent
Applications: Industrial defoamer in aqueous systems; adhesive manufacture; ink manufacture; latex processing; soap manufacture; starch processing; paint (water based) manufacture
Form: Fluid; Color: White; Odor: Slight sorbic acid odor
Properties: Solubility: Good in water; Density: 8.4 lb/gal; Sp. gr.: 1.01; Visc.: 300 max. cps; Stability: Good
Composition: 30% active ingred.
Toxicity: May cause eye irritation
Containers: 5 gal cubitainers; 55 gal drums

AF 66. General Electric
Chemical Description: Filled polydimethy siloxane
Category: Antifoam agent
Applications: Antifoam agent in nonaqueous industrial applications; adhesive, ink and paint manufacture; resin polymerization; petroleum processing
Form: Fluid; Color: Grayish white Odor: None
Properties: Solubility: Sol. in aliphatic, aromatic, chlorinated solvents, negligible sol. in water; Density: 8.4 lb/gal; Sp. gr.: 1.01; Visc.: 500 max (Cstks.); Flash point: 600 F min. (open cup); Stability: Indefinite
Composition: 100% active ingred.
Toxicity: May cause eye irritation
Containers: 5 gal containers, 55 gal fiberpak drums

AF 72. General Electric
Chemical Description: Polydimethylsiloxane
Category: Antifoam agent, nonionic surfactant
Applications: Antifoam in food processing; adhesive, ink (water-based), paint (water-based), and soap manufacture; latex processing; manufacture or processing of fruit, instant coffee, instant foods, jams, jelly, juice, pickles, rice, soft drinks, sauces; sugar refining
Form: Fluid; Color: White; Odor: Slight sorbic acid odor
Properties: Solubility: Good in water; Density: 8.4 lb/gal; Sp. gr.: 1.01; Visc.: 1000 max. cps; Stability: Good
Composition: 30% active ingred.
Toxicity: May cause eye irritation
Containers: 5 gal cubitainers, 55 gal drums

AF 75. General Electric
Chemical Description: Polydimethylsiloxane
Category: Antifoam agent, nonionic surfactant
Applications: Food grade antifoam; chemical processing of adhesives, ink, latex, soap, starch; textiles and paper industry—paper

coating and manufacture, textile dyeing, finishing, sizing and wood pulping; processing or manufacture of fruit, instant coffee, jams, juice, pickles, rice, sauce, soft drinks, syrup, potato, sugar refining; leather finishing; metalworking
Form: Fluid; Color: White; Odor: Slight sorbic acid odor
Properties: Solubility: Good in water; Density: 8.4 lb/gal; Sp. gr.: 1.02; Visc.: 2500 cps; Stability: Good
Composition: 10% active ingred.
Toxicity: May cause eye irritation
Containers: 55 gal fiberpak drums

Agrilan A. Lankro Chem. Ltd.
Chemical Description: Blend oil soluble sulfonates and polyoxyethylene ethers
Category: Emulsifier, anionic/nonionic
Applications: Emulsifier for agricultural products, solvents, oils, etc.
Form: Viscous liquid; Color: Brown; Odor: Faint
Properties: Solubility: Turbid water-sol. @ 20 C; Sp. gr.: 1.062 @ 20 C; Visc.: 2400 cs @ 20 C; Flash Pt.: > 300 F (COC); Pour Pt.: < 0 C; pH: 4.5–6.5 (1% aq.)
Composition: 100% active ingred.
Containers: Store in SS, glass, resin or polythene-lined equipment
Warnings and Caution: Avoid prolonged contact with concentrated product; keep from strong oxidizing and reducing agents (potentially explosive)

Agrilan AC. Lankro Chem. Ltd.
Chemical Description: Blend oil soluble sulfonates and polyoxyethylene ethers
Category: Emulsifier, anionic/nonionic
Applications: Emulsifier for agricultural products (pesticides and herbicide systems), oils and aromatic solvents in degreaser formulations
Form: Viscous liquid; Color: Amber; Odor: Alcoholic
Properties: Solubility: Water-sol. @ 20 C; Sp. gr.: 1.052 @ 20 C; Visc.: 500 cs @ 20 C; Flash Pt.: > 120 F (Abel CC); Pour Pt.: 17 C; pH: 4.5–6.5 (1% aq.)
Composition: 90% active ingred.
Toxicity: Avoid prolonged contact with concentrated form
Containers: Store in SS, glass, resin, or polythene-lined equipment
Warnings and Caution: Contains flammable solvent; do not mix with strong oxidizing or reducing agents (potentially explosive)

Agrilan BA. Lankro Chem. Ltd.
Chemical Description: Blend oil soluble sulfonates and polyoxyethylene ethers
Category: Emulsifier, anionic/nonionic
Applications: Emulsifier for pesticides, herbicides, oils and aromatic solvents in degreaser formulations
Form: Viscous liquid; Color: Brown; Odor: Alcoholic
Properties: Solubility: Turbid wa-

ter-sol. @ 20 C; Sp. gr.: 1.034 @ 20 C; Visc.: 3100 cs @ 20 C; Flash Pt.: > 90 F (Abel CC); Pour Pt.: < 0 C; pH: 4.5–6.5 (1% aq.)
Composition: 80% active ingred.
Toxicity: Avoid prolonged contact with concentrated form
Containers: Store in SS, glass, resin, or polythene-lined equipment
Warnings and Caution: Contains flammable solvent; do not mix with strong oxidizing or reducing agents (potentially explosive)

Agrilan BM. Lankro Chem. Ltd.
Chemical Description: Blend of oil soluble sulfonates and polyoxyethylene ethers
Category: Emulsifier, anionic/nonionic
Applications: Emulsifier for pesticides, herbicides, oils and aromatic solvents in degreaser formulations
Form: Liquid; Color: Amber; Odor: Characteristic
Properties: Solubility: Turbid water-sol. @ 20 C; Sp. gr.: 1.022 @ 20 C; Visc.: 196 cs @ 20 C; Flash Pt.: 78 F (Abel CC); Pour Pt.: < 0 C; pH: 5.0–6.5 (1% aq.)
Composition: 70% active ingred.
Toxicity: Avoid prolonged contact with concentrated form
Containers: Store in SS, glass, resin, or polythene-lined equipment
Warnings and Caution: Contains flammable solvent; do not mix with strong oxidizing or reducing agents (potentially explosive)

Agrimul 26-B. Diamond Shamrock Process Chemicals Div.
Chemical Description: Nonionic surfactant
Category: Emulsifier, nonionic
Applications: Emulsifier for citrus and other agricultural spray oils
Form: Liquid; Color: Amber
Properties: Density: 8.3 lb/gal; Visc.: 95 cps; Pour Pt.: 15 C
Composition: 100% conc.

Agrimul 70-A. Diamond Shamrock Process Chemicals Div.
Chemical Description: Alkyl aryl polyether alcohol
Category: Emulsifier, nonionic
Applications: Agricultural emulsifier for chlorinated hydrocarbons
Form: Liquid; Color: Amber
Properties: Density: 8.8 lb/gal; Visc.: 8850 cps; Pour Pt.: 0 C
Composition: 100% conc.

Agrimul A-300. Diamond Shamrock Process Chemicals Div.
Chemical Description: Aromatic sulfonate oxide condensate blend
Category: Emulsifier, anionic/nonionic
Applications: Agricultural emulsifier for chlorinated and phosphated insecticide concentrates
Form: Liquid; Color: Amber
Properties: Density: 8.9 lb/gal; Visc.: 1050 cps; Pour Pt.: 5 C
Composition: 100% conc.

Ahco A-117. ICI United States
Chemical Description: Alkylaryl sulfonate

Category: Emulsifier, anionic surfactant
Applications: Emulsifier: dispersing waxes, pigments, resins; Detergent in textile, leather, laundering industries; Wetting agent for dyeing and finishing
Form: Liquid; Color: Amber; Odor: Faint
Properties: Solubility: Isopropanol @ 25 C; Sp. gr.: ≈ 1.0; Visc.: ≈ 7000 cps
Composition: 100% active ingred.
Reaction: Alkaline
Toxicity: Irritating to eyes, skin
References: TB 628

Ahco AB-100. ICI United States
Chemical Description: Formulated surfactant
Category: Emulsifier, anionic-nonionic blend
Applications: Dye carrier, solvent emulsifier
Form: Liquid; Color: Amber; Odor: Solvent
Properties: Solubility: Isopropanol @ 25 C; Sp. gr.: ≈ 1.04; Visc.: 1900 cs; Flash Pt.: ≈ 64 F; Pour Pt.: ≈ –15 F
Composition: Proprietary % active
Toxicity: Poisonous; strong irritant to eyes and skin; avoid inhalation
Warnings and Cautions: Poisonous if ingested; flammable
References: TB 633

Ahco AB-118. ICI United States
Chemical Description: Formulated surfactant; anionic-nonionic blend
Category: Emulsifier
Applications: Dye carrier; solvent emulsifier
Form: Liquid; Color: Amber; Odor: Solvent
Properties: Solubility: Isopropanol @ 25 C; Sp. gr.: ≈ 1.0; Visc.: 300 cs; Flash Pt.: ≈ 100 F; Pour Pt.: ≈ 20 F
Composition: Proprietary % active
Reaction: Alkaline
Toxicity: Poisonous; strong irritant to eyes and skin; avoid inhalation
Warnings and Cautions: Poisonous if ingested; flammable
References: TB 643

Ahco AB-120. ICI United States
Chemical Description: Formulated surfactant
Category: Emulsifier; anionic
Applications: Dye carrier; solvent emulsifier
Form: Liquid; Color: Amber; Odor: Solvent
Properties: Water @ 25 C; Sp. gr.: ≈ 1.0; Visc.: 600 cps; Flash Pt.: ≈ 80 F; Pour Pt.: ≈ 40 F; Stability: Good
Composition: Proprietary % active
Reaction: Alkaline
Toxicity: Poisonous; strong irritant to skin and eyes; avoid inhalation
Warnings and Cautions: Poisonous if ingested; flammable
References: TB 645

Ahco AB-135. ICI United States
Chemical Description: Formulated surfactant
Category: Emulsifier, anionic-nonionic blend
Applications: Dye carrier; solvent emulsifier
Form: Liquid; Color: Amber; Odor: Solvent

Properties: Sp. gr.: ≈ 1.0; Visc.: ≈ 100 cps; Flash Pt.: ≈ 90 F; Pour Pt.: ≈ 25 F; Stability: Good
Composition: Proprietary % active
Reaction: Alkaline
Toxicity: Poisionous if ingested; strong irritant to skin and eyes
Warnings and Cautions: Poisonous if ingested; flammable
References: TB 647

Ahco AB-146. ICI United States
Chemical Description: Formulated surfactant
Category: Emulsifier; anionic
Applications: Dye carrier; solvent emulsifiers
Form: Liquid; Color: Amber; Odor: Solvent
Properties: Solubility: Isopropanol @ 25 C; Sp. gr.: ≈ 1.0; Visc.: 1000 cps; Stability: Good
Composition: Proprietary % active
Reaction: Alkaline
Toxicity: Poisonous; strong irritant to skin and eyes; avoid inhalation
Warnings and Cautions: Poisonous if ingested; flammable
References: TB 648

Ahco AB-160. ICI United States
Chemical Description: Formulated surfactant
Category: Emulsifier; anionic
Applications: Dye carrier; solvent emulsifier
Form: Liquid; Color: Amber; Odor: Solvent
Properties: Solubility: Isopropanol @ 25 C; Sp. gr.: ≈ 1; Visc.: 750 cs; Stability: Good
Composition: Proprietary % active
Reaction: Alkaline
Toxicity: Poisonous; strong irritant to skin and eyes; avoid inhalation
Warnings and Caution: Poisonous if ingested; flammable
References: TB 644

Ahco AB-228. ICI United States
Chemical Description: Formulated surfactant
Category: Emulsifier; anionic
Applications: Dye carrier; solvent emulsifier
Form: Liquid; Color: Amber; Odor: Faint
Properties: Solubility: Water @ 25 C; Sp. gr.: ≈ 1.0; Visc.: 2600 cs; Stability: Good
Composition: 100% active ingred.
Reaction: Neutral
Toxicity: Strong irritant to eyes; irritant to skin
Warnings and Cautions: Avoid eye contact
References: TB 646

Ahco AJ-110. ICI United States
Chemical Description: Sulfated castor oil
Category: Wetting agent, lubricant, penetrant, dispersant, emulsifier, anionic surfactant
Applications: Dye bath agent; detergent–textile, leather industries; wetting agent in desizing/boiling off cottons, bleaching, re-wetting dried skins, tanning, finishing, sizing; fat liquoring
Form: Liquid; Color: Amber; Odor: Castor oil
Properties: Solubility: Water @ 25 C; Sp. gr.: ≈ 1.05
Composition: 50% active ingred., 50% water

Reaction: Alkaline
Toxicity: Eye irritant
Warnings and Caution: Avoid contact with eyes
References: TB 582

Akwilox 133. Swift Chem.
Chemical Description: Brominated oil
Category: Food additive
Applications: Viscosity adjustment for soft drinks
Form: Liquid; Color: Light amber; Odor: Bland; Taste: Bland
Properties: Sp. gr.: 1.33; Stability: 200 hr min.
Composition: 100% active ingred.
Containers: 600 lb drum; 65 lb pail

Alacsan 7LUF. Alcolac
Chemical Description: Modified alkyl dimethyl benzyl ammonium chloride
Category: Detergent; cationic
Applications: Toiletries, household cleaners, textile and plastics industries, diaper rash formulations, medicated soaps, protective hand creams, vaginal preparations, deodorants, after-shave lotions
Form: Viscous liquid; Color: 5 max (Gardner)
Properties: Strong attraction to electronegative surfaces; good solubilizing properties; high foam capacity
Composition: 50% active ingred.

Albacer. Glycol Chem.
Applications: Synthetic waxes and plastic lubricants
Form: Very hard solid; Color: Cream
Properties: Sp. gr.: 1.05 @ 20 C; M.P.: 75–85 C; Flash Pt.: ≈ 196 C

Alcojet. Alconox, Inc.
Chemical Description: Blend of sodium metasilicate, sodium carbonate, polyoxyethylene ester of mixed fatty and resin acids
Category: Detergent, nonionic
Applications: In mechanical washers; stain removal for laboratory and hospital glassware, metal instruments, etc.
Form: Powder
Properties: Solubility: Water; Biodegradable
Containers: 5 lb tubette; 25, 50 lb carton, 100, 300 lb drums

Alcolec Series. American Lecithin
Chemical Description: Phosphatides
Category: Wetting agent, emulsifier, release agent, nonionic
Applications: Foods, cosmetics, paints, pharmaceuticals
Form: Liquid, powder; Color: Amber
Properties: Solubility: Most fat solvents except acetone; Sp. gr.: 1.03; Visc.: 10,000 cps; Stability: Good
Composition: 60–100% active ingredient; other components: fatty oil
Reaction: pH 6.6
Containers: 460 lb drums

Alcolec 439-C. American Lecithin
Chemical Description: Water-dispersible lecithin
Category: Wetting agent, emulsifier, nonionic

Applications: Water-base paint emulsifier
Form: Liquid; Color: Amber; Odor: Bland
Properties: Solubility: Most fat solvents except acetone; Sp. gr.: 1.03; Visc.: 7000 cps; Stability: Good
Composition: 100% active ingred.
Containers: 460 lb drums

Alcolec Granules. American Lecithin
Chemical Description: Soy phosphatides, 95%
Category: Emulsifier, stabilizer, nonionic
Applications: Diet supplement; cosmetics; pharmaceuticals
Form: Beads; Color: Amber; Odor: Bland
Properties: Sp. gr.: 0.5
Composition: 95+% active ingred.
Containers: 200 lb drums

Alcolec Powder. American Lecithin
Chemical Description: Soy phosphatides, 95%
Category: Wetting agent, emulsifier, stabilizer, nonionic
Applications: Diet supplements, foods, cosmetics, pharmaceuticals
Form: Powder
Properties: Sp. gr.: 0.6
Composition: 100% active ingred.
Containers: 200 lb drums

Alcolec Z-3. American Lecithin
Chemical Description: Hydroxylated lecithin
Category: Wetting agent, emulsifier, nonionic
Applications: Foods
Form: Liquid
Properties: Solubility: Most fat solvents except acetone; Visc.: 8000 cps

Alconate 80. Witco Chem., Sonneborne Div.
Category: Emulsifier, wetting agent
Form: Viscous liquid; Color: Dark brown
Properties: Sp. gr.: 1.1 @ 60 F
Composition: 77% active ingred.; other components: mineral oil (15%), water (8.0%)
Reaction: Acid
Toxicity: Nontoxic with respect to Federal Hazardous Substance Labelling Act; may cause eye burn
Containers: Drum, T/W, T/C
Warnings and Caution: May cause eye burn

Alconox. Alconox, Inc.
Chemical Description: Blend of alkyl aryl sulfonates, lauryl alcohol sulfates, phosphates, carbonates
Category: Detergent with wetting, sequestering and synergistic agents, anionic
Applications: Detergent for removal of soil, grit, blood, fats from glass, metal, plastic, rubber; ultrasonic cleaners
Form: Powder; Color: Colorless; Odor: None
Properties: Solubility: Complete in water; Biodegradable; Anticorrosive properties
Containers: Drums, Cartons, boxes
References: Technical Bulletin #1

Alcopol AH New. Allied Colloids Ltd.
Chemical Description: Mono ethanolamine salt of fatty acid ester
Category: Dispersant, anionic
Applications: Dispersant esp. useful in nonaqueous systems; dry cleaning aid; good water carrying properties
Form: Liquid
Composition: 99% conc.

Alcopol OD. Allied Colloids Ltd.
Chemical Description: Sodium ditredecyl sulfosuccinate
Category: Dispersant, wetting agent, emulsifier, anionic
Form: Liquid
Composition: 60% conc.

Alcopol OB. Allied Colloids Ltd.
Chemical Description: Sodium deisobutyl sulfosuccinate
Category and Applications: See Alcopol FL
Form: Liquid
Composition: 45% conc.

Alcopol OS. Allied Colloids Ltd.
Chemical Description: Sodium dehexyl sulfosuccinate
Category and Applications: See Alcopol FL
Form: Liquid
Composition: 60% conc.

Alcotab. Alconox, Inc.
Chemical Description: Blend of alkyl aryl sulfonates, lauryl alcohol sulfates, phosphates, carbonates
Category: Detergent with wetting, sequestering and synergistic agents, anionic
Application: Cleaners in syphon-type pipette or test-tube rinsers
Form: Tablet
Properties: Biodegradable
References: Technical Bulletin #3

Aldo 8. Glyco Chem.
Same as Aldo TC in Cosmetic Grade

Aldo 10. Glyco Chem.
Same as Aldo TGTC in cosmetic grade

Aldo 12. Glyco Chem.
Same as Aldo TCL in cosmetic grade

Aldo DGDO. Glyco Chem.
Chemical Description: Decaglycerol decaoleate
Form: Liquid; Color: Amber
Properties: HLB: 3.0; Sap. Value: 165–185

Aldo HGDS. Glyco Chem.
Chemical Description: Hexaglycerol distearate
Form: Solid; Color: Tan
Properties: M.P.: 53–57 C; HLB: 4.0; Sap. Value: 105–125

Aldo HMO. Glyco Chem.
Chemical Description: Glycerol monooleate, high mono.
Form: Semi-liquid; Color: Yellow
Properties: M.P.: $<$ 20 C; HLB: 3.0; Sap. Value: 145–155

Aldo HMS. Glyco Chem.
Chemical Description: Glycerol monostearate, high mono.
Form: White beads

Properties: M.P.: 61–68; HLB: 2.8; Sap. Value: 165–175

Aldo LS. Glyco Chem.
Chemical Description: Glycerol lactostearate
Form: Solid; Color: White
Properties: M.P.: 47–54 C; HLB: 9.0; Sap. Value: 245–260

Aldo MC. Glyco Chem.
Chemical Description: Glycerol monococoate
Category: Emulsifier, nonionic
Form: Soft solid; Color: Cream
Properties: Solubility: Sol. in ethanol, ethyl acetate, toluol, naphtha, mineral and vegetable oils; disp. in water; Sp. gr.: 0.97; M.P.: 20–26 C; HLB: 6.8; Sapon. Value: 180–190; Acid No.: < 6; Iodine Value: < 10; pH: 7.5–8.5 (5% aq.)
Composition: 100% conc.
References: Bulletin E-100

Aldo MLD. Glyco Chem.
Chemical Description: Glycerol monolaurate (disp.)
Category: Emulsifier, nonionic
Form: Soft solid; Color: Cream
Properties: Solubility: Sol. in ethanol, ethyl acetate, toluol, naphtha, mineral and vegetable oils; disp. in water; Sp. gr.: 0.97; M.P.: 21–26 C; HLB: 6.8; Sapon. Value: 185–195; Acid No.: < 5; Iodine Value: < 9; pH: 7.5–8.5 (5% aq.)
Composition: 100% conc.
References: Bulletin E-100

Aldo MO. Glyco Chem.
Chemical Description: Glycerol monooleate
Category: Emulsifier, nonionic
Applications: Baking, ice creams, cosmetics and industrial processes
Form: Soft solid; Color: Yellow
Properties: Solubility: Sol. in ethyl acetate; miscible in certain proportions with methanol, ethanol, acetone, toluol, naphtha, mineral and vegetable oils; Sp. gr.: 0.95; M.P.: < 25 C; HLB: 3.4; Sapon. Value: 170–180; Acid No.: < 5; Iodine Value: 68–73; pH: 4.5–6.5
Composition: 100% conc.
References: Bulletin E-100

Aldo MOD. Glyco Chem.
Chemical Description: Glycerol monooleate (disp.); available in animal, vegetable, food and Kosher grades
Category: Emulsifier, nonionic
Applications: Baking, ice creams, cosmetics, industrial processes
Form: Liquid; Color: Yellow
Properties: Solubility: Sol. in toluol, naphtha, mineral and vegetable oils; disp. in water; Sp. gr.: 0.95; M.P.: < 0 C; HLB: 5.0; Sapon. Value: 141–147; Acid No.: < 8; Iodine Value: 65–69; pH: 8.5–9.5 (5% aq.)
Composition: 100% conc.
References: Bulletin E-100

Aldo MO Tech. Glyco Chem.
Chemical Description: Glycerol monooleate, tech. grade
Category: Emulsifier, nonionic
Form: Soft solid; Color: Amber
Properties: Solubility: Sol. in ethyl acetate; miscible in certain proportions with methanol, ethanol, ace-

tone, toluol, naphtha, mineral and vegetable oils; Sp. gr.: 0.95; M.P.: < 25 C; HLB: 3.4; Sapon. Value: 170–180; Acid No.: < 5; Iodine Value: 68–73; pH: 4.5–6.5
Composition: 100% conc.
References: Bulletin E-100

Aldo MR. Glyco Chem.
Chemical Description: Glycerol monoricinoleate (disp.)
Category: Emulsifier, nonionic
Form: Liquid; Color: Yellow
Properties: Solubility: Sol. in methanol, ethanol, ethyl acetate, toluol, vegetable oil; disp. in water; Sp. gr.: 1.02; M.P.: < -8 C; HLB: 5.6; Sapon. Value: 122–132; Acid No.: < 4; Iodine Value: 66–69; pH: 8.3–9.3
Composition: 100% conc.
References: Bulletin E-100

Aldo MS. Glyco Chem.
Chemical Description: Glycerol monostearate
Category: Emulsifier, nonionic
Applications: Baking, ice creams, cosmetics, industrial processes
Form: Beads; Color: White
Properties: Solubility: Sol. hot in methanol, ethanol, toluol, naphtha, mineral and vegetable oils; disp. hot in water; Sp. gr.: 0.97; M.P.: 57–61 C; HLB: 3.6; Sapon. Value: 158–165; Acid No.: < 3; Iodine Value: < 5; pH: 7.6–8.6 (3% aq.)
Composition: 100% conc.
References: Bulletin E-100

Aldo MS Industrial. Glyco Chem.
Chemical Description: Glycerol monostearate (industrial grade)
Category: Emulsifier, nonionic
Form: Flakes; Color: Tan
Properties: Solubility: Sol. hot in methanol, ethanol, toluol, naphtha, mineral and vegetable oils; disp. hot in water; Sp. gr.: 0.97; M.P.: 55–62 C; HLB: 3.6; Sapon. Value: 157–177; Acid No.: < 5; Iodine Value: < 5; pH: 6.5–8.5 (5% aq.)
Composition: 100% conc.
References: Bulletin E-100

Aldo MSA. Glyco Chem.
Chemical Description: Glycerol monostearate, acid stable
Form: White beads
Properties: M.P.: 51–58 C; HLB: 11.0; Sap. Value: 90–100

Aldo MSC. Glyco Chem.
Chemical Description: Glycerol monostearate (cosmetic)
Category: Emulsifier, nonionic
Form: Flakes; Color: White
Properties: Solubility: Sol. hot in methanol, ethanol, toluol, naphtha, mineral and vegetable oils; disp. hot in water; Sp. gr.: 0.97; M.P.: 55–58 C; HLB: 3.5; Sapon. Value: 168–176; Acid No.: < 6; Iodine Value: < 3; pH: 4.0–7.0 (5% aq.)
Composition: 100% conc.
References: Bulletin E-100

Aldo MSD. Glyco Chem.
Chemical Description: Glycerol monostearate, disp.
Category: Emulsifier, nonionic

Applications: Baking, ice creams, cosmetics, industrial processes
Form: Beads; Color: White
Properties: Solubility: Sol. hot in methanol, ethanol, toluol, naphtha, mineral and vegetable oils; disp. hot in water; Sp. gr.: 0.97; M.P.: 56–60 C; HLB: 5.8; Sapon. Value: 140–150; Acid No.: < 3.5; Iodine Value: < 5; pH: 9.2–10.2 (3% aq.)
Composition: 100% conc.
References: Bulletin E-100

Aldo MSLG. Glyco Chem.
Chemical Description: Glycerol monostearate, low glycerine
Form: White beads
Properties: M.P.: 58–62 C; HLB: 3.3; Sap. Value: 165–175

Aldo TC. Glyco Chem.
Chemical Description: Liq. Fract. esters, edible coconut oil
Form: Straw. liquid
Properties: M.P./Solid Pt.: < – 5 C; HLB: 1.4; Sap. Value: 335–360

Aldo TCL. Glyco Chem.
Chemical Description: Liq. fract. esters, edible coconut oil
Form: Straw. liquid
Properties: M.P./Solid Pt.: < 0 C; HLB: 1.4; Sap. Value: 300–325

Aldo TGMS. Glyco Chem.
Chemical Description: Triglycerol monostearate
Form: Tan beads
Properties: M.P.: 54–58 C; HLB: 4.0; Sap. Value: 120–140

Aldo TGTC. Glyco Chem.
Chemical Description: Liq. fract. esters, edible coconut oil
Form: Straw. liquid
Properties: M.P./Solid Pt.: < 0 C; HLB: 4.0; Sap. Value: 250–275

Aldo TO. Glyco Chem.
Chemical Description: Glycerol trioleate
Form: Yellow liquid
Properties: M.P.: < 0 C; HLB: 0.8; Sap. Value: 188–195

Aldosperse AA 40/60. Glyco Chem.
Chemical Description: 40% Aldosperse MS-20, 60% Aldo HMS
Form: White beads
Properties: M.P.: 40–60 C; HLB: 8.0; Sap. Value: 120–130

Aldosperse MS-20. Glyco Chem.
Chemical Description: POE 20 glycerol monostearate
Form: Soft solid; Color: White
Properties: M.P.: 30–40 C; HLB: 13.1; Sap. Value: 65–75

Aldosperse O-20. Glyco Chem.
Chemical Description: 80% glycerol monostearate, 20% POE 20 sorbitan monooleate
Form: Straw. beads
Properties: M.P.: 55–65 C; HLB: 5.2; Sap. Value: 140–150

Aldosperse TS-20. Glyco Chem.
Chemical Description: 80% glycerol monostearate, 20% POE 20 sorbitan tristearate
Form: Straw. beads
Properties: M.P.: 55–65 C; HLB: 4.3; Sap. Value: 150–160

Aldosperse TS-40. Glyco Chem.
Chemical Description: 60% glycerol monostearate, 40% POE 20 sorbitan tristearate
Form: Straw. beads
Properties: M.P.: 40–60 C; HLB: 8.0; Sap. Value: 120–130

Alfol 4 Alcohol. Continental Oil Co.
Applications: Intermediate for amines, fatty acid esters
Form: Liquid; Color: Colorless; Odor: Characteristic, penetrating, sweet
Properties: Refractive Index: 1.3993; Solubility: Limited sol. in water; Density: 6.76; Sp. gr.: 0.809; Visc.: 3.4; M.P. –128 C; B.P.: 117.7 C
Composition: 99.3% active ingred.; 0.07% water

Alfol 6 Alcohol. Continental Oil Co.
Form: Liquid; Color: Colorless
Properties: Refractive Index: 1.4179; Density: 6.83; Sp. gr.: 0.823; Visc.: 5.5; M.P.: –49; B.P.: 315
Composition: 98.5% active ingred.; 0.15% max water

Alfol 8 Alcohol. Continental Oil Co.
Category: Wetting agent, emulsifier
Applications: Raw material for preparation of acetate esters, stabilizers, mercaptans, caprylic acid, octaldehyde; intermediate for surfactants
Form: Liquid; Color: Colorless; Odor: Mild aromatic
Properties: Solubility: Immiscible: water; Miscible: alcohol, chloroform, ether, benzene; Density: 6.86; Sp. gr.: 0.823; M.P. 5; B.P.: 383
Composition: 99.2% active ingred.; 0.03% water
Containers: SS, heresite-lined steel or aluminum or dry carbon steel tanks

Alfol 10 Alcohol. Continental Oil Co.
Category: Emulsifier
Applications: Chemical intermediate for surfactants, alcohol sulfates, ethoxylates, ether sulfates; foam builders
Form: Liquid; Color: Colorless; Odor: Sweet, fat-like
Properties: Solubility: Alcohol, ether, benzene, glacial acetic acid; insol. water; Density: 6.90; Sp. gr.: 0.828; M.P.: 44; B.P.: 448
Composition: 99% active ingred.; 0.04% water
Containers: Dry carbon steel; SS; heresite-lined steel or aluminum

Alfol 12 Alcohol. Continental Oil Co.
Form: Liquid; Color: Colorless
Properties: Sp. gr.: 0.831; M.P.: 75; B.P. 498
Composition: 99.1% active ingred.
Containers: Dry carbon steel; SS, epoxy or heresite-lined steel, aluminum
Warnings and Caution: Avoid inhaling vapors from the hot liquid; wash body parts contacting the alcohol with water; combustible

Alfol 14 Alcohol. Continental Oil Co.
Form: Solid, crystalline; Color: White; Odor: Typical fatty alco-

hol
Properties: Sp. gr.: 0.824; M.P.: 99; B.P. 552
Composition: 99% active ingred.
Reaction: Typical of compounds with the OH functional group
Containers: Dry carbon steel, SS, epoxy or heresite-lined steel, aluminum
Warnings and Caution: Avoid inhaling vapors from the hot liquid; wash body parts contacting the alcohol with water; combustible; treat with same precautions as with high boiling hydrocarbons

Alfol 16 Alcohol. Continental Oil Co.
Form: Waxy solid; Color: White; Odor: Typical fatty alcohol
Properties: Density: 6.77; Sp. gr.: 0.813; Visc.: 6.77; M.P. 117; B.P.: 604
Composition: 98.9% active ingred.
Containers: Dry carbon steel; aluminum, epoxy or heresite-lined steel
Warnings and Caution: Avoid inhaling vapors form the hot liquid; Wash body parts contacting the alcohol with water; combustible; treat with same precautions as high boiling hydrocarbons

Alfol 18 Alcohol. Continental Oil Co.
Applications: Surfactant range alcohol
Form: Waxy solid; Color: White; Odor: Typical fatty alcohol
Properties: Density: 6.71; Sp. gr.: 0.8075; Visc.: 13.5; M.P.: 135; B.P.: 640
Composition: 98.7% active ingred.
Reaction: Typical of compounds with the –OH functional group
Containers: Aluminum, SS, epoxy or heresite-lined steel. dry carbon steel
Warnings and Caution: Avoid inhaling vapors from the hot liquid; Wash body parts contacting the alcohol with water; combustible; treat with same precautions as high boiling hydrocarbons

Alfol 20 Alcohol. Continental Oil Co.
Chemical Description: Blend of C_2O and higher even-carbon-number primary linear alcohols
Category: Lubricant, defoamer
Applications: Lubricants for plastics, textiles, metals; ethoxylation and sulfation to prepare emulsifier and surfactants; esters; defoamers for paper, latex, other aqueous slurries; waxes; chemical intermediate
Form: Solid; Color: Off white
Properties: Density: 6.80; Sp. gr.: 0.817; M.P.: 131

Alfol 610 Alcohol. Continental Oil Co.
Category: Intermediate
Applications: Intermediate for plasticizers and surfactants; chemical specialty as emulsifier and foam builders; mercerizing agent in textile industry; intermediate for aldehydes, acids, amines, halides and mercaptans, lubricating oil additive
Form: Liquid; Color: Colorless; Odor: Mild, sweet

Properties: Solubility: Insol. water, sol. alcohol, ether, benzene; Density: 6.92; Sp. gr.: 0.829; Visc.: 11.0; M.P.: –8; B.P.: 350
Composition: 99% active ingred.
Warnings and Caution: Avoid inhaling vapors from the hot liquid; Wash body parts contacting the alcohol with water; combustible; treat with same precautions as high boiling hydrocarbons

Alfol 810 Alcohol. Continental Oil Co.
Category: Intermediate
Applications: High-quality derivatives; intermediate for plasticizers, surfactants; chemical specialty as emulsifier and foam builder; mercerizing agent in textiles; lubricating oil additive; chemical intermediate for manufacture of halides, amines, aldehydes, acids and mercaptans
Form: Liquid; Color: Clear colorless; Odor: Mild, sweet
Properties: Solubility: Sol. alcohol, ether, benzene–almost insol. water; Density: 6.93; Sp. gr.: 0.831; Visc.: 8.9; M.P.: 7; B.P.: 400
Composition: 99% active ingred.
Containers: SS, aluminum, heresite-lined steel; dry carbon steel

Alfol 1012 Alcohol. Continental Oil Co.
Category: Intermediate
Applications: Surfactant intermediate
Form: Liquid; Color: Colorless; Odor: Sweet
Properties: Density: 7.0; Sp. gr.: 0.834; Visc.: 15.0; M.P.: 35–40; B.P.: 427
Composition: 99% active ingred.
Reaction: Typical of compounds with the –OH functional group
Containers: SS, aluminum, epoxy-lined or heresite-lined steel tanks; dry mild steel

Alfol 1214 Alcohol. Continental Oil Co.
Category: Surfactant intermediate
Form: Liquid; Color: Clear, colorless; Odor: Sweet, typical fatty alcohol
Properties: Density: 7.0; Sp. gr.: 0.838; Visc.: 14.3 (120); M.P.: 73–75; B.P.: 518
Composition: 99% active ingred.
Reaction: Typical of compounds with the –OH functional group
Containers: SS, aluminum, heresite-lined steel; dry carbon steel
Warnings and Caution: Avoid inhaling vapor from the hot liquid; wash body parts contacting the alcohol with water; combustible; treat with same precautions as high boiling hydrocarbons

Alfol 1216 Alcohol. Continental Oil Co.
Category: Surfactant intermediate
Form: Liquid; Color: Clear colorless; Odor: Typical fatty alcohol
Properties: Density: 7.0; Sp. gr.: 0.830; Visc.: 15.5 (120); M.P.: 64–70; B.P.: 514
Composition: 98.5% active ingred.
Reaction: Typical of compounds with the –OH functional group
Containers: See Alfol 1214 Alcohol

Warnings and Caution: See Alfol 1214 Alcohol

Alfol 1218 Alcohol. Continental Oil Co.
Category: Surfactant intermediate
Form: Solid; Color: White
Properties: Density: 6.85; Sp. gr.: 0.825; Visc.: 15.5 (120); M.P.: 99–101; B.P.: 524
Composition: 100% active ingred.
Reaction: Typical of compounds with the –OH functional group
Containers: See Alfol 1214 Alcohol
Warnings and Caution: See Alfol 1214 Alcohol

Alfol 1412 Alcohol. Continental Oil Co.
Category: Surfactant intermediate
Form: Soft, waxy solid; Color: White; Odor: Typical fatty alcohol
Properties: Density: 7.0; Sp. gr.: 0.838; Visc.: 16.7 (100); M.P. 72–75; B.P.: 560
Composition: 99.4% active ingred.
Reaction: Typical of compounds with the –OH functional group
Containers: See Alfol 1214 Alcohol
Warnings and Caution: See Alfol 1214 Alcohol

Alfol 1618 Alcohol. Continental Oil Co.
Category: Surfactant intermediate
Form: Waxy solid; Color: White; Odor: Typical fatty alcohol
Properties: Density: 6.81; Sp. gr.: 0.820; Visc.: 22 (120); M.P.: 110–115; B.P.: 609
Composition: 98.7% active ingred.
Reaction: Typical of compounds with the –OH functional group
Containers: See Alfol 1214 Alcohol
Warnings and Caution: See Alfol 1214 Alcohol

Alfol 1620 Alcohol. Continental Oil Co.
Category: Surfactant intermediate
Form: Waxy solid; Color: White; Odor: Typical fatty alcohol
Properties: Density: 6.81; Sp. gr.: 0.817; Visc.: 13.5 (160); M.P.: 110–113; B.P.: 605
Composition: 98.5% active ingred.
Reaction: Typical of compounds with the –OH functional group
Containers: See Alfol 1214 Alcohol
Warnings and Caution: See Alfol 1214 Alcohol

Alfol C22-Alcohol. Continental Oil Co.
Chemical Description: Blend of C_{22} and higher even-carbon-number primary linear alcohols
Category: Lubricant, defoamer
Applications: Lubricant for plastics, textiles, metals; ethoxylation and sulfation to prepare emulsifiers and surfactants; esters; defoamers for paper, latex and other aqueous slurries; waxes; chemical intermediate
Form: Solid; Color: Off white
Properties: Solubility: nil in water; Density: 6.81; Sp. gr.: 0.817; M.P.: 132

Alfonic 1012-60. Conoco Chem.
Chemical Description: Ethoxylated

linear alcohols (60% E.O.)
Category: Detergent, wetting agent, emulsifier, foaming agent, intermediate, dispersant, nonionic
Applications: Industrial–metal cleaners, textile and leather processing agents, agricultural formulations, pulp and paper processing, maintenance cleaners, coatings; Household–light-duty dishwashing liquids. heavy-duty (liquid or dry) laundry products; all-purpose cleaners, car washing compounds
Form: Liquid; Color: 20 (APHA); Odor: Sweet
Properties: Density: 8.2; Sp. gr.: 0.98; Visc.: 20.5 cps @ 100; M.P.: 36; B.P.: 214; Stability: Good in presence of anionics, cationics, acids, alkalies, other builders; compatible with most oxiding and bleaching agents in textile industry; Surface Tension: 26 dynes/cm @ 0.1%; Biodegradable
Composition: 99.8% active ingred.
Toxicity: Relatively nontoxic; in conc. form may produce eye or skin irritation
Containers: aluminum, SS, phenolic coated tanks
Warnings and Caution: If it contacts the body, flush with water; combustible; treat with same precautions as hydrocarbon; foam or dry chemicals to extinguish fires

Alfonic 1412-40 Ethoxylate. Conoco Chem.
ChemicalDescription: Ethoxylated linear alcohols (40% E.O.)
Category: Nonionic
Applications: Raw material for manufacture of sodium, ammonium, potassium, and mono-, di-, or triethanolamine ethoxysulfates; can be further processed to yield ethoxyphosphates
Form: Liquid; Color: Clear; Odor: Slight
Properties: Density: 7.64; Sp. gr.: 0.91; Visc.: 52; M.P. 41; Biodegradable
Composition: 100% active ingred.
Toxicity: Relatively nontoxic; in conc. form may cause skin or eye irritation
Warnings and Caution: If contacts body, flush exposed area with water

Alfonic 1412-60. Conoco Chem.
ChemicalDescription: Ethoxylated linear alcohols (60% E.O.)
Category: Detergent, wetting agent, emulsifier, foaming agent, nonionic surfactant
Applications: Industrial–metal cleaners, textile and leather processing agents, agricultural formulations, pulp and paper processing, maintenance cleaners, coatings; Household–heavy-duty (liquid or dry) laundry products
Form: Liquid; Color: Cloudy; Odor: Mild, sweet; Viscous at room temperature
Properties: Density: 8.2; Sp. gr.: 0.98; Visc.: 33 cps @ 100; Stability: Good in presence of anionic and cationic compounds, acids, alkalies, other detergent builders; Biodegradable
Composition: 100% active ingred.
Toxicity: Relatively nontoxic; in conc. form could cause skin or eye irritation

Warnings and Caution: Should not be used with strong oxidizing agents such as conc. hydrogen peroxide (potentially explosive); if contacts body, flush exposed area with water

Alfonic 1412-A Ether Sulfate. Conoco Chem.
Chemical Description: Sulfated, ethoxylated alcohol derivatives
Category: Detergent, wetting agent, emulsifier, foaming agent
Applications: Household–light-duty dishwashing liquids, bubble baths, fine fabric cleaner, wool washing liquids, upholstery and rug cleaners, cosmetic cleansers, all-purpose cleaners; Industrial–emulsion polymerization, petroleum production, wallboard foaming agent, paper processing, agricultural formulations, maintenance cleaners
Form: Liquid; Color: Clear; Odor: Mild
Properties: Density: 8.4; Visc.: 76 cps @ 100; Biodegradable
Composition: 58% active ingred.; other components: water
Toxicity: Moderate irritation to eyes; nontoxic skin, oral, inhalation tests
Containers: 316 SS
Warnings and Caution: If contacts body, flush exposed area with water

Alfonic 1412-S Ether Sulfate. Conoco Chem.
Chemical Description: Sulfated ethoxylated alcohol derivatives
Category: Detergent, emulsifier, wetting and foaming agent, anionic
Applications: Household–light-duty dishwashing liquids, bubble baths, fine fabric and wool cleaners, upholstery and rug cleaners, cosmetic cleansers, all-purpose cleaners; Industrial–emulsion polymerization, petroleum production, wallboard foaming agent, paper processing, agricultural formulations, maintenance cleaners
Form: Liquid; Color: Clear; Odor: Mild
Properties: Density: 8.8; Visc.: 64 cps @ 100; Biodegradable
Composition: 38% active ingred.; 0.2% water; Na_2CO_3 added as buffer
Toxicity: Moderate eye irritation; nontoxic by skin, oral, inhalation tests
Containers: 316 SS
Warnings and Caution: If it contacts body, flush exposed area with water

Alfonic 1618-65. Conoco Chem.
Chemical Description: Ethoxylated linear alcohols (65% E.O.)
Category: Detergent, emulsifier, foaming and wetting agent; nonionic
Applications: Industrial–metal cleaners, textile and leather processing, agricultural formulations, pulp and paper processing, maintenance cleaners, coatings; Household–heavy-duty (liquid or dry) laundry products
Form: Solid; Color: Near white; Odor: Nil
Properties: Solubility: Good in water, ethanol, glycol ether (mon-

obutyl), isopropyl alcohol, xylene, perchlorethyl; slight in kerosene; insol. mineral oil; Density: 8.2; Sp. gr.: 0.98; Visc.: 81 cps @ 100; M.P.: 88; B.P.: 234; Biodegradable
Composition: 100% active ingred.
Reaction: Compatible and stable in presence of anionics and cationics, acids, alkalies, other detergent builders; compatible with most oxidizing and bleaching agents at normal use concentrations
Toxicity: Relatively nontoxic; may produce skin or eye irritation
Containers: Aluminum, SS, phenolic coated tanks
Warnings and Caution: If contacts body, flush exposed area with water; treat with same precautions as high boiling hydrocarbon; with strong oxidizing agents (e.g., conc. hydrogen peroxide) potential explosive mixtures can result

Alipal CD-128. GAF Corp.
Chemical Description: Sulfate-ester-type surfactant based on a linear hydrophobe
Category: Foaming agent, wetting agent, anionic
Applications: Use in all-purpose detergents and concentrates; foaming agent for air drilling of oil and gas wells; wetting agent for conc. caustic sol'ns.
Form: Liquid; Color: Straw
Properties: Solubility: Good in electrolyte systems; Sp. gr.: 1.056; Pour Pt.: –5 C; Stability: Good in high electrolyte, alkaline
Composition: 56% min. active ingredient
References: Data sheet

Alipal CO-433. GAF Corp.
Chemical Description: Sodium salt of a sulfated alkylphenoxypoly (ethyleneoxy)-ethanol
Category: Detergent, wetting, dispersing, emulsifying and foaming agents; anionic
Applications: Liquid all-purpose cleaners; light-duty dishwashing detergents; textile wet processing; cosmetics; emulsion polymerization and post-stabilization; pesticide emulsifiers
Form: Clear liquid; Color: 4 max (Varnish); Odor: Mild aromatic
Properties: Solubility: Water sol.; Density: 8.9 lb/gal; Sp. gr.: 1.065; Visc.: 2500 cps; Stability: Good; Surface Tension: 32 dynes/cm (1% conc.)
Composition: 28% active ingred.
References: Tech. Bulletin 6220-001

Alipal CO-436. GAF Corp.
Chemical Description: Ammonium salt of a sulfated alkylphenoxypoly(ethyleneoxy)-ethanol
Category and Applications: See Alipal CO-433
Form: Clear liquid; Color: 5 max. (Varnish); Odor: Alcoholic
Properties: Solubility: Water sol.; Density: 8.9 lb/gal; Sp. gr.: 1.065; Visc.: 100 cps; Stability: Good; Surface Tension: 34 dynes/cm (1% sol'n.)
Composition: 58% active ingred.
Toxicity: Eye irritant

Alipal EO-526. GAF Corp.
Chemical Description: Sodium salt of a sulfated alkylphenoxypoly

(ethyleneoxy)ethanol
Category and Applications: See Alipal CO-433
Form: Hazy liquid; Color: 10 max (Varnish); Odor: Alcoholic
Properties: Solubility: Water sol.; Density: 9.3 lb/gal; Sp. gr.: 1.115; Visc.: 400 cps; Stability: Good; Surface Tension: 35 dynes/cm (1% sol'n.)
Composition: 58% active ingred.

Alipal EP-110. GAF Corp.
Chemical Description: Ammonium salt of sulfated alkylphenoxypoly (ethyleneoxy) ethanol
Category: Emulsifier
Applications: Emulsion polymerization yielding latices with outstanding multivalent ion stability, water-spot resistance, freedom from grain and coagulum, film clarity and uniformity
Form: Liquid (slight haze); Color: Pale yellow
Properties: Sp. gr.: 1.04; Visc.: 91.0 cks; Pour Pt.: 0 C; Surface Tension: 38.3 dynes/cm (1%)
Composition: 30% active ingred.
Toxicity: Moderate eye irritation

Alipal EP-115. GAF Corp.
Chemical Description: Ammonium salt of a sulfated alkylphenoxypoly (ethyleneoxy) ethanol
Form: Clear viscous liquid; Color: 3 max (VCS)
Composition: 30% active ingred.

Alipal EP-120. GAF Corp.
Chemical Description: Ammonium salt of sulfated alkylphenoxypoly (ethyleneoxy) ethanol
Category: Emulsifier
Applications: See Alipal EP-110
Form: Liquid (slight haze); Color: Pale yellow
Properties: Sp. gr.: 1.06; Visc.: 111 cks; Pour Pt.: 8 C; Surface Tension: 40 dynes/cm (1%)

Alkadet LD-50. Alkaril Chem.
Chemical Description: Blend of surfactants
Category: Detergent, anionic
Applications: Light-duty liquid detergent concentrate for dishwashing and general purpose, esp. scouring metal surfaces, all flooring surfaces
Form: Clear liquid; Color: Amber
Properties: Visc.: 500 cps; Cloud Pt.: < 30 F
Composition: 48% active ingred.

Alkadet LD50-Q. Alkaril Chem.
Chemical Description: Compounded liquid detergent concentrate
Category: Detergent, anionic
Applications: Liquid detergent concentrate for dishwashing and general purpose cleaning esp. flooring surfaces, lacquered, painted surfaces
Form: Clear liquid; Color: Green
Properties: Sp. gr.: 1.075; Visc.: 100–130 cps
Composition: 50% active ingred.

Alkadet LD-60. Alkaril Chem.
Chemical Description: Blend of surfactants
Category and Applications: See Alkadet LD-50
Form: Clear liquid; Color: Amber
Properties: Sp. gr.: 1.05; Visc.:

≈ 100 cps; Cloud Pt.: 7 C max.
Composition: 58–60% active

Alkameen C-5. Alkaril Chem.
Chemical Description: Tertiary amines; ethoxylates of primary tallow amines
Category: Emulsifier, detergent, wetting agent, cationic
Applications: Emulsifier–fat liquoring; making of cutting and soluble oils; dispersing waxes, pigments, resins; printing; Wetting agent–desizing and boiling off cottons; dyeing; tanning; insecticides, herbicides, fungicides; Detergents–textiles; skins; leather; dyeing
Form: Liquid; Color: Amber; Odor: Typical
Properties: Solubility: Insol. water
Composition: 100% active ingred.
Reaction: Alkaline

Alkameen C-10. Alkaril Chem.
Chemical Description: Tertiary amines; ethoxylates of primary tallow amines
Category: Emulsifier, detergent, wetting agent, cationic
Applications: Emulsifier–fat liquoring, making of cutting and soluble oils; Wetting Agents–pickling; electroplating; insecticides, herbicides, fungicides; Detergents–textiles; mines and minerals; cosmetics
Form: Liquid; Color: Amber; Odor: Typical
Properties: Solubility: Insol. water
Composition: 100% active ingred.
Reaction: Alkaline

Alkameen HT-2. Alkaril Chem.
Chemical Description: Tertiary amines; ethoxylates of hydrogenated primary tallow amines
Category: Emulsifier, cationic
Applications: Emulsifier–dispersing waxes, pigments, resins; Wetting agent–rotary washer, waterproofing; Detergents–textiles
Form: Solid; Color: Amber; Odor: Typical
Properties: Solubility: Insol. water
Composition: 100% active ingred.
Reaction: Alkaline

Alkameen HT-5. Alkaril Chem.
Chemical Description: See Alkameen HT-2
Category: Detergent, wetting agent, emulsifier, cationic
Applications: Emulsifier–fat liquoring; Wetting agent–insecticides, herbicides, fungicides; Detergent–textiles
Form: Solid; Color: Amber; Odor: Typical
Properties: Solubility: Insol. water
Composition: 100% active ingred.
Reaction: Alkaline

Alkameen HT-12. Alkaril Chem.
Chemical Description: See Alkameen HT-2
Category: Emulsifier, detergent, wetting agent, cationic
Applications: Emulsifier–fat liquoring; dispersing waxes, pigments, resins; Detergent–textiles
Form: Solid; Color: Amber; Odor: Typical
Properties: Solubility: Insol. water
Composition: 100% active ingred.
Reaction: Alkaline

Alkameen HT-15. Alkaril Chem.

Chemical Description: See Alkameen HT-2
Category: Detergent, emulsifier, wetting agent, cationic
Applications: Emulsifer–printing; Wetting agent–dyeing; Detergent–textiles, dyeing
Form: Solid; Color: Amber; Odor: Typical
Composition: 100% active ingred.
Reaction: Alkaline

Alkameen HT-25. Alkaril Chem.
Chemical Description: See Alkameen HT-2
Category: Detergent, wetting agent, emulsifier, cationic
Applications: Emulsifier–dispersing waxes, pigments, resins; preparation of emulsified paints; Detergents–textiles; mines and mining; paints
Form: Solid; Color: Yellow; Odor: Typical
Composition: 100% active ingred.
Reaction: Alkaline

Alkameen HT-30. Alkaril Chem.
Chemical Description: See Alkameen HT-2
Category: Emulsifier, wetting agent, cationic
Applications: Emulsifier–dispersing waxes, pigments, resins; preparation of emulsified paints; Detergents–textiles, mines and minerals, paints
Form: Solid; Color: Tan; Odor: Typical
Composition: 100% active ingred.
Reaction: Alkaline

Alkamide 1002. Alkaril Chem.
Chemical Description: 2:1 Coconut Alkanolamide
Category: Detergent, Nonionic-anionic surfactant
Applications: Hair shampoo, bubble bath, specialty detergents
Form: Liquid; Color: Straw-color
Properties: Solubility: Good in water; thickening and foam stabilizing properties
Composition: 100% active ingred.
References: Technical Data # 15

Alkamide 2104. Alkaril Chem.
Chemical Description: 2:1 Coconut Diethanolamide
Category: Detergent, foam booster, nonionic-anionic
Applications: Detergent–textiles, household, cosmetics, hair shampoos, bubble bath, cleaners of street, building, hospitals; Emulsifier–making of cutting and soluble oils
Form: Viscous liquid; Color: Amber; Odor: Low
Properties: Solubility: Good in water; thickening and foam stabilizing properties
Composition: 100% active ingred.
References: Technical Data # 11

Alkamide 2106. Alkaril Chem.
Chemical Description: 2:1 Coconut Alkanolamide
Category: Detergent, nonionic-anionic
Applications: Compounded with alkaline salts for industrial and household floor cleaning formulations
Form: Viscous liquid; Color: Amber

Properties: Water-soluble; thickening and foam stabilizing properties; good wetting and soil suspension properties
Composition: 100% active ingred.
References: Technical Data # 12

Alkamide 2112. Alkaril Chem.
Chemical Description: 2:1 Fatty acid alkanolamide, sulfonate blend
Category: Detergent, nonionic-anionic
Applications: Heavy-duty detergent base; base for highly alkaline wax strippers and degreasers; cleaners of street, building, hospitals
Form: Viscous liquid; Color: Brown; Odor: Low
Properties: Solubility: Good in water; Cloud Pt.: 45 F max.
Composition: 100% active ingred.
Reaction: Alkaline
References: Technical Data

Alkamide 2122. Alkaril Chem.
Chemical Description: 2:1 alkanolamide sulfonate concentrate
Category: Detergent, wetting, dispersing and air entraining agent, nonionic-anionic
Applications: Air entraining agent in construction materials, e.g., prestressed concrete, concrete brick, pipe and block; light-duty liquid detergents
Form: Viscous liquid; Color: Amber
Properties: Solubility: Good in water; Cloud Pt.: 45 F max.; High foam stabilizing properties
Composition: 100% active ingred.
References: Technical Data # 16

Alkamide 2124. Alkaril Chem.
Chemical Description: 2:1 Lauric diethanolamide
Category: Detergent, wetting agent, emulsifier, foaming agent, nonionic-anionic surfactant
Applications: Hair shampoo concentrate, liquid detergent, bubble bath formulations, specialty emulsifiers
Form: Viscous liquid; Color: Amber
Properties: Solubility: Good in water; Sp. gr.: 1.0–1.05; Good thickening and foam stabilizing action
Composition: 100% active ingred.
References: Technical Data # 13

Alkamide 2204. Alkaril Chem.
Chemical Description: 2:1 Coconut diethanolamide
Category: Detergent, nonionic-anionic
Applications: Heavy-duty detergent base, household use, cleaners of street, buildings, hospital; rust-inhibitor
Form: Viscous liquid; Color: Amber; Odor: Low
Properties: Solubility: Good in water
Composition: 100% active ingred.
Containers: Steel tanks or drums may be used due to anti-rusting properties
References: Technical Data # 14

Alkamide CDE. Alkaril Chem.
Chemical Description: Coconut diethanolamide
Category: Detergent, emulsifier, stabilizer, nonionic surfactant

Applications: Foam booster, viscosity builder; Detergent–textiles, skins, leather, laundering, household, cosmetics, street, buildings and hospital cleaners; Emulsifiers–making of cutting and soluble oils; printing
Form: Liquid; Color: Light amber; Odor: Low
Properties: Solubility: In detergent systems, aromatic and chlorinated aliphatic solvents, petroleum solvents, mineral oils; Sp. gr.: 1.04 @ 20 C
Composition: 90% active ingred.; other components: free diethanolamine (7% max.), free fatty acid (0.5% max.), and amide ester (4% max.)
Reaction: Alkaline
References: Technical Data #4

Alkamide CDO. Alkaril Chem.
Chemical Description: Coconut diethanolamide
Category: Detergent, nonionic
Applications: Foam booster, viscosity builder; Detergent–textiles, skins, leather, laundering, household, cosmetics, street, building and hospital cleaners; Wetting agent–removal of spots, rotary washer; Emulsufier–making of cutting and soluble oils; dispersing waxes, pigments, resins
Form: Liquid; Color: Amber; Odor: Low
Properties: Solubility: In aromatic and chlorinated aliphatic solvents, most petroleum solvents, mineral oils; Sp. gr.: 1.04 @ 20 C
Composition: 82% active ingred.; other components: free diethanolamine (7.0% max.), free fatty acid (0.5% max.), amide ester (4% max.)
References: Technical Data #3

Alkamide CL63. Alkaril Chem.
Chemical Description: Coconut-lauric diethanolamine
Category: Detergent, emulsifier, nonionic
Applications: Toiletry preparations–shampoos, bubble bath (as active skin conditioner and foam stabilizer); cleaning formulations–rug shampoo, all-purpose cleaners thickening agents
Form: Liquid; Color: Clear to cloudy at room temperature
Properties: Solubility: In detergent systems, aromatic and chlorinated aliphatic solvents, petroleum solvents, mineral oils; Sp. gr.: 1.03–1.05
Composition: 92% min. active ingred.; other components: free diethanolamine (7% max.); free fatty acid (0.5% max.), amide ester (4 % max.)
References: Technical Data #2

Alkamide CME. Alkaril Chem.
Chemical Description: Coconut monoethanolamides
Category: Detergent, foam booster, viscosity builder, opacifier
Applications: Detergents–laundering, household, cosmetics; street, buildings, hospital cleaners; pharmaceuticals
Form: Waxy solid; Color: White; Odor: Low
Properties: Solubility: In detergent systems, aromatic or chlorinated

aliphatic solvents, petroleum solvents, mineral oils; M.P.: 56 C
Composition: 95% active ingred.; other components: free monoethanolamine (2% max.), free fatty acid (0.5% max.)
Reaction: Neutral
References: Technical Data # 5

Alkamide CMO. Alkaril Chem.
Chemical Description: Coconut monoethanolamide
Category: Detergent, nonionic
Applications: Detergent–laundering, household, cosmetics, pharmaceuticals
Form: Waxy solid; Color: Cream
Properties: Solubility: In detergent systems, aromatic or chlorinated aliphatic solvents, petroleum solvents, mineral oils; M.P.: 50 C
Composition: 90% active ingred.; other components: free monoethanolamine (2% max.), free fatty acid (1.0% max.)
References: Technical Data # 5

Alkamide HTDE. Alkaril Chem.
Chemical Description: Modified alkanolamide of higher fatty acids
Category: Detergent, thickener, viscosity builder for kerosene, etc.
Applications: Detergent–textiles, skins, papermaking, household, cleaners of street, buildings, hospitals
Form: Waxy solid; Color: Cream; Odor: Low
Properties: M.P.: 55 C
Composition: 90% active ingred.; other components: free diethanolamine (8% max.), free fatty acid (as oleic) (1% max.)
Reaction: Alkaline
References: Technical Data # 9

Alkamide HTME. Alkaril Chem.
Chemical Description: Monoethanolamide of higher fatty acids
Category: Detergent, opacifier, thickener for cosmetics; nonionic
Form: Waxy solid; Color: Cream; Odor: Low
Properties: M.P.: 85–90 C
Composition: 90% active ingred.; other components: free monoethanolamine (1.5 max.)
Reaction: Alkaline
References: Technical Data # 10

Alkamide L7DE. Alkaril Chem.
Chemical Description: Lauric-myristic diethanolamine
Category: Detergent, foam booster, viscosity builder, nonionic
Applications: Superfatting and thickening agent for toiletry preparations–shampoos, bubble bath; cleaning formulations
Form: Paste (liquid above 25 C); Color: White; Odor: Low
Properties: Solubility: In detergent systems, aromatic and chlorinated aliphatic solvents, petroleum solvents, mineral oils; Sp. gr.: 1.03–1.05; M.P.: 15 C
Composition: 92% active ingred.; other components: free diethanolamine (7% max.), free fatty acid (0.5% max.), amide ester (4.0% max.)
Reaction: Alkaline
References: Technical Data # 2

Alkamide L7DE-BT. Alkaril Chem.
Chemical Description: Liquefied

form of modified lauric-myristic diethanolamide
Category: Detergent, foam stabilizer, viscosity builder
Applications: Shampoo concentrates
Form: Liquid; Color: Straw; Odor: Bland
Properties: Sp. gr.: 0.99; Visc.: 275 cps @ 30 C; F.P.: 0 C; Cloud Pt.: 0 C max.
Composition: 84% active ingred.
Reaction: Alkaline
References: Technical Data sheet

Alkamide L9DE. Alkaril Chem.
Chemical Description: Lauric diethanolamine
Category: Detergent, foam stabilizer, viscosity builder, nonionic
Applications: Toiletry applications–shampoos, bubble bath (as active skin conditioner and foam stabilizer); cleaning formulations–rug shampoo, all-purpose cleaners; thickening agents
Form: Waxy solid (liquid > 25 C); Color: White; Odor: Low
Properties: Solubility: In detergent systems, aromatic and chlorinated aliphatic solvents, petroleum solvents, mineral oils; Sp. gr.: 1.04; F.P.: 25 C
Composition: 92% active ingred.; free diethanolamine (7% max.); free fatty acid (0.5% max.); amide ester (4.0% max.)
Reaction: Alkaline
References: Technical Data # 2

Alkamide L9ME. Alkaril Chem.
Chemical Description: Lauric monoethanolamide
Category: Detergent, foam stabilizer, viscosity builder
Applications: Detergent–laundering, household, cosmetics (active skin protector in technical rinses, barrier creams, hand cleaners, etc.), cleaners of street, buildings, hospital; pharmaceuticals
Form: Waxy solid; Color: White; Odor: Low
Properties: Solubility: In detergent systems, aromatic and chlorinated aliphatic solvents, petroleum solvents, mineral oils; M.P.: 60 C
Composition: 95% active ingred.; other components: free monoethanolamine (2% max.), free fatty acid (0.5% max.)
Reaction: Alkaline
References: Technical Data # 1

Alkamide RDO. Alkaril Chem.
Chemical Description: Ricinoleic diethanolamide
Category: Emulsifier, lubricant, nonionic
Applications: Making of cutting and soluble oils; textiles, skins, leather, household; street, buildings, hospital cleaners; antiblocking and mold release agent
Form: Clear liquid; Color: Brown; Odor: Low
Composition: 100% active ingred.
Reaction: Alkaline
References: Technical Data # 7

Alkamide SDO. Alkaril Chem.
Chemical Description: Modified oleic diethanolamide
Category: Foam stabilizer, viscosity builder
Applications: Textiles, skins, laun-

dering, leather, household, cosmetics (shampoo and bubble bath as an active skin conditioner), street, buildings, hospital cleaners; cleaning hangings; making of cutting and soluble oils; corrosion inhibitor
Form: Viscous liquid; Color: Amber; Odor: Low
Properties: Solubility: In detergent systems, aromatic and chlorinated aliphatic solvents, petroleum solvents, mineral oils; Sp. gr.: 1.04
Composition: 80% active ingred.; other components: free diethanolamine (12% max.), free fatty acid (0.5% max.)
Reaction: Alkaline
References: Technical Data # 6

Alkamidox CME-2. Alkaril Chem.
Chemical Description: Coconut monoethanolamide ethoxylate containing 2 moles ethylene oxide
Category: Detergent, foam stabilizer, nonionic
Applications: Dishwashing detergents, heavy duty liquid laundry detergents, cosmetic applications
Form: Solid; Color: Cream
Properties: Solubility: Soluble in ethanol, xylene; insol. kerosene, mineral oil; Stability: Unstable in aqueous systems above pH 12, stable to strong alkalies
Composition: 100% active ingred.
References: Technical Data sheet

Alkamidox CME-5. Alkaril Chem.
Chemical Description: Coconut monoethanolamide ethoxylate containing 5 moles ethylene oxide
Category: Detergent, foam stabilizer, nonionic
Applications: Dishwashing detergents, heavy duty liquid laundry detergents, cosmetic applications
Form: Liquid; Color: Straw
Properties: Solubility: In water, ethanol, xylene; insol. kerosene, mineral oil; Stability: Unstable in aqueous systems, stable to strong alkalies
Composition: 100% active ingred.
References: Technical Data sheet

Alkaminox C2. Alkaril Chem.
Chemical Description: Tertiary amine with one coconut derived alkyl group and two hydroxyethyl groups
Category: Antistatic agent
Applications: Antistatic agent and corrosion inhibitor in steam generating and circulating systems
Form: Clear liquid; Color: Amber
Properties: Solubility: Sol. in most organic solvents, mineral acid; insol. water; Sp. gr.: 0.87
References: Technical Data #18-1

Alkaminox T2. Alkaril Chem.
Chemical Description: Tertiary amine with one tallow fatty group and two oxyethylene groups
Category: Emulsifier, dispersant
Applications: Emulsifier, dispersant, internal antistat for polyethylene, antistat agent for synthetic fibers
Form: Paste; Color: 8 max. (Gardner)
Properties: Solubility: Sol. in most organic solvents, mineral acid; insol. water; Sp. gr.: 0.92
References: Technical Data #18-4

Alkaminox T-5. Alkaril Chem.
Chemical Description: Tertiary amine with one tallow fatty group and two polyoxyethylene groups
Category: Emulsifier, dispersant
Applications: Emulsifier and dispersant in textile industry; dispersing agent for pigments in coatings; corrosion inhibitor in steam generating and circulating systems
Form: Clear liquid; Color: 8 max. (Gardner)
Properties: Solubility: Sol. in most organic solvents, mineral acid; insol. water; Sp. gr.: 0.97
References: Technical Data #18-5

Alkaminox T-15. Alkaril Chem.
Chemical Description: Tertiary amine with one tallow fatty group and two polyoxyethylene groups
Category: Emulsifier, dispersant
Applications: Emulsifier and dispersing agent in textile industry; emulsifier for wax emulsions
Form: Clear liquid; Color: 8 max. (Gardner)
Properties: Solubility: Sol. in most organic solvents, mineral acid and water; Sp. gr.: 1.03
References: Technical Data #18-7

Alkamox C20. Alkaril Chem.
Chemical Description: Amine oxide of ethoxylated coco amine
Category: Detergent, wetting agent, foam booster
Applications: Foam booster and stabilizer for latex foam for carpets and carpet backing
Form: Liquid; Color: Yellow; Odor: Faint
Properties: F.P.: <0 C
Composition: 33% active ingred.; other components: water
Reaction: Neutral

Alkamox CAPO. Alkaril Chem.
Chemical Description: Coco amido propyl dimethyl amine oxide
Category: Detergent, wetting agent, foaming agent and foam stabilizer
Applications: Foaming agent and foam stabilizer for shampoos, bath products, fine fabric cleaners
Form: Liquid; Color: Colorless; Odor: Faint
Properties: Solubility: Complete in water; F.P.: <0 C
Composition: 30% active ingred.; other components: water
Reaction: Neutral

Alkamox LO. Alkaril Chem.
Chemical Description: Amine oxide of lauryl dimethyl amine
Category: Detergent, wetting agent, foaming agent and foam stabilizer, antistat agent, nonionic-cationic
Applications: Foaming agent and foam stabilizer for shampoos and detergents; foam rubber; electroplating; paper coating; as a pour point depressant for mineral oils
Form: Liquid; Color: Colorless; Odor: Faint
Properties: Solubility: Excellent in water; F.P.: <0 C
Composition: 30% active ingred.; other components: water
Reaction: Neutral
References: Technical Data #19-2

Alkamuls Series. Alkaril Chem.
Chemical Description: Polyethylene glycol esters

Category: Emulsifier, dispersing agents, nonionic
Applications: General emulsifiers for O/W emulsions, dispersing agents for cosmetic and industrial use including: textiles; leather; papermaking; paints; varnishes; foods; household; dyeing; finishing; sizing; coating; waterproofing; insecticides, herbicides, fungicides; treatment of grains and forage; making of cutting and soluble oils; dispersing waxes, pigments, resins; printing; dispersing dry powders; preparation of emulsified paints
Form, Properties, Composition: see individual listings
Reaction: Neutral

Alkamuls 400-DL. Alkaril Chem.
Chemical Description: PEG 400 dilaurate
Form: Liquid
Properties: HLB: 10.0
Composition: 100% active ingred.

Alkamuls 400-DO. Alkaril Chem.
Chemical Description: Polyethylene glycol 400 dioleate
Category: Emulsifier, wetting agents, nonionic
Applications: Emulsifiers for defoamers, cosmetic creams, agricultural oils; thickeners in lotions; opacifiers for shampoos; finishing agents for textile fibers
Form: Clear Liquid; Color: Amber
Properties: Sp. gr.: 0.97–0.99; HLB: 7.2
Composition: 100% active ingred.
References: Technical Data 24-5

Alkamuls 400-DS. Alkaril Chem.
Chemical Description: Polyethylene glycol 400 distearate
Category and Applications: See Alkamuls 400-DO
Form: Waxy flakes; Color: White
Properties: HLB: 7.8; Solidification Pt.: 95 F typ
Composition: 100% active ingred.

Alkamuls 400-MC. Alkaril Chem.
Chemical Description: Polyethylene glycol 400 monococoate
Category and Applications: See Alkamuls 400-DO
Form: Liquid
Properties: HLB: 13.9
Composition: 100% active ingred.

Alkamuls 400-ML. Alkaril Chem.
Chemical Description: Polyethylene glycol 400 monolaurate
Category and Applications: See Alkamuls 400-DO
Form: Clear liquid; Color: Amber
Properties: Sp. gr.: 1.02–1.04; Cloud Pt.: 50 F max.; HLB: 12.8
Composition: 100% active ingred.
References: Technical Data 24-1

Alkamuls 400-MO. Alkaril Chem.
Chemical Description: Polyethylene glycol 400 monooleate
Category and Applications: See Alkamuls 400-DO; also used in dairy industry as an emulsifier to wet out and emulsify powder milk substitute for feeding of calves
Form: Clear liquid; Color: Amber
Properties: Sp. gr.: 1.01–1.03; HLB: 11.0; Solidification Pt.: $<$ 40 F
Composition: 100% active ingred.
References: Technical Data 24–2

Alkamuls 400-MS. Alkaril Chem.

Chemical Description: Polyethylene glycol 400 monostearate
Category and Applications: See Alkamuls 400-DO
Form: Paste
Properties: HLB: 11.2
Composition: 100% active ingred.

Alkamuls 600-DO. Alkaril Chem.
Chemical Description: Polyethylene glycol 600 dioleate
Category and Applications: See Alkamuls 400-DO
Form: Liquid; Color: Amber
Properties: Sp. gr.: 1.01–1.03; Solidification Pt.: 15 C typical
Composition: 100% active ingred.
References: Technical Data 24-11

Alkamuls 600-ML. Alkaril Chem.
Chemical Description: Polyethylene glycol 600 monolaurate
Form: Liquid
Properties: HLB: 14.6
Composition: 100% active

Alkamuls 600-MO. Alkaril Chem.
Chemical Description: Polyethylene glycol 600 monooleate
Form: Liquid
Properties: HLB: 13.0
Composition: 100% active

Alkamuls 600-MS. Alkaril Chem.
Chemical Description: Polyethylene glycol 600 monostearate
Form: Solid
Properties: HLB: 13.2
Composition: 100% active

Alkamuls DEG Series. Alkaril Chem.
Chemical Description: See individual listings
Category: Emulsifier, dispersing agents, nonionic
Applications: Detergent–skins, leather, papermaking, transport; Wetting agent–tanning, coating, insecticides, fungicides, herbicides; Emulsifiers–making of cutting and soluble oils; dispersing waxes, pigments, resins; printing; preparation of emulsified paints
Composition: 100% active ingred.
Reaction: Neutral

Alkamuls DEG-DT. Alkaril Chem.
Chemical Description: DEG Ditallate
Category: Dispersing agent, emulsifier
Form: Liquid; Color: Brown; Odor: Faint

Alkamuls DEG-ML. Alkaril Chem.
Chemical Description: DEG monolaurate
Category: Emulsifier
Form: Liquid; Color: Brown; Odor: Faint
Properties: HLB: 6.5

Alkamuls DEG-MO. Alkaril Chem.
Chemical Description: DEG monooleate
Category: Lubricant
Form: Liquid; Color: Brown; Odor: Faint
Properties: HLB: 4.7

Alkamuls DEG-MS. Alkaril Chem.
Chemical Description: DEG monostearate
Category: Opacifier, emulsifier, plasticizer
Form: Liquid; Color: Brown; Odor: Faint
Properties: HLB: 4.7

Alkamuls GML. Alkaril Chem.
Chemical Description: Glycerol monolaurate
Category: Emulsifier, nonionic
Applications: Textiles, leather, foods
Form: Paste
Properties: HLB: 3.0
Composition: 100% active
Reaction: Neutral

Alkamuls GMO. Alkaril Chem.
Chemical Description: Glycerol monooleate
Category and Applications: See Alkamuls 400-DO
Form: Liquid; Color: Light yellow
Properties: M.P.: 25 C max. HLB: 3.4
Composition: 100% active ingred.
References: Technical Data 24-33

Alkamuls GMR. Alkaril Chem.
Chemical Description: Glycerol monoricinoleate
Category: Emulsifier, stabilizer, nonionic
Applications: Textiles, leather, foods
Form: Liquid
Composition: 100% active
Reaction: Neutral

Alkamuls GMS. Alkaril Chem.
Chemical Description: Glycerol monostearate
Category and Applications: See Alkamuls 400-DO
Form: Solid
Properties: HLB: 3.4

Alkamuls PEMO. Alkaril Chem.
Chemical Description: Pentaerythritol monooleate
Category: Corrosion preventive in lubricants
Applications: Prevents corrosion in automotive and steel parts by depositing as a film on exposed parts
Form: Liquid; Color: Brown; Odor: Low
Composition: 100% active ingred.
Reaction: Neutral

Alkamuls PSML-20. Alkaril Chem.
Chemical Description: Ethoxylated sorbitan (20) monolaurate
Category: Emulsifier, solubilizer, nonionic surfactant
Applications: General purpose water-soluble surfactants, used as O/W emulsifiers
Form: Liquid; Color: Yellow
Properties: HLB: 16.7
Composition: 97% active ingred.; 3% max. water
References: Technical Data 25-11

Alkamuls PSMO-5. Alkaril Chem.
Chemical Description: Polyoxyethylene sorbitan (5) monooleate
Category: Emulsifier, nonionic surfactant
Applications: Preparation of emulsified paints; insecticides, herbicides, fungicides; foods; cosmetics
Form: Liquid; Color: Amber; Odor: Typical
Properties: Solubility: Water insol.; HLB: 10
Composition: 98.5% active ingred.
Reaction: Neutral
References: Technical Data 25-15

Alkamuls PSMO-20. Alkaril Chem.
Chemical Description: Polyoxyethylene sorbitan (20) monooleate

Category: Wetting agent, emulsifier, nonionic
Applications: Preparation of emulsified paints; insecticides; herbicides; fungicides; textiles; foods; cosmetics
Form: Liquid; Color: Yellow; Odor: Typical
Properties: Solubility: Water sol.; HLB: 15
Composition: 97% active ingred.
Reaction: Neutral
References: Technical Data 25-14

Alkamuls PSMS-4. Alkaril Chem.
Chemical Description: Polyoxyethylene sorbitan (4) monostearate
Category: Emulsifier, nonionic
Applications: Preparation of emulsified paints; insecticides, herbicides, fungicides; textiles; foods; cosmetics
Form: Solid; Color: Tan; Odor: Typical
Properties: Solubility: Water insol.; HLB: 9.6
Composition: 97% active
Reaction: Neutral
References: Technical Data 25-13

Alkamuls PSMS-20. Alkaril Chem.
Chemical Description: Polyoxyethylene sorbitan (20) monostearate
Category: Wetting agent, emulsifier, nonionic
Applications: Preparation of emulsified paints; insecticides, herbicides, fungicides; foods; cosmetics
Form: Viscous liquid, gel on standing; Color: Yellow; Odor: Typical
Properties: Solubility: Water sol.; HLB: 14.9
Composition: 97% active ingred.
Reaction: Neutral
References: Technical Data 25-12

Alkamuls PSTO-20. Alkaril Chem.
Chemical Description: Polyoxyethylene sorbitan (20) trioleate
Category: Emulsifier, nonionic
Applications: Preparation of emulsified paints; insecticides, herbicides, fungicides; foods; cosmetics
Form: Liquid; Color: Amber; Odor: Typical
Properties: Solubility: Water insol.; HLB: 11
Composition: 97% active ingred.
Reaction: Neutral

Alkamuls PSTS-20. Alkaril Chem.
Chemical Description: Polyoxyethylene sorbitan (20) tristearate
Category: Wetting agent, emulsifier, nonionic
Applications: Preparation of emulsified paints; insecticides, herbicides, fungicides; foods; cosmetics
Form: Solid; Color: Tan; Odor: Typical
Properties: Solubility: Water insol.; HLB: 10.5
Composition: 97% active ingred.
Reaction: Neutral
References: Technical Data 25-17

Alkamuls SML. Alkaril Chem.
Chemical Description: Sorbitan monolaurate
Category: Emulsifier, nonionic
Applications: Fat liquoring, dyeing, insecticides, herbicides, fungicides, textiles, foods, cosmetics
Form: Liquid (may form a sediment on standing which redissolves on warming); Color: Amber; Odor: Typical

Properties: Solubility: Water insol., moderately soluble in most alcohols, vegetable oils, mineral oils; Sp. gr.: 1.05 @ 60 F; HLB: 8.5
Composition: 97% active ingred.
Reaction: Neutral
References: Technical Data 25-1

Alkamuls SMO. Alkaril Chem.
Chemical Description: Sorbitan monooleate
Category: Wetting agent, emulsifier, nonionic
Applications: Making of cutting and soluble oils; dispersing waxes, pigments, resins; dyeing; insecticides, herbicides, fungicides; textiles; foods; cosmetics
Form: Liquid; Color: Amber; Odor: Typical
Properties: Solubility: Water insol., soluble in most vegetable and mineral oils, not soluble in glycol; Sp. gr.: 1.0; HLB: 4.3
Composition: 97% active ingred.
Reaction: Neutral
References: Technical Data 25-3

Alkamuls SMS. Alkaril Chem.
Chemical Description: Sorbitan monostearate
Category: Emulsifier, nonionic
Applications: Foods; cosmetics; dispersing waxes, pigments, resins
Form: Flakes; Color: Cream to tan; Odor: Typical
Properties: Solubility: Water insol.; HLB: 4.7
Composition: 98.5% active ingred.
Reaction: Neutral
References: Technical Data 25-2

Alkamuls STO. Alkaril Chem.
Category: Emulsifier, nonionic
Applications: Textiles; foods; cosmetics; insecticides, herbicides, fungicides
Form: Liquid; Color: Amber; Odor: Typical
Properties: Solubility: Water insol.; HLB: 1.8
Composition: 100% active ingred.
Reaction: Neutral

Alkamuls STS. Alkaril Chem.
Category: Emulsifier
Applications: Foods; cosmetics; dispersing waxes, pigments, resins
Form: Solid; Color: Tan; Odor: Typical
Properties: Solubility: Water insol.; HLB: 2.1
Composition: 100% active ingred.
Reaction: Neutral

Alkanol 189-S. DuPont Co.
Chemical Description: Sodium alkyl sulfonate
Category: Wetting agent, detergent, penetrant, anionic
Applications: Metal cleaning and pickling; bottle washing; general detergent applications
Form: Liquid; Color: Reddish-brown
Properties: Visc.: 30 cps @ 80 F; Stability: Good in acid, neutral and alkaline media; Surface Tension: 31.2 dynes/cm (1% sol'n. @ 90 F)
Composition: 31.5% active ingred.
References: Data sheet # 3

Alkanol A-CN. DuPont Co.
Chemical Description: Ethylene oxide condensate
Category: Antiprecipitant, dyeing

assistant, dye solubilizer, contrast agent, nonionic surfactant
Applications: Use in application of dyes to styled carpeting; aids dye solubility and exhaustion of two types of dye in same bath; aids level absorption of dyes
Form: Liquid; Color: Light amber
Properties: Solubility: > 50% in water; Density: 8.2 lb/gal @ 25 C; Stability: Indefinite in storage; Surface Tension: 38 dynes/cm (1% conc. @ 25 C)
References: Data sheet

Alkanol CNR. DuPont
Chemical Description: Amphoteric surfactant
Category: Dye assistant
Applications: Dye assistant for carpeting with both cationic- and anionic-dyeable nylon fibers; inhibits cross-staining; enhances level absorption of dyes
Form: Clear liquid; Color: Amber
Properties: Solubility: Miscible with water; Density: 8.61 lb/gal; Sp. gr.: 1.033
References: Bulletin

Alkanol DOA. DuPont Co.
Chemical Description: Amine polymer solution
Category: Dispersant
Applications: General dispersant for nonaqueous use; dispersant for drycleaning systems; paint manufacture; for use in reboilers in chemical manufacturing; to reduce sludge formation in storage of nonaqueous products; in quenching oils to prevent sludging
Form: Viscous liquid; Color: Light amber
Properties: Solubility: Complete in kerosene, hydrocarbons; insol. in acetone; Density: 7.4 lb/gal @ 25 C; Visc.: 540 cs @ 25 C; Flash Pt. (C.O.C.): 68 C; Pour Pt.: < 29 C; Stability: Excellent thermal stability
Toxicity: Acute oral toxicity, skin irritant
References: Data sheet

Alkanol DW. DuPont Co.
Chemical Description: Sodium alkylaryl sulfonate
Category: Emulsifier, wetting agent, detergent, anionic
Applications: Emulsifier for textile cleaning solvents; scouring, bleaching and dyeing textiles; defibering of waste paper; tar removal from wool; rewetting of textiles and paper mill felts; penetrant and spreader for shoe finishes; bottle washing; acid and alkaline cleaners; dairy and paint cleaners
Form: Liquid; Color: Yellow; Odor: Alcoholic
Properties: Solubility: Miscible in water; Density: 8.9 lb/gal; Cloud Pt.: 5 C (max.); Stability: Good in acid, alkali media, and storage
Composition: 28% active ingred.
References: Data sheet # 2

Alkanol ND. DuPont Co.
Chemical Description: Alkyl diaryl sulfonate
Category: Dyeing assistant, anionic surfactant
Applications: Dyeing assistant, aids leveling of acid dyes on nylon
Form: Liquid; Color: Yellow; Odor: Terpene

Properties: Solubility: Miscible with water; Density: 9.6 lb/gal; Visc.: Slightly viscous; Stability: Good in strong acids, alkalies, and in storage
Composition: 45% active ingred.
References: Data sheet

Alkanol S. DuPont Co.
Chemical Description: Sodium tetrahydronaphthalene sulfonate
Category: Dispersant, stabilizer, solubilizing agent, anionic
Applications: As dispersing, stabilizing and solubilizing agent in emulsions systems–textile printing compositions, cosmetic lotions and creams
Form: Small flakes; Color: Light cream
Properties: Solubility: Water sol. approx. 20%; Density: 2.6 lb/gal
Composition: 98% active ingred.
References: Data sheet # 3A

Alkanol WXN. DuPont Co.
Chemical Description: Sodium alkylaryl sulfonate
Category: Wetting, rewetting agent, dyeing assistant, anionic
Applications: Textile dyeing assistant and rewetting agent; leveling of acid dyes on corrugating medium and calender stains on boxboard; rewetting dried paper broke; rag cooking; dust control in coal piles
Form: Liquid; Color: Light yellow; Odor: Alcoholic
Properties: Solubility: (in water) miscible; Density: 8.6 lb/gal; Cloud Pt.: 0 C max.; Stability: Good in acid and alkali media, and storage; Surface Tension: 26 dynes/cm (1% sol'n. @ 25 C)
Composition: 30% active ingred.
References: Data sheet # 4

Alkanol XC. DuPont Co.
Chemical Description: Sodium alkylnaphthalene sulfonate
Category: Wetting agent, dispersant, penetrant, anionic
Applications: Rapid wetting and dispersing where excess foaming is undesirable and detergency is unnecessary; bleaching and dyeing of textiles, leather, paper; paper and paper mill felts; reduction of shrinkage in ceramics manufacture; dry colors manufacture
Form: Powder; Color: Light cream
Properties: Solubility: Soluble in ethyl alcohol, acetone, benzene, 8–10% in water @ 25 C; Sp. gr.: 0.55; Stability: Good in acid, alkali media and in storage
References: Data sheet # 1

Alkaphos 3. Alkaril Chem.
Chemical Description: Free acid of complex organic phosphate esters
Category: Detergent, wetting agent, emulsifier, solubilizer, antistatic agent, lubricants, anionic
Applications: Detergent–textiles, skins, leather, laundering esp. dry cleaning detergents, household and street, buildings, hospitals cleaners; esp. in high alkaline systems; Wetting agent–desizing and boiling off cottons; peroxide bleaching; finishing agent; insecticides, herbicides, fungicides; removing starches, wall paper; cleaning hangings; Emulsifer–Kier boil-

ing; making of cutting and soluble oils; printing
Form: Viscous liquid; Color: Pale yellow; Odor: Low
Properties: Solubility:Sol. in water and most aromatic and chlorinated solvents; more oil-sol. than Alkaphos-6 and -10; Sp. gr.: 1.025; Visc.: 863 cps @ 75 F; Stability: Excellent in acid and alkali media
Composition: 100% active ingred.
Reaction: Acid
References: Technical Data 45

Alkaphos 6. Alkaril Chem.
Chemical Description: Free acid of complex organic phosphate esters
Category: Detergent, wetting agent, emulsifier, solubilizer, antistatic agent, lubricant, anionic surfactant
Applications: See Alkaphos 3
Form: Viscous liquid; Color: Pale yellow; Odor: Low
Properties: Solubility: Sol. in water and most aromatic and chlorinated solvents; more oil-sol. than Alkaphos-10 and more hydrophilic than Alkaphos-3; Sp. gr.: 1.055; Visc.: 1045 @ 75 F; Stability: Excellent in acid and alkali media; Good rust inhibition properties
Composition: 100% active ingred.
Reaction: Acid
References: Technical Data 45

Alkaphos 10. Alkaril Chem.
Chemical Description: Free acid of complex organic phosphate esters
Category: Detergent, wetting agent, emulsifier, solubilizer, antistatic agent, lubricant, anionic
Applications: See Alkaphos 3; also water-repellent finish for cottons
Form: Viscous liquid; Color: Pale yellow; Odor: Low
Properties: Solubility: Sol. in water and most aromatic and chlorinated solvents; Sp. gr.: 1.085; Visc.: 2100 cps @ 75 F; Stability: Excellent in acid and alkali media; Good rust inhibition; Excellent cloud pt. depressant for nonionics
Composition: 100% active ingred.
Reaction: Acid
References: Technical Data 45

Alkaphos K 380. Alkaril Chem.
Chemical Description: Potassium salt of organic phosphate ester
Category: Solubilizer, hydrotropic surfactant, anionic
Applications: Solubilizer for built liquid detergents; liquid rinse aids to increase the solubility and raise cloud pt. of low foam nonionics
Form: Clear liquid
Properties: Cloud Pt.: < 0 C
Composition: 44% active ingred.
References: Technical Data 47

Alkaphos QS. Alkaril Chem.
Chemical Description: Free acid of complex organic phosphate ester
Category: Detergent, wetting agent, emulsifier, hydrotropic surfactant, solubilizer, anionic
Applications: Detergent–textiles, skins, laundering, household and street, building, hospital cleaners, esp. high alkaline and high electrolyte content cleaning preparations; Emulsifier–printing
Form: Liquid; Color: Colorless; Odor: Bland

Properties: Solubility: Excellent in water, caustic soda liquid and caustic potash liquid; Cloud Pt.: 68 C (1% sol'n. in 15% NaOH)
Composition: 90% active ingred.
Reaction: Alkaline
References: Technical Data 46

Alkapol PEG 400. Alkaril Chem.
Chemical Description: Polyoxyethylene glycol with average m.w. of 400
Category: Intermediate, plasticizer, solvent, lubricant
Applications: Intermediate for production of PEG-fatty acid esters; solvent or co-solvent for cosmetics, pharmaceuticals
Form: Liquid; Color: White
Properties: Solubility: Water sol.; Sp. gr.: 1.13 (20/15.5 C); F.P.: 4–8 C; Stability: Good physical chemical stability
References: Technical Data sheet

Alkapol PEG 600. Alkaril Chem.
Chemical Description: Polyoxyethylene glycol with average m.w. 600
Category: Intermediate, plasticizer, solvent, lubricant
Applications: Intermediate for production of PEG-fatty acid esters; solvent or co-solvent for cosmetics, pharmaceuticals
Form: Liquid; Color: White
Properties: Solubility: Water sol.; Sp. gr.: 1.13 (30/15.5 C); F.P.: 20–25 C; Stability: Good physical chemical stability
References: Technical Data sheet

Alkaquat C. Alkaril Chem.
Chemical Description: Dicoco imidazolinium quaternary
Category: Dyestuff retardant
Applications: Textiles processing
Form: Liquid; Color: Amber; Odor: Of solvent
Properties: Solubility: Dispersible in water; F.P.: < 0 C
Composition: 75% active ingred.; other components: isopropanol
Reaction: Neutral

Alkaquat DMB-451. Alkaril Chem.
Chemical Description: N-alkyl (C_{12}-C_{14}-C_{16}) methyl benzyl ammonium chloride
Category: Wetting agent, emulsifier, biocide, disinfectant, cationic
Applications: Bacteriocidal and cleaning formulations for beverage industry, dairy industry (plant equipment, milk tank trucks, milk pails cleaners), food processing; water treatment (algae control in swimming pools, water cooling systems); slimicides and mold control in paper industries; general disinfection and sanitization for public washrooms, hospitals, laundires; pest control; preservatives; antidandruff rinses
Form: Liquid; Color: Pale yellow (working solutions are odorless, colorless)
Properties: Solubility: Sol. in water, ethanol, acetone, many aliphatic solvents; Sp. gr.: 0.96; Stability: Good; Surface Tension: 33 dynes/cm @ 1% sol'n.
Composition: 50% min. active ingred.; other components: ethanol (10%), water (40% approx.)
Toxicity: Skin burns, severe eye

damage at concentration supplied
References: Technical Data sheet

Alkaquat O. Alkaril Chem.
Chemical Description: Oleic imidazoline quaternary
Category: Softening agent, cationic
Applications: Liquid form of Alkaquat T for use in high concentrations of fabric softeners; textiles; leather; laundering; papermaking; household; cosmetics
Form: Liquid; Color: Amber; Odor: Of solvent
Properties: Sp. gr.: 0.95; F.P.: $<$ 0 C
Composition: 75% active ingred.; other components: isopropanol
Reaction: Neutral
References: Technical Data

Alkaquat ST. Alkaril Chem.
Chemical Description: Modified dialkyl imidazolinium quaternary based on stearyl derivative
Category: Fabric softener, antistat, cationic
Applications: Fabric softener and antistat for textile fibers, esp. for direct application to cloth; manufacture of sheet fabric softeners for dryer use
Form: Solid; Color: Cream; Odor: Isopropanol solvent
Properties: Solubility: Isopropanol solvent; M.P.: 45 C
Composition: 75% active ingred.; other components: isopropanol
Reaction: Neutral
References: Technical Data 71

Alkaquat T. Alkaril Chem.
Chemical Description: Ditallow imidazolinium quaternary
Category: Softening agent, cationic
Applications: Fabric softener for household and commercial laundry use and for textile finishing operations; papermaking; cosmetics
Form: Slightly viscous fluid; Color: Pale yellow; Odor: Typical
Properties: Solubility: Water insol.; Sp. gr.: 0.95
Composition: 74–76% active ingred.; other components: isopropanol
Reaction: Neutral
References: Technical Data 70

Alkaquest EDTA. Alkaril Chem.
Chemical Description: Tetrasodium ethylene diamine tetra acetate, aqueous solution
Category: Chelating agent
Applications: General purpose chelating agent; complexes calcium, magnesium, iron, etc., through wide pH range
Form: Liquid; Color: Straw
Properties: m.w.: 380 (anhydrous); Sp. gr.: 1.280; Chelation value: 10 (calcium, gms. of calcium carbonate sequestered by 100 gms. of product)
Composition: 37.5% active ingred.
References: Technical Data 60

Alkasurf ALS. Alkaril Chem.
Chemical Description: Ammonium lauryl sulfate
Category: Detergent, wetting agent, emulsifier, anionic
Applications: Hair shampoo and other cosmetics; household use
Form: Liquid; Color: Light; Odor: Typical

Properties: Solubility: Water sol.
Composition: 27% active ingred.; other components: water
Reaction: Neutral

Alkasurf CA. Alkaril Chem.
Chemical Description: Calcium alkaryl sulfonate
Category: Emulsifier
Applications: Emulsifier for agricultural oils; insecticides, herbicides, fungicides; also in combination with ethoxylated nonionics in self-dispersing liquids
Form: Liquid; Color: Amber; Odor: Typical
Properties: Solubility: Water insol.; HLB: 10.5
Composition: 60% active ingred.; other components: buthanol
Reaction: Neutral
References: Technical Data 44

Alkasurf CBX16. Alkaril Chem.
Chemical Description: Fatty alcohol polyglycol ether carboxylic salt
Category: Detergent, wetting agent, anionic
Applications: Detergent with improved resistance to alkali/electrolytes; textiles; laundering; household; street, buildings, hospitals cleaners; desizing and boiling off cottons
Form: Liquid; Color: Light yellow; Odor: None
Properties: Solubility: Good in alkali; F.P.: < 0 C
Composition: 70% active ingred.
Reaction: Neutral
References: Technical Data 40

Alkasurf CBX16 Acid. Alkaril Chem.
Chemical Description: Fatty alcohol polyglycol ether carboxylic acid
Category: Detergent, wetting agent, anionic
Applications: Detergent with improved resistance to alkali/electrolytes; textiles; laundering; household; street, buildings, hospitals cleaners; desizing and boiling off cottons
Form: Liquid; Color: Light yellow; Odor: None
Properties: Solubility: Good in alkali
Composition: 100% active ingred.
Reaction: Acid
References: Technical Data 40

Alkasurf CO-25. Alkaril Chem.
Chemical Description: Ethoxylated castor oil
Category: Detergent, wetting agent, emulsifier, nonionic
Applications: Textiles; cosmetics
Form: Liquid; Color: Yellow; Odor: Typical
Composition: 100% active ingred.
Reaction: Neutral

Alkasurf DLS. Alkaril Chem.
Chemical Description: Diethanolamine lauryl sulfate
Category: Detergent, wetting agent, emulsifier
Applications: Hair shampoos, cosmetics, household use
Form: Liquid; Color: Light; Odor: Typical
Properties: Solubility: Water sol.
Composition: 35% active ingred.; other components: water

Reaction: Neutral

Alkasurf EA-60. Alkaril Chem.
Chemical Description: Ammonium lauryl ether sulfate
Category: Detergent, wetting agent, emulsifier, anionic
Applications: General purpose surfactant, esp. good quality dishwashing detergent, shampoo, bubble bath; also laundering, dyeing, household, cosmetic, cleaners of street, buildings, hospitals; wetting agents–dyeing, spot removal
Form: Clear slightly viscous liquid; Color: Light amber; Odor: Typical
Properties: Solubility: Water sol.; Sp. gr.: 1.04 @ 20 C; Cloud Pt.: $<$ 40 F
Composition: 58% active ingred.; other components: water, ethanol
Reaction: Neutral
References: Tech. Data

Alkasurf ES-60. Alkaril Chem.
Chemical Description: Sodium lauryl ether sulfate
Category: Detergent, emulsifier, wetting agent, anionic
Applications: General purpose surfactant esp. dishwashing detergent, shampoo, bubble bath; also textiles; laundering; dyeing; household; cosmetics; street, buildings, hospitals cleaners; dyeing and spot removal as wetting agent
Composition: 58% active ingred.; other components: water, ethanol
Reaction: Neutral
References: Tech. Data

Alkasurf IPAM. Alkaril Chem.
Chemical Description: Linear dodecyl benzene isopropyl amine sulfonate
Category: Emulsifier
Applications: Oil-soluble emulsifier, dry cleaning surfactant, emulsion degreasers, fuel oil additives, textiles, transport, insecticides, fungicides, herbicides
Form: Clear viscous liquid; Color: Pale yellow
Properties: Solubility: Water insol., sol. in most commonly used solvents; Sp. gr.: 1.06
Composition: 95% active ingred.
Reaction: Neutral
References: Tech. Data

Alkasurf LA-3. Alkaril Chem.
Chemical Description: Linear fatty alcohol ethoxylate
Category: Emulsifier, nonionic
Applications: Emulsifier and surfactant intermediate esp. in agricultural concentrates, dry cleaning soaps, degreasing compounds, wool lubricants, rubber based paints; making of cutting and sol. oils; dispersing waxes, pigments, resins
Form: Liquid; Color: Transluscent to opaque; Odor:
Properties: Solubility: Water insol.; Sp. gr.: 0.9 (25/25 C); Visc.: 19 (100 F); Cloud Pt.: 58–62 C; Stability: Good; HLB: 7.8
Composition: 100% active ingred.
Reaction: Neutral
References: Tech. Data sheet 21-1

Alkasurf LA-7. Alkaril Chem.
Chemical Description: Linear fatty alcohol ethoxylate
Category: Detergent, emulsifier,

wetting agent, nonionic
Applications: General purpose detergent–household, industrial cleaners; textiles; laundering; papermaking; paints, varnishes; transport; Wetting agent–caustic and mineral acid cleaners; dyeing; rewetting dried skins; pickling; electroplating; rotary washer; wetting pulp in beater; removing starches, wall paper; rewetting agent for paper towels, tissues; Emulsifier–making of cutting and soluble oils; dispersing waxes, pigments, resins; printing
Form: Liquid; Color: Light; Odor:
Properties: Solubility: Water sol.; Sp. gr.: 0.96 (50/25 C); Visc.: 32 cps @ 100 F; M.P.: 11–15 C; HLB: 12; Cloud Pt.: 44–48 C; Flash Pt.: 440 F (COC); Pour Pt.: 70 F; Stability: Good; Biodegradable
Composition: 100% active ingred.
Reaction: Neutral
References: Tech. Data sheet 21-2

Alkasurf LA-9. Alkaril Chem.
Chemical Description: Linear fatty alcohol ethoxylate
Category: Detergent, wetting agent, emulsifier, nonionic
Applications: General purpose surfactant–household and industrial cleaners; textiles; laundering; dyeing; papermaking; paints; varnishes; transport; cosmetics; Wetting agents–dyeing; rewetting dried skins; pickling; electroplating; rotary washer; wetting pulp in beater; caustic and mineral acid cleaners; Emulsifier–making of cutting and sol. oils; dispersing waxes, pigments, resins; printing
Form: Semisolid; Color: Light; Odor: Typical
Properties: Solubility:Water sol.; Sp. gr.: 1.01 (20 C); HLB: 13.1; Stability: Good; Biodegradable
Composition: 100% active ingred.
Reaction: Neutral
References: Tech. Data sheet 21-3

Alkasurf LA-9 85%. Alkaril Chem.
Chemical Description: See Alkasurf LA-9
Category and Applications: See Alkasurf LA-9
Form: Liquid form of Alkasurf LA-9
Composition: 85% active ingred.

Alkasurf LA-12. Alkaril Chem.
Chemical Description: Linear fatty alcohol ethoxylate
Category: Detergent, wetting agent, emulsifier, nonionic
Applications: See Alkasurf LA-9; also kier boiling; O/W emulsifier and solubilizer for topical cosmetic applications
Form: White solid; Odor: Low
Properties: Solubility: Water sol.; Sp. gr.: 1.0 (50/25 C); HLB: 14.4
Composition: 100% active ingred.
Reaction: Neutral

Alkasurf LA-15. Alkaril Chem.
Chemical Description: Linear fatty alcohol ethoxylate
Category: Emulsifier, detergent, nonionic
Applications: O/W emulsifier and solubilizer for topical cosmetic applications (has low sol. in nonpolar solvents); household and industrial cleaners (esp. high alka-

linity, high temperature cleaners); emulsion polymerization systems
Form: White solid
Properties: Biodegradable
Composition: 100% active ingred.
References: Tech. Data sheet 21-5

Alkasurf LA-16. Alkaril Chem.
Chemical Description: Straight chain polyoxyethylene fatty alcohol condensate
Category: Detergent, emulsifier, wetting agent, nonionic
Applications: Detergent—textiles, skins, leather, laundering, dyeing, papermaking, transport, household, cosmetics, industrial cleaners; Wetting agent—desizing and boiling off cottons; carbonizing woolens; dyeing; rewetting dried skins; tanning; pickling; electroplating; rotary washer; removing starches, wall paper; Emulsifer—making of cutting and sol. oils; dispersing waxes, pigments, resins; printing
Form: Clear colorless liquid; Odor: Low
Properties: Solubility: Water sol.; Sp. gr.: 1.0 (20/20 C); Visc.: 267 SUS (20 C); Biodegradable
Composition: 100% active ingred.
Reaction: Neutral
References: Tech. Data sheet 22-1

Alkasurf LA Acid. Alkaril Chem.
Chemical Description: High active sulfonic acid produced from the continuous SO_3 sulfonation of linear alkylate
Category: Detergent, emulsifier, penetrating and foaming agent
Applications: Dishwashing and heavy-duty liquid, powdered and paste detergents; car shampoos; bubble baths; degreasers and acid cleaners; dry cleaning detergents; metal cleaners; emulsion polymerization; agricultural emulsifiers
Form: Viscous liquid; Color: Brown
Properties: Stability: Good in hard and soft water, strong alkalies and strong acids; Biodegradable
Composition: 96% active ingred.
References: Tech. Data Sheet 41

Alkasurf LAN-23. Alkaril Chem.
Chemical Description: Linear fatty alcohol ethoxylate
Category: Emulsifier, nonionic surfactant
Applications: O/W emulsifier esp. topical cosmetic applications
Form: Solid; Color: Almost colorless; Odor: Characteristic
Properties: Biodegradable
Composition: 100% active ingred.
References: Tech. Data sheet 21-11

Alkasurf MLS. Alkaril Chem.
Chemical Description: Monoethanolamine lauryl sulfate
Category: Detergent, wetting agent
Applications: Surfactant for clear viscous shampoos; wax strippers where lower ammonia odor is required
Form: Liquid; Color: Pale yellow; Odor: Low
Composition: 32% active ingred.
Reaction: Neutral

Alkasurf NP-4. Alkaril Chem.

Chemical Description: Nonyl phenol ethoxylate
Category: Emulsifier, dispersant, intermediate, nonionic
Applications: Emulsifier for mineral oils, kerosene, chlorinated hydrocarbons; co-emulsifier with water soluble surfactants; oil soluble dispersant, surfactant intermediate; household, industrial, cosmetic use; fat liquoring; making of cutting and sol. oils; dispersing waxes, pigments, resins
Form: Liquid; Color: Light; Odor: Low
Properties: Solubility: Water insol.; Sp. gr.: 1.02; HLB: 9; Stability: Complete in acid or alkaline media
Composition: 100% active ingred.
Reaction: Neutral
References: Tech. Data sheet 23-1

Alkasurf NP-5. Alkaril Chem.
Chemical Description: Nonyl phenol ethoxylate
Category: Emulsifier, co-emulsifier, oil soluble dispersant, nonionic
Applications: See Alkasurf NP-4; also printing applications; preparation of emulsified paints
Form: Liquid; Color: Light; Odor: Low
Properties: Solubility: Water insol.; Sp. gr.: 1.03; HLB: 10; Stability: Complete in acid or alkaline media
Composition: 100% active ingred.
Reaction: Neutral
References: Tech. Data sheet 23-2

Alkasurf NP-6. Alkaril Chem.
Chemical Description: Nonyl phenol ethoxylate
Category: Emulsifier, co-emulsifier, oil soluble dispersant, nonionic
Applications: Household and industrial cleaners; intermediate; plasticizer and antistatic agent for plastics; emulsifier for mineral oils; insecticides, fungicides, herbicides; fat liquoring; making of cutting and sol. oils; dispersing waxes, pigments, resins; printing; preparation of emulsified paints
Form: Liquid; Color: Light; Odor: Characteristic
Properties: Solubility: Water insol.; Sp. gr.: 1.04; HLB: 11.0; Stability: Good
Composition: 100% active ingred.
Reaction: Neutral
References: Tech. Data sheet 23-3

Alkasurf NP-8. Alkaril Chem.
Chemical Description: Nonyl phenol ethoxylate
Category: Detergent, emulsifier, wetting agent
Applications: Low foaming general purpose detergent and wetting agent; household and industrial cleaners; textiles; leather; laundering; dyeing; papermaking; paints; cosmetics; dyeing; rewetting dried skins; electroplating; rotary washer; wetting pulp in beater; Kier boiling; fat liquoring; making of cutting and soluble oils; dispersing waxes, pigments, resins; printing

Form: Liquid; Color: Light; Odor: Low
Properties: Solubility: Water sol.; Sp. gr.: 1.05; HLB: 12.0; Stability:

Composition: 100% active ingred.
Reaction: Neutral
References: Tech. Data sheet 23-4

Alkasurf NP-9. Alkaril Chem.
Chemical Description: Nonyl phenol ethoxylate
Category: Detergent, wetting agent, emulsifier, nonionic
Applications: Detergent–household and industrial cleaners; textiles; skins; leather; laundering; dyeing; papermaking; paints; varnishes; cosmetics; Wetting agent–dyeing; rewetting dried skins; tanning; spot removal; pickling; electroplating; rotary washer; wetting pulp in beater; sizing; insecticides, fungicides, herbicides; Emulsifer–Kier boiling; fat liquoring; making of cutting and sol. oils; dispersing waxes, pigments, resins; printing; making of bitumen emulsions
Form: Liquid; Color: Light; Odor: Low
Properties: Solubility: Water sol.; Sp. gr.: 1.06; HLB: 13.4; Stability: Good
Composition: 100% active ingred.
Reaction: Neutral
References: Tech. Data sheet 23-5

Alkasurf NP-10. Alkaril Chem.
Chemical Description: Nonyl phenol ethoxylate
Category: Detergent, emulsifier, wetting agent, nonionic
Applications: Household and industrial cleaners; textiles; skins; leather; laundering; dyeing; papermaking; paints; varnishes; cosmetics; desizing and boiling off cottons; carbonizing woolens; rewetting dried skins; spot removal; pickling; electroplating; rotary washer; wetting pulp in beater; sizing; making of bitumen emulsions
Form: Liquid; Color: Light; Odor: Low
Properties: Solubility: Water sol.; Sp. gr.: 1.06; HLB: 13.5; Stability: Good
Composition: 100% active ingred.
Reaction: Neutral
References: Tech. Data sheet 23-6

Alkasurf NP-11. Alkaril Chem.
Chemical Description: Nonyl phenol ethoxylate
Category: Detergent, emulsifier, wetting agent, nonionic
Applications: Household and industrial cleaners; textiles; skins; leather; laundering; dyeing; paints; varnishes; cosmetics; dyeing; rewetting dried skins; spot removal; pickling; electroplating; rotary washer; wetting pulp in beater; removing starches, wall paper; cleaning hangings; dispersing waxes, pigments, resins; making of bitumen emulsions
Form: Liquid; Color: Light; Odor: Low
Properties: Solubility: Water sol.; Sp. gr.: 1.06; HLB: 13.8; Stability: Good
Composition: 100% active ingred.
Reaction: Neutral
References: Tech. Data sheet 23-7

Alkasurf NP-15. Alkaril Chem.
Chemical Description: Nonyl phenol ethoxylate

Category: Detergent, wetting agent, nonionic
Applications: High temperature general purpose detergent, compatible with higher conc. of electrolytes than lower NP series; textiles; skins; leather; laundering; dyeing; paints; varnishes; household; cosmetics; street, building, hospital cleaners; bleaching; pickling; electroplating; rotary washer; dispersing waxes, pigments, resins; making of bitumen emulsions
Form: Paste; Color: Light; Odor: Low
Properties: Solubility: Water sol.; Sp. gr.: 1.07; HLB: 15; Stability: Good
Composition: 100% active ingred.
Reaction: Neutral
References: Tech. Data sheet 23-8

Alkasurf NP-40. Alkaril Chem.
Chemical Description: Nonyl phenol ethoxylate with ≈ 40 moles of ethylene oxide
Category: Wetting agent, emulsifier, stabilizer, dispersant, penetrant, nonionic
Applications: Polymer emulsifier, high temperature applications; textiles; dyeing; paints; dispersing waxes, pigments, resins; printing; preparation of emulsified paints
Form: Liquid; Color: Light; Odor: Low
Properties: Solubility: Excellent water sol.; F.P.: 50–60 F
Composition: 70% active ingred.
Reaction: Neutral
References: Tech. Data sheet 23-11

Alkasurf OP-5. Alkaril Chem.
Chemical Description: Octyl phenoxy polyethoxy ethanol containing 5 moles of ethylene oxide
Category: Emulsifier, nonionic surfactant
Applications: Improves detergency, dispersability, wetting action of solvent based systems; emulsion type cleaners; co-emulsifier
Form: Liquid; Color: Light; Odor: Characteristic
Properties: Solubility: Oil sol.; Sp. gr.: 1.05 (15/15 C); HLB: 10.4
Composition: 100% active ingred.
References: Tech. Data sheet 20-2

Alkasurf OP-8. Alkaril Chem.
Chemical Description: Octyl phenoxy polyethoxy ethanol containing 7-8 moles of ethylene oxide
Category: Emulsifier, detergent, wetting agent, nonionic
Applications: Household and industrial cleaning and laundering, low foam level; agricultural pesticides; textiles; dyeing, cement, papermaking; paints; varnishes; transport; spot removal; pickling; electroplating; rotary washer
Form: Liquid; Color: Light; Odor: Low
Properties: Solubility: Water sol.; Sp. gr.: 1.06 (15/15 C); HLB: 12.4; Stability: Good
Composition: 100% active ingred.
Reaction: Neutral
References: Tech. Data sheet 20-3

Alkasurf OP-10. Alkaril Chem.
Chemical Description: Octyl phenoxy polyethoxy ethanol containing ≈ 10 moles of ethylene oxide
Category: Detergent, emulsifier, wetting agent, nonionic

Applications: Household and industrial cleaners; textiles; laundering; dyeing; paints; varnishes; transport; rewetting dried skins; spot removal; pickling; electroplating; rotary washer; caustic and mineral acid cleaners; dispersing waxes, pigments, resins; printing
Form: Liquid; Color: Light; Odor: Low
Properties: Solubility: Water sol.; Sp. gr.: 1.1 (15/15 C); HLB: 13.5; Stability: Good
Composition: 100% active ingred.
Reaction: Neutral
References: Tech. Data sheet 20-4

Alkasurf OP-12. Alkaril Chem.
Chemical Description: Octyl phenoxy polyethoxy ethanol containing 12-13 moles of ethylene oxide
Category: Detergent, emulsifier, wetting agent, nonionic
Applications: Household and industrial cleaners; textiles; laundering; dyeing; paints; varnishes; transport; spot removal; pickling; electroplating; rotary washer; removing starches, wall paper; cleaning hangings; wetting agent in caustic and mineral acid cleaners
Form: Liquid; Color: Light; Odor: Low
Properties: Solubility: Water sol.; Sp. gr.: 1.07 (25/25 C); HLB: 14.6; Stability: Good
Composition: 100% active ingred.
Reaction: Neutral
References: Tech. Data sheet 20-5

Alkasurf OP-30 70%. Alkaril Chem.
Chemical Description: Octyl phenoxy polyethoxy ethanol containing 30 moles of ethylene oxide
Category: Detergent, emulsifier, wetting agent, nonionic
Applications: Used where maximum water solubility is required at normal and high temperature and conc. electrolyte sol'ns.; stabilizer for synthetic latices
Form: Liquid; Color: Light; Odor: Low
Properties: Solubility: Excellent water sol.; Sp. gr.: 1.1; HLB: 17.3
Composition: 70% active ingred.
References: Tech. Data sheet 20-7

Alkasurf OP-40. Alkaril Chem.
Chemical Description: Octyl phenoxy polyethoxy ethanol containing 40 moles of ethylene oxide
Category: Detergent, emulsifier, wetting agent, nonionic
Applications: See Alkasurf OP-30; textiles; laundering; dyeing; household; industrial cleaners; pickling; electroplating; rotary washer
Form: Liquid; Color: Light; Odor: Low
Properties: Solubility: Excellent water sol.; Sp. gr.: 1.1; HLB: 18; Stability: Good
Composition: 70% active ingred.; other components: water
Reaction: Neutral
References: Tech. Data sheet 20-8

Alkasurf SS-DA-6. Alkaril Chem.
Chemical Description: Sodium sulfosuccinate ester of an ethoxylated C_{10} to C_{12} straight-chain alcohol mixture
Category: Detergent, foaming and wetting agent, anionic

Applications: Emulsion polymerization of vinyl acetate and acrylates
Form: Liquid; Color: Pale yellow; Odor: Bland
Properties: Solubility: Water sol.; Sp. gr.: 1.1; F.P. < 0 C
Composition: 30% active ingred.; other components: water
Reaction: Neutral
References: Tech. Data sheet

Alkasurf SS-L9DE. Alkaril Chem.
Chemical Description: Sodium lauryl alkylolamide sulfosuccinate
Category: Detergent, anionic
Applications: Low irritation surfactant for toiletries; household use
Form: Liquid; Color: Light yellow; Odor: Bland
Properties: Solubility: Excellent water sol.; Sp. gr.: 1.17; Stability: Good; Biodegradable
Composition: 40% active ingred.; other components: water
Reaction: Neutral
References: Tech. Data sheet 32

Alkasurf SS-L9ME. Alkaril Chem.
Chemical Description: Sodium lauryl alkylolamide sulfosuccinate
Category: Detergent, anionic
Applications: High foaming surfactant for toiletry and carpet shampoos; laundering; cleaners of street, buildings, hospitals
Form: Viscous liquid; Color: Cream white; Odor: Bland
Properties: Sp. gr.: 1.00; Stability: Good; Biodegradable
Composition: 40% active ingred.
Reaction: Neutral
References: Tech. Data sheet 33

Alkasurf SS-LA-3. Alkaril Chem.
Chemical Description: Sodium polyglycol ether sulfosuccinate
Category: Detergent, emulsifier, wetting agent, anionic
Applications: Cosmetics esp. hair shampoos, bubble bath; emulsion polymerization; household and industrial cleaners
Form: Clear liquid; Color: Light; Odor: Typical
Properties: Solubility: Water sol.; Sp. gr.: 1.10
Composition: 30% active ingred.; other components: water
Reaction: Neutral
References: Tech. Data sheet 31

Alkasurf SS-O-60. Alkaril Chem.
Chemical Description: Sodium dioctyl sulfosuccinate
Category: Wetting agent, anionic
Applications: Textile industry–desizing, dyeing and printing, wetting and penetrating; dry cleaning; agriculture–insecticides, fungicides, treatment of grains and forage; activation of straw manures; cosmetics; paints; emulsion polymerization; rewetting dried skins; tanning; finishing; electroplating; abatement of dust; wetting pulp in beater; removing starches, wall paper; cleaning hangings
Form: Clear colorless liquid
Properties: Solubility: Water 1%; most organic liquids; Sp. gr.: 1.07; Surface Tension: 29 dynes/cm (0.1% active in water)
Composition: 60% active ingred.; other components: ≈ 13% ethanol, ≈ 27% water
Reaction: Neutral

Toxicity: Eye irritant
References: Tech. Data sheet 34

Alkasurf SS-O-75. Alkaril Chem.
Chemical Description: Sodium dioctyl sulfosuccinate
Category and Applications: Powerful wetting agent for aqueous and organic systems; see Alkasurf SS-O-60
Form: Clear colorless liquid; Odor: Typical
Properties: Solubility: 1% in water; soluble in most organic liquids; Sp. gr.: 1.09; F.P.: <–20 C; Surface Tension: 29 dynes/cm (0.1% active in dist. water)
Composition: 75% active ingred.; other components: ≈ 16% water, ≈ 9% ethanol
Reaction: Neutral
Toxicity: Eye irritant
References: Tech. Data sheet 35

Alkasurf SS-OA. Alkaril Chem.
Chemical Description: Disodium N-oleyl sulfosuccinamate
Category: Foaming agent
Applications: Foaming agent for rubber latex foam; emulsifying, dispersing, solubilizing, foam stabilizing action on soaps, etc.
Form: Liquid; Color: Amber; Odor: Low
Properties: Solubility: Water sol.; Visc.: 25–30 cps @ 70 F; F.P.: < 0 C
Composition: 31% active ingred.; other components: water
Reaction: Alkaline
References: Tech. Data sheet 37

Alkasurf SS-O-ME. Alkaril Chem.
Chemical Description: Fatty alkanolamide sulfosuccinate
Category: Emulsifier, anionic
Applications: Emulsifier for use in synthetic latices and other polymer emulsions
Form: Liquid
Composition: 40% conc.

Alkasurf SS-TA. Alkaril Chem.
Chemical Description: Disodium N-octadecyl sulfosuccinamate
Category: Detergent, foaming agent
Applications: Foaming agent for rubber latex foam; heavy duty alkaline cleaner; cement, brick, tile cleaners; emulsification of waxes for oil and wax polishes; emulsion polymerization of vinyl acetate
Form: Paste; Color: Cream; Odor: Low
Properties: Solubility: Poor in water
Composition: 35% active ingred.; other components: water
Reaction: Alkaline
References: Tech. Data sheet 36

Alkasurf T. Alkaril Chem.
Chemical Description: Cosmetic grade triethanolamine linear alkylate sulfonate
Category: Detergent, emulsifier, wetting agent
Applications: Hair shampoo, bubble bath, other cosmetic, household use
Form: Clear liquid; Color: Straw; Odor: Typical
Properties: Solubility: Complete in water; Visc.: 2500 cps max.
Composition: 60% active ingred.; other components: water
Reaction: Neutral

References: Tech. Data sheet 43

Alkasurf TDA-7.5. Alkaril Chem.
Chemical Description: Tridecyl alcohol ethoxylate
Category: Emulsifier, wetting agent, nonionic
Applications: Foam builder and solubilizer for alkyl aryl sulfates, etc.; emulsifier and co-emulsifier for O/W systems; wetting and rewetting agent in various industrial applications; chemical specialty manufacturing
Form: Opaque liquid; Color: Almost water white; Odor: Mild
Properties: Solubility: Water sol.; Sp. gr.: 1.00 (20/20 C)
Composition: 100% active ingred.
References: Tech. Data sheet

Alkasurf TDA-12. Alkaril Chem.
Chemical Description: Tridecyl alcohol ethoxylate
Category: Emulsifier, detergent, nonionic
Applications: Emulsifier and co-emulsifier for O/W systems; leveling agent; solubilizer; high-temperature detergent
Form: White solid; Odor: Mild
Properties: Solubility: Complete in water; M.P.: ≈ 27 C
Composition: 100% active ingred.
References: Tech. Data sheet

Alkasurf TDA-15. Alkaril Chem.
Chemical Description: Tridecyl alcohol ethoxylate
Category: Emulsifier, detergent, nonionic
Applications: Emulsifier and co-emulsifier for O/W systems; leveling agent, solubilizer, high-temperature detergent
Form: White solid; Odor: Mild
Properties: Solubility: Complete in water; M.P.: ≈ 30 C
Composition: 100% active ingred.
References: Tech. Data sheet

Alkasurf TLS. Alkaril Chem.
Chemical Description: Triethanolamine lauryl sulfate
Category: Detergent, emulsifier, wetting agent, anionic
Applications: Hair shampoos, bubble baths, and other toiletries where low cloud point properties are critical; household uses
Form: Liquid; Color: Light amber; Odor: Characteristic
Properties: Solubility: Water sol.; Sp. gr.: 1.02 (20/20 C); Cloud Pt.: 0 C
Composition: 40% active ingred.
Reaction: Neutral
References: Tech. Data sheet 26

Alkasurf WADX. Alkaril Chem.
Chemical Description: Modified diethanolamine lauryl sulfate
Category: Detergent, anionic
Applications: Cosmetic and pharmaceutical cleaning preparations
Form: Clear to slightly hazy viscous liquid; Color: Pale straw; Odor: Bland
Properties: Water sol.
Composition: 30% active ingred.
References: Tech. Data sheet 29

Alkasurf WAQ. Alkaril Chem.
Chemical Description: Sodium lauryl sulfate
Category: Detergent, emulsifier,

wetting agent, anionic
Applications: Hair shampoo, hand lotions, bubble bath, other cosmetics; institutional detergents, sanitizers, floor cleaners, rug and upholstery shampoos; laundry and dry cleaning uses
Form: Liquid; Color: White pearlescent; Odor: Typical
Properties: Solubility: Water sol.
Composition: 30% active ingred.
Reaction: Neutral
References: Tech. Data sheet 27

Alkasurf WHC. Alkaril Chem.
Chemical Description: Blend of emulsifiers suited to compounding waterless hand cleaner
Category: Wetting agent
Applications: Waterless hand cleaner base
Form: Clear viscous liquid; Color: Amber; Odor: Typical
Composition: 100% active ingred.
Reaction: Alkaline
References: Tech. Data sheet 65

Alkaterge-E. IMC Chem.
Chemical Description: Oxazoline-type compound
Category: Detergent, emulsifier, wetting agent
Applications: Salts and soaps; paper, textiles, metal cleaners; emulsion stabilizer; acid acceptor; pigment grinding and dispersion; antifoam agent; antioxidant
Form: Clear liquid; Color: 15 max. (Gardner)
Properties: Solubility: Sol. in most organic liquids, slight sol, in water; Density: 7.74 lb/gal; Sp. gr.: 0.9; Visc.: 155 cp @ 25 C; F.P.: ≈ – 31 C; Flash Pt.: > 212 F; Surface Tension: 40 dynes/cm (0.001% aqueous sol'n.)
Composition: 79% active
Containers: Can, drums
References: NP Data sheet 4B

Alkaterge-T. IMC Chem.
Chemical Description: Oxazoline-type compound
Category: Emulsifier, dispersant
Applications: Corrosion inhibitor; emulsifying agent; pigment dispersion
Form: Waxy solid; Color: Buff to brown
Properties: Solubility: Highly sol. in aromatic hydrocarbons; Flash Pt.: > 200 F (Tag closed cup); Surface Tension: 30.4 dynes/cm
Toxicity: Skin irritant
Warnings and Cautions: Avoid prolonged or repeated exposure to skin
References: NP Data sheet 4A

Alkateric BC. Alkaril Chem.
Chemical Description: Alkyl dimethyl betaine
Category: Detergent
Applications: Low irritation shampoo base and hair conditioning additive; coupling agent for quaternary based shampoos; amphoteric surfactant; dyeing assistant in textile industry; household, cosmetic use; cleaners of street, buildings, hospitals
Form: Liquid; Color: Pale straw; Odor: Low
Properties: Solubility: Water sol.; F.P.: <0 C; pH: 7.5–8.5

Composition: 31% active ingred.
Reaction: Neutral
References: Tech. Data sheet

Alkateric CAB. Alkaril Chem.
Chemical Description: Betaine of coconut alkylolamide
Category: Detergent, amphoteric foaming agent
Applications: Foaming/coupling agent for shampoos; household, cosmetic use; cleaners of street, buildings, hospitals
Form: Liquid; Color: Amber; Odor: Bland
Properties: Solubility: Water complete
Composition: 32% active ingred.; other component: water
Reaction: Neutral

Alkateric CIB. Alkaril Chem.
Category: Detergent
Applications: Amphoteric foaming agent and surfactant; baby shampoo; household and cosmetic use; cleaners of street, buildings, hospitals
Form: Liquid; Color: Amber; Odor: Bland
Properties: Sp. gr.: 1.1
Composition: 40% active ingred.; other components: water
Reaction: Alkaline
Toxicity: Low eye irritation

Alkateric STB. Alkaril Chem.
Chemical Description: Stearyl betaine
Category: Detergent, amphoteric surfactant/softener
Applications: Detergent, amphoteric surfactant/softener unaffected by hard water; household, cosmetic use; cleaners of street, buildings, hospitals
Form: Liquid; Color: Pale yellow
Composition: 50% active ingred.; other components: water/alcohol
Reaction: Neutral

Alkateric TB. Alkaril Chem.
Chemical Description: Tallow betaine
Category: Detergent; amphoteric surfactant/softener
Applications: Household, cosmetic use; cleaners of street, buildings and hospitals
Form: Liquid; Color: Amber
Composition: 70% active ingred.; other components: water/alcohol
Reaction: Neutral

Alkatrope SX-40. Alkaril Chem.
Chemical Description: Sodium xylene sulfonate
Category: Hydrotrope, solubilizer and coupling agent
Applications: Used to dissolve detergents in electrolyte sol'ns.; used to lower cloud point and viscosity of detergent systems
Form: Clear liquid; Color: Light
Composition: 41% active ingred.
Containers: 550-lb bung-type drums

Alkazine C. Alkaril Chem.
Chemical Description: Coconut hydroxyethyl imidazoline
Category: Emulsifier, antistat
Applications: Textiles, leather, household
Form: Paste; Color: Brown; Odor: Amine
Properties: Solubility: Sol. in aqueous acid; F.P.: 20 C; Stability:

Not stable to alkali
Composition: 85% active ingred.
Reaction: Alkaline
References: Tech. Data sheet 80

Alkazine O. Alkaril Chem.
Chemical Description: Oleic hydroxyethyl imidazoline
Category: Emulsifier, corrosion inhibitor, antistat agent
Applications: Emulsifier; asphalt emulsions; corrosion inhibitor; acid cleaners; solvent degreasers; textile antistat; leather and household uses
Form: Liquid; Color: Brown; Odor: Amine
Properties: Solubility: Sol. in aqueous acid and organics; F.P.: 5 C; Stability: Not stable in alkaline medium
Composition: 85% active ingred.
Reaction: Alkaline
References: Tech. Data sheet 81

Alkazine ST. Alkaril Chem.
Chemical Description: Stearic hydroxyethyl imidazoline
Category: Emulsifier, dispersant
Applications: Wax emulsions; corrosion inhibitor; household formulations; dispersing waxes, pigments, resins
Form: Waxy solid; Color: Tan; Odor: Amine
Properties: Solubility: Sol. in acid and organic systems
Composition: 85% active ingred.
Reaction: Alkaline
References: Tech. Data sheet 83

Alkazine TO. Alkaril Chem.
Chemical Description: Tall oil hydroxyethyl imidazoline
Category: Emulsifier, cationic
Applications: Asphalt emulsions; corrosion inhibitor–oil burning systems, pickling bath operations; dispersing waxes, pigments, resins; making of bitumen emulsions
Form: Liquid; Color: Brown; Odor: Amine
Properties: Solubility: Sol. in acid systems; Stability: Not stable to alkali
Composition: 80% active ingred.
Reaction: Alkaline
References: Tech. Data sheet 82

Alkylate A215. Monsanto Co.
Chemical Description: Alkyl benzene
Category: Intermediate, detergent
Applications: Intermediates for the manufacture of anionic liquid and solid detergents
Form: Liquid; Color: Colorless
Properties: m.w.: 231–241; Density: 7.2 lb/gal; Sp. gr.: 0.855–0.870; Biodegradable
References: Data sheet

Alkylate A225. Monsanto Co.
Chemical Description: Alkylbenzene
Category: Intermediate
Applications: See Alkylate A215
Form: See Alkylate A215
Properties: m.w.: 238–248; Density: 7.2 lb/gal; Sp. gr.: 0.855–0.870; Biodegradable

Alkylate A230. Monsanto Co.
Chemical Description: Alkylbenzene
Category: Intermediate

Applications and Form: See Alkylate A215
Properties: m.w.: 258–266; Density: 7.2 lb/gal; Sp. gr.: 0.855–0.870; Biodegradable

Alkylate H230H. Monsanto Co.
Chemical Description, Category, Applications, Form: See Alkylate H230L
Properties: m.w.: 360 (avg.); Ref. Index: 1.495–1.515; Sp. gr.: 0.890–0.910; Visc.: 140–220 SUS @ 100 F; I.B.P.: 290 C min.

Alkylate H230L. Monsanto Chem.
Chemical Description: By-product of alkylbenzene manufacture
Category: Intermediate
Applications: Anionic surfactant intermediates for formulating liquid and solid detergents
Form: Liquid; Color: Yellow
Properties: m.w.: 335 (avg.); Ref. Index: 1.495–1.515; Sp. gr.: 0.880–0.910; Visc.: 100–160 SUS @ 100 F; I.B.P.: 290 C min.

Alkylene Series. Hart Prod.
Category: Detergent–all except 8, 10, 11, 12; Wetting agent–17-24; Emulsifier–41-49
Applications: Detergents–Textiles; skins; leather; laundering; dyeing; metallurgy; mines and minerals; papermaking; transport; household; cosmetics; street, building, hospital cleaners; Wetting agents–Desizing and boiling off cottons; carbonizing woolens; bleaching; dyeing; rewetting dried skins; tanning; finishing; spot removal; Emulsifiers–Kier boiling; fat liquoring; making of cutting and soluble oils; dispersing waxes, pigments, resins; printing; dispersing dry powders; preparation of emulsified paints; making of bitumen emulsions
Form: Liquid; Color: Colorless; Odor: Practically odorless
Properties: Solubility: Water and organic solvents
Composition: Varies with grade; some in series are 100% active
Reaction: Neutral
Containers: Drums

Alox 100. Alox Corp.
Applications: Source of synthetic acids; manufacture of corrosion inhibitors
Form: Gel, paste; Color: Tan; Odor: Mild
Properties: Solubility: Nonpolar solvents; Density: 7.6–7.7 lb/gal; Sp. gr.: 0.92; M.P.: 100 F
Composition: 100% active ingred.
Reaction: Acid

Alox 111. Alox Corp.
Category: Mold release agent
Applications: Base for concrete mold release; rust preventive for metal forms
Form: Liquid; Color: Tan; Odor: Mild
Properties; Solubility: Nonpolar solvents; Density: 7.6–7.7 lb/gal; Sp. gr.: 0.92; Flash Pt.: 310 F min. (O.C.); Pour Pt.: 65 F max.
Composition: 100% active ingred.
Reaction: Slightly acid

Alox 162. Alox Corp.

Category: Lubricity agent, corrosion inhibitor
Applications: Lubricity agent in mineral oil sol'n.; railroad journal oils; making of cutting and sol. oils
Form: Paste; Color: Brown; Odor: Mild
Properties: Solubility: Nonpolar solvents; Density: 7.3 lb/gal; Sp. gr.: 0.88; M.P.: 86–89 F
Composition: 100% active ingred.
Reaction: Slightly acid
References: Bulletin

Alox 164. Alox Corp.
Category: Detergent, corrosion inhibitor
Applications: Base for crankcase oils for improved anti-wear, detergency, solvency, lubricity, antirust
Form: Liquid; Color: Brown; Odor: Mild
Properties: Solubility: Nonpolar solvents; Visc.: 35–45 SUS @ 100 F; Flash Pt.: 150 F min.; Pour Pt.: 60 F
Composition: 100% active ingred.
Reaction: Slightly acid

Alox 318F. Alox Corp.
Category: Corrosion inhibitor
Applications: Oil or solvent corrosion inhibitor, ferrous and nonferrous surfaces
Form: Gel, paste; Color: Brown; Odor: Mild
Properties: Nonpolar solvents; Density: 7.7 lb/gal @ 60 F; Sp. gr.: 0.91; M.P.: 100 F; Flash Pt.: 350 min. (O.C.)
Composition: 100% active ingred.
Reaction: Slightly acid
Containers: 15, 30, 55-gal containers, tankcar
References: Bulletin

Alox 319F. Alox Corp.
Category: Corrosion inhibitor
Applications: Oil or solvent corrosion inhibitor; making of cutting and sol. oils
Form: Paste; Color: Brown; Odor: Mild
Properties: Solubility: Nonpolar solvents; Density: 7.7 lb/gal @ 60 F; Sp. gr.: 0.91; M.P.: 100 F min.; Flash Pt.: 350 F min. (O.C.)
Composition: 100% active ingred.
Reaction: Acid
Toxicity: $LD_{50} > 5$ g/kg (Rats)
References: Bulletin

Alox 350. Alox Corp.
Chemical Description: Esters of an oxidized petroleum fraction
Category: Lubricity agent, corrosion inhibitor
Applications: Making of cutting and soluble oils
Form: Paste; Color: Brown; Odor: Mild
Properties: Solubility: Nonpolar solvents; Density: 7.7 lb/gal; Sp. gr.: 0.91; M.P.: 100 F
Reaction: Acid
Toxicity: $LD_{50} > 5$g/kg (Rat)
Containers: Drums, tank cars

Alox 400L. Alox Corp.
Applications: Source of synthetic acids; manufacture of corrosion inhibitors
Form: Liquid; Color: Dark; Odor: Mild
Properties: Solubility: Nonpolar solvents; Sp. gr.: 0.9; Pour Pt.: 10

F
Composition: 100% active ingred.
Reaction: Acid

Alox 436. Alox Corp.
Category: Mold release
Applications: Base for concrete mold release lubricants; corrosion inhibitor
Form: Dark liquid
Properties: Density: 7.7 lb/gal; Flash Pt.: 280 F; Stability: Good

Alox 436A. Alox Corp.
Category and Applications: See Alox 436–lower cost version of Alox 436
Form: Dark liquid; Odor: Mild
Properties: Solubility: Nonpolar solvents; Sp. gr.: 0.91
Composition: 100% active ingred.
Reaction: Slightly acid

Alox 488. Alox Corp.
Chemical Description: Mixture from oxidation of petroleum fractions
Category: Corrosion inhibitor, lubricant
Applications: Upper cylinder lubricant; additive for industrial burner fuels
Form: Liquid; Color: Dark; Odor: Mild
Properties: Solubility: Nonpolar solvents; Density: 7.3 lb/gal @ 60 F; Sp. gr.: 0.88; Visc.: 140-190 SUS @ 100 F; Flash Pt.: > 230 F
Composition: 100% active ingred.
Reaction: Slightly acid
References: Data sheet

Alox 490. Alox Corp.
Applications: Upper cylinder lubricant; engine scavenging additive
Properties: Density: 7.2 lb/gal; Sp. gr.: 0.86 @ 15.6 C; Flash Pt.: 200 F (O.C.)

Alox 502. Alox Corp.
Chemical Description: Oxygenated hydrocarbon partially neutralized with barium
Category: Corrosion inhibitor
Applications: Base for corrosion inhibitors; preservative lubricants
Form: Color: Dark; Odor: Mild
Properties: Solubility: Nonpolar solvents; Density: 8.1 lb/gal; Sp. gr.: 0.97; Flash Pt.: 390 F
Composition: 100% active
Reaction: Neutral
References: Data sheet

Alox 575. Alox Chem.
Chemical Description: Oxygenated hydrocarbon with barium and a sodium petroleum sulfonate
Category: Emulsifier, corrosion inhibitor
Applications: Oil soluble corrosion inhibitor (emulsifiable); making of cutting and soluble oils
Form: Liquid, gel; Color: Tan; Odor: Mild
Properties: Solubility: Oils; Density: 8.5 lb/gal; Sp. gr.: 1.02 @ 15.6 C; Flash Pt.: 400 F; Pour Pt. 80 F
Composition: 100% active ingred.
Reaction: Slightly acid
References: Bulletin

Alox 600. Alox Chem.
Chemical Description: Mixture of high m.w. organic acids, lactones and esters from petroleum
Category: Source of synthetic ac-

ids, manufacture of corrosion inhibitors; chemical intermediate
Applications: Rust preventives, metal-working compounds, automobile undercoating
Form: Solid; Color: Dark; Odor: Mild
Properties: Solubility: Nonpolar solvents; Density: 7.9 lb/gal @ 75 F; Sp. gr.: 0.95 @ 75 F; M.P.: 115 F; Flash Pt.: 345 F; Stability: Good
Composition: 100% active ingred.
Reaction: Acid
Toxicity: $LD_{50} > 5$ g/kg (Rat)
Containers: Drums, tankcars
References: Bulletin

Alox 604. Alox Corp.
Chemical Description: Oxygenated hydrocarbon partially neutralized with calcium
Category: Corrosion inhibitor
Applications: Base for soft film cut back rust preventives
Form: Paste; Color: Dark; Odor: Mild
Properties: Solubility: Nonpolar solvents; Density: 7.99 lb/gal; Sp. gr.: 0.96 @ 15.6 C; M.P.: 220 F; Flash Pt.: 400 F; Pour Pt.: 150 F
Composition: 100% active ingred.
Containers: Drums
References: Product bulletin

Alox 606. Alox Corp.
Category: Corrosion inhibitor
Applications: Outdoor rust preventive for protection of metal surfaces, with good salt resistance
Form: Solid; Color: Dark; Odor: Mild
Properties: Solubility: Nonpolar solvents; Density: 8.2 lb/gal; Sp. gr.: 0.98; M.P.: 250 F
Composition: 100% active ingred.
Reaction: Neutral
References: Tech. Data sheet

Alox 607. Alox Corp.
Category and Applications: See Alox 606; Alox 607 is a higher melting version of 606
Properties: Solubility: Nonpolar solvents; Density: 8.3 lb/gal; Sp. gr.: 0.98; M.P. 280 F
Composition: 100% active
Reaction: Neutral
References: Tech. Data sheet

Alox 707. Alox Corp.
Chemical Description: Mixture of organic acids, lactones and esters
Category: Corrosion inhibitor
Applications: Long term protection of steel during outdoor storage, with good salt resistance; undercoatings for automobiles
Form: Solid
Properties: Density: 8.8 lb/gal; M.P.: 185 F min.; Flash Pt.: 410 F (O.C.)
Containers: 100, 400 lb drums
References: Product bulletin

Alox 800. Alox Corp.
Category: Source of synthetic acids
Applications: Manufacture of corrosion inhibitors
Form: Liquid; Color: Dark; Odor: Mild
Properties: Solubility: Nonpolar solvents; Sp. gr.: 0.86; Pour Pt.: –25 F
Composition: 100% active ingred.
Reaction: Acid

Alox 853. Alox Corp.
Category: Corrosion inhibitor
Applications: Corrosion inhibitor esp. metals protection; fuel additive; anti-freeze additive
Properties: Solubility: Water insol.; soluble in ethanol, isopropanol; Density: 7.1 lb/gal; Sp. gr.: 0.85 @ 60 F; Flash Pt.: 155 F min. (O.C.); Pour Pt.: –25 F max.
Containers: Drums, tankcar

Alox 901. Alox Corp.
Category: Source of synthetic acids
Applications: Manufacture of corrosion inhibitors
Form: Liquid; Color: Dark; Odor: Mild
Properties: Solubility: Nonpolar solvents; Sp. gr.: 0.93; Pour Pt.: 30 F
Composition: 100% active ingred.
Reaction: Acid

Alox 904. Alox Corp.
Category: Corrosion inhibitor
Applications: Base for high flash fingerprint remover; corrosion inhibitor
Form: Gel, paste; Color: Dark
Properties: Solubility: Nonpolar solvents; Density: 8.16 lb/gal; Sp. gr.: 0.98 @ 77 F; Flash Pt.: 295 F (C.O.C.); Pour Pt.: 65 F
Composition: 100% active ingred.
Reaction: Slightly acid
References: Technical Data sheet

Alox 940. Alox Corp.
Chemical Description: Oxygenated hydrocarbon partially neutralized with calcium and a sodium petroleum sulfonate
Category: Corrosion inhibitor
Applications: Oil and solvent rust preventive base
Form: Viscous liquid; Color: Dark; Odor: Mild
Properties: Density: 8.0 lb/gal; Sp. gr.: 0.97; Flash Pt.: 300 F min. (O.C.)
Composition: 100% active ingred.
Reaction: Slightly acid
Containers: Drums, tankcars
References: Product bulletin

Alox 1509H. Alox Corp.
Chemical Description: Concentrated sol'n. of an inhibited resin in mineral spirits
Applications: Protective film for metal surfaces
Properties: Solubility: Gasoline, mineral spirits, aromatic and chlorinated solvents; Density: 8.2 lb/gal @ 60 F; Flash Pt.: 102 F; Pour Pt.: 40 F
References: Tech. data sheet

Alox 1516. Alox Corp.
Category: Corrosion inhibitor
Applications: Water soluble corrosion inhibitor for circulating cooling systems
Form: Viscous liquid; Color: Dark
Properties: Solubility: Water sol.; Density: 8.7 lb/gal; Sp. gr.: 1.04 @ 127 F; Flash Pt.: 220 F (O.C.); Pour Pt.: 30 F
References: Product bulletin

Alox 1516B. Alox Corp.
Category: Corrosion inhibitor
Applications: Water soluble corrosion inhibitor
Form: Viscous liquid; Color: Dark; Odor: Mild

Properties: Solubility: Water sol.; Sp. gr.: 1.04; Pour Pt.: 30 F
Composition: 100% active ingred.
Reaction: Neutral, amine salt

Alox 1643. Alox Corp.
Applications: Additive for industrial burner fuels promoting more efficient burning, better heat transfer, sludge suspension
Form: Liquid; Color: Brown
Properties: Density: 7.12 lb/gal
References: Tech. data sheet

Alox 1680. Alox Corp.
Chemical Description: Oxygenated hydrocarbon containing minor amount of a phosphatide
Category: Corrosion inhibitor
Applications: Penetrating oil additive
Form: Liquid; Color: Dark; Odor: Mild
Properties: Solubility: Nonpolar solvents; Density: 7.6 lb/gal; @ 60 F; Sp. gr.: 0.92 (60/60 F); Visc. 220 SUS @ 100 F; Flash Pt. : 350 F
Composition: 100% active ingred.
Containers: 15, 30, 55 gal drums and in bulk
References: Product bulletin

Alox 1689. Alox Corp.
Chemical Description: Ester of an oxygenated hydrocarbon
Category: Emulsifier, nonionic
Applications: Lubricity agent and nonionic emulsifier; making of cutting and soluble oils; rolling oil additive for stainless steel sheets and titanium sheets
Form: Solid; Color: Dark; Odor: Mild
Properties: Solubility: Nonpolar solvents; Density: 7.7 lb/gal; Sp. gr.: 0.93; M.P.: 115 F; Flash Pt.: 400 F (O.C.)
Composition: 100% active ingred.
Reaction: Slightly acid
References: Product bulletin

Alox 1727. Alox Corp.
Category: Emulsifier, corrosion inhibitor
Applications: Nonstaining slushing oil additive rust preventive for steel
Form: Viscous liquid; Color: Dark; Odor: Mild
Properties: Solubility: Nonpolar solvents; Density: 8.0 lb/gal; Sp. gr.: 0.98; Flash Pt.: 350 F
Composition: 100% active ingred.
Reaction: Slightly acid
References: Product bulletin

Alox 1805. Alox Corp.
Category: Corrosion inhibitor, anionic
Applications: Water soluble corrosion inhibitor; hydraulic and brake fluids; grinding coolants and metal cleaning compositions
Form: Viscous liquid; Color: Dark; Odor: Mild
Properties: Solubility: Water sol.; Density: 9.2 lb/gal; Sp. gr.: 1.1 @ 60 F; Flash Pt.: 212 F; Pour Pt.: 0 F; Surface Tension: 39 dynes/cm @ 0.1% conc.
Composition: 75% active ingred.; other components: 25% water
Reaction: Neutral; amine salt
Toxicity: Mild irritant to skin and eyes
Warnings and Cautions: Contains

para tertiary butyl benzoic acid
References: Tech. data sheet

Alox 1832. Alox Corp.
Category: Corrosion inhibitor
Applications: Oil soluble corrosion inhibitor for low viscosity, low temperature oils (not for use in internal combustion engines)
Properties: Density: 8.3 lb/gal; Sp. gr.: 0.99–1.0 @ 60 F; Flash Pt.: 440 F min.
References: Tech. data sheet

Alox 1843. Alox Corp.
Chemical Description: Amine salt of an organic acid
Category: Corrosion inhibitor
Applications: Water soluble corrosion inhibitor; low foaming surfactant
Form: Liquid; Color: Pale yellow; Odor: Mild
Properties: Solubility: Water, alcohols, glycols; Density: 9.3 lb/gal @ 60 F; Sp. gr.: 1.12 @ 60 F; Visc.: 138 cs @ 100 F; Pour Pt.: 15 F; Biodegradability: 80%
Composition: 75% active ingred.; other components: water (25%)
Reaction: Neutral; amine salt
Toxicity: Mild irritant to skin and eyes
Warnings and Caution: Contains para tertiary butyl benzoic acid
References: Product bulletin

Alox 1846. Alox Corp.
Category: Corrosion inhibitor
Applications: Corrosion inhibitor for jet fuels
Form: Liquid; Color: Tan; Odor: Mild
Properties: Solubility: Nonpolar solvents; Density: 7.39 lb/gal; Sp. gr.: 0.89; Flash Pt.: 175 F (C.O.C.); Pour Pt.: 20 F
Composition: 100% active ingred.
Reaction: Slightly acid
References: Tech. data sheet

Alox 2000. Alox Corp.
Category: Corrosion inhibitor
Applications: Base for corrosion inhibitor
Form: Viscous liquid; Color: Dark; Odor: Mild
Properties: Solubility: Nonpolar solvents; Density: 7.96 lb/gal; Sp. gr.: 0.95; Flash Pt.: 380 F; Pour Pt.: 40 F
Composition: 100% active ingred.
Reaction: Slightly acid
References: Product bulletin

Alox 2019A. Alox Corp.
Category: Corrosion inhibitor
Applications: Additive for fingerprint remover
Form: Liquid; Color: Dark; Odor: Mild
Properties: Solubility: Nonpolar solvents; Density: 7.9 lb/gal; Sp. gr.: 0.94; Visc.: 365 SUS @ 100 F; Flash Pt.: 137 F (C.C.)
Composition: 100% active ingred.
Reaction: Acid
References: Product bulletin

Alox 2028. Alox Corp.
Chemical Description: Blend of inhibitors including the barium soap of an oxygenated hydrocarbon
Category: Corrosion inhibitor
Applications: Base for nonemulsifiable, nonstaining, water displac-

ing rust preventive
Form: Paste; Color: Dark; Odor: Mild
Properties: Solubility: Nonpolar solvents; Density: 7.9 lb/gal; Sp. gr.: 0.95; M.P.: 95 F; Flash Pt.: 300 F (O.C.)
Composition: 100% active ingred.
Reaction: Neutral
References: Product bulletin

Alox 2116A. Alox Corp.
Chemical Description: Partial barium soap of an oxygenated hydrocarbon
Category: Corrosion inhibitor
Applications: Corrosion inhibitor for grease
Form: Paste; Color: Dark; Odor: Mild
Properties: Solubility: Nonpolar solvents; Sp. gr.: 0.97; M.P.: 86 F; Flash Pt.: 355 F (O.C.)
Composition: 100% active ingred.
Reaction: Acid
References: Tech. data sheet

Alox 2138F. Alox Corp.
Category: Corrosion inhibitor
Applications: Bullet lubricant used alone or with beeswax
Form: Solid; Color: Dark; Odor: Mild
Properties: Solubility: Nonpolar solvents; Density: 7.9 lb/gal @ 76 F; Sp. gr.: 0.95; M.P.: 160 F; Flash Pt.: 350 F
Reaction: Acid
Containers: Drums, bulk
References: Tech. data sheet

Alox 2140S. Alox Corp.
Category: Corrosion inhibitor
Applications: Additive for production of preservative lubricants and soft solvent cut-back rust preventives; automotive rustproofing with high resistance to salt spray
Form: Dark colored rubbery
Properties: Density: 7.9 lb/gal; Sp. gr.: 0.95; M.P.: 280 F min.; Flash Pt.: 490 F min.
References: Product bulletin

Alox 2140S-70. Alox Corp.
See Alox 2140S
Composition: 70% Alox 2140S, 30% mineral spirits

Alox 2162. Alox Corp.
Chemical Description: Barium soap of an oxygenated hydrocarbon
Category: Corrosion inhibitor
Applications: Base for thin film rust preventive for limited salt spray protection
Form: Paste; Color: Dark; Odor: Mild
Properties: Solubility: Nonpolar solvents; Density: 8.0 lb/gal @ 60 F; Sp. gr.: 0.96; M.P.: 95 F; Flash Pt.: 300 F
Composition: 100% active ingred.
Reaction: Slightly acid
References: Product bulletin

Alox 2200. Alox Corp.
Category: Corrosion inhibitor
Applications: Base for corrosion inhibitors
Form: Viscous liquid; Color: Brown; Odor: Mild
Properties: Solubility: Nonpolar solvents; Density: 8.39 lb/gal; Sp. gr.: 1.01; Flash Pt.: > 375 F

Composition: 100% active ingred.
References: Tech. data sheet

Alox 2201. Alox Corp.
Category: Corrosion inhibitor
Applications: Manufacture of water displacing, general purpose lubricating oil (low temperature)
Form: Liquid; Color: Dark; Odor: Mild
Properties: Solubility: Nonpolar solvents; Density: 8.0; Sp. gr.: 0.96; Flash Pt.: > 350 F (O.C.)
Composition: 100% active ingred.
Reaction: Slightly acid
References: Tech. data sheet

Alox 2211K. Alox Corp.
Category: Corrosion inhibitor
Applications: Thixotropic rust preventive depositing a thin film for long time protection from humidity and salt spray
Form: Film type–dry
Properties: Density: 7.9 lb/gal; Sp. gr.: 0.95; M.P.: ≈ 205 F; Flash Pt.: 350 F min.
References: Tech. data sheet

Alox 2213. Alox Corp.
Category: Corrosion inhibitor
Properties: Density: 7.7 lb/gal; Sp. gr.: 0.93; M.P.: 110–120 F; Flash Pt.: > 350 F
References: Tech. data sheet

Alox 2213A. Alox Corp.
Category: Corrosion inhibitor
Applications: Base for thin soft film water displacing rust preventive for household and industrial use; limited resistance to salt spray and acids; lubricant and penetrant
Form: Paste; Color: Dark; Odor: Mild
Properties: Solubility: Nonpolar solvents; Density: 7.75–8.08 lb/gal; Sp. gr.: 0.95; M.P.: 100 F; Flash Pt.: 300 F
Composition: 100% active ingred.
Reaction: Acid
Containers: Drums, bulk
References: Tech. data sheet

Alox 2268. Alox Corp.
Category: Corrosion inhibitor
Applications: Water emulsifiable corrosion inhibitor and lubricity agent; thin waxy film for protection of steel during extended indoor storage; excellent antiwear and lubricity characteristics
Form: Waxy soft solid; Color: Brown; Odor: Mild
Properties: Solubility: Water (emulsion); Density: 7.6 lb/gal; Sp. gr.: 0.92; M.P.: 135 F; Flash Pt.: > 275 F
Composition: 100% active ingred.
Reaction: Acid
Containers: Drums, bulk
References: Tech. data sheet

Alrodyne 6104. Ciba-Geigy
Chemical Description: Polyoxyalkylene fatty ester polymer
Category: Nonionic emulsifier
Applications: Nonionic emulsifier used with aromatic or aliphatic solvents and various toxicants
Form: Viscous liquid; Color: Dark amber

Alromine RA. Ciba-Geigy
Chemical Description: Fatty acid amide
Applications: Used in hair condi-

tioners and antistats
Form: Paste; Color: Off-white

Alrosol B. Ciba-Geigy
Chemical Description: Fatty acid amide
Applications: Used in liquid soaps and detergents for car washes, metal cleaners, dishwashing and hard surface cleaners
Form: Viscous liquid; Color: Amber

Alrosol C. Ciba-Geigy
Chemical Description: Fatty acid amide
Category: Foam booster, wetting agent
Applications: Used in hand soaps, hair and rug shampoos, metal cleaners
Form: Viscous liquid; Color: Light amber

Alrosol Conc. Ciba-Geigy
Chemical Description: Fatty acid amide
Category: Foam stabilizer
Applications: Foam stabilizer for shampoos, liquid soaps, shaving creams; alkaline hand and metal cleaners, degreasers, spot removers
Form: Clear viscous liquid; Color: Yellow

Alrosol O. Ciba-Geigy
Chemical Description: Fatty acid amide
Category: Emulsifier
Applications: Emulsifier for hydrocarbons, pigments, abrasives at pH 3–5; making of cutting oils; produces stable emulsions in presence of salt
Form: Viscous liquid; Color: Amber

Alrosperse 11P Flake. Ciba-Geigy
Chemical Description: Fatty acid amide
Category: Nonionic emulsifier, lubricant
Applications: Hair rinses, in hand modifiers for textiles, spreading agent in paste waxes and polishes
Form: Waxy flake; Color: Off white

Alrosperse 100. Ciba-Geigy
Chemical Description: Blend of nonionic and cationic surfactants
Category: Emulsifier, solubilizing agent
Applications: Solubilizing agent and emulsifier for nonaqueous media; in solvent-based metal cleaners; as dispersant in fuel oil and dry cleaning detergents
Form: Liquid; Color: Dark amber

Alscoap SS-90. Toho Chem.
Category: Detergent
Applications: Formulated hair shampoo; skins
Form: Paste; Color: Pale yellow
Composition: 90% active ingred.; other components: water

Amacarrier CE. AC&C
Chemical Description: Perc-biphenyl-benzoate
Category: Carrier, nonionic
Applications: Jet dyeing polyester with level dyeing, high-color yield, clean machines
Form: Clear liquid; Color: Pale yellow
Properties: Solubility: Self emul-

sifiable; Density: 10 lb/gal; Stability: Indefinite; Emulsifiability: 100%
Composition: 100% active ingred.
Reaction: Neutral, pH 8
Containers: 450 lb drums
References: Product bulletin

Amacarrier DDQ. AC&C
Category: Dye carrier, slightly anionic
Applications: Dye carrier for Quiana; atmospheric; mixed fiber dyeing
Form: Liquid; Color: Clear amber
Properties: Solubility: Dispersible in water; Stability: Indefinite; Emulsifiability: 100%
Composition: 100% active ingred.
Reaction: Neutral, pH 8
References: Product bulletin

Amacarrier HB. AC&C
Chemical Description: Self-emulsifying aromatic ester (benzoate)
Category: Dye carrier, nonionic-anionic
Applications: Dyeing polyester and triacetate; atmospheric conditions
Form: Liquid; Color: Clear amber
Properties: Solubility: Water sol.; Sp. gr.: 1.036; Stability: Indefinite; Emulsifiability: 100%; Biodegradable
Composition: 100% active ingred. (benzoate)
Reaction: Neutral, pH 8
Containers: 450 lb drums
References: Product bulletin

Amacarrier L. AC&C
Chemical Description: Blend TCB/biphenyl
Category: Dye carrier, nonionic
Applications: Polyester dyeing, atmospheric carrier
Form: Liquid; Color: Clear amber
Properties: Solubility: Self-emulsifiable at 100–120 F; Stability: Indefinite; Emulsifiability: 100%
Composition: 100% active ingred.
Reaction: Neutral, pH 8
Containers: 450 lb drums
References: Product bulletin

Amacarrier LBP. AC&C
Chemical Description: Liquid biphenyl
Category: Dye carrier, slightly anionic
Applications: Atmospheric carrier for polyester dyeing, mixed fiber dyeing
Form: Liquid; Color: Clear amber; Odor: Low
Properties: Solubility: Dispersible in warm water; Stability: Indefinite; Emulsifiability: 100%
Composition: 100% active ingred.
Reaction: Neutral, pH 8
References: Product bulletin

Amafix PAM. AC&C
Category: Dye fixative, anionic
Applications: Color fixation of acid dyes on nylon, dyed and printed fabrics
Form: Liquid; Color: Tan
Properties: Solubility: 100% water sol.; Density: 8.9 lb/gal; Sp. gr.: 1.07; Stability: Good; pH: 8
Composition: 25% active ingred.
Reaction: Neutral
Containers: 450 lb drums
References: Product bulletin

Amaglide 53. AC&C

Category: Dyebath lubricant, nonionic
Applications: Dyebath lubricant for synthetic and natural fabrics, esp. to control formation of crack marks on synthetic fabrics
Form: Liquid; Color: Milky white
Properties: Solubility: Water dispersible; Density: 8.3 lb/gal; Stability: Good; Emulsifiability: 100%
Composition: 30% active ingred.
Reaction: Neutral, pH 8
Containers: 450 lb drums
References: Product bulletin

Amamigrant PA. AC&C
Chemical Description: Polymeric
Category: Dyeing assistant, nonionic
Applications: Dye migration control in continuous dyeing
Form: Viscous liquid; Color: White
Properties: Solubility: 100% water sol.; Sp. gr.: 1.01; Visc.: 1000 cps (Brookfield); Stability: Good
Reaction: Neutral, pH 8
Containers: 450 lb drums
References: Product bulletin

Amaquest. AC&C
Chemical Description: EDTA-HEDTA
Category: Sequestering agent
Applications: Sequestering metals; Kier boiling; fat liquoring; bleaching; dyeing; finishing; insecticides, herbicides, fungicides; textiles; skins; leather; papermaking
Form: Liquid; Color: Clear
Properties: Solubility: 100%; Stability: Good
Containers: 450 lb drums
References: Product bulletin

Amascour. AC&C
Category: Detergent, wetting agent, anionic
Applications: Scouring agents for textiles; dyeing; desizing and boiling off cottons; bleaching; tanning; spot removal; Kier boiling; fat liquoring
Form: Liquid; Color: Clear white
Properties: Solubility: 100% water sol.; Stability: Indefinite
Reaction: Neutral, pH 8
Containers: 450 lb drums
References: Product bulletin

Amasoft 16-7. AC&C
Category: Fabric softener, cationic
Applications: Softener and antistat for synthetic fibers; softener for natural fibers; finishing
Form: Liquid; Color: Clear
Properties: Solubility: 100%; Density: 8.3 lb/gal
Containers: 450 lb drums
References: Product bulletin

Amasperse N. AC&C
Category: Anionic dispersing agent
Applications: Synthetic fiber dyeing; super dispersing agent for disperse color; tanning; sizing; textiles; skins; leather; metallurgy; mines and minerals; printing; dispersing dry powders; making of bitumen emulsions
Form: Liquid; Color: Clear
Properties: Solubility: 100% water sol.; Density: 9.7 lb/gal; Stability: Indefinite
Composition: 40% active ingred.
Containers: 450 lb drums
References: Product bulletin

Amawet AMR. AC&C
Chemical Description: Potassium salt of a long chain ester
Category: Wetting agent, anionic
Applications: Textile scouring, dyeing and finishing; specialty wetting agents; Kier boiling; continuous peroxide bleaching; vat dyeing; sulfur dyeing
Form: Clear liquid; Color: Light amber
Properties: Solubility: 100% water sol.; Stability: Good; pH: 8
Containers: 450 lb drums
References: Product bulletin

Amawet DF. AC&C
Category: Dye assistant
Applications: Nonfoaming auxiliary to increase flow of liquor in pump circulating machines; high temperature use
Form: Slightly viscous liquid; Color: Clear yellow
Properties: Density: 8.6 lb/gal; Sp. gr.: 1.03; Stability: Good
References: Product bulletin

Amawet LF. AC&C
Chemical Description: Exthoxylated linear alcohol
Category: Wetting, emulsifying, dispersing and penetrating agents; slightly anionic
Form: Water white liquid; Odor: Faint fatty alcohol
Properties: Biodegradable; low foaming
Reaction: Essentially neutral
References: Product bulletin

Amawet P. AC&C
Chemical Description: Sulfated fatty ester
Category: Dye assistant, anionic
Applications: Textile dyeing auxiliary aiding level dyeing; package, beam, cheese dyeing of cotton; preparation of dyeing of wool and synthetic knit, woven piece goods
Form: Liquid; Color: Dark amber
Properties: Solubility: Water sol; pH: 6.5
References: Product bulletin

Ambidex. Cincinnati Milacron Chem.
Category: Wetting agent, emulsifier, cationic
Applications: Wetting agent and emulsifier for oils and asphalts
Form: Viscous liquid; Color: Dark brown; Odor: Amine
Properties: Solubility: Oil sol.; Density: 8.14 lb/gal @ 60 F; Sp. gr.: 0.985 @ 60 F; Visc.: 6600 SUS @ 100 F; Flash Pt.: 450 F; Pour Pt.: 40 F
Composition: 100% active ingred.
Containers: 55 gal drums, tankcar
References: Tech. bulletin

Ambidex-PB. Cincinnati Milacron Chem.
Category: Wetting agent, emulsifier, cationic
Applications: Wetting agent for oils and asphalts
Form: Viscous liquid; Color: Dark brown; Odor: Slight amine
Properties: Solubility: Oil sol.; Density: 8.2 lb/gal; Sp. gr.: 0.985 @ 60 F; Visc.: 660 SUS @ 140 F; Flash Pt.: 370 F; Pour Pt.: 72 F
Composition: 100% active ingred.
Containers: 55-gal drums, tankcar
References: Tech. bulletin

Amerchol Series. Amerchol, a Unit of CPC International
Chemical Description: Sterol extracts
Category: Emulsifier, stabilizer, plasticizer, nonionic
Applications: Aerosols, baby toiletries, creams and lotions, dermatological vehicles, hair products, lotion detergents, pigmented makeup items
References: Products bulletin

Amerchol 400. Amerchol, a Unit of CPC International
Category and Applications: See Amerchol Series; Least effective emulsifier; adjust payoff of anhydrous makeups; suitable for medium to heavy creams and lotions
Form: Soft solid; Color: Yellow-ivory
Properties: M.P.: 48–52 C

Amerchol BL. Amerchol, a Unit of CPC International
Category and Applications: See Amerchol Series; Highest ester content of the series; ideal for heavy creams and lotions
Form: Semisolid; Color: Yellow-amber

Amerchol C. Amerchol, a Unit of CPC International
Category and Applications: See Amerchol Series; Moderate ester content; suitable for medium to heavy creams and lotions
Form: Soft solid; Color: Pale yellow-ivory
Properties: M.P.: 40–46 C

Amerchol CAB. Amerchol, a Unit of CPC International
Category and Applications: See Amerchol Series; Contains no esters or free fatty acids; hypoallergenic preparations; medium to heavy creams and lotions
Form: Soft solid; Color: Pale ivory
Properties: M.P.: 40–46 C

Amerchol H-9. Amerchol, a Unit of CPC International
Category and Applications: See Amerchol Series; Moderate ester content; spreads easily but retains consistency at body temp.; ideal for dermatologicals and heavy creams and lotions
Form: Soft solid; Color: Pale yellow
Properties: M.P.: 55–62 C

Amerchol L-99. Amerchol, a Unit of CPC International
Category and Applications: See Amerchol Series; Thin fluid form; economical alternative for less critical emulsions; light lotions to heavy creams
Form: Oily liquid; Color: Pale yellow; Odor: Faint, characteristic sterol
Properties: Sp. gr.: 0.856–0.866

Amerchol L-101. Amerchol, a Unit of CPC International
Category and Applications: See Amerchol Series; Most efficient stabilizer for emulsified makeups; plasticizer for hair sprays; hypoallergenic products; ideal for light lotions to heavy creams
Form: Oily liquid; Color: Pale yellow; Odor: Faint, characteristic sterol

Properties: Sp. gr.: 0.84–0.86

Amerchol L-500. Amerchol, a Unit of CPC International
Category and Applications: See Amerchol Series; Ideal for low oil/high wax content systems; economical conc. extract for light lotions to heavy creams
Form: Thick liquid to semisolid; Color: Amber; Odor: Characteristic sterol

Amerchol RC. Amerchol, a Unit of CPC International
Category and Applications: See Amerchol Series; High triterpene content; least effective emulsifier; shade developer for cosmetic pigments; ideal for anhydrous and emulsified makeups
Form: Soft solid; Color: Ivory
Properties: M.P.: 55–64 C

Amerchol Polysorbate 80. Amerchol, a Unit of CPC International
Chemical Description: Polyoxyethylene (20) sorbitan monooleate
Category: Emulsifier, solubilizer, nonionic
Applications: Used in cosmetic, toiletry, and pharmaceutical products
Form: Liquid
Properties: Solubility: Sol. in water and alcohol; HLB: 15.0
Composition: 100% active ingred.

Ameroxol OE-2. Amerchol, a Unit of CPC International
Chemical Description: Oleyl alcohol ethoxylate
Category: Nonionic emulsifier and solubilizer
Applications: W/O emulsifier; creams and lotions; ideal spreading and dispersing agent for bath oils
Form: Clear liquid; Color: Pale straw; Odor: Bland
Properties: Solubility: Sol. in mineral oil, isopropyl esters, anhydrous ethanol
References: Products bulletin

Ameroxol OE-10. Amerchol, a Unit of CPC International
Chemical Description and Category: See Amerchol OE-2
Applications: O/W emulsifier and solubilizer, esp. for fragrance products
Form: Semisolid; Color: White; Odor: Bland
Properties: Solubility: Sol. in water, ethanol, hydroalcoholics, glycols; Cloud Pt.: 47–55 C
References: Products bulletin

Ameroxol OE-20. Amerchol, a Unit of CPC International
Chemical Description and Category: See Amerchol OE-2
Applications: O/W emulsifier and solubilizer with soft afterfeel
Form: Waxy solid; Color: White; Odor: Bland
Properties: Solubility: Sol. in water, ethanol, hydroalcoholics, glycols; Cloud Pt.: 87–93 C

Amidan. Grindsted Products
Chemical Description: Distilled monoglycerides integrated with other food ingredients
Category: Emulsifier
Applications: Dough conditioner in bread, rolls, bread improvers

Amidox C. Stepan Chem.
Chemical Description: Ethoxylat-

ed coconut fatty acids monoethanol-amides
Category: Detergent, emulsifier
Applications: Shampoos, dishwashing detergents
Form: Liquid to paste
Composition: 100% active ingred.

Amine C. Ciba-Geigy
Chemical Description: Imidazoline derivative
Category: Emulsifier, dispersant, detergent, cationic
Applications: Acid cleaners, antistat agents for textiles, plastics; paints; corrosion inhibitor; making of cutting and rust preventive oils
Form: Yellow liquid
Properties: Solubility: Sol. in polar, organic solvents, hydrocarbons, relatively insol. water; Stability: Good
Composition: 100% active ingred.
Toxicity: Very irritating to skin and eyes
References: Tech. data

Amine O. Ciba-Geigy
Chemical Description: Oleyl imidazoline
Category: Emulsifier, dispersant, detergent, antistat, cationic
Applications: Antistat agent for textiles, plastics, paints; in acid cleaners; in corrosion inhibitor formulations; lubricating oils; asphalts
Form: Amber liquid
Properties: See Amine C
Composition: 100% active ingred.
Toxicity: Very irritating to skin and eyes
References: Tech. data

Amine S. Ciba-Geigy
Chemical Description: Stearyl imidazoline
Category: Emulsifier, dispersant, detergent, antistat, cationic
Applications: Antistat for textiles, paints; in acid cleaners; in corrosion inhibitors; emulsion waxes
Form: Tan wax
Properties: See Amine C
Composition: 100% active ingred.
Toxicity: Very irritating to skin and eyes
References: Tech. data

Amine T. Ciba-Geigy
Chemical Description: Tall oil fatty acid imidazoline
Category and Applications: See Amine O
Form: Dark amber liquid
Composition: 100% active ingred.
Toxicity: Very irritating to skin and eyes
References: Tech. data

Aminogen DG Liquid. . AC&C
Category: Foaming agent
Applications: Foam control for dyeing in combination with Aminogen K or SD; aids solubility of dyestuff
Form: Clear liquid
Properties: Density: 8.64 lb/gal; Stability: Stable and pourable at 25 F; High foam level; pH: 8.8
Composition: 100% active ingred.
References: Product bulletin

Aminogen EB. AC&C
Category: Dye assistant, anionic-nonionic
Applications: Energy saving dispersion, leveling antiprecipitant for

low temperature rapid dyeing of nylon stock, yarn, woven, knits, carpets
Form: Liquid; Color: Clear amber
Properties: Solubility: 100% water sol.; Stability: Indefinite
Reaction: Neutral, pH 7
Containers: 450 lb drums
References: Product bulletin

Aminogen K-LF. AC&C
Category: Dye assistant
Applications: Continuous dyeing of nylon yarn, carpet; maximize color value
Form: Light yellow liquid
Properties: Density: 8.14 lb/gal; Stability: Stable and pourable @ 25 F; Low foam level; pH: 9.0
Composition: 100% active ingred.
References: Product bulletin

Aminogen P-LF. AC&C
Category: Dye assistant
Applications: Printing of nylons; maximize color value
Form: Light yellow liquid
Properties: Density: 8.58 lb/gal; Stability: Stable and pourable @ 25 F; Moderate foam level
Composition: 100% active ingred.
References: Product bulletin

Aminogen R. AC&C
Category: Dye assistant
Applications: Energy saving low temperature dyeing of wool; aids rapid dyeing and dye penetration
Form: Liquid; Color: Clear
Properties: Solubility: 100% water sol.; Density: 8.7 lb/gal; Stability: Good ; pH 6.0–6.5
Containers: 450 lb drums
References: Product bulletin

Aminogen SD-LF. AC&C
Category: Dye assistant
Applications: Space dyeing; flat bed printing; yarn dyeing by pad application; nylon; carpet dyeing
Form: Light yellow liquid
Properties: Solubility: 100%; Density: 8.58 lb/gal; Stability: Stable and pourable @ 25 F; Low foam level; pH: 9.7
Composition: 100% active ingred.
Containers: 450 lb drums
References: Product bulletin

Amisoft C-T-12. Ajinomoto U.S.A.
Chemical Description: Monotriethanolamine *n*-cocoyl-*l*-glutamate
Category: Emulsifier
Applications: Cosmetics
Form: Liquid
Composition: 20% active ingred.

Amisoft GS-11. Ajinomoto U.S.A.
Chemical Description: Monosodium *n*-cocoyl-*l*-glutamate
Category: Emulsifier
Applications: Cosmetics
Form: Powder
Composition: 100% active ingred.

Amisoft HS-11. Ajinomoto U.S.A.
Chemical Description: Monosodium *n*-hydrogenated-tallowyl-*l*-glutamate
Category: Emulsifier
Applications: Cosmetics
Form: Powder
Composition: 100% active ingred.

Amisoft HT 12. Ajinomoto U.S.A.
Chemical Description: Monotriethanolamine *n*-hydrogenated-tallowyl-*l*-glutamate
Category: Emulsifier

Application: Cosmetics
Form: Liquid
Composition: 20% active ingred.

Amisoft LS11. Ajinomoto U.S.A.
Chemical Description: Monosodium *n*-lauroryl-*l*-glutamate
Category: Emulsifier
Applications: Cosmetics
Form: Powder
Composition: 100% active ingred.

Amisoft LT-12. Ajinomoto U.S.A.
Chemical Description: Monotriethanolamine *n*-lauroyl-*l*-glutamate
Category: Emulsifier
Applications: Cosmetics
Form: Liquid
Composition: 20% active ingred.

Ammonyx 4. Onyx Chem. Co.
Chemical Description: Stearyl dimethyl benzyl ammonium chloride
Category: Emulsifier, conditioner, softener, emollient, cationic
Applications: Hair rinses, skin creams and lotions
Form: Paste
Properties: Sp. gr.: 0.99 (25/20 C); Flash Pt.: > 200 F
Composition: 17–19% quaternary; 24–28% solids

Ammonyx 4B. Onyx Chem. Co.
Chemical Description: Stearyl dimethyl benzyl ammonium chloride
Category: Emulsifier, conditioner, softener, emollient, cationic
Applications: Hair rinses, skin creams and lotions
Form: Paste
Properties: Sp. gr.: 0.99 (25/20 C); Flash Pt.: > 200 F
Composition: 16–18% quaternary; 24–26% solids

Ammonyx 485. Onyx Chem. Co.
Chemical Description: Stearyl dimethyl benzyl ammonium chloride
Category: Emulsifier, conditioner, softener, emollient, cationic
Applications: Hair rinses, skin creams and lotions
Form: Powder
Properties: Sp. gr.: 0.45 (25/20 C); Flash Pt.: 122 F
Composition: 85% min. quaternary; 87% min. solids

Ammonyx 490. Onyx Chem. Co.
Chemical Description: Stearyl dimethyl benzyl ammonium chloride
Category: Emulsifier, conditioner, softener, emollient, cationic
Applications: Hair rinses, skin creams and lotions
Form: Powder
Properties: Sp. gr.: 0.47 (25/20 C); Flash Pt.: 130 F
Composition: 90% min. quaternary; 92 min. solids

Ammonyx 856. Onyx Chem. Co.
Chemical Description: Tallow dimethyl benzyl ammonium chloride
Category: Dispersant, suspending agent, deemulsifier, cationic
Applications: Industrial grade pigment dispersant and suspending agent; deemulsifier for solvent extraction processes

Form: Paste
Properties: Sp. gr. 0.90 (25/20 C); Flash Pt.: 64 F
Composition: 75% min. quaternary; 77% min. solids

Ammonyx 4002. Onyx Chem. Co.
Chemical Description: Stearyl dimethyl benzyl ammonium chloride
Category: Emulsifier, conditioner, softener, emollient, cationic
Applications: Hair rinses, skin creams and lotions
Form: Powder
Properties: Sp. gr.: 0.52 (25/20 C); Flash Pt.: 170 F
Composition: 94% min. quaternary; 96% min. solids

Ammonyx 4080. Onyx Chem. Co.
Chemical Description: 1-Methyl-1-tallow amido-ethyl-2-heptadecyl imidazolinium methosulfate
Category: Softener, cationic
Applications: Liquid fabric softener for home and industrial use; formulated over wide viscosity range
Form: Liquid
Properties: Sp. gr.: 0.97 (26/20 C); Flash Pt.: 68 F
Composition: 74–76% quaternary; 81% max. solids

Ammonyx CDO. Onyx Chem. Co.
Chemical Description: Coco amido propyl dimethyl amine oxide
Category: Wetting, foaming agent, foam stabilizer, conditioner, nonionic
Applications: Bubble baths, bath oils, dishwash concentrates, hair coloring systems, softeners and cleansers
Form: Liquid
Properties: Sp. gr.: 1.02 (25/20 C); Flash Pt.: > 200 F
Composition: 30% amine oxide

Ammonyx CETAC. Onyx Chem.
Chemical Description: Cetyl trimethyl ammonium chloride
Category: Emulsifier, conditioner, softener, emollient, cationic
Applications: Hair rinses, skin creams and lotions
Form: Liquid
Properties: Sp. gr.: 0.93 (25/20 C); Flash Pt.: > 200 F
Composition: 24–26% quaternary; 25% min. solids

Ammonyx CO. Onyx Chem. Co.
Chemical Description: Cetyl dimethyl amine oxide
Category: Conditioner, nonionic
Applications: Household, cosmetic and janitorial products
Form: Liquid
Properties: Sp. gr.: 0.96 (25/20 C); Flash Pt.: > 200 F; Stability: Stable in high electrolyte concentrations
Composition: 30% amine oxide

Ammonyx DMCD-40. Onyx Chem.
Chemical Description: Lauryl dimethyl amine oxide
Category: Wetting, foaming agent, foam stabilizer, nonionic
Applications: Cosmetic, household and janitorial products
Form: Liquid
Properties: Sp. gr.: 0.91 (25/20 C); Flash Pt.: 86 F
Composition: 40–42% amine oxide

Ammonyx DME. Onyx Chem. Co.
Chemical Description: Cetyl dimethyl benzyl ammonium chloride
Category: Emulsifier, wetting agent, cationic
Applications: Personal care products
Form: Paste
Properties: Sp. gr.: 0.96 (25/20 C); Flash Pt.: 52 F
Composition: 75% min. quaternary; 77% min. solids

Ammonyx KP. Onyx Chem. Co.
Chemical Description: Oleyl dimethyl benzyl ammonium chloride
Category: Conditioner, antistat, cationic
Applications: Clear hair rinses
Form: Liquid
Properties: Sp. gr.: 0.98 (25/20 C); Flash Pt.: > 200 F
Composition: 50% min. quaternary; 52% min. solids

Ammonyx LO. Onyx Chem. Co.
Chemical Description: Lauryl dimethyl amine oxide
Category: Wetting, foaming agent, foam stabilizer, grease emulsifier, nonionic
Applications: Shampoos, bath products, fine fabric cleaners, shaving creams, lotions
Form: Liquid
Properties: Sp. gr.: 0.96 (25/20 C); Flash Pt.: > 200 F
Composition: 30% amine oxide

Ammonyx MCO. Onyx Chem. Co.
Chemical Description: Myristyl/cetyl dimethyl amine oxide
Category: Wetting and foaming agent, foam stabilizer, nonionic
Applications: Cosmetic, household and janitorial products
Form: Liquid
Properties: Sp. gr.: 0.96 (25/20 C); Flash Pt.: > 200 F
Composition: 30% amine oxide

Ammonyx MO. Onyx Chem. Co.
Chemical Description: Myristyl dimethyl amine oxide
Category: Wetting and foaming agent, foam stabilizer, nonionic
Applications: Cosmetic, household and janitorial products
Form: Liquid
Properties: Sp. gr.: 0.96 (25/20 C); Flash Pt.: > 200 F
Composition: 30% amine oxide

Ammonyx SO. Onyx Chem. Co.
Chemical Description: Stearyl dimethyl amine oxide
Category: Conditioner, emulsifier, nonionic
Applications: Cream rinses, lotions, liquid bath products
Form: Paste
Properties: Sp. gr.: 0.99 (25/20 C); Flash Pt.: > 200 F
Composition: 25% amine oxide

Ammonyx T. Onyx Chem. Co.
Chemical Description: Cetyl dimethyl benzyl ammonium chloride
Category: Emulsifier, wetting agent, cationic
Applications: In specialty latex dispersions as emulsifier and wetting agent

Form: Liquid
Properties: Sp. gr.: 0.99 (25/20 C); Flash Pt.: > 200 F
Composition: 25% min. quaternary; 25% min. solids

AMP. IMC Chem.–NP Div.
Chemical Description: 2-amino-2-methyl-1-propanol
Category: Emulsifier, catalyst, anionic
Applications: As raw materials in polymer applications, e.g. oxazoline structures (as corrosion inhibitors, antifoam compounds, synthetic waxes); in coatings as pigment dispersants; in emulsions for cosmetics, cleaners, industrial insecticides, asphalt emulsions; in boiler-water treatment; as catalysts; in formaldehyde scavenging
Form: Solid; Color: 20 (APHA)
Properties: m.w.: 89.14; Density: 7.78 lb/gal; Sp. gr.: 0.928 (40/40C); Visc.: 102 cp @ 30 C; M.P.: 30 C; B.P.:165 C; Flash Pt.: 172F
Composition: 100% active ingred.
Toxicity: Skin irritant; causes eye burns
Containers: 55 gal drums or bulk; store in iron or steel
Warnings and Caution: Corrosive to copper, brass and aluminum; combustible; protective goggles and gloves should be worn
References: TDS No. 10

AMP-95. IMC Chem.–NP Div.
Chemical Description: 2-amino-2-methyl-1-propanol
Category and Applications: See AMP
Form: Liquid; Color: Colorless; Odor: Low
Properties: m.w.: 89.14; Density: 7.85 lb/gal; Sp. gr.: 0.942 (25/25C); Visc.: 147 cp; F.P.: –2 C; Flash Pt.: 182 F (Tag closed cup); Surface Tension: 36–38 dynes/cm
Composition: 95% active ingred.
Toxicity: Skin irritant, causes eye burns
Containers: See AMP
Warnings and Caution: See AMP
References: TDS No. 10A

Ampho B11-34. Armstrong Chem.
Chemical Description: Complex cocobetaine derived from dimethyl amine propyl amine
Category: High foaming amphoteric surfactant with outstanding foam stability; wetting agent
Applications: Cosmetic formulations
Form: Clear liquid; Color: Light yellow; Odor: Mild
Properties: Solubility: Solv. water; Density: 8 lb/gal; Sp. gr.: 0.96
Composition: 35% active ingred.
Reaction: Compatible with anionic, cationic and nonionic detergents; compatible with acid and alkaline systems

Ampho T-35. Armstrong Chem.
Chemical Description: Potassium salt of fatty amine polyethoxy carboxylate
Category: Detergent, amphoteric
Applications: Heavy duty surface cleaners; chelating agent; solubilizer for difficult materials; stabilizer; dispersant and penetrant; corrosion inhibitors; metal-to-metal lubricant; skin softening in dish-

washing compounds; laundry detergents
Form: Liquid; Color: Amber; Odor: Mild
Properties: Solubility: Water and polar solvents; Density: 8.7 lb/gal; Sp. gr.: 1.05 @ 80 F; Stability: Stable to acid or alkali; Biodegradable
Composition: 35% active ingred.
Reaction: Compatible with alkali or acid systems

Ampholan U 203. Diamond Shamrock
Chemical Description: Sodium salt of an amphoteric surfactant
Category: Detergent, wetting agent
Applications: Foaming agent for synthetic latex; effective in alkaline solution
Form: Liquid
Composition: 30% active ingred.
Containers: Drums

Ampholan U203. Lankro Chem. Ltd. Surfactants Div.
Chemical Description: Amphoteric surfactant with carboxyl and amino groups
Category: Detergent, foaming and stabilizing agent, dispersant, amphoteric
Applications: Industrial cleaners and detergents esp. in high electrolyte concentrations; auzilliary foaming and stabilizing agent in synthetic rubber latices in carpet backing industry; shampoo formulations; germicides; in textile industry as dispersant for pigments in manufacture of synthetic fibers
Form: Liquid; Color: Pale amber; Odor: Characteristic
Properties: Solubility: Water-sol. @ 20 C; Sp. gr.: 1.035 @ 20 C; Visc.: 132 cs @ 20 C; Flash Pt.: > 200 F (COC); Pour Pt.: < 0 C; pH: 6.0–8.0 (1% aq.)
Composition: 30% active ingred.
Reaction: Forms a cationic quaternary ammonium salt in acidic sol'ns.; forms anionic carboxylate salt in alkaline conditions
Containers: Store in SS, glass, resin or polythene lined equipment
Warnings and Caution: Avoid prolonged contact with concentrated product; should not be mixed with strong oxidizing or reducing agents (potentially explosive)

Anionyx 12S. Onyx Chem. Co.
Chemical Description: Disodium monooleamido diglycol sulfosuccinate
Category: Detergent, anionic
Applications: Low-irritation shampoos, bubble baths, dishwash detergents
Form: Liquid
Properties: Sp. gr.: 1.05 (25/20 C); Flash Pt.: 138 F
Composition: 20% active

Antara LB-400. GAF Chem. Prod.
Chemical Description: Fatty alcohol polyethyleneoxy phosphate ester acid
Category: Lubricant, anionic
Applications: Lubricant, extreme-pressure additive, rust inhibitor with good antiwear, antiweld, wetting, detergency, dispersancy, emulsification properties; chain

lubricants, circulating oils, cutting fluids, drawing compounds, diesel oils, greases, hydraulic fluids, rolling oils, rust preventives, steel-mill slushing oils, turbine oils, gear oils, glass cutting and polishing lubricants
Form: Opaque viscous liquid; Color: Amber
Properties: Solubility: Oil and water sol. and in aliphatic and aromatic solvents; Sp. gr.: 1.04
Composition: 99% active ingred.
Toxicity: Skin and eye irritant
References: Data sheet 450-65

Antara LE-500. GAF Chem. Prod.
Chemical Description: Free acid of a complex organic phosphate ester
Category and Applications: See Antara LB-400
Form: Clear viscous liquid; Color: Yellow
Properties: Solubility: Oil and water sol. and in aliphatic and aromatic solvents; Sp. gr.: 1.11

Antara LE-700. GAF Chem. Prod.
Chemical Description: Free acid of a complex organic phosphate ester
Category and Applications: See Antara LB-400
Form: Clear viscous liquid; Color: Pale yellow
Properties: Solubility: Water sol. and aromatic solvents; Sp. gr.: 1.12
Composition: 100% active

Antara LF-200. GAF Chem. Prod.
Chemical Description: Complex organic phosphate ester free acid
Category and Applications: See Antara LB-400
Form: Slight cloudy viscous liquid; Color: 8 (VCS)
Composition: 100% active ingred.

Antara LK-500. GAF Chem. Prod.
Chemical Description: Complex phosphate ester of an ethoxylated linear alcohol
Category: Low-foaming lubricant
Applications: Lubricant for either oil or synthetic water-based metalworking fluids, with excellent extreme-pressure, anti-wear and rust inhibition properties
Form: Slightly viscous, clear liquid Color: Pale yellow
Properties: Solubility: Sol. in xylene, ethanol, ethylene glycol, carbon tetrachloride, butyl cellosolve, water; Sp. gr.: 1.1; Visc.: 21 centistokes @ 100 C; Biodegradable
Composition: 100% active ingred.

Antara LM-400. GAF Chem. Prod.
Chemical Description: Free acid of a complex organic phosphate ester
Category and Applications: See Antara LB-400
Form: Clear liquid; Color:Yellow
Properties: Solubility: Oil sol. and aliphatic and aromatic solvents; Sp. gr.: 1.05
Composition: 100% active ingred.

Antara LM-600. GAF Chem. Prod.
Chemical Description: Free acid of a complex organic phosphate ester
Category and Applications: See Antara LB-400
Form: Clear viscous liquid; Color: Yellow

Properties: Solubility: Oil and water sol. and in aliphatic and aromatic solvents; Sp. gr.: 1.06
Composition: 100% active ingred.

Antara LP-700. GAF Chem. Prod.
Chemical Description: Free acid of a complex organic phosphate ester
Category and Applications: See Antara LB-400
Form: Clear viscous liquid; Color: Pale yellow
Properties: Solubility: Water, aromatic solvents; Sp. gr.: 1.20
Composition: 100% active ingred.

Antara LS-500. GAF Chem. Prod.
Chemical Description: Free acid of a complex organic phosphate ester
Category and Applications: See Antara LB-400
Form: Clear viscous liquid; Color: Pale yellow
Properties: Solubility: Oil and water sol. and in aliphatic and aromatic solvents; Sp. gr.: 1.00
Composition: 100% active ingred.

Antaron FC-34. GAF Chem. Prod.
Chemical Description: Complex fatty amido compound (fatty group derived from coconut oil)
Category: Detergent, wetting agent, emulsifier, dispersant, emollient, amphoteric surfactant
Applications: In textile processing, esp. fulling of woolens; leather degreasing; fat liquoring; household and industrial liquid-cleaners and shampoos (rugs, upholstery, floors, cars, white-wall tires); steel pickling; steam cleaning of bright aluminum; aerosol formulations; toiletries
Form: Clear semiviscous liquid; Color: Amber
Properties: Solubility: Water-sol. esp. high electrolyte content; Sp. gr.: 1.09; Stability: Good stability in acid, alkaline, or electrolyte systems; Surface Tension: 32.0 dynes/cm (0.0155% @ 20 C)
Composition: 37% active ingred.
Reaction: Compatible with wide range of anionic, cationic and nonionic surfactants
Toxicity: Skin irritant; causes eye burn
Containers: 10 lb pkg; 50, 500 lb drums; bulk avail. in tank cars
References: Bulletin 2303-004

Antaron MC-44. GAF Chem. Prod.
Chemical Description: Dicarboxylic coco imidazoline, sodium salt
Category: Emulsifier, solubilizer, coupling agent, amphoteric
Applications: Nonirritating shampoos, skin cleansers, cosmetics; industrial and household cleaners
Form: Viscous liquid; Color: Amber
Composition: 38% active ingred.

Antaron PC-37. GAF Chem. Prod.
Chemical Description: Sulfated fatty polyoxyethylene quaternary nitrogen compound
Category: Amphoteric shampoo surfactant
Applications: Cosmetics esp. shampoos due to skin and eye mildness, ease of formulation, hair manageability, broad surfactant and resin

compatibility
Form: Clear slightly viscous liquid; Color: Amber
Properties: Solubility (10% conc.): miscible–water, ethanol; Sp. gr.: 1.15; Visc.: 1100–1200 cps (Brookfield, 25C); Stability: Excellent > pH 5.5, in hard water and strong bases; Surface Tension: 39 dynes/cm (0.1%); Biodegradable
Composition: 75% active ingred.; 25% water
Reaction: Compatible with anionic, cationic, nonionics
Toxicity: Very mild to skin and eyes
References: Bulletin 9655-055

Antarox BL-225. GAF Chem. Prod.
Chemical Description: Modified linear aliphatic polyether
Category: Low-foaming nonionic surfactant with excellent detergency and wetting properties
Applications: Commercial dishwashing formulations; alkaline metal cleaners; machine floor-scrubbers; desizing baths; continuous bleaching, dyeing, finishing of textiles
Form: Clear liquid; Color: Slightly yellow; Odor: Bland
Properties: Solubility: Water and organic solvents except paraffins; Sp. gr.: 0.99; Visc.: 40 cps; Cloud Pt.: 25–29 C; Flash Pt.: 190 C (COC); Pour Pt.: < –18 C; Stability: Good to acids and alkalis; Surface Tension: 30 dynes/cm (0.1%); Biodegradable
Composition: 100% active ingred.
Toxicity: Eye irritant
References: Bulletin 9643-003

Antarox BL-330. GAF Chem. Prod.
Chemical Description: Aliphatic polyether
Category: Detergent, wetting agent, nonionic
Applications: Low foaming detergent and wetting agent; rinse aids, metal cleaning, mechanical dishwashing
Form: Liquid
Composition: 95% active ingred.

Antarox BL-344. GAF Chem. Prod.
Chemical Description: Aliphatic polyether
Category: Detergent, nonionic
Applications: High-temperature detergent; nondiscoloring dedusters for caustic powders
Form: Liquid
Properties: Stability: Good at high temperatures
Composition: 90% active ingred.

Antarox BL-240. GAF Chem. Prod.
Chemical Description: Modified linear aliphatic polyether
Category: See Antarox BL-225
Applications: High-temperature applications; metal spray cleaning; textile wet processing; rinse aid
Form: Liquid; Color: Slightly yellow; Odor: Bland
Properties: Solubility: More water sol. than Antarox BL-225; Sp. gr.: 0.99; Visc.: 35 cps; Cloud Pt.: 38–42 C; Flash Pt.: 124 C; Pour Pt.: < –21 C; Surface Tension: 28 dynes/cm (0.1%); Biodegradable
Composition: 100% active ingred.
Toxicity: Eye irritant

References: Bulletin 9643-003

Antarox G-200. GAF Chem. Prod.
Chemical Description: Amino amide from mixed fatty acids and mixed polyamines reacted with ethylene oxide
Category: Dispersant, nonionic
Applications: Dispersant substantive to ferrous metals; viscose spin bath additive (tire cords, textile filaments); deposit and sludge control in cellophane mfg., casting solutions, etc.; steelmaking process
Form: Opaque, viscous liquid; Color: Brown
Properties: Refractive Index: 1.48; Cloud Pt.: 64–68 C
Composition: 100% active ingred.

Antarox LF-330. GAF Chem. Prod.
Chemical Description: Aliphatic polyether
Category: Detergent, wetting agent, nonionic
Applications: Low-foaming surfactant for rinse aid concentrates; metal spray cleaners; machine dishwashing; textile wetting and scouring; aircraft, bottle, dairy, and food plant cleaning; laundry detergents; paper-machine-felt washing; pulp deresination; steam cleaning
Form: Clear liquid; Color: Pale yellow; Odor: Bland
Properties: Solubility: Water and most organic solvents except paraffins; Visc.: 29 centistokes; Cloud Pt.: 30 C; Pour Pt.: 3 C; Stability: Excellent to acid and strong alkali; Surface Tension: 29 dynes/cm (0.1%)
Composition: 95% active ingred.
Reaction: Slightly acid
Toxicity: Eye irritant
References: Bulletin 7543-097

Antarox LF-344. GAF Chem. Prod.
Chemical Description: Aliphatic polyether
Category and Applications: See Antarox LF-330. LF-344 has better solubility at high temperatures
Form: Clear liquid; Color: Pale yellow; Odor: Bland
Properties: Solubility: Water and most organic solvents except paraffins; Visc.: 61 centistokes; Cloud Pt.: 44 C; Pour Pt.: 6 C; Surface Tension: 33 dynes/cm (0.1%)
Composition: 90% active ingred.
Reaction: Slightly acid
Toxicity: Eye irritant
References: Bulletin 7543-097

Antiprex 461. Allied Colloids Inc.
Chemical Description: Polyacrylic acid
Category: Antiscalant
Applications: Antiscalant for use in industrial water systems, e.g., cooling towers, boilers, heat exchangers
Form: Liquid; Color: Light amber
Properties: m.w.: low; Sp. gr.: 1.26; Visc.: 400–700 cps
Composition: 50% active ingred.
Containers: 55 gal steel drums or bulk
Warnings and Caution: Keep away from skin and eyes
References: Technical data sheet

Antiprex A. Allied Colloids Inc.
Chemical Description: Acrylate polymer
Category: Scale inhibitor, dispersant, anionic

Applications: Scale inhibitor for use in aqueous systems, e.g., recirculating cooling systems, evaporators, boilers, in the chemical, mineral and oil industries
Form: Liquid; Color: Pale yellow; Odor: Slight
Properties: m.w.: 2–4000; Solubility: Water sol.; Sp. gr.: 1.3; Visc.: 1400 cp; Stability: Stable to high temperatures > 500 F
Composition: 45% active ingred.
Containers: 55 gal steel drums
References: Bulletin GID101

Apex Level Dye #2123. Apex
Category: Detergent
Applications: Solvent detergent blend; formulated for single bath scouring and dyeing of polyester fabrics
Form: Liquid
Properties: Solubility: Emulsifies in water
Containers: 55 gal drums

Apex Mersapex #2251. Apex
Category: Detergent
Applications: Scouring agent; removal of oily soils from synthetic fabrics and blends; dyeing applications
Form: Gel; Color: Water white
Properties: Solubility: In water
Containers: 55 gal drums

Apex Mersynol #450. Apex
Category: Detergent, wetting agent
Applications: Scouring and wetting agent; formulated for Kier boiling and bleaching; desizing and boiling
Properties: Solubility: Water
Containers: 55 gal drums

Apex Mersynol #4150. Apex
Category: Wetting agent, dispersant
Applications: Wetting and dispersing agent in Kier scouring and bleaching of cotton; dispersing and leveling assistant in dyeing; desizing and boiling off cottons
Form: Liquid; Color: Clear
Containers: 55 gal drums

Apex Pentrapex #200. Apex
Category: Detergent, wetting agent
Applications: Low foaming scouring agent for Kier processing; effective in baths containing high concentrations of salts and alkalis
Form: Liquid; Color: Amber
Containers: 55 gal drums

Apex Pentrapex #1923. Apex
Category: Wetting agent, penetrating agent
Applications: Wetting agent effective in solutions containing high concentrations of salts; wetting and penetrating agent effective and stable in Apex flame retardants
Form: Liquid; Color: Yellow
Properties: Solubility: Miscible with water
Containers: 55 gal drums
Warnings and Caution: Rinse well with water if spilled on skin or clothing

Apex Tarem #20. Apex
Category: Detergent, wetting a-

gent
Applications: Allows single bath scouring and dyeing of cellulosic/synthetic blends; applicable in pressure dyeing equipment
Form: Liquid
Properties: Solubility: Water dispersible
Containers: 55 gal drums

Apex Tarem #1498. Apex
Category: Detergent
Applications: Solvent based surfactant; solvent scour for cellulosics, synthetics and blends; bleaching and dyeing
Form: Liquid
Properties: Solubility: Forms emulsion in water
Containers: 55 gal drums

Apexomine #1639. Apex
Category: Detergent
Applications: Detergent; Kier scouring assistant; scouring agent for natural fibers; permits single bath scouring and dyeing of nylon or polyester fabrics; desizing and boiling off cottons; bleaching and dyeing
Containers: 55 gal drums

Apexomine #1771. Apex
Category: Detergent
Applications: Low foaming solvent scour; solvent emulsion for fabrics containing high concentrations of oils, grease, etc.
Form: Liquid; Color: Slightly yellow
Properties: Solubility: Emulsifies in water
Containers: 55 gal drums
Warnings and Caution: If spilled on skin or clothing, rinse well with water

Apexscour #1962. Apex
Category: Detergent
Applications: Continuous scouring agent
Form: Liquid
Properties: Solubility: Water
Containers: 55 gal drums

Apexscour #2590. Apex
Category: Detergent, wetting agent
Applications: Continuous scouring agent; economical anionic/nonionic surfactant for continuous scouring of printed fabrics
Form: Liquid
Containers: 55 gal drums

Aremul AIS. Arol Chem. Prod.
Category: Detergent, emulsifier
Applications: Industrial detergent base, emulsifier for aliphatic solvents; base for solvent degreasing compounds, dry cleaning soaps, waterless hand cleaners, emulsions of mineral oil and chlorinated solvents; Detergent–Textiles, skins, laundering, household, industrial; Wetting agent–Removing starches, wall paper; Emulsifier–Kier boiling, dispersing waxes, pigments, resins
Form: Liquid; Color: Amber; Odor: Mild
Properties: Solubility: Sol. in various aliphatic and chlorinated solvents; Sp. gr.: 1.03; Stability: Excellent
Composition: 90–95% active ingred.; other components: amine

salt of an alkyl aryl sulfonic acid
Reaction: Acid
Toxicity: Avoid ingestion; otherwise considered nontoxic
Containers: 55 gal drums
Warnings and Cautions: Avoid contact with eyes
References: Technical data sheet

Aremul DCL. Arol Chem. Prod.
Chemical Description: Blend of anionic sulfonates
Category: Detergent, emulsifier
Applications: Basic emulsifier for carrier solvents in polyester dyeing; dye leveler; wetting agent–dyeing and finishing; detergent–textiles, skins, laundering, cleaners of street, buildings, hospitals; dispersing waxes, pigments, resins
Form: Liquid; Color: Dark amber; Odor: Mild
Properties: Solubility: Miscible with water and ODCB, TCB, butyl benzoate and methyl salicylate; Sp. gr.: 1.04; Stability: Stable in presence of acids, anionic and nonionic dyeing assistants
Composition: 55–60% active ingred.; other components: blend of anionic sulfonates
Reaction: Acid
Toxicity: Avoid ingestion; otherwise considered nontoxic
Containers: 55 gal drums
Warnings and Caution: Avoid contact with eyes
References: Technical data sheet

Aremul V-91. Arol Chem. Prod.
Chemical Description: Blend of alkyl aryl sulfonates, modified nonionics and organic coupling agents
Category: Detergent, emulsifier
Applications: Concentrated emulsifier; synthetic detergent base; base for solvent type scouring, degreasing and cleaning agents; Detergent–textiles, household, street, buildings, hospitals cleaners; Emulsifier–Kier boiling; making of cutting and soluble oils; dispersing waxes, pigments, resins
Form: Liquid; Color: Dark amber; Odor: Mild
Properties: Solubility: Sol. in warm water, readily sol. in aliphatic solvents; Stability: Stable under mild acid and alkaline conditions up to boil and to hard water conditions
Composition: 90–95% active ingred.
Reaction: Acid
Toxicity: Avoid ingestion, otherwise considered nontoxic
Containers: 55 gal drums
Warnings and Caution: Avoid contact with eyes
References: Technical data sheet

Aremul X-91. Arol Chem. Prod.
Chemical Description: Blend of alkyl aryl sulfonates, modified nonionics and organic coupling agents
Category: Detergent, emulsifier
Applications: Concentrated emulsifier; synthetic detergent base; base for scouring agents, dye dispersant and scouring agent, degreaser, cleaner and emulsifier for fuel oil treatment; Detergent–textiles, skins, leather, laundering, dyeing, household, cosmetics,

cleaners of street, buildings, hospitals; Emulsifier–Kier boiling, dispersing waxes, pigments, resins
Form: Liquid; Color: Amber; Odor: Mild
Properties: Solubility: Sol. in warm water and aromatic solvents; Sp. gr.: 1.03; Stability: Stable to mild acids and alkalines up to boil, stable under hard water conditions
Composition: 90–95% active ingred.
Reaction: Acid
Toxicity: Avoid ingestion; otherwise considered nontoxic
Containers: 55 gal drums
Warnings and Caution: Avoid contact with eyes
References: Technical data sheet

Arizona DR-22. Arizona Chem.
Category: Emulsifier, detergent, wetting agent
Applications: Emulsifier for SBR rubber production; used to make Arizona disproportionated rosin soaps (DRS-40, 42, 43, 44)
Properties: Saponification No.: 165; Softening Pt.: 66 C

Arizona DRS-40. Arizona Chem.
Chemical Description: Potassium soap of disproportionated tall oil rosin
Category: Emulsifier, detergent, wetting agent, anionic
Applications: Emulsifier for SBR rubber production; Detergent–mines and minerals
Form: Paste; Color: Tan; Odor: Bland
Properties: Solubility: Water-dispersible; Density: 9.11 lb/gal @ 75 C; Sp. gr.: 1.1; Visc.: 500 cps @ 75 C; Stability: Good
Composition: 80% solids; DRS-40 contains an electrolyte to prevent stratification

Arizona DRS-42. Arizona Chem.
Chemical Description: Potassium soap of disproportionated tall oil rosin
Category and Applications: See Arizona DRS-40
Form: Paste; Color: Tan; Odor: Bland
Properties: Solubility: Water-dispersible; Density: 9.11 lb/gal @ 75 C; Sp. gr.: 1.1; Visc.: 800 cps @ 75 C; Stability: Good; Lower Acid No. and higher visc. than DRS-40
Composition: 80% solids

Arizona DRS-43. Arizona Chem.
Chemical Description: Sodium salt of disproportionated tall oil rosin
Category and Applications: See Arizona DRS-40
Form: Paste; Color: Tan; Odor: Bland
Properties: Solubility: Water-dispersible; Density: 8.94 lb/gal @ 75 C; Sp. gr.: 1.1; Visc.: 1700 cps @ 75 C; Stability: Good
Composition: 71% solids

Arizona DRS-44. Arizona Chem.
Chemical Description: Sodium salt of disproportionated tall oil rosin
Category and Applications: See Arizona DRS-40
Form: Paste; Color: Tan; Odor: Bland

Properties: Solubility: Water-dispersible; Density: 8.99 lb/gal @ 75 C; Sp. gr.: 1.1; Visc.: 750 cps @ 75 C; Stability: Good
Composition: 75% solids

Arkomon SO. American Hoechst Corp.
Chemical Description: Oleoyl sarcoside
Category: Intermediate, anionic
Applications: Intermediate in manufacture of anticorrosive oil; amine salts as emulsifiers for cutting oil emulsion
Form: Liquid

Arkopal N040. American Hoechst Corp.
Chemical Description: Nonyl phenoxy poly (ethyleneoxy) ethanol
Category: Detergent, wetting agent, emulsifier, nonionic
Applications: General industrial use
Form: Liquid
Composition: 100% conc.

Arkopal N060. American Hoechst Corp.
Chemical Description: Nonyl phenoxy poly (ethyleneoxy) ethanol
Category and Applications: See Arkopal N040
Form: Liquid
Composition: 100% conc.

Arkopal N080. American Hoechst Corp.
Chemical Description: Nonyl phenoxy poly (ethyleneoxy) ethanol
Category and Applications: See Arkopal N040
Form: Liquid
Composition: 100% conc.

Arkopal N090. American Hoechst Corp.
Chemical
Chemical Description: Nonyl phenoxy poly (ethyleneoxy) ethanol
Category and Applications: See Arkopal N040
Form: Liquid
Composition: 100% conc.

Arkopal N100. American Hoechst Corp.
Chemical Description: Nonyl phenoxy poly (ethyleneoxy) ethanol
Category and Applications: See Arkopal N040
Form: Liquid
Composition: 100% conc.

Arkopal N110. American Hoechst Corp.
Chemical Description: Nonyl phenoxy poly (ethyleneoxy) ethanol
Category and Applications: See Arkopal N040
Form: Liquid
Composition: 100% conc.

Arkopal N130. American Hoechst Corp.
Chemical Description: Nonyl phenoxy poly (ethyleneoxy) ethanol
Category and Applications: See

Arkopal N040
Form: Paste
Composition: 100% conc.

Arkopal N150. American Hoechst Corp.
Chemical Description: Nonyl phenoxy poly (ethyleneoxy) ethanol
Category and Applications: See Arkopal N040
Form: Paste
Composition: 100% conc.

Arlacel 20. I.C.I. United States
Chemical Description: Sorbitan monolaurate
Category: Emulsifier, nonionic
Form: Liquid; Color: Yellow
Properties: Solubility: Sol. in methanol, ethanol, mineral, cottonseed and corn oils, ethylene glycol; Sp. gr.: 1.0; Visc.: 4250 cps; HLB: 8.6; Flash Pt.: > 300 F; Fire Pt.: > 300 F
Composition: 100% active ingred.

Arlacel 40. I.C.I. United States
Chemical Description: Sorbitan monopalmitate
Category: Nonionic emulsifier
Form: Cream beads
Properties: Solubility: Sol. in isopropanol; Sp. gr.: ≈ 1; HLB: 6.7; Flash Pt.: > 300 F; Fire Pt.: > 300 F
Composition: 100% active ingred.

Arlacel 60. I.C.I. United States
Chemical Description: Sorbitan monostearate
Category: Nonionic surfactant
Applications: Cosmetic use
Form: Waxy beads; Color: Cream
Properties: Solubility: Sol. in isopropanol; HLB: 4.7; Flash Pt.: > 300 F; Fire Pt.: > 300 F; Pour Pt.: ≈ 53 C
Composition: 100% active ingred.
Containers: 51 gal fiber lever pkg.

Arlacel 60. I.C.I. United States
Chemical Description: Sorbitan monostearate
Category and Applications: See Arlacel C
Form: Solid
Properties: HLB: 4.7
Composition: 100% conc.

Arlacel 80. I.C.I. United States
Chemical Description: Sorbitan monooleate
Category: Nonionic surfactant
Form: Oil liquid; Color: Yellow
Properties: Solubility: Sol. in isopropanol, mineral and cottonseed oils; Sp. gr.: ≈ 1; Visc.: ≈ 1900 cps; HLB: 4.3; Flash Pt.: > 300 F; Fire Pt.: > 300 F
Composition: 100% active ingred.
Containers: 55 gal steel drum

Arlacel 83. I.C.I. United States
Chemical Description: Sorbitan sesquioleate
Category: and Applications: See Arlacel C. Arlacel 83 is cosmetic and pharmaceutical grade of Arlacel C
Category: Nonionic surfactant
Form: Oil liquid; Color: Yellow
Properties: Solubility: Sol. in

mineral and cottonseed oils, ethanol, isopropanol; Sp. gr.: ≈ 1; Visc.: ≈ 1500 cps; HLB: 3.7; Flash Pt.: > 300 F; Fire Pt.: > 300 F
Composition: 100% active ingred.

Arlacel 85. I.C.I. United States
Chemical Description: Sorbitan trioleate
Category: Nonionic surfactant
Form: Oily liquid; Color: Yellow
Properties: Solubility: Sol. in isopropyl alcohol, mineral oil, cottonseed oil, corn oil; Sp. gr.: ≈ 0.95; Visc.: ≈ 250 cps; HLB: 1.8; Flash Pt.: > 300 F; Fire Pt.: > 300 F
Composition: 100% active ingred.

Arlacel 165. I.C.I. United States
Chemical Description: Glycerol monostearate and polyoxethylene stearate blend
Category: Nonionic surfactant, emulsifier
Applications: Cosmetics field
Form: Beads; Color: White; Odor: Bland
Properties: Solubility: Dispersible in water; HLB: 11; Flash Pt.: > 300 F; Fire Pt.: > 300 F; Pour Pt.: ≈ 48 C; Stability: Acid stable; Self-emulsifying
Composition: 100% active ingred.

Arlacel 186. I.C.I. United States
Chemical Description: Mono- and diglycerides with 0.02% BHA and 0.01% citric acid in propylene glycol added as a preservative
Category: Nonionic surfactant, emulsifier
Form: Clear liquid; Color: Pale yellow
Properties: Solubility: Sol. in ethanol, isopropanol, cottonseed oil and mineral; Sp. gr.: ≈ 1.0; Visc.: ≈ 150 cps; HLB: 2.8; Flash Pt.: > 300 F; Fire Pt.: > 300 F
Composition: 100% active ingred.

Arlacel 481. I.C.I. United States
Chemical Description: Mixture of fatty acid esters
Category: Emulsifier, nonionic
Applications: Cosmetic creams and lotions
Form: Waxy
Composition: 100% conc.

Arlacel 986. I.C.I. United States
Chemical Description: Formulated product–mixture of fatty acid esters
Category and Applications: See Arlacel 481
Form: Solid
Composition: 100% conc.

Arlacel 987. I.C.I. United States
Chemical Description: Sorbitan mono fatty acid ester
Category and Applications: See Arlacel 481
Form: Liquid
Composition: 100% conc.

Arlacel 988. I.C.I. United States
Chemical Description: Polyoxyethylene glycerine sorbitan fatty acid ester
Category and Applications: See

Arlacel 481
Form: Solid
Composition: 100% conc.

Arlacel 989. I.C.I. United States
Chemical Description: Polyoxyethylene fatty glyceride
Category and Applications: See Arlacel 481
Form: Solid
Composition: 100% conc.

Arlacel C. I.C.I. United States
Chemical Description: Sorbitan sesquioleate
Category: Nonionic surfactant, W/O emulsifier
Form: Oily liquid; Color: Amber
Properties: Solubility: Sol. in mineral oil, cottonseed oil, ethanol, isopropanol; Sp. gr.: ≈ 1.0; Visc.: ≈ 1000 cps; HLB: 3.7; Flash Pt.: > 300 F; Fire Pt.: > 300 F
Composition: 100% active ingred.

Arlamol E. I.C.I. United States
Chemical Description: Polyoxypropylene (15) stearyl ether
Category: Emollient solvent
Form: Oily liquid
Properties: Solubility: Sol. in mineral oil, isopropyl esters, cottonseed oil, ethanol, isopropanol, hexadecyl alcohol; Sp. gr.: ≈ 0.95; Visc.: ≈ 90 cps; Pour Pt.: Typically < 0 C
Composition: 100% active ingred.

Arlasolve 200. I.C.I. United States
Chemical Description: Polyoxyethylene (20) isohexadecyl ether
Category: Nonionic surfactant, emulsifier, solubilizer
Applications: Cosmetics
Form: Soft white waxy solid
Properties: Solubility: Water, propylene glycol, ethanol; Sp. gr.: ≈ 1.0; HLB: 15.7; Flash Pt.: > 230 F; Fire Pt.: > 230 F
Composition: 100% active ingred.
Toxicity: Eye irritant

Arlatone 285. I.C.I. United States
Chemical Description: Polyoxyethylene fatty glyceride
Category: Coupling agent, solubilizer, nonionic
Applications: Cosmetics esp. for coupling and solubilizing perfumes and fragrances
Form: Semifluid
Composition: 100% conc.

Arlatone 289. I.C.I. United States
Chemical Description: Polyoxyethylene fatty glyceride
Category and Applications: See Arlatone 285
Form: Waxy
Composition: 100% conc.

Arlatone 970. I.C.I. United States
Chemical Description: Polyoxyethylene sorbitan fatty acid ester
Category and Applications: See Arlatone 285
Form: Liquid
Composition: 100% conc.

Arlatone 983. I.C.I. United States
Chemical Description: Polyoxyethylene fatty acid ester
Category: Emulsifier, nonionic
Applications: Emulsifier for oil/water creams and lotions
Form: Solid
Composition: 100% conc.

Arlatone B. I.C.I. United States
Chemical Description: Mixture of polyoxyethylene sorbitan fatty acid esters and alkyl aryl alcohols
Category: Emulsifier, nonionic
Applications: Blooming bath oils, waterless hand cleaners, lotion formulas, anhydrous aerosol foams based on mineral oil or deodorized kerosene; other cosmetic cleansing products
Form: Clear liquid; Color: Amber
Properties: Solubility: Miscible with mineral oil, deodorized kerosene, isopropyl myristate, isopropyl palmitate, hexadecyl alcohol, ethyl alcohol, isopropyl alcohol, cottonseed oil; Sp. gr.: ≈ 1; Visc.: ≈ 400 cp; HLB: 9; Flash Pt.: 300F; Pour Pt.: 0 F
Composition: 100% active ingred.

Arlatone G. I.C.I. United States
Chemical Description: Polyoxyethylene fatty glyceride
Category: Nonionic surfactant
Applications: For use in formulating clear gels
Form: Yellow viscous liquid to soft paste
Properties: Solubility: Sol. in water, ethanol and isopropanol; Sp. gr.: ≈ 1.0; HLB: 10.8; Flash Pt.: > 300 F; Fire Pt.: > 300 F
Composition: 100% active ingred.

Arlatone T. I.C.I. United States
Chemical Description: Polyoxyethylene (40) sorbitol septaoleate
Category: Emulsifier, solubilizer, antistat, lubricant, nonionic
Applications: General purpose water-soluble hydrophilic surfactants used as O/W emulsifiers, esp. with corresponding "Span" products; also in textile industry as antistat and fiber lubricant
Form: Liquid; Color: Yellow
Properties: Solubility: Sol. in vegetable oils, mineral oil, isopropyl myristate, isopropyl palmitate; dispersible in water; Sp. gr.: ≈ 1; Visc.: ≈ 175 cps; HLB: 9; Flash Pt.: > 300 F; Fire Pt.: > 300 F
Composition: 100% active ingred.

Arlex. I.C.I United States
Chemical Description: Aqueous solution of sorbitol and its anhydrides, and oxypropylene ether (83%)
Category: Polyhydric alcohol (polyol)
Applications: Industrial humectant
Form: Clear viscous liquid; Odor: Essentially odor free
Properties: Solubility: Sol. in water, alcohol, glycerin, propylene glycol; Sp. gr.: ≈ 1.32; Visc.: ≈ 1700 cps; Flash Pt.: > 300 F; Fire Pt.: > 300 F
Composition: 83% active ingred.
Containers: 55 gal closed-head steel drum

Armac 18D. Armak
Chemical Description: Octadecylamine acetate–distilled grade
Category: Wetting agent, emulsifier, cationic, corrosion inhibitor
Applications: Water-soluble cationic reagent for pigment flushing, froth flotation of minerals, flocculation, corrosion inhibition
Form: Flake; Color: White

Properties: Solubility: Water-sol.; Sp. gr.: 0.827 @ 80 C; M.P.: 85 C; Flash Pt.: 320 F (COC); Fire Pt.: 340 F (COC)
Composition: 100% active ingred.
Toxicity: Irritating to skin and eyes
Containers: 50 lb cardboard boxes, stored in dry place away from heat
Warnings and Caution: Protective goggles and gloves worn when handling
References: Bulletin 76-30

Armac C. Armak
Chemical Description: Cocoamine acetate
Category: See Armac 18D
Applications: See Armac 18D; also as stripping agent
Form: Solid; Color: Yellow
Properties: m.w.: 260; Solubility: Water-sol.; Sp. gr.: 0.852 @ 80 C; M.P.: 50 C; Flash Pt.: 285 F (COC); Fire Pt.: 300 F (COC)
Containers: 55 gal drum

Armac HT. Armak
Chemical Description: (hydrogenated-tallow)amine acetate
Category and Applications: See Armac 18D
Form: Solid; Color: White
Properties: Sp. gr.: 0.820 @ 80 C; M.P.: 60 C; Flash Pt.: 325 F (COC); Fire Pt.: 340 F (COC)
Composition: 100% active ingred.

Armac OD. Akzo Chemie
Chemical Description: Oleylamine acetate
Category: Corrosion inhibitor, flotation agent
Applications: Mineral flotation agent in mining industry; corrosion inhibitor
Form: Solid; Color: 13 max (Gardner)
Properties: M.P.: 55 C; Flash Pt.: 170 C (COC)
Composition: 95% min. active ingred.; 4% max. secondary amine
Containers: 172 kg net open head drums

Armac T. Akzo Chemie
Chemical Description: Tallowaamine acetate
Category: Flotation agent, corrosion inhibitor, lubricant
Applications: Mineral flotation agent in mining industry; corrosion inhibitor and lubricant in metal processing
Form: Solid; Color: 10 max. (Gardner)
Properties: Sp. gr.: 0.87 @ 60 C; M.P.: 45–61 C; Flash Pt.: 168 C (COC)
Composition: 94% min active ingred.; 5% max. secondary amine
Toxicity: Skin irritant, severe eye irritant
Containers: 172 kg net open head drums
Warnings and Caution: Avoid contact with strong oxidizing agents and strong mineral acids

Armeen Series. Armak

Chemical Description: Aliphatic amines (primary, secondary, or tertiary) with long-chain alkyl groups derived from fatty acids derived from coconut oil, soybean

oil and tallow
Category: Chemical intermediates, emulsifier, wetting agent
Applications: Preparation of amine salts of pesticide chemicals; synthesis of fuel oil and gasoline additives; amine oxides and quaternary ammonium salts for cosmetic formulations; amines and their salts as collectors in mineral flotation; salts and derivatives as asphalt emulsifiers, textile finishing agents, pigment dispersants, antistatic agents, corrosion inhibitors, metal polishes, antifoaming agents (textiles)
Form and Properties: See individual Armeen listings
Reaction: Armeen amines form salts with acids, amides with acylating agents, complexes with heavy metals, Schiff bases with aromatic aldehydes, ethoxylated tertiary amines with alkylene oxides and certain aldehydes, quaternary ammonium compounds with alkylating agents; ionic type: cationic
Toxicity: Severe burns to skin and eyes when in prolonged contact
Containers: 55-gal drums, tankcars
References: Product data bulletin 78-5

Armeen 2C. Armak
Chemical Description: Dicocoamine (secondary amine)
Form: Solid
Properties: Solubility: Sol. in chloroform, slightly sol. in isopropanol, toluene, carbon tetrachloride, kerosene; Sp. gr.: 0.793 @ 60/40 C; Visc.: 49.1 SSU @ 60 C; M.P.: 104–117 F; Flash Pt.: 350 F; Fire Pt.: 410 F; Pour Pt.: 80 F

Armeen 2HT. Armak
Chemical Description: Di(hydrogenated tallow)amine (secondary amine)
Applications: Corrosion inhibitor and pour point depressant for petroleum-refined products
Form: Solid
Properties: Solubility: Sol. in chloroform, slightly sol. in toluene, carbon tetrachloride, kerosene; Sp. gr.: 0.803 @ 60/4 C; Visc.: 63.9 SSU @ 70 C; M.P.: 140–149 F; Flash Pt.: 435 F; Fire Pt.: 480 F; Pour Pt.: 115 F

Armeen 8D. Armak
Chemical Description: Octylamine (primary amine)
Form: Liquid
Properties: Solubility: Sol. in methanol, ethanol, acetone, isopropanol, chloroform, toluene, carbon tetrachloride, kerosene; Sp. gr.: 0.793 @ 25/4 C; Visc.: 34.4 SSU @ 25 C; M.P.: 10–21 F; Flash Pt.: 145 F; Fire Pt.: 160 F; Pour Pt.: 15 F

Armeen 12D. Armak
Chemical Description: Dodecylamine (primary amine)
Applications: Lubricants (metal treatment industry)
Form: Liquid
Properties: Solubility: See Armeen 8D; Sp. gr.: 0.801 @ 25/4 C; Visc.: 42.2 SSU @ 25 C; Melting Range: 79–86 F; Flash Pt.: 235 F;

Fire Pt.: 250 F; Pour Pt.: 80 F

Armeen 16D. Armak
Chemical Description: Hexadecylamine (primary amine)
Form: Solid
Properties: Solubility: Sol. in methanol, ethanol, isopropanol, chloroform, toluene, carbon tetrachloride, slightly sol. in acetone, kerosene; Sp. gr.: 0.789 @ 60/4 C; Visc.: 37.5 SSU @ 55 C; Melting Range: 100–118 F; Flash Pt.: 285 F; Fire Pt.: 325 F; Pour Pt.: 100 F

Armeen 18. Armak
Chemical Description: Octadecylamine (primary amine)
Applications: Anticaking agent
Form: Solid
Properties: Solubility: Sol. in ethanol, isopropanol, chloroform, toluene, carbon tetrachloride, slightly sol. in acetone, kerosene; Sp. gr.: 0.792 @ 60/4 C; Visc.: 45.6 SSU @ 55 C; Melting Range: 122–133 F; Flash Pt.: 320 F; Fire Pt.: 365 F; Pour Pt.: 115 F

Armeen 18D. Armak
Chemical Description: Octadecylamine (primary amine)
Form: Solid
Properties: Solubility: Sol. in ethanol, isopropanol, chloroform, toluene, carbon tetrachloride, slightly sol. in methanol, kerosene; Sp. gr.: 0.791 @ 60/4 C; Visc.: 43.7 SSU @ 45 C; Melting Range: 122–133F; Flash Pt.: 300 F; Fire Pt.: 350 F; Pour Pt.: 110 F

Armeen C. Armak
Chemical Description: Cocoamine (primary amine)
Applications: Stripping agent for paints
Form: Liquid
Properties: Solubility: Sol. in methanol, ethanol, acetone, isopropanol, chloroform, toluene, carbon tetrachloride, kerosene; Sp. gr.: 0.805 @ 25/4 C; Visc.: 44.2 SSU @ 35 C; Melting Range: 54–59 F; Flash Pt.: 240 F; Fire Pt.: 270 F; Pour Pt.: 45 F

Armeen CD. Armak
Chemical Description: Cocoamine (primary amine)
Form: Liquid
Properties: Solubility: See Armeen C; Sp. gr.: 0.804 @ 25/4 C; Visc.: 43.0 SSU @ 35 C; Melting Range: 57–63 F; Flash Pt.: 230 F; Fire Pt.: 250 F; Pour Pt.: 55 F

Armeen DM12D. Akzo Chemie
Chemical Description: Dodecyldimethyl amine
Category: Chemical intermediate, cationic
Applications: Raw material for surfactants (e.g., amine oxides and quats)
Form: Liquid; Color: Yellow; Odor: Amine
Properties: Solubility: Water insol.; Sp. gr.: 0.78; Visc.: 3.30 cs; F.P.: −15 C; B.P.: 80–115 C @ 3 mm Hg.; Flash Pt.: 122 C (COC); Iodine Value: 3 max.
Composition: 98% active ingred.; 1% max. primary amine; 1% max. secondary amine

Reaction: Alkaline
Toxicity: Skin irritant, severe eye irritant
Containers: 200 L. bung type steel drums
Warnings and Caution: Avoid contact with strong oxidizing agents

Armeen DM14D. Akzo Chemie
Chemical Description: Tetradecyldimethyl amine
Category: Chemical intermediate, cationic
Applications: Raw material for manufacture of pharmaceutical quaternaries
Form: Liquid; Color: Yellow; Odor: Amine
Properties: Solubility: Insol. in water; Sp. gr.: 0.79; Visc.: 4.70 cs; F.P.: –8 C; B.P.: 100–125 C @ 3 mm Hg.; Flash Pt.: 28 C (COC); Iodine Value: 3 max.
Composition: 98% active ingred.; 1% max. prim. amine; 1% max. sec. amine
Reaction: Alkaline
Toxicity: Skin irritant, severe eye irritant
Containers: 200 L. bung type steel drums
Warnings and Caution: Avoid contact with strong oxidizing agents

Armeen DM16D. Akzo Chemie
Chemical Description: Hexadecyldimethyl amine
Category: Chemical intermediate, cationic
Applications: Raw material for surfactants (e.g., amine oxides and quaternaries)
Form: Liquid; Color: Yellow; Odor: Amine
Properties: Solubility: Insol. in water; Sp. gr.: 0.80; Visc.: 6.75 cs; F.P.: 8 C; B.P.: 100–136 C @ 3 mm Hg.; Flash Pt.: 150 C (COC); Iodine Value: 3 max.
Composition: 98% active ingred.; 1% max. prim. amine; 1% max. sec. amine
Reaction: Alkaline
Toxicity: Skin irritant, severe eye irritant

Armeen DM18D. Akzo Chemie
Chemical Description: Octadecyldimethyl amine
Category: Chemical intermediate, cationic
Applications: Raw material for surfactants (e.g., amine oxides and quaternaries)
Form: Liquid; Color: Yellow; Odor: Amine
Properties: Solubility: Insol. in water; Sp. gr.: 0.79; F.P.: 20 C; B.P.: 145–160 C @ 3 mm Hg.; Flash Pt.: 178 C (COC); Iodine Value: 3 max.
Composition: 98% active ingred.; 1% max. prim. amine; 1% max. sec. amine
Reaction: Alkaline
Toxicity: Skin irritant, severe eye irritant
Containers: 200 L. bung type steel drums
Warnings and Caution: Avoid contact with strong oxidizing agents and strong acids

Armeen DMAD. Akzo Chemie
Chemical Description: Alkyl (C_{10}–

C_{18}) dimethyl amine
Category: Chemical intermediate
Applications: Raw material for surfactants (e.g., quaternaries)
Form: Liquid; Color: Yellow; Odor: Amine
Properties: Solubility: Insol. in water; Sp. gr.: 0.79; Visc.: 4.44 cs; B.P.: 45–150 C @ 3 mm Hg.; Flash Pt.: 130 C (COC); Iodine Value: 3 max.
Composition: 98% active ingred.; 1% max. prim. amine; 1% max. sec. amine
Reaction: Alkaline
Toxicity: Skin irritant, severe eye irritant
Containers: 200 L. bung type steel drums
Warnings and Caution: Avoid contact with strong oxidizing agents

Armeen DMCD. Akzo Chemie
Chemical Description: Alkyl (C_8–C_{18}) dimethyl amine
Category: Chemical intermediate
Applications: Raw material for surfactants (e.g., amine oxides and quaternaries)
Form: Liquid; Color: Yellow; Odor: Amine
Properties: Solubility: Insol. in water; Sp. gr.: 0.79; Visc.: 4.20 cs; B.P.: 42–150 C @ 3 mm Hg.; Flash Pt.: 108 C (COC); Iodine Value: 3 max.
Composition: 98% active ingred.; 1% max. prim. amine; 1% max. sec. amine
Reaction: Alkaline
Toxicity: Skin irritant, severe eye irritant
Containers: 200 L. bung type steel drums
Warnings and Caution: Avoid contact with strong oxidizing agents

Armeen DMHTD. Akzo Chemie
Chemical Description: Hydrogenated tallow dimethyl amine
Category: Chemical intermediate
Applications: Raw material for surfactants (e.g., amine oxide, quaternaries)
Form: Liquid; Color: Yellow; Odor: Amine
Properties: Solubility: Insol. in water; Sp. gr.: 0.80; Visc.: 8.75 cs; F.P.: 18 C; B.P.: 100–155 C; Flash Pt.: 162 C (COC); Iodine Value: 3 max.
Composition: 98% active ingred.; 1% max. prim. amine; 1% max. sec. amine
Reaction: Alkaline
Toxicity: Skin irritant, severe eye irritant
Containers: 200 L. bung type steel drums
Warnings and Caution: Avoid contact with strong oxidizing agents

Armeen DMMCD. Akzo Chemie
Chemical Description: Alkyl (C_{12}–C_{14}) dimethyl amine
Category: Chemical intermediate
Applications: Raw material for surfactants (e.g., amine oxides, quaternaries)
Form: Liquid; Color: Yellow; Odor: Amine
Properties: Solubility: Insol. in water; Sp. gr.: 0.79; Visc.: 3.95 cs; F.P.: –19 C; B.P.: 90–125 C;

Flash Pt.: 125 C (COC); Iodine Value: 3 max.
Composition: 98% active ingred.; 1% max. prim. amine; 1% max. sec. amine
Reaction: Alkaline
Toxicity: Skin irritant, severe eye irritant
Containers: 200 L. bung type steel drums
Warnings and Caution: Avoid contact with strong oxidizing agents

Armeen HT. Armak
Chemical Description: (hydrogenated-tallow)amine (primary amine)
Form: Solid; Color: 9 (Gardner)
Properties: Solubility: Sol. in methanol, ethanol, isopropanol, chloroform, toluene, carbon tetrachloride, slightly sol. in kerosene; Sp. gr.: 0.795 @ 60/4 C; Visc.: 47.5 SSU @ 55 C; Melting Range: 79–136 F; Cloud Pt.: 115 F; Flash Pt.: 320 F; Fire Pt.: 365 F; Pour Pt.: 110 F

Armeen HTD. Armak
Chemical Description: (hydrogenated-tallow) amine (primary amine)
Form: Solid; Color: 2 (Gardner)
Properties: Solubility: See Armeen HT; Sp. gr.: 0.794 @ 60/4C; Visc.: 44.1 SSU @ 55 C; Melting Range: 70–120 F; Cloud Pt.: 110 F; Flash Pt.: 315 F; Fire Pt.: 360 F; Pour Pt.: 100 F

Armeen L-10. Armak
Chemical Description: Beta-amine (straight chain, saturated, petroleum derived chemicals with an amino functional group attached to an interior carbon atom)
Category: Cationic surfactant
Form: Clear liquid; Color: 1 (Gardner)
Properties: Solubility: Sol. in isopropyl alcohol, mineral oil, kerosene; Sp. gr.: 0.787; F.P.: –50 F; Cloud Pt.: – 50 F; Flash Pt.: 85 C; Fire Pt.: 94 C; Biodegradable
References: Bulletin 74-13

Armeen L-11. Armak
Chemical Description: See Armeen L-10
Form: Clear liquid; Color: 2 (Gardner)
Properties: Solubility: Sol. in isopropyl alcohols, mineral oil, kerosene; Sp. gr.: 0.802; M.P.: –20 F; Cloud Pt.: –20 F; Flash Pt.: 100C; Fire Pt.: 115 C; Biodegradable
References: Bulletin 74-13

Armeen L-15. Armak
Chemical Description: See Armeen L-10
Form: Clear liquid; Color: 5 (Gardner)
Properties: Solubility: Sol. in isopropyl alcohol, mineral oil, kerosene; Sp. gr.: 0.813; Cloud Pt.: 54 F; Flash Pt.: 157 C; Fire Pt.: 175 C; Pour Pt.: 50 F
References: Bulletin 74-13

Armeen N-CMD. Akzo Chemie
Chemical Description: *N*-coco morpholine
Category: Catalyst

Applications: Catalyst in polyurethane foams
Form: Liquid; Color: 3 (Gardner)
Properties: Solubility: Sol. in acetone, IPA; Sp. gr.: 0.87; Visc.: 50 SSU @ 55 C; M.P.: −29/−10 C; Flash Pt.: 155 C (COC); Pour Pt.: −10 C
Composition: 97% min. active ingred.
Toxicity: Skin irritant, severe eye irritant
Warnings and Caution: Avoid contact with strong oxidizing agents

Armeen O. Armak
Chemical Description: Oleylamine (primary amine)
Form: Paste; Color: 8 (Gardner)
Properties: Solubility: Sol. in acetone, methanol, ethanol, isopropanol, chloroform, toluene, carbon tetrachloride, kerosene, white mineral oil; Sp. gr.: 0.820 @ 38/4 C; Visc.: 57.0 SSU @ 25 C; Melting Range: 50–68 F; Flash Pt.: 320 F; Fire Pt.: 360 F

Armeen OD. Akzo Chemie
Chemical Description: Oleylamine
Category: Wetting agent
Applications: Lube oil additive (anticorrosive)
Form: Paste; Color: 2 (Gardner)
Properties: Solubility: Insol. in water; Sp. gr.: 0.79 @ 60 C; Visc.: 56.6 SSU; M.P.: 8–18 C; Flash Pt.: 154 C (COC); Iodine Value: 78 min.
Composition: 98% active ingred.; 3% max. sec. amine
Reaction: Alkaline
Toxicity: Skin irritant, severe eye irritant
Containers: 200 L. bung type steel drums
Warnings and Caution: Avoid contact with strong mineral acids, strong oxidizing agents

Armeen SD. Armak
Chemical Description: Soyaamine (primary amine)
Form: Paste; Color: 3 (Gardner)
Properties: Solubility: Sol. in methanol, ethanol, acetone, isopropanol, chloroform, toluene, carbon tetrachloride; Sp. gr.: 0.81 @ 38/4 C; Visc.: 46.2 SSU @ 35 C; Melting Range: 81–86 F; Cloud Pt.: 85 F; Flash Pt.: 305 F; Fire Pt.: 345 F; Pour Pt.: 70 F

Armeen SZ. Akzo Chemie
Chemical Description: Sodium salt of *N*-coco β-amino butyric acid
Category: Detergent, wetting agent, stabilizer, corrosion inhibitor
Applications: Paints; dispersion and softening aid for pigments; dyeing and levelling agents; fungicides (paint films); stabilizer for styrene-butadiene based paints and other latex based compositions; corrosion inhibitor
Form: Liquid; Color: Yellow
Properties: Solubility: Sol. in water; Sp. gr.: 0.96; Visc.: 19 cps @ 55 C; Flash Pt.: 40 C (Abel-Pensky closed cup); pH: 9.9–10.8
Composition: 39–41% active in-

gred.; 18% isopropylalcohol; 6% max. free amine; 40% water
Reaction: Alkaline
Toxicity: Moderate skin irritant, severe eye irritant
Containers: 200 L. bung type steel drums
Warnings and Caution: Avoid contact with strong oxidizing agents/strong acids

Armeen T. Armak
Chemical Description: Tallowamine (primary amine)
Form: Paste; Color: 11 (Gardner)
Properties: Solubility: Sol. in isopropanol, methanol, ethanol, chloroform, toluene, carbon tetrachloride; Sp. gr.: 0.813 @ 38/4 C; Visc.: 47.0 SSU @ 35 C; Melting Range: 64–117 F; Cloud Pt.: 100 F; Flash Pt.: 315 F; Fire Pt.: 355 F; Pour Pt.: 70 F
Composition: 95% active ingred.
Toxicity: Corrosive to skin and eyes
Containers: 55-gal open-head drums; tank cars, tank wagons
Warnings and Caution: Protective goggles and gloves
References: Bulletin 77–7

Armeen TD. Armak
Chemical Description: Tallowamine (primary amine)
Form: Paste; Color: 2 (Gardner)
Properties: Solubility: Same as Armeen T; Sp. gr.: 0.812 @ 38/4C; Visc.: 45.2 SSU @ 35 C; Melting Range: 64–118 F; Cloud Pt.: 102 F; Flash Pt.: 305 F; Fire Pt.: 345 F; Pour Pt.: 70 F
Composition: 98% active ingred.

Armeen Z. Armak
Chemical Description: N-coco-beta-aminobutyric acid
Applications: Cosmetic; pigment softening and dispersing agent; freeze-thaw stabilizer for latex paints; lubricating aid in printing inks; dispersing agent in latex rubber reclamation; cooling tower corrosion inhibitor
Form: Pumpable slurry; Color: 8 (Gardner)
Properties: Solubility: Sol. in water, isopropanol; Sp. gr.: 0.98 @ 25/4 C; Visc.: 247 SSU @ 25 C; Flash Pt.: Foamed to 210 F; Pour Pt.: 65 F
Reaction: Amphoteric; compatible with cationic, anionic and nonionic surfactants

Armeen Z-9. Akzo Chemie
Chemical Description: *N*-coco β-amine butyric acid
Category: Emulsifier
Applications: Pigment dispersion aid
Form: Liquid; Color: Yellow
Properties: Solubility: Sol. in water; Sp. gr.: 0.90; Flash Pt.: 33 C (Abel-Pensky closed cup)
Composition: 46–50% active ingred.; 10% isopropylalcohol; 6% max. free amine; 40% water
Reaction: Neutral
Toxicity: Skin irritant, severe eye irritant
Containers: 200 L. bung type steel drums
Warnings and Caution: Avoid contact with strong oxidizing agents

Armid 18. Armak
Chemical Description: Octadecanamide
Category: Antiblock agent, stabilizers, dispersants
Applications: Antiblock, slip agents in manufacturing operations, in coatings and films; in manufacture of food packaging materials; as additive for printing inks, dyes, carbon paper coatings, fusible coatings for glassware and ceramics due to good viscosity control and improvement in dye solubility; foam viscosity stabilizers and boosters for detergents; raw materials for synthetic waxes; water repellents for textiles; pigment dispersing aids; thickening agents for paints and varnish removers
Form: Flake; Color: 7 (Gardner)
Properties: M.P.: 99–109 C; Flash Pt.: 225 C; Fire Pt.: 250 C
Composition: 90% active ingred.
Containers: Cardboard cartons, 55-gal steel drums
References: Bulletin 73–5

Armid C. Armak
Chemical Description: Cocoamide
Category and Applications: See Armid 18
Form: Flake; Color: 10 (Gardner)
Properties: M.P.: 85 C; Flash Pt.: 174 C; Fire Pt.: 185 C
Composition: 90% active ingred.
Containers: Cardboard cartons, 55-gal steel drums
References: Bulletin 73–5

Armid E. Akzo Chemie
Chemical Description: Erucamide
Category: Plastic additive
Applications: Slip and antiblocking agent
Form: Solid, flakes; Color: Yellow
Properties: Solubility: Insol. in water; M.P.: 79 C; Flash Pt.: 230 C (COC); Iodine Value: 72–80
Composition: 90% active ingred.; 3.5% free fatty acid
Containers: 50 kg. fiberboard keg
Warnings and Caution: Avoid contact with strong oxidizing agents

Armid HT. Armak
Chemical Description: (hydrogenated-tallow)amide
Category and Applications: See Armid 18
Form: Flake, powder; Color: 7 (Gardner)
Properties: M.P.: 98–103 C; Flash Pt.: 225 C; Fire Pt.: 250 C
Composition: 90% active ingred.
Containers: Cardboard cartons, 55-gal steel drums
References: Bulletin 73–5

Armid O. Armak
Chemical Description: Oleamide
Category and Applications: See Armid 18; also as release agent in cosmetics; penetrant in paper manufacture
Form: Flake, solid pellet; Color: 7 (Gardner); Odor: Bland
Properties: M.P.: 68 C; Flash Pt.: 207 C; Fire Pt.: 228 C
Composition: 90% active
Containers: Cardboard cartons, 55-gal steel drums
References: Bulletin 73–5

Armix 146. Arjay, Inc.
Chemical Description: Blended

adjuvant
Category: Anionic surfactant
Applications: Specialty surfactant to enhance the effectiveness of MSMA formulations; used for DSMA and high caustic industrial detergent cleaners
Form: Liquid
Composition: 100% conc.

Armoflo 48. Armak
Category: Conditioning agent
Applications: Conditioning agent to control caking and/or dusting, esp. effective for ammonium sulfate, ammonium nitrate, diammonium phosphate, urea; also in treatment of thermosetting plastics, detergents, insecticides
Form: Liquid; Color: 10 (Gardner)
Properties: Sp. gr.: 0.854 @ 60 C; Visc.: 69.1 SSU @ 60 C; M.P.: –4 C

Armoflo 65. Armak
Category and Applications: See Armoflo 48
Form: Liquid; Color: 12 (Gardner)
Properties: Sp. gr.: 0.824; Visc.: 52.2 SSU @ 60 C; M.P.: 5 C

Armoflo 66. Armak
Category and Applications: See Armoflo 48
Form: Semiliquid
Properties: Sp. gr.: 0.826 @ 60 C; Visc.: 59.0 SSU @ 60 C; M.P.: 38C

Armogloss 100. Armak
Applications: Concentrated liquid polishes for use in auto car washers, with excellent rinse beading to decrease drying time
Form: Liquid; Color: Clear yellow
Properties: Sp. gr.: 0.88; Visc.: 500 cp; Pour Pt.: 35 F

Armogloss 123. Armak
Applications: See Armogloss 100
Form: Liquid; Clear golden
Properties: Sp. gr.: 0.85; Visc.: 100 cp; Flash Pt.: > 200 F; Pour Pt.: 22 F

Armohib 25. Armak
Chemical Description: Amine-based
Category: Corrosion inhibitor, cationic
Applications: Metal cleaning and pickling; plant cleaning; corrosion inhibitor in sulfuric, phosphoric, sulfamic and citric acids; oil well acidizing
Form: Liquid; Color: Amber
Properties: Solubility: Readily sol. in acids; Density: 7.4 lb/gal; Sp. gr.: 0.89; Cloud Pt.: 15 F; Flash Pt.: 70 F (COC); Pour Pt.: < 0 F
Composition: 50% active ingred.
Toxicity: Irritant to skin and eyes
Containers: 55 gal steel drums with bungs; 5 and 1 gal cans
Warnings and Caution: Goggles and gloves worn
References: Bulletin 73–9

Armohib 28. Armak
Chemical Description: Amine-based
Category: Corrosion inhibitor, cationic
Applications: Metal cleaning and pickling; corrosion inhibitor in hydrochloric acid and HCl/HF

mixtures
Form: Liquid; Color: Amber
Properties: Solubility: Readily sol. in acids; Density: 7.69 lb/gal; Sp. gr.: 0.95; Flash Pt.: 90 F (COC); Pour Pt.: 52 F
Composition: 75% active ingred.
Toxicity: Irritant to skin and eyes
Containers: 55 gal steel drums with bungs, 5 and 1 gal cans
Warnings and Caution: Goggles and gloves
References: Bulletin 73-9

Armohib 31. Armak
Category: Corrosion inhibitor, cationic
Applications: Corrosion inhibitor in sulfuric, phosphoric, sulfamic and citric acids; metal cleaning and pickling
Form: Liquid; Color: Amber
Properties: Solubility: Readily sol. in acids; Density: 8.66 lb/gal; Sp. gr.: 1.042; Cloud Pt.: 28 F; Flash Pt.: 340 F (COC); Pour Pt.: 24 F
Composition: 100% active ingred.
Toxicity: Irritant to skin and eyes
Containers: 55 gal steel drums with bungs; 5 and 1 gal cans
Warnings and Caution: Goggles and gloves
References: Bulletin 73-9

Armosoft L. Armak
Chemical Description: Ditallow dimethyl ammonium chloride
Category: Fabric softener, cationic
Applications: Household and commercial fabric softener with antistatic, lubricating, bacteriostatic properties
Form: Hazy to clear liquid; Color: 6 (Gardner)
Properties: Density: 7.09 lb/gal; Sp. gr.: 0.85 @ 100 F; Visc.: 45.5 cps @ 120 F; Pour Pt.: 70–75 F
Composition: 75% active ingred.
Toxicity: Skin and eye irritant
Containers: 55 gal steel drums, tank trucks and tank cars
Warnings and Caution: Flammable; gloves and goggles should be worn
References: Bulletin 75-21

Armotan MO. Akzo Chemie Italia
Chemical Description: Sorbitan monooleate
Category: Emulsifier
Applications: Cosmetic and pharmaceutical preparations; making of cutting and soluble oils
Form: Liquid; Color: 8 max. (Gardner)
Properties: Sp. gr.: 1.01 @ 20 C; Visc.: 9.5–11 poise; Pour Pt.: ⩽ −12 C
Composition: > 99% active ingred.
Reaction: Neutral

Armotan MS. Akzo Chemie Italia
Chemical Description: Sorbitan monostearate
Category: Emulsifier
Applications: Cosmetic and pharmaceutical preparations
Form: Needle-like; Color: Cream
Properties: M.P.: 51–58 C
Composition: > 99% active ingred.
Containers: 25 kg bags

Armotan PMD20. Akzo Chemie Italia
Chemical Description: Polysor-

bate 80
Category: Wetting agent, emulsifier
Applications: Cosmetic and pharmaceutical preparations; rewetting dried skins
Form: Liquid; Color: 7 max (Gardner)
Properties: Sp. gr.: 1.02 @ 20 C; Visc.: 345–445 cs; Pour Pt.: –16 C
Reaction: Neutral

Arneel 18D. Armak
Chemical Description: Stearyl nitrile
Category: Detergent, wetting agent, rust inhibitor
Applications: Metal-wetting with mineral oils; plasticizer for synthetic rubbers and plastics
Form: Color: 6 (Gardner)
Properties: F.P.: 39 C; B.P.: 330–360 C

Arneel C. Armak
Chemical Description: Coco nitrile
Category and Applications: See Arneel 18D

Arneel OD. Armak
Chemical Description: Oleyl nitrile
Category and Applications: See Arneel 18D
Form: Color: 3 (Gardner)
Properties: Sp. gr.: 0.83; F.P.: 5C; B.P.: > 325 C

Arneel T. Armak
Chemical Description: Tallow nitrile
Category and Applications: See Arneel 18D
Properties: F.P.: 4 C; B.P.: 330–360 C

Arodet 25 S. Arol Chem. Prod.
Chemical Description: Blend of amine condensate and alkyl aryl sulfonates
Category: Detergent, wetting agent, emulsifier
Applications: Textile fulling agent, industrial detergent base; base for hair shampoos, car washes, glass cleaners, textile detergents, all purpose cleaners; detergent–textiles, skins, laundering, foods, household, cleaners of street, buildings, hospitals
Form: Liquid; Color: Amber; Odor: Mild
Properties: Solubility: Complete in water; Sp. gr.: 0.98; Stability: Excellent
Composition: 100% active ingred.
Reaction: Alkaline
Toxicity: Avoid ingestion and eye contact
Containers: 55 gal drums
References: Technical data sheet

Arodet 60 T Soft. Arol Chem.
Chemical Description: Triethanolamine salt of linear alkyl sulfonate
Category: Detergent, wetting agent, emulsifier
Applications: High foaming, liquid detergent concentrate and coupling agent; shampoos, liquid dishwashing detergents, bubble baths, auto shampoos; textiles; skins; laundering; foods; household; cosmetics; street, building, hospital cleaners

Form: Liquid; Color: Pale yellow; Odor: Mild
Properties: Solubility: Complete in water; Sp. gr.: 1.09; Stability: Excellent
Composition: 60% active ingred.
Reaction: Alkaline
Toxicity: Avoid ingestion and contact with eyes
Containers: 55 gal drums
References: Technical data sheet

Arodet AA-350. Arol Chem. Prod.
Chemical Description: Alkyl aryl sulfonate with glycol coupling agents
Category: Detergent, wetting agent, dyeing assistant
Applications: Scouring and wetting agent; pre- and after-scouring agent, leveler and retarder, dye dispersant; finishing agent; emulsifier–dispersing waxes, pigments, resins
Form: Liquid; Color: Pale yellow; Odor: Mild
Properties: Solubility: Readily miscible in water at ambient temperature; Sp. gr.: 1.04; Stability: Excellent
Composition: 35% active ingred.
Reaction: Acid
Toxicity: Avoid ingestion and contact with eyes
Containers: 55 gal drums
References: Technical data sheet

Arodet AN-100. Arol Chem. Prod.
Chemical Description: Modified nonionic derivative
Category: Detergent, wetting agent, emulsifier, penetrant, leveling agent, scouring agent, dye dispersant
Applications: Detergent–textiles, dyeing; Wetting agent–desizing and boiling off cottons, dyeing, finishing; Emulsifier–Kier boiling
Form: Liquid; Color: Water white to pale yellow; Odor: Mildly alcoholic
Properties: Solubility: Complete in water; Sp. gr.: 1.11; Stability: Excellent
Composition: 68–70% active ingred.
Reaction: Acid
Toxicity: Avoid ingestion and contact with eyes
Containers: 55 gal drums
References: Technical data sheet

Arodet BN-100. Arol Chem. Prod.
Chemical Description: Ethoxylated alcohol
Category: Detergent, wetting agent, emulsifier, dyeing assistant
Applications: Detergent–textiles, dyeing; Wetting agent–dyeing, finishing; Emulsifier–for solvents, oils, waxes; dispersing waxes, pigments, resins
Form: Liquid; Color: Clear; Odor: Mild
Properties: Solubility: Sol. in lukewarm water; Stability: Excellent
Composition: 100% active ingred.
Reaction: Acid
Toxicity: Avoid ingestion and contact with eyes
Containers: 55 gal drums
References: Technical data sheet

Arodet F. Arol Chem. Prod.
Chemical Description: Fatty amide derivative

Category: Detergent, wetting agent
Applications: Scouring agent, Kier boiling, bleaching and dyeing assistant; all purpose textile surfactant; desizing and boiling off cottons
Form:
Form: Gel; Color: Pale straw; Odor: Mildly perfumed
Properties: Solubility: Readily sol. in warm water; Sp. gr.: 1.07; Stability: Excellent
Composition: 14–16% active ingred.
Reaction: Acid
Toxicity: Avoid ingestion and contact with eyes
Containers: 55 gal drums
References: Technical data sheet

Arodet L-70 M. Arol Chem. Prod.
Chemical Description: Ethylene oxide derivative
Category: Detergent, wetting agent, emulsifier, dye assistant, nonionic
Applications: All purpose detergent with high cloud point; Detergent–textiles, skins, leather, laundering, dyeing, housekhold, cleaners of street, buildings, hospitals; Wetting agent–desizing and boiling off cottons, bleaching, dyeing, finishing, spot removal; Emulsifier–Kier boiling, dispersing waxes, pigments, resins
Form: Liquid; Color: Water white; Odor: Mildly alcoholic
Properties: Solubility: Water sol.; Sp. gr.: 1.03; Stability: Stable to acids and alkalies due to nonionic character, stable in hard water conditions
Toxicity: Avoid ingestion and contact with eyes
Containers: 55 gal drums
References: Technical data sheet

Arodet RA. Arol Chem. Prod. Co.
Chemical Description: Sulfated modified alcohol derivative
Category: Detergent, scouring agent, penetrant, dyeing assistant
Applications: Textiles, laundering
Form: Liquid; Color: Pale amber; Odor: Mild
Properties: Solubility: Readily sol. in warm water; Stability: Stable to hard water conditions, hypochlorite and peroxide bleaching solutions, alkalis and mild acid conditions
Composition: 32–35% active ingred.
Reaction: Acid
Toxicity: Avoid ingestion and contact with eyes
Containers: 55 gal drums
References: Technical data sheet

Arodet SC Special. Arol Chem.
Chemical Description: Mixture of alcohol sulfates, alkyl aryl sulfonates, amine condensates and sequesterants
Category: Detergent, emulsifier, wetting agent
Applications: General purpose cleaning agent; rug shampoo base; hair shampoos; liquid car washes; bar glass cleaners; textiles; laundering; foods; household; cosmetics; cleaners of street, buildings, hospitals; wetting agent–removal of starches, wall paper

Form: Liquid; Color: Clear; Odor: Mild
Properties: Solubility: Readily sol. in water over wide temperature range; Sp. gr.: 1.09; Stability: Excellent
Reaction: Alkaline
Toxicity: Avoid ingestion and contact with eyes
Containers: 55 gal drums
References: Technical data sheet

Arodet TX-7. Arol Chem. Prod.
Chemical Description: Modified nonionic blend
Category: Detergent, wetting agent, emulsifier, nonionic
Applications: Detergent–skins, leather, textiles, household and commercial uses; Wetting agent–bleaching, dyeing, finishing, sizing; Emulsifier– for Varsol and after scour to improve crock resistance of polyester fibers
Form: Liquid; Color: Amber; Odor: Mildly alcoholic
Properties: Solubility: Readily sol. in water; Sp. gr.: 1.01; Stability: Excellent
Composition: 50–55% active
Reaction: Acid
Toxicity: Avoid ingestion and contact with eyes
Containers: 55 gal drums
References: Technical data sheet

Arofos 200 Concentrate. Arol
Chemical Description: Phosphate ester blend
Category: Detergent, wetting agent, emulsifier, penetrant, dye leveling agent
Applications: Kier boiling; dispersing waxes, pigments, resins
Form: Liquid; Color: Water white to pale yellow; Odor: Mild
Properties: Solubility: Readily sol. in water; Sp. gr.: 1.08; Stability: Excellent
Composition: 85% active ingred.
Reaction: Mild alkaline
Toxicity: Avoid ingestion and contact with eyes
Containers: 55 gal drums
References: Technical data sheet

Arofos 326. Arol Chem. Prod. Co.
Chemical Description: Poly phosphorylated surfactant
Category: Detergent, emulsifier, wetting agent
Applications: Detergent base for alkaline cleaners, textiles, dyeing, household, cleaners of street, buildings, hospitals; Excellent emulsifier for chlorinated and aromatic solvents in steam jenny cleaners, all purpose cleaners, alkaline cleaners
Form: Liquid; Color: Pale yellow; Odor: Mild
Properties: Solubility: Complete in water; Sp. gr.: 1.09; Stability: Excellent
Composition: 99+% active ingred.
Reaction: Acid
Toxicity: Irritant to skin, fingers, eyes
Containers: 55 gal drums
Warnings and Caution: Avoid contact with eyes and skin; product is an acid
References: Technical data sheet

Arofos TD-19 Special. Arol Chem.
Chemical Description: Phosphory-

lated ethylene oxide adduct in acid form
Category: Detergent, emulsifier, wetting agent
Applications: Phosphoric acid thickener; used in hair preparations, cosmetics, heavy-duty cleansers, floor/wax strippers, metal processing, textile wetting, scouring and emulsifying agent; household and commercial detergents; wetting agent–bleaching and dyeing; emulsifier–dispersing waxes, pigments, resins and printing applications
Form: Gel; Color: Light amber; Odor: Mild
Properties: Solubility: Excellent in aromatic and chlorinated solvents, completely sol. in water; Sp. gr.: 1.10; Stability: Excellent
Composition: 98% active ingred.
Reaction: Alkaline
Toxicity: Avoid ingestion and contact with eyes
Containers: 55 gal drums
References: Technical data sheet

Arol Defoamer JT. Arol Chem.
Chemical Description: Complex alkyl ethoxylate blend
Category: Defoamer
Applications: Nonsilicone defoamer; dye dispersant; leveling agent; defoamer for jet machine dyeing; dispersing waxes, pigments, resins; preparation of emulsified paints
Form: Liquid; Color: Water white; Odor: Mild
Properties: Solubility: Readily dispersible at ambient temperatues in water; Stability: Excellent
Composition: 100% active ingred.
Reaction: Acid
Toxicity: Avoid ingestion and contact with eyes
Containers: 55 gal drums
References: Technical data sheet

Arolterge 100 M. Arol Chem. Prod.
Chemical Description: Blend of fatty acid alkanolamides and phosphate esters
Category: Detergent, emulsifier, wetting agent
Applications: Industrial detergent base or used by itself; floor stripper cleaner; auto engine cleaners, tanks, garage floors; textiles; removal of starches, wall paper
Form: Liquid; Color: Dark amber; Odor: Mild
Properties: Solubility: Complete in water; Sp. gr.: 1.0; Stability: Excellent
Composition: 100% active ingred.
Reaction: Alkaline
Toxicity: Avoid ingestion and contact with eyes
Containers: 55 gal drums
References: Technical data sheet

Aromox 18/12. Armak
Category and Applications: See Aromox C/12
Form: Soft paste; Color: 2 (Gardner)
Properties: Solubility: Solv. aqueous isopropanol; Sp. gr.: 0.927 @ 100 F; Visc.: 54 cps @ 30 C; Flash Pt.: 130 F; Pour Pt.: > 120 F
Composition: 50% active ingred.
Containers: See Aromox C/12
Warnings and Caution: Protective clothing, goggles, gloves should be worn

References: Bulletin 74-21

Aromox C/12. Armak
Chemical Description: bis(2-hydroxyethyl)cocoamine oxide
Category: Wetting agent, emulsifier, stabilizer, foaming agent, antistat, cationic
Applications: Detergent additive, shampoos, cosmetics, textiles, metal plating, petroleum additives, paper, plastics, rubber, agricultural chemicals, paints and dyes, lubricant compounding, chemical specialties
Form: Clear liquid; Color: 2 (Gardner)
Properties: Solubility: Solv. aqueous isopropanol; Sp. gr.: 0.949; Visc.: 52 cp; Cloud Pt.: 18 F; Flash Pt.: 82 F; Fire Pt.: 175 F; Pour Pt.: 0 F; Surface Tension: 33 dynes/cm (0.1%); Biodegradable
Composition: 50% active ingred.
Containers: SS, glass or fiberglass reinforced polyester tanks
Warnings and Caution: Protective clothing, gloves, goggles should be worn
References: Bulletin 74-21

Aromox C/12-W. Armak
Chemical Description: bis(2-hydroxyethyl)cocoamine oxide (water-based)
Category and Applications: See Aromox C/12
Form: Clear liquid; Color: 2 (Gardner)
Properties: Solubility: Solv. water; Sp. gr.: 0.997; Visc.: 2097 cp (Brookfield #2 spindle); Flash Pt.: > 212 F; Pour Pt.: 35 F; Surface Tension: 30.8 dynes/cm (0.1%); Biodegradable
Composition: 40% active ingred.
Containers: See Aromox C/12
Warnings and Caution: Protective clothing, goggles, gloves should be worn
References: Bulletin 74-21

Aromox CD/12. Armak
Chemical Description: bis(2-hydroxyethyl)cocoamine oxide
Category and Applications: See Aromox C/12
Form: Clear liquid; Color: 1 (Gardner)
Properties: Solubility: Solv. aqueous isopropanol; Sp. gr.: 0.949; Visc.: 52 cp; Cloud Pt.: 18 F; Flash Pt.: < 80 F; Fire Pt.: 175 F; Pour Pt.: 0 F; Biodegradable
Composition: 50% active ingred.
Containers: See Aromox C/12
Warnings and Caution: Protective clothing, goggles, gloves should be worn
References: Armak catalogue

Aromox DM14D-W. Akzo Chemie
Chemical Description: Tetradecyldimethylamine oxide
Category: Detergent
Applications: Thickener of sodium hypochlorite; household and cosmetic preparations
Form: Liquid; Color: 1 max. (Gardner)
Properties: Sp. gr.: 0.96; M.P.: –3 C; Flash Pt.: Nonflammable; pH: 7–8
Composition: 30% min. active ingred.; 70% water; 0.5% max. free amine; 0.1% max. free peroxide

Reaction: Neutral
Containers: Bung-type polythene drum; net 200 kg
Warnings and Caution: Avoid contact with heavy metals, salts of Fe, Cu, Ni, etc.

Aromox DM16. Armak
Chemical Description: Dimethylhexadecylamine oxide
Category and Applications: See Aromox C/12
Form: Clear liquid; Color: 1 (Gardner)
Properties: Solubility: Solv. aqueous isopropanol; Sp. gr.: 0.885; Visc.: 19 cp; Cloud Pt.: 44 F; Flash Pt.: 80 F; Fire Pt.: 183 F; Pour Pt.: 0 F; Surface Tension: 31.6 dynes/cm (0.1%); Biodegradable
Composition: 40% active ingred.
Containers: See Aromox C/12
Warnings and Caution: Protective clothing, goggles, gloves should be worn
References: Bulletin 74-21

Aromox DM16D-W. Akzo Chemie
Chemical Description: Hexadecyldimethylamine oxide
Category: Detergent, emollient
Applications: Emollient in cosmetics
Form: Liquid/paste; Color: 1 max. (Gardner)
Properties: Sp. gr.: 0.94 @ 60 C; M.P.: 33 C; Flash Pt.: Nonflammable; pH: 7–8
Composition: 25% min. active ingred.; 75% water; 0.5% max. free amine; 0.1% max. free peroxide
Reaction: Neutral
Containers: 100 kg net open-head polythene drums
Warnings and Caution: Avoid contact with heavy metals, salts of Fe, Cu, Ni, etc.

Aromox DM18D-W. Akzo Chemie
Chemical Description: Octadecyldimethylamine oxide
Category: Detergent, emollient
Applications: Emollient in cosmetics
Form: Paste
Properties: Solubility: Water-sol.; pH: 7–8
Composition: 25% min. active ingred.; 75% water; 0.5% max. free amine; 0.1% max. free peroxide
Reaction: Neutral
Containers: 100 kg net open-head polythene drums
Warnings and Caution: Avoid contact with heavy metals, salts, Fe, Cu, Ni, etc.

Aromox DMC. Armak
Chemical Description: Dimethylcocoamine oxide
Category and Applications: See Aromox C/12
Form: Clear liquid; Color: 1 (Gardner)
Properties: Solubility: Solv. aqueous isopropanol; Sp. gr.: 0.89; Visc.: 18 cp; Cloud Pt.: 6 F; Flash Pt.: < 80 F; Fire Pt.: 100 F; Pour Pt.: < –25 F; Surface Tension: 33.1 dynes/cm (0.1%); Biodegrad.
Composition: 40% active ingred.
Containers: See Aromox C/12
Warnings and Caution: Protective clothing, goggles, gloves should be worn

References: Bulletin 74-21

Aromox DMCD. Akzo Chemie
Chemical Description: Alkyl (C_8–C_{18})dimethylamine oxide
Category: Detergent
Applications: Thickener of NaOCl; household and cosmetic preparations
Form: Liquid; Color: 1 max. (Gardner)
Properties: Sp. gr.: 0.89; Flash Pt.: 21 C (Abel-Pensky closed cup); pH: 6–9
Composition: 39% min. active ingred.; 30% water; 30% isopropylalcohol; 1.5% max. free amine; 0.35% max. free oxide
Containers: 100 kg net open-head polythene drums
Warnings and Caution: Avoid contact with heavy metals, salts of Fe, Cu, Ni, etc.

Aromox DMC-W. Armak
Chemical Description: Dimethylcocoamine oxide (water-based)
Category and Applications: See Aromox C/12
Form: Clear liquid; Color: 1 (Gardner)
Properties: Solubility: Solv. water; Sp. gr.: 0.971; Visc.: 17 cp; Cloud Pt.: 34 F; Flash Pt.: > 212 F; Pour Pt.: 35 F; Surface Tension: 32.5 dynes/cm (0.1%); Biodegradable
Composition: 30% active ingred.
Containers: See Aromox C/12
Warnings and Caution: Protective clothing, goggles, gloves should be worn
References: Bulletin 74-21

Aromox DMHT. Armak
Chemical Description: Dimethyl (hydrogenated-tallow)amine oxide
Category and Applications: See Aromox C/12
Form: Clear liquid; Color: 1 (Gardner)
Properties: Solubility: Aqueous isopropanol; Sp. gr.: 0.886; Visc.: 32 cp; Cloud Pt.: 64 F; Flash Pt.: < 85 F; Fire Pt.: 105 F; Pour Pt.: 35 F; Surface Tension: 36.2 dynes/cm (0.1%)
Composition: 40% active ingred.
Containers: See Aromox C/12
Warnings and Caution: Protective clothing, goggles, gloves should be worn
References: Bulletin 74-21

Aromox DMHTD. Akzo Chemie
Chemical Description: Hydrogenated tallow dimethylamine oxide
Category: Detergent
Applications: Cosmetics (shampoos); household preparations
Form: Color: 1 max. (Gardner)
Properties: Sp. gr.: 0.89; Flash Pt.: 29 C (PMCC)
Composition: 39% min. active ingred.; 1.5% max. free amine; 30% water; 30% isopropylalcohol; 0.34% free peroxide
Toxicity: Severe skin irritation; minimal eye irritation
Containers: 100 kg net open-head polythene drums
Warnings and Caution: Avoid contact with heavy metals, salts of Fe, Cu, Ni, etc.

Aromox DMMC-W. Armak
Chemical Description: Dimethyl-

cocoamine oxide (middle-cut coco fraction)–water-based
Category and Applications: See Aromox C/12
Form: Clear liquid; Color: 1 (Gardner)
Properties: Solubility: Solv. water; Sp. gr.: 0.96; Visc.: 90 cp; Cloud Pt.: 22 F; Flash Pt.: >212F; Pour Pt.: 36 F; Surface Tension: 31.2 (0.1%)
Composition: 30% active ingred.
Containers: See Aromox C/12
Warnings and Caution: Protective clothing, goggles, gloves should be worn
References: Bulletin 74-21

Aromox T/12. Armak
Chemical Description: bis(2-hydroxyethyl)tallow amine oxide
Category and Applications: See Aromox C/12
Form: Clear liquid; Color: 4 (Gardner)
Properties: Solubility: Solv. aqueous isopropanol; Sp. gr.: 0.94; Visc.: 77 cp; Cloud Pt.: 60 F; Flash Pt.: 90 F; Fire Pt.: 180 F; Pour Pt.: 55 F; Surface Tension: 33.0 dynes/cm (0.1%)
Composition: 50% active ingred.
Containers: See Aromox C/12
Warnings and Caution: Protective clothing, goggles, gloves should be worn
References: Bulletin 74-21

Arosurf AA-22. Sherex
Category: Emulsifier, cationic
Applications: Asphalt mixing-grade emulsions

Arosurf AA-23. Sherex
Chemical Description: Diamine type
Category: Emulsifier, cationic
Applications: Asphalt rapid-set and mixing-grade emulsions

Arosurf AA-54. Sherex
Chemical Description: Modified diamine type
Category: Emulsifier, cationic
Applications: Asphalt rapid-set and mixing-grade emulsions
Form: Liquid

Arosurf AA-55. Sherex
Chemical Description: Modified diamine type
Category: Emulsifier, cationic
Applications: Asphalt rapid-set and mixing-grade emulsions
Form: Liquid

Arosurf AA-57. Sherex
Category: Emulsifier, cationic
Applications: Asphalt slow-set emulsions

Arosurf AA-60. Sherex
Category: Emulsifier, cationic
Applications: Asphalt rapid-set and mixing-grade emulsions

Arosurf AA-76. Sherex
Category: Emulsifier, cationic
Applications: Asphalt slurry-seal applications

Arosurf TA-100. Sherex
Chemical Description: Dimethyl distearyl ammonium chloride

Category: Softener, cationic
Applications: Creme rinse concentrate for hair care; fabric softener for home and commercial launderies, textile processing
Form: Powder
Composition: 97% solids content

Arowet NF. Arol Chem. Prod.
Category: Wetting agent, anionic
Applications: Nonfoaming type wetting agent with good dye leveling properties; dyeing and bleaching applications
Form: Liquid; Color: Reddish brown
Properties: Solubility: Readily sol. in water; Sp. gr.: 1.04; Stability: Compatible under mild acid and alkaline conditions to boil; stable to hard water conditions
Reaction: Neutral
Toxicity: Avoid ingestion and contact with eyes
Containers: 55 gal drums
References: Technical data sheet

Arowet S. Arol Chem. Prod.
Chemical Description: Blend of sulfonated esters
Category: Wetting agent
Applications: Wetting and rewetting agent; excellent for pasting-out dyestuffs, good leveling properties; desizing and boiling off cottons; bleaching and dyeing
Form: Liquid; Color: Amber; Odor: Mild
Properties: Solubility: Readily sol. in water; Sp. gr.: 1.03; Stability: Excellent
Toxicity: Avoid ingestion and contact with eyes
Containers: 55 gal drums
References: Technical data sheet

Arquad 2C-75. Armak
Chemical Description: Dimethyl dicocoammonium chloride
Category and Applications: See Arquad 12-33
Form: Semiliquid; Color: 7 (Gardner)
Properties: m.w.: 447; Solubility: Solv. aqueous isopropanol; Sp. gr.: 0.89; HLB: 11.4; Flash Pt.: < 80 F; Pour Pt.: 10 F; Surface Tension: 30 dynes/cm (0.1%)
Composition: 75% active ingred.
Toxicity: Skin irritant, severe eye irritant
Containers: See Arquad 12-33
Warnings and Caution: Flammable; protective clothing, goggles, gloves should be worn
References: Bulletins 76-15–17

Arquad 2HT-75. Armak
Chemical Description: Dimethyldi-(hydrogenated-tallow)ammonium chloride
Category and Applications: See Arquad 12-33; also as bacteriostat
Form: Soft paste; Color: White
Properties: m.w. (of active): 573; Solubility: Solv. aqueous isopropanol; Density: 7.22 lb/gal; Sp. gr.: 0.87; Visc.: 47.5 cps @ 120 F; F.P.: 95 F; HLB: 9.7; Flash Pt.: 112 F; Pour Pt.: 90–100 F; Surface Tension: 37 dynes/cm (0.1%)
Composition: 75% active ingred.
Toxicity: Skin and eye irritant
Containers: See Arquad 12-33

Warnings and Caution: Combustible; protective gloves and glasses should be worn
References: Bulletins 76-15-17

Arquad 12-33. Armak
Chemical Description: Trimethyldodecylammonium chloride
Category: Emulsifier, foaming, wetting, dispersion agents, corrosion inhibitor, antistat, cationic
Applications: Antistatic agents, corrosion inhibitors, fabric softeners, hair conditioners, combing aids, latex foaming agents, leather softeners, metal cleaning and polishing emulsions, mineral flotation agents, liquid starch stabilizers, water and sewage treatment flocculants
Form: Liquid; Color: 7 (Gardner)
Properties: m.w. (of active): 263; Solubility: Solv. aqueous isopropanol; Sp. gr.: 0.98; F.P.: 5 F; HLB: 17.1; Flash Pt.: 140 F; Stability: Good; Biodegradable
Composition: 33% active ingred.
Toxicity: Skin irritant, severe eye irritant
Containers: 55-gal nonreturnable epoxy-phenolic lined drums; bulk available SS tank trucks
Warnings and Cautions: Flammable; protective clothing, goggles, gloves should be worn
References: Bulletins 76-15-17

Arquad 12-50. Armak
Chemical Description: Trimethyldodecylammonium chloride
Category and Applications: See Arquad 12-33
Form: Liquid; Color: 1 (Gardner)
Properties: m.w. (of active): 263; Solubility: Solv. aqueous isopropanol; Sp. gr.: 0.89; F.P.: 13 F; HLB: 17.1; Flash Pt.: < 80 F; Stability: Good; Surface Tension: 33 dynes/cm (0.1% conc.); Biodegradable
Composition: 50% active ingred.
Toxicity: Skin irritant, severe eye irritant
Containers: See Arquad 12-33
Warnings and Caution: Flammable; protective clothing, goggles, gloves should be worn
References: Bulletins 76-15-17

Arquad 16-29. Armak
Chemical Description: Trimethylhexadecylammonium chloride
Category and Applications: See Arquad 12-33
Form: Liquid; Color: 6 (Gardner)
Properties: m.w.: 319; Solubility: Solv. aqueous isopropanol; Sp. gr.: 0.96; F.P.: 61 F; HLB: 15.8; Flash Pt.: > 212 F; Stability: Good; Biodegradable
Composition: 29% active aqueous
Toxicity: Skin irritant, severe eye irritant
Containers: See Arquad 12-33
Warnings and Caution: Corrosive; protective goggles, gloves should be worn
References: Bulletins 76-15-17

Arquad 16-50. Armak
Chemical Description: Trimethylhexadecylammonium chloride
Category and Applications: See Arquad 12-33
Form: Liquid; Color: 6 (Gardner)

Properties: m.w. (of active): 319; Solubility: Solv. aqueous isopropanol; Sp. gr.: 0.88; HLB: 15.8; Flash Pt.: $<$ 80 F; Stability: Good; Surface Tension: 34 dynes/cm (0.1%); Biodegradable
Composition: 50% active ingred.
Toxicity: Skin irritant, severe eye irritant
Containers: See Arquad 12-33
Warnings and Caution: Flammable; protective clothing, goggles, gloves should be worn
References: Bulletins 76-15–17

Arquad 18-50. Armak
Chemical Description: Trimethyloctadecylammonium chloride
Category and Applications: See Arquad 12-33; also as dye-leveling agents, viscosity stabilizers, fuel oil additives, in lubricant compounding
Form: Liquid; Color: 7 (Gardner)
Properties: m.w. (of active): 347; Solubility: Solv. aqueous isopropanol; Sp. gr.: 0.88; HLB: 15.7; Flash Pt.: $<$ 80 F; Stability: Good; Surface Tension: 34 dynes/cm (0.1%); Biodegradable
Composition: 50% active ingred.
Toxicity: Skin irritant, severe eye irritant
Containers: See Arquad 12-33
Warnings and Cautions: Flammable; protective clothing, goggles, gloves, should be worn
References: Bulletins 76-15–17

Arquad B-50 USP. Akzo Chemie
Chemical Description: Alkyl (C_8–C_{18}) dimethylbenzyl ammonium chloride USP
Category: Disinfectant, germicide
Form: Liquid; Color: 2 max. (Gardner)
Properties: Sp. gr.: 0.98
Composition: 50% active ingred.
Reaction: Neutral
Toxicity: Skin irritant, severe eye irritant
Containers: 100 kg net open-head polythene drums
Warnings and Caution: Avoid contact with strong oxidizing agents, anionics; kills bacteria in biological disposal systems

Arquad B-90 USP. Akzo Chemie
Chemical Description: Alkyl (C_8–C_{18}) dimethylbenzyl ammonium chloride USP
Category: Disinfectant
Applications: Disinfectant, germicide, fungicide
Form: Viscous mass @25 C; Color: 6 max. (Gardner)
Composition: 90% active ingred.
Toxicity: Skin irritant, severe eye irritant
Containers: 100 kg net open-head polythene drums
Warnings and Caution: Avoid contact with strong oxidizing agents, anionics; kills bacteria in biological disposal systems

Arquad B-100. Armak
Chemical Description: Dimethylalkylbenzyl ammonium chloride
Category: Bacteriocide
Applications: EPA approved bacteriocide, sanitizer for schools, hospitals, food processing plants

Form: Liquid; Color: 2 (Gardner)
Properties: m.w.: 380; Solubility: Solv. aqueous isopropanol; Flash Pt.: < 80 F
Composition: 50% active ingred.
Toxicity: Irritating to skin and eyes
Warnings and Caution: Flammable, protective gloves and goggles should be worn
References: Bulletins 76-1, 76-15–17

Arquad C-33. Armak
Chemical Description: Trimethylcocoammonium chloride
Category and Applications: See Arquad 12-33
Form: Liquid; Color: 7 (Gardner)
Properties: m.w. (of active): 278; Solubility: Solv. aqueous isopropanol; Sp. gr.: 0.98; F.P.: 25 F; HLB: 16.5; Stability: Good; Biodegradable
Composition: 33% active ingred.
Toxicity: Skin irritant, severe eye irritant
Containers: See Arquad 12-33
Warnings and Caution: Corrosive; protective clothing, goggles, gloves should be worn
References: Bulletins 76-15–17

Arquad C-33-W. Akzo Chemie
Chemical Description: Cocotrimethyl ammonium chloride
Category: Emulsion-break retardant in cosmetics
Form: Liquid
Properties: Sp. gr.: 0.96; F.P.: –3 C; Flash Pt.: Nonflammable; pH: 5–8
Composition: 32–25% active

Arquad C-50. Armak
Chemical Description: Trimethylcocoammonium chloride
Category and Applications: See Arquad 12-33
Form: Liquid; Color: 7 (Gardner)
Properties: m.w. (of active): 278; Solubility: Solv. aqueous isopropanol; Sp. gr.: 0.89; F.P.: 5 F; HLB: 16.5; Flash Pt.: < 80 F; Stability: Good; Surface Tension: 31 dynes/cm (0.1%); Biodegradable
Composition: 50% active ingred.
Toxicity: Skin irritant, severe eye irritant
Containers: See Arquad 12-33
Warnings and Caution: Flammable; protective clothing, goggles, gloves should be worn
References: Bulletins 76-15–17

Arquad DM14B-90. Akzo Chemie
Chemical Description: Tetradecyldimethylbenzyl ammonium chloride
Category: Bactericide, fungicide, germicide, disinfectant
Form: Powder; Color: White
Composition: 90% active ingred.
Toxicity: Skin irritant, severe eye irritant
Containers: 50 kg net fiberboard kegs
Warnings and Caution: Avoid contact with strong oxidizing agents, anionics; product will kill bacteria in biological disposal systems

Arquad DM18B-90. Akzo Chemie

Chemical Description: Octadecyldimethylbenzyl ammonium chloride
Category: Conditioner, emulsifier, softener
Applications: Cosmetics and pharmaceuticals; conditioner/softener in after shampoo creams
Form: Powder; Color: White
Properties: Solubility: 100% water sol.; pH: 5–8 (1% sol'n.)
Composition: 90% min. active
Toxicity: Skin irritant, severe eye irritant
Containers: 50 kg fiberboard kegs
Warnings and Caution: Avoid contact with strong oxidizing agents, anionics; product will kill bacteria in biological disposal systems

Arquad DMHTB-75. Akzo Chemie
Chemical Description: Hydrogenated tallow dimethylbenzyl ammonium chloride
Category: Bactericide, disinfectant, softening agent
Applications: Disinfection; textile softening agent
Form: Soft paste; Color: 4 max. (Gardner)
Properties: F.P.: 60 C
Composition: 75% active ingred.
Reaction: Neutral
Toxicity: Skin irritant, severe eye irritant
Containers: 50 kg net fiberboard kegs
Warnings and Caution: Avoid contact with strong oxidizing agents, anionics; product will kill bacteria in biological disposal systems

Arquad DMMCB-50. Akzo Chemie
Chemical Description: Alkyl (C_{12}–C_{14}, C_{16}) dimethyl benzyl ammonium chloride
Category: Antistat, flocculant, emulsifier, softener, corrosion inhibitor
Applications: Cosmetics industry; soil stabilization; antistatic agent in textile industry; flocculant; fabric softener; corrosion inhibitor; fungicides, bactericides
Form: Liquid
Composition: 50% active

Arquad DMMCB-75. Akzo Chemie
Chemical Description: Alkyl (C_{12}–C_{14}) dimethylbenzyl ammonium chloride
Category: Disinfectant, bacteri-
Cloud Pt.: –36 F; Flash Pt.: < 80 F; Biodegradable
Toxicity: Skin and eye irritant
Warnings and Caution: Flammable, protective gloves and goggles should be worn

Arquad L-11. Armak
Chemical Description: Quarternized beta-amines
Category: Emulsifier, cationic
Properties: Solubility: Sol. in water; Sp. gr.: 0.885; M.P.: –90F;
cide, germicide, fungicide
Form: Liquid; Color: 2 max. (Gardner)
Properties: Sp. gr.: 0.98; Flash Pt.: 37 C
Composition: 50% active ingred.
Toxicity: Skin irritant, severe eye irritant

Containers: 100 kg net open-head polythene drums
Warnings and Caution: Avoid contact with strong oxidizing agents, anionics; product will kill bacteria in biological disposal systems

Arquad L-15. Armak
Chemical Description: Quarternized beta-amines
Category: Emulsifier, cationic
Form: Clear liquid; Color: 9 (Gardner)
Properties: Solubility: Water-sol.; Sp. gr.: 0.877; M.P.: 46 F; Cloud Pt.: 42 F; Flash Pt.: 100 F
Composition: 50% active ingred.

Arquad S-50. Armak
Chemical Description: Trimethylsoyaammonium chloride
Category and Applications: See Arquad 12-33
Applications: Used in butumen emulsions; to control growth of organism in spray water for air conditioning; algae and slime control agents in water systems
Form: Liquid; Color: 8 max. (Gardner)
Properties: Sp. gr.: 0.89; F.P.: 20 C; pH: 5–8
Composition: 49–52% active
Toxicity: Skin irritant, severe eye irritant
Containers: 180 kg net bung type steel drums
Warnings and Caution: Avoid contact with strong oxidizng agents, anioncs; avoid temperatures > 70 C product will kill bacteria in biological disposal systems

References: Bulletins 76-15–17

Arquad S-2C-50. Akzo Chemie
Chemical Description: 1:1 mixture of oleyltrimethyl ammonium chloride and dicocodimethyl ammonium chloride
Category: Bactericide, wetting agent, corrosion inhibitor
Applications: Bactericide; oil recovery as wetting agent
Form: Liquid; Color: 10 max. (Gardner)
Properties: Sp. gr.: 9.87; F.P.: –15 C; pH: 5–8
Composition: 50–55% active
Containers: 172 kg net bung type drums

Arquad T-50. Armak
Chemical Description: Trimethyltallowammonium chloride
Category and Applications: See Arquad 12-33
Applications: In manufacture of tetramycin (oxytetracycline) antibiotics
Form: Liquid; Color: 8 max. (Gardner)
Form: Liquid; Color: 8 (Gardner)
Properties: m.w.: 340; Solubility: Solv. aqueous isopropanol; Sp. gr.: 0.88; HLB: 14.2; Flash Pt.: < 80 F; Stability: Good; Biodegradable
Composition: 50% active ingred.
Toxicity: Skin irritant, severe eye irritant
Containers: 180 kg net bung type steel drums
Warnings and Caution: Avoid contact with strong oxidizing agents,

anionics; avoid temperatures > 70 C; will kill bacteria in biological disposal systems Flammable; protective clothing, goggles, gloves should be worn
References: Bulletins 76-15–17

Arquad T-2C-50. Armak
Chemical Description: 1:1 mixture of trimethyltallowammonium chloride and dimethyldicocoammonium chloride
Category and Applications: See Arquad 12-33
Form: Liquid; Color: 8 (Gardner)
Properties: m.w. (of active): 394; Solubility: Solv. aqueous isopropanol; Sp. gr.: 0.87; HLB: 13.0; Flash Pt.: < 80 F
Composition: 50% active ingred.
Toxicity: Skin irritant, severe eye irritant
Containers: See Arquad 12-33
Warnings and Caution: Flammable, protective gloves and goggles should be worn
References: Bulletins 76-15–17

Arquest 665-S. Arjay, Inc.
Chemical Description: Phosphate ester
Category: Scale inhibitor, anionic
Applications: Scale inhibitor for calcium carbonate and calcium sulfate
Form: Liquid

Arquest 669-S. Arjay, Inc.
Chemical Description: Acrylic polymer
Category: Scale inhibitor, anionic
Applications: Controls barium sulfate, calcium sulfate, and calcium carbonate scale
Form: Liquid

Arquest 674-S. Arjay, Inc.
Chemical Description: Phosphonate
Category: Scale inhibitor, sequestering and deflocculating agent, anionic
Form: Liquid

Arquest 675-S. Arjay, Inc.
Chemical Description: Phosphonate
Category: Scale inhibitor, anionic
Applications: Scale inhibitor for calcium carbonate, calcium sulfate, and barium sulfonate
Form: Liquid

Arsul 458-S. Arjay, Inc.
Chemical Description: Alcohol ether sulfate
Category: Detergent, foaming agent, anionic
Applications: Used in light duty dishwash detergents, all purpose cleaners, bubble baths, upholstery and rug cleaners; emulsion polymerization; foaming agent for wall board manufacturing, air drilling, foam markers
Form: Liquid
Composition: 60% conc.

Arsul DDB. Arjay, Inc.
Chemical Description: Branched chain dodecyl benzene sulfonic acid
Category: Chemical intermediate,

anionic
Applications: Intermediate for reaction with different alkali bases to form the metal or amine salts; sulfonate derivatives used as emulsifiers, dispersants, cleaning compounds, emulsion breakers, and corrosion inhibitors
Form: Liquid
Composition: 90% conc.

Arsul LAS. Arjay, Inc.
Chemical Description: Biodegradable linear dodecyl benzene sulfonic acid
Category and Applications: See Arsul DDB
Form: Liquid
Composition: 98% conc.

Arsul WAQ. Arjay, Inc.
Chemical Description: Sodium lauryl sulfate
Category: Anionic surfactant
Applications: Used in applications requiring good foam stability; used in emulsion polymerization and cosmetic formulations
Form: Liquid
Composition: 30% conc.

Artodan. Grinsted Prod.
Chemical Description: Sodium stearyl-2-lactylate
Category: Emulsifier, anionic
Applications: Food emulsifier, dough conditioner, starch complexing agents for antistaling, structure improvement, increased volume in bread
Form: Powder
Properties: HLB: 9
Composition: 100% active ingred.

Artrads 6524, 6923, 7522. Arjay, Inc.
Chemical Description: Dimer trimer polybasic acids
Category: Intermediate
Applications: Intermediate for reaction with polyfunctional materials, e.g., diols, diamines polyols, and polyamines, to produce good corrosion inhibitors
Form: Liquid
Composition: 100% conc.

Arylan CA. Lankro Chem. Ltd.
Chemical Description: Straighter chain dodecyl benzene sulfonate, calcium salt
Category: Emulsifier, anionic
Applications: Emulsifier for certain pesticidal and herbicidal preparations; hydrocarbon solvents and waxes; oils and aromatic solvents in degreaser formulations
Form: Viscous liquid; Odor: Alcoholic
Properties: Solubility: Water-insol.; Sp. gr.: 1.012 @ 20 C; Visc.: 3025 cs @ 20 C; Flash Pt.: 94 F (Abel CC); Pour Pt.: < 0 C; pH: 5–8 (1% aq.)
Composition: 70% active ingred.
Toxicity: Avoid prolonged contact with concentrated form
Containers: Store in SS, glass, resin, or polythene-lined equipment
Warnings and Caution: Contains flammable solvent; do not mix with strong oxidizing or reducing agents (potentially explosive)

Arylan DA 36. Lankro Chem. Ltd.

Chemical Description: Blended liquid detergent with nonionic foam stabilizer
Category: Detergent, anionic/nonionic
Applications: Concentrated blend of liquid detergents. Fully biodegradable.
Form: Liquid;
Properties: Biodegradable
Composition: 36% Conc.

Arylan DT. Lankro Chem. Ltd.
Chemical Description: Specially blended liquid detergent concentrate based on alkyl aryl sulfonates with foam stabilizers
Category: Detergent
Applications: High active detergents, hard surface cleaners, upholstery cleaners, industrial cleaners
Form: Clear liquid; Color: Light amber; Odor: Mild
Properties: Sp. gr.: 1.05 @ 20 C; Visc.: 330 cs @ 20 C; Flash Pt.: > 200 F (COC); Pour Pt.: < 0 C; Stability: Good; pH: 6–7.5 (1% aq.)
Composition: 30% active ingred.
Containers: Store in SS, glass, resin or polythene-lined equipment
Warnings and Caution: Avoid prolonged contact with concentrated form; do not mix with strong oxidizing or reducing agents (potentially explosive)

Arylan HAL. Lankro Chem. Ltd.
Chemical Description: Specially blended liquid detergent concentrate based on alkyl aryl sulfonates with foam stabilizers
Category: Detergent
Applications: High active detergents, hard surface cleaners, upholstery cleaners
Form: Clear liquid; Color: Amber; Odor: Mild, alcoholic
Properties: Sp. gr.: 1.04 @ 20 C; Visc.: 300 cs @ 20 C; Flash Pt.: 90 F (Abel CC); Pour Pt.: < 0 C; Stability: Good; pH: 6.5–7.5 (1% aq.)
Composition: 87% active ingred.
Containers: Store in SS, glass, resin or polythene-lined equipment
Warnings and Caution: Avoid prolonged contact with concentrated form; do not mix with strong oxidizing or reducing agents (potentially explosive)

Arylan LQ. Lankro Chem. Ltd.
Chemical Description: Specially blended liquid detergent concentrate based on alkyl aryl sulfonates with foam stabilizers
Category: Detergent
Applications: High active detergents, hard surface cleaners, upholstery cleaners, industrial cleaners
Form: Clear liquid; Color: Light amber; Odor: Mild
Properties: Sp. gr.: 1.053 @ 20 C; Visc.: 350 cs @ 20 C; Flash Pt.: > 200 F (COC); Pour Pt.: < 0 C; Stability: Good; pH: 7–9 (1% aq.)
Composition: 40% active ingred.
Containers: SS, glass, resin or polythene lined equipment
Warnings and Caution: Avoid prolonged contact with concentrated form; do not mix with strong oxidizing or reducing agents (potentially explosive)

Arylan PWS. Lankro Chem. Ltd.
Chemical Description: Amine alkylaryl sulfonate
Category: Emulsifier
Applications: Emulsifier for mineral oils, kerosene, waxes and chlorinated solvents
Form: Liquid
Properties: Visc.: 4200 cs @ 20 C
Composition: 90% active ingred.
Reaction: Acid

Arylan S Acid. Diamond Shamrock Europe/Process Chemicals Div.
Chemical Description: Dodecyl benzene sulfonic acid (branched chain).
Category: Intermediary, anionic
Applications: Intermediary for preparation of anionic surfactants
Form: Liquid
Composition: 96% Conc.

Arylan S90 Flake. Lankro Chem. Ltd.
Chemical Description: Sodium dodecyl benzene sulfonate flake (branched chain)
Category: Wetting agent, emulsifier, anionic
Applications: Primary emulsifier in the emulsion polymerization of PVA: wetting agent for wettable powders
Form: Liquid
Composition: 95% Conc.

Arylan S Flake. Lankro Chem. Ltd.
Chemical Description: Sodium dodecylbenzene sulfoante
Category: Wetting agent, emulsifier
Applications: Emulsion polymerization; wettable powders
Form: Flake; Color: Cream
Composition: 82% active ingred.
Reaction: Neutral

Arylan SBC 25. Lankro Chem. Ltd.
Chemical Description: Straighter chain dodecyl benzene sulfonate, sodium salt
Category: Detergent, emulsifier, anionic
Applications: Liquid detergent base and primary emulsifier for emulsion polymerization
Form: Liquid
Composition: 25% Conc.

Arylan SBC Acid. Lankro Chem. Ltd.
Chemical Description: Straighter chain dodecyl benzene sulfonic acid
Category: Detergent, anionic
Applications: Biodegradable intermediate for detergents
Form: Liquid
Properties: Biodegradable
Composition: 97% Conc.

Arylan SC 30. Lankro Chem. Ltd.
Chemical Description: Sodium dodecylbenzene sulfonate
Category: Detergent, emulsifier
Applications: Base for liquid detergents; primary emulsifier for emulsion polymerization
Form: Liquid

Composition: 30% active ingred.
Reaction: Neutral

Arylan SC Acid. Lankro Chem. Ltd.
Chemical Description: Dodecylbenzene sulfonic acid
Category: Intermediate
Applications: Intermediate for detergents, esp. liquid detergents
Form: Liquid; Color: Brown
Properties: Visc.: 1800 cs @ 20 C
Composition: 97% active ingred.
Reaction: Acid
Toxicity: Strong organic acid, hence irritating to eyes and skin
Containers: Resin-lined drums

Arylan SNS. Diamond Shamrock
Chemical Description: Sodium naphthalene sulfonic acid formaldehyde condensate
Category: Dispersing agent
Applications: Dispersing solid particles in aqueous suspensions
Form: Powder; Color: Buff
Composition: 90% active ingred.
Reaction: Neutral

Arylan SO60 Acid. Lankro Chem. Ltd.
Chemical Description: Branched chain tridecyl benzene sulfonic acid
Category: Detergent, emulsifier, wetting agent, emulsion polymerization aids
Applications: Base surfactant for heavy duty powder detergent formulations
Form: Viscous liquid; Color: Brown; Odor: Characteristic
Properties: Solubility: Water-sol. @ 20 C; Sp. gr.: 1.045 @ 20 C; Visc.: > 25,000 cs @ 20 C; Flash Pt.: > 250 F (COC); Pour Pt.: < 0 C; pH: 2.0 (1% aq.)
Composition: > 96% active ingred.
Toxicity: Corrosive organic acid; extreme eye and skin irritant
Containers: Store in SS, resin, polythene or glass-lined equipment
Warnings and Caution: Protective goggles, gloves and clothing should be worn; use with adequate ventilation (may evolve sulfur dioxide odor); in case of fire, toxic sulfur dioxide fumes may be produced

Arylan SX Flake. Lankro Chem. Ltd.
Chemical Description: Straighter chain dodecyl benzene sulfonate flake, sodium salt
Category: Detergent, wetting agent, anionic
Applications: Base for detergents; wetting agent for powders
Form: Liquid
Composition: 75% Conc.

Arylan TE/C. Lankro Chem. Ltd.
Chemical Description: Straighter chain dodecyl benzene sulfonate (TEA salt)
Category: Detergent, emulsifier, wetting agent, emulsion polymerization aids, anionic
Applications: Specialty wax emulsifier; co-emulsifier with an Ethylan nonionic for the emulsification of phenolics and chlorinated

solvents
Form: Clear liquid; Odor: Mild
Properties: Solubility: Water-sol. @ 20 C; Sp. gr.: 1.060 @ 20 C; Visc.: 900 cs @ 20 C; Flash Pt.: > 200 F (COC); Pour Pt.: < 0 C; Stability: Good in alkaline, mild acid, low electrolyte concentration and in hard or soft waters; pH: 7.0–8.5 (1% aq.); Biodegradable
Composition: 40% active ingred.
Containers: Store in SS, glass, resin, or polythene-lined equipment
Warnings and Caution: Avoid prolonged contact with concentrated form; in case of fire, toxic fumes of sulfur dioxide may be produced

Arylene M40. Hart Chem.
Chemical Description: Dioctyl sulfosuccinate
Category: Wetting agent, anionic
Applications: Scouring of cellulosics and in desizing process
Form: Liquid; Color: Clear
Properties: Refractive Index: 1.3932; Density: 1.03; Visc.: 50 cps
Composition: 40% active ingred.
Reaction: Neutral
Toxicity: Skin and eye irritant
Containers: T/L/Drums
Warnings and Caution: Avoid contact with skin and eyes

Arylene M60. Hart Chem.
Chemical Description: Dioctyl sulfosuccinate
Category: Wetting agent
Applications: Scouring of cellulosics and in desizing processes
Form: Liquid; Color: Clear
Properties: Refractive Index: 1.3932; Density: 1.06; Visc.: 50 cps
Composition: 60% active ingred.
Toxicity: Skin and eye irritant
Containers: T/L/Drums
Warnings and Caution: Avoid contact with eyes and skin

Arzoline 215. Arjay, Inc.
Chemical Description: Alkyl cyclic amido amine
Category: Corrosion inhibitor base, cationic
Applications: Corrosion inhibitor base; may be further modified to produce oil soluble, water dispersible, or water soluble inhibitors
Form: Liquid
Composition: 100% conc.

Asco-66. Berol Kemi AB
Chemical Description: Polyamine derivative
Category: Antistripping agent
Applications: Antistripping agent in road construction, jointsealing compounds, undersealing compounds for cars; promotes passive and active adhesion between binder and aggregate
Form: Liquid; Color: Dark; Odor: Ammoniacal
Properties: Solubility: Sol. in asphalt, oils and many solvents; Flash Pt.: > 150 C (PM); Stability: Good heat stability
Reaction: Alkaline
Toxicity: May cause skin and eye injuries

Containers: 200 kg net drums
Warnings and Caution: Protective face shields, goggles should be worn

Asol. Lucas Meyer
Chemical Description: Lecithin fraction
Category: Antispatter agent, release agent, emulsifier, nonionic
Applications: Antispatter agent, release agent for margarine; emulsifier for cosmetics, pharmaceuticals
Form: Liquid
Composition: 40–100% conc.

Aston AP Concentrate. Onyx Chem. Co.
Chemical Description: Cationic polyamine
Category: Antistatic agent, softener, cationic
Applications: Water and solvent soluble antistatic agent and softener for industrial and textile use
Form: Liquid
Properties: Solubility: Water and solvent sol.; Sp. gr.: 1.04 (25/20 C); Flash Pt.: > 200 F
Composition: 100% solids

Aston BO. Onyx Chem. Co.
Chemical Description: Special cationic
Category: Antistat, cationic
Applications: Antistatic agent for rug shampoos which reduces resoiling after shampooing
Form: Liquid
Properties: Sp. gr.: 0.99 (25/20 C); Flash Pt.: > 200 F
Composition: 25% min. solids

Atcopen SD. Bostik South
Chemical Description: Alkyl phosphate
Category: Wetting agent, anionic
Applications: Low foaming, surfactant wetting agent, stable in acid and alkaline solvents
Form: Liquid
Composition: 78% conc.

Atcowet W. Bostik South
Chemical Description: Dioctyl sodium sulfosuccinate
Category: Wetting and rewetting agent, anionic
Applications: Textile wetting and rewetting applications
Form: Liquid
Composition: 75% conc.

Atlas G-263. I.C.I. United States
Chemical Description: Cetyl ethyl morpholinium ethosulfate (35% aqueous sol'n.)
Category: Antistatic agent, cationic
Applications: Textile specialties
Form: Liquid
Properties: HLB: 30.0
Composition: 35% active ingred.

Atlas G-263 Anhydrous. ICI United States
Chemical Description: Cetyl ethyl morpholinium ethosulfate
Category and Applications: See Atlas G-263
Form: Waxy solid; Color: Tan to brown; Odor: Fruity
Properties: Solubility: Sol. in water and ethanol forming cloudy solutions
Composition: 99% active ingred.

Atlas G-271. I.C.I. United States
Chemical Description: Soya ethyl morpholinium ethosulfate (35% aqueous solution)
Category: Deodorant, cosmetic specialty, cationic
Form: Liquid
Properties: HLB: 30.0
Composition: 35% active ingred.

Atlas G-695. I.C.I. United States
Chemical Description: Mono- and diglycerides of edible fats or oils
Category: Food emulsifier
Form: Paste; Odor and Taste: Bland
Properties: Solubility: Sol. in lower alcohols, vegetable and mineral oils; Sp. gr.: 0.96; Visc.: 528 cps; HLB: ≈ 3; Clear Pt.: 40 C; Flash Pt.: > 300 F; Fire Pt.: > 300 F

Atlas G-711. I.C.I. Unites States
Chemical Description: Alkyl aryl sulfonate
Category: Detergent, anionic
Applications: Dry cleaning detergent
Form: Liquid; Color: Light yellow
Properties: Solubility: Sol. in chlorinated solvents, ethylene glycol, lower alcohols, aniline and petroleum solvents; Sp. gr.: ≈ 1; Visc.: ≈ 5500 cps; HLB: 11.7; Flash Pt.: > 300 F; Fire Pt.: > 300 F
Composition: 100% active ingred.

Atlas G-750. I.C.I. United States
Chemical Description: 80% sodium sorbitol borate
Category: Water-sol. resin
Form: Viscous liquid; Color: Light yellow
Properties: Refractive Index: 1.45–1.456; Solubility: Sol. in water; Sp. gr.: 1.375–1.390; Visc.: ≈ 4500 cp

Atlas G-950. I.C.I. United States
Chemical Description: Sorbide dioleate
Category: Nonionic surfactant
Form: Liquid; Color: Orange-red
Properties: Solubility: Dispersible in water; sol. in aromatic and aliphatic hydrocarbons, vegetable oils; Sp. gr.: ≈ 1; Visc.: ≈ 250 cps; HLB: 2.7; Flash Pt.: > 300 F; Fire Pt.: > 300 F
Containers: 55 gal closed head steel drum

Atlas G-991. I.C.I. United States
Chemical Description: Mono- and diglycerides with 0.02% BHA and 0.01% citric acid added as preservatives
Category: Food emulsifier
Form: Ivory white, plastic solid
Properties: Solubility: Sol. above its melting point in vegetable oils and mineral oil; M.P.: ≈ 118 F; HLB: 2.8; Flash Pt.: > 300 F; Fire Pt.: > 300 F

Atlas G-1045. I.C.I. United States
Chemical Description: Polyoxyethylene sorbitol laurate
Category: Emulsifier, nonionic
Applications: Fiber lubricant; general purpose emulsifier, esp. for insecticides
Form: Oily liquid; Color: Light yellow

Properties: Solubility: Dispersible with haze in water; sol. in toluene, mineral spirits, acetone, diethyl ether, dioxane, cellsolve, ethyl acetate, aniline; Sp. gr.: ≈ 1; Visc.: ≈ 175 cp; HLB: 11.5; Flash Pt.: > 300 F; Fire Pt.: > 300 F
Composition: 100% active ingred.

Atlas G-1052. I.C.I. United States
Chemical Description: Polyoxyethylene sorbitol distearate
Category: Nonionic surfactant
Form: Wax; Color: Cream
Properties: Solubility: Sol. in isopropanol; HLB: 8.5; Flash Pt.: > 300 F

Atlas G-1069. I.C.I. United States
Chemical Description: Polyoxyethylene sorbitol ester of mixed fatty acids
Category: Nonionic surfactant
Form: Liquid; Color: Amber
Properties: Solubility: Dispersible in water, sol. in isopropanol, most vegetable oils, hydrocarbon solvent; Sp. gr.: ≈ 1.0; Visc.: ≈ 200 cps; HLB: 9.4; Pour Pt.: ≈ 45 F

Atlas G-1086. I.C.I. United States
Chemical Description: Polyoxyethylene sorbitol hexaoleate
Category: Nonionic emulsifier
Form: Oily liquid; Color: Yellow
Properties: Solubility: Sol. in toluene, mineral spirits, many vegetable oils, acetone, carbon tetrachloride, cellosolve, methanol, lower alcohols, aniline; Sp. gr.: ≈ 1; Visc.: ≈ 200 cps; HLB: 10.2; Flash Pt.: > 300 F; Fire Pt.: > 300 F
Composition: 100% active ingred.

Atlas G-1096. I.C.I. United States
Chemical Description: Polyoxyethylene (50) sorbitol hexaoleate
Category: Nonionic surfactant
Form: Liquid; Color: Amber
Properties: Solubility: Sol. in vegetable oils, acetone, cellosolve, lower alcohols, some aromatic solvents, carbon tetrachloride; Sp. gr.: ≈ 1.0; Visc.: ≈ 220 cps; HLB: 11.4; Flash Pt.: > 300 F; Fire Pt.: > 300 F
Composition: 100% active ingred.

Atlas G-1186. I.C.I. United States
Chemical Description: Polyoxyethylene sorbitol oleate
Category: Nonionic surfactant
Applications: Fiber lubricant; general purpose emulsifier, esp. for insecticides
Form: Oily liquid; Color: Amber
Properties: Solubility: Sol. in lower alcohols, vegetable and mineral oils; Sp. gr.: ≈ 1; Visc.: ≈ 230 cps; HLB: 10.2; Flash Pt.: >300 F; Fire Pt.: > 300 F
Composition: 100% active ingred.

Atlas G-1223. I.C.I. United States
Chemical Description: Polyoxyethylene esters of mixed fatty and resin acids with 0.025% Antifoam added
Category: Detergent, nonionic
Form: Oily liquid; Color: Light amber
Properties: Solubility: Sol. in water, diethyl ether, dioxane, cellosolve, ethyl acetate, lower alcohols, toluene, acetone; Sp. gr.: ≈ 1.1; Visc.: ≈ 350 cps; HLB: 13.8;

Flash Pt.: > 300 F; Fire Pt.: > 300 F

Atlas G-1236. I.C.I. United States
Category: Blend of nonionic lubricants, emulsifiers, and rewetters
Form: Liquid; Color: Clear amber
Properties: Solubility: Dispersible in water; sol. in alcohol, chloroform; Visc.: ≈ 600 cps; Flash Pt.: > 300 F; Fire Pt.: 300 F; Pour Pt.: ≈ 0 C

Atlas G-1292. I.C.I. United States
Chemical Description: Polyoxyethylene fatty glyceride
Category: Nonionic surfactant
Form: Pale yellow liquid to soft paste
Properties: Solubility: Sol. in water, ethanol, isopropanol; Sp. gr.: ≈ 1.0; HLB: 10.8; Flash Pt.: > 300 F; Fire Pt.: > 300 F
Containers: 55 gal closed head steel drums

Atlas G-1295. I.C.I. United States
Chemical Description: Polyoxyethylene fatty glyceride
Category: Nonionic surfactant
Form: White to cream colored, hard, waxy solid
Properties: Solubility: Sol. in water, lower alcohols, acetone, and chloroform; Sp. gr.: ≈ 1.05 @ 60 C; HLB: 17.5; Flash Pt.: > 300 F; Fire Pt.: > 300 F

Atlas G-1300. I.C.I. United States
Chemical Description: Polyoxyethylene glyceride ester
Category: Emulsifier, nonionic
Applications: Emulsifiers for use in textile chemical formulations
Form: Hard waxy solid; Color: Tan
Properties: Solubility: Sol. in water, lower alcohols, acetone, chloroform; HLB: ≈ 18.1; Flash Pt.: > 300 F; Fire Pt.: > 300 F

Atlas G-1441. I.C.I. United States
Chemical Description: Polyoxyethylene (40) sorbitol lanolin derivative
Category: Emulsifier, nonionic
Applications: Cosmetic use
Form: Solid; Color: Red brown
Properties: Solubility: Sol. in water (with haze), isopropanol; Sp. gr.: ≈ 1; HLB: 14; Flash Pt.: > 300 F; Fire Pt.: > 300 F; Pour Pt.: ≈ 33 C

Atlas G-1471. I.C.I. United States
Chemical Description: Polyoxyethylene (75) lanolin derivative
Category: Nonionic surfactant, emulsifiers
Applications: Cosmetic use
Form: Soft paste; Color: Amber
Properties: Solubility: Sol. in isopropyl alcohol, ethylene glycol; HLB: 16; Flash Pt.: > 300 F; Fire Pt.: > 300 F

Atlas G1556. I.C.I. United States
Chemical Description: Blend of nonionic waxy esters of polyhydric alcohols
Form: Waxy solid; Color: Light tan
Properties: Solubility: Dispersible in hot water, sol. in carbon tetrachloride, chloroform, anhydrous methanol; Sp. gr.: ≈ 1.05; Flash

Pt.: > 300 F; Fire Pt.: > 300 F; Pour Pt.: > 43 C

Atlas G-1564. I.C.I. United States
Chemical Description: Blend of nonionic waxy esters of polyhydric alcohols
Category: Textile lubricant and softener
Form: Waxy granules; Color: Light ivory
Properties: Solubility: Sol. in hot carbon tetrachloride, chloroform, anhydrous methanol; dispersible in warm water; Sp. gr.: ≈ 1.04; Flash Pt.: > 300 F; Pour Pt.: ≈ 50 C
Composition: 100% active ingred.

Atlas G-1565. I.C.I. United States
Chemical Description: Blend of nonionic waxy esters of polyhydric alcohols
Category: Textile lubricant and softener
Form: Waxy solid; Color: Light ivory
Properties: Solubility: Dispersible in warm water; sol. in hot carbon tetrachloride, chloroform, anhydrous methanol; Sp. gr.: ≈ 1.05; HLB: 10.1; Flash Pt.: > 300 F; Fire Pt.: > 300 F; Pour Pt.: ≈ 50 C
Composition: 100% active ingred.

Atlas G-1649. I.C.I. United States
Chemical Description: Polyoxyethylene alkyl aryl ether
Category: Nonionic surfactant
Form: Waxy solid; Color: Cream
Properties: Solubility: Sol. in water, toluol, methanol, isopropanol, carbon tetrachloride; Sp. gr.: ≈ 1.1; HLB: 16.0; Flash Pt.: > 300 F; Fire Pt.: > 300 F

Atlas G-1702. I.C.I. United States
Chemical Description: Polyoxyethylene (6) sorbitol beeswax derivative
Category: Nonionic surfactant
Form: Waxy solid; Color: Tan
Properties: Solubility: Slightly sol. in vegetable oils; HLB: 5; Flash Pt.: > 300 F; Fire Pt.: > 300 F

Atlas G-1726. I.C.I. United States
Chemical Description: Polyoxyethylene (20) sorbitol beeswax derivative
Category: Nonionc surfactant
Form: Waxy solid; Color: Tan
Properties: Solubility: Slightly sol. in vegetable oils; HLB: 5; Flash Pt.: > 300 F; Fire Pt.: > 300 F

Atlas G-1790. I.C.I. United States
Chemical Description: Polyoxyethylene (20) lanolin derivative
Category: Nonionic, emulsifier
Applications: Cosmetic use
Form: Solid; Color: Brown
Properties: Solubility: Sol. in water; HLB: 11; Flash Pt.: > 300 F; Fire Pt.: > 300 F

Atlas G-1795. I.C.I. United States
Chemical Description: Polyoxyethylene (50) lanolin derivative
Category: Nonionic emulsifier
Form: Waxy solid; Color: Tan
Properties: Solubility: Sol. in warm water and ethanol; Sp. gr.: ≈ 1.15; HLB: 17.0; Flash Pt.: > 300 F; Fire Pt.: > 300 F

Atlas G-2079. I.C.I. United States
Chemical Description: Polyoxyethylene (20) palmitate
Category: Nonionic surfactant
Form: Waxy solid; Color: White
Properties: Solubility: Sol. in water and ethanol; HLB: 15.5; Flash Pt.: > 300 F; Fire Pt.: > 300 F

Atlas G-2090. I.C.I. United States
Chemical Description: Blend of polyoxyethylene fatty amine and polyoxyethylene sorbitol oleate
Category: Emulsifier blend
Applications: Detergent, emulsifier, detergent specialty
Form: Mobile liquid; Color: Red-brown
Properties: Solubility: Dispersible in water; sol. in isopropanol, xylene, mineral spirits, kerosene; Sp. gr.: ≈ 0.95; Visc.: ≈ 140 cps; HLB: 12.5; Flash Pt.: > 300 F; Fire Pt.: > 300 F
Composition: 100% active ingred.

Atlas G-2109. I.C.I. United States
Chemical Description: Polyoxyethylene coconut fatty acid
Category: Lubricant, nonionic
Applications: Heat resistant textile lubricant
Form: Liquid
Composition: 100% active ingred.

Atlas G2127. I.C.I. United States
Chemical Description: Polyoxyethylene laurate
Category: Nonionic surfactant
Form: Liquid; Color: Light yellow
Properties: Solubility: Sol. in isopropyl alcohol; dispersible in water and many vegetable oils; Sp. gr.: ≈ 1.0; HLB: 12.8; Flash Pt.: > 300 F

Atlas G2143. I.C.I. United States
Chemical Description: Polyoxyethylene (10) monoleate
Category: Nonionic emulsifier
Form: Liquid; Color: Light yellow
Properties: Solubility: Dispersible in water; sol. in most mineral oils, lower alcohols; Flash Pt.: > 300 F; Pour Pt.: ≈ −10 C

Atlas G-2151. I.C.I. United States
Chemical Description: Polyoxyethylene (30) stearate
Category: Nonionic surfactant
Form: Waxy solid; Color: Ivory
Properties: Solubility: Sol. in propylene glycol, ethylene glycol, water; HLB: 16; Flash Pt.: > 300 F; Fire Pt.: > 300 F

Atlas G-2160. I.C.I. United States
Chemical Description: Polyoxyethylene (25) propylene glycol monostearate
Category: Nonionic surfactant
Form: Semisolid; Color: Light tan
Properties: Solubility: Sol. in water, methanol and ethanol; HLB: 16; Flash Pt.: > 300 F; Fire Pt.: > 300 F

Atlas G-2162. I.C.I. United States
Chemical Description: Polyoxyethylene (25) propylene glycol stearate
Category: Nonionic surfactant
Form: Semisolid; Color: Cream

Properties: Solubility: Sol. in water, methanol, ethanol; HLB: 16; Flash Pt.: > 300 F; Fire Pt.: > 300 F
Composition: 100% active ingred.

Atlas G-2183. I.C.I. United States
Chemical Description: Propylene glycol monopalmitate
Category: Food emulsifier
Form: Waxy solid; Color: Ivory
Properties: Solubility: Sol. in vegetable oils, common organic solvents; M.P.: ≈ 87 F; HLB: 3.2

Atlas G-2185. I.C.I. United States
Chemical Description: Propylene glycol monooleate
Category: Food emulsifier
Form: Liquid; Color: Yellow
Properties: Solubility: Sol. in vegetable oils, common organic solvents; HLB: 3.2; Congealing Pt.: 40 F or less

Atlas G-2198. I.C.I. United States
Chemical Description: Polyoxyethylene (40) stearate
Category: Nonionic surfactant
Form: Waxy solid; Color: Light tan
Properties: Solubility: Sol. in water, acetone, ether, dioxane, cellosolve, carbon tetrachloride, lower alcohols, ethyl acetate, aniline and many organic solvents; HLB: 16.9; Flash Pt.: > 300 F; Fire Pt.: > 300 F; Pour Pt.: ≈ 38 C

Atlas G-2240. I.C.I. United States
Chemical Description: Polyoxyethylene (6) sorbitol
Category: Polyhydric alcohol (polyol)
Form: Syrupy pale amber transparent liquid; Odor: Faint bland
Properties: Solubility: Sol. in water, sorbitol sol'ns., ethanol, methanol, cellosolve, dioxane, aniline; Sp. gr.: ≈ 1.2; Visc.: ≈ 1100 cp; Flash Pt.: > 300 F; Fire Pt.: > 300 F

Atlas G-2320. I.C.I. United States
Chemical Description: Polyoxyethylene (20) sorbitol
Category: Polyhydric alcohol
Form: Liquid; Color: Light
Properties: Solubility: Sol. in water, lower alcohols, ethylene glycol, cellosolve, dioxane, aniline; Sp. gr.: ≈ 1.15; Visc.: ≈ 400 cps; Flash Pt.: > 300 F; Fire Pt.: > 300 F

Atlas G-2401. I.C.I. United States
Chemical Description: Aqueous solution of hydroxypropyl sorbitol
Category: Polyhydric alcohol
Form: Liquid; Color: Straw; Taste: Slightly bitter
Properties: Refractive Index: ≈ 1.475; Solubility: Sol. in water, lower alcohols, propylene glycol, ethylene glycol, cellosolve; Sp. gr.: ≈ 1.25; Visc.: ≈ 2000 cps

Atlas G-2402. I.C.I. United States
Chemical Description: Hydroxy propyl sorbitol
Category: Polyhydric alcohol
Form: Liquid; Color: Straw
Properties: Solubility: Sol. in water, lower alcohols, propylene glycol, ethylene glycol, cellosolve; Sp. gr.: ≈ 1.3; Visc.: ≈ 12,000 cps;

Flash Pt.: > 300 F; Fire Pt.: > 300 F

Atlas G-2684. I.C.I. United States
Chemical Description: Blend of sorbitan monooleate and polyoxyethylene esters of mixed fatty and rosin acids
Category: Detergent, nonionic
Form: Liquid; Color: Amber
Properties: Solubility: Dispersible in water; sol. in acetone, toluene, dioxane, carbon tetrachloride, cyclohexanol, aniline, mineral oils; Sp. gr.: ≈ 1; Visc.: ≈ 500 cps; HLB: 7.8; Flash Pt.: > 300 F; Fire Pt.: > 300 F; Pour Pt.: ≈ 15 F

Atlas G-3300. I.C.I. United States
Chemical Description: Alkyl aryl sulfonate
Category: Anionic surfactant
Applications: Dispersant, pigments
Form: Liquid; Color: Amber
Properties: Solubility: Sol. at low conc. in water and hard water; sol. in acetone, carbon tetrachloride, cellosolve, diethyl ether, dioxane, ethylene glycol, methanol, aniline, mineral spirits, kerosene, toluene; Sp. gr.: ≈ 1.0; Visc.: ≈ 6000 cps; HLB: 11.7; Flash Pt.: > 300 F; Fire Pt.: > 300 F
Composition: 90% active ingred.

Atlas G-3634A. I.C.I. United States
Chemical Description: Quaternary ammonium derivative
Category: Antistat
Applications: Liquid textile antistatic agent; textile softener
Form: Liquid; Color: Light amber
Properties: Solubility: Sol. in water, lower alcohols, acetone, propylene glycol, carbon tetrachloride, perchlorethylene; Sp. gr.: ≈ 1.05; Visc.: ≈ 2500 cps; HLB: 18.5; Flash Pt.: > 300 F
Containers: 55 gal closed head steel drums

Atlas G-3684. I.C.I. United States
Chemical Description: Polyoxyethylene fatty amine
Category: Polyoxyethylene alkyl amine surfactant
Form: Liquid; Color: Amber
Properties: Solubility: Sol. in acetone, lower alcohols, 8% sulfuric acid, 6% sodium hydroxide, water; Sp. gr.: ≈ 1.03; Visc.: ≈ 150 cps; Flash Pt.: > 300 F; Fire Pt.: > 300 F; Pour Pt.: ≈ 20 F

Atlas G-3780A. I.C.I. United States
Chemical Description: Polyoxyethylene alkyl amine
Category: Surfactant
Form: Liquid (may become hazy); Color: Amber
Properties: Solubility: Sol. in water, lower alcohols, acetone, ethyl acetate, ethylene glycol; Sp. gr.: ≈ 1.04; Visc.: ≈ 250 cps; HLB: ≈ 15.5; Flash Pt.: > 300 F

Atlas G-3801. I.C.I. United States
Chemical Description: Polyoxyethylene (2) alkyl alcohol (antioxidant added)
Category: Nonionic emulsifier
Form: Pourable liquid; Color: White
Properties: Sp. gr.: ≈ 0.89; Visc.: ≈ 60 cps; HLB: 5.3; Stability:

Alkalai stable

Atlas G-3811. I.C.I. United States
Chemical Description: Polyoxyethylene (10) alkyl alcohol (antioxidant added)
Category: Nonionic emulsifier
Form: Pourable liquid; Color: White
Properties: Sp. gr.: ≈ 0.992; Visc.: ≈ 415 cps; HLB: 12.9; Stability: Alkalai stable

Atlas G-3821. I.C.I. United States
Chemical Description: Polyoxyethylene (20) alkyl alcohol (antioxidant added)
Category: Nonionic emulsifier
Form: Soft wax
Properties: Sp. gr.: ≈ 1.04; HLB: 15.7; Pour Pt.: ≈ 32 C; Stability: Alkali stable

Atlas G-7166P. I.C.I. United States
Chemical Description: Polyoxyethylene (16) sorbitan tristearate
Category: Nonionic surfactant
Form: Waxy solid; Color: Tan
Properties: Solubility: Dispersible in warm water, tending to gel at higher concentrations; dispersible in most organic solvents, in vegetable and mineral oils; Sp. gr.: ≈ 1.05; HLB: ≈ 10

Atlas G-7596H. I.C.I. United States
Chemical Description: Polyoxyethylene sorbitan laurate
Category: Nonionic emulsifier
Form: Oily liquid; Color: Orange
Properties: Solubility: Sol. in water; Sp. gr.: ≈ 1.1; Visc.: ≈ 400 cps; HLB: 14.3; Flash Pt.: > 300 F; Fire Pt.: > 300 F

Atlas G-7596-J. I.C.I. United States
Chemical Description: Polyoxyethylene (10) sorbitan monolaurate
Category: Detergent, nonionic
Form: Liquid; Color: Yellow
Properties: Solubility: Sol. in water, acetone, dioxane, cellosolve, lower alcohols, ethyl acetate, aniline; Sp. gr.: ≈ 1.1; Visc.: ≈ 400 cps; HLB: 14.9; Flash Pt.: > 300 F; Fire Pt.: > 300 F

Atlas G-9046T. I.C.I. United States
Chemical Description: Polyoxyethylene (20) mannitan monolaurate
Category: Nonionic surfactant
Form: Oily liquid; Color: Yellow
Properties: Solubility: Sol. in water, ethanol, methanol, propylene glycol; Sp. gr.: ≈ 1.1; HLB: 16.7; Flash Pt.: > 300 F; Fire Pt.: > 300 F

Atlox 775. I.C.I. United States
Category: Nonionic emulsifier
Applications: Emulsifier for agricultural spray oils
Form: Liquid
Composition: 100% active ingred.

Atlox 1045A. I.C.I. United States
Chemical Description: Polyoxyethylene sorbitol oleate-laurate
Category: Nonionic agricultural surfactant
Form: Liquid; Color: Yellow
Properties: Solubility: Sol. in most organic solvents and pesticides; dispersible in water; Sp. gr.:

≈ 1.0; Visc.: 200 cps; HLB: 200 cps; HLB: 11.4; Flash Pt.: > 300F

Atlox 1087. I.C.I. United States
Chemical Description: Polyoxyethylene sorbitol septaoleate
Category: Nonionic emulsifier
Form: Liquid; Color: Yellow
Properties: Solubility: Sol. in most vegetable and mineral oil; dispersible in water; Sp. gr.: ≈ 1; Visc.: 175 cps; HLB: 9.2; Flash Pt.: > 300 F; Fire Pt.: > 300 F

Atlox 1196. I.C.I. United States
Chemical Description: Polyoxyethylene sorbitol oleate
Category: Nonionic emulsifier
Form: Oily liquid; Color: Amber
Properties: Solubility: Dispersible in water; sol. in xylol, methanol, acetone, most agricultural concentrates; Sp. gr.: ≈ 1; Visc.: ≈ 450 cp; HLB: 11.4; Flash Pt.: ≈ 475 F; Fire Pt.: ≈ 600 F

Atlox 1256. I.C.I. United States
Chemical Description: Polyoxyethylene sorbitol esters of mixed fatty and resin acids
Category: Nonionic surfactant
Form: Liquid; Color: Dark amber
Properties: Solubility: Sol. in most organic solvents and pesticide formulations; dispersible in water; Sp. gr.: ≈ 1.0; Visc.: ≈ 1000 cps; HLB: 9.7; Flash Pt.: > 300 F; Fire Pt.: > 300 F

Atlox 2081 B. Atlas Chemical Industries
Chemical Description: Polyoxyethylene sorbitan esters of fatty and resin acids–biodegradable alkylaryl sulfonate blend
Category: Emulsifier
Applications: Surfactants used with all types of agricultural emulsifiable concentrates
Form: Liquid
Properties: Biodegradable

Atlox 3300 B. Atlas Chemical Industries
Chemical Description: Polyoxyethylene sorbitan esters of fatty and resin acids; alkyl aryl sulfonate blend
Category: Anionic surfactants
Applications: Surfactants used with all types of agricultural emulsifiable concentrates
Form: Liquid
Properties: Biodegradable

Atlox 3335. I.C.I. United States
Chemical Description: Alkyl aryl sulfonate blended with polyoxyethylene sorbitan ester of tall oil
Category: Agricultural surfactant
Form: Liquid; Color: Amber
Properties: Solubility: Dispersible in water; sol. in most aromatic solvents and pesticide formulations; Sp. gr.: ≈ 1.1 @ 20 C; Visc.: ≈ 2200 cps; HLB: 13; Flash Pt.: ≈ 160 F

Atlox 3335 B. Atlas Chemical Industries
Chemical Description: Polyoxyethylene sorbitan esters of fatty and resin acids; alkyl aryl sulfonate blend
Category: Anionic surfactants
Applications: Surfactants used

with all types of agricultural emulsifiable concentrates
Form: Liquid
Properties: Biodegradable

Atlox 3386 B. Atlas Chemical Industries
Chemical Description: Polyoxyethylene sorbitol fatty esters; alkyl aryl sulfonate blend
Category: Anionic surfactants
Applications: Surfactants used with all types of agricultural emulsifiable concentrates
Form: Liquid
Properties: Biodegradable

Atlox 3387. I.C.I. United States
Chemical Description: Anionic-nonionic surfactant blend
Category: Agricultural surfactant
Form: Viscous liquid; Color: Amber
Properties: Solubility: Sol. in most aromatic solvents and pesticide formulations; Sp. gr.: ≈ 1 @ 20 C; Visc.: ≈ 4000 cps

Atlox 3387 B. Atlas Chemical Industries
Chemical Description: Polyoxyethylene glyceride ester; alkyl aryl sulfonate blend
Category: Anionic surfactants
Applications: Surfactants used with all types of agricultural emulsifiable concentrates
Form: Liquid
Properties: Biodegradable

Atlox 3400 B. Atlas Chemical Industries
Chemical Description: Polyoxyethylene alkyl phenol; alkyl aryl sulfonate blend
Category: Anionic surfactants
Applications: Surfactants used with all types of agricultural emulsifiable concentrates
Form: Liquid
Properties: Biodegradable

Atlox 3403F. I.C.I. United States
Chemical Description: Anionic/nonionic blend
Category: Emulsifier
Applications: Pesticide emulsifier
Form: Liquid; Color: Amber
Properties: Solubility: Sol. in most aromatic solvents used in pesticide formulations; dispersible in water; Density: ≈ 8.4 lb/gal @ 20 C; Sp. gr.: ≈ 1.01 @ 20 C; Visc.: ≈ 100 cp @ 77 F; Flash Pt.: ≈ 58 F; Pour Pt.: ≈ 25 F
Toxicity: Eye irritant; poison
Warnings and Caution: Flammable liquid

Atlox 3404F. I.C.I. United States
Chemical Description: Anionic/nonionic blend
Category: Emulsifier
Applications: Pesticide emulsifier
Form: Liquid; Color: Amber
Properties: Solubility: Sol. in kerosene and most aromatic solvents used in pesticide formulations; dispersible in water; Density: ≈ 8.7 lb/gal; Sp. gr.: ≈ 1.04; Visc.: ≈ 1400 cp @ 77 F; Flash Pt.: ≈ 64 F; Pour Pt.: ≈ –15 F
Toxicity: Poison; skin and eye irritant
Warnings and Caution: Flammable liquid

Atlox 3406F. I.C.I. United States
Chemical Description: Anionic/nonionic blend
Category: Emulsifier
Applications: Pesticide emulsifier
Form: Liquid; Color: Amber
Properties: Solubility: Sol. in most aromatic solvents used in pesticide formulations; dispersible in water; Density: ≈ 8.6 lb/gal @ 20 C; Sp. gr.: ≈ 1.03 @ 20 C; Visc.: ≈ 200 cp @ 77 F; Flash Pt.: ≈ 61 F; Pour Pt.: ≈ 20 F
Toxicity: Poison; eye irritant
Warnings and Caution: Flammable

Atlox 3407F. I.C.I. United States
Chemical Description: Anionic/nonionic blend
Category: Pesticide emulsifier
Form: Liquid; Color: Amber
Properties: Solubility: Sol. in most aromatic solvents used in pesticide formulations; dispersible in water; Sp. gr.: ≈ 1.03 @ 20/20 C; Visc.: ≈ 250 cp @ 70 F; Flash Pt.: ≈ 59 F; Pour Pt.: ≈ 10 F
Toxicity: Poisonous if ingested; eye contact and repeated skin exposure can cause chemical burns
Warnings and Caution: Flammable liquid; goggles and gloves should be worn

Atlox 3408F. I.C.I. United States
Chemical Description: Anionic/nonionic blend
Category: Pesticide emulsifier
Form: Liquid; Color: Amber
Properties: Solubility: Sol. in most aromatic solvents used in pesticide formulations; dispersible in water; Density: ≈ 8.5 lb/gal @ 20 C; Sp. gr.: ≈ 1.02 @ 20/20C; Visc.: ≈ 150 cp @ 77 F; Flash Pt.: ≈ 57 F; Pour Pt.: ≈ 20 F
Toxicity: Poisonous if ingested; eye contact and repeated skin exposure can cause chemical burns
Warnings and Caution: Goggles and gloves should be worn; avoid inhalation. Flammable

Atlox 3409F. I.C.I. United States
Chemical Description: Anionic/nonionic blend
Category: Pesticide emulsifier
Form: Liquid; Color: Amber
Properties: Solubility: Sol. in most aromatic solvents used in pesticide formulations; dispersible in water; Density: ≈ 8.4 lb/gal @ 20C; Sp. gr.: ≈ 1.01 @ 20 C; Visc.: ≈ 150 cp @ 77 F; Flash Pt.: ≈ 54 F; Pour Pt.: ≈ 40 F
Toxicity: Poisonous if ingested; eye irritant
Warnings and Caution: Flammable liquid;

Atlox 3414. I.C.I. United States
Chemical Description: Anionic/nonionic blend
Category: Emulsifier
Form: Viscous liquid
Properties: Solubility: Sol. in kerosene, fuel oil, xylene, heavy aromatic naptha; Sp. gr.: ≈ 1.03; Flash Pt.: > 240 F

Atlox 3421F. I.C.I. United States
Chemical Description: Anionic/nonionic surfactant blend
Category: Emulsifier
Applications: Agricultural emulsifiers for improved concentrate

ageing; esp. effective with organophosphate pesticides
Form: Hazy liquid
Properties: Solubility: Sol. in most armoatic solvents and pesticide formulations; dispersible in water; Sp. gr.: 1.04; Visc.: 250 cp; Flash Pt.: 48 F; Pour Pt.: 55 F
Warnings and Caution: Flammable

Atlox 3422F. I.C.I. United States
Chemical Description: Anionic/nonionic surfactant blend
Category: Emulsifier
Applications: See Atlox 3421F
Form: Hazy liquid
Properties: Solubility: Sol. in most aromatic solvents and pesticide formulations; dispersible in water; Sp. gr.: 1.04; Visc.: 900 cp; Flash Pt.: 58 F; Pour Pt.: 50 F
Warnings and Caution: Flammable

Atlox 3450F. I.C.I. United States
Chemical Description: Anionic/nonionic blend
Category: Emulsifier
Applications: Agricultural surfactant; emulsifier for aerial spray formulations that are applied at dilution rates as low as 1:1; in more dilute sprays for ground application;emulsifies high-poundage formulations
Form: Hazy liquid; Color: Dark brown
Properties: Solubility: Sol. in most aromatic solvents and pesticide formulations; dispersible in water; Density: 8.7 lb/gal; Sp. gr.: ≈ 1.04; Visc.: ≈ 700 cp; Flash Pt.: ≈ 120 F; Pour Pt.: ≈ 60 F; Stability: Good storage stability
Warnings and Caution: Combustible

Atlox 3453F. I.C.I. United States
Chemical Description: Nonionic/anionic blend
Category: Emulsifier
Applications: General purpose emulsifiers for pesticide concentrates where high flash point is required
Form: Liquid
Composition: 100% active ingred.

Atlox 3454F. I.C.I. United States
See Atlox 3453F

Atlox 3455F. I.C.I. United States
See Atlox 3453F

Atlox 3459F. I.C.I. United States
Chemical Description: Anionic/nonionic surfactant blend
Category: Emulsifier
Applications: Agricultural emulsifiers for improved concentrate ageing; sometimes required for optimum emulsification of extremely high poundage phosphate formulas
Form: Clear liquid
Properties: Solubility: Sol. in most aromatic solvents and pesticide formulations; sol. in water; Sp. gr.: 1.05; Visc.: 500 cp; Flash Pt.: 64 F; Pour Pt.: 15 F
Warnings and Caution: Flammable

Atlox 4851 B. Atlas Chemical Industries
Chemical Description: Polyoxyethylene triglyceride alkyl aryl sulfonate blend
Category: Anionic surfactants

Applications: Surfactants used with all types of agricultural emulsifiable concentrates
Form: Liquid
Properties: Biodegradable

Atlox 4853 B. Atlas Chemical Industries
Chemical Description: Blend of ionic and nonionic surfactants
Category: Anionic, nonionic
Applications: Surfactants used with all types of agricultural emulsifiable concentrates
Form: Liquid
Properties: Biodegradable

Atlox 4855 B. Atlas Chemical Industries
Chemical Description: Polyoxyethylene triglyceride; alkyl aryl sulfonate blend
Category: Anionic surfactants
Applications: Surfactants used with all types of agricultural emulsifiable concentrates
Form: Liquid
Properties: Biodegradable

Atlox 4856 B. Atlas Chemical Industries
Chemical Description: Polyoxyethylene alcohol alkyl aryl sulfonate blend
Category: Anionic surfactants
Applications: Surfactants used with all types of agricultural emulsifiable concentrates
Form: Liquid
Properties: Biodegradable

Atlox 4857 B. Atlas Chemical Industries
Chemical Description: Polyoxyethylene alkyl phenol; alkyl aryl sulfonate blend
Category: Anionic surfactants
Applications: Surfactants used with all types of agricultural emulsifiable concentrates
Form: Liquid
Properties: Biodegradable

Atlox 4858 B. Atlas Chemical Industries
Chemical Description: Polyoxyethylene sorbitan esters of fatty and resin acids; alkyl aryl sulfonate blend
Category: Anionic surfactants
Applications: Surfactants used with all types of agricultural emulsifiable concentrates
Form: Liquid
Properties: Biodegradable
Composition: 100% conc.

Atlox 4860. Atlas Chemical Industries
Chemical Description: Polyoxyethylene alkyl ether and fatty acids; alkyl aryl sulfonate blend
Category: Anionic surfactants
Applications: Surfactants used with all types of agricultural emulsifiable concentrates
Form: Powder
Properties: Biodegradable
Composition: 100% conc.

Atlox 4862. Atlas Chemical Industries
Chemical Description: Anionic suspending agent
Category: Wetting/suspending a-

gent, anionic
Applications: Agricultural wettable powders
Form: Powder
Composition: 100% conc.

Atlox 4868B. Atlas Chemical Industries
Chemical Description: Polyoxyethylene triglyceride/alkyl aryl sulfonate blend
Category: Wetting/suspending agent
Applications: Agricultural wettable powders
Form: Liquid
Properties: Biodegradable

Atlox 4870 B. Atlas Chemical Industries
Chemical Description: Polyethylene sorbitan esters of fatty and resin acids; biodegradable alkyl aryl sulfonate blend
Category: Wetting/suspending agent
Applications: Agricultural wettable powders
Form: Liquid
Properties: Biodegradable

Atlox 4991. Atlas Chemical Industries
Chemical Description: Polyoxyethylene alkyl aryl ether
Category: Nonionic surfactant
Form: Liquid
Composition: 100% conc.

Atlox 4995. Atlas Chemical Industries
Category: Wetting/suspending agent, nonionic
Applications: Agricultural wettable powders
Form: Powder
Composition: 100% conc.

Atlox 4996 B. Atlas Chemical Industries
Category: Suspending/wetting agent, anionic
Applications: Agricultural wettable powders
Form: Powder

Atlox 8916 P. Atlas Chemical Industries
Chemical Description: Polyoxyethylene esters of mixed fatty and resin acids
Category: Nonionic surfactant
Form: Liquid
Composition: 100% conc.

Atlox 8916P. I.C.I. United States
Chemical Description: Polyoxyethylene (16) sorbitan ester of tall oil
Category: Nonionic surfactant
Applications: Agricultural surfactant
Form: Liquid; Color: Amber
Properties: Solubility: Sol. in water and most aromatic solvents and pesticide formulations; Sp. gr.: ≈ 1.1; Visc.: ≈ 1320 cps; HLB: 14.6; Flash Pt.: > 300 F

Atlox 8916PF. I.C.I. United States
Chemical Description: Alkoxylated polyol fatty acid ester
Category: Emulsifier
Applications: Agricultural surfactant; pesticides

Form: Liquid; Color: Amber
Properties: Solubility: Sol. with haze in most aromatic solvents used in pesticide formulations; gels upon addition to water, sol. in water after heating; Density: ≈ 9.0 lb/gal; Sp. gr.: ≈ 1.08; Visc.: ≈ 500 cp; Flash Pt.: > 200 F; Pour Pt.: ≈ –20 F
Warnings and Caution: Combustible liquid

Atlox 8916TF. I.C.I. United States
Chemical Description: Alkoxylated polyol fatty acid ester
Category: Emulsifier
Applications: Agricultural surfactant; pesticides
Form: Liquid; Color: Amber
Properties: Solubility: Sol. in most aromatic solvents used in pesticide formulations; sol. with haze in high aromatic naphtha, sol. without haze in xylene; sol. in water after heating; Density: ≈ 8.3 lb/gal; Sp. gr.: ≈ 1.0; Visc.: ≈ 500 cp; Flash Pt.: > 200 F; Pour Pt.: ≈ –20 F
Warnings and Caution: Combustible liquid

Atmer 507. I.C.I. United States
Chemical Description: Aliphatic polyether
Category: Nonionic surfactant
Form: Liquid; Color: Amber
Properties: Solubility: Dispersible in water; sol. in acetone, ethylene acetate, dioxane, cellosolve, lower alcohols; Sp. gr.: ≈ 1.1; Visc.: ≈ 300 cps; Flash Pt.: > 300 F; Fire Pt.: > 300 F

Atmer 508. I.C.I. United States
Chemical Description: Aromatic polyether
Category: Nonionic surfactant
Form: Liquid; Color: Light
Properties: Solubility: Sol. in lower alcohols, cellosolve, many petroleum solvents; Sp. gr.: ≈ 1.0; Visc.: ≈ 220 cps; Flash Pt.: > 300 F; Fire Pt.: > 300 F

Atmer 665. I.C.I. United States
Chemical Description: Nonionic ester
Category: Nonionic surfactant
Form: Waxy beads; Color: Tan
Properties: Solubility: Sol. in isopropanol; HLB: 2.1; Flash Pt.: > 300 F; Fire Pt.: > 300 F; Pour Pt.: ≈ 53 C

Atmos 150. I.C.I. United States
Chemical Description: Mono- and diglycerides
Category: Food emulsifier
Form: Ivory white beads; Odor and Taste: Bland
Properties: Solubility: Sol. above its melting point in vegetable oils, mineral oil and isopropyl alcohol; M.P.: ≈ 140 F; HLB: 3.2; Flash Pt.: > 300 F; Fire Pt.: > 300 F
Containers: 47 gal fiber lever pak

Atmos 300. I.C.I. United States
Chemical Description: Mono- and diglycerides with 0.02% BHA and 0.01% citric acid in propylene glycol added as preservatives
Category: Food emulsifier, nonionic
Form: Clear liquid; Color: Yellow

Properties: Solubility: Sol. in cottonseed oil, mineral oil, isopropyl alcohol; Sp. gr.: ≈ 0.96; Visc.: ≈ 130 cps; HLB: 2.8; Flash Pt.: > 300 F; Fire Pt.: > 300 F
Composition: 100% active ingred.

Atmos 729. I.C.I. United States
Chemical Description: Hydrated mono- and diglycerides, polysorbate 60 (11.3%), and sorbitan monostearate (11.3%); with phosphoric acid, sodium propionate and sodium benzoate as preservatives
Category: Food emulsifier
Applications: Used in various cake types for improved batter aeration
Form: Creamy white soft plastic
Properties: Solubility: Disperses in water; Flash Pt.: > 300 F; Fire Pt.: > 300 F
Composition: 45% active ingred.

Atmos 758. I.C.I. United States
Chemical Description: Hydrated sorbitan, monostearate (22.5%), polysorbate 60 (13.5%), and mono- and diglycerides, with phosphoric acid, sodium propionate and sodium benzoate added as preservatives
Category: Food emulsifier
Applications: Cake doughnut emulsifier for improved batter aeration
Form: Creamy white soft plastic
Properties: Solubility: Disperses in water; Flash Pt.: > 300 F; Fire Pt.: > 300 F
Composition: 45% active ingred.

Atmos 2462. I.C.I. United States
Chemical Description: Hydrated mono- and diglycerides, polysorbate 60 (13%), sorbitan monostearate (13%) and propylene glycol; with phosphoric acid, sodium propionate and sodium benzoate added as preservatives
Category: Food emulsifier
Applications: Cake emulsifier for improved batter aeration in various cake types
Form: Creamy white soft plastic
Properties: Solubility: Disperses in water; HLB: 7.4; Flash Pt.: > 300 F; Fire Pt.: > 300 F
Composition: 39% active ingred.

Atmul 80. I.C.I. United States
Chemical Description: Mono- and diglycerides with 0.02% BHA and 0.01% citric acid added as preservatives
Category: Food emulsifier, nonionic
Applications: Food shortenings applications
Form: Ivory white plastic solid
Properties: Solubility: Sol. above its melting point in isopropyl alcohol, vegetable oils, mineral oil; M.P.: ≈ 115 F; HLB: 2.8; Flash Pt.: > 300 F; Fire Pt.: > 300 F
Composition: 100% active ingred.

Atmul 84. I.C.I. United States
Chemical Description: Mono- and diglycerides
Category: Food emulsifier
Form: Ivory white beads or flakes; Odor and Taste: Bland
Properties: Solubility: Sol. above

its melting point in alcohol, vegetable oils, mineral oil; M.P.: ≈ 140 F; HLB: 2.8; Flash Pt.: > 300 F; Fire Pt.: > 300 F

Atmul 84K. I.C.I. United States
Chemical Description: Mono- and diglycerides
Category: Food emulsifier
Applications: Certified as Kosher
Form: Ivory-white flakes; Odor and Taste: Bland
Properties: Solubility: Sol. above its melting point in alcohol, vegetable oils, mineral oil; M.P.: ≈ 144 F; HLB: 2.8; Flash Pt.: > 300 F; Fire Pt.: > 300 F

Atmul 86K. I.C.I. United States
Chemical Description: Mono- and diglycerides with 0.02% BHA and 0.01% citric acid added as preservatives
Category: Food emulsifier
Applications: For Kosher use
Form: Ivory white plastic solid
Properties: Solubility: Sol. above its melting point in isopropyl alcohol, vegetable oils, mineral oil; M.P.: ≈ 117 F; HLB: 2.8; Flash Pt.: > 300 F; Fire Pt.: > 300 F

Atmul 122. I.C.I. United States
Chemical Description: Mono- and diglycerides with 0.02% BHA and 0.01% citric acid added as preservatives
Category: Food emulsifier
Form: Ivory white plastic solid
Properties: Solubility: Sol. above its melting point in isopropyl alcohol, vegetable oil, mineral oil; M.P.: ≈ 126 F; HLB: 3.5; Flash Pt.: > 300 F; Fire Pt.: > 300 F

Atmul 124. I.C.I. United States
Chemical Description: Mono- and diglycerides
Category: Food emulsifier
Form: Ivory white flakes; Odor and Taste: Bland
Properties: Solubility: Sol. above its melting point in isopropyl alcohol, vegetable oils, mineral oil; M.P.: ≈ 140 F; HLB: 3.5; Flash Pt.: > 300 F; Fire Pt.: > 300 F
Containers: 50 lb bags

Atmul 500. I.C.I. United States
Chemical Description: Mono- and diglycerides with 0.02% BHA and 0.01% citric acid added as preservatives
Category: Food emulsifier
Form: Ivory white plastic solid
Properties: Solubility: Sol. above its melting point in isopropyl alcohol, vegetable oils, mineral oils; M.P.: ≈ 132 F; HLB: 3.5; Flash Pt.: > 300 F; Fire Pt.: > 300 F

Atmul 651K. I.C.I. United States
Chemical Description: Mono- and diglycerides with 0.02% BHA and 0.01% citric acid added as preservatives
Category: Food emulsifier
Applications: For Kosher use
Form: Ivory white plastic solid
Properties: Solubility: Sol. above its melting point in isopropyl alcohol, vegetable oils, mineral oil; M.P.: ≈ 133 F; HLB: 3.5; Flash Pt.: > 300 F; Fire Pt.: > 300 F

Atmul 700H. I.C.I. United States

Chemical Description: Hydrated mono- and diglycerides with polysorbate 60 (1%), and sodium propionate and lactic acid added as preservatives
Category: Food emulsifier
Applications: Softener for yeast-raised bakery products
Form: Soft plastic; Color: Creamy white
Properties: Solubility: Dispersible in water; HLB: 4.0; Flash Pt.: > 300 F; Fire Pt.: > 300 F
Composition: Total monoglycerides: ≈ 25%

Atolex AST/2. Standard Chem.
Category: Antistat, nonionic
Applications: Antistatic liquid and fiber lubricant for textile industry

Atolex AST/3. Standard Chem.
Category: Antistat, cationic
Applications: Textile antistatic agent
Form: Paste

Atolex BT. Standard Chem.
Category: Dyebath auxiliary
Applications: Backtanning agent particularly for acid dyes on nylon

Atolex DA/25. Standard Chem.
Category: Dyebath auxiliary
Applications: Anionic levelling and dispersing agent

Atolex DCP/2. Standard Chem.
Category: Dyebath auxiliary
Applications: Dye carrier for high temperature pressure dyeing of polyester

Atolex LDA/40. Standard Chem.
Category: Dyebath auxiliary, nonionic
Applications: Levelling and dispersing agent

Atolex LS/3. Standard Chem.
Chemical Description: Sulphated fatty alcohol
Category: Detergent and wetting agent, nonionic/anionic
Form: Liquid

Atolex NI/100B. Standard Chem.
Category: Detergent, wetting agent, nonionic
Applications: Concentrated detergent and wetting agent for textiles
Form: Liquid

Atolex PFC. Standard Chem.
Category: Lubricant
Applications: Cationic polyethylene emulsion substantive lubricant and softener for textile industry

Atolex PFN. Standard Chem.
Category: Lubricant
Applications: Nonionic polyethylene emulsion lubricant and softener

Atolex PMC. Standard Chem.
Category: Lubricant, cationic
Applications: Modified cationic polyethylene emulsion substantive lubricant and softener also applicable in dyebath

Atpet 80. I.C.I. United States
Chemical Description: Sorbitan

monooleate
Category: Nonionic rust inhibitor
Form: Liquid; Color: Amber
Properties: Solubility: Dispersible in water; sol. in most petroleum oils, most vegetable oils, lower alcohols and chlorinated solvents; Sp. gr.: ≈ 1; Visc.: ≈ 1200 cps; HLB: 4.3; Flash Pt.: > 300 F; Fire Pt.: > 300 F; Pour Pt.: ≈ 0 F
Composition: 100% active ingred.

Atpet 100. I.C.I. United States
Chemical Description: Sorbitan fatty acid ester
Category: Nonionic rust inhibitor
Form: Viscous liquid (may gel on standing); Color: Red brown
Properties: Solubility: Sol. in most mineral oils and vegetable oils, lower alcohols, chlorinated solvents; dispersible in water; Sp. gr.: ≈ 1.0; Visc.: ≈ 900 cps; HLB: 4.3; Flash Pt.: > 300 F; Fire Pt.: > 300 F

Atpet 200. I.C.I. United States
Chemical Description: Sorbitan fatty acid ester
Category: Nonionic rust inhibitor
Form: Liquid; Color: Reddish brown
Properties: Solubility: Sol. in most mineral oils, vegetable oils, lower alcohols, chlorinated solvents; dispersible in water; Sp. gr.: ≈ 1.0; Visc.: ≈ 1200 cps; HLB: 4.3; Flash Pt.: > 300 F; Fire Pt.: > 300 F; Pour Pt.: ≈ 0 F
Containers: 55 gal closed head steel drums

Atplus 300F. I.C.I. United States
Chemical Description: Nonionic surfactants with coupling agents
Category: Agricultural surfactant, nonionic
Applications: Surfactant portion of crop oil concentrate adjuvant for pesticides
Form: Liquid; Color: Amber
Properties: Solubility: Sol. in most aromatic solvents used in pesticide formulations; dispersible in water; sol. in agricultural grade paraffin oil at 3 to 50%; Density: ≈ 8.6 lb/gal @ 20 C; Sp. gr.: ≈ 1.03 @ 20 C; Visc.: ≈ 400 cp; Flash Pt.: > 200 F; Pour Pt.: ≈ 15 F
Composition: 92% solids
Warnings and Caution: Combustible

Atplus 400. I.C.I. United States
Category: Agricultural surfactant, anionic
Form: Liquid; Color: Pale yellow
Properties: Solubility: Sol. in water; Density: ≈ 9 lb/gal @ 20 C; Sp. gr.: ≈ 1.08; Visc.: < 100 cps @ 20 C

Atplus 401. I.C.I. United States
Chemical Description: Anionic surfactant with coupling agents
Category: Agricultural surfactant, wetting agent and adjuvant
Applications: For use with monosodium acid methane arsonate solutions and other herbicides containing MSMA; also with high salt concentration formulations; low phytotoxicity to crops
Form: Liquid; Color: Colorless or light yellow
Properties: Solubility: Disperses

readily without heat; Sp. gr.: ≈ 1.21; Visc.: ≈ 31 centistokes; Flash Pt.: ≈ 118 F; Pour Pt.: ≈ –20 F
Toxicity: May cause skin and eye irritation on contact
Containers: 55 gal closed head steel drums; tank wagons; tank cars available

Atplus 411F. I.C.I. United States
Chemical Description: Oil/surfactant concentrate
Category: Agricultural surfactant
Applications: Widely used with many herbicides and harvest aid chemicals; shows activity with pre-emergent herbicides

Atplus 522. I.C.I. United States
Chemical Description: Formulated product
Category: Agricultural surfactant
Applications: For use in sodium chlorate solutions; improves wetting properties by reduction of surface tension
Form: Liquid; Color: Amber
Properties: Solubility: Disperses readily, sol. in water; Density: ≈ 10.0 lb/gal @ 20 C; Sp. gr.: ≈ 1.2 @ 20 C; Visc.: ≈ 35 centistokes; Flash Pt.: ≈ 115 F; Pour Pt.: < 32 F
Toxicity: May cause eye and skin irritation on contact
Containers: 55 gal closed head steel drums; tank wagons and tank cars available

Atplus 526. I.C.I. United States
Chemical Description: Concentrated blend of fatty acids, fatty acid esters and alkoxylated polyhydric alcohol fatty acid esters
Category: Agricultural surfactant, nonionic
Applications: Spray tank adjuvant for improving the wetting, spreading and sticking properties of agricultural sprays; specifically developed for repackaging by formulators
Form: Liquid

Atplus 535. I.C.I. United States
Category: Agricultural surfactant; anionic
Applications: Stabilizer for liquid fertilizer-pesticide mixtures; for repackaging/resale
Form: Light amber liquid
Properties: Solubility: Complete in liquid fertilizers; Density: 10.5 lb/gal; Sp. gr.: ≈ 1.25; Visc.: ≈ 40 cps; Flash Pt.: > 200 F

Atplus 540. I.C.I. United States
Chemical Description: Alkoxylated alcohol ether
Category: Agricultural foaming agent
Applications: Spray tank additive for short life foam with a stability of 15 to 30 min.
Form: Liquid; Color: Amber
Properties: Density: ≈ 8.1 lb/gal; Sp. gr.: ≈ 0.97; Visc.: ≈ 30 cps; Flash Pt.: ≈ 65 F; Pour Pt.: ≈ 20 F
Composition: 75% alkoxylated alcohol ether; 25% constituents ineffective as foaming agents

Atplus 555. I.C.I. United States
Chemical Description: Nonionic

surfactant and coupling agents
Category: Agricultural surfactant, nonionic
Applications: Very effective with post-emergent herbicides; powerful wetting agent

Atplus 989. I.C.I. United States
Chemical Description: Nonionic surfactant and coupling agents
Category: Agricultural surfactant
Applications: Shows activity with pre-emergent Caporal®

Atplus 1034. I.C.I. United States
Chemical Description: Aromatic sulfates and polyhydroxy compounds
Category: Agricultural foaming agent
Applications: Spray tank additive for long life foam with stability of 2 to 24 hours
Form: Liquid; Color: Amber
Properties: Density: ≈ 10 lb/gal; Sp. gr.: ≈ 1.2; Visc.: ≈ 140 cps; Flash Pt.: > 200 F; Pour Pt.: ≈ −35 F

Atplus 1364. I.C.I. United States
Chemical Description: Nonionic surfactant with sticker and coupler
Category: Agricultural surfactant
Applications: Used with post-emergent WP pesticides

Atplus 1399. I.C.I. United States
Chemical Description: Nonionic surfactant and coupling agents
Category: Agricultural surfactant
Applications: Active with some pre-emergent herbicides; medium HLB

Atplus 1400. I.C.I. United States
Chemical Description: Nonionic surfactant with coupling agents
Category: Agricultural surfactant
Applications: Medium-high HLB; for activation, reduction of crop injury, pre-emergent herbicides

Atplus 1401. I.C.I. United States
Chemical Description: Nonionic surfactant and coupling agents
Category: Agricultural surfactant
Applications: High HLB; used with growth regulators; increases pesticide soil movement

Atplus 1402. I.C.I. United States
Chemical Description: Nonionic surfactant with coupling agents
Category: Agricultural surfactant
Applications: High HLB; similar to Atplus 1401 but more effective in some cases

Atplus 1403. I.C.I. United States
Chemical Description: Anionic surfactant and coupling agents
Category: Agricultural surfactant
Applications: Wetting agent, activation; used with post-emergent herbicides

Atsurf 2801. I.C.I. United States
Chemical Description: Polyol oleate
Category: Industrial grade surfactant, emulsifier, nonionic
Applications: Polyethylene, carnauba, and paraffin wax emulsions for paper coatings, floor polishes, furniture polishes, fiberboard and particle board production; silicone oil emulsions for mold release a-

gents, polishes, antifoam agents; perchloroethylene, mineral spirits, kerosene emulsions for metal cleaning; mineral oil emulsions for metalworking
Form: Liquid; Color: Amber
Properties: Solubility: Sol. in isopropanol, perchloroethylene, xylene, cottonseed oil, paraffinic oil; Density: 1.00 g/ml; Visc.: 379 cs; HLB: 5; Pour Pt.: −39 F
Composition: 100% active ingred.

Atsurf 2802. I.C.I. United States
Chemical Description: Polyol trioleate
Category and Applications: See Atsurf 2801
Form: Liquid; Color: Amber
Properties: Solubility: Sol. in perchloroethylene, xylene, cottonseed and paraffinic oils; Density: 0.94 g/ml; Visc.: 204 cs; HLB: 2.5; Pour Pt.: < −35 F
Composition: 100% active ingred.

Atsurf 2803. I.C.I. United States
Chemical Description: Polyol stearate
Category and Applications: See Atsurf 2801
Form: Solid; Color: Light tan
Properties: Solubility: Sol. in isopropanol (1%), perchloroethylene, xylene; Density: 1.06 g/ml; Visc.: 165 cs @ 160 F; HLB: 5; Pour Pt.: 125 F
Composition: 100% active ingred.

Atsurf 2821. I.C.I. United States
Chemical Description: Ethoxylated polyol oleate
Category and Applications: See Atsurf 2801
Form: Liquid; Color: Amber
Properties: Solubility: Sol. in water, isopropanol, perchloroethylene (10%), xylene (1%), cottonseed oil (1%); Density: 1.06 g/ml; Visc.: 846 cs; HLB: 16; Pour Pt.: 55 F
Composition: 100% active ingred.

Atsurf 2822. I.C.I. United States
Chemical Description: Ethoxylated polyol trioleate
Category and Applications: See Atsurf 2801
Form: Liquid; Color: Amber
Properties: Solubility: Sol. in isopropanol, perchloroethylene, xylene, cottonseed and paraffinic oils; Density: 1.02 g/ml; Visc.: 349 cs; HLB: 11; Pour Pt.: 15 F
Composition: 100% active ingred.

Atsurf 2823. I.C.I. United States
Chemical Description: Ethoxylated polyol stearate
Category and Applications: See Atsurf 2801
Form: Semisolid; Color: Amber
Properties: Solubility: Sol. in water, isopropanol, xylene (1%), cottonseed oil (1%); Density: 1.08 g/ml; Visc.: 50 cs @ 160 F; HLB: 16; Pour Pt.: 90 F
Composition: 100% active ingred.

Avirol 200. Henkel Inc.
Chemical Description: Special ammonium lauryl sulfate
Category: Foaming agent, dispersant
Applications: Foaming and dispersing agent for dry wall manu-

facture; foaming agent and suspension aid in rock drilling; foaming agent for fire fighting; dispersing agent for dyes and pigments
Form: Liquid; Color: Light amber; Odor: Typical
Properties: Visc.: 1000 cps; Cloud Pt.: 10 C max.
Composition: 28% fatty alcohol sulfate; 0.5% ammonium chloride; 1.5% ammonium sulfate

Avirol 300. Henkel Inc.
Chemical Description: TEA lauryl sulfate
Applications: Base for shampoos, bubble baths, and other cosmetic preparations for cleansing purposes
Form: Liquid; Color: Light amber; Odor: Pleasant, mild
Properties: Visc.: 500 cps max.
Composition: 39–41% fatty alcohol sulfate

AXS. Conoco Chemical
Chemical Description: Ammonium xylene sulfonate
Applications: Hydrotrope for detergent formulations
Form: Liquid; Color: Near colorless; Odor: Slight ammonia
Properties: Density: 9.5; Sp. gr.: 1.141; Visc.: 4.8; M.P.: 14; B.P.: 212
Composition: 42% active ingred.; other components: water

B

Barium Petronate 50S Neutral. Witco Chem. Sonneborn Div.
Applications: Additive for lube and industrial oils, fuels, specialty oils and greases
Form: Color: Dark brown
Properties: Sp. gr.: 1.0 @ 60 F
Composition: 45% active ingred.; 54% mineral oil; 1% water
Toxicity: May cause eye burns
Containers: Drums, tankwagons, tankcars

Barium Petronate 50-S Neutral. Witco Chem. Sonneborn Div.
Chemical Description: Synthetic sulfonate
Category: Lubricant
Applications: Additives for lube and industrial oils, fuel oil, undercoating and greases
Properties: Visc.: 100 SSU
Composition: 50–52% barium sulfonate

Barium Petronate B-70. Witco Chem. Sonneborn Div.
Applications: Lube and industrial oil and fuel additive
Form: Viscous liquid; Color: Dark brown
Properties: Sp. gr.: 1.0 @ 60 F
Composition: 46% active ingred.; 53.5% mineral oil; 0.5% water
Reaction: Acid
Toxicity: May cause eye burn
Containers: Drums, tankwagons, tank cars

Barium Petronate Basic. Witco Chem. Sonneborn Div.
Chemical Description: Barium carbonate salt of hydrocarbon sulfonic acid
Category: Lubricant
Applications: Additives for lube and industrial oils and fuels, specialty oils, greases
Properties: Visc.: 250 SSU
Composition: 40–45% barium sulfonate

Base 11-T. Cincinnati Milacron Chem. Inc.
Category: Emulsifier
Applications: Preparation of o/w emulsions for metal cutting, rust proofing and quenching

Form: Liquid; Color: Brown
Properties: Solubility: Oil sol.; Density: 8.7 lb/gal @ 60 F; Sp. gr.: 1.05 @ 60 F; Visc.: 190 SUS @ 210 F
Composition: 100% active ingred.
Containers: 55 gal drums

Berol 02. Berol Kemi AB
Chemical Description: Alkyl phenol ethylene oxide adduct
Category: Emulsifier, nonionic
Applications: Emulsifier for waxes, oils, solvents; emulsion polymerization
Form: Clear liquid; Color: < 200 (Hazen)
Properties: Solubility: Sol. in ethanol, xylene, trichloroethylene, light fuel oil; disp. in water, paraffin oil, white spirit; Density: 1.04 g/cm^3 @ 20 C; Visc.: ≈ 350 cps @ 20 C; HLB: 10.9; Cloud Pt.: 60–70 C; Clear Pt.: < 0 C; Flash Pt.: ≫ 100 C; Pour Pt.: < 0 C; Stability: Stable to weak acids, bases, oxidizing and reducing agents; Surface Tension: 31 dynes/cm; Biodegradable in normal biological sewage purification conditions; pH: 6–7 (1% aq.)
Composition: 100% active ingred.
Toxicity: Degreasing effect on skin in concentrated form
Containers: 200 kg net drums

Berol 07. Berol Kemi AB
Chemical Description: Fatty alcohol ethoxylate
Category: Detergent, equalizing and dispersing agent, nonionic
Applications: Low-foam washing machine detergents in powder form; textile industry for high-temperature washing; equalizing and dispersing agent for dyes
Form: Wax; Color: White
Properties: Solubility: Sol. in water, ethanol, xylene, trichloroethylene; disp. in paraffin oil, white spirit, light fuel oil; Density: 1.09 g/cm^3; Visc.: ≈ 50 cps @ 50 C; M.P.: 36–38 C; HLB: 15.0; Cloud Pt.: 74–75 C; Flash Pt.: ≫ 100 C; Stability: Stable to weak acids, bases, oxidizing and reducing agents; Surface Tension: 40 dynes/cm; Biodegradable; pH: 6–7 (1% aq.)
Composition: 100% active ingred.
Toxicity: Degreasing effect on skin in concentrated form
Containers: 100 kg net drums

Berol 08 Powder. Berol Kemi AB
Chemical Description: Fatty alcohol ethoxylate
Category: Detergent, dispersant, nonionic
Applications: Low foaming powder detergents; in its melted form as dispersant for other powders with very small particle size (to form prills)
Form: Powder; Color: White
Properties: Solubility: Sol. in water, ethanol, trichloroethylene; disp. in xylene; Density: 0.55–0.65 g/cm^3; Visc.: ≈ 200 cps (@ 70 C); M.P.: 52–54 C; HLB: 18.7; Cloud Pt.: 76–77 C; Flash Pt.: ≫ 100 C; Stability: Stable to weak acids, bases, oxidizing and reducing agents; Surface Tension: 48 dynes/cm; Biodegradable; pH: 6–7 (1% aq.)

Composition: 100% active ingred.
Toxicity: Degreasing effect on skin in concentrated form
Containers: 25 kg net sacks

Berol 09. Berol Kemi AB
Chemical Description: Alkyl phenol ethylene oxide adduct
Category: Detergent, wetting agent, emulsifier, dispersant, nonionic
Applications: Detergent; textile wetting and cleaning agent for natural and synthetic fibers, esp. for washing virgin wool; emulsifier for waxes, oil, solvents; emulsion polymerization; dispersant for pigments and tinting pastes in paints and varnishes; emulsifier for paint resins
Form: Clear liquid; Color: < 150 (Hazen)
Properties: Solubility: Sol. in water, ethanol, xylene, trichloroethylene; disp. in paraffin oil, white spirit, light fuel oil; Density: 1.05 g/cm^3 @ 20 C; Visc.: ≈ 350 cps @ 20 C; HLB: 13.3; Cloud Pt.: 52–58 C; Clear Pt.: ≈ 7 C; Flash Pt.: ⩾ 100 C; Pour Pt.: ≈ 5 C; Stability: Stable to weak acids, bases, oxidizing and reducing agents; Surface Tension: 33 dynes/cm; Biodegradable under normal biological sewage purification conditions; pH: 6–7 (1% aq.)
Composition: 100% active ingred.
Toxicity: Degreasing effect on skin in concentrated form
Containers: 200 kg net drums

Berol 17. Berol Kemi AB
Chemical Description: Resin acid ethylene oxide adduct
Category: Lubricant, emulsifier, nonionic
Applications: Textile industry as a spinning oil for natural and synthetic fibers; with alkyl aryl sulphonates as emulsifier for biocides and aromatic solvents
Form: Clear liquid; Color: 15 (Gardner)
Properties: Solubility: Sol. in water, ethanol, trichloroethylene, xylene; disp. in acetone, propylene glycol; Density: 1.11 g/cm^3 @ 20 C; Visc.: 50 cps @ 20 C; HLB: 14; Cloud Pt.: 45 C (1% in 10% NaCl sol'n.); Clear Pt.: ≈ 17 C; Flash Pt.: ⩾ 100 C; Pour Pt.: ≈ 15 C; Stability: Stable to weak acids and bases; Surface Tension: 38 dynes/cm; pH: 7.5–8.5 (1% aq.)
Composition: 100% active ingred.
Toxicity: Degreasing effect on skin in concentrated form

Berol 26. Berol Kemi AB
Chemical Description: Alkyl phenol ethylene oxide adduct
Category: Emulsifier, detergent additive, base for sulfation, nonionic
Applications: Emulsifier for waxes, oils, solvents; detergent additive to improve degreasing properties; base for sulfation to produce ether sulfate for dishwashing liquids
Form: Clear liquid; Color: < 250 (Hazen)
Properties: Solubility: Sol. in ethanol, xylene, trichloroethylene, paraffin oil, white spirit, light fuel oil; disp. in water; Density: 1.02

g/cm^3 @ 20 C; Visc.: ≈ 350 cps @ 20 C; HLB: 8.9; Cloud Pt.: 42–46 C; Clear Pt.: < 0 C; Flash Pt.: ≫ 100 C; Pour Pt.: < 0 C; Stability: Stable to weak acids, bases, oxidizing and reducing agents; Biodegradable in normal conditions in biological sewage purification; pH: 6–7 (1% aq. disp.)
Composition: 100% active ingred.
Toxicity: Degreasing effect on skin in concentrated form
Containers: 200 kg net drums

Berol 048. Berol Kemi AB
Chemical Description: Alcohol ethylene oxide adduct
Category: Wetting agent, detergent, nonionic
Applications: Wetting and cleaning agent in industrial cleaners containing strong acids and alkalis; pickling and phosphating of metals; rust removal; degreaser for cleaning bath for solution reuse; textile wetting agent (adhesive removal, bleaching); wetting and dispersing talc and magnesium stearate
Form: Clear liquid; Color: < 100 (Hazen)
Properties: Solubility: Sol. in water and ethanol; disp. in propylene glycol, benzene, trichloroethylene, paraffin oil, white spirit; Density: 1.02 g/cm^3 @ 20 C; Visc.: 170 cps @ 20 C; HLB: 13.5; Cloud Pt.: ≈ 70 C; Clear Pt.: ≈ 4 C; Flash Pt.: ≫ 100 C; Pour Pt.: ≈ 0 C; Stability: Stable to strong acids and bases; Surface Tension: 27 dynes/cm; pH: 6–7 (1% aq.)
Composition: 85% active ingred.
Toxicity: Degreasing effect on skin in concentrated form

Berol 049. Berol Kemi AB
Chemical Description: n-primary alcohol ethylene oxide adduct
Category: Detergent, nonionic
Applications: Liquid cleaners, dishwashing liquids, general purpose high foam cleaners
Form: Clear liquid; Color: < 100 (Hazen)
Properties: Solubility: Sol. in water, ethanol, xylene, trichloroethylene, light fuel oil, white spirit (hazy); Density: 0.99 g/cm^3 @ 20 C; Visc.: ≈ 40 cps @ 20 C; HLB: 12.0; Cloud Pt.: 55–60 C; Clear Pt.: ≈ 15 C; Flash Pt.: ≫ 100 C; Pour Pt.: ≈ 7 C; Stability: Stable to weak acids, bases, oxidizing and reducing agents; Surface Tension: ≈ 30 dynes/cm; Biodegradable; pH: 6–7 (1% aq.)
Composition: 100% active ingred.
Toxicity: Degreasing effect on skin in concentrated form

Berol 050. Berol Kemi AB
Chemical Description: Fatty alcohol ethoxylate
Category: Sulfonation raw material, emulsifier, nonionic
Applications: Sulfonation raw material to produce alkyl ether sulfates for hair shampoos, bubble baths, washing-up liquids, etc.; emulsifier for waxes, oils, solvents
Form: Clear liquid; Color: < 100 (Hazen)
Properties: Solubility: Sol. in ethanol, xylene, trichloroethylene, paraffin oil, white spirit, light fuel

oil; disp. in water; Density: 0.93 g/cm^3 @ 20 C; Visc.: 30 cps @ 20 C; HLB: 8.4; Cloud Pt.: 48–52 C; Clear Pt.: ≈ 20 C; Flash Pt.: ≫ 100 C; Pour Pt.: ≈ 0 C; Stability: Stable to weak acids, bases, oxidizing and reducing agents; Biodegradable; pH: 6–7 (1% aq. disp.)
Composition: 100% active ingred.
Toxicity: Degreasing effect on skin in concentrated form
Containers: 185 kg net drums

Berol 055. Berol Kemi AB
Chemical Description: n-primary alcohol ethylene oxide adduct
Category: Sulfonation raw material,emulsifier, nonionic
Applications: Sulfonation raw material to produce alkyl ether sulfates for hair shampoos, bubble baths, etc.; emulsifier for waxes, oils, solvents
Form: Clear liquid; Color: < 100 (Hazen)
Properties: Solubility: Sol. in ethanol, xylene, trichloroethylene, paraffin oil, white spirit, light fuel oil; disp. in water; Density: 0.90 g/cm^3 @ 20 C; Visc.: ≈ 30 cps @ 20 C; HLB: 6.5; Cloud Pt.: 35–40 C; Clear Pt.: ≈ 15 C; Flash Pt.: ≫ 100 C; Pour Pt.: ≈ 0 C; Stability: Stable to weak acids, bases, oxidizing and reducing agents; Biodegradable; pH: 6–7 (1% aq. disp.)
Composition: 100% active ingred.
Toxicity: Degreasing effect on skin in concentrated form
Containers: 185 kg net drums

Berol 057. Berol Kemi AB
Chemical Description: Fatty alcohol ethoxylate
Category: Emulsifier, nonionic
Applications: Emulsion polymerization, alone or in combination with less hydrophilic nonionics or with anionic surfactants to obtain a very small particle size
Form: Soft wax; Color: < 300 (Hazen)
Properties: Solubility: Sol. in water, ethanol, xylene, trichloroethylene; disp. in paraffin oil, white spirit; Density: 1.00 g/cm^3 @ 60 C; Visc.: ≈ 60 cps @ 40 C; HLB: 15.5; Cloud Pt.: 74–76 C (1% in 10% NaCl sol'n.); Flash Pt.: ≫ 100 C; Pour Pt.: ≈ 30 C; Stability: Stable to weak acids, bases, oxidizing and reducing agents; Surface Tension: 39 dynes/cm; Biodegradable; pH: 6–7 (1% aq.)
Composition: 100% active ingred.
Toxicity: Degreasing effect on skin; otherwise nontoxic
Containers: 200 kg net drums

Berol 059. Berol Kemi AB
Chemical Description: n-primary alcohol ethylene oxide adduct
Category: Detergent, nonionic
Applications: Liquid cleaners, dishwashing liquids, etc.
Form: Clear liquid; Color: < 100 (Hazen)
Properties: Solubility: Sol. in water, ethanol, xylene, trichloroethylene, light fuel oil, white spirit (hazy); Density: 0.97 g/cm^3 @ 20 C; Visc.: ≈ 35 cps; HLB: 11.3; Cloud Pt.: 35–40 C (1% aq.) @ 20 C; Clear Pt.: ≈ 13 C; Flash Pt.: ≫ 100 C; Pour Pt.: ≈ 5 C; Stability: Stable to weak acid, bases, oxidiz-

ing and reducing agents; Surface Tension: ≈ 29 dynes/cm; Biodegradable; pH: 6–7 (1% aq.)
Composition: 100% active ingred.
Toxicity: Degreasing effect on skin in concentrated form

Berol 065. Berol Kemi AB
Chemical Description: Fatty alcohol ethoxylate
Category: Detergent, nonionic
Applications: Low-foam washing machine detergents, esp. excellent grease and pigment removal; detergents for spray drying
Form: Soft wax; Color: White
Properties: Solubility: Sol. in water, ethanol, xylene, trichloroethylene; disp. in paraffin oil, white spirit, light fuel oil; Density: 0.98 g/cm^3 @ 40 C; Visc.: ≈ 30 cps @ 50 C; HLB: 12.6; Cloud Pt.: 78–82 C (1% aq.); Clear Pt.: ≈ 44 C; Flash Pt.: ≫ 100 C; Pour Pt.: ≈ 33 C; Stability: Stable to weak acids, bases, oxidizing and reducing agents; Surface Tension: 34 dynes/cm; Biodegradable; pH: 6–7 (1% aq.)
Composition: 100% active ingred.
Toxicity: Degreasing effect on skin in concentrated form
Containers: 200 kg net wt drums

Berol 79. Berol Kemi AB
Category: Emulsifier, nonionic
Applications: In emulsifiable solvents for cleaning, degreasing auto parts, workshop floors, car bodies, etc.; antistatic and emulsifiable spinning oil for textile industry
Form: Clear liquid; Color: Yellow
Properties: Solubility: Sol. in ethanol, benzene, trichloroethylene, paraffin oil, light mineral oil; disp. in water; Density: 0.97 g/cm^3 @ 20 C; Pour Pt.: 0 C
Composition: 100% active ingred.
Containers: 200 kg net drums

Berol 081 Flakes. Berol Kemi AB
Chemical Description: Fatty alcohol ethoxylate
Category: Detergent, emulsifier, dispersant, stabilizer, nonionic
Applications: Low foam detergents in powder form; emulsifier for waxes; dispersant for pigments; latice stabilizer
Form: Flakes; Color: White
Properties: Solubility: Sol. in water, ethanol, trichloroethylene; disp. in xylene; Density: 0.55–0.65 g/cm^3; M.P.: 46–50 C; HLB: 16.7; Cloud Pt.: 75–76 C (1% in 10% NaCl sol'n.); Flash Pt.: ≫ 100 C; Stability: Stable to weak acids, bases, oxidizing and reducing agents; Surface Tension: 43 dynes/cm; Biodegradable; pH: 6–7 (1% aq.)
Composition: 100% active ingred.
Toxicity: Degreasing effect on skin in concentrated form
Containers: 25 kg net sacks

Berol 081 Powder. Berol Kemi AB
Chemical Description: Fatty alcohol ethoxylate
Category: Detergent, emulsifier, dispersant, stabilizer, nonionic
Applications: Low foam powder detergents; emulsifier for waxes; dispersing agent for pigments; latice stabilizer
Form: Powder; Color: White

Properties: Solubility: Sol. in water, ethanol, trichloroethylene; disp. in xylene; Density: 0.55–0.65 g/cm^3; M.P.: 46–50 C; HLB: 16.7; Cloud Pt.: 75–76 C; Flash Pt.: $\gg$ 100 C; Stability: Stable to weak acids, bases, oxidizing and reducing agents; Surface Tension: 43 dynes/cm; Biodegradable; pH: 6–7 (1% aq.)
Composition: 100% active ingred.
Toxicity: Degreasing effect on skin in concentrated form
Containers: 25 kg net sacks

Berol 087. Berol Kemi AB
Chemical Description: n-primary alcohol surfactant
Category: Detergent, nonionic
Applications: Low foam detergent for machine dishwashing, high pressure spraying and brushing; combined with a phosphate ester to increase the solubility at high temperatures and strong electrolyte concentrations
Form: Clear liquid; Color: 75 (Hazen)
Properties: Solubility: Sol. in water, ethanol, trichloroethylene; disp. in xylene; Density: 0.997 g/cm^3 @ 20 C; Visc.: 85 cps @ 20 C; HLB: $\approx$ 11.5; Cloud Pt.: 42 C; Clear Pt.: 8 C; Flash Pt.: $\gg$ 100 C; Pour Pt.: –1 C; Stability: Stable to weak acids, bases, oxidizing and reducing agents; Surface Tension: 31.9 dynes/cm; Biodegradable under normal sewage treatment conditions; pH: 5–7 (1% aq.)
Composition: 100% active ingred.
Toxicity: Degreasing effect on skin in concentrated form

Berol 106. Berol Kemi AB
Chemical Description: Alkyl glyceride ethylene oxide adduct
Category: Emulsifier, nonionic
Applications: O/W emulsifier; in combination with more hydrophobic surfactants for biocides and pesticides
Form: Cloudy liquid; Color: $<$ 4 (Gardner)
Properties: Solubility: Sol. in water, ethanol, propylene glycol, xylene, trichloroethylene; Density: 1.055 g/cm^3; Visc.: $\approx$ 650 cps; HLB: $\approx$ 15; Cloud Pt.: 54–58 C; Clear Pt.: $\approx$ 25 C; Flash Pt.: $\gg$ 100 C; Pour Pt.: $\approx$ 15 C; Stability: Stable to weak acids and bases; Surface Tension: 44 dynes/cm; Biodegradable; pH: 5–7 (1% aq.)
Composition: 100% active ingred.
Toxicity: Degreasing effect on skin in concentrated form

Berol 121. Berol Kemi AB
Chemical Description: Fatty alcohol ethoxylate
Category: Detergent, nonionic
Applications: Liquid detergents, dishwashing liquids, general purpose high foam cleaners (in combination with anionics, e.g., alkyl aryl sulfonate or alkyl ether sulfate)
Form: Clear/opal liquid; Color: $<$ 100 (Hazen)
Properties: Solubility: Sol. in ethanol, white spirit, light fuel oil; opaquely sol. in water and xylene; disp. in trichloroethylene, paraffin oil; Density: 0.97 g/cm^3 @ 20 C; HLB: 11.0; Cloud Pt.: 75–80 C (in 15% n-propanol); Clear Pt.: $\approx$

26 C; Flash Pt.: ≫ 100 C; Pour Pt.: ≈ 20 C; Stability: Stable to weak acids, bases, oxidizing and reducing agents; Surface Tension: 29 dynes/cm; Biodegradable; pH: 6–7 (1% aq.)
Composition: 97.5% active ingred.
Toxicity: Degreasing effect on skin in concentrated form
Containers: 200 kg net drums

Berol 151. Berol Kemi AB
Chemical Description: Fatty acid alkanolamide type based on lauric/myristic acid diethanolamide
Category: Foam stabilizer and modifier, dispersant, nonionic
Applications: Foam stabilizer and booster in hair shampoos, bubble baths, cleaners, etc. in combination with alkyl sulfates and alkyl ether sulfates; viscosity booster; dispersant for calcium soaps, liquid hand soap
Form: Clear liquid; Color: ⩽ 200 (Hazen)
Properties: Solubility: Sol. in ethanol, propylene glycol, xylene; disp. in water, white spirit, paraffin oil; Density: 0.98 g/cm^3 @ 20 C; Visc.: 1100 cps @ 20 C; HLB: ≈ 12; Clear Pt.: ≈ 3 C; Flash Pt.: ≫ 100 C; Pour Pt.: ≈ –5 C; Stability: Stable to weak acids and bases; Surface Tension: 29.5 dynes/cm; Biodegradable; pH: ≈ 9.5 (1% aq.)
Composition: 100% active ingred.
Toxicity: Degreasing effect on skin in concentrated form

Berol 160. Berol Kemi AB
Chemical Description: Fatty alcohol ethoxylate
Category: Detergent, emulsifier, nonionic
Applications: Liquid detergents; emulsifier for waxes, oils, solvents
Form: Clear liquid; Color: < 100 (Hazen)
Properties: Solubility: Sol. in water, ethanol, xylene, trichloroethylene, light fuel oil; disp. in white spirit; Density: 0.98 g/cm^3 @ 20 C; Visc.: ≈ 50 cps @ 20 C; HLB: 11.8; Cloud Pt.: 40–50 C; Clear Pt.: ≈ 19 C; Flash Pt.: ≫ 100 C; Pour Pt.: ≈ 15 C; Stability: Stable to weak acids, bases, oxidizing and reducing agents; Surface Tension: 29 dynes/cm; Biodegradable; pH: 6–7 (1% aq.)
Composition: 100% active ingred.
Toxicity: Degreasing effect on skin in concentrated form
Containers: 200 kg net drums

Berol 173. Berol Kemi AB
Chemical Description: Fatty alcohol ethoxylate
Category: Detergent, nonionic
Applications: Low foaming machine washing detergents; pigment and fatty soil removal over wide temperature range for natural and synthetic fabrics; spray dried and mechanically mixed detergent powders
Form: Paste; Color: White
Properties: Solubility: Sol. in xylene, trichloroethylene, light fuel oil; opaquely sol. in water and ethanol; Density: 0.99 g/cm^3 @ 20 C; Visc.: ≈ 30 cps @ 50 C; HLB: 12.5; Cloud Pt.: 65–70 C (1% aq.); Clear Pt.: ≈ 44 C; Flash Pt.: ≫ 100 C; Pour Pt.: ≈ 25 C;

Stability: Stable to weak acids, bases, oxidizing and reducing agents; Surface Tension: 33 dynes/cm (0.1% sol'n.); pH: 6–7 (1% aq.); Easily biodegradable
Composition: 100% active ingred.
Toxicity: Degreasing effect on skin in concentrated form
Containers: 200 kg net drums

Berol 185. Berol Kemi AB
Chemical Description: Modified fatty alcohol ethoxylate
Category: Detergent, nonionic
Applications: Concentrated liquid detergents; dishwashing liquids and general purpose cleaners with high foaming and cleaning; strong pigment and fatty soil removal; textile washing processes
Form: Clear liquid; Color: Colorless; Odor: Less tendency than similar products to unpleasant odor > 30 C
Properties: Solubility: Sol. in water, ethanol; disp. in xylene, trichloroethylene, white spirit, light fuel oil; Density: 1.02 g/cm^3 @ 20 C; Visc.: ≈ 100 cps @ 20 C; HLB: 13.5; Cloud Pt.: 65–70 C; Clear Pt.: < 5 C; Flash Pt.: ≫ 100 C; Pour Pt.: < 0 C; Stability: Stable to weak acids, bases, oxidizing and reducing agents; Surface Tension: 31.5 dynes/cm (0.1% sol'n.); pH: 6–7 (1% aq.); Easily biodegradable
Composition: 90% active ingred.
Toxicity: Degreasing effect on skin in concentrated form
Containers: 200 kg net drums

Berol 186. Berol Kemi AB
Chemical Description: Modified fatty alcohol ethoxylate
Category: Detergent, nonionic
Applications: Concentrated liquid detergents; dishwashing liquids and general purpose cleaners with high foaming and cleaning; strong pigment and fatty soil removal; textile washing processes
Form: Clear/opal liquid; Color: < 100 (Hazen)
Properties: Solubility: Sol. in water, ethanol, xylene, trichloroethylene, white spirit, light fuel oil: Density: 1.01 g/cm^3 @ 20 C; Visc: ≈ 100 cps @ 20 C; HLB: 13.5; Cloud Pt.: 65–70 C (1% aq.); Clear Pt.: ≈ 18 C; Flash Pt.: ≫ 100 C; Pour Pt.: ≈ 10 C; Stability: Stable to weak acids, bases, oxidizing and reducing agents; Surface Tension: 31.5 dynes/cm (0.1% sol'n.); pH: 6–7 (1% aq.); Biodegradable
Composition: 100% active ingred.
Toxicity: Degreasing effect on skin in concentrated form
Containers: 200 kg net drums

Berol 196. Berol Kemi AB
Chemical Description: Alkyl glyceride ethylene oxide adduct type
Category: Emulsifier, nonionic
Applications: Emulsifier in connection with foodstuffs; superfatting agent in bath oils, hand cleaners
Form: Clear liquid; Color: ⩽5 Gardner;
Properties: Solubility: Sol. in ethanol, propylene glycol, trichloroethylene, white spirity; opaquely sol. in water· Density: 1.020 g/cm^3 @ 20 C; Visc.: ≈ 500 cps @

20 C; HLB: 9; Cloud Pt.: ≈ 58 C; Clear Pt.: < 0 C; Flash Pt.: ≫ 100 C; Pour Pt.: – 15 C; Stability: Stable to weak acids and bases; pH: 5–7 (1% aq.); Biodegradable
Composition: 100% active ingred.
Toxicity: Degreasing effect on skin in concentrated form

Berol 252. Berol Kemi AB
Chemical Description: Alkyl phenol ethylene oxide adduct type
Category: Emulsifier, nonionic
Applications: Emulsifier in emulsion polymerization; combined with anionic surfactants such as alkyl aryl sulfonates, alkyl and alkyl ether sulfates
Form: Clear liquid; Color: Max. 200 (Hazen)
Properties: Solubility: Sol. in water, ethanol; Density: 1.09 g/cm^3 @ 20 C; Visc.: 1000 cps @ 20 C; Cloud Pt.: 77 C; Clear Pt.: ≈ 17 C; Flash Pt.: ≫ 100 C; Pour Pt.: ≈ 12 C; Stability: Stable to weak acids, bases, oxidizing and reducing agents; Surface Tension: 41 dynes/cm; pH: 6–7 (1% aq.); Biodegradable
Composition: 70% active ingred.
Toxicity: Degreasing effect on skin in concentrated form

Berol 259. Berol Kemi AB
Chemical Description: Alkyl phenol ethylene oxide adduct type
Category: Stabilizer, nonionic
Applications: Foaming in industrial processes; intermediary product of lubricating oil additives and as a stabilizer for rubber
Form: Clear liquid; Color: ⩽ 500 (Hazen)
Properties: Solubility: Sol. in ethanol, xylene, trichloroethylene, paraffin oil, white spirit, light fuel oil; dispersible in water; Density: 1.01 g/cm^3 @ 20 C; Visc.: ≈ 550 cps @ 20 C; HLB: 5.7; Cloud Pt.: 50–54 C; Clear Pt.: < 0 C; Flash Pt.: ≫ 100 C; Pour Pt.: < 0 C; Stability: Stable to weak acids, bases, oxidizing and reducing agents; pH: 6–7; Biodegradable
Composition: 100% active ingred.
Toxicity: Degreasing effect on skin in concentrated form
Containers: 200 kg net drums

Berol 263. Berol Kemi AB
Chemical Description: Alkyl phenol ethylene oxide adduct type
Category: Detergent, nonionic
Applications: Detergents and cleaners; dispersing agent for pigments
Form: Soft wax; Color: < 200 (Hazen)
Properties: Solubility: Sol. in water, ethanol, xylene, trichloroethylene; dispersible in paraffin oil, white spirit; Density: 1.06 g/cm^3 @ 30 C; Visc.: ≈ 325 cps @ 25 C; HLB: 14.7; Cloud Pt.: 92–94 C (1% aq. sol'n.); Clear Pt.: ≈ 24 C; Flash Pt.: ≫ 100 C; Pour Pt.: ≈ 20 C; Stability: Stable to weak acids, bases, oxidizing and reducing agents; Surface Tension: 36 dynes/cm (0.1% sol'n.); pH: 6–7 (1% aq.); Biodegradable
Composition: 100% active ingred.
Toxicity: Degreasing effect on skin in concentrated form

Berol 267. Berol Kemi AB
Chemical Description: Alkyl phenol ethylene oxide adduct type
Category: Emulsifier, detergent, nonionic
Applications: Used as an emulsifier for waxes, oils and solvents; dishwashing compounds; washing of virgin wool
Form: Clear liquid; Color: < 200 (Hazen)
Properties: Solubility: Sol. in ethanol, xylene, trichloroethylene; disp. in water, paraffin oil, white spirit, light fuel oil; Density: 1.04 g/cm^3 @ 20 C; Visc.: $\approx$ 350 cps @ 20 C; HLB: 12.3; Cloud Pt.: $\approx$ 10 C (1% aq.); Clear Pt.: < 0 C; Flash Pt.: $\gg 100$ C; Pour Pt.: < 0 C; Stability: Stable to weak acids, bases, oxidizing and reducing agents; Surface Tension: 31 dynes/cm; pH: 6–7; (1% aq.); Biodegradable
Composition: 100% active ingred.
Toxicity: Degreasing effect on skin in concentrated form
Containers: 200 kg net drums

Berol 269. Berol Kemi AB
Chemical Description: Dialkyl phenol ethylene oxide adduct type
Category: Emulsifier, nonionic
Applications: Emulsifier for heavy paraffin and aromatic oils
Form: Opal liquid; Color: < 500 (Hazen)
Properties: Solubility: Sol. in ethanol, xylene, trichloroethylene, light fuel oil; disp. in water and white spirit; Density: 1.02 g/cm^3 @ 20 C; Visc.: $\approx$ 550 cps @ 20 C; HLB: 10.5; Cloud Pt.: $\approx$ 60 C (in 15% *n*-propanol); Flash Pt.: $\gg$ 100 C; Pour Pt.: < 0 C; Stability: Stable to weak acids, bases, oxidizing and reducing agents; pH: 6–7 (1% aq.)
Composition: 100% active ingred.
Toxicity: Degreasing effect on skin in concentrated form
Containers: 200 kg net drums

Berol 272. Berol Kemi AB
Chemical Description: Dialkyl phenol ethylene oxide adduct type
Category: Detergent, nonionic
Applications: Low-foaming washing machine detergents in powder form; good stability for spray drying
Form: Paste; Color: Light yellow
Properties: Solubility: Sol. in water, ethanol, xylene, trichloroethylene; disp. in paraffin oil; Density: 1.04 g/cm^3 @ 20 C; Visc.: $\approx$ 300 cps @ 35 C; M.P.: $\approx$ 34 C; HLB: 13.4; Cloud Pt.: $\approx$ 60 C (1% aq.); Flash Pt.: $\gg$ 100 C; Pour Pt.: $\approx$ 23 C; Stability: Stable to weak acids, bases, oxidizing and reducing agents; Surface Tension: 30 dynes/cm (0.1% sol'n.); pH: 6–7 (1% aq.)
Composition: 100% active ingred.
Toxicity: Degreasing effect on skin in concentrated form
Containers: 200 kg net drums

Berol 274. Berol Kemi AB
Chemical Description: Alkyl phenol ethylene oxide adduct type
Category: Emulsifier, nonionic
Applications: Emulsifier in emulsion polymerization; increases sta-

bility of the latice
Form: Soft wax; Color: < 200 (Hazen)
Properties: Solubility: Sol. in water, ethanol, xylene, trichloroethylene; Density: 1.07 g/cm^3 @ 50 C; Visc.: ≈ 200 cps @ 50 C; HLB: ≈ 17; Cloud Pt.: 77 C (1% in 10% NaCl sol'n.); Clear Pt.: ≈ 44 C; Flash Pt.: ≫ 100 C; Pour Pt.: ≈ 41 C; Stability: Stable to weak acids, bases, oxidizing and reducing agents; Surface Tension: 41 dynes/cm (0.1% sol'n.); pH: 6–7 (1% aq.); Easily biodegradable under normal biological sewage purification conditions
Composition: 100% active ingred.
Toxicity: Degreasing effect on skin in concentrated form

Berol 277. Berol Kemi AB
Chemical Description: Alkyl phenol ethylene oxide adduct type
Category: Emulsifier, nonionic
Applications: Emulsion polymerization; increases stability of latice
Form: Clear liquid; Color: < 200 (Hazen)
Properties: Solubility: Sol. in water and ethanol; Density: 1.09 g/cm^3 @ 20 C; Visc.: 1000 cps @ 20 C; HLB: 17; Cloud Pt.: 77 C (1% in 10/ NaCl sol'n.); Clear Pt.: ≈ 17 C; Flash Pt.: ≫ 100 C; Pour Pt.: ≈ 12 C; Stability: Stable to weak acids, bases, oxidizing and reducing agents; Surface Tension: 41 dynes/cm (0.1% sol'n.); pH: 6–7 (1% aq.); Easily biodegradable under normal biological sewage purification conditions
Composition: 70% active ingred.
Toxicity: Degreasing effect on skin in concentrated form

Berol 278. Berol Kemi AB
Chemical Description: Alkyl phenol ethylene oxide adduct type
Category: Emulsifier, nonionic
Applications: Emulsion polymerization; combined with anionics to obtain a very small particle size
Form: Clear liquid; Color: < 200 (Hazen)
Properties: Solubility: Sol. in water and ethanol; Density: 1.08 g/cm^3 @ 20 C; Visc.: ≈ 625 cps @ 20 C; HLB: 15.2; Cloud Pt.: 97–99 C (1% aq.); Clear Pt.: ≈ 1 C; Flash Pt.: ≫ 100 C; Pour Pt.: ≈ 0 C; Stability: Stable to weak acids, bases, oxidizing and reducing agents; Surface Tension: 37 dynes/cm (0.1% sol'n.); pH: 6–7 (1% aq.); Easily biodegradable under normal biological sewage purification conditions
Composition: 80% active ingred.
Toxicity: Degreasing effect on skin in concentrated form
Containers: 200 kg net drums

Berol 281. Berol Kemi AB
Chemical Description: Alkyl phenol ethylene oxide type
Category: Emulsifier, nonionic
Applications: Emulsion polymerization; combined with anionics to obtain a very small particle size
Form: Clear viscous liquid; Color: < 200 (Hazen)
Properties: Solubility: Sol. in water and ethanol; Density: 1.09 g/cm^3 @ 20 C; Visc.: ≈ 625 cps @ 20 C; HLB: 16.0; Cloud Pt.: 73 C

(1% in 10% NaCl sol'n.); Clear Pt.: ≈ 14 C; Flash Pt.: ≫ 100 C; Pour Pt.: ≈ 7 C; Stability: Stable to weak acids, bases, oxidizing and reducing agents; Surface Tension: 39 dynes/cm (0.1% sol'n.); pH: 6–7 (1% aq.); Easily biodegradable under normal biological sewage purification conditions
Composition: 80% active ingred.
Toxicity: Degreasing effect on skin in concentrated form
Containers: 200 kg net drums

Berol 282. Berol Kemi AB
Chemical Description: Alkyl phenol ethylene oxide adduct type
Category: Emulsifier, nonionic
Applications: Emulsion polymerization; with anionics to obtain a very small particle size; increases stability of the latice
Form: Soft paste; Color: < 200 (Hazen)
Properties: Solubility: Sol. in water and ethanol; Density: 1.09 g/cm^3 @ 30 C; Visc.: 625 cps @ 20 C; HLB: 17; Cloud Pt.: 77 C (1% in 10% NaCl sol'n.); Clear Pt.: ≈ 24 C; Flash Pt.: ≫ 100 C; Pour Pt.: ≈ 21 C; Stability: Stable to weak acids, bases, oxidizing and reducing agents; Surface Tension: 41 dynes/cm (0.1% sol'n.); pH: 6–7 (1% aq.); Easily biodegradable under normal biological sewage purification conditions
Composition: 80% active ingred.
Toxicity: Degreasing effect on skin in concentrated form

Berol 291. Berol Kemi AB
Chemical Description: Alkyl phenol ethylene oxide adduct type
Category: Emulsifier, nonionic
Applications: Emulsion polymerization; with anionics to obtain a very small particle size; increase stability of the latice
Form: Hard wax; Color: < 200 (Hazen)
Properties: Solubility: Sol. in water, ethanol, xylene, trichloroethylene; Density: ≈ 1.07 g/cm^3 @ 60 C; Visc.: ≈ 300 cps @ 60 C; HLB: 18.2; Cloud Pt.: 77 C (1% in 10% NaCl sol'n.); Clear Pt.: ≈ 53 C; Flash Pt.: ≫ 100 C; Pour Pt.: ≈ 51 C; Stability: Stable to weak acids, bases, oxidizing and reducing agents; Surface Tension: 45 cm; pH: 6-7 (1% aq.)
Composition: 100% active ingred.
Toxicity: Degreasing effect on skin in concentrated form
Containers: 100 kg net drums

Berol 292. Berol Kemi AB
Chemical Description: Alkyl phenol ethylene oxide adduct type
Category: Emulsifier nonionic
Applications: Emulsion polymerization; with anionics to obtain a very small particle size
Form: Soft wax; Color: < 200 (Hazen)
Properties: Solubility: Sol. in water, ethanol, xylene, trichloroethylene, disp. in white spirit; Density: 1.07 g/cm^3 @ 40 C; Visc.: ≈ 200 cps @ 40 C; HLB: 16.0; Cloud Pt.: 73 C (1% in 10% NaCl sol'n'); Clear Pt.: ≈ 35 C; Flash Pt.: ≫ 100 C; Pour Pt.: ≈ 33 C; Stability: Stable to weak acids, bases, oxidizing and reducing a-

gents; Surface Tension: 39 dynes/cm; pH: 6–7 (1% aq.); Biodegradable
Composition: 100% active ingred.
Toxicity: Degreasing effect on skin in concentrated form
Containers: 100 kg net drums

Berol 295. Berol Kemi AB
Chemical Description: Alkyl phenol ethylene oxide adduct type
Category: Emulsifier, nonionic
Applications: Emulsion polymerization; with anionics to obtain a very small particle size; increases stability of the latice
Form: Clear liquid; Color: < 300 (Hazen)
Properties: Solubility: Sol. in water, ethanol; Density: 1.10 g/cm^3 @ 20 C Visc.: 1500 cps @ 20 C; HLB: 17.8; Cloud Pt.: 77 C (1% in 10% NaCl sol'n.); Clear Pt.: ≈ 19 C; Flash Pt.: ≫ 100 C; Pour Pt.: ≈ 14 C; Stability: Stable to weak acids, bases, oxidizing and reducing agents; Surface tension: 43 dynes/cm; pH: 5–7 (1% aq.)
Composition: 70% active ingred.
Toxicity: Degreasing effect on skin in concentrated form
Containers: 200 kg net drums

Berol 296. Berol Kemi AB
Chemical Description: Alkyl phenol ethylene oxide adduct type
Category: Emulsifier, nonionic
Applications: Emulsion polymerization; with anionics to obtain a very small particle size; biocides and petroleum solvents
Form: Soft paste; Color: < 200 (Hazen)
Properties: Solubility: Sol. in water, ethanol, xylene, trichloroethylene; disp. in light fuel oil; Density: 1.07 g/cm^3 @ 30 C; Visc.: ≈ 375 cps @ 30 C; HLB: 15.2; Cloud Pt.: 97–99 C (1% aq.); Clear Pt.: ≈ 27 C; Flash Pt.: ≫ 100 C; Pour Pt.: ≈ 26 C; Stability: Stable to weak acids, bases, oxidizing and reducing agents; Surface tension: 37 dynes/cm; pH: 6–7 (1% aq.); Biodegradable
Composition: 100% active ingred.
Toxicity: Degreasing effect on skin in concentrated form
Containers: 200 kg net drums

Berol 297. Berol Kemi AB
Chemical Description: Alkyl phenol ethylene oxide adduct type
Category: Emulsifier, nonionic
Applications: Used for various types of waxes, oils and solvents
Form: Clear liquid; Color: < 250 (Hazen)
Properties: Solubility: Sol. in ethanol, xylene, trichloroethylene, paraffin oil, white spirit, light fuel oil; disp. in water; Density: 1.03 g/cm^3 @ 20 C; Visc.: ≈ 350 cps @ 20 C; HLB: 9.5; Cloud Pt.: 48–54 C; Clear Pt.: < 0 C; Flash Pt.: ≫ 100 C; Pour Pt.: < 0 C; Stability: Stable to weak acids, bases, oxidizing and reducing agents; pH: 6–7 (1% aq.); Biodegradable
Composition: 100% active ingred.
Toxicity: Degreasing effect on skin in concentrated form

Berol 305. Berol Kemi AB
Chemical Description: Amine oxide

Category: Foam controller, nonionic
Form: Clear liquid; Color: Pale yellow
Properties: Solubility: Sol. in water, ethanol, acetone; Density: 0.96 @ 20 C; Cloud Pt.: 95 C (9% NaOH sol'n.); Clear Pt.: - 1 C; Pour Pt.: 0 C; Stability: Stable to acid and alkaline solutions; Surface Tension: 31 dynes/cm
Composition: 28% active ingred.
Containers: 200 kg net drums

Berol 370. Berol Kemi AB
Chemical Description: Ethylene oxide/propylene oxide block polymer
Category: Detergent, rinse aid additive, nonionic
Applications: Low foaming detergent for use with strong mechanical action, spraying under pressure, e.g., machine dishwashing detergents, alkaline degreasers; rinse aid additive for dishwashers, car cleaners; windscreen washer fluids
Form: Clear liquid; Color: < 150 (Hazen)
Properties: m.w.: ≈ 1400; Solubility: Sol. in ethanol, xylene, trichloroethylene, white spirit; opaquely sol. in water; Density: 1.05 g/cm^3 @ 20 C; Visc.: ≈ 230 cps @ 20 C; Cloud Pt.: 30–34 C (1% aq.); Clear Pt.: < –20 C; Flash Pt.: ≫ 100 C; Pour Pt.: < –10 C; Stability: Stable to weak acids, bases, oxidizing, and reducing agents; Surface Tension: 41 dynes/cm (0.1% sol'n.); pH: 5–7 (1% aq.)
Composition: 100% active ingred.
Toxicity: Degreasing effect on skin in concentrated form
Containers: 200 kg net drums

Berol 371. Berol Kemi AB
Chemical Description: Ethylene oxide/propylene oxide block polymer
Category: Detergent, antistat, emollient, solvent, nonionic
Applications: Low foaming detergent for use with strong mechanical action or spraying under pressure; textile industry as water-sol. spinning oil, antistat agent (esp. for synthetic fibers); cosmetics as emollient and solvent for deodorants and antiperspirants
Form: Soft paste; Color: < 150 (Hazen)
Properties: m.w.: ≈ 2900; Solubility: Sol. in water, ethanol, xylene, trichloroethylene; Density: 1.05 g/cm^3 @ 20 C; Visc.: ≈ 700 cps @ 30 C; Cloud Pt.: 62–66 C (1% aq.); Clear Pt.: ≈ 30 C; Flash Pt.: ≫ 100 C; Pour Pt.: ≈ 28 C; Stability: Stable to weak acids, bases, oxidizing and reducing agents; Surface Tension: 38 dynes/cm (0.1% sol'n.); pH: 5–7 (1% aq.)
Composition: 100% active ingred.
Toxicity: Degreasing effect on skin in concentrated form
Containers: 200 kg net drums

Berol 372 Flakes. Berol Kemi AB
Chemical Description: Ethylene oxide/propylene oxide block polymer
Category: Emulsifier, emollient, nonionic
Applications: Emulsifier in emul-

sion polymerization; improves storage and freeze/thaw stabilities in finished latice; emollient for deodorants and antiperspirants; produces material in block form, e.g., perfume, insect repellents, etc.
Form: Flakes; Color: White
Properties: m.w.: ≈ 9000; Solubility: Sol. in water, ethanol, trichloroethylene; Density: 0.55–0.65 g/cm^3; M.P.: ≈ 50 C; Cloud Pt.: ≈ 70 C (1% in 10% NaCl sol'n.); Flash Pt.: ≫ 100 C; Stability: Stable to weak acids, bases, oxidizing and reducing agents; Surface Tension: 42 dynes/cm (0.1% sol'n.); pH: 5–7 (1% aq.)
Composition: 100% active ingred.
Toxicity: Degreasing effect on skin in concentrated form
Containers: 25 kg net sacks

Berol 374. Berol Kemi AB
Chemical Description: Ethylene oxide/propylene oxide block polymer
Category: Emulsifier, foam depressant/detergent, nonionic
Applications: Low foaming emulsifier in emulsion polymerization esp. suited to manufacture of paint, etc.; foam depressant/detergent in foodstuffs industry
Form: Clear liquid; Color: < 100 (Hazen)
Properties: m.w.: ≈ 2200; Solubility: Sol. in ethanol, xylene, trichloroethylene, white spirit; disp. in water; Density: 1.05 g/cm^3 @ 20 C; Visc.: ≈ 450 cps @ 20 C; Cloud Pt.: 24–26 C (1% aq.); Clear Pt.: < –20 C; Flash Pt.: ≫ 100 C; Pour Pt.: < –10 C; Stability: Stable to weak acids, bases, oxidizing and reducing agents; Surface Tension: 40 dynes/cm (0.1% sol'n.); pH: 5–7 (1% aq.)
Composition: 100% active ingred.
Toxicity: Degreasing effect on skin in concentrated form
Containers: 200 kg net drums

Berol 452. Berol Kemi AB
Chemical Description: Alkyl ether sulfate type consisting of the sodium salt of sulfated technical lauryl alcohol polyglycol ether
Category: Wetting and foaming agent, emulsifier, anionic
Applications: Good foaming surfactant with strong degreasing action; in combination with less strongly degreasing surfactants to produce hair shampoos, foam baths, manual dishwashing compounds, general detergents; emulsion polymerization of vinyl chloride, vinyl acetate, acrylic acids, etc.
Form: Clear liquid; Color: < 100 (Hazen)
Properties: Solubility: Sol. in water, ethanol, propylene glycol; disp. in acetone; Density: 1.04 g/cm^3 @ 20 C; Visc.: ≈ 200 cps @ 20 C; Clear Pt.: < 12 C; Flash Pt.: ≫ 100 C; Stability: Stable to weak bases and very weak acids; Surface Tension: 37 dynes/cm (0.1% sol'n.); pH: 6.5–8.0 (1% aq.); Biodegradable
Composition: 27–29% active
Toxicity: Degreasing effect on skin in concentrated form

Berol 463. Berol Kemi AB

Chemical Description: Alkyl sulfate type consisting of the diethanolamine salt of sulfated technical lauryl alcohol
Category: Foaming agent, anionic
Applications: Hair shampoos, bath foams, hand cleaners
Form: Clear liquid; Color: < 100 (Hazen)
Properties: Solubility: Sol. in water, ethanol, propylene glycol, acetone; Density: 1.050 g/cm^3 @ 20 C; Visc.: ≈ 1000 cps @ 20 C; Clear Pt.: < 15 C; Flash Pt.: ≫ 100 C; Stability: Stable to weak bases and very weak acids; Surface Tension: ≈ 30 dynes/cm (0.1% sol'n.); pH: 6.5–7.5 (1% aq.); Biodegradable
Composition: 33–35% active
Toxicity: Degreasing effect on skin in concentrated form

Berol 472. Berol Kemi AB
Chemical Description: Alkyl sulfate type consisting of sodium cetyl sulfate
Category: Emulsifier, dispersant, anionic
Applications: Emulsion polymerization of vinyl chloride, vinyl acetate, styrene/butadiene, acrylics, etc.; dispersing agent for pigments in manufacture of latex based paints and adhesives
Form: Cloudy liquid; Color: < 150 (Hazen)
Properties: Solubility: Sol. in propylene glycol; disp. in water, ethanol; Density: 1.055 g/cm^3 @ 20 C; Visc.: ≈ 20 cps @ 20 C; Flash Pt.: ≫ 100 C; Stability: Stable to weak bases and very weak acids; Surface Tension: 37 dynes/cm (0.1% sol'n.); pH: 7–8 (1% aq.); Biodegradable
Composition: 32 ± 1% active
Toxicity: Degreasing effect on skin in concentrated form

Berol 473. Berol Kemi AB
Chemical Description: Nonylphenol ether sulfate type
Category: Detergent, anionic
Applications: Detergent with good fatty soil removal, foaming properties for industrial detergents, dishwashing liquids
Form: Clear liquid; Color: < 100 (Hazen)
Properties: Density: 1.075 g/cm^3 @ 20 C; Visc.: 205 cps @ 20 C; Clear Pt.: ≈ 2 C; Flash Pt.: ≫ 100 C; Pour Pt.: ≈ 2 C; Stability: Stable to weak bases and very weak acids; pH: ≈ 3.9; Biodegradable
Composition: ≈ 30% active ingred.
Toxicity: Degreasing effect on skin in concentrated form

Berol 474. Berol Kemi AB
Chemical Description: Alkyl sulfate type consisting of the sodium salt of sulfated technical lauryl alcohol
Category: Emulsifier, anionic
Applications: Emulsion polymerization of vinyl chloride and styrene/butadiene; in combination with other surfactants in polymerization of acrylic acids and vinyl acetate
Form: Soft paste; Color: < 50 (Hazen)
Properties: Solubility: Sol. in water, ethanol, propylene glycol;

Density: 1.030 g/cm^3 @ 20 C; Clear Pt.: ≈ 25 C; Flash Pt.: ≫ 100 C; Stability: Stable to weak bases and very weak acids; Surface Tension: 25 dynes/cm (0.1% sol'n.); pH: 7–8 (1% aq.); Biodegradable
Composition: 39–41% active
Toxicity: Degreasing effect on skin in concentrated form

Berol 475. Berok Kemi AB
Chemical Description: Alkyl ether sulfate type consisting of the sodium salt of sulfated *n*-primary alcohol polyglycol ether
Category: Degreaser, dispersant, anionic
Applications: Cleaner with good pigment removal and balanced degreasing; esp. for washing fur coats, leathers, sensitive woollen materials; spray-dried household detergents; cream hair shampoos, hand cleaners; dispersant for water-based pigment pastes
Form: Soft paste; Color: < 300 (Hazen)
Properties: Solubility: Sol. in propylene glycol and water (slightly opal sol'n.); disp. in ethanol, trichloroethylene, xylene; Density: 0.97 g/cm^3 @ 20 C; Flash Pt.: ≫ 100 C; Stability: Stable to weak bases and very weak acids; Surface Tension: 37 dynes/cm (0.1% sol'n.); pH: 6.5–8.0 (1% aq.); Biodegradable
Composition: 39–41% active
Toxicity: Degreasing effect on skin in concentrated form

Berol 480. Berol Kemi AB
Chemical Description: Alkyl sulfate type consisting of the triethanol amine salt of sulfated technical lauryl alcohol
Category: Detergent, foaming agent, anionic
Applications: Hair shampoos, bath foams, hand cleaners
Form: Clear liquid; Color: < 100 (Hazen)
Properties: Solubility: Sol. in water, ethanol, propylene glycol; Density: 1.04 g/cm^3 @ 20 C; Visc.: 50 cps @ 20 C; Clear Pt.: 12 C; Flash Pt.: ≫ 100 C; Stability: Stable to weak bases and very weak acids; Surface Tension: 26 dynes/cm (0.1% sol'n.); pH: 6.5–7.5 (1% aq.); Biodegradable
Composition: 39–41% active
Toxicity: Degreasing effect on skin in concentrated form

Berol 484. Berol Kemi AB
Chemical Description: Alkyl sulfate type consisting of mono ethanol amine salt of sulfated technical lauryl alcohol
Category: Foaming, anionic
Applications: Hair shampoos, bath foams, hand cleaners
Form: Clear liquid; Color: ⩽ 200 (Hazen)
Properties: Solubility: Sol. in water, propylene glycol, acetone; disp. in ethanol; Density: 1.020 g/cm^3 @ 20 C; Visc.: 150–350 cps @ 20 C; Clear Pt.: ⩽ 15 C; Flash Pt.: ≫ 100 C; Stability: Stable to weak and strong bases, very weak acids, oxidizing and reducing agents; Surface Tension: 33 dynes/cm; pH: 6.5–7.5 (1% aq.); Bio-

degradable
Composition: 32–34% active ingred.
Toxicity: Degreasing effect on skin in concentrated form

Berol 490. Berol Kemi AB
Chemical Description: Sodium alkyl aryl sulfonate
Category: Wetting agent, foaming and emulsifier, anionic
Applications: Because of its good wetting, foaming and emulsifying properties, improved emulsifying power can be achieved when combined with other surfactants
Form: Clear liquid; Color: < 300 (Hazen)
Properties: Solubility: Sol. in water, ethanol, propylene glycol, acetone; disp. in light mineral oil, white spirit; Density: 1.037 g/cm^3 @ 20 C; Visc.: ≈ 250 cps @ 20 C; Clear Pt.: < 18 C; Flash Pt.: ≫ 100 C; Pour Pt.: ≈ 5 C; Stability: Stable to high concentrations of acids and bases; Surface Tension: 35 dynes/cm; pH: 6–8 (1% aq.); Biodegradable
Composition: 24–26% active ingred.
Toxicity: Skin and eye irritant

Berol 493. Berol Kemi AB
Chemical Description: Sodium alkyl aryl sulfonate
Category: Detergent, foaming, wetting agent, anionic
Applications: High foaming and effective washing and cleaning agents, used in textile industry as a general wetting and penetration agent
Form: Clear liquid; Color: < 150 (Hazen)
Properties: Solubility: Sol. in water, ethanol, propylene glycol; disp. in acetone, trichloroethylene; Density: 1.100 g/cm^3 @ 20 C; Visc.: ≈ 5000 cps @ 20 C; Clear Pt.: < 15 C; Flash Pt.: ≫ 100 C; Stability: Stable to strong acids and bases; Surface tension: 35 dynes/cm; pH: 6–8 (1% aq.); Biodegradable
Composition: 34–36% active

Berol 496. Berol Kemi AB
Chemical Description: Linear Sodium alkyl aryl sulfonate
Category: Detergent, foaming, wetting, penetrant, emulsifier, anionic
Applications: Good cleaning effect, foaming and wetting power; used in textile industry as a general wetting and penetration agent; emulsifier for polymerization
Form: Soft paste; Color: 150 (Hazen)
Properties: Solubility: Sol. in water, ethanol, propylene glycol, acetone; disp. in trichloroethylene, spindle oil, white spirit; Density: 1.060 g/cm^3 @ 20 C; Clear Pt.: ≈ 0 C; Flash Pt.: ≫ 100 C; Stability: Stable to strong acids and bases; Surface tension: 35 dynes/cm; pH: 6–8; Biodegradable
Composition: 80% active ingred.
Toxicity: Degreasing effect on skin in concentrated form

Berol 513. Berol Kemi AB
Chemical Description: Alkyl ether phosphate ester and an *n*-primary

alcohol ethylene adduct type
Category: Detergent, anionic/nonionic
Applications: Good cleaning effect with corrosion-inhibiting qualities;
Form: Paste; Color: < 5 Gardner
Properties: Solubility: Sol. in ethanol, acetone, xylene, propylene glycol, white spirit; disp. in water, trichloroethylene; Density: 1.050 g/cm^3 @ 30 C; Visc.: ≈ 330 cps @ 30 C; Flash Pt.: ≫ 100 C; Stability: Stable at pH 8–12; Surface tension: 34.4 dynes/cm; Biodegradable
Composition: ≈ 100% active ingred.
Warnings and Caution: Avoid contact with skin

Berol 516. Berol Kemi AB
Chemical Description: Mixture of an alkyl ether phosphate ester and an *n*-primary alcohol ethylene oxide adduct
Category: Detergent, corrosion inhibitor, anionic/nonionic
Applications: Detergents and cleaning agents with builders which may cause corrosion to soft metals; washing machines
Form: Paste; Color: < 5 (Gardner)
Properties: Solubility: Sol. in ethanol, acetone, xylene, propylene glycol, white spirit; disp. in water; Density: 1.050 g/cm^3 @ 30 C; Visc.: ≈ 330 cps @ 30 C; Clear Pt.: ≈ 40 C; Flash Pt.: ≫ 100 C; Pour Pt.: ≈ 30 C; Stability: Stable at pH 8–12 in its neutralized form; limited stability in presence of water; Surface Tension: 34.4 dynes/cm (0.1% sol'n.); pH: 2.3 (1% aq.); Biodegradable
Composition: 100% active ingred.
Toxicity: Acid–avoid contact with skin

Berol 518. Berol Kemi AB
Chemical Description: Mixture of an alkyl ether phosphate ester and an *n*-primary alcohol ethylene oxide adduct
Category: Detergent, foam controller, anionic/nonionic
Applications: Machine detergents, spray-dried detergents
Form: Soft wax; Color: < 5 (Gardner)
Properties: Solubility: Sol. in ethanol, acetone, xylene, trichloroethylene, white spirit, light fuel oil, propylene glycol (slightly opal sol'n); disp. in water; Density: 1.010 g/cm^3 @ 50 C; Visc.: 1500 cps @ 35 C, 125 cps @ 50 C; Clear Pt.: ≈ 35 C; Flash Pt.: ≫ 100 C; Pour Pt.: ≈ 35 C; Stability: Stable at pH 8–12 in neutralized form; limited stability in presence of water; Surface Tension: 38 dynes/cm (0.1% sol'n.); pH: 2.3 (1% aq.); Biodegradable
Composition: 100% active ingred.
Toxicity: Acid–avoid contact with skin

Berol 521. Berol Kemi AB
Chemical Description: Water sol'n. of potassium alkyl phosphate esters
Category: Corrosion inhibitor, solubilizer, anionic
Applications: Corrosion inhibitor; solubilizes nonionic surfactants in

presence of high electrolyte conc.; detergents and cleaning agents with builders which can corrode soft metals; liquid and powder products
Form: Clear liquid; Color: < 300 (Hazen)
Properties: Solubility: Sol. in water, propylene glycol: Density: 1.145 g/cm³ @ 20 C; Visc.: 315 cps @ 20 C; Clear Pt.: 25 C; Flash Pt.: ≫ 100 C; Pour Pt.: –10 C; Stability: Stable at pH 8–12; pH: 7–8 (1% aq.); Biodegradable
Composition: ≈ 40% active
Toxicity: Causes burning and eye irritation; avoid skin contact

Berol 522. Berol Kemi AB
Chemical Description: Water sol'n. of a potassium alkyl phosphate ester
Category: Solubilizer, anionic
Applications: Solubilizes nonionic surfactants in high electrolyte conc. (esp. for nonyl phenol ethylene oxide condensates); household and commercial liquid laundry detergents; industrial cleaning
Form: Clear liquid; Color: < 300 (Hazen)
Properties: Solubility: Sol. in water, propylene glycol; Density: < 1.245 g/cm³ @20 C; Visc.: < 125 cps @ 20 C; Clear Pt.: < 0 C; Flash Pt.: ≫ 100 C; Pour Pt.: < –10 C; Stability: Stable at pH 8–12; Surface Tension: 31 dynes/cm (0.1% conc.); pH: < 8–9 (1% aq.); Biodegradable
Composition: 40–42% active
Toxicity: Causes burning and eye irritation; avoid contact with skin

Berol 525. Berol Kemi AB
Chemical Description: Mixture of an alkyl ether phosphate ester and a straight-chain primary alcohol ethylene oxide adduct
Category: Detergent, solubilizer, anionic/nonionic
Applications: Solubilizes nonionic surfactants in presence of inorganic salts; liquid alkaline industrial cleaners; household and institutional slightly alkaline detergents; spray-dried detergents
Form: Paste; Color: < 150 (Hazen)
Properties: Solubility: Sol. in water, ethanol, acetone, propylene glycol (slightly opal), trichloroethylene, xylene, white spirit; Density: 1.060 g/cm³; Visc.: 1500–2000 cps; Clear Pt.: ≈ 30 C; Flash Pt.: ≫ 100 C; Pour Pt.: ≈ 23 C; Stability: Stable at pH 8–12 in neutralized form; limited stability in presence of water; Surface Tension: 33 dynes/cm (0.1% conc.); pH: 2.3 (1% aq.); Biodegradable
Composition: 100% active ingred.
Toxicity: Acid–avoid skin contact

Berol 548. Berol Kemi AB
Chemical Description: Amphoteric surfactant
Category: Detergent, wetting agent, amphoteric
Applications: Industrial detergents
Form: Clear liquid; Color: 50 (Hazen)
Properties: Solubility: Sol. in 30% caustic sol'ns.; Visc.: 70 cps @ 20 C; Clear Pt.: –5 C; Flash Pt.: > 100 C; Surface Tension: 28 dynes/cm (0.1% conc.); pH: 9.3 (1% aq.)
Composition: 35% active ingred.

Toxicity: Slightly corrosive; degreasing effect on skin in concentrated form
Containers: 200 liter trauser drum

Berol 563. Berol Kemi AB
Chemical Description: Alkyl polyglycol ether ammonium methyl sulfate
Category: Detergent, cationic
Applications: Alkaline degreasing and cleaning agents with good solubility and stability in alkaline conditions
Form: Clear liquid; Color: ⩽ 7 (Gardner)
Properties: Solubility: Sol. in water, ethanol, propylene glycol, xylene; Density: 1.110 g/cm^3 @ 20 C; Visc.: 500 cps @ 20 C; Cloud Pt.: ≈ 85 C (1% in 9% NaOH); Clear Pt.: 18 C; Flash Pt.: ⩾ 100 C; Pour Pt.: 10 C; Stability: Stable to weak acids, bases; Surface Tension: 43 dynes/cm (0.1% conc.); pH: 7–9 (1% aq.)
Composition: 100% active ingred.
Toxicity: Degreasing effect on skin in concentrated form

Berol 594. Berol Kemi AB
Chemical Description: Hydroxyethyl-2-alkyl imidazolin type
Category: Adhesion aid, dispersing agent, wetting and foaming agent, corrosion inhibitor
Applications: Strong cationic activity in acid sol'ns. and oils and solvents; temporary corrosion inhibitor; aids adhesion in solvent-based paints and varnishes, emulsions of oils, waxes and asphalt; dispersant for pigments; in acid cleaning agents
Form: Clear liquid; Color: ⩽ 12 (Gardner)
Properties: m.w.: ≈ 360; Solubility: Sol. in ethanol, xylene, trichloroethylene, white spirit, paraffin oil and dilute sulfuric, hydrochloric, phosphoric and acetic acids; disp. in water; Density: 0.950 g/cm^3 @ 20 C; Visc.: 300 cps @ 25 C; Clear Pt.: ⩽ –10 C; Flash Pt.: ⩾ 100 C; Pour Pt.: < 0 C; Stability: Stable to dilute acids; Surface Tension: 32 dynes/cm (0.1% sol'n.); pH: 10–11 (1% aq.)
Composition: 100% active ingred.
Toxicity: Can burn and irritate skin
Warnings and Caution: Protective clothing and goggles should be worn

Berol 716. Berol Kemi AB
Chemical Description: Alkyl phenol ethylene oxide adduct
Category: Detergent, nonionic
Applications: Low foaming machine washing detergents in powder form; good pigment and grease removal; spray drying
Form: Cloudy liquid; Color: < 500 (Hazen)
Properties: Solubility: Sol. in ethanol, xylene, trichloroethylene, paraffin oil (opaque); disp. in water, white spirit, light fuel oil; Density: 1.04 g/cm^3 @ 20 C; Visc.: ≈ 800 cps @ 20 C; HLB: 13; Cloud Pt.: 80–85 C (in 10% *n*-propanol); Clear Pt.: ≈ 35 C; Flash Pt.: ⩾ 100 C; Pour Pt.: ≈ 20 C; Stability: Stable to weak acids, bases, oxi-

dizing and reducing agents; pH: 6–7 (1% aq.)
Composition: 100% active ingred.
Toxicity: Degreasing effect on skin in concentrated form
Containers: 200 kg net drums

Berol 730. Berol Kemi AB
Chemical Description: Polypropylene glycols
Category: Antifoaming agent, lubricant, release agent, hydraulic fluid lubricant
Applications: Substitute for different oils; dust adhesive fluid in air filters
Form: Clear liquid; Color: < 100 (Hazen)
Properties: m.w.: 1200; Solubility: Sol. in mineral spirit, acetone, xylene, methanol, carbontetrachloride @ 25 C (5% sol'n); sol. in water @ 25 C (< 2% sol'n.); Density: 1.0 g/cm^3 @ 25 C; Visc.: 160 cps @ 25 C; Flash Pt.: > 200 C; Pour Pt.: – 40 C
Composition: 100% active ingred.
Containers: 200 kg net drums

Berol 733. Berol Kemi AB
Chemical Description: Potassium salt of a phosphated alkyl phenol ethylene oxide adduct
Category: Detergent, anionic
Applications: Solubilizes nonionic surfactants in presence of inorganic salts; liquid alkaline industrial cleaners; household and institutional slightly alkaline detergents; spray-dried detergents
Form: Clear liquid; Color: < 100 (Hazen)
Properties: Solubility: Sol. in water, 6% sodium hydroxide sol'n., 35% tetra-potassium pyrophosphate sol'n., 50% sodium metasilicate sol'n., propylene glycol; disp. in acetone; Density: 1.050 g/cm^3 @ 20 C; Visc.: ≈ 400 cps @ 20 C; Clear Pt.: – 5 C; Flash Pt.: ≫ 100 C; Pour Pt.: –7 C; Stability: Stable at pH 8–12; Surface Tension: 41 dynes/cm; pH: 9–10 (1% aq.); Biodegradable
Composition: 39–41% active
Toxicity: Degreasing effect on skin in concentrated form

Berol 740. Berol Kemi AB
Chemical Description: Polypropylene glycols
Category: Antifoaming agent, lubricant, release agent, hydraulic fluid lubricant
Applications: Substitute for different oils; dust adhesive fluid in air filters
Form: Clear liquid; Color: < 100 (Hazen)
Properties: m.w.: 1800; Solubility: Sol. in mineral spirit, acetone, xylene, methanol, carbontetrachloride @ 25 C (5% sol'n.); sol. in water @ 25 C (< 0.1% sol'n.); Density: 1.0 g/cm^3 @ 25 C; Visc.: 200 cps @ 25 C; Flash Pt.: > 200 C; Pour Pt.: – 35 C
Composition: 100% active ingred.
Containers: 200 kg net drums

Berol 752. Berol Kemi AB
Chemical Description: Surfactant
Category: Foam stabilizer, anionic
Applications: Stabilization of mechanically whipped PVC foam; uniform and even distribution of

the size of the air pores
Form: Viscous liquid; Color: < 7 Gardner
Properties: Solubility: Disp. in water, propylene glycol, dioctylphthalate, tributylphosphate; Density: 1.070 g/cm^3; Visc.: ≈ 10,000 cps @ 25 C; Flash Pt.: ⩾ 100 C; Pour Pt.: ≈ 25 C; Stability: Stable to weak acids and alkalies; pH: 6–8 (1% sol'n. in 1:1 water/ethanol)
Composition: 95–100% active
Toxicity: Degreasing effect on skin in concentrated form

Berol 755. Berol Kemi AB
Chemical Description: Linear alkyl-aryl-sulfonate and of a *n*-primary alcohol ethylene oxide adduct
Category: Detergent, anionic/nonionic
Applications: Combination giving excellent detergency and relatively strong and stable foaming
Form: Clear liquid; Color: ⩽ 5 Gardner
Properties: Solubility: Sol. in water, ethanol, isopropanol, xylene; disp. in white spirit, light fuel oil; Density: 1.050 g/cm^3 @ 20 C; Visc.: 150 cps @ 25 C; Clear Pt.: ≈ 0 C; Flash Pt.: ⩾ 100 C; Pour Pt.: ≈ –10 C; Stability: Stable to weak acids and bases; Surface Tension: 36 dynes/cm; pH: 6–7; Biodegradable
Composition: 60% active ingred.
Toxicity: Degreasing effect on skin in concentrated form

Berol 767. Berol Kemi AB
Chemical Description: Sodium salt of an alkyl ether sulfate and an alkyl phenol ethoxylate
Category: Emulsifier, anionic/nonionic
Applications: Emulsion polymerization of vinyl acetate and vinyl acetate/acrylate
Form: Clear liquid; Color: Max. 100 (Hazen)
Properties: Solubility: Sol. in water, ethanol, acetone, propylene glycol; disp. in light fuel oil; Density: 1.060 g/cm^3 @ 20 C; Visc.: 230 cps @ 20 C; Clear Pt.: below 0 C; Flash Pt.: ⩾ 100 C; Pour Pt.: below 0 C; Stability: Stable to weak bases and very weak acids; Surface tension: 45 dynes/cm; pH: 6–7 (1% aq.)
Composition: 34–36% active
Toxicity: Degreasing effect on skin in concentrated form

Berol 777. Berol Kemi AB
Chemical Description: Linear alkyl aryl sulfonate and alkyl phenol ethylene oxide adduct
Category: Detergent, anionic/nonionic
Applications: Excellent detergency and relatively strong foaming; used in textile industry as a general wetting and cleaning agent; excellent pigment dispersing properties;
Form: Clear liquid; Color: ⩽ 250 (Hazen)
Properties: Solubility: Sol. in water, propylene glycol; disp. in ethanol, acetone, trichloroethylene, xylene; Density: 1.055 g/cm^3 @ 20 C; Visc.: 425 cps @ 20 C; Clear Pt.: ≈ 1 C; Flash Pt.: ⩾ 100 C; Pour Pt.: ≈ 0 C; Stability: Stable

to weak acids and bases; Surface tension: ≈ 38 dynes/cm; pH: 6–7 (1% aq.); Biodegradable
Composition: 34% active ingred.
Toxicity: Degreasing effect on skin in concentrated form

Berol 779. Berol Kemi AB
Category: Anionic surfactant
Applications: Solvent degreasing
Form: Clear liquid; Color: < 5 (Gardner)
Properties: Solubility: Sol. in water (opal sol'n.), ethanol, xylene, trichloroethylene, white spirit, light fuel oil; Density: 0.98 g/cm^3 @ 20 C; Visc.: ≈ 150 cps @ 20 C; Clear Pt.: < –20 C; Flash Pt.: 70 C; Pour Pt.: < –20 C; Stability: Stable to moderate concentrations of weak acids and bases; pH: 5.5–6.5 (1% aq.); Biodegradable
Composition: 100% active ingred.
Toxicity: Degreasing effect on skin in concentrated form

Berol 782. Berol Kemi AB
Chemical Description: Blended emulsifier
Category: Emulsifer
Applications: Dissolve large quantities of grease, oil, asphalt
Form: Clear liquid; Color: 12 Gardner
Properties: Solubility: Sol. in ethyl alcohol, xylene, trichloroethylene, white spirit, light fuel oil; Density: 0.943 g/cm^3; Visc.: 60 cps @ 25 C; Clear Pt.: –30 C; Flash Pt.: 72 C; Pour Pt.: –30 C; pH: 6 ± 0.1; Biodegradable
Composition: 100% active ingred.
Containers: 195 kg net drums
Warnings and Caution: Skin and eye irritant

Berol 786. Berol Kemi AB
Chemical Description: Oil-in-water emulsion
Category: Emulsifier
Applications: Corrosion inhibitor
Form: Liquid; Color: Yellow-white
Properties: Density: 0.994 g/cm^3 @ 20 C; Visc.: 100 cps @ 20 C; Flash Pt.: > 100 C; Pour Pt.: 0 C;
Containers: 200 kg net drums

Berol 788. Berol Kemi AB
Chemical Description: Linear alkyl aryl sulfonate and an alkyl phenol ethylene oxide adduct
Category: Detergent, anionic/nonionic
Applications: Grease and pigment removing properties and relatively strong foaming
Form: Clear liquid; Color: 500 (Hazen)
Properties: Solubility: Sol. in water, ethanol, propylene glycol; disp. in acetone, trichloroethylene, white spirit; Density: 1.050 g/cm^3 @ 20 C; Visc.: 450 cps @ 20 C; Clear Pt.: ≈ 5 C; Flash Pt.: ≫ 100 C; Pour Pt.: ≈ 0 C; Stability: Stable to weak acids and bases; Surface tension: 37 dynes/cm; pH: 6–7; Biodegradable
Composition: 30% active ingred.
Toxicity: Degreasing effect on skin in concentrated form

Berol 792. Berol Kemi AB
Category: Emulsifier, nonionic
Applications: Emulsifier for oils

and solvents; high penetrating and excellent degreasing and cleaning properties for engine parts, workshop floors, car bodies, etc.
Form: Clear liquid; Color: Yellow
Properties: Solubility: Sol. in ethanol, paraffin oil, light mineral oil; disp. in water, benzene, trichloroethylene; Density: 1.02 g/cm^3 @ 20 C; Pour Pt.: < -10 C
Containers: 200 kg net drums

Berol 808. Berol Kemi AB
Category: Detergent, softener, antistat, anionic/cationic/nonionic
Applications: Textile softening and antistatic agent; liquid light duty detergents; post additive to powder formulations
Form: Clear liquid; Color: Yellow
Properties: Solubility: Sol. in ethanol, propylene glycol, white spirit, light fuel oil; opaquely sol. in water; Density: 1.013 g/cm^3; Visc.: 105 cps; Cloud Pt.: $>$ 90 C (1% aq.); Clear Pt.: $\approx$ 2 C; Flash Pt.: 60 C; Flow Pt.: –4 C; Stability: Stable to weak acids and bases; Surface Tension: 32 dynes/cm (0.1%); pH: 5.5 (1% aq.); Biodegradable under normal sewage treatment conditions
Composition: 80% active ingred.
Toxicity: Causes burning and eye irritation; degreasing effect on skin in concentrated form

Berol 809. Berol Kemi AB
Chemical Description: Combined anionic/cationic/nonionic surfactant
Category: Softener, antistat, detergent, anionic/cationic/nonionic
Applications: Textile softening and antistatic agent; light duty detergents
Form: Clear liquid; Color: Light yellow
Properties: Solubility: Sol. in water, ethanol, white spirit; Density: 1.04 g/cm^3; Visc.: 170 cps; Cloud Pt.: $>$ 55 C (opal, 1% aq.); Clear Pt.: 1 C; Flash Pt.: $\gg$ 100 C; Pour Pt.: –4 C; Stability: Stable to weak acids and bases; Surface Tension: 31.5 dynes/cm (0.1% sol'n.); pH: 6.3 (1% aq.); Biodegradable under normal sewage treatment conditions
Composition: 80% active ingred.
Toxicity: Causes burning and eye irritation; degreasing effect on skin in concentrated form

Berol 822. Berol Kemi AB
Chemical Description: Alkyl aryl sulfonate type
Category: Emulsifier, anionic
Applications: Emulsifier of biocide (pesticide) concentrates, and some highly aromatic and chlorinated solvents
Form: Clear viscous liquid; Color: $<$ 10 (Gardner)
Properties: Solubility: Sol. in ethanol, trichloroethylene, paraffin oil, xylene, white spirit, light fuel oil; disp. in water; Density: 1.03 g/cm^3 @ 20 C; Visc.: 1200 cps @ 20 C; Flash Pt.: $\approx$ 35 C; Pour Pt.: $\approx$ 5 C; Stability: Stable to high concentrations of strong acids and bases; pH: 5–8 (1% aq.); Biodegradable
Composition: 60–62% active; solvent: butanol

Toxicity: Degreasing effect on skin in concentrated form
Containers: 200 kg net drums

Berol Dinonylphenol. Berol Kemi AB
Chemical Description: Dinonylphenol
Category: Chemical intermediate
Applications: Reaction products include salts with mono- and polyvalent metal ions, ethers, resins with good oil solubility, polyoxyethylene derivatives (used as emulsifiers, detergents and dispersants); as component in lubricating oils, corrosion inhibitors, stabilizer for rubber, plasticizer for polymers and resins; hydrophobe for nonionic surfactants
Form: Viscous liquid; Color: Yellow; Odor: Phenol-like
Properties: Ref. Index: 1.4985; Solubility: Sol. in ethanol, acetone, trichloroethylene, butyl glycol, white spirit, mineral oil; Density: 0.911 g/cm^3 @ 20 C; Visc.: ≈ 400 cps @ 40 C; B.P.: 326–334 C (5%); Flash Pt.: > 500 C; Pour Pt.: ≈ 0 C; Hydroxyl No.: 165
Toxicity: Severe skin and eye irritant; respiratory irritant on prolonged exposure to vapor
Containers: Store in stainless steel or aluminum
Warnings and Caution: Protective clothing, goggles and gloves should be worn

Berol EGA 07. Berol Kemi AB
Chemical Description: Fatty alcohol ethoxylate
Category: Dye levelling and dispersing agent, nonionic
Applications: Textile dye levelling and dispersing agent; protective colloid; stripping of vat dyes
Form: Clear liquid; Color: < 150 (Hazen)
Properties: Solubility: Sol. in water, ethanol, xylene, trichloroethylene; disp. in paraffin oil, white spirit, light fuel oil; Density: 1.01 g/cm^3 @ 20 C; Visc.: ≈ 10 cps @ 20 C; HLB: 15.0; Cloud Pt.: 74–76 C (5% in 10% NaCl sol'n.); Clear Pt.: < 0 C; Flash Pt.: ≫ 100 C; Pour Pt.: < 0 C; Stability: Stable to weak acids, bases, oxidizing and reducing agents; Surface Tension: 40 dynes/cm (0.1%); pH: 6–7 (5% aq.); Biodegradable
Composition: 20% active ingred.
Toxicity: Degreasing effect on skin in concentrated form
Containers: 200 kg net drums

Berol EGA 192. Berol Kemi AB
Chemical Description: Alkyl glyceride ethylene oxide adduct type
Category: Equalizing and dispersing agent, nonionic
Applications: Textile dye equalizing and dispersing agent
Form: Clear liquid; Color: < 3 (Gardner)
Properties: Solubility: Sol. in water, ethanol, acetone; disp. in propylene glycol, xylene; Density: 1.080 g/cm^3 @ 20 C; Visc.: 120 cps @ 20 C; HLB: ≈ 18; Cloud Pt.: 62 C (1% in 10% NaCl sol'n.); Clear Pt.: ≈ 14 C; Flash Pt.: ≫ 100 C; Pour Pt.: ≈ 13 C; Stability: Stable to weak acids and bases; Surface Tension: 42 dynes/cm;

pH: 6–7 (1% aq.); Biodegradable
Composition: 85% active ingred.
Toxicity: Degreasing effect on skin in concentrated form

Berol Fintex 10. Berol Kemi AB
Chemical Description: Fatty acid ethylene oxide adduct type
Category: Softener, antistat, equalizing agent, emulsifier, nonionic
Applications: Textile softener and antistat esp. on cotton and rayon; adhesive bath additive; emulsifier for greases and oils (under certain conditions with foodstuffs)
Form: Wax; Color: White
Properties: Solubility: Sol. in ethanol, trichloroethylene, xylene; disp. in water, propylene glycol, acetone, white spirit; Density: 1.010 g/cm^3 @ 36 C; Visc.: ≈ 80 cps @ 36 C; HLB: ≈ 12; Cloud Pt.: 75–80 C; Clear Pt.: ≈ 38 C; Flash Pt.: ⩾ 100 C; Pour Pt.: ≈ 35 C; Stability: Stable to weak acids and bases; Surface Tension: 42 dynes/cm (0.1%); pH: 6–7 (1% aq.); Biodegradable
Composition: 100% active ingred.
Toxicity: Degreasing effect on skin in concentrated form

Berol Fintex 41. Berol Kemi AB
Category: Softener, cationic
Applications: Textile softener esp. for cotton and rayon; in anticrease treatment with resins; adhesive bath additive
Form: Paste; Color: Yellowish
Properties: Density: 0.900–1.000 g/cm^3; Flash Pt.: ⩾ 100 C; Stability: Stable to weak acids and bases; pH: ≈ 9 (1% aq.)
Composition: 50% active ingred.
Toxicity: Degreasing effect on skin in concentrated form

Berol Fintex 42. Berol Kemi AB
Chemical Description: Weakly cationic surfactant
Category: Softener, antistat, cationic
Applications: Textile softener for natural and synthetic fibers, esp. for cotton and rayon; in anticrease treatment with resins; adhesive bath additive
Form: Liquid dispersion; Color: Light yellow
Properties: Density: 1.000 g/cm^3 @ 20 C; Visc.: ≈ 500 cps @ 20 C; Flash Pt.: ⩾ 100 C; Pour Pt.: ≈ 0 C; Stability: Stable to weak acids and bases; pH: ≈ 9 (1% aq.)
Composition: 25% active ingred.
Toxicity: Degreasing effect on skin in concentrated form

Berol Fintex 573. Berol Kemi AB
Chemical Description: Cationic surfactant consisting of alkyl polyglycol ether ammonium ethyl sulfate and a nonionic wetting agent
Category: Antistat, bactericide, cationic
Applications: Antistatic agent for synthetic textiles and plastics; bactericide for gram-postive bacteria
Form: Clear liquid; Color: ⩽ 7 (Gardner)
Properties: Solubility: Sol. in water, ethanol, propylene glycol,

acetone, trichloroethylene, xylene, white spirit; Density: 1.050 g/cm^3 @ 20 C; Visc.: 750 cps @ 20 C; Cloud Pt.: ≈ 70 C (1% in 10% NaCl); Clear Pt.: 10 C; Flash Pt.: ≫ 100 C; Pour Pt.: ≈ 2 C; Stability: Stable to weak acids and bases; Surface Tension: 35 dynes/cm (0.1% sol'n.); pH: 7.0 ± 0.5 (1% aq.)
Composition: 100% active ingred.
Toxicity: Degreasing effect on skin in concentrated form

Berol Fintex 577. Berol Kemi AB
Chemical Description: Quaternary ammonium compound
Category: Softener, rewetting agent, antistat, cationic
Applications: Softening, antistatic and rewetting agent for natural and synthetic textiles; hair cream rinses; antibacterial properties
Form: Clear liquid; Color: < 300 (Hazen)
Properties: Solubility: Disp. in water (hot or cold); sol. in ethanol, xylene, propylene glycol, white spirit; Density: 0.94 g/cm^3 @ 20 C; Visc.: 100 cps @ 20 C; Clear Pt.: < 15 C; Flash Pt.: 25 C; Pour Pt.: ≈ 5 C; Stability: Stable to weak acids and bases; moderately stable to strong acids and bases; Surface Tension: 32 dynes/cm; pH: 6–7 (1% aq.)
Composition: ≈ 70% active
Toxicity: Degreasing effect on skin in concentrated form

Berol Fintex 713. Berol Kemi AB
Chemical Description: Highly concentrated emulsion of polyethylene wax
Category: Finishing agent, nonionic
Applications: Finishing agent for cotton and rayon to improve tear-, wear- and crease-resistance; softening agent; textile filler
Form: Milky-white emulsion
Properties: Density: 0.980 g/cm^3 @ 20 C; Visc.: ≈ 100 cps @ 20 C; Flash Pt.: ≫ 100 C; Stability: Stable to acids and bases; pH: ≈ 9 (2% emulsion)
Composition: 45% active ingred.
Toxicity: Degreasing effect on skin in concentrated form

Berol Fintex 755. Berol Kemi AB
Chemical Description: Mixture of anionic (linear alkyl aryl sulfonate) and nonionic (*n*-primary alcohol) surfactants
Category: Wetting agent, dispersing agent, detergent, anionic/nonionic
Applications: Textile wetting, cleaning, dispersing agents; levelling agent in dyeing
Form: Clear liquid; Color: ⩽ 5 (Gardner)
Properties: Solubility: Sol. in water, ethanol, isopropanol, trichloroethylene (opal), xylene; Density: 1.050 g/cm^3 @ 20 C; Visc.: 150 cps @ 25 C; Clear Pt.: 0 C; Flash Pt.: ≫ 100 C; Pour Pt.: –10 C; Stability: Stable to weak acids and bases; Surface Tension: 36 dynes/cm (0.1% sol'n.); pH: 6–7 (1% aq.); Biodegradable
Composition: 70% active ingred.
Toxicity: Degreasing effect on skin in concentrated form

Berol Fintex 776. Berol Kemi AB
Chemical Description: Fatty alcohol ethoxylate
Category: Detergent, nonionic
Applications: Low foam detergent for textile industry for washing and dyeing/washing in Jet or HP machines
Form: Clear liquid; Color: < 100 (Hazen)
Properties: Solubility: Sol. in water, ethanol, xylene, trichloroethylene, white spirit, light fuel oil; Density: 1.00 g/cm^3 @ 20 C; Visc.: ≈ 50 cps @ 20 C; Cloud Pt.: ≈ 45 C; Clear Pt.: < 5 C; Flash Pt.: > 100 C; Pour Pt.: < 0 C; Stability: Stable to weak acids, bases, oxidizing and reducing agents; Surface Tension: 32 dynes/cm (0.1% sol'n.); pH: 6–7 (1% aq.); Biodegradable
Composition: 90% active ingred.
Toxicity: Degreasing effect on skin in concentrated form
Containers: 200 kg net drums

Berol Lanco. Berol Kemi AB
Chemical Description: Alkyl phenol ethylene oxide adduct type
Category: Detergent, wetting agent, emulsifier, dispersant, nonionic
Applications: Textile wetting and cleaning agent for natural and synthetic fibers; esp. for washing of virgin wool; emulsifier for waxes, oils, solvents, and paint resins; emulsion polymerization; dispersant for pigment and tinting pastes in paint industry
Form: Clear liquid; Color: < 150 (Hazen)
Properties: Solubility: Sol. in water, ethanol, xylene, trichloroethylene; Density: 1.05 g/cm^3 @ 20 C; Visc.: ≈ 350 cps @ 20 C; HLB: 13.3; Cloud Pt.: 52–58 C (1% aq.); Clear Pt.: ≈ 7 C; Flash Pt.: ≫ 100 C; Pour Pt.: ≈ 5 C; Stability: Stable to weak acids, bases, oxidizing and reducing agents; Surface Tension: 33 dynes/cm; pH: 6–7 (1% aq.); Biodegradable under normal biological sewage purification conditions
Composition: 100% active
Toxicity: Degreasing effect on skin in concentrated form
Containers: 200 kg net drums

Berol Nonylphenol. Berol Kemi AB
Chemical Description: Nonylphenol
Category: Raw material
Applications: Raw material for detergents, dispersants, wetting agents, additives for lubricating oils, corrosion inhibitors, chemicals for the rubber industry, plasticizers, stabilizers for plastics, germicides, insecticides, oil-soluble resins, tanning agents, dyestuffs and pharmaceutical preparations
Form: Viscous liquid; Color: Almost colorless; Odor: Faint phenol
Properties: m.w.: 220; Ref. Index: 1.513; Solubility: Sol. in ethanol, acetone, trichloroethylene, butyl glycol, white spirit, mineral oil; Density: 0.949 g/cm^3 @ 20 C; Visc.: ≈ 2750 cps @ 20 C; Flash Pt.: 145 C; Pour Pt.: < 0 C; Hydroxyl No. 252
Reaction: Reacts with ethylene oxide to form polyoxyethylene

adducts; with hexamethylenetetramine, etc. to form polymers; can be sulfonated, halogenated, nitrated, etc. to yield various products
Toxicity: Severe irritation and burning to skin, eyes, etc.; respiratory irritation on prolonged exposure to vapor
Warnings and Caution: Protective clothing, goggles, gloves should be worn

Berol WASC. Berol Kemi AB
Chemical Description: Alkyl phenol ethylene oxide adduct
Category: Detergent, rewetting agent, dispersant, nonionic
Applications: Household and industrial detergents, dishwashing liquids; textile washing agent; rewetting agent for cotton and cellulose; dispersing agent for pigment pastes in paint and varnish industry
Form: Clear liquid; Color: < 150 (Hazen)
Properties: Solubility: Sol. in water, ethanol; disp. in xylene, trichloroethylene, paraffin oil, white spirit; Density: 1.06 g/cm^3 @ 20 C; Visc.: ≈ 375 cps @ 20 C; HLB: 14.1; Cloud Pt.: 75–80 C (1% aq.); Clear Pt.: ≈ 5 C; Flash Pt.: ≫ 100 C; Pour Pt.: ≈ 0 C; Stability: Stable to weak acids, bases, oxidizing and reducing agents; Surface Tension: 34 dynes/cm (0.1% sol'n.); pH: 6–7 (1% aq.); Biodegradable under normal biological sewage purification conditions
Composition: 95% active ingred.
Toxicity: Degreasing effect on skin in concentrated form
Containers: 200 kg net drums

Biopal NR-20. GAF Corp.
Chemical Description: Nonylphenoxy poly (ethyleneoxy) ethanol–iodine complex
Form: Viscous free flowing liquid; Color: Dark, reddish brown

Biopal VRO-20. GAF Corp.
Chemical Description: Nonylphenoxypoly(ethyleneoxy)-ethanol iodine complex providing 20% available iodine
Category: Iodophor concentrate
Applications: Iodine complexed with a nonionic surfactant for formulation of sanitizing rinses, germicidal rinses and cleaners, sanitizing cleaners, detergent-sanitizers, disinfecting agents
Form: Pourable liquid; Color: Dark brown
Properties: Solubility: Water-sol.; Sp. gr.: 1.34; Visc.: 1700 cps; Pour Pt.: 10.4 C; Solidification Pt.: 5.2 C; Stability: Excellent (storage)
Warnings and Caution: Mixing with concentrated forms of oxidizing or reducing agents may be hazardous

Bio Soft D-35X. Stepan Chem.
Chemical Description: Sodium linear alkylate sulfonates
Category: Detergent with good foaming, wetting, emulsifying properties
Applications: Liquid detergents
Form: Liquid; Color: Light straw

Properties: Solubility: Water-sol.; Density: 8.9 lb/gal; Cloud Pt.: 10 C; Stability: Good; Biodegradable
Composition: 34% min. active ingred.
Reaction: Excellent compatability with other anionics, nonionics and alkaline builders

Bio Soft D-40. Stepan Chem.
Chemical Description: Sodium linear alkylate sulfonate
Category: Detergent with good foaming, wetting and emulsifying properties
Applications: Liquid detergent
Form: Clear liquid; Odor: Bland
Properties: Solubility: Water-sol.; Density: 8.7 lb/gal; Cloud Pt.: 29 C; Biodegradable
Composition: 40% active ingred.
Reaction: Compatible with other anionics, nonionics and alkaline builders

Bio Soft D-60. Stepan Chem.
Chemical Description: Linear sodium alkyl aryl sulfonate
Category: Detergent, foaming and wetting agent
Form: Slurry; Color: Light
Properties: Solubility:Water-sol.; Surface Tension: 37.4 dynes/cm (0.1% conc.); Biodegradable
Composition: 50% active ingred.; 21% water

Bio Soft D-62. Stepan Chem.
Chemical Description: Sodium linear alkylate sulfonates
Category: Detergent with good foaming, wetting, emulsifying properties
Form: Slurry; Color: Straw yellow; Odor: Bland
Properties: Solubility: Water sol.; Density: 9.1 lb/gal; Biodegradable
Composition: 60% active ingred.
Reaction: Compatible with other anionics, nonionics and alkaline builders

Bio Soft EA-8. Stepan Chem.
Chemical Description: Linear alkoxypolyoxethylene ethanols
Category: Detergent, nonionic
Applications: Chemical specialties: heavy duty laundry detergents, agricultural emulsions, sanitizers, metal cleaning and pickling, shampoo ingredient; Textiles: bleaching and dyeing assistant, Kier boiling, scouring of wool and cotton, soaping and washing, wool carbonizing; Petroleum: drilling muds, oil well acidizing; Paper: de-inking, de-pitching, felt washing, re-wetting
Form: Hazy white liquid
Properties: Solubility: Sol. in various organic solvents; Density: 8.42 lb/gal; Sp. gr.: 1.01; Cloud Pt.: 46 C; Stability: Stable to acid or alkaline hydrolysis, electrolytes, oxidizing or reducing agents, cationics, anionics and hard water; Biodegradable
Composition: 100% active ingred.

Bio Soft EA-10. Stepan Chem.
Chemical Description: Linear alkoxypolyoxethylene ethanols
Category and Applications: See Bio Soft EA-8
Form: Hazy, white liquid

Properties: Solubility: Sol. in various organic solvents, oil, water; Density: 8.42 lb/gal; Sp. gr.: 1.01; Cloud Pt.: 58.5 C; Stability: See Bio Soft EA-8; Biodegradable
Composition: 100% active ingred.

Bio Soft HD-100. Stepan Chem.
Chemical Description: Blend of nonionic and anionic surfactants
Category: Intermediate
Applications: Intermediate for the formulation of built and unbuilt liquid laundry detergents
Form: Clear liquid; Color: Light yellow
Properties: Solubility:Water-sol.; Density: 8.9 lb/gal; Visc.: 600 cps; Pour Pt.: 5 C
Composition: 100% active ingred.

Bio Soft LD-32. Stepan Chem.
Chemical Description: Anionic-nonionic blend with sodium cation
Category: Detergent
Applications: Liquid dishwashing detergent concentrate; bubble baths, car wash detergent, rug shampoos, textile scouring, aircraft cleaners, air entraining agent in concrete
Form: Clear liquid; Color: Pale straw
Properties: Solubility: Water-sol.; Density: 9.23 lb/gal; Visc.: 350 cps; Cloud Pt.: 5 C; Biodegradable
Composition: 60% active ingred.

Bio Soft LD-68. Stepan Chem.
Category: Detergent
Applications: Liquid detergent concentrate for dishwashing, car wash, fine fabrics
Form: Clear liquid; Color: Pale straw
Properties: Density: 9.0 lb/gal; Visc.: 800 cps; Biodegradable
Composition: 60% solids contains no phosphates

Bio Soft LD-70. Stepan Chem.
Chemical Description: Sodium olefin sulfonate
Category: Detergent, foaming agent, nonionic
Applications: Shampoos, bubble baths, dishwashing, car washing, fine fabric detergents
Form: Clear liquid; Color: Amber
Properties: Density: 8.7 lb/gal; Visc.: 300 cps; Cloud Pt.: $>$ 40 F; Biodegradable
Composition: 35% active ingred.; contains no phosphates

Bio Soft LD-80. Stepan Chem.
Chemical Description: Sodium olefin sulfonate
Category: Detergent
Applications: Dishwashing detergent concentrate
Form: Liquid; Color: Amber
Properties: Solubility: Water-sol.; Density: 8.8 lb/gal; Visc.: 225 cps; Cloud Pt.: $<$ 5 C; Biodegradable
Composition: 41% solids

Bio Soft LD-95. Stepan Chem.
Chemical Description: Anionic-nonionic blend with sodium cation
Applications: Liquid dishwashing detergent concentrate
Form: Clear liquid; Color: Pale straw
Properties: Solubility: Water-sol.; Density: 9.11 lb/gal; Visc.: 600

cps; Cloud Pt.: 5 C; Biodegradable
Composition: 60% active ingred.

•

Bio Soft N-21. Stepan Chem.
Category: Detergent
Applications: Liquid dishwashing detergent concentrate
Form: Clear liquid; Color: Amber
Properties: Solubility: Water-sol.; Density: 8.6 lb/gal; Visc.: 220 cps; Biodegradable
Composition: 60% active ingred.

Bio Soft N-300. Stepan Chem.
Chemical Description: Triethanolamine linear alkyl aryl sulfonate
Category: Detergent, wetting and foaming agent, anionic
Applications: Liquid dishwashing, car washing detergent, shampoos, textile scouring agents, metal cleaners, latex paints
Form: Clear liquid; Color: Pale yellow
Properties: Solubility: Water-sol.; Density: 9.0 lb/gal; Sp. gr.: 1.08; Visc.: 3200 cps; Surface Tension: 38.8 dynes/cm (0.1% conc.); Biodegradable
Composition: 60% active ingred.

Bio Soft S-100. Stepan Chem.
Chemical Description: Linear alkyl aryl sulfonic acid
Category: Intermediate
Applications: Intermediate for formulation of built detergents by dry neutralization
Form: Dark viscous liquid
Properties: Solubility: Water-sol.; Sp. gr.: 1.04; Biodegradable
Composition: 97% active ingred.; 0.5% sulfuric acid; 1.8% unsulfonated oil

Bio Terge AS-40. Stepan Chem.
Chemical Description: 40% aqueous sol'n. of sodium alpha olefin sulfonate
Category: Detergent, foaming agent
Applications: Mild liquid dishwash, car wash; commercial and industrial formulations
Form: Liquid; Color: Yellow
Properties: Solubility: Water-sol.; Density: 8.9 lb/gal; Biodegradable
Composition: 40% active ingred.; contains no phosphates

Bio Terge AS-90F. Stepan Chem.
Chemical Description: Sodium alpha olefin
Category: Detergent
Applications: Powdered detergents
Form: Cream flake
Properties: Solubility: Water-sol.; Biodegradable
Composition: 90% active ingred.

Blancol. GAF Corp.
Chemical Description: Sodium naphthalenesulfonate-formaldehyde condensate
Category: Thinning, dispersing, dye-leveling agents; anionic
Applications: Paper industry for slime control, reduction in two-sidedness, improved retention of fillers or fines, improved sizing, lowered viscosity of coatings
Form: Free-flowing coarse granules; Color: Tan to brown; Odor: None
Properties: Solubility: Sol. in warm or cold water; Density:

0.77–0.82 g/ml; Stability: Good
Composition: 88–90% active
Containers: C/L bags; 10 lb pkgs

Blancol N. GAF Corp.
Chemical Description: Sodium naphthalenesulfonate - formaldehyde condensate
Category and Applications: See Blancol
Form: Free-flowing granular powder; Color: Tan to brown; Odor: None
Properties: Solubility: Sol. in warm or cold water; Density: 0.65–0.75 g/ml; Stability: Good
Composition: 88–90% active
Containers: C/L bags, 10 lb pkgs

Brij 30. I.C.I. United States
Chemical Description: Polyoxyethylene (4) lauryl ether
Category: Nonionic surfactant
Form: Liquid; Color: Colorless to light yellow (may become hazy or form a precipitate)
Properties: Solubility: Sol. in alcohol, propylene glycol, cottonseed oil; Sp. gr.: ≈ 0.95; Visc.: ≈ 30 cps; HLB: 9.7; Flash Pt.: > 300 F; Fire Pt.: > 300 F

Brij 30 SP. I.C.I. United States
Chemical Description: Polyoxyethylene (4) lauryl ether with 0.01% BHA and 0.005% citric acid added as preservatives
Category: Nonionic surfactant
Form: Liquid; Color: Colorless to light yellow
Properties: Solubility: Sol. in alcohol, propylene glycol, cottonseed oil; Sp. gr.: ≈ 0.95; Visc.: ≈ 30 cps; HLB: 9.5; Flash Pt.: > 300 F; Fire Pt.: > 300 F

Brij 35. I.C.I. United States
Chemical Description: Polyoxyethylene (23) lauryl ether
Category: Nonionic surfactant
Form: Waxy solid; Color: White
Properties: Solubility: Sol. in water, alcohol, propylene glycol; Sp. gr.: ≈ 1.05; HLB: 16.9; Flash Pt.: > 300 F; Fire Pt.: > 300 F; Pour Pt.: ≈ 33 C

Brij 35 SP. I.C.I. United States
Chemical Description: Polyoxyethylene (23) lauryl ether with 0.01% BHA and 0.005% citric acid added as preservatives
Category: Nonionic surfactant
Form: Waxy solid; Color: White
Properties: Solubility: Sol. in water, alcohol, propylene glycol; Sp. gr.: ≈ 1.05; HLB: 16.9; Flash Pt.: > 300 F; Fire Pt.: > 300 F; Pour Pt.: ≈ 33 C

Brij 52. I.C.I. United States
Chemical Description: Polyoxyethylene (2) cetyl ether (antioxidants added)
Category: Nonionic surfactant
Form: Waxy solid; Color: White
Properties: Solubility: Sol. in alcohol, cottonseed oil, mineral oil; HLB: 5.3; Flash Pt.: > 300 F; Fire Pt.: > 300 F; Pour Pt.: ≈ 33 C

Brij 56. I.C.I. United States
Chemical Description: Polyoxyethylene (10) cetyl ether (antioxidants added)
Category: Nonionic surfactant

Form: Waxy solid; Color: White
Properties: Solubility: Sol. in alcohol; HLB: 12.9; Flash Pt.: > 300 F; Fire Pt.: > 300 F; Pour Pt.: ≈ 31 C

Brij 58. I.C.I. United States
Chemical Description: Polyoxyethylene (20) cetyl ether with 0.01% BHA and 0.005% citric acid added as preservatives
Category: Nonionic surfactant
Form: Waxy solid; Color: White
Properties: Solubility: Sol. in water and alcohol; HLB: 15.7; Flash Pt.: > 300 F; Fire Pt.: > 300 F; Pour Pt.: ≈ 38 C
Containers: 55 gal open head steel drum

Brij 72. I.C.I. United States
Chemical Description: Polyoxyethylene (2) stearyl ether with 0.01% BHA and 0.005% citric acid added as preservatives
Category: Nonionic surfactant
Form: Waxy solid; Color: White
Properties: Solubility: Sol. in alcohol, cottonseed oil; HLB: 4.9; Flash Pt.: > 300 F; Fire Pt.: > 300 F; Pour Pt.: ≈ 43 C
Containers: 55 gal open head steel drum

Brij 76. I.C.I. United States
Chemical Description: Polyoxyethylene (10) stearyl ether with 0.01% BHA and 0.005% citric acid added as preservatives
Category: Nonionic surfactant
Form: Waxy solid; Color: White
Properties: Solubility: Sol. in ethanol, propylene glycol; HLB: 12.4; Flash Pt.: > 300 F; Fire Pt.: > 300 F; Pour Pt.: ≈ 38 C

Brij 78. I.C.I. United States
Chemical Description: Polyoxyethylene (20) stearyl ether with 0.01% BHA and 0.005% citric acid added as preservatives
Category: Nonionic surfactant
Form: Waxy solid; Color: White
Properties: Solubility: Sol. in alcohol; HLB: 15.3; Flash Pt.: > 300 F; Fire Pt.: > 300 F; Pour Pt.: ≈ 38 C

Brij 92. I.C.I. United States
Chemical Description: Polyoxyethylene (2) oleyl ether with 0.01% BHA and 0.005% citric acid added as preservatives
Category: Nonionic surfactant
Form: Liquid; Color: Yellow
Properties: Solubility: Sol. in alcohol, propylene glycol, cottonseed and mineral oil; Visc.: ≈ 30 cps; HLB: 4.9; Flash Pt.: > 300 F; Fire Pt.: > 300 F; Pour Pt.: ≈ 10 C

Brij 93. I.C.I. United States
Chemical Description: Polyoxyethylene (2) oleyl ether with 0.01% BHA and 0.005% citric acid added as preservatives
Category: Nonionic surfactant
Form: Liquid; Color: Pale yellow
Properties: Solubility: Sol. in alcohol, propylene glycol, cottonseed and mineral oils; Visc.: ≈ 30 cps; HLB: 4.9; Flash Pt.: > 300 F; Fire Pt.: > 300 F; Pour Pt.: ≈ 10 C

Brij 96. I.C.I. United States
Chemical Description: Polyoxyethylene (10) oleyl ether (antioxidants added)
Category: Nonionic surfactant
Form: Yellow liquid with some solids
Properties: Solubility: Sol. in water and alcohol; Visc.: ≈ 100 cps; HLB: 12.4; Flash Pt.: > 300 F; Fire Pt.: > 300 F; Pour Pt.: ≈ 16 C

Brij 97. I.C.I. United States
Chemical Description: Polyoxyethylene (10) oleyl ether with 0.01% BHA and 0.005% citric acid added as preservatives
Category: Nonionic surfactant
Form: Liquid; Color: Pale yellow
Properties: Solubility: Sol. in water and alcohol; Visc.: ≈ 100 cps; HLB: 12.4; Flash Pt.: > 300 F; Fire Pt.: > 300 F; Pour Pt.: ≈ 16 C
Containers: 55 gal closed head steel drum

Brij 98. I.C.I. United States
Chemical Description: Polyoxyethylene (20) oleyl ether (antioxidants added)
Category: Nonionic surfactant
Form: Soft waxy solid; Color: Cream
Properties: Solubility: Sol. in water, alcohol, propylene glycol; HLB: 15.3; Flash Pt.: > 300 F; Fire Pt.: > 300 F; Pour Pt.: ≈ 30 C

Brij 99. I.C.I. United States
Chemical Description: Polyoxyethylene (20) oleyl ether (antioxidants added)
Category: Nonionic surfactant
Form: Solid; Color: Cream
Properties: Solubility: Sol. in water, alcohol, propylene glycol; HLB: 15.3; Flash Pt.: > 300 F; Fire Pt.: > 300 F; Pour Pt.: ≈ 33 C

BTC 50, 50 USP. Onyx Chem. Co.
Chemical Description: *N*-alkyl dimethyl benzyl ammonium chlorides
Category: Disinfectant, cationic
Applications: Hard surface disinfection, sanitization and deodorization
Form: Liquid
Properties: Sp. gr.: 0.96 (25/20 C); Flash Pt.: 126 F
Composition: 50% min. quaternary

BTC 65, 65 USP. Onyx Chem. Co.
Chemical Description: *N*-alkyl dimethyl benzyl ammonium chlorides
Category: Disinfectant, cationic
Applications: Hard surface disinfection, sanitization and deodorization
Form: Liquid
Properties: Sp. gr.: 0.97 (25/20 C); Flash Pt.: 132 F
Composition: 50% min. quaternary

BTC 100. Onyx Chem. Co.
Chemical Description: *N*-alkyl dimethyl benzyl ammonium chlorides
Category: Disinfectant, cationic

Applications: Hard surface disinfection, sanitization and deodorization
Form: Solid
Properties: Sp. gr.: 0.95 (25/20 C); Flash Pt.: > 200 F
Composition: 95% min. quaternary

BTC 776. Onyx Chem. Co.
Chemical Description: *N*-alkyl dimethyl benzyl and dialkyl methyl benzyl ammonium chlorides
Category: Disinfectant, cationic
Applications: Algaecidal and slimicidal agents for swimming pool and industrial water treatment
Form: Liquid
Properties: Sp. gr.: 0.96 (25/20 C); Flash Pt.: 102 F
Composition: 50% min. quaternary

BTC 812. Onyx Chem. Co.
Chemical Description: *N*-octyl dodecyl dimethyl ammonium chlorides
Category: Disinfectant, cationic
Applications: Hard surface disinfection and sanitization fungicide
Form: Liquid
Properties: Sp. gr.: 0.93 (25/20 C); Flash Pt.: 126 F
Composition: 50% min. quaternary

BTC 824. Onyx Chem. Co.
Chemical Description: *N*-alkyl dimethyl benzyl ammonium chlorides
Category: Disinfectant, cationic
Applications: Hard surface disinfection and sanitization, algaecide in swimming pool and industrial water treatment
Form: Liquid
Properties: Sp. gr.: 0.96 (25/20 C); Flash Pt.: 120 F
Composition: 50% min. quaternary

BTC 824 P-100. Onyx Chem. Co.
Chemical Description: *N*-tetradecyl dimethyl benzyl ammonium chloride monohydrate
Category: Disinfectant, cationic
Applications: Nondusting water soluble powder for manufacturing disinfectants, sanitizers, and deodorizers
Form: Powder
Properties: Sp. gr.: 0.40 (25/20 C); Flash Pt.: > 200 F
Composition: 95% min. quaternary

BTC 835. Onyx Chem. Co.
Chemical Description: *N*-alkyl dimethyl benzyl ammonium chlorides
Category: Disinfectant, cationic
Applications: Hard surface disinfection and sanitization, algaecide in swimming pool and industrial water treatment
Form: Liquid
Properties: Sp. gr.: 0.97 (25/20 C); Flash Pt.: 130 F
Composition: 50% min. quaternary

BTC 1100. Onyx Chem. Co.
Chemical Description: *N*-alkyl dimethyl 1-naphthylmethyl ammonium chlorides

Category: Disinfectant, cationic
Applications: High active nondusting powder used as a disinfectant, sanitizer, algaecide, and deodorizer
Form: Powder
Properties: Sp. gr.: 0.50 (25/20 C); Flash Pt.: > 200 F
Composition: 99% min. quaternary

BTC 2125. Onyx Chem. Co.
Chemical Description: *N*-alkyl dimethyl benzyl and dimethyl ethylbenzyl ammonium chlorides
Category: Disinfectant, cationic
Applications: Hard surface disinfection and sanitization, algaecide in swimming pool and industrial water treatment
Form: Liquid
Properties: Sp. gr.: 0.97 (25/20 C); Flash Pt.: > 200 F
Composition: 50% min. quaternary

BTC 2125 80%. Onyx Chem. Co.
Chemical Description: *N*-alkyl dimethyl benzyl and dimethyl ethylbenzyl ammonium chlorides
Category: Disinfectant, cationic
Applications: Hard surface disinfection and sanitization, algaecide in swimming pool and industrial water treatment
Form: Liquid
Properties: Sp. gr.: 0.94 (25/20 C); Flash Pt.: 130 F
Composition: 80% min. quaternary

BTC 2125M. Onyx Chem. Co.
Chemical Description: *N*-alkyl dimethyl benzyl and dimethyl ethylbenzyl ammonium chlorides
Category: Disinfectant, cationic
Applications: Used in disinfectant and sanitizer formulations
Form: Liquid
Properties: Sp. gr.: 0.97 (25/20 C); Flash Pt.: > 200 F
Composition: 50% min. quaternary

BTC 2125M 80%. Onyx Chem. Co.
Chemical Description: *N*-alkyl dimethyl benzyl and dimethyl ethylbenzyl ammonium chlorides
Category: Disinfectant, cationic
Applications: Used in disinfectant and sanitizer formulations
Form: Liquid
Properties: Sp. gr.: 0.94 (25/20 C); Flash Pt.: 130 F
Composition: 80% min. quaternary

BTC 2125 P-40. Onyx Chem. Co.
Chemical Description: *N*-alkyl dimethyl benzyl and dimethyl ethylbenzyl ammonium chlorides
Category: Disinfectant, cationic
Applications: Hard surface disinfection and sanitization, algaecide in swimming pool and industrial water treatment
Form: Powder
Properties: Sp. gr.: 0.55 (25/20 C); Flash Pt.: > 200 F
Composition: 40% min. quaternary

BTC 2565. Onyx Chem. Co.
Chemical Description: *N*-alkyl dimethyl benzyl ammonium chlo-

rides
Category: Disinfectant, cationic
Applications: Algaecidal and slimicidal agents for swimming pool and industrial water treatment
Form: Liquid
Properties: Sp. gr.: 0.96 (25/20 C); Flash Pt.: 110 F
Composition: 50% min quaternary

BTC 8248. Onyx Chem. Co.
Chemical Description: *N*-alkyl dimethyl benzyl ammonium chlorides
Category: Disinfectant, cationic
Applications: Hard surface disinfection and sanitization, algaecide in swimming pool and industrial water treatment
Form: Liquid
Properties: Sp. gr.: 0.95 (25/20 C); Flash Pt.: 110 F
Composition: 80% min. quaternary

BTC 8249. Onyx Chem. Co.
Chemical Description: *N*-alkyl dimethyl benzyl ammonium chlorides
Category: Disinfectant, cationic
Applications: Hard surface disinfection and sanitization, algaecide in swimming pool and industrial water treatment
Form: Liquid
Properties: Sp. gr.: 0.94 (25/20 C); Flash Pt.: 118 F
Composition: 90% min. quaternary

BTC E-8358. Onyx Chem. Co.
Chemical Description: *N*-alkyl dimethyl benzyl ammonium chlorides
Category: Disinfectant, cationic
Applications: Hard surface disinfection and sanitization, algaecide in swimming pool and industrial water treatment
Form: Liquid
Properties: Sp. gr.: 0.93 (25/20 C); Flash Pt.: 100 F
Composition: 80% min. quaternary

BTCO 1010. Onyx Chem. Co.
Chemical Description: *N*-didecyl dimethyl ammonium chloride
Category: Disinfectant, cationic
Applications: Hard surface disinfection and sanitization fungicide
Form: Liquid
Properties: Sp. gr.: 0.89 (25/20 C); Flash Pt.: 86 F
Composition: 50% min. quaternary

Bubble Breaker 259. Witco Chem.
Chemical Description: Blend of organic, nonsilicone compounds
Category: Defoamer
Applications: Manufacture of water-based paper coatings, textile-processing formulations, paints, adhesives, and inks

Bubble Breaker 260. Witco Chem.
Chemical Description: Blend of organic, nonsilicone compounds
Category: Defoamer
Applications: Latex manufacturing operations including polymerization, transfer and monomer stripping; formulation of agricultural chemical products

Bubble Breaker 613-M. Witco Chem.
Chemical Description: Blend of organic, nonsilicone compounds
Category: Defoamer
Applications: Processing of beet sugar; formulation of agricultural chemical products

Bubble Breaker 622. Witco Chem.
Chemical Description: Blend of organic, nonsilicone compounds
Category and Applications: See Bubble Breaker 613-M

Bubble Breaker 730. Witco Chem.
Chemical Description: Blend of organic, nonsilicone compounds
Category and Applications: See Bubble Breaker 259

Bubble Breaker 737. Witco Chem.
Chemical Description: Blend of organic, nonsilicone compounds
Category and Applications: See Bubble Breaker 259

Bubble Breaker 746. Witco Chem.
Chemical Description: Blend of organic, nonsilicone compounds
Category: Defoamer
Applications: Manufacture of water-based paper coatings, textile-processing formulations, paints, adhesives, inks; formulation of agricultural chemical products

Bubble Breaker 748. Witco Chem.
Chemical Description: Blend of organic, nonsilicone compounds
Category and Applications: See Bubble Breaker 746

Bubble Breaker 776. Witco Chem.
Chemical Description: Dispersion of reacted silica in hydrocarbon solvent
Category: Defoamer
Applications: Latex manufacturing operations including polymerization, transfer, monomer stripping and effluent treatment; formulation of water-based paints and adhesives

Bubble Breaker 900. Witco Chem.
Chemical Description: Blend of organic, nonsilicone compounds
Category and Applications: See Bubble Breaker 746

Bubble Breaker 913. Witco Chem.
Chemical Description: Blend of organic, nonsilicone compounds
Category and Applications: See Bubble Breaker 259

Bubble Breaker 917. Witco Chem.
Chemical Description: Blend of organic, nonsilicone compounds
Category: Defoamer
Applications: Formulation of semi- and high-gloss latex paints, agricultural chemical products

Bubble Breaker 1840X. Witco Chem.
Chemical Description: Sodium salt of sulfonated fatty acid
Category: Defoamer
Applications: Defoamer for "wet-process" production of phosphoric acid

Bubble Breaker 3009-F. Witco Chem.
Chemical Description: Blend of organic, nonsilicone compounds
Category: Defoamer
Applications: Defoamer for wash water of sliced potatoes

Bubble Breaker 3017-A. Witco Chem.
Chemical Description: Dispersion of reacted silica in hydrocarbon solvent
Category and Applications: See Bubble Breaker 776

Bubble Breaker 3056-A. Witco Chem.
Chemical Description: Dispersion of reacted silica in hydrocarbon solvent
Category: Defoamer
Applications: Latex manufacturing operations including polymerization, transfer, monomer stripping and effluent treatment; formulation of water-based paints and adhesives, water-based drilling fluids for the petroleum industry

Bubble Breaker 3073-7. Witco Chem.
Chemical Description: Blend of organic, nonsilicone compounds
Category and Applications: See Bubble Breaker D

Bubble Breaker D. Witco Chem.
Chemical Description: Blend of organic, nonsilicone compounds
Category: Defoamer
Applications: For water-based drilling fluids for the petroleum industry

Bubble Breaker PR. Witco Chem.
Chemical Description: Sodium salt of sulfonated fatty acid
Category and Applications: See Bubble Breaker 1840X

C

Calcium Petronate. Witco Chem. Sonneborn Div.
Category: Detergent
Applications: Detergent and rust inhibitor component in lube oil additives, in rust preventive formulations and emulsifier for w/o systems
Form: Viscous liquid; Color: Dark brown
Properties: Solubility: Sol. in organic solvents; Sp. gr.: 0.98
Composition: 44% active ingred; 55% mineral oil; 1.0% water
Reaction: Acid
Toxicity: May cause eye burn
Containers: Drums, tank wagons, tank cars

Calsolene Oil HSA. I.C.I. United States
Chemical Description: Sulfated fatty acid ester
Category: Anionic surfactant
Form: Clear liquid (may be turbid at room temp.); Color: Gold yellow
Properties: Solubility: Dissolves in hard and soft water and in salt solutions; Flash Pt.: 200 F (open cup)
Toxicity: Causes eye burns and skin irritation
Warnings and Caution: Avoid contact; use with forced ventilation

Carbonox. NL Baroid/NL Industries
Category: Thinner, dispersant
Applications: Thinner and dispersant used to control viscosity and gel strengths of drilling fluids, to aid in filtration control, to emulsify oil into water in emulsion muds
Containers: Multiwall paper bags containing 50 lb

Cardene. Cincinnati Milacron
Category: Wetting agent
Applications: Secondary surfactant improving o/w emulsions
Form: Liquid; Color: Light amber
Properties: Solubility: Oil sol.; Sp. gr.: 0.979 @ 60 F; Visc.: 2200 SUS @ 100 F
Composition: 100% active ingred.
Containers: 55 gal drums

Carnapol 77X. Reilly-Whiteman Inc.
Chemical Description: Fatty dia-

mide
Category: Release agent, nonionic
Applications: High melting point wax for use as release agents and wax finishes
Form: Solid
Composition: 100% active

Carsamide 7644. Carson Chem.
Chemical Description: Modified coconut diethanolamide
Category: Emulsifier, thickening agent; nonionic/anionic
Applications: Controlled suds detergent, textile fulling and scouring agent
Form: Liquid; Color: 6 (Gardner)
Properties: Solubility: Sol. in water, alcohol, chlorinated and/or aromatic hydrocarbons; dispersible in mineral oil, kerosene, natural fats and oils, mineral spirits; Density: 8.3 lb/gal

Carsamide C-3. Carson Chem.
Category: Intermediate, emulsifier
Applications: Intermediate for liquid detergent formulation; car and hair shampoos; improves the foaming, detergency and hard water resistance of soaps; solvent cleaners and dry cleaning formulations; emulsifier for aromatic solvents
Containers: 55 gal steel drums

Carsamide C-7944. Carson Chem.
Form: Amber liquid free of foreign matter

Carsamide CA. Carson Chem.
Chemical Description: High activity coco diethanolamide
Category: Detergent, emulsifier, dispersant, wetting agent, nonionic
Applications: Industrial, institutional, cosmetic and household cleaners
Properties: Solubility: Sol. in water, alcohol, chlorinated hydrocarbons, aromatic hydrocarbons, "Polysolve," "Cellosolves," "Carbitols" and natural fats and oils; Density: 8.3 lb/gal; Sp. gr.: 1.0; Biodegradable
Composition: 100% active ingred.

Carsamide LE. Carson Chem.
Chemical Description: Super coco fatty diethanolamide
Form: Liquid; Color: Light amber

Carsamide SAC. Carson Chem.
Chemical Description: Coco super diethanolamide
Category: Nonionic detergent, emulsifier, dispersant, wetting agent, viscosity builder
Applications: Detergent for industrial, institutional, cosmetic and household cleaners; foam stabilizer for shampoos, bubble baths, detergents
Form: Viscous liquid; Color: Pale amber; Odor: Mild, fatty
Properties: Solubility: Sol. in water, alcohol, chlorinated and aromatic hydrocarbons, "Polysolvs," "Cellosolves," "Carbitols," natural fats and oils; Density: 8.2 lb/gal
Composition: 100% active ingred.

Carsamide SAL-7. Carson Chem.
Chemical Description: High activity lauric diethanolamide
Category: Nonionic detergent,

emulsifier, foaming agent and foam stabilizer
Applications: Foam stabilizer, thickener, detergent and foaming agent in household, institutional and industrial cleaners; car washes, rug and upholstery cleaners, shampoos, bubble baths, lotions, creams
Form: Solid; Color: 2 (Gardner); Odor: Mild, fatty
Properties: Solubility: Sol. in alcohol, chlorinated and aromatic hydrocarbons; Density: 8.2 lb/gal; M.P.: 104–114 F; Biodegradable
Containers: 55 gal open top steel drums

Carsamide SAL-9. Carson Chem.
Chemical Description: High activity lauric diethanolamine
Category: Detergent, emulsifier foaming agent, nonionic
Applications: Foam stabilizer thickener, detergent and foaming agent in household, institutional and industrial cleaners, car washes rug and upholstery cleaners, shampoos, bubble baths, lotions, creams
Form: Solid; Color: 2 (Gardner); Odor: Mild, fatty
Properties: Solubility: Sol. in alcohol, chlorinated and aromatic hydrocarbons; Density: 8.1 lb/gal; Biodegradable
Containers: 55 gal open top steel drums

Carsamide SAL-82. Carson Chem.
Form: Clear yellow liquid free from foreign matter
Properties: Visc.: 500 cps max.

Carsamine 2414. Carson Chem.
Chemical Description: A mixture of aliphatic and heterocyclic mono and diamines
Properties: Solubility: Complete in water; Density: 9.0 lb/gal; Boiling range: 242–310 C; decomposes at 310 C

Carsamine CB. Carson Chem.
Form: Clear liquid; Color: Amber
Properties: Density: 7.6 lb/gal

Carsamine Oxide LM. Carson Chem.
Chemical Description: Lauryl/myristyl-dimethyl amine oxide
Category: Foaming and conditioning agent
Applications: Foaming and conditioning agent in household detergents, industrial detergents, shampoos, bubble baths; foam stabilizer; wetting agent in concentrated electrolyte sol'ns.; mildly cationic under acidic conditions
Form: Clear liquid; Color: 2 (Gardner)
Composition: 29% active ingred.
Reaction: Compatible with anionic and nonionic detergents

Carsemol A-500. Carson Chem.
Chemical Description: Ethyl palmitate, ethyl myristate, ethyl stearate and ethyl pelargonate
Applications: Emollient creams and lotions, moisturizers, toners, eye make-up, hair products, lipsticks, nail enamels, bath products
Properties: Sp. gr.: 0.865; Cloud Pt.: 11 C

Carsodet TD-7C. Carson Chem.
Chemical Description: Trideceth-7

carboxylic acid supplied as free carboxylic acid
Applications: Neutralized with other materials to produce mild anionic surfactants; topical cleansing products; shampoos
Form: Colorless liquid
Properties: Density 8.5 lb/gal; pH: 1 (as is)

Carsofoam 211. Carson Chem.
Chemical Description: Blend of sulfonates, nonionics, and coconut derived alkanolamides
Category: Detergent concentrate
Applications: All-purpose cleaners, solvent emulsion cleaners, paint strippers, high foam light-duty liquids; base for aircraft cleaners, dairy cleaners, dishwashing liquids, car washes, metal cleaners, laundry cleaners, textile processing
Form: Liquid; Color: Amber; Odor: Mild
Properties: Density: 8.8 lb/gal; Clear Pt.: 1 C; pH: 8.2 ± 0.2; Biodegradable
Composition: 100% active ingred.

Carsofoam MS Concentrate. Carson Chem.
Category: Detergent
Applications: Shampoo concentrate esp. for mild shampoos and baby shampoos
Form: Hazy liquid; Color: Light yellow
Properties: pH ≅ 7.3 (10% sol'n.)
Composition: 50% active ingred.; 2% sodium chloride

Carsofoam PS-1. Carson Chem.
Category: Shampoo concentrate
Applications: Manufacture of liquid cream and pearlized shampoos
Form: Pearlized lotion; Color: White; Odor: Mild, pleasant
Composition: ≈ 38% active

Carsofoam PS-3. Carson Chem.
Category: Shampoo concentrate
Applications: Manufacture of liquid pearlized and cream shampoos
Form: Pearlized lotion; Color: White; Odor: Mild, pleasant
Properties: pH ≅ 7 (10% sol'n.)
Composition: ≈ 38% active

Carsofoam SC. Carson Chem.
Chemical Description: Sodium laureth sulfate, cocamide DEA, DEA lauryl sulfate and tetrasodium EDTA
Category: Detergent concentrate
Applications: High quality shampoos and bubble baths
Form: Thick, flowable gel; Color: 4 (Gardner)
Properties: pH ≅ 7 (10% sol'n.)
Composition: 65% solids

Carsofoam T-60-L. Carson Chem.
Chemical Description: Triethanolamine salt of linear alkyl benzene sulfonic acid
Category: High foaming detergent, wetting agent
Applications: Liquid cleaning compounds for cosmetic, industrial and institutional usage
Form: Clear, viscous liquid; Color: < 5 (Gardner); Odor: Bland
Properties: Density: 9.0 lb/gal; pH: 7.0–7.6 (as is); Biodegradable

Composition: 60% solids
Containers: 55 gal lined steel or liquipak drums

Carsofos 4326. Carson Chem.
Chemical Description: Anionic/nonionic blend
Category: Anionic detergent, wetting agent, emulsifier, penetrant
Applications: All-purpose cleaners
Form: Liquid; Color: Pale amber; Odor: Mild, fatty
Properties: Density: 8.25 lb/gal; Sp. gr.: 0.989; pH: 7.5–8.5 (as is)
Composition: 92% solids

Carsofos N. Carson Chem.
Applications: Formulation of aqueous degreasers
Form: Clear viscous liquid; Color: 5 (Gardner)
Properties: Density: 8.39 lb/gal; Visc.: 300 cps (10% sol'n. in water); pH: 9.2 (1% sol'n.)

Carsofos NP-6. Carson Chem.
Form: Viscous liquid; Color: Amber, 5 (Gardner)
Properties: Acid Value: 105; pH: 2.5 (1% aq.)

Carsofos NP-9. Carson Chem.
Form: Slightly hazy viscous liquid; Color: Yellow, 4 (Gardner)
Properties: Acid Value: 120; pH: 2.0 (1% aq.)

Carsonam 3. Carson Chem.
Category: Amphoteric surfactant, detergent, wetting agent, emulsifier, foaming and coupling agents
Applications: Shampoos, lotions, antidandruff shampoos and germicides; chemical specialties such as liquid hand cleaners, rug and upholstery shampoos, dishwashing compounds
Form: Liquid; Color: 5
Properties: Density: 8.45 lb/gal; pH: 5.5–7.5 (10% sol'n.); Biodegradable
Composition: 30% active ingred; 64% water; 6% max. chloride

Carsonam 32. Carson Chem.
Chemical Description: Coco amido betaine (industrial grade)
Applications: Industrial use
Form: Clear liquid; Color: Amber
Properties: Solubility: Solvent–Proprietary ethanol; pH: 6.0–7.0 (as is)
Composition: 58–62% solids; 4% max. sodium chloride

Carsonam 33-S. Carson Chem.
Category and Applications: See Carsonam 3
Form: Liquid; Color: 6 max.
Properties: Density: 8.45 lb/gal; pH: 5.5–7.5 (10% sol'n.); Biodegradable
Composition: 30% active ingred.; 63% water; 7% max. chloride

Carsonam 3147. Carson Chem.
Category and Applications: See Carsonam 3
Form: Liquid; Color: 5 max.
Properties: Density: 8.45 lb/gal; Biodegradable
Composition: 33% active ingred.; 64% water; 2% max. chloride

Carsonam BCW. Carson Chem.
Chemical Description: Coco fatty

betaine
Category: Foaming and conditioning agent
Applications: Used in household and industrial detergents; also hair shampoos, bubble baths
Form: Liquid; Color: 2 (Gardner)
Properties: Density: 8.5 lb/gal; Sp. gr.: 1.044; pH: 8.0 (as is); Biodegradable
Composition: 44% solids; 56% water; 8% sodium chloride

Carsonam C. Carson Chem.
Chemical Description: Coco imidazoline betaine
Category: Emulsifier, detergent, sequestrant, mild germistat and fungistat, amphoteric
Applications: Human, rug, upholstery shampoos; heavy duty liquid cleaners, dishwashing compounds, liquid steam cleaners
Form: Liquid; Color: 5 (Gardner)
Properties: pH: 9.7 (as is); Biodegradable
Composition: 37% solids content

Carsonam C-SF. Carson Chem.
Chemical Description: Coco imidazoline betaine–salt free
Category and Applications: See Carsonam C
Form: Liquid; Color: 5 (Gardner)
Properties: pH: 9.7 (as is); Biodegradable
Composition: 37% solids content

Carsonam C-SPCL. Carson Chem.
Chemical Description: Coco imidazoline betaine
Category and Applications: See Carsonam C
Form: Color: 7 (Gardner)
Properties: pH: 6.0 (as is); Biodegradable
Composition: 48% solids content

Carsonam DC. Carson Chem.
Chemical Description: Coco imidazoline betaine dicarboxylate
Category and Applications: See Carsonam C
Form: Gel; Color: 9 (Gardner)
Properties: pH: 8.5 (5% sol'n.); Biodegradable
Composition: 50% solids; 11% sodium chloride; 2.4% nitrogen

Carsonam DC-SF. Carson Chem.
Chemical Description: Coco imidazoline betaine dicarboxylate–salt free
Category and Applications: See Carsonam C
Form: Gel; Color: 8 (Gardner)
Properties: pH: 9.6 (10% sol'n.); Biodegradable
Composition: 38% solids; 2.5% nitrogen

Carsonam DC-70%-SF. Carson Chem.
Chemical Description: Coco imidazoline betaine dicarboxylate–salt free
Category and Applications: See Carsonam C
Form: Color: 11 (Gardner)
Properties: pH: 9.6 (10% sol'n.); Biodegradable
Composition: 70% solids; 4.35% nitrogen

Carsonam L. Carson Chem.
Chemical Description: Lauryl

imidazoline betaine
Category and Applications: See Carsonam C
Form: Liquid; Color: 6 (Gardner)
Properties: Sp. gr.: 1.1; pH: 9.3 (as is); Biodegradable
Composition: 43% solids; 6.5% sodium chloride; 2.6% nitrogen

Carsonol ALS. Carson Chem.
Chemical Description: Ammonium lauryl sulfate
Category: Detergent with high foam, good wetting and emulsifying properties, anionic
Applications: Cosmetics–shampoos, bubble baths, shaving and cleansing creams, skin lotions; Chemical Specialties–industrial cleaners, rug and upholstery cleaners, liquid household detergents, vegetable scrubbing, emulsion polymerization, automobile and pet shampoos
Form: Viscous liquid; Color and Odor: Low
Properties: Solubility: Water sol.; Density: 8.4 lb/gal; Stability: Good > pH 4; pH: 6.5–7.0 (10% sol'n.); Biodegradable
Composition: 30% active ingred.

Carsonol ALS Special. Carson Chem.
Chemical Description: Ammonium lauryl sulfate
Category and Applications: See Carsonol ALS
Form: Clear liquid
Properties: Solubility: Water sol.; Density: 8.4 lb/gal; Stability: Good > pH 4; pH: 6.5–7.0 (10% sol'n.); Biodegradable
Composition: 30% active ingred.

Carsonol BD. Carson Chem.
Chemical Description: Blend of carboxybetaine, sulfate and amine groups
Category: Detergent, foaming agent, anionic/amphoteric
Applications: Shampoo concentrate with good hair manageability, foaming, cleansing and hard water stability; bubble baths
Form: Moderately viscous liquid; Color: Light amber
Properties: Solubility: Excellent water sol.; Cloud Pt.: 50 F; pH: 7.5–8.5
Composition: 30–32% active

Carsonol DLS. Carson Chem.
Chemical Description: Diethanolamine lauryl sulfate
Category and Applications: See Carsonol ALS
Form: Clear liquid; Color and Odor: Low
Properties: Solubility: Water sol.; Density: 8.49 lb/gal; pH: 7.5–8.5 (10% sol'n.); Biodegradable
Composition: 35% active ingred.

Carsonol ILS. Carson Chem.
Chemical Description: Mixed isopropanolamine lauryl sulfate
Category and Applications: See Carsonol ALS
Form: Viscous liquid; Color and Odor: Low
Properties: Solubility: Water sol.; Density: 8.63 lb/gal; pH: 7.0–7.4 (10% sol'n.); Biodegradable
Composition: 75% active ingred.

Carsonol MLS. Carson Chem.
Chemical Description: Magnesium lauryl sulfate
Category and Applications: See Carsonol ALS
Form: Viscous liquid; Color and Odor: Low
Properties: Solubility: Water sol.; Density: 8.57 lb/gal; pH: 6.5–7.5 (10% sol'n.); Biodegradable
Composition: 30% active ingred.

Carsonol SES-A. Carson Chem.
Chemical Description: Ammonium lauryl ether sulfate
Category: Anionic surfactants with excellent foaming in hard and soft water
Applications: Shampoos, bubble baths, liquid detergents, automobile shampoos, household chemical specialties, industrial foaming agents
Form: Liquid
Properties: Density: 8.4 lb/gal; pH: 6.5–7.0 (10% sol'n.); Biodegradable
Composition: 60% active ingred.

Carsonol SES-S. Carson Chem.
Chemical Description: Sodium lauryl ether sulfate
Category and Applications: See Carsonol SES-A
Form: Liquid
Properties: Density: 8.8 lb/gal; Stability: Good in acidic and alkaline media; pH: 7.0–8.0 (10% sol'n.); Biodegradable
Composition: 60% active ingred.

Carsonol SHS. Carson Chem.
Chemical Description: Sodium 2-ethylhexyl sulfate
Category: Low foam detergent, wetting agent, penetrant
Applications: Used in caustic solutions for peeling of fruits and vegetables; very stable to high electrolyte and in acid and basic solutions
Form: Clear liquid; Color: 3 (Gardner)
Properties: Density: 9.6 lb/gal; Sp. gr.: 1.15; Visc.: 35 cps; pH: 10.3 (10% sol'n.)
Composition: 40% active ingred.
Containers: 480 lb net

Carsonol SLES. Carson Chem.
Chemical Description: Sodium lauryl ether sulfate
Category and Applications: See Carsonol SES-A
Form: Liquid
Properties: Density: 8.7 lb/gal; Stability: Good in acidic and alkaline media; pH: 7.5–8.5 (10% sol'n.); Biodegradable
Composition: 30% active ingred.

Carsonol SLES-2. Carson Chem.
Chemical Description: Sodium lauryl ether sulfate
Category and Applications: See Carsonol SES-A
Form: Liquid
Properties: Density: 8.63 lb/gal; Stability: Good in acidic and alkaline media; pH: 7.0–8.0 (10% sol'n.); Biodegradable
Composition: 30% active ingred.

Carsonol SLS. Carson Chem.
Chemical Description: Sodium lauryl sulfate

Category and Applications: See Carsonol ALS
Form: Clear liquid; Color and Odor: Low
Properties: Solubility: Water sol.; Density: 8.7 lb/gal; pH: 7.5–8.5 (10% sol'n.); Biodegradable
Composition: 30% active ingred.

Carsonol SLS Paste B. Carson Chem.
Chemical Description: Sodium lauryl sulfate
Category and Applications: See Carsonol ALS
Form: Paste; Color and Odor: Low
Properties: Solubility: Water sol.; Density: 8.8 lb/gal; pH: 8.4–8.8 (10% sol'n.); Biodegradable
Composition: 30% active ingred.

Carsonol SLS Special. Carson Chem.
Chemical Description: Sodium lauryl sulfate
Category and Applications: See Carsonol ALS
Form: Liquid; Color and Odor: Low
Properties: See Carsonol SLS
Composition: 30% active ingred.

Carsonol TLS. Carson Chem.
Chemical Description: Triethanolamine lauryl sulfate
Category and Applications: See Carsonol ALS
Form: Clear liquid; Color and Odor: Low
Properties: Solubility: Water sol.; Density: 8.7 lb/gal; pH: 7.0–7.5 (10% sol'n.); Biodegradable
Composition: 40% active ingred.

Carsonon L-985. Carson Chem.
Chemical Description: 9–10 mole ethoxylated linear primary alcohol
Category: Wetting agent, detergent, emulsifier, dispersant, nonionic
Applications: Maintenance and institutional cleaners; textile processing; metal cleaning; paper and paint industries
Form: Liquid; Color: Colorless
Properties: Solubility: Sol. in acid and alkaline solutions; Density: 8.32 lb/gal; Cloud Pt.: ≈ 137 F; Pour Pt.: 39–43 F; Solidification Pt.: 30–34 F; Stability: Stable in acid and alkaline solutions; pH: neutral (3% sol'n.); Biodegradable
Composition: 85% active ingred.

Carsonon LF-5. Carson Chem.
Chemical Description: Linear alcohol alkoxylate
Category: Low foam surfactant, nonionic
Applications: Pulp and paper additives, mechanical dishwashing, metal cleaning, dairy equipment cleaning
Form: Color: 100 (APHA); Odor: Mild
Properties: Solubility: Sol. in aromatic solvents, polar solvents, water below cloud pt. and in the common chlorinated solvents; Sp. gr.: 1.02 (60 F/60 F); Visc.: 125 cts; Cloud Pt.: 43–49 C; Flash Pt.: 460 F (COC); Stability: Stable to dry alkalis; Surface Tension: 33.5 dynes/cm (0.1% conc.); pH: 5.5–7.0 (1% aq.); Biodegradable

Carsonon LF-46. Carson Chem.
Chemical Description: Polyoxyalkalene glycol
Category: Low foam surfactant, nonionic
Applications: Low foam use as pulp and paper additives, mechanical dishwashing, metal cleaning, dairy equipment cleaning
Form: Color: 1 (Gardner); Odor: Mild
Properties: Solubility: Sol. in aromatic solvents, polar solvents, water below cloud pt., and in the common chlorinated solvents; Visc.: 200 cps; Cloud Pt.: 44 C; Flash Pt.: > 230 F; pH: 5.5–7.0 (1% aq.); Biodegradable

Carsonon N-4. Carson Chem.
Chemical Description: 4-mole ethoxylated nonylphenol
Category: Emulsifier, detergent, dispersant, nonionic
Applications: Oil soluble emulsifier; oil soluble detergent and dispersing agent for petroleum oils; intermediate in the synthesis of high foaming, water soluble sulfate esters; drycleaning detergent
Form: Liquid; Color: Pale
Properties: Solubility: Sol. in kerosene, Stoddard solvent, alcohol, chlorinated and aromatic solvents; Density: 8.6 lb/gal; Sp. gr.: 1.02; Visc.: 175–250 cps; Flash Pt.: 500–600 F; Fire Pt.: 500–600 F; Pour Pt.: -5 ± 2 F; Solidification Pt.: -20 ± 2 F
Composition: 100% active ingred.

Carsonon N-6. Carson Chem.
Chemical Description: 6-mole ethoxylated nonylphenol
Category: Emulsifier, wetting agent, nonionic
Applications: Used for foam control in household detergent, industrial cleaning compounds; low foaming wetting agent; used with Carsonon N-9 to improve oil solubility and detergency in solvent systems
Form: Liquid; Color: Pale; Odor: Mild, pleasant
Properties: Solubility: Sol. in alcohols, aromatic, chlorinated and aliphatic solvents, fatty oil; dispersible in water; Density: 8.68 lb/gal; Visc.: 180–240; Cloud Pt.: 90–102 F; Flash Pt.: 500 F; Solidification Pt.: –33 to –29 F
Composition: 100% active ingred.

Carsonon N-8. Carson Chem.
Chemical Description: 8–8½ mole ethoxylated nonylphenol
Category: Wetting agent, detergent, emulsifier, dispersant, nonionic
Applications: Household and industrial cleaners, textile processing, paper industry, wetting agent for use with mineral acids and corrosion inhibitors
Form: Liquid; Color: Pale; Odor: Mild, pleasant
Properties: Solubility: Sol. in water, alcohol, chlorinated and aromatic solvents; Density: 8.75 lb/gal; Visc.: 200–300 cps; Cloud Pt.: 82–86 F; Flash Pt.: ≈ 435 F; Fire Pt.: ≈ 465 F; Pour Pt.: 41–45 F; Solidification Pt.: 30–34 F; pH: neutral (1% sol'n.)
Composition: 100% active ingred.

Carsonon N-9. Carson Chem.
Chemical Description: 9–10 mole ethoxylated nonylphenol
Category: Wetting agent, detergent, emulsifier, dispersant, nonionic
Applications: Household and industrial cleaners; textile processing; paper industry; wetting agent for use with mineral acids and corrosion inhibitors
Form: Liquid; Color: Pale; Odor: Mild, pleasant
Properties: Solubility: Sol. in water, alcohol, chlorinated and aromatic solvents; Density: 8.83 lb/gal; Visc.: 180–250 cps; Flash Pt.: ≈ 540 F; Fire Pt.: ≈ 600 F; Cloud Pt.: ≈ 130 F; Pour Pt.: 35–39 F; Solidification Pt.: 24–28 F; pH: neutral (1% sol'n.)
Composition: 100% active ingred.

Carsonon N-10. Carson Chem.
Chemical Description: 10–11 mole ethoxylated nonylphenol
Category: Wetting agent, detergent, emulsifier, dispersant, nonionic
Applications: Household and industrial cleaners, textile processing, paper industry, wetting agent for use with mineral acids and corrosion inhibitors
Form: Liquid; Color: Pale; Odor: Mild, pleasant
Properties: Solubility: Sol. in water, alcohol, chlorinated and aromatic solvents; Density: 8.84 lb/gal; Visc.: 180–250 cps; Cloud Pt.: 143–146 F; Flash Pt.: ≈ 540 F; Fire Pt.: ≈ 600 F; Pour Pt.: 35–39 F; Solidification Pt.: 24–28 F; pH: neutral (1% sol'n.)
Composition: 100% active ingred.

Carsonon N-30. Carson Chem.
Chemical Description: 30 mole ethoxylated nonylphenol
Category: Wetting agent, detergent, emulsifier, nonionic
Applications: Emulsion polymerization; emulsifier for fats, oils, waxes and essential oils
Form: Solid; Odor: Mild, pleasant
Properties: Density: 9.75 lb/gal; Sp. gr.: 1.17; Cloud Pt.: Clear @ 212 F (1% sol'n.); Solidification Pt.: 104 ± 2 F; pH: neutral (1% sol'n.)
Composition: 100% active ingred.

Carsonon N-30, 70%. Carson Chem.
Category and Applications: See Carsonon N-30
Form: Liquid to semisolid; Color: Pale yellow
Properties: Sp. gr.: 1.09; Cloud Pt.: 167–171 F; pH: 6.0–7.0 (3% sol'n.)
Composition: 70% active ingred.; 30% water

Carsonon N-40, 70%. Carson Chem.
Form: Semisolid; Color: White to yellow
Properties: Cloud Pt.: Clear @ 212 F (1% sol'n.); pH: 6.0–7.0 (3% sol'n.)
Composition: 70% active, 30% water

Carsonon N-50, 70%. Carson Chem.
Form: Semisolid; Color: White to yellow
Properties: Cloud Pt.: Clear @ 212 F (1% sol'n.); pH: 6.0–7.0

(3% sol'n.)
Composition: 70% active ingred.; 30% water

Carsonon N-100, 70%. Carson Chem.
Chemical Description: Polyethoxylated nonylphenol
Category: Emulsifier, stabilizer
Applications: Effective at high temperatures and in concentrated electrolyte solutions; very high water solubility
Form: Semisolid; Color: White to off white
Properties: Solubility: Very high water sol.; Cloud Pt.: Clear @ 212 F (1% sol'n.); pH: 6.0–7.0 (3% sol'n.)
Composition: 70% active, 30% water

Carsonon ND-317. Carson Chem.
Chemical Description: Polyethoxylated nonylphenol
Category: Nonionic surfactant
Applications: Base for solid toilet bowl cleaners
Form: Opaque, viscous liquid; Color: 2 (Gardner)
Properties: Sp. gr.: 1.11; M.P.: High; Cloud Pt.: 162 ± 2 F (1% in 10% NaCl sol'n.)

Carsonon P-30M. Carson Chem.
Form: Color: 5 max (Molten)
Properties: Cloud Pt.: 30–32 ml H_2O; pH: 7.5–8.0 (3% aq.)
Composition: 1% max water

Carsonon TD-10. Carson Chem.
Chemical Description: 10 mole ethoxylated tridecyl alcohol
Category: Emulsifier, detergent, wetting agent, foam builder, solubilizer, nonionic
Applications: Household industrial cleaners; textile processing; paper industry; emulsifier for essential oils, aromatic solvents, fats; wetting agent for use with mineral acids and corrosion inhibitors
Form: Liquid; Color: 1 (Gardner)
Properties: Solubility: Water sol.; Density: 8.5 lb/gal; HLB: 13.7; Cloud Pt.: 162–166 F (1% in distilled water); Flash Pt.: > 300 F (COC); Pour Pt.: 67 F; pH: 5.0–7.0 (3% aq.)
Composition: 100% active ingred.

Carsonon TD-11. Carson Chem.
Chemical Description: 11 mole ethoxylated tridecyl alcohol; available as 100% material and 70% solution in water
Category: Detergent, emulsifier, wetting agent, nonionic
Applications: Household and industrial cleaners; textile processing; paper industry; emulsifier for essential oils, aromatic solvents, fats; wetting agent for use with mineral acids and corrosion inhibitors; high foaming properties; high temperature detergency
Form: Waxy solid; Color: 1 (Gardner)
Properties: Density: 8.5 lb/gal; HLB: 14.6; Cloud Pt.: 137–141 F (1% in 10% NaCl sol'n.); Flash Pt.: > 300 F (COC); Pour Pt.: 92 F; pH: 5.0–7.0 (3% aq.)
Composition: Available as 100% material or 70% solution in water
Reaction: Compatible with electrolytes, strong alkalis

Carsoquat 621. Carson Chem.
Chemical Description: n-alkyl dimethyl benzyl ammonium chloride
Category: Corrosion inhibitor, cationic surfactant
Form: Liquid; Color: Almost water-white to pale yellow; Odor: Mild
Properties: Solubility: Miscible with water, lower alcohols, most polar solvents; Density: 8 lb/gal; Sp. gr.: 0.96; Stability: Good
Composition: 50% active
Reaction: Compatible with nonionic and amphoteric detergents
Containers: 55 gal drums
Warnings and Cautions: Avoid contact with eyes, skin, clothing; wash affected areas with plenty of soap and water

Carsoquat 621 (80%). Carson Chem.
Chemical Description: Alkyl dimethyl ammonium chloride
Category and Applications: See Carsoquat 621
Form: Clear liquid; Color: Straw
Properties: m.w.: 361; pH: 7–8 (1% sol'n.)
Composition: 80% active ingred.; 10% water; 10% alcohol

Carsoquat 816-C. Carson Chem.
Chemical Description: Blended cetearyl alcohol, PEG-40 castor oil and stearalkonium chloride
Category: Creme rinse concentrate
Applications: Finished creme rinse requires only dispersion in water; compatible with hydrolized animal protein, lanolin, lanolin derivatives
Form: White waxy flakes; Odor: Mild, fatty
Properties: Solubility: Disperses readily in water; pH: 6.3 (5% disp.)
Composition: 96% min active

Carsoquat CTM-29. Carson Chem.
Chemical Description: Cetyl trimethyl ammonium chloride
Category: Cationic surfactant
Applications: Conditioners in specialty hair preparations; coagulating agent in the manufacture of antibiotics and other pharmaceuticals
Form: Liquid; Color: Almost clear to pale yellow
Properties: m.w.: 319; Density: 8.1 lb/gal; Sp. gr.: 0.968; pH: 7.0–9.0 (10% active)
Composition: 29% active ingred.; 71% water
Reaction: Compatible with nonionic and cationic polymers, organic chelates, most polar solvents and inorganic detergent builders

Carsoquat CTM-429. Carson Chem.
Chemical Description: Cetyl trimethyl ammonium chloride
Category and Applications: See Carsoquat CTM-29
Form and Properties: Identical to Carsoquat CTM-29 except for pH: 3.5–4.0 (2% active)
Composition: 29% active; 71% water

Carsoquat SDQ-25. Carson Chem.
Chemical Description: Stearyl dimethyl benzyl ammonium chloride
Category: Cationic surfactant
Applications: Softener, creme hair

rinse, hair conditioner; for cosmetics and natural fibers
Form: Thick creamy paste; Color: White; Odor: Mild and sweet
Properties: pH: 3.0–4.0 (1% disp.)
Composition: 20–21 % active

Carsoquat SDQ-85. Carson Chem.
Chemical Description: Stearyl dimethyl benzyl ammonium chloride
Category: Hair conditioner and softener, cationic
Form: Waxy flake; Color: White
Properties: M.P.: ≈ 140 F; pH: 5–7 (0.5 sol'n.)
Composition: 85% active ingred.

Carsosoft S-75. Carson Chem.
Chemical Description: Alkyamidoethyl alkyl imidazolinium methyl methosulfate
Category: Fabric softener base, cationic
Applications: Fabric softener base with lubricity, softness and antistatic and antilinting properties, excellent rewet properties and nonyellowing
Form: Liquid; Color: 4 (Gardner)
Properties: m.w.: 720; Stability: Excellent thermal stability; pH: 5–7 (5% sol'n.)
Composition: 75% active ingred.; diluent: isopropanol

Carsosoft S-90. Carson Chem.
Form: Semisold; Color: Yellow
Properties: m.w.: 723; pH: 5.0–7.0 (5% aq.)
Composition: 85–88% active

Carsosoft S-90-M. Carson Chem.
Chemical Description: Modified alkyamidoethyl alkyl imidazolinium methyl methosulfate
Category and Applications: See Carsosoft S-75
Form: Liquid; Color: 4 (Gardner)
Properties: Solubility: Readily dispersible in warm and cold water; Stability: Excellent thermal stability; pH: 5.0–7.0 (5% sol'n.)
Composition: 88–91% active ingred.; diluent: isopropanol

Carsosoft T-90. Carson Chem.
Chemical Description: Quaternary ditallow diamido
Category: Fabric softener, cationic
Applications: Home and commercial laundry fabric softener with excellent softening, rewet properties, nonyellowing, antistat and antilint properties
Form: Liquid; Color: Yellow
Properties: pH: 5–7 (5% sol'n.)
Composition: 88–90% active ingred.; Diluent: isopropanol

Carsosoft V-90. Carson Chem.
Chemical Description: Dihydrogenated tallow dimethyl ammonium methosulfate
Form: Paste; Color: White, 4 (Gardner)
Properties: pH: 5.5 (10% in 1:1 IPA/aq.)
Composition: 91% solids

Carsosoft V-100. Carson Chem.
Chemical Description: Dihydrogenated tallow dimethyl ammonium sulfate
Form: Flake or powder; Color: White
Properties: pH: 5.5–6.5 (10% in

1:1 IPA/aq.)
Composition: 95% solids

Carsosulf SXS-Liquid. Carson Chem.
Chemical Description: Sodium xylene sulfonate
Category: Coupling agent, solubilizer, anionic
Applications: Nonflammable coupling agent and solubilizer in liquid detergent systems; esp. for coupling inorganics into cleaning compounds
Form: Liquid
Properties: Sp. gr.: 1.18–1.22; pH: 7.5–10.5 (as is)
Composition: 40% active ingred.

Carsosulf UL-100 Acid. Carson Chem.
Chemical Description: Linear alkyl benzene sulfonic acid
Properties: m.w.: 316; Density: 8.84 lb/gal; Sp. gr.: 1.06; Biodegradable
Composition: 97 ± 1% active
Containers: 55 gal steel lined drum

Cartaretin F-4. Sandoz Colors and Chem.
Category: Cationic polymer
Applications: Conditioning agent for hair and skin; as wetting agent for sizing
Form: Liquid; Color: Light yellow; Odor: Natural
Properties: Solubility: Sol. in water and alcohol @ 20 C; Sp. gr.: 1.07; Visc.: 400 cps; F.P.: –4 C; B.P.: 220 F; Stability: Very stable
Composition: 30% active ingred.
Containers: 500 lb fiber drum
References: Bulletin 7-159/78

Cartaretin F-8. Sandoz Colors and Chem.
Category: Cationic polymer
Applications: Conditioning agent for hair and skin; wetting agent in sizing
Form: Liquid; Color: Amber; Odor: Natural
Properties: Solubility: Sol. in water and alcohol @ 20 C; Sp. gr.: 1.07; Visc.: 800 cps; F.P.: –4 C; B.P.: 220 F; Stability: Very stable
Composition: 30% active ingred.
Containers: 500 lb fiber drum
References: Bulletin 7-159/78
Patents: U.S. 3,632,559

Casul 70 HF. Thompson-Hayward Chemical Co.
Chemical Description: Calcium salt of a branched chain dodecylbenzene alkylate
Category: Emulsifier, anionic
Applications: Co-emulsifier for oil-in-water and water-in-oil formulations
Form: Liquid; Color: Amber
Properties: Density: 8.50 lb/gal; Sp. gr.: 1.02; Visc.: 18,400 cps; Flash Pt.: 104 F (PMCC); Pour Pt.: 46 F; pH: 5.0–7.0
Composition: 70% active ingred.
Containers: 55 gal. drums

Cation Softener X. Onyx Chem. Co.
Chemical Description: Alkyl imidazoline derivative
Category: Softener, lubricant, mordant
Applications: Softener, lubricant and glass fiber mordant with good substantivity

Form: Paste
Properties: Sp. gr.: 0.92 (25/20 C); Flash Pt.: 60 F

CC-16. NL Baroid Petroleum Services
Chemical Description: Sodium salt of a humic acid material
Category: Drilling fluid additive, emulsifier, gel dispersing agent, anionic
Applications: Thinner and emulsifier for oil-in-water emulsion muds esp. in normal-to-high pH drilling fluids
Form: Free flowing powder
Properties: Solubility: Sol. in water-based drilling fluids; Stability: Stable at high temperature
Composition: 100% active ingred.
Containers: 50 lb multiwall paper bags

CC-603. PVO International
Chemical Description: Blend of carrageenan, guar gum and dextrose
Category: Stabilizer
Applications: Food grade stabilizer for use in whipped toppings, fillings, coffee whiteners

Centrol Series. Central Soya Co.
Chemical Description: Natural soybean lecithin
Category: Wetting agent, emulsifier, emollient, softening agent, lubricant, solubilizer
Applications: Cosmetics; food processing; pharmaceuticals; coatings manufacture; plastic and rubber industry; glass and ceramics processing; paper and printing industry; masonry and asphalt products; petroleum industry as specialty oils and greases; metal processing; pesticides; textiles and leather
Form: Liquid
Composition: 100% active ingred.

Centrolene Series. Central Soya Co.
Chemical Description: Modified soybean lecithin
Category: Wetting agent, emulsifier, release agent, dispersing agent, lubricant, anticorrosive agent, emollient
Applications: Food processing; cosmetics; pharmaceuticals; coatings manufacture; plastic and rubber industries; glass and ceramics; paper and printing; masonry and asphalt products; petroleum industry as specialty greases and oils; metal processing; textiles and leather
Form: Liquid
Composition: 100% active ingred.
Toxicity: Food approved and nonfood approved

Centrolex Series. Central Soya Co.
Chemical Description: Refined "oil free" phosphatides (soybean lecithin)
Category: Wetting agent, emulsifier, dietary supplement, dispersant, emollient
Applications: Food processing; cosmetics; pharmaceuticals; plastics and rubber industries
Form: Granular
Composition: 100% active ingred.

Centromix Series. Central Soya Co.
Chemical Description: Standard

soybean lecithin with selected surfactants
Category: Emulsifier, wetting agent, dispersant, penetrant
Applications: Food processing; cosmetics; coatings manufacture and finishes; plastic and rubber industries; glass and ceramic processing; paper and printing; masonry and asphalt; metal processing; pesticides; adhesives; textiles and leather
Form: Liquid, paste
Composition: 100% active ingred.

Centrophil Series. Central Soya Co.
Chemical Description: Refined soybean lecithin with selected carriers
Category: Wetting agent, emulsifier, dispersant, release agent, dietary supplement
Applications: Food processing; cosmetics; pharmaceuticals; plastics and rubber industries; masonry and asphalt products; metal processing
Form: Liquid, solid, other
Composition: 100% active ingred.
Toxicity: Most are food approved; some for industrial use only

Ceramol. Aceto Chem.
Chemical Description: Partially sulfonated fatty alcohol from hydrogenated vegetable oil
Category: Emulsifier
Applications: Cosmetics
Form: Flake

Ceranine AT. Sandoz Ltd.
Chemical Description: Fatty acid derivative
Category: Softener, antistatic, cationic
Applications: Softener and antistatic agent for synthetic fibers
Form: Paste

Ceranine AW. Sandoz Ltd.
Chemical Description: Fatty acid derivative
Category: Softener, antistatic, cationic
Applications: Softener and antistatic agent for synthetic fibers
Form: Paste/solid

Ceranine HC. Sandoz Ltd.
Chemical Description: Fatty acid derivative
Category: Softener, cationic
Applications: Softener for all fibers
Form: Paste

Ceranine HC Hi Conc. Sandoz Colors and Chem.
Category: Emulsifier
Applications: Skin care products
Form: Solid; Color: Tan; Odor: Slight
Properties: Solubility: Dispersible in water; M.P.: 55–65 C; Stability: Good
Composition: 100% active ingred.
Containers: 200 lb fiber drum

Ceranine HCA. Sandoz Colors and Chem.
Chemical Description: A fatty amide derivative
Category: Emulsifier, cationic
Applications: Skin care products; softener for textiles
Form: Granule; Color: Cream;

Odor: Slight
Properties: Solubility: Dispersible in water; M.P.: 71–81 C; Stability: Good
Composition: 100% active ingred.
Containers: 250 lb fiber drum

Ceranine HCL. Sandoz Ltd.
Chemical Description: Fatty acid derivative
Category: Softener, cationic
Applications: Softener for all fibers
Form: Liquid/flakes

Ceranine HCS. Sandoz Ltd.
Chemical Description: Fatty acid derivative
Category: Softener, cationic
Applications: Softener for all fibers
Form: Liquid

Ceranine L. Sandoz Ltd.
Chemical Description: Polyethylene dispersion
Category: Softener, nonionic
Applications: Softener for cellulosic fibers and their blends with synthetic fibers, especially for high-grade finishing
Form: Liquid

Ceranine PAP. Sandoz Ltd.
Chemical Description: Polymerization product
Category: Softener, cationic
Applications: Permanent softener for polyamide fibers
Form: Liquid

Ceranine PNA. Sandoz Ltd.
Chemical Description: Fatty acid derivative
Category: Softener, cationic
Applications: Softener for synthetic fibers
Form: Paste/powder

Ceranine PNL. Sandoz Ltd.
Chemical Description: Substituted fatty acid amide
Category: Softener, cationic
Applications: Softener for synthetic fibers
Form: Liquid

Ceranine PNS. Sandoz Ltd.
Chemical Description: Fatty acid derivative
Category: Softener, cationic
Applications: Softener for all fibers, especially for synthetic fibers
Form: Flakes/paste

Ceranine RW. Sandoz Ltd.
Chemical Description: Fatty acid derivative
Category: Softener, cationic
Applications: Softener with rewetting properties for cellulosic fibers and their blends with synthetic fibers
Form: Liquid/paste

Ceranine RWN. Sandoz Ltd.
Chemical Description: Fatty acid derivative
Category: Softener, cationic
Applications: Softener with rewetting properties for cellulosic fibers and their blends with synthetic fibers
Form: Liquid

Ceranine SG. Sandoz Ltd.
Chemical Description: Fatty acid derivative
Category: Softener, nonionic/anionic
Applications: Softener for all fibers, particularly for scrooping
Form: Paste/solid

Ceranine SIT. Sandoz Ltd.
Chemical Description: Fatty acid derivative/silicones
Category: Softener, cationic
Applications: Softener for all fibers with antistatic effect
Form: Paste/solid

Ceranine SWP. Sandoz Ltd.
Chemical Description: Fatty acid derivative
Category: Softener, cationic
Applications: Permanent softener for polyester and polyacrylic fibers
Form: Liquid

Ceranine VR. Sandoz Ltd.
Chemical Description: Fatty alcohol sulfates
Category: Softener, anionic
Applications: Softener for all fibers
Form: Paste

Ceranine VS. Sandoz Ltd.
Chemical Description: Fatty acid condensation product in aqueous dispersion
Category: Softener, cationic
Applications: Softener with hydrophobic properties for cellulosic fibers and their blends with synthetic fibers
Form: Liquid

Ceranine WN. Sandoz Ltd.
Chemical Description: Fatty acid ester
Category: Softener, nonionic
Applications: Softener for cellulosic and synthetic fibers
Form: Liquid/flakes

Cerol A Liquid. Sandoz Colors and Chem.
Chemical Description: Cationic zirconium/wax emulsion
Category: Water repellent, cationic
Applications: Water repellent for use on cotton and polyester/cotton blends

Cerol C. Sandoz Ltd.
Chemical Description: Fatty acid chrom-complex
Category: Waterproof agent, cationic
Applications: Waterproofing agent, fast to washing and dry cleaning
Form: Liquid

Cerol EWL. Sandoz Ltd.
Chemical Description: Modified synthetic resin
Category: Waterproof agent, cationic
Applications: Waterproofing agent, wash fast
Form: Liquid

Cerol ZN. Sandoz Ltd.
Chemical Description: Paraffin dispersion containing zirconium
Category: Waterproof agent, cationic
Applications: Waterproofing agent
Form: Liquid

Charlab Condensate-K. Catawba-Charlab, Inc.
Chemical Description: Alkanolamide
Category: Wetting agent, emulsifier, nonionic
Applications: Wetting agent, emulsifier, scouring agent
Form: Liquid; Color: Amber
Properties: Solubility: Water sol.; Sp. gr.: 1.01
Composition: 100% active ingred.

Charlab Leveler AT Special. Catawba-Charlab, Inc.
Chemical Description: Sodium hydrocarbon sulfonate
Category: Detergent, emulsifier, anionic
Applications: Textile dyeing-leveling agent, softener, lubricant
Form: Solid; Color: Amber
Properties: Solubility: Sol. in water @ 160 F; Sp. gr.: 1.0; M.P.: 110–125 F
Composition: 80% active ingred.

Charlab Leveler DSL. Catawba-Charlab, Inc.
Chemical Description: Sodium hydrocarbon sulfonate
Category: Detergent, emulsifier, anionic
Applications: Textile leveling agent for dyeing, softener, dispersant
Form: Paste; Color: Light brown
Properties: Solubility: Sol. in hot water @ 160 F; Density: 1.03 lb/gal
Composition: 40% active ingred.

Charlab LPC. Catawba-Charlab Inc.
Chemical Description: Lauryl pyridinium chloride
Category: Cationic surfactant
Form: Solid; Color: Light brown; Odor: Pyridine
Properties: Solubility: Water sol.; F.P.: 45 C
Composition: 97–100% active ingred.; other components: small amount pyridine; lauryl alcohol; lauryl ether
Toxicity: Skin irritant; solid material or powder may cause a rash with certain individuals

Chemannate C-60B. Chemform Corp.
Chemical Description: Calcium salt of dodecyl benzene sulfonic acid
Category: Anionic base for emulsifiers
Form: Viscous liquid; Color: Amber
Properties: Solubility: Oil sol.; Density: 8.45 lb/gal; Visc.: 600 cps @ 20 C (Brookfield); Flash Pt.: > 100 F; pH: 4–7 (5% in 25% isopropanol)
Composition: 60% active ingred.

Chemannate C-70B. Chemform Corp.
Chemical Description: Calcium salt of dodecyl benzene sulfonic acid
Category: Emulsifier base, anionic
Applications: Agricultural chemicals
Form: Viscous liquid; Color: Dark amber
Properties: Visc.: 3500 cps @ 140 F; Flash Pt.: > 100 F; pH:

3.0–6.0 (5% in 25% IPA)
Composition: 70% active ingred.

Chemcol 14-Z. Chemform Corp.
Chemical Description: Anionic/nonionic blend
Category: Emulsifier, anionic/nonionic
Applications: Emulsifier for aromatic solvent aquatic weed control esp. for xylene or aromatic hydrocarbon-xylene solvents
Form: Color: Light amber
Properties: Solubility: Sol. in xylene, miscible with water; Pour Pt.: < 25 F; Stability: Good
Containers: 55 gal drums

Chemcol 610. Chemform Corp.
Category: Emulsifier
Applications: Low foaming, oil sol. emulsifier for agricultural sprays, esp. petroleum oils for citrus sprays
Form: Clear liquid; Color: < 1 (Gardner); Odor: Mild
Properties: Solubility: Oil sol.; Pour Pt.: < –10 F; pH: 6–8 (1% sol'n.)
Containers: 55 gal drums; tank cars and tank trucks

Chemcol Agwet 110. Chemform Corp.
Chemical Description: Alkyl aryl polyoxyethylene ether
Category: Agricultural surfactant, nonionic
Applications: Reduces surface tension, improves wetting and plant coverage of herbicide and pesticide sprays
Form: Clear liquid
Properties: Solubility: Water sol.; Density: 8.2 lb/gal; Flash Pt.: near 80 F
Composition: 80% active ingred.; 20% isopropyl alcohol
Toxicity: Skin and eye irritant
Containers: Bulk and drum
Warnings and Caution: Avoid skin and eye contact; flammable

Chemcol Agwet 115-M. Chemform Corp.
Chemical Description: Alkyl polyoxyethylene glycol ether
Category: Wetting agent, nonionic
Applications: Agricultural surfactant to improve wetting properties, reduce surface tension, improve plant coverage of pesticide and herbicide sprays
Form: Liquid; Color: Water white to pale yellow
Properties: Solubility: Water sol.; Density: 7.8 lb/gal; Flash Pt.: 108 F; Biodegradable
Composition: 80% active ingred.; 20% methyl propanol
Toxicity: Skin and eye irritant
Containers: Bulk and 55 gal drums

Chemcol CO-28. Chemform Corp.
Chemical Description: Polyethoxylated triglyceride (28-mole ethylene oxide adduct of castor oil)
Category: Emulsifier, nonionic
Applications: Textiles and agricultural chemicals
Form: Liquid; Color: Pale yellow
Properties: Density: 8.6 lb/gal; Cloud Pt.: < 25 C; Pour Pt.: < 0 C; pH: 6–8
Composition: 100% active ingred.

Chemcol CO-30. Chemform Corp.
Chemical Description: Polyethoxylated triglyceride (reaction of a castor oil with 30 moles of ethylene oxide
Category and Applications: See Chemcol CO-28
Form: Liquid; Color: Yellow
Properties: Density: 8.7 lb/gal; HLB: 11.7; Cloud Pt.: > 100 C; Pour Pt.: < 0 C; pH: 6–8
Composition: 100% active ingred.

Chemcol CO-40. Chemform Corp.
Chemical Description: Polyethoxylated triglyceride (40-mole ethylene oxide adduct of castor oil)
Category: Emulsifier, lubricant, dispersant, antistat, nonionic
Applications: Agricultural chemicals; textiles; promotes dye leveling
Form: Viscous liquid; Color: Yellow
Properties: Density: 8.8–8.9 lb/gal; HLB: 13.1; Cloud Pt.: > 100 C; Pour Pt.: 10 C; pH: 6.5–7.5
Composition: 97% active ingred.

Chemcol HE-1. Chemform Corp.
Chemical Description: Anionic/nonionic blended surfactant
Category: Agricultural emulsifiers, anionic/nonionic
Applications: Improves soft water performance of emulsifiable concentrates; used to couple polar materials into nonpolar formulations
Form: Slightly viscous liquid; Color: Amber
Properties: Density: 8.3 lb/gal; Visc.: 80 cps @ 20 C; Flash Pt.: > 100 F; Pour Pt.: < 5 C
Toxicity: Skin and eye irritant on prolonged contact
Containers: Bulk and 55 gal drums

Chemcol HE-2. Chemform Corp.
Category: Agricultural emulsifier, nonionic/anionic
Applications: Improves hard water performance of many toxicant formulations
Form: Viscous liquid; Color: Light amber
Properties: Density: 8.6 lb/gal; Visc.: 146 cps @ 20 C; Flash Pt.: > 100 F; Pour Pt.: 17 C; pH: 4–7 (5% in 25% IPA)
Toxicity: Skin and eye irritant on prolonged contact
Containers: Bulk and 55 gal drums

Chemcol HE-8. Chemform Corp.
Category: Agricultural emulsifier, nonionic/anionic
Applications: Improves spontaneity and emulsion stability of organo-phosphate insecticides
Form: Viscous liquid; Color: Light amber
Properties: Density: 8 lb/gal; Visc.: 1850 cps @ 20 C; Flash Pt.: > 100 F; Pour Pt.: 15 C; pH: 4–7 (5% in 25% IPA)
Toxicity: Skin and eye irritant on prolonged contact
Containers: Bulk and 55 gal drums

Chemcol JL-60. Chemform Corp.
Chemical Description: Polyoxyethylene ether alcohol
Category: Detergent, dispersant, scouring agent, leveling agent, nonionic

Applications: Cosmetics, textiles
Form: Cloudy liquid; Color: White
Properties: Density: 8.2 lb/gal; Cloud Pt.: < 25 C; Pour Pt.: 4.5–7.5 C; pH: 6–8
Composition: 100% active ingred.

Chemcol JL-120. Chemform Corp.
Chemical Description: Polyoxyethylene ether alcohol
Category: Detergent, dispersant, scouring and leveling agents, nonionic
Applications: Textiles, detergents, cosmetics, agricultural chemicals
Form: Semisolid; Color: White
Properties: Density: 8.4 lb/gal; Pour Pt.: Pourable liquid @ 50 C; pH: 6–8
Composition: 100% active ingred.

Chemcol JL-180. Chemform Corp.
Chemical Description: Polyoxyethylene ether alcohol
Category: Detergent, wetting agent, dispersant, scouring and leveling agent, nonionic
Applications: Textiles
Form: White waxy solid
Properties: Cloud Pt.: > 100 C; Pour Pt.: Pourable liquid @ 50 C; pH: 6–8
Composition: 100% active ingred.

Chemcol MS-910. Chemform Corp.
Category: Agricultural chemical, nonionic/anionic
Applications: Miscible surfactant for organic arsenicals
Form: Clear liquid; Color: < 6 (Gardner); Odor: Mild, pleasant
Properties: Density: 9.1 lb/gal; Sp. gr.: 1.09 @ 20 C; pH: 5.5–6.5 (10% sol'n.)
Composition: 100% active ingred.
Toxicity: Skin and eye irritant on prolonged contact
Containers: Bulk and drum; corrosive to ferrous surfaces; store in stainless steel, fiberglass lined tanks

Chemcol NPE-40. Chemform Corp.
Chemical Description: Nonyl phenol ethoxylate (the reaction product of nonyl phenol with 4 moles of ethylene oxide)
Category: Detergent, stabilizer, coemulsifier, dispersant, nonionic
Applications: Textile, household cleaners, agricultural chemicals, petroleum oil dispersant
Form: Liquid; Color: Pale yellow
Properties: Solubility: Oil sol.; Density: 8.5 lb/gal; HLB: 8.8; Cloud Pt.: Insoluble (1% in water); Flash Pt.: > 400 F (TOC); Pour Pt.: < 0 C; pH: 6–8 (5% in 25% IPA)
Composition: 100% active ingred.

Chemcol NPE-60. Chemform Corp.
Chemical Description: Nonyl phenol ethoxylate (reaction product of nonyl phenol with 6 moles of ethylene oxide)
Category and Applications: See Chemcol NPE-40
Form: Liquid; Color: Pale yellow
Properties: Solubility: Water sol.; Density: 8.7 lb/gal; HLB: 10.8; Cloud Pt.: Insoluble (1% in water); Flash Pt.: > 400 F (TOC); Pour Pt.: < 10 C; pH: 6–8 (5% in 25% IPA)
Composition: 100% active ingred.

Chemcol NPE-100. Chemform Corp.
Chemical Description: Nonyl phenol ethoxylate (reaction of nonyl phenol with 10 moles ethylene oxide)
Category and Applications: See Chemcol NPE-40
Form: Liquid; Color: Colorless to pale yellow
Properties: HLB: 13.4; Cloud Pt.: 58–64 C (1% in water); pH: 6–8 (5% in 25% IPA)
Composition: 100% active ingred.

Chemcol NPE-140. Chemform Corp.
Chemical Description: Nonyl phenol ethoxylate (reaction product of nonyl phenol with 14 moles of ethylene oxide)
Category and Applications: See Chemcol NPE-40
Form: Viscous liquid to off-white solid
Properties: Density: 8.9 lb/gal; HLB: 14.7; Cloud Pt.: 92–96 C (1% in water); Flash Pt.: > 450 F (TOC); Pour Pt.: ≈ 24 C; pH: 3.5–4.5 (10% sol'n.)
Composition: 100% active ingred.

Chemcol NPE-200. Chemform Corp.
Chemical Description: Nonyl phenol ethoxylate (reaction product of nonyl phenol with 20 moles of ethylene oxide)
Category and Applications: See Chemcol NPE-40
Form: Off-white solid
Properties: Density: Solid; HLB: 16.0; Cloud Pt.: 67–73 C (1% in 10% NaCl); Flash Pt.: > 450 F (TOC); Pour Pt.: ≈ 36 C; pH: 3.5–4.5 (10% sol'n.)
Composition: 100% active ingred.

Chemcol NPE-400. Chemform Corp.
Chemical Description: Nonyl phenol ethoxylate (reaction product of nonyl phenol with 40 moles of ethylene oxide)
Category and Applications: See Chemcol NPE-40
Form: Off-white solid
Properties: Density: Solid; HLB: 17.7; Cloud Pt.: > 100 C (1% in water); Flash Pt.: > 450 F (TOC); Pour Pt.: ≈ 48 C; pH: 3.5–4.5 (10% sol'n.)
Composition: 100% active ingred.

Chemcol PE-34. Chemform Corp.
Chemical Description: Blended anionic/nonionic surfactant
Category: Emulsifier, anionic/nonionic
Applications: Emulsifier for agricultural chemicals esp. Propanil and where mesityl oxide is the principal solvent; used in soft and hard waters
Form: Viscous liquid; Color: Dark amber; Odor: Mild, aromatic
Properties: Density: 8.9 lb/gal @ 40 C; Sp. gr.: 1.067 @ 40 C; Visc.: 550 cps @ 40 C; Pour Pt.: < 0 C; pH: 3.5–4.5 (10% sol'n.)
Composition: 100% active ingred.
Toxicity: Skin and eye irritant on prolonged contact
Containers: Bulk or drum quantities; store in phenolic lined equipment
Warnings and Caution: Flush with water in case of contact with skin or eyes; flammable liquid

Chemcol PS-160. Chemform Corp.
Chemical Description: Complex phosphate ester of an ethylene oxide adduct
Category: Detergent, emulsifier
Applications: Agricultural chemicals; emulsion polymerization; good electrolyte tolerance
Form: Clear liquid; Color: Straw; Odor: Mild
Properties: Density: 9.2 lb/gal; Sp. gr.: 1.12; Pour Pt.: < 0 C
Composition: 100% active ingred.

Chemcol T-10. Chemform Corp.
Chemical Description: Ethoxylated fatty amine (reaction of tallow-amine with 10 moles of ethylene oxide)
Category: Emulsifier, antistat, dispersant, cationic
Applications: Textiles
Form: Liquid; Color: 8 (Gardner)
Properties: Density: 8.2 lb/gal; Cloud Pt.: 80 ± 10 C; Pour Pt.: 54 F
Composition: 100% active ingred.

Chemcol T-15. Chemform Corp.
Chemical Description: Ethoxylated fatty amine (reaction of tallow-amine with 15 moles of ethylene oxide)
Category: Emulsifier, antistat, dispersant, cationic
Applications: Textiles
Form: Liquid; Color: Amber
Properties: Density: 8.5 lb/gal; Cloud Pt.: > 100 C; Pour Pt.: < 5 C; pH: 11.74
Composition: 100% active ingred.

Chemcol T-20. Chemform Chem.
Chemical Description: Ethoxylated fatty amine (reaction of tallow-amine with 20 moles of ethylene oxide)
Category: Emulsifier, antistat, dispersant, cationic
Applications: Textiles
Form: Liquid; Color: Amber
Properties: Density: 8.7 lb/gal; Cloud Pt.: > 100 C; Pour Pt.: 10 C; pH: 11.59
Composition: 100% active ingred.

Chemcol T-25. Chemform Corp.
Chemical Description: Ethoxylated fatty amine (reaction of tallow-amine with 25 moles of ethylene oxide)
Category: Emulsifier, cationic
Applications: Textiles
Form: Solid; Color: 6 (Gardner)
Properties: Density: 8.8 lb/gal; Cloud Pt.: > 100 C; Pour Pt.: 110 F
Composition: 100% active ingred.

Chemcol TO-10. Chemform Corp.
Chemical Description: Polyethoxylated mixed rosin and fatty acid (reaction of tall-oil with 10 moles of ethylene oxide)
Category: Emulsifier, lubricant, dye assistant, nonionic
Applications: Textiles
Form: Liquid; Color: Yellow
Properties: Sp. gr.: 0.872; Pour Pt.: < 50 F; Saponification No.: 77 ± 5
Composition: 100% active ingred.

Chemcol TO-15. Chemform Corp.
Chemical Description: Polyethoxylated mixed rosin and fatty acid

(reaction of tall-oil with 15 moles of ethylene oxide)
Category: Emulsifier, lubricant, dye assistant, nonionic
Applications: Textiles
Form: Solid; Color: 8 (Gardner)
Properties: Sp. gr.: 0.866; Pour Pt.: 80 F; Saponification No.: 59 ± 5; pH: 6–8
Composition: 100% active ingred.

Chemcol TO-16. Chemform Corp.
Chemical Description: Polyethoxylated mixed rosin and fatty acid (reaction of tall-oil with 16 moles of ethylene oxide)
Category: Emulsifier, lubricant, dye assistant, nonionic
Applications: Textiles
Form: Liquid; Color: Light amber
Properties: Density: 9 lb/gal; HLB: 13.2; Pour Pt.: 68 F; pH: 7–8
Composition: 97% active ingred.

Chemcol VCC-5. Chemform Corp.
Chemical Description: Anionic/nonionic blend of metal sulfonates and polyethoxylates
Category: Emulsifier, anionic/nonionic
Applications: Multi-purpose pesticide emulsifier
Form: Liquid; Color: Amber
Properties: Density: 8.49 lb/gal; Sp. gr.: 1.019; Flash Pt.: 170 F (TOC); pH: 4–6 (5% sol'n. in 3:1 water:IPA)
Composition: 100% active ingred.
Toxicity: Avoid contact with skin and eyes; flush with water in case of contact
Containers: Store and transfer using carbon steel equipment
Warnings and Caution: Contains flammable solvents

Chemical Base 6532. Sandoz Colors and Chem.
Chemical Description: Stearamido ethyl diethalamine
Category: Emulsifier, cationic
Applications: Conditioning agent for skin and hair
Form: Solid; Color: Yellowish tan; Odor: Slight
Properties: M.P.: 35–45 C; Stability: Good
Composition: 100% active ingred.
Containers: 250 lb fiber drum
References: Bulletin 7-254/78

Chemstick 56. Chemform Corp.
Applications: Spray sticker designed to increase the effectiveness of wettable powder and emulsifiable concentrate for agricultural sprays
Form: Liquid; Color: Amber
Properties: Density: 7.56 lb/gal

Cholesteryl Acetate. Van Schuppen chemie
Chemical Description: Acetic ester of a high quality cholesterol
Category: Raw material
Applications: As a starting material for the synthesis of various steroid derivatives, esp. vitamin D_3; in liquid crystal compositions
Form: Microcrystalline powder; Color: White or faintly yellow; Odor: Odorless
Properties: Solubility: Sol. in acetic acid, alcohol, chloroform, dioxane, ether, ethyl acetate, hexane, pyridine; insol. in water; M.P.:

111.5–115 C; Acid No.: 0.3 max.
References: Product information No. 3A

Cholesteryl Adipate. Van Schuppen chemie
Applications: Production of thermochromic liquid crystals and temperature-insensitive liquid crystals for industrial and medical fields and in displays; as a starting material for pharmaceuticals (e.g., vitamin D_3 and hormones) and for cosmetic preparations
Form: Crystalline solids; Color: White to faint yellow
Properties: M.P.: 184 C

Cholesteryl Benzoate. Van Schuppen chemie
Applications: See Cholesteryl Adipate
Form: Crystalline solids; Color: White to faint yellow
Properties: M.P.: 149 C

Cholesterol B.P. Van Schuppen chemie
Category: Raw material, emulsifier
Applications: A basic material for many pharmaceuticals, e.g., vitamin D_3, steroid hormones and for liquid crystals; also as a natural emulsifier in cosmetic preparations (characteristic emolleint feel)
Form: Powder; Color: White; Odor: Almost odorless
Properties: Solubility: Sol. in acetic acid, acetone, alcohol, benzene, chloroform, dioxane, ether, ethyl acetate, hexane, pyridine, toluene and vegetable oils; insol. in water; M.P.: 145–149 C; Acid No.: 0.3 max.
References: Product information No. 1 B

Cholesteryl Butoxy Ethyl Carbonate. Van Schuppen chemie
Applications: See Cholesteryl adipate
Form: Crystalline solids; Color: White to faint yellow
Properties: M.P.: 65 C

Cholesteryl Butyrate. Van Schuppen
Applications: See Cholesteryl adipate
Form: Crystalline solids; Color: White to faint yellow
Properties: M.P.: 97 C

Cholesterol Techn. Van Schuppen Chemie
Category: Raw material, emulsifier, nonionic
Applications: Raw material for alkoxylated compounds; W/O emulsifier for cosmetics
Form: Powder; Color: White to faintly yellow; Odor: Odorless
Properties: Solubility: Sol. in isopropyl myristate, isostearic acid, hexadecyl alcohol, chlorinated hydrocarbons and common organic solvents; insol. in water; M.P.: 135–140 C; Acid No.: 1.0 max.
References: Product information No. 1D

Cholesterol USP. Van Schuppen chemie
Category: Raw material, emulsifier
Applications: Raw material for

pharmaceuticals, e.g., vitamin D_3, steroid hormones, for liquid crystals; natural emulsifier in cosmetic preparations (emollient feel)
Form: Crystalline powder; Color: White; Odor: Almost odorless
Properties: Solubility: Sol. in acetic acid, acetone, alcohol, benzene, chloroform, dioxane, ether, ethyl acetate, hexane, pyridine, toluene and vegetable oils; insol. in water; M.P.: 147–150 C; Acid No.: 0.3 max.
References: Product information No. 1A

Cindet 201. Cindet Chemicals
Chemical Description: Blend of ethoxylated alcohols and aryl sulfonates
Category: Detergent, dyeing assistant, dispersing agent, penetrant, anionic
Applications: Detergent for scouring wool, cotton, synthetics; in dyeing procedures at neutral or alkaline pH
Form: Liquid; Color: Light amber
Properties: Stability: Stable in hard water, alkali media and moderate concentrations of acids; pH: 7.5–8.5 (1% sol'n.)
Composition: 25% active ingred.

Cindet 542. Cindet Chemicals
Chemical Description: Polyoxyethylene compound
Category: Dye assistant, scouring and soaping off agent, detergent, retarding and leveling agent, nonionic
Applications: Preparation, dyeing and finishing of textiles; wet processing of natural and synthetic fibers; retarding and leveling agent in dyeing of polyamide, polyester, acrylic fibers; detergent as afterscour; thermosol dye assistant
Form: Liquid; Color: Water white
Properties: Stability: Stable in hard water and acid and alkali media; Fast wetting speed; medium foaming; excellent rinsability; pH: 6.5–7.5 (1% sol'n.)
Composition: 50% active ingred.
References: Bulletin TB-270-2

Cindet AL. Cindet Chemicals
Chemical Description: Sodium alcohol ether sulfate
Category: Dye assistant, dispersant, penetrant, scouring agent, anionic
Applications: Scouring agent for wool, cotton, synthetics; dyeing procedures at neutral or alkaline pH; desizing
Form: Liquid; Color: Light amber
Properties: Stability: Stable to alkali media; Very free rinsing, fast wetting speed, medium foaming; pH: 7.0–8.0 (1% sol'n.)
Composition: 60% active ingred.

Cindet AP. Cindet Chemicals
Chemical Description: Modified blend of fatty nonionic and sulfonated hydrocarbons
Category: Wetting agent, scouring agent, afterscour, anionic
Applications: Wetting agent for causticizing and bleach operations; improves wetting speed and levelness in hydrogen peroxide bleach systems; soaping-off aid on con-

tinuous dyeing operations
Form: Transparent liquid; Color: Amber
Properties: Solubility: Water sol.; pH: 7.5–8.5 (1% sol'n.)

Cindet C-12. Cindet Chemicals
Chemical Description: Coconut amine condensate
Category: Detergent, emulsifier, wetting agent, scouring agent, lubricant, fulling agent, dyeing assistant, anionic
Applications: Scouring baths, fulling formulations for wool, scouring agent and detergent as afterscour
Form: Viscous liquid; Color: Amber
Properties: Stability: Stable in alkaline media; Medium wetting and foaming properties; good rinsability; pH: 8.5–9.5 (1% sol'n.)
Composition: 99% active ingred.
References: Bulletin TB-270-10

Cindet GE. Cindet Chemicals
Chemical Description: Isopropylamine salt of dodecyl benzene sulfonic acid
Category: Emulsifier, anionic
Applications: Emulsifier in polyester and triacetate dye carriers; emulsifier with varsol, xylene, etc. in solvent scours
Form: Viscous liquid; Color: Amber
Properties: Solubility: Sol. in low concentrations in water, completely sol. in many solvents; Medium foaming; pH: 7.2–7.8 (1% sol'n.)
Composition: 94% active ingred.

Cindet JS. Cindet Chemicals
Chemical Description: Polyoxyethylene compound
Category: Dye assistant, scouring agent, retarding and leveling agent, soaping off agent, detergent, nonionic
Applications: Nonfoaming surfactant in scouring and dyeing in the Gaston County Jet Machine; desizing woven goods; retarding and leveling agent in dyeing of polyamide, polyester, acrylic fibers; afterscour
Form: Liquid; Color: Light amber
Properties: Stability: Stable in acid and alkaline media; Excellent rinsability; pH: 6.5–7.5 (1% sol'n.)

Cindet K-1. Cindet Chemicals
Chemical Description: Blend of synthetic detergents, phosphates, and other alkalis
Category: Detergent, scouring agent, anionic
Applications: Scouring agent for cotton, rayon, nylon, polyester, polyolefin fibers; kier boiling of cotton prior to bleaching; free rinsing; in hard water conditions
Form: Powder; Color: White
Properties: Stability: Stable in alkaline media; Medium wetting speed and foaming; excellent rinsability; pH: 10–11 (1% sol'n.)
Composition: 100% active ingred.

Cindet NAK. Cindet Chemicals
Chemical Description: Blend of organic compounds
Category: Detergent; fulling, scouring and soaping agent; dye dispersant, anionic

Applications: Dye assistant as detergent, dispersant, fulling, scouring and soaping agent; removes dye carriers after dyeing hydrophobic fibers; oil, grease, soil removal
Form: Viscous liquid; Color: Light amber
Properties: Stability: Stable to alkali, hard water and peroxides; low foaming, medium wetting speed, excellent rinsability; pH: 7.5–8.5 (1% sol'n.)
Composition: 50% active ingred.
References: Bulletin TB-270-5

Cindet R. Cindet Chemicals
Chemical Description: Fatty polyoxethylene compound
Category: Dyeing assistant, scouring agent, retarder, detergent as afterscour, emulsifier, nonionic
Applications: Preparation, dyeing, and finishing of textiles; used in neutral, acid or alkaline media for wet processing of natural or synthetic fibers; oil and grease emulsifier; soaping off after dyeing
Form: Liquid; Color: Light amber
Properties: Stability: Stable in acid and alkaline media; Medium wetting speed, medium to low foaming, excellent rinsability; pH: 6.5–7.5 (1% sol'n.)
Composition: 70% active ingred.
References: Bulletin TB-270-1

Cinwet A-35. Cindet Chemicals
Chemical Description: Blend of anionic and nonionic surfactants
Category: Detergent, dyeing assistant, dispersant, penetrant
Applications: Detergent of superior wetting speed for scouring wool, cotton, synthetics; oil and grease removal
Form: Clear liquid; Color: Slight yellow
Properties: Stability: Stable in acid and alkaline media in moderate concentration; Fast wetting speed, medium foaming, excellent rinsability; pH: 7–8 (1% sol'n.)
Composition: 35% active ingred.

Cinwet BOS. Cindet Chemicals
Chemical Description: Sulfated propyl oleate
Category: Dyeing assistant, dispersant, leveling agent, wetting agent, lubricant, anionic
Applications: Dispersant and leveling agent in dyeing of polyester and polyamide; wetting agent in bleaching and dyeing of cellulosic fibers; coning emulsions; rewetting properties
Form: Liquid; Color: Amber
Properties: Stability: Stable in alkaline media and moderately acid media; Medium to fast wetting speed; low foaming; excellent rinsability; pH: 6–7 (1% sol'n.)
Composition: 65% active ingred.

Cithrol 2ML. Croda Chemicals
Chemical Description: Polyethylene glycol 200 monolaurate
Category: Wetting agent, emulsifier, detergent, thickening agent, solubilizer, dispersant, softening agent, lubricant, antistat
Applications: Cosmetics, textiles, glass fiber, metal treatment
Properties: m.w.: 200; HLB: 8.8; Saponification Value: 158–168

Cithrol 4ML. Croda Chemicals
Chemical Description: Polyethylene glycol 400 monolaurate
Category: Emulsifier, dye assistant, penetrant, detergent, nonionic
Applications: Component of paper and paperboard; in closures for food containers; as viscosity stabilizer and antistat in plastics industry; emulsifier for DDT and Chordane insecticides; cotton and rayon dyeing; scouring, desizing and dewashing of textiles
Properties: m.w.: 400; Solubility: Water sol.; HLB: 19; Saponification Value: 92–98

Cithrol 6ML. Croda Chemicals
Chemical Description: Polyethylene glycol 600 monolaurate
Category: See Cithrol 2ML
Applications: See Cithrol 2ML; also particularly suited for preparation of clear cod oil concentrates for dispersion in water to form fat liquor emulsions for leather treatment
Properties: m.w.: 600; HLB: 15.8; Saponification Value: 64–74

Cithrol 10ML. Croda Chemicals
Chemical Description: Polyethylene glycol 1000 monolaurate
Category and Applications: See Cithrol 2ML; esp. as emulsifier for oleic acid and colloidal silica; low foam detergent for mechanical washing machines; in leather industry as emulsifier and penetrant in straight sperm oil emulsions
Properties: m.w.: 1000; Solubility: Sol. in water, alcohol, polar solvents; HLB: 18; Saponification Value: 51–55

Cithrol 60ML. Croda Chemicals
Chemical Description: Polyethylene glycol 6000 monolaurate
Category and Applications: See Cithrol 2ML
Properties: m.w.: 6000; HLB: 19.2; Saponification Value: 8–13

Cithrol 2MO. Croda Chemicals
Chemical Description: Polyethylene glycol 200 monooleate
Category: Wetting agent, penetrant, detergent, emulsifier, solubilizer, thickening agent, dispersant, textile auxiliary, lubricant
Applications: O/W emulsifier for cosmetics; textile industry as dyeing, dispersing and softening agents; lubricants in metal working processes; emulsifier for soluble oils and insecticides
Properties: m.w.: 200; HLB: 6.2; Saponification Value: 110–112

Cithrol 4MO. Croda Chemicals
Chemical Description: Polyethylene glycol 400 monooleate
Category and Applications: See Cithrol 2MO
Properties: m.w.: 400; HLB: 11.4; Saponification Value: 85–93

Cithrol 6MO. Croda Chemicals
Chemical Description: Polyethylene glycol 600 monooleate
Category and Applications: See Cithrol 2MO
Properties: m.w.: 600; HLB: 13.1;

Saponification Value: 65–75

Cithrol 10MO. Croda Chemicals
Chemical Description: Polyethylene glycol 1000 monooleate
Category and Applications: See Cithrol 2MO
Properties: m.w.: 1000; HLB: 16.2; Saponification Value: 35–40

Cithrol 40MO. Croda Chemicals
Chemical Description: Polyethylene glycol 4000 monooleate
Category and Applications: See Cithrol 2MO
Properties: m.w.: 4000; HLB: 18.7; Saponification Value: 10–15

Cithrol 60MO. Croda Chemicals
Chemical Description: Polyethylene glycol 6000 monoleate
Category and Applications: See Cithrol 2MO
Properties: m.w.: 6000; HLB: 19.0; Saponification Value: 5–10

Cithrol 2MS. Croda Chemicals
Chemical Description: Polyethylene glycol 200 monostearate
Category: Nonionic emulsifier, detergent, wetting agent, solubilizer, thickening agent, anti-frothing agent, foaming agent
Applications: O/W and W/O emulsifiers; cosmetics as O/W emulsifiers and wetting agents; solubilizers and thickeners in perfumery; emulsifiers for insecticides; esp. useful in preparing emulsions of high electrolyte content
Properties: m.w.: 200; M.P.: 39–41 C; HLB: 6.3; Saponification Value: 110–120

Cithrol 3MS. Croda Chemicals
Chemical Description: Polyethylene glycol 300 monostearate
Category and Applications: See Cithrol 2MS
Properties: m.w.: 300; M.P.: 30–33 C; HLB: 10.7; Saponification Value: 95–105

Cithrol 4MS. Croda Chemicals
Chemical Description: Polyethylene glycol 400 monostearate
Category and Applications: See Cithrol 2MS
Properties: m.w.: 400; M.P.: 31–34 C; HLB: 11.0; Saponification Value: 95–105

Cithrol 6MS. Croda Chemicals
Chemical Description: Polyethylene glycol 600 monostearate
Category and Applications: See Cithrol 2MS
Properties: m.w.: 600; M.P.: 33–35 C; HLB: 14.0; Saponification Value: 68–76

Cithrol 10MS. Croda Chemicals
Chemical Description: Polyethylene glycol 1000 monostearate
Category and Applications: See Cithrol 2MS
Properties: m.w.: 1000; M.P.: 34–40 C; HLB: 16.0; Saponification Value: 36–50

Cithrol 15MS. Croda Chemicals
Chemical Description: Polyethylene glycol 1500 monostearate
Category and Applications: See

Cithrol 2MS
Properties: m.w.: 1500; HLB: 17.0; Saponification Value: 30–55

Cithrol 40MS. Croda Chemicals
Chemical Description: Polyethylene glycol 4000 monostearate
Category and Applications: See Cithrol 2MS
Properties: m.w.: 4000; HLB: 18.8; Saponification Value: 12–17

Cithrol 'A'. Croda Chemicals
Chemical Description: Poloxyethylene glycol oleate
Category: Emulsifier, dispersant, antistat, nonionic
Applications: Antistatic agent for textile processing; emulsions of mineral and vegetable oils, waxes, and fatty acids; emulsion-type metal cleaners, polishers, cutting oils; emulsions for leather and paper trades; emulsions of casein for the lacquer industry; oil emulsions for rubber latex; rag pulling oils and wool lubricants
Form: Liquid; Color: Amber; Odor: Slight agreeable
Properties: Solubility: Sol. in white spirit and liquid paraffin; completely sol. in xylol, ethyl acetate, methylated spirits, oleic acid; dispersible in water; Sp. gr.: 1.013; Flash Pt.: ≈ 215 C; Saponification Value: 85–93; pH: 6.5–7.5 (10% sol'n.)
Containers: Mild steel drums

Cithrol GMS A/S ES 0743. Croda Chemicals
Chemical Description: Glyceryl monostearate
Category: Emulsifier
Applications: Emulsifier and opacifier for cosmetics, esp. antiperspirants based on astringent aluminum salts, acidic skin creams, hair rinses and bleaches; base for pharmaceutical creams and ointments
Form: Color: 3 (Gardner)
Properties: M.P.: ≈ 52 C; Saponification Value: ≈ 96; Stability: Stable in acids, solutions of electrolytes and with anionics and cationics

Clindrol 101CG. Clintwood Chem.
Chemical Description: Alkanolamide
Category: Emulsifier, detergent, thickener, nonionic
Applications: Emulsifier of aliphatic hydrocarbons; for hydrocarbon propellents and aerosol packaged products; household uses as detergent
Form: Liquid; Color: Dark amber; Odor: Typical alkanolamide
Properties: Solubility: Sol. in alcohols, glycols, ketones, esters, aromatic and aliphatic hydrocarbons; Sp. gr.: 0.99; F.P.: 6 C
Composition: 100% active ingred.
Reaction: Alkaline
Toxicity: Skin and eye irritant
Containers: 55 gal tight head steel drums (450# net each)

Clindrol 101LI. Clintwood Chem.
Chemical Description: Lauric isopropanolamide
Category: Foam stabilizer and

thickener, nonionic
Applications: Foam stabilizer and thickener in rug shampoo, hand cleaners, hair shampoo; thickener for alpha olefin sulfonate solutions
Form: Solid flakes; Color: Tan
Properties: Solubility: Sol. in alcohols, glycols, ketones, esters, aromatic and chlorinated hydrocarbons; F.P.: 53–60 C; Stability: Stable
Composition: 100% active ingred.
Reaction: Alkaline
Toxicity: Skin and eye irritant
Containers: 55 gal open head fiber drums 200# net each

Clindrol 200CG. Clintwood Chem.
Chemical Description: Coconut oil alkanolamide
Category: Detergent, wetting agent, foam stabilizer, thickener, nonionic
Applications: Detergent; general purpose cleaners, metal cleaners, hard surface cleaners for household use, textiles; Kier boiling
Form: Liquid; Color: Dark amber; Odor: Typical alkanolamide
Properties: Solubility: Sol. in water, alcohols, glycols, ketones, aromatic and chlorinated hydrocarbons; Sp. gr.: 1.00; F.P.: –8 C
Composition: 100% active ingred.
Reaction: Alkaline
Toxicity: Skin and eye irritant
Containers: 55 gal tight head steel drums 450# net each

Clindrol 200 CGN. Clintwood Chem.
Chemical Description: Coconut oil diethanolamide
Category: Detergent, wetting agent, emulsifier, foam stabilizer, thickener, nonionic/anionic
Applications: Built hard surface cleaners, hand cleaners, metal degreasers, burnishing compounds, floor cleaners, wax strippers; textiles, laundering and household uses; desizing and boiling off cottons; Kier boiling, making of cutting and soluble oils, rinseable degreasers
Form: Liquid; Color: Dark amber; Odor: Typical alkanolamide
Properties: Solubility: Sol. in alcohols, water, ketones, aromatic and chlorinated hydrocarbons; Sp. gr.: 0.99; F.P.: 0 C
Composition: 100% active ingred.
Reaction: Alkaline
Toxicity: Skin and eye irritant
Containers: 55 gal tight head steel drums 450# net each

Clindrol 200HC. Clintwood Chem.
Chemical Description: Hydrogenated coconut fatty acid alkanolamide
Category: Detergent, emulsifier, foam stabilizer, thickener, nonionic
Applications: Hair shampoos, metal cleaners; household and cosmetic uses
Form: Liquid; Color: Straw; Odor: Typical alkanolamide
Properties: Solubility: Sol. in alcohols, water, aromatic and chlorinated hydrocarbons, ketones; Sp. gr.: 1.02; F.P.: 14–15 C
Composition: 100% active ingred.
Reaction: Alkaline
Toxicity: Skin and eye irritant

Containers: 55 gal tight head steel drums 450# net each

Clindrol 200L. Clintwood Chem.
Chemical Description: Lauric acid diethanolamide
Category: Detergent, wetting agent, emulsifier, foam stabilizer and thickener, nonionic
Applications: Foam stabilizer and thickener in hair shampoos, bubble bath and liquid dishwash detergent; hard surface cleaners; household and cosmetic uses
Form: Paste; Color: White; Odor: Typical alkanolamide
Properties: Solubility: Sol. in alcohols, water, ketones, glycols, aromatic hydrocarbons; Sp. gr.: 1.02; M.P.: 25–30 C
Composition: 100% active ingred.
Reaction: Alkaline
Toxicity: Skin and eye irritant
Containers: 55 gal open head steel drums 425# net each

Clindrol 200-O. Clintwood Chem.
Chemical Description: Diethanolamide of oleic acid
Category: Emulsifier, thickening agent, nonionic
Applications: Emulsifier for higher boiling aliphatic hydrocarbons; rinseable degreasers, fire safe fuels, thickener for amphoteric surfactants (shampoos, bubble baths); making of cutting and soluble oils; household and cosmetic uses
Form: Liquid; Color: Amber
Properties: Solubility: Sol. in alcohols, glycols, aliphatic and aromatic hydrocarbons, ketones, chlorinated hydrocarbons; Sp. gr.: 0.99; F.P.: –8 C
Composition: 100% active ingred.
Reaction: Alkaline
Toxicity: Skin and eye irritant
Containers: 55 gal tight head steel drums 450# net each

Clindrol 200MS. Clintwood Chem.
Chemical Description: Alkanolamide
Category: Pearl producing agent
Applications: Hair shampoos, hand cleansers, skin care products; household and cosmetics uses
Form: Flakes; Color: Tan; Odor: Ammoniacal
Properties: Solubility: Sol. in alcohols, glycols, esters, ketones, aromatic hydrocarbons; Sp. gr.: 0.998; M.P.: 98 C
Composition: 100% active ingred.
Reaction: Alkaline
Toxicity: Skin and eye irritants
Containers: 55 gal open head fiber drums 200# net each

Clindrol 200RC. Clintwood Chem.
Chemical Description: Coconut fatty acid alkanolamide
Category: Detergent, wetting agent, emulsifier, solubilizing agent, nonionic
Applications: Solubilizing agent, emulsifier in insecticide compositions, dye solvent for use in inks in printing industry, foam stabilizer in hair shampoos and other cosmetics; household detergents
Form: Liquid; Color: Amber
Properties: Solubility: Sol. in alcohols, glycols, water, ketones, aromatic and chlorinated hydrocarbons; Sp. gr.: 1.02

Composition: 100% active ingred.
Reaction: Alkaline
Toxicity: Skin and eye irritant
Containers: 55 gal tight head steel drums, 450# net each

Clindrol 200S. Clintwood Chem.
Chemical Description: Stearic acid derived alkanolamide
Category: Emulsifier, thickener and dispersing agent, nonionic
Applications: Emulsifier for skin, hair care products, other cosmetics (creams and lotions); hand cleaners, waterless hand cleaners; gel shampoos; porcelain cleaners; abrasive metal cleaners
Form: Solid; Color: White
Properties: Solubility: Sol. in alcohols, glycols, ketones, esters, aromatic and chlorinated hydrocarbons; M.P.: 36–40 C
Composition: 100% active ingred.
Reaction: Alkaline
Toxicity: Skin and eye irritant
Containers: 55 gal open head steel drums, 400# net each

Clindrol 201CGN. Clintwood Chem.
Chemical Description: Coconut alkanolamide
Category: Detergent, emulsifier, wetting agent, foam stabilizer, thickener
Applications: Built hard surface cleaners, hand cleaners, metal degreasers, burnishing compounds, floor cleaners, wax strippers; textiles, laundering, household uses; desizing and boiling off cottons; Kier boiling; making of cutting and soluble oils; rinseable degreasers
Form: Liquid; Color: Dark amber; Odor: Typical alkanolamide
Properties: Solubility: Sol. in alcohols, water, ketones, aromatic and chlorinated hydrocarbons; Sp. gr.: 0.99; F.P.: 0 C
Composition: 100% active ingred.
Reaction: Alkaline
Toxicity: Skin and eye irritant
Containers: 55 gal tight head steel drums, 450# net each

Clindrol 202CGN. Clintwood Chem.
Chemical Description: Coconut oil alkanolamide
Category: Detergent, emulsifier, wetting agent, foam stabilizer and thickener, nonionic/anionic
Applications: Built hard surface cleaners, hand cleaners, metal degreasers, burnishing compounds, floor cleaners, wax strippers, steam cleaners; textiles, laundering, transport and household uses as detergents; desizing and boiling off cottons; Kier boiling; making of cutting and soluble oils; rinseable degreasers
Form: Liquid; Color: Dark amber; Odor: Typical alkanolamide
Properties: Solubility: Sol. in alcohols, water, ketones, aromatic and chlorinated hydrocarbons; Sp. gr.: 0.99; F.P.: 0 C
Composition: 100% active ingred.
Reaction: Alkaline
Toxicity: Skin and eye irritant
Containers: 55 gal tight head steel drums, 450# net each

Clindrol 203CG. Clintwood Chem.
Category: Detergent, wetting a-

gent, foam stabilizer, thickener
Applications: Detergent, general purpose cleaners, metal cleaners, hard surface cleaners, cleaner-sanitizers with quaternary ammonium compounds; textiles, household and cosmetic uses; Kier boiling
Form: Liquid; Color: Straw to light amber; Odor: Typical alkanolamide
Properties: Solubility: Sol. in alcohols, water, glycols, ketones, aromatic and chlorinated hydrocarbons; Sp. gr.: 1.00; F.P.: –8 C
Composition: 100% active ingred.
Reaction: Alkaline
Toxicity: Skin and eye irritant
Containers: 55 gal tight head steel drums 450# net each

Clindrol 204CGN. Clintwood Chem.
Chemical Description: Coconut oil alkanolamide
Category: Detergent, emulsifier, wetting agent, foam stabilizer and thickener
Applications: Built hard surface cleaners, hand cleaners, metal degreasers, burnishing compounds, floor cleaners, wax strippers; textiles, laundering and household uses as detergent; desizing and boiling off cottons; Kier boiling; making of cutting and soluble oils; rinseable degreasers
Form: Liquid; Color: Straw to light amber; Odor: Typical alkanolamide
Properties: Solubility: Sol. in alcohols, water, ketones, aromatic and chlorinated hydrocarbons; Sp. gr.: 0.99; F.P.: 0 C
Composition: 100% active ingred.
Reaction: Alkaline
Toxicity: Skin and eye irritant
Containers: 55 gal tight head steel drums, 450# net each

Clindrol 206CGN. Clintwood Chem.
Chemical Description: Coconut alkanolamide
Category: Detergent, emulsifier, wetting agent, conveyor chain lubricant, anionic
Applications: Hard surface cleaner and fabric cleaner; household and industrial hard surface cleaners, car wash, dishwash, food machinery cleaner, fine fabric cleaner; detegent in transport and cosmetics industries; wetting agent for removal of spots, insecticides, herbicides, fungicides
Form: Liquid; Color: Dark amber
Properties: Solubility: Sol. in water, alcohols, glycols, esters, ketones, aromatic and chlorinated hydrocarbons; Sp. gr.: 1.02; F.P.: –5 C
Composition: 100% active ingred.
Reaction: Alkaline
Toxicity: Skin and eye irritant
Containers: 55 gal tight head steel drums, 450# net each

Clindrol 207CGN. Clintwood Chem.
Chemical Description: Coconut alkanolamide
Category: Detergent, emulsifier, wetting agent, conveyor chain lubricant, anionic
Applications: Household and industrial hard surface cleaners, car wash, dishwash, food machinery cleaner, fine fabric cleaner, con-

veyor chain lubricant; detergent in laundering, foods, transport, cosmetics industries; as wetting agent in removal of spots and in insecticides, herbicides, fungicides
Form: Liquid; Color: Amber
Properties: Solubility: Sol. in water, alcohols, glycols, esters, ketones, aromatic and chlorinated hydrocarbons; Sp. gr.: 1.02; F.P.: –5 C
Composition: 100% active ingred.
Reaction: Alkaline
Toxicity: Skin and eye irritant
Containers: 55 gal tight head steel drums, 450# net each

Clindrol 209CGN. Clintwood Chem.
Chemical Description: Coconut alkanolamide
Category: Detergent, wetting agent, emulsifier, solubilizer, nonionic/anionic
Applications: Solubilizing agent and emulsifier for pesticides; agricultural wetting agent; household detergents; degreasers
Form: Liquid; Color: Amber
Properties: Solubility: Sol. in water, alcohols, glycols, glycol ethers, aromatic and chlorinated hydrocarbons; Sp. gr.: 1.06; M.P.: –7 to –3 C
Composition: 100% active ingred.
Reaction: Alkaline
Toxicity: Skin and eye irritant
Containers: 55 gal tight head steel drums, 450# net each

Clindrol 210CGN. Clintwood Chem.
Chemical Description: Coconut alkanolamide
Category: Detergent, wetting agent, emulsifier, conveyor chain lubricant, nonionic/anionic
Applications: Conveyor chain lubricants, rinseable degreasers, making of cutting and soluble oils
Form: Liquid; Color: Amber
Properties: Solubility: Sol. in water, alcohols, glycols, esters, ketones, aromatic and aliphatic hydrocarbons; Sp. gr.: 0.99; M.P.: 0–5 C
Composition: 100% active ingred.
Reaction: Alkaline
Toxicity: Skin and eye irritant
Containers: 55 gal tight head steel drums, 450# net each

Clindrol 868. Clintwood Chem.
Chemical Description: Fatty acid alkanolamide
Category: Emulsifier, thickener, nonionic
Applications: Gel shampoos and hand cleaners; household and cosmetic applications
Form: Solid; Color: White
Properties: Solubility: Sol. in alcohols, ketones, chlorinated and aromatic solvents; M.P.: 58–63 C
Composition: 100% active ingred.
Reaction: Alkaline
Toxicity: Skin and eye irritant
Containers: Cubitainers, 40# net each

Clindrol SDG. Clintwood Chem.
Chemical Description: Diethylene glycol monostearate
Category: Emulsifier, opacifying agent, nonionic
Applications: Auxiliary emulsifier and opacifier in skin and hair care products

Form: Solid; Color: White
Properties: Solubility: Sol. in alcohols, glycols, chlorinated, aromatic and aliphatic hydrocarbons; Sp. gr.: 1.01; M.P.: 42–45 C
Composition: 100% active ingred.
Reaction: Neutral
Containers: 40 lb net cubitainers

Clindrol SEG. Clintwood Chem.
Chemical Description: Ethylene glycol monostearate
Category: Opacifier and pearl producing agent, nonionic
Applications: Hair shampoos, liquid dishwash detergent, hair and skin care products, hand cleaners
Form: Flakes; Color: White; Odor: Slightly fatty
Properties: Solubility: Sol. in alcohols, glycols, esters, ketones, aromatic, aliphatic and chlorinated hydrocarbons; Sp. gr.: 0.998; M.P.: 52–60 C
Composition: 100% active ingred.
Reaction: Neutral
Containers: 55 gal open head fiber drums, 200# net each

Clindrol Superamide 100C. Clintwood Chem.
Chemical Description: Coconut diethanolamide
Category: Foam stabilizer and thickener, nonionic
Applications: Hair shampoos, bubble bath, liquid dishwash detergent, hand cleaners
Form: Liquid; Color: Straw
Properties: Solubility: Sol. in alcohols, glycols, ketones, aromatic and chlorinated hydrocarbons; Sp. gr.: 0.99; F.P.: 8 C; Stability: Stable
Composition: 100% active ingred.
Reaction: Alkaline
Toxicity: Skin and eye irritant
Containers: 55 gal open head drums, 425# net each

Clindrol Superamide 100CG. Clintwood Chem.
Chemical Description: Coconut diethanolamide
Category: Foam stabilizer and thickener, nonionic
Applications: Foam stabilizer and thickener in hair shampoos, bubble bath, liquid dishwash detergents, hand cleaners
Form: Liquid; Color: Straw to amber
Properties: Solubility: Sol. in alcohols, glycols, water, aromatic and chlorinated hydrocarbons; Sp. gr.: 0.99; F.P.: –3 C
Composition: 100% active ingred.
Reaction: Alkaline
Toxicity: Skin and eye irritant
Containers: 55 gal tight head steel drums, 450# net each

Clindrol Superamide 100L. Clintwood Chem.
Chemical Description: Lauric diethanolamide
Category: Foam stabilizer and thickener, nonionic
Applications: Foam stabilizer in aqueous systems; in shampoos, bubble bath, dishwash detergent; thickener in gel shampoos
Form: Solid; Color: White; Odor: Typical alkanolamide
Properties: Solubility: Sol. in alco-

hols, ketones, chlorinated and aromatic solvents; Sp. gr.: 0.99; M.P.: 42–46 C; Stability: Stable
Composition: 100% active ingred.
Reaction: Alkaline
Toxicity: Skin and eye irritant
Containers: 55 gal open head steel drums, 435# net; cubitainers, 40# net

Clindrol Superamide 100LM. Clintwood Chem.
Chemical Description: Diethanolamide of 70% lauric–30% myristic acid
Category: Foam stabilizer and thickener, nonionic
Applications: Foam stabilizer in aqueous systems; for hair shampoos, bubble bath, liquid dishwash detergent; thickener in gel shampoos
Form: Paste, solid; Color: White
Properties: Solubility: Sol. in alcohols, ketones, chlorinated and aromatic solvents; Sp. gr.: 0.99; M.P.: 29–33 C; Stability: Stable
Composition: 100% active ingred.
Reaction: Alkaline
Toxicity: Skin and eye irritant
Containers: 55 gal open head steel drums, 425# net each

Clindrol 100MC. Clintwood Chem.
Chemical Description: Monoethanolamide of coconut fatty acid
Category: Foam stabilizer, nonionic
Applications: Foam stabilizer in shampoo and liquid dishwash detergent
Form: Flakes; Color: Tan; Odor: Ammoniacal
Properties: Solubility: Sol. in ethanol, isopropanol, aliphatic and aromatic hydrocarbons; Sp. gr.: 1.0; M.P.: 63 C
Composition: 100% active ingred.
Reaction: Alkaline
Toxicity: Skin and eye irritant
Containers: 55 gal open head fiber drums, 200# net each

CM-88. Georgia-Pacific
Chemical Description: Aqueous alkaline solution of a lignoamine
Category: Modifier, cationic
Applications: Replaces 40 to 60% of a primary emulsifier thus providing a lower cost emulsifier system
Form: Solid
Properties: Density: 1.25 g/ml; Solidification: 46%

CMTC-T. Hercules Inc.
Chemical Description: Sodium carboxymethylcellulose
Category: Cellulose gum, ionic
Applications: Used as a synthetic detergent to prevent redeposition of soil that has been removed from fabrics, built soaps, laundry sizes, imparting soil resistance to cotton carpeting, adhesions to cellulose yarns, binding and adhesive properties that make it useful alone in combination with other ingredients as an adhesive
Form: Powder
Properties: Solubility: Sol. in water
Containers: 50 lbs/bag

Cobee 76. PVO International
Chemical Description: Refined,

bleached, deodorized coconut oil
Category: Emollient, superfatting agent, clouding agent
Applications: Food industry as frying oil, ice cream coatings, cracker spray, chewy candies; hair conditioners; pharmaceutical vehicles
Form: Soft solid; Color: 20/2 (Lovibond); Odor and Taste: None
Properties: M.P.: 76–80 F

Cobee 92. PVO International
Chemical Description: Refined, bleached, deodorized, hydrogenated coconut oil
Category: Emollient, superfatting agent, clouding agent
Applications: Cream fillings, nonchewy candies, icings, sandwich fillings, snack roasting, frozen dessert, peanut roasting; vegetable dairy industry; soaps and shampoos as emollients and superfatting agents
Form: Solid; Color: 20/2 (Lovibond); Odor and Taste: None
Properties: M.P.: 98–101 F

Cobee 110. PVO International
Chemical Description: Refined, bleached, deodorized, hydrogenated coconut oil
Category and Applications: See Cobee 92
Form: Solid; Color: 20/2 (Lovibond); Odor and Taste: None
Properties: M.P.: 110–114 F

Complemix 50. American Cyanamid Co.
Category: Wetting agent
Applications: Food and beverage;
Form: Liquid; Color: Nearly colorless
Properties: Sp. gr.: 0.93; B.P.: 173 F
Composition: 50% active ingred.; other components: beverage grade ethanol
Toxicity: Eye irritant

Complemix 100. American Cyanamid Co.
Category: Wetting agent
Applications: Food and beverage
Form: Solid; Color: White
Properties: Solubility: 1.5 g in 100 ml water @ 25 C; Sp. gr.: 1.1
Composition: 100% active ingred.
Toxicity: Eye irritant
Warnings and Caution: Avoid eye contact

Compound 8-S. DuPont Co.
Chemical Description: Complex alkylaryl sulfonate
Category: Dispersing agent, anionic
Applications: Nonfoaming dispersant for disperse- and vat-type dyes; application to clays; dispersing agent for carbon black
Form: Liquid
Properties: Solubility: Water-sol.; Density: 9.6 lb/gal; Sp. gr.: 1.152; Visc.: Readily pourable; Stability: Freezing affects but does not damage physical form; thaw and mix before using
References: Bulletin

Compound 9264B. PVO Internat'l.
Chemical Description: Ethoxylated glycerol monostearate

Category: Softener, nonionic
Form: Solid

Conco AAS-98S. Continental Chem. Co.
Chemical Description: Dodecyl benzene sulfonic acid
Category: Detergent intermediate, anionic
Applications: For manufacturing detergents by neutralizing dodecyl benzene sulfonic acid; manufacture of shampoos, detergent, metal cleaners
Form: Liquid; Color: Brown
Properties: Solubility: Water sol.; Sp. gr.: 1.04; Biodegradable
Composition: 98+% active ingred.; other components: water and dodecyl benzene
Reaction: Acid
Containers: 55 gal open head drums (450 lb net), tankwagons or tankcars
Warnings and Caution: Acid is corrosive (salts of the acid are noncorrosive)
References: Technical bulletin

Conco ATR-98S. Continental Chem. Co.
Chemical Description: Tridecyl benzene sulfonic acid
Category: Detergent intermediate, anionic
Applications: Detergent intermediate; neutralized with wide range of alkalies
Form: Liquid
Properties: Sp. gr.: 1.04; Biodegradable
Composition: 98+% active ingred.
Reaction: Acid
Containers: 55 gal drums
Warnings and Caution: Acid is corrosive
References: Technical bulletin

Condanol 42/FAS. Dutton and Reinisch Ltd.
Chemical Description: Surfactant blend
Category: Detergent
Applications: Shampoo base
Form: Liquid; Color: Pale yellow; Odor: Faint, typical
Properties; Sp. gr.: ≈ 1.0; Visc.: Medium
Composition: 25% active ingred.; other components: water
Reaction: Neutral
Containers: Lacquered drums

Condanol B 2003. Dutton and Reinisch Ltd.
Chemical Description: Tetra-sodium (1,2-dicarboxyethyl)–N-alkyl sulfosuccinamide
Category: Detergent, emulsifier, foaming agent, foam stabilizer, solubilizer, anionic
Applications: Industrial and metal cleaners; dishwashing liquids, household and general purpose cleaners; carpet and upholstery shampoos; soldering aid; reduction of gel formation; dispersion of pigments; emulsifier for wax and oil; cosmetic preparations
Form: Clear liquid; Color: Amber
Properties: Solubility: Sol. in water and electrolytes; Stability: Good esp. in presence of inorganic salts
Composition: 35% solids

Condanol CMS. Dutton and Reinisch Ltd.

Chemical Description: Disodium coconut sulphosuccinamate
Category: Emulsifier, foaming agent
Form: Liquid; Color: Pale yellow; Odor: Slight typical
Properties: Sp. gr.: ≈ 1.0
Composition: 40% active ingred.; other components: water
Reaction: Slight alkaline
Containers: Lacquered drums

Condanol CS 30. Dutton and Reinisch Ltd.
Chemical Description: Surfactant blend
Category: Detergent
Applications: Carpet shampoo base
Form: Liquid; Color: Pale yellow; Odor: Of formalin
Properties: Sp. gr.: ≈ 1.0; Visc.: Low
Composition: 80% active ingred.; other components: water
Reaction: Slight acid
Containers: Lacquered drums

Condanol DLS 35. Dutton and Reinisch Ltd.
Chemical Description: Diethanolamine lauryl sulfate
Category: Detergent
Applications: Shampoo base
Form: Liquid; Color: Pale yellow; Odor: Slight typical
Properties: Sp. gr.: ≈ 1.0; Visc.: Low
Composition: 35% active ingred.; other components: water
Reaction: Neutral
Containers: Lacquered drums

Condanol Emulsifier BW. Rewo Chem.
Chemical Description: Modified nonionic fatty acid derivative
Category: Emulsifier, nonionic
Applications: Cleansing liquids, hand cleaners, cleaning of oily and greasy surfaces
Form: Clear viscous liquid; Color: Yellow
Properties: Solubility: Oil sol.
Composition: 97% solids; 3% water

Condanol Emulsifier BWA. Dutton and Reinisch Ltd.
Chemical Description: Modified nonionic fatty acid derivative
Category: Emulsifier, nonionic
Applications: Metal cleaning formulations as emulsifier for engine cleaners, solvent degreasers; emulsifying paraffin wax; emulsifier for cold cleaners to de-wax cars
Form: Clear viscous liquid; Color: Yellow
Composition: 97% solids; 3% water

Condanol FBR. Dutton and Reinisch Ltd.
Chemical Description: Detergent combination
Category: Detergent, anionic
Applications: Raw material for weakly alkaline household and general-purpose cleaners; removal of latex based floor waxes and polishes; corrosion-inhibitors
Form: Clear viscous liquid
Properties: Density: 0.98 g/cm^3 @ 20 C; Visc.: 350 cp (Brookfield spindle 3)
Composition: 98% solids; 2% water

Condanol FLM. Dutton and Reinisch Ltd.
Category: Anionic foam fire-extinguisher base
Applications: Raw material for low-density foam fire-extinguisher
Form: Clear liquid; Color: Light yellow
Properties: Density @ 20 C: 1.0–1.1 g/cm^3; Clear Pt.: –12 C; Biodegradable

Condanol M 5047. Dutton and Reinisch Ltd.
Chemical Description: Detergent combination
Category: Detergent
Applications: Surface cleaner preconcentrate
Form: Clear liquid; Color: Colorless
Composition: 45% solids content

Condanol MLS 35. Dutton and Reinisch Ltd.
Chemical Description: Monoethanolamine lauryl sulfate
Category: Detergent
Applications: Shampoo base
Form: Liquid; Color: Pale yellow; Odor: Slight typical
Properties: Sp. gr.: ≈ 1.0; Visc.: High
Composition: 35% active ingred.; other components: water
Reaction: Neutral
Containers: Lacquered drums

Condanol NL2. Dutton and Reinisch Ltd.
Chemical Description: Sodium lauryl alcohol polyglycol ether sulfate (approx. 2 mole ethylene oxide)
Category: Detergent
Applications: Shampoo base
Form: Liquid; Color: Water white; Odor: Slight typical
Properties: Sp. gr.: ≈ 1.0; Visc.: Medium
Composition: 30% active ingred.; other components: water
Reaction: Neutral
Containers: Lacquered drums

Condanol NL3. Dutton and Reinisch Ltd.
Chemical Description: Sodium lauryl polyglycol ether sulfate (3 moles ethylene oxide)
Category: Detergent
Applications: Shampoo base
Form: Liquid; Color: Water white; Odor: Slight typical
Properties: Sp. gr.: ≈ 1.0; Visc.: Low
Composition: 30% active ingred.; other components: water
Reaction: Neutral
Containers: Lacquered drums

Condanol NL 3, 28%. Dutton and Reinisch Ltd.
Chemical Description: Sodium lauryl ether sulfate
Category: Detergent, foaming agent, anionic
Applications: Cleaning formulations, high quality shampoos and foam bath preparations
Form: Clear liquid; Color: Almost colorless; Odor: Low
Properties: Visc.: 50–150 cp; Stability: Good temperature stability
Composition: 29% active ingred.

Condanol NLS 28. Dutton and Reinisch Ltd.
Chemical Description: Sodium lauryl sulfate

Category: Detergent, emulsifier, anionic
Applications: Washing and cleaning formulations; cosmetic preparations: foam baths and shampoos; emulsion polymerization
Form: Liquid; Color: Light yellow
Properties: m.w.: 300
Composition: 28% active ingred.

Condanol NLS 30. Dutton and Reinisch Ltd.
Chemical Description: Sodium lauryl sulfate
Category: Detergent
Applications: Shampoo base, emulsion polymerization
Form: Paste; Color: Off white; Odor: Slight typical
Properties: Sp. gr.: ≈ 1.0
Composition: 30% active ingred.; other components: water
Reaction: Alkaline
Containers: Lacquered drums

Condanol NLS 90. Dutton and Reinisch Ltd.
Chemical Description: Sodium lauryl sulfate
Category and Applications: See Condanol NLS 28
Form: Powder; Color: White to light yellow
Composition: 85% active ingred.

Condanol OFR. Dutton and Reinisch Ltd.
Category: Surface cleaner
Applications: Cleaning of synthetic car seats and furniture coverings, white-walled tires, baths, sinks, table-tops, floors, mirrors
Form: Clear, colorless liquid
Composition: 16% solids content

Condanol S 1590. Dutton and Reinisch Ltd.
Chemical Description: Detergent combination
Category: Detergent, anionic
Applications: Wash-active preconcentrate; dishwashing liquids
Form: Clear viscous liquid; Color: Yellow
Properties: Stability: Temperature stable down to 3 C
Composition: 44% solids content

Condanol S 1605. Dutton and Reinisch Ltd.
Chemical Description: Detergent combination
Category: Detergent, anionic; good wetting and emulsifying properties; foam stability
Applications: Liquid washing, rinsing, cleaning agents for household and industrial uses; skin-protecting properties
Form: Clear viscous liquid; Color: Yellow
Composition: 48% solids content

Condanol SB3OL. Dutton and Reinisch Ltd.
Chemical Description: Surfactant blend
Category: Detergent
Applications: Shampoo base, cream shampoos
Form: Liquid; Color: Pale yellow; Odor: Nil
Properties: Sp. gr.: ≈ 1.0; Visc.: Medium
Composition: 30% active ingred.; other components: water
Reaction: Neutral
Containers: Lacquered drums

Condanol SBA/1. Dutton and Reinisch Ltd.
Chemical Description: Sodium salt of the monosulfosuccinate of lauryl alcohol polyglycol ether (short chain)
Category: Detergent
Applications: Hair shampoos, foam baths, metal cleaning
Form: Paste; Color: Off white; Odor: Slight typical
Properties: Sp. gr.: ≈ 1.0
Composition: 40% active ingred.; other components: water
Reaction: Slight acid
Containers: Lacquered drums

Condanol SBD060. Dutton and Reinisch Ltd.
Chemical Description: Dioctyl sodium sulfosuccinate based on 2-ethyl hexanol
Category: Wetting agent
Applications: Super wetting agent with low foaming properties
Form: Liquid; Color: Pale yellow; Odor: Strong typical
Properties: Sp. gr.: ≈ 1.0; Visc.: Medium
Composition: 60% active ingred.; other components: water and isopropanol
Reaction: Neutral
Containers: Lacquered drums

Condanol SB-DO 70. Dutton and Reinisch Ltd.
Chemical Description: Sodium diisooctyl sulfosuccinate
Category: Wetting agent, solubilizer, anionic
Applications: Wetting agent in cosmetic preparations and household cleaners; cleaning lotions; nail-varnish removers; shampoos; foam baths; dishwashing agents; window and glass cleaners technical applications: metal cleaning, anti-smear agent for car windscreens; paper, textile and paint industries; pickling agents; dye industry; dry-cleaning process
Form: Almost colorless liquid
Properties: Stability: Good in pH range 3–9; Surface Tension: 30 dynes/cm (0.1% sol'n. @ 20 C); Biodegradable
Composition: 64% active ingred.

Condanol SBF 12. Dutton and Reinisch Ltd.
Chemical Description: Disodium lauryl alcohol sulfosuccinate
Category: Detergent, foam-improver, anionic
Applications: Cosmetic preparations; carpet and upholstery shampoos; synthetic bar soap formulations
Form: White paste
Properties: Biodegradable
Composition: 39% solids content

Condanol SBF 12-Powder. Dutton and Reinisch Ltd.
Chemical Description: Disodium lauryl alcohol sulfosuccinate
Category: Detergent, anionic
Applications: Synthetic soaps, normal soap bars, dry shampoos, enzyme washing powders, moisture sensitive technical products; used to improve the spray-drying properties of softer materials
Form: Powder; Color: White or light yellow

Properties: Solubility: Water-sol.; Density: 250 g/l; Stability: Good in weakly acid and weakly alkaline media; Biodegradable
Composition: 95% solids content

Condanol SBFA/3. Dutton and Reinisch Ltd.
Chemical Description: Sodium salt of the monosulfosuccinate of lauryl alcohol polyglycol ether (medium chain)
Category: Detergent
Applications: Mild hair shampoos and foam baths
Form: Liquid; Color: Water white
Properties: Sp. gr.: ≈ 1.0; Visc.: Low
Composition: 40% active ingred.; other components: water
Reaction: Neutral
Containers: Lacquered drums

Condanol SBFA 30, 40%. Dutton and Reinisch Ltd.
Chemical Description: Disodium lauryl alcohol polyglycol-ether sulfosuccinate
Category: Detergent raw material, anionic
Applications: Mild detergent base with good foaming properties; high quality shampoos, baby preparations, intimate hygiene preparations
Form: Clear liquid; Color: Almost colorless
Properties: Visc.: 100–200 cp; Biodegradable
Composition: 39% solids content

Condanol SBIL/3. Dutton and Reinisch Ltd.
Chemical Description: Sodium salt of the monosulfosuccinate of lauric isopropanolamide polyglycol ether (medium chain)
Category: Detergent
Applications: Hair shampoos
Form: Gel; Color: Pale yellow; Odor: Typical slight
Properties: Sp. gr.: ≈ 1.0; Visc.: High
Composition: 50% active ingred.; other components: water
Reaction: Slight acid
Containers: Lacquered drums

Condanol SBL. Dutton and Reinisch Ltd.
Chemical Description: Sodium salt of the monosulfosuccinate of lauric monoethanolamide
Category: Detergent, wetting agent
Applications: Carpet shampoo base; fire fighting formulations
Form: Liquid, paste; Color: Pale yellow; Odor: Slight typical
Properties: Sp. gr.: ≈ 1.0; Visc.: Low
Composition: 35% active ingred.; other components: water
Reaction: Slight acid
Containers: Lacquered drums

Condanol SBL 203. Dutton and Reinisch Ltd.
Chemical Description: Disodium lauric acid ethanolamide sulfosuccinate
Category: Detergent, anionic
Applications: Mild detergent, foam improver; cosmetics; aids spray-drying; carpet and uphols-

tery shampoos; synthetic soaps
Form: Turbid liquid; Color: Yellow
Properties: Biodegradable
Composition: 39% solids content

Condanol SBV. Dutton and Reinisch Ltd.
Chemical Description: Modified sulfosuccinate
Category: Detergent, foaming agent, anionic
Applications: Mild detergent, nonirritant shampoos and foam baths, rinsing and cleaning agents with skin-protecting properties
Form: Clear liquid; Color: Light yellow
Properties: Visc.: 2500 cp; Biodegradable
Composition: 28% solids content

Condanol SBZ. Dutton and Reinisch Ltd.
Chemical Description: Sodium salt of the monosulfosuccinate of lauric isopropanolamide polyglycol ether (medium chain)
Category: Detergent
Applications: Hair shampoos
Form: Gel; Color: Pale yellow; Odor: Slight typical
Properties: Sp. gr.: ≈ 1.0; Visc.: Medium
Composition: 50% active ingred.; other components: water
Reaction: Neutral
Containers: Lacquered drums

Condanol SCK 2040. Dutton and Reinisch Ltd.
Chemical Description: Detergent combination
Category: Detergent raw material, anionic
Applications: Preparation of liquid cleaning agents, shampoos
Form: Clear viscous liquid; Color: Yellow
Properties: Visc.: 600 cp; Biodegradable
Composition: 39% solids content

Condanol TLS 40. Dutton and Reinisch Ltd.
Chemical Description: Triethanolamine lauryl sulfate
Category: Detergent
Applications: Shampoo base
Form: Liquid; Color: Pale yellow; Odor: Slight typical
Properties: Sp. gr.: ≈ 1.0; Visc.: Low
Composition: 45% active ingred.; other components: water
Reaction: Neutral
Containers: Lacquered drums

Condanol TMS. Dutton and Reinisch Ltd.
Chemical Description: Disodium alkyl sulfosuccinamide
Category: Emulsifier, wetting agent, foam-additive; anionic
Applications: Foam-additive for natural and synthetic lattices; in carpet backing, foam mattresses, furniture foam blocks; emulsion polymerization
Form: Soft creamy paste or liquid
Properties: Stability: Stable in alkaline and acidic media; Biodegradable
Composition: 35% solids content

Condanol TS 25. Dutton and Reinisch Ltd.
Chemical Description: Sulfosuc-

cinate-base detergent combination
Category: Detergent, anionic
Applications: Concentrate for foam cleaners for coarse-fibered textiles (carpets, upholstery, car-seat covers)
Form: Liquid; Color: Light yellow
Properties: Clear Pt.: 8 C; Biodegradable
Composition: 25% solids content

Condasol Sulfonic Acid K. Dutton and Reinisch Ltd.
Chemical Description: Straight chain dodecylbenzene sulfonic acid
Category: Anionic surfactant
Applications: Dishwashing, washing and cleaning formulations
Form: Viscous liquid; Color: Brown
Properties: 90% biodegradable
Composition: 97% sulfonic acid; 1% sulfuric acid; 2% unsulfated matter

Condasol T 65. Rewo Chem.
Chemical Description: p-toluene sulfonic acid, 65%
Category: Catalyst, electrolyte, intermediate
Applications: Catalyst for preparation of esters, for acid hydrolysis, polymerization, dehydration, alkylation reactions; electrolyte; intermediate product in preparation of cresol
Form: Clear, pale liquid; Odor: Characteristic
Composition: 65% sulfonic acid; 1.5% sulfuric acid; 0.2% toluene; 0.3% sulfone

Warnings and Caution: Strong acid; rubber gloves and goggles should be worn

Condasol TKS 90/F. Dutton and Reinisch Ltd.
Chemical Description: Triammonium dodecylbenzenesulfonate
Category: Raw material for detergent formulations; anionic
Applications: Raw material for production of concentrated cleansing agents, steam-jet and high pressure cleaners, defatting agents, car shampoos; chain greasing; cleaning steel ball bearings
Form: Viscous liquid; Color: Dark amber
Properties: Biodegradable
Composition: 85% active ingred.

Condensate 4476. Proctor Chemical Company, Inc.
Chemical Description: Amine condensate of fatty acid
Category: Wetting agent, detergent, high foamer
Applications: Substitute for 100% coconut amine condensate, condensate PC
Form: Liquid; Color: 6–8 (Gardner)
Properties: Sp. gr.: 1.020
Composition: 88–90% active

Condensate PC. Proctor Chemical Company, Inc.
Chemical Description: Coconut-diethanolamine condensation product
Category: Wetting and rewetting agent, Detergent, high foamer, thickener, emulsifier
Applications: Excellent detergen-

cy on fabrics, scouring piece goods and yarns after naphthol dyeing to remove excess dyestuff adhering to the surface of the fiber, liquid dishwashing formulations, thickener even at concentrations of 5% to 10%, oil soluble emulsifier for preparing self-emulsifiable oil systems, formulating sanitary supply chemicals
Form: Liquid; Color: Amber
Composition: 100% active ingred.
Containers: 480 lb drum

Corax. Diamond Shamrock
Chemcial Description: Wax
Applications: Rubber lubricant for O-rings
Form: Solid
Composition: 100% active ingred.

Corona Lanolin. Croda Chem. Ltd.
Chemical Description: Anhydrous lanolin
Category: Conditioning emollient, moisturizing agent, co-solvent, plasticizer, W/O emulsifier, superfatting agent, wetting/dispersing agent
Applications: Pharmaceutical bases and salves, skin and hair care preparations, make-up cosmetics, soaps, protective creams

Coronet Lanolin. Croda Chem. Ltd.
Chemical Description: Super refined cosmetic grade lanolin
Category and Applications: See Corona lanolin
Form: Minimum odor, color and batch variation in this grade
Properties: Slightly higher M.P. than USP and BP standards

Corrosion Inhibitor 500. Alkaril Chem. Ltd.
Category: Corrosion inhibitor
Applications: Used where corrosion is caused by hydrochloric acid, organic acids and carbon dioxide; hydrochloric acid industrial cleaning products and pickling baths
Form: Clear liquid; Color: Deep amber
Properties: Solubility: Sol. in distilled water, aliphatic and aromatic solvents, 15% hydrochloric acid, 15% sulfuric acid, 30% phosphoric acid
References: Technical bulletin 101

Corrosion Inhibitor 600. Alkaril Chem. Ltd.
Category: Corrosion inhibitor
Applications: Filming type corrosion inhibitor when applied to steel; used in concentrated hydrochloric, phosphoric and sulfuric acids; concentrated mineral acid industrial cleaning products for pickling bath and drain cleaner applications
Form: Clear liquid; Color: Amber
Properties: Solubility: Sol. in hydrochloric, phosphoric and sulfuric acids; pH: 6.0-7.0
References: Technical bulletin 102

CPA-Alpha. Dixo Co. Inc.
Chemical Description: Anionic/nonionic detergent mixture
Category: Detergent, wetting agent, emulsifier
Applications: Detergent for dry-

cleaning solvent use to improve or accelerate soil removal; "prespotting" material in drycleaning; emulsifier for water and drycleaning solvents
Form: Liquid; Color: Very light yellow amber; Odor: Slight
Properties: Solubility: Completely sol. in drycleaning solvents
Composition: 100% active ingred.
Containers: 1, 5, 15, 30, and 55 gal containers

Cremophor A 6. BASF AG
Chemical Description: Ethoxylated saturated fatty alcohol
Category: Emulsifer, nonionic
Applications: Emulsifying agents for ointments, creams, liquid emulsions, gels and other cosmetic and pharmaceutical preparations
Form: Wax; Color: White
Properties: Ref. Index: 1.4420–1.4424 @ 60 C; Solubility: Sol. in: water, alcohol; Density: 0.896–0.906 g/cm^3 @ 60 C; HLB: 10–12; Sapon. Value: < 3; Acid No.: < 1; Hydroxyl No.: 115–135; Iodine Value: < 1; Stability: Stable to acids, bases and salts; Emulsifiability: Oil-in-water types of emulsion; pH: 6–7
Composition: 100% active ingred.
References: Technical data sheet M 5551 e

Cremophor A 11. BASF AG
Chemical Description: Ethoxylated saturated fatty alcohol
Category: Emulsifier, nonionic
Applications: Emulsifying agents for ointments, creams, liquid emulsions, gels and other cosmetic and pharmaceutical preparations
Form: Wax; Color: White
Properties: Ref. Index: 1.4464–1.4474 @ 60 C; Solubility: Sol. in: water, alcohol; Density: 0.964–0.968 g/cm^3 @ 60 C; HLB: 12–14; Sapon. Value: < 1; Acid No.: < 1; Hydroxyl No.: 70–80; Iodine Value: < 1; Stability: Stable to acids, bases and salts; Emulsifiability: Oil-in-water types of emulsion; pH: 6–7
Composition: 100% active ingred.
References: Technical Data sheet M 5551 e

Cremophor A 25. BASF AG
Chemical Description: Ethoxylated saturated fatty alcohol
Category: Emulsifier, nonionic
Applications: Emulsifying agents for ointments, creams, liquid emulsions, gels and other cosmetic and pharmaceutical preparations;
Form: Powder; Color: White
Properties: Ref. Index: 1.4512–1.4520 @ 60 C; Solubility: Sol. in: water, alcohol; Density: 1.020–1.028 g/cm^3 @ 60 C; HLB: 15–17; Sapon. Value: < 3; Acid No.: < 1; Hydroxyl No.: 35–45; Iodine Value: < 1; Stability: Stable to acids, bases and salts; Emulsifiability: Oil-in-water types of emulsions; pH: 5–7
Composition: 100% active ingred.
References: Technical Data sheet M 5551 e

Cremophor EL. BASF AG
Chemical Description: Fatty acid

esters of glycerolpolyglycols and polyglycols
Category: Solubilizer, emulsifier,, nonionic
Applications: Pharmaceuticals, cosmetics, veterinary medicine, and animal feeding stuffs; used for producing aqueous preparations of hydrophobic substances
Form: Liquid: Color: Pale yellow
Properties: Ref. Index: ≈ 1.471; Solubility: Sol. in water, ethyl alcohol, *n*-propyl alcohol, isopropyl alcohol, ethyl acetate, chloroform, carbon tetrachloride, trichlorethylene, benzene, toluene and xylene; Density: 1.05–1.06 ml; Visc.: 700–850 cps; HLB: 12–14; Sapon. Value: 65–70; Acid No.: < 2; Hydroxyl No.: 65–78; Iodine Value: 28–32; Stability: Stable to electrolytes, e.g., acids and salts; Emulsifiability: Fatty soluble vitamins A, D, E and K; ethereal oils or perfume oils; pH: 6–8 (10% aq.)
Composition: 100% active ingred.
Warnings and Caution: Should be stored in closed containers and protected from light
References: Technical data sheet M 1714 e

Cremophor RH 40. BASF AG
Chemical Description: Hydrophobic consisting of fatty acid glycerol polyglycol esters and fatty acid polyglycol esters
Category: Solubilizer, emulsifier, nonionic
Applications: Solubilization of essential oils and perfumery synthetics
Form: Paste; Color: White
Properties: Ref. Index: 1.453–1.457 @ 60 C; Solubility: Sol. in water, ethanol, isopropyl alcohol, *n*-propanol, ethyl acetate, chloroform, carbon tetrachloride, trichloroethylene, benzene, toluene and xylene; HLB: 14–16; Sapon Value: 50–60; Acid No.: < 1; Hydroxyl No.: 60–80; Iodine Value: < 1; Stability: Stable to aqueous alcohol; Emulsifiability: hydrophobic substances; pH: 6–7 (10% aq. sol'n.)
Composition: 100% active ingred.
Warnings and Caution: Drums or tanks should be kept tightly closed or microorganisms may grow
References: Technical data sheet M 5629 e

Cremophor RH 60. BASF
Chemical Description: Hydrophobic consisting of fatty acid glycerol polyglycol esters and fatty acid polyglycol esters
Category: Solubilizer, emulsifier, nonionic
Applications: Solubilization of essential oils and perfumery synthetics
Form: Paste; Color: White
Properties: Ref. Index: 1.453–1.457 @ 60 C; Solubility: Sol. in water, ethanol, isopropyl alcohol, *n*-propanol, ethyl acetate chloroform, carbon tetrachloride, trichloroethylene, benzene, toluene and xylene; HLB: 15–17; Sapon. Value: 40–50; Acid No.: < 1; Hydroxyl No.: 50–70; Iodine Value: < 1; Stability: Stable to aqueous alcohol; Emulsifiability:

Hydrophobic substances; pH: 6–7 (10% aq. sol'n.)
Composition: 100% active ingred.
Warnings and Caution: Drums or tanks should be kept tightly closed or microorganisms may grow
References: Technical data sheet M 5629 e

Cremophor RH 410. BASF
Chemical Description: Hydrophobic consisting of fatty acid glycerol polyglycol esters and fatty acid polyglycol esters
Category: Solubilizer, emulsifier, nonionic
Applications: Solubilization of essential oils and perfumery synthetics
Form: Liquid
Properties: Ref. Index: 1.457–1.462; Solubility: Sol. in water, ethanol, isopropyl alcohol, *n*-propanol, ethyl acetate, chloroform, carbon tetrachloride, trichloroethylene, benzene, toluene and xylene; Visc.: 1500–2000 cps; HLB: 14–16; Sapon. Value: 45–55; Acid No.: < 1; Iodine Value: < 1; Stability: Stable to aqueous alcohol; Emulsifiability: Hydrophobic substances; pH: 6–7 (10% aq.)
Composition: 90% active ingred.
Warnings and Caution: Drums or tanks should be kept tightly closed or microorganisms may grow
References: Technical data sheet M 5629 e

Cremophor RH 455. BASF AG
Chemical Description: Hydrophobic consisting of fatty acid glycerol polyglycol esters and fatty acid polyglycol esters
Category: Solubilizer, emulsifier, nonionic
Applications: Solubilization of essential oils and perfumery synthetics
Form: Liquid
Properties: Ref. Index: 1.459–1.464; Solubility: Sol. in water, ethanol, isopropyl alcohol, *n*-propanol, ethyl acetate, chloroform, carbon tetrachloride, trichloroethylene, benzene, toluene and xylene; Visc.: 1000–1500 cps; HLB: 14–16; Sapon. Value: 45–55; Acid No.: < 1; Iodine Value: < 1; Stability: Stable to aqueous alcohol; Emulsifiability: Hydrophobic substances; pH: 6–7 (10% aq.)
Composition: 90% active ingred.
Warnings and Caution: Drums or tanks should be kept tightly closed or microorganisms may grow
References: Technical data sheet M 5629 e

Cremophor S 9. BASF AG
Chemical Description: Ester of a polyethylene glycol with stearic acid
Category: Emulsifier, nonionic
Applications: Oil-in-water type emulsions; suspension stabilizer; thickening agent; good lubricating and anti-tack effect
Form: Solid; Color: Yellowish white
Properties: Ref. Index: ≈ 1.446 @ 60 C; Solubility: Sol. in water, ethyl alcohol, isopropyl alcohol,

n-propyl alcohol, acetone, ethyl acetate, chloroform, benzene, dimethyl phthalate, castor oil, cod liver oil and oleic acid; Density: ≈ 0.97 g/cm^3 @ 60 C; HLB: ≈ 12; Sapon. Value: 88–98; Acid No.: 2; Hydroxyl No.: 80–105; Iodine Value: < 2; Stability: Stable to polyethylene glycols and other inert bases, electrolytes; Emulsifiability: Fatty oils, greases, vaseline, solid and liquid paraffin, beeswax, silicone oils, fatty alcohols
Composition: 100% active ingred.
References: Technical data sheet M 5542 e

Cremophor WO-A. BASF AG
Chemical Description: Fatty alcohol glycerol ether
Category: Emulsifier, nonionic
Applications: Water-in-oil emulsion-based cosmetics and pharmaceuticals
Form: Liquid; Color: Yellowish
Properties: Solubility: Sol. in mineral and native oils and fats; Sp. gr.: 0.950–0.980 g/cm^3 @ 20 C; HLB: 4.8–5.8; Sapon. Value: < 5; Acid No.: < 0.1; Hydroxyl No.: 230–300; Iodine Value: 50–60; Stability: Stable to acids, bases and electrolytes
Composition: 100% active ingred.
References: Technical data sheet M 5614 e

Crestalans. Croda Chem.
Chemical Description: Derivatives of fractionated liquid lanolin and isopropyl esters
Category: Emollient, moisturizing agent, conditioners, plasticizers, super-fatting agents
Applications: Skin preparations, baby creams, lotions and oils, skin make-up, hair preparations; plasticizers for aerosol hair sprays and lacquers; super-fatting agents for nail varnish removers and similar cosmetic solvent compositions
Form: Clear liquid; Color: Pale yellow; Odor: Faint sterol

Crill Series. Croda Chem.
Chemical Description: Partial esters of sorbitan
Category: Nonionic surfactant, emulsifier, dispersant, co-solvent, wetting agents, antifoams, viscosity reducers, mold release agents, internal slip agents, antiblock agents
Applications: Cosmetics, food and food packaging, insecticides and herbicides, leather treatment, metal working fluids, oil slick dispersants, paints and inks, pharmaceuticals, plastics, polishes, textiles

Crill 1. Croda Chem.
Chemical Description: Sorbitan monolaurate
Category: Emulsifier, wetting and pigment dispersing aid, antifoam, corrosion inhibitor, lubricant, antistatic agent
Applications: Auxiliary emulsifier in cosmetics and pharmaceutical oil in water systems; antifoam in food and protein processing (e.g., milk, molasses, sugar and yeast); manufacture of paper board for food packaging; used in coating polyolefin films; dry cleaning

systems; leather treatment; metal and furniture polishes; pesticide emulsions; textile processing
Form: Clear viscous liquid; Color: Pale yellow
Properties: Solubility: Sol. in ethyl alcohol, oleyl alcohol, mineral oil; HLB: 8.6; Sap. Value: 158–170

Crill 2. Croda Chem.
Chemical Description: Sorbitan monopalmitate
Category and Applications: See Crill Series; may be used in many of the applications designated to Crill 3; leaves nongreasy residue
Form: Hard waxy solid; Color: Pale tan
Properties: Solubility: Partly sol. in propylene glycol, ethyl and oleyl alcohols, olive oil, oleic acid; M.P.: 46 C; HLB: 6.7; Sap. Value: 140–150

Crill 3. Croda Chem.
Chemical Description: Sorbitan monostearate
Category: Emulsifier, lubricant, antistat
Applications: O/W emulsions; cosmetic and pharmaceutical creams and lotions; silicone and wax polishes; insecticides, herbicides; metal cleaners; buffing compounds; textile lubricants; antistatic agents; food applications
Form: Hard waxy solid; Color: Pale tan; Odor: Low
Properties: Solubility: Partly sol. in oleyl alcohol, olive oil, oleic acid; M.P.: 51–54 C; HLB: 4.7; Sap. Value: 147–157

Crill 4. Croda Chem.
Chemical Description: Sorbitan monooleate
Category: Emulsifier, wetting agent, dispersant, lubricant, coupling agent, antifoam
Applications: Industrial W/O emulsifiers; cosmetic and pharmaceutical applications (as wetting agents and dispersants for medicaments and fillers, pigment dispersing aid in lipstick, eye makeup); aids dispersion for aerosol powder systems; aerosol polishes; insecticidal sprays; inks and surface coatings; metal working lubricants; cutting oils; textile lubricants; dry cleaning operations; food process antifoam; oil slick dispersant
Form: Viscous liquid; Color: Amber
Properties: Solubility: Sol. in ethyl and oleyl alcohols, mineral oil, IPM, olive oil, oleic acid; HLB: 4.3; Sap. Value: 147–160

Crill 6. Croda Chem.
Chemical Description: Sorbitan monoisostearate (based on a branched chain fatty acid; completely saturated ester)
Category: W/O emulsifier, wetting agent, pigment dispersant
Form: Viscous liquid; Color: Pale yellow
Properties: Solubility: Sol. in mineral oil, olive oil, partly sol. in oleyl alcohol, IPM; HLB: 4.7; Sap. Value: 143–153

Crill 35. Croda Chem.
Chemical Description: Sorbitan

tristearate
Category and Applications: See Crill 3
Form: Hard waxy solid; Color: Pale tan
Properties: Solubility: Partly sol. in oleyl alcohol, mineral oil, IPM, olive oil, oleic acid; M.P.: 48 C; HLB: 2.1; Sap. Value: 176–188

Crill 43. Croda Chem.
Chemical Description: Sorbitan sesquioleate
Category and Applications: See Crill 4
Form: Viscous liquid; Color: Amber
Properties: Solubility: Sol. in oleyl alcohol, mineral oil, olive oil, oleic acid; HLB: 3.7; Sap. Value: 149–160

Crill 45. Croda Chem.
Chemical Description: Sorbitan trioleate
Category and Applications: See Crill 4
Form: Viscous liquid; Color: Amber
Properties: Solubility: Sol. in oleyl alcohol, mineral oil, IPM, olive oil, oleic acid; HLB: 1.8; Sap. Value: 172–186

Crillet Series. Croda Chem.
Chemical Description: Polyethoxylated sorbitan esters
Category: Nonionic surfactants; emulsifier, wetting agent, solubilizer; often combined with a member of the Crill range in emulsification systems
Applications: Cosmetics, food and food packaging, household products, insecticides and herbicides, metal working fluids, paints and inks, pharmaceuticals, polishes and cleaners, polymers, textiles

Crillet 1. Croda Chem.
Chemical Description: POE 20 sorbitan monolaurate
Category: Solubilizer, emulsifier
Applications: Solubilizer for essential oils and perfumes in cosmetics; emulsifier and dispersant for pharmaceuticals (e.g., menthol, phenols); cosmetic creams and lotions; animal feedstuffs
Form: Clear liquid; Color: Yellow; Odor: Low
Properties: Solubility: Sol. in water, ethyl alcohol, oleyl alcohol, oleic acid; HLB: 16.7; Sap. Value: 40–51; Surface Tension: 38.5 dynes/cm (0.1% sol'n. @ 25 C)

Crillet 2. Croda Chem.
Chemical Description: POE 20 sorbitan monopalmitate
Category: Emulsifier, dispersant, penetrant, levelling agent, lubricant, antistat
Applications: Similar to Crillet 3 with slightly higher HLB
Form: Pasty liquid; Color: Yellow amber
Properties: Solubility: Sol. in water, ethyl alcohol, oleyl alcohol, oleic acid; HLB: 15.6; Sap. Value: 43–49; Surface Tension: 41.5 dynes/cm (0.1% @ 25 C)

Crillet 3. Croda Chem.
Chemical Description: POE 20 sorbitan monostearate

Category: Emulsifier, dispersant, penetrant, levelling agent, lubricant, antistat
Applications: As emulsifier, alone or with Crill 3, for cosmetics and pharmaceuticals; for solvents and polish ingredients; dispersant for insecticides, herbicides, cattle dyes; textile dyeing as penetrant and levelling agent; lubricant/antistat in synthetic fiber processing
Form: Yellow liquid gels to soft solid on cooling
Properties: Solubility: Sol. in ethyl and oleyl alcohol, oleic acid; partly sol. in water, isopropyl myristate, xylene, trichlorethylene; HLB: 14.9; Sap. Value: 45–55; Surface Tension: 42.5 dynes/cm (0.1% @ 25 C)

Crillet 4. Croda Chem.
Chemical Description: POE 20 sorbitan monooleate
Category: Emulsifier, dispersant, solubilizer, lubricant, detergent, antistat, wetting agent
Applications: Cosmetics and pharmaceuticals, often used with appropriate Crill; emulsifier for solvents, waxes, silicones, aerosol and nonaerosol polishes; oil slick dispersant, marine tank cleaners; synthetic animal feedstuffs, hand cleaners; insecticides, herbicides; leather degreasing; veterinary products
Form: Clear liquid; Color: Yellow amber; Odor: Faint characteristic fatty
Properties: Solubility: Sol. in water, ethyl and oleyl alcohols, oleic acid; partly sol. in isopropyl myristate, xylene; HLB:15.0; Sap. Value: 45–55; Surface Tension: 42.5 dynes/cm (0.1% @ 25 C)

Crillet 6. Croda Chem.
Chemical Description: POE 20 sorbitan monoisostearate
Category and Applications: See Crillet Series; Solubilizing characteristics similar to Crillet 3; low temperature liquidity and superior color, odor and chemical stability
Form: Clear liquid; Color: Yellow
Properties: Solubility: Sol. in water, ethyl and oleyl alcohols, oleic acid, xylene, trichlor-ethylene; HLB: 14.9; Sap. Value: 40–50; Surface Tension: 38.6 dynes/cm (0.1% @ 25 C)

Crillet 11. Croda Chem.
Chemical Description: POE 4 sorbitan monolaurate
Category and Applications: See Crillet 1
Form: Clear liquid; Color: Yellow
Properties: Solubility: Sol. in ethyl and oleyl alcohols, oleic acid, kerosene; partly sol. in water, isopropyl myristate, olive oil, xylene, trichlor-ethylene; HLB: 13.3; Sap. Value: 100–115; Surface Tension: 34.7 dynes/cm (0.1% @ 25 C)

Crillet 31. Croda Chem.
Chemical Description: POE 4 sorbitan monostearate
Category and Applications: See Crillet 3
Form: Waxy solid; Color: Cream pale yellow
Properties: Solubility: Sol. in oleyl

alcohol, oleic acid; HLB: 9.6; Sap. Value: 98–113; Surface Tension: 41.5 dynes/cm (0.1% @ 25 C)

Crillet 35. Croda Chem.
Chemical Description: POE 20 sorbitan tristearate
Category and Applications: See Crillet 3
Form: Waxy solid; Color: Cream/buff
Properties: Solubility: Sol. in ethyl and oleyl alcohols, oleic acid, trichlor-ethylene; partly sol. in water, mineral oil, isopropyl myristate, olive oil, kerosene, xylene; HLB: 10.5; Sap. Value: 88–98; Surface Tension: 42.5 dynes/cm (0.1% @ 25 C)

Crillet 41. Croda Chem.
Chemical Description: POE 5 sorbitan monooleate
Category and Applications: See Crillet 4
Form: Clear liquid; Color: Yellow amber
Properties: Solubility: Sol. in ethyl alcohol, oleyl alcohol, isopropyl myristate, oleic acid, kerosene, butyl stearate; HLB: 10.0; Sap. Value: 96–104; Surface Tension: 37 dynes/cm (0.1% @ 25 C)

Crillet 45. Croda Chem.
Chemical Description: POE 20 sorbitan trioleate
Category and Applications: See Crillet 4
Form: Clear viscous liquid; Color: Amber
Properties: Solubility: Sol. in ethyl and oleyl alcohols, isopropyl myristate, oleic acid, kerosene, trichlor-ethylene, butyl stearate; HLB: 11.0; Sap. Value: 82–95; Surface Tension: 41 dynes/cm (0.1% @ 25 C)

Crodafos Series. Croda Chem.
Chemical Description: Alkyl ether phosphates; available in acid form or neutralized with appropriate base
Category: Anionic surfactants, detergents, wetting agents, coupling agents, corrosion inhibitors, thickeners, foam stabilizers, lubricants, antistats
Applications: Detergents and cleaners (hard surface, floor, industrial, steam cleaners, dry cleaning process); metal working fluids, cutting oils, lubricating oils, soluble cutting oils; grinding fluids; anticorrosion agent; antistatic agent for various polymers; textiles (processing of synthetic fibers, fabric softening, dye baths, bleaching and scouring operations); herbicides, pesticides and fertilizers; emulsifiers for solvents and industrial specialties
Form: Many and varied
Composition: 99% min. active
Warnings and Caution: Treat Crodafos acid phosphates as acids; avoid eye and skin contact; flush with water in case of contact

Crodafos 25D Series. Croda Chem.
Chemical Description: Alkyl ether phosphates based on synthetic C12/C15 alcohols; available with 2, 5, and 10 moles ethylene oxide and as Crodafos 25D TEA salt

Category and Applications: See Crodafos Series
Form: Liquid to soft paste; Color: Colorless to pale yellow
Properties: pH: 1.8–2.5 (3% sol'n. @ 20 C)
Composition: 99% min. active

Crodafos CS Series. Croda Chem.
Chemical Description: Alkyl ether phosphates based on ceto stearyl alcohol; available with 2, 5, and 10 moles ethylene oxide and as Crodafos CS TEA salt
Category and Applications: See Crodafos Series
Form: Hard wax; Color: White to buff
Properties: pH: 1.8–2.5 (3% sol'n. @ 20 C)
Composition: 99% min. active

Crodafos ID Series. Croda Chem.
Chemical Description: Alkyl ether phosphates based on isodecanol; available with 2, 5, and 10 moles ethylene oxide and as Crodafos ID TEA salts
Category and Applications: See Crodafos Series
Form: Fluid to paste; Color: Colorless to pale yellow
Properties: pH: 1.8–2.5 (3% sol'n. @ 20 C)
Composition: 99% min. active

Crodafos N Series. Croda Chem.
Chemical Description: Phosphated oleyl polyether; available as Crodafos N3 based on Volpo N3 and Crodafos N10 based on Volpo N10; available as the free acids and as neutralized (salt) versions
Category: Emulsifier, solubilizer, detergent, antistat, anionic
Applications: Emulsifier esp. for use at low concentrations; cosmetics, pharmaceutical and toiletry preparations (e.g., bubble bath, shampoo, waterless hand cleaners)
Form: Paste
Composition: 100% active ingred.

Crodafos O Series. Croda Chem.
Chemical Description: Alkyl ether phosphates based on oleyl alcohol; members differ in polyether chain length with 2, 5, 10 moles of ethylene oxide respectively (Crodafos O.2 acid, Crodafos O.5 acid, Crodafos O.10 acid)
Also available neutralized by triethanolamine as Crodafos O.2 TEA salt, etc.
Category and Applications: See Crodafos Series
Form: Slightly viscous liquid to soft paste; Color: Pale yellow to amber
Properties: pH: 1.8–2.5 (3% sol'n. @ 20 C)
Composition: 99% active ingred.

Crodafos SG. Croda Chem.
Chemical Description: Phosphated ester derived from an alkoxylate of cosmetic grade cetyl alcohol
Category: Emulsifier, antistat, corrosion inhibitor, antigelling agent, anionic
Applications: Shampoos, cream rinses, soap-free cleansing products; depilatories, hair straighteners and other alkaline cosmetics;

corrosion inhibitors and antigelling agents in soluble salt aerosol antiperspirants
Form: Liquid
Composition: 100% active ingred.

Crodafos T Series. Croda Chem.
Chemical Description: Alkyl ether phosphates based on tridecanol; available with 2, 5, and 10 moles ethylene oxide and as Crodafos T TEA salts
Category and Applications: See Crodafos Series
Form: Liquid to viscous liquid; Color: Colorless to pale yellow
Properties: pH: 1.8–2.5 (3% sol'n. @ 20 C)
Composition: 99% min. active

Crodalan AWS. Croda Chem.
Chemical Description: Acetylated ethoxylated lanolin alcohol derivative
Category: Emollient, super-fatting agent, conditioning agent, O/W emulsifier, solubilizing agent, nonionic
Applications: Shampoos, foam baths, bubble baths, skin creams/lotions, make-up preparations, shaving compositions
Form: Liquid; Color: Golden; Odor: Faint fatty
Composition: 100% active ingred.

Crodalan IPL. Croda Chem.
Chemical Description: Isopropyl esters based on fractionated lanolin acids; available in liquid or solid form
Category: Emollient, wetting agent
Applications: Bath oils, skin, hair and make-up cosmetics
Form: Thin opalescent liquid or soft paste liquifying at skin temperature; Color: Pale yellow

Crodalan LA. Croda Chem.
Chemical Description: Acetylated lanolin alcohols
Category: Emollient, penetrant, plasticizer, nonionic
Applications: Cosmetics (hair sprays and lacquers, skin preparations, baby products, hair preparations, make-up); pharmaceuticals; as plasticizer in polymeric substances and high melting point fats
Form: Thin mobile liquid; Color: Clear; Odor: Odorless
Properties: Visc.: Low
Composition: 100% active ingred.

Crodamol BS. Croda Chem.
Chemical Description: Butyl stearate
Category: Emollient, plasticizer, wetting agent
Applications: Cosmetics: Emollient without oiliness in lotions and creams; plasticizer in hair sprays and nail varnish (also improves gloss); nail varnish removers; wetting agent for pigments and dyes in lipsticks
Form: Liquid; Color: Water white

Crodamol CAP. Croda Chem.
Chemical Description: Blend of branched chain esters
Category: Emollient
Applications: Skin care preparations as emollient with good water

repellency, reduced stickiness
Form: Color: Water white; Odor: Low

Crodamol CL. Croda Chem.
Chemical Description: Cetyl lactate
Category: Emollient
Applications: Emollient with good solubility in aqueous alcoholic solutions; dry emollient in make-up and skin care products; in preparations packed in polystyrene
Form: Solid; Color: Colorless to pale straw
Properties: Solubility: Good in aqueous alcoholic solutions; sol. in various glycols

Crodamol CSL. Croda Chem.
Chemical Description: Ceto stearyl lactate
Category: Emollient
Applications: Emollient, esp. in compositions packaged in styrene
Properties: Similar to Crodamol ML with slightly higher M.P.

Crodamol DA. Croda Chem.
Chemical Description: Diisopropyl adipate
Category: Super-fatting agent, solvent for cosmetics, plasticizer, emollient
Applications: Solvent for aromatic oils and perfumes, lipstick dyestuffs, after shave lotions, colognes, skin fresheners; super-fatting agent for aqueous alcoholic lotions and nail varnish removers; plasticizer in nail varnishes and hair sprays; skin care preparations and shampoos as emollient
Form: Liquid; Color: Water white; Odor: Neutral
Properties: Solubility: Sol. in aqueous alcoholic solutions; miscible with mineral oils, vegetable oils, synthetic esters, acetoglycerides, triacetin and silicone oils of low viscosity; Visc.: Very low

Crodamol DO. Croda Chem.
Chemical Description: Decyl oleate
Category: Emollient, lubricant, penetrant
Applications: Cosmetics and pharmaceuticals; leaves nonsticky dry film on skin; in lipsticks as a partial substitute for castor oil
Form: Liquid; Color: Pale straw; Odor: Mild characteristic; Taste: Bland

Crodamol ICS. Croda Chem.
Chemical Description: Isocetyl stearate
Category: Emollient, lubricant, spreading agent
Applications: Controls viscosity and crystallization and reduces rheological changes in emulsion systems; bath oils, skin creams and lotions, hair-dressing, make-up cosmetics; cosmetics packaged in styrene
Form: Liquid; Color: Water white
Properties: Visc.: Low; Liquid sets at ≈ 0 C

Crodamol IPM. Croda Chem.
Chemical Description: Isopropyl myristate
Category: Spreading agent, emol-

lient, co-solvent
Applications: Vehicle and co-solvent for cosmetic raw materials; bath oils, aerosol antiperspirants, deodorants, hygiene sprays, sunscreens, lipsticks, pre-shave preparations, skin care, make-up, and hair care cosmetics
Form: Liquid; Color: Water white; Odor: Odorless

Crodamol LL. Croda Chem.
Chemical Description: Lauryl lactate
Category: Emollient
Applications: Promotes spreading and wetting characteristics of cosmetics; leaves thin dry film on skin and hair; antiperspirants, after shaves, make-up, hair and skin care products
Form: Liquid; Color: Colorless to pale straw; Odor: Faint bland
Properties: Solubility: Sol. in highly polar materials such as glycols; does not attack styrene

Crodamol ML. Croda Chem.
Chemical Description: Myristyl lactate
Category: Emollient
Applications: Improves smoothness and gloss in lipsticks without oiliness; tack reducer in antiperspirants; hair care preparations
Form: Liquid; Color: White to pale yellow
Properties: Solubility: Sol. in aqueous alcoholic solutions; essentially lipophilic but also sol. in a variety of glycols; does not attack polystyrene

Crodamol MM. Croda Chem.
Chemical Description: Myristyl myristate
Applications: Substitute for spermaceti wax and/or beeswax in cosmetic and pharmaceutical formulations; hair creams, cold creams, cleansing creams, suppositories, lipsticks, eye make-up
Form: Waxy solid; Color: Creamy white

Crodamol OP. Croda Chem.
Chemical Description: 2 ethyl hexyl palmitate
Category: Lubricant
Applications: Aids prevention of insensible perspiration and skin respiration; used in make-up, skin and hair care preparations and to modify the occlusiveness of creams, ointments and oils
Form: Liquid; Color: Water white; Odor: Mild
Properties: Low order of reaction with polystyrene packaging materials

Crodamol PC. Croda Chem.
Chemical Description: Propylene glycol diester of selected coconut fatty acids
Category: Emollient, spreading agent
Applications: Cosmetic systems esp. bath oils, antiperspirants, lipsticks, skin and hair care preparations
Form: Liquid; Odor: Slight bland
Properties: Visc.: Low; Cloud Pt.: Very low

Crodamol SS. Croda Chem.

Chemical Description: Synthetic spermaceti NF; essentially a mixture of myristyl myristate and cetostearyl palmitate
Category: Emollient
Applications: Gives body and emollience to cosmetic and pharmaceutical formulations
Form: Crystalline solid; Color: Almost white

Crodamol W. Croda Chem.
Chemical Description: Blend of straight chain saturated esters
Category: Emollient and water repellent
Applications: Cosmetics and toiletries; melts rapidly on application to skin giving cooling effect; water repellent and nongreasy emollient film on application to skin
Form: Waxy solid; Color: White; Odor: Mild
Properties: M.P.: Just below skin temperature

Crodasinic LS35. Croda Chem.
Chemical Description: Sodium N-lauroyl sarcosinate
Category: Foaming agent, wetting agent, detergent, lubricant, antistat, corrosion inhibitor, bacteriostat, penetrant, anionic
Applications: For use in slightly acidic conditions; as inhibitor for hexokinase, used in toothpaste, mouth washes, other dental preparations; corrosion inhibitor in aqueous systems; pharmaceuticals, depilatories, shampoos, shaving preparations, toilet soaps; surgical scrubs; alkali cleaners, carpet shampoos; cutting oils; emulsion polymerization; food packaging; textile scouring and fulling of wool; household and industrial cleaners
Form: Clear liquid
Properties: Solubility: Water sol.; Stability: Stable to alkaline hydrolysis, under moderately acid conditions at normal temperatures; Biodegradable
Composition: 35% active ingred. (also available as 30% active)

Crodateric C. Croda Chem.
Chemical Description: Derived from coconut fatty acid
Category: Amphoteric surfactants
Properties: m.w.: 360
Composition: 50% active aqueous solution

Crodateric Cy. Croda Chem.
Chemical Description: Derived from caprylic acid
Category: Amphoteric surfactants
Properties: m.w.: 284
Composition: 50% active aqueous solution

Crodateric L. Croda Chem.
Chemical Description: Derived from lauric acid
Category: Amphoteric surfactants
Properties: m.w.: 340
Composition: 50% active aqueous solution

Crodateric O. Croda Chem.
Chemical Description: Derived from oleic acid
Category: Amphoteric surfactants
Properties: m.w.: 417

Composition: 50% active aqueous solution

Crodateric O.100. Croda Chem.
Chemical Description: Derived from oleic acid
Category: Amphoteric surfactants
Properties: m.w.: 417
Composition: 100% active

Crodateric S. Croda Chem.
Chemical Description: Derived from stearic acid
Category: Amphoteric surfactants
Properties: m.w.: 410
Composition: 50% active aqueous solution

Crodazoline C. Croda Chem.
Chemical Description: Fatty acid imidazoline
Form: Color: 9–10 (Gardner)
Properties: m.w.: 288; M.P.: 42 C; Surface Tension: 32.5 (1%/0.36%); pH: 11.5 (10% aq. disp.)
Composition: 90% min. imidazoline

Crodazoline Cy. Croda Chem.
Chemical Description: Fatty acid imidazoline
Form: Color: 10–11 (Gardner)
Properties: m.w.: 212; M.P.: 45 C; Surface Tension: 27.8 (1%/0.36%); pH: 12 (10% aq. disp.)
Composition: 90% min. imidazoline

Crodazoline H. Croda Chem.
Chemical Description: Fatty acid imidazoline
Form: Color: 8–9 (Gardner)
Properties: m.w.: 383; M.P.: 57 C; Surface Tension: 43 (1%/0.36%); pH: 11.5 (10% aq. disp.)
Composition: 90% min. imidazoline

Crodazoline L. Croda Chem.
Chemical Description: Fatty acid imidazoline
Form: Color: 4–5 (Gardner)
Properties: m.w.: 268; M.P.: 40 C; Surface Tension: 37.2 (1%/0.36%); pH: 11.3 (10% aq. disp.)
Composition: 90% min. imidazoline

Crodazoline M. Croda Chem.
Chemical Description: Fatty acid imidazoline
Form: Color: 9–10 (Gardner)
Properties: m.w.: 297; M.P.: 51 C; Surface Tension: 36.5 (1%/0.36%); pH: 11.5 (10% aq. disp.)
Composition: 90% min. imidazoline

Crodazoline O. Croda Chem.
Chemical Description: Fatty acid imidazoline
Form: Color: 6–7 (Gardner)
Properties: m.w.: 345; Pour Pt.: –12 C; Surface Tension: 26.5 (1%/0.36%); pH: 11.1 (10% aq. disp.)
Composition: 90% min. imidazoline

Crodazoline S. Croda Chem.
Chemical Description: Fatty acid imidazoline
Form: Color: 4–5 (Gardner)
Properties: m.w.: 338; M.P.: 52 C; Surface Tension: 43.2 (1%/0.36%); pH: 11.0 (10% aq. disp.)

Composition: 90% min. imidazoline

Crodazoline T. Croda Chem.
Chemical Description: Fatty acid imidazoline
Form: Color: 10–11 (Gardner)
Properties: m.w.: 350; Pour Pt.: –10 C; Surface Tension: 36 (1%/0.36%); pH: 11.2 (10% aq. disp.)
Composition: 90% min. imidazoline

Crodazoline TW. Croda Chem.
Chemical Description: Fatty acid imidazoline
Form: Color: 7–8 (Gardner)
Properties: m.w.: 338; M.P.: 49 C; Surface Tension: 38 (1%/0.36%); pH: 11.1 (10% aq. disp.)
Composition: 90% min. imidazoline

Croderol G7000. Croda Chem.
Chemical Description: Glycerine
Category: Humectant, moisturizing agent, diluent, plasticizer
Applications: Diluent and plasticizer for many polar materials; as humectant, moisturizer in cosmetics and pharmaceuticals (skin and hair care preparations, shaving compositions, etc.)

Crodesta. Croda Chem.
Chemical Description: Sucrose fatty acid esters based on stearic/palmitic acid; the monoester content varies in the series
Category: Emulsifier
Applications: Cosmetic emulsifier, esp. for lotion compositions; eye, lip, face make-up and baby products due to mildness to skin and eyes

Crodet L4. Croda Chem.
Chemical Description: Lauric acid POE 4 (4 moles of ethylene oxide)
Category: Emulsifier, wetting agent, solubilizer, dispersant, nonionic
Applications: O/W emulsifiers for cosmetics and pharmaceutical creams, lotions and ointments; solubilizer for perfumes or aqueous alcoholic preparations (bath essences, colognes, after shave lotions); L4 is useful as plasticizer for polymers and resins used in hair setting lotions and sprays
Form: Liquid; Color: Pale straw
Properties: Solubility: Sol. in ethyl and oleyl alcohols, oleic acid, ceto stearyl alcohol; HLB: 9.8; Sap. Value: 138–150; Stability: Stable to moderately strong acids and alkalis, to oxidation under normal conditions, in presence of many electrolytes

Crodet L8. Croda Chem.
Chemical Description: Lauric acid POE 8 (8 moles of ethylene oxide)
Category and Applications: See Crodet L4; L8 esp. used as solubilizer for perfumes and plasticizer for hair sprays
Form: Liquid; Color: Pale straw
Properties: Solubility: Sol. in water, ethyl and oleyl alcohols, oleic acid, ceto stearyl alcohol; HLB: 12.7; Sap. Value: 95–106; Stability: Stable to moderately strong acids and alkalis, to oxidation under normal conditions, in pres-

ence of many electrolytes

Crodet L12. Crodet Chem.
Chemical Description: Lauric acid POE 12
Category and Applications: See Crodet L4; L12 esp. as solubilizer for perfumes; dispersant in aqueous systems
Form: Liquid; Color: Pale straw
Properties: Solubility: Sol. in water, ethyl and oleyl alcohols, oleic acid, ceto stearyl alcohol; HLB: 14.5; Sap. Value: 72–82; Stability: Stable to moderately strong acids and alkalis, to oxidation under normal conditions, in presence of many electrolytes

Crodet L24. Crodet Chem.
Chemical Description: Lauric acid POE 24
Category and Applications: See Crodet L4; L24 esp. as solubilizer for perfumes
Form: Soft paste; Color: Off white
Properties: Solubility: Sol. in water, ethyl and ceto stearyl alcohols; HLB: 16.8; Sap. Value: 42–48; Stability: Stable to moderately strong acids and alkalis, to oxidation under normal conditions, in presence of many electrolytes

Crodet L40. Croda Chem.
Chemical Description: Lauric acid POE 40
Category and Applications: See Crodet L4
Form: Waxy solid; Color: Off white
Properties: Solubility: Sol. in water, ethyl and ceto stearyl alcohols; HLB: 17.9; Sap. Value: 26–31; Stability: Stable to moderately strong acids and alkalis, to oxidation under normal conditions, in presence of many electrolytes

Crodet L100. Croda Chem.
Chemical Description: Lauric acid POE 100
Category and Applications: See Crodet L4
Form: Waxy solid; Color: Pale yellow
Properties: Solubility: Sol. in water, ethyl and ceto stearyl alcohols; HLB: 19.1; Sap. Value: 11–15; Stability: Stable to moderately strong acids and alkalis, to oxidation under normal conditions, to presence of many electrolytes

Crodet S4. Croda Chem.
Chemical Description: Stearic acid POE 4 (4 moles of ethylene oxide)
Category and Applications: See Crodet L4; also as auxiliary surfactants to optimize HLB
Form: Soft paste; Color: Off white
Properties: Solubility: Sol. in ethyl and oleyl alcohols, oleic acid, ceto stearyl alcohol; essentially sol. in arachis oil and isoparaffinic solvent; HLB: 7.7; Sap. Value: 117–129; Stability: Stable to moderately strong acids and alkalis, to oxidation under normal conditions, and to many electrolytes

Crodet S8. Croda Chem.
Chemical Description: Stearic acid POE 8

Category and Applications: See Crodet L4
Form: Soft paste; Color: Off white
Properties: Solubility: Sol. in ethyl and oleyl alcohols, oleic acid, ceto stearyl alcohol; essentially sol. in arachis oil and isoparaffinic solvent; HLB: 10.8; Sap. Value: 84–94; Stability: Stable to moderately strong acids and alkalis, oxidation under normal conditions, and to many electrolytes

Crodet S12. Croda Chem.
Chemical Description: Stearic acid POE 12
Category and Applications: See Crodet L4
Form: Waxy solid; Color: Off white
Properties: Solubility: Sol. in ethyl and oleyl alcohols, oleic acid, ceto stearyl alcohol; essentially sol. in water; HLB: 13.4; Sap. Value: 65–75; Stability: Stable to moderately strong acids and alkalis, oxidation under normal conditions, to many electrolytes

Crodet S24. Croda Chem.
Chemical Description: Stearic acid POE 24
Category and Applications: See Crodet L4
Form: Waxy solid; Color: Off white
Properties: Solubility: Sol. in water, ethyl and ceto stearyl alcohols; partly sol. in arachis oil, oleyl alcohol, isopropyl myristate, oleic acid; HLB: 15.8; Sap. Value: 38–47; Stability: Stable to moderately strong acids and alkalis, oxidation under normal conditions, to many electrolytes

Crodet S40. Croda Chem.
Chemical Description: Stearic acid POE 40
Category and Applications: See Crodet L4
Form: Waxy solid; Color: Off white
Properties: Solubility: Sol. in water, ethyl and ceto stearyl alcohols; partly sol. in oleic acid, propylene glycol; HLB: 16.7; Sap. Value: 23–30; Stability: Stable to moderately strong acids and alkalis, oxidation under normal conditions, to many electrolytes

Crodet S100. Croda Chem.
Chemical Description: Stearic acid POE 100
Category and Applications: See Crodet L4
Form: Waxy solid; Color: Off white
Properties: Solubility: Sol. in water, ethyl and ceto stearyl alcohols; partly sol. in propylene glycol; HLB: 18.8; Sap. Value: 10–14; Stability: Stable to moderately strong acids and alkalis, oxidation under normal conditions, to many electrolytes

Crodex A. Croda Chem.
Chemical Description: Prepared from ceto stearyl alcohol and sodium lauryl sulfate
Category: Emulsifying wax BP, anionic
Applications: Pharmaceutical and

cosmetic uses; yields emulsions with dry emollience; sunscreens; cold creams; hair creams; depilatory; lubricating creams; insect repellent
Form: Waxy solid; Color: Almost white; Odor: Faint characteristic
Properties: Stability: Stable to weak acids and alkalis and some metal salts

Crodex C. Croda Chem.
Chemical Description: Prepared from ceto stearyl alcohol and Cetrimide BP
Category: Cationic emulsifying wax BPC; bacteriacides
Applications: Pharmaceuticals and cosmetics; germicidal creams, ointments and lotions, hair conditioning preparations, baby creams, skin care products
Form: Waxy solid; Color: Almost white; Odor: Faint characteristic

Crodex N. Croda Chem.
Chemical Description: Prepared from ceto stearyl alcohol and Cetomacrogol 1000 BPC
Category: Nonionic emulsifying wax BP; wetting agent and penetrant
Applications: Emulsifier for most emollient materials in cosmetics and pharmaceuticals
Form: Waxy solid; Color: Almost white; Odor: Faint characteristic

Croduret 10. Croda Chem.
Chemical Description: POE 10 hydrogenated castor oil
Category: Emulsifier, solubilizer, emollient, superfatting agent, detergent, nonionic
Applications: Cosmetics (shampoos, foam baths, skin care products; solubilizer for Vitamins A and D); Textiles (spin finish additives, emulsifiers, lubricants, and antistats); water soluble metal working fluids; emulsion polymerization; insecticides, herbicides; household detergent formulations
Form: Liquid; Color: Straw
Properties: Solubility: Sol. in oleyl alcohol, naphtha, methyl ethyl ketone, oleic acid, trichloroethylene; HLB: 6.3; Cloud Pt.: < 20 C; Sap. Value: 120–130; Surface Tension: 37 dynes/cm (0.1%)

Croduret 30. Croda Chem.
Chemical Description: POE 30 hydrogenated castor oil
Category and Applications: See Croduret 10
Form: Liquid; Color: Pale straw
Properties: Solubility: Sol. in water, ethyl and oleyl alcohols, naphtha, methyl ethyl ketone, oleic acid, trichloroethylene; HLB: 11.6; Cloud Pt.: 48 C; Sap. Value: 70–80; Surface Tension: 46 (0.1% sol'n.)

Croduret 40. Croda Chem.
Chemical Description: POE 40 hydrogenated castor oil
Category and Applications: See Croduret 10
Form: Viscous paste; Color: Off white
Properties: Solubility: Sol. in water, ethyl alcohol, naphtha, methyl ethyl ketone, oleic acid, trichloroethylene; HLB: 13.0; Cloud

Pt.: 62 C; Sap. Value: 60–65; Surface Tension: 46 dynes/cm (0.1% sol'n.)

Croduret 60. Croda Chem.
Chemical Description: POE 60 hydrogenated castor oil
Category and Applications: See Croduret 10
Form: Stiff paste; Color: Off white
Properties: Solubility: Sol. in water, ethyl alcohol, methyl ethyl ketone, oleic acid, trichloroethylene; HLB: 14.7; Cloud Pt.: 71 C; Sap. Value: 45–50; Surface Tension: 47.5 dynes/cm (0.1% sol'n.)

Croduret 100. Croda Chem.
Chemical Description: POE 100 hydrogenated castor oil
Category and Applications: See Croduret 10
Form: Waxy solid; Color: Off white
Properties: Solubility: Sol. in water, ethyl alcohol, methyl ethyl ketone, trichloroethylene; HLB: 16.5; Cloud Pt.: 65 C; Sap. Value: 25–35; Surface Tension: 46.2 dynes/cm (0.1% sol'n.)

Crolactil CSL. Croda Chemicals
Chemical Description: Acyl lactylates
Category: Emulsifier
Applications: Water-in-oil emulsions, ingredients of food products
Form: Powder; Color: Off-white
Properties: HLB: 5.0; Sapon. Value: 175–250; Acid No.: 50–86

Crolactil SISL. Croda Chemicals
Chemical Description: Acyl lactylates
Category: Emulsifier
Applications: Oil-in-water emulsions, ingredients of food products
Form: Liquid; Color: Yellow
Properties: HLB: 6.5; Sapon. Value: 210–280; Acid No.: 60–90

Crolactil SSL. Croda Chemicals
Chemical Description: Acyl lactylates
Category: Emulsifier
Applications: Oil-in-water emulsions, ingredients of food products
Form: Powder; Color: Off-white
Properties: HLB: 6.5; Sapon. Value: 210–280; Acid No.: 60–90

Crotein A, C, O. Croda Chem.
Chemical Description: Refined collagen hydrolysates (differing in molecular weights)
Category: Conditioning agent, foam stabilizer and booster, dye leveling agent
Applications: Substantive to hair and skin; protective action in hair care products for dyed, bleached, waved hair; hair conditioning and texturizing; foam stabilizers and boosters in shampoos; dye leveler in hair dyes and bleaches; cold waves and straighteners; hair sprays, rinses, sets; hand and body lotions, make-up, treatment creams, moisturizers; nail hardeners; depilatories
Form: Powder; Color and Odor: Very low
Composition: 100% active ingred.

Crotein CCT. Croda Chem.
Chemical Description: Triethanolamine salt of a complex protein coconut fatty acid ester
Category: Detergent, foaming agent
Applications: Skin cleansers, hair and body shampoos; reduces irritation of other detergents

Crotein Q. Croda Chem.
Chemical Description: Quaternary ammonium derivative of a hydrolyzed collagen protein
Category: Hair conditioning agent, cationic
Applications: Substantive to hair; hair conditioners and shampoos with reduction of split ends, superior gloss, improved manageability and texture
Properties: Solubility: Water sol.

p-Cumyl phenol 7.5 mole ethoxylate (developmental product). Monsanto Industrial Chem.
Category: Wetting agent, emulsifier
Applications: Emulsifier in solvent cleaners, in metal cleaning compounds
Form: Liquid; Color: Light yellow; Odor: Mild phenolic
Properties: Ref. Index: 1.5146 @ 60 C; Solubility: Sol. in water; Sp. gr.: 1.1121 @ 20 C; Visc.: 370 cps; Cloud Pt.: 48.8 C; Flash Pt.: 565 F; Pour Pt.: –1 F; Stability: Good; Surface Tension: 39.1 dynes/cm
Composition: 99.5% active ingred.; other components: 0.4% water max.; 0.1% ash max.
Reaction: Neutral
Toxicity: Moderate skin and eye irritant
sol.; Sp. gr.: 1.1191 @ 20 C; Visc.: 370 cps; Cloud Pt.: 94 C; Flash Pt.: 565 F; Pour Pt.: 56 F; Stability: Good; Surface Tension: 38.9 dynes/cm
Composition: 99.5% active ingred.; other components: 0.4% water max.; 0.1% ash max.
Reaction: Neutral
Toxicity: Mild eye irritant

p-Cumyl phenol 10 mole ethoxylate (developmental product). Monsanto Industrial Chem.
Category: Wetting agent, emulsifier
Applications: Emulsifier in solvent cleaners, in hard surface and general industrial cleaners
Form: Liquid; Color: Light yellow; Odor: Mild phenolic
Properties: Refractive Index: 1.4990 @ 60 C; Solubility: Sol. in water; Sp. gr.: 1.1141 @ 20 C; Visc.: 410 cps; Cloud Pt.: 74.6 C; Flash Pt.: 545 F; Pour Pt.: 26 F; Stability: Good; Surface Tension: 43.0 dynes/cm
Composition: 99.5% active ingred.; other components: 0.4% water max.; 0.1% ash max.
Reaction: Neutral
Toxicity: Mild eye irritant

p-Cumyl phenol 15 mole ethoxylate (developmental product) Monsanto Industrial Chem.
Category: Wetting agent, emulsifier
Applications: Emulsifier, in sol-

vent cleaners, in hard surface, metal and general industrial cleaners
Form: Liquid; Color: Light yellow; Odor: Mild phenolic
Properties: Refractive Index: 1.4878 @ 60 C; Solubility: Water-

Cyclochem 326A. Cyclo Chem. Corp.
Chemical Description: A synthetic replacement for natural beeswax
Category: Emulsifer, nonionic
Applications: Its plasticity, emulsion forming power and behavior with solvents are indistinguishable from the natural product; particularly suited to both w/o and o/w emulsions including hair creams, cold creams, lotions and lipsticks
Form: Flakes; Color: Off-white
Properties: M. P.: 51–54 C; Acid No.: 25–27

Cyclochem CL. Cyclo Chem. Corp.
Chemical Description: Cetyl lactate
Category: Penetrant
Applications: To promote penetration and feel, useful in imparting creaminess to lipstick and conditioning effect to cream mascara, imparts a luxurious feel to bath oils and lotions
Form: Solid; Color: Off-white
Properties: M. P.: 24 C; Sapon. Value: 180; Acid No.: 3.5 max.

Cyclochem CP. Cyclo Chem. Corp.
Chemical Description: Cetyl palmitate
Category: Ester
Applications: Ester developed for use as a substitute for spermaceti in cold creams, lipsticks and topical cosmetics
Form: Flakes; Color: Off-white
Properties: M. P.: 50 C; Sapon. Value: 114; Acid No.: 3.0 max.

Cyclochem EGDS. Cyclo Chem. Corp.
Chemical Description: Ethylene glycol distearate pure
Category: Thickener
Applications: Used in shampoos and cosmetic lotions as a thickener and to provide optimum opacity
Form: Flakes; Color: White
Properties: M. P.: 61 C; Sapon. Value: 195; Acid No.: 20.0 max.

Cyclochem EGMS. Cyclo Chem. Corp.
Chemical Description: Ethylene glycol monostearate pure
Category: Additive
Applications: Widely used to add viscosity, pearlescence and opacity to lotions and shampoos
Form: Flakes; Color: White
Properties: M. P.: 56 C; Sapon. Value: 182; Acid No.: 4.0 max.

Cyclochem EM 324. Cyclo Chem. Corp.
Chemical Description: A blend of natural fatty alcohols, fatty acids and other derivatives
Category: Emulsifier, anionic
Applications: Self-emulsifying; useful in formulating o/w hair and skin conditioning creams; tolerates up to 5% organic acids such as citric and lactic acids

Form: Flakes; Color: Off-white
Properties: M. P.: 47–50 C; Acid No.: 93–98

Cyclochem EM 560. Cyclo Chem. Corp.
Chemical Description: A blend of natural fatty alcohols and fatty esters
Category: Emulsifier, anionic
Applications: Self-emulsifying; useful for o/w creams on the acid side
Form: Wax; Color: Off-white
Properties: M. P.: 62–65 C; Acid No.: 1.0 max.

Cyclochem GMO. Cyclo Chem. Corp.
Chemical Description: Glyceryl monooleate
Category: Emulsifier, emollient, opacifier, lubricant
Applications: Effective water-in-oil emulsifier, emollient and opacifier, provides lubricity to pre- and aftershave lotions and cosmetic systems
Form: Liquid; Color: Amber
Properties: M.P.: 9 C; Sapon. Value: 170; Acid No.: 9.0 max.

Cyclochem GMS. Cyclo Chem. Corp.
Chemical Description: Glyceryl monostearate pure
Category: Emulsifier
Applications: A strongly lipophilic emulsifier, easily solubilized by PEG esters, enhances feel and appearance of formulations
Form: Flakes; Color: White
Properties: M. P.: 56 C; Sapon. Value: 170; Acid No.: 5.0 max.

Cyclochem GMS 21. Cyclo Chem. Corp.
Chemical Description: Glyceryl monostearate acid stable self emulsifying
Category: Electrolyte tolerant, emulsifier, emollient, anionic
Applications: High electrolyte tolerance, emulsifier and emollient for creams, lotions and related toiletry products
Form: Flakes; Color: Off-white
Properties: M. P.: 52 C; Sapon. Value: 157; Acid No.: 18.0 max.

Cyclochem GMS 165. Cyclo Chem. Corp.
Chemical Description: Glyceryl monostearate acid stable self emulsifying
Category: Emulsifier, electrolyte tolerant
Applications: A principle emulsifier in o/w creams and lotions, exhibits exceptional electrolyte tolerance and functions well in the acid pH range
Form: Flakes; Color: Off-white
Properties: M. P.: 55 C; Sapon. Value: 93; Acid No.: 2.0 max.

Cyclochem GTIS. Cyclo Chem. Corp.
Chemical Description: Isostearyl triglyceride
Category: Emollient, co-emulsifier, softener
Applications: A super emollient, its liquid nature allows it to be easily incorporated into creams and lotions; as a co-emulsifier, it

provides good texture in creams; promotes softness and a smooth feeling to the skin in creams and lotions
Form: Liquid; Color: Yellow
Properties: M. P.: < 0 C; Sapon. Value: 192; Acid No.: 8.0 max.

Cyclochem GTL. Cyclo Chem. Corp.
Chemical Description: Lauric/myristic triglyceride
Category: Emollient
Applications: To promote emolliency and an oily feel to cream and lotions
Form: Liquid/paste
Properties: M. P.: 27 C; Sapon. Value: 245; Acid No.: 30.0 max.

Cyclochem INEO. Cyclo Chem. Corp.
Chemical Description: Isostearyl neopentanoate
Category: Penetrant
Applications: Highly acid and alkali stable ester with rapid penetration that promotes a dry feel to the skin
Form: Liquid; Color: Off-white
Properties: M. P.: < 0 C; Sapon. Value: 149; Acid No.: 2.0 max.

Cyclomide LIPA. Cyclo Chem. Corp.
Chemical Description: Lauric monoisopropanolamide
Category: Viscosity builder, foam stabilizer
Applications: Builds viscosity, stabilizes foam and is mild to the skin
Form: Flakes; Color: Off-white
Composition: 100% active, ≈ 95% amide content

Cyclochem LVL. Cyclo Chem. Corp.
Chemical Description: Lauryl lactate
Category: Penetrant, emollient, pigment dispersant
Applications: To provide rapid non-oily penetration; important component of cake powders, lotions and creams; emollient and pigment dispersant for liquid makeups; reduces tackiness of aluminum salt anti-perspirants
Form: Liquid; Color: Off-white
Properties: M. P.: 4 C; Sapon. Value: 199; Acid No.: 3.0 max.

Cyclochem ML. Cyclo Chem. Corp.
Chemical Description: Myristyl lactate
Category: Melting ester, penetrant, emollient, pigment dispersant
Applications: Slightly higher melting ester with similar properties to Cyclochem LVL
Form: Liquid; Color: Off-white
Properties: M. P.: 12 C; Sapon. Value: 172; Acid No.: 3.0 max.

Cyclochem MM/M. Cyclo Chem. Corp.
Chemical Description: Myristyl myristate
Category: Fatty acid esters
Applications: With its melt point at body temperature, it imparts a soft, velvety feel which persists after rinsing; at low levels (under 1%) it provides infinite spreading so as to prevent the

watery feel of hand and body lotions high in water content
Form: Flakes; Color: Off-white
Properties: M. P.: 38 C; Sapon. Value: 130; Acid No.: 3.0 max.

Cyclochem MST. Cyclo Chem. Corp.
Chemical Description: Myristyl Stearate
Category: Fatty acid esters
Applications: To impart soft, velvety feel to emulsions where a stiffer emulsion with superior opacity is desired
Form: Flakes; Color: Off-white
Properties: M. P.: 45 C; Sapon. Value: 115; Acid No.: 5.0 max.

Cyclochem PEG 400 DS. Cyclo Chem. Corp.
Chemical Description: Polyethylene glycol 400 distearate
Category: Fatty acid esters, nonionic
Applications: These esters offer the formulator varied hydrophilic and lipophilic properties
Form: Soft solid; Color: White
Properties: M. P.: 33 C; Sapon. Value: 123; Acid No.: 10.0 max.

Cyclochem PEG 400 ML. Cyclo Chem. Corp.
Chemical Description: Polyethylene glycol 400 monolaurate
Category: Fatty acid esters, nonionic
Applications: These esters offer the formulator varied hydrophilic and lipophilic properties
Form: Liquid; Color: Light straw
Properties: M. P.: 10 C; Sapon. Value: 90; Acid No.: 5.0 max.

Cyclochem PEG 400 MO. Cyclo Chem. Corp.
Chemical Description: Polyethylene glycol 400 monooleate
Category: Fatty acid esters, nonionic
Applications: These esters offer the formulator varied hydrophilic and lipophilic properties
Form: Liquid; Color: Light straw
Properties: M. P.: 10 C; Sapon. Value: 85; Acid No.: 5.0 max.

Cyclochem PETO. Cyclo Chem. Corp.
Chemical Description: Pentaerythritol tetraoleate
Category: Fatty acid esters, nonionic
Applications: Unique liquid ester with good high temperature performance, imparts gloss in lipsticks and assists in the development of the 'wet look'
Form: Liquid; Color: Yellow
Properties: M. P.: < 0 C; Sapon. Value: 187; Acid No.: 5.0 max.

Cyclochem PETS. Cyclo Chem. Corp.
Chemical Description: Pentaerythritol tetrastearate
Category: Wax, nonionic
Applications: Synthetic, hard, high melting, neutral wax that provides high heat stability in lubricants; domestic substitute for expensive neutral waxes in polishes, coatings and various hot-melt compositions and in cosmetic products
Form: Flakes; Color: Off-white

Properties: M. P.: 57 C; Sapon. Value: 190; Acid No.: 9.0 max.

Cyclochem PGMS. Cyclo Chem. Corp.
Chemical Description: Propylene glycol monostearate, pure
Category: Emulsifier, occlusive agent, nonionic
Applications: An auxiliary emulsifier for lotions and make-up preparations, a good occlusive agent for reducing skin moisture loss
Form: Soft solid; Color: Off-white
Properties: M. P.: 36 C; Sapon. Value: 186; Acid No.: 5.0 max.

Cyclochem POL. Cyclo Chem. Corp.
Chemical Description: A blend of higher fatty alcohol and ethoxylated adducts
Category: Emulsifier, nonionic
Applications: Self-emulsifying; product of choice for creams and lotions containing high quantities of alkaline ingredients
Form: Wax; Color: Off-white
Properties: M. P.: 48–52 C; Acid No.: 2.0 max.

Cyclochem SPS. Cyclo Chem. Corp.
Chemical Description: Synthetic spermaceti
Category: Fatty acid esters, nonionic
Applications: Chemically and physically identical to the natural product; provides sheen and smooth feel to creams; gloss to stick makeup products; at 0.5% in aerosol talc, it reduces billowing and improves talc adhesion
Form: Flakes; Color: Off-white
Properties: M. P.: 46 C; Sapon. Value: 115; Acid No.: 3.0 max.

Cyclochem SS. Cyclo Chem. Corp.
Chemical Description: Stearyl stearate
Category: Fatty acid esters
Applications: A component of many natural waxes which provides a 'feel' on skin and hair that is less oily than mineral oil and less drying than isopropyl myristate
Form: Flakes; Color: Off-white
Properties: M. P.: 53 C; Sapon. Value: 110; Acid No.: 5.0 max.

Cyclogol Cetyl Alcohol NF. Cyclo Chem. Corp.
Chemical Description: *n*-hexadecanol
Category: Emollient carrier, viscosity builder, nonionic
Applications: Emollient carrier and viscosity builder for creams, lotions and other cosmetic emulsions
Form: Flakes; Color: Off-white
Properties: M. P.: 45–50 C; Acid No.: nil

Cyclogol Cetyl-Stearyl Alcohol. Cyclo Chem. Corp.
Chemical Description: *n*-hexadecanol and *n*-octadecanol
Category: Fatty alcohol, nonionic
Applications: Combines the viscosity effect of stearyl with the skin feel of cetyl in lotions and cosmetic emulsions
Form: Flakes; Color: Off-white
Properties: M. P.: 48–53 C; Acid

No.: nil

Cyclogol EL. Cyclo Chem. Corp.
Chemical Description: Ethoxylated fatty alcohol
Category: Solubilizer, nonionic
Applications: Valuable in formulating water soluble bath oils; solubilizer of perfume oils, super-fatting agents and microbicides
Form: Paste; Color: White
Properties: M. P.: 28–33 C; Acid No.: 2.0 max.

Cyclogol LTH. Cyclo Chem. Corp.
Chemical Description: Ethoxylated linear alcohol
Category: Detergent, dispersant, nonionic
Applications: Excellent for emulsifying oils, for degreasing and for engine cleaners
Form: Clear liquid
Properties: M. P.: 22–24 C; Acid No.: nil

Cyclogol NI. Cyclo Chem. Corp.
Chemical Description: A blend of higher fatty alcohol and ethoxylated adducts
Category: Emulsifier, nonionic
Applications: Useful in preparing o/w ointments, creams and lotions containing hard to solubilize additives such as thioglycolates, antihistamines and polyvalent salts
Form: Flakes; Color: White
Properties: M. P.: 49–50 C; Acid No.: 1.0 max.

Cyclogol Stearyl Alcohol. Cyclo Chem. Corp.
Chemical Description: *n*-octadecanol
Category: Viscosity builder, nonionic
Applications: Viscosity builder for creams, lotions and other cosmetic emulsions
Form: Flakes; Color: Off-white
Properties: M. P.: 56–60 C; Acid No.: nil

Cyclomide AC 28. Cyclo Chem. Corp.
Chemical Description: Fatty acid condensate
Category: Anti-corrosive mineral oil additive
Applications: Anti-corrosive mineral oil additive for cutting fluids and soluble cutting oil compositions
Form: Liquid
Composition: 100% active

Cyclomide BWA/M. Cyclo Chem. Corp.
Chemical Description: Fatty alkanolamide blends
Category: Emulsifier
Applications: Effective o/w emulsifier system for use as an engine degreaser
Form: Liquid
Composition: 100% active

Cyclomide C212. Cyclo Chem. Corp.
Chemical Description: Coconut monoethanolamide
Category: Additive
Applications: Effective additive for soap or synthetic based washing powders to promote and stabilize lather, increase detergent power and impart skin mildness

Form: Flakes; Color: Cream
Composition: 100% active, ≈ 95% amide content

Cyclomide CODI. Cyclo Chem. Corp.
Chemical Description: Cocamidopropyl dimethylamine
Category: Cationic emulsions
Applications: Base for preparing cationic emulsions for use in creams, lotions and hair rinse formulations
Form: Liquid; Color: Yellow
Composition: 100% active, ≈ 98% amide content

Cyclomide DC212. Cyclo Chem. Corp.
Chemical Description: 2:1 Coconut diethanolamide
Category: Viscosity booster, foaming agent, detergency
Applications: Used in cosmetic and laundry products to boost viscosity, foam quality and detergency
Form: Liquid; Color: Amber
Composition: 100% active

Cyclomide DC212M. Cyclo Chem. Corp.
Chemical Description: 2:1 Modified coconut diethanolamide
Category: Cleaners, solventless degreasers and wax strippers
Applications: For hard surface cleaners, solventless degreasers and wax strippers where high alkali compatibility is a necessity
Form: Liquid; Color: Amber
Composition: 100% active

Cyclomide DC212/S. Cyclo Chem. Corp.
Chemical Description: 1:1 Coconut diethanolamide
Category: Foam stabilizer, viscosity booster
Applications: Foam stabilizing and viscosity boosting when low cloud point is essential
Form: Liquid
Composition: 100% active, ≈ 85% amide content

Cyclomide DC212/SE. Cyclo Chem. Corp.
Chemical Description: 1:1 Coconut diethanolamide
Category: Thickening and foam stabilization
Applications: Maximizes thickening and foam stabilization in combination with alkyl aryl sulfonates or lauryl sulfates in shampoos and detergent formulations
Form: Liquid
Composition: 100% active, ≈ 95% amide content

Cyclomide DL203. Cyclo Chem. Corp.
Chemical Description: 2:1 Lauric diethanolamide
Category: Stabilizer
Applications: Outstanding power to stabilize lather of shampoos and liquid detergents
Form: Off-white liquid/White paste
Composition: 100% active

Cyclomide DL203/S. Cyclo Chem. Corp.
Chemical Description: 1:1 Lauric

diethanolamide
Category: Foam and detergency booster
Applications: High quality foam booster for cosmetic products, boosts detergency and contributes super-fatty effect
Form: Crystalline solid; Color: Off-white
Composition: 100% active, ≈ 95% amide content

Cyclomide DL207/S. Cyclo Chem. Corp.
Chemical Description: 1:1 Lauric/myristic diethanolamide
Category: Foam and viscosity booster, wetting agent
Applications: Maximum mildness combined with outstanding foam and viscosity boosting, fastest wetting action of series plus improved super-fatting effect
Form: Off-white liquid/White paste
Composition: 100% active, ≈ 95% amide content

Cyclomide DO280/S. Cyclo Chem. Corp.
Chemical Description: 1:1 Oleic diethanolamide
Category: Thickening and super-fatting agent, emulsifier
Applications: Excellent thickening and super-fatting agent for lotion shampoos, good auxiliary emulsifier
Form: Liquid
Composition: 100% active, ≈ 85% amide content

Cyclomide DOTS. Cyclo Chem. Corp.
Chemical Description: 1:1 Linoleic diethanolamide
Applications: Used as a viscosity builder, effective as an emulsifier and as a pigment dispersant
Form: Liquid; Color: Amber
Composition: 100% active, ≈ 95% amide content

Cyclomide DP240/S. Cyclo Chem. Corp.
Chemical Description: 1:1 Palm kernel diethanolamide
Category: Viscosity builder, foam stabilizer
Applications: Liquid viscosity builder and foam stabilizer for organic formulations
Form: Liquid
Composition: 100% active, ≈ 85% amide content

Cyclomide IODI. Cyclo Chem. Corp.
Chemical Description: Isostearamidopropyl dimethylamine
Category: Cationic emulsions
Applications: Base for preparing cationic emulsions, promotes substantivity and adhesion of other additives to the skin, enhances protein substantivity
Form: Clear liquid
Composition: 100% active, ≈ 98% amide content

Cyclomide KD. Cyclo Chem. Corp.

Chemical Description: 1:1 Coconut diethanolamide
Category: Viscosity builder, foam stabilizer
Applications: Superior to other 1:1 superamides as a viscosity builder while providing excellent foam stabilization
Form: Liquid; Color: Pale Yellow
Composition: 100% active, ≈ 95% amide content

Cyclomide L203. Cyclo Chem. Corp.
Chemical Description: Lauric monoethanolamide
Category: High viscosity and foam stabilizer
Applications: Promotes high viscosity with superior foam stability
Form: Flakes; Color: Off-white
Composition: 100% active, ≈ 95% amide content

Cyclomide Pinemulse. Cyclo Chem. Corp.
Chemical Description: Modified coconut polydiethanolamide
Category: Emulsifier, coupler
Applications: Energy saving emulsifier and coupler, designed for formulating extremely stable pine oil cleaners without the aid of heat
Form: Liquid; Color: Amber
Composition: 100% active

Cyclomide S280. Cyclo Chem. Corp.
Chemical Description: Stearic monoethanolamide
Category: Skin protectant
Applications: Outstanding skin protectant in toilet bars, creams, lotions and pastes
Form: Flakes; Color: Off-white
Composition: 100% active, ≈ 95% amide content

Cyclomide SODI. Cyclo Chem. Corp.
Chemical Description: Stearamidopropyl dimethylamine
Category: Cationic emulsions
Applications: Solid material for preparing cationic emulsions, unique effect produced when neutralized with water soluble acids
Form: Solid; Color: Off-white
Composition: 100% active, ≈ 98% amide content

Cyclophos P4. Cyclo Chem. Corp.
Chemical Description: Aliphatic
Category: Acid ester, anionic
Applications: Hydrophilic phosphate ester that imparts detergency, moderate foaming and good coupling properties in powdered machine dishwashing products and steam carpet cleaners; excellent surface tension depressant
Form: Clear thin liquid
Composition: 98% active

Cyclophos PB12. Cyclo Chem. Corp.
Chemical Description: Aromatic
Category: Acid ester, anionic
Applications: Developed for metal

cleaning; highly stable under prolonged use in soak tank cleaning; retains cleaning power under high soil loads
Form: Clear liquid
Composition: 98% active

Cyclophos PE5. Cyclo Chem Corp.
Chemical Description: Aromatic
Category: Acid ester, anionic
Applications: Hydrotrope that is dispersible in either oil or water systems; promotes detergency in dry cleaning, concrete cleaners, and powdered laundry formulations; effective at elevated temperatures, in solubilizing nonionic
Form: Clear liquid
Composition: 98% active

Cyclophos PE9. Cyclo Chem. Corp.
Chemical Description: Aromatic
Category: Acid ester, anionic
Applications: High in electrolyte tolerance; used to promote detergency and viscosity in heavy duty, all purpose liquid formulations and built steam cleaners
Form: Clear liquid
Composition: 99% active

Cyclophos PE30. Cyclo Chem. Corp.
Chemical Description: Aromatic
Category: Acid ester, anionic
Applications: Useful in emulsion polymerization to promote clarity and stability of cast acrylic and PVA films; latex stabilizer; for oil emulsion printing pastes; dyeing assistant
Form: Paste; Color: Off-white
Composition: 90% active

Cyclophos PL3. Cyclo Chem. Corp.
Chemical Description: Aliphatic
Category: Acid ester, anionic
Applications: An effective emulsifier for cosmetic oils, creams and waterless hand cleaners; a lubricity additive in soluble oils and a rust inhibitor; used in lubricating and rolling oils and water-based cutting fluids
Form: Liquid; Color: Pale yellow
Composition: 99% active

Cyclophos PL6A. Cyclo Chem. Corp.
Chemical Description: Aliphatic
Category: Acid ester, anionic
Applications: Compatible in concentrated electrolyte solutions; excellent coupler for nonionics in liquid alkaline systems; stable and non-coloring on flake caustic; moderate foamer
Form: Liquid; Color: Pale yellow
Composition: 98% active

Cyclophos PL31. Cyclo Chem. Corp.
Chemical Description: Aliphatic
Category: Acid ester, anionic
Applications: Used as coupling agent in extreme alkali systems; additive for the improvement of lubricity and the loadbearing properties of oil-based lubricants along with emulsification of cutting oils
Form: Clear liquid
Composition: 98% active

Cyclophos TE12. Cyclo Chem. Corp.
Chemical Description: Aromatic

Category: Acid ester, anionic
Applications: Highly water soluble phosphate ester that is an excellent emulsifier and solubilizer of nonionics in high electrolyte solutions; excellent alkali stability provides hard surface detergency for heavy duty cleaning formulations, household detergents, floor cleaning concentrates and metal cleaning products
Form: Liquid; Color: Amber
Composition: 98% active

Cyclopol SB-5. Cyclo Chem. Corp.
Chemical Description: Sulfosuccinate of ethoxylated lanolin alcohol
Category: Moisturizing agent, anionic
Applications: Ideal for shampoos, foam baths, skin cleansers, skin creams, hair sprays, dishwashing liquids
Form: Liquid; Color: Yellow
Composition: 48–50% active

Cyclopol SBCA198. Cyclo Chem. Corp.
Chemical Description: Sulfosuccinamate of mixed fatty amine
Category: Liquid surface active agent, anionic
Applications: Liquid surface active agent useful in latex systems; produces copious, very stable foam
Form: Liquid; Color: Yellow
Composition: 33–35% active

Cyclopol SBCP. Cyclo Chem. Corp.
Chemical Description: Sulfosuccinate of coconut alkanolamide
Category: Anionic
Applications: Functioning in a wide viscosity range, it promotes low irritation in mild shampoos, bubble baths and skin cleansers; compatible with anionics, nonionics, and amphoterics
Form: Liquid; Color: Yellow
Composition: 39–41% active

Cyclopol SBDO. Cyclo Chem. Corp.
Chemical Description: Dioctyl sulfosuccinate
Category: Reducing agent, anionic
Applications: Outstanding surface tension reducing agent
Form: Liquid
Composition: 68–70% active

Cyclopol SBFA30. Cyclo Chem. Corp.
Chemical Description: Sulfosuccinate of ethoxylated lauryl alcohol
Category: Raw material
Applications: Mild, non-irritating raw material for bubble baths, shampoos and general cleansing agents
Form: Liquid
Composition: 38–40% active

Cyclopol SBG 212. Cyclo Chem. Corp.
Chemical Description: Sulfosuccinate of disodium monococamide
Category: Base, foaming agent
Applications: Mild, non-irritating base for shampoos and skin products; a high foamer that is an easily handled liquid without use of heat
Form: Liquid; Color: Yellow
Composition: 29–31% active

Cyclopol SBG 280. Cyclo Chem. Corp.
Chemical Description: Sulfosuccinate of disodium mono-oleamide
Category: Foaming agent, anionic
Applications: Excellent surfactant that is non-irritating to the eyes and skin; produces good foam structure, viscosity control, and is compatible with anionic, nonionic and amphoteric surfactants
Form: Liquid; Color: Yellow
Composition: 29–31% active

Cyclopol SBG 280/T. Cyclo Chem. Corp.
Chemical Description: Sulfosuccinate of TEA-monooleamide
Category: Base, anti-irritant, anionic
Applications: High active, extremely mild, non-irritating base for skin products and shampoos; anti-irritant when combined with other surfactants
Form: Liquid; Color: Yellow
Composition: 39–41% active

Cyclopol SBL203. Cyclo Chem. Corp.
Chemical Description: Sulfosuccinate of lauric acid monoethanolamide
Category: Raw material, anionic
Applications: Raw material for carpet shampoos; when combined with sodium lauryl sulfate produces brittle, tack free residue; mild ingredient for bubble baths and shampoos
Form: Liquid; Color: Yellow
Composition: 38–40% active

Cyclopol SBR-3. Cyclo Chem. Corp.
Chemical Description: Sulfosuccinate of higher fatty acid monoethanolamide
Category: Skin-protective, anti-irritant surfactant
Applications: Eliminates or reduces skin irritation caused by other surfactants; recommended for skin cleansing compositions, dishwashing liquids, toiletries containing surfactants
Form: Liquid; Color: Yellow
Composition: 38–40% active

Cyclopol SBU-185. Cyclo Chem. Corp.
Chemical Description: Sulfosuccinate of undecylenic acid alkanolamide
Category: Fungicidal and bacteriostatic additive, anti-dandruff agent, anionic
Applications: Used in shampoos, body lotions, foot preparations and similar detergent compositions
Form: Liquid; Color: Yellow
Composition: 49–51% active

Cycloryl 21. Cyclo Chem. Corp.
Chemical Description: Sodium lauryl sulfate cosmetic grade
Category: Base, anionic
Applications: Low cost base for cream and liquid shampoos and specialty cleaners where fast solubility and free rinsing are sought
Form: Liquid/paste; Color: Pale Yellow
Composition: 28–30% active

Cycloryl 21 LS. Cyclo Chem. Corp.
Chemical Description: Sodium lauryl sulfate low salt
Category: Base, anionic
Applications: Low chloride base for use in shampoos, hand cleaners and detergents
Form: Liquid; Color: Pale Yellow
Composition: 28–30% active

Cycloryl 21 SP. Cyclo Chem. Corp.
Chemical Description: Sodium lauryl sulfate low cloud point
Category: Flowable process aid, anionic
Applications: Flowable process aid for the preparation of compounds, frothing agents, and detergents
Form: Liquid; Color: Yellow
Composition: 28–30% active

Cycloryl ABSA. Cyclo Chem. Corp.
Chemical Description: Linear alkyl aryl sulfonic acid
Category: Detergent, anionic
Applications: Hard surface cleaners, built detergents and shampoo system
Form: Liquid; Color: Amber
Properties: Biodegradable
Composition: 97% min. active

Cycloryl DA. Cyclo Chem. Corp.
Chemical Description: Diethanolamine lauryl sulfate
Category: Base, anionic
Applications: Base for preparation of clear, mild shampoos
Form: Liquid; Color: Yellow
Composition: 34–36% active

Cycloryl DCA. Cyclo Chem. Corp.
Chemical Description: Modified coconut polydiethanolamide
Category: Detergent, anionic
Applications: Concentrated detergent for human and pet shampoos; used to build viscosity and to solubilize difficult ingredients
Form: Liquid; Color: Amber
Composition: 98–100% active

Cycloryl DDB40. Cyclo Chem. Corp.
Chemical Description: Sodium alkyl aryl sulfonate
Category: Base, anionic
Applications: Base for medium duty detergents, cleaners and strippers
Form: Liquid; Color: Yellow
Composition: 38–40% active

Cycloryl DDB60T. Cyclo Chem. Corp.
Chemical Description: Triethanolamine alkyl aryl sulfonate
Category: Base, anionic
Applications: High active base for cleaners and shampoos
Form: Liquid; Color: Yellow
Composition: 58–60% active

Cycloryl JB-8. Cyclo Chem. Corp.
Chemical Description: Sodium lauryl ether sulfate plus special additives
Category: Base, anionic
Applications: Base for foaming bath gels and liquids
Form: Liquid; Color: Amber
Composition: 33–36% active

Cycloryl LDC. Cyclo Chem. Corp.
Chemical Description: Ammonium lauryl ether sulfate plus sulfonate and amide
Category: Base, anionic
Applications: Base for light duty detergent and liquid dishwashing compounds; car washing applications
Form: Liquid; Color: Amber
Composition: 60–62% active

Cycloryl M1. Cyclo Chem. Corp.
Chemical Description: Built pearlized sodium lauryl ether sulfate
Category: Base, anionic
Applications: Easy to use base for luxurious pearly cream shampoos having high flash foam
Form: Creamy liquid; Color: White
Composition: 34–36% active

Cycloryl MA. Cyclo Chem. Corp.
Chemical Description: Ammonium lauryl sulfate
Category: Base, anionic
Applications: Base for shampoos, bubble baths, and light duty detergents
Form: Liquid; Color: Pale Yellow
Composition: 28–30% active

Cycloryl MA 330. Cyclo Chem. Corp.
Chemical Description: Ammonium lauryl ether sulfate
Category: Base, degreasing agent, anionic
Applications: Base for shampoos and light duty detergents; grease cutting and oil solubilization to formulations
Form: Liquid; Color: Pale Yellow
Composition: 27–29% active

Cycloryl MA 360. Cyclo Chem. Corp.
Chemical Description: Ammonium lauryl ether sulfate
Category: Base, degreasing agent, anionic
Applications: High active base for compounding light duty dish detergents, fine fabric cleaners; superior grease cutting
Form: Liquid; Color: Pale Yellow
Composition: 58–60% active

Cycloryl ME 60. Cyclo Chem. Corp.
Chemical Description: Sodium myristyl ether sulfate
Category: Base, anionic
Applications: High active concentrated base for consumer or professional shampoos, foaming bath oils, bubble baths and cosmetic emulsions
Form: Liquid; Color: Pale Yellow
Composition: 58–60% active

Cycloryl ME 361. Cyclo Chem. Corp.
Chemical Description: Ammonium myristyl ether sulfate
Category: Base, anionic
Applications: High active concentrated base for acid shampoos and mild high foaming cleansing products
Form: Liquid; Color: Pale Yellow
Composition: 58–60% active

Cycloryl MG. Cyclo Chem. Corp.
Chemical Description: Magnesium

lauryl sulfate
Category: Base, anionic
Applications: Carpet, upholstery shampoo base having very low cloud point; forms tack free residue
Form: Liquid; Color: Pale Yellow
Composition: 28–30% active

Cycloryl MS. Cyclo Chem. Corp.
Chemical Description: Sodium myristyl sulfate
Category: Mild emulsifier effective, anionic
Applications: Mild emulsifier effective at very low levels; provides detergency to cleansing creams; excellent for fine fabric detergents
Form: Paste; Color: White
Composition: 25–27% active

Cycloryl NA. Cyclo Chem. Corp.
Chemical Description: Sodium lauryl ether sulfate
Category: Base, anionic
Applications: Base for shampoos and light duty detergents; superior electrolyte sensitivity, and good hard water stability to formulations
Form: Liquid; Color: Pale Yellow
Composition: 27–29% active

Cycloryl NA-2. Cyclo Chem. Corp.
Chemical Description: Sodium lauryl ether sulfate
Category: Base, anionic
Applications: Base for clear liquid and colored shampoos, bubble baths and light duty detergents
Form: Liquid; Color: Clear
Composition: 24–26% active

Cycloryl NA 61. Cyclo Chem. Corp.
Chemical Description: Sodium lauryl ether sulfate
Category: Base, anionic
Applications: High active base for high foaming detergents, dishwashes, car washes, textile scours and shampoos
Form: Liquid; Color: Pale Yellow
Composition: 58–60% active

Cycloryl NA61 CG. Cyclo Chem. Corp.
Chemical Description: Sodium lauryl ether sulfate
Category: Base, anionic
Applications: Cosmetics where color and high activity are important
Form: Liquid
Composition: 58–60% active

Cycloryl NWC. Cyclo Chem. Corp.
Chemical Description: Sodium lauryl ether sulfate plus builders
Category: Base, anionic
Applications: Base for shampoos and bubble baths
Form: Liquid; Color: Amber
Properties: Biodegradable
Composition: 62–64% active

Cycloryl SET. Cyclo Chem. Corp.
Chemical Description: Ethoxylated primary alcohol plus additives
Category: Base, anionic
Applications: High active base for superior heavy duty liquid laundry detergent
Form: Liquid
Composition: 79–81% active

Cycloryl TD. Cyclo Chem. Corp.
Chemical Description: Triethanolamine lauryl ether sulfate
Category: Base, anionic
Applications: Base for viscous shampoos with high salt tolerance
Form: Liquid; Color: Pale Yellow
Composition: 39–41% active

Cycloryl TEALS. Cyclo Chem. Corp.
Chemical Description: Triethanolamine lauryl sarcosinate
Category: Foaming agent, detergent, anionic
Applications: High flash foam, mild detergent for shampoos, bubble baths; performs well in hard water applications
Form: Liquid
Composition: 39–41% active

Cycloryl TS30. Cyclo Chem. Corp.
Chemical Description: Sodium lauryl sulfate plus additves
Category: Detergent, anionic
Applications: Carpet shampoo concentrate
Form: Liquid; Color: Amber
Composition: 28–30% active

Cycloryl WAT. Cyclo Chem. Corp.
Chemical Description: Triethanolamine lauryl sulfate
Category: Base, anionic
Applications: Base for light colored clear shampoos, bubble baths and other cleaning preparations where a creamy, mild lather is desired
Form: Liquid; Color: Pale Yellow
Composition: 38–40% active

Cycloryl WAT-I. Cyclo Chem. Corp.
Chemical Description: Triamine lauryl sulfate
Category: Base, anionic
Applications: Base for highly concentrated foaming toiletry products; useful where water content must be minimized
Form: Liquid
Composition: 86–90% active

Cycloryl XL. Cyclo Chem. Corp.
Chemical Description: Amphoteric modified lauryl sulfate
Category: Base, anionic
Applications: Base for providing conditioning and low irritation shampoos
Form: Liquid; Color: Pale Yellow
Composition: 36–38% active

Cyclosheen 100. Cyclo Chem. Corp.
Chemical Description: A blend of alkanolamides and opacifiers
Category: Emulsifier, nonionic
Applications: Used to impart stable pearl, to improve lathering and to increase viscosity of liquid shampoos and cleansers
Form: Soft wax; Color: Off-white
Properties: M. P.: 58–60 C; Acid No.: 2.0 max.

Cycloteric BET C-30. Cyclo Chem. Corp.
Chemical Description: Betaine amphoteric
Category: Foaming agent
Applications: Mild, high foaming surfactant; a broad range of viscosity control is provided when used with anionics; for promoting foam and detergency over a wide

pH range
Form: Liquid; Color: Yellow
Properties: Stability: Stable in acid and alkaline media; pH: 4.5–5.5
Composition: 29–31% active

Cycloteric BET I-30. Cyclo Chem. Corp.
Chemical Description: Betaine amphoteric
Category: Conditioning agent
Applications: Outstanding conditioning agent that provides an excellent viscosity building effect in compounded systems
Form: Liquid; Color: Yellow
Properties: pH: 5.5–6.5
Composition: 29–31% active

Cycloteric BET O-30. Cyclo Chem Corp.
Chemical Description: Betaine amphoteric
Category: Foaming, conditioning, and thickening agents
Applications: Foam stabilizer, conditioner, and thickening agent for hair and skin preparations
Form: Liquid; Color: Yellow
Properties: pH: 5.5–6.5
Composition: 29–31% active

Cycloteric DC-SF. Cyclo Chem. Corp.
Chemical Description: Dicarboxylate amphoteric
Category: Cleansing agent
Applications: High tolerance to alkali and other electrolytes
Form: Liquid; Color: Clear
Properties: pH: 8.0–9.0
Composition: 38–40% active

Cycloteric MV-SF. Cyclo Chem. Corp.
Chemical Description: Monocarboxylate amphoteric
Category: Low foaming, wetting agent
Applications: A powerful, low foaming wetting agent valuable in compounding highly acidic and highly alkaline cleansers where fast action and rust inhibition are sought
Form: Liquid; Color: Clear
Properties: pH: 9.0–11.0
Composition: 38–40% active

Cycloton CT 100. Cyclo Chem. Corp.
Chemical Description: Fatty quaternary compound combined with compatible emulsifiers
Category: Emulsifier, cationic
Applications: An excellent creme rinse/conditioner is produced when the cationic wax is diluted
Form: Flakes; Color: White
Properties: M. P.: 76–77 C; Acid No.: 1.0 max.

Cyncal 80%. Hilton-Davis
Chemical Description: Blend of alkyl dimethyl benzyl ammonium chlorides
Category: Antimicrobial agent, antistat, cationic
Applications: Disinfectant, sanitizer, alone or in combination with detergents for use in food, beverage processing plants, food service establishments, hospitals, industrial plants and farm use; fabric

antistatic agent; water treatment; good hard water tolerance
Form: Liquid; Color: Light yellow; Odor: Mild, pleasant
Properties: m.w.: 359 (of active); Solubility: Freely sol. in water, lower alcohols, ketones, glycols; Density: 7.8 lb/gal; Sp. gr.: 0.94; Stability: Crystallizes after prolonged storage at low temperature (returns to homogeneous state on warming to room temperature)
Composition: 80% active ingred.
Reaction: Compatible with nonionics, inorganic detergent builders, organic chelates, most polar solvents
Toxicity: Skin irritant; causes eye damage
Warnings and Caution: Corrosive. Goggles and gloves should be worn. Harmful or fatal if swallowed. Avoid contamination of food
References: Tech bulletin A-2-9/72

D

Dapral. Akzo Chemie
Chemical Description: Phosphate derivatives
Category: Detergent, wetting agent, emulsifier
Applications: Detergent–textiles, laundering, dyeing, household and cosmetics; as wetting agent in dyeing; emulsifiers for making of cutting and soluble oils; dispersing waxes, pigments, resins
Form: Liquid, paste

Darvan No. 1. R. T. Vanderbilt Co.
Chemical Description: Complex polymerized organic salts of sulfonic acids of alkylaryl type
Category: Dispersing agent
Applications: General purpose dispersant esp. for good color; in adhesives, as defoaming agents in paperboard, in gasket sealers for food packaging; agricultural sprays; aids ball and tube mill grinding; cosmetic emulsions and suspensions esp. liquid make-up bases, eye make-up, calamine and talc lotions; household polishes; for barium sulfate suspensions in medical research and x-ray work; leather tanning processes
Form: Granular; Color: Buff
Properties: Solubility: 40% sol. in water; Density: 1.55 mg/m^3; Stability: Stable to freezing, mild acid and alkali; pH: 8.0–10.5 (1% sol'n.)

Darvan No. 2. R. T. Vanderbilt Co.
Chemical Description: Complex polymerized organic salts of sulfonic acids of alkylaryl type
Category: Dispersing agent, emulsifier
Applications: Pelleting aid in animal feed; agricultural sprays; aids in ball and tube mill grinding; O/W emulsions; cosmetic emulsions and suspensions esp. liquid make-up bases, eye make-up, calamine and talc lotions; household polishes; pharmaceutical tablets containing insolubles; in microbiology
Form: Powder; Color: Dark brown
Properties: Solubility: 25% water sol.; Density: 1.59 mg/m^3; Stability: Stable to freezing, mild

acid and alkali; pH: 7.0–8.5 (1% sol'n.)

Darvan No. 7. R. T. Vanderbilt Co.
Chemical Description: Polyelectrolyte dispersing agent in aqueous solution
Category: Dispersant
Applications: Nondiscoloring dispersant for preparing high solids dispersions; agricultural sprays; aids in ball and tube mill grinding; cosmetic emulsions esp. liquid make-up bases, eye make-up, calamine and talc lotions; household polishes; used by museums and universities in cleaning artifacts; in rubber and ceramic fields to disperse TiO_2
Form: Liquid; Color: Water white
Properties: Solubility: Very water sol.; Density: 1.16 mg/m^3; Stability: Stable to freezing, mild acid and alkali; pH: 9.5–10.5
Composition: 25% solids
References: Bulletin No. 118

Daxad 11. W.R. Grace & Co.
Chemical Description: Sodium salt of polymerized alkyl naphthalene sulfonic acids
Category: Dispersing agent
Applications: Agricultural chemicals; mastics, caulks and sealants; pigment slurries and dispersions; reduces slurry viscosity; emulsification properties
Form: Fine powder; Color: Buff
Properties: Density: 48 lb/ft^3; Surface Tension: 71 dynes/cm (1% sol'n. @ 20 C)

Daxad 11G. W.R. Grace & Co.
Chemical Description: Sodium salt of polymerized alkyl naphthalene sulfonic acids
Category: Dispersing agent
Applications: Agricultural chemicals; inks; mastics, caulks and sealants; synthetic polymers; emulsification; paper industry; polymerization; slurry viscosity reduction; wetting action toward carbon black
Form: Granules; Color: Buff
Properties: Density: 35 lb/ft^3; Stability: Stable to mild acids and alkalies; Surface Tension: 71 dynes/cm (1% @ 20 C)
Composition: 87% active ingred.; 4.5% max. moisture content; 8.5% max. inert solubles
References: Bulletin DX-72-1

Daxad 11KLS. W.R. Grace & Co.
Chemical Description: Potassium salt of polymerized alkyl naphthalene sulfonic acids
Category: Dispersant
Applications: Dyes, dyestuffs; inks; latex paints; pigment slurries and dispersions; latex viscosity reduction; textile treatment
Form: Granules; Color: Light buff
Properties: Density: 35 lb/ft^3; Surface Tension: 71 dynes/cm (1% @ 20 C)

Daxad 13. W.R. Grace & Co.
Chemical Description: Sodium salt of polymerized alkyl naphthalene sulfonic acids
Category: Dispersent

Applications: Disperses inorganic and organic pigments; paper coating dispersion; paper mill slime control
Form: Powder; Color: Amber
Properties: Solubility: Very sol. in water systems; Density: 47 lb/ft^3; Surface Tension: 66 dynes/cm (1% sol'n. @ 72 F); pH: 9.5 (1% sol'n. @ 72 F)

Daxad 15. W.R. Grace & Co.
Chemical Description: Sodium salt of polymerized alkyl naphthalene sulfonic acids
Category: Dispersant
Applications: Agricultural chemicals; ceramics; concrete admixtures; dyes; cements; industrial cleaners; inks; water treatment chemicals
Form: Powder; Color: Amber
Properties: Density: 47 lb/ft^3; Surface Tension: 70 dynes/cm (1% sol'n. @ 20 C)

Daxad 16. W.R. Grace & Co.
Chemical Description: Sodium salts of polymerized alkyl naphthalene sulfonic acids
Category: Dispersant
Applications: Agricultural chemicals; concrete admixtures; dyes; pigment slurries and dispersions; synthetic polymers; synthetic tanning agents; water treatment chemicals
Form: Transparent liquid; Color: Dark brown
Properties: Density: 10.4 lb/gal.; Stability: Stable to mild acids and alkalies; Surface Tension: 70 dynes/cm (1% sol'n. @ 20 C); pH: 9–10 (1% sol'n.)
Composition: 42.5% active ingred.
References: Bulletin DX-66-2

Daxad 17. W.R. Grace & Co.
Chemical Description: Sodium salts of polymerized alkyl naphthalene sulfonic acids
Category: Dispersant
Applications: Agricultural chemicals; ceramics; concretes and cements; dyes; industrial cleaners; inks; mastics; pigment slurries; synthetic polymers; synthetic tanning agents; water treatment chemicals
Form: Granules; Color: Amber
Properties: Density: 35 lb/ft^3; Surface Tension: 70 dynes/cm (1% sol'n. @ 20 C)

Daxad 19. W.R. Grace & Co.
Chemical Description: Sodium salts of polymerized alkyl naphthalene sulfonic acids
Category: Dispersant
Applications: Concretes and cements; gypsum wallboard; pigment slurries; synthetic tanning agents
Form: Powder; Color: Amber
Properties: Solubility: Very sol. in water systems; Density: 47 lb/ft^3; Surface Tension: 71 dynes/cm (1% sol'n. @ 20 C); pH: 9.0 (1% sol'n. @ 72 F)

Daxad 21. W.R. Grace & Co.
Chemical Description: Mono-calcium salt of polymerized aryl alkyl

sulfonic acids
Category: Dispersant
Applications: Agricultural chemicals; concretes and cements; dyes; gypsum wallboard; pigment slurries; water treatment chemicals
Form: Powder; Color: Brown
Properties: Density: 30 lb/ft^3; Stability: Stable to mild acids and alkalies; Surface Tension: 62 dynes/cm (1% sol'n. @ 20 C); pH: 5–8 (1% sol'n. @ 72 F)
Composition: 92% active ingred.
References: Bulletin DX-71-1

Daxad 23. W.R. Grace & Co.
Chemical Description: Sodium salts of polymerized substituted benzoid alkyl sulfonic acids
Category: Dispersant
Applications: Agricultural chemicals; concretes and cements; dyes; linoleum pastes; pigment slurries; water treatment chemicals
Form: Powder; Color: Dark brown
Properties: Density: 37 lb/ft^3; Stability: Stable to mild acids and alkalies; Surface Tension: 56 dynes/cm (1% sol'n. @ 20 C); pH: 7.0–8.5 (1% sol'n. @ 72 F)
Composition: 84.5% active min.
References: Bulletin DX-63-2

Daxad 27. W.R. Grace & Co.
Chemical Description: Sodium salts of polymerized substituted benzoid alkyl sulfonic acids formulated with an inorganic suspending agent
Category: Dispersant
Applications: Long lasting suspensions of larger size particles; agricultural chemicals; pigment slurries and dispersions
Form: Free-flowing powder; Color: Light gray-brown
Properties: Density: 48 lb/ft^3; Stability: Stable to mild acids and alkalies; Surface Tension: 62 dynes/cm (1% sol'n. @ 20 C); pH: 6–8 (30% aq. susp.)
Composition: 84.5% active min.
References: Bulletin DX-62-2

Daxad 30. W.R. Grace & Co.
Chemical Description: Sodium salt of a carboxylated polyelectrolyte
Category: Dispersant
Applications: Rapid dispersion of pigments in aqueous solutions; paint formulations; emulsion polymerization; large particle suspensions; agricultural chemicals; ceramics; industrial cleaners; inks; latex paints; mastics; water treatment chemicals
Form: Liquid; Color: Clear to slightly opalescent
Properties: Solubility: Sol. in water systems; Density: 9.6 lb/gal; Sp. gr.: 1.15; Visc.: 75 cps (Brookfield #2 spindle); F.P.: < -5 C; Stability: Stable over wide pH range; Surface Tension: 70 dynes/cm (1% sol'n. @ 20 C); pH: 10.0
Composition: 25% solids

Daxad 30-30. W.R. Grace & Co.
Chemical Description: Sodium salt of a carboxylated polyelectrolyte
Category: Dispersant
Applications: Rapid dispersion of pigments in aqueous solutions;

paint formulations; emulsion polymerization; large particle suspensions; agricultural chemicals; ceramics; industrial cleaners; inks; latex paints; mastics; water treatment chemicals
Form: Liquid; Color: Clear to slightly opalescent
Properties: Solubility: Sol. in water systems; Density: 9.9 lb/gal; Sp. gr.: 1.21; Visc.: 1 cps max. (Brookfield #2 spindle); F.P.: < −5 C; Stability: Stable over wide pH range; Surface Tension: 70 dynes/cm (1% sol'n. @ 20 C); pH: 10
Composition: 30% solids content

Daxad 30S. W.R. Grace & Co.
Chemical Description: Sodium salt of a carboxylated polyelectrolyte
Category: Dispersant
Applications: Rapid dispersion of pigments in aqueous solutions; agricultural chemicals; industrial cleaners; mastics; pigment slurries; water treatment chemicals
Form: Fine powder; Color: White
Properties: Solubility: Sol. in water systems; Density: 32 lb/ft^3; Stability: Stable over wide pH range; Surface Tension: 70 dynes/cm (1% sol'n. @ 20 C)
Composition: 90% active ingred.
References: Bulletin DX-61-2

Daxad 31. W.R. Grace & Co.
Chemical Description: Sodium salt of a carboxylated polyelectrolyte
Category: Dispersant
Applications: Latex paints and coatings; dispersion of organic and inorganic pigments
Form: Clear liquid; Color: Yellow
Properties: Solubility: Sol. in water systems; Density: 9.2 lb/gal; Sp. gr.: 1.11; Visc.: 30 cps (Brookfield #2 spindle); Surface Tension: 50 dynes/cm (1% sol'n. @ 25 C)
Composition: 25% solids content

Daxad 32. W.R. Grace & Co.
Chemical Description: Ammonium salt of carboxylated polyelectrolyte
Category: Dispersant
Applications: Ceramics; pigment slurries and dispersions
Form: Clear to slightly opalescent fluid; Color: Amber
Properties: Solubility: Very sol. in water systems; Density: 9.1–9.4 lb/gal; Sp. gr.: 1.11; Visc.: 75 cps (Brookfield #2 spindle); Surface Tension: 70 dynes/cm (1% sol'n. @ 20 C); pH: 7.5–9.0
Composition: 25% solids content
References: Bulletin DX-60-2

DDBSA 99-b. Monsanto Ind. Chem.
Chemical Description: Dodecylbenzene sulfonic acid
Applications: Used in low salt liquid detergents; general purpose neutralization reactions; spray dried products
Properties: m.w.: 316 avg.; Sp. gr.: 1.0;
Composition: 98% active ingred.
Toxicity: Causes burns; harmful if absorbed through skin
Containers: 450 lb net, 55 gal metal drums; 5000 gal tanktrucks;

8000 and 10,000 gal tankcars
Warnings and Caution: Do not get in eyes, on skin or clothing; avoid breathing vapor or mist; in case of contact flush thoroughly with water for 15 min; call physician

Deceresol Surfactant CW. American Cyanamid Co.
Category: Wetting agent, modified nonionic
Applications: Wetting of various textiles during processing; dyeing; desizing and boiling off cottons; bleaching; sizing; Kier boiling; papermaking; wet out of fiber mulch for dyeing process; desizing of starch-based sizes
Form: Liquid; Color: Clear
Properties: Solubility: Dissolves readily with stirring; Density: 8.8 lb/gal; Sp. gr.: 1.06; Stability: Excellent
Composition: 92% active ingred.; other components: water and 2-ethyl hexanol
Reaction: Modified nonionic
Toxicity: Severe eye irritant; drying to skin
Warnings and Caution: Rubber gloves and goggles should be worn
References: Bulletin 292

Deceresol Surfactant NI Conc. American Cyanamid Co.
Category: Detergent, emulsifier, wetting agent, nonionic
Applications: Reactant finishing of textiles, desizing, scouring, bleaching, dyeing, printing, carbonizing woolens; Kier boiling; making of cutting and soluble oils; dispersing waxes, pigments, resins
Form: Liquid; Color: Clear
Properties: Solubility: Dissolves readily with stirring in cold water; less sol. in water above 140 F; Density: 8.8 lb/gal; Sp. gr.: 1.05; Stability: Excellent; stable to freezing and thawing; pH: 6–7
Composition: 100% active ingred.
Reaction: Nonionic
Toxicity: Nontoxic in normal usage; avoid skin and eye contact; severe eye irritant and drying to skin
Containers: 470 lb containers
Warnings and Caution: Wear rubber gloves and goggles when handling
References: Bulletin 182

Deceresol Surfactant OT 75%. American Cyanamid Co.
Category: Wetting agent, anionic
Applications: Wetting out fabric; desizing; scouring and bleaching; dyeing and printing; finishing
Form: Viscous liquid; Color: Water white; Odor: Slight hexanol
Properties: Solubility: Completely as dilute solution; Density: 8.9 lb/gal; Sp. gr.: 1.082; Visc.: 172 cp; F.P.: 30 F; B.P.: ≈ 200 F; Stability: Excellent; Biodegradable in 24 h; pH: 5.6 (1% sol'n.)
Composition: 75% active ingred.; other components: 6–8% alcohol
Toxicity: Nontoxic in normal usage; severe eye irritant and drying to skin
Containers: 475 lb containers
Warnings and Caution: Wear rub-

ber gloves and goggles when handling

Deceresol Surfactant OT Special 70%. American Cyanamid Co.
Category and Applications: See Deceresol Surfactant OT 75%
Form: Thin liquid; Color: Colorless; Odor: Slight
Properties: Solubility: Dissolves with agitation in warm water; Density: 8.8 lb/gal; Sp. gr.: 1.05; Stability: Good; pH: 5.7
Composition: 70% active ingred.; other components: water and 2-ethyl hexanol
Reaction: Slightly acid
Toxicity: Nontoxic in normal usage; severe eye irritant; drying to skin
Containers: 475 lb containers
Warnings and Caution: Wear rubber gloves and goggles when handling
References: Bulletin 263

Deceresol Surfactant P Special. American Cyanamid Co.
Category: Detergent
Applications: Textile detergent
Form: Liquid; Color: Light amber
Properties: Solubility: Infinite in water; Density: 8.5 lb/gal; Stability: Stable to freeze/thaw
Composition: 18% active ingred.
Reaction: Neutral
Containers: 450 lb containers
References: Bulletin 183

Defoamer 347D. Hart Chem.
Chemical Description: Anionic specialty formulated product
Category: Defoamer
Applications: Pulp and paper effluent defoamer
Form: Paste; Color: Amber
Properties: Solubility: Fair emulsion
Composition: 53% active ingred.
Reaction: Acid
Toxicity: Skin and eye irritant
Containers: Drums, T/L
Warnings and Caution: Avoid contact with eyes and skin

Defoamer 543. Hart Chem.
Chemical Description: Nonionic specialty formulated product
Category: Defoamer
Applications: Paper machine and industrial effluent defoamer
Form: Liquid; Color: Yellow
Properties: Refractive Index: 1.4550; Solubility: Fair emulsion; Density: 0.874; Visc.: 31 cps
Composition: 100% active ingred.
Reaction: Neutral
Toxicity: Skin and eye irritant
Containers: Drums, T/L
Warnings and Caution: Avoid contact with skin and eyes

Defoamer 2000 Series. Hart Chem.
Chemical Description: Nonionic specialty formulated products
Category: Defoamer
Applications: Defoamers for brownstock, screen room, effluent and bleach plant
Form: Liquid; Color: Cream
Properties: Density: 0.870; Visc.: 700 cps
Composition: 100% active ingred.
Toxicity: Skin and eye irritant
Containers: Drums, T/T, T/C

Defoamer 4201. Hart Chem.
Chemical Description: Nonionic specialty formulated product
Category: Defoamer
Applications: Brownstock defoamer
Form: Liquid; Color: White
Properties: Solubility: Dispersible Density: 0.87; Visc.: 450 cps
Composition: 100% active ingred.
Toxicity: Skin and eye irritant
Containers: Drums, T/T, T/C
Warnings and Caution: Avoid contact with skin and eyes

Defoamer 4203. Hart Chem.
Chemical Description: Nonionic specialty formulated product
Category: Defoamer
Applications: Brown stock defoamer
Form: Liquid; Color: White
Properties: Solubility: Dispersible; Sp. gr.: 0.86; Visc.: 450 cps
Composition: 100% active ingred.
Toxicity: Skin and eye irritant
Containers: Drums, T/T, T/C
Warnings and Caution: Avoid contact with skin and eyes

Defoamer 4205. Hart Chem.
Chemical Description: Nonionic specialty formulated product
Category: Defoamer
Applications: Screen room and effluent defoamer
Form: Liquid; Color: White
Properties: Solubility: Fair emulsion; Sp. gr.: 0.87; Visc.: 150 cps
Composition: 100% active ingred.
Toxicity: Skin and eye irritant
Containers: Drums, T/T, T/C
Warnings and Caution: Avoid contact with skin and eyes

Defoamer 4418. Hart Chem.
Chemical Description: Nonionic specialty formulated product
Category: Defoamer
Applications: Board machine defoamer
Form: Liquid; Color: White
Properties: Solubility: Dispersible; Sp. gr.: 0.87; Visc.: 350 cps
Composition: 100% active ingred.
Toxicity: Skin and eye irritant
Containers: Drums, T/T, T/C
Warnings and Caution: Avoid contact with skin and eyes

Defoamer 4440. Hart Chem.
Chemical Description: Nonionic specialty formulated product
Category: Defoamer
Applications: Cold stock screen room defoamer
Form: Liquid; Color: White
Properties: Solubility: Dispersible; Sp. gr.: 0.86
Composition: 100% active ingred.
Toxicity: Skin and eye irritant
Containers: Drums, T/T, T/C
Warnings and Caution: Avoid contact with skin and eyes

Defoamer C5B. Hart Chem.
Chemical Description: Nonionic specialty formulated product
Category: Defoamer
Applications: Board machine and industrial effluent defoamer
Form: Liquid; Color: Cream
Properties: Solubility: Fair emulsion; Sp. gr.: 0.86; Visc.: 400 cps

Composition: 100% active ingred.
Toxicity: Skin and eye irritant
Containers: Drums, T/T, T/C
Warnings and Caution: Avoid contact with skin and eyes

Defoamer HF-2. Hart Chem.
Chemical Description: Nonionic specialty formulated product
Category: Defoamer
Applications: Paper machine defoamer
Form: Paste; Color: White
Properties: Solubility: Stable emulsion; Visc.: 6000 cps
Composition: 30% active ingred.
Toxicity: Skin and eye irritant
Containers: 45 IG Drums
Warnings and Caution: Avoid contact with skin and eyes

Defoamer LH-192. Hart Chem.
Chemical Description: Nonionic specialty formulated product
Category: Defoamer
Applications: Liquid phosphate rock defoamer
Form: Liquid; Color: Brown
Properties: Solubility: Dispersible; Visc.: 140 cps
Composition: 100% active ingred.
Reaction: Acid
Toxicity: Skin and eye irritant
Containers: T/L, T/T
Warnings and Caution: Avoid contact with skin and eyes

Defoamer PBR-2. Hart Chem.
Chemical Description: Nonionic specialty formulated product
Category: Defoamer
Applications: Control of foam problems in linerboard and corrugated mills
Form: Liquid; Color: White
Properties: Solubility: Dispersible; Sp. gr.: 0.86; Visc.: 450 cps
Composition: 100% active ingred.
Toxicity: Skin and eye irritant
Containers: Drums, T/L, mini bulk
Warnings and Caution: Avoid contact with skin and eyes

Defoamer PM. Hart Chem.
Chemical Description: Nonionic specialty formulated product
Category: Defoamer
Applications: Paper machine defoamer
Form: Liquid; Color: Amber
Properties: Refractive Index: 1.4864; Solubility: Fair emulsion; Sp. gr.: 0.88; Visc.: 40 cps
Composition: 100% active ingred.
Reaction: Neutral
Toxicity: Skin and eye irritant
Containers: Drums, T/L
Warnings and Caution: Avoid contact with skin and eyes

Defoamer S. Hart Chem.
Chemical Description: Nonionic silicone emulsion
Category: Defoamer
Applications: Multipurpose silicone defoaming agent for industrial and textile
Form: Viscous liquid; Color: White
Properties: Solubility: Readily dispersible; Sp. gr.: 0.988; Visc.: 1500 cps
Reaction: Acid
Toxicity: Skin and eye irritant
Containers: Drums, T/L
Warnings and Caution: Avoid contact with skin and eyes

Delvet 68. Diamond Shamrock
Category: Nonionic surfactant
Applications: Chlorinated paraffin flame retardant additive
Form: Fluid
Composition: 68% active ingred.

Delvet 70. Diamond Shamrock
Category: Nonionic surfactant
Applications: Chlorinated paraffin flame retardant additive
Form: Fluid
Composition: 68% active ingred.

Demulfer Series. Toho Chem.
Category: Demulsifier
Applications: Demulsifier for crude oil production
Form: Liquid

Denlube 33-B. Graden Chem.
Chemical Description: Ammonium stearate dispersion
Category: Foam rubber additive
Applications: Frothing aid and foam stabilizer in acrylic and SBR latex systems
Form: Liquid; Color: White; Odor: Ammoniacal
Properties: Density: 8.0 lb/gal; Stability: Unstable above 100 F when heated; pH: 10.0 ± 1.0 (2% sol'n.)
Composition: 30% active
Containers: 55 gal drums
Warnings and Caution: Solidifies on freezing but regains normal fluidity on thawing; should be stored at room temperature and kept tightly covered to avoid loss of ammonia

Denlube 40. Graden Chem.
Chemical Description: High density polyethylene emulsion
Category: Lubricant, nonionic
Applications: By lubricating the coating, it improves flow and leveling characteristics, making possible smooth even application; reduces cracking of the dry coating and provides roll release
Form: Liquid; Color: Amber
Properties: Density: 8.4 lb/gal; pH: 8.5–9.0 (1% sol'n.)
Composition: 22% active
Containers: 55 gal drums
Warnings and Caution: Freezing seriously damages the product; containers should be kept tightly closed when not in use to prevent evaporation loss

Denlube 50 NFG. Graden Chem.
Chemical Description: Scale wax emulsion
Category: Foam rubber additive
Applications: Added to latex during foaming to preserve the resilience of the foam rubber article; it lubricates the inside surfaces of the cells in the foam producing an anti-blocking effect
Form: Liquid; Color: White
Composition: 50% active
Containers: 55 gal drum
Warnings and Caution: Freezing seriously damages the product; protect from cold in transit and storage

Denlube 1025-KA. Graden Chem.
Chemical Description: A blend of higher molecular weight sulfonated

aliphatic compounds
Category: Antiblocking agent
Applications: Strip-stock lubricant during rubber processing; it also acts as a processing aid
Form: Paste; Color: Cream
Properties: Solubility: Disp. in all proportions in water at 180 F; pH: 8.5–9.0 (1% sol'n.)
Composition: 80% active
Containers: 450 lb net drums

Denphos P-610. Graden Chem.
Chemical Description: Phosphate ester
Category: Detergent, emulsifier, wetting agent, anionic
Applications: Used as the sole detergent and wetting agent because of its stability and solubility; low foaming product recommended for industrial and household cleaners
Form: Liquid; Color: Water-white to pale yellow; Odor: Mild
Properties: Solubility: Sol. in aqueous solutions containing high concentrations of alkaline builders; Sp. gr.: 1.155; pH: 1.5–2.5 (2% sol'n.)
Composition: 100% active
Containers: 500 lb net drums

Densol 284. Graden Chem.
Chemical Description: Highly sulfated fatty acid
Category: Stabilizer for natural and synthetic latex, anionic
Applications: Protects natural and synthetic latices from coagulation during high speed stirring or pumping
Form: Liquid; Color: Dark amber; Odor: Fatty
Properties: Stable to strong acids, such as dilute acetic or hydrochloric acid
Containers: 55 gal drums

Densol 1010. Graden Chem.
Chemical Description: Modified sodium polyacrylate
Category: Thickener, plasticizer, thinning compound
Applications: Extensively used in rug and upholstery backing adhesives; stabilizer for latex compounding systems
Form: Gel; Color: Light amber
Properties: Density: 8.5 lb/gal; pH: 10 (2% sol'n.)
Composition: 10% active
Containers: 55 gal drums

Densol 6920. Graden Chem.
Chemical Description: Synthetic replacement for sulfated castor oil
Category: Emulsifier, anionic
Applications: A good emulsifier for mineral oil as well as effective dispersant or emulsifier for colors in oil systems; prevents hard settling of pigments during storage
Form: Liquid
Properties: Sp. gr.: 0.972 @ 60 F; pH: 5.4 (10% sol'n.)
Composition: 75% active
Containers: 55 gal drums

Densol 6930. Graden Chem.
Chemical Description: Synthetic replacement for sulfated sperm oil
Category: Emulsifier
Properties: pH: 5.4 (10% sol'n.)

Containers: 55 gal drum

Densol P-82. Graden Chem.
Chemical Description: Sulfated oleic ester
Category: Wetting agent, emulsifier
Applications: Excellent wetting agent; rewetting agent for lath paper and corrugating medium, .009 paper; performs well in up to 1% sodium hydroxide solution; readily dissolves in water from 60 F to boiling; soluble in a number of organic solvents
Form: Liquid; Color: Amber
Properties: Density: 8.57 lb/gal; Sp. gr.: 1.0288; pH: 5.5 (2% sol'n.)
Composition: 63% active
Containers: 55 gal drums

Denwet CM. Graden Chem.
Chemical Description: Sodium dioctyl sulfosuccinate
Category: Wetting agent, anionic
Applications: Reduces surface and interfacial tension; aids in wetting, dispersing, emulsifying, penetrating and solubilizing; used in emulsion polymerization of latex, particularly polyvinylacetate
Form: Clear liquid
Properties: Density: 9.0 lb/gal; pH: 7.0 ± 0.5 (2% sol'n.)
Composition: 75 ± 1% active
Containers: 55 gal liquapak drum

Denwet RG-7. Graden Chem.
Chemical Description: Sulfonated alkyl ester
Category: Wetting agent, anionic
Applications: Reduces surface and interfacial tension; aids in wetting, dispersing, emulsifying, penetrating and solubilizing; drycleaning detergent which does not absorb on fabrics and develop rancid odors; contributes significantly to the adhesion of paint films to various substrates, especially shiny or glossy surfaces; used in oil solutions of herbicides and pesticides as a wetting and spreading agent; readily soluble in a wide range of solvents which makes it useful as a coupling agent or solubilizer for emulsifiers of lower solubility
Form: Clear liquid
Properties: Density: 9.0 lb/gal; pH: 7.0 ± 0.5 (2% sol'n.)
Composition: 75 ± 1% active
Containers: 55 gal liquapak drum

Depuma. American Color & Chem.
Chemical Description: Silicone type antifoaming agent
Category: Nonionic antifoaming agent
Applications: Textiles and general defoaming needs for water systems, in acid and alkaline liquors; skins, laundering, leather, dyeing, metallurgy, mines and minerals and papermaking industires; bleaching, dyeing, tanning, finishing, sizing, coating, electroplating processes; rotary washer; waterpaint calsomines, adhesives; insecticides, herbicides, fungicides; removing starches, wall paper; Kier boiling; fat liquoring; printing; making of bitumen emulsions
Form: Pourable viscous emulsion; Color: Pearly white
Properties: Solubility: Water dis-

persible; Density: 8.2 lb/gal; Stability: Indefinitely stable at room temperature; keep from freezing
Composition: Silicone
Containers: 450 lb drums

Depuma OB. American Color & Chem.
Chemical Description: Organic, nonsilicone type antifoam
Category: Nonionic antifoaming agent
Applications: Foam control for textiles in water systems; used in skins, leather, dyeing, metallurgy, mines and minerals, papermaking and paints industries; desizing and boiling off cottons; bleaching, dyeing, tanning, finishing, sizing, coating processes; water-paint calsomines, adhesives; insecticides, herbicides, fungicides; Kier boiling; fat liquoring; making of cutting and soluble oils; dispersing waxes, pigments, resins; printing; preparation of emulsified paints, bitumen emulsions
Form: Liquid; Color: White
Properties: Solubility: Sol. in most organic solvents; emulsifies in warm water; Stability: Stable; Emulsifiability: 100%; Biodegradable
Composition: 100% active ingred.
Reaction: Neutral, pH 7
Containers: 450 lb drums

Deriphat. General Mills Chem.
Category: Detergent, wetting agent, emulsifier
Applications: Detergents: textiles, leather, household, cosmetic (shampoos), cleaners of streets, buildings, hospitals; as wetting agent in bleaching process
Form: Liquid, powder, flake

DET. Swastik
Chemical Description: Alkyl aryl sulfonate sodium salt built with S.T.P.P. and other alkaline builders
Category: Detergent
Applications: Suitable for hand washing cottons, woolens, and synthetic fabrics
Form: Powder; Color: White
Properties: pH: 10.5 (1% sol'n.)

DET Liquid. Swastik
Chemical Description: Alkyl aryl sulfonate plus polyoxyethylene alcohol ether sulfate plus additives
Category: Cleansing, detergent
Applications: Liquid household cleanser, recommended for cleansing crockery, stainless steel utensils, tiles, linoleum, plastic wares
Form: Clear liquid
Properties: pH: 8.0–9.5 (1% sol'n.)

DET-Washmatic. Swastik
Chemical Description: Alkyl aryl sulfonate sodium salt
Category: Detergent
Applications: Suitable for use in washing machines
Form: Spray-dried powder; Color: Blue
Properties: pH: 10.5 (1% sol'n.)

Det-O-Jet. Alconox Inc.
Category: Detergent
Applications: Low sudsing deter-

gent with good wetting, emulsifying and penetrating properties for mechanical and ultrasonic washers; removing grease, heavy soils and stains from porcelain, glass, electronic parts
Form: Liquid
Properties: Solubility: Sol. in hot or cold water, soft or hard water; Biodegradable
Composition: Potassium hydroxide, silicate of soda and chlorine
Reaction: Highly alkaline
Toxicity: Skin and eye irritant
Containers: 1 gal container, 15 and 55 gal drums
Warnings and Caution: Avoid contact with skin, eyes, clothing; flush immediately with water in case of contact; do not ingest
References: Bulletin # 6

Detergent CR. Arol Chem.
Chemical Description: Fatty ethanolamine condensate
Category: Detergent, wetting agent, emulsifier
Applications: Textile scouring, emulsifying, wetting, penetrating and leveling agent; general purpose wetting, rewetting, penetrating and leveling agent; soaping-off prints; heavy-duty scouring; thickener in cosmetics; base for bubble baths, hand soaps; laundering, foods and household uses as detergent; Kier boiling; dispersing pigments, waxes, resins; printing
Form: Liquid; Color: Pale amber; Odor: Mild
Properties: Solubility: Readily sol. in water over wide temperature range; Sp. gr.: 1.01; Stability: Excellent
Composition: 100% active ingred.
Reaction: Alkaline
Toxicity: Avoid ingestion; otherwise considered nontoxic
Containers: 55 gal drums
Warnings and Caution: Avoid contact with eyes

Detergent Honey. Dixo Co. Inc.
Category: Detergent, emulsifier, wetting agent
Applications: Drycleaning detergent; emulsifier for a drycleaning solvent and water; wetting agent in drycleaning stain removal
Form: Liquid; Color: Yellow-amber; Odor: Sweet
Properties: Solubility: Sol. in common drycleaning solvents
Containers: 1, 5, 15, 30 and 55 gal metal containers

Dilasoft RW. Sandoz Ltd.
Chemical Description: Fatty acid derivative modified
Category: Softener, cationic/nonionic
Applications: Softening agent with wetting properties for cellulosic fibers
Form: Liquid

Dilasoft TF. Sandoz Ltd.
Chemical Description: Sulfonate
Category: Softening/filling agent, anionic
Applications: Softening/filling agent
Form: Liquid

Dion 3-800. Diamond Shamrock
Chemical Description: Mercaptan
Applications: Rapid-cure additive for epoxy resins
Form: Liquid
Composition: 100% active ingred.

Dion Emulsifier 26. Diamond Shamrock
Category: Emulsifier, nonionic
Applications: Emulsifier for epoxy resins
Form: Solid
Composition: 100% active ingred.

Dion Emulsifier 37. Diamond Shamrock
Category: Emulsifier, nonionic
Applications: Emulsifier for epoxy resins
Form: Liquid
Composition: 50% active ingred.

Dispersant 3000E. Hart Chem.
Chemical Description: Nonionic specialty formulated product
Category: Dispersant
Applications: Nonionic pitch dispersant
Form: Liquid; Color: Yellow
Properties: Solubility: Clear soluble; Sp. gr.: 1.00
Composition: 85% active ingred.
Reaction: Neutral
Toxicity: Skin and eye irritant
Containers: Drums, T/L, T/T, T/C
Warnings and Caution: Avoid contact with skin and eyes

Dispersant 3050B. Hart Chem.
Chemical Description: Nonionic specialty formulated product
Category: Dispersant
Applications: Nonionic pitch dispersant
Form: Liquid; Color: Yellow
Properties: Solubility: Dispersible; Sp. gr.: 1.03
Composition: 48% active ingred.
Reaction: Neutral
Toxicity: Skin and eye irritant
Containers: Drums, T/L, T/T, T/C
Warnings and Caution: Avoid contact with skin and eyes

Dispersol ACA. I.C.I. United States
Category: Dispersant
Applications: Dispersing agent for dyes, level dyeing assistant; dispersing waxes, pigments, resins; printing; dispersing dry powders
Form: Free-flowing powder; Color: Tan; Odor: Faint
Properties: Solubility: Water sol.; Sp. gr.: ≈ 1.0
Composition: 100% active ingred.
Reaction: Alkaline
References: TB 302R

Dispersol CWLA. I.C.I. United States
Category: Dyeing assistant, mildly cationic
Applications: Dye assistant for nylon or wool, esp. for acid dyes; afterscour
Form: Liquid; Color: Amber; Odor: Faint
Properties: Solubility: Water sol.; Sp. gr.: ≈ 1.0
Composition: Proprietary % active
Reaction: Acid
Toxicity: Causes eye burns
Warnings and Caution: Wear goggles; avoid inhaling vapors or

mists; in case of eye contact, flush immediately with water
References: TB 589

Dispex A40. Allied Colloids
Chemical Description: Ammonium salt of a polymeric carboxylic acid
Category: Deflocculant
Applications: Ceramic industry, glaze manufacture, stabilization and for preparation of clay slips for casting and spray-drying; sanitaryware; tableware
Form: Mobile liquid; Color: Pale yellow
Properties: Solubility: Miscible with water; Sp. gr.: 1.16; Visc.: 3.5–5.5 cs @ 15%; Stability: Good storage stability; imparts viscosity stability to products even at high temperature and under alkaline conditions; pH: 7.5–8.5
Composition: 40% active ingred.
Containers: 55 gal or 5 gal nonreturnable steel drums
References: Data sheet GID/317.4

Dispex N40. Allied Colloids
Chemical Description: Sodium salt of a polymeric carboxylic acid
Category and Applications: See Dispex A40
Form: Mobile liquid; Color: Pale yellow
Properties: Solubility: Miscible with water; Sp. gr.: 1.30; Visc.: 15–25 cs @ 25%; Stability: Good storage stability; imparts viscosity stability to products even at high temperature and under alkaline conditions; pH: 7.0–7.5
Composition: 40% active ingred.
Containers: 55 gal or 5 gal nonreturnable steel drums
References: Data sheet GID/317.4

DMAMP. IMC Chem. Group
Chemical Description: 2-dimethylamino-2-methyl-1-propanol
Category: Emulsifier, corrosion inhibitor, solubilizer, catalyst
Applications: Emulsifier for waxes in combination with an unsaturated fatty acid; vapor-phase corrosion inhibitor in steam-condensate systems; neutralizing agent in aqueous systems; solubilizer for resin systems; catalyst for production of polyurethane foams; curing agent for liquid epoxy resins; stabilizer; as a raw material for chemical synthesis
Form: Essentially anhydrous
Properties: Solubility: Sol. in water, organic solvents such as methanol, butanol, acetone, ethyl acetate, benzene, hexane, mineral oil, castor oil; Density: 7.5 lb/gal; Sp. gr.: 0.9; Visc.: A–A_2 (Gardner); F.P.: 19 C; B.P.: 160 C (@ 760 mm Hg); Flash Pt.: 135 F (TOC)
Composition: 100% active ingred.
Reaction: Alkaline
Toxicity: Causes skin and eye burns
Containers: 1, 5, 55 gal drums; tank cars; tank trucks; store in iron or steel
Warnings and Caution: Goggles, gloves, protective clothing should be worn; do not ingest; in case of contact, flush eyes or skin with plenty of water; combustible liquid
References: NP Series TDS No. 18

DMAMP-80. IMC Chem. Group
Chemical Description: 2-dimethylamino-2-methyl-1-propanol
Category and Applications: See DMAMP
Form: Liquid
Properties: Density: 7.9 lb/gal; Sp. gr.: 0.95; Visc.: A–A_2 (Gardner); F.P.: –20 C; B.P.: ≈ 98 C; Flash Pt.: 150 F (TOC)
Composition: 80% active ingred.; 20% water
Reaction: Highly alkaline
Toxicity: Causes skin and eye burns
Containers: 1, 5, 55 gal drums; tank cars; tank trucks; store in iron or steel
Warnings and Caution: Goggles, gloves, protective clothing should be worn; do not ingest; in case of contact, flush with plenty of water; combustible liquid
References: NP Series TDS No. 18

Dobane 45. Shell Chemicals
Chemical Description: Narrow cut linear alkylbenzene
Category: Detergent
Applications: Base material for preparation of linear alkylbenzene sulfonate; suitable for applications in powder detergent formulations
Form: Clear liquid
Properties: Ref. Index: 1.4830; Solubility: Not miscible with water; Density: 0.859 kg/l @ 20 C; Visc.: 13 cs @ 20 C; M.P.: –69 C; Flash Pt.: 172 C; Stability: Stable over a wide range of temperature; Biodegradable: 90%
Toxicity: Eye irritation and repeated or prolonged contact can cause skin irritation
Containers: Mild steel or stainless steel 182 kg drums
References: Data sheet 2A 116

Dobane 055. Shell Chemicals
Chemical Description: Linear alkylbenzene
Category: Detergent, anionic after sulfonation
Applications: Base material for preparation of linear alkylbenzene sulfonate; high foaming household powder detergent
Form: Clear liquid
Properties: Ref. Index: 1.4840; Solubility: Not miscible with water; Density: 0.860 kg/l @ 20 C; Visc.: 10 cs @ 20 C; M.P.: < –70 C; Flash Pt.: 150 C; Stability: Stable over a wide range of temperature; Biodegradable: 90%
Composition: 100% active ingred.
Toxicity: Eye irritation and repeated or prolonged contact can cause skin irritation
Containers: Mild steel or stainless steel 182 kg drums
References: Data sheet 2A 115

Dobane 80. Shell Chemicals
Chemical Description: Narrow cut linear alkylbenzene
Category: Detergent
Applications: Base material for preparation of linear alkylbenzene sulfonate; intended for use in a mixture with a higher molecular mass alkylbenzene when it endows the derived sulfonate with improved solubility and cold stability characteristics

Form: Clear liquid
Properties: Ref. Index: 1.4860; Solubility: Not miscible with water; Density: 0.862 kg/l @ 20 C; Visc.: 5 cs @ 20 C; M.P.: < –70 C; Stability: Stable over a wide range of temperature; Biodegradable: 90%
Toxicity: Eye irritation and repeated or prolonged contact can cause skin irritation
Containers: Mild steel or stainless steel 182 kg drums
References: Data sheet 2A 110

Dobane 83. Shell Chemicals
Chemical Description: Linear alkylbenzene
Category: Detergent, anionic after sulfonation
Applications: Base material for preparation of linear alkylbenzene sulfonate; light duty liquids where the sulfonate gives excellent solubility and cold stability characteristics
Form: Clear liquid
Properties: Ref. Index: 1.4855; Solubility: Note miscible with water; Density: 0.861 kg/l @ 20 C; Visc.: 7 cs @ 20 C; M.P.: < –70 C; Flash Pt.: 138 C; Stability: Stable over a wide range of temperature; Biodegradable: 90%
Composition: 100% active
Toxicity: Eye irritation and repeated or prolonged contact can cause skin irritation
Containers: Mild steel or stainless steel 182 kg drums
References: Data sheet 2A 111

Dobane 102. Shell Chemicals
Chemical Description: Narrow cut linear alkylbenzene
Category: Detergent
Applications: Base material for preparation of linear alkylbenzene sulfonate; designed for applications in premium quality dishwashing liquids where it provides optimum performance and ease of formulation
Form: Clear liquid
Properties: Ref. Index: 1.4850; Solubility: Not miscible with water; Density: 0.861 kg/l @ 20 C; Visc.: 7 cs @ 20 C; M.P.: < –70 C; Flash Pt.: 140 C; Stability: Stable over a wide range of temperature; Biodegradable: 90%
Toxicity: Eye irritation and repeated or prolonged contact can cause skin irritation
Containers: Mild steel or stainless steel 210 liter drums
References: Data sheet 2A 112

Dobane 123. Shell Chemicals
Chemical Description: Narrow cut linear alkylbenzene
Category: Detergent
Applications: Base material for preparation of linear alkylbenzene sulfonate; recommended for applications in household powder detergents
Form: Clear liquid
Properties: Ref. Index: 1.4835; Solubility: Note miscible with water; Density: 0.860 kg/l @ 20 C; Visc.: 11 cs @ 20 C; M.P.: < –70 C; Flash Pt.: 160 C; Stability: Stable over a wide range of temperature; Biodegradable: 90%
Toxicity: Eye irritation and re-

peated or prolonged contact can cause skin irritation
Containers: Mild steel or stainless steel 210 liter drums
References: Data sheet 2A 114

Dobane 945. Shell Chemicals
Chemical Description: Broad cut linear alkylbenzene
Category: Detergent
Applications: Base material for preparation of linear alkylbenzene sulfonate; high foaming household powder detergents; suitable as a biodegradable replacement for branched chain alkylbenzene
Form: Clear liquid
Properties: Ref. Index: 1.4835; Solubility: Not miscible with water; Density: 0.860 kg/l @ 20 C; Visc.: 11 cs @ 20 C; M.P.: –70 C; Flash Pt.: 142 C; Stability: Stable over a wide range of temperature; Biodegradable: 90%
Toxicity: Eye irritation and repeated or prolonged contact can cause skin irritation
Containers: Mild steel or stainless steel 182 kg drums
References: Data sheet 2A 117

Dobane HP. Shell Chemicals
Chemical Description: Broad cut linear alkylbenzene
Category: Detergent
Applications: Base material for preparation of linear alkylbenzene sulfonate; high foaming household powder detergents and light duty liquids
Form: Clear liquid
Properties: Ref. Index: 1.4845; Solubility: Not miscible with water; Density: 0.861 kg/l @ 20 C; Visc.: 9 cs @ 20 C; M.P.: < –70 C; Flash Pt.: 140 C; Stability: Stable over a wide range of temperature; Biodegradable: 90%
Toxicity: Eye irritation and repeated or prolonged contact can cause skin irritation
peated or prolonged contact can cause skin irritation
Containers: Mild steel or stainless steel 182 kg drums
References: Data sheet 2A 113

Dobanic Acid 83. Shell Chemicals
Chemical Description: Linear alkylbenzene sulfonic acid
Category: Detergent, anionic
Containers: Mild steel or stainless steel 182 kg drums
References: Data Sheet 2A 118

Dobane JN. Shell Chemicals
Chemical Description: Linear alkylbenzene
Category: Detergent, anionic after sulfonation
Applications: Base material for preparation of linear alkylbenzene sulfonate; preparation of all types of controlled foaming and high foaming household powders and light duty liquids
Form: Clear liquid
Properties: Ref. Index: 1.4850; Solubility: Not miscible with water; Density: 0.861 kg/l @ 20 C; Visc.: 8 cs @ 20 C; M.P.: < –70 C; Flash Pt.: 144 C; Stability: Stable over a wide range of temperature; Biodegradable: 90%
Composition: 100% active
Toxicity: Eye irritation and re-

Applications: A component of dishwashing and general cleaning liquids where it enables excellent cold stability characteristics
Form: Liquid; Color: Brown; Odor: Sulfur dioxide
Properties: Solubility: Miscible in water up to 40%; Density: 1.06 kg/l @ 20 C; Visc.: 1550 cs @ 20 C; M.P.: −10 to −12 C; Biodegradable: > 90%
Composition: 97% active
Toxicity: Eye irritation and repeated or prolonged contact can cause skin irritation
Containers: Mild steel 222 kg drums
Warnings and Caution: When handling product, always wear safety goggles/face shield to protect the eyes and face against splashes; wear rubber/plastic gloves, protective apron
References: Data sheet 2A 120

Dobanic Acid 102. Shell Chemicals
Chemical Description: Linear alkylbenzene sulfonic acid
Category: Detergent
Applications: A component for premium quality dishwashing liquids where it provides optimum performance with excellent cold stability charactertistics
Form: Liquid; Color: Brown; Odor: Sulfur dioxide
Properties: Solubility: Miscible in water up to 40%; Density: 1.06 kg/l @ 20 C; Visc.: 1800 cs @ 20 C; M.P.: −10 to −12 C; Biodegradable: > 90%
Composition: 97% active
Toxicity: Eye irritation and repeated or prolonged contact can cause skin irritation
Containers: Mild steel 222 kg drums
Warnings and Caution: When handling product, always wear safety goggles/face shield to protect the eyes and face against splashes; wear rubber/plastic gloves, protective apron
References: Data sheet 2A 121

Dobanic Acid JN. Shell Chemicals
Chemical Description: Linear alkylbenzene sulfonic acid
Category: Detergent, anionic
Applications: A component of all types of high foaming and controlled foaming household powders and light duty liquids
Form: Liquid; Color: Brown; Odor: Sulfur dioxide
Properties: Solubility: Miscible in water up to 40%; Density: 1.05 kg/l @ 20 C; Visc.: 1600 cs @ 20 C; M.P.: −10 to −12 C; Biodegradable: > 90%
Composition: 97% active
Toxicity: Eye irritation and repeated or prolonged contact can cause skin irritation
Containers: Mild steel 222 kg drums
Warnings and Caution: When handling product, always wear safety goggles/face shield to protect the eyes and face against splashes; wear rubber/plastic gloves, protective apron
References: Data sheet 2A 122

Dobanol 23. Shell Chemicals
Chemical Description: A narrow cut blend of primary alcohols of high purity
Category: Base material
Applications: Base material for preparation of alcohol ethoxylates, sulfates, ethoxysulfates
Form: Liquid; Odor: Mild
Properties: Ref. Index: 1.4420; Solubility: Not miscible with water; Density: 0.831 kg/l; Visc.: 14 cs @ 40 C; M. P.: 17–19 C; Flash Pt.: 132 C; Acid No.: < 0.1; Hydroxyl No.: 290 ± 5; Iodine Value: 0.1; Biodegradable: $> 95\%$
Composition: 100% active
Toxicity: Eye irritation and repeated or prolonged contact can cause skin irritation
Containers: Mild steel 168 kg drums
References: Data sheet 2A 131

Dobanol 23-2. Shell Chemicals
Chemical Description: Primary alcohol ethoxylate
Category: Detergent, nonionic
Applications: Intermediate for production of primary alcohol ethoxysulfates as components of shampoos, bubble baths, and other toiletries
Form: Clear liquid
Properties: Density: 0.91 kg/l @ 20 C; Visc.: 15 cs @ 40 C; M.P.: 4–5 C; HLB: 6.2; Flash Pt.: 148 C; Acid No.: 0.05–0.2; Hydroxyl No.: 199 ± 5; Biodegradable: $> 95\%$
Composition: 100% active
Toxicity: Eye irritation and repeated or prolonged contact can cause skin irritation
Containers: Mild steel 190 kg drums
Warnings and Caution: Storage temperature should not exceed 50 C
References: Data sheet 2A 144

Dobanol 23-6.5. Shell Chemicals
Chemical Description: Primary alcohol ethoxylate
Category: Detergent, nonionic
Applications: Used in high performance clothes washing powders and in light duty liquids
Form: Slurry; Color: White
Properties: Solubility: Miscible in all proportions but high viscosity gels may be formed at mid-range concentrations: Density: 0.96 kg/l @ 40 C; Visc.: 26 cs @ 40 C; M.P.: 17–19 C; HLB: 11.9; Cloud Pt.: 43 ± 3 C (5% aq. sol'n.); Flash Pt.: 172 C; Acid No.: 0.05–0.2; Hydroxyl No.: 117 ± 5; pH: 6.8 (0.5% aq. sol'n.)
Composition: 100% active
Toxicity: Eye irritation and repeated or prolonged contact can cause skin irritation
Containers: Mild steel 190 kg drums
Warnings and Caution: Storage temperature should not exceed 50 C
References: Data sheet 2A 145

Dobanol 25. Shell Chemicals
Chemical Description: A blend of primary alcohols of high purity
Category: Base material
Applications: Base material for

preparation of alcohol ethoxylates, sulfates, and ethoxysulfates
Form: Liquid; Odor: Mild
Properties: Ref. Index: 1.4440; Solubility: Not miscible with water; Density: 0.831 kg/l; Visc.: 14 cs @ 40 C; M.P.: 20–22 C; Acid No.: < 0.1; Hydroxyl No.: 270 ± 5; Iodine Value: < 0.1; Biodegradable
Composition: 100% active
Toxicity: Eye irritation and repeated contact can cause skin irritation
Containers: Mild steel 168 kg drums
Warnings and Caution: Storage temperature should not exceed 45 C
References: Data sheet 2A 133

Dobanol 25-3. Shell Chemicals
Chemical Description: Primary alcohol ethoxylate
Category: Detergent, nonionic
Applications: Intermediate for production of primary alcohol ethoxysulfates as components of high performance dishwashing liquids
Form: Clear liquid
Properties: Density: 0.93 kg/l @ 20 C; Visc.: 18 cs @ 40 C; M.P.: 5–6 C; HLB: 7.8; Flash Pt.: 162 C; Acid No.: 0.05–0.2; Hydroxyl No.: 166 ± 6; Biodegradable: > 95%
HLB: 7.8; Flash Pt.: 162 C; Acid No.: 0.05–0.2; Hydroxyl No.: 166 ± 6; Biodegradable: > 95%
Composition: 100% active
Toxicity: Eye irritation and repeated or prolonged contact can cause skin irritation
Containers: Mild steel 195 kg drums
Warnings and Caution: Storage temperature should not exceed 50 C
References: Data sheet 2A 147

Dobanol 25-3A/27 EU. Shell Chem.
Chemical Description: Aqueous-ethanol sol'n. of primary alcohol ethoxysulfate (ammonia cation)
Category: Detergent
Applications: Component of higher performance light duty liquid formulations which may require ethanol and urea as solubilizing additives
Form: Clear liquid
Properties: Solubility: Miscible in all proportions; Density: 1.00 kg/l @ 20 C; Visc.: 30 cs @ 20 C; Flash Pt.: 33 C; pH: 7.0–9.0 (5% aq. sol'n.); Biodegradable: > 95%
Composition: 25.5–27.5% active
Toxicity: Eye irritation and repeated or prolonged contact can cause skin irritation
Containers: Mild steel 215 kg drums
Warnings and Caution: Storage temperature should not exceed 40 C; the vapor is inflammable; use flame-proofed equipment and ensure storage and handling areas are free of naked lights and sparks
References: Data sheet 2A 164

Dobanol 25-3A/60. Shell Chem.
Chemical Description: Aqueous-ethanol concentrate of primary alcohol ethoxysulfate (ammonia

cation)
Category: Detergent, anionic
Applications: Component of high active matter, high performance light duty liquid formulations
Form: Liquid
Properties: Solubility: Miscible in all proportions but high viscosity gels may be formed at mid-range concentrations; Density: 1.04 kg/l @ 20 C; Visc.: 60 cs @ 20 C; Cloud Pt.: 8 C; Clear Pt.: 10 C; Flash Pt.: 26 C; pH: 6.5–8.0 (5% aq. sol'n.); Biodegradable: > 95%
Composition: 59–61% active
Toxicity: Eye irritation and repeated or prolonged contact can cause skin irritation
Containers: Mild steel 218 kg drums
Warnings and Caution: Storage temperature should not exceed 40 C; the vapor is inflammable; use flame-proofed equipment and ensure storage and handling areas are free of naked lights and sparks
References: Data sheet 2A 162

Dobanol 25-3S/27. Shell Chemicals
Chemical Description: Aqueous solution of primary aclohol ethoxysulfate (sodium cation)
Category: Detergent, anionic
Applications: Component of light duty liquid detergents
Form: Clear liquid
Properties: Solubility: Miscible in all proportions; Density: 1.05 kg/l @ 20 C; Visc.: 60 cs @ 20 C; Cloud Pt.: 1 C; Clear Pt.: 14 C; pH: 6.5–8.0 (5% aq. sol'n.); Biodegradable: > 95%
Composition: > 26.0% active
Toxicity: Eye irritation and repeated or prlonged contact can cause skin irritation
Containers: Mild steel 218 kg drums
Warnings and Caution: Storage temperature should not exceed 40 C
References: Data sheet 2A 161

Dobanol 25-3S/60. Shell Chem.
Chemical Description: Aqueous-ethanol concentrate of primary alcohol ethoxysulfate (sodium cation)
Category: Detergent, anionic
Applications: Component of high active matter, high performance, light, duty, liquid formulations
Form: Liquid
Properties: Solubility: Miscible in all proportions but high viscosity gels may be formed at mid-range concentrations; Density: 1.05 kg/l @ 20 C; Visc.: 60 cs @ 20 C; Flash Pt.: 25 C; pH: 6.5–8.0 (5% aq. sol'n.); Biodegradable: > 95%
Composition: 58–60% active
Toxicity: Eye irritation and repeated or prolonged contact can cause skin irritation
Containers: Mild steel 220 kg drums
Warnings and Caution: Storage temperature should not exceed 40 C; the vapor is inflammable; use flame-proofed equipment and ensure storage and handling areas are free of naked lights and sparks
References: Data sheet 2A 163

Dobanol 25-3S/70. Shell Chem.
Chemical Description: Aqueous concentrate of primary alcohol ethoxysulfate (sodium cation)
Category: Detergent
Applications: Component of high active matter, high performance, light duty, liquid formulations
Form: Clear gelled liquid
Properties: Solubility: Miscible in all proportions but high viscosity gels may be formed at mid-range concentrations; Density: 1.08 kg/l @ 20 C; Visc.: 5000 cP @ 20 C at shear rate 10 reciprocal s; pH: 7.5–9.5 (5% aq. sol'n.); Biodegradable: > 95%
Composition: 68–72% active
Toxicity: Eye irritation and repeated or prolonged contact can cause skin irritation
Containers: Mild steel drums
Warnings and Caution: Storage temperature should not exceed 40 C
References: Data sheet 2A 165

Dobanol 25-7. Shell Chemicals
Chemical Description: Primary alcohol ethoxylate
Category: Detergent, nonionic
Applications: Used in high performance clothes washing powders
Form: Paste/solid; Color: White
Properties: Solubility: Miscible in all proportions but high viscosity gels may be formed at mid-range concentrations; Density: 0.97 kg/l @ 40 C; Visc.: 31 cs @ 40 C; M.P.: 21–23 C; HLB: 12.0; Cloud Pt.: 51 ± 3 C (5% aq. sol'n.); Flash Pt.: 180 C; Acid No.: 0.05–0.2; Hydroxyl No. 109 ± 5; pH: 6.8 (0.5% aq. sol'n.)
Composition: 100% active
Toxicity: Eye irritation and repeated or prolonged contact can cause skin irritation
Containers: Mild steel 190 kg drums
Warnings and Caution: Storage temperature should not exceed 50 C
References: Data sheet 2A 148

Dobanol 25-9. Shell Chemicals
Chemical Description: Primary alcohol ethoxylate
Category: Detergent, nonionic
Applications: Used in high performance clothes washing powders
Form: Solid; Color: White
Properties: Solubility: Miscible in all proportions but high viscosity gels may be formed at mid-range concentrations; Density: 0.98 kg/l @ 40 C; Visc.: 40 cs @ 40 C; M.P.: 25–28 C; HLB: 13.2; Cloud Pt.: 79 ± 4 (5% aq. sol'n.); Flash Pt.: 190 C; Acid No.: 0.05–0.2; Hydroxyl No.: 92 ± 4; pH: 6.8 (0.5% aq. sol'n.)
Composition: 100% active
Toxicity: Eye irritation and repeated or prolonged contact can cause skin irritation
Containers: Mild steel (open top) 204 kg drums
Warnings and Caution: Storage temperature should not exceed 50 C
References: Data sheet 2A 149

Dobanol 45. Shell Chemicals
Chemical Description: A narrow cut blend of primary alcohols of high purity
Category: Base material
Applications: Base material for preparation of alcohol ethoxylates, sulfates, ethoxysulfates
Form: Liquid; Odor: Mild
Properties: Solubility: Not miscible with water; Density: 0.824 kg/l @ 40 C; Visc.: 17 cs @ 40 C; M.P.: 29–31 C; Flash Pt.: 152 C; Acid No.: < 0.1; Hydroxyl No.: 255 ± 5; Iodine Value: 0.1; Biodegradable
Composition: 100% active ingred.
Toxicity: Eye irritation and repeated or prolonged contact can cause skin irritation
Containers: Mild steel 168 kg drums
Warnings and Caution: Storage temperature should not exceed 45 C
References: Data sheet 2A 134

Dobanol 45-7. Shell Chemicals
Chemical Description: Primary alcohol ethoxylate
Category: Detergent, nonionic
Applications: Used in high performance clothes washing powders
Form: Solid; Color: White
Properties: Solubility: Miscible in all proportions but high viscosity gels may be formed at mid-range concentrations; Density: 0.96 kg/l @ 50 C; Visc.: 33 cs @ 40 C; M.P.: 22–24 C; HLB: 11.6; Cloud Pt.: 46 ± 3 C (5% aq. sol'n.); Flash Pt.: 185 C; Acid No.: 0.05–0.2; Hydroxyl No.: 106 ± 5; pH: 6.8 (0.5% aq. sol'n.)
Composition: 100% active
Toxicity: Eye irritation and repeated or prolonged contact can cause skin irritation
Containers: Mild steel 195 kg drums
Warnings and Caution: Storage temperature should not exceed 50 C
References: Data sheet 2A 150

Dobanol 45-11. Shell Chemicals
Chemical Description: Primary alcohol ethoxylate
Category: Detergent, nonionic
Applications: Used in high performance clothes washing powders
Form: Solid; Color: White
Properties: Solubility: Miscible in all proportions but high viscosity gels may be formed at mid-range concentrations; Density: 0.99 kg/l @ 50 C; Visc.: 47 cs @ 40 C; M.P.: 28–31 C; HLB: 13.7; Cloud Pt.: 88 ± 4 C (5% aq. sol'n.); Flash Pt.: 200 C; Acid No.: 0.05–0.2; Hydroxyl No.: 80 ± 4; pH: 6.8 (0.5% aq. sol'n.)
Composition: 100% active
Toxicity: Eye irritation and repeated or prolonged contact can cause skin irritation
Containers: Mild steel 197 kg drums
Warnings and Caution: Storage temperature should not exceed 50 C
References: Data sheet 2A 151

Dobanol 45-18. Shell Chemicals
Chemical Description: Primary alcohol ethoxylate
Category: Detergent, nonionic
Applications: Used in high performance clothes washing powders
Form: Solid; Color: White
Properties: Solubility: Miscible in all proportions but high viscosity gels may be formed at mid-range concentrations; Density: 1.02 kg/l @ 50 C; Visc.: 42 cs @ 60 C; M.P.: 35–39 C; HLB: 15.7; Cloud Pt.: > 100 C (5% aq. sol'n.); Flash Pt.: 200 C; Acid No.: 0.05–0.2; Hydroxyl No.: 55 ± 4; pH: 6.8 (0.5% aq. sol'n.)
Composition: 100% active
Toxicity: Eye irritation and repeated or prolonged contact can cause skin irritation
Containers: Mild steel drums
Warnings and Caution: Storage temperature should not exceed 50 C
References: Data sheet 2A 152

Dobanol 91. Shell Chemicals
Chemical Description: A blend of primary alcohols of high purity
Category: Base material
Applications: Base material for preparation of alcohol ethoxylates, sulfates, ethoxysulfates
Properties: Ref. Index: 1.4370; Solubility: Not miscible with water; Density: 0.832 kg/l; Visc.: 9 cs @ 40 C; M.P.: –8 to –10 C; Flash Pt.: 107 C (PMCC); Acid No.: < 0.1; Hydroxyl No.: 350 ± 5; Iodine Value: 0.1; Biodegradable
Composition: 100% active
Toxicity: Eye irritation and repeated or prolonged contact can cause skin irritation
Containers: Mild steel 175 kg drums
Warnings and Caution: Storage temperature should not exceed 45 C
References: Data sheet 2A 130

Dobanol 91-2.5. Shell Chemicals
Chemical Description: Primary alcohol ethoxylate
Category: Chemical intermediate, nonionic
Applications: Used for the production of primary alcohol ethoxysulfates; component of emulsifier formulations
Form: Clear liquid
Properties: Density: 0.93 kg/l @ 20 C; Visc.: 11 cs @ 40 C; M. P.: –20 to –22 C; HLB: 8.1; Flash Pt.: 120 C (PMCC); Acid No.: 0.05-0.2; Hydroxyl No.: 208 ± 6; Biodegradable: > 95%
Composition: 100% active
Toxicity: Eye irritation and repeated or prolonged contact can cause skin irritation
Containers: Mild steel 195 kg drums
Warnings and Caution: Storage temperature should not exceed 50 C
References: Data sheet 2A 140

Dobanol 91-5. Shell Chemicals
Chemical Description: Primary alcohol ethoxylate
Category: Wetting agent, emulsi-

fier, nonionic
Applications: Component of hard surface liquid cleaners
Form: Clear liquid
Properties: Solubility: Miscible in all proportions; Density: 0.98 kg/l @ 20 C; Visc.: 18 cs @ 40 C; M. P.: 3–6 C; HLB: 11.6; Cloud Pt.: 36 ± 3 (5% aq. sol'n.); Flash Pt.: 146 C (PMCC); Acid No.: 0.05–0.2; Hydroxyl No.: 148 ± 4; pH: 6.8 (0.05% aq. sol'n.)
Composition: 100% active
Toxicity: Eye irritation and repeated or prolonged contact can cause skin irritation
Containers: Mild steel 200 kg drums
Warnings and Caution: Storage temperature should not exceed 50 C
References: Data sheet 2A 141

Dobanol 91-6. Shell Chemicals
Chemical Description: Primary alcohol ethoxylate
Category: Detergent, wetting agent, nonionic
Applications: Detergent and wetting agent in textile processing, degreasing, metal cleaning, pulp deresination
Form: Clear liquid
Properties: Solubility: Miscible in all proportions; Density: 0.99 kg/l @ 20 C; Visc.: 21 cs @ 40 C; M. P.: 6–9 C; HLB: 12.5; Cloud Pt.: 56 ± 3 (5% aq. sol'n.); Flash Pt.: 150 C (PMCC); Acid No.: 0.05–0.2; Hydroxyl No.: 133 ± 6; pH: 6.8 (0.5% aq. sol'n.)
Composition: 100% active
Toxicity: Eye irritation and repeated or prolonged contact can cause skin irritation
Containers: Mild steel 203 kg drums
Warnings and Caution: Storage temperature should not exceed 50 C
References: Data sheet 2A 142

Dobanol 91-8. Shell Chemicals
Chemical Description: Primary alcohol ethoxylate
Category: Detergent, nonionic
Applications: Component of dishwashing liquids in blends with other surfactants
Form: Liquid/slurry
Properties: Solubility: Miscible in all proportions but high viscosity gels may be formed at mid-range concentrations; Density: 1.00 kg/l @ 20 C; Visc.: 28 cs @ 40 C; M. P.: 14–17 C; HLB: 13.7; Cloud Pt.: 81.5 ± 3.5 (5% aq. sol'n.); Flash Pt.: 160 C (PMCC); Acid No.: 0.05–0.2; Hydroxyl No.: 110 ± 5; pH: 6.8 (0.5% aq. sol'n.)
Composition: 100% active
Toxicity: Eye irritation and repeated or prolonged contact can cause skin irritation
Containers: Mild steel 200 kg drums
Warnings and Caution: Storage temperature should not exceed 50 C
References: Data sheet 2A 143

Dobanol 235. Shell Chemicals
Chemical Description: A blend of primary alcohols of high purity
Category: Base material

Applications: Base material for preparation of alcohol ethoxylates, sulfates, ethoxysulfates
Properties: Ref. Index: 1.4430; Solubility: Not miscible with water; Density: 0.831 kg/l; Visc.: 14 cs @ 40 C; M.P.: 18–20 C; Flash Pt.: 134 C (PMCC); Acid No.: $<$ 0.1; Hydroxyl No.: 280 ± 5; Iodine Value: 0.1; Biodegradable
Toxicity: Eye irritation and repeated or prolonged contact can cause skin irritation
Containers: Mild steel drums
Warnings and Caution: Storage temperature should not exceed 45 C
References: Data sheet 2A 132

Dobanol 235-3. Shell Chemicals
Chemical Description: Primary alcohol ethoxylate
Category: Chemical intermediate
Applications: Used for production of primary alcohol ethoxysulfate as component of high performance dishwashing liquids and toiletries
Form: Clear liquid
Properties: Density: 0.92 kg/l @ 20 C; Visc.: 17 cs @ 40 C; M. P.: 4–5 C; HLB: 7.8; Flash Pt.: 160 C (PMCC); Acid No.: 0.05–0.2; Hydroxyl No.: 172 ± 6; Biodegradable: $>$ 95%
Composition: 100% active
Toxicity: Eye irritation and repeated or prolonged contact can cause skin irritation
Containers: Mild steel drums
Warnings and Caution: Storage temperature should not exceed 50 C
References: Data sheet 2A 146

Doittol 891. Diamond Shamrock
Category: Anionic surfactant
Applications: Rubber anti-blocking agent
Form: Liquid
Composition: 20% active ingred.

Doittol K-21. Diamond Shamrock
Category: Anionic surfactant
Applications: Rubber anti-blocking agent
Form: Liquid
Composition: 25% active ingred.

Dow Corning 190 Surfactant. Dow Corning
Chemical Description: Silicone-glycol copolymer
Category: Silicone surfactant
Applications: Used for producing flexible slab stock (urethane) foam; has high processing latitude; good efficiency
Form: Low-viscosity liquid; Color: 2 (Gardner)
Properties: Sp. gr.: 1.035; Visc.: 1500 cst; Flash Pt.: 250 F (CCC); Stability: Good in water and amine premixes
Containers: 40 and 441 lb containers net weight
References: Bulletin 22-315

Dow Corning 193 Surfactant. Dow Corning
Chemical Description: Nonhydrolyzable silicone glycol copolymer
Applications: For producing all

types of rigid urethane foam; to manufacture board, laminates, foam in place, spray foam and shoe soles
Form: Clear viscous liquid; Color: 2 (Gardner)
Properties: Refractive Index: 1.45; Sp. gr.: 1.07; Visc.: 465 cs; F.P.: 50 F; Flash Pt.: 300 F (COC); Pour Pt.: 52 F; Stability: Good premix stability; becomes hazy < 77 F; solidifies to soft wax < 50 F
Containers: 40, 441 lb containers
References: Bulletin 22-316A

Dow Corning 196 Surfactant. Dow Corning Corp.
Chemical Description: Silicone glycol copolymer
Category: Profoamer
Applications: Used for producing low-density, flexible-slab urethane foams; with high efficiency and excellent processing latitude
Form: Low viscosity liquid; Color: < 4 (Gardner)
Properties: Solubility: Sol. in water, polyol, fluorocarbon or water-amine streams; Sp. gr.: 1.035; Visc.: 2000 cst; Flash Pt.: 225 F (open cup)
Toxicity: Mild eye irritant; moderate skin irritant on prolonged contact
Containers: 441 lb drums; 2000 and 4000 gal tank trucks
References: Bulletin 22-291

Dow Corning 197 Surfactant. Dow Corning Corp.
Chemical Description: Nonhydrolyzable silicone
Category: Profoamer
Applications: Profoamer for high- and low-density rigid urethane foams with good system compatibility, low freeze point, excellent emulsifying properties
Form: Thin liquid; Color: Clear yellow
Properties: Refractive Index: 1.446; Sp. gr.: 1.05; Visc.: 300 cst; Flash Pt.: 120 F
Toxicity: Skin and eye irritant; prolonged inhalation of vapors may be harmful
Containers: 40 and 440 lb containers
Warnings and Caution: Combustible; keep away from heat and flame. Use with adequate ventilation. Avoid prolonged breathing of vapors and eye and skin contact
References: Bulletin 22-249a

Dow Corning 198 Surfactant. Dow Corning Corp.
Chemical Description: Silicone glycol copolymer
Applications: High processing latitude surfactant for high- and low-density flexible urethane foams, hot molding and filled foam operations
Form: Low viscosity liquid; Color: < 4 (Gardner)
Properties: Solubility: Sol. in water, polyol, fluorocarbon, or water-amine streams; Sp. gr.: 1.035; Visc.: 1500 cst; Flash Pt.: > 150F (open cup)
Toxicity: Slight eye irritant; prolonged skin contact may cause moderate irritation

Containers: 441 lb drums, 2000 and 4000 gal tank trucks
References: Bulletin 22-301

Dow Corning 470A Fluid. Dow Corning
Chemical Description: Silicone glycol copolymer
Category: Nonionic emulsifier, emulsifier additive, wetting agent, surface tension depressant
Applications: Used to make powders more wettable and dispersible; used to treat gels, thickeners, clays, powdered gums, dyes, pigments and water-dispersible powders; emulsifier additive; polish additive; glass polish ingredient; soap and detergent additive; cosmetic and aerosol shave ingred.
Form: Liquid; Color: Medium straw
Properties: Refractive Index: 1.45; Solubility: Water sol. @ < 160 F; Sp. gr.: 1.075; Visc.: 275 cst; Flash Pt.: 500 F
Composition: 100% solids
Containers: 1 gal pails, 5, 55 gal drums
References: Bulletin 22-205

Dow Corning 1248 Fluid (formerly Q4-3558). Dow Corning Corp.
Chemical Description: Silicone polycarbinols (graft copolymer type)
Applications: Improves properties of organics such as release, abrasion resistance, lubricity, water repellency, softening and flexibility, low- and high-temperature flexibility; improves properties of urethanes
Form: Liquid; Color: Clear
Properties: Refractive Index: 1.4110; Solubility: Sol. in ethanol, isopropanol, methylene-chloride, n-hexane, toluene (@ 10%); Visc.: 160 cs; Flash Pt.: 250 F; Pour Pt.: –130 F
Containers: 5 and 55 gal containers
References: Product Information

Dow Corning 1250 Surfactant. Dow Corning Corp.
Chemical Description: Silicone in solvent (xylene)
Category: Silicone profoamer
Applications: Mechanical frothing of vinyl plastisols into a uniform fine-celled foam with better water resistance; allow foam fusing at low temperatures
Form: Thin liquid; Color: Very light straw
Properties: Sp. gr.: 1.00; Visc.: 6 cst; Flash Pt.: 85 F (open cup)
Composition: 50% silicone content
Toxicity: Skin irritant on prolonged contact; avoid prolonged breathing of vapors
Containers: 8, 40, 440 lb containers
Warnings and Caution: Solvent is combustible; keep away from heat and open flame; use with adequate ventilation. Avoid prolonged contact with skin and breathing of vapors
References: Bulletin 22-008c
Patent: U.S. 3,511,788

Dow Corning 1251 Surfactant. Dow

Corning Corp.
Chemical Description: Silicone in solvent (xylene)
Category and Applications: See Dow Corning 1250 Surfactant
Form: Thin liquid; Color: Very light straw
Properties: Sp. gr.: 1.00; Visc.: 4 cst; Flash Pt.: 85 F (open cup)
Composition: 50% silicone content
Toxicity: Skin irritant on prolonged contact
Containers: 8, 40, 440 lb containers
Warnings and Caution: Combustible solvent; keep away from heat and open flame; use with adequate ventilation; avoid prolonged contact with skin or breathing of vapors
Patent: U.S. 3,511,788

Dow Corning 1252 Surfactant. Dow Corning Corp.
Chemical Description: Silicones in solvent (texanol isobutyrate)
Category and Applications: See Dow Corning 1250 Surfactant
Form: Thin liquid; Color: Straw
Properties: Sp. gr.: 1.06; Visc.: 250 cs; Flash Pt.: 135 F (open cup)
Composition: 50% silicone content
Toxicity: Skin irritant on prolonged content
Containers: 8, 40, 440 lb containers
Warnings and Caution: Combustible solvent; keep away from heat and open flame; use with adequate ventilation. Avoid prolonged contact with skin or inhalation of vapors
References: Bulletin 22-008c
Patent: U.S. 3,511,788

Dow Corning 1253 Surfactant. Dow Corning Corp.
Chemical Description: Silicones in solvent (texanol isobutyrate)
Category and Applications: See Dow Corning 1250 Surfactant
Form: Thin liquid; Color: Straw
Properties: Sp. gr.: 1.07; Visc.: 250 cs; Flash Pt.: 135 F
Composition: 50% silicone content
Toxicity: Skin irritant on prolonged contact
Containers: 8, 40, 440 lb containers
Warnings and Caution: Combustible solvent; keep away from heat and open flame. Avoid prolonged breathing or contact with skin
References: Bulletin 22-008c
Patent: U.S. 3,511,788

Dow Corning FF-400. Dow Corning Corp.
Chemical Description: Organosiloxane
Category: Fiber finish
Applications: Textile finishes with high heat stability, high lubrication at low add-on; filament spin finishes; staple fiber finishes; knitting yarn lubricants; coning oil ingredients; texturizing finishes; additives to organic finishes; thread finishes
Form: Fluid; Color: Light amber

Properties: Solubility: Sol. in polar solvents such as isopropanol, low m.w. alcohols, glycol ethers; water-dispersible; Sp. gr.: 1.022; Visc.: 300 cs; Flash Pt.: 177 C (open cup)
Composition: 100% active ingred.
References: Bulletin 25-008

Dow Corning FF-412. Dow Corning Corp.
Chemical Description: Organopolysiloxane
Category and Applications: See Dow Corning FF-400
Form: Fluid; Color: Light amber
Properties: Solubility: Sol. in low m.w. alcohols, glycol ethers; dispersible in water; Sp. gr.: 1.040; Visc.: 200 cs; Flash Pt.: 177 C (open cup)
Composition: 100% active ingred.
References: Bulletin 25-008

Dow Corning FF-414. Dow Corning Corp.
Chemical Description: Organopolysiloxanes
Category and Applications: See Dow Corning FF-400
Form: Fluid; Color: Light amber
Properties: Solubility: Sol. in water, alcohols, glycol ether; Sp. gr.: 1.070; Visc.: 465 cs; Flash Pt.: 177 C (open cup)
Composition: 100% active ingred.
References: Bulletin 25-008

Dow Corning Q2-8026 Fluid. Dow Corning Corp.
Chemical Description: Silicone polycarbinols (graft copolymer type)
Applications: See Dow Corning 1248 Fluid
Form: Liquid; Color: Clear straw
Properties: Refractive Index; 1.45; Solubility: Sol. in water, ethanol, isopropanol, methylene-chloride, toluene; dispersible in n-hexane (@ 10%); Visc.: 465 cs; Flash Pt.: 400 F; Pour Pt.: 60 F
Toxicity: Slight eye irritant on direct contact
Containers: 5 and 55 gal containers

Dow Corning Q4-3557 Fluid. Dow Corning Corp.
Chemical Description: Silicone polycarbinols (graft copolymer type)
Applications: See Dow Corning 1248 Fluid
Form: Liquid; Color: Clear straw
Properties: Refractive Index: 1.43; Solubility: Sol. in ethanol, isopropanol, methylene-chloride, n-hexane, toluene; dispersible in mineral oil (@ 10%); Visc.: 40 cs; Flash Pt.: > 200 F; Pour Pt.: −80 F
Containers: 5 and 55 gal containers

Dow Corning Q4-3667 Fluid. Dow Corning Corp.
Chemical Description: Silicone polycarbinols (ABA type)
Applications: See Dow Corning 1248 Fluid
Form: Liquid; Color: Clear straw
Properties: Refractive Index: 1.44; Solubility: Sol. in ethanol, isopropanol, methylene-chloride, toluene; dispersible in water, n-hexane

(@ 10%); Visc.: 320 cs; Flash Pt.: 285 F; Pour Pt.: 65 F
Containers: 5 and 55 gal containers

Dowfax 2A1. Dow Corning Corp.
Chemical Description: Sodium dodecyl diphenyl ether disulfonate
Category: Detergent, emulsifier, wetting agent, solubilizer
Applications: Emulsion polymerization, dyeing assistant; primary emulsifier for SB & ABS polymers; leveling agent for acid dyeing of nylon; phenolic solubilizer for disinfectants; as detergent for dyeing, mines and minerals, cleaners of street, buildings, hospitals; wetting agent in electroplating; preparation of emulsified paints, bitumen emulsions
Form: Liquid; Color: Amber
Properties: Sp. gr.: 1.16; Visc.: 145 cps @ 0.10%; Surface Tension: 31 dynes/cm @ 0.10%
Composition: 45% active ingred.
Toxicity: Causes eye burns; may cause skin irritation; toxic to fish
Containers: 55 gal drum and bulk

Dowfax 2A0. Dow
Chemical Description: Dodecyl diphenyl-ether disulfonic acid
Category: Intermediate
Applications: Used to prepare non-sodium salts of Dowfax 2A1 for special applications
Form: Liquid; Color: Black
Properties: Sp. gr.: 1.11; Visc.: 510 cps; Surface Tension: 30 dynes/cm @ 0.10%
Composition: 40% active ingred.
Reaction: Acid
Toxicity: Irritant to skin, hair, fingers, eyes
Containers: 55 gal drum and bulk
Warnings and Caution: Strong acid—causes burns, toxic to fish

Dowfax 3B2. Dow Corning Corp.
Chemical Description: Sodium n-decyl diphenyl-ether disulfonate
Category: Detergent, wetting agent, emulsifier, solubilizer
Applications: Pesticide spray adjuvant, esp. for methane arsonate herbicides; phenolic solubilization for disinfectants; detergent in strong alkaline/acid cleaners, cement, transport and industrial cleaners; preparation of emulsified paints
Form: Liquid; Color: Red-brown
Properties: Sp. gr.: 1.17; Visc.: 120 cps; Surface Tension: 38 dynes/cm @ 0.10%
Composition: 45% active ingred.
Toxicity: Causes eye burns; may cause skin irritation; toxic to fish
Containers: 22 gal drum and bulk

Dowfax 9N6. Dow Chem. Europe
Category: Detergent, emulsifier, wetting agent, dispersing agent
Applications: Household wax and polishes; solvent based cleaners; emulsifier and wetting agent for pesticides; emulsifier for metal cleaners, drycleaning "soap," household-type wax and polish coemulsifier; detergent in metallurgy and household products; dispersing waxes, pigments, resins
Form: Liquid; Color: Colorless
Properties: Refractive Index:

1.486; Solubility: Water insol.; Sp. gr.: 1.04; Visc.: 220 cst; Stability: Good
Composition: 100% active ingred.
Reaction: Neutral
Toxicity: Slightly irritating to eyes
Containers: Bulk, drums
Warnings and Caution: Avoid eye contact and ingestion

Dowfax 9N9. Dow Chem. Europe
Category: Detergent, emulsifier, wetting agent
Applications: Household liquid cleaners; industrial detergent; leather and laundering industries; emulsifier for latex paints; textile processing aid; paper industry processing aid; fat liquoring; preparation of emulsified paints, bitumen emulsions; wetting agent for desizing and boiling off cottons, carbonizing woolens, abatement of dust, wetting pulp in beater, insecticides
Form: Liquid; Color: Slightly yellow
Properties: Refractive Index: 1.480; Solubility: Water sol.; Sp. gr.: 1.06; Visc.: 270 cst; Stability: Stable; Surface Tension: 33 dynes/cm
Composition: 100% active ingred.
Reaction: Neutral
Toxicity: Slightly irritating to eyes
Containers: Bulk, drrums
Warnings and Caution: Avoid eye contact and ingestion

Dowfax 9N30. Dow Chem. Europe
Category: Detergent, emulsifier, wetting agent
Applications: Heavy duty detergents; detergent and wetting agent in system involving near boiling or boiling temperatures and/or high concentrations of electrolytes; dispersing agent; preparation of emulsified paints, bitumen emulsions
Form: Solid; Color: White
Properties: Ref. Index: 1.470; Solubility: Water sol.; Sp. gr.: 1.05; M.P.: 49 C; Stability: Stable; Surface Tension: 42 dynes/cm
Composition: 100% active ingred.
Reaction: Neutral
Toxicity: Slight eye irritant
Containers: Bulk, drums
Warnings and Caution: Avoid eye contact and ingestion

Dowfax 10-5. Dow Chem. Europe
Category: Detergent, wetting agent, dispersing agent
Applications: Liquid detergents for household and industrial use; dry cleaning formulations; textiles, leather, laundering applications as detergents; desizing and boiling off cottons, removal of spots, abatement of dust, insecticides as wetting agent
Form: Liquid; Color: Colorless
Properties: Refractive Index: 1.439; Solubility: Water sol.; Sp. gr.: 0.97; Visc.: 29 cst; M.P.: 5 C; Stability: Stable; Surface Tension: 28 dynes/cm
Composition: 100% active ingred.
Reaction: Neutral
Toxicity: Slight eye irritant
Containers: Bulk, drums
Warnings and Caution: Avoid eye

contact and ingestion

Dowfax 10-6. Dow Chem. Europe
Category: Detergent, wetting agent
Applications: Liquid detergents for household and industrial use; dry cleaning formulations; textiles, leather, laundering uses as detergents; desizing and boiling off cottons, removal of spots, abatement of dust, insecticides as wetting agents
Form: Liquid; Color: Colorless
Properties: Refractive Index: 1.443; Solubility: Water sol.; Sp. gr.: 0.98; Visc.: 50 cst; M.P.: 8 C; Stability: Stable; Surface Tension: 28 dynes/cm
Composition: 100% active ingred.
Reaction: Neutral
Toxicity: Slight eye irritant
Containers: Bulk, drums
Warnings and Caution: Avoid eye contact and ingestion

Dowfax 12-3. Dow Chem. Europe
Category: Detergent, emulsifier, raw material for ethersulfates
Applications: Raw material for cosmetic grade ethersulfates for shampoos and bubble bath; quench oil additive; automotive engine cleaner; mineral oil emulsions
Form: Liquid; Color: Colorless
Properties: Refractive Index: 1.439; Solubility: Water insol.; Sp. gr.: 0.93; Visc.: 30 cst; Stability: Stable
Composition: 100% active ingred.
Reaction: Neutral
Toxicity: Slight eye irritant
Containers: Bulk, drums
Warnings and Caution: Avoid eye contact and ingestion

Dowfax 14-3. Dow Chem. Europe
Category: Detergent, emulsifier
Applications: Raw material for detergent grade ethersulfates for liquid detergents, quench oil additive, automotive engine cleaner, mineral oil emulsions
Form: Liquid; Color: Colorless
Properties: Refractive Index: 1.438; Solubility: Water insol.; Sp. gr.: 0.925; Visc.: 35 cst; Stability: Stable
Composition: 100% active ingred.
Reaction: Neutral
Toxicity: Slight eye irritant
Containers: Bulk, drums
Warnings and Caution: Avoid eye contact and ingestion

Dowfax 14-12. Dow Chem. Europe
Category: Detergent, wetting agent
Applications: Heavy duty detergents; as detergent and wetting agent in systems involving boiling or near boiling temperatures and/or high concentrations of electrolytes; as detergents in textiles, laundering, household applications
Form: Paste; Color: Slightly yellow
Properties: Refractive Index: 1.449; Solubility: Water sol.; Sp. gr.: 0.98; M.P.: 43 C; Stability: Stable; Surface Tension: 32 dynes/cm
Composition: 100% active ingred.
Reaction: Neutral
Toxicity: Slight eye irritant

Containers: Bulk, drums
Warnings and Caution: Avoid eye contact and ingestion

Dowfax 18-11. Dow Chem. Europe
Category: Detergent, wetting agent, emulsifier
Applications: Heavy duty detergents; formulations of low foaming heavy duty detergent for textiles, laundering and household uses; emulsifier for cosmetics and pharmaceuticals
Form: Solid; Color: White
Properties: Refractive Index: 1.450; Solubility: Water sol.; Sp. gr.: 0.97; M.P.: 33 C; Stability: Stable; Surface Tension: 36 dynes/cm
Composition: 100% active ingred.
Reaction: Neutral
Toxicity: Slight eye irritant
Containers: Bulk, drums
Warnings and Caution: Avoid eye contact and ingestion

Dowfax 18-25. Dow Chem. Europe
Category: Detergent
Applications: Heavy duty detergents; detergent and wetting agent in systems involving near boiling or boiling temperatures and/or high concentrations of electrolytes; detergents for textiles, laundering and household uses
Form: Solid; Color: White
Properties: Refractive Index: 1.453; Solubility: Water sol.; Sp. gr.: 1.01; M.P.: 50 C; Stability: Stable; Surface Tension: 45 dynes/cm
Reaction: Neutral
Toxicity: Slight eye irritant
Containers: Bulk, drums
Warnings and Caution: Avoid eye contact and ingestion

Dowfax 63N10. Dow Chem. Europe
Category: Detergent, defoamer, demulsifier
Applications: As detergent for dishwashing machines; heavy duty liquid detergents; defoamer for anionic detergent slurries, for breaking oil emulsions; as detergent for textiles, laundering, foods, household uses; dispersing waxes, pigments, resins
Form: Liquid; Color: Colorless
Properties: Solubility: Slightly water soluble; Sp. gr.: 1.02; Visc.: 340 cst; Stability: Stable; Surface Tension: 42 dynes/cm
Composition: 100% active ingred.
Reaction: Neutral
Toxicity: Slight eye irritant
Containers: Bulk, drums
Warnings and Caution: Avoid eye contact and ingestion

Dowfax 70N14D. Dow Chem. Europe
Category: Demulsifier
Applications: Crude oil demulsifier; break of crude oil emulsions
Form: Liquid; Color: Slightly yellow
Properties: Solubility: Slightly sol. in water; dispersible in water; Sp. gr.: 0.95; Visc.: 25 cst; F.P.: –50 C; B.P.: 64 C; Stability: Stable
Composition: 65% active ingred.
Reaction: Neutral
Toxicity: To be determined; eye irritant

Containers: Bulk
Warnings and Caution: Fire hazard; vapor-air mixture can be explosive

Dresinate 81 Liquid Soap. Hercules
Chemical Description: Modified aqueous solution of sodium soap
Category: Emulsifier ingredient
Applications: Emulsifier ingred. and/or stabilizer in solvent cleaners and soluble oils for metal working, disinfectants, oil-well drilling muds, drawing and grinding compounds; as plasticizing agents in specialty coatings
Form: Liquid; Color: Pale
Properties: Solubility: Dilutable with water and aqueous solutions of alcohols and glycols; Density: 8.7 lb/gal @ 21 C; Visc.: 5.7 poises @ 140 F
Composition: 87% solids content
Containers: Bung-type and open-head drums (465 lb net wt)
References: PD No. 7246

Dresinate 90 Liquid Soap. Hercules
Chemical Description: Modified aqueous solution of potassium soap
Category and Applications: See Dresinate 81
Form: Liquid; Color: Pale
Properties: Solubility: Dilutable with water and aqueous solutions of alcohols and glycols; Density: 9.0 lb/gal @ 21 C; Visc.: 14.7 poises @ 140 F
Composition: 90% solids content
References: PD No. 7246

Dresinate 91 Liquid Soap. Hercules
Chemical Description: Modified aqueous solution of potassium soap
Category and Applications: See Dresinate 81
Form: Liquid; Color: Pale
Properties: Solubility: Dilutable with water and aqueous solutions of alcohols and glycols; Density: 8.8 lb/gal @ 21 C; Visc.: 5.3 poises @ 140 F
Composition: 88% solids content
References: PD No. 7246

Dresinate 92 Liquid Soap. Hercules
Chemical Description: Modified aqueous solution of potassium soap
Category and Applications: See Dresinate 81
Form: Liquid; Color: Dark
Properties: Solubility: Dilutable with water and aqueous solutions of alcohols and glycols; Density: 9.0 lb/gal @ 21 C; Visc.: 9.2 poises @ 140 F
Composition: 88% solids
References: PD No. 7246

Dresinate 93 Liquid Soap. Hercules
Chemical Description: Modified aqueous solution of potassium soap
Category and Applications: See Dresinate 81
Form: Liquid; Color: Dark
Properties: Solubility: Dilutable with water and aqueous solutions of alcohols and glycols; Density: 9.5 lb/gal @ 21 C; Visc.: 7.3 poises @ 140 F
Composition: 87% solids content
References: PD No. 7246

Dresinate 94 Liquid Soap. Hercules
Chemical Description: Modified aqueous solution of potassium soap
Category and Applications: See Dresinate 81
Form: Liquid; Color: Pale
Properties: Solubility: Dilutable with water and aqueous solutions of alcohols and glycols; Density: 9.3 lb/gal @ 21 C; Visc.: 12.4 poises @ 140 F
Composition: 87% solids
References: PD No. 7246

Dresinate 95 Liquid Soap. Hercules
Chemical Description: Modified aqueous solution of potassium soap
Category and Applications: See Dresinate 81
Form: Liquid; Color: Dark
Properties: Solubility: Dilutable with water and aqueous solutions of alcohols and glycols; Density: 9.1 lb/gal; Visc.: 4.5 poises @ 140 F
Composition: 88% solids
References: PD No. 7246

Dresinate 214 Paste Soap. Hercules
Chemical Description: Potassium soap of a pale modified rosin
Category: Emulsifier, wetting agent, foaming agent, dispersant
Applications: Used in the manufacture of adhesives, polymer emulsions, synthetic latices; as sizing agents; in foaming and flotation operations; for pigment wetting and dispersing
Form: Paste
Properties: Solubility: Dilutable with water and aqueous solutions of alcohols and glycols; Density: 9.22 lb/gal @ 60 C; Visc.: 23 poises @ 140 F; Stability: Tends to separate or stratify on prolonged standing
Composition: 80% solids content
Containers: Open-head metal drums (475 lb net wt)
References: PD No. 7247-1

Dresinate 515 Paste Soap. Hercules
Chemical Description: Potassium soap of a pale modified rosin
Category and Applications: See Dresinate 214; Dresinate 515 is a nonstratifying soap recommended to replace Dresinate 214
Form: Paste
Properties: Solubility: Dilutable with water and aqueous solutions of alcohols and glycols; Density: 9.24 lb/gal @ 60 C; Visc.: 64 poises @ 140 F
Composition: 80% solids
Containers: Open-head metal drums (475 lb net wt)
References: PD No. 7247-1

Dresinate 731 Paste Soap. Hercules
Chemical Description: Sodium soap of a pale modified rosin
Category and Applications: See Dresinate 214
Form: Paste
Properties: Solubility: Dilutable with water and aqueous solutions of alcohols and glycols; Density: 9.1 lb/gal @ 60 C; Visc.: 11 poises @ 140 F
Composition: 70% solids
Containers: Open-head metal drums (475 lb net wt)

References: PD No. 7247-1

Dresinate TX Dry Powder Soap. Hercules
Chemical Description: Sodium soap of a tall oil
Category and Applications: See Dresinate X
Form: Free-flowing powder; Color: Dark brown
Properties: Solubility: Sol. in water and aqueous solutions of alcohols and glycols; Density: 26 lb/ ft^3
Composition: 96% min. solids
Containers: Multiwall moisture vapor-resistant paper bags (50 lb net wt)
References: PD No. 7248-1

Dresinate X Dry Powder Soap. Hercules
Chemical Description: Sodium soap of a pale rosin
Category: Emulsifier, detergent aid, dispersant, stabilizer
Applications: As ingredients of highly built alkaline metal cleaners and other industrial cleaners, paintbrush cleaners, driveway cleaners, household cleaners, scouring powders; as dispersants for pigments, in the production of resinated colors; as stabilizers for synthetic rubber latices, conditioners for sulfur dusts; as emulsifiers for vegetable and mineral oils, pine oil and asphalt
Form: Free-flowing powder; Color: Pale cream
Properties: Solubility: Sol. in water and aqueous solutions of alcohols and glycols; Density: 26 lb/ ft^3
Composition: 95% min. solids
Containers: Multiwall moisture vapor-resistant paper bags (50 lb net wt)
References: PD No. 7248-1

Dresinate XX Dry Powder Soap. Hercules
Chemical Description: Sodium soap of a dark rosin
Category and Applications: See Dresinate X
Form: Free-flowing powder; Color: Light brown
Properties: Solubility: Sol. in water and aqueous solutions of alcohols and glycols; Density: 28 lb/ ft^3
Composition: 96% min. solids
Containers: Multiwall moisture vapor-resistant paper bags (50 lb net wt)
References: PD No. 7248-1

Drew BRX-354. Drew Produtos Quimicos Ltda.
Chemical Description: Coco amide propyl amine oxide
Category: Detergent, nonionic
Applications: Bubble bath and acid shampoos formulations
Form: Liquid
Composition: 30% active

Drewclad 2. PVO International
Category: Mothproofing agent
Applications: Permanent treatment for wool; for spray-on application in acid systems to raw stock dyed wool

Form: Liquid
Properties: Stability: Very stable to strong acids; Emulsifiability: Self-emulsifiable
Composition: 20% active ingred.

Drewclad 33. PVO International
Category: Mothproofing agent
Applications: Permanent treatment for wool; for use in the dye bath; resistant to extraction by laundering or drycleaning
Form: Liquid
Composition: 20% active ingred.

Drewclad C/F. PVO International
Category: Mothproofing agent
Applications: Permanent treatment for wool; stable and resistant to extraction
Form: Liquid

Drewfax 220. Drew Produtos Quimicos Ltda.
Chemical Description: Sodium alkil suf.
Category: Wetting agent, emulsifier, anionic
Applications: Wetting agent and emulsifier for polymerization
Form: Liquid
Composition: 29–30% active

Drewfax 250. Drew Produtos Quimicos Ltda.
Chemical Description: Sodium alkil suf.
Category: Wetting agent, anionic
Applications: Wetting agent for pesticides
Form: Powder
Composition: 90% active

Drewfax 502. Drew Produtos Quimicos Ltda.
Chemical Description: Sulfosuccinate derivative
Category: Wetting agent, emulsifier, anionic and nonionic
Applications: Wetting agent and emulsifier for polymerization and flowable pesticides
Form: Liquid

Drewfax 528. Drew Produtos Quimicos Ltda.
Chemical Description: Dioctyl ester of sodium sulfosuccinate
Category: Wetting agent, emulsifier, anionic
Applications: Wetting agent and emulsifier for polymerization
Form: Paste
Composition: 70% active

Drewfax 539. Drew Produtos Quimicos Ltda.
Chemical Description: Sulfosuccinate derivative
Category: Wetting agent, emulsifier, anionic and nonionic
Applications: Wetting agent and emulsifier for polymerization and flowable pesticides
Form: Liquid

Drewfax 549. Drew Produtos Quimicos Ltda.
Chemical Description: Sulfosuccinate derivative
Category: Wetting agent, emulsifier, anionic
Applications: Wetting agent and emulsifier for polymerization and flowable pesticides
Form: Liquid

Drewfax 550. Drew Produtos Quimicos Ltda.
Chemical Description: Sulfosuccinate derivative
Category: Wetting agent, emulsifier, anionic
Applications: Wetting agent and emulsifier for polymerization and flowable pesticides
Form: Liquid

Drewfax 560. Drew Produtos Quimicos Ltda.
Chemical Description: Sulfosuccinate derivative
Category: Wetting agent, emulsifier, anionic
Applications: Wetting agent and emulsifier for polymerization and flowable pesticides
Form: Powder

Drewfax 577. Drew Produtos Quimicos Ltda.
Chemical Description: Sulfosuccinate derivative
Category: Wetting agent, emulsifier, anionic
Applications: Wetting agent and emulsifier for polymerization and flowable pesticides
Form: Powder

Drewlate 30. PVO International
Chemical Description: Acetylated tartrated monoglyceride of vegetable fat
Category: Food emulsifier
Form: Syrupy liquid; Color: Amber

Drewlene 10. PVO International
Chemical Description: Propylene glycol monostearate
Category: Food emulsifier
Applications: Used to increase whipping ability and aeration in cake batters, icings and toppings
Form: Solid, flake or bead; Color: Yellow
Properties: M.P.: 110 F; HLB: 3.4; Sapon. Value: 170–190

Drewmulse 3-1-O. PVO International
Chemical Description: Triglycerol monooleate
Category: Nonethoxylated emulsifiers, solubilizers, dispersants, nonionic
Applications: W/O and O/W emulsions; emollient creams and lotions; clear emulsions; pharmaceutical vehicles; cosmetics; FDA approved for internal use (degraded by the body to glycerine and fatty acid)
Form: Liquid; Color: 8 (Gardner)
Properties: Solubility: Sol. in isopropanol, peanut oil, mineral oil; dispersible in water; HLB: 7.0; Sap. Value: 125–150
References: PVO brochure

Drewmulse 3-1-S. PVO International
Chemical Description: Triglycerol monostearate
Category and Applications: See Drewmulse 3-1-O
Form: Solid; Color: 10 (Gardner)
Properties: Solubility: Dispersible in water, isopropanol, propylene glycol, peanut oil, mineral oil; HLB: 7.0; Sap. Value: 120–140
References: PVO brochures

Drewmulse 6-2-S. PVO International
Chemical Description: Hexaglycerol distearate
Category and Applications: See Drewmulse 3-1-O
Form: Solid; Color: 13 (Gardner)
Properties: Solubility: Sol. hot in isopropanol, peanut and mineral oils; disp. in propylene glycol; HLB: 8.0; Sap. Value: 105–125
References: PVO brochures

Drewmulse 10. PVO International
Chemical Description: Glycerol mono fat from soya oil
Category: Lipophilic emulsifier
Applications: Various food industry uses as emulsifier, dispersing aid, antistaling agent; improves stability and shelf life, chewing or flow properties; aids aeration; dough conditioners; antistick agent
Form: Plastic; Color: Cream
Properties: M.P.: 118 F; HLB: 2.8; Sap. Value: 170–190
Composition: 40% alpha mono
References: PVO brochures

Drewmulse 10-4-O. PVO International
Chemical Description: Decaglycerol tetraoleate
Category and Applications: See Drewmulse 3-1-O
Form: Liquid; Color: 8 (Gardner)
Properties: Solubility: Sol. in isopropanol, peanut and mineral oils; disp. in water; HLB: 6.0; Sap. Value: 125–145
References: PVO brochures

Drewmulse 10-8-O. PVO International
Chemical Description: Decaglycerol octaoleate
Category and Applications: See Drewmulse 3-1-O
Form: Liquid; Color: 8 (Gardner)
Properties: Solubility: Sol. in peanut and mineral oils; HLB: 4.0; Sap. Value: 155–175
References: PVO brochures

Drewmulse 10-10-O. PVO International
Chemical Description: Decaglycerol decaoleate
Category and Applications: See Drewmulse 3-1-O
Form: Liquid; Color: 8 (Gardner)
Properties: Solubility: Sol. in peanut and mineral oils; HLB: 3.0; Sap. Value: 165–185
References: PVO brochures

Drewmulse 10-10-S. PVO International
Chemical Description: Decaglycerol decastearate
Category and Applications: See Drewmulse 3-1-O
Form: Solid; Color: 8 (Gardner)
Properties: Solubility: Sol. hot in peanut and mineral oils; slightly sol. in propylene glycol; HLB: 3.0; Sap. Value: 160–180
References: PVO brochures

Drewmulse 15. PVO International
Chemical Description: Glycerol mono cottonseed oil
Category and Applications: See Drewmulse 10
Form: Soft solid; Color: Cream

Properties: M.P.: 110 F
Composition: 40% alpha mono
References: PVO brochures

Drewmulse 20/200/V. PVO International
Chemical Description: Glycerol monostearate
Category and Applications: See Drewmulse 10
Form: Beads; Color: Cream
Properties: M.P.: 140 F
Composition: 40% alpha mono
References: PVO brochures

Drewmulse 21. PVO International
Chemical Description: Glycerol monostearate from soya oil
Category and Applications: See Drewmulse 10
Form: Bead or flake; Color: Cream
Properties: HLB: 8.4; Sap. Value: 160–180
Composition: 30% alpha mono
References: PVO brochures

Drewmulse 30/900. PVO International
Chemical Description: Glycerol monostearate
Category and Applications: See Drewmulse 10
Form: Beads; Color: Cream
Properties: M.P.: 138 F
Composition: 52% alpha mono
References: PVO brochures

Drewmulse 50. PVO International
Chemical Description: Glycerol mono fat
Category and Applications: See Drewmulse 10
Form: Cream; Color: Cream
Composition: 52% alpha mono
References: PVO brochures

Drewmulse 55. PVO International
Chemical Description: Glycerol mono shortening from tallow and soya
Category and Applications: See Drewmulse 10
Form: Plastic; Color: Cream
Properties: Tallow: M.P.: 130–135 F; HLB: 3.4; Sap. Value: 165–185; Soya: M.P.: 116–122 F; HLB: 2.4
Composition: 52% alpha mono
References: PVO brochures

Drewmulse 70. Drew Produtos Quimicos Ltda.
Chemical Description: Blend of glycerol monoesters
Category: Emulsifier, nonionic
Applications: Food emulsifier, baking additives, margarine, and shortening emulsifiers
Form: Liquid
Composition: 13.5–19% active

Drewmulse 75. Drew Produtos Quimicos Ltda.
Chemical Description: Sorbitan ester

Category: Emulsifier, nonionic
Applications: See Drewmulse 70
Form: Paste
Composition: 100% active

Drewmulse 85. PVO International
Chemical Description: Glycerol monooleate from tallow and soya
Category and Applications: See Drewmulse 10
Form: Semiliquid; Color: Light yellow
Properties: HLB: 3.4; Sap. Value: 160–180
Composition: 40% alpha mono
References: PVO brochure

Drewmulse 200. PVO International
Chemical Description: Glycerol monostearate from soya and tallow
Category and Applications: See Drewmulse 10
Form: Beads; Color: Cream
Properties: M.P.: 135–145 F; HLB: 2.8; Sap. Value: 154–170 (soya), 163–177 (tallow)
Composition: 40% alpha mono
References: PVO brochures

Drewmulse 300. PVO International
Chemical Description: Glycerol monostearate (80%) and polysorbate (20%) blend from soya and tallow oils
Category and Applications: See Drewmulse 10
Form: Beads; Color: Cream
Properties: M.P.: ≈ 135 F; HLB: 4.3
Composition: 30% alpha mono
References: PVO brochures

Drewmulse 365. Drew Productos Quimicos Ltda.
Chemical Description: Glycerol monostearate (67%) and polysorbate 65 (33%) blend from soya and tallow oils
Category and Applications: See Drewmulse 10
Form: Beads; Color: Cream
Properties: M.P.: ≈ 136 F; HLB: 5.3
Composition: 25–30% alpha mono
References: PVO brochures

Drewmulse 700. PVO International
Chemical Description: Glycerol monostearate (80%) and polysorbate 80 (20%) blend from soya and tallow oils
Category and Applications: See Drewmulse 10
Form: Beads; Color: Cream
Properties: M.P.: ≈ 135 F; HLB: 5.2
Composition: 30% alpha mono
References: PVO brochures

Drewmulse 900. PVO International
Chemical Description: Glycerol monostearate
Category and Applications: See Drewmulse 10
Form: Beads; Color: Cream
Properties: M.P.: ≈ 138 F; HLB: 3.8; Sap. Value: 160–180
Composition: 52% alpha mono
References: PVO brochures

Drewmulse 1128. PVO International
Chemical Description: Glycerol monostearate and polyoxyethylene propylene glycol stearate

Category and Applications: See Drewmulse CNO
Form: Flake; Color: 2 (Gardner)
Properties: HLB: 8.4; Stability: Acid stable; Emulsifiability: Self-emulsifying
Composition: 20% mono

Drewmulse 1129. PVO International
Chemical Description: Glycerol monostearate and polyoxyethylene stearate
Category and Applications: See Drewmulse CNO
Form: Flake
Properties: HLB: 11.0; Stability: Acid stable; Emulsifiability: Self-emulsifying
Composition: 20% mono

Drewmulse CNO. PVO International
Chemical Description: Glycerol monococate
Category: Emulsifier, stabilizer, opacifier, viscosity builder
Applications: Emulsifier, stabilizer and viscosity builder in cosmetics and pharmaceuticals; opacifier; bases for creams and lotions; cream rinses; hair conditioners
Form: Solid; Color: 4 (Gardner)
Properties: HLB: 3.8
Composition: 50% mono

Drewmulse DGMS. Drew Produtos Quimicos Ltda.
Chemical Description: Diethylene glycol monostearate
Category: Emulsifier, opacifier and pearlescence agents, nonionic
Applications: Emulsifier for W/O and O/W systems; opacifier and pearlescence agents for shampoos, additives for pigment milling
Form: Flakes
Composition: 85% active

Drewmulse EGDS. Drew Produtos Quimicos Ltda.
Chemical Description: Ethylene glycol distearate
Category and Applications: See Drewmulse DGMS
Form: Flakes
Composition: 48% active

Drewmulse EGMS. Drew Produtos Quimicos Ltda.
Chemical Description: Ethylene glycol monostearate
Category and Applications: See Drewmulse DGMS
Form: Flakes
Composition: 48% active

Drewmulse GMC-8. PVO International
Chemical Description: Glycerol monocaprylate-caprate
Category and Applications: See Drewmulse CNO; also for food applications (see Drewmulse 10)
Form: Liquid; Color: 3 (Gardner)
Properties: HLB: 8.3
Composition: 75% mono

Drewmulse GMO. PVO International
Chemical Description: Glycerol monooleate
Category and Applications: See

Drewmulse CNO
Form: Liquid; Color: 6 (Gardner)
Properties: HLB: 3.4
Composition: 40% mono

Drewmulse GMRO. Drew Produtos Quimicos Ltda.
Chemical Description: Glycerol monoricinoleate
Category and Applications: See Drewmulse DGMS
Form: Liquid

Drewmulse GMS. Drew Produtos Quimicos Ltda.
Chemical Description: Glycerol monostearate
Category and Applications: See Drewmulse DGMS
Form: Flakes
Composition: 40–45% active

Drewmulse HM 100. PVO International
Chemical Description: Glycerol monostearate and polyoxyethylene stearate
Category and Applications: See Drewmulse CNO; self-emulsifying and acid stable; esp. effective in antiperspirant formulations
Form: Flake; Color: 2 (Gardner)
Properties: HLB: 8.4
Composition: 24% mono

Drewmulse M. Drew Produtos Quimicos Ltda.
Chemical Description: Blend of glycerol monoesters
Category: Emulsifier, nonionic
Applications: See Drewmulse 70
Form: Paste
Composition: 32% active

Drewmulse PGML. PVO International
Chemical Description: Propylene glycol monolaurate
Category: Chemical specialty
Form: Liquid

Drewmulse PGMS. Drew Produtos Quimicos Ltda.
Chemical Description: Propylene glycol monostearate
Category and Applications: See Drewmulse DGMS
Form: Flakes
Composition: 48% active

Drewmulse POE-SML. PVO International
Chemical Description: Polyoxyethylene 20 sorbitan monolaurate
Category: Nonionic emulsifiers, solubilizers, dispersants, wetting agents, detergents, viscosity control agents
Applications: W/O emulsifiers for cosmetics and pharmaceuticals; emollient creams and lotions; shave creams; fragrance and flavor solubilizers; shampoos; topical ointments; vitamins; germicides in aqueous systems
Form: Liquid
Properties: HLB: 16.7; Sap. Value: 40–51

Drewmulse POE-SMO. PVO International
Chemical Description: Polyoxyethylene 20 sorbitan monooleate
Category and Applications: See

Drewmulse POE-SML
Form: Liquid
Properties: HLB: 15.0; Sap. Value: 45–55

Drewmulse POE-SMS. PVO International
Chemical Description: Polyoxyethylene 20 sorbitan monostearate
Category and Applications: See Drewmulse POE-SML
Form: Solid
Properties: HLB: 14.9; Sap. Value: 45–55

Drewmulse POE-STS. PVO International
Chemical Description: Polyoxyethylene 20 sorbitan tristearate
Category and Applications: See Drewmulse POE-SML
Form: Solid
Properties: HLB: 10.5; Sap. Value: 88–98

Drewmulse PNO. Drew Produtos Quimicos Ltda.
Chemical Description: Blend of glycerol monoesters
Category: Emulsifier, nonionic
Applications: See Drewmulse 70
Form: Paste
Composition: 38–42% active

Drewmulse SML. PVO International
Chemical Description: Sorbitan monolaurate
Category and Applications: See Drewmulse POE-SML
Form: Liquid
Properties: HLB: 8.6; Sap. Value: 159–171

Drewmulse SMO. PVO International
Chemical Description: Sorbitan monooleate
Category and Applications: See Drewmulse POE-SML
Form: Liquid
Properties: HLB: 4.3; Sap. Value: 144–156

Drewmulse SMS. PVO International
Chemical Description: Sorbitan monostearate
Category and Applications: See Drewmulse POE-SML
Form: Solid
Properties: HLB: 4.7; Sap. Value: 147–157

Drewmulse STS. PVO International
Chemical Description: Sorbitan tristearate
Category and Applications: See Drewmulse POE-SML
Form: Solid
Properties: HLB: 2.1; Sap. Value: 170–190

Drewmulse TP. PVO International
Chemical Description: Glycerol monostearate (triple pressed)
Category and Applications: See Drewmulse CNO
Form: Flake; Color: 4 (Gardner)
Properties: HLB: 3.8
Composition: 40% mono

Drewmulse V. PVO International
Chemical Description: Glycerol monostearate (vegetable)

Category and Applications: See Drewmulse CNO
Form: Bead; Color: 6 (Gardner)
Properties: HLB: 3.8
Composition: 40% mono

Drewmulse V-SE. PVO International
Chemical Description: Glycerol monostearate (self-emulsifying)
Category and Applications: See Drewmule CNO
Form: Bead; Color: 6 (Gardner)
Properties: HLB: 8.4
Composition: 30% mono

Drewplast 017. PVO International
Category: Plastic additive, antistatic and slip agent
Applications: Antistatic and slip agent for PVC and polyolefin resin; esp. for PVC food wrap; FDA listed GRAS
Properties: Stability: Excellent heat stability

Drewplast 029. PVO International
Category: Plastic additive, antistatic agent
Applications: Antistatic agent for polyolefins and polystyrene; food grade, accepted by FDA as direct food additive

Drewplast 030. PVO International
Category: Plastic additive, antifogging agent, lubricant
Applications: Antifogging agent for PVC meat and produce wrap; internal lubricant
Properties: Stability: Excellent heat stability

Drewplast 050. Drew Produtos Quimicos Ltda.
Chemical Description: Glycerol ester
Category: Plasticizer, nonionic
Form: Paste

Drewplast 051. PVO International
Category: Plastic additive, lubricant
Applications: Antistatic internal lubricant with antifogging properties in PVC and polyolefins; FDA listed GRAS
Properties: Stability: Excellent heat stability

Drewplast 100. PVO International
Category: Plastic additive, antifogging agent
Applications: Economical antifogging agent for PVC food wrap; improves machinability of film or sheet in automatic wrapping machines; FDA approved

Drewplast 111. Drew Produtos Quimicos Ltda.
Chemical Description: Fatty ester
Category: Plasticizer, nonionic
Form: Liquid

Drewplast 175. Drew Produtos Quimicos Ltda.
Chemical Description: Fatty ester
Category: Plasticizer, nonionic
Form: Liquid

Drewpol 3-1-O. PVO International
Chemical Description: Triglycerol monooleate
Category: Food emulsifier

Applications: The Drewpols have various functions as food additives such as solubilizer for flavors and colors; improves dispersibility, shelf life, texture, volume, emulsion stability; provides gloss in confectionary coatings; imparts aeration, dryness; foam control
Form: Liquid; Color: Amber
Properties: HLB: 7.0; Sap. Value: 120–160

Drewpol 3-1-S. PVO International
Chemical Description: Triglycerol monostearate
Category and Applications: See Drewpol 3-1-O
Form: Beads; Color: Amber
Properties: HLB: 7.0; Sap. Value: 120–140

Drewpol 3-1-SH. PVO International
Chemical Description: Triglycerol mono shortening (Drewmulse 8731-S)
Category and Applications: See Drewpol 3-1-O
Form: Thick liquid; Color: Amber
Properties: HLB: 7.4; Sap. Value: 110–140

Drewpol 6-2-S. PVO International
Chemical Description: Hexaglycerol distearate
Category and Applications: See Drewpol 3-1-O
Form: Beads; Color: Amber
Properties: HLB: 4.0; Sap. Value: 100–130

Drewpol 10-4-O. PVO International
Chemical Description: Decaglycerol tetraoleate
Category and Applications: See Drewpol 3-1-O
Form: Liquid; Color: Amber
Properties: HLB: 6.0; Sap. Value: 125–145

Drewpol 10-8-O. PVO International
Chemical Description: Decaglycerol octaoleate
Category and Applications: See Drewpol 3-1-O
Form: Liquid; Color: Amber
Properties: HLB: 4.0; Sap. Value: 150–180

Drewpol 10-10-O. PVO International
Chemical Description: Decaglycerol decaoleate
Category and Applications: See Drewpol 3-1-O
Form: Liquid; Color: Amber
Properties: HLB: 3.0; Sap. Value: 160–190

Drewpol 10-10-S. PVO International
Chemical Description: Decaglycerol decastearate
Category and Applications: See Drewpol 3-1-O
Form: Solid; Color: Tan
Properties: HLB: 3.0; Sap. Value: 150–190

Drewpon 40. Drew Produtos Quimicos Ltda.
Chemical Description: Triethanolamine lauryl sulfate
Category: Emulsifier, detergent, wetting agent, anionic

Applications: Liquid and emulsion shampoos, bubble baths, rug shampoo, household cleaning products, emulsion polymerization, wetting agents, pigment dispersants, dishwashing detergents
Form: Liquid
Composition: 39–42% active

Drewpon 40LS. Drew Produtos Quimicos Ltda.
Chemical Description: Triethanolamine lauryl sulfate
Category: Emulsifier, detergent, wetting agent, anionic
Applications: See Drewpon 40
Form: Liquid
Composition: 38–40% active

Drewpon 100. Drew Produtos Quimicos Ltda.
Chemical Description: Sodium lauryl sulfate
Category: Detergent, wetting agent, anionic
Applications: Used as detergent, wetting dispersant for tooth pastes, shampoos, bubble baths, pesticides
Form: Powder
Composition: 90% active

Drewpon AC-50. Drew Produtos Quimicos Ltda.
Chemical Description: Amphoteric imidazoline
Category: Detergent, amphoteric
Applications: Detergent for non-irritating shampoos
Form: Liquid
Composition: 38% active

Drewpon AS. Drew Produtos Quimicos Ltda.
Chemical Description: Sodium lauryl sulfate
Category: Detergent, anionic
Applications: Detergent for non-irritating shampoos
Form: Paste
Composition: 28–30% active

Drewpon CN. Drew Produtos Quimicos Ltda.
Chemical Description: Sodium lauryl sulfate
Category: Detergent, anionic
Applications: Detergent for non-irritating shampoos
Form: Liquid/paste
Composition: 28-30% active

Drewpon EBO. Drew Produtos Quimicos Ltda.
Chemical Description: Sodium lauryl ether sulfate
Category: Detergent, anionic
Applications: Detergent for non-irritating shampoos
Form: Liquid
Composition: 26–28% active

Drewpon ECM. Drew Produtos Quimicos Ltda.
Chemical Description: Monoethanolamine lauryl ether sulfate
Category: Detergent, anionic
Applications: Detergent for non-irritating shampoos
Form: Liquid
Composition: 30–32% active

Drewpon EDT. Drew Produtos Quimicos Ltda.
Chemical Description: Triethanol-

amine lauryl ether sulfate
Category: Detergent, anionic
Applications: Detergent for non-irritating shampoos
Form: Liquid
Composition: 34–36% active

Drewpon EG. Drew Produtos Quimicos Ltda.
Chemical Description: Sodium lauryl ether sulfate
Category: Detergent, anionic
Applications: Used for pearlescent shampoos
Form: Liquid
Composition: 22% active

Drewpon EKZ. Drew Produtos Quimicos Ltda.
Chemical Description: Ammonium lauryl ether sulfate
Category: Detergent, anionic
Applications: Detergent for non-irritating shampoos
Form: Liquid
Composition: 23–25% active

Drewpon ELB. Drew Produtos Quimicos Ltda.
Chemical Description: Sodium lauryl ether sulfate
Category: Detergent, anionic
Applications: Detergent for non-irritating shampoos
Form: 27–28% active

Drewpon ESG. Drew Produtos Quimicos Ltda.
Chemical Description: Magnesium lauryl ether sulfate
Category: Detergent, anionic
Applications: Detergent for non-irritating shampoos
Form: Liquid
Composition: 26–28% active

Drewpon ETH. Drew Produtos Quimicos Ltda.
Chemical Description: Sodium lauryl ether sulfate
Category: Detergent, anionic
Applications: Detergent for non-irritating shampoos
Form: Liquid
Composition: 26–28% active

Drewpon EVE. Drew Produtos Quimicos Ltda.
Chemical Description: Sodium lauryl ether sulfate
Category: Detergent, anionic
Applications: Detergent for non-irritating shampoos
Form: Liquid
Composition: 26–28% active

Drewpon MEA. Drew Produtos Quimicos Ltda.
Chemical Description: Monoethanolamine lauryl sulfate
Category: Detergent, anionic
Applications: Detergent for non-irritating shampoos
Form: Liquid
Composition: 32–34% active

Drewpon MG. Drew Produtos Quimicos Ltda.
Chemical Description: Magnesium lauryl sulfate
Category: Detergent, anionic
Applications: Detergent for non-irritating shampoos
Form: Liquid
Composition: 28–30% active

Drewpon NH. Drew Produtos Quimicos Ltda.
Chemical Description: Ammonium lauryl sulfate
Category: Detergent, anionic
Applications: Detergent for non-irritating shampoos
Form: Liquid
Composition: 27–29% active

Drewpon NHAC. Drew Produtos Quimicos Ltda.
Chemical Description: Ammonium lauryl sulfate
Category: Detergent, anionic
Applications: Detergent for non-irritating shampoos
Form: Liquid
Composition: 37.5% active

Drewpone 60. PVO International
Chemical Description: Polyoxyethyiene 20 sorbitan monostearate
Category: Food emulsifier
Applications: Hydrophilic food emulsifiers generally used with more lipophilic emulsifiers; the Drewpones have various food applications including as foaming agents, dough conditioners; to improve volume, aeration, texture, gloss, dispersiblity (for color and flavor, in vitamin products), dryness, stability
Form: Liquid to soft solid; Color: Yellow
Properties: HLB: 14.9; Sap. Value: 45–55
Composition: 65–69.5% ETO

Drewpone 65. PVO International
Chemical Description: Polyoxyethylene 20 sorbitan tristearate
Category and Applications: See Drewpone 60
Form: Soft solid; Color: Amber
Properties: HLB: 10.5; Sap. Value: 88–98
Composition: 46–50% ETO

Drewpone 80. PVO International
Chemical Description: Polyoxyethylene 20 sorbitan monooleate
Category and Applications: See Drewpone 60
Form: Liquid; Color: Yellow
Properties: HLB: 15.0; Sap. Value: 45–55
Composition: 65–69.5% ETO

Drewsorb 60. PVO International
Chemical Description: Sorbitan monostearate
Category: Food emulsifier
Applications: Used by itself or with hydrophilic emulsifiers in food applications
Form: Beads; Color: Cream
Properties: Solubility: Dispersible in water; HLB: 4.7; Sap. Value: 147–157

Druspin 1223. PVO International
Chemical Description: Ester
Category: Textile lubricant
Applications: Lubricant to minimize breakage in worsted carding; used for ordinary filet wire in the card clothing
Form: Liquid
Properties: Visc.: Low; Stability: Good
Composition: 100% active ingred.

Druspin 1400. PVO International
Chemical Description: Ester
Category: Textile lubricant
Applications: Lubricant to minimize breakage in worsted carding; used for metallic wire in the card clothing
Form: Liquid
Properties: Visc.: Low; Stability: Good
Composition: 100% active ingred.

Druspin LO-VIS. PVO International
Chemical Description: Blended product
Category: Textile lubricant, antistat, emulsifier
Applications: Low viscosity precarding lubricant for worsted processing; esp. when nonionic scouring in absence of added electrolyte results in static charged wool
Form: Liquid
Properties: Visc.: Low
Composition: 100% active ingred.

Druspin Supreme. PVO International
Chemical Description: Blend of mineral oil, emulsifiers, lubricants, antistats
Category: Emulsifier, lubricant, antistatic agent
Applications: Antistatic worsted oil with excellent emulsion stability; imparts good "spinability," lubrication, antistatic protection
Form: Liquid
Composition: 100% active ingred.

Drustat 75. PVO International
Chemical Description: Complex quarternized fatty amide
Category: Cationic antistatic agent
Applications: Antistatic agent for all synthetic fibers, imparting softness, lubricity
Form: Liquid; Color: Amber
Composition: 100% active ingred.
Reaction: Compatible with silicone water repellent systems

Drustat T. PVO International
Chemical Description: Quarternized fatty amide
Category: Cationic antistatic agent
Applications: Antistatic agent on wool and all synthetic fibers (e.g., polyamide, polyester, acrylics, acetate, dynel); nondiscoloring to fabric; resistant to removal by washing
Form: Paste; Color: Cream
Properties: Solubility: Readily water dispersible
Composition: 50% active
Reaction: Compatible with aminoplast resin systems

Dugista. Van Schuppen
Chemical Description: Blend of de-oiled soybean lecithin, primary dried yeast and wheat germs
Category: Dietary supplement
Applications: Dietary supplement in correcting and preventing arterial diseases; general stimulant of activity; natural sources of choline, inositol, phosphatides, B-vitamins, vitamin E, amino acids and minerals
Form: Granular; Color: Light yel-

low; Odor and Taste: Bland
Composition: 65–70% phosphatides; 12% min. protein; 12–15% carbohydrates; 3–4% vegetable oil; 2.5–3% phosphorus; 2–2.5% nitrogen; 2.5–3% tricalciumphosphate
References: Product Information No. 62

Dulectin Granular. Van Schuppen
Chemical Description: De-oiled soybean lecithin (complex mixture of acetone-insoluble phosphatides from soybeans)
Category: Dietary supplement
Applications: Dietary supplement in correcting and preventing heart and arterial diseases; generally as a stimulant of activity
Form: Granular; Color: Light yellow-brown; Odor and Taste: Bland
Properties: Solubility: Dispersible in water; sol. in animal, mineral and vegetable oils and fats, ether and various hydrocarbons
Composition: 97% min. acetone-insoluble phosphatides
References: Product information No. 61A

Dulectin Powdered. Van Schuppen
Chemical Description: De-oiled soybean lecithin (complex mixture of acetone-insoluble phosphatides from soybeans)
Category: Food additive, dietary supplement
Applications: In margarine, macaroni and baking industries as colloidal, interfacial, emulsifying, softening and antioxidant agents; in dried full cream milk powder; as a dietary supplement in correcting and preventing heart and arterial diseases; activity stimulant
Form: Dry, free-flowing powder; Color: Light yellow; Odor and Taste: Bland
Properties: Solubility: Dispersible in water; sol. in animal, mineral and vegetable oils and fats, ether and various hydrocarbons
Composition: 97% min. acetone-insoluble phosphatides
References: Product information No. 61 B

Duofol T. Hart Chem.
Chemical Description: Anionic sulfated ester
Category: Wetting agent
Applications: Sulfonated ester used for wetting, dispersing and finishing
Form: Liquid; Color: Amber
Properties: Solubility: Clear sol. 5% water; Sp. gr.: 1.068; Visc.: 70 cps
Composition: 55% active ingred.
Reaction: Neutral
Toxicity: Skin and eye irritant
Containers: Drums, T/L
Warnings and Caution: Avoid contact with skin and eyes

Duomac C. Akzo Chemie
Chemical Description: Coco-1,3-diaminopropane diacetate
Category: Corrosion inhibitor, bactericide in drilling fluids
Properties: Solubility: 100% in isopropanol; Sparingly sol. in water @ 60 C; 100% in C_2H_5OH @

30 C; Sp. gr.: 0.85 @ 25 C; M. P.: 20–24 C
Composition: 89% active
Containers: Net 172 kg open head drum
Warnings and Caution: Avoid contact with strong oxidizing agents, acids and bases

Duomac T. Armak
Chemical Description: N-tallow-1,3-propanediamine diacetate
Category and Applications: The Duomac diamine acetate salts are water-soluble derivatives of the corresponding diamines; used when a water-soluble cationic reagent is required for pigment flushing, froth flotation of minerals, flocculation
Form: Paste; Color: Brown; Odor: Vinegar
Properties: m.w.: 480; Solubility: Sol. in water, chloroform, ethanol, isopropanol; Sp. gr.: 0.892 @ 90 C; M.P.: 181 F; B.P.: > 300 C; Flash Pt.: 335 F (COC)
Composition: 85% Duomeen T as the acetate
Toxicity: Corrosive to skin and eyes; causes severe burns on prolonged contact
Warnings and Caution: Protective goggles and gloves must be worn
References: Bulletin 76-19

Duomeen C. Armak
Chemical Description: N-coco-1,3-propanediamine
Category: Organic bifunctional bases, corrosion inhibitors, fuel oil additives, flotation agents, cationic
Applications: Form strongly bonded films on surfaces of metals, textiles, plastics; as corrosion inhibitors, multi-purpose gasoline and fuel oil additives, flotation agents, epoxy curing agents; in preparation of amine salts of some herbicides
Form: Liquid; Color: Dark amber; Odor: Ammonia
Properties: m.w.: 276; Solubility: Sol. in naphtha, mineral oil, isopropanol; Sp. gr.: 0.836; M.P.: 71 F; B.P.: > 300 C; Flash Pt.: 295 F (COC)
Composition: 43% min. primary amine; 43% min. secondary amine
Reaction: Readily neutralized to form either water- or oil-soluble mono- or diacid salts
Toxicity: Corrosive to skin and eyes; causes severe burns on prolonged contact
Warnings and Caution: Protective goggles and gloves must be worn
References: Bulletin 76-19

Duomeen C Special. Armak
Chemical Description: N-coco-1,3-propanediamine
Category and Applications: See Duomeen C
Form: Liquid; Color: Yellow; Odor: Ammonia
Properties: m.w.: 276; Solubility: Sol. in naphtha, mineral oil, isopropanol; Sp. gr.: 0.836; M.P.: 80 F; B.P.: > 300 C; Flash Pt.: 295 F (COC)
Composition: 45% min. primary amine; 45% min. secondary amine
Reaction: Readily neutralized to

form either water- or oil-soluble mono- or diacid salts
Toxicity: Corrosive to skin and eyes; causes severe burns on prolonged contact
Warnings and Caution: Protective goggles and gloves must be worn
References: Bulletin 76-19

Duomeen L-11. Armak
Chemical Description: N-s-alkyl-1,3-propanediamine
Category and Applications: See Duomeen C
Form: Liquid; Color: Yellow; Odor: Ammonia
Properties: m.w.: 300; Solubility: Sol. in mineral oil, isopropanol; Sp. gr.: 0.841; M.P.: –32 F; B.P.: > 300 C; Flash Pt.: 226 F (COC)
Toxicity: Corrosive to skin and eyes; causes severe burns on prolonged contact
Warnings and Caution: Protective goggles and gloves must be worn
References: Bulletin 76-19

Duomeen L-15. Armak
Chemical Description: N-s-alkyl-1,3-propanediamine
Category and Applications: See Duomeen C
Form: Liquid; Color: Yellow; Odor: Ammonia
Properties: m.w.: 383; Solubility: Sol. in mineral oil, isopropanol; Sp. gr.: 0.842; M.P.: 44 F; B.P.: > 300 C; Flash Pt.: 266 F (COC)
Toxicity: Corrosive to skin and eyes; causes severe burns on prolonged contact
Warnings and Caution: Protective goggles and gloves must be worn
References: Bulletin 76-19

Duomeen O. Armak
Chemical Description: N-oleyl-1,3-propanediamine
Category and Applications: See Duomeen C
Form: Paste; Color: Yellow; Odor: Ammonia
Properties: m.w.: 350; Solubility: Sol. in naphtha @ 35 C, isopropanol; Sp. gr.: 0.841; M.P.: 59 F; B.P.: > 300 C; Flash Pt.: 370 F (COC)
Composition: 43% min. primary amine; 43% min. secondary amine
Reaction: Readily neutralized to form either water- or oil-soluble mono- or diacid salts
Toxicity: Corrosive to skin and eyes; causes severe burns on prolonged contact
Warnings and Caution: Protective goggles and gloves must be worn
References: Bulletin 76-19

Duomeen T. Armak
Chemical Description: N-tallow-1,3-propanediamine
Category and Applications: See Duomeen C; also as textile finishing agent, as dispersant for inorganic pigments in paints, as larvicidal oil additive
Form: Paste; Color: Dark amber; Odor: Ammonia
Properties: m.w.: 350; Solubility: Sol. in naphtha @ 35 C, isopropanol; Sp. gr.: 0.841; M.P.: 115 F; B.P.: > 300 C; Flash Pt.: 370 F (COC)
Composition: 43% min. primary

amine; 43% min. secondary amine
Reaction: Readily neutralized to form either water- or oil-soluble mono- or diacid salts
Toxicity: Corrosive to skin and eyes; causes severe burns on prolonged contact
Warnings and Caution: Protective goggles and gloves must be worn
References: Bulletin 76-19

Duoquad T-50. Akzo Chemie
Chemical Description: N, N, N′, N′, N′ - pentamethyl-*n*-tallow-1,3-propanadiammonium dichloride
Category: Detergent, corrosion inhibitor, metal cleaner, cationic
Form: Liquid; Color: Max. 13 (Gardner)
Properties: Sp. gr.: 0.90; F.P.: –20 C; Flash Pt.: 13 C; pH: 6–8
Composition: 48–52% active
Toxicity: Skin irritant, severe eye irritant
Containers: Net 180 kg drums
Warnings and Caution: Avoid contact with strong oxidizing agents and overexposure

Duomeen T Special. Armak
Chemical Description: N-tallow-1,3-propanediamine
Category and Applications: See Duomeen C
Form: Paste; Color: Dark amber; Odor: Ammonia
Properties: m.w.: 330; Solubility: Sol. in naphtha, isopropanol; Sp. gr.: 0.841; M.P.: 115 F; B.P.: > 300 C; Flash Pt.: 370 F (COC)
Composition: 43% min. primary amine; 43% min. secondary amine
Reaction: Readily neutralized to form either water- or oil-soluble mono- or diacid salts
Toxicity: Corrosive to skin and eyes; causes severe burns on prolonged contact
Warnings and Caution: Protective goggles and gloves must be worn
References: Bulletin 76-19

Duomeen TDO. Armak
Chemical Description: N-tallow-1,3-propanediamine dioleate
Category and Applications: See Duomeen C; also in metal treatment as a film and boundary lubricant, a metal drawing additive, corrosion inhibitor; in paint industry as a dispersant for inorganic and organic pigments; added to primer paints to insure adhesion to metallic surfaces and displace moisture
Form: Semisolid; Color: Dark; Odor: Ammonia
Properties: m.w.: 924; Solubility: Sol. in naphtha, mineral oil, isopropanol; Sp. gr.: 0.865 @ 65 C; M.P.: 77 F; B.P.: > 300 C; Flash Pt.: 455 F (COC)
Composition: 33% Duomeen T
Toxicity: Corrosive to skin and eyes; causes severe burns on prolonged contact
Warnings and Caution: Protective goggles and gloves must be worn
References: Bulletin 76-19

Duomeen TX. Akzo Chemie
Chemical Description: Tallow-1,3-diaminopropane
Category: Emulsifier, base material

Applications: Base material for ethloxylated products; emulsifier for bitumen emulsions
Form: Solid; Color: 7 (Gardner); Odor: Ammoniacal
Properties: Solubility: Insoluble in water @ 25 C; Sp. gr.: 0.84; M.P.: 46–55 C; Flash Pt.: 180 C; Iodine Value: 30–40
Composition: 92% min. active
Toxicity: Skin irritant, severe eye irritant
Containers: Net 165 kg open head steel drums
Warnings and Caution: Avoid contact with strong oxidizing agents and strong acids

Du Pont Retarder LAN. Du Pont
Chemical Description: Alkyl trimethylammonium halide
Category: Dye retardant, stripping agent, emulsifier, cationic
Applications: Dye retarding and stripping agent with azoic dyes; esp. in application of cationic dyes to acrylic fibers
Form: Liquid; Color: Dark red-brown
Properties: Solubility: Miscible with water; Stability: Stable to acids, alkalis and in storage; Surface Tension: 43.4 dynes/cm (0.1% sol'n.)
Composition: 20% active ingred.
References: Data Sheet No. 20

Duponol 80. Du Pont
Chemical Description: Sodium n-octyl sulfate technical
Category: Wetting agent, dispersant, penetrant, softening agent, emulsifier, anionic
Applications: Textiles, leather and metal processes such as mineral dyeing, leather tanning, soldering fluxes, metal pickling and cleaning compositions; highly surface active in electrolyte solutions of moderate to high concentrations
Form: Liquid; Color: Light yellow
Properties: m.w.: 232; Solubility: Sol. in presence of many multivalent ions; Visc.: 30 cps @ 27 C; Surface Tension: 51.7 dynes/cm (0.1% sol'n.)
Composition: 33–35% active
References: DS No. 9

Duponol C. Du Pont
Chemical Description: Sodium lauryl sulfate USP
Category: Emulsifier, dispersant, detergent, wetting agent, anionic
Applications: Cosmetic, dental and medical preparations; O/W emulsions; emollient creams, vanishing creams, hand lotions, brushless shaving creams, cold creams
Form: Spray-dried beads; Color: White; Odor: Mildly fatty
Properties: Solubility: 10–15 g in 100 ml water @ 25 C; Density: 3.3 lb/gal; Stability: Stable to alkali and in normal storage; hydrolyzes in strong acid solutions
Composition: 90–96% sodium lauryl sulfate
Containers: 51-gal Leverpak (150 lb net wt)
Warnings and Caution: Store in well-closed containers; corrosive to most metals

Duponol D Paste. Du Pont
Chemical Description: Sodium lauryl oleyl sulfate technical
Category: Detergent, emulsifier, wetting agent, dispersant, anionic
Applications: Textile industry for heavy duty scouring and dyeing; dispersing agent for dyes; scouring of shearlings; leather degreasing; scouring and conditioning of paper mill felts; in metal cleaning compositions
Form: Paste; Color: Light yellow
Properties: m.w.: 328; Surface Tension: 30.5 dynes/cm (0.1% sol'n.)
Composition: 38% active ingred.
References: DS No. 8

Duponol EP. Du Pont
Chemical Description: Diethanolamine lauryl sulfate (technical)
Category: Foaming agent, detergent, wetting agent, emulsifier, anionic
Applications: Cosmetics esp. clear liquid shampoos and pearlescent liquid cream lotion shampoos; as detergent in textile scouring esp. for fabrics with "Lycra" spandex yarn
Form: Slightly viscous liquid; Color: Very pale golden; Odor: Bland and clean
Properties: Solubility: Sol. in water in all proportions; in polar solvents with some electrolyte precipitation; Density: 8.4 lb/gal; Visc.: 50–150 cps @ 27 C; Cloud Pt.: 5 C; Stability: Stable to freezing, heat (150 F), uv light, aging
Composition: 33–36% active
Warnings and Caution: Some corrosion on iron, steel, aluminum
References: D-2955

Duponol G. Du Pont
Chemical Description: Fatty alcohol amine sulfate
Category: Emulsifier, wetting agent, anionic
Applications: Powerful oil- and wax-soluble emulsifier; wetting agent in water and aqueous electrolyte solutions; at low concentrations can be added to flammable solvents to increase the electrical conductance and reduce static-caused fire hazards
Form: Viscous translucent paste; Color: Yellow; Odor: Mild, fatty, penetrating
Properties: Solubility: Sol. in ethyl alcohol, ethyl acetate, Stoddard solvent, perchlorethylene, Cellosolve, toluene, mineral oil; at least 50% sol. in water; Density: 8.2 lb/gal; Stability: Stable to alkalis and in normal storage; hydrolyzes in strong acid; Surface Tension: 29.6 dynes/cm (0.1% sol'n.)
Composition: 92% active ingred.
References: Data sheet No. 12

Duponol LS Paste. Du Pont
Chemical Description: Sodium oleyl sulfate technical
Category: Scouring agent, rewetting agent, anionic
Applications: Scouring agent in textile and leather industries; as rewetting agent for paper; in washing paper mill felts
Form: Paste; Color: Light tan

Properties: m.w.: 355; Surface Tension: 37.8 dynes/cm (0.1% sol'n.)
Composition: 26% active ingred.
References: Data Sheet No. 13

Duponol ME Dry Surface Active Agent. Du Pont
Chemical Description: Sodium lauryl sulfate, technical (manufactured from synthetic fatty alcohols)
Category: Detergent, wetting agent, foaming agent, emulsifier, anionic
Applications: Detergent esp. effective for greasy soils; general industrial applications
Form: Powder; Color: White to medium cream
Properties: m.w.: 302; Solubility: 30% sol. in water; active ingred. sol. in methanol and ethanol; Density: 3.3 lb/gal; Cloud Pt.: 70 F (1% aqueous sol'n.); Stability: Stable to alkalis and in normal storage; hydrolyzes in acids; Surface Tension: 35 dynes/cm (0.1% sol'n.)
Composition: 90–96% active
References: Data Sheet No. 28

Duponol QC. Du Pont
Chemical Description: Based on sodium lauryl sulfate
Category: Shampoo base, anionic
Applications: Highly effective for clear liquid, liquid cream lotion and paste shampoos with excellent foaming properties
Form: Slightly viscous liquid; Color: Essentially water white; Odor: Bland, clean; characteristic lauryl alcohol sulfate
Properties: Solubility: Sol. in water in all proportions, in polar solvents with some electrolyte precipitation; Visc.: $<$ 500 cps @ 80 F; Cloud Pt.: $\approx$ 65 F; Stability: Stable to freezing, uv light, aging; stable to heat 150 F for several hours
Composition: 29–30% active
References: Brochure D-3077

Duponol RA. Du Pont
Chemical Description: Fortified sodium ether-alcohol sulfate
Category: Detergent, dispersant, emulsifier, penetrating and leveling agent, wetting agent, anionic
Applications: Textile processing as dispersing, emulsifying, penetrating and leveling agents; as wetting agent in aqueous solutions at elevated temperatures; in scouring and dyeing of fabrics and yarns of natural and man-made fibers
Form: Clear liquid; Color: Light amber
Properties: Solubility: Miscible with water in all proportions; Density: 8.7 lb/gal; Sp. gr.: 1.04; Visc.: 178 cps (Brookfield @ 80 F); Cloud Pt.: 68 F; Flash Pt.: $>$ 212 F (TOC); Stability: Stable to alkalis and in normal storage; stable to low concentrations in acid media; Surface Tension: 33 dynes/cm (0.1% sol'n.)
References: Brochure D-5205

Duponol WA Dry Surface Active Agent. Du Pont
Chemical Description: Sodium lauryl sulfate surfactant

Category: Detergent, emulsifier, wetting agent, penetrant, anionic
Applications: Detergent esp. effective on greasy soils; for dry mixing with alkaline detergent builders and general industrial applications; reduces surface tension of aqueous systems; produces dense, stable foam
Form: Powder; Color: White
Properties: Solubility: Sol. up to 40% product conc. in water; the active ingred. is sol. in highly polar solvents (e.g., methanol, ethanol); Density: 2.6 lb/gal; Cloud Pt.: 37 F (1% sol'n.); Stability: Stable to alkalis, in storage; hydrolyzes in acid conditions; Surface Tension: 33 dynes/cm (0.1% sol'n.)
Composition: 44% sodium alkyl sulfate

Duponol WA Paste. Du Pont
Chemical Description: Alcohol sulfate product
Category and Applications: See Duponol WAQA

Duponol WAQA. Du Pont
Chemical Description: Alcohol sulfate product
Category: Shampoo base
Applications: Shampoo formulations esp. liquid cream lotion, solid cream or paste shampoos; controlled degreasing action, superior cleansing and foaming

Duponol WAQE. Du Pont
Chemical Description: Sodium lauryl sulfate technical
Category: Emulsifier, anionic
Applications: Emulsion polymerization systems esp. preparation of plastomeric, elastomeric latices and dry resins
Form: Liquid; Color: Pale yellow
Properties: m.w.: 302; Solubility: Sol. in water in all proportions; sol. in polar solvents with some electrolyte precipitation; Density: 8.6 lb/gal; Visc.: 500 cps max. @ 27 C; Cloud Pt.: ≈ 65 F; Stability: Stable to freezing, aging; stable to heat (150 F) for several hours
Composition: 29–30.5% active
Containers: Perma-Plex drums; tank trucks, tank cars
References: Brochure D-2989

Duponol WN. Du Pont
Chemical Description: Sodium octyl/decyl sulfate, technical
Category: Wetting and penetrating agent, anionic
Applications: Wetting agent and penetrant esp. in aqueous solutions of high electrolyte content; leather tanning, electroplating, oil well acidizing, dyeing of vat colors, caustic impregnation of warps, wool chlorination, acid fulling, cleaning of fur pelts
Form: Liquid; Color: Pale yellow
Properties: m.w.: 246; Visc.: 30 cps @ 27 C; Surface Tension: 52.5 dynes/cm (0.1% sol'n.)
Composition: 33–35% active
References: Data Sheet No. 19

Duponol XL. Du Pont
Category: Shampoo surfactant, anionic-amphoteric

Applications: Shampoo formulations with exceptional foaming power; substantive to hair and skin; controlled degreasing
Form: Moderately viscous liquid; Color: Golden; Odor: Faintly spicy
Properties: Solubility: Sol. in water in all proportions; in polar solvents with some electrolyte precipitation; Visc.: 2500–4500 cps @ 80 F; Cloud Pt.: ≈ 50 F; Clear Pt.: 50–55 F; Stability: Stable to freezing, heat (150 F); to uv light, to aging
Composition: 36% solids content

Dur-Em 104K. Durkee Ind. Foods
Chemical Description: Mono- and diglycerides from hydrogenated vegetable oil with BHA and BHT, citric acid
Category: Food emulsifier, nonionic
Applications: Bakery, prepared mixes, margarines, convenience foods, frozen desserts; products designated K are available in kosher form
Form: Plastic; Color: Cream
Properties: M.P.: 115–121 F
Composition: 100% active ingred.

Dur-Em 107K. Durkee Ind. Foods
Chemical Description: Mono- and diglycerides from hydrogenated vegetable oil with BHA, BHT and citric acid
Category and Applications: See Dur-Em 104K
Form: Bead; Color: White
Properties: M.P.: 140–145 F
Composition: 100% active ingred.
Containers: 200 lb drums, bulk

Dur-Em 114K. Durkee Ind. Foods
Chemical Description: Mono- and diglycerides from hydrogenated vegetable oil with BHA, BHT and citric acid
Category and Applications: See Dur-Em 104K
Form: Plastic; Color: Cream
Properties: M.P.: 110–120 F
Composition: 100% active ingred.
Containers: 50 lb carton, bulk

Dur-Em 117K. Durkee Ind. Foods
Chemical Description: Mono- and diglycerides from hydrogenated vegetable oil with citric acid
Category and Applications: See Dur-Em 104K; also in chewing gums
Form: Beads, flakes; Color: White
Properties: M.P.: 145–150 F
Containers: 50 lb bags, 200 lb drums, bulk

Dur-Em 127K. Durkee Ind. Foods
Chemical Description: Mono- and diglycerides from hydrogenated vegetable oil with citric acid
Category and Applications: Peanut butter stabilizer
Form: Beads; Color: White
Properties: M.P.: 140–145 F
Composition: 100% active ingred.
Containers: 50 lb carton, bulk

Dur-Em 136K. Durkee Ind. Foods
Chemical Description: Mono- and diglycerides from hydrogenated

vegetable oils with BHA, BHT and citric acid
Category and Applications: See Dur-Em 104K; also for vegetable dairy systems
Form: Plastic
Composition: 100% active ingred.

Dur-Em 187. Durkee Ind. Foods
Chemical Description: Mono- and diglycerides from hydrogenated animal fats with citric acid
Category and Applications: See Dur-Em 104K
Form: Bead or flake; Color: White
Properties: M.P.: 140–145 F
Composition: 100% active ingred.
Containers: Bulk only

Dur-Em 195. Durkee Ind. Foods
Chemical Description: Mono- and diglycerides from hydrogenated animal and vegetable fats with BHA, BHT, citric acid
Category and Applications: Dry mix systems, dog foods, prepared mixes, continuous and conventional breads, cakes and sweet doughs
Form: Plastic; Color: Cream
Properties: M.P.: 123–129 F
Composition: 100% active ingred.
Containers: Bulk only

Dur-Em 204K. Durkee Ind. Foods
Chemical Description: Mono- and diglycerides from hydrogenated vegetable oil with BHA, BHT, citric acid
Category and Applications: See Dur-Em 104K
Form: Plastic; Color: Cream
Properties: M.P.: 120–130 F
Composition: 100% active ingred.
Containers: 50 lb carton, bulk

Dur-Em 207-E. Durkee Ind. Foods
Chemical Description: Mono- and diglycerides with citric acid
Category and Applications: For use directly in continuous and conventional breads, sweet doughs; emulsifier hydrate for breads; cosmetic/pharmaceutical uses
Form: Powder; Color: White
Properties: M.P.: 140–146 F; Emulsifiability: Self-emulsifying
Composition: 100% active ingred.
Containers: 50 lb carton

Dur-Em 207K. Durkee Ind. Foods
Chemical Description: Mono- and diglycerides from hydrogenated vegetable oil with citric acid
Category and Applications: See Dur-Em 104K
Form: Bead or plastic; Color: White
Properties: M.P.: 140–146 F
Composition: 100% active ingred.
Containers: 200 lb drum, 50 lb bag, bulk

Dur-Em 227K. Durkee Ind. Foods
Chemical Description: Mono- and diglycerides from hydrogenated vegetable oil with citric acid
Category and Applications: See Dur-Em 104K
Form: Bead
Composition: 100% active ingred.

Dur-Em 287. Durkee Ind. Foods
Chemical Description: Mono- and

diglycerides from hydrogenated animal fat with citric acid
Category and Applications: See Dur-Em 104K
Form: Plastic
Composition: 100% active ingred.

Dur-Em 287E. Durkee Ind. Foods
Chemical Description: Mono- and diglycerides from hydrogenated animal fat with citric acid
Category: Nonionic food emulsifier
Applications: Conventional and continuous breads, sweet doughs; emulsifier hydrate for breads and nonfat cakes
Form: Powder; Color: White
Properties: M.P.: 140–146 F
Composition: 100% active ingred.

Dur-Em 295. Durkee Ind. Foods
Chemical Description: Mono- and diglycerides from hydrogenated animal and vegetable fats with and without BHA, BHT and citric acid
Category and Applications: See Dur-Em 104K
Form: Plastic; Color: Cream
Properties: M.P.: 123–129 F
Composition: 100% active ingred.

Duratex. Durkee Ind. Foods
Chemical Description: Partially hydrogenated cottonseed oil
Category: Lubricant, release agent
Applications: Lubricant for tablets; release agent; candy dusting and coating for hydroscopic materials
Form: Powder
Properties: M.P.: 142–148 F
Containers: 50 lb carton

Durfax 60K. Durkee Ind. Foods
Chemical Description: Polyoxyethylene 20 sorbitan monostearate
Category: Nonionic food emulsifier
Applications: O/W emulsions in food industry
Form: Viscous liquid; Color: Yellow
Properties: Solubility: Sol. in water, and cottonseed oil (at low levels); HLB: 14.9
Composition: 100% active ingred.
Containers: 470 lb packaging, bulk

Durfax 65K. Durkee Ind. Foods
Chemical Description: Polyoxyethylene 20 sorbitan tristearate
Category and Applications: See Durfax 60K
Form: Solid; Color: Tan
Properties: Solubility: Sol. at low levels in cottonseed oil; dispersible in water
Composition: 100% active ingred.
Containers: 470 lb packaging, bulk

Durfax 80K. Durkee Ind. Foods
Chemical Description: Polyoxyethylene 20 sorbitan monooleate
Category and Applications: See Durfax 60K
Form: Liquid; Color: Yellow
Properties: Solubility: Sol. in water and at low levels in cottonseed oils; HLB: 15.9
Composition: 100% active ingred.
Containers: 470 lb packaging, bulk

Durfax EOM K. Durkee Ind. Foods
Chemical Description: Ethoxylated mono- and diglycerides
Category: Food emulsifier, nonionic
Applications: Dough conditioner in yeast-raised baked goods; cakes, icings
Form: Paste
Properties: M.P.: 80–85 F; HLB: 13.1
Composition: 100% active ingred.

Durkee 07. Durkee Ind. Foods
Chemical Description: Partially hydrogenated cottonseed oil
Category: Food additive
Applications: Peanut butter stabilizer, breads, shortenings, chewing gum, caramel coatings
Form: Flake or Beads
Properties: M.P.: 142–148 F
Containers: 200 lb drum, 50 lb bag, bulk

Durkee 17. Durkee Ind. Foods
Chemical Description: Partially hydrogenated soybean oil
Category: Food additive
Applications: Shortenings
Form: Flake
Properties: M.P.: 152–158 F
Containers: 50 lb bag, bulk

Durkee 27. Durkee Ind. Foods
Chemical Description: Partially hydrogenated palm oil
Category: Food additive
Applications: Peanut butter stabilizer
Form: Bead
Properties: M.P.: 136–143 F
Containers: 50 lb carton, bulk

Durlac 100WK. Durkee Ind. Foods
Chemical Description: Glyceryl lactoesters of fatty acids from hydrolyzed vegetable oil
Category: Food emulsifier, nonionic
Applications: Prepared mixes, whipped toppings, other vegetable dairy systems; cakes, shortenings
Form: Flake
Properties: M.P.: 115–130 F
Composition: 100% active
Containers: 50 lb bag

Durlac 200K. Durkee Ind. Foods
Chemical Description: Glyceryl lactoesters of fatty acids from vegetable oil
Category and Applications: See Durlac 100WK; also for cosmetic and pharmaceutical use
Form: Solid
Properties: M.P.: 97–103 F
Composition: 100% active ingred.
Containers: 50 lb carton, bulk

Durlac 300K. Durkee Ind. Foods
Chemical Description: Lactylated fatty acid esters of glycerin and propylene glycol (PGLS)
Category: Food emulsifier, nonionic
Applications: Vegetable dairy systems, cakes, cake mixes, aerating agent, cosmetic and pharmaceutical uses
Form: Flake; Color: Cream
Properties: M.P.: 95–105 F
Composition: 100% active ingred.
Containers: 50 lb carton, bulk

Durpeg 400MO. Durkee Ind. Foods
Chemical Description: Mono- and dioleate esters with polyethylene glycol from animal and vegetable sources
Category: Food emulsifier, nonionic
Applications: In calf milk replacers, processing aid in animal feeds
Form: Semi-solid; Color: Amber
Properties: HLB: 11.1
Composition: 100% active ingred.
Containers: 450 lb packaging

Durpro 107K. Durkee Ind. Foods
Chemical Description: Propylene glycol, mono- and diesters, and mono- and diglycerides with citric acid
Category: Food emulsifier, nonionic
Applications: Cake and cake mix emulsifier
Form: Flake; Color: Cream
Properties: M.P.: 117–125; HLB: 2.5
Composition: 100% active ingred.
Containers: 50 lb carton

Durpro 187. Durkee Ind. Foods
Chemical Description: Propylene glycol, mono- and diesters, and mono- and diglycerides
Category: Food emulsifier, nonionic
Applications: Cake and cake mix emulsifier
Form: Flake; Color: Cream
Properties: M.P.: 117–125 F; HLB: 2.5
Composition: 100% active ingred.

Durtan 60K. Durkee Ind. Foods
Chemical Description: Sorbitan monostearate
Category: Food emulsifier, nonionic
Applications: Cakes and cake mixes; chocolate compound coatings as gloss-enhancers; vegetable dairy systems
Form: Beads; Color: Cream
Properties: M.P.: 121–127 F; HLB: 4.7
Composition: 100% active ingred.

Dusoran MD. Van Schuppen
Chemical Description: Lanolin alcohols, distilled
Category: Stabilizer, softener, emulsifier, nonionic
Applications: When mixed with paraffins, it is a choice emulsifier for the preparation of water-in-oil emulsions; in absorption bases, nourishing creams (moisturizing and conditioning), emollient creams and lotions, hair creams, cleansing preparations
Form: Waxy solid; Color: Light yellow
Properties: Solubility: Sol. in mineral oils, abs. alcohol, alcohol 95%, chloroform, ether, light petroleum, toluene; Sapon. Value: 8.0; Acid No.: 2.0; Hydroxyl No.: 130–150
Composition: 100% active
References: Data sheet 33

Dusoran NG. Van Schuppen
Chemical Description: Wool and lanolin alcohols
Category: Stabilizer, softener, emulsifier
Applications: When mixed with paraffins, it is a choice emulsifier

for the preparation of water-in-oil emulsions; in absorption bases, nourishing creams (moisturizing and conditioning), lotions, hair creams, cleansing preparations, lipstick bases
Form: Solid; Color: Tan; Odor: Faint
Properties: Solubility: Sol. in mineral oils, abs. alcohol, alcohol 95%, chloroform, toluene; Sapon. Value: 10; Acid No.: 3.0; Hydroxyl No.: 120–180
References: Data sheet 31 B

Dusoran NO. Van Schuppen
Chemical Description: Wool and lanolin alcohols
Category: Stabilizer, softener, emulsifier
Applications: When mixed with paraffins, it is a choice emulsifier for the preparation of water-in-oil emulsions; in absorption bases, nourishing creams (moisturizing and conditioning), lotions, hair creams, cleansing preparations, lipstick bases
Form: Solid; Color: Light brown; Odor: Faint
Properties: Solubility: Sol. in mineral oils, abs. alcohol, alcohol 95%, chloroform, toluene; Sapon. Value: 10; Acid No.: 2; Hydroxyl No.: 120–180
References: Data sheet 31 A

Dyesperse DC. Hart Chem.
Chemical Description: Anionic naphthalene formaldehyde sulfonate
Category: Dispersant
Applications: Dispersant for disperse and acid dyes
Form: Liquid; Color: Amber
Properties: Refractive Index: 1.4028; Solubility: Clear sol.; Sp. gr.: 1.138; Visc.: 10 cps
Composition: 33% active ingred.
Reaction: Neutral
Toxicity: Skin and eye irritant
Containers: Drums, T/L
Warnings and Caution: Avoid contact with skin and eyes

Dymsol 12-03-A. Diamond Shamrock
Category: Anionic emulsifier
Applications: Biodegradable primary or secondary emulsifier for emulsion polymerization
Form: Liquid
Composition: 59% active ingred.

Dymsol 31-P. Diamond Shamrock
Category: Anionic surfactant, stabilizer
Applications: Biodegradable mechanical and chemical stabilizer for latex
Form: Liquid
Properties: Biodegradable
Composition: 55% active ingred.

Dymsol 36-A. Diamond Shamrock
Category: Anionic emulsifier
Applications: Biodegradable emulsifier for vinyl chloride monomer
Form: Liquid
Composition: 24.5% active ingred.

Dymsol 38-C. Diamond Shamrock
Category: Anionic emulsifier
Applications: Biodegradable secondary emulsifier to cut down

coagulum in SBR production
Form: Liquid
Composition: 70% active ingred.

Dymsol 2031. Diamond Shamrock
Category: Anionic emulsifier
Applications: Biodegradable primary or secondary emulsifier for emulsion polymerization
Form: Liquid
Composition: 60% active ingred.

Dymsol B. Diamond Shamrock
Category: Nonionic surfactant
Applications: Rubber anti-blocking agent
Form: Dispersion
Composition: 45% active ingred.

Dymsol L. Diamond Shamrock
Category: Nonionic stabilizer
Applications: Biodegradable stabilizer giving good wetting, controlled viscosity and sharp coagulation point
Form: Liquid
Composition: 100% active ingred.

Dymsol LP. Diamond Shamrock
Category: Anionic emulsifier
Applications: Biodegradable primary emulsifier for large particle size latices
Form: Liquid
Properties: Biodegradable
Composition: 33% active ingred.

Dymsol N. Diamond Shamrock
Category: Nonionic stabilizer
Applications: Biodegradable chemical stabilizer for latex
Form: Liquid
Composition: 97.5% active ingred.

Dymsol S. Diamond Shamrock
Category and Applications: Replacement for sulfated sperm oil fatliquors; tanning leather
Form: Liquid; Color: Greenish brown
Containers: 55 gal lined open-head steel drums
References: Bulletin LE-55 Rev.

Dymsol TXM. Diamond Shamrock
Category: Lubricant, nonionic
Applications: Polyester lubricant liquid for plastics industry
Form: Liquid
Composition: 100% active ingred.

Dymsol VD. Diamond Shamrock
Category: Anionic surfactant
Applications: Latex thickener
Form: Liquid
Composition: 15% active ingred.

Dyqex. Georgia-Pacific
Chemical Description: Sodium lignosulfonate
Category: Dye dispersant extender
Applications: Compatible with primary dispersants and contributes the additional advantages of low viscosity in use and of exceptional redispersibility of dye powders
Form: Powder; Color: Light

E

EC-25K. Durkee Ind. Foods
Chemical Description: Propylene glycol mono- and diesters of fats and fatty acids, mono- and diglycerides, partially hydrogenated soybean oil with lecithin, BHA, citric acid
Category: Food emulsifier, nonionic
Applications: Cakes, mixes; used with all-purpose shortenings for baker's cakes or with emulsified shortenings for very lean cakes
Form: Plastic; Color: Light straw
Properties: M.P.: 90–100 F; HLB: 2.6
Composition: 100% active ingred.
Containers: 50 lb carton

EC-117MK. Durkee Ind. Foods
Chemical Description: Mono- and diglycerides and lactylic esters of fatty acids
Category: Food emulsifier, nonionic
Applications: Emulsified system for freeze-thaw coffee whitener, aerating composition, foam mat drying
Form: Beads; Color: Cream to light tan
Properties: M.P.: 123–133 F; HLB: 4.6
Composition: 100% active ingred.
Containers: 50 lb carton

Ecconol 61. Essential Chem.
Chemical Description: Foam-stabilized anionic-nonionic complex of amide sulfonate
Category: Detergent
Applications: Dishwash and carwash products; high-sudsing cleaners
Form: Liquid; Color: Amber
Properties: Solubility: Water sol.; Biodegradable
Composition: 67% active ingred.

Ecconol 66. Essential Chem.
Chemical Description: Fatty amide condensate
Category: Detergent, wetting agent, thickener, emulsifier, nonionic
Applications: Cleaning compounds and lubricants; liquid detergents
Form: Liquid; Color: Amber
Properties: Biodegradable

Composition: 100% active ingred.

Ecconol 606. Essential Chem.
Chemical Description: Organic alkylaryl sulfonate
Category: Detergent, wetting agent, anionic
Applications: Cleaning compounds; detergency, foaming, wetting, surface tension reducing properties for liquid formulations
Form: Liquid; Color: Amber
Properties: Biodegradable
Composition: 60% active ingred.

Ecconol 628. Essential Chem.
Chemical Description: Coconut oil amide
Category: Detergent, wetting agent, nonionic
Applications: Cleaning agents, thickener, foam stabilizer
Form: Liquid
Properties: Biodegradable
Composition: 100% active ingred.

Ecconol B. Essential Chem.
Chemical Description: Fatty alkanolamide
Category: Detergent, wetting agent, emulsifier, nonionic
Applications: Cleaning agents, balanced water-sol. base detergent for floor and wall cleaners, heavy-duty formulations
Form: Liquid
Properties: Biodegradable
Composition: 90% active ingred.

Ekaline G-80 Flakes. Sandoz
Category: Detergent, wetting agent, emulsifier, dispersant
Applications: General purpose scouring, emulsifying and leveling agent; promotes disperse dye migration on filiment nylon carpets; scouring/dyeing of wool for removal of residual grease and leveling of dyes; afterscouring disperse dyeings and prints esp. polyester
Form: Flakes; Color: White
Properties: Solubility: 20% in water
Composition: 100% active ingred.
Reaction: Neutral

Ekaline G-80 Liquid. Sandoz
Category: Detergent, wetting agent, emulsifier, dispersant
Applications: General purpose scouring, emulsifying and leveling agent; promotes disperse dye migration on filement nylon carpets; scouring/dyeing of wool for removal of residual grease and leveling of dyes; afterscouring disperse dyeings and prints, esp. polyester
Form: Liquid; Color: Colorless
Properties: Solubility: Miscible with water; Sp. gr.: 1.02
Composition: 20% active ingred.
Reaction: Acid

Elfan 200 Pulver. Akzo Chemie
Chemical Description: Laurylalcoholsulfate sodium salt (basis: native fatty alcohol C_{12})
Category: Detergent
Applications: Cosmetics; tooth pastes
Form: Powder, beads; Color: White
Properties: Density: 220 g/l; Stability: In storage, protect against humidity

Composition: 90% active ingred.
Reaction: Neutral
Warnings and Caution: Avoid contact with acids because of hydrolysis

Elfan 240 T/S. Akzo Chemie
Chemical Description: Laurylalcohol sulfate, triethanolamine salt (basis: synthetic fatty alcohol C_{12}/C_{14})
Category: Detergent
Applications: Skins and cosmetics applications; shampoos
Form: Liquid; Color: Yellowish
Properties: Solubility: Sol. in water; Sp. gr.: 1.04; Visc.: 150 cps @ 20 C; Stability: No special storage precautions
Composition: 40% active ingred.; 2% max. unsulfated parts
Reaction: Neutral
Containers: 20 ton containers; 125 kg drums
Warnings and Caution: Prevent contact with acids because of hydrolysis

Elfan 260 S. Akzo Chemie
Chemical Description: Lauryl alcohol sulfate sodium salt (basis: synthetic fatty alcohol C_{12}/C_{16})
Category: Detergent, emulsifier
Applications: Textiles, skins, cosmetics uses as detergents (esp. hairwashing, pasty cleaners); emulsion polymerization
Form: Liquid, paste; Color: Yellowish
Properties: Solubility: Sol. in water @ 30 C min.; Sp. gr.: 1.04; Visc.: 200 cps @ 30 C; F.P.: 15 C
Composition: 30% active ingred.; 1.2% max. Na_2SO_4; 1.2% max. unsulfated parts
Reaction: Neutral
Containers: 125 kg drums, bulk
Warnings and Caution: Prevent contact with acids because of hydrolysis

Elfan 280. Akzo Chemie
Chemical Description: Coco fatty alcohol sulfate sodium salt from native fatty alcohol C_{12}/C_{18}
Category: Detergent, anionic
Applications: General purpose heavy duty detergents for textiles, skins, leather, laundering, dyeing, metallurgy, mines and minerals, cement, papermaking, paints, varnish, foods, transport, household and cosmetic uses; hand cleaning
Form: Paste; Color: Yellowish
Properties: Solubility: 700 g/l water @ 50 C; Sp. gr.: 1.05; Visc.: 8000 cps @ 50 C; pH: 8.5 ± 1
Composition: 42% active ingred.; other components: 1%max. inorganic salts, 1.5% max. unsulfated parts
Reaction: Acid
Containers: 125 kg drums
Warnings and Caution: Prevent contact with acids because of hydrolysis

Elfan 680. Akzo Chemie
Chemical Description: Oleyl-cetyl alcohol sulfate sodium salt from oleyl-cetyl alcohol
Category: Detergent, anionic
Applications: Powder detergents, pasty cleaning agents and detergents; textile uses
Form: Paste; Color: Yellow-brown

Properties: Solubility: < 10 g/l water @ 20 C; Sp. gr.: 1.05; M.P.: 40 C; Stability: No special precautions for storage; pH: 11 ± 1
Composition: 50% active ingred.; other components: 3% max. Na_2SO_4, 1.5% max. NaCl, 7% max. unsulfated parts
Reaction: Alkaline
Containers: 125 kg drums
Warnings and Caution: Prevent contact with acids because of hydrolysis

Elfan A. Akzo Chemie
Chemical Description: Betain-type surfactant
Category: Wetting agent, emulsifier, amphoteric
Applications: Cosmetics
Form: Liquid

Elfan A 432. Akzo Chemie
Chemical Description: Amphoteric compound of the betain type from lauric-myristic acid
Category: Detergent, amphoteric
Applications: Cosmetics and disinfecting cosmetics; baby shampoos, bubble baths
Form: Liquid; Color: Yellowish
Properties: Solubility: Water-sol.; Sp. gr.: 1.04; F.P.: < 0 C; Stability: Protect against frost; Isoelectric Pt.: 6.5; pH: 5 ± 1
Composition: 30% active ingred.; other components: 5% inorganic salt
Reaction: Acid
Containers: 20 ton containers; 125 kg drums

Elfan KT 550. Akzo Chemie
Chemical Description: Coco-tallow fatty alcohol sulfate sodium salt, mixing ratio 50:50 from native fatty alcohol C_{12}/C_{18}
Category: Detergent, anionic
Applications: Heavy duty detergent in powder form, washing and cleaning pastes; textiles, skins, leather, laundering, dyeing, mines and minerals, cement, papermaking, paints, foods, transport, household and cosmetic applications
Form: Paste; Color: Yellowish
Properties: Solubility: Poor sol. in water; Sp. gr.: 1.05; Visc.: 10,000 cps @ 50 C; Stability: No special precautions for storage; pH: 8 ± 0.5
Composition: 42% active ingred.; other components: 1% max. Na_2SO4, 0.4% max. NaCl; 1.5% max. unsulfated parts
Reaction: Alkaline
Containers: 125 kg drums
Warnings and Caution: Prevent contact with acids because of hydrolysis

Elfan L 310. Akzo Chemie
Chemical Description: Ethylene glycol distearate
Category: Additive for shampoos and bubble baths, anionic
Applications: Additive for shampoos and bubble baths (pearlescent effect)
Form: Solid, flakes; Color: Light yellow
Properties: Solubility: Dispersible in water; Sp. gr.: 0.96; M.P.: 64–66 C; Stability: No special precau-

tions; Sapon. Value: 200 max.; OH No.: ca. 30; Acid No.: 6 max.
Reaction: Neutral
Containers: Paperbags 25 kg

Elfan NS. Akzo Chemie
Chemical Description: Lauryl ether sulfates
Category: Detergent, wetting agent
Applications: Dishwashing detergents, shampoos; textiles, laundering, dyeing, papermaking, household, cosmetics uses
Form: Liquid or paste
Composition: 30 or 70% active

Elfan NS 242. Akzo Chemie
Chemical Description: Lauryl ether sulfate sodium salt from native fatty alcohol C_{12}/C_{14} with ca. 2.5 mol EO
Category: Detergent, wetting agent
Applications: Liquid shampoos, bubble baths, dishwashing liquids; textiles, skins, household and cosmetic applications and in fire-extinguishers
Form: Paste; Color: Yellowish
Properties: Solubility: 400 g/l water; Visc.: 30,000 cps (@ 20 C), 15,000 cps (@ 40 C); F.P.: < 0 C; Stability: Store between 0–50 C; pH: 7.5 ± 0.5
Composition: 70% active ingred.; other components: 0.6% max. Na_2SO_4, 2% max. unsulfated parts
Reaction: Neutral
Containers: 20 ton containers
Warnings and Caution: Prevent contact with acids because of hydrolysis

Elfan NS 243 S. Akzo Chemie
Chemical Description: Lauryl ether sulfate sodium salt from synthetic fatty alcohol C_{12}/C_{14} with ca. 3 mol. EO
Category: Detergent, wetting agent, anionic
Applications: Liquid shampoos, bubble baths, dishwashing liquids; textiles, skins, household, cosmetic applications and in fire extinguishers
Form: Liquid; Color: Clear
Properties: Solubility: Water sol.; Sp. gr.: 1.04; Visc.: 50 cps@20 C; F.P.: < 0 C; Stability: Store between 0 and 50 C; pH: 7 ± 0.5
Composition: 28% active ingred.; other components: 0.3% max. Na_2SO_4; 0.8% unsulfated parts
Reaction: Neutral
Containers: 20 ton containers
Warnings and Caution: Prevent contact with acids because of hydrolysis

Elfan NS 243 S Conc. Akzo Chemie
Chemical Description: Lauryl ether sulfate sodium salt from synthetic fatty alcohol C_{12}/C_{14} with ca. 3 mol. EO
Category: Detergent, wetting agent, anionic
Applications: Liquid shampoos, bubble baths, dishwashing liquids; textiles, skins, household and cosmetic applications and in fire extinguishers
Form: Paste; Color: Yellowish
Properties: 400 g/l water; Sp. gr.: 1.12; Visc.: 30,000 cps (@ 20 C), 15,000 cps (@ 40 C); F.P.: < 0 C;

Stability: Store between 0 and 50 C; pH: 8.5 ± 1
Composition: 70% active ingred.; other components: 0.6% max. Na_2SO_4, 2% max. unsulfated parts
Reaction: Weakly alkaline
Containers: 20 ton containers
Warnings and Caution: Prevent contact with acids because of hydrolysis

Elfan NS 252 S. Akzo Chemie
Chemical Description: Lauryl ether sulfate sodium salt from synthetic fatty alcohol C_{12}/C_{15} with ca. 2.5 mol. EO
Category: Detergent, wetting agent, anionic
Applications: Liquid shampoos, bubble baths, dishwashing liquids; textiles, skins, household and cosmetic applications and in fire-extinguishers
Form: Liquid; Color: Yellowish
Properties: Solubility: Water sol.; Sp. gr.: 1.05; Visc.: 150 cps @ 20 C; F.P.: $<$ 0 C; Stability: Store between 0 and 50 C; pH: 7 ± 0.5
Composition: 28% active ingred.; other components: 0.3% max. Na_2SO_4, 0.8% max. unsulfated parts
Reaction: Neutral
Containers: 20 ton containers
Warnings and Caution: Prevent contact with acids because of hydrolysis

Elfan NS 252 S Conc. Akzo Chemie
Chemical Description: Lauryl ether sulfate sodium salt from synthetic fatty alcohol C_{12}/C_{15} with ca. 2.5 mol. EO
Category: Detergent, wetting agent, anionic
Applications: Liquid shampoos, bubble baths, dishwashing liquids; textiles, skins, household and cosmetic uses and in fire extinguishers
Form: Paste; Color: Yellowish
Properties: Solubility: 400 g/l water; Sp. gr.: 1.11; Visc.: 30,000 cps (@ 20 C), 15,000 cps (@ 40 C); F.P.: $<$ 0 C; Stability: Store between 0 and 50 C; pH: 8.5 ± 1
Composition: 70% active ingred.; other components: 0.5% max. Na_2SO_4, 2% max. unsulfated parts
Reaction: Alkaline
Containers: 20 ton containers
Warnings and Caution: Prevent contact with acids because of hydrolysis

Elfan NS 423 SH. Akzo Chemie
Chemical Description: Lauryl ether sulfate sodium salt from synthetic fatty alcohol C_{12}/C_{14} with increased amount of C_{14} with ca. 3 mol. EO
Category: Detergent, anionic
Applications: Dishwashing liquids; textiles, skins, household detergents
Form: Liquid; Color: Yellowish
Properties: Solubility: Water-sol.; Visc.: 100 cps @ 20 C; F.P.: $<$ 0 C; Stability: Store between 0 and 50 C; pH: 7.5 ± 0.5
Composition: 28% active ingred.; other components: 0.5% max. Na_2SO_4, 0.8% max. unsulfated parts
Reaction: Neutral
Container: 20 ton containers
Warnings and Caution: Prevent

contact with acids because of hydrolysis

Elfan NS 423 SH Conc. Akzo Chemie
Chemical Description: Lauryl ether sulfate sodium salt from synthetic fatty alcohol C_{12}/C_{14} with increased amount of C_{14} with ca. 3 mol. EO
Category: Detergent, anionic
Applications: Textiles, skins, household detergents esp. dishwashing liquids
Form: Paste; Color: Yellowish
Properties: Solubility: 400 g/l water; Sp. gr.: 1.10; Visc.: 30,000 cps (@ 20 C), 15,000 cps (@ 40 C); F.P.: < 0 C; Stability: Store between 0 and 50 C; pH: 8.5 ± 1
Composition: 70% active ingred.; other components: 0.6% max. Na_2SO_4, 2% max. unsulfated parts
Reaction: Alkaline
Containers: 20 ton containers
Warnings and Caution: Prevent contact with acids because of hydrolysis

Elfan OS. Akzo Chemie
Chemical Description: Olefin sulfonates
Category: Detergent, wetting agents, emulsifier
Applications: Dishwashing detergents, shampoos; textiles, laundering, dyeing, household and cosmetic detergents; making of cutting and soluble oils
Form: Liquid
Composition: 30–40 % actives

Elfan OS 46. Akzo Chemie
Chemical Description: Alpha-olefin sulfonate sodium salt from alpha-olefin C_{14}/C_{16}
Category: Detergent, anionic
Applications: Shampoos, bubble baths, dishwashing liquids, liquid and pasty cleaning agents; skins, household and cosmetic cleaners
Form: Liquid; Color: Yellow
Properties: Solubility: Water sol.; Sp. gr.: 1.05; Visc.: 250 cps @ 20 C; F.P.: < 0 C; Stability: Store between 10 and 40 C; pH: 8 ± 1
Composition: 37% active ingred.; other components: 1% max. Na_2SO_4, 1.5% max. unsulfated parts
Reaction: Alkaline
Containers: 20 ton containers

Elfan PM 600. Akzo Chemie
Chemical Description: Surfactant concentrate from fatty alcohol sulfates with increased amount of lauric alcohol and special additives for the skin
Category: Detergent, anionic
Applications: Household cleaning, light duty detergents
Form: Paste; Color: Light yellow
Properties: Solubility: Poor sol. in water; Sp. gr.: 1.06; Stability: No special precautions: pH: 8.5 ± 1
Composition: 60% active ingred.; other components: 2% min. alkylol-amide; 6% max. inorganic salts; 7% max. unsulfated parts
Reaction: Alkaline
Containers: 125 kg drums

Elfan SG 35. Akzo Chemie
Chemical Description: Silver gloss concentrate from ether sulfate

and ethylene glycol distearate
Category: Detergent, anionic
Applications: Cosmetics; shampoo additive with silver gloss; base for pearlescent shampoos and bubble bath
Form: Paste; Color: White
Properties: Solubility: Dispersible in water; Sp. gr.: 1.04; Visc.: 10,000 cps @ 20 C; Stability: Store between 10 and 40 C; pH: 8.5 ± 1
Composition: 37% active ingred.; other components: water
Reaction: Alkaline
Containers: 125 kg drums

Elfan SP 500. Akzo Chemie
Chemical Description: Surfactant concentrate from ether sulfate and alkyl benzene sulfonate with a nonionic surfactant
Category: Detergent, anionic
Applications: Household cleaners, all purpose cleaners, liquid dishwashing
Form: Liquid; Color: Yellowish
Properties: Solubility: 250 g/l water; Sp. gr.: 1.05; Visc.: 5000 cps @ 20 C; F.P.: < 0 C; Stability: Protect against frost; pH: 7 ± 0.5
Composition: 50% active ingred.; other components: water
Reaction: Neutral
Containers: 20 ton containers, 125 kg drums

Elfan WA. Akzo Chemie
Chemical Description: Dodecylbenzene sulfonate sodium salt from n-dodecylbenzene
Category: Detergent, emulsifier, anionic
Applications: Heavy duty detergent, dishwashing liquids; textiles and household cleaning; emulsion polymerization
Form: Paste; Color: Yellowish
Properties: Solubility: 250 g/l water @ 20 C; Sp. gr.: 1.08; F.P.: < 0 C; M.P.: 5 C; Stability: Store in tanks between 80–90 C; pH: 9 ± 1
Composition: 50% active ingred.; other components: 1% max. Na_2SO_4, 1% max. unsulfated parts
Reaction: Alkaline
Containers: 20 ton tanks, 125 kg drums

Elfan WA Pulver. Akzo Chemie
Chemical Description: Dodecylbenzene sulfonate sodium salt from n-dodecylbenzene
Category: Detergent
Applications: Heavy duty detergent for household and textiles uses
Form: Powder, beads; Color: Light-yellowish
Properties: Solubility: 100 g/l water @ 20 C; Density: 300 g/l; Stability: Hygroscopic; store dry; pH: 8–9
Composition: 80% active ingred.; other components: 2% max. Na_2SO_4, 18% Na_2CO_3, 1% max. unsulfated parts
Reaction: Alkaline
Containers: Paperbag 25 kg

Elfan WA Sulfosäure. Akzo Chemie
Chemical Description: Dodecylbenzene sulfonic acid from n-dodecylbenzene
Category: Raw material

Applications: Raw material for inorganic and organic neutralisates
Form: Liquid; Color: Dark brown; Odor: SO_3
Properties: Solubility: Water sol.; Sp. gr.: 1.06; Visc.: 500 cps @ 20 C; Flash Pt.: 150 C (Pensky-Martens, closed cup); Stability: Store preferaby at room temperature; Acid No.: 185 max.
Composition: 98% active ingred.; other components: 1.5% max. H_2SO_4, 1.8% unsulfated parts
Reaction: Acid
Toxicity: Skin and eye irritant
Containers: 20 ton containers
Warnings and Caution: Etching and corrosive to metals; severely irritating liquid

Elfan WAT. Akzo Chemie
Chemical Description: Dodecylbenzene sulfonate triethanolamine salt from n-dodecylbenzene
Category: Detergent, emulsifier, anionic
Applications: Liquid detergents for textile and household uses; emulsifier for insecticides
Form: Liquid; Color: Yellow-brown
Properties: Solubility: Water sol.; Sp. gr.: 1.06; Visc.: 1000 cps @ 20 C; F.P.: $<$ 0 C; Stability: No special precautions for storage; pH: 6 ± 0.5
Composition: 50% active ingred.; other components: 0.8% max. unsulfated parts
Reaction: Acid
Containers: 20 ton containers, 125 kg drums

Elfanol. Akzo Chemie
Chemical Description: Semiesters of succinic acid with ethoxylated fatty alcohols, etc.
Category: Detergent, wetting agent
Applications: Shampoos, hand cleaners; textiles, laundering, dyeing, papermaking, household and cosmetic cleaners; as wetting agents for bleaching, dyeing, coating
Form: Liquid, paste
Composition: 40–60% actives

Elfanol 510. Akzo Chemie
Chemical Description: Sulfosuccinic acid monoester of a fatty acid alkylol amide sodium salt; basis: tallow fatty acid monoethanolamide
Category: Detergent, anionic
Applications: Light duty detergents and Syndet bars; detergents for textiles, skins, leather, laundering, dyeing, metallurgy, mining, cement, papermaking, paints, foods, transport, household and cosmetic industries
Form: Paste; Color: Light yellow
Properties: Solubility: Poor water sol.; Sp. gr.: 1.06; Stability: Not stable @ pH > 8 and pH < 6; pH: 6.5 ± 1
Composition: 50% active ingred.; other components: water
Reaction: Acid
Containers: 125 kg drums

Elfanol 616. Akzo Chemie
Chemical Description: Sulfosuccinic acid monoester of an ethoxylated fatty alcohol sodium salt;

basis: synthetic fatty alcohol C_{12}/C_{14} with ca. 3 mol. EO
Category: Detergent, anionic
Applications: Cosmetics, shampoos, bubble and baby baths
Form: Liquid; Color: Clear
Properties: Solubility: Water sol.; Sp. gr.: 1.05; Visc.: 100 cps @ 20 C; F.P.: 0 C; Stability: Not stable at pH > 8 or pH < 6; pH: 7
Composition: 30% active ingred.; other components: 0.05% max. Na_2SO_4
Reaction: Neutral
Containers: 125 kg drums, 2 ton containers

Elfanol 850. Akzo Chemie
Chemical Description: Sulfosuccinic acid monoester of an ethoxylated fatty acid alkylolamide sodium salt; basis: ethoxylated coco fatty acid monoethanol amide
Category: Detergent, anionic
Applications: Hair shampoos, baby shampoos, bubble baths, other cosmetics
Form: Liquid; Color: Yellow-brown
Properties: Solubility: Poor water sol.; Sp. gr.: 1.06; F.P.: 14 C; Stability: Not stable at pH > 8 or pH < 6; pH: 7
Composition: 45% active ingred.
Reaction: Neutral
Containers: 125 kg drums

Elfanol 883. Akzo Chemie
Chemical Description: Sulfosuccinic acid diisooctylester sodium salt from 2-ethylhexanol
Category: Wetting agent, anionic
Applications: Wetting agent for technical processes
Form: Liquid; Color: Yellowish
Properties: Solubility: 15 g/l water @ 20 C; Sp. gr.: 1.08; Visc.: 50 cps @ 20 C; F.P.: < 0 C; Flash Pt.: 25 C (Pensky-Martens, closed cup); Stability: Not stable at pH > 8 or pH < 6; pH: 7
Composition: 63% active ingred.; other components: 6% isopropyl-alcohol; other: water
Reaction: Neutral
Containers: 125 kg drums
Warnings and Caution: Product contains isopropyl alcohol

Elfapur. Akzo Chemie
Chemical Description: Elfapur n ethoxylated alkyl phenols. Elfapur ethoxylated amines, amides, alcohols
Category: Detergent, wetting agent, emulsifier
Applications: Detergent formulations for textiles, laundering, dyeing, household, cosmetics industries; as wetting agents in bleaching, dyeing, waterproofing; as emulsifier for making of cutting and soluble oils, dispersing waxes, pigments and resins
Form: Liquid, paste

Elfapur KA 45. Akzo Chemie
Chemical Description: Coco fatty acid amide polyglycolether from coco fatty acid monoethanolamide
Category: Detergent, nonionic
Applications: Detergent for textiles, household esp. liquid dishwashing

Form: Liquid; Color: Yellowish
Properties: Solubility: 300 g/l water @ 20 C; Sp. gr.: 1.05; F.P.: 10–15 C; Cloud Pt.: 74–76 C (1% in water); Flash Pt.: 180 C (Pensky-Martens closed cup); Stability: No special precautions
Composition: 88% active ingred.; other components: 3% max. PEG
Reaction: Neutral
Containers: 125 kg drums, 20 ton containers

Elfapur LM 25. Akzo Chemie
Chemical Description: Lauryl alcohol polyglycol ether from native fatty alcohol C_{12}/C_{14}
Category: Raw material
Applications: Raw material for anionic surfactants esp. ethersulfates
Form: Liquid; Color: Yellowish
Properties: Solubility: Poor water sol.; Sp. gr.: 1.01; F.P.: 5 C; Cloud Pt.: 56–60 C (10% in 25% butyldiglycol); Flash Pt.: 180 C (P-M closed cup); OH No.: 187 ± 2; Stability: No special precautions
Composition: 97% active ingred.; other components: 3% max. PEG
Reaction: Neutral
Containers: 125 kg drums

Elfapur LM 30 S. Akzo Chemie
Chemical Description: Lauryl alcohol polyglycol ether from synthetic fatty alcohol C_{12}/C_{14}
Category: Raw material
Applications: Raw material for anionic surfactants esp. ethersulfates
Form: Liquid; Color: Yellowish
Properties: Solubility: Poor water sol.; F.P.: 5 C; Cloud Pt.: 60–64 C (10% in 25% butyl diglycol); OH No.: 170 ± 2; Stability: No special precautions
Composition: 97% active ingred.; other components: 3% max. PEG
Reaction: Neutral
Containers: 125 kg drums

Elfapur LM 75 S. Akzo Chemie
Chemical Description: Lauryl alcohol polyglycol ether from synthetic fatty alcohol C_{12}/C_{14}
Category: Detergent, nonionic
Applications: Detergent for textiles and household uses
Form: Liquid; Color: Yellowish
Properties: Solubility: 300 g/l water @ 20 C; Sp. gr.: 1.00; F.P.: 10 C; M.P.: 10–15 C; Cloud Pt.: 60–62 C (1% in water); Stability: No special precautions
Composition: 87% active ingred.; other components: 3% max. PEG, 10% water; also available water-free
Reaction: Neutral
Containers: 125 kg drums

Elfapur LP 25 S. Akzo Chemie
Chemical Description: Lauryl alcohol polyglycol ether from synthetic fatty alcohol C_{12}/C_{15}
Category: Raw material
Applications: Raw material for anionic surfactants esp. ethersulfates
Form: Liquid; Color: Yellowish
Properties: Solubility: Poor water sol.; Sp. gr.: 1.01; F.P.: 5 C;

Cloud Pt.: 50–54 C (10% in 25% butyl diglycol); Flash Pt.: 180 C (P-M closed cup); OH No.: 179 ± 2; Stability: No special precautions
Composition: 97% active ingred.; other components: 3% PEG
Reaction: Neutral
Containers: 125 kg drums

Elfapur ML 30 SH. Akzo Chemie
Chemical Description: Lauryl alcohol polyglycol ether from synthetic fatty alcohol C_{12}/C_{14} with increased amount of C_{14}
Category: Raw material
Applications: Raw material for anionic surfactants esp. ethersulfates
Form: Liquid; Color: Yellowish
Properties: Solubility: Poor water sol.; Sp. gr.: 1.01; F.P.: 5 C; Cloud Pt.: 62–66 C (10% in 25% butyl diglycol); Flash Pt.: 180 C (P-M closed cup); OH No. 165 ± 2; Stability: No special precautions
Composition: 97% active ingred.; other components: 3% max. PEG
Reaction: Neutral
Containers: 125 kg drums

Elfapur N 50. Akzo Chemie
Chemical Description: Nonylphenol polyglycol ether
Category: Detergent, emulsifier, nonionic
Applications: Detergent for textiles, household uses
Form: Liquid; Color: Yellowish
Properties: Solubility: Poor water sol.; Sp. gr.: 1.05; F.P.: –5 C; Cloud Pt.: 60–62 C (10% in 25% butyl diglycol); Flash Pt.: 90 C (P-M closed cup); Stability: No special precautions
Composition: 97% active ingred.; other components: 3% max. PEG
Reaction: Neutral
Containers: 125 kg drums, 20 ton containers

Elfapur N 70. Akzo Chemie
Chemical Description: Nonylphenol polyglycol ether
Category: Detergent, emulsifier
Applications: Household and textile applications
Form: Liquid; Color: Yellowish; Gel at 50–70% active ingred.
Properties: Solubility: 300 g/l water @ 20 C; Sp. gr.: 1.05; F.P.: 0 C; Cloud Pt.: 22–24 C (1% in water); Flash Pt.: 103 C (P-M, closed cup); Stability: No special precautions
Composition: 97% active ingred.; other components: 3% max. PEG
Reaction: Neutral
Containers: 125 kg drums, 20 ton containers

Elfapur N 90. Akzo Chemie
Chemical Description: Nonylphenol polyglycol ether
Category: Detergent, emulsifier, nonionic
Applications: Textiles and household applications
Form: Liquid; Color: Yellowish; Gel at 50–70% active ingred.
Properties: Solubility: 300 g/l water @ 20 C; Sp. gr.: 1.06; F.P.: 15 C; Cloud Pt.: 60–62 C (1% in water); Flash Pt.: 114 C (P-M, closed cup); Stability: No special

precautions
Composition: 97% active ingred.; other components: 3% max. PEG
Reaction: Neutral
Containers: 125 kg drums, 20 ton containers

Elfapur N 120. Akzo Chemie
Chemical Description: Nonylphenol polyglycol ether
Category: Detergent, emulsifier
Applications: Textile and household detergents; liquid dishwashing and all purpose cleaners
Form: Liquid; Color: Yellowish; Gel at 50–70% active ingred.
Properties: Solubility: 300 g/l water @ 20 C; Sp. gr.: 1.07; M.P.: 13–18 C; Cloud Pt.: 88–90 C (1% in water); Flash Pt.: 102 C (P-M closed cup); Stability:No special precautions
Composition: 97% active ingred.; 3% max. PEG
Reaction: Neutral
Containers: 125 kg drums, 20 ton containers

Elfapur N 150. Akzo Chemie
Chemical Description: Nonylphenol polyglycol ether
Category: Detergent, emulsifier, nonionic
Applications: Textiles, household cleaning
Form: Paste; Color: Yellowish; Gel at 40–70% active ingred.
Properties: Solubility: 400 g/l water @ 20 C; Sp. gr.: 1.07; M.P.: 25–30 C; Cloud Pt.: 64–66 C (1% in 10% NaCl); Flash Pt.: 94 C (P-M, closed cup); Stability: No special precautions
Composition: 97% active ingred.; other components: 3% max. PEG
Reaction: Neutral
Containers: 20 ton containers, 125 kg drums

Elfapur O 80. Akzo Chemie
Chemical Description: Oleyl-cetyl alcohol polyglycol ether from oleyl-cetyl alcohol
Category: Detergent, nonionic
Applications: Textiles industry; foam controlled heavy duty detergent
Form: Liquid; Color: Yellow-brown; Gel at 15–90% active
Properties: Solubility: 400 g/l water @ 20 C; Sp. gr.: 1.05; M.P.: 25–30 C; Cloud Pt.: 44–46 C (1% in water); Iodine No.: 22; Stability: No special precautions
Composition: 97% active ingred.; other components: 3% max. PEG
Reaction: Neutral
Containers: 125 kg drums, 20 ton containers

Elfapur O 160. Akzo Chemie
Chemical Description: Oleyl-cetyl alcohol polyglycol ether
Category: Detergent
Applications: Textiles industry
Form: Paste; Color: Yellowish; Gel at 30–80% active ingred.
Properties: Solubility: 400 g/l water @ 20 C; Sp. gr.: 1.06; M.P.: 29–35 C; Cloud Pt.: 64–66 C (1% in 10% NaCl); Stability: No special precautions
Composition: 97% active ingred.; other components: 3% max. PEG

Reaction: Neutral
Containers: 125 kg drums, 20 ton containers

Elfapur T 110. Akzo Chemie
Chemical Description: Tallow fatty alcohol polyglycol ether
Category: Detergent, nonionic
Applications: Foam controlled heavy duty detergent powders; textile use
Form: Paste; Color: Yellowish; Gel at 30–80% active ingred.
Properties: Solubility: 300 g/l water @ 20 C; Sp. gr.: 1.05; M.P.: 35–40 C; Cloud Pt.: 77–79 C (1% in water); Flash Pt.: 125 C (P-M, closed cup); Stability: 70 C max. for containers
Composition: 97% active ingred.; other components: 3% max. PEG
Reaction: Neutral
Containers: 125 kg drums, 20 ton containers

Elfapur T 250 Pulver. Akzo Chemie
Chemical Description: Tallow fatty alcohol polyglycol ether
Category: Detergent, nonionic
Applications: Foam controlled heavy duty detergent, scouring powder; textiles industry
Form: Powder, beads; gel at active ingred. > 25%; Color: White
Properties: Solubility: 250 g/l water @ 20 C; Density: 300 g/l; Cloud Pt.: 77–79 C (1% in 10% NaCl); Stability: No special precautions
Composition: 98% active ingred.; Other components: 2% max. PEG
Reaction: Neutral
Containers: 25 kg paperbags

Elfugin ANT. Sandoz Ltd.
Chemical Description: Phosphoric ester
Category: Antistatic, anionic
Applications: Antistatic agent
Form: Liquid

Elfugin KNP. Sandoz Ltd.
Chemical Description: Amine derivative
Category: Antistatic, nonionic/cationic
Applications: Antistatic agent
Form: Liquid

Elfugin V. Sandoz Ltd.
Chemical Description: Ethylene oxide adduct
Category: Antistatic, amphoteric
Applications: Antistatic agent for synthetic fibers
Form: Liquid/paste

Eltesol AX 40. Albright & Wilson Ltd. Marchon Div.
Chemical Description: Ammonium xylene sulfonate
Category: Hydrotrope, anionic
Applications: In manufacture of heavy duty detergent powders to reduce viscosity of the slurry before spray-drying; as cloud point depressants in formulating light duty liquid detergents; as solubilizers and coupling agents in heavy duty detergent formulation
Form: Liquid; Color: Pale yellow
Properties: pH: 7–8.5
Composition: 41% active ingred.; other components: 1.5% max. inorganic sulfate; balance water
Reaction: Alkaline

Eltesol PT 93. Albright & Wilson Ltd. Marchon Div.
Chemical Description: Potassium toluene sulfonate
Category and Applications: See Eltesol AX 40
Form: Powder; Color: Off white
Properties: pH: 9–10.5
Composition: 93% min. active ingred.; other components: 4.5% max. inorganic sulfate; balance water
Reaction: Alkaline

Eltesol PX 93. Albright & Wilson Ltd. Marchon Div.
Chemical Description: Potassium xylene sulfonate
Category and Applications: See Eltesol AX 40
Form: Powder; Color: Off white
Properties: pH: 9–10.5
Composition: 93% min. active ingred.; 4.5% max. inorganic sulfate; balance water
Reaction: Alkaline

Eltesol ST 34. Albright & Wilson Ltd. Marchon Div.
Chemical Description: Sodium toluene sulfonate
Category and Applications: See Eltesol AX 40
Form: Liquid; Color: Pale yellow
Properties: pH: 7–10
Composition: 34% active ingred.; other components: 4.5% max. inorganic sulfate; balance water
Reaction: Alkaline

Eltesol ST 90. Albright & Wilson Ltd. Marchon Div.
Chemical Description: Sodium toluene sulfonate
Category and Applications: See Eltesol AX 40
Form: Powder; Color: Off white (also available as pellets)
Properties: pH: 9–10.5
Composition: 89% min. active ingred.; other components: 8.5% max. inorganic sulfate; balance water
Reaction: Alkaline

Eltesol SX 30. Albright & Wilson Ltd. Marchon Div.
Chemical Description: Sodium xylene sulfonate
Category and Applications: See Eltesol AX 40
Form: Liquid; Color: Pale yellow
Properties: pH: 7–10
Composition: 30% active ingred.; other components: 1.5% max. inorganic sulfate; balance water
Reaction: Alkaline

Eltesol SX 93. Albright & Wilson Ltd. Marchon Div.
Chemical Description: Sodium xylene sulfonate
Category and Applications: See Eltesol AX 40
Form: Powder; Color: Off white (also available in pellet form)
Properties: pH: 9–10.5
Composition: 93% min. active ingred.; other components: 4.5% max. inorganic sulfate; balance water
Reaction: Alkaline

Eltesol TA65. Albright & Wilson Ltd. Marchon Div.
Chemical Description: Toluene sulfonic acid
Category: Intermediate and catalyst
Applications: As hardening agents in plastics industry; as activators for nicotine insecticides; for metal descaling and in resin-bound sand castings; as catalysts in the preparation of esters, in acetylation
Form: Clear liquid; Color: Light amber
Properties: Density: 1.2 g/cc; Visc.: 9–12 cs
Composition: 65.5% sulfonic acid

Eltesol TA96. Albright & Wilson Ltd. Marchon Div.
Chemical Description: Toluene sulfonic acid
Category and Applications: See Eltesol TA65
Form: Thick crystalline liquid; Color: Brown
Properties: Density: 1.3 g/cc @ 50 C
Composition: 97% sulfonic acid

Eltesol XA65. Albright & Wilson Ltd. Marchon Div.
Chemical Description: Xylene sulfonic acids
Category and Applications: See Eltesol TA65
Form: Clear liquid; Color: Light amber
Properties: Density: 1.2 g/cc; Visc.: 9–12 cs
Composition: 65% sulfonic acid

Eltesol XA/M65. Albright & Wilson Ltd. Marchon Div.
Chemical Description: Xylene sulfonic acid
Category and Applications: See Eltesol TA65
Form: Slightly hazy liquid; Color: Light amber
Properties: Density: 1.2 g/cc; Visc.: 9–12 cs
Composition: 65% sulfonic acid

Emcol 27-21. Witco Chem.
Chemical Description: Blend of oil soluble sulfonate and polyoxyethylene ethers
Category: Emulsifier, anionic/nonionic
Applications: Emulsifier for Ortho-Dichlorobenzene
Form: Liquid
Composition: 100% active ingred.

Emcol 27-21N. Witco Chem.
Chemical Description: Blend of oil soluble sulfonate and polyoxyethylene ethers
Category: Emulsifier, anionic/nonionic
Applications: Emulsifier for Ortho-Dichlorobenzene
Form: Liquid
Composition: 100% active ingred.

Emcol 64-6. Witco Chem.
Chemical Description: Blend of nonionic surfactants
Category: Emulsifier
Applications: For pesticide emulsifiable concentrations
Form: Liquid

Composition: 100% active ingred.

Emcol 64-6A. Witco Chem.
Chemical Description: Blend of nonionic surfactants
Category: Emulsifier
Applications: For pesticide emulsifiable concentrations
Form: Liquid
Composition: 100% active ingred.

Emcol 64-45D. Witco Chem.
Chemical Description: Blend of anionic and nonionic surfactants
Category: Emulsifier
Applications: For pesticide emulsifiable concentrates
Form: Liquid
Composition: 100% active ingred.

Emcol 4100M. Witco Chem. Organics Div.
Chemical Description: Disodium sulfosuccinate
Category: Foaming agent, cleansing agent, anionic
Applications: Bubble bath and shampoo formulations; carpet and upholstery cleaners; textiles
Form: Liquid
Properties: Solubility: Completely water sol.

Emcol 4161L. Witco Chem. Organics Div.
Chemical Description: Disodium sulfosuccinate
Category: Foaming agent, detergent
Applications: Bubble bath and shampoo formulations; cleansing agents for cosmetics and toiletries

Emcol 4300. Witco Chem. Organics Div.
Chemical Description: Disodium sulfosuccinate
Category: Foaming agent, detergent
Applications: Bubble bath and shampoo formulations; cosmetics and toiletries cleansing agent

Emcol 4500. Witco Chem. Organics Div.
Chemical Description: Sodium dioctyl sulfosuccinate
Category: Wetting agent, detergent, dispersant
Applications: Cosmetics and toiletries; textiles
Properties: Solubility: Completely water sol.

Emcol 4560. Witco Chem. Organics Div.
Chemical Description: Modified sodium dioctyl sulfosuccinate
Category: Detergent and water solubilizer
Applications: Drycleaning

Emcol 4600. Witco Chem. Organics Div.
Chemical Description: Sodium bis (tridecyl) sulfosuccinate
Category: Detergent, foam modifier, wetting agent, dispersant
Applications: Drycleaning; aqueous aerosol formulations; dispersing agent for pigments in oil-based systems

Emcol AC 62-36. Witco Chem.
Chemical Description: Blend of anionic and nonionic surfactants
Category: Emulsifier
Applications: Emulsifier for arsenic based herbicides
Form: Liquid
Composition: 100% active ingred.

Emcol AC 62-37. Witco Chem.
Chemical Description: Blend of anionic and nonionic surfactants
Category and Applications: See Emcol AC 62-36
Form: Liquid
Composition: 100% active ingred.

Emcol AC 62-38. Witco Chem.
Chemical Description: Blend of anionic and nonionic surfactants
Category and Applications: See Emcol AC 62-36
Form: Liquid
Composition: 100% active ingred.

Emcol AD 5-06. Witco Chem.
Chemical Description: Blend of oil soluble sulfonate and polyoxyethylene ethers
Category: Emulsifier, anionic/nonionic
Applications: For pesticide emulsifiable concentrates
Form: Liquid
Composition: 100% active ingred.

Emcol AD 6-89A. Witco Chem.
Chemical Description: Blend of oil soluble sulfonate and polyoxyethylene ethers
Category: Emulsifier, anionic/nonionic
Applications: For pesticide emulsifiable concentrates
Form: Liquid
Composition: 100% active ingred.

Emcol AG 4-48N. Witco Chem.
Chemical Description: Blend of oil soluble sulfonate and polyoxyethylene ethers
Category: Emulsifier, anionic/nonionic
Applications: For pesticide concentrates
Form: Liquid
Composition: 100% active ingred.

Emcol AK 16-11. Witco Chem.
Chemical Description: Blend of nonionic surfactants
Category: Emulsifier
Applications: Emulsifier for the direct solubilization of DDVP in water
Form: Liquid
Composition: 100% active ingred.

Emcol AK 16-11N. Witco Chem.
Chemical Description: Blend of nonionic surfactants
Category: Emulsifier
Applications: Emulsifier for the direct solubilization of DDVP in water
Form: Liquid
Composition: 100% active ingred.

Emcol AK 16-95. Witco Chem.
Chemical Description: Blend of nonionic surfactants
Category: Emulsifier
Applications: For pesticide emulsifiable concentrates
Form: Liquid
Composition: 100% active ingred.

Emcol AK 16-97N. Witco Chem.
Chemical Description: Blend of oil soluble sulfonate and polyoxyethylene ethers
Category: Emulsifier, anionic/nonionic
Applications: For pesticide emulsifiable concentrates
Form: Liquid
Composition: 100% active ingred.

Emcol AK 18-35. Witco Chem.
Chemical Description: Blend of oil soluble sulfonate and polyoxyethylene ethers .
Category: Emulsifier, anionic/nonionic
Applications: For pesticide emulsifiable concentrates
Form: Liquid
Composition: 100% active ingred.

Emcol AK 18-72A. Witco Chem.
Chemical Description: Blend of anionic and nonionic surfactants
Category: Emulsifier
Applications: Specific emulsifier for DDVP emulsifiable concentrates
Form: Liquid
Composition: 100% active ingred.

Emcol AL 26-43. Witco Chem.
Chemical Description: Blend of anionic and nonionic surfactants
Category: Emulsifier
Applications: For pesticide emulsifiable concentrates
Form: Liquid
Composition: 100% active ingred.

Emcol AL 69-49. Witco Chem.
Chemical Description: Blend of nonionic surfactants
Category: Emulsifier
Applications: Emulsifier for pesticide emulsifiable concentrates where anionic surfactants would decompose the toxicant, e.g., for dimethoate
Form: Liquid
Composition: 100% active ingred.

Emcol AP 08-21. Witco Chem.
Chemical Description: Blend of anionic and nonionic surfactants
Category: Emulsifier
Applications: Emulsifier for emulsifiable concentrates of highly concentrated pesticide blends
Form: Liquid
Composition: 100% active ingred.

Emcol AP 08-22. Witco Chem.
Chemical Description: Blend of anionic and nonionic surfactants
Category: Emulsifier
Applications: Emulsifier for emulsifiable concentrates of highly concentrated pesticide blends
Form: Liquid
Composition: 100% active ingred.

Emcol AP 08-22C. Witco Chem.
Chemical Description: Blend of anionic and nonionic surfactants
Category: Emulsifier
Applications: Emulsifier for emulsifiable concentrates of highly concentrated pesticide blends
Form: Liquid
Composition: 100% active ingred.

Emcol AP 08-34A. Witco Chem.
Chemical Description: Blend of anionic and nonionic surfactants

Category: Emulsifier
Applications: Emulsifier for pesticide emulsifiable concentrates
Form: Liquid
Composition: 100% active ingred.

Emcol CC-9. Witco Chem. Organics Div.
Chemical Description: Polypropoxylated quaternary ammonium chloride
Category: Dispersant, emulsifier, conditioner, detergent, antistat, wetting agent, penetrant, lubricant, cationic
Applications: Cosmetica and toiletries; industrial surfactant; textile industry; antistat and detergent for cosmetic aerosols; pigment dispersant in aqueous systems
Form: Liquid
Composition: 100% conc.

Emcol CC-36. Witco Chem. Organics Div.
Chemical Description: Polypropoxylated quaternary ammonium chloride
Category: Dispersant, antistat, detergent, emulsifier, conditioner, wetting agent, penetrant, lubricant, cationic
Applications: Cosmetics and toiletries; textile industry; aerosol formulations; pigment dispersant in aq. or nonaq. systems
Form: Liquid
Composition: 100% conc.

Emcol CC-37-18. Witco Chem. Organics Div.
Chemical Description: Coco betaine
Category: Foaming agent and stabilizer, viscosity modifier, amphoteric
Applications: Foaming agent and stabilizer for cosmetics and toiletries; viscosity modifier for cationic, anionic or nonionic formulations
Form: Liquid

Emcol CC-42. Witco Chem. Organics Div.
Chemical Description: Polypropoxylated quaternary ammonium chlorides
Category: Emulsifier, conditioner, dispersant, detergent, antistat, cationic
Applications: Drycleaning detergent; industrial and textile surfactant; cosmetics and toiletries
Form: Liquid

Emcol CC-55. Witco Chem. Organics Div.
Chemical Description: Polypropoxylated quaternary ammonium acetate
Category: Experimental cationic surfactant
Applications: Industrial, cosmetics and toiletries applications

Emcol CC-57. Witco Chem. Organics Div.
Chemical Description: Polypropoxylated quaternary ammonium phosphate
Category: Experimental cationic surfactant; detergent, antistat
Applications: Drycleaning surfactant, industrial uses, cosmetics and toiletries

Emcol CS-136, CS-143, CS-151, and CS-165. Witco Chem.
Chemical Description: Organic phosphate ethers
Category: Detergents, emulsifiers, dispersants, solubilizers, anionic
Applications: Series range from highly lipophilic to highly hydrophilic for use as detergents, emulsifiers, dispersants and solubilizers for nonionic surfactants
Form: Liquid

Emcol CS-1361. Witco Chem.
Chemical Description: Sodium salt of complex organo-phosphate ester
Applications: Used for scale control
Form: Clear liquid
Properties: Solubility: Sol. in mineral oil, kerosene, heavy aromatic naphtha, xylene, perchlorethylene, ethanol, carbon tetrachloride, water; dispersible in 10% caustic soda; Sp. gr.: 1.10; Pour Pt.: 40 F; Acid No.: 5 (to pH 9.5); Surface Tension: 31.9 dynes/cm (0.05%); pH: 5 (3% aq.)
Composition: 10% max. moisture

Emcol D 24-25. Witco Chem.
Chemical Description: Calcium sulfonate
Category: Emulsifier
Applications: Oil-in-water emulsifier for industrial use

Emcol E-607L. Witco Chem. Organics Div.
Chemical Description: Lapyrium chloride
Category: Emulsifier, conditioner, emollient, foaming, cleansing and substantive agent, cationic
Applications: Cosmetics and toiletries; industrial applications; cosmetic aerosol formulations
Form: Powder
Composition: 90% conc.

Emcol E-607S. Witco Chem. Organics Div.
Chemical Description: Stearoyl analog of Emcol E-607L
Category: Emulsifier, conditioner, emollient, detergent, substantive agent, cationic
Applications: Cosmetics and toiletries; aerosol formulations
Form: Powder
Composition: 90% conc.

Emcol H-400X. Witco Chem.
Chemical Description: Anionic/nonionic blend
Form: Slightly viscous liquid; Color: Clear to dark amber
Properties: Sp. gr.: 1.03 @ 20 C; pH: 6–7 (3% disp.)

Emcol H-470X. Witco Chem.
Chemical Description: Blend of oil soluble sulfonate and polyoxyethylene ethers
Category: Emulsifier
Applications: Emulsifier for pesticide emulsifiable concentrates
Form: Liquid
Composition: 100% active ingred.

Emcol N-83T. Witco Chem.
Chemical Description: Blend of oil soluble sulfonate and polyoxyethylene ethers
Category: Emulsifier

Applications: Emulsifier for pesticide emulsifiable concentrates
Form: Liquid
Composition: 100% active ingred.

Emcol N-85T. Witco Chem.
Chemical Description: Blend of oil soluble sulfonates and polyoxyethylene ethers
Category: Emulsifier
Applications: Emulsifier for pesticide emulsifiable concentrates
Form: Liquid
Composition: 100% active ingred.

Emcol N-140B. Witco Chem.
Chemical Description: Blend of oil soluble sulfonates and polyoxyethylene ethers
Category: Emulsifier
Applications: Emulsifier for pesticide emulsifiable concentrates
Form: Hazy liquid; Color: Dark amber
Properties: Sp. gr.: 1.01–1.03; pH: 6–7
Composition: 100% active ingred.

Emcol N-141B. Witco Chem.
Chemical Description: Blend of oil soluble sulfonates and polyoxyethylene ethers
Category: Emulsifier
Applications: Emulsifier for pesticide emulsifiable concentrates
Form: Hazy liquid; Color: Dark amber
Properties: Sp. gr.: 1.01–1.03; pH: 6–7
Composition: 100% active ingred.

Emcol N-300B. Witco Chem.
Chemical Description: Blend of oil soluble sulfonates and polyoxyethylene ethers
Category: Emulsifier, anionic/nonionic
Applications: Emulsifier for pesticide emulsifiable concentrates bal. toward high anionic content
Form: Slightly viscous liquid; Color: Clear to dark amber
Properties: Sp. gr.: 1.03 @ 20 C; pH: 6–7
Composition: 100% active ingred.

Emcol N-369. Witco Chem.
Chemical Description: Blend of oil soluble sulfonates and polyoxyethylene ethers
Category: Emulsifier
Applications: Emulsifier for pesticide emulsifiable concentrates
Form: Liquid
Composition: 100% active ingred.

Emcol N-469. Witco Chem.
Chemical Description: Blend of oil soluble sulfonates and polyoxyethylene ethers
Category: Emulsifier
Applications: Emulsifier for pesticide emulsifiable concentrates
Form: Liquid
Composition: 100% active ingred.

Emcol N-500B. Witco Chem.
Chemical Description: Blend of oil soluble sulfonates and polyoxyethylene ethers
Category: Emulsifier, anionic/nonionic
Applications: Emulsifier for pesticide emulsifiable concentrates balanced toward high nonionic con-

tent
Form: Slightly viscous liquid; Color: Clear to dark amber
Properties: pH: 6–7
Composition: 100% active ingred.

Emcol N-710. Witco Chem.
Chemical Description: Blend of oil soluble sulfonates and polyoxyethylene ethers
Category: Emulsifier
Applications: Emulsifier for pesticide emulsifiable concentrates
Form: Liquid
Composition: 100% active ingred.

Emcol N-712. Witco Chem.
Chemical Description: Blend of oil soluble sulfonates and polyoxyethylene ethers
Category: Emulsifier
Applications: Emulsifier for pesticide emulsifiable concentrates
Form: Liquid
Composition: 100% active ingred.

Emcol N-714. Witco Chem.
Chemical Description: Blend of oil soluble sulfonates and polyoxyethylene ethers
Category: Emulsifier
Applications: Emulsifier for pesticide emulsifiable concentrates
Form: Liquid
Composition: 100% active ingred.

Emcol P10-20 BU. Witco Chem.
Chemical Description: Calcium alkylarylsulfonate
Category: Intermediate
Applications: Intermediate for agricultural emulsifiers
Form: Viscous liquid; Color: Amber
Properties: Solubility: Oil sol.; Sp. gr.: 1.01–1.03; pH: 6.0–7.5 (3% in 20% isopropanol)
Composition: 68–70% active

Emcol P-1045. Witco Chem.
Chemical Description: Amine salt of alkyl-aryl sulfonic acid
Category: Wetting agent, dispersant, anionic
Applications: Wetting agent in oil-based systems, e.g., dry cleaning, floor wax removal, emulsion polymerization; sludge dispersant in fuel oil
Form: Liquid
Properties: Solubility: Oil sol., water dispersible
Composition: 90% active ingred.

Emcol P-1049. Witco Chem.
Chemical Description: Amine salt of alkyl-aryl sulfonic acid
Category: Anionic surfactant
Applications: Exceptionally water soluble surfactant for hard-to-clean applications
Form: Liquid
Properties: Solubility: Excellent water sol.
Composition: 60% active ingred.

Emcol P-1052. Witco Chem.
Category: Intermediate, anionic
Applications: Intermediate for agricultural emulsifier
Form: Viscous liquid; Color: Dark brown
Properties: pH: 6–6.4
Composition: 71–72% active

Emcol P-1059. Witco Chem.
Chemical Description: Amine salt of alkyl-aryl sulfonic acid
Category: Anionic surfactant
Applications: Surfactant for O/W emulsions, cleaning and degreasing
Form: Liquid
Composition: 90% active ingred.

Emcol P-1059B. Witco Chem.
Chemical Description: Amine salt of alkyl-aryl sulfonic acid
Category: Anionic surfactant
Applications: Dry cleaning surfactant; water solubilizer
Form: Liquid
Composition: 90% active ingred.

Emcol P-1073. Witco Chem.
Chemical Description: Amine salt of alkyl-aryl sulfonic acid
Category: Anionic surfactant
Applications: For viscosity control in invert emulsion drilling mud
Form: Viscous cloudy liquid; Color: Dark brown
Properties: Solubility: Oil sol.
Composition: 90% active ingred.

Emcol PS-331. Witco Chem.
Chemical Description: Organic phosphate ester
Category: Anionic surfactant
Applications: Surfactant for water based systems, e.g., detergents, hydrotropes, emulsion polymerization
Form: Liquid

Emcol PS-400. Witco Chem.
Chemical Description: Organic phosphate ester
Category: Anionic surfactant
Applications: Surfactant for water based systems, e.g., detergents, hydrotropes, emulsion polymerization
Form: Liquid

Emcol PS-415. Witco Chem.
Chemical Description: Organic phosphate ester
Form: Liquid
Category: Anionic surfactant
Applications: Surfactant for water based systems, e.g., detergents, hydrotropes, emulsion polymerization

Emcol TS-230. Witco Chem.
Chemical Description: Organic phosphate ester
Category: Dispersant, lubricant, emulsifier
Applications: Dispersant and lubricant in oil drilling muds; in water-based fluids during drilling and post-completion well treatment; O/W emulsifier; flow control agent
Form: Clear liquid; Color: Amber
Properties: Solubility: Sol. in xylene, ethanol, carbon tetrachloride, water, 10% caustic soda; Sp. gr.: 1.22; Pour Pt.: < 50 F; Acid No.: 170 (to pH 9.5); Surface Tension: 49.6 dynes/cm (0.05%); pH: 2 (3% aq.)

Emcol ZM-1117. Witco Chem.
Chemical Description: Blend of oil soluble surfactants and polyoxyethylene ethers
Category: Emulsifier, anionic/nonionic

Applications: Emulsifier for pesticide emulsifiable concentrates
Form: Liquid
Composition: 100% active ingred.

Emcon E. Fanning Chem.
Chemical Description: Mixture of glycerides, phosphatides, and egg-derived sterols
Category: Emulsifier, nonionic
Applications: Cosmetic/pharmaceutical emulsions; excellent W/O emulsifier to form bases with water absorbing properties and add emollience to formulations; fat liquoring; leather, foods, household applications
Form: Liquid; Color: Yellow; Odor: Bland
Properties: Solubility: Mineral oil, vegetable oil, most organic solvents; Density: 9.95
Composition: 100% active ingred.
Reaction: Neutral
Toxicity: Slightly irritating; excellent emollient
Containers: Steel drums

Emphos CS-141. Witco Chem.
Chemical Description: Organic phosphate ester
Category: Emulsifier, lubricant, wetting agent, dispersant, solubilizer, corrosion inhibitor, antistatic agent, anionic
Applications: Hard-surface cleaners, emulsion polymerization, textile wet processing, dry cleaning detergents, cosmetic creams and lotions, metal degreasing, paper making, pesticide formulation, corrosion inhibitors, drilling mud fluids
Form: Clear viscous liquid
Properties: Solubility: Sol. in heavy aromatic naphtha, xylene, perchlorethylene, ethanol, carbon tetrachloride, water, 10% caustic soda; partially sol. in kerosene; Sp. gr.: 1.12; Pour Pt.: > 40 F; Acid No.: 110 (to pH 9.5); Surface Tension: 30.6 dynes/cm (0.05% sol'n.); pH: 2 (3% aq.)
Warnings and Caution: Acid–protect with rubber gloves and goggles while handling
References: Bulletin 234-CS141

Emphos CS-143. Witco Chem.
Chemical Description: Phosphoric acid ester
Category: Emulsifier for Kelthane
Form: Clear liquid
Properties: Acid No.: 150 mg KOH/g (to pH 9.5); pH: 2.5–3 (3% aq.)
Reaction: Acid

Emphos CS-151. Witco Chem.
Chemical Description: Organic phosphate ester
Category: Emulsifier, lubricant, wetting agent, dispersant, solubilizer, corrosion inhibitor, antistat, anionic
Applications: Hard surface cleaners, emulsion polymerization, textile wet processing, dry cleaning detergents, cosmetic creams and lotions, metal degreasing, paper making, pesticide formulation, corrosion inhibitors, drilling mud fluids
Form: Clear viscous liquid
Properties: Solubility: Sol. in heavy aromatic naphtha, xylene,

perchlorethylene, ethanol, carbon tetrachloride, water, 10% caustic soda; partially sol. in kerosene; Sp. gr.: 1.11; Pour Pt.: > 50 F; Acid No.: 95 (to pH 9.5); Surface Tension: 36.8 dynes/cm (0.05%); pH: 2 (3% aq.)
Reaction: Acid
Warnings and Caution: Acid; protect during handling with rubber gloves and goggles
References: Bulletin 234-CS151

Emerest 2301. Emery Industries
Chemical Description: Methyl oleate
Category: Lubricant, defoamer, floatation agent, plasticizer, nonionic
Applications: As components in mold release agents, defoamers, flotation agents, plasticizers for cellulosic plastics, needle lubricants, bases for industrial lubricants; when sulfated, as wetting, rewetting and dye leveling agents for textiles and leather
Form: Liquid; Color: < 6 (Gardner)
Properties: Solubility: Sol. in IPA, toluol, mineral oil; Density: 7.3 lb/gal; Visc.: 5 cs @ 38 C; F.P.: –16 C; Flash Pt.: 177 C; Sapon. Value: 195; Acid No.: 4
Composition: 100% active ingred.

Emerest 2302. Emery Industries
Chemical Description: Propyl oleate
Category and Applications: See Emerest 2301
Form: Liquid; Color: 7 max. (Gardner)
Properties: Solubility: Sol. in IPA, toluol, mineral oil; Density: 7.2 lb/gal; Visc.: 6 cs @ 38 C; F.P.: –20 C; Sapon. Value: 175; Acid No.: 6
Composition: 100% active ingred.

Emerest 2314. Emery Industries
Chemical Description: Isopropyl myristate
Category: Emollient, lubricant, nonionic
Applications: Cosmetic emollient; sewing thread lubricant
Form: Liquid; Color: 1 max. (Gardner)
Properties: Solubility: Sol. in IPA, toluol, mineral oil; Density: 7.1 lb/gal; Visc.: 6 cs @ 38 C; F.P.: –5 C; Flash Pt.: 152 C; Sapon. Value: 210; Acid No.: 1
Composition: 100% active ingred.

Emerest 2316. Emery Industries
Chemical Description: Isopropyl palmitaite (from high purity palmitic acid)
Category: Lubricant, emollient, nonionic
Applications: Lubricant for synthetic fibers; cosmetic emollient
Form: Liquid; Color: 1 max. (Gardner)
Properties: Solubility: Sol. in IPA, toluol, mineral oil; Density: 7.1 lb/gal; Visc.: 8 cs @ 38 C; F.P.: 14 C; Flash Pt.: 163 C; Sapon. Value: 188; Acid No.: 1
Composition: 100% active ingred.

Emerest 2326. Emery Industries
Chemical Description: Butyl stearate
Category: Lubricant, nonionic
Applications: Used in aluminum foil rolling, wire drawing; in products used in synthetic fiber spin finishing, fiber and yarn processing
Form: Liquid; Color: 1 max. (Gardner)
Properties: Solubility: Sol. in IPA, toluol, mineral oil; Density: 7.2 lb/gal; Visc.: 10 cs; F.P.: 20 C; Flash Pt.: 188 .C; Sapon. Value: 172; Acid No.: 1
Composition: 100% active ingred.

Emerest 2350. Emery Industries
Chemical Description: Ethylene glycol monostearate
Category: Opacifying and pearlescing agent, nonionic
Applications: Liquid cosmetic and detergent compounds as opacifier and pearlescing agent
Composition: 100% active ingred.

Emerest 2355. Emery Industries
Chemical Description: Ethylene glycol distearate
Category: Opacifier, nonionic
Applications: Liquid detergent and cosmetic products as opacifier
Form: Beads; Color: 1 (Gardner)
Properties: Solubility: Sol. in IPA, toluol, mineral oil; F.P.: 50 C; HLB: 2.1; Sapon. Value: 185; Acid No.: 8
Form: Flake; Color: 4 (Gardner)
Properties: Solubility: Dispersible in mineral oil; F.P.: 62 C; HLB: 1.2; Sapon. Value: 195; Acid No.: 6
Composition: 100% active ingred.

Emerest 2381. Emery Industries
Chemical Description: Propylene glycol monostearate
Category: Pearlescing and spreading agent, anionic
Applications: Imparts pearlescence to cosmetic and dishwashing products; as spreading agent in hand lotions
Form: Beads; Color: 2 (Gardner)
Properties: Solubility: Sol. in IPA; dispersible in water, mineral oil, toluol (hot); F.P.: 40 C; HLB: 4.0; Sapon. Value: 170; Acid No.: 20
Composition: 100% active ingred.

Emerest 2400. Emery Industries
Chemical Description: Glycerol monostearate
Category: Emulsifier, nonionic
Applications: In hand creams and other cosmetics; in industrial lubricants and lubricant softeners for textiles
Form: Beads; Color: 1 (Gardner)
Properties: Solubility: Sol. in IPA; sol. hot in toluol, mineral oil; F.P.: 57.5 C; Sapon. Value: 173; Acid No. 3; Iodine Value: 0.5
Composition: 100% active ingred.

Emerest 2401. Emery Industries
Chemical Description: Glycerol monostearate
Category: Emulsifier, nonionic
Applications: Technical grade of Emerest 2400 with similar functions

Form: Beads; Color: 2 (Gardner)
Properties: Solubility: Sol. in IPA; sol. hot in toluol, mineral oil; F.P.: 58 C; Sapon. Value: 153; Acid No.: 3; Iodine Value: 1
Composition: 100% active ingred.

Emerest 2407. Emery Industries
Chemical Description: Glycerol monostearate, SE
Category: Emulsifier, modified anionic
Applications: Cosmetic creams and lotions
Form: Beads; Color: 3 (Gardner)
Properties: Solubility: Sol. in IPA, sol. hot in toluol, mineral oil; F.P.: 58 C; Sapon. Value: 153; Acid No.: 20; Iodine Value: 1; Emulsifiability: Self-emulsifying
Composition: 100% active ingred.

Emerest 2421. Emery Industries
Chemical Description: Glycerol monooleate
Category: Emulsifier, nonionic
Applications: As a component in mold release agents, a vehicle for agricultural insecticides, an anti-icing fuel additive, a rust preventive additive for compounded oils; in textiles as a lubricant component in synthetic fiber spin finishes
Form: Liquid; Color: 11 (Gardner)
Properties: Solubility: Sol. in toluol, mineral oil; Density: 7.9 lb/gal; Visc.: 91 cs @ 38 C; F.P.: 6 C; Flash Pt.: 235 C; Sapon. Value: 170; Acid No.: 6; Iodine Value: 73
Composition: 100% active ingred.

Emerest 2423. Emery Industries
Chemical Description: Glycerol trioleate
Category: Lubricant, emulsifier, nonionic
Applications: Called synthetic olive oil; when emulsified, as lubricant for metals, leather, textiles; sulfated GTO used as a softener for leathers and textiles
Form: Liquid; Color: 3 (Gardner)
Properties: Solubility: Sol. in toluol, mineral oil; Density: 7.6 lb/gal; Visc.: 47 cs; F.P.: 9 C; Flash Pt.: 293 C; Sapon. Value: 197; Acid No.: 5; Iodine Value: 85
Composition: 100% active ingred.

Emerest 2620. Emery Industries
Chemical Description: PEG-200 monolaurate
Category: Emulsifier, coupling agent, defoamer, nonionic
Applications: Defoamer in water base coatings; viscosity depressant in vinyl plastisols; viscosity control additive in hair rinse formulations; paper softener
Form: Liquid; Color: 1 (Gardner)
Properties: Solubility: Dispersible in water, mineral oil, butyl stearate, glycerol trioleate, perchloroethylene, Stod. solvent; Density: 8.2 lb/gal; Visc.: 40 cs; HLB: 9.8; Cloud Pt.: < 25 C
Composition: 100% active ingred.

Emerest 2622. Emery Industries
Chemical Description: PEG-200 dilaurate
Category: Coemulsifier, lubricant, nonionic

Applications: Coemulsifier and lubricant in self-emulsifiable textile and industrial oils; mold release agent; viscosity control agent; in specialty paper coatings
Form: Liquid; Color: 2 (Gardner)
Properties: Solubility: Dispersible in water; sol. in mineral oil, butyl stearate, glycerol trioleate, perchloroethylene, Stod. solvent; Density: 8.0 lb/gal; Visc.: 35 cs; HLB: 6.8; Cloud Pt.: < 25 C
Composition: 100% active ingred.

Emerest 2640. Emery Industries
Chemical Description: PEG-400 monostearate
Category: Emulsifier, lubricant, nonionic
Applications: Emulsifier of oils and fats in the manufacture of fluid or paste emulsions for industrial lubricants, consumer products, textile lubricants and softeners; as thickening agent and stabilizer for starch coatings on paper; as water-dispersible paper size; lubricant for channeling wire through conduit
Form: Soft waxy solid; Color: < 1 (Gardner)
Properties: Solubility: Dispersible in water, mineral oil, butyl stearate, glycerol trioleate, perchloroethylene, Stod. solvent; M.P.: 32 C; HLB: 11.7; Cloud Pt.: < 25 C
Composition: 100% active ingred.

Emerest 2642. Emery Industries
Chemical Description: PEG-400 distearate
Category: Emulsifier, thickener, nonionic
Applications: Cosmetic and industrial emulsions
Form: Soft solid; Color: 2 (Gardner)
Properties: Solubility: Sol. in mineral oil, butyl stearate, glycerol trioleate, perchloroethylene, Stod. solvent; dispersible in water; M.P.: 36 C; HLB: 8.1; Cloud Pt.: < 25 C
Composition: 100% active ingred.

Emerest 2646. Emery Industries
Chemical Description: PEG-400 monooleate
Category: Emulsifier, nonionic
Applications: Balanced emulsifier for oil-based soluble cutting oils and other specialty industrial lubricants; industrial degreasers; viscosity stabilizer for vinyl plastisols; used in textile softeners and lubricant bases; as component in drycleaning formulations
Form: Liquid; Color: 2 (Gardner)
Properties: Solubility: Dispersible in water, butyl stearate, glycerol trioleate, perchloroethylene, Stod. solvent; Density: 8.5 lb/gal; Visc.: 100 cs; HLB: 11.7; Cloud Pt.: < 25 C
Composition: 100% active ingred.

Emerest 2647. Emery Industries
Chemical Description: PEG-400 sesquioleate
Category: Emulsifier, nonionic
Applications: Oil and fat emulsifier in industrial and textile lubricants

Form: Liquid; Color: 2 (Gardner)
Properties: Solubility: Sol. in mineral oil, butyl stearate, glycerol trioleate, perchloroethylene; disp. in water, Stod. solvent; Density: 8.2 lb/gal; Visc.: 75 cs; HLB: 9.2; Cloud Pt.: < 25 C
Composition: 100% active ingred.

Emerest 2648. Emery Industries
Chemical Description: PEG-400 dioleate
Category: Emulsifier, solubilizer, nonionic
Applications: Lipophilic emulsifier and solubilizer for mineral oils, fats, and solvents; emulsifier of kerosene in agricultural and pesticide sprays; latex paints, metalworking fluids, solvents, specialty and industrial lubricants
Composition: 100% active ingred.

Emerest 2650. Emery Industries
Chemical Description: PEG-400 monolaurate
Category: Wetting agent, defoamer, leveling agent, dispersant, solubilizer, nonionic
Applications: Wetting agent, defoamer and leveling agent in latex paints; dispersant for pigment grinding; as solubilizer for oils and solvents; as antiblock agent in vinyls; in textile fiber processing
Form: Liquid; Color: 1 (Gardner)
Properties: Solubility: Sol. in water; disp. in butyl stearate, glycerol trioleate, perchloroethylene, Stod. solvent; Density: 8.6 lb/gal; Visc.: 100 cs; HLB: 13.1; Cloud Pt.: 33 C
Composition: 100% active ingred.

Emerest 2652. Emery Industries
Chemical Description: PEG-400 dilaurate
Category: Emulsifier, nonionic
Applications: Oil-soluble emulsifier, used as softener, lubricant, release agent for paper; coupler and lubricant in synthetic fiber spin finishes in textile industry
Form: Liquid; Color: 2 (Gardner)
Properties: Solubility: Sol. in mineral oil, butyl stearate, glycerol trioleate, perchloroethylene, Stod. solvent; disp. in water; Density: 8.3 lb/gal; Visc.: 65 cs; HLB: 10.4; Cloud Pt.: < 25 C
Composition: 100% active ingred.

Emerest 2660. Emery Industries
Chemical Description: PEG-600 monooleate
Category: Emulsifier, nonionic
Applications: Water-soluble emulsifier and detergent; specialty lubricants; dye leveling agent in textiles; component in systems for the acid washing of printed circuit boards
Form: Liquid; Color: 2 (Gardner)
Properties: Solubility: Sol. in water; disp. in butyl stearate, glycerol trioleate, Stod. solvent; Density: 8.7 lb/gal; Visc.: 115 cs; HLB: 13.6; Cloud Pt.: 47 C
Composition: 100% active ingred.

Emery 6705. Emery Industries
Chemical Description: Phenoxyethanol
Category: Solvent

Applications: Solvent for nitrocellulose, cellulose acetate, ethylcellulose, and vinyl, alkyl and ester type resins; solvent in printing inks; in special formulations for removing industrial finishes and coatings
Form: Liquid; Color: 1 (Gardner)
Properties: Solubility: Sol. in perchloroethylene, Stod. solvent; disp. in butyl stearate; Density: 9.1 lb/gal; Visc.: 21 cs
Composition: 100% active ingred.

Emery 6748. Emery Industries
Chemical Description: Amphoteric betaine surfactant
Applications: Shampoos, hair products, bath products, skin cleaners, mild baby shampoos, other specialty cosmetics
Form: Liquid; Color: 4 (Gardner)
Properties: Solubility: Sol. in water; disp. in glycerol trioleate; Density: 8.7 lb/gal; Stability: Stable over wide pH range; high tolerance to hard water and electrolytes; high foaming
Composition: 30% active ingred.

Emery 6752. Emery Industries
Chemical Description: Monocarboxylate coconut imidazolinium derivative
Category: Amphoteric surfactant
Applications: Cleaning compounds and shampoos; viscosity improver and foam stabilizer in combination with anionic or nonionic surfactants; detergent in alkaline systems and steam cleaning compounds
Form: Liquid; Color: 9 (Gardner)
Properties: Solubility: Sol. in water; disp. in glycerol trioleate; Density: 8.7 lb/gal; Stability: Stable over wide pH range; high tolerance to hard water and electrolytes; isoelectric point = 7
Composition: 38% active ingred.

Emery 6849. Emery Industries
Chemical Description: Linear POE alcohol
Applications: Special purpose product used in solid built toilet bowl cleaners
Form: Liquid; Color: 1 (Gardner)
Properties: Solubility: Water sol.; Density: 8.2 lb/gal; Cloud Pt.: > 100 C
Composition: 65% active ingred.

Emgard 2035. Emery Industries
Category: Nonionic surfactant
Applications: Used in surfactant systems for soluble citrus spray oils
Form: Liquid; Color: 4 (Gardner)
Properties: Solubility: Sol. in mineral oil, glycerol trioleate; disp. in butyl stearate, perchloroethylene, Stod. solvent; Density: 8.1 lb/gal; HLB: 8.5
Composition: 100% active ingred.

Emgard 2051. Emery Industries
Category: Emulsifier
Applications: Emulsifier for aromatic weed oils
Form: Liquid; Color: 6 (Gardner)
Properties: Solubility: Sol. in water, glycerol trioleate; disp. in min-

eral oil, butyl stearate, Stod. solvent; Density: 8.9 lb/gal; HLB: 15.0
Composition: 90% active ingred.

Emgard 2063. Emery Industries
Category: Penetrant
Applications: Agricultural formulations
Form: Liquid; Color: 1 (Gardner)
Properties: Solubility: Sol. in water, butyl stearate, glycerol trioleate, Stod. solvent; Density: 8.2 lb/gal; Cloud Pt.: < 25 C
Composition: 100% active ingred.

Emgard 2066. Emery Industries
Category: Spreader/activator
Applications: Agricultural surfactant systems
Form: Liquid; Color: 1 (Gardner)
Properties: Solubility: Sol. in water, butyl stearate, glycerol; disp. hot in mineral oil, Stoddard solvent; Density: 8.7 lb/gal; Cloud Pt.: 14 C
Composition: 100% active ingred.

Emgard 2067. Emery Industries
Category: Spreader base
Applications: Contains isopropanol as a water coupling agent
Form: Liquid; Color: 1 (Gardner)
Properties: Solubility: Sol. in water, mineral oil, butyl stearate, glycerol trioleate, Stod. solvent
Composition: 80% active ingred.

Emid 6500. Emery Industries
Chemical Description: Coconut monoethanolamide
Category: Foam stabilizer, thickener, nonionic
Applications: Shampoos, liquid detergents, rug cleaners
Form: Flake; Color: 4 (Gardner)
Properties: Solubility: Sol. in mineral oil, butyl stearate, glycerol trioleate, perchloroethylene, Stod. solvent; disp. in water; M.P.: 71 C
Composition: 100% active ingred.

Emid 6510. Emery Industries
Chemical Description: Lauric super diethanolamide
Category: Foam booster, stabilizer, thickener in combination with anionic surfactants; nonionic
Applications: Shampoos, bubble baths, general purpose detergents
Form: Solid; Color: 2 (Gardner)
Properties: Solubility: Sol. in perchloroethylene, Stod. solvent; disp. in water, mineral oil, butyl stearate, glycerol trioleate; M.P.: 42 C
Composition: 100% active ingred.

Emid 6511. Emery Industries
Chemical Description: Lauric super diethanolamide
Category: Emulsifier, nonionic
Applications: Heavy duty, built, liquid formulations
Form: Solid; Color: 2 (Gardner)
Properties: Solubility: Sol. in perchloroethylene, Stod. solvent; disp. in water, mineral oil, butyl stearate, glycerol trioleate; M.P.: 40 C
Composition: 100% active ingred.

Emid 6514. Emery Industries
Chemical Description: Coco super

diethanolamide
Category: Foam stabilizer, thickener, detergent, nonionic
Applications: Liquid cleaning compounds
Form: Liquid; Color: 3 (Gardner)
Properties: Solubility: Sol. in water, butyl stearate, glycerol trioleate, perchloroethylene, Stod. solvent; disp. in mineral oil; Density: 8.2 lb/gal
Composition: 100% active ingred.

Emid 6515. Emery Industries
Chemical Description: Coco oil super diethanolamide
Category: Foam booster and stabilizer, nonionic
Applications: Boosts and stabilizes foam; inhibits redeposition of soils
Form: Liquid; Color: 3 (Gardner)
Properties: Solubility: Sol. in water, butyl stearate, glycerol trioleate, perchloroethylene, Stod. solvent; disp. in mineral oil; Density: 8.3 lb/gal
Composition: 100% active ingred.

Emid 6533. Emery Industries
Chemical Description: Modified coco diethanolamide
Category: Emulsifier, thickening agent, foamer
Applications: Emulsifier, thickening agent, moderate foamer for controlled suds detergents; fulling and scouring agent for textiles
Form: Liquid; Color: 4 (Gardner)
Properties: Solubility: Sol. in water, perchloroethylene, Stod. solvent; disp. in mineral oil, butyl stearate, glycerol trioleate; Density: 8.3 lb/gal
Composition: 100% active ingred.

Emid 6534. Emery Industries
Chemical Description: Modified coco diethanolamide
Category: Thickener, foam stabilizer, emulsifier, nonionic
Applications: Shampoos; fatty acid soaps in cosmetic formulations
Form: Liquid; Color: 4 (Gardner)
Properties: Solubility: Sol. in water, perchloroethylene, Stod. solvent; disp. in mineral oil, butyl stearate, glycerol trioleate; Density: 8.5 lb/gal
Composition: 100% active ingred.

Emid 6538. Emery Industries
Chemical Description: Modified coco diethanolamide
Category: Thickener, foam stabilizer, nonionic
Applications: Built liquid detergents
Form: Liquid; Color: 4 (Gardner)
Properties: Solubility: Sol. in water, perchloroethylene, Stod. solvent; disp. in mineral oil, butyl stearate, glycerol trioleate; Density: 8.3 lb/gal
Composition: 100% active ingred.

Emid 6541. Emery Industries
Chemical Description: Lauric diethanolamide
Category: Emulsifier, foam booster, stabilizer, nonionic
Applications: Shampoos, general purpose cleaners

Form: Paste; Color: 2 (Gardner)
Properties: Solubility: Sol. in water, perchloroethylene, Stod. solvent; Density: 8.3 lb/gal
Composition: 100% active ingred.

Emid 6543. Emery Industries
Chemical Description: Stearic-oleic diethanolamide
Category: Coemulsifier, nonionic
Applications: Hydrophobic coemulsifier for mineral oils, fatty esters, and various solvents
Form: Liquid; Color: 4 (Gardner)
Properties: Solubility: Sol. in water, perchloroethylene, Stod. solvent; disp. in mineral oil, butyl stearate, glycerol trioleate; Density: 8.5 lb/gal
Composition: 100% active ingred.

Emid 6544. Emery Industries
Chemical Description: Capric diethanolamide
Category: Foam booster, wetting agent, detergency builder, nonionic
Applications: Used in salt and alkali systems; dedusting agent for alkaline powders
Properties: Solubility: Sol. in butyl stearate, perchloroethylene, Stod. solvent; disp. in water, mineral oil, glycerol trioleate; M.P.: 30 C
Composition: 100% active ingred.

Emid 6545. Emery Industries
Chemical Description: Oleic diethanolamide
Category: Emulsifier, foam suppresant, nonionic
Applications: Emulsifier for mineral oils in formulating antistatic fiber processing aids and yarn lubricants; foam suppresant in dye carrier and solvent emulsions
Form: Liquid; Color: 7 (Gardner)
Properties: Solubility: Sol. in mineral oil, butyl stearate, glycerol trioleate, perchloroethylene, Stod. solvent; disp. in water; Density: 7.8 lb/gal
Composition: 100% active ingred.

Emkabase. Emkay Chem. Co.
Chemical Description: Naphthenic, amino sulfonate blend
Category: Emulsifier, anionic
Applications: Emulsifier for mineral oil
Form: Liquid

Emkagen. Emkay Chem. Co.
Chemical Description: Amide amine condensate
Category: Detergent
Form: Liquid

Emkalane. Emkay Chem. Co.
Category: Dyeing assistant
Applications: Dyeing assistant for wools and blends

Emkanol. Emkay Chem. Co.
Category: Mercerizing assistants

Emkanyl. Emkay Chem. Co.
Category: Dyeing assistant
Applications: Dyeing assistant for hosiery

Emkapol. Emkay Chem. Co.
Applications: Soap products

Form: Paste

Emkapon. Emkay Chem. Co.
Chemical Description: Amide sulfonates
Category: Detergent blend

Emkatex. Emkay Chem.
Applications: Dyestuff dispersing condensate

Empicol AL 30. Albright & Wilson Ltd.
Chemical Description: Ammonium salt of sulfated narrow cut lauryl alcohol
Category: Detergent, anionic
Applications: Shampoos, carpet shampoos
Form: Liquid/paste; Color: Amber
Properties: m.w.: 291; Visc.: 14,000 cs @ 20 C; Stability: May become pastes or hazy liquids under cold storage; becomes clear liquid on warming; pH: 6.5 ± 0.5 (5% sol'n.)
Composition: 29% active ingred.

Empicol CHC 30. Albright & Wilson Ltd. Marchon Div.
Chemical Description: Sodium cetyl oleyl sulfate
Category: Detergent, anionic
Applications: Multi purpose textile detergent with good wetting, dispersing and emulsifying properties; wool scouring
Form: Soft paste; Color: Cream
Properties: Solubility: Readily sol. in warm water; Stability: Chemically stable when heated under alkaline conditions and in acid solutions; pH: 8.5 min. (2% sol'n.)
Composition: 30% active ingred.
Reaction: Alkaline

Empicol DLS. Albright & Wilson Ltd. Marchon Div.
Chemical Description: Diethanolamine lauryl sulfate
Category: Detergent, anionic
Applications: Liquid and lotion shampoos; mild to hair and skin
Form: Clear liquid; Color: Straw
Properties: m.w.: 379 (of active); Density: 1.05 g/cc; Visc.: 500 cs max. @ 20 C; Cloud Pt.: < 0 C; pH: 7–8 (1% aq.)
Composition: 34% active ingred.

Empicol EL. Albright & Wilson Ltd. Marchon Div.
Chemical Description: Concentrated blend of anionic and nonionic surfactants
Category: Detergent
Applications: Manufacture of liquid shampoos
Form: Viscous liquid; Color: Amber
Properties: Sp. gr.: 1.025 @ 20 C; Visc.: Visc. approx. doubles at 50% sol'n.; 586 cs (conc. @ 20 C); 1143 cs (50% sol'n. @ 20 C); Cloud Pt.: –2 C
Composition: 50% active ingred.

Empicol EMD. Albright & Wilson Ltd. Marchon Div.
Chemical Description: Blend of natural, straight chain fatty alcohol
Category: Detergent
Applications: Very mild detergent

for shampoos, children's and baby products with good foaming, hard water performance
Form: Liquid; Color: Pale straw; Odor: Almost odorless
Properties: Solubility: Completely sol. in cold water; Density: 1.05 g/cc @ 20 C; Visc.: < 250 cs @ 20 C; Cloud Pt.: < 0 C; Stability: Resistant to hydrolysis in strong alkalis; decomposed by strong acids; pH: 6.5–7.5 (5% sol'n.); Biodegradable
Composition: 30% active ingred.

Empicol ESB 3. Albright & Wilson Ltd. Marchon Div.
Chemical Description: Sodium salt of sulfated ethoxylated lauryl alcohol
Category: Shampoo base, anionic
Applications: Liquid and lotion shampoo base with good foaming properties
Form: Liquid; Color: Colorless
Properties: Solubility: Water sol.; Density: 1.0; Visc.: ≈ 3000 cs @ 20 C; Cloud Pt.: 0 C; Stability: Stable to alkalis, to acids under certain conditions; pH: 6.5–7.5 (5% sol'n.)
Composition: 27.5% active ingred.

Empicol ESB 3/S. Albright & Wilson Ltd. Marchon Div.
Chemical Description: Sodium salt of sulfated ethoxylated fatty alcohol
Category: Shampoo base
Applications: Raw material for liquid and lotion shampoos, body shampoos and bubble baths with good foaming properties, clarity
Form: Liquid; Color: Colorless
Properties: Solubility: Excellent; Density: ≈ 1.01 g/cc; Visc.: 2600 cs @ 20 C; Cloud Pt.: < 0 C; Biodegradable
Composition: 25% sodium lauryl ether sulfate
Reaction: Neutral (5% sol'n.)

Empicol ESB 30. Albright & Wilson Ltd. Marchon Div.
Chemical Description: Sodium salt of sulfated ethoxylated lauryl alcohol
Category: Shampoo base
Applications: Liquid and lotion shampoo base with good foaming properties
Form: Liquid; Color: Colorless
Properties: m.w.: 384 (of active); Solubility: Water sol.; Density: ≈ 1.0; Visc.: ≈ 20,000 cs @ 20 C; Cloud Pt.: 0 C; Stability: Stable to alkalis, to acids under certain conditions; pH: 6.5–7.5 (5% sol'n.)
Composition: 27.5% active ingred.

Empicol ESB 50. Albright & Wilson Ltd. Marchon Div.
Chemical Description: Sodium salt of sulfated ethoxylated lauryl alcohol
Category: Shampoo base
Applications: Liquid and lotion shampoo base with good foaming properties
Form: Cloudy liquid; Color: Pale straw
Properties: m.w.: 384 (of active); Solubility: Slowly sol. in cold water; may tend to form a gel; Den-

sity: ≈ 1.1; Visc.: Pourable liquid; Stability: Stable to alkalis, to acids under certain conditions; pH: 6.5–7.5
Composition: 50% active ingred.

Empicol LZG30. Albright & Wilson Ltd. Marchon Div.
Chemical Description: Sodium salt of sulfated medium cut lauryl alcohol
Category: Anionic surfactant, wetting agent
Applications: Industrial grade anionic surfactants; wetting agents with low inorganic salt content
Form: Paste; Color: Cream
Composition: 30% active ingred.
Reaction: Alkaline

Empicol LZGV. Albright & Wilson Ltd. Marchon Div.
Chemical Description: Sodium salt of sulfated medium cut lauryl alcohol
Category: Anionic surfactant, wetting agent
Applications: Industrial grade anionic surfactants; wetting agents with low inorganic salt content
Form: Needle; Color: Cream
Composition: 85% active ingred.
Reaction: Alkaline

Empicol MD. Albright & Wilson Ltd. Marchon Div.
Chemical Description: Blend of natural, straight chain fatty alcohol
Category: Detergent
Applications: Very mild detergent for shampoos, children's and baby products with good foaming, hard water performance
Form: Liquid; Color: Pale straw; Odor: Almost odorless
Properties: Solubility: Completely sol. in cold water; Density: 1.05 g/cc @ 20 C; Visc.: < 250 cs @ 20 C; Cloud Pt.: < 0 C; Stability: Resistant to hydrolysis in strong alkalis; decomposed by strong acids; pH: 6.5–7.5; Biodegradable
Composition: 30% active ingred.

Empicol ML 26. Albright & Wilson Ltd. Marchon Div.
Chemical Description: Magnesium salt of sulfated narrow-cut lauryl alcohol
Category: Detergent, anionic
Applications: In manufacture of shampoos, toothpastes
Form: Liquid; Color: Pale straw
Properties: Solubility: Miscible with water; may form slightly hazy solution due to residual alkalinity; add small amount of citric acid to form clear solution; Density: ≈ 1.0 g/cc; Visc.: ≈ 150 cs
Composition: 26% active ingred.
Reaction: Alkaline

Empicol TAS90. Albright & Wilson Ltd. Marchon Div.
Chemical Description: Sodium tallow alcohol sulfate
Category and Applications: See Empicol TAS80V; also as high temperature fabric scouring agent; enhances luster of calendered fabrics due to wax and oil emulsifying powers; additive during sizing
Form: Powder; Color: Cream

Properties: pH: 10
Composition: 90% active ingred.
Reaction: Alkaline

Empicol TC 34. Albright & Wilson Ltd. Marchon Div.
Chemical Description: Ammonium/triethanolamine salt of sulfated narrow cut lauryl alcohol
Category: Detergent, anionic
Applications: Shampoos
Form: Clear liquid; Color: Amber
Properties: m.w.: 317; Visc.: 27,000 cs @ 20 C; Stability: May become pastes or hazy liquid under cold storage; becomes clear liquid on warming; pH: 6.5 ± 0.3 (5% sol'n.)
Composition: 31.5% active ingred.

Empicol TCR. Albright & Wilson Ltd. Marchon Div.
Chemical Description: Ammonium/triethanolamine salt of sulfated narrow cut lauryl alcohol
Category: Detergent, foaming agent, anionic
Applications: Shampoos, liquid detergents
Form: Clear liquid; Color: Golden yellow
Properties: m.w.: 317; Visc.: 12,500 cs @ 20 C; Stability: May become pastes or hazy liquid under cold storage; becomes clear liquid on warming; pH: 6.5 ± 0.5 (5% sol'n.)
Composition: 21.5% active ingred.

Empicol TDL 75. Albright & Wilson Ltd. Marchon Div.
Chemical Description: Diethanolamine/triethanolamine salt of sulfated lauryl alcohol
Category: Shampoo raw material
Applications: Raw material for shampoos, bubble baths, bath oils, etc. with excellent detergency, wetting and foaming properties, good solubility; mild to skin and scalp
Form: Soft paste; Color: Yellow to amber
Properties: m.w.: 410 (of active); Solubility: Readily sol. in water; Density: ≈ 1.1 g/cc; pH: 7.5 ± 0.2 (2% sol'n.)
Composition: 75% active ingred.; 6% max. ethanol; 4.5% max. free fatty material; 0.5% max. chloride ion; 2.5% max. sulfate ion

Empicol TL 40. Albright & Wilson Ltd. Marchon Div.
Chemical Description: Triethanolamine salt of sulfated narrow-cut lauryl alcohol
Category: Detergent
Applications: Manufacture of liquid and lotion shampoos
Form: Clear liquid; Color: Very pale yellow
Properties: Solubility: Miscible with water; Density: 1.025 g/cc; Visc.: 300 cs @ 20 C; Stability: Stable to acids and alkalis under certain conditions; pH: 6.8–7.1 (5% sol'n.)
Composition: 40% triethanolamine lauryl sulfate

Empicol TL40/S. Albright & Wilson Ltd. Marchon Div.
Chemical Description: Triethanolamine salt of sulfated fatty alcohol
Category: Detergent, foaming a-

gent
Applications: Liquid and lotion shampoo base with good detergency, foaming power, solubility and viscosity characteristics
Form: Clear liquid; Color: Very pale yellow
Properties: Density: ≈ 1.025 g/cc
Composition: 38% active ingred.
Reaction: Neutral

Empicol TLP. Albright & Wilson Ltd. Marchon Div.
Chemical Description: Triethanolamine salt of sulfated narrow-cut lauryl alcohol
Category: Detergent
Applications: Manufacture of liquid and lotion shampoos; alkylolamide acts as a foam booster and stabilizer and viscosity modifier and lends some superfatting properties
Form: Clear liquid; Color: Golden yellow
Properties: Solubility: Miscible with water; Density: 1.025 g/cc; Visc.: 1000 cs @ 20 C; pH: 6.8–7.1 (5% sol'n.)
Composition: 40.5% triethanolamine lauryl sulfate; 1.0% alkylolamide

Empicol TLR. Albright & Wilson Ltd. Marchon Div.
Chemical Description: Triethanolamine salt of sulfated narrow-cut lauryl alcohol
Category: Detergent
Applications: Manufacture of liquid and lotion shampoos; alkylolamide acts as a foam booster and stabilizer and viscosity modifier and lends some superfatting properties
Form: Clear liquid; Color: Golden yellow
Properties: Solubility: Miscible with water; Density: 1.025 g/cc; Visc.: 4000 cs @ 20 C; pH: 6.5–7.2 (5% sol'n.)
Composition: 35.5% triethanolamine lauryl sulfate; 3% min. alkylolamide

Empicol WA. Albright & Wilson Ltd. Marchon Div.
Chemical Description: Sodium salt of sulfated narrow-cut lauryl alcohol
Category: Industrial surfactant
Applications: Industrial uses, e.g., preparation of rubber foams for carpet backing
Form: Clear liquid; Color: Pale yellow
Properties: m.w.: 292 (of active); Sp. gr.: ≈ 1.04; pH: 8.5–9.5 (5% aq.)
Composition: 27% active ingred.; other components: 5% solubilizer; 66.5% water

Empicol XC35. Albright & Wilson Ltd. Marchon Div.
Chemical Description: Blend of anionic and nonionic components
Category: Shampoo base
Applications: Ingredient for liquid cream (lotion) shampoos; formulations should contain a preservative
Form: Viscous pearly liquid; Odor: Almost odorless
Reaction: Neutral

Empicol XC 35/S. Albright & Wilson Ltd. Marchon Div.
Chemical Description: Blend of anionic and nonionic components
Category: Shampoo base
Applications: Ingredient for pearly lotion shampoos; formulations should contain a preservative
Form: White liquid; Odor: Almost odorless
Properties: Stability: Good
Reaction: Neutral

Empicol XDB. Albright & Wilson Ltd. Marchon Div.
Chemical Description: Detergent complex
Category: Detergent
Applications: Detergent, foaming and conditioning agent for shampoos; substantive to hair and skin
Form: Clear liquid; Color: Golden
Properties: Visc.: ≈ 4000 cs; Cloud Pt.: < 0 C; pH: 8 ± 0.5 (5% sol'n.)
Composition: 30% solids content; 70% water content
Reaction: Alkaline

Empicol XM 17. Albright & Wilson Ltd.
Category: Detergent, anionic
Applications: Carpet shampoo double concentrate
Form: Liquid; Color: Pale amber
Properties: Visc.: ≈ 3000 cs @ 20 C; Cloud Pt.: 5 C; pH: 6–7 (5% sol'n.)
Composition: 21% anionic active ingred.; 4% foam booster

Empicol XT 45. Albright & Wilson Ltd. Marchon Div.
Chemical Description: Blend of anionic and nonionic components
Category: Shampoo base
Applications: Active ingredient for lacquer removing hair shampoo
Form: Mobile liquid; Color: Pale yellow
Properties: Sp. gr.: ≈ 1.05 @ 20 C; Visc.: 100 cs max. @ 20 C; Cloud Pt.: < 0 C; pH: 9.0–9.5 (5% aq.)
Composition: 26–27% active ingred.
Reaction: Alkaline

Empicryl APD. Albright & Wilson Ltd. Marchon Div.
Chemical Description: Aqueous solution of the disodium salt of a copolymer of maleic anhydride and diisobutylene
Category: Dispersant
Applications: Pigment dispersant for water based paints; pigment stabilizer; imparts high gloss to paints
Form: Clear liquid; Color: Pale yellow
Properties: Sp. gr.: 1.12 @ 18 C; Visc.: 16 cs; F.P.: –3 C; pH: 12.0–12.7
Composition: 22% min. diisobutylene disodium maleate copolymers; 3% max. sodium maleate
Reaction: Alkaline

Empicryl APD/B. Albright & Wilson Ltd. Marchon Div.
Chemical Description: Aqeuous solution of the disodium salt of a copolymer of maleic anhydride and diisobutylene
Category and Applications: See Empicryl APD

Form and Properties: Identical to Empicryl APD but with a lower pH (concentrate) of 10.0–10.5

Empicryl LM. Albright & Wilson Ltd. Marchon Div.
Chemical Description: Lauryl/myristyl methacrylate monomer
Applications: Used for polymerization to yield polymers in solution or emulsion form for surface coatings for textiles, leather, etc.; stabilized with Topanol A against premature polymerization
Form: Color: 100 max. (Hazen)
Properties: Sapon. Value: 191–218
Composition: 62% min. lauryl methacrylate; 22% min. myristyl methacrylate

Empicryl TM. Albright & Wilson Ltd. Marchon Div.
Chemical Description: Ceto-stearyl methacrylate monomer
Applications: See Empicryl LM
Form: Color: 200 max. (Hazen)
Properties: Sapon. Value: 152–178
Composition: 57% min. stearyl methacrylate; 25% min. cetyl methacrylate

Empigen BAC. Albright & Wilson Ltd. Marchon Div.
Chemical Description: Benzalkonium chloride solution B.P.
Category: Disinfectant, cationic
Applications: Disinfectant for dairy, food processing, restaurant, brewing and bottling industries; medical and pharmaceutical applications
Form: Liquid; Color: Pale yellow; Odor: Aromatic
Properties: Solubility: Miscible with water, alcohol and acetone
Composition: 50% active ingred.; 0.2% max. sulfated ash
Reaction: Incompatible with soap and other anionic active surfactants and with alkali hydroxides

Empigen BB. Albright & Wilson Ltd. Marchon Div.
Chemical Description: Alkyl dimethyl betaine
Category: Foam booster/stabilizer, wetting agent, thickening agent
Applications: Ingredient for shampoos and detergents, esp. baby shampoos due to mildness; static control
Form: Liquid; Color: Almost water white
Properties: Stability: Excellent acid and alkali stability
Composition: 30% betaine

Empilan BQ 100. Albright & Wilson Ltd. Marchon Div.
Chemical Description: Polyethylene glycol monoester of a fatty acid
Category: Emulsifier, wetting agent, antistat, nonionic
Applications: Emulsifier for insecticides, spindle oils, etc; industrial wetting agent; scouring agent; antifoam in laundry works, in glue manufacture and in paper coating mixes; ingred. in hand cleaning jellies; antistat for PVC; turbine oil additive
Form: Liquid; Color: Dark brown;

Odor: Faint, typical
Properties: m.w.: 400 (of glycol base); Solubility: Dispersible in water; yields clear solutions with ethanol, xylene, oleic acid; yields turbid solutions with white spirit, mineral oil, perchlorethylene, kerosene, naphtha; Sp. gr.: 1.017 @ 20 C; Visc.: 130 cps; M.P.: ≈ −1 C
Composition: 100% active ingred.

Empilan CDE. Albright & Wilson Ltd. Marchon Div.
Chemical Description: Coconut diethanolamide
Category: Foam stabilizer, solubilizer, detergent, nonionic
Applications: Shampoos, liquid detergent formulations
Form: Liquid/paste; Color: Golden yellow
Composition: 100% active ingred.

Empilan CME. Albright & Wilson Ltd. Marchon Div.
Chemical Description: Coconut monoethanolamide
Category: Foam booster/stabilizer; viscosity modifier, nonionic
Applications: Foam booster/stabilizers for liquid and powder detergents; viscosity modifier for liquid detergents, dishwashers, hair shampoos, carpet shampoos; as bases for manufacture of ethoxylated derivatives
Form: Waxy flake; Color: Cream
Properties: Solubility: Dispersible in hot water; Density: 0.4 g/cc; Stability: Stable to acids and alkalis under certain conditions
Composition: 94.2% active ingred.
Reaction: Alkaline

Empilan CDX. Albright & Wilson Ltd. Marchon Div.
Chemical Description: Coconut diethanolamide
Category and Applications: See Empilan CDE
Form: Liquid/paste; Color: Straw
Composition: 25% free amine

Empilan DL 40. Albright & Wilson Ltd. Marchon Div.
Chemical Description: Fatty alcohol ethoxylate
Category: Emulsifier, wetting, dye leveling and dispersing agents, nonionic
Applications: Textile dye leveling and dispersing agent for cellulosic, polyester, acrylic fibers, wool and nylon; stripping of vat dyeings using sodium bisulfite and caustic soda; dyeing and printing with azoic dyes; bleaching of textiles; desizing baths; antistat; emulsifier for fatty acids and mineral oils for fiber lubricants; wetting and stabilizing agent for synthetic rubber latices
Form: Clear liquid; Color: Pale yellow
Properties: Solubility: Miscible with water; Density: 1.046 g/cc; Cloud Pt.: > 100 C in distilled water; Stability: Yields neutral solutions stable to acids, alkalis and most electrolytes at the concentrations found in textile industry
Composition: 40% active ingred.

Empilan DL 100. Albright & Wilson Ltd. Marchon Div.

Chemical Description: Fatty alcohol ethoxylate
Category and Applications: See Empilan DL 40
Form: Stiff waxy solid
Properties: Solubility: Readily sol. in warm water; M.P.: ≈ 36 C; Cloud Pt.: > 100 C (1% in distilled water)
Composition: 100% active ingred.

Empilan EGMS. Albright & Wilson Ltd. Marchon Div.
Chemical Description: Ethylene glycol monostearate
Category: Opacifier, pearling agent, emulsifier, stabilizer
Applications: Opacifier/pearling agent and emulsifying/stabilizing agent in liquid cream (lotion) and cream shampoos; oil-soluble emollient in hand lotions and creams
Form: Hard wax-like flake; Color: White
Properties: Solubility: Oil-soluble; M.P.: ≈ 56 C
Composition: 65% diester; 30% monoester; 3.5% max. free fatty acid

Empilan GMS LSE 32. Albright & Wilson Ltd. Marchon Div.
Chemical Description: Glyceryl monostearate, self-emulsifying, food grade
Category: Emulsifier, nonionic
Form: Wax-like powder; Color: White
Properties: m.w.: 358 (of monoglyceride); M.P.: 55–60 C; pH: 7–9 (10% aq.)
Composition: 32.5% min. monoglyceride

Empilan GMS NSE 32. Albright & Wilson Ltd. Marchon Div.
Chemical Description: Glyceryl monostearate, food grade
Category: Nonionic emulsifier
Form: Wax-like powder; Color: White
Properties: M.P.: 54–59 C; pH: 6–8 (10% aq.)
Composition: 32.5% min. monoglyceride

Empilan GMS NSE 40. Albright & Wilson Ltd. Marchon Div.
Chemical Description: Glyceryl monostearate, food grade
Category: Nonionic emulsifier
Form: Wax-like powder; Color: White
Properties: M.P.: 53–58 C; pH: 6–8 (10% aq.)
Composition: 38% min. monoglyceride

Empilan GMS SE 32. Albright & Wilson Ltd. Marchon Div.
Chemical Description: Glyceryl monostearate, self-emulsifying
Category: Nonionic emulsifier
Form: Wax-like powder; Color: White
Properties: M.P.: 55–60 C; pH: 7–9 (10% aq.)
Composition: 32.5% min. monoglyceride

Empilan GMS SE 40. Albright & Wilson Ltd. Marchon Div.
Chemical Description: Glyceryl monostearate
Category: Nonionic emulsifier
Form: Wax-like powder; Color: White

Properties: M.P.: 56–61 C; pH: 7–9 (10% aq.)
Composition: 38% min. monoglyceride

Empilan KA3. Albright & Wilson Ltd. Marchon Div.
Chemical Description: Fatty alcohol ethylene oxide condensates
Category: Emulsifier, detergent, dispersant, wetting agent, nonionic
Applications: For textiles as scouring and wetting agents, as emulsifiers in self emulsifiable spinning oils, as dye leveling and dispersing agents; for metal processing as wetting agents in alkaline cleaners, electrolyte and pickling baths, as additives in cutting and rolling oils; for paper industry as wetting and deinking assistants; for paint industry as emulsifiers in acrylic/vinyl acetate emulsions; in insecticides and pesticides
Form: Clear liquid; Color: Pale straw (@ 20 C)
Properties: Solubility: Sol. in mineral oils and other organic solvents; HLB: 8.9; Cloud Pt.: Insoluble; Stability: Good; Biodegradable
Composition: 100% active ingred.
Reaction: Neutral or slightly alkaline

Empilan KA5. Albright & Wilson Ltd. Marchon Div.
Chemical Description: Fatty alcohol ethylene oxide condensate
Category and Applications: See Empilan KA3
Form: Clear liquid; Color: Pale straw (@ 20 C)
Properties: HLB: 11.5; Cloud Pt.: 37 C (1% aq. sol'n.); Stability: Good; Biodegradable
Composition: 100% active ingred.
Reaction: Neutral or slightly alkaline

Empilan KA590. Albright & Wilson Ltd. Marchon Div.
Chemical Description: Fatty alcohol ethylene oxide condensate
Category and Applications: See Empilan KA3
Form: Clear liquid; Color: Water white
Properties: Density: 9.8 lb/gal; HLB: 11.4; Cloud Pt.: 95 F; Pour Pt.: 12 F; Biodegradable
Composition: 90% active ingred.
Reaction: Neutral or slightly alkaline; compatible with wide range of anionics, cationics, acids, alkalis, detergent builders and oxidizing and bleaching agents in textile industry in normal use
Warnings and Caution: Avoid high concentrations Empilan KA590 with high concentration of strong oxidizing agent; instability and explosion may result

Empilan KA880. Albright & Wilson Ltd. Marchon Div.
Chemical Description: Fatty alcohol ethylene oxide condensate
Category and Applications: See Empilan KA3
Form: Clear liquid; Color: Pale straw (@ 20 C)
Properties: HLB: 13.6; Cloud Pt.: 81 C; Stability: Good; Biodegradable
Composition: 80% active ingred.

Reaction: Neutral or slightly alkaline

Empilan KA1080. Albright & Wilson Ltd. Marchon Div.
Chemical Description: Fatty alcohol ethylene oxide condensate
Category and Applications: See Empilan KA3
Form: Clear liquid; Color: Pale straw (@ 20 C)
Properties: HLB: 14.6; Cloud Pt.: 95 C; Stability: Good; Biodegradable
Composition: 80% active ingred.
Reaction: Neutral or slightly alkaline

Empilan KB 2. Albright & Wilson Ltd. Marchon Div.
Chemical Description: Ethylene oxide condensate of narrow-cut lauryl alcohol
Category: Emulsifier, foam booster, super-fatting agents, nonionic
Applications: For sulfation process to yield salts of lauryl ether sulfates and phosphates; as emulsifiers, foam boosters and superfatting agents in detergents and emulsifying systems
Form: Liquid; Color: Almost colorless
Properties: Solubility: Insol. in cold water; Density: ≈ 0.9; Visc.: 25 cs; Stability: Stable to acids and alkalis
Composition: 100% active ingred.

Empilan KB 3. Albright & Wilson Ltd. Marchon Div.
Chemical Description: Ethylene oxide condensates of narrow-cut lauryl alcohol
Category and Applications: See Empilan KB 2
Form: Liquid; Color: Almost colorless
Properties: Solubility: Insol. in cold water; Density: ≈ 0.9; Visc.: 25 cs; Stability: Stable to acids and alkalis; pH: 6–7 (5% sol'n.)
Composition: 100% active ingred.

Empilan KL 6. Albright & Wilson Ltd. Marchon Div.
Chemical Description: Ethylene oxide condensate of cetyl/oleyl alcohol
Category: Emulsifier, detergent, wetting agent
Applications: Emulsifier for mineral oils in cosmetics; detergents and wetting agents in textile industry
Form: Soft paste; Color: Cream
Properties: Solubility: Dispersible in cold water; Density: 0.9 g/cc; M.P.: 35 C; Stability: Stable to acids and alkalis; pH: 8–10
Composition: 100% active ingred.
Reaction: Alkaline

Empilan KL 10. Albright & Wilson Ltd. Marchon Div.
Chemical Description: Ethylene oxide condensate of cetyl/oleyl alcohol
Category and Applications: See Empilan KL 6
Form: Paste; Color: Cream
Properties: Solubility: Sol. in cold water; Density: 0.9 g/cc; M.P.: 40 C; Stability: Stable to acids and alkalis; pH: 8–10

Composition: 100% active ingred.
Reaction: Alkaline

Empilan KL 20. Albright & Wilson Ltd. Marchon Div.
Chemical Description: Ethylene oxide condensate of cetyl/oleyl alcohol
Category and Applications: See Empilan KL 6
Form: Soft solid; Color: Cream
Properties: Solubility: Sol. in cold water; Density: 0.9 g/cc; M.P.: 42 C; Stability: Stable to acids and alkalis
Composition: 100% active ingred.
Reaction: Alkaline; pH = 8–10

Empilan KM 11. Albright & Wilson Ltd. Marchon Div.
Chemical Description: Polyethanoxy alkyl ether
Category: Emulsifier, nonionic
Applications: Emulsifier of waxes and certain fatty alcohols; foam control agent in synthetic heavy duty detergents; soap additive
Form: Paste; Color: Colorless to pale straw
Properties: Setting Pt.: ≈ 30 C; Hydroxyl Value: 73
Composition: 100% active ingred.

Empilan KM 20. Albright & Wilson Ltd. Marchon Div.
Chemical Description: Polyethanoxy alkyl ether
Category: Emulsifier, nonionic
Applications: Emulsifier of waxes and certain fatty alcohols; foam control agent in synthetic heavy duty detergents; toilet bowl cleaners
Form: Waxy flake/block; Color: Colorless to pale straw
Properties: Setting Pt.: ≈ 38 C; Hydroxyl Value: 49

Empilan KM 50. Albright & Wilson Ltd. Marchon Div.
Chemical Description: Polyethanoxy alkyl ether
Category and Applications: See Empilan KM 20
Form: Waxy flake/block (also available in powder form); Color: Colorless to pale straw
Properties: Setting Pt.: ≈ 40 C; Hydroxyl Value: 23

Empilan LDE. Albright & Wilson Ltd. Marchon Div.
Chemical Description: Lauric diethanolamide
Category and Applications: See Empilan CDE
Form: Solid; Color: Straw
Composition: 6% max. free amine

Empilan LIS. Albright & Wilson Ltd. Marchon Div.
Chemical Description: Lauric monoisopropanolamide
Category and Applications: See Empilan CME
Form: Waxy flake; Color: Cream
Properties: Solubility: Dispersible in hot water; Density: 0.4 g/cc; Stability: Stable to acids and alkalis under certain conditions
Composition: 87.5% active ingred.

Empilan LP2. Albright & Wilson Ltd. Marchon Div.
Chemical Description: Ethylene oxide condensate of Empilan CME

Category: Foam and detergent booster, nonionic
Applications: Liquid detergent additive as foam stabilizer, detergent booster, and solubilizer
Form: Soft paste; Color: Creamy yellow; Odor: Faint characteristic
Properties: Sapon. Value: 14 max. (mg KOH/g); Stability: Stable to acids and alkalis
Composition: 100% active ingred.
Reaction: Slightly alkaline

Empilan MAA. Albright & Wilson Ltd. Marchon Div.
Chemical Description: Ethylene oxide condensate of Empilan CME
Category and Applications: See Empilan LP2
Form: Soft paste; Color: Creamy yellow; Odor: Faint characteristic
Properties: Sapon. Value: 15 max. (mg KOH/g); Stability: Stable to acids and alkalis
Composition: 100% active ingred.
Reaction: Slightly alkaline

Empilan NP Series. Albright & Wilson Ltd. Marchon Div.
Chemical Description: Nonyl phenol ethoxylate
Category: Wetting agent, detergent, emulsifier, nonionics
Applications: Empilan NP4: emulsifiers; when sulfated yields anionic surfactants for liquid detergents; Empilan NP9 and NP12: detergent and cleaning systems; Empilan NP15: in rubber industry for latex preparation; Empilan NP20 and NP30: high temperature alkaline detergents; other series general applications: in concrete mixing, in metal polishing preparations, as demulsifiers for crude petroleum emulsions, as solubilizers for certain insecticides
Composition: 100% active ingred.

Empilan NP9. Albright & Wilson Ltd. Marchon Div.
Chemical Description: Nonyl phenol ethoxylate
Category and Applications: See Empilan NP Series
Form: Soft paste/liquid; Color: Pale straw
Properties: Density: 1.0 g/cc; Visc.: 300 cs; Cloud Pt.: 55 C (1% in water); Hydroxyl Value: 90 (mg KOH/g)
Composition: 100% active ingred.
Reaction: Neutral or slightly alkaline

Empimin KSN 27. Albright & Wilson Ltd. Marchon Div.
Chemical Description: Sodium lauryl ether sulfate based on a narrow-cut lauryl alcohol, industrial grade
Category: Detergent ingredient, anionic
Applications: Used in liquid detergents in conjunction with sodium alkyl benzene sulfonate to yield product with mild "feel" and good foaming properties
Form: Clear or slightly turbid liquid; Color: Pale straw
Properties: Visc.: 25–100 cs; Stability: Good sol. and stability under alkaline conditions; Biodegradable
Composition: 27% active ingred.

Empimin KSN 60. Albright & Wilson Ltd. Marchon Div.
Chemical Description: Sodium lauryl ether sulfate based on narrow-cut lauryl alcohol; with 10% ethanol as solubilizer
Category and Applications: See Empimin KSN 27
Form: Translucent liquid; Color: Pale amber
Properties: Visc.: 2500–3000 cs; Stability: Stable and sol. under alkaline conditions; Biodegradable
Composition: 60% active ingred.

Empimin LF. Albright & Wilson Ltd. Marchon Div.
Chemical Description: Blend of foaming agents
Category: Foaming agent
Applications: Foaming agent for use with noncarboxylated SBR latices in manufacture of foamed latex carpet backing
Form: Clear liquid; Color: Yellow
Properties: Visc.: 25 cs @ 20 C; Cloud Pt.: −3 C; Stability: Good; pH: 8 (1% aq.)
Composition: 35% active ingred.
Reaction: Alkaline

Empimin MHH. Albright & Wilson Ltd.
Chemical Description: Disodium N-lauryl sulfosuccinamate
Category: Foaming agent, anionic
Applications: Foaming agent for rubber latices, suggested for high density carpet
Form: Liquid; Color: Pale cream
Properties: pH: 8 (5% aq.)
Composition: 40% active ingred.
Reaction: Alkaline

Empimin MKK. Albright & Wilson Ltd.
Chemical Description: Disodium N-stearyl sulfosuccinamate
Category: Foaming agent, anionic
Applications: Foaming agent for rubber latices
Form: Soft paste; Color: Pale cream
Properties: pH: 8 (5% aq.)
Composition: 35% active ingred.
Reaction: Alkaline

Empimin MSS. Albright & Wilson Ltd.
Chemical Description: Diammonium N-lauryl sulfosuccinamate
Category: Foaming agent, anionic
Applications: Foaming agent for rubber latices, suggested as fast foaming agent for low-density backings
Form: Liquid; Color: Pale cream
Properties: pH: 8 (5% aq.)
Composition: 40% active ingred.
Reaction: Alkaline

Empimin MTT. Albright & Wilson Ltd.
Chemical Description: Disodium N-oleyl sulfosuccinamate
Category: Foaming agent, anionic
Applications: Foaming agent for rubber latices, suggested for high density carpet and for foaming of carboxylated SBR latices
Form: Liquid; Color: Pale cream
Properties: pH: 8 (5% aq.)
Composition: 40% active ingred.

Reaction: Alkaline

Empiphos 4KP. Albright & Wilson Ltd. Marchon Div.
Chemical Description: Tetra potassium pyrophosphate
Category: Detergent builder
Applications: Builder for heavy duty liquid detergents; as clarifying agent in liquid soaps; manufacture of synthetic rubber; as a buffer in electroplating baths; as a pigment dispersant and stabilizer in emulsion paints; in high pressure boiler water treatment to reduce scale
Properties: Solubility: 191 g/100 g water @ 25 C; Density: 75 lb/ft^3
Reaction: Alkaline; pH = 10.3 (1% sol'n.)

Empiwax SK. Albright & Wilson Ltd.
Chemical Description: Based on cetyl/stearyl alcohol
Category: Emulsifier, anionic
Applications: Self-emulsifying wax as O/W emulsifier for pharmaceuitcal and toilet preparations and ointments
Form: Flake; Color: White
Properties: Stability: 10% emulsion in water is stable

Empiwax SK/BP. Albright & Wilson Ltd. Marchon Div.
Chemical Description: Based on cetyl/stearyl alcohol
Category: Emulsifier, anionic
Applications: Self-emulsifying wax as O/W emulsifier for pharmaceutical and toilet preparations and ointments; esp. for use in penicillin cream bases; complies with B.P. specs for emulsifying wax
Form: Flake; Color: White
Properties: Stability: 10% emulsion in water is stable

Emsorb 2500. Emery Industries
Chemical Description: Sorbitan monooleate
Category: Emulsifier, coupling agent, lubricant, softener, nonionic
Applications: Textile fibers and leather; can be used with water-sol. emulsifiers in formulating clear emulsifiable concentrates of petroleum oils and waxes, natural fats, alkyl esters; cosmetics and household products
Form: Liquid; Color: 8 (Gardner)
Properties: Solubility: Sol. in mineral oil, butyl stearate, perchloroethylene, Stod. solvent; disp. in glycerol trioleate; Density: 8.3 lb/gal; Visc.: 1000 cs; HLB: 4.6
Composition: 100% conc.

Emsorb 2502. Emery Industries
Chemical Description: Sorbitan sesquioleate
Category: Emulsifier, coupling agent, nonionic
Applications: Primary emulsifier for W/O systems; coemulsifier for O/W systems for mineral oil, fats, waxes; coupling agent; household aerosols; cosmetics; industrial oils; textile oils
Form: Liquid; Color: 6 (Gardner)
Properties: Solubility: Sol. in mineral oil, butyl stearate, glycerol trioleate, perchloroethylene, Stod.

solvent; water insol.; Density: 8.2 lb/gal; Visc.: 1500 cs; HLB: 3.7
Composition: 100% conc.

Emsorb 2503. Emery Industries
Chemical Description: Sorbitan trioleate
Category: Coupling agent, emulsifier, nonionic
Applications: Processing and softening of textiles and leather
Form: Liquid; Color: 7 (Gardner)
Properties: Solubility: Sol. in mineral oil, butyl stearate, glycerol trioleate, perchloroethylene, Stod. solvent; Density: 7.9 lb/gal; Visc.: 200 cs; HLB: 2.7
Composition: 100% active ingred.

Emsorb 2505. Emery Industries
Chemical Description: Sorbitan monostearate
Category: Emulsifier, nonionic
Applications: Hydrophobic emulsifier for emulsions of mineral oils, fats, waxes and silicones; coemulsifier for industrial oils, household products and cosmetics; textile lubricant; with Emsorb 6905 in paraffin wax emulsions and silicone defoamers for paper and textile processing
Form: Solid; Color: 4 (Gardner)
Properties: Solubility: Disp. in butyl stearate, glycerol trioleate, perchloroethylene; water insol.; M.P.: 54 C; HLB: 5.4
Composition: 100% conc.

Emsorb 2507. Emery Industries
Chemical Description: Sorbitan tristearate
Category: Emulsifier, nonionic
Applications: Hydrophobic emulsifier similar to Emsorb 2503; also as a fiber-to-metal textile lubricant
Form: Waxy solid; Color: 4 (Gardner)
Properties: Solubility: Sol. in butyl stearate, glycerol trioleate, perchloroethylene, Stod. solvent; disp. in mineral oil; water insol.; M.P.: 54 C; HLB: 2.4
Composition: 100% conc.

Emsorb 2510. Emery Industries
Chemical Description: Sorbitan monopalmitate
Category: Emulsifier, nonionic
Applications: Oil soluble waxy emulsifier for cosmetics and household products; fiber-to-metal lubricant for synthetic and cellulosic fibers
Form: Solid; Color: 8 (Gardner)
Properties: Solubility: Sol. in Stod. solvent; disp. in butyl stearate; M.P.: 48 C; HLB: 6.6
Composition: 100% conc.

Emsorb 2515. Emery Industries
Chemical Description: Sorbitan monolaurate
Category: Emulsifier, nonionic
Applications: Water dispersible emulsifier with antifoam properties for oils, fats, solvents in household specialties, industrial oils, cosmetics and emulsion polymerization
Form: Liquid; Color: 7 (Gardner)
Properties: Solubility: Sol. in mineral oil, butyl stearate, glycerol trioleate, perchloroethylene, Stod. solvent; dispersible in water; Density: 8.8 lb/gal; Visc.: 4500 cs;

HLB: 7.6
Composition: 100% conc.

Emsorb 6900. Emery Industries
Chemical Description: POE (20) sorbitan monooleate
Category: Emulsifier, dispersant, solubilizer, nonionic
Applications: Hydrophilic emulsifier and coemulsifier for petroleum oils, fats, solvents and waxes in cosmetics, household products, industrial lubricants, textile dye carriers; as dispersant for pigments in coatings; as solubilizer for oils and fragrances; emulsifier for aliphatic alcohols in tobacco sucker control concentrates
Form: Liquid; Color: 5 (Gardner)
Properties: Solubility: Sol. in water; disp. in glycerol trioleate; Density: 9.0 lb/gal; Visc.: 400 cs; HLB: 15.1; Cloud Pt.: > 100 C
Composition: 100% conc.

Emsorb 6901. Emery Industries
Chemical Description: POE (5) sorbitan monooleate
Category: Emulsifier, lubricant, nonionic
Applications: O/W emulsifier and lubricant with mineral and vegetable oils in industrial lubricants; textile lubricants and softeners
Form: Liquid; Color: 6 (Gardner)
Properties: Solubility: Sol. in glycerol trioleate, Stod. solvent; disp. in water and mineral oil; Density: 8.5 lb/gal; Visc.: 450 cs; HLB: 10.1; Cloud Pt.: < 25 C
Composition: 100% conc.

Emsorb 6903. Emery Industries
Chemical Description: POE (20) sorbitan trioleate
Category: Emulsifier, lubricant, nonionic
Applications: O/W emulsifier for petroleum oils, fats, waxes and alkyl esters; lubricant for metals, textiles, leather; in soluble oils for metal processing and finishing; in textile processing and finishing; glass fiber lubricants; automotive lubricant additives
Form: Liquid; Color: 6 (Gardner)
Properties: Solubility: Sol. in butyl stearate, glycerol trioleate, Stod. solvent; disp. in water, mineral oil and perchloroethylene; Density: 8.6 lb/gal; Visc.: 300 cs; HLB: 10.9; Cloud Pt.: < 25 C
Composition: 95% active ingred.

Emsorb 6905. Emery Industries
Chemical Description: POE (20) sorbitan monostearate
Category: Emulsifier, lubricant, nonionic
Applications: O/W emulsifier for mineral oil, fats, waxes; fiber-to-metal textile lubricant; coemulsifier with Emsorb 2505 in paraffin wax emulsions for textiles, paper and wallboard coatings; cosmetic and household products
Form: Waxy semisolid; Color: 3 (Gardner)
Properties: Solubility: Sol. in water, Stod. solvent; disp. in perchloroethylene; Density: 8.9 lb/gal; Visc.: 500 cs; HLB: 15.2; Cloud Pt.: > 100 C
Composition: 100% conc.

Emsorb 6906. Emery Industries
Chemical Description: POE (3) sorbitan monostearate
Category: Emulsifier, lubricant, nonionic
Applications: W/O emulsifier in household formulations; fiber-to-metal lubricant for synthetic and cellulosic fibers and yarns
Form: Solid; Color: 3 (Gardner)
Properties: Solubility: Sol. in Stod. solvent; disp. in water, mineral oil, butyl stearate, glycerol trioleate, perchloroethylene; M.P.: 39 C; HLB: 9.0; Cloud Pt.: < 25 C
Composition: 100% conc.

Emsorb 6907. Emery Industries
Chemical Description: POE (20) sorbitan tristearate
Category: Emulsifier, lubricant, nonionic
Applications: O/W emulsifier for petroleum oils and natural fats; textile processing and finishing compounds as lubricant and softener
Form: Waxy solid; Color: 5 (Gardner)
Properties: Solubility: Sol. in mineral oil, butyl stearate, glycerol trioleate, Stod. solvent; disp. in water, perchloroethylene; M.P.: 34 C; HLB: 11.0; Cloud Pt.: < 25 C
Composition: 100% conc.

Emsorb 6910. Emery Industries
Chemical Description: POE (20) sorbitan monopalmitate
Category: Emulsifier, lubricant, nonionic
Applications: O/W emulsifier and lubricant; similar functions as Emsorb 6905 with lower viscosity and greater water solubility
Form: Viscous liquid; Color: 6 (Gardner)
Properties: Solubility: Sol. in water and glycerol trioleate; disp. in perchloroethylene; Density: 9.2 lb/gal; Visc.: 600 cs; HLB: 15.8; Cloud Pt.: > 100 C
Composition: 100% conc.

Emsorb 6915. Emery Industries
Chemical Description: POE (20) sorbitan monolaurate
Category: Emulsifier, solubilizer, nonionic
Applications: O/W emulsifier and solubilizer of petroleum oils, solvents and fats, essential oils, fragrances; cosmetic creams and lotions; as viscosity modifier in shampoos; in textile industry as emulsifier for dye carriers, as an antistatic scrooping agent in spin finishes, as a fiber processing aid
Form: Liquid; Color: 6 (Gardner)
Properties: Solubility: High water sol.; also sol. in glycerol trioleate; disp. in perchloroethylene; Density: 9.2 lb/gal; Visc.: 400 cs; HLB: 16.5; Cloud Pt.: > 100 C
Composition: 100% conc.

Emsorb 6916. Emery Industries
Chemical Description: POE (4) sorbitan monolaurate
Category: Emulsifier
Applications: O/W emulsifier in emulsion polymerization of PVC
Form: Liquid; Color: 7 (Gardner)
Properties: Solubility: Disp. in

water, butyl stearate, perchloroethylene and Stod. solvent; Density: 9.0 lb/gal; Visc.: 700 cs; HLB: 12.1; Cloud Pt.: < 25 C

Emulan A. BASF AG
Chemical Description: Oleic acid oxyethylate
Category: Emulsifier, nonionic
Applications: O/W emulsifier for cutting oils, rolling oils, wiredrawing oils, formwork release oils, washable oils, cold cleaners, metal cleaning and metal polishing emulsions
Form: Liquid; Color: Yellowish; Odor: Characteristic
Properties: Solubility: Sol. in mineral oils, fatty oils, fatty acids, polar organic solvents; emulsifiable in water; Sp. gr.: ≈ 0.89 (20 C); Visc.: ≈ 120 mPa•s @ 20 C; HLB: ≈ 11; Flash Pt.: ≈ 210 C; Solidification Pt.: ≈ –20 C; Sapon. Value: ≈ 110; Acid No.: 0–1; pH: 6–7.5 (1% aq.)
Composition: 100% active
Warnings and Caution: Avoid prolonged contact with skin
References: Technical leaflet M 5220 e

Emulan AF. BASF AG
Chemical Description: Fatty alcohol oxyethylate
Category: Emulsifier, nonionic
Applications: Paraffin wax emulsions, mineral oil emulsions
Form: Soft wax; Color: White
Properties: Solubility: Sol. in mineral oil and polar organic media; Sp. gr.: ≈ 0.91 @ 50 C; M.P.: ≈ 42 C; HLB: ≈ 11; Flash Pt.: ≈ 190 C; Sapon. Value: Almost 0; Acid No.: Almost 0; pH: 6–7.5 (1% aq.)
Composition: 100% active
Warnings and Caution: Avoid prolonged contact with skin
References: Technical leaflet M 5220 e

Emulan EL. BASF AG
Chemical Description: Castor oil oxyethylate
Category: Emulsifier, nonionic
Applications: Emulsifying agent in fatty acids, fatty oils, organic solvents, e.g., plastics, resins, dyes
Form: Liquid; Color: Pale yellow; Odor: Characteristic
Properties: Solubility: Sol. in water and polar organic media; Sp. gr.: ≈ 1.06 @ 30 C: Visc.: 200–500 mPa•s @ 30 C; HLB: ≈ 14; Flash Pt.: ≈ 280 C; Solidification Pt.: 15–25 C; Sapon. Value: ≈ 60; Acid No.: 0–1; pH: 6–7.5 (1% aq.)
Composition: 100% active
Warnings and Caution: Avoid prolonged contact with skin
References: Technical leaflet M 5220 e

Emulan FJ. BASF AG
Chemical Description: Fatty acid alkylolamide
Category: Emulsifier, nonionic
Applications: W/O emulsifier for impregnation, lubrication, polishing and cleaning; O/W emulsifier for anticorrosive component in metal working

Form: Liquid; Color: Brownish; Odor: Faint
Properties: Solubility: Sol. in mineral oils, fatty oils and mixtures; emulsifiable in water; Sp. gr.: ≈ 0.96 @ 20 C; Visc.: ≈ 200 mPa•s @ 20 C; HLB: 3–5 (W/O type); ≈ 8.5 (O/W type); Solidification Pt.: ≈ −20 C; Sapon. Value: ≈ 150; Acid No.: ≈ 2; pH: ≈ 9 (1% aq.)
Composition: 100% active
Warnings and Caution: Avoid prolonged contact with skin
References: Technical leaflet M 5220 e

Emulan FM. BASF AG
Chemical Description: Triethanolamine monooleic ester
Category: Emulsifier, nonionic
Applications: W/O emulsifier for impregnation, lubrication, polishing and cleaning; O/W emulsifier for polishes; rust-preventing additive for metal working and corrosion protection
Form: Liquid; Color: Brown; Odor: Characteristic
Properties: Solubility: Sol. in mineral oils, fatty oils and mixtures; Sp. gr.: ≈ 0.95 @ 20 C; Visc.: ≈ 200 mPa•s @ 20 C; HLB: 3–5 (W/O type); ≈ 9.5 (O/W type); Flash Pt.: ≈ 240 C; Solidification Pt.: ≈ −5 C; Sapon. Value: ≈ 135; Acid No.: 0–5; pH: ≈ 10 (1% aq.)
Composition: 100% active
Warnings and Caution: Avoid prolonged contact with skin
References: Technical leaflet M 5220 e

Emulan OC. BASF AG
Chemical Description: Fatty alcohol oxyethylate
Category: Emulsifier, nonionic
Applications: O/W emulsifier for dry bright emulsions; stabilizing agent for emulsions and dispersions
Form: Soft wax; Color: White or pale yellowish
Properties: Solubility: Sol. in water, fatty acids, waxes, and polar organic solvents; Sp. gr.: ≈ 1.03 @ 50 C; M.P.: 35–40 C; HLB: ≈ 17; Flash Pt.: ≈ 210 C; Solidification Pt.: ≈ 34 C; Sapon. Value: Almost 0; Acid No.: Almost 0; pH: 6–7.5 (1% aq.)
Composition: 100% active
Warnings and Caution: Avoid prolonged contact with skin
References: Technical leaflet M 5220 e

Emulan OG. BASF AG
Chemical Description: Fatty alcohol oxyethylate
Category: Emulsifier, dispersant, stabilizer, nonionic
Applications: O/W emulsifier for fatty acids, waxes, polar organic solvents; dispersant for solid substances, in grinding dyes; stabilizer, e.g., for hydraulic and anticorrosion oils
Form: Waxy powder in bead form
Properties: Solubility: Sol. in water and polar organic media; Density: ≈ 0.6 kg/l; M.P.: 45–50 C; HLB: ≈ 17; Flash Pt.: ≈ 240 C; Sapon. Value: Almost 0; Acid No.:

Almost 0; pH: 6–7.5 (1% aq.)
Composition: 100% active
Warnings and Caution: Avoid prolonged contact with skin
References: Technical leaflet M 5220 e

Emulan OSN. BASF AG
Chemical Description: Fatty alcohol oxyethylate
Category: Emulsifier, nonionic
Applications: O/W emulsifier for dry bright wax emulsions
Form: Soft wax; Color: White or pale yellowish
Properties: Solubility: Sol. in water, fatty acids, waxes, polar organic solvents; Sp. gr.: ≈ 1.03 @ 50 C; M.P.: 35–40 C; HLB: ≈ 17; Flash Pt.: ≈ 230 C; Solidification Pt.: ≈ 34 C; Sapon. Value: Almost 0; Acid No.: Almost 0; pH: 6–7 (1% aq.)
Composition: 100% active
Warnings and Caution: Avoid prolonged contact with skin
References: Technical leaflet M 5220 e

Emulan OU. BASF AG
Chemical Description: Fatty alcohol oxyethylate
Category: Emulsifier, dispersant, stabilizer, nonionic
Applications: O/W emulsifier for fatty acids, waxes, polar organic solvents; dispersing of solid substances, in grinding processes; stabilizing of emulsions and dispersions
Form: Soft wax; Color: White or pale yellowish
Properties: Solubility: Sol. in water, fatty acids, molten greases and ester waxes, polar organic solvents; Sp. gr.: ≈ 1.02 @ 50 C; M.P.: ≈ 40 C; HLB: ≈ 17; Flash Pt.: ≈ 210 C; Solidification Pt.: ≈ 35 C; Sapon. Value: Almost 0; Acid No.: Almost 0; pH: 6–7.5 (1% aq.)
Composition: 100% active
Warnings and Caution: Avoid prolonged contact with skin
References: Technical leaflet M 5220 e

Emulan P. BASF AG
Chemical Description: Fatty alcohol oxyethylate
Category: Emulsifier, nonionic
Applications: Emulsions of highly refined mineral oil fractions, e.g., paraffin oil, washable oils, cold cleaners, other solvent-based cleaning agents
Form: Clear liquid; Color: Colorless
Properties: Solubility: Sol. in paraffin oil and other mineral oils; emulsifiable in water; Sp. gr.: ≈ 0.91 @ 20 C; Visc.: ≈ 400 mPa•s @ 20 C; HLB: 11–12; Flash Pt.: ≈ 150 C; Solidification Pt.: ≈ 10 C; Sapon. Value: Almost 0; Acid No.: Almost 0; pH: 6–7.5 (1% aq.)
Composition: 100% active
Warnings and Caution: Avoid prolonged contact with skin
References: Technical leaflet M 5220 e

Emulan PO. BASF AG
Chemical Description: Alkylphe-

nol oxyethylate
Category: Emulsifier, nonionic
Applications: Cold cleaners and other solvent-based cleaners, formwork release oils, drilling oils
Form: Clear liquid; Color: Colorless to pale yellowish
Properties: Solubility: Sol. in spindle and other mineral oils, fatty oils and polar organic solvents; emulsifiable in water; Sp. gr.: ≈ 1.04 @ 20 C; Visc.: ≈ 400 mPa•s @ 20 C; HLB: ≈ 10–12; Flash Pt.: ≈ 240 C; Solidification Pt.: ≈ –15 C; Sapon. Value: Almost 0; Acid No.: Almost 0; pH: 6–7.5 (1% aq.)
Composition: 100% active
Warnings and Caution: Avoid prolonged contact with skin
References: Technical leaflet M 5220 e

Emulan SH. BASF AG
Chemical Description: Nitrogenous carboxylic acid
Category: Salts of Emulan SH as emulsifiers, wetting agents, anticorrosive agents, anionic
Applications: Salts of Emulan SH as emulsifiers; as anti-corrosive agents in aqueous hydraulic liquids, anticorrosion emulsions and oils; in aq. and solvent-based cleaning agents and metal degreasers, e.g., industrial and workshop cleaners
Form: Clear to slightly turbid liquid; Color: Brownish; Odor; Faint
Properties: Solubility: In unneutralized form, sol. in mineral oils, fatty oils, organic solvents; when neutralized with alkali or amine, sol. in water; Sp. gr.: ≈ 0.95 @ 20 C; Visc.: ≈ 2500 mPa•s @ 20 C; Flash Pt.: ≈ 225 C; Solidification Pt.: ≈ –10 C; Sapon. Value: ≈ 160; Acid No.: ≈ 158; pH: ≈ 4 (1% aq.)
Composition: 100% active
Warnings and Caution: Avoid prolonged contact with skin
References: Technical leaflet M 5220 e

Emulan SH 85. BASF AG
Chemical Description: Nitrogenous carboxylic acid
Category and Applications: See Emulan SH
Form: Clear fluid liquid; Color: Brownish
Properties: Solubility: In unneutralized form, sol. in mineral oils, fatty oils, polar organic solvents; when neutralized with alkali or amine, sol. in water; Sp. gr.: ≈ 0.94 @ 20 C; Visc.: ≈ 250 mPa•s @ 20 C; Flash Pt.: ≈ 90 C; Solidification Pt.: ≈ –15 C; Sapon. Value: ≈ 135; Acid No.: ≈ 127; pH: ≈ 4 (1% aq.)
Composition: 85% active; 12% isobutanol; 3% water
Warnings and Caution: Avoid prolonged contact with skin
References: Technical leaflet M 5220 e

Emulbon Series. Toho Chem.
Category: Emulsifier
Applications: Sizing assistant, lubricant
Form: Liquid

Composition: 100% conc.

Emulgade F. Henkel Canada
Chemical Description: Blend of Dehydag Wax O and Dehydag Wax E with a nonionic emulsifying agent based on ethoxylated fatty acid
Applications: Self-emulsifying raw material for cosmetic creams and fluid emulsions
Composition: 95% fatty matter

Emulgade F Special. Henkel Canada
Chemical Description: Blend of Dehydag Was O with a nonionic emulsifying agent based on ethoxylated fatty acid
Category: Raw material, nonionic
Applications: Self-emulsifying raw material for fluid emulsions of the oil-in-water type; cosmetic creams and emulsions; raw material for processing cationics
Composition: 99% fatty matter

Emulphogene BC-420. GAF Corp.
Chemical Description: Tridecyloxypoly (ethyleneoxy) ethanol with relatively low ethylene oxide content
Category: Emulsifier, detergent, dispersant, intermediate, nonionic
Applications: Intermediate in manufacture of high-foaming anionic surfactants of the water-sol. sulfate ester type; oil-sol. detergent and dispersant for petroleum oils
Form: Cloudy, slightly viscous but pourable liquid; Color: # 2 max. (VCS @ 50 C); Odor: Mild, pleasant
Properties: Solubility: Insol. in water, kerosene; sol. in toluene; Sp. gr.: 0.92–0.94; Cloud Pt.: Insol.; Flash Pt.: > 200 F (P-M, closed cup); Fire Pt.: > 165 C; Pour Pt.: –7 to –1 C; Stability: Stable to sulfuric acid and alkalis; pH: 6.5–8.5 (10% sol'n.)
Composition: 100% active ingred.

Emulphogene BC-610. GAF Corp.
Chemical Description: Tridecyloxypoly (ethyleneoxy) ethanol
Category: Detergent, wetting agent, emulsifier, nonionic
Applications: Low-foaming detergent and wetting agent for mechanical dishwashing formulations and spray-type alkaline cleaners; in pulp and paper industry; as wool-scouring agent with degreasing action at low temperatures; emulsifier for silicones and petroleum oils
Form: Cloudy, slightly viscous pourable liquid; Color: # 2 max. (VCS @ 50 C); Odor: Mild, pleasant
Properties: Solubility: Sol. in methanol, ethanol, acetone, toluene; dispersible in water and mineral oil; Sp. gr.: 0.97–0.99; Cloud Pt.: Dispersible (1% sol'n.) in distilled water; Flash Pt.: > 200 F (P-M, closed cup); Fire Pt.: > 165 C; Pour Pt.: 4.5–7.5 C; Stability: Stable to sulfuric acid and alkalis; pH: 6–8 (10% sol'n.)
Composition: 100% active ingred.

Emulphogene BC-630. GAF Corp.
Chemical Description: Tridecyloxypoly (ethyleneoxy) ethanol
Category and Applications: See

Emulphogene BC-610; slightly more hydrophilic than BC-610
Form: Cloudy, slightly viscous but pourable liquid; Color: # 2 max. (VCS @ 50 C); Odor: Mild, pleasant
Properties: Solubility: Sol. in methanol, ethanol, acetone, toluene; partially sol. in tetrachloroethylene; Sp. gr.: ≈ 0.99; Cloud Pt.: 50–60 C (1% sol'n.); Flash Pt.: > 200 F (P-M, closed cup); Fire Pt.: > 165 C; Pour Pt.: 15–20 C; Stability: Stable to sulfuric acid, alkalis and metallic ions; pH: 6–8 (10% sol'n.)
Composition: 100% active ingred.

Emulphogene BC-720. GAF Corp.
Chemical Description: Tridecyloxypoly (ethyleneoxy) ethanol with relatively high ethylene oxide content
Category: Foam stabilizer, detergent, solubilizer, wetting and rewetting agent, emulsifier, nonionic
Applications: Foam stabilizer and detergent booster for alkylarylsulfonates; solubilizer for salt-free alkylarylsulfonates; light- and heavy-duty detergent formulations; textile detergent in neutral, acid and alkaline media; paper industry for cleaning paper machine felts, rewetting paper towels; wetting agent for mineral acids and corrosion inhibitors; emulsifier and solubilizer for aromatic solvents, fats, essential oils, waxes
Form: Opaque, viscous pourable liquid; Color: # 2 max. (VCS @ 50 C); Odor: Mild, pleasant
Properties: Solubility: Sol. in water, alcohol, acetone, toluene; partially sol. in tetrachloroethylene; Sp. gr.: 1.00–1.03 (@ 40 C); Cloud Pt.: 58–62 C (1% sol'n.); Flash Pt.: > 200 F (P-M, closed cup); Fire Pt.: > 165 C; Pour Pt.: ≈ 20 C; Stability: Stable to sulfuric acid, alkalis, metallic ions; pH: 6–8 (10% sol'n.)
Composition: 100% active ingred.

Emulphogene BC-840. GAF Corp.
Chemical Description: Tridecyloxypoly (ethyleneoxy) ethanol with relatively high ethylene oxide content
Category: Foam stabilizer, detergent, solubilizer, wetting agent, emulsifier, nonionic
Applications: Foam stabilizer and detergent booster for alkylarylsulfonates; solubilizer for salt-free alkylarylsulfonates; in light- and heavy-duty detergents; detergent and wetting agent for use with strong alkalis, high temperatures; wetting agent in high concentrations of electrolytes; stabilizer and emulsifier for synthetic latices; emulsifier and solubilizer for aromatic solvents, fats, essential oils, waxes
Form: Paste; Color: # 2 max. (VCS @ 50 C); Odor: Mild, pleasant
Properties: Solubility: Sol. in water, alcohol, acetone, toluene; partially sol. in tetrachloroethylene; Sp. gr.: 1.01–1.04 (@ 40 C); Cloud Pt.: > 95 C (1% sol'n.); Flash Pt.: > 200 F (P-M, closed cup); Fire Pt.: > 165 C; Pour Pt.: 33–40 C; Stability: Stable to sulfuric acid,

alkalis, metallic ions; pH: 6–8 (10% sol'n.)
Composition: 100% active ingred.

Emulphogene TB-970. GAF Corp.
Chemical Description: Linear aliphatic ethoxylate
Category: Detergent, dispersant, textile wet processing aid, nonionic
Applications: Commercial and household laundry detergents; wall and floor cleaners; dairy cleaners and sanitizers; dye leveling, scouring and other textile wet processing; automatic bowl cleaners
Form: Waxy solid and free-flowing flakes; Color: Off-white
Properties: Solubility: Sol. in water, ethanol, carbon tetrachloride, "Cellosolve," xylene (warm) (10% in 90% solvent); Cloud Pt.: > 100 C (1% in distilled water); Solidification Pt.: 50 C min.; Neutralization No.: 0.5 max.; Surface Tension: 50.8 dynes/cm (0.1% conc.)
Composition: 100% conc.
References: Data sheet 9643-027

Emulphor EL-620. GAF Corp.
Chemical Description: Polyoxyethylated vegetable oil
Category: Emulsifier, dispersant, antistat, lubricant, solubilizer, wetting agent, nonionic
Applications: In water-emulsion paints based on vinyl, alkyd or acrylic latices; in flexible and rigid urethane foams; in polyester resin emulsions as binders for glass mats and preforms; for pickle-degreasing, fatliquoring and dyeing of leather; to solubilize basic and acid dyestuffs in paper industry for beater and calender operations; in textile industry as emulsifiers, dyeing assistants, antistats for acrylic fibers, solubilizers; in liquid wax polishes for floors, furniture, automobiles, shoes, etc.; in pesticides and herbicides; magnetic coupling fluids; as emulsifiers for cosmetic and pharmaceutical preparations
Form: Clear liquid; Color: Light brown; Odor: Mild oily
Properties: Solubility: Sol. in water, acetone, carbon tetrachloride, ethanol, vegetable oil, ether, toluene, butyl Cellosolve, xylene, methanol; Density: 8.7–8.8 lb/gal; Sp. gr.: 1.04–1.05; Visc.: 600–1000 cps; Cloud Pt.: 40–45 C (1% aq. sol'n.); Flash Pt.: 291–295 C; Fire Pt.: 322–326 C; Neutralization No.: 0.5 max.; Stability: Stable to metallic ions and weak alkali; partially stable to 1% acid; Surface Tension: 41 dynes/cm (0.1% sol'n.)
Composition: 100% active ingred.
References: Booklet 9655-030

Emulphor EL-719. GAF Corp.
Chemical Description: Polyoxyethylated vegetable oil
Category and Applications: See Emulphor EL-620
Form: Clear liquid; Color: Yellow; Odor: Oily
Properties: Solubility: Sol. in water, acetone, carbon tetrachloride, ethanol, vegetable oil, ether, toluene, butyl Cellosolve, xylene,

methanol; Density: 8.9–9.0 lb/gal; Sp. gr.: 1.06–1.07; Visc.: 500–800 cps; Cloud Pt.: 83–88 C (1% aq. sol'n.); Flash Pt.: 275–279 C; Fire Pt.: 328–332 C; Neutralization No.: 0.5 max.; Stability: Stable to metallic ions, weak alkali; partially stable to 1% acid; Surface Tension: 38 dynes/cm (0.1% sol'n.)
Composition: 96% min. active ingred.
References: Booklet 9655=030

Emulphor EL-980. GAF Corp.
Chemical Description: Polyoxyethylated vegetable oil
Category: Emulsifier, antistat, lubricant, nonionic
Applications: Spin-finish additives esp. for high temperature use; in over-spray compositions; dyeing assistants
Form: Waxy solid; Color: 5 max. (VCS)
Properties: Solubility: Readily sol. in water; Neutralization No.: 0.5 max.
Composition: 100% active ingred.
References: Data sheet 9643-014R

Emulphor EL-985. GAF Corp.
Chemical Description: Polyoxyethylated vegetable oil
Category and Applications: See Emulphor EL-980
Form: Viscous liquid; Color: 4 max. (VCS)
Properties: Solubility: Readily sol. in water; Neutralization No.: 0.5 max.
Composition: 50% active ingred.
References: Data sheet 9643-014R

Emulphor ON-870. GAF Corp.
Chemical Description: Polyoxyethylated oleyl alcohol
Category: Emulsifier, dispersant, stabilizer, solubilizer, wetting agent, nonionic
Applications: To prevent premature coagulation and stabilize natural or synthetic latices in rubber-based paints, adhesives, etc.; in textile industry as emulsifiers in acid degreasing, in dyeing wool/acrylic blends, as dispersing and stabilizing agent in haphthol dyeing; in leather degreasing, dyeing and fatliquoring; as emulsifiers for cosmetic and pharmaceutical creams and lotions; as solubilizer for essential oils and drugs; as wetting agent in metal cleaners with mineral acids and corrosion inhibitors
Form: Solid wax; Color: White
Properties: Solubility: Sol. in water, xylene, ethanol, ethylene glycol, butyl Cellosolve; Sp. gr.: 1.04; Cloud Pt.: > 100 C; Flash Pt.: > 93 C (P-M, closed cup); Pour Pt.: 46 C; Solidification Pt.: 43 C; Surface Tension: 37 dynes/cm (0.1% conc.)
Composition: 100% active ingred.
Toxicity: In absence of specific tests, regard as probable eye irritant
References: Data sheet 9643-023

Emulphor ON-877. GAF Corp.
Chemical Description: Polyoxyethylated oleyl alcohol
Category and Applications: See

Emulphor ON-870
Form: Pourable, viscous liquid; Color: 3 max. (VCS)
Properties: Solubility: Sol. in water, ethanol, ethylene glycol, butyl Cellosolve; Sp. gr.: 1.06; Cloud Pt.: 95–100 C (1% sol'n.); Flash Pt.: > 93 C (P-M, closed cup); Pour Pt.: 15 C; Solidification Pt.: 9 C; Surface Tension: 40.5 dynes/cm (0.1% conc.)
Composition: 70% active ingred.
References: Data sheet 9643-023

Emulphor VN-430. GAF Corp.
Chemical Description: Polyoxyethylated oleic acid
Category: Emulsifier, lubricant, nonionic
Applications: In textile industry as lubricant for wool, in cotton and rayon finishing, as dyeing aids; in leather processing to degrease pickled skins, improve fatliquoring, increase tensile strength of chrome side leathers; as oil-sol. surfactant in cutting oils, degreasing solvents and cleaners; as W/O emulsifiers for cold creams, lotions, hair-grooming preparations and other cosmetics
Form: Clear liquid; Color: Dark amber
Properties: Solubility: Dispersible in water; sol. in xylene, white mineral oil, ethanol, butyl Cellosolve; Sp. gr.: ≈ 1.0; Cloud Pt.: < 10 C (1% sol'n.); Flash Pt.: > 93 C (P-M, closed cup); Pour Pt.: 13 C; Solidification Pt.: 10 C; Surface Tension: 32 dynes/cm (0.1% conc.); pH: 7–7.5
Composition: 100% active ingred.
Reaction: Neutral
References: Bulletin 9643-022

Emulphor VT-650. GAF Corp.
Chemical Description: Ethoxylated stearic acid
Category: Emulsifier, lubricant, nonionic
Applications: Emulsifier for silicone and nonsilicone defoamers, low-foaming emulsifier/lubricant for dye carriers, emulsifier additive for mineral oil, alkyl esters and triglycerides; softener and lubricant for textile processing esp. as additive in oversprays and in spin finishes; dye assistant
Form: Soft wax; Color: 2 (VCS); Odor: Slight acetic acid
Properties: Ref. Index: ≈ 1.451; Solubility: Sol. in xylene, kerosene, ethanol, carbon tetrachloride, butyl Cellosolve and perchloroethylene; Density: 8.3 lb/gal; Sp. gr.: 0.99 (@ 50 C); Visc.: ≈ 9 cs (@ 100 C); Flash Pt.: 162 F (CC); Solidification Pt.: 32 C; Sapon. Value: 83–94; Neutralization No.: 0.5 max.
Composition: 100% active ingred.
References: Data sheet 9643-030

Emulsan O. Reilly-Whiteman Inc.
Chemical Description: Polyethylene glycol 400 monooleate
Category: Emulsifier, nonionic
Applications: Same as Emulsan K but with better oil solubility
Form: Liquid
Composition: 99% active ingred.

Emulsifier WHC. Stepan Chem. Co.
Category: Emulsifier

Applications: Waterless hand cleaner formulations
Form: Liquid; Color: Light amber
Properties: Solubility: Dispersible in water; sol. in kerosene; Density: 8.25 lb/gal; pH: 8 (1% aq.)

Emulso E-309. Toho Chem.
Category: Emulsifier
Applications: Oil spill dispersant
Form: Liquid
Properties: Sp. gr.: 0.789; Visc.: 4.1 cst (@ 30 C)

Emulton AA. Shamtan Lubricants & Chem. Ltd.
Chemical Description: Fatty acid-alkanolamine condensate
Category: Emulsifier, detergent, thickener, nonionic
Composition: 100% active ingred.

Emulton G. Shamtan Lubricants & Chem. Ltd.
Chemical Description: Fatty acid-glycerol esters
Category: Emulsifier, stabilizer, nonionic
Form: Solid or liquid
Composition: 100% active ingred.

Emulton PG. Shamtan Lubricants & Chem. Ltd.
Chemical Description: Fatty acid-polyglycol esters
Category: Emulsifier, lubricant, plasticizer, wetting agent, softener, defoamer, nonionic
Form: Solid or liquid
Composition: 100% active ingred.

Emulton PS. Shamtan Lubricants & Chem. Ltd.
Chemical Description: Petroleum sulfonates from sodium, calcium, barium and ammonium salts
Category: Emulsifiers, rust preventives and oil additives, anionic
Composition: ≈ 60% active in mineral oil

Emulvis. C.P. Hall Co.
Chemical Description: Polyoxyethylene stearate
Category: Viscosity builder, solubility retarder, nonionic
Applications: Viscosity builder in cosmetics (shampoos, hair rinses, skin cleaners, etc.); solubility retarder for gum arabic, methyl cellulose, natural gums and resins, in timed release tablets and coatings, soap bars; antiblock agent for decals; evaporation control for farm ponds; viscosity improver for inks; thickener in pet shampoos; melt point control in suppositories
Form: Flake; Color: White to cream; Odor: Bland
Properties: Solubility: 50% sol. in 70% isopropyl alcohol; 3% sol. in water (greater water sol. at high temperatures); M.P.: 53–58 C; pH: 4–6 (3% aq.)
Reaction: Acid

Endet. Swastik
Chemical Description: Heavy duty detergent powder containing proteolytic enzyme
Category: Detergent
Applications: Household heavy duty washing powder esp. for dislodging stubborn proteinous stains
Form: Spray dried powder (white)

and colored enzyme beads
Properties: pH: 9.5 (1% sol'n. in distilled water)

Epicol TAS80V. Albright & Wilson Ltd. Marchon Div.
Chemical Description: Sodium tallow alcohol sulfate
Category: Detergent
Applications: Heavy duty detergent formulations with softening effects; controlled foam formulations; esp. useful in 50/50 blend with sodium alkyl benzene sulfonates; ore flotation agent; wetting and emulsifying agents
Form: Needles; Color: Cream
Properties: pH: 10
Composition: 80% active ingred.
Reaction: Alkaline;

Epoxol 5-2E. Swift Chem. Co.
Chemical Description: Epoxidized octyl tallate
Category: Stabilizer
Applications: High oxirane stabilizing plasticizers with low volatility; produces hardness in semi-rigid vinyl formulations; extrusion aid; minimizes air release problems in plastisols
Form: Color: 1 (Gardner)
Properties: Sp. gr.: 0.9244; Visc.: A (Gardner); Sapon. Value: 139.8; Acid No.: 0.5
References: Booklet 3MDP7

Epoxol 7-4. Swift Chem. Co.
Chemical Description: Epoxidized soybean oil
Category: Plasticizer
Applications: High oxirane stabilizing plasticizer; FDA approved for food packaging materials
Form: Color: 70 (Cobalt-platinum); Odor: Low; Taste: Low
Properties: Ref. Index: 1.4705; Solubility: Miscible with esters, ketones, heptane, ethyl ether, vinyl plasticizers, aromatic and chlorinated hydrocarbons; Density: 8.28 lb/gal; Sp. gr.: 0.994; Visc.: 314 cps; Flash Pt.: 590 F (COC); Pour Pt.: 25 F; Sapon. Value: 178.1; Acid No.: 0.10; Hydroxyl No.: 6.5
Containers: Drums, tank wagons
References: Booklet 3MDP7

Epoxol 8-2B. Swift Chem. Co.
Chemical Description: Epoxidized butyl esters of linseed oil fatty acids
Category: Plasticizer
Applications: High oxirane stabilizing plasticizers with low viscosity, low volatility, excellent heat and light stabilization; FDA approved for food packaging materials
Form: Color: 1 (Gardner)
Properties: Sp. gr.: 0.965; Visc.: A (Gardner); Sapon. Value: 150; Acid No.: 0.25–0.50
References: Booklet 3MDP7

Epoxol 9-5. Swift Chem. Co.
Chemical Description: Epoxidized linseed oil
Category: Plasticizer
Applications: High oxirane stabilizing plasticizer; FDA approved for food packaging materials
Form: Color: 70 (Cobalt-platinum); Odor: Low; Taste: Low
Properties: Ref. Index: 1.4715;

Solubility: Miscible with esters, ketones, heptane, ethyl ether, vinyl plasticizers, aromatic and chlorinated hydrocarbons; Density: 8.58 lb/gal; Sp. gr.: 1.03; Visc.: 619 cps; Flash Pt.: 595 F (COC); Pour Pt.: 30 F; Sapon. Value: 172; Acid No.: 0.12; Hydroxyl No.: 12
References: Booklet 3MDP7

Equex AEM. Procter & Gamble
Chemical Description: Ammonium lauryl ether sulfate with alkanolamide added
Category: Detergent, foamer, anionic
Applications: Concentrate for foaming bubble baths, low pH shampoos, etc.
Form: Clear liquid; Color: Light amber; Odor: Sweet, pleasant
Properties: Density: 8.2 lb/gal; Sp. gr.: 0.98; Visc.:60 cps (Brookfield); Cloud Pt.: ≈ 40 F; Solidification Pt.: ≈ –10 F; pH: 6.3 (as is)
Composition: 37.5% active ingred.; other components: 9% alkanolamide; 1.2% petroleum ether extract; 20% denatured ethyl alcohol

Equex S. Procter & Gamble
Chemical Description: Sodium lauryl sulfate, cosmetic grade
Category: Detergent, emulsifier, anionic
Applications: Shampoos, liquid cleaners, specialty products
Form: Liquid
Properties: Density: 8.7 lb/gal; Sp. gr.: 1.04; Visc.: 50–200 cps (Brookfield); Cloud Pt.: ≈ 54 F; Solidification Pt.: ≈ 35 F; pH: 7.8 (10% sol'n.)
Composition: 29% active ingred.

Equex SP. Procter & Gamble
Chemical Description: Sodium lauryl sulfate, polymerization grade
Category: Emulsifier, anionic
Applications: Emulsion polymerization
Form: Clear liquid; Color: Pale yellow; Odor: Clean, pleasant
Properties: Density: 8.7 lb/gal; Sp. gr.: 1.04; Visc.: ≈ 90 cps (Brookfield @ 80 F); Cloud Pt.: ≈ 55 F; Solidification Pt.: ≈ 34 F; pH: 8 (10% sol'n.)
Composition: 29.8% active ingred.

Equex STM. Procter & Gamble
Chemical Description: Sodium and triethanolamine lauryl sulfate blend with alkanolamide added
Category: Detergent, foamer, anionic
Applications: Concentrate for shampoos, cleaners, specialty
Form: Thick, honey-like fluid; Color: Opaque white to straw; Odor: Clean, pleasant
Properties: Density: 8.7 lb/gal; Sp. gr.: 1.04; Visc.: 9000 cps (Brookfield); Cloud Pt.: 52–55 F; Solidification Pt.: 32–34 F; pH: 8 (as is)
Composition: 22.5% active ingred.; other components: 7.5% alkanolamide

Equex SW. Procter & Gamble
Chemical Description: Sodium

lauryl sulfate
Category: Detergent, wetting agent, emulsifier, anionic
Applications: Cream shampoos, paste cleaners, rug shampoos
Form: Thick, honey-like fluid; Color: Pale yellow; Odor: Clean, pleasant
Properties: Density: 8.8 lb/gal; Sp. gr.: 1.05; Visc.: ≈ 40,000 cps (Brookfield @ 80 F); Cloud Pt.: ≈ 77 F; Solidification Pt.: ≈ 55 F; pH: 7.8 (10% sol'n.)
Composition: 28.5% active ingred.

Equex T. Procter & Gamble
Chemical Description: Triethanolamine lauryl sulfate
Category: Detergent, anionic
Applications: Shampoos
Form: Clear liquid; Color: Water white to pale yellow; Odor: Clean, pleasant
Properties: Density: 8.7 lb/gal; Sp. gr.: 1.04; Visc.: ≈ 50 cps (Brookfield @ 80 F); Cloud Pt.: ≈ 28 F; Solidification Pt.: ≈ 18 F; pH: 7.3 (10% sol'n.)
Composition: 40% active ingred.

ESI-Terge 40% Coconut Oil Soap. Emulsion Systems
Chemical Description: Cochin oil soap
Category: Detergent, anionic
Applications: Liquid hand soap
Form: Liquid; Color: Straw; Odor: Coconut
Properties: Solubility: Water sol.; Density: 8.5 lb/gal; Visc.: Water-like; F.P.: 320 F
Composition: 40% active ingred.
Reaction: Alkaline
Containers: Drums

ESI-Terge 320. Emulsion Systems
Chemical Description: Phosphated nonylphenoxy polyethoxy ethanol
Category: Detergent, anionic
Applications: Detergent, viscosity aid, self-coupling wetting agent in synthetic cleaners, hard surface cleaners, wax strippers and compounds with high caustic concentrations
Form: Liquid; Color: Light straw; Odor: Mild
Properties: Solubility: Water sol.; Density: 9.34 lb/gal; Acid No.: 100 ± 10
Composition: 100% active ingred.
Containers: Drums
Warnings and Cautions: Avoid skin contact
References: Bulletin G.P. 5

ESI-Terge 330. Emulsion Systems
Chemical Description: Phosphated glycol ester
Category: Detergent, wetting agent, anionic
Applications: Self-coupling wetting agent for high alkaline hard surface cleaner; low to intermediate viscosity detergents; steam cleaners; cement cleaners; wax strippers; oven cleaners
Form: Liquid; Color: Light straw; Odor: Mild
Properties: Solubility: Water sol.; Density: 9.18 lb/gal; Acid No.: 285 ± 25
Composition: 100% active ingred.
Reaction: Acid

Containers: Drums
Warnings and Caution: Avoid skin contact
References: Bulletin GP-6

ESI-Terge B-15. Emulsion Systems
Chemical Description: 2:1 coconut diethanolamide
Category: Detergent, wetting agent, nonionic
Applications: Detergent for mild all purpose cleaners; detergent foam builder and thickener for all purpose cleaners
Form: Liquid; Color: Amber
Properties: Solubility: Water sol.; Density: 8.4 lb/gal; Sp. gr.: 1.01; pH: 9.8–10.8 (1% sol'n.); Biodegradable
Composition: 100% active ingred.
Reaction: Alkaline
Containers: 55 gal drums
References: Bulletin GP-3

ESI-Terge C-5. Emulsion Systems
Chemical Description: Modified coconut diethanolamide
Category: Detergent, wetting agent, nonionic
Applications: Detergent for mild all purpose cleaners; detergent foam builder and thickener
Form: Liquid; Color: Amber
Properties: Solubility: Water sol.; Density: 8.1 lb/gal; Sp. gr.: 0.97; pH: 8.5–9.5 (1% sol'n.); Biodegradable
Composition: 100% active ingred.
Reaction: Alkaline
Containers: 55 gal drums
References: Bulletin GP-3

ESI-Terge HA-20. Emulsion Systems
Chemical Description: Modified amine condensate
Category: Detergent base, anionic/nonionic
Applications: Self-coupling detergent for high and low pH cleaners compatible with glycol ethers, high viscosity all purpose cleaners; high concentrate high alkali cleaners
Form: Liquid; Color: Amber
Properties: Solubility: Water and some solvents; Density: 8.64 lb/gal; Acid No.: 60 ± 10; pH: 7.5–8.5 (1% sol'n.)
Composition: 100% active ingred.
Reaction: Alkaline
Containers: 55 gal drums
References: Bulletin GP-7

ESI-Terge RT-61. Emulsion Systems
Applications: Grease and oil emulsifiers; detergent formulations
Form: Liquid; Color: Water clear
Properties: Density: 8 lb/gal; Visc.: 45 cps (LV #1 spindle); pH: 7.5 (1% sol'n.)
Composition: 90% active ingred.

ESI-Terge S-10. Emulsion Systems
Chemical Description: 1:1 coconut diethanolamide
Category: Detergent, emulsifier, nonionic
Applications: Foam stabilizer, thickener; liquid dishwashing and car washing detergent
Form: Liquid; Color: Light straw
Properties: Density: 8.15 lb/gal; Sp. gr.: 0.98; pH: 9–10; Biodegradable
Composition: 100% active ingred.

Reaction: Alkaline
Containers: 55 gal drums
References: Bulletin GP-3

ESI-Terge T-60. Emulsion Systems
Chemical Description: Triethanolamine dodecyl benzene sulfonate
Category: Detergent, wetting agent, foam stabilizer, anionic
Applications: Car washing and dishwashing detergent, synthetic hand soap
Form: Liquid; Color: Amber
Properties: Density: 9.1 lb/gal; Sp. gr.: 1.09; F.P.: 320 F; pH: 6.5–7.5 (1% sol'n.); Biodegradable
Composition: 60% active ingred.
Reaction: Neutral
Containers: 500 lb net wt drums
References: Bulletin GP-4

Estrasan 1-A. Reilly-Whiteman Inc.
Chemical Description: Methyl ricinoleate
Category: Lubricity aid, nonionic
Applications: Enhances lubricity and wettability of petroleum oils for metals
Form: Liquid
Composition: 100% active

Estrex 1G58. Swift Chem.
Chemical Description: Glycerol esters
Category: Lubricant
Applications: Ironer's wax
Form: Beads, flakes; Color: White
Properties: M.P.: 57–62 C; pH: 6–7
Composition: 100% active ingred.
Containers: Drums

Ethoduomeen T/13. Akzo Chemie
Chemical Description: N,N′,N′-tris (2-hydroxyethyl)-N-tallow-1,3 diamminopropane
Category: Emulsifier
Applications: Making of bitumen emulsions, dispersing waxes
Form: Liquid; Color: 18 max. (Gardner)
Properties: Sp. gr.: 0.95; F.P.: 20 C; M.P.: 17 C; B.P.: 150 C min.; Flash Pt.: 204 C (COC)
Composition: 95% min. active ingred.; other components: 2% max. primary and secondary amines; 1% max. moisture
Toxicity: Skin irritant, severe eye irritant
Containers: 200 lb bung type steel drums, net 180 kg
Warnings and Caution: Avoid contact with strong oxidizing agents

Ethoduomeen T/20. Akzo Chemie
Chemical Description: N,N′,N′-polyoxyethylene(10)-N-tallow-1,3-diaminopropane
Category: Wetting agent
Applications: Coating preparation on paperboard
Form: Liquid; Color: 18 max. (Gardner)
Properties: Sp. gr.: 0.99; F.P.: < 25 C; B.P.: 150 C min.; Flash Pt.: 90 C min. (P-M, closed cup)
Composition: 95% min. active ingred.; 2% max. primary and sec. amines; 1% max. moisture
Toxicity: Skin irritant, severe eye irritant
Containers: 200 lb bung type steel drums, net 180 kg
Warnings and Caution: Avoid contact with strong oxidizing agents

Ethoduomeen T/25. Akzo Chemie
Chemical Description: N,N′,N′-polyoxyethylene (15)-N-tallow-1,3-diaminopropane
Category: Corrosion inhibitor
Applications: Corrosion inhibitor in water treatment chemicals in secondary oil recovery
Form: Liquid; Color: 18 min. (Gardner)
Properties: Sp. gr.: 1.02; M.P.: 25 C max. Flash Pt.: 238 C (COC)
Composition: 95% active ingred.; other components: 2% max. prim. and sec. amines; 1% max. moisture
Toxicity: Skin irritant, severe eye irritant
Containers: Bung type drum, net 180 kg
Warnings and Caution: Avoid contact with strong oxidizing agents

Ethoduomeen TD/13. Armak
Chemical Description: Ethoxylated diamine from tallow fatty acid
Category and Applications: See Ethoduomeen T/13
Form: Clear liquid; Color: 4 max. (Gardner)
Properties: Sp. gr.: 0.95; Surface Tension: 34.2 dynes/cm (0.1% sol'n.)
References: Bulletin 73-7, 78-1

Ethoduquad T/20. Armak
Chemical Description: Polyethoxylated quaternary ammonium salt from tallow fatty acid
Category and Applications: See Ethoquad C/12
Form: Clear liquid
Properties: pH: 6–9 (10% aq.)
Composition: 72% min. active
References: Bulletin 73-7

Ethofat 60/15. Armak
Chemical Description: Ethylene oxide condensate of stearic acid
Category and Applications: See Ethofat C/15
Form: Soft paste; Color: 8 max. (Gardner)
Properties: Solubility: Sol. in acetone, isopropanol, carbon tetrachloride, dioxane, benzene; disperses in water; Sp. gr.: 1.01; Sapon. Value: 110–120; Surface Tension: 39 dynes/cm (0.1% sol'n.); pH: 6.0–7.5 (10% aq.)
References: Bulletin No. 73-7

Ethofat 60/20. Armak
Chemical Description: Ethylene oxide condensate of stearic acid
Category and Applications: See Ethofat C/15
Form: Clear liquid; Color: 8 max. (Gardner)
Properties: Solubility: Sol. in acetone (< 80 C); isopropanol (< 50 C), carbon tetrachloride (< 75 C), dioxane (< 75 C), benzene; disperses in water; Sp. gr.: 1.02; Sapon. Value: 70–80; Surface Tension: 36 dynes/cm (0.1% sol'n.); pH: 6.0–7.5 (10% aq.)
Containers: 55 gal unlined steel drums
References: Bulletin No. 73-7, 78-1

Ethofat 60/25. Armak Co.
Chemical Description: Ethylene

oxide condensate of stearic acid
Category and Applications: See Ethofat C/15
Form: Paste; Color: 11 max. (Gardner)
Properties: Sapon. Value: 55–65; pH: 6.0–7.5 (10% aq.)
References: Bulletin No. 73-7

Ethofat 142/20. Armak
Chemical Description: Ethylene oxide condensate of 4% rosin fatty acid
Category and Applications: See Ethofat C/15
Form: Clear liquid; Color: 11 max. (Gardner)
Properties: Sp. gr.: 1.03; Sapon. Value: 72–82; Surface Tension: 39.5 dynes/cm (0.1% sol'n.); pH: 6.0–7.5 (10% aq.)
References: Bulletin 73-7

Ethofat 242/25. Armak
Chemical Description: Ethylene oxide condensate of 50% rosin fatty acids
Category and Applications: See Ethofat C/15
Form: Clear liquid; Color: 12 max. (Gardner)
Properties: Solubility: Sol. in acetone, isopropanol, carbon tetrachloride, dioxane, benzene, water (viscous sol'n.); Sp. gr.: 1.08; Sapon. Value: 55–65; Surface Tension: 42 dynes/cm (0.1% sol'n.); pH: 6.0–7.5 (10% aq.)
Composition: 100% conc.
Containers: 55 gal unlined steel drums
References: Bulletin 73-7

Ethofat C/15. Armak
Chemical Description: Ethylene oxide condensate of coco fatty acid
Category: Emulsifier, detergent, wetting agent, dispersant, suspending agent, nonionic
Applications: Textile, cosmetic, agricultural chemicals, metal working, leather treating, petroleum production, paints, etc.
Form: Clear liquid; Color: 9 (Gardner)
Properties: Solubility: Sol. in acetone, isopropanol, carbon tetrachloride, dioxane, benzene; disperses in water < 70 C; Sp. gr.: 1.00; HLB: 10.6; Sapon. Value: 120–130; Surface Tension: 33 dynes/cm (0.1% sol'n.); pH: 6–7.5 (10% aq.)
Composition: 100% conc.
References: Bulletin No. 73-7

Ethofat C/25. Armak
Chemical Description: Ethylene oxide condensate of coco fatty acid
Category and Applications: See Ethofat C/15
Form: Paste; Color: 8 (Gardner)
Properties: Sapon. Value: 60–70; pH: 6.0–7.5 (10% aq.)
References: Bulletin No. 73-7

Ethofat O/15. Armak
Chemical Description: Ethylene oxide condensate of oleic acid
Category and Applications: See Ethofat C/15
Form: Clear liquid; Color: 9 max. (Gardner)
Properties: Solubility: Sol. in ace-

tone (> 25 C), isopropanol, carbon tetrachloride (cloudy), dioxane, benzene; disperses in water; Sp. gr.: 0.99; HLB: 8.6. Sapon. Value: 110–120; Surface Tension: 35 dynes/cm (0.1% sol'n.); pH: 6.0–7.5 (10% aq.)
Composition: 100% conc.
References: Bulletin No. 73-7

Ethofat O/20. Armak
Chemical Description: Ethylene oxide condensate of oleic acid
Category and Applications: See Ethofat C/15
Form: Clear liquid; Color: 9 max. (Gardner)
Properties: Solubility: Sol. in acetone (> 50 C), isopropanol, carbon tetrachloride, dioxane, Stoddard solvent (> 25 C), benzene, water (> 75 C); Sp. gr.: 1.03; Sapon. Value: 75–85; Surface Tension: 41 dynes/cm (0.1% sol'n.)
Composition: 100% conc.
Reaction: pH = 6.0–7.5 (10% aq. sol'n.)
References: Bulletin No. 73-7

Ethomeen 18/12. Armak
Chemical Description: Ethoxylated tertiary amine from stearic acid
Category and Applications: See Ethomeen C/12
Form: Solid; Color: 7 max. (Gardner)
Properties: Solubility: Sol. in acetone (< 25 C), isopropanol, carbon tetrachloride, benzene; insol. water; Sp. gr.: 0.96
Composition: 95% tertiary amine
Toxicity: Corrosive, skin irritant, severe eye irritant
References: Bulletin 73-7, 78-1

Ethomeen 18/15. Armak
Chemical Description: Ethoxylated tertiary amine from stearic acid
Category and Applications: See Ethomeen C/12
Form: Liquid to paste; Color: 8 max. (Gardner)
Properties: Solubility: Sol. in acetone (cloudy @ < 50 C), isopropanol (< 75 C), carbon tetrachloride (cloudy), dioxane (< 75 C), Stoddard solvent (cloudy), benzene; forms gel in water; Sp. gr.: 0.98; Surface Tension: 34 dynes/cm (0.1% sol'n.)
Toxicity: Corrosive, skin irritant, severe eye irritant
References: Bulletin 73-7, 78-1

Ethomeen 18/20. Armak
Chemical Description: Ethoxylated tertiary amine from stearic acid
Category and Applications: See Ethomeen C/12
Form: Clear liquid; Color: 8 max. (Gardner)
Properties: Sp. gr.: 1.02; Surface Tension: 40 dynes/cm (0.1% sol'n.)
Toxicity: Skin irritant, severe eye irritant
References: Bulletin 73-7, 78-1

Ethomeen 18/25. Armak Co.
Chemical Description: Ethoxylated tertiary amine from stearic acid

Category and Applications: See Ethomeen C/12
Form: Clear liquid; Color: 8 max. (Gardner)
Properties: Sp. gr.: 1.04; Surface Tension: 43 dynes/cm (0.1% sol'n.)
References: Bulletin 73-7, 78-1

Ethomeen 18/60. Armak
Chemical Description: Ethoxylated tertiary amine from stearic acid
Category and Applications: See Ethomeen C/12
Form: Solid; Color: White
Properties: Solubility: Sol. in acetone (< 100 C), isopropanol (< 100 C), carbon tetrachloride, dioxane (< 100 C), benzene (< 100 C), water; Sp. gr.: 1.12; Surface Tension: 49 dynes/cm (0.1% sol'n.)
Toxicity: Mild skin and eye irritant
References: Bulletin 73=7, 78-1

Ethomeen C/12. Armak
Chemical Description: Ethoxylated tertiary amine from coco fatty acid
Category: Emulsifier, dispersant, cationic
Applications: Textile processing
Form: Clear liquid; Color: 6 max. (Gardner)
Properties: Solubility: Sol. in acetone, isopropanol, carbon tetrachloride, Stoddard solvent, benzene; forms gel in water; Sp. gr.: 0.87
Composition: 95% min. tertiary amine
Toxicity: Skin irritant, severe eye irritant
References: Bulletin 73-7

Ethomeen C/15. Armak
Chemical Description: Ethoxylated tertiary amine from coco fatty acid
Category and Applications: See Ethomeen C/12
Form: Clear liquid; Color: 6 max. (Gardner)
Properties: Solubility: Sol. in acetone (> 25 C), isopropanol, carbon tetrachloride (> 25 C), dioxane, Stoddard solvent (> 10 C), benzene, water (cloudy); Sp. gr.: 0.98; HLB: 13.9; Surface Tension: 33 dynes/ cm (0.1% sol'n.)
Toxicity: Skin irritant, severe eye irritant
References: Bulletin 73-7, 78-1

Ethomeen C/20. Armak
Chemical Description: Ethoxylated tertiary amine from coco fatty acid
Category and Applications: See Ethomeen C/12
Form: Clear liquid; Color: 11 max. (Gardner)
Properties: Solubility: Sol. in acetone, isopropanol, carbon tetrachloride (> 25 C), dioxane, Stoddard solvent (> 75 C), benzene, water (cloudy); Sp. gr.: 1.02; Surface Tension: 39 dynes/cm (0.1% sol'n.)
Toxicity: Skin irritant, severe eye irritant
References: Bulletin 73-7, 78-1

Ethomeen C/25. Armak
Chemical Description: Ethoxyl-

ated tertiary amine from coco fatty acid
Category and Applications: See Ethomeen C/12
Form: Clear liquid; Color: 12 max. (Gardner)
Properties: Solubility: Sol. in acetone, isopropanol, carbon tetrachloride (> 25 C), dioxane, Stoddard solvent (> 80 C), benzene, water; Sp. gr.: 1.04; Surface Tension: 41 dynes/cm (0.1% sol'n.)
Toxicity: Skin irritant, severe eye irritant
References: Bulletin 73-7, 78-1

Ethomeen O/12. Armak
Chemical Description: Ethoxylated tertiary amine from oleic acid
Category and Applications: See Ethomeen C/12
Form: Clear liquid; Color: 8 max. (Gardner)
Properties: Sp. gr.: 0.90; Surface Tension: 31.5 dynes/cm (0.1% sol'n.)
Toxicity: Skin irritant, severe eye irritant
References: Bulletin 73-7, 78-1

Ethomeen O/15. Armak
Chemical Description: Ethoxylated tertiary amine from oleic acid
Category and Applications: See Ethomeen C/12
Form: Clear liquid; Color: 8 max. (Gardner)
Properties: Sp. gr.: 0.96; Surface Tension: 35.3 dynes/cm (0.1% sol'n.)
Toxicity: Skin irritant, severe eye irritant
References: Bulletin 73-7, 78-1

Ethomeen O/25. Armak
Chemical Description: Ethoxylated tertiary amines from oleic acid
Category and Applications: See Ethomeen C/12
Form: Paste; Color: 8 max. (Gardner)
Toxicity: Skin irritant, severe eye irritant
References: Bulletin 73-7, 78-1

Ethomeen S/12. Armak
Chemical Description: Ethoxylated tertiary amine from soya fatty acid
Category and Applications: See Ethomeen C/12
Form: Clear heavy liquid; Color: 14 max. (Gardner)
Properties: Solubility: Sol. in acetate, isopropanol, carbon tetrachloride, Stoddard solvent, benzene; insol. water; Sp. gr.: 0.91; Surface Tension: 31.4 dynes/cm (0.1% sol'n.)
Composition: 95% tertiary amine
Toxicity: Skin irritant, severe eye irritant
References: Bulletin 73-7, 78-1

Ethomeen S/15. Armak
Chemical Description: Ethoxylated tertiary amine from soya fatty acid
Category and Applications: See Ethomeen C/12
Form: Clear liquid; Color: 14 max. (Gardner)
Properties: Solubility: Sol. in ace-

tone (cloudy @ > 50 C), isopropanol, carbon tetrachloride (> 25 C), dioxane, Stoddard solvent (> 75 C), benzene; forms gel or dispersion in water; Sp. gr.: 0.95; Surface Tension: 33 dynes/cm (0.1% sol'n.)
Toxicity: Skin irritant, severe eye irritant
References: Bulletin 73-7, 78-1

Ethomeen S/20. Armak
Chemical Description: Ethoxylated tertiary amine from soya fatty acid
Category and Applications: See Ethomeen C/12
Form: Clear liquid; Color: 14 max. (Gardner)
Properties: Solubility: Sol. in acetone (cloudy), isopropanol, carbon tetrachloride (> 25 C), dioxane, Stoddard solvent (> 50 C), benzene, water (forms viscous sol'n.); Sp. gr.: 1.02; Surface Tension: 40 dynes/cm (0.1% sol'n.)
Toxicity: Skin irritant, severe eye irritant
References: Bulletin 73-7, 78-1

Ethomeen S/25. Armak
Chemical Description: Ethoxylated tertiary amine from soya fatty acid
Category and Applications: See Ethomeen C/12
Form: Clear liquid; Color: 18 max. (Gardner)
Properties: Solubility: Sol. in acetone (cloudy), isopropanol, carbon tetrachloride (> 25 C), dioxane, Stoddard solvent (> 75 C), benzene, water; Sp. gr.: 1.04; Surface Tension: 43 dynes/cm (0.1% sol'n.)
Toxicity: Skin irritant, severe eye irritant
References: Bulletin 73-7, 78-1

Ethomeen T/12. Armak
Chemical Description: Ethoxylated tertiary amine from tallow fatty acid
Category and Applications: See Ethomeen C/12
Form: Paste; Color: 8 max. (Gardner)
Properties: Solubility: Sol. in acetone, isopropanol, carbon tetrachloride, Stoddard solvent, benzene, water (> 50 C); Sp. gr.: 0.92
Composition: 95% tertiary amine
Toxicity: Corrosive, skin irritant, severe eye irritant
References: Bulletin 73-7, 78-1

Ethomeen T/15. Armak
Chemical Description: Ethoxylated tertiary amine from tallow fatty acid
Category and Applications: See Ethomeen C/12
Form: Clear liquid; Color: 8 max. (Gardner)
Properties: Solubility: Forms cloudy sol'ns. in acetone, isopropanol, carbon tetrachloride, dioxane, Stoddard solvent, benzene; forms gel in water @ 30–80 C; Sp. gr.: 0.97; Surface Tension: 34 dynes/cm (0.1% sol'n.)
Toxicity: Corrosive, skin irritant, severe eye irritant
References: Bulletin 73-7, 78-1

Ethomeen T/25. Armak
Chemical Description: Ethoxylated tertiary amine from tallow fatty acid
Category and Applications: See Ethomeen C/12
Form: Clear liquid; Color: 8 max. (Gardner)
Properties: Solubility: Forms cloudy sol'ns. in acetone, isopropanol, carbon tetrachloride, dioxane, benzene (< 100 C), water; Sp. gr.: 1.03; Surface Tension: 41 dynes/cm (0.1% sol'n.)
Toxicity: Skin irritant, severe eye irritant
References: Bulletin 73-7, 78-1

Ethomeen TD/15. Armak
Chemical Description: Ethoxylated tertiary amine from tallow fatty acid
Category and Applications: See Ethomeen C/12
Form: Clear liquid; Color: 4 max. (Gardner)
Properties: Sp. gr.: 0.97; Surface Tension: 35.8 dynes/cm (0.1% sol'n.)
References: Bulletin 73-7, 78-1

Ethomeen TD/25. Armak
Chemical Description: Ethoxylated tertiary amine from tallow fatty acid
Category and Applications: See Ethomeen C/12
Form: Clear liquid; Color: 4 max. (Gardner)
Properties: Sp. gr.: 1.03; Surface Tension: 43.5 dynes/cm (0.1% sol'n.)
Toxicity: Skin irritant, severe eye irritant
References: Bulletin 73-7, 78-1

Ethomid HT/15. Armak
Chemical Description: Ethoxylated amide from hyd. tallow acid
Category: Emulsifier, dispersant, detergent, nonionic
Form: Solid; Color: 10 max. (Gardner)
Properties: Solubility: Sol. in isopropanol (< 15 C); forms dispersions in acetone (> 25 C), carbon tetrachloride (> 25 C), dioxane, Stoddard solvent (> 30 C), benzene (> 50 C), water; Sp. gr.: 1.03; Hydroxyl No.: 95–110; Surface Tension: 37 dynes/cm (0.1% sol'n.)
Composition: 100% conc.
References: Bulletin 73-7

Ethomid HT/60. Armak
Chemical Description: Ethoxylated amide from hyd. tallow acid
Category: See Ethomid HT/15
Form: Hard solid; Color: 11 max. (Gardner)
Properties: Solubility: Sol. in isopropanol (< 75 C), carbon tetrachloride (< 75 C), dioxane (< 75 C), benzene (< 80 C), water (< 75 C); Sp. gr.: 1.14; Hydroxyl Value: 40–50; Surface Tension: 47 dynes/cm (0.1% sol'n.)
Reaction: Neutral
References: Bulletin 73-7

Ethomid O/15. Armak
Chemical Description: Ethoxylated amide from oleic acid
Category: See Ethomid HT/15

Form: Liquid; Color: 12 max. (Gardner)
Properties: Solubility: Sol. in isopropanol (< 50 C), carbon tetrachloride (< 25 C), dioxane (< 25 C); forms gel in water @ 50–80 C; Sp. gr.: 1.0; Hydroxyl No.: 100–120; Surface Tension: 35 dynes/cm (0.1% sol'n.)
Reaction: Neutral
References: Bulletin 73-7

Ethoquad 18/12. Armak
Chemical Description: Polyethoxylated quaternary ammonium salt from stearic acid
Category and Applications: See Ethoquad C/12
Form: Paste; Color: 6 max. (Gardner)
Properties: Solubility: Sol. in acetone, isopropanol (< 25 C), benzene, water; Sp. gr.: 0.919; Surface Tension: 40.2 dynes/cm (0.1% sol'n.); pH: 7–8 (10% aq.)
Composition: 70% min. active
References: Bulletin 73-7

Ethoquad 18/25. Armak
Chemical Description: Polyethoxylated quaternary ammonium salt from stearic acid
Category and Applications: See Ethoquad C/12
Form: Clear liquid; Color: 11 max. (Gardner)
Properties: Solubility: Sol. in acetone, isopropanol, carbon tetrachloride, dioxane, benzene, water; Sp. gr.: 1.058; Surface Tension: 50.1 dynes/cm (0.1% sol'n.); pH: 6–9 (10% aq.)
Composition: 95% min. active
References: Bulletin 73-7

Ethoquad C/12. Armak
Chemical Description: Polyethoxylated quaternary ammonium salt from coco fatty acid
Category: Antistatic agents, emulsifiers, dyeing assistants, cationic
Applications: Textile industry, electroplating bath additives
Form: Clear liquid; Color: 9 max. (Gardner)
Properties: Solubility: Sol. in acetone, isopropanol, carbon tetrachloride, dioxane (> 15 C), Stoddard solvent (> 50 C), benzene, water; Sp. gr.: 0.969; Stability: Stable to acid and alkali; Surface Tension: 35.4 dynes/cm (0.1% sol'n.); pH: 7–8 (10% aq.)
Composition: 75% min. active
References: Bulletin 73-7

Ethoquad C/25. Armak
Chemical Description: Polyethoxylated quaternary ammonium salt from coco fatty acid
Category and Applications: See Ethoquad C/12
Form: Clear liquid; Color: 11 max. (Gardner)
Properties: Solubility: Sol. in acetone, isopropanol, carbon tetrachloride (> 30 C), dioxane, benzene, water; Sp. gr.: 1.071; Stability: Stable to acid and alkali; Surface Tension: 43.4 dynes/cm (0.1% sol'n.)
Composition: 95% min. active
References: Bulletin 73-7

Ethoquad O/12. Armak
Chemical Description: Polyethox-

ylated quaternary ammonium salt from oleic acid
Category and Applications: See Ethoquad C/12
Form: Clear liquid; Color: 9 max. (Gardner)
Properties: Solubility: Sol. in acetone, isopropanol, carbon tetrachloride, dioxane, Stoddard solvent (> 10 C), benzene, water; Sp. gr.: 0.932; Surface Tension: 40.3 dynes/cm (0.1% sol'n.); pH: 7–8 (10% aq.)
Composition: 72% min. active
References: Bulletin 73-7

Ethoquad O/25. Armak
Chemical Description: Polyethoxylated quaternary ammonium salt from oleic acid
Category and Applications: See Ethoquad C/12
Form: Clear liquid; Color: 11 max. (Gardner)
Properties: Solubility: Sol. in acetone, isopropanol, carbon tetrachloride, dioxane, benzene, water; Sp. gr.: 1.062; Surface Tension: 40.8 dynes/cm (0.1% sol'n.); pH: 7–9 (10% aq.)
Composition: 95% min. active
References: Bulletin 73-7

Ethosperse G-26. Glycol Chem.
Chemical Description: POE 26 glycerol
Category and Applications: See Ethosperse LA-4
Form: Liquid; Color: Pale straw
Properties: Solubility: Sol. in water, methanol, ethanol, acetone, ethyl acetate, toluol; Sp. gr.: 1.12 (@ 38 C); Visc.: 150 cps (@ 38 C); Acid No.: 2 max.; Hydroxyl No.: 133–142

Ethosperse LA-4. Glyco Chem.
Chemical Description: POE 4 ether lauryl alcohol
Category: Emulsifier, nonionic
Applications: Cosmetic, pharmaceutical and industrial uses
Form: Liquid; Color: Pale straw
Properties: Solubility: Sol. in methanol, ethanol, acetone, ethyl acetate. toluol, mineral oil; miscible with water, vegetable oil in certain proportions; Sp. gr.: 0.95; Visc.: 30 cps; HLB: 9.5; Acid No.: 2 max.; Hydroxyl No.: 145–160
Composition: 100% active ingred.

Ethosperse LA-12. Glyco Chem.
Chemical Description: POE 12 ether lauryl alcohol
Category and Applications: See Ethosperse LA-4
Form: Turbid liquid to soft paste
Properties: Solubility: Sol. in methanol, ethanol, acetone, ethyl acetate, toluol; miscible with water in certain proportions and in vegetable oil if hot; Sp. gr.: 1.10; Visc.: 1000 cps; HLB: 4.8; Acid No.: 2 max.; Hydroxyl No. 72–82
Composition: 100% active ingred.

Ethosperse SL-20. Glyco Chem.
Chemical Description: POE 20 sorbitol
Category and Applications: See Ethosperse LA-4
Form: Liquid; Color: Light yellow
Properties: Solubility: Sol. in water, methanol, ethanol, acetone;

Sp. gr.: 1.16; Visc.: 460 cps; HLB: 16.6; Acid No.: 0.5 max.; Hydroxyl No.: 385–430
Composition: 100% active ingred.

Ethosperse TDA-6. Glyco Chem.
Chemical Description: POE 6 ether tridecyl alcohol
Category and Applications: See Ethosperse LA-4
Form: Liquid; Color: Pale straw
Properties: Solubility: Sol. in methanol, ethanol, acetone, ethyl acetate, toluol; dispersible in water; Sp. gr.: 0.98; Visc.: 80 cps; HLB: 11.0; Acid No.: 1 max.; Hydroxyl No.: 118–133
Composition: 100% active ingred.

Ethylan 20. Lankro Chem. Ltd.
Chemical Description: Nonyl phenol ethoxylate
Category: Detergent, wetting agent, stabilizer, emulsifier, nonionic
Applications: Detergent, foam stabilizer and booster; pesticide formulations; essential oil solubilizer; latex stabilization; emulsification of waxes and polyester resin
Form: Waxy solid; Color: White; Odor: Negligible
Properties: Solubility: Water-sol.; Sp. gr.: 1.065 @ 40 C; Visc.: 160 cs @ 20 C; HLB: 16.0; Cloud Pt.: > 100 C (1% aq.); Flash Pt.: > 400 F (COC); Pour Pt.: 30 C; Stability: Stable to electrolytes, hard water, extremes of pH and temperature; pH: 6–8 (1% aq.)
Composition: 100% active ingred.
Containers: 200 kg net mild-steel drums or bulk
Warnings and Caution: Avoid prolonged contact with concentrated form; spillages may be slippery; do not mix with strong oxidizing or reducing agents (potentially explosive)
References: Bulletin SI:13 FM678

Ethylan 44. Lankro Chem. Ltd.
Chemical Description: Nonyl phenol ethoxylate
Category: Emulsifier, nonionic
Applications: Emulsifier for hydrocarbons, silicones, in formulations of degreasers and hand cleaning gels
Form: Clear liquid; Color: Straw; Odor: Negligible
Properties: Solubility: Water-insol.; Sp. gr.: 1.023 @ 20 C; Visc.: 340 cs @ 20 C; HLB: 9; Flash Pt.: > 400 F (COC); Pour Pt.: < 0 C; Stability: Stable to electrolytes, hard water, extremes of pH and temperature; pH: 6–8 (1% aq.)
Composition: 100% active ingred.
Containers: 200 kg net mild-steel drums or bulk
Warnings and Caution: Avoid prolonged contact with concentrated form; spillages may be slippery; do not mix with strong oxidizing or reducing agents (potentially explosive)
References: Bulletin SI:13 FM678

Ethylan 55. Lankro Chem. Ltd.
Chemical Description: Nonyl phenol ethoxylate
Category and Applications: See Ethylan 44
Form: Clear liquid; Color: Straw
Properties: Solubility: Water-in-

sol.; Sp. gr.: 1.035 @ 20 C; Visc.: 290 cs @ 20 C; HLB: 10.5; Flash Pt.: > 400 F (COC); Pour Pt.: < 0 C; Stability: Stable to electrolytes, hard water, extremes of pH and temperature; pH: 6–8 (1% aq.)
Composition: 100% active ingred.
Containers: 200 kg net mild-steel drums or bulk
Warnings and Caution: Avoid prolonged contact with concentrated form; spillages may be slippery; do not mix with strong oxidizing or reducing agents (potentially explosive)
References: Bulletin SI:13 FM678

Ethylan 77. Lankro Chem. Ltd.
Chemical Description: Nonyl phenol ethoxylate
Category and Applications: See Ethylan 44
Form: Clear liquid; Color: Pale straw; Odor: Negligible
Properties: Solubility: Water-disp.; Sp. gr.: 1.038 @ 20 C; Visc.: 360 cs @ 20 C; HLB: 10.9; Cloud Pt.: < 2 C (1% aq.); Flash Pt.: > 400 F (COC); Pour Pt.: < 0 C; Stability: Stable to electrolytes, hard water, extremes of pH and temperature; pH: 6–8 (1% aq.)
Composition: 100% active ingred.
Containers: 200 kg net mild-steel drums or bulk
Warnings and Caution: Avoid prolonged contact with concentrated form; spillages may be slippery; do not mix with strong oxidizing or reducing agents (potentially explosive)
References: Bulletin SI:13 FM678

Ethylan 172. Lankro Chem. Ltd.
Chemical Description: Cetyl, oleyl alcohol ethoxylate
Category: Emulsifier, nonionic
Applications: Emulsifying mineral oils and hydrophobic waxes
Form: Liquid; Color: Lt. amber
Properties: HLB: 7.0
Composition: 100% active ingred.
Reaction: Neutral

Ethylan A2. Lankro Chem. Ltd.
Chemical Description: PEG 200 oleate
Category: Emulsifier, antifoam agent, nonionic
Applications: Kerosene and mineral oil emulsifier, antifoam
Form: Liquid; Color: Lt. amber
Properties: Sp. gr.: 0.976; Visc.: 67 cs @ 20 C; HLB: 7.0
Composition: 100% active ingred.
Reaction: Neutral

Ethylan A3. Lankro Chem. Ltd.
Chemical Description: PEG 300 oleate
Category: Emulsifier, antifoam
Applications: Emulsifier for mineral and vegetable oils, antifoam, dispersion aid
Form: Liquid; Color: Lt. amber
Properties: Sp. gr.: 1.000; Visc.: 100 cs @ 20 C; HLB 8.9
Composition: 100% active ingred.
Reaction: Neutral

Ethylan A4. Lankro Chem. Ltd.
Chemical Description: PEG 400 oleate
Category: Emulsifier, antistatic, nonionic
Applications: Oil emulsifier, anti-

stat for plastics, dispersion aid
Form: Liquid; Color: Lt. amber
Properties: Sp. gr.: 1.017; Visc.: 130 cs @ 20 C; HLB: 10.3
Composition: 100% active ingred.
Reaction: Neutral

Ethylan A6. Lankro Chem. Ltd.
Chemical Description: PEG 600 oleate
Category: Dispersing agent, nonionic
Applications: Oil emulsifier and antistat for plastics, pigment dispersant
Form: Liquid; Color: Lt. amber
Properties: Sp. gr.: 1.037; Visc.: 160 cs @ 20 C; HLB: 12.3
Composition: 100% active ingred.
Reaction: Neutral

Ethylan A15. Lankro Chem. Ltd.
Chemical Description: Coconut diethanolamide
Category: Emulsifier, nonionic
Applications: Industrial degreasing, metal cleaning applications, emulsifying solvents and polyethylene waxes
Form: Clear Liquid; Color: Straw
Properties: Sp. gr.: 1.010; Visc.: 880 cs @ 20 C
Composition: 99% active ingred.
Reaction: Alkaline

Ethylan BCP. Lankro Chem. Ltd.
Chemical Description: Nonyl phenol ethoxylate
Category: Detergent, wetting agent, emulsifier, nonionic
Applications: Textile scouring agent, dry cleaning detergent; wetting and emulsifying agent for oils and aromatic solvents
Form: Clear colorless liquid; Odor: Negligible
Properties: Solubility: Water-sol.; Sp. gr.: 1.058 @ 20 C; Visc.: 320 cs @ 20 C; HLB: 12.9; Cloud Pt.: 50 C (1% aq.); Flash Pt.: > 400 F (COC); Pour Pt.: 0 C; Stability: Stable to electrolytes, hard water, extremes of pH and temperature; pH: 6–8 (1% aq.)
Composition: 100% active ingred.
Containers: 200 kg net mild-steel drums or bulk
Warnings and Caution: Avoid prolonged contact with concentrated form; spillages may be slippery; do not mix with strong oxidizing or reducing agents (potentially explosive)
References: Bulletin SI:13 FM678

Ethylan BV. Lankro Chem. Ltd.
Chemical Description: Nonyl phenol ethoxylate
Category: Foam stabilizer, emulsifier, solubilizer, nonionic
Applications: Foam stabilizer and booster; emulsifier of solvents and oleins, in pesticides; as emulsifier and solubilizer for essential oils and perfumes; emulsion polymerization
Form: Soft paste; Color: White; Odor: Negligible
Properties: Solubility: Water-sol.; Sp. gr.: 1.083 @ 20 C; Visc.: 380 cs @ 20 C; HLB: 14.5; Cloud Pt.: 90 C (1% aq.); Flash Pt.: > 400 F (COC); Pour Pt.: 20 C; Stability: Stable to electrolytes, hard water, extremes of pH and temperature; pH: 6–8 (1% aq.)

Composition: 100% active ingred.
Containers: 200 kg net mild-steel drums or bulk
Warnings and Caution: Avoid prolonged contact with concentrated form; spillages may be slippery; do not mix with strong oxidizing or reducing agents (potentially explosive)
References: Bulletin SI:13 FM678

Ethylan BZA. Lankro Chem. Ltd.
Chemical Description: Modified alkylphenol ethoxylate
Category: Detergent, nonionic
Applications: Low-foam nonionic with high stability on solid caustic and good food soil defoaming characteristics for mechanical dishwashing; metal cleaning
Form: Liquid
Properties: Cloud Pt.: 18 C
Composition: 100% active ingred.

Ethylan C 40 AH. Lankro Chem. Ltd.
Chemical Description: PEG ester of unsaturated fatty acid
Category: Emulsifier, nonionic
Applications: Emulsifier for chlorinated solvents, olein and wax, pesticides; dye leveling agent; low-foaming dispersing aid; perfumery solubilizer
Form: Hazy liquid; Color: Yellow
Properties: Sp. gr.: 1.054; Visc.: 320 cs @ 40 C; HLB: 13.5
Composition: 100% active ingred.
Reaction: Neutral

Ethylan C 160. Lankro Chem. Ltd.
Chemical Description: PEG unsaturated fatty acid ester
Category: Emulsifier, dispersant, solubilizer
Applications: Olein and wax emulsifier; perfumery; low-foaming dispersant, solubilizer
Form: Waxy solid; Color: Cream
Properties: HLB: 17.6
Composition: 100% active ingred.

Ethylan CD913. Lankro Chem. Ltd.
Chemical Description: Synthetic lower fraction primary alcohol ethylene oxide condensate
Category: Detergent, emulsifier, wetting agent, solubilizer, nonionic
Applications: W/O and O/W emulsification; co-emulsifier for mineral oils and waxes; emulsification of alkyd resins and paraffin hydrocarbons
Form: Clear liquid; Color: Colorless
Properties: Solubility: Water insol.; Sp. gr.: 0.936 @ 20 C; Visc.: 27 cs @ 20 C; HLB: 8.8; Flash Pt.: 290 F (COC); Pour Pt.: < 0 C; pH: 6–8 (1% aq.)
Composition: 97% active
Containers: 200 kg net mild-steel drums or bulk
Warnings and Caution: Avoid prolonged contact with concentrated form; spillages may be slippery; do not mix with strong oxidizing or reducing agents (potentially explosive)
References: Bulletin SI: 8FM678

Ethylan CD916. Lankro Chem. Ltd.

Chemical Description: Synthetic lower fraction primary alcohol ethylene oxide condensate
Category: Emulsifier, detergent, wetting agent, nonionic
Applications: O/W emulsification of paraffin waxes and hydrocarbon solvents; bases for toxicant emulsifiers and co-emulsifiers for mineral oils; scouring and wetting agents; textile antistatic agents; specialty cleaners
Form: Clear liquid; Color: Colorless; Odor: Characteristic fatty alcohol
Properties: Solubility: Water-sol.; Sp. gr.: 0.993 @ 20 C; Visc.: 53 cs @ 20 C; HLB: 12.8; Cloud Pt.: 58 C; Flash Pt.: 355 F (COC); Pour Pt.: 0 C; pH: 6–8 (1% aq.)
Composition: 96% active
Containers: 200 kg net mild-steel drums or bulk
Warnings and Caution: Avoid prolonged contact with concentrated form; spillages may be slippery; do not mix with strong oxidizing or reducing agents (potentially explosive)
References: Bulletin SI: 8FM678

Ethylan CD919. Lankro Chem. Ltd.
Chemical Description: Synthetic lower fraction primary alcohol ethylene oxide condensate
Category: Emulsifier, detergent, wetting agent, solubilizer, nonionic
Applications: O/W emulsifier; wetting agents and detergents, sudsing agents for liquid detergents; scouring and dye levelling agents; co-emulsifiers for aromatic solvents, natural and synthetic waxes and toxicants; detergent sanitizers; essential oil solubilizers
Form: Clear liquid; Color: Colorless; Odor: Characteristic fatty alcohol
Properties: Solubility: Water-sol.; Sp. gr.: 1.008 @ 20 C; Visc.: 77 cs @ 20 C; HLB: 14.4; Cloud Pt.: 82 C; Flash Pt.: 385 F (COC); Pour Pt.: 7 C; pH: 6–8 (1% aq.)
Composition: 97% active
Containers: 200 kg net mild-steel drums or bulk
Warnings and Caution: Avoid prolonged contact with concentrated form; spillages may be slippery; do not mix with strong oxidizing or reducing agents (potentially explosive)
References: Bulletin SI: 8FM678

Ethylan CD9112. Lankro Chem. Ltd.
Chemical Description: Synthetic lower fraction primary alcohol ethylene oxide condensate
Category and Applications: See Ethylan CD919
Form: Waxy solid; Color: White; Odor: Characteristic fatty alcohol
Properties: Solubility: Water-sol.; Sp. gr.: 1.031 @ 40 C; Visc.: 56 cs @ 40 C; HLB: 15.5; Cloud Pt.: 100 C; Flash Pt.: > 400 F (COC); Pour Pt.: 23 C; pH: 6–8 (1% aq.)
Composition: 100% active
Containers: 200 kg net mild-steel drums or bulk
Warnings and Caution: Avoid prolonged contact with concentrated form; spillages may be slippery;

do not mix with strong oxidizing or reducing agents (potentially explosive)
References: Bulletin SI: 8FM678

Ethylan CD9130. Lankro Chem. Ltd.
Chemical Description: Synthetic lower fraction primary alcohol ethylene oxide condensate
Category: Wetting agent, emulsifier, nonionic
Applications: O/W emulsification; emulsifier for synethetic waxes; wetting agent at high temperature; detergent sanitizers; toilet block component
Form: Waxy solid; Color: White; Odor: Characteristic fatty alcohol
Properties: Solubility: Water-sol.; Visc.: 62 cs @ 60 C; HLB: 17.8; Cloud Pt.: > 100 C; Flash Pt.: > 400 F (COC); Pour Pt.: 45 C; pH: 6–8 (1% aq.)
Composition: 100% active ingred.
Containers: 200 kg net mild-steel drums or bulk
Warnings and Caution: Avoid prolonged contact with concentrated form; spillages may be slippery; do not mix with strong oxidizing or reducing agents (potentially explosive)
References: Bulletin SI: 8FM678

Ethylan CF71. Lankro Chem. Ltd.
Chemical Description: Coconut fatty acid ester
Category: Emulsifier, lubricant, nonionic
Applications: Oil emulsifier; emulsifier for cosmetics and pharmaceuticals; fiber lubricant
Form: Liquid; Color: Pale straw
Properties: HLB: 14.0
Composition: 100% active ingred.

Ethylan CH. Lankro Chem. Ltd.
Chemical Description: Ethoxylated coconut fatty acid alkylolamide
Category: Foam stabilizer, co-emulsifier, nonionic
Applications: Foam stabilizer in liquid detergents, general purpose or hard surface cleaners and in shampoos; co-emulsifier for chlorinated solvents
Form: Clear liquid; Color: Amber; Odor: Mild
Properties: Solubility: Water-sol.; Sp. gr.: 1.025 @ 20 C; Visc.: 270 cs @ 20 C; Cloud Pt.: 80 C (1% aq.); Flash Pt.: > 300 F (COC); Pour Pt.: 14 C; Stability: Good in presence of hard water and electrolytes; pH: 7.0–9.0 (1% aq.)
Composition: 100% active ingred.
Reaction: Neutral
Warnings and Caution: Avoid prolonged contact with concentrated form; spillages may be slippery; do not mix with strong oxidizing or reducing agents (potentially explosive)
References: Bulletin SI:12FM678

Ethylan CPG 630. Lankro Chem. Ltd.
Chemical Description: Modified alcohol ethoxylate
Category: Detergent, wetting agent, nonionic
Applications: Low foaming wet-

ting agent and detergent for rinse aids, machine dishwashing, metal cleaning, spray cleaning and bottle washing

Form: Clear liquid; Color: Colorless; Odor: Mild

Properties: Solubility: Water-sol.; Sp. gr.: 0.978 @ 20 C; Visc.: 65 cs @ 20 C; Cloud Pt.: 38 C (1% aq.); Flash Pt.: > 350 F (COC); Pour Pt.: 1 C; Stability: Stable to alkalis in aq. sol'n. up to boil, to acids except in presence of strong oxidizing agents, to chlorine releasing agents; pH: 6–7.5 (1% aq.)

Composition: 99% active ingred.

Reaction: Neutral

Containers: 200 kg net mild steel drums or bulk

Warnings and Caution: Avoid prolonged contact with concentrated form; spillages may be slippery; do not mix with strong oxidizing or reducing agents (potentially explosive)

References: Bulletin SI:10 FM678

Ethylan CPG 660. Lankro Chem. Ltd.

Chemical Description: Modified alcohol ethoxylate

Category and Applications: See Ethylan CPG 630

Form: Clear liquid; Color: Colorless; Odor: Mild

Properties: Solubility: Water-sol.; Sp. gr.: 0.984 @ 20 C; Visc.: 77 cs @ 20 C; Cloud Pt.: 29 C (1% aq.); Flash Pt.: > 350 F (COC); Pour Pt.: < –10 C; Stability: Stable to alkalis in aq. sol'n. up to boil, to acids except in presence of strong oxidizing agents, to chlorine releasing agents; pH: 6–7.5 (1% aq.)

Composition: 99% active ingred.

Reaction: Neutral

Containers: 200 kg net mild steel drums or bulk

Warnings and Caution: Avoid prolonged contact with concentrated form; spillages may be slippery; do not mix with strong oxidizing or reducing agents (potentially explosive)

References: Bulletin SI:10 FM678

Ethylan CPG 745. Lankro Chem. Ltd.

Chemical Description: Modified alcohol ethoxylate

Category and Applications: See Ethylan CPG 630

Form: Clear liquid; Color: Colorless; Odor: Mild

Properties: Solubility: Water-sol.; Sp. gr.: 0.985 @ 20 C; Visc.: 84 cs @ 20 C; Cloud Pt.: 37 C (1% aq.); Flash Pt.: > 350 F (COC); Pour Pt.: –6 C; Stability: Stable to alkalis in aq. sol'n. up to boil, to acids except in presence of strong oxidizing agents, to chlorine releasing agents; pH: 6–7.5 (1% aq.)

Composition: 99% active ingred.

Reaction: Neutral

Containers: 200 kg net mild steel drums or bulk

Warnings and Caution: Avoid prolonged contact with concentrated form; spillages may be slippery; do not mix with strong oxidizing or reducing agents (potentially explosive)

References: Bulletin SI.10 FM678

Ethylan CPG 945. Lankro Chem. Ltd.

Chemical Description: Modified alcohol ethoxylate

Category and Applications: See Ethylan CPG 630

Form: Clear liquid; Color: Colorless; Odor: Mild

Properties: Solubility: Water-sol.; Sp. gr.: 1.005 @ 20 C; Visc.: 88 cs @ 20 C; Cloud Pt.: 47 C (1% aq.); Flash Pt.: > 350 F (COC); Pour Pt.: 2 C; Stability: Stable to alkalis in aq. sol'n. up to boil, to acids except in presence of strong oxidizing agents, to chlorine releasing agents; pH: 6–7.5 (1% aq.)

Composition: 99% active ingred.

Reaction: Neutral

Containers: 200 kg net mild steel drums or bulk

Warnings and Caution: Avoid prolonged contact with concentrated form; spillages may be slippery· do not mix with strong oxidizing or reducing agents (potentially explosive)

References: Bulletin SI:10 FM678

Ethylan CRS. Lankro Chem. Ltd.

Chemical Description: Ethoxylated coconut fatty acid alkylolamide

Category: Emulsifier, foam stabilizer, nonionic

Applications: Foam stabilizer in liquid detergent, general purpose or hard surface cleaners and in shampoos; co-emulsifier for chlorinated solvents

Form: Clear liquid; Color: Amber; Odor: Mild

Properties: Solubility: Water-sol.; Sp. gr.: 1.026 @ 20 C; Visc.: 250 cs @ 20 C; Flash Pt.: > 300 F (COC); Pour Pt.: 10 C; Stability: Good in presence of hard water and electrolytes; pH: 7–9 (1% aq.)

Composition: 100% active ingred.

Reaction: Neutral

Warnings and Caution: Avoid prolonged contact with concentrated form; spillages may be slippery; do not mix with strong oxidizing or reducing agents (potentially explosive)

References: Bulletin SI:12FM678

Ethylan D252. Lankro Chem. Ltd.

Chemical Description: Synthetic primary alcohol ethoxylate

Category: Emulsifier, nonionic

Applications: w/o and o/w emulsifier for mineral oils and waxes; emulsification of alkyd resins and paraffin hydrocarbons

Form: Clear liquid; Color: Colorless; Odor: Faint

Properties: Solubility: Oil-sol.; Sp. gr.: 0.903 @ 20 C; Visc.: 31 cs @ 20 C; HLB: 5.6; Flash Pt.: 310 F (COC); Pour Pt.: 5 C; pH: 6–8 (1% aq.)

Composition: 100% active ingred.

Reaction: Neutral

Containers: 200 kg net mild steel drums or bulk

Warnings and Caution: Avoid prolonged contact with concentrated form; spillages may be slippery; do not mix with strong oxidizing or reducing agents (potentially explosive)

References: Bulletin SI: 7FM678

Ethylan D253. Lankro Chem. Ltd.
Chemical Description: Synthetic primary alcohol ethoxylate
Category and Applications: See Ethylan D252; also as viscosity depressant for PVC plastisols; intermediate for sulfation
Form: Clear liquid; Color: Colorless; Odor: Faint
Properties: Solubility: Oil-sol.; Sp. gr.: 0.930 @ 20 C; Visc.: 36 cs @ 20 C; HLB: 7.8; Flash Pt.: 335 F (COC); Pour Pt.: 3 C; pH: 6–8 (1% aq.)
Composition: 100% active ingred.
Reaction: Neutral
Containers: 200 kg net mild steel drums or bulk
Warnings and Caution: Avoid prolonged contact with concentrated form; spillages may be slippery; do not mix with strong oxidizing or reducing agents (potentially explosive)
References: Bulletin SI: 7FM678

Ethylan D254. Lankro Chem. Ltd.
Chemical Description: Synthetic primary alcohol ethoxylate
Category and Applications: See Ethylan D252; also as dry cleaning assistant
Form: Clear liquid; Color: Colorless; Odor: Faint
Properties: Solubility: Oil-sol.; Sp. gr.: 0.948 @ 20 C; Visc.: 57 cs @ 20 C; HLB: 9.8; Flash Pt.: 360 F (COC); Pour Pt.: 5 C; pH: 6–8 (1% aq.)
Composition: 97% active ingred.
Reaction: Neutral
Containers: 200 kg net mild steel drums or bulk
Warnings and Caution: Avoid prolonged contact with concentrated form; spillages may be slippery; do not mix with strong oxidizing or reducing agents (potentially explosive)
References: Bulletin SI: 7FM678

Ethylan D256. Lankro Chem. Ltd.
Chemical Description: Synthetic primary alcohol ethoxylate
Category: Emulsifier, detergent, wetting agent, nonionic
Applications: o/w emulsification of paraffin waxes and hydrocarbon solvents; bases for toxicant emulsifiers and co-emulsifiers for mineral oils; scouring and wetting agents; textile antistatic agents; specialty cleaners
Form: Hazy liquid; Odor: Faint
Properties: Solubility: Water-sol.; Sp. gr.: 0.972 @ 20 C; Visc.: 72 cs @ 20 C; HLB: 11.4; Cloud Pt.: 43 C (1% aq.); Flash Pt.: 375 F (COC); Pour Pt.: 15 C; pH: 6–8 (1% aq.)
Composition: 100% active ingred.
Reaction: Neutral
Containers: 200 kg net mild steel drums or bulk
Warnings and Caution: Avoid prolonged contact with concentrated form; spillages may be slippery; do not mix with strong oxidizing or reducing agents (potentially explosive)
References: Bulletin SI: 7FM678

Ethylan D257. Lankro Chem. Ltd.
Chemical Description: Synthetic primary alcohol ethoxylate

Category and Applications: See Ethylan D256
Form: Hazy liquid; Odor: Faint
Properties: Solubility: Water-sol.; Sp. gr.: 0.974 @ 40 C; Visc.: 31 cs @ 40 C; HLB: 12.2; Cloud Pt.: 49 C (1% aq.); Flash Pt.: 385 F (COC); Pour Pt.: 21 C; pH: 6–8 (1% aq.)
Composition: 100% active ingred.
Reaction: Neutral
Containers: 200 kg net mild steel drums or bulk
Warnings and Caution: Avoid prolonged contact with concentrated form; spillages may be slippery; do not mix with strong oxidizing or reducing agents (potentially explosive)
References: Bulletin SI: 7FM678

Ethylan D259. Lankro Chem. Ltd.
Chemical Description: Synthetic primary alcohol ethoxylate
Category: Detergent, wetting agent, emulsifier, solubilizer, nonionic
Applications: Liquid formulations; solubilizing wax, essential oils
Form: Solid; Color: White
Properties: Solubility: Water-sol.; Sp. gr.: 0.988; Visc.: 36 cs @ 40 C; Cloud Pt.: 76 C
Composition: 100% active ingred.

Ethylan D2512. Lankro Chem. Ltd.
Chemical Description: Synthetic primary alcohol ethoxylate
Category: Emulsifier, detergent, wetting agent, solubilizer, nonionic
Applications: o/w emulsification; detergent, wetting and sudsing agents for liquid detergents; scouring and dye levelling agents; coemulsifiers for aromatic solvents, natural and synthetic waxes and toxicants; detergent sanitizers; essential oil solubilizers
Form: Waxy solid; Color: White; Odor: Faint
Properties: Solubility: Water-sol.; Sp. gr.: 1.001 @ 40 C; Visc.: 46 cs @ 40 C; HLB: 14.2; Cloud Pt.: 92 C (1% aq.); Flash Pt.: > 400 F (COC); Pour Pt.: 29 C; pH: 6–8 (1% aq.)
Composition: 100% active ingred.
Reaction: Neutral
Containers: 200 kg net mild steel drums or bulk
Warnings and Caution: Avoid prolonged contact with concentrated form; spillages may be slippery; do not mix with strong oxidizing or reducing agents (potentially explosive)
References: Bulletin SI: 7FM678

Ethylan D2560. Lankro Chem. Ltd.
Chemical Description: Synthetic primary alcohol ethoxylate
Category: Wetting agent, emulsifier, nonionic
Applications: Latex additive; toilet block component; o/w emulsification; wetting at elevated temperatures
Form: Flake; Color: White; Odor: Faint
Properties: Solubility: Water-sol.; Visc.: 120 cs @ 60 C; HLB: 18.6; Cloud Pt.: > 100 C (1% aq.); Flash Pt.: > 400 F (COC); Pour Pt.: 45 C; pH: 6–8 (1% aq.)
Composition: 100% active ingred.

Reaction: Neutral
Containers: 200 kg net mild steel drums or bulk
Warnings and Caution: Avoid prolonged contact with concentrated form; spillages may be slippery; do not mix with strong oxidizing or reducing agents (potentially explosive)
References: Bulletin SI: 7FM678

Ethylan DP. Lankro Chem. Ltd.
Chemical Description: Nonyl phenol ethoxylate
Category: Foam stabilizer, emulsifier, solubilizer, nonionic
Applications: Foam stabilizer and booster; emulsifier of solvents and oleins, in pesticides; as emulsifier and solubilizer for essential oils and perfumes; emulsion polymerization
Form: Slightly hazy, viscous liquid; Odor: Negligible
Properties: Solubility: Water-sol.; Sp. gr.: 1.068 @ 20 C; Visc.: 400 cs @ 20 C; HLB: 14.0; Cloud Pt.: 80 C (1% aq.); Flash Pt.: >400 F (COC); Pour Pt.: 13 C; Stability: Stable to electrolytes, hard water, extremes of pH and temperature; pH: 6–8 (1% aq.)
Composition: 100% active ingred.
Containers: 200 kg net mild-steel drums or bulk
Warnings and Caution: Avoid prolonged contact with concentrated form; spillages may be slippery; do not mix with strong oxidizing or reducing agents (potentially explosive)
References: Bulletin SI:13 FM678

Ethylan ENTX. Lankro Chem.Ltd.
Chemical Description: Alkylphenol ethoxylate
Category: Emulsifier, plasticizer, nonionic
Applications: Mineral oil and wax emulsifier; mastic plasticizer
Form: Liquid; Color: Pale straw; Odor: Faint
Properties: Solubility: Oil-sol.; Sp. gr.: 1.006 @ 20 C; Visc.: 440 cs @ 20 C; HLB: 9.0; Flash Pt.: >350 F (COC); Pour Pt.: 2 C; Stability: Good in presence of acids, alkalis and electrolytes; pH: 6.0–7.5 (1% aq.)
Composition: 100% active ingred.
Reaction: Neutral
Containers: 200 kg net mild steel drums or bulk
Warnings and Caution: Avoid prolonged contact with concentrated form; spillages may be slippery; do not mix with strong oxidizing or reducing agents (potentially explosive)
References: Bulletin SI:14FM678

Ethylan GD. Lankro Chem. Ltd.
Chemical Description: Special fatty acid diethanolamide
Category: Emulsifier, antistat, lubricant, nonionic
Applications: Emulsifier for mineral oil, cutting oils, dewatering fluids, metal degreasing agents
Form: Clear liquid; Color: Amber; Odor: Mild
Properties: Solubility: Excellent oil-sol.; Sp. gr.: 0.982 @ 20 C; Visc.: 1200 cs @ 20 C; Flash Pt.: > 300 F (COC); Pour Pt.: 10 C; Stability: Good in presence of

hard water and electrolytes; pH: 8–11 (1% aq.)
Composition: 75% active ingred.
Reaction: Alkaline
Warnings and Caution: Avoid prolonged contact with concentrated form; spillages may be slippery; do not mix with strong oxidizing or reducing agents (potentially explosive)
References: Bulletin SI:12FM678

Ethylan GMF. Lankro Chem. Ltd.
Chemical Description: Alkylphenol ethoxylate
Category: Wetting agent, emulsifier, nonionic
Applications: Emusifier for hydrocarbon solvents; emulsion polymerization
Form: Clear liquid; Color: Straw; Odor: Faint
Properties: Solubility: Water-sol.; Sp. gr.: 1.050 @ 20 C; Visc.: 330 cs @ 20 C; HLB: 12.5; Cloud Pt.: 47 C (1% aq.); Flash Pt.: > 350 F (COC); Pour Pt.: 7 C; Stability: Good in presence of acids, alkalis and electrolytes; pH: 6–7.5 (1% aq.)
Composition: 100% active ingred.
Reaction: Neutral
Containers: 200 kg net mild steel drums or bulk
Warnings and Caution: Avoid prolonged contact with concentrated form; spillages may be slippery; do not mix with strong oxidizing or reducing agents (potentially explosive)
References: Bulletin SI:14FM678

Ethylan HA. Lankro Chem. Ltd.
Chemical Description: Nonyl phenol ethoxylate
Category and Applications: See Ethylan 20
Form: Waxy solid; Color: White; Odor: Negligible
Properties: Solubility: Water-sol.; Sp. gr.: 1.064 @ 60 C; Visc.: 120 cs @ 60 C; HLB: 17.4; Cloud Pt.: > 100 C (1% aq.); Flash Pt.: > 400 F (COC); Pour Pt.: 43 C; Stability: Stable to electrolytes, hard water, extremes of pH and temperature; pH: 6–8 (1% aq.)
Composition: 100% active ingred.
Containers: 200 kg net mild-steel drums or bulk
Warnings and Caution: Avoid prolonged contact with concentrated form; spillages may be slippery; do not mix with strong oxidizing or reducing agents (potentially explosive)
References: Bulletin SI:13 FM678

Ethylan HP. Lankro Chem. Ltd.
Chemical Description: Nonyl phenol ethoxylate
Category and Applications: See Ethylan 20
Form: Waxy solid; Color: White; Odor: Negligible
Properties: Solubility: Water-sol.; Sp. gr.: 1.072 @ 40 C; Visc.: 78 cs @ 60 C; HLB: 16.6; Cloud Pt.: > 100 C (1% aq.); Flash Pt.: > 400 F (COC); Pour Pt.: 36 C; Stability: Stable to electrolytes, hard water, extremes of pH and temperature; pH: 6–8 (1% aq.)
Composition: 100% active ingred.
Containers: 200 kg net mild-steel drums or bulk

Warnings and Caution: Avoid prolonged contact with concentrated form; spillages may be slippery; do not mix with strong oxidizing or reducing agents (potentially explosive)
References: Bulletin SI:13 FM678

Ethylan KELD. Lankro Chem. Ltd.
Chemical Description: Modified fatty alkylolamide
Category: Emulsifier, nonionic
Applications: Wax , kerosene and olein emulsification; solvent degreasing
Form: Clear liquid; Color: Amber; Odor: Mild
Properties: Solubility: Disp. in water; Sp. gr.: 1.006 @ 20 C; Visc.: 630 cs @ 20 C; Flash Pt.: > 300 F (COC); Pour Pt.: < 0 C; Stability: Good in presence of hard water and electrolytes; pH: 8–10 (1% aq.)
Composition: 100% active ingred.
Warnings and Caution: Avoid prolonged contact with concentrated form; spillages may be slippery; do not mix with strong oxidizing or reducing agents (potentially explosive)
References: Bulletin SI:12 FM678

Ethylan KEO. Lankro Chem. Ltd.
Chemical Description: Nonyl phenol ethoxylate
Category and Applications: See Ethylan BCP
Form: Clear colorless liquid; Odor: Negligible
Properties: Solubility: Water-sol.; Sp. gr.: 1.060 @ 20 C; Visc.: 340 cs @ 20 C; HLB: 13.0; Cloud Pt.: 54 C (1% aq.); Flash Pt.: > 400 F (COC); Pour Pt.: 3 C; Stability: Stable to electrolytes, hard water, extremes of pH and temperature; pH: 6–8 (1% aq.)
Composition: 100% active ingred.
Containers: 200 kg net mild-steel drums or bulk
Warnings and Caution: Avoid prolonged contact with concentrated form; spillages may be slippery; do not mix with strong oxidizing or reducing agents (potentially explosive)
References: Bulletin SI:13 FM678

Ethylan L10. Lankro Chem. Ltd.
Chemical Description: PEG 1000 laurate
Category: Emulsifier
Applications: Cosmetics and pharmaceuticals
Form: Solid; Color: White
Properties: Sp. gr.: 1.056; Visc.: 76 cs @ 40 C
Reaction: Neutral

Ethylan LD. Lankro Chem. Ltd.
Chemical Description: Coconut diethanolamide
Category: Emulsifier, foam stabilizer, antistat, nonionic
Applications: Foam stabilizer in liquid detergents, general purpose or hard surface cleaners and in shampoos; emulsifier in hand cleaning gels; antistat for plastics; kerosene and oil emulsifier
Form: Clear liquid; Color: Straw; Odor: Mild
Properties: Solubility: Disp. in water; Sp. gr.: 0.981 @ 20 C; Visc.: 960 cs @ 20 C; Flash Pt.: > 350 F

(COC); Pour Pt.: 15 C; Stability: Good in presence of hard water and electrolytes; pH: 8.5–10.5 (1% aq.)
Composition: 90% active ingred.
Warnings and Caution: Avoid prolonged contact with concentrated form; spillages may be slippery; do not mix with strong oxidizing or reducing agents (potentially explosive)
References: Bulletin SI:12FM678

Ethylan LDS. Lankro Chem. Ltd.
Chemical Description: Coconut diethanolamide
Category: Foam stabilizer, emulsifier, nonionic
Applications: Foam stabilizer in liquid detergents with cold temperature stability, general purpose or hard surface cleaners and in shampoos; kerosene and oil emulsifier
Form: Clear liquid; Color: Straw; Odor: Mild
Properties: Solubility: Disp. in water; Sp. gr.: 0.990 @ 20 C; Visc.: 810 cs @ 20 C; Flash Pt.: > 350 F (COC); Pour Pt.: 10 C; Stability: Good in presence of hard water and electrolytes; pH: 8.5–10.5 (1% aq.)
Composition: 87% active ingred.
Warnings and Caution: Avoid prolonged contact with concentrated form; spillages may be slippery; do not mix with strong oxidizing or reducing agents (potentially explosive)
References: Bulletin SI:12FM678

Ethylan LM. Lankro Chem. Ltd.
Chemical Description: Coconut monoalkanolamide
Category: Emulsifier, antistat, foam stabilizer, nonionic
Applications: Additive for hard surface cleaners; foam stabilizer for liquid detergents
Form: Flake; Color: Pale straw; Odor: Mild
Properties: Solubility: Water-insol.; Sp. gr.: 0.909 @ 80 C; Visc.: 40 cs @ 80 C; Flash Pt.: > 350 F (COC); Pour Pt.: 63 C; Stability: Good in presence of hard water and electrolytes; pH: 8.5–10.5 (1% aq.)
Composition: 94% active ingred.
Warnings and Caution: Avoid prolonged contact with concentrated form; spillages may be slippery; do not mix with strong oxidizing or reducing agents (potentially explosive)
References: Bulletin SI:12FM678

Ethylan LM2. Lankro Chem. Ltd.
Chemical Description: Ethoxylated coconut fatty acid alkylolamide
Category: Foam stabilizer, emulsifier, nonionic
Applications: Detergent formulations
Form: Soft paste; Color: Off-white; Odor: Mild
Properties: Solubility: Water-sol.; Sp. gr.: 0.963 @ 40 C; Visc.: 108 cs @ 40 C; Flash Pt.: > 300 F (COC); Pour Pt.: 2 C; Stability: Good in presence of hard water and electrolytes; pH: 7.0–9.0 (1% aq.)
Composition: 100% active ingred.

Warnings and Caution: Avoid prolonged contact with concentrated form; spillages may be slippery; do not mix with strong oxidizing or reducing agents (potentially explosive)
References: Bulletin SI:12FM678

Ethylan MLD. Lankro Chem. Ltd.
Chemical Description: Lauric diethanolamide
Category: Foam booster and stabilizer, antistat, nonionic
Applications: Foam booster and stabilizer in toiletries and detergent formulations (e.g., shampoos, bubble baths); antistat for plastics
Form: Waxy flake; Color: Pale cream; Odor: Negligible
Properties: Solubility: Disp. in water; Visc.: 130 cs @ 60 C; Flash Pt.: > 300 F (COC); Pour Pt.: 40 C ; Stability: Good in presence of hard water and electrolytes; pH: 8–9 (1% aq.)
Composition: 95% active ingred.
Reaction: Alkaline
Warnings and Caution: Avoid prolonged contact with concentrated form; spillages may be slippery; do not mix with strong oxidizing or reducing agents (potentially explosive)
References: Bulletin SI:12FM678

Ethylan N30. Lankro Chem. Ltd.
Chemical Description: Nonyl phenol ethoxylate
Category and Applications: See Ethylan 20
Form: Waxy solid; Color: White; Odor: Negligible
Properties: Solubility: Water-sol.; Sp. gr.: 1.064 @ 60 C; Visc.: 90 cs @ 60 C; HLB: 17.0; Cloud Pt.: > 100 C (1% aq.); Flash Pt.: > 400 F (COC); Pour Pt.: 39 C; Stability: Stable to electrolytes, hard water, extremes of pH and temperature; pH: 6–8 (1% aq.)
Composition: 100% active ingred.
Containers: 200 kg net mild-steel drums or bulk
Warnings and Caution: Avoid prolonged contact with concentrated form; spillages may be slippery; do not mix with strong oxidizing or reducing agents (potentially explosive)
References: Bulletin SI:13 FM678

Ethylan N50. Lankro Chem. Ltd.
Chemical Description: Nonyl phenol ethoxylate
Category and Applications: See Ethylan 20
Form: Waxy solid; Color: White; Odor: Negligible
Properties: Solubility: Water-sol.; Sp. gr.: 1.073 @ 60 C; Visc.: 135 cs @ 60 C; HLB: 18.2; Cloud Pt.: > 100 C (1% aq.); Flash Pt.: > 400 F (COC); Pour Pt.: 43 C; Stability: Stable to electrolytes, hard water, extremes of pH and temperature; pH: 6–8 (1% aq.)
Composition: 100% active ingred.
Containers: 200 kg net mild-steel drums or bulk
Warnings and Caution: Avoid prolonged contact with concentrated form; spillages may be slippery; do not mix with strong oxidizing or reducing agents (potentially explosive)
References: Bulletin SI:13 FM678

Ethylan N92. Lankro Chem. Ltd.
Chemical Description: Nonyl phenol ethoxylate
Category and Applications: See Ethylan 20
Form: Waxy solid; Color: White; Odor: Negligible
Properties: Solubility: Water-sol.; Sp. gr.: 1.078 @ 60 C; Visc.: 340 cs @ 60 C; HLB: 19.0; Cloud Pt.: > 100 C (1% aq.); Flash Pt.: > 400 F (COC); Pour Pt.: 50 C; Stability: Stable to electrolytes, hard water, extremes of pH and temperature; pH: 6–8 (1% aq.)
Composition: 100% active ingred.
Containers: 200 kg net mild-steel drums or bulk
Warnings and Caution: Avoid prolonged contact with concentrated form; spillages may be slippery; do not mix with strong oxidizing or reducing agents (potentially explosive)
References: Bulletin SI:13 FM678

Ethylan NP 1. Lankro Chem. Ltd.
Chemical Description: Nonyl phenol ethoxylate
Category: Defoamer, nonionic
Applications: Defoaming applications, oil emulsifier
Form: Clear liquid; Color: Straw
Properties: Solubility: Oil-sol.; Sp. gr.: 0.987 @ 20 C; Visc.: 650 cs @ 20 C; HLB: 4.5; Flash Pt.: 300 F (COC); Pour Pt.: < 0 C; Stability: Stable to electrolytes, hard water, and extremes of pH and temperature; pH: 6–8 (1% aq.)
Composition: 100% active ingred.
Containers: 200 kg net mild-steel drums or bulk
Warnings and Caution: Avoid prolonged contact with concentrated form; spillages may be slippery; do not mix with strong oxidizing or reducing agents (potentially explosive)
References: Bulletin SI:13 FM678

Ethylan PQ. Lankro Chem. Ltd.
Chemical Description: Modified alkylphenol ethoxylate
Category: Solubilizer, nonionic
Applications: Manufacutre of iodophor
Form: Water-white liquid (clear @ 30 C); Odor: Faint
Properties: Solubility: Water-sol.; Sp. gr.: 1.045 @ 40 C; Visc.: 110 cs @ 40 C; HLB: 14.2; Cloud Pt.: 83 C (1% aq.); Flash Pt.: > 350 F (COC); Pour Pt.: 15 C; Stability: Good in presence of acids, alkalis and electrolytes pH: 6.0–7.5 (1% aq.)
Composition: 98% active ingred.
Reaction: Neutral
Containers: 200 kg net mild steel drums or bulk
Warnings and Caution: Avoid prolonged contact with concentrated form; spillages may be slippery; do not mix with strong oxidizing or reducing agents (potentially explosive)
References: Bulletin SI:14 FM678

Ethylan OE. Lankro Chem. Ltd.
Chemical Description: Cetyl/oleyl alcohol ethoxylate
Category: Emulsifier, solubilizer, stabilizer, leveling agent, nonionic
Applications: Emulsifier for fatty acids, alcohols and waxes; essen-

tial oil solubilizer; latex stabilizer; in the manufacture of polishes; emulsion polymerization aid; dye leveling agent
Form: Waxy solid; Color: Cream; Odor: Negligible
Properties: Solubility: Water-sol.; Sp. gr.: 1.009 @ 40 C; Visc.: 67 cs @ 40 C; HLB: 14; Cloud Pt.: 90 C (1% aq.); Flash Pt.: > 350 F (COC); Pour Pt.: 31 C; Stability: Good in presence of acids, alkalis and electrolytes; pH: 6–8 (1% aq.)
Composition: 100% active ingred.
Reaction: Neutral
Containers: 200 kg net mild steel drums or bulk
Warnings and Caution: Avoid prolonged contact with concentrated form; spillages may be slippery; do not mix with strong oxidizing or reducing agents (potentially explosive)
References: Bulletin SI:9 FM678

Ethylan R. Lankro Chem. Ltd.
Chemical Description: Cetyl/oleyl alcohol ethoxylate
Category and Applications: See Ethylan OE
Form: Waxy solid; Color: Cream; Odor: Negligible
Properties: Solubility: Water-sol.; Sp. gr.: 1.026 @ 40 C; Visc.: 150 cs @ 40 C; HLB: 15.4; Cloud Pt.: > 100 C (1% aq.); Flash Pt.: > 350 F (COC); Pour Pt.: 36 C; Stability: Good in presence of acids, alkalis and electrolytes; pH: 6–8 (1% aq.)
Composition: 100% active ingred.
Reaction: Neutral
Containers: 200 kg net mild steel drums or bulk
Warnings and Caution: Avoid prolonged contact with concentrated form; spillages may be slippery; do not mix with strong oxidizing or reducing agents (potentially explosive)
References: Bulletin SI:9 FM678

Ethylan TC. Lankro Chem. Ltd.
Chemical Description: Coconut fatty amine ethoxylate
Category: Wetting agent, nonionic
Applications: Wetting agents in acid or alkaline formulations for metal cleaning and stripping of surface coatings; fiber antistat; cosmetics
Form: Liquid; Color: Dark amber; Odor: Mild fatty amine
Properties: Solubility: Water-sol.; Sp. gr.: 1.040 @ 20 C; Visc.: 140 cs @ 20 C; HLB: 15; Flash Pt.: > 300 F (COC); Pour Pt.: < 0 C; Stability: Good in presence of acids, alkalis and high electrolyte concentrations; pH: 8.5–10.0 (1% aq.)
Composition: 100% active ingred.
Reaction: Alkaline
Containers: 200 kg net mild steel drums or bulk
Warnings and Caution: Avoid prolonged contact with concentrated form; spillages may be slippery; do not mix with strong oxidizing or reducing agents (potentially explosive)
References: Bulletin SI:15 FM678

Ethylan TCO. Lankro Chem. Ltd.
Chemical Description: Complex

amine oxide
Category: Stabilizer, foam booster, emulsifier, nonionic
Applications: Foam and suds stabilizer in light and heavy duty liquid detergents; cosmetic formulations; foam stabilizer in no-gel carpet backing and latex foam odor control agent
Form: Clear to hazy liquid; Color: Straw; Odor: Mild fatty amine
Properties: Solubility: Water-sol.; Sp. gr.: 1.008 @ 20 C; Visc.: 159 cs @ 20 C; Flash Pt.: > 300 F (COC); Pour Pt.: 0 C; Stability: Good in presence of acids, alkalis, electrolytes; pH: 6–8 (1% aq.)
Composition: 40% active ingred.
Reaction: Neutral
Containers: 200 kg net mild steel drums or bulk
Warnings and Caution: Avoid prolonged contact with concentrated form; spillages may be slippery; do not mix with strong oxidizing or reducing agents (potentially explosive)
References: Bulletin SI:15FM678

Ethylan TF. Lankro Chem. Ltd.
Chemical Description: Ethoxylated coconut fatty amine
Category: Emulsifier, antistat, nonionic
Applications: Oil emulsifier with anticorrosive properties; antistat for synthetic fibers, with polystyrene; formulation of wax polishes
Form: Liquid; Color: Light amber; Odor: Mild fatty amine
Properties: Solubility: Disp. in water; Sp. gr.: 0.912 @ 20 C; Visc.; 140 cs @ 20 C; HLB: 6; Flash Pt.: > 300 F (COC); Pour Pt.: < 0 C; Stability: Good in presence of acids, alkalis and electrolytes; pH: 8.5–10.0 (1% aq.)
Composition: 100% active ingred.
Toxicity: Slightly toxic; skin and eye irritant
Containers: 200 kg net mild steel drums or bulk
Warnings and Caution: Avoid prolonged contact with concentrated form; spillages may be slippery; do not mix with strong oxidizing or reducing agents (potentially explosive)
References: Bulletin SI:15FM678

Ethylan TH-2. Lankro Chem. Ltd.
Chemical Description: Hydrogenated fatty amine ethoxylate
Category: Emulsifier, anticorrosive agent, antistat, nonionic
Applications: Oil emulsifier with anticorrosive properties; antistat for synthetic fibers and polystyrene
Form: Solid; Color: Pale brown; Odor: Mild fatty amine
Properties: Solubility: Disp. in water; Sp. gr.: 0.878 @ 60 C; Visc.: 33 cs @ 60 C; HLB: 5; Flash Pt.: > 300 F (COC); Pour Pt.: 40 C; Stability: Good in presence of acids, alkalis and electrolytes; pH: 8.5–10.0 (1% aq.)
Composition: 100% active ingred.
Toxicity: Slightly toxic; skin and eye irritant
Containers: 200 kg net mild steel drums or bulk
Warnings and Caution: Avoid prolonged contact with concentrated form; spillages may be slippery;

do not mix with strong oxidizing or reducing agents (potentially explosive)
References: Bulletin SI:15 FM678

Ethylan TN-10. Lankro Chem. Ltd.
Chemical Description: Coconut fatty amine ethoxylate
Category: Emulsifier, nonionic
Applications: Cosmetics manufacture, hair-conditioner formulations; oil emulsifier with anticorrosive properties; antistat for synthetic fibers; formulation of wax polishes
Form: Liquid; Color: Dark amber; Odor: Mild fatty amine
Properties: Solubility: Water-sol.; Sp. gr.: 1.015 @ 20 C; Visc.: 174 cs @ 20 C; HLB: 14; Flash Pt.: > 300 F (COC); Pour Pt.: < 0 C; Stability: Good in acids, alkalis and electrolytes; pH: 8.5–10.0 (1% aq.)
Composition: 100% active ingred.
Containers: 200 kg net mild steel drums or bulk
Warnings and Caution: Avoid prolonged contact with concentrated form; spillages may be slippery; do not mix with strong oxidizing or reducing agents (potentially explosive)
References: Bulletin SI:15 FM678

Ethylan TT-15. Lankro Chem. Ltd.
Chemical Description: Tallow amine ethoxylate
Category and Applications: See Ethylan TC
Form: Paste; Color: Pale brown; Odor: Mild fatty amine
Properties: Solubility: Water-sol.; Sp. gr.: 1.030 @ 20 C; Visc.: 252 cs @ 20 C; HLB: 14; Flash Pt.: > 300 F (COC); Pour Pt.: 0 C; Stability: Good in presence of acids, alkalis and electrolytes; pH: 8.5–10.0 (1% aq.)
Composition: 100% active ingred.
Containers: 200 kg net mild steel drums or bulk
Warnings and Caution: Avoid prolonged contact with concentrated form; spillages may be slippery; do not mix with strong oxidizing or reducing agents (potentially explosive)
References: Bulletin SI:15 FM678

Ethylan TU. Lankro Chem. Ltd.
Chemical Description: Nonyl phenol ethoxylate
Category: Detergent, wetting and emulsifying agent, nonionic
Applications: Textile scouring agent, dry cleaning detergent; wetting and emulsifying agent for oils and aromatic solvents
Form: Clear liquid; Color: Pale straw; Odor: Negligible
Properties: Solubility: Water-sol.; Sp. gr.: 1.055 @ 20 C; Visc.: 340 cs @ 20 C; HLB: 12.2; Cloud Pt.: 32 C (1% aq.); Flash Pt.: > 400 F (COC); Pour Pt.: < 0 C; Stability: Stable to electrolytes, hard water, extremes of pH and temperature; pH: 6–8 (1% aq.)
Composition: 100% active ingred.
Containers: 200 kg net mild-steel drums or bulk
Warnings and Caution: Avoid prolonged contact with concentrated form; spillages may be slippery;

do not mix with strong oxidizing or reducing agents (potentially explosive)
References: Bulletin SI:13 FM678

Ethylan VPK. Lankro Chem. Ltd.
Chemical Description: PEG unsaturated fatty acid ester
Category: Emulsifier, lubricant, nonionic
Applications: Textile oil emulsifier, fiber lubricant
Form: Clear liquid; Color: Pale straw
Properties: Sp. gr.: 1.043; Visc.: 120 cs @ 20 C; HLB: 12.4
Composition: 100% active ingred.
Reaction: Neutral

Etocas 10. Croda Chem. Ltd.
Chemical Description: POE 10 castor oil
Category: Emulsifier, solubilizer, lubricant, emollient, superfatting agent, antistat, softener, detergents, nonionic
Applications: Cosmetics (perfume and essential oil solubilizers; shampoos, foam baths; skin care preparations); textile surfactants (lubricants, antistats, spin finish additives, softeners, tufting aids, dye leveling agents for synthetic fiber processing); water soluble metal working fluids; emulsion polymerization; detergents and emulsifiers in insecticides, herbicides and household products
Form: Liquid; Color: Pale yellow
Properties: Solubility: Sol. in ethyl alcohol, oleyl alcohol, naphtha, methyl ethyl ketone, oleic acid, trichloroethylene; part sol. in butyl stearate, arachis oil; disp. in water; HLB: 6.3; Cloud Pt.: < 20 C (1% in 10% brine); Sapon. Value: 120–130; Acid No.: 1.0 max.; Hydroxyl No.: 120–140; Iodine Value: 55–60; Surface Tension: Insol. (0.1 deionized water); pH: 6.0–7.5
Composition: 97% active ingred.

Etocas 35. Croda Chem. Ltd.
Chemical Description: POE 35 castor oil
Category and Applications: See Etocas 10
Form: Liquid; Color: Pale yellow
Properties: Solubility: Sol. in water, ethyl and oleyl alcohols, naphtha, methyl ethyl ketone, oleic acid, trichloroethylene; part sol. in butyl stearate and arachis oil; HLB: 12.5; Cloud Pt.: 35–40 C; Sapon. Value: 62–72; Acid No.: 1.0 max.; Hydroxyl Value: 75–90; Iodine Value: 30–35; Surface Tension: 41.5 dynes/cm (0.1 deionized water); pH: 6.0–7.5

Etocas 40. Croda Chem. Ltd.
Chemical Description: POE 40 castor oil
Category and Applications: See Etocas 10
Form: Liquid; Color: Pale yellow
Properties: Solubility: Sol. in water, ethyl and oleyl alcohols, naphtha, methyl ethyl ketone, oleic acid, trichloroethylene; part. sol. in butyl stearate; HLB: 13.0; Cloud Pt.: 50 C; Sapon. Value: 60–65; Acid No.: 1.0 max.; Hy-

droxyl No.: 60–75; Iodine Value: 28–32; pH: 6.0–7.5
Composition: 97% active ingred.

Etocas 60. Croda Chem. Ltd.
Chemical Description: POE 60 castor oil
Category and Applications: See Etocas 10
Form: Soft paste; Color: Pale yellow
Properties: Solubility: Sol. in water, ethyl alcohol, naphtha, methyl ethyl ketone, oleic acid, trichloroethylene; part sol. in oleyl alcohol, butyl stearate; HLB: 14.7; Cloud Pt.: 60 C; Sapon. Value: 45–50; Acid No.: 1.0 max.; Hydroxyl No.: 50–60; Iodine Value: 22–26; Surface Tension: 43.2 dynes/cm (0.1 deionized water); pH: 6.0–7.5

Etocas 100. Croda Chem. Ltd.
Chemical Description: POE 100 castor oil
Category and Applications: See Etocas 10
Form: Waxy solid; Color: Pale yellow
Properties: Solubility: Sol. in water, ethyl alcohol, naphtha, methyl ethyl ketone, oleic acid, trichloroethylene; part sol. in oleyl alcohol; HLB: 16.5; Cloud Pt.: 66 C; Sapon. Value: 25–35; Acid No.: 1.0 max.; Hydroxyl Value: 35–45; Iodine Value: 14–18; Surface Tension: 41.6 dynes/cm (0.1 deionized water); pH: 6.0–7.5
Composition: 97% active ingred.

Eureka 102. Atlas Refinery, Inc.
Chemical Description: Sulfated castor oil
Category: Emulsifier, detergent, anionic
Applications: Topping oil for suede leather; grinding aid in pigment dispersions; plasticizer in finish coatings
Form: Liquid; Color: Light
Properties: pH: 7.6
Composition: 72% active ingred.
Reaction: Neutral

Eureka 392. Atlas Refinery, Inc.
Chemical Description: Sulfated fatty acid from vegetable oil
Category: Emulsifier, anionic
Applications: Carrier for refined oils; base for solvent fatliquor systems
Form: Liquid
Properties: pH: 5.6
Composition: 75% active ingred.

Eureka 400. Atlas Refinery, Inc.
Chemical Description: Sulfonated oil based on refined sperm oil
Category: Emulsifier, lubricant, softener, anionic
Applications: Leather processing
Form: Liquid; Color: Light
Properties: Stability: Good over entire pH range; pH: 4.5
Composition: 87% active ingred.

Eureka E-2. Atlas Refinery, Inc.
Chemical Description: Cationic derivative of a fatty acid diamine condensate
Category: Emulsifier, lubricant, softener, cationic
Applications: Emulsifier for raw

oils (e.g., neatsfoot, mineral)
Form: Heavy viscous liquid
Properties: Density: 8.32 lb/gal @ 60 F; Sp. gr.: 0.999 @ 60 F; pH: 6–6.5 (10% sol'n.)
Composition: 100% active ingred.

Exp. Surfactant XD-8390. Dow
Category: Emulsifier, wetting agent
Applications: Emulsion polymerization; primary emulsifier for SB and ABS polymers; provides stability and dispersibility for flowable pesticide formulations
Form: Liquid; Color: Red-brown
Properties: Sp. gr.: 1.11 @ 20 C; Visc.: 10 cps
Composition: 35% active ingred.
Toxicity: Acute oral, skin and eye irritant
Containers: 55 gal drums, bulk
Warnings and Caution: Burns the eyes; may cause skin irritation; toxic to fish

F

Flexricin 9. NL Industries, Ind. Chem. Div.
Chemical Description: Propylene glycol monoricinoleate
Category: Wetting agent, solvent, plasticizier, nonionic
Applications: Wetting agent, dye solvent, wax plasticizer; textile, household and cosmetic applications; rewetting dried skins
Form: Liquid; Color: 2+ (Gardner)
Properties: Solubility: Sol. in toluene, butyl acetate, MEK, ethanol; Sp. gr.: 0.96; Visc.: 3 stokes
Composition: 100% active ingred.
Containers: 55 gal drums
Warnings and Caution: Keep containers closed; avoid contact with oxidizing materials

Flexricin 13. NL Industries, Ind. Chem. Div.
Chemical Description: Glyceryl monoricinoleate
Category: Wetting agent, plasticizer, release agent, antifoam agent, nonionic
Applications: Wetting agent, wax plasticizer, plasticizer and mold release agent for rubber polymers, antifoam agent; textile, household and cosmetic applications; rewetting dried skins
Form: Liquid; Color: 2 min. (Gardner)
Properties: Solubility: Sol. in toluene, butyl acetate, MEK, ethanol; Sp. gr.: 0.985; Visc.: 8.8 stokes
Composition: 100% active ingred.
Containers: 55 gal drums
Warnings and Caution: Keep containers closed; avoid contact with oxidizing materials

Flexricin 15. NL Industries, Ind. Chem. Div.
Chemical Description: Ethylene glycol monoricinoleate
Category: Wetting agent, plasticizer, nonionic
Applications: Textiles, household and cosmetic uses; rewetting dried skins
Form: Liquid; Color: 4 (Gardner)
Properties: Solubility: Sol. in toluene, butyl acetate, MEK, ethanol; Sp. gr.: 0.965; Visc.: 3.9 stokes
Containers: 55 gal drums
Warnings and Caution: Keep con-

tainers closed; avoid contact with oxidizing materials

Fluilan. Croda Chem. Ltd.
Chemical Description: Pure liquid lanolin from anhydrous lanolin BP/USP
Category: Conditioning agent, emollient, moisturizer, penetrant, superfatting agent, dispersant, emulsifier, nonionic
Applications: Skin care cosmetics, make-up
Form: Viscous liquid; Color: Golden; Odor: Faint characteristic lanolin
Composition: 100% active ingred.

Fluilanol. Croda Chem. Ltd.
Chemical Description: Liquid lanolin
Category: Emulsifier, emollient, dispersant, wetting agent, superfatting agent, nonionic
Applications: Make-up cosmetics, bath oils; improves emulsion stability, dispersing and spreading properties; wetting agent for pigments in lipophilic media; superfatting agent for soaps and detergents
Form: Liquid; Color: Pale; Odor: Faint sterol
Properties: Self-emulsifying
Composition: 100% active ingred.

Fluorad FC-93. 3M Co.
Chemical Description: Ammonium perfluoroalkyl sulfonate
Category: Wetting agent, anionic
Applications: Wetting agent in etching solutions in manufacture of semi-conductor devices; active in acid and alkaline media
Form: Clear liquid
Properties: Solubility: Sol. in mixtures of alcohol and water; only slightly sol. in water; Sp. gr.: 1.1; Flash Pt.: 81 F (PMCC); Stability: Excellent chemical and thermal stability; pH: 7 (1% aq.)
Composition: 25% active ingred.; other components: 73% water, 27% isopropyl alcohol
Reaction: Neutral
Warnings and Caution: Flammable
References: Bulletin Y-ITD93 (79.05)BPH

Fluorad FC-95. 3M Co.
Chemical Description: Potassium perfluoroalkyl sulfonate
Category: Wetting and foaming agent, anionic
Applications: Chrome plating baths; conversion coatings; metal and plastic etchants; acid pickling baths; alkaline cleaning systems; for systems requiring low surface tension and dense foam
Form: Free-flowing powder
Properties: Solubility: Low sol. in water and most inorganic solvents; recommended with co-solvent such as DGMBE; M.P.: Decomposes @ 390 C; Stability: Stable to strongly oxidizing and reducing media; chemically and thermally stable; Surface Tension: 22 dynes/cm (0.1 g/100 g water); pH: 7–8 (0.1% aq.)
Composition: 100% active ingred.
Warnings and Caution: Avoid eye contact, prolonged breathing; use with adequate ventilation
References: Bulletin Y-ISTDI

(75-.2)NPI

Fluorad FC-98. 3M Co.
Chemical Description: Potassium perfluoroalkyl sulfonate
Category and Applications: See Fluorad FC-95; FC-98 is slightly less surface active and produces less stable and less dense foam than FC-95
Form: Free-flowing powder
Properties: Solubility: Low sol. in water and most inorganic solvents; recommended with co-solvent such as DGMBE; M.P.: Decomposes @ 390 C; Stability: Stable to acid and oxidizing systems; chemically and thermally stable; Surface Tension: 40 dynes/cm (0.1 g/100 g water); pH: 6–8 (0.1% aq.)
Composition: 100% active ingred.
Warnings and Caution: Avoid eye contact, prolonged breathing; use with adequate ventilation
References: Bulletin Y-ISTDI (75-.2)NPI

Fluorad FC-99. 3M Co.
Chemical Description: Amine perfluoroalkyl sulfonate sol'n. in water
Category: Wetting and foaming agent, anionic
Applications: Chrome plating baths, metal and plastic etchants, acid cleaning and pickling baths
Form: Clear liquid; Color: Yellow to amber
Properties: Solubility: Water miscible; Sp. gr.: 1.1; Stability: Chemically and thermally stable; active in acidic systems; Surface Tension: 22 dynes/cm (0.4 g/100 g dist. water); pH: 6–7 (as is)
Composition: 25% active ingred.; other components: water
Toxicity: Considered minimally irritating
References: Bulletin Y-ITD99(79.03)R1

Fluorad FC-120. 3M Co.
Chemical Description: Anionic fluorochemical
Category: Wetting and leveling agent, anionic
Applications: Aqueous coating systems; clear polishes based on alkali soluble polymers
Form: Low viscosity liquid
Properties: Sp. gr.: 1.06; Flash Pt.: > 38 C (TCC); pH: 8.5–9.5
Composition: 25% active ingred. in 37.5% ethylene glycol monobutyl ether and 37.5% water
Toxicity: Minor eye and skin irritant
Warnings and Caution: Avoid prolonged breathing of vapor; goggles and gloves recommended
References: Bulletin Y-ITD120 (68.05)R1

Fluorad FC-126. 3M Co.
Chemical Description: Anionic fluorochemical
Category: Anionic surfactant
Form: Free-flowing powder
Applications: Excellent surface activity in alkaline media
Properties: Solubility: Sol. in water, acetone, methyl alcohol; M.P.: Decomposes @ 120 C; Surface Tension: 44 dynes/cm (0.001 conc.); pH: 4–6

Composition: 100% active ingred.

Fluorad FC-128. 3M Co.
Chemical Description: Potassium fluorinated alkyl carboxylate
Category: Wetting, spreading and leveling agent, anionic
Applications: Excellent leveling in alkaline media; good foaming; water-based coatings; alkaline cleaners; leak detection solutions
Form: Powder
Properties: Solubility: Forms gel in water; Surface Tension: 17 dynes/cm (0.1 g/100 g dist. water); pH: 7–8 (0.1% aq.)
Composition: 100% active ingred.
Warnings and Caution: Avoid contact with skin, eyes; avoid prolonged breathing of vapors
References: Bulletin Y-ISTDJ (75-.2)NPI

Fluorad FC-129. 3M Co.
Chemical Description: Potassium fluorinated alkyl carboxylate
Category: Wetting and leveling agent, anionic
Applications: Floor polish, alkaline cleaners
Form: Liquid; Color: Amber
Properties: Solubility: Limited sol. in aq. acids and bases; stronger sol. in organic polar liquids (enhanced by solvents in FC-129); Sp. gr.: 1.3; Flash Pt.: 115 F (PMCC); Surface Tension: 17 dynes/cm (0.2 g/100 g dist. water); pH: 7–8 (0.1% solids in aq. sol'n.)
Composition: 50% active ingred.; other components: 28% water, 18% butyl Cellosolve, 4% ethyl alcohol
Warnings and Caution: Flammable; keep tightly closed; avoid contact with skin, eyes; avoid prolonged breathing
References: Bulletin Y-ITD129 (29.05)R1

Fluorad FC-134. 3M Co.
Chemical Description: Fluorinated alkyl quaternary ammonium iodides
Category: Wetting, spreading and leveling agent, cationic
Applications: Floor polishes, photographic development, corrosion inhibitors, leak detector sol'ns.; excellent activity in acid, basic and neutral media
Form: Powder
Properties: Solubility: Forms gel in water; Surface Tension: 17 dynes/cm (0.1 g/100 g dist. water); pH: 3–4 (0.1% aq.)
Composition: 100% active ingred.
Warnings and Caution: Avoid prolonged breathing, skin and eye contact

Fluorad FC-135. 3M Co.
Chemical Description: Fluorinated alkyl quaternary ammonium iodides
Category: Wetting and leveling agent, cationic
Applications: Cleaning and polishing products; substantive to many metal surfaces
Form: Liquid; Color: Dark amber
Properties: Solubility: Low sol. in most nonpolar organic solvents; limited sol. in highly polar solvents, aq. acids and bases; Sp. gr.: 1.2; Flash Pt.: 69 F (PMCC); Sur-

face Tension: 17 dynes/cm (0.2 g/100 g dist. water); pH: 3–4 (0.1% solids in aq. sol'n.)
Composition: 50% active solids; 34% isopropyl alcohol; 16% water
Toxicity: Slight skin and eye irritant
Warnings and Caution: Flammable; keep from heat, sparks and flame; protect skin and eyes from contact
References: Bulletin Y-ITD135 (59.025)R1

Fluorad FC-143. 3M Co.
Chemical Description: Ammonium perfluoroalkyl carboxylate
Category: Emulsifier, wetting agent, anionic
Applications: Fluropolymer emulsions; wetting agent in alkaline media
Form: Free-flowing powder
Properties: Solubility: Sol. in water, methyl alcohol, acetone; Density: 0.6–0.7 g/cc; M.P.: Decomposes @ 130 C; Stability: Excellent chemical and thermal stability and in acid, alkaline, oxidizing and reducing media; Surface Tension: 44 dynes/cm (0.1 g/100 g dist. water); pH: 4–6 (0.5 aq. sol'n.)
Composition: 100% active ingred.
Warnings and Caution: Avoid prolonged breathing of dust or fumes; use with adequate ventilation
References: Bulletin Y-ISTDH (65.2)R

Fluorad FC-170-C. 3M Co.
Chemical Description: Fluorinated alkyl polyoxethylene ethanol
Category: Wetting, spreading and leveling agent, nonionic
Applications: Floor polish formulations; good activity in water and acid solutions
Form: Liquid; Color: Amber
Properties: Solubility: Sol. in hydrochloric acid, methyl alcohol, ethyl alcohol, isopropyl alcohol; Sp. gr.: 1.32; Cloud Pt.: < 0 C; Flash Pt.: > 220 F (PMCC); Surface Tension: 20 dynes/cm (0.1 g/100 g dist. water); pH: 7–8 (1% aq.)
Composition: 95% active ingred.
References: Bulletin Y-ITD-170C (38.05)R1

Fluorad FC-171. 3M Co.
Chemical Description: Fluorinated alkyl alkoxylate
Category: Wetting, spreading and leveling agent, nonionic
Applications: Floor polishes, cleaners, high solids coatings and water reducible coatings
Form: Liquid; Color: Amber
Properties: Solubility: 20% sol. in butyl Cellosolve, Cellosolve acetate, toluene, methyl ethyl ketone, dimethyl formamide, isopropyl alcohol, dibutyl phthalate; forms cloudy sol'n. in water; Sp. gr.: 1.4; Flash Pt.: > 220 F (PMCC); Pour Pt.: ≈ 20 F; Surface Tension: 20 dynes/cm (0.1 g/100 g dist. water)
Composition: 100% active ingred.
Toxicity: Minor eye irritant
References: Bulletin Y-ITD171

Fluorad FC-430. 3M Co.
Chemical Description: Fluorinated alkyl ester
Category: Wetting, leveling,

spreading and flow control agents, nonionic
Applications: Water-based coatings and polymer systems; solvent-based coatings; soldering systems
Form: Viscous liquid; Color: Straw to amber
Properties: Ref. Index: 1.445; Solubility: > 20 g of solid/100 ml of solvent for: dist. water, Cellosolve acetate, methyl ethyl ketone, toluene, dimethylformamide; Sp. gr.: 1.15; Visc.: 15,000 cp (Brookfield spindle # 3 @ 6 rpm); Flash Pt.: > 300 F (TOC); Surface Tension: 30.3 dynes/cm (0.1 g/100 ml dist. water)
Composition: 100% active ingred.
Warnings and Caution: Avoid prolonged breathing of vapors and eye and skin contact
References: Bulletin Y-ITD-430-1 (114-.2)NPI

Fluorad FC-431. 3M Co.
Chemical Description: Fluorinated alkyl ester
Category and Applications: See Fluorad FC-430
Form: Thin liquid; Color: Straw to amber
Properties: Ref. Index: 1.406; Solubility: > 20 g of solid/100 ml of: butyl Cellosolve, Cellosolve acetate, methyl ethyl ketone, toluene, Solvesso 100, dimethylformamide, dibutyl phthalate; Sp. gr.: 1.05; Visc.: 200 cp (Brookfield spindle # 1 @ 60 rpm); Flash Pt.: 40 F (TOC); Surface Tension: 36.6 dynes/cm (0.1 g/100 ml dist. water)
Composition: 50% active solids in ethyl acetate
Warnings and Caution: Flammable solvent; keep away from heat, sparks or flame; avoid prolonged breathing and skin and eye contact
References: Bulletin Y-ITD-430 (114-.2)NPI

Fluorad FC-432. 3M Co.
Chemical Description: Fluorinated alkyl esters
Category: Nonionic surfactant
Applications: Active in low polarity organic solvents; forms hydrocarbon liquids; solvent-based coatings; oil-based insecticides, grease removers, penetrating oils
Form: Low viscosity liquid
Properties: Ref. Index: 1.40; Solubility: Sol. in perchloroethylene, toluene, benzene, heptane; Sp. gr.: 0.78; Flash Pt.: 19.4 F (TCC); Surface Tension: 20.5 dynes/cm (2 g/100 g ethyl acetate)
Composition: 25% active in heptane
Warnings and Caution: Flammable solvent; keep from heat, sparks or flame; avoid prolonged breathing and skin contact
References: Bulletin Y-ISTDG (65.2)R

Fluorad FC-740. 3M Co.
Chemical Description: Fluorinated alkyl ester
Category: Well stimulation additive, nonionic
Applications: Well stimulation additive active in organic solvents; produces stable foams in low po-

larity hydrocarbon liquids
Form: Liquid; Color: Amber
Properties: Sp. gr.: 1.01; Visc.: 100 cps; Flash Pt.: > 100 F (PMCC); Pour Pt.: 10 F; Surface Tension: 24.1 dynes/cm (0.2% FC-740 in kerosene @ 22 C)
Composition: 50% solids in S.C. (Central Solvents and Chemicals Co.) Solvent 150
Toxicity: Mild skin and eye irritant
Warnings and Caution: Keep away from heat, sparks or flame; avoid prolonged skin contact

Fluorad FC-750. 3M Co.
Chemical Description: Fluorinated alkyl quaternary ammonium iodides
Category: Well stimulation additive, cationic
Applications: Oil and gas well stimulation fluids; active in acid, neutral and alkaline media; offers lowest surface tension for aq. acids; enhances activity of organic corrosion inhibitors
Form: Liquid; Color: Dark
Properties: Sp. gr.: 1.2; Flash Pt.: 53 F (PMCC); Pour Pt.: 32–40 F; Surface Tension: 17.5 dynes/cm (100 ppm in deionized water)
Composition: 50% solids in 2/1 isopropanol/water
Toxicity: Minor skin and eye irritant
Warnings and Caution: Keep away from flames; protect eyes and skin from contact
References: Bulletin Y-ITD750 (99.05)II

Fluorad FC-760. 3M Co.
Chemical Description: Fluorinated alkyl alkoxylates
Category: Well stimulation additive, nonionic
Applications: Oil and gas well stimulation fluids; useful at low levels to reduce surface tension of fluids used for acidizing, fracturing and water flooding; active in acid, neutral or alkaline sol'ns.
Form: Liquid; Color: Amber
Properties: Sp. gr.: 1.4; Flash Pt.: > 220 F (PMCC); Pour Pt.: 20 F; Surface Tension: 19 dynes/cm (100 ppm in deionized water)
Toxicity: Minimal eye irritant
References: Bulletin Y-ITD760 (99.05)II

Foamaster 206-A. Diamond Shamrock Europe
Chemical Description: Nonsilicone type, based on chemically modified oils and fats
Category: Defoamer
Applications: Thermosol dyeing, exhaust dyeing, printing for textile industry
Form: Liquid
Properties: Solubility: Readily dispersible in water to form emulsions
Composition: 100% active ingred.

Foamaster 267-A. Diamond Shamrock Europe
Chemical Description: Nonsilicone type, based on chemically modified oils and fats
Category: Defoamer, nonionic
Applications: In textile industry

for thermosol and exhaust dyeing, printing; defoamer for most coatings based on synthetic latices
Form: Liquid
Properties: Solubility: Readily dispersible in water to form emulsions
Composition: 100% active ingred.

Foamaster 335-A. Diamond Shamrock Europe
Chemical Description: Nonsilicone type based on chemically modified oils and fats
Category: Defoamer
Applications: Textile industry for thermosol and exhaust dyeing, printing
Form: Liquid
Properties: Solubility: Readily dispersible in water to form emulsions
Composition: 100% active ingred.

Foamaster 1407-50. Diamond Shamrock Europe
Category: Defoamer, nonionic
Form: Fluidized paste
Properties: Solubility: Excellent water dispersibility

Foamaster 1719-A. Diamond Shamrock Europe
Category: Defoamer, nonionic
Applications: Soluble defoamer for use in gelatine
Form: Liquid
Composition: 50% active ingred.

Foamaster 8034. Diamond Shamrock Europe
Category: Defoamer, nonionic
Applications: Paints based on fine particle size acrylic emulsions; suggested for coating applications where standard defoamers are not effective; foam control in manufacture and in finished adhesives; rubber and latex latices
Form: Liquid
Composition: 100% active ingred.

Foamaster B. Diamond Shamrock Europe Ltd.
Category: Defoamer, nonionic
Applications: Predispersible defoamer for latex systems (paints, synthetic rubber and adhesives)
Form: Liquid
Composition: 100% active ingred.

Foamaster DD-72. Diamond Shamrock Europe Ltd.
Category: Defoamer, nonionic
Applications: Emulsion paints, wallpaper coatings, water-based printing inks)
Form: Liquid
Composition: 100% active ingred.

Foamaster DF-177-F. Diamond Shamrock Europe
Category: Foam suppressor, nonionic
Applications: Easily dispersible foam suppressor for coating and adhesive formulations
Form: Liquid
Composition: 100% active ingred.

Foamaster DF-178. Diamond Shamrock Europe
Category: Foam suppressor, nonionic
Applications: Coating and adhe-

sive formulations
Form: Liquid
Composition: 100% active ingred.

Foamaster DF-198-L. Diamond Shamrock Europe
Category: Defoamer, nonionic
Applications: Defoamer for paints based on fine particle size acrylic emulsions
Form: Liquid
Composition: 100% active ingred.

Foamaster DNH-1. Diamond Shamrock Europe
Category: Defoamer, nonionic
Applications: Defoamer for emulsion paints and polymers
Form: Liquid
Composition: 100% active ingred.

Foamaster G. Diamond Shamrock Europe
Category: Defoamer, nonionic
Applications: Dispersible latex stripping defoamer
Form: Liquid
Composition: 100% active ingred.

Foamaster JMY. Diamond Shamrock Europe
Category: Foam control agent, nonionic
Applications: Defoamers for adhesives, rubber and latex latices
Form: Liquid
Composition: 100% active ingred.

Foamaster KF-99. Diamond Shamrock Europe
Chemical Description: Nonsilicone types, based on chemically modified oils and fats
Category: Defoamer
Applications: Thermosol dyeing, exhaust dyeing, printing for textile industry
Form: Liquid
Properties: Solubility: Readily dispersible in water to form emulsions
Composition: 100% active ingred.

Foamaster NDW. Diamond Shamrock Europe
Category: Defoamer, nonionic
Applications: Defoamer for water-based paints, adhesive and cement screeds; coatings based on starch, protein or synthetic latex adhesives; rubber and latex latices
Form: Liquid
Composition: 100% active ingred.

Foamaster NS-20. Diamond Shamrock Europe
Category: Defoamer, nonionic
Applications: Silicone-free defoamer for coating or paper stock, air knife coatings
Form: Liquid
Composition: 100% active ingred.

Foamaster NXZ. Diamond Shamrock Europe
Category: Defoamer, nonionic
Applications: Defoamer for synthetic latices, paints adhesives; blade and roll coatings; emulsifiable latex stripping defoamer; for foam control in manufacture and in finished adhesive
Form: Liquid
Composition: 100% active ingred.

Foamaster P. Diamond Shamrock Europe
Category: Defoamer, nonionic
Applications: Foam control in polyvinyl alcohol systems (gives clear film) in adhesive industry
Form: Liquid
Composition: 35% active ingred.

Foamaster PD-1. Diamond Shamrock Europe
Category: Defoamer, nonionic
Applications: Defoamer for dry mixes, adhesives and cement screeds, joint cements, spackles, etc.
Form: Powder
Composition: 67% active ingred.

Foamaster TDB. Diamond Shamrock Europe
Category: Defoamer, nonionic
Applications: Foam control agent in dry adhesives, joint cements, spackles; rubber and latex latices
Form: Liquid
Composition: 100% active ingred.

Foamaster TDB-1. Diamond Shamrock Europe
Category: Defoamer, nonionic
Applications: Defoamer for water-soluble resins
Form: Liquid
Composition: 100% active ingred.

Foamaster V. Diamond Shamrock Europe
Category: Defoamer, nonionic
Applications: Dispersible latex stripping defoamer
Form: Liquid
Composition: 100% active ingred.

Foamaster VL. Diamond Shamrock Europe
Category: Defoamer, nonionic
Applications: Emulsifiable latex monomer stripping defoamer; latex and emulsion paints
Form: Liquid
Composition: 100% active ingred.

ENCYCLOPEDIA OF SURFACTANTS

VOLUME II

Compiled by

Michael and Irene Ash

Chemical Publishing Co.
New York, N.Y.

Chemical Publishing Co., Inc.

ISBN 0-8206-0287-6

Printed in the United States of America

PREFACE

This comprehensive, modern encyclopedia of surfactant products has been compiled with the cooperation of the major national and international surfactant manufacturers in order to give the user quick and thorough information on each tradename product that is included. Information from thousands of brochures, technical bulletins, and data sheets has been gleaned in order to make a complete profile on each surfactant product available from one source—this three volume compendium. Though sometimes limited by the information provided by the manufacturers, in most instances, the chemical ingredients, properties, form, applications, etc. of each product are discussed in detail.

The use of surfactants pervades almost every chemical industrial concern—including textiles, paints, pesticides, foods, cosmetics, pharmaceuticals, etc. We hope that because of the universality of surfactants, this reference work will be able to serve as a useful tool to the many chemical engineers and salespeople here and abroad.

We want to extend our thanks to Roberta Dakan for her dedication and tireless efforts in helping make this encyclopedia as accurate and consistent as possible.

When the temperature at which properties such as viscosity, density, surface tension, solubilities, etc., is not included, a standard temperature of 25 C is to be assumed. The information in this publication is reliable to the best of our knowledge. We would be glad to be informed of any errors or omissions so that these can be integrated into subsequent editions of this encyclopedia.

M. and I. Ash

CONTRIBUTORS

Akzo Chemie bv
Stationsstraat 48
P.O. Box 247
3800 AE-Amersfoort
The Netherlands

Akzo Chemie Italia SpA.
Via Vismara
20020 Arese
Milano, Italy

Albright & Wilson (Australia) Ltd.
Box 4544
610 St. Kilda Rd.
GPO Melbourne 3001
Australia

Albright & Wilson Ltd.
Detergent Sector
P.O. Box 15
White Haven, Cumberland
CA 28 9QQ, England

Alkaril Chemicals Ltd.
3265 Wolfedale Road
Mississauga, Ont., Canada

Amerchol Corp.
(Unit of CPC International Inc.)
P.O. Box 351
Talmadge Road
Edison, N.J. 08902

American Can Co.
Chemical Products Dept.
American Lane
Greenwich, Conn. 06830

American Hoechst Corp.
Industrial Chemicals Div.
Rt. 202–206 North
Somerville, N.J. 08876

Armak–Industrial Chemicals
P.O. Box 1805
Chicago, Ill. 60690

Arol Chem. Prod. Co.
649 Ferry St.
Newark, N.J. 07105

BASF Wyandotte Corp.
Textile Colors and Chem. Dept.
100 Cherry Hill Road
Parsippany, N.J. 07054

BASF Aktiengesellschaft
Carl-Bosch-Str. 36
6700 Ludwigshafen
Germany

Bofors Lakeway
5025 Evanston Ave.
Muskegon, Mich. 49443

Chemische Werke Huls AG
Postfach 1320
D-4370 Marl 1, West Germany

Ciba-Geigy Corp.
Dyestuff & Chemicals Div.
P.O. Box 11422
Greensboro, N.C. 27409

Coastal Industries,
Carlstadt, N.J.

Colloids, Inc.
394 Frelinghuysen Ave.
Newark, N.Y. 07114

Conoco Chemicals
Continental Oil Co.
5 Greenway Plaza East
P.O. Box 2197
Houston, TX 77001

Croda Chemicals Ltd.
Cowick Hall
Snaith Goole
North Humberside DN14 9AA
England

Crompton & Knowles Corp.
Dyes & Chemical Div.
Route 208
Fair Lawn, N.J. 07410

Crown Zellerbach Corp.
Chem. Products Div.
P.O. Box 4266
Vancouver (Orchards), Wash. 98662

Diamond Shamrock Corp.
Process Chemicals Div.
350 Mt. Kemble Ave.
Morristown, N.J. 07960

Du Pont de Nemours, E.I. & Co.
Specialty Chemicals & Products
Nemours Bldg.
Wilmington, DE 19898

Durkee Industrial Foods Group/
SCM Corp.
900 Union Commerce Bldg.
Cleveland, Ohio 44115

Dutton & Reinisch Ltd.
Crown House, London Road
Morden, Surrey
SM45DU, England

Dynamit Nobel AG
2 Haberstrasse,
D-5210 Troisdorf-Oberlar,
West Germany

Eastman Chemical Products Inc.
Health & Nutrition Div.
P.O. Box 431
Kingsport, TN 37662

Elektrochemische Fabrik Kempen
GmbH
P.O. Box 21
4152 Kempen, West Germany

Emery Industries, Inc.
Malmstrom Div.
1501 W. Elizabeth Ave.
Linden, N.J. 07036

Fanning Chem. Co., Inc.
357 West Erie St.
Chicago, Ill. 60610

GAF Corp.
Chemical Products
140 W. 51st St.
New York, N.Y. 10020

Georgia-Kaolin Co.
1185 Mary St.
Elizabeth, N.J. 07207

Georgia-Pacific Corp.
Bellingham Div.
P.O. Box 1236
Bellingham, Wash. 98225

Glyco Chemicals, Inc.
51 Weaver St.
Greenwich, Conn. 06830

W.R. Grace & Co.
Organic Chemicals Div.
55 Hayden Ave.
Lexington, MA 02173

Graden Chemical Co., Inc.
426 Bryan St.
Havertown, Pa. 19083

Grindsted Products
Grindstedvaerket A/S
38 Edwin Rahrs Vej
DK-8220 Brabrand
Denmark

A. Gross & Co.
Millmaster Onyx Group
P.O. Box 818
Newark, N.J. 07101

C.P. Hall Co.
7300 S. Central Ave.
Chicago, Ill. 60638

Hamblet & Hayes Co.
P.O. Box 730
Colonial Road
Salem, Mass. 01970

Hart Chemical Ltd.
Guelph, Canada N1H 6K8

Hart Products Corp.
173 Sussex St.
Jersey City, N.J. 07302

Henkel Inc.
Chemical Specialties Div.
1301 Jefferson St.
Hoboken, N.J. 07030

Hercules Inc.
Organics Dept.
Hercules Tower
910 Market St.
Wilmington, Del. 19894

Heterene Chemical Co., Inc.
792 21 Ave.
Paterson, N.J. 07513

High Point Chem. Corp.
609 Taylor St.
P.O. Box 2316
High Point, N.C. 27261

Humko Sheffield Chem.
Div. Kraft Inc.
P.O. Box 398
Memphis, TN 38101

ICI United States Inc.
New Murphy Rd. & Concord Pike
Wilmington, Del. 19897

Inolex Corp.
Personal Care Div.
4221 S. Western Blvd.
Chicago, Ill. 60609

Interstab Chemicals, Inc.
500 Jersey Ave.
New Brunswick, N.J. 08903

Isochem Corp.
99 Cook St.
Lincoln, RI 02865

Lancaster Chemical Co.
Div. of AZS Corp.
660 Frelinghuysen Ave.
Newark, N.J. 07144

Lankro Chemicals Ltd.
Eccles
Manchester M30 OBH
England

Lilachem
37-33 Rue de la Loi
1040 Brussels, Belgium

Lipo Chemicals, Inc.
207 19th Ave.
Paterson, N.J. 07504

Lonza Inc.
22-10 Route 208
Fair Lawn, N.J. 07410

Manchem Ltd.
Aston New Rd.
Manchester M11 4AT
England

Manro Products Ltd.
Bridge St.
Stalybridge Cheshire SK15 1PH
England

Marathon Morco Co.
4401 Park Ave.
Dickinson, Texas 77539

Marlowe-Van Loan Corp.
P.O. Box 1851
High Point, N.C. 27261

Mars Chemical Corp.
762 Marietta Blvd. N.W.
Atlanta, Ga. 30318

Mazer Chemicals Inc.
3938 Porett Dr.
Gurnee, Ill. 60031

Miranol Chem. Co., Inc.
277 Colt St.
Irvington, N.J. 07111

Mona Industries, Inc.
76 E. 24th St.
Paterson, N.J. 07524

Monsanto Co.
800 N. Lindbergh Blvd.
St. Louis, Mo. 63166

Nease Chemical Co.
P.O. Box 221
State College, Pa. 16801

Niacet Corp.
47th St. and Pine Ave.
Niagara Falls, N.Y. 14302

Nikko Chemicals Co., Ltd.
1-4-8 Nihonbashi-Bakurocho
Chuo-ku, Tokyo, Japan

Nippon Nyukazai Co., Ltd.
19-9, 3-chome, Ginza,
Chuo-ku, Tokyo 104, Japan

Nippon Oils & Fats Co., Ltd.
No. 5-1 chome, Yurakucho,
Chiyoda-Ku, Tokyo, Japan

Onyx Chemical Co.
Millmaster Onyx Group
190 Warren St.
Jersey City, N.J. 07302

Proctor & Gamble Co.
Sharonwood Technical Center
Cincinnati, Ohio 45217

PVO International Inc.
416 Division St.
Boonton, N.J. 07005

Reed Ltd.,
Chemical Div.–Lignin Products
P.O. Box 2025
Quebec, P.Q., Canada

R.I.T.A. Corp.
P.O. Box 556
Crystal Lake, Ill. 60014

Robinson-Wagner Co., Inc.
628 Waverly Ave.
Mamaroneck, N.Y. 10543

Rohm and Haas Co.
Independence Mall West
Philadelphia, Pa. 19105

Ronsheim & Moore Ltd.
Ings Lane, Castleford,
Yorkshire, WF10 2JT England

Sandoz Colors & Chemicals
Hanover, N.J. 07936

Shell Chemical Co.
One Shell Plaza
Houston, TX 77002

Stepan Chemical Co.
Edens & Winnetka Rds.
Northfield, Ill. 60093

Swastik Household & Industrial
Products Ltd.
Shaghibag House
13 Walchand Hirachand Marg
Bellard Estate
Bombay 400 038 India

The Theobald Industries
P.O. Box 72
Harrison, N.J. 07029

Tokaiseiyu Ind. Co., Ltd.
Yamadahigashimachi 2-Chome 67
Higashiku,
Nagoya, Japan

Toho Chemical Industry Co., Ltd.
1-14-9 Nihonbashi
Kakigara-cho, Chuo-ku
Tokyo, Japan

Van Schuppen Chemie
Nieuweweg, Veenendaal
Netherlands

Werner G. Smith
Cleveland, Ohio

Westbrook Lanolin Co.
Argonaut Works
Laisterdyke, Bradford
Yorks BD4 8AU, England

Westvaco-Polychemicals
P.O. Box 5207
N. Charleston, S.C. 29406

Witco Chemical Corp.
Organics Div. & Sonneborn Div.
277 Park Ave.
New York, N.Y. 10017

Witco Chemical Ltd.
P.O. Box 975
Haifa 31000
Israel

ABBREVIATIONS

@	at
aq.	aqueous
avg..	average
Be	Baumé
BHA	butylated hydroxyanisole
B.P.	boiling point
C	degrees Centigrade
cc.	cubic centimeter(s)
CC	Closed cup
cm	centimeter(s)
COC	Cleveland open cup
conc..	concentration, concentrated
cps	centipoise(s)
cs or cst	centistoke(s)
disp.	dispersion, dispersible
dist.	distilled
EO	ethylene oxide
esp..	especially
ETO	ethylene oxide
F	degrees Fahrenheit
F.P.	freezing point
ft^3	cubic feet
g	gram(s)
gal	gallon(s)
HLB	hydrophilic lipophilic balance
ingred..	ingredient(s)
insol..	insoluble
IPA	isopropyl alcohol
kg.	kilogram(s)
l.	liter(s)
lb	pound(s)
max.	maximum
min.	minimum
ml	milliliter(s)
mm.	millimeter(s)
mN	millinewton
M.P.	melting point
m.w.	molecular weight

no. . . . . number
OC . . . . Open cup
O/W . . . . oil in water
PEG . . . . polyethylene glycol
pH . . . . hydrogen ion concentration
PM . . . . Pensky-Martens
PMCC . . . . Pensky-Martens closed cup
POE . . . . polyoxyethylated
pt. . . . . point
ref. . . . . refractive
sapon. . . . . saponification
SBR . . . . styrene-butadiene resin
sol. . . . . soluble
solid. . . . . solidification
sol'n. . . . . solution
sp. gr. . . . . specific gravity
Stod. . . . . Stoddard
STPP . . . . sodium tripolyphosphate
SUS . . . . Saybolt Universal Seconds
T/C. . . . . tankcar
TCC . . . . Taggart closed cup
tech. . . . . technical
TGA . . . . triglycollamic acid
T/L. . . . . tank load
TOC . . . . Taggart open cup
TSP . . . . trisodium phosphate
T/T. . . . . tank truck
T/W . . . . tank wagon
USP . . . . United States Pharmacopeia
visc. . . . . viscosity
W/O . . . . water in oil
wt . . . . weight
% . . . . percent
≈ . . . . approximately
< . . . . less than
> . . . . greater than
⩽ . . . . less than or equal to
⩾ . . . . greater than or equal to
. . . . number

The following products which are included in this volume, are registered with the U.S. Patent Office.

Gafac®
Gafamide®
Gafstat®
Glucamate®
Glucate®
Hallcomid®
Hyamine®
Hyonic®
Imwitor®
Kessco®
Kieralon®
Klearfac®
Lantrol®
Laventin®
Leophen®
Lexaine®
Lexate®
Lexemul®
Lexol®
Lignosite®
Lomar®
Lusynton®
Lutensit®
Lutensol®
Lytron®
Marlamid®
Marlazin®
Marlican®
Marlinat®
Marlipal®
Marlon®
Marlophen®
Marlophor®
Marlopon®
Marlosol®
Marlowet®
Marlox®
Maypon®
Merpol®
Miglyol®
Miramine®
Miranol®
Monawet®
Monflor®
Monopol®
Morco®
Myrj®
Myvacet®
Myvaplex®
Myvatex®
Myverol®
Nekal®
Nekanil®
Neodol®
Neofat®
Niaproof®
Nopalcol®

G

Gafac BH-650. GAF Corp.
Chemical Description: Free acid of complex organic phosphate ester
Category: Emulsifier, detergent, anionic
Applications: Pesticide emulsifiers with high electrolyte tolerance; industrial and household detergent, hard-surface cleaners
Form: Clear viscous liquid
Properties: Solubility: Sol. in butyl Cellosolve, ethanol, water; Density: 11.4 lb/gal; Sp. gr.: 1.36–1.38 (@ 50 C); Pour Pt.: < 5 C; Acid No.: 370–390; Stability: Stable to alkaline and neutral media; pH: < 3 (10% sol'n.)
Composition: 100% active ingred.
Warnings and Caution: Protect skin and eyes from contact
References: Bulletin 2303-034

Gafac BI-729. GAF Corp.
Chemical Description: Free acid of complex organic phosphate ester
Category: Emulsifier, detergent, anionic
Applications: Pesticide emulsifier, household and industrial detergents for high-solids formulations
Form: Clear viscous liquid
Properties: Solubility: Sol. in butyl Cellosolve, ethanol, water; Density: 10.3 lb/gal; Sp. gr.: 1.23–1.25 (@ 50 C); Pour Pt.: < 0 C; Acid No.: 250–270 (as 100%); Stability: Stable to alkalis and neutral media; pH: < 2.5 (10% sol'n.)
Composition: 89% active ingred.
Reaction: Acid
Warnings and Caution: Protect skin and eyes from contact
References: Bulletin 2303-034

Gafac BI-750. GAF Corp.
Chemical Description: Free acid of complex organic phosphate ester
Category and Applications: See Gafac BI-729
Form: Clear viscous liquid
Properties: Solubility: Sol. in butyl Cellosolve, ethanol, water; Density: 10.6 lb/gal; Sp. gr.: 1.26–1.28 (@ 50 C); Pour Pt.: < 0 C; Acid No.: 250–275; Stability: Sta-

ble to alkaline and neutral media; pH: < 2.5 (10% sol'n.)
Composition: 100% active ingred.
Reaction: Acid
Warnings and Caution: Protect skin and eyes from contact
References: Bulletin 2303-034

Gafac GB-520. GAF Corp.
Chemical Description: Partial sodium salt of a complex organic phosphate ester
Category: Textile antistat, lubricant, emulsifier, anionic
Applications: Formulation of synthetic fiber finishes, esp. heavy denier types (e.g., nylon tire cord)
Form: Hazy viscous liquid
Properties: Solubility: Sol. in xylene, butyl Cellosolve, perchloroethylene, ethanol; sol. (hazy) in mineral oil, kerosene, Stod. solvent, cottonseed oil; dispersible in water; Density: 8.6 lb/gal; Sp. gr.: 1.03–1.04; Pour Pt.: 9 C; pH: 2.5–4.5 (10% sol'n.)
Composition: 98% active ingred.
Reaction: Acid
References: Bulletin 2303-034

Gafac LO-529. GAF Corp.
Chemical Description: Partial sodium salt of a complex organic phosphate ester
Category: Detergent, anionic
Applications: Household and industrial detergent formulations esp. in concentrated alkali and salt media
Form: Clear viscous liquid
Properties: Solubility: Sol. in kerosene, Stod. solvent, cottonseed oil, xylene, butyl Cellosolve, perchloroethylene, water; sol. (hazy) in ethanol; Density: 9.1 lb/gal; Sp. gr.: 1.05–1.15; Pour Pt.: 6 C; pH: 5–6 (10% sol'n.)
Composition: 88% active ingred.
Reaction: Acid
References: Bulletin 2303-034

Gafac MC-470. GAF Corp.
Chemical Description: Partial sodium salt of a complex phosphate ester
Category: Detergent, emulsifier, textile lubricant, antistat, softener, anionic
Applications: Household and industrial detergents, textile industry
Form: Clear viscous liquid
Properties: Solubility: Sol. in mineral oil, kerosene, xylene, perchloroethylene; dispersible in butyl Cellosolve, ethanol, water; Density: 8.6 lb/gal; Sp. gr.: 1.02–1.04; Pour Pt.: 0 C; pH: 5–6.5 (10% sol'n.)
Composition: 95% active ingred.
References: Bulletin 2303-034

Gafac PE-510. GAF Corp.
Chemical Description: Free acid of complex organic phosphate ester
Category: Emulsifier, detergent, anionic
Applications: Pesticide emulsifier, household, industrial and drycleaning detergent, emulsion polymerization
Form: Clear to slightly hazy viscous liquid

Properties: Solubility: Sol. in kerosene, Stod. solvent, xylene, butyl Cellosolve, perchloroethylene, ethanol; disp. in water and cottonseed oil; Density: 9 lb/gal; Sp. gr.: 1.08–1.09; Pour Pt.: 20 C; Acid No.: 49–59; Stability: Stable to alkaline and neutral media; pH: < 2.5 (10% sol'n.)
Composition: 100% active ingred.
Reaction: Acid
Warnings and Caution: Protect skin and eyes from contact
References: Bulletin 2303-034

Gafac RA-600. GAF Corp.
Chemical Description: Free acid of complex organic phosphate ester
Category: Detergent, dispersant, wetting agent, anionic
Applications: Household and industrial detergents esp. for hard surface cleaning; in textile wet processing (boil off, bleaching, etc.) as detergents, dispersants and wetting agents
Form: Clear to slightly hazy viscous liquid
Properties: Solubility: Sol. in kerosene, xylene, butyl Cellosolve, perchloroethylene, ethanol, water; disp. in mineral oil; Density: 8.9 lb/gal; Sp. gr.: 1.06–1.08; Pour Pt.: < 0 C; Acid No.: 100–115; Stability: Stable to alkaline and neutral media; pH: < 2.5 (10% sol'n.)
Composition: 98.5% active ingred.
Reaction: Acid
Warnings and Caution: Protect skin and eyes from contact
References: Bulletin 2303-034

Gafac RB-400. GAF Corp.
Chemical Description: Free acid of a complex organic phosphate ester
Category: Lubricant, antistat, emulsifier, anionic
Applications: Fiber and metal lubricants, antistats and emulsifiers
Form: Opaque, viscous liquid (separates on standing)
Properties: Solubility: Sol. in mineral oil, kerosene, Stod. solvent, xylene, butyl Cellosolve, perchloroethylene, ethanol; disp. in water; Density: 8.6 lb/gal; Sp. gr.: 1.03–1.04; Pour Pt.: 18 C; Acid No.: 80–90; Stability: Stable to alkaline and neutral media; pH: < 2.5 (10% sol'n.)
Composition: 98% active ingred.
Reaction: Acid
Warnings and Caution: Protect skin and eyes from contact
References: Bulletin 2303-034

Gafac RD-510. GAF Corp.
Chemical Description: Free acid of complex organic phosphate ester
Category: Emulsifier, antistat, lubricant, solubilizer, anionic
Applications: Fiber and metal lubricant; esp. in the formulation of low denier synthetic fiber finishes (i.e., polyesters, polypropylene); cosmetic emulsifier and solubilizer
Form: Clear viscous liquid
Properties: Solubility: Sol. in mineral oil, kerosene, xylene, butyl Cellosolve, perchloroethylene, ethanol; disp. in water; Density:

8.8 lb/gal; Sp. gr.: 1.05–1.06; Pour Pt.: 13 C; Acid No.: 100–115; Stability: Stable to alkaline and neutral media; pH: < 2.5 (10% sol'n.)
Composition: 98% active ingred.
Reaction: Acid
Warnings and Caution: Protect skin and eyes from contact
References: Bulletin 2303-034

Gafac RE-410. GAF Corp.
Chemical Description: Free acid of a complex organic phosphate ester
Category: Detergent, emulsifier, wetting agent, dispersant, antistat, lubricant, anionic
Applications: Pesticide emulsifier with high electrolyte tolerance; drycleaning detergent (excellent antisoil redeposition in Stod. solvent and perchloroethylene systems); textile wet processing (boil off, bleaching, etc.) for cottons and synthetics; fiber and metal lubricants, antistats and emulsifiers
Form: Clear viscous liquid
Properties: Solubility: Sol. in kerosene, Stod. solvent, cottonseed oil, xylene, butyl Cellosolve, perchloroethylene, ethanol; disp. in water; Density: 9.1 lb/gal; Sp. gr.: 1.00–1.20; Pour Pt.: 18 C; Acid No.: 85–100; Stability: Stable to alkaline and neutral media; pH: < 2.5 (10% sol'n.)
Composition: 100% active ingred.
Reaction: Acid
Warnings and Caution: Protect skin and eyes from contact
References: Bulletin 2303-034

Gafac RE-610. GAF Corp.
Chemical Description: Free acid of a complex organic phosphate ester
Category and Applications: See Gafac RE-410; also in household and industrial detergents; water-repellant fabric finishes; emulsion polymerization
Form: Slightly hazy viscous liquid
Properties: Solubility: Sol. in xylene, butyl Cellosolve, perchloroethylene, ethanol, water; disp. in mineral oil; Density: 9.2 lb/gal; Sp. gr.: 1.10–1.12; Pour Pt.: < 0 C; Acid No.: 62–72; Stability: Stable to alkaline and neutral media; pH: < 2.5 (10% sol'n.)
Composition: 100% active ingred.
Reaction: Acid
Warnings and Caution: Protect skin and eyes from contact
References: Bulletin 2303-034

Gafac RE-870. GAF Corp.
Chemical Description: Free acid of a complex organic phosphate ester
Category and Applications: See Gafac RE-410; also as cosmetic emulsifier and solubilizer and in emulsion polymerization process
Form: Soft, waxy paste
Properties: Solubility: Sol. in xylene, butyl Cellosolve, perchloroethylene (hazy), ethanol, water; Density: 9.6 lb/gal; Sp. gr.: 1.10–1.20 (@ 50 C); Pour Pt.: 29 C; Acid No.: 80–95; pH: < 2.5 (10% sol'n.)
Composition: 99% active ingred.
Reaction: Acid
Warnings and Caution: Protect

skin and eyes from contact
References: Bulletin 2303-034

Gafac RE-960. GAF Corp.
Chemical Description: Free acid of a complex organic phosphate ester
Category and Applications: See Gafac RE-410; also in emulsion polymerization
Form: Soft, waxy paste
Properties: Solubility: Sol. in cottonseed oil, butyl Cellosolve, ethanol, water; sol. (hazy) in xylene, perchloroethylene; Density: 9.8 lb/gal; Sp. gr.: 1.17–1.18 (@ 50 C); Pour Pt.: 20 C; Acid No.: 98–110; Stability: Stable to alkaline and neutral media; pH: < 2.5 (10% sol'n.)
Composition: 90% active ingred.
Reaction: Acid
Warnings and Caution: Protect skin and eyes from contact
References: Bulletin 2303-034

Gafac RK-500. GAF Corp.
Chemical Description: Free acid of a complex organic phosphate ester
Category: Lubricant, antistat, emulsifier, anionic
Applications: Fiber and metal treatment
Form: Clear, viscous liquid
Properties: Solubility: Sol. in xylene, butyl Cellosolve, perchloroethylene, ethanol, water; disp. in mineral oil and kerosene; Density: 9.1 lb/gal; Sp. gr.: 1.05–1.15; Pour Pt.: < 0 C; Acid No.: 132–145; Stability: Stable to alkaline and neutral media; pH: < 2.5 (10% sol'n.)
Composition: 100% active ingred.
Reaction: Acid
Warnings and Caution: Protect skin and eyes from contact
References: Bulletin 2303-034

Gafac RM-410. GAF Corp.
Chemical Description: Free acid of a complex organic phosphate ester
Category: Detergent, emulsifier, anionic
Applicaitons: Drycleaning detergent; pesticide emulsifier
Form: Slightly hazy, viscous liquid
Properties: Solubility: Sol. in mineral oil, kerosene, Stod. solvent, cottonseed oil, xylene, butyl Cellosolve, perchloroethylene, ethanol; disp. in water; Density: 8.8 lb/gal; Sp. gr.: 1.05–1.07; Pour Pt.: 19 C; Acid No.: 51–64; Stability: Stable to alkaline and neutral media; pH: < 2.5 (10% sol'n.)
Composition: 100% active ingred.
Reaction: Acid
Warnings and Caution: Protect skin and eyes from contact
References: Bulletin 2303-034

Gafac RM-510. GAF Corp.
Chemical Description: Free acid of a complex organic phosphate ester
Category: Emulsifier, anionic
Applications: Pesticide emulsifier; cosmetic emulsifier and solubilizer
Form: Hazy viscous liquid
Properties: Solubility: Sol. in kerosene, Stod. solvent, cottonseed oil, xylene, butyl Cellosolve, perchloroethylene, ethanol; disp. in water and mineral oil; Density: 8.8 lb/gal; Sp. gr.: 1.05–1.07;

Pour Pt.: 5 C; Acid No.: 45–55; Stability: Stable to alkaline and neutral media; pH: < 2.5 (10% sol'n.)
Composition: 100% active ingred.
Reaction: Acid
Warnings and Caution: Protect skin and eyes from contact
References: Bulletin 2303-034

Gafac RM-710. GAF Corp.
Chemical Description: Free acid of a complex organic phosphate ester
Category: Emulsifier, solubilizer, anionic
Applications: Cosmetic emulsifier and solubilizer; low skin and eye irritation
Form: Hazy, viscous liquid
Properties: Solubility: Sol. in xylene (hazy), butyl Cellosolve, perchloroethylene, ethanol, water; Density: 8.9 lb/gal; Sp. gr.: 1.06–1.08; Pour Pt.: 15 C; Acid No.: 33–41; Stability: Stable to alkaline and neutral media; pH: < 3.0 (10% sol'n.)
Composition: 100% active ingred.
Reaction: Acid
Warnings and Caution: Protect skin and eyes from contact
References: Bulletin 2303-034

Gafac RO-310. GAF Corp.
Chemical Description: Free acid of a complex organic phosphate ester
Category: Emulsifier, anionic
Applications: Cosmetic emulsifier for lotions, creams and clear oil-water hair grooming products
Form: Clear, viscous liquid
Properties: Solubility: Sol. in mineral oil, kerosene, cottonseed oil, xylene, butyl Cellosolve, perchloroethylene, ethanol; disp. in water; Density: 8.4 lb/gal; Sp. gr.: 1.00–1.02; Pour Pt.: < 0 C; Acid No.: 73–80; pH: < 2.5 (10% sol'n.)
Composition: 98% active ingred.
Reaction: Acid
Warnings and Caution: Protect skin and eyes from contact
References: Bulletin 2303-034

Gafac RO-660. GAF Corp.
Chemical Description: Free acid of a complex organic phosphate ester
Category and Applications: See Gafac RO-310; more hydrophilic version of RO-310
Form: Soft, waxy paste
Properties: Solubility: Sol. in mineral oil, kerosene, Stod. solvent, xylene, butyl Cellosolve, perchloroethylene, ethanol, water; Density: 8.6 lb/gal; Sp. gr.: 1.03–1.04; Pour Pt.: 25 C; Acid No.: 52–58; Stability: Stable to alkaline and neutral media; pH: < 2.5 (10% sol'n.)
Composition: 98% active ingred.
Reaction: Acid
Warnings and Caution: Protect skin and eyes from contact
References: Bulletin 2303-034

Gafac RP-710. GAF Corp.
Chemical Description: Free acid of a complex organic phosphate ester
Category: Detergent, anionic
Applications: Household and industrial detergents, hard surface cleaners

Form: Slightly hazy, viscous liquid
Properties: Solubility: Sol. in xylene, butyl Cellosolve, perchloroethylene, ethanol, water; Density: 10.2 lb/gal; Sp. gr.: 1.22–1.23; Pour Pt.: –7 C; Acid No.: 85–110; pH: < 2.5 (10% sol'n.)
Composition: 100% active ingred.
Reaction: Acid
Warnings and Caution: Protect skin and eyes from contact
References: Bulletin 2303-034

Gafac RS-410. GAF Corp.
Chemical Description: Free acid of a complex organic phosphate ester
Category: Emulsifier, detergent, dispersant, wetting agent, antistat, lubricant, anionic
Applications: Pesticide emulsifier; drycleaning detergent; textile wet processing; fiber and metal treatment
Form: Hazy, viscous liquid
Properties: Solubility: Sol. in mineral oil, kerosene, Stod. solvent, cottonseed oil, xylene (hazy), butyl Cellosolve, perchloroethylene, ethanol; disp. in water; Density: 8.6 lb/gal; Sp. gr.: 1.03–1.04; Pour Pt.: –15 C; Acid No.: 95–115; pH: < 2.5 (10% sol'n.)
Composition: 100% active ingred.
Reaction: Acid
Warnings and Caution: Protect skin and eyes from contact
References: Bulletin 2303-034

Gafac RS-610. GAF Corp.
Chemical Description: Free acid of a complex organic phosphate ester
Category and Applications: See Gafac RS-410
Form: Hazy, viscous liquid
Properties: Solubility: Sol. in kerosene, Stod. solvent, cottonseed oil, xylene, butyl Cellosolve, perchloroethylene, ethanol; disp. in water; Density: 8.7 lb/gal; Sp. gr.: 1.04–1.06; Pour Pt.: < 0 C; Acid No.: 75–85; Stability: Stable to alkaline and neutral media; pH: < 2.5 (10% sol'n.)
Composition: 100% active ingred.
Reaction: Acid
Warnings and Caution: Protect skin and eyes from contact
References: Bulletin 2303-034

Gafac RS-710. GAF Corp.
Chemical Description: Free acid of a complex organic phosphate ester
Category and Applications: See Gafac RS-410
Form: Opaque, viscous liquid
Properties: Solubility: Sol. in Stod. solvent, cottonseed oil, xylene, butyl Cellosolve, perchloroethylene, ethanol and water; Density: 8.7 lb/gal; Sp. gr.: 1.04–1.06; Pour Pt.: 18 C; Acid No.: 58–70; Stability: Stable to alkaline and neutral media; pH: < 2.5 (10% sol'n.)
Composition: 100% active ingred.
Reaction: Acid
Warnings and Caution: Protect skin and eyes from contact
References: Bulletin 2303-034

Gafamide CDD-518 Alkanolamide. GAF Corp.
Chemical Description: Coconut oil-diethanolamine condensate

Category: Foam stabilizer, viscosity builder, detergent, wetting agent, anionic
Applications: Shampoos, bubble baths, dishwashing liquids, heavy-duty liquid laundry detergents, waterless hand cleaners, floor cleaners, other industrial cleaning applications
Form: Clear viscous liquid
Properties: Solubility: Sol. in water, ethanol, perchloroethylene, ethylene glycol, butyl Cellosolve, butyl ether, propylene glycol; Sp. gr.: ≈ 1.0; Visc.: 1050 cst (Ostwald-Fenske); pH: 10 (1% sol'n.)
Composition: 84% active ingred.
Reaction: Alkaline
References: Bulletin 9643-017

Gafstat S. GAF Corp.
Chemical Description: Modified nitrogen compound
Category: Textile finishing agent, cationic
Applications: Textile finishing agent, antistat for natural and synthetic fibers and fabrics; good heat stability
Form: Clear free-flowing liquid; Color: Amber
Properties: Solubility: Readily sol. in water to form clear sol'n.; pH: 6–7 (10% in dist. water)
Composition: 24% min. solids
Toxicity: Skin and eye irritant
Containers: 55 gal open head steel drums with polyethylene liners; 450 lb net wt
Warnings and Caution: Avoid contact with eyes, skin or clothing

Gantrez AN. GAF Corp.
Chemical Description: Poly (methyl vinyl ether/maleic anhydride)
Category: Dispersing, coupling, stabilizing agents, thickener, emulsifier, solubilizer, corrosion inhibitor, film former, antistat, anionic
Applications: Adhesives and coatings (paints, glues, finishes, inks, etc.); agriculture industry in pesticides, herbicides, aerosols, soil conditioners; chemical processing in polymerization, latices, water treatment; in acid bowl cleaners, built detergents, hand cleaners; for cosmetics and toiletries (hair sprays, shampoos, detergent bars); in fire extinguishing foams; tanning of leather; in paper industry; pharmaceutical products (bacteriostatic agents, spray bandages, tablet coating, etc.); in photoreproduction processes; pigment grinding for paints; in textiles for spandex, cotton, rayon, nylon, polyester, and glass-fiber finishes, printing pastes, etc.; waxes and polishes; ceramics, asbestos and steel processing
Form: Fluffy powder; Color: White
Properties: Solubility: Sol. in water over entire pH range and in several organic solvents (alcohols, phenols, pyridine, aldehydes, ketones, lactams and lower aliphatic esters); Density: 0.32 g/cc; Sp. gr.: 1.37; Softening Pt. Range: 200–225 C; Stability: Stable to acid and alkalis; pH (free acid–5% aq. sol'n.) ≈ 2
Composition: 100% conc.

References: Technical bulletin 7543-017

Gardilene IPA. Albright & Wilson (Australia) Ltd.
Chemical Description: Isopropylamine alkyl benzene sulfonate based on an SO_3 sulfonated linear alkylate, in a solvent system
Category: Emulsifier, dispersant, anionic
Applications: Emulsifier, drycleaning assistant, and dispersing agent
Form: Clear, viscous liquid; Color: Amber
Properties: pH: 6.5 (aq. disp.)
Composition: 85% active ingred.
Containers: 200 kg net lined closed head mild steel drum
Warnings and Caution: Avoid prolonged contact with skin; store away from naked flame–vapors are inflammable
References: Data sheet SD67-678

Gardilene S25L. Albright & Wilson (Australia) Ltd.
Chemical Description: Aq. sol'n. of sodium alkyl benzene sulfonate based on an SO_3 sulfonated linear "soft" alkylate
Category: Wetting, dispersing, and foaming agent, emulsifier, anionic
Applications: Formulating specialty cleaners; as emulsifier in emulsion polymerization processes
Form: Clear, viscous liquid; Color: Pale yellow
Properties: Visc.: 3300 cps (20 C); Clear Pt.: < 6 C; pH: 7.2 (10% sol'n.)
Composition: 25% active ingred.
Containers: 200 kg net lined open head mild steel drum with double bungs in lid
Warnings and Caution: Avoid prolonged contact with skin; spillages are slippery
References: Data sheet SD69-378

Gardinol WA Paste. Ronsheim & Moore Ltd.
Chemical Description: Sodium salt of sulfated broad cut fatty alcohol
Category: Detergent, emulsifier, anionic
Applications: Manufacture of cream and lotion shampoos; mild detergent for woolens and fine fabrics; emulsifier for cosmetic emulsions
Form: Stiff paste; Color: White to cream
Composition: 40% conc.

Gardiquat 1450. Albright & Wilson (Australia) Ltd.
Chemical Description: Alkyl dimethyl benzyl ammonium chloride (complies with USP specifications)
Category: Germicide, deodorant, algaecide, slimicide
Form: Clear liquid; Color: Almost water white (@ 20 C)
Properties: Cloud Pt.: 1 C max. (quick cool); pH: 7.0 ± 0.5 (1% sol'n.)
Composition: 50% active ingred.
Containers: 180 kilo net lined closed head mild steel drum
Warnings and Caution: Avoid contact with skin; store away from naked flame–flammable
References: Data sheet SD71-678

Gardiquat 1480. Albright & Wilson (Australia) Ltd.
Chemical Description: Alkyl dimethyl benzyl ammonium chloride (complies with USP specifications)
Category: Germicide, deodorant, algaecide, slimicide
Form: Clear liquid; Color: Almost water white
Properties: Sp. gr.: 0.945; Flash Pt.: 32 C (PMCC); Cloud Pt.: 1 C max. (quick cool); pH: 7.0 ± 0.5 (1% sol'n.)
Composition: 80% active ingred.
Containers: 180 kilo net lined closed head mild steel drum
Warnings and Caution: Avoid contact with skin; store away from naked flame—flammable
References: Data sheet SD71A-379

Gardiquat SV480. Albright & Wilson (Australia) Ltd.
Chemical Description: Alkyl dimethyl benzyl ammonium chloride
Category: Germicide, deodorant, algaecide, slimicide
Form: Clear liquid; Color: Water white (at normal ambient temperatures)
Properties: Sp. gr.: 0.929; Visc.: 106 s (No. 4 cup); Cloud Pt.: < 0 C (quick cool); Flash Pt.: 29 C (PMCC); pH: 7.2 (1% sol'n.)
Composition: 80% active ingred.
Containers: 180 kilo net lined closed head mild steel drum
Warnings and Caution: Avoid contact with skin; store away from naked flame—flammable liquid
References: Data sheet SD 71B-180

GC-44-14. PVO International
Chemical Description: Dipentaerythritol ester
Category: Lubricant
Applications: Functional fluids for aviation, automotive, and military formulated lubricants and gas turbine engine oils
Form: Liquid
Properties: Sp. gr.: 1.01; Visc.: 56 cs @ 100 F; Flash Pt.: 565 F; Fire Pt.: 625 F; Pour Pt.: < –80 F

Geitol RC-100. Tokaiseiyu Ind. Co., Ltd.
Chemical Description: Mixture of nonionic agent and specific builders
Category: Detergent, nonionic
Applications: Reduction cleaning agent for polyester fibers
Form: Powder
Composition: 100% conc.

Genamine C-020. American Hoechst Corp.
Chemical Description: Coconut fatty amine oxethylate
Category: Raw material, cationic
Applications: Used as starting materials for the chemo-technical industry, e.g., in manufacture of dyeing and textile auxiliaries, mineral oil additives, plant protection agents and insecticides and pesticides, cosmetic bases, adhesives
Form: Clear liquid
Properties: Ref. Index: 1.4559 @

50 C; Solubility: Forms clear sol'n. (@ 1:10) in mineral oil, benzol, tetra, glycol; turbid sol'n. in water (10 g/l); Sp. gr.: 0.895 (50 C); Visc.: 28.8 cps (50 C); Flash Pt.: 188 C (Marcusson); Solidifying Pt.: 2 C; Stability: Generally stable; unaffected by hard water; Surface Tension: 26 dynes/cm (1g/l @ 20 C); pH: 9–10 (10 g/l)
Composition: 100% active ingred.
Toxicity: Corrosive effect on skin and mucous membranes

Genamine C-050. American Hoeschst Corp.
Chemical Description: Coconut fatty amine oxethylate
Category and Applications: See Genamine C-020
Form: Clear liquid
Properties: Ref. Index.: 1.4575 @ 50 C; Solubility: Forms clear sol'n. in water (@ 10 g/l) and in mineral oil, benzol, tetra, and glycol (@ 1:10); Sp. gr.: 0.950 (50 C); Visc.: 30.4 cps (50 C); Flash Pt.: 224 C (Marcusson); Solidifying Pt.: −17 C; Stability: Generally stable; unaffected by hard water; Surface Tension: 32 dynes/cm (1 g/l @ 20 C); pH: 9–10 (10 g/l)
Composition: 100% active ingred.
Toxicity: Corrosive to skin and mucous membranes

Genamine C-080. American Hoecsht Corp.
Chemical Description: Coconut fatty amine oxethylate
Category and Applications: See Genamine C-020
Form: Clear liquid
Properties: Ref. Index: 1.4582 @ 50 C; Solubility: Forms clear sol'n. in water (10 g/l), and in benzol, tetra, and glycol (1:10); turbid in mineral oil (1:10); Sp. gr.: 0.988 (50 C); Visc.: 35.2 cps (50 C); Flash Pt.: 249 C (Marcusson); Solidifying Pt.: −18 C; Stability: Generally stable, unaffected by hard water; Surface Tension: 37 dynes/cm (1 g/l @ 20 C); pH: 9–10 (10 g/l)
Composition: 100% active ingred.
Toxicity: Corrosive to skin and mucous membranes

Genamine C-100. American Hoechst Corp.
Chemical Description: Coconut fatty amine oxethylate
Category and Applications: See Genamine C-020
Form: Clear liquid
Properties: Ref. Index.: 1.4590 (50 C); Solubility: Forms clear sol'n. in water (10 g/l), and in benzol, tetra, glycol (1:10); turbid in mineral oil (1:10); Sp. gr.: 1.000 (50 C); Visc.: 37.0 cps (50 C); Flash Pt.: 273 C (Marcusson); Solidifying Pt.: −18 C; Stability: Generally stable, unaffected by hard water; Surface Tension: 40 dynes/cm (1 g/l @ 20 C); pH: 9–10 (10 g/l)
Composition: 100% active ingred.
Toxicity: Corrosive to skin and mucous membranes

Genamine C-150. American Hoechst Corp.
Chemical Description: Coconut fatty amine oxethylate

Category and Applications: See Genamine C-020
Form: Clear liquid
Properties: Ref. Index: 1.4597 @ 50 C; Solubility: Forms clear sol'n. in water (10 g/l) and in benzol, tetra, glycol (1:10); Sp. gr.: 1.027 (50 C); Visc.: 45.5 cps (50 C); Flash Pt.: 287 C (Marcusson); Solidifying Pt.: –17 C; Stability: Generally stable, unaffected by hard water; Surface Tension: 44 dynes/cm (1 g/l @ 20 C); pH: 9–10 (10 g/l)
Composition: 100% active ingred.
Toxicity: Corrosive to skin and mucous membranes

Genamine C-200. American Hoechst Corp.
Chemical Description: Coconut fatty amine oxethylate
Category and Applications: See Genamine C-020
Form: Clear liquid
Properties: Ref. Index.: 1.4600 (50 C); Solubility: Forms clear sol'n. in water (10 g/l), and in benzol, tetra, and glycol (1:10); Sp. gr.: 1.043 (50 C); Visc.: 55.0 cps (50 C); Flash Pt.: 295 C (Marcusson); Solidifying Pt.: –5 C; Stability: Generally stable, unaffected by hard water; Surface Tension: 45 dynes/cm (1 g/l @ 20 C); pH: 9–10 (10 g/l)
Composition: 100% active ingred.
Toxicity: Corrosive to skin and mucous membranes

Genamine C-250. American Hoechst Corp.
Chemical Description: Coconut fatty amine oxethylate
Category and Applications: See Genamine C-020
Form: Turbid liquid
Properties: Ref. Index: 1.4600 (50 C); Solubility: Forms clear sol'n. in water (10 g/l), and in benzol, tetra, and glycol (1:10); Sp. gr.: 1.059 (50 C); Visc.: 65.2 cps (50 C); Flash Pt.: 298 C (Marcusson); Solidifying Pt.: 5 C; Stability: Generally stable, unaffected by hard water; Surface Tension: 47 dynes/cm (1 g/l @ 20 C); pH: 9–10 (10 g/l)
Composition: 100% active ingred.
Toxicity: Corrosive to skin and mucous membranes

Genamine O-020. American Hoechst Corp.
Chemical Description: Oleylamine oxethylate
Category and Applications: See Genamine C-020
Form: Turbid liquid
Properties: Ref. Index: 1.4632 (50 C); Solubility: Forms clear sol'n. in mineral oil, benzol, and tetra (1:10); turbid sol'n. in glycol (1:10); Sp. gr.: 0.883 (50 C); Visc.: 37.4 cps (50 C); Flash Pt.: 220 C (Marcusson); Solidifying Pt.: –2 C; Stability: Generally stable, unaffected by hard water; pH: 9–10 (10 g/l)
Composition: 100% active ingred.
Toxicity: Corrosive to skin and mucous membranes

Genamine O-050. American Hoechst Corp.
Chemical Description: Oleylamine

oxethylate
Category and Applications: See Genamine C-020
Form: Clear liquid
Properties: Ref. Index: 1.4630 (50 C); Solubility: Forms clear sol'n. in mineral oil, benzol, tetra, and glycol (1:10); turbid sol'n. in water (10 g/l); Sp. gr.: 0.956 (50 C); Visc.: 45.3 cps (50 C); Flash Pt.: 269 C (Marcusson); Solidifying Pt.: –10 C; Stability: Generally stable, unaffected by hard water; Surface Tension: 42 dynes/cm (1 g/l @ 20 C); pH: 9–10 (10 g/l)
Composition: 100% active ingred.
Toxicity: Corrosive to skin and mucous membranes

Genamine O-080. American Hoechst Corp.
Chemical Description: Oleylamine oxethylate
Category and Applications: See Genamine C-020
Form: Clear liquid
Properties: Ref. Index: 1.4630 (50 C); Solubility: Forms clear sol'n. in water (10 g/l), and in mineral oil, benzol, tetra, and glycol (1:10); Sp. gr.: 0.983 (50 C); Visc.: 49.6 cps (50 C); Flash Pt.: 282 C (Marcusson); Solidifying Pt.: –19 C; Stability: Generally stable, unaffected by hard water; Surface Tension: 39 dynes/cm (1 g/l @ 20 C); pH: 9–10 (10 g/l)
Composition: 100% active ingred.
Toxicity: Corrosive to skin and mucous membranes

Genamine O-100. American Hoechst Corp.
Chemical Description: Oleylamine oxethylate
Category and Applications: See Genamine C-020
Form: Turbid liquid
Properties: Ref. Index.: 1.4628 (50 C); Solubility: Forms clear sol'n. in water (10 g/l), and in benzol, tetra, and glycol (1:10); Sp. gr.: 0.995 (50 C); Visc.: 52.1 cps (50 C); Flash Pt.: 288 C (Marcusson); Solidifying Pt.: –19 C; Stability: Generally stable, unaffected by hard water; Surface Tension: 41 dynes/cm (1 g/l @ 20 C); pH: 9–10 (10 g/l)
Composition: 100% active ingred.
Toxicity: Corrosive to skin and mucous membranes

Genamine O-150. American Hoechst Corp.
Chemical Description: Oleylamine oxethylate
Category and Applications: See Genamine C-020
Form: Turbid liquid
Properties: Ref. Index: 1.4623 (50 C); Solubility: Forms clear sol'n. in water (10 g/l), and in benzol, tetra, and glycol (1:10); Sp. gr.: 1.024 (50 C); Visc.: 65.1 cps (50 C); Flash Pt.: 307 C (Marcusson); Solidifying Pt.: –6 C; Stability: Generally stable, unaffected by hard water; Surface Tension: 45 dynes/cm (1 g/l @ 20 C); pH: 9–10 (10 g/l)
Composition: 100% active ingred.
Toxicity: Corrosive to skin and mucous membranes

Genamine O-200. American Hoechst Corp.
Chemical Description: Oleylamine oxethylate
Category and Applications: See Genamine C-020
Form: Paste
Properties: Ref. Index: 1.4623 (50 C); Solubility: Forms clear sol'n. in water (10 g/l), and in benzol, tetra, and glycol (1:10); Sp. gr.: 1.063 (50 C); Visc.: 75.9 cps (50 C); Flash Pt.: 314 C (Marcusson); Stability: Generally stable, unaffected by hard water; Surface Tension: 48 dynes/cm (1 g/l @ 20 C); pH: 9–10 (10 g/l)
Composition: 100% active ingred.
Toxicity: Corrosive to skin and mucous membranes

Genamine O-250. American Hoechst Corp.
Chemical Description: Oleylamine oxethylate
Category and Applications: See Genamine C-020
Form: Wax
Properties: Ref. Index: 1.4620 (50 C); Solubility: Forms clear sol'n. in water (10 g/l), and in benzol, tetra, and glycol (1:10); Sp. gr.: 1.080 (50 C); Visc.: 89.2 cps (50 C); Flash Pt.: 317 C (Marcusson); Stability: Generally stable, unaffected by hard water; Surface Tension: 51 dynes/cm (1 g/l @ 20 C); pH: 9–10 (10 g/l)
Composition: 100% active ingred.
Toxicity: Corrosive to skin and mucous membranes

Genamine S-020. American Hoecsht Corp.
Chemical Description: Stearylamine oxethylate
Category and Applications: See Genamine C-020
Form: Wax
Properties: Ref. Index: 1.4582 (50 C); Solubility: Forms clear sol'n. in mineral oil, benzol, and tetra (1:10); turbid sol'n. in water (10 g/l) and glycol (1:10); Sp. gr.: 0.889 (50 C); Visc.: 40.6 cps (50 C); Flash Pt.: 219 C (Marcusson); Stability: Generally stable, unaffected by hard water; pH: 9–10 (10 g/l)
Composition: 100% active ingred.
Toxicity: Corrosive to skin and mucous membranes

Genamine S-050. American Hoechst Corp.
Chemical Description: Stearylamine oxethylate
Category and Applications: See Genamine C-020
Form: Clear liquid
Properties: Ref. Index: 1.4591 (50 C); Solubility: Forms clear sol'n. in mineral oil, benzol, tetra, and glycol (1:10); turbid sol'n. in water (10 g/l); Sp. gr.: 0.943 (50 C); Visc.: 39.0 cps (50 C); Flash Pt.: 258 C (Marcusson); Solidifying Pt.: 10 C; Stability: Generally stable, unaffected by hard water; Surface Tension: 35 dynes/cm (1 g/l @ 20 C); pH: 9–10 (10 g/l)
Composition: 100% active ingred.
Toxicity: Corrosive to skin and mucous membranes

Genamine S-080. American Hoecsht Corp.
Chemical Description: Stearylamine oxethylàte
Category and Applications: See Genamine C-020
Form: Turbid liquid
Properties: Ref. Index: 1.4597 (50 C); Solubility: Forms clear sol'n. in water (10 g/l), and in mineral oil, benzol, tetra, and glycol (1:10); Sp. gr.: 0.974 (50 C); Visc.: 43.3 cps (50 C); Flash Pt.: 273 C (Marcusson); Solidifying Pt.: 0 C; Stability: Generally stable, unaffected by hard water; Surface Tension: 39 dynes/cm (1 g/l @ 20 C); pH: 9–10 (10 g/l)
Composition: 100% active ingred.
Toxicity: Corrosive to skin and mucous membranes

Genamine S-100. American Hoechst Corp.
Chemical Description: Stearylamine oxethylate
Category and Applications: See Genamine C-020
Form: Turbid liquid
Properties: Ref. Index: 1.4598 (50 C); Solubility: Forms clear sol'n. in water (10 g/l), and in mineral oil, benzol, tetra, and glycol (1:10); Sp. gr.: 0.987 (50 C); Visc.: 46.1 cps (50 C); Flash Pt.: 277 C (Marcusson); Solidifying Pt.: –5 C; Stability: Generally stable, unaffected by hard water; Surface Tension: 42 dynes/cm (1 g/l @ 20 C); pH: 9–10 (10 g/l)
Composition: 100% active ingred.
Toxicity: Corrosive to skin and mucous membranes

Genamine S-150. American Hoechst Corp.
Chemical Description: Stearylamine oxethylate
Category and Applications: See Genamine C-020
Form: Turbid liquid
Properties: Ref. Index: 1.4600 (50 C); Solubility: Forms clear sol'n. in water (10 g/l), and in benzol, tetra, and glycol (1:10); Sp. gr.: 1.011 (50 C); Visc.: 55.6 cps (50 C); Flash Pt.: 280 C (Marcusson); Solidifying Pt.: –7 ·C; Stability: Generally stable, unaffected by hard water; Surface Tension: 44 dynes/cm (1 g/l @ 20 C); pH: 9–10 (10 g/l)
Composition: 100% active ingred.
Toxicity: Corrosive to skin and mucous membranes

Genamine S-200. American Hoechst Corp.
Chemical Description: Stearylamine oxethylate
Category and Applications: See Genamine C-020
Form: Turbid liquid
Properties: Ref. Index: 1.4606 (50 C); Solubility: Forms clear sol'n. in water (10 g/l), and in benzol, tetra, glycol (1:10); turbid sol'n. in mineral oil (1:10); Sp. gr.: 1.033 (50 C); Visc.: 67.8 cps (50 C); Flash Pt.: 281 C (Marcusson); Solidifying Pt.: –10 C; Stability: Generally stable, unaffected by hard water; Surface Tension: 47 dynes/cm (1 g/l @ 20 C); pH: 9–10 (10 g/l)
Composition: 100% active ingred.

Toxicity: Corrosive to skin and mucous membranes

Genamine S-250. American Hoechst Corp.
Chemical Description: Stearylamine oxethylate
Category and Applications: See Genamine C-020
Form: Paste
Properties: Ref. Index: 1.4606 (50 C); Solubility: Forms clear sol'n. in water (10 g/l), and in benzol, tetra, and glycol (1:10); Sp. gr.: 1.050 (50 C); Visc.: 81.8 cps (50 C); Flash Pt.: 282 C (Marcusson); Stability: Generally stable, unaffected by hard water; Surface Tension: 49 dynes/cm (1 g/l @ 20 C); pH: 9–10 (10 g/l)
Composition: 100% active ingred.
Toxicity: Corrosive to skin and mucous membranes

Genapol AMS. American Hoechst Corp.
Chemical Description: Acylaminopolyglycol ether sulfate triethanolamine salt
Category: Detergent, foaming agent, anionic
Applications: Top-grade cosmetics cleansers, e.g., baby shampoo, bath preparations, feminine hygiene products, conditioning shampoos; lime soap dispersant
Form: Clear, low viscosity liquid; Color: Yellow; Odor: Weak
Properties: Density: ≈ 1.0 g/cm^3 @ 20 C; Visc.: 200 cps max. @ 20 C; pH: 6.5–8.0 (1% active sol'n. @ 20 C); Biodegradable
Composition: ≈ 40% active

Genapol C-020. American Hoechst Corp.
Chemical Description: Coconut fatty alcohol polyglycol ether
Category: Raw material, nonionic
Applications: Starting materials for the chemo-technical industry, e.g., in manufacture of textile, leather, and paper auxiliaries, detergents, emulsifiers, cosmetics, plant protection agents, etc.
Form: Turbid liquid
Properties: Ref. Index: 1.4412 (50 C); Solubility: Forms clear sol'n. in mineral oil, benzene, tetra, and glycol (1:10); turbid sol'n. in water (10 g/l); Sp. gr.: 0.892 (50 C); Visc.: 10.6 cps (50 C); Flash Pt.: 170 C (Marcusson); Solidifying Pt.: 12 C; Surface Tension: 34 dynes/cm (1 g/l @ 20 C)
Composition: 100% active ingred.
Reaction: Neutral

Genapol C-050. American Hoechst Corp.
Chemical Description: Coconut fatty alcohol polyglycol ether
Category and Applications: See Genapol C-020
Form: Turbid liquid
Properties: Ref. Index: 1.4465 (50 C); Solubility: Forms clear sol'n. in mineral oil, benzene, and tetra (1:10); turbid sol'n. in water (10 g/l); Sp. gr.: 0.952; Visc.: 17.6 cps (50 C); Flash Pt.: 201 C (Marcusson); Solidifying Pt.: 11 C; Surface Tension: 32 dynes/cm (1 g/l @ 20 C)
Composition: 100% active ingred.
Reaction: Neutral

Genapol C-080. American Hoechst Corp.
Chemical Description: Coconut fatty alcohol polyglycol ether
Category and Applications: See Genapol C-020
Form: Paste
Properties: Ref. Index: 1.4498 (50 C); Solubility: Forms clear sol'n. in water (10 g/l), and in benzene, tetra, and glycol (1:10); turbid sol'n. in mineral oil (1:10); Sp. gr.: 0.979 (50 C); Visc.: 25.4 cps (50 C); Flash Pt.: 246 C (Marcusson); Surface Tension: 36 dynes/cm (1 g/l @ 20 C)
Composition: 100% active ingred.
Reaction: Neutral.

Genapol C-100. American Hoechst Corp.
Chemical Description: Coconut fatty alcohol polyglycol ether
Category and Applications: See Genapol C-020
Form: Paste
Properties: Ref. Index: 1.4503 (50 C); Solubility: Forms clear sol'n. in water (10 g/l), and in benzene, tetra, and glycol (1:10); turbid sol'n. in mineral oil (1:10); Sp. gr.: 0.990 (50 C); Visc.: 30.4 cps (50 C); Flash Pt.: 251 C (Marcusson); Surface Tension: 38 dynes/cm (1 g/l @ 20 C)
Composition: 100% active ingred.
Reaction: Neutral

Genapol C-150. American Hoechst Corp.
Chemical Description: Coconut fatty alcohol polyglycol ether
Category and Applications: See Genapol C-020
Form: Wax
Properties: Ref. Index: 1.4535 (50 C); Solubility: Forms clear sol'n. in water (10 g/l), and in benzene and tetra (1:10); turbid sol'n. in glycol (1:10); Sp. gr.: 1.027 (50 C); Visc.: 45.8 cps (50 C); Flash Pt.: 260 C (Marcusson); Surface Tension: 43 dynes/cm (1 g/l @ 20 C)
Composition: 100% active ingred.
Reaction: Neutral

Genapol C-200. American Hoechst Corp.
Chemical Description: Coconut fatty alcohol polyglycol ether
Category and Applications: See Genapol C-020
Form: Wax
Properties: Ref. Index: 1.4550 (50 C); Solubility: Forms clear sol'n. in water (10 g/l), and in benzene and tetra (1:10); turbid sol'n. in glycol (1:10); Sp. gr.: 1.032 (50 C); Visc.: 61.5 cps (50 C); Flash Pt.: 264 C (Marcusson); Surface Tension: 44 dynes/cm (1 g/l @ 20 C)
Composition: 100% active ingred.
Reaction: Neutral

Genapol C-250. American Hoechst Corp.
Chemical Description: Coconut fatty alcohol polyglycol ether
Category and Applications: See Genapol C-020
Form: Wax
Properties: Ref. Index: 1.4564

(50 C); Solubility: Forms clear sol'n. in water (10 g/l), and in benzene and tetra (1'10); turbid sol'n. in glycol (1:10); Sp. gr.: 1.055 (50 C); Visc.: 80.6 cps (50 C); Flash Pt.: 267 C (Marcusson); Surface Tension: 47 dynes/cm (1 g/l @ 20 C)
Composition: 100% active ingred.
Reaction: Neutral

Genapol CRT 40. American Hoechst Corp.
Chemical Description: Alkyl sulfate triethanolamine ammonium salt
Category: Detergent, anionic
Applications: Detergent raw material for cosmetics industry; used in clear liquid shampoos and body cleansing agents; good foaming and mild cleansing
Form: Clear liquid; Color: Pale yellow; Odor: Slight
Properties: m.w.: 422 (mean); Solubility: Sol. in water; miscible with other anionic surfactants; Density: 1.0–1.05 g/cm^3 @ 20 C; Visc.: 100 cps max. @ 20 C; pH: 7.0 ± 0.5 (1% aq. @ 20 C); Biodegradable
Composition: 40.5 ± 1.5% alkyl sulfate

Genapol LRO Liquid. American Hoechst Corp.
Chemical Description: Alkyl diglycol ether sulfate sodium salt based on natural fatty alcohols
Category: Foaming agent, detergent, anionic
Applications: Cosmetics products, shampoos, bubble baths; degreasing properties; lime soap dispersant
Form: Clear liquid; Color: Pale yellow; Odor: Slight
Properties: m.w.: 382 (avg.); Solubility: Miscible with water in any ratio; Density: ≈ 1.05 g/cm^3 @ 20 C; Visc.: 100 cps max. @ 20 C; pH: 6.5–8.0 (1% aq. @ 20 C); Biodegradable
Composition: 27.0 ± 0.5% alkylether sulfate

Genapol LRO Paste. American Hoechst Corp.
Chemical Description: Alkyl diglycol ether sulfate sodium salt based on natural fatty alcohols
Category: Detergent, foaming agent, anionic
Applications: Cosmetics products, hair shampoo, bubble bath, shower preparations, hand cleansing agent
Form: Mobile paste; Color: Slightly yellowish; Odor: Slight
Properties: m.w.: 382 (avg.); Solubility: Miscible with water in any ratio; Density: ≈ 1.0 g/cm^3; pH: 7.2 ± 0.6 (1% aq. @ 20 C); Biodegradable
Composition: 68.0 ± 1.5% alkyl ether sulfonate

Genapol O-020. American Hoechst Corp.
Chemical Description: Oleyl alcohol polyglycol ether
Category and Applications: See Genapol C-020
Form: Clear liquid
Properties: Ref. Index: 1.4510 (50 C); Solubility: Forms clear

sol'n. in mineral oil, benzene, and tetra (1:10); turbid in water (10 g/l); Sp. gr.: 0.894 (50 C); Visc.: 12.4 cps (50 C); Flash Pt.: 186 C (Marcusson); Solidifying Pt.: 2 C
Composition: 100% active ingred.
Reaction: Neutral

Genapol O-050. American Hoechst Corp.
Chemical Description: Oleyl alcohol polyglycol ether
Category and Applications: See Genapol C-020
Form: Turbid liquid
Properties: Ref. Index: 1.4530 (50 C); Solubility: Forms clear sol'n. in benzene and tetra (1:10); turbid sol'n. in water (10 g/l) and mineral oil (1:10); Sp. gr.: 0.936 (50 C); Visc.: 18.4 cps (50 C); Flash Pt.: 225 C (Marcusson); Solidifying Pt.: 1 C; Surface Tension: 54 dynes/cm (1 g/l @ 20 C)
Composition: 100% active ingred.
Reaction: Neutral

Genapol O-080. American Hoechst Corp.
Chemical Description: Oleyl alcohol polyglycol ether
Category and Applications: See Genapol C-020
Form: Turbid liquid
Properties: Ref. Index: 1.4538 (50 C); Solubility: Forms clear sol'n. in water (10 g/l), and in benzene, tetra, and glycol (1:10); turbid sol'n. in mineral oil (1:10); Sp. gr.: 0.960 (50 C); Visc.: 25.1 cps (50 C); Flash Pt.: 246 C (Marcusson); Solidifying Pt.: 17 C; Surface Tension: 44 dynes/cm (1 g/l @ 20 C)
Composition: 100% active ingred.
Reaction: Neutral

Genapol O-100. American Hoechst Corp.
Chemical Description: Oleyl alcohol polyglycol ether
Category and Applications: See Genapol C-020
Form: Paste
Properties: Ref. Index: 1.4558 (50 C); Solubility: Forms clear sol'n. in water (10 g/l), and in benzene, tetra, and glycol (1:10); turbid sol'n. in mineral oil (1:10); Sp. gr.: 0.989 (50 C); Visc.: 33.0 cps (50 C); Flash Pt.: 260 C (Marcusson); Solidifying Pt.: 19 C; Surface Tension: 41 dynes/cm (1 g/l @ 20 C)
Composition: 100% active ingred.
Reaction: Neutral

Genapol O-120. American Hoechst Corp.
Chemical Description: Oleyl alcohol polyglycol ether
Category and Applications: See Genapol C-020
Form: Paste
Properties: Ref. Index: 1.4560 (50 C); Solubility: Forms clear sol'n. in water (10 g/l), and in benzene, tetra, and glycol (1:10); turbid sol'n. in mineral oil; Sp. gr.: 1.000 (50 C); Visc.: 42.5 cps (50 C); Flash Pt.: 265 C (Marcusson); Surface Tension: 42 dynes/cm (1 g/l @ 20 C)
Composition: 100% active ingred.
Reaction: Neutral

Genapol O-150. American Hoechst Corp.
Chemical Description: Oleyl alcohol polyglycol ether
Category and Applications: See Genapol C-020
Form: Wax
Properties: Ref. Index: 1.4570 (50 C); Solubility: Forms clear sol'n. in water (10 g/l), and in benzene and tetra (1:10); turbid sol'n. in mineral oil and glycol (1:10); Sp. gr.: 1.020 (50 C); Visc.: 49 1 cps (50 C); Flash Pt.: 271 C (Marcusson); Surface Tension: 45 dynes/cm (1 g/l @ 20 C)
Composition: 100% active ingred.
Reaction: Neutral

Genapol O-200. American Hoechst Corp.
Chemical Description: Oleyl alcohol polyglycol ether
Category and Applications: See Genapol C-020
Form: Wax
Properties: Ref. Index: 1.4580 (50 C); Solubility: Forms clear sol'n. in water (10 g/l), and in benzene and tetra (1:10); turbid sol'n. in mineral oil and glycol (1:10); Sp. gr.: 1.037 (50 C); Visc.: 65.9 cps (50 C); Flash Pt.: 278 C (Marcusson); Surface Tension: 47 dynes/cm (1 g/l @ 20 C)
Composition: 100% active ingred.
Reaction: Neutral

Genapol O-230. American Hoechst Corp.
Chemical Description: Oleyl alcohol polyglycol ether
Category and Applications: See Genapol C-020
Form: Wax
Properties: Ref. Index: 1.4582 (50 C); Solubility: Forms clear sol'n. in water (10 g/l), and in benzene and tetra (1:10); turbid sol'n. in mineral oil and glycol (1:10); Sp. gr.: 1.042 (50 C); Visc.: 79.5 cps (50 C); Flash Pt.: 279 C (Marcusson); Surface Tension: 47 dynes/cm (1 g/l @ 20 C)
Composition: 100% active ingred.
Reaction: Neutral

Genapol O-250. American Hoechst Corp.
Chemical Description: Oleyl alcohol polyglycol ether
Category and Applications: See Genapol C-020
Form: Wax
Properties: Ref. Index: 1.4587 (50 C); Solubility: Forms clear sol'n. in water (10 g/l) and in benzene and tetra (1:10); turbid sol'n. in mineral oil and glycol (1:10); Sp. gr.: 1.042 (50 C); Visc.: 88.1 cps (50 C); Flash Pt.: 280 C (Marcusson); Surface Tension: 48 dynes/cm (1 g/l @ 20 C)
Composition: 100% active ingred.
Reaction: Neutral

Genapol PGM Conc. American Hoechst Corp.
Chemical Description: Dispersion of nonionic pearl-lustering substances in an alkyl ether sulfate sodium salt with viscosity-controlling and foam-stabilizing additives
Category: Pearl-luster concentrate, anionic

Applications: Used in pearl luster shampoos, bubble baths, and other cosmetics
Form: Homogeneous, fluid paste; Color: White; Odor: Slight
Properties: Solubility: Miscible with water in all ratios; Density: ≈ 1.0 g/cm^3; pH: 7.2 ± 0.8% (1% active); Biodegradable
Composition: 40.0 ± 1.0% active

Genapol PN 30. American Hoechst Corp.
Chemical Description: Ethylene oxide/propylene oxide crossed linked blocked polymer
Category: Lubricant, nonionic
Applications: Lubricant for formulating water soluble metalworking fluids in combination with corrosion inhibitors; low foaming properties make it suitable for production of controlled low foam detergents, wetting agents, and cleaners
Form: Liquid; Color: Yellow
Properties: Ref. Index: ≈ 1.448 (50 C); Solubility: Cloudy in water; Sp. gr.: ≈ 1.01 (50 C); Cloud Pt.: 30 ± 1 C (5 g in 25 ml of 24% butyl diglycol sol'n.); Solidification Pt.: < –10 C; Stability: Stable to hard water, metal salts, acids, and alkali
Composition: 100% active ingred.

Genapol S-020. American Hoechst Corp.
Chemical Description: Stearyl alcohol polyglycol ether
Category and Applications: See Genapol C-020
Form: Wax
Properties: Ref. Index: 1.4458 (50 C); Solubility: Forms clear sol'n. in mineral oil, benzene, and tetra (1:10); turbid in water (10 g/l) and glycol (1:10); Sp. gr.: 0.890 (50 C); Visc.: 15.2 cps (50 C); Flash Pt.: 198 C (Marcusson); Surface Tension: 50 dynes/cm (1 g/l @ 20 C)
Composition: 100% active ingred.
Reaction: Neutral

Genapol S-050. American Hoechst Corp.
Chemical Description: Stearyl alcohol polyglycol ether
Category and Applications: See Genapol C-020
Form: Wax
Properties: Ref. Index: 1.4480 (50 C); Solubility: Forms clear sol'n. in mineral oil, benzene, and tetra (1:10); turbid in water (10 g/l) and glycol (1:10); Sp. gr.: 0.934 (50 C); Visc.: 21.2 cps (50 C); Flash Pt.: 224 C (Marcusson); Surface Tension: 50 dynes/cm (1 g/l @ 20 C)
Composition: 100% active ingred.
Reaction: Neutral

Genapol S-080. American Hoechst Corp.
Chemical Description: Stearyl alcohol polyglycol ether
Category and Applications: See Genapol C-020
Form: Paste
Properties: Ref. Index: 1.4510 (50 C); Solubility: Forms clear sol'n. in benzene, tetra (1:10); turbid sol'n. in water (10 g/l), gly-

col and mineral oil (1:10); Sp. gr.: 0.971 (50 C); Visc.: 28.4 cps (50 C); Flash Pt.: 228 C (Marcusson); Surface Tension: 44 dynes/cm (1 g/l @ 20 C)
Composition: 100% active ingred.
Reaction: Neutral

Genapol S-100. American Hoechst Corp.
Chemical Description: Stearyl alcohol polyglycol ether
Category and Applications: See Genapol C-020
Form: Paste
Properties: Ref. Index: 1.4512 (50 C); Solubility: Forms clear sol'n. in benzene and tetra (1:10); turbid sol'n. in water (10 g/l), and mineral oil and glycol (1:10); Sp. gr.: 0.978 (50 C); Visc.: 34.0 cps (50 C); Flash Pt.: 239 C (Marcusson); Surface Tension: 45 dynes/cm (1 g/l @ 20 C)
Composition: 100% active ingred.
Reaction: Neutral

Genapol S-150. American Hoechst Corp.
Chemical Description: Stearyl alcohol polyglycol ether
Category and Applications: See Genapol C-020
Form: Wax
Properties: Ref. Index: 1.4540 (50 C); Solubility: Forms clear sol'n. in water (10 g/l), and in benzene and tetra (1:10); turbid sol'n. in mineral oil and glycol (1:10); Sp. gr.: 1.009 (50 C); Visc.: 50.3 cps (50 C); Flash Pt.: 266 C (Marcusson); Surface Tension: 47 dynes/cm (1 g/l @ 20 C)
Composition: 100% active ingred.
Reaction: Neutral

Genapol S-200. American Hoechst Corp.
Chemical Description: Stearyl alcohol polyglycol ether
Category and Applications: See Genapol C-020
Form: Wax
Properties: Ref. Index: 1.4555 (50 C); Solubility: Forms clear sol'n. in water (10 g/l), and in benzene and tetra (1:10); turbid sol'n. in mineral oil and glycol (1:10); Sp. gr.: 1.031 (50 C); Visc.: 67.4 cps (50 C); Flash Pt.: 267 C (Marcusson); Surface Tension: 48 dynes/cm (1 g/l @ 20 C)
Composition: 100% active ingred.
Reaction: Neutral

Genapol S-250. American Hoechst Corp.
Chemical Description: Stearyl alcohol polyglycol ether
Category and Applications: See Genapol C-020
Form: Wax
Properties: Ref. Index: 1.4570 (50 C); Solubility: Forms clear sol'n. in water (10 g/l) and in benzene and tetra (1:10); turbid sol'n. in mineral oil and glycol (1:10); Sp. gr.: 1.049 (50 C); Visc.: 106 cps (50 C); Flash Pt.: 273 C (Marcusson); Surface Tension: 49 dynes/cm (1 g/l @ 20 C)
Composition: 100% active ingred.
Reaction: Neutral

Generol 122. Henkel Inc.
Chemical Description: Soya sterol
Category: Emollient, emulsifier, stabilizer, nonionic
Applications: Emollient, auxiliary or primary emulsifier, emulsion stabilizer, and viscosity modifier for cosmetics
Form: Waxy flake; Color: White
Properties: HLB: 3.0
Composition: 90% conc.

Generol 122 E 5. Henkel Inc.
Chemical Description: PEG-5 soya sterol
Category: Emollient, emulsifier, stabilizer, nonionic
Applications: Substantive to hair, emollient and emulsifier, modifier and emulsion stabilizer
Form: Soft waxy amorphous solid
Properties: HLB: 5.0
Composition: 100% conc.

Generol 122 E 10. Henkel Inc.
Chemical Description: PEG-10 soya sterol
Category: Emollient, emulsifier, stabilizer, nonionic
Applications: Emollient, auxiliary emulsifier, substantive to hair, emulsion stabilizer, and viscosity modifier
Form: Soft wax; Color: Light amber
Properties: HLB: 12.0
Composition: 100% conc.

Generol 122 E 16. Henkel Inc.
Chemical Description: PEG-16 soya sterol
Category: Emollient, solubilizer, wetting agent, nonionic
Applications: Pigment wetter, emollient, solubilizer
Form: Hard wax; Color: Ivory
Properties: HLB: 15.0
Composition: 100% conc.

Generol 122 E 25. Henkel Inc.
Chemical Description: PEG-25 soya sterol
Category: Solubilizer, emollient, wetting and deflocculating agent, nonionic
Applications: Solubilizer and emollient; pigment wetter and deflocculating agent
Form: Hard wax; Color: Ivory
Properties: HLB: 17.0
Composition: 100% conc.

Glucamate SSE-20. Amerchol Corp.
Chemical Description: POE (20) methyl glucoside sesquistearate
Category: O/W emulsifier, nonionic
Applications: O/W emulsifier; used with Glucate SS in O/W emulsions to provide viscosity stability, recovery from work, mildness to skin and eyes at low concentrations
Form: Soft solid; Color: Pale yellow
Properties: Solubility: Sol. in water, isopropanol, ethanol, castor oil, corn oil; HLB: 15.3; Cloud Pt.: 74 C (1% in 5% NaCl); Flash Pt.: 570 F (OC); Fire Pt.: 630 F (OC); Sapon. Value: 47; Acid No.: < 1; Hydroxyl No.: 97; Iodine Value: < 1; pH: 6.5 (10% aq.)
Composition: 100% conc.

Glucate SS. Amerchol Corp.
Chemical Description: Methyl glucoside sesquistearate
Category: Emulsifier, nonionic
Applications: W/O emulsifier; used with Glucamate SSE-20 for O/W emulsions to provide viscosity stability, mildness on skin and eyes, recovery from work at low concentrations
Form: Flakes; Color: Off white
Properties: Solubility: Sol. in warm isopropanol; miscible with common oil phase ingredients; water insol; M.P.: 51 C; HLB: 6.4; Flash Pt.: 530 F (OC); Fire Pt.: 545 F (OC); Sapon. Value: 136; Acid No.: 5.5; Hydroxyl No.: 270; Iodine Value; < 1; pH: 5.5 (10% in 1:1 IPA:water)
Composition: 100% conc.

Glycerol Mono Stearate. Graden Chem. & Equipment
Form: Flake; Color: 40Y/4R max. (Lovibond)
Properties: Solubility: Insol. in water; M.P.: 53–57 C; Sapon. Value: 167–177; Iodine Value: $<$ 1
Containers: 50 lb multiwall bags

Glycomul L. Glyco Chem.
Chemical Description: Sorbitan mono laurate
Category: Emulsifier, nonionic
Applications: Edible, cosmetic, pharmaceutical and industrial uses
Form: Liquid; Color: Amber
Properties: Solubility: Sol. in methanol, ethanol, naphtha; miscible in certain proportions with acetone, ethyl acetate, toluol, mineral oil, vegetable oil; Sp. gr.: $\approx$ 1.0; Visc.: 4500 cps; HLB: 8.6; Sapon. Value: 157–171; Acid No.: 7 max.; Hydroxyl No.: 330–360
Composition: 100% conc.
References: Bulletin G-100

Glycomul LC. Glyco Chem.
Chemical Description: Sorbitan mono laurate, cosmetic grade
Category: Emulsifier, nonionic
Applications: Cosmetics
Form: Liquid; Color: Light amber
Properties: Solubility: Sol. in methanol, ethanol, naphtha; disp. in water; miscible in certain proportions with acetone, ethyl acetate, toluol, mineral and vegetable oils; Sp. gr.: $\approx$ 1.0; Visc.: 5000 cps; HLB: 8.6; Sapon. Value: 154–163; Acid No.: 10 max.; Hydroxyl No.: 324–356
Composition: 100% conc.
References: Bulletin G-100

Glycomul MA. Glyco Chem.
Chemical Description: Sorbitan ester mixed fatty acids
Category and Applications: See Glycomul L
Form: Liquid; Color: Dark amber
Properties: Solubility: Sol. in methanol, ethanol, acetone, toluol, naphtha, mineral and vegetable oils; Sp. gr.: $\approx$ 1.0; Visc.: $\approx$ 1000 cps; Sapon. Value: 139–156; Acid No.: 3.5–7.5; Hydroxyl No.: 161–183
Composition: 100% conc.
References: Bulletin G-100

Glycomul O. Glyco Chem.
Chemical Description: Sorbitan mono oleate
Category and Applications: See Glycomul L
Form: Liquid; Color: Amber
Properties: Solubility: Sol. in ethyl acetate, mineral and vegetable oils; miscible in certain proportions with methanol, ethanol, toluol, naphtha; disp. in water; Sp. gr.: ≈ 1.0; Visc.: 1000 cps; HLB: 4.3; Sapon. Value: 148–161; Acid No.: 7.5 max.; Hydroxyl No.: 193–209
Composition: 100% conc.
References: Bulletin G-100

Glycomul P. Glyco Chem.
Chemical Description: Sorbitan mono palmitate
Category and Applications: See Glycomul L
Form: Beads; Color: Cream
Properties: Solubility: Sol. in vegetable oil; sol. hot in ethyl acetate; poorly sol. hot in ethanol, acetone, toluol, mineral oil; HLB: 6.7; Sapon. Value: 139–150; Acid No.: 7.5 max.; Hydroxyl No.: 272–306
Composition: 100% conc.
References: Bulletin G-100

Glycomul S. Glyco Chem.
Chemical Description: Sorbitan mono stearate (also available in vegetable grade and kosher grade)
Category and Applications: See Glycomul L
Form: Beads; Color: Cream
Properties: Solubility: Sol. in vegetable oil; poorly sol. hot in ethanol, ethyl acetate, toluol, mineral oil; HLB: 4.7; Sapon. Value: 146–158; Acid No.: 10 max.; Hydroxyl No.: 236–260
Composition: 100% conc.
References: Bulletin G-100

Glycomul SOC. Glyco Chem.
Chemical Description: Sorbitan sesqui oleate
Category and Applications: See Glycomul L
Form: Liquid; Color: Amber
Properties: Solubility: Sol. in methanol, ethanol, ethyl acetate; miscible with mineral oil in certain proportions; Sp. gr.: ≈ 1.0; Visc.: 1000 cps; HLB: 4.0; Sapon. Value: 149–166; Acid No.: 7.5–11.5; Hydroxyl No.: 184–206
Composition: 100% conc.
References: Bulletin G-100

Glycomul TAO. Glyco Chem.
Chemical Description: Sorbitan ester mixed resin and fatty acids
Category and Applications: See Glycomul L
Form: Liquid; Color: Amber
Properties: Solubility: Sol. in methanol, ethanol, toluol, naphtha, mineral and vegetable oils; disp. in water; Sp. gr.: ≈ 1.02; Visc.: 1300 cps; HLB: 4.3; Sapon. Value: 135–148; Acid No.: 4–7; Hydroxyl No.: 180–208
Composition: 100% conc.
References: Bulletin G-100

Glycomul TO. Glyco Chem.
Chemical Description: Sorbitan trioleate
Category and Applications: See Glycomul L

Form: Oily liquid; Color: Amber
Properties: Solubility: Sol. in ethyl acetate, toluol, naphtha, mineral and vegetable oils; disp. in water; Sp. gr.: ≈ 0.95; Visc.: 200 cps; HLB: 1.8; Sapon. Value: 171–185; Acid No.: 13.5 max.; Hydroxyl No.: 58–69
Composition: 100% conc.
References: Bulletin G-100

Glycomul TS. Glyco Chem.
Chemical Description: Sorbitan tristearate
Category and Applications: See Glycomul L
Form: Beads; Color: Light tan
Properties: Solubility: Poorly sol. in ethyl acetate, toluol; disp. in acetone, naphtha, mineral and vegetable oils; HLB: 2.1; Sapon. Value: 175–190; Acid No.: 15 max.; Hydroxyl No.: 65–80
Composition: 100% conc.
References: Bulletin G-100

Glycosperse L-20. Glyco Chem.
Chemical Description: POE 20 sorbitan mono laurate
Category: Emulsifier, nonionic
Applications: Emulsifier for food, cosmetic and pharmaceutical uses
Form: Liquid; Color: Yellow
Properties: Solubility: Sol. in water, methanol, ethanol, acetone, ethyl acetate; miscible with toluol and vegetable oil in certain proportions; Sp. gr.: ≈ 1.1; Visc.: 400 cps; HLB: 16.7; Sapon. Value: 39–52; Acid No.: 2 max.; Hydroxyl No.: 96–114
Composition: 100% conc.
References: Bulletin G-100

Glycosperse L-20X. Glyco Chem.
Chemical Description: POE 20 sorbitan mono laurate, anhydrous
Category and Applications: See Glycosperse L-20
Form: Liquid; Color: Yellow
Properties: Solubility: Sol. in water, methanol, ethanol, acetone, ethyl acetate; miscible with toluol and vegetable oil in certain proportions; Sp. gr.: 1.1; Visc.: 400 cps; HLB: 16.7; Sapon. Value: 39–52; Acid No.: 2 max.; Hydroxyl No.: 100–117
Composition: 100% conc.
References: Bulletin G-100

Glycosperse O-5. Glyco Chem.
Chemical Description: POE 5 sorbitan mono oleate
Category and Applications: See Glycosperse L-20
Form: Liquid (gels on standing); Color: Amber
Properties: Solubility: Sol. in ethanol, methanol, ethyl acetate, mineral oil; disp. in water; miscible with acetone, toluol and naphtha in certain proportions; Sp. gr.: ≈ 1.0; Visc.: 450 cps; HLB: 10.0; Sapon. Value: 95–105; Acid No.: 2 max.; Hydroxyl No.: 136–152
Composition: 100% conc.
References: Bulletin G-100

Glycosperse O-20. Glyco Chem.
Chemical Description: POE 20 sorbitan mono oleate
Category and Applications: See Glycosperse L-20
Form: Liquid; Color: Yellow
Properties: Solubility: Sol. in water, methanol, ethanol, ethyl ace-

tate, toluol, vegetable oil; partly sol. in acetone; Sp. gr.: ≈ 1.0; Visc.: 400 cps; HLB: 15; Sapon. Value: 44–56; Acid No.: 2 max.; Hydroxyl No.: 64–81
Composition: 100% conc.
References: Bulletin G-100

Glycosperse O-20 VEG. Glyco Chem.
Chemical Description: POE 20 sorbitan mono oleate, vegetable grade
Category and Applications: See Glycosperse L-20
Form: Liquid; Color: Yellow
Properties: Solubility: Sol. in water, methanol, ethanol, ethyl acetate, toluol, vegetable oil; partly sol. in acetone; Sp. gr.: 1.0; Visc.: 400 cps; Sapon. Value: 44–56; Acid No.: 2 max.; Hydroxyl No.: 64–81
Composition: 100% conc.
References: Bulletin G-100

Glycosperse O-20X. Glyco Chem.
Chemical Description: POE 20 sorbitan monooleate, anhydrous
Category and Applications: See Glycosperse L-20
Form: Liquid; Color: Yellow
Properties: Solubility: See Glycosperse O-20 VEG; Sp. gr.: 1.0; Visc.: 600 cps; Sapon. Value: 44–56; Acid No.: 2 max.; Hydroxyl No.: 64–81
Composition: 100% conc. (0.2% max. water)
References: Bulletin G-100

Glycosperse P-20. Glyco Chem.
Chemical Description: POE 20 sorbitan mono palmitate
Category and Applications: See Glycosperse L-20
Form: Liquid; Color: Yellow
Properties: Solubility: Sol. in water, methanol, ethanol, acetone, ethyl acetate; miscible with toluol, naphtha and vegetable oil in certain proportions; Sp. gr.: 1.0; Visc.: 550 cps; HLB: 15.6; Sapon. Value: 40–53; Acid No.: 2 max.; Hydroxyl No.: 91–107
Composition: 100% conc.
References: Bulletin G-100

Glycosperse S-20. Glyco Chem.
Chemical Description: POE 20 sorbitan mono stearate
Category and Applications: See Glycosperse L-20
Form: Liquid; Color: Yellow
Properties: Solubility: Sol. in water, ethyl acetate, toluol; miscible with methanol, ethanol, acetone and naphtha in certain proportions; Sp. gr.: 1.1; HLB: 14.9; Sapon. Value: 44–56; Acid No.: 2 max.; Hydroxyl No.: 80–96
Composition: 100% conc.
References: Bulletin G-100

Glycosperse TO-20. Glyco Chem.
Chemical Description: POE 20 sorbitan tri oleate
Category and Applications: See Glycosperse L-20
Form: Liquid (gels on standing); Color: Yellow
Properties: Solubility: Sol. in ethanol, methanol, ethyl acetate; partly sol. in mineral oil; disp. in water; miscible with acetone, toluol, naphtha and vegetable oil in certain proportions; Sp. gr.: 1.0;

Visc.: 300 cps; HLB: 11.0; Sapon. Value: 82–95; Acid No.: 2 max.; Hydroxyl No.: 39–52
Composition: 95% conc.
References: Bulletin G-100

Glycosperse TS-20. Glyco Chem.
Chemical Description: POE 20 sorbitan tri stearate
Category and Applications: See Glycosperse L-20
Form: Waxy solid; Color: Tan
Properties: Solubility: Sol. in ethanol, methanol, acetone, ethyl acetate, naphtha, mineral and vegetable oils; disp. in water and toluol; Sp. gr.: 1.05; HLB: 10.5; Sapon. Value: 88–98; Acid No.: 2 max.; Hydroxyl No.: 44–60
Composition: 100% conc.
References: Bulletin G-100

Golden Dawn Grade 1. Westbrook Lanolin Co.
Chemical Description: Anhydrous lanolin
Category: Emollient, emulsifier, nonionic
Applications: Emollient, W/O emulsifier, ointment base, hair conditioner, wax crystal inhibitor, lipstick binder
Form: Soft wax; Color: Pale

Golden Dawn Grade 2. Westbrook Lanolin Co.
Chemical Description: Anhydrous lanolin
Category and Applications: See Golden Dawn Grade 1; Grade 2 has slightly darker color
Form: Soft wax

Golden Dawn Superfine. Westbrook Lanolin Co.
Chemical Description: Anhydrous lanolin of finest quality and palest color
Category: Emollient, emulsifier, nonionic
Applications: Emollient, w/o emulsifier, ointment bases, hair conditioners, wax crystal inhibitors, lipstick binder
Form: Soft wax
Properties: HLB: 4.0
Composition: 100% conc.

Golden Fleece DF. Westbrook Lanolin Co.
Chemical Description: Extra refined anhydrous lanolin (detergent-free)
Category and Applications: See Golden Dawn Grade 1
Form: Soft solid
Properties: HLB: 4.5
Composition: 100% conc.

Golden Fleece P-80. Westbrook Lanolin Co.
Chemical Description: Anhydrous lanolin (trace pesticides reduced by 80%)
Category and Applications: See Golden Dawn Grade 1

Golden Fleece P-95. Westbrook Lanolin Co.
Chemical Description: Anhydrous lanolin (trace pesticides reduced by 95%, low in natural free alcohols)
Category and Applications: See Golden Dawn Grade 1

Golden Fleece RA. Westbrook Lanolin Co.
Chemical Description: Anhydrous lanolin (detergent-free and low in natural free alcohols)
Category and Applications: See Golden Dawn Grade 1
Form: Soft solid
Composition: 100% conc.

Graden Butyl Oleate. Graden Chem. & Equipment
Chemical Description: Butyl oleate
Category: Plasticizer
Applications: Low-temperature plasticizer for natural and synthetic elastomers
Form: Liquid; Color: 2 max. (Gardner); Odor: Mild fatty
Properties: Ref. Index: 1.451; Solubility: Sol. in acetone, benzol, carbon tetrachloride, isopropanol, mineral oil, castor oil, peanut oil; Density: 7.2 lb/gal; Sp. gr.: 0.864–0.877; F.P.: –10 C; B.P.: 190–220 C; Flash Pt.: 375 F (COC); Acidity: 1% max.; Iodine Value: 70–80
Containers: 55 gal steel drum; 400 lb net wt

Graden Butyl Stearate. Graden Chem. & Equipment
Form: Oily liquid; Color: 35 max. APHA; Odor: Practically none
Properties: Ref. Index: 1.442; Solubility: Sol. in acetone, benzol, ethyl acetate, carbon tetrachloride, isopropanol, heptane, castor oil, peanut oil, mineral oil; partly sol. in methanol; Density: 7.1 lb/gal; Sp. gr.: 0.853–0.859; Visc.: 7 cps; F.P.: 19 C; B.P.: 196.5 C; Flash Pt.: 369 F (COC); Acidity: 0.5% max.; Iodine Value: 6 max.
Containers: 55 gal steel drum (bung type), 380 lb net wt

Gradol 250. Graden Chem. Co.
Chemical Description: Sodium salt of a carboxylated polyelectrolyte
Category: Dispersing agent, anionic
Applications: Dispersant for pigments in latex paint formulations with minimum foaming; viscosity depressant for latex, clay slurries and pigment dispersions
Form: Clear liquid; Color: 40 max. (APHA)
Properties: Solubility: Sol. in water systems; Density: 9.6 lb/gal; F.P.: $<$ –5 C; Stability: Stable over wide pH range; Surface Tension: 69–71 dynes/cm (1% sol'n. @ 72 F); pH: 9.5–10.5 (1% sol'n.)
Composition: 25% active ingred.
Containers: 55 gal Liquipak

Gradol 250A. Graden Chem. Co.
Chemical Description: Ammonium salt of a carboxylated polyelectrolyte
Category: Dispersing agent, anionic
Applications: Dispersant in ceramics industry and where sodium or potassium salts may not be tolerated
Form: Clear liquid; Color: Amber
Properties: Solubility: Sol. in water systems; Density: 9.1–9.4 lb/gal; pH: 7.5–9.0

Composition: 25% active ingred.
Containers: 55 gal nonreturnable Liquipak

Gradol 300. Graden Chem. Co.
Chemical Description: Sodium salt of a carboxylated polyelectrolyte
Category: Dispersant, anionic
Applications: Nonfoaming dispersing agent for pigments, fillers and other insol. fine particles in water; improves hiding qualities, increases color development and solids loading, decreases sedimentation rates, lowers viscosity in pigmented latex systems
Form: Liquid; Color: Light
Properties: Solubility: Sol. in aq. systems; Sp. gr.: 1.21; pH: 10 (5% sol'n.)
Composition: 30% solids content
Reaction: Alkaline
Containers: 55 gal Liquipak drums
Warnings and Caution: Protect from freezing

Gradonic N-95. Graden Chem. Co.
Chemical Description: Nonyl phenol plus 9.5 moles of ethylene oxide
Category: Detergent, wetting agent, dispersant, emulsifier, nonionic
Applications: Detergent formulations alone or with anionic surfactants and alkaline salts with good foaming and rapid wetting
Form: Liquid; Color: Almost colorless; Odor: Mild
Properties: Solubility: Sol. in water, dilute inorganic salt and caustic sol'ns., in mineral acid sol'ns.; Cloud Pt.: 130 F (5% sol'n.); Stability: Stable to hydrolysis in acid or alkali; pH: 7 (2% sol'n.)
Composition: 100% active ingred.
Reaction: Neutral

Grillocin. R.I.T.A. Corp.
Category: Odor absorbent, nonionic
Applications: Deodorants, body sprays, room sprays, antiperspirants
Form: Liquid; Color: Golden; Odor: None
Properties: Sp. gr.: 1.083; Stability: Good
Composition: 100% active ingred.

Grindtek AML 60. Grindtek
Chemical Description: Acetylated lauric/myristic fatty acid ester of glycerol
Category: Lubricant, plasticizer, cosolvent
Applications: Lubricant and plasticizer for plastics and coatings; cosolvent for polar additives in low polarity systems (e.g., metal cleaning products)
Form: Liquid; Color: Light tan
Properties: Solubility: Sol. in ethanol, peanut oil, toluene, white spirit, paraffin oil; partially sol. warmed in propylene glycol; HLB: 2.1

Grindtek AMOS 90. Grindtek
Chemical Description: Acetylated oleic/stearic/palmitic fatty acid esters of glycerol

Category and Applications: See Grindtek AML 60
Form: Liquid; Color: Yellow
Properties: Solubility: Sol. in ethanol, peanut oil, toluene, white spirit, paraffin oil; partly sol. warmed in propylene glycol; HLB: 1.8

Grindtek CA-P. Grindtek
Chemical Description: Citric acid ester of glycerol monostearate (from stearic/palmitic fatty acids)
Category: Emulsifier
Applications: O/W emulsifier
Form: Powder; Color: Cream
Properties: Solubility: Sol. warmed in toluene, white spirit; partly sol. warmed in propylene glycol; HLB: 11 (value estimated from dispersibility in water at 5%)

Grindtek DAT-L. Grindtek
Chemical Description: Diacetylated tartaric acid ester of glycerol (oleic/stearic) fatty acid esters
Category: Emulsifier
Applications: O/W emulsifier, esp. when partially neutralized in situ with hydroxides or amines
Form: Viscous liquid; Color: Yellow-brown
Properties: Solubility: Sol. in ethanol, toluene; sol. warmed in peanut oil; disp. in water; HLB: 7 (value estimated from dispersibility in water at 5%)

Grindtek DAT-S. Grindtek
Chemical Description: Diacetylated tartaric acid esters of glycerol (stearic/palmitic) fatty acid esters
Category and Applications: See Grindtek DAT-L
Form: Flake; Color: Cream
Properties: Solubility: Sol. in toluene; sol. warmed in white spirit; partly sol. warmed in peanut oil; HLB: 8 (value estimated from dispersibility in water at 5%)

Grindtek DAT-VO. Grindtek
Chemical Description: Diacetylated tartaric acid esters of glycerol (oleic/stearic) fatty acid esters
Category: Emulsifier
Applications: W/O emulsifier
Form: Liquid; Color: Yellow-brown
Properties: Solubility: Sol. in peanut oil, toluene, white spirit; disp. in water and propylene glycol; HLB: 6 (value estimated from dispersibility in water at 5%)

Grindtek FAL 1. Grindtek
Chemical Description: Stearic/palmitic fatty acid ester of lactylic acid
Category: Emulsifier
Applications: O/W emulsifier
Form: Powder; Color: Cream
Properties: Solubility: Sol. warm in ethanol, propylene glycol, toluene, white spirit, paraffin oil; HLB: 10 (value estimated from dispersibility in water at 5%)

Grindtek FAL 2. Grindtek
Chemical Description: See Grindtek FAL 1
Category: Emulsifier
Applications: W/O emulsifier
Form: Flake; Color: Cream

Properties: Solubility: Sol. warm in toluene, white spirit; partly sol. warm in peanut oil; HLB: 2 (value estimated from dispersibility in water at 5%)

Grindtek FAL 3. Grindtek
Chemical Description: See Grindtek FAL 1
Category: Emulsifier
Applications: O/W emulsifier when partially neutralized in situ with hydroxides or amines
Form: Block; Color: Tan
Properties: Solubility: Sol. in toluene; sol. warm in ethanol, propylene glycol, peanut oil; partly sol. warm in white spirit, paraffin oil; HLB: 2 (value estimated from dispersibility in water at 5%)

Grindtek FAL 4. Grindtek
Chemical Description: Caprylic fatty acid ester of lactylic acid and its salt
Category: Emulsifier
Applications: O/W emulsifier with expected fungistatic properties
Form: Plastic; Color: Yellow/brown
Properties: Solubility: Sol. in ethanol, propylene glycol, toluene; sol. warm in water, white spirit; HLB: > 13 (value estimated from dispersibility in water at 5%)

Grindtek LA 15. Grindtek
Chemical Description: Lactic acid ester of glycerol monostearate (from stearic/palmitic fatty acid)
Applications: Component in W/O and O/W creams
Form: Flake; Color: Whitish
Properties: Solubility: Sol. warm in peanut oil, toluene, white spirit; partly sol. in paraffin oil; HLB: 5.1

Grindtek LA 30. Grindtek
Chemical Description: Lactic acid ester of glycerol monostearate (from stearic/palmitic fatty acid)
Applications: Component in W/O and O/W creams
Form: Block; Color: Whitish
Properties: Solubility: Sol. warm in ethanol, peanut oil, toluene, white spirit, paraffin oil; partly sol. warm in propylene glycol; HLB: 8.2

Grindtek ML 90. Grindtek
Chemical Description: Fatty acid ester of glycerol (from lauric fatty acid)
Category: Lubricant, antistat, antifogging agent
Applications: Components in W/O and O/W creams; lubricants for processing of polyvinylchloride, polystyrene and copolymers of styrene; antistat and antifogging agent in polyolefins; antimicrobial effects reported
Form: Block; Color: Whitish
Properties: Solubility: Sol. in ethanol; sol. warm in propylene glycol, toluene; partly sol. in white spirit; HLB: 5.3

Grindtek MM 90. Grindtek
Chemical Description: Myristic fatty acid ester of glycerol
Category: Lubricant, antistat, antifogging agent

Applications: Components in W/O and O/W creams; lubricant for processing of polyvinylchloride, polystyrene, and copolymers of styrene; antistat and antifogging agent in polyolefins
Form: Block; Color: Whitish
Properties: Solubility: Sol. warm in ethanol, propylene glycol, toluene; HLB: 5.0

Grindtek MOL 90. Grindtek
Chemical Description: Oleic/linoleic fatty acid ester of glycerol
Category and Applications: See Grindtek MM 90
Form: Semisolid; Color: Whitish
Properties: Solubility: Sol. in ethanol, toluene; sol. warm in propylene glycol, peanut oil, white spirit, paraffin oil; HLB: 4.3

Grindtek MOP 90. Grindtek
Chemical Description: Oleic/palmitic fatty acid ester of glycerol
Category and Applications: See Grindtek MM 90
Form: Block; Color: Whitish
Properties: Solubility: Sol. warm in ethanol, propylene glycol, toluene, white spirit; HLB: 4.3

Grindtek MSP 32-6. Grindtek
Chemical Description: Stearic/palmitic fatty acid ester of glycerol
Category and Applications: See Grindtek MM 90
Form: Powder; Color: Whitish
Properties: Solubility: Disp. in water; disp. warm in toluene, white spirit; HLB: 3.7

Grindtek MSP 40. Grindtek
Chemical Description: Stearic/palmitic fatty acid ester of glycerol
Category and Applications: See Grindtek MM 90
Form: Powder; Color: Whitish
Properties: Solubility: Sol. warm in toluene; disp. warm in white spirit; HLB: 2.8

Grindtek MSP 40F. Grindtek
Chemical Description: Stearic/palmitic fatty acid ester of glycerol
Category and Applications: See Grindtek MM 90
Form: Fine powder; Color: Whitish
Properties: Solubility: Sol. warm in toluene; disp. warm in white spirit; HLB: 2.8

Grindtek MSP 52. Grindtek
Chemical Description: Stearic/palmitic fatty acid ester of glycerol
Category and Applications: See Grindtek MM 90
Form: Powder; Color: Whitish
Properties: Solubility: Sol. warm in toluene; disp. warm in white spirit; HLB: 3.8

Grindtek MSP 90. Grindtek
Chemical Description: Stearic/palmitic fatty acid ester of glycerol
Category and Applications: See Grindtek MM 90
Form: Powder; Color: Whitish
Properties: Solubility: Sol. warm in ethanol, toluene; HLB: 4.3

Grindtek PGE 25. Grindtek
Chemical Description: Stearic/palmitic/oleic/linoleic fatty acid ester

of polyglycerol
Category: Emulsifier, antifogging agent
Applications: O/W emulsifier; antifogging agent for plastics
Form: Liquid; Color: Yellow-brown
Properties: Solubility: Sol. in ethanol; sol. warm in peanut oil, toluene, paraffin oil; disp. in water, propylene glycol; HLB: 5.5

Grindtek PGE 55. Grindtek
Chemical Description: Stearic/palmitic fatty acid ester of polyglycerol
Category and Applications: See Grindtek PGE 25
Form: Powder; Color: Whitish
Properties: Solubility: Sol. warm in toluene, paraffin oil; partly sol. warm in ethanol, peanut oil, white spirit; HLB: 6.8

Grindtek PGE 55-6. Grindtek
Chemical Description: Stearic/palmitic fatty acid ester of polyglycerol
Category and Applications: See Grindtek PGE 25
Form: Block; Color: Tan
Properties: Solubility: Sol. warm in ethanol; partly sol. warm in propylene glycol, toluene, white spirit; HLB: 7.4

Grindtek PGE-DSO. Grindtek
Chemical Description: Oleic/linoleic fatty acid ester of polyglycerol
Category: Emulsifier, antifogging agent
Applications: O/W emulsifier, anti fogging agent for plastics; effective for W/O emulsification with vegetable oils
Form: Viscous liquid; Color: Yellow-brown
Properties: Solubility: Sol. in toluene; sol. warm in peanut oil, white spirit, paraffin oil (partially); HLB: 2.8

Grindtek PGMS 90. Grindtek
Chemical Description: Stearic/palmitic fatty acid ester of propylene glycol.
Category: Emulsifier
Applications: W/O emulsifier
Form: Block; Color: Whitish
Properties: Solubility: Sol. in ethanol, toluene; sol. warm in propylene glycol, peanut oil, white spirit, paraffin oil; HLB: 3.5

Grindtek PK 60. Grindtek
Chemical Description: Lauric/myristic fatty acid ester of glycerol
Category: Lubricant, antistat, antifogging agent
Applications: Components in W/O and O/W creams; lubricants for processing of polyvinylchloride, polystyrene and copolymers of styrene; antistat and antifogging agents in polyolefins
Form: Block; Color: Whitish
Properties: Solubility: Sol. in ethanol; sol. warm in peanut oil, toluene, paraffin oil; partly sol. in propylene glycol; HLB: 4.1

Grindtek SMS. Grindtek
Chemical Description: Stearic/palmitic fatty acid ester of glycerol

Category: Emulsifier
Applications: O/W and W/O emulsifier
Form: Powder; Color: Tan
Properties: Solubility: Sol. warm in toluene; partly sol. warm in ethanol, white spirit, paraffin oil; HLB: 5.2

Grindtek STS. Grindtek
Chemical Description: Stearic/palmitic fatty acid ester of glycerol
Category: Emulsifier
Applications: O/W and W/O emulsifier
Form: Powder; Color: Tan
Properties: Solubility: Sol. warm in toluene, white spirit; partly sol. warm in ethanol, peanut oil, paraffin oil; HLB: 2.3

Grocor 1000. A. Gross & Co.
Chemical Description: Glycerol trioleate
Category: Wetting agent, emulsifier, nonionic
Applications: Sulfated product is a surfactant for textiles, leathers; desizing and boiling off cottons; sizing; rewetting dried skins; printing
Form: Liquid; Color: 6 max. (Gardner); Odor: Mild
Properties: Ref. Index: 1.4670; Solubility: Sol. in oils and hydrocarbons; Sp. gr.: 0.9145; Visc.: 190–205 saybolt seconds @ 100 F; Cloud Pt.: 18 F max.; Stability: Good
Composition: 97% min. active
Reaction: Neutral
Containers: SS drums or aluminum tank cars

Grocor 1200. A. Gross & Co.
Chemical Description: Glycerol trioleate
Category: Wetting agent, emulsifier, nonionic
Applications: Surfactant for textiles and leathers; desizing and boiling off cottons; sizing; rewetting dried skins; printing
Form: Liquid; Color: Amber; Odor: Mild
Properties: Solubility: Sol. in oils and hydrocarbons; Sp. gr.: 0.945; Visc.: 190–205 SUS @ 100 F; Cloud Pt.: 5 F max.
Composition: 97% min. active
Reaction: Neutral
Containers: SS drums, aluminum tank cars and tank wagons

Grocor 2000. A. Gross & Co.
Chemical Description: Glycerol monooleate
Category: Wetting agent, emulsifier, penetrant, nonionic
Applications: Textile finishing compounds; specialty lubricants; vinyl stabilizers; odorless paints base
Form: Liquid; Color: Pale yellow; Odor: Mild
Properties: Ref. Index: 1.4689; Solubility: Sol. in oils and hydrocarbons; Sp. gr.: 0.949; Visc.: 375 SUS @ 100 F; Cloud Pt.: 5 F
Composition: 97.5% active ingred.
Reaction: Neutral
Containers: SS drums

Grocor 2500. A. Gross & Co.
Chemical Description: Glycerol dioleate

Category: Emulsifier, thickener, stabilizer, nonionic
Applications: Auxiliary emulsifier and emulsion stabilizer for cosmetics, textiles, paints
Form: Liquid; Color: Yellow; Odor: Mild
Properties: Solubility: Sol. in oils and hydrocarbons; Sp. gr.: .93; Visc.: 90 cps; F.P.: 0 C
Composition: 95% min. active
Reaction: Neutral
Containers: SS drums

Grocor 3000. A. Gross & Co.
Chemical Description: Propyl oleate
Category: Wetting agent, stabilizer, nonionic
Applications: Chemical specialties, textiles (desizing and boiling off cottons, sizing), leather finishing, cosmetics, wetting pulp in beater
Form: Liquid; Color: Light yellow; Odor: Mild
Properties: Solubility: Sol. in IPA, oils, etc.; Sp. gr.: 0.89; F.P.: 0 C
Composition: 97% min. active
Reaction: Neutral
Containers: Bulk or drums

Grocor 4000. A. Gross & Co.
Chemical Description: Butyl oleate Cat
Category: Wetting agent, plasticizer, nonionic
Applications: Textile wetting agent, finishing, wetting pulp in beater; plasticizer for chloroprene and vinyl chloride; leather treatment
Form: Liquid; Color: Slightly yellow
Properties: Ref. Index; 1.45; Solubility: Sol. in IPA, hydrocarbons, mineral oil; Sp. gr.: 0.86; F.P.: –20 C; B.P.: 170 C
Composition: 97% min. active
Reaction: Neutral
Containers: Bulk or drums

Grocor 5221 SE. A. Gross & Co.
Chemical Description: Diglycol stearate SE
Category: Wetting agent, emulsifier, plasticizer, lubricant, opacifying agent, anionic
Applications: Textile lubricant and finishing; bodying and opacifying agent for cosmetics
Form: Solid, beads, flake; Color: White
Properties: Solubility: Sol. in hydrocarbons; disp. hot in water; Sp. gr.: 0.96; Softening Pt.: 46–56 C
Composition: 100% active
Reaction: Neutral
Containers: Bulk or drums

Grocor 5410. A. Gross & Co.
Chemical Description: Butyl stearate
Category: Plasticizer, lubricant, defoamer, nonionic
Applications: Lubricant and spinning compound for textiles; paper defoamer; cutting oils for metals; finishing; wetting pulp in beater
Form: Liquid; Color: Water white; Odor: Mild
Properties: Ref. Index: 1.4328; Solubility: Sol. in IPA, toluol, oils; Sp. gr.: 0.855; Visc.: 10 cs; F.P.: 20 C

Composition: 98% min. active
Reaction: Neutral
Containers: Bulk or drums

Grocor 5500. A. Gross & Co.
Chemical Description: Glycerol monostearate
Category: Emulsifier, thickener, opacifier, nonionic
Applications: Opacifier in cosmetics; lubricant softener for textiles; dispersant for dry powders
Form: Solid, powder, beads, flake; Color: White; Odor: Mild
Properties: Solubility: Sol. in IPA; Sp. gr.: 0.97; M.P.: 56–59 C
Composition: 100% active ingred.
Reaction: Neutral
Containers: Bulk, drums

Grocor 5510. A. Gross & Co.
Chemical Description: Butyl stearate
Category: Wetting agent, plasticizer, lubricant, nonionic
Applications: Lubricating, softening and spinning compounds for the textile industry; defoamer for paper and paperboard; lubricants and cutting oils for metals; plasticizer for chlorinated rubbers; used in cosmetics and coatings
Form: Liquid; Color: Water white; Odor: Mild
Properties: Ref. Index: 1.4328; Solubility: Sol. in IPA and toluol; Sp. gr.: 0.855; Visc.: 10 cs; F.P.: 20 C
Composition: 98% min. active
Reaction: Neutral
Containers: Bulk or drums

Grocor 5721. A. Gross & Co.
Chemical Description: Tridecyl stearate
Category: Wetting agent, emulsifier, lubricant, nonionic
Applications: Dye carrier and textile lubricant; wetting agent for pigments and for coating dispersing dry powders
Form: Liquid; Color: Water white
Properties: Sp. gr.: 0.86; F.P.: 6 C
Composition: 100% conc.
Reaction: Neutral

Grocor 5820 G. A. Gross & Co.
Chemical Description: 2-hexyldecyl stearate
Category: Wetting agent, emollient, lubricant, nonionic
Applications: Emollient, lubricant and intermediate for cosmetics, textiles, printing and paints
Form: Liquid; Color: Water white; Odor: Mild
Properties: Sp. gr.: 0.853; Visc.: 90 SUS @ 100 F; F.P.: –7 C
Composition: 98% min. active
Reaction: Neutral
Toxicity: May cause eye irritation
Containers: Bulk or drums

Grocor 6000. A. Gross & Co.
Chemical Description: Glycerol monostearate
Category: Emulsifier, lubricant, plasticizer, suspending agent, nonionic
Applications: Lubricant and plasticizer for textile, rubber, paper, cosmetic and chemical specialties *industries*
Form: Solid, beads; Color: White;

Odor: Mild
Properties: Solubility: Sol. in IPA; Sp. gr.: 0.91; M.P.: 58 C
Reaction: Neutral
Containers: Paper bags, drums, or bulk

Grocor 6000 E. A. Gross & Co.
Chemical Description: Food grade glycerol monostearate
Category: Emulsifier, stabilizer, plasticizer, nonionic
Applications: Emulsifier, stabilizer, plasticizer, antitack agent for candy, chewing gum, ice cream and other foods; also edible lubricant
Form: Solid, beads; Color: White; Odor: Mild
Properties: Ref. Index: 0.910; Solubility: Sol. in IPA; M.P.: 58 C
Composition: 100% conc.
Reaction: Neutral
Containers: Bulk, drums, bags

Grocor 6000 SE. A. Gross & Co.
Chemical Description: Glycerol monostearate, self-emulsifying
Category: Emulsifier, lubricant, plasticizer, anionic
Applications: Cosmetics, textile, rubber, plastic specialty industries
Form: Solid, beads; Color: White; Odor: Bland
Properties: Solubility: Disp. in water; sol. in IPA; Sp. gr.: 0.92; M.P.: 58 C
Composition: 100% conc.
Reaction: Acid
Containers: Bags or drums

Grocor 8002. A. Gross & Co.
Chemical Description: Methyl oleate
Category: Wetting agent, lubricant, plasticizer, nonionic
Applications: Textiles, chemical specialties, cosmetics, leather, plastics, dyeing, sizing
Form: Liquid; Color: Lt. yellow; Odor: Mild
Properties: Solubility: Sol. in IPA; Sp. gr.: 0.872; Cloud Pt.: 0 C
Composition: 98% min. active
Reaction: Neutral
Containers: Bulk or drums

Grocor 8008. A. Gross & Co.
Chemical Description: Methyl oleate
Category: Wetting agent, lubricant, plasticizer, nonionic
Applications: Textiles, plastics, chemical specialties, cosmetics, leather, dyeing
Form: Liquid; Color: Light yellow
Properties: Solubility: Sol. in IPA; Sp. gr.: 0.872; Cloud Pt.: 0 C
Composition: 98% min. active
Reaction: Neutral
Containers: Bulk or drums

Grocor 8041. A. Gross & Co.
Chemical Description: Methyl tallowate
Category: Detergent, lubricant, plasticizer, nonionic
Applications: Textiles, leather and wire drawing
Form: Liquid; Color: Water white
Properties: F.P.: 15 C max.

Composition: 98% min. active
Reaction: Neutral
Containers: Bulk or drums

Grocor 8058. A. Gross & Co.
Chemical Description: Methyl stearate
Category: Detergent, lubricant, plasticizer, nonionic
Applications: Textiles, leather, wire drawing
Form: Liquid, gel; Color: Water white
Properties: Solubility: Sol. in IPA; Sp. gr.: 0.860; F.P.: 32 C max.
Composition: 98% min. active
Reaction: Neutral
Containers: Bulk or drums

Grocor PEG 300. A. Gross & Co.
Chemical Description: Monooleate
Category: Detergent, wetting agent, emulsifier, lubricant, nonionic
Applications: Textile lubricant, dyeing and finishing, paper, cosmetics, leather and chemical specialties
Form: Liquid; Color: Light yellow; Odor: Mild
Properties: Sp. gr.: 0.97; Visc.: 170 SUS @ 100 F
Composition: 98% min. active
Reaction: Neutral
Containers: Bulk or drums

Grotex 1120. A. Gross & Co.
Chemical Description: Self-emulsifying softener
Category: Detergent, wetting agent, lubricant, thickener, nonionic
Applications: Self-emulsifying softener, suspending agent in textiles
Form: Solid, beads; Color: White; Odor: Mild
Properties: Solubility: Dispersible in water; Sp. gr.: 0.92; M.P.: 52 C
Composition: 100% conc.
Reaction: Slightly acid
Containers: Fiber drums

H

Hallcomid M-6. C. P. Hall Co.
Chemical Description: N,N-Dimethyl caproamide
Category: Detergent, de-emulsifier, penetrant, solubilizer, plasticizer, softening agent
Applications: Solvents used for insecticides, plasticizers, softening agents, cleaning aids, pain removers, metal cleaners, gas absorber, de-icer, herbicides, corrosion inhibitor, degassing agent, antifoaming agent, adhesives, agricultural and automotive specialties, coatings, cosmetics, dispersions, drilling muds, textiles, germicides
Form: Liquid; Color: Colorless to very pale yellow; Odor: Low
Properties: m.w.: 143; Solubility: Completely soluble in common organic solvents, hydrocarbons, plasticizers, fatty acids, waxes, vegetable oils, esters, water; Visc.: 7.0 cps; F.P.: −40 C; B.P.: 83–89 C @ 3 mm Hg; Flash Pt.: 190 F; Fire Pt.: 208 F
Composition: 95% N,N-Dimethyl amide

Hallcomid M 8-10. C.P. Hall Co.
Chemical Description: N,N-dimethyl caprylamide-capramide
Category: Solvent, corrosion inhibitor, nonionic
Applications: Mutual solvent for polar and nonpolar ingredients in cosmetics, cleaners; corrosion inhibitor
Form: Liquid; Color: Colorless to very pale yellow
Properties: m.w.: ≈ 171–199; Solubility: 0.17 g/100 g water; sol. in most common organic solvents, hydrocarbons, plasticizers, fatty acids, and vegetable oils; Sp. gr.: 0.880; Visc.: 10.5 cps; F.P.: −21 C; B.P.: 115–170 C (3 mm Hg); Flash Pt.: 245 F
Composition: 95% conc.
References: Tech. data 827921

Hallcomid M-12. C.P. Hall Co.
Chemical Description: N,N-dimethyl lauramide
Category: Hydrotrope, solubilizer, corrosion inhibitor, nonionic
Applications: Hydrotrope; solubilizer for variety of polar and non-

polar emulsifiers, dyes, perfumes; corrosion inhibitor
Form: Liquid; Color: Colorless to very pale yellow
Properties: m.w.: ≈ 227; Solubility: < 0.2 g/100 g water @ 20 C; sol. in most common organic solvents, hydrocarbons, plasticizers, fatty acids, and vegetable oils; Sp. gr.: 0.872; Visc.: 14.5 cps; F.P.: 14–16 C; B.P.: 147–163 C (3 mm Hg); HLB: 8.0; Flash Pt.: 292 F
Composition: 95% conc.
References: Tech. data 827922

Hallcomid M-18. C.P. Hall Co.
Chemical Description: N,N-dimethyl stearamide
Category: Solubilizer, corrosion inhibitor, nonionic
Applications: Solubilizes easily, improves viscosity of shampoos and controls foam; corrosion inhibitor
Form: Solid; Color: Colorless to very pale yellow
Properties: m.w.: ≈ 312; Solubility: < 0.1 g/100 g water @ 20 C; sol. in most common organic solvents, hydrocarbons, plasticizers, fatty acids, and vegetable oils; Sp. gr.: 0.851; Visc.: 13.5 cps (40 C); F.P.: 38–40 C; B.P.: 194–214 C (3 mm Hg); HLB: 7.0; Flash Pt.: 399 F
Composition: 95% conc.
References: Tech. data 827923

Hallcomid M 18-OL. C.P. Hall Co.
Chemical Description: N,N-dimethyl oleamide
Category: Solubilizer, solvent, dispersant, wetting agent, nonionic
Form: Liquid; Color: Slightly yellow
Properties: m.w.: ≈ 310; Solubility: < 0.1 g/100 g water @ 20 C; sol. in moxt common organic solvents, hydrocarbons, plasticizers, fatty acids, and vegetable oils; Sp. gr.: 0.875; Visc.: 18.5 cps; F.P.: –8 C; B.P.: 196–211 C (3 mm Hg); HLB: 7.0; Flash Pt.: 410 F
Composition: 95% conc.
References: Tech. data 827924

Halvopon FS. Albright & Wilson
Chemical Description: Specially formulated catalyst
Category: Fat splitter
Applications: Used for the splitting operation yielding a good quality fatty acid
Form: Liquid; Color: Brown; Odor: Typical

Hamposyl C. W.R. Grace & Co.
Chemical Description: Cocoyl sarcosine
Category: Detergent, wetting and foaming agents, anionic
Applications: Used for hair and rug shampoos, cosmetics and compatible with many cationics
Form: Liquid; Color: 3 (Gardner)
Properties: Solubility: Sol. in most organic solvents; Sp. gr.: 0.97–0.99; M.P.: 23–26 C; Biodegradable
Composition: 94% active
Reaction: Acid
Containers: 55 gal drums
References: Hamposyl brochure

Hamposyl C-30. W.R. Grace & Co.
Chemical Description: Sodium cocoyl sarcosine
Category: Detergent, wetting and foaming agents, anionic
Applications: Used for hair and rug shampoos, cosmetics and compatible with many cationics
Form: Liquid; Color: 100 (APHA)
Properties: Solubility: Miscible in water; Sp. gr.: 1.02–1.03; Visc.: 30 cps; F. P.: –1 C; Surface Tension: 30 dynes/cm; Biodegradable
Composition: 30% active
Reaction: Neutral
Toxicity: 14,000 mg/kilo of body weight
Containers: 55 gal drums
References: Hamposyl brochure

Hamposyl L. W. R. Grace & Co.
Chemical Description: Layroyl sarcosine
Category: Detergent, wetting and foaming agents, anionic
Applications: Used for hair and rug shampoos, cosmetics and compatible with many cationics
Form: Solid; Color: 2 (Gardner)
Properties: Solubility: Sol. in most organic solvents; Sp. gr.: 0.97–0.99; M.P.: 34–37 C; Biodegradable
Composition: 94% active
Reaction: Acid
Containers: 55 gal drums
References: Hamposyl brochure

Hamposyl L-30. W. R. Grace & Co.
Chemical Description: Sodium lauroyl sarcosine
Category: Detergent, wetting and foaming agents, anionic
Applications: Used for hair and rug shampoos, cosmetics and compatible with many cationics
Form: Liquid; Color: 60 (APHA)
Properties: Solubility: Miscible in water; Sp. gr.: 1.02–1.03; Visc.: 30 cps; F. P.: –1 C; Surface Tension: 30 dynes/cm; Biodegradable
Composition: 30% active
Reaction: Neutral
Toxicity: LD 50 15,000 mg/kilo of body weight
Containers: 55 gal drums
References: Hamposyl brochure

Hamposyl L-95. W. R. Grace & Co.
Chemical Description: Sodium lauroyl sarcosine
Category: Detergent, wetting and foaming agents, anionic
Applications: Used for hair and rug shampoos, cosmetics, toothpaste, and compatible with many cationics
Form: Powder; Color: White
Properties: Solubility: Sol. in water; Density: 25 lbs/ft^3; Biodegradable
Composition: 94% active
Reaction: Neutral
Containers: 55 gal drums
References: Hamposyl brochure

Hamposyl M. W.R. Grace & Co.
Chemical Description: Myristoyl sarcosine
Category: Detergent, wetting and foaming agents, anionic
Applications: Used for hair and rug shampoos, cosmetics and compatible with many cationics
Form: Solid; Color: 2 (Gardner)

Properties: Solubility: Sol. in most organic solvents; Sp. gr.: 0.97–0.99; M.P.: 48–53 C; Biodegradable
Composition: 94% active
Reaction: Acid
Containers: 55 gal drums
References: Hamposyl brochure

Hamposyl M-30. W.R. Grace & Co.
Chemical Description: Sodium myristoyl sarcosine
Category: Detergent, wetting and foaming agents, emulsifier, anionic
Applications: Used for hair and rug shampoos, cosmetics and compatible with many cationics
Form: Liquid; Color: 80 (ALPHA)
Properties: Solubility: Miscible in water; Sp. gr.: 1.02–1.03; Visc.: 30 cps; F. P.: −1 C; Surface Tension: 30 dynes/cm; Biodegradable
Composition: 30% active
Reaction: Neutral
Containers: 55 gal drums
References: Hamposyl brochure

Hamposyl O. W.R. Grace & Co.
Chemical Description: Oleoyl sarcosine
Category: Detergent, wetting and foaming agents, anionic
Applications: Used for hair and rug shampoos, cosmetics, corrosion inhibitors and compatible with many cationics
Form: Liquid; Color: 4 (Gardner)
Properties: Solubility: Sol. in most organic solvents; Sp. gr.: 0.95–0.97; Visc.: 250 cps; Biodegradable
Composition: 94% active
Reaction: Acid
Containers: 55 gal drums
References: Hamposyl brochure

Hamposyl S. W.R. Grace & Co.
Chemical Description: Stearoyl sarcosine
Category: Detergent, wetting and foaming agents, anionic
Applications: Used for hair and rug shampoos, cosmetics and compatible with many cationics
Form: Solid; Color: 4 (Gardner)
Properties: Solubility: Sol. in most organic solvents; Sp. gr.: 0.96–0.98; M.P.: 53–58 C; Biodegradable
Composition: 94% active
Reaction: Acid
Containers: 55 gal drums
References: Hamposyl brochure

Haroil SCO-65. Graden Chem.
Chemical Description: Sulfated castor oil
Category: Superfatting agent, emulsifier, plasticizer, anionic
Applications: A superfatting agent for cosmetic creams, soaps; a good emulsifier for use in cosmetic formulations; used as a plasticizer for animal glue, starch and casein adhesives
Form: Liquid; Color: Amber
Properties: pH: 7.0 ± 0.5 (2% aq. sol'n.)
Composition: 50% active
Containers: 55 gal drums

Haroil SCO-7525. Graden Chem.
Chemical Description: Sulfated castor oil
Category: Antisagging, antisettling

agent, anionic
Applications: Will prevent sagging of paint film and hard settling of pigments during storage; improves the brushing properties of paints; used in water-based shingle stains to prevent settling
Form: Liquid; Color: Amber
Properties: Density: 1.025 @ 68 F
Composition: 75% active
Containers: 55 gal drums

Harol D. Graden Chem.
Chemical Description: Highly polymerized naphthalene sulfonate
Category: Stabilizer, viscosity depressant, dispersing agent, anionic
Applications: Used in latex and casein paper coating compounds; good dispersing agent for carbon black, clay dispersions and calcium carbonate dispersions
Form: Powder
Properties: Density: 47 lb/ft^3; Surface Tension: 71 dynes/cm (1% sol'n. @ 20 C); pH: 9 (1% sol'n. @ 20 C)
Composition: 83–85% active

Harol RG-71. Graden Chem.
Chemical Description: Sodium salt of polymerized alkyl naphthalene sulfonic acid
Category: Dye retardant, leveler, pitch dispersant, anionic
Applications: Provides freedom from dusting and faster solubility than powdered dispersants of this type
Form: Powder
Properties: Density: 35–37 lb/ft^3; Stability: Stable in the presence of mild acids and alkalies; Surface Tension: 70–72 dynes/cm (1% sol'n.) @ 72 F; pH: 7.8–10.4 (1% sol'n.) @ 72 F
Compositon: 95% active

Hartaine CB-40. Hart Products
Chemical Description: Cocobetaine
Category: Detergent, wetting agent, emulsifier
Applications: Detergent–textiles; dyeing; household; cosmetics; Wetting agent–desizing and boiling off cottons; bleaching; removal of spots; Emulsifier–dispersing waxes, pigments, resins; printing; dispersing dry powders
Form: Liquid; Color: Very light
Properties: Solubility: Sol. in water; Density: 8.85 lb/gal; Sp. gr.: 1.06
Composition: 40% active
Containers: Drums

Hartamide AD. Hart Products
Chemical Description: Coconut alkanolamide
Category: Detergent, wetting agent, emulsifier, thickener, nonionic
Applications: Detergent–textiles; laundering; dyeing; household; cosmetics; cleaners of street, buildings, hospitals; Wetting agent–dyeing; Emulsifier–making of cutting and soluble oils; dispersing waxes, pigments, resins; Thickener
Form: Liquid; Color: Amber;

Odor: Fatty
Properties: Solubility: Sol. in water
Composition: 100% active
Containers: Drums

Hartamide CE 80. Hart Products
Chemical Description: Lauric-myristic super amide
Category: Detergent, wetting agent, emulsifier, thickener, nonionic
Applications: Detergent–textiles; laundering; dyeing; household; cosmetics; cleaners of street, buildings, hospitals; Wetting agent–dyeing; Emulsifier–making of cutting and soluble oils; dispersing waxes, pigments, resins; Thickener
Form: Liquid; Color: Light amber; Odor: Fatty
Properties: Solubility: Sol. in water
Composition: 100% active
Reaction: Acid
Containers: Drums

Hartamide CE-90. Hart Products
Chemical Description: High purity coconut fatty acid alkanolamide
Category: Detergent, wetting agent, emulsifier, thickener, nonionic
Applications: Detergent–textiles; laundering; dyeing; household; cosmetics; Wetting agent–dyeing; Emulsifier–making of cutting and soluble oils; dispersing waxes, pigments, resins; Thickener
Form: Liquid; Color: Light amber; Odor: Fatty
Properties: Solubility: Sol. in water
Composition: 100% active
Reaction: Acid
Containers: Drums

Hartamide LDA 70. Hart Chem.
Chemical Description: Lauric diethanolamide
Category: Detergent, nonionic
Applications: Foam stabilizer and viscosity regulator for liquid and powdered detergent systems
Form: Solid; Color: Clear
Composition: 100% active
Containers: Drums/T/L
Warnings and Caution: Avoid contact with skin and eyes

Hartamide LDA 90. Hart Chem.
Chemical Description: Lauric diethanolamide
Category: Detergent, nonionic
Applications: Foam stabilizer and viscosity regulator for liquid and powdered detergent systems
Form: Solid; Color: Clear
Composition: 100% active
Containers: Drums/T/L
Warnings and Caution: Avoid contact with skin and eyes

Hartamide LE-90. Hart Products
Chemical Description: High purity lauric alkanolamide
Category: Detergent, wetting agent, emulsifier, thickener, nonionic
Applications: Detergent–textiles; laundering; dyeing; household, cosmetics; Wetting agent–dyeing; Emulsifier–making of cutting and soluble oils; dispersing waxes, pig-

ments, resins; Thickener
Form: Solid; Color: Light; Odor: Fatty
Properties: Solubility: Sol. in water
Composition: 100% active
Reaction: Acid
Containers: Drums

Hartamide LM 11D. Hart Products
Chemical Description: High purity lauric myristic alkanolamide
Category: Detergent, wetting agent, emulsifier, thickener, nonionic
Applications: Detergent–textiles; laundering; dyeing; household; cosmetics; Wetting agent–dyeing; Emulsifier–making of cutting and soluble oils; dispersing waxes, pigments, resins; Thickener
Form: Gel/Solid; Color: Light; Odor: Fatty
Properties: Solubility: Sol. in water
Composition: 100% active
Reaction: Acid
Containers: Drums

Hartamide OD. Hart Chem.
Chemical Description: Coconut diethanolamide
Category: Detergent, nonionic
Applications: Foam stabilizer and viscosity regulator for liquid and powdered detergent systems
Form: Liquid; Color: Yellow
Properties: Density: 1.02 lb/gal; Visc.: 900 cps
Composition: 100% active
Reaction: Alkaline
Containers: Drums/T/L
Warnings and Caution: Avoid contact with skin and eyes

Hartamide OS. Hart Products
Chemical Description: Coco-oleic super amide
Category: Detergent, wetting agent, emulsifier, nonionic
Applications: Detergent–textiles; laundering; dyeing; household; cosmetics; Wetting agent–dyeing; Emulsifier–making of cutting and soluble oils; dispersing waxes, pigments, resins
Form: Liquid; Color: Light amber; Odor: Fatty
Properties: Solubility: Sol. in water
Composition: 100% active
Reaction: Acid
Containers: Drums

Hartenol LAS-30. Hart Products
Chemical Description: Sodium lauryl sulfate
Category: Detergent, wetting agent, emulsifier, anionic
Applications: Detergent–textiles; skins; leather; laundering; dyeing; cosmetics; cleaners of street, buildings, hospitals; Wetting agent–desizing and boiling off cottons; carbonizing woolens; bleaching; dyeing; rewetting dried skins; rotary washer; abatement of dust; removing starches, wall paper; cleaning hangings; Emulsifier–dispersing waxes, pigments, resins; printing; dispersing dry powders

Form: Liquid; Color: Very light
Composition: 30% active
Containers: Drums

Hartenol LES-60. Hart Products
Chemical Description: Sol lauryl sulfate
Category: Detergent, wetting agent, emulsifier, anionic
Applications: Detergent–textiles; dyeing; household; cosmetics; cleaners of street, buildings, hospitals; Wetting agent–desizing and boiling off cottons; bleaching; dyeing; Emulsifier–dispersing waxes, pigments, resins; dispersing dry powders
Form: Liquid; Color: Light
Composition: 60% active
Reaction: Acid
Containers: Drums

Hartex San Q 50. Hart Chem.
Chemical Description: Lauryl dimethylbenzyl ammonium chloride
Category: Sanitizer, cationic
Applications: Germicide, sanitizer, and disinfectant; very efficient in the presence of hard water salts
Form: Liquid; Color: Amber
Properties: Ref. Index: 1.4412; Density: 0.980 lb/gal; Visc.: 60 cps
Composition: 50% active
Reaction: Neutral
Containers: Drums/T/L
Warnings and Caution: Avoid contact with skin and eyes

Hartex V63. Hart Chem.
Chemical Description: Sulfonated castor oil
Category: Penetrant, plasticizer, lubricant, emulsifier, anionic
Applications: Emulsifier for terpineols, mineral and fatty acid
Form: Liquid; Color: Amber
Properties: Ref. Index: 1.4448; Density: 1.056 lb/gal; Visc.: 520 cps
Composition: 70% active
Containers: Drums/T/L
Warnings and Caution: Avoid contact with skin and eyes

Hartex V64. Hart Chem.
Chemical Description: Sulfonated castor oil
Category: Penetrant, plasticizer, lubricant, emulsifier, anionic
Applications: Emulsifier for terpineols, mineral and fatty acid
Form: Liquid; Color: Amber
Properties: Density: 1.02 lb/gal; Visc.: 250 cps
Composition: 45% active
Containers: Drums/T/L
Warnings and Caution: Avoid contact with skin and eyes

Hartofix 2X. Hart Chem.
Chemical Description: Cationic resin
Category: Dye auxilaries, cationic
Applications: Fixative for direct dyes; improves the fastness of most substantive colors to water and soaps
Form: Liquid; Color: Clear
Properties: Ref. Index: 1.4406; Density: 1.17 lb/gal; Visc.: 20 cps
Composition: 40% active
Reaction: Acid

Containers: Drums/T/L
Warnings and Caution: Avoid contact with skin and eyes

Hartolite. Croda Chem. Ltd.
Chemical Description: Fraction of lanolin alcohols
Category: Emollient, skin conditioner, moisturizing agent, nonionic
Applications: Highly recommended as the primary emulsifier in W/O creams based on hydrocarbon oils and waxes
Form: Soft waxy solid; Color: Pale Yellow
Properties: Stability: Stable to W/O
Composition: 100% active ingred.

Hartolon 368. Hart Chem.
Chemical Description: Fatty acid esters
Category: Softener, nonionic
Applications: Effective on cotton corduroy and cotton yarn where increased suppleness is needed
Form: Liquid; Color: Amber
Properties: Density: 0.9 lb/gal
Composition: 95% active ingred.
Reaction: Neutral
Containers: Drums/T/L T/T
Warnings and Caution: Avoid contact with skin and eyes

Hartolon AL. Hart Chem.
Chemical Description: Fatty acid condensate
Category: Softener, nonionic
Applications: Provides maximum softening features coupled with excellent antistatic properties on all yarns
Form: Paste; Color: White
Composition: 95% active ingred.
Reaction: Neutral
Containers: Drums/T/L T/T
Warnings and Caution: Avoid contact with skin eyes

Hartolon NA. Hart Chem.
Chemical Description: Blend of saturated esters and alkyl condensates
Category: Softener, nonionic
Applications: Provides good lubricity and antistatic properties with exceptional resistance to cross staining of dyed polyester/cotton blends
Form: Paste; Color: White
Composition: 25% active ingred.
Reaction: Neutral
Containers: Drums/T/L T/T
Warnings and Caution: Avoid contact with skin and eyes

Hartonyl L531. Hart Chem.
Chemical Description: Ethoxylated fatty amine
Category: Levelling aid, dispersing agent, cationic
Applications: Levelling aid for dyeing polyamides; migrating levelling and dispersing agent for acid and disperse dye stuffs; increases contrast and clarity between nylon fibers of different affinities; effective in both continuous and winch dyeing of carpets
Form: Liquid; Color: Yellow
Properties: Ref. Index: 1.407
Composition: 45% active ingred.
Reaction: Alkaline

Containers: Drums/T/L
Warnings and Caution: Avoid contact with skin and eyes

Hartonyl L535. Hart Chem.
Chemical Description: Ethoxylated fatty amine
Category: Dyeing assistant, antiprecipitant, dispersant, cationic
Applications: Dyeing assistant for levelling and dispersing of acid and disperse dye stuffs; promotes contrast with regular and deep dyeable nylon
Form: Liquid; Color: Yellow
Properties: Density: 1.025 lb/gal
Composition: 33% active ingred.
Reaction: Acid
Containers: Drums/T/L
Warnings and Caution: Avoid contact with skin and eyes

Hartonyl L537. Hart Chem.
Chemical Description: Ethoxylated fatty amine
Category: Levelling and retarding agents, cationic
Applications: Levelling and retarding agent for acid dyes; used for dyeing of regular/deep dyeable nylon carpet; effective in both continuous and winch dyeing
Form: Liquid; Color: Yellow
Properties: Ref. Index: 1.412; Density: 0.999 lb/gal
Composition: 60% active ingred.
Reaction: Alkaline
Containers: Drums/T/L
Warnings and Caution: Avoid contact with skin and eyes

Hartonyl W504. Hart Chem.
Chemical Description: Modified linear alcohol ethoxylate
Category: Wetting and foaming aid, nonionic
Applications: Wetting and foaming aid used in the continuous dyeing of carpets by the Kuster dyeing system; compatible with all dye stuff classes over a wide pH range; retains wetting properties even at high temperatures
Form: Liquid; Color: Clear
Composition: 85% active ingred.
Reaction: Neutral
Containers: Drums/T/L
Warnings and Caution: Avoid contact with skin and eyes

Hartopol 4270. Hart Chem.
Chemical Description: Polyoxyalkylene glycols
Category: Detergent, emulsifier, nonionic
Applications: Rinse aids, latex coatings, textile penetrant, felt washing and cutting fluids
Form: Liquid; Color: Clear
Properties: Ref. Index: 1.458; Visc.: 425 cps
Composition: 100% active ingred.
Reaction: Neutral
Containers: Drums/T/L T/T
Warnings and Caution: Avoid contact with skin and eyes

Hartopol 4277. Hart Chem.
Chemical Description: Polyoxyalkylene glycols
Category: Detergent, emulsifier, nonionic
Applications: Rinse aids, polyurethane, oil demulsifier
Form: Liquid; Color: Clear
Properties: Ref. Index: 1.460; Visc.: 275 cps

Composition: 100% active ingred.
Reaction: Neutral
Containers: Drums/T/L T/T
Warnings and Caution: Avoid contact with skin and eyes

Hartopol 4278. Hart Chem.
Chemical Description: Polyoxyalkylene glycols
Category: Detergent, emulsifier, nonionic
Applications: Cosmetics and pharmaceuticals, pulp and paper drainage, emulsion polymerization
Form: Liquid; Color: Clear
Properties: Ref. Index: 1.458; Density: 1.05 lb/gal; Visc.: 540 cps
Composition: 100% active ingred.
Reaction: Neutral
Containers: Drums/T/L T/T
Warnings and Caution: Avoid contact with skin and eyes

Hartopol 4299. Hart Chem.
Chemical Description: Polyoxyalkylene glycols
Category: Detergent, emulsifier, nonionic
Applications: Pulp and paper foam control, water treating, emulsion polymerization, felt washing
Form: Liquid; Color: Clear
Properties: Ref. Index: 1.454; Sp. gr.: 1.03; Visc.: 400 cps
Composition: 100% active ingred.
Reaction: Neutral
Containers: Drums/T/L T/T
Warnings and Caution: Avoid contact with skin and eyes

Hartopol LF-1. Hart Chem.
Chemical Description: Polyoxyalkylene propylene glycol condensate
Category: Detergent, emulsifier, nonionic
Applications: Low foaming for powdered dish detergents, rinse aids, and general industrial detergent applications
Form: Liquid; Color: Clear
Properties: Ref. Index: 1.452; Sp. gr.: 1.01; Visc.: 320 cps
Composition: 100% active ingred.
Reaction: Neutral
Containers: Drums/T/L T/T
Warnings and Caution: Avoid contact with skin and eyes

Hartopol LF-2. Hart Chem.
Chemical Description: Polyoxyalkylene propylene glycols condensate
Category: Detergent, emulsifier, nonionic
Applications: Low foaming for powdered dish detergents, rinse aids, and general industrial detergent applications
Form: Liquid; Color: Clear
Properties: Ref. Index: 1.458; Sp. gr.: 1.035; Visc.: 400 cps
Composition: 100% active ingred.
Reaction: Neutral
Containers: Drums/bulk
Warnings and Caution: Avoid contact with skin and eyes

Hartopol LF-4. Hart Chem.
Chemical Description: Polyoxyalkylene propylene glycol condensate
Category: Detergent, wetting agent, emulsifier, nonionic
Applications: Designed for use in automatic dishwashing detergents

Form: Liquid; Color: Clear
Properties: Sp. gr.: 1.035; Visc.: 375 cps
Composition: 100% active ingred.
Reaction: Neutral
Containers: Drums/T/T
Warnings and Caution: Avoid contact with skin and eyes

Hartopon FS. Hart Chem.
Chemical Description: Modified linear alcohol ethoxylate
Category: Detergent, nonionic
Applications: Detergent used for soaping off of dyestuffs; excellent scouring and wetting properties in both acid and alkaline solutions; effective emulsifying and soil suspending agent; may be used on a variety of natural and synthetic fibers
Form: Liquid; Color: Clear
Properties: Density: 1.04 lb/gal; Biodegradable
Composition: 75% active ingred.
Reaction: Neutral
Containers: Drums/T/L
Warnings and Caution: Avoid contact with skin and eyes

Hartosoft 75. Hart Chem.
Chemical Description: Quaternary imidazoline
Category: Antistatic, cationic
Applications: A nonyellowing cationic softener with good antistatic properties
Form: Liquid; Color: White
Properties: Density: 0.989; Visc.: 530 cps
Composition: 15% active ingred.
Reaction: Acid
Containers: Drums/T/L
Warnings and Caution: Avoid contact with skin and eyes

Hartosoft 171. Hart Chem.
Chemical Description: Blend of esters and quaternary amines
Category: Softener, cationic
Applications: Softener used to obtain a full and lofty hand
Form: Liquid; Color: White
Properties: Visc.: 1400 cps
Composition: 15% active ingred.
Reaction: Acid
Containers: Drums/T/L
Warnings and Caution: Avoid contact with skin and eyes

Hartosoft ORF. Hart Chem.
Chemical Description: Modified fatty nitrogenous compound
Category: Softener, cationic
Applications: Textile softener; generally applied to acrylic yarn
Form: Paste; Color: White
Composition: 100% active ingred.
Reaction: Acid
Containers: Drums/T/L
Warnings and Caution: Avoid contact with skin and eyes

Hartosoft TAF. Hart Chem.
Chemical Description: High molecular weight polyamine
Category: Softener, cationic
Form: Liquid; Color: Cream
Composition: 18% active ingred.
Reaction: Acid
Containers: Drums/T/L
Warnings and Caution: Avoid contact with skin and eyes

Hartotrope AXS 40. Hart Chem.
Chemical Description: Xylene sulfonate, ammonium salt
Category: Detergent, anionic
Applications: Solubilizer and cloud point depressant for light duty and built liquid detergent systems
Form: Liquid; Color: Clear
Properties: Density: 1.125 lb/gal
Composition: 40% active ingred.
Reaction: Neutral
Containers: Drums/T/T
Warnings and Caution: Avoid contact with skin and eyes

Hartotrope KTS 50. Hart Chem.
Chemical Description: Toluene sulfonate, potassium salt
Category: Detergent, anionic
Applications: See Hartotrope AXS 40
Form: Liquid; Color: Clear
Properties: Density: 1.24 lb/gal
Composition: 50% active ingred.
Reaction: Neutral
Containers: Drums/T/T
Warnings and Caution: Avoid contact with skin and eyes

Hartotrope STS 40. Hart Chem.
Chemical Description: Toluene sulfonate, sodium salt
Category: Detergent, anionic
Applications: See Hartotrope AXS 40
Form: Liquid; Color: Clear
Properties: Density: 1.19 lb/gal
Composition: 40% active ingred.
Reaction: Neutral
Containers: Drums/T/T
Warnings and Caution: Avoid contact with skin and eyes

Hartotrope STS Powder. Hart Chem.
Chemical Description: Toluene sulfonate, sodium salt
Category: Detergent, anionic
Applications: Solubilizer and cloud point depressant for light duty and built liquid detergent systems; anticaking agent in powdered detergents
Form: Powder; Color: White
Composition: 93% active ingred.
Reaction: Neutral
Containers: Bags
Warnings and Caution: Avoid contact with skin and eyes

Hartotrope SXS 40. Hart Chem.
Chemical Description: Xylene sulfonate, sodium salt
Category: Detergent, anionic
Applications: See Hartotrope AXS 40
Form: Liquid; Color: Clear
Properties: Density: 1.19 lb/gal
Composition: 40% active ingred.
Reaction: Neutral
Containers: Drums/T/T
Warnings and Caution: Avoid contact with skin and eyes

Hartox DMCD. Hart Products
Chemical Description: Cocodimethylamine oxide
Category: Detergent, wetting agent, cationic
Applications: Detergent–textiles; dyeing; cosmetics; Wetting agent–bleaching; dyeing
Form: Liquid
Composition: 40% active ingred.
Reaction: Neutral
Containers: Drums

Hercules CMC. Hercules
Chemical Description: Sodium carboxymethylcellulose
Category: Thickener, binder, stabilizer, protective colloid, suspending agent, anionic
Applications: Used in food systems, cosmetics, pharmaceuticals, paper products, adhesives, ceramics
Form: Liquid
Properties: Ref. Index: 1.3355 @ 2% sol'n.; Solubility: Sol. in cold or hot water; Density: 0.75 g/ml; Sp. gr.: 1.0068 @ 2% sol'n.; Surface Tension: 71 dynes/cm (1% sol'n.); pH: 7.5 @ 2% sol'n.
Composition: 99.5% active ingred.
References: Hercules brochure

Hercules AR 150. Hercules
Chemical Description: Polyethylene glycol ester
Category: Low-foaming, detergent, nonionic
Applications: Compatible with pine oil, quaternaries, and other germicidal agents; can be formulated with sulfate, alkaline, phosphate, silicate, and carbonate salts; can be used with most other nonionic, anionic, and cationic surfactants
Form: Liquid/Soft wax; Color: 10 (Gardner)
Properties: Solubility: Sol. in water, isopropanol, benzene, toluene, ester solvents; Visc.: 300 cps @ 38 C; Cloud Pt.: 60 C (1% sol'n.); Stability: Stable to acids and alkalies; Surface tension: 39 dynes/cm @ 0.01% in distilled water; pH: 9 (1% sol'n.)
References: Data sheet 767-4

Hetamide LA. Heterene Chem.
Chemical Description: Lauric acid
Category: Foam stabilizer, emulsifier, viscosity builder, detergent, nonionic
Applications: Used for car wash shampoos, bottle cleaning, cosmetics, dry cleaning soaps, degreasing compounds, dispersants, strippers, hand cleaners, floor cleaners, lubricants
Form: Liquid @ 20 C
Properties: Sp. gr.: 1.00 @ 20 C; Acid No.: 10–14; pH: 9–10 (1% aq. sol'n.)
Composition: 100% active ingred.

Hetamide MC. Heterene Chem.
Chemical Description: Coconut fatty
Category: Foam stabilizer, emulsifier, viscosity builder, detergent, nonionic
Applications: See Hetamide LA
Form: Liquid @ 20 C
Properties: Sp. gr.: 0.98 @ 20 C; pH: 9.5–10.5 (1% aq. sol'n.)
Composition: 100% active ingred.

Hetamide ML. Heterene Chem.
Chemical Description: Lauric acid
Category: Foam stabilizer, emulsifier, viscosity builder, detergent, nonionic
Applications: See Hetamide LA
Form: Solid @ 20 C
Properties: Sp. gr.: 0.96 @ 20 C; M.P.: 35 C
Composition: 100% active

Hetamide MMC. Heterene Chem.
Chemical Description: Mixed fatty acids
Category: Foam stabilizer, emulsifier, viscosity builder, detergent, nonionic
Applications: See Hetamide LA
Properties: Sp. gr.: 1.01 @ 20 C; pH: 9–10.5 (1% aq. sol'n.)
Composition: 100% active ingred.

Hetamide OC. Heterene Chem.
Chemical Description: Mixed fatty acids
Category: Foam stabilizer, emulsifier, viscosity builder, detergent, nonionic
Applications: See Hetamide LA
Form: Liquid @ 20 C
Properties: Sp. gr.: 1.01 @ 20 C; Acid No.: 40–45%; pH: 9–10 (1% aq. sol'n.)
Composition: 100% active ingred.

Hetamide RC. Heterene Chem.
Chemical Description: Coconut oil
Category: Foam stabilizer, emulsifier, viscosity builder, detergent, nonionic
Applications: See Hetamide LA
Form: Liquid @ 20 C
Properties: Sp. gr.: 1.01 @ 20 C; pH: 9–10 (1% aq. sol'n.)
Composition: 100% active ingred.

Hetoxamate FA5. Heterene Chem.
Chemical Description: Ethoxylated compounds based on PEG 200 monotallate
Category: Detergent, emulsifier, lubricant, softener, nonionic
Applications: Cosmetic emulsifiers for creams and lotions, hand cleaners, wax and oil emulsifiers, emollients; textile emulsifiers, scouring agents, dyeing assistants; metal cleaning agents, leather processing (fat liquoring)
Form: Liquid
Properties: Solubility: Sol. in isopropanol, mineral oil; disp. in water; HLB: 8.4; Sapon. Value: 100–120; Acid No.: 2.0

Hetoxamate FA20. Heterene Chem.
Chemical Description: Ethoxylated compounds based on PEG 200 monotallate
Category: Detergent, emulsifier, lubricant, softener, nonionic
Applications: See Hetoxamate FA5
Form: Semisolid
Properties: Solubility: Sol. in water, isopropanol; disp. in mineral oil; HLB: 14.9; Sapon. Value: 50–60; Acid No.: 2.0

Hetoxamate LA5. Heterene Chem.
Chemical Description: Ethoxylated compounds based on PEG 200 monolaurate
Category: Detergent, emulsifier, lubricant, softener, nonionic
Applications: See Hetoxamate FA5
Form: Liquid
Properties: Solubility: Sol. in isopropanol; disp. in water; partly sol. in mineral oil; HLB: 9.8; Sapon. Value: 125–145; Acid No.: 2.0

Hetoxamate LA9. Heterene Chem.
Chemical Description: Ethoxylated compounds based on PEG 400

monolaurate
Category: Detergent, emulsifier, lubricant, softener, nonionic
Applications: See Hetoxamate FA5
Form: Liquid
Properties: Solubility: Sol. in water, isopropanol; HLB: 13.1; Sapon. Value: 89–96; Acid No.: 2.0

Hetoxamate MO5. Heterene Chem.
Chemical Description: Ethoxylated compounds based on PEG 200 mono oleate
Category: Detergent, emulsifier, lubricant, softener, nonionic
Applications: See Hetoxamate FA5
Form: Liquid
Properties: Solubility: Sol. in isopropanol, disp. in water; HLB: 8.0; Sapon. Value: 115–125; Acid No.: 2.0

Hetoxamate MO9. Heterene Chem.
Chemical Description: Ethoxylated compounds based on PEG 400 mono oleate
Category: Detergent, emulsifier, lubricant, softener, nonionic
Applications: See Hetoxamate FA5
Form: Liquid
Properties: Solubility: Sol. in isopropanol; disp. in water; HLB: 11.4; Sapon. Value: 80–88; Acid No.: 2.0

Hetoxamate MO15. Heterene Chem.
Chemical Description: Ethoxylated compounds based on PEG 600 mono oleate
Category: Detergent, emulsifier, lubricant, softener, nonionic
Applications: See Hetoxamate FA5
Form: Liquid
Properties: Solubility: Sol. in water, isopropanol; HLB: 13.5; Sapon. Value: 60–70; Acid No.: 2.0

Hetoxamate SA5. Heterene Chem.
Chemical Description: Ethoxylated compounds based on PEG 200 monostearate
Category: Detergent, emulsifier, lubricant, softener, nonionic
Applications: See Hetoxamate FA5
Form: Solid
Properties: Solubility: Sol. in isopropanol; dispersible hot in water; HLB: 8.0; Sapon. Value: 120–130; Acid No.: 2.0

Hetoxamate SA9. Heterene Chem.
Chemical Description: Ethoxylated compounds based on PEG 400 monostearate
Category: Detergent, emulsifier, lubricant, softener, nonionic
Applications: See Hetoxamate FA5
Form: Solid
Properties: Solubility: Sol. in isopropanol; dispersible hot in water; HLB: 11.5; Sapon. Value: 84–93; Acid No.: 2.0

Hetoxamate SA35. Heterene Chem.
Chemical Description: Ethoxylated compounds based on PEG 1540 monostearate

Category: Detergent, emulsifier, lubricant, softener, nonionic
Applications: See Hetoxamate FA5
Form: Solid
Properties: Solubility: Sol. in water, isopropanol; HLB: 16.9; Sapon. Value: 24–34; Acid No.: 2.0

Hetoxamate SA90. Heterene Chem.
Chemical Description: Ethoxylated compounds based on PEG 4000 monostearate
Category: Detergent, emulsifier, lubricant, softener, nonionic
Applications: See Hetoxamate FA5
Form: Solid
Properties: Solubility: Sol. in water, isopropanol; HLB: 18.6; Sapon. Value: 10–18; Acid No.: 2.0

Hetoxamine C2. Heterene Chem.
Chemical Description: Ethoxylated compounds based on coco
Category: Emulsifier, desizing agent, softener, antistatic agent, cationic
Applications: Emulsifier in the agricultural, waxes and oils, leather processing, and metal cleaning industries; desizing agent, softener, antistatic agent, water repellant and wet spinning assistant in the textile industries
Form: Liquid
Properties: m.w.: 285; Solubility: Sol. in isopropanol, mineral oil; gel in water

Hetoxamine C5. Heterene Chem.
Chemical Description: Ethoxylated compounds based on coco
Category: Emulsifier, desizing agent, softener, antistatic agent, cationic
Applications: See Hetoxamine C2
Form: Liquid
Properties: m.w.: 425; Solubility: Sol. in water, isopropanol, mineral oil

Hetoxamine C15. Heterene Chem.
Chemical Description: Ethoxylated compounds based on coco
Category: Emulsifier, desizing agent, softener, antistatic agent, cationic
Applications: See Hetoxamine C2
Form: Liquid
Properties: m.w.: 860; Solubility: Sol. in water, isopropanol

Hetoxamine O2. Heterene Chem.
Chemical Description: Ethoxylated compounds based on oleyl
Category: Emulsifier, desizing agent, softener, antistatic agent, nonionic
Applications: See Hetoxamine C2
Form: Liquid
Properties: m.w.: 350; Solubility: Sol. in isopropanol, mineral oil

Hetoxamine O5. Heterene Chem.
Chemical Description: Ethoxylated compounds based on oleyl
Category: Emulsifier, desizing agent, softener, antistatic agent, nonionic
Applications: See Hetoxamine C2
Form: Liquid
Properties: m.w.: 492; Solubility: Sol. in isopropanol, mineral oil

Hetoxamine O15. Heterene Chem.
Chemical Description: Ethoxylated compounds based on oleyl
Category: Emulsifier, desizing agent, softener, antistatic agent, nonionic
Applications: See Hetoxamine C2
Form: Liquid
Properties: m.w.: 930; Solubility: Sol. in water, isopropanol

Hetoxamine S2. Heterene Chem.
Chemical Description: Ethoxylated compounds based on soya
Category: Emulsifier, desizing agent, softener, antistatic agent, cationic
Applications: See Hetoxamine C2
Form: Liquid
Properties: m.w.: 350; Solubility: Sol. in isopropanol, mineral oil

Hetoxamine S5. Heterene Chem.
Chemical Description: Ethoxylated compounds based on soya
Category: Emulsifier, desizing agent, softener, antistatic agent, cationic
Applications: See Hetoxamine C2
Form: Liquid
Properties: m.w.: 480; Solubility: Sol. in isopropanol; partyl sol. in mineral oil; gel in water

Hetoxamine ST2. Heterene Chem.
Chemical Description: Ethoxylated compounds based on stearyl
Category: Emulsifier, desizing agent, softener, antistatic agent, nonionic
Applications: See Hetoxamine C2
Form: Solid
Properties: m.w.: 388; Solubility: Sol. in isopropanol, mineral oil

Hetoxamine ST5. Heterene Chem.
Chemical Description: Ethoxylated compounds based on stearyl
Category: Emulsifier, desizing agent, softener, antistatic agent, nonionic
Applications: See Hetoxamine C2
Form: Solid
Properties: m.w.: 520; Solubility: Sol. in isopropanol, mineral oil

Hetoxamine T2. Heterene Chem.
Chemical Description: Ethoxylated compounds based on tallow
Category: Emulsifier, desizing agent, softener, antistatic agent, cationic
Applications: See Hetoxamine C2
Form: Liquid
Properties: m.w.: 350; Solubility: Sol. in isopropanol, mineral oil

Hetoxamine T5. Heterene Chem.
Chemical Description: Ethoxylated compounds based on tallow
Category: Emulsifier, desizing agent, softener, antistatic agent, cationic
Applications: See Hetoxamine C2
Form: Liquid/paste
Properties: m.w.: 490; Solubility: Sol. in water, isopropanol, mineral oil

Hetoxamine T15. Heterene Chem.
Chemical Description: Ethoxylated compounds based on tallow
Category: Emulsifier, desizing a-

gent, softener, antistatic agent, cationic
Applications: See Hetoxamine C2
Form: Liquid/paste
Properties: m.w.: 925; Solubility: Sol. in water, isopropanol

Hetoxamine T20. Heterene Chem.
Chemical Description: Ethoxylated compounds based on tallow
Category: Emulsifier, desizing agent, softener, antistatic agent, nonionic
Applications: See Hetoxamine C2
Form: Liquid/paste
Properties: m.w.: 1150; Solubility: Sol. in water, isopropanol

Hetoxide BN-13. Heterene Chem.
Chemical Description: Ethoxylated compound based on B-napthol
Category: Emollient, emulsifier, nonionic
Applications: Used in cosmetics as perfume solubilizers, emollients, emulsifiers, lotions, viscosity control agents, lubricants, dispersants; used in textiles as scouring agents, dyeing assistants, dye carriers; used for household cleaning compounds, metal treating agents, metal plating, intermediate for chemicals
Form: Paste @ 30 C
Properties: Solubility: Sol. in water, isopropanol, mineral oil; Acid No.: 1.0; Hydroxyl No.: 73–83

Hetoxide BP-3. Heterene Chem.
Chemical Description: Ethoxylated compound based on modified butanol
Category: Emollient, emulsifier, nonionic
Applications: See Hetoxide BN-13
Form: Liquid @ 30 C
Properties: Solubility: Sol. in water, isopropanol; Hydroxyl No.: 245–265

Hetoxide BY-3. Heterene Chem.
Chemical Description: Ethoxylated compound based on butyndiol
Category: Emollient, emulsifier, nonionic
Applications: See Hetoxide BN-13
Form: Liquid @ 30 C
Properties: Solubility: Sol. in water, isopropanol

Hetoxide C2. Heterene Chem.
Chemical Description: Ethoxylated compound based on castor oil
Category: Emollient, emulsifier, nonionic
Applications: See Hetoxide BN-13
Form: Liquid @ 30 C
Properties: Solubility: Sol. in isopropanol, mineral oil; HLB: 4.3; Sapon. Value: 155–170; Acid No.: 1.0

Hetoxide C9. Heterene Chem.
Chemical Description: Ethoxylated compound based on castor oil
Category: Emollient, emulsifier, nonionic
Applications: See Hetoxide BN-13
Form: Liquid @ 30 C
Properties: Solubility: Sol. in isopropanol, mineral oil; HLB: 11.2; Sapon. Value: 120–136; Acid No.: 1.0

Hetoxide C15. Heterene Chem.
Chemical Description: Ethoxylated compound based on castor oil
Category: Emollient, emulsifier, nonionic
Applications: See Hetoxide BN-13
Form: Liquid @ 30 C
Properties: Solubility: Sol. in isopropanol, mineral oil; HLB: 13.6; Sapon. Value: 95–105; Acid No.: 1.0

Hetoxide C25. Heterene Chem.
Chemical Description: Ethoxylated compound based on castor oil
Category: Emollient, emulsifier, nonionic
Applications: See Hetoxide BN-13
Form: Liquid @ 30 C
Properties: Solubility: Sol. in isopropanol, mineral oil; HLB: 15.6; Sapon. Value: 74–82; Acid No.: 1.0

Hetoxide C40. Heterene Chem.
Chemical Description: Ethoxylated compound based on castor oil
Category: Emollient, emulsifier, nonionic
Applications: See Hetoxide BN-13
Form: Paste @ 30 C
Properties: Solubility: Sol. in water, isopropanol; HLB: 16.9; Sapon. Value: 55–65; Acid No.: 1.0

Hetoxide C200. Heterene Chem.
Chemical Description: Ethoxylated compound based on castor oil
Category: Emollient, emulsifier, nonionic
Applications: See Hetoxide BN-13
Form: Solid @ 30 C
Properties: Solubility: Sol. in water, isopropanol; Sapon. Value: 16–18; Acid No.: 1.0

Hetoxide DNP-4. Heterene Chem.
Chemical Description: Ethoxylated compound based on dinonyl phenol
Category: Emollient, emulsifier, nonionic
Applications: See Hetoxide BN-13
Form: Liquid @ 30 C
Properties: Solubility: Sol. in isopropanol, mineral oil; disp. in water; HLB: 6.6; Hydroxyl No.: 120-140

Hetoxide DNP9.6. Heterene Chem.
Chemical Description: Ethoxylated compound based on dinonyl phenol
Category: Emollient, emulsifier, nonionic
Applications: See Hetoxide BN-13
Form: Liquid @ 30 C
Properties: Solubility: Sol. in isopropanol, mineral oil; disp. in water; HLB: 10.7; Hydroxyl No.: 65–75

Hetoxide G7. Heterene Chem.
Chemical Description: Ethoxylated compound based on glycerine
Category: Emollient, emulsifier, nonionic
Applications: See Hetoxide BN-13
Form: Liquid @ 30 C
Properties: Solubility: Sol. in water, isopropanol; Acid No.: 1.0; Hydroxyl No.: 400–428

Hetoxide HC-16. Heterene Chem.
Chemical Description: Ethoxylated compound based on hydrogen-

ated castor oil
Category: Emollient, emulsifier, nonionic
Applications: See Hetoxide BN-13
Form: Liquid @ 30 C
Properties: Solubility: Sol. in isopropanol, mineral oil; HLB: 11.3; Sapon. Value: 85–95; Acid No.: 1.0

Hetoxide MTG. Heterene Chem.
Chemical Description: Ethoxylated compound based on alcohol
Category: Emollient, emulsifier, nonionic
Applications: See Hetoxide BN-13
Form: Liquid @ 30 C
Properties: Solubility: Sol. in water, isopropanol, mineral oil; Acid No.: 1.0; Hydroxyl No.: 290–330

Hetoxide NP-4. Heterene Chem.
Chemical Description: Ethoxylated compound based on nonyl phenol
Category: Emollient, emulsifier, nonionic
Applications: See Hetoxide BN-13
Form: Liquid @ 30 C
Properties: Solubility: Sol. in isopropanol, mineral oil; HLB: 7.6; Hydroxyl No.: 135–140

Hetoxide NP-9. Heterene Chem.
Chemical Description: Ethoxylated compound based on nonyl phenol
Category: Emollient, emulsifier, nonionic
Applications: See Hetoxide BN-13
Form: Liquid @ 30 C
Properties: Solubility: Sol. in water, isopropanol; HLB: 13.0;Hydroxyl No.: 85–95

Hetoxide NP-40. Heterene Chem.
Chemical Description: Ethoxylated compound based on nonyl phenol
Category: Emollient, emulsifier, nonionic
Applications: See Hetoxide BN-13
Form: Liquid @ 30 C
Properties: Solubility: Sol. in water, isopropanol; HLB: 17.0

Hetoxol CA 2. Heterene Chem.
Chemical Description: Ethoxylated compounds based on cetyl
Category: Detergent, emulsifier, leveling agent, intermediate, nonionic
Applications: Used for cosmetic shampoos, lotions, bath oils, wax, oil emulsifers, textile emulsifiers, scouring agents, dye leveling agents, dye assistants; intermediates for phosphating or sulfating agents, household formulations, silicone emulsification, surfactants
Form: Solid
Properties: Solubility: Sol. in isopropanol, mineral oil; HLB: 5.3; Hydroxyl No.: 160–180

Hetoxol CA20. Heterene Chem.
Chemical Description: Ethoxylated compounds based on cetyl
Category: Detergent, emulsifier, leveling agent, intermediate, nonionic
Applications: See Hetoxol CA 2
Form: Solid

Properties: Solubility: Sol. in water, isopropanol; HLB: 15.7; Hydroxyl No.: 45–60

Hetoxol CS9. Heterene Chem.
Chemical Description: Ethoxylated compounds based on cetyl stearyl
Category: Detergent, emulsifier, leveling agent, intermediate, nonionic
Applications: See Hetoxol CA 2
Form: Solid
Properties: Solubility: Sol. in water, isopropanol; HLB: 12.0; Hydroxyl No.: 80–90

Hetoxol CS20. Heterene Chem.
Chemical Description: Ethoxylated compounds based on cetyl stearyl
Category: Detergent, emulsifier, leveling agent, intermediate, nonionic
Applications: See Hetoxol CA 2
Form: Solid
Properties: Solubility: Sol. in water, isopropanol; HLB: 15.4; Hydroxyl No.: 50–70

Hetoxol CS30. Heterene Chem.
Chemical Description: Ethoxylated compounds based on cetyl stearyl
Category: Detergent, emulsifier, leveling agent, intermediate, nonionic
Applications: See Hetoxol CA 2
Form: Solid
Properties: Solubility: Sol. in water, isopropanol; HLB: 16.7; Hydroxyl No.: 40–52

Hetoxol L3N. Heterene Chem.
Chemical Description: Ethoxylated compounds based on lauryl
Category: Detergent, emulsifier, leveling agent, intermediate, nonionic
Applications: See Hetoxol CA 2
Form: Liquid
Properties: Solubility: Sol. in isopropanol, mineral oil; HLB: 7.9; Hydroxyl No.: 170–176

Hetoxol L4N. Heterene Chem.
Chemical Description: Ethoxylated compounds based on lauryl
Category: Detergent, emulsifier, leveling agent, intermediate, nonionic
Applications: See Hetoxol CA 2
Form: Liquid
Properties: Solubility: Sol. in isopropanol, mineral oil; HLB: 9.7; Hydroxyl No.: 145–165

Hetoxol L9N. Heterene Chem.
Chemical Description: Ethoxylated compounds based on lauryl
Category: Detergent, emulsifier, leveling agent, intermediate, nonionic
Applications: See Hetoxol CA 2
Form: Liquid
Properties: Solubility: Sol. in water, isopropanol; HLB: 11.8; Hydroxyl No.: 90–110

Hetoxol L23N. Heterene Chem.
Chemical Description: Ethoxylated compounds based on lauryl
Category: Detergent, emulsifier, leveling agent, intermediate, nonionic

Applications: See Hetoxol CA 2
Form: Solid
Properties: Solubility: Sol. in water, isopropanol; HLB: 16.9; Hydroxyl No.: 40–55

Hetoxol LS9. Heterene Chem.
Chemical Description: Ethoxylated compounds based on lauryl
Category: Detergent, emulsifier, leveling agent, intermediate, nonionic
Applications: See Hetoxol CA 2
Form: Solid
Properties: Solubility: Sol. in water, isopropanol; HLB: 13.3; Hydroxyl No.: 90–100

Hetoxol OL2. Heterene Chem.
Chemical Description: Ethoxylated compounds based on oleyl
Category: Detergent, emulsifier, leveling agent, intermediate, nonionic
Applications: See Hetoxol CA 2
Form: Liquid
Properties: Solubility: Sol. in isopropanol, mineral oil; HLB: 4.9; Hydroxyl No.: 152–162

Hetoxol OL4. Heterene Chem.
Chemical Description: Ethoxylated compounds based on oleyl
Category: Detergent, emulsifier, leveling agent, intermediate, nonionic
Applications: See Hetoxol CA 2
Form: Liquid
Properties: Solubility: Sol. in isopropanol, mineral oil; HLB: 8.0; Hydroxyl No.: 120–135

Hetoxol OL5. Heterene Chem.
Chemical Description: Ethoxylated compounds based on oleyl
Category: Detergent, emulsifier, leveling agent, intermediate, nonionic
Applications: See Hetoxol CA 2
Form: Liquid
Properties: Solubility: Sol. in water, isopropanol; HLB: 9.1; Hydroxyl No.: 110–120

Hetoxol OL23. Heterene Chem.
Chemical Description: Ethoxylated compounds based on oleyl
Category: Detergent, emulsifier, leveling agent, intermediate, nonionic
Applications: See Hetoxol CA 2
Form: Semisolid
Properties: Solubility: Sol. in water, isopropanol; HLB: 15.9; Hydroxyl No.: 47–62

Hetoxol OL40. Heterene Chem.
Chemical Description: Ethoxylated compounds based on oleyl
Category: Detergent, emulsifier, leveling agent, intermediate, nonionic
Applications: See Hetoxol CA 2
Form: Solid
Properties: Solubility: Sol. in water, isopropanol; HLB: 17.4; Hydroxyl No.: 30–45

Hetoxol STA2. Heterene Chem.
Chemical Description: Ethoxylated compounds based on stearyl
Category: Detergent, emulsifier, leveling agent, intermediate, nonionic

Applications: See Hetoxol CA 2
Form: Solid
Properties: Solubility: Sol. in isopropanol; HLB: 4.9; Hydroxyl No.: 145–165

Hetoxol TD3. Heterene Chem.
Chemical Description: Ethoxylated compounds based on tridecyl
Category: Detergent, emulsifier, leveling agent, intermediate, nonionic
Applications: See Hetoxol CA 2
Form: Liquid
Properties: Solubility: Sol. in isopropanol, mineral oil; HLB: 7.9; Hydroxyl No.: 165–175

Hetoxol TD6. Heterene Chem.
Chemical Description: Ethoxylated compounds based on tridecyl
Category: Detergent, emulsifier, leveling agent, intermediate, nonionic
Applications: See Hetoxol CA 2
Form: Liquid
Properties: Solubility: Sol. in water, isopropanol; HLB: 11.3; Hydroxyl No.: 115–125

Hetoxol TD12. Heterene Chem.
Chemical Description: Ethoxylated compounds based on tridecyl
Category: Detergent, emulsifier, leveling agent, intermediate, nonionic
Applications: See Hetoxol CA 2
Form: Liquid
Properties: Solubility: Sol. in water, isopropanol; HLB: 14.5; Hydroxyl No.: 72–87

Hetsorb L-10. Heterene Chem.
Chemical Description: Ethoxylated sorbitan esters based on monolaurate
Category: Detergent, emulsifier, lubricant
Applications: Emulsifier for oils, fats, solvents, waxes, lubricants, viscosity control, cosmetic formulations
Form: Liquid
Properties: Solubility: Sol. in water, isopropanol; HLB: 8.4; Sapon. Value: 66–76; Hydroxyl No.: 150–170

Hetsorb L-20. Heterene Chem.
Chemical Description: Ethoxylated sorbitan esters based on monolaurate polysorbate 20
Category: Detergent, emulsifier, lubricant
Applications: Emulsifier for oils, fats, solvents, waxes, lubricants, viscosity control, cosmetic formulations
Form: Liquid
Properties: Solubility: Sol. in water, isopropanol; HLB: 16.7; Sapon. Value: 40–50; Hydroxyl No.: 95–108

Hetsorb O-20. Heterene Chem.
Chemical Description: Ethoxylated sorbitan esters based on mono oleate polysorbate 80
Category: Detergent, emulsifier, lubricant
Applications: Emulsifier for oils, fats, solvents, waxes, lubricants, viscosity control, cosmetic formulations

Hetsorb TS-20. Heterene Chem.
Chemical Description: Ethoxylated sorbitan esters based on tristearate polysorbate 65
Category: Detergent, emulsifier, lubricant
Applications: Emulsifier for oils, fats, solvents, waxes, lubricants, viscosity control, cosmetic formulations
Form: Solid
Properties: Solubility: Sol. in isopropanol, mineral oil (when heated); disp. in water; HLB: 10.5; Sapon. Value: 88–98; Hydroxyl No.: 44–60
Form: Liquid
Properties: Solubility: Sol. in water, isopropanol; HLB: 15.0; Sapon. Value: 45–55; Hydroxyl No.: 65–80

Hetsorb P-20. Heterene Chem.
Chemical Description: Ethoxylated sorbitan esters based on monopalmatate
Category: Detergent, emulsifier, lubricant
Applications: Emulsifier for oils, fats, solvents, waxes, lubricants, viscosity control, cosmetic formulations
Form: Liquid
Properties: Solubility: Sol. in water, isopropanol; misc. in mineral oil @ certain %; HLB: 15.6; Sapon. Value: 43–49; Hydroxyl No.: 89–105

Hetsorb S-20. Heterene Chem.
Chemical Description: Ethoxylated sorbitan esters based on monostearate polysorbate 60
Category: Detergent, emulsifier, lubricant
Applications: Emulsifier for oils, fats, solvents, waxes, lubricants, viscosity control, cosmetic formulations
Form: Liquid
Properties: Solubility: Sol. in water; misc. in isopropanol; HLB: 14.9; Sapon. Value: 45–55; Hydroxyl No.: 80–96

Hetsulf 40. Heterene Chem.
Chemical Description: Sodium dodecylbenzene sulfonate
Category: Wetting agent, emulsifier, dispersant, anionic
Applications: Intermediate, emulsifier, base for neutralized surfactant
Form: Paste
Properties: pH: 7.5–8.5 (5% aq. sol'n.); Biodegradable
Composition: 38% active ingred.

Hetsulf 40X. Heterene Chem.
Chemical Description: Sodium dodecylbenzene sulfonate
Category: Wetting agent, emulsifier, dispersant, anionic
Applications: Liquid formulation syndet, detergent
Form: Liquid
Properties: pH: 7.0–8.0 (5% aq. sol'n.); Biodegradable
Composition: 35% active ingred.

Hetsulf 50A. Heterene Chem.
Chemical Description: Ammonium dodecylbenzene sulfonate
Category: Wetting agent, emulsi-

fier, dispersant, anionic
Applications: Light duty detergent formulations, detergent
Form: Slurry
Properties: pH: 6.0–7.0 (5% aq. sol'n.); Biodegradable
Composition: 48% active ingred.

Hetsulf 60S. Heterene Chem.
Chemical Description: Sodium dodecylbenzene sulfonate
Category: Wetting agent, emulsifier, dispersant, anionic
Applications: Base for formulated products, light and heavy duty type
Form: Slurry
Properties: pH: 7.0–8.0 (5% aq. sol'n.); Biodegradable
Composition: 57% active ingred.

Hetsulf 60T. Heterene Chem.
Chemical Description: Amine dodecylbenzene sulfonate
Category: Wetting agent, emulsifier, dispersant, anionic
Applications: Cosmetic, bubble bath formulas, shampoos
Form: Liquid
Properties: pH: 7.0–7.5 (5% aq. sol'n.); Biodegradable
Composition: 52% active ingred.

Hetsulf Acid. Heterene Chem.
Chemical Description: Dodecylbenzene sulfonic acid
Category: Wetting agent, emulsifier, dispersant, anionic
Applications: Intermediate, emulsifier, base for neutralized surfactant
Form: Liquid
Properties: Density: 1.05; Biodegradable
Composition: 97% active ingred.

Hetsulf IPA. Heterene Chem.
Chemical Description: Amine dodecylbenzene sulfonate
Category: Wetting agent, emulsifier, dispersant, anionic
Form: Liquid
Properties: pH: 5.0–6.5 (5% aq. sol'n.); Biodegradable

2E Hexyl Palmitate. Akzo Chemie
Category: Emollient
Applications: Cosmetic preparations
Form: Liquid
Properties: Ref. Index: 1.440; Sp. gr.: 0.850; F. P.: 0.0 C
Composition: >93% active ingred.
Reaction: Neutral
Containers: 180 kg steel drums

Hi-Brite. Coastal Industries
Category: Detergent
Applications: Cleaners of street, buildings, hospitals, etc.
Form: Liquid; Color: Pink; Odor: Bland
Properties: Solubility: Sol. in water; Density: 8.60 lb/gal; Sp. gr.: 1.030
Composition: 17% active ingred.
Reaction: Alkaline
Containers: Poly bottles

Hi-Tex. Coastal Industries
Category: Detergent
Applications: Used for laundering; rug shampooing

Form: Liquid; Color: Amber
Properties: Solubility: Sol. in water; Density: 8.45 lb/gal; Sp. gr.: 1.012
Composition: 14% active ingred.
Reaction: Alkaline
Containers: Poly

Hi-Tex AC. Coastal Industries
Category: Detergent
Applications: Cleaners of street, buildings, hospitals, etc.
Form: Liquid; Color: Amber; Odor: Mild
Properties: Solubility: Sol. in water; Density: 8.36 lb/gal; Sp. gr.: 1.001; Stability: Good
Composition: 12% active ingred.
Reaction: Acid
Containers: Poly and poly-lined drums

Hipochem ADN. High Point Chem.
Chemical Description: Quaternary ammonium complex
Category: Dyeing assistant, nonionic
Applications: Leveling agent of acid dyes on nylon
Form: Liquid; Color: Amber; Odor: Mild alcohol
Properties: Solubility: Sol. in water; Density: 8.73 lb/gal; Sp. gr.: 1.048; Stability: Excellent
Composition: 50% active ingred.
Reaction: Acid (slightly)
Containers: Bulk or drums

Hipochem AM-99. High Point Chem.
Chemical Description: Super amide condensate
Category: Detergent, wetting agent, emulsifier, foam builder, nonionic
Applications: Multipurpose detergent and emulsifier
Form: Liquid; Color: Amber; Odor: Mild amine
Properties: Solubility: Sol. in water; Density: 8.89 lb/gal; Sp. gr.: 1.06; Stability: Excellent
Composition: 99% active ingred.
Reaction: Alkaline
Containers: Bulk or drums

Hipochem AMC. High Point Chem.
Chemical Description: Amine condensate
Category: Detergent, wetting agent, emulsifier, foam builder, nonionic
Applications: Multipurpose detergent and emulsifier
Form: Liquid; Color: Dark brown; Odor: Mild amine
Properties: Solubility: Sol. in water; Density: 8.16 lb/gal; Sp. gr.: .98
Composition: 99–100% active
Reaction: Alkaline
Containers: Bulk or drums

Hipochem C-95. High Point Chem.
Chemical Description: Quaternary ammonium compound
Category: Dyeing assistant, cationic
Applications: Retarder for cationic dyes on acrylics
Form: Liquid; Color: Amber; Odor: Mild alcohol
Properties: Solubility: Sol. in water at all temperatures; Density: 8.29 lb/gal; Sp. gr.: .995; Stability: Excellent

Composition: 50% active ingred.
Reaction: Acid
Containers: Bulk or drums

Hipochem D2. High Point Chem.
Chemical Description: Quaternary ammonium compound
Category: Dyeing assistant, cationic
Applications: Retarder for basic dyes on acrylics
Form: Liquid; Color: Light amber
Properties: Solubility: Sol. in water @ 25 C (10% sol'n.); Density: 8.08 lb/gal; Sp. gr.: 0.97 @ 22 C; Stability: Excellent
Composition: 40% active ingred.
Reaction: Neutral
Containers: Bulk or drums

Hipochem Dispersol GTO. High Point Chem.
Chemical Description: Sulfated glycerol trioleate
Category: Detergent, wetting agent, emulsifier, dyeing assistant, anionic
Applications: Multipurpose dyeing assistant, detergent, and emulsifier
Form: Liquid; Color: Dark amber; Odor: Mild sulfate
Properties: Solubility: Sol. in water at all temperatures; Density: 8.66 lb/gal; Sp. gr.: 1.04; Stability: Excellent
Composition: 55% active ingred.
Reaction: Acid
Containers: Bulk or drums

Hipochem Dispersol SB. High Point Chem.
Chemical Description: Sulfated butyl oleate
Category: Detergent, wetting agent, emulsifier, dyeing assistant, anionic
Applications: Multipurpose dyeing assistant
Form: Liquid; Color: Dark amber; Odor: Mild sulfate
Properties: Solubility: Disp. in water at all temperatures; Density: 8.41 lb/gal; Sp. gr.: 1.01; Stability: Excellent
Composition: 55% active ingred.
Reaction: Acid
Containers: Bulk or drums

Hipochem Dispersol SCO. High Point Chem.
Chemical Description: Sulfated castor oil
Category: Detergent, wetting agent, emulsifier, dyeing assistant, anionic
Applications: Multipurpose dyeing assistant
Form: Liquid; Color: Dark amber; Odor: Mild sulfate
Properties: Solubility: Disp. in water at all temperatures; Density: 8.33 lb/gal; Sp. gr.: 1.0; Stability: Excellent
Composition: 70% active ingred.
Reaction: Acid
Containers: Bulk or drums

Hipochem EFK. High Point Chem.
Chemical Description: Aliphatic sulfate
Category: Detergent
Applications: Textile lubricant, substantive to cellulosic fibers
Form: Paste; Color: White

Properties: Solubility: Sol. in water @ 25 C (10% sol'n.); Stability: Excellent
Composition: 40% active ingred.
Reaction: Neutral
Containers: Bulk or drums

Hipochem EK-6. High Point Chem.
Chemical Description: Long chain phosphated alcohol
Category: Wetting agent, anionic
Applications: Wetting agent for desizing, dyeing, bleaching; low foaming
Form: Liquid; Color: Light amber; Odor: Mild alcohol
Properties: Solubility: Sol. in water @ 120 F; Density: 8.70 lb/gal; Sp. gr.: 1.044; Stability: Excellent
Composition: 60% active ingred.
Reaction: Neutral
Containers: Bulk or drums

Hipochem EK-18. High Point Chem.
Chemical Description: Sodium dioctyl sulfosuccinate
Category: Wetting agent, anionic
Applications: Super fast wetting agent for cellulosics; excellent rewetting; suitable for desizing, bleaching, scouring, etc.
Form: Liquid; Color: Light amber; Odor: Alcohol (slightly)
Properties: Solubility: Sol. in water @ 25 C (10% sol'n.); Density: 8.91 lb/gal; Sp. gr.: 1.07; Stability: Excellent
Composition: 75% active ingred.
Reaction: Neutral
Containers: Bulk or drums

Hipochem LCA. High Point Chem.
Chemical Description: Sulfonated oleyl alcohol
Category: Detergent, lubricant, anionic
Applications: Leveling; a textile dye leveling agent
Form: Paste; Color: White; Odor: Mild sulfate
Properties: Solubility: Disp. in water (10% sol'n.); Stability: Excellent
Composition: 40% active ingred.
Reaction: Acid
Containers: Drums

Hipochem LH-Soap. High Point Chem.
Chemical Description: Coconut oil soap
Category: Detergent, wetting agent, emulsifier, anionic
Applications: Developer and brightener
Form: Liquid; Color: Amber; Odor: Mild fatty
Properties: Solubility: Sol. in water at all temperatures; Density: 7.77 lb/gal; Sp. gr.: 0.93; Stability: Excellent
Composition: 40% active ingred.
Reaction: Alkaline
Containers: Bulk or drums

Hipochem M-51. High Point Chem.
Chemical Description: Non-surface active quaternary ammonium compound
Category: Dyeing assistant, cationic
Applications: Dyeing assistant; migrating agent for cationic dyes on acrylics
Form: Liquid; Color: White
Properties: Solubility: Sol. in water at all temperatures; Density:

8.86 lb/gal; Sp. gr.: 1.064; Stability: Excellent
Composition: 60% active ingred.
Reaction: Acid
Containers: Bulk or drums

Hipochem No. 40-L. High Point Chem.
Chemical Description: A modified ABS
Category: Detergent, wetting agent, emulsifier, dyeing penetrant, anionic
Applications: Wetting, penetrant; a fast wetting agent
Form: Liquid; Color: Dark amber
Properties: Solubility: Sol. in water at all temperatures; Density: 8.33 lb/gal; Sp. gr.: 1.0; Stability: Excellent
Composition: 40% active ingred.
Reaction: Alkaline
Containers: Bulk or drums

Hipochem No. 641. High Point Chem.
Chemical Description: Alkyl diaryl sulfonate
Category: Dyeing assistant, anionic
Applications: Dyeing assistant for polyamide fibers
Form: Liquid; Color: Light brown
Properties: Solubility: Sol. in water at all temperatures; Density: 9.70 lb/gal; Sp. gr.: 1.16; Stability: Excellent
Composition: 45% active ingred.
Reaction: Alkaline
Containers: Bulk or drums

Hipochem PE-25. High Point Chem.
Chemical Description: Emulsified polyethylene
Category: Softener, cationic
Applications: Polyethylene softener
Form: Liquid; Color: Amber; Odor: Bland waxy
Properties: Solubility: Sol. in water at all temperatures; Density: 8.25 lb/gal; Sp. gr.: 0.99; Stability: Excellent
Composition: 25% active ingred.
Reaction: Alkaline
Containers: Bulk or drums

Hipochem PNO-11. High Point Chem.
Chemical Description: Phosphate ester derivative
Category: Detergent, anionic
Applications: Wool-dyeing assistant and reserving agent in multifiber cross dyeings
Form: Liquid; Color: Light amber; Odor: Alcoholic (slightly)
Properties: Solubility: Sol. in water @ 25 C (10% sol'n.); Density: 8.83 lb/gal; Sp. gr.: 1.06 @ 22 C; Stability: Excellent
Composition: 90% active ingred.
Reaction: Neutral
Containers: Bulk or drums

Hipochem Retarder CJ. High Point Chem.
Chemical Description: Quaternary ammonium compound
Category: Dyeing assistant, cationic
Applications: Retarder for cationic dyes on acrylic fibers; dye retarder for jet machines
Form: Liquid; Color: Amber; Odor: Mild amine
Properties: Solubility: Sol. in wa-

ter @ 25 C (10% sol'n.); Density: 8.21 lb/gal; Sp. gr.: 0.985; Stability: Excellent
Composition: 40% active ingred.
Reaction: Acid
Containers: Bulk or drums

Homodan. Grindsted Products
Chemical Description: Special emulsifiers and blends
Category; Plasticizer, nonionic
Applications: Improves plasticity of the margarine and produces a flakier puff pastry
Form: Flakes, liquids, pills, solids
Composition: 100% active ingred.

Hostacerin DGO. American Hoechst Corp.
Chemical Description: Oleic acid polyglycerol ester
Category: Emulsifier, nonionic
Applications: Emulsifier for manufacture of W/O emulsions for cosmetics and pharmaceuticals, e.g., night cream, baby cream
Form: Liquid; Color: Yellow; Odor: Weakly fatty
Properties: Solubility: Sol. in most organic solvents, paraffin oil, and neutral oil; sol. in warm fatty alcohols, fatty acid esters, and their mixtures; Density: ≈ 0.97 g/cm^3 @ 20 C; Visc.: 600 ± 200 cps @ 20 C; Flash Pt.: ≈ 250 C; Setting Pt.: ≈ 0 C; Sapon. No.: 165 ± 15; Acid No.: 1 max.; Iodine Value: 70 ± 5
Composition: 100% active

Hostacerin O-5. American Hoechst Corp.
Chemical Description: Alkyl polyglycol ether
Category: Emulsifier, nonionic
Applications: Emulsifier, esp. paraffin emulsifier; used in preparation of cosmetics creams and lotions
Form: Slightly cloudy liquid; Color: Yellowish
Properties: Ref. Index: ≈ 1.463 (20 C); Solubility: Dissolves to a clear sol'n. in all liquid or molten hydrocarbons, fatty alcohols, fatty acids, fatty esters and their mixtures; readily sol. in most organic solvents; dissolves to cloudy sol'n. in water; Sp. gr.: 0.953 ± 0.005 (20 C); Visc.: 67 ± 3 cps (20 C); Flash Pt.: ≈ 240 C (Marcusson); Sapon. No.: 1; Acid No.: 1
Composition: 100% active
Reaction: Neutral

Hostacerin T-3. American Hoechst Corp.
Chemical Description: Alkyl polyglycolether
Category: Ointment base, emulsifier, nonionic
Applications: Manufacture of ointments, creams, and liquid emulsions; additive for shampoos, soaps, and other preparations for body hygiene; superfatting agent
Form: Soft, waxlike substance; Color: White
Properties: Ref. Index: ≈ 1.446; Solubility: Sol. when warmed in all hydrocarbons, fatty alcohols; Sp. gr.: 0.905 (50 C); Visc.: 15 ± 3 cps (50 C); Cloud Pt.: ≈ 54 C (in butyl diglycol); Flash Pt.: ≈ 220 C; Sapon. No.: 1 max.; Acid

No.: 1 max.
Composition: 100% active
Reaction: Neutral

Hostacor BM. American Hoechst Corp.
Chemical Description: Alkyl sulfonamidocarboxylic acid ester 70%
Category: Corrosion inhibitor
Applications: Emulsifier and corrosion inhibitor in coolants containing mineral oil alone or in combination with anionic and nonionic emulsifiers, corrosion inhibitors, solubilizer mineral oil and water
Form: Clear liquid; Color: Brown
Properties: Solubility: Colloidal—5% in water @ 340 ppm; Sp. gr.: 1.1 (20 C); Flame Pt.: 302 F; Pour Pt.: 14 F; pH: ≈ 9 (1% sol'n.)

Hostacor BS. American Hoechst Corp.
Chemical Description: Aryl sulfonamidocarboxylic acid ester 70%
Category: Corrosion inhibitor
Applications: Corrosion inhibitor in cooling water systems and in coolants free of mineral oil (alone or in combination with alkanolamines, glycols, and/or blocked polymers, and water)
Form: Clear liquid; Color: Yellow
Properties: Solubility: Clear sol. 5% in water @ 340 ppm); Sp. gr.: 1.1 (20 C); Flame Pt.: 302 F; Pour Pt.: 14 F; pH: ≈ 9 (1% sol'n.)

Hostacor DT. American Hoechst Corp.
Chemical Description: Fatty acid alkanolamide
Category: Corrosion inhibitor
Applications: Used in metalworking fluids, esp. water-sol. cooling lubricants based on mineral oil
Form: Oily liquid; Color: Brown
Properties: Ref. Index: ≈ 1.48 (20 C); Solubility: Dissolves to turbid sol'n. in water, dissolves to clear sol'n. in mineral oils; Sp. gr.: ≈ 1.01 (20 C); Flash Pt.: ≈ 210 C (Marcusson); Stability: Sufficiently stable to water hardness constituents; pH: ≈ 9 (1% aq.)

Hostacor H Liquid. Ameican Hoechst Corp.
Chemical Description: Arylsulfonamidocarboxylic acid
Category: Chemical intermediate, anionic
Applications: Intermediate for the manufacture of corrosion inhibitors for aq. systems; water-sol. salts of Hostacor H Liquid are used in mineral oil-free metalworking fluids alone or in combination with water-sol. lubricants
Form: Viscous liquid; Color: Yellowish
Properties: Solubility: Sol. in ethanol and glycol; insol. in water; Sp. gr.: 1.17 ± 0.05 (122 F); Solidification Pt.: ≈ 3 F; Acid No.: 161 ± 10

Hostacor KS 1. American Hoechst Corp.
Chemical Description: Alkanolamine salt of a nitrogen-containing condensation product

Category: Corrosion inhibitor, lubricant, anionic
Applications: Used alone or in combination with other water-sol. corrosion inhibitors and lubricants as cooling and machining auxiliaries for metalworking, esp. for machining of ferrous metals
Form: Clear, viscous liquid; Color: Pale brown
Properties: Solubility: Dissolves to clear sol'n. in water, glycols, alkanolamines; insol. in mineral oils; Sp. gr.: 1.16 ± 0.02 (20 C); Flash Pt.: > 200 C (Marcusson); Setting Pt.: < –20 C; Stability: Largely unaffected by water hardness constituents; pH: 8.0–8.5 (1% aq.)
Reaction: Weakly alkaline

Hostacor R. American Hoechst Corp.
Chemical Description: Alkaline earth salt of a nitrogen-containing condensation product
Category: Corrosion inhibitor
Applications: Corrosion inhibitor for the preparation of mineral oil-based rust-inhibiting oils; improves lubricant properties of straight mineral oils (used in nonemulsified cutting oils); used in lubricating oils with other active substances
Form: Clear liquid of oily consistency; Color: Dark brown
Properties: Solubility: Sol. in most mineral oils and organic solvents; insol. in water; Sp. gr.: 0.99–1.00 (20 C); Flash Pt.: 140 C (Marcusson); Setting Pt.: –10 C

Hostapal BV Conc. American Hoechst Corp.
Chemical Description: Alkyl aryl polyglycol ether sulfate, sodium salt
Category: Dispersing agent, wetting agent, emulsifier, anionic
Applications: Dispersant for organic and inorganic pigment preparations; excellent wetting and emulsifying properties
Form: Gelatinous paste; Color: Yellowish brown
Properties: Solubility: Miscible with water in any proportion; Density: 9.10 lb/gal (20 C); Stability: Stable to boiling, hard water, neutral salts, and alkali; compatible with anionic and nonionic substances
Composition: 50 ± 1% active
Reaction: Neutral to weakly alkaline

Hostapal CVS. American Hoechst Corp.
Chemical Description: Alkylphenol polyglycol ether
Category: Wetting agent, nonionic
Applications: Wetting agent for organic and inorganic pigment preparations; promotes good wetting and flow characteristics, allows increased pigment loads and reduced milling time; for use with anionic dispersing agents
Form: Viscous liquid; Color: Slightly yellow
Properties: Density: 8.85 lb/gal; Cloud Pt.: 52–56 C (1% aq.)
Composition: 100% active ingred.
Reaction: Neutral to weakly alkaline

Hostapur SAS 30. American Hoechst Corp.
Chemical Description: Secondary n-alkane sulfonate sodium salt
Category: Detergent, wetting agent, anionic
Applications: Detergent bases for manufacture of washing-up liquids, detergents, shampoos, bubble baths, textile auxiliaries, leather finishing agents; good skin compatibility, excellent wetting properties over wide temperature and pH range in hard and soft water
Form: Clear liquid; Color: Colorless to weakly yellow; Odor: Virtually odorless
Properties: m.w.: 328 (avg.); Sp. gr.: 1.048 (20 C); Visc.: 400 cps (20 C); Biodegradable
Composition: 30 ± 0.3% active
Reaction: Neutral to weakly alkaline

Hostapur SAS 60.. American Hoechst Corp.
Chemical Description: Secondary n-alkane sulfonate, sodium salt
Category and Applications: See Hostapur SAS 30
Form: Soft paste; Color: Yellowish; Odor: Virtually odorless
Properties: m.w.: 328 (avg.); Sp. gr.: 1.087 (20 C); Visc.: 26 cps (20 C); Biodegradable
Composition: 60 ± 0.5% active
Reaction: Neutral to weakly alkaline

Hostapur SAS-93. American Hoechst Corp.
Chemical Description: Secondary n-alkane sulfonate sodium salt
Category and Applications: See Hostapur SAS 30
Form: Flake
Properties: Biodegradable
Composition: 93% active ingred.

Hyamine 10-X. Rohm and Haas
Chemical Description: Diisobutylcresoxyethoxyethyl dimethyl benzyl ammonium chloride
Category: Disinfectant, cationic
Applications: Germicide, disinfectant, and sanitizer in restaurant and pharmaceutical uses
Form: Crystals; Color: White
Properties: Density: 27.5 lb/ft^3; Visc.: Solid; Surface Tension: 40 dynes/cm (0.01% aq. sol'n.)
Composition: 100% active ingred.
References: Rohm and Haas handbook CS-16

Hyamine 1622. Rohm and Haas
Chemical Description: Diisobutylphenoxyethoxyethyl dimethylbenzyl ammonium chloride
Category: Disinfectant, cationic
Applications: Germicide, disinfectant, and sanitizer in restaurant and pharmaceutical uses
Form: Liquid; Color: Light amber
Properties: Density: 8.6 lb/gal; Flash Pt.: 110 F (TOC); Pour Pt.: 25 F
Composition: 50% active ingred.
References: Rohm and Haas handbook CS-16

Hyamine 2389. Rohm and Haas
Chemical Description: Methyldodecylbenzyl trimethyl ammonium chloride (80%), methyldodecylxylene bis (trimethyl ammon-

ium chloride) (20%)
Category: Disinfectant, cationic
Applications: Germicide, disinfectant, and sanitizer in restaurant and pharmaceutical uses
Form: Liquid; Color: Light amber
Properties: Density: 8.3 lb/gal; Visc.: 360 cps; Flash Pt.: > 200 F (TOC); Pour Pt.: 25 F; Surface Tension: 43 dynes/cm (0.01% aq. sol'n.)
Composition: 50% active ingred.
References: Rohm and Haas handbook CS-16

Hyamine 3500. Rohm and Haas
Chemical Description: *n*-Alkyl (50% C_{14}, 40% C_{12}, 10% C_{16}) dimethyl benzyl ammonium chloride
Category: Disinfectant, cationic
Applications: Germicide, disinfectant, and sanitizer in restaurant and pharmaceutical uses
Form: Liquid; Color: 2–5 VCS
Properties: Density: 8.0 lb/gal; Visc.: 42 cps; Flash Pt.: 105 F (TOC); Pour Pt.: 15 F; Surface Tension: 40 dynes/cm (0.01% aq. sol'n.)
Composition: 50% active ingred.
References: Rohm and Haas handbook CS-16

Hybase C-300. Witco Chemical Sonneborn Div.
Chemical Description: Highly basic oil-soluble calcium sulfonate
Category: Detergent, rust inhibitor
Applications: Marine or stationary diesels requiring separate cylinder-type lubrication; corrosion inhibiting additive
Form: Liquid; Color: Dark brown
Properties: Solubility: Sol. in organic solvents at all temperatures; Density: 9.4 lb/gal; Sp. gr.: 1.13 @ 60 F; Visc.: 800 sus @ 210 F; Flash Pt.: 380 F
Composition: 29% active ingred.
Reaction: Acid
Containers: Drums, T/W, T/C
Warnings and Caution: May cause eye burn
References: Data sheet S-3

Hybase M-300. Witco Chemical Sonneborn Div.
Chemical Description: Highly basic oil-soluble magnesium sulfonate
Category: Lubricant, corrosive inhibitor
Applications: Marine or stationary diesels requiring separate cylinder-type lubrication; corrosion inhibiting additive
Form: Liquid; Color: Dark brown; Odor: Petroleum
Properties: Solubility: Sol. in organic solvents at all temperatures; Density: 9.3 lb/gal; Sp. gr.: 1.12 @ 60 F; Visc.: 300 cs @ 210 F; Flash Pt.: 380 F
Composition: 28% active ingred.
Reaction: Alkaline, acid
Containers: Drums, T/W, T/C
Warnings and Caution: May cause eye burn
References: Data sheet S-1

Hybase M-400. Witco Chemical Sonneborn Div.
Chemical Description: Highly basic oil-soluble magnesium sulfonate
Category: Lubricant, corrosive inhibitor

Applications: Marine or stationary diesels requiring separate cylinder-type lubrication; corrosion inhibiting additive
Form: Liquid; Color: Dark brown
Properties: Solubility: Sol. in organic solvents at all temperatures; Density: 9.6 lb/gal; Sp. gr.: 1.15 @ 60 F; Visc.: 300 cs @ 210 F; Flash Pt.: 375 F (COC)
Composition: 27% active ingred.
Reaction: Alkaline, acid
Containers: Drums, T/W, T/C
Warnings and Caution: May cause eye burn
References: Data sheet S-2

Hydrotopic Compound. Akzo Chemie
Category: Hydrotopic agent
Applications: Used for all-purpose cleaners
Form: Liquid; Color: Yellow
Properties: Solubility: Sol. in water at all temperatures; Sp. gr.: 1.02; Visc.: 10 cps @ 20 C; F. P.: <0 C
Composition: 40% active ingred.
Reaction: Acid
Containers: 125 kg drums

Hymolon K-90. Hart Chem.
Chemical Description: Coconut diethanolamide
Category: Detergent, emulsifier, nonionic
Applications: Used for household detergents; scouring agent for wool, cotton, and polyester
Form: Liquid; Color: Yellow
Properties: Density: 1.04 lb/gal
Composition: 91% active ingred.

Hyonic PE-40. Diamond Shamrock
Chemical Description: Nonyl phenol
Category: Emulsifier, nonionic
Applications: Emulsifier for solvent cleaning compounds and waterless hand cleaners; low foaming emulsifier for silicone products; solvent based metal cleaners; detergent for petroleum oils; viscosity reducer for plastisols; intermediate for productions of anionic surfactants
Form: Liquid; Color: Clear
Properties: Solubility: Dispersible; Density: 8.5 lb/gal; Cloud Pt.: 0 C (1% aq. sol'n.); Pour Pt.: –29 C; Surface Tension: 28 dynes/cm (0.01% sol'n.); pH: 7.0 (1% aq. sol'n.)
Composition: >99% active ingred.
Containers: 55 gal steel drums or bulk
References: Diamond Shamrock brochure

Hyonic PE-90. Diamond Shamrock
Chemical Description: Nonyl phenol
Category: Detergent, wetting agent, co-emulsifier, nonionic
Applications: Detergent for household and industrial cleaning compounds; detergent-penetrant for acid cleaning solutions; alkaline salt cleaners; degreasing compounds; detergent for all phases of textile processing; wetting agent for paper towelling or tissue; paper deinking
Form: Liquid; Color: Clear
Properties: Solubility: Forms clear

Reaction: Alkaline
Containers: Drums, T/L
Warnings and Caution: Avoid contact with skin and eyes
solution; Density: 8.8 lb/gal; Cloud Pt.: 54 C (1% aq. sol'n.); Pour Pt.: 2 C; Surface Tension: 31 dynes/cm (0.01% sol'n.); pH: 7.0 (1% aq. sol'n.)
Composition: >99% active ingred.
Containers: 55 gal steel drums or bulk
References: Diamond Shamrock brochure

Hyonic PE-100. Diamond Shamrock
Chemical Description: Nonyl phenol
Category: Detergent, wetting agent, base, penetrant, emulsifier, stabilizer, rewetting agent, nonionic
Applications: Base for liquid dishwashing detergents; detergent for low temperature laundry preparations; wetting agent in mineral acid and alkali solutions; base of household and industrial formulated cleaners; penetrant for corrosion inhibitors; emulsifier adjuvant in liquid formulations; stabilizer for latex preparations; rewetting agent for paper towelling and tissue; hide soaking
Form: Liquid; Color: Clear
Properties: Solubility: Clear; Density: 8.9 lb/gal; Cloud Pt.: 68 C (1% aq. sol'n.); Pour Pt.: 4 C; Surface Tension: 32 dynes/cm (0.01% sol'n.); pH: 7.0 (1% aq. sol'n.)
Composition: >99% active ingred.
Containers: 55 gal steel drums or bulk
References: Diamond Shamrock brochure

Hyonic PE-120. Diamond Shamrock
Chemical Description: Nonyl phenol
Category: Detergent, penetrant, wetting agent, co-emulsifier, nonionic
Applications: Detergent formulations for high temperatures; detergent-penetrant in concentrated solutions of electrolytes; wetting agent for caustic solutions; coemulsifier for oil-in-water emulsions
Form: Liquid/Semi-solid; clear liquid melt
Properties: Solubility: Forms clear solutions; Density: 8.9 lb/gal; Cloud Pt.: 91 C (1% aq. sol'n.); Pour Pt.: 15 C; Surface Tension: 36 dynes/cm (0.01% sol'n.); pH: 7.0 (1% aq. sol'n.)
Composition: >99% active ingred.
Containers: 55 gal steel drums or bulk
References: Diamond Shamrock brochure

Hyonic PE-240. Diamond Shamrock
Chemical Description: Alkyl phenoxy polyoxyethylene ethanol
Category: Wetting agent, foaming agent, nonionic
Form: Liquid; Color: Clear
Properties: Solubility: Forms clear solution; Density: 8.8 lb/gal; Cloud Pt.: 25 C; Pour Pt.: –9 C
Composition: 100% active ingred.

Hyonic PE-250. Diamond Shamrock
Chemical Description: Alkyl phenoxy polyoxyethylene ethanol
Category: Wetting agent, foaming agent, nonionic
Form: Liquid; Color: Clear
Properties: Solubility: Forms clear solution; Density: 8.9 lb/gal; Cloud Pt.: 52 C; Pour Pt.: 5 C
Composition: 100% active ingred.

Hyonic PE-260. Diamond Shamrock
Chemical Description: Alkyl phenoxy polyoxyethylene ethanol
Category: Wetting agent, foaming agent, nonionic
Form: Liquid; Color: Clear
Properties: Solubility: Forms clear liquid; Density: 8.9 lb/gal; Cloud Pt.: 65 C; Pour Pt.: 7 C
Composition: 100% active ingred.

Hyonic PE-360. Diamond Shamrock
Chemical Description: Tridecyl alcohol
Category: Detergent, emulsifier, wetting agent, nonionic
Form: Semi-solid
Properties: Solubility: Dispersible; Density: 8.1 lb/gal; Cloud Pt.: 0 C; Pour Pt.: 18 C

Hystrene 3675. Humko Sheffield Chem.
Chemical Description: 75% dimer acid
Category and Applications: Imparts flexibility into polymeric systems (used in polyesters, polyamides, polyurethanes, polyureas, and epoxy systems); corrosion inhibitors; metal-working lubricants; adhesives; inks; surface coatings
Form: Liquid; Color: 9 max. (Gardner)
Properties: Visc.: 9000 cs; Sapon. No.: 191–199; Acid No.: 189–197
Composition: 100% conc.

Hystrene 3675C. Humko Sheffield Chem.
Chemical Description: 75% dimer acid and 3% monomer
Category and Applications: See Hystrene 3675
Form: Liquid; Color: 9 max. (Gardner)
Properties: Visc.: 7500 cs; Sapon. No.: 191–199; Acid No.: 189–197
Composition: 100% conc.

Hystrene 3675CS. Humko Sheffield Chem.
Chemical Description: 75% dimer acid and 3% monomer
Category and Applications: See Hystrene 3675
Form: Liquid; Color: 8 max. (Gardner)
Properties: Visc.: 12,000 cs; Sapon. No.: 196–203; Acid No.: 194–201
Composition: 100% conc.

Hystrene 3675CX. Humko Sheffield Chem.
Chemical Description: 75% dimer acid and 3% monomer
Category and Applications: See Hystrene 3675
Form: Liquid; Color: 9 max. (Gardner)
Properties: Visc.: 8000 cs; Sapon. No.: 193–201; Acid No.: 192–200
Composition: 100% conc.

Hystrene 3675X. Humko Sheffield Chem.
Chemical Description: 75% dimer acid
Category and Applications: See Hystrene 3675
Form: Liquid; Color: 9 max. (Gardner)
Properties: Visc.: 9300 cs; Sapon. No.: 193–201; Acid No.: 192–200
Composition: 100% conc.

Hystrene 3680. Humko Sheffield Chem.
Chemical Description: 80% dimer acid
Category and Applications: See Hystrene 3675
Form: Liquid; Color: 8 max. (Gardner)
Properties: Visc.: 8500 cs; Sapon. No.: 191–199; Acid No.: 190–197
Composition: 100% conc.

Hystrene 3680S. Humko Sheffield Chem.
Chemical Description: 80% dimer acid
Category and Applications: See Hystrene 3675
Form: Liquid; Color: 8 max. (Gardner)
Properties: Visc.: 14,000 cs; Sapon. No.: 196–203; Acid No.: 194–201
Composition: 100% conc.

Hystrene 3680X. Humko Sheffield Chem.
Chemical Description: 80% dimer acid
Category and Applications: See Hystrene 3675
Form: Liquid; Color: 8 max. (Gardner)
Properties: Visc.: 8300 cs; Sapon. No.: 196–203; Acid No.: 194–201
Composition: 100% conc.

Hystrene 3695. Humko Sheffield Chem.
Chemical Description: 95% dimer acid
Category and Applications: See Hystrene 3675
Form: Liquid; Color: 6 max. (Gardner)
Properties: Visc.: 8000 cs; Sapon. No.: 198–202; Acid No.: 194–198
Composition: 100% conc.

Hystrene 3695S. Humko Sheffield Chem.
Chemical Description: 95% dimer acid
Category and Applications: See Hystrene 3675
Form: Liquid; Color: 7 max. (Gardner)
Properties: Visc.: 9500 cs; Sapon. No.: 195–206; Acid No.: 194–202
Composition: 100% conc.

Hystrene 3695X. Humko Sheffield Chem.
Chemical Description: 95% dimer acid
Category and Applications: See Hystrene 3675
Form: Liquid; Color: 7 max. (Gardner)
Properties: Visc.: 7800 cs; Sapon. No.: 196–202; Acid No.: 194–199
Composition: 100% conc.

Hystrene 4516. Humko Sheffield Chem.
Chemical Description: 45% palmitic fatty acid
Category: Lubricant, textile auxiliary, emulsifier, plasticizer, intermediate
Applications: Highest purity, saturated fatty acids used in cosmetics, shampoos, synthetic lubricants, textile auxiliaries, pharmaceuticals, metallic soaps, emulsifiers, greases, plasticizers, enamels, and chemical intermediates with excellent stability
Form: Solid
Properties: Sapon. No.: 207–212; Acid No.: 206–211
Composition: 100% conc.

Hystrene 5016. Humko Sheffield Chem.
Chemical Description: Triple pressed grade fatty acid
Category and Applications: See Hystrene 4516
Form: Solids
Properties: Sapon. No.: 208–211; Acid No.: 207–210
Composition: 100% conc.

Hystrene 5460. Humko Sheffield Chem.
Chemical Description: Trimer acid
Category and Applications: See Hystrene 3675
Form: Liquid
Properties: Visc.: 40,000 cs; Sapon. No.: 190–198; Acid No.: 182–190
Composition: 100% conc.

Hystrene 7018. Humko Sheffield Chem.
Chemical Description: 70% stearic fatty acid
Category and Applications: See Hystrene 4516
Form: Solid
Properties: Sapon. No.: 200–206; Acid No.: 199–205
Composition: 100% conc.

Hystrene 7022. Humko Sheffield Chem.
Chemical Description: 70% arachidic and behenic fatty acid
Category and Applications: See Hystrene 4516
Form: Solid
Properties: Sapon. No.: 172–182; Acid No.: 171–181
Composition: 100% conc.

Hystrene 8016. Humko Sheffield Chem.
Chemical Description: 80% palmitic fatty acid
Category and Applications: See Hystrene 4516
Form: Solid
Properties: Sapon. No.: 214–218; Acid No.: 213–217
Composition: 100% conc.

Hystrene 8018. Humko Sheffield Chem.
Chemical Description: 80% stearic fatty acid
Category and Applications: See Hystrene 4516
Form: Solid
Properties: Sapon. No.: 199–204; Acid No.: 198–203
Composition: 100% conc.

Hystrene 9008. Humko Sheffield Chem.
Chemical Description: 90% caprylic fatty acid
Category and Applications: See Hystrene 4516
Form: Liquid
Properties: Sapon. No.: 379–388; Acid No.: 377–385
Composition: 100% conc.

Hystrene 9010. Humko Sheffield Chem.
Chemical Description: 90% capric fatty acid
Category and Applications: See Hystrene 4516
Form: Solid
Properties: Sapon. No.: 318–328; Acid No.: 315–325
Composition: 100% conc.

Hystrene 9014. Humko Sheffield Chem.
Chemical Description: 90% myristic fatty acid
Category and Applications: See Hystrene 4516
Form: Solid
Properties: Sapon. No.: 241–244; Acid No.: 240–243
Composition: 100% conc.

Hystrene 9016. Humko Sheffield Chem.
Chemical Description: 90% palmitic fatty acid
Category and Applications: See Hystrene 4516
Form: Solid
Properties: Sapon. No.: 217-221; Acid No.: 216–220
Composition: 100% conc.

Hystrene 9018. Humko Sheffield Chem.
Chemical Description: 90% stearic fatty acid
Category and Applications: See Hystrene 4516
Form: Solid
Properties: Sapon. No.: 197–202; Acid No.:196–201
Composition: 100% conc.

Hystrene 9022. Humko Sheffield Chem.
Chemical Description: 90% arachidic and behenic fatty acid
Category and Applications: See Hystrene 4516
Form: Solid
Properties: Sapon. No.: 169–179; Acid No.: 168–178
Composition: 100% conc.

Hystrene 9512. Humko Sheffield Chem.
Chemical Description: 95% lauric fatty acid
Category and Applications: See Hystrene 4516
Form: Solid
Properties: Sapon. No.: 277–282; Acid No.: 276–281
Composition: 100% conc.

Hystrene 9718. Humko Sheffield Chem.
Chemical Description: 92% plus stearic fatty acid
Category and Applications: See Hystrene 4516
Form: Solid
Properties: Sapon. No.: 196–201; Acid No.: 195–200
Composition: 100% conc.

I

ICE #2FK. Durkee
Chemical Description: 80/20 blend of monoglycerides and polysorbate 80 derived from vegetable fat
Category: Emulsifier, nonionic
Applications: Used with stabilizers in frozen desserts for overrun, dryness
Form: Bead; Color: Cream
Properties: M.P.: 138–144 F; HLB: 5.2
Composition: 100% conc.

ICE #10K. Durkee
Chemical Description: 80/20 blend of mono- and diglycerides and polysorbate 65 derived from vegetable fats
Category: Emulsifier, nonionic
Applications: Used with stabilizers in frozen desserts for overrun, dryness
Form: Bead; Color: Light ivory
Properties: M.P.: 141–144 F; HLB: 5.9
Composition: 100% conc.

ICE #12K. Durkee
Chemical Description: 80/20 blend of mono- and diglycerides and polysorbate 65 derived from vegetable fats
Category: Emulsifier, nonionic
Applications: Used with stabilizers in frozen desserts for overrun, dryness
Form: Bead; Color: Cream
Properties: M.P.: 140–144 F; HLB: 4.1
Composition: 100% conc.

Idet 5L. Swastik Household and Industrial Products
Chemical Description: Sodium salt of alkyl aryl sulfonate
Category: Wetting agent, detergent, anionic
Form: Clear liquid; Color: Yellow
Properties: pH: 6.8 (1% sol'n.)
Composition: 17% conc.

Idet 5L SP. NF. Swastik Household and Industrial Products
Chemical Description: Specially formulated alkyl aryl sulfonate
Category: Wetting, dispersing, penetrating agent, anionic
Applications: Detergent and wetting agent for wet processes in

textile industry
Form: Clear viscous liquid; Color: Yellow
Properties: pH: 6.8 (1% sol'n.)
Composition: 25% conc.

Idet 10. Swastik Household and Industrial Products
Chemical Description: Alkyl aryl sulfonate sodium salt
Category: Wetting agent, detergent, anionic
Applications: Wetting agent in textile wet process, e.g., desizing, Kier-boiling, scouring, sizing of warp yarn (cotton), boiling; boiling and wetting out in the jigger, bleaching, dyeing, finishing, etc.
Form: Spray-dried powder
Properties: pH: 7.0–8.5 (1% aq.)
Composition: 37% active ingred.

Idet 20-P. Swastik Household and Industrial Products
Chemical Description: Alkyl aryl sulfonate sodium salt
Category: Wetting agent, detergent, anionic
Applications: Wetting agent in formulations of water-dispersible powders, such as DDT, BHC, COC, etc.; detergent in household scouring powder, dishwashing powder
Form: Spray-dried powder
Properties: pH: 7.5–9.0 (1% aq.)
Composition: 76% active ingred.

Igepal CA-210. GAF Corp.
Chemical Description: Octylphenoxypoly(ethyleneoxy)ethanol
Category: Emulsifier, nonionic
Applications: Emulsifier for nonpolar hydrocarbon solvents and oils, e.g., heptane and mineral oil, in solvent emulsion cleaners and drycleaning systems; pesticides; floor polish
Form: Liquid; Color: Yellow; Odor: Aromatic
Properties: Solubility: Sol. in low-visc. white mineral oil, deodorized kerosene, Stoddard solvent, xyylene, butyl Cellosolve, perchloroethylene, ethanol; Sp. gr.: 0.99; Visc.: 500–800 cps; HLB: 4.8; Cloud Pt.: Insol. (1% sol'n.); Flash Pt.: > 200 F (PMCC); Pour Pt.: 23 ± 2 F; Solidification Pt.: 18 ± 2 F; Stability: Stable to acids, alkalis and dilute sol'ns. of many oxidizing and reducing agents; Surface Tension: Insol. (0.01% sol'n.)
Composition: 100% active; 25% ethylene oxide
Toxicity: Severe eye irritant
Warnings and Caution: Contact with concentrated oxidizing or reducing agents may be hazardous (explosive)
References: Bulletin 2303-020

Igepal CA-420. GAF Corp.
Chemical Description: Octylphenoxypoly(ethyleneoxy)ethanol
Category: Emulsifier, nonionic
Applications: Emulsifier for nonpolar hydrocarbon solvents, e.g., heptane and mineral oil, in solvent emulsion cleaners and drycleaning systems; pesticides; floor polish
Form: Liquid; Color: Pale yellow; Odor: Aromatic

Properties: Solubility: Sol. in deodorized kerosene, Stoddard solvent, xylene, butyl Cellosolve, perchloroethylene, ethanol; Sp. gr.: 1.02; Visc.: 310–400 cps; HLB: 8.0; Cloud Pt.: Insol. (1% sol'n.); Flash Pt.: > 200 F; Pour Pt.: −10 ± 2 F; Solidification Pt.: −15 ± 2 F; Stability: Stable to acids, alkalis, dilute sol'ns. of many oxidizing and reducing agents; Surface Tension: Insol. (0.01% sol'n.)
Composition: 100% active; 40% ethylene oxide content
Toxicity: Severe eye irritant
Warnings and Caution: Contact with concentrated oxidizing or reducing agents may be hazardous (explosive)
References: Bulletin 2303-020

Igepal CA-520. GAF Corp.
Chemical Description: Octylphenoxypoly(ethyleneoxy)ethanol
Category: Emulsifier, detergent, wetting agent, nonionic
Applications: Emulsifier for nonpolar hydrocarbon solvents and oils, e.g., heptane and mineral oil, in solvent emulsion cleaners and drycleaning systems; automotive gasoline anti-icing additive; pesticides; floor polishes
Form: Liquid; Color: Pale yellow; Odor: Aromatic
Properties: Solubility: Sol. in Stoddard solvent, xylene, butyl Cellosolve, perchloroethylene, ethanol; disp. in water; Sp. gr.: 1.04; Visc.: 250–280 cps; HLB: 10.0; Cloud Pt.: Insol. (1% sol'n.); Flash Pt.: > 200 F (PMCC); Pour Pt.: −15 ± 2 F; Solidification Pt.: −20 ± 2 F; Stability: Stable to acids, alkalis, dilute sol'ns. of many oxidizing and reducing agents; Surface Tension: 30 dynes/cm (0.01% sol'n.)
Composition: 100% active; 50% ethylene oxide content
Toxicity: Severe eye irritant
Warnings and Caution: Contact with concentrated oxidizing or reducing agents may be hazardous (explosive)
References: Bulletin 2303-020

Igepal CA-620. GAF Corp.
Chemical Description: Octylphenoxypoly(ethyleneoxy)ethanol
Category: Detergent, wetting agent, emulsifier, nonionic
Applications: Used in detergent compounding and aqueous processing in textile and paper industries; detergents in industrial metal cleaners, floor cleaners, detergent-sanitizers, waterless hand cleaners, household and industrial controlled-foam detergents; emulsifier for polar solvents, e.g., xylene
Form: Liquid; Color: Pale yellow; Odor: Aromatic
Properties: Solubility: Sol. in xylene, butyl Cellosolve, perchloroethylene, ethanol, ethylene glycol, water; Sp. gr.: 1.05; Visc.: 240–260 cps; HLB: 12.0; Cloud Pt.: 70–75 F (1% sol'n.); Flash Pt.: > 200 F (PMCC); Pour Pt.: 15 ± 2 F; Solidification Pt.: 10 ± 2 F; Stability: Stable to acids, alkalis, dilute sol'ns. of many oxidizing and

reducing agents; Surface Tension: 30 dynes/cm (0.01% sol'n.)
Composition: 100% active; 60% ethylene oxide content
Toxicity: Severe eye irritant
Warnings and Caution: Contact with concentrated oxidizing or reducing agents may be hazardous (explosive)
References: Bulletin 2303-020

Igepal CA-630. GAF Corp.
Chemical Description: Octylphenoxypoly(ethyleneoxy)ethanol
Category: Wetting agent, detergent, emulsifier, nonionic
Applications: Used in detergent compounding and aqueous processing in textile and paper industries; detergents in industrial metal cleaners, floor cleaners, detergent-sanitizers, waterless hand cleaners; emulsifier for polar solvents, e.g., xylene
Form: Liquid; Color: Pale yellow; Odor: Aromatic
Properties: Solubility: Sol. in xylene, butyl Cellosolve, perchloroethylene, ethanol, ethylene glycol, water; Sp. gr.: 1.06; Visc.: 230–260 cps; HLB: 13.0; Cloud Pt.: 146–153 (1% sol'n.); Flash Pt.: > 200 F (PMCC); Pour Pt.: 45 ± 2 F; Solidification Pt.: 40 ± 2 F; Stability: Stable to acids, alkalis, dilute sol'ns. of many oxidizing and reducing agents; Surface Tension: 31 dynes/cm (0.01% sol'n.)
Composition: 100% active; 65% ethylene oxide
Toxicity: Severe eye irritant
Warnings and Caution: Contact with concentrated oxidizing or reducing agents may be hazardous (explosive)
References: Bulletin 2303-020

Igepal CA-720. GAF Corp.
Chemical Description: Octylphenoxypoly(ethyleneoxy)ethanol
Category: Detergent, wetting agent, emulsifier, nonionic
Applications: Hard-surface detergent with aq. solubility at high temperatures, avoids film formation and soil redeposition in hot sol'ns., has good rinsability; used in hot spray, soak, and steam-cleaning systems; electrolytic cleaning and metal pickling operations
Form: Dispersed, opaque liquid; Odor: Aromatic
Properties: Solubility: Sol. in xylene, butyl Cellosolve, perchloroethylene, ethanol, ethylene glycol, water; Sp. gr.: 1.07; Visc.: 70–76 cps @ 50 C; HLB: 14.6; Cloud Pt.: 187–194 F (1% sol'n.); Flash Pt.: > 200 F (PMCC); Pour Pt.: 69 ± 2 F; Solidification Pt.: 64 ± 2 F; Stability: Stable to acids, alkalis, dilute sol'ns. of many oxidizing and reducing agents; Surface Tension: 32 dynes/cm (0.01% sol'n.)
Composition: 100% active; 73% ethylene oxide content
Toxicity: Severe eye irritant
Warnings and Caution: Contact with concentrated oxidizing or reducing agents may be hazardous (explosive)
References: Bulletin 2303-020

Igepal CA-887. GAF Corp.
Chemical Description: Octylphenoxypoly(ethyleneoxy)ethanol
Category: Emulsifier, stabilizer, nonionic
Applications: Primary emulsifier for vinyl acetate and acrylate polymerization; post-stabilizer for synthetic latices; dyeing assistant; emulsifier for fats and waxes
Form: Liquid; Color: Pale yellow; Odor: Aromatic
Properties: Sp. gr.: 1.10; HLB: 17.4; Cloud Pt.: Clear at 212 F (1% sol'n.); Flash Pt.: > 200 F (PMCC); Pour Pt.: 36 ± 2 F; Solidification Pt.: 31 ± 2 F; Stability: Stable to acids, alkalis, dilute sol'ns. of many oxidizing and reducing agents; Surface Tension: 39 dynes/cm (0.01% sol'n.)
Composition: 70% active; 87% ethylene oxide content
Toxicity: Severe eye irritant
Warnings and Caution: Contact with concentrated oxidizing or reducing agents may be hazardous (explosive)
References: Bulletin 2303-020

Igepal CA-890. GAF Corp.
Chemical Description: Octylphenoxypoly(ethyleneoxy)ethanol
Category: Emulsifier, stabilizer, nonionic
Applications: Primary emulsifier for vinyl acetate and acrylate polymerization; post-stabilizer for synthetic latices; dyeing assitant; emulsifier for fats and waxes
Form: Wax; Color: Off-white; Odor: Aromatic
Properties: Solubility: Sol. in butyl Cellosolve, ethanol, water; Sp. gr.: 1.08 @ 50 C; Visc.: Solid; HLB: 18.0; Cloud Pt.: Clear at 212 F (1% sol'n.); Flash Pt.: > 200 F (PMCC); Pour Pt.: 115 ± 2 F; Solidification Pt.: 110 ± 2 F; Stability: Stable to acids, alkalis, dilute sol'ns. of many oxidizing and reducing agents; Surface Tension: 42 dynes/cm (0.01% sol'n.)
Composition: 100% active; 90% ethylene oxide content
Toxicity: Severe eye irritant
Warnings and Caution: Contact with concentrated oxidizing or reducing agents may be hazardous (explosive)
References: Bulletin 2303-020

Igepal CA-897. GAF Corp.
Chemical Description: Octylphenoxypoly(ethyleneoxy)ethanol
Category: Emulsifier, stabilizer, nonionic
Applications: Primary emulsifier for vinyl acetate and acrylate polymerization; post-stabilizer for synthetic latices; dyeing assistant; emulsifier for fats and waxes
Form: Liquid; Color: Pale yellow; Odor: Aromatic
Properties: Sp. gr.: 1.10; HLB: 18.0; Cloud Pt.: Clear at 212 F (1% sol'n.); Flash Pt.: > 200 F (PMCC); Pour Pt.: 25 ± 2 F; Solidification Pt.: 20 ± 2 F; Stability: Stable to acids, alkalis, dilute sol'ns. of many oxidizing and reducing agents; Surface Tension: 48 dynes/cm (0.01% sol'n.)
Composition: 70% active; 90%

ethylene oxide content
Warnings and Caution: Contact with concentrated oxidizing or reducing agents may be hazardous (explosive)
References: Bulletin 2303-020

Igepal CO-210. GAF Corp.
Chemical Description: Nonylphenoxypoly(ethyleneoxy)ethanol
Category: Foam stabilizer, intermediate, defoamer, coemulsifier, detergent, dispersant, nonionic
Applications: Oil-soluble surfactant and intermediate; foam stabilizer (in low concentrations) for high-foaming detergents; defoamer (in higher concentrations) for low-foaming and cold-water detergents; coemulsifier in nonionic blends; emulsion stabilizer; detergent and dispersant for petroleum oils
Form: Liquid; Color: Yellow; Odor: Aromatic
Properties: Solubility: Sol. in low-visc. white mineral oil, deodorized kerosene, Stoddard solvent, corn oil, heavy aromatic naphtha, xylene, butyl Cellosolve, dibutyl phthalate, perchloroethylene, ethylene dichloride, ethanol; Sp. gr.: 0.99; Visc.: 300–400 cps; HLB: 4.6; Flash Pt.: > 200 F (PMCC); Pour Pt.: –3 ± 2 F; Solidification Pt.: –8 ± 2 F; Stability: Stable to acids, alkalis, dilute sol'ns. of many oxidizing and reducing agents
Composition: 100% active; 23% ethylene oxide
Warnings and Caution: Contact with concentrated oxidizing or reducing agents may be explosive
References: Bulletin 9643-019

Igepal CO-430. GAF Corp.
Chemical Description; Nonylphenoxypoly(ethyleneoxy)ethanol
Category: Coemulsifier, plasticizer, antistat, stabilizer, intermediate, detergent, dispersant, corrosion inhibitor, nonionic
Applications: Oil-sol. coemulsifier with water-sol. CO-850 and CO-880; plasticizer and antistat for polyvinyl acetate; freeze-thaw stabilizer for latex emulsions; intermediate for synthesis of high-foaming, water-sol. sulfate esters; detergent and dispersant for petroleum oils; ashless corrosion inhibitor for two-cycle engine oils
Form: Liquid; Color: Pale yellow; Odor: Aromatic
Properties: Solubility: Sol. in deodorized kerosene, Stoddard solvent, corn oil, heavy aromatic naphtha, xylene, butyl Cellosolve, dibutyl phthalate, perchloroethylene, ethylene dichloride, ethanol; Sp. gr.: 1.02; Visc.: 250–325 cps; HLB: 8.8; Flash Pt.: > 200 F (PMCC); Pour Pt.: –15 ± 2 F; Solidification Pt.: –20 ± 2 F; Stability: Stable to acids, alkalis, dilute sol'ns. of many oxidizing and reducing agents
Composition: 100% active; 44% ethylene oxide
Warnings and Caution: Contact with concentrated oxidizing or reducing agents may be explosive
References: Bulletin 9643-019

Igepal CO-520. GAF Corp.
Chemical Description: Nonylphenoxypoly(ethyleneoxy)ethanol
Category: Intermediate, emulsi-

fier, coupling agent, deicing fluid, rust inhibitor, dispersant, nonionic
Applications: Oil-sol. surfactant with HLB on border of oil and water solubility; intermediate for anionic surfactants; deicing fluid for jet aircraft fuels and automotive gasoline; rust inhibitor in home oil-storage tanks; dispersant for petroleum oils
Form: Liquid; Color: Pale yellow; Odor: Aromatic
Properties: Solubility: Sol. in Stoddard solvent, corn oil, heavy aromatic naphtha, xylene, butyl Cellosolve, dibutyl phthalate, perchloroethylene, ethylene dichloride, ethanol; Sp. gr.: 1.03; Visc.: 240–300 cps; HLB: 10.0; Flash Pt.: > 200 F (PMCC); Pour Pt.: −24 ± 2 F; Solidification Pt.: −29 ± 2 F; Stability: Stable to acids, alkalis, dilute sol'ns. of many oxidizing and reducing agents; Surface Tension: 30 dynes/cm (0.01% sol'n.)
Composition: 100% active; 50% ethylene oxide
Warnings and Caution: Contact with concentrated oxidizing or reducing agents may be explosive
References: Bulletin 9643-019

Igepal CO-530. GAF Corp.
Chemical Description: Nonylphenoxypoly(ethyleneoxy)ethanol
Category: Emulsifier, detergent, dispersant, nonionic
Applications: Emulsifier for silicones, agricultural chemicals, mineral oils; detergent and dispersant for petroleum oils; deinking of paper
Form: Liquid; Color: Pale yellow; Odor: Aromatic
Properties: Solubility: Sol. in Stoddard solvent, corn oil, heavy aromatic naphtha, xylene, butyl Cellosolve, dibutyl phthalate, perchloroethylene, ethylene dichloride, ethanol; Sp. gr.: 1.04; Visc.: 230–300 cps; HLB: 10.8; Flash Pt.: > 200 F (PMCC); Pour Pt.: −26 ± 2 F; Solidification Pt.: −31 ± 2 F; Stability: Stable to acids, alkalis, dilute sol'ns. of many oxidizing and reducing agents; Surface Tension: ≈ 28 dynes/cm (0.01% sol'n.)
Composition: 100% active; 54% ethylene oxide
Warnings and Caution: Contact with concentrated oxidizing or reducing agents may be explosive
References: Bulletin 9643-019

Igepal CO-610. GAF Corp.
Chemical Description: Nonylphenoxypoly(ethyleneoxy)ethanol
Category: Detergent, wetting agent, emulsifier, nonionic
Applications: Water-sol., low-foaming surfactant with properties of Igepal CO-630
Form: Liquid; Color: Pale yellow; Odor: Aromatic
Properties: Solubility: Sol. in corn oil, heavy aromatic naphtha, xylene, butyl Cellosolve, dibutyl phthalate, perchloroethylene, ethylene dichloride, ethanol, ethylene glycol, water; Sp. gr.: 1.05; Visc.: 230–290 cps; HLB: 12.2; Cloud Pt.: 72–82 F (1% sol'n.); Flash Pt.: > 200 F (PMCC); Pour Pt.: 37 ±

2 F; Solidification Pt.: 32 ± 2 F; Stability: Stable to acids, alkalis, dilute sol'ns. of many oxidizing and reducing agents; Surface Tension: 30 dynes/cm (0.01% sol'n.)
Composition: 100% active; 61% ethylene oxide
Warnings and Caution: Contact with concentrated oxidizing or reducing agents may be explosive
References: Bulletin 9643-019

Igepal CO-630. GAF Corp.
Chemical Description: Nonylphenoxypoly(ethyleneoxy)ethanol
Category: Detergent, wetting agent, penetrant, emulsifier, nonionic
Applications: Textile processing (detergent for wool, cotton, synthetic fibers; in cotton finishing; formulation of water-repellent finishes; wetting agent in warp sizing and carbonizing; penetrant in continuous peroxide bleaching); cleaning paper-machine felts; rewetting agent for paper towels and tissues; wetting agent in hide soaking; fatliquoring; household and industrial cleaners (laundry detergents, dairy and food plant cleaners, santizing rinses, etc.); water-based paints; metal processing with mineral acids and corrosion inhibitors; emulsion cleaning
Form: Liquid; Color: Almost colorless; Odor: Aromatic
Properties: Solubility: Sol. in corn oil, heavy aromatic naphtha, xylene, butyl Cellosolve, dibutyl phthalate, perchloroethylene, ethylene dichloride, ethanol, ethylene glycol, water; Sp. gr.: 1.06; Visc.: 225–300 cps; HLB: 13.0; Cloud Pt.: 126–133 F (1% sol'n.); Flash Pt.: > 200 F (PMCC); Pour Pt.: 31 ± 2 F; Solidification Pt.: 26 ± 2 F; Stability: Stable to acids, alkalis, dilute sol'ns. of many oxidizing and reducing agents; Surface Tension: 31 dynes/cm (0.01% sol'n.)
Composition: 100% active; 65% ethylene oxide
Warnings and Caution: Contact with concentrated oxidizing or reducing agents may be explosive
References: Bulletin 9643-019

Igepal CO-660. GAF Corp.
Chemical Description: Nonylphenoxypoly(ethyleneoxy)ethanol
Category and Applications: Similar to, but more soluble than Igepal CO-630
Form: Liquid; Color: Pale yellow; Odor: Aromatic
Properties: Solubility: Sol. in corn oil, heavy aromatic naphtha, xylene, butyl Cellosolve, dibutyl phthalate, perchloroethylene, ethylene dichloride, ethanol, water; Sp. gr.: 1.06; Visc.: 225–275 cps; HLB: 13.2; Cloud Pt.: 140–149 F (1% sol'n.); Flash Pt.: > 200 F (PMCC); Pour Pt.: 46 ± 2 F; Solidification Pt.: 41 ± 2 F; Stability: Stable to acids, alkalis, dilute sol'ns. of many oxidizing and reducing agents; Surface Tension: 31 dynes/cm (0.01% sol'n.)
Composition: 100% active; 66% ethylene oxide
Warnings and Caution: Contact

with concentrated oxidizing or reducing agents may be explosive
References: Bulletin 9643-019

Igepal CO-710. GAF Corp.
Chemical Description: Nonylphenoxypoly(ethyleneoxy)ethanol
Category and Applications: Similar to, but more sol. at slightly elevated temperatures than, Igepal CO-630; also scouring agent for raw wool under any pH conditions
Form: Liquid; Color: Pale yellow; Odor: Aromatic
Properties: Solubility: Sol. in heavy aromatic naphtha, xylene, butyl Cellosolve, dibutyl phthalate, perchloroethylene, ethylene dichloride, ethanol, ethylene glycol, water; Sp. gr.: 1.06; Visc.: 240–300 cps; HLB: 13.6; Cloud Pt.: 158–165 F (1% sol'n.); Flash Pt.: > 200 F (PMCC); Pour Pt.: 49 ± 2 F; Solidification Pt.: 45 ± 2 F; Stability: Stable to acids, alkalis, dilute sol'ns. of many oxidizing and reducing agents; Surface Tension: 32 dynes/cm (0.01% sol'n.)
Composition: 100% active; 68% ethylene oxide
Warnings and Caution: Contact with concentrated oxidizing or reducing agents may be explosive
References: Bulletin 9643-019

Igepal CO-720. GAF Corp.
Chemical Description: Nonylphenoxypoly(ethyleneoxy)ethanol
Category and Applications: See Igepal CO-710
Form: Dispersed opaque liquid; Odor: Aromatic
Properties: Solubility: Sol. in corn oil, heavy aromatic naphtha, xylene, butyl Cellosolve, dibutyl phthalate, perchloroethylene, ethylene dichloride, ethanol, ethylene glycol, water; Sp. gr.: 1.06; Visc.: 260–340 cps; Flash Pt.: > 200 F (PMCC); Pour Pt.: 62 ± 2 F; Solidification Pt.: 57 ± 2 F; Stability: Stable to acids, alkalis, dilute sol'ns. of many oxidizing and reducing agents; Surface Tension: 34 dynes/cm (0.01% sol'n.)
Composition: 100% active; 71% ethylene oxide
Warnings and Caution: Contact with concentrated oxidizing or reducing agents may be explosive
References: Bulletin 9643-019

Igepal CO-730. GAF Corp.
Chemical Description: Nonylphenoxypoly(ethyleneoxy)ethanol
Category: Detergent, dispersant, emulsifier, penetrant, wetting agent, nonionic
Applications: General detergent and dispersant esp. in high-temperature applications; emulsifier for fats, oils, waxes; penetrant and wetting agent in caustic formulations, e.g., alkaline metal cleaners, bottle washing fluids; surfactant for use with high electrolyte concentrations, in presence of mineral acids and corrosion inhibitors
Form: Dispersed liquid; Color: Yellow; Odor: Aromatic
Properties: Solubility: Sol. in heavy aromatic naphtha, xylene, butyl Cellosolve, dibutyl phthalate; perchloroethylene, ethylene

dichloride, ethanol, ethylene glycol, water; Sp. gr.: 1.07; Visc.: 450–550 cps; HLB: 15.0; Cloud Pt.: 203–212 F (1% sol'n.); Flash Pt.: > 200 F (PMCC); Pour Pt.: 71 ± 2 F; Solidification Pt.: 68 ± 2 F; Stability: Stable to acids, alkalis, dilute sol'ns. of many oxidizing and reducing agents; Surface Tension: 36 dynes/cm (0.01% sol'n.)
Composition: 100% active; 75% ethylene oxide
Warnings and Caution: Contact with concentrated oxidizing or reducing agents may be explosive
References: Bulletin 9643-019

Igepal CO-850. GAF Corp.
Chemical Description: Nonylphenoxypoly(ethyleneoxy)ethanol
Category: Detergent, wetting agent, dispersant, emulsifier, demulsifier, stabilizer, intermediate, nonionic
Applications: Water-sol. detergent, dispersant and wetting agent at elevated temperatures and high electrolyte concentrations; emulsifier for fats, oils, waxes, solvents; demulsifier for crude petroleum oil emulsions; stabilizer for synthetic latices; emulsifier for polyester resins; ingredient of silicone emulsions used in glass mold-release agents
Form: Wax; Color: Pale yellow; Odor: Aromatic
Properties: Solubility: Sol. in heavy aromatic naphtha, xylene, butyl Cellosolve, dibutyl phthalate, perchloroethylene, ethylene dichloride, ethanol, ethylene glycol, water; Sp. gr.: 1.08 @ 50 C; Visc.: Solid; HLB: 16.0; Cloud Pt.: Clear at 212 F (1% sol'n.); Flash Pt.: > 200 F (PMCC); Pour Pt.: 91 ± 2 F; Solidification Pt.: 86 ± 2 F; Stability: Stable to acids, alkalis, dilute sol'ns. of many oxidizing and reducing agents; Surface Tension: 39 dynes/cm (0.01% sol'n.)
Composition: 100% active; 80% ethylene oxide
Warnings and Caution: Contact with concentrated oxidizing or reducing agents may be explosive
References: Bulletin 9643-019

Igepal CO-880. GAF Corp.
Chemical Description: Nonylphenoxypoly(ethyleneoxy)ethanol
Category: Detergent, solubilizer, nonionic
Applications: Detergent for high-temperature scouring of textiles in pressure equipment; solubilizer for chlordane, toxaphene, kerosene, essential oils; other uses similar to Igepal CO-850
Form: Wax; Color: Pale yellow; Odor: Aromatic
Properties: Solubility: Sol. in xylene, butyl Cellosolve, ethylene dichloride, ethanol, ethylene glycol, water; Sp. gr.: 1.08 @ 50 C; Visc.: Solid; HLB: 17.2; Cloud Pt.: Clear at 212 F (1% sol'n.); Flash Pt.: > 200 F (PMCC); Pour Pt.: 109 ± 2 F; Solidification Pt.: 104 ± 2 F; Stability: Stable to acids, alkalis, dilute sol'ns. of many oxidizing and reducing agents; Surface Tension: 43 dynes/cm (0.01% sol'n.)
Composition: 100% active; 86%

ethylene oxide
Warnings and Caution: Contact with concentrated oxidizing or reducing agents may be explosive
References: Bulletin 9643-019

Igepal CO-887. GAF Corp.
Chemical Description: Nonylphenoxypoly(ethyleneoxy)ethanol
Category and Applications: See Igepal CO-880
Form: Liquid; Color: Pale yellow; Odor: Aromatic
Properties: Sp. gr.: 1.09; HLB: 17.2; Cloud Pt.: Clear at 212 F (1% sol'n.); Flash Pt.: > 200 F (PMCC); Pour Pt.: 34 ± 2 F; Solidification Pt.: 25 ± 2 F; Stability: Stable to acids, alkalis, dilute sol'ns. of many oxidizing and reducing agents
Composition: 70% active; 86% ethylene oxide
Warnings and Caution: Contact with concentrated oxidizing or reducing agents may be explosive
References: Bulletin 9643-019

Igepal CO-890. GAF Corp.
Chemical Description: Nonylphenoxypoly(ethyleneoxy)ethanol
Category: Emulsifier, stabilizer, wetting agent, dyeing assistant, nonionic
Applications: Highly water-sol. emulsifiers and stabilizers effective at high temperatures; wetting agents in concentrated electrolyte sol'ns.; stabilizers for synthetic latices; emulsifiers and stabilizers in floor waxes and polishes; polymerization emulsifiers for vinyl acetate and acrylic emulsions; dyeing assistants for azoic and acid dyestuffs
Form: Wax; Color: Off-white; Odor: Aromatic
Properties: Solubility: Sol. in ethylene dichloride, ethanol, water; Sp. gr.: 1.09 @ 50 C; HLB: 17.8; Cloud Pt.: Clear at 212 F (1% sol'n.); Flash Pt.: > 200 F (PMCC); Pour Pt.: 112 ± 2 F; Solidification Pt.: 106 ± 2 F; Stability: Stable to acids, alkalis, dilute sol'ns. of many oxidizing and reducing agents;
Composition: 100% active; 89% ethylene oxide
Warnings and Caution: Contact with concentrated oxidizing or reducing agents may be explosive
References: Bulletin 9643-019

Igepal CO-897. GAF Corp.
Chemical Description: Nonylphenoxypoly(ethyleneoxy)ethanol
Category and Applications: See Igepal CO-890
Form: Liquid; Color: Pale yellow; Odor: Aromatic
Properties: Sp. gr.: 1.10; HLB: 17.8; Cloud Pt.: Clear at 212 F (1% sol'n.); Flash Pt.: > 200 F (PMCC); Pour Pt.: 46 ± 2 F; Solidification Pt.: 41 ± 2 F; Stability: Stable to acids, alkalis, dilute sol'ns. of many oxidizing and reducing agents
Composition: 70% active; 89% ethylene oxide
Warnings and Caution: Contact with concentrated oxidizing or reducing agents may be explosive
References: Bulletin 9643-019

Igepal CO-970. GAF Corp.
Chemical Description: Nonylphenoxypoly(ethyleneoxy)ethanol
Category and Applications: See Igepal CO-890
Form: Wax; Color: Off-white; Odor: Aromatic
Properties: Solubility: Sol. in ethylene dichloride, ethanol, water; Sp. gr.: 1.10 @ 50 C; HLB: 18.2; Cloud Pt.: Clear at 212 F (1% sol'n.); Flash Pt.: > 200 F (PMCC); Pour Pt.: 114 ± 2 F; Solidification Pt.: 108 ± 2 F; Stability: Stable to acids, alkalis, dilute sol'ns. of many oxidizing and reducing agents
Composition: 100% active; 91% ethylene oxide
Warnings and Caution: Contact with concentrated oxidizing or reducing agents may be explosive
References: Bulletin 9643-019

with concentrated oxidizing or re-

Igepal CO-977. GAF Corp.
Chemical Description: Nonylphenoxypoly(ethyleneoxy)ethanol
Category and Applications: See Igepal CO-890
Form: Liquid; Color: Pale yellow; Odor: Aromatic
Properties: Sp. gr.: 1.10; HLB: 18.2; Cloud Pt.: Clear at 212 F (1% sol'n.); Flash Pt.: > 200 F (PMCC); Pour Pt.: 52 ± 2 F; Solidification Pt.: 46 ± 2 F; Stability: Stable to acids, alkalis, dilute sol'ns. of many oxidizing and reducing agents
Warnings and Caution: Contact with concentrated oxidizing or reducing agents may be explosive
References: Bulletin 9643-019

Igepal CO-990. GAF Corp.
Chemical Description: Nonylphenoxypoly(ethyleneoxy)ethanol
Category and Applications: See Igepal CO-890
Form: Wax; Color: Off-white; Odor: Aromatic
Properties: Solubility: Sol. in ethylene dichloride, ethanol, water; Sp. gr.: 1.12 @ 50 C; HLB: 19.0; Cloud Pt.: Clear at 212 F (1% sol'n.); Flash Pt.: > 200 F (PMCC); Pour Pt.: 122 ± 2 F; Solidification Pt.: 116 ± 2 F; Stability: Stable to acids, alkalis, dilute sol'ns. of many oxidizing and reducing agents
Composition: 100% active; 95% ethylene oxide
Warnings and Caution: Contact with concentrated oxidizing or reducing agents may be explosive
References: Bulletin 9643-019

Igepal CO-997. GAF Corp.
Chemical Description: Nonylphenoxypoly(ethyleneoxy)ethanol
Category and Applications: See Igepal CO-890
Form: Liquid; Color: Pale yellow; Odor: Aromatic
Properties: Sp. gr.: 1.11; HLB: 19.0; Cloud Pt.: Clear at 212 F (1% sol'n.); Flash Pt.: > 200 F (PMCC); Pour Pt.: 68 ± 2 F; Solidification Pt.: 63 ± 2 F; Stability: Stable to acids, alkalis, dilute sol'ns. of many oxidizing and re-

ducing agents
Composition: 70% active; 95% ethylene oxide
Warnings and Caution: Contact with concentrated oxidizing or reducing agents may be explosive
References: Bulletin 9643-019

Igepal CTA-639. GAF Corp.
Chemical Description: Modified alkylphenoxypoly(ethyleneoxy)-ethanol
Category: Detergent, wetting agent, emulsifier, nonionic
Applications: Emulsifier in rubber latex emulsion paints esp. when a calcium-insensitive emulsifier is required
Form: Free-flowing, slightly viscous liquid
Properties: Solubility (1%): Sol. in water, xylene, ethanol, carbon tetrachloride, ethylene glycol, ethylene glycol monobutyl ether; partly sol. in Stoddard solvent; Sp. gr.: 1.06; Visc.: 387 cps; Cloud Pt.: 70–78 C (1% aq.); Flash Pt.: 249 C (COC); Stability: Stable to 3% H_2SO_4; Surface Tension: 29.8 dynes/cm (0.1% aq.); pH: 6 (1% sol'n.)
Composition: 100% active ingred.
References: Bulletin 7543-127

Igepal DM-430. GAF Corp.
Chemical Description: Dialkylphenoxypoly(ethyleneoxy)ethanol
Category: Emulsifier, nonionic
Applications: W/O emulsifier for nonpolar solvents and oils, e.g., hexane, heptane, mineral oil; emulsification of agricultural chemicals, in emulsion polymerization, in fatliquoring of leather, etc.
Form: Slightly hazy liquid; Color: Yellow; Odor: Aromatic
Properties: Solubility: Sol. in corn oil, heavy aromatic naphtha, xylene, monobutyl glycol ether, perchloroethylene, ethanol; partly sol. in Stoddard solvent; Density: 8.3 lb/gal; Sp. gr.: 0.995; Visc.: 13 cps @ 100 C; HLB: 9.4; Flash Pt.: 500–520 F (COC); Pour Pt.: –7 C; Stability: Stable to acids, alkalis, dilute sol'ns. of many oxidizing and reducing agents
Composition: 100% active
Toxicity: Eye irritant
Warnings and Caution: Contact with concentrated oxidizing or reducing agents may be explosive
References: Bulletin 2303-018

Igepal DM-530. GAF Corp.
Chemical Description: Dialkylphenoxypoly(ethyleneoxy)ethanol
Category: Emulsifier, nonionic
Applications: Emulsifier for naphthenic oils used in emulsifiable cutting oils; acid bowl-cleaner emulsions; textile finishing oils; agricultural emulsions
Form: Slightly hazy liquid; Color: Yellow; Odor: Aromatic
Properties: Solubility: Sol. in corn oil, heavy aromatic naphtha, xylene, monobutyl glycol ether, perchloroethylene, ethanol; partly sol. in Stoddard solvent, water; Density: 8.43 lb/gal; Sp. gr.: 1.010; Visc.: 15 cps @ 100 C; HLB: 10.6; Flash Pt.: 500–520 F (COC); Pour

Pt.: 0 C; Stability: Stable to acids, alkalis, dilute sol'ns. of many oxidizing and reducing agents; Surface Tension: 29 dynes/cm (0.01% sol'n.)
Composition: 100% active
Toxicity: Eye irritant
Warnings and Caution: Contact with concentrated oxidizing or reducing agents may be explosive
References: Bulletin 2303-018

Igepal DM-710. GAF Corp.
Chemical Description: Dialkylphenoxypoly(ethyleneoxy)ethanol
Category: Detergent, emulsifier, nonionic
Applications: Detergents esp. for cellulosic and synthetic fibers; for built, controlled-suds laundering compounds; industrial metal cleaners; acid cleaners; floor cleaners; detergent-sanitizers; leather degreasing; paper-mill felt washing; latex emulsification; emulsifiable pesticides
Form: Opaque liquid; Color: Pale yellow; Odor: Aromatic
Properties: Solubility: Sol. in corn oil, heavy aromatic naphtha, xylene, monobutyl glycol ether, perchloroethylene, ethanol, water; partly sol. in Stoddard solvent; Density: 8.72 lb/gal; Sp. gr.: 1.045; Visc.: 19 cps @ 100 C; Cloud Pt.: 48–52 C (1% sol'n.); Flash Pt.: 500–520 F (COC); Pour Pt.: 18 C; Stability: Stable to acids, alkalis, dilute sol'ns. of many oxidizing and reducing agents; Surface Tension: 29 dynes/cm (0.01% sol'n.)
Composition: 100% active
Toxicity: Eye irritant
Warnings and Caution: Contact with concentrated oxidizing or reducing agents may be explosive
References: Bulletin 2303-018

Igepal DM-730. GAF Corp.
Chemical Description: Dialkylphenoxypoly(ethyleneoxy)ethanol
Category and Applications: See Igepal DM-710
Form: Paste; Color: Pale yellow; Odor: Aromatic
Properties: Solubility: Sol. in xylene, monobutyl glycol ether, ethanol, ethylene glycol, water; partly sol. in corn oil, heavy aromatic naphtha, perchloroethylene; Density: 8.75 lb/gal; Sp. gr.: 1.049; Visc.: 25 cps @ 100 C; Cloud Pt.: > 100 C (1% sol'n.); Flash Pt.: 500–520 F (COC); Pour Pt.: 25 C; Stability: Stable to acids, alkalis, dilute sol'ns. of many oxidizing and reducing agents; Surface Tension: 33 dynes/cm (0.01% sol'n.)
Composition: 100% active
Toxicity: Eye irritant
Warnings and Caution: Contact with concentrated oxidizing or reducing agents may be explosive
References: Bulletin 2303-018

Igepal DM-880. GAF Corp.
Chemical Description: Dialkylphenoxypoly(ethyleneoxy)ethanol
Category: Emulsifier, solubilizer, nonionic
Applications: Primary emulsifier in emulsion polymerization and

latex stabilization; solubilizer for essential oils; emulsifier for fatty acids, polar solvents, and oils; as surfactant in high-temperature acid and alkali cleaners
Form: Wax; Color: Pale yellow; Odor: Aromatic
Properties: Solubility: Sol. in xylene, monobutyl glycol ether, ethanol, ethylene glycol, water; Sp. gr.: 1.050 @ 50 C; Visc.: 55 cps @ 100 C; Cloud Pt.: > 100 C (1% sol'n.); Flash Pt.: 500–520 F (COC); Pour Pt.: 47 C; Stability: Stable to acids, alkalis, dilute sol'ns. of many oxidizing and reducing agents; Surface Tension: 38 dynes/cm (0.01% sol'n.)
Composition: 100% active
Toxicity: Eye irritant
Warnings and Caution: Contact with concentrated oxidizing or reducing agents may be explosive
References: Bulletin 2303-018

Igepal DM-970. GAF Corp.
Chemical Description: Dialkylphenoxypoly(ethyleneoxy)ethanol
Category: Detergent, dispersant, wetting agent, stabilizer, nonionic
Applications: Cotton detergent and soil dispersant in commercial laundry formulations and in controlled-suds, household and laundry products; textile scouring agent and dye levelling aid; hard-surface detergent (e.g., wall and floor cleaners, dairy cleaners and sanitizers, marine-fuel-tank detergents for use with salt water, and truck and railroad car cleaners; cosmetic bars and cleansing creams; insecticide formulations; in paper coatings as a stabilizer for latices and dispersant for high-solids clay pigments; oil well drilling aid; in asphalt emulsifiers; latex stabilizer in emulsion paints
Form: Free-flowing flakes; Color: White
Properties: Solubility: Sol. in water, aromatic solvents, methanol, ethanol; Sp. gr.: ≈ 1.05 @ 80 C; HLB: 19.0; Cloud Pt.: > 100 C (1% aq.); Solidification Pt.: ≈ 54 C; Stability: Stable to storage
Composition: 100% active ingred.
Reaction: Neutral
Containers: 50 lb bags
Warnings and Caution: Contact with concentrated oxidizing or reducing agents may be explosive
References: Bulletin 2303-009

Igepal RC-520. GAF Corp.
Chemical Description: Dodecylphenoxypoly(ethyleneoxy)ethanol
Category: Rewetting agent, emulsifier, nonionic
Applications: Rewetting agent for paper, paper towels, tissues; emulsifier for vegetable and mineral oils; textile wetting agent for desizing and dyeing operations of cotton and rayon fabrics
Form: Slightly viscous liquid; Color: 2 max (VCS)
Properties: Solubility: Sol. in most hydrocarbon and polar solvents; disp. in water; Sp. gr.: 1.03; Visc.: 360 cps (Ostwald-Fenske); Pour Pt.: −10 C; Stability: Stable to storage at high temperatures; unaffected by hard water, ionic surfactants, and acids or alkalis
Composition: 100% active ingred.

Containers: Bulk and closed-head steel drums
References: Bulletin 2303-010

Igepal RC-620. GAF Corp.
Chemical Description: Dodecylphenoxypoly(ethyleneoxy)ethanol
Category: Detergent, wetting agent, emulsifier, dispersant, nonionic
Applications: All-purpose detergent and wetting agent with good emulsifying and dispersing properties; acid-cleaning; detergent-sanitizing; grease-cutting formulations; emulsifier and wetter in agricultural compounds, rug shampoos, white-wall tire cleaners; dedusting agent
Form: Liquid; Color: 200 max (APHA)
Properties: HLB: 10.0; Cloud Pt.: 38–42 C (1% aq.); Stability: Stable to high-temperature storage; unaffected by hard water, ionic surfactants, acids, alkalis
Composition: 100% active ingred.

Igepal RC-630. GAF Corp.
Chemical Description: Dodecylphenoxypoly(ethyleneoxy)ethanol
Category: Detergent, wetting agent, emulsifier, dispersant, nonionic
Applications: Industrial cleaners, heavy-duty liquid household detergents and detergent tablets
Form: Liquid; Color: 200 max (APHA)
Properties: HLB: 12.7; Cloud Pt.: 62–66 C (1% aq.)
Composition: 100% active ingred.

Igepon AM-78. GAF Corp.
Chemical Description: Myristic acid, 2-sulfoethyl ester, sodium salt
Category: Detergent, anionic
Applications: Detergent with good compatibility with hard water, mildness, high foam for bubble baths, cosmetic cleansing bars, shampoos, sudsing lotions and creams, syndet/soap bars
Form: Powder
Properties: Solubility: 0.01 g/100 ml; Surface Tension: 32.5 dynes/cm (0.05%)
Composition: 83% active ingred.

Igepon CN-42. GAF Corp.
Chemical Description: Sodium-N-cyclohexyl-N-palmitoyl taurate
Category: Detergent, dispersant, stabilizer, anionic
Applications: Low foaming detergent and dispersant for mechanical dishwasher and industrial cleaner formulations; stabilizer for synthetic rubber emulsions, esp. GR-S latices; finishing agent for tow yarns
Form: Soft firm paste; Color: Light tan; Odor: Fatty
Properties: m.w.: 467.2; Solubility: Readily sol. in water with moderate stirring; pH: 8.0–9.5 (10% sol'n.)
Composition: 23% min. active

Igepon T-33. GAF Corp.
Chemical Description: Sodium-N-methyl-N-oleoyl taurate
Category: Detergent, wetting agent, dispersant, anionic
Applications: General-purpose sur-

factant; textile surfactant for washing piece goods, raw stock, yarns at all temperatures and pH conditions; soaping prints, vat, and naphthol dyed fibers; assistant in Kier boiling, bleaching, dyeing, wetting, textile finishing; in industrial detergents, e.g., rug shampoos, liquid industrial cleaners, alkaline bottle-washing compounds, metal cleaners; in the manufacture of corrugating media from semichemical or kraft pulp (overcomes "air-sizing")
Form: Nearly clear, free-flowing liquid; Color: Pale yellow; Odor: Perfumed, alcoholic
Properties: m.w.: 425; Solubility: Sol. in water; pH: 6.5–8.0 (10% sol'n.)
Composition: 32% min. active

Igepon T-43. GAF Corp.
Chemical Description: Sodium-N-methyl-N-oleoyl taurate
Category and Applications: See Igepon T-33
Form: Viscous heterogeneous liquid, slurry; Color: Milky white; Odor: Fatty
Properties: m.w.: 425; Solubility: Readily sol. in water > 54 C; slowly sol. @ 25 C; pH: 6.5–8.0 (5% sol'n.)
Composition: 33% min. active

Igepon T-51. GAF Corp.
Chemical Description: Sodium-N-methyl-N-oleoyl taurate
Category and Applications: See Igepon T-33
Form: Gel; Color: Light amber; Odor: Perfumed
Properties: m.w.: 425; Solubility: Readily sol. in water @ 54 C; pH: 6.5–8.0 (10% sol'n.)
Composition: 13.9% min. active

Igepon T-77. GAF Corp.
Chemical Description: Sodium-N-methyl-N-oleoyl taurate
Category and Applications: See Igepon T-33
Form: Flakes; Color: Cream; Odor: Fatty
Properties: m.w.: 425; Solubility: Slowly sol. in water @ 54 C; pH: 6.5–8.0 (2% sol'n.)
Composition: 67% min. active

Igepon TC-42. GAF Corp.
Chemical Description: Sodium-N-methyl-N-"coconut oil acid" taurate
Category: Detergent, foaming agent, dispersant, anionic
Applications: Chemically stable detergent with good foaming, lathering, and dispersing properties for use in detergent bars, shampoos, bubble baths, and other cosmetic preparations
Form: Soft, smooth paste; Color: White
Properties: m.w.: 363; Stability: Good; pH: 7.0–8.5 (10% sol'n.)
Composition: 24% min. active

Igepon TK-32. GAF Corp.
Chemical Description: Sodium-N-methyl-N-"tall oil acid" taurate
Category: Detergent, suspending and dispersing agent, anionic
Applications: Precipitation inhibitor for organic and inorganic salts of Ba, Ca, Sr; effective at less than

stoichiometric concentrations; in the petroleum industry to prevent scale formation in oil-well tubing and surface flow lines
Form: Clear liquid; Color: 8 max. (VCS)
Properties: m.w.: 439; Solubility: Readily sol. in water; pH: 8–10 (10% sol'n.)
Composition: 20% min. active

Igepon TM-43. GAF Corp.
Chemical Description: Sodium N-methyl-N-oleoyl taurate
Category and Applications: See Igepon AM-78
Form: Slurry
Properties: Solubility: 20 g/100 ml; Surface Tension: 36.8 dynes/cm (0.05%); Stability: Good resistance to hydrolysis over wide pH range
Composition: 33% active ingred.

Igepon TN-74. GAF Corp.
Chemical Description: Sodium-N-methyl-N-acyl taurate
Category: Detergent, suspending and dispersing agent, anionic
Applications: Detergent with suspending and dispersing properties; herbicide and insecticide dispersant
Form: Fine powder; Color: White; Odor: Fatty
Properties: m.w.: 406; Solubility: Readily sol. in water > 43 C; pH: 6.5–8.0 (2% sol'n.)
Composition: 44% min. active

Imerol DU. Sandoz Ltd.
Chemical Description: Ethylene oxide adduct
Category: Softener, antistatic, nonionic
Applications: Dry cleaning assistant, softener, antistatic agent
Form: Liquid

Imerol NCP. Sandoz Ltd.
Chemical Description: Sulfonate containing solvents
Category: Wetting agent, anionic
Applications: Wetting agent
Form: Liquid

Imerol NCP Liquid. Sandoz Ltd.
Chemical Description: Sulfonate containing solvents
Category: Emulsifier, dispersant, wetting agent, anionic
Applications: Used with Lyogen V (U) to convert basic dyes to dispersed form directly in the print paste
Form: Liquid; Color: Brown
Properties: Solubility: Good in print pastes; Sp. gr.: 1.09
Composition: 57% active ingred.

Imerol SS Liquid. Sandoz Ltd.
Category: Detergent
Applications: Solvent scouring of all fibers; preparation of cotton/polyester; scouring and dyeing of nylon hosiery; scouring synthetic fibers; scouring tricot on beams; tar and oil spot removal on cotton
Form: Liquid; Color: Yellowish
Properties: Solubility: Miscible with water; Sp. gr.: 0.874
Composition: 88% active ingred.

Imerol VLF Liquid. Sandoz Ltd.
Category: Detergent
Applications: Solvent/surfactant

combination for use in polyester dyeing; prescouring polyester knits and other synthetics in Gaston County jets; one-step scouring/dyeing of polyester where carrier free dyeing is preferred
Form: Liquid; Color: Yellow
Properties: Solubility: Disp. in water @ 70–100 F; Sp. gr.: 0.94
Composition: 100% active ingred.
Reaction: Alkaline

Imerol XN. Sandoz Ltd.
Chemical Description: Anionic modified polyglycol ether
Category: Scouring agent, weakly anionic
Applications: Odorless solvent-sol. scouring and stain removing agent
Form: Liquid

Imwitor 780 K. Dyanamit Nobel Aktiengesellschaft
Chemical Description: Modified partial glyceride of isostearic acid
Category: Emulsifier, nonionic
Applications: W/O emulsifier for cosmetic and pharmaceutical preparations
Form: Med. visc. liquid; Color: Yellowish; Odor: Slight characteristic
Properties: Solubility: Sol. in chloroform, light petroleum, benzene, alcohol (96%), 1,6-hexandiol, liquid paraffin, vegetable and animal oils; Sp. gr.: 0.96–0.98 @ 20 C; Visc.: 700–900 cps; HLB: 3.7; Sapon. No.: 240–260; Acid No.: 5 max.; Iodine Value: 7 max.
Composition: 100% conc.
Containers: 25 kg net plastic containers

Imwitor 900 K. Dynamit Nobel Aktiengesellschaft
Chemical Description: Glycerylmonodiester of palmitic and stearic acid
Category: Stabilizer, emulsifier, nonionic
Applications: Auxiliary in emulsification of fats for W/O emulsions; emulsion stabilizer; thickener; cosmetic preparations and production of lipsticks
Form: Waxy solid; Color: White to cream; Odor: Slightly fatty
Properties: Solubility: Sol. in all fats, oils, waxes, chloroform, ether, ethanol (@ 60 C); M.P.: 56–60 C; HLB: 3.8; Sapon. No.: 160–176; Acid No.: 3 max.; Iodine Value: 3 max.
Composition: 42–48% monoglycerides
Containers: 25 kg net cartons–powder or flakes

Imwitor 960 K. Dynamit Nobel Aktiengesellschaft
Chemical Description: Partial glyceride mixture of natural palmitic and stearic acid with an anionic emulsifier
Category: Emulsifier, anionic
Applications: Emulsifier and ointment base for pharmaceutical and cosmetic creams of the O/W type; self-emulsifying
Form: Hard waxy solid; Color: White; Odor: Typical
Properties: Solubility: Sol. in all fats, oils, waxes, in chloroform, benzene, ether, ethanol (96% @ 60 C); M.P.: 56–60 C; Sapon. No.:

155–170; Acid No.: 3 max.; Iodine Value: 3 max.
Composition: 30–40% monoester content
Containers: 20 kg net cartons

Indulin W-1. Westvaco, Polychemicals Dept.
Chemical Description: Amine derivative of pine lignin
Category: Emulsifier, retarding agent, cationic
Applications: Emulsifier in asphalt emulsions; retarding agent in cement
Form: Free-flowing powder; Color: Brown
Properties: Solubility: Sol. in water ⩾ pH 8.3 and ⩽ pH 4.2; Density: 22 lb/ft^3 (loose); Visc.: ≈ 100 cps @ 20% solids, 10,000 cps @ 30% solids; Surface Tension: 47 dynes (1% aq.); pH: 10.5 (2% aq.)
Composition: 100% conc.
Containers: 40 lb net multiwall paper bags

Indu-Sol 238. Alkaril Chem.
Category: Wetting agent, suspending agent, lubricant, rust inhibitor
Applications: Concentrated, synthetic, nonpetroleum type cutting fluid and coolant for ferrous metals and aluminum; in grinding operations and general machining operations
Form: Translucent liquid; Color: Greenish-red
Properties: Density: 11 lb/gal (Imp.); Flash Pt.: None; Fire Pt.: None; Stability: Excellent @ 5%; pH: 9.2–9.6 @ 5%; Biodegradable

Industrene 105. Humko Sheffield Chem.
Chemical Description: Oleic fatty acid
Category and Applications: Used in alkyd resins, rubber compounding, mineral-flotation adjuvants, foam depressants, caulking compounds, buffing compounds, water repellents, paper manufacture, polishes, soaps, abrasives, cutting oils, candles, crayons, emulsifiers
Form: Liquid; Color: 6 max. (Gardner)
Properties: Sapon. No.: 197–205; Acid No.: 195–204
Composition: 100% conc.

Industrene 138. Humko Sheffield Chem.
Chemical Description: Distilled cotton fatty acid
Category and Applications: See Industrene 105
Form: Liquid; Color: 7 max. (Gardner)
Properties: Sapon. No.: 197–205; Acid No.: 196–204
Composition: 100% conc.

Industrene 143. Humko Sheffield Chem.
Chemical Description: Distilled animal fatty acid
Category and Applications: See Industrene 105
Form: Paste; Color: 5 max. (Gardner)
Properties: Sapon. No.: 203–207; Acid No.: 202–206
Composition: 100% conc.

Industrene 154-BG. Humko Sheffield Chem.
Chemical Description: Hydrogenated buffing grade fatty acid
Category and Applications: See Industrene105
Form: Solid; Color: 9 max. (Gardner)
Properties: Sapon. No.: 191–203; Acid No.: 190–202
Composition: 100% conc.

Industrene 205. Humko Sheffield Chem.
Chemical Description: White oleic fatty acid
Category and Applications: See Industrene 105
Form: Liquid
Properties: Sapon. No.: 197–205; Acid No.: 195–204
Composition: 100% conc.

Industrene 225. Humko Sheffield Chem.
Chemical Description: Distilled soya alkyd fatty acid
Category and Applications: See Industrene 105
Form: Liquid; Color: 3 max. (Gardner)
Properties: Sapon. No.: 197–203; Acid No.: 195–201
Composition: 100% conc.

Industrene 226. Humko Sheffield Chem.
Chemical Description: Distilled soya alkyd fatty acid
Category and Applications: See Industrene 105
Form: Liquid; Color: 4 max. (Gardner)
Properties: Sapon. No.: 199–204; Acid No.: 198–203
Composition: 100% conc.

Industrene 232. Humko Sheffield Chem.
Chemical Description: Distilled corn-type fatty acid
Category and Applications: See Industrene 105
Form: Liquid; Color: 6 max. (Gardner)
Properties: Sapon. No.: 197–205; Acid No.: 196–204
Composition: 100% conc.

Industrene 239. Humko Sheffield Chem.
Chemical Description: Distilled cotton fatty acid
Category and Applications: See Industrene 105
Form: Liquid; Color: 5 max. (Gardner)
Properties: Sapon. No.: 200–206; Acid No.: 199–205
Composition: 100% conc.

Industrene 325. Humko Sheffield Chem.
Chemical Description: Distilled coconut fatty acid
Category and Applications: See Industrene 105
Form: Paste
Properties: Sapon. No.: 260–280; Acid No.: 258–278
Composition: 100% conc.

Industrene 328. Humko Sheffield Chem.
Chemical Description: Stripped coconut fatty acid

Category and Applications: See Industrene 105
Form: Paste
Properties: Sapon. No.: 254–261; Acid No.: 253–260
Composition: 100% conc.

Industrene 333. Humko Sheffield Chem.
Chemical Description: Hydrogenated stripped coconut fatty acid
Category and Applications: See Industrene 105
Form: Paste
Properties: Sapon. No.: 251–265; Acid No.: 250–264
Composition: 100% conc.

Industrene 365. Humko Sheffield Chem.
Chemical Description: 50% caprylic acid
Category and Applications: See Industrene 105
Properties: Sapon. No.: 346–361; Acid No.: 345–360
Composition: 100% conc.

Industrene 3022. Humko Sheffield Chem.
Chemical Description: Hydrogenated fish fatty acid
Category and Applications: See Industrene 105
Form: Solid
Properties: Sapon. No.: 196–204; Acid No.: 195–203
Composition: 100% conc.

Industrene 4022. Humko Sheffield Chem.
Chemical Description: 40% arachidic and behenic fatty acid
Category and Applications: See Industrene 105
Form: Solid
Properties: Sapon. No.: 191–199; Acid No.: 190–198
Composition: 100% conc.

Industrene 4516. Humko Sheffield Chem.
Chemical Description: 45% palmitic fatty acid
Category and Applications: See Industrene 105
Form: Solid
Properties: Sapon. No.: 206–212; Acid No.: 205–211
Composition: 100% conc.

Industrene 4518. Humko Sheffield Chem.
Chemical Description: Single pressed grade fatty acid
Category and Applications: See Industrene 105
Form: Solid; Color: 3 max. (Gardner)
Properties: Sapon. No.: 205–212; Acid No.: 204–211
Composition: 100% conc.

Industrene 5016. Humko Sheffield Chem.
Chemical Description: Double pressed grade fatty acid
Category and Applications: See Industrene 105
Form: Solid
Properties: Sapon. No.: 208–211; Acid No.:207–210
Composition: 100% conc.

Industrene 5022. Humko Sheffield Chem.
Chemical Description: 50% ara-

chidic and behenic fatty acid
Category and Applications: See Industrene 105
Form: Solid; Color: 7 max. (Gardner)
Properties: Sapon. No.: 178–188; Acid No.: 177–187
Composition: 100% conc.

Industrene 7018. Humko Sheffield Chem.
Chemical Description: 70% stearic fatty acid
Category and Applications: See Industrene 105
Form: Solid
Properties: Sapon. No.: 201–208; Acid No.: 200–207
Composition: 100% conc.

Industrene 8518. Humko Sheffield Chem.
Chemical Description: 85% stearic fatty acid
Category and Applications: See Industrene 105
Form: Solid
Properties: Sapon. No.: 198–205; Acid No.: 197–204
Composition: 100% conc.

Industrene 9018. Humko Sheffield Chem.
Chemical Description: 90% stearic fatty acid
Category and Applications: See Industrene 105
Form: Solid
Properties: Sapon. No.: 197–202; Acid No.: 196–201
Composition: 100% conc.

Industrene B. Humko Sheffield Chem.
Chemical Description: Hydrogenated mixed fatty acid
Category and Applications: See Industrene 105
Form: Solid
Properties: Sapon. No.: 200–208; Acid No.: 199–207
Composition: 100% conc.

Industrene R. Humko Sheffield Chem.
Chemical Description: Hydrogenated rubber grade fatty acid
Category and Applications: See Industrene 105
Form: Solid; Color: 10 max. (Gardner)
Properties: Sapon. No.: 194–214; Acid No.: 193–213
Composition: 100% conc.

Interwet 33. Interstab Chem., Inc.
Category: Emulsifier, penetrant, nonionic
Applications: Water-borne formulations; netting and penetrating agent; coating; water-paint calsomines, adhesives; insecticides, herbicides, fungicides; dispersing waxes, pigments, resins; preparation of emulsified paints; making of bitumen emulsions
Form: Liquid; Color: Yellow; Odor: Mild fatty acid
Properties: Solubility: Disp. in water; sol. in alcohols, acetones, toluol (at room temperature); Sp. gr.: 1.01; Visc.: 85 cps (C Gardner-Holdt); F.P.: ≈ 2 C; B.P.: 290 C; HLB: 11.5; Stability: Very stable; aq. sol'ns. stable to high conc. of salts, inc. Ca, Ba, Mg
Composition: 100% active ingred.
Reaction: Neutral
Toxicity: Prolonged contact may

cause slight irritation to skin and eyes
Containers: Unlined steel drums

Intralan Salt HA. Crompton and Knowles
Chemical Description: Neutralized naphthalene condensate
Category: Dispersing and leveling agent, anionic
Applications: Dispersant for disperse dyes; leveling agent for dyeing of synthetic fibers; dyeing auxiliary
Form: Powder
Properties: Solubility: Water-sol.
Reaction: Neutral

Intralan Salt N. Crompton and Knowles
Chemical Description: Ethylene oxide condensate
Category: Dye assistant, leveling agent, penetrant, nonionic
Applications: Dye assistant for wool/pan blends and for Intrachrome and Intralan dyes; leveling agent for 1:2 premetalized acid dyes
Form: Liquid; Color: Yellow
Properties: Solubility: Sol. in boiling water
Reaction: Neutral

Intraphasol PC. Crompton and Knowles
Chemical Description: Sulfuric acid ester salt of polyglycol ether compound
Category: Wetting agent, emulsifier, dispersant, anionic
Applications: Printing of polyamide and polyester carpets
Form: Liquid; Color: Amber; Odor: Alcoholic
Properties: Solubility: Water sol.; Density: 8.7 lb/gal
Reaction: Neutral

Intratex A. Crompton and Knowles
Chemical Description: Complex amino condensate
Category: Leveling agent, amphoteric
Applications: Nonfoaming leveling agent for dyeing wool and p.a.; leveling agent for premetalized, chrome, acid, and selected reactive dyes
Form: Liquid; Color: Tan
Properties: Solubility: Water sol.; Density: 9 lb/gal
Reaction: Neutral

Intratex AN. Crompton and Knowles
Chemical Description: Ethylene oxide condensate
Category: Leveling agent, nonionic
Applications: Wool dyeing; dye stripping; overcomes tippy dyeing of wool; reduces bleeding of prints in afterwash
Form: Liquid
Properties: Solubility: Water sol.
Reaction: Neutral

Intratex B. Crompton and Knowles
Chemical Description: Polyglycol ether derivative
Category: Leveling agent, detergent, amphoteric
Applications: Level dyeing of wool, esp. low temperature wool dyeing with reactive dyes; wool

scouring
Form: Liquid; Color: Tan
Properties: Solubility: Water sol.; Density: 8.7 lb/gal
Reaction: Neutral

Intratex C. Crompton and Knowles
Chemical Description: Complex amino condensate
Category: Leveling agent, amphoteric
Applications: Dyeing; derivative agent; dyeing wool stock, tops, yarn
Form: Liquid; Color: Clear
Properties: Density: 8 lb/gal
Reaction: Neutral

Intratex OR. Crompton and Knowles
Chemical Description: Low molecular polyamide, nonionic
Category: Leveling agent, nonionic
Applications: Vat dyeing assistant; dyeing cellulosics with vats
Form: Liquid; Color: Yellow brown
Properties: Solubility: Water sol.; Density: 8.6 lb/gal; Sp. gr.: 1.04
Reaction: Neutral

Intratex POK. Crompton and Knowles
Chemical Description: Sulfonated benzimioazole derivative
Category: Dispersant, leveling agent, anionic
Applications: Leveling agent for vat and disperse dyes; dyeing cellulosics with vats and acetate, nylon with disperse dyes
Form: Liquid; Color: Yellow brown
Properties: Solubility: Water sol.
Reaction: Neutral

Intratex W New. Crompton and Knowles
Chemical Description: Ethylene oxide condensate
Category: Leveling agent, penetrant, nonionic
Applications: Wool dyeing; compatibility of anionic/cationic dyes; wool dyeing with milling dyes
Form: Liquid; Color: Yellow
Properties: Solubility: Water sol.; Density: 8.5 lb/gal
Reaction: Neutral

Intravon AN. Crompton and Knowles
Chemical Description: Fatty acid derivative
Category: Detergent, emulsifier, nonionic
Applications: Scouring and emulsifying agent for wool, cotton, synthetics, and blends; detergent in laundering, dyeing, household, and cosmetics applications
Form: Liquid; Color: Amber
Properties: Solubility: Water sol.; Density: 8.4 lb/gal; Sp. gr.: 1.01
Reaction: Alkaline

Intravon JET. Crompton and Knowles
Chemical Description: Blended surfactants
Category: Detergent, emulsifier, nonionic/anionic
Applications: Low foaming scouring agent for textiles in jet machines and other foam generating

equipment
Form: Liquid; Color: Yellow brown
Properties: Solubility: Water sol.
Reaction: Alkaline

Intravon JF. Crompton and Knowles
Chemical Description: Ethylene oxide condensate
Category: Detergent, emulsifier, dispersant, nonionic
Applications: Scouring of synthetic fibers, wool, cotton; Kier boiling assistant; detergent in textile, laundering, dyeing, household, and cosmetics applications; desizing and boiling off cottons
Form: Liquid; Color: Yellow
Properties: Solubility: Water sol.; Density: 8.4 lb/gal
Reaction: Neutral

Intravon JU. Crompton and Knowles
Chemical Description: Ethylene oxide condensate
Category: Detergent, wetting agent, emulsifier, penetrant, nonionic
Applications: Multipurpose surfactant; textile scouring, emulsifying, wetting; detergent in textiles, skins, leather, laundering, dyeing, household, industrial, and cosmetics applications; wetting agent in bleaching, dyeing, rewetting dried skins, finishing; emulsifier in Kier boiling
Form: Liquid; Color: Pale yellow
Properties: Solubility: Water sol.; Density: 8.5 lb/gal
Reaction: Neutral

Intravon SO. Crompton and Knowles
Chemical Description: Blend of surfactants
Category: Detergent, wetting agent, emulsifier, anionic/nonionic
Applications: Scouring and dyeing textiles; laundering and household applications as detergent
Form: Liquid; Color: Yellow
Properties: Solubility: Water sol.; Density: 8.6 lb/gal
Reaction: Neutral

Invermul. NL Industries/Baroid Div.
Chemical Description: Modified calcium salts of higher organic acids
Category: Emulsifier, anionic
Applications: Emulsifier for W/O emulsion drilling fluids; stabilizes the emulsion, aids suspension properties, and aids in reducing filtration
Form: Powder
Composition: 100% active ingred.

I.P.M. Akzo Chemie Italia
Chemical Description: Isopropyl myristate
Category: Emollient, nonionic
Applications: Cosmetics preparations
Form: Liquid
Properties: Sp. gr.: 0.850; Visc.: 6.2–6.4 cs @ 20 C; F.P.: 7 C
Composition: > 99% active
Reaction: Neutral
Containers: 180 kg steel drums

Iscolan. Croda Italiana S.R.L.
Chemical Description: Lanolin ester

Category: Emollient, wetting and spreading agent, nonionic
Applications: Skin, make-up, and hair care cosmetics; excellent non-greasy emollient, esp. suited in nonocclusive cosmetics
Form: Slightly opalescent liquid; Color: Pale yellow
Properties: Lipophilic

Isochemstrip 215. Isochem Corp.
Category: Stripping agent
Applications: Polyurethane and epoxy stripper
Properties: Nonvolatile, nontoxic, water soluble

Isokleen. Isochem Corp.
Category: Detergent, dispersant, solubilizer
Applications: Alkaline concentrated cleaner; high dispersing action and good flushing off powers for most organic or natural resin, coatings, colors, or adhesives; solubilizer for natural or synthetic latices, organisols, acrylics, acetates, and caseinates; removes caked on gums, resins, incl. vinyl, starches, latex, greases, soot, oils; alkaline booster for other cleaners
Form: Dry powder
Properties: Solubility: Water sol.
Composition: 100% active

Isomul Extra. Isochem Corp.
Chemical Description: Alkanolamide
Category: Detergent, emulsifier, nonionic
Applications: Base for solubilizing waxes, solvents, oils, etc.; base for formulated metal and textile cleaners; good deinker
Form: Liquid
Properties: Biodegradable
Composition: 100% active ingred.

Isopropyl Lanolate. van Schuppen Chemie
Chemical Description: Isopropyl ester of normal, branched chain and hydroxy acids of lanolin
Category: Emollient, dispersant, wetting agent
Applications: In nourishing creams (moisturizing and conditioning), emollient creams and lotions, cleansing preparations, etc.; dispersing and wetting agent for pigmented products
Form: Nontacky soft solid; Color: Yellowish-brown; Odor: Practically odorless
Properties: Solubility: Sol. in hydrocarbons, chlorinated hydrocarbons, isopropyl myristate, isopropyl palmitate, ethyl alcohol, isopropyl alcohol, mineral and vegetable oils, and other common formulating materials; Sapon. Value: 130–160; Acid No.: 18.0 max.; Hydroxyl Value: 45–65
Composition: 100% active ingred.
References: Product information No. 29

Isoslushoff X-5. Isochem Corp.
Chemical Description: Phenolic base
Category: Detergent
Applications: Low cost, high strength cleaner where biodegradability is not important

Isoslushoff X-M-50. Isochem Corp.
Chemical Description: Alkaline alkanolamide

Category: Detergent, solubilizer, nonionic
Applications: Cleaner, esp. for removing encrusted paints and resins from process equipment and floors
Form: Liquid
Properties: Nonvolatile, nonflammable

Isoslushoff X-NP-25. Isochem Corp.
Chemical Description: Alkaline alkanolamide
Category: Detergent, nonionic
Applications: High potency cleaner for paper, resins, and polyethylenes
Form: Liquid; Odor: None
Properties: Solubility: Water dilutable; Biodegradable

Isoslushoff X-NP-30. Isochem Corp.
Chemical Description: Alkaline-free type of Isoslushoff X-NP-25
Category: Detergent, nonionic
Applications: Used on glass-lined equipment or wherever alkali is objectionable; low-foaming with high dissolving powers; good for polyethylene, acrylics, most synthetic adhesives and coatings
Properties: Nonflammable; Biodegradable

Isosolvent #7. Isochem Corp.
Category: Solvent
Applications: Nonvolatile, water-disp. replacement for acetone, ketone, xylol, and methyl ethyl ketone; neutral solvent for metal or plastic cleaning prior to bonding, encapsulation, or painting
Form: Liquid; Odor: Free from objectionable fumes and odors

Isosolvent #33. Isochem Corp.
Category: Stripping and solubilizing agent
Applications: Stripping and solubilizing agent for polyurethane and other plastic films and foams; equipment cleanups and process equipment
Properties: Low volatility, nontoxic, nonflammable

Isothan Q-75. Onyx Chem. Co.
Chemical Description: n-alkyl isoquinolinium bromides
Category: Fungicide, cationic
Applications: Used as a fungicide in agricultural products, shampoos, and hair rinses
Form: Liquid
Properties: Sp. gr.: 1.08 (25/20 C); Flash Pt.: 92 F
Composition: 75% min. quaternary

JK

Juniorlan. Emery Ind./ Malmstrom Div.
Chemical Description: Combination of lanolin and fatty acid esters
Category: Lanolin substitute, nonionic
Form: Soft wax
Composition: 100% conc.

Kadif 50 Flakes. Witco Chem. Ltd.
Chemical Description: Sodium dodecylbenzene sulfonate
Category: Wetting and penetrating agent, detergent
Applications: Dry-blended cleaning compounds; general-purpose wetting and penetrating agent in wallpaper removal and dye leveling; dishwashing detergent, washing compound for farm produce; with phosphates, alkalies, and other builders for industrial and institutional cleaners, laundry detergents, metal, floor, and scouring cleaners, upholstery shampoo and rug cleaning compounds
Form: Flakes; Color: Light; Odor: Practically odorless
Properties: Solubility: Water sol.; pH: 8.0 (1% aq.)
Composition: 50% active ingred.
Containers: 15 kg net paper bags
Warnings and Caution: Store in a cool, dry place; avoid moisture pickup

Kalex Acids. Hart Chem.
Chemical Description: Ethylene diamine tetraacetic acid
Category: Sequestering agent, anionic
Applications: Used in systems where the sodium ion is undesirable for preparation of amine or other alkali metal salts
Form: Powder; Color: White
Properties: Solubility: Insol.
Composition: 99% active ingred.
Reaction: Acid
Toxicity: Skin and eye irritant
Containers: Kits, T/C
Warnings and Caution: Avoid contact with skin and eyes

Kalex Liquid 50%. Hart Chem.
Chemical Description: Tetrasodium ethylenediamine tetraace-

tate
Category: Sequestering agent, anionic
Applications: Complexes calcium, magnesium, and most other common metals through a wide pH range
Form: Liquid; Color: Yellow
Properties: Solubility: Clear sol.
Composition: 50% active ingred.
Reaction: Alkaline
Toxicity: Skin and eye irritant
Containers: Drums, T/T
Warnings and Caution: Avoid contact with skin and eyes

Kalex OH. Hart Chem.
Chemical Description: Disodium dihydroxyethyl ethylenediamine diacetate
Category: Sequestering agent, anionic
Applications: Sequesters calcium and magnesium and may be used for control of iron in the pH range 8.0–10.5
Form: Liquid
Properties: Solubility: Clear sol.
Composition: 42% active ingred.
Reaction: Alkaline
Toxicity: Skin and eye irritant
Containers: Drums, T/T
Warnings and Caution: Avoid contact with skin and eyes

Kalex Penta. Hart Chem.
Chemical Description: Pentasodium diethylenetriamine pentaacetate
Category: Sequestering agent, anionic
Applications: Sequestering agent used when slightly higher metal chelate stability is required
Form: Liquid; Color: Yellow
Properties: Solubility: Clear sol.; Stability: Stable in presence of strong oxidizing agents
Composition: 40% conc.
Reaction: Alkaline
Toxicity: Skin and eye irritant
Containers: Drums, T/T
Warnings and Caution: Avoid contact with skin and eyes

Kalex Powder FC. Hart Chem.
Chemical Description: Tetrasodium ethylenediamine tetraacetate dihydrate
Category: Sequestering agent, anionic
Applications: Powdered form of Kalex Liquid 50; advantageous when formulating a dry product
Form: Powder; Color: Yellow
Properties: Solubility: Clear sol.
Composition: 82% conc.
Reaction: Alkaline
Toxicity: Skin and eye irritant
Containers: Kits, T/L
Warnings and Caution: Avoid contact with skin and eyes

Kar-Kote. Coastal Industries, Inc.
Category: Detergent
Applications: Washing painted surfaces on cars, trucks, and buses
Form: Powder; Color: Tan; Odor: Bland
Properties: Stability: Good
Reaction: Alkaline
Containers: Fiber drums

Katapol PN-430. GAF Corp.
Chemical Description: Polyoxyethylated alkylamine
Category: Emulsifier, corrosion

inhibitor, cationic
Applications: Emulsifier for mineral oils, highly absorbed on most solid surfaces; acid corrosion inhibitor for ferrous alloys
Form: Free flowing, slightly viscous liquid; Color: Brown
Properties: Solubility: Sol. in kerosene (10%) with slight haze; pH: 8.0–10.0 (10% aq. disp.)
Composition: 99% active ingred.

Katapol PN-730. GAF Corp.
Chemical Description: Polyoxyethylated (15) tallow amine (distilled)
Category: Emulsifier, dye assistant, softener, antistat, lubricant, cationic
Applications: Water soluble emulsifier for oils; dye assistant for textiles and leather; textile softening agent, antistat, and fiber lubricant
Form: Clear, free flowing liquid; Color: Amber
Properties: Solubility: Sol. in water
Composition: 99% min. active

Katapol PN-810. GAF Corp.
Chemical Description: Polyoxyethylated (20) tallow amine (distilled)
Category and Applications: See Katapol PN-730; used where a nonhydrophilic product is required
Form: Clear, free flowing liquid (may cloud below 25 C); Color: Amber
Composition: 99% min. active

Kelig 32. American Can Co.
Chemical Description: Lignosulfonate and modified sugar acids
Category: Sequestrant, dispersant, anionic
Applications: Lubricant removal from zinc phosphate coatings; formulation of cooling water treatment compounds and alkaline cleaners; scale control; oil well cement retardant; good under alkaline conditions
Form: Powder
Composition: 100% conc.

Kemamide B. Humko Sheffield Chem.
Chemical Description: Behenamide
Category and Applications: Slip and antiblock agent in plastic films, esp. for polyolefins; foam stabilizer; in printing inks; as wiredrawing lubricant; in wax compositions; in water repellents for textiles
Form: Solid; Color: 5 max. (Gardner)
Properties: m.w.: 312; Solubility: Insol. in water; slightly sol. in organic solvents at room temperature; M.P.: 98–108 C; Acid No.: 4.0 max.; Iodine Value: 4.0 max.
Composition: 100% conc.
Reaction: Neutral

Kemamide E. Humko Sheffield Chem.
Chemical Description: Erucamide
Category and Applications: See Kemamide B

Form: Solid; Color: 5 max. (Gardner)
Properties: m.w.: 335; Solubility: Insol. in water; slightly sol. in organic solvents at room temperature; M.P.: 76–86 C; Acid No.: 4.0 max.; Iodine Value: 70–80
Composition: 100% conc.
Reaction: Neutral

Kemamide S. Humko Sheffield Chem.
Chemical Description: Stearamide
Category and Applications: See Kemamide B
Form: Solid; Color: 5 max. (Gardner)
Properties: m.w.: 278; Solubility: Insol. in water; slightly sol. in organic solvents at room temperature; M.P.: 98–108; Acid No.: 4.0 max.; Iodine Value: 3.0 max.
Composition: 100% conc.
Reaction: Neutral

Kemamide U. Humko Sheffield Chem.
Chemical Description: Oleamide
Category and Applications: See Kemamide B
Form: Solid; Color: 5 max. (Gardner)
Properties: m.w.: 275; Solubility: Insol. in water; slightly sol. in organic solvents at room temperature; M.P.: 68–78 C; Acid No.: 4.0 max.; Iodine Value: 72–90
Composition: 100% conc.
Reaction: Neutral

Kemamide W 10. Humko Sheffield Chem.
Chemical Description: N,N′-ethylenebis (tall oil amide)
Category: Lubricant
Applications: Synthetic waxes; as a process lubricant for styrenated plastics; as a mold-release agent for phenolics; antiblock agent for polyolefins; in asphalts, paper coatings, rubber; in potting and dipping compounds for electrical components; defoamer in pulp mill operations
Form: Solid; Color: 6 max. (Gardner)
Properties: Solubility: Insol. in water; slightly sol. in organic solvents at room temperature; M.P.: ≈ 110 C; Acid No.: 10 max.
Composition: 100% conc.
Reaction: Neutral

Kemamide W 20. Humko Sheffield Chem.
Chemical Description: N,N′-ethylenebisoleamide
Category and Applications: See Kemamide W 10
Form: Solid; Color: 6 max. (Gardner)
Properties: Solubility: Insol. in water; slightly sol. in organic solvents at room temperature; M.P.: ≈ 120 C; Acid No.: 10 max.
Composition: 100% conc.
Reaction: Neutral

Kemamide W 35. Humko Sheffield Chem.
Chemical Description: N,N′-ethylenebisstearamide
Category and Applications: See Kemamide W 10
Form: Solid; Color: 6 max. (Gardner)

Properties: Solubility: Insol. in water; slightly sol. in organic solvents at room temperature; M.P.: ≈ 140 C; Acid No.: 10 max.
Composition: 100% conc.
Reaction: Neutral

Kemamide W 40. Humko Sheffield Chem.
Chemical Description: N,N′-ethylenebisstearamide
Category and Applications: See Kemamide W 10
Form: Solid; Color: 3 max. (Gardner)
Properties: Solubility: Insol. in water; slightly sol. in organic solvents at room temperature: M.P.: ≈ 140 C; Acid No.: 10 max.
Composition: 100% conc.
Reaction: Neutral

Kemamide W 45. Humko Sheffield Chem.
Chemical Description: N,N′-ethylenebisstearamide
Category and Applications: See Kemamide W 10
Form: Solid; Color: 3 max. (Gardner)
Properties: Solubility: Insol. in water; slightly sol. in organic solvents at room temperature; M.P.: ≈ 145 C; Acid No.: 10 max.
Composition: 100% conc.
Reaction: Neutral

Kemamine BQ-9702C. Humko Sheffield Chem.
Chemical Description: Dimethyl hydrogenated tallow benzyl quaternary ammonium chloride
Category: Germicide, antistat, softener, emulsifier, corrosion inhibitor, cationic
Applications: Germicides, sanitizers, slimicides, antistats, textile softening agents, dyeing aids, corrosion inhibitors, emulsifiers
Form: Liquid; Color: 4 max. (Gardner)
Properties: m.w.: 420; Solubility: Disp. in water; pH: 9 max. (5% sol'n.)
Composition: 75% active ingred.

Kemamine BQ-9742C. Humko Sheffield Chem.
Chemical Description: Dimethyl tallow benzyl quaternary ammonium chloride
Category and Applications: See Kemamine BQ-9702C
Form: Paste; Color: 6 max. (Gardner)
Properties: m.w.: 420; Solubility: Disp. in water; pH: 9 max. (5% sol'n.)
Composition: 75% active ingred.

Kemamine D-150. Humko Sheffield Chem.
Chemical Description: N-50% arachidyl-behenyl 1,3-propylenediamine
Category: Detergent, bactericide, corrosion inhibitor, cationic
Applications: Gasoline detergents, bactericides, corrosion inhibitors in petroleum production, epoxy hardeners, dispersing aids, cationic asphalt systems
Form: Solid; Color: 5 max. (Gardner)
Properties: Iodine Value: 10.0 max.
Composition: 88% conc.

Kemamine D-190. Humko Sheffield Chem.
Chemical Description: N-90% arachidyl-behenyl 1,3-propylenediamine
Category and Applications: See Kemamine D-150
Form: Solid; Color: 5 max. (Gardner)
Properties: Iodine Value: 10.0 max.
Composition: 88% conc.

Kemamine D-650. Humko Sheffield Chem.
Chemical Description: N-coconut 1,3-propylenediamine
Category and Applications: See Kemamine D-150
Form: Paste; Color: 6 max. (Gardner)
Properties: Iodine Value: 12.0 max.
Composition: 88% conc.

Kemamine D-970. Humko Sheffield Chem.
Chemical Description: N-hydrogenated tallow 1,3-propylenediamine
Category and Applications: See Kemamine D-150
Form: Solid; Color: 6 max. (Gardner)
Properties: Iodine Value: 3.0 max.
Composition: 88% conc.

Kemamine D-974. Humko Sheffield Chem.
Chemical Description: N-tallow 1,3-propylenediamine
Category and Applications: See Kemamine D-150
Form: Solid; Color: 12 max. (Gardner)
Properties: Iodine Value: 32 min.
Composition: 88% conc.

Kemamine D-989. Humko Sheffield Chem.
Chemical Description: N-oleyl 1,3-propylenediamine
Category and Applications: See Kemamine D-150
Form: Liquid; Color: 7 max. (Gardner)
Properties: Iodine Value: 60 min.
Composition: 88% conc.

Kemamine DP Series. Humko Sheffield Chem.
Chemical Description: Dimer amines
Category: Corrosion inhibitor, intermediate
Applications: Corrosion inhibitors for petroleum-processing equipment; as intermediates, extenders, and cross-linking agents in high-polymer systems

Kemamine P-150. Humko Sheffield Chem.
Chemical Description: 50% arachidyl-behenyl primary amine
Category: Emulsifier, flotation agent, dispersant, intermediate, cationic
Applications: Acid scavenger, corrosion inhibitor, fuel oil additive, mold release for rubber and plastics, dispersing and flushing agent; ethoxylated amine derivatives used as emulsifiers, spinning aids, and lubricants in metal-working

oils
Form: Solid; Color: 3 max. (Gardner)
Properties: Solubility: Sol. in common organic solvents; Iodine Value: 5.0 max.
Composition: 93% conc.

Kemamine P-150D. Humko Sheffield Chem.
Chemical Description: Distilled 50% arachidyl-behenyl primary amine
Category and Applications: See Kemamine P-150
Form: Solid; Color: 1 max. (Gardner)
Properties: Solubility: Sol. in common organic solvents; Iodine Value: 5.0 max.
Composition: 97% conc.

Kemamine P-190. Humko Sheffield Chem.
Chemical Description: 90% arachidyl-behenyl primary amine
Category and Applications: See Kemamine P-150
Form: Solid; Color: 3 max. (Gardner)
Properties: Solubility: Sol. in common organic solvents; Iodine Value: 5.0 max.
Composition: 93% conc.

Kemamine P-190D. Humko Sheffield Chem.
Chemical Description: Distilled 90% arachidyl-behenyl primary amine
Category and Applications: See Kemamine P-150
Form: Solid; Color: 1 max. (Gardner)
Properties: Solubility: Sol. in common organic solvents; Iodine Value: 5.0 max.
Composition: 97% conc.

Kemamine P-650. Humko Sheffield Chem.
Chemical Description: Technical coconut primary amine
Category and Applications: See Kemamine P-150
Form: Liquid; Color: 3 max. (Gardner)
Properties: Solubility: Sol. in common organic solvents; Iodine Value: 12.0 max.
Composition: 93% conc.

Kemamine P-650D. Humko Sheffield Chem.
Chemical Description: Distilled coconut primary amine
Category and Applications: See Kemamine P-150
Form: Liquid; Color: 1 max. (Gardner)
Properties: Solubility: Sol. in common organic solvents; Iodine Value: 12.0 max.
Composition: 97% conc.

Kemamine P-880. Humko Sheffield Chem.
Chemical Description: Technical palmityl primary amine
Category and Applications: See Kemamine P-150
Form: Solid; Color: 3 max. (Gardner)
Properties: Solubility: Sol. in common organic solvents; Iodine Value: 2.0 max.
Composition: 93% conc.

Kemamine P-880D. Humko Sheffield Chem.
Chemical Description: Distilled palmityl primary amine
Category and Applications: See Kemamine P-150
Form: Solid; Color: 1 max. (Gardner)
Properties: Solubility: Sol. in common organic solvents; Iodine Value: 2.0 max.
Composition: 97% conc.

Kemamine P-970. Humko Sheffield Chem.
Chemical Description: Technical hydrogenated tallow primary amine
Category and Applications: See Kemamine P-150
Form: Solid; Color: 3 max. (Gardner)
Properties: Solubility: Sol. in common organic solvents; Iodine Value: 3.0 max.
Composition: 93% conc.

Kemamine P-970D. Humko Sheffield Chem.
Chemical Description: Distilled hydrogenated tallow primary amine
Category and Applications: See Kemamine P-150
Form: Solid; Color: 1 max. (Gardner)
Properties: Solubility: Sol. in common organic solvents; Iodine Value: 3.0 max.
Composition: 97% conc.

Kemamine P-974. Humko Sheffield Chem.
Chemical Description: Technical tallow primary amine
Category and Applications: See Kemamine P-150
Form: Paste; Color: 3 max. (Gardner)
Properties: Solubility: Sol. in common organic solvents; Iodine Value: 38 min.
Composition: 93% conc.

Kemamine P-974D. Humko Sheffield Chem.
Chemical Description: Distilled tallow primary amine
Category and Applications: See Kemamine P-150
Form: Paste; Color: 1 max. (Gardner)
Properties: Solubility: Sol. in common organic solvents; Iodine Value: 38 min.
Composition: 97% conc.

Kemamine P-989. Humko Sheffield Chem.
Chemical Description: Technical oleyl primary amine
Category and Applications: See Kemamine P-150
Form: Liquid; Color: 3 max. (Gardner)
Properties: Solubility: Sol. in common organic solvents; Iodine Value: 70 min.
Composition: 93% conc.

Kemamine P-989D. Humko Sheffield Chem.
Chemical Description: Distilled oleyl primary amine
Category and Applications: See Kemamine P-150
Form: Liquid; Color: 1 max.

(Gardner)
Properties: Solubility: Sol. in common organic solvents; Iodine Value: 70 min.
Composition: 97% conc.

Kemamine P-990. Humko Sheffield Chem.
Chemical Description: Technical stearyl primary amine
Category and Applications: See Kemamine P-150
Form: Solid; Color: 3 max. (Gardner)
Properties: Solubility: Sol. in common organic solvents; Iodine Value: 2.0 max.
Composition: 93% conc.

Kemamine P-990D. Humko Sheffield Chem.
Chemical Description: Distilled stearyl primary amine
Category and Applications: See Kemamine P-150
Form: Solid; Color: 1 max. (Gardner)
Properties: Solubility: Sol. in common organic solvents; Iodine Value: 2.0 max.
Composition: 97% conc.

Kemamine P-997. Humko Sheffield Chem.
Chemical Description: Technical soya primary amine
Category and Applications: See Kemamine P-150
Form: Paste; Color: 3 max. (Gardner)
Properties: Solubility: Sol. in common organic solvents; Iodine Value: 90 min.
Composition: 93% conc.

Kemamine P-997D. Humko Sheffield Chem.
Chemical Description: Distilled soya primary amine
Category and Applications: See Kemamine P-150
Form: Paste; Color: 2 max. (Gardner)
Properties: Solubility: Sol. in common organic solvents; Iodine Value: 90 min.
Composition: 97% conc.

Kemamine Q-1902C. Humko Sheffield Chem.
Chemical Description: Dimethyl di-90% arachidyl-behenyl quaternary ammonium chloride
Category and Applications: See Kemamine BQ-9702C
Form: Solid; Color: 4 max. (Gardner)
Properties: m.w.: 680; Solubility: Disp. in water; pH: 9 max. (5% sol'n.)
Composition: 75% active ingred.

Kemamine Q-6502C. Humko Sheffield Chem.
Chemical Description: Dimethyl di-coconut quaternary ammonium chloride
Category and Applications: See Kemamine BQ-9702C
Form: Liquid; Color: 4 max. (Gardner)
Properties: m.w.: 465; Solubility: Disp. in water; pH: 9 max. (5% sol'n.)
Composition: 75% active ingred.

Kemamine Q-9702C. Humko Sheffield Chem.
Chemical Description: Dimethyl

di-hydrogenated tallow quaternary ammonium chloride
Category and Applications: See Kemamine BQ-9702C
Form: Solid; Color: 2 max. (Gardner)
Properties: m.w.: 575; Solubility: Disp. in water; pH: 9 max. (5% sol'n.)
Composition: 75% active ingred.

Kemamine S-190. Humko Sheffield Chem.
Chemical Description: Di-90% arachidyl-behenyl secondary amine
Category: Chemical intermediate, cationic
Applications: Chemical intermediate for oil additives and textile auxiliaries
Form: Solid; Color: 1 max. (Gardner)
Properties: Iodine Value: 5.0 max.
Composition: 90% conc.

Kemamine S-970. Humko Sheffield Chem.
Chemical Description: Di-hydrogenated tallow secondary amine
Category and Applications: See Kemamine S-190
Form: Solid; Color: 1 max. (Gardner)
Properties: Iodine Value: 3.0 max.
Composition: 90% conc.

Kemamine T-1902D. Humko Sheffield Chem.
Chemical Description: Distilled dimethyl-90% arachidyl-behenyl tertiary amine
Category: Chemical intermediate, catalyst, cationic
Applications: Intermediate for quaternary ammonium derivatives formed by using an alkylating agent; acid scavenger in petroleum products; epoxy hardener; catalyst in manufacture of flexible polyurethane foams
Form: Liquid; Color: 1 max. (Gardner)
Composition: 95% conc.

Kemamine T-6501. Humko Sheffield Chem.
Chemical Description: Methyl dicoconut tertiary amine
Category and Applications: See Kemamine T-1902D
Form: Liquid; Color: 3 max. (Gardner)
Composition: 95% conc.

Kemamine T-6502D. Humko Sheffield Chem.
Chemical Description: Distilled dimethyl coconut tertiary amine
Category and Applications: See Kemamine T-1902D
Form: Color: 1 max. (Gardner)
Composition: 95% conc.

Kemamine T-9701. Humko Sheffield Chem.
Chemical Description: Methyl dihydrogenated tallow tertiary amine
Category and Applications: See Kemamine T-1902D
Form: Liquid; Color: 3 max. (Gardner)
Composition: 95% conc.

Kemamine T-9702D. Humko Sheffield Chem.
Chemical Description: Distilled dimethyl hydrogenated tallow tertiary amine
Category and Applications: See Kemamine T-1902D
Form: Liquid; Color: 1 max. (Gardner)
Composition: 95% conc.

Kemamine T-9742D. Humko Sheffield Chem.
Chemical Description: Distilled dimethyl tallow tertiary amine
Category and Applications: See Kemamine T-1902D
Form: Liquid; Color: 1 max. (Gardner)
Composition: 95% conc.

Kemamine T-9902D. Humko Sheffield Chem.
Chemical Description: Distilled dimethyl stearyl tertiary amine
Category and Applications: See Kemamine T-1902D
Form: Liquid; Color: 1 max. (Gardner)
Composition: 95% conc.

Kemamine T-9972D. Humko Sheffield Chem.
Chemical Description: Distilled dimethyl soya tertiary amine
Category and Applications: See Kemamine T-1902D
Form: Liquid; Color: 1 max. (Gardner)
Composition: 95% conc.

Kemester 105. Humko Sheffield Chem.
Chemical Description: Methyl oleate
Category: Intermediate, solvent
Applications: Intermediate in production of superamides; in metalworking lubricants; as a specialized solvent
Form: Liquid; Color: 2 max. (Gardner)
Properties: Sapon. No.: 194–203; Acid No.: 4.0 max.
Composition: 100% conc.

Kemester 205. Humko Sheffield Chem.
Chemical Description: Methyl oleate (white)
Category and Applications: See Kemester 105
Form: Liquid; Color: 1 max. (Gardner)
Properties: Sapon. No.: 194–203; Acid No.: 4.0 max.
Composition: 100% conc.

Kemester 226. Humko Sheffield Chem.
Chemical Description: Methyl soyate
Category and Applications: See Kemester 105
Form: Liquid; Color: 4 max. (Gardner)
Properties: Sapon. No.: 195–200; Acid No.: 4.0 max.
Composition: 100% conc.

Kemester 325. Humko Sheffield Chem.
Chemical Description: Methyl cocoate

Category and Applications: See Kemester 105
Form: Liquid; Color: 1 max. (Gardner)
Properties: Sapon. No.: 246–256; Acid No.: 4.0 max.
Composition: 100% conc.

Kemester EGDS. Humko Sheffield Chem.
Chemical Description: Ethylene glycol distearate
Category and Applications: See Kemester 105
Form: Solid; Color: 2 max. (Gardner)
Properties: Sapon. No.: 190–200; Acid No.: 6.0 max.; Hydroxyl No.: 33–43
Composition: 100% conc.

Kessco 639. Armak Co.
Chemical Description: Blend of isopropyl myristate and isopropyl palmitate
Category: Emollient, lubricant, solvent, nonionic
Applications: Cosmetics and toiletries formulations
Form: Liquid; Color: 30 (APHA)
Properties: Ref. Index: 1.436; Density: 7.1 lb/gal; Sp. gr.: 0.849–0.855; Visc.: 5.9 cps; F.P.: 7 C; B.P.: 170 C @ 4 mm; Flash Pt.: 305 F (COC); Acid Value: 1.0 max.; Iodine Value: 1.0

Kessco 653. Armak Co.
Chemical Description: Synthetic spermaceti wax
Category: Synthetic spermaceti wax, viscosity booster, nonionic
Applications: In pharmaceutical and cosmetic products to increase viscosity or raise the melting point of ointments and creams; in cold creams
Form: Glistening flakes; Color: White
Properties: Solubility: Sol. in boiling alcohol, ether, chloroform, other waxes, oils, hydrocarbons; insol. in water; M.P.: 51–55 C; Saponification Value: 109–117; Acid No.: 2.0 max.; Iodine Value: 1.0 max.
Containers: 200 lb net wt fiber drum
References: Bulletin #77-4

Kessco Acetin. Armak Co.
Chemical Description: Glycerol ester of fatty acid
Category: Emulsifier, stabilizer, emollient, nonionic
Applications: Cosmetic formulations; Auxiliary emulsifier; emulsion stabilizer; opacifying, emollient, and bodying effect in cosmetics
Form: Liquid; Color: 75 max. (APHA)
Properties: Density: 9.9 lb/gal; Sp. gr.: 1.188–1.198 (25/20 C); Visc.: 66.8 cps; F.P.: –30 C; Flash Pt.: 305 F (COC); Acidity: 0.5% (as acetic)

Kessco Alcohol Ester–Butoxyethyl Oleate. Armak Co.
Category: Plasticizer, nonionic
Applications: Specialty plasticizer in rubber and polyvinyl chloride

products
Form: Liquid; Color: 250 (APHA)
Properties: Ref. Index: 1.449; Density: 7.4 lb/gal; Sp. gr.: 0.881–0.889; Visc.: 10 cps; F.P.: –37 C; B.P.: 220 C @ 4 mm; Acidity: 0.2% (as acetic); Iodine Value: 70

Kessco Alcohol Ester–Butoxyethyl Stearate. Armak Co.
Category and Applications: See Kessco Alcohol Ester–butoxyethyl oleate
Form: Liquid; Color: 60 (APHA)
Properties: Ref. Index: 1.444; Density: 7.3 lb/gal; Sp. gr.: 0.874–0.880; Visc.: 11 cps; F.P.: 12 C; B.P.: 218 C @ 4 mm; Flash Pt.: 380 F (COC); Acidity: 0.1% (as acetic); Iodine Value: 5.0

Kessco Alcohol Ester–Butyl Oleate. Armak Co.
Category: Emollient, lubricant, solvent, plasticizer, solubilizer, nonionic
Applications: Emollient in creams and lotions; plasticizer in aerosol hair sprays; cosolvent in nonaqueous preparations; dye solubilizer in lipstick; in synthetic fiber spin finishes; specialty plasticizer in rubber and polyvinyl chloride products
Form: Liquid; Color: 200 (APHA)
Properties: Ref. Index: 1.450; Density: 7.2 lb/gal; Sp. gr.: 0.861–0.869 (25/20 C); Visc.: 8 cps; F.P.: < –15 C; B.P.: 185 C @ 4 mm; Flash Pt.: 380 F (COC); Acidity: 0.1% (as acetic)

Kessco Alcohol Ester–Butyl Stearate Cosmetic. Armak Co.
Category: Emollient, lubricant, solvent, plasticizer, solubilizer, nonionic
Applications: Emollient in cosmetic creams and lotions; plasticizer in aerosol hair sprays; cosolvent in nonaqueous preparations; dye solubilizer in lipstick; in synthetic fiber spin finishes
Form: Liquid; Color: 20 (APHA)
Properties: Ref. Index: 1.442; Density: 7.1 lb/gal; Sp. gr.: 0.850–0.860 (25/20 C); Visc.: 7 cps; F.P.: 19 C; B.P.: 200 C @ 4 mm; Flash Pt.: 370 F (COC); Acid No.: 1.0; Iodine Value: 0.5 max.

Kessco Alcohol Ester–Butyl Stearate Distilled. Armak Co.
Category and Applications: See Kessco Alcohol Ester–Butyl Stearate Cosmetic
Form: Liquid; Color: 100 (APHA)
Properties: Ref. Index: 1.442; Density: 7.1 lb/gal; Sp. gr.: 0.853–0.859; Visc.: 7 cps; F.P.: 18 C; B.P.: 200 C @ 4 mm; Flash Pt.: 370 F (COC); Acidity: 0.5% (as acetic); Iodine Value: 6.0 max.

Kessco Alcohol Ester–Dibutoxyethyl Phthalate. Armak Co.
Category: Plasticizer, nonionic
Applications: Specialty plasticizer in rubber and polyvinyl chloride products
Form: Liquid; Color: 100 (APHA)
Properties: Ref. Index: 1.482; Density: 8.8 lb/gal; Sp. gr.: 1.054–

1.060 (25/20 C); Visc.: 30 cps; F.P.: –55 C; Flash Pt.: 400 F (COC); Acidity: 0.3% (as acetic)

Kessco Alcohol Ester–Isobutyl Palmitate. Armak Co.
Category: Emollient, lubricant, nonionic
Applications: In synthetic fiber spin finishes
Form: Liquid; Color: 60 (APHA)
Properties: Ref. Index: 1.437; Density: 7.1 lb/gal; Sp. gr.: 0.850–0.856 (25/20 C); Visc.: 6.7 cps; F.P.: 13 C; B.P.: 170 C @ 4 mm; Flash Pt.: 325 F (COC); Acidity: 0.5% (as palmitic); Iodine Value: 1.0 max.

Kessco Alcohol Ester–Isocetyl Stearate. Armak Co.
Category: Emollient, nonionic
Applications: Emollient with advantages of liquid form, low freezing point, low volatility, a high degree of saturation, and a relatively high molecular weight
Form: Liquid; Color: 200 (APHA)
Properties: Ref. Index: 1.452; Density: 7.1 lb/gal; Sp. gr.: 0.853–0.859 (25/20 C); Visc.: 32 cps; F.P.: 0 C; Flash Pt.: 450 F (COC); Acid Value: 3.0 max.; Iodine Value: 5.0 max.

Kessco Alcohol Ester–Isopropyl Myristate. Armak Co.
Category: Emollient, lubricant, solvent, plasticizer, nonionic
Applications: Cosmetic creams and lotions; in electric preshave products and bath oil formulations; solvent for silicone oils, lanolin, sun-screening agents, resorcinol, methyl salicylate, undecylenic acid, and benzyl benzoate; plasticizer in aerosol hair sprays; cosolvent in nonaqueous preparations; dye solubilizer in lipstick
Form: Liquid; Color: 20 (APHA)
Properties: Ref. Index: 1.433; Density: 7.1 lb/gal; Sp. gr.: 0.849–0.855 (25/20 C); Visc.: 4.8 cps; F.P.: –3 C; B.P.: 160 C @ 4 mm; Flash Pt.: 305 F (COC); Acid No.: 1.0 max.; Iodine Value: 1.0 max.

Kessco Alcohol Ester–Isopropyl Palmitate. Armak Co.
Category and Applications: See Kessco Alcohol Ester–Isopropyl Myristate
Form: Liquid; Color: 20 (APHA)
Properties: Ref. Index: 1.437; Density: 7.1 lb/gal; Sp. gr.: 0.849–0.855 (25/20 C); Visc.: 6.7 cps; F.P.: 13 C; B.P.: 170 C @ 4 mm; Flash Pt.: 325 F (COC); Acid No.: 1.0 max.; Iodine Value: 1.0 max.

Kessco Diacetin. Armak Co.
Chemical Description: Glycerol ester of fatty acid
Category and Applications: See Kessco Acetin
Form: Liquid
Properties: Density: 9.8 lb/gal; Sp. gr.: 1.173–1.183 (25/20 C); Visc.: 36 cps; F.P.: ≈ –35 C; Flash Pt.: 295 F (COC); Acidity: 0.5% (as acetic)

Kessco Diethylene Glycol Monostearate. Armak Co.
Chemical Description: Glycol ester
Category: Bodying and opacifying agents, nonionic
Applications: Cosmetic creams and lotions
Form: Flakes
Properties: M.P.: 44.5–47.5 C; HLB: 4.3; Flash Pt.: 395 F (COC); Acid No.: 5.0 max.; Iodine Value: 0.5 max.

Kessco Diglycol Stearate Neutral. Armak Co.
Chemical Description: Glycol ester
Category and Applications: See Kessco Diethylene Glycol Monostearate
Form: Flakes
Properties: M.P.: 42–48 C; Flash Pt.: 360 F (COC); Acid No.: 103 max.; Iodine Value: 7.0 max.

Kessco Diglycol Stearate S.E. Armak Co.
Chemical Description: Glycol ester
Category: Bodying and opacifying agent, anionic
Applications: Cosmetic creams and lotions
Form: Flakes
Properties: M.P.: 48–53 C; Flash Pt.: 345 F (COC); Acid No.: 103 max.; Iodine Value: 7.0 max.

Kessco Ethylene Glycol Distearate. Armak Co.
Chemical Description: Glycol ester
Category: Opacifying agent, nonionic
Applications: Opacifying agent; sufficiently insol. in high concentrations of detergents to provide a high degree of opacity to liquid cream shampoos, etc.
Form: Flakes
Properties: M.P.: 60–63 C; Flash Pt.: 390 F (COC); Acid No.: 15.0 max.; Iodine Value: 0.5 max.

Kessco Ethylene Glycol Monostearate. Armak Co.
Chemical Description: Glycol ester
Category and Applications: See Kessco Ethylene Glycol Distearate
Form: Flakes
Properties: M.P.: 56–60 C; HLB: 2.9; Flash Pt.: 390 F (COC); Acid No.: 2.0 max.; Iodine Value: 0.5 max.

Kessco Ethylene Glycol Monostearate 70. Armak Co.
Chemical Description: Glycol ester; high mono-ester content
Category and Applications: See Kessco Ethylene Glycol Distearate
Form: Flakes
Properties: M.P.:52–56 C; Flash Pt.: 370 F (COC); Acid No.:2.0 max.; Iodine Value: 0.5 max.

Kessco Glycerol Dioleate. Armak Co.
Chemical Description: Glycerol ester of fatty acid
Category and Applications: See Kessco Acetin
Form: Liquid; Color: Yellow to

amber
Properties: Density: 7.7 lb/gal; Sp. gr.: 0.923–0.929; Visc.: 90 cps; F.P.: 0 C; HLB: 2.9; Flash Pt.: 520 F (COC); Acid No.: 5.0 max.; Iodine Value: 82 max.

Kessco Glycerol Monolaurate. Armak Co.
Chemical Description: Glycerol ester of fatty acid
Category and Applications: See Kessco Acetin
Form: Solid; Color: White
Properties: M.P.: 53.9 C; Flash Pt.: 425 F (COC); Acid No.: 5.0 max.; Iodine Value: 1.0 max.

Kessco Glycerol Monooleate. Armak Co.
Chemical Description: Glycerol ester of fatty acid
Category and Applications: See Kessco Acetin; also as W/O emulsifier
Form: Liquid, solids; Color: Yellow
Properties: Density: 7.9 lb/gal; Sp. gr.: 0.945–0.953; Visc.: 204 cps; HLB: 3.8; Flash Pt.: 435 F (COC); Acid No.: 5.0 max.; Iodine Value: 77 max.

Kessco Glycerol Monostearate 860. Armak Co.
Chemical Description: Glycerol ester of fatty acid
Category and Applications: See Kessco Acetin; also in tablet coatings in pharmaceuticals
Form: Flakes; Color: White
Properties: M.P.: 58.5–61.5; HLB: 3.8 Flash Pt.: 450 F (COC); Acid No.: 3.0 max.; Iodine Value: 2.0 max.

Kessco Glycerol Monostearate DH-1. Armak Co.
Chemical Description: Glycerol ester of fatty acid
Category and Applications: See Kessco Acetin; also as tablet coatings in pharmaceuticals
Form: Flakes; Color: White
Properties: M.P.: 55–58 C; HLB: 3.8; Flash Pt.: 410 F (COC); Acid No.: 5.0 max.; Iodine Value: 6.0 max.

Kessco Glycerol Monostearate Pure. Armak Co.
Chemical Description: Glycerol ester of fatty acid
Category and Applications: See Kessco Acetin; also as tablet coatings in pharmaceuticals
Form: Flakes; Color: White
Properties: M.P.: 56.5–58.5; HLB: 3.8; Flash Pt.: 410 F (COC); Acid No.: 3.0 max.; Iodine Value: 0.5 max.

Kessco Glycerol Monostearate SE. Armak Co.
Chemical Description: Glycerol ester of fatty acid; acid stable and self-emulsifying
Category and Applications: See Kessco Acetin; also as tablet coatings in pharmaceuticals, and as primary emulsifiers
Form: Flakes; Color: White to cream

Properties: M.P.: 54–58 C; Acid No.: 3.0 max.; Stability: Acid stable; Emulsifiability: Self-emulsifying

Kessco PEG Series. Armak Co.
Chemical Description: Polyethylene glycol ester
Category: Nonionic surfactant, thickening agent, solubilizer
Applications: Versatile and stable nonionic surfactant for use in cosmetic, pharmaceutical, food, agricultural, plastic, and other industries; thickener for shampoos and cream rinses; solubilizer in bath oils and fragrance compositions; in ointments and nonaqueous hair preparations to provide washability
Form and Properties: See individual listings
Containers: 450 lb steel drums
References: Bulletin 76-14

Kessco PEG 200 Dilaurate. Armak Co.
Category and Applications: See Kessco PEG Series
Form: Liquid; Color: Light yellow
Properties: Solubility: Sol. in isopropyl alcohol, acetone, carbon tetrachloride, ethyl acetate, toluol, isopropyl myristate, white oil; disp. in water; Density: 7.9 lb/gal; Sp. gr.: 0.951; F.P.: $<$ 9 C; HLB: 5.9; Flash Pt.: 460 F (COC); Fire Pt.: 510 F; Sapon. Value: 176–186; Acid No.: 10.0 max.; Iodine No.: 12.5 max.

Kessco PEG 200 Monolaurate. Armak Co.
Category and Applications: See Kessco PEG Series
Form: Liquid; Color: Light yellow
Properties: Solubility: Sol. in isopropyl alcohol, acetone, carbon tetrachloride, ethyl acetate, toluol, isopropyl myristate; disp. in water; Density: 8.2 lb/gal; Sp. gr.: 0.985; F.P.: $<$ 5 C; HLB: 9.8; Flash Pt.: 385 F (COC); Fire Pt.: 420 F; Sapon. Value: 132–142; Acid No.: 5.0 max.; Iodine No.: 9.5 max.; pH: 4.5 (3% disp. @ 20 C)

Kessco PEG 200 Dioleate. Armak Co.
Category and Applications: See Kessco PEG Series
Form: Liquid; Color: Light amber
Properties: Solubility: Sol. in naphtha, kerosene, isopropyl alcohol, acetone, carbon tetrachloride, ethyl acetate, toluol, isopropyl myristate, peanut oil, white oil; disp. in Na_2SO_4 and water; Density: 7.9 lb/gal; Sp. gr.: 0.942; F.P.: < -15 C; HLB: 6.0; Flash Pt.: 545 F (COC); Fire Pt.: 625 F; Sapon. Value: 148–158; Acid No.: 10.0 max.; Iodine No.: 70.0; pH: 5.0 (3% disp.)

Kessco PEG 200 Monooleate. Armak Co.
Category and Applications: See Kessco PEG Series
Form: Liquid; Color: Light amber

Properties: Solubility: Sol. in isopropyl alcohol, acetone, carbon tetrachloride, ethyl acetate, toluol; disp. in Na_2SO_4 and water; Density: 8.1 lb/gal; Sp. gr.: 0.973; F.P.: < −15 C; HLB: 8.0; Flash Pt.: 395 F (COC); Fire Pt.: Sapon. Value: 115–124; Acid No.: 5.0 max.; Iodine No.: 56.0 max.; pH: 5.0 (3% disp.)

Kessco PEG 200 Distearate. Armak Co.
Category and Applications: See Kessco PEG Series
Form: Soft solid; Color: White to cream
Properties: Solubility: Sol. in naphtha, kerosene, isopropyl alcohol, acetone, carbon tetrachloride, ethyl acetate, toluol, isopropyl myristate, peanut oil, white oil; disp. hot in water; Sp. gr.: 0.9060 @ 65 C; M.P.: 34 C; HLB: 5.0; Flash Pt.: 475 F (COC); Fire Pt.: 525 F; Sapon. Value: 153–162; Acid No.: 10.0 max.; Iodine No.: 0.5 max.; pH: 5.0 (3% disp.)

Kessco PEG 200 Monostearate. Armak Co.
Category and Applications: See Kessco PEG Series
Form: Soft solid; Color: White to cream
Properties: Solubility: Sol. in isopropyl alcohol, acetone, carbon tetrachloride, ethyl acetate, toluol, isopropyl myristate, peanut oil; disp. in Na_2SO_4 (5%); disp. hot in water; Sp. gr.: 0.9360 @ 65 C; M.P.: 31 C; HLB: 7.9; Flash Pt.: 410 F (COC); Fire Pt.: 450 F; Sapon. Value: 120–129; Acid No.: 5.0 max.; Iodine No.: 0.5 max.; pH: 5.0 (3% disp.)

Kessco PEG 300 Dilaurate. Armak Co.
Category and Applications: See Kessco PEG Series
Form: Liquid; Color: Light yellow
Properties: Solubility: Sol. in naphtha, isopropyl alcohol, acetone, carbon tetrachloride, ethyl acetate, toluol, isopropyl myristate; partly sol. in white oil, kerosene; disp. in water; Density: 8.1 lb/gal; Sp. gr.: 0.975; F.P.: < 13 C; HLB: 9.8; Flash Pt.: 475 F (COC); Fire Pt.: 550 F; Sapon. Value: 148–158; Acid No.: 10.0 max.; Iodine No.: 10.0 max.

Kessco PEG 300 Monolaurate. Armak Co.
Category and Applications: See Kessco PEG Series
Form: Liquid; Color: Light yellow
Properties: Solubility: Sol. in isopropyl alcohol, acetone, carbon tetrachloride, ethyl acetate, toluol; partly sol. in isopropyl myristate; disp. in water; Density: 8.4 lb/gal; Sp. gr.: 1.011; F.P.: < 8 C; HLB: 11.4; Flash Pt.: 445 F (COC); Fire Pt.: 500 F; Sapon. Value: 104–114; Acid No.: 5.0 max.; Iodine No.: 7.5 max.; pH: 4.5 (3% disp. @ 20 C)

Kessco PEG 300 Dioleate. Armak Co.
Category and Applications: See Kessco PEG Series

Form: Liquid; Color: Light amber
Properties: Solubility: Sol. in isopropyl alcohol, acetone, carbon tetrachloride, ethyl acetate, toluol, isopropyl myristate, peanut oil; partly sol. in naphtha, kerosene, white oil; disp. in water; Density: 8.0 lb/gal; Sp. gr.: 0.962; F.P.: < −5 C; HLB: 7.2; Flash Pt.: 510 F (COC); Fire Pt.: 585 F; Sapon. Value: 128–137; Acid No.: 10.0 max.; Iodine No.: 61.0 max.; pH: 5.0 (3% disp.)

Kessco PEG 300 Monooleate. Armak Co.
Category and Applications: See Kessco PEG Series
Form: Liquid; Color: Light amber
Properties: Solubility: Sol. in isopropyl alcohol, acetone, carbon tetrachloride, ethyl acetate, toluol; disp. in water; Density: 8.3 lb/gal; Sp. gr.: 0.998; F.P.: < −5 C; HLB: 9.6; Flash Pt.: 450 F (COC); Sapon. Value: 94–102; Acid No.: 5.0 max.; Iodine No.: 45.0 max.; pH: 5.0 (3% disp.)

Kessco PEG 300 Distearate. Armak Co.
Category and Applications: See Kessco PEG Series
Form: Soft solid; Color: White to cream
Properties: Solubility: Sol. in naphtha, kerosene, isopropyl alcohol, acetone, carbon tetrachloride, ethyl acetate, toluol, isopropyl myristate, peanut oil, white oil; disp. hot in water; M.P.: 32 C; HLB: 6.5; Sapon. Value: 130–139; Acid No.: 10.0 max.; Iodine No.: 0.5 max.; pH: 5.0 (3% disp.)

Kessco PEG 300 Monostearate. Armak Co.
Category and Applications: See Kessco PEG Series
Form: Soft solid; Color: White to cream
Properties: Solubility: Sol. in isopropyl alcohol, acetone, carbon tetrachloride, ethyl acetate, toluol, isopropyl myristate, peanut oil; disp. hot in water; Sp. gr.: 0.9660 @ 65 C; M.P.: 28 C; HLB: 9.7; Flash Pt.: 475 F (COC); Fire Pt.: 520 F; Sapon. Value: 97–105; Acid No.: 5.0 max.; Iodine No.: 0.5 max.; pH: 5.0 (3% disp.)

Kessco PEG 400 Dilaurate. Armak Co.
Category and Applications: See Kessco PEG Series
Form: Liquid; Color: Light yellow
Properties: Solubility: Sol. in naphtha, isopropyl alcohol, acetone, carbon tetrachloride, ethyl acetate, toluol, isopropyl myristate, peanut oil; partly sol. in kerosene; disp. in water; Density: 8.3 lb/gal; Sp. gr.: 0.990; F.P.: 18 C; HLB: 9.8; Flash Pt.: 480 F (COC);; Fire Pt.: 555 F; Sapon. Value: 127–137; Acid No.: 10.0 max.; Iodine No.: 9.5 max.

Kessco PEG 400 Monolaurate. Armak Co.
Category and Applications: See Kessco PEG Series

Form: Liquid; Color: Light Yellow
Properties: Solubility: Sol. in isopropyl alcohol, acetone, carbon tetrachloride, ethyl acetate, toluol, water; Density: 8.6 lb/gal; Sp. gr.: 1.028; F.P.: 12 C; HLB: 13.1; Flash Pt.: 475 F (COC); Fire Pt.: 535 F; Sapon. Value: 86–96; Acid No.: 5.0 max.; Iodine No.: 6.5 max.

Kessco PEG 400 Dioleate. Armak Co.
Category and Applications: See Kessco PEG Series
Form: Liquid; Color: Light amber
Properties: Solubility: Sol. in naphtha, isopropyl alcohol, acetone, carbon tetrachloride, ethyl acetate, toluol, isopropyl myristate, peanut oil; partly sol. in kerosene, white oil; disp. in Na_2SO_4 (5%) and water; Density: 8.1 lb/gal; Sp. gr.: 0.977; F.P.: < 7 C; HLB: 8.5; Flash Pt.: 520 F (COC); Fire Pt.: 635 F; Sapon. Value: 113–122; Acid No.: 10.0 max.; Iodine No.: 55.0 max.; pH: 5.0 (3% disp.)

Kessco PEG 400 Monooleate. Armak Co.
Category and Applications: See Kessco PEG Series
Form: Liquid; Color: Light amber
Properties: Solubility: Sol. in isopropyl alcohol, acetone, carbon tetrachloride, ethyl acetate, toluol; disp. in Na_2SO_4 (5%) and water; Density: 8.4 lb/gal; Sp. gr.: 1.013; F.P.: < 10 C; HLB: 11.4; Flash Pt.: 510 F (COC); Fire Pt.: 585 F; Sapon. Value: 80–89; Acid No.: 5.0 max.; Iodine No.: 39.0 max.; pH: 5.0 (3% disp.)

Kessco PEG 400 Distearate. Armak Co.
Category and Applications: See Kessco PEG Series
Form: Soft solid; Color: White to cream
Properties: Solubility: Sol. in naphtha, isopropyl alcohol, acetone, carbon tetrachloride, ethyl acetate, toluol, isopropyl myristate, peanut oil, white oil; disp. hot in water; partly sol. in kerosene; Sp. gr.: 0.9390 @ 65 C; M.P.: 36 C; HLB: 8.0; Flash Pt.: 500 F (COC); Fire Pt.: 545 F; Sapon. Value: 115–124; Acid No.: 10.0 max.; Iodine No.: 0.5 max.; pH: 5.0 (3% disp.)

Kessco PEG 400 Monostearate. Armak Co.
Category and Applications: See Kessco PEG Series
Form: Soft solid; Color: White to cream
Properties: Solubility: Sol. in isopropyl alcohol, acetone, carbon tetrachloride, ethyl acetate, toluol; partly sol. in isopropyl myristate; disp. hot in water; Sp. gr.: 0.9780 @ 65 C; M.P.: 32 C; HLB: 11.6; Flash Pt.: 480 F (COC); Fire Pt.: 525 F; Sapon. Value: 83–92; Acid No.: 5.0 max.; Iodine No.: 0.5 max.; pH: 5.0 (3% disp.)

Kessco PEG 600 Dilaurate. Armak Co.
Category and Applications: See Kessco PEG Series

Form: Soft solid; Color: Cream
Properties: Solubility: Sol. in isopropyl alcohol, acetone, carbon tetrachloride, ethyl acetate, toluol, isopropyl myristate; partly sol. in naphtha; disp. in water; Sp. gr.: 0.9820 @ 65 C; F.P.: 24 C; HLB: 12.2; Flash Pt.: 465 F (COC); Fire Pt.: 530 F; Sapon. Value: 102–112; Acid No.: 10.0 max.; Iodine No.: 7.5 max.

Kessco PEG 600 Monolaurate. Armak Co.
Category and Applications: See Kessco PEG Series
Form: Liquid; Color: Light yellow
Properties: Solubility: Sol. in Na_2SO_4, isopropyl alcohol, acetone, carbon tetrachloride, ethyl acetate, toluol, water; Density: 8.8 lb/gal; Sp. gr.: 1.050; F.P.: 23 C; HLB: 14.9; Flash Pt.: 475 F (COC); Fire Pt.: 555 F; Sapon. Value: 64–74; Acid No.: 5.0 max.; Iodine No.: 5.0 max.

Kessco PEG 600 Dioleate. Armak Co.
Category and Applications: See Kessco PEG Series
Form: Liquid; Color: Light amber
Properties: Solubility: Sol. in isopropyl alcohol, acetone, carbon tetrachloride, ethyl acetate, toluol, isopropyl myristate, peanut oil; partly sol. in naphtha; disp. in water; Density: 8.3 lb/gal; Sp. gr.: 1.001; F.P.: 19 C; HLB: 10.5; Flash Pt.: 495 F (COC); Fire Pt.: 615 F; Sapon. Value: 92–102; Acid No.: 10.0 max.; Iodine No.: 45.0 max.; pH: 5.0 (3% disp.)

Kessco PEG 600 Monooleate. Armak Co.
Category and Applications: See Kessco PEG Series
Form: Liquid; Color: Light amber
Properties: Solubility: Sol. in isopropyl alcohol, acetone, carbon tetrachloride, ethyl acetate, toluol, water; disp. in Na_2SO_4 (5%); Density: 8.7 lb/gal; Sp. gr.: 1.037; F.P.: 23 C; HLB: 13.5; Flash Pt.: 525 F (COC); Fire Pt.: 620 F; Sapon. Value: 60–69; Acid No.: 5.0 max.; Iodine No.: 30.0 max.; pH: 5.0 (3% disp.)

Kessco PEG 600 Distearate. Armak Co.
Category and Applications: See Kessco PEG Series
Form: Soft solid; Color: White to cream
Properties: Solubility: Sol. in isopropyl alcohol, acetone, carbon tetrachloride, ethyl acetate, toluol, isopropyl myristate, peanut oil; partly sol. in naphtha, kerosene; disp. hot in water; Sp. gr.: 0.9670 @ 65 C; M.P.: 39 C; HLB: 10.6; Flash Pt.: 490 F (COC); Fire Pt.: 530 F; Sapon. Value: 93–102; Acid No.: 10.0 max.; Iodine No.: 0.5 max.; pH: 5.0 (3% disp.)

Kessco PEG 600 Monostearate. Armak Co.
Category and Applications: See Kessco PEG Series
Form: Soft solid; Color: White to cream
Properties: Solubility: Sol. in isopropyl alcohol, acetone, carbon tetrachloride, ethyl acetate, toluol,

water; Sp. gr.: 1.000 @ 65 C; M.P.: 37 C; HLB: 13.6; Flash Pt.: 480 F (COC); Fire Pt.: 550 F; Sapon. Value: 61–70; Acid No.: 5.0 max.; Iodine No.: 0.25 max.; pH: 5.0 (3% disp.)

Kessco PEG 1000 Dilaurate. Armak Co.
Category and Applications: See Kessco PEG Series
Form: Soft solid; Color: Cream
Properties: Solubility: Sol. in isopropyl alcohol, acetone, carbon tetrachloride, ethyl acetate, toluol, isopropyl myristate, water; Sp. gr.: 1.015 @ 65 C; F.P.: 38 C; HLB: 14.5; Flash Pt.: 475 F (COC); Fire Pt.: 495 F; Sapon. Value: 68–78; Acid No.: 10.0 max.; Iodine No.: 5.5 max.

Kessco PEG 1000 Monolaurate. Armak Co.
Category and Applications: See Kessco PEG Series
Form: Soft solid; Color: Cream
Properties: Solubility: Sol. in propylene glycol (hot), Na_2SO_4, isopropyl alcohol, acetone, carbon tetrachloride, ethyl acetate, toluol, water; Sp. gr.: 1.035 @ 65 C; F.P.: 40 C; HLB: 16.5; Flash Pt.: 490 F (COC); Fire Pt.: 505 F; Sapon. Value: 41–51; Acid No.: 5.0 max.; Iodine No.: 3.5 max.

Kessco PEG 1000 Dioleate. Armak Co.
Category and Applications: See Kessco PEG Series
Form: Soft solid; Color: Cream
Properties: Solubility: Sol. in isopropyl alcohol, acetone, carbon tetrachloride, ethyl acetate, toluol, water; disp. in Na_2SO_4 (5%); Sp. gr.: 1.005 @ 65 C; F.P.: 37 C; HLB: 13.1; Flash Pt.: 505 F (COC); Fire Pt.: 615 F; Sapon. Value: 64–74; Acid No.: 10.0 max.; Iodine No.: 33.0 max.; pH: 5.0 (3% disp.)

Kessco PEG 1000 Monooleate. Armak Co.
Category and Applications: See Kessco PEG Series
Form: Soft solid; Color: Cream
Properties: Solubility: Sol. in Na_2SO_4 (5%), isopropyl alcohol, acetone, carbon tetrachloride, ethyl acetate, toluol, water; sol. hot in propylene glycol; Sp. gr.: 1.035 @ 65 C; F.P.: 39 C; HLB: 15.4; Flash Pt.: 515 F (COC); Fire Pt.: 595 F; Sapon. Value: 40–49; Acid No.: 5.0 max.; Iodine No.: 21.0 max.; pH: 5.0 (3% disp.)

Kessco PEG 1000 Distearate. Armak Co.
Category and Applications: See Kessco PEG Series
Form: Wax; Color: Cream
Properties: Solubility: Sol. in isopropyl alcohol, acetone, carbon tetrachloride, ethyl acetate, toluol, water; partly sol. in isopropyl myristate and peanut oil; Sp. gr.: 1.005 @ 65 C; M.P.: 40 C; HLB: 12.3; Flash Pt.: 485 F (COC); Fire Pt.: 510 F; Sapon. Value: 65–74; Acid No.: 10.0 max.; Iodine No.: 0.25 max.; pH: 5.0 (3% disp.)

Kessco PEG 1000 Monostearate. Armak Co.
Category and Applications: See Kessco PEG Series
Form: Wax; Color: Cream
Properties: Solubility: Sol. in Na_2SO_4 (5%), isopropyl alcohol, acetone, carbon tetrachloride, ethyl acetate, toluol, water; Sp. gr.: 1.030 @ 65 C; M.P.: 41 C; HLB: 15.6; Flash Pt.: 475 F (COC); Fire Pt.: 495 F; Sapon. Value: 40–48; Acid No.: 5.0 max.; Iodine No.: 0.25 max.; pH: 5.0 (3% disp.)

Kessco PEG 1540 Dilaurate. Armak Co.
Category and Applications: See Kessco PEG Series
Form: Wax; Color: Cream
Properties: Solubility: Sol. in water, Na_2SO_4 (5%); sol. hot in propylene glycol, isopropyl alcohol, acetone, carbon tetrachloride, ethyl acetate, toluol; Sp. gr.: 1.04 @ 65 C; F.P.: 42 C; HLB: 15.7; Flash Pt.: 450 F (COC); Fire Pt.: 515 F; Sapon. Value: 48–56; Acid No.: 10.0 max.; Iodine Number: 4.0 max.; pH: 4.5 (3% disp. @ 20 C)

Kessco PEG 1540 Monolaurate. Armak Co.
Category and Applications: See Kessco PEG Series
Form: Wax; Color: Cream
Properties: Solubility: Sol. in water, Na_2SO_4 (5%); sol. hot in propylene glycol, isopropyl alcohol, acetone, carbon tetrachloride, ethyl acetate, toluol; Sp. gr.: 1.06 @ 65 C; F.P.: 46 C; HLB: 17.6; Flash Pt.: 445 F (COC); Fire Pt.: 520 F; Sapon. Value: 26–36; Acid No.: 5.0 max.; Iodine No.: 2.5 max.; pH: 4.5 (3% disp. @ 20 C)

Kessco PEG 1540 Dioleate. Armak Co.
Category and Applications: See Kessco PEG Series
Form: Wax; Color: Cream
Properties: Solubility: Sol. in propylene glycol, Na_2SO_4, water; sol. hot in isopropyl alcohol, acetone, carbon tetrachloride, ethyl acetate, toluol; Sp. gr.: 1.025 @ 65 C; F.P.: 44 C; HLB: 15.0; Flash Pt.: 480 F (COC); Fire Pt.: 560 F; Sapon. Value: 45–55; Acid No.: 10.0 max.; Iodine No.: 24.5 max.; pH: 5.0 (3% disp.)

Kessco PEG 1540 Monooleate. Armak Co.
Category and Applications: See Kessco PEG Series
Form: Wax; Color: Cream
Properties: Solubility: Sol. in propylene glycol, Na_2SO_4, water; sol. hot in isopropyl alcohol, acetone, carbon tetrachloride, ethyl acetate, toluol; Sp. gr.: 1.050 @ 65 C; F.P.: 45 C; HLB: 17.0; Flash Pt.: 520 F (COC); Fire Pt.: 545 F; Sapon. Value: 28–37; Acid No.: 5.0 max.; Iodine No.: 14.5 max.; pH: 5.0 (3% disp.)

Kessco PEG 1540 Distearate. Armak Co.
Category and Applications: See Kessco PEG Series

Form: Wax; Color: Cream
Properties: Solubility: Sol. in isopropyl alcohol, acetone, carbon tetrachloride, ethyl acetate, toluol, water; partly sol. in peanut oil; Sp. gr.: 1.015 @ 65 C; M.P.: 45 C; HLB: 14.8; Flash Pt.: 490 F (COC); Fire Pt.: 540 F; Sapon. Value: 49–58; Acid No.: 10.0 max.; Iodine No.: 0.25 max.; pH: 5.0 (3% disp.)

Kessco PEG 1540 Monostearate. Armak Co.
Category and Applications: See Kessco PEG Series
Form: Wax; Color: Cream
Properties: Solubility: Sol. in Na_2SO_4 (5%), isopropyl alcohol, acetone, carbon tetrachloride, ethyl acetate, toluol, water; Sp. gr.: 1.050 @ 65 C; M.P.: 47 C; HLB: 17.3; Flash Pt.: 495 F (COC); Fire Pt.: 540 F; Sapon. Value: 27–36; Acid No.: 5.0 max.; Iodine No.: 0.25 max.; pH: 5.0 (3% disp.)

Kessco PEG 4000 Dilaurate. Armak Co.
Category and Applications: See Kessco PEG Series
Form: Wax; Color: Cream
Properties: Solubility: Sol. in water, Na_2SO_4 (5%); sol. hot in propylene glycol, isopropyl alcohol, acetone, carbon tetrachloride, ethyl acetate, toluol; Sp. gr.: 1.065 @ 65 C; F.P.: 52 C; HLB: 17.6; Flash Pt.: 495 F (COC); Fire Pt.: 515 F; Sapon. Value: 20–30; Acid No.: 5.0 max.; Iodine No.: 2.0 max.; pH: 4.5 (3% disp. @ 20 C)

Kessco PEG 4000 Monolaurate. Armak Co.
Category and Applications: See Kessco PEG Series
Form: Wax; Color: Cream
Properties: Solubility: Sol. in water, Na_2SO_4 (5%); sol. hot in propylene glycol, isopropyl alcohol, acetone, carbon tetrachloride, ethyl acetate, toluol; Sp. gr.: 1.075 @ 65 C; F.P.: 55 C; HLB: 18.8; Flash Pt.: 515 F (COC); Fire Pt.: 565 F; Sapon.Value: 9–18; Acid No.: 5.0 max.; Iodine No.: 1.0 max.; pH: 4.5 (3% disp. @ 20 C)

Kessco PEG 4000 Dioleate. Armak Co.
Category and Applications: See Kessco PEG Series
Form: Wax; Color: Cream
Properties: Solubility: Sol. in propylene glycol, Na_2SO_4, water; sol. hot in isopropyl alcohol, acetone, carbon tetrachloride, ethyl acetate, toluol; Sp. gr.: 1.060 @ 65 C; F.P.: 49 C; HLB: 17.8; Flash Pt.: 500 F (COC); Fire Pt.: 520 F; Sapon. Value: 19–27; Acid No.: 5.0 max.; Iodine No.: 11.0 max.; pH: 5.0 (3% disp.)

Kessco PEG 4000 Monooleate. Armak Co.
Category and Applications: See Kessco PEG Series
Form: Wax; Color: Cream
Properties: Solubility: Sol. in propylene glycol, Na_2SO_4, water; sol. hot in isopropyl alcohol, acetone, carbon tetrachloride, ethyl acetate, toluol; Sp. gr.: 1.075 @ 65 C;

F.P.: 55 C; HLB: 18.3; Flash Pt.: 495 F (COC); Fire Pt.: 515 F; Sapon. Value: 10–18; Acid No.: 5.0 max.; Iodine No.: 6.0 max.; pH: 5.0 (3% disp.)

Kessco PEG 4000 Distearate. Armak Co.
Category and Applications: See Kessco PEG Series
Form: Wax; Color: Cream
Properties: Solubility: Sol. in Na_2SO_4 (5%), isopropyl alcohol, acetone, carbon tetrachloride, ethyl acetate, toluol, water; Sp. gr.: 1.060 @ 65 C; M.P.: 51 C; HLB: 17.3; Flash Pt.: 515 F (COC); Fire Pt.: 545 F; Sapon. Value: 19–27; Acid No.: 5.0 max.; Iodine No.: 0.2 max.; pH: 5.0 (3% disp.)

Kessco PEG 4000 Monostearate. Armak Co.
Category and Applications: See Kessco PEG Series
Form: Wax; Color: Cream
Properties: Solubility: Sol. in Na_2SO_4 (5%), isopropyl alcohol, acetone, carbon tetrachloride, ethyl acetate, toluol, water; Sp. gr.: 1.075 @ 65 C; M.P.: 56 C; HLB: 18.6; Flash Pt.: 465 F (COC); Fire Pt.: 520 F; Sapon. Value: 10–18; Acid No.: 5.0 max.; Iodine No.: 0.1 max.; pH: 5.0 (3% disp.)

Kessco PEG 6000 Dilaurate. Armak Co.
Category and Applications: See Kessco PEG Series
Form: Wax; Color: Cream
Properties: Solubility: Sol. in water, Na_2SO_4 (5%); sol. hot in propylene glycol, isopropyl alcohol, acetone, carbon tetrachloride, ethyl acetate, toluol; Sp. gr.: 1.077 @ 65 C; F.P.: 57 C; HLB: 18.7; Flash Pt.: 435 F (COC); Fire Pt.: 495 F; Sapon. Value: 12–20; Acid No.: 9.0 max.; Iodine No.: 1.5 max.; pH: 4.5 (3% disp. @ 20 C)

Kessco PEG 6000 Monolaurate. Armak Co.
Category and Applications: See Kessco PEG Series
Form: Wax; Color: Cream
Properties: Solubility: Sol. in water, Na_2SO_4 (5%); sol. hot in propylene glycol, isopropyl alcohol, acetone, carbon tetrachloride, ethyl acetate, toluol; F.P.: 61 C; HLB: 19.2; Sapon. Value: 7–13; Acid No.: 5.0 max.; Iodine No.: 1.0 max.; pH: 4.5 (3% disp. @ 20 C)

Kessco PEG 6000 Dioleate. Armak Co.
Category and Applications: See Kessco PEG Series
Form: Wax; Color: Cream
Properties: Solubility: Sol. in propylene glycol, Na_2SO_4 (5%), water; sol. hot in isopropyl alcohol, acetone, carbon tetrachloride, ethyl acetate, toluol; Sp. gr.: 1.070 @ 65 C; F.P.: 56 C; HLB: 18.3; Flash Pt.: 500 F (COC); Fire Pt.: 525 F; Sapon. Value: 13–21; Acid No.: 9.0 max.; Iodine No.: 7.5 max.; pH: 5.0 (3% disp.)

Kessco PEG 6000 Monooleate. Armak Co.
Category and Applications: See Kessco PEG Series
Form: Wax; Color: Cream
Properties: Solubility: Sol. in propylene glycol, Na_2SO_4 (5%), water; sol. hot in isopropyl alcohol, acetone, carbon tetrachloride, ethyl acetate, toluol; Sp. gr.: 1.085 @ 65 C; F.P.: 59 C; HLB: 19.0; Flash Pt.: 470 F (COC); Fire Pt.: 495 F; Sapon. Value: 7–13; Acid No.: 5.0 max.; Iodine No.: 4.3 max.; pH: 5.0 (3% disp.)

Kessco PEG 6000 Distearate. Armak Co.
Category and Applications: See Kessco PEG Series
Form: Wax; Color: Cream
Properties: Solubility: Sol. in propylene glycol, Na_2SO_4 (5%), isopropyl alcohol, acetone, carbon tetrachloride, ethyl acetate, toluol; partly sol. in water; Sp. gr.: 1.075 @ 65 C; M.P.: 55 C; HLB: 18.4; Flash Pt.: 475 F (COC); Fire Pt.: 525 F; Sapon. Value: 14–20; Acid No.: 9.0 max.; Iodine No.: 0.1 max.; pH: 5.0 (3% disp.)

Kessco PEG 6000 Monostearate. Armak Co.
Category and Applications: See Kessco PEG Series
Form: Wax; Color: Cream
Properties: Solubility: Sol. in propylene glycol, Na_2SO_4 (5%), isopropyl alcohol, acetone, carbon tetrachloride, ethyl acetate, toluol, water; Sp. gr.: 1.080 @ 65 C; M.P.: 61 C; HLB: 18.8; Flash Pt.: 480 F (COC); Fire Pt.: 525 F; Sapon. Value: 7–13; Acid No.: 5.0 max.; Iodine No.: 0.1 max.; pH: 5.0 (3% disp.)

Kessco Propylene Glycol Monostearate Pure. Armak Co.
Chemical Description: Glycol ester
Category: Auxiliary emulsifier, opacifier
Applications: Hand care products, makeup preparations, solidified perfume compositions; melting pt. is close to body temperature so it improves application properties of creams and lotions on the skin
Form: Flakes
Properties: M.P.: 33.5–38.5 C; Flash Pt.: 390 F (COC); Acid No.: 3.0 max.; Iodine Value: 0.5 max.

Kessco Synthetic Spermaceti N.F. Armak Co.
Chemical Description: Synthetic spermaceti wax meeting N.F. specifications
Applications: For use in raising the viscosity or melting point of face creams, cold creams, lipsticks, hand lotions, body lotions, shaving creams, etc.
Form: Flakes; Color: White
Properties: Solubility: Sol. in boiling alcohol, ether, chloroform, other waxes, oils, hydrocarbons; insol. in water; M.P.: 43–47 C; Saponification Value: 109–117; Acid No.: 2.0 max.; Iodine Value: 1.0 max.

Containers: 200 lb net wt fiber drum
References: Bulletin #77-5

Kessco Triacetin. Armak Co.
Chemical Description: Glycerol ester of fatty acid
Category and Applications: See Kessco Acetin; also as solvent for essential oils
Form: Oily liquid; Color: 10 max. (APHA)
Properties: Ref. Index: 1.429; Solubility: Sol. in benzol, ethyl acetate, methyl ethyl ketone, isopropanol, methanol, propylene glycol; partly sol. in castor oil; solubility in water: 6.1%; Density: 9.6 lb/gal; Sp. gr.: 1.151–1.158 (25/20 C); Visc.: 16 cps; F.P.: –50 C (gel); Flash Pt.: 290 F (COC); Acidity: 0.005% max. (as acetic)
Containers: 55 gal bung type steel drum
References: Bulletin # 76-3

Kesscolin. Armak Co.
Category: Lanolin replacement, nonionic
Applications: Lanolin replacement water absorption base; base for conditioning and moisturizing creams and lotions; nonallergenic ointment bases
Form: Paste; Color: 125 (APHA–hazy)
Properties: Visc.: 55.3 cps @ 122 F; M.P.: 47.0 C; Flash Pt.: 425 F; Fire Pt.: 445 F; Iodine Value: 8.25; Stability: Good
References: Bulletin No. 73-21

Kesscowax B. Armak Co.
Category: Beeswax substitute
Applications: Replacement for natural beeswax in beeswax-borax cosmetic cold cream formulations
Form: Wax; Color: White
Properties: M.P.: 64–68 C; Sapon. Value: 197 ± 5; Acid No.: 12–30
Toxicity: Slight eye irritant
References: Bulletin 78-3

Kieralon B. BASF Aktiengesellschaft
Chemical Description: Mixture of nonionic and anionic surfactants
Category: Detergent, wetting agent, dispersant, anionic
Applications: Boil-off and Kier boiling of cellulosic fibers in long liquors; bleaching of cellulosic fibers; aftersoaping of dyed and printed cellulosic fibers and acrylics; precleansing of synthetic fibers
Form: Low visc. liquid; Color: Brownish
Properties: Solubility: Readily sol. in cold and warm water; Density: ≈ 1.03 g/ml; Setting Pt.: ≈ –3 C; Stability: Stable to hard water, acids, and reduction or oxidation bleaching agents in the concentrations normally used for textile fibers
Toxicity: Skin and eye irritant on prolonged contact with concentrated form
References: Leaflet M 1761 e

Kieralon B Highly Conc. BASF Aktiengesellschaft
Chemical Description: Mixture of nonionic and anionic surfactants
Category and Applications: See

Kieralon B
Form: Liquid; Color: Brownish
Properties: Solubility: Readily sol. in warm water; Density: ≈ 1.05 g/ml; Visc.: Higher visc. than Kieralon B; Setting Pt.: ≈ –16 C; Stability: Stable to hard water, acids, and reduction or oxidation bleaching agents in the concentrations normally used for textile fibers
Toxicity: Skin and eye irritant on prolonged contact with concentrated form
References: Leaflet M 1761 e

Kieralon D. BASF Aktiengesellschaft
Chemcial Description: Mixture of surfactants with affinity for dyes and a dye-dissolving effect
Category: Detergent, nonionic
Applications: Detergent for afterwashing prints, colored woven fabrics and dyed shades; detaches nonfixed dye and prevents it from going on to unprinted or paler areas in scouring operation; improves the wet fastness of deep shades
Form: Clear, viscous liquid; Color: Colorless to weakly yellow
Properties: Solubility: Readily sol. by stirring into warm water; Density: ≈ 1.01 g/ml; Pour Pt.: 6–7 C; Stability: Stable to hard water, alkalis, acids, reducing agents, and peroxides in concentrations normally used in afterwashing
Reaction: Weakly alkaline
Toxicity: Severe skin and eye irritant on prolonged contact
References: Leaflet M 5451 e

Kieralon JET. BASF Aktiengesellschaft
Chemical Description: Emulsion of nonionic components with minimum foam
Category: Wetting agent, detergent, nonionic
Applications: Wetting agent and detergent with minimum foam for textile finishing; in processes with strong liquor turbulence (e.g., jet dyeing machines); precleansing, boil-off, or bleaching in long liquors; aftersoaping of dyed shades; as foam inhibitor
Form: Viscous liquid; Color: White
Properties: Density: ≈ 1 g/ml @ 20 C; Pour Pt.: ≈ 3 C; Stability: Stable to the usual electrolyte and bleaching agent concentrations in the recommended fields up to 20 ml/l NaOH 38 Be.
Warnings and Caution: Protect eyes from splashes of undiluted product

Kieralon KB. BASF Aktiengesellschaft
Chemical Description: Mixture of anionic and nonionic components
Category: Detergent, wetting agent, anionic
Applications: Electrolyte-resistant detergent and wetting agent for textiles, esp. in alkaline liquors containing electrolytes; boiling-off of cotton and its blends with synthetic fibers; aftersoaping of prints and dyed shades on cellulosic fibers; wetting and scouring in bleaching cellulosic fibers with peroxide, chlorite, or hypochlorite
Form: Clear liquid of low viscosi-

ty; Color: Yellowish
Properties: Solubility: Dissolves readily in cold water; Density: ≈ 1.08 g/ml; Pour Pt.: ≈ –5 C; Stability: Stable to hard water, alkalis, acids, and to reduction and oxidation bleaches in normal concentrations for treating textile fibers; esp. stable in caustic liquors and in the presence of salts
Toxicity: Skin and eye irritant on prolonged contact with concentrated product
References: Leaflet M 5666 e

Kieralon OL. BASF Aktiengesellschaft
Chemical Description: Mixture of nonionic and anionic components
Category: Detergent, anionic
Applications: Specialty product for removing oily, fatty, and waxy impurities and for washing-out spin finishes and water-soluble sizes; in high concentrations, as stain-remover
Form: Transparent, pourable paste; Color: Light yellow
Properties: Solubility: Readily sol. in warm water; Stability: Stable to hard water, acids, and reduction or oxidation bleaches in normal concentrations for textile fibers
Toxicity: Skin and eye irritant on prolonged contact with concentrated product
References: Leaflet M 5131 e

Kilfoam. Arol Chem. Prod. Co.
Chemical Description: Emulsion of silicones, emulsifiers, and other stabilizers
Category: Defoamer
Applications: Silicone-based defoamer for textile and industrial foaming problems; emulsifier for dispersing waxes, pigments, resins, in preparation of emulsified paints
Form: Liquid; Color: Milky white; Odor: Mild
Properties: Solubility: Readily disp. in water at ambient temperatures; Sp. gr.: 0.97; Stability: Excellent
Reaction: Neutral
Containers: 55 gal drums
Warnings and Caution: Avoid ingestion and contact with eyes

Kito 703. Hamblet & Hayes Co.
Chemical Description: Linear alcohol, ethoxylated polyhydric
Category: Wetting agent, emulsifier, detergent. defoamer, nonionic
Applications: Rewetting in papermaking; adaptable in many detergent, wetting, and emulsifying processes; wetting agent in bleaching, sizing, and coating
Form: Liquid; Color: Colorless; Odor: Mild
Properties: Solubility: Sol. in oils and organic media; water-insol.; Sp. gr.: 0.924; Visc.: 37 cs @ 20 C; F.P.: –32 C; HLB: 9.0; Flash Pt.: 370 F; Biodegradable
Composition: 100% active ingred.
Reaction: Neutral
Toxicity: Slight skin irritant; severe eye irritant; defats tissues
Containers: 55 gal and bulk
Warnings and Caution: Protect eyes; avoid prolonged skin contact

Kito 709. Hamblet & Hayes Co.
Chemical Description: Linear alcohol, ethoxylated polyhydric
Category: Detergent, wetting agent, emulsifier, nonionic
Applications: See Kito 703; also for wetting pulp in beater
Form: Liquid; Color: Colorless; Odor: Mild
Properties: Solubility: Sol. in water, various organic media; Sp. gr.: 1.006; Visc.: 86 cs @ 20 C; F.P.: 13 C; HLB: 13.3; Flash Pt.: 470; Surface Tension: 29 dynes/cm; Biodegradable
Composition: 100% active ingred.
Reaction: Neutral
Toxicity: Slight skin irritant; severe eye irritant; defats tissue
Containers: 55 gal and bulk
Warnings and Caution: Protect eyes; avoid prolonged skin contact

Klearfac AA 040. BASF Wyandotte Corp.
Chemical Description: Free acid of an organic phosphate ester
Category: Emulsifier, solubilizer, detergent, antistat, anionic
Applications: Emulsifier and solubilizer for nonionics in high electrolyte and strongly alkaline solutions; heavy-duty cleaning formulations (airplane and railroad car cleaners, bottle washing compounds, household detergents, wall and floor cleaner concentrates, paint and wax strippers, solvent cleaners for dry cleaning and dedusting); metal cleaning formulations, alkaline soak tank cleaning, and electrolytic metal cleaning; in textile processing (Kier boiling, scouring, bleaching, soaping, print washing operations; antistat for synthetic yarns, primary emulsifier for textile lubricants); in emulsion polymerization of acrylic monomers, in vinyl acetate and vinyl acetate-acrylic systems, pesticides, herbicides, insecticides, and liquid fertilizer ingredients; antistat in plastics
Form: Clear liquid; Color: Light yellow
Properties: Solubility: > 10% sol. in acetone, ethyl alcohol, ethylene glycol, propylene glycol; easily disp. in xylene, carbon tetrachloride, kerosene, water; Sp. gr.: 1.112; Visc.: 240 cps (Brookfield); Cloud Pt.: 2 C (1% aq.); Pour Pt.: 55 F; Stability: Good in high electrolyte and alkaline media; Surface Tension: 43.4 dynes/cm (0.1% aq.); pH: 2.1 (1% aq.); Biodegradable
Composition: 60% active ingred.
Toxicity: Acidic—causes irritation on contact with skin or mucous membranes
References: Bulletin 0-722R

Klearfac AA 270. BASF Wyandotte Corp.
Chemical Description: Free acid of an organic phosphate ester
Category and Applications: See Klearfac AA 270
Form: Clear liquid; Color: Light yellow
Properties: Solubility: > 10% sol. in xylene, acetone, ethyl alcohol,

ethylene glycol, propylene glycol, carbon tetrachloride, water; Sp. gr.: 1.165; Visc.: 5900 cps (Brookfield); Cloud Pt.: 95 C (1% aq.); Pour Pt.: 25 F; Stability: Good in high electrolyte and alkaline media; Surface Tension: 42.3 dynes/cm (0.1% aq.); pH: 2.2 (1% aq.); Biodegradable
Composition: 90% active ingred.
Toxicity: Acidic–causes irritation on contact with skin or mucous membranes
References: Bulletin 0-722R

Klearfac AA 420. BASF Wyandotte Corp.
Chemical Description: Free acid of an organic phosphate ester
Category and Applications: See Klearfac AA 040
Form: Clear liquid; Color: Light yellow
Properties: Solubility: > 10% sol. in xylene, acetone, ethyl alcohol, ethylene glycol, propylene glycol, carbon tetrachloride, water; Sp. gr.: 1.118; Visc.: 2950 cps (Brookfield); Cloud Pt.: 67 C (1% aq.); Pour Pt.: 20 F; Stability: Good in high electrolyte and alkaline media; Surface Tension: 38.0 dynes/cm (0.1% aq.); pH: 2.1 (1% aq.); Biodegradable
Composition: 90% active ingred.
Toxicity: Acidic–causes irritation on contact with skin or mucous membranes
References: Bulletin 0-722R

Kling/Catimuls. Lancaster Chem. Co.
Category: Wetting agent, emulsifier
Applications: Asphalt emulsions; asphalt antistripping; waterproofing; making of bitumen emulsions
Form: Liquid; Color: Dark brown
Properties: Solubility: Sol. in petroleum solvents; Sp. gr.: 0.98–1.02; B.P.: 260 F; Stability: Indefinite at room temperature
Composition: 100% active ingred.
Reaction: Acid
Toxicity: Slight dermatitic effects due to free amine
Containers: Bulk and 55 gal drums
Warnings and Caution: Wash with copious amount of water if contacted with skin

Klucel. Hercules
Chemical Description: Hydroxypropylcellulose
Category: Thickener, nonionic
Applications: Acts as thickener, protective colloid binder, stabilizer, and suspending agent in shampoos, creams, lotions, esp. water-solvent mixtures
Form: Powder; Color: White; Odor: None
Properties: Solubility: Sol. in water and polar organic solvents; Visc.: 10–2000 cps (1% sol'n.); Softening Pt.: 110–180 C; Stability: Good; Surface Tension: 43.8 dynes/cm
Composition: 99.5% active
Containers: 100 lb drums

L

Lactodan. Grindsted Prod.
Chemical Description: Lactic acid esters of monoglycerides
Category: Emulsifier, nonionic
Applications: Food emulsifier for shortening, cake improvers, toppings, dessert powders; improves aeration and foam stabilization; enhances whippability in whipped toppings
Form: Solid, flakes
Properties: HLB: 3.5
Composition: 100% active ingred.

Lakeway 100-CA. Bofors Lakeway
Chemical Description: Coco diethanolamide
Category: Foam booster, thickener, stabilizer, nonionic
Applications: Shampoos, bubble baths, light duty liquids
Form: Liquid; Color: Light straw
Properties: Density: 8.3 lb/gal; Cloud Pt.: ≈ –3 C; pH: 10 (10%)
Composition: 90% conc.

Lakeway 101-10. Bofors Lakeway
Chemical Description: Sodium lauryl sulfate
Category: Detergent, anionic
Applications: Personal care products, hair shampoo, bubble bath, rug shampoos
Form: Liquid; Color: White to light yellow
Properties: Density: 8.5 lb/gal; Visc.: 350 cps (Brookfield); Cloud Pt.: ≈ 18 C; pH: 8.0 (10%)
Composition: 29% conc.

Lakeway 101-10P,N. Bofors Lakeway
Chemical Description: Dry sodium lauryl sulfate
Applications: Emulsion polymerization; plant and printing ink dispersions; personal care products; lubricant
Form: Powder or needle; Color: White
Properties: pH: 10 (1%)
Composition: 90% conc.

Lakeway 101-11. Bofors Lakeway
Chemical Description: Sodium lauryl sulfate
Category: Detergent, anionic
Applications: For emulsion polymerization
Form: Liquid; Color: White to

light yellow
Properties: Density: 8.5 lb/gal; Visc.: 250 cps (Brookfield); Cloud Pt.: ≈ 15 C; pH: 8.0 (10%)
Composition: 29% conc.

Lakeway 101-19. Bofors Lakeway
Chemical Description: Sodium lauryl sulfate
Category and Applications: See Lakeway 101-10 (101-19 is high actives grade)
Form: Liquid; Color: White to light yellow
Properties: Density: 8.6 lb/gal; Visc.: 100 cps (Brookfield); Cloud Pt.: ≈ 19 C; pH: 8.0 (10%)
Composition: 45% conc.

Lakeway 101-20. Bofors Lakeway
Chemical Description: Ammonium lauryl sulfate
Category: Detergent, anionic
Applications: Personal care products, bubble baths, industrial processing, fire fighting foams
Form: Liquid; Color: White to light yellow
Properties: Density: 8.5 lb/gal; Visc.: 1000 cps (Brookfield); Cloud Pt.: ≈ 14 C; pH: 7.0 (10%)
Composition: 29% conc.

Lakeway 101-30. Bofors Lakeway
Chemical Description: Triethanolamine lauryl sulfate
Category: Detergent, anionic
Applications: Low cloud point, low irritancy for personal care products
Form: Liquid; Color: White to light yellow
Properties: Density: 8.6 lb/gal; Visc.: 120 cps (Brookfield); Cloud Pt.: ≈ 3 C; pH: 8.0 (10%)
Composition: 40% conc.

Lakeway 106R. Bofors Lakeway
Chemical Description: Formulated concentrate
Category: Rug shampoo concentrate, anionic
Applications: High actives rug shampoo for low cloud point–dry foam
Form: Liquid; Color: White to light yellow
Properties: Density: 8.6 lb/gal; Visc.: 265 cps (Brookfield); Cloud Pt.: ≈ 0 C; pH: 7.5 (10%)
Composition: 30% conc.

Lakeway 120. Bofors Lakeway
Chemical Description: Formulated concentrate
Category: Rug shampoo concentrate, anionic
Applications: Anti-resoil dry residue shampoo
Form: Liquid; Color: Milky white to light yellow
Properties: Density: 8.9 lb/gal; Visc.: 150 cps (Brookfield); pH: 7.0 (10%)
Composition: 30% conc.

Lakeway 201-10. Bofors Lakeway
Chemical Description: Sodium lauryl polyether sulfate
Category: Detergent, anionic
Applications: High actives, low cloud point detergent for shampoos, personal care products

Form: Liquid; Color: Straw
Properties: Density: 8.8 lb/gal; Visc.: 285 cps (Brookfield); Cloud Pt.: ≈ 13 C; pH: 7.5 (10%)
Composition: 58% conc.

Lakeway 201-11. Bofors Lakeway
Chemical Description: Sodium lauryl polyether sulfate
Category and Applications: Lower actives version of Lakeway 201-10
Form: Liquid; Color: Straw
Properties: Density: 8.5 lb/gal; Visc.: 100 cps (Brookfield); Cloud Pt.: ≈ 0 C; pH: 8.0 (10%)
Composition: 29% conc.

Lakeway 201-20. Bofors Lakeway
Chemical Description: Ammonium lauryl polyether sulfate
Category: Detergent, anionic
Applications: High actives detergent for dishwash, laundry, shampoos, personal care products
Form: Liquid; Color: Straw
Properties: Density: 8.4 lb/gal; Visc.: 75 cps (Brookfield); Cloud Pt.: ≈ 0 C; pH: 7.2 (10%)
Composition: 58% conc.

Lakeway 201-29. Bofors Lakeway
Chemical Description: Ammonium lauryl polyether sulfate
Category and Applications: Lower actives version of Lakeway 201-20
Form: Liquid; Color: Straw
Properties: Density: 8.4 lb/gal; Visc.: 60 cps (Brookfield); Cloud Pt.: ≈ 0 C; pH: 6.5 (10%)
Composition: 29% conc.

Lakeway 301-10. Bofors Lakeway
Chemical Description: Sodium alpha olefin sulfonate
Category: Detergent, anionic
Applications: Low cost, high sudsing detergent for hair shampoo, bubble bath, personal care products, dishwash, rug shampoos, industrial processing products
Form: Liquid; Color: Light yellow
Properties: Density: 8.9 lb/gal; Visc.: 75 cps (Brookfield); Cloud Pt.: ≈ 2 C; pH: 8.5 (10%)
Composition: 39% conc.

Lakeway 301-10 F. Bofors Lakeway
Chemical Description: Dry sodium alpha olefin sulfonate
Category: Detergent, anionic
Applications: High actives flake form for dry bubble baths, metal cleaning
Form: Flake (also available as powder); Color: White to off white
Properties: Density: 2.9 lb/gal; pH: 7.0 (10%)
Composition: 90% conc.

Lakeway 302. Bofors Lakeway
Chemical Description: Formulated concentrate
Category: Detergent, anionic
Applications: Liquid hand dishwash and car wash base
Form: Liquid; Color: Straw
Properties: Density: 9.3 lb/gal; Visc.: 710 cps (Brookfield); Cloud Pt.: ≈ < −6 C; pH: 6.8 (10%)
Composition: 60% conc.

Lakeway 303. Bofors Lakeway
Chemical Description: Formulated concentrate
Category: Detergent, anionic
Applications: Liquid hand dish-

wash concentrate, liquid industrial hand soap
Form: Lotionized liquid; Color: Pink
Properties: Density: 9.3 lb/gal; Visc.: 640 cps (Brookfield); Cloud Pt.: ≈ < -6 C; pH: 6.8 (10%)
Composition: 60% conc.

Lakeway 401-M. Bofors Lakeway
Chemical Description: Formulated concentrate
Category: Detergent, anionic
Applications: Base concentrate for hair shampoo, bubble bath; voluminous sudsing, premium grade base; superior viscosity building
Form: Liquid; Color: Straw
Properties: Density: 8.6 lb/gal; Visc.: 5000 cps (Brookfield); Cloud Pt.: ≈ 6 C; pH: 9.4 (10%)
Composition: 39% conc.

Lakeway 402. Bofors Lakeway
Chemical Description: Formulated anionic blend
Applications: Personal care products, hair shampoo, bubble bath, rug shampoo
Form: Liquid; Color: White to yellow
Properties: Density: 8.7 lb/gal; Visc.: 100 cps (Brookfield); Cloud Pt.: ≈ 10 C; pH: 8.0 (10%)
Composition: 40% conc.

Lakeway 500 Series. Bofors Lakeway
Chemical Description: Formulated concentrate
Category: Foaming agent, anionic
Applications: Standard, high actives, noncorrosives types available for gypsum board production
Form: Liquids
Properties: Density: 8.5 lb/gal; pH: 7.0 (10%)
Composition: 28–60% conc.

Lakeway I-100. Bofors Lakeway
Chemical Description: Formulated concentrate
Category: Rug shampoo concentrate, anionic
Applications: High actives regular grade rug shampoo, fine fabrics shampoo
Form: Liquid; Color: White to light yellow
Properties: Density: 8.6 lb/gal; Visc.: 240 cps (Brookfield); Cloud Pt.: ≈ 9 C; pH: 7.2 (10%)
Composition: 31% conc.

Lakeway I-101. Bofors Lakeway
Chemical Description: Formulated concentrate
Category: Rug shampoo concentrate, anionic
Applications: High actives rug shampoo for aerosol packaging
Form: Liquid; Color: White to light yellow
Properties: Density: 8.6 lb/gal; Visc.: 65 cps (Brookfield); Cloud Pt.: ≈ 2 C; pH: 7.2 (10%)
Composition: 31% conc.

Lakeway I-108. Bofors Lakeway
Chemical Description: Formulated concentrate
Category: Rug shampoo concentrate, anionic

Applications: Dry foamer rug and upholstery shampoo with dry residue
Form: Liquid; Color: White to light yellow
Properties: Density: 8.5 lb/gal; Visc.: 150 cps (Brookfield); Cloud Pt.: ≈ 18 C; pH: 7.0 (10%)
Composition: 24% conc.

Lakeway I-200. Bofors Lakeway
Chemical Description: Formulated concentrate
Category: Upholstery shampoo concentrate, anionic
Applications: Mild, dry foamer upholstery shampoo
Form: Liquid; Color: White to light yellow
Properties: Density: 8.5 lb/gal; Visc.: 100 cps (Brookfield); Cloud Pt.: ≈ 12 C; pH: 7.8 (10%)
Composition: 21% conc.

Lakeway SA. Bofors Lakeway
Chemical Description: Linear dodecylbenzene sulfonic acid
Category: Detergent, intermediate, anionic
Applications: Low cost high sudsing detergent base for dishwash, carwash, industrial cleaners
Form: Viscous fluid; Color: Dark amber
Properties: Density: 8.7 lb/gal; Visc.: 1440 cps (Brookfield); Acid No.: 182
Composition: 98% conc.

Lakeway SXS. Bofors Lakeway
Chemical Description: Sodium xylene sulfonate
Category: Hydrotrope, solubilizer, anionic
Applications: Hydrotrope for solubilizing, used in low cloud dishwash, shampoos, metal cleaning, etc.
Form: Low viscosity liquid; Color: Light yellow
Properties: Density: 10.0 lb/gal; Visc.: 10 cps (Brookfield); Cloud Pt.: < –5 C; pH: 8.5 (as is)
Composition: 41% conc.

Lan-Aqua-Sol. Fanning Chem. Co.
Chemical Description: Ethoxylated lanolins
Category: Emulsifier, anionic
Applications: Cosmetics/pharmaceutical emulsions; as a primary emulsifier and to add emollience to skin and hair care formulations; household detergent applications
Form: Gel, solid; Color: Yellow; Odor: Trace
Properties: Solubility: Sol. in water, alcohol; Sp. gr.: 1.00–1.03; M.P.: 45–51 C
Composition: 50 and 100% active
Containers: Steel drums

Lanbritol Wax N21. Ronsheim & Moore Ltd.
Chemical Description: Blend of cetyl and stearyl alcohols and a nonionic emulsifier which is a cetyl/oleyl alcohol ether of polyethylene glycol
Category: Emulsifier, nonionic
Applications: Self-emulsifying wax produces excellent thin stable O/W emulsions for pharmaceuticals, cosmetics and hair preparations

Form: Waxy solid; Color: White
Properties: M.P.: 45–53 C; Acid Value: 0.5; Biodegradable
Composition: 100% conc.

Lancare. Diamond Shamrock Europe/Process Chem. Div./Lancare Hygiene Prod.
Category: Anionic/nonionic surfactant
Applications: Range of formulated shampoos, bubble baths, mild abrasive cleaners
Form: Liquid

Lanethyl. Croda Chem. Ltd.
Chemical Description: Alcohol extract of lanolin alcohols
Category: Emulsifier, plasticizer, conditioning agent, nonionic
Applications: Plasticizer or film modifier for hair setting resins; W/O emulsifier; used in alcohol-based systems, hair sprays and lacquers
Form: Brittle wax; Color: Amber
Properties: Solubility: Sol. in ethyl alcohol; Lipophilic
Composition: 100% conc.

Laneto-50. R.I.T.A. Chem. Corp.
Chemical Description: Ethoxylated USP lanolin
Category: Emollient, lubricant, emulsifier, plasticizer, nonionic
Applications: Water emulsifier for cosmetics and pharmaceuticals; shampoos; waterless hand cleaners
Form: Liquid; Color: Yellow
Properties: Solubility: Sol. in water; Sp. gr.: 1.00–1.10
Composition: 50% conc.

Laneto-100. R.I.T.A. Chem. Corp.
Chemical Description: Ethoxylated USP lanolin
Category and Applications: See Laneto-50; also as solubilizer
Form: Solid; Color: Yellow
Properties: Solubility: Water sol.; M.P.: 48–52 C
Composition: 100% conc.

Lanette Wax CAT. Ronsheim & Moore Ltd.
Chemical Description: Blend of fatty alcohols with cationic and nonionic emulsifiers
Category: Emulsifier, cationic
Applications: Self-emulsifying wax for use in toiletry and pharmaceutical preparations, esp. in manufacture of "Hairaid" type of hair conditioners; also in germicidal hand creams and lotions
Form: Waxy solid
Composition: 100% conc.

Lanette Wax SX. Ronsheim & Moore Ltd.
Chemical Description: Blend of sodium alkyl sulfate and cetyl/stearyl alcohol
Category: Emulsifier, anionic
Applications: Self-emulsifying wax producing stable O/W emulsions for pharmaceutical and cosmetic creams, lotions, ointments
Form: Waxy flakes; Color: Cream to pale yellow; Odor: Faint characteristic
Properties: Solubility: Partially sol. in alcohol; almost insol. in water forming an emulsion; M.P.: ≈

50 C; Acid Value: 0.5
Composition: 100% conc.
Reaction: Neutral

Lanette Wax S.X.B.P. Ronsheim & Moore Ltd.
Chemical Description: Same as Lanette Wax SX but complying to B.P. specifications
Category and Applications: See Lanette Wax SX

Lanex. Croda Chem. Ltd.
Chemical Description: Alcohol extract of pure lanolin
Category: Super-fatting agent, emollient, conditioning agent, film modifier, plasticizer, nonionic
Applications: Alcohol sol. emollient, plasticizer, and conditioner in alcohol-based cosmetics, hair sprays and lacquers, preshave lotions
Form: Viscous liquid; Color: Amber
Properties: Solubility: Sol. in ethyl alcohol; Lipophilic
Composition: 100% conc.

Lanexol AWS. Croda Chem. Ltd.
Chemical Description: Alkoxylated liquid lanolin
Category: Emollient, lubricant, plasticizer, super-fatting agent, emulsifier, nonionic
Applications: O/W emulsifier; emollient and lubricant for alcoholic and aqueous compositions; plasticizer for hair sprays, lacquers, and wave set lotions; super-fatting agent in solvent systems based on alcohol or acetone; in shampoos, after shaves, skin creams and lotions, hair rinses, dyes, and conditioners
Form: Viscous liquid; Color: Amber
Properties: Solubility: Sol. in water, ethanol, and mixtures thereof; Cloud Pt.: 65–80 C (1% aq.); Pour Pt.: 13 C max.; Sapon. Value: 10–20; Acid No.: 2.0 max.; Hydroxyl No.: 30–45; Iodine Value: 9–17
Composition: 97% conc.

Lanfrax. Emery Industries/Malmstrom Div.
Chemical Description: Solid lanolin fraction
Category: Stabilizer, emollient, emulsifier, nonionic
Applications: Cosmetic emulsion stabilizer; W/O emulsifier and waxing agent; O/W auxiliary emulsifier; slip reducing agent in floor finishing compounds; floor polishes; furniture polishes
Form: Wax
Composition: 100% conc.

Lanfrax WS 55. Emery Ind./Malmstrom Div.
Chemical Description: Wool wax polyglycol ester
Category: Emulsifier, nonionic
Applications: Cosmetic O/W emulsifier in hair tonics, hair sprays
Form: Wax
Composition: 100% conc.

Lankro Mud-Aids. Diamond Shamrock Europe/Process Chem. Div.
Applications: A series of general workability aids for drilling muds incl. mud surfactants and defoamers

Lankro Mud Detergents. Diamond Shamrock Europe/Process Chem. Div.
Category: Anionic or nonionic detergent
Applications: A series of detergents for oil well drilling mud applications

Lankro Mud-Emuls. Diamond Shamrock Europe/Process Chem. Div.
Category: Emulsifier
Applications: A series of W/O and O/W emulsifiers for water-based and invert muds

Lankrolan SHR-3. Diamond Shamrock Europe/Process Chem. Div.
Chemical Description: Organic thiosulfate
Category: Shrink resist agent, anionic
Applications: Shrink resist agent for wool
Form: Liquid
Composition: 42% conc.

Lankromul OSD. Lankro Chem. Ltd.
Chemical Description: Emulsifier in aliphatic hydrocarbon solvent
Category: Emulsifier, dispersant
Applications: Emulsifier and dispersant for spillages of crude oil, fuel oils, or refined mineral oils with low toxicity on marine life; beach cleaning; cleaning of harbor walls, and oil storage tanks
Form: Clear liquid; Color: Straw; Odor: Mild, fatty
Properties: Solubility: Disp. in water; Sp. gr.: 0.808 @ 10 C; Visc.: 3.5 cs @ 20 C; Flash Pt.: 88 C (COC); Pour Pt.: < -20 C
Reaction: Neutral
Toxicity: Skin irritant on prolonged contact with concentrated form
Containers: Mild steel drums
Warnings and Caution: Inflammable solvent—keep away from heat, flames, and sparks; protective gloves, goggles, and clothing should be worn

Lankropearl T. Diamond Shamrock Europe/Process Chem. Div.
Chemical Description: Blend of pearlizing agents in an anionic base
Category: Surfactant/pearlizing agent, anionic/nonionic
Applications: Surfactant/pearlizing agent for toiletries
Form: Liquid
Composition: 40% conc.

Lankropol ADF. Lankro Chem. Ltd.
Chemical Description: Modified sulfosuccinamate
Category: Foaming agent, anionic
Applications: Used in foamed aq. polymer systems, e.g., in mechanical foaming of polymer coatings in the textile, leather, and carpet industries; upholstery, draperies, and household fabric coatings
Form: Clear liquid; Color: Light brown; Odor: Characteristic
Properties: Solubility: Sol. in water @ 20 C; Sp. gr.: 1.078 @ 20 C; Visc.: 11 cs @ 20 C; Flash Pt.: $>$

200 F (COC); Pour Pt.: −5 C; pH: 6.0–8.0 (1% aq.)
Composition: 35% active ingred.
Toxicity: Skin irritant on prolonged contact with concentrated form
Warnings and Caution: Protective gloves and goggles should be worn; spillages may be slippery; in a fire may produce toxic fumes of sulfur dioxide

Lankropol ATE. Lankro Chem. Ltd.
Chemical Description: Tetrasodium N-(1,2-dicarboxyethyl)-N-octadecyl sulfosuccinamate
Category: Emulsifier, stabilizer, foaming agent, solubilizer
Applications: Primary emulsifier and mechanical stabilizer for emulsion polymers with low critical micelle concentration; auxiliary foaming agent for latex foam carpet backings; solubilizing agent with high salt tolerance for industrial and household cleaning applications; cosmetics and agricultural formulations
Form: Slightly hazy liquid; Color: Amber; Odor: Ethanolic
Properties: Sp. gr.: 1.119 @ 20 C; Visc.: 36 cs @ 20 C; Flash Pt.: 73 F (Abel CC); Pour Pt.: 3 C; pH: 7.0–8.5 (1% aq.)
Composition: 35% active ingred.
Toxicity: Skin irritant on prolonged contact with concentrated form
Warnings and Caution: Protective gloves and goggles should be worn; spillages may be slippery; contains ethanol–keep away from heat, flame, and sparks

Lankropol KMA. Lankro Chem. Ltd.
Chemical Description: Sodium dihexyl sulfosuccinate (contains ethanol)
Category: Wetting agent, solubilizer, anionic
Applications: Wetting agent esp. in sol'ns. of electrolytes; solubilizer for soaps; emulsion polymerization aid in preparation of high solids fluid latices, esp. with styrene and its copolymers
Form: Hazy liquid; Color: Pale straw; Odor: Ethanolic
Properties: Solubility: Sol. in water @ 20 C; Sp. gr.: 1.082 @ 20 C; Visc.: 31 cs @ 20 C; Flash Pt.: 91 F (Abel CC); Pour Pt.: $<$ 0 C; pH: 6.0–7.5 (1% aq.)
Composition: 60% active ingred.
Reaction: Neutral
Toxicity: Skin irritant on prolonged contact with concentrated form
Containers: Stainless steel, glass, resin, or polythene-lined equipment
Warnings and Caution: Protective gloves and goggles should be worn; spillages may be slippery; in a fire may produce toxic fumes of sulfur dioxide; contains ethanol–keep away from heat, flames, and sparks
References: Bulletin SI:18 FM678

Lankropol KO Special. Lankro Chem. Ltd.
Chemical Description: Sodium di-

octyl sulfosuccinate (contains mineral oil)
Category: Wetting agent, emulsifier, water carrier, anionic
Applications: Solvent emulsifier; water carrier in drycleaning formulations; dewatering aid; emulsifier for mineral oil with nonionics where improved high temperature emulsion stability is achieved
Form: Clear liquid; Color: Pale straw; Odor: Mild
Properties: Solubility: Disp. in water @ 20 C; Sp. gr.: 1.050 @ 20 C; Visc.: 1500 cs @ 20 C; Flash Pt.: > 200 F (COC); Pour Pt.: < 0 C; pH: 6–7 (1% aq.)
Composition: 60% active ingred.
Reaction: Neutral
Toxicity: Skin irritant on prolonged contact with concentrated form
Containers: Stainless steel, glass, resin, or polythene-lined equipment
Warnings and Caution: Protective gloves and goggles should be worn; spillages may be slippery; in a fire may produce toxic fumes of sulfur dioxide
References: Bulletin SI:18 FM678

Lankropol KO2. Lankro Chem. Ltd.
Chemical Description: Sodium dioctyl sulfosuccinate (contains ethanol)
Category: Wetting agent, anionic
Applications: Wetting agent; emulsion polymerization aid
Form: Clear liquid; Color: Pale straw; Odor: Ethanolic
Properties: Solubility: Sol. up to 0.5% in water @ 20 C; Sp. gr.: 1.050 @ 20 C; Visc.: 35 cs @ 20 C; Flash Pt.: 81 F (Abel CC); Pour Pt.: < 0 C; pH: 6.0–7.0 (1% aq.)
Composition: 60% active ingred.
Reaction: Neutral
Toxicity: Skin irritant on prolonged contact with concentrated form
Containers: Stainless steel, glass, resin, or polythene-lined equipment
Warnings and Caution: Protective gloves and goggles should be worn; spillages may be slippery; in a fire may produce toxic fumes of sulfur dioxide; contains ethanol–keep away from heat, flames, and sparks
References: Bulletin SI:18 FM678

Lankropol ODS/LS. Lankro Chem. Ltd.
Chemical Description: Disodium N-octadecyl sulfosuccinamate
Category: Foaming agent, anionic
Applications: Foaming agent for aq. polymer dispersions, esp. for high solids styrene-butadiene latex secondary foam carpet backing
Form: Hazy liquid; Color: Light amber; Odor: Mild
Properties: Solubility: Sol. in water @ 20 C; Sp. gr.: 1.082 @ 20 C; Visc.: 12 cs @ 20 C; Flash Pt.: > 200 F (COC); Pour Pt.: –5 C; pH: 7.0–9.0 (1% aq.)
Composition: 35% active ingred.
Reaction: Neutral
Toxicity: Skin irritant on prolonged contact with concentrated form

Warnings and Caution: Protective gloves and goggles should be worn; spillages may be slippery; in a fire may produce toxic fumes of sulfur dioxide

Lankropol ODS/PT. Lankro Chem. Ltd.
Chemical Description: Disodium N-octadecyl sulfosuccinamate
Category and Applications: See Lankropol ODS/LS
Form: Paste/liquid; Odor: Mild
Properties: Sp. gr.: 1.066 @ 20 C; Visc.: 5400 cs @ 20 C; Flash Pt.: > 200 F (COC); Pour Pt.: 4 C; pH: 7.0–9.0 (1% aq.)
Composition: 35% active ingred.
Toxicity: Skin irritant on prolonged contact with concentrated form
Warnings and Caution: Protective gloves and goggles should be worn; spillages may be slippery; in a fire may produce toxic fumes of sulfur dioxide

Lankropol OPA. Lankro Chem. Ltd.
Chemical Description: Potassium salt of a fatty acid sulfonate
Category: Wetting agent, detergent, dispersant, anionic
Applications: Surfactant with good ionic stability, high surface activity, and low foaming properties; in metal industry as alkaline cleaners, steam cleaners, acid pickling; household specialties (toilet bowl cleaners, oven cleaners, wax strippers); textiles for Kier boiling, mercerizing, dye leveling; in emulsion polymerization of styrene and vinyl acetate monomers
Form: Clear liquid; Color: Dark amber; Odor: Mild fatty
Properties: Solubility: Sol. in water @ 20 C; Sp. gr.: 1.125 @ 20 C; Visc.: 270 cs @ 20 C; Flash Pt.: > 200 F (COC); Pour Pt.: < 0 C; pH: 5.5–6.0 (1% aq.); Biodegradable
Composition: 50% active ingred.
Toxicity: Skin irritant on prolonged contact with concentrated form
Containers: Stainless steel, glass, resin, or polythene-lined equipment
Warnings and Caution: Protective gloves and goggles should be worn; spillages may be slippery; in a fire may produce toxic fumes of sulfur dioxide
References: Bulletin SI:17 FM678

Lankropol WA. Lankro Chem. Ltd.
Chemical Description: Ammonium salt of a sulfated monoester of fatty acid
Category: Wetting agent, detergent, dispersant, anionic
Applications: Low foaming wetting agent in textile industry; dispersion aid in paint industry; solvent emulsifier in degreasing formulations; emulsifier in herbicides
Form: Clear liquid; Color: Dark amber; Odor: Characteristic
Properties: Solubility: Sol. in water @ 20 C; Sp. gr.: 1.030 @ 20 C; Visc.: 120 cs @ 20 C; Flash Pt.: > 200 F (COC); Pour Pt.: < 0 C; pH: 6.0–7.0 (1% aq.); Biodegradable

Composition: 50% active ingred.
Toxicity: Skin irritant on prolonged contact with concentrated form
Containers: Stainless steel, glass, resin, or polythene-lined equipment
Warnings and Caution: Protective gloves and goggles should be worn; spillages may be slippery; in a fire may produce toxic fumes of sulfur dioxide
References: Bulletin SI:17 FM678

Lankropol WN. Lanrko Chem. Ltd.
Chemical Description: Sodium salt of a sulfated monoester of fatty acid
Category and Applications: See Lankropol WA; also esp. recommended in textile mercerizing operations
Form: Clear liquid; Color: Dark amber; Odor: Characteristic
Properties: Solubility: Sol. in water @ 20 C; Sp. gr.: 1.050 @ 20 C; Visc.: 75 cs @ 20 C; Flash Pt.: > 200 F (COC); Pour Pt.: < 0 C; pH: 6.0–7.0 (1% aq.); Biodegradable
Composition: 50% active ingred.
Toxicity: Skin irritant on prolonged contact with concentrated form
Containers: Stainless steel, glass, resin, or polythene-lined equipment
Warnings and Caution: Protective gloves and goggles should be worn; spillages may be slippery; in a fire may produce toxic fumes of sulfur dioxide
References: Bulletin SI:17 FM678

Lankrosol SXS-30. Lankro Chem. Ltd.
Chemical Description: Sodium xylene sulfonate
Category: Solubilizer, viscosity modifier, hydrotrope, anionic
Applications: Hydrotrope and viscosity modifier for high active liquid detergents; solubilizer for anionic surfactants in high electrolyte sol'ns., e.g., hard surface cleaners; dye leveling agent in the dyeing of nylon
Form: Clear liquid; Color: Pale straw; Odor: Negligible
Properties: Solubility: Sol. in water @ 20 C; Sp. gr.: 1.128 @ 20 C; Visc.: 2.7 cs @ 20 C; Flash Pt.: > 200 F (COC); Pour Pt.: < 0 C; pH: 7.0 (1% aq.)
Composition: 30% active ingred.
Toxicity: Skin irritant on prolonged contact with concentrated form
Containers: Stainless steel, glass, resin, or polyethylene lined equipment
Warnings and Caution: Protective gloves and goggles should be worn;
References: Bulletin SI:20 FM678

Lanogel 21. Robinson-Wagner Co.
Chemical Description: Polyoxyethylene derivative of lanolin
Category: Emollient, emulsifier, dispersant, wetting agent, solubilizer, foam stabilizer, nonionic
Applications: Cosmetics and pharmaceuticals incl. shampoos, hair conditioners and rinses, shaving creams, soaps, liquid detergents,

antiperspirants, facial tissues
Form: Gel; Color: 3 max. (ASTM)
Properties: Solubility: Disp. in water; Sapon. Value: 12–24; Iodine Value: 12 max. (Hanus); Stability: Stable in hard water; resistant to wide temperature and pH ranges, and to oxidation
Composition: 50% active ingred.

Lanogel 31. Robinson-Wagner Co.
Chemical Description: Polyoxyethylene derivative of lanolin
Category and Applications: See Lanogel 21
Form: Gel; Color: 3 max. (ASTM)
Properties: Solubility: Partly sol. in water; Sapon. Value: 9–18; Iodine Value: 10 max.; Stability: Stable in hard water; resistant to wide temperature and pH ranges, and to oxidation
Composition: 50% active ingred.

Lanogel 41. Robinson-Wagner Co.
Chemical Description: Polyoxyethylene derivative of lanolin
Category and Applications: See Lanogel 21
Form: Gel; Color: 3 max. (ASTM)
Properties: Solubility: Sol. in water; Sapon. Value: 6–15; Iodine Value: 8 max. (Hanus); Stability: Stable in hard water; resistant to wide temperature and pH ranges, and to oxidation
Composition: 50% active ingred.

Lanogel 61. Robinson-Wagner Co.
Chemical Description: Polyoxyethylene derivative of lanolin
Category and Applications: See Lanogel 21
Form: Gel; Color: 3 max. (ASTM)
Properties: Solubility: Sol. in water and ethyl alcohol/water sol'ns.; Sapon. Value: 4–12; Iodine Value: 6 max. (Hanus); Stability: Stable in hard water; resistant to wide temperature and pH ranges, and to oxidation
Composition: 50% active ingred.

Lanolic Acid. Croda Chem. Ltd.
Chemical Description: Distilled lanolin acids
Category: Soaps ingredient, nonionic
Applications: Ingredient in soaps; Lanolic acid and its soaps possess good dry emollience; in skin, hair, make-up cosmetics; additives for shaving foams/creams; soaps used in aerosols
Form: Soft solid; Color: Pale yellow; Odor: Faint characteristic
Properties: Solubility: Highly sol. in most cosmetics raw materials; excellent cosolvent properties; Lipophilic
Composition: 100% conc.

Lanolin Alcohols LG. Van Schuppen Chemie
Chemical Description: Complex mixture of high molecular normal and branched chain monohydric alcohols and sterols produced by saponification of lanolin
Category: Emulsifier, stabilizer, emollient, nonionic
Applications: Neutral wool grease substitute in technical applications; auxiliary agent in textile lubricants (batching oils), leather preservatives (fatliquoring), anti-

corrosion compounds, polishing materials, emulsion paints, wood preservatives (water-repellent and antiblooming agent); in cosmetics; in waterproofing agents for textiles and paper
Form: Soft, bleached wax; Color: Tan
Properties: Solubility: Sol. in hydrocarbons, chlorinated hydrocarbons, vegetable and mineral oils; insol. in water; Sapon. Value: 20–40; Acid No.: 6.0 max.; Hydroxyl No.: 85–105
Composition: 100% conc.
References: Product information No. 36 B

Lanolin Alcohols LO. Van Schuppen Chemie
Chemical Description: Complex mixture of high molecular normal and branched chain monohydric alcohols and sterols produced by saponification of lanolin
Category and Applications: See Lanolin Alcohols LG
Form: Soft solid, unbleached, technical quality; Color: Brown
Properties: Solubility: Sol. in hydrocarbons, chlorinated hydrocarbons, vegetable and mineral oils; insol. in water; Sapon. Value: 20–40; Acid No.: 6.0 max.; Hydroxyl No.: 90–110
Composition: 100% conc.
References: Product information No. 36 A

Lanolin Alcohols THG. Van Schuppen Chemie
Chemical Description: Complex mixture of high molecular normal and branched chain monohydric alcohols and sterols produced by saponification of lanolin
Category: Emulsifier, stabilizer, emollient, nonionic
Applications: In cosmetic preparations, esp. hair dyes; a substitute for "wool alcohols B.P." in technical applications
Form: Solid, bleached; Color: Tan
Properties: Solubility: Sol. in hydrocarbons, chlorinated hydrocarbons, vegetable and mineral oils; insol. in water; Sapon. Value: 20.0 max.; Acid No.: 6.0 max.; Hydroxyl No.: 105–125
Composition: 100% conc.
References: Product information No. 35 D

Lanolin Alcohols THO. Van Schuppen Chemie
Chemical Description: Complex mixture of high molecular normal and branched chain monohydric alcohols and sterols produced by saponification of lanolin
Category and Applications: See Lanolin Alcohols THG
Form: Solid, unbleached; Color: Brown
Properties: Solubility: Sol. in hydrocarbons, chlorinated hydrocarbons, vegetable and mineral oils; insol. in water; Sapon. Value: 20.0 max.; Acid No.: 6.0 max.; Hydroxyl No.: 105–125
Composition: 100% conc.
References: Product information No. 35 C

Lanolin Fatty Acids G. Van Schuppen Chemie
Chemical Description: Lanolin fatty acid
Category: Emusifier, anionic
Applications: Due to excellent emulsifying and oil binding properties of their metal soaps, used in lubricating greases (esp. in calcium, sodium, and lithium base greases); in solid and liquid polishing compositions, emulsion polishes, anticorrosion compounds, etc.
Form: Wax, bleached quality
Properties: Solubility: Sol. in mineral oils and many organic solvents, e.g., benzene, toluene, trichloroethylene, white spirit; Sapon. Value: 145–165; Acid No.: 120–135; Hydroxyl No.: 60–80; Iodine Value: 10.0 max.
Composition: 100% conc.
References: Product information No. 42 A

Lanolin Fatty Acids O. Van Schuppen Chemie
Chemical Description: Lanolin fatty acid
Category and Applications: See Lanolin Fatty Acids G
Form: Wax, unbleached, technical quality; Color: Brown; Odor: Characteristic
Properties: Solubility: Sol. in mineral oils and many organic solvents, e.g., benzene, toluene, trichloroethylene, white spirit; Sapon. Value: 140–160; Acid No.: 120–135; Hydroxyl No.: 65–90; Iodine Value: 11.0 max.
Composition: 100% conc.
References: Product information No. 41 A

Lanolin (Hydrous Wool Fat). Westbrook Lanolin Co.
Chemical Description: W/O emulsion containing 30% water
Category and Applications: Emollient, emulsifier, ointment base, hair conditioner, wax crystal inhibitor, lipstick binder

Lanosterol. Van Schuppen Chemie
Chemical Description: Lanolin derivative; mixture of wool wax triterpenoids
Category: Emulsifier, raw material, nonionic
Applications: Auxiliary emulsifier in cosmetics, esp. in lipsticks and cosmetic creams; component of impregnated paper electric condensors and other uses in electrical engineering; starting material for the synthesis of tetracyclic triterpenoid derivatives
Form: Crystalline powder; Color: White to faintly yellow; Odor: Almost odorless
Properties: Solubility: Easily sol. in dioxane, ether, ethyl acetate; sol. in acetone, chlorinated hydrocarbons, open-chain and cyclic hydrocarbons; slightly sol. in ethanol and methanol; insol. in water; M.P.: 138–145 C; Acid No.: 0.5 max.
Composition: 100% conc.
References: Product information No. 4

Lanpol 5. Croda Chem. Ltd.
Chemical Description: Polyethoxylated lanolin acid
Category: Emulsifier, solubilizer, wetting agent, solvent, nonionic
Applications: O/W emulsifier; nonsticky, useful in emulsions of low solids contents; solvents for bromo acid dyes in lipsticks; skin care and make-up products
Form: Pasty solid; Color: Pale yellow
Properties: HLB: 7.5; Sapon. Value: 105–120; Acid No.: 10 max.; Hydroxyl No.: 110–125; Iodine Value: 11 max.; pH: 4–7 (3%)
Composition: 97% conc.

Lanpol 10. Croda Chem. Ltd.
Chemical Description: Polyethoxylated lanolin acid
Category and Applications: See Lanpol 5
Form: Pasty solid; Color: Pale yellow
Properties: HLB: 10.9; Sapon. Value: 60–80; Acid No.: 10 max.; Hydroxyl No.: 60–80; Iodine Value: 7 max.; pH: 4–7 (3%)
Composition: 97% conc.

Lanpol 20. Croda Chem. Ltd.
Chemical Description: Polyethoxylated lanolin acid
Category and Applications: See Lanpol 5
Form: Pasty solid; Color: Pale yellow
Properties: HLB: 14.1; Sapon. Value: 45–60; Acid No.: 10 max.; Hydroxyl No.: 50–75; Iodine Value: 5 max.; pH: 4–7 (3%)
Composition: 97% conc.

Lanquell 206. Lankro Chem. Ltd.
Chemical Description: Polyglycol based
Category: Antifoam
Applications: Low toxicity antifoam for beet sugar processing, yeast production, fermentation processes, e.g., antibiotic and enzyme production, and in the manufacture of citric acid; brewing industry; manufacture of paper for use in food packaging
Form: Liquid; Color: Pale yellow; Odor: Faint
Properties: Solubility: Disp. in water @ 20 C; Sp. gr.: 0.984 @ 20 C; Visc.: 180 cs @ 20 C; Flash Pt.: > 150 F (Abel CC); Pour Pt.: –28 C; pH: 7.0 (1% aq.)
Composition: 92% active ingred.
Containers: Mild steel containers; take care to exclude further moisture
Warnings and Caution: Contains alcohol–keep away from flame, spark, or heat; spillages may be slippery; do not mix with strong oxidizing or reducing agents (potentially explosive)
References: Bulletin SI:21 FM678

Lanquell 217. Lankro Chem. Ltd.
Chemical Description: Polyglycol based
Category and Applications: See Lanquell 206
Form: Slightly hazy liquid; Color: Pale yellow; Odor: Faint
Properties: Solubility: Disp. in water @ 20 C; Sp. gr.: 1.005 @ 20 C; Visc.: 355 cs @ 20 C; Flash Pt.: > 200 F (Abel CC); Pour Pt.: –13 C; pH: 7.0 (1% aq.)

Composition: 98% active ingred.
Containers: Mild steel containers; take care to exclude further moisture
Warnings and Caution: Spillages may be slippery; do not mix with strong oxidizing or reducing agents (potentially explosive)
References: Bulletin SI:21 FM678

Lantrol. Emery Ind./Malmstrom Div.
Chemical Description: Liquid lanolin fraction
Category: Emollient, nonionic
Applications: Skin and hair emollient which emulsifies with water; in cosmetic creams, lotions, shampoos, lipsticks, moisturizing formulations
Form: Viscous liquid
Properties: HLB: 9.0
Composition: 100% conc.

Lantrol AWS. Emery Ind./Malmstrom Div.
Chemical Description: Ethoxylated lanolin
Category: Emulsifier, emollient, nonionic
Applications: Water and alcohol sol. lanolin oil; auxiliary O/W emulsifier with a soft, silky feel; hair sprays, alcoholic lotions, shampoos, foam baths, shower foams
Form: Viscous liquid
Properties: Solubility: Sol. in water and alcohol; HLB: 18.0
Composition: 100% conc.

Lathanol LAL Flake. Stepan Chem. Co.
Chemical Description: Sodium lauryl sulfoacetate
Category: Emulsifier, wetting agent, detergent, foaming agent, anionic
Applications: Cosmetics, dentifrices, bubble baths, synthetic soap products, light duty detergents, shampoos
Form: Flake; Color: Whitish; Odor: Faint lauryl alcohol; Taste: Very slightly acrid
Properties: Solubility: Water @ 25 C: 3.5 g/100 ml; Density: 0.38 g/ml (untamped); Stability: Stable to water hardness and at pH 5.0–7.3; Surface Tension: 30.1 dynes/cm (0.1% conc.); pH: 5.0–7.5 (5% conc.)
Composition: 70% active ingred.
Reaction: Neutral
Warnings and Caution: Avoid inhalation of dust

Lathanol LAL Powder . Stepan Chem. Co.
Chemical Description: Sodium lauryl sulfoacetate
Category and Applications: See Lathanol LAL Flake
Form: Powder; Color: Whitish; Odor: Faint lauryl alcohol; Taste: Very slightly acrid
Properties: Solubility: Water sol. @ 25 C: 3.5 g/100 ml; Density: 0.85 g/ml (tamped); Stability: Stable to water hardness and at pH 5.0–7.3; Surface Tension: 30.1 dynes/cm (0.1% conc.); pH: 5.0–7.5 (5% conc.)

Composition: 70% active ingred.
Reaction: Neutral
Warnings and Caution: Avoid inhalation of dust

Laurex 16/18. Albright & Wilson
Chemical Description: Primary fatty alcohol derived from naturally occurring oils and fats
Category: Raw material
Applications: Raw material for manufacture of sulfated derivatives for use as detergent active matter (e.g., cetyl stearyl alcohol sulfate, cetyl stearyl alcohol ether sulfate); raw material in the manufacture of additives (alkyl methacrylates) for lubricating oils
Form: Solid; Color: White
Properties: M.P.: 50–54 C; Sapon. Value: 1.5; Acid No.: 0.3; Hydroxyl No.: 204–216; Iodine Value: 1.5

Laurex 16/18D. Albright & Wilson
Chemical Description: Primary fatty alcohols derived from naturally occurring oils and fats; double distilled grade
Category: Raw material
Applications: Raw material for manufacture of sulfated derivatives for use as detergent active matter (e.g., cetyl stearyl alcohol sulfate, cetyl stearyl alcohol ether sulfate)
Form: Solid; Color: White
Properties: M.P.: 50–54 C; Sapon. Value: 0.8; Acid No.: 0.1; Hydroxyl No.: 208–227; Iodine Value: 1.0

Laurex CH. Albright & Wilson
Chemical Description: Primary fatty alcohol
Category: Superfatting agent, raw material
Applications: Superfatting agent in shampoos; raw material for sulfation, ethoxylation, etc.
Form: Soft solid; Color: White
Properties: M.P.: 18–23 C; Sapon. Value: 1.5; Acid No.: 0.3; Hydroxyl No.: 284–303; Iodine Value: 0.6

Laurex CS. Albright & Wilson
Chemical Description: Primary fatty alcohol
Category and Applications: See Laurex CH
Form: Flake or solid; Color: White
Properties: M.P.: 50–54 C; Sapon. Value: 1.5; Acid No.: 0.3; Hydroxyl No.: 204–216; Iodine Value: 1.5

Laurex L1. Albright & Wilson
Chemical Description: Primary fatty alcohol
Category and Applications: See Laurex CH
Form: Soft solid; Color: White
Properties: M.P.: 20–23 C; Sapon. Value: 0.5; Acid No.: 0.1; Hydroxyl No.: 265–280; Iodine Value: 0.5

Laurex NC. Albright & Wilson
Chemical Description: Primary fatty alcohol
Category and Applications: See Laurex CH
Form: Soft solid; Color: White

Properties: M.P.: 20–23 C; Sapon. Value: 0.5; Acid No.: 0.1; Hydroxyl No.: 285–297; Iodine Value: 0.5

Laurex PKH. Albright & Wilson
Chemical Description: Primary fatty alcohol
Category and Applications: See Laurex CH
Form: Soft solid; Color: White
Properties: M.P.: 18–23 C; Sapon. Value: 1.5; Acid No.: 0.3; Hydroxyl No.: 270–282; Iodine Value: 0.6

Lauridit KD. Akzo Chemie GmbH
Chemical Description: Coco fatty acid diethanolamide
Category: Detergent, emulsifier, nonionic
Applications: Cosmetic and household applications; viscosity increasing additive for shampoos and bubble baths; liquid cleaning and dishwashing agents; emulsifier for mineral and fatty oils
Form: Liquid; Color: Yellowish
Properties: Solubility: 40 g/l water @ 20 C; Sp. gr.: 1.02; M.P.: 14–15 C; Flash Pt.: 104 C (PMCC); Acid No.: 4 max.; Stability: Below room temperature, product is a paste; pH: 9
Composition: 90% active ingred.
Reaction: Alkaline
Containers: 125 kg drums

Lauridit KDG. Akzo Chemie GmbH
Chemical Description: Coco fatty acid diethanolamide
Category and Applications: See Lauridit KD
Form: Liquid; Color: Light yellow
Properties: Ref. Index: 1.472; Solubility: 800 g/l water @ 20 C; Sp. gr.: 1.02; Flash Pt.: 114 C (PMCC); Acid No.: 1.5 max.; pH: 9.5
Composition: 85% active ingred.
Reaction: Alkaline
Containers: 20 tons; 125 kg drums

Lauridit KM. Akzo Chemie GmbH
Chemical Description: Coco fatty acid monoethanolamide
Category: Detergent, nonionic
Applications: Detergent for textile, household, and cosmetic applications; foam stabilizer for high foaming heavy duty detergents, shampoos, bubble baths
Form: Flakes; Color: Yellowish
Properties: Solubility: Poor water sol.; Sp. gr.: 0.97; M.P.: 65–70 C; Sapon. No.: 14 max.; Acid No.: 4 max.; Ester No.: 12 max.; pH: 9
Composition: 93% active ingred.
Reaction: Alkaline
Containers: 25 kg paper bags

Lauridit LM. Akzo Chemie GmbH
Chemical Description: Lauric acid monoethanolamide
Category and Applications: See Lauridit KM
Form: Flakes; Color: Yellowish
Properties: Solubility: Poor water sol.; Sp. gr.: 1.01 @ 80 C; M.P.: 80–84 C; Sapon. No.: 20 max.; Acid No.: 4 max.; Stability: Protect from high temperature; pH: 9

Composition: 93% active ingred.
Reaction: Alkaline
Containers: 25 kg paper bags

Lauridit OD. Akzo Chemie GmbH
Chemical Description: Oleic acid diethanolamide
Category and Applications: See Lauridit KD
Form: Liquid; Color: Yellow-brown
Properties: Solubility: 100 g/l water @ 20 C; Sp. gr.: 0.97; F.P.: 0 C; Flash Pt.: 100 C (PMCC); Acid No.: 4 max.; pH: 9
Composition: 85% active ingred.
Reaction: Alkaline
Containers: 20 tons; 125 kg drums

Lauridit PD. Akzo Chemie GmbH
Chemical Description: Fatty acid polydiethanolamide/fatty acid diethanolamide 1:2
Category: Detergent, emulsifier
Applications: Household detergents; corrosion inhibitor for cutting oils and additive for acid cleaners
Form: Liquid; Color: Brownish
Properties: Solubility: Sol. in water; Sp. gr.: 1.02; Flash Pt.: 105 C (PMCC); Acid No.: 16 max.; pH: 9
Composition: 65% active ingred.
Reaction: Alkaline
Toxicity: Bacterial toxicity against *Pseudomonas putida*: 16 mg/l
Containers: 125 kg drums

Laventin CW. BASF Aktiengesellschaft
Chemical Description: Alkanol polyglycol ether
Category: Wetting agent, detergent, nonionic
Applications: Water-free wetting agent and detergent with good emulsifying and dispersing effect on oil and fatty contaminants for all textile fibers; used in oxidative desizing, precleansing, scouring, reductive and oxidative bleaching, afterwashing, etc.
Form: Clear, low viscosity liquid; Color: Colorless to slightly yellow
Properties: Solubility: Dissolves readily when stirred into water; Density: 0.99 g/ml; Stability: Stable to hard water, alkalis, acids, reductive and oxidizing bleaching agents in normal concentrations for textile treatment; pH: 6.5–7.5 (1% sol'n.); Biodegradable
Toxicity: Skin and eye irritant on prolonged contact with concentrated form
References: Leaflet M 5358 e

Laventin W. BASF Aktiengesellschaft
Chemical Description: Oxyethylated fatty acid derivative
Category: Detergent, nonionic
Applications: Water-free detergent for washing all types of fibers, esp. in acid media; used for backwashing of wool slubbing, scouring wool, acid washing, milling
Form: Viscous liquid or semisolid paste that becomes solid at low temperatures; Color: Yellow-brown
Properties: Solubility: Water sol.; Stability: Stable to hard water, alkalis, acids, and to reductive and

oxidative bleaches in normal concentrations for textile treatment
Reaction: Slightly alkaline
Toxicity: Skin and eye irritant on prolonged contact with concentrated form
References: Leaflet M 5164 e

Lecidan. Grindsted Prod.
Chemical Description: Blends of monoglycerides and special soya lecithin
Category: Emulsifier, nonionic
Applications: Food emulsifier for frying margarine; imparts excellent frying properties and minimizes spattering
Form: Pills
Properties: HLB: 5.0
Composition: 100% active ingred.

Leophen BN. BASF Aktiengesellschaft
Chemical Description: Alcohol sulfonate
Category: Wetting agent, anionic
Applications: Foam-free wetting agent in mercerizing cotton yarn and piece goods
Form: Clear, low viscosity liquid; Color: Yellow-brown
Properties: Solubility: Miscible with water and dilute caustic soda; Density: ≈ 1.09 g/ml; Solidification Pt.: ≈ –15 C; Stability: Stable in mercerizing liquors up to 30 Be; Biodegradable
Toxicity: Eye irritant
References: Leaflet M 1157 e

Leophen LG. BASF Aktiengesellschaft
Chemical Description: Mixture of sodium salts of aliphatic sulfonic acids
Category: Wetting agent
Applications: Cresol-free, nonfoaming, alkali-stable wetting agent for use in caustic baths up to 25 Be, allowing material to be treated without previous boil-off; in one-bath pigment dyeing and resin finishing
Form: Almost clear, slightly viscous liquid; Color: Yellowish
Reaction: Almost neutral
References: Leaflet M 1619 e

Leophen M. BASF Aktiengesellschaft
Chemical Description: Neutral phosphoric acid ester with nonionic emulsifiers
Category: Wetting agent, antifoam
Applications: Wetting agent in bleaching, dyeing, and finishing of cellulosic fibers, synthetic fibers, and wool in the form of loose stock, yarn, and piece goods; foam-inhibiting and air-displacing effect; detergent for piece goods of synthetics in jet dyeing machines
Form: Liquid; Color: Colorless to weakly yellow
Properties: Solubility: Readily miscible with water; Density: ≈ 1.0 g/ml; Solidification Pt.: < –15 C; Stability: Stable in hard water, in medium acid to strongly alkaline pH ranges; stable to oxidation and reduction bleaching agents; Emulsifiability: Self-emulsifying
Toxicity: Skin, eye, and respira-

tory tract irritant; avoid inhaling vapors
References: Leaflet M 5675 e

Leophen ML. BASF Aktiengesellschaft
Chemical Description: Specific alkylsulfate
Category: Wetting agent
Applications: Wetting agent for mercerizing and caustic-treating yarn, fabrics, or knitgoods with minimum foam formation
Form: Liquid; Color: Yellowish-brown
Properties: Density: ≈ 1 g/ml @ 20 C; Pour Pt.: ≈ –3 C
Toxicity: Eye irritant

Leophen RA. BASF Aktiengesellschaft
Chemical Description: Concentrated grade of the sodium salt of a sulfosuccinate
Category: Wetting and rewetting agent, anionic
Applications: Splittable wetting agent for dyeing in circulating-liquor systems, for continuous dyeing of cellulosic fibers by the pigment-pad process, for the chlorine-peroxide bleach, and the continuous chlorination of wool; as rewetting agent imparts absorptivity to cellulosics in bleaching
Form: Liquid; Color: Yellowish
Properties: Solubility: Sol. in hot water; 5% sol'ns. are viscous and turbid but become clearer when diluted; Stability: Resistant to chlorine, hard water, and weak acids; split slowly by weak alkalis, and more rapidly by strong alkalis; pH: 5.5–7.0 (1% aq.)
Toxicity: Attacks skin and (eye) mucous membranes; avoid inhalation of vapors from hot liquors
References: Leaflet M 5281 e

Leophen RBD. BASF Aktiengesellschaft
Chemical Description: Dialkyl sulfimide
Category: Wetting and rewetting agent, anionic
Applications: Low-foaming rapid wetting agent for use in bleaching and dyeing; rewetting agent for cellulosic and synthetic fibers
Form: Clear, viscous liquid; Color: Golden yellow
Properties: Solubility: Readily sol. in water at all temperatures; Density: ≈ 1.1 g/ml; Setting Pt.: ≈ –7 C
Toxicity: Skin and eye irritant on prolonged or repeated contact
References: Leaflet M 5624 e

Leophen U. BASF Aktiengesellschaft
Chemical Description: Combination of nonionic and anionic surfactants
Category: Wetting agent, detergent, dispersant
Applications: Versatile wetting agent/detergent with excellent stability to electrolytes and rapid wetting over wide pH range; used in desizing, treating with alkali, and bleaching with hypochlorite, peroxide, or chlorite
Form: Paste; Color: Yellowish

Properties: Solubility: Readily sol. in cold and hot water; Stability: Stable to electrolytes; pH: 7.5–8 (10 g/l)
Warnings and Caution: Wear protective goggles; avoid prolonged contact with skin
References: Leaflet M 5127 e

Levelan AO192. Lankro Chem. Ltd.
Chemical Description: Fatty alcohol ethoxylate
Category: Leveling agent, nonionic
Applications: Leveling agent for direct dyestuffs
Form: Clear liquid; Odor: Faint
Properties: Solubility: Sol. in water; Sp. gr.: 1.018 @ 20 C; Visc.: 13 cs @ 20 C; Cloud Pt.: > 100 C (1% active aq.); Flash Pt.: > 200 F (COC); Pour Pt.: < 0 C; Stability: Stable to acids, alkalis, and electrolytes; effective in hard and soft water; pH: 6.0–8.0 (1% aq.)
Composition: 20% active ingred.
Toxicity: Skin irritant on prolonged contact with concentrated form
Containers: Stainless steel, glass, resin, or polythene lined equipment
Warnings and Caution: Wear protective gloves and goggles; spillages may be slippery; do not mix with strong oxidizing or reducing agents (potentially explosive)
References: Bulletin SI: 16 FM678

Levelan P148. Lankro Chem. Ltd.
Chemical Description: Nonyl phenol ethylene oxide condensate
Category: Emulsifier, stabilizer, nonionic
Applications: Primary emulsifier in emulsion polymer industry; latex stabilization
Form: Clear liquid; Odor: Faint
Properties: Solubility: Sol. in water; Sp. gr.: 1.069 @ 20 C; Visc.: 490 cs @ 20 C; Cloud Pt.: > 100 C (1% aq.); Flash Pt.: > 200 F (COC); Pour Pt.: < 0 C; Stability: Stable to acids, alkalis, and electrolytes; effective in hard and soft water; pH: 6.0–8.0 (1% aq.)
Composition: 80% active ingred.
Toxicity: Skin irritant on prolonged contact with concentrated form
Containers: Stainless steel, glass, resin, or polythene lined equipment
Warnings and Caution: Wear protective gloves and goggles; spillages may be slippery; do not mix with strong oxidizing or reducing agents (potentially explosive)
References: Bulletin SI:16 FM678

Levelan P208. Lankro Chem. Ltd.
Chemical Description: Nonyl phenol ethoxylate
Category: Wetting agent, stabilizer, emulsifier, nonionic
Applications: Wetting agent at high temperature and high electrolyte concentration in textile industry; latex stabilizer and emulsifier in emulsion polymer industry
Form: Clear liquid; Odor: Faint
Properties: Solubility: Sol. in water; Sp. gr.: 1.086 @ 20 C; Visc.: 460 cs @ 20 C; Cloud Pt.: > 100 C (1% aq.); Flash Pt.: > 200 F

(COC); Pour Pt.: < 0 C; Stability: Stable to acids, alkalis, electrolytes; effective in hard and soft water; pH: 6.0–8.0 (1% aq.)
Composition: 80% active ingred.
Toxicity: Skin irritant on prolonged contact with concentrated form
Containers: Stainless steel, glass, resin, or polythene lined equipment
Warnings and Caution: Wear protective gloves and goggles; spillages may be slippery; do not mix with strong oxidizing or reducing agents (potentially explosive)
References: Bulletin SI:16 FM678

Levelan P307. Lankro Chem. Ltd.
Chemical Description: Alkylphenol ethoxylate
Category: Wetting agent
Applications: Wetting agent for high temperature applications
Form: Liquid
Composition: 70% active ingred.

Levelan P357. Lankro Chem. Ltd.
Chemical Description: Nonyl phenol ethylene oxide condensate
Category and Applications: See Levelan P208
Form: Clear liquid; Odor: Faint
Properties: Solubility: Sol. in water; Sp. gr.: 1.090 @ 20 C; Visc.: 960 cs @ 20 C; Cloud Pt.: > 100 C (1% aq.); Flash Pt.: > 200 F (COC); Pour Pt.: 2 C; Stability: Stable to acids, alkalis, electrolytes; effective in hard and soft water; pH: 6.0–8.0 (1% aq.)
Composition: 70% active ingred.
Toxicity: Skin irritant on prolonged contact with concentrated form
Containers: Stainless steel, glass, resin, or polythene lined equipment
Warnings and Caution: Wear protective gloves and goggles; spillages may be slippery; do not mix with strong oxidizing or reducing agents (potentially explosive)
References: Bulletin SI:16 FM678

Lexaine C. Inolex Corp.
Chemical Description: Cocamidopropyl betaine
Category: Viscosity builder, foam booster, thickener, amphoteric
Applications: Specialty shampoos, e.g., baby, acid balance, gel type; hair products, bath products, skin cleansers, disinfectants; thickener in foaming and dilutable conditioners
Form: Liquid; Color 4 max. (Gardner)
Properties: Sp. gr.: 1.044; F.P.: 0 C; pH: 4.5–5.5
Composition: 30% active ingred.
Containers: 55 gal drums

Lexaine CG-30. Inolex Corp.
Chemical Description: Cocoamidoalkyl betaine
Category: Amphoteric surfactant
Applications: Light duty liquid detergent systems, in shampoos and other hair care products, bath products, skin cleansers; forms high viscosity systems
Form: Clear, nonviscous liquid; Color: Straw to light amber; Odor:

Bland
Properties: pH: 6.5–8.0
Composition: 29% active ingred.
Containers: 400 lb net polyethylene-lined fiber drums

Lexaine CG-50. Inolex Corp.
Chemical Description: Cocamidopropyl betaine derived by direct synthesis from coconut oil
Category: Amphoteric surfactant
Applications: See Lexaine CG-30; CG-50 is more concentrated with less salt content and can be used to formulate high active light duty detergents with good cold stability
Form: Clear viscous liquid; Color: Straw to light amber; Odor: Bland
Properties: pH: 7.5 (10% aq.)
Composition: 50% active ingred.
Containers: 450 lb net polyethylene-lined fiber drums

Lexaine IBC-70. Inolex Corp.
Chemical Description: Blend of DEA lauryl sulfate, cocamidopropyl betaine, and propylene glycol
Category: Shampoos concentrate, amphoteric/anionic
Applications: High active, ionically balanced, mild shampoo concentrate in nonaqueous liquid form
Form: Slightly hazy liquid; Color: Amber
Properties: Density: 8.8 lb/gal; Visc.: 1850 cps; F.P.: –4 C; pH: 6.0–7.0 (10% aq.)
Composition: 70% active ingred.
Containers: 400 lb net polyethylene-lined fiber drums

Lexate CL-60. Inolex Corp.
Chemical Description: Blend of cationic and intronium surfactants
Category: Conditioner concentrate
Applications: Hair conditioner concentrate with good viscosity after dilution
Form: Liquid; Color: 7 (Gardner); Odor: Mild, fatty
Properties: Sp. gr.: 1.01; Visc.: 150 cps; Cloud Pt.: $<$ 5 C
Composition: 75% active ingred.
Reaction: Alkaline
Toxicity: Protect skin and eyes from contact

Lexate CRC. Inolex Corp.
Chemical Description: Blend of stearamidopropyl dimethylamine, glycol stearate, and ceteth-2
Category: Conditioner, emulsifier, cationic
Applications: Cream rinse concentrate and conditioner
Form: Waxy flake; Color: Cream to tan; Odor: Mild, fatty amino
Properties: Solubility: Easily melted and dispersed in water; M.P.: 49–53 C
Composition: 100% active ingred.
Containers: 85 lb net fiber drum
Warnings and Caution: Avoid contact with eyes; store below 35 C

Lexate IL. Inolex Corp.
Chemical Description: Lanolin and mineral oil
Category: Emollient, nonionic
Applications: Lanolin additive in W/O emulsions; cleansing and conditioning lotions; as superfatting agent in soaps; emollient additive

in spreading or dispersible bath oils
Form: Paste; Color: Straw; Odor: Less odor than lanolin
Properties: M.P.: 35 C; Sapon. Value: 66; Acid Value: 1; Hydroxyl Value: 37; Iodine Value: 16
Composition: 100% active ingred.
Containers: 30 lb net plastic pail

Lexate PX. Inolex Corp.
Chemical Description: Blend of petrolatum, lanolin, and ozokerite
Category: Emulsifier, emollient
Applications: O/W emulsifier for incorporating hydrophobic additives; auxiliary emulsifier to aid stability and consistency in O/W and W/O emulsions; lanolin cream base for night creams, cold creams, cleansing creams, hair creams, topical pharmaceuticals; conditioning additive in soaps and shaving cream
Form: Semisolid; Color: Faintly yellow; Odor: Low
Properties: M.P.: 41 C; Sapon. Value: 1.0; Acid Value: 0.3; Hydroxyl Value: 20; Iodine Value: 16; Emulsifiability: Self-emulsifying
Containers: 375 lb net open head steel drums

Lexate TA. Inolex Corp.
Chemical Description: Blend of glyceryl stearate, isopropyl myristate, and stearyl stearate
Category: Emollient, emulsifier
Applications: Auxiliary lipophilic emulsifier in W/O emulsions; emollient in cosmetic creams and lotions
Form: Waxy solid; Color: White; Odor: Odorless–bland, fatty
Properties: M.P.: 45–50 C; Sapon. Value: 159–169; Acid Value: 5–9; Iodine Value: 2.0 max.
Containers: 30 lb net plastic pail

Lexate TL. Inolex Corp.
Chemical Description: Blend of glyceryl stearate, butyl stearate, and stearyl stearate
Category: Emollient, emulsifier
Applications: Superfatting agent in milled bar soap; enhances barrier properties in creams and lotions; auxiliary lipophilic emulsifier in W/O systems
Form: Waxy solid; Color: White; Odor: Mild, fatty
Properties: M.P.: 50–55 C; Sapon. Value: 145–152; Acid Value: 3 max.; Iodine Value: 2.0 max.
Containers: 30 lb net plastic pail

Lexemul 55G. Inolex Corp.
Chemical Description: Glyceryl monostearate
Category: Nonionic surfactant
Applications: Used in cosmetics and topical pharmaceuticals; improves lubricity and gloss in cosmetic creams and lotions
Form: Flakes; Color: White; Odor: Mild fatty
Properties: M.P.: 55–59 C; Sapon. Value: 160–170; Acid Value: 2.0 max.; Iodine Value: 1.0 max.; pH: 5.5 (3% aq. suspension)
Composition: 100% conc.
Containers: 100 lb net fiber drums

Lexemul 530. Inolex Corp.
Chemical Description: Glycerol monostearate
Category: Emulsifier, anionic modfied
Applications: Primary emulsifier in O/W systems, esp. for cosmetic creams and lotions, hair dressings, shaving creams; not for use with cationics or in products containing acid or acid-reacting salts
Form: Flakes; Color: White to cream; Odor: Low
Properties: Solubility: Sol. @ 60 C in carbon tetrachloride, ethyl acetate, isopropanol, mineral oil; disp. in water @ 60 C; M.P.: 56 C; Sapon. Value: 146–154; Acid Value: 20.0 max.; Iodine Value: 3.0 max.; Stability: Good; Emusifiability: Self-emulsifying; pH: 8.5 (3% aq. disp.)
Composition: 100% conc.
Reaction: Slightly basic
Containers: 100 lb net fiber drum

Lexemul 561. Inolex Corp.
Chemical Description: Glyceryl monostearate, cosmetic grade
Category: Emulsifier, nonionic
Applications: Emulsifier for all types of O/W cream or lotion systems; hand creams, antiperspirants, bleach lotions, etc.
Form: Brittle flakes; Color: White to light cream; Odor: Mild ethoxylate
Properties: Solubility: Disp. in water @ 60 C; Sapon. Value: 90–100; Acid Value: 2 max.; Iodine Value: 1.5 max.; Self-emulsifying; pH: 4.3 (3% aq. disp.)
Composition: 100% conc.
Containers: 100 lb net fiber drum
Warnings and Caution: Avoid storage above 50 C

Lexemul AR. Inolex Corp.
Chemical Description: Glyceryl monostearate
Category: Emulsifier, stabilizer, opacifier, emollient, cationic
Applications: Emulsifier in cosmetics, toiletries, deodorants and antiperspirants, and topical pharmaceuticals
Form: Flakes; Color: White to cream; Odor: Mild fatty
Properties: Solubility: Disp. in water @ 60 C; M.P.: 60 C; Sapon. Value: 166–174; Acid Value: 25–31; Iodine Value: 3.0 max.; Stability: Acid-stable; Emulsifiability: Self-emulsifying
Composition: 100% conc.
Containers: 100 lb net fiber drums

Lexemul AS. Inolex Corp.
Chemical Description: Glyceryl monostearate, self-emulsifying
Category: Emulsifier, stabilizer, opacifier, emollient, anionic
Applications: Cosmetics, toiletries, topical pharmaceuticals; used over wide pH range; electrolyte tolerant
Form: Flakes; Color: White to cream; Odor: Mild fatty
Properties: Solubility: Disp. in water @ 60 C; M.P.: 60 C; Sapon. Value: 153–162; Acid Value: 14–18; Iodine Value: 3.0 max.; Emulsifiability: Self-emulsifying
Composition: 100% conc.
Containers: 100 lb net fiber drums

Lexol 60. Inolex Corp.
Chemical Description: Mixed isopropyl esters of myristic, palmitic, and stearic acids
Category: Emollient, nonionic
Applications: Lower cost substitute for isopropyl myristate; emollient oils, cosmetics, topical pharmaceuticals; bath oils, electric preshaving lotions, aerosol antiperspirants
Form: Clear liquid; Color: Colorless; Odor: Odorless
Properties: Solubility: Sol. in acetone, benzene, carbon tetrachloride, castor oil, chloroform, ethanol, heptane, isopropanol; insol. in glycerol, propylene glycol, water; Density: 7.1 lb/gal; Sp. gr.: 0.852; F.P.: 7 C; Sapon. Value: 190–198; Iodine Value: 1.0 max.; Biodegradable
Composition: 100% conc.
Containers: 55 gal steel drums

Lexol IPM. Inolex Corp.
Chemical Description: Isopropyl myristate
Category: Emollient, solvent, penetrant, cloud point depressant
Applications: Cosmetics, bath oils, electric preshave lotions
Form: Clear liquid; Color: Colorless; Odor: Essentially odorless
Properties: m.w. (theoretical): 270; Ref. Index: 1.433; Solubility: Sol. in acetone, benzene, carbon tetrachloride, castor oil, chloroform, ethanol, heptane, isopropanol; Density: 7.1 lb/gal; Sp. gr.: 0.847–0.854; Visc.: 4.8 cp; F.P.: 3 C; B.P.: 170 C @ 4 mm; Flash Pt.: 305 F; Sapon. Value: 202–212; Iodine Value: 1.0 max.; Stability: Good oxidative stability
Containers: 55 gal steel drums

Lexol IPP. Inolex Corp.
Chemical Description: Isopropyl palmitate
Category: Emollient, solubilizer
Applications: Emollient in conditioning cosmetics; solubilizer for many cosmetic and topical pharmaceuticals
Form: Clear liquid; Color: Colorless; Odor: None
Properties: m.w. (theoretical): 298; Ref. Index: 1.437; Solubility: Sol. in acetone, benzene, carbon tetrachloride, castor oil, chloroform, ethanol, heptane, isopropanol; Density: 7.1 lb/gal; Sp. gr.: 0.850–0.855; Visc.: 7 cps; F.P.: 12 C; B.P.: 172 C @ 4 mm; Flash Pt.: 335 F; Sapon. Value: 182–191; Iodine Value: 1.0 max.; Stability: Good oxidative stability
Containers: 55 gal steel drum

Lignosite 260. Georgia-Pacific/Bellingham Div.
Chemical Description: Modified lignosulfonate
Category: Emulsifier, dispersant, anionic
Applications: Dispersing pigments, insecticides, finely divided solids; O/W emulsifier
Form: Fine powder; Color: Tan
Composition: 80% lignosulfonates as sodium and calcium salts; 15% carbohydrate-derived material; 5% other
Containers: 50-lb bags

Lignosite 431. Georgia-Pacific/Bellingham Div.
Chemical Description: Sodium-based lignosulfonate
Category: Dispersant, anionic
Applications: Dispersion of pigments; formulating wettable powders; manufacture of wax emulsions; formulating water-treatment compounds and industrial cleaners
Form: Fine powder; Color: Brown; also available as 50% sol'n.
Properties: Solubility: Dissolves in hot or cold water; Density: ≈ 10.5 lb/gal (50% sol'n.); pH: 4.0 (10% sol'n.)
Composition: 80% sodium lignin sulfonates; 15% carbohydrate-derived material; 5% miscellaneous
Reaction: Slightly acidic

Lignosite 458. Georgia-Pacific/Bellingham Div.
Chemical Description: Sodium lignosulfonate
Category: Dispersant, anionic
Applications: Dispersion of pigments; manufacture of concrete admixtures and wax emulsions; formulating wettable powders, water-treatment compounds, boiler-treating compounds, and industrial cleaners
Form: Fine powder; Color: Brown; also available in 46–48% solids sol'n.
Properties: Solubility: Dissolves in hot or cold water; Density: Powder avg. bulk: 23 lb/ft^3; Liquid: ≈ 10.3 lb/gal (46% sol'n.)
Composition: Powder: 80% sodium lignosulfonate; Liquid: 46% solids

Lignosite 823. Georgia-Pacific/Bellingham Div.
Chemical Description Sodium lignosulfonate
Category: Dispersant, anionic
Applications: Dispersant for fine solids in aq. systems; manufacture of wax emulsions; formulating wettable powders; dispersion of dyestuffs and pigments
Form: Solubility: Dissolves in cold or hot water; Density: Avg. bulk 23 lb/ft^3; pH: 7–8 (10% aq.)
Composition: 70% sodium lignin sulfonates; 15% carbohydrate-derived material; 15% miscellaneous
Reaction: Neutral

Lignosite 854. Georgia-Pacific/Bellingham Div.
Chemical Description: Sodium lignosulfonate
Category: Dispersant, anionic
Applications: Dispersion of pigments; manufacture of concrete admixtures and wax emulsions; formulating wettable powders, water-treatment compounds, and industrial cleaners
Form: Fine powder; Color: Brown; also available as 50% sol'n.
Properties: Solubility: Dissolves in hot or cold water; Density: Powder avg. bulk: 23 lb/ft^3; Liquid: ≈ 10.5 lb/gal (50% sol'n.); pH: 5.0 (10% aq.)
Composition: 80% sodium lignin

sulfoantes; 15% carbohydrate-derived material; 5% miscellaneous
Reaction: Slightly acidic

Lignosite 1840. Georgia-Pacific/Bellingham Div.
Chemical Description: Calcium base lignosulfonate
Category: Extender, dispersant
Applications: Extender for adhesive and binder systems; dispersant for manufacture of wax emulsions where light color is important
Form: Fine powder; Color: Light; also available as 50% solids sol'n.
Properties: Solubility: Dissolves in hot or cold water; Density: Powder avg. bulk: 23 lb/ft^3; Liquid: ≈ 10.5 lb/gal (50% sol'n.); pH: 4.5 (10% aq.)
Composition: 70% calcium lignin sulfonates; 24% carbohydrate-derived material; 6% miscellaneous
Reaction: Slightly acidic

Lignosite FML. Georgia-Pacific/Bellingham Div.
Chemical Description: Aqueous sol'n. of a ferromagnetic iron lignosulfonate
Category: Ferromagnetic fluid
Applications: Ferromagnetic fluid; used in separation processes to adsorb to certain clays, minerals, and flocculating agents to enhance their separation by application of a magnetic field
Form: Liquid
Properties: Density: 10.1 lb/gal; Visc.: 60 cps (Brookfield); Stability: Stable to dilution without loss of magnetic properties; stable at pH 2–13; good salt tolerance
Composition: 35% solids content

Lignosite GPM-38. Georgia-Pacific/Bellingham Div.
Chemical Description: Modified sodium lignosulfonate
Form: Powder; also available as 38–40% solids sol'n.
Properties: Solubility: Dissolves in hot or cold water; pH: 10 (10% aq.)
Composition: 80% modified sodium lignosulfonate; 6.2% sulfur; 10% sodium
Reaction: Alkaline

Lignosol AXD. Reed Ltd.
Chemical Description: Spray dried acidic desugarized and partially desulfonated sodium lignosulfonate
Category: Wetting agent, emulsifier, dispersant, anionic
Applications: Dispersing waxes, pigments, resins; wetting agent for insecticides, herbicides, fungicides; sludge conditioner for boiler and cooling water applications
Form: Powder; Color: Brown
Properties: Solubility: Water sol.; Density: 28–32 lb/ft^3 bulk; Stability: Up to 250 C; pH: 4.9 (27% sol'n.)
Composition: 95% active ingred.
Reaction: Acid
Containers: 50 lb paper bags
Warnings and Caution: Organic powder—avoid contact with oxidizing agents

Lignosol B. Reed Ltd.
Chemical Description: Calcium lignosulfonate
Category: Wetting agent, emulsifier, dispersant, binder, anionic
Applications: Binding agent for refractories, bricks, foundry cores; linoleum cement; grouting and soil stabilization in construction; tanning and filling; insecticides, herbicides, fungicides
Form: Liquid; Color: Black
Properties: Sp. gr.: 1.25; Visc.: 200 cps @ 80 F; F.P.: –10 C; B.P.: 105 C
Composition: 50% active ingred.
Reaction: Acid
Containers: Drums or bulk

Lignosol BD. Reed Ltd.
Chemical Description: Spray dried calcium lignosulfonate
Category: Wetting agent, emulsifier, binder, anionic
Applications: Binder; water reduction and extrusion aid in brick and refractory manufacture; grouting and soil stabilization in construction; foundry core binder; gypsum wallboard additive; plating bath dispersant; tanning agent; stabilizer for wax emulsions; insecticides
Form: Powder; Color: Yellow
Properties: Solubility: Complete in water; Density: 28–32 lb/ft^3 bulk; Stability: Up to 250 C; pH: 4.5 (27% sol'n.)
Composition: 95% active ingred.
Reaction: Acid
Containers: 50 lb paper bags

Lignosol D-10. Reed Ltd.
Chemical Description: Sodium lignosulfonate
Category: Wetting agent, dispersant, anionic
Applications: Dispersant for disperse dyes
Form: Powder; Color: Brown
Properties: Solubility: Water sol.; Stability: Up to 250 C
Composition: 95% active ingred.
Reaction: Acid
Containers: 50 lb paper bags

Lignosol D-30. Reed Ltd.
Chemical Description: Sodium lignosulfonate
Category: Wetting agent, dispersant, anionic
Applications: Primary dispersant for disperse dyes
Form: Powder; Color: Brown
Properties: Solubility: Complete in water; Stability: Up to 250 C
Composition: 95% active ingred.
Reaction: Alkaline

Lignosol DXD. Reed Ltd.
Chemical Description: Spray dried alkaline desugarized and partially desulfonated sodium lignosulfonate
Category: Wetting agent, emulsifier, dispersant, anionic
Applications: Dispersant in dyestuffs; emulsion stabilizer for asphalt emulsions; dispersant and wetting agent in industrial cleaners; insecticides, herbicides, fungicides

Form: Powder; Color: Black
Properties: Solubility: Complete in water; Density: 30–32 lb/ft^3 bulk; Stability: Up to 250 C; pH: 9.0 (27% sol'n.)
Composition: 95% active ingred.
Reaction: Alkaline

Lignosol FTA. Reed Ltd.
Chemical Description: Spray dried fractionated, desugarized, and desulfonated sodium lignosulfonate
Category: Wetting agent, dispersant, anionic
Applications: Primary dispersant for disperse dyes
Form: Powder; Color: Brown
Properties: Solubility: Complete in water; Density: 30–32 lb/ft^3 bulk; Stability: Up to 250 C; pH: 9.5 (27% sol'n.)
Composition: 95% active ingred.
Reaction: Alkaline

Lignosol HCX. Reed Ltd.
Chemical Description: Alkaline sugar-free sodium lignosulfonate
Category: Wetting agent, dispersant, anionic
Applications: Insecticides, herbicides, fungicides
Form: Powder; Color: Brown
Properties: Solubility: Water sol.; Stability: Up to 250 C
Composition: 95% active ingred.
Reaction: Alkaline

Lignosol Humectant F. Reed Ltd.
Chemical Description: Liquid low molecular weight sodium lignosulfonate
Category: Chelating agent
Applications: Complexing metal ions
Form: Liquid

Lignosol R30. Reed Ltd.
Chemical Description: Spray dried desugarized calcium sodium lignosulfonate
Category: Slime control agent
Applications: Slime control agent in mining
Form: Powder

Lignosol SF. Reed Ltd.
Chemical Description: Spray dried desugarized calcium lignosulfonate
Category: Dispersant, anionic
Applications: Dispersant for Portland cement, concrete; water reduction in gypsum wallboard; retanning agent
Form: Powder; Color: Brown
Properties: Solubility: Water sol.; Density: 28–32 lb/ft^3 bulk; Stability: Up to 250 C; pH: 6.0 (27% sol'n.)
Composition: 95% active ingred.
Reaction: Acid

Lignosol SFL. Reed Ltd.
Chemical Description: 46% solids desugarized calcium lignosulfonate
Category: Dispersant, anionic
Applications: Dispersant for Portland cement, concrete
Form: Liquid; Color: Black
Properties: Solubility: Complete in water; Sp. gr.: 1.24; F.P.: −10 C; B.P.: 100 C
Composition: 48% active ingred.
Reaction: Acid

Lignosol SFX. Reed Ltd.
Chemical Description: Spray dried desugarized sodium lignosulfonate
Category: Dispersant, anionic
Applications: Concrete admixtures; sludge conditioner for boiler and cooling water applications; dyestuff dispersant and extender
Form: Powder; Color: Brown
Properties: Density: 28–32 lb/ft^3 bulk; pH: 6.8 (27% sol'n.)
Composition: 95% active ingred.

Lignosol SFX-65. Reed Ltd.
Chemical Description: Spray dried fractionated and desugarized lignosulfonate
Form: Powder; Color: Yellow
Properties: Density: 28–32 lb/ft^3 bulk; pH: 6.0 (27% sol'n.)
Composition: 95% active ingred.

Lignosol TS. Reed Ltd.
Chemical Description: 49% solids ammonium lignosulfonate
Category: Wetting agent, emulsifier, anionic
Applications: Slurry water reducers and grinding aids in cement manufacture; tanning, retanning, and filling; electroplating; dispersing waxes, pigments, resins
Form: Liquid; Color: Black
Properties: Solubility: Complete in water; Sp. gr.: 1.23; F.P.: –10 C
Composition: 47% active ingred.
Reaction: Acid

Lignosol TSD. Reed Ltd.
Chemical Description: Spray dried ammonium lignosulfonate
Category: Wetting agent, emulsifier, dispersant, anionic
Applications: Leather retanning agent; plating bath dispersant; slurry water reducer and grinding aid in cement manufacture; carbon black pelletizing; phenolformaldehyde extender; slurry explosives and flotation aid
Form: Powder; Color: Yellow
Properties: Solubility: Water sol.; Density: 28–32 lb/ft^3 bulk; Stability: Up to 250 C; pH: 4.5 (27% sol'n.)
Composition: 95% active ingred.
Reaction: Acid

Lignosol TSF. Reed Ltd.
Chemical Description: Spray dried desugarized ammonium lignosulfonate
Category: Dispersant
Applications: Insecticides, fungicides, and herbicides
Form: Powder; Color: Brown
Properties: Density: 28–32 lb/ft^3 bulk; pH: 4.3 (27% sol'n.)
Composition: 95% active ingred.

Lignosol WT. Reed Ltd.
Chemical Description: Spray dried low molecular weight, desugarized lignosulfonate
Category: Wetting agent, dispersant, anionic
Applications: Sludge conditioner, boiler feed water treatment
Form: Powder; Color: Yellow
Properties: Solubility: Water sol.; Stability: Up to 300 C
Composition: 95% active ingred.
Reaction: Alkaline

Lignosol X. Reed Ltd.
Chemical Description: 45% solids sodium lignosulfonate
Category: Wetting agent, emulsifier, anionic
Applications: Tanning and retanning agent; dye leveling; wax emulsion; slurry water reducer and grinding aid
Form: Liquid; Color: Black
Properties: Solubility: Water sol.; Sp. gr.: 1.24; F.P.: –5 C; B.P.: 105 C
Composition: 47% active ingred.
Reaction: Neutral

Lignosol XD. Reed Ltd.
Chemical Description: Spray dried sodium lignosulfonate
Category: Wetting agent, emulsifier, anionic
Applications: Retanning leather; water treatment; wax emulsions stabilizer; dyeing; pickling; dispersing waxes, pigments, resins
Form: Powder; Color: Yellow
Properties: Solubility: Water sol.; Density: 28–32 lb/ft^3 bulk; Stability: Up to 250 C; pH: 6.5 (27% sol'n.)
Composition: 95% active ingred.
Reaction: Neutral

Lignosol XD-65. Reed Ltd.
Chemical Description: Spray dried fractionated sodium lignosulfonate
Category: Flotation aid
Applications: Slurry explosives and flotation aid in mining industry
Form: Powder; Color: Yellow
Properties: Density: 28–32 lb/ft^3 bulk; pH: 6.8 (27% sol'n.)
Composition: 95% active ingred.
Reaction: Neutral

Lilafilm CA. Lilachim S.A.
Applications: Additive for water treatment

Lilaflot. Lilachim S.A.
Applications: Series of flotation reagents for minerals

Lilamac 140. Lilachim S.A.
Chemical Description: Acetate amine salt of Lilamin 140
Category: Flotation aid, cationic
Applications: Aids for crude oil extraction, soil stabilization, wax emulsions, ore flotation
Form: Solid; Color: 8 max. (Gardner)
Properties: m.w.: ≈ 330; M.P.: 60 C; Acid Value: 165–175; Iodine Value: 3 max.
Composition: 95% active
Toxicity: Skin and eye irritant
Warnings and Caution: Wear protective goggles, gloves, clothing, and preventive ointment

Lilamac 160. Lilachim S.A.
Chemical Description: Acetate amine salt of Lilamin 160
Category and Applications: See Lilamac 140
Form: Solid; Color: 8 max. (Gardner)
Properties: m.w.: ≈ 280; M.P.: 38 C; Acid Value: 195–205; Iodine Value: 12 max.
Composition: 95% active ingred.
Toxicity: Skin and eye irritant

Warnings and Caution: Wear protective goggles, gloves, clothing, and preventive ointment

Lilamac 170. Lilachim S.A.
Chemical Description: Acetate amine salt of Lilamin 170
Category and Applications: See Lilamac 140
Form: Solid; Color: 8 max. (Gardner)
Properties: m.w.: ≈ 332; M.P.: 45 C; Acid Value: 165–175; Iodine Value: 35 min.
Composition: 95% active ingred.
Toxicity: Skin and eye irritant
Warnings and Caution: Wear protective goggles, gloves, clothing, and preventive ointment

Lilamin Series. Lilachim S.A.
Chemical Description: Series of primary, secondary, and tertiary amines
Category: Emulsifier, anticorrosive, lubricant, bactericidal, dispersant, cationic
Applications: Additives for crude oil extraction, drilling aids; anti-stripping aids; adhesion aids in road building; gasoline and lubricant additives; improves pigment dispersion in printing inks; pigment and filler dispersants in natural and synthetic rubbers
Form: Liquid to solid
Toxicity: Skin and eye irritant
Warnings and Caution: Wear protective goggles, gloves, clothing, and preventive ointment

Lilamin 101 D. Lilachim S.A.
Chemical Description: Arachidyl behenyl primary amine (distilled)
Category and Applications: See Lilamin Series; primary amines also as mold release agents for hard rubber
Form: Flakes; Color: 3 max. (Gardner)
Properties: m.w.: ≈ 295; Iodine Value: 3 max.
Composition: 97% active ingred.

Lilamin 115. Lilachim S.A.
Chemical Description: Soyabean primary amine; also available in distilled grade (Lilamin 115 D)
Category and Applications: See Lilamin 101
Form: Liquid; Color: 5 max. (Gardner)
Properties: m.w.: ≈ 270; Iodine Value: 90 min.
Composition: 95% active ingred.

Lilamin 140. Lilachim S.A.
Chemical Description: Hydrogenated tallow primary amine; also available in distilled grade (Lilamin 140 D)
Category and Applications: See Lilamin 101; also for coating clays and esp. good as mold release agent for hard rubber
Form: Solid; Color: 3 max. (Gardner)
Properties: m.w.: ≈ 266; Sp. gr.: 0.795 @ 60 C; Visc.: 4.40 cps @ 60 C; M.P.: 46 C; Flash Pt.: 158 C (OC); Iodine Value: 3 max.
Composition: 95% active ingred.

Lilamin 142. Lilachim S.A.
Chemical Description: Stearyl primary amine; also available in distilled grade (Lilamin 142 D)

Category and Applications: See Lilamin 101
Form: Solid; Color: 3 max. (Gardner)
Properties: m.w.: ≈ 273; Iodine Value: 3 max.
Composition: 95% active ingred.

Lilamin 151. Lilachim S.A.
Chemical Description: Tall primary amine
Category and Applications: See Lilamin 101
Form: Liquid; Color: 5 max. (Gardner)
Properties: m.w.: ≈ 270; Iodine Value: 100 min.
Composition: 92% active ingred.

Lilamin 160. Lilachim S.A.
Chemical Description: Coconut primary amine; also available in distilled grade (Lilamin 160 D)
Category and Applications: See Lilamin 101
Form: Liquid; Color: 3 max. (Gardner)
Properties: m.w.: ≈ 205; Sp. gr.: 0.781 @ 60 C; Visc.: 2.75 cps @ 60 C; M.P.: 16 C; Flash Pt.: 112 C (OC); Iodine Value: 13 max.
Composition: 95% active ingred.

Lilamin 163. Lilachim S.A.
Chemical Description: Lauryl primary amine; also available in distilled grade (Lilamin 163 D)
Category and Applications: See Lilamin 101; 163 D also used in fungicides
Form: Liquid/solid; Color: 3 max. (Gardner)
Properties: m.w.: ≈ 190; Sp. gr.: 0.777 @ 60 C; Visc.: 1.80 cps @ 60 C; M.P.: 25 C; Flash Pt.: 110 C (OC); Iodine Value: 2 max.
Composition: 95% active ingred.

Lilamin 170. Lilachim S.A.
Chemical Description: Tallow primary amine; also available in distilled grade (Lilamin 170 D)
Category and Applications: See Lilamin 101; also for coating clays and as emulsifier for creams and beauty products
Form: Paste; Color: 3 max. (Gardner)
Properties: m.w.: ≈ 266; Sp. gr.: 0.799 @ 60 C; Visc.: 3.35 cps @ 60 C; M.P.: 32 C; Flash Pt.: 155 C (OC); Iodine Value: 38 min.
Composition: 95% active ingred.

Lilamin 172. Lilachim S.A.
Chemical Description: Oleyl primary amine; also available in distilled grade (Lilamin 172 D)
Category and Applications: See Lilamin 101
Form: Liquid; Color: 3 max. (Gardner)
Properties: m.w.: ≈ 270; Sp. gr.: 0.802 @ 60 C; Visc.: 3.20 cps @ 60 C; M.P.: 19 C; Flash Pt.: 155 C (OC); Iodine Value: 70 min.
Composition: 95% active ingred.

Lilamin 308 D. Lilachim S.A.
Chemical Description: Caprylyl dimethyl tertiary amine (distilled)
Category and Applications: See Lilamin Series; tertiary amines also as polymerization catalysts in

epoxy resins and polyurethane foams, and extraction reagents for uranium, vanadium, cobalt, etc.
Form: Liquid; Color: 2 max. (Gardner)
Properties: m.w.: ≈ 165; Iodine Value: 1 max.
Composition: 95% active ingred.

Lilamin 310 D. Lilachim S.A.
Chemical Description: Capryl dimethyl tertiary amine (distilled)
Category and Applications: See Lilamin 308 D
Form: Liquid; Color: 2 max. (Gardner)
Properties: m.w.: ≈ 195; Iodine Value: 1 max.
Composition: 95% active ingred.

Lilamin 312 D. Lilachim S.A.
Chemical Description: Lauryl dimethyl tertiary amine (distilled)
Category and Applications: See Lilamin 308 D
Form: Liquid; Color: 2 max. (Gardner)
Properties: m.w.: ≈ 220; Iodine Value: 1 max.
Composition: 95% active ingred.

Lilamin 314 D. Lilachim S.A.
Chemical Description: Myristyl dimethyl tertiary amine (distilled)
Category and Applications: See Lilamin 308 D
Form: Liquid; Color: 2 max. (Gardner)
Properties: m.w.: ≈ 250; Iodine Value: 2 max.
Composition: 95% active ingred.

Lilamin 316 D. Lilachim S.A.
Chemical Description: Palmityl dimethyl tertiary amine (distilled)
Category and Applications: See Lilamin 308 D
Form: Liquid; Color: 2 max. (Gardner)
Properties: m.w.: ≈ 280; Iodine Value: 2 max.
Composition: 95% active ingred.

Lilamin 342 D. Lilachim S.A.
Chemical Description: Stearyl dimethyl tertiary amine (distilled)
Category and Applications: See Lilamin 308 D
Form: Liquid; Color: 2 max. (Gardner)
Properties: m.w.: ≈ 305; Flash Pt.: 164 C (OC); Iodine Value: 2 max.
Composition: 95% active ingred.

Lilamin 343. Lilachim S.A.
Chemical Description: Dihydrogenated tallow methyl tertiary amine
Category and Applications: See Lilamin Series
Form: Solid; Color: 2 max. (Gardner)
Properties: m.w.: ≈ 525; M.P.: 30 C; Iodine Value: 3 max.
Composition: 95% active ingred.

Lilamin 345 D. Lilachim S.A.
Chemical Description: Hydrogenated tallow-dimethyl tertiary amine (distilled)
Category and Applications: See Lilamin 308 D
Form: Liquid; Color: 2 max. (Gardner)

Properties: m.w.: ≈ 295; Flash Pt.: 160 D (OC); Iodine Value: 3 max.
Composition: 95% active ingred.

Lilamin 363. Lilachim S.A.
Chemical Description: Trilauryl tertiary amine
Category and Applications: See Lilamin Series
Form: Liquid; Color: 2 max. (Gardner)
Properties: m.w.: ≈ 529
Composition: 95% active ingred.

Lilamin 364. Lilachim S.A.
Chemical Description: Tri (caprylyl capryl) tertiary amine
Category and Applications: See Lilamin Series
Form: Liquid; Color: 2 max. (Gardner)
Properties: m.w.: ≈ 390
Composition: 95% active ingred.

Lilamin 367 D. Lilachim S.A.
Chemical Description: Coconut dimethyl tertiary amine (distilled)
Category and Applications: See Lilamin 308 D
Form: Liquid; Color: 2 max. (Gardner)
Properties: m.w.: ≈ 230; Flash Pt.: 109 C (OC); Iodine Value: 15 max.
Composition: 95% active ingred.

Lilamin 368. Lilachim S.A.
Chemical Description: Tri (caprylyl capryl/lauryl) tertiary amine
Category and Applications: See Lilamin Series
Form: Liquid; Color: 2 max. (Gardner)
Properties: m.w.: ≈ 436
Composition: 95% active ingred.

Lilamin 369. Lilachim S.A.
Chemical Description: Dicoco methyl tertiary amine
Category and Applications: See Lilamin Series
Form: Liquid; Color: 2 max. (Gardner)
Properties: m.w.: ≈ 390; Iodine Value: 15 max.
Composition: 95% active ingred.

Lilamin 372 D. Lilachim S.A.
Chemical Description: Oleyl dimethyl tertiary amine (distilled)
Category and Applications: See Lilamin 308 D
Form: Liquid; Color: 2 max. (Gardner)
Properties: m.w.: ≈ 300; Iodine Value: 75 min.
Composition: 95% active ingred.

Lilamin 381. Lilachim S.A.
Chemical Description: Tri isooctyl tertiary amine
Category and Applications: See Lilamin Series
Form: Liquid; Color: 2 max. (Gardner)
Properties: m.w.: ≈ 362
Composition: 95% active ingred.

Lilamin 382. Lilachim S.A.
Chemical Description: Tri isodecyl tertiary amine
Category and Applications: See Lilamin Series

Form: Liquid; Color: 2 max. (Gardner)
Properties: m.w.: ≈ 450
Composition: 95% active ingred.

Lilamin 383. Lilachim S.A.
Chemical Description: Tri tridecyl tertiary amine
Category and Applications: See Lilamin Series
Form: Liquid; Color: 2 max. (Gardner)
Properties: m.w.: ≈ 567
Composition: 95% active ingred.

Lilamin 540. Lilachim S.A.
Chemical Description: Hydrogenated tallow diamine
Category and Applications: See Lilamin Series; diamines also as polymerization catalysts in epoxy resins, polyurethane foams
Form: Solid; Color: 5 max. (Gardner)
Properties: m.w.: ≈ 340; Iodine Value: 3 max.
Composition: 85% active ingred.

Lilamin 560. Lilachim S.A.
Chemical Description: Coconut diamine
Category and Applications: See Lilamin 540; also as corrosion inhibitor in crude oil distillation; pigment wetting agent in surface coatings
Form: Paste; Color: 6 max. (Gardner)
Properties: m.w.: ≈ 265; Sp. gr.: 0.812 @ 60 C; Visc.: 3.85 cps @ 60 C; M.P.: 25 C; Flash Pt.: ± 125 C (OC); Iodine Value: 10 max.
Composition: 85% active ingred.

Lilamin 570. Lilachim S.A.
Chemical Description: Tallow diamine
Category and Applications: See Lilamin 560; also in rapid-set emulsions in road building; corrosion inhibitor in laminating, wire-drawing, and cutting oils
Form: Paste; Color: 5 max. (Gardner)
Properties: m.w.: ≈ 340; Sp. gr.: 0.820 @ 60 C; Visc.: 4.05 cps @ 60 C; M.P.: 42 C; Flash Pt.: ± 190 C (OC); Iodine Value: 30–40
Composition: 85% active ingred.

Lilamin 570 DO. Lilachim S.A.
Chemical Description: Dioleate of Lilamin 570
Category and Applications: See Lilamin Series; also as pigment wetting aids, grinding aids, corrosion inhibitor for primers
Form: Paste; Color: 12 max. (Gardner)
Properties: m.w.: ≈ 885; Acid Value: 120; Iodine Value: 60 min.
Composition: 95% active ingred.

Lilamin 572. Lilachim S.A.
Chemical Description: Oleyl diamine
Category and Applications: See Lilamin 540
Form: Liquid; Color: 6 max. (Gardner)
Properties: m.w.: ≈ 350; Sp. gr.: 0.815 @ 60 C; Visc.: 3.95 cps @ 60 C; M.P.: 27 C; Flash Pt.: ± 195 C (OC); Iodine Value: 60 min.
Composition: 85% active ingred.

Lilamin AC. Lilachim S.A.
Chemical Description: Amines blend
Category: Anticaking aids, cationic
Applications: Anticaking aids and dust inhibitors for fertilizer manufacture
Form: Solid and liquid

Lilamin EO. Lilachim S.A.
Chemical Description: Ethoxylated amine derivatives
Category: Emulsifier, corrosion inhibitor, antistat, cationic
Applications: Crude oil extraction and drilling aids; gas additive; antistat for vinyl and plastics; hair care lotions; electrostatic paints; textile sizing aids, dye-fixing aids; corrosion inhibitor and lubricant in metallurgy, cutting oils

Lilamin LS. Lilachim S.A.
Applications: Series of fuel oil additives

Lilaminox. Lilachim S.A.
Chemical Description: Tertiary amine oxides
Applications: Gasoline additives; shampoos bases

Lilastat. Lilachim S.A.
Category: Antistatic agents
Applications: Antistatics for vinyl and polyamide plastics

Lipal 3 TD. PVO International
Chemical Description: Ethoxylated fatty ethers based on polyoxyethylene 3 tridecyl
Category: Emulsifier, solubilizer, wetting agent, dispersant, nonionic
Applications: See Lipal 4 LA
Form: Liquid; Color: 1 (Gardner)
Properties: Solubility: Sol. in isopropanol, propylene glycol, peanut oil, mineral; disp. in water; HLB: 8.0 ± 1; Cloud Pt.: 40–43 C (1% sol'n.); Hydroxyl No.: 165–175; Stability: Stable over a wide pH range
Composition: 100% active ingred.

Lipal 4 LA. PVO International
Chemical Description: Ethoxylated fatty ethers based on polyoxyethylene 4 lauryl
Category: Emulsifier, solubilizer, wetting agent, dispersant, nonionic
Applications: Compatability with both cationic and anionic compounds; broad range of hydrophilic and lipophilic properties; excellent emulsifier and solubilizer; used in emollient creams and lotions; make-up bases, hair preparations, clear gels, antiperspirants, ointment bases, shave creams, germicidal skin cleansers, quick breaking foams
Form: Liquid; Color: 1 (Gardner)
Properties: Solubility: Sol. in isopropanol, propylene glycol, peanut oil; disp. in water, mineral oil; HLB: 9.7 ± 1; Cloud Pt.: 52–53 C (1% sol'n.); Hydroxyl No.: 150–170; Stability: Stable over a wide pH range
Composition: 100% active ingred.

Lipal 5 L. PVO International
Chemical Description: Polyethylene glycol ester based on polyoxyethylene 5 laurate
Category: Emulsifier, solubilizer, wetting agent, dispersant, nonionic
Applications: See Lipal 300 W
Form: Liquid @ 20 C; Color: 2 (Gardner)
Properties: Solubility: Sol. in isopropanol; slightly sol. in peanut oil; disp. in water; HLB: 9.8 ± 1; Sapon. Value: 135–150; Acid No.: 5; Iodine Value: 8; Stability: Stable over a wide pH range
Composition: 100% active ingred.

Lipal 5 OA. PVO International
Chemical Description: Ethoxylated fatty ethers based on polyoxyethylene 5 oleyl
Category: Emulsifier, solubilizer, wetting agent, dispersant, nonionic
Applications: See Lipal 4 LA
Form: Liquid; Color: 3 (Gardner)
Properties: Solubility: Sol. in isopropanol, propylene glycol, peanut oil, mineral oil; HLB: 9.0 ± 1; Cloud Pt.: 57–60 C (1% sol'n.); Hydroxyl No.: 117–132; Stability: Stable over a wide pH range
Composition: 100% active ingred.

Lipal 5 S. PVO International
Chemical Description: Polyethylene glycol ester based on polyoxyethylene 5 stearate
Category: Emulsifier, solubilizer, wetting agent, dispersant, nonionic
Applications: See Lipal 300 W
Form: Solid @ 20 C; Color: 4 (Gardner)
Properties: Solubility: Sol. in isopropanol; slightly sol. in peanut oil; disp. in water; HLB: 6.8 ± 1; Sapon. Value: 105–120; Acid No.: 2; Iodine Value: 8; Stability: Stable over a wide pH range
Composition: 100% active ingred.

Lipal 6 TD. PVO International
Chemical Description: Ethoxylated fatty ethers based on polyoxyethylene 6 tridecyl
Category: Emulsifier, solubilizer, wetting agent, dispersant, nonionic
Applications: See Lipal 4 LA
Form: Liquid; Color: 1 (Gardner)
Properties: Solubility: Sol. in water, isopropanol, propylene glycol, mineral oil; HLB: 11.0 ± 1; Cloud Pt.: 66–69 C (1% sol'n.); Hydroxyl No.: 110–125; Stability: Stable over a wide pH range
Composition: 100% active ingred.

Lipal 9 C. PVO International
Chemical Description: Polyethylene glycol ester based on polyoxyethylene 9 castor oil
Category: Emulsifier, solubilizer, wetting agent, dispersant, nonionic
Applications: See Lipal 300 W
Form: Liquid @ 20 C; Color: 4 (Gardner)
Properties: Solubility: Sol. in isopropanol; disp. in water; HLB: 8.0 ± 1; Sapon. Value: 120–136; Acid

No.: 2; Iodine Value: 55–65; Stability: Stable over a wide pH range
Composition: 100% active ingred.

Lipal 9 L. PVO International
Chemical Description: Polyethylene glycol ester based on polyoxyethylene 9 laurate
Category: Emulsifier, solubilizer, wetting agent, dispersant, nonionic
Applications: See Lipal 300 W
Form: Transitional @ 20 C; Color: 2 (Gardner)
Properties: Solubility: Sol. in water, isopropanol; slightly sol. in peanut oil; HLB: 13.0 ± 1; Sapon. Value: 89–104; Acid No.: 5; Iodine Value: 6; Stability: Stable over a wide pH range
Composition: 100% active ingred.

Lipal 9 LA. PVO International
Chemical Description: Ethoxylated fatty ethers based on polyoxyethylene 9 lauryl
Category: Emulsifier, solubilizer, wetting agent, dispersant, nonionic
Applications: See Lipal 4 LA
Form: Liquid; Color: 1 (Gardner)
Properties: Solubility: Sol. in water, isopropanol, propylene glycol; HLB: 13.6 ± 1; Cloud Pt.: 73–76 C (1% sol'n.); Hydroxyl No.: 90–110; Stability: Stable over a wide pH range
Composition: 100% active ingred.

Lipal 9 N. PVO International
Chemical Description: Ethoxylated fatty ethers based on polyoxyethylene 9 nonyl phenyl
Category: Emulsifier, solubilizer, wetting agent, dispersant, nonionic
Applications: See Lipal 4 LA
Form: Liquid; Color: 1 (Gardner)
Properties: Solubility: Sol. in water, isopropanol, propylene glycol; HLB: 15.0 ± 1; Cloud Pt.: 52–56 C (1% sol'n.); Hydroxyl No.: 85–100; Stability: Stable over a wide pH range
Composition: 100% active ingred.

Lipal 9 OL. PVO International
Chemical Description: Polyethylene glycol ester based on polyoxyethylene 9 oleate
Category: Emulsifier, solubilizer, wetting agent, dispersant, nonionic
Applications: See Lipal 300 W
Form: Liquid @ 20 C; Color: 4 (Gardner)
Properties: Solubility: Sol. in isopropanol; slightly sol. in peanut oil; disp. in water; HLB: 12.0 ± 1; Sapon. Value: 84–100; Acid No.: 4; Iodine Value: 45; Stability: Stable over a wide pH range
Composition: 100% active ingred.

Lipal 10 OA. PVO International
Chemical Description: Ethoxylated fatty ethers based on polyoxyethylene 10 oleyl
Category: Emulsifier, solubilizer, wetting agent, dispersant, nonionic
Applications: See Lipal 4 LA
Form: Transitional; Color: 3 (Gardner)
Properties: Solubility: Sol. in wa-

ter, isopropanol; HLB: 12.4 ± 1; Cloud Pt.: 48–52 C (1% sol'n.); Hydroxyl No.: 80–95; Stability: Stable over a wide pH range
Composition: 100% active ingred.

Lipal 10 TD. PVO International
Chemical Description: Ethoxylated fatty ethers based on polyoxyethylene 10 tridecyl
Category: Emulsifier, solubilizer, wetting agent, dispersant, nonionic
Applications: See Lipal 4 LA
Form: Solid; Color: 1 (Gardner)
Properties: Solubility: Sol. in water, isopropanol, propylene glycol; HLB: 13.8 ± 1; Cloud Pt.: 75–85 C (1% sol'n.); Hydroxyl No.: 80–95; Stability: Stable over a wide pH range
Composition: 100% active ingred.

Lipal 12 LA. PVO International
Chemical Description: Ethoxylated fatty ethers based on polyoxyethylene 12 lauryl
Category: Emulsifier, solubilizer, wetting agent, dispersant, nonionic
Applications: See Lipal 4 LA
Form: Solid; Color: 1 (Gardner)
Properties: Solubility: Sol. in water, isopropanol, propylene glycol; HLB: 14.8 ± 1; Cloud Pt.: 65–68 C (1% sol'n.); Hydroxyl No.: 72–87; Stability: Stable over a wide pH range
Composition: 100% active ingred.

Lipal 15 CSA. PVO International
Chemical Description: Ethoxylated fatty ethers based on polyoxyethylene 15 cetyl/stearyl
Category: Emulsifier, solubilizer, wetting agent, dispersant, nonionic
Applications: See Lipal 4 LA
Form: Solid; Color: 1 (Gardner)
Properties: Solubility: Sol. in water, isopropanol, propylene glycol; HLB: 14.4 ± 1; Cloud Pt.: 67–70 C (1% sol'n.); Hydroxyl No.: 56–68; Stability: Stable over a wide pH range
Composition: 100% active ingred.

Lipal 15 T. PVO International
Chemical Description: Polyethylene glycol ester based on polyoxyethylene 15 tallate
Category: Emulsifier, solubilizer, wetting agent, dispersant, nonionic
Applications: See Lipal 300 W
Form: Liquid @ 20 C; Color: 12 (Gardner)
Properties: Solubility: Sol. in water, isopropanol; HLB: 10.5 ± 1; Sapon. Value: 25–40; Acid No.: 2; Iodine Value: 30–45; Stability: Stable over a wide pH range
Composition: 100% active ingred.

Lipal 20 OA. PVO International
Chemical Description: Ethoxylated fatty ethers based on polyoxyethylene 20 oleyl
Category: Emulsifier, solubilizer, wetting agent, dispersant, nonionic
Applications: See Lipal 4 LA
Form: Solid; Color: 3 (Gardner)
Properties: Solubility: Sol. in water, isopropanol; HLB: 15.4 ± 1;

Cloud Pt.: 67–72 C (1% sol'n.); Hydroxyl No.: 47–62; Stability: Stable over a wide pH range
Composition: 100% active ingred.

Lipal 20 SA. PVO International
Chemical Description: Ethoxylated fatty ethers based on polyoxyethylene 20 stearyl
Category: Emulsifier, solubilizer, wetting agent, dispersant, nonionic
Applications: See Lipal 4 LA
Form: Solid; Color: 1 (Gardner)
Properties: Solubility: Sol. in water, isopropanol, propylene glycol; HLB: 15.3 ± 1; Cloud Pt.: 70–75 C (1% sol'n.); Hydroxyl No.: 50–70; Stability: Stable over a wide pH range
Composition: 100% active ingred.

Lipal 23 LA. PVO International
Chemical Description: Ethoxylated fatty ethers based on polyoxyethylene 23 lauryl
Category: Emulsifier, solubilizer, wetting agent, dispersant, nonionic
Applications: See Lipal 4 LA
Form: Solid; Color: 1 (Gardner)
Properties: Solubility: Sol. in water, isopropanol, propylene glycol; HLB: 16.9 ± 1; Cloud Pt.: 75–78 C (1% sol'n.); Hydroxyl No.: 40–55 Stability: Stable over a wide pH range
Composition: 100% active ingred.

Lipal 25 C. PVO International
Chemical Description: Polyethylene glycol ester based on polyoxyethylene 25 castor oil
Category: Emulsifier, solubilizer, wetting agent, dispersant, nonionic
Applications: See Lipal 300 W
Form: Liquid @ 20 C; Color: 4 (Gardner)
Properties: Solubility: Sol. in water, isopropanol; HLB: 14.5 ± 1; Sapon Value: 76–88; Acid No.: 2; Iodine Value: 35–45; Stability: Stable over a wide pH range
Composition: 100% active ingred.

Lipal 25 S. PVO International
Chemical Description: Polyethylene glycol ester based on polyoxyethylene 25 stearate
Category: Emulsifier, solubilizer, wetting agent, dispersant, nonionic
Applications: See Lipal 300 W
Form: Solid @ 20 C; Color: 4 (Gardner)
Properties: Solubility: Sol. in water, isopropanol, propylene glycol; slightly sol. in peanut oil; HLB: 14.0 ± 1; Sapon. Value: 35–47; Acid No.: 2; Iodine Value: 2; Stability: Stable over a wide pH range
Composition: 100% active ingred.

Lipal 30 SA. PVO International
Chemical Description: Ethoxylated fatty ethers based on polyoxyethylene 30 stearyl
Category: Emulsifier, solubilizer, wetting agent, dispersant, nonionic
Applications: See Lipal 4 LA
Form: Solid; Color: 1 (Gardner)

Properties: Solubility: Sol. in water, isopropanol, propylene glycol; HLB: 17.0 ± 1; Cloud Pt.: 76–78 C (1% sol'n.); Hydroxyl No.: 28–43; Stability: Stable over a wide pH range
Composition: 100% active ingred.

Lipal 39 S. PVO International
Chemical Description: Polyethylene glycol ester based on polyoxyl 40 stearate USP
Category: Emulsifier, solubilizer, wetting agent, dispersant, nonionic
Applications: See Lipal 300 W
Form: Solid @ 20 C; Color: 4 (Gardner)
Properties: Solubility: Sol. in water, isopropanol; disp. in propylene glycol, peanut oil; HLB: 16.9 ± 1; Sapon. Value: 23–35; Acid No.: 2; Iodine Value: 2; Stability: Stable over a wide pH range
Composition: 100% active ingred.

Lipal 50 OA. PVO International
Chemical Description: Ethoxylated fatty ethers based on polyoxyethylene 50 oleyl
Category: Emulsifier, solubilizer, wetting agent, dispersant, nonionic
Applications: See Lipal 4 LA
Form: Solid; Color: 3 (Gardner)
Properties: Solubility: Sol. in water, isopropanol, propylene glycol; HLB: 18.5 ± 1; Cloud Pt.: 75–78 C (1% sol'n.); Hydroxyl No.: 29–33; Stability: Stable over a wide pH range
Composition: 100% active ingred.

Lipal 50 S. PVO International
Chemical Description: Polyethylene glycol ester based on polyoxyethylene 50 stearate
Category: Emulsifier, solubilizer, wetting agent, dispersant, nonionic
Applications: See Lipal 300 W
Form: Solid @ 20 C; Color: 2 (Gardner)
Properties: Solubility: Sol. in water, isopropanol; HLB: 17.9 ± 1; Sapon. Value: 18–30; Acid No.: 1; Iodine Value: 1; Stability: Stable over a wide pH range
Composition: 100% active ingred..

Lipal 52 C. PVO International
Chemical Description: Polyethylene glycol ester based on polyoxyethylene 52 castor oil
Category: Emulsifier, solubilizer, wetting agent, dispersant, nonionic
Applications: See Lipal 300 W
Form: Solid @ 20 C; Color: 4 (Gardner)
Properties: Solubility: Sol. in water, isopropanol, propylene glycol; HLB: 16.0 ± 1; Sapon. Value: 46–58; Acid No.: 2; Iodine Value: 20–30; Stability: Stable over a wide pH range

Lipal 300 DL. PVO International
Chemical Description: Polyethylene glycol ester based on PEG 300 dilaurate
Category: Emulsifier, solubilizer, wetting agent, dispersant, nonionic
Applications: See Lipal 300 W
Form: Liquid @ 20 C; Color: 1

(Gardner)
Properties: Solubility: Sol. in isopropanol, peanut oil, mineral oil; disp. in water; HLB: 6.3 ± 1; Sapon. Value: 148–162; Acid No.: 8; Iodine Value: 6; Stability: Stable over a wide pH range
Composition: 100% active ingred.

Lipal 300 W. PVO International
Chemical Description: Polyethylene glycol ester based on PEG monooleate
Category: Emulsifier, solubilizer, wetting agent, dispersant, nonionic
Applications: Compatability with both cationic and anionic compounds; broad range of hydrophilic and lipophilic properties; excellent emulsifier and solubilizer; used in emollient creams and lotions, shampoos, make-up bases, shave creams, germicidal skin cleansers, hair preparations, clear gels
Form: Liquid @ 20 C; Color: 5 (Gardner)
Properties: Solubility: Sol. in isopropanol, mineral oil; slightly sol. in peanut oil; disp. in water; HLB: 9.2 ± 1; Sapon. Value: 102–115; Acid No.: 6; Iodine Value: 55; Stability: Stable over a wide pH range
Composition: 100% active ingred.

Lipal 400 DL. PVO International
Chemical Description: Polyethylene glycol ester based on PEG 400 dilaurate
Category: Emulsifier, solubilizer, wetting agent, dispersant, nonionic
Applications: See Lipal 300 W
Form: Liquid @ 20 C; Color: 4 (Gardner)
Properties: Solubility: Sol. in isopropanol, peanut oil, mineral oil; disp. in water; HLB: 10.4 ± 1; Sapon. Value: 125–140; Acid No.: 10; Iodine Value: 8; Stability: Stable over a wide pH range
Composition: 100% active ingred.

Lipal 400 DS. PVO International
Chemical Description: Polyethylene glycol ester based on PEG 400 distearate
Category: Emulsifier, solubilizer, wetting agent, dispersant, nonionic
Applications: See Lipal 300 W
Form: Solid @ 20 C; Color: 4 (Gardner)
Properties: Solubility: Sol. in ispropanol, peanut oil, mineral oil; disp. hot in water; HLB: 7.2 ± 1; Sapon. Value: 122–132; Iodine Value: 3; Stability: Stable over a wide pH range
Composition: 100% active ingred.

Lipal 400 DW. PVO International
Chemical Description: Polyethylene glycol ester based on 400 dioleate
Category: Emulsifier, solubilizer, wetting agent, dispersant, nonionic
Applications: See Lipal 400 W
Form: Liquid
Composition: 100% active ingred.

Lipal 400 OL. PVO International
Chemical Description: Polyethylene glycol ester based on PEG 400 monooleate
Category: Emulsifier, solubilizer, wetting agent, dispersant, nonionic
Applications: See Lipal 300 W
Form: Liquid @ 20 C; Color: 13 (Gardner)
Properties: Solubility: Sol. in isopropanol; slightly sol. in peanut oil; disp. in water; HLB: 6.8 ± 1; Sapon. Value: 83–88; Acid No.: 6; Iodine Value: 47; Stability: Stable over a wide pH range
Composition: 100% active ingred.

Lipal 400 S. PVO International
Chemical Description: Polyethylene glycol ester based on PEG 400 monostearate
Category: Emulsifier, solubilizer, wetting agent, dispersant, nonionic
Applications: See Lipal 300 W
Form: Solid @ 20 C; Color: 1 (Gardner)
Properties: Solubility: Sol. in isopropanol; disp. in water; HLB: 11.3 ± 1; Sapon. Value: 82–92; Acid No.: 5; Iodine Value: 1; Stability: Stable over a wide pH range
Composition: 100% active ingred.

Lipal 400 W. PVO International
Chemical Description: Polyethylene glycol ester based on 400 monooleate
Category: Emulsifier, solubilizer, wetting agent, dispersant, nonionic
Applications: Emulsifier for W/O and O/W systems; used for cosmetic and pharmaceutical preparations
Form: Liquid
Composition: 100% active ingred.

Lipal 600 S. PVO International
Chemical Description: Polyethylene glycol ester based on 600 monostearate
Category: Emulsifier, solubilizer, wetting agent, dispersant, nonionic
Applications: See Lipal 400 W
Form: Solid
Composition: 100% active ingred.

Lipal 600 W. PVO International
Chemical Description: Polyethylene glycol ester based on 600 monooleate
Category: Emulsifier, solubilizer, wetting agent, dispersant, nonionic
Applications: See Lipal 400 W
Form: Liquid
Composition: 100% active ingred.

Lipal 610. PVO International
Chemical Description: Ethoxylated fatty ethers based on polyoxyethylene 8 tridecyl
Category: Emulsifier, solubilizer, wetting agent, dispersant, nonionic
Applications: See Lipal 4 LA
Form: Liquid; Color: 1 (Gardner)
Properties: Solubility: Sol. in water, isopropanol, propylene glycol; HLB: 12.5 ± 1; Cloud Pt.: 42–46 C

(1% sol'n.); Hydroxyl No.: 85–100; Stability: Stable over a wide pH range
Composition: 85% active ingred.

Lipal CE 38. PVO International
Chemical Description: Ethoxylated polyol POE 5 coconut ester
Category: Emulsifier, solubilizer, wetting agent, coupling agent, dispersant, nonionic
Applications: Compatability with both cationic and anionic compounds; broad range of hydrophilic and lipophilic properties; excellent emulsifier and solubilizer; used in emollient creams and lotions, shampoos, shave creams, bath oil spreading agents, hair preparations
Form: Transitional; Color: 3 red (Lovibond)
Properties: Solubility: Sol. in isopropanol, carnation mineral oil, trichloroethylene, alcohol; HLB: 7.6 ± 1; Sapon. Value: 130–150; Stability: Extremely stable; not effected by electrolytes
Composition: 100% active ingred.

Lipal CE 55. PVO International
Chemical Description: Ethoxylated polyol POE 11 coconut ester
Category: Emulsifier, solubilizer, wetting agent, coupling agent, dispersant, nonionic
Applications: See Lipal CE 38
Form: Transitional; Color: 3 red (Lovibond)
Properties: Solubility: Sol. in isopropanol, carnation mineral oil, trichloroethylene, alcohol; HLB: 11.0 ± 1; Sapon. Value: 90–115; Stability: Extremely stable; not effected by electrolytes
Composition: 100% active ingred.

Lipal CE 64. PVO International
Chemical Description: Ethoxylated polyol POE 16 coconut ester
Category: Emulsifier, solubilizer, wetting agent, coupling agent, dispersant, nonionic
Applications: See Lipal CE 38
Form: Transitional; Color: 3 red (Lovibond)
Properties: Solubility: Sol. in water, isopropanol, trichloroethylene, carnation mineral oil, alcohol; HLB: 12.8 ± 1; Sapon. Value: 70–100; Stability: Extremely stable; not effected by electrolytes
Composition: 100% active ingred.

Lipal CE 71. PVO International
Chemical Description: Ethoxylated polyol POE 21 coconut ester
Category: Emulsifier, solubilizer, wetting agent, coupling agent, dispersant, nonionic
Applications: See Lipal CE 38
Form: Transitional; Color: 3 red (Lovibond)
Properties: Solubility: Sol. in water, isopropanol, trichloroethylene, alcohol; HLB: 14.2 ± 1; Sapon. Value: 50–80; Stability: Extremely stable; not effected by electrolytes
Composition: 100% active ingred.

Lipal OE 55. PVO International
Chemical Description: Ethoxylated polyol POE 12 safflower ester
Category: Emulsifier, solubilizer,

wetting agent, coupling agent, dispersant, nonionic
Applications: See Lipal CE 38
Form: Liquid; Color: 6 red (Lovibond)
Properties: Solubility: Sol. in water, isopropanol, trichloroethylene, carnation mineral oil, alcohol; HLB: 11.0 ± 1; Sapon. Value: 70–90; Stability: Extremely stable; not effected by electrolytes

Lipal OE 64. PVO International
Chemical Description: Ethoxylated polyol POE 17 safflower ester
Category: Emulsifier, solubilizer, wetting agent, coupling agent, dispersant, nonionic
Applications: See Lipal CE 38
Form: Liquid; Color: 6 red (Lovibond)
Properties: Solubility: Sol. in water, isopropanol, trichloroethylene, carnation mineral oil, alcohol; HLB: 12.8 ± 1; Sapon. Value: 55–75; Stability: Extremely stable; not effected by electrolytes

Lipal OE 70. PVO International
Chemical Description: Ethoxylated polyol POE 23 safflower ester
Category: Emulsifier, solubilizer, wetting agent, coupling agent, dispersant, nonionic
Applications: See Lipal CE 38
Form: Liquid; Color: 6 red (Lovibond)
Properties: Solubility: Sol. in water, isopropanol, trichloroethylene, carnation mineral oil, alcohol; HLB: 14.0 ± 1; Sapon. Value: 42–62; Stability: Extremely stable; not effected by electrolytes

Lipal TE 43. PVO International
Chemical Description: Ethoxylated polyol POE 8 tallow ester
Category: Emulsifier, solubilizer, wetting agent, coupling agent, dispersant, nonionic
Applications: See Lipal CE 38
Form: Liquid; Color: 6 red (Lovibond)
Properties: Solubility: Sol. in isopropanol, trichloroethylene, alcohol, carnation mineral oil; HLB: 8.6 ± 1; Sapon. Value: 90–115; Stability: Extremely stable; not effected by electrolytes
Composition: 100% active ingred.

Lipal TE 55. PVO International
Chemical Description: Ethoxylated polyol POE 12 tallow ester
Category: Emulsifier, solubilizer, wetting agent, coupling agent, dispersant, nonionic
Applications: See Lipal CE 38
Form: Transitional; Color: 6 red (Lovibond)
Properties: Solubility: Sol. in water, isopropanol, trichloroethylene, carnation mineral oil, alcohol; HLB: 11.0 ± 1; Sapon. Value: 70–100; Stability: Extremely stable; not effected by electrolytes
Composition: 100% active ingred.

Lipal TE 70. PVO International
Chemical Description: Ethoxylated polyol POE 22 tallow ester
Category: Emulsifier, solubilizer, wetting agent, coupling agent, dispersant, nonionic
Applications: See Lipal CE 38
Form: Soft solid; Color: 6 red (Lovibond)

Properties: Solubility: Sol. in water, isopropanol, trichloroethylene, carnation mineral oil, alcohol; HLB: 14.0 ± 1; Sapon. Value: 40–65; Stability: Extremely stable; not effected by electrolytes
Composition: 100% active ingred.

Lipal TE 76. PVO International
Chemical Description: Ethoxylated polyol POE 30 tallow ester
Category: Emulsifier, solubilizer, wetting agent, coupling agent, dispersant, nonionic
Applications: See Lipal CE 38
Form: Soft solid; Color: 6 red (Lovibond)
Properties: Solubility: Sol. in water, isopropanol, trichloroethylene, carnation mineral oil, alcohol; HLB: 15.2 ± 1; Sapon. Value: 30–60; Stability: Extremely stable; not effected by electrolytes
Composition: 100% active ingred.

Lipamide S. Lipo Chemicals Inc.
Chemical Description: 1:1 diethanolamide of stearic acid
Category: Emulsifier, thickener, emulsion stabilizer, lubricant, nonionic
Applications: Skin and hair care creams and lotions, shampoos, personal washing liquids, antiperspirants, and cold waves
Form: Wax; Color: Off-white; Odor: Mild, waxy
Properties: Solubility: Miscible hot with common oil phase ingred. and most organic solvents; disp. in hot water; M.P.: 41–46 C; Acid Value: 12–18; Stability: Stable over a broad pH range; excellent electrolyte tolerance; pH: 9–10.5 (1% aq.)
Composition: 100% active ingred.

Lipo DGLS. Lipo Chemicals Inc.
Chemical Description: PEG-2 laurate SE
Category and Applications: See Lipopeg 4-L
Form: Liquid; Color: Yellow
Properties: HLB: 8.3 ± 1; Sapon. Value: 160–170; Acid Value: 4 max.
Composition: 100% active ingred.

Lipo GMS 450. Lipo Chemicals Inc.
Chemical Description: Glyceryl stearate
Category: Emulsifier, emollient, opacifier, viscosity builder, nonionic
Applications: General purpose emulsifiers; emollient, opacifier, and viscosity builder in creams and lotions
Form: Bead or flake; Color: White
Properties: HLB: 3.6 ± 1; Sapon. Value: 169–182; Acid Value: 5 max.
Composition: 100% active ingred.

Lipo GMS 470. Lipo Chemicals Inc.
Chemical Description: Glyceryl stearate SE
Category and Applications: See Lipo GMS 450
Form: Bead or flake; Color: White
Properties: HLB: 5.8 ± 1; Sapon. Value: 138–152; Acid Value: 5 max.
Composition: 100% active ingred.

Lipocol B. Lipo Chemicals Inc.
Chemical Description: PEG-9 stearate and PEG-9 laurate
Category: Spreading agent, emulsifier, dispersant, lubricant, nonionic
Applications: Bath oils, creams, lotions
Form: Solid wax; Color: White to off-white
Properties: HLB: 11.0 ± 1; Sapon. Value: 94–104; Acid Value: 5 max.
Composition: 100% active ingred.

Lipocol C-2. Lipo Chemicals Inc.
Chemical Description: Ceteth-2
Category: Emulsifier, defoamer, wetting agent, solubilizer, conditioning agent, nonionic
Applications: Acid and alkaline stable nonionic surfactant for use in antiperspirants, depilatories, creams, lotions, and pigment dispersions
Form: Solid wax; Color: White
Properties: HLB: 5.3 ± 1; Acid Value: 1 max.; Hydroxyl Value: 160–180
Composition: 100% active ingred.

Lipocol C-10. Lipo Chemicals Inc.
Chemical Description: Ceteth-10
Category and Applications: See Lipocol C-2
Form: Solid wax; Color: White
Properties: HLB: 12.9 ± 1; Acid Value: 1 max.; Hydroxyl Value: 75–90
Composition: 100% active ingred.

Lipocol C-20. Lipo Chemicals Inc.
Chemical Description: Ceteth-20
Category and Applications: See Lipocol C-2
Form: Solid wax; Color: White
Properties: HLB: 15.7 ± 1; Acid Value: 2 max.; Hydroxyl Value: 50–58
Composition: 100% active ingred.

Lipocol L-1. Lipo Chemicals Inc.
Chemical Description: Laureth-1
Category and Applications: See Lipocol C-2
Form: Liquid; Color: Colorless
Properties: HLB: 3.6 ± 1; Acid Value: 2 max.; Hydroxyl Value: 231–243
Composition: 100% active ingred.

Lipocol L-4. Lipo Chemicals Inc.
Chemical Description: Laureth-4
Category and Applications: See Lipocol C-2
Form: Liquid; Color: Colorless
Properties: HLB: 9.7 ± 1; Acid Value: 0.5 max.; Hydroxyl Value: 145–160
Composition: 100% active ingred.

Lipocol L-12. Lipo Chemicals Inc.
Chemical Description: Laureth-12
Category and Applications: See Lipocol C-2
Form: Solid wax; Color: White
Properties: HLB: 14.5 ± 1; Acid Value: 1 max.; Hydroxyl Value: 72–87
Composition: 100% active ingred.

Lipocol L-23. Lipo Chemicals Inc.
Chemical Description: Laureth-23
Category and Applications: See Lipocol C-2
Form: Solid wax; Color: White
Properties: HLB: 16.9 ± 1; Acid Value: 0.5 max.; Hydroxyl Value: 42–52
Composition: 100% active ingred.

Lipocol O-2. Lipo Chemicals Inc.
Chemical Description: Oleth-10
Category and Applications: See Lipocol C-2
Form: Liquid; Color: Yellow
Properties: HLB: 4.9 ± 1; Acid Value: 1 max.; Hydroxyl Value: 160–180
Composition: 100% active ingred.

Lipocol O-10. Lipo Chemicals Inc.
Chemical Description: Oleth-10
Category and Applications: See Lipocol C-2
Form: Liquid; Color: Yellow
Properties: HLB: 12.4 ± 1; Acid Value: 2 max.; Hydroxyl Value: 74–84
Composition: 100% active ingred.

Lipocol S-2. Lipo Chemicals Inc.
Chemical Description: Steareth-2
Category and Applications: See Lipocol C-2
Form: Solid wax; Color: White
Properties: HLB: 4.9 ± 1; Acid Value: 1 max.; Hydroxyl Value: 155–165
Composition: 100% active ingred.

Lipocol S-10. Lipo Chemicals Inc.
Chemical Description: Steareth-10
Category and Applications: See Lipocol C-2
Form: Solid wax; Color: White
Properties: HLB: 12.4 ± 1; Acid Value: 1 max.; Hydroxyl Value: 75–90
Composition: 100% active ingred.

Lipocol S-20. Lipo Chemicals Inc.
Chemical Description: Steareth-20
Category and Applications: See Lipocol C-2
Form: Solid wax; Color: White
Properties: HLB: 15.3 ± 1; Acid Value: 1 max.; Hydroxyl Value: 45–60
Composition: 100% active ingred.

Lipocol SC-15. Lipo Chemicals Inc.
Chemical Description: Ceteareth-15
Category and Applications: See Lipocol C-2
Form: Solid wax; Color: White
Properties: HLB: 14.3 ± 1; Acid Value: 2 max.; Hydroxyl Value: 50–65
Composition: 100% active ingred.

Lipocol TD-12. Lipo Chemicals Inc.
Chemical Description: Trideceth-12
Category and Applications: See Lipocol C-2
Form: Paste; Color: White
Properties: HLB: 14.6 ± 1; Acid Value: 1 max.; Hydroxyl Value: 70–85
Composition: 100% active ingred.

Lipodan. Grindsted Prod.
Chemical Description: Distilled monoglyceride specialties
Category: Emulsifier, nonionic
Applications: Food emulsifier for recombined and filled cream; improves aeration, enhances whippability, stabilizes cream during storage
Form: Flakes
Composition: 100% active ingred.

Lipolan 31. Lipo Chemicals Inc.
Chemical Description: 24 mole polyethylene glycol ether of the lanolin alcohol complex formed by the controlled hydrogenation of lanolin
Category: Emulsifier, solubilizer, nonionic
Applications: O/W emulsifier; cosmetics, toiletries, and topical pharmaceuticals; emollient and conditioner; skin and hair care products, makeup bases, antiperspirants, depilatories, cream rinses and bleaches; solubilizer for shampoos, hair grooms, aftershaves, and skin fresheners
Form: Waxy solid; Color: Cream; Odor: Bland, characteristic
Properties: Solubility: Sol. in water and ethanol; insol. in mineral oil; Sapon. Value: 8 max.; Acid Value: 2 max.; Hydroxyl Value: 35–55; Stability: Stable at high and low pH; good electrolyte tolerance
Composition: 100% active ingred.
Containers: 450 lb net open head steel drums

Lipomulse 165. Lipo Chemicals Inc.
Chemical Description: Glyceryl stearate and PEG-100 stearate
Category and Applications: See Lipo GMS 450
Form: Bead or flake; Color: White
Properties: HLB: 11.0 ± 1; Sapon. Value: 90–100; Acid Value: 2 max.
Composition: 100% active ingred.

Liponate CL. Lipo Chemicals Inc.
Chemical Description: Cetyl lactate
Category: Emollient, thickener, viscosity control agent
Applications: Emollient for obtaining desired feel and penetration in a finished cream, lotion, or bath preparation; also used for thickening and viscosity control
Form: Soft solid to liquid; Color: White
Properties: Sapon. Value: 174–195; Acid Value: 3 max.

Liponate GC. Lipo Chemicals Inc.
Chemical Description: Caprylic/capric triglyceride
Category and Applications: See Liponate CL
Form: Liquid; Color: Colorless
Properties: Sapon. Value: 325–355; Acid Value: 0.1 max.

Liponate IPM. Lipo Chemicals Inc.
Chemical Description: Isopropyl myristate
Category and Applications: See Liponate CL

Form: Liquid; Color: Colorless
Properties: Sapon. Value: 202–211; Acid Value: 2 max.

Liponate IPP. Lipo Chemicals Inc.
Chemical Description: Isopropyl palmitate
Category and Applications: See Liponate CL
Form: Liquid; Color: Colorless
Properties: Sapon. Value: 183–190; Acid Value: 2 max.

Liponate ML. Lipo Chemicals Inc.
Chemical Description: Myristyl lactate
Category and Applications: See Liponate CL
Form: Liquid to solid; Color: White to yellow
Properties: Sapon. Value: 165–182; Acid Value: 5 max.

Liponate MM. Lipo Chemicals Inc.
Chemical Description: Myristyl myristate
Category and Applications: See Liponate CL
Form: Solid wax; Color: White
Properties: Sapon. Value: 120–135; Acid Value: 5 max.

Liponate PC. Lipo Chemicals Inc.
Chemical Description: Propylene glycol dicaprylate/dicaprate
Category and Applications: See Liponate CL
Form: Liquid; Color: Colorless
Properties: Sapon. Value: 315–335; Acid Value: 0.1 max.

Liponate PO-4. Lipo Chemicals Inc.
Chemical Description: Pentaerythritol tetraoleate
Category and Applications: See Liponate CL
Form: Liquid; Color: Yellow
Properties: Sapon. Value: 185–195; Acid Value: 10 max.

Liponate PS-4. Lipo Chemicals Inc.
Chemical Description: Pentaerythritol tetrastearate
Category and Applications: See Liponate CL
Form: Flake; Color: White
Properties: Sapon. Value: 183–198; Acid Value: 10 max.

Liponate SPS. Lipo Chemicals Inc.
Chemical Description: Synthetic spermaceti
Category and Applications: See Liponate CL
Form: Flake; Color: White to cream
Properties: Sapon. Value: 109–120; Acid Value: 5 max.

Liponate SS. Lipo Chemicals Inc.
Chemical Description: Stearyl stearate
Category and Applications: See Liponate CL
Form: Flake; Color: Off-white
Properties: Sapon. Value: 103–117; Acid Value: 5 max.

Liponic EG-1. Lipo Chemicals Inc.
Chemical Description: 26 mole ethylene oxide adduct of glycerin
Category: Humectant, lubricant, nonionic

Applications: Preelectric shaves, aftershaves, skin fresheners; humectant in creams and lotions; dispersing medium for makeup pigments; plasticizer for hair setting and styling resins; foam modifier and slip enhancer in soap, detergent systems, and shaving creams
Form: Clear to slightly hazy viscous liquid; Color: Colorless; Odor: Practically odorless
Properties: Solubility: Sol. in water, alcohol, acetone, and ethyl acetate; Acid Value: 0.5 max.; Hydroxyl Value: 128–138
Composition: 100% active ingred.
Containers: 500 lb net closed head steel drums

Lipopeg 4-L. Lipo Chemicals Inc.
Chemical Description: PEG-8 laurate
Category: Spreading agent, emulsifier, dispersant, lubricant, nonionic
Applications: Mild fatty esters for use in bath oils, creams, lotions
Form: Liquid; Color: Yellow
Properties: HLB: 13.0 ± 1; Sapon. Value: 90–100; Acid Value: 5 max.
Composition: 100% active ingred.

Lipopeg 4-S. Lipo Chemicals Inc.
Chemical Description: PEG-8 stearate
Category and Applications: See Lipopeg 4-L
Form: Paste; Color: Cream
Properties: HLB: 11.2 ± 1; Sapon. Value: 80–90; Acid Value: 5 max.
Composition: 100% active ingred.

Lipopeg 6-L. Lipo Chemicals Inc.
Chemical Description: PEG-12 laurate
Category and Applications: See Lipopeg 4-L
Form: Liquid; Color: Yellow
Properties: HLB: 14.6 ± 1; Sapon. Value: 65–76; Acid Value: 5 max.
Composition: 100% active ingred.

Lipopeg 39-S. Lipo Chemicals Inc.
Chemical Description: PEG-40 stearate
Category and Applications: See Lipopeg 4-L
Form: Solid wax; Color: White
Properties: HLB: 16.9 ± 1; Sapon. Value: 23–35; Acid Value: 1 max.
Composition: 100% active ingred.

Lipopeg 100-S. Lipo Chemicals Inc.
Chemical Description: PEG-100 stearate
Category and Applications: See Lipopeg 4-L
Form: Flake or bead; Color: Tan
Properties: HLB: 18.8 ± 1; Sapon. Value: 9–20; Acid Value: 1 max.
Composition: 100% active ingred.

Lipopeg 6000-DS. Lipo Chemicals Inc.
Chemical Description: PEG-150 distearate
Category and Applications: See Lipopeg 4-L
Form: Flake; Color: Off-white
Properties: HLB: 18.4 ± 1; Sapon. Value: 12–20; Acid Value: 10 max.
Composition: 100% active ingred.

Liposorb L. Lipo Chemicals Inc.
Chemical Description: Sorbitan laurate
Category: Emulsifier, thickener, lubricant, antistat, nonionic
Applications: All-purpose lipophilic surfactants used in conjunction with the polyoxyethylated Liposorb series
Form: Liquid; Color: Amber
Properties: HLB: 8.6 ± 1; Sapon. Value: 158–170; Hydroxyl Value: 330–360
Composition: 100% active ingred.

Liposorb L-10. Lipo Chemicals Inc.
Chemical Description: PEG-10 sorbitan laurate
Category: Emulsifier, lubricant, antistat, nonionic
Applications: All-purpose hydrophilic surfactants used for solubilizing oils and in conjunction with Liposorb esters
Form: Liquid; Color: Yellow
Properties: HLB: 14.9 ± 1; Sapon. Value: 66–76; Hydroxyl Value: 150–170
Composition: 100% active ingred.

Liposorb L-20. Lipo Chemicals Inc.
Chemical Description: Polysorbate 20
Category and Applications: See Liposorb L-10
Form: Liquid; Color: Yellow
Properties: HLB: 16.7 ± 1; Sapon. Value: 40–50; Hydroxyl Value: 96–108
Composition: 100% active ingred.

Liposorb O. Lipo Chemicals Inc.
Chemical Description: Sorbitan oleate
Category and Applications: See Liposorb L
Form: Liquid; Color: Amber
Properties: HLB: 4.3 ± 1; Sapon. Value: 145–160; Hydroxyl Value: 193–210
Composition: 100% active ingred.

Liposorb O-20. Lipo Chemicals Inc.
Chemical Description: Polysorbate 80
Category and Applications: See Liposorb L-10
Form: Liquid; Color: Yellow
Properties: HLB: 15.0 ± 1; Sapon. Value: 45–55; Hydroxyl Value: 65–80
Composition: 100% active ingred.

Liposorb P. Lipo Chemicals Inc.
Chemical Description: Sorbitan palmitate
Category and Applications: See Liposorb L
Form: Beads or flakes; Color: Tan
Properties: HLB: 6.7 ± 1; Sapon. Value: 139–151; Hydroxyl Value: 272–306
Composition: 100% active ingred.

Liposorb P-20. Lipo Chemicals Inc.
Chemical Description: Polysorbate 40
Category and Applications: See Liposorb L-10
Form: Liquid; Color: Yellow
Properties: HLB: 15.6 ± 1; Sapon.

Value: 40–53; Hydroxyl Value: 90–107
Composition: 100% active ingred.

Liposorb S. Lipo Chemicals Inc.
Chemical Description: Sorbitan stearate
Category and Applications: See Liposorb L
Form: Beads or flakes; Color: Cream
Properties: HLB: 4.7 ± 1; Sapon. Value: 147–157; Hydroxyl Value: 235–260
Composition: 100% active ingred.

Liposorb S-20. Lipo Chemicals Inc.
Chemical Description: Polysorbate 60
Category and Applications: See Liposorb L-10
Form: Paste; Color: Yellow
Properties: HLB: 14.9 ± 1; Sapon. Value: 45–55; Hydroxyl Value: 81--96
Composition: 100% active ingred.

Liposorb SQO. Lipo Chemicals Inc.
Chemical Description: Sorbitan sesquioleate
Category and Applications: See Liposorb L
Form: Liquid; Color: Amber
Properties: HLB: 3.7 ± 1; Sapon. Value: 149–160; Hydroxyl Value: 188–210
Composition: 100% active ingred.

Liposorb TO. Lipo Chemicals Inc.
Chemical Description: Sorbitan trioleate
Category and Applications: See Liposorb L
Form: Liquid; Color: Amber
Properties: HLB: 1.8 ± 1; Sapon. Value: 171–185; Hydroxyl Value: 58–69
Composition: 100% active ingred.

Liposorb TO-20. Lipo Chemicals Inc.
Chemical Description: Polysorbate 85
Category and Applications: See Liposorb L-10
Form: Liquid; Color: Yellow
Properties: HLB: 11.0 ± 1; Sapon. Value: 82–95; Hydroxyl Value: 39–52
Composition: 100% active ingred.

Liposorb TS. Lipo Chemicals Inc.
Chemical Description: Sorbitan tristearate
Category and Applications: See Liposorb L
Form: Beads or flakes; Color: Cream
Properties: HLB: 2.1 ± 1; Sapon. Value: 175–190; Hydroxyl Value: 65–80
Composition: 100% active ingred.

Liposorb TS-20. Lipo Chemicals Inc.
Chemical Description: Polysorbate 65
Category and Applications: See Liposorb L-10
Form: Solid wax; Color: Tan
Properties: HLB: 10.5 ± 1; Sapon. Value: 88–98; Hydroxyl Value: 44–60
Composition: 100% active ingred.

Lipovol A. Lipo Chemicals Inc.
Chemical Description: Avocado oil
Category: Emollient, lubricant
Applications: Natural emollient oil for use as lubricant and emollient in skin products, hair-care products, makeups, fine soaps, bath oils, and anhydrous systems
Form: Oil; Color: Yellow to green
Properties: Sapon. Value: 177–198; Acid Value: 3 max.; Iodine Value: 65–95

Lipovol ALM. Lipo Chemicals Inc.
Chemical Description: Sweet almond oil
Category and Applications: See Lipovol A
Form: Oil; Color: Yellow
Properties: Sapon. Value: 185–200; Acid Value: 2 max.; Iodine Value: 95–115

Lipovol CP. Lipo Chemicals Inc.
Chemical Description: Cherry pit oil
Category and Applications: See Lipovol A
Form: Oil; Color: Light amber
Properties: Sapon. Value: 182–202; Acid Value: 1 max.; Iodine Value: 95–115

Lipovol G. Lipo Chemicals Inc.
Chemical Description: Grape seed oil
Category and Applications: See Lipovol A
Form: Oil; Color: Yellow amber
Properties: Sapon. Value: 183–205; Iodine Value: 132–152

Lipovol J. Lipo Chemicals Inc.
Chemical Description: Jojoba oil
Category and Applications: See Lipovol A
Form: Oil; Color: Yellow
Properties: Sapon. Value: 85–110; Acid Value: 5 max.; Iodine Value: 75–95

Lipovol P. Lipo Chemicals Inc.
Chemical Description: Apricot kernel oil
Category and Applications: See Lipovol A
Form: Oil; Color: Yellow
Properties: Sapon. Value: 185–195; Acid Value: 1 max.; Iodine Value: 90–115

Lipovol SES. Lipo Chemicals Inc.
Chemical Description: Sesame oil
Category and Applications: See Lipovol A
Form: Oil; Color: Yellow
Properties: Sapon. Value: 188–195; Acid Value: 0.2 max.; Iodine Value: 103–116

Lipovol SO. Lipo Chemicals Inc.
Chemical Description: Hybrid safflower oil
Category and Applications: See Lipovol A
Form: Oil; Color: Yellow
Properties: Sapon. Value: 184–196; Acid Value: 1 max.; Iodine Value: 90–105

Lipovol WGO. Lipo Chemicals Inc.
Chemical Description: Wheat germ oil
Category and Applications: See Lipovol A

Form: Oil; Color: Yellow
Properties: Sapon. Value: 183–195; Acid Value: 2 max.; Iodine Value: 130–140

Lipowax C. Lipo Chemicals Inc.
Chemical Description: N,N′-ethylene-bis-stearamide
Category: Lubricant, antistat, nonionic
Applications: Internal lubricant in thermoplastic and thermosetting resins; imparts antitack, antiblocking, and antistatic properties; used in vinyls, polystyrene, polyethylene, ABS; mold release, lubricant, antiblock, and detackifying agent in synthetic rubbers; surface-finishing agent in hard rubber; in paints and lacquers to aid pigment suspension and salt spray resistance; FDA-approved for use in food packaging materials
Form: Prill, powder, or atomized powder; Color: Light cream
Properties: Solubility: Sol. in hot toluene, naphtha, kerosene, and turpentine; M.P.: 140–145 C; Flash Pt.: ≈ 285 C (COC); Acid Value: 10 max.; Stability: Resistant to water, acids, and alkalis
Composition: 100% active ingred.
Containers: 50 lb net bags and 200 lb net fiber drums (prill); 160 lb net fiber drums (powder and atomized powder)

Lipowax D. Lipo Chemicals Inc.
Chemical Description: High molecular wt saturated fatty alcohols and their ethylene oxide adducts
Category: Emulsifying wax, nonionic
Applications: O/W self-emulsifying wax, esp. for oil phases having HLB values of 9–13; skin and hair care creams and loitons; antiperspirants, depilatories, cream rinses, and opacified hair dyes and bleaches
Form: Waxy solid; Color: White to off-white; Odor: Bland, characteristic
Properties: Solubility: Sol. in alcohol; miscible warm with most oil phase ingred.; disp. in warm water; M.P.: 46–55 C; HLB: 11; Sapon. Value: 2 max.; Acid Value: 1 max.; Iodine Value: 2 max.; Stability: Stable over pH range normally found in cosmetics and toiletries; good electrolyte tolerance
Composition: 100% active ingred.

Lipowax P. Lipo Chemicals Inc.
Chemical Description: Balanced blend of lipophilic high molecular wt fatty alcohols and hydrophilic ethoxylated products
Category: Emulsifying wax, nonionic
Applications: O/W self-emulsifying wax for neutral and mildly acidic and alkaline pH systems; emulsifies emollient oils and waxes, esp. at HLB 7–11; hair and skin care creams and lotions; acid-mantle preparations, cream rinses, and antiperspirants
Form: Waxy solid; Color: Creamy white; Odor: Bland characteristic
Properties: Solubility: Sol. in alcohol; miscible warm with most oil phase ingred.; disp. in warm water; M.P.: 48–52 C; HLB: 9; Sa-

pon. Value: 14 max.; Hydroxyl Value: 178–192
Composition: 100% active ingred.

Lomar D. Diamond Shamrock
Chemical Description: Highly polymerized naphthalene sulfonate
Category: Dispersant, anionic
Applications: Dispersing agent for carbon black pigments in latex systems
Form: Powder; Color: Clear amber
Properties: pH: 9.3
Composition: 84% active ingred.
Containers: 50 or 75 lb paper bags or 250 lb net wt fiber drum
References: Diamond Shamrock product bulletin

Lomar LS. Diamond Shamrock
Chemical Description: Sodium salt of a condensed mono naphthalene sulfonic acid
Category: Dispersant, anionic
Applications: Dispersant in agricultural formulations for low concentration of electrolytes
Form: Powder
Composition: 95% active ingred.

Lomar PL. Diamond Shamrock
Chemical Description: Sodium salt of a condensed mono naphthalene sulfonic acid
Category: Dispersant, emulsifier, anionic
Applications: Recommended in any application involving dispersion of solid particles
Form: Liquid; Color: Amber
Properties: Solubility: Clear (10% aq. sol'n.); pH: 9.5 (10% aq. sol'n.)
Composition: 46% active ingred.
Containers: 55 gal fiber drum
References: Diamond Shamrock product bulletin

Lomar PW. Diamond Shamrock
Chemical Description: Sodium neutralized, condensed naphthalene sulfonic acid
Category: Dispersant, anionic
Applications: Dispersant of solid particles in aqueous media; reduces the viscosity of solids when dispersed in water; valuable aid in ceramics, especially in molding operations; viscosity suppressant in the cold styrene-butadiene rubber process; used as a dispersant in gypsum board, emulsion paints; successfully disperses and eliminates pitch from paper pulp; viscosity reducer for pigment slurries; dispersant and stabilizer in the synthetic rubber polymerization process; in wet milling, it ruptures agglomerates and improves the grinding action
Form: Powder; Color: Tan
Properties: Solubility: Sol. in hard or soft water; pH: 9.5 (10% aq. sol'n.)
Composition: 87% active ingred.
Containers: 50 or 75 lb paper bags or 250 lb net wt drums
References: Diamond Shamrock product bulletin

Lomar PWA. Diamond Shamrock
Chemical Description: Ammonia neutralized, condensed naphthalene sulfonic acid
Category: Ammonia neutralized

dispersant, viscosity depressant, anionic
Applications: Promotes higher slurry solids, particularly for molding and extruding in ceramic operations; highly recommended as a dispersant for emulsion paints; extensively used as a dispersant, stabilizer and viscosity depressant in the emulsion polymerization of synthetic elastomers, especially SBR; efficient viscosity reducer for pigment slurries; easily added to dry adhesives, joining compounds, spackles and sealants; it improves working properties and in-place water resistance
Form: Powder; Color: Light Tan
Properties: Solubility: Sol. in hard or soft water; pH: 7.2 (10% aq. sol'n.)
Composition: 92% active ingred.
Containers: 55 gal fiber drums or paper bags
References: Diamond Shamrock product bulletin

Lonzaine C. Lonza Inc.
Chemical Description: Cocoamido betaine
Category: Foaming agent, conditioner, viscosity booster, wetting agent, amphoteric
Applications: Cosmetic/toiletries incl. shampoos and hair conditioners; alkaline and acid cleaners; light and heavy duty detergents; metal finishing; electroplating; textile finishing and dyeing; lime soap dispersing; disinfectant/sanitizer cleaners and antibacterial scrubs
Form: Nonviscous liquid; Color: 2.0 max. (Gardner); Odor: Characteristic
Properties: Stability: Good tolerance to high levels of detergent builders, electrolytes, alkali, and acids; Surface Tension: 33.6 dynes/cm (0.1%); pH: 4.5–5.5 (10% sol'n.) ; Biodegradable
Composition: 30% actives

Lonzaine CO. Lonza Inc.
Chemical Description: Cocoamido betaine
Category and Applications: See Lonzaine C
Form: Nonviscous liquid; Color: 2.0 max. (Gardner)
Properties: Density: 8.77 lb/gal; Sp. gr.: 1.051; Stability: Good tolerance to high levels of detergent builders, electrolytes, alkali, and acids; Surface Tension: 33.9 dynes/cm (0.1%); pH: 6–8 (10% sol'n.); Biodegradable
Composition: 30% actives

Lonzest 143-S. Lonza Inc.
Chemical Description: Myristyl propionate
Category: Emollient, penetrant, solvent, nonionic
Applications: Emollient, penetrant and spreading agent in cosmetics; perfume solvent in formulation of bath oils, antiperspirants, spray deodorants, preshave lotions, sunburn and tanning lotions; also in waterless hand cleaners, makeup and lipstick removers, facial cleansers
Form: Clear liquid; Color: 5–10

(APHA)
Properties: Ref. Index: 1.433; Solubility: Sol. in acetone, carbon tetrachloride, castor oil, cottonseed oil, ethanol, ethyl acetate, heptane, isopropanol, lanolin oil, peanut oil, silicone oils, toluene; Sp. gr.: 0.852; Visc.: 5 cps; F.P.: –5 C
Composition: 100% active ingred.

Lonzest PEG-4DO. Lonza Inc.
Chemical Description: Polyethylene glycol 400 dioleate
Category: Detergent, lubricant, dispersant, anticorrosive, emulsifier, defoamer, nonionic
Applications: Solvent emulsions, degreasers, dry cleaning detergents; lubricants and wool oil additives for textiles; rust preventative oils, degreasers, and cutting oils for metals; lubricants, sludge dispersants, anticorrosive and water emulsifier for turbine oils in petroleum industry; insecticide/agricultural sprays; fur and leather treatment; defoamer for glue and proteins
Form: Liquid; Color: Yellow
Properties: Solubility: Sol. in mineral oil, isopropanol, toluol; disp. in water; Sp. gr.: 0.98; M.P.: 0 C; HLB: 9.9; Sapon. No.: 120–130; Acid No.: 5–10; Iodine Value: 45–55
Composition: 100% conc.

Lonzest PEG 4-L. Lonza Inc.
Chemical Description: Polyethylene glycol 400 monolaurate
Category: Dispersant, rewetting agent, stabilizer, nonionic
Applications: Polish removers for cosmetics and pharmaceuticals; in deodorant sprays; clarifying agent for rosin soaps, lime soaps; rewetting agent for sanforizing, wool dyeing, acetate dye dispersant in textile industry; rewetting agent and softener for toweling and tissues in paper industry; emulsion paints, inks, latex stabilizers; pigment dispersant; antiblocking agent for cellulose acetate, vinyls, protein, rubber, and waxes; in animal feed formulations
Form: Liquid; Color: Yellow
Properties: Solubility: Sol. in water, isopropanol, toluol; Sp. gr.: 1.03; M.P.: 10 C; HLB: 13.1; Sapon. No.: 90–100; Acid No.: 5–10; Iodine Value: 7–8
Composition: 100% conc.

Lonzest PEG-4-O. Lonza Inc.
Chemical Description: Polyethylene glycol 400 monooleate
Category: Detergent, emulsifier, nonionic
Applications: Polish remover in cosmetics and pharmaceuticals; solvent emulsions; degreasers, dry cleaning detergents; rust preventative oils and degreasers for metals; in wax emulsions
Form: Liquid; Color: Yellow
Properties: Solubility: Sol. in isopropanol, toluol; partly sol. in mineral oil; disp. in water; Sp. gr.: 1.01; M.P.: 0 C; HLB: 11.4; Sapon. No.: 86–96; Acid No.: 5–10; Iodine Value: 31–40
Composition: 100% conc.

Lonzol LA-300. Lonza Inc.
Chemical Description: Ammonium lauryl sulfate
Category: Foaming agent, emulsifier, detergent, anionic
Applications: Personal care products incl. shampoos, bubble baths with good cleansing and mildness; industrial and household products incl. rug and upholstery shampoos, all purpose cleaners, emulsion polymerization
Form: Viscous liquid
Properties: pH: 6.0–7.0 (10% sol'n.)
Composition: 27–30% active

Lonzol LS-300. Lonza Inc.
Chemical Description: Sodium lauryl sulfate
Category and Applications: See Lonzol LA-300
Form: Clear liquid
Properties: pH: 7.3–8.5 (10% sol'n.)
Composition: 28–30% active

Lonzol LT-400. Lonza Inc.
Chemical Description: Triethanolamine lauryl sulfate
Category and Applications: See Lonzol LA-300
Form: Clear liquid
Properties: pH: 7.0–8.0 (10% sol'n.)
Composition: 38–42% active

Loraderm 8120. Dutton & Reinisch Ltd.
Chemical Description: Special triglyceride
Category: Neutral oil base, solubilizer
Applications: Neutral oil base for cosmetic creams and lotions; solubilizer for active ingredients
Form: Liquid
Composition: 100% active ingred.

Loraderm S 1333. Dutton & Reinisch Ltd.
Chemical Description: Di-sodium sulfosuccinate of a specially selected vegetable fatty acid
Category: Detergent, anionic
Applications: Used for washing-up liquids; liquid and powder detergents; all types of shampoos; foam bath; skin creams; face lotions; make-up foundations; all types of cleaners; drilling- and cutting-oils
Form: Liquid; Color: Amber
Properties: pH: 6.5–7.5 (5% solids)
Composition: 40% active ingred.

Loraderm SB/LAN 5. Dutton & Reinisch Ltd.
Chemical Description: Wool wax alcohol sulfosuccinate
Category: Detergent, anionic
Applications: Moisturizing skin protective agent for creams, lotions, and cleansing formulations
Form: Liquid; Color: Brown
Properties: Solubility: slightly sol. in alcohols, glycols and other solvents; pH: 6.5–7.5 (5% solids); biodegradable
Composition: 50% active ingred.

Loraderm SBLAN/5. Dutton & Reinisch Ltd.
Chemical Description: Sodium salt of the sulfosuccinate of wool wax alcohol

Category: Detergent, anionic
Applications: Moisturizing, skin-protective agent for creams, lotions, and cleansing formulations
Form: Paste; Color: Yellow; Odor: Typical (slight)
Properties: Sp. gr.: ≈ 1.0
Composition: 50% active ingred.
Reaction: Acid (slightly)
Containers: Lacquered drums

Loraderm SBLAN/E. Dutton & Reinisch Ltd.
Chemical Description: Sodium salt of the mono-sulfosuccinate of wool wax alcohol (pure grade)
Category: Detergent, anionic
Applications: Moisturizing, skin-protective agent for creams, lotions, and cleansing formulations
Form: Paste; Color: Yellow; Odor: Typical (slight)
Composition: 50% active ingred.
Reaction: Acid (slightly)
Containers: Lacquered drums

Loramine AC 28. Dutton & Reinisch Ltd.
Chemical Description: Groundnut fatty acid polydialkanolamide
Category: Detergent, emulsifier, nonionic
Applications: Used as an anti-corrosion additive for emulsifiable and nonemulsifiable mineral-oil-based metal-working lubricants, including drilling-, cutting-, and grinding-oils
Form: Liquid; Color: Brown
Properties: Density: 0.98 g/cm^3 @ 20 C; Sp. gr.: ≈ 1.0; Visc.: medium; Acid No.: 14–16.5
Composition: 100% active ingred.
Reaction: Alkaline
Containers: Mild steel drums

Loramine B 204. Dutton & Reinisch Ltd.
Chemical Description: Coconut fatty acid alkylamido propyldimethylamine oxide
Category: Foam booster, detergent, softener, nonionic
Applications: Highly effective foam booster in many wash-active formulations; increases volume of foam; improves foam stability; has excellent emulsifying and calcium soap dispersing properties; in shampoos it produces a good conditioning effect on the hair
Form: Liquid; Color: Pale yellow
Properties: pH: 5–7 (10% solids); Biodegradable
Composition: 35% active ingred.

Loramine C 212. Dutton & Reinisch Ltd.
Chemical Description: Coconut fatty acid monoethanolamide
Category: Detergent, nonionic
Applications: Can be incorporated in all types of detergent products to provide: improvement of foam quantity, stability and texture; increased viscosity; superfatting effect/skin protection; improved detergency; stabilizer of emulsions
Form: Flakes; Color: Off-white; Odor: Amidic (slight)
Properties: M.P.: ≈ 73 C
Composition: 100% active ingred.
Reaction: Alkaline
Containers: Paper sacks

Loramine DAS. Dutton & Reinisch Ltd.
Chemical Description: Special surfactant
Category: Disinfectant cleaning agent
Form: Liquid
Composition: 50% active ingred.

Loramine DC 212. Dutton & Reinisch Ltd.
Chemical Description: Specialized fatty acid polydiethanolamide
Category: Lubricating additive, corrosion inhibitor, nonionic
Form: Liquid
Composition: 100% active ingred.

Loramine DC 212/S. Dutton & Reinisch Ltd.
Chemical Description: Coconut fatty acid diethanolamide
Category: Detergent, nonionic
Applications: Foam stabilizer and viscosity modifier used for shampoos and detergents
Form: Liquid; Color: Light yellow; Odor: Typical (slight)
Properties: Density: 0.99 g/cm^3 @ 20 C; Visc.: Medium
Composition: 100% active ingred.
Reaction: Alkaline
Containers: Mild steel drums

Loramine DC 212/SE. Dutton & Reinisch Ltd.
Chemical Description: Coconut fatty acid diethanolamine
Category: Detergent, foam stabilizer, detergent builder, nonionic
Applications: Cosmetic shampoos and bath preparations including personal hygiene toiletries
Form: Liquid
Composition: 100% active ingred.

Loramine DC 220/SE. Dutton & Reinisch Ltd.
Chemical Description: Coconut fatty acid diethanolamide
Category: Detergent, nonionic
Applications: Thickening agent, foam stabilizer, superfatting agent
Form: Soft paste
Composition: 100% active ingred.

Loramine DC 280. Dutton & Reinisch Ltd.
Chemical Description. Oleic acid polydiethanolamide
Category: Detergent, nonionic
Applications: Wide range of liquid and paste products in the cosmetic, cleaning-agent, and detergent industry
Form: Liquid; Color: Amber; Odor: Typical (slight)
Properties: Sp. gr.: ≈ 1.0; Visc.: Medium
Composition: 100% active ingred.
Reaction: Alkaline

Loramine DL 203. Dutton & Reinisch Ltd.
Chemical Description: Lauric acid polydialkanolamide
Category: Detergent, nonionic
Applications: Increases foam quality, quantity, and stability; increases viscosity; good superfatting; improves washing power; stabilization of emulsions; fixation of perfumes

Form: Wax; Color: White
Properties: Density: 0.97 g/cm^3 @ 20 C; Sp. gr.: ≈ 1.0; Visc.: High
Composition: 100% active ingred.
Reaction: Alkaline

Loramine DL 203/S. Dutton & Reinisch Ltd.
Chemical Description: Lauric acid diethanolamide
Category: Detergent, nonionic
Applications: Foam stabilizer and viscosity modifier for shampoos and detergents
Form: Solid; Color: Off-white; Odor: Typical (slight)
Properties: M.P.: ≈ 40 C
Composition: 100% active ingred.
Reaction: Alkaline
Containers: Mild steel drums

Loramine DL 240. Dutton & Reinisch Ltd.
Chemical Description: Coconut fatty acid polydiethanolamide
Category: Detergent, nonionic
Applications: Can be added to liquid and paste cosmetic preparations and all types of cleaning- and washing-agents to produce the following improvements: increased viscosity; improved foam quality, quantity, and stability; stabilization of emulsions; superfatting; improved washing power
Form: Liquid: Color: Yellow; Odor: Typical (faint)
Properties: Visc.: Medium
Composition: 100% active ingred.
Reaction: Alkaline
Containers: Mild steel drums

Loramine DLM/SE. Dutton & Reinisch Ltd.
Chemical Description: Fatty acid diethanolamide
Category: Detergent, nonionic
Applications: Thickening and superfatting raw material for cosmetic preparations
Form: Wax
Composition: 100% active ingred.

Loramine DO 280/S. Dutton & Reinisch Ltd.
Chemical Description: Groundnut fatty acid diethanolamide
Category: Detergent, nonionic
Applications: Foam stabilizer and viscosity modifier for shampoos and detergents
Form: Liquid; Color: Pale brown; Odor: Typical (slight)
Properties: Visc.: Medium
Composition: 100% active ingred.
Reaction: Alkaline
Containers: Mild steel drums

Loramine DO 280/SE. Dutton & Reinisch Ltd.
Chemical Description: Oleic acid diethanolamide
Category: Detergent, nonionic
Applications: Recommended for cosmetic products for which improved skin-compatibility, foam quality, and viscosity characteristics are required
Form: Liquid; Color: Honey
Properties: Density: 0.96 g/cm^3 @ 20 C
Composition: 100% active ingred.

Loramine DU 185. Dutton & Reinisch Ltd.
Chemical Description: Undecylenic acid polydialkanolamide
Category: Detergent, emulsifier, nonionic
Applications: Used as bacteriocide, thickening, foam stabilizing agent in shampoos
Form: Liquid; Color: Yellow
Properties: Solubility: Sol. in water, oil, alcohols, glycols; Sp. gr.: ≈ 1.0; Visc.: High
Composition: 100% active ingred.
Reaction: Acid
Containers: Lacquered drums

Loramine IM-BT. Dutton & Reinisch Ltd.
Chemical Description: Imidazoline derivate
Category: Adhesive additive, corrosion inhibitor, emulsifier, cationic
Applications: Adhesive additive for bitumen; adhesive agent and emulsifier in normal acidic bitumen emulsions; rust inhibitor in acidic media
Form: Liquid; Color: Green-brown, 13 max. (Gardner); Odor: Amine
Properties: Density: 0.94 g/cm^3 @ 20 C
Composition: 100% active ingred.
Reaction: Alkaline
Toxicity: Avoid prolonged contact with skin
Containers: Iron containers

Loramine IM-CA. Dutton & Reinisch Ltd.
Chemical Description: 1-Hydroxyethyl-2-heptadecenyl imidazoline
Category: Corrosion inhibitor, emulsifier, penetrant, wetting agent, additive, cationic
Applications: Leather industry—used in pickling baths as a corrosion inhibitor; Metal-working industry—used as an additive for corrosion inhibiting oils, in colling-fluids, and in cutting oils; small quantities will improve in the penetration and binding power of paint or dyes in the priming of wood and metal surfaces; wetting agent for carbonization baths; additive for the preparation of glass fiber
Form: Wax; Color: Yellow
Properties: m.w.: 285; Solubility: Sol. in most polar and nonpolar solvents; disp. in water; Stability: Stable in neutral and acidic media
Composition: 100% active ingred.
Reaction: Alkaline
Toxicity: Avoid prolonged contact with skin
Containers: Iron containers
Warnings and Caution: Should be stored in air-tight containers as it is a tertiary amine and can react with carbon dioxide

Loramine IM-OA. Dutton & Reinisch Ltd.
Chemical Description: 1-Hydroxyethyl-2-heptadecenyl imidazoline
Category: Rust inhibitor, adhesive agent, coupling agent, emulsifier, detergent, antistat, raw material, cationic
Applications: Can be used whenever inhibition of corrosion and/or an improved adhesive effect is re-

quired; in heating oils, fatty acid combination provides good rust prevention; used as a coupling agent where rust prevention is required; in pickling baths, used as an inhibitor to reduce attack on the steel or iron surface and to reduce hydrogen brittleness; excellent cationic emulsifier for the emulsification of oils and bitumens; acidic solutions can be prepared which are suitable for the formulations of metal cleaners, toilet bowl cleaners, rust removers; antistatic effect when dissolved in aqueous and alcoholic solutions; raw material for quaternization reactions; suitable for the flotation of phosphates, potassium, feldspar, iron, and other ores; used in the paint industry for pigment flushing and to improve coupling during priming of surfaces
Form: Liquid; Color: Amber
Properties: m.w.: 345; Solubility: Sol. in most polar and nonpolar solvents; disp. in water; Density: 0.94 g/cm^3 @ 20 C; Visc.: 800 cps; Stability: High thermal stability; pH: 10–11 (1% aq. sol'n.)
Composition: 100% active ingred.
Reaction: Alkaline
Toxicity: Avoid prolonged contact with skin
Containers: Iron containers
Warnings and Caution: This product will tend to crystallize at a rate dependent on length and temperature of storage

Loramine IM-OD. Dutton & Reinisch Ltd.
Chemical Description: 1-hydroxyethyl-2-alkyl-imidazoline
Category: Corrosion inhibitor, hydrophobing agent
Form: Liquid
Composition: 100% active ingred.

Loramine IPE 280. Dutton & Reinisch Ltd.
Chemical Description: Oleic monoisopropanolamide
Category: Detergent, nonionic
Applications: Skin-protecting additive for liquid and paste products; in addition to its excellent super-fatting properties, it helps to produce a fine-structured and creamy foam
Form: Semisolid; Color: Light yellow; Odor: Typical (slight)
Composition: 100% active ingred.
Reaction: Alkaline
Containers: Lacquered drums

Loramine IPL 203. Dutton & Reinisch Ltd.
Chemical Description: Lauric monoisopropanolamide
Category: Detergent, nonionic
Applications: Additive for solid and paste end-products, whose properties are as follows: increased foam quality and stability; increased viscosity; good superfatting; improved washing power; stabilization of emulsions
Form: Flakes; Color: White; Odor: Typical (slight)
Properties: M.P.: ∿ 55 C
Composition: 100% active ingred.
Reaction: Alkaline
Containers: Paper sacks

Loramine IPP 240. Dutton & Reinisch Ltd.
Chemical Description: Coconut fatty acid monoisopropanolamide
Category: Detergent, nonionic
Applications: See Loramine IPL 203
Form: Flakes; Color: White; Odor: Typical (slight)
Properties: M.P.: ∿ 50 C
Composition: 100% active ingred.
Reaction: Alkaline
Containers: Mild steel drums

Loramine L203. Dutton & Reinisch Ltd.
Chemical Description: Lauric monoethanolamide
Category: Detergent, nonionic
Applications: Suitable for all types of solid, paste, and powder detergent preparations; improvement of foam texture, volume, and stability; increased viscosity; superfatting effect/skin protection; fixation of perfumes; stabilization of emulsions; improvement of detergency
Form: Flakes; Color: White; Odor: Typical (slight)
Properties: M.P.: ∿ 84 C
Composition: 100% active ingred.
Reaction: Alkaline
Containers: Paper sacks

Loramine L 408. Dutton & Reinisch Ltd.
Chemical Description: Lauryl dimethylamineoxide
Category: Detergent, nonionic
Applications: Foam booster for shampoos and foam baths, hair conditioner
Form: Liquid
Composition: 30% active ingred.

Loramine MRC. Dutton & Reinisch Ltd.
Chemical Description: Mixture of fatty acid alkanolamides
Category: Corrosion inhibitor, detergent
Applications: Corrosion inhibitor; aluminum cleaners; increases the power of cleaning formulations
Form: Liquid
Composition: 100% active ingred.

Loramine RE. Dutton & Reinisch Ltd.
Chemical Description: Unadulterated combination of nonionic fatty acid blends
Category: Anticorrosion additive, emulsifier, nonionic
Applications: Additive for emulsifiable metal-working lubricants such as drilling-, cutting-, and grinding oils; inhibits corrosion; good emulsifying power; mild to the skin; increases lubricity; low toxicity
Form: Liquid; Color: Dark brown
Properties: Density: 0.98 g/cm^3 @ 20 C; Biodegradable
Composition: 100% active ingred.

Loramine S 280. Dutton & Reinisch Ltd.
Chemical Description: Stearic acid monoethanolamide
Category: Detergent, nonionic
Applications: Excellent super-fatting effect; increased mildness to

the skin; improved foam quality; increased viscosity; improved washing power; stabilization of emulsions
Form: Flakes; Color: White; Odor: Typical (slight)
Properties: M.P.: ∿ 92 C
Composition: 100% active ingred.
Reaction: Alkaline
Containers: Paper sacks

Loramine SBU 185. Dutton & Reinisch Ltd.
Chemical Description: Sodium salt of the mono-sulfosuccinate of undecylenic acid monoethanolamide
Category: Fungicide, detergent, foaming agent, anionic
Applications: Exhibits exceptional antiseborrheoic and antiallergic properties; can be combined with all anionic, nonionic, and amphoteric products without any reduction in its activity; used in foambaths, shampoos, skin-cleaners, foot baths, washing agents
Form: Liquid; Color: Pale yellow; Odor: Typical (slight)
Properties: Solubility: Sol. in alcohol, glycol; Sp. gr.: ∿ 1.0; Visc.: Medium; Stability: Stable at room temperature; pH: 6.5–7.5 (5% solids); Biodegradable
Composition: 50% active ingred.
Reaction: Acid (slight)
Containers: Lacquered drums

Loramine U 185. Dutton & Reinisch Ltd.
Chemical Description: Monoethanolamide of undecylenic acid
Category: Fungicide, detergent, nonionic
Applications: Improves foam quality and stability, super-fatting and increases viscosity; fully compatible with all anionic, cationic, amphoteric, and other nonionic surface-active agents
Form: Flakes; Color: Yellow; Odor: Typical (strong)
Properties: Solubility: Sol. in alcohol, glycol; disp. in warm water; M. P.: ∿ 50 C
Composition: 100% active ingred.
Reaction: Alkaline

Lorapal C 6. Dutton & Reinisch Ltd.
Chemical Description: Coconut amide polyglycolether
Category: Dispersant, wetting agent, foaming agent, nonionic
Applications: Excellent calcium soap dispersing and wetting properties; used advantageously as a solvent for perfumes in shampoo and foam-bath formulations; improves foam quality when used in combination with other surface-active raw materials; can be easily combined with all anionic, amphoteric, cationic, and nonionic materials; can be used for the formulation of spray- and immersion bath-cleaners
Form: Liquid; Color: Amber
Properties: Cloud Pt.: 80–90 C (2% in distilled water); Stability: Stable in dilute acid and alkaline solutions; pH: 8–10 (1% solids); Biodegradable
Composition: 100% active ingred.

Lorapal HV 5. Dutton & Reinisch Ltd.
Chemical Description: Nonylphenol polyglycolether
Category: Emulsifier, raw material, wetting agent, degreasing agent, nonionic
Applications: Used as an emulsifier for preparations of all sorts of emulsions and insecticides; raw material for rinsing, washing, and cleaning agents of all types; wetting agent in the paper and cellulose industry; raw material for textile treatment; degreasing agent in pickling and alkaline immersion-bath cleaners
Form: Liquid; Color: 2 max. (Gardner)
Properties: Solubility: Sol. in mineral oil, alcohol, ketone, toluene, xylene, chlorinated hydrocarbons; Cloud Pt.: 59–62 C (10% in 25% aq. butyl diglycol); Stability: Stable in the presence of dilute acids and alkalis, in hard water, and oxidizing agents; pH: 5–7 (1% solids)
Composition: 100% active ingred.

Lorapal HV 8. Dutton & Reinisch Ltd.
Chemical Description: Nonylphenol polyglycolether
Category: Emulsifier, raw material, wetting agent, degreasing agent, nonionic
Applications: See Lorapal HV 5
Form: Liquid; Color: 2 max. (Gardner)
Properties: Solubility: Sol. in mineral oil, alcohol, ketone, toluene, xylene, chlorinated hydrocarbons; Cloud Pt.: 41–45 C (10% in 25% aq. butyl diglycol); Stability: Stable in the presence of dilute acids and alkalis, in hard water, and oxidizing agents; pH: 5–7 (1% solids)
Composition: 100% active ingred.

Lorapal HV 9. Dutton & Reinisch Ltd.
Chemical Description: Nonylphenol polyglycolether
Category: Emulsifier, raw material, wetting agent, degreasing agent, nonionic
Applications: See Lorapal HV 5
Form: Liquid; Color: 2 max. (Gardner)
Properties: Solubility: Sol. in mineral oil, alcohol, ketone, toluene, xylene, chlorinated hydrocarbons; Cloud Pt.: 59–62 C (2% in distilled water); Stability: Stable in the presence of dilute acids and alkalis, in hard water, and oxidizing agents; pH: 5–7 (1% solids)
Composition: 100% active ingred.

Lorapal HV 10. Dutton & Reinisch Ltd.
Chemical Description: Nonylphenol polyglycolether
Category: Emulsifier, raw material, wetting agent, degreasing agent, nonionic
Applications: See Lorapal HV 5
Form: Liquid; Color: 2 max. (Gardner)
Properties: Solubility: Sol. in mineral oil, alcohol, ketone, toluene, xylene, chlorinated hydrocarbons; Cloud Pt.: 70–73 C (2%

in distilled water); Stability: Stable in the presence of dilute acids and alkalis, in hard water, and oxidizing agents; pH: 5--7 (1% solids)
Composition: 100% active ingred.

Lorapal HV 14. Dutton & Reinisch Ltd.
Chemical Description: Nonylphenol polyglycolether
Category: Emulsifier, raw material, wetting agent, degreasing agent, nonionic
Applications: See Lorapal HV 5
Form: Soft wax; Color: White
Properties: Solubility: Sol. in mineral oil, alcohol, ketone, toluene, xylene, chlorinated hydrocarbons; Cloud Pt.: 60–63 C (2% in 10% salt sol'n.); Stability: Stable in the presence of dilute acids and alkalis, in hard water, and oxidizing agents; pH: 5–7 (1% solids)
Composition: 100% active ingred.

Lorapal HV 25. Dutton & Reinisch Ltd.
Chemical Description: Nonylphenol polyglycolether
Category: Emulsifier, raw material, wetting agent, degreasing agent, nonionic
Applications: See Lorapal HV 5
Form: Wax; Color: White
Properties: Solubility: Sol. in mineral oil, alcohol, ketone, toluene, xylene, chlorinated hydrocarbons; Cloud Pt.: 73–76 C (2% in 10% salt sol'n.); Stability: Stable in the presence of dilute acids and alkalis, in hard water, and oxidizing agents; pH: 5–7 (1% solids)
Composition: 100% active ingred.

Lorapal LA 3. Dutton & Reinisch Ltd.
Chemical Description: Fatty alcohol polyglycolether
Category: Emulsifier, coupling agent, raw material, nonionic
Applications: Because of its low level of ethoxylation, this product has a definite hydrophobic character; ideal raw material for the production of various ether sulfates with excellent foaming properties; can be easily incorporated into sulfosuccinate-based shampoos to increase the viscosity and to provide a mild super-fatting effect
Form: Liquid; Color: Clear
Properties: Solubility: Sol. in mineral oil and many organic solvents; Hydroxyl No.: 165–180; pH: 5--7 (1% solids); Biodegradable

Lorapal M 365. Dutton & Reinisch Ltd.
Chemical Description: Rolling- and hydraulic-oil additive
Category: Raw material, high-pressure additive, lubricant, corrosion inhibitor, nonionic
Applications: Used as an additive for water-soluble metal working liquids; for rolling- and hydraulic-oil bathes on polyglycolether derivatives; raw material for metalworking preparations, such as screw-thread cutting lubricants, ball-bearing polishes; widely used as a component in many rolling-oil formulations; compatible with anionic, cationic, amphoteric, and other nonionic products over a

wide pH range
Form: Liquid; Color: Amber
Properties: Solubility: Sol. in water; Cloud Pt.: 37–43 C (2% solids in distilled water); pH: 7–9 (1% solids); Biodegradable
Composition: 100% active ingred.

Lorapal O 8. Dutton & Reinisch Ltd.
Chemical Description: Oleic acid amide polyglycolether
Category: Wetting and dispersing agent, emulsifier, detergent, nonionic
Applications: Excellent wetting and calcium soap dispersing agent; good oil-in-water emulsifier; can be combined with all anionic, cationic, amphoteric, and nonionic materials; suitable for machine washing formulations
Form: Liquid; Color: Brown
Properties: Solubility: Sol. in water; fairly sol. in alcohol, ketones, esters, chlorinated hydrocarbons, benzene, fatty oils; sparingly sol. in toluene, xylene; Cloud Pt.: 75–85 C; Stability: Stable in dilute acid and alkaline solutions; pH: 8–10 (1% solids); Biodegradable
Composition: 100% active ingred.

Lorapal TA 11. Dutton & Reinisch Ltd.
Chemical Description: Fatty alcohol polyglycolether
Category: Wash-active base, nonionic
Applications: Has excellent wetting-, dispersing-, emulsifying-, cleaning-power
Form: Soft wax; Color: White
Properties: Solubility: Sol. in water and many organic solvents; Cloud Pt.: 69–73 C (2% in distilled water); Stability: Stable in dilute acids, alkalines, hard water solutions; pH: 5–7 (1% solids); Biodegradable

Lorapal TA 25. Dutton & Reinisch Ltd.
Chemical Description: Fatty alcohol polyglycolether
Category: Wash-active base, nonionic
Applications: See Lorapal TA 11
Form: Powder; Color: White
Properties: Solubility: Sol. in water and many organic solvents; Cloud Pt.: 75–90 C (2% in 10% salt sol'n.); Stability: Stable in dilute acids, alkalines, hard water solutions; pH: 5–7 (1% solids); Biodegradable

Lorapal TA 50. Dutton & Reinisch Ltd.
Chemical Description: Fatty alcohol polyglycolether
Form: Powder; Color: White
Properties: Cloud Pt.: 72–76 (2% in 10% salt sol'n.); pH: 5–7 (1% solids); Biodegradable

Loraphat E 1027. Dutton & Reinisch Ltd.
Chemical Description: Alkylphenol polyglycolether phosphate
Category: Corrosion inhibitor, emulsifier, dispersant, wetting agent, anionic
Applications: Used in pure oil for

cutting, rolling, and extruding metals; reduces friction, prevents wear, cleans and has antistatic properties
Form: Liquid; Color: Yellow
Properties: Acid No.: 100 ± 5; Stability: Stable in neutral and alkaline solutions
Composition: 100% active ingred.

Loraphat EAK 8190. Dutton & Reinisch Ltd.
Chemical Description: Fatty alcohol polyglycolether phosphate
Category: Corrosion inhibitor, emulsifier, dispersant, wetting agent, anionic
Applications: See Loraphat E 1027
Form: Liquid; Color: Yellow
Properties: Solubility: Disp. in water; Acid No.: 150 ±10; Stability: Stable in neutral and alkaline solutions; Biodegradable
Composition: 100% active ingred.

Lorapon AM-2C. Dutton & Reinisch Ltd.
Chemical Description: Coconut based ampholyte
Category: Detergent, dispersant, foam booster, wetting agent, amphoteric
Applications: Stabilizes the foam, and acts as a calcium soap dispersant when used in hard water; used in toiletries, mainly in mild shampoo and foam-bath preparations, and especially in baby products; compatible with all anionic, cationic, and nonionic wash active substances
Form: Liquid; Color: Golden-yellow
Properties: Biodegradable
Composition: 49% active ingred.

Lorapon AM-2L. Dutton & Reinisch Ltd.
Chemical Description: Imidazoline-based ampholyte
Category: Detergent, amphoteric
Applications: Mild foam baths, shampoos, person hygiene toiletries
Form: Liquid
Composition: 50% active ingred.

Lorapon AM-B 13. Dutton & Reinisch Ltd.
Chemical Description: Alkyl amido betaine
Category: Detergent, foam booster, amphoteric
Applications: Suitable for the formulation of shampoos, foam baths, and other toiletries; can be combined with anionic wash active agents to give clear solutions; has a strong foam stabilizing effect even in the presence of oil and soap; compatible with all anionic, cationic, nonionic, and other amphoteric wash active substances
Form: Liquid; Color: Yellow
Properties: pH: 4–6 (10% solids); Biodegradable
Composition: 30% active ingred.

Lorapon AM-B 13/T. Dutton & Reinisch Ltd.
Chemical Description: Alkyl amido betaine
Category: Detergent
Applications: Alkaline and acid

cleaning agents, car shampoos, metal cleaners
Form: Liquid
Composition: 35% active ingred.

Lorapon AM-CA. Dutton & Reinisch Ltd.
Chemical Description: Coconut based ampholyte
Category: Detergent, foaming agent, amphoteric
Applications: Excellent foaming properties which are almost unaffected by the presence of salts and calcium soaps; highly suitable for the formulation of shampoos and for improving the soap compatibility of foam baths
Form: Liquid; Color: Clear yellow
Properties: pH: 8–8.5 (10% solids)
Composition: 30% active ingred.

Lorapon AM-DML. Dutton & Reinisch Ltd.
Chemical Description: Dimethyl lauryl amino-betaine
Category: Detergent, amphoteric
Applications: Baby shampoos, hard-surface cleaners, steam-jet cleaners, pickling agents
Form: Liquid
Composition: 40% active ingred.

Lorapon AM-G 30. Dutton & Reinisch Ltd.
Chemical Description: Imidazoline-based betaine, modified
Category: Detergent
Applications: Baby and child care cosmetic formulations
Form: Liquid
Composition: 30% active ingred.

Lorapon AM-KSF. Dutton & Reinisch Ltd.
Chemical Description: Imidazoline-based ampholyte, salt free
Category: Detergent, amphoteric
Applications: Baby and child care cosmetic products, personal hygiene toiletries
Form: Liquid
Composition: 70% active ingred.

Lorapon AM-LL. Dutton & Reinisch Ltd.
Chemical Description: Linear alkyl-amido amphoteric
Category: Detergent, amphoteric
Applications: High skin compatible amphoteric for cosmetics
Form: Liquid
Composition: 30% active ingred.

Lorapon AM-V. Dutton & Reinisch Ltd.
Chemical Description: Short-chain fatty-acid based ampholyte
Category: Detergent, wetting agent, amphoteric
Applications: Exhibits excellent wetting properties in acid and in alkaline media; high tolerance towards the presence of large amounts of electrolytes; suitable for the preparation of clear stable cleaning agents with a high salt content; used as a wetting agent for all types of acidic and alkaline cleaners; reduces surface tension enabling more rapid and complete penetration of the electrolyte
Form: Liquid; Color: Dark amber
Properties: pH; 10–12 (10% solids)
Composition: 28% active ingred.

Loraquat B 41. Dutton & Reinisch Ltd.
Chemical Description: Alkyl triethanol ammonium chloride
Category: Disinfectant, cationic
Form: Liquid
Composition: 50% active ingred.

Loraquat B 50. Dutton & Reinisch Ltd.
Chemical Description: Lauryl dimethyl benzyl ammonium chloride
Category: Disinfectant
Applications: Disinfectant for all types of cleaners, for dairies, and other food manufacturing industries; algacide
Form: Liquid
Composition: 50% active ingred.

Loraquat CPEM. Dutton & Reinisch Ltd.
Chemical Description: Coco-pentaethoxy methyl-ammonium methosulfate
Category: Softener, emulsifier, antistatic agent
Applications: Hair conditioner for shampoos; emulsifier for polymerization processes; antistatic agent
Form: Liquid
Composition: 100% active ingred.

Loraquat CR 3099. Dutton & Reinisch Ltd.
Chemical Description: Di-fatty acid ester dimethyl ammonium metho-sulfate
Category: Softener, dry-cleaning agent, textile and leather auxiliary
Form: Liquid
Composition: 100% active ingred.

Loraquat M 5040. Dutton & Reinisch Ltd.
Chemical Description: 1-methyl-2 alkyl-3 alkylamidoethyl-imidazolinium-methosulfate
Category: Softener, detergent, antistatic agent, cationic
Applications: Textile softening rinses for domestic use and commercial laundry use; textile finishing; effective softening agent for paper; textile softener with disinfectant properties; additive for hair conditioning rinses
Form: Liquid; Color: Amber
Properties: Solubility: Disp. in cold water; Stability: Stable in mild acid or alkaline media; pH: 4–5 (1% solids)
Composition: 75% active ingred.

Loraquat M 5040 H. Dutton & Reinisch Ltd.
Chemical Description: Quaternary imidazolinium derivative
Category: Softener
Applications: Softener for textiles
Form: Soft paste
Composition: 75% active ingred.

Loraquat QA 100. Dutton & Reinisch Ltd.
Chemical Description: Alkyl dimethylbenzyl ammonium saccharinate
Category: Disinfectant, detergent, cationic
Applications: Creams, salves, oral hygiene, pharmaceuticals, oral sprays, toothpastes, skin disinfection, hair shampoos, household

deodorants, shoe insoles, paper handkerchiefs, paints, deodorants, antiperspirants, body sprays, disinfectants, hair lotions, plastics
Form: Crystals; Color: White
Properties: Solubility: Sol. in alcohol, glycol, glycol ethers, ketones, carbon tetrachloride; M. P.: 40–50 C; Stability: Stable below 220 C
Composition: 94% active ingred.

Loraquat W 595. Dutton & Reinisch Ltd.
Chemical Description: Combination of cationic dialkyl derivatives
Category: Softener, cationic
Applications: Fabric softener
Form: Soft paste
Composition: 75% active ingred.

Loraquat W 7500. Dutton & Reinisch Ltd.
Chemical Description: Quaternary imidazolinium derivative
Category: Softener
Applications: Fabric softener
Form: Liquid
Composition: 75% active ingred.

Loraquat W 7525. Dutton & Reinisch Ltd.
Chemical Description: Combination of cationic dialkyl derivatives
Category: Softener, cationic
Applications: Fabric softener
Form: Liquid
Composition: 75% active ingred.

LSF–54. Witco Chemical Ltd.
Chemical Description: Synthetic
Category: Detergent, wetting agent, anionic
Applications: High flame resistance, good fluidity, gas-tight foam blanket; prevents reignition; can act as wetting agent
Form: Foam
Properties: Sp. gr.: 1.032 @ 20 C; Visc.: 5.0 cs @ 20 C; F. P.: 5 C; pH: 6.5–8.0
Containers: 25 kg net plastic cans

Lusynton A. BASF AG
Chemical Description: Mixture of organic and inorganic substances
Category: Stabilizer, Corrosion inhibitor
Applications: Excellent chlorite stabilizer with good resistance to hard water and acid with good wetting and detergent effects
Form: Powder; Color: Yellow/white
Properties: Solubility: Sol. in warm water; pH: 9.6 (1% aq. sol'n.)

Lutensit A-EP. BASF AG
Chemical Description: Acid phosphate ester of a fatty alcohol alkoxylate
Category: Wetting, dispersing, solubilizing, and cleaning agent, anionic
Applications: Suitable for liquid and powder-form industrial and household cleaning agents, such as high-pressure cleaners for motor cars, floor polishes, wax-removing and solvent-based cold cleaner formulations
Form: Liquid; Color: Yellowish
Properties: Solubility: Sol. in water, 5% caustic soda, 5% hydro-

chloric acid, 5% common salt solution, isopropanol, white spirit, trichlorethylene, xylene; Density: 1.07 g/ml @ 20 C; Surface Tension: 39 dynes/cm @ 20 C (1 g/l water); pH: 2 (0.1% water); Biodegradable
Composition: 100% active ingred.
References: Technical information paper M 5389 e

Lutensit A-ES. BASF AG
Chemical Description: Alkyl phenol ether sulfate, sodium salt
Category: Wetting agent, dispersing agent, anionic
Applications: Suitable for cleaning agents to be used in engineering, industry, and households; good emulsifying capacity for methylene chloride; can be combined with nonionic water-soluble Lutensol types
Form: Liquid; Color: Yellowish
Properties: Solubility: Sol. in water, 5% caustic soda, 5% hydrochloric acid, 5% common salt solution, isopropanol; Density: 1.09 g/ml @ 20 C; Stability: Stable to hard water; Surface Tension: 31 dynes/cm @ 20 C (1 g/l water); pH: 7–9 (0.1% water); Biodegradable
Composition: 40% active ingred.
References: Technical information paper M 5389 e

Lutensit A-FK. BASF AG
Chemical Description: Fatty acid condensation product, sodium salt
Category: Wetting agent, dispersing agent, anionic
Applications: Suitable for cleaning agents to be used in engineering, industry, and households; wetting and dispersing agent that is low foaming and even has an antifoam effect; can be combined with nonionic wetting agents and emulsifiers; good emulsifying capacity for perchlorethylene and high aromatics
Form: Liquid; Color: Yellowish
Properties: Solubility: Sol. in water, trichlorethylene, xylene; Density: 1.00 g/ml @ 20 C; Pour Pt.: 5 C; Surface Tension: 36 dynes/cm @ 20 C (1 g/l water); pH: 8–9.5 (0.1% water); Biodegradable
Composition: 55% active ingred.
References: Technical information paper M 5389 e

Lutensit A-LBA. BASF AG
Chemical Description: Alkyl benzene sulfonate, alkanolamine salt
Category: Wetting agent, dispersing agent, cleaning agent, anionic
Applications: Suitable for cleaning agents to be used in engineering, industry, and households; useful component in rinsing and cleaning formulations and water-based all-purpose cleaners; used as an emulsifying component for various technical emulsions; emulsifying capacity for chlorinated hydrocarbons (methylene chloride, perchlorethylene, trichlorethylene and for aromatics
Form: Liquid; Color: Yellowish
Properties: Solubility: Sol. in water, 5% caustic soda, 5% hydro-

chloric acid, 5% common salt solution, isopropanol, white spirit, trichlorethylene, xylene; Density: 1.08 g/ml @ 20 C; Pour Pt.: –5 C; Surface Tension: 33 dynes/cm @ 20 C (1 g/l water); pH: 7–9 (0.1% water); Biodegradable
Composition: 55% active ingred.
References: Technical information paper M 5389 e

Lutensit A–PS. BASF AG
Chemical Description: Alkane sulfonate, sodium salt
Category: Wetting agent, dispersing agent, cleansing agent, anionic
Applications: Low-foaming wetting, dispersing, and cleansing agent for industrial and household cleaners with a high electrolyte content, for alkaline household cleaning agents, for alkaline and acid metal cleaning formulations and steam-jet cleaning agents; good foam behavior in acid solutions
Form: Liquid; Color: Yellowish
Properties: Solubility: Sol. in water, 5% caustic soda; 5% hydrochloric acid, 5% common salt solution; Density: 1.2 g/ml @ 20 C; Surface Tension: 34 dynes/cm @ 20 C (1 g/l water); pH: 7–8 (0.1% water); Biodegradable
Composition: 65% active ingred.
Toxicity: Prolong contact may cause skin irritation
References: Technical information paper M 5389 e

Lutensit AN 10. BASF AG
Chemical Description: Mixtures of various nonionic alkoxylates and an anionic, acidic, readily neutralizable surfactant
Category: Degreasing agent, anionic/nonionic
Applications: Suitable for the manufacture of aqueous degreasing agents for use at room or elevated temperatures
Form: Liquid; Color: Yellowish
Properties: Density: 0.97 g/cm^3 @ 20 C; Flash Pt.: 120 C; Pour Pt.: <-15 C; Acid No.: 195 mg KOH/g; Surface Tension: 31 mN/m @ 20 C (1 g/l sodium salt); pH: 4–5 (0.1% aq. sol'n.)
Composition: 100% active ingred.
Toxicity: Prolong contact with skin and eyes may cause irritation
References: Technical information Paper 2522 e

Lutensit AN 20. BASF AG
Chemical Description: Mixtures of various nonionic alkoxylates and an anionic, acidic, readily neutralizable surfactant
Category: Cold cleaner, anionic/nonionic
Applications: Used for the formulation of readily de-emulsifiable solvent-type cold cleaners
Form: Liquid; Color: Yellowish
Properties: Density: 0.96 g/cm^3 @ 20 C; Flash Pt.: 120 C; Pour Pt.: <-15 C; Acid No.: 195 mg KOH/g; Surface Tension: 36 mN/m @ 20 C (1 g/l sodium salt); pH: 3.5–4.5 (0.1% aq. sol'n.)
Composition: 100% active ingred.
Toxicity: Prolong contact may cause skin and eye irritation

References: Technical information paper 2522 e

Lutensit AS 3330. BASF AG
Chemical Description: Sodium salt of sulfated oxoalcohol polyglycol ether
Category: Foaming agent, anionic
Applications: High foaming surfactant for the light chemical and cosmetics industry
Form: Liquid
Composition: 28% active ingred.

Lutensit AS 3334. BASF AG
Chemical Description: Sodium salt of sulfated fatty alcohol polyglycol ether
Category: Foaming agent, anionic
Applications: High foaming surfactant for the light chemical and cosmetics industry
Form: Liquid
Composition: 28% active ingred.

Lutensit K-HP. BASF AG
Chemical Description: Fatty acid condensation product
Category: Detergent, cationic
Applications: Hydrophobic rinse aid for car washing
Form: Liquid
Composition: 100% active ingred.

Lutensit K-LC. BASF AG
Chemical Description: Dimethyl $C_{12}C_{14}$ fatty alkyl benzyl ammonium chloride
Category: Biocidal, wetting and dispersing agent, cationic
Applications: Suitable for the production of disinfectant cleaners for the beverages and foodstuffs industries, trading concerns and the household
Form: Liquid; Color: Clear
Properties: Solubility: Sol. in water, 5% sodium hydroxide, 5% hydrochloric acid, 5% sodium chloride, alcohol, chlorinated hydrocarbons; slightly sol. in oil and white spirit; Density: 0.98 g/cm^3 @ 23 C; Visc.: 80 mPa s @ 23 C; Cloud Pt.: −4 C; Surface Tension: 37 mN/m @ 23 C (1 g/l); pH: 6 (1% aq. sol'n.); Biodegradable: 80%
Composition: 50% active ingred.
Toxicity: Prolonged contact with skin may cause irritation
Containers: Plastic, glass, or coated metal
References: Technical information paper 2593 e

Lutensit K-LC 80. BASF AG
Chemical Description: Dimethyl $C_{12}C_{14}$ fatty alkyl benzyl ammonium chloride
Category: Biocidal, wetting and dispersing agent, cationic
Applications: Suitable for the production of disinfectant cleaners for the beverages and foodstuffs industries, trading concerns and the household
Form: Liquid; Color: Yellowish
Properties: Solubility: Sol. in water, 5% sodium hydroxide, 5% hydrochloric acid, 5% sodium chloride, alcohol, xylene, chlorinated hydrocarbons; slightly sol. in oil and white spirit; Density: 0.98 g/cm^3 @ 23 C; Visc.: 500

mPa s @ 23 C; Cloud Pt.: −10 C; Surface Tension: 37 mN/m @ 23 C (1 g/l); pH: 6 (0.1% aq. sol'n.); Biodegradable: 80%
Composition: 80% active ingred.
Toxicity: Prolonged contact with skin may cause irritation
Containers: Plastic, glass, or coated metal
References: Technical information paper 2593 e

Lutensit K–OC. BASF AG
Chemical Description: Dimethylalkylbenzylammoniumchloride
Category: Biocidal, wetting and dispersing agent, cationic
Applications: Suitable for the production of disinfectant cleaners for the beverages and foodstuffs industries, trading concerns and the household
Form: Liquid; Color: Yellowish
Properties: Solubility: Sol. in water, 5% sodium hydroxide, alcohol; slightly sol. in oil and white spirit; Density: 0.98 g/cm^3 @ 23 C; Visc.: 100 mPa s @ 23 C; Cloud Pt.: −3 C; Surface Tension: 31 mN/m @ 23 C (1 g/l); pH: 6 (0.1% aq. sol'n.); Biodegradable: 80%
Composition: 50% active ingred.
Toxicity: Prolonged contact with skin may cause irritation
Containers: Plastic, glass, or coated metal
References: Technical information paper 2593 e

Lutensit K–TI. BASF AG
Chemical Description: Iodine surfactant complex
Category: Disinfectant
Applications: Biocidal surfactant for disinfectant cleaners
Form: Liquid

Lutensol A 7. BASF AG
Chemical Description: $C_{16}C_{18}$ fatty alcohol ethoxylated with 7 EO
Category: Detergent, nonionic
Applications: Water-soluble surfactants for the production of detergents and cleaning agents, and for the chemicals processing industry
Form: Liquid
Composition: 100% active ingred.

Lutensol A 8. BASF AG
Chemical Description: Coconut oil alcohol ethoxylate
Category: Detergent, wetting agent, dispersing agent, emulsifying agent, nonionic
Applications: Suitable for the production of detergents and cleaning agents; can be combined with other nonionic, anionic, or cationic surfactants and auxiliaries
Form: Liquid; Color: Clear
Properties: Solubility: Sol. in water, 5% caustic soda, 5% hydrochloric acid, 5% common salt solution, mineral oil, gasoline, chlorinated hydrocarbon; Cloud Pt.: ≈ 55 C (0.1% water); Solidification Pt.: ≈ 10 C; Stability: Stable to hard water; pH: 6.0–7.5 (1% water)
Composition: ≈ 90% active ingred.
References: Technical information paper M 5459 e

Lutensol A 80. BASF AG
Chemical Description: Tallow alcohol ethoxylate

Category: Detergent, wetting agent, dispersing agent, emulsifying agent, nonionic
Applications: Suitable for the production of detergents and cleaning agents; can be combined with other nonionic, anionic, or cationic surfactants and auxiliaries
Form: Waxy powder; Color: White or light yellowish
Properties: Solubility: Sol. in water, 5% caustic soda, 5% hydrochloric acid, 5% common salt solution, alcohol, ketones, benzene, xylene, chlorinated hydrocarbon; Cloud Pt.: > 100 C (1% water); Solidification Pt.: ≈ 40 C; Stability: Stable to hard water; pH: 6.0–7.5 (1% water)
Composition: 100% active ingred.
References: Technical information paper M 5459 e

Lutensol AO 3. BASF AG
Chemical Description: Straight chained synthetic fatty alcohol with 3 EO
Category: Detergent, emulsifying and dispersing agent, nonionic
Applications: Mainly used in the cleaners and detergents industry for the production of household and industrial detergents; can be combined with other nonionic, anionic, or cationic surfactants; very compatible with dyes, protective colloids, and swelling agents; mainly used as an emulsifying agent and surfactant in specific formulations; very efficient in removing greasy stains from fabrics
Form: Liquid; Color: Clear (dull)
Properties: Solubility: Sol. in petroleum fractions, alcohols, aromatic hydrocarbons, chlorinated hydrocarbons; slightly sol. in water, 5% hydrochloric acid, 5% sodium chloride; Density: 0.93 g/cm^3 @ 20 C; Visc.: 40 mPa s @ 20 C; Flash Pt.: 130 C (DIN 51376); Solidification Pt.: < 5 C; Hydroxyl No.: 160 (mg KOH/g); pH: 7 (1% water); Biodegradable: > 80%
Composition: 100% active ingred.
References: Technical information paper 2588 e

Lutensol AO 5. BASF AG
Chemical Description: Straight chained synthetic fatty alcohol with 5 EO
Category: Detergent, emulsifying and dispersing agent, nonionic
Applications: See Lutensol AO 3
Form: Liquid; Color: Clear (dull)
Properties: Solubility: Sol. in petroleum fractions, alcohols, aromatic hydrocarbons, chlorinated hydrocarbons; slightly sol. in water, 5% sodium hydroxide, 5% hydrochloric acid, 5% sodium chloride; Density: 0.96 g/cm^3 @ 20 C; Visc.: 40 mPa s @ 20 C; Flash Pt.: 150 C (DIN 51376); Solidification Pt.: < 5 C; Hydroxyl No.: 130 (mg KOH/g); pH: 7 (1% water); Biodegradable: > 80%
Composition: 100% active ingred.
References: Technical information paper 2588 e

Lutensol AO 7. BASF AG
Chemical Description: Straight chained synthetic fatty alcohol with 7 EO

Category: Detergent, emulsifying and dispersing agent, nonionic
Applications: See Lutensol AO 3
Form: Liquid; Color: Clear (dull)
Properties: Solubility: Sol. in water, 5% hydrochloric acid, petroleum fractions, alcohols, aromatic hydrocarbons, chlorinated hydrocarbons; slightly sol. in 5% sodium hydroxide, 5% sodium chloride; Density: 0.98 g/cm^3 @ 20 C; Visc.: 120 mPa s @ 20 C; Cloud Pt.: 43 C (1 g of surfactant + 100 g of water); Flash Pt.: 190 C (DIN 51376); Solidification Pt.: 13 C; Hydroxyl No.: 110 (mg KOH/g); pH: 7 (1% water); Biodegradable: > 80%
Composition: 100% active ingred.
References: Technical information paper 2588 e

Lutensol AO 8. BASF AG
Chemical Description: Straight chained synthetic fatty alcohol with 8 EO
Category: Detergent, emulsifying and dispersing agent, nonionic
Applications: See Lutensol AO 3
Form: Liquid; Color: Clear (dull)
Properties: Solubility: Sol. in water, 5% sodium hydroxide, 5% hydrochloric acid, 5% sodium chloride, petroleum fractions, alcohols, aromatic hydrocarbons, chlorinated hydrocarbons; Density: 0.96 g/cm^3 @ 60 C; Visc.: 30 mPa s @ 60 C; Cloud Pt.: 52 C (1 g of surfactant + 100 g of water); Flash Pt.: 200 C (DIN 51376); Solidification Pt.: 14 C; Hydroxyl No.: 100 (mg KOH/g); pH: 7 (1% water); Biodegradable: > 80%
Composition: 100% active ingred.
References: Technical information paper 2588 e

Lutensol AO 10. BASF AG
Chemical Description: Straight chained synthetic fatty alcohol with 10 EO
Category: Detergent, emulsifying and dispersing agent, nonionic
Applications: See Lutensol AO 3
Form: Soft paste; Color: White
Properties: Solubility: Sol. in water, 5% sodium hydroxide, 5% hydrochloric acid, 5% sodium chloride, alcohols, aromatic hydrocarbons, chlorinated hydrocarbons; Density: 0.98 g/cm^3 @ 60 C; Visc.: 40 cps @ 60 C; Cloud Pt.: 80 C (1 g of surfactant + 100 g of water); Flash Pt.: 200 C, DIN 51376; Solidification Pt.: 20 C; Hydroxyl No.: 85 (mg KOH/g); pH: 7 (1% water); Biodegradable: > 80%
Composition: 100% active ingred.
References: Technical information paper 2588 e

Lutensol AO 11. BASF AG
Chemical Description: Straight chained synthetic fatty alcohol with 11 EO
Category: Detergent, emulsifying and dispersing agent, nonionic
Applications: See Lutensol AO 3
Form: Soft paste; Color: White
Properties: Solubility: Sol. in water, 5% sodium hydroxide, 5% hydrochloric acid, 5% sodium chloride, alcohols, aromatic hydrocar-

bons, chlorinated hydrocarbons; Density: 0.99 g/cm^3 @ 60 C; Visc.: 40 cps @ 60 C; Cloud Pt.: 86 C (1 g of surfactant + 100 g of water); Flash Pt.: 200 C, DIN 51376; Solidification Pt.: 20 C; Hydroxyl No.: 80 (mg KOH/g); pH: 7 (1% water); Biodegradable: > 80%
Composition: 100% active ingred.
References: Technical information paper 2588 e

Lutensol AO 12. BASF AG
Chemical Description: Straight chained synthetic fatty alcohol with 10 EO
Category: Detergent, emulsifying and dispersing agent, nonionic
Applications: See Lutensol AO 3
Form: Solid paste; Color: White
Properties: Solubility: Sol. in water, 5% sodium hydroxide, 5% hydrochloric acid, 5% sodium chloride, alcohols, aromatic hydrocarbons, chlorinated hydrocarbons; Density: 1.00 g/cm^3 @ 60 C; Visc.: 40 cps @ 60 C; Cloud Pt.: 91 C (1 g of surfactant + 100 g of water); Flash Pt.: 200 C, DIN 51376; Solidification Pt.: 21 C; Hydroxyl No.: 75 (mg KOH/g); pH: 7 (1% water); Biodegradable: > 80%
Composition: 100% active ingred.
References: Technical information paper 2588 e

Lutensol AO 109. BASF AG
Chemical Description: Straight chained synthetic fatty alcohol with 10 EO
Category: Detergent, emulsifying and dispersing agent, nonionic
Applications: See Lutensol AO 3
Form: Liquid; Color: Clear (dull)
Properties: Solubility: Sol. in water, 5% sodium hydroxide, 5% hydrochloric acid, 5% sodium chloride, petroleum fractions, alcohols, chlorinated hydrocarbons; Density: 1.03 g/cm^3 @ 20 C; Visc.: 200 cps @ 20 C; Cloud Pt.: 80 (1 g of surfactant + 100 g of water); Solidification Pt.: 13 C; Hydroxyl No.: 85 (mg KOH/g); pH: 7 (1% water); Biodegradable: > 80%
Composition: 90% active ingred.
References: Technical information paper 2588 e

Lutensol AP 6. BASF AG
Chemical Description: Alkylphenol with 6 EO
Category: Detergent, wetting, dispersing, and emulsifying agents, nonionic
Applications: Highly efficient components of detergents and cleaning agents for industry and household; good emulsifying effect on various oils and solvents; very suitable for dispersing solids formed by chemical reaction, precipitation, or coagulation; extremely useful for numerous surface treatments and finishes and in chemical and physical conversion processes; highly efficient wetting, washing, and levelling agents which do not increase the water absorption of the leather and fur skins; highly efficient wetting, emulsifying, and dispersing agents in the paper industry; good surfactant effect in the production and processing

aqueous dye and paint systems
Form: Liquid; Color: Yellowish
Properties: Solubility: Sol. in mineral oils, alcohols, ketones, aromatic hydrocarbons, chlorinated hydrocarbons; Sp. gr.: 1.05 @ 20 C; Visc.: 400 cps @ 20 C; Solidification Pt.: < –15 C; Surface Tension: 31 dynes/cm @ 20 C
Composition: 100% active ingred.
References: Technical information paper M 5054 e

Lutensol AP 7. BASF AG
Chemical Description: Alkylphenol with 7 EO
Category: Detergent, wetting, dispersing, and emulsifying agents, nonionic
Applications: See Lutensol AP 6
Form: Liquid; Color: Clear
Properties: Solubility: Sol. in mineral oils, alcohols, ketones, aromatic hydrocarbons, chlorinated hydrocarbons; Sp. gr.: 1.04 @ 20 C; Visc.: 350 cps @ 20 C; Solidification Pt.: < –15 C; Surface Tension: 31 dynes/cm @ 20 C
Composition: 100% active ingred.
References: Technical information paper M 5054 e

Lutensol AP 8. BASF AG
Chemical Description: Alkylphenol with 8 EO
Category: Detergent, wetting, dispersing, and emulsifying agents, nonionic
Applications: See Lutensol AP 6
Form: Liquid; Color: Clear
Properties: Solubility: Sol. in water, 5% hydrochloric acid, 5% common salt solution, alcohols, ketones, aromatic hydrocarbons, chlorinated hydrocarbons; Sp. gr.: 1.05 @ 20 C; Visc.: 350 cps @ 20 C; Surface Tension: 31 dynes/cm @ 20 C
Composition: 100% active ingred.
References: Technical information paper M 5054 e

Lutensol AP 9. BASF AG
Chemical Description: Alkylphenol with 9 EO
Category: Detergent, wetting, dispersing, and emulsifying agents, nonionic
Applications: See Lutensol AP 6
Form: Liquid; Color: Clear
Properties: Solubility: Sol. in water, 5% hydrochloric acid, 5% common salt solution, alcohols, ketones, aromatic hydrocarbons, chlorinated hydrocarbons; Sp. gr.: 1.05 @ 20 C; Visc.: 350 cps @ 20 C; Solidification Pt.: 5 C; Surface Tension: 31 dynes/cm @ 20 C
Composition: 100% active ingred.
References: Technical information paper M 5054 e

Lutensol AP 10. BASF AG
Chemical Description: Alkylphenol with 10 EO
Category: Detergent, wetting, dispersing, and emulsifying agents, nonionic
Applications: See Lutensol AP 6
Form: Liquid; Color: Clear
Properties: Solubility: Sol. in water, 5% hydrochloric acid, 5% common salt solution, alcohols, ketones, aromatic hydrocarbons,

chlorinated hydrocarbons; Sp. gr.: 1.06 @ 20 C; Visc.: 350 cps @ 20 C; Solidification Pt.: 10 C; Surface Tension: 32 dynes/cm @ 20 C
Composition: 100% active ingred.
References: Technical information paper M 5054 e

Lutensol AP 14. BASF AG
Chemical Description: Alkylphenol with 14 EO
Category: Detergent, wetting, dispersing, and emulsifying agents, nonionic
Applications: See Lutensol AP 6
Form: Liquid; Color: Clear
Properties: Solubility: Sol. in water, 5% caustic soda, 5% hydrochloric acid, 5% common salt solution, alcohols, ketones, aromatic hydrocarbons, chlorinated hydrocarbons; Sp. gr.: 1.05 @ 50 C; Visc.: 250 cps @ 30 C; Solidification Pt.: 20–22 C; Surface Tension: 37 dynes/cm @ 20 C
Composition: 100% active ingred.
References: Technical information paper M 5054 e

Lutensol AP 20. BASF AG
Chemical Description: Alkylphenol with 20 EO
Category: Detergent, wetting, dispersing, and emulsifying agents, nonionic
Applications: See Lutensol AP 6
Form: Soft wax; Color: White
Properties: Solubility: Sol. in water, 5% caustic soda, 5% hydrochloric acid, 5% common salt solution, alcohols, ketones, aromatic hydrocarbons, chlorinated hydrocarbons; Sp. gr.: 1.06 @ 50 C; M. P.: 35–40 C; Surface Tension: 42 dynes/cm @ 20 C
Composition: 100% active ingred.
References: Technical information paper M 5054 e

Lutensol AP 30. BASF AG
Chemical Description: Alkylphenol with 30 EO
Category: Detergent, wetting, dispersing, and emulsifying agents, nonionic
Applications: See Lutensol AP 6
Form: Wax
Composition: 100% active ingred.

Lutensol AT 11. BASF AG
Chemical Description: Saturated $C_{16}C_{18}$ alcohol with 11 EO
Category: Detergent, dispersant, wetting agent, emulsifier, nonionic
Applications: Water-sol. surfactant for chemical industry and production of detergents; cleansing components in aq. and solvent-based aq. formulations; emulsifier components; industrial and household detergents; limesoap dispersant
Form: Paste; Color: White to pale yellowish
Properties: Solubility: Sol. in distilled water; partly sol. in white spirit; clear in benzene, chlorinated hydrocarbons, ethanol, isopropanol; Visc.: ≈ 30 cps (60 C); Cloud Pt.: ≈ 86 C (0.5% aq.); Solidification Pt.: ≈ 30 C; pH: Practically neutral (1% sol'n.)
Composition: 100% active ingred.
References: Technical leaflet M 5117 e

Lutensol AT 18. BASF AG
Chemical Description: Saturated $C_{16}C_{18}$ alcohol with 18 EO
Category and Applications: See Lutensol AT 11
Form: Paste
Composition: 100% active ingred.

Lutensol AT 25. BASF AG
Chemical Description: Saturated $C_{16}C_{18}$ alcohol with 25 EO
Category and Applications: See Lutensol AT 11
Form: Powder; Color: White
Properties: Solubility: Clear in distilled water, benzene, chlorinated hydrocarbons, ethanol, isopropanol; Visc.: ≈ 70 cps (60 C); Cloud Pt.: > 100 C (0.5% aq.); Solidification Pt.: ≈ 37 C; pH: Practically neutral (1% aq.)
Composition: 100% active ingred.
References: Technical leaflet M 5117 e

Lutensol AT 50. BASF AG
Chemical Description: Saturated $C_{16}C_{18}$ alcohol with 50 EO
Category and Applications: See Lutensol AT 11
Form: Powder; Color: White
Properties: Solubility: Clear in distilled water, benzene, chlorinated hydrocarbons, ethanol; partly sol. in isopropanol; Visc.: ≈ 125 cps (60 C); Cloud Pt.: > 100 C (0.5% aq.); Solidification Pt.: ≈ 38 C; pH: Practically neutral (1% aq.)
Composition: 100% active ingred.
References: Technical leaflet M 5117 e

Lutensol ED 140. BASF AG
Chemical Description: Compound formed by the addition of propylenoxide to ethylene diamine, followed by the addition of ethylene oxide
Category: Antistat, detergent, dispersant, defoamer, wetting agent, gelling agent, solubilizer, emulsifier, demulsifier, lubricant, nonionic
Applications: Used as material for dye leveling agents, dispersing agents, defoamers, foam controllers, mold releasing agents, antistats, wetting agents, emulsion polymerization; ED 140 used in household and industrial rinse aids; solubilizer; low-foaming surfactant
Form: Clear liquid; Color: Yellowish
Properties: Solubility: 10% is sol. in water, 5% caustic soada, 5% hydrochloric acid, benzene, ethanol, isopropyl alcohol, trichloroethylene; Density: ≈ 1.05 g/cm^3; Visc.: ≈ 750 cps; Cloud Pt.: ≈ 80 C (1% aq.); Pour Pt.: ≈ –20 C; Surface Tension: ≈ 44 mN/m (1 g/l @ 20 C); pH: ≈ 9 (5% aq.)
Composition: 100% active ingred.
References: TI/P 2524 e

Lutensol ED 310. BASF AG
Chemical Description: See Lutensol ED 140
Category and Applications: See Lutensol ED 140; also foam suppressor in alkaline industrial cleaners, e.g., spray-applied degreasers; demulsifier for crude oil/water emulsions; lubricant in metalwork-

ing for water-sol. coolants
Form: Clear to slightly dull liquid; Color: Yellowish
Properties: Solubility: 10% is sol. in benzene, ethanol, isopropyl alcohol; Density: ≈ 1.00 g/cm^3; Visc.: ≈ 800 cps; Cloud Pt.: ≈ 20 C (1% aq.); Pour Pt.: ≈ −25 C; Surface Tension: ≈ 33 mN/m (1 g/l @ 20 C); pH: ≈ 9 (5% aq.)
Composition: 100% active ingred.
References: TI/P 2524 e

Lutensol ED 370. BASF AG
Chemical Description: See Lutensol ED 140
Category and Applications: See Lutensol ED 140; also in alkaline dairy cleaners, powder or solid cleaners, e.g., glass rinse tablets; dust binding; emulsion polymers
Form: Fine powder; Color: White to yellowish
Properties: Solubility: 10% is sol. in water, 5% hydrochloric acid, ethanol, isopropyl alcohol, trichloroethylene; Density: ≈ 1.05 g/cm^3; Visc.: ≈ 600 cps; M.P.: ≈ 50 C; Cloud Pt.: > 100 C (1% aq.); Surface Tension: ≈ 44 mN/m (1 g/l @ 20 C); pH: ≈ 8 (5% aq.)
Composition: 100% active ingred.
References: TI/P 2524 e

Lutensol ED 610. BASF AG
Chemical Description: See Lutensol ED 140
Category and Applications: See Lutensol ED 140; foam suppressor for other surfactants, e.g., in bottle-washing lines; additive for coolants used in metalworking
Form: Dull liquid; Color: Colorless to yellowish
Properties: Solubility: 10% is sol. in benzene, isopropyl alcohol, ethanol; Density: ≈ 1.02 g/cm^3; Visc.: ≈ 1300 cps; Cloud Pt.: ≈ 15 C (1% aq.); Pour Pt.: ≈ −25 C; Surface Tension: ≈ 33 mN/m (1 g/l @ 20 C); pH: ≈ 8 (5% aq.)
Composition: 100% active ingred.
References: TI/P 2524 e

Lutensol FA 12. BASF AG
Chemical Description: Oleyl amine ethoxylate
Category: Emulsifier, dispersant, detergent, wetting agent, nonionic
Applications: Acidic degreasing agents; components in heavy-duty and other detergents, liquid and pasty cleaning agents, shampoos
Form: Liquid; Color: Yellowish
Properties: Solubility: Forms clear 10% sol'ns. in distilled water, 5% caustic soda, 5% hydrochloric acid, 5% common salt sol'n., alcohol; Density: ≈ 1.02 g/cm^3 (20 C); Visc.: ≈ 200 cps (20 C); Surface Tension: ≈ 35 dynes/cm (1 g/l distilled water @ 20 C); pH: ≈ 8 (1% aq.); Biodegradability: > 80%
Composition: 100% active ingred.
References: Technical leaflet M 5592 e

Lutensol FSA 10. BASF AG
Chemical Description: Oleic acid amide ethoxylate
Category: Emulsifier, dispersant, detergent, wetting agent, emulsifier, nonionic
Applications: Good emulsifying

effect on chlorinated and aromatic hydrocarbons; mineral oil emulsions; powder and liquid fine detergents and in hand cleaners; auxiliary for fur dressing
Form: Liquid with some sediment; Color: Reddish
Properties: Solubility: Forms clear 10% sol'ns. in distilled water, 5% caustic soda, 5% hydrochloric acid, 5% common salt sol'n., and alcohol; Density: ≈ 1.01 (60 C); Visc.: ≈ 70 cps (60 C); Cloud Pt.: ≈ 80 C (1% aq.); Surface Tension: ≈ 35 dynes/cm (1 g/l water @ 20 C); pH: ≈ 8 (1% aq.); Biodegradability: > 80%
Composition: 100% active ingred.
References: Technical leaflet M 5592 e

Lutensol LF 400. BASF AG
Chemical Description: Alkoxylated straight chain alcohol
Category: Low-foaming detergent, wetting and dispersing agent, nonionic
Applications: Acidic rinse aids; formulations for dish washing machines and industrial glass cleaners; alkaline metal cleaning, e.g., in dairy and brewery cleaners; antifoam in production and processing of powder-form cleaners
Form: Clear liquid; Color: Colorless
Properties: Ref. Index: 1.45; Solubility: 10% is sol. in water, 5% hydrochloric acid, white spirit, spindle oil, benzene, ethanol, isopropanol, trichloroethylene; Density: ≈ 0.97 g/cm^3 (20 C); Visc.: ≈ 60 cps; Cloud Pt.: ≈ 31 C (1% aq.); Setting Pt.: ≈ –2 C; Surface Tension: ≈ 30 dynes/cm (0.1% aq.); pH: ≈ 7 (5% aq.)
Composition: 100% active ingred.
Toxicity: Skin and eye irritant on prolonged contact
References: Technical leaflet M 5544 e

Lutensol LF 401. BASF AG
Chemical Description: Alkoxylated straight chain alcohol
Category: Low foaming detergent, wetting and dispersing agent, nonionic
Applications: Heavy-duty detergents, general-purpose cleaners, specialty cleaners; low phosphate formulations; acid pickling baths; alkaline metal cleaners
Form: Clear liquid; Color: Colorless
Properties: Ref. Index: 1.46; Solubility: 10% is sol. in water, 5% hydrochloric acid, white spirit, benzene, ethanol, isopropanol, trichloroethylene; Density: ≈ 1.05 g/cm^3 (20 C); Visc.: ≈ 330 cps; Cloud Pt.: ≈ 77 C (1% aq.); Setting Pt.: ≈ 7 C; Surface Tension: ≈ 40 dynes/cm (0.1% aq.); pH: ≈ 7 (5% aq.)
Composition: 100% actve ingred.
Toxicity: Skin and eye irritant on prolonged contact
References: Technical leaflet M 5544 e

Lutensol LF 600. BASF AG
Chemical Description: Alkoxylated straight chain alcohol

Category: Low foaming detergent, wetting and dispersing agent, nonionic
Applications: Suitable for cleaners based on di- and triphosphates, e.g., brush and high-pressure washing of motorcars; other hard surface cleaners, e.g., garage and workshop floors, plastic surfaces, and metal parts
Form: Clear liquid; Color: Colorless
Properties: Ref. Index: 1.46; Solubility: 10% is sol. in water, 5% hydrochloric acid, white spirit, spindle oil, benzene, ethanol, isopropanol, trichloroethylene; Density: ≈ 0.97 g/cm^3 (20 C); Visc.: ≈ 110 cps; Cloud Pt.: ≈ 52 C (1% aq.); Setting Pt.: −10 C; Surface Tension: ≈ 40 dynes/cm (0.1% aq.); pH: ≈ 7 (5% aq.)
Composition: 100% active ingred.
Toxicity: Skin and eye irritant on prolonged contact
References: Technical leaflet M 5544 e

Lutensol LF 700. BASF AG
Chemical Description: Alkoxylated straight chain alcohol
Category: Low foaming detergent, wetting and dispersing agent, nonionic
Applications: Low foaming, good wetting surfactant for dish washing and rinse aids, in dairy and brewery cleaners, and in egg washing systems; foam regulator in household detergents, e.g., light duties, and in industrial cleaners
Form: Clear liquid; Color: Slightly yellowish
Properties: Ref. Index: 1.45; Solubility: 10% is sol. in 5% hydrochloric acid, white spirit, spindle oil, benzene, ethanol, isopropanol, and trichloroethylene; Density: ≈ 0.96 g/cm^3 (20 C); Visc.: ≈ 110 cps; Setting Pt.: ≈ 6 C; Surface Tension: ≈ 34 dynes/cm (0.1% aq.); pH: ≈ 7 (5% aq.)
Composition: 100% active ingred.
Toxicity: Skin and eye irritant on prolonged contact
References: Technical leaflet M 5544 e

Lutensol LF 711. BASF AG
Chemical Description: Alkoxylated straight chain alcohol
Category: Low foaming detergent, wetting and dispersing agent, nonionic
Applications: Very acid-stable with good wetting properties; used in formulations for dish washing machines, rinse aids, and in acid and alkaline metal cleaning baths; binding of dust, esp. coal dust
Form: Clear liquid; Color: Colorless
Properties: Ref. Index: 1.45; Solubility: 10% is sol. in water, 5% hydrochloric acid, white spirit (opaque), spindle oil, benzene, ethanol, isopropanol, trichloroethylene; Density: ≈ 1.0 g/cm^3 (20 C); Visc.: ≈ 110 cps; Cloud Pt.: ≈ 36 C (1% aq.); Setting Pt.: ≈ −25 C; Surface Tension: ≈ 28 dynes/cm (0.1% aq.); pH: ≈ 7 (5% aq.)
Composition: 100% active ingred.

Toxicity: Skin and eye irritant on prolonged contact
References: Technical leaflet M 5544 e

Lutensol LF 1300. BASF AG
Chemical Description: Alkoxylated straight chain alcohol
Category: Low foaming detergent, wetting and dispersing agent, nonionic
Applications: See Lutensol LF 700; lower foaming than LF 700
Form: Clear liquid; Color: Slightly yellowish
Properties: Ref. Index: 1.45; Solubility: 10% is sol. in white spirit, spindle oil, benzene, ethanol, isopropanol, trichloroethylene; Density: ≈ 0.97 g/cm^3 (20 C); Visc.: ≈ 140 cps; Setting Pt.: ≈ −6 C; Surface Tension: ≈ 35 dynes/cm (0.1% aq.); pH: ≈ 7 (5% aq.)
Composition: 100% active ingred.
Toxicity: Skin and eye irritant on prolonged contact
References: Technical leaflet M 5544 e

Lutensol LSV. BASF AG
Chemical Description: Alkoxylated straight chain alcohol
Category: Low foaming detergent, wetting and dispersing agent, nonionic
Applications: Suitable for formulation of weakly alkaline or acid additives to be used in strongly alkaline baths for bottle washing; used alone or in combination with other low foamers as additives with wetting and foam-suppressing effect, esp. with phosphorus compounds that disperse lime and prevent scale formation
Form: Clear liquid; Color: Slightly yellowish
Properties: Ref. Index: 1.45; Solubility: 10% is sol. in benzene, 5% hydrochloric acid, ethanol, isopropanol, trichloroethylene; 10% is sol. opaque in water, 5% caustic soda, white spirit, and spindle oil; Density: ≈ 1.04 g/cm^3 (20 C); Visc.: ≈ 500 cps; Cloud Pt.: ≈ 30 C (1% aq.); Setting Pt.: ≈ −15 C; Surface Tension: ≈ 42 dynes/cm (0.1% aq.); pH: 8–9 (5% aq.)
Composition: 100% active ingred.
Toxicity: Skin and eye irritant on prolonged contact
References: Technical leaflet M 5544 e

Lutensol LT 30. BASF AG
Chemical Description: Specially alkoxylated fatty alcohol
Category: Nonionic surfactant
Applications: Surfactant for acid and alkaline low-foaming cleaners
Form: Liquid
Composition: 100% active ingred.

Lutensol ON 30 D. BASF AG
Chemical Description: Synthetic fatty alcohol with 3 EO
Category: Nonionic surfactant
Applications: Surfactant for chemical industry and for production of detergents
Form: Liquid
Composition: 100% active ingred.

Lutensol ON 50 D. BASF AG
Chemical Description: Synthetic fatty alcohol with 5 EO
Category and Applications: See Lutensol ON 30 D
Form: Liquid
Composition: 100% active ingred.

Lutensol ON 60 D. BASF AG
Chemical Description: Synthetic fatty alcohol with 6 EO
Category and Applications: See Lutensol ON 30 C
Form: Liquid
Composition: 100% active ingred.

Lutensol ON 70 D. BASF AG
Chemical Description: Synthetic fatty alcohol with 7 EO
Category and Applications: See Lutensol ON 30 D
Form: Liquid
Composition: 100% active ingred.

Lutensol ON 110 D. BASF AG
Chemical Description: Synthetic fatty alcohol with 11 EO
Category and Applications: See Lutensol ON 30 D
Form: Paste
Composition: 100% active ingred.

Lutensol ON 300 D. BASF AG
Chemical Description: Synthetic fatty alcohol with 30 EO
Category and Applications: See Lutensol ON 30 D
Form: Solid
Composition: 100% active ingred.

Lyogen AFS Liquid. Sandoz
Category: Detergent
Applications: Scouring, leveling, and finishing of nylon hosiery; dyeing of nylon hosiery with disperse dyes or dyeboarding machines
Form: Liquid; Color: Brownish
Properties: Solubility: Turbid in water; Sp. gr.: 1.01
Composition: 85–86% active
Reaction: Acid

Lyogen BE Liquid. Sandoz
Category: Dyeing assistant
Applications: Neutral dyeing of nylon to produce uniform pick-up of dyes from fiber to fiber (barré); barré covering agent in the dyeing of nylon at a neutral pH, such as nylon carpets with Nylosan C dyes
Form: Liquid; Color: Yellow
Properties: Solubility: 55% in water @ 70 F; Sp. gr.: 1.16
Composition: 45% active ingred.
Reaction: Alkaline

Lyogen DFT Liquid. Sandoz
Category: Detergent, wetting agent, emulsifier, dispersant
Applications: Multipurpose dyeing assistant for polyester; trimer depressant for polyester; leveling agent for disperse dyes on polyester; possesses antiprecipitation properties
Form: Liquid; Color: Yellow
Properties: Solubility: Miscible with water; Sp. gr.: 1.05
Composition: 30% active ingred.
Reaction: Alkaline

Lyogen F Liquid. Sandoz
Category: Leveling agent, dyeing assistant

Applications: Dyeing of wool in all forms, esp. with reactive dyes; low foaming dyeing assistant for dyeing wool to prevent unevenness between top and bottom, increase dye yield, and dyeing below the boil
Form: Liquid; Color: Brown
Properties: Solubility: Miscible with water; Sp. gr.: 1.06
Composition: 85% active ingred.
Reaction: Neutral

Lyogen KF Liquid. Sandoz
Category: Detergent, wetting agent, emulsifier, dispersant
Applications: Used with vat dyes for improved levelness and penetration
Form: Liquid; Color: Yellow
Properties: Sp. gr.: 1.05
Composition: 17% active ingred.
Reaction: Neutral

Lyogen MS Liquid. Sandoz
Category: Dyeing assistant
Applications: Leveling agent for dyeing wool and nylon; dyeing wool and nylon with acid, acid milling, and premetallized dyes to ensure even dyeing; dyeing modified polypropylene fibers with acid milling and 1:2 metal complex dyes
Form: Liquid; Color: Yellow
Properties: Solubility: Miscible with water; Sp. gr.: 1.08
Composition: 33% active ingred.
Reaction: Alkaline

Lyogen NL Liquid. Sandoz
Category: Dyeing assistant
Applications: Dyeing nylon with acid dyes, esp. tricot and hosiery on becks and paddle machines for barré coverage and dyestuff migration; dyeing nylon automotive carpets with premetallized dyes
Form: Liquid; Color: Amber
Properties: Solubility: Miscible with water; Sp. gr.: 1.05
Composition: 34% active ingred.
Reaction: Alkaline

Lyogen P Liquid. Sandoz
Category: Dyeing assistant
Applications: Dyeing nylon on the acid side (Nylosan, Lanasyn dyes) to promote barré coverage and migration
Form: Liquid; Color: Yellow
Properties: Sp. gr.: 1.09
Composition: 35% active ingred.
Reaction: Neutral

Lyogen PAA Liquid. Sandoz
Category: Antiprecipitant
Applications: Dyeing acrylic fibers and blends with wool and nylon where cationic and anionic dyes are used simultaneously to prevent coprecipitation
Form: Liquid; Color: Amber
Properties: Solubility: Miscible with water
Composition: 43% active ingred.
Reaction: Neutral

Lyogen SF Liquid. Sandoz
Category: Dyeing assistant
Applications: Dyeing wool, silk, and nylon with acid dyes; for promoting even dyeing (leveling) with acid dyes without prolonged boiling; leveling unlevel wool batches

Form: Liquid; Color: Yellow
Properties: Sp. gr.: 1.03
Composition: 25% active ingred.
Reaction: Alkaline

Lyogen SMK-40 Liquid. Sandoz
Category and Applications: See Lyogen SMK Paste
Form: Liquid; Color: Amber
Properties: Solubility: Miscible with water; Sp. gr.: 1.03
Composition: 20% active ingred.
Reaction: Alkaline

Lyogen SMK Paste. Sandoz
Category: Dye leveling agent, antiprecipitant
Applications: Antiprecipitating agent for acid/basic or acid/disperse/basic dyes (wool/acrylic, acrylic/nylon, etc.); esp. in beck dyeing carpets, upholstery, and knitgoods
Form: Gel, paste; Color: Brown
Properties: Solubility: Miscible with water @ 90 F
Composition: 49.5% active
Reaction: Alkaline

Lyogen V (U) Liquid. Sandoz
Category: Detergent, wetting agent
Applications: Surfactant and leveling agent for scouring and printing of polyester and nylon; low temperature beck scouring of texturized polyester knits; printing texturized nylon fabrics and carpets to ensure dye fixation and eliminate frostiness; continuous dyeing of wool stuffings and tops
Form: Liquid; Color: Amber
Properties: Solubility: Water disp.; Sp. gr.: 1.04
Reaction: Alkaline

Lyogen WD Liquid. Sandoz
Category: Dyeing assistant
Applications: Leveling agent for dyeing of wool and cellulose with wool and direct colors except 1:2 metal complex dyes, to ensure even dyeings
Form: Liquid; Color: Amber
Properties: Solubility: Miscible with water; Sp. gr.: 1.05
Composition: 40% active ingred.
Reaction: Alkaline

Lytron 614 Latex. Monsanto Plastics and Resins Co.
Chemical Description: Modified polystyrene latex
Category: Opacifier, anionic
Applications: Opacifier for liquid detergents, esp. liquid dishwashing detergents and all-purpose cleaners incl. low solids and/or high alcohol content systems
Form: Low viscosity liquid; Color: White
Properties: Sp. gr.: ≈ 1.03; Visc.: 100 cps max.; Stability: Stable in acid, neutral or alkaline systems; pH: 5.5 ± 0.3
Composition: 40% active ingred.
References: Data sheet No. 6450

Lytron Latex 621. Monsanto Plastics and Resins Co.
Chemical Description: Modified polystyrene latex

Category: Opacifier, anionic
Applications: See Lytron 614 Latex
Form: Low viscosity liquid; Color: White
Properties: Sp. gr.: ≈ 1.03; Visc.: 100 cps max.; Stability: Stable in acid, neutral, or alkaline systems; pH: 5.5 ± 0.3
Composition: 40% active ingred.
References: Data sheet No. 6232A

M

Macol 660. Mazer Chem. Inc.
Chemical Description: N-butoxy polyoxyethylene polyoxypropylene glycol
Category: Defoamer, lubricant, intermediate, nonionic
Applications: Defoamer, hydraulic fluid, heat transfer fluid, metal working fluid, intermediate, mold release agent, rubber lubricant
Form: Clear viscous liquid; Color: 100 max. (APHA)
Properties: Ref. Index: 1.459; Solubility: Sol. in acetone, propylene glycol, oleic acid, castor oil, water, toluene, isopropanol; Density: 8.72 lb/gal; Sp. gr.: 1.047; Visc.: 660 s (Saybolt @ 100 F); Flash Pt.: 430 F (COC); Pour Pt.: –40 F; Acid No.: 0–1; Stability: Excellent heat stability; tolerant to acid and alkali systems; pH: 5–7

Macol 3520. Mazer Chem. Inc.
Chemical Description: N-butoxy polyoxyethylene polyoxypropylene glycol
Category and Applications: See Macol 660
Form: Clear viscous liquid; Color: 100 max. (APHA)
Properties: Ref. Index: 1.461; Solubility: Sol. in acetone, propylene glycol, oleic acid, castor oil, water, toluene, isopropanol; Density: 8.75 lb/gal; Sp. gr.: 1.050; Visc.: 3520 s (Saybolt @ 100 F); Flash Pt.: 430 F (COC); Pour Pt.: –20 F; Acid No.: 0–1; Stability: Excellent heat stability; tolerant to acid and alkali systems; pH: 5–7

Macol 5100. Mazer Chem. Inc.
Chemical Description: N-butoxy polyoxyethylene polyoxypropylene glycol
Category and Applications: See Macol 660
Form: Clear viscous liquid; Color: 100 max. (APHA)
Properties: Ref. Index: 1.462; Solubility: Sol. in acetone, propylene glycol, oleic acid, castor oil, water, toluene, isopropanol; Density: 8.75 lb/gal; Sp. gr.: 1.050; Visc.: 5100 s (Saybolt @ 100 F);

Flash Pt.: 430 F (COC); Pour Pt.: −20 F; Acid No.: 0–1; Stability: Excellent heat stability; tolerant to acid and alkali systems; pH: 5–7

Macol LA-4. Mazer Chem. Inc.
Chemical Description: POE (4) lauryl ether
Category: Detergent, wetting agent, emulsifier, nonionic
Applications: Dispersants, solubilizers, and coupling agents for cosmetics, textiles, metalworking lubricants, household products, industrial applications
Form: Colorless Liquid
Properties: Solubility: Sol. in IPA, mineral oil; HLB: 9.7; Hydroxyl Value: 145–165; Stability: Stable to hydrolysis by strong acids and alkalis
Composition: 100% conc.

Macol LA-9. Mazer Chem. Inc.
Chemical Description: POE (9) lauryl ether
Category and Applications: See Macol LA-4
Form: Colorless liquid
Properties: Solubility: Sol. in water, IPA; HLB: 11.8; Hydroxyl Value: 90–110; Stability: Stable to hydrolysis by strong acids and alkalis

Macol LA-12. Mazer Chem. Inc.
Chemical Description: POE (12) lauryl ether
Category and Applications: See Macol LA-4
Form: Solid; Color: White
Properties: Solubility: Sol. in water and IPA; HLB: 14.5; Hydroxyl Value: 70–80; Stability: Stable to hydrolysis by strong acids and alkalis
Composition: 100% conc.

Macol LA-23. Mazer Chem. Inc.
Chemical Description: POE (23) lauryl ether
Category and Applications: See Macol LA-4
Form: Solid; Color: White
Properties: Solubility: Sol. in water and IPA; M.P.: 32 C; HLB: 16.9; Hydroxyl Value: 40–55; Stability: Stable to hydrolysis by strong acids and alkalis
Composition: 100% conc.

Macol OA-2. Mazer Chem. Inc.
Chemical Description: POE (2) oleyl ether
Category and Applications: See Macol LA-4
Form: Colorless liquid
Properties: Solubility: Sol. in IPA, mineral oil; HLB: 4.9; Hydroxyl Value: 152–162; Stability: Stable to hydrolysis by strong acids and alkalis
Composition: 100% conc.

Macol OA-4. Mazer Chem. Inc.
Chemical Description: POE (4) oleyl ether
Category and Applications: See Macol LA-4
Form: Colorless liquid
Properties: Solubility: Sol. in IPA and mineral oil; HLB: 8.0; Hydroxyl Value: 120–136
Composition: 100% conc.

Macol OA-10. Mazer Chem. Inc.
Chemical Description: POE (10) oleyl ether
Category and Applications: See Macol LA-4
Form: Solid; Color: Cream
Properties: Solubility: Sol. in water, IPA; disp. in mineral oil; M.P.: 16 C; HLB: 12.4; Hydroxyl Value: 80–95; Stability: Stable to hydrolysis by strong acids and alkalis
Composition: 100% conc.

Macol OA-20. Mazer Chem. Inc.
Chemical Description: POE (20) oleyl ether
Category and Applications: See Macol LA-4
Form: Solid; Color: Cream
Properties: Solubility: Sol. in water, IPA; M.P.: 30 C; HLB: 15.3; Hydroxyl Value: 50–65; Stability: Stable to hydrolysis by strong acids and alkalis
Composition: 100% conc.

Macol SA-2. Mazer Chem. Inc.
Chemical Description: POE (2) stearyl ether
Category and Applications: See Macol LA-4
Form: Solid; Color: White
Properties: Solubility: Sol. in IPA; M.P.: 43 C; HLB: 4.9; Hydroxyl Value: 145–165; Stability: Stable to hydrolysis by strong acids and alkalis
Composition: 100% conc.

Macol SA-5. Mazer Chem. Inc.
Chemical Description: POE (5) stearyl ether
Category and Applications: See Macol LA-4
Form: Solid; Color: White
Properties: Solubility: Sol. in IPA; HLB: 8.9; Hydroxyl Value: 110–125; Stability: Stable to hydrolysis by strong acids and alkalis
Composition: 100% conc.

Macol SA-10. Mazer Chem. Inc.
Chemical Description: POE (10) stearyl ether
Category and Applications: See Macol LA-4
Form: Solid; Color: White
Properties: Solubility: Sol. in IPA; M.P.: 38 C; HLB: 12.4; Hydroxyl Value: 75–90; Stability: Stable to hydrolysis by strong acids and alkalis
Composition: 100% conc.

Macol SA-15. Mazer Chem. Inc.
Chemical Description: POE (15) stearyl ether
Category and Applications: See Macol LA-4
Form: Solid; Color: White
Properties: Solubility: Sol. in water and IPA; M.P.: 38 C; HLB: 14.2; Hydroxyl Value: 57–72; Stability: Stable to hydrolysis by strong acids and alkalis
Composition: 100% conc.

Macol SA-20. Mazer Chem. Inc.
Chemical Description: POE (20) stearyl ether
Category and Applications: See Macol LA-4
Form: Solid; Color: White
Properties: Solubility: Sol. in wa-

ter and IPA; M.P.: 38 C; HLB: 15.3; Hydroxyl Value: 45–60; Stability: Stable to hydrolysis by strong acids and alkalis
Composition: 100% conc.

Macol SA-40. Mazer Chem. Inc.
Chemical Description: POE (40) stearyl ether
Category and Applications: See Macol LA-4
Form: Solid; Color: White
Properties: Solubility: Sol. in water and IPA; M.P.: 40 C; HLB: 17.4; Hydroxyl Value: 25–40; Stability: Stable to hydrolysis by strong acids and alkalis
Composition: 100% conc.

Mafo 13. Mazer Chem. Inc.
Chemical Description: Potassium salt of complex n-stearyl amino acid
Category: Emulsifier, wetting agent, corrosion inhibitor, solubilizer, lubricant, emollient, amphoteric
Applications: Solubilizer for difficult materials, e.g., phenolic germicides; wetting, dispersing, and penetrating on hard surfaces; corrosion protection to metal containers; dairy chain and metal-to-metal lubricant; emulsifier in cutting fluids based on vegetable oils; emollient effect on skin
Form: Clear liquid; Color: Amber
Properties: Solubility: Sol. in water and most polar solvents; Sp. gr.: 1.015; Visc.: 140–200 cps (Brookfield); B.P.: 230 F; Flash Pt.: 230 F; Pour Pt.: 10 C; Stability: Stable to acid or alkali in wide pH range; tolerates hard water; pH: 10–11 (100%); Biodegradable
Composition: 70% active ingred.
Toxicity: Contact with skin and eyes may cause irritation

Mafo C-12. Mazer Chem. Inc.
Chemical Description: Potassium salt of complex n-coco amino acid
Category: Amphoteric surfactant
Applications: Hard surface cleaners, agriculture emulsions, cosmetics, steam cleaners, dishwashing, etc.
Form: Clear liquid; Color: Light yellow
Properties: Solubility (@ 5% conc.): Sol. in 18% KOH, 7% NaOH, 15% HCl, 10% trisodium phosphate; sol. in sea water @ pH 4.0 and 8.0; disp. at pH 9–10; Density: 8.47 lb/gal; Sp. gr.: 1.02; Flash Pt.: 212 F (COC); pH: 10–11
Composition: 29–31% solids

Makon 4. Stepan Chem. Co.
Chemical Description: Alkylphenoxy polyoxyethylene ethanol
Category: Detergent, emulsifier, nonionic
Applications: Chemical specialties, e.g., leather finishing, cosmetic and agriculture emulsions, industrial and metal cleaners, metal pickling, detergent intermediate, perfume solubilizer, shampoo ingredient; textile industry as dyeing and bleaching assistants, in Kier boiling, scouring of wool, cotton, and synthetics, wool carbonizing,

lubricant emulsions; in paper industry for deinking, felt washing, rewetting; in petroleum industry for drilling muds, oil well acidizing, and secondary oil recovery
Form: Liquid; Color: Light straw
Properties: Solubility: Oil sol.; Density: 8.5 lb/gal; Visc.: 260 cps; Pour Pt.: –20 C; Solidification Pt.: –28.9 C; pH: 7.7 (1% sol'n.)
Composition: 100% active ingred.

Makon 6. Stepan Chem. Co.
Chemical Description: Alkylphenoxy polyoxyethylene ethanol
Category and Applications: See Makon 4
Form: Liquid; Color: Light straw
Properties: Solubility: Disp. in water; Density: 8.67 lb/gal; Visc.: 255 cps; Pour Pt.: –29 C; Solidification Pt.: –34.4 C; pH: 7.9 (1% sol'n.)
Composition: 100% active ingred.

Makon 8. Stepan Chem. Co.
Chemical Description: Alkylphenoxy polyoxyethylene ethanol
Category and Applications: See Makon 4
Form: Liquid; Color: Light straw
Properties: Solubility: Sol. in water at room temperature; disp. at higher temperatures; Density: 8.76 lb/gal; Visc.: 205 cps; Cloud Pt.: 24 C (1% sol'n.); Pour Pt.: –5 C; Solidification Pt.: 0 C; pH: 7.0 (1% sol'n.)
Composition: 100% active ingred.

Makon 10. Stepan Chem. Co.
Chemical Description: Alkylphenoxy polyoxyethylene ethanol
Category and Applications: See Makon 4
Form: Liquid; Color: Light straw
Properties: Solubility: Water sol.; Density: 8.85 lb/gal; Visc.: 235 cps; Cloud Pt.: 54 C (1% sol'n.); Pour Pt.: 2.8 C; Solidification Pt.: 4 C; pH: 8.2 (1% sol'n.)
Composition: 100% active ingred.

Makon 12. Stepan Chem. Co.
Chemical Description: Alkylphenoxy polyoxyethylene ethanol
Category and Applications: See Makon 4
Form: Liquid; Color: Light straw
Properties: Solubility: Water sol.; Density: 8.9 lb/gal; Visc.: 300 cps; Cloud Pt.: 81 C (1% sol'n.); Pour Pt.: 12.2 C; Solidification Pt.: 10 C; pH: 7.2 (1% sol'n.)
Composition: 100% active ingred.

Makon 14. Stepan Chem. Co.
Chemical Description: Alkylphenoxy polyoxyethylene ethanol
Category and Applications: See Makon 4
Form: Liquid; Color: Light straw
Properties: Solubility: Water sol.; Density: 8.9 lb/gal; Visc.: 520 cps; Cloud Pt.: 94 C (1% sol'n.); Pour Pt.: 18.9 C; Solidification Pt.: 17.2 C; pH: 7.2 (1% sol'n.)
Composition: 100% active ingred.

Makon 20. Stepan Chem. Co.
Chemical Description: Alkylphenoxy polyoxyethylene ethanol
Category and Applications: See Makon 4
Form: Waxy solid; Color: Off-white

Properties: Solubility: Water sol.; Density: 9.0 lb/gal; Cloud Pt.: Clear at 100 C; Pour Pt.: 30 C; Solidification Pt.: 30 C; pH: 9.0 (1% sol'n.)
Composition: 100% active ingred.

Makon 30. Stepan Chem. Co.
Chemical Description: Alkylphenoxy polyoxyethylene ethanol
Category and Applications: See Makon 4
Form: Waxy solid; Color: Off-white
Properties: Solubility: Sol. in water; Density: 9.0 lb/gal; Cloud Pt.: Clear at 100 C; Pour Pt.: 40 C; Solidification Pt.: 40 C; pH: 8.5 (1% sol'n.)
Composition: 100% active ingred.

Makon NF-5. Stepan Chem. Co.
Chemical Description: Polyethoxylated aliphatic base
Category: Detergent, wetting agent, nonionic
Applications: Low foaming wetting agent for mechanical dishwashing, industrial metal cleaning, bottle and spray washing, textile wetting and scouring, oil field additives, emulsion polymerization, agricultural spray adjuvant
Form: Liquid; Color: Colorless to pale straw; Odor: Bland
Properties: Sp. gr.: 1.021; Pour Pt.: < 40 F; pH: 9.0
Composition: 97+% active ingred.

Makon NF-12. Stepan Chem. Co.
Chemical Description: Polyethoxylated aliphatic base
Category and Applications: See Makon NF-5
Form: Liquid; Color: Colorless to pale straw; Odor: Bland
Properties: Sp. gr.: 0.995; Pour Pt.: < 15 F; pH: 9.0
Composition: 100% active ingred.

Makon OP-9. Stepan Chem. Co.
Chemical Description: Octylphenol-based with 9–10 moles ethylene oxide
Category: Detergent, nonionic
Applications: Hard surface cleaners, textile and paper industries
Form: Clear liquid; Color: Pale straw
Properties: m.w.: 628 ± 10; Solubility: Water sol.; Visc.: 200–300 cps (as is); HLB: 13.5; Cloud Pt.: 65 ± 2 (1% aq.); Pour Pt.: 45 F; Hydroxyl Value: 89 ± 5; pH: 7.5–8.5 (5% in 1:1 water:IPA)
Composition: 100% active ingred.

Manoxol IB. Manchem Ltd.
Chemical Description: Sodium di-(iso-butyl) sulfosuccinate
Category: Wetting agent, anionic
Applications: Dispersion, agricul-
Category: Wetting agent, anionic
Applications: Dispersion, agriculture, battery separators, dry cleaning and laundry, dust control, emulsion polymerization, leather, metal technology, mold release, oil additives, paint, pharmaceutical, photography, printing ink, textiles
Properties: m.w.: 332; Solubility: 76.0 g/100 ml in water; Sp. gr.: 1.06 (60% sol'n.); Surface Tension:

49.4 dynes/cm (0.1% @ 20 C)
Composition: 100% solid
Containers: Multiply paper sacks with polythene liners

Manoxol MA. Manchem Ltd.
Chemical Description: Sodium di-(methyl-amyl) sulfosuccinate
Category and Applications: See Manoxol IB
Form: Mobile liquid
Properties: Solubility: 34.3 g/100 ml in water @ 25 C; Sp. gr.: 1.06 (60% sol'n.); Surface Tension: 49.1 dynes/cm (0.1% @ 20 C)
Composition: 60% active ingred.
Toxicity: Eye irritant; degreasing to skin
Containers: 10 gal drums with polythene inner containers
Warnings and Caution: Avoid skin and eye contact, inhalation and ingestion; combustible

Manoxol N. Manchem Ltd.
Chemical Description: Sodium dinonyl sulfosuccinate
Category and Applications: See Manoxol IB
Properties: m.w.: 472; Solubility: 11.5 g/100 ml water (@ 60 C); Surface Tension: 27.3 dynes/cm (0.1% sol'n. @ 20 C)
Composition: 100% solid and 60% water/alcohol sol'n.
Containers: Sol'n. grades packed in 10 gal drums with polythene inner containers, solid grades packed in multiply paper sacks with polythene liners

Manoxol OT. Manchem Ltd.
Chemical Description: Dioctyl sodium sulfosuccinate
Category and Applications: See Manoxol IB
Form: Waxy lumps; Color: Off-white
Properties: m.w.: 444; Solubility: Very sol. in most solvents, oils, and organic media; ≈ 1.5% sol. in water @ 25 C; Sp. gr.: ≈ 1.1; Stability: Hydrolyzes in acid and alkaline sol'ns.; generally used at pH 1–10; Surface Tension: 30 dynes/cm (0.1% @ 20 C)
Composition: 100% active ingred.
Toxicity: Eye irritant; degreasing to skin
Containers: Multiply paper sacks with polythene liners
Warnings and Caution: Avoid eye and skin contact, inhalation, and ingestion; combustible; burning produces toxic fumes

Manoxol OT 60%. Manchem Ltd.
Chemical Description: Solution of Manoxol OT (60%), industrial methylated spirits (≈ 15%), and water (≈ 25%)
Category and Applications: See Manoxol IB
Form: Liquid; Color: Water white
Properties: Solubility: Disperses readily in water; Sp. gr.: ≈ 1.05; Flash Pt.: ≈ 80 F; Stability: Hydrolyzes in acid and alkaline sol'ns.; generally used at pH 1–10
Composition: 60% active ingred.
Toxicity: Eye irritant; degreasing to skin
Containers: 10 gal drums with

polythene inner containers
Warnings and Caution: Avoid skin and eye contact, inhalation and ingestion; combustible

Manoxol OT/B. Manchem Ltd.
Chemical Description: 85% Manoxol OT and 15% sodium benzoate
Category and Applications: See Manoxol IB; OT/B used for making wettable powders
Form: Powder
Composition: 85% active ingred.

Manoxol OT/P. Manchem Ltd.
Chemical Description: Dioctyl sodium sulfosuccinate of high purity
Category and Applications: Pharmaceutical grade of Manoxol OT

Manro ALES 27. Manro Products Ltd.
Chemical Description: Ammonium lauryl ether sulfate
Applications: Shampoos, bubble baths
Form: Clear liquid; Color: Pale yellow; Odor: Mild
Properties: Sp. gr.: 1.03 @ 20 C; Visc.: 3000 cps @ 20 C; Biodegradable
Composition: 27% active ingred.
Containers: 45 gal drums or road tankers

Manro ALES 60. Manro Products Ltd.
Chemical Description: Ammonium synthetic C_{13}–C_{15} fatty alcohol ether sulfate
Applications: Liquid detergents, cleaning products, fire fighting foams, and other products where high foaming is required
Form: Slightly hazy mobile gel; Color: Amber; Odor: Slight alcoholic
Properties: Sp. gr.: 1.033 @ 20 C; pH: 7–8 (10% aq.); Biodegradable
Composition: 60% active ingred.
Containers: 45 gal drums or road tankers

Manro ALS 30. Manro Products Ltd.
Chemical Description: Ammonium lauryl sulfate
Applications: Cosmetics and toiletries, esp. shampoos
Form: Clear to slightly hazy, viscous liquid; Color: Pale yellow; Odor: Mild
Properties: Sp. gr.: 1.02 @ 20 C; Visc.: 5000 cps @ 20 C; pH: 6.0–7.0 (10% aq.); Biodegradable
Composition: 28% min. active
Containers: 45 gal drums or road tankers

Manro ALS 30S. Manro Products Ltd.
Chemical Description: Ammonium alkyl sulfate
Applications: Cosmetics and toiletries, esp. shampoos
Form: Clear to slightly hazy, viscous liquid; Color: Pale yellow; Odor: Negligible
Properties: Sp. gr.: 1.03 @ 20 C; Visc.: 5000 cps @ 20 C; pH: 6–7 (10% aq.); Biodegradable

Composition: 28% min. active
Containers: 45 gal drums or road tankers

Manro BA. Manro Products Ltd.
Chemical Description: Dodecyl benzene sulfonic acid based on a broad cut, straight chain, alkyl benzene
Applications: Heavy and light duty detergents, emulsion polymerization, etc.
Form: Viscous liquid; Color: Dark brown; Odor: Characteristic
Properties: Sp. gr.: 1.04 @ 20 C; Visc.: 1500 cps @ 20 C; Biodegradable
Composition: 95% min. active
Containers: 45 gal drums or road tankers

Manro BA 25. Manro Products Ltd.
Chemical Description: Sodium salt of SO_3 sulfonated broad cut dodecyl benzene of the predominantly straight chain type
Applications: Liquid detergents, cleaning products, emulsion polymerization systems
Form: Clear liquid; Color: Pale amber; Odor: Mild
Properties: Sp. gr.: 1.04 @ 20 C; Visc.: 250 cps @ 20 C; pH: 7–8 (10% aq.); Biodegradable
Composition: 25% min. active
Containers: 45 gal drums or road tankers

Manro BA 30. Manro Products Ltd.
Chemical Description: Sodium salt of SO_3 sulfonated broad cut dodecyl benzene of the predominantly straight chain type
Applications: Liquid detergents, cleaning products, emulsion polymerization
Form: Clear liquid; Color: Pale amber; Odor: Mild
Properties: Sp. gr.: 1.05 @ 20 C; Visc.: 1000 cps @ 20 C; pH: 7–8 (10% aq.); Biodegradable
Composition: 29% min. active
Containers: 45 gal drums or road tankers

Manro BES 27. Manro Products Ltd.
Chemical Description: Sodium alkyl ether sulfate
Category: Foaming agent, detergent, emulsifier
Applications: Cosmetics, toiletries, detergent, and industrial products
Form: Mobile liquid; Color: Pale yellow; Odor: Mild
Properties: Sp. gr.: 1.04 @ 20 C; Visc.: 200 cps @ 20 C; pH: 6.5–7.5 (10% aq.); Biodegradable
Composition: 26.5% min. active
Containers: 45 gal drums or road tankers

Manro BES 60. Manro Products Ltd.
Chemical Description: Sodium synthetic C_{13}–C_{15} fatty alcohol ether sulfate
Category: Foaming agent, detergent, emulsifier
Applications: Liquid detergents, cleaning products, fire fighting foams
Form: Slightly hazy, mobile gel; Color: Amber; Odor: Slight al-

coholic
Properties: Sp. gr.: 1.090 @ 20 C; pH: 7–8 (10% aq.); Biodegradable
Composition: 60% active ingred.
Containers: 45 gal drums or road tankers

Manro BES 70. Manro Products Ltd.
Chemical Description: Sodium alkyl ether sulfate
Category: Foaming agent, detergent, emulsifier, dispersant, intermediate, anionic
Applications: Cosmetics esp. shampoos and bubble baths; foam stabilizer in liquid detergents; primary detergent and foaming agent in industrial and household cleaning products; fire fighting foams; lime soap dispersant; emulsifier for chlorinated solvents
Form: Mobile gel; Color: Pale yellow; Odor: Mild
Properties: Sp. gr.: 1.05 @ 20 C; Stability: Good in high electrolyte concs., mild alkaline media; pH: 7–9 (10% aq.); Biodegradable
Composition: 67.5% min. active
Toxicity: Degreasing to skin on prolonged contact
Containers: 45 gal drums or road tanker

Manro BSC. Manro Products Ltd.
Chemical Description: Technical grade of benzene sulfonyl chloride
Category: Chemical intermediate
Applications: Used in the manufacture of dyestuffs, etc.
Form: Clear liquid; Color: Water white
Composition: 95% min. active
Containers: 45 gal drums

Manro CD. Manro Products Ltd.
Chemical Description: Coconut fatty acid diethanolamide
Category: Foam stabilizer, solubilizer, nonionic
Applications: Liquid detergents, shampoos, bubble baths with good mildness, detergency and suspending power
Form: Clear viscous liquid; Color: Golden yellow; Odor: Mild
Properties: Sp. gr.: 0.982 @ 20 C; Visc.: 950 cps @ 20 C; Set Pt.: 20 C max.; Biodegradable
Composition: 90% active ingred.
Containers: 45 gal drums or road tankers

Manro CDS. Manro Products Ltd.
Chemical Description: Coconut fatty acid diethanolamide
Category: Foam stabilizer, solubilizer
Applications: Liquid detergents, shampoos, bubble baths
Form: Clear, viscous liquid; Color: Golden yellow; Odor: Mild
Properties: Sp. gr.: 0.990 @ 20 C; Visc.: 800 cps @ 20 C; Set Pt.: 10 C max.; Biodegradable
Composition: 85% active ingred.
Containers: 45 gal drums or road tankers

Manro CDX. Manro Products Ltd.
Chemical Description: Coconut fatty acid diethanolamide
Category and Applications: See

Manro CD
Form: Clear, viscous liquid; Color: Golden yellow; Odor: Mild
Properties: Sp. gr.: 1.01 @ 20 C; Visc.: 950 cps @ 20 C; Biodegradable
Composition: 70% active ingred.
Containers: 45 gal drums or road tankers

Manro CMEA. Manro Products Ltd.
Chemical Description: Coconut fatty acid monoethanolamide
Category: Foam booster, foam stabilizer, nonionic
Applications: Liquid and powder detergent systems
Form: Waxy flake; Color: Cream; Odor: Mild
Properties: Biodegradable
Composition: 94% min. active
Containers: 25 kilo polypropylene bags

Manro D Paste. Manro Products Ltd.
Chemical Description: Sodium cetyl oleyl alcohol sulfate
Applications: Wool scouring and Kier boiling in textile industry
Form: Paste; Color: Cream; Odor: Characteristic
Properties: Sp. gr.: 1.08 @ 40 C; pH: 7.0–9.5 (10% aq.)
Composition: 29% active ingred.
Containers: 45 gal drums

Manro DB 30 S. Manro Products Ltd.
Chemical Description: Blend of anionic/nonionic surfactants
Category: Detergent, foaming agent
Applications: Liquid detergents, hard surface cleaners
Form: Clear, viscous liquid; Color: Golden yellow; Odor: Negligible
Properties: Sp. gr.: 1.048 @ 20 C; Visc.: 500 cps @ 20 C; Cloud Pt.: 0 C max.; pH: 6.2–6.8
Composition: 30% active ingred.
Containers: 45 gal drums or road tankers

Manro DB 45. Manro Products Ltd.
Chemical Description: Anionic/nonionic surfactant blend
Category: Detergent, foaming agent
Applications: Liquid detergents, hard surface cleaners
Form: Clear, viscous liquid; Color: Golden yellow; Odor: Negligible
Properties: Sp. gr.: 1.05 @ 20 C; Visc.: 1200 cps @ 20 C; Cloud Pt.: 5 C max.; pH: 6.5–7.5 (10% aq.)
Composition: 45% active ingred.
Containers: 45 gal drums or road tankers

Manro DL 28. Manro Products Ltd.
Chemical Description: Sodium alkyl sulfate
Applications: Foamable precoats and no-gel foam compounds in latex compounding industry
Form: Clear liquid; Color: Pale yellow; Odor: Mild, characteristic
Properties: Sp. gr.: 1.05 @ 20 C; Visc.: 50 cps @ 20 C; Cloud Pt.: 0 C max.; Stability: Good low

temperature stability; pH: 7–8 (10% aq.); Biodegradable
Composition: 28% min. active
Containers: 45 gal drums or road tankers

Manro DP 60. Manro Products Ltd.
Chemical Description: Formulated product
Category: Detergent
Applications: Nonscratch cream cleanser for use on stainless steel sinks, laminated work tops, cookers, baths, wash basins, vitreous enamel, plastics, paintwork
Form: Viscous liquid; Color: White; Odor: Negligible
Properties: Sp. gr.: 1.47 @ 20 C; Visc.: 3200 cps @ 20 C; pH: 10.5 (10% aq.)
Composition: 60% solids content
Containers: 200 liter drums or road tankers

Manro DS 35. Manro Products Ltd.
Chemical Description: Sodium alkyl sulfate
Category: Solubilizer, wetting agent
Applications: Industrial detergents
Form: Clear liquid; Color: Pale yellow; Odor: Characteristic
Properties: Sp. gr.: 1.07 @ 20 C; Visc.: 50 cps max. @ 20 C; pH: 7–8 (10% aq.)
Composition: 35% active ingred.
Containers: 45 gal drums or road tankers

Manro FCM 95LV. Manro Products Ltd.
Chemical Description: Acid compound based on an aromatic sulfonic acid
Category: Catalyst, curing agent
Applications: Catalyst in production of resin bound sand castings, esters, e.g., phthalates; curing agent in production of resins
Form: Liquid; Color: Dark brown; Odor: Slight, aromatic
Properties: Sp. gr.: 1.28 @ 20 C; Visc.: 1100 cps
Composition: 91% min. active
Containers: 45 gal drums or road tankers

Manro FCM 100. Manro Products Ltd.
Chemical Description: Acid compound based on an aromatic sulfonic acid
Category and Applications: See Manro FCM 95LV
Form: Clear liquid; Color: Light brown; Odor: Slight
Properties: Sp. gr.: 1.24 @ 20 C; Visc.: 175 cps @ 20 C
Composition: 90% active ingred.
Containers: 45 gal drums or road tanker

Manro HA. Manro Products Ltd.
Chemical Description: Dodecyl benzene sulfonic acid
Applications: Heavy and light duty detergents
Form: Viscous liquid; Color: Dark brown; Odor: Characteristic
Properties: Sp. gr.: 1.06 @ 20 C; Visc.: 18,000 cps @ 20 C
Composition: 95% min. active
Containers: 45 gal drums or road tankers

Manro HCS. Manro Products Ltd.
Chemical Description: Isopropylamine salt of SO_3 sulfonated dodecyl benzene
Category: Emulsifier
Applications: Emulsifier in the manufacture of kerosene-based hand cleaning preparations
Form: Clear viscous liquid; Color: Amber; Odor: Mild
Properties: Sp. gr.: 1.03 @ 20 C; Visc.: 3000 cps @ 20 C; pH: 5-6 (2% aq.); Biodegradable
Composition: 89.5% min. active
Containers: 45 gal drums or road tankers

Manro KOL 20. Manro Products Ltd.
Chemical Description: Potassium oleate sol'n.
Category: Foaming agent
Applications: Gel foam systems, emulsion polymerization
Form: Clear liquid; Color: Amber; Odor: Mild
Properties: Sp. gr.: 1.02 @ 20 C; Visc.: 130 cps @ 20 C; pH: 11.5-12.5 (10% aq.)
Composition: 20% solids content
Containers: 45 gal drums or road tankers

Manro KOL 35. Manro Products Ltd.
Chemical Description: Potassium oleate sol'n.
Category: Foaming agent
Applications: Gel foam systems, emulsion polymerization for the latex industry
Form: Clear, viscous liquid; Color: Amber
Properties: Sp. gr.: 1.04 @ 20 C; Visc.: 1500 cps @ 20 C; pH: 11.5-12.5 (10% aq.)
Composition: 35% solids content
Containers: 45 gal drums or road tankers

Manro KXS 40. Manro Products Ltd.
Chemical Description: 40% aq. sol'n. of potassium xylene sulfonate
Category: Solubilizing agent
Form: Clear liquid; Color: Pale yellow; Odor: Negligible
Properties: Sp. gr.: 1.18 @ 40 C; pH: 7.0-10.0 (10% aq.)
Composition: 39% min. active
Containers: 45 gal drums or road tankers

Manro MA 35. Manro Products Ltd.
Chemical Description: 35% disodium octadecyl sulfosuccinamate
Applications: Used in No Gel foam systems based on high solids, noncarboxylated SBR latex, or natural rubber latex
Form: Fluid; Color: Light cream; Odor: Mild, characteristic
Properties: Sp. gr.: 1.05 @ 20 C; pH: 7.5-9.5 (10% aq.)
Composition: 35% solids content
Containers: 45 gal drums or road tankers

Manro ML 33. Manro Products Ltd.
Chemical Description: 33% monoethanolamine lauryl sulfate
Applications: Cosmetics and toile-

tries, esp. shampoos, bubble baths
Form: Clear to slightly hazy, viscous liquid; Color: Pale yellow; Odor: Mild
Properties: Sp. gr.: 1.03 @ 20 C; Visc.: 7500 cps @ 20 C; pH: 6.3–7.3 (10% aq.); Biodegradable
Composition: 32% min. active
Containers: 45 gal drums or road tankers

Manro ML 33S. Manro Products Ltd.
Chemical Description: Monoethanolamine alkyl sulfate
Applications: Cosmetics and toiletries, esp. shampoos, bubble baths
Form: Clear to slightly hazy, viscous liquid; Color: Pale yellow; Odor: Negligible
Properties: Sp. gr.: 1.03 @ 20 C; Visc.: 7500 cps @ 20 C; pH: 6.3–7.3 (10% aq.)
Composition: 32% min. active
Containers: 45 gal drums or road tankers

Manro NA. Manro Products Ltd.
Chemical Description: Dodecyl benzene sulfonic acid
Category: Raw material
Applications: Suitable for neutralization to serve as a base for liquid heavy and light duty detergents
Form: Viscous liquid; Color: Dark brown; Odor: Characteristic
Properties: Sp. gr.: 1.06 @ 20 C; Visc.: 2000 cps @ 20 C
Composition: 96% min. active
Containers: 45 gal drums or road tankers

Manro NEC. Manro Products Ltd.
Chemical Description: 27% active sodium lauryl ether sulfate
Category: Foaming agent
Applications: Shampoos, bubble baths
Form: Clear liquid; Color: Water white; Odor: Mild
Properties: Sp. gr.: 1.05 @ 20 C; Visc.: 3000 cps @ 20 C; pH: 6.0–7.0 (10% aq.); Biodegradable
Composition: 27% min. active
Containers: 45 gal drums or road tankers

Manro NEC 70. Manro Products Ltd.
Chemical Description: 70% active sodium lauryl ether sulfate
Category and Applications: See Manro NEC
Form: Mobile gel; Color: Pale yellow; Odor: Mild
Properties: Sp. gr.: 1.05 @ 20 C; pH: 7–8 (10% aq.); Biodegradable
Composition: 68% min. active
Containers: 45 gal drums or road tankers

Manro PC 35N. Manro Products Ltd.
Chemical Description: Anionic/nonionic blend
Category: Shampoo concentrate
Applications: Pearlized lotion concentrate for shampoos, bubble baths
Form: Viscous liquid with pearly effect; Color: White; Odor: Negligible
Properties: Sp. gr.: 1.00 @ 20 C; Visc.: 16,000 cps @ 20 C; pH:

6.5–7.0 (10% aq.)
Composition: 35% active ingred.
Containers: 45 gal drums or road tankers

Manro PTSA 65 E. Manro Products Ltd.
Chemical Description: Toluene sulfonic acid
Category: Catalyst
Applications: Catalyst in production of resin bound sand castings and in the manufacture of esters
Form: Clear liquid; Color: Amber; Odor: Slight
Properties: Sp. gr.: 1.24 @ 20 C; Visc.: ≈ 10 cps @ 20 C
Composition: 65% active ingred.
Containers: 45 gal drums or road tankers

Manro PTSA 65 H. Manro Products Ltd.
Chemical Description: Toluene sulfonic acid
Category and Applications: See Manro PTSA 65 E
Form: Clear liquid; Color: Amber; Odor: Slight
Properties: Sp. gr.: 1.24 @ 20 C; Visc.: ≈ 10 cps @ 20 C
Composition: 62.5% active
Containers: 45 gal drums or road tankers

Manro PTSA 65 LS. Manro Products Ltd.
Chemical Description: Toluene sulfonic acid
Category and Applications: See Manro PTSA 65 E
Form: Clear liquid; Color: Amber; Odor: Slight
Properties: Sp. gr.: 1.24 @ 20 C; Visc.: ≈ 10 cps @ 20 C
Composition: 65% active ingred.
Containers: 45 gal drums or road tankers

Manro PTSA Crystals. Manro Products Ltd.
Chemical Description: Para toluene sulfonic acid monohydrate
Category: Catalyst, curing agent
Applications: Catalyst in production of esters, e.g., phthalate ester plasticizers; curing agent for coating resins such as epoxies, aminoplasts, etc., and in adhesive systems; in manufacture of dyes and pigments; to reduce stoving temperatures of car finishes
Form: Crystals; Color: White to pink
Composition: 97% min. active
Containers: 25 kg polypropylene bags

Manro PTSC. Manro Products Ltd.
Chemical Description: Para toluene sulfonyl chloride, technical grade
Category: Chemical intermediate
Applications: Chemical intermediate in preparation of derivatives for pharmaceuticals; raw material and processing aid in dyestuffs industry
Form: Crystalline solid; Color: White
Properties: Set Pt.: 64.5 C min.
Composition: 98% min. active
Containers: 100 kg polythene-lined fiberboard kegs

Manro SDBS 30. Manro Products Ltd.
Chemical Description: Sodium salt of SO_3 sulfonated narrow cut dodecyl benzene
Applications: Liquid detergents and cleaning products; emulsion polymerization
Form: Clear, viscous liquid; Color: Amber; Odor: Mild
Properties: Sp. gr.: 1.05 @ 30 C; Visc.: 1000 cps @ 30 C; pH: 7–9 (10% aq.); Biodegradable
Composition: 29% min. active
Containers: 45 gal drums or road tankers

Manro SIOS. Manro Products Ltd.
Chemical Description: 40% aq. sol'n. of sodium iso-octyl sulfate
Category: Wetting agent
Applications: Wetting agent in industrial detergents
Form: Clear liquid; Color: Pale amber; Odor: Characteristic
Properties: Sp. gr.: 1.12 @ 20 C; pH: 7–8 (10% aq.)
Composition: 38% min. active
Containers: 45 gal drums or road tankers

Manro SLS 28. Manro Products Ltd.
Chemical Description: 28% active sodium lauryl sulfate
Applications: Cosmetics, esp. shampoos, toiletries, emulsion polymerization
Form: Clear liquid; Color: Very pale yellow; Odor: Mild
Properties: Sp. gr.: 1.04 @ 20 C; Visc.: 200 cps @ 20 C; pH: 7–8 (10% aq.); Biodegradable
Composition: 28% min. active
Containers: 45 gal drums or road tankers

Manro SLS 28S. Manro Products Ltd.
Chemical Description: Sodium alkyl sulfate
Category and Applications: See Manro SLS 28
Form: Clear liquid; Color: Very pale yellow; Odor: Negligible
Properties: Sp. gr.: 1.04 @ 20 C; Visc.: 7000 cps @ 20 C; pH: 7–8 (10% aq.); Biodegradable
Composition: 28% min. active
Containers: 45 gal drums or road tankers

Manro STS 40. Manro Products Ltd.
Chemical Description: 40% aq. sol'n. of sodium toluene sulfonate
Category: Solubilizing agent
Applications: Solubilizer; improves cold temperature stability; decreases viscosity of finished liquid detergent
Form: Clear liquid; Color: Pale yellow; Odor: Negligible
Properties: Sp. gr.: 1.19 @ 20 C; pH: 7–10 (10% aq.)
Composition: 39.5% min. active
Containers: 45 gal drums or road tankers

Manro SXS 30. Manro Products Ltd.
Chemical Description: 30% aq. sol'n. of sodium xylene sulfonate

Category and Applications: See Manro STS 40
Form: Clear liquid; Color: Pale yellow; Odor: Negligible
Properties: Sp. gr.: 1.15 @ 20 C; pH: 8–10 (10% aq.)
Composition: 29% min. active
Containers: 45 gal drums or road tankers

Manro SXS 40. Manro Products Ltd.
Chemical Description: 40% aq. sol'n. of sodium xylene sulfonate
Category and Applications: See Manro STS 40
Form: Clear liquid; Color: Pale yellow
Properties: Sp. gr.: 1.17 @ 30 C; pH: 7–10 (10% aq.)
Composition: 39% min. active
Containers: 45 gal drums or road tankers

Manro TDBS 60. Manro Products Ltd.
Chemical Description: Triethanolamine salt of SO_3 sulfonated dodecyl benzene
Applications: Detergents, car shampoos
Form: Clear viscous liquid; Color: Amber; Odor: Slight, alcoholic
Properties: Sp. gr.: 1.06 @ 20 C; Visc.: 400 cps @ 20 C; pH: 6.5–7.5 (10% aq.)
Composition: 60% active ingred.
Containers: 45 gal drums or road tankers

Manro TL 40. Manro Products Ltd.
Chemical Description: 40% active triethanolamine lauryl sulfate
Applications: Shampoos, bubble baths
Form: Clear liquid; Color: Pale yellow; Odor: Mild
Properties: Sp. gr.: 1.025 @ 20 C; Visc.: 100 cps @ 20 C; pH: 6.5–7.0 (10% aq.); Biodegradable
Composition: 40% min. active
Containers: 45 gal drums or road tankers

Manro TL 40S. Manro Products Ltd.
Chemical Description: Triethanolamine alkyl sulfate
Applications: Shampoos, bubble baths
Form: Clear liquid; Color: Pale yellow; Odor: Negligible
Properties: Sp. gr.: 1.04 @ 20 C; Visc.: 300 cps @ 20 C; pH: 6.5–7.0 (10% aq.)
Composition: 37% min. active
Containers: 45 gal drums or road tankers

Mapeg 200 DL. Mazer Chem. Inc.
Chemical Description: PEG 200 dilaurate
Category: Emulsifier, dispersant, nonionic
Applications: Metalworking lubricants, fiber lubricants and softeners, solubilizers, defoamers, antistats, cosmetics, pharmaceuticals, chemical intermediates; used in neutral or moderately acidic or alkaline systems
Form: Clear liquid; Color: Yellow
Properties: Solubility: Sol. in isopropanol, toluol, soybean and mineral oils; disp. in water; Sp.

gr.: 0.95; M.P.: 10 C; HLB: 6.8; Sapon. No.: 176–192; Acid No.: 10.0 max.; Iodine No.: 12 max.
Composition: 100% conc.

Mapeg 200 ML. Mazer Chem. Inc.
Chemical Description: PEG 200 monolaurate
Category and Applications: See Mapeg 200 DL
Form: Clear liquid; Color: Yellow
Properties: Solubility: Sol. in isopropanol, toluol; partly sol. in soybean oil and mineral oil; disp. in water; Sp. gr.: 0.99; M.P.: 5 C; HLB: 9.8; Sapon. No.: 139–159; Acid No.: 5.0 max.; Iodine No.: 10 max.
Composition: 100% conc.

Mapeg 200 DO. Mazer Chem. Inc.
Chemical Description: PEG 200 dioleate
Category and Applications: See Mapeg 200 DL
Form: Liquid; Color: Yellow
Properties: Solubility: Sol. in isopropanol, toluol, soybean and mineral oils; disp. in water; Sp. gr.: 0.95; M.P.: < –10 C; HLB: 6.0; Sapon. No.: 148–158; Acid No.: 10.0 max.; Iodine No.: 70 max.
Composition: 100% conc.

Mapeg 200 MO. Mazer Chem. Inc.
Chemical Description: PEG 200 monooleate
Category and Applications: See Mapeg 200 DL
Form: Liquid; Color: Yellow
Properties: Solubility: Sol. in isopropanol, toluol, soybean oil; disp. in water; Sp. gr.: 0.98; M.P.: < –10 C; HLB: 8.0; Sapon. No.: 115–125; Acid No.: 5.0 max.; Iodine No.: 56 max.
Composition: 100% conc.

Mapeg 200 DS. Mazer Chem. Inc.
Chemical Description: PEG 200 distearate
Category and Applications: See Mapeg 200 DL
Form: Solid; Color: White
Properties: Solubility: Sol. in isopropanol, toluol, soybean and mineral oils; disp. hot in water; M.P.: 34 C; HLB: 4.7; Sapon. No.: 155–165; Acid No.: 10.0 max.; Iodine No.: 1.0 max.
Composition: 100% conc.

Mapeg 200 MS. Mazer Chem. Inc.
Chemical Description: PEG 200 monostearate
Category and Applications: See Mapeg 200 DL
Form: Solid; Color: White
Properties: Solubility: Sol. in isopropanol, toluol, soybean oil; disp. hot in water; M.P.: 33 C; HLB: 8.0; Sapon. No.: 120–130; Acid No.: 5.0 max.; Iodine No.: 1.0 max.
Composition: 100% conc.

Mapeg 400 DL. Mazer Chem. Inc.
Chemical Description: PEG 400 dilaurate
Category and Applications: See Mapeg 200 DL
Form: Liquid; Color: Light yellow
Properties: Solubility: Sol. in iso-

propanol, toluol, soybean oil; disp. in water; Sp. gr.: 0.98; M.P.: 18 C; HLB: 9.8; Sapon. No.: 130–140; Acid No.: 10.0 max.; Iodine No.: 9 max.
Composition: 100% conc.

Mapeg 400 ML. Mazer Chem. Inc.
Chemical Description: PEG 400 monolaurate
Category and Applications: See Mapeg 200 DL
Form: Liquid; Color: Light yellow
Properties: Solubility: Sol. in water, isopropanol, toluol; Sp. gr.: 1.01; M.P.: 12 C; HLB: 13.1; Sap. No.: 89–96; Acid No.: 5.0 max.; Iodine No.: 8 max.
Composition: 100% conc.

Mapeg 400 DO. Mazer Chem. Inc.
Chemical Description: PEG 400 dioleate
Category and Applications: See Mapeg 200 DL
Form: Liquid; Color: Yellow
Properties: Solubility: Sol. in isopropanol, toluol, soybean oil; partly sol. in mineral oil; disp. in water; Sp. gr.: 0.98; M.P.: < 7 C; HLB: 8.5; Sapon. No.: 114–122; Acid No.: 10.0 max.; Iodine No.: 55 max.
Composition: 100% conc.

Mapeg 400 MO. Mazer Chem. Inc.
Chemical Description: PEG 400 monooleate
Category and Applications: See Mapeg 200 DL
Form: Liquid; Color: Yellow
Properties: Solubility: Sol. in isopropanol, toluol; partly sol. in soybean oil; disp. in water; Sp. gr.: 1.01; M.P.: < 10 C; HLB: 11.4; Sapon. No.: 80–88; Acid No.: 5.0 max.; Iodine No.: 40 max.
Composition: 100% conc.

Mapeg 400 DS. Mazer Chem. Inc.
Chemical Description: PEG 400 distearate
Category and Applications: See Mapeg 200 DL
Form: Solid; Color: White
Properties: Solubility: Sol. in isopropanol, toluol, soybean and mineral oils; disp. hot in water; M.P.: 36 C; HLB: 8.1; Sapon. No.: 116–125; Acid No.: 10.0 max.; Iodine No.: 1.0 max.
Composition: 100% conc.

Mapeg 400 MS. Mazer Chem. Inc.
Chemical Description: PEG 400 monostearate
Category and Applications: See Mapeg 200 DL
Form: Solid; Color: White
Properties: Solubility: Sol. in isopropanol, toluol, soybean oil; disp. hot in water; M.P.: 33; HLB: 11.5; Sapon. No.: 84–93; Acid No.: 5.0 max.; Iodine No.: 1.0 max.
Composition: 100% conc.

Mapeg 600 DO. Mazer Chem. Inc.
Chemical Description: PEG 600 dioleate
Category and Applications: See Mapeg 200 DL
Form: Liquid; Color: Yellow
Properties: Solubility: Sol. in iso-

propanol, toluol, soybean oil; disp. in water; Sp. gr.: 1.00; M.P.: 20 C; HLB: 10.5; Sapon. No.: 92–102; Acid No.: 10.0 max.; Iodine No.: 45 max.
Composition: 100% conc.

Mapeg 600 MO. Mazer Chem. Inc.
Chemical Description: PEG 600 monooleate
Category and Applications: See Mapeg 200 DL
Form: Liquid; Color: Yellow
Properties: Solubility: Sol. in water, isopropanol, toluol; partly sol. in soybean oil; Sp. gr.: 1.03; M.P.: 25 C; HLB: 13.5; Sapon. No.: 60–70; Acid No.: 5.0 max.; Iodine No.: 29 max.
Composition: 100% conc.

Mapeg 600 DS. Mazer Chem. Inc.
Chemical Description: PEG 600 distearate
Category and Applications: See Mapeg 200 DL
Form: Solid or flake; Color: White
Properties: Solubility: Sol. in isopropanol, toluol, soybean oil; disp. hot in water; M.P.: 41 C; HLB: 10.6; Sapon. No.: 94–104; Acid No.: 10.0 max.; Iodine No.: 1.0 max.
Composition: 100% conc.

Mapeg 600 MS. Mazer Chem. Inc.
Chemical Description: PEG 600 monostearate
Category and Applications: See Mapeg 200 DL
Form: Solid; Color: White
Properties: Solubility: Sol. in water, isopropanol, toluol, propylene glycol, soybean oil; disp. hot in mineral oil; M.P.: 36 C; HLB: 13.6; Sapon. No.: 62–70; Acid No.: 5.0 max.; Iodine No.: 0.5 max.
Composition: 100% conc.

Mapeg 1000 MS. Mazer Chem. Inc.
Chemical Description: PEG 1000 monostearate
Category and Applications: See Mapeg 200 DL
Form: Solid or flake; Color: White
Properties: Solubility: Sol. in water, isopropanol, toluol, propylene glycol; M.P.: 42 C; HLB: 15.7; Sapon. No.: 41–49; Acid No.: 5.0 max.; Iodine No.: 0.5 max.
Composition: 100% conc.

Mapeg 1540 DS. Mazer Chem. Inc.
Chemical Description: PEG 1540 distearate
Category and Applications: See Mapeg 200 DL
Form: Solid or flake; Color: White
Properties: Solubility: Sol. in water, isopropanol, toluol, propylene glycol; partly sol. in soybean oil; M.P.: 45 C; HLB: 14.8; Sapon. No.: 49–58; Acid No.: 10.0 max.; Iodine No.: 1.0 max.
Composition: 100% conc.

Mapeg 6000 DS. Mazer Chem. Inc.
Chemical Description: PEG 6000 distearate
Category and Applications: See Mapeg 200 DL
Form: Solid or flake; Color: White

Properties: Solubility: Sol. in water, toluol, propylene glycol, isopropanol; M.P.: 55 C; HLB: 18.4; Sapon. No.: 14–20; Acid No.: 9.0 max.; Iodine No.: 0.1 max.
Composition: 100% conc.

Mapeg EGDS. Mazer Chem. Inc.
Chemical Description: Ethylene glycol distearate
Category and Applications: See Mapeg 200 DL
Form: Solid or flake; Color: White to cream
Properties: Solubility: Sol. in isopropanol, toluol, soybean and mineral oils; M.P.: 63 C; HLB: 1.4; Sapon. No.: 190–199; Acid No.: 6.0 max.; Iodine No.: 1.0 max.
Composition: 100% conc.

Mapeg EGMS. Mazer Chem. Inc.
Chemical Description: Ethylene glycol monostearate
Category and Applications: See Mapeg 200 DL
Form: Solid or flake; Color: White to cream
Properties: Solubility: Sol. in isopropanol, toluol, soybean and mineral oils; M.P.: 56 C; HLB: 2.4; Sapon. No.: 180–188; Acid No.: 4.0 max.; Iodine No.: 0.5 max.
Composition: 100% conc.

Mapeg S-40. Mazer Chem. Inc.
Chemical Description: PEG 1760 monostearate
Category and Applications: See Mapeg 200 DL
Form: Solid or flake; Color: White
Properties: Solubility: Sol. in water, isopropanol, toluol, propylene glycol; M.P.: 48 C; HLB: 17.4; Sapon. No.: 25–35; Acid No.: 1.0 max.; Iodine No.: 0.5 max.
Composition: 100% conc.

Mapeg TAO-15. Mazer Chem. Inc.
Chemical Description: PEG 660 monotallate
Category and Applications: See Mapeg 200 DL
Form: Clear liquid; Color: Amber
Properties: Solubility: Sol. in water, isopropanol, toluol; Sp. gr.: 1.05; M.P.: 20 C; HLB: 13.8; Sapon. No.: 55–65; Acid No.: 5.0 max.; Iodine No.: 42 max.
Composition: 100% conc.

Maphos 17. Mazer Chem. Inc.
Chemical Description: Aromatic phosphate ester
Category: Emulsifier, anionic
Applications: Emulsion polymerization, esp. acrylic and polyvinyl acetate
Form: Clear viscous liquid; Color: Yellow
Properties: Solubility: Sol. in water @ pH 2.0 and 9.5 (@ 5%); clear to 90 C in 16% NaOH (@ 1%); clear to boiling in 10% STPP and 5% TSP (@ 1%); Sp. gr.: 1.110; Pour Pt.: < 0 C; Stability: Good in neutral and alkaline media; pH: 1.5–2.5 (1% aq.)
Composition: 68.5–71.5% surfactant content
Toxicity: Acid—eye and skin irritant on prolonged contact
Warnings and Caution: Wear protective goggles and gloves

Maphos 60. Mazer Chem. Inc.
Chemical Description: Aliphatic phosphate ester
Category: Hydrotrope, detergent, anticorrosive agent, anionic
Applications: Hydrotropes with good detergency and wetting; hard surface cleaners; dedusting applications
Form: Clear viscous liquid; Color: Light yellow
Properties: Solubility: Sol. in distilled water @ pH 2.0 and 9.5 (@ 5%), perchloroethylene (@ 10%), xylene (@ 10%); clear to boiling in 10% STPP and 5% TSP (@ 1%); clear to 30 C in 16% NaOH (@ 1%); Sp. gr.: 1.08; Pour Pt.: < 0 C; Stability: Good in neutral and alkaline media; pH: 1.5–2.5 (1% aq.)
Composition: 100% surfactant content
Toxicity: Acid—eye and skin irritant on prolonged contact
Warnings and Caution: Wear protective goggles and gloves

Maphos 66. Mazer Chem. Inc.
Chemical Description: Potassium salt of phosphate ester
Category and Applications: See Maphos 60
Form: Clear liquid; Color: Colorless
Properties: Solubility: Sol. in distilled water @ pH 9.5 (@ 5%); clear to boiling in 10% STPP, 5% TSP, and 16% NaOH @ 1% conc.; Sp. gr.: 1.18; Pour Pt.: < 0 C; Stability: Good in neutral and alkaline media; pH: 9.0-9.5 (1% aq.)
Composition: 45–47% surfactant content
Toxicity: Defatting to skin

Maphos 76. Mazer Chem. Inc.
Chemical Description: Aromatic phosphate ester
Category: Detergent, antistat, anticorrosive agent, emulsifier, anionic
Applications: Drycleaning formulations; hard surface cleaners; emulsion polymerization, esp. acrylic and polyvinyl acetate (approved for food packaging)
Form: Clear viscous liquid; Color: Yellow
Properties: Solubility: Sol. in distilled water @ pH 9.5 (@ 5%), in perchloroethylene (@ 10%), in xylene (@ 10%); clear to boiling in 10% STPP and 5% TSP (@ 1%); Sp. gr.: 1.09; Pour Pt.: < 0 C; Stability: Good in neutral and alkaline media; pH: 1.5–2.5 (1% aq.)
Composition: 100% active ingred.
Toxicity: Acid—eye and skin irritant on prolonged contact
Warnings and Caution: Wear protective goggles and gloves

Maphos 76 NA. Mazer Chem. Inc.
Chemical Description: Partial sodium salt of aromatic phosphate ester
Category: Detergent, antistat, anticorrosive agent, anionic
Applications: Drycleaning formulations
Form: Clear liquid; Color: Light yellow
Properties: Solubility: Sol. in dis-

tilled water @ pH 2.0 and 9.5 (@ 5%), in perchloroethylene (@10%), in xylene (@ 10%); clear to boiling in 5% TSP and 10% STPP (@ 1%); Sp. gr.: 1.08; Pour Pt.: 8 C; Stability: Good in neutral and alkaline media; pH: 5.0–6.0 (1% aq.)
Composition: 87–89% active
Toxicity: Defatting to skin

Maphos 91. Mazer Chem. Inc.
Chemical Description: Aromatic phosphate ester
Category and Applications: See Maphos 76; also used in dedusting applications and lubricant additives (greases, synthetic cutting oils, chain belt lubricants, rust preventatives)
Form: Clear viscous liquid; Color: Yellow
Properties: Solubility: Sol. in distilled water @ pH 2.0 and 9.5 (@ 5%), in perchloroethylene (@10%), in xylene (@ 10%); clear to boiling in 10% STPP and 5% TSP (@ 1%); clear to 30 C in 16% NaOH (@ 1%); Sp. gr.: 1.10; Pour Pt.: 2 C; Stability: Good in neutral and alkaline media; pH: 1.5–2.5 (1% aq.)
Composition: 100% active ingred.
Toxicity: Acid–eye and skin irritant on prolonged contact
Warnings and Caution: Wear protective goggles and gloves

Maphos 8078. Mazer Chem. Inc.
Chemical Description: Aromatic phosphate ester
Category: Lubricant additive, anionic
Applications: Greases, synthetic cutting oils, drawing compounds, chain belt lubricants, grease oils, rust preventatives
Form: Clear liquid; Color: Amber
Properties: Solubility: Sol. in xylene (@ 10%); clear to boiling in 10% STPP and 5% TSP (@ 1%); sol. with slight haze in distilled water (@ 5%); Sp. gr.: 1.11; Pour Pt.: 5 C; Stability: Good in neutral and alkaline media; pH: 1.5–2.5 (1% aq.)
Composition: 100% active ingred.
Toxicity: Acid–eye and skin irritant on prolonged contact
Warnings and Caution: Wear protective goggles and gloves

Maphos 8135. Mazer Chem. Inc.
Chemical Description: Aromatic phosphate ester
Category: Hydrotrope, lubricant, anionic
Applications: Hydrotrope, lubricant additive (greases, synthetic cutting oils, drawing compounds, etc.)
Form: Viscous liquid; Color: Light amber
Properties: Solubility: Sol. in distilled water @ pH 2.0 and 9.5 (@ 5%); clear to boiling in 5% TSP (@ 1%); clear to 25 C in 16% NaOH (@ 1%); hazy at 95 C in 10% STPP (@ 1%); Sp. gr.: 1.07; Pour Pt.: 2 C; Stability: Good in neutral and alkaline media; pH: 1.5–2.5 (1% aq.)
Composition: 100% active ingred.
Toxicity: Acid–eye and skin irri-

tant on prolonged contact
Warnings and Caution: Wear protective goggles and gloves

Maphos L-4. Mazer Chem. Inc.
Chemical Description: Aromatic phosphate ester
Category and Applications: See Maphos 8078
Form: Clear viscous liquid; Color: Yellow
Properties: Solubility: Sol. in perchloroethylene, mineral oil, and xylene (@ 10%); Sp. gr.: 1.09; Pour Pt.: < 0 C; Stability: Good in neutral and alkaline media; pH: 1.5–2.5 (1% aq.)
Composition: 100% active ingred.
Toxicity: Acid–eye and skin irritant on prolonged contact
Warnings and Caution: Wear protective goggles and gloves

Maphos L-6. Mazer Chem. Inc.
Chemical Description: Aromatic phosphate ester
Category: Detergent, antistat, anticorrosive agent, lubricant, anionic
Applications: Drycleaning formulations; lubricant additives
Form: Viscous liquid; Color: Yellow
Properties: Solubility: Sol. @ 10% conc. in perchloroethylene, mineral oil, and xylene; Sp. gr.: 1.05; Pour Pt.: 8 C; Stability: Good in neutral and alkaline media; pH: 1.5–2.5 (1% aq.)
Composition: 100% active ingred.
Toxicity: Acid–eye and skin irritant on prolonged contact
Warnings and Caution: Wear protective goggles and gloves

Maphos L-10. Mazer Chem. Inc.
Chemical Description: Aliphatic phosphate ester
Category and Applications: See Maphos L-4
Form: Clear viscous liquid; Color: Light yellow
Properties: Solubility: Sol. in distilled water @ pH 9.5 (@ 5%), and @ 10% in perchloroethylene, mineral oil, and xylene; Sp. gr.: 1.045; Pour Pt.: < 0 C; Stability: Good in neutral and alkaline media; pH: 1.5–2.5 (1% aq.)
Composition: 100% active ingred.
Toxicity: Acid–eye and skin irritant on prolonged contact
Warnings and Caution: Wear protective goggles and gloves

Maphos L-12. Mazer Chem. Inc.
Chemical Description: Aliphatic phosphate ester
Category: Detergent, lubricant, emulsifier, anionic
Applications: Hard surface cleaners; lubricant additives; emulsion polymerization esp. acrylic and polyvinyl acetate (approved for food packaging)
Form: Clear liquid; Color: Light yellow
Properties: Solubility: Sol. in distilled water @ pH 9.5 (@ 5%), in mineral oil and xylene (@ 10%); clear to boiling in 10% STPP and 5% TSP @ 1%; Sp. gr.: 1.01; Pour Pt.: 12 C; Stability: Good in neutral and alkaline media; pH: 1.5–2.5 (1% aq.)
Composition: 100% conc.

Maphos L-13. Mazer Chem. Inc.
Chemical Description: Aliphatic phosphate ester
Category and Applications: See Maphos 8078
Form: Clear liquid; Color: Light yellow amber
Properties: Solubility: Sol. in distilled water @ pH 9.5 (@ 5%), and in perchloroethylene, mineral oil, and xylene @ 10%; clear to boiling in 5% TSP (@ 1%); Sp. gr.: 1.015; Pour Pt.: 3 C; Stability: Good in neutral and alkaline media; pH: 1.5–2.5 (1% aq.)
Composition: 100% active ingred.
Toxicity: Acid–eye and skin irritant on prolonged contact
Warnings and Caution: Wear protective goggles and gloves

Maphos L-22. Mazer Chem. Inc.
Chemical Description: Aliphatic phosphate ester
Category and Applications: See Maphos 8078
Form: Clear liquid; Color: Light yellow
Properties: Solubility: Sol. in distilled water @ pH 9.5 (@ 5%), and in perchloroethylene, mineral oil, and xylene @ 10%; clear to boiling in 5% TSP @ 1%; Sp. gr.: 1.03; Pour Pt.: 17 C; Stability: Good in neutral and alkaline media; pH: 1.5–2.5 (1% aq.)
Composition: 100% active ingred.
Toxicity: Acid–eye and skin irritant on prolonged contact
Warnings and Caution: Wear protective goggles and gloves

Maprofix 563. Onyx Chem. Co.
Chemical Description: Sodium lauryl sulfate
Category: Detergent, foamer, anionic
Applications: Recrystallized, high purity dentifrice (pharmaceutical grade) approved by the FDA for use in toothpaste, mouthwashes, food additives, fruit washing compounds
Form: Powder or granule
Properties: Sp. gr.: 0.66 (25/20 C); Flash Pt.: > 200 F
Composition: 97% min. active

Maprofix DFS. Onyx Chem. Co.
Chemical Description: Formulated anionic blend of sodium lauryl sulfate and an amine oxide
Category: Detergent, anionic
Applications: Good foaming, minimum wetting and low cloud point surfactant for rug and upholstery shampoos, hard surface cleaners
Form: Liquid
Properties: Sp. gr.: 0.98 (25/20 C); Flash Pt.: 140 F; pH: 7.5–9.5
Composition: 27.5% min. solids

Maprofix ES. Onyx Chem. Co.
Chemical Description: Sodium lauryl ether sulfate (3.0 moles of ETO)
Category: Detergent, foamer, anionic
Applications: Clear liquid shampoos, bubble baths, cosmetics; esp. mild with low cloud point; easily adjusted to desired viscosities
Form: Liquid

Properties: Sp. gr.: 1.05 (25/20 C); Flash Pt.: > 200 F
Composition: 27% min. active

Maprofix ESY. Onyx Chem. Co.
Chemical Description: Sodium lauryl ether sulfate (1.0 mole of ETO)
Category and Applications: See Maprofix ES
Form: Liquid
Properties: Sp. gr.: 1.04 (25/20 C); Flash Pt.: > 200 F
Composition: 24.5–25.5% active

Maprofix LCP. Onyx Chem. Co.
Chemical Description: Sodium lauryl sulfate
Category: Foaming and wetting agent, anionic
Applications: Emulsion polymerization; light duty industrial and household formulations; low cloud point
Form: Liquid
Properties: Sp. gr.: 1.05 (25/20 C); Flash Pt.: > 200 F
Composition: 28–30% active

Maprofix LES-60A. Onyx Chem. Co.
Chemical Description: Ammonium lauryl sulfate (3.0 moles of ETO)
Category: Detergent, anionic
Applications: Used in light duty liquid detergents together with linear alkyl aryl sulfonates, esp. for low pH formulations, nonalkaline shampoo concentrates, dishwash detergents
Form: Liquid
Properties: Sp. gr.: 1.06 (25/20 C); Flash Pt.: 82 F
Composition: 57–60% active

Maprofix LES-60C. Onyx Chem. Co.
Chemical Description: Sodium lauryl ether sulfate (3.0 moles of ETO)
Category: Detergent, anionic
Applications: Used in light duty liquid detergents together with linear alkyl aryl sulfonates, esp. for shampoo concentrates, bubble baths, dishwash detergents
Form: Liquid
Properties: Sp. gr.: 1.06 (25/20 C); Flash Pt.: 82 F
Composition: 56–59% active

Maprofix LK (USP). Onyx Chem. Co.
Chemical Description: Sodium lauryl sulfate
Category: Foaming agent, dispersant, detergent, anionic
Applications: Powdered detergent formulations for household and personal care products; industrial pigment dispersant
Form: Powder
Properties: Sp. gr.: 0.35 (25/20 C); Flash Pt.: > 200 F
Composition: 90% min. active

Maprofix MG. Onyx Chem. Co.
Chemical Description: Magnesium lauryl sulfate
Category: Wetting agent, anionic
Applications: Rug and upholstery shampoos; low viscosity, low cloud

point with fairly high electrolyte tolerance
Form: Liquid
Properties: Sp. gr.: 1.02 (25/20 C); Flash Pt.: > 200 F
Composition: 27–29% active

Maprofix NH. Onyx Chem. Co.
Chemical Description: Ammonium lauryl sulfate
Category: Detergent, foaming agent, anionic
Applications: Nonalkaline shampoos, bubble baths, and mild detergent cleaners below pH 7.0
Form: Viscous liquid
Properties: Sp. gr.: 1.01 (25/20 C); Flash Pt.: > 200 F
Composition: 27–30% active

Maprofix NHL. Onyx Chem. Co.
Chemical Description: Ammonium lauryl sulfate
Category: Foaming agent, anionic
Applications: Low viscosity industrial foamer
Form: Liquid
Properties: Sp. gr.: 1.02 (25/20 C); Flash Pt.: > 200 F
Composition: 27–30% active

Maprofix NHL-22. Onyx Chem. Co.
Chemical Description: Ammonium lauryl sulfate
Category and Applications: Low viscosity version of Maprofix NH
Form: Liquid
Properties: Sp. gr.: 1.01 (25/20 C); Flash Pt.: > 200 F
Composition: 27–28% active

Maprofix RS. Onyx Chem. Co.
Chemical Description: Formulated anionic blend of sodium lauryl sulfate and sodium lauroyl sarcosinate
Category: Foaming and wetting agent, anionic
Applications: Low cloud point, high foaming, and mimimum wetting for rug and upholstery shampoo concentrates; leaves a dry residue for easy removal
Form: Liquid
Properties: Sp. gr.: 1.04 (25/20 C); Flash Pt.: > 200 F; pH: 7.5–9.5
Composition: 28.0% min. solids

Maprofix RSN. Onyx Chem. Co.
Chemical Description: Formulated anionic blend of ammonium lauryl sulfate and ammonium lauroyl sarcosinate
Category: Foaming agent, detergent, anionic
Applications: High foaming, minimum wetting, low cloud point, and excellent detergency for nonalkaline rug and upholstery shampoo concentrates; leaves a dry residue for easy removal
Form: Liquid
Properties: Sp. gr.: 1.01 (25/20 C); Flash Pt.: > 200 F; pH: 6.5–7.5
Composition: 24.5% min. solids

Maprofix SP. Onyx Chem. Co.
Chemical Description: Anionic/amine oxide blend
Category: Anionic surfactant
Applications: High foaming blend for formulating acid and alkaline personal care products; easily built to high viscosities

Form: Liquid
Properties: Sp. gr.: 1.03 (25/20 C); Flash Pt.: > 200 F; pH: 7.5–9.0
Composition: 24.5–27.5% solids

Maprofix TLS. Onyx Chem. Co.
Chemical Description: Triethanolamine lauryl sulfate
Category: Detergent, anionic
Applications: High active detergent for shampoo concentrates
Form: Paste
Properties: Sp. gr.: 1.05 (25/20 C); Flash Pt.: 88 F
Composition: 79% min. active

Maprofix TLS-65. Onyx Chem. Co.
Chemical Description: Triethanolamine lauryl sulfate
Category and Applications: See Maprofix TLS
Form: Paste
Properties: Sp. gr.: 1.06 (25/20 C); Flash Pt.: > 200 F
Composition: 64–67% active

Maprofix TLS-500. Onyx Chem. Co.
Chemical Description: Triethanolamine lauryl sulfate
Category: Detergent, foaming agent, anionic
Applications: Cosmetics and industrial formulations with low cloud point, mildness, low color
Form: Liquid; Color: Very light
Properties: Sp. gr.: 1.05 (25/20 C); Cloud Pt.: 0 C; Flash Pt.: > 200 F
Composition: 40% min. active

Maprofix TLS-513. Onyx Chem. Co.
Chemical Description: Triethanolamine lauryl sulfate
Category and Applications: See Maprofix TLS-500
Form: Liquid; Color: Very light
Properties: Sp. gr.: 1.05 (25/20 C); Cloud Pt.: 10 C; Flash Pt.: > 200 F
Composition: 39% min. active

Maprofix WA Paste. Onyx Chem. Co.
Chemical Description: Sodium lauryl sulfate
Category: Detergent, anionic
Applications: High active industrial and cosmetic detergent for rug shampoos and hair shampoo concentrates
Form: Paste
Properties: Sp. gr.: 1.09 (25/20 C); Flash Pt.: > 200 F
Composition: 41.5% min. active

Maprofix WAC. Onyx Chem. Co.
Chemical Description: Sodium lauryl sulfate
Category: Detergent, foamer, anionic
Applications: Shampoos, bubble baths, detergent specialties
Form: Liquid; Color: Very light
Properties: Sp. gr.: 1.04 (25/20 C); Flash Pt.: > 200 F
Composition: 28–30% active

Maprofix WAC-LA. Onyx Chem. Co.
Chemical Description: Sodium lauryl sulfate
Category: Detergent, anionic
Applications: Personal care and industrial formulations with low viscosity and low salt content;

emulsion polymerization
Form: Liquid; Color: Water white
Properties: Sp. gr.: 1.04 (25/20 C); Flash Pt.: > 200 F
Composition: 28–30% active

Maprolyte 101. Onyx Chem. Co.
Chemical Description: Anionic/amphoteric blend
Category: Detergent, conditioner, anionic/amphoteric
Applications: Mild blend with good detergency and high foam in hard water used in shampoos and bubble baths for conditioning properties
Form: Liquid
Properties: Sp. gr.: 1.03 (25/20 C); Flash Pt.: > 200 F; pH: 7.5–8.5
Composition: 33–37% solids

Maprolyte C. Onyx Chem. Co.
Chemical Description: Cocoamidopropyl betaine
Category: Detergent, wetting and foaming agent, stabilizer, nonionic
Applications: Shampoos, bubble baths
Form: Liquid
Properties: Sp. gr.: 1.04 (25/20 C); Flash Pt.: > 200 F
Composition: 33–37% solids

Maprolyte LX. Onyx Chem. Co.
Chemical Description: Anionic/amphoteric blend of DEA lauryl sulfate and DEA lauramino propionate
Category: Foaming agent, conditioner, anionic/amphoteric
Applications: Mild concentrate for shampoos, body soaps, bubble baths; good foaming power and substantivity to skin and hair
Form: Liquid
Properties: Sp. gr.: 1.03 (25/20 C); Flash Pt.: > 200 F; pH: 7.5–8.5
Composition: 33–37% solids

Maprosyl 30. Onyx Chem. Co.
Chemical Description: Sodium n-lauroyl sarcosinate
Category: Detergent, wetting and foaming agent, anionic
Applications: Personal care and household detergent products, low cloud point, extreme mildness; good tolerance to many cationic compounds
Form: Liquid
Properties: Sp. gr.: 1.03 (25/20 C); Flash Pt.: > 200 F
Composition: 29.5–30.5% active

Marabond 21. American Can Co.
Chemical Description: Lignosulfonate
Applications: Oil well cement retardation

Maracarb N-1. American Can Co.
Chemical Description: Salts of low molecular weight lignosulfonic acids
Category: Modifying agent, dispersant, humectant, anionic
Applications: Modifying agent, dispersant, and humectant in dyestuff pastes; industrial cleaners esp. pulp mill slime control and alkaline metal cleaners; plant foliar spray; chelates metal ions
Form: Powder and liquid
Properties: pH: 8.1–9.0 (powder–

3% sol'n.), 7.0–8.0 (liquid–3% sol'n.)
Composition: 94% solids (powder), 51–53% solids (liquid)
References: Bulletin No. 131

Maracell E. American Can Co.
Chemical Description: Partially desulfonated sodium lignosulfonate
Category: Scale inhibitor
Applications: Scale inhibitor, antifoam synergist for boiler water treatment and antifouling treatment of cooling water; industrial cleaners; manufacture of paste for lead acid storage batteries
Form: Powder
Properties: Solubility: Water sol.; pH: 10.8 (3% sol'n.)
Containers: 50 lb net multiwall paper bag

Maracon MC. American Can Co.
Chemical Description: Lignosulfonate
Category: Concrete additive
Applications: Low cost retarder imparting more air (than Maracon R) to concrete mix
Form: Liquid or powder

Maracon N. American Can Co.
Category: Concrete additive
Applications: Water-reducing and strength-improving admixture for concrete; for use in all ready-mixed structural concrete
Form: Liquid or powder

Maracon R. American Can Co.
Category: Concrete additive
Applications: Water-reducing admixture; retards setting and increases strength; used during hot weather pours and in large mass concrete structures such as dams or bridges
Form: Liquid or powder

Maracon T. American Can Co.
Category: Concrete additive
Applications: Used in prestressed concrete; initially retards, then accelerates setting; improves bond of concrete to steel
Form: Liquid or powder

Marasperse C-21. American Can Co.
Chemical Description: Highly purified calcium lignosulfonate
Category: Dispersant, anionic
Applications: Emulsion stabilizer; agriculture chemical formulations; gypsum board additive; industrial cleaners; clay deflocculants in the manufacture of brick, tile, refractories, pottery, and porcelain ware; oil well drilling muds
Form: Powder; Color: Brown
Properties: Solubility: 100% water sol.; Density: 35–40 lb/ft^3 (bulk); Surface Tension: 49.4 dynes/cm (1% sol'n.); pH: 7.0–8.2 (3% sol'n.)
Composition: 100% conc.
Containers: 50 lb net multiwall paper bags
References: Bulletin No. 131

Marasperse CB. American Can Co.
Chemical Description: Lignosulfonate
Category: Dispersant, stabilizer, anionic
Applications: O/W emulsion stabilizer; manufacture of disperse dyes

for dyeing acetate and polyesters; agriculture chemical formulations; gypsum board additive; industrial cleaners; clay deflocculants in ceramics industry; carbon black slurries
Form: Powder; Color: Black
Properties: Solubility: 100% water sol.; Density: 43–47 lb/ft^3 (bulk); Surface Tension: 51.4 dynes/cm (1% sol'n.); pH: 8.5–9.2 (3% sol'n.)
Containers: 50 lb net multiwall paper bags
References: Bulletin No. 131

Marasperse N-22. American Can Co.
Chemical Description: Sodium lignosulfonate
Category: Dispersant, stabilizer, anionic
Applications: O/W emulsion stabilizer; manufacture of disperse dyes for dyeing acetate and polyesters; dispersant and sequestering agent in cooling water treatments; agricultural chemical formulations; gypsum board additive; industrial cleaners;
Form: Powder; Color: Brown
Properties: Solubility: 100% water sol.; Density: 35–40 lb/ft^3 (bulk); Surface Tension: 52.8 dynes/cm (1% sol'n.); pH: 7.5–8.5 (3% sol'n.)
Composition: 100% conc.
Containers: 50 lb net multiwall paper bags
References: Bulletin No. 131

Maratan 22. American Can Co.
Chemical Description: Sodium-fractionated lignosulfonate with extra high tan purity
Category: Tanning agent
Applications: Tanning and retanning agents; replaces up to 50% of normal requirement for vegetable tans; used alone or with vegetable tans to retan over chrome tanned leather
Properties: pH: 3.2–3.8 (5% sol'n.)

Maratan SN. American Can Co.
Chemical Description: Sodium lignosulfonate
Category: Tanning agent auxiliary
Applications: Fillers, buffers, and neutralizers in tanning
Properties: pH: 3.2–3.4 (5% sol'n.)

Maratan SNV. American Can Co.
Chemical Description: Polyphenolic-sodium lignosulfonate
Category and Applications: See Maratan 22
Properties: pH: 3.2–3.8 (5% sol'n.)

Marlamid A 18. Chemische Werke Huls AG
Chemical Description: Fatty acid alkylolamide
Category: Intermediate, cationic
Applications: Intermediate for finishing and softening agents in textile industry
Form: Liquid
Composition: 100% active ingred.

Marlamid AS 18. Chemische Werke Huls AG
Chemical Description: Fatty acid alkylolamide
Category: Raw material, cationic
Applications: Raw material for the manufacture of fabric soften-

ers, so-called fabric softener rinses
Form: Liquid
Composition: 100% active ingred.

Marlamid D 1218. Chemische Werke Huls AG
Chemical Description: Coconut diethanolamide
Category: Foam stabilizer, superfatting agent, nonionic
Applications: Stabilizes the foam of wash liquors, improves the soil-suspending power and has a superfatting effect
Form: Liquid
Composition: 100% active ingred.

Marlamid D 1885. Chemische Werke Huls AG
Chemical Description: Oleic acid diethanolamide
Category and Applications: See Marlamid D 1218
Form: Liquid
Composition: 100% active ingred.

Marlamid DF 1218. Chemische Werke Huls AG
Chemical Description: Coconut diethanolamide
Category and Applications: See Marlamid D 1218
Form: Liquid
Composition: 100% active ingred.

Marlamid KL. Chemische Werke Huls AG
Chemical Description: Fatty acid alkylolamide of high molecular weight
Category: Nonionic surfactant
Applications: Gives pearlescent appearance to liquid and paste detergents
Form: Solid
Composition: 100% active ingred.

Marlamid KLA. Chemische Werke Huls AG
Chemical Description: Fatty acid alkylolamide of high molecular weight
Category and Applications: See Marlamid KL
Form: Liquid
Composition: 40% active ingred.

Marlamid M 1218. Chemische Werke Huls AG
Chemical Description: Coconut monoethanolamide
Category and Applications: See Marlamid D 1218
Form: Solid
Composition: 100% active ingred.

Marlamid O 18. Chemische Werke Huls AG
Chemical Description: Fatty acid alkylolamide
Category and Applications: See Marlamid A 18
Form: Liquid
Composition: 100% active ingred.

Marlamid OS 18. Chemische Werke Huls AG
Chemical Description: Fatty acid alkylolamide
Category and Applications: See Marlamid AS 18
Form: Liquid
Composition: 100% active ingred.

Marlazin 7265. Chemische Werke Huls AG
Chemical Description: Combination of cationic agents
Category and Applications: See Marlazin OK 1
Form: Solid
Composition: 100% active ingred.

Marlazin 8567. Chemische Werke Huls AG
Chemical Description: Aromatic amine
Category: Cationic surfactant
Applications: Special product for the production of floor cleaners and so-called polymer strippers
Form: Liquid
Composition: 100% active ingred.

Marlazin L 10. Chemische Werke Huls AG
Chemical Description: Fatty amine polyglycol ether
Category: Nonionic surfactant
Applications: Used as component in alkali- and acid-resistant industrial liquid and paste cleaners, as well as in the production of dye and fiber auxiliaries
Form: Liquid
Composition: 100% active ingred.

Marlazin L 410. Chemische Werke Huls AG
Chemical Description: Fatty amine polyglycol ether
Category and Applications: See Marlazin L 10
Form: Liquid
Composition: 50% active ingred.

Marlazin OK 1. Chemische Werke Huls AG
Chemical Description: Basic fatty acid condensation product
Category: Cationic surfactant
Applications: Used in the manufacture of products having water-repellent properties for use on smooth and painted surfaces, e.g., window cleaners, car washing and rinsing agents
Form: Solid
Composition: 100% active ingred.

Marlazin OL 20. Chemische Werke Huls AG
Chemical Description: Fatty amine polyglycol ether
Category and Applications: See Marlazin L 10
Form: Liquid
Composition: 100% active ingred.

Marlazin S 10. Chemische Werke Huls AG
Chemical Description: Fatty amine polyglycol ether
Category and Applications: See Marlazin L 10
Form: Liquid
Composition: 100% active ingred.

Marlazin T 10. Chemische Werke Huls AG
Chemical Description: Fatty amine polyglycol ether
Category and Applications: See Marlazin L 10
Form: Liquid
Composition: 100% active ingred.

Marlazin T 410. Chemische Werke Huls AG
Chemical Description: Fatty amine polyglycol ether
Category: Cationic surfactant
Applications: See Marlazin L 10
Form: Liquid
Composition: 50% active ingred.

Marlican. Chemische Werke Huls AG
Chemical Description: Straight-chain dodecylbenzene
Category: Anionic after sulfonation
Applications: Starting product for the production of alkyl benzene sulfonates
Form: Liquid
Composition: 100% active ingred.
References: Brochure No. 1007/2

Marlinat DF 8. Chemische Werke Huls AG
Chemical Description: Dioctyl ester of sodium sulfosuccinic acid
Category: Wetting agent, anionic
Applications: Highly active wetting agent for the textile, paint, and paper industries; suitable for the manufacture of cleaners for glass, dishes, and metal; improves wetting power of cosmetic preparations
Form: Liquid
Composition: 70% active ingred.

Marlinat HA 12. Chemische Werke Huls AG
Chemical Description: Sodium salt of lauric acid monoethanolamide sulfosuccinate
Category: Detergent, anionic
Applications: Detergent raw material having a gentle action for the production of cleaners of sensitive textiles such as carpets, upholstery, and clothing; possesses good cleaning power, pronounced foaming properties, and, on drying, leaves nonhygroscopic residues that can be readily removed by vacuum cleaning; good skin compatibility makes it suitable for manufacture of cosmetic detergents
Form: Paste
Composition: 45% active ingred.

Marlinat SRN 30. Chemische Werke Huls AG
Chemical Description: Blend of anionics
Category and Applications: See Marlinat HA 12
Form: Liquid
Composition: 30% active ingred.

Marlipal 34/30, 34/50, 34/60, 34/70, 34/90, 34/100, 34/110, 34/120, 34/140, 34/200, 34/300. Chemische Werke Huls AG
Chemical Description: C_{13}/C_{14}-oxoalcohol polyglycol ethers
Category: Nonionic surfactants
Applications: Nonionic surfactants in powder and liquid detergents for household and industry; emulsifiers and wetting agents in textile industry
Form: Liquid to solid
Composition: 100% active ingred.
References: Brochure No. 1058/2

Marlipal 124, 129, 1218/5, 1618/11, 1618/25, 1618/40, 1850/5, 1850/8, 1850/10, 1850/30, 1850/80. Chemische Werke Huls AG
Chemical Description: Fatty alcohol polyglycol ethers
Category: Nonionic surfactants
Applications: These grades are soluble in oil and water; depending on the length of their polyglycol ether groups and the type of fatty alcohols used, they possess outstanding dispersing, wetting, detergent, cleaning, soil-suspending, and homogenizing powers
Form: Liquid to wax
Composition: 100% active ingred.

Marlipal BS. Chemische Werke Huls AG
Chemical Description: Fatty acid polyglycol ester
Category: Nonionic surfactant
Applications: Nonionic product having a superfatting and viscosity-increasing effect, for producing liquid and paste detergents, esp. for hair shampoos and similar cosmetic preparations
Form: Wax
Composition: 100% active ingred.

Marlipal FS. Chemische Werke Huls AG
Chemical Description: Fatty acid polyglycol ester
Category and Applications: See Marlipal BS
Form: Liquid
Composition: 100% active ingred.

Marlipal KE, KF, MG, ML, SU. Chemische Werke Huls AG
Chemical Description: Fatty alcohol polyglycol ethers
Category: Nonionic surfactants
Applications: Nonionic detergent raw materials employed as base materials and as additives for detergents and dishwashing agents, in liquid, paste, and powder form; also used in manufacture of tablet soaps
Form: Liquids to solid
Composition: 100% active ingred.

Marlon A 350, A 360, A 365, A 375. Chemische Werke Huls AG
Chemical Description: Sodium dodecylbenzene sulfonates
Category: Anionic surfactants
Applications: High-quality low-salt alkylbenzene sulfonates for the manufacture of liquid and paste detergents and cleaners
Form: Liquid to paste
Composition: 50–75% active
References: Brochure No. 1064/2

Marlon A 390, A 396, ARL. Chemische Werke Huls AG
Chemical Description: Sodium dodecylbenzene sulfonates
Category: Anionic surfactants
Applications: Highly concentrated alkylbenzene sulfonate powders for the manufacture of general-purpose detergent and cleaning powders for domestic, commercial, and industrial use

Form: Powder
Composition: 90–96% active
References: Brochure No. 1064/2

Marlon AFM 40, AFM 43, AFM 50 N. Chemische Werke Huls AG
Chemical Description: Blends of sodium dodecylbenzene sulfonate, alkyl ether sulfate, and nonionic
Category: Anionic/nonionic
Applications: Homogeneous, highly viscous, well-balanced liquid blends of anionic and nonionic detergents; readily soluble, highly foaming grades gentle to the skin, particularly designed as bases for dishwashing and cleaning agents as well as for industrial cleaners
Form: Viscous liquid
Composition: 40–50% active
References: Brochure No. 1064/2

Marlon AFR. Chemische Werke Huls AG
Chemical Description: Modified sodium dodecylbenzene sulfonate
Category: Anionic
Applications: Readily, homogeneous special formulation of Marlon A intended as a base material for dishwashing agents, cosmetic detergents, and car shampoos
Form: Liquid
Composition: 30% active ingred.
References: Brochure No. 1064/2

Marlon AM 80. Chemische Werke Huls AG
Chemical Description: Sodium dodecylbenzene sulfonate/nonionic blend
Category: Anionic/nonionic
Applications: Versatile paste and liquid formulations
Form: Liquid
Composition: 80% active ingred.
References: Brochure No. 1064/2

Marlon AMX. Chemische Werke Huls AG
Chemical Description: Amine salt of dodecylbenzene sulfonic acid
Category: Anionic
Applications: Base material for the production of drycleaning detergents, degreasing agents for the metal industry, and floor cleaners
Form: Liquid
Composition: 90% active ingred.

Marlon AS_3. Chemische Werke Huls AG
Chemical Description: Straight chain dodecylbenzene sulfonic acid
Category: Anionic surfactant
Applications: Intermediate for the manufacture of anionic surfactants
Form: Liquid
Composition: 100% active ingred.
References: Brochure No. 1053/2

Marlophen 83, 84, 85, 86, 86 S, 87, 88, 89, 810, 812, 814, 820, 825, 850. Chemische Werke Huls AG
Chemical Description: Nonylphenolpolyglycol ethers
Category: Detergent, wetting and dispersing agent, nonionic
Applications: These are soluble in oil and water; depending on the length of the polyglycolether groups, they possess marked wetting, detergent, dispersing, and

homogenizing powers; Marlophen 83 is also used as a starting material for sulfation
Form: Liquid to wax
Composition: 100% active ingred.
References: Brochure No. 1063/2

Marlophen DNP 16, DNP 18, DNP 30. Chemische Werke Huls AG
Chemical Description: Dinonylphenol polyglycol ethers
Category: Raw material, dispersant, nonionic
Applications: Low-foaming raw materials for textile and paper auxiliaries, with marked dispersing powers
Form: Wax
Composition: 100% active ingred.

Marlophen P 1. Chemische Werke Huls AG
Chemical Description: Ethoxylated phenol
Category: Solvent, solubilizer, nonionic
Applications: Solvent and solubilizer for organic materials
Form: Liquid
Composition: 100% active ingred.

Marlophen X. Chemische Werke Huls AG
Chemical Description: Alkylphenol polyglycol ether
Category: Wetting agent, nonionic
Applications: Wetting agent for use in binders for coal and rock dust as well as pigments, and as an additive in concrete manufacture
Form: Liquid
Composition: 100% active ingred.

Marlophor AS-Acid, CS-Acid, DS-Acid, NP 5-Acid, NP 7-Acid, NP 20-Acid. Chemische Werke Huls AG
Chemical Description: Partial phosphate ester as acid
Category: Anionic surfactant
Applications: Base components for the manufacture of acid cleaners for metal, glass, and ceramics; cleaners for motor vehicles, engines, sanitary installations, structural facings, and windows; these products are outstanding for their noncorrosive properties, their ability to take a good polish, marked rust removing capacity, and their inertness to common paint finishes
Form: Liquid to solid
Composition: 100% active ingred.

Marlophor FC-Acid. Chemische Werke Huls AG
Chemical Description: Partial phosphate ester as acid
Category: Wetting agent, anionic
Applications: Low-foam wetting agent for alkaline sol'ns., used principally in the manufacture of agents for automatic dishwashing machines and industrial cleaners
Form: Liquid
Composition: 100% active ingred.

Marlophor FC-Sodium Salt, N 3-Sodium Salt. Chemische Werke Huls AG
Chemical Description: Partial phosphate ester as salt
Category and Applications: See Marlophor FC-Acid
Form: Liquid
Composition: 100% active ingred.

Marlophor ND. Chemische Werke Huls AG
Chemical Description: Partial phosphate ester
Category: Wetting agent, antistat, anionic
Applications: Water- and oil-sol. wetting agent for the textile and paper industries, as well as an antistatic agent for natural and synthetic fibers and for use in drycleaning detergents
Form: Liquid
Composition: 100% active ingred.

Marlophor T 10-Acid. Chemische Werke Huls AG
Chemical Description: Partial phosphate ester as acid
Category: Anionic
Applications: Special liquid formulation for the production of drycleaning detergents with an antistatic effect
Form: Liquid
Composition: 100% active ingred.

Marlophor T 10-Sodium Salt. Chemische Werke Huls AG
Chemical Description: Partial phosphate ester as salt
Category and Applications: See Marlophor T 10-Acid
Form: Liquid
Composition: 100% active ingred.

Marlopon ADS 50. Chemische Werke Huls AG
Chemical Description: Diethanolamine dodecylbenzene sulfonate
Category: Raw material, anionic
Applications: Raw material for liquid phosphate-containing detergents and cleaners, incl. those with an alkaline reaction
Form: Fluid paste
Composition: 50% active ingred.

Marlopon AMS 60. Chemische Werke Huls AG
Chemical Description: Amine dodecylbenzene sulfonate/nonionic blend
Category: Anionic/nonionic surfactant
Applications: Used for detergents, dishwashing agents and cleaners
Form: Fluid paste
Composition: 60% active ingred.

Marlopon AT 50. Chemische Werke Huls AG
Chemical Description: Triethanolamine dodecylbenzene sulfonate
Category: Anionic surfactant
Applications: Neutral detergent base for the manufacture of cosmetic detergents, dishwashing agents, car shampoos, etc.
Form: Liquid
Composition: 50% active ingred.

Marlosol 183, 189, 1820, 1825, OL 7, OL 8, OL 10, OL 15, OL 20. Chemische Werke Huls AG
Chemical Description: Fatty acid polyglycol esters
Category: Raw materials, nonionic
Applications: Raw materials for finishing agents in the synthetic fiber industry
Form: Liquid to wax
Composition: 100% active ingred.

Marlowet 4508. Chemische Werke Huls AG
Chemical Description: Organic ammonium salt
Category: Emulsifier, anionic
Applications: Emulsifier for white spirit, kerosene; for environmentally compatible cold cleaners
Form: Liquid
Composition: 100% active ingred.
References: Brochure No. 1049/2

Marlowet 4603. Chemische Werke Huls AG
Chemical Description: Carboxylic acid polyglycol ester
Category: Emulsifier, nonionic
Applications: Emulsifier for mineral oils; for W/O emulsions, textile finishes
Form: Liquid
Composition: 100% active ingred.
References: Brochure No. 1049/2

Marlowet 4702. Chemische Werke Huls AG
Chemical Description: Carboxylic acid polyglycol ester
Category: Emulsifier, nonionic
Applications: Emulsifier for mineral oils; for shuttering oils, metal working
Form: Liquid
Composition: 100% active ingred.
References: Brochure No. 1049/2

Marlowet 4703. Chemische Werke Huls AG
Chemical Description: Carboxylic acid polyglycol ester
Category: Emulsifier, nonionic
Applications: Emulsifier for solvents, mineral oils; for metal cleaning, metal working
Form: Liquid
Composition: 100% active ingred.
References: Brochure No. 1049/2

Marlowet 4800. Chemische Werke Huls AG
Chemical Description: Alkyl polyglycol ether
Category: Emulsifier, nonionic
Applications: Emulsifier for waxes, oleic acid; for dry-bright emulsion polishes, textile lubricants, furniture polishes, leather care agents
Form: Wax
Composition: 100% active ingred.
References: Brochure No. 1049/2

Marlowet 4857. Chemische Werke Huls AG
Chemical Description: Alkyl polyglycol ether
Category: Emulsifier, nonionic
Applications: Emulsifier for paraffins, low melting waxes, silicone oils; for textile, wood, and paper impregnation, mold release agents, furniture polish
Form: Paste
Composition: 100% active ingred.
References: Brochure No. 1049/2

Marlowet 4862. Chemische Werke Huls AG
Chemical Description: Alkyl polyglycol ether
Category: Emulsifier, nonionic
Applications: Emulsifier for higher melting waxes, hard paraffins; for furniture polishes, dry-bright

emulsion polishes
Form: Flakes
Composition: 100% active ingred.
References: Brochure No. 1049/2

Marlowet 5001. Chemische Werke Huls AG
Chemical Description: Alkyl polyglycol ether
Category: Emulsifier, nonionic
Applications: Emulsifier for white oils; for textile auxiliaries, cosmetic preparations, textile lubricants and finishes
Form: Liquid
Composition: 100% active ingred.
References: Brochure No. 1049/2

Marlowet 5100. Chemische Werke Huls AG
Chemical Description: Alkylene oxide adduct of aliphatic diamine
Category: Emulsifier, nonionic
Applications: Emulsifier for lubricating aids; for metal working, foam control
Form: Liquid
Composition: 100% active ingred.
References: Brochure No. 1049/2

Marlowet 5300. Chemische Werke Huls AG
Chemical Description: Partially exterified phosphoric acid
Category: Emulsifier, anionic
Applications: Emulsifier for solvents, mineral oils; for metal working and metal cleaning
Form: Liquid
Composition: 100% active ingred.
References: Brochure No. 1049/2

Marlowet 5301. Chemische Werke Huls AG
Chemical Description: Partially exterified phosphoric acid
Category: Emulsifier, anionic
Applications: Emulsifier for mineral oils; for metal working
Form: Liquid
Composition: 100% active ingred.
References: Brochure No. 1049/2

Marlowet 5305. Chemische Werke Huls AG
Chemical Desccription: Partially exterified phosphoric acid
Category and Applications: See Marlowet 5301
Form: Liquid
Composition: 100% active ingred.
References: Brochure No. 1049/2

Marlowet 5306. Chemische Werke Huls AG
Chemical Description: Partially exterified phosphoric acid
Category and Applications: See Marlowet 5301
Form: Liquid
Composition: 100% active ingred.
References: Brochure No. 1049/2

Marlowet 5311. Chemische Werke Huls AG
Chemical Description: Partially exterified phosphoric acid
Category: Emulsifier, anionic
Applications: Emulsifier for chlorinated hydrocarbons; for paint removers
Form: Liquid
Composition: 100% active ingred.
References: Brochure No. 1049/2

Marlowet 5320. Chemische Werke Huls AG
Chemical Description: Partially exterified phosphoric acid
Category and Applications: See Marlowet 5301
Form: Liquid
Composition: 100% active ingred.
References: Brochure No. 1049/2

Marlowet 5400. Chemische Werke Huls AG
Chemical Description: Alkylamine polyglycol ether
Category: Emulsifier, nonionic
Applications: Emulsifier for mineral and paraffin oils; for car wash rinses, furniture polishes
Form: Liquid
Composition: 100% active ingred.
References: Brochure No. 1049/2

Marlowet 5436. Chemische Werke Huls AG
Chemical Description: Carboxylic acid alkylolamine ester
Category: Emulsifier, nonionic
Applications: Emulsifier for mineral oils; for metal working, corrosion protection
Form: Liquid
Composition: 100% active ingred.
References: Brochure No. 1049/2

Marlowet 5439. Chemische Werke Huls AG
Chemical Description: Carboxylic acid alkylolamide
Category and Applications: See Marlowet 5436
Form: Liquid
Composition: 100% active ingred.
References: Brochure No. 1049/2

Marlowet 5440. Chemische Werke Huls AG
Chemical Description: Substituted imidazoline
Category: Emulsifier, cationic
Applications: Emulsifier for mineral oils; for corrosion protection, car wash rinses
Form: Liquid
Composition: 100% active ingred.
References: Brochure No. 1049/2

Marlowet 5480. Chemische Werke Huls AG
Chemical Description: Carboxylic acid alkylolamide
Category: Emulsifier, nonionic
Applications: Emulsifier for mineral oils; for metal working, corrosion protection
Form: Liquid
Composition: 100% active ingred.
References: Brochure No. 1049/2

Marlowet 5529. Chemische Werke Huls AG
Chemical Description: Mixture of alkoxylated amines and carboxylic acids
Category: Emulsifier, cationic
Applications: Emulsifier for white spirit, petrol; for fast-separating cold cleaners
Form: Liquid
Composition: 100% active ingred.
References: Brochure No. 1049/2

Marlowet 5600. Chemische Werke Huls AG
Chemical Description: Amine salt of carboxylic acid
Category: Emulsifier, anionic
Applications: Emulsifier for wax-

es; for leather care agents, wood and paper impregnation agents
Form: Solid
Composition: 100% active ingred.
References: Brochure No. 1049/2

Marlowet 5606. Chemische Werke Huls AG
Chemical Description: Amine salt of carboxylic acid
Category: Emulsifier, anionic
Applications: Emulsifier for paraffin waxes; for manufacture of water-repellent chipboard
Form: Wax
Composition: 100% active ingred.
References: Brochure No. 1049/2

Marlowet 5609. Chemische Werke Huls AG
Chemical Description: Amine salt of carboxylic acid
Category: Emulsifier, anionic
Applications: Emulsifier for white spirit, trichlorethylene; for dewatering agents, cavity preserving agents
Form: Liquid
Composition: 100% active ingred.
References: Brochure No. 1049/2

Marlowet 5622. Chemische Werke Huls AG
Chemical Description: Amine salt of carboxylic acid
Category: Emulsifier, anionic
Applications: Emulsifier for mineral oils, solvents; for metal working, corrosion protection, floor care
Form: Liquid
Composition: 100% active ingred.
References: Brochure No. 1049/2

Marlowet BIK. Chemische Werke Huls AG
Chemical Description: Mixture of carboxylic acid amine salt and alkyl polyglycol ether
Category: Emulsifier, anionic/nonionic
Applications: Emulsifier for mineral oils and hydrocarbons, bitumen; for metal working, road construction
Form: Liquid
Composition: 100% active ingred.
References: Brochure No. 1049/2

Marlowet BL. Chemische Werke Huls AG
Chemical Description: Alkyl polyglycol ether
Category: Emulsifier, nonionic
Applications: Emulsifier for spindle oils, esp. white oils; for textile lubricants
Form: Liquid
Composition: 100% active ingred.
References: Brochure No. 1049/2

Marlowet EF. Chemische Werke Huls AG
Chemical Description: Mixture of carboxylic acid polyglycol esters
Category: Emulsifier, nonionic
Applications: Emulsifier for fatty acids (esp. oleic acid), solvents, ethyl hexandiol, and similar active agents; for textile lubricants, pest control, cosmetic preparations
Form: Paste
Composition: 100% active ingred.
References: Brochure No. 1049/2

Marlowet FOX. Chemische Werke Huls AG
Chemical Description: Alkyl polyglycol ether
Category: Emulsifier, nonionic
Applications: Emulsifier for higher melting waxes and oleic acid; for textile lubricants, car and furniture polishes, leather care, drybright emulsion polishes
Form: Wax
Composition: 100% active ingred.
References: Brochure No. 1049/2

Marlowet GFN. Chemische Werke Huls AG
Chemical Description: Alkyl polyglycol ether
Category: Emulsifier, nonionic
Applications: Emulsifier for higher melting waxes and ester waxes; for furniture and floor care, drybright emulsion polishes
Form: Paste
Composition: 100% active ingred.
References: Brochure No. 1049/2

Marlowet GFW. Chemische Werke Huls AG
Chemical Description: Alkyl polyglycol ether
Category: Emulsifier, nonionic
Applications: Emulsifier for higher melting waxes and hard paraffins; for furniture polishes, drybright emulsion polishes
Form: Flake
Composition: 100% active ingred.
References: Brochure No. 1049/2

Marlowet IHF. Chemische Werke Huls AG
Chemical Description: Mixture of n-alkyl benzene sulfonate, alkyl polyglycol ether, and solvent
Category: Emulsifier, nonionic
Applications: Emulsifier for insecticides, herbicides; for pesticides, cleaning agents
Form: Liquid
Composition: 100% active ingred.
References: Brochure No. 1049/2

Marlowet ISM. Chemische Werke Huls AG
Chemical Description: Alkyl polyglycol ether
Category: Emulsifier, nonionic
Applications: Emulsifier for solvents, esp. solvent naphtha and aromatic hydrocarbons; for pesticides, cold cleaners
Form: Liquid
Composition: 100% active ingred.
References: Brochure No. 1049/2

Marlowet LVX. Chemische Werke Huls AG
Chemical Description: Mixture of carboxylic acid polyglycol esters
Category: Emulsifier, nonionic
Applications: Emulsifier for animal and vegetable oils; for leather auxiliaries, textile lubricants
Form: Liquid
Composition: 100% active ingred.
References: Brochure No. 1049/2

Marlowet MA. Chemische Werke Huls AG
Chemical Description: Mixture of carboxylic acid polyglycol esters and alkylolamides
Category: Emulsifier, nonionic
Applications: Emulsifier for min-

eral oils, esp. spindle oils; for metal working (rolling oils), textile lubricants, solvents
Form: Liquid
Composition: 100% active ingred.
References: Brochure No. 1049/2

Marlowet NF. Chemische Werke Huls AG
Chemical Description: Mixture of carboxylic acid polyglycol esters
Category: Emulsifier, nonionic
Applications: Emulsifier for animal and vegetable oils, chloroparaffins; for leather, auxiliaries, textile lubricants
Form: Liquid
Composition: 100% active ingred.
References: Brochure No. 1049/2

Marlowet OFA. Chemische Werke Huls AG
Chemical Description: Mixture of n-alkyl benzene sulfonate, carboxylic acid polyglycol esters, and alkyl polyglycol ethers
Category: Emulsifier, anionic/nonionic
Applications: Emulsifier for oleic acid, solvents, esp. chlorinated and aromatic hydrocarbons and kerosene; for mixed oil and fat lubricants, pesticides, metal cleaning
Form: Liquid
Composition: 100% active ingred.
References: Brochure No. 1049/2

Marlowet OFW. Chemische Werke Huls AG
Chemical Description: Mixture of carboxylic acid polyglycol esters and alkyl polyglycol ethers
Category: Emulsifier, nonionic
Applications: Emulsifier for mineral oils, esp. white oils; for textile lubricants, metal working
Form: Liquid
Composition: 100% active ingred.
References: Brochure No. 1049/2

Marlowet OTS. Chemische Werke Huls AG
Chemical Description: Carboxylic acid polyglycol esters
Category: Emulsifier, nonionic
Applications: Emulsifier for mineral oils, esp. spindle oils; for metal working and cleaning, textile lubricants
Form: Liquid
Composition: 100% active ingred.
References: Brochure No. 1049/2

Marlowet PA. Chemische Werke Huls AG
Chemical Description: Alkyl polyglycol ether
Category: Nonionic surfactant
Applications: Mixture for paraffin; for textile and paper impregnation, mold release agents, furniture polishes
Form: Paste
Composition: 100% active ingred.
References: Brochure No. 1049/2

Marlowet PW. Chemische Werke Huls AG
Chemical Description: Alkyl polyglycol ether
Category: Emulsifier, nonionic
Applications: Emulsifier for paraffin and low melting waxes; for

textile and paper impregnation, mold release agents, furniture polishes
Form: Paste
Composition: 100% active ingred.
References: Brochure No. 1049/2

Marlowet RA. Chemische Werke Huls AG
Chemical Description: Mixture of n-alkylbenzene sulfonate and carboxylic acid polyglycol esters
Category: Emulsifier, anionic/nonionic
Applications: Emulsifier for essential oils; for cosmetic preparations
Form: Liquid
Composition: 100% active ingred.
References: Brochure No. 1049/2

Marlowet RNP. Chemische Werke Huls AG
Chemical Description: Mixture of carboxylic acid polyglycol esters, alkylaryl polyglycol ether, and a glycol derivative
Category: Emulsifier, nonionic
Applications: Emulsifier for essential oils, for cosmetic preparations
Form: Liquid
Composition: 100% active ingred.
References: Brochure No. 1049/2

Marlowet SAF. Chemische Werke Huls AG
Chemical Description: Mixture of alkyl polyglycol esters
Category: Emulsifier, nonionic
Applications: Emulsifier for paraffinic mineral oils, esp. white oils; for textile lubricants
Form: Liquid
Composition: 100% active ingred.
References: Brochure No. 1049/2

Marlowet SLM. Chemische Werke Huls AG
Chemical Description: Mixture of carboxylic acid polyglycol esters and alkylolamides
Category: Emulsifier, nonionic
Applications: Emulsifier for mineral oils, spindle oils, solvents; for textile lubricants
Form: Liquid
Composition: 100% active ingred.
References: Brochure No. 1049/2

Marlowet SLS. Chemische Werke Huls AG
Chemical Description: Carboxylic acid polyglycol ester
Category: Emulsifier, nonionic
Applications: Emulsifier for mineral oils, white spirit, bitumen; for loom oils, textile lubricants, cold cleaners, road construction
Form: Liquid
Composition: 100% active ingred.
References: Brochure No. 1049/2

Marlowet SW. Chemische Werke Huls AG
Chemical Description: Mixture of carboxylic acid polyglycol esters and alkyl polyglycol ether
Category: Emulsifier, nonionic
Applications: Emulsifier for mineral oils; for textile lubricants, metal working
Form: Liquid
Composition: 100% active ingred.
References: Brochure No. 1049/2

Marlowet SWN. Chemische Werke Huls AG
Chemical Description: Mixture of carboxylic acid polyglycol esters and alkyl polyglycol ethers
Category: Emulsifier, nonionic
Applications: Emulsifier for spindle and white oils, solvents; for easily washed-out textile lubricants, metal cleaning
Form: Liquid
Composition: 100% active ingred.
References: Brochure No. 1049/2

Marlowet T. Chemische Werke Huls AG
Chemical Description: Amine salt of carboxylic acid
Category: Emulsifier, anionic
Applications: Emulsifier for mineral oils, white spirit, bitumen; for metal working, road construction
Form: Paste
Composition: 100% active ingred.
References: Brochure No. 1049/2

Marlowet TM. Chemische Werke Huls AG
Chemical Description: Alkylaryl polyglycol ether
Category: Emulsifier, nonionic
Applications: Emulsifier for mineral oils; for metal working, cold cleaners
Form: Liquid
Composition: 100% active ingred.
References: Brochure No. 1049/2

Marlowet WOE. Chemische Werke Huls AG
Chemical Description: Alkyl polyglycol ether
Category: Emulsifier, nonionic
Applications: Emulsifier for paraffinic mineral oils, esp. white oils; for textile lubricants
Form: Liquid
Composition: 100% active ingred.
References: Brochure No. 1049/2

Marlowet WSD. Chemische Werke Huls AG
Chemical Description: Mixture of alkyl and alkylaryl polyglycol ethers
Category: Emulsifier, nonionic
Applications: Emulsifier for mineral oils, esp. white oils; for textile lubricants
Form: Liquid
Composition: 100% active ingred.
References: Brochure No. 1049/2

Marlox LM 25/30, LM 75/30, LP 90/20. Chemische Werke Huls AG
Chemical Description: Polypropylene glycol/ethylene oxide addition products
Category: Nonionic surfactant
Applications: Components for liquid, paste, and powder detergents and cleaning agents, esp. in low-foam dishwashing agents
Form: Liquid
Composition: 100% active ingred.

Marlox NP 109. Chemische Werke Huls AG
Chemical Description: Alkylphenol/alkylene oxide addition product
Category: Nonionic surfactant
Applications: Components having good wetting power for low-foam

formulations
Form: Liquid
Composition: 100% active ingred.

Marlox S 58. Chemische Werke Huls AG
Chemical Description: Fatty alcohol/alkylene oxide addition products
Category: Detergent, nonionic
Applications: For low-foam formulations and in combination with other detergent raw materials
Form: Liquid
Composition: 100% active ingred.

Marquest 100. Mars Chem. Corp.
Chemical Description: Tetrasodium salt of ethylenediaminetetraacetic acid
Category: Chelating agent, stabilizer, anionic
Applications: Sanitary supply and industrial cleaners; clarification of soap and detergent sol'ns.; keeps metal cleaning and anodizing baths cleaner; clarification and stabilization of cosmetics
Form: Clear liquid; Color: Very pale yellow
Properties: Density: 11.1 lb/gal; Sp. gr.: 1.3; Stability: Good in hot water, acids, alkalis, or electrolytes; pH: 11–12 (10% sol'n.)
Composition: 38% min. active

Mars AMLS. Mars Chem. Corp.
Chemical Description: Ammonium lauryl sulfate
Cateogry: Foaming agent, detergent, wetting agent, emulsifier, anionic
Applications: Cosmetics—hair shampoo, bubble bath, and as emulsifier in cleansing creams; sanitary supply and industrial cleaners—dry foam carpet shampoo, upholstery cleaner, car wash, dishwash, other high foam detergent formulations
Form: Clear liquid; Color: Pale yellow
Properties: Density: 8.5 lb/gal; Sp. gr.: 1.02; Visc.: 600 cps max.; Stability: Good to anionic and nonionic surfactants, hard water, alkali, acids, lower alcohols, and electrolytes; pH: 7.0–7.5 (10% sol'n.)
Composition: 30% active ingred.

Mars Anionic 68. Mars Chem. Corp.
Chemical Description: Blend of linear sodium alkyl aryl sulfonate and nonionic
Category: Detergent, anionic
Applications: Sanitary supply and industrial cleaners—all purpose cleaners, car wash, dishwash detergents, drum dried and spray dried cleaners, other light to medium duty liquid detergents
Form: Clear gel; Color: Medium amber
Properties: Solubility: Sol. in water (incl. calcium and magnesium hardness), lower alcohols, anionic, and nonionic surfactants; Density: 8.8 lb/gal; Sp. gr.: 1.05; pH: 6.8–7.2 (10% sol'n.); Biodegradable
Composition: 68% active ingred.

Mars Defoamer A-69. Mars Chem. Corp.
Chemical Description: Blend of silicone and nonsilicone defoam-

ing components
Category: Defoamer, nonionic
Applications: Defoamer in waste treatment plants, for recovery tanks in carpet extraction cleaners and automatic scrubbers; in textile wet processing incl. preparation, dyeing, and finishing; in agricultural spray application and washing operations such as egg washing
Form: Opaque emulsion; Color: Off white; Odor: Bouquet
Properties: Solubility: Sol. in water; Density: 8.2 lb/gal; Sp. gr.: 0.98; Visc.: 400 cps; pH: 7.5–8.5 (10% sol'n.); Biodegradable
Composition: 11% active ingred.

Mars Defoamer D-10. Mars Chem. Corp.
Chemical Description: Silicone emulsion
Category: Defoamer
Applications: Defoamer in sanitary supply and industrial cleaners (formulated nonfoaming cleaners, automatic scrubber and carpet extractor cleaning, waste treatment plants and lift stations); in textile wet processing incl. preparation, dyeing, and finishing; in paper processing incl. bleaching and coating operations; in agricultural sprays and washing operations
Form: Opaque emulsion; Color: Off white
Properties: Solubility: Disp. in water; Density: 8.3 lb/gal; Sp. gr.: 1.0; Visc.: 200–800 cps; Stability: Stable in acid, alkaline, or neutral systems; pH: 7.5–8.5 (10% sol'n.); Biodegradable
Composition: 10% min. active

Mars Defoamer E-3. Mars Chem. Corp.
Chemical Description: Silicone emulsion
Category: Defoamer, nonionic
Applications: Silicone base for manufacture of defoamers; defoamer in textile wet processing, extraction cleaning, lift station, and waste treatment plants, in nonfoaming cleaners, in wet processing in the pulp and paper industry
Form: Opaque emulsion; Color: Off white
Properties: Solubility: Disp. in water; Density: 8.1 lb/gal; Sp. gr.: 0.97; Visc.: 400 cps; pH: 7–8; Biodegradable
Composition: 30% active ingred.

Mars Defoamer G-34. Mars Chem. Corp.
Chemical Description: Nonsilicone emulsion
Category: Defoamer
Applications: Defoamer for textile wet processing, for sanitary supply and industrial chemicals, and in the pulp and paper industry where silicone defoamers are undesirable
Form: Opaque emulsion; Color: White
Properties: Solubility: Disp. in water; Density: 7.2 lb/gal; Sp. gr.:

0.86; Visc.: 250 cps; pH: 6.5–7.5 (2% sol'n.)
Composition: 85% active ingred.

Mars Emulsifier C. Mars Chem. Corp.
Chemical Description: Complex difatty quaternary
Category: Emulsifier, antistat, lubricant, cationic
Applications: Emulsification of solvents, mineral oils, and waxes; sanitary supply and industrial chemicals (auto spray waxes and wash-and-wax formulations, emulsifiable mop oil, solvent degreasers); textile dry cleaning antistats, fabric lubricants and softeners, dye levelers
Form: Clear liquid; Color: Pale amber; Odor: Mild, fatty
Properties: Solubility: Sol. in mineral oil, kerosene, xylene, mineral spirits, alcohols; disp. in water; Density: 7.8 lb/gal; Sp. gr.: 0.93; Visc.: 200 cps; Pour Pt.: 10 C; Stability: Stable in presence of nonionic and cationic surfactants; pH: 4.5–5.5 (10% sol'n.); Biodegradable
Composition: 74% min. active

Mars ES-60. Mars Chem. Corp.
Chemical Description: 60% active sodium lauryl ether sulfate
Category: Detergent, foaming agent, anionic
Applications: Cosmetics (hair shampoos, bubble baths); sanitary supply and industrial cleaners (dish detergents, car wash, foaming degreasers, and other applications where foam and grease removal are necessary)
Form: Clear liquid; Color: Light amber
Properties: Density: 8.8 lb/gal; Sp. gr.: 1.04; Visc.: 64 cs @ 100 F; Stability: Stable to moderate concentrations of acid or alkali, in hard water aq. systems, in hydrophilic solvent-based systems; pH: 8.0–9.0 (5% sol'n.); Biodegradable
Composition: 58–60% active

Mars NI-100. Mars Chemical Corp.
Chemical Description: Polyethylene glycol ether
Category: Wetting and rewetting agent, dispersant, foaming agent, emulsifier, nonionic
Applications: Cosmetics; sanitary supply and industrial cleaners as a detergent, wetting agent, and emulsifier; agricultural dormant oil and insecticide sprays; latex and polymer emulsifier, detergent, and wetting agent in textile, paper, leather, compounding, and paint industries; as wetting agent with mineral acids and corrosion inhibitors in metal processing industry
Form: Clear, slightly viscous liquid; Color: Pale yellow
Properties: Solubility: Sol. in water, alcohols, glycols, certain solvents (however, certain concs. in water at room temperature will form gels); Density: 8.75 lb/gal; Sp. gr.: 1.06; Visc.: 30.0 cps; HLB: 13.0; Cloud Pt.: 54 ± 2 C (2% sol'n.); Pour Pt.: 15 C; Stability: Stable to acid, alkalis, and

most electrolytes in typical aq. sol'ns.; pH: 7–8 (10% sol'n.); Biodegradable
Composition: 100% active ingred.

Mars NI-233. Mars Chem. Corp.
Chemical Description: Ethoxylated alcohol
Category: Emulsifier, nonionic
Applications: Oil-sol. emulsifier for bath oils, and cream and lotion cleansing creams in cosmetics industry; sanitary supply and industrial cleaners (mop treatments, solvent-based motor cleaners, carburetor and parts cleaners, fuel oil additives); agricultural dormant oil spray concentrates, insecticides, and herbicides
Form: Clear liquid; Color: Colorless
Properties: Solubility: Sol. in hexylene and propylene glycol, methanol, ethanol, isopropanol, mono butyl ethers, perchloroethylene, heavy aromatic naphtha, kerosene, mineral oil, xylene, toluene; disp. in water; Density: 7.7 lb/gal; Sp. gr.: 0.93; Visc.: 40 cps; HLB: 8.0; pH: 7–8 (10% sol'n.); Biodegradable
Composition: 100% active ingred.

Mars PA-400. Mars Chem. Corp.
Chemical Description: Free acid form of a phosphated polyether alcohol
Category: Detergent, coupling agent, corrosion inhibitor, anionic
Applications: Sanitary supply and industrial cleaners—aq. and powdered steam cleaning compounds, heavy duty degreasers, and other highly built detergents; coupling agent for certain surfactants; anticorrosive effect @ pH $\geqslant$ 9.5
Form: Viscous liquid; Color: Yellow
Properties: Solubility: Sol. in xylene, toluene trichlorethylene, carbon tetrachloride, most glycols and short chain alcohols, monobutyl ethers; sol. in aq. sol'ns. @ pH 1–14; disp. in mineral oil, kerosene; Density: 9.15 lb/gal; Sp. gr.: 1.1; Visc.: 6000 cps; Pour Pt.: $<$ 0 C; Stability: Excellent tolerance to alkali and alkaline builders in built detergent formulations; pH: 1.0–2.0 (10% sol'n.); Biodegradable
Composition: 100% active ingred.

Mars Pro Conc I. Mars Chem. Corp.
Chemical Description: Anionic/nonionic surfactant blend with foam stabilizer and water softening agents
Category: Detergent, foaming agent, foam stabilizer, water softener, anionic/nonionic
Applications: Sanitary supply and industrial cleaners—dish detergent, car wash, hand soap, neutral and all purpose cleaners, butyl degreaser cleaners for pressure washer, butyl degreasers and wax strippers
Form: Clear liquid; Color: Amber
Properties: Density: 9.1 lb/gal; Sp. gr.: 1.09; Visc.: 500 cps; pH: 8.5–9.5 (10% sol'n.)
Composition: 64–66% solids

Mars Pro Conc II. Mars Chem. Corp.
Chemical Description: Anionic/nonionic surfactant blend
Category: Detergent, anionic/nonionic
Applications: Dry foam carpet and upholstery shampoos, animal shampoo, liquid waterless hand soap, car wash, dishwash; high foaming with good detergency in high electrolyte and low temperature conditions; cosmetics and shampoos
Form: Clear liquid; Color: Light yellow
Properties: Solubility: Sol. in water; Density: 8.5 lb/gal; Sp. gr.: 1.02; Visc.: 1000 cps; Cloud Pt.: 15 C; pH: 7–8 (10% sol'n.); Biodegradable
Composition: 36–38% solids

Mars SA-98-S. Mars Chem. Corp.
Chemical Description: Linear alkyl aryl sulfonic acid
Category: Detergent intermediate, wetting agent, anionic
Applications: Sanitary supply, institutional, and janitorial cleaners and cosmetics; detergent intermediate (neutralized sodium salt for all-purpose cleaners; isopropylamine salt as emulsifier and dispersant in solvent cleaning compounds, etc.)
Form: Viscous liquid; Color: Reddish amber
Properties: Solubility: Completely miscible with water; Density: 8.8 lb/gal; Sp. gr.: 1.06; Acid Value: 185; Biodegradable
Composition: 98% active ingred.

Mars SLS-30. Mars Chem. Corp.
Chemical Description: Sodium lauryl sulfate
Category: Foaming agent, detergent, wetting agent, emulsifier, anionic
Applications: Cosmetics–hair shampoo, bubble bath, and as emulsifier in cleansing creams and cosmetics; sanitary supply and industrial cleaners–dry foam carpet shampoo, upholstery cleaner, car wash, dishwash, and other high foam detergent formulations
Form: Clear liquid; Color: Pale yellow
Properties: Density: 8.6 lb/gal; Sp. gr.: 1.04; Visc.: 550 cps; Cloud Pt.: 14 C max.; Stability: Good to anionic and nonionic surfactants, hard water, alkali, acids, lower alcohols, and electrolytes; pH: 7.5–9.0 (10% sol'n.); Biodegradable
Composition: 29% active ingred.

Mars SLS-A95. Mars Chem. Corp.
Chemical Description: Sodium lauryl sulfate
Category and Applications: See Mars SLS-30
Form: Clear liquid; Color: Yellow
Properties: Density: 8.6 lb/gal; Sp. gr.: 1.03; Visc.: 40 cps; Cloud Pt.: 18 C max.; Stability: Good to anionic and nonionic surfactants, hard water, alkali, acids, lower alcohols, and electrolytes; pH: 7.0–8.5 (10% sol'n.); Biodegradable
Composition: 30% solids

Mars SXS. Mars Chem. Corp.
Chemical Description: Sodium xylene sulfate
Category: Solubilizer, coupling agent, hydrotrope
Applications: As a hydrotrope for coupling builder with detergents and as a cloud point depressant in cosmetics and sanitary supply industries
Form: Clear liquid; Color: Colorless; Odor: Mild alcohol
Properties: Density: 9.8 lb/gal; Sp. gr.: 1.18; Visc.: < 20 cps; Stability: Compatible with most inorganic phosphates and silicates, acid, and alkali; pH: 8–9 (10% sol'n.)
Composition: 40% active

Mars TLS-40. Mars Chem. Corp.
Chemical Description: Triethanolamine lauryl sulfate
Category: Detergent, foaming agent, wetter, emulsifier, anionic
Applications: Cosmetic industry–hair shampoo, bubble bath, aerosol shaving cream, and as emulsifier in cleansing creams and cosmetics, esp. where mildness is required; sanitary supply and industrial cleaners–dry foam carpet shampoo, upholstery cleaner, car wash, dishwash, and other high foam detergent formulations
Form: Clear liquid; Color: Pale yellow
Properties: Density: 8.7 lb/gal; Sp. gr.: 1.04; Visc.: 500 cps; Cloud Pt.: 0 C; Stability: Stable in acidic and alkaline systems @ pH 4–12, hard water aq. systems, hydrophilic solvent-based systems, and systems containing anionic and/or nonionic surfactants; pH: 6.5–7.5 (10% sol'n.)
Composition: 40% active ingred.

Marsamid 10. Mars Chem. Corp.
Chemical Description: Modified 2:1 coconut diethanolamide with blend of amide, free fatty acid, and free diethanolamine
Category: Detergent, coupler, emulsifier, viscosity builder, anionic/nonionic
Applications: Sanitary supply and industrial cleaners–general purpose household and heavy duty detergents, floor strippers and cleaners, car wash, white sidewall cleaner, and oil and solvent emulsifier; solvent emulsifier in pesticide formulations
Form: Clear liquid; Color: Dark amber
Properties: Density: 8.4 lb/gal; Sp. gr.: 1.01; Visc.: 1900 cps; Pour Pt.: 10 C min.; Acid Value: 35–45; Stability: Excellent in highly alkaline systems; pH: 8.5–9.5 (5% sol'n.); Biodegradable
Composition: 100% solids content

Marsamid 40. Mars Chem. Corp.
Chemical Description: 2:1 coconut diethanolamide
Category: Foam stabilizer, viscosity builder, wetting agent, detergent, nonionic
Applications: Sanitary supply and industrial cleaners–general purpose cleaners and degreasers and

in combination with other surfactants to produce formulations with special properties, e.g., foam, viscosity, and corrosion inhibition
Form: Clear liquid; Color: Amber
Properties: Solubility: Sol. in water (in all concs.), polar organic solvents, and many hydrocarbons and chlorinated organic compounds; Density: 8.55 lb/gal; Sp. gr.: 1.025; Visc.: 1450 cps; Acid Value: < 10; Stability: Good in alkaline systems; pH: 9.5–10.5 (2% sol'n.); Biodegradable
Composition: 99+% solids
Reaction: Slightly alkaline

Marsamid 50. Mars Chem. Corp.
Chemical Description: 1:1 coconut diethanolamide with high amide content of 85%
Category: Viscosity builder, foam builder and stabilizer, nonionic
Applications: Cosmetic industry—shampoos, bubble baths, cosmetics; sanitary supply and industrial cleaners—light duty detergents, dishwash, floor cleaners, car wash, white sidewall tire cleaners, waterless hand soap
Form: Viscous liquid; Color: Pale yellow
Properties: Solubility: Disp. in water; sol. in alcohols, glycols, carbon tetrachloride, dimethyl sulfoxide, Cellosolve-type ethers, Stoddard solvent, pine oil; Density: 8.21 lb/gal; Visc.: 1500 cps; Acid Value: < 10; pH: 9.0–9.5 (2% sol'n.); Biodegradable
Composition: 100% solids

Marsoft 75C. Mars Chem. Corp.
Category: Fabric softener, cationic
Applications: Commercial and household laundry softener; all fabric softener for textile industry; in paper industry, used to impregnate paper for home laundry use, and as paper additive for softening
Form: Clear liquid; Color: Amber; Odor: Mild, fatty
Properties: Solubility: Disp. in water; Density: 7.8 lb/gal; Sp. gr.: 0.93; Visc.: 200 cps; Pour Pt.: 10 C; Acid No.: 5; pH: 3.5–6.5; Biodegradable
Composition: 74% active ingred.

Marsonol SF-C2. Mars Chem. Corp.
Chemical Description: Di-sodium salt of a dicarboxylated imidazoline derived from coconut fatty acid and amine condensate
Category: Foaming agent and stabilizer, detergent, wetting agent, emulsifier, amphoteric
Applications: Industrial cleaners—heavy duty liquid cleaners, steam cleaner, aluminum brightener, automotive cleaner, white wall tire cleaner, dishwash, foaming wax stripper, heavy duty spray degreaser; cosmetics—bath oil, waterless hand soap, conditioning shampoos
Form: Clear liquid; Color: Amber
Properties: Solubility: Sol. in water; Density: 8.9 lb/gal; Sp. gr.: 1.07; Visc.: 50 cps; Stability: Stable to very high levels of electrolytes and/or alkalis, and to acid

and base; pH: 9.6 (20% @ 30 C); Biodegradable
Composition: 38–40% active

Marvanol 55% SPO. Marlowe-Van Loan
Chemical Description: Sulfated propyl oleate
Category: Detergent, wetting agent, anionic
Applications: Wetting and leveling agent
Form: Liquid; Color: Amber
Composition: 55% active ingred.

Marvanol 60% SBO. Marlowe-Van Loan
Chemical Description: Sulfated butyl oleate
Category: Detergent, wetting agent, emulsifier, anionic
Applications: Wetting and leveling agent
Form: Liquid; Color: Amber
Composition: 60% active ingred.

Marvanol 75% SCO. Marlowe-Van Loan
Chemical Description: Sulfated castor oil
Category: Detergent, wetting agent
Applications: Dyeing assistant and lubricant; finishing operations
Form: Liquid; Color: Amber
Composition: 75% active ingred.

Marvanol CO. Marlowe-Van Loan
Chemical Description: Ethoxylated phenol
Category: Detergent, emulsifier, wetting agent
Applications: Textiles detergent; wetting agent in finishing operations
Form: Liquid; Color: Colorless
Composition: 99+ % active

Marvanol DC. Marlowe-Van Loan
Chemical Description: Ethoxylated alcohol
Category: Wetting agent
Applications: Textile fiber wetting and dyeing
Form: Liquid; Color: Colorless
Composition: 35% active ingred.

Marvanol GC. Marlowe-Van Loan
Chemical Description: Sulfo-succinamide
Category: Wetting agent, anionic
Applications: Textile fiber wetting and dye leveling
Form: Gel; Color: Amber
Composition: 65% active ingred.

Marvanol Penetrant 35. Marlowe-Van Loan
Chemical Description: Amine neutralized sulfonic acid
Category: Detergent, wetting agent, anionic
Applications: Textile dyeing and finishing leveling and retarding agent for acid and neutral acid dyes
Form: Liquid; Color: Amber
Composition: 40% active ingred.
Reaction: Neutral

Marvanol RE-1274. Marlowe-Van Loan
Chemical Description: 1:1 coconut fatty amide
Category: Emulsifier, cationic

Applications: Viscosity builder, dye leveler
Form: Liquid; Color: Amber
Properties: Sp. gr.: 0.950
Composition: 97% active ingred.

Marvanol RE-1281. Marlowe-Van Loan
Chemical Description: 2:1 coconut fatty amide
Category: Emulsifier, cationic
Applications: Viscosity builder, dye leveler
Form: Liquid; Color: Amber
Properties: Sp. gr.: 1.02
Composition: 98% active ingred.

Marvanol Solvent Scour 34. Marlowe-Van Loan
Chemical Description: Solvents and detergents
Category: Detergent, anionic
Applications: Solvent scour
Form: Liquid; Color: Amber
Composition: 100% active ingred.

Marvanscour KW. Marlowe-Van Loan
Chemical Description: Solvents and detergents
Category: Detergent, nonionic
Applications: Prescour and/or afterscour
Form: Liquid; Color: Straw
Composition: 83% active ingred.

Marvantex BS. Marlowe-Van Loan
Chemical Description: Alkaline built blend
Category: Detergent, anionic
Applications: Alkaline scour for dye stripping
Form: Powder; Color: Yellow to white
Composition: 100% active ingred.
Reaction: Alkaline

Marvantex GS. Marlowe-Van Loan
Chemical Description: Alkaline scour
Category: Detergent, anionic
Applications: Oil, grease, and graphite removal
Form: Powder
Composition: 100% active ingred.
Reaction: Alkaline

Mayodan. Grindsted Prod.
Chemical Description: Stabilizer blends
Category: Emulsifier
Applications: Food emulsifier for mayonnaise and salad dressings; improves emulsification, regulates viscosity, and enhances flavor

Maypon 4C. Stepan Chem. Co.
Chemical Description: Potassium salt polypeptide coco fatty acid condensate
Category: Detergent, anionic
Applications: Personal care products; highly refined, general purpose cleansers with good foaming properties and mildness
Form: Clear to slightly hazy liquid; Color: Amber
Properties: Solubility: Miscible with water in all proportions; Visc.: 500 cps max.; pH: 6.7–7.3 (as is)
Composition: 34–36% solids

Maypon 4CT. Stepan Chem. Co.
Chemical Description: Triethanolamine salt polypeptide coco fatty acid condensate
Category: Detergent, anionic
Applications: Organic form similar to Maypon 4C with lower cloud pt.
Form: Clear to slightly hazy liquid; Color: Light amber
Properties: Solubility: Miscible with water in all proportions; Visc.: 150 cps max.; pH: 6.7–7.3
Composition: 38–40% solids

Maypon K. Stepan Chem. Co.
Chemical Description: Sodium salt polypeptide soya fatty acid concentrate
Category: Detergent, anionic
Applications: For industrial cleaners; mildness helps to prevent dermatitis
Form: Clear liquid; Color: Amber
Properties: Solubility: Miscible with water at 50 C in all proportions; Sp. gr.: 1.12–1.14; Visc.: 6000 cps max.; pH: 8.5–9.5 (as is)
Composition: 42–44% solids

Maypon UD. Stepan Chem. Co.
Chemical Description: Potassium salt polypeptide undecylenic fatty acid condensate
Category: Detergent, anionic
Applications: Personal care products; antifungal properties; good for dermatological preparations
Form: Clear liquid; Color: Dark amber
Properties: Solubility: Sol. in alcohol, glycerin, glycols, other polyols; miscible with water in all proportions; Sp. gr.: 1.06–1.09; Visc.: 100 cps max.; pH: 6.3–6.8
Composition: 34–38% solids

Mazamide 25. Mazer Chem. Inc.
Chemical Description: 2:1 coconut sulfonic acid
Category and Applications: See Mazamide 65
Form: Liquid
Properties: Sp. gr.: 1.053; Biodegradable

Mazamide 65. Mazer Chem. Inc.
Chemical Description: 2:1 mixed fatty acid
Category: Emulsifier, detergent, nonionic/anionic
Applications: See Mazamide 70
Form: Liquid
Properties: Sp. gr.: 0.99–1.02; pH: 9.0–10.0; Biodegradable
Toxicity: Defatting to skin

Mazamide 65CZ. Mazer Chem. Inc.
Chemical Description: 2:1 mixed fatty acid
Category and Applications: See Mazamide 65
Form: Liquid
Properties: Sp. gr.: 0.99–1.02; pH: 9.0–10.0; Biodegradable
Toxicity: Defatting to skin

Mazamide 66. Mazer Chem. Inc.
Chemical Description: 2:1 mixed fatty acid
Category and Applications: See Mazamide 65
Form: Liquid
Properties: Sp. gr.: 0.99–1.02; pH: 9.0–10.0; Biodegradable
Toxicity: Defatting to skin

Mazamide 70. Mazer Chem. Inc.
Chemical Description: 2:1 coconut oil alkanolamide
Category: Emulsifier, detergent, nonionic
Applications: Hard surface cleaners, dishwash liquids; as emulsifiers and detergents in metalworking fluids/synthetic coolants, waterless hand cleaners and automotive specialties; can be neutralized with fatty acids to make amine soaps which are themselves anionic surfactants
Form: Liquid
Properties: Sp. gr.: 0.99–1.02; pH: 9.0–10.0; Biodegradable
Toxicity: Defatting to skin

Mazamide 80. Mazer Chem. Inc.
Chemical Description: 1:1 coconut diethanol superamide
Category and Applications: See Mazamide LS-196; also used in hard surface wetters and cleaners
Form: Liquid
Properties: Sp. gr.: 0.98–1.00; pH: 9.0–10.5; Biodegradable

Mazamide C-5. Mazer Chem. Inc.
Chemical Description: POE 6 coco monoethanolamide
Category and Applications: See Mazamide SS-10
Form: Liquid
Properties: Sp. gr.: 1.0–1.1; pH: 9.5–10.5; Biodegradable
Toxicity: Defatting to skin

Mazamide CA-20. Mazer Chem. Inc.
Chemical Description: 2:1 coco fatty acid
Category and Applications: See Mazamide 65
Form: Liquid
Properties: Sp. gr.: 0.99–1.02; pH: 9.0–10.0; Biodegradable
Toxicity: Defatting to skin

Mazamide CS-148. Mazer Chem. Inc.
Chemical Description: 1:1 coconut oil alkanolamide
Category and Applications: See Mazamide LS-196
Form: Liquid
Properties: Sp. gr.: 0.98–1.00; pH: 9.0–10.5; Biodegradable

Mazamide L-5. Mazer Chem. Inc.
Chemical Description: POE 6 lauric monoethanolamide
Category and Applications: See Mazamide SS-10
Form: Solid
Properties: Sp. gr.: 1.0–1.1; pH: 9.5–10.5; Biodegradable
Toxicity: Defatting to skin

Mazamide LM-20. Mazer Chem. Inc.
Chemical Description: 2:1 coco fatty acid
Category and Applications: See Mazamide 65
Form: Liquid
Properties: Sp. gr.: 0.99–1.01; pH: 9.2–10.0; Biodegradable
Toxicity: Defatting to skin

Mazamide LS-173. Mazer Chem. Inc.
Chemical Description: 1:1 lauric myristic diethanol superamide
Category and Applications: See

Mazamide LS-196
Form: Solid
Properties: Sp. gr.: 0.98–1.00; pH: 9.0–10.5; Biodegradable
Toxicity: Defatting to skin

Mazamide LS-196. Mazer Chem. Inc.
Chemical Description: 1:1 lauric alkanolamide
Category: Thickener, foam stabilizer, emulsifier, nonionic
Applications: Cosmetic/toiletry formulations; thicken, stabilize foam structure, consume less acid to facilitate pH adjustment, better odor
Form: Solid
Properties: Sp. gr.: 0.98–1.00; pH: 9.0–10.5; Biodegradable
Toxicity: Defatting to skin

Mazamide O-20. Mazer Chem. Inc.
Chemical Description: 2: 1 oleic
Category and Applications: See Mazamide SS-10
Form: Liquid
Properties: Sp. gr.: 0.99–1.01; pH: 9.0–10.0; Biodegradable
Toxicity: Defatting to skin

Mazamide SS-10. Mazer Chem. Inc.
Chemical Description: 1:1 soybean oil alkanolamide
Category: Emulsifier, lubricant, rust inhibitor, buffing compound, nonionic
Form: Liquid
Properties: Sp. gr.: 0.98–1.00; pH: 9.0–10.5; Biodegradable
Toxicity: Defatting to skin

Mazamide TO-10. Mazer Chem. Inc.
Chemical Description: 1:1 tall oil alkanolamide
Category: Emulsifier, nonionic
Form: Liquid
Properties: Sp. gr.: 0.98–1.00; pH: 9.0–10.0; Biodegradable
Toxicity: Degreasing to skin

Mazamide TO-20. Mazer Chem. Inc.
Chemical Description: 2:1 tall oil
Category and Applications: See Mazamide 70
Form: Liquid
Properties: Sp. gr.: 0.99–1.01; pH: 9.0–10.0; Biodegradabe
Toxicity: Defatting to skin

Mazawet 77. Mazer Chem. Inc.
Chemical Description: Alkyl polyoxyalkylene ether
Category: Wetting agent, low foam surfactant above its cloud pt., moderate foamer below its cloud pt., nonionic
Applications: Used in textile processing, spray dried detergents, emulsion polymerization, dry cleaning surfactant systems, hard surface cleaners, deresination of sulfite pulp
Form: Color: 1 (Gardner)
Properties: m.w.: 426 avg.; Solubility: 5% is sol. in toluene, perchloroethylene, Stoddard solvent, mineral seal oil, propylene glycol, water, sodium hydroxide (1% aq.), potassium hydroxide (3% aq.), sodium chloride (5% aq.), STPP (1% aq.), TKPP (1% aq.); Sp. gr.: 0.965; Cloud Pt.: 45 C (1% aq.); Flash Pt.: 340 F (PMCC);

Pour Pt.: 11 C; Hydroxyl Value: 15; Surface Tension: 28.1 dynes/ cm (0.1%); pH: 6.5 (3% aq.)
Composition: 100% active ingred.

Mazeen C 2. Mazer Chem. Inc.
Chemical Description: Polyoxyethylene (2) coco amine
Category: Emulsifier, antistat, rewetting agent, lubricant, cationic
Applications: Insecticides and herbicides; antistatic agents; grease additives; textile lubricants; emulsifiers for lubricants, water-based inks, and cosmetics
Form: Liquid; Color: 11 (Gardner)
Properties: m.w.: 285; Solubility: Sol. in benzene, acetone, isopropanol, mineral oil; forms gel in water; Sp. gr.: 0.874; Surface Tension: 28 dynes/ cm (0.1% sol'n.)
Composition: 100% conc.

Mazeen C 5. Mazer Chem. Inc.
Chemical Description: Polyoxyethylene (5) coco amine
Category and Applications: See Mazeen C 2
Form: Liquid; Color: 11 (Gardner)
Properties: m.w.: 425; Solubility: Sol. in water, benzene, acetone, isopropanol, mineral oil; Sp. gr.: 0.976; Surface Tension: 33 dynes/ cm (0.1% sol'n.)
Composition: 100% conc.

Mazeen C 10. Mazer Chem. Inc.
Chemical Description: Polyoxyethylene (10) coco amine
Category and Applications: See Mazeen C 2
Form: Liquid; Color: 11 (Gardner)
Properties: m.w.: 645; Solubility: Sol. in water, benzene, acetone, isopropanol; Sp. gr.: 1.017; Surface Tension: 39 dynes/cm (0.1% sol'n.)
Composition: 100% conc.

Mazeen C 15. Mazer Chem. Inc.
Chemical Description: Polyoxyethylene (15) coco amine
Category and Applications: See Mazeen C 2
Form: Liquid; Color: 9 (Gardner)
Properties: m.w.: 860; Solubility: Sol. in water, benzene, acetone, isopropanol; Sp. gr.: 1.042; Surface Tension: 41 dynes/cm (0.1% sol'n.)
Composition: 100% conc.

Mazeen S 2. Mazer Chem. Inc.
Chemical Description: Polyoxyethylene (2) soy amine
Category and Applications: See Mazeen C 2
Form: Liquid; Color: 14 (Gardner)
Properties: m.w.: 350; Solubility: Sol. in benzene, acetone, isopropanol, mineral oil; Sp. gr.: 0.911; Surface Tension: 26 dynes/cm (0.1% sol'n.)
Composition: 100% conc.

Mazeen S 5. Mazer Chem. Inc.
Chemical Description: Polyoxyethylene (5) soy amine
Category and Applications: See Mazeen C 2
Form: Liquid; Color: 14 (Gardner)
Properties: m.w.: 480; Solubility: Sol. in benzene, isopropanol; partly sol. in acetone and mineral oil;

forms gel in water; Sp. gr.: 0.951; Surface Tension: 33 dynes/cm (0.1% sol'n.)
Composition: 100% conc.

Mazeen S 10. Mazer Chem. Inc.
Chemical Description: Polyoxyethylene (10) soy amine
Category and Applications: See Mazeen C 2
Form: Liquid;Color: 14 (Gardner)
Properties: m.w.: 710; Solubility: Sol. in water, benzene, acetone, isopropanol; partly sol. in mineral oil; Sp. gr.: 1.020; Surface Tension: 40 dynes/cm (0.1% sol'n.)
Composition: 100% conc.

Mazeen S 15. Mazer Chem. Inc.
Chemical Description: Polyoxyethylene (15) soy amine
Category and Applications: See Mazeen C 2
Form: Liquid;Color: 18 (Gardner)
Properties: m.w.: 930; Solubility: Sol. in water, benzene, acetone, isopropanol; Sp. gr.: 1.040; Surface Tension: 43 dynes/cm (0.1% sol'n.)
Composition: 100% conc.

Mazeen T 2. Mazer Chem. Inc.
Chemical Description: Polyoxyethylene (2) mixed amine
Category and Applications: See Mazeen C 2
Form: Liquid;Color: 11 (Gardner)
Properties: m.w.: 350; Solubility: Sol. in benzene, acetone, isopropanol, mineral oil; Sp. gr.: 0.916; Surface Tension: 29 dynes/cm (0.1% sol'n.)
Composition: 100% conc.

Mazeen T 5. Mazer Chem. Inc.
Chemical Description: Polyoxyethylene (5) mixed amine
Category and Applications: See Mazeen C 2
Form: Liquid-paste; Color: 12 (Gardner)
Properties: m.w.: 480; Solubility: Sol. in benzene, acetone, isopropanol, mineral oil; gels in water; Sp. gr.: 0.966; Surface Tension: 34 dynes/cm (0.1% sol'n.)
Composition: 100% conc.

Mazeen T 15. Mazer Chem. Inc.
Chemical Description: Polyoxyethylene (15) mixed amine
Category and Applications: See Mazeen C 2
Form: Liquid; Color: 18 (Gardner)
Properties: m.w.: 925; Solubility: Sol. in water, benzene, acetone, isopropanol; Sp. gr.: 1.028; Surface Tension: 41 dynes/cm (0.1% sol'n.)
Composition: 100% conc.

Mazol 80 MG. Mazer Chem. Inc.
Chemical Description: Ethoxylated monoglyceride
Category: Emulsifier
Applications: Emulsifier for cakes and icings; whipped toppings
Form: Liquid; Color: Yellow
Properties: Solubility: Sol. in ethanol; partly sol. in soybean oil; disp. in propylene glycol; M.P.: 25–27 C; HLB: ≈ 11.0; Sapon. No.: 65–75; Acid No.: 2.0 max.; Iodine Value: 1

Mazol 300. Mazer Chem. Inc.
Chemical Description: Glycerol monooleate
Category: Antifoam agent, coemulsifier, nonionic
Applications: Antifoam agent for sugar and protein processing, coemulsifier with T-Maz 60 or T-Maz 80
Form: Liquid; Color: Yellow
Properties: Solubility: Sol. in propylene glycol, soybean oil; disp. in ethanol; HLB: ≈ 3.0; Sapon. No.: 150–160; Acid No.: 5.0 max.; Iodine Value: 90
Composition: 100% conc.

Mazol GMO. Mazer Chem. Inc.
Chemical Description: Glycerol monooleate
Category and Applications: See Mazol 300
Form: Liquid; Color: Yellow
Properties: Solubility: Sol. in soybean oil; disp. in ethanol, propylene glycol; HLB: ≈ 2.4; Sapon. No.: 167–177; Acid No.: 5.0 max.; Iodine Value: 90
Composition: 100% conc.

Mazol PGMS. Mazer Chem. Inc.
Chemical Description: Propylene glycol monostearate
Category: Coemulsifier, dispersant, nonionic
Applications: Coemulsifier for edible oil and shortenings; dispersing aid for nondairy creamers
Form: Solid; Color: White-cream
Properties: Solubility: Sol. in propylene glycol; partly sol. in soybean oil; disp. in ethanol; M.P.: 39–46 C; HLB: ≈ 3.4; Sapon. No.: 170–180; Acid No.: 5.0 max.; Iodine Value: 2.0
Composition: 100% conc.

Mazol PGO-31. Mazer Chem. Inc.
Chemical Description: Triglycerol monooleate
Category: Emulsifier, nonionic
Applications: Emulsifier for dietetic, bakery, and confectionary food products; in cosmetics, toiletries, pharmaceuticals, lubricants, mold release compounds, plasticizers for synthetic fibers and plastics
Form: Liquid; Color: 8 max. (Gardner)
Properties: HLB: 6.2; Sapon. Value: 125–150; Acid Value: 4 max.; Iodine Value: 78 max.
Composition: 100% conc.

Mazol PGO-62. Mazer Chem. Inc.
Chemical Description: Hexaglycerol dioleate
Category and Applications: See Mazol PGO-31
Form: Liquid; Color: 10 max. (Gardner)
Properties: HLB: 8.5; Sapon. Value: 105–125; Acid Value: 6 max.; Iodine Value: 75 max.

Mazol PGO-66. Mazer Chem. Inc.
Chemical Description: Hexaglycerol hexaoleate
Category and Applications: See Mazol PGO-31
Form: Liquid; Color: 8 max. (Gardner)

Properties: HLB: 3.0; Sapon. Value: 160–180; Acid Value: 6 max.; Iodine Value: 75 max.

Mazol PGO-104. Mazer Chem. Inc.
Chemical Description: Decaglycerol tetraoleate
Category and Applications: See Mazol PGO-31
Form: Liquid; Color: 8 max. (Gardner)
Properties: HLB: 6.2; Sapon. Value: 125–150; Acid Value: 8 max.; Iodine Value: 60 max.

Mazol PGO-108. Mazer Chem. Inc.
Chemical Description: Decaglycerol octaoleate
Category and Applications: See Mazol PGO-31
Form: Liquid; Color: 8 max. (Gardner)
Properties: HLB: 3.4; Sapon. Value: 155–175; Acid Value: 10 max.; Iodine Value: 75 max.

Mazol PGO-1010. Mazer Chem. Inc.
Chemical Description: Decaglycerol decaoleate
Category and Applications: See Mazol PGO-31
Form: Liquid; Color: 8 max. (Gardner)
Properties: HLB: 2.5; Sapon. Value: 165–185; Acid Value: 10 max.; Iodine Value: 80 max.

Mazol PGS-31. Mazer Chem. Inc.
Chemical Description: Triglycerol monostearate
Category and Applications: See Mazol PGO-31
Form: Solid; Color: 8 max. (Gardner)
Properties: HLB: 6.2; Sapon. Value: 120–145; Acid Value: 6 max.; Iodine Value: 3 max.

Mazol PGS-62. Mazer Chem. Inc.
Chemical Description: Hexaglycerol distearate
Category and Applications: See Mazol PGO-31
Form: Solid; Color: 10 max. (Gardner)
Properties: HLB: 8.5; Sapon. Value: 105–125; Acid Value: 6 max.; Iodine Value: 3 max.

Mazol PGSH-31. Mazer Chem. Inc.
Chemical Description: Triglycerol mono shortening
Category and Applications: See Mazol PGO-31
Form: Semisolid; Color: 10 max. (Gardner)
Properties: HLB: 6.0; Sapon. Value: 115–140; Acid Value: 6 max.; Iodine Value: 40 max.

Mazol PGSH-62. Mazer Chem. Inc.
Chemical Description: Hexaglycerol di shortening
Category and Applications: See Mazol PGO-31
Form: Semisolid; Color: 10 max. (Gardner)
Properties: HLB: 8.3; Sapon. Value: 100–125; Acid Value: 6 max.; Iodine Value: 40 max.

Mazoline OA. Mazer Chem. Inc.
Chemical Description: Oleyl-based imidazoline
Category: Emulsifier

Applications: Mazoline OA or its acid salts are emulsifiers for preparing W/O emulsions; enhances corrosion inhibition and lubricity; used in metalworking
Properties: Solubility: Disp. freely in nonpolar solvents such as kerosene and mineral oils; lipophilic

Mazon R.I. #4. Mazer Chem. Inc.
Chemical Description: Amine-based corrosion inhibitor
Category: Corrosion inhibitor, nonionic
Applications: Synthetic and semi-synthetic metalworking coolants for cast iron, steel, and other ferrous metals; effective at low concentrations; contributes to lubrication
Form: Clear to hazy liquid; Color: Light
Properties: Solubility: Sol. in water and glycols; Sp. gr.: 1.245; Visc.: ≈ 6000 cps; Flash Pt.: 290 F (P-M); pH: 9.6 (1% aq.)

Mazon R.I. #325. Mazer Chem. Inc.
Category: Corrosion inhibitor, nonionic
Applications: Corrosion inhibitor for ferrous metals; coolants and metalworking fluids; resists microbial deterioration
Form: Clear liquid; Color: Light
Properties: Solubility: Sol. in water and lower alcohols; Sp. gr.: 1.130; Visc.: 50 cps; Flash Pt.: 425 F (P-M); Stability: Good; pH: 9.74 (as is)
Composition: 30% water

Mazu DF-100S. Mazer Chem. Inc.
Chemical Description: 100% silicone defoamer compound
Category: Defoamer, nonionic
Applications: Food grade defoamer for aq. systems
Form: Syrupy; Color: Translucent light gray
Properties: Solubility: Disp. in aromatic and chlorinated hydrocarbons; slightly in aliphatic hydrocarbons with slight agitation; Sp. gr.: 1.01
Composition: 100% active ingred.

Mazu DF-110S. Mazer Chem. Inc.
Chemical Description: 10% silicone emulsion defoamer
Category and Applications: See Mazu DF-100S
Form: Light cream; Color: White
Properties: Solubility: Instantly disp. in cold water; Sp. gr.: 1.00
Composition: 10% active ingred.

Mazu DF-130S. Mazer Chem. Inc.
Chemical Description: 30% silicone emulsion defoamer
Category and Applications: See Mazu DF-100S
Form: Medium cream; Color: White
Properties: Solubility: Instantly disp. in cold water; Sp. gr.: 1.01
Composition: 30% active ingred.

Mazu DF 200S. Mazer Chem. Inc.
Chemical Description: 100% silicone compound
Category: Antifoam agent
Applications: Antifoam in nonaq. food and industrial applications—

insecticides, latex manufacture, pharmaceuticals, resin polymerization, ink processing, adhesive manufacturing, food processing
Form: Liquid; Odor: None
Properties: Solubility: Sol. in aliphatic chlorinated and aromatic solvents; Density: 8.34 lb/gal; Sp. gr.: 1.000; Visc.: 2000 cs; Flash Pt.: 310 F (OC)
Composition: 100% active ingred.
Toxicity: Contact with eyes or skin may cause irritation
Containers: 8, 40, 440 lb net wt containers

Mazu DF 210S. Mazer Chem. Inc.
Chemical Description: 10% silicone emulsion
Category: Antifoam agent, nonionic
Applications: Antifoam in food processing, brine sol'ns., dehydrating and evaporating, fermentation processes, abrasive slurries, adhesives, latex emulsions, cutting oils, commercial cleaning compounds, insecticides and pesticides
Form: Liquid; Color: White
Properties: Solubility: Sol. in water @ >90%; Sp. gr.: 1.0; Visc.: 300–500 cs; B.P.: 212 F; Flash Pt.: > 300 F (OC); Stability: Good; pH: 7.0
Composition: 10% active ingred.
Toxicity: May cause temporary eye discomfort
Containers: 40 and 440 lb net wt containers

Mazu DF 230S. Mazer Chem. Inc.
Chemical Description: 30% silicone emulsion
Category: Antifoam agent, nonionic
Applications: Antifoam agent in adhesive manufacturing, chemical processing, latex processing, fruit and juice processing, paint and soap manufacture, starch, textile, and paper processing, metalworking, petroleum processing
Form: Liquid; Color: White
Properties: Solubility: Sol. in water > 90%; Density: 8.34 lb/gal; Sp. gr.: 1.00; Visc.: 1000–2000 cs; B.P.: 212 F; Flash Pt.: > 300 F (OC); Stability: Good; pH: 7.0
Composition: 30% active ingred.
Toxicity: May cause temporary eye discomfort
Containers: 40 and 440 lb net wt containers

Medialan KA Conc. American Hoechst Corp.
Chemical Description: Based on fatty acid sarcoside sodium salt
Category: Detergent, foaming agent, anionic
Applications: Additive to top quality mild toilet preparations and hair shampoos; may be used in cream shampoos in combination with other surfactants
Form: Soft paste; Color: Yellowish; Odor: Weak
Properties: m.w.: 310 ± 5 (avg.); Solubility: Miscible with water in any ratio; Biodegradable
Composition: 63 ± 2% active

Medialan KF. American Hoechst Corp.
Chemical Description: Salt of the condensation product of saturated

fatty acids of medium chain length and sarcosine
Category: Detergent, anionic
Applications: Preparation of clear, liquid shampoos and body lotions, and cream shampoos alone or in combination with other surfactants
Form: Clear, low viscosity liquid; Color: Yellowish
Properties: Solubility: Miscible with water in any proportion
Composition: 40% active ingred.
Reaction: Practically neutral

Melatex 295K. Tokaiseiyu Ind. Co., Ltd.
Chemical Description: Polyacrylic emulsion
Category: Anionic surfactant
Applications: Used for stiff-finishing of synthetic fibers
Form: Liquid
Composition: 45% conc.

Melatex BM-600. Tokaiseiyu Ind. Co., Ltd.
Chemical Description: Polyacrylic emulsion
Category: Anionic surfactant
Applications: Used for stiff-finishing of synthetic and natural fibers
Form: Liquid
Composition: 45% conc.

Mercerol GVC-65 Liquid. Sandoz Ltd.
Category: Wetting agent, mercerizing agent
Applications: Mercerizing penetrant for efficient wetout in caustic soda saturator
Form: Liquid; Color: Yellowish
Properties: Solubility: Disp. in water; Sp. gr.: 1.03
Composition: 65% active ingred.
Reaction: Alkaline

Mercerol NL. Sandoz Ltd.
Chemical Description: Sulfonate, cresol-free, containing solvents
Category: Wetting agent, anionic
Applications: Wetting agent for causticizing baths
Form: Liquid

Mercerol QW. Sandoz Ltd.
Chemical Description: Sulfonate, cresol-free, containing solvents
Category and Applications: See Mercerol NL
Form: Liquid

Mercol 15. Theobald Industries
Chemical Description: Sodium dodecylbenzene sulfonate
Category: Detergent, emulsifier, wetting agent, anionic
Applications: Surfactant component in variety of cleaning agents for textiles, skins, leather, laundering, dyeing, metallurgy, papermaking, paints, foods, transport, household, cosmetics, industrial; for wetting pulp in beater; in water and fire proofing compounds; in insecticides; for removing starches, wall paper; dispersing waxes, pigments, resins
Form: Light density beads; Color: White; Odor: Neutral
Properties: Solubility: Water sol.; Density: 0.15 g/cc; pH: 7.0–8.5; Biodegradable
Composition: 40% active ingred.
Reaction: Alkaline

Toxicity: Eye irritant
Containers: Bags/drums
Warnings and Caution: Avoid prolonged exposure to dust; do not ingest

Mercol 25. Theobald Industries
Chemical Description: Sodium dodecylbenzene sulfonate
Category and Applications: See Mercol 15
Form: Med. density beads; Color: White; Odor: Neutral
Properties: Solubility: Water sol.; Density: 0.25 g/cc; pH: 7.0–8.5; Biodegradable
Composition: 40% active ingred.
Reaction:
Toxicity: Eye irritant
Containers: Bags/drums
Warnings and Caution: Avoid prolonged exposure to dust; do not ingest

Mercol 30. Theobald Industries
Chemical Description: Sodium dodecylbenzene sulfonate
Category and Applications: See Mercol 15
Form: High density beads; Color: White; Odor: Neutral
Properties: Solubility: Water sol.; Density: 0.30 g/cc; pH: 7.0–8.5; Biodegradable
Composition: 40% active ingred.
Reaction: Alkaline
Toxicity: Eye irritant
Containers: Bags/drums
Warnings and Caution: Avoid prolonged exposure to dust; do not ingest

Mercol Special. Theobald Industries
Chemical Description: Sodium dodecylbenzene sulfonate
Category and Applications: See Mercol 15
Form: Beads; Odor: Neutral
Properties: Solubility: Water sol.; Density: 0.30 g/cc; pH: 7.0–8.5; Biodegradable
Composition: 20% active ingred.
Reaction: Alkaline
Toxicity: Eye irritant
Containers: Bags/drums
Warnings and Caution: Avoid prolonged exposure to dust; do not ingest

Merpalyt L-15-1. Elektrochemische Fabrik Kempen
Chemical Description: Alkylamide betaine
Category: Ampholyte
Applications: Cosmetic and shampoo formulations; alcalic and acid cleaning formulations
Form: Liquid; Color: Yellow
Composition: 45% active ingred.
Containers: 120 kg net poly drums

Merpasol DIO 60. Elektrochemische Fabrik Kempen
Chemical Description: Di-i-octylsulfosuccinate, sodium salt
Category: Wetting agent, anionic
Applications: Rapid wetting agent for textile and paper industries
Form: Viscous liquid; Color: Light
Composition: 60% active ingred.
Containers: 125 kg net drums

Merpasol DIO 75. Elektrochemische Fabrik Kempen
Chemical Description: Di-i-octyl-

sulfosuccinate, sodium salt
Category and Applications: See Merpasol DIO 60
Form: Viscous liquid; Color: Light
Composition: 75% active ingred.
Containers: 125 kg net drums

Merpinal D 40. Elektrochemische Fabrik Kempen
Chemical Description: Sodium decyl sulfate
Category: Wetting agent, anionic
Applications: Low-foaming wetting agent for alcalic metal baths; degreasing and cleaning of hard surfaces
Form: Viscous liquid; Color: White
Properties: Biodegradable
Composition: 40% active ingred.
Containers: Coated 200 kg net iron drums

Merpinal LM 40. Elektrochemische Fabrik Kempen
Chemical Description: Narrow cut lauryl sodium sulfate
Category: Detergent, anionic
Applications: Liquid household detergents, hair shampoo, and foam bath preparations
Form: Paste; Color: White
Composition: 40% active ingred.
Containers: 125 kg net drums

Merpinal LM 50. Elektrochemische Fabrik Kempen
Chemical Description: Sodium lauryl sulfate $C_{12}-C_{14}$
Category: Detergent, anionic
Applications: High-foaming paste and powder household and textile detergents
Form: Paste; Color: White
Properties: Biodegradable
Composition: 50% active ingred.
Containers: 200 kg net iron drums

Merpinal LMT 40. Elektrochemische Fabrik Kempen
Chemical Description: Narrow cut lauryl triethanolamine sulfate
Category and Applications: See Merpinal LM 40
Form: Viscous liquid; Color: Light
Composition: 40% active ingred.
Containers: 125 kg net drums

Merpinal LS 40. Elektrochemische Fabrik Kempen
Chemical Description: Broad cut lauryl sodium sulfate
Category: Detergent, anionic
Applications: Paste and powder household and textile detergents
Form: Soft paste; Color: White
Composition: 40% active ingred.
Containers: 200 kg net steel drums

Merpinal LS 50. Elektrochemische Fabrik Kempen
Chemical Description: Sodium lauryl sulfate $C_{12}-C_{18}$
Category: Detergent, anionic
Applications: High-foaming paste and powder household and textile detergents
Form: Soft paste; Color: White
Properties: Biodegradable
Composition: 50% active ingred.
Containers: 200 kg net iron drums

Merpinal Q 101. Elektrochemische Fabrik Kempen
Chemical Description: Sodium octadecyl sulfate

Category: Detergent, anionic
Applications: Paste and powder detergents
Form: Paste; Color: White
Properties: Biodegradable
Composition: 30% active ingred.
Containers: Coated 200 kg net iron drums

Merpinal Q 147. Elektrochemische Fabrik Kempen
Chemical Description: Sodium 2-ethylhexyl sulfate
Category: Wetting agent, anionic
Applications: Low-foaming wetting agent for alcalic metal baths; degreasing and cleaning of hard surfaces
Form: Clear liquid; Color: Yellowish
Properties: Biodegradable
Composition: 40% active ingred.
Containers: Coated 200 kg net iron drums

Merpinal T 35. Elektrochemische Fabrik Kempen
Chemical Description: Tallow fatty alcohol sodium sulfate
Category: Detergent, anionic
Applications: Paste and powder detergents
Form: Paste; Color: Yellowish-white
Composition: 35% active ingred.
Containers: 200 kg net steel drums

Merpinamid KD 11. Elektrochemische Fabrik Kempen
Chemical Description: Broad cut lauryl diethanolamide
Category: Superfatting agent, foam stabilizer, nonionic
Applications: Superfatting agent, foam stabilizer, shampoos and liquid cleaning and washing preparations for household and industry
Form: Viscous liquid; Color: Yellowish
Containers: 200 kg net steel drums

Merpinamid KD 12. Elektrochemische Fabrik Kempen
Chemical Description: Broad cut lauryl diethanolamide
Category and Applications: See Merpinamid KD 11
Form: Viscous liquid; Color: Yellowish
Containers: 200 kg net steel drums

Merpinamid KM. Elektrochemische Fabrik Kempen
Chemical Description: Broad cut lauryl monoethanolamide
Category and Applications: See Merpinamid KD 11
Form: Soft paste; Color: Light-brownish
Containers: 190 kg net steel drums

Merpinamid LD/E. Elektrochemische Fabrik Kempen
Chemical Description: Narrow cut lauryl diethanolamide
Category and Applications: See Merpinamid KD 11
Form: Viscous liquid; Color: Yellowish
Containers: 200 kg net steel drums

Merpinamid LM. Elektrochemische Fabrik Kempen
Chemical Description: Narrow cut lauryl monoethanolamide
Category and Applications: See Merpinamid KD 11
Form: Flakes; Color: Yellowish-white
Containers: 40 kg net bags

Merpinamid LMIPA. Elektrochemische Fabrik Kempen
Chemical Description: Lauric fatty acid isopropanolamide
Category: Superfatting agent, stabilizer, detergent, nonionic
Applications: Superfatting agent, foam stabilizer, shampoo and liquid cleaner; washing preparation for household and industry
Form: Wax; Color: White
Containers: 200 kg net iron drums

Merpinamid LSD. Elektrochemische Fabrik Kempen
Chemical Description: Broad cut lauryl polydiethanolamide
Category and Applications: See Merpinamid KD 11
Form: Viscous liquid; Color: Yellowish
Containers: 200 kg net steel drums

Merpinamid LSD/E. Elektrochemische Fabrik Kempen
Chemical Description: Broad cut lauryl diethanolamide
Category and Applications: See Merpinamid KD 11
Form: Viscous liquid; Color: Yellowish
Containers: 200 kg net steel drums

Merpinamid LSM. Elektrochemische Fabrik Kempen
Chemical Description: Broad cut lauryl monoethanolamide
Category and Applications: See Merpinamid KD 11
Form: Flakes; Color: Yellowish-white
Containers: 40 kg net bags

Merpinamid OD. Elektrochemische Fabrik Kempen
Chemical Description: Oleic acid-diethanolamide
Category: Superfatting agent, stabilizer, detergent, nonionic
Applications: Superfatting agent, foam stabilizer, shampoo and liquid cleaner; washing preparation for household and industry
Form: Viscous liquid; Color: Yellow
Containers: 200 kg net iron drums

Merpinol CK 30. Elektrochemische Fabrik Kempen
Chemical Description: Protein coconut fatty acid condensate–kalium salt
Applications: Protein product for shampoos and skin and body care products
Form: Viscous liquid; Color: Yellow-brown
Composition: 30% active ingred.
Containers: 120 kg net poly drums

Merpinol S. Elektrochemische Fabrik Kempen
Chemical Description: Protein oleic acid condensate–sodium salt
Applications: Protein product used as textile auxiliary
Form: High-viscous liquid; Color: Light brown
Composition: 25% active ingred.
Containers: 120 kg net poly drums

Merpiphos A 30/50. Elektrochemische Fabrik Kempen
Chemical Description: Alkyl ether addition compound–sodium salt
Category: Wetting agent, anionic
Applications: Wetting agent for alcalic metal baths; mercerizing auxiliary
Form: Liquid; Color: Light
Composition: 100% active ingred.
Containers: 120 kg net poly drums

Merpiphos BP 80. Elektrochemische Fabrik Kempen
Chemical Description: Organic phosphoric acid ester
Category: Wetting agent, emulsifier, anionic
Applications: Universal wetting agent; emulsifier for drilling oils
Form: Liquid; Color: Light yellow
Composition: 100% active ingred.
Containers: 120 kg net poly drums

Merpiphos EP. Elektrochemische Fabrik Kempen
Chemical Description: Organic phosphoric acid ester
Category: Emulsifier, antistat, anionic
Applications: Antistatic textile auxiliary; drycleaning agent; pesticide emulsifier
Form: Liquid; Color: Brown
Composition: 100% active ingred.
Containers: 60 kg net poly drums

Meripiphos GP. Elektrochemische Fabrik Kempen
Chemical Description: Organic phosphoric acid ester
Category: Emulsifier, detergent, anionic
Applications: Emulsifier; alcalic industrial cleaning agent; drycleaning product; pesticide emulsifier
Form: Liquid; Color: Yellowish
Composition: 100% active ingred.
Containers: 60 kg net poly drums

Meripiphos LP. Elektrochemische Fabrik Kempen
Chemical Description: Organic phosphoric acid ester
Category and Applications: See Meriphos EP
Form: Liquid; Color: Light brown
Composition: 100% active ingred.
Containers: 60 kg net poly drums

Meripiphos MP. Elektrochemische Fabrik Kempen
Chemical Description: Organic phosphoric acid ester
Category: Emulsifier, detergent, lubricant, anionic
Applications: Textile cleaning agent, emulsifier, lubricant for synthetic fibers
Form: Paste; Color: Yellow
Composition: 100% active ingred.
Containers: 120 kg net poly drums

Meripiphos PP. Elektrochemische Fabrik Kempen
Chemical Description: Organic phosphoric acid ester
Category: Hydrotrope, detergent, anionic
Applications: Hydrotrope; liquid cleaner with high anorganic content
Form: Liquid; Color: Light yellow
Composition: 100% active ingred.
Containers: 60 kg net poly drums

Meripiphos RP. Elektrochemische Fabrik Kempen
Chemical Description: Organic phosphoric acid ester
Category and Applications: See Meriphos PP
Form: Liquid; Color: Light yellow
Composition: 100% active ingred.
Containers: 60 kg net poly drums

Meripiphos SP. Elektrochemische Fabrik Kempen
Chemical Description: Organic phosphoric acid ester
Category: Hydrotrope, anionic
Applications: Hydrotrope; for slurries; active agent in sprayed washing powders
Form: Liquid; Color: Light brown
Composition: 100% active ingred.
Containers: 120 kg net poly drums

Meripiphos TP. Elektrochemische Fabrik Kempen
Chemical Description: Organic phosphoric acid ester
Category and Applications: See Meriphos SP
Form: Liquid; Color: Light brown
Composition: 100% active ingred.
Containers: 120 kg net poly drums

Meripiphos UP. Elektrochemische Fabrik Kempen
Chemical Description: Organic phosphoric acid ester
Category and Applications: See Meriphos EP
Form: Liquid; Color: Light yellow
Composition: 100% active ingred.
Containers: 60 kg net poly drums

Meripiphos VP. Elektrochemische Fabrik Kempen
Chemical Description: Organic phosphoric acid ester
Category and Applications: See Meriphos MP
Form: Pasty liquid; Color: Brown
Composition: 100% active ingred.
Containers: 120 kg net poly drums

Merpisap A 497. Elektrochemische Fabrik Kempen
Chemical Description: Blend of anionic active and nonionic basic products
Category: Detergent, anionic
Applications: Liquid washing and cleaning agents for household and industry
Form: Pumpable paste; Color: Light-brownish
Composition: 97% active ingred.
Containers: 200 kg net steel drums

Merpisap AC 70. Elektrochemische Fabrik Kempen
Chemical Description: Dodecyl benzene calcium sulfonate
Category: Emulsifier, anionic
Applications: Insecticide formulations

Form: Viscous oil; Color: Yellowish
Composition: 70% active ingred.
Containers: 200 kg net steel drums

Merpisap AE 50. Elektrochemische Fabrik Kempen
Chemical Description: Dodecyl benzene sodium sulfonate
Category: Detergent, anionic
Applications: Liquid, paste, and powder detergents and cleaning agents for household and industry
Form: Pumpable paste; Color: White
Composition: 50% active ingred.
Containers: 125 kg net drums

Merpisap AE 60. Elektrochemische Fabrik Kempen
Chemical Description: Linear sodium dodecyl benzene sulfonate
Category: Detergent, anionic
Applications: Liquid, paste, and powder detergents and cleaning agents for household and industry; textile auxiliary
Form: Paste; Color: White
Properties: Biodegradable
Composition: 60% active ingred.
Containers: 120 kg net poly drums

Merpisap AE 70. Elektrochemische Fabrik Kempen
Chemical Description: Dodecyl benzene sodium sulfonate
Category and Applications: See Merpisap AE 50
Form: Paste; Color: White
Composition: 70% active ingred.
Containers: 200 kg net drums

Merpisap AN 90. Elektrochemische Fabrik Kempen
Chemical Description: Blend of anionic active and nonionic basic products
Category: Detergent, anionic
Applications: Liquid washing and cleaning agents for household and industry
Form: Soft paste; Color: Yellowish
Composition: 90% active ingred.
Containers: 200 kg net steel drums

Merpisap AP 85 W. Elektrochemische Fabrik Kempen
Chemical Description: Alkylaryl sodium sulfonate salt
Category: Detergent, wetting agent, anionic
Applications: Powdery detergents for household and industry, scouring and wetting agents
Form: Fine powder; Color: White
Composition: 85% active ingred.
Containers: 40 kg net bags

Merpisap AP 90. Elektrochemische Fabrik Kempen
Chemical Description: Linear sodium dodecyl benzene sulfonate
Category: Detergent, wetting agent, anionic
Applications: Powder detergents for household and industry; scouring and wetting agent; textile auxiliary
Form: Free-flowing powder; Color: White
Properties: Biodegradable
Composition: 88% active ingred.
Containers: 40 kg net poly bags

Merpisap AP 90 P. Elektrochemische Fabrik Kempen
Chemical Description: Linear sodium dodecyl benzene sulfonate
Category and Applications: See Merpisap AP 90
Form: Free-flowing powder with anticaking agent; Color: White
Properties: Biodegradable
Composition: 85% active ingred.
Containers: 40 kg net poly bags

Merpisap AS 98. Elektrochemische Fabrik Kempen
Chemical Description: Linear dodecyl benzene sulfonic acid
Category: Intermediate, anionic
Applications: Basic product for liquid, paste, and powder detergents, wetting, and cleansing formulations
Form: Viscous liquid; Color: Dark brownish
Properties: Biodegradable
Composition: 98% active ingred.
Containers: 210 kg net iron drums

Merpisap AT 50. Elektrochemische Fabrik Kempen
Chemical Description: Linear triethanolamine dodecyl benzene sulfonate
Category: Detergent, anionic
Applications: Liquid washing and cleaning agents for household and industry; textile auxiliary; cosmetic industry
Form: Pumpable paste; Color: Yellow
Properties: Biodegradable
Composition: 50% active ingred.
Containers: 200 kg net iron drums

Merpisap AT 90. Elektrochemische Fabrik Kempen
Chemical Description: Linear triethanolamine dodecyl benzene sulfonate
Category and Applications: See Merpisap AT 50
Form: Pumpable paste; Color: Yellow
Properties: Biodegradable
Composition: 90% active ingred.
Containers: 200 kg net iron drums

Merpisap D 397. Elektrochemische Fabrik Kempen
Chemical Description: Blend of anionic active and nonionic basic products
Category: Detergent, anionic
Applications: Liquid washing and cleaning agents for household and industry
Form: Pumpable paste; Color: Light brownish
Composition: 97% active ingred.
Containers: 200 kg net steel drums

Merpisap DC 70. Elektrochemische Fabrik Kempen
Chemical Description: A calcium sulfonate
Category: Emulsifier, anionic
Applications: Emulsifier for insecticide formulations
Form: Viscous oil; Color: Light brownish
Composition: 70% active ingred.
Containers: 200 kg net steel drums

Merpisap DE 70. Elektrochemische Fabrik Kempen
Chemical Description: Sodium tetrapropylene benzene sulfonate

Category: Detergent, anionic
Applications: Liquid, paste, and powder detergents and cleaning agents for household and industry; textile auxiliary
Form: Paste; Color: White
Composition: 70% active ingred.
Containers: 120 kg net poly drums or road tankers

Merpisap DN 90. Elektrochemische Fabrik Kempen
Chemical Description: Blend of anionic active and nonionic basic products
Category and Applications: See Merpisap D 397
Form: Paste; Color: Light
Composition: 90% active ingred.
Containers: 200 kg net steel drums

Merpisap DP 82. Elektrochemische Fabrik Kempen
Chemical Description: Sodium tetrapropylene benzene sulfonate
Category and Applications: See Merpisap AP 90
Form: Free-flowing powder; Color: White
Composition: 83% active ingred.
Containers: 40 kg net poly bags

Merpisap DT 90. Elektrochemische Fabrik Kempen
Chemical Description: Triethanolamine tetrapropylene benzene sulfonate
Category and Applications: See Merpisap AT 50
Form: Pumpable paste; Color: Yellow
Composition: 90% active ingred.
Containers: 200 kg net iron drums

Merpisap F-9-3. Elektrochemische Fabrik Kempen
Chemical Description: Blend of anionic and nonionic products
Category: Detergent, anionic
Applications: Semifinished liquid washing and cleaning agents for household and industry
Form: Low-viscous opaque liquid; Color: Light
Properties: Biodegradable
Composition: 50% active ingred.
Containers: 120 kg net plastic drum

Merpisap KH 30. Elektrochemische Fabrik Kempen
Chemical Description: Liquified linear sodium dodecyl benzene sulfonate
Category: Detergent, anionic
Applications: Textile auxiliary; liquid washing and cleaning agent for household and industry
Form: Pumpable paste; Color: Yellow
Properties: Biodegradable
Composition: 30% active ingred.
Containers: 200 kg net iron drums or road tankers

Merpisap KO 30. Elektrochemische Fabrik Kempen
Chemical Description: Blend of linear sodium dodecyl benzene sulfonate and nonionic products
Category and Applications: See Merpisap F-9-3
Form: Pumpable paste; Color: Yellow
Properties: Biodegradable

Composition: 30% active ingred.
Containers: 200 kg net iron drums or road tankers

Merpisap TS 98. Elektrochemische Fabrik Kempen
Chemical Description: Tetrapropylene benzene sulfonic acid
Category and Applications: See Merpisap AS 98
Form: Viscous liquid; Color: Dark brownish
Composition: 98% active ingred.
Containers: 210 kg net iron drums

Merpol C. Du Pont Co.
Chemical Description: Long-chain alcohol sulfate
Category: Wetting agent, detergent, penetrant, emulsifier, anionic
Applications: Textile industry–scouring and spot-cleaning, dye assistant, Kier boiling and bleaching assistant; cleaning of mill equipment; leather industry wetting and penetrating skins; emulsifier in fatliquoring; paper industry–making of paper pulps; cleaning boilers; removing ink
Form: Liquid; Color: Light amber
Properties: Solubility: Miscible with water in all proportions; Density: 7.5 lb/gal; pH: 8 (1% sol'n.)

Merpol CH-196. Du Pont Co.
Chemical Description: Modified ethylene oxide condensate
Category: Detergent, wetting agent, nonionic
Applications: Wet processing of all fibers; preparation of fabrics with higher levels of oils and waxes such as knit goods, napped goods, synthetic blends; stain removal; detergent in batch operations such as beck or jig scouring, and in continuous range operations
Form: Clear liquid; Color: Colorless; Odor: Aromatic
Properties: Solubility: Disp. in water in all proportions; Density: 8.1 lb/gal; Sp. gr.: 0.973; Visc.: 24 cps (Brookfield @ 80 F); HLB: 13.0; Stability: Stable in acid media; separates in sol'n. of strong alkalis; Surface Tension: 27.2 dynes/cm (0.1 g/100 ml conc.); pH: 6–8

Merpol DA. Du Pont Co.
Chemical Description: Ethylene oxide condensate
Category: Leveling agent, dyeing assistant, nonionic
Applications: Leveling agent on disperse, acid, premetalized, and cationic dyes on hydrophobic fibers and wool; low foaming
Form: Slightly viscous, clear liquid; Color: Light yellow; Odor: Alcoholic
Properties: Solubility: Sol. in water > 50%; sol. in ethyl alcohol, isopropyl alcohol, ethylene glycol, Cellosolve, acetone, methyl ethyl ketone, oleic acid; Density: 8.5 lb/gal; Visc.: 110 cps @ 27 C; Cloud Pt.: 10 C (lower), > 100 C (upper, 1% sol'n.); Clear Pt.: 12 C; Stability: Stable to acids, alkalis, and most heavy metal ions; Surface Tension: 50 dynes/cm (0.1% conc.); pH: 9 (1% sol'n.)
Composition: 60% active ingred.

Merpol HCS. Du Pont Co.
Chemical Description: Ethylene oxide condensate sol'n.
Category: Wetting agent, penetrant, antistat, leveling agent, detergent, dyeing assistant, emulsifier, stabilizer, nonionic
Applications: Used in textile finishing operations, dyeing, scouring; rug and upholstery cleaning; stabilizer for rubber latices and in water emulsion paints; in cutting oils, polishing waxes; dispersant for pigments in paints and inks; in permanent wave lotions, medicinal ointments, antiperspirant cream formulations in cosmetics industry
Form: Clear, viscous liquid; Color: Light yellow; Odor: Mild, aromatic
Properties: Solubility: 40% sol. in water; sol. in organic solvents that are miscible with water; Density: 8.63 lb/gal; M.P.: Viscous paste @ 32 F; HLB: 15.5; Cloud Pt.: 43 F; Flash Pt.: > 235 F; Stability: Stable to acids, alkalis, and most heavy metal ions; Surface Tension: 42.9 dynes/cm (0.1% conc.); pH: 6–8 (10% sol'n.)
Composition: 60% active ingred.

Merpol HCW. Du Pont Co.
Chemical Description: Higher fatty alcohol ethylene oxide condensate
Category: Emulsifier, leveling agent, stabilizer, wetting agent, dispersant, detergent, nonionic
Applications: Cold water make-up version of Merpol HCS; dyebath additive for all fibers
Form: Clear, viscous liquid; Odor: Mild, alcoholic
Properties: Solubility: Water sol.; Density: 8.32 lb/gal; F.P.: < 3 F; Cloud Pt.: > 212 F; Flash Pt.: > 80 F; pH: 6–7 (10% sol'n.)
Composition: 30% active ingred.

Merpol LF-H. Du Pont Co.
Chemical Description: Polyether
Category: Detergent, wetting agent, dispersant, emulsifier, dye assistant, lubricant, nonionic
Applications: Nonfoam wetting, scouring, and dyeing assistant for textiles; jet dyer applictions; low-foam detergent esp. with alkaline builders; wetting agent for nylon carpet print pastes
Form: Clear liquid; Color: Pale yellow; Odor: Mild
Properties: Solubility: 40% max. sol. in water below 70 F; Density: 7.96 lb/gal; Visc.: 28 cps (Brookfield @ 80 F); Cloud Pt.: 78 F (upper, 1% aq.); Flash Pt.: > 200 F (TCC); Stability: Store betw. 35 and 90 F; may freeze below 35 F and may separate above 90 F; pH: 6–8; Biodegradable

Merpol OJS. Du Pont Co.
Chemical Description: Ethylene oxide condensate
Category: Wetting and dispersing agent, emulsifier, antistat, detergent, nonionic
Applications: Dyeing assistant, leveling agent for textiles, antistatic finish for synthetic fiber fabrics; detergent esp. with alka-

line builders
Form: Clear liquid; Color: Slightly yellow; Odor: Mild, fatty
Properties: Solubility: Sol. in water; sol. or disp. in alcohol, ketones, aliphatic and aromatic hydrocarbons; Density: 8.04 lb/gal; Sp. gr.: 0.9648; Visc.: 46 cps (Brookfield @ 27 C); HLB: 12.7; Cloud Pt.: 30 F; Flash Pt.: 167 F (TOC); Stability: Stable to acid, alkali, and most heavy metal ions; Surface Tension: 33.5 dynes/cm (0.1%); pH: 6–8
Composition: 60% active ingred.

Merpol SE. Du Pont Co.
Chemical Description: Long-chain fatty alcohol ethylene oxide condensate
Category: Wetting and rewetting agent, emulsifier, dispersant, nonionic
Applications: Low-foam general purpose wetting agent
Form: Clear liquid; Color: Colorless to pale yellow
Properties: Solubility: 0.1% sol. in water (cloudy disp. at higher conc.); Density: 8.1 lb/gal; HLB: 10.5; Stability: Stable to acids, alkalis, salts; pH: 6.2 (0.1% sol'n.)
Composition: 95% active ingred.

Merpol SH. Du Pont Co.
Chemical Description: Long chain alcohol-ethylene oxide condensate
Category: Detergent, wetting agent, nonionic
Applications: Textile processing—preparation of fabrics for dyeing or finishing; detergent in continuous and jig scouring operations; wetting agent in soft and hard water under acidic and alkaline conditions; wetting agent in desize bath
Form: Clear liquid; Color: None; Odor: Alcoholic
Properties: Solubility: Miscible with water in all proportions up to 110 F; sol. or disp. in ethyl and isopropyl alcohols, acetone, methyl ethyl ketone, ethylene glycol, glycerin, and oleic acid; Density: 8.0 lb/gal; Visc.: 24 cps (Brookfield @ 27 C); HLB: 12.9; Cloud Pt.: $<$ 40 F; Flash Pt.: 92 F (TCC); Surface Tension: 27.4 dynes/cm (0.1 g/100 ml conc.); pH: 5–7 (10% aq.); Biodegradable
Composition: 50% active ingred.

Merpoxal LM 3028. Elektrochemische Fabrik Kempen
Chemical Description: Sodium lauryl ether sulfate
Category: Detergent, emulsifier, anionic
Applications: Cosmetic, hair shampoo, and foam bath preparations; liquid washing, cleaning, and shampooing agents for household and industry; emulsifier
Form: Water-clear liquid
Properties: Biodegradable
Composition: 28% active ingred.
Containers: 120 kg net poly drums

Merpoxal LM 3030. Elektrochemische Fabrik Kempen
Chemical Description: Sodium sulfate, lauryl narrow cut
Category: Anionic surfactant

Applications: Hair shampoo and foam-bath preparations
Form: Viscous liquid; Color: Waer-white
Composition: 30% active ingred.
Containers: 210 kg net steel drums

Merpoxal LM 3070. Elektrochemische Fabrik Kempen
Chemical Description: Sodium lauryl ether sulfate
Category and Applications: See Merpoxal LM 3028
Form: Water-clear liquid
Properties: Biodegradable
Composition: 70% active ingred.
Containers: 120 kg net poly drums

Merpoxal NOA 4080. Elektrochemische Fabrik Kempen
Chemical Description: Ammonium alkyl aryl ether sulfate
Category: Detergent, emulsifier, anionic
Applications: Liquid washing, cleaning, and shampooing agents for household and industry; emulsifier
Form: Soft paste; Color: Yellow
Composition: 80% active ingred.
Containers: 200 kg net iron drums

Merpoxal TR 4060. Elektrochemische Fabrik Kempen
Chemical Description: Sodium-i-tridecyl ether sulfates
Category and Applications: See Merpoxal LM 3028
Form: Soft paste; Color: Yellow
Composition: 60% active ingred.
Containers: 200 kg net iron drums

Merpoxen DL 52. Elektrochemische Fabrik Kempen
Chemical Description: Alkyl polyglycol ether 5.2 mol EO
Category: Wetting agent, nonionic
Applications: Wetting agent; basic product for liquid and powder cleaning and washing preparations
Form: Clear, viscous liquid
Properties: Turbidity Pt.: 33–35 C (1% aq.)
Composition: 100% active ingred.
Containers: 200 kg net iron drums

Merpoxen DNP 160. Elektrochemische Fabrik Kempen
Chemical Description: Dinonyl phenol ethoxylate
Category: Emulsifier, detergent, wetting agent, nonionic
Applications: Emulsifier, liquid and powder cleaning, washing and wetting agents
Form: Viscous liquid; Color: Yellowish
Composition: 100% active ingred.
Containers: 200 kg net steel drums

Merpoxen KM 45. Elektrochemische Fabrik Kempen
Chemical Description: Coconut fatty acid monoethanolamide-ethoxylate
Category: Stabilizer, nonionic
Applications: Foam stabilizer for shampoos and liquid detergents
Form: Liquid-pasty substance; Color: Yellowish
Composition: 100% active ingred.
Containers: 200 kg net iron drums

Merpoxen KM 60. Elektrochemische Fabrik Kempen
Chemical Description: Coconut fatty acid monoethanolamide-ethoxylate
Category and Applications: See Merpoxen KM 45
Form: Liquid; Color: Yellow
Composition: 100% active ingred.
Containers: 200 kg net iron drums

Merpoxen LM 20. Elektrochemische Fabrik Kempen
Chemical Description: C_{12}–C_{14} alcohol polyglycol ether 2 mol EO
Category: Nonionic intermediate
Applications: Basic product for ether sulfates
Form: Clear liquid
Properties: Turbidity Pt.: 27–28 C (5 g in 20 ml 25% butylglycol sol'n.)
Composition: 100% active ingred.
Containers: 190 kg net iron drums

Merpoxen LM 30. Elektrochemische Fabrik Kempen
Chemical Description: C_{12}–C_{14} alcohol polyglycol ether 3 mol EO
Category and Applications: See Merpoxen LM 20
Form: Clear liquid
Properties: Turbidity Pt.: 44–45 C (5 g in 20 ml 25% butylglycol sol'n.)
Composition: 100% active ingred.
Containers: 190 kg net iron drums

Merpoxen LP 60. Elektrochemische Fabrik Kempen
Chemical Description: C_{12}–C_{15} alcohol polyglycol ether (synth.)
Category: Detergent, emulsifier, nonionic
Applications: Cleaning agent, emulsifier, textile auxiliary
Form: Clear, opaque liquid
Properties: Turbidity Pt.: 44–46 C (1% aq.)
Composition: 100% active ingred.
Containers: 190 kg net iron drums

Merpoxen LP 70. Elektrochemische Fabrik Kempen
Chemical Description: C_{12}–C_{15} alcohol polyglycol ether (synth.)
Category and Applications: See Merpoxen LP 60
Form: Clear, opaque liquid
Properties: Turbidity Pt.: 84–86 C (1% aq.)
Composition: 100% active ingred.
Containers: 190 kg net iron drums

Merpoxen LP 109. Elektrochemische Fabrik Kempen
Chemical Description: C_{12}–C_{15} alcohol polyglycol ether (synth.)
Category: Detergent, nonionic
Applications: Textile auxiliary; washing agent
Form: Clear, opaque liquid
Properties: Turbidity Pt.: 72–74 C (1 g in 100 ml 10% NaCl sol'n.)
Composition: 100% active ingred.
Containers: 190 kg net iron drums

Merpoxen LP 120. Elektrochemische Fabrik Kempen
Chemical Description: C_{12}–C_{15} alcohol polyglycol ether (synth.)
Category and Applications: See Merpoxen LP 109
Form: Soft paste; Color: White

Properties: Turbidity Pt.: 86–88 C (1% aq.)
Composition: 100% active ingred.
Containers: 190 kg net iron drums

Merpoxen NO 15. Elektrochemische Fabrik Kempen
Chemical Description: Nonylphenol 1.5 mol EO
Category: Emulsifier, defoamer, nonionic
Form: Viscous liquid; Color: Water-white
Composition: 100% active ingred.
Containers: 200 kg net iron drums

Merpoxen NO 40. Elektrochemische Fabrik Kempen
Chemical Description: Nonylphenol 4 mol EO
Category: Emulsifier, intermediate, nonionic
Applications: Emulsifier; basic product for ether sulfates
Form: Viscous liquid; Color: Water-white
Properties: Turbidity Pt.: 36–38 C (5 g in 20 ml 25% butylglycol sol'n.)
Composition: 100% active ingred.
Containers: 200 kg net iron drums

Merpoxen NO 60. Elektrochemische Fabrik Kempen
Chemical Description: Nonylphenol 6 mol EO
Category: Emulsifier, wetting agent, detergent, nonionic
Applications: Emulsifier; textile auxiliary; wetting, washing, and cleaning agent; powder cleaner
Form: Viscous liquid; Color: Water-white
Properties: Turbidity Pt.: 59–61 C (5 g in 20 ml 25% butylglycol sol'n.)
Composition: 100% active ingred.
Containers: 200 kg net iron drums

Merpoxen NO 80. Elektrochemische Fabrik Kempen
Chemical Description: Nonylphenol 8 mol EO
Category and Applications: See Merpoxen NO 60
Form: Viscous liquid; Color: Water-white
Properties: Turbidity Pt.: 44–46 C (1% aq.)
Composition: 100% active ingred.
Containers: 200 kg net iron drums

Merpoxen NO 90. Elektrochemische Fabrik Kempen
Chemical Description: Nonylphenol ethoxylate
Category and Applications: See Merpoxen NO 60
Form: Viscous liquid; Color: Water-white
Properties: Turbidity Pt.: 52–54 C
Composition: 100% active ingred.
Containers: 200 kg net steel drums

Merpoxen NO 95. Elektrochemische Fabrik Kempen
Chemical Description: Nonylphenol 9.5 mol EO
Category and Applications: See Merpoxen NO 60
Form: Viscous liquid; Color: Water-white
Properties: Turbidity Pt.: 56–58 C (1% aq.)

Composition: 100% active ingred.
Containers: 200 kg net iron drums or road tanker

Merpoxen NO 100. Elektrochemische Fabrik Kempen
Chemical Description: Nonylphenol ethoxylate
Category and Applications: See Merpoxen NO 60
Form: Viscous liquid; Color: Water-white
Properties: Turbidity Pt.: 60–62 C
Composition: 100% active ingred.
Containers: 200 kg net steel drums

Merpoxen NO 110. Elektrochemische Fabrik Kempen
Chemical Description: Nonyl phenol ethoxylate
Category and Applications: See Merpoxen NO 60
Form: Viscous liquid; Color: Water-white
Properties: Turbidity Pt.: 72–74 C
Composition: 100% active ingred.
Containers: 200 kg net steel drums

Merpoxen NO 120. Elektrochemische Fabrik Kempen
Chemical Description: Nonylphenol ethoxylate
Category and Applications: See Merpoxen NO 60
Form: Soft paste; Color: White
Properties: Turbidity Pt.: 83–85 C
Composition: 100% active ingred.
Containers: 200 kg net steel drums

Merpoxen NO 140. Elektrochemische Fabrik Kempen
Chemical Description: Nonylphenol ethoxylate
Category and Applications: See Merpoxen NO 60
Form: Soft paste; Color: White
Properties: Turbidity Pt.: 90–92 C
Composition: 100% active ingred.
Containers: 200 kg net steel drums

Merpoxen NO 150. Elektrochemische Fabrik Kempen
Chemical Description: Nonylphenol 15 mol EO
Category: Emulsifier, detergent, nonionic
Applications: Emulsifier; liquid, paste, and powder washing and cleaning agents
Form: Soft wax; Color: White
Properties: Turbidity Pt.: 94–96 C (1% aq.)
Composition: 100% active ingred.
Containers: 200 kg net iron drums

Merpoxen NO 200. Elektrochemische Fabrik Kempen
Chemical Description: Nonylphenol 20 mol EO
Category and Applications: See Merpoxen NO 150
Form: Wax; Color: White
Properties: Turbidity Pt.: 71–73 C (1 g in 100 ml 10% NaCl sol'n.)
Composition: 100% active ingred.
Containers: 200 kg net iron drums

Merpoxen NO 300. Elektrochemische Fabrik Kempen
Chemical Description: Nonylphenol 30 mol EO
Category and Applications: See Merpoxen NO 150
Form: Wax; Color: White
Properties: Turbidity Pt.: 73–75 C

(1 g in 100 ml 10% NaCl sol'n.)
Composition: 100% active ingred.
Containers: 200 kg net iron drums

Merpoxen OA 150. Elektrochemische Fabrik Kempen
Chemical Description: Alkyl amine ethoxylate
Category: Emulsifier, leveling and softening agent, nonionic
Applications: Used in textile industry
Form: Soft paste; Color: Light brownish
Composition: 100% active ingred.
Containers: 200 kg net steel drums

Merpoxen OFS 60. Elektrochemische Fabrik Kempen
Chemical Description: Oleic fatty acid ethoxylate, 6 mol EO
Category: Emulsifier, nonionic
Applications: Emulsifier for textile and detergents industries
Form: Liquid; Color: Yellowish
Composition: 100% active ingred.
Containers: 200 kg net iron drums

Merpoxen OFS 80. Elektrochemische Fabrik Kempen
Chemical Description: Oleic fatty acid ethoxylate, 8 mol EO
Category and Applications: See Merpoxen OFS 60
Form: Liquid; Color: Yellowish
Composition: 100% active ingred.
Containers: 200 kg net iron drums

Merpoxen PEG 200. Elektrochemische Fabrik Kempen
Chemical Description: Polyethylene glycol, mol wt 200
Category: Intermediate, softener, lubricant, nonionic
Applications: Basic product for the manufacture of fatty acid ethers; cosmetic preparations; paper softener; lubricant for hydraulic liquids
Form: Clear liquid
Composition: 100% active ingred.
Containers: 200 kg net iron drums

Merpoxen PEG 300. Elektrochemische Fabrik Kempen
Chemical Description: Polyethylene glycol, mol wt 300
Category and Applications: See Merpoxen PEG 200
Form: Clear liquid
Composition: 100% active ingred.
Containers: 200 kg net iron drums

Merpoxen PEG 400. Elektrochemische Fabrik Kempen
Chemical Description: Polyethylene glycol, mol wt 400
Category and Applications: See Merpoxen PEG 200
Form: Clear liquid
Composition: 100% active ingred.
Containers: 200 kg net iron drums

Merpoxen PEG 600. Elektrochemische Fabrik Kempen
Chemical Description: Polyethylene glycol, mol wt 600
Category: Softener, lubricant, nonionic
Applications: Cosmetic preparations; paper softener; lubricant for hydraulic liquids
Form: Clear paste
Composition: 100% active ingred.
Containers: 200 kg net iron drums

Merpoxen PEG 800. Elektrochemische Fabrik Kempen
Chemical Description: Polyethylene glycol, mol wt 800
Category and Applications: See Merpoxen PEG 800
Form: Clear paste
Composition: 100% active ingred.
Containers: 200 kg net iron drums

Merpoxen PEG 1000. Elektrochemische Fabrik Kempen
Chemical Description: Polyethylene glycol, mol wt 1000
Category and Applications: See Merpoxen PEG 800
Form: Wax; Color: White
Composition: 100% active ingred.
Containers: 200 kg net iron drums

Merpoxen PEG 1500 S. Elektrochemische Fabrik Kempen
Chemical Description: Polyethylene glycol, mol wt 1500
Category and Applications: See Merpoxen PEG 800
Form: Flakes; Color: White
Composition: 100% active ingred.
Containers: 50 kg net poly bags

Merpoxen RO 200. Elektrochemische Fabrik Kempen
Chemical Description: Castor oil polyglycol ether, 20 mol EO
Category: Emulsifier, nonionic
Applications: Emulsifier, textile auxiliary
Form: Viscous liquid; Color: Yellow
Composition: 100% active ingred.
Containers: 200 kg net iron drums

Merpoxen RO 300. Elektrochemische Fabrik Kempen
Chemical Description: Castor oil polyglycol ether, 30 mol EO
Category and Applications: See Merpoxen RO 200
Form: Wax; Color: White
Composition: 100% active ingred.
Containers: 200 kg net iron drums

Merpoxen RO 400. Elektrochemische Fabrik Kempen
Chemical Description: Castor oil polyglycol ether, 40 mol EO
Category and Applications: See Merpoxen RO 200
Form: Wax; Color: White
Composition: 100% active ingred.
Containers: 200 kg net iron drums

Merpoxen S-5-96/10. Elektrochemische Fabrik Kempen
Chemical Description: Alkylamine polyalkylenoxide ether
Category: Detergent, lubricant, nonionic
Applications: Low foaming resp. nonfoaming washing and cleaning agent; cooling lubricant; cosmetic industry
Form: Opaque liquid; Color: Light yellow
Properties: Turbidity Pt.: 25–28 C (1% aq.)
Composition: 100% active ingred.
Containers: 200 kg net poly drums

Merpoxen S-5-96/20. Elektrochemische Fabrik Kempen
Chemical Description: Alkylamine polyalkylenoxide ether
Category and Applications: See

Merpoxen S-5-96/10
Form: Opaque, viscous liquid; Color: Light yellow
Properties: Turbidity Pt.: 26–29 C (1% aq.)
Composition: 100% active ingred.
Containers: 200 kg net poly drums

Merpoxen S-5-96/40. Elektrochemische Fabrik Kempen
Chemical Description: Alkylamine polyalkylenoxide ether
Category and Applications: See Merpoxen S-5-96/10
Form: Soft paste; Color: Light yellow
Properties: Turbidity Pt.: 32–35 C (1% aq.)
Composition: 100% active ingred.
Containers: 200 kg net poly drums

Merpoxen S-5-96/70. Elektrochemische Fabrik Kempen
Chemical Description: Alkylamine polyalkylenoxide ether
Category and Applications: See Merpoxen S-5-96/10
Form: Paste; Color: Light yellow
Properties: Turbidity Pt.: 100 C (1% aq.)
Composition: 100% active ingred.
Containers: 200 kg net poly drums

Merpoxen SFS 60. Elektrochemische Fabrik Kempen
Chemical Description: Stearic fatty acid ethoxylate
Category: Emulsifier, nonionic
Applications: In textile auxiliaries and as an emulsifier
Form: Soft paste; Color: White
Composition: 100% active ingred.
Containers: 200 kg net steel drums

Merpoxen SFS 80. Elektrochemische Fabrik Kempen
Chemical Description: Stearic fatty acid ethoxylate, 8 mol EO
Category and Applications: See Merpoxen SFS 60
Form: Soft paste; Color: White
Composition: 100% active ingred.
Containers: 110 kg net poly drums

Merpoxen SFS 120. Elektrochemische Fabrik Kempen
Chemical Description: Stearic fatty acid ethoxylate
Category and Applications: See Merpoxen SFS 60
Form: Wax-like substance; Color: White
Composition: 100% active ingred.
Containers: 200 kg net steel drums

Merpoxen SP 100. Elektrochemische Fabrik Kempen
Chemical Description: Oleyl-cetyl alcohol polyglycol ether, 10 mol EO
Category: Emulsifier, intermediate, nonionic
Applications: Emulsifier; basic product for powder detergents
Form: Waxy paste; Color: White
Properties: Turbidity Pt.: 72–74 C (1% aq.)
Composition: 100% active ingred.
Containers: 190 kg net iron drums

Merpoxen SP 200. Elektrochemische Fabrik Kempen
Chemical Description: Oleyl-cetyl alcohol polyglycol ether, 20 mol EO
Category: Emulsifier, detergent,

nonionic
Applications: Emulsifier; egalizer for the textile industry; washing agents
Form: Waxy paste; Color: White
Properties: Turbidity Pt.: 72–74 C (1 g in 100 ml 10% NaCl sol'n.)
Composition: 100% active ingred.
Containers: 200 kg net iron drums

Merpoxen SP 230. Elektrochemische Fabrik Kempen
Chemical Description: Oleyl-cetyl alcohol polyglycol ether, 23 mol EO
Category and Applications: See Merpoxen SP 200
Form: Waxy paste; Color: White
Properties: Turbidity Pt.: 75–76 C (1 g in 100 ml 10% NaCl sol'n.)
Composition: 100% active ingred.
Containers: 200 kg net iron drums

Merpoxen STA 150. Elektrochemische Fabrik Kempen
Chemical Description: Alkyl amine ethoxylate
Category: Emulsifier, leveling and softening agent, nonionic
Applications: Emulsifier, leveling and softening agent in textile industry
Form: Soft paste; Color: Light brownish
Composition: 100% active ingred.
Containers: 200 kg net steel drums

Merpoxen STA 200. Elektrochemische Fabrik Kempen
Chemical Description: Alkyl amine ethoxylate
Category and Applications: See Merpoxen STA 150
Form: Soft wax; Color: Light brownish
Composition: 100% active ingred.
Containers: 200 kg net steel drums

Merpoxen STA 250. Elektrochemische Fabrik Kempen
Chemical Description: Alkyl amine ethoxylate
Category and Applications: See Merpoxen STA 150
Form: Wax; Color: Light brownish
Composition: 100% active ingred.
Containers: 200 kg net steel drums

Merpoxen T-4-8. Elektrochemische Fabrik Kempen
Chemical Description: I-tridecylalcohol polyglycol ether (modified)
Category: Wetting agent, nonionic
Applications: Low-foaming wetting agent
Form: Clear liquid; Color: Whitish
Properties: Turbidity Pt.: 28–29 C (1% aq.)
Composition: 100% active ingred.
Containers: 200 kg net iron drums

Merpoxen T-5-5. Elektrochemische Fabrik Kempen
Chemical Description: Alkyl polyglycol ether (modified)
Category and Applications: See Merpoxen T-4-8
Form: Clear liquid; Color: Whitish
Properties: Turbidity Pt.: 34–36 C (1% aq.)
Composition: 100% active ingred.
Containers: 200 kg net iron drums

Merpoxen T-5-7. Elektrochemische Fabrik Kempen
Chemical Description: Alkyl polyglycol ether (modified)
Category and Applications: See Merpoxen T-4-8
Form: Clear liquid; Color: Whitish
Properties: Turbidity Pt.: 11–14 C (1% aq.)
Composition: 100% active ingred.
Containers: 200 kg net iron drums

Merpoxen T 110. Elektrochemische Fabrik Kempen
Chemical Description: Tallow fatty polyglycol ether, 11 mol EO
Category: Detergent, emulsifier, nonionic
Applications: Basic product for powder detergents and cleansers; emulsifiers for paraffins
Form: Waxy paste; Color: White
Properties: Turbidity Pt.: 81–82 C (1% aq.)
Composition: 100% active ingred.
Containers: 190 kg net iron drums

Merpoxen T 180. Elektrochemische Fabrik Kempen
Chemical Description: Tallow fatty polyglycol ether, 18 mol EO
Category and Applications: See Merpoxen T 110
Form: Waxy paste; Color: White
Properties: Turbidity Pt.: 73–75 C (1 g in 100 ml 10% NaCl sol'n.)
Composition: 100% active ingred.
Containers: 200 kg net iron drums

Merpoxen T 250. Elektrochemische Fabrik Kempen
Chemical Description: Tallow fatty polyglycol ether, 25 mol EO
Category: Detergent, nonionic
Applications: Basic product for powder detergents and cleansers
Form: Wax (also available as powder or flakes); Color: White
Properties: Turbidity Pt.: 75–77 C (1 g in 100 ml 10% NaCl sol'n.)
Composition: 100% active ingred.
Containers: 200 kg net iron drums or 50 kg net poly bags

Merpoxen T 500. Elektrochemische Fabrik Kempen
Chemical Description: Tallow fatty polyglycol ether, 50 mol EO
Category and Applications: See Merpoxen T 250
Form: Wax (also available as powder or flakes); Color: White
Properties: Turbidity Pt.: 74–75 C (1 g in 100 ml 10% NaCl sol'n.)
Composition: 100% active ingred.
Containers: 200 kg net iron drums or 50 kg net poly bags

Merpoxen T 800. Elektrochemische Fabrik Kempen
Chemical Description: Tallow fatty polyglycol ether, 80 mol EO
Category and Applications: See Merpoxen T 250
Form: Wax (also available as powder or flakes); Color: White
Properties: Turbidity Pt.: 71–73 C (1 g in 100 ml 10% NaCl sol'n.)
Composition: 100% active ingred.
Containers: 200 kg net iron drums or 50 kg net poly bags

Merpoxen TA 120. Elektrochemische Fabrik Kempen
Chemical Description: Tall oil fatty acid ethoxylate

Category: Detergent, emulsifier, nonionic
Applications: Washing and lathering agents in the textile industry; emulsifier
Form: Viscous liquid; Color: Yellowish
Composition: 100% active ingred.
Containers: 200 kg net steel drums

Merpoxen TR 40. Elektrochemische Fabrik Kempen
Chemical Description: I-tridecyl alcohol polyglycol ether, 40 mol EO
Category: Intermediate, emulsifier, wetting agent, nonionic
Applications: Basic product for ether sulfates; emulsifier for wettable powders; wetting agent for textile industry; for washing and cleansing formulations
Form: Viscous liquid; Color: Water-white
Properties: Turbidity Pt.: 42–44 C (5 g in 20 ml 25% butylglycol sol'n.)
Composition: 100% active ingred.
Containers: Coated 200 kg net iron drums

Merpoxen TR 70. Elektrochemische Fabrik Kempen
Chemical Description: I-tridecyl alcohol polyglycol ether, 7 mol EO
Category and Applications: See Merpoxen TR 40
Form: Viscous liquid; Color: Whitish
Properties: Turbidity Pt.: 69–70 C (5 g in 20 ml 25% butylglycol sol'n.)
Composition: 100% active ingred.
Containers: Coated 200 kg net iron drums

Merpoxen TR 80. Elektrochemische Fabrik Kempen
Chemical Description: I-tridecyl polyglycol ether
Category and Applications: See Merpoxen TR 40
Form: Soft paste; Color: White
Composition: 100% active ingred.
Containers: 200 kg net steel drums

Merpoxen TR 100. Elektrochemische Fabrik Kempen
Chemical Description: I-tridecyl alcohol polyglycol ether, 10 mol EO
Category and Applications: See Merpoxen TR 40
Form: Soft paste; Color: White
Properties: Turbidity Pt.: 70–72 C (1% aq.)
Composition: 100% active ingred.
Containers: Coated 200 kg net iron drums

Merpoxen UN 22. Elektrochemische Fabrik Kempen
Chemical Description: Alkyl C_9–C_{11} alcohol polyglycol ether, 2.2 mol EO
Category: Defoamer, intermediate, nonionic
Applications: Defoamer; basic product for the production of ether sulfates
Form: Clear, viscous liquid
Properties: Turbidity Pt.: 35–37 C

(5 g in 20 ml 25% butylglycol sol'n.)
Composition: 100% active ingred.
Containers: 190 kg net iron drums

Merpoxen UN 40. Elektrochemische Fabrik Kempen
Chemical Description: Alkyl C_9–C_{11} alcohol polyglycol ether, 4 mol EO
Category: Intermediate, nonionic
Applications: Basic product for liquid and powder washing and cleaning agent
Form: Clear, viscous liquid
Properties: Turbidity Pt.: 59–61 C (5 g in 20 ml 25% butylglycol sol'n.)
Composition: 100% active ingred.
Containers: 190 kg net iron drums

Merpoxen UN 50. Elektrochemische Fabrik Kempen
Chemical Description: Alkyl C_9–C_{11} alcohol polyglycol ether 5 mol EO
Category and Applications: See Merpoxen UN 40
Form: Clear, viscous liquid
Properties: Turbidity Pt.: 33–35 C (1% aq.)
Composition: 100% active ingred.
Containers: 190 kg net iron drums

Merpoxen UN 65. Elektrochemische Fabrik Kempen
Chemical Description: Alkyl C_9–C_{11} alcohol polyglycol ether, 6.5 mol EO
Category and Applications: See Merpoxen UN 40
Form: Clear, viscous liquid
Properties: Turbidity Pt.: 56–58 C (1% aq.)
Composition: 100% active ingred.
Containers: 190 kg net iron drums

Microlan. Croda Italiana S.R.L.
Chemical Description: Lanolin adsorbed onto silicate base
Category: Superfatting agent, nonionic
Applications: Superfatting agent for powder and pigmented compositions
Form: Powder; Color: Off white

Miglyol 810. Dynamit Nobel Chem.
Chemical Description: Caprylic/capric triglyceride
Category: Emollient, nonionic
Applications: Low viscosity, oxidation stable, and skin-compatible oil for cosmetics and pharmaceuticals, incl. oral products, as vehicle and solvent for injection products, topical ointments, creams, and lotions, suppositories, dietetic products
Form: Liquid; Color: 3 max. (Gardner); Odor and Taste: Neutral
Properties: Ref. Index: 1.4485–1.4505; Solubility: Sol. @ 20 C in diethyl ether, petroleum ether, chloroform, isopropanol, toluene; Sp. gr.: 0.94–0.95 @ 20 C; Visc.: 27–30 cps @ 20 C; Cloud Pt.: ≈ 0 C; Sapon. No.: 340–360; Acid Value: 0.1 max.; Iodine Value: 0.5 max.; Biodegradable
Reaction: Neutral

Miglyol 812. Dynamit Nobel Chem.
Chemical Description: Caprylic/caprictriglyceride

Category and Applications: See Miglyol 810
Form: Liquid; Color: Almost colorless; Odor and Taste: Neutral
Properties: m.w.: ≈ 510; Ref. Index: 1.448–1.450; Solubility: Sol. in ether, benzene, toluene, xylene, petroleum ether, light liquid paraffin, chloroform, carbon tetrachloride, acetone, methyl ethyl ketone; Sp. gr.: 0.940–0.950 @ 20 C; Visc.: 28–32 cps @ 20 C; Sapon. No.: 345–350; Acid Value: 0.1 max.; Iodine Value: 1 max.; Biodegradable
Reaction: Neutral

Miglyol 818. Dynamit Nobel Chem.
Chemical Description: Triglyceride of fractionated coconut oil fatty acids with 5% linoleic acid
Category and Applications: See Miglyol 810
Form: Liquid; Color: 4.5 max. (Gardner); Odor and Taste: None
Properties: Ref. Index: 1.4500–1.4525; Solubility: Sol. @ 20 C in diethyl ether, petroleum ether, chloroform, isopropanol, toluene, 96% alcohol; Sp. gr.: 0.93–0.95 @ 20 C; Visc.: 30–33 cps; Cloud Pt.: ≈ 10 C; Sapon. No.: 315–335; Acid Value: 0.2 max.; Iodine Value: 10 max.; Biodegradable
Reaction: Neutral

Miglyol 840. Dynamit Nobel Chem.
Chemical Description: Propylene glycol diester of saturated vegetable fatty acids
Category and Applications: See Miglyol 810
Form: Liquid; Color: 3.0 max. (Gardner); Odor and Taste: None
Properties: Ref. Index: 1.4400–1.4420; Solubility: Sol. @ 20 C in diethyl ether, petroleum ether, chloroform, isopropanol, toluene, 96% alcohol; Sp. gr.: 0.91–0.92 @ 20 C; Visc.: 9–12 cps @ 20 C; Cloud Pt.: ≈ −30 C; Sapon. No.: 320–340; Acid Value: 0.1 max.; Iodine Value: 0.5 max.; Biodegradable
Reaction: Neutral

Miglyol 840-Gel. Dynamit Nobel Chem.
Category: Stabilizer, emollient
Applications: Anhydrous preparations (ointments, make up, lipgloss); improves consistency and thermal stability in cosmetics and topical pharmaceuticals; emulsion stabilizers; base for W/O creams in combination with suitable emulsifiers
Form: Paste; Color: Cream to light brown; Odor: Practically odorless
Properties: Sapon. Value: 290–310; Acid Value: 0.5 max.; Hydroxyl Value: 7 max.; Iodine Value: 2 max; Stability: Stable to oxidation and below 0 C; gel does not melt
Reaction: Neutral

Miglyol-Gel. Dynamit Nobel Chem.
Chemical Description: Caprylic/caprictriglyceride and stearalkonium hectorite
Category: Stabilizer, nonionic
Applications: Improves consistency and thermostability in cosmet-

ic and pharmaceutical creams; emulsion stabilizer; as a base for W/O and O/W creams when combined with suitable emulsifiers; does not melt at even 100 C
Form: Paste; Color: Cream to light brown; Odor: Nearly odorless
Properties: Sapon. Value: 295–315; Acid No.: 0.5 max.; Hydroxyl No.: 5 max.; Iodine Value: 1.5 max.
Composition: 100% conc.
Reaction: Neutral

Mill Oil Remover BP. Proctor Chem. Co.
Chemical Description: Emulsified solvents
Applications: Preparation and boil off of greige cotton or cotton/polyester goods; removes natural waxes and pectins from cotton, mill grease and spin finish oils from polyester; used in padding, continuous bleaching, beck, and jig methods
Form: Clear liquid; Color: Amber
Properties: Solubility: Forms stable emulsion in cold or hot water; pH: 10.0; Stability: Compatible with moderate amounts of caustic soda (about 1%)
Reaction: Alkaline

Minemal 350. Nippon Nyukazai Co., Ltd.
Category: Emulsifier, nonionic/anionic
Applications: Emulsifier for machine oil
Form: Liquid

Mineral Colloid MO. Georgia Kaolin Co.
Chemical Description: Colloidal hydrated aluminum silicate
Category: Emulsifier
Applications: Dispersing waxes, pigments, resins; preparation of emulsified paints
Form: Powder; Color: Buff; Odor: None
Properties: Ref. Index: 1.56; Sp. gr.: 2.59; Visc.: High; Stability: Good
Reaction: Alkaline
Containers: 50 lb bags

Miramine CC. Miranol Chem. Co.
Chemical Description: 1-(2-hydroxyethyl)-2-norcoco-2-imidazoline
Category: Corrosion inhibitor, lubricant
Applications: Corrosion inhibitor in cutting oils; lubricant for metal surfaces; used in acid shampoos; formulations with mineral spirits emulsify grease and oil
Form: Liquid; Color: Amber; Odor: Bland to faintly ammoniacal
Properties: Solubility: Sol. in many organic solvents; disp. in water; Sp. gr.: 0.940; Biodegradable
Composition: 99% active ingred.
Toxicity: Severe skin and eye irritant

Miramine OC. Miranol Chem. Co.
Chemical Description: 1-(2-hydroxyethyl)-2-[(Z)-8-heptadecyl]-2-imidazoline
Category: Detergent, thickener, corrosion inhibitor, cationic

Applications: Industrial cleaners and detergents; emulsifies kerosene, mineral oil, and other nonpolar solvents; thickens hydrochloric acid; corrosion inhibitor in metal treating and cutting fluids
Form: Liquid to pasty solid; Color: Amber; Odor: Ammoniacal
Properties: Solubility: Oil sol. and water disp.; Sp. gr.: 0.930; Biodegradable
Composition: 99% min. active
Reaction: Alkaline
Toxicity: Severe skin and eye irritant
Warnings and Caution: Wear protective goggles and gloves

Miramine SC. Miranol Chem. Co.
Chemical Description: 1-(2-hydroxyethyl)-2-norsoya-2-imidazoline
Category: Emulsifier, corrosion inhibitor, lubricant
Applications: Emulsifier and coemulsifier in solvent/water systems; corrosion inhibitor in cutting oils; lubricant for metal surfaces
Form: Liquid; Color: Amber; Odor: Ammoniacal
Properties: Solubility: Sol. in many organic solvents; disp. in water; Sp. gr.: 0.930; Biodegradable
Composition: 99% active ingred.
Toxicity: Severe skin and eye irritant

Miranol 2MCA Modified. Miranol Chem. Co.
Chemical Description: Amphoteric salt–lauryl sulfate salt of a dicarboxylated cocoyl imidazoline derivative
Applications: Formulating shampoos which are nonirritating to skin and eyes and noneye stinging
Form: Clear liquid; Color: Light amber
Properties: Solubility: Sol. in water; pH: 8.0–8.5 (20% aq.)
Composition: 46–48% solids

Miranol 2MCAS Modified. Miranol Chem. Co.
Chemical Description: Mixture of the lauryl sulfate and laureth-3-sulfate salts of a dicarboxylated cocoyl imidazoline derivative
Category: Amphoteric surfactant
Applications: Formulating shampoos which are nonirritating to skin and eyes and noneye stinging
Form: Clear liquid; Color: Light yellow
Properties: Solubility: Sol. in water; pH: 7.5–8.0
Composition: 45.5–48.0% solids

Miranol 2MCT Modified. Miranol Chem. Co.
Chemical Description: Trideceth-3-sulfate salt of a dicarboxylated cocoyl imidazoline derivative
Category: Base detergent, amphoteric
Applications: Base detergent for nonirritating and noneye stinging cleansing and make-up removal products of the clear liquid, lotion, and paste types
Form: Clear liquid; Color: Light amber

Properties: Solubility: Sol. in water; pH: 7.5–8.0
Composition: 48.5–51.5% solids

Miranol 2MHT Modified. Miranol Chem. Co.
Chemical Description: Trideceth-3-sulfate salt of a dicarboxylated lauryl imidazoline derivative
Category: Detergent base, amphoteric
Applications: Uses similar to Miranol 2MCT Modified; may be used at higher levels; greater mildness
Form: Clear liquid; Color: Light amber
Properties: Solubility: Sol. in water; pH: 8.0–8.5 (20% aq.)
Composition: 48–50% solids

Miranol C2M Anhydrous Acid. Miranol Chem. Co.
Chemical Description: Acidic form of Miranol C2M Conc.
Category: Detergent, wetting agent, amphoteric
Applications: High foaming surfactant for dispersion on caustic soda and on powder mixes, metal cleaning, etc.; leveling agent in tin plating from acid baths
Form: Stiff gel; Color: Amber
Properties: Solubility: Sol. in water and alcohol; slightly sol. in nonpolar organic solvents; pH: ≈ 4.6 (50% sol'n.)
Composition: 100% active ingred.

Miranol C2M Conc. Miranol Chem. Co.
Chemical Description: Dicarboxylic coconut derivative, sodium salt
Category: Detergent, amphoteric
Applications: Nonirritating shampoos, skin cleansers, make up removers; emulsifier, solubilizer, and stabilizer in pharmaceuticals; hand soaps; household and industrial cleaners; detoxification of other surfactants
Form: Clear, viscous liquid
Properties: Solubility: Sol. in water; pH: 8.0–8.5 (20% aq.)
Composition: 38% active

Miranol C2M-SF 70%. Miranol Chem. Co.
Chemical Description: Disodium salt of a dicarboxylated imidazoline derivative
Category and Applications: See Miranol C2M-SF
Form: Paste at room temperature; pumps and pours readily @ 50 C
Properties: pH: 9.0–10.0 (10% aq.); Biodegradable
Composition: 70% active ingred.

Miranol C2M-SF Conc. Miranol Chem. Co.
Chemical Description: Disodium salt of a dicarboxylated imidazoline derivative; salt free version of Miranol C2M Conc.
Category: Detergent, amphoteric
Applications: Used for extra heavy-duty liquid cleaning compounds, steam cleaners, pressure cleaners, metal cleaners, etc.
Form: Clear liquid; Color: Amber
Properties: Solubility: Sol. in water and alcohol; pH: 9.4–9.8; Biodegradable
Composition: 39% active ingred.

Miranol CM Conc. Miranol Chem. Co.
Chemical Description: Monocarboxylated cocoyl imidazoline derivative
Category: Detergent, wetting and foaming agent, sequestrant, emulsifier, dispersant, germicidal, amphoteric
Applications: Detergent with good foaming, wetting, sequestration, emulsifying properties; lime soap dissolving; mild germicidal properties; mild to skin and eyes; useful where alkali and electrolyte tolerance of no more than 20% is required; viscosity builder; suspends particulate soil and emulsified grease
Form: Viscous liquid; Color: Light amber
Properties: pH: 9.0–9.5 (new process), 12.0–13.0 (old process); Biodegradable
Composition: 37% conc.

Miranol CM-SF Conc. Miranol Chem. Co.
Chemical Description: Sodium salt of a monocarboxylate cocoyl-imidazoline derivative, salt free
Category and Applications: See Miranol CM Conc.
Form: Clear liquid; Color: Light amber
Properties: Solubility: Sol. in water and alcohol; pH: 9.5–10.5; Biodegradable
Composition: 36–38% solids

Miranol CS Conc. Miranol Chem. Co.
Chemical Description: Sulfonated coconut derivative, sodium salt
Category: Detergent, wetting agent, corrosion inhibitor, emulsifier, sequestrant, foaming agent, amphoteric
Applications: Household and industrial uses–cosmetic shampoos, bubble baths, skin cleansers; cold water wool and fine fabric detergent with fabric softening properties; metal polishes; specialty laundry detergents; steel pickling as acid corrosion inhibitor and wetting agent
Form: Liquid
Properties: Stability: Stable @ all pH; compatible with high conc. of electrolytes and alkalies, and with quaternary ammonium salts; Biodegradable
Composition: 38% conc.

Miranol DM. Miranol Chem. Co.
Chemical Description: Sodium salt of a monocarboxylated stearyl imidazoline derivative
Category: Antistat, softener, lubricant, amphoteric
Applications: General antistat and softener on animal, cellulosic and synthetic fibers; imparts a full, smooth hand to all textiles; softens yarns, bleached or dyes piece goods, and hosiery; wool lubricant; household softener
Form: Creamy paste; Color: White; readily pourable at 60 C
Properties: pH: 5.4–6.0 @ 65 C; Biodegradable
Composition: 25–27% solids

Miranol DS. Miranol Chem. Co.
Chemical Description: Sulfonated stearyl derivative, sodium salt

Category: Detergent, amphoteric
Applications: Laundry compounds, household and dishwash products, shampoos, germicidal shampoos, hand soaps; metal processing
Form: Paste
Composition: 21% conc.

Miranol H2M Conc. Miranol Chem. Co.
Chemical Description: Dicarboxylated lauryl imidazoline derivative
Category and Applications: Uses similar to Miranol C2M Conc.; H2M Conc. recommended where foaming under a load of greasy soil is important; formulations have slightly lower viscosities
Form: Viscous liquid; Color: Light amber
Properties: Solubility: Sol. in water; pH: 8.0–8.5 (20% aq.); Biodegradable
Composition: 49–51% solids

Miranol H2M-SF 70%. Miranol Chem. Co.
Chemical Description: Disodium salt of a dicarboxylated lauryl imidazoline derivative
Category and Applications: See Miranol H2M-SF Conc.
Form: Paste at room temperature; pours and pumps readily at 50 C
Composition: 70% active ingred.

Miranol H2M-SF Conc. Miranol Chem. Co.
Chemical Description: Disodium salt of a dicarboxylated lauryl imidazoline derivative, salt free
Category and Applications: Used in place of Miranol C2M-SF Conc. where higher quality of a pure lauric derivative is required, as in cosmetic formulations
Form: Clear liquid; Color: Light amber
Properties: Solubility: Sol. in water and alcohol; pH: 9.2–9.8; Biodegradable
Composition: 38–40% solids

Miranol HM Conc. Miranol Chem. Co.
Chemical Description: Monocarboxylated lauryl imidazoline derivative
Category and Applictions: See Miranol CM Conc.
Form: Viscous liquid; Color: Light amber
Properties: Solubility: Water sol.; pH: 9.0–9.5; Biodegradable
Composition: 43–45% solids

Miranol HS Conc. Miranol Chem. Co.
Chemical Description: Sulfonated lauric derivative, sodium salt
Category: Detergent, wetting agent, foaming agent, emulsifier, sequestrant, amphoteric
Applications: General surfactant; sequesters heavy metals; disperses and solubilizes lime-soap
Form: Liquid
Composition: 38% conc.

Miranol J2M Anhydrous Acid. Miranol Chem. Co.
Chemical Description: Anhydrous acid of Miranol J2M Conc.

Category and Applications: See Miranol JEM Anhydrous Acid
Form: Stiff gel; Color: Amber
Properties: Solubility: Sol. in water and alcohol; slightly sol. in nonpolar organic solvents; pH: ≈ 4.6 (50% sol'n.)
Composition: 100% active ingred.

Miranol J2M Conc. Miranol Chem. Co.
Chemical Description: Sodium salt of a dicarboxylated caprylyl imidazoline derivative
Category and Applications: See Miranol JEM Conc.; slightly better rinsing properties than JEM Conc.; also used in emulsion polymerization of synthetic rubbers and resins
Form: Clear liquid
Properties: pH: 8.2–8.6; Biodegradable
Composition: 48–50% solids

Miranol J2M-SF Conc. Miranol Chem. Co.
Chemical Description: Dicarboxylated caprylyl imidazoline derivative, salt free
Category and Applications: Salt free version of Miranol J2M Conc.
Form: Liquid
Properties: pH: 8.8–9.3; Biodegradable
Composition: 38–40% solids

Miranol JEM Anhydrous Acid. Miranol Chem. Co.
Chemical Description: Anhydrous acid of Miranol JEM Conc.
Category: Detergent, wetting, amphoteric
Applications: Low foaming surfactant for dispersion on caustic soda and on powder mixes, metal cleaning, etc.
Form: Stiff gel; Color: Amber
Properties: Solubility: Sol. in water and alcohol; slightly sol. in nonpolar organic solvents; pH: ≈ 4.6 (50% sol'n.)
Composition: 100% active ingred.

Miranol JEM Conc. Miranol Chem. Co.
Chemical Description: Sodium salt of a dicarboxylated caprylic and ethylhexoic imidazoline derivative
Category: Wetting agent, penetrant, amphoteric
Applications: Wetting agent in 3–22% caustic soda sol'ns.; low foaming; used in bottle washing compounds and wax stripping formulations, for alkaline degreasing, mercerizing and Kier boiling, electrolytic degreasing, fruit and vegetable lye peeling, and all applications where surface tension reduction is important
Form: Clear liquid
Properties: Stability: Stable in caustic soda sol'ns.; retains solubility and wetting in high electrolyte and acid sol'ns.; Surface Tension: 30.5 dynes/cm (0.12% in water); pH: 10.0–10.5; Biodegradable
Composition: 33–35% solids

Miranol JS Conc. Miranol Chem. Co.
Chemical Description: Sulfonated caprylic derivative, sodium salt

Category: Wetting agent, corrosion inhibitor, amphoteric
Applications: Pickling acids
Form: Liquid
Properties: Stability: Good; high tolerance for inorganic salts; Biodegradable
Composition: 40% conc.

Miranol L2M-SF Conc. Miranol Chem. Co.
Chemical Description: Disodium dicarboxylated imidazoline derivative, salt free
Category and Applications: Powerful foamer with good oil emulsifying properties; similar alkali and electrolyte tolerances to Miranol CM Conc.
Form: Liquid
Properties: pH: 10–10.5; Biodegradable
Composition: 38–40% solids

Miranol LM-SF Conc. Miranol Chem. Co.
Chemical Description: Monocarboxylated linoleic imidazoline derivative, salt free
Category and Applications: Similar to Miranol L2M-SF Conc., but more hydrophobic; suggested as lubricant for ferrous and nonferrous metals
Form: Viscous liquid which sets to a pasty gel on aging at room temperature; pourable at 60 C
Properties: pH: 9.5–10.5; Biodegradable
Composition: 36–38% solids

Miranol MHT. Miranol Chem. Co.
Chemical Description: Trideceth-3-sulfate salt of a monocarboxylated lauryl imidazoline derivative
Category: Detergent base, amphoteric
Applications: Base detergent for nonirritating and noneye stinging foaming hair and skin cleansing products
Form: Clear liquid; Color: Light amber
Properties: Solubility: Sol. in water; pH: 9.4–10.2
Composition: 33.5–36.5% solids

Miranol OM-SF Conc. Miranol Chem. Co.
Chemical Description: Sodium salt of a monocarboxylated oleyl imidazoline derivative, salt free
Category and Applications: See Miranol CM Conc.
Form: Clear liquid; Color: Light amber
Properties: Solubility: Sol. in water and alcohol; pH: 9.8–10.2; Biodegradable
Composition: 36–38% solids

Miranol OS. Miranol Chem. Co.
Chemical Description: Sulfonated oleic derivative, sodium salt
Category and Applications: See Miranol HS Conc.
Form: Paste
Composition: 21% conc.

Miranol S2M Conc. Miranol Chem. Co.
Chemical Description: Sodium salt of a dicarboxylated capryl

imidazoline derivative
Category and Applications: See Miranol CM Conc.; S2M Conc. retains good foaming properties at isoelectric point area and in the presence of soil or grease
Form: Clear liquid
Properties: pH: 8.1–8.3; Biodegradable
Composition: 50–52% solids

Miranol S2M-SF Conc. Miranol Chem. Co.
Chemical Description: Sodium salt of a dicarboxylated capryl imidazoline derivative, salt free
Category and Applications: See Miranol S2M Conc.
Form: Liquid
Properties: pH: 8.8–9.2; Biodegradable
Composition: 38–40% solids

Miranol SM Conc. Miranol Chem. Co.
Chemical Description: Sodium salt of a monocarboxylated capryl imidazoline derivative
Category: Wetting agent, foaming agent, detergent, amphoteric
Applications: Medicated and germicidal shampoos and handsoaps, surgical soaps, rug and upholstery shampoos; in emulsion polymerization
Form: Clear liquid
Properties: pH: 9.0–9.5; Biodegradable
Composition: 40–42% solids

Mirapol A-15. Miranol Chem. Co.
Chemical Description: Polyquaternary ammonium chloride
Category: Softening, conditioning, and surface modifying agent, cationic
Applications: Used in cream rinses, conditioning-type shampoos
Form: Viscous liquid; Color: Amber
Properties: m.w.: 2260 (avg.); Solubility: Dissolves readily in water in all proportions; pH: 8.5
Composition: 64% active ingred.

Mirataine CB. Miranol Chem. Co.
Chemical Description: Aq. sol'n. of the cocoamidoalkyl betaine
Category: Amphoteric surfactant
Applications: Mild, high foaming base for shampoo formulations
Form: Liquid; Color: Pale
Properties: Density: 8.71 lb/gal; Sp. gr.: 1.045; Cloud Pt.: Clear to freeze; pH: 8.5
Composition: 35% active ingred.

Mirataine CBS. Miranol Chem. Co.
Chemical Description: Aq. sol'n. of cocoamidoalkylammonium sulfonate
Category: Amphoteric surfactant
Applications: Mild, high foaming base for mild shampoo formulations
Form: Liquid
Properties: Density: 9.1 lb/gal; Cloud Pt.: ⩽ −12 C; pH: 8.2
Composition: 50% active ingred.

Mirataine CEB. Miranol Chem. Co.
Chemical Description: Aq. sol'n. of the cocoamidoalkyl betaine
Category: Amphoteric surfactant
Applications: Mild, high foaming

base for shampoo formulations
Form: Liquid; Color: Pale
Properties: Density: 8.75 lb/gal; Sp. gr.: 1.05; Cloud Pt.: Clear to freeze; pH: 8.5; Biodegradable
Composition: 40% active ingred.
Toxicity: May cause eye and skin irritation

Mirataine T2C. Miranol Chem. Co.
Chemical Description: Aq. sol'n. of disodium n-tallow-3, 3′ iminodipropionate
Category: Detergent, solubilizer, amphoteric
Applications: Moderate foaming surfactant with good detergency and solubilizing properties; substantive to skin, hair, leather, and textile fibers contributing lubricity and antistatic properties; used in industrial acid cleaners; compatible with high levels of alkalies and inorganic builders
Form: Liquid
Properties: Solubility: Clear (30% solids in 1% NaOH); Surface Tension: 31.6 dynes/cm (1% aq); pH: 11.5 (as is); Biodegradable
Composition: 35% active ingred.

Modicol L. Diamond Shamrock/ Process Chem. Div.
Chemical Description: Polyethylene glycol fatty ester
Category: Coagulant, wetting and dispersing agent, nonionic
Applications: Wetting and viscosity control in natural and synthetic latexes and resin emulsions; additive in grinding pastes for latex compounding; aids compounding of high solids synthetic latex
Form: Clear liquid; Color: Yellow
Properties: Solubility: Sol. in ethanol, acetone, ethyl acetate, toluol, vegetable oil, water; partly sol. in mineral oil; Density: 8.6 lb/gal; Acid Value: 1.0; Stability: Good; pH: 7.5 (2% sol'n.)
Composition: 99% active ingred.
Containers: 55 gal tighthead steel drums

Modicol S. Diamond Shamrock/ Process Chem. Div.
Chemical Description: Sulfonated fatty product
Category: Stabilizer, anionic
Applications: In rubber industry to stabilize natural and synthetic latexes during compounding, storage, and application; ensures mechanical and chemical stability; prevents premature coagulation during high-speed agitation, acidification, or pigmentation
Form: Liquid; Color: Dark reddish-brown; Odor: Fatty
Properties: Solubility: Completely sol. in water and soluble alcohols; hazy sol'n. in 4% sodium hydroxide, in toluol, and in mineral and vegetable oils; Sp. gr.: 1.1; Stability: Stable in 2% hydrochloric acid; pH: 7.5 (2% sol'n.)
Composition: 60% active ingred.
Containers: 45 gal Liquapak drums

Monacor 39. Mona Industries Inc.
Chemical Description: Amidoester carboxylic acid derivative
Category: Corrosion inhibitor, anionic

Applications: Ashless, oil-sol. corrosion inhibitor for use in slushing oils, rolling oils, lubricating oils, etc.
Form: Viscous liquid; Color: Amber
Properties: Solubility @ 10%: Sol. in isopropanol, butyl carbitol, mineral oil, benzene, perchloroethylene; disp. in water (separates on standing); Density: 8.1 lb/gal; Sp. gr.: 0.97 g/cc; Acid No.: 70
Composition: 100% active ingred.
References: Tech. bulletin no. 251

Monacor BE. Mona Industries Inc.
Category: Corrosion inhibitor
Applications: Nonfoaming, water-sol. corrosion inhibitor for cast iron, steel, and other ferrous metals; for use in metal working lubricants, cooling towers, etc.
Form: Clear, viscous liquid
Properties: Solubility: Water-sol.; Density: 9.24 lb/gal; Sp. gr.: 1.107; pH: 9.1 (1%)
Composition: 100% active ingred.
References: Tech. bulletin no. 252

Monafax 785. Mona Industries Inc.
Chemical Description: Mixture of mono and di-phosphate esters
Category: Emulsifier, lubricant, antistat, detergent, corrosion inhibitor, anionic
Applications: Emulsifier for vinyl acetate, acrylates, and SBR; hard surface cleaners with good tolerance in concentrated alkali and salt sol'ns.; herbicide and insecticide emulsifiers, esp. in fertilizers; in water-sol. metal working lubricants; dry cleaning detergents (good antisoil redeposition and antistatic properties)
Form: Clear to slightly hazy viscous liquid
Properties: Solubility: Sol. in water, ethanol, perchloroethylene, ethylene glycol mono butyl ether, xylene; Density: ≈ 9.2 lb/gal; Sp. gr.: 1.115; Acid Value: 70 ± 3 (@ pH 5.0–5.5; pH: < 2.5 (10% sol'n.)
Composition: 100% active ingred.
References: Tech. bulletin no. 290c

Monafax 786. Mona Industries Inc.
Chemical Description: Mixture of mono and di-phosphate esters
Category: Emulsifier, lubricant, antistat, detergent, corrosion inhibitor, anionic
Applications: Herbicide and insecticide emulsifier, esp. in fertilizers; hard surface cleaners with good tolerance in conc. alkali and salt sol'ns.; in water-sol. metal working lubricants; dry cleaning detergents (good antisoil redeposition and antistatic properties)
Form: Clear to slightly hazy viscous liquid
Properties: Solubility @ 10%: Sol. in ethanol, perchloroethylene, ethylene glycol mono butyl ether, Stoddard solvent, xylene; disp. in water and cottonseed oil; Density: ≈ 9.0 lb/gal; Sp. gr.: 1.085; Acid Value: 57 ± 3 (@ pH 5.0–5.5); pH: < 2.5 (10% sol'n.)
Composition: 100% active ingred.
References: Tech. bulletin no. 290c

Monafax 794. Mona Industries Inc.
Chemical Description: Mixture of mono and di-phosphate esters
Category: Emulsifier, anionic
Applications: Primary emulsifier for vinyl acetate, acrylates, and SBR; permits preparation of high solids polymers with heat and light stability; corrosion inhibitor
Form: Soft, waxy paste
Properties: Solubility @ 10%: Sol. in water, ethanol, perchloroethylene (hazy), ethylene glycol mono butyl ether, cottonseed oil, and xylene (hazy); Density: ≈ 9.8 lb/gal; Sp. gr.: 1.175 (50 C); Acid Value: 98 ± 5 (@ pH 5.0–5.5); pH: < 2.5 (10% sol'n.)
Composition: 90% active ingred.
References: Tech. bulletin no. 290c

Monafax 831. Mona Industries Inc.
Chemical Description: Mixture of mono and di-phosphate esters
Category: Emulsifier, lubricant, antistat, detergent, corrosion inhibitor, anionic
Applications: Hard surface cleaners with moderate foaming, good wetting, esp. in alkaline and high electrolyte concentrations; soak tank cleaners, heavy duty alkaline all purpose cleaners, and steam cleaners
Form: Clear viscous liquid
Properties: Solubility @ 10%: Sol. in water, ethanol, perchloroethylene, ethylene glycol mono butyl ether, mineral oil, kerosene, Stoddard solvent, cottonseed oil, and xylene; Density: ≈ 9.0 lb/gal; Sp. gr.: 1.08; Acid Value: 110 ± 3 (@ pH 5.0–5.5); pH: < 2.5 (10% sol'n.)
Composition: 100% active ingred.
References: Tech. bulletin no. 290c

Monafax 872. Mona Industries Inc.
Chemical Description: Organic phosphate ester, potassium salt
Category: Hydrotrope, solubilizer, anionic
Applications: Hydrotrope and solubilizer for conventional and low-foam nonionics in alkaline and high electrolyte systems; liquid hand dishwashes, all purpose cleaners
Form: Clear liquid; Color: Light straw
Properties: Solubility @ 10%: Sol. in water; disp. in ethanol, ethylene glycol mono butyl ether, and cottonseed oil; Density: ≈ 11.24 lb/gal; Sp. gr.: 1.35; Acid No.: 20 ± 2 (@ pH 9.5–9.8); pH: 7–9 (10% sol'n.)
Composition: 50% active ingred.
References: Tech. bulletins no. 290c, 291a

Monafax 939. Mona Industries Inc.
Chemical Description: Alcohol phosphate ester in the acid form
Category: Lubricant, anionic
Applications: Used in metal working lubricants and synthetic cutting fluids; provides low coefficient of friction, rust inhibition, and reduced surface tension; high phosphorous content; used in wa-

ter and oil based formulations
Form: Clear liquid; Color: Amber
Properties: m.w.: 297 (avg.); Solubility: Sol. in water at neutral and alkaline pH; sol. @ 1% and 10% in ethanol, chlorinated hydrocarbons, aromatic hydrocarbons, mineral spirits, kerosene, and mineral oil; Density: 8.6 lb/gal; Acid Value: 220 ± 3 (1:1 isopropanol/water); pH: 2.5 max. (10% sol'n. IPA/water)
Composition: 100% active ingred.
Toxicity: Very mild skin irritant and mild ocular irritant
References: Tech. bulletin no. 293 a and b

Monafax H-15. Mona Industries Inc.
Chemical Description: Mixture of mono and di-phosphate esters
Category: Emulsifier, lubricant, antistat, detergent, corrosion inhibitor, anionic
Applications: Emulsifier for herbicides and insecticides, esp. in fertilizer sol'ns.; high electrolyte tolerance
Form: Hazy, viscous liquid
Properties: Solubility @ 10%: Sol. in water, ethanol, perchloroethylene, ethylene glycol mono butyl ether, xylene (hazy); disp. in Stoddard solvent; Density: ≈ 8.9 lb/gal; Sp. gr.: 1.07; Acid Value: 40 ± 3 (@ pH 5.0–5.5); pH: < 2.5 (10% sol'n.)
Composition: 100% active ingred.
References: Tech. bulletin no. 290c

Monafax L-10. Mona Industries Inc.
Chemical Description: Mixture of mono and di-phosphate esters
Category: Emulsifier, lubricant, antistat, detergent, corrosion inhibitor, anionic
Applications: Herbicide and insecticide emulsifier, esp. in fertilzer sol'ns.; in drycleaning detergents (good antisoil redeposition and antistatic properties); high electrolyte tolerance
Form: Slightly hazy, viscous liquid
Properties: Solubility @ 10%: Sol. in ethanol, perchloroethylene, ethylene glycol mono butyl ether, kerosene, Stoddard solvent, cottonseed oil, and xylene; disp. in water and mineral oil; Density: ≈ 8.8 lb/gal; Sp. gr.: 1.06; Acid Value: 48 ± 3 (@ pH 5.0–5.5); pH: < 2.5 (10% sol'n.)
Composition: 100% active ingred.
References: Tech. bulletin no. 290c

Monalube 29-78. Mona Industries Inc.
Chemical Description: Alkanolamide
Category: Lubricant, corrosion inhibitor, nonionic
Applications: Synthetic, water-sol. lubricant for metal working with good rust preventive power; drawing and cutting oils for iron, steel, stainless steel, aluminum, etc.; in fiber lubricants and textile specialties
Form: Liquid
Properties: Solubility: Water-sol.;

Acid No.: 22–32
Composition: 100% active ingred.
References: Tech. bulletin no. 701b

Monalube 780. Mona Industries Inc.
Chemical Description: Modified alkanolamide
Category: Lubricant, corrosion inhibitor, nonionic/anionic
Applications: Synthetic, water-disp. copper drawing lubricant producing minimal copper soap and minimal foam; corrosion inhibitor
Form: Liquid; Color: Amber
Properties: Solubility: Disp. in water; Density: 8.2 lb/gal; Acid No.: 7.5; Stability: Good
Composition: 100% active ingred.
References: Tech. bulletin no. 704

Monamate CPA-40%. Mona Industries Inc.
Chemical Description: Sulfosuccinate half ester of an alkanolamide, di-sodium salt
Category: Anionic surfactant
Applications: High foaming non-irritating surfactant for shampoos, bubble baths, skin cleansers, rug shampoos, etc.
Form: Liquid
Properties: Acid No.: 5–12; Stability: Stable @ pH 4–8; good heat stability at its normal pH; Biodegradable
Composition: 40% active ingred.
References: Tech. bulletin no. 275B

Monamate OPA-30. Mona Industries Inc.
Chemical Description: Sulfosuccinate half ester of an alkanolamide, di-sodium salt
Category: Anionic surfactant
Applications: High foaming, non-irritating surfactant for shampoos, bubble baths, skin cleansers
Form: Liquid; Color: Light yellow
Properties: Density: 8.85 lb/gal; Sp. gr.: 1.06; Stability: Stable @ pH 4–8; good heat stability at its normal pH; pH: 5.6 (as is); Biodegradable
Composition: 30% solids
References: Tech. bulletin no. 272

Monamid Series. Mona Industries Inc.
Chemical Description: Fatty acid-alkanolamides
Category: Foam boosters and stabilizers, emulsifiers, detergents, wetting agents, corrosion inhibitors, viscosity builders, lubricants, dispersants
Applications: Cosmetics, bubble baths, shampoos, drycleaning detergents, dishwashing detergents, metal cleaners, hand soaps, waterless hand cleaners, textile detergents, fiber and hair conditioners, lubricants, rust inhibitors, rug shampoos, metal cutting fluids, dustless soap powders, metal polishes, emulsifiable waxes, pigment dispersions, agricultural sprays, fuel oil additives, leather and fur preparations
Form and Properties: See individual entries

Monamid 7-100. Mona Industries Inc.
Chemical Description: 1:1 coconut diethanolamide
Category: Nonionic
Applications: See Monamid Series
Form: Liquid; Color: 6 max. (Gvcs-33)
Properties: Solubility @ 10%: Sol. ethyl alcohol, chlorinated and aromatic hydrocarbons, mineral spirits, kerosene, white mineral oil, natural oils and fats; disp. in water; Density: 8.00 lb/gal; Sp. gr.: 0.96 (20 C); Acid No.: 0–2; pH: 8–9 (10% sol'n.); Biodegradable
Composition: 100% active ingred.
Containers: 450 lb net tighthead steel drums
References: Tech. bulletin 240 D

Monamid 7-153 CS. Mona Industries Inc.
Chemical Description: Modified coconut diethanolamide
Category: Nonionic/anionic
Applications: See Monamid Series
Form: Liquid; Color: 3 max. (Gvcs-33)
Properties: Solubility @ 10%: Sol. in ethyl alcohol, chlorinated and aromatic hydrocarbons, mineral spirits, kerosene, white mineral oil; disp. in water and natural oils and fats; Density: 8.75 lb/gal; Sp. gr.: 1.05 (20 C); Acid No.: 0–2; pH: 8.5–9.5 (10% sol'n.); Biodegradable
Composition: 100% active ingred.
Containers: 480 lb net Liquipak
References: Tech. bulletin 240 D

Monamid 15-70W. Mona Industries Inc.
Chemical Description: 1:1 linoleic diethanolamide
Category: Nonionic
Applications: See Monamid Series
Form: Liquid; Color: 11 max. (Gvcs-33)
Properties: Solubility @ 10%: Sol. in ethyl alcohol, chlorinated and aromatic hydrocarbons, mineral spirits, kerosene, natural oils and fats; disp. in white mineral oil; gels in water; Density: 8.00 lb/gal; Sp. gr.: 0.96 (20 C); Acid No.: 0–0–1; pH: 10–11 (10% sol'n.); Biodegradable
Composition: 100% active ingred.
Containers: 440 lb net tighthead steel drum
References: Tech. bulletin 240 D

Monamid 150-AD. Mona Industries Inc.
Chemical Description: 1:1 coconut diethanolamide
Category: Nonionic
Applications: See Monamid Series
Form: Liquid; Color: 11 max. (Gvcs-33)
Properties: Solubility @ 10%: Sol. in water, ethyl alcohol, chlorinated hydrocarbons, aromatic hydrocarbons, natural oils and fats; Density: 8.25 lb/gal; Sp. gr.: 0.99; Acid No.: 0–3; pH; 9.8–10.8 (10% sol'n.); Biodegradable
Composition: 100% active ingred.
Containers: 450 lb net tighthead steel drums
References: Tech. bulletin 240 D

Monamid 150-ADD. Mona Industries Inc.
Chemical Description: 1:1 coconut diethanolamide
Category: Nonionic
Applications: See Monamid Series
Form: Liquid; Color: 4 max. (Gvcs-33)
Properties: Solubility @ 10%: Sol. in water, ethyl alcohol, chlorinated hydrocarbons, aromatic hydrocarbons, natural oils and fats; Density: 8.30 lb/gal; Sp. gr.: 1.00; Acid No.: 0–3; pH: 10–11 (10% sol'n.); Biodegradable
Composition: 100% active ingred.
Containers: 450 lb net tighthead steel drums
References: Tech. bulletin 240 D

Monamid 150-ADY. Mona Industries Inc.
Chemical Description: 1:1 mixed fatty acid-diethanolamide
Category: Nonionic
Applications: See Monamid Series
Form: Liquid; Color: 10 max. (Gvcs-33)
Properties: Solubility @ 10%: Sol. in ethyl alcohol, chlorinated and aromatic hydrocarbons, mineral spirits, kerosene; disp. in white mineral oil, natural oils and fats; gels in water; Density: 8.20 lb/gal; Sp. gr.: 0.97 (20 C); Acid No.: 0–1; pH: 10–11 (10% sol'n.); Biodegradable
Composition: 100% active ingred.
Containers: 450 lb net open head steel drums
References: Tech. bulletin 240 D

Monamid 150-CW. Mona Industries Inc.
Chemical Description: 1:1 capric-diethanolamide
Category: Nonionic
Applications: See Monamid Series
Form: Liquid (crystallizes on aging); Color: 7 max. (Gvcs-33)
Properties: Solubility @ 10%: Sol. water, ethyl alcohol, chlorinated and aromatic hydrocarbons, natural oils and fats; Density: 8.25 lb/gal; Sp. gr.: 0.99 (20 C); Acid No.: 0–2; pH: 10.3–11.3 (10% sol'n.); Biodegradable
Composition: 100% active ingred.
Containers: 450 lb net Liquipak
References: Tech. bulletin 240 D

Monamid 150-DR. Mona Industries Inc.
Chemical Description: 1:1 coconut diethanolamide
Category: Nonionic
Applications: See Monamid Series
Form: Liquid; Color: 5 max. (Gvcs-33)
Properties: Solubility @ 10%: Sol. in ethyl alcohol, mineral spirits, kerosene, white mineral oil, natural oils and fats; disp. in water, chlorinated and aromatic hydrocarbons; Density: 8.20 lb/gal; Sp. gr.: 0.98 (20 C); Acid No.: 0–5; pH: 8.5–9.5 (10% sol'n.); Biodegradable
Composition: 100% active ingred.
Containers: 450 lb net open head steel drums
References: Tech. bulletin 240 D

Monamid 150-GLT. Mona Industries Inc.
Chemical Description: Modified lauric diethanolamide
Category: Thickener and foam stabilizer, nonionic
Applications: Cosmetic and toiletry products
Form: Clear liquid (may become hazy or crystallize at cold temperatures); Color: Amber
Properties: Acid No.: 1 max.; pH: 9.0–10.5 (1% sol'n.)
Composition: 100% active ingred.
References: Tech. bulletin no. 260

Monamid 150-IS. Mona Industries Inc.
Chemical Description: 1:1 isostearic diethanolamide
Category: Nonionic
Applications: See Monamid Series
Form: Liquid; Color: 6 max. (Gvcs-33)
Properties: Solubility @ 10%: Sol. in ethyl alcohol, chlorinated and aromatic hydrocarbons, mineral spirits, kerosene, white mineral oil, natural oils and fats; disp. in water; Density: 8.00 lb/gal; Sp. gr.: 0.96 (20 C); Acid No.: 5–10; pH: 8.8–9.8 (10% sol'n.); Biodegradable
Composition: 100% active ingred.
Containers: 440 lb net tighthead steel drums
References: Tech. bulletin 240 D

Monamid 150-LMW-C. Mona Industries Inc.
Chemical Description: 70/30 lauric myristic diethanolamide
Category: Nonionic
Applications: See Monamid Series
Form: Solid; Color: 3 max. (Gvcs-33)
Properties: Solubility @ 10%: Sol. in ethyl alcohol, chlorinated and aromatic hydrocarbons, mineral spirits; disp. in kerosene, white mineral oil, natural oils and fats; gels in water; Density: 8.20 lb/gal; Sp. gr.: 0.98 (40 C); Acid No.: 0-1; pH: 10.2–11.2 (10% sol'n.); Biodegradable
Composition: 100% active ingred.
Containers: 450 lb net open head steel drums
References: Tech. bulletin 240 D

Monamid 150-LW. Mona Industries Inc.
Chemical Description: 1:1 lauric diethanolamide
Category: Nonionic
Applications: See Monamid Series
Form: Solid; Color: 3 max. (Gvcs-33)
Properties: Solubility @ 10%: Sol. in ethyl alcohol, chlorinated and aromatic hydrocarbons; disp. in water, mineral spirits, kerosene, white mineral oil, natural oils and fats; Density: 8.20 lb/gal; Sp. gr.: 0.98 (40 C); Acid No.: 0–1; pH: 10–11 (10% sol'n.); Biodegradable
Composition: 100% active ingred.
Containers: 450 lb net open head steel drums
References: Tech. bulletin 240 D

Monamid 150-LWA. Mona Industries Inc.
Chemical Description: 1:1 lauric diethanolamide

Category: Nonionic
Applications: See Monamid Series
Form: Solid; Color: 3 max. (Gvcs-33)
Properties: Solubility @ 10%: Sol. ethyl alcohol, chlorinated and aromatic hydrocarbons; disp. in water, mineral spirits, kerosene, white mineral oil, natural oils and fats; Density: 8.20 lb/gal; Sp. gr.: 0.98 (40 C); Acid No.: 0–1; pH: 9.5–10.5 (10% sol'n.); Biodegradable
Composition: 100% active ingred.
Containers: 450 lb net open head steel drums
References: Tech. bulletin 240 D

Monamid 150-MW. Mona Industries Inc.
Chemical Description: 1:1 myristic diethanolamide
Category: Nonionic
Applications: See Monamid Series
Form: Solid; Color: 2 max. (Gvcs-33)
Properties: Solubility @ 10%: Sol. in ethyl alcohol, chlorinated and aromatic hydrocarbons; disp. in water; gels in natural oils and fats; Density: 8.20 lb/gal; Sp. gr.: 0.98 (45 C); Acid No.: 0–3; pH: 9.5–10.5 (10% sol'n.); Biodegradable
Composition: 100% active ingred.
Containers: 100 lb net fiberboard drum
References: Tech. bulletin 240 D

Monamid 716. Mona Industries Inc.
Chemical Description: Modified lauric diethanolamide
Category: Foam booster and stabilizer, viscosity builder, solubilizer, coupler, wetting agent, nonionic
Applications: Shampoos, bubble baths, and liquid detergent formulations; extremely mild
Form: Clear liquid; Color: Light amber
Properties: Solubility @ 10%: Sol. in water, ethyl alcohol, chlorinated and aromatic hydrocarbons, natural oils and fats; Density: 8.2 lb/gal; Sp. gr.: 0.98 (20 C); Cloud Pt.: < 1 C; Acid No.: 0–3; pH: 10.0–11.0 (10% sol'n.); Biodegradable
Composition: 100% active ingred.
Containers: 450 lb net tighthead steel drums
References: Tech. bulletin 267d, 240 D

Monamid 718. Mona Industries Inc.
Chemical Description: Stearic diethanolamide
Category: Nonionic
Applications: See Monamid Series
Form: Solid; Color: 4 max. (Gvcs-33)
Properties: Solubility @ 10%: Sol. in ethyl alcohol, aromatic hydrocarbons; disp. in chlorinated hydrocarbons; gels in water; Density: 8.00 lb/gal; Sp. gr.: 0.96 (45 C); Acid No.: 21 ± 3; pH: 9.3–10.3 (10% sol'n.); Biodegradable
Composition: 100% active ingred.
Containers: 450 lb net open head steel drums
References: Tech. bulletin 240 D

Monamid 770. Mona Industries Inc.
Chemical Description: Modified coconut diethanolamide
Category: Nonionic
Applications: See Monamid Series
Form: Liquid; Color: 4 max. (Gvcs-33)
Properties: Solubility @ 10%: Sol. in water, ethyl alcohol, chlorinated and aromatic hydrocarbons, natural oils and fats; Density: 8.25 lb/gal; Sp. gr.: 0.99 (20 C); Acid No.: 0–1; pH: 9.2–10.2 (10% sol'n.); Biodegradable
Composition: 85% active ingred.
Containers: 450 lb net open head steel drums
References: Tech. bulletin 240 D

Monamid 982. Mona Industries Inc.
Chemical Description: 1:1 mixed fatty acid diethanolamide
Category: Foam booster and stabilizer, viscosity builder
Applications: Shampoos, bubble baths, skin cleansers, household and industrial liquid detergents
Form: Clear liquid; Color: Amber
Properties: Density: ≈ 8.16 lb/gal; Acid No.: 1 max.
Composition: 100% active ingred.
References: Tech. bulletin no. 282

Monamid 1007. Mona Industries Inc.
Chemical Description: 1:1 mixed fatty acid diethanolamide
Category and Applications: See Monamid 982
Form: Clear liquid; Color: Amber
Properties: Density: ≈ 8.16 lb/gal; Acid No.: 1 max.
Composition: 100% active ingred.
References: Tech. bulletin no. 282

Monamid CMA. Mona Industries Inc.
Chemical Description: 1:1 coconut monoethanolamide
Category: Nonionic
Applications: See Monamid Series
Form: Granular; Color: Tan
Properties: Solubility @ 10% and 25 C: Sol. in ethyl alcohol; disp. in water, chlorinated and aromatic hydrocarbons, mineral spirits, kerosene, white mineral oil, natural oils and fats; Solid. Pt.: 63 ± 2 C; Acid No.: 0–1; pH: 9.4–10.8 (10% sol'n.); Biodegradable
Composition: 100% active ingred.
Containers: 200 lb net fiberboard drums
References: Tech. bulletin 240 D

Monamid LIPA. Mona Industries Inc.
Chemical Description: 1:1 lauric monoisopropanolamide
Category: Nonionic
Applications: See Monamid Series
Form: Granular; Color: Tan
Properties: Solubility @ 10% and 25 C: Sol. in ethyl alcohol; disp. in water, kerosene, white mineral oil, natural oils and fats; Solid. Pt.: 55 ± 3 C; Acid No.: 0–1; pH: 10.3–11.3 (10% sol'n.); Biodegradable
Composition: 100% active ingred.
Containers: 175 lb net fiberboard drums
References: Tech. bulletin 240 D

Monamid LMA. Mona Industries Inc.
Chemical Description: 1:1 lauric monoethanolamide
Category: Nonionic
Applications: See Monamid Series
Form: Granular; Color: Tan
Properties: Solubility @ 10% and 25 C: Sol. in ethyl alcohol; disp. in water, aromatic hydrocarbons, kerosene, white mineral oil, natural oils and fats; Solid. Pt.: 80 ± 2 C; Acid No.: 0–1; pH: 10–11 (10% sol'n.); Biodegradable
Composition: 100% active ingred.
Containers: 200 lb net fiberboard drums
References: Tech. bulletin 240 D

Monamid LMMA. Mona Industries Inc.
Chemical Description: 1:1 lauric myristic monoethanolamide
Category: Nonionic
Applications: See Monamid Series
Form: Granular; Color: Tan
Properties: Solubility @ 10% and 25 C: Sol. in ethyl alcohol; disp. in water, aromatic hydrocarbons, kerosene, white mineral oil, natural oils and fats; Solid. Pt.: 80 ± 2 C; Acid No.: 0–1; pH: 9.7–10.7 (10% sol'n.); Biodegradable
Composition: 100% active ingred.
Containers: 200 lb net fiberboard drums
References: Tech. bulletin 240 D

Monamid R31-42. Mona Industries Inc.
Chemical Description: 70/30 lauric myristic diethanolamide
Category: Nonionic
Applications: See Monamid Series
Form: Liquid; Color: 2 max. (Gvcs-33)
Properties: Solubility @ 10%: Sol. in water, ethyl alcohol, chlorinated and aromatic hydrocarbons; disp. in natural oils and fats; Density: 8.25 lb/gal; Sp. gr.: 0.99 (20 C); Acid No.: 0–1; pH: 10–11 (10% sol'n.); Biodegradable
Composition: 80% active ingred.
Containers: 450 lb net open head steel drums
References: Tech. bulletin 240 D

Monamid S. Mona Industries Inc.
Chemical Description: 1:1 stearic monoethanolamide
Category: Nonionic
Applications: See Monamid Series
Form: Granular; Color: Tan
Properties: Solubility @ 10% and 25 C: Disp. in water, chlorinated and aromatic hydrocarbons, mineral spirits, kerosene, white mineral oil, natural oils and fats; Solid. Pt.: 87 ± 2 C; Acid No.: 0–1; pH: 9.5–11.0 (10% sol'n.); Biodegradable
Composition: 100% active ingred.
Containers: 200 lb net fiberboard drums
References: Tech. bulletin 240 D

Monamine Series. Mona Industries Inc.
Chemical Description: Fatty acid-alkanolamide
Category: Foam boosters and stabilizers, emulsifiers, detergents, wetting agents, corrosion inhibi-

tors, viscosity builders, lubricants, dispersants
Applications: Cosmetics, bubble baths, shampoos, drycleaning detergents, dishwashing detergents, metal cleaners, hand soaps, waterless hand cleaners, textile detergents, fiber and hair conditioners, lubricants, rust inhibitors, rug shampoos, metal cutting fluids, dustless soap powders, metal polishes, emulsifiable waxes, pigment dispersions, agricultural sprays, fuel oil additives, leather and fur preparations
Form and Properties: See individual entries
References: Tech. bulletin 240 D

Monamine 779. Mona Industries Inc.
Chemical Description: Amidoether sulfate complex
Category: Foaming agent, viscosity builder, detergent, wetting and penetrating agent, surface tension reducer, nonionic/anionic
Applications: Shampoos, bubble baths, foaming bath oils, body and skin cleansers; household and industrial cleaners; good hard water and soap tolerance and stability with alkali
Form: Clear viscous liquid; Color: Light yellow
Properties: Solubility @ 10%: Sol. in water, polar solvents, alcohol, glycols, fatty acid esters, chlorinated and aromatic hydrocarbons; Density: 8.75 lb/gal; Sp. gr.: 1.05 ± 0.1; Cloud Pt.: 2 C; Stability: Stable to alkali; Surface Tension: 29 dynes/cm (0.1% conc.); pH: 9.25 ± 0.5 (10% sol'n.); Biodegradable
Composition: 100% active ingred.
References: Tech. bulletin no. 274a

Monamine 853. Mona Industries Inc.
Chemical Description: Modified alkanolamide
Category: Detergent
Applications: Gel and lotion type waterless hand cleaners with good water rinsability and heavy soil detergency
Form: Clear liquid; Color: Amber
Properties: Density: 8.6 lb/gal; Acid No.: 8 max.
Composition: 100% active ingred.
References: Tech. bulletin 246a

Monamine 928. Mona Industries Inc.
Chemical Description: Modified long-chain alkanolamide
Category: Emulsifier, lubricant, corrosion inhibitor, nonionic
Applications: Synthetic coolants and metal working fluids; secondary emulsifier in conventional soluble oil systems
Form: Paste; Color: Cream
Properties: Density: 8.0 lb/gal (50 C); Acid No.: 5 max.
Composition: 100% active ingred.
References: Tech. bulletin 265

Monamine 929. Mona Industries Inc.
Chemical Description: Modified long-chain alkanolamide

Category: Emulsifier, lubricant, corrosion inhibitor, nonionic/anionic
Applications: See Monamine 928
Form: Clear to opaque liquid; Color: Amber
Properties: Density: 8.3 lb/gal; Acid No.: 56 ± 3
Composition: 100% active ingred.
References: Tech. bulletin 265

Monamine AA-100. Mona Industries Inc.
Chemical Description: 1:2 distilled coconut diethanolamide
Category: Nonionic/anionic
Applications: See Monamine Series
Form: Liquid; Color: 7 max. (Gvcs-33)
Properties: Solubility @ 10%: Sol. in water, ethyl alcohol, chlorinated and aromatic hydrocarbons; Density: 8.30 lb/gal; Sp. gr.: 1.00 (20 C); Acid No.: 28–32; pH: 9.5–10.5 (10% sol'n.); Biodegradable
Composition: 100% active ingred.
Containers: 450 lb net tighthead steel drums
References: Tech. bulletin 240 D

Monamine AC-100. Mona Industries Inc.
Chemical Description: 1:2 mixed fatty acid diethanolamide
Category: Nonionic/anionic
Applications: See Monamine Series
Form: Liquid; Color: 13 max. (Gvcs-33)
Properties: Solubility @ 10%: Sol. in ethyl alcohol, chlorinated and aromatic hydrocarbons; gels in water; Density: 8.30 lb/gal; Sp. gr.: 1.00 (20 C); Acid No.: 22–32; pH: 9.5–10.5 (10% sol'n.); Biodegradable
Composition: 100% active ingred.
Containers: 450 lb net open head steel drums
References: Tech. bulletin 240 D

Monamine ACO-100. Mona Industries Inc.
Chemical Description: 1:2 lauric diethanolamide
Category: Nonionic/anionic
Applications: See Monamine Series
Form: Paste; Color: 5 max. (Gvcs-33)
Properties: Solubility @ 10%: Sol. in water, ethyl alcohol, chlorinated and aromatic hydrocarbons; Density: 8.40 lb/gal; Sp. gr.: 1.01 (20 C); Acid No.: 10–14; pH: 9.5–10.5 (10% sol'n.); Biodegradable
Composition: 100% active ingred.
Containers: 450 lb net open head steel drums
References: Tech. bulletin 240 D

Monamine AD-100. Mona Industries Inc.
Chemical Description: 1:2 coconut diethanolamide
Category: Nonionic
Applications: See Monamine Series
Form: Liquid; Color: 11 max. (Gvcs-33)
Properties: Solubility @ 10%: Sol. in water, ethyl alcohol, chlorinated hydrocarbons; disp. in aromat-

ic hydrocarbons; Density: 8.50 lb/gal; Sp. gr.: 1.02 (20 C); Acid No.: 2–8; pH: 9.5–10.5 (10% sol'n.); Biodegradable
Composition: 100% active ingred.
Containers: 450 lb net tighthead steel drums
References: Tech. bulletin 240 D

Monamine ADD-100. Mona Industries Inc.
Chemical Description: 1:2 coconut diethanolamide
Category: Nonionic
Applications: See Monamine Series
Form: Liquid; Color: 3 max. (Gvcs-33)
Properties: Solubility @ 10%: Sol. in water, ethyl alcohol, chlorinated hydrocarbons; disp. in aromatic hydrocarbons; Density: 8.50 lb/gal; Sp. gr.: 1.02 (20 C); Acid No.: 2–6; pH: 9.5–10.5 (10% sol'n.); Biodegradable
Composition: 100% active ingred.
Containers: 450 lb net tighthead steel drums
References: Tech. bulletin 240 D

Monamine ADD-100 LE. Mona Industries Inc.
Chemical Description: 1:2 coconut diethanolamide
Category: Nonionic
Applications: See Monamine Series
Form: Liquid; Color: 3 max. (Gvcs-33)
Properties: Solubility @ 10%: Sol. in water, ethyl alcohol, chlorinated and aromatic hydrocarbons; Density: 8.50 lb/gal; Sp. gr.: 1.02 (20 C); Acid No.: 0–2; pH: 10.5–11.5 (10% sol'n.); Biodegradable
Composition: 100% active ingred.
Containers: 450 lb net tighthead steel drums
References: Tech. bulletin 240 D

Monamine ADDS-100. Mona Industries Inc.
Chemical Description: 1:2 mixed fatty acid diethanolamide
Category: Nonionic/anionic
Applications: See Monamine Series
Form: Liquid; Color: 7 max. (Gvcs-33)
Properties: Solubility @ 10%: Sol. in water, ethyl alcohol, chlorinated and aromatic hydrocarbons; Density: 8.30 lb/gal; Sp. gr.: 1.00 (20 C); Acid No.: 48–52; pH: 9.0–10.0 (10% sol'n.); Biodegradable
Composition: 100% active ingred.
Containers: 450 lb net tighthead steel drums
References: Tech. bulletin 240 D

Monamine ADS-100. Mona Industries Inc.
Chemical Description: 1:2 mixed fatty acid diethanolamide
Category: Nonionic/anionic
Applications: See Monamine Series
Form: Liquid; Color: 11 max. (Gvcs-33)
Properties: Solubility @ 10%: Sol. in water, ethyl alcohol, chlorinated and aromatic hydrocarbons; Density: 8.30 lb/gal; Sp. gr.: 1.00 (20 C); Acid No.: 48–52; pH: 9.0–

10.0 (10% sol'n.); Biodegradable
Composition: 100% active ingred.
Containers: 450 lb net tighthead steel drums
References: Tech. bulletin 240 D

Monamine ADY-100. Mona Industries Inc.
Chemical Description: 1:2 mixed fatty acid diethanolamide
Category: Nonionic
Applications: See Monamine Series
Form: Liquid; Color: 9 max. (Gvcs-33)
Properties: Solubility @ 10%: Sol. in ethyl alcohol, mineral spirits, kerosene; disp. in water, chlorinated and aromatic hydrocarbons; Density: 8.20 lb/gal; Sp. gr.: 0.98 (20 C); Acid No.: 0–2; pH: 10.5–11.5 (10% sol'n.); Biodegradable
Composition: 100% active ingred.
Containers: 450 lb net open head steel drums
References: Tech. bulletin 240 D

Monamine AF-100. Mona Industries Inc.
Chemical Description: 1:2 mixed fatty acid diethanolamide
Category: Nonionic/anionic
Applications: See Monamine Series
Form: Liquid; Color: 12 max. (Gvcs-33)
Properties: Solubility @ 10%: Sol. in water, ethyl alcohol, chlorinated and aromatic hydrocarbons, mineral spirits; disp. in kerosene, white mineral oil, natural oils and fats; Density: 8.25 lb/gal; Sp. gr.: 0.99 (20 C); Acid No.: 38–42; pH: 9.0–10.0 (10% sol'n.); Biodegradable
Composition: 100% active ingred.
Containers: 450 lb net tighthead steel drums
References: Tech. bulletin 240 D

Monamine ALX-80 SS. Mona Industries Inc.
Chemical Description: 1:2 modified coconut diethanolamide
Category: Nonionic/anionic
Applications: See Monamine Series
Form: Liquid; Color: 6 max. (Gvcs-33)
Properties: Solubility @ 10%: Sol. in water, ethyl alcohol; Density: 8.75 lb/gal; Sp. gr.: 1.05 (20 C); Acid No.: 52–60; pH: 8.5–9.5 (10% sol'n.); Biodegradable
Composition: 80% active ingred.
Containers: 480 lb net open head steel drums
References: Tech. bulletin 240 D

Monamine ALX-100 S. Mona Industries Inc.
Chemical Description: 1:2 modified coconut diethanolamide
Category: Nonionic/anionic
Applications: See Monamine Series
Form: Liquid; Color: 8 max. (Gvcs-33)
Properties: Solubility @ 10%: Sol. in water, ethyl alcohol, chlorinated and aromatic hydrocarbons; Density: 8.85 lb/gal; Sp. gr.: 1.06 (20 C); Acid No.: 62–70; pH: 8.5–

9.5 (10% sol'n.); Biodegradable
Composition: 100% active ingred.
Containers: 480 lb net open head steel drums

Monamine ARA-100. Mona Industries Inc.
Chemical Description: 1:2 coconut diethanolamide
Category: Nonionic/anionic
Applications: See Monamine Series
Form: Liquid; Color: 7 max. (Gvcs-33)
Properties: Solubility @ 10%: Sol. in water, ethyl alcohol, chlorinated and aromatic hydrocarbons; Density: 8.30 lb/gal; Sp. gr.: 1.00 (20 C); Acid No.: 28–32; pH: 9.5–10.5 (10% sol'n.); Biodegradable
Composition: 100% active ingred.
Containers: 450 lb net tighthead steel drums
References: Tech. bulletin 240 D

Monamine CF-100 M. Mona Industries Inc.
Chemical Description: 1:2 mixed fatty acid diethanolamide
Category: Nonionic/anionic
Applications: See Monamine Series
Form: Liquid; Color: 11 max. (Gvcs-33)
Properties: Solubility @ 10%: Sol. in water, ethyl alcohol, chlorinated and aromatic hydrocarbons; Density: 8.40 lb/gal; Sp. gr.: 1.01 (20 C); Acid No.: 56–64; pH: 8.5–9.5 (10% sol'n.); Biodegradable
Composition: 100% active ingred.
Containers: 450 lb net tighthead steel drums
References: Tech. bulletin 240 D

Monamine I-76. Mona Industries Inc.
Chemical Description: 1:2 coconut diethanolamide
Category: Nonionic/anionic
Applications: See Monamine Series
Form: Liquid; Color: 12 max. (Gvcs-33)
Properties: Solubility @ 10%: Sol. in water, ethyl alcohol, chlorinated and aromatic hydrocarbons; Density: 8.30 lb/gal; Sp. gr.: 1.00 (20 C); Acid No.: 45–55; pH: 8.5–9.5 (10% sol'n.); Biodegradable
Composition: 100% active ingred.
Containers: 450 lb net tighthead steel drums
References: Tech. bulletin 240 D

Monamine LM-100. Mona Industries Inc.
Chemical Description: 70/30 lauric myristic diethanolamide
Category: Nonionic/anionic
Applications: See Monamine Series
Form: Liquid; Color: 4 max. (Gvcs-33)
Properties: Solubility @ 10%: Sol. in water, ethyl alcohol, chlorinated and aromatic hydrocarbons; Density: 8.50 lb/gal; Sp. gr.: 1.02 (20 C); Acid No.: 18–23; pH: 9.5–10.5 (10% sol'n.); Biodegradable
Composition: 100% active ingred.
Containers: 450 lb net open head steel drum
References: Tech. bulletin 240 D

Monamine R8-26. Mona Industries Inc.
Chemical Description: 1:2 mixed fatty acid diethanolamide
Category: Nonionic/anionic
Applications: See Monamine Series
Form: Liquid; Color: 11 max. (Gvcs-33)
Properties: Solubility @ 10%: Sol. in water, ethyl alcohol, chlorinated and aromatic hydrocarbons; Density: 8.25 lb/gal; Sp. gr.: 0.99 (20 C); Acid No.: 75–85; pH: 9.0–10.0 (10% sol'n.); Biodegradable
Composition: 100% active ingred.
Containers: 450 lb net open head steel drums
References: Tech. bulletin 240 D

Monamine T-100. Mona Industries Inc.
Chemical Description: 1:2 tall oil diethanolamide
Category: Nonionic/anionic
Applications: See Monamine Series
Form: Liquid; Color: 10 max. (Gvcs-33)
Properties: Solubility @ 10%: Sol. in ethyl alcohol, chlorinated and aromatic hydrocarbons, mineral spirits, kerosene; disp. in water; Density: 8.10 lb/gal; Sp. gr.: 0.97 (20 C); Acid No.: 10–16; pH: 10.0–11.0 (10% sol'n.); Biodegradable
Composition: 100% active ingred.
Containers: 450 lb net tighthead steel drums
References: Tech. bulletin 240 D

Monamulse CI. Mona Industries Inc.
Chemical Description: Modified imidazoline
Category: Corrosion inhibitor, lubricant, penetrant, anionic
Applications: Corrosion inhibitor at low concentrations in greases and oil-based lubrication systems; improves water resistance, lubricating and penetrating properties, and emulsion stability
Form: Clear liquid; Color: Amber
Properties: Solubility @ 10%: Sol. in alcohol, chlorinated and aromatic hydrocarbons, mineral spirits, kerosene, white mineral oil, natural oils and fats; Sp. gr.: 0.959 ± .005; Acid No.: 0–2
Composition: 100% active ingred.
References: Tech. bulletin 703a

Monaquat ISIES. Mona Industries Inc.
Chemical Description: Quaternized heptadecyl imidazoline
Category: Antistat, lubricant, softener, corrosion inhibitor, cationic
Applications: In cosmetics industry as a base for cream rinses, and in hair sprays and hair grooms; in industrial and textile applications used in synthetic coolants and lubricants, plastics, textile softeners, and fiber lubricants
Form: Liquid; Color: Amber
Properties: Solubility @ 10%: Sol. in water, ethanol, butyl Cellosolve, aromatic, chlorinated, and fluorinated hydrocarbons; Sp. gr.: 1.03;

pH: 6.9 (10% sol'n.); Biodegradable
Composition: 100% active ingred.
References: Tech. bulletin 902a

Monaterge 85. Mona Industries Inc.
Chemical Description: Fatty acid amido complex
Category: Detergent, wetting agent, emulsifier, nonionic/anionic
Applications: Alkaline cleaners, e.g., heavy duty all purpose cleaners, wax strippers, floor and wall cleaners, white wall tire cleaners, drain cleaners, liquid laundry detergents; emulsifier for chlorinated solvents
Form: Clear liquid; Color: Amber
Properties: Solubility: Sol. in water, ethyl alcohol, aromatic hydrocarbons, natural oils and fats; Density: 8.5 lb/gal; Sp. gr.: 1.02; Acid No.: 40; Stability: Stable in systems containing high concentrations of sodium hydroxide, potassium hydroxide, sodium and potassium phosphates, sodium silicates, and sodium carbonate; Surface Tension: 27.9 dynes/cm (0.05% conc.); pH: 8.9 (10% sol'n.); Biodegradable
Composition: 85% active ingred.
References: Tech. bulletin 231

Monaterge LF-945. Mona Industries Inc.
Chemical Description: Modified imidazoline amphoteric
Category: Detergent, amphoteric/nonionic
Applications: Low-foaming detergent for heavy duty alkaline and acid cleaners at low temperatures, e.g., spray tank cleaners, steam cleaners, soak tank cleaners
Form: Clear to hazy liquid; Color: Light yellow
Properties: Density: 8.3 lb/gal; Sp. gr.: 1.00; pH: 10.0–11.0 (10% sol'n.)
Composition: 100% active ingred.
References: Tech. bulletin 230a

Monateric 805. Mona Industries Inc.
Chemical Description: Amphoteric salt
Category: Amphoteric/anionic surfactant
Applications: High-foaming, very mild surfactant for shampoos, hair conditioners, skin care products
Form: Viscous liquid; Color: Light amber
Properties: Density: 9.16 lb/gal; Sp. gr.: 1.10; Acid No.: 18 ± 4; pH: 7.8 ± 0.5 (10% sol'n.)
Composition: 40 ± 1% active
References: Tech. bulletin 956c

Monateric 811. Mona Industries Inc.
Chemical Description: Amphoteric based on alkyl imidazoline
Category: Corrosion inhibitor, detergent, wetting agent, amphoteric
Applications: Corrosion inhibitor in aq. systems; noncorrosive cleaners and industrial detergents; good tolerance to high electrolyte systems
Form: Liquid; Color: Amber
Properties: Solubility: Water-sol.; Density: 8.7 lb/gal; Sp. gr.: 1.04;

Acid No.: Nil; Stability: Clear in 10% NaOH; pH: 10.4 (10% sol'n.)
Composition: 50% active ingred.
References: Tech. bulletin 955a

Monateric 1000. Mona Industries Inc.
Chemical Description: Amphoteric based on alkyl imidazoline
Category: Corrosion inhibitor, detergent, wetting agent, amphoteric
Applications: Corrosion inhibitor in aq. systems; primary surfactant in metal cleaning, soak tank cleaners, etc.; replacement for sodium nitrite in cutting fluids, synthetic lubricants, metal cleaners, burnishing compounds, liquid detergents
Form: Liquid; Color: Light amber
Properties: Density: 8.8 lb/gal; Sp. gr.: 1.05; pH: 11.2 (10% sol'n.)
Composition: 50% active ingred.
References: Tech. bulletin 962

Monateric ADA. Mona Industries Inc.
Chemical Description: Coco-amido betaine
Applications: Air drilling, foam drilling, foam blanketing; air entraining agent for cement, gypsum, wallboard, etc.; high foaming with good salt tolerance
Form: Clear liquid; Color: Amber
Properties: pH: 7.9 (as is)
Composition: 38% solids
References: Tech. bulletin 963

Monateric CA-35%. Mona Industries Inc.
Chemical Description: Amphoteric based on coconut imidazoline
Category: Detergent, wetting agent, emulsifier, dispersant, amphoteric
Applications: Cosmetic, household, and industrial products, incl. shampoos, bubble baths, skin cleansers, hair conditioners, cream rinses, hair dye formulations, medium to high foaming floor cleaners, metal cleaners, and other high electrolyte industrial detergent systems
Form: Liquid; Color: Amber
Properties: m.w.: 360; Density: 8.5 lb/gal; Sp. gr.: 1.02; Acid No.: 50 ± 5; Stability: Stable to high concs. of acids and alkalis; compatible with high electrolyte sol'ns.; Surface Tension: 29.5 dynes/cm (0.1% conc.); Biodegradable
Composition: 35% active ingred.
References: Tech. bulletin 952a

Monateric CAB. Mona Industries Inc.
Chemical Description: Coco-amido betaine
Category: Amphoteric surfactant
Applications: Very mild, high foaming base surfactant for shampoos, bubble baths, hair conditioners, etc.; good hard water tolerance
Form: Clear liquid; Color: Yellow
Properties: pH: 7.0 ± 0.5 (as is)
Composition: 30% active ingred.
References: Tech. bulletin 959a

Monateric CDX-38. Mona Industries Inc.
Chemical Description: Imidazoline-derived dicarboxylic acid am-

photeric
Category: Detergent, foaming agent, amphoteric
Applications: Mild base surfactant for baby shampoos, daily use shampoos, and skin cleansers
Form: Viscous liquid; Color: Light amber
Properties: Density: 9.8 lb/gal; Sp. gr.: 1.18; Acid No.: 15.0; pH: 8.4 (10% as is); Biodegradable
Composition: 38% active ingred.
References: Tech. bulletin 961

Monateric CEM-38%. Mona Industries Inc.
Chemical Description: Sodium salt of a dicarboxyethyl fatty acid derived imidazoline
Category: Detergent, wetting agent, emulsifier, dispersant, solubilizer, hydrotrope, amphoteric
Applications: Liquid detergent systems, esp. high electrolyte and alkaline systems
Form: Clear to hazy liquid; Color: Amber
Properties: Density: 8.75 lb/gal; Acid No.: 32 ± 3; Surface Tension: 35 dynes/cm (0.1% conc.); pH: 8.5 ± 0.5 (10% as is); Biodegradable
Composition: 38% active ingred.
References: Tech. bulletin 957b

Monateric CM-36S. Mona Industries Inc.
Chemical Description: Coco imidazoline-based amphoteric
Category: Foaming agent, emulsifier, detergent, wetting agent, amphoteric
Applications: High foaming, low irritation surfactant for cosmetics industry; as fulling agent in textile industry; in industrial cleaners; good pH stability and electrolyte tolerance
Form: Clear liquid; Color: Yellow
Properties: Solubility: Water-sol.; Stability: Good pH stability and electrolyte tolerance; Surface Tension: 32.0 dynes/cm (0.015% conc. @ 20 C); pH: 12.8 (as is)
Composition: 36% active ingred.
References: Tech. bulletin 964

Monateric CSH-32. Mona Industries Inc.
Chemical Description: Sodium salt of a dicarboxymethyl fatty acid derived imidazoline
Category: Amphoteric surfactant
Applications: High foaming, nonirritating surfactant for baby shampoos and "daily use" shampoos
Form: Clear liquid; Color: Yellow
Properties: Density: 9.4 lb/gal; Sp. gr.: 1.13; Visc.: ≈ 2000 cps; F.P.: < 0 C; Cloud Pt.: < 0 C; pH: 8.4 (10%); Biodegradable
Composition: 32% active ingred.
References: Tech. bulletin 960

Monateric Cy Na-50%. Mona Industries Inc.
Chemical Description: Sodium salt of a capryl imidazoline derivative
Category: Detergent, emulsifier, wetting agent, amphoteric
Applications: Low to moderate foaming surfactant for use in concentrated electrolyte systems;

used in bottle washing compounds, wax strippers, steam cleaners, industrial cleaners, all-purpose cleaners, alkaline degreasing, kier boiling, mercerizing, acid-pickling baths
Form: Liquid; Color: Amber
Properties: Acid No.: 3 max.; Stability: Stable to acid and alkaline systems and in a wide range of concentrated electrolyte systems; pH: 10.5 ± 0.5 (10% as is); Biodegradable
Composition: 50% active ingred.
References: Tech. bulletin 953a

Monateric ISA-35%. Mona Industries Inc.
Chemical Description: Isostearic imidazoline
Category: Amphoteric surfactant
Applications: Contributes lubricity and viscosity building properties to cosmetic and industrial products; medium foaming and very mild
Form: Clear to hazy flowable gel; Color: Amber
Properties: Solubility @ 10%: Sol. in water, alcohol, glycol ethers, and alkanolamines; disp. in mineral oil and natural oils and fats; Acid No.: 40–50 (isopropanol); Stability: Stable to high concentrations of acids and alkalis; compatible with high electrolyte sol'ns.; Surface Tension: 30 dynes/cm (0.1% conc.); pH: 5–6 (10% sol'n.); Biodegradable
Composition: 35% active ingred.
References: Tech. bulletin 954B

Monateric LF-100. Mona Industries Inc.
Chemical Description: 2-alkyl imidazoline amphoteric
Category: Detergent, amphoteric
Applications: Low foaming detergent for use in either low or high temperature alkaline and acid cleaners; "scrub-vac" cleaners, e.g., indoor-outdoor carpet cleaners or hard surface cleaners; heavy duty all-purpose industrial cleaners, esp. with high caustic conc.
Form: Liquid; Color: Amber
Properties: m.w.: 290; Solubility: Readily sol. in aq. acid systems; neutralizes slowly in alkaline systems; Density: 8.7 lb/gal; Sp. gr.: 1.04; pH: 12.0 ± 0.5 (10% sol'n.); Biodegradable
Composition: 99% min. active
References: Tech. bulletin 958a

Monateric LF Na-50. Mona Industries Inc.
Chemical Description: Sodium salt of Monateric LF-100
Category and Applications: Similar to Monateric LF-100 with greater solubility permitting more rapid incorporation into liquid alkaline cleaners
Form: Liquid; Color: Amber
Properties: Density: 9.1 lb/gal; Sp. gr.: 1.09; Stability: Stable in highly alkaline and acid systems; pH: 11.0 ± 0.5 (10% sol'n.); Biodegradable
Composition: 50% active ingred.
References: Tech. bulletin 958a

Monawet MB-45. Mona Industries Inc.
Chemical Description: Sodium-diisobutyl sulfosuccinate
Category: Wetting, dispersing, emulsifying, penetrating, and solubilizing agent; reduces surface and interfacial tension; anionic
Applications: Used in emulsion polymerization of styrene/butadiene for rug backing
Form: Clear liquid; Color: Colorless
Properties: m.w.: 332; Solubility: Very sol. in water; fair sol. in polar and nonpolar solvents; Density: 9.3 lb/gal; Sp. gr.: 1.12; Cloud Pt.: 13 C; Flash Pt.: 215 F (PMCC); Acid No.: 2.5 max.; Stability: Stable @ pH 2–10; Surface Tension: 54 dynes/cm (0.1%); pH: 6 ± 1 (as is)
Composition: 45% active ingred.
Toxicity: Eye irritant; defatting to skin
Containers: Polyethylene-lined fiber containers
References: Tech. bulletin 212F

Monawet MM-80. Mona Industries Inc.
Chemical Description: Sodium-dihexyl sulfosuccinate
Category: See Monawet MB-45
Applications: Emulsion and suspension polymerization; water-sol. pesticidal sprays; shampoo formulations; soap and detergent products and final rinse aid; auxiliary stabilizer and emulsifier in latex paints; indirect food additive
Form: Clear liquid; Color: Colorless
Properties: m.w.: 388; Solubility: 33 g/100 g water @ 20 C; good sol. in polar and nonpolar solvents; Density: 9.2 lb/gal; Sp. gr.: 1.10; Cloud Pt.: $<$ 0 C; Flash Pt.: 110 F (PMCC); Acid No.: 2.5 max.; Stability: Stable @ pH 2–10; Surface Tension: 46 dynes/cm (0.1%); pH: 6 ± 1 (as is)
Composition: 80% active ingred.
Toxicity: Eye irritant; defatting to skin
Containers: Polyethylene-lined fiber containers
References: Tech. bulletin 212F

Monawet MO65-150. Mona Industries Inc.
Chemical Description: Sodium dioctyl sulfosuccinate, anhydrous
Category: Wetting, penetrating, and spreading agent, emulsifier, anionic
Applications: Oil well rig cleaner; penetrants, lubricants, and rust inhibitors; drycleaning detergents and spot removers; solvent cleaners and strippers
Form: Clear liquid; Color: Colorless
Properties: Density: 8.75 lb/gal; Sp. gr.: 1.05; Flash Pt.: 42–43 C (COC); Acid No.: 2.5 max.; pH: 5.5 ± 1.0 (10% sol'n.)
Composition: 65% active ingred.
References: Tech. bulletin 222

Monawet MO-70. Mona Industries Inc.
Chemical Description: Sodium-dioctyl sulfosuccinate

Category: See Monawet MB-45
Applications: Emulsion and suspension polymerization; textile desizing and printing and dyeing processes; fertilizers, insecticides, and fungicides; cosmetic creams, lotions, bath oils, shampoos; drycleaning systems; indirect food additives
Form: Clear liquid; Color: Colorless
Properties: m.w.: 444; Solubility: Very good sol. in polar and nonpolar solvents; Density: 9.0 lb/gal; Sp. gr.: 1.08; Cloud Pt.: $<$ −5 C; Flash Pt.: 325 F (PMCC); Acid No.: 2.5 max.; Stability: Stable @ pH 2–10; Surface Tension: 29 dynes/cm (0.1%); pH: 6 ± 1 (as is)
Composition: 70% active ingred.
Toxicity: Eye irritant; defatting to skin
Containers: Polyethylene-lined fiber containers
References: Tech. bulletin 212F

Monawet MO-70E. Mona Industries Inc.
Chemical Description: Sodium-dioctyl sulfosuccinate
Category: See Monawet MB-45
Applications: Emulsion and suspension polymerization; textile desizing, and printing and dyeing processes; fertilizers, insecticides, and fungicides; cosmetic creams, lotions, bath oils, shampoos; indirect food additives; window and glass cleaners; auxiliary stabilizer and emulsifier in latex paints; cleaning agent in meat and poultry packaging plants
Form: Clear liquid; Color: Colorless
Properties: m.w.: 444; Solubility: Very good sol. in polar and nonpolar solvents; Density: 9.0 lb/gal; Sp. gr.: 1.08; Cloud Pt.: $\leq$ −5 C; Flash Pt.: 82 F (PMCC); Acid No.: 2.5 max.; Stability: Stable @ pH 2–10; Surface Tension: 29 dynes/cm (0.1%); pH: 6 ± 1 (as is)
Composition: 70% active ingred.
Toxicity: Eye irritant; defatting to skin
Containers: Polyethylene-lined fiber containers
References: Tech. bulletin 212F

Monawet MO-70R. Mona Industries Inc.
Chemical Description: Sodium-dioctyl sulfosuccinate
Category: See Monawet MB-45
Applications: Textile desizing, printing and dyeing processes; fertilizers, insecticides, and fungicides; cosmetic creams, lotions, bath oils, shampoos; drycleaning systems; indirect food additives
Form: Clear liquid; Color: Colorless
Properties: m.w.: 444; Solubility: Very good sol. in polar and nonpolar solvents; Density: 8.8 lb/gal; Sp. gr.: 1.06; Cloud Pt.: $<$ −5 C; Flash Pt.: 280 F (PMCC); Acid No.: 2.5 max.; Stability: Stable @ pH 2–10; Surface Tension: 29 dynes/cm (0.1%); pH: 6 ± 1 (as is)
Composition: 70% active ingred.

Toxicity: Eye irritant; defatting to skin
Containers: Polyethylene-lined fiber containers
References: Tech. bulletin 212F

Monawet MO-75E. Mona Industries Inc.
Chemical Description: Sodium dioctyl sulfosuccinate
Category: See Monawet MB-45
Applications: Textile desizing, printing and dyeing processes; fertilizers, insecticides, and fungicides; cosmetic creams, lotions, bath oils, shampoos; indirect food additives
Form: Clear liquid; Color: Colorless
Properties: m.w.: 444; Solubility: Very good in polar and nonpolar solvents; Density: 9.0 lb/gal; Sp. gr.: 1.08; Cloud Pt.: < –5 C; Flash Pt.: 80 F (PMCC); Acid No.: 2.5 max.; Stability: Stable @ pH 2–10; Surface Tension: 29 dynes/cm (0.1%); pH: 6 ± 1 (as is)
Composition: 75% active ingred.
Toxicity: Eye irritant; defatting to skin
Containers: Polyethylene-lined fiber containers
References: Tech. bulletin 212F

Monawet MO-84R2W. Mona Industries Inc.
Chemical Description: Sodium-dioctyl sulfosuccinate
Category: See Monawet MB-45
Applications: Textile desizing, printing and dyeing processes; fertilizers, insecticides, and fungicides; cosmetic creams, lotions, bath oils, shampoos; preparation of nonaq. pigment dispersions for printing inks; indirect food additives
Form: Viscous liquid; Color: Light yellow
Properties: m.w.: 444; Solubility: Very good in polar and nonpolar solvents; Density: 9.2 lb/gal; Sp. gr.: 1.10; Cloud Pt.: < –10 C; Flash Pt.: 223 F (PMCC); Acid No.: 2.5 max.; Stability: Stable @ pH 2–10; Surface Tension: 29 dynes/cm (0.1%); pH: 5.5 ± 1.0 (10%)
Composition: 83% active ingred.
Toxicity: Eye irritant; defatting to skin
Containers: Polyethylene-lined fiber containers
References: Tech. bulletin 212F

Monawet MO-85P. Mona Industries Inc.
Chemical Description: Sodium dioctyl sulfosuccinate
Category: Wetting agent, anionic
Applications: Wetting agent for wettable powders, pigments, etc.
Form: Fine powder; Color: White
Properties: Solubility: 0.85% sol. in water @ room temperature; Surface Tension: 26 dynes/cm (1% aq.); pH: 6.5 (1%)
Composition: 85% active ingred.
References: Tech. bulletin 223

Monawet MT-70. Mona Industries Inc.
Chemical Description: Sodium-ditridecyl sulfosuccinate

Category: See Monawet MB-45
Applications: Emulsion and suspension polymerization; indirect food additives
Form: Clear liquid; Color: Light straw
Properties: m.w.: 584; Solubility: Excellent in polar and nonpolar solvents; Density: 8.5 lb/gal; Sp. gr.: 1.02; Cloud Pt.: −2 C; Flash Pt.: 230 F (PMCC); Acid No.: 2.5 max.; Stability: Stable @ pH 2–10; Surface Tension: 29 dynes/cm (0.1%); pH: 6 ± 1 (as is)
Composition: 70% active ingred.
Toxicity: Eye irritant; defatting to skin
Containers: Polyethylene-lined fiber containers
References: Tech. bulletin 212F

Monawet MT-70E. Mona Industries Inc.
Chemical Description: Sodium-ditridecyl sulfosuccinate
Category and Applications: See Monawet MT-70
Form: Clear liquid; Color: Light straw
Properties: m.w.: 584; Solubility: Excellent in polar and nonpolar solvents; Density: 8.4 lb/gal; Sp. gr.: 1.01; Cloud Pt.: −15 C; Flash Pt.: 86 F (PMCC); Acid No.: 2.5 max.; Stability: Stable @ pH 2–10; Surface Tension: 29 dynes/cm (0.1%); pH: 6 ± 1 (as is)
Composition: 70% active ingred.
Toxicity: Eye irritant, defatting to skin
Containers: Polyethylene-lined fiber containers
References: Tech. bulletin 212F

Monawet MT-80H2W. Mona Industries Inc.
Chemical Description: Sodium-ditridecyl sulfosuccinate
Category: See Monawet MB-45
Applications: Preparation of nonaq. pigment dispersions used in printing inks; indirect food additives
Form: Viscous liquid; Color: Light yellow
Properties: m.w.: 584; Solubility: Excellent in polar and nonpolar solvents; Density: 8.5 lb/gal; Sp. gr.: 1.02; Cloud Pt.: < 0 C; Flash Pt.: 225 F (PMCC); Acid No.: 2.5 max.; Stability: Stable @ pH 2–10; Surface Tension: 29 dynes/cm (0.1%); pH: 5.5 ± 1.0 (10%)
Composition: 80% active ingred.
Toxicity: Eye irritant; defatting to skin
Containers: Polyethylene-lined fiber containers
References: Tech. bulletin 212F

Monawet SNO-35. Mona Industries Inc.
Chemical Description: Tetrasodium N-(1, 2-dicarboxyethyl) N-alkyl (C_{18}) sulfosuccinamate
Category: Wetting agent, solubilizer, emulsifier, dispersant, viscosity depressant, mild detergent, anionic
Applications: Industrial detergents, agricultural, cosmetic, and textile products; specialty emulsions or dispersions for polymerization systems
Form: Clear liquid; Color: Light amber

Properties: m.w.: 653 (avg.); Solubility: Sol. in water; good sol. in high electrolyte salt sol'ns.; Density: 9.5 lb/gal; Sp. gr.: 1.12–1.16; Visc.: 16–18 s (#2 Zahn cup); F.P.: 45 ± 5 F; Acid No.: 2.0 max.; Iodine No.: 0.5 max.; Stability: High alkaline stability; high calcium tolerance; Surface Tension: 43 dynes/cm (0.1%); pH: 7–8 (as is); Biodegradable
Composition: 35% solids
Containers: 55 gal Liquipaks or bulk tankwagons
References: Tech. bulletin 285 D

Monawet TD-30. Mona Industries Inc.
Chemical Description: Half ester of sulfosuccinic acid based on an ethoxylated fatty alcohol
Category: Anionic surfactant
Applications: Used in emulsion polymerization to develop high molecular wt, high solids polymers with improved freeze-thaw stability; foam and emulsifying properties make it useful in cosmetic and fine fabric detergents; textile wet processing; low surface tension properties
Form: Liquid; Color: Light yellow
Properties: Density: 9.0 lb/gal; Sp. gr.: 1.08; Visc.: 45 cps (Brookfield no.3 spindle @ 100 rpm); Cloud Pt.: –3 C; Stability: Stable to hard water, other multivalent ions, e.g., iron, and to acids; Surface Tension: 32.9 dynes/cm (0.1% conc.); pH: 5–6 (as is)
Composition: 30% active ingred.
References: Tech. bulletin 220

Monazoline Series. Mona Industries Inc.
Chemical Description: 1-hydroxyethyl-2 alkylimidazolines
Category: Wetting agent, emulsifier, detergent, thickener, corrosion inhibitor, antistat, softener, bactericide, cationic
Applications: Agricultural sprays, fungicides, and herbicides; emulsifiers for nonpolar liquids; paint industry pigment flushing operations; corrosion inhibitors and thickening agents in petroleum industry; antistatic additives in plastic film processing; textile antistats and softeners for cotton, wool, and rayon; dye assistants flocculation; flotation agents; leather processing
Form: Liquid
Properties: Solubility: Sol. in oil; disp. in water; Biodegradable
References: Tech. bulletin 280E

Monazoline C. Mona Industries Inc.
Chemical Description: 1-hydroxyethyl-2 alkylimidazoline based on coconut fatty acid
Category and Applications: See Monazoline Series; also in acid dairy cleaners, as corrosion inhibitor in latex primer paints, in oil well acidifying and secondary recovery operations, in cutting and quenching oils
Form: Liquid (may crystallize on aging); Color: Amber
Properties: m.w.: 282; Solubility @ 10%: Sol. in ethyl alcohol, chlorinated hydrocarbons, mineral oil, toluene, kerosene, mineral spirits,

vegetable oil; disp. in water; Density: 7.75 lb/gal; Sp. gr.: 0.93; Acid No.: 1 max.; pH: 10.5–12 (10% disp.); Biodegradable
Composition: 90% min. imidazoline
Containers: 55 gal open head steel drums
References: Tech. bulletin 280E

Monazoline CY. Mona Industries Inc.
Chemical Description: 1-hydroxyethyl-2 alkylimidazoline based on caprylic acid
Category and Applications: See Monazoline Series
Form: Liquid (may crystallize on aging); Color: Amber
Properties: m.w.: 212; Solubility @ 10%: Sol. in ethyl alcohol, chlorinated hydrocarbons, toluene, vegetable oil; Density: 8.25 lb/gal; Sp. gr.: 0.99; Acid No.: 1 max.; pH: 10.5–12 (10% disp.); Biodegradable
Composition: 90% min. imidazoline
Containers: 55 gal closed head steel drums
References: Tech. bulletin 280E

Monazoline O. Mona Industries Inc.
Chemical Description: 1-hydroxyethyl-2 alkylimidazoline based on oleic acid
Category and Applications: See Monazoline Series; also in phenothiazine (agricultural) preparations, acid dairy cleaners, emulsion cleaners for greases and oils, rinse aids for automatic car washing, thickener for oil-based paints and sealants, in alkyd paints and printing inks, fuel oil additive for sludge dispersion and corrosion control, acid fulling agent and penetrant in textile carbonizing
Form: Liquid (may crystallize on aging); Color: Amber
Properties: m.w.: 345; Solubility @ 10%: Sol. in ethyl alcohol, chlorinated hydrocarbons, mineral oil, toluene, kerosene, mineral spirits, vegetable oil; disp. in water; Density: 7.66 lb/gal; Sp. gr.: 0.92; Acid No.: 1 max.; pH: 10–11.5 (10% disp.); Biodegradable
Composition: 90% min. imidazoline
Containers: 55 gal closed head steel drums
References: Tech. bulletin 280E

Monazoline T. Mona Industries Inc.
Chemical Description: 1-hydroxyethyl-2 alkylimidazoline based on tall oil fatty acid
Category and Applications: See Monazoline Series; also aids gravel to asphalt bonding in road construction; rinse aids for automatic car washing; thickeners for oil-based paints and sealants; printing ink additive; protective metal coatings
Form: Liquid (may crystallize on aging); Color: Amber
Properties: m.w.: 350; Solubility @ 10%: Sol. in ethyl alcohol, chlorinated hydrocarbons, mineral oil, toluene, kerosene, mineral spirits, vegetable oil; disp. in water; Den-

sity: 7.75 lb/gal; Sp. gr.: 0.93; Acid No.: 1 max.; pH: 10–11.5 (10% disp.); Biodegradable
Composition: 90% min. imidazoline
Containers: 55 gal closed head steel drums
References: Tech. bulletin 280E

Monflor 31. I.C.I. United States
Chemical Description: Fluorochemical surfactant
Category: Wetting agent, emulsifier, anionic
Applications: Industrial cleaning, chemical specialties, floor polishes, suspension/emulsion polymerization, ink/paint pigments, emulsification, metal treatment, wetting agents and rinse aids
Form: Liquid
Properties: Density: 1.15 g/cc; Decomposition Temperature: 250 C (TGA)
Composition: 30% active ingred.
Toxicity: Skin irritant on prolonged contact with concentrated solutions
Warnings and Caution: Flammable

Monflor 32. I.C.I. United States
Chemical Description: Fluorochemical surfactants
Category: Leveling and wetting agent, dispersant, emulsifier, anionic
Applications: Ink, paint pigments
Form: Liquid
Properties: Density: 1.09 g/cc; Decomposition Temp.: 225 C (TGA)
Composition: 30% active ingred.
Toxicity: Skin irritant on prolonged contact with concentrated form; severe eye irritant
Warnings and Caution: Combustible

Monflor 51. I.C.I. United States
Chemical Description: Fluorochemical surfactants
Category: Dispersant, emulsifier, foam stabilizer, leveling agent, pigment antiflotation agent, spreading and wetting agent, nonionic
Applications: Mold release, industrial cleaning, chemical specialties, floor polishes, plastics additives, suspension/emulsion polymerization, photographic film emulsions, ink/paint pigments, emulsification, wetting agents and rinse aids
Form: Waxy solid
Properties: Solubility: Very sol. in methanol, isopropyl alcohol, ethyl alcohol, acetone, methyl ethyl ketone, 1,1,1-trichloroethane, water; sol. in toluene; Density: 1.25 g/cc; Decomposition Temp.: 220 C (TGA)
Composition: 100% active ingred.
Toxicity: Skin irritant on prolonged contact with concentrated form

Monflor 52. I.C.I. United States
Chemical Description: Fluorochemical surfactants
Category: Dispersant, emulsifier, foam stabilizer, leveling agent, wetting and spreading agent, nonionic
Applications: Mold release, chemical specialties, plastics additives,

photographic film emulsions, ink/paint pigments, emulsification, wetting agents and rinse aids
Form: Viscous liquid
Properties: Solubility: Very sol. in methanol, isopropyl alcohol, ethyl alcohol, acetone, methyl ethyl ketone, 1,1,1-trichloroethane, toluene, water; Density: 1.32 g/cc; Decomposition Temp.: 200 C (TGA)
Composition: 100% active ingred.
Toxicity: Skin irritant on prolonged contact with concentrated form

Monflor 53. I.C.I. United States
Chemical Description: Fluorochemical surfactant
Category: Wetting agent, emulsifier, nonionic
Applications: Photographic film emulsions, emulsification
Form: Oily liquid
Properties: Solubility: Very sol. in n-heptane, methanol, isopropyl alcohol, ethyl alcohol, acetone, methyl ethyl ketone, perchloroethylene, 1,1,1-trichloroethane, toluene; Density: 1.14 g/cc; Decomposition Temp.: 220 C (TGA)
Composition: 100% active ingred.
Toxicity: Skin irritant on prolonged contact with concentrated form

Monflor 72. I.C.I. United States
Chemical Description: Fluorochemical surfactant
Category: Wetting agent, emulsifier, cationic
Applications: Chemical specialties, floor polishes, plastics additives, ink/paint pigments
Form: Liquid
Properties: Density: 1.18 g/cc; Decomposition Temp. 225 C (TGA)
Composition: 30% active ingred.
Toxicity: Skin irritant on prolonged contact with concentrated form
Warnings and Caution: Combustible

Monflor 91. I.C.I. United States
Chemical Description: Fluorochemical surfactant
Category: Wetting agent, emulsifier, anionic
Applications: Mold release, metal treatment
Form: Liquid
Properties: Density: 1.6 g/cc; Decomposition Temp.: 230 C (TGA)
Composition: 10% active ingred.
Toxicity: Contact with hot glowing surfaces may produce corrosive acid fumes— avoid breathing vapors

Monolan 1206/2. Lankro Chem. Ltd.
Chemical Description: Ethylene oxide-propylene oxide block polymer
Category: Antifoam, nonionic
Applications: Antifoam additive for detergents and wetting agents; used in synthetic lubricants
Form: Clear liquid; Color: Colorless; Odor: Faint
Properties: Solubility: Insol. in

water; Sp. gr.: 1.003 (20 C); Visc.: 445 cs (20 C); Flash Pt.: > 400 F (COC); Pour Pt.: < 0 C; Stability: Stable to electrolytes, acids, alkalis, and hard water; pH: 6.0–8.0 (1% aq.)
Composition: 100% active ingred.
Toxicity: Skin irritant on prolonged contact with concentrated state
Containers: 200 kg net mild steel drums or bulk
Warnings and Caution: Spillages may be slippery; do not mix with strong oxidizing or reducing agents—potentially explosive
References: Bulletin SI:22 FM678

Monolan 2000 E/12. Lankro Chem. Ltd.
Chemical Description: Ethylene oxide-propylene oxide block polymer
Category: Detergent, wetting agent, antifoam, nonionic
Applications: Low-foaming formulations; also used as antifoam
Form: Clear liquid; Color: Colorless; Odor: Faint
Properties: Solubility: Insol. in water; Sp. gr.: 1.020 (20 C); Visc.: 390 cs (20 C); Flash Pt.: > 400 F (COC); Pour Pt.: < 0 C; Stability: Stable to electrolytes, acids, alkalis, and hard water; pH: 6–8 (1% aq.)
Composition: 100% active ingred.
Toxicity: Skin irritant on prolonged contact with concentrated state
Containers: 200 kg net mild steel drums or bulk
Warnings and Caution: Spillages may be slippery; do not mix with strong oxidizing or reducing agents—potentially explosive
References: Bulletin SI:22 FM678

Monolan 2500 E/30. Lankro Chem. Ltd.
Chemical Description: Ethylene oxide-propylene oxide block polymer
Category: Detergent, wetting agent, nonionic
Applications: Low-foaming formulations; emulsion polymerization; metal cleaning; resin plasticizers; latex stabilization; textile processing aids
Form: Clear liquid; Color: Colorless; Odor: Faint
Properties: Solubility: Sol. in water; Sp. gr.: 1.043 (20 C); Visc.: 650 cs (20 C); Cloud Pt.: 35 C (1% aq.); Flash Pt.: > 400 F (COC); Pour Pt.: < 0 C; Stability: Stable to electrolytes, acids, alkalis, and hard water; pH: 6–8 (1% aq.)
Composition: 100% active ingred.
Reaction: Neutral
Toxicity: Skin irritant on prolonged contact with concentrated state
Containers: 200 kg net mild steel drums or bulk
Warnings and Caution: Spillages may be slippery; do not mix with strong oxidizing or reducing agents—potentially explosive
References: Bulletin SI:22 FM678

Monolan 3000 E/60. Lankro Chem. Ltd.
Chemical Description: Ethylene oxide-propylene oxide block polymer
Category: Detergent, wetting agent, antifoam, nonionic
Applications: Low-foaming industrial and domestic detergents; emulsion polymerization; metal cleaning; resin plasticizers; latex stabilization; textile processing aids
Form: Solid; Color: White; Odor: Faint
Properties: Solubility: Sol. in water; Sp. gr.: 1.052 (20 C); Visc.: 355 cs (40 C); Cloud Pt.: 92 C (1% aq.); Flash Pt.: > 400 F (COC): Pour Pt.: 31 C; Stability: Stable to electrolytes, acids, alkalis, and hard water; pH: 6–8 (1% aq.)
Composition: 100% active ingred.
Toxicity: Skin irritant on prolonged contact with concentrated state
Containers: 200 kg net mild steel drums or bulk
Warnings and Caution: Spillages may be slippery; do not mix with strong oxidizing or reducing agents–potentially explosive
References: Bulletin SI:22 FM678

Monolan 8000 E/80. Lankro Chem. Ltd.
Chemical Description: Ethylene oxide-propylene oxide block polymer
Category: Detergent, wetting agent, emulsifier, nonionic
Applications: Low-foaming industrial and domestic detergents; emulsion polymerization; metal cleaning; resin plasticizer; latex stabilizer; textile processing aids
Form: Flake; Color: White; Odor: Faint
Properties: Solubility: Sol. in water; Sp. gr.: 1.070 (20 C); Visc.: 1100 cs (60 C); Cloud Pt.: 100 C (1% aq.); Flash Pt.: > 400 F (COC); Pour Pt.: 49 C; Stability: Stable to electrolytes, acids, alkalis, and hard water; pH: 6–8 (1% aq.)
Composition: 100% active ingred.
Reaction: Neutral
Toxicity: Skin irritant on prolonged contact with concentrated state
Containers: 200 kg net mild steel drums or bulk
Warnings and Caution: Spillages may be slippery; do not mix with strong oxidizing or reducing agents–potentially explosive
References: Bulletin SI:22 FM678

Monolan M. Lankro Chem. Ltd.
Chemical Description: Ethylene oxide-propylene oxide block polymer
Category: Emulsifier, nonionic
Applications: Low-foaming herbicide and pesticide concentrate formulation; low-foam emulsifier
Form: Clear liquid; Color: Colorless; Odor: Faint
Properties: Solubility: Sol. in water; Sp. gr.: 1.020 (20 C); Visc.: 470 cs (20 C); Cloud Pt.: 24 C (1% aq.); Flash Pt.: > 400 F

(COC); Pour Pt.: < 0 C; Stability: Stable to electrolytes, acids, alkalis, and hard water; pH: 6–8 (1% aq.)
Composition: 100% active ingred.
Reaction: Neutral
Toxicity: Skin irritant on prolonged contact with concentrated state
Containers: 200 kg net mild steel drums or bulk
Warnings and Caution: Spillages may be slippery; do not mix with strong oxidizing or reducing agents–potentially explosive
References: Bulletin SI:22 FM678

Monolan P222. Lankro Chem. Ltd.
Chemical Description: Ethylene oxide-propylene oxide block polymer
Category: Antifoam, nonionic
Applications: Antifoam additive for detergents and wetting agents; synthetic lubricants
Form: Clear liquid; Color: Colorless; Odor: Faint
Properties: Solubility: Insol. in water; Sp. gr.: 1.004 (20 C); Visc.: 440 cs (20 C); Flash Pt.: > 400 F (COC); Pour Pt.: < 0 C; Stability: Stable to electrolytes, acids, alkalis, and hard water; pH: 6–8 (1% aq.)
Composition: 100% active ingred.
Reaction: Neutral
Toxicity: Skin irritant on prolonged contact with concentrated state
Containers: 200 kg net mild steel drums or bulk
Warnings and Caution: Spillages may be slippery; do not mix with strong oxidizing or reducing agents–potentially explosive
References: Bulletin SI:22 FM678

Monolan PB. Lankro Chem. Ltd.
Chemical Description: Complex ethylene oxide-propylene oxide copolymer
Category: Wetting agent, rinse aid, nonionic
Applications: Low-foam detergents and rinse aids; agricultural sprays; antifoams and defoamers; dye leveling agents; car shampoos; window cleaning liquids; pigment dispersants; yarn lubricants
Form: Liquid; Color: Colorless; Odor: Faint
Properties: Solubility: Disp. in water; Sp. gr.: 1.020 (20 C); Visc.: 680 cs (20 C); Cloud Pt.: 19 C (1% aq.); Flash Pt.: > 400 F (COC); Pour Pt.: < 0 C; Stability: Stable to electrolytes, acids, alkalis, and hard water; pH: 6–8 (1% aq.)
Composition: 100% active ingred.
Reaction: Neutral
Toxicity: Skin irritant on prolonged contact with concentrated state
Containers: 200 kg net mild steel drums or bulk
Warnings and Caution: Spillages may be slippery; do not mix with strong oxidizing or reducing agents–potentially explosive
References: Bulletin SI:22 FM678

Monolan PC. Lankro Chem. Ltd.
Chemical Description: Complex ethylene oxide-propylene oxide

copolymer
Category and Applications: See Monolan PB
Form: Liquid; Color: Colorless; Odor: Faint
Properties: Solubility: Sol. in water; Sp. gr.: 1.040 (20 C); Visc.: 800 cs (20 C); Cloud Pt.: 44 (1% aq.); Flash Pt.: > 400 F (COC); Pour Pt.: 8 C; Stability: Stable to electrolytes, acids, alkalis, and hard water; pH: 6–8 (1% aq.)
Composition: 100% active ingred.
Reaction: Neutral
Toxicity: Skin irritant on prolonged contact with concentrated state
Containers: 200 kg net mild steel drums or bulk
Warnings and Caution: Spillages may be slippery; do not mix with strong oxidizing or reducing agents—potentially explosive
References: Bulletin SI:22 FM678

Monolan PM7. Lankro Chem. Ltd.
Chemical Description: Ethylene oxide-propylene oxide copolymer on an organic substrate
Category: Detergent, wetting agent, rinse aid, nonionic
Applications: Low foam detergent and wetting agent for use in bottle washing and automatic dishwashing formulations
Form: Clear liquid; Color: Colorless; Odor: Faint
Properties: Solubility: Sol. in water; Sp. gr.: 1.050 (20 C); Visc.: 710 cs (20 C); Cloud Pt.: 41 C (1% aq.); Flash Pt.: > 400 F (COC); Pour Pt.: < 0 C; Stability: Stable to electrolytes, acids, alkalis, and hard water; pH: 6–8 (1% aq.)
Composition: 100% active ingred.
Reaction: Neutral
Toxicity: Skin irritant on prolonged contact with concentrated state
Containers: 200 kg net mild steel drums or bulk
Warnings and Caution: Spillages may be slippery; do not mix with strong oxidizing or reducing agents—potentially explosive
References: Bulletin SI:22 FM678

Monolan PPG440. Lankro Chem. Ltd.
Chemical Description: Polypropylene glycol
Category: Lubricant, antistat, plasticizer, nonionic
Applications: Cosolvents in dyestuff, ink, resin, and rubber industries; cosmetic preparations; oil-free lubricants; hydraulic fluids; intermediates in surfactants and plastic production
Form: Liquid; Color: Colorless; Odor: Faint
Properties: m.w.: 400 (avg.); Solubility: Sol. in water; Sp. gr.: 1.010 (20 C); Visc.: 80 cs (20 C); Flash Pt.: > 450 F (COC); Pour Pt.: < 0 C; Stability: Chemically inert, good thermal stability
Composition: 100% active ingred.
Toxicity: Skin irritant on prolonged contact with concentrated state
Containers: 200 kg net mild steel drums or bulk

Warnings and Caution: Spillages may be slippery; do not mix with strong oxidizing or reducing agents–potentially explosive
References: Bulletin SI:23 FM678

Monolan PPG1100. Lankro Chem. Ltd.
Chemical Description: Polypropylene glycol
Category and Applications: See Monolan PPG440
Form: Liquid; Color: Colorless; Odor: Faint
Properties: m.w.: 1000 (avg.); Solubility: Insol. in water; Sp. gr.: 1.005 (20 C); Visc.: 180 cs (20 C); Flash Pt.: > 450 F (COC); Pour Pt.: < 0 C; Stability: Chemically inert, good thermal stability
Composition: 100% active ingred.
Toxicity: Skin irritant on prolonged contact with concentrated state
Containers: 200 kg net mild steel drums or bulk
Warnings and Caution: Spillages may be slippery; do not mix with strong oxidizing or reducing agents–potentially explosive
References: Bulletin SI:23 FM678

Monolan PPG2200. Lankro Chem. Ltd.
Chemical Description: Polypropylene glycol
Category and Applications: See Monolan PPG440
Form: Liquid; Color: Colorless; Odor: Faint
Properties: m.w.: 2000 (avg.); Solubility: Insol. in water; Sp. gr.: 1.004 (20 C); Visc.: 450 cs (20 C); Flash Pt.: > 450 F (COC); Pour Pt.: < 0 C; Stability: Chemically inert, good thermal stability
Composition: 100% active ingred.
Toxicity: Skin irritant on prolonged contact with concentrated state
Containers: 200 kg net mild steel drums or bulk
Warnings and Caution: Spillages may be slippery; do not mix with strong oxidizing or reducing agents–potentially explosive
References: Bulletin SI:23 FM678

Monolan PT. Lankro Chem. Ltd.
Chemical Description: Complex ethylene oxide-propylene oxide copolymer
Category and Applications: See Monolan PB
Form: Liquid; Color: Colorless; Odor: Faint
Properties: Solubility: Sol. in water; Sp. gr.: 1.037 (20 C); Visc.: 750 cs (20 C); Cloud Pt.: 28 C (1% aq.); Flash Pt.: > 400 F (COC); Pour Pt.: < 0 C; Stability: Stable to electrolytes, acids, alkalis, and hard water; pH: 6–8 (1% aq.)
Composition: 100% active ingred.
Reaction: Neutral
Toxicity: Skin irritant on prolonged contact with concentrated state
Containers: 200 kg net mild steel drums or bulk
Warnings and Caution: Spillages may be slippery; do not mix with

strong oxidizing or reducing agents—potentially explosive
References: Bulletin SI:22 FM678

Monopol Oil 75. GAF Corp.
Chemical Description: Specially sulfated castor oil
Category: Wetting agent, penetrant, leveling agent, softener, anionic
Applications: Softener/lubricant or softener additive for cellulosic fibers; wetter, penetrant, and leveler for dyeing direct, developed, sulfur, and vat dyes on cellulosic and cellulosic blends
Form: Clear liquid; Color: Light yellow
Properties: Sp. gr.: 1.05; Stability: Stable to acids, alkalis, and electrolytes
Composition: 75% active ingred.

Morco H-62 Natural. Marathon Morco Co.
Chemical Description: Sodium petroleum sulfonate
Applications: Drilling mud compounds, drycleaning compounds, fuel oil additives, lube oil additive manufacture, metal working compounds, ore flotation, pigment dispersion, rust preventives
Properties: m.w.: 520; Solubility: Oil-sol.; Density: 8.45 lb/gal (60 F); Visc.: 300 cs (210 F)
Composition: 62% sulfonate
Containers: 55 gal open head drums

Morco H-70 Natural. Marathon Morco Co.
Chemical Description: Sodium petroleum sulfonate
Category: Dispersant, anionic
Applications: Drilling mud compounds, drycleaning compounds, fuel oil additives, lube oil additive manufacture, metal working compounds, ore flotation, pigment dispersion, rust preventives
Form: Liquid
Properties: m.w.: 520; Solubility: Oil-sol.; Density: 8.60 lb/gal (60 F); Visc.: 850 cs (210 F)
Composition: 70% sulfonate
Containers: 55 gal open head drums

Morco HM-62 Natural. Marathon Morco Co.
Chemical Description: Sodium petroleum sulfonate
Applications: Degreasing compounds, demulsifying crude oil, drilling mud compounds, drycleaning compounds, fuel oil additives, lube oil additive manufacture, metal working compounds, ore flotation, pigment dispersion, rust preventives
Properties: m.w.: 490; Solubility: Oil-sol.; Density: 8.48 lb/gal (60 F); Visc.: 260 cs (210 F)
Composition: 62.0% sulfonate
Containers: 55 gal open head drums

Morco L-62 Natural. Marathon Morco Co.
Chemical Description: Sodium petroleum sulfonate

Applications: Degreasing compounds, demulsifying crude oil, dry cleaning compounds, fuel oil additives, metal working compounds, soluble oils
Properties: m.w.: 430; Solubility: Oil-sol.; Density: 8.60 lb/gal (60 F); Visc.: 150 cs (210 F)
Composition: 62% sulfonate
Containers: 55 gal open head drums

Morco L-70 Natural. Marathon Morco Co.
Chemical Description: Sodium petroleum sulfonate
Category: Emulsifier, wetting agent, anionic
Applications: Degreasing compounds, demulsifying crude oil, drycleaning compounds, fuel oil additives, metal working compounds, soluble oils
Form: Liquid
Properties: m.w.: 430; Solubility: Oil-sol.; Density: 8.80 lb/gal (60 F); Visc.: 400 cs (210 F)
Composition: 70% sulfonate
Containers: 55 gal open head drums

Morco M-62 Natural. Marathon Morco Co.
Chemical Description: Sodium petroleum sulfonate
Applications: Degreasing compounds, demulsifying crude oil, drilling mud compounds, drycleaning compounds, fuel oil additives, lube oil additive manufacture, metal working compounds, ore flotation, soluble oils
Properties: m.w.: 465; Solubility: Oil-sol.; Density: 8.50 lb/gal (60 F); Visc.: 225 cs (210 F)
Composition: 62% sulfonate
Containers: 55 gal open head drums

Morco M-70 Natural. Marathon Morco Co.
Chemical Description: Sodium petroleum sulfonate
Category: Emulsifier, dispersant, anionic
Applications: Degreasing compounds, demulsifying crude oil, drilling mud compounds, drycleaning compounds, fuel oil additives, lube oil additive manufacture, metal working compounds, ore flotation, soluble oils
Form: Liquid
Properties: m.w.: 465; Solubility: Oil-sol.; Density: 8.65 lb/gal (60 F); Visc.: 500 cs (210 F)
Composition: 70% sulfonate
Containers: 55 gal open head drums

Musloid 815D. Colloids Inc.
Chemical Description: Fatty acid ethoxylate
Category: Dispersant, stabilizer, leveling agent, nonionic
Applications: Low-foaming surfactant providing auxiliary dispersing, stabilizing, and leveling action in emulsion paint systems
Form: Viscous liquid; Color: Amber
Properties: Sp. gr.: 0.95 (70 F); Acid No.: 4.5
Composition: 100% active ingred.

Musloid 815M. Colloids Inc.
Chemical Description: Fatty acid ethoxylate
Category and Applications: See Musloid 815D
Form: Viscous liquid; Color: Amber
Properties: Sp. gr.: 1.01 (70 F); Acid No.: 1.2
Composition: 100% active ingred.

Musloid 815S. Colloids Inc.
Chemical Description: Fatty acid ethoxylate
Category and Applications: See Musloid 815D
Form: Viscous liquid; Color: Amber
Properties: Sp. gr.: 0.98 (70 F); Acid No.: 4.0
Composition: 100% active ingred.

Myrj 45. I.C.I. United States
Chemical Description: Polyoxyethylene 8 stearate
Category: Emulsifier, nonionic
Applications: General purpose O/W emulsifiers for cosmetics, pharmaceuticals, etc.
Form: Soft waxy solid; Color: Cream
Properties: Solubility: Sol. in isopropanol, methanol, ethanol; disp. in water; Sp. gr.: ≈ 1.0; HLB: 11.1; Flash Pt.: > 300 F; Pour Pt.: ≈ 28 C; Sapon. No.: 82–95; Acid No.: 2.0 max.; Hydroxyl No.: 87–105
Composition: 100% conc.

Myrj 52. I.C.I. United States
Chemical Description: Polyoxyethylene 40 stearate
Category and Applications: See Myrj 45
Form: Waxy solid; Color: Ivory
Properties: Solubility: Sol. in water, acetone, ether, methanol, and ethanol; Sp. gr.: ≈ 1.1; HLB: 16.9; Flash Pt.: > 300 F; Pour Pt.: ≈ 38 C; Sapon No.: 25–35; Acid No.: 1.0 max.; Hydroxyl No.: 27–40
Composition: 100% active ingred.
Containers: 30 gal fiber lever pak

Myrj 52S. I.C.I. United States
Chemical Description: Polyoxyethylene 40 stearate
Category and Applications: See Myrj 45
Form: Waxy granular solid; Color: White
Properties: Solubility: Sol. in water, toluol, acetone, ether, dioxane, cellosolve, carbon tetrachloride, methanol. ethanol, ethyl acetate, aniline, and 5% aq. sol'ns. of sulfuric acid, sodium sulfate, and aluminum chloride; Sp. gr.: ≈ 1.1; HLB: 16.9; Flash Pt.: > 300 F; Pour Pt.: 38 C; Sapon. No.: 25–35; Acid No.: 1.5 max.; Hydroxyl No.: 25–37
Composition: 100% conc.

Myrj 53. I.C.I. United States
Chemical Description: Polyoxyethylene 50 stearate
Category and Applications: See Myrj 45
Form: Solid; Color: Cream
Properties: Solubility: Sol. in water, ethanol, and isopropanol;

HLB: 17.9; Pour Pt.: ≈ 42 C; Sapon. No.: 20–28; Acid No.: 2.0 max.; Hydroxyl No.: 23–35
Composition: 100% conc.

Myvacet 5-07. Eastman Chemical Products
Chemical Description: Distilled acetylated monoglycerides
Category: Emulsifier, nonionic
Applications: Confectionery products, cosmetics and pharmaceuticals, dry topping mix, food coatings, gravies, sauces, syrups
Form: Waxy solid
Properties: Solubility: Major portion sol. in most common organic solvents, but some crystals form even at low concentrations; M.P.: 41–46 C; HLB: 3.8–4.0; Sapon. Value: 278–292; Acid Value: 3.0 max.; Hydroxyl Value: 132–153; Iodine Value: 0.5–3.0
Composition: 100% active ingred.
Containers: 400 lb fiber drums

Myvacet 7-00. Eastman Chemical Products
Chemical Description: Distilled acetylated monoglycerides
Category: Emulsifier, nonionic
Applications: Confectionery products, cosmetics and pharmaceuticals, food coatings
Form: Waxy solid
Properties: Solubility: Major portion sol. in most common organic solvents, but some crystals form even at low concentrations; M.P.: 37–40 C; HLB: 3.8–4.0; Sapon. Value: 316–331; Acid Value: 3.0 max.; Hydroxyl Value: 80–96; Iodine Value: 0.5–3.0
Composition: 100% active ingred.
Containers: 400 lb fiber drums

Myvacet 7-07K. Eastman Chemical Products
Chemical Description: Distilled acetylated monoglycerides
Category: Emulsifier, nonionic
Applications: Confectionery products, cosmetics and pharmaceuticals, food coatings
Form: Waxy solid
Properties: Solubility: Major portion sol. in most common organic solvents, but some crystals form even at low concentrations; M.P.: 37–40 C; HLB: 3.8–4.0; Sapon. Value: 316–331; Acid Value: 3.0 max.; Hydroxyl Value: 80–96; Iodine Value: 0.5–3.0
Composition: 100% active ingred.
Containers: 400 lb fiber drums

Myvacet 9-40. Eastman Chemical Products
Chemical Description: Distilled acetylated monoglycerides
Category: Emulsifier, nonionic
Applications: Antidusting applications, cosmetics and pharmaceuticals, defoaming, lubricants, plasticizers, release coatings, slab dressing, slip agents, wax-softening uses
Form: Clear liquid
Properties: Solubility: Sol. in all common organic solvents, 80% w/w aq. ethanol, and vegetable and mineral oils; M.P.: 4–12 C; HLB: 3.8–4.0; Sapon. Value: 375–385; Acid Value: 3.0 max.; Hy-

droxyl Value: 0–15; Iodine Value: 43
Composition: 100% active ingred.
Containers: 40 lb steel pail and 425 lb bung-in-head steel drum

Myvacet 9-45. Eastman Chemical Products
Chemical Description: Distilled acetylated monoglycerides
Category and Applications: See Myvacet 9-40
Form: Clear liquid
Properties: Solubility: Sol. in all common organic solvents, 80% w/w aq. ethanol, and vegetable and mineral oils; M.P.: 4–12 C; HLB: 3.8–4.0; Sapon. Value: 375–385; Acid Value: 3.0 max.; Hydroxyl Value: 0–15; Iodine Value: 48
Composition: 100% active ingred.
Containers: 40 lb steel pail and 425 lb bung-in-head steel drum

Myvaplex 600. Eastman Chemical Products
Chemical Description: Food-grade concentrated glyceryl monostearate
Category: Emulsifier, nonionic
Applications: Improves texture and appearance of macaroni and cereal products
Form: Powder
Properties: Sp. gr.: 0.92 (80 C); Congeal Pt.: 70 C; Clear Pt.: 78 C; Acid Value: 4.0 max.; Iodine Value: 5.0 max.
Composition: 90% min. monoester content
Containers: 100 lb fiber drum

Myvatem Dispersing Agent. Eastman Chemical Products
Chemical Description: Diacetyl tartaric acid esters of distilled monoglycerides
Category: Emulsifier, dispersant
Applications: Baked goods
Form: Sticky viscous semisolid
Containers: 425 lb open head steel drum

Myvatex 3-50. Eastman Chemical Products
Chemical Description: Distilled glyceryl monostearate and distilled propylene glycol monostearate
Category: Emulsifier, nonionic
Applications: Canned puddings, foam mat drying, fruit purees, whipped desserts
Form: Small beads
Properties: Congeal Pt.: ≈ 56 C; Clear Pt.: 60 C; Acid Value: 3.0 max.
Composition: 90% max. monoester content
Containers: 100 and 250 lb fiber drums

Myvatex 8-20. Eastman Chemical Products
Chemical Description: Distilled monoglycerides with 20% by wt of hydrogenated soybean oil
Category: Emulsifier, nonionic
Applications: Coffee whiteners
Form: Small beads
Properties: Congeal Pt.: ≈ 60 C; Clear Pt.: 68 C
Composition: 70% max. mono-

ester content
Containers: 100 and 250 lb fiber drums

Myvatex 8-20E. Eastman Chemical Products
Chemical Description: Distilled monoglycerides with 20% by wt of hydrogenated soybean oil
Category: Emulsifier, nonionic
Applications: Ice cream and frozen desserts; whipped toppings, puddings
Form: Powder or small beads
Properties: Congeal Pt.: ≈ 58 C; Clear Pt.: 63 C; Acid Value: 3.0 max.
Composition: 70% max. monoester content
Containers: 100 and 250 lb fiber drums

Myvatex 25-00, 25-07. Eastman Chemical Products
Chemical Description: Distilled monoglycerides, lecithin, and water
Category: Emulsifier, nonionic
Applications: Bread and sweet goods shortening
Form: Soft solid
Composition: 25% active ingred.
Containers: 400 lb fiber drums

Myvatex 40-06S. Eastman Chemical Products
Chemical Description: Myvatex 3-50, lactylic stearate, and water
Category: Emulsifier, nonionic
Applications: Cakes
Form: Soft solid
Properties: HLB: 3.8–4.0
Composition: 25% active ingred.
Containers: 400 lb fiber drums

Myvatex Peanut Butter Stabilizer. Eastman Chemical Products
Chemical Description: Distilled glyceryl monostearate and fully hydrogenated palm oil, in equal parts
Category: Emulsifier, stabilizer, nonionic
Applications: Peanut butter
Form: Small beads
Properties: Congeal Pt.: ≈ 62 C; Clear Pt.: 71 C; Acid Value: 3.0 max.
Composition: 45% max. monoester content
Containers: 100 and 250 lb fiber drums

Myvatex Super DŌ. Eastman Chemical Products
Chemical Description: Succinylated monoglycerides and distilled monoglycerides
Category: Emulsifier, nonionic
Applications: Bread, buns, baked goods
Form: Powder
Properties: Congeal Pt.: ≈ 54 C; Acid Value: 70–80
Composition: 35–40% monoester content
Containers: 100 lb fiber drums

Myverol 18-00. Eastman Chemical Products
Chemical Description: Distilled monoglycerides derived from hydrogenated lard

Category: Emulsifier, nonionic
Applications: Baked goods, confectionery products, cosmetics, dehydrated potatoes, drum-dried products, foam mat drying, margarine, peanut butter, whipped desserts and toppings
Form: Small beads
Properties: Sp. gr.: 0.91 (80 C); Congeal Pt.: 68 C; Clear Pt.: 73 C; Acid Value: 4.0 max.; Iodine Value: 5.0 max.
Composition: 100% active
Containers: 250 lb fiber drums

Myverol 18-04. Eastman Chemical Products
Chemical Description: Distilled monoglycerides derived from hydrogenated palm oil
Category and Applications: See Myverol 18-00
Form: Small beads
Properties: Sp. gr.: 0.94 (80 C); Congeal Pt.: 66 C; Clear Pt.: 71.5 C; Acid Value: 4.0 max.; Iodine Value: 5.0 max.
Composition: 100% active
Containers: 250 lb fiber drums

Myverol 18-06. Eastman Chemical Products
Chemical Description: Distilled monglycerides derived from hydrogenated vegetable oils
Category and Applications: See Myverol 18-00
Form: Small beads
Properties: Sp. gr.: 0.92 (80 C); Congeal Pt.: 70 C; Clear Pt.: 78 C; Acid Value: 4.0 max.; Iodine Value: 5.0 max.
Composition: 100% active
Containers: 250 lb fiber drums

Myverol 18-07. Eastman Chemical Products
Chemical Description: Distilled monoglycerides derived from hydrogenated vegetable oils
Category and Applications: See Myverol 18-00
Form: Small beads
Properties: Sp. gr.: 0.92 (80 C); Congeal Pt.: 67 C; Clear Pt.: 76 C; Acid Value: 4.0 max.; Iodine Value: 5.0 max.
Composition: 100% active
Containers: 250 lb fiber drums

Myverol 18-30. Eastman Chemical Products
Chemical Description: Distilled monoglycerides derived from edible tallow
Category: Emulsifier, nonionic
Applications: Baked goods, cake shortenings, coffee whiteners, confectionery products, cosmetics, icing and cream-filling shortenings, margarine, whipped desserts and toppings
Form: Plastic
Properties: Sp. gr.: 0.92 (80 C); Congeal Pt.: 58 C; Clear Pt.: 62 C; Acid Value: 4.0 max.
Composition: 100% active
Containers: 50 lb cube

Myverol 18-35. Eastman Chemical Products
Chemical Description: Distilled

monoglycerides derived from refined palm oil
Category and Applications: See Myverol 18-30
Form: Plastic
Properties: Sp. gr.: 0.94 (80 C); Congeal Pt.: 58 C; Clear Pt.: 63 C; Acid Value: 4.0 max.
Composition: 100% active
Containers: 50 lb cube and 400 lb fiber drum

Myverol 18-40. Eastman Chemical Products
Chemical Description: Distilled monoglycerides derived from lard
Category and Applications: See Myverol 18-30
Form: Plastic
Properties: Sp. gr.: 0.91 (80 C); Congeal Pt.: 54 C; Clear Pt.: 60 C; Acid Value: 4.0 max.
Composition: 100% active
Containers: 50 lb cube and 400 lb fiber drum

Myverol 18-50K. Eastman Chemical Products
Chemical Description: Distilled monoglycerides
Category and Applications: See Myverol 18-30
Form: Plastic
Properties: Sp. gr.: 0.91 (80 C); Congeal Pt.: 49 C; Clear Pt.: 59 C; Acid Value: 4.0 max.
Composition: 100% active
Containers: 50 lb cube and 400 lb fiber drum

Myverol 18-85. Eastman Chemical Products
Chemical Description: Distilled monoglycerides derived from cottonseed oil
Category and Applications: See Myverol 18-30
Form: Soft plastic
Properties: Sp. gr.: 0.92 (80 C); Congeal Pt.: 46 C; Clear Pt.: 56 C; Acid Value: 4.0 max.
Composition: 100% active
Containers: 50 lb cube and 400 lb fiber drum

Myverol 18-98. Eastman Chemical Products
Chemical Description: Distilled monoglycerides derived from safflower oil
Category: Emulsifier, nonionic
Applications: Diet margarines, icing and cream-filling shortenings
Form: Soft plastic
Properties: Sp. gr.: 0.90 (80 C); Congeal Pt.: 25 C; Clear Pt.: 41 C; Acid Value: 4.0 max.
Composition: 100% active
Containers: 425 lb open head steel drum

N

Na-Cumene Sulfonate 40. Chemische Werke Huls AG
Chemical Description: Sodium cumene sulfonate
Category: Hydrotrope, anionic
Applications: Hydrotrope; solution aid for liquid detergents and cleaners, as well as for washing powder preparations (slurries)
Form: Liquid
Composition: 40% active ingred.
References: Brochure No. 1066/2

Na-Cumene Sulfonate Powder. Chemische Werke Huls AG
Chemical Description: Sodium cumene sulfonate
Category and Applications: See Na-Cumene Sulfonate 40
Form: Powder
Composition: 96% active ingred.
References: Brochure No. 1066/2

Na-Toluene Sulfonate 30, Na-Toluene Sulfonate 40. Chemische Werke Huls AG
Chemical Description: Sodium toluene sulfonate
Category and Applications: See Na-Cumene Sulfonate 40
Form: Liquid
Composition: 30–40% active ingred.
References: Brochure No. 1066/2

Nacconol 35SL. Stepan Chemical Co.
Chemical Description: Linear alkylate sulfonate
Category: Detergent, wetting and foaming agent, emulsifier, dispersant, penetrant, anionic
Applications: Textile wet processing and dyeing; rug and upholstery shampoo and laundering; liquid acid and alkali dairy cleaners; paper and leather industries; domestic cleaners; cosmetics (hair shampoos, liquid bubble baths); washing vehicles, fruits, vegetables; curing aid in fertilizers
Form: Clear liquid; Color: Light straw; Odor: Faint
Properties: Solubility: Water-sol.; Density: 9.12 lb/gal (20 C); Sp. gr.: 1.11 (20 C); Visc.: 103 cps; Cloud Pt.: 0 C max.; Stability: Stable to acids, alkalis, water hard-

ness, oxidizing and reducing agents; Surface Tension: 41.8 dynes/cm (0.1% conc.); pH: 6.0–7.5; Biodegradability: 90% min.
Composition: 35–38% active
Containers: 55 gal Liquipack drums

Nacconol 40F. Stepan Chemical Co.
Chemical Description: Sodium alkyl aryl sulfonate
Category: Wetting and foaming agent, dispersant, detergent, anionic
Applications: Used in cement clinker grinding operation, in dishwashing and all purpose cleaners, wet processing and dyeing of textiles, electroplating, herbicides and insecticides, kier boiling, laundry, metal cleaning, metal working, ore flotation, pickling baths
Form: Free flowing flakes; Color: White; Odor: Faint
Properties: Solubility: Water-sol.; Density: 18.7–22.0 lb/ft^3 (untamped); Stability: Stable to mineral and organic acids, dilute alkalis, and oxidizing and reducing agents; Surface Tension: 31 dynes/cm (0.1% conc.); pH: 6.4–7.6 (1% aq.); Biodegradability: 90% min.
Composition: 40% active ingred.
Containers: 55 gal Leverpak drums with an aluminum barrier

Nacconol 90F. Stepan Chemical Co.
Chemical Description: Sodium alkyl aryl sulfonate based on a linear alkylate
Category: Wetting agent, emulsifier, dispersant, foaming agent, detergent, anionic
Applications: Wettable insecticide powders, pesticidal sprays, fertilizers; liquid or solid light and heavy duty cleaners, shampoos; textile wet processing and dyeing; antipitting agent in nickel plating; emulsion polymerization for latex, rubber, plastics; in firefighting; gypsum board; laundry; metal cleaning; mining; paper; pickling baths;
Form: Flakes; Color: Light cream; Odor: Faint
Properties: Solubility: Water-sol.; Density: 22.0 lb/ft^3 (untamped); Stability: Stable to mineral and organic acids, dilute alkalis, and reducing agents; oxidizing agents may cause some darkening; Surface Tension: 35 dynes/cm (0.1% conc.); pH: 6.0–7.5 (10% aq.); Biodegradability: 90% min.
Composition: 90% active ingred.
Warnings and Caution: Store in a cool, dry place

Nalkylene 500. Conoco Chem.
Chemical Description: Linear alkylbenzene
Category: Surfactant intermediate
Applications: Intermediate for production of linear alkylbenzene sulfonate, for formulation of biodegradable, high-foaming, detergent products
Form: Liquid; Color: Saybolt 30+; Odor: Odorless
Properties: F.P.: < -70 F

Composition: 100% active ingred.
Containers: Carbon steel
Warnings and Caution: Observe normal precautions for handling hydrocarbons

Nalkylene 550. Conoco Chem.
Chemical Description: Linear alkylbenzene
Category: Intermediate
Applications: Intermediate to biodegradable linear alkylbenzene sulfonate for many uses as anionic surfactant in heavy and light duty detergents
Form: Liquid; Color: Saybolt 30+; Odor: Odorless
Properties: Sp. gr.: 0.862; M.P.: < -90 F
Composition: 100% active ingred.
Containers: Carbon steel
Warnings and Caution: Observe normal precautions for handling hydrocarbons; keep away from open flames

Nalkylene 515. Conoco Chem.
Chemical Description: Linear alkylbenzene
Category and Applications: See Nalkylene 500
Form: Oily liquid; Color: Water white; Odor: Odorless
Properties: Sp. gr.: 0.861; M.P.: < -90 F
Containers: Carbon steel
Warnings and Caution: Observe normal precautions for handling hydrocarbons

Nalkylene 600. Conoco Chem.
Chemical Description: Linear alkylbenzene
Category: Surfactant intermediate
Applications: Intermediate to production of biodegradable sulfonic acids and slurries for heavy duty detergent formulations
Form: Liquid; Color: Saybolt 30+; Odor: Odorless
Properties: Sp. gr.: 0.865; Visc.: 46.9 SSU @ 100 F; M.P.: < -90 F; B.P.: 580
Composition: 100% active ingred.
Containers: Carbon steel
Warnings and Caution: Observe normal precautions for handling hydrocarbons

Nansa 1192. Albright & Wilson/Marchon Div.
Category: Foaming agent
Applications: For use in high speed mixing and foaming of gypsum slurries for plaster board production
Form: Clear liquid; Color: Golden yellow
Properties: Density: 1.03 g/cc; Visc.: 290 cs (20 C); Cloud Pt.: < 0 C; pH: 8 (5% sol'n.)

Nansa AS 40. Albright & Wilson/Marchon Div.
Chemical Description: Ammonium salt of SO_3-sulfonated dodecyl benzene
Category: Detergent
Applications: Formulation of domestic and industrial liquid deter-

gents, e.g., dishwashing formulations, car shampoos, general cleaning formulations
Form: Opaque gel; Color: Golden yellow
Properties: pH: 6.0–6.5 (2% aq.)
Composition: 40% active ingred.

Nansa BXS. Albright & Wilson/Marchon Div.
Chemical Description: Sodium alkyl naphthalene sulfonate
Category: Wetting and dispersing agent, anionic
Applications: For colloidal systems without detergent properties; agricultural dispersant for pesticides, fertilizers, and sulfur; leather tanning liquors; stabilizer in latex formulations; paper and pulp additive; pigment dispersant in cements, ceramics, clays, dyestuffs; metal cleaning and plating; textile processing; preserving fluids
Form: Free flowing powder; Color: Cream
Properties: Stability: Very stable; wetting power unaffected by dilute acids, alkalis, divalent salts, hard water, or chlorine; pH: 7.5–9.5 (1% sol'n.)
Composition: 38% min. active

Nansa HS40 Soft. Albright & Wilson/Marchon Div.
Chemical Description: Sodium alkyl benzene sulfonate
Category: Wetting agent, dispersant, anionic
Applications: Dry mix detergent compounds and scouring powders; industrial applications and as a light duty cleanser for automobiles, floors, walls, and other hard surfaces
Form: Free flowing coarse powder; Color: Cream
Properties: Density: 0.55 g/cm^3 (bulk); pH: 8.5 ± 1.0 (1% sol'n.); Biodegradability: > 80%
Composition: 38 ± 3% active
Containers: 25 kg net multiply paper bag with polythene liner
Warnings and Caution: Avoid prolonged contact with skin and inhalation of dust
References: Data sheet SD80-678

Nansa HS80P. Albright & Wilson/Marchon Div.
Chemical Description: Sodium alkyl benzene sulfonate
Category: Wetting and dispersing agent, emulsifier
Applications: Used in abrasive scouring powders and in quality detergent compounds for domestic and industrial use
Form: Free flowing powder; Color: Cream
Properties: Density: 0.52 g/cm^3 (bulk, loose); pH: 8.5 ± 1.0 (1% sol'n.); Biodegradability: > 80%
Composition: 80 ± 3.0% active
Containers: 25 kg net multiply paper bags
Warnings and Caution: Avoid prolonged contact with skin and inhalation of dust
References: Data sheet SD81A-678

Nansa HS80 Soft. Albright & Wilson/Marchon Div.
Chemical Description: Sodium alkyl benzene sulfonate
Category: Wetting and dispersing agent, anionic
Applications: In dry-mix laundry detergent powders and heavy duty detergent compounds; scouring powders; insecticide sprays; metal pickling sol'ns.
Form: Free flowing coarse powder; Color: Cream
Properties: Density: 0.55 g/cm^3 (bulk); pH: 8.5 ± 1.0 (1% sol'n.); Biodegradability: > 80%
Composition: 78 ± 3% active
Containers: 25 kg net multiply paper bag with polythene liner
Warnings and Caution: Avoid prolonged contact with skin and inhalation of dust
References: Data sheet SD81-678

Nansa HS 85/S. Albright & Wilson/Marchon Div.
Chemical Description: Sodium alkyl benzene sulfonate
Category and Applications: See Nansa HS 80/S
Form: Flake; Color: Cream
Properties: Density: 0.5 g/cc (as packed); pH: 9.0–11.0 (1% aq.); Biodegradable
Composition: 85 ± 3% active

Nansa HSA/L. Albright & Wilson/Marchon Div.
Chemical Description: Alkyl benzene sulfonic acid, sulfuric acid sulfonated, based on a linear "soft" alkylate
Category: Detergent, anionic
Applications: Basic ingredient in low active light duty liquid detergents
Form: Viscous liquid; Color: Dark brown
Properties: Acid Value: 220 ± 10; Biodegradability: > 80%
Composition: 86% min. active
Containers: 205 kg net closed head mild steel drums
Warnings and Caution: Acidic and corrosive–handle with care; store in a cool place
References: Data sheet SD79-678

Nansa LES 42. Albright & Wilson/Marchon Div.
Chemical Description: Blended detergent concentrate
Category: Detergent concentrate
Applications: Dishwashing and light duty clothes washing formulations
Form: Clear liquid; Color: Golden
Properties: Visc.: ≈ 250 cs (20 C); Cloud Pt.: < 0 C (quick cool); Surface Tension: 30 dynes/cm (1 g/l); pH: 6.2–6.8 (5% sol'n.)
Composition: 37% active ingred.

Nansa MA 30. Albright & Wilson/Marchon Div.
Chemical Description: Blend of sodium dodecyl benzene sulfonate and ethoxylated nonionic surfactants
Category: Detergent, anionic/nonionic
Applications: Base for liquid detergents for dishwashing and other hard surface cleansing operations

Form: Liquid; Color: Golden
Properties: Sp. gr.: 1.046; Visc.: ≈ 500 cs (20 C); Cloud Pt.: < 0 C (quick cool); pH: 6.5–7.0 (5% aq.)
Composition: 30% active ingred.

Nansa S40/S. Albright & Wilson/Marchon Div.
Chemical Description: Spray-dried sodium alkyl benzene sulfonate
Category: Detergent, wetting agent, anionic
Applications: Light duty detergent, or in formulation of heavy duty detergents, special dairy, metal, floor, vehicle, and bottle cleaners; wetter in insecticides, metal pickling, printing inks, paper processing
Form: Free-flowing powder; Color: White/pale cream
Properties: pH: 8.0–9.0 (1% aq.)
Composition: 40% active ingred.

Nansa SBA. Albright & Wilson/Marchon Div.
Chemical Description: SO_3-sulfonated dodecyl benzene sulfonic acid
Category: Detergent, intermediate, anionic
Applications: Manufacture of liquid detergents, soapless detergent powders; increases biological softness in end product
Form: Liquid; Color: Dark brown
Properties: Visc.: ≈ 20,000 cs (20 C); Stability: Stable to acids and alkalis
Composition: 96% active ingred.

Nansa SL 30 Liquid. Albright & Wilson/Marchon Div.
Chemical Description: Sodium salt of SO_3-sulfonated dodecyl benzene sulfonic acid
Category: Anionic surfactant
Applications: Detergent formulations
Form: Liquid; Color: Pale yellow
Properties: Visc.: 2500–3500 cs (20 C); Cloud Pt.: 15 C; Stability: Stable to acids and alkalis; pH: 7.0–7.5 (2% sol'n.)
Composition: 30% active ingred.

Nansa SS 30 Liquid. Albright & Wilson/Marchon Div.
Chemical Description: Sodium salt of SO_3-sulfonated dodecyl benzene sulfonic acid
Category and Applications: See Nansa SL 30 Liquid
Form: Liquid; Color: Golden
Properties: Visc.: ≈ 250 cs (20 C); Cloud Pt.: 0 C; Stability: Stable to acids and alkalis; pH: 6.0–6.8 (2% sol'n.)
Composition: 30% min. active

Nansa SS 60 Paste. Albright & Wilson/Marchon Div.
Chemical Description: Sodium salt of SO_3-sulfonated dodecyl benzene sulfonic acid
Category and Applications: See Nansa SL 30 Liquid
Form: Paste; Color: Cream
Properties: Stability: Stable to acids and alkalis; pH: 7.0–9.0 (2% sol'n.)
Composition: 60% min. active

Nansa SSA. Albright & Wilson/Marchon Div.
Chemical Description: SO_3-sulfonated dodecyl benzene sulfonic acid
Category and Applications: See Nansa SBA
Form: Liquid; Color: Dark brown
Properties: Visc.: ≈ 2000 cs (20 C); Stability: Stable to acids and alkalis
Composition: 96% active ingred.

Nansa SSAL. Albright & Wilson/Marchon Div.
Chemical Description: Alkyl benzene sulfonic acid, SO_3-sulfonated, based on a linear "soft" alkylate
Category: Anionic surfactant
Applications: Basic ingredient in light duty liquid detergents and industrial/institutional cleaners; in the manufacture of wetting, emulsifying, and foaming agents
Form: Viscous liquid; Color: Dark brown
Properties: Acid Value: 183 ± 5; Biodegradability: > 80%
Composition: 94% min. active
Containers: 205 kg net closed head mild steel drums
Warnings and Caution: Acidic and corrosive–handle with care; store in a cool place
References: Data sheet SD82-678

Nansa TS 60. Albright & Wilson/Marchon Div.
Chemical Description: Triethanolamine salt of SO_3-sulfonated dodecyl benzene
Category: Detergent, emulsifier, dispersant, anionic
Applications: Formulation of liquid detergents with low cloud points, dishwashing, bubble bath, car shampoo formulations, and in heavy duty liquid detergents; emulsifier for carnauba wax and pine oil derivatives or in emulsion polymerization reactions; as desizing agent for synthetic fibers; pigment dispersing agent
Form: Viscous liquid; Color: Medium brown
Properties: Sp. gr.: 1.09; Visc.: ≈ 9000 cs (20 C); Cloud Pt.: < 0 C (quick cool, 50% sol'n.); pH: 6.6–7.0 (2% aq.)
Composition: 60% active ingred.

Nansa UCA/S. Albright & Wilson/Marchon Div.
Chemical Description: Spray-dried built powders based on sodium alkyl benzene sulfonate
Category: Detergent, anionic
Applications: Domestic and industrial detergent powders containing organic and inorganic builders
Form: Free-flowing powder; Color: Blue
Properties: Solubility: Sol. in hot or cold water; Biodegradable
Composition: 17% active ingred.

Nansa UCP/S. Albright & Wilson/Marchon Div.
Chemical Description: Spray-dried built powders based on sodium alkyl benzene sulfonate
Category and Applications: See Nansa UCA/S
Form: Free-flowing powder; Col-

or: Blue
Properties: Solubility: Sol. in hot or cold water; Biodegradable
Composition: 15% active ingred.

Nansa YS 94. Albright & Wilson/Marchon Div.
Chemical Description: Isopropylamine salt of SO_3-sulfonated dodecyl benzene
Category: Emulsifier, coupling agent
Applications: Emulsifier for solvent-based hand cleaners; in emulsifiable solvent degreasing; as coupling agent for water in charge detergent systems
Form: Viscous liquid; Color: Amber
Properties: Visc.: ≈ 10,000 cs (20 C); Cloud Pt.: < 0 C (quick cool); pH: 5.0–8.0 (2% aq.)
Composition: 94% active ingred.

Nansol. Albright & Wilson/Marchon Div.
Chemical Description: Blend of sodium dodecyl benzene sulfonate and sodium alkyl ether sulfate
Category: Detergent, anionic
Applications: Base for production of liquid detergents for dishwashing, and hard surface cleansing operations; good foaming properties
Form: Liquid; Color: Golden
Properties: Sp. gr.: 1.05 (20 C); Visc.: ≈ 400 cs (20 C); pH: 6.5 ± 0.5; Biodegradable
Composition: 30% active ingred.

Naxonate 4AX. Nease Chem. Co.
Chemical Description: Ammonium xylene sulfonate, 40% sol'n.
Category: Hydrotrope, anionic
Applications: Formulating heavy duty liquid, powder, and dishwashing detergents; solubilizing inks, organic agents for electroplating baths, organic dyestuffs, cresols in germicidal formulations, of pine oil in water, of polymers; stabilizing leather stain sol'ns. and sol'ns. of alkanolamides; viscosity control in animal glues, foam sol'ns.; maintains homogenity of cutting and drawing compounds
Form: Liquid; Color: 50 max. (Klett)
Properties: pH: 6.5–8.5 (as is)
Composition: 40% min. active
Containers: 55 gal lined drums or tankwagons

Naxonate 4KT. Nease Chem. Co.
Chemical Description: Potassium toluene sulfonate, 40% sol'n.
Category and Applications: See Naxonate 4AX
Form: Liquid; Color: 50 max. (Klett)
Properties: pH: 7–9 (as is)
Composition: 40% min. active
Containers: 55 gal lined drums or tankwagons

Naxonate 4L. Nease Chem. Co.
Chemical Description: Sodium xylenesulfonate, 40% sol'n.
Category: Hydrotrope, anionic
Applications: Formulating heavy duty liquid, powder, and dishwash-

ing detergents; solubilizing inks, organic agents for electroplating baths, organic dyestuffs, cresols in germicidal formulations, of pine oil in water, of polymers; stabilizing leather stain sol'ns. and sol'ns. of alkanolamides; viscosity control in animal glues, foam sol'ns.; maintains homogenity of cutting and drawing compounds
Form: Liquid; Color: 50 max. (Klett)
Properties: pH: 7–9 (as is)
Composition: 40% min. active
Containers: 55 gal lined drums or tankwagons

Naxonate 4ST. Nease Chem. Co.
Chemical Description: Sodium toluene sulfonate, 40% sol'n.
Category and Applications: See Naxonate 4AX
Form: Liquid; Color: 50 max. (Klett)
Properties: pH: 7–9 (as is)
Composition: 40% min. active
Containers: 55 gal lined drums or tankwagons

Naxonate 5KT. Nease Chem. Co.
Chemical Description: Potassium toluene sulfonate, 50% sol'n.
Category and Applications: See Naxonate 4AX
Form: Liquid; Color: 50 max. (Klett)
Properties: pH: 7–9 (as is)
Composition: 50% min. active
Containers: Tankwagons only

Naxonate 5L. Nease Chem. Co.
Chemical Description: Sodium xylene sulfonate, 50% sol'n.
Category and Applications: See Naxonate 4AX
Form: Liquid; Color: 60 max. (Klett)
Properties: pH: 7–9 (as is)
Composition: 50% min. active
Containers: Tankwagons only

Naxonate 6AC. Nease Chem. Co.
Chemical Description: Ammonium cumene sulfonate, 60% sol'n.
Category and Applications: See Naxonate 4AX
Form: Liquid; Color: 80 max. (Klett)
Properties: pH: 7–8 (as is)
Composiiton: 60% min. active
Containers: 55 gal lined drums or tankwagons

Naxonate 45SC. Nease Chem. Co.
Chemical Description: Sodium cumene sulfonate, 45% sol'n.
Category and Applications: See Naxonate 4AX
Form: Liquid; Color: 75 max. (Klett)
Properties: pH: 7–9 (as is)
Composition: 44% min. active
Containers: 55 gal lined drums or tankwagons

Naxonate AX. Nease Chem. Co.
Chemical Description: Ammonium xylene sulfonate
Category and Applications: See Naxonate 4AX

Form: Powder; Color: Cream
Composition: 95% min. active
Containers: 200 leverpak

Naxonate G. Nease Chem. Co.
Chemical Description: Sodium xylene sulfonate
Category and Applications: See Naxonate 4AX
Form: Powder; Color: Cream
Composition: 95% min. active
Containers: 200 leverpak

Naxonate KT. Nease Chem. Co.
Chemical Description: Potassium toluene sulfonate
Category and Applications: See Naxonate 4AX
Form: Powder; Color: Cream
Composition: 95% min. active
Containers: 200 leverpak

Naxonate SC. Nease Chem. Co.
Chemical Description: Sodium cumene sulfonate
Category and Applications: See Naxonate 4AX
Form: Powder; Color: Cream
Composition: 95% min. active
Containers: 125 leverpak

Naxonate ST. Nease Chem. Co.
Chemical Description: Sodium toluene sulfonate
Category and Applications: See Naxonate 4AX
Form: Powder; Color: Cream
Composition: 95% min. active
Containers: 175 leverpak

Nekal BA-77. GAF Corp.
Chemical Description: Sodium alkyl naphthalene sulfonate
Category: Wetting agent, dispersant, stabilizer without detergent properties; anionic
Applications: Leather dyeing, textile processing, formulation of lakes, paints, inks, pesticides, synthetic latex emulsions; prevents coagulation of synthetic rubbers
Form: Free-flowing powder; Color: Tan
Properties: m.w.: 314; pH: 8–10 (5% sol'n.)
Composition: 75% min. active

Nekal BX-78. GAF Corp.
Chemical Description: Sodium alkyl naphthalene sulfonate
Category and Applications: See Nekal BX Conc. Paste
Form: Free-flowing powder; Color: Cream to tan
Properties: m.w.: 326; Solubility: Readily sol. in water @ 60 C, slowly sol. @ 25 C; pH: 6.0–8.0 (5% sol'n.)
Composition: 70% min. active

Nekal BX Conc. Paste. BASF AG
Chemical Description: Sodium salt of an alkyl naphthalene sulfonic acid
Category: Emulsifier, wetting and dispersing agent, foaming agent, anionic
Applications: Water-sol. emulsifier for mineral oils, liquid and solid fats and fatty acids, waxes, paraffin waxes, solvents, etc.; auxiliary

in grinding processes in dyes industry, in the production of plant protective agents, and in the processing of building materials; pigment manufacture; production of aq. pigment pastes and emulsion paints
Form: High viscosity paste; Color: Yellowish
Properties: Solubility: Water-sol.; Stability: Adequate stability to hard water and electrolytes; pH: 10–12 (5% aq.)
Composition: ≈ 60% active
Toxicity: Irritates skin and inflames (eyes) mucous membranes on prolonged contact
References: Tech. leaflet M 5216 e

Nekal BX Dry. BASF AG
Chemical Description: Sodium salt of an alkyl naphthalene sulfonic acid
Category and Applications: See Nekal BX Conc. Paste
Form: Hygroscopic powder; Color: Pale
Properties: Solubility: Dissolves readily in water at low concentrations; Stability: Adequate stability to hard water and electrolytes for most applications; pH: 9–11 (5% aq.)
Composition: ≈ 65% active
Toxicity: Irritates skin and inflames (eyes) mucous membranes on prolonged contact
References: Tech. leaflet M 5216 e

Nekal NF. GAF Corp.
Chemical Description: Alkyl naphthalene sulfonate
Category: Wetting agent, dispersant, penetrant, anionic
Applications: Used for continuous pad dyeing of polyester/cotton, cotton, and rayon fabrics with direct, sulfur, and vat dyes; low foaming and lowers the surface tension of the pad bath permitting use of high speeds in padding operation; scouring of materials; leveling of dyes in exhaust systems
Form: Clear liquid; Color: Amber
Properties: Solubility: Miscible with water in all proportions; Sp. gr.: 1.04; Stability: Good stability to acid, alkali, and hard water; pH: 6–8
Reaction: Slightly alkaline (1%)

Nekal WS-21. GAF Corp.
Chemical Description: Sulfonated aliphatic polyester
Category: Wetting and rewetting agent, anionic
Applications: Textile processing–wetting in high-speed continuous processing and long liquor work, sizing and desizing, in bleach liquors, in raw stock processing, package and beam dyeing, resin finishes; rewetting prior to dyeing and printing, in compressive shrinking, finishing, and before mercerizing
Form: Slightly viscous liquid; Color: Water-white; Odor: Slightly alcoholic
Properties: Solubility: Sol. in water in all proportions; Sp. gr.: 0.9–1.0; Stability: Stable in hard and soft water, and in moderately acid or alkaline aq. sol'ns.; pH: 6.0–

7.5 (10% sol'n.)
Composition: 48% conc.
References: Tech. bulletin 7543-135

Nekal WS-25. GAF Corp.
Chemical Description: Sulfonated aliphatic polyester
Category and Applications: See Nekal WS-21
Form: Slightly viscous liquid; Color: Water-white; Odor: Slightly alcoholic
Properties: Solubility: Sol. in water in all proportions; Sp. gr.: 1.0–1.1; Stability: Stable in hard and soft water and in moderately acid or alkaline aq. sol'ns.; pH: 6.0–7.5 (10% sol'n.)
Composition: 55–60% conc.
References: Tech. bulletin 7543-135

Nekal WT-27. GAF Corp.
Chemical Description: Sodium dioctyl sulfosuccinate
Category: Wetting and rewetting agent, anionic
Applications: Wetting for padding and long liquor dyeing, sizing and desizing, raw-stock processing, package and beam dyeing, resin finishes; rewetting prior to dyeing and printing, and in compressive shrinking, mercerizing, and finishing; paper rewetting; also used in antifogging compositions, battery separators, drycleaning detergents, electroplating, emulsion polymerization, glass cleaning, wallpaper removers, firefighting; effective at very low concentrations
Form: Clear liquid; Color: Water-white
Properties: Solubility: Readily sol. in water up to 1.2% solids @ 20 C and 5.5% @ 70 C; sol. to some degree in other polar and nonpolar organic solvents; Stability: Good at pH 2–9; pH: 5.0–6.0 (1% sol'n.)
Composition: 70% active ingred.
References: Tech. bulletin 7543-168

Nekanil 907. BASF AG
Chemical Description: Low-ethoxylated alkyl phenol
Category: Detergent, wetting agent, nonionic
Applications: In textile industry for scouring of wool in-the-grease, scouring of wool and its blends in the form of yarn and piece goods, milling
Composition: 100% active ingred.
Toxicity: Skin and eye irritant on prolonged contact
References: Tech. leaflet M 5440 e

Nekanil 910. BASF AG
Chemical Description: Alkyl phenol adduct with 9–10 moles of EO
Category: Detergent, wetting and dispersing agent, nonionic
Applications: Textile industry for scouring of wool in-the-grease, precleaning of cellulosic and synthetic fibers, afterwashing of dyed shades and prints; wetting agent in desizing and in carbonizing liquors; esp. useful above 60 C
Composition: 100% active ingred.
Toxicity: Skin and eye irritant on prolonged contact
References: Tech. bulletin M 5440 e

Nekanil LN. BASF AG
Chemical Description: Low-ethoxylated alkyl phenol
Category: Detergent, wetting agent, emulsifier, nonionic
Applications: In textile industry for scouring of wool in-the-grease, scouring of wool and its blends as yarns and piece goods, milling, wetting in carbonizing liquors and in desizing; removal of oil stains; esp. effective below 60 C and in weakly alkaline liquor
Composition: 100% active ingred.
Toxicity: Skin and eye irritant on prolonged contact
References: Tech. leaflet M 5440 e

Neobee 18. PVO International Inc.
Chemical Description: Highly refined and deodorized hybrid safflower oil
Category: Emollient, solubilizer, solvent
Applications: Emollient oil for cosmetics and pharmaceuticals; used for cooking and frying oil
Form: Liquid; Odor and Taste: Bland
Properties: Solubility: Sol. in alcohol, mineral oil, acetone; Sp. gr.: 0.915; Visc.: 70 cps; Setting Pt.: ≈ –18 C; Sapon. Value: 190–200; Iodine Value: 95; Stability: Stable to oxidation; Surface Tension: 31.6 dynes/cm

Neobee 62. PVO International Inc.
Chemical Description: Soya stearine
Category: Solubilizer, stabilizer
Applications: Food applications—used to adjust melting points of fats; peanut butter stabilizer; fast setting for icings and coatings
Form: Flakes
Properties: Setting Pt.: 60 C; Sapon. Value: 189-195; Iodine Value: 3.0

Neobee 1053. PVO International Inc.
Chemical Description: Coconut oil derived triglycerides
Category: Diluent, clouding agent
Applications: Diluent vehicle for flavoring, medicinals, colorings, etc.; clouding agent in beverages; coatings for collagen casing used in sausage production
Form: Liquid; Odor and Taste: None
Properties: Solubility: Sol. in alcohol, mineral oil, acetone; Sp. gr.: 0.930–0.960; Setting Pt.: –2 C; Sapon. Value: 325–355; Iodine Value: 0.5

Neobee 1054. PVO International Inc.
Chemical Description: Propylene glycol diester of coconut fatty acids
Category: Cosolvent
Applications: Carrier for flavors and colors; cosolvent for cosmetics and pharmaceuticals; very low setting point
Form: Liquid
Properties: Solubility: Sol. in alcohol, mineral oil, acetone; Sp. gr.: 0.910–0.923; Setting Pt.: –20 C; Sapon. Value: 315–335; Iodine Value: 0.5

Neobee 1062. PVO International Inc.
Chemical Description: Coconut oil-derived triglycerides
Category: Cosolvent
Applications: Carrier for fat-sol. vitamins, medicinals, and ointments in food, cosmetic, and pharmaceutical industries; slabbing oils; polishing agent for candy pans, hard gums, and sugar coated pieces replacing mineral oil
Form: Liquid
Properties: Solubility: Sol. in mineral oil and acetone; Sp. gr.: 0.925–0.945; Setting Pt.: 5 C; Sapon. Value: 295–315; Iodine Value: 8.0

Neobee 1223 (Neobee O). PVO International Inc.
Chemical Description: Coconut oil-derived triglycerides
Category: Carrier, spreading agent, penetrant
Applications: Carrier for fat-sol. vitamins, medicinals, and ointments; slabbing oils; polishing agent for candy pans, hard gums, and sugar coated pieces replacing mineral oil; spreading and penetrating agent for pharmaceutical vehicles
Form: Liquid; Odor: Practically odorless
Properties: Solubility: Sol. in mineral oil and acetone; Sp. gr.: 0.938; Visc.: 30 cps; Setting Pt.: ≈ –4 C; Sapon. Value: 300–315; Iodine Value: 8; Stability: Stable to oxidation; Surface Tension: 32.0 dynes/cm

Neobee 1400 Oil (Neobee M-5). PVO International Inc.
Chemical Description: Caprylic/capric triglycerides
Category: Diluent, emollient
Applications: Diluent vehicle for flavoring, medicinals, colorings, etc.; clouding agent in beverages; coatings for collagen casing used in sausage production; cosmetic and pharmaceutical production
Form: Liquid; Odor and Taste: None
Properties: Solubility: Sol. in alcohol, mineral oil, acetone; Sp. gr.: 0.930–0.960; Visc.: 23 cps; Setting Pt.: ≈ –5 C; Sapon. Value: 335–360; Iodine Value: 0.5; Stability: Stable to oxidation; Surface Tension: 32.3 dynes/cm

Neobee M-5. PVO International Inc.
See Neobee 1400

Neobee M-20. PVO International Inc.
Chemical Description: Propylene glycol dicaprylate/dicaprate
Category: Emollient, carrier
Applications: Emollient oil for cosmetics and pharmaceuticals with excellent deposition on skin; carrier for flavors and colors in foods
Form: Liquid
Properties: Solubility: Sol. in alcohol, mineral oil, acetone; Sp. gr.: 0.920; Visc.: 9 cps; Setting Pt.: ≈ –20 C; Sapon. Value: 315–335; Iodine Value: 0.5; Surface Tension: 31.0 dynes/cm

Neobee O. PVO International Inc.
See Neobee 1223

Neodol 23-3. Shell Chem. Co.
Chemical Description: $C_{12}-C_{13}$ alcohol (3EO) ethoxylate
Category: Detergent intermediate
Applications: Preparation of sulfates for high-foaming liquid detergents
Form: Clear to slightly hazy liquid; Color: 50 max. (APHA); Odor: Mild
Properties: m.w.: 310–342; Density: 7.7 lb/gal; Sp. gr.: 0.925; Visc.: 19 cs (100 F); M.P.: 5–6 C; HLB: 8.1; Flash Pt.: 300 F (PMCC); Pour Pt.: 4 C; Hydroxyl No.: 172
Composition: 100% active ingred.
Toxicity: Severe eye irritant, and skin irritant on prolonged contact with conc. form
References: Tech. bulletin SC: 89-76

Neodol 23-6.5. Shell Chem. Co.
Chemical Description: $C_{12}-C_{13}$ alcohol (6.5 EO) ethoxylate
Category: Detergent, wetting agent, nonionic
Applications: Formulation of high-foaming liquid detergents, e.g., manual dishwashing liquids; textile wetting agent; metal cleaning; paper, paint, and leather processing
Form: Liquid; Color: Practically colorless; Odor: Mild
Properties: m.w.: 484; Density: 8.08 lb/gal (100 F); Sp. gr.: 0.963 (50/25 C); Visc.: 29.5 cs (100 F); M.P.: 11–15 C; HLB: 12.0; Cloud Pt.: 45 C (1% aq.); Flash Pt.: 330 F (PMCC); Pour Pt.: 16 C; Hydroxyl No.: 116; pH: 6.0 (1% aq.); Biodegradable
Composition: 100% active ingred.
Toxicity: Severe eye irritant, and skin irritant on prolonged contact with conc. form
References: Tech. bulletin SC: 90-76

Neodol 25-3. Shell Chem. Co.
Chemical Description: $C_{12}-C_{15}$ alcohol (3 EO) ethoxylate
Category and Applications: See Neodol 23-3
Form: Clear to slightly hazy liquid; Color: Colorless; Odor: Mild
Properties: m.w.: 320–350; Density: 7.70 lb/gal; Sp. gr.: 0.925; Visc.: 19 cs (100 F); M.P.: 5–6 C; HLB: 7.9; Flash Pt.: 315 F (PMCC); Pour Pt.: 0 C; Hydroxyl No.: 167
Composition: 100% active ingred.
Toxicity: Severe eye irritant, and skin irritant on prolonged contact with conc. form
References: Tech. bulletin SC:86-76

Neodol 25-3A. Shell Chem. Co.
Chemical Description: $C_{12}-C_{15}$ alcohol (3 EO) ethoxysulfate, ammonium salt
Category: Detergent, anionic
Applications: Formulation of high-foaming liquid detergents, e.g., manual dishwashing, light duty liquids, etc.
Form: Clear, viscous liquid; Color: Light; Odor: Mild, ethanol only

Properties: m.w.: 432 (avg.); Density: 8.50 lb/gal (60 F); Sp. gr.: 1.02; Visc.: 45 cs (100 F); Flash Pt.: 74 F (PMCC); pH: 7.3; Biodegradable
Composition: 59% active ingred.
Toxicity: Severe eye irritant, and skin irritant on prolonged contact with conc. form
References: Tech. bulletin SC:91-77

Neodol 25-3S. Shell Chem. Co.
Chemical Description: $C_{12}-C_{15}$ alcohol (3 EO) ethoxysulfate, sodium salt
Category and Applications: See Neodol 25-3A
Form: Viscous liquid; Color: Light; Odor: Mild, ethanol only
Properties: m.w.: 437 (avg., of actives); Density: 8.76 lb/gal (60 F); Sp. gr.: 1.02; Visc.: 45 cs (100 F); Flash Pt.: 73 F (PMCC); pH: 7.5–9.0 (as is)
Composition: 59% active ingred.
Toxicity: Severe eye irritant, and skin irritant on prolonged contact with conc. form
Warnings and Caution: Corrosive to mild steel
References: Tech. bulletin SC: 92-76

Neodol 25-7. Shell Chem. Co.
Chemical Description: $C_{12}-C_{15}$ alcohol (7 EO) ethoxylate
Category: Detergent, nonionic
Applications: Formulation of heavy-duty controlled foaming detergents; general maintenance; textile processing; metal cleaning; paper, paint, and leather processing
Form: Paste-like; Color: Practically colorless; Odor: Mild
Properties: m.w.: 519; Density: 8.12 lb/gal (100 F); Sp. gr.: 0.958 (50/25 C); Visc.: 34 cs (100 F); M.P.: 12–17 C; HLB: 12.2; Cloud Pt.: 50 C (1% aq.); Flash Pt.: 350 F (PMCC); Pour Pt.: 21 C; Hydroxyl No.: 108; pH: 6.0 ± 0.5 (1% aq.)
Composition: 100% active ingred.
Toxicity: Severe eye irritant, and skin irritant on prolonged contact with conc. form
References: Tech. bulletin SC: 87-76

Neodol 25-9. Shell Chem. Co.
Chemical Description: $C_{12}-C_{15}$ alcohol (9 EO) ethoxylate
Category and Applications: See Neodol 25-7
Form: Paste-like; Color: Practically colorless; Odor: Mild
Properties: m.w.: 610; Density: 8.25 lb/gal (100 F); Sp. gr.: 0.982 (50/25 C); Visc.: 41 cs (100 F); M.P.: 25–28 C; HLB: 13.3; Cloud Pt.: 74 C (1% aq.); Flash Pt.: 370 F (PMCC); Pour Pt.: 24 C; Hydroxyl No.: 92; pH: 6.0 ± 0.5 (1% aq.)
Composition: 100% active ingred.
Toxicity: Severe eye irritant and skin irritant on prolonged contact with conc. form
References: Tech. bulletin SC:88-76

Neodol 25-12. Shell Chem. Co.
Chemical Description: $C_{12}-C_{15}$ alcohol (12 EO) ethoxylate
Category and Applications: See Neodol 25-7
Form: Paste-like; Color: Practically colorless; Odor: Mild
Properties: m.w.: 729; Density: 8.40 lb/gal (100 F); Sp. gr.: 1.003 (50/25 C); Visc.: 53 cs (100 F); M.P.: 30–33 C; HLB: 14.4; Cloud Pt.: 97 C (1% aq.); Flash Pt.: 380 F (PMCC); Pour Pt.: 27 C; Hydroxyl No.: 75; pH: 6.0 ± 0.5 (1% aq.)
Composition: 100% active ingred.
Toxicity: Severe eye irritant, and skin irritant on prolonged contact with conc. form
References: Tech. bulletin SC:131:76

Neodol 91-2.5. Shell Chem. Co.
Chemical Description: C_9-C_{11} primary alcohol 2.5 mole ethoxylate
Category: Emulsifier, wetting agent, intermediate, nonionic
Applications: Oil-sol. emulsifier and wetting agent; its sulfates are high-foaming rapidly biodegradable anionic surfactants for industrial processing aids, high-foaming maintenance and hard surface cleaners with high salt tolerance
Form: Liquid; Color: Practically colorless; Odor: Mild
Properties: m.w.: 270 (avg.); Solubility: Sol. in a wide range of oxygenated and hydrocarbon solvents; partial sol. in water; Density: 7.73 lb/gal; Sp. gr.: 0.934; Visc.: 12 cs (100 F); HLB: 8.1; Flash Pt.: 280 F (PMCC); Pour Pt.: −12 C; Hydroxyl No.: 208; pH: 6.0 (1% aq. disp.); Rapidly biodegradable
Composition: 100% active ingred.
Toxicity: Severe eye irritant, and skin irritant on prolonged contact with conc. form

Neodol 91-6. Shell Chem. Co.
Chemical Description: C_9-C_{11} primary alcohol 6 mole ethoxylate
Category: Nonionic surfactant, dispersant, wetting agent
Applications: General purpose surfactant; used in maintenance, janitorial, transportation, and other hard surface cleaners, in laundry products, and for wet processing of textiles, pulp and paper, and leather
Form: Liquid; Color: Practically colorless
Properties: m.w.: 424 (avg.); Density: 8.24 lb/gal; Sp. gr.: 0.991; Visc.: 23 cs (100 F); M.P.: 6–9 C; HLB: 12.5; Cloud Pt.: 52 C (1% sol'n.); Flash Pt.: 334 F (PMCC); Pour Pt.: 7 C; Hydroxyl No.: 132; pH: 6.0 (1% aq.); Rapidly biodegradable
Composition: 100% active ingred.
Toxicity: Severe eye irritant, and skin irritant on prolonged contact with conc. form

Neodol 91-8. Shell Chem. Co.
Chemical Description: C_9-C_{11} primary alcohol 8.4 mole ethoxylate
Category: Nonionic surfactant,

wetting agent
Applications: General purpose surfactant with good wetting and high cloud pt.; used in maintenance, janitorial, transportation, and other hard surface cleaners, in laundry products, and for wet processing of textiles, pulp and paper, and leather
Form: Liquid (may solidify to a soft paste at low ambient temps.); Color: Practically colorless
Properties: m.w.: 529 (avg.); Solubility: Readily disp. in cool water; Density: 8.42 lb/gal; Sp. gr.: 1.002; Visc.: 30 cs (100 F); M.P.: 14–17 C; HLB: 14.0; Cloud Pt.: 80 C (1% sol'n.); Flash Pt.: 349 F (PMCC); Pour Pt.: 16 C; Hydroxyl No.: 106; pH: 6.0 (1% aq.); Rapidly biodegradable
Composition: 100% active ingred.
Toxicity: Severe eye irritant, and skin irritant on prolonged contact with conc. form

Neoscoa 203C. Toho Chemical Industry
Chemical Description: Polyoxyethylene alkyl ether
Category: Detergent, nonionic
Applications: Scouring agent for cotton, wool, and synthetic fibers
Form: Liquid
Composition: 52% active

Neoscoa 363. Toho Chemical Industry
Chemical Description: Blend of nonionics
Category: Detergent, nonionic
Applications: Scouring agent for raw wool and textiles; metal cleaning
Form: Liquid
Composition: 100% active

Neoscoa 500C. Toho Chemical Industry
Chemical Description: Blend of nonionics and anionics
Category: Detergent, nonionic/anionic
Applications: Scouring agent for cotton, wool, and synthetic fibers
Form: Liquid
Composition: 50% active

Neoscoa FS-100. Toho Chemical Industry
Chemical Description: Nonionic blend
Applications: Deinking agent for paper production by flotation method
Form: Liquid

Neoscoa GF 3C. Toho Chemical Industry
Chemical Description: Blend of polyoxyethylene alkylaryl ether and polyoxyethylene alkylaryl ether phosphate
Category: Detergent, wetting agent, nonionic/anionic
Applications: Scouring and wetting agent for cotton and chemical and synthetic fibers
Form: Liquid
Composition: 80% active

Neoscoa PRA-60. Toho Chemical Industry
Chemical Description: Specially blended nonionics

Category: Detergent, nonionic/anionic
Applications: Aftersoaping agent esp. for polyester fibers
Form: Liquid
Composition: 54% active

Neoscoa SS-10. Toho Chemical Industry
Chemical Description: Nonionic blend
Applications: Deinking agent for paper production by washing method
Form: Liquid

Neoscoa TH-102. Toho Chemical Industry
Chemical Description: Specially blended nonionics
Category: Detergent, nonionic/anionic
Applications: Dyeing machine cleaner for disperse dyeing
Form: Liquid
Composition: 50% active

Neospinol 264. Toho Chem. Industry Co., Ltd.
Chemical Description: Blend of special anionics, nonionics, and mineral oil
Category: Dispersant
Applications: Dispersant for sulfur in rayon production
Form: Liquid
Composition: 100% conc.

Neospinol 358. Toho Chem. Industry Co., Ltd.
Chemical Description: Special nonionic complex (nitrogen compound)
Category and Applications: See Neospinol 264
Form: Liquid
Composition: 100% conc.

Neustrene 045. Humko Sheffield Chem.
Chemical Description: Hydrogenated fish glycerides
Category: Lubricant, intermediate, emulsifier
Applications: Lubricating greases, pharmaceutical intermediates, synthetic waxes, textile lubricants, mold release agents, buffing compounds, emulsifying agents
Form: Solid
Properties: Sapon. No.: 188–201; Acid No.: 6.0 max.
Composition: 100% conc.

Neustrene 053. Humko Sheffield Chem.
Chemical Description: Hydrogenated fish glycerides
Category and Applications: See Neustrene 045
Form: Solid
Properties: Sapon. No.: 186–201; Acid No.: 5.0 max.
Composition: 100% conc.

Neustrene 059. Humko Sheffield Chem.
Chemical Description: Hydrogenated tallow glycerides
Category and Applications: See Neustrene 045
Form: Solid; Color: 5 max. (Gardner)

Properties: Sapon. No.: 193–205; Acid No.: 10.0 max.
Composition: 100% conc.

Neustrene 060. Humko Sheffield Chem.
Chemical Description: Refined hydrogenated tallow glycerides
Category and Applications: See Neustrene 045
Form: Solid
Properties: Sapon. No.: 193–205; Acid No.: 2.0 max.
Composition: 100% conc.

Neustrene 064. Humko Sheffield Chem.
Chemical Description: Hydrogenated soya glycerides
Category and Applications: See Neustrene 045
Form: Solid
Properties: Sapon. No.: 188–200; Acid No.: 4.0 max.
Composition: 100% conc.

Neutronyx 600. Onyx Chem. Co.
Chemical Description: Nonylphenol polyglycol ether (9.5 moles EO)
Category: Detergent, dispersant, emulsifier, wetting agent, nonionic
Applications: Used in detergent disinfectants and sanitizers, household detergents, dishwash, fine fabric, and woolen cleaners; esp. good in hard water
Form: Liquid
Properties: Sp. gr.: 1.05; Cloud Pt.: 54 C; Flash Pt.: > 200 F; Stability: Stable in acid and alkaline systems
Composition: 100% conc.

Neutronyx 656. Onyx Chem. Co.
Chemical Description: Nonylphenol polyglycol ether (11.0 moles EO)
Category and Applications: See Neutronyx 600
Form: Liquid
Properties: Sp. gr.: 1.05; Cloud Pt.: 71 C; Flash Pt.: > 200 F; Stability: Stable in acid and alkaline systems
Composition: 100% conc.

Neutronyx S-60. Onyx Chem. Co.
Chemical Description: Ammonium nonylphenol polyglycol ether sulfate
Category: Detergent, foaming agent, anionic
Applications: Used in disinfectant hand cleaner formulations, light duty dishwash, carwash, and glass cleaners; high foaming
Form: Liquid
Composition: 60% conc.

Newcol 20. Nippon Nyukazai
Chemical Description: Sorbitan laurate
Category: Antistat, emulsifier, corrosion inhibitor, lubricant, solubilizer, nonionic
Applications: Extreme pressure lubricant; emulsion solubilizer
Form: Solid
Composition: 100% conc.

Newcol 25. Nippon Nyukazai
Chemical Description: Polyoxyethylene sorbitan laurate
Category and Applications: See

Newcol 20
Form: Liquid
Composition: 100% conc.

Newcol 40. Nippon Nyukazai
Chemical Description: Sorbitan palmitate
Category and Applications: See Newcol 20
Form: Solid
Composition: 100% conc.

Newcol 45. Nippon Nyukazai
Chemical Description: Polyoxyethylene sorbitan palmitate
Category and Applications: See Newcol 20
Form: Solid
Composition: 100% conc.

Newcol 60. Nippon Nyukazai
Chemical Description: Sorbitan stearate
Category and Applications: See Newcol 20
Form: Solid
Composition: 100% conc.

Newcol 65. Nippon Nyukazai
Chemical Description: Polyoxyethylene sorbitan stearate
Category and Applications: See Newcol 20
Form: Solid
Composition: 100% conc.

Newcol 80, 3-80. Nippon Nyukazai
Chemical Description: Sorbitan oleate
Category and Applications: See Newcol 20
Form: Liquid
Composition: 100% conc.

Newcol 85. Nippon Nyukazai
Chemical Description: Polyoxyethylene sorbitan oleate
Category and Applications: See Newcol 20
Form: Liquid
Composition: 100% conc.

Newcol 150. Nippon Nyukazai
Chemical Description: Polyoxyethylene laurate
Category: Emulsifier, dispersant, lubricant, nonionic
Form: Liquid
Composition: 100% conc.

Newcol 170. Nippon Nyukazai
Chemical Description: Polyoxyethylene oleate
Category: Emulsifier, dispersant, lubricant, nonionic
Form: Liquid
Composition: 100% conc.

Newcol 180. Nippon Nyukazai
Chemical Description: Polyoxyethylene stearate
Category: Emulsifier, dispersant, lubricant, nonionic
Form: Solid
Composition: 100% conc.

Newcol 180T. Nippon Nyukazai
Category: Nonionic surfactant
Applications: Used in emulsion polymerization
Form: Solid
Composition: 100% conc.

Newcol 210. Nippon Nyukazai
Chemical Description: NH_4 dodecylbenzene sulfonate
Category: Wetting agent, emulsi-

fier, detergent, anionic
Form: Liquid
Composition: 50% conc.

Newcol 271A. Nippon Nyukazai
Chemical Description: Na alkyl diphenyl ether di sulfonate
Category: Detergent, emulsifier, anionic
Applications: Used in emulsion polymerization
Form: Liquid
Composition: 45% conc.

Newcol 290K. Nippon Nyukazai
Chemical Description: Na alkyl sulfosuccinate
Category: Wetting and dispersing agent, solubilizer, penetrant, anionic
Form: Liquid
Composition: 75% conc.

Newcol 290M. Nippon Nyukazai
Chemical Description: Na alkyl sulfosuccinate
Category: See Newcol 290K
Form: Liquid
Composition: 75% conc.

Newcol 290P. Nippon Nyukazai
Chemical Description: Na alkyl sulfosuccinate
Category: See Newcol 290K
Form: Liquid
Composition: 30% conc.

Newcol 405. Nippon Nyukazai
Chemical Description: Polyoxyethylene lauryl amine
Category: Corrosion inhibitor, intermediate, nonionic
Applications: Intermediate for textile industry
Form: Liquid
Composition: 100% conc.

Newcol 410. Nippon Nyukazai
Chemical Description: Polyoxyethylene lauryl amine
Category and Applications: See Newcol 405
Form: Solid
Composition: 100% conc.

Newcol 420. Nippon Nyukazai
Chemical Description: Polyoxyethylene lauryl amine
Category and Applications: See Newcol 405
Form: Solid
Composition: 100% conc.

Newcol 506. Nippon Nyukazai
Chemical Description: Polyoxyethylene nonylphenyl ether
Category: Nonionic surfactant
Applications: Used in emulsion polymerization
Form: Solid
Composition: 100% conc.

Newcol 508. Nippon Nyukazai
Chemical Description: Block copolymer of ethylene oxide and propylene oxide
Category: Detergent, nonionic
Applications: Low-foaming detergent
Form: Solid
Composition: 100% conc.

Newcol 560. Nippon Nyukazai
Chemical Description: Polyoxyethylene nonylphenyl ether
Category: Wetting agent, pene-

trant, spreading and dispersing agent, emulsifier, nonionic
Applications: Base materials of detergents; emulsifier of agricultural chemicals, machine oils, etc.
Form: Liquid
Composition: 100% conc.

Newcol 560SF. Nippon Nyukazai
Chemical Description: NH_4 polyoxyethylene nonylphenyl ether sulfate
Category and Applications: See Newcol 560SN
Form: Liquid
Composition: 50% conc.

Newcol 560SN. Nippon Nyukazai
Chemical Description: Na polyoxyethylene nonylphenyl ether sulfate
Category: Emulsifier, wetting agent, penetrant, detergent, anionic
Applications: Emulsion polymerization
Form: Liquid
Composition: 30% conc.

Newcol 561H. Nippon Nyukazai
Chemical Description: Polyoxyethylene nonylphenyl ether
Category: Antifoaming agent, mold lubricant, nonionic
Form: Liquid
Composition: 100% conc.

Newcol 562. Nippon Nyukazai
Chemical Description: Polyoxyethylene nonylphenyl ether
Category and Applications: See Newcol 560
Form: Liquid
Composition: 100% conc.

Newcol 564. Nippon Nyukazai
Chemical Description: Polyoxyethylene nonylphenyl ether
Category and Applications: See Newcol 560
Form: Liquid
Composition: 100% conc.

Newcol 565. Nippon Nyukazai
Chemical Description: Polyoxyethylene nonylphenyl ether
Category and Applications: See Newcol 560
Form: Liquid
Composition: 100% conc.

Newcol 565FH. Nippon Nyukazai
Chemical Description: Block copolymer of ethylene oxide and propylene oxide
Category and Applications: See Newcol 508
Form: Liquid
Composition: 100% conc.

Newcol 566. Nippon Nyukazai
Chemical Description: Polyoxyethylene nonylphenyl ether
Category and Applications: See Newcol 560
Form: Liquid
Composition: 100% conc.

Newcol 568. Nippon Nyukazai
Chemical Description: Polyoxyethylene nonylphenyl ether
Category: Nonionic surfactant
Applications: Used in emulsion polymerization
Form: Solid
Composition: 100% conc.

Newcol 569E. Nippon Nyukazai
Chemical Description: Block copolymer of ethylene oxide and propylene oxide
Category and Applications: See Newcol 508
Form: Paste
Composition: 100% conc.

Newcol 607. Nippon Nyukazai
Chemical Description: Polyoxyethylene alkyl aryl ether
Category: Emulsifier, solubilizer, detergent, nonionic
Applications: Emulsion solubilizer, emulsion polymerization
Form: Liquid
Composition: 100% conc.

Newcol 610. Nippon Nyukazai
Chemical Description: Polyoxyethylene alkyl aryl ether
Category and Applications: See Newcol 607
Form: Solid
Composition: 100% conc.

Newcol 614. Nippon Nyukazai
Chemical Description: Polyoxyethylene alkyl aryl ether
Category and Applications: See Newcol 607
Form: Solid
Composition: 100% conc.

Newcol 623. Nippon Nyukazai
Chemical Description: Polyoxyethylene alkyl aryl ether
Category and Applications: See Newcol 607
Form: Solid
Composition: 100% conc.

Newcol 704. Nippon Nyukazai
Chemical Description: Polyoxyethylene alkyl aryl ether
Category and Applications: See Newcol 607
Form: Liquid
Composition: 100% conc.

Newcol 707. Nippon Nyukazai
Chemical Description: Polyoxyethylene alkyl aryl ether
Category and Applications: See Newcol 607
Form: Liquid
Composition: 100% conc.

Newcol 710. Nippon Nyukazai
Chemical Description: Polyoxyethylene alkyl aryl ether
Category and Applications: See Newcol 607
Form: Solid
Composition: 100% conc.

Newcol 714. Nippon Nyukazai
Chemical Description: Polyoxyethylene alkyl aryl ether
Category and Applications: See Newcol 607
Form: Solid
Composition: 100% conc.

Newcol 723. Nippon Nyukazai
Chemical Description: Polyoxyethylene alkyl aryl ether
Category and Applications: See Newcol 607
Form: Solid
Composition: 100% conc.

Newcol 860. Nippon Nyukazai
Chemical Description: Polyoxyethylene nonylphenyl ether derivative
Category: Detergent, nonionic
Applications: Low-foaming detergent

Newcol 861SE. Nippon Nyukazai
Chemical Description: Na octylphenoxyethoxyethyl sulfonate
Category and Applications: See Newcol 560SN
Form: Paste
Composition: 30% conc.

Newcol 862. Nippon Nyukazai
Chemical Description: Polyoxyethylene nonylphenyl ether derivative
Category: Detergent, nonionic
Applications: Low-foaming detergent
Form: Paste
Composition: 100% conc.

Newcol 864. Nippon Nyukazai
Chemical Description: Polyoxyethylene octylphenyl ether
Category: Antifoaming agent, mold lubricant, nonionic
Form: Liquid
Composition: 100% conc.

Newcol 865. Nippon Nyukazai
Chemical Description: Polyoxyethylene octylphenyl ether
Category: Wetting, penetrating, spreading, and dispersing agent, emulsifier, nonionic
Applications: Base materials of detergents; emulsifier of agricultural chemicals, machine oils, etc.
Form: Liquid
Composition: 100% conc.

Newcol 1010. Nippon Nyukazai
Chemical Description: Polyoxyethylene octyl ether
Category: Stabilizer, nonionic
Applications: Used as stabilizer of latex
Form: Solid
Composition: 100% conc.

Newcol 1020. Nippon Nyukazai
Chemical Description: Polyoxyethylene octyl ether
Category: Stabilizer, nonionic
Applications: Used as stabilizer of latex
Form: Solid
Composition: 100% conc.

Newcol 1100. Nippon Nyukazai
Chemical Description: Polyoxyethylene lauryl ether
Category and Applications: See Newcol 1010
Form: Solid
Composition: 100% conc.

Newcol 1105. Nippon Nyukazai
Chemical Description: Polyoxyethylene octyl ether
Category: Emusifier, lubricant, detergent, nonionic
Applications: Dyeing additive agent, emulsifier of mineral oil
Form: Liquid
Composition: 100% conc.

Newcol 1110. Nippon Nyukazai
Chemical Description: Polyoxyethylene lauryl ether
Category and Applications: See Newcol 1010
Form: Solid
Composition: 100% conc.

Newcol 1120. Nippon Nyukazai
Chemical Description: Polyoxyethylene lauryl ether
Category: Emulsifier, nonionic
Applications: Emulsifier of mineral oil; dyeing additive agent, etc.
Form: Solid
Composition: 100% conc.

Newcol 1200. Nippon Nyukazai
Chemical Description: Polyoxyethylene oleyl ether
Category and Applications: See Newcol 1120
Form: Solid
Composition: 100% conc.

Newcol 1203. Nippon Nyukazai
Chemical Description: Polyoxyethylene lauryl ether
Category and Applications: See Newcol 1120
Form: Liquid
Composition: 100% conc.

Newcol 1204. Nippon Nyukazai
Chemical Description: Polyoxyethylene oleyl ether
Category and Applications: See Newcol 1120
Form: Liquid
Composition: 100% conc.

Newcol 1208. Nippon Nyukazai
Chemical Description: Polyoxyethylene oleyl ether
Category and Applications: See Newcol 1120
Form: Solid
Composition: 100% conc.

Newcol 1210. Nippon Nyukazai
Chemical Description: Polyoxyethylene oleyl ether
Category and Applications: See Newcol 1120
Form: Solid
Composition: 100% conc.

Newcol 1305. Nippon Nyukazai
Chemical Description: Polyoxyethylene tridecyl ether
Category: Emulsifier, lubricant, nonionic
Applications: Emulsifier of silicone; dyeing additive agent
Form: Solid
Composition: 100% conc.

Newcol 1305SN, 1310SN. Nippon Nyukazai
Chemical Description: Na polyoxyethylene tridecyl ether sulfate
Category and Applications: See Newcol 560SN
Form: Liquid
Composition: 30% conc.

Newcol 1310. Nippon Nyukazai
Chemical Description: Polyoxyethylene tridecyl ether
Category and Applications: See Newcol 1305
Form: Solid
Composition: 100% conc.

Newcol 1515. Nippon Nyukazai
Chemical Description: Polyoxyethylene tridecyl ether
Category and Applications: See Newcol 1305
Form: Liquid
Composition: 100% conc.

Newcol 1525. Nippon Nyukazai
Chemical Description: Polyoxyethylene tridecyl ether
Category and Applications: See Newcol 1305
Form: Liquid
Composition: 100% conc.

Newcol 1545. Nippon Nyukazai
Chemical Description: Polyoxyethylene alkyl ether
Category and Applications: See Newcol 1305
Form: Solid
Composition: 100% conc.

Newcol 1610. Nippon Nyukazai
Chemical Description: Polyoxyethylene cetyl ether
Category: Emulsifier, stabilizer, nonionic
Applications: Emulsifier of paraffin wax; latex stabilizer
Form: Solid
Composition: 100% conc.

Newcol 1620. Nippon Nyukazai
Chemical Description: Polyoxyethylene cetyl ether
Category and Applications: See Newcol 1610
Form: Solid
Composition: 100% conc.

Newcol 1807. Nippon Nyukazai
Chemical Description: Polyoxyethylene stearyl ether
Category and Applications: See Newcol 1610
Form: Solid
Composition: 100% conc.

Newcol 1820. Nippon Nyukazai
Chemical Description: Polyoxyethylene stearyl ether
Category and Applications: See Newcol 1610
Form: Solid
Composition: 100% conc.

Newcol B4, B10. Nippon Nyukazai
Chemical Description: Polyoxyethylene naphthyl ether
Category: Dispersant, suspending agent, diffusion agent, nonionic
Form: Liquid
Composition: 100% conc.

Newcol B18. Nippon Nyukazai
Chemical Description: Polyoxyethylene naphthyl ether
Category: See Newcol B4
Form: Solid
Composition: 100% conc.

Niaproof 4. Niacet Corp.
Chemical Description: Sodium tetradecyl sulfate
Category: Detergent, wetting agent, emulsifier, anionic
Applications: Adhesives and sealants, coatings, photo chemicals, photo emulsion applications, emulsion polymerization
Form: Liquid; Color: Essentially colorless

Properties: Solubility: Miscible with water @ 20 C; Density: 8.58 lb/gal (20 C); Sp. gr.: 1.031 (20/20 C); Surface Tension: 47 dynes/cm (0.1% aq.); pH: 8.5 (0.1% aq.)
Composition: 27% active ingred.
Toxicity: Moderate oral and skin penetration toxicity
Containers: 55 gal tight steel composite drums
Warnings and Caution: Avoid contact with skin or eyes; avoid breathing vapor; store above 60 F to avoid separation

Niaproof 7. Niacet Corp.
Chemical Description: Sodium heptadecyl sulfate
Category: Detergent, wetting agent, emulsifier, anionic
Applications: Adhesives, metal finishing, leather treating, crystal growth control
Form: Liquid; Color: Essentially colorless
Properties: Solubility: Miscible with water @ 20 C; Density: 8.82 lb/gal (20 C); Sp. gr.: 1.060 (20/20 C); Surface Tension: 34 dynes/cm (0.1% aq.); pH: 5.3 (0.1% aq.)
Composition: 26% active ingred.
Toxicity: Moderate oral and skin penetration toxicity
Containers: 55 gal tight steel composite drums
Warnings and Caution: Avoid contact with skin or eyes; avoid breathing vapor; store above 60 F to avoid separation

Niaproof 08. Niacet Corp.
Chemical Description: Sodium ethylhexyl sulfate
Category: Detergent, wetting agent, emulsifier, anionic
Applications: Metal cleaning, electroplating, photo chemicals, industrial cleaners, caustic vegetable peeling, adhesives, emulsion polymerization
Form: Liquid; Color: Essentially colorless
Properties: Solubility: Miscible with water @ 20 C; Density: 9.23 lb/gal (20 C); Sp. gr.: 1.109 (20/20 C); Surface Tension: 63 dynes/cm (0.1% aq.); pH: 7.3 (0.1% aq.)
Composition: 40% active ingred.
Toxicity: Moderate oral and skin penetration toxicity
Containers: 55 gal tight polyethylene and steel composite drums
Warnings and Caution: Avoid contact with skin or eyes, avoid breathing vapor; store above 60 F to avoid separation

Nikkol BL-1SY. Nikko Chem. Co., Ltd.
Chemical Description: Ethylene glycol dodecyl ether
Category: Nonionic
Applications: 98% pure, homogeneous compound for use in research on ether-type nonionics, e.g., studies of surface and interfacial phenomena, physical, biological, and surfactant properties, and as control standard in chromatography, etc.
Properties: m.w.: 230.4; Ref. Index: 1.4413 (30 C); Sp. gr.: 0.8681; M.P.: 19.5 C; B.P.: 135 C (1 mmHg)

Nikkol BL-2SY. Nikko Chem. Co., Ltd.
Chemical Description: Diethylene glycol dodecyl ether
Category and Applications: See Nikkol BL-1SY
Properties: m.w.: 274.4; Ref. Index: 1.4441 (30 C); Sp. gr.: 0.8996; M.P.: 17.6 C; B.P.: 164 C (1 mmHg)

Nikkol BL-3SY. Nikko Chem. Co., Ltd.
Chemical Description: Triethylene glycol dodecyl ether
Category and Applications: See Nikkol BL-1SY
Properties: m.w.: 318.5; Ref. Index: 1.4464 (30 C); Sp. gr.: 0.9216; M.P.: 16.0 C; B.P.: 194–197 C (1.5 mmHg)

Nikkol BL-4SY. Nikko Chem. Co., Ltd.
Chemical Description: Tetraethylene glycol dodecyl ether
Category and Applications: See Nikkol BL-1SY
Properties: m.w.: 362.6; Ref. Index: 1.4485 (30 C); Sp. gr.: 0.9442; M.P.: 18.7 C; B.P.: 214–215 (2 mmHg)

Nikkol BL-5SY. Nikko Chem. Co., Ltd.
Chemical Description: Pentaethylene glycol dodecyl ether
Category and Applications: See Nikkol BL-1SY
Properties: m.w.: 406.6; Ref. Index: 1.4503 (30 C); Sp. gr.: 0.9586; M.P.: 22.8 C; B.P.: 225–229 (2 mmHg); Cloud Pt.: 30.5 C (0.1%); Surface Tension: 29.6 dynes/cm

Nikkol BL-6SY. Nikko Chem. Co., Ltd.
Chemical Description: Hexaethylene glycol dodecyl ether
Category and Applications: See Nikkol BL-1SY
Properties: m.w.: 450.6; Ref. Index: 1.4520 (30 C); Sp. gr.: 0.9756; M.P.: 24.8 C; Cloud Pt.: 55.0 C (0.1%); Surface Tension: 31.5 dynes/cm

Nikkol BL-7SY. Nikko Chem. Co., Ltd.
Chemical Description: Heptaethylene glycol dodecyl ether
Category and Applications: See Nikkol BL-1SY
Properties: m.w.: 494.6; Ref. Index: 1.4533 (30 C); Sp. gr.: 0.9837; M.P.: 26.1 C; Cloud Pt.: 70.0 C (0.1%); Surface Tension: 33.4 dynes/cm

Nikkol BL-8SY. Nikko Chem. Co., Ltd.
Chemical Description: Octaethylene glycol dodecyl ether
Category and Applications: See Nikkol BL-1SY
Properties: m.w.: 538.7; Ref. Index: 1.4542 (30 C); Sp. gr.: 0.9951; M.P.: 31.4 C; Cloud Pt.: 79.0 C (0.1%); Surface Tension: 34.6 dynes/cm

Nikkol BPS-5. Nikko Chem. Co., Ltd.
Chemical Description: POE (5)-phytosterol
Category: Emulsifier, solubilizer, dispersant, emollient, foam stabilizer,
Applications: Used in basic cosmetics, hair-care toiletries, make-up cosmetics, and pharmaceuticals as emulsifiers for O/W and W/O compounds, solubilizers, viscosity and consistency adjusters, foam stabilizers, conditioner, harmonizing agent, emollient, etc.
Form: Paste; Color: White to pale yellow
Properties: Solubility @ 25 C and 5% conc.: Sol. in propylene glycol, ethanol, 2-octyldecanol, olive oil; partially sol. in water, liquid paraffin, isocetyl myristate, and isocetyl isostearate; HLB: 9.5; Acid Value: 0.25 max.; Hydroxyl Value: 95.8; pH: 4.8 (5%)

Nikkol BPS-10. Nikko Chem. Co., Ltd.
Chemical Description: POE (10)-phytosterol
Category and Applications: See Nikkol BPS-5
Form: Paste or solid; Color: White to pale yellow
Properties: Solubility @ 25 C and 5% conc.: Sol. in water, propylene glycol, ethanol, 2-octyldecanol, olive oil; HLB: 12.5; Acid Value: 0.18 max.; Hydroxyl Value: 67.3; pH: 5.2 (5%)

Nikkol BPS-15. Nikko Chem. Co., Ltd.
Chemical Description: POE (15)-phytosterol
Category and Applications: See Nikkol BPS-5
Form: Wax-like solid; Color: White to pale yellow
Properties: Solubility @ 25 C and 5% conc.: Sol. in water, propylene glycol, ethanol, 2-octyldecanol; HLB: 15.0; Acid Value: 0.06 max.; Hydroxyl Value: 52.8; pH: 5.0 (5%)

Nikkol BPS-20. Nikko Chem. Co., Ltd.
Chemical Description: POE (20)-phytosterol
Category and Applications: See Nikkol BPS-5
Form: Wax-like solid; Color: White to pale yellow
Properties: Solubility @ 25 C and 5% conc.: Sol. in water, propylene glycol, ethanol, 2-octyldecanol; HLB: 15.5; Acid Value: 0.07 max.; Hydroxyl Value: 43.9; pH: 5.5 (5%)

Nikkol BPS-25. Nikko Chem. Co., Ltd.
Chemical Description: POE (25)-phytosterol
Category and Applications: See Nikkol BPS-5
Form: Wax-like solid; Color: White to pale yellow
Properties: Solubility @ 25 C and

5% conc.: Sol. in water, propylene glycol, ethanol; HLB: 18.0; Acid Value: 0.11 max.; Hydroxyl Value: 40.3; pH: 5.7 (5%)

Nikkol BPS-30. Nikko Chem. Co., Ltd.
Chemical Description: POE (30)-phytosterol
Category and Applications: See Nikkol BPS-5
Form: Wax-like solid; Color: White to pale yellow
Properties: Solubility @ 25 C and 5% conc.: Sol. in water, propylene glycol, ethanol; HLB: 18.0; Acid Value: 0.09 max.; Hydroxyl Value: 37.0; pH: 5.7 (5%)

Nikkol GO-430. Nikko Chem. Co., Ltd.
Chemical Description: POE (30) sorbitol tetraoleate
Category: Emulsifier, solubilizer, superfatting agent, nonionic
Applications: Used in drug and cosmetics industries for emulsifying cream, lotion, bath toiletries, ointments; for solubilizing cosmetic liquors, tonics; for superfatting shampoo, rinse, cleansing cream, make-up; for composing oil phase of make-up preparations; also for emulsion polymerization of plastics, emulsification of silicone, resins, waxes, cutting oils, agricultural chemicals, and printing inks
Form: Liquid; Color: Pale yellow
Properties: Ref. Index: 1.4727; Solubility: Sol. in ethanol, PEG 400, ethyl acetate, xylene, kerosene; partially sol. in water and propylene glycol; Sp. gr.: 1.048; HLB: 11.5; Congealing Pt.: −2.8 C
Composition: 100% conc.

Nikkol GO-440. Nikko Chem. Co., Ltd.
Chemical Description: POE (40) sorbitol tetraoleate
Category and Applications: See Nikkol GO-430
Form: Liquid; Color: Pale yellow
Properties: Ref. Index: 1.4718; Solubility: Sol. in ethanol, PEG 400, ethyl acetate, xylene; sol. warm in kerosene; partially sol. in water and propylene glycol; Sp. gr.: 1.054; HLB: 12.5; Congealing Pt.: −6.9 C
Composition: 100% conc.

Nikkol GO-460. Nikko Chem. Co., Ltd.
Chemical Description: POE (60) sorbitol tetraoleate
Category and Applications: See Nikkol GO-430
Form: Liquid; Color: Pale yellow
Properties: Ref. Index: 1.4719; Solubility: Sol. in water, ethanol, PEG 400, ethyl acetate, olive oil, xylene; partially sol. in propylene glycol; Sp. gr.: 1.060; HLB: 14.0; Congealing Pt.: 3.5 C
Composition: 100% conc.

Nikkol GS-460. Nikko Chem. Co., Ltd.
Chemical Description: POE (60) sorbitol tetrastearate
Category and Applications: See Nikkol GO-430

Form: Liquid; Color: Pale yellow
Properties: Solubility: Sol. in water, ethanol, ethyl acetate, xylene; sol. warm in olive oil; partially sol. in PEG 400; M.P.: 35 C; HLB: 13.0
Composition: 100% conc.

Nikkol TS-10. Nikko Chem. Co., Ltd.
Chemical Description: Polyoxyethylene sorbitan monostearate
Category: Emulsifier, solubilizer
Applications: Emulsifier for use with Nikkol SS-10, SO-10, etc. for O/W products, cosmetics, pharmaceuticals, e.g., vanishing cream, emollient, shaving, deodorant, protective, hand creams, hair dressings, sun tan lotion, lip pomade, ointment base; also for some W/O-type products, and as a solubilizer; used for paraffin wax, lacquer, solvent emulsion, and in rinsing or polishing agents and paint
Form: Viscous liquid; Color: Pale yellow to yellow
Properties: Solubility: Sol. in water, ethanol, ethyl acetate, toluene; Sapon. Value: 43–49; Acid Value: 1.5 max.; Hydroxyl Value: 84–100; pH: 5.7–7.7 (5%)
Composition: 100% conc.
Containers: 15 kg net petroleum can

Nilo VON. Sandoz Ltd.
Chemical Description: Ethylene oxide adduct
Category: Emulsifier, nonionic
Applications: Emulsifier for mineral oils
Form: Liquid

Ninate 401. Stepan Chem. Co.
Chemical Description: Calcium alkylbenzene sulfonate
Category: Emulsifier, anionic
Applications: Used in combination with nonionics to form emulsifier systems for pesticide formulations, and in self-dispersing liquids
Form: Liquid; Color: Dark amber
Properties: Solubility: Sol. in kerosene, xylene, and aromatic naphtha; insol. in water and mineral oil; Sp. gr.: 1.00; Visc.: ≈ 2500 cps; Flash Pt.: ≈ 108 F (Setaflash closed tester); pH: 4.5–5.5 (5% in 50% aq. isopropanol)
Composition: 65% active ingred.

Ninate 401-HF. Stepan Chem. Co.
Chemical Description: Calcium alkylbenzene sulfonate
Category and Applications: See Ninate 401; 401-HF recommended for use where higher flash is required
Form: Liquid; Color: Dark amber
Properties: Solubility: Sol. in kerosene, xylene, and heavy aromatic naphtha; insol. in water and mineral oil; Sp. gr.: 1.02; Visc.: ≈ 2500 cps; Flash Pt.: ≈ 123 F (Setaflash closed tester); pH: 4.5–5.5 (5% in 50% aq. isopropanol)
Composition: 60% active ingred.

Ninate 411. Stepan Chem. Co.
Chemical Description: Alkylamine dodecylbenzene sulfonate
Category: Detergent, emulsifier, anionic
Applications: In emulsifiable kero-

sene formulations for removing grease and carbon from engine blocks, metal parts, garage floors; in water-displacing oils for dry cleaning, etc.; causes kerosene or mineral oil to dissolve clearly in water for mop treating, hand cleaners; fuel oil additives
Form: Viscous liquid; Color: Yellow
Properties: m.w.: 385; Solubility: Sol. in kerosene, deodorized kerosene, Stoddard solvent, chlorinated and aromatic solvents, alcohol, and pine oil; insol. in mineral oils; forms a clear sol'n. in water below 1% conc., disp. in water at higher concs.; Density: 8.5 lb/gal; Congealing Pt.: 30 F; Surface Tension: 30 dynes/cm (0.1% aq.); pH: 3.6 (20% in 1:1 isopropanol/water)
Composition: 95% active ingred.

Ninol 128 Extra. Stepan Chem. Co.
Chemical Description: Alkylolamide from refined coconut oil fatty acid
Category and Applications: See Ninol AA-62 Extra
Form: Liquid; Color: Yellow
Properties: Density: 8.3 lb/gal
Composition: 100% conc.

Ninol 201. Stepan Chem. Co.
Chemical Description: Alkylolamide from oleic fatty acid
Category: Emulsifier, nonionic
Applications: Improves viscosity control in personal care and liquid detergent products; emulsifier in industrial lubricant systems
Form: Liquid
Properties: Solubility: Sol. in oils; disp. in water; Density: 8.23 lb/gal; pH: 9.5–10.5 (as is)

Ninol 1281. Stepan Chem. Co.
Chemical Description: Specialty alkylolamide
Category: Detergent, nonionic
Applications: Floor cleaners and all purpose detergent formulations with good soil suspension, high viscosity, rust inhibition, and controlled foam and wetting; compatible with alkalies
Form: Liquid; Color: Amber
Properties: Density: 8.5 lb/gal; pH: 9.0 (1% sol'n.)
Composition: 100% active ingred.

Ninol 1285. Stepan Chem. Co.
Chemical Description: Specialty alkylolamide
Category and Applications: See Ninol 1281
Form: Viscous liquid; Color: Straw
Properties: Density: 8.35 lb/gal; pH: 9.0 (1% sol'n.)
Composition: 100% active ingred.

Ninol 2012 Extra. Stepan Chem. Co.
Chemical Description: Alkylolamide from coconut fatty acid
Category and Applications: See Ninol AA-62 Extra
Form: Liquid; Color: Yellow
Properties: Density: 8.2 lb/gal
Composition: 100% conc.

Ninol AA-62. Stepan Chem. Co.
Chemical Description: Alkylolamide from lauric fatty acid
Category: Foam booster and stabilizer, nonionic
Applications: Improves viscosity in shampoos and liquid detergent formulations
Form: Paste
Properties: Solubility: Sol. in water; Density: 8.34 lb/gal; pH: 9.0–10.5 (5% aq.)
Composition: 100% active ingred.

Ninol AA-62 Extra. Stepan Chem. Co.
Chemical Description: Alkylolamide from lauric fatty acid
Category: Foam booster and stabilizer, detergent, thickener, nonionic
Applications: Shampoos, bubble baths, dishwashing detergents in combination with anionics
Form: Wax; Color: White
Properties: Density: 8.2 lb/gal
Composition: 100% conc.

Ninol HDA-7. Stepan Chem. Co.
Chemical Description: Specialty alkylolamide
Category and Applications: See Ninol 1281
Form: Liquid; Color: Amber
Properties: Density: 8.34 lb/gal; pH: 9.5 (1% sol'n.)
Composition: 93% active ingred.

Ninol P-616. Stepan Chem. Co.
See Ninol AA-62 Extra

Ninol P-621. Stepan Chem. Co.
Chemical Description: Alkylolamide from lauric/myristic fatty acid
Category and Applications: See Ninol AA-62 Extra
Form: Paste; Color: White
Properties: Density: 8.2 lb/gal
Composition: 100% conc.

Ninol P-650. Stepan Chem. Co.
See Ninol 2012 Extra

Nissan Amine AB. Nippon Oil and Fats Co., Ltd.
Chemical Description: Primary octadecylamine
Category: Anticorrosive agent, germicide, wetting agent, mold releasing agent, softener, emulsifier, dispersant, intermediate
Applications: Water treatment, ore flotation, pigment flushing aid, fabric softener, grease additive, intermediate for synthesis of various organic compounds, asphalt antistripping and concrete waterproofing, rubber reclaiming and mold releasing, textile treatment, anticaking agent in agriculture and ceramics
Form: Waxy solid or flake; Color: White
Properties: Solidifying Pt.: 47–53 C; Iodine Value: 2 max.
Composition: 98% min. primary amine
Containers: 15 kg paper bag, 13 kg can, 160 kg drum

Nissan Amine ABT. Nippon Oil and Fats Co., Ltd.
Chemical Description: Primary

tallow-hydrogenated-alkylamine
Category and Applications: See Nissan Amine AB
Form: Waxy solid or flake; Color: White
Properties: Solidifying Pt.: 40–46 C; Iodine Value: 3 max.
Composition: 98% min. primary amine
Containers: 15 kg paper bag, 14 kg can, 160 kg drum

Nissan Amine ABT_2. Nippon Oil and Fats Co., Ltd.
Chemical Description: Primary tallow-alkylamine
Category and Applications: See Nissan Amine AB
Form: Waxy solid; Color: White
Properties: Solidifying Pt.: 30–40 C; Iodine Value: 35 min.
Composition: 98% min. primary amine
Containers: 13 kg can, 160 kg drum

Nissan Amine BB. Nippon Oil and Fats Co., Ltd.
Chemical Description: Primary dodecylamine
Category and Applications: See Nissan Amine AB
Form: Liquid in summer, solid in winter; Color: 120 max. (APHA)
Properties: Solidifying Pt.: 24–28 C; Iodine Value: 1 max.
Composition: 98% min. primary amine
Containers: 13 kg can, 160 kg drum

Nissan Amine DT. Nippon Oil and Fats Co., Ltd.
Chemical Description: Tallow-alkyl propylenediamine
Category: Emulsifier, corrosion inhibitor, wetting agent, dispersant
Applications: Asphalt emulsifier, water treatment, pigment flushing and dispersing, ore flotation, gasoline, grease, and fuel oil additive
Form: Solid; Color: Yellowish brown
Properties: Solidifying Pt.: 25–34 C; Iodine Value: 10–50
Containers: 160 kg drum

Nissan Amine DTH. Nippon Oil and Fats Co., Ltd.
Chemical Description: Tallow-hydrogenated-alkyl propylene diamine
Category and Applications: See Nissan Amine DT
Form: Flake; Color: Yellowish brown
Properties: Iodine Value: 3 max.
Containers: 16 kg paper bag

Nissan Amine FB. Nippon Oil and Fats Co., Ltd.
Chemical Description: Primary coco-alkylamine
Category and Applications: See Nissan Amine AB
Form: Liquid in summer, solid in winter; Color: 120 max. (APHA)
Properties: Solidifying Pt.: 16–22 C; Iodine Value: 12 max.
Composition: 98% min. primary amine
Containers: 13 kg can, 160 kg drum

Nissan Amine MB. Nippon Oil and Fats Co., Ltd.
Chemical Description: Primary tetradecylamine
Category and Applications: See Nissan Amine AB
Form: Waxy solid; Color: White
Properties: Solidifying Pt.: 31–35 C; Iodine Value: 1 max.
Composition: 98% min. primary amine
Containers: 13 kg can, 160 kg drum

Nissan Amine OB. Nippon Oil and Fats Co., Ltd.
Chemical Description: Primary oleylamine
Category and Applications: See Nissan Amine AB
Form: liquid in summer, solid in winter; Color: 140 max. (APHA)
Properties: Solidifying Pt.: 30 C max.; Iodine Value: 60 min.
Composition: 98% min. primary amine
Containers: 13 kg can, 160 kg drum

Nissan Amine PB. Nippon Oil and Fats Co., Ltd.
Chemical Description: Primary hexadecylamine
Category and Applications: See Nissan Amine AB
Form: Waxy solid; Color: White
Properties: Solidifying Pt.: 38–46 C; Iodine Value: 1 max.
Composition: 98% min. primary amine
Containers: 13 kg can, 160 kg drum

Nissan Anon BF. Nippon Oil and Fats Co., Ltd.
Chemical Description: Dimethyl cocoalkyl betaine
Category: Antistatic agent, softener, germicide, amphoteric
Applications: Asphalt antistripping agent; textile treatment; plastics antistat; germicide in foods and cleaning industries
Form: Liquid; Color: Light yellow
Composition: 25% min. active
Containers: 17 kg Atron can

Nissan Anon BL. Nippon Oil and Fats Co., Ltd.
Chemical Description: Dimethyl dodecyl betaine
Category: Antistatic agent, softener, germicide, amphoteric
Applications: Textile industry; rinse aid in cosmetics; germicide in foods and cleaning applications; extraction aid in fermentation
Form: Liquid; Color: Light yellow
Properties: pH: 4.5–6.5
Composition: 30–33% min. active
Containers: 17 kg Atron can

Nissan Anon LG. Nippon Oil and Fats Co., Ltd.
Chemical Description: Alkyl di-(aminoethyl)glycine
Category: Antistatic agent, softener, germicide, amphoteric
Applications: Water treatment; duckweed killer; germicide in textiles and cleaning industries; extraction aid for fermentation
Form: Liquid; Color: Light yel-

low
Properties: pH: 8.5–10.5
Composition: 30% min. active
Containers: 17 kg Atron can

Nissan Asphasol No. 20. Nippon Oils and Fats Co., Ltd.
Chemical Description: Cationic diamine
Category: Emulsifier, cationic
Applications: Asphalt emulsifier for penetrating and mixing processes
Form: Semisolid; Color: Yellow brown
Properties: Solubility: Sol. in aq. sol'n. of organic acid, e.g., acetic acid and inorganic acid; insol. in water; Acid Value: 5 max.
Toxicity: Skin irritant
Containers: 14 kg net tin can or 160 kg net iron drum

Nissan Asphasol No. 100. Nippon Oils and Fats Co., Ltd.
Chemical Description: Cationic diamine
Category: Emulsifier, cationic
Applications: Asphalt emulsifier for mixing process
Form: Liquid; Color: Reddish brown
Properties: Solubility: Dissolves easily in aq. sol'n. of organic acid, e.g., acetic acid, and inorganic acid; insol. in water; Solidifying Pt.: ≈ 20 C
Toxicity: Skin irritant
Containers: 18 kg net tin can or 180 kg net iron drum

Nissan Cation AB. Nippon Oil and Fats Co., Ltd.
Chemical Description: Octadecyl trimethyl ammonium chloride
Category: Germicide, dispersant, coagulant, antistat, softener, cationic
Applications: Germicide in water treatment, petroleum, paper, foods, and textile industries; duckweed killer; antistat in plastics, pulp and paper industries; rinse agent; pigment coating agent
Form: Liquid; Color: Light yellow
Composition: 23% min. active
Containers: 16 kg Atron can, 170 kg drum

Nissan Cation ABT-350. Nippon Oil and Fats Co., Ltd.
Chemical Description: Tallow-hydrogenated alkyl trimethyl ammonium chloride
Category and Applications: See Nissan Cation AB
Form: Liquid; Color: Light yellow
Composition: 33–37% min. active
Containers: 15 kg Atron can

Nissan Cation ABT-500. Nippon Oil and Fats Co., Ltd.
Chemical Description: Tallow-hydrogenated alkyl trimethyl ammonium chloride
Category and Applications: See Nissan Cation AB
Form: Liquid; Color: Light yellow
Composition: 50% min. active
Containers: 15 kg Atron can

Nissan Cation AR-4. Nippon Oil and Fats Co., Ltd.
Chemical Description: Alkyl imidazoline chloride
Category and Applications: See Nissan Cation AB
Form: Liquid; Color: Yellowish brown
Composition: 35% min. active
Containers: 14 kg Atron can

Nissan Cation BB. Nippon Oil and Fats Co., Ltd.
Chemical Description: Dodecyl trimethyl ammonium chloride
Category and Applications: See Nissan Cation AB
Form: Liquid; Color: Light yellow
Composition: 30% min. active
Containers: 18 kg Atron can

Nissan Cation F_2-10R. Nippon Oil and Fats Co., Ltd.
Chemical Description: Coco-alkyl dimethyl benzyl ammonium chloride
Category and Applications: See Nissan Cation AB
Form: Liquid; Color: Colorless
Composition: 9.5% min. active
Containers: 18 kg Atron can

Nissan Cation F_2-20R. Nippon Oil and Fats Co., Ltd.
Chemical Description: Coco-alkyl dimethyl benzyl ammonium chloride
Category and Applications: See Nissan Cation AB
Form: Liquid; Color: Colorless
Composition: 19–21% active
Containers: 10 kg poly bottle, 180 kg drum

Nissan Cation F_2-40E. Nippon Oil and Fats Co., Ltd.
Chemical Description: Coco-alkyl dimethyl benzyl ammonium chloride
Category and Applications: See Nissan Cation AB
Form: Liquid; Color: Colorless
Composition: 40–44% active
Containers: 17 kg Atron can, 100 kg drum

Nissan Cation F_2-50. Nippon Oil and Fats Co., Ltd.
Chemical Description: Coco-alkyl dimethyl benzyl ammonium chloride
Category and Applications: See Nissan Cation AB
Form: Liquid; Color: Light yellow
Composition: 50% min. active
Containers: 17 kg Atron can, 100 kg drum

Nissan Cation FB. Nippon Oil and Fats Co., Ltd.
Chemical Description: Coco-alkyl trimethyl ammonium chloride
Category and Applications: See Nissan Cation AB
Form: Liquid; Color: Light yellow
Composition: 30% min. active
Containers: 17 kg Atron can, 100 kg and 200 kg drums

Nissan Cation FB-500. Nippon Oil and Fats Co., Ltd.
Chemical Description: Coco-alkyl trimethyl ammonium chloride
Category and Applications: See Nissan Cation AB
Form: Liquid; Color: Light yellow
Composition: 50% min. active
Containers: 15 kg Atron can, 170 kg drum

Nissan Cation L-207. Nippon Oil and Fats Co., Ltd.
Chemical Description: Polyoxyethylene dodecyl monomethylammonium chloride
Category and Applications: See Nissan Cation AB
Form: Viscous liquid; Color: Yellowish brown
Composition: 90% min. active
Containers: 18 kg Atron can

Nissan Cation M_2-100. Nippon Oil and Fats Co., Ltd.
Chemical Description: Tetradecyl dimethyl benzyl ammonium chloride
Category and Applications: See Nissan Cation AB
Form: Crystalline powder; Color: White or light yellow
Composition: 85% min. active
Containers: 10 kg paper box

Nissan Cation MA. Nippon Oil and Fats Co., Ltd.
Chemical Description: Tetradecylamine acetate
Category: Corrosion inhibitor, germicide, antistatic agent, cationic
Applications: Anticorrosive and germicide in water treatment and petroleum industry; duckweed killer; pigment dispersant; antistat, softener, germicide for textiles; ore flotation; rubber mold releasing; powder coating agent
Form: Flake; Color: White or light yellow
Properties: Solidifying Pt.: 60–72 C
Containers: 20 kg paper bag

Nissan Cation PB-40. Nippon Oil and Fats Co., Ltd.
Chemical Description: Hexadecyl trimethyl ammonium chloride
Category and Applications: See Nissan Cation AB
Form: Liquid; Color: Light yellow
Composition: 40% min. active
Containers: 15 kg Atron can, 180 kg drum

Nissan Cation PB-300. Nippon Oil and Fats Co., Ltd.
Chemical Description: Hexadecyl trimethyl ammonium chloride
Category and Applications: See Nissan Cation AB
Form: Liquid; Color: Light yellow
Composition: 27% min. active
Containers: 16 kg Atron can, 170 kg drum

Nissan Cation S_2-100. Nippon Oil and Fats Co., Ltd.
Chemical Description: Octadecyl dimethyl benzyl ammonium chlo-

ride
Category and Applications: See Nissan Cation AB
Form: Crystalline powder; Color: White or light yellow
Composition: 85% min. active
Containers: 10 kg paper box

Nissan Cation SA. Nippon Oil and Fats Co., Ltd.
Chemical Description: Octadecylamine acetate
Category and Applications: See Nissan Cation MA
Form: Flake; Color: White or light yellow
Properties: Solidifying Pt.: 72–82 C
Containers: 15 kg paper box

Nissan Dispanol 16A. Nippon Oil and Fats Co., Ltd.
Chemical Description: Polyoxyethylene alkyl ether
Category: Emulsifier, nonionic
Applications: Water-disp. emulsifier for paraffin
Form: Liquid; Color: 120 max. (APHA)
Properties: Solubility: Sol. in methanol, xylene, kerosene; sol. (cloudy) in ethyl ether; disp. in water; HLB: 10.7; Cloud Pt.: 0 C max. (1% aq.)

Nissan Dispanol LS-100. Nippon Oil and Fats Co., Ltd.
Chemical Description: Polyoxyalkylene alkyl aryl ether
Category: Detergent, nonionic
Applications: Low-foam detergent for synthetic fibers
Form: Liquid; Color: 50 max. (APHA)
Properties: Solubility: Sol. in methanol, ethyl ether; sol. (cloudy) in water; disp. in xylene; Cloud Pt.: 23–29 C (1% aq.)

Nissan Dispanol N-100. Nippon Oil and Fats Co., Ltd.
Chemical Description: Polyoxyethylene alkyl aryl ether
Category: Detergent, nonionic
Applications: Water-sol. detergent for animal cellulosic, and synthetic fibers
Form: Liquid; Color: 120 max. (APHA)
Properties: Solubility: Sol. in water, methanol, ethyl ether, xylene; HLB: 12.5; Cloud Pt.: 33–41 C (1% aq.)

Nissan Dispanol TOC. Nippon Oil and Fats Co., Ltd.
Chemical Description: Polyoxyethylene alkyl ether
Category: Detergent, wetting agent, emulsifier, nonionic
Applications: Water-sol. surfactant, esp. as detergent for wool fiber
Form: Liquid; Color: 120 max. (APHA)
Properties: Solubility: Sol. in water, methanol, ethyl ether; disp. in xylene; HLB: 12.0; Cloud Pt.: 45–53 C (1% aq.)

Nissan Monogly I. Nippon Oil and Fats Co., Ltd.
Chemical Description: Glycerol monostearate, self-emulsifying

Category: Emulsifier, nonionic
Applications: Emulsifier for creams and lotions of the O/W type
Form: Flake; Color: White
Properties: Solubility: Disp. in water; M.P.: 55–63 C; HLB: 11.0; Acid Value: 5 max.
Composition: 37% min. monoglyceride content

Nissan Monogly M. Nippon Oil and Fats Co., Ltd.
Chemical Description: Glycerol monostearate
Category: Emulsifier, nonionic
Applications: Cosmetics and pharmaceuticals
Form: Flake; Color: White
Properties: M.P.: 53–61 C; HLB: 3.0; Acid Value: 3 max.
Composition: 40% min. monoglyceride content

Nissan Nonion DS-60HN. Nippon Oil and Fats Co., Ltd.
Chemical Description: PEG distearate
Category: Emulsifier, thickener, nonionic
Applications: Cosmetic applications, textile, and other industrial uses
Form: Solid
Properties: Solubility: Sol. in water, tetrachloromethan, methanol; sol. warm in diethylene glycol; M.P.: 54–62 C; HLB: 19.0; Acid Value: 2 max.
Composition: 100% conc.

Nissan Nonion E-205. Nippon Oil and Fats Co., Ltd.
Chemical Description: Polyoxyethylene oleyl ether
Category: Emulsifier, dispersant, detergent, wetting agent, nonionic
Applications: Textile processing, cosmetics, metal working compounds, agricultural preparations, and industrial cleaners
Form: Liquid; Color: 140 max. (APHA)
Properties: Solubility: Sol. in xylene, kerosene, soybean oil, ether, tetrachloromethan, methanol, diethylene glycol; sol. warm in liquid paraffin; disp. in water; HLB: 9.0; Cloud Pt.: 0 C (1% aq.)
Composition: 100% conc.

Nissan Nonion E-215. Nippon Oil and Fats Co., Ltd.
Chemical Description: Polyoxyethylene oleyl ether
Category and Applications: See Nissan Nonion E-205
Form: Semisolid; Color: 140 max. (APHA)
Properties: Solubility: Sol. in water, liquid paraffin, methanol, diethylene glycol; sol. (cloudy) in xylene, kerosene, soybean oil, ether, tetrachloromethan; HLB: 14.2; Cloud Pt.: 95 C min. (1% aq.)
Composition: 100% conc.

Nissan Nonion E-230. Nippon Oil and Fats Co., Ltd.
Chemical Description: Polyoxyethylene oleyl ether
Category and Applications: See

Nissan Nonion E-205
Form: Solid; Color: 140 max. (APHA)
Properties: Solubility: Sol. in water, methanol, diethylene glycol; sol. (cloudy) in xylene, ether, tetrachloromethan; HLB: 16.6; Cloud Pt.: 100 C min. (1% aq.); Solid. Pt.: 37–45 C
Composition: 100% conc.

Nissan Nonion HS Series. Nippon Oil and Fats Co., Ltd.
Chemical Description: Polyoxyethylene octylphenyl ether
Category: Detergent, emulsifier, nonionic
Applications: Household and industrial detergents and cleaners, textile processing, paints, printing inks, agricultural preparations, and latex polymerization
Form and Properties: See individual listings
Composition: 100% conc.

Nissan Nonion HS-204.5. Nippon Oil and Fats Co., Ltd.
Form: Liquid; Color: 120 max. (APHA)
Properties: Solubility: Sol. in xylene, kerosene, soybean oil, ether, tetrachloromethan, methanol, diethylene glycol; disp. in water; HLB: 9.8; Cloud Pt.: 0 C max. (1% aq.); Solid. Pt.: −10 C max.

Nissan Nonion HS-206. Nippon Oil and Fats Co., Ltd.
Form: Liquid; Color: 120 max. (APHA)
Properties: Solubility: Sol. in xylene, kerosene, soybean oil, ether, tetrachloromethan, methanol, diethylene glycol; disp. in water; HLB: 11.2; Solid. Pt.: 0 C max.

Nissan Nonion HS-208. Nippon Oil and Fats Co., Ltd.
Form: Liquid; Color: 120 max. (APHA)
Properties: Solubility: Sol. in xylene, ether, tetrachloromethan, methanol, diethylene glycol; disp. in water and soybean oil; HLB: 12.6; Cloud Pt.: 42–46 C (1% aq.); Solid. Pt.: 0 C max.

Nissan Nonion HS-210. Nippon Oil and Fats Co., Ltd.
Form: Liquid; Color: 120 max. (APHA)
Properties: Solubility: Sol. in water, xylene, ether, tetrachloromethan, methanol, diethylene glycol; HLB: 13.3; Cloud Pt.: 70–78 C (1% aq.); Solid. Pt.: −1–10 C

Nissan Nonion HS-215. Nippon Oil and Fats Co., Ltd.
Form: Liquid; Color: 120 max. (APHA)
Properties: Solubility: Sol. in water, xylene, ether, tetrachloromethan, methanol, diethylene glycol; HLB: 15.0; Cloud Pt.: 91–99 C (1% aq.); Solid. Pt.: 19–26 C

Nissan Nonion HS-220. Nippon Oil and Fats Co., Ltd.
Form: Semisolid; Color: 120 max. (APHA)
Properties: Solubility: Sol. in water, methanol, diethylene glycol;

sol. (cloudy) in xylene, ether, and tetrachloromethan; HLB: 16.2; Cloud Pt.: 100 C min. (1% aq.); Solid. Pt.: 24–32 C

Nissan Nonion HS-240. Nippon Oil and Fats Co., Ltd.
Form: Solid; Color: 120 max. (APHA)
Properties: Solubility: Sol. in water, methanol, diethylene glycol; HLB: 17.9; Cloud Pt.: 100 C min. (1% aq.); Solid. Pt.: 39–47 C

Nissan Nonion K-202. Nippon Oil and Fats Co., Ltd.
Chemical Description: Polyoxyethylene lauryl ether
Category and Applications: See Nissan Nonion E-205
Form: Liquid; Color: 200 max. (APHA)
Properties: Solubility: Sol. in xylene, ether, tetrachloromethan, methanol, diethylene glycol; disp. in kerosene, liquid paraffin, soybean oil; HLB: 6.0; Cloud Pt.: 0 C max. (1% aq.)
Composition: 100% conc.

Nissan Nonion K-203. Nippon Oil and Fats Co., Ltd.
Chemical Description: Polyoxyethylene lauryl ether
Category and Applications: See Nissan Nonion E-205
Form: Liquid; Color: 200 max. (APHA)
Properties: Solubility: Sol. in xylene, ether, tetrachloromethan, methanol, diethylene glycol; disp. in kerosene, liquid paraffin, soybean oil; HLB: 8.0; Cloud Pt.: 0 C max. (1% aq.)
Composition: 100% conc.

Nissan Nonion K-204. Nippon Oil and Fats Co., Ltd.
Chemical Description: Polyoxyethylene lauryl ether
Category and Applications: See Nissan Nonion E-205
Form: Semisolid; Color: 200 max. (APHA)
Properties: Solubility: Sol. in xylene, ether, tetrachloromethan, methanol, diethylene glycol; sol. warm in kerosene; disp. in water, liquid paraffin, soybean oil; HLB: 9.2; Cloud Pt.: 0 C max. (1% aq.); Solid. Pt.: 5–21 C
Composition: 100% conc.

Nissan Nonion K-207. Nippon Oil and Fats Co., Ltd.
Chemical Description: Polyoxyethylene lauryl ether
Category and Applications: See Nissan Nonion E-205
Form: Semisolid; Color: 50 max. (APHA)
Properties: Solubility: Sol. in water, xylene, methanol, diethylene glycol; sol. warm in ether; disp. in liquid paraffin, tetrachloromethan; HLB: 12.1; Cloud Pt.: 55–63 C (1% aq.); Solid. Pt.: 20–35 C
Composition: 100% conc.

Nissan Nonion K-211. Nippon Oil and Fats Co., Ltd.
Chemical Description: Polyoxyethylene lauryl ether
Category and Applications: See

Nissan Nonion E-205
Form: Semisolid; Color: 50 max. (APHA)
Properties: Solubility: Sol. in water, xylene, methanol, diethylene glycol; disp. in liquid paraffin, tetrachloromethan; HLB: 14.1; Cloud Pt.: 90–98 C (1% aq.); Solid. Pt.: 25–40 C
Composition: 100% conc.

Nissan Nonion K-215. Nippon Oil and Fats Co., Ltd.
Chemical Description: Polyoxyethylene lauryl ether
Category and Applications: See Nissan Nonion E-205
Form: Semisolid; Color: 140 max. (APHA)
Properties: Solubility: Sol. in water, xylene, methanol, diethylene glycol; disp. in liquid paraffin, tetrachloromethan; HLB: 15.2; Cloud Pt.: 100 C min. (1% aq.); Solid. Pt.: 23–33 C
Composition: 100% conc.

Nissan Nonion K-220. Nippon Oil and Fats Co., Ltd.
Chemical Description: Polyoxyethylene lauryl ether
Category and Applications: See Nissan Nonion E-205
Form: Solid; Color: 120 max. (APHA)
Properties: Solubility: Sol. in water, xylene, methanol, diethylene glycol; disp. in liquid paraffin, tetrachloromethan; HLB: 16.2; Cloud Pt.: 100 C min. (1% aq.); Solid. Pt.: 29–39 C
Composition: 100% conc.

Nissan Nonion K-230. Nippon Oil and Fats Co., Ltd.
Chemical Description: Polyoxyethylene lauryl ether
Category and Applications: See Nissan Nonion E-205
Form: Semisolid; Color: 120 max. (APHA)
Properties: Solubility: Sol. in water, xylene, methanol, diethylene glycol; disp. in liquid paraffin, tetrachloromethan; HLB: 17.3; Cloud Pt.: 100 C min. (1% aq.); Solid. Pt.: 37–45 C
Composition: 100% conc.

Nissan Nonion L-2. Nippon Oil and Fats Co., Ltd.
Chemical Description: PEG monolaurate
Category: Emulsifier, nonionic
Applications: General emulsifier for textile, leather, and other industrial uses
Form: Liquid; Color: 4 max. (Gardner)
Properties: Solubility: Sol. in methanol; sol. warm in kerosene; disp. in water; HLB: 9.5; Acid Value: 1 max.
Composition: 100% conc.

Nissan Nonion L-4. Nippon Oil and Fats Co., Ltd.
Chemical Description: PEG monolaurate
Category and Applications: See Nissan Nonion L-2
Form: Liquid; Color: 4 max. (Gardner)

Properties: Solubility: Sol. in water, ether, methanol; HLB: 13.1; Acid Value: 2 max.
Composition: 100% conc.

Nissan Nonion LP-20·R. Nippon Oil and Fats Co., Ltd.
Chemical Description: Sorbitan monolaurate
Category: Emulsifier, stabilizer, thickener, lubricant, softener, nonionic
Applications: Oil-sol. emulsifier for cosmetic, pharmaceutical, and food applications; as O/W emulsion stabilizer and thickener, esp. in combination with LT-221, ST-221, and OT-221; also in textile industry as fiber lubricant and softener
Form: Oily liquid; Color: 5 max. (Gardner)
Properties: Solubility: Sol. in methanol, ethanol, acetone, xylene, ethyl ether, kerosene; disp. in water; HLB: 8.6; Solid. Pt.: 10 C max.; Acid Value: 7 max.
Composition: 100% conc.

Nissan Nonion LT-221. Nippon Oil and Fats Co., Ltd.
Chemical Description: Polyoxyethylene sorbitan monolaurate
Category: Emulsifier, nonionic
Applications: Water-sol. emulsifier for cosmetic, pharmaceutical, and food applications; as O/W emulsifiers, esp. in conjunction with LP-20·R, PP-40·R, SP-60·R, and OP series
Form: Oily liquid; Color: 6 max. (Gardner)
Properties: Solubility: Sol. in water, methanol, ethanol, acetone, xylene, ethyl ether, tetrachloromethan, ethylene glycol; HLB: 16.7; Solid. Pt.: 0 C max.
Composition: 100% conc.

Nissan Nonion NS Series. Nippon Oil and Fats Co., Ltd.
Chemical Description: Polyoxyethylene nonylphenyl ether
Category: Detergent, emulsifier, nonionic
Applications: Household and industrial detergents and cleaners, textile processing, paints, printing inks, agricultural preparations, and latex polymerization
Form and Properties: See individual listings
Composition: 100% conc.

Nissan Nonion NS-202. Nippon Oil and Fats Co., Ltd.
Form: Liquid; Color: 120 max. (APHA)
Properties: Solubility: Sol. in xylene, kerosene, soybean oil, ether, tetrachloromethan, methanol, diethylene glycol; disp. in water; HLB: 5.7; Cloud Pt.: 0 C max. (1% aq.); Solid. Pt.: –10 C max.

Nissan Nonion NS-204.5. Nippon Oil and Fats Co., Ltd.
Form: Liquid; Color: 120 max. (APHA)
Properties: Solubility: Sol. in xylene, kerosene, soybean oil, ether, tetrachloromethan, methanol, diethylene glycol; disp. in water; HLB: 9.5; Cloud Pt.: 0 C max. (1% aq.); Solid. Pt.: –10 C max.

Nissan Nonion NS-206. Nippon Oil and Fats Co., Ltd.
Form: Liquid; Color: 120 max. (APHA)
Properties: Solubility: Sol. in xylene, ether, tetrachloromethan, methanol, diethylene glycol; disp. in water and soybean oil; HLB: 10.9; Cloud Pt.: 0 C max. (1% aq.); Solid. Pt.: –10 C max.

Nissan Nonion NS-208.5. Nippon Oil and Fats Co., Ltd.
Form: Liquid; Color: 120 max. (APHA)
Properties: Solubility: Sol. in xylene, ether, tetrachloromethan, methanol, diethylene glycol; disp. in water; HLB: 12.6; Cloud Pt.: 33–41 C (1% aq.)

Nissan Nonion NS-210. Nippon Oil and Fats Co., Ltd.
Form: Liquid; Color: 120 max. (APHA)
Properties: Solubility: Sol. in water, xylene, ether, tetrachloromethan, methanol, diethylene glycol; HLB: 13.3; Cloud Pt.: 46–54 C (1% aq.)

Nissan Nonion NS-212. Nippon Oil and Fats Co., Ltd.
Form: Liquid; Color: 120 max. (APHA)
Properties: Solubility: Sol. in water, xylene, ether, tetrachloromethan, methanol, diethylene glycol; HLB: 14.1; Cloud Pt.: 73–84 C (1% aq.)

Nissan Nonion NS-215. Nippon Oil and Fats Co., Ltd.
Form: Liquid; Color: 120 max. (APHA)
Properties: Solubility: Sol. in water, xylene, ether, methanol, diethylene glycol; sol. (cloudy) in tetrachloromethan; HLB: 15.0; Cloud Pt.: 89–97 C (1% aq.); Solid. Pt.: 12–22 C

Nissan Nonion NS-220. Nippon Oil and Fats Co., Ltd.
Form: Semisolid; Color: 120 max. (APHA)
Properties: Solubility: Sol. in water, xylene, ether, methanol, diethylene glycol; HLB: 16.0; Cloud Pt.: 100 C min. (1% aq.); Solid. Pt.: 22–32 C

Nissan Nonion NS-230. Nippon Oil and Fats Co., Ltd.
Form: Solid; Color: 120 max. (APHA)
Properties: Solubility: Sol. in water, xylene, methanol, diethylene glycol; sol. (cloudy) in ether; HLB: 17.1; Cloud Pt.: 100 C min. (1% aq.); Solid. Pt.: 34–42 C

Nissan Nonion NS-240. Nippon Oil and Fats Co., Ltd.
Form: Solid; Color: 120 max. (APHA)
Properties: Solubility: Sol. in water, methanol, diethylene glycol; HLB: 17.8; Cloud Pt.: 100 C min. (1% aq.); Solid. Pt.: 40–46 C

Nissan Nonion NS-270. Nippon Oil and Fats Co., Ltd.
Form: Solid; Color: 120 max. (APHA)
Properties: Solubility: Sol. in water, methanol, diethylene glycol; HLB: 18.7; Cloud Pt.: 100 C min. (1% aq.); Solid. Pt.: 47–53 C

Nissan Nonion O-2. Nippon Oil and Fats Co., Ltd.
Chemical Description: PEG monooleate
Category and Applications: See Nissan Nonion L-2
Form: Liquid; Color: 10 max. (Gardner)
Properties: Solubility: Sol. in kerosene, tetrachloromethan, methanol; disp. in water, xylene, liquid paraffin, soybean oil, and ether; HLB: 7.9; Acid Value: 2 max.
Composition: 100% conc.

Nissan Nonion O-3. Nippon Oil and Fats Co., Ltd.
Chemical Description: PEG monooleate
Category and Applications: See Nissan Nonion L-2
Form: Liquid; Color: 10 max. (Gardner)
Properties: Solubility: Sol. in kerosene, tetrachloromethan, methanol; disp. in water, xylene, liquid paraffin, soybean oil, ether; HLB: 10.3; Acid Value: 2 max.
Composition: 100% conc.

Nissan Nonion O-4. Nippon Oil and Fats Co., Ltd.
Chemical Description: PEG monooleate
Category and Applications: See Nissan Nonion L-2
Form: Liquid; Color: 9 max. (Gardner)
Properties: Solubility: Sol. in xylene, kerosene, tetrachloromethan, methanol; sol. warm in ether; disp. in water and soybean oil; HLB: 11.5; Acid Value: 2 max.
Composition: 100% conc.

Nissan Nonion O-6. Nippon Oil and Fats Co., Ltd.
Chemical Description: PEG monooleate
Category and Applications: See Nissan Nonion L-2
Form: Liquid; Color: 9 max. (Gardner)
Properties: Solubility: Sol. in water, xylene, kerosene, ether, tetrachloromethan, methanol; HLB: 13.5; Acid Value: 2 max.
Composition: 100% conc.

Nissan Nonion OP-80·R. Nippon Oil and Fats Co., Ltd.
Chemical Description: Sorbitan monooleate
Category and Applications: See Nissan Nonion LP-20·R.
Form: Oily liquid; Color: 9 max. (Gardner)
Properties: Solubility: Oil-sol.; HLB: 4.3; Solid. Pt.: 0 C max.; Acid Value: 7 max.
Composition: 100% conc.

Nissan Nonion OP-83•RAT. Nippon Oil and Fats Co., Ltd.
Chemical Description: Sorbitan sesquioleate
Category and Applications: See Nissan Nonion LP-20•R
Form: Oily liquid; Color: 9 max. (Gardner)
Properties: Solubility: Sol. in ethanol, acetone, xylene, ethyl ether, tetrachloromethan, kerosene; slightly sol. in methanol; disp. warm in water; HLB: 3.7; Solid. Pt.: 0 C max.; Acid Value: 7 max.
Composition: 100% conc.

Nissan Nonion OP-85•R. Nippon Oil and Fats Co., Ltd.
Chemical Description: Sorbitan trioleate
Category and Applications: See Nissan Nonion LP-20•R
Form: Oily liquid; Color: 9 max. (Gardner)
Properties: Solubility: Oil-sol.; HLB: 1.8; Solid. Pt.: 0 C max.; Acid Value: 7 max.
Composition: 100% conc.

Nissan Nonion OT-221. Nippon Oil and Fats Co., Ltd.
Chemical Description: Polyoxyethylene sorbitan monooleate
Category and Applications: See Nissan Nonion LT-221
Form: Oily liquid; Color: 6 max. (Gardner)
Properties: Solubility: Sol. in water, ethanol, acetone, xylene, ethyl ether, tetrachloromethan; disp. in methanol; HLB: 15.0; Solid. Pt.: 0 C max.
Composition: 100% conc.

Nissan Nonion P-208. Nippon Oil and Fats Co., Ltd.
Chemical Description: Polyoxyethylene cetyl ether
Category and Applications: See Nissan Nonion E-205
Form: Liquid; Color: 120 max. (APHA)
Properties: Solubility: Sol. in water, xylene, methanol, diethylene glycol; HLB: 11.9; Cloud Pt.: 43–53 C (1% aq.); Solid. Pt.: 19–29 C
Composition: 100% conc.

Nissan Nonion P-210. Nippon Oil and Fats Co., Ltd.
Chemical Description: Polyoxyethylene cetyl ether
Category and Applications: See Nissan Nonion E-205
Form: Semisolid; Color: 120 max. (APHA)
Properties: Solubility: Sol. in water, xylene, methanol, diethylene glycol; HLB: 12.9; Cloud Pt.: 64–74 C (1% aq.); Solid. Pt.: 23–33 C
Composition: 100% conc.

Nissan Nonion P-213. Nippon Oil and Fats Co., Ltd.
Chemical Description: Polyoxyethylene cetyl ether
Category and Applications: See Nissan Nonion E-205
Form: Semisolid; Color: 120 max. (APHA)
Properties: Solubility: Sol. in water, kerosene, soybean oil, ether,

methanol, diethylene glycol; disp. in xylene and tetrachloromethan; HLB: 14.1; Cloud Pt.: 87–97 C (1% aq.); Solid. Pt.: 28–38 C
Composition: 100% conc.

Nissan Nonion PP-40•R. Nippon Oil and Fats Co., Ltd.
Chemical Description: Sorbitan monopalmitate
Category and Applications: See Nissan Nonion LP-20•R
Form: Waxy solid; Color: 7 max. (Gardner)
Properties: Solubility: Oil-sol.; HLB: 6.7; Solid. Pt.: 45–51 C; Acid Value: 7 max.
Composition: 100% conc.

Nissan Nonion PT-221. Nippon Oil and Fats Co., Ltd.
Chemical Description: Polyoxyethylene sorbitan monopalmitate
Category and Applications: See Nissan Nonion LT-221
Form: Oily liquid; Color: 8 max. (Gardner)
Properties: Solubility: Sol. in water, methanol, ethanol, acetone, xylene, ethyl ether, tetrachloromethan, ethylene glycol; HLB: 15.3; Solid. Pt.: 17–27 C
Composition: 100% conc.

Nissan Nonion S-2. Nippon Oil and Fats Co., Ltd.
Chemical Description: PEG monostearate
Category and Applications: See Nissan Nonion DS-60HN
Form: Semisolid; Color: 4 max. (Gardner)
Properties: Solubility: Sol. in methanol, tetrachloromethan; sol. warm in xylene; sol. (cloudy) in liquid paraffin, soybean oil; disp. in water, kerosene, and ether; M.P.: 33–41 C; HLB: 8.0; Acid Value: 1 max.
Composition: 100% conc.

Nissan Nonion S-4. Nippon Oil and Fats Co., Ltd.
Chemical Description: PEG monostearate
Category and Applications: See Nissan Nonion DS-60HN
Form: Semisolid; Color: 4 max. (Gardner)
Properties: Solubility: Sol. in kerosene, ether, tetrachloromethan, methanol; sol. warm in xylene; disp. in water; M.P.: 30–40 C; HLB: 11.6; Acid Value: 2 max.
Composition: 100% conc.

Nissan Nonion S-6. Nippon Oil and Fats Co., Ltd.
Chemical Description: PEG monostearate
Category and Applications: See Nissan Nonion DS-60HN
Form: Semisolid; Color: 4 max. (Gardner)
Properties: Solubility: Sol. in water, soybean oil, tetrachloromethan, methanol; sol. warm in xylene, ether; sol. (cloudy) in kerosene; M.P.: 35–41 C; HLB: 13.6; Acid Value: 2 max.
Composition: 100% conc.

Nissan Nonion S-10. Nippon Oil and Fats Co., Ltd.
Chemical Description: PEG mono-

stearate
Category and Applications: See Nissan Nonion DS-60HN
Form: Semisolid; Color: 4 max. (Gardner)
Properties: Solubility: Sol. in water, xylene, soybean oil, ether, tetrachloromethan, methanol, diethylene glycol; M.P.: 38–44 C; HLB: 15.2; Acid Value: 2 max.
Composition: 100% conc.

Nissan Nonion S-15. Nippon Oil and Fats Co., Ltd.
Chemical Description: PEG monostearate
Category and Applications: See Nissan Nonion DS-60HN
Form: Semisolid; Color: 4 max. (Gardner)
Properties: Solubility: Sol. in kerosene, ether, tetrachloromethan, methanol; sol. warm in xylene; disp. in water; M.P.: 39–45 C; HLB: 12.8; Acid Value: 2 max.
Composition: 100% conc.

Nissan Nonion S-15.4. Nippon Oil and Fats Co., Ltd.
Chemical Description: PEG monostearate
Category and Applications: See Nissan Nonion DS-60HN
Form: Semisolid; Color: 4 max. (Gardner)
Properties: Solubility: Sol. in water, xylene, ether, tetrachloromethan, methanol, diethylene glycol; M.P.: 42–48 C; HLB: 16.7; Acid Value: 2 max.
Composition: 100% conc.

Nissan Nonion S-40. Nippon Oil and Fats Co., Ltd.
Chemical Description: PEG monostearate
Category and Applications: See Nissan Nonion DS-60HN
Form: Semisolid; Color: 6 max. (Gardner)
Properties: Solubility: Sol. in water, xylene, ether, tetrachloromethan, methanol, diethylene glycol; M.P.: 49–55 C; HLB: 18.2; Acid Value: 2 max.
Composition: 100% conc.

Nissan Nonion S-207. Nippon Oil and Fats Co., Ltd.
Chemical Description: Polyoxyethylene stearyl ether
Category and Applications: See Nissan Nonion E-205
Form: Semisolid; Color: 120 max. (APHA)
Properties: Solubility: Sol. in xylene, kerosene, liquid paraffin, soybean oil, tetrachloromethan, methanol, diethylene glycol; sol. (cloudy) in ether; disp. in water; HLB: 10.7; Cloud Pt.: 0 C max. (1% aq.); Solid. Pt.: 30–36 C
Composition: 100% conc.

Nissan Nonion S-215. Nippon Oil and Fats Co., Ltd.
Chemical Description: Polyoxyethylene stearyl ether
Category and Applications: See Nissan Nonion E-205
Form: Semisolid; Color: 120 max. (APHA)
Properties: Solubility: Sol. in water, xylene, kerosene, soybean oil,

ether, tetrachloromethan, methanol, diethylene glycol; HLB: 14.2; Cloud Pt.: 100 C min. (1% aq.); Solid. Pt.: 35–41 C
Composition: 100% conc.

Nissan Nonion S-220. Nippon Oil and Fats Co., Ltd.
Chemical Description: Polyoxyethylene stearyl ether
Category and Applications: See Nissan Nonion E-205
Form: Semisolid; Color: 120 max. (APHA)
Properties: Solubility: Sol. in water, xylene, kerosene, soybean oil, ether, tetrachloromethan, methanol, diethylene glycol; HLB: 15.3; Cloud Pt.: 100 C min. (1% aq.); Solid Pt.: 38–44 C
Composition: 100% conc.

Nissan Nonion SP-60•R. Nippon Oil and Fats Co., Ltd.
Chemical Description: Sorbitan monostearate
Category and Applications: See Nissan Nonion LP-20•R
Form: Waxy solid; Color: 5 max. (Gardner)
Properties: Solubility: Sol. warm in methanol, ethanol, xylene, tetrachloromethan, kerosene; slightly sol. in ethyl ether; disp. warm in water; HLB: 4.7; Solid. Pt.: 49–55 C; Acid Value: 7 max.
Composition: 100% conc.

Nissan Nonion ST-221. Nippon Oil and Fats Co., Ltd.
Chemical Description: Polyoxyethylene sorbitan monostearate
Category and Applications: See Nissan Nonion LT-221
Form: Oily liquid; Color: 5 max. (Gardner)
Properties: Solubility: Sol. in water, methanol, ethanol, acetone, xylene, ethyl ether, tetrachloromethan, kerosene; HLB: 14.9; Solid. Pt.: 20–30 C
Composition: 100% conc.

Nissan Nymeen DT-203. Nippon Oil and Fats Co., Ltd.
Chemical Description: Polyoxyethylene tallow-alkylpropylene diamine
Category: Dispersant, wetting agent, emulsifier, softener, antistat, nonionic
Applications: Pigment wetting and dispersing; detergent, emulsifier, and acid cleaner additive for metals; textile antistatic agent and auxiliary
Form: Liquid; Color: Light brown
Properties: Cloud Pt.: 0 C max.; Amine Value: 237 ± 10
Containers: 16 kg can, 180 kg drum

Nissan Nymeen DT-208. Nippon Oil and Fats Co., Ltd.
Chemical Description: Polyoxyethylene tallow-alkylpropylene diamine
Category and Applications: See Nissan Nymeen DT-203
Form: Liquid; Color: Light brown
Properties: Cloud Pt.: 100 C min.; Amine Value: 163 ± 17
Containers: 18 kg can

Nissan Nymeen L-201. Nippon Oil and Fats Co., Ltd.
Chemical Description: Oxyethylene dodecylamine
Category and Applications: See Nissan Nymeen DT-203
Form: Liquid; Color: Colorless or light yellow
Properties: Cloud Pt.: 0 C max.; Amine Value: 239 ± 7
Containers: 15 kg can, 160 kg can

Nissan Nymeen L-202. Nippon Oil and Fats Co., Ltd.
Chemical Description: Polyoxyethylene dodecylamine
Category and Applications: See Nissan Nymeen DT-203
Form: Liquid; Color: Light yellow
Properties: Cloud Pt.: 0 C max.; Amine Value: 202 ± 10
Containers: 15 kg can, 170 kg drum

Nissan Nymeen L-207. Nippon Oil and Fats Co., Ltd.
Chemical Description: Polyoxyethylene dodecylamine
Category and Applications: See Nissan Nymeen DT-203
Form: Liquid; Color: Light brown
Properties: Cloud Pt.: 80 C min.; Amine Value: 113 ± 6
Containers: 18 kg can

Nissan Nymeen S-202. Nippon Oil and Fats Co., Ltd.
Chemical Description: Polyoxyethylene octadecylamine
Category and Applications: See Nissan Nymeen DT-203
Form: Solid; Color: Yellowish-brown
Properties: Cloud Pt.: 0 C max. Amine Value: 159 ± 7
Containers: 180 kg drum

Nissan Nymeen S-204. Nippon Oil and Fats Co., Ltd.
Chemical Description: Polyoxyethylene octadecylamine
Category and Applications: See Nissan Nymeen DT-203
Form: Solid; Color: Yellowish-brown
Properties: Cloud Pt.: 0 C max.; Amine Value: 127 ± 7
Containers: 180 kg drum

Nissan Nymeen S-210. Nippon Oil and Fats Co., Ltd.
Chemical Description: Polyoxyethylene octadecylamine
Category and Applications: See Nissan Nymeen DT-203
Form: Liquid in summer, solid in winter; Color: Light brown
Properties: Cloud Pt.: 95 C min.; Amine Value: 80 ± 5
Containers: 18 kg can

Nissan Nymeen S-215. Nippon Oil and Fats Co., Ltd.
Chemical Description: Polyoxyethylene octadecylamine
Category and Applications: See Nissan Nymeen DT-203
Form: Solid; Color: Yellowish-brown
Properties: Cloud Pt.: 100 C min.; Amine Value: 62 ± 4
Containers: 18 kg can, 200 kg drum

Nissan Nymeen S-220. Nippon Oil and Fats Co., Ltd.
Chemical Description: Polyoxyethylene octadecylamine
Category and Applications: See Nissan Nymeen DT-203
Form: Solid; Color: Yellowish-brown
Properties: Cloud Pt.: 100 C min.; Amine Value: 51 ± 4
Containers: 18 kg can

Nissan Nymeen T_2-206. Nippon Oil and Fats Co., Ltd.
Chemical Description: Polyoxyethylene tallow-alkylamine
Category and Applications: See Nissan Nymeen DT-203
Form: Liquid; Color: Light brown
Properties: Cloud Pt.: 95 C min.; Amine Value: 106 ± 5
Containers: 17 kg can

Nissan Nymeen T_2-210. Nippon Oil and Fats Co., Ltd.
Chemical Description: Polyoxyethylene tallow-alkylamine
Category and Applications: See Nissan Nymeen DT-203
Form: Liquid; Color: Light brown
Properties: Cloud Pt.: 95 C min.; Amine Value: 80 ± 7
Containers: 18 kg can, 200 kg drum

Nissan Nymeen T_2-230. Nippon Oil and Fats Co., Ltd.
Chemical Description: Polyoxyethylene tallow-alkylamine
Category and Applications: See Nissan Nymeen DT-203
Form: Solid; Color: Light brown
Properties: Cloud Pt.: 100 C min.; Amine Value: 36 ± 4
Containers: 18 kg can, 200 kg drum

Nissan Nymeen T_2-260. Nippon Oil and Fats Co., Ltd.
Chemical Description: Polyoxyethylene tallow-alkylamine
Category and Applications: See Nissan Nymeen DT-203
Form: Solid; Color: Light brown
Properties: Cloud Pt.: 100 C min.; Amine Value: 19 ± 4
Containers: 18 kg can

Nissan Plonon 102. Nippon Oil and Fats Co., Ltd.
Chemical Description: Polyoxyethylene-polyoxypropylene ether
Category: Emulsifier, solubilizer, dispersant, detergent, antifoaming agent, wetting agent, nonionic
Applications: Detergent for metal surfaces; emulsion polymerization; antifoaming agent in fermentation industry, for CO_2 absorbing liquid, for acrylonitrile manufacture, aq. sol'n. of polymer, etc.; in paper coating color process
Form: Liquid; Color, Odor, and Taste: None
Properties: m.w.: 1250 (avg.); Solubility @ 20 C: Sol. in water, ethyl alcohol, aceton, benzol, dioxane; Visc.: 105 cs (37.8 C); HLB: 7; Cloud Pt.: 22 C (10% aq.); Stability: Stable to acid, alkali, peroxide, and metallic ion; Surface Tension: 41.1 dynes/cm^2 (0.1% conc.)
Composition: 100% conc.

Nissan Plonon 104. Nippon Oil and Fats Co., Ltd.
Chemical Description: Polyoxyethylene-polyoxypropylene ether
Category and Applications: See Nissan Plonon 102
Form: Liquid; Color, Odor, and Taste: None
Properties: m.w.: 1650 (avg.); Solubility @ 20 C: Sol. in water, ethyl alcohol, aceton, benzol, dioxane; Visc.: 150 cs (37.8 C); HLB: 15; Cloud Pt.: 64 C (10% aq.); Solid. Pt.: 18 C; Stability: Stable to acid, alkali, peroxide, and metallic ion; Surface Tension: 43.5 dynes/cm^2 (0.1% conc.)
Composition: 100% conc.

Nissan Plonon 108. Nippon Oil and Fats Co., Ltd.
Chemical Description: Polyoxyethylene-polyoxypropylene ether
Category and Applications: See Nissan Plonon 102
Properties: m.w.: 4000 (avg.); HLB: 30.5; Cloud Pt.: 100 C min. (10% aq.); Solid. Pt.: 45 C; Stability: Stable to acid, alkali, peroxide, and metallic ion

Nissan Plonon 171. Nippon Oil and Fats Co., Ltd.
Chemical Description: Polyoxyethylene-polyoxypropylene ether
Category and Applications: See Nissan Plonon 102
Properties: m.w.: 2000 (avg.); Solubility: Sol. in ethyl alcohol, aceton, benzol, dioxane; Visc.: 160 cs (37.8 C); HLB: 3; Cloud Pt.: 24 C (1% aq.); Stability: Stable to acid, alkali, peroxide, and metallic ion; Surface Tension: 39.5 dynes/cm^2 (0.1% conc.)

Nissan Plonon 172. Nippon Oil and Fats Co., Ltd.
Chemical Description: Polyoxyethylene-polyoxypropylene ether
Category and Applications: See Nissan Plonon 102
Properties: m.w.: 2400 (avg.); Solubility @ 20 C: Sol. in ethyl alcohol, aceton, benzol, dioxane; Visc.: 210 cs (37.8 C); HLB: 7; Cloud Pt.: 28 C (1% aq.); Stability: Stable to acid, alkali, peroxide, and metallic ion

Nissan Plonon 201. Nippon Oil and Fats Co., Ltd.
Chemical Description: Polyoxyethylene-polyoxypropylene ether
Category and Applications: See Nissan Plonon 102
Form: Liquid; Color, Odor, and Taste: None
Properties: m.w.: 2200 (avg.); Solubility @ 20 C: Sol. in ethyl alcohol, aceton, benzol, dioxane; Visc.: 190 cs (37.8 C); HLB: 3; Cloud Pt.: 21 C (1% aq.); Stability: Stable to acid, alkali, peroxide, and metallic ions; Surface Tension: 38.0 dynes/cm^2 (0.1% conc.)
Composition: 100% conc.

Nissan Plonon 204. Nippon Oil and Fats Co., Ltd.
Chemical Description: Polyoxyethylene-polyoxypropylene ether
Category and Applications: See

Nissan Plonon 102
Form: Paste; Color, Odor, and Taste: None
Properties: m.w.: 3400 (avg.); Solubility @ 20 C; Sol. in water, ethyl alcohol; Visc.: 370 cs (37.8 C); HLB: 13.5; Cloud Pt.: 64 C (10% aq.); Solid. Pt.: 21 C; Stability: Stable to acid, alkali, peroxide, and metallic ion; Surface Tension: 39.8 dynes/cm^2 (0.1% conc.)
Composition: 100% conc.

Nissan Plonon 208. Nippon Oil and Fats Co., Ltd.
Chemical Description: Polyoxyethylene-polyoxypropylene ether
Category and Applications: See Nissan Plonon 102
Form: Paste; Color, Odor, and Taste: None
Properties: m.w.: 8000 (avg.); Solubility @ 20 C: Sol. in water; HLB: 28; Cloud Pt.: 100 C min. (10% aq.); Solid. Pt.: 50 C; Stability: Stable to acid, alkali, peroxide, and metallic ion; Surface Tension: 44.1 dynes/cm^2 (0.1% conc.)
Composition: 100% conc.

Nissan Tertiary Amine AB. Nippon Oil and Fats Co., Ltd.
Chemical Description: Octadecyldimethylamine
Category: Intermediate, corrosion inhibitor, germicide, catalyst
Applications: Intermediate for various surfactants; viscosity index improver for lubricating oil; curing catalyst for epoxy resin
Form: Liquid or half-solid; Color: Light yellow
Properties: Iodine Value: 2 max.
Composition: 95% min. tertiary amine
Containers: 13 kg can, 160 kg drum

Nissan Tertiary Amine ABT. Nippon Oil and Fats Co., Ltd.
Chemical Description: Tallow-hydrogenated alkyl dimethylamine
Category and Applications: See Nissan Tertiary Amine AB
Form: Liquid or half-solid; Color: Light yellow
Properties: Iodine Value: 3 max.
Composition: 95% min. tertiary amine
Containers: 13 kg can, 160 kg drum

Nissan Tertiary Amine BB. Nippon Oil and Fats Co., Ltd.
Chemical Description: Dodecyldimethylamine
Category and Applications: See Nissan Tertiary Amine AB
Form: Liquid; Color: Light yellow
Properties: Iodine Value: 1 max.
Composition: 95% min. tertiary amine
Containers: 13 kg can, 160 kg drum

Nissan Tertiary Amine FB. Nippon Oil and Fats Co., Ltd.
Chemical Description: Coco-alkyl dimethylamine
Category and Applications: See Nissan Tertiary Amine AB
Form: Liquid; Color: Light yellow
Properties: Iodine Value: 12 max.
Composition: 95% min. tertiary

amine
Containers: 13 kg can, 160 kg drum

Nissan Tertiary Amine MB. Nippon Oil and Fats Co., Ltd.
Chemical Description: Tetradecyldimethylamine
Category and Applications: See Nissan Tertiary Amine AB
Form: Liquid or half-solid; Color: Light yellow
Properties: Iodine Value: 1 max.
Composition: 95% min. tertiary amine
Containers: 13 kg can, 160 kg drum

Nissan Tertiary Amine PB. Nippon Oil and Fats Co., Ltd.
Chemical Description: Hexadecyldimethylamine
Category and Applications: See Nissan Tertiary Amine AB
Form: Liquid or half-solid; Color: Light yellow
Properties: Iodine Value: 1 max.
Composition: 95% min. tertiary amine
Containers: 13 kg can, 160 kg drum

Nonasol 3922. Hart Chem. Ltd.
Chemical Description: Blend of sulfates, sulfonates, and amide
Category: Detergent, anionic
Applications: Wide variety of cleaning compounds, light duty liquid dishwashing detergent
Form: Liquid; Color: Yellow
Properties: Sp. gr.: 1.03
Composition: 60% active ingred.
Reaction: Neutral
Containers: Drums, T/T
Warnings and Caution: Avoid contact with skin and eyes

Nonasol N4AS. Hart Chem. Ltd.
Chemical Description: Sulfated linear alcohol ethoxylate ammonium salt
Category: Detergent, anionic
Applications: High quality dishwashing detergent; high foaming, good wetting and detergency
Form: Liquid; Color: Yellow
Properties: Sp. gr.: 1.02; Biodegradable
Composition: 58% active ingred.
Containers: Drums, T/T
Warnings and Caution: Avoid contact with skin and eyes

Nonasol N4SS. Hart Chem. Ltd.
Chemical Description: Sulfated linear alcohol ethoxylate, sodium salt
Category: Detergent, anionic
Applications: Base for high foaming liquid detergents; good wetting and detergency
Form: Liquid; Color: Yellow
Properties: Sp. gr.: 1.04; Biodegradable
Composition: 60% active ingred.
Containers: Drums, T/T
Warnings and Caution: Avoid contact with skin and eyes

Nonionic Emulsifier WGS. Werner G. Smith, Inc.
Chemical Description: Alkyl ether ester
Category: Emulsifier, nonionic

Applications: Low foaming emulsifier for oils and emulsion systems; helps solubilize lanolin and degras; making of cutting and soluble oils, metal working lubricants
Form: Liquid; Color: 10–12 (Gardner)
Properties: Solubility: Sol. in paraffinic oils; disp. in water; Density: ≈ 8.07 lb/gal; Sp. gr.: 0.965–0.980; HLB: 10; Cloud Pt.: 50–60 F; Flash Pt.: 400 F; Sapon. No.: 60–70; Acid No.: 10–15; Iodine No.: 40–45; pH: 6–7 (1% sol'n. in tap water); Biodegradable
Composition: 100% active ingred.
Containers: 55-gal drums

Nonipol T-28. Toho Chemical Industry Co., Ltd.
Chemical Description: Nonionics and mineral oil
Category: Nonionic surfactant
Applications: Spinning oil for wool
Form: Liquid
Composition: 100% active ingred.

Nonisol 100. Ciba-Geigy Corp.
Chemical Description: PEG 400 monolaurate
Category: Emulsifier, solubilizing agent, wetting agent, thickener, nonionic
Applications: Cosmetics, textiles, hand cleaners; used in spreading agents
Form: Liquid; Color: Yellow
Properties: Solubility: Water-sol.; insol. in kerosene and xylene
Composition: 100% conc.

Nonisol 210. Ciba-Geigy Corp.
Chemical Description: PEG 400 dioleate
Category: Emulsifier, dispersant, lubricant, nonionic
Applications: Used to emulsify toxicants, to disperse sludge in fuel oil; as textile lubricant
Form: Liquid; Color: Light amber
Properties: Solubility: Sol. in kerosene
Composition: 100% conc.

Nonisol 300. Ciba-Geigy Corp.
Chemical Description: PEG 400 monostearate
Category: Emulsifier, wetting agent, stabilizer, nonionic
Applications: Base for ointments, creams, suppositories, hair dressings, and liquid makeup; starch stabilizer
Form: Paste; Color: White
Properties: Solubility: Sol. in xylene; disp. in water; M.P.: 30 C
Composition: 100% conc.

Nopalcol Series. Diamond Shamrock/Process Chem. Div.
Chemical Description: Polyethylene glycol fatty esters
Category: Emulsifier, plasticizer, lubricant, wetting agent, dispersant, binding and thickening agent, nonionic
Applications: Cosmetics, distillation defoamers, dry cleaning agents, leather industry (pigment dispersant), mineral oil emulsions, paper industry (softener, dye dispersant, lubricant, leveling a-

gent, etc.), solvents, textiles (leveler, dispersant, lubricant, antistat), wall tile mastic
Form: Liquid to solid
Properties: Stability: Stable to acid and alkali at room temperature; may hydrolyze at higher temperatures or in conc. acid and alkaline sol'ns.
Composition: 100% conc.

Nopco RDY. Diamond Shamrock/ Process Chem. Div.
Chemical Description: Nonyl phenol (50 mol EO)
Category: Nonionic surfactant
Applications: Agriculture
Form: Solid
Properties: Cloud Pt.: 100 C; Pour Pt.: 46 C

Nopcogen 22-O. Diamond Shamrock/Process Chem. Div.
Chemical Description: Imidazoline derived from oleic acid
Category: Emulsifier, wetting agent, corrosion inhibitor, cationic
Applications: Emulsifier for mineral oils, diesel oil, and kerosene; textile and paper oil emulsions; preparation of acid detergents for dairy and bottling plants, acid-based porcelain and metal cleaning compounds; antimicrobial properties in sanitizing applications; asphalt bonding aid; demulsification; corrosion inhibition
Form: Viscous liquid; Color: Amber
Properties: Solubility: 1.6 g/100 g water; 37.5 g/100 g 5% acetic acid; 20.2 g/100 g 5% hydrochloric acid (forms thick, foamy sol'ns.); miscible with pine oil and mineral oil; Cloud Pt.: 0 C
Composition: 90% active ingred.
References: Prod. bulletin GEN-10 Rev.

Nopcosant K. Diamond Shamrock/ Process Chem. Div.
Chemical Description: Anionic polymer
Category: Dispersant, anionic
Applications: Low foaming dispersant for latex systems; disperses pigments and fillers, increases solids loading, lowers viscosity, and decreases sedimentation rates
Form: Clear liquid
Properties: Solubility: Miscible with water in all proportions; Density: 10.9 lb/gal; pH: 12 (5% sol'n.)
Composition: 34% active ingred.
Containers: 55 gal openhead steel drums with a rigid polyethylene insert
References: Prod. bulletin PAD-22 Rev.

Nopcosant L. Diamond Shamrock/ Process Chem. Div.
Chemical Description: Sulfated naphthalene
Category: Dispersant, anionic
Applications: Nonfoaming dispersant for latex systems; latex paints, coatings for paper, textiles, and leather
Form: Liquid; Color: Light
Properties: Solubility: Water-sol.; Density: 9.8 lb/gal; pH: 9.6 (5% sol'n.)

Composition: 25% solids
Containers: 55 gal openhead fiber drums with polyethylene liner
References: Prod. bulletin PAD-32 Rev.

Nopcosperse 44. Diamond Shamrock
Chemical Description: Polyeletrolyte
Category: Dispersant, anionic
Applications: Dispersant for paint systems with high efficiency, good gloss retention, viscosity stability, color development, low foaming; compatible with reactive pigments; needs no additional wetting or dispersing agents in the grind
Form: Clear liquid; Color: Light amber
Properties: Solubility: Readily sol. in water and glycols; Density: 10.2 lb/gal; pH: 7.8
Composition: 35% active ingred.

Norlig. American Can Co.
Chemical Description: Calcium and sodium lignosulfonate
Category: Binder, dispersant, anionic
Applications: Manufacture of clay products and refractory brick; binder for briquets and animal feedstuff pellets; manufacture of linoleum mastics; soil stabilizer and dust control agent in treatment of unsurfaced roads
Form: Liquid or powder
Composition: 50–60% conc. (liquids); 100% conc. (powders)
Containers: Tank trucks, tank cars, or 55 gal drums (liquids); 50 lb net multiwall paper bags (powders)
References: Bulletin No. 131

Novol. Croda Chem. Ltd.
Chemical Description: Super refined oleyl alcohol
Category: Emollient, stabilizer, superfatting agent, suspending aid
Applications: Cosmetics products, e.g., lipsticks (solvent properties, improves glide and slip), all-purpose emollient and cleansing creams and lotions, hair grooming and conditioning creams and lotions, sunscreen preparations, antiperspirant and deodorant creams and lotions, bath oil preparations
Form: Liquid; Color: Low; Odor: Mild
Properties: Solubility: Miscible with fat, oil, and wax mixtures; Stability: Good, resistant to oxidation on aging

Nylomine Assistant DN. I.C.I. United States
Chemical Description: Blend of surfactants
Category: Dyeing assistant, anionic
Applications: Leveling agent for dyeing polyamide fibers with acid dyes; eliminates barré, provides scouring, leveling, and lubricity
Form: Liquid; Color: Yellow
Properties: Solubility: Sol. in water; Sp. gr.: ≈ 1.0; F.P.: ≈ 0 C; B.P.: ≈ 100 C; Stability: Good; compatible with mineral acids, alkalis, salts; pH: 6.0–7.5 (1% sol'n.)

Composition: ≈ 55% active
Reaction: Alkaline
Toxicity: May cause eye burns and skin irritation on prolonged contact with conc. form; mists can cause respiratory and eye irritation
References: Tech. bulletin 278R

O

OHlan. Amerchol, a Unit of CPC International
Chemical Description: Hydroxylated lanolin
Category: Emulsifier, wetting agent, dispersant, emollient, nonionic
Applications: Primary W/O emulsifier; auxiliary emulsifier and stabilizer for O/W systems; pigment wetting and dispersing for makeup products; adds tack and gloss, and raises the melting point of wax-based cosmetic formulations; superfatting emollient and conditioner in soaps, shaving creams; dermatological vehicles; hair grooms; effective at low concs. and produces stable emulsions
Form: Waxy solid; Color: Yellow-amber to light tan; Odor: Mild
Properties: Solubility: Miscible with common oil phase ingred.; sol. at low levels in castor oil; M.P.: 39–46 C; HLB: 4.0; Sapon. Value: 95–110; Acid Value: 10 max.; Hydroxyl Value: 60–80 (Stetzler); Iodine Value: 15–23
Composition: 100% conc.

Onamine 12. Onyx Chem. Corp.
Chemical Description: n-dodecyl dimethyl amine
Category: Intermediate, cationic
Applications: Intermediate used in the synthesis of quaternaries, amine oxides, oil and grease additives
Form: Liquid
Properties: Sp. gr.: 0.78 (25/20 C); Flash Pt.: > 200 F
Composition: 100% conc.

Onamine 14. Onyx Chem. Corp.
Chemical Description: n-tetradecyl dimethyl amine
Category and Applications: See Onamine 12
Form: Liquid
Properties: Sp. gr.: 0.80 (25/20 C); Flash Pt.: > 200 F
Composition: 100% conc.

Onamine 16. Onyx Chem. Corp.
Chemical Description: n-hexadecyl dimethyl amine
Category and Applications: See Onamine 12
Form: Liquid

Properties: Sp. gr.: 0.80 (25/20 C); Flash Pt.: > 200 F
Composition: 100% conc.

Onamine 18. Onyx Chem. Corp.
Chemical Description: n-stearyl dimethyl amine
Category and Applications: See Onamine 12
Form: Liquid
Properties: Sp. gr.: 0.80 (25/20 C); Flash Pt.: > 200 F
Composition: 100% conc.

Onamine 65. Onyx Chem. Corp.
Chemical Description: n-alkyl dimethyl amine
Category and Applications: See Onamine 12
Form: Liquid
Properties: Sp. gr.: 0.79 (25/20 C); Flash Pt.: > 200 F
Composition: 100% conc.

Onamine 835. Onyx Chem. Corp.
Chemical Description: n-alkyl dimethyl amine
Category and Applications: See Onamine 12
Form: Liquid
Properties: Sp. gr.: 0.79 (25/20 C); Flash Pt.: > 200 F
Composition: 100% conc.

Onamine 1214. Onyx Chem. Corp.
Chemical Description: n-alkyl dimethyl amine
Category and Applications: See Onamine 12
Form: Liquid
Properties: Sp. gr.: 0.79 (25/20 C); Flash Pt.: > 200 F
Composition: 100% conc.

Onamine 1416. Onyx Chem. Corp.
Chemical Description: n-alkyl dimethyl amine
Category and Applications: See Onamine 12
Form: Liquid
Properties: Sp. gr.: 0.79 (25/20 C); Flash Pt.: > 200 F
Composition: 100% conc.

Onyxide 75. Onyx Chem. Corp.
Chemical Description: n-alkenyl dimethyl ethyl ammonium bromides
Category: Algaecide, cationic
Applications: Used in recirculating water systems, swimming pools, and humidifiers
Form: Paste
Properties: Sp. gr.: 0.95; Flash Pt.: 68 F
Composition: 75% min. active

Onyxide 172. Onyx Chem. Corp.
Chemical Description: n-alkyl ammonium cyclohexylsulfamate
Category: Antifungal agent, preservative, cationic
Applications: An organic, nonvolatile, antifungal agent and preservative for latex emulsions, paints, adhesives, coated fabrics, and cutting oils
Form: Liquid
Properties: Sp. gr.: 1.02; Flash Pt.: > 200 F
Composition: 77–81% active

Onyxide 200. Onyx Chem. Corp.
Chemical Description: Hexahydro-1.3.5-tris(2-hydroxethyl)-s-triazine
Category: Preservative, nonionic

Applications: Preservative for soluble cutting fluids and coolants
Form: Liquid
Properties: Sp. gr.: 1.15; Flash Pt.: > 200 F
Composition: 78.5% min. active

Onyxide 3300. Onyx Chem. Corp.
Chemical Description: n-alkyl dimethyl benzyl ammonium saccharinate
Category: Germicide, cationic
Applications: Toiletry and pharmaceutical formulations
Form: Powder
Properties: Sp. gr.: 0.38; Flash Pt.: > 200 F
Composition: 95% min. active

Onyxol 42. Onyx Chem. Corp.
Chemical Description: Stearic fatty acid
Category: Softener, dye carrier, opacifying and thickening agent
Applications: Used in the textile industry as a softener and dye carrier; in cosmetic preparations as a pearlizing, opacifying, and thickening agent
Form: Wax
Properties: Sp. gr.: 0.97; Flash Pt.: > 200 F; Acid No.: 18–24; Alkali No.: 30–70

Onyxol 336. Onyx Chem. Corp.
Chemical Description: Lauric fatty acid diethanolamide
Category: Foaming, wetting, and dispersing agent, nonionic
Applications: Personal care products; exhibits a stable dense foam with excellent wetting and dispersing properties
Form: Liquid
Properties: Sp. gr.: 1.01 (25/20 C); Flash Pt.: > 200 F; Acid No.: 18–24; Alkali No.: 130–165
Composition: 97% conc.

Onyxol 345. Onyx Chem. Corp.
Chemical Description: Coconut fatty acid diethanolamide
Category: Foam stabilizer, wetting agent, dispersant, thickening agent, nonionic
Applications: Cosmetic formulations; low eye irritation
Form: Liquid
Properties: Sp. gr.: 1.00 (25/20 C); Flash Pt.: > 200 F; Acid No.: 8–13; Alkali No.: 130–165
Composition: 97% conc.

Onyxol WW. Onyx Chem. Corp.
Chemical Description: Mixed fatty acid diethanolamide
Applications: Used in heavy and light duty detergents; allows compounder flexibility to tailor its performance by neutralizing the excess diethanolamine
Form: Liquid
Properties: Sp. gr.: 0.99 (25/20 C); Flash Pt.: > 200 F; Acid No.: 18–24; Alkali No.: 130–165

Orzan A. Crown Zellerbach
Chemical Description: Ammonium lignin sulfonate
Category: Dispersant, anionic
Applications: Dispersing-reducing slurry viscosity for fine mineral particles; used for skins, leather, metallurgy, animal feed, tanning,

electroplating, abatement of dust, insecticides, fungicides, and herbicides, dispersing waxes, pigments, and resins, dispersing dry powders
Form: Powder; Color: Brown
Properties: Solubility: 55% in water; Sp. gr.: 1.21 (50% aq.); Visc.: 300 cps (50% aq.)
Composition: 95% conc.
Reaction: Slight acid
Toxicity: Mild eye irritant
Containers: Bags, drums, or bulk

Orzan S. Crown Zellerbach
Chemical Description: Sodium lignin sulfonate
Category: Dispersant, anionic
Applications: Dispersing or reducing slurry viscosity for fine mineral particles; used for dyeing, metallurgy, cement, insecticides, herbicides, and fungicides, dispersing waxes, pigments, resins, dry powders, making of bitumen emulsions
Form: Powder; Color: Brown
Properties: Solubility: 60% in water; Sp. gr.: 1.264 (50% aq.); Visc.: 250 cps (50% aq.); Stability: Stable
Composition: 95% conc.
Reaction: Neutral
Toxicity: Slight eye and skin irritant
Containers: Bags, drums, and bulk

Ospin Salt ON. Tokaiseiyu Ind. Co., Ltd.
Chemical Description: Quaternary ammonium salt
Category: Migrator, cationic
Applications: Migrating agent for dyeing polyacrylnitrile fibers with cationic dyes
Form: Liquid
Composition: 50% conc.

Ospin TAN. Tokaiseiyu Ind. Co., Ltd.
Chemical Description: Quaternary ammonium salt
Category: Retarder, cationic
Applications: Retarder for polyacrylnitrile fibers dyeing with cationic dyes
Form: Liquid
Composition: 50% conc.

ENCYCLOPEDIA OF SURFACTANTS

VOLUME III

Compiled by

Michael and Irene Ash

Chemical Publishing Co., Inc.
New York, N.Y.

Chemical Publishing Co., Inc.

ISBN 0-8206-0289-2

Printed in the United States of America

PREFACE

This comprehensive, modern encyclopedia of surfactant products has been compiled with the cooperation of the major national and international surfactant manufacturers in order to give the user quick and thorough information on each tradename product that is included. Information from thousands of brochures, technical bulletins, and data sheets has been gleaned in order to make a complete profile on each surfactant product available from one source–this three volume compendium. Though sometimes limited by the information provided by the manufacturers, in most instances, the chemical ingredients, properties, form, applications, etc. of each product are discussed in detail.

The use of surfactants pervades almost every chemical industrial concern–including textiles, paints, pesticides, foods, cosmetics, pharmaceuticals, etc. We hope that because of the universality of surfactants, this reference work will be able to serve as a useful tool to the many chemical engineers and salespeople here and abroad.

We want to extend our thanks to Roberta Dakan for her dedication and tireless efforts in helping make this encyclopedia as accurate and consistent as possible.

When the temperature at which properties such as viscosity, density, surface tension, solubilities, etc. is not included, a standard temperature of 25 C is to be assumed. The information in this publication is reliable to the best of our knowledge. We would be glad to be informed of any errors or omissions so that these can be integrated into subsequent editions of this encyclopedia.

M. and I. Ash

CONTRIBUTORS

Air Products and Chemicals Inc.
Allentown, Pa. 18105

Alcolac Inc.
3440 Fairfield Rd.
Baltimore, Md. 21226

Amerchol Corp.
(Unit of CPC International Inc.)
P.O. Box 351
Talmadge Road
Edison, N.J. 08902

American Color & Chem. Corp.
11400 Westinghouse Blvd.
Charlotte, N.C. 28232

American Cyanamid Co.
Fine Chemicals Dept.
Pearl River, N.Y. 10965

American Cyanamid Co.
Industrial Chemicals Div.
Berdan Ave.
Wayne, N.J. 07470

American Cyanamid Co.
Organic Chemicals Div.
Bound Brook, N.J. 08805

Armak –Industrial Chemicals
P.O. Box 1805
Chicago, Ill. 60690

Armstrong Chem. Co., Inc.
1530 South Jackson St.
Janesville, Wis. 53545

Ashland Chemical Co.
5200 Paul G. Blazer Memorial Pkwy.
Dublin, Ohio 43017

Atlas Chem.
See ICI United States Inc.

AZS Chemical Co.
Division of AZS Corp.
762 Marietta Blvd. N.W.
Atlanta, Ga. 30318

BASF Wyandotte Corp.
Industrial Chemicals Group
1609 Biddle Ave.
Wyandotte, Mich. 48192

BASF Wayndotte Corp.
Textile Colors and Chem. Dept.
100 Cherry Hill Road
Parsippany, N.J. 07054

BASF AG
Carl-Bosch-Str. 36
6700 Ludwigshafen
Germany

Berol Kemi AB
S-444 01 Stenungsund 1
Sweden

Canada Packers Ltd.
Chemical Div.
2200 St. Clair Ave. W.
Toronto, Ontario MGN-1K4,
Canada

Carson Chemicals, Inc.
2779 East El Presidio
Long Beach, Calif. 90810

Ciba-Geigy Corp.
Dyestuff & Chemicals Div.
P.O. Box 11422
Greensboro, N.C. 27409

Cincinnati Milacron Chem. Inc.
West St.
Cincinnati, Ohio 45215

Cities Service Co.
ICD, Cities Service Bldg.
3445 Peachtree Rd. N.E.
Atlanta, Ga. 30326

Colloids, Inc.
394 Frelinghuysen Ave.
Newark, N.Y. 07114

Conoco Chemicals
Continental Oil Co.
5 Greenway Plaza East
P.O. Box 2197
Houston, TX 77001

Continental Oil Co.
See Conoco

Croda Chemicals Ltd.
Cowick Hall
Snaith Goole
North Humberside DN14 9AA
England

Croda Inc.
51 Madison Ave.
New York, N.Y. 10010

Dexter Chemical Corp.
845 Edgewater Rd.
Bronx, N.Y. 10474

Diamond Shamrock Corp.
Process Chemicals Div.
350 Mt. Kemble Ave.
Morristown, N.J. 07960

Drew Productos Quimicos Ltds.
Rua Sampalo Viana, 425
Sao Paulo, Brazil

Du Pont de Nemours, E.I. & Co.
Specialty Chemicals & Products
Nemours Bldg.
Wilmington, DE 19898

Durkee Industrial Foods Group/
SCM Corp.
900 Union Commerce Bldg.
Cleveland, Ohio 44115

Dutton & Reinisch Ltd.
Crown House, London Road
Morden, Surrey
SM45DU, England

Dynamit Nobel AG
2 Haberstrasse,
D-5210 Troisdorf-Oberlar
West Germany

Elektrochemische Fabrik
Kempen GmbH
P.O. Box 21
4152 Kempen, West Germany

Emery Industries, Inc.
Malmstrom Div.
1501 W. Elizabeth Ave.
Linden, N.J. 07036

Formosa Chem. & Fibre Corp.
1 Nanking East Road
Sec. 2, Taipei
Taiwan, China

GAF Corp.
Chemical Products
140 W. 51st St.
New York, N.Y. 10020

General Electric Co.
Silicone Prod. Dept.
Waterford, N.Y. 12188

Glyco Chemicals, Inc.
51 Weaver St.
Greenwich, Conn. 06830

Graden Chemical Co., Inc.
426 Bryan St.
Havertown, Pa. 19083

Grindsted Products
Grindstedvaerket A/S
38 Edwin Rahrs Vej
DK-8220 Brabrand
Denmark

Hamblet & Hayes Co.
P.O. Box 730
Colonial Road
Salem, Mass. 01970

Hart Chemical Ltd.
Guelph, Canada N1H 6K8

Hart Products Corp.
173 Sussex St.
Jersey City, N.J. 07302

Henkel Inc.
Chemical Specialties Div.
1301 Jefferson St.
Hoboken, N.J. 07030

Henkel Argentina S.A.
Avda. E. Madero Piso 14
1106 Capital Federal
Argentina

Hilton-Davis Chemical Co.
2235 Langdon Farm Rd.
Cincinnati, Ohio 45237

ICI United States Inc.
New Murphy Rd. & Concord Pike
Wilmington, Del. 19897

ICI Australia Petrochemicals Ltd.
ICI House
1 Nicholson St.
Melbourne 3001 Australia

Intex Products, Inc.
P.O. Box 6648
Greenville, S.C. 29606

ITT Rayonier Inc.
605 3rd Ave.
New York, N.Y. 10016

Jefferson Chemical Co., Inc.
P.O. Box 4128
Austin, Texas 78765

Lankro Chemicals Ltd.
Eccles
Manchester M30 OBH
England

Lilachem
37-33 Rue de la Loi
1040 Brussels, Belgium

Marathon Morco Co.
4401 Park Ave.
Dickinson, Texas 77539

Mars Chemical Corp.
762 Marietta Blvd. N.W.
Atlanta, Ga. 30318

Mazer Chemicals Inc.
3938 Porett Dr.
Gurnee, Ill. 60031

Monsanto Co.
800 N. Lindbergh Blvd.
St. Louis, Mo. 63166

Morton Chemical
Div. of Morton-Norwich Prod., Inc.
1275 Lake Ave.
Woodstock, IL 60098

Nikko Chemicals Co., Ltd.
1-4-8 Nihonbashi-Bakurocho
Chuo-ku, Tokyo, Japan

Nippon Nyukazai Co., Ltd.
19-9, 3-chome, Ginza,
Chuo-ku, Tokyo 104, Japan

Nippon Oils & Fats Co., Ltd.
No. 5-1 chome, Yurakucho,
Chiyoda-Ku, Tokyo, Japan

NL Industries
Industrial Chemicals Div.
P.O. Box 700
Hightstown, N.J. 08520

NL Industries
NL Baroid
P.O. Box 1675
2404 Southwest Freeway
Houston, TX 77001

Oleofina SA
33 Rue de la Loi
1040 Brussels, Belgium

Aarhus Oliefabrik A/S
DK-1800 Aarhus C. Danmark

Onyx Chemical Co.
Millmaster Onyx Group
190 Warren St.
Jersey City, N.J. 07302

Patco Cosmetic Products
C.J. Paterson Co.
3947 Broadway
Kansas City, MO 64111

Penreco Div. of Pennzoil Co.
106 South Main St.
Butler, Pa. 16001

Pfau's Sons Co., Inc.
P.O. Box 7
Jeffersonville, Ind. 47130

Pro-Chem, Inc.
16536 Broadway
Cleveland, Ohio 44137

Proctor Chem. Co.
P.O. Box 399
Salisbury, N.C. 28144

PVO International Inc.
416 Division St.
Boonton, N.J. 07005

Quad Chemical Corp.
2779 East El Presidio
Long Beach, CA 90810

Reed Ltd.
Chemical Div.—Lignin Products
P.O. Box 2025
Quebec, P.Q., Canada

R.I.T.A. Corp.
P.O. Box 556
Crystal Lake, Ill. 60014

Robeco Chemicals Inc.
99 Park Ave.
New York, N.Y. 10016

Robinson-Wagner Co., Inc.
628 Waverly Ave.
Mamaroneck, N.Y. 10543

Rohm and Haas Co.
Independence Mall West
Philadelphia, Pa. 19105

Ruetgers-Nease Chem. Co., Inc.
Box 221
State College, Pa. 16801

Sandoz Colors & Chemicals
Hanover, N.J. 07936

Sandoz Ltd.
CH-4002
Basle, Switzerland

Scher Chemicals
1 Styertowne Rd.
Clifton, N.J. 07012

Shell Chemical Co.
One Shell Plaza
Houston, TX 77002

Sherex Chemical Co., Inc.
P.O. Box 646
Dublin, Ohio 43017

Stauffer Chemical Co.
Westport, Conn. 06880

Standard Chemical Co.
Mill Lane
Cheadle, Chesire SK8 2NX
England

Stepan Chemical Co.
Edens & Winnetka Rds.
Northfield, Ill. 60093

Swastik Household & Industrial
Products Ltd.
Shaghibag House
13 Walchand Hirachand Marg
Bellard Estate
Bombay 400 038 India

Swift Chemical Co.
383 Orenda Road
Bramaleg, Ontario, Canada

Tanatex Chem. Co.
Div. of Sybron Corp.
P.O. Box 125
Wellford, S.C. 29385

Tennessee Corp.
See Cities Service Co.

The Theobald Industries
P.O. Box 72
Harrison, N.J. 07029

Thompson-Hayward Chem. Co.
5200 Speaker Rd.
Kansas City, Kansas 66106

Toho Chemical Industry Co., Ltd.
1-14-9 Nihonbashi
Kakigara-cho, Chuo-ku
Tokyo, Japan

Tokaiseiyu Ind. Co., Ltd.
Yamadahigashimachi 2-Chome 67
Higashiku, Nagoya,
Japan

Union Carbide Corp.
Chemicals & Plastics
270 Park Ave.
New York, N.Y. 10017

Uniroyal Chemicals
Div. of Uniroyal Inc.
Elm St.
Naugatuck, Conn. 06770

Valchem
Chemical Division of United Merchants & Manufacturers Inc.
Langley, S.C. 29834

Van Schuppen Chemie
Nieuweweg, Veenendaal
Netherlands

Westvaco-Polychemicals
P.O. Box 5207
N. Charleston, S.C. 29406

Witco Chemical Corp.
Organics Div. & Sonneborn Div.
277 Park Ave.
New York, N.Y. 10017

Witco Chemical Ltd.
P.O. Box 975
Haifa 31000
Israel

ABBREVIATIONS

aq. . . . aqueous
avg.. . . . average
B.P. . . . boiling point
C . . . degrees Centigrade
cc. . . . cubic centimeter(s)
CC . . . closed cup
cm . . . centimeter(s)
CMC . . . sodium carboxy methyl cellulose
COC . . . Cleveland open cup
conc.. . . . concentration, concentrated
cps . . . centipoise(s)
cs or cst. . . . centistoke(s)
DDVP . . . 2,2-dichlorovinyl dimethyl phosphate
DEA . . . diethanolamine
disp. . . . dispersion, dispersible
dist. . . . distilled
EO . . . ethylene oxide
esp.. . . . especially
F . . . degrees Fahrenheit
F.P. . . . freezing point
ft . . . foot, feet
g . . . gram(s)
gal . . . gallon(s)
HLB . . . hydrophilic lipophilic balance
ingred.. . . . ingredient(s)
IPA. . . . isopropyl alcohol
IPM . . . isopropyl myristate
kg. . . . kilogram(s)
l. . . . liter(s)
lb . . . pound(s)
max . . . maximum
MEA. . . . monoethanolamine
MEK. . . . methyl ethyl ketone
min. . . . minimum
MIPA . . . monoisopropanolamine
ml . . . milliliter(s)
mm. . . . millimeter(s)
M.P. . . . melting point
mfg. . . . manufacture, manufacturing

m.w. . . . . molecular weight
NF . . . . National Formulary
no. . . . . number
OC . . . . open cup
O/W . . . . oil in water
PEG . . . . polyethylene glycol
pH . . . . hydrogen ion concentration
PM . . . . Pensky-Martens
PMCC . . . . .Pensky-Martens closed cup
POE . . . . .polyoxyethylated
pt. . . . . point
PVC . . . . polyvinyl chloride
PVP . . . . polyvinylpyrrolidone
quat.. . . . . quaternary
ref.. . . . . refractive
reg.. . . . . regular
sapon.. . . . . saponification
sol.. . . . . .soluble
solid.. . . . . .solidification
sol'n.. . . . . solution
sp. gr. . . . . .specific gravity
SSU . . . . Saybolt Seconds Universal
Stod.. . . . . Stoddard
SUS . . . . Saybolt Universal Seconds
T/C. . . . . tank car
TCC . . . . Taggart closed cup
tech.. . . . . .technical
T/L. . . . . tank load
TOC . . . . Taggart open cup
T/T. . . . . .tank truck
T/W . . . . tank wagon
USP . . . . United States Pharmacopeia
visc. . . . . viscosity
W/O . . . . water in oil
wt . . . . weight
@ . . . . at
% . . . . .percent
≈. . . . . approximately
<. . . . . less than
>. . . . . .greater than
⩽. . . . . less than or equal to
⩾. . . . . greater than or equal to
. . . . number

The following products, which are included in this volume, are registered with the U.S. Patent Office.

Palegal®
Peregal®
Pluracol®
Plurafac®
Pluriol®
Primasol®
Priminox®
Proxel®
Pycal®
Reax®
Redicote®
Renex®
Ritachol®
Robane®
Robuoy®
Roccal®
Sarkosyl®
Scripset®
Sellogen®
Separol®
Sipex®
Sipon®
Siponate®
Siponic®
Sodasil®
Softigen®
Softisan®
Solricin®
Solulan®
Solusol®
Sontex®
Spermwax®
Sponto®
Standamul®
Standapol®
Starfol®
Steinapon®
Steol® 4N
Stepanate® X
Stepanol® WAC
Sterox®
Strodex®
Sulfosil®
Sulframin®
Super-Pro® #5A
Surfac®
Surfactol®
Surfonic®
Surfynol®
Synthrapol®
Tamol®
Tanapon®
Tanaterge®
Tandem®
T-Det®
Tergitol®
Tetronic®
Texpon®
T-Mulz®
Triton®
Tryfac®
Tween®
Tween-Mos®
Ultra Blend®
Ultra Sulfate®
Unamide®
Uniperol®
Uniquat®
Valdet®
Varine® C
Varion® CADG
Varsulf®
Veegum®
Victawet®
Witbreak®
Witcamide®
Witcamine® 210
Zelec®
Zonyl®

P

Palegal A. BASF AG

Chemical Description: Mixture of nonionic and anionic products
Category: Dyeing assistant, nonionic/anionic
Applications: Auxiliary for high-temperature dyeing of polyester fibers with disperse dyes by the Palegal process; leveling, detergent, and dispersing effects
Form: Viscous liquid; Color: Brown
Properties: Solubility: Readily sol. in water in all proportions; Stability: Stable to hard water and acids
Reaction: Almost neutral (aq. sol'ns.)
Toxicity: Skin irritant on prolonged contact with concentrated form
Warnings and Caution: Use goggles when taking product out of container; store at room temperature
References: Tech. leaflet M 5363 e

Palegal B. BASF AG

Chemical Description: Mixture of nonionic, mostly aliphatic compounds
Category: Leveling agent, dye assistant, nonionic
Applications: Leveling auxiliary for dyeing polyester fibers under high-temperature conditions; improves the leveling-out of disperse dyes and reduces marking of variations in the polyester fiber material
Form: Viscous liquid; Color: Light brown
Properties: Stability: Resistant to hard water and acids
Toxicity: Skin irritant; may cause inflammation of the mucous membranes of the eyes
Warnings and Caution: Avoid contact with skin; wear goggles when handling concentrated product
References: Tech. leaflet M 5530 e

Palegal HT. BASF AG

Chemical Description: Mixture of aliphatic oxyethylation products
Category: Leveling agent, dye assistant, nonionic
Applications: Low-foaming leveling agent for dyeing polyester fi-

bers under high-temperature conditions; improves levelness and penetration of polyester piece goods in beam-dyeing machines with an uneven rate of liquor flow; used for shading under high-temperature conditions
Form: Viscous liquid; Color: Brown
Properties: Solubility: Dissolves in cold water with stirring (forms a milky emulsion); Stability: Resistant to hard water and acids
Reaction: Neutral
Toxicity: Concentrated form may cause irritation of the mucous membranes, and of the skin on prolonged contact
Warnings and Caution: Wear goggles when handling concentrated product
References: Tech. leaflet M 5593 e

Palegal SF. BASF AG
Chemical Description: Mixture of nonionic, mainly aliphatic compounds
Category: Leveling agent, dye assistant, nonionic
Applications: Leveling auxiliary for dyeing polyester fibers under high-temperature conditions; ensures that disperse dyes go evenly onto the fiber; promotes migration in the high-temperature phase: reduces marking of variations in affinity due to the textile material; nonfoaming and suitable for use in jet dyeing machines that are not completely flooded
Form: Viscous liquid; Color: Yellowish
Properties: Solubility: Forms stable emulsion when stirred into cold water; Stability: Unaffected by hard water and acids
Warnings and Caution: Avoid prolonged contact with skin in conc. form
References: Tech. leaflet M 5694 e

Paracol H. Nippon Nyukazai Co., Ltd.
Category: Nonionic/anionic
Applications: Organic phosphate insecticides
Form: Liquid
Composition: 100% conc.

Paracol L. Nippon Nyukazai Co., Ltd.
Category: Nonionic/anionic
Applications: Organic phosphate insecticides
Form: Liquid
Composition: 100% conc.

Paracol M. Nippon Nyukazai Co., Ltd.
Category: Nonionic/anionic
Applications: Organic phosphate insecticides
Form: Liquid
Composition: 100% conc.

Paracol OP. Nippon Nyukazai Co., Ltd.
Chemical Description: Alkyl phosphate
Category: Anticorrosion agent, emulsion breaker, anionic
Form: Liquid
Composition: 100% conc.

Paracol SV. Nippon Nyukazai Co., Ltd.
Category: Emulsifier, nonionic/anionic
Applications: Emulsifier for organic phosphate insecticides (DDVP)
Form: Liquid
Composition: 100% conc.

Pationic 122A. Patco Cosmetic Prod./C.J. Patterson Co.
Chemical Description: Sodium capryl lactylate
Category: Anionic surfactant
Applications: Surfactant with microbial inhibitor properties
Form: Clear viscous liquid; Color: Amber
Properties: Solubility @ 1% conc. and 25 C: Sol. in propylene glycol, IPM, IPA; sol. (hazy) in distilled water; Visc.: 24 cps (2% in distilled water); HLB: 11.3; Sapon. Value: 235–265; Acid Value: 65–85; Surface Tension: 24.72 dynes/cm (0.1% conc.); pH: 5.25 (2% aq.)
Composition: 100% conc.
Toxicity: Moderate skin irritant in conc. form
Warnings and Caution: Store under cool, dry conditions
References: Bulletin No. 100, 101, 117

Pationic 138C. Patco Cosmetic Prod./C.J. Patterson Co.
Chemical Description: Sodium lauroyl lactylate
Category: Detergent, conditioner, viscosity builder, emulsifier, anionic
Applications: Cleansing, conditioning agent in shampoos; viscosity builder; lipophilic emulsifier; self-emulsifying wax
Form: Waxy solid; Color: Beige
Properties: Solubility @ 1% conc. and 25 C: Sol. (hazy) in dist. water; disp. in IPM; Visc.: 19 cps (2% in dist. water); M.P.: 55–59 C; HLB: 14.4; Sapon. Value: 175–205; Acid Value: 50–70; Surface Tension: 24.30 dynes/cm (0.1% conc.); pH: 6.90 (2% aq.)
Composition: 100% conc.
Toxicity: Minor skin irritant in conc. form
Warnings and Caution: Store in cool, dry place
References: Bulletin No.: 101, 116

Pationic 145A. Patco Cosmetic Prod./C.J. Patterson Co.
Chemical Description: Sodium stearoÿl-1-lactylate
Category: Emulsifier, anionic
Applications: Cosmetics products
Form: Flaked solid; Color: White
Properties: Solubility @ 1% conc. and 25 C: Disp. in dist. water; Visc.: 185 cps (2% in dist. water); M.P.: 57–60 C; HLB: 8.3; Sapon. Value: 169–190; Acid Value: 60–80; Surface Tension: 28.87 dynes/cm (0.1% conc.); pH: 5.70 (2% aq.)
Composition: 100% conc.
Toxicity: Minor skin irritant in conc. form

Warnings and Caution: Store in a cool, dry place
References: Bulletin No. 101, 115

Pationic CSL. Patco Cosmetic Prod./C.J. Patterson Co.
Chemical Description: Calcium stearoyl-2-lactylate
Category: Emulsifier, anionic
Applications: Emulsifier and protein complexer for cosmetics
Form: Powder; Color: Light tan
Properties: Solubility @ 1% conc. and 25 C: Disp. in mineral oil, propylene glycol, IPA; Visc.: 85 cps (2% in mineral oil); M.P.: 40–45 C; HLB: 5.0; Sapon. No.: 195–230; Acid Value: 50–65; Surface Tension: 37.00 dynes/cm (0.1% conc.); pH: 4.90 (2% aq.)
Composition: 100% conc.
Warnings and Caution: Store in a cool, dry place; avoid exposure to temp. > 32 C
References: Bulletin No. 101, 113

Pationic ISL. Patco Cosmetic Prod./C.J. Patterson Co.
Chemical Description: Sodium isostearoyl-2-lactylate
Category: Anionic surfactant
Applications: Cosmetics products
Form: Clear viscous liquid; Color: Straw or honey
Properties: Solubility @ 1% conc. and 25 C: Sol. in mineral oil, propylene glycol, IPM, IPA; disp. in dist. water; HLB: 5.9; Sapon. Value: 205–225; Acid Value: 60–70; Surface Tension: 26.28 dynes/cm (0.1% conc.); pH: 6.30 (2% aq.)
Composition: 100% active ingred.
Toxicity: Minor skin irritant in conc. form
Warnings and Caution: Store in a cool, dry place
References: Bulletin No. 101, 120

Pationic NMF. Patco Cosmetic Prod./C.J. Patterson Co.
Chemical Description: Blend of straight and branched chain aliphatic esters and alcohol
Category: Emollient, moisturizer, nonionic
Applications: Cosmetics products
Form: Clear liquid; Odor: Bland
Properties: Ref. Index: 1.4456 (20 C); Solubility: Oil-sol.; Density: 7.3 lb/gal; Sp. gr.: 0.888; Visc.: 40 cps; Sapon. Value: 170–190; Acid Value: 2.0 max.
Composition: 100% conc.
Containers: 400 lb net steel drum, 35 lb net plastic pack
References: Bulletin No. 173

Pationic SSL. Patco Cosmetic Prod./C.J. Patterson Co.
Chemical Description: Sodium stearoyl-2-lactylate
Category: Emulsifier, viscosity builder, anionic
Applications: Emulsifier, viscosity builder, protein complexer for cosmetics products
Form: Powder; Color: Light tan
Properties: Solubility @ 1% conc. and 25 C: Disp. in dist. water and mineral oil; Visc.: 175 cps (2% in dist. water); M.P.: 41–46 C; HLB: 6.5; Sapon. Value: 210–235; Acid Value: 60–70; Surface Tension: 32.34 dynes/cm (0.1%

conc.); pH: 5.95 (2% aq.)
Composition: 100% conc.
Warnings and Caution: Store in a cool, dry place; avoid exposure to temp. > 32 C
References: Bulletin No. 101, 112

Patlac IL. Patco Cosmetic Prod./ C.J. Patterson Co.
Chemical Description: Isostearyl lactate
Category: Nonionic surfactant
Applications: Cosmetic products
Form: Low viscosity liquid; Color: Pale yellow
Properties: Solubility @ 1% conc. and 25 C: Sol. in ethanol (3A), propylene glycol, mineral oil, IPM; Sp. gr.: 0.92; HLB: 10.3 ± 1.0; pH: 4.5 (2% aq.)
Composition: 100% conc.
References: Bulletin No. 210

Pave 100. Cincinnati Milacron
Chemical Description: Complex amide amine
Category: Asphalt additive, wetting and bonding aid
Applications: Used in asphalt cement, cutbacks, and emulsions to produce effective bond with wet or dry aggregate for all types of bituminous paving; wetting and bonding aid for asphalt; increases antistripping properties of asphalt
Form: Viscous liquid; Color: Dark brown
Properties: Density: 8.2 lb/gal (60 F); Sp. gr.: 0.985 (60 F); Visc.: 6600 SUS (100 F); Flash Pt.: 450 F (COC); Fire Pt.: 485 F (COC); Pour Pt.: 40 F (ASTM); Stability: Completely heat stable in asphalt; chemically stable in storage or in asphalt
Composition: 100% active ingred.
Containers: 440 lb net drums, tankcars, and tanktrucks

Pave 192. Cincinnati Milacron
Chemical Description: Complex amide amine
Category and Applications: See Pave 100
Form: Liquid; Color: Dark brown
Properties: Density: 7.8 lb/gal (60 F); Sp. gr.: 0.934 (60 F); Visc.: 1104 SUS (100 F); Flash Pt.: 420 F (COC); Fire Pt.: 445 F (COC); Pour Pt.: 10 F (ASTM); Stability: Completely heat stable in asphalt; chemically stable in storage or in asphalt
Composition: 100% active ingred.
Containers: 420 lb net drums, tankcars, or tanktrucks

Pave Bond. Cincinnati Milacron
Category: Coating and bonding additive, emulsifier
Applications: Coating and bonding additive in asphalt cement, cutbacks, and emulsions; improves initial bonding and inhibiting of water stripping; emulsifier for cutback asphalt emulsions
Form: Viscous liquid; Color: Dark brown
Properties: Density: 7.95 lb/gal (100 F); Sp. gr.: 0.953 (100/60 F); Visc.: 2290 SUS (100 F); Flash Pt.: 370 F (COC); Fire Pt.: 400 F (COC); Pour Pt.: 55 F (ASTM D-97)

Composition: 100% active ingred.
Containers: 400 lb net drums, tankcars, tanktrucks
Warnings and Caution: Wear safety glasses and protective gloves in well-ventilated area

Pave Bond LP. Cincinnati Milacron
Chemical Description: Complex amide amine
Category: Coating additive
Applications: Increases coating and bonding properties in asphalt cement and cutbacks, and in stockpile mixes; inhibits water stripping; may be used with hot asphalt cements
Form: Liquid; Color: Dark brown
Properties: Density: 7.68 lb/gal; Sp. gr.: 0.921 (77/60 F); Visc.: 466 cs; Flash Pt.: 315 F (COC); Fire Pt.: 350 F (COC); Pour Pt.: 15 F (ASTM D-97)
Composition: 100% active ingred.
Reaction: Alkaline
Toxicity: Corrosive to eyes and skin; harmful if inhaled
Containers: 400 lb net drums, tankcars, tanktrucks
Warnings and Caution: Wear safety glasses and protective gloves; use in well-ventilated area

Pave Bond Special. Cincinnati Milacron
Chemical Description: Complex amide amine
Category and Applications: See Pave Bond LP
Form: Liquid; Color: Dark brown
Properties: Density: 8.04 lb/gal (100 F); Sp. gr.: 0.963 (100/60 F); Visc.: 92.3 cs (100 F); Flash Pt.: 310 F (COC); Pour Pt.: 23 F (ASTM D-97)
Composition: 100% active ingred.
Reaction: Alkaline
Toxicity: Corrosive to skin and eyes; harmful if inhaled
Containers: 400 lb net drums, tankcars, tanktrucks
Warnings and Caution: Wear safety glasses and protective gloves; use in well-ventilated area; ground all equipment while transferring, blending, or pumping

PB-92. Reed Ltd.
Chemical Description: Modified calcium lignosulfate
Category: Wetting agent, emulsifier, dispersant, anionic
Applications: Oil well mud thinner
Form: Solid; Color: Yellow
Properties: Solubility: Sol. in water; Stability: Up to 300 C
Composition: 95% active ingred.
Reaction: Alkaline

Peacock. Pfau's Sons Co., Inc.
Category: Emulsifier
Applications: Tanning, fat liquoring, making of cutting and soluble oils
Form: Liquid; Color: Various; Odor: Characteristic
Properties: Sp. gr.: 0.910–0.920; Visc.: 200–220 cps
Reaction: Neutral
Containers: Drums, bulk

Pegnol HA-120. Toho Chemical Industry Co., Ltd.
Chemical Description: Polyoxy-

ethylene alkyl amine
Category: Emulsifier
Form: Liquid
Composition: 100% active ingred.

Pegosperse 50 MS. Glyco Chem.
Chemical Description: Ethylene glycol monostearate
Category: Emulsifier, dispersant, nonionic
Applications: General emulsifier for O/W emulsions
Form: Flakes; Color: White
Properties: Solubility: Sol. hot in methanol, ethanol, acetone, ethyl acetate, toluol, naphtha, mineral and vegetable oils; Sp. gr.: 0.96; M.P.: 55–60 C; HLB: 0.9 ± 0.5; Sapon. Value: 180–187; Acid Value: < 5; Iodine Value: < 2; pH: 4.0–6.0 (3% aq. disp.)
Composition: 100% conc.

Pegosperse 100 L. Glyco Chem.
Chemical Description: Diethylene glycol monolaurate
Category and Applications: See Pegosperse 50 MS
Form: Liquid; Color: Straw
Properties: Solubility: Sol. in ethanol, toluol, naphtha, mineral oil; miscible with methanol, acetone, ethyl acetate, vegetable oil; disp. in water; Sp. gr.: 0.97; HLB: 7.4 ± 0.5; Solid. Pt.: < 13 C; Sapon. Value: 160–170; Acid Value: < 4 max.; Iodine Value: < 9; pH: 8–10 (5% aq. disp.)
Composition: 100% conc.

Pegosperse 100 ML. Glyco Chem.
Chemical Description: Diethylene glycol monolaurate
Category and Applications: See Pegosperse 50 MS
Form: Liquid; Color: Straw
Properties: Solubility: Sol. in methanol, ethanol, acetone, ethyl acetate; partly sol. in naphtha; miscible hot with mineral and vegetable oils; Sp. gr.: 0.96; HLB: 6.0 ± 0.5; Solid. Pt.: 16–19 C; Sapon. Value: 180–190; Acid Value: < 4 max.; Iodine Value: < 10; pH: 3.0–6.0 (5% aq. disp.)
Composition: 100% conc.

Pegosperse 100 MR. Glyco Chem.
Chemical Description: Diethylene glycol monoricinoleate
Category and Applications: See Pegosperse 50 MS
Form: Liquid; Color: Amber
Properties: Solubility: Sol. in methanol, ethanol, acetone, ethyl acetate; partly sol. in mineral oil; miscible with toluol, naphtha, vegetable oil; Sp. gr.: 0.98; HLB: 4.8 ± 0.5; Solid. Pt.: < 0 C; Sapon. Value: 140–150; Acid Value: < 10; Iodine Value: 65–71; pH: 3–5 (5% aq. disp.)
Composition: 100% conc.

Pegosperse 100 O. Glyco Chem.
Chemical Description: Diethylene glycol oleate
Category and Applications: See Pegosperse 50 MS
Form: Liquid; Color: Amber
Properties: Solubility: Sol. in methanol, ethanol, toluol, naphtha; miscible with acetone, ethyl acetate, mineral and vegetable oils; disp. in water; Sp. gr.: 0.93;

HLB: 3.5 ± 0.5; Solid. Pt.: < 0 C; Sapon. Value: 160–175; Acid Value: 80–95; Iodine Value: 70–80; pH: 8–9 (5% aq. disp.)
Composition: 100% conc.

Pegosperse 100 S. Glyco Chem.
Chemical Description: Diethylene glycol stearate
Category and Applications: See Pegosperse 50 MS
Form: Beads; Color: White
Properties: Solubility: Sol. hot in methanol, ethanol, toluol, naphtha, mineral and vegetable oils; miscible hot with acetone and ethyl acetate; disp. hot in water; Sp. gr.: 0.96; M.P.: 46–54 C; HLB: 3.8 ± 0.5; Sapon. Value: 165–175; Acid Value: 95–105; Iodine Value: < 2; pH: 6.8–7.5 (3% aq. disp.)
Composition: 100% conc.

Pegosperse 200 ML. Glyco Chem.
Chemical Description: Polyethylene glycol 200 monolaurate
Category and Applications: See Pegosperse 50 MS
Form: Liquid; Color: Yellow
Properties: Solubility: Sol. in methanol, ethanol, acetone, ethyl acetate, toluol; miscible with water, naphtha, mineral and vegetable oils; Sp. gr.: 0.99; HLB: 8.6 ± 0.5; Solid. Pt.: < 5 C; Sapon. Value: 149–159; Acid Value: < 5; Iodine Value: < 9; pH: 4.0–6.5 (5% aq. disp.)
Composition: 100% conc.

Pegosperse 400 DL. Glyco Chem.
Chemical Description: Polyethylene glycol 400 dilaurate
Category and Applications: See Pegosperse 50 MS
Form: Liquid; Color: Yellow
Properties: Solubility: Sol. in methanol, ethanol, acetone, ethyl acetate, toluol, naphtha, mineral and vegetable oils; Sp. gr.: 0.99; HLB: 10.0 ± 0.5; Solid. Pt.: 5–12 C; Sapon. Value: 127–140; Acid Value: < 5; Iodine Value: < 7; pH: 3–5 (5% aq. disp.)
Composition: 100% conc.

Pegosperse 400 DO. Glyco Chem.
Chemical Description: Polyethylene glycol 400 dioleate
Category and Applications: See Pegosperse 50 MS
Form: Liquid; Color: Amber
Properties: Solubility: Sol. in methanol, ethanol, ethyl acetate, toluol, naphtha, mineral and vegetable oils; disp. hot in water; Sp. gr.: 0.97; HLB: 7.2 ± 0.5; Solid. Pt.: < 0 C; Sapon. Value: 115–125; Acid Value: < 10; Iodine Value: 52–57; pH: 4.5–6.5 (5% aq. disp.)
Composition: 100% conc.

Pegosperse 400 DS. Glyco Chem.
Chemical Description: Polyethylene glycol 400 distearate
Category and Applications: See Pegosperse 50 MS
Form: Soft solid; Color: Cream
Properties: Solubility: Sol. in ethanol, ethyl acetate, toluol, naphtha, mineral oil; sol. hot in methanol and vegetable oil; disp. hot in water; Sp. gr.: 0.98; M.P.: 29–37 C; HLB: 7.8 ± 0.5; Sapon. Value: 115–125; Acid Value: <

10; Iodine Value: < 1; pH: 4.0–6.5 (5% aq. disp.)
Composition: 100% conc.

Pegosperse 400 DTR. Glyco Chem.
Chemical Description: Polyethylene glycol 400 di tri ricinoleate
Category and Applications: See Pegosperse 50 MS
Form: Liquid; Color: Amber
Properties: Solubility: Sol. in acetone, ethyl acetate, toluol, naphtha, mineral and vegetable oils; partly sol. in ethanol; disp. in water; Sp. gr.: 0.95–0.97; HLB: 1.8 ± 0.5; Solid. Pt.: < 10 C; Sapon. Value: 156–166; Acid Value: < 12; Iodine Value: 75–82; pH: 3.5–5.0 (5% aq. disp.)
Composition: 100% conc.

Pegosperse 400 MC. Glyco Chem.
Chemical Description: Polyethylene glycol 400 monococoate
Category and Applications: See Pegosperse 50 MS
Form: Liquid; Color: Yellow
Properties: Solubility: Sol. in methanol, ethanol, acetone, ethyl acetate, toluol, vegetable oil; sol. hot in mineral oil; miscible with water and naphtha; Sp. gr.: 1.03; HLB: 13.9 ± 0.5; Solid. Pt.: < 6 C; Sapon. Value: 88–100; Acid Value: < 4; Iodine Value: < 5; pH: 4–6 (5% aq. disp.)
Composition: 100% conc.

Pegosperse 400 ML. Glyco Chem.
Chemical Description: Polyethylene glycol 400 monolaurate
Category and Applications: See Pegosperse 50 MS
Form: Liquid; Color: Straw
Properties: Solubility: Sol. in methanol, ethanol, acetone, ethyl acetate, toluol, vegetable oil; sol. hot in mineral oil; miscible with water and naphtha; Sp. gr.: 1.03; HLB: 13.9 ± 0.5; Solid. Pt.: < 7 C; Sapon. Value: 90–100; Acid Value: < 3; Iodine Value: < 4; pH: 4.0–6.0 (5% aq. disp.)
Composition: 100% conc.

Pegosperse 400 MO. Glyco Chem.
Chemical Description: Polyethylene glycol 400 monooleate
Category and Applications: See Pegosperse 50 MS
Form: Liquid; Color: Amber
Properties: Solubility: Sol. in methanol, ethanol, acetone, ethyl acetate, toluol; sol. hot in naphtha; partly sol. hot in mineral and vegetable oils; disp. in water; Sp. gr.: 1.01; HLB: 11.0 ± 0.5; Solid. Pt.: < 0 C; Sapon. Value: 80–88; Acid Value: < 5; Iodine Value: 36–38; pH: 4–6 (5% aq. disp.)
Composition: 100% conc.

Pegosperse 400 MOT. Glyco Chem.
Chemical Description: Polyethylene glycol 400 monooleate (tallate)
Category and Applications: See Pegosperse 50 MS
Form: Liquid; Color: Yellow
Properties: Solubility: Sol. in methanol, ethanol, acetone, ethyl acetate, toluol; sol. hot in naphtha; partly sol. hot in mineral and vegetable oils; disp. in water; Sp. gr.:

1.02; HLB: 11.0 ± 0.5; Solid. Pt.: < 0 C; Sapon. Value: 83–89; Acid Value: < 5–8; Iodine Value: 54–59; pH: 3.5–5.0 (5% aq. disp.)
Composition: 100% conc.

Pegosperse 400 MS. Glyco Chem.
Chemical Description: Polyethylene glycol 400 monostearate
Category and Applications: See Pegosperse 50 MS
Form: Soft solid; Color: White
Properties: Solubility: Sol. in methanol, ethanol, acetone, ethyl acetate, toluol; sol. hot in naphtha, mineral and vegetable oils; disp. hot in water; Sp. gr.: 1.0; M.P.: 30 C min.; HLB: 11.2 ± 0.5; Sapon. Value: 83–94; Acid Value: < 3; Iodine Value: < 1; pH: 3.5–6.0 (5% aq. disp.)
Composition: 100% conc.

Pegosperse 600 ML. Glyco Chem.
Chemical Description: Polyethylene glycol 600 monolaurate
Category and Applications: See Pegosperse 50 MS
Form: Liquid; Color: Yellow
Properties: Solubility: Sol. in water, methanol, ethanol, acetone, ethyl acetate; partly sol. in mineral oil; sol. hot in naphtha; miscible with toluol and vegetalbe oil; Sp. gr.: 1.01; HLB: 14.6 ± 0.5; Solid. Pt.: 16–21 C; Sapon. Value: 65–75; Acid Value: < 1; Iodine Value: < 4; pH: 6–8 (5% aq. disp.)
Composition: 100% conc.

Pegosperse 600 MS. Glyco Chem.
Chemical Description: Polyethylene glycol 600 monostearate
Category and Applications: See Pegosperse 50 MS
Form: Soft solid; Color: White
Properties: Solubility: Sol. in methanol, ethanol, acetone, ethyl acetate, toluol; sol. hot in vegetable oil; disp. hot in water; Sp. gr.: 1.01; M.P.: 27–32 C; HLB: 13.2 ± 0.5; Sapon. Value: 72–78; Acid Value: < 8.5; Iodine Value: < 1; pH: 3–5 (5% aq. disp.)
Composition: 100% conc.

Pegosperse 700 TO. Glyco Chem.
Chemical Description: Polyethylene glycol 700 monotallate
Category and Applications: See See Pegosperse 50 MS
Form: Liquid; Color: Amber
Properties: Solubility: Sol. in water, methanol, ethanol, acetone, ethyl acetate, toluol; sol. hot in vegetable oil; Sp. gr.: 1.10; HLB: 13.5 ± 0.5; Solid. Pt.: < 15 C; Sapon. Value: 35–45; Acid Value: < 2; Iodine Value: 38–43; pH: 6–8 (5% aq. disp.)
Composition: 100% conc.

Pegosperse 1000 MS. Glyco Chem.
Chemical Description: Polyethylene glycol 1000 monostearate
Category and Applications: See Pegosperse 50 MS
Form: Solid; Color: Cream
Properties: Solubility: Sol. in methanol, ethanol, acetone, ethyl acetate, toluol; sol. hot in naphtha, vegetable oil; partly sol. in mineral oil; disp. hot in water; Sp. gr.: 1.02; M.P.: 37–43 C; HLB: 15.2 ± 0.5; Sapon. Value: 45–55;

Acid Value: < 3; Iodine Value: < 1; pH: 3–5 (5% aq. disp.)
Composition: 100% conc.

Pegosperse 1500 DO. Glyco Chem.
Chemical Description: Polyethylene glycol 1500 dioleate
Category and Applications: See Pegosperse 50 MS
Form: Soft solid; Color: Amber
Properties: Solubility: Sol. in methanol, ethanol, acetone, ethyl acetate, toluol; sol. hot in vegetable oil; disp. hot in water; Sp. gr.: 1.05; M.P.: 30–38 C; HLB: 7.8 ± 0.5; Sapon. Value: 104–112; Acid Value: < 11; Iodine Value: 65–72; pH: 3.5–5.0 (5% aq. disp.)
Composition: 100% conc.

Pegosperse 1500 MS. Glyco Chem.
Chemical Description: Polyethylene glycol 1500 monostearate
Category and Applications: See Pegosperse 50 MS
Form: Waxy solid; Color: Cream
Properties: Solubility: Sol. in ethanol, ethyl acetate, toluol, vegetable oil; sol. hot in methanol, naphtha; miscible with water; Sp. gr.: 1.05; M.P.: 27–31 C; HLB: 13.8 ± 0.5; Sapon. Value: 57–67; Acid Value: < 3.5; Iodine Value: < 1; pH: 3–5 (5% aq. disp.)
Composition: 100% conc.

Pegosperse 4000 MS. Glyco Chem.
Chemical Description: Polyethylene glycol 4000 monostearate
Category and Applications: See Pegosperse 50 MS
Form: Waxy solid; Color: Cream
Properties: Solubility: Sol. hot in water, methanol, ethanol, acetone, ethyl acetate, toluol, vegetable oil; partly sol. hot in mineral oil; disp. in naphtha; Sp. gr.: 1.10; M.P.: 54–61 C; HLB: 18.0 ± 0.5; Sapon. Value: 17–22; Acid Value: < 6; Iodine Value: < 1; pH: 3–5 (5% aq. disp.)
Composition: 100% conc.

Pegosperse 9000 CO. Glyco Chem.
Chemical Description: Polyethylene glycol 9000 castor oil
Category and Applications: See Pegosperse 50 MS
Form: Solid; Color: Tan
Properties: Solubility: Sol. in water, methanol, ethanol, acetone; Sp. gr.: 1.08; M.P.: 41–44 C; HLB: 18.1 ± 0.5; Sapon. Value: 14–20; Acid Value: < 1; Iodine Value: < 10; pH: 6.0–7.0 (3% aq. disp.)
Composition: 100% conc.

Pegosperse MFE. Glyco Chem.
Chemical Description: Mixed polyethylene glycol esters
Category and Applications: See Pegosperse 50 MS
Form: Soft solid; Color: Brown
Properties: Solubility: Sol. hot in methanol, ethanol, vegetable oil; partly sol. hot in mineral oil; disp. hot in water; Sp. gr.: 1.05; M.P.: 46–54 C; HLB: 13.0 ± 0.5; Sapon. Value: 60–70; Acid Value: < 10; Iodine Value: 39–42; pH: 3–5 (5% aq. disp.)
Composition: 100% conc.

Peltex. Reed Ltd.
Chemical Description: Modified ferro-chromium lignosulfate
Category: Wetting agent, emulsifier, dispersant, anionic
Applications: Oil well drilling mud thinner (gypsum, sodium, and lime-based muds)
Form: Powder; Color: Black
Properties: Solubility: Water-sol.; Density: 0.5; Stability: Up to 300 C
Composition: 95% active ingred.
Reaction: Acid
Toxicity: Toxic due to chromium ions

Penetron Conc. Hart Products
Chemical Description: Sodium dioctylsulfosuccinate
Category: Detergent, wetting agent, emulsifier
Applications: Textile wetting and rewetting; wallpaper removal; skins; leather; laundering; dyeing; papermaking; household and industrial detergents; bleaching; cleaning hangings; dispersing waxes, pigments, resins, dry powders; printing; preparation of emulsified paints
Form: Liquid; Color: Colorless; Odor: Alcoholic
Properties: Solubility: Sol. in water and polar organic solvents; Sp. gr.: 1.05
Composition: 70% active ingred.
Reaction: Neutral
Containers: Drums

Pentex 40. Colloids, Inc.
Chemical Description: Alkylaryl sulfonate
Category: Wetting agent, penetrant, emulsifier, anionic
Applications: Textile and industrial processing; textile scouring processes; improves dye uniformity on velvets and tight weaves; aids rewetting of leather, paper, and paper-mill felts; improves pick-up of warp-yarn sizing sol'ns.
Form: Liquid; Color: Red-brown
Properties: Sp. gr.: 1.15; Pour Pt.: 0 C; Stability: Stable in acid and alkaline sol'ns.; pH: 9.5 (5%)

Pentex 99. Colloids, Inc.
Chemical Description: Liquid form of dioctyl sodium sulfosuccinate
Category: Wetting agent, detergent, dispersant, anionic
Applications: Textile scouring and dispersant for dyes; paper rewetting and felt washing surfactant; wetting agent in cosmetic creams and lotions; detergent additive in dry cleaning fluids, glass cleaning compounds, dishwashing compounds; wallpaper removers; emulsion polymerization; water-based paint formulations
Form: Clear liquid
Properties: Sp. gr.: 1.08–1.13 (70 F); Visc.: 500–1000 cps (70 F); Pour Pt.: 40 F; Stability: Stable @ pH 7–10; effective in hard and soft waters up to boiling temp.
Composition: 75% conc.

Peregal O. GAF Corp.
Chemical Description: Polyoxyethylated fatty alcohol
Category: Leveling agent, pene-

trant, dye assistant, nonionic
Applications: Leveling agent for selected vat dyes and for direct dyes; stripping assistant for vat dyes
Form: Clear liquid; Color: Yellowish-brown
Properties: Solubility: Readily miscible with water at all concs.; Sp. gr.: 1.02; Stability: Unaffected by hard water, acids, alkalis, and salts

Peregal OK. GAF Corp.
Chemical Description: Methyl polyethanol quaternary amine
Category: Textile auxiliary, leveling and stripping agent, cationic
Applications: Promotes level dyeing and improves penetration, esp. for light shades; leveling, retarding, and stripping agent for vat dyestuff on rayon and cotton fibers
Form: Slightly cloudy liquid; Color: Pale amber
Properties: Solubility: Readily miscible with water at all concs.; Sp. gr.: 1.03; pH: 7.6

Peregal ST. GAF Corp.
Chemical Description: Polyvinylpyrrolidone
Category: Dispersant, stripping assistant, detergent, cationic
Applications: Stripping assistant for cotton and rayon for vat, sulfur, or direct dyes; suitable for use in long liquors, package machines, or jigs; fulling soap additive; rag-stripping assistant in paper industry; detergent-assistant and soil-suspending agent for washing cellulosic and animal-fiber yarns
Form: Liquid; Color: 5 max. (VCS)
Properties: Solubility: Readily sol. in water; pH: 5–7 (5% aq.)
Composition: 30–32% active

Perlankrol ACM-2. Lankro Chem. Ltd.
Chemical Description: Fatty alkylolamide ether sulfate
Category: Base
Applications: Preparation of high-foaming shampoos and toiletries
Composition: 30% active ingred.

Perlankrol ADP3. Lankro Chem. Ltd.
Chemical Description: Sodium salt of fatty alcohol ether sulfate
Category: Ingredient, anionic
Applications: Base for preparation of high-foaming bubble baths and shampoos, and other toiletries
Form: Clear liquid; Color: Water-white; Odor: Mild
Properties: Solubility: Water-sol.; Sp. gr.: 1.048 (20 C); Visc.: 28 cs; Flash Pt.: > 200 F (COC); Pour Pt.: < 0 C; Stability: Stable in conc. electrolyte sol'ns.; pH: 7.0–8.5 (1% aq.); Biodegradable
Composition: 27% active ingred.
Reaction: Neutral
Toxicity: May cause irritation on prolonged contact with conc. form
Containers: Stainless steel, glass, resin, or polythene-lined equipment
Warnings and Caution: Wear protective gloves and goggles; spillages

may be slippery; in a fire, toxic fumes of sulfur dioxide may be produced
References: Bulletin SI:24 FM678

Perlankrol ASC2. Lankro Chem. Ltd.
Chemical Description: Sodium salt of fatty alcohol ether sulfate
Category and Applications: See Perlankrol ADP3
Form: Clear liquid; Color: Water-white; Odor: Mild
Properties: Solubility: Water-sol.; Sp. gr.: 1.047 (20 C); Visc.: 1000 cs; Flash Pt.: > 200 F; Pour Pt.: < 0 C; Stability: Stable in conc. electrolyte sol'ns.; pH: 6.5–7.5 (1% aq.); Biodegradable
Composition: 27.5% active
Reaction: Neutral
Toxicity: May cause irritation on prolonged contact with conc. form
Containers: Stainless steel, glass, resin, or polythene-lined equipment
Warnings and Caution: Wear protective gloves and goggles; spillages may be slippery; in a fire, toxic fumes of sulfur dioxide may be produced
References: Bulletin SI:24 FM678

Perlankrol ATL40. Lankro Chem. Ltd.
Chemical Description: Triethanolamine salt of narrow cut lauryl sulfate
Category and Applications: See Perlankrol ADP3
Form: Liquid; Color: Pale straw; Odor: Mild
Properties: Solubility: Water-sol.; Sp. gr.: 1.048 (20 C); Visc.: 60 cs; Flash Pt.: > 200 F (COC); Pour Pt.: < 0 C; Stability: Stable in conc. electrolyte sol'ns.; pH: 7–8 (1% aq.); Biodegradable
Composition: 40% active ingred.
Reaction: Neutral
Toxicity: May cause irritation on prolonged contact with conc. form
Containers: Stainless steel, glass, resin, or polythene-lined equipment
Warnings and Caution: Wear protective gloves and goggles; spillages may be slippery; in a fire, toxic fumes of sulfur dioxide may be produced
References: Bulletin SI:24 FM678

Perlankrol DAF25. Lankro Chem. Ltd.
Chemical Description: Ammonium salt of synthetic primary alcohol sulfate
Category: Foaming agent, base, anionic
Applications: Formulation of toiletries and carpet shampoos
Form: Liquid; Color: Pale straw
Composition: 25% active ingred.

Perlankrol DGS. Lankro Chem. Ltd.
Chemical Description: Sodium primary alcohol sulfate
Category: Solubilizer, detergent, textile auxiliary, foaming agent, emulsifier, anionic
Applications: Detergent bases in

household and industrial cleaners; foaming agent in synthetic latex industry; primary emulsifier in emulsion polymerization; this product is available on a made-to-order basis only
Form: Slurry; Color: White; Odor: Faint fatty alcohol
Properties: Solubility: Water-sol.; Sp. gr.: 1.041 (20 C); Visc.: 220 cs (40 C); Flash Pt.: > 200 F (COC); Pour Pt.: 12 C; Stability: Stable to most electrolytes and in hard water; tolerant to mildly alkaline sol'ns. at ambient temp.; pH: 7.0–8.5 (1% aq.); Biodegradable
Composition: 28% active ingred.
Toxicity: May cause irritation on prolonged contact with conc. form
Containers: Stainless steel, glass, resin, or polythene-lined equipment
Warnings and Caution: Wear protective gloves and goggles; spillages may be slippery; in a fire, toxic fumes of sulfur dioxide may be produced
References: Bulletin SI:25 FM678

Perlankrol DSA. Lankro Chem. Ltd.
Chemical Description: Sodium salt of synthetic primary alcohol sulfate
Category and Applications: See Perlankrol DGS
Form: Slurry; Color: White; Odor: Faint fatty alcohol
Properties: Solubility: Water-sol.; Sp. gr.: 1.046 (20 C); Visc.: 300 cs (40 C); Flash Pt.: > 200 F (COC); Pour Pt.: 10 C; Stability: Stable to most electrolytes and in hard water; tolerant to mildly alkaline sol'ns. at ambient temps.; pH: 7.0–8.5 (1% aq.); Biodegradable
Composition: 28% active ingred.
Toxicity: May cause irritation on prolonged contact with conc. form
Containers: Stainless steel, glass, resin, or polythene-lined equipment
Warnings and Caution: Wear protective gloves and goggles; spillages may be slippery; in a fire, toxic fumes of sulfur dioxide may be produced
References: Bulletin SI:25 FM678

Perlankrol EAD-60. Lankro Chem. Ltd.
Chemical Description: Ammonium salt of synthetic primary alcohol ether sulfate
Category: Foam booster and stabilizer, wetting agent, detergent, dispersant, emulsifier, anionic
Applications: Liquid detergents, industrial and household cleaners; foaming agent for fire fighting foam compounds; dispersant and emulsifier for chlorinated solvents; good wetting and foaming in neutral and mildly alkaline conditions; good lime soap dispersion power
Form: Clear mobile gel; Color: Straw; Odor: Faint alcoholic
Properties: Solubility: Water-sol.; Sp. gr.: 1.033 (20 C); Visc.: 240

cs (20 C); Flash Pt.: 85 F (Abel CC); Pour Pt.: < 0 C; Stability: Stable to high electrolyte concs. and hard water; tolerant to mildly alkaline sol'ns. at ambient temps.; pH: 7–9 (1% aq.); Biodegradable
Composition: 60% active ingred.
Reaction: Neutral
Toxicity: May cause irritation on prolonged contact with conc. form
Containers: Stainless steel, glass, resin, or polythene-lined equipment
Warnings and Caution: Wear protective gloves and goggles; spillages may be slippery; flammable—keep from heat, naked flames, and sparks; in a fire, toxic fumes of sulfur dioxide may be produced
References: Bulletin SI:26 FM678

Perlankrol ESD. Lankro Chem. Ltd.
Chemical Description: Sodium salt of synthetic primary alcohol ether sulfate
Category and Applications: See Perlankrol EAD-60
Form: Clear liquid; Color: Water-white; Odor: Faint
Properties: Solubility: Water-sol.; Sp. gr.: 1.053 (20 C); Visc.: 130 cs (20 C); Flash Pt.: > 200 F (COC); Pour Pt.: < 0 C; Stability: Stable to high electrolyte concs. and hard water; tolerant to mildly alkaline sol'ns. at ambient temps.; pH: 7–9 (1% aq.); Biodegradable
Composition: 27% active ingred.
Reaction: Neutral
Toxicity: May cause irritation on prolonged contact with conc. form
Containers: Stainless steel, glass, resin, or polythene-lined equipment
Warnings and Caution: Wear protective gloves and goggles; spillages may be slippery; in a fire, toxic fumes of sulfur dioxide may be produced
References: Bulletin SI:26 FM678

Perlankrol ESD-60. Lankro Chem. Ltd.
Chemical Description: Sodium salt of synthetic primary alcohol ether sulfate
Category and Applications: See Perlankrol EAD-60
Form: Clear to hazy mobile gel; Color: Water-white; Odor: Faint alcoholic
Properties: Solubility: Water-sol.; Sp. gr.: 1.090 (20 C); Visc.: 210 cs (20 C); Flash Pt.: 90 F (Abel CC); Pour Pt.: < 0 C; Stability: Stable to high electrolyte conc. and hard water; tolerant to mildly alkaline sol'ns. at ambient temps.; pH: 7–9 (1% aq.); Biodegradable
Composition: 60% active ingred.
Reaction: Neutral
Toxicity: May cause irritation on prolonged contact with conc. form
Containers: Stainless steel, glass, resin, or polythene-lined equipment
Warnings and Caution: Wear protective gloves and goggles; spillages may be slippery; flammable—keep

from heat, naked flames, and sparks; in a fire, toxic fumes of sulfur dioxide may be produced
References: Bulletin SI:26 FM678

Perlankrol ESS-25. Lankro Chem. Ltd.
Chemical Description: Sodium salt of synthetic primary alcohol ether sulfate
Category and Applications: See Perlankrol EAD-60
Form: Clear liquid; Color: Water-white; Odor: Faint
Properties: Solubility: Water-sol.; Sp. gr.: 1.040 (20 C); Visc.: 65 cs (20 C); Flash Pt.: > 200 F (COC); Pour Pt.: < 0 C; Stability: Stable to high electrolyte conc. and hard water; tolerant to mildly alkaline sol'ns. at ambient temps.; pH: 7–9 (1% aq.); Biodegradable
Composition: 25% active ingred.
Toxicity: May cause irritation on prolonged contact with conc. form
Containers: Stainless steel, glass, resin, or polythene-lined equipment
Warnings and Caution: Wear protective gloves and goggles; spillages may be slippery; in a fire, toxic fumes of sulfur dioxide may be produced
References: Bulletin SI:26 FM678

Perlankrol FB-25. Lankro Chem. Ltd.
Chemical Description: Sodium salt of modified alkyl phenol ether sulfate
Category: Emulsifier, anionic
Applications: Primary emulsifier for emulsion polymerization
Form: Clear liquid; Color: Straw; Odor: Faint alcoholic
Properties: Solubility: Water-sol.; Sp. gr.: 1.038 (20 C); Visc.: 48 cs (20 C); Flash Pt.: 86 F (Abel CC); Pour Pt.: –3 C; Stability: Stable in conc. electrolyte sol'ns. and hard water; tolerant to mildly alkaline sol'ns. at ambient temps.; pH: 6.0–8.5 (1% aq.)
Composition: 25% active ingred.
Reaction: Neutral
Toxicity: May cause irritation on prolonged contact with conc. form
Containers: Stainless steel, glass, resin, or polythene-lined equipment
Warnings and Caution: Wear protective gloves and goggles; spillages may be slippery; flammable–keep from heat, naked flames, and sparks; in a fire, toxic fumes of sulfur dioxide may be produced
References: Bulletin SI:27 FM678

Perlankrol FD-63. Lankro Chem. Ltd.
Chemical Description: Ammonium salt of alkyl phenol ether sulfate
Category and Applications: See Perlankrol FB-25
Form: Clear liquid; Color: Straw; Odor: Faint alcoholic
Properties: Solubility: Water-sol.; Sp. gr.: 1.056 (20 C); Visc.: 106 cs (20 C); Flash Pt.: 80 F (Abel CC); Pour Pt.: < 0 C; Stability: Stable in conc. electrolyte sol'ns.

and hard water; tolerant to mildly alkaline sol'ns. at ambient temps.; pH: 6.0–8.5 (1% aq.)
Composition: 63% active ingred.
Reaction: Neutral
Toxicity: May cause irritation on prolonged contact with conc. form
Containers: Stainless steel, glass, resin, or polythene-lined equipment
Warnings and Caution: Wear protective gloves and goggles; spillages may be slippery; flammable–keep from heat, naked flames, and sparks; in a fire, toxic fumes of sulfur dioxide may be produced; do not blend with strongly alkaline material (ammoniacal odor develops)
References: Bulletin SI:27 FM678

Perlankrol FF. Lankro Chem. Ltd.
Chemical Description: Ammonium salt of alkyl phenol ether sulfate
Category: Foam booster, stabilizing agent, anionic
Applications: Industrial detergents
Form: Hazy viscous liquid; Color: Amber; Odor: Faint ammoniacal
Properties: Solubility: Water-sol.; Sp. gr.: 1.115 (20 C); Visc.: 4200 cs (20 C); Flash Pt.: 84 F (Abel CC); Pour Pt.: 5 C; Stability: Stable in conc. electrolyte sol'ns. and hard water; tolerant to mildly alkaline sol'ns. at ambient temps.; pH: 6.0–8.5 (1% aq.)
Composition: 90% active ingred.
Reaction: Neutral
Toxicity: May cause irritation on prolonged contact with conc. form
Containers: Stainless steel, glass, resin, or polythene-lined equipment
Warnings and Caution: Wear protective gloves and goggles; spillages may be slippery; flammable–keep from heat, naked flames, and sparks; in a fire, toxic fumes of sulfur dioxide may be produced; do not blend with strongly alkaline material (ammoniacal odor develops)
References: Bulletin SI:27 FM678

Perlankrol FN-65. Lankro Chem. Ltd.
Chemical Description: Sodium salt of alkyl phenol ether sulfate
Category: Detergent base, foam stabilizer, emulsifier, anionic
Applications: Specialty detergent formulations; primary emulsifier in emulsion polymerization
Form: Clear liquid; Color: Straw; Odor: Faint alcoholic
Properties: Solubility: Water-sol.; Sp. gr.: 1.110 (20 C); Visc.: 200 cs (20 C); Flash Pt.: 85 F (Abel CC); Pour Pt.: $<$ 0 C; Stability: Stable in conc. electrolyte sol'ns. and hard water; tolerant to mildly alkaline sol'ns. at ambient temps.; pH: 6.0–8.5 (1% aq.)
Composition: 65% active ingred.
Reaction: Neutral
Toxicity: May cause irritation on prolonged contact with conc. form
Containers: Stainless steel, glass,

resin, or polythene-lined equipment
Warnings and Caution: Wear protective gloves and goggles; spillages may be slippery; flammable–keep from heat, naked flames, and sparks; in a fire, toxic fumes of sulfur dioxide may be produced
References: Bulletin SI:27 FM678

Perlankrol O. Lankro Chem. Ltd.
Chemical Description: Sodium alkyl sulfate
Category: Clarifying and stabilizing agent, wetting agent, viscosity modifier, anionic
Applications: Detergent and cleaning formulations; industrial detergents and hard surface cleaners
Form: Clear liquid; Color: Light amber; Odor: Characteristic
Properties: Solubility: Water-sol.; Sp. gr.: 1.120 (20 C); Visc.: 44 cs (20 C); Flash Pt.: > 200 F (COC); Pour Pt.: < 0 C; Stability: Stable to most electrolytes and in hard water; tolerant to mildly alkaline sol'ns. at ambient temps.; pH: 7.0–8.5 (1% aq.); Biodegradable
Composition: 40% active ingred.
Reaction: Neutral
Toxicity: May cause irritation on prolonged contact with conc. form
Containers: Stainless steel, glass, resin, or polythene-lined equipment
Warnings and Caution: Wear protective goggles and gloves; spillages may be slippery; in a fire, toxic fumes of sulfur dioxide may be produced
References: Bulletin SI:25 FM678

Perlankrol PA Conc. Lankro Chem. Ltd.
Chemical Description: Ammonium salt of alkyl phenol ether sulfate
Category: Detergent base, foam stabilizer, foaming agent, emulsifier, anionic
Applications: Liquid detergent formulations; emulsifier for cresylic acid
Form: Hazy viscous liquid; Color: Amber; Odor: Faint ammoniacal
Properties: Solubility: Water-sol.; Sp. gr.: 1.110 (20 C); Visc.: 6000 cs (20 C); Flash Pt.: 86 F (Abel CC); Pour Pt.: 7 C; Stability: Stable in conc. electrolyte sol'ns. and hard water; tolerant to mildly alkaline sol'ns. at ambient temps.; pH: 6.0–8.5 (1% aq.)
Composition: 90% active ingred.
Toxicity: May cause irritation on prolonged contact with conc. form
Containers: Stainless steel, glass, resin, or polythene-lined equipment
Warnings and Caution: Wear protective goggles and gloves; spillages may be slippery; flammable–keep from heat, naked flames, and sparks; in a fire, toxic fumes of sulfur dioxide may be produced; do not blend with strongly alkaline material (ammoniacal odor develops)
References: Bulletin SI:27 FM678

Perlankrol RN-75. Lankro Chem. Ltd.
Chemical Description: Sodium salt of alkyl phenol ether sulfate
Category: Foaming and wetting agent, emulsifier, anionic
Applications: Specialty foaming and wetting agent at high electrolyte concentrations; primary emulsifier in emulsion polymerization
Form: Clear liquid; Color: Straw; Odor: Faint alcoholic
Properties: Solubility: Water-sol.; Sp. gr.: 1.120 (20 C); Visc.: 470 cs (20 C); Flash Pt.: 88 F (Abel CC); Pour Pt.: < 0 C; Stability: Stable in conc. electrolyte sol'ns. and hard water; tolerant to mildly alkaline sol'ns. at ambient temps.; pH: 6.0–8.5 (1% aq.)
Composition: 75% active ingred.
Reaction: Neutral
Toxicity: May cause irritation on prolonged contact with conc. form
Containers: Stainless steel, glass, resin, or polythene-lined equipment
Warnings and Caution: Wear protective goggles and gloves; spillages may be slippery; flammable–keep from heat, naked flames, and sparks; in a fire, toxic fumes of sulfur dioxide may be produced
References: Bulletin SI:27 FM678

Perlankrol SN. Lankro Chem. Ltd.
Chemical Description: Sodium salt of alkyl phenol ether sulfate
Category: Foam booster and stabilizer, wetting agent, emulsifier, anionic
Applications: Industrial detergent formulations; wetting agent, esp. at high electrolyte concentrations; primary emulsifier in emulsion polymerization
Form: Clear liquid; Color: Water-white; Odor: Faint
Properties: Solubility: Water-sol.; Sp. gr.: 1.054 (20 C); Visc.: 48 cs (20 C); Flash Pt.: > 200 F (COC); Pour Pt.: < 0 C; Stability: Stable in conc. electrolyte sol'ns. and hard water; tolerant to mildly alkaline sol'ns. at ambient temps.; pH: 6.0–8.5 (1% aq.)
Composition: 30% active ingred.
Reaction: Neutral
Toxicity: May cause irritation on prolonged contact with conc. form
Containers: Stainless steel, glass, resin, or polythene-lined equipment
Warnings and Caution: Wear protective goggles and gloves; spillages may be slippery; in a fire, toxic fumes of sulfur dioxide may be produced
References: Bulletin SI:27 FM678

Perlankrol TM1. Lankro Chem. Ltd.
Chemical Description: Sodium salt of fatty amide ether sulfate
Category and Applications: See Perlankrol ADP3
Form: Clear liquid; Color: Pale yellow; Odor: Mild
Properties: Solubility: Water-sol.; Sp. gr.: 1.045 (20 C); Visc.: 715 cs; Flash Pt.: > 200 F (COC); Pour Pt.: < 0 C; Stability: Stable

in conc. electrolyte sol'ns.; pH: 7–8 (1% aq.); Biodegradable
Composition: 30% active ingred.
Reaction: Neutral
Toxicity: May cause irritation on prolonged contact with conc. form
Containers: Stainless steel, glass, resin, or polythene-lined equipment
Warnings and Caution: Wear protective goggles and gloves; spillages may be slippery; in a fire, toxic fumes of sulfur dioxide may be produced
References: Bulletin SI:24 FM678

Petromixes. Witco Chem./Sonneborn Div.
Chemical Description: Blend of sodium sulfonates, auxiliary fatty acid soaps, coupling agents, and antifoaming agents
Category: Emulsifier, anionic
Applications: Soluble oil base and degreaser; forms stable, nonfoaming O/W emulsions for use as cutting oils, textile processing oils, and metal degreasers
Form: Liquid; Color: Dark brown
Properties: Solubility: Sol. in organic solvents; Sp. gr.: 1.0 (60 F)
Reaction: Acid
Toxicity: May cause eye burn
Containers: Drums, T/W, T/C

Petronate 25C, 25H. Witco Chem./Sonneborn Div.
Chemical Description: Calcium petroleum sulfonate
Category: Detergent, rust inhibitor, emulsifier, anionic
Applications: Detergent and rust inhibitor component in lube oil additives, in rust preventive formulations, and as emulsifier for W/O systems
Form: Liquid; Color: 3.5 (ASTM, dilute)
Properties: Density: 8.2 lb/gal; Visc.: 170 SSU (100 C); Flash Pt.: 360 F; Fire Pt.: 390 F; Hydroxyl No.: 21 min.
Composition: 43–46% calcium sulfonate

Petronate CR. Witco Chem./Sonneborn Div.
Chemical Description: Sodium petroleum sulfonate
Category: Wetting agent, emulsifier, dispersant, rust preventative, anionic
Form: Viscous liquid; Color: Dark brown; Odor: Petroleum
Properties: Solubility: Sol. in all organic solvents; Density: 8.5 lb/gal; Sp. gr.: 1.02 (60 F); Visc.: 175 s (Furol, 100 C); Flash Pt.: 380 F; Fire Pt.: 415 F
Composition: 61–63% active
Reaction: Alkaline
Toxicity: May cause eye burns
Containers: Drums, T/W, T/C

Petronate HL. Witco Chem./Sonneborn Div.
Chemical Description: Sodium petroleum sulfonate
Category: Emulsifier, dispersing and wetting agent, anionic
Form: Viscous liquid; Color: Dark brown; Odor: Petroleum
Properties: Solubility: Sol. in oil,

organic solvents; Density: 8.5 lb/gal; Sp. gr.: 1.1 (60 F); Visc.: 135 s (Furol, 100 C); Flash Pt.: 380 F; Fire Pt.: 415 F
Composition: 61–63% active
Reaction: Alkaline
Toxicity: May cause eye burn
Containers: Drums, T/W, T/C

Petronate HMW. Witco Chem./Sonneborn Div.
Chemical Description: Calcium petroleum sulfonate
Category: Detergent, rust inhibitor
Applications: Lubricating oil additives, rust preventive formulations
Properties: Density: 8.2 lb/gal; Visc.: 200 SSU (100 C); Flash Pt.: 350 F; Fire Pt.: 380 F; Hydroxyl No.: 20 min.
Composition: 45–46% calcium sulfonate

Petronate K. Witco Chem./Sonneborn Div.
Chemical Description: Sodium petroleum sulfonate
Category: Wetting agent, emulsifier, dispersant, anionic
Applications: Emulsifier and wetting agent in dry cleaning soaps, textile processing, and other applications demanding light color
Form: Viscous liquid; Color: Light brown
Properties: Solubility: Oil-sol.; Density: 8.3 lb/gal; Sp. gr.: 1.1 (60 F); Visc.: 85 s (Furol, 100 C); Flash Pt.: 380 F; Fire Pt.: 415 F
Composition: 61–63% active
Reaction: Alkaline
Toxicity: May cause eye burn
Containers: Drums, T/W, T/C

Petronate L. Witco Chem./Sonneborn Div.
Chemical Description: Sodium petroleum sulfonate
Category: Wetting agent, emulsifier, dispersant, anionic
Applications: Emulsifier and wetting agent for oil and solvent systems
Form: Viscous liquid; Color: Dark brown; Odor: Petroleum
Properties: Solubility: Sol. in oil, organic solvents; Density: 8.5 lb/gal; Sp. gr.: 1.1 (60 F); Visc.: 125 s (Furol, 100 C); Flash Pt.: 380 F; Fire Pt.: 415 F
Composition: 61–63% active
Reaction: Alkaline
Toxicity: May cause eye burn
Containers: Drums, T/W, T/C

Petronate Basic. Witco Chem./Sonneborn Div.
Chemical Description: Barium petroleum sulfonate
Category: Oil additive, anionic
Applications: Additives for lube and industrial oils and fuels, specialty oils, greases
Form: Liquid; Color: 7.0 (ASTM, dilute)
Properties: Visc.: 250 SSU (100 C); Hydroxyl No.: 65 min.
Composition: 40–45% barium sulfonate

Petronate Neutral (50-S). Witco Chem./Sonneborn Div.
Chemical Description: Barium

petroleum sulfonate
Category: Oil additive, anionic
Applications: Additive for lube and industrial oils, fuel oil, undercoating, and greases
Form: Liquid; Color: 3.5 (ASTM, dilute)
Properties: Visc.: 100 SSU (100 C); Hydroxyl No.: 5.0 max.
Composition: 50–52% barium sulfonate

Petrosul 545. Penreco
Chemical Description: Sodium petroleum sulfonate
Category: Emulsifier, anionic
Applications: Demulsifying agent, flotation agent, fuel oil and grease additives, leather oils, metal working oils and metal cleaners, textile oils
Form: Viscous liquid; Color: 3.5 max. (ASTM-D-1500)
Properties: m.w.: 460–475 (avg.); Solubility: Oil-sol.; Density: 8.3 lb/gal; Stability: Heat stable to 400 F
Composition: 50% active ingred.
Containers: 55 gal steel drums, tank cars, transports

Petrosul 550. Penreco
Chemical Description: Sodium petroleum sulfonate
Category: Solubilizer, rust preventative, dispersant, anionic
Applications: Demulsifying agent, fuel oil and grease additives, lubricating oil additive manufacture, metal working oils and metal cleaners, rust preventatives
Form: Viscous liquid; Color: 3.5 max. (ASTM-D-1500)
Properties: m.w.: 505–520 (avg.); Solubility: Oil-sol.; Density: 8.3 lb/gal; Stability: Heat stable up to 400 F
Composition: 50% active ingred.
Containers: 55 gal drums, tank cars, transports

Petrosul 742. Penreco
Chemical Description: Sodium petroleum sulfonate
Category: Emulsifier, solubilizer, anionic
Applications: Textile and dry cleaning formulations; fat splitting; solubilization; demulsifying agent; flotation agent; fuel oil additives; leather oils; pigment grinding agents
Form: Viscous liquid; Color: 3.5 max. (ASTM-D-1500)
Properties: m.w.: 415–430 (avg.); Solubility: Oil-sol.; Density: 8.4 lb/gal; Stability: Heat stable up to 400 F
Composition: 62% active ingred.
Containers: 55 gal steel drums, tank cars, transports

Petrosul 744CL. Penreco
Chemical Description: Sodium petroleum sulfonate
Category: Emulsifier, solubilizing agent, anionic
Applications: Demulsifying agent, dry cleaning detergents, fat splitting agent, flotation agent, fuel oil additives, grease additives, leather oils, metal working oils, metal cleaners, pigment grinding agents, textile oils; esp. where a light

colored emulsifier is needed
Form: Viscous liquid; Color: 5 max. (ASTM-D-1500)
Properties: m.w.: 440–450 (avg.); Solubility: Oil-sol.; Density: 8.6 lb/gal; Stability: Heat stable up to 400 F
Composition: 62% active ingred.
Containers: 55 gal steel drums, tank cars, transports

Petrosul 745. Penreco
Chemical Description: Sodium petroleum sulfonate
Category: Emulsifier, anionic
Applications: Demulsifying agent, dry cleaning detergents, fat splitting agent, flotation agent, fuel oil additives, leather oils, pigment grinding aids, textile oils
Form: Viscous liquid; Color: 3.5 max. (ASTM-D-1500)
Properties: m.w.: 460–475 (avg.); Solubility: Oil-sol.; Density: 8.6 lb/gal; Stability: Heat stable up to 400 F
Composition: 62% active ingred.
Containers: 55 gal steel drums, tank cars, transports

Petrosul 750. Penreco
Chemical Description: Sodium petroleum sulfonate
Category: Dispersant, rust preventative, anionic
Applications: Demulsifying agent, fuel oil and grease additives, lubricating oil additive manufacture, metal working oils, metal cleaners, rust preventives
Form: Viscous liquid; Color: 3.5 max. (ASTM-D-1500)
Properties: m.w.: 505–520 (avg.); Solubility: Oil-sol.; Density: 8.6 lb/gal; Stability: Heat stable up to 400 F
Composition: 62% active ingred.
Containers: 55 gal steel drums, tank cars, transports

Petrosul Neutral Barium. Penreco
Chemical Description: Barium petroleum sulfonate
Category: Dispersant, rust preventative, anionic
Applications: Fuel oil and grease additives, rust preventives
Form: Viscous liquid; Color: 5 max. (ASTM-D-1500)
Properties: m.w.: 1100–1140 (avg.); Solubility: Oil-sol.; Density: 8.2 lb/gal; Stability: Heat stable up to 400 F
Composition: 40% active ingred.
Containers: 55 gal steel drums, tank cars, transports

Petrosul Neutral Calcium. Penreco
Chemical Description: Calcium petroleum sulfonate
Category: Dispersant, rust preventative, anionic
Applications: Fuel oil and grease additives; rust preventatives
Form: Viscous liquid; Color: 5 max. (ASTM-D-1500)
Properties: m.w.: 1000–1040 (avg.); Solubility: Oil-sol.; Density: 8.0 lb/gal; Stability: Heat stable up to 400 F
Composition: 40% active ingred.
Containers: 55 gal steel drums, tank cars, transports

Petrowet R. Du Pont Co.
Chemical Description: Sodium alkyl sulfonate
Category: Wetting agent, detergent, dispersant, penetrant, anionic
Applications: Petroleum industry–acidizing oil wells; metal working industry–pickling, preflux, and plating baths, degreasing, component in soldering flux; textile industry–vat color pigment padding, causticizing cotton, carbonizing wool; paper industry–parchmentizing paper; equipment cleaning–boiler cleaning, cleaning sugar mill rolls, refrigeration equipment; ceramic cleaning–marble, tile, stone, terrazzo cleaning sol'ns
Form: Clear liquid; Color: Yellow to amber; Odor: Alcoholic
Properties: Solubility: Sol. in water, 15% HCl, 10% H_2SO_4, 25% NaOH; Density: 9 lb/gal; Visc.: 15 cps (27 C); Gel Pt.: 21–23 F; Flash Pt.: 196 F (OC); Stability: Stable in high concs. of inorganic salts; effective surface and interfacial tension reduction in strong organic acids, nonoxidizing mineral acids, and strongly alkaline sol'ns.; Surface Tension: 47.1 dynes/cm (0.1% conc. @ 80 F); pH: 4.0–5.5 (1% aq.)
Composition: 22% active ingred.

Phospholan ALF-5. Lankro Chem. Ltd.
Chemical Description: Complex phosphate ester from fatty alcohol base
Category: Foam limiter, anionic (in neutralized form)
Applications: Foam limiter for use with anionic surfactants
Form: Solid; Color: Off-white; Odor: Mild
Properties: Solubility: Insol. in water; Visc.: 120 cs (80 C); Flash Pt.: > 300 F (COC); Pour Pt.: 65 C; Stability: Excellent stability in high concs. of acid, alkali, and electrolytes; pH: 2.0–3.0 (1% aq.)
Composition: 100% active ingred.
Toxicity: Corrosive organic acid–extremely irritating to skin and eyes
Containers: Stainless steel, glass, resin, or polythene-lined equipment
Warnings and Caution: Wear protective goggles and gloves

Phospholan KPE-4. Lankro Chem. Ltd.
Chemical Description: Potassium salt of complex phosphate ester
Category: Hydrotrope, anionic
Applications: Solubilization of low foaming surfactants into highly built liquids, e.g., bottle washing, low foam alkaline cleaning, liquid machine dishwashing detergents, other industrial and domestic detergents and cleaners
Form: Liquid; Color: Pale straw; Odor: Mild
Properties: Solubility: Full sol. in water @ 20 C; Sp. gr.: 1.294 (20 C); Visc.: 180 cs (20 C); Flash Pt.: > 200 F (COC); Pour Pt.: < 0 C; Stability: Stable under highly alkaline and acidic conditions; pH: 6.0–8.0 (1% aq.)

Composition: 65% active ingred.
Reaction: Neutral
Toxicity: May cause irritation on prolonged contact with conc. form
Containers: Stainless steel, glass, resin, or polythene-lined equipment
Warnings and Caution: Wear protective goggles and gloves; spillages may be slippery; do not mix with strong oxidizing or reducing agents (potentially explosive)
References: Bulletin SI:29 FM678

Phospholan PDB-3. Lankro Chem. Ltd.
Chemical Description: Complex phosphate ester based on fatty alcohol ethoxylate
Category: Wetting agent, detergent, emulsifier, lubricant, anionic
Applications: Detergent component, esp. for alkaline cleaners and heavy duty industrial powders and liquid detergents; kier boiling of cottons and in mercerizing baths; dyestuff carrier and emulsifier for phenols; dry cleaning applications; electrolytic metal cleaning formulations; emulsifier in emulsion polymerization and for pesticides and herbicides; specialty waterless hand cleaners; metal working lubricants with anticorrosive properties for cutting fluid formulations and in grinding, rolling, and drawing; stripping of surface coatings; production of iodophors; oil emulsifier and antistat for degreasing of synthetic fibers; esp. useful in conc. alkali, acid, and electrolyte sol'ns.
Form: Liquid; Color: Pale straw; Odor: Mild
Properties: Solubility: Disp. in water @ 20 C; Sp. gr.: 1.000 (20 C); Visc.: 2400 cs (20 C); Flash Pt.: > 300 F (COC); Pour Pt.: 10 C; Stability: Excellent stability in high concs. of acid, alkali, and electrolytes; pH: 2.0–3.0 (1% aq.)
Composition: 100% active ingred.
Reaction: Acid
Toxicity: Corrosive organic acid—extremely irritating to skin and eyes
Containers: Stainless steel, glass, resin, or polythene-lined equipment
Warnings and Caution: Wear protective goggles and gloves
References: Bulletin SI:28 FM678

Phospholan PNP-9. Lankro Chem. Ltd.
Chemical Description: Complex phosphate ester based on nonyl phenol ethoxylate, acid form
Category and Applications: See Phospholan PDB-3
Form: Viscous liquid; Color: Dark amber; Odor: Mild
Properties: Solubility: Sol. in water @ 20 C; Sp. gr.: 1.200 (20 C); Visc.: 9400 cs (40 C); Flash Pt.: > 300 F (COC); Pour Pt.: 15 C; Stability: Excellent stability in high concs. of acid, alkali, and electrolytes; pH: 2.0–3.0 (1% aq.)
Composition: 100% active ingred.
Reaction: Acid
Toxicity: Corrosive organic acid—

extremely irritating to skin and eyes
Containers: Stainless steel, glass, resin, or polythene-lined equipment
Warnings and Caution: Wear protective goggles and gloves
References: Bulletin SI;28 FM678

Pluracol E-200. BASF Wyandotte Corp.
Chemical Description: Polyethylene glycol
Category: Intermediate, lubricant
Applications: Intermediate for preparation of nonionic surfactants, etc.; base, coating, vehicle, extender, and coupling agent for pharmaceutical and cosmetic lotions, creams, ointments, suppositories, and tablets; stabilizer, solvent, and coupler in antiperspirants, hair rinses, etc.; electropolishing baths for metals; lubricant for metal extruding, drawing; rubber industry lubricant; wood treatment; textile conditioning and antistatic agents, sizing agents, softeners, and lubricants; inks
Form: Clear liquid; Color: Colorless
Properties: Solubility: Sol. in water and most organic solvents except aliphatic hydrocarbons; Density: 9.4 lb/gal; Sp. gr.: 1.12; Visc.: 4.36 cs (210 F); F.P.: –48 C; Flash Pt.: 360 F; Surface Tension: 57.2 dynes/cm (1% sol'n.); pH: 6.5 (5% aq.)
Containers: 5 and 55 gal drums, tank cars, tank trucks

Pluracol E-300. BASF Wyandotte Corp.
Chemical Description: Polyethylene glycol
Category and Applications: See Pluracol E-200; also as coating, binder, plasticizing agent, lubricant, dispersant in food tablets and preparations
Form: Clear liquid; Color: Colorless
Properties: Solubility: Sol. in water and most organic solvents except aliphatic hydrocarbons; Density: 9.4 lb/gal; Sp. gr.: 1.12; Visc.: 5.75 cs (210 F); F.P.: –22 C; Flash Pt.: 410 F; Surface Tension: 62.9 dynes/cm (1% sol'n.); pH: 5.7 (5% aq.)
Containers: 5 and 55 gal drums, tank cars, tank trucks

Pluracol E-400. BASF Wyandotte Corp.
Chemical Description: Polyethylene glycol
Category and Applications: See Pluracol E-300
Form: Clear liquid; Color: Colorless
Properties: Solubility: Sol. in water and most organic solvents except aliphatic hydrocarbons; Density: 9.4 lb/gal; Sp. gr.: 1.12; Visc.: 7.39 cs (210 F); F.P.: 2.5 C; Flash Pt.: 460 F; Surface Tension: 66.6 dynes/cm (1% sol'n.); pH: 6.2 (5% aq.)
Containers: 5 and 55 gal drums, tank cars, tank trucks

Pluracol E-600. BASF Wyandotte Corp.
Chemical Description: Polyethylene glycol
Category and Applications: See Pluracol E-300
Form: Clear liquid; Color: Colorless
Properties: Solubility: Sol. in water and most organic solvents except aliphatic hydrocarbons; Density: 9.4 lb/gal; Sp. gr.: 1.12; Visc.: 10.83 cs (210 F); F.P.: 16.5 C; Flash Pt.: 480 F; Surface Tension: 65.2 dynes/cm (1% sol'n.); pH: 5.3 (5% aq.)
Containers: 5 and 55 gal drums, tank cars, tank trucks

Pluracol E-1500. BASF Wyandotte Corp.
Chemical Description: Polyethylene glycol
Category and Applications: See Pluracol E-300
Form: Waxy solid; Color: White
Properties: Solubility: Sol. in water and most organic solvents except aliphatic hydrocarbons; Density: 10.0 lb/gal; Sp. gr.: 1.20; M.P.: 46.0–47.5 C; Flash Pt.: > 490 F; Surface Tension: 62.8 dynes/cm (1% sol'n.); pH: 6.7 (5% aq.)
Containers: 55 gal drums, tank cars, tank trucks

Pluracol E-4000. BASF Wyandotte Corp.
Chemical Description: Polyethylene glycol
Category and Applications: See Pluracol E-300
Form: Waxy solid; Color: White
Properties: Solubility: Sol. in water and most organic solvents except aliphatic hydrocarbons; Density: 10.0 lb/gal; Sp. gr.: 1.20; M.P.: 59.5 C; Flash Pt.: > 490 F; Surface Tension: 61.9 dynes/cm (1% sol'n.); pH: 6.7 (5% aq.)
Containers: 55 gal drums, tank cars, tank trucks

Pluracol E-6000. BASF Wyandotte Corp.
Chemical Description: Polyethylene glycol
Category and Applications: See Pluracol E-200
Form: Waxy solid; Color: White
Properties: Solubility: Sol. in water and most organic solvents except aliphatic hydrocarbons; Density: 10.0 lb/gal; Sp. gr.: 1.21; M.P.: 61.0 C; Flash Pt.: > 490 F; Surface Tension: 62.1 dynes/cm (1% sol'n.); pH: 6.7 (5% aq.)
Containers: 55 gal drums, tank cars, tank trucks

Pluracol P-410. BASF Wyandotte Corp.
Chemical Description: Polypropylene glycol
Category: Intermediate, antifoam agent, lubricant, plasticizer
Applications: Chemical intermediate; antifoam in fermentation and in paint formulations; antiblooming agent for pentachlorophenol treated wood; binder and lubricant for ceramics; lubricant for the cold reduction of magne-

sium and aluminum; plasticizer of resin-treated papers; preparation of polyurethane foams; hydraulic and grinding fluids; ore flotation; Kosher approved
Form: Liquid; Color: Water-white; Odor: Slight ether-like
Properties: m.w.: 425; Solubility: Completely water-sol.; Density: 8.36 lb/gal; Sp. gr.: 1.005; Visc.: 35 cps (100 F); Flash Pt.: > 400 F (PMCC); Pour Pt.: –35 F; pH: 6–7
Toxicity: Slight eye irritant
Containers: 55 gal drums, tank cars

Pluracol P-710. BASF Wyandotte Corp.
Chemical Description: Polypropylene glycol
Category and Applications: See Pluracol P-410
Form: Liquid; Color: Water-white; Odor: Slight ether-like
Properties: m.w.: 775; Solubility: Moderately sol. in water; Density: 8.35 lb/gal; Sp. gr.: 1.004; Visc.: 65 cps (100 F); Flash Pt.: > 400 F (PMCC); Pour Pt.: –35 F; pH: 6–7
Toxicity: Slight eye irritant
Containers: 55 gal drums, tank cars

Pluracol P-1010. BASF Wyandotte Corp.
Chemical Description: Polypropylene glycol
Category and Applications: See Pluracol P-410
Form: Liquid; Color: Water-white; Odor: Slight ether-like
Properties: m.w.: 1050; Solubility: Practically insol. in water @ room temp.; Density: 8.38 lb/gal; Sp. gr.: 1.007; Visc.: 80 cps (100 F); Flash Pt.: > 400 F (PMCC); Pour Pt.: –35 F; pH: 6–7
Toxicity: Slight eye irritant
Containers: 55 gal drums, tank cars

Pluracol P-2010. BASF Wyandotte Corp.
Chemical Description: Polypropylene glycol
Category and Applications: See Pluracol P-410
Form: Liquid; Color: Water-white; Odor: Slight ether-like
Properties: m.w.: 2000; Solubility: Practically insol. in water @ room temp.; Density: 8.34 lb/gal; Sp. gr.: 1.002; Visc.: 175 cps (100 F); Flash Pt.: > 400 F (PMCC); Pour Pt.: –35 F; pH: 6–7
Toxicity: Slight eye irritant
Containers: 55 gal drums, tank cars

Pluracol P-3010. BASF Wyandotte Corp.
Chemical Description: Polypropylene glycol
Category and Applications: See Pluracol P-410
Form: Liquid; Color: Water-white; Odor: Slight ether-like
Properties: m.w.: 3000; Solubility: Practically insol. in water @ room temp.; Density: 8.32 lb/gal; Sp. gr.: 1.00; Visc.: 295 cps (100

F); Flash Pt.: > 400 F (PMCC); Pour Pt.: –20 F; pH: 6–7
Toxicity: Slight eye irritant
Containers: 55 gal drums, tank cars

Pluracol P-4010. BASF Wyandotte Corp.
Chemical Description: Polypropylene glycol
Category and Applications: See Pluracol P-410
Form: Liquid; Color: Water-white; Odor: Slight ether-like
Properties: m.w.: 4000; Solubility: Practically insol. in water @ room temp.; Density: 8.33 lb/gal; Sp. gr.: 1.00; Visc.: 550 cps (100 F); Flash Pt.: > 400 F (PMCC); Pour Pt.: –20 F; pH: 6–7
Toxicity: Slight eye irritant
Containers: 55 gal drums, tank cars

Pluracol W-170. BASF Wyandotte Corp.
Chemical Description: Polyalkoxylated polyglycol
Category: Defoaming agent, lubricant
Applications: Components in demulsifying and wetting formulations; brake fluids and metalworking fluids; rubber lubricant; textile applications; fiber lubricant; defoamers for hot and cold applications, food processing, chemical processing; cosmetic formulations
Form: Liquid; Color: 50 max. (APHA)
Properties: Solubility: Sol. in water, alcohols, ketones, esters, benzene, toluene, glycol ethers, chlorinated solvents; slightly sol. in gasoline and petroleum oils; Density: 8.58 lb/gal; Sp. gr.: 1.03 (20/20 C); Visc.: 160–180 SUS (100 F); Cloud Pt.: 73 C (1%); Flash Pt.: 360 F (OC); Pour Pt.: –45 F; Stability: Excellent thermal and chemical stability; pH: 5.5–7.0 (10% aq.)
Containers: 55 gal steel drums, tank cars, tank trucks

Pluracol W352ON. BASF Wyandotte Corp.
Chemical Description: Polyalkoxylated polyglycol
Category and Applications: See Pluracol W-170
Form: Viscous liquid; Color: 40 max. (APHA)
Properties: Solubility: Sol. in water, alcohols, ketones, esters, benzene, toluene, glycol ethers, chlorinated solvents; slightly sol. in gasoline and petroleum oils; Density: 8.83 lb/gal; Sp. gr.: 1.06 (20/20 C); Visc.: 3520 SUS (100 F); Cloud Pt.: 57 C (1%); Flash Pt.: 437 F (OC); Pour Pt.: –20 F; Stability: Excellent thermal and chemical stability; pH: 6.0–7.5 (10% aq.)
Containers: 55 gal steel drums, tank cars, tank trucks

Pluracol W352ON-RL. BASF Wyandotte Corp.
Chemical Description: Polyalkoxylated polyglycol
Category and Applications: See Pluracol W-170

Form: Viscous liquid; Color: 200 max. (APHA)
Properties: Solubility: Sol. in water, alcohols, ketones, esters, benzene, toluene, glycol ethers, chlorinated solvents; slightly sol. in gasoline and petroleum oils; Density: 8.87 lb/gal; Sp. gr.: 1.065 (20/20 C); Visc.: 1752–2500 SUS (100 F); Cloud Pt.: 55.5 C (1%); Flash Pt.: 440 F (OC); Pour Pt.: 28.4 F; Stability: Excellent thermal and chemical stability; pH: 6.0–7.5 (10% aq.)
Containers: 55 gal steel drums, tank cars, tank trucks

Pluracol W-660. BASF Wyandotte Corp.
Chemical Description: Polyalkoxylated polyglycol
Category and Applications: See Pluracol W-170
Form: Viscous liquid; Color: 50 max. (APHA)
Properties: Solubility: Sol. in water, alcohols, ketones, esters, benzene, toluene, glycol ethers, chlorinated solvents; slightly sol. in gasoline and petroleum oils; Density: 8.79 lb/gal; Sp. gr.: 1.055 (20/20 C); Visc.: 660 SUS (100 F); Cloud Pt.: 60.5 C (1%); Flash Pt.: 440 F (OC); Pour Pt.: –34 F; Stability: Excellent thermal and chemical stability; pH: 5.5–7.5 (10% aq.)
Containers: 55 gal steel drums, tank cars, tank trucks

Pluracol W-2000. BASF Wyandotte Corp.
Chemical Description: Polyalkoxylated polyglycol
Category and Applications: See Pluracol W-170
Form: Viscous liquid; Color: 50 max. (APHA)
Properties: Solubility: Sol. in water, alcohols, ketones, esters, benzene, toluene, glycol ethers, chlorinated solvents; slightly sol. in gasoline and petroleum oils; Density: 8.83 lb/gal; Sp. gr.: 1.06 (20/20 C); Visc.: 2000 SUS (100 F); Cloud Pt.: 57.0 C (1%); Flash Pt.: 440 F (OC); Pour Pt.: –25 F; Stability: Excellent thermal and chemical stability; pH: 5.5–7.5 (10% aq.)
Containers: 55 gal steel drum, tank car, tank truck

Pluracol W5100. BASF Wyandotte Corp.
Chemical Description: Polyalkoxylated polyglycol
Category and Applications: See Pluracol W-170
Form: Viscous liquid; Color: 40 max. (APHA)
Properties: Solubility: Sol. in water, alcohols, ketones, esters, benzene, toluene, glycol ethers, chlorinated solvents; slightly sol. in gasoline and petroleum oils; Density: 8.83 lb/gal; Sp. gr.: 1.06 (20/20 C); Visc.: 5100 SUS (100 F); Cloud Pt.: 55 C (1%); Flash Pt.: 437 F (OC); Pour Pt.: –20 F; Stability: Excellent thermal and chemical stability; pH: 5.5–7.0 (10% aq.)
Containers: 55 gal steel drums, tank cars, tank trucks

Pluracol W5100N. BASF Wyandotte Corp.
Chemical Description: Identical to Pluracol W-5100 except W-5100N contains ≈ 0.25% potassium acetate
Form and Properties: See Pluracol W5100

Pluradot HA-410. BASF Wyandotte/Ind. Chem. Group
Chemical Description: Trifunctional polyoxyalkylene glycol
Category: Emulsifier, solubilizer, deduster, wetting agent, demulsifier, detergent, dispersant, nonionic
Applications: Low foaming and low pour pt. surfactant for hard surface cleaning, machine dishwashing, etc.
Form: Liquid
Properties: m.w.: 3200; Sp. gr.: 1.03 (26/26 C); Visc.: 450 cps; Cloud Pt.: 25 C (1%); Pour Pt.: −15 F; Surface Tension: 36 dynes/cm (0.1%)
Composition: 100% conc.

Pluradot HA-420. BASF Wyandotte/Ind. Chem. Group
Chemical Description: Trifunctional polyoxyalkylene glycol
Category and Applications: See Pluradot HA-410
Form: Liquid
Properties: m.w.: 3600; Sp. gr.: 1.04 (26/26 C); Visc.: 565 cps; Cloud Pt.: 28 C (1%); Pour Pt.: −10 F; Surface Tension: 37 dynes/cm (0.1%)
Composition: 100% conc.

Pluradot HA-430. BASF Wyandotte/Ind. Chem. Group
Chemical Description: Trifunctional polyoxyalkylene glycol
Category and Applications: See Pluradot HA-410
Form: Liquid
Properties: m.w.: 3900; Sp. gr.: 1.05 (26/26 C); Visc.: 700 cps; Cloud Pt.: 43 C (1%); Pour Pt.: 20 C; Surface Tension: 38 dynes/cm (0.1%)
Composition: 100% conc.

Pluradot HA-433. BASF Wyandotte/Ind. Chem. Group
Chemical Description: Trifunctional polyoxyalkylene glycol
Category and Applications: See Pluradot HA-410
Form: Liquid
Properties: m.w.: 3900; Solubility: Not completely soluble; Sp. gr.: 1.05 (26/26 C); Visc.: 700 cps; Pour Pt.: 20 F
Composition: 100% conc.

Pluradot HA-440. BASF Wyandotte/Ind. Chem. Group
Chemical Description: Trifunctional polyoxyalkylene glycol
Category and Applications: See Pluradot HA-410
Form: Liquid
Properties: m.w.: 4450; Sp. gr.: 1.05 (26/26 C); Visc.: 750 cps; Cloud Pt.: 54 C (1%); Pour Pt.: 25 C; Surface Tension: 39 dynes/cm (0.1% conc.)
Composition: 100% conc.

Pluradot HA-450. BASF Wyandotte/Ind. Chem. Group
Chemical Description: Trifunctional polyoxyalkylene glycol
Category and Applications: See Pluradot HA-410
Form: Paste
Properties: m.w.: 5700; Sp. gr.: 1.06 (26/26 C); Cloud Pt.: 74 C (1%); Pour Pt.: 60 F; Surface Tension: 41 dynes/cm (0.1%)
Composition: 100% conc.

Pluradot HA-510. BASF Wyandotte/Ind. Chem. Group
Chemical Description: Trifunctional polyoxyalkylene glycol
Category and Applications: See Pluradot HA-410
Form: Liquid
Properties: m.w.: 4600; Sp. gr.: 1.03 (26/26 C); Visc.: 710 cps; Cloud Pt.: 24 C (1%); Pour Pt.: −15 F; Surface Tension: 35 dynes/cm (0.1%, not completely sol.)
Composition: 100% conc.

Pluradot HA-520. BASF Wyandotte/Ind. Chem. Group
Chemical Description: Trifunctional polyoxyalkylene glycol
Category and Applications: See Pluradot HA-410
Form: Liquid
Properties: m.w.: 5000; Sp. gr.: 1.03 (26/26 C); Visc.: 760 cps; Cloud Pt.: 27 C (1%); Pour Pt.: 5 F; Surface Tension: 36 dynes/cm (0.1%)
Composition: 100% conc.

Pluradot HA-530. BASF Wyandotte/Ind. Chem. Group
Chemical Description: Trifunctional polyoxyalkylene glycol
Category and Applications: See Pluradot HA-410
Form: Liquid
Properties: m.w.:5300; Sp. gr.: 1.04 (26/26 C); Visc.: 850 cps; Cloud Pt.: 42 C (1%); Pour Pt.: 30 F; Surface Tension: 37 dynes/cm (0.1%)
Composition: 100% conc.

Pluradot HA-540. BASF Wyandotte/Ind. Chem. Group
Chemical Description: Trifunctional polyoxyalkylene glycol
Category and Applications: See Pluradot HA-410
Form: Liquid
Properties: m.w.: 6000; Sp. gr.: 1.05 (26/26 C); Visc.: 1240 cps; Cloud Pt.: 54 C (1%); Pour Pt.: 40 F; Surface Tension: 38 dynes/cm (0.1%)
Composition: 100% conc.

Pluradot HA-550. BASF Wyandotte/Ind. Chem. Group
Chemical Description: Trifunctional polyoxyalkylene glycol
Category and Applications: See Pluradot HA-410
Form: Paste
Properties: m.w.: 7500; Sp. gr.: 1.06 (26/26 C); Cloud Pt.: 77 C (1%); Pour Pt.: 75 F; Surface Tension: 39 dynes/cm (0.1%)
Composition: 100% conc.

Plurafac A-24. BASF Wyandotte/ Ind. Chem. Group
Chemical Description: Straight chain, primary aliphatic specially oxyalkylated alcohol
Category: Detergent, nonionic
Applications: Commercial and home cleaning formulations, e.g., heavy-duty, light-duty, all purpose liquid, powder, and tablet products
Form: Liquid; Color: Colorless
Properties: Ref. Index: 1.4501; Solubility: Sol. in methyl, ethyl, and isopropyl alcohols, acetone, methyl ethyl ketone, butyl Cellosolve, propylene glycol, chloroform, perchloroethylene, toluene, xylene, kerosene; sol. < 10% in water; Density: 7.7 lb/gal; Sp. gr.: 0.927; Visc.: 27 cps; HLB: 6; Cloud Pt.: 0 C (1%); Flash Pt.: 400 F; Fire Pt.: 405 F; Pour Pt.: 56 F; Surface Tension: 29.4 dynes/cm (insol. @ 0.1%); pH: 6–7 (1% sol'n.); Biodegradable
Composition: 100% conc.
Reaction: Neutral
Toxicity: Mild skin irritant

Plurafac A-27. BASF Wyandotte/ Ind. Chem. Group
Chemical Description: Straight chain, primary aliphatic specially oxyalkylated alcohol
Category and Applications: See Plurafac A-24
Form: Paste; Color: White
Properties: Solubility: Sol. in water, methyl alcohol, ethyl alcohol, acetone, methyl ethyl ketone, butyl Cellosolve, propylene glycol, chloroform, toluene, xylene; sol. < 10% in isopropyl alcohol, ethylene glycol, and perchloroethylene; Density: 8.4 lb/gal (50/25 C); HLB: 14.5; Cloud Pt.: 100 C (1%); Flash Pt.: 460 F; Pour Pt.: 75 F; Surface Tension: 31.0 dynes/cm (0.1%); pH: 6–7 (1% sol'n.); Biodegradable
Composition: 100% conc.
Reaction: Neutral
Toxicity: Mild skin irritant

Plurafac A-38. BASF Wyandotte/ Ind. Chem. Group
Chemical Description: Straight chain, primary aliphatic specially oxyalkylated alcohol
Category and Applications: See Plurafac A-24
Form: Solid; Color: White
Properties: Solubility: Sol. in water, methyl, ethyl, and isopropyl alcohols, acetone, methyl ethyl ketone, butyl Cellosolve, chloroform, perchloroethylene, toluene, xylene; M.P.: 114 F; HLB: 19; Cloud Pt.: > 100 C (1%); Flash Pt.: 480 F; Fire Pt.: 570 F; Surface Tension: 43.6 dynes/cm (0.1%); pH: 6–7; Biodegradable
Composition: 100% conc.
Reaction: Neutral
Toxicity: Mild skin irritant

Plurafac A-39. BASF Wyandotte/ Ind. Chem. Group
Chemical Description: Straight chain, primary aliphatic specially oxyalkylated alcohol
Category and Applications: See Plurafac A-24

Form: Solid; Color: White
Properties: Solubility: Sol. in water, chloroform; sol. < 10% in methyl and ethyl alcohols, acetone, methyl ethyl ketone, toluene, xylene; M.P.: 132 F; HLB: 20; Cloud Pt.: > 100 C (1%); Flash Pt.: 480 F; Fire Pt.: 570 F; Surface Tension: 53.2 dynes/cm (0.1%); pH: 6–7; Biodegradable
Composition: 100% conc.
Reaction: Neutral
Toxicity: Mild skin irritant

Plurafac A-46. BASF Wyandotte/ Ind. Chem. Group
Chemical Description: Straight chain, primary aliphatic specially oxyalkylated alcohol
Category and Applications: See Plurafac A-24
Form: Paste; Color: White
Properties: Solubility: Sol. in water, methyl, ethyl, and isopropyl alcohols, acetone, methyl ethyl ketone, butyl Cellosolve, chloroform, perchloroethylene, toluene, xylene; Density: 8.2 lb/gal (50/25 C); HLB: 13; Cloud Pt.: 72 C (1%); Flash Pt.: 475 F; Pour Pt.: 75 F; Surface Tension: 32.5 dynes/cm (0.1%); pH: 6–7 (1% sol'n.); Biodegradable
Composition: 100% conc.
Reaction: Neutral
Toxicity: Mild skin irritant

Plurafac B-25-5. BASF Wyandotte/Ind. Chem. Group
Chemical Description: Straight chain, primary aliphatic specially oxyalkylated alcohol
Category and Applications: See Plurafac A-24
Form: Clear liquid; Color: Colorless
Properties: Ref. Index: 1.4570; Solubility: Sol. in water, methyl, ethyl, and isopropyl alcohols, acetone, methyl ethyl ketone, butyl Cellosolve, propylene glycol, chloroform, perchloroethylene, toluene, xylene; sol. < 10% in propylene glycol and kerosene; Density: 8.2 lb/gal; Sp. gr.: 1.002; Visc.: 100 cps; HLB: 12; Cloud Pt.: 65 C (1%); Pour Pt.: 9 F; Surface Tension: 36.4 dynes/cm (0.1%); pH: 6–7 (1%); Biodegradable
Composition: 100% conc.
Reaction: Neutral
Toxicity: Mild skin irritant

Plurafac B-26. BASF Wyandotte/ Ind. Chem. Group
Chemical Description: Straight chain, primary aliphatic specially oxyalkylated alcohol
Category and Applications: See Plurafac A-24
Form: Clear liquid; Color: Colorless
Properties: Ref. Index: 1.4587; Solubility: Sol. in water, methyl, ethyl, and isopropyl alcohols, acetone, methyl ethyl ketone, butyl Cellosolve, chloroform, perchloroethylene, toluene, xylene; sol. < 10% in propylene glycol; Density: 8.5 lb/gal; Sp. gr.: 1.020; Visc.: 114 cps; HLB: 14; Cloud Pt.: 73 C (1%); Flash Pt.: 475 F; Fire Pt.: 550 F; Pour Pt.: 35 F; Surface Tension: 36.3 dynes/cm (0.1%);

pH: 6–7 (1% sol'n.); Biodegradable
Composition: 100% conc.
Reaction: Neutral
Toxicity: Mild skin irritant

Plurafac C-17. BASF Wyandotte/ Ind. Chem. Group
Chemical Description: Straight chain, primary aliphatic specially oxyalkylated alcohol
Category and Applications: See Plurafac A-24
Form: Opaque liquid; Color: White
Properties: Ref. Index: 1.4526; Solubility: Sol. in water, methyl, ethyl, and isopropyl alcohols, acetone, methyl ethyl ketone, butyl Cellosolve, ethylene glycol, propylene glycol, chloroform, perchloroethylene, toluene, xylene; Density: 8.6 lb/gal; Sp. gr.: 1.030; Visc.: 100 cps; HLB: 16; Cloud Pt.: 81 C (1%); Flash Pt.: 520 F; Fire Pt.: 535 F; Pour Pt.: 56 F; Surface Tension: 34.4 dynes/cm (0.1%); pH: 6–7 (1% sol'n.); Biodegradable
Composition: 100% conc.
Reaction: Neutral
Toxicity: Mild skin irritant

Plurafac D-25. BASF Wyandotte/ Ind. Chem. Group
Chemical Description: Straight chain, primary aliphatic specially oxyalkylated alcohol
Category and Applications: See Plurafac A-24
Form: Slightly cloudy liquid
Properties: Ref. Index: 1.4560; Solubility: Sol. in water, methyl, ethyl, and isopropyl alcohols, acetone, methyl ethyl ketone, butyl Cellosolve, propylene glycol, chloroform, perchloroethylene, toluene, xylene; Density: 8.4 lb/gal; Sp. gr.: 1.007; Visc.: 95 cps; HLB: 11; Cloud Pt.: 59 C (1%); Flash Pt.: 465 F; Fire Pt.: 515 F; Pour Pt.: 0 F; Surface Tension: 34.3 dynes/cm (0.1%); pH: 6–7 (1% sol'n.); Biodegradable
Composition: 100% conc.
Reaction: Neutral
Toxicity: Mild skin irritant

Plurafac RA-20. BASF Wyandotte/ Ind. Chem. Group
Chemical Description: Straight chain, primary aliphatic specially oxyalkylated alcohol
Category: Detergent, dispersant, wetting agent, emulsifier, deduster, nonionic
Applications: Esp. useful for low foam applications, e.g., in rinse aids, and dishwashing products
Form: Clear liquid; Color: Colorless
Properties: Ref. Index: 1.4530; Solubility: Sol. in water, methyl, ethyl, and isopropyl alcohols, acetone, methyl ethyl ketone, butyl Cellosolve, propylene glycol, chloroform, perchloroethylene, toluene, xylene, kerosene; Density: 8.3 lb/gal; Sp. gr.: 0.992; Visc.: 83 cps; HLB: 10; Cloud Pt.: 45 C (1%); Flash Pt.: 475 F; Fire Pt.: 530 F; Pour Pt.: 45 F; Surface Tension: 32.8 dynes/cm (0.1%); pH: 6–7 (1% sol'n.); Biodegradable

Composition: 100% conc.
Reaction: Neutral
Toxicity: Mild skin irritant

Plurafac RA-30. BASF Wyandotte/ Ind. Chem. Group
Chemical Description: Straight chain, primary aliphatic specially oxyalkylated alcohol
Category and Applications: See Plurafac RA-20
Form: Clear liquid; Color: Colorless
Properties: Ref. Index: 1.4534; Solubility: Sol. in water, methyl, ethyl, and isopropyl alcohols, acetone, methyl ethyl ketone, butyl Cellosolve, propylene glycol, chloroform, perchloroethylene, toluene, xylene, kerosene; Density: 8.1 lb/gal; Sp. gr.: 0.973; Visc.: 56 cps; HLB: 9; Cloud Pt.: 35 C (1%); Flash Pt.: 455 F; Fire Pt.: 520 F; Pour Pt.: 50 F; Surface Tension: 30.6 dynes/cm (0.1%); pH: 6–7 (1% sol'n.); Biodegradable
Composition: 100% conc.
Reaction: Neutral
Toxicity: Mild skin irritant

Plurafac RA-40. BASF Wyandotte/ Ind. Chem. Group
Chemical Description: Straight chain, primary aliphatic specially oxyalkylated alcohol
Category and Applications: See Plurafac RA-20
Form: Clear liquid; Color: Colorless
Properties: Ref. Index: 1.4510; Solubility: Sol. in water, methyl, ethyl, and isopropyl alcohols, acetone, methyl ethyl ketone, butyl Cellosolve, chloroform, perchloroethylene, toluene, xylene, kerosene; Density: 8.1 lb/gal; Sp. gr.: 0.970; Visc.: 91 cps; HLB: 7; Cloud Pt.: 25 C (1%); Flash Pt.: 437 F; Fire Pt.: 525 F; Pour Pt.: –15 F; Surface Tension: 30.5 dynes/cm (0.1%); pH: 6–7 (1% sol'n.); Biodegradable
Composition: 100% conc.
Reaction: Neutral
Toxicity: Mild skin irritant

Plurafac RA-43. BASF Wyandotte/ Ind. Chem. Group
Chemical Description: Straight chain, primary aliphatic specially oxyalkylated alcohol
Category and Applications: See Plurafac RA-20
Form: Opaque liquid; Color: White
Properties: Density: 7.9 lb/gal; Sp. gr.: 0.950; Visc.: 120 cps; HLB: 7; Cloud Pt.: 24 C (1%); Flash Pt.: 437 F; Fire Pt.: 525 F; Pour Pt.: 21 F; Surface Tension: 30.5 dynes/cm (0.1%); pH: 3–4 (1% sol'n.); Biodegradable
Composition: 100% conc.
Toxicity: Mild skin irritant

Plurafac T-55. BASF Wyandotte Corp.
Chemical Description: Oxyalkylated primary aliphatic alcohol
Category: Detergent, emulsifier, nonionic
Applications: Agricultural spray adjuvants, emulsifiers, heavy duty

laundry detergents, general purpose cleaners, car wash detergents, degreasers
Form: Liquid; Color: Pale yellow
Properties: Solubility: Water-sol.; Sp. gr.: 1.012; Visc.: 164 cps; Cloud Pt.: 57 C (1%); Pour Pt.: 34 F; Surface Tension: 36.2 dynes/cm (0.1% conc.); Biodegradable
Toxicity: Eye irritant, severe skin irritant

Pluriol E 200. BASF AG
Chemical Description: Polyethylene glycol
Category: Solubilizer, impregnating agent, plasticizer, humectant, lubricant, dispersant, mold-release agent, flow improver, thermal and hydraulic fluid, organic intermediate, nonionic
Applications: Detergents and cleaners; dye and pigment dispersant; in inks–Indian, flexographic, etc.; textile and coatings industry; coloring ceramics; softener in paper industry; plasticizer in adhesives industry and in production of cellulose film; ceramics and metalworking lubricant
Form: Clear liquid; Color: Colorless
Properties: m.w.: 200; Solubility: Water-sol.; Density: ≈ 1.12 g/cc (20 C); Visc.: 55–65 cs (20 C); M.P.: < –40 C; Flash Pt.: > 150 C; Stability: Good heat stability in absence of oxygen; resistant to moderately strong acids, alkalis, and salt sol'ns.; pH: 6.0–7.5 (1% aq.)
Composition: 100% conc.

Pluriol E 300. BASF AG
Chemical Description: Polyethylene glycol
Category and Applications: See Pluriol E 200
Form: Clear liquid; Color: Colorless
Properties: m.w.: 300; Solubility: Water-sol.; Density: ≈ 1.13 g/cc (20 C); Visc.: 75–95 cs (20 C); M.P.: –20 to –10 C; Flash Pt.: > 150 C; Stability: Good heat stability in absence of oxygen; resistant to moderately strong acids, alkalis, and salt sol'ns.; pH: 6.0–7.5 (1% aq.)
Composition: 100% conc.

Pluriol E 400. BASF AG
Chemical Description: Polyethylene glycol
Category and Applications: See Pluriol E 200
Form: Clear liquid; Color: Colorless
Properties: m.w.: 400; Solubility: Water-sol.; Density: ≈ 1.13 g/cc (20 C); Visc.: 105–120 cs (20 C); M.P.: –10 to 10 C; Flash Pt.: > 150 C; Stability: Good heat stability in absence of oxygen; resistant to moderately strong acids, alkalis, and salt sol'ns.; pH: 6.0–7.5 (1% aq.)
Composition: 100% conc.

Pluriol E 600. BASF AG
Chemical Description: Polyethylene glycol

Category and Applications: See Pluriol E 200
Form: Semisolid
Properties: m.w.: 600; Solubility: Water-sol.; Density: ≈ 1.15 g/cc (20 C); Visc.: 8–11 cs (99 C); M.P.: 10–20 C; Flash Pt.: > 220 C; Stability: Good heat stability in absence of oxygen; resistant to moderately strong acids, alkalis, and salt sol'ns.; pH: 6.0–7.5 (1% aq.)
Composition: 100% conc.

Pluriol E 1500. BASF AG
Chemical Description: Polyethylene glycol
Category: Solubilizer, plasticizer, humectant, lubricant, dispersant, binder, nonionic
Applications: Binders and hardeners in toilets and hard soaps; dispersing dyes and pigments; in inks–Indian, flexographic, etc.; textile and coatings industries; coloring ceramics; paper industry softener; plasticizer in adhesives industry and in the production of cellulose film; ceramics and metalworking lubricants; auxiliaries for copper and nickel electroplating baths, electrolytic polishing of steel
Form: Fine powder; Color: White
Properties: m.w.: 1500; Solubility: Water-sol.; Density: ≈ 0.6 kg/l (bulk); Visc.: 25–35 cs (99 C); M.P.: ≈ 45 C; Flash Pt.: > 230 C; Solid. Pt.: ≈ 35 C; Stability: Good heat stability in absence of oxygen; resistant to moderately strong acids, alkalis, and salt sol'ns.; pH: 6.0–7.5 (1% aq.)
Composition: 100% conc.

Pluriol E 4000. BASF AG
Chemical Description: Polyethylene glycol
Category and Applications: See Pluriol E 1500
Form: Fine powder; Color: White
Properties: m.w.: 4000; Solubility: Water-sol.; Density: ≈ 0.6 kg/l (bulk); Visc.: 80–130 cs (99 C); M.P.: ≈ 55 C; Flash Pt.: > 240 C; Solid. Pt.: ≈ 43 C; Stability: Good heat stability in absence of oxygen; resistance to moderately strong acids, alkalis, and salt sol'ns.; pH: 6.0–7.5 (1% aq.)
Composition: 100% conc.

Pluriol E 6000. BASF AG
Chemical Description: Polyethylene glycol
Category and Applications: See Pluriol E 1500
Form: Fine powder; Color: White
Properties: m.w.: 6000; Solubility: Water-sol.; Density: ≈ 0.6 kg/l (bulk); Visc.: ≈ 300 cs (99 C); M.P.: ≈ 60 C; Flash Pt.: > 240 C; Solid. Pt.: ≈ 45 C; Stability: Good heat stability in absence of oxygen; resistant to moderately strong acids, alkalis, and salt sol'ns.; pH: 6.0–7.5 (1% aq.)
Composition: 100% conc.

Pluriol E 9000. BASF AG
Chemical Description: Polyethyl-

ene glycol
Category and Applications: See Pluriol E 1500; also used in rubber processing; as mold release agent in production of latex and rubber
Form: Fine powder; Color: White
Properties: m.w.: 9000; Solubility: Water-sol.; Density: ≈ 0.6 kg/l (bulk); Visc.: ≈ 1900 cs (99 C); M.P.: ≈ 65 C; Flash Pt.: > 240 C; Solid. Pt.: ≈ 46 C; Stability: Good heat stability in absence of oxygen; resistant to moderately strong acids, alkalis, and salt sol'ns.; pH: 6.0–7.5 (1% aq.)
Composition: 100% conc.

Pluriol P 600. BASF AG
Chemical Description: Polypropylene glycol
Category: Lubricant, mold release agent, antifoaming agent, solubilizer, intermediate, nonionic
Applications: Additives for synthetic oils, water-sol. cutting fluids, rolling oils, and hydraulic fluids; high-temp. lubricants; internal and external lubricants for rubber; antifoams in paints and hydraulic fluids and in the processing of synthetic rubber and latex; heat transmission fluids; refrigerants, coolants; consistency improvers and solubilizers; intermediate which can enter into esterification, etherification, and polycondensation reactions
Form: Clear liquid; Color: Colorless
Properties: m.w.: ≈ 600; Solubility: Miscible with water, ethanol, toluene, trichloroethylene, oil; Density: 1.0 g/cc (20 C); Visc.: ≈ 130 cs (20 C); Flash Pt.: ≈ 216 C; Pour Pt.: ≈ –43 C; pH: 6.5–7.5 (1% aq.)
Composition: 100% conc.
References: Tech. information TI/P 2561 e

Pluriol P 900. BASF AG
Chemical Description: Polypropylene glycol
Category and Applications: See Pluriol P 600; also in quick-break solvent-type cold cleaners
Form: Clear liquid; Color: Colorless
Properties: m.w.: ≈ 900; Solubility: Miscible with ethanol, toluene, trichloroethylene, most oils, 5% hydrochloric acid; forms sol'ns. of about 2% or less in water; Density: 1.0 g/cc (20 C); Visc.: ≈ 180 cs (20 C); Cloud Pt.: 33 C (1% aq.); Flash Pt.: ≈ 220 C; Pour Pt.: ≈ –38 C; pH: 6.5–7.5 (1% aq.)
Composition: 100% conc.
References: Tech. information TI/P 2561 e

Pluriol P 2000. BASF AG
Chemical Description: Polypropylene glycol
Category and Applications: See Pluriol P 600
Form: Clear liquid; Color: Colorless
Properties: m.w.: ≈ 2000; Solubility: Miscible with ethanol, tolu-

ene, trichloroethylene, most oils; Density: 1.0 g/cc (20 C); Visc.: ≈ 440 cs (20 C); Flash Pt.: ≈ 222 C; Pour Pt.: ≈ –35 C; pH: 6.5–7.5 (1% aq.)
Composition: 100% conc.
References: Tech. information TI/P 2561 e

Pluriol PE 3100. BASF AG

Chemical Description: Propylene oxide/ethylene oxide block polymer
Category: Dispersant, defoamer, nonionic
Applications: Dispersing agent for colorants; defoamer in various applications
Form: Clear liquid; Color: Colorless
Properties: m.w.: ≈ 1100; Ref. Index: ≈ 1.45; Solubility @ 10%: Sol. in ethanol, 2-propanol, toluene, tetrachloroethylene, mineral spirit; Density: ≈ 1.02 g/cc; Visc.: ≈ 175 cps; Cloud Pt.: ≈ 40 C (1% aq.); Pour Pt.: ≈ –20 C; Stability: Stable in acidic media; resistant to alkalis within limits; Surface Tension: ≈ 45 dynes/cm (0.1% @ 20 C); pH: ≈ 7 (5% aq.); Biodegradability: < 80%
Composition: 100% conc.
Toxicity: Skin and (eye) mucous membrane irritant on prolonged contact with conc. form
Containers: Stainless steel, aluminum, or iron coated with a phenolic resin
References: TI/P 2596 e

Pluriol PE 4300. BASF AG

Chemical Description: Propylene oxide/ethylene oxide block polymer
Category: Low-foaming surfactant, nonionic
Applications: Dishwashing detergents and rinse aids
Form: Clear liquid (undergoes reversible opalescence on cooling); Color: Colorless
Properties: m.w.: ≈ 1700; Ref. Index: ≈ 1.45; Solubility @ 10%: Sol. in water, 10% hydrochloric acid, ethanol, 2-propanol, tetrachloroethylene; Density: ≈ 1.04 g/cc; Visc.: ≈ 300 cps; Cloud Pt.: ≈ 42 C (1% aq.); Pour Pt.: ≈ –20 C; Stability: Stable in acidic media; resistant to alkalis within limits; Surface Tension: ≈ 43 dynes/cm (0.1% @ 20 C); pH: ≈ 7 (5% aq.); Biodegradability: < 80%
Composition: 100% conc.
Toxicity: Skin and (eye) mucous membrane irritant on prolonged contact with conc. form
Containers: Stainless steel, aluminum, or iron coated with a phenolic resin
References: TI/P 2596 e

Pluriol PE 6100. BASF AG

Chemical Description: Propylene oxide/ethylene oxide block polymer
Category: Defoamer, nonionic
Applications: Defoamer in household dishwashing machines, bottle-

washing plants, metal cleaners, boiler feedwater, acid dyebaths, antifreeze
Form: Clear liquid (undergoes reversible opalescence on cooling); Color: Colorless
Properties: m.w.: ≈ 2000; Ref. Index: ≈ 1.45; Solubility @ 10%: Sol. in ethanol, 2-propanol, toluene, tetrachloroethylene, mineral spirit; Density: ≈ 1.02 g/cc; Visc.: ≈ 350 cps; Cloud Pt.: ≈ 24 C (1% aq.); Pour Pt.: ≈ –20 C; Stability: Stable in acidic media; resistant to alkalis within limits; Surface Tension: ≈ 44 dynes/cm (0.1% @ 20 C); pH: ≈ 7 (5% aq.); Biodegradability: < 80%
Composition: 100% conc.
Toxicity: Skin and (eye) mucous membrane irritant on prolonged contact with conc. form
Containers: Stainless steel, aluminum, or iron coated with a phenolic resin
References: TI/P 2596 e

Pluriol PE 6101. BASF AG
Chemical Description: Propylene oxide/ethylene oxide block polymer
Category: Defoamer, nonionic
Applications: Rinse aids; higher cloud pt. than PE 6100
Form: Clear to slightly dull liquid
Properties: m.w.: ≈ 2000; Ref. Index: ≈ 1.45; Solubility @ 10%: Sol. in 10% hydrochloric acid, ethanol, 2-propanol, toluene, tetrachloroethylene, mineral spirit; Density: ≈ 1.02 g/cc; Visc.: ≈ 350 cps; Cloud Pt.: ≈ 28 C (1% aq.); Pour Pt.: ≈ –10 C; Stability: Stable in acidic media; resistant to alkalis within limits; Surface Tension: ≈ 44 dynes/cm (0.1% @ 20 C); pH: ≈ 7 (5% aq.); Biodegradability: < 80%
Composition: 100% conc.
Toxicity: Skin and (eye) mucous membrane irritant on prolonged contact with conc. form
Containers: Stainless steel, aluminum, or iron coated with a phenolic resin
References: TI/P 2596 e

Pluriol PE 6200. BASF AG
Chemical Description: Propylene oxide/ethylene oxide block polymer
Category: Low-foaming wetting agent, emulsifier, demulsifier, nonionic
Applications: Mechanical cleaning processes; household rinse aids; phosphatizing baths; emulsifier in polymerization processes, e.g., production of vinylacryl polymers; demulsifier in splitting crude oil emulsions
Form: Clear liquid (undergoes reversible opalescence on cooling); Color: Colorless
Properties: m.w.: ≈ 2500; Ref. Index: ≈ 1.45; Solubility @ 10%: Sol. in water, 10% hydrochloric acid, ethanol, 2-propanol; sol. (opaque) in toluene and tetrachloroethylene; Density: ≈ 1.04 g/cc; Visc.: ≈ 500 cps; Cloud Pt.: ≈ 32 C (1% aq.); Pour Pt.: ≈ –14 C;

Stability: Stable in acidic media; resistant to alkalis within limits; Surface Tension: ≈ 44 dynes/cm (0.1% @ 20 C); pH: ≈ 7 (5% aq.); Biodegradability: < 80%
Composition: 100% conc.
Toxicity: Skin and (eye) mucous membrane irritant on prolonged contact with conc. form
Containers: Stainless steel, aluminum, or iron coated with a phenolic resin
References: TI/P 2596 e

Pluriol PE 6400. BASF AG

Chemical Description: Propylene oxide/ethylene oxide block polymer
Category: Detergent, dispersant, lubricant, emulsifier, nonionic
Applications: Low-foaming detergent for cleaning liquors subjected to severe mechanical shear, e.g., in dishwashing machines, bottle-washing lines, dairy cleaners; disperses sparingly sol. calcium and magnesium salts; in metalworking as a coolant and lubricant for grinding, drilling, and cutting; emulsion polymerization
Form: Slightly dull liquid
Properties: m.w.: ≈ 3000; Ref. Index: ≈ 1.45; Solubility @ 10%: Sol. in water, 10% hydrochloric acid, ethanol, 2-propanol, toluene, tetrachloroethylene; Density: ≈ 1.05 g/cc; Visc.: ≈ 850 cps; Cloud Pt.: ≈ 59 C (1% aq.); Pour Pt.: ≈ 16 C; Stability: Stable in acidic media; resistant to alkalis within limits; Surface Tension: ≈ 44 dynes/cm (0.1% @ 20 C); pH: ≈ 7 (5% aq.); Biodegradability: < 80%
Composition: 100% conc.
Toxicity: Skin and (eye) mucous membrane irritant on prolonged contact with conc. form
Containers: Stainless steel, aluminum, or iron coated with a phenolic resin
References: TI/P 2596 e

Pluriol PE 6800. BASF AG

Chemical Description: Propylene oxide/ethylene oxide block polymer
Category: Detergent, dust binder, dispersant, solubilizer, emulsifier, nonionic
Applications: Powder detergents for household and industry; dust binder in powder formulations; disperses sparingly sol. calcium and magnesium salts in metal pickling baths; solubilizes essential oils in cosmetics; production of emulsion polymers
Form: Fine powder; Color: White
Properties: m.w.: ≈ 8500; Ref. Index: ≈ 1.45 (70 C); Solubility @ 10%: Sol. in water, 10% hydrochloric acid, ethanol, toluene; Density: ≈ 1.06 g/cc (70 C); Cloud Pt.: > 100 C (1% aq.); Stability: Stable in acidic media; resistant to alkalis within limits; Surface Tension: ≈ 48 dynes/cm (0.1% @ 20 C); pH: ≈ 7 (5% aq.); Biodegradability: < 80%
Composition: 100% conc.
Toxicity: Skin and (eye) mucous

membrane irritant on prolonged contact with conc. form
Containers: Stainless steel, aluminum, or iron coated with a phenolic resin
References: TI/P 2596 e

Pluriol PE 8100. BASF AG
Chemical Description: Propylene oxide/ethylene oxide block polymer
Category: Defoamer, surfactant, nonionic
Applications: Defoamer and low-foaming surfactant for cleaning processes involving high rates of mechanical shear, e.g., dishwashing detergents, cleaners for breweries and dairies, rinse aids
Form: Clear to slightly dull liquid; Color: Colorless
Properties: m.w.: ≈ 2600; Ref. Index: ≈ 1.45 (70 C); Solubility @ 10%: Sol. in 2-propanol, toluene, tetrachloroethylene; Density: ≈ 1.04 g/cc; Visc.: ≈ 600 cps; Cloud Pt.: ≈ 18 C (1% aq.); Pour Pt.: ≈ −30 C; Stability: Stable in acidic media; resistant to alkalis within limits; Surface Tension: ≈ 35 dynes/cm (0.1% @ 20 C); pH: ≈ 7 (5% aq.); Biodegradability: < 80%
Composition: 100% conc.
Toxicity: Skin and (eye) mucous membrane irritant on prolonged contact with conc. form
Containers: Stainless steel, aluminum, or iron coated with a phenolic resin
References: TI/P 2596 e

Pluriol PE 9200. BASF AG
Chemical Description: Propylene oxide/ethylene oxide block polymer
Category: Wetting agent, nonionic
Applications: Wetting agent for cleaning processes; surfactant in dishwashing machine detergents for the household and canteens; in acidic rinse aids; imparts attractive high gloss to glasses and dishes
Form: Clear to slightly dull liquid
Properties: m.w.: ≈ 3650; Ref. Index: ≈ 1.45 (70 C); Solubility @ 10%: Sol. in ethanol, 2-propanol, toluene, tetrachloroethylene; sol. (opaque) in water and 10% hydrochloric acid; Density: ≈ 1.03 g/cc; Visc.: ≈ 700 cps; Cloud Pt.: ≈ 22 C (1% aq.); Pour Pt.: ≈ −3 C; Stability: Stable in acidic media; resistant to alkalis within limits; Surface Tension: ≈ 35 dynes/cm (0.1% @ 20 C); pH: ≈ 7 (5% aq.); Biodegradability: < 80%
Composition: 100% conc.
Toxicity: Skin and (eye) mucous membrane irritant on prolonged contact with conc. form
Containers: Stainless steel, aluminum, or iron coated with a phenolic resin
References: TI/P 2596 e

Pluriol PE 9400. BASF AG
Chemical Description: Propylene oxide/ethylene oxide block polymer
Category: Emulsifier, dispersant, nonionic

Applications: Emulsifier in the oil industry and the production of aq. dispersions; in cleaning liquors subjected to severe mechanical shear, e.g., in dishwashing machines, bottle-washing lines, and dairy cleaners; disperses sparingly sol. calcium and magnesium salts
Form: Waxy; Color: White
Properties: m.w.: ≈ 4600; Ref. Index: ≈ 1.45 (70 C); Solubility @ 10%: Sol. in 10% hydrochloric acid, ethanol, 2-propanol, toluene, tetrachloroethylene; sol. (opaque) in water; Density: ≈ 1.03 g/cc; Stability: Stable in acidic media; resistant to alkalis within limits; Surface Tension: ≈ 40 dynes/cm (0.1% @ 20 C); pH: ≈ 7 (5% aq.); Biodegradability: < 80%
Composition: 100% conc.
Toxicity: Skin and (eye) mucous membrane irritant on prolonged contact with conc. form
Containers: Stainless steel, aluminum, or iron coated with a phenolic resin
References: TI/P 2596 e

Pluriol PE 10100. BASF AG
Chemical Description: Propylene oxide/ethylene oxide block polymer
Category: Wetting and antifoam agent, nonionic
Applications: Cleaners and rinse aids for dishwashing machines
Form: Clear to slightly dull liquid
Properties: m.w.: ≈ 3550; Ref. Index: ≈ 1.45; Solubility @ 10%: Sol. in 10% hydrochloric acid, ethanol, 2-propanol, toluene, tetrachloroethylene, mineral spirit; Density: ≈ 1.02 g/cc; Visc.: ≈ 1000 cps; Cloud Pt.: ≈ 15 C (1% aq.); Pour Pt.: ≈ –30 C; Stability: Stable in acidic media; resistant to alkalis within limits; Surface Tension: ≈ 35 dynes/cm (0.1% @ 20 C); pH: ≈ 7 (5% aq.); Biodegradability: < 80%
Composition: 100% conc.
Toxicity: Skin and (eye) mucous membrane irritant on prolonged contact with conc. form
Containers: Stainless steel, aluminum, or iron coated with a phenolic resin
References: TI/P 2596 e

Pluriol PE 10500. BASF AG
Chemical Description: Propylene oxide/ethylene oxide block polymer
Category: Emulsifier, nonionic
Applications: Used together with other Pluriol PE types in household and industrial cleaners; in emulsion polymerization; emulsifier for agrochemicals
Form: Waxy product; Color: White
Properties: m.w.: ≈ 6500; Ref. Index: ≈ 1.45 (70 C); Solubility @ 10%: Sol. in water, 10% hydrochloric acid, ethanol, 2-propanol; sol. (opaque) in toluene and tetrachloroethylene; Density: ≈ 1.03 g/cc; Cloud Pt.: > 100 C (1% aq.); Stability: Stable in acidic media; resistant to alkalis within limits; Surface Tension: ≈ 36 dynes/cm

(0.1% @ 20 C); pH: ≈ 7 (5% aq.); Biodegradability: < 80%
Composition: 100% conc.
Toxicity: Skin and (eye) mucous membrane irritant on prolonged contact with conc. form
Containers: Stainless steel, aluminum, or iron coated with a phenolic resin
References: TI/P 2596 e

Pluriol RPE 2540. BASF AG
Chemical Description: Polyethylene-polypropylene glycol
Category: Nonionic surfactant
Applications: For defoaming and foam controlled detergents
Form: Liquid
Properties: Density: ≈ 1.0 g/cc; Visc.: ≈ 1500 cps (20 C); Surface Tension: ≈ 37 dynes/cm (1 g/l @ 20 C); pH: ≈ 9 (5% aq.)
Composition: 100% conc.

Pluriol RPE 3110. BASF AG
Chemical Description: Polyethylene-polypropylene glycol
Category and Applications: See Pluriol RPE 2540
Form: Liquid
Composition: 100% conc.

Pluronic 10R5. BASF Wyandotte/ Ind. Chem. Group
Chemical Description: Block copolymer formed by sequential addition of ethylene and propylene oxide to an ethylene glycol base
Category: Emulsifier, wetting agent, detergent, binder, stabilizer, plasticizer, lubricant, solubilizer, dispersant, dedusting and viscosity control agent, defoamer, nonionic
Applications: Hard surface detergents; rinse aids; automatic dishwashing; textile processing; cosmetics; pharmaceuticals; pulp and paper industry; petroleum industry (crude oil demulsifier, pipeline additives); agricultural products; in iodophors; water treating and steam generating systems; fermentation processes; as intermediates; cutting and grinding fluids
Form: Liquid
Properties: m.w.: 1970 (avg.); Ref. Index: 1.4587; Solubility (> 10 g/ 100 ml solvent): Sol. in water, propylene glycol, xylene, isopropanol, ethyl acetate, perchloroethylene, isopropyl myristate; Sp. gr.: 1.058; Visc.: 400 cps; HLB: 21.0; Cloud Pt.: 69 C (1% aq.); Flash Pt.: > 450 F (COC); Pour Pt.: 15 C; Stability: Stable over wide pH range; effective in hard or soft water; Surface Tension: 50.9 dynes/ cm (0.1% conc.)
Composition: 100% active ingred.
Reaction: Neutral
Toxicity: Slight eye irritant
Containers: 470 lb net steel drums, tank cars, tank trucks

Pluronic 10R8. BASF Wyandotte/ Ind. Chem. Group
Chemical Description: Block copolymer formed by sequential addition of ethylene and propylene oxide to an ethylene glycol base
Category and Applications: See Pluronic 10R5

Form: Flakable solid
Properties: m.w.: 5000 (avg.); Solubility (> 10 g/100 ml solvent): Sol. in water, xylene, ethyl acetate; Sp. gr.: 1.062 (77 C); M.P.: 46 C; HLB: 33.0; Cloud Pt.: 99 C (1% aq.); Flash Pt.: > 450 F (COC); Stability: Stable over wide pH range; effective in hard water; Surface Tension: 54.1 dynes/cm (0.1% conc.)
Composition: 100% active ingred.
Reaction: Neutral
Toxicity: Slight eye irritant
Containers: 250 lb net Leverpak drums, tank cars, tank trucks

Pluronic 17R1. BASF Wyandotte/ Ind. Chem. Group
Chemical Description: Block copolymer formed by sequential addition of ethylene and propylene oxide to an ethylene glycol base
Category and Applications: See Pluronic 10R5; also for foam control in paper sizing operations
Form: Liquid
Properties: m.w.: 1950 (avg.); Ref. Index: 1.4516; Solubility (> 10 g/ 100 ml solvent): Sol. in xylene, isopropanol, ethyl acetate, mineral oil, perchloroethylene, isopropyl myristate, trichlorotrifluoroethane; Sp. gr.: 1.018; Visc.: 300 cps; HLB: 4.0; Cloud Pt.: 32 C (1% aq.); Flash Pt.: > 450 F (COC); Pour Pt.: –27 C; Stability: Stable over wide pH range; effective in hard water; Surface Tension: 33.0 dynes/cm (0.1% conc.)
Composition: 100% active ingred.
Reaction: Neutral
Toxicity: Minimal eye irritant
Containers: 460 lb net steel drums, tank car, tank truck

Pluronic 17R2. BASF Wyandotte/ Ind. Chem. Group
Chemical Description: Block copolymer formed by sequential addition of ethylene and propylene oxide to an ethylene glycol base
Category and Applications: See Pluronic 10R5
Form: Liquid
Properties: m.w.: 2100 (avg.); Ref. Index: 1.4535; Solubility (> 10 g/ 100 ml solvent): Sol. in water, xylene, isopropanol, ethyl acetate, perchloroethylene, isopropyl myristate, trichlorotrifluoroethane; Sp. gr.: 1.030; Visc.: 350 cps; HLB: 8.0; Cloud Pt.: 39 C (1% aq.); Flash Pt.: > 450 F (COC); Pour Pt.: –25 C; Stability: Stable over wide pH; effective in hard water; Surface Tension: 41.9 dynes/cm (0.1% conc.)
Composition: 100% active ingred.
Reaaction: Neutral
Containers: 460 lb net steel drums, tank cars, tank trucks

Pluronic 17R4. BASF Wyandotte/ Ind. Chem. Group
Chemical Description: Block copolymer formed by sequential addition of ethylene and propylene oxide to an ethylene glycol base
Category and Applications: See Pluronic 10R5
Form: Liquid
Properties: m.w.: 2700 (avg.); Ref. Index: 1.4572; Solubility (> 10 g/

100 ml solvent): Sol. in water, propylene glycol, xylene, isopropanol, ethyl acetate, perchloroethylene, isopropyl myristate, trichlorotrifluoroethane; Sp. gr.: 1.048; Visc.: 560 cps; HLB: 16.0; Cloud Pt.: 47 C (1% aq.); Flash Pt.: > 450 F (COC); Pour Pt.: 18 C; Stability: Stable over wide pH range; effective in hard water; Surface Tension: 44.1 dynes/cm (0.1% conc.)
Composition: 100% active ingred.
Reaction: Neutral
Toxicity: Slight eye irritant
Containers: 470 lb net steel drums, tank cars, tank trucks

Pluronic 17R8. BASF Wyandotte/ Ind. Chem. Group
Chemical Description: Block copolymer formed by sequential addition of ethylene and propylene oxide to an ethylene glycol base
Category and Applications: See Pluronic 10R5; esp. in dry toilet bowl cleaners, dye levelers, solubilizer of water-insol. drugs, stick-type cosmetics, soap bars, dispersant in deinking operations
Form: Flakable solid
Properties: m.w.: 7500 (avg.); Solubility (> 10 g/100 ml solvent); Sol. in water, ethyl acetate, isopropyl myristate; Sp. gr.: 1.064 (77 C); M.P.: 53 C; HLB: 32.0; Cloud Pt.: 81 C (1% aq.); Flash Pt.: > 450 F (COC); Stability: Stable over wide pH range; effective in hard water; Surface Tension: 47.3 dynes/cm (0.1% conc.)
Composition: 100% active ingred.
Reaction: Neutral
Containers: 250 lb net Leverpak drums, tank cars, tank trucks

Pluronic 25R1. BASF Wyandotte/ Ind. Chem. Group
Chemical Description: Block copolymer formed by sequential addition of ethylene and propylene oxide to an ethylene glycol base
Category and Applications: See Pluronic 10R5; esp. for foam control in paper sizing operations and antifreeze
Form: Liquid
Properties: m.w.: 2800 (avg.); Ref. Index: 1.4521; Solubility (> 10 g/100 ml solvent): Sol. in xylene, isopropanol, ethyl acetate, perchloroethylene, isopropyl myristate, trichlorotrifluoroethane; Sp. gr.: 1.017; Visc.: 460 cps; HLB: 2.3; Cloud Pt.: 28 C (1% aq.); Flash Pt.: > 450 F (COC); Pour Pt.: –27 C; Stability: Stable over wide pH range; effective in hard water; Surface Tension: 36.3 dynes/cm (0.1% conc.)
Composition: 100% active ingred.
Reaction: Neutral
Toxicity: Slight eye irritant
Containers: 460 lb net steel drums, tank cars, tank trucks

Pluronic 25R2. BASF Wyandotte/ Ind. Chem. Group
Chemical Description: Block copolymer formed by sequential addition of ethylene and propylene oxide to an ethylene glycol base
Category and Applications: See Pluronic 10R5; esp. good wetting

and rinsing properties with low foaming; lubricant and leveling agent for paper coating
Form: Liquid
Properties: m.w.: 3120 (avg.); Ref. Index: 1.4541; Solubility (> 10 g/100 ml solvent): Sol. in xylene, isopropanol, ethyl acetate, perchloroethylene, isopropyl myristate, trichlorotrifluoroethane; Sp. gr.: 1.039; Visc.: 680 cps; HLB: 6.3; Cloud Pt.: 33 C (1% aq.); Flash Pt.: > 450 F (COC); Pour Pt.: –5 C; Stability: Stable over wide pH; effective in hard water; Surface Tension: 37.5 dynes/cm (0.1% conc.)
Composition: 100% active ingred.
Reaction: Neutral
Toxicity: Slight eye irritant
Containers: 460 lb net steel drums, tank cars, tank trucks

Pluronic 25R4. BASF Wyandotte/Ind. Chem. Group
Chemical Description: Block copolymer formed by sequential addition of ethylene and propylene oxide to an ethylene glycol base
Category and Applications: See Pluronic 10R5
Form: Liquid
Properties: m.w.: 3800 (avg.); Ref. Index: 1.4574; Solubility (> 10 g/100 ml solvent): Sol. in water, propylene glycol, xylene, isopropanol, ethyl acetate, perchloroethylene, isopropyl myristate, trichlorotrifluoroethane; Sp. gr.: 1.046; Visc.: 1110 cps; HLB: 14.3; Cloud Pt.: 40 C (1% aq.); Flash Pt.: > 450 F (COC); Pour Pt.: 25 C; Stability: Stable over wide pH; effective in hard water; Surface Tension: 40.9 dynes/cm (0.1% conc.)
Composition: 100% active ingred.
Reaction: Neutral
Toxicity: Slight eye irritant
Containers: 470 lb net steel drums, tank cars, tank trucks

Pluronic 25R5. BASF Wyandotte/Ind. Chem. Group
Chemical Description: Block copolymer formed by sequential addition of ethylene and propylene oxide to an ethylene glycol base
Category and Applications: See Pluronic 10R5; also thickener for cosmetic pastes and creams
Form: Paste
Properties: m.w.: 4500 (avg.); Solubility (> 10 g/100 ml solvent): Sol. in water, propylene glycol, xylene, isopropanol, ethyl acetate, perchloroethylene, isopropyl myristate, trichlorotrifluoroethane; Sp. gr.: 1.036 (60 C); M.P.: 33 C; HLB: 18.3; Cloud Pt.: 44 C (1% aq.); Flash Pt.: > 450 F (COC); Stability: Stable over wide pH; effective in hard water; Surface Tension: 43.5 dynes/cm (0.1% conc.)
Composition: 100% active ingred.
Reaction: Neutral
Containers: 470 lb net steel drums, tank cars, tank trucks

Pluronic 25R8. BASF Wyandotte/Ind. Chem. Group
Chemical Description: Block copolymer formed by sequential ad-

dition of ethylene and propylene oxide to an ethylene glycol base
Category and Applications: See Pluronic 10R5; esp. for dry toilet bowl cleaners, dye levelers for wool, cotton, and synthetics, solubilizer for water-insol. drugs, thickener for cosmetic pastes and creams, stick-type cosmetics, soap bars, deinking operations, felt washing operations
Form: Flakable solid
Properties: m.w.: 9000 (avg.); Solubility (> 10 g/100 ml solvent): Sol. in water; Sp. gr.: 1.062; M.P.: 56 C; HLB: 30.3; Cloud Pt.: 80 C (1% aq.); Flash Pt.: > 450 F (COC); Stability: Stable over wide pH; effective in hard water; Surface Tension: 46.1 dynes/cm (0.1% conc.)
Composition: 100% active ingred.
Reaction: Neutral
Toxicity: Minimal eye irritant
Containers: 250 lb net Leverpak drums, tank cars, tank trucks

Pluronic 31R1. BASF Wyandotte/ Ind. Chem. Group
Chemical Description: Block copolymer formed by sequential addition of ethylene and propylene oxide to an ethylene glycol base
Category and Applications: See Pluronic 10R5; esp. for floating bath oils, foam control in antifreeze
Form: Liquid
Properties: m.w.: 3200 (avg.); Ref. Index: 1.4522; Solubility (>10 g/100 ml solvent): Sol. in xylene, isopropanol, ethyl acetate, perchloroethylene, isopropyl myristate, trichlorotrifluoroethane; Sp. gr.: 1.018; Visc.: 578 cps; HLB: 1.7; Cloud Pt.: 25 C (1% aq.); Flash Pt.: > 450 F (COC); Pour Pt.: –25 C; Stability: Stable over wide pH; effective in hard water; Surface Tension: 34.1 dynes/cm (0.1% conc.)
Composition: 100% active ingred.
Reaction: Neutral
Toxicity: Minimal eye irritant
Containers: 460 lb net steel drums, tank cars, tank trucks

Pluronic 31R2. BASF Wyandotte/ Ind. Chem. Group
Chemical Description: Block copolymer formed by sequential addition of ethylene and propylene oxide to an ethylene glycol base
Category and Applications: See Pluronic 10R5; also as paper coating color additives, in deinking operations, and felt washing operations
Form: Liquid
Properties: m.w.: 3400 (avg.); Ref. Index: 1.4542; Solubility (>10 g/100 ml solvent): Sol. in water, propylene glycol, xylene, isopropanol, ethyl acetate, perchloroethylene, isopropyl myristate, trichlorotrifluoroethane; Sp. gr.: 1.030; Visc.: 818 cps; HLB: 5.7; Cloud Pt.: 30 C (1% aq.); Flash Pt.: > 450 F (COC); Pour Pt.: 9 C; Stability: Stable over wide pH; effective in hard water; Surface Tension: 38.9 dynes/cm (0.1% conc.)
Composition: 100% active ingred.

Reaction: Neutral
Toxicity: Minimal eye irritant
Containers: 460 lb net steel drums, tank cars, tank trucks

Pluronic 31R4. BASF Wyandotte/ Ind. Chem. Group
Chemical Description: Block copolymer formed by sequential addition of ethylene and propylene oxide to an ethylene glycol base
Category and Applications: See Pluronic 10R5
Form: Paste
Properties: m.w.: 4300 (avg.); Solubility (> 10 g/100 ml solvent): Sol. in xylene, isopropanol, ethyl acetate, perchloroethylene, isopropyl myristate, trichlorotrifluoroethane; Sp. gr.: 1.028 (60 C); M.P.: 26 C; HLB: 13.7; Cloud Pt.: 31 C (1% aq.); Flash Pt.: > 450 F (COC); Pour Pt.: 26 C; Stability: Stable over wide pH; effective in hard water; Surface Tension: 41.2 dynes/cm (0.1% conc.)
Composition: 100% active ingred.
Reaction: Neutral
Toxicity: Slight eye irritant
Containers: 470 lb net steel drums, tank cars, tank trucks

Pluronic F38. BASF Wyandotte/ Ind. Chem. Group
Chemical Description: Polyol formed by addition of propylene oxide to the two hydroxyl groups of a propylene glycol initiator
Category: Wetting agent, emulsifier, demulsifier, foam control agent and/or low to moderate foaming agent, viscosity control agent, dispersant, antistatic agent, deduster, dye leveler, gelling agent, nonionic
Applications: Agricultural chemicals; cosmetics and personal care products; pharmaceuticals; metal cleaning; pulp and paper industry; rinse aids; textile processing–spin finishes, dyeing assistant, scouring aid, dye leveler, lubricant; water treatment
Form: Prilled
Properties: m.w.: 5000 (avg.); Solubility (> 10 g/100 ml solvent): Sol. in ethyl alcohol, propylene glycol, water, toluene; Density: 8.9 lb/gal (77 C); Sp. gr.: 1.07 (77 C); Visc.: 260 cps (77 C); M.P.: 48 C; HLB: 30.5; Cloud Pt.: > 100 C (1% aq.); Flash Pt.: 505 F (COC); Stability: Compatible with soaps, CMC, common alkaline salts; Surface Tension: 52.2 dynes/ cm (0.1% conc.)
Composition: 100% active ingred.
Containers: 55 gal drums, tank cars, tank trucks

Pluronic F68. BASF Wyandotte/ Ind. Chem. Group
Chemical Description: Polyol formed by addition of propylene oxide to the two hydroxyl groups of a propylene glycol initiator
Category and Applications: See Pluronic F38
Form: Prilled
Properties: m.w.: 8350 (avg.); Solubility (> 10 g/100 ml solvent): Sol. in ethyl alcohol, water; Density: 8.8 lb/gal (77 C); Sp. gr.: 1.06 (77 C); Visc.: 1000 cps (77 C);

M.P.: 52 C; HLB: 29.0; Cloud Pt.: > 100 C (1% aq.); Flash Pt.: 500 F (COC); Stability: Compatible w/ soaps, CMC, common alkaline salts; Surface Tension: 50.3 dynes/cm (0.1% conc.)
Composition: 100% active ingred.
Containers: 55 gal drums, tank cars, tank trucks

Pluronic F68LF. BASF Wyandotte/Ind. Chem. Group
Chemical Description: Polyol formed by addition of propylene oxide to the two hydroxyl groups of a propylene glycol initiator
Category and Applications: See Pluronic F38
Form: Prilled
Properties: m.w.: 7560 (avg.); Solubility (> 10 g/100 ml solvent): Sol. in ethyl alcohol, toluene; Density: 8.8 lb/gal (77 C); Sp. gr.: 1.06 (77 C); Visc.: 850 cps (77 C); M.P.: 50 C; Cloud Pt.: 32 C (1% aq.); Stability: Compatible with soaps, CMC, common alkaline salts; Surface Tension: 43.7 dynes/cm (0.1% conc.)
Composition: 100% active ingred.
Containers: 55 gal drums, tank cars, tank trucks

Pluronic F77. BASF Wyandotte/Ind. Chem. Group
Chemical Description: Polyol formed by addition of propylene oxide to the two hydroxyl groups of a propylene glycol initiator
Category and Applications: See Pluronic F38
Form: Prilled
Properties: m.w.: 6600 (avg.); Solubility (> 10 g/100 ml solvent): Sol. in ethyl alcohol, water, toluene; Density: 8.7 lb/gal (77 C); Sp. gr.: 1.04 (77 C); Visc.: 415 cps (77 C); M.P.: 48 C; HLB: 24.5; Cloud Pt.: >100 C (1% aq.); Flash Pt.: 485 F (COC); Stability: Compatible with soaps, CMC, common alkaline salts; Surface Tension: 47.0 dynes/cm (0.1% conc.)
Composition: 100% active ingred.
Containers: 55 gal drums, tank cars, tank trucks

Pluronic F87. BASF Wyandotte/Ind. Chem. Group
Chemical Description: Polyol formed by addition of propylene oxide to the two hydroxyl groups of a propylene glycol initiator
Category and Applications: See Pluronic F38
Form: Prilled
Properties: m.w.: 7700 (avg.); Solubility (> 10 g/100 ml solvent): Sol. in ethyl alcohol, water, toluene; Density: 8.7 lb/gal (77 C); Sp. gr.: 1.04 (77 C); Visc.: 700 cps (77 C); M.P.: 49 C; HLB: 24.0; Cloud Pt.: > 100 C (1% aq.); Flash Pt.: 472 F (COC); Stability: Compatible with soaps, CMC, common alkaline salts; Surface Tension: 44.0 dynes/cm (0.1% conc.)
Composition: 100% active ingred.
Containers: 55 gal drums, tank cars, tank trucks

Pluronic F88. BASF Wyandotte/Ind. Chem. Group
Chemical Description: Polyol formed by addition of propylene

oxide to the two hydroxyl groups of a propylene glycol initiator
Category and Applications: See Pluronic F38
Form: Prilled
Properties: m.w.: 10,800 (avg.); Solubility (> 10 g/100 ml solvent): Sol. in ethyl alcohol and water; Density: 8.8 lb/gal (77 C); Sp. gr.: 1.06 (77 C); Visc.: 2300 cps (77 C); M.P.: 54 C; HLB: 28.0; Cloud Pt.: >100 C (1% aq.); Stability: Compatible with soaps, CMC, common alkaline salts; Surface Tension: 48.5 dynes/cm (0.1% conc.)
Composition: 100% active ingred.
Containers: 55 gal drums, tank cars, tank trucks

Pluronic F98. BASF Wyandotte/ Ind. Chem. Group
Chemical Description: Polyol formed by addition of propylene oxide to the two hydroxyl groups of a propylene glycol initiator
Category and Applications: See Pluronic F38
Form: Prilled
Properties: m.w.: 13,000 (avg.); Solubility (> 10 g/100 ml solvent): Sol. in ethyl alcohol, water, perchloroethylene; Density: 8.8 lb/gal (77 C); Sp. gr.: 1.06 (77 C); Visc.: 2700 cps (77 C); M.P.: 55 C; HLB: 27.5; Cloud Pt.: > 100 C (1% aq.); Flash Pt.: 491 F (COC); Stability: Compatible with soaps, CMC, common alkaline salts; Surface Tension: 43.0 dynes/ cm (0.1% conc.)
Composition: 100% active ingred.
Containers: 55 gal drums, tank cars, tank trucks

Pluronic F108. BASF Wyandotte/ Ind. Chem. Group
Chemical Description: Polyol formed by addition of propylene oxide to the two hydroxyl groups of a propylene glycol initiator
Category and Applications: See Pluronic F38
Form: Prilled
Properties: m.w.: 14,000 (avg.); Solubility (> 10 g/100 ml solvent): Sol. in ethyl alcohol, water; Density: 8.8 lb/gal (77 C); Sp. gr.: 1.06 (77 C); Visc.: 8000 cps (77 C); M.P.: 57 C; HLB: 27.0; Cloud Pt.: > 100 C (1% aq.); Flash Pt.: 495 F (COC); Stability: Compatible with soaps, CMC, common alkaline salts; Surface Tension: 41.2 dynes/cm (0.1% conc.)
Composition: 100% active ingred.
Containers: 55 gal drums, tank cars, tank trucks

Pluronic F127. BASF Wyandotte/ Ind. Chem. Group
Chemical Description: Polyol formed by addition of propylene oxide to the two hydroxyl groups of a propylene glycol initiator
Category and Applications: See Pluronic F38
Form: Prilled
Properties: m.w.: 12,500 (avg.); Solubility (> 10 g/100 ml solvent): Sol. in ethyl alcohol, water, toluene, perchloroethylene; Den-

sity: 8.8 lb/gal (77 C); Sp. gr.: 1.05 (77 C); Visc.: 3100 cps (77 C); M.P.: 56 C; HLB: 22.0; Cloud Pt.: > 100 C (1% aq.); Stability: Compatible with soaps, CMC, common alkaline salts; Surface Tension: 40.6 dynes/cm (0.1% conc.)
Composition: 100% active ingred.
Containers: 55 gal drums, tank cars, tank trucks

Pluronic L31. BASF Wyandotte/ Ind. Chem. Group
Chemical Description: Polyol formed by addition of propylene oxide to the two hydroxyl groups of a propylene glycol initiator
Category and Applications: See Pluronic F38
Form: Liquid
Properties: m.w.: 1100 (avg.); Ref. Index: 1.4515; Solubility (> 10 g/ 100 ml solvent): Sol. in ethyl alcohol, propylene glycol, water, toluene, xylene, perchloroethylene; Density: 8.5 lb/gal; Sp. gr.: 1.02; Visc.: 165 cps; HLB: 4.5; Cloud Pt.: 37 C (1% aq.); Flash Pt.: 439 F (COC); Pour Pt.: −32 C; Stability: Compatible with soaps, CMC, common alkaline salts; Surface Tension: 46.9 dynes/ cm (0.1% conc.)
Composition: 100% active ingred.
Containers: 55 gal drums, tank cars, tank trucks

Pluronic L35. BASF Wyandotte/ Ind. Chem. Group
Chemical Description: Polyol formed by addition of propylene oxide to the two hydroxyl groups of a propylene glycol initiator
Category and Applications: See Pluronic F38
Form: Liquid
Properties: m.w.: 1900 (avg.); Ref. Index: 1.4593; Solubility (> 10 g/ 100 ml solvent): Sol. in ethyl alcohol, propylene glycol, water, toluene, xylene, perchloroethylene; Density: 8.8 lb/gal; Sp. gr.: 1.06; Visc.: 340 cps; HLB: 18.5; Cloud Pt.: 77 C (1% aq.); Pour Pt.: 7 C; Stability: Compatible with soaps, CMC, common alkaline salts; Surface Tension: 48.8 dynes/cm (0.1% conc.)
Composition: 100% active ingred.
Containers: 55 gal drums, tank cars, tank trucks

Pluronic L42. BASF Wyandotte/ Ind. Chem. Group
Chemical Description: Polyol formed by addition of propylene oxide to the two hydroxyl groups of a propylene glycol initiator
Category and Applications: See Pluronic F38
Form: Liquid
Properties: m.w.: 1630 (avg.); Ref. Index: 1.4541; Solubility (> 10 g/ 100 ml solvent): Sol. in ethyl alcohol, propylene glycol, water, toluene, xylene, perchloroethylene; Density: 8.6 lb/gal; Sp. gr.: 1.03; Visc.: 250 cps; HLB: 8.0; Cloud Pt.: 37 C (1% aq.); Flash Pt.: 450 F (COC); Pour Pt.: −26 C; Stability: Compatible with soaps, CMC, common alkaline salts; Surface Tension: 46.5 dynes/ cm (0.1% conc.)

Composition: 100% active ingred.
Containers: 55 gal drums, tank cars, tank trucks

Pluronic L43. BASF Wyandotte/ Ind. Chem. Group
Chemical Description: Polyol formed by addition of propylene oxide to the two hydroxyl groups of a propylene glycol initiator
Category and Applications: See Pluronic F38
Form: Liquid
Properties: m.w.: 1850 (avg.); Ref. Index: 1.4563; Solubility (> 10 g/ 100 ml solvent): Sol. in ethyl alcohol, propylene glycol, water, toluene, xylene, perchloroethylene; Density: 8.7 lb/gal; Sp. gr.: 1.04; Visc.: 310 cps; HLB: 12.0; Cloud Pt.: 42 C (1% aq.); Pour Pt.: −1 C; Stability: Compatible with soaps, CMC, alkaline salts; Surface Tension: 47.3 dynes/cm (0.1% conc.)
Composition: 100% active ingred.
Containers: 55 gal drums, tank cars, tank trucks

Pluronic L44. BASF Wyandotte/ Ind. Chem. Group
Chemical Description: Polyol formed by addition of propylene oxide to the two hydroxyl groups of a propylene glycol initiator
Category and Applications: See Pluronic F38
Form: Liquid
Properties: m.w.: 2200 (avg.); Ref. Index: 1.4580; Solubility (> 10 g/ 100 ml solvent): Sol. in ethyl alcohol, propylene glycol, water, toluene, xylene, perchloroethylene; Density: 8.8 lb/gal; Sp. gr.: 1.05; Visc.: 440 cps; HLB: 16.0; Cloud Pt.: 65 C (1% aq.); Flash Pt.: 464 F (COC); Pour Pt.: 16 C; Stability: Compatible with soaps, CMC, common alkaline salts; Surface Tension: 45.3 dynes/cm (0.1% conc.)
Composition: 100% active ingred.
Containers: 55 gal drums, tank cars, tank trucks

Pluronic L61. BASF Wyandotte/ Ind. Chem. Group
Chemical Description: Polyol formed by addition of propylene oxide to the two hydroxyl groups of a propylene glycol initiator
Category and Applications: See Pluronic F38
Form: Liquid
Properties: m.w.: 2000 (avg.); Ref. Index: 1.4520; Solubility (> 10 g/ 100 ml solvent): Sol. in ethyl alcohol, toluene, xylene, perchloroethylene; Density: 8.4 lb/gal; Sp. gr.: 1.01; Visc.: 285 cps; HLB: 3.0; Cloud Pt.: 24 C (1% aq.); Flash Pt.: 455 F (COC); Pour Pt.: −29 C; Stability: Compatible with soaps, CMC, common alkaline salts
Composition: 100% active ingred.
Containers: 55 gal drums, tank cars, tank trucks

Pluronic L62. BASF Wyandotte/ Ind. Chem. Group
Chemical Description: Polyol formed by addition of propylene oxide to the two hydroxyl groups

of a propylene glycol initiator
Category and Applications: See Pluronic F38
Form: Liquid
Properties: m.w.: 2500 (avg.); Ref. Index: 1.4550; Solubility (>10 g/100 ml solvent): Sol. in ethyl alcohol, propylene glycol, water, toluene, xylene, perchloroethylene; Density: 8.6 lb/gal; Sp. gr.: 1.03; Visc.: 400 cps; HLB: 7.0; Cloud Pt.: 32 C (1% aq.); Flash Pt.: 466 F (COC); Pour Pt.: –4 C; Stability: Compatible with soaps, CMC, common alkaline salts; Surface Tension: 42.8 dynes/cm (0.1% conc.)
Composition: 100% active ingred.
Containers: 55 gal drums, tank cars, tank trucks

Pluronic L62D. BASF Wyandotte/Ind. Chem. Group
Chemical Description: Polyol formed by addition of propylene oxide to the two hydroxyl groups of a propylene glycol initiator
Category and Applications: See Pluronic F38
Form: Liquid
Properties: m.w.: 2350 (avg.); Ref. Index: 1.4557; Solubility (>10 g/100 ml solvent): Sol. in ethyl alcohol, propylene glycol, toluene, xylene, perchloroethylene; Density: 8.7 lb/gal; Sp. gr.: 1.04; Visc.: 385 cps; HLB: 7.0; Cloud Pt.: 35 (1% aq.); Flash Pt.: 455 F (COC); Pour Pt.: –1 C; Stability: Compatible with soaps, CMC, common alkaline salts; Surface Tension: 43.0 dynes/cm (0.1% conc.)
Composition: 100% active ingred.
Containers: 55 gal drums, tank cars, tank trucks

Pluronic L62LF. BASF Wyandotte/Ind. Chem. Group
Chemical Description: Polyol formed by addition of propylene oxide to the two hydroxyl groups of a propylene glycol initiator
Category and Applications: See Pluronic F38
Form: Liquid
Properties: m.w.: 2450 (avg.); Ref. Index: 1.4546; Solubility (>10 g/100 ml solvent): Sol. in ethyl alcohol, propylene glycol, water, toluene, xylene, perchloroethylene; Density: 8.6 lb/gal; Sp. gr.: 1.03; Visc.: 400 cps; HLB: 6.6; Cloud Pt.: 28 C (1% aq.); Pour Pt.: –10 C; Stability: Compatible with soaps, CMC, common alkaline salts; Surface Tension: 38.6 dynes/cm (0.1% conc.)
Composition: 100% active ingred.
Containers: 55 gal drums, tank cars, tank trucks

Pluronic L63. BASF Wyandotte/Ind. Chem. Group
Chemical Description: Polyol formed by addition of propylene oxide to the two hydroxyl groups of a propylene glycol initiator
Category and Applications: See Pluronic F38
Form: Liquid
Properties: m.w.: 2650 (avg.); Ref. Index: 1.4562; Solubility (>10 g/100 ml solvent): Sol. in ethyl alcohol, propylene glycol, water,

toluene, xylene, perchloroethylene; Density: 8.7 lb/gal; Sp. gr.: 1.04; Visc.: 490 cps; HLB: 11.0; Cloud Pt.: 34 C (1% aq.); Pour Pt.: 10 C; Stability: Compatible with soaps, CMC, common alkaline salts; Surface Tension: 43.3 dynes/cm (0.1% conc.)
Composition: 100% active ingred.
Containers: 55 gal drums, tank cars, tank trucks

Pluronic L64. BASF Wyandotte/ Ind. Chem. Group
Chemical Description: Polyol formed by addition of propylene oxide to the two hydroxyl groups of a propylene glycol initiator
Category and Applications: See Pluronic F38
Form: Liquid
Properties: m.w.: 2900 (avg.); Ref. Index: 1.4575; Solubility (>10 g/ 100 ml solvent): Sol. in ethyl alcohol, propylene glycol, water, toluene, xylene, perchloroethylene; Density: 8.8 lb/gal; Sp. gr.: 1.05; Visc.: 550 cps; HLB: 15.0; Cloud Pt.: 61 C (1% aq.); Flash Pt.: 485 F (COC); Pour Pt.: 16 C; Stability: Compatible with soaps, CMC, common alkaline salts; Surface Tension: 43.2 dynes/cm (0.1% conc.)
Composition: 100% active ingred.
Containers: 55 gal drums, tank cars, tank trucks

Pluronic L72. BASF Wyandotte/ Ind. Chem. Group
Chemical Description: Polyol formed by addition of propylene oxide to the two hydroxyl groups of a propylene glycol initiator
Category and Applications: See Pluronic F38
Form: Liquid
Properties: m.w.: 2750 (avg.); Ref. Index: 1.4542; Solubility (>10 g/ 100 ml solvent): Sol. in ethyl alcohol, water, toluene, xylene, perchloroethylene; Density: 8.6 lb/ gal; Sp. gr.: 1.03; Visc.: 510 cps; HLB: 6.5; Cloud Pt.: 25 C (1% aq.); Flash Pt.: 442 F (COC); Pour Pt.: –7 C; Stability: Compatible with soaps, CMC, common alkaline salts; Surface Tension: 39.0 dynes/cm (0.1% conc.)
Composition: 100% active ingred.
Containers: 55 gal drums, tank cars, tank trucks

Pluronic L81. BASF Wyandotte/ Ind. Chem. Group
Chemical Description: Polyol formed by addition of propylene oxide to the two hydroxyl groups of a propylene glycol initiator
Category and Applications: See Pluronic F38
Form: Liquid
Properties: m.w.: 2750 (avg.); Ref. Index: 1.4526; Solubility (>10 g/ 100 ml solvent): Sol. in ethyl alcohol, toluene, xylene, perchloroethylene; Density: 8.5 lb/gal; Sp. gr.: 1.02; Visc.: 475 cps; HLB: 2.0; Cloud Pt.: 20 C (1% aq.); Pour Pt.: –37 C; Stability: Compatible with soaps, CMC, common alkaline salts
Composition: 100% active ingred.
Containers: 55 gal drums, tank cars, tank trucks

Pluronic L92. BASF Wyandotte/ Ind. Chem. Group
Chemical Description: Polyol formed by addition of propylene oxide to the two hydroxyl groups of a propylene glycol initiator
Category and Applications: See Pluronic F38
Form: Liquid
Properties: m.w.: 3650 (avg.); Ref. Index: 1.4547; Solubility (>10 g/ 100 ml solvent): Sol. in ethyl alcohol, toluene, xylene, perchloroethylene; Density: 8.6 lb/gal; Sp. gr.: 1.03; Visc.: 700 cps; HLB: 5.5; Cloud Pt.: 26 C (1% aq.); Flash Pt.: 445 F (COC); Pour Pt.: 7 C; Stability: Compatible with soaps, CMC, common alkaline salts; Surface Tension: 35.9 dynes/ cm (0.1% conc.)
Composition: 100% active ingred.
Containers: 55 gal drums, tank cars, tank trucks

Pluronic L101. BASF Wyandotte/ Ind. Chem. Group
Chemical Description: Polyol formed by addition of propylene oxide to the two hydroxyl groups of a propylene glycol initiator
Category and Applications: See Pluronic F38
Form: Liquid
Properties: m.w.: 3800 (avg.); Ref. Index: 1.4524; Solubility (>10 g/ 100 ml solvent): Sol. in ethyl alcohol, toluene, xylene, perchloroethylene; Density: 8.5 lb/gal; Sp. gr.: 1.02; Visc.: 800 cps; HLB: 1.0; Cloud Pt.: 15 C (1% aq.); Pour Pt.: –23 C; Stability: Compatible with soaps, CMC, common alkaline salts
Composition: 100% active ingred.
Containers: 55 gal drums, tank cars, tank trucks

Pluronic L121. BASF Wyandotte/ Ind. Chem. Group
Chemical Description: Polyol formed by addition of propylene oxide to the two hydroxyl groups of a propylene glycol initiator
Category and Applications: See Pluronic F38
Form: Liquid
Properties: m.w.: 4400 (avg.); Ref. Index: 1.4527; Solubility (>10 g/ 100 ml solvent): Sol. in ethyl alcohol, toluene, xylene, perchloroethylene; Density: 8.4 lb/gal; Sp. gr.: 1.01; Visc.: 1200 cps; HLB: 5.0; Cloud Pt.: 14 C (1% aq.); Pour Pt.: 5 C; Stability: Compatible with soaps, CMC, common alkaline salts; Surface Tension: 33.0 dynes/cm (0.1% conc)
Composition: 100% active ingred.
Containers: 55 gal drums, tank cars, tank trucks

Pluronic L122. BASF Wyandotte/ Ind. Chem. Group
Chemical Description: Polyol formed by addition of propylene oxide to the two hydroxyl groups of a propylene glycol initiator
Category and Applications: See Pluronic F38
Form: Liquid
Properties: m.w.:5000 (avg.); Ref. Index: 1.4558; Solubility (>10 g/ 100 ml solvent): Sol. in ethyl al-

cohol, toluene, xylene, perchloroethylene; Density: 8.6 lb/gal; Sp. gr.: 1.03; Visc.: 1750 cps; HLB: 4.0; Cloud Pt.: 19 C (1% aq.); Flash Pt.: 490 F (COC); Pour Pt.: 20 C; Stability: Compatible with soaps, CMC, common alkaline salts; Surface Tension: 33.0 dynes/cm (0.1% conc.)
Composition: 100% active ingred.
Containers: 55 gal drums, tank cars, tank trucks

Pluronic P65. BASF Wyandotte/Ind. Chem. Group
Chemical Description: Polyol formed by addition of propylene oxide to the two hydroxyl groups of a propylene glycol initiator
Category and Applications: See Pluronic F38
Form: Paste
Properties: m.w.: 3400 (avg.); Solubility (> 10 g/100 ml solvent): Sol. in ethyl alcohol, propylene glycol, water, toluene, xylene, perchloroethylene; Density: 8.8 lb/gal (60 C); Sp. gr.: 1.06 (60 C); Visc.: 180 cps (60 C); M.P.: 30 C; HLB: 17.0; Cloud Pt.: 82 C (1% aq.); Pour Pt.: 27 C; Stability: Compatible with soaps, CMC, common alkaline salts; Surface Tension: 46.3 dynes/cm (0.1% conc.)
Composition: 100% active ingred.
Containers: 55 gal drums, tank cars, tank trucks

Pluronic P75. BASF Wyandotte/Ind. Chem. Group
Chemical Description: Polyol formed by addition of propylene oxide to the two hydroxyl groups of a propylene glycol initiator
Category and Applications: See Pluronic F38
Form: Paste
Properties: m.w.: 4150 (avg.); Solubility (> 10 g/100 ml solvent): Sol. in ethyl alcohol, propylene glycol, water, toluene, xylene, perchloroethylene; Density: 8.8 lb/gal (60 C); Sp. gr.: 1.06 (60 C); Visc.: 250 cps (60 C); M.P.: 34 C; HLB: 16.5; Cloud Pt.: 82 C (1% aq.); Pour Pt.: 27 C; Stability: Compatible with soaps, CMC, common alkaline salts; Surface Tension: 42.8 dynes/cm (0.1% conc.)
Composition: 100% active ingred.
Containers: 55 gal drums, tank cars, tank trucks

Pluronic P84. BASF Wyandotte/Ind. Chem. Group
Chemical Description: Polyol formed by addition of propylene oxide to the two hydroxyl groups of a propylene glycol initiator
Category and Applications: See Pluronic F38
Form: Paste
Properties: m.w.: 4200 (avg.); Solubility (> 10 g/100 ml solvent): Sol. in ethyl alcohol, propylene glycol, water, toluene, xylene, perchloroethylene; Density: 8.6 lb/gal (60 C); Sp. gr.: 1.03 (60 C); Visc.: 265 cps (60 C); M.P.: 34 C; HLB: 14.0; Cloud Pt.: 74 C (1% aq.); Flash Pt.: 442 F (COC); Pour Pt.: 18 C; Stability: Compat-

ible with soaps, CMC, common alkaline salts; Surface Tension: 42.0 dynes/cm (0.1% conc.)
Composition: 100% active ingred.
Containers: 55 gal drums, tank cars, tank trucks

Pluronic P85. BASF Wyandotte/ Ind. Chem. Group
Chemical Description: Polyol formed by addition of propylene oxide to the two hydroxyl groups of a propylene glycol initiator
Category and Applications: See Pluronic F38
Form: Paste
Properties: m.w.: 4600 (avg.); Solubility (> 10 g/100 ml solvent): Sol. in ethyl alcohol, propylene glycol, water, toluene, xylene, perchloroethylene; Density: 8.7 lb/gal (60 C); Sp. gr.: 1.04 (60 C); Visc.: 310 cps (60 C); M.P.: 40 C; HLB: 16.0; Cloud Pt.: 85 C (1% aq.); Pour Pt.: 29 C; Stability: Compatible with soaps, CMC, common alkaline salts; Surface Tension: 42.5 dynes/cm (0.1% conc.)
Composition: 100% active ingred.
Containers: 55 gal drums, tank cars, tank trucks

Pluronic P94. BASF Wyandotte/ Ind. Chem. Group
Chemical Description: Polyol formed by addition of propylene oxide to the two hydroxyl groups of a propylene glycol initiator
Category and Applications: See Pluronic F38
Form: Paste
Properties: m.w.: 5100 (avg.); Solubility (> 10 g/100 ml solvent): Sol. in ethyl alcohol, propylene glycol, water, toluene, xylene, perchloroethylene; Density: 8.7 lb/gal (60 C); Sp. gr.: 1.04 (60 C); Visc.: 110 cps (60 C); M.P.: 37 C; HLB: 13.5; Cloud Pt.: 76 C (1% aq.); Pour Pt.: 33 C; Stability: Compatible with soaps, CMC, common alkaline salts; Surface Tension: 39.6 dynes/cm (0.1% conc.)
Composition: 100% active ingred.
Containers: 55 gal drums, tank cars, tank trucks

Pluronic P103. BASF Wyandotte/ Ind. Chem. Group
Chemical Description: Polyol formed by addition of propylene oxide to the two hydroxyl groups of a propylene glycol initiator
Category and Applications: See Pluronic F38
Form: Paste
Properties: m.w.: 4950 (avg.); Solubility (> 10 g/100 ml solvent): Sol. in ethyl alcohol, water, toluene, xylene, perchloroethylene; Density: 8.7 lb/gal (60 C); Sp. gr.: 1.04 (60 C); Visc.: 285 cps (60 C); M.P.: 30 C; HLB: 9.0; Cloud Pt.: 86 C (1% aq.); Pour Pt.: 21 C; Stability: Compatible with soaps, CMC, common alkaline salts; Surface Tension: 34.4 dynes/cm (0.1% conc.)
Composition: 100% active ingred.
Containers: 55 gal drums, tank cars, tank trucks

Pluronic P104. BASF Wyandotte/Ind. Chem. Group
Chemical Description: Polyol formed by addition of propylene oxide to the two hydroxyl groups of a propylene glycol initiator
Category and Applications: See Pluronic F38
Form: Paste
Properties: m.w.: 5850 (avg.); Solubility (> 10 g/100 ml solvent): Sol. in ethyl alcohol, water, toluene, xylene, perchloroethylene; Density: 8.7 lb/gal (60 C); Sp. gr.: 1.04 (60 C); Visc.: 550 cps (60 C); M.P.: 37.5 C; HLB: 13.0; Cloud Pt.: 81 C (1% aq.); Flash Pt.: 448 F (COC); Pour Pt.: 32 C; Stability: Compatible with soaps, CMC, common alkaline salts; Surface Tension: 33.1 dynes/cm (0.1% conc.)
Composition: 100% active ingred.
Containers: 55 gal drums, tank cars, tank trucks

Pluronic P105. BASF Wyandotte/Ind. Chem. Group
Chemical Description: Polyol formed by addition of propylene oxide to the two hydroxyl groups of a propylene glycol initiator
Category and Applications: See Pluronic F38
Form: Paste
Properties: m.w.: 6500 (avg.); Solubility (> 10 g/100 ml solvent): Sol. in ethyl alcohol, water, toluene, xylene, perchloroethylene; Density: 8.8 lb/gal (60 C); Sp. gr.: 1.05 (60 C); Visc.: 800 cps (60 C); M.P.: 42 C; HLB: 15.0; Cloud Pt.: 91 C (1% aq.); Pour Pt.: 35 C; Stability: Compatible with soaps, CMC, common alkaline salts; Surface Tension: 39.1 dynes/cm (0.1% conc.)
Composition: 100% active ingred.
Containers: 55 gal drums, tank cars, tank trucks

Pluronic P123. BASF Wyandotte/Ind. Chem. Group
Chemical Description: Polyol formed by addition of propylene oxide to the two hydroxyl groups of a propylene glycol initiator
Category and Applications: See Pluronic F38
Form: Paste
Properties: m.w.: 5750 (avg.); Solubility (> 10 g/100 ml solvent): Sol. in ethyl alcohol, water, toluene, xylene, perchloroethylene; Density: 8.5 lb/gal (60 C); Sp. gr.: 1.02 (60 C); Visc.: 350 cps (60 C); M.P.: 31 C; HLB: 8.0; Cloud Pt.: 90 C (1% aq.); Pour Pt.: 31 C; Stability: Compatible with soaps, CMC, common alkaline salts; Surface Tension: 34.1 dynes/cm (0.1% conc.)
Composition: 100% active ingred.
Containers: 55 gal drums, tank cars, tank trucks

Pogol 200. Hart Chem. Ltd.
Chemical Description: Polyethylene glycol
Category: Solubilizer, nonionic
Applications: Pharmaceuticals and cosmetics for the preparation of ointments, lotions, and tablet binder

Form: Liquid; Color: Clear
Properties: Density: 1.14
Composition: 100% active ingred.
Reaction: Neutral
Toxicity: Skin and eye irritant
Containers: Drums, T/T
Warnings and Caution: Avoid contact with skin and eyes

Pogol 300. Hart Chem. Ltd.
Chemical Description: Polyethylene glycol
Category: Solubilizer, dispersant, emulsifier, nonionic
Applications: Agricultural herbicides and pesticides
Form: Liquid; Color: Clear
Properties: Density: 1.13
Composition: 100% active ingred.
Reaction: Neutral
Toxicity: Skin and eye irritant
Containers: Drums, bulk
Warnings and Caution: Avoid contact with skin and eyes

Pogol 400. Hart Chem. Ltd.
Chemical Description: Polyethylene glycol
Category: Emulsifier, solubilizer, dispersant, nonionic
Applications: Agricultural herbicides and pesticides
Form: Liquid; Color: Clear
Properties: Density: 1.12; Visc.: 92 cps
Composition: 100% active ingred.
Reaction: Neutral
Toxicity: Skin and eye irritant
Containers: Drums, T/T
Warnings and Caution: Avoid contact with skin and eyes

Pogol 400 USP. Hart Chem. Ltd.
Chemical Description: Polyethylene glycol, USP grade
Category: Solubilizer, lubricant, nonionic
Applications: Mold release agent and lubricant for both natural and synthetic products
Form: Liquid; Color: Clear
Properties: Density: 1.12; Visc.: 92 cps
Composition: 100% active ingred.
Reaction: Neutral
Toxicity: Skin and eye irritant
Containers: Drums, T/T
Warnings and Caution: Avoid contact with skin and eyes

Pogol 600. Hart Chem. Ltd.
Chemical Description: Polyethylene glycol
Category: Solubilizer, lubricant, nonionic
Applications: For lubricity in paper coating mixes; also as antisticking agents and oil resistors
Form: Paste; Color: Clear
Properties: Ref. Index: 1.4699; Density: 1.13; Visc.: 130 cps
Composition: 100% active ingred.
Reaction: Neutral
Toxicity: Skin and eye irritant
Containers: Drums, T/T
Warnings and Caution: Avoid contact with skin and eyes

Pogol 1540. Hart Chem. Ltd.
Chemical Description: Polyethylene glycol
Category: Solubilizer, nonionic
Applications: Base for the manufacture of fatty acid esters, emul-

sifiers, and dispersants
Form: Solid; Color: White
Composition: 100% active ingred.
Reaction: Neutral
Toxicity: Skin and eye irritant
Containers: Drums, T/T
Warnings and Caution: Avoid contact with skin and eyes

Pogol 1570. Hart Chem. Ltd.
Chemical Description: Polyethylene glycol
Category: Emulsifier, nonionic
Applications: Base for the manufacture of fatty acid esters, emulsifiers, and dispersants
Form: Liquid; Color: Clear
Properties: Ref. Index: 1.4360; Density: 1.112; Visc.: 110 cps
Composition: 70% active ingred.
Reaction: Neutral
Toxicity: Skin and eye irritant
Containers: Drums, T/T
Warnings and Caution: Avoid contact with skin and eyes

Polawax. Croda Chem. Ltd.
Chemical Description: Emulsifying wax
Category: Emulsifier, nonionic
Applications: O/W emulsifier, self-bodying, for use in basic emulsions, cosmetic formulations, e.g., skin creams and lotions, hair creams, conditioning rinses, baby preparations, make-up, aerosol formulations, pharmaceutical preparations
Form: Wax
Properties: Drop Pt.: 47–51 C; Sapon. Value: 7.8–15.0; Acid Val-

solubilizer, wetting agent, nonionic
Applications: W/O emulsifiers; water and alcohol soluble emollients; used for cosmetics, esp. in skin care creams and lotions, make-up cosmetics, and alcoholic lotions; bromo dye solvents in lipsticks
Form: Soft pastes to waxy solids
Properties: HLB (respectively): 7.3, 10.7, 12.7, 14.0, and 16.4
Composition: 100% conc.

Polylube GK. Hart Chem. Ltd.
Chemical Description: Polyoxyalkylene condensate
Category: Lubricant, nonionic
Applications: Water-sol. synthetic fiber lubricant, antistatic agent; also effective on wool
Form: Liquid
Properties: Ref. Index: 1.467; Sp. gr.: 1.09; Visc.: 330 cps
Composition: 100% active ingred.
Reaction: Neutral
Toxicity: Skin and eye irritant
Containers: Drums, T/T
Warnings and Caution: Avoid contact with skin and eyes

Polylube NPP. Hart Chem. Ltd.
Chemical Description: Blend of polyoxyalkylene and esters
Category: Lubricant, nonionic
Applications: Lubricant for processing of synthetic fiber blends on high-speed, spinning equipment
Form: Liquid; Color: Yellow
Properties: Sp. gr.: 0.99; Visc.: 230 cps
Composition: 87% active ingred.

Reaction: Neutral
Toxicity: Skin and eye irritant
Containers: Drums, T/T
Warnings and Caution: Avoid contact with skin and eyes

Polylube SC. Hart Chem. Ltd.
Chemical Description: Fatty acid condensate
Category: Lubricant, nonionic
Applications: Lubricant for processing of polyester fiber blends on high-speed, spinning equipment
Form: Liquid; Color: Clear
Properties: Sp. gr.: 1.02; Visc.: 110 cps
Composition: 87% active ingred.
Reaction: Neutral
Toxicity: Skin and eye irritant
Containers: Drums, T/L
Warnings and Caution: Avoid contact with skin and eyes

Polylube WS. Hart Chem. Ltd.
Chemical Description: Polyalkylene glycol ether
Category: Lubricant
Applications: Nonyellowing lubricant for spinning nylon and wool yarns
Form: Liquid; Color: Clear
Properties: Ref. Index: 1.4642; Sp. gr.: 1.086; Visc.: 500 cps
Composition: 100% active ingred.
Reaction: Neutral
Toxicity: Skin and eye irritant
Containers: Drums, T/T
Warnings and Caution: Avoid contact with skin and eyes

Polypeptide 37. Stepan Chem. Co.
Chemical Description: Sodium salt of protein hydrolysate
Category: Anionic surfactant
Applications: General use in hair and skin conditioning preparations, e.g., hair straighteners and waving sol'ns., setting lotions, shampoos
Form: Clear liquid; Color: Dark amber
Properties: Solubility: Miscible with water in all proportions; Sp. gr.: 1.20–1.23; Visc.: 50 cps max.; pH: 7.0–7.4
Composition: 48–50% solids

Polypeptide AAS 20%. Stepan Chem. Co.
Chemical Description: Ammonium salt of protein hydrolysate in ethanol
Category: Mildly cationic
Applications: For aerosol hair sprays and conditioners, and skin care preparations (provides a silky-soft feel)
Form: Clear liquid; Color: Amber; Odor: Alcoholic
Properties: Solubility: Sol. in anhydrous ethyl alcohol, isopropanol, propylene glycol, hexylene glycol, water; Sp. gr.: 0.87–0.88; pH: 5.5–6.5 (10% aq.)
Composition: 20% min. solids

Polypeptide LSN. Stepan Chem. Co.
Chemical Description: Ammonium salt of protein hydrolysate
Category: Anionic surfactant
Applications: Low ash, low salt content to insure clear, bright film formation; used for hair and skin care products

Form: Clear liquid; Color: Amber
Properties: Solubility: Miscible with water in all proportions; 20 parts in 60% alcohol; Visc.: 100 cps max.; pH: 6.6–7.4
Composition: 40–42% solids

Polypeptide LSN Anhydrous. Stepan Chem. Co.
Chemical Description: Ammonium salt of protein hydrolysate
Category: Anionic surfactant
Applications: Low salt content insures clear, bright film formation; substantive to hair; personal care products
Form: Fluffy, free-flowing powder (spray-dried); Color: White
Properties: Solubility: Miscible with water in all proportions; dissolves readily; Density: 20–21 lb/ft^3 (bulk)
Composition: 94% min. solids

Polypeptide SF. Stepan Chem. Co.
Chemical Description: Partial ammonium salt of protein hydrolysate
Category: Anionic surfactant
Applications: High purity product for the most demanding skin and hair applications; very low odor, color, salt, and ash
Form: Clear liquid; Color: Amber
Properties: Solubility: Miscible with water in all proportions; pH: 6.5–7.0
Composition: 56–58% solids

Polystep A-2. Stepan Chem. Co.
Chemical Description: Sodium xylene sulfonate
Category: Hydrotrope
Form: Liquid; Color: Pale to colorless
Composition: 40% active ingred.

Polystep A-4. Stepan Chem. Co.
Chemical Description: Sodium linear alkyl (C_{12}) benzene sulfonate
Category: Emulsifier, anionic
Applications: Low salt emulsifier for emulsion polymerization
Form: Turbid liquid; Color: Yellow
Composition: 50% active ingred.

Polystep A-6. Stepan Chem. Co.
Chemical Description: Sodium linear alkyl (C_{12}) benzene sulfonate
Applications: Emulsion polymerization
Form: Viscous slurry; Color: Yellow
Composition: 58% active ingred.

Polystep A-7. Stepan Chem. Co.
Chemical Description: Sodium linear alkyl (C_{12}) benzene sulfonate
Applications: Emulsion polymerization
Form: Slurry; Color: Yellow
Composition: 39% active ingred.

Polystep A-9. Stepan Chem. Co.
Chemical Description: Triethanolamine linear alkyl (C_{12}) benzene sulfonate
Applications: Emulsion polymerization
Form: Clear liquid; Color: Yellow
Composition: 60% active ingred.

Polystep A-10. Stepan Chem. Co.
Chemical Description: Calcium branched alkyl benzene sulfonate
Applications: Emulsion polymerization
Form: Viscous liquid; Color: Dark
Properties: Solubility: Oil-sol.
Composition: 65% active ingred.

Polystep A-11. Stepan Chem. Co.
Chemical Description: Isopropyl amine branched alkyl benzene sulfonate
Applications: Emulsion polymerization
Form: Clear, viscous liquid; Color: Pale
Properties: Solubility: Oil-sol.
Composition: 88% active ingred.

Polystep A-12. Stepan Chem. Co.
Chemical Description: Potassium branched alkyl (C_{15}) benzene sulfonate
Applications: Emulsion polymerization
Form: Clear, viscous liquid; Color: Pale
Composition: 47% active ingred.

Polystep A-13. Stepan Chem. Co.
Chemical Description: Linear alkyl (C_{12}) benzene sulfonic acid
Applications: Emulsion polymerization
Form: Viscous liquid
Composition: 97% active ingred.

Polystep A-15. Stepan Chem. Co.
Chemical Description: Sodium linear alkyl (C_{12}) benzene sulfonate
Category: Emulsifier, anionic
Applications: Alcohol-free emulsifier for emulsion polymerization
Form: Clear to hazy liquid; Color: Pale
Composition: 22% active ingred.

Polystep A-16. Stepan Chem. Co.
Chemical Description: Sodium branched alkyl (C_{12}) benzene sulfonate
Category: Emulsifier, anionic
Applications: Pumpable product for styrene polymerization
Form: Turbid liquid; Color: Pale
Composition: 30% active ingred.

Polystep A-17. Stepan Chem. Co.
Chemical Description: High purity branched dodecylbenzene sulfonic acid
Category: Intermediate, emulsifier
Applications: Intermediate for the production of sodium salts (wetting agent and high foamer); neutralized acid as emulsifier; salts as emulsifiers for SBR latex and can be "isolated" to coagulate the rubber; improves properties in emulsion polymerization
Form: Viscous liquid
Properties: Solubility: Acid and its salts are sol. in water; calcium and amine salts sol. in oil; Density: 9.1 lb/gal; Sp. gr.: 1.09; Visc.: 9000 cps (30 C)
Composition: 97.4% active ingred.
Toxicity: Can cause chemical burns if spilled on skin
Containers: 475 lb net full open head, lined steel drums
Warnings and Caution: Wear protective goggles or face mask

Polystep A-18. Stepan Chem. Co.
Chemical Description: Sodium linear alkenyl (C_{14}, C_{16}) sulfonate
Applications: For vinyl chloride, vinylidene chloride, SBR polymerization
Form: Clear liquid; Color: Pale
Composition: 40% active ingred.

Polystep B-1. Stepan Chem. Co.
Chemical Description: Ammonium alkyl phenol ethoxylate sulfate (4 moles EO)
Category: Emulsifier, anionic
Applications: Emulsion polymerization
Form: Viscous liquid; Color: Yellow
Composition: 59% active ingred.

Polystep B-3. Stepan Chem. Co.
Chemical Description: Sodium lauryl sulfate
Applications: High purity surfactant for emulsion polymerization
Form: Powder; Color: White
Composition: 97.5% active ingred.

Polystep B-5. Stepan Chem. Co.
Chemical Description: Sodium lauryl sulfate
Category: Emulsifier, anionic
Applications: Carefully controlled polymer grade emulsifier
Form: Clear liquid; Color: Pale
Composition: 29% active ingred.

Polystep B-6. Stepan Chem. Co.
Chemical Description: Triethanolamine lauryl sulfate
Applications: Emulsion polymerization
Form: Clear liquid; Color: Pale
Composition: 40% active ingred.

Polystep B-7. Stepan Chem. Co.
Chemical Description: Ammonium lauryl sulfate
Applications: Emulsion polymerization; provides better water resistance in coatings
Form: Clear liquid; Color: Pale
Composition: 28% active ingred.

Polystep B-8. Stepan Chem. Co.
Chemical Description: Diethanolamine lauryl sulfate
Applications: Emulsion polymerization
Form: Clear liquid; Color: Pale
Composition: 33% active ingred.

Polystep B-9. Stepan Chem. Co.
Chemical Description: Magnesium lauryl sulfate
Applications: Emulsion polymerization; improves stability of latex
Form: Clear liquid; Color: Pale
Composition: 28% active ingred.

Polystep B-10. Stepan Chem. Co.
Chemical Description: Sodium lauryl ethoxylate sulfate (4 moles EO)
Applications: Ethanol-free product for emulsion polymerization
Form: Clear liquid; Color: Pale
Composition: 28% active ingred.

Polystep B-11. Stepan Chem. Co.
Chemical Description: Ammonium lauryl ethoxylate sulfate (4 moles EO)

Category: Emulsifier, anionic
Applications: High active emulsifier for emulsion polymerization
Form: Clear liquid
Composition: 59% active ingred.

Polystep B-12. Stepan Chem. Co.
Chemical Description: Sodium lauryl ethoxylate sulfate (4 moles EO)
Category: Emulsifier, anionic
Applications: Emulsion polymerization
Form: Clear liquid; Color: Pale
Composition: 60% active ingred.

Polystep B-13. Stepan Chem. Co.
Chemical Description: Ammonium alkyl polyether ethoxylate sulfate
Category: Emulsifier
Applications: Versatile emulsifier for emulsion polymerization; insensitive to polyvalent cations
Form: Clear liquid; Color: Pale
Composition: 60% active ingred.

Polystep B-14. Stepan Chem. Co.
Chemical Description: Sodium alkyl polyether ethoxylate sulfate
Category: Emulsifier
Applications: Emulsion polymerization; insensitive to polyvalent cations
Form: Clear liquid; Color: Pale
Composition: 60% active ingred.

Polystep B-16. Stepan Chem. Co.
Chemical Description: Sodium lauryl ethoxylate sulfate
Applications: Emulsion polymerization
Form: Clear liquid; Color: Pale
Composition: 25% active ingred.

Polystep B-19. Stepan Chem. Co.
Chemical Description: Sodium lauryl ethoxylate sulfate (30 moles EO)
Category: Emulsifier, anionic
Applications: Emulsifier for floor polish polymers
Form: Clear liquid; Color: Pale
Composition: 26% active ingred.

Polystep B-20. Stepan Chem. Co.
Chemical Description: Ammonium lauryl ethoxylate sulfate (30 moles EO)
Category: Emulsifier, anionic
Applications: Emulsifier for floor finish latexes; good water resistance
Form: Clear liquid; Color: Pale
Composition: 26% active ingred.

Polystep B-21. Stepan Chem. Co.
Chemical Description: Sodium lauryl ethoxylate sulfate (12 moles EO)
Applications: For high carboxyl latex
Form: Clear liquid; Color: Pale
Composition: 28% active ingred.

Polystep B-22. Stepan Chem. Co.
Chemical Description: Ammonium lauryl ethoxylate sulfate (12 moles EO)
Category: Emulsifier, anionic
Applications: Emulsifier for acrylic copolymers
Form: Clear liquid; Color: Pale
Composition: 28% active ingred.

Polystep B-23. Stepan Chem. Co.
Chemical Description: Sodium lauryl ethoxylate sulfate (12 moles EO)
Applications: High active version of Polystep B-21; more oil soluble
Form: Hazy liquid; Color: Amber
Composition: 60% active ingred.

Polystep B-24. Stepan Chem. Co.
Chemical Description: Sodium lauryl sulfate
Applications: Emulsion polymerization; low cloud point
Form: Liquid; Color: Pale
Composition: 28% active ingred.

Polystep B-25. Stepan Chem. Co.
Chemical Description: Sodium decyl sulfate
Applications: Emulsion polymerization; high surface tension, low foam, hydrophilic
Form: Clear liquid; Color: Pale
Composition: 38% active ingred.

Polystep B-26. Stepan Chem. Co.
Chemical Description: Sodium cetyl-stearyl sulfate
Applications: Emulsion polymerization; Ca, Mg, H ion sensitive; less hydrophilic
Form: Paste; Color: Pale
Composition: 25% active ingred.

Polystep B-27. Stepan Chem. Co.
Chemical Description: Sodium alkyl phenol ethoxylate sulfate (4 moles EO)
Category: Emulsifier
Applications: Versatile emulsifier; esp. for acrylics and SBR
Form: Clear liquid; Color: Pale
Composition: 28% active ingred.

Polystep B-28. Stepan Chem. Co.
Chemical Description: Sodium alkyl polyether ethoxylate sulfate
Applications: Emulsion polymerization; low foam, low viscosity, high surface tension
Form: Clear liquid; Color: Amber
Composition: 40% active ingred.

Polystep F-1. Stepan Chem. Co.
Chemical Description: Nonyl phenol ethoxylate (4 moles EO)
Category: Dispersant, nonionic
Applications: Pigment dispersant; low foaming
Form: Clear liquid; Color: Pale
Properties: Solubility: Oil-sol.
Composition: 100% active ingred.

Polystep F-2. Stepan Chem. Co.
Chemical Description: Nonyl phenol ethoxylate (6 moles EO)
Category: Dispersant, nonionic
Applications: Pigment dispersant
Form: Clear liquid; Color: Pale
Properties: Solubility: Oil-sol.
Composition: 100% active ingred.

Polystep F-3. Stepan Chem. Co.
Chemical Description: Nonyl phenol ethoxylate (8 moles EO)
Category: Nonionic
Form: Clear liquid; Color: Pale
Properties: Solubility: Borderline water-sol.
Composition: 100% active ingred.

Polystep F-4. Stepan Chem. Co.
Chemical Description: Nonyl phenol ethoxylate (10 moles EO)
Category: Emulsifier, nonionic
Applications: Emulsion polymerization
Form: Clear liquid; Color: Pale
Composition: 100% active ingred.

Polystep F-5. Stepan Chem. Co.
Chemical Description: Nonyl phenol ethoxylate (12 moles EO)
Category: Emulsifier, stabilizer, nonionic
Applications: Emulsion polymerization
Form: Clear to turbid liquid
Composition: 100% active ingred.

Polystep F-6. Stepan Chem. Co.
Chemical Description: Nonyl phenol ethoxylate (14 moles EO)
Category: Nonionic
Applications: Emulsion polymerization; excellent with anionics
Form: Clear to turbid liquid
Composition: 100% active ingred.

Polystep F-7. Stepan Chem. Co.
Chemical Description: Nonyl phenol ethoxylate (15 moles EO)
Applications: Emulsion polymerization
Form: Turbid, viscous liquid
Composition: 100% active ingred.

Polystep F-8. Stepan Chem. Co.
Chemical Description: Nonyl phenol ethoxylate (20 moles EO)
Applications: For straight nonionic systems
Form: Clear to hazy liquid or gel
Composition: 70% active ingred.

Polystep F-9. Stepan Chem. Co.
Chemical Description: Nonyl phenol ethoxylate (30 moles EO)
Category: Emulsifier, nonionic
Applications: For acrylic systems
Form: Clear to hazy liquid or gel
Composition: 70% active ingred.

Polystep F-10. Stepan Chem. Co.
Chemical Description: Nonyl phenol ethoxylate (40 moles EO)
Category: Emulsifier, nonionic
Applications: Very hydrophilic emulsifier for acrylic systems
Form: Clear to hazy liquid or gel
Composition: 70% active ingred.

Polystep F-12. Stepan Chem. Co.
Chemical Description: Lauryl alcohol ethoxylate (12 moles EO)
Category: Nonionic
Applications: For preemulsification and post stabilization in latex systems
Form: Paste; Color: White
Composition: 100% active ingred.

Polystep F-13. Stepan Chem. Co.
Chemical Description: Lauryl alcohol ethoxylate (30 moles EO)
Category: Nonionic surfactant
Applications: Versatile hydrophilic surfactant
Form: Paste; Color: White
Composition: 100% active ingred.

Polystep J-2. Stepan Chem. Co.
Chemical Description: Anionic/nonionic blend
Category: Emulsifier
Applications: Emulsifier for coat-

ings and adhesive polymers
Form: Clear to hazy liquid
Composition: 70% active ingred.

Polystep J-3. Stepan Chem. Co.
Chemical Description: Anionic/nonionic blend
Category: Emulsifier
Applications: Emulsifier for vinyl acetate and acrylics
Form: Clear liquid
Composition: 28% active ingred.

Polytard. Westvaco/Polychemicals Dept.
Chemical Description: Sulfonic acid derivative of Kraft lignin
Category: Specialty cement additive
Applications: Extends setting time in mortars; aids air-entraining agents in masonry cements
Form: Powder

Polyquart H. Henkel Inc.
Chemical Description: PEG-15 tallow polyamine
Category: Antistat, pseudo-cationic
Applications: Pseudo-cationic for use in hair and skin applications; antistatic properties leaves hair manageable, glossy, and easy to style; offers good wet and dry compatibility; no build-up on hair shaft with repeated applications; compatible with anionics and cationics; skin care applications at low levels, esp. in O/W nonionic emulsions where it contributes to smooth texture, reduction of oiliness, and improved rub-out; completely nonirritating to skin and eyes
Form: Clear, viscous liquid; Color: Light yellow
Composition: 50% conc.

Polyquart H-7102. Henkel Inc.
Chemical Description: PEG-15 coco polyamine and stearalkonium chloride
Category and Applications: See Polyquart H; the quaternary in Polyquart H-7102 increases the antistatic qualities; ideal for "acid-pH systems"
Form: Clear, viscous liquid; Color: Light yellow
Composition: 50% conc.

Polywet AX-4. Uniroyal Chem.
Category: Emulsifier
Applications: Manufacture of acrylate latices
Form: Liquid; Color: Amber
Properties: Sp. gr.: 1.12; Visc.: 12,000 cps (20 C); F.P.: 30 F; pH: 5–6
Composition: 40% active ingred.
Reaction: Acid
Containers: Drum or bulk
Warnings and Caution: Treat as mild acid—wear gloves, goggles, etc.; wash spills with water; absorb large spills with sand, etc.

Polywet KX-3. Uniroyal Chem.
Chemical Description: Potassium salt of a functionalized oligomer
Category: Emulsifier, anionic
Applications: Manufacture of styrene-butadiene latices
Form: Liquid; Color: Amber;

Odor: Mild thiol
Properties: Solubility: Complete in water; Sp. gr.: 1.19; Visc.: 380 cps; F.P.: 15 C; HLB: 16; Pour Pt.: 5 F; Surface Tension: 47 dynes/cm (0.1% conc.); pH: 7–8
Composition: 46% active ingred.
Reaction: Neutral
Containers: Drums or bulk
Warnings and Caution: Wear eye protection; wash spills or splashes with water; absorb large spills with sand, etc.

Polywet KX-4. Uniroyal Chem.
Chemical Description: Potassium salt of a functionalized oligomer
Category: Emulsifier, anionic
Applications: Manufacture of acrylate latices
Form: Liquid; Color: Amber; Odor: Mild thiol
Properties: Solubility: Complete in water; Sp. gr.: 1.177; Visc.: 5000 cps (15 C); F.P.: 20 F; HLB: 17, Pour Pt.: 15 F; Surface Tension: 47 dynes/cm (0.1% conc.); pH: 7–8
Composition: 40% active ingred.
Reaction: Neutral
Containers: Drums or bulk
Warnings and Caution: Wear eye protection; wash spills or splashes with water; absorb large spills with sand, etc.

Polywet XK-5. Uniroyal Chem.
Chemical Description: Potassium salt of a functionalized oligomer
Form: Liquid; Color: Yellow to amber
Properties: Solubility: Complete in water; Sp. gr.: 1.19; Visc.: 1200 cps; HLB: 26; Pour Pt.: 15 F; Surface Tension: 38 dynes/cm (0.1% conc.); pH: 6.5–8.0
Composition: 45 ± 1% solids

Polywet ND-1. Uniroyal Chem.
Chemical Description: Sodium salt of a polyfunctional oligomer
Category: Dispersant, anionic
Applications: Dispersant for minerals; nonfoaming dispersant for pigments and fillers in paint, paper, latex compounding, and water treatment applications; esp. in TiO_2 and ZnO dispersions which are not affected by mercury biocides or heat; useful in conductive paper coatings
Form: Liquid; Color: Slight yellow
Properties: Sp. gr.: 1.13; Visc.: 35 cps; F.P.: 15 F; pH: 6.8–7.8
Composition: 25% active ingred.
Reaction: Neutral
Toxicity: Mild eye irritant
Containers: Drums or bulk
Warnings and Caution: Wear eye protection; wash spills and splashes with water

Polywet ND-2. Uniroyal Chem.
Chemical Description: Sodium salt of a functionalized oligomer
Category: Dispersant, anionic
Applications: Nonfoaming dispersant for pigments, minerals, extenders, fillers in aq. systems; used in latex paints, esp. gloss latex enamels; scale inhibitor for water treatment; complies with FDA regulations governing coatings, adhesives,

paper and paperboard, and boiler water
Form: Liquid; Color: Colorless to slight yellow
Properties: Sp. gr.: 1.16; Visc.: 32 cps; F.P.:15 F; pH: 6.8–7.8
Composition: 25% active ingred.
Toxicity: Mild eye irritant
Containers: Drums or bulk
Warnings and Caution: Wear eye protection; wash spills and splashes with water

Primasol FP. BASF AG
Chemical Description: Mixture of sodium salts of aliphatic sulfo acids
Category: Textile auxiliary, wetting agent, penetrant, anionic
Applications: Auxiliary that promotes penetration of untreated cotton goods which are difficult to wet; used in continuous and batchwise dyeing of loom-state cotton with vat, sulfur, naphthol, and reactive dyes
Form: Low viscosity liquid; Color: Yellowish-brown
Properties: Solubility: Readily sol. in water; Stability: Very stable to electrolytes and large amounts of alkali in padding liquors and dyebaths; stable to hard water and acid liquors
Reaction: Weakly alkaline
Toxicity: Skin irritant on prolonged contact with conc. form
Warnings and Caution: Wear protective goggles; use with good ventilation
References: Tech. leaflet M 1477 e

Primasol KW. BASF AG
Chemical Description: Mixture of anionic and nonionic compounds, polyfunctional, synergistic
Category: Textile auxiliary, wetting agent, detergent, stabilizer, anionic/nonionic
Applications: Dyeing auxiliary for continuous dyeing of wool, polyamide, and polyester fibers; used for continuous dyeing and vigoureux printing of slubbing and tow, continuous dyeing of textile floor coverings, and space dyeing; ensures rapid penetration and uniform dispersion of the padding liquor, acts as a detergent, and improves solubility of dyes and stabilizes the disp. of sparingly sol. dyes
Form: Liquid; Color: Yellowish-brown
Properties: Solubility: Dissolves readily in cold or warm water; Stability: Resistant to acids, alkalis, and electrolytes in certain concentrations; pH: 6–6.5
Reaction: Slightly acid to almost neutral
Toxicity: Causes inflammation of skin on prolonged contact
References: Tech. leaflet M 5143 e

Primasol PAC. BASF AG
Chemical Description: Mixture of high-boiling, low-volatile esters and ethers with nonionic surfactants
Category: Textile auxiliary, wetting agent
Applications: Auxiliary for continuous or batch-wise dyeing of

acrylic fibers with selected cationic dyes by the pad-steam process; wets and accelerates diffusion of dyes into fiber; used in dyeing slubbing, tow, fabrics, carpets, nonwovens, plushes, and high-bulk yarns
Form: Clear, slightly viscous liquid; Color: Almost colorless
Properties: Solubility: Readily sol. in water
Reaction: Neutral
Toxicity: Attacks mucous membranes of the eyes; vapors may cause dangerous inflammation of the mucous membranes when inhaled
Warnings and Caution: Protect eyes; avoid inhalation of vapors
References: Tech. leaflet M 5420 e

Primasol SD. BASF AG
Chemical Description: Mixture of oxyethylation products
Category: Textile auxiliary, stabilizer, wetting agent, nonionic
Applications: Auxiliary for the space dyeing of polyamide and acrylic carpet yarns with acid, metal complex, and basic dyes; improves solubility of dyes, stabilizes padding liquors and print pastes, and promotes fixation rate of dyes in steaming; ensures rapid wetting and uniform dispersion of the padding liquor on the fiber surface
Form: Liquid; Color: Colorless or light yellowish
Properties: Stability: Stable to alkalis, acids, or salts when not in excessively high concs.; pH: 6.5–7.5 (10% sol'n.)
Toxicity: Eye and skin irritant on prolonged contact
Warnings and Caution: Wear protective eye goggles; avoid prolonged contact with skin
References: Tech. leaflet M 5455 e

Priminox T-25. Rohm and Haas Co.
Chemical Description: Condensate with 25 moles EO bound to an alkyl chain
Category: Emulsifier, nonionic
Form: Solid; Color: 11 (VCS)
Properties: Density: 8.6 lb/gal; HLB: 15.6; Cloud Pt.: > 100 C (1% sol'n.); Flash Pt.: > 350 F (TOC); Surface Tension: 29 dynes/cm (0.01%)
Composition: 100% active ingred.

Printac CN-230. Toho Chem. Industry Co., Ltd.
Category: Wetting agent
Applications: Leveling and penetrating agent in acrylic fabrics printing
Form: Liquid
Composition: 50% active ingred.

Procetyl 10. Croda Chem. Ltd.
Chemical Description: POP 10 cetyl ether
Category: Emollient, lubricant, cosolvent, plasticizer, superfatting agent, penetrant, wetting agent, spreading agent, nonionic
Applications: Cosmetics, skin make up, hair preparations, bath oils, alcoholic and aq. alcoholic compositions; emollient and valve lubricant in aerosol antiperspi-

rants and deodorants
Form: Clear liquid; Color: 150 max. (APHA); Odor: Faint, characteristic sweet
Properties: Solubility: Excellent aq. alcoholic solubility; Acid Value: 1 max.; Hydroxyl Value: 75–85; Iodine Value: 2 max.; Stability: Stable to oxidation; useful in both acidic and alkaline media; pH: 6.0–7.5 (3% disp.)
Composition: 100% active ingred.

Procetyl 20. Croda Chem. Ltd.
Chemical Description: POP 20 cetyl ether
Category and Applications: See Procetyl 10
Form: Clear liquid; Color: 150 max. (APHA); Odor: Faint, characteristic sweet
Properties: Solubility: Excellent aq. alcoholic solubility; Acid Value: 1 max.; Hydroxyl Value: 45–55; Iodine Value: 2 max.; Stability: Stable to oxidation; useful in both acidic and alkaline media; pH: 6.0–7.5 (3% disp.)
Composition: 100% active ingred.

Procetyl 30. Croda Chem. Ltd.
Chemical Description: POP 30 cetyl ether
Category and Applications: See Procetyl 10
Form: Clear liquid; Color: 150 max. (APHA); Odor: Faint, characteristic sweet
Properties: Solubility: Excellent aq. alcoholic solubility; Acid Value: 1 max.; Hydroxyl Value: 35–45; Iodine Value: 2 max.; Stability: Stable to oxidation; useful in both acidic and alkaline media; pH: 6.0–7.5 (3% disp.)
Composition: 100% active ingred.

Procetyl 50. Croda Chem. Ltd.
Chemical Description: POP 50 cetyl ether
Category and Applications: See Procetyl 10
Form: Clear liquid; Color: 150 max (APHA); Odor: Faint, characteristic sweet
Properties: Solubility: Excellent aq. alcoholic solubility; Acid Value: 1 max.; Hydroxyl Value: 20–30; Iodine Value: 2 max.; Stability: Stable to oxidation; useful in both acidic and alkaline media; pH: 6-7.5 (3% disp.)
Composition: 100% active ingred.

Procetyl AWS. Croda Chem. Ltd.
Chemical Description: Alkoxylated cetyl ether
Category: Emollient, solubilizer, emulsifier, plasticizer, nonionic
Applications: Emollient and solubilizer in aq. and aq. alcoholic systems, i.e., bath oils and essences, aerosol antiperspirants
Form: Slightly turbid, oily liquid; Color: Water white; Odor: Characteristic sweet
Properties: Solubility: Water-sol.; hydrophilic
Composition: 100% active ingred.

Prochols. Croda Chem. Ltd.
Chemical Description: Propoxylated distilled lanolin alcohols, differing in degree of propoxylation

(5, 10, 20, and 50 moles)
Category: Emollient, cosolvent
Applications: Alcohol-sol. emollients; possess good aq. alcoholic compatibility; used in all cosmetics, esp. skin and bath oils, skin care products, make-up cosmetics
Form: Liquids; Color: Amber
Properties: Solubility: Alcohol-sol.

Product BCO. DuPont
Chemical Description: Cetyl betaine
Category: Wetting agent, detergent, emulsifier, dispersant, softener, leveling and rewetting agent, dyeing assistant, antistat; amphoteric surfactant
Applications: Dyeing applications; softening agent for textiles; leveling and rewetting agent in the paper industry; dyeing assistant and degreasing agent in the leather industry; antistatic agent on plastic films
Form: Clear liquid; Color: Brown; Odor: Mild
Properties: Solubility: Miscible in water; Density: 8.4 lb/gal; Visc.: Free-flowing; Stability: Good; Surface Tension: 27.8 dynes/cm @ 32 C (0.1% conc.)
Composition: 33% solids

Produkt B 2045. Croda Chem. Ltd.
Chemical Description: Alkenyl succinic anhydride
Category: Corrosion inhibitor, hardener
Applications: Corrosion inhibitor additive for nonaq. systems; hardener in cosmetics and pharmaceuticals, esp. for topical applications; forms liquid emulsions
Form: Solid; Color: Colorless; Odor: Practically odorless
Properties: M.P.: 55–63 C; Sapon. Value: 2 max.; Acid Value: 1 max.; Iodine Value: 2 max. (Hanus); Stability: Stable to acidic and alkaline media
Composition: 100% conc.

Prosol 5. Hart Chem. Ltd.
Chemical Description: Mineral oil and fatty acid condensate
Category: Lubricant, nonionic
Applications: Mineral oil lubricant for processing of wool and wool synthetic blends on rigorous mechanical equipment
Form: Liquid; Color: Amber
Properties: Ref. Index: 1.4913; Sp. gr.: 0.922; Visc.: 70 cps; Stability: Forms stable emulsion
Reaction: Acid
Toxicity: Skin and eye irritant
Containers: Drums, T/L
Warnings and Caution: Avoid contact with skin and eyes

Prosol 518. Hart Chem. Ltd.
Chemical Description: Refined mineral oil and emulsifiers
Category: Lubricant, nonionic
Applications: High performance mineral oil lubricant for processing of wool and wool synthetic blends
Form: Liquid; Color: Amber
Properties: Ref. Index: 1.471; Sp. gr.: 0.895; Visc.: 65 cps; Stability: Forms stable emul-

sion
Composition: 100% active ingred.
Reaction: Acid
Toxicity: Skin and eye irritant
Containers: Drums, T/L
Warnings and Caution: Avoid contact with skin and eyes

Prosol C70. Hart Chem. Ltd.
Chemical Description: Hydrocarbons and polyether glycol
Category: Lubricant, nonionic
dener for epoxy resins
Form: Viscous liquid
Composition: 100% conc.

Produkt B 3010. Croda Chem. Ltd.
Chemical Description: Disodium alkenyl succinate
Category: Corrosion inhibitor
Applications: Rust-inhibiting additive for aq. systems; corrosion inhibitor for aerosol packaging
Form: Viscous liquid
Composition: 40% active ingred.

Produkt B 3032. Croda Chem. Ltd.
Chemical Description: Trialkanolammonium alkenyl succinate
Category: Rust inhibitor
Applications: Rust-inhibiting additive for aq. systems
Form: Viscous liquid
Composition: 40% active ingred.

Promulgen D. Robinson-Wagner Co.
Chemical Description: Polyethylene glycol ether complexes of fatty alcohols
Category: Emulsifier, stabilizer, gelling agent, nonionic
Applications: O/W emulsifier, emollient, and stabilizer for cosmetics and pharmaceuticals, esp. for topical application; forms cream emulsions
Form: Solid; Color: Colorless; Odor: Practically odorless
Properties: M.P.: 47–55 C; Sapon. Value: 2 max.; Acid Value: 1 max.; Iodine Value: 2 max. (Hanus); Stability: Stable to acidic or alkaline media
Composition: 100% conc.

Promulgen G. Robinson-Wagner Co.
Chemical Description: Polyethylene glycol ether complex of fatty alcohols
Category: Emulsifier, stabilizer, gelling agent, nonionic
Applications: O/W emulsifier, emollient, and stabilizer for cos-
Applications: Nonyellowing coning oil for processing of both texturized filament nylon and polyester yarns
Form: Liquid; Color: Yellow
Properties: Sp. gr.: 0.848; Visc.: 35 cps; Stability: Forms stable emulsion
Composition: 100% active ingred.
Reaction: Alkaline
Toxicity: Skin and eye irritant
Containers: Drums, T/L
Warnings and Caution: Avoid contact with skin and eyes

Prostearyl 15. Croda Chem. Ltd.
Chemical Description: POP 15 stearyl ether
Category: Emollient
Applications: Cosmetics

Protodye 71. Proctor Chem. Co.
Category: Emulsifier, dyeing assistant, detergent, nonionic
Applications: Emulsifier for Varsol, a solvent for pretreatment of synthetic fibers before dyeing; forms emulsion with Varsol which is stable to caustic soda and organic sequestering agents; dyeing assistant; detergent in conjunction with alkalis such as soda ash, trisodium or tetrasodium pyrophosphate

Protodye NS. Proctor Chem. Co.
Category: Dyeing assistant, scouring agent
Applications: Improves scouring and level dyeing by causing nylon and polyester fabrics to sink during beck scouring and dyeing; low foaming; suitable for pressure jets; scouring operations
Form: Liquid
Properties: Solubility: Disperses in cold water

Protodye QNF. Proctor Chem. Co.
Chemical Description: Naphthalene sulfonate, sodium salt
Category: Wetting and dispersing agent
Applications: Good wetting action and dispersion properties for dyes, pigments, latices; wetting agent for all cotton and rayon processing; pasting agent for direct, sulfur, and vat dyestuffs; wool carbonizing
Form: Liquid; Color: Light yellow
Properties: Solubility: Water-sol.; Stability: Excellent stability in strong acids or alkalis
Composition: 16% active ingred.

Protopan. Proctor Chem. Co.
Chemical Description: Ethylene oxide condensate
Category: Detergent, dyeing assistant, wetting agent, emulsifier, anionic
Applications: Textile scouring, dyeing, and preparation work; penetrant in desizing operations; in solvent/desizing systems for preparing polyester/cotton; Kier boiling of cotton; open width scouring; in chrome/acetic acid oxidizing bath of vat color; rinses freely from the cloth
Form: Clear liquid; Color: Straw
Properties: Solubility: Completely sol. in cold water with the absence of a cloud pt.; Stability: Good stability with practical concs. of acid or alkali, and enzymes; pH: ≈ 7.5 (1% sol'n.); Biodegradable
Composition: 70% conc.

Protosan EP. Proctor Chem. Co.
Chemical Description: Alkyl benzene sulfonate
Category: Wetting and rewetting agent, dyeing assistant, anionic
Applications: Nonfoaming surfactant for sanforizing, vat and sulfur dyeing, and goods which are difficult to rewet; good penetrating and level dyeing properties
Form: Liquid
Properties: Stability: Very stable to all conc. of acids, alkalis, and salts, except for mercerizing-

strength caustic acid
Composition: 42% conc.
Containers: 480 lb drums

Protowet 35. Proctor Chem. Co.
Chemical Description: Complex alcohol phosphate
Category: Detergent, dyeing assistant, anionic
Applications: Cotton preparation, e.g., in acid steeping, Kier boiling, peroxide or chlorine bleaching, open width boil off; scouring of synthetics in jig, pressure machines, or dye beck; as a dyeing assistant in pigment pad vat dyeing of cotton, continuous vat dyeing, vat dyeing in the beck or jig, direct, developed, acid, and basic dyestuffs; soaping off vats, naphthols, etc.
Form: Clear liquid; Color: Straw
Properties: Solubility: Completely sol. in hot or cold water; Stability: Compatible with caustic, sulfuric, or acetic acid as high as 5%; stable with metal salts; pH: ≈ 7 (1% sol'n.)
Containers: 450 lb net drums

Protowet 60. Proctor Chem. Co.
Chemical Description: Complex alcohol phosphate
Category and Applications: See Protowet 35
Form: Clear liquid; Color: Straw
Properties: Solubility: Complete in hot or cold water; Stability: Compatible with caustic, sulfuric or acetic acid as high as 5%; stable with metal salts; pH: ≈ 7 (1% sol'n.)
Composition: 60% active ingred.
Containers: ≈ 450 lb net drums

Protowet 3072. Proctor Chem. Co.
Chemical Description: Phosphated alkyl ester, free acid
Category: Chemical intermediate
Applications: Amine neutralized forms are used as antistats in spin finishes with good effects on synthetic fibers, e.g., acetate, polyester, nylon, and acrylics
Form: Liquid; Color: Light brown
Properties: Solubility: Insol. in water as is; the triethanolamine salt is clearly sol. in water; pH: 6.5 (1% sol'n. of triethanolamine neutralized form)
Composition: 100% active ingred.

Protowet 3193. Proctor Chem. Co.
Chemical Description: Emulsified solvent
Category: Detergent, anionic
Applications: Used to clean finishing and dye pad rolls; as scouring agent for removal of fats and oils from fabrics made from synthetic or natural fibers
Form: Liquid; Color: Light yellow
Properties: pH: ≈ 5–5.5
Composition: Solvent–aliphatic hydrocarbon

Protowet BPX. Proctor Chem. Co.
Chemical Description: Emulsified solvents
Applications: Preparation and boil off of greige cotton or cotton/polyester goods; removes the natural waxes and pectins from cot-

ton, mill grease and spin finish oils from polyester
Form: Clear liquid; Color: Amber
Properties: Solubility: Forms stable emulsion in cold or hot water; Stability: Compatible with moderate amounts of caustic soda (about 1%); pH: 10.0 (3% sol'n.)

Protowet C. Proctor Chem. Co.
Chemical Description: Sulfonated fatty acid ester
Category: Wetting and rewetting agent, emulsifier, anionic
Applications: Textile wet processing, including preshrinking, dye penetration of heavy fabrics, package dyeing assistant, package dyed yarns for knitting trade, slasher dyeing, warp sizing, preparation of wettable insecticide powders
Form: Clear oil; Color: Amber
Properties: Solubility: Readily dissolves in water from 60 F to the boil (wetting efficiency greater at the higher temperatures); sol. in many organic solvents, e.g., Stoddard solvent; Stability: Stable to acids at pH down to 1.0, alkalis not exceeding concentrations of 1% sol'ns., and in salt concs. of 700–750 ppm; Biodegradable
Composition: 60% conc.

Protowet D-75. Proctor Chem. Co.
Chemical Description: Sulfonated aliphatic polyester
Category: Wetting and rewetting agent, emulsifier, anionic
Applications: Textile wetting and rewetting, effective in small concs.; rewetting agent before sanforizing and continuous dyeing; wetting out and penetrating fabrics before tub, jig, or package dyeings; emulsifier for insol. fatty products; also for metal cleaning, pigment dispersion, glass cleaning, dry cleaning compounds
Form: Slightly viscous liquid; Color: Very light; (also available in the form of a 25% gel)
Properties: Stability: Should be used in a nearly neutral bath
Composition: 75% active ingred.
Containers: 480 lb net drums

Protowet D-75 S. Proctor Chem. Co.
Chemical Description: Sulfonated aliphatic polyester
Category and Applications: See Protowet D-75; D-75 S is stabilized to enable its use in the presence of resin catalyst
Form: Slightly viscous liquid; Color: Very light (also available in the form of a 25% gel)
Properties: Stability: Should be used in a nearly neutral bath
Composition: 75% active ingred.
Containers: 480 lb net drums

Protowet E-4. Proctor Chem. Co.
Category: Wetting agent, emulsifier, penetrant, dyeing assistant, nonionic
Applications: Wetting and emulsifying agent with the property of non-rewetting, e.g., useful in water repellent for stain and soil reactant finishes; penetrant for resin baths; stabilizes finishing

baths; dyeing assistants; in fiber finish components; prescour on Jet machines, becks, and other batch equipment; good for continuous scouring, desizing and dyeing, and afterwashing resin treated goods

Protowet FHL. Proctor Chem. Co.
Chemical Description: Sodium salt of sulfated glycerides
Category: Emulsifier, lubricant, dyeing assistant, scouring agent, anionic
Applications: Emulsifier and lubricant in select spin finish formulations; emulsifies mineral oil, butyl stearate, coconut oil; dyebath lubricant to prevent chafing marks and crowsfeet in batch type operations; dyeing assistant to help paste up dyes, esp. Naphthol dyes; scouring of beams and packages where too much wetting can cause channeling of the dye liquor
Form: Liquid; Color: Medium amber
Properties: Solubility: Complete—forming translucent sol'n.; pH: 5–6 (1% sol'n.); Biodegradable
Composition: 75% active ingred.

Protowet MO. Proctor Chem. Co.
Chemical Description: Solvent base product
Category: Detergent
Applications: Removing mill oils, waxes, and grease from greige goods; scour in reworking finished goods; used in a prepad and in the boil off
Form: Liquid; Color: Straw
Properties: Stability: Stable in 4% caustic soda at the boil

Protowet NRW. Proctor Chem. Co.
Chemical Description: Phosphated alcohol
Category: Penetrant, emulsifier, wetting agent, scouring agent, dyeing assistant, anionic
Applications: Scouring cottons or synthetics in acid or alkaline sol'ns.; wetting agent during desizing; assistant in Kier boiling, open width boil off, or continuous bleaching; scouring in the beck, pressure machine, or jig; dyeing assistant with vats, sulfurs, directs, and on basic colors; soaping off vats, developed, and naphthol dyed goods; soaping off prints
Form: Clear liquid; Color: Light yellow
Properties: Solubility: Excellent in hot or cold water; pH: ≈ 7 (1% sol'n.)
Composition: 30% active ingred.; also available in 100% active free acid form

Protowet PTB. Proctor Chem. Co.
Category: Scouring agent, anionic
Applications: Scouring agent, esp. for polyester and its blends; removes coning oils, fiber lubricants, etc. from carpet fibers; used in batch scouring operations, e.g., becks, package machines, or beam machines, and in continuous scouring
Form: Clear liquid; Color: Slightly yellow

Protowet RW. Proctor Chem. Co.
Chemical Description: Sulfonated fatty acid ester
Category: Rewetting agent, emulsifier, anionic
Applications: Rewetting, esp. on heavy tightly woven fabrics that have been dyed with sulfur colors or Indo carbon black; for preshrinking, dye penetration of heavy fabrics, package dyed yarns for knitting trade; emulsifier for many organic solvents
Form: Clear oil; Color: Amber
Properties: Solubility: Readily dissolves in water from 60 F to the boil; sol. in many organic solvents, e.g., Stoddard solvent; Stability: Stable in acids at pH down to 1.0, in alkalis not exceeding 1% concs., and in salts of concs. 700–750 ppm
Composition: 62% active ingred.

Protowet TJ. Proctor Chem. Co.
Chemical Description: Phosphated alcohol
Category: Emulsifier, dispersant, wetting agent, detergent, anionic
Applications: Preparation of polyester or polyester blends, removes fiber lubricants and coning oils prior to dyeing; used in desizing, bleaching, and scouring of cotton or polyester/cotton blends; afterscouring on polyester yarns where alkaline hydro sol'ns. are used; as detergent in afterwash of precured resin finished goods
Form: Liquid; Color: Pale yellow
Properties: Solubility: Sol. in cold water in all proportions; Sp. gr.: 1.10–1.20; Stability: Stable in caustic soda sol'ns. up to 8%; compatible with enzymes used in desizing
Composition: 85% active ingred.

Protowet XL. Proctor Chem. Co.
Chemical Description: Sulfonated fatty acid ester
Category and Applications: See Protowet C
Form: Clear oil; Color: Amber
Properties: Solubility: Readily dissolves in water from 60 F to the boil; sol. in many organic solvents, e.g., Stoddard solvent; Stability: Stable in acids at pH down to 1.0, in alkalis not exceeding 1% sol'ns., and in salts in concs. of 700–750 ppm; pH: ≈ 6
Composition: 62% active ingred.

Proxel CRL. I.C.I. United States
Chemical Description: 1,2-benzisothiazolin-3-one sol'n.
Category: Microbiostat preservative
Form: Liquid, noncrystallizing above −10 C; Color: Dark brown; Odor: Amine-like
Properties: Solubility: Sol. in water; Sp. gr.: ≈ 1.12; Visc.: ≈ 15 cps; Stability: Aq. sol'ns. are stable in presence of electrolytes at pH down to ≈ 4.5; pH: 9.5–11.5
Composition: 30–35% active
Containers: 45 lb pail, 506 lb drum

Pycal 94. I.C.I. United States
Chemical Description: Polyoxyethylene aryl ether
Category: Plasticizer

Form: Liquid; Color: Pale yellow
Properties: Ref. Index: ≈ 1.502; Solubility: Sol. in water, ethers, alcohols, ketones, lower aliphatic esters, and aromatic hydrocarbon solvents; Sp. gr.: ≈ 1.1; Visc.: ≈ 50 cps; Flash Pt.: > 300 F; Fire Pt.: > 300 F; Acid No.: 2.0 max.; Hydroxyl No.: 203–218; pH: 4–6 (50% aq.)

Pyronate 40. Witco Chem./Sonneborn Div.
Chemical Description: Sodium petroleum sulfonate
Category: Wetting agent, dispersant, anionic
Applications: Oil flotation; froth flotation of metallic and nonmetallic ores and whenever demulsification properties are required
Form: Liquid; Color: Dark
Properties: Solubility: Water-sol.; Density: 9.6 lb/gal; Sp. gr.: 1.18 (60 F)
Composition: 42% active ingred.; other components: 18% mineral oil, 40% water
Reaction: Alkaline
Toxicity: May cause eye burn
Containers: Drums, T/W, T/C

Q

Quad Chemical Caramel Color. Quad Chem. Corp.
Applications: Single strength, acid-proof caramel color
Properties: Solubility: 60% in ethanol; Density: 10.99 lb/gal (60 F); Sp. gr.: 1.3182 (60 F); Visc.: 275 cps (80 F); pH: 2.9 (as is)

Quad Chemical Caramel Color. Quad Chem. Corp.
Applications: Double strength, acid-proof caramel color
Properties: Solubility: 70% in ethanol; Density: 10.42 lb/gal (60 F); Sp. gr.: 1.2500 (60 F); Visc.: 200 cps (80 F); pH: 2.9 (as is)

Quad Opacifier #35. Carson Chem.
Chemical Description: Polymer emulsion
Category: Opacifying agent
Applications: Opacifying water-based liquids; used in household ammonia, liquid laundry detergent, light duty dishwashing detergent, acid bowl cleaners, fabric softeners, cream hair conditioners
Properties: Density: 8.5 lb/gal; Stability: Good stability in alkaline and acid media, and wide range of surfactants; pH: 2–3
Composition: 34–36% solids

Quadrilan AT. Lankro Chem. Ltd.
Chemical Description: Quaternized fatty amine ethoxylate
Category: Antistatic agent, cationic
Applications: Antistat for PVC and other polymers
Form: Clear liquid; Color: Amber; Odor: Characteristic
Properties: Solubility: Sol. in water @ 20 C; Sp. gr.: 1.081 (20 C); Visc.: 400 cs (20 C); Flash Pt.: $>$ 300 F (COC); Pour Pt.: $<$ 0 C; pH: 6–9 (1% aq.); Biodegradable
Composition: 100% active ingred.
Reaction: Neutral
Toxicity: May cause irritation on prolonged contact with conc. form
Containers: Stainless steel, glass, polythene, or resin lined equipment
Warnings and Caution: Wear protective gloves and goggles; spillages

may be slippery; do not mix with strong oxidizing or reducing agents–potentially explosive
References: Bulletin SI:30 FM678

Quadrilan BC. Lankro Chem. Ltd.
Chemical Description: Alkyl benzyl dimethyl ammonium chloride sol'n. (BP grade)
Category: Germicide, emulsifier, cationic
Applications: Germicide in disinfectants and sanitizers; emulsifier in the dyeing industry
Form: Liquid; Color: Pale yellow; Odor: Mild, characteristic
Properties: Solubility: Sol. in water @ 20 C; Sp. gr.: 0.988 (20 C); Visc.: 120 cs (20 C); Flash Pt.: > 200 F (COC); Pour Pt.: < 0 C; pH: 7–8 (1% aq.); Biodegradable
Composition: 50% active ingred.
Reaction: Neutral
Toxicity: May cause irritation on prolonged contact with conc. form
Containers: Stainless steel, glass, polythene, or resin lined equipment
Warnings and Caution: Wear protective gloves and goggles; spillages may be slippery; do not mix with strong oxidizing or reducing agents–potentially explosive
References: Bulletin SI:30 FM678

Quadrilan OD-75. Lankro Chem. Ltd.
Chemical Description: Quaternary ammonium compound
Category: Softener, antistatic agent, cationic
Applications: Textiles
Form: Stiff paste; Color: White; Odor: Isopropanol
Properties: Solubility: Disp. in water @ 20 C; Sp. gr.: 0.870 (40 C); Visc.: 55 cs (50 C); Flash Pt.: 60 F (COC); Pour Pt.: 35–40 C; pH: 6.0–9.0 (1% aq.); Biodegradable
Composition: 75% active ingred.
Reaction: Neutral
Toxicity: May cause irritation on prolonged contact with conc. form
Containers: Stainless steel, glass, polythene, or resin lined equipment
Warnings and Caution: Wear protective gloves and goggles; spillages may be slippery; do not mix with strong oxidizing or reducing agents–potentially explosive; contains isopropanol–keep from heat, naked flames, and other sources of ignition
References: Bulletin SI:30 FM678

Quadrilan SK. Lankro Chem. Ltd.
Chemical Description: Quaternized fatty amine ethoxylate
Category: Antistatic agent, cationic
Applications: Polymer antistatic agent
Form: Liquid
Composition: 100% conc.

Quaternary O. Ciba-Geigy
Chemical Description: Quaternized oleyl imidazoline
Category: Detergent, wetting agent, penetrant, cationic

Applications: Penetrant for strong acids, solvents, germicides, fungistats; used in acid cleaners and corrosion inhibitor formulations; ore flotation; asphalt wetting
Form: Liquid; Color: Amber
Composition: 100% conc.

Quell-Oil C1. Lankro Chem. Ltd.
Category: Dispersant
Applications: Oil spill dispersant concentrate which is dilutable in water before use; very low toxicity; cleans oil from beaches; also used for spillages on dry land
Form: Clear liquid; Color: Light amber; Odor: Mild, fatty
Properties: Solubility: Readily disp. in water; Sp. gr.: 0.970 (30 C); Visc.: 43 cs (20 C); Flash Pt.: 101 C (COC); Pour Pt.: < -20 C
Reaction: Neutral
Containers: Stored in mild steel tanks and drums, and in pillow tanks
Warnings and Caution: Wear protective goggles and gloves during spraying; avoid inhalation of spray

Quell-Oil QSC. Lankro Chem. Ltd.
Applications: For cleaning oil deposits from harbor walls, piers, promenades, jetties, and other vertical surfaces, esp. for removal of oil subjected to a long period of weathering; applied by spray technique
Form: Clear liquid; Color: Straw; Odor: Mild
Properties: Solubility: Forms a gel in water (facilitates adhesion to surfaces); Sp. gr.: 0.808 (30 C); Visc.: 4.5 cs (20 C); Flash Pt.: $\approx$ 88 C (COC); Pour Pt.: < -20 C
Containers: Stored in mild steel drums and containers
Warnings and Caution: Contains an inflammable solvent–keep from heat, naked flames, and sparks; wear protective goggles and gloves; avoid inhalation of spray

Querton 14 BCL. Lilachim S.A.
Chemical Description: Myristyl dimethyl benzyl quaternary ammonium chloride
Category: Cationic surfactant
Form: Solid
Composition: 100% conc.

Querton 16 BCL. Lilachim S.A.
Chemical Description: Palmityl dimethyl benzyl quaternary ammonium chloride
Category: Cationic surfactant
Form: Solid
Composition: 100% conc.

Querton 28 CL. Lilachim S.A.
Chemical Description: Dioctyl dimethyl quaternary ammonium chloride
Category: Cationic surfactant
Form: Liquid; Color: 3 max. (Gardner)
Properties: m.w.: $\approx$ 320
Composition: 50% conc.

Querton 210 CL. Lilachim S.A.
Chemical Description: Didecyl dimethyl quaternary ammonium chloride
Category: Cationic surfactant
Form: Liquid; Color: 3 max. (Gardner)

Properties: m.w.: ≈ 360
Composition: 45% conc.

Querton HBG BCL. Lilachim S.A.
Chemical Description: Hydrogenated tallow dimethyl benzyl quaternary ammonium chloride
Category: Cationic surfactant
Form: Solid or liquid; Color: 3 max. (Gardner)
Properties: Iodine Value: 3; pH: 7 (5% aq.)
Composition: 100% conc. (solid); 75% conc. (liquid)

Querton IL 49. Lilachim S.A.
Chemical Description: Trialkyl benzyl quaternary ammonium chloride
Category: Cationic surfactant
Form: Liquid; Color: 3 max. (Gardner)
Properties: m.w.: ≈ 420; pH: 7 (5% aq.)
Composition: 50% conc.

Querton KK BCL. Lilachim S.A.
Chemical Description: Coconut dimethyl benzyl quaternary ammonium chloride
Category: Cationic surfactant
Form: Liquid, Color: 3 max. (Gardner)
Properties: pH: 7 (5% aq.)
Composition: 50% conc.

R

RAC-100. Tokaiseiyu Ind. Co., Ltd.
Chemical Description: Polyamine derivative
Category: Lubricant, softening agent, cationic
Applications: Lubricating and softening agent for various fibers
Form: Liquid

Radia 7040. Oleofina S.A.
Chemical Description: n-butyl oleate
Category: Chemical intermediate, lubricant
Applications: Chemical synthesis; carbon source in antibiotic culture broths; lubricity improvers in mineral oils; formulation of cutting oils, lamination oils, textile oils, rust inhibitors; textile and leather industry
Properties: m.w.: 327 (avg.); Ref. Index: 1.4518; Solubility: Sol. in most solvents and in mineral and vegetable oils; Sp. gr.: 0.854 (37.8 C); Visc.: 5.30 cps (37.8 C); Cloud Pt.: –15 C; Flash Pt.: 195 C (COC); Surface Tension: 31.5 dynes/cm

Radia 7051. Oleofina S.A.
Chemical Description: n-Butyl stearate
Category: Chemical intermediate, lubricant, plasticizer
Applications: Chemical synthesis; lubricants in mineral, cutting, lamination, textile oils, and rust inhibitors, textile and leather applications
Properties: m.w.: 325 (avg.); Ref. Index: 1.4432; Solubility: Sol. in most solvents, vegetable and mineral oils; Sp. gr.: 0.844 (37.8 C); Visc.: 6.10 cps (37.8 C); Cloud Pt.: 22.5 C; Flash Pt.: 200 C (COC); Surface Tension: 32.00 dynes/cm

Radia 7060. Oleofina S.A.
Chemical Description: Methyl oleate
Category: Chemical intermediate, lubricant
Applications: Chemical synthesis; lubricity improvers in mineral oils; formulation of cutting oils, lamination oils, textile oils, rust inhibitors; textile and leather industry

Properties: m.w.: 286 (avg.); Ref. Index.: 1.4513; Solubility: Sol. in most solvents and in mineral and vegetable oils; Sp. gr.: 0.862 (37.8 C); Visc.: 4.10 cps (37.8 C); Cloud Pt.: −11 C; Flash Pt.: 178 C (COC); Surface Tension: 31.50 dynes/cm

Radia 7070. Oleofina S.A.
Chemical Description: n-Butyl myristate
Category: Chemical intermediate, lubricant
Applications: Chemical synthesis; lubricants in mineral, cutting, lamination, textile oils, and rust inhibitors; textile and leather applications
Properties: m.w.: 282 (avg.); Ref. Index: 1.4388; Solubility: Sol. in most solvents, vegetable and mineral oils; Sp. gr.: 0.845 (37.8 C); Visc.: 4.10 cps (37.8 C); Cloud Pt.: 0 C; Flash Pt.: 172 C (COC); Surface Tension: 30.00 dynes/cm

Radia 7104. Oleofina S.A.
Chemical Dèscription: Glycerol capromyristate
Category: Lubricant, chemical intermediate
Applications: Formulation of cutting oils, lamination oils, textile oils, corrosion inhibitors; chemical synthesis
Properties: m.w.: 603 (avg.); Solubility: Sol. in trichlorethylene; sol. (cloudy) in hexane, benzene, isopropanol, mineral oil; Sp. gr.: 0.881 (98.9 C); Visc.: 6.25 cps (98.9 C); M.P.: 36 C; Flash Pt.: 216 C (COC)

Radia 7110. Oleofina S.A.
Chemical Description: Methyl stearate
Category and Applications: See Radia 7070
Properties: m.w.: 285 (avg.); Solubility: Sol. in most solvents, vegetable and mineral oils; M.P.: 31.0 C; Flash Pt.: 166 C (COC)

Radia 7118. Oleofina S.A.
Chemical Description: Methyl laurate
Category and Applications: See Radia 7070
Properties: m.w.: 214 (avg.); Ref. Index: 1.4322; Solubility: Sol. in most solvents, vegetable and mineral oils; Sp. gr.: 0.855 (37.8 C); Visc.: 2.30 cps (37.8 C); Cloud Pt.: 1.5 C; Flash Pt.: 132 C (COC)

Radia 7120. Oleofina S.A.
Chemical Description: Methyl palmitate
Category and Applications: See Radia 7070
Properties: m.w.: 267 (avg.); Solubility: Sol. in most solvents, vegetable and mineral oils; Sp. gr.: 0.855 (37.8 C); Visc.: 4.30 cps (37.8 C); M.P.: 27.5 C; Flash Pt.: 149 C (COC)

Radia 7127. Oleofina S.A.
Chemical Description: Ethylhexyl laurate
Category and Applications: See Radia 7070
Properties: m.w.: 330 (avg.); Ref. Index: 1.4422; Solubility: Sol. in

most solvents, vegetable and mineral oils; Sp. gr.: 0.846 (37.8 C); Visc.: 4.50 cps (37.8 C); Cloud Pt.: −25 C; Flash Pt.: 171 C (COC); Surface Tension: 30.50 dynes/cm

Radia 7129. Oleofina S.A.
Chemical Description: Ethylhexyl palmitate
Category and Applications: See Radia 7070
Properties: m.w.: 357 (avg.); Ref. Index: 1.4467; Solubility: Sol. in most solvents, vegetable and mineral oils; Sp. gr.: 0.847 (37.8 C); Visc.: 7.60 cps (37.8 C); Cloud Pt.: −1 C; Flash Pt.: 206 C (COC); Surface Tension: 31.00 dynes/cm

Radia 7130. Oleofina S.A.
Chemical Description: Isooctyl stearate
Category and Applications: See Radia 7070; also as plasticizer in paper industry
Properties: m.w.: 377 (avg.); Ref. Index: 1.4479; Solubility: Sol. in most solvents, vegetable and mineral oils; Sp. gr.: 0.846 (37.8 C); Visc.: 9.00 cps (37.8 C); Cloud Pt.: 12.5 C; Flash Pt.: 220 C (COC)

Radia 7161. Oleofina S.A.
Chemical Description: Glycerol trioleate
Category and Applications: See Radia 7104; also as carbon source in antibiotic culture broths
Properties: m.w.: 880 (avg.); Ref. Index: 1.4685; Solubility: Sol. in hexane, benzene, isopropanol, trichlorethylene, mineral and vegetable oils; Sp. gr.: 0.908 (37.8 C); Visc.: 49.20 cps (37.8 C); Cloud Pt.: −9 C; Flash Pt.: 274 C (COC)

Radia 7163. Oleofina S.A.
Chemical Description: Glycerol trioleate
Category and Applications: See Radia 7161
Properties: m.w.: 880 (avg.); Ref. Index: 1.4688; Solubility: Sol. in hexane, benzene, isopropanol, trichlorethylene, mineral and vegetable oils; Sp. gr.: 0.906 (37.8 C); Visc.: 42.30 (37.8 C); Cloud Pt.: −12 C; Flash Pt.: 283 C (COC)

Radia 7171. Oleofina S.A.
Chemical Description: Pentaerythritol tetraoleate
Category and Applications: See Radia 7104
Properties: m.w.: 1160 (avg.); Ref. Index: 1.4733; Solubility @ 10%: Sol. in trichlorethylene, mineral oil; sol. (cloudy) in hexane, benzene, isopropanol, vegetable oil; Sp. gr.: 0.913 (37.8 C); Visc.: 70.80 cps (37.8 C); Cloud Pt.: −9.5 C; Flash Pt.: 302 C (COC)

Radia 7176. Oleofina S.A.
Chemical Description: Pentaerythritol tetrastearate
Category and Applications: See Radia 7104
Properties: m.w.: 1140 (avg.); Solubility @ 10%: Sol. in trichlor-

ethylene; Sp. gr.: 0.868 (98.9 C); Visc.: 14.50 cps (98.9 C); M.P.: 64.5 C; Flash Pt.: 290 C

Radia 7190. Oleofina S.A.
Chemical Description: Isopropyl myristate
Category and Applications: See Radia 7070; also as emollient, plasticizer, solubilizer of active components in cosmetics and pharmaceuticals
Properties: m.w.: 271 (avg.); Ref. Index: 1.4345; Solubility: Sol. in most solvents, vegetable and mineral oils; Sp. gr.: 0.840 (37.8 C); Visc.: 3.50 cps (37.8 C); Cloud Pt.: –3.5 C; Flash Pt.: 157 C (COC); Surface Tension: 27.50 dynes/cm

Radia 7194. Oleofina S.A.
Chemical Description: Diisopropyl adipate
Category and Applications: See Radia 7190
Properties: m.w.: 230 (avg.); Ref. Index: 1.4237; Solubility: Sol. in most solvents, vegetable and mineral oils; Sp. gr.: 0.938 (37.8 C); Visc.: 2.80 cps (37.8 C); Cloud Pt.: –14.5 C; Flash Pt.: 131 C (COC); Surface Tension: 30.00 dynes/cm

Radia 7195. Oleofina S.A.
Chemical Description: Isopropyl stearate
Category and Applications: See Radia 7190
Properties: m.w.: 315 (avg.); Ref. Index: 1.4382; Solubility: Sol. in most solvents, vegetable and mineral oils; Sp. gr.: 0.842 (37.8 C); Visc.: 5.10 cps (37.8 C); Cloud Pt.: 18.5 C; Flash Pt.: 175 C (COC); Surface Tension: 30.50 dynes/cm

Radia 7197. Oleofina S.A.
Chemical Description: Diisobutyl adipate
Category and Applications: See Radia 7070
Properties: m.w.: 258 (avg.); Ref. Index: 1.4300; Solubility: Sol. in most solvents, vegetable and mineral oils; Sp. gr.: 0.939 (37.8 C); Visc.: 3.90 cps (37.8 C); Cloud Pt.: –30 C; Flash Pt.: 153 C (COC); Surface Tension: 30.70 dynes/cm

Radia 7200. Oleofina S.A.
Chemical Description: Isopropyl palmitate
Category and Applications: See Radia 7190
Properties: m.w.: 300 (avg.); Ref. Index: 1.4373; Solubility: Sol. in most solvents, vegetable and mineral oils; Sp. gr.: 0.843 (37.8 C); Visc.: 5.05 cps (37.8 C); Cloud Pt.: 11.5 C; Flash Pt.: 167 C (COC); Surface Tension: 30.50 dynes/cm

Radia 7230. Oleofina S.A.
Chemical Description: Isobutyl oleate
Category and Applications: See Radia 7040
Properties: m.w.: 327 (avg.); Ref. Index: 1.4499; Solubility: Sol. in

most solvents, vegetable and mineral oils; Sp. gr.: 0.854 (37.8 C); Visc.: 5.60 cps (37.8 C); Cloud Pt.: –17 C; Flash Pt.: 191 C (COC); Surface Tension: 31.50 dynes/cm

Radia 7231. Oleofina S.A.
Chemical Description: Isopropyl oleate
Category and Applications: See Radia 7060
Properties: m.w.: 313 (avg.); Ref. Index: 1.4479; Solubility: Sol. in most solvents, vegetable and mineral oils; Sp. gr.: 0.853 (37.8 C); Visc.: 5.05 cps (37.8 C); Cloud Pt.: –17 C; Flash Pt.: 169 C (COC); Surface Tension: 31.00 dynes/cm

Radia 7240. Oleofina S.A.
Chemical Description: Isobutyl stearate
Category and Applications: See Radia 7070; also as gloss promoters in special lacquers, and as plasticizers in paper industry
Properties: m.w.: 325 (avg.); Ref. Index: 1.4421; Solubility: Sol. in most solvents, vegetable and mineral oils; Sp. gr.: 0.844 (37.8 C); Visc.: 6.20 cps (37.8 C); Cloud Pt.: 17 C; Flash Pt.: 181 C (COC); Surface Tension: 30.20 dynes/cm

Radia 7303. Oleofina S.A.
Chemical Description: Glycerol trioleate
Category and Applications: See Radia 7161
Properties: m.w.: 880 (avg.); Ref. Index: 1.4709; Solubility: Sol. in hexane, benzene, isopropanol, trichlorethylene, mineral and vegetable oils; Sp. gr.: 0.910 (37.8 C); Visc.: 45.20 cps (37.8 C); Cloud Pt.: 5 C; Flash Pt.: 275 C (COC)

Radia 7330. Oleofina S.A.
Chemical Description: Isooctyl oleate
Category and Applications: See Radia 7060
Properties: m.w.: 373 (avg.); Ref. Index: 1.4542; Solubility: Sol. in most solvents, vegetable and mineral oils; Sp. gr.: 0.857 (37.8 C); Visc.: 7.50 cps (37.8 C); Cloud Pt.: –20 C; Flash Pt.: 208 C (COC); Surface Tension: 31.50 dynes/cm

Radia 7331. Oleofina S.A.
Chemical Description: Ethylhexyl oleate
Category and Applications: See Radia 7060
Properties: m.w.: 373 (avg.); Ref. Index: 1.4538; Solubility: Sol. in most solvents, vegetable and mineral oils; Sp. gr.: 0.857 (37.8 C); Visc.: 7.50 cps (37.8 C); Cloud Pt.: –23 C; Flash Pt.: 193 C (COC); Surface Tension: 32.50 dynes/cm

Radia 7363. Oleofina S.A.
Chemical Description: Glycerol trioleate
Category and Applications: See Radia 7161
Properties: m.w.: 880 (avg.); Ref.

Index: 1.4691; Solubility: Sol. in hexane, benzene, isopropanol, trichlorethylene, mineral and vegetable oils; Sp. gr.: 0.906 (37.8 C); Visc.: 41.20 cps (37.8 C); Cloud Pt.: –13 C; Flash Pt.: 289 C (COC)

Radia 7370. Oleofina S.A.
Chemical Description: Trimethylpropane trioleate
Category and Applications: See Radia 7104
Properties: m.w.: 900 (avg.); Ref. Index: 1.4712; Solubility @ 10%: Sol. in hexane, benzene, isopropanol, trichlorethylene, mineral and vegetable oils; Sp. gr.: 0.904 (37.8 C); Visc.: 45.90 cps (37.8 C); Cloud Pt.: –24 C; Flash Pt.: 307 C (COC)

Radia 7500. Oleofina S.A.
Chemical Description: Cetyl palmitate
Category and Applications: See Radia 7070; also used for wax formulation due to its high melting point
Properties: m.w.: 484 (avg.); Solubility @ 10%: Sol. in benzene and trichlorethylene; Sp. gr.: 0.805 (98.9 C); Visc.: 4.30 (98.9 C); M.P.: 49 C; Flash Pt.: 214 C (COC)

Radia 7501. Oleofina S.A.
Chemical Description: Cetyl stearate
Category and Applications: See Radia 7500
Properties: m.w.: 534 (avg.); Solubility @ 10%: Sol. in trichlorethylene; Sp. gr.: 0.807 (98.9 C); Visc.: 4.60 cps (98.9 C); M.P.: 56 C; Flash Pt.: 216 C (COC)

Radia 7510. Oleofina S.A.
Chemical Description: Isononyl stearate
Category and Applications: See Radia 7070
Properties: m.w.: 380 (avg.); Ref. Index: 1.4493; Solubility: Sol. in most solvents, vegetable and mineral oils; Sp. gr.: 0.851 (37.8 C); Visc.: 11.30 cps (37.8 C); Cloud Pt.: 11 C; Flash Pt.: 201 C (COC); Surface Tension: 31.50 dynes/cm

Radia 7514. Oleofina S.A.
Chemical Description: Pentaerythritol tetrabehenate
Category and Applications: See Radia 7104
Properties: m.w.: 1300 (avg.); Solubility @ 10%: Sol. (cloudy) @ 75 C in trichlorethylene, mineral oil; Sp. gr.: 0.860 (98.9 C); Visc.: 17.20 cps (98.9 C); M.P.: 73.5 C; Flash Pt.: 289 C (COC)

Radiamuls 125. Oleofina S.A.
Chemical Description: Sorbitan monolaurate
Category: Food emulsifier, dispersant, antistaling agent, spray drying aid, solubilizer, dryness improver, whipping aid, nonionic
Applications: Yeasted baked goods, bakery products, calf milk replacers, carbonated beverages, coffee whiteners, confectioners coating and chocolate, dairy prod-

ucts, soups (dry mixed), sugar refining, vitamin oil compounds, whipped toppings
Form: Liquid
Properties: HLB: ≈ 6.8; Acid Value: 7 max.; Iodine Value: 10 max.
Containers: 190 kg net bung drums
References: PA/12A/1276

Radiamuls 135. Oleofina S.A.
Chemical Description: Sorbitan monopalmitate
Category and Applications: See Radiamuls 125
Form: Powder or flake
Properties: M.P.: ≈ 50 C; HLB: ≈ 5.3; Acid Value: 7 max.; Iodine Value: 1 max.
Containers: 190 kg net bung drums
References: PA/12A/1276

Radiamuls 137. Oleofina S.A.
Chemical Description: Sorbitan POE-20-monolaurate
Category and Applications: See Radiamuls 125
Form: Liquid
Properties: HLB: ≈ 16; Acid Value: 2 max.; Iodine Value: 5 max.
Containers: 190 kg net bung drums
References: PA/12A/1276

Radiamuls 141. Oleofina S.A.
Chemical Description: Glycerol monostearate, self-emulsifying
Category: Emulsifier, nonionic
Applications: Food emulsifier; used for binding of water, regulation of fat crystallization, and complexation of amylose in ice cream and frozen desserts, dairy products, calf milk replacers, bakery and pastry products, confectionery compounds, dressings, sauces, soups
Form: Powder or flakes
Properties: m.w.: 509 (avg.); Solubility @ 10%: Sol. @ 75 C in isopropanol, trichlorethylene, mineral oil; sol. (cloudy) @ 75 C in hexane and vegetable oil; Sp. gr.: 0.902 (98.9 C); Visc.: 20.10 cps (98.9 C); M.P.: 58 C; HLB: 3.5 avg.; Flash Pt.: 198 C (COC); Sapon. Value: 145–160; Iodine Value: 1 max.
Composition: ≈ 40% monoglycerides
Containers: 25 kg net bags or bulk
References: PA/8aA/1176

Radiamuls 142. Oleofina S.A.
Chemical Description: Glycerol monostearate
Category and Applications: See Radiamuls 141
Form: Powder or flakes
Properties: m.w.: 515 (avg.); Solubility @ 10%: Sol. @ 75 C in benzene, isopropanol, trichlorethylene, mineral and vegetable oils; Sp. gr.: 0.885 (98.9 C); Visc.: 12.00 cps (98.9 C); M.P.: 58 C; HLB: 3.0 avg.; Flash Pt.: 233 C (COC); Sapon. Value: 169–176; Iodine Value: 1 max.
Composition: ≈ 40% monoglycerides
Containers: 25 kg net bags or bulk
References: PA/8aA/1176

Radiamuls 143. Oleofina S.A.
Chemical Description: Vegetable glycol monostearate
Category and Applications: See Radiamuls 141
Form: Powder or flakes
Properties: M.P.: 56–60 C; HLB: ≈ 3.0; Sapon. Value: 165–180; Iodine Value: 1 max.
Composition: ≈ 40% monoglycerides
Containers: 25 kg net bags or bulk
References: PA/8aA/1176

Radiamuls 145. Oleofina S.A.
Chemical Description: Sorbitan monostearate
Category and Applications: See Radiamuls 125
Form: Powder or flakes
Properties: M.P.: 50–60 C; HLB: ≈ 5; Acid Value: 7 max.; Iodine Value: 60–75
Containers: 25 kg net multiply paperbags
References: PA/12A/1276

Radiamuls 147. Oleofina S.A.
Chemical Description: Sorbitan POE-20-monostearate
Category and Applications: See Radiamuls 125
Form: Liquid
Properties: HLB: ≈ 15; Acid Value: 2 max.; Iodine Value: 1 max.
Containers: 190 kg net bung drums
References: PA/12A/1276

Radiamuls 152. Oleofina S.A.
Chemical Description: Glycerol monooleate
Category and Applications: See Radiamuls 141
Form: Liquid
Properties: HLB: ≈ 2.8; Cloud Pt.: 10 C max.; Sapon. Value: 165–178; Iodine Value: 72–77
Composition: ≈ 40% monoglycerides
Containers: 190 kg net bung drums or bulk
References: PA/8aA/1176

Radiamuls 155. Oleofina S.A.
Chemical Description: Sorbitan monooleate
Category and Applications: See Radiamuls 125
Form: Liquid
Properties: HLB: ≈ 4.7; Acid Value: 7 max.; Iodine Value: 1 max.
Containers: 190 kg net bung drums
References: PA/12A/1276

Radiamuls 157. Oleofina S.A.
Chemical Description: Sorbitan POE-20-monooleate
Category and Applications: See Radiamuls 125
Form: Liquid
Properties: HLB: ≈ 15.7; Acid Value: 2 max.; Iodine Value: 20–26
Containers: 190 kg net bung drums
References: PA/12A/1276

Radiamuls 341. Oleofina S.A.
Chemical Description: Glycerol monostearate, self-emulsifying
Category and Applications: See Radiamuls 141
Form: Powder or flakes
Properties: M.P.: 56–60 C; HLB: ≈ 5; Sapon. Value: 164–174; Iodine Value: 2 max.
Composition: ≈ 40% monoglycerides
Containers: 25 kg net bags or bulk
References: PA/8aA/1176

Radiamuls 345. Oleofina S.A.
Chemical Description: Sorbitan tristearate
Category and Applications: See Radiamuls 125
Form: Powder or flakes
Properties: M.P.: 50–60 C; HLB: ≈ 2.7; Acid Value: 17 max.; Iodine Value: 1 max.
Containers: 25 kg net multiply paperbags
References: PA/12A/1276

Radiamuls 600. Oleofina S.A.
Chemical Description: Glycerol monostearate
Category and Applications: See Radiamuls 141
Form: Powder or flakes
Properties: m.w.: 460 (avg.); Solubility @ 10%: Sol. @ 75 C in hexane, benzene, isopropanol, trichlorethylene, mineral and vegetable oils; Sp. gr.: 0.893 (98.9 C); Visc.: 13.40 cps (98.9 C); M.P.: 59 C; HLB; 3.3 avg.; Flash Pt.: 223 C (COC); Sapon. Value: 165–175; Iodine Value: 1 max.
Composition: 60% min. monoglycerides
Containers: 25 kg net bags or bulk
References: PA/8aA/1176

Radiamuls 601. Oleofina S.A.
Chemical Description: Glycerol monostearate
Category and Applications: See Radiamuls 141
Form: Powder or flakes
Properties: m.w.: 460 (avg.); Solubility @ 10%: Sol. @ 75 C in hexane, benzene, isopropanol, trichlorethylene, mineral and vegetable oils; Sp. gr.: 0.900 (98.9 C); Visc.: 10.70 cps (98.9 C); M.P.: 52 C; HLB: 3.3 avg.; Flash Pt.: 217 C (COC); Sapon. Value:164–174; Iodine Value: 50–55
Composition: 60% min. monoglycerides
Containers: 25 kg net bags or bulk
References: PA/8aA/1176

Radiamuls 602. Oleofina S.A.
Chemical Description: Glycerol monosoyate
Category and Applications: See Radiamuls 141
Form: Liquid
Properties: m.w.: 464 (avg.); Ref. Index: 1.475; Solubility @ 10%: Sol. in hexane, benzene, isopropanol, mineral and vegetable oils; Sp. gr.: 0.951 (37.8 C); Visc.: 102.30 cps (37.8 C); HLB: 3.1 avg.; Cloud Pt.: 17.5 C; Flash Pt.: 212 C (COC); Sapon. Value: 165–175; Iodine Value: 110–115
Composition: 60% min. mono-

glycerides
Containers: 190 kg net drums or bulk
References: PA/8aA/1176

Radiamuls 900. Oleofina S.A.
Chemical Description: Glycerol monostearate
Category and Applications: See Radiamuls 141
Form: Powder or flakes
Properties: M.P.: ≈ 70 C; HLB: ≈ 4.6; Sapon. Value: 155–170; Iodine Value: 1 max.
Composition: 90% min. monoglycerides
Containers: 25 kg net bags or bulk
References: PA/8aA/1176

Radianol 7106. Oleofina S.A.
Chemical Description: Glycerol tricaprate/caprylate
Category: Emollient, plasticizer, solubilizer, food additive
Applications: Used in the food and feed industry, and in cosmetics and pharmaceuticals
Form: Low viscosity oil
Properties: m.w.: 518 (avg.); Ref. Index: 1.4402; Solubility @ 10%: Sol. in hexane, benzene, isopropanol, trichlorethylene, mineral and vegetable oils; Sp. gr.: 0.933 (37.8 C); Visc.: 15.90 cps (37.8 C); Cloud Pt.: < –30 C; Flash Pt.: 178 C (COC)

Radianol 7376. Oleofina S.A.
Chemical Description: Glycerol triheptanoate
Category and Applications: See Radianol 7106; also used as butter tracers
Form: Low viscosity oil
Properties: m.w.: 432 (avg.); Ref. Index: 1.4444; Solubility @ 10%: Sol. in hexane, benzene, isopropanol, trichlorethylene, mineral and vegetable oils; Sp. gr.: .0952 (37.8 C); Visc.: 10.10 cps (37.8 C); Cloud Pt.: < –30 C; Flash Pt.: 207 C (COC)

Radiasurf 7000. Oleofina S.A.
Chemical Description: POE-20-glycerol monostearate
Category: Chemical intermediate, emulsifier, detergent, lubricant, wetting agent, nonionic
Applications: Chemical synthesis; O/W or W/O cosmetic and pharmaceutical emulsions; pearlescent shampoo formulations; detergency and cleaning products; formulation of cutting oils, lamination oils, textile oils, rust inhibitors; pigment wetting, faster grinding, and improved gloss in paints, lacquers, and printing inks; lubricant and mold release in plastics
Form: Liquid
Properties: m.w.: 1395 (avg.); Ref. Index: 1.4672; Solubility @ 10%: Sol. in trichlorethylene; sol. (cloudy) in hexane, benzene, isopropanol, mineral and vegetable oils; disp. in water; Sp. gr.: 1.032 (37.8 C); Visc.: 113.30 cps (37.8 C); HLB: 13.0 avg.; Cloud Pt.: 27.5 C; Flash Pt.: 244 C (COC)

Radiasurf 7125. Oleofina S.A.
Chemical Description: Sorbitan monolaurate
Category: Emulsifier, descouring

aid, antistat, anticorrosive agent, surfatting and bodying aid, antifog aid, dispersant, detergent, nonionic
Applications: Emulsion of solvents; cutting oils; textile lubricant additives; leather auxiliaries; cosmetics and pharmaceuticals; plastics; pesticides; dry cleaning formulations; paper additives; pigment dispersants; detergent additives; anticorrosion additives for pipelines; cleaners for metallic surfaces; additives to improve grinding of concrete
Form: Liquid; Color: Amber
Properties: m.w.: 450 (avg.); Ref. Index: 1.4718; Solubility @ 10%: Sol. in hexane, trichlorethylene, mineral oil; sol. (cloudy) in benzene, isopropanol, vegetable oil; disp. in water; Sp. gr.: 1.025 (37.8 C); Visc.: 1193.0 cps (37.8 C); HLB: 6.8 avg.; Cloud Pt.: 23 C; Flash Pt.: 198 C (COC); Sapon. Value: 155–175; Acid Value: 7 max.; Iodine Value: 10 max.
Containers: 190 kg net bung drums or bulk
References: Spec/1176/7A6

Radiasurf 7135. Oleofina S.A.
Chemical Description: Sorbitan monopalmitate
Category and Applications: See Radiasurf 7125
Form: Solid; Color: White
Properties: m.w.: 500 (avg.); Solubility @ 10%: Sol. (cloudy) in isopropanol, trichlorethylene, mineral oil; disp. in water; Sp. gr.: 0.943 (98.9 C); Visc.: 51.70 cps (98.9 C); M.P.: 54.5 C; HLB: 5.3 avg.; Flash Pt.: 214 C (COC); Sapon. Value: 140–160; Acid Value: 7 max.; Iodine Value: 1 max.
Containers: 25 kg net multiply paper bags or bulk

Radiasurf 7137. Oleofina S.A.
Chemical Description: Sorbitan POE(20)-monolaurate
Category and Applications: See Radiasurf 7125
Form: Liquid; Color: Amber
Properties: m.w.: 1340 (avg.); Ref. Index: 1.4712; Solubility @ 10%: Sol. in water, benzene, isopropanol, trichlorethylene; sol. (cloudy) in mineral and vegetable oils; Sp. gr.: 1.095 (37.8 C); Visc.: 172.3 cps (37.8 C); HLB: 16 avg.; Cloud Pt.: −10 C; Flash Pt.: 248 C (COC); Sapon. Value: 40–55; Acid Value: 2 max.; Iodine Value: 5 max.
Containers: 190 kg net bung drums or bulk
References: Spec/1176/7A6

Radiasurf 7145. Oleofina S.A.
Chemical Description: Sorbitan monostearate
Category and Applications: See Radiasurf 7125
Form: Solid; Color: White
Properties: m.w.: 510 (avg.); Solubility @ 10%: Sol. (cloudy) in trichlorethylene; disp. in water; Sp. gr.: 0.943 (98.9 C); Visc.: 48.20 cps (98.9 C); M.P.: 58 C; HLB: 5 avg.; Flash Pt.: 232 C (COC); Sapon. Value: 146–158; Acid Value: 7 max.; Iodine Value:

1 max.
Containers: 25 kg net multiply paper bags or bulk
References: Spec/1176/7A6

Radiasurf 7146. Oleofina S.A.
Chemical Description: Glycerol mono-12-hydroxystearate
Category: Chemical intermediate, emulsifier, detergent, wetting agent, lubricant, nonionic
Applications: Chemical synthesis; cosmetics and pharmaceuticals; pearlescent shampoo formulations; detergency and cleaning products; cutting, lamination, textile oils and rust inhibitors; pigment wetting, grinding, and improved gloss in paints, lacquers, printing inks; paper industry; lubricant and mold release in plastics; textile and leather processing
Form: Solid
Properties: m.w.: 550 (avg.); Solubility @ 10%: Sol. @ 75 C in benzene, isopropanol, trichlorethylene; Sp. gr.: 0.925 (98.9 C); Visc.: 25.80 cps (98.9 C); M.P.: 70 C; HLB: 1.9 avg.; Flash Pt.: 260 C (COC)
Composition: 40% monoester content

Radiasurf 7147. Oleofina S.A.
Chemical Description: Sorbitan POE(20)-monostearate
Category and Applications: See Radiasurf 7125
Form: Liquid; Color: Amber
Properties: m.w.: 1400 (avg.); Ref. Index: 1.4694; Solubility @ 10%: Sol. in trichlorethylene, mineral oil; sol. (cloudy) in water, benzene, isopropanol, vegetable oil; Sp. gr.: 1.068 (37.8 C); Visc.: 202.6 cps (37.8 C); HLB: 15 avg.; Cloud Pt.: 21.5 C; Flash Pt.: 260 C (COC); Sapon. Value: 44–56; Acid Value: 2 max.; Iodine Value: 1 max.
Containers: 190 kg net bung drums or bulk
References: Spec/1176/7A6

Radiasurf 7151. Oleofina S.A.
Chemical Description: Glycerol monooleate, self-emulsifying
Category: Chemical intermediate, emulsifier, detergent, wetting agent, lubricant, nonionic
Applications: Chemical synthesis; cosmetics and pharmaceuticals; detergency and cleaning products; cutting, lamination, textile oils and rust inhibitors; pigment wetting, grinding, and improved gloss in paints, lacquers, and printing inks; paper industry; lubricant and mold release in plastics; textile and leather processing
Form: Liquid
Properties: m.w.: 515 (avg.); Ref. Index: 1.4696; Solubility @ 10%: Sol. in isopropanol, mineral oil; sol. (cloudy) in hexane, benzene, trichlorethylene, vegetable oil; Sp. gr.: 0.947 (37.8 C); Visc.: 137.70 cps (37.8 C); HLB: 3.3 avg.; Cloud Pt.: 12.0 C; Flash Pt.: 208 C (COC)
Composition: 40% monoester content

Radiasurf 7152. Oleofina S.A.
Chemical Description: Glycerol monooleate
Category and Applications: See Radiasurf 7151
Form: Liquid
Properties: m.w.: 521 (avg.); Ref. Index: 1.4697; Solubility @ 10%: Sol. in hexane, benzene, isopropanol, mineral oil; sol. (cloudy) in trichlorethylene and vegetable oil; Sp. gr.: 0.934 (37.8 C); Visc.: 77.10 cps (37.8 C); HLB: 2.8 avg.; Cloud Pt.: 7.5 C; Flash Pt.: 232 C (COC)
Composition: 40% monoester content

Radiasurf 7153. Oleofina S.A.
Chemical Description: Glycerol monoricinoleate
Category and Applications: See Radiasurf 7151; also antistatic and antifogging additive in plastics
Form: Liquid
Properties: m.w.: 554 (avg.); Ref. Index: 1.4774; Solubility @ 10%: Sol. in benzene, isopropanol; sol. (cloudy) in trichlorethylene, mineral and vegetable oils; Sp. gr.: 0.982 (37.8 C); Visc.: 356.20 cps (37.8 C); HLB: 2.5 avg., Cloud Pt.: −16.0 C; Flash Pt.: 229 C (COC)
Composition: 40% monoester content

Radiasurf 7155. Oleofina S.A.
Chemical Description: Sorbitan monooleate
Category and Applications: See Radiasurf 7125
Form: Liquid; Color: Amber
Properties: m.w.: 510 (avg.); Ref. Index: 1.4778; Solubility @ 10%: Sol. in hexane, isopropanol, trichlorethylene, mineral and vegetable oils; Sp. gr.: 0.987 (37.8 C); Visc.: 395.2 cps (37.8 C); HLB: 4.7 avg.; Cloud Pt.: 7 C; Flash Pt.: 249 C (COC); Sapon. Value: 145–161; Acid Value: 7 max.; Iodine Value: 60–75
Containers: 190 kg net bung drums or bulk
References: Spec/1176/7A6

Radiasurf 7157. Oleofina S.A.
Chemical Description: Sorbitan POE(20)-monooleate
Category and Applications: See Radiasurf 7125
Form: Liquid; Color: Amber
Properties: m.w.: 1400 (avg.); Ref. Index: 1.4726; Solubility @ 10%: Sol. in water, benzene, trichlorethylene; sol. (cloudy) in isopropanol, mineral and vegetable oils; Sp. gr.: 1.071 (37.8 C); Visc.: 240.4 cps (37.8 C); HLB: 15.7 avg.; Cloud Pt.: 5.5 C; Flash Pt.: 287 C (COC); Sapon. Value: 44–56; Acid Value: 2 max.; Iodine Value: 20–26
Containers: 190 kg net bung drums or bulk
References: Spec/1176/7A6

Radiasurf 7175. Oleofina S.A.
Chemical Description: Pentaerythritol monostearate
Category and Applications: See Radiasurf 7146

Form: Solid
Properties: m.w.: 675 (avg.); Solubility @ 10%: Sol. in trichlorethylene; Sp. gr.: 0.886 (98.9 C); Visc.: 17.60 cps (98.9 C); M.P.: 53.0 C; HLB: 2.3 avg.; Flash Pt.: 259 C (COC)

Radiasurf 7196. Oleofina S.A.
Chemical Description: Propylene glycol monomyristate
Category: Wetting aid, lubricant, opacifier, antistat, dispersant, W/O emulgent, scouring and detergent aid, defoamer, plasticizer, rust inhibitor, nonionic
Applications: Cosmetics and pharmaceuticals, lubricating oils, cutting oils, textile and leather aids, pigment grinding in paints and printing inks, latex and emulsion paints, plastics, waxes and maintenance products, insecticides
Form: Liquid; Color: White to amber
Properties: m.w.: 378 (avg.); Ref. Index: 1.4379; Solubility @ 10%: Sol. in benzene, isopropanol, trichlorethylene, vegetable oil; sol. (cloudy) in hexane, mineral oil; disp. in water; Sp. gr.: 0.902 (37.8 C); Visc.: 17.30 cps (37.8 C); HLB: 3.9 avg.; Cloud Pt.: 16 C; Flash Pt.: 145 C (COC); Sapon. Value: 190–205; Acid Value: 5 max.
Containers: 190 kg net bung drums or bulk
References: Spec/1176/7A9

Radiasurf 7201. Oleofina S.A.
Chemical Description: Propylene glycol monostearate
Category and Applications: See Radiasurf 7196
Form: Solid; Color: White to cream
Properties: m.w.: 457 (avg.); Solubility @ 10%: Sol. (cloudy) in hexane, trichlorethylene, mineral oil; Sp. gr.: 0.849 (98.9 C); Visc.: 4.50 cps (98.9 C); M.P.: 39 C; HLB: 1.9 avg.; Flash Pt.: 206 C (COC); Sapon. Value: ≈ 177; Acid Value: 2 max.; Iodine Value: 1 max.
Containers: 190 kg net bung drums or bulk
References: Spec/1176/7A9

Radiasurf 7206. Oleofina S.A.
Chemical Description: Propylene glycol monooleate
Category and Applications: See Radiasurf 7196
Form: Liquid; Color: Amber
Properties: m.w.: 460 (avg.); Ref. Index: 1.4606; Solubility @ 10%: Sol. in benzene, isopropanol, trichlorethylene, vegetable oil; sol. (cloudy) in hexane, mineral oil; Sp. gr.: 0.900 (37.8 C); Visc.: 18.80 cps (37.8 C); HLB: 2.3 avg.; Cloud Pt.: –9 C; Flash Pt.: 157 C (COC); Sapon. Value: 172–180; Acid Value: 2 max.; Iodine Value: 74–80
Containers: 190 kg net bung drums or bulk
References: Spec/1176/7A9

Radiasurf 7269. Oleofina S.A.
Chemical Description: Ethylene glycol distearate
Category and Applications: See

Radiasurf 7196
Form: Solid, Color: White
Properties: Solubility @ 10%: Sol. @ 75 C in trichlorethylene; sol. (cloudy) @ 75 C in hexane, benzene, isopropanol, mineral and vegetable oils; M.P.: ≈ 65 C; HLB: ≈ 1.5; Sapon. Value: 190–200; Acid Value: 3 max.; Iodine Value: 1 max.
Containers: 25 kg net multiply paper bags or bulk
References: Spec/1176/7A9

Radiasurf 7270. Oleofina S.A.
Chemical Description: Ethylene glycol monostearate
Category and Applications: See Radiasurf 7196
Form: Solid; Color: White
Properties: m.w.: 445 (avg.); Solubility @ 10%: Sol. @ 75 C in benzene, isopropanol, trichlorethylene, vegetable oil; Sp. gr.: 0.851 (98.9 C); Visc.: 4.80 cps (98.9 C); M.P.: 61 C; HLB: 1.8 avg.; Flash Pt.: 214 C (COC); Sapon. Value: 180–190; Acid Value: 2 max.; Iodine Value: 1 max.
Containers: 25 kg net multiply paper bags or bulk
References: Spec/1176/7A9

Radiasurf 7345. Oleofina S.A.
Chemical Description: Sorbitan tristearate
Category and Applications: See Radiasurf 7125
Form: Solid
Properties: m.w.: 900 (avg.); Solubility @ 10%: Sol. @ 75 C in trichlorethylene, mineral and vegetable oils; Sp. gr.: 0.893 (98.9 C); Visc.: 19.50 cps (98.9 C); M.P.: 60 C; HLB: 2.7 avg.; Flash Pt.: 253 C (COC)

Radiasurf 7372. Oleofina S.A.
Chemical Description: Trimethylolpropane monooleate
Category and Applications: See Radiasurf 7151
Form: Liquid
Properties: m.w.: 514 (avg.); Ref. Index: 1.4705; Solubility @ 10%: Sol. in hexane, benzene, isopropanol, trichlorethylene, mineral and vegetable oils; Sp. gr.: 0.935 (37.8 C); Visc.: 90 cps (37.8 C); HLB: 4.4 avg.; Cloud Pt.: –18.5 C; Flash Pt.: 224 C (COC)

Radiasurf 7400. Oleofina S.A.
Chemical Description: Diethylene glycol monooleate
Category and Applications: See Radiasurf 7196
Form: Liquid; Color: Amber
Properties: m.w.: 478 (avg.); Ref. Index: 1.4642; Solubility @ 10%: Sol. in benzene, isopropanol, trichlorethylene; sol. (cloudy) in hexane, mineral and vegetable oils; disp. in water; Sp. gr.: 0.929 (37.8 C); Visc.: 58.50 cps (37.8 C); HLB: 3.8 avg.; Cloud Pt.: –6.5 C; Flash Pt.: 180 C (COC); Sapon. Value: 155–168; Acid Value: 4 max.; Iodine Value: 65–75
Containers: 190 kg net bung drums or bulk
References: Spec/1176/7A9

Radiasurf 7402. Oleofina S.A.
Chemical Description: PEG 200 monooleate
Category: Wetting aid, lubricant, opacifier, antistat, dispersant, O/W emulgent, scouring and detergent aid, defoamer, plasticizer, rust inhibitor, viscosity modifier, antifog aid, nonionic
Applications: Cosmetics and pharmaceuticals; lubricating oils; cutting oils; textile and leather aids; pigment grinding aids in paints and printing inks; latex and emulsion paints; plastics; waxes and maintenance products; glass fiber; insecticides; silicones
Form: Liquid; Color: Amber
Properties: m.w.: 512 (avg.); Ref. Index: 1.4645; Solubility @ 10%: Sol. in benzene, isopropyl, trichlorethylene; sol. (cloudy) in hexane, mineral and vegetable oils; disp. in water; Sp. gr.: 0.962 (37.8 C); Visc.: 34.00 cps (37.8 C); HLB: 8.4 avg.; Cloud Pt.: –10 C; Flash Pt.: 218 C (COC); Sapon. Value: 105–125; Acid Value: < 3; Iodine Value: 50–60; Surface Tension: 32.00 dynes/cm (0.1% aq.)
Containers: 190 kg net bung drums or bulk
References: Spec/1176/7A11

Radiasurf 7403. Oleofina S.A.
Chemical Description: PEG 400 monooleate
Category and Applications: See Radiasurf 7402
Form: Liquid; Color: Amber
Properties: m.w.: 728 (avg.); Ref. Index: 1.4672; Solubility @ 10%: Sol. in benzene, isopropanol, trichlorethylene; sol. (cloudy) in hexane, mineral and vegetable oils; disp. in water; Sp. gr.: 1.007 (37.8 C); Visc.: 49.50 cps (37.8 C); HLB: 11.7 avg.; Cloud Pt.: –1 C; Flash Pt.: 261 C (COC); Sapon. Value: 75–90; Acid Value: < 3; Iodine Value: 34–42; Surface Tension: 34.00 dynes/cm (0.1% aq.)
Containers: 190 kg net bung drums or bulk
References: Spec/1176/7A11

Radiasurf 7404. Oleofina S.A.
Chemical Description: PEG 600 monooleate
Category and Applications: See Radiasurf 7402
Form: Liquid; Color: Amber
Properties: m.w.: 920 (avg.); Ref. Index: 1.4669; Solubility @ 10%: Sol. in water, benzene, isopropanol, trichlorethylene; sol. (cloudy) in hexane, mineral and vegetable oils; Sp. gr.: 1.030 (37.8 C); Visc.: 66.30 cps (37.8 C); HLB: 13.2 avg.; Cloud Pt.: 19 C; Flash Pt.: 254 C (COC); Sapon. Value: 60–75; Acid Value: < 3; Iodine Value: 27–37; Surface Tension: 37.00 dynes/cm (0.1% aq.)
Containers: 190 kg net bung drums or bulk
References: Spec/1176/7A11

Radiasurf 7410. Oleofina S.A.
Chemical Description: Diethylene glycol monostearate
Category and Applications: See Radiasurf 7196

Form: Solid; Color: White
Properties: m.w.: 475 (avg.); Solubility @ 10%: Sol. (cloudy) @ 75 C in hexane, benzene, isopropanol, trichlorethylene, mineral and vegetable oils; Sp. gr.: 0.873 (98.9 C); Visc.: 5.95 cps (98.9 C); M.P.: 51.5 C; HLB: 3.5 avg.; Flash Pt.: 191 C (COC); Sapon. Value: 160–175; Acid Value: 3 max.; Iodine Value: 1 max.
Containers: 25 kg net multiply paper bags or bulk
References: Spec/1176/7A9

Radiasurf 7411. Oleofina S.A.
Chemical Description: Diethylene glycol monostearate, self-emulsifying
Category and Applications: See Radiasurf 7196
Form: Solid; Color: White
Properties: m.w.: 470 (avg.); Solubility @ 10%: Sol. @ 75 C in trichlorethylene; sol. (cloudy) @ 75 C in hexane, benzene, isopropanol, mineral and vegetable oils; Sp. gr.: 0.878 (98.9 C); Visc.: 5.90 cps (98.9 C); M.P.: 52 C; HLB: 4.1 avg.; Flash Pt.: 176 C (COC); Sapon. Value: 160–170; Acid Value: 3 max.; Iodine Value: 1 max.
Containers: 25 kg net multiply paper bags or bulk
References: Spec/1176/7A9

Radiasurf 7412. Oleofina S.A.
Chemical Description: PEG 200 monostearate
Category and Applications: See Radiasurf 7402
Form: Paste; Color: White
Properties: m.w.: 522 (avg.); Solubility @ 10%: Sol. in benzene, isopropanol, trichlorethylene; sol. (cloudy) in hexane, mineral oil; Sp. gr.: 0.913 (98.9 C); Visc.: 6.70 cps (98.9 C); M.P.: 36 C; HLB: 7.5 avg.; Flash Pt.: 186 C (COC); Sapon. Value: 120–135; Acid Value: < 3; Iodine Value: 1 max.
Containers: 190 kg net bung drums or bulk
References: Spec/1176/7A11

Radiasurf 7413. Oleofina S.A.
Chemical Description: PEG 400 monostearate
Category and Applications: See Radiasurf 7402
Form: Paste; Color: White
Properties: m.w.: 722 (avg.); Solubility @ 10%: Sol. in benzene, trichlorethylene, mineral oil; sol. (cloudy) in hexane; Sp. gr.: 0.951 (98.9 C); Visc.: 9.45 cps (98.9 C); M.P.: 34 C; HLB: 11.9 avg.; Flash Pt.: 248 C (COC); Sapon. Value: 75–90; Acid Value: < 3; Iodine Value: 1 max.
Containers: 190 kg net bung drums or bulk
References: Spec/1176/7A11

Radiasurf 7414. Oleofina S.A.
Chemical Description: PEG 600 monostearate
Category and Applications: See Radiasurf 7402
Form: Paste; Color: White
Properties: m.w.: 906 (avg.); Solubility @ 10%: Sol. in benzene,

isopropanol, trichlorethylene; sol. (cloudy) in hexane, mineral and vegetable oils; Sp. gr.: 0.981 (98.9 C); Visc.: 12.60 cps (98.9 C); M.P.: 38 C; HLB: 13.3 avg.; Flash Pt.: 241 C (COC); Sapon. Value: 60–75; Acid Value: < 3; Iodine Value: 1 max.
Containers: 190 kg net bung drums or bulk
References: Spec/1176/7A11

Radiasurf 7417. Oleofina S.A.
Chemical Description: PEG 1500 monostearate
Category and Applications: See Radiasurf 7402
Form: Solid; Color: White
Properties: m.w.: 1812 (avg.); Solubility @ 10%: Sol. in benzene, trichlorethylene; Sp. gr.: 1.023 (98.9 C); Visc.: 28.40 cps (98.9 C); M.P.: 47 C; HLB: 16.5 avg.; Flash Pt.: 248 C (COC); Sapon. Value: 30–40; Acid Value: < 3; Iodine Value: 1 max.; Surface Tension: 46.50 dynes/cm (0.1% aq.)
Containers: 25 kg net multiply paper bags or bulk
References: Spec/1176/7A11

Radiasurf 7420. Oleofina S.A.
Chemical Description: Diethylene glycol monolaurate
Category and Applications: See Radiasurf 7196
Form: Paste; Color: White
Properties: m.w.: 352 (avg.); Solubility @ 10%: Sol. in benzene, isopropanol, trichlorethylene; sol. (cloudy) in hexane, mineral and vegetable oils; Sp. gr.: 0.942 (37.8 C); Visc.: 15.20 cps (37.8 C); HLB: 5.7 avg.; Cloud Pt.: 28.5 C; Flash Pt.: 164 C (COC); Sapon. Value: 195–205; Acid Value: 3 max.
Containers: 190 kg net bung drums or bulk
References: Spec/1176/7A9

Radiasurf 7421. Oleofina S.A.
Chemical Description: Diethylene glycol monolaurate, self-emulsifying
Category and Applications: See Radiasurf 7196
Form: Liquid; Color: White
Properties: m.w.: 346 (avg.); Solubility @ 10%: Sol. in trichlorethylene; sol. (cloudy) in hexane, benzene, isopropanol, mineral and vegetable oils; Sp. gr.: 0.901 (98.9 C); Visc.: 3.60 cps (98.9 C); HLB: 6.3 avg.; Cloud Pt.: 29.5 C; Flash Pt.: 173 C (COC); Sapon. Value: 183–193; Acid Value: 3 max.; Iodine Value: 10 max.
Containers: 190 kg net bung drums or bulk
References: Spec/1176/7A9

Radiasurf 7422. Oleofina S.A.
Chemical Description: PEG 200 monolaurate
Category and Applications: See Radiasurf 7402
Form: Liquid; Color: White
Properties: m.w.: 379 (avg.); Ref. Index: 1.4542; Solubility @ 10%: Sol. in benzene, isopropanol, trichlorethylene; sol. (cloudy) in hexane, mineral and vegetable

oils; disp. in water; Sp. gr.: 0.983 (37.8 C); Visc.: 23.40 cps (37.8 C); HLB: 9.6 avg.; Cloud Pt.: –3 C; Flash Pt.: 188 C (COC); Sapon. Value: 140–150; Acid Value: < 3; Iodine Value: 1 max.; Surface Tension: 30.00 dynes/cm (0.1% aq.)
Containers: 190 kg net bung drums or bulk
References: Spec/1176/7A11

Radiasurf 7423. Oleofina S.A.
Chemical Description: PEG 400 monolaurate
Category and Applications: See Radiasurf 7402
Form: Liquid; Color: White
Properties: m.w.: 610 (avg.); Ref. Index: 1.4596; Solubility @ 10%: Sol. in benzene, isopropanol, trichlorethylene; sol. (cloudy) in water, hexane, mineral and vegetable oils; Sp. gr.: 1.023 (37.8 C); Visc.: 41.70 cps (37.8 C); HLB: 13.2 avg.; Cloud Pt.: 4 C; Flash Pt.: 238 C (COC); Sapon. Value: 90–100; Acid Value: < 3; Iodine Value: 1 max.; Surface Tension: 31.00 dynes/cm (0.1% aq.)
Containers: 190 kg net bung drums or bulk
References: Spec/1176/7A11

Radiasurf 7431. Oleofina S.A.
Chemical Description: PEG 300 monooleate
Category and Applications: See Radiasurf 7402
Form: Liquid; Color: Amber
Properties: Solubility @ 10%: Sol. in benzene, isopropanol, trichlorethylene; sol. (cloudy) in hexane, mineral and vegetable oils; disp. in water; HLB: ≈ 9.5; Cloud Pt.: ≈ –8 C; Sapon. Value: 95–115; Acid Value: < 3; Iodine Value: 40–50
Containers: 190 kg net bung drums or bulk
References: Spec/1176/7A11

Radiasurf 7432. Oleofina S.A.
Chemical Description: PEG 300 monostearate
Category and Applications: See Radiasurf 7402
Form: Paste; Color: White
Properties: Solubility @ 10%: Sol. in benzene, isopropanol, trichlorethylene, mineral oil; sol. (cloudy) in hexane; M.P.: 28–35 C; HLB: ≈ 9.7; Sapon. Value: 95–115; Acid Value: < 3; Iodine Value: 1 max.
Containers: 190 kg net bung drums or bulk
References: Spec/1176/7A11

Radiasurf 7443. Oleofina S.A.
Chemical Description: PEG 400 dioleate
Category and Applications: See Radiasurf 7402
Form: Liquid; Color: Amber
Properties: m.w.: 911 (avg.); Ref. Index: 1.4655; Solubility @ 10%: Sol. in trichlorethylene, mineral and vegetable oils; sol. (cloudy) in hexane, benzene, isopropanol; disp. in water; Sp. gr.: 0.962 (37.8 C); Visc.: 47.00 cps (37.8 C); HLB: 7.4 avg.; Cloud Pt.: –3 C; Flash Pt.: 248 C (COC); Sapon. Value: 120–130; Acid Value: < 5; Iodine Value: 50–60

Containers: 190 kg net bung drums or bulk
References: Spec/1176/7A11

Radiasurf 7453. Oleofina S.A.
Chemical Description: PEG 400 distearate
Category and Applications: See Radiasurf 7402
Form: Paste; Color: White
Properties: m.w.: 901 (avg.); Solubility @ 10%: Sol. in benzene, isopropanol, trichlorethylene; Sp. gr.: 0.920 (98.9 C); Visc.: 9.85 cps (98.9 C); M.P.: 39 C; HLB: 7.7 avg.; Flash Pt.: 242 C (COC); Sapon. Value: 120–130; Acid Value: < 5; Iodine Value: 1 max.
Containers: 190 kg net bung drums or bulk
References: Spec/1176/7A11

Radiasurf 7454. Oleofina S.A.
Chemical Description: PEG 600 distearate
Category and Applications: See Radiasurf 7402
Form: Solid; Color: White
Properties: m.w.: 1101 (avg.); Solubility @ 10%: Sol. in benzene, trichlorethylene; sol. (cloudy) in hexane, isopropanol, mineral oil; Sp. gr.: 0.940 (98.9 C); Visc.: 12.20 cps (98.9 C); M.P.: 40 C; HLB: 9.7 avg.; Flash Pt.: 249 C; Sapon. Value: 100–110; Acid Value: < 5; Iodine Value: 1 max.
Containers: 25 kg net multiply paper bags or bulk
References: Spec/1176/7A11

Radiasyn 7174. Oleofina S.A.
Chemical Description: Pentaerythritol tetra-C_5/C_9
Applications: Synthetic ester lubricants
Form: Liquid
Properties: m.w.: 580 (avg.); Ref. Index: 1.4543; Solubility @ 10%: Sol. in hexane, benzene, isopropanol, trichlorethylene, mineral and vegetable oils; Sp. gr.: 0.971 (37.8 C); Visc.: 29.50 cps (37.8 C); Cloud Pt.: –35 C; Flash Pt.: 275 C (COC)

Radiasyn 7177. Oleofina S.A.
Chemical Description: Pentaerythritol tetra-C_7
Applications: See Radiasyn 7174
Form: Liquid
Properties: m.w.: 580 (avg.); Ref. Index: 1.4526; Solubility @ 10%: Sol. in hexane, benzene, isopropanol, trichlorethylene, mineral and vegetable oils; Sp. gr.: 0.971 (37.8 C); Visc.: 24.20 cps (37.8 C); Cloud Pt.: –33 C; Flash Pt.: 251 C (COC)

Radiasyn 7178. Oleofina S.A.
Chemical Description: Pentaerythritol tetra-C_8/C_{10}
Applications: See Radiasyn 7174
Form: Liquid
Properties: m.w.: 690 (avg.); Ref. Index: 1.4546; Solubility @ 10%: Sol. in hexane, benzene, isopropanol, trichlorethylene, mineral and vegetable oils; Sp. gr.: 0.945 (37.8 C); Visc.: 31.50 cps (37.8 C); Cloud Pt.: –2 C; Flash Pt.: 282 C (COC)

Radiasyn 7364. Oleofina S.A.
Chemical Description: Trimethylolpropane tri-C_5/C_9
Applications: See Radiasyn 7174
Form: Liquid
Properties: m.w.: 465 (avg.); Ref. Index: 1.4511; Solubility @ 10%: Sol. in hexane, benzene, isopropanol, trichlorethylene, mineral and vegetable oils; Sp. gr.: 0.951 (37.8 C); Visc.: 15.40 cps (37.8 C); Cloud Pt.: < -35 C; Flash Pt.: 255 C (COC)

Radiasyn 7367. Oleofina S.A.
Chemical Description: Trimethylolpropane tri-C_7
Applications: See Radiasyn 7174
Form: Liquid
Properties: m.w.: 465 (avg.); Ref. Index: 1.4514; Solubility @ 10%: Sol. in hexane, benzene, isopropanol, trichlorethylene, mineral and vegetable oils; Sp. gr.: 0.950 (37.8 C); Visc.: 14.70 cps (37.8 C); Cloud Pt.: < -35 C; Flash Pt.: 223 C (COC)

Radiasyn 7368. Oleofina S.A.
Chemical Description: Trimethylolpropane tri-C_8/C_{10}
Applications: See Radiasyn 7174
Form: Liquid
Properties: m.w.: 550 (avg.); Ref. Index: 1.4527; Solubility @ 10%: Sol. in hexane, benzene, isopropanol, trichlorethylene, mineral and vegetable oils; Sp. gr.: 0.932 (37.8 C); Visc.: 22.30 cps (37.8 C); Cloud Pt.: < -35 C; Flash Pt.: 248 C (COC)

Raylig. ITT Rayonier, Inc.
Chemical Description: Lignin derivative
Category: Dispersant, anionic
Applications: Deflocculation and viscosity control in water dispersions of solids
Form: Liquid; Color: Brown; Odor: Pleasant, vanilla-like
Properties: Solubility: Water-sol.; Sp. gr.: 1.23; Visc.: 500 cps; F.P.: 0 C; B.P.: 100 C; Stability: Stable; pH: 5.5
Composition: 50% active ingred.
Reaction: Acid
Toxicity: Mild eye irritant
Containers: Tank trucks, cars

Reax 15B. Westvaco/Polychemicals Dept.
Chemical Description: Sodium salt of highly modified Kraft pine lignin
Category: Dispersant, sequestrant, anionic
Applications: Treatment of industrial water supplies to sequester alkaline earth and heavy metals; dyestuff dispersant; dispersant for organic pesticides in wettable powder formulations; additive in negative plates for lead-acid storage batteries
Form: Free-flowing powder
Properties: Solubility: Readily sol.; pH: 10.5
Composition: 100% conc.
References: Tech. bulletin 81025

Reax 65AE. Westvaco/Polychemicals Dept.
Chemical Description: Solution of

saponified rosin and blended fatty acids
Category: Air entrainment aid, plasticizer
Applications: Masonry and portland cements
Form: Liquid
Composition: 34% solids

Reax 65D. Westvaco/Polychemicals Dept.
Category: Air-entraining agent, plasticizer
Applications: Dry air-entraining systems in portland or masonry cement manufacture; plasticizer and air-entraining agent in stucco and plastic cements
Form: Free-flowing powder

Reax 77. Westvaco/Polychemicals Dept.
Chemical Description: Amine acetates and sulfonated polyhydroxyethers
Category: Grinding aid, flow promoter, anionic
Applications: Production of portland cements; provides balance between improved grinding efficiency and flow properties
Form: Liquid
Properties: Sp. gr.: 1.200; Visc.: 40 cps; Stability: Stable
Composition: 70% active ingred.

Reax 77XF. Westvaco/Polychemicals Dept.
Chemical Description: Amine acetates and polyhydroxy alcohols
Category: Flow promoter, grinding aid, anionic
Applications: Portland cement production; provides extra flowability at normal dosages
Form: Liquid
Properties: Sp. gr.: 1.220; Visc.: 75 cps; Stability: Stable

Reax 88B. Westvaco/Polychemicals Dept.
Chemical Description: Sugar-free sodium based sulfonate of modified Kraft pine lignin
Category: Dispersant, viscosity reducer, anionic
Applications: Used in aq. systems where low pH, high electrolyte content, or high dispersed solids ratio necessitates a high-solubility dispersant; high solids flowable liquid pesticides; pesticides and plant foods in high analysis liquid fertilizers; viscosity reducer in liquid dye formulations
Form: Free-flowing powder (also available as 35% solids aq. sol'n.)
Properties: Solubility: Easily sol.; Visc.: 95; pH: 11.5 (aq. sol'ns.)
Composition: 100% active ingred.

Reax G-1. Westvaco/Polychemicals Dept.
Chemical Description: Polyglycol-based with a combination of selected aliphatic hydroxyl functional groups
Category: Grinding aid, flow promoter, anionic
Applications: Portland cements
Form: Liquid
Properties: Solubility: Easily handled and readily dilutable in water; Sp. gr.: 1.125; Visc.: 50 cps

Reax G-2. Westvaco/Polychemicals Dept.
Chemical Description: Glycol-based with a combination of amines, aliphatic and aromatic hydroxyls, carboxylic acids, and sulfonates
Category: Flow promoter, grinding aid, anionic
Applications: Used in portland cements where unusual flow problems are not encountered
Form: Liquid
Properties: Sp. gr.: 1.110; Visc.: 30 cps

Reax SR-1. Westvaco/Polychemicals Dept.
Chemical Description: Sodium salt of sulfonated modified Kraft lignin
Category: Dispersant, anionic
Applications: Dyestuff dispersant for textile industry; does not reduce or stain fibers and maintains capacity for high temperature stabilization
Form: Free-flowing powder; Color: Brown
Properties: Solubility: Virtually no solubility in most organic solvents; increased sol. with the addition of water, in water-miscible solvents, e.g., DMF, ethylene glycol, methyl alcohol; Density: 24 lb/ft^3 (bulk, loose); Surface Tension: 55.8 dynes (1% aq.); pH: 9.9 (2% aq.)
Composition: 100% conc.
Containers: 50 lb net multiwall paper bags
References: Bulletin 58959, 58990

Reax SR-7. Westvaco/Polychemicals Dept.
Chemical Description: Sodium salt of sulfonated modified Kraft lignin
Category and Applications: See Reax SR-1; SR-7 recommended for azo dyes which have electronegative groups ortho to the azo link
Form: Free-flowing powder; Color: Brown
Properties: Surface Tension: 58.8 dynes (1% aq.); pH: 6.7 (2% aq.)
Composition: 100% conc.
References: Bulletin 58990

Recodan. Grindsted Prod.
Chemical Description: Integrated emulsifiers and stabilizers
Category: Emulsifier, stabilizer, nonionic
Applications: Improves fat dispersion, increases heat stability, and enhances palatability in sterilized or pasteurized, recombined, filled, imitation, chocolate, and flavored milk products
Form: Powder
Composition: 100% conc.

Redicote 80-S. Armak Co.
Chemical Description: Amine-based
Category: Emulsifier, adhesion and antistripping agent, cationic
Applications: Antistripping agents in cutbacks, asphalt cements; produces water-resistant film of asphalt on inferior grades of aggregates, uniform and firmer coating of each stone

Form: Liquid; Color: Dark
Properties: Density: 7.83 lb/gal; Sp. gr.: 0.940; Visc.: 130 cps (60 C); Flash Pt.: > 180 C (PM); Pour Pt.: 4.4 C; Stability: Excellent heat stability
Reaction: Alkaline
Toxicity: Severe skin and eye irritants
Containers: Drums, truck load, or tank truck
Warnings and Caution: Alkaline corrosive liquid—wear protective goggles and gloves
References: Bulletin 78-7

Redicote 81-S. Armak Co.
Chemical Description: Amine-based
Category and Applications: See Redicote 80-S
Form: Opaque liquid; Color: Brown
Properties: Density: 7.92 lb/gal; Sp. gr.: 0.953; Visc.: 134 cps (60 C); Flash Pt.: 89.0 C (PM); Pour Pt.: 4.4 C; Stability: Excellent heat stability
Reaction: Alkaline
Toxicity: Severe skin and eye irritant
Containers: Drums, truck load, or tank truck
Warnings and Caution: Alkaline corrosive liquid—wear protective goggles and gloves
References: Bulletin 78-7

Redicote 82-S. Armak Co.
Chemical Description: Amine-based
Category and Applications: See Redicote 80-S
Form: Opaque liquid; Color: Brown
Properties: Density: 7.00 lb/gal; Sp. gr.: 0.841; Visc.: 56.3 cps (60 C); Flash Pt.: 89.0 C (PM); Pour Pt.: 0.3 C; Stability: Excellent heat stability
Reaction: Alkaline
Toxicity: Severe skin and eye irritant
Containers: Drums, truck load, or tank truck
Warnings and Caution: Alkaline corrosive liquid—wear protective goggles and gloves
References: Bulletin 78-7

Redicote 4442 Emulsion Additive. Armak Co.
Category: Emulsion additive, nonionic
Applications: Improves emulsion properties, e.g., increases viscosity, improves shear, storage, and dilution stabilities, prevents graininess, improves fine aggregate coating, gives thixotropic properties, etc.
Form: Free-flowing powder
Properties: Solubility: Easily disp. in water; disp. in isopropanol, naphtha; Stability: Sensitive to pH—do not add directly to acidified systems or use in systems with < pH 2
Containers: 100 lb bags or truck loads
References: Bulletin No. 75-8

Redicote AP. Armak Co.
Chemical Description: Amine-based asphalt emulsifier

Category: Emulsifier, cationic
Applications: Additive to hot paving asphalt prior to emulsification for improved emulsion settlement properties and increased viscosity
Properties: Density: 7.99 lb/gal; Sp. gr.: 0.957; Visc.: 205 cps (60 C); Pour Pt.: 10 C
Reaction: Alkaline
Toxicity: Severe skin and eye irritant
Containers: Drums, truck load, or tank truck
Warnings and Caution: Alkaline corrosive liquid—wear protective goggles and gloves
References: Bulletin No. 76-7

Redicote E-54. Armak Co.
Chemical Description: Amine-based asphalt emulsifiers
Category: Emulsifier, anionic
Applications: Emulsifier for high quality RS-grade or MS-grade emulsions
Form: Hazy liquid
Properties: Density: 8.59 lb/gal; Sp. gr.: 1.03; Visc.: 2700 SSU (22 C); Flash Pt.: 122 C (TOC); Pour Pt.: 15.5 C
Reaction: Neutral
Containers: Drums or bulk
References: Bulletin No. 74-17

Redicote E-55. Armak Co.
Chemical Description: Amine-based asphalt emulsifier
Category: Emulsifier, cationic
Applications: Slow set and mixing grade asphalt emulsions; for slurry seal and cold mix stabilization
Form: Liquid; Color: Dark
Properties: Density: 8.369 lb/gal; Sp. gr.: 1.003; Visc.: 220 SSU; Flash Pt.: 185 C; Pour Pt.: −7 C
Containers: Drums or bulk
References: Bulletin No. 75-5

Redicote E-56. Amrak Co.
Chemical Description: Amine-based asphalt emulsifier
Category: Emulsifier, anionic
Applications: Rapid setting asphalt emulsions; also for MS grades with addition of a suitable distillate
Form: Viscous liquid; Color: Dark
Properties: Solubility: Rapidly sol. in water; Density: 9.1 lb/gal; Sp. gr.: 1.092; Visc.: 2700 SSU (60 C); Flash Pt.: 196 C; Pour Pt.: −12 C
Containers: Drums or bulk
Warnings and Caution: Quite corrosive when diluted; care should be taken not to add water to conc. form—may cause considerable heat and dangerous splashing
References: Bulletin No. 75-6

Redicote E-57. Armak Co.
Chemical Description: Amine-based asphalt emulsifier
Category: Emulsifier, anionic
Applications: Slow setting emulsions; for dense mixing, fog sealing, mulching slurry seal; with a pH change, produces an emulsion with many RS and MS properties; stone coating abilities at lower conc.
Form: Viscous liquid; Color: Dark

Properties: Solubility: Water-sol.; Density: 10.50 lb/gal; Sp. gr.: 1.260; Visc.: 307 SSU (60 C); Flash Pt.: 230 C (COC); Pour Pt.: −11 C
Containers: Drums or bulk
References: Bulletin No. 75-7

Redicote E-62. Armak Co.
Chemical Description: Amine-based asphalt emulsifier
Category: Emulsifier, anionic
Applications: Preparation of high float rapid and medium setting grade emulsions, and rapid and medium setting emulsions
Form: Viscous liquid; Color: Black
Properties: Density: 8.26 lb/gal; Sp. gr.: 0.990; Visc.: 745 SSU (60 C); Flash Pt.: 100 C (PM); Pour Pt.: 10 C
Containers: Drums or bulk
Warnings and Caution: Organic acid—wear protective goggles and gloves
References: Bulletin 77-6

Redicote E-63. Armak Co.
Chemical Description: Amine-based asphalt emulsifier
Category: Emulsifier, cationic
Applications: Cationic slow-setting emulsions; produces solvent-free emulsions for the coating and handling of dense aggregate and fine-grained soil mixes
Form: Liquid
Properties: Density: 9.73 lb/gal; Sp. gr.: 1.166; Visc.: 230 cps (60 C); Pour Pt.: 1.7
Toxicity: Severe skin and eye irritant
Containers: Drums or bulk
Warnings and Caution: Acid corrosive liquid—wear protective goggles and gloves
References: Bulletin No. 76-8

Redicote E-64. Armak Co.
Category: Emulsifier, cationic
Applications: Rapid-setting cationic asphalt emulsions with good stability and stone-coating properties; can be used in coarse mixing grade formulations
Form: Clear liquid; Color: Yellow
Properties: Solubility: Readily sol. in acidified cold water; Density: 6.91 lb/gal; Sp. gr.: 0.831; Visc.: 3.6 cps (60 C); Flash Pt.: $<$ 21 C (PM); Pour Pt.: 4.3 C; Stability: Stable after many freeze/thaw cycles
Toxicity: Severe skin and eye irritant
Containers: Drums or bulk
Warnings and Caution: Alkaline corrosive liquid—wear protective goggles and gloves; flammable—keep from heat, sparks, and open flame
References: Bulletin 77-8

Redicote E-65. Armak Co.
Category: Emulsifier, cationic
Applications: Slow-setting asphalt emulsions; does not require addition of acid; produces solvent-free emulsions for the coating and handling of dense aggregate and fine-grained soil mixes
Form: Hazy liquid; Color: Amber
Properties: Density: 9.002 lb/gal; Sp. gr.: 1.083; Visc.: 54.2 cps (60

C); Flash Pt.: 32 C (PM); Pour Pt.: –18.9 C
Reaction: Neutral
Toxicity: Severe skin and eye irritant
Containers: Drums or bulk
Warnings and Caution: Acid corrosive liquid—wear protective goggles and gloves
References: Bulletin 77-9

Remanol 1300. Elektrochemische Fabrik Kempen
Chemical Description: Collagen protein hydrolystae mol wt 1300
Applications: Protein product; skin-compatibility improving component for shampoos, etc.
Form: Low-viscous liquid; Color: Brown
Composition: 50% active ingred.
Containers: 120 kg net poly drums

Renex 20. I.C.I. United States
Chemical Description: POE ester of mixed fatty acid and heterocyclic resin acids, 16 mols EO
Category: Detergent, nonionic
Applications: Detergent with low foam, synergism with alkyl aryl sulfonates and with soaps in hard water, dedusting action, and good spot removing on hydrophobic fibers; used in home laundry, textile scouring, spotting agent, rug shampoo
Form: Clear liquid; Color: Amber
Properties: Solubility: Sol. in water, lower alcohols, toluene, acetone, diethyl ether, dioxane, cellosolve, ethyl acetate, aniline; disp. in carbon tetrachloride, ethylene glycol; Sp. gr.: ≈ 1.1; Visc.: ≈ 350 cps; HLB: 13.8; Cloud Pt.: ≈ 126 F (1% aq.); Flash Pt.: > 300 F; Fire Pt.: > 300 F; Pour Pt.: ≈ 55 F; Acid No.: 2.0 max.; Hydroxyl No.: 57–67; Surface Tension: 41 dynes/ cm (0.01% sol'n.)
Composition: 100% active ingred.

Renex 22. I.C.I. United States
Chemical Description: POE ester of mixed fatty and resin acids, 12 mols EO
Category and Applications: Similar to Renex 20 with lower foam; recommended for mechanical dishwashing, metal cleaning formulas where spray application is used
Form: Liquid; Color: Amber
Properties: Solubility: Disp. in water; sol. in acetone, ethyl acetate, dioxane, cellosolve, aniline, ethyl alcohol; Sp. gr.: ≈ 1.1; Visc.: ≈ 300 cps; HLB: 12.2; Cloud Pt.: ≈ 32 F (1% aq.); Flash Pt.: > 300 F; Fire Pt.: > 300 F; Pour Pt.: ≈ 35 F; Acid No.: 2.0 max.; Hydroxyl No.: 68–81; Surface Tension: 39 dynes/cm (0.01% sol'n.)

Renex 25. I.C.I. United States
Chemical Description: Renex 20 and urea
Category: Detergent, nonionic
Applications: Convenient for formulating powdered or granulated cleaning products; used in home laundry, rug shampoos
Form: Free-flowing powder; Color: Tan

Properties: Solubility: Sol. in water; slightly sol. in lower alcohols, cellosolve; HLB: 13.5; Cloud Pt.: 124 F (1% sol'n.); Flash Pt.: > 300 F; Surface Tension: 39 dynes/cm (0.01% sol'n.); pH: ≈ 7 (1% aq.)
Composition: 50% active ingred.

Renex 26. I.C.I. United States
Chemical Description: Ethoxylated fatty acid ester
Category: Detergent, nonionic
Form: Liquid; Color: Amber
Properties: Solubility: Sol. in water, petroleum solvent, and cleaners' naphtha; miscible with isopropanol; Visc.: ≈ 600 cps; Flash Pt.: > 300 F; Fire Pt.: > 300 F; Acid No.: 3.0 max.; Hydroxyl No.: 75–95

Renex 30. I.C.I. United States
Chemical Description: POE (12) tridecyl ether
Category: Detergent, wetting agent, nonionic
Applications: Good detergency and wetting action at low concentrations in hard water, alkaline, and acid sol'ns.; used for raw wool scouring and carbonizing, metal cleaning, bottle washing, home laundry
Form: Clear to hazy liquid (precipitates on standing); Color: Colorless
Properties: Solubility: Sol. in water, lower alcohols, xylene, butyl Cellosolve, carbon tetrachloride, ethylene glycol, propylene glycol; slightly sol. in vegetable oils; Sp. gr.: ≈ 1.0; Visc.: ≈ 60 cps; HLB: 14.5; Cloud Pt.: ≈ 183 F (1% aq.); Flash Pt.: > 300 F; Fire Pt.: > 300 F; Pour Pt.: ≈ 55 F; Acid No.: 1.0 max.; Hydroxyl No.: 75–85; Surface Tension: 28 dynes/cm (0.01% sol'n.); pH: ≈ 6 (1% in dist. water)
Composition: 100% conc.
Toxicity: Causes eye irritation
Warnings and Caution: Avoid contact with eyes

Renex 31. I.C.I. United States
Chemical Description: POE (15) tridecyl ether
Category: Detergent, wetting agent, nonionic
Applications: Good wetting with high foam and high cloud point; used for liquid hand dishwashing formulations and metal cleaning; good in high salt concs., e.g., sea water
Form: Soft paste; Color: White
Properties: Solubility: Sol. in water, propylene glycol, lower alcohols, butyl Cellosolve, carbon tetrachloride, xylene; slightly sol. in perchloroethylene; Sp. gr.: ≈ 1.0; Visc.: ≈ 130 cps; HLB: 15.4; Cloud Pt.: ≈ 210 F (1% aq.); Flash Pt.: > 300 F; Fire Pt.: > 300 F; Pour Pt.: ≈ 61 F; Acid No.: 1.0 max.; Hydroxyl No.: 60–74; Surface Tension: 34 dynes/cm (0.01% sol'n.)
Composition: 100% conc.
Containers: 55 gal open head steel drums

Renex 35. I.C.I. United States
Chemical Description: Urea complex of POE (12) tridecyl ether
Category and Applications: See Renex 30; convenient for formulating powdered or granulated cleaning products
Form: Free-flowing powder; Color: Ivory white
Properties: Solubility: Sol. in water, ethylene glycol, lower alcohols; M.P.: ≈ 257 F; HLB: 14.5; Cloud Pt.: ≈ 180 F (1% aq.); Flash Pt.: > 300 F; Fire Pt.: > 300 F; Surface Tension: 33 dynes/cm (0.01% sol'n.); pH: ≈ 7 (1% aq.)
Composition: 50% conc.
Containers: 51 gal fiber lever pak

Renex 36. I.C.I. United States
Chemical Description: POE (6) tridecyl ether
Category: Detergent, dispersant, wetting agent, nonionic
Applications: Lowest foamer of this series with greatest resistance to discoloration by solid caustic soda; for use in strongly alkaline cleaners, e.g., in metal cleaning and bottle washing; dispersant for solids and paint pigments; dishwashing
Form: Cloudy to clear, thin oily liquid; Color: Colorless
Properties: Solubility: Sol. in acetone, carbon tetrachloride, ethylacetate, cellosolve, ethanol, toluene, aniline; disp. (cloudy) in water and 5% aq. sol'ns. of sulfuric acid, hydrochloric acid, and sodium hydroxide; Sp. gr.: ≈ 1.0; Visc.: ≈ 80 cps; HLB: 11.4; Cloud Pt.: < 32 F (1% aq.); Flash Pt.: > 300 F; Fire Pt.: > 300 F; Acid No.: 1.0 max.; Hydroxyl No.: 118–133; Surface Tension: 27 dynes/cm (0.01% sol'n.); pH: ≈ 6 (1% in dist. water)
Composition: 100% conc.

Renex 647. Atlas Chemical Industries NV
Chemical Description: POE nonyl phenol ether
Category: Emulsifier, detergent, nonionic
Form: Liquid
Composition: 100% conc.

Renex 648. I.C.I. United States
Chemical Description: POE (5) nonyl phenol ether
Category: Detergent, nonionic
Applications: Solvent and emulsion cleaning applications, , metal cleaning
Form: Clear liquid
Properties: Solubility @ 10%: Sol. in propylene glycol, perchloroethylene, isopropyl alcohol, cottonseed oil; self-emulsifying in water; Sp. gr.: 1.03; Visc.: 223 cps; HLB: 10; Flash Pt.: > 200 F; Pour Pt.: –20 F; Stability: Very good in acids, alkalis, and in hard water
Composition: 100% active ingred.

Renex 649. Atlas Chemical Industries NV
Chemical Description: POE nonyl phenol ether
Category: Emulsifier, detergent, nonionic
Form: Solid
Composition: 100% conc.

Renex 650. I.C.I. United States
Chemical Description: POE (30) nonyl phenol ether
Category: Detergent, emulsifier, stabilizer, nonionic
Applications: High foam and high cloud pt. surfactant for detergent use at high temperatures and salt concentrations, emulsification, and latex stabilization
Form: Solid; Color: White
Properties: Solubility @ 10%: Sol. in water, propylene glycol, isopropyl alcohol, xylene; Sp. gr.: 1.15; HLB: 17.1; Cloud Pt.: Clear @ 212 F; Flash Pt.: > 300 F; Pour Pt.: 91 F; Stability: Very good in acids, alkalis, and hard water; Surface Tension: 42 dynes/cm (0.01% sol'n.)
Composition: 100% active ingred.

Renex 678. I.C.I. United States
Chemical Description: POE (15) nonyl phenol ether
Category: Detergent, wetting agent, nonionic
Applications: High cloud point and high foaming for use at high temperatures and salt concs.; used in dishwashing
Form: Liquid; Color: Clear
Properties: Solubility @ 10%: Sol. in water, propylene glycol, perchloroethylene, isopropyl alcohol, xylene; Sp. gr.: 1.08; Visc.: 319 cps; HLB: 15; Cloud Pt.: 211 F (1% sol'n.); Flash Pt.: > 300 F; Pour Pt.: 69 F; Stability: Very good in acids, alkalis, and hard water; Surface Tension: 33 dynes/cm (0.01% sol'n.)
Composition: 100% active ingred.

Renex 679. Atlas Chemical Industries NV
Chemical Description: POE nonyl phenol ether
Category: Emulsifier, detergent, nonionic
Form: Liquid
Composition: 100% conc.

Renex 688. I.C.I. United States
Chemical Description: POE (8) nonyl phenol ether
Category: Detergent, wetting agent, nonionic
Applications: Low foaming surfactant for dishwashing, textile scouring
Form: Liquid
Properties: Solubility @ 10%: Sol. in propylene glycol, perchloroethylene, isopropyl alcohol, xylene; sol. (hazy) in water and cottonseed oil; Sp. gr.: 1.05; Visc.: 240 cps; HLB: 12.3; Cloud Pt.: 87 F (1% sol'n.); Flash Pt.: > 300 F; Pour Pt.: 12 F; Stability: Very good in acids, alkalis, and hard water; Surface Tension: 30 dynes/cm (0.01% sol'n.)
Composition: 100% active ingred.

Renex 690. I.C.I. United States
Chemical Description: POE (10) nonyl phenol ether
Category: Detergent, wetting agent, nonionic
Applications: General purpose detergents for home and commercial laundry products, textile scouring, wetting, dyeing and desizing, acid metal cleaners, sanitizers, and hard surface cleaning
Form: Liquid

Properties: Solubility @ 10%: Sol. in propylene glycol, perchloroethylene, isopropyl alcohol, xylene; sol. (hazy) in water, cottonseed oil; Sp. gr.: 1.06; Visc.: 260 cps; HLB: 13.3; Cloud Pt.: 150 F (1% sol'n.); Flash Pt.: > 300 F; Pour Pt.: 33 F; Stability: Very good in acids, alkalis, and hard water; Surface Tension: 31 dynes/cm (0.01% sol'n.)
Composition: 100% active ingred.

Renex 697. I.C.I. United States
Chemical Description: POE (6) nonyl phenol ether
Category and Applications: See Renex 648
Form: Clear liquid
Properties: Solubility @ 10%: Sol. in propylene glycol, perchloroethylene, isopropyl alcohol, xylene, cottonseed oil; self-emulsifying in water; Sp. gr.: 1.04; Visc.: 250 cps; HLB: 10.9; Cloud Pt.: < 32 F (1% sol'n.); Flash Pt.: > 300 F; Pour Pt.: –21 F; Stability: Very good in acids, alkalis, and in hard water; Surface Tension: 30 dynes/cm (0.01% sol'n.)
Composition: 100% active ingred.

Renex 698. I.C.I. United States
Chemical Description: POE (9–9.5) nonyl phenol ether
Category and Applications: See Renex 690
Form: Clear liquid
Properties: Solubility @ 10%: Sol. in propylene glycol, perchloroethylene, isopropyl alcohol, xylene; sol. (hazy) in water, cottonseed oil; Sp. gr.: 1.06; Visc.: 245 cps; HLB: 13; Cloud Pt.: 129 F (1% sol'n.); Flash Pt.: > 300 F; Pour Pt.: 28 F; Stability: Very good in acids, alkalis, and in hard water; Surface Tension: 30 dynes/cm (0.01% sol'n.)
Composition: 100% active ingred.

Renex 702, 703, 704, 707, 711, 714, 720. Atlas Chemical Industries NV
Chemical Description: POE alkyl ether
Category: Emulsifier, detergent, nonionic
Applications: General purpose detergents, emulsifiers
Form: Liquid
Properties: Biodegradable
Composition: 100% conc.

Renex 751. Atlas Chemical Industries NV
Chemcial Description: POE octyl phenol ether
Category: Lubricant, emulsifier, nonionic
Applications: Textile processing
Form: Liquid
Composition: 100% conc.

Retarder CA. Hart Chem. Ltd.
Chemical Description: N-alkyl dimethylbenzyl ammonium chloride
Category: Retarding agent, cationic
Applications: Retarding agent for acrylic fiber dyeing with cationic dyestuff; allows maximum retarding and leveling at high temps.
Form: Liquid; Color: Clear

Properties: Sp. gr.: 0.98
Composition: 40% active ingred.
Reaction: Alkaline
Toxicity: Skin and eye irritant
Containers: Drums, T/L
Warnings and Caution: Avoid contact with skin and eyes

Rewocid DU 185. Dutton & Reinisch Ltd.
Chemical Description: Undecylenic acid polydiethanolamide
Category: Detergent, emulsifier, nonionic
Applications: Used as bacteriocide, thickening, foam stabilizing agent in shampoos
Form: Viscous liquid; Color: Yellow
Properties: Solubility: Sol. in water, oil, alcohols, glycols; Sp. gr.: ≈ 1.0; Visc.: High
Composition: 100% active ingred.
Reaction: Acid
Containers: Lacquered drums

Rewocid SBU 185. Dutton & Reinisch Ltd.
Chemical Description: Sodium salt of the monosulfosuccinate of undecylenic acid monoethanolamide
Category: Fungicide, detergent, foaming agent, anionic
Applications: Exhibits exceptional antiseborrheoic and antiallergic properties; can be combined with all anionic, nonionic, and amphoteric products without any reduction in its acitivity; used in foambaths, shampoos, skin-cleaners, foot baths, washing agents
Form: Powder; Color: Pale yellow; Odor: Typical (slight)
Properties: Solubility: Sol. in alcohol, glycol; Sp. gr.: ≈ 1.0; Stability: Stable at room temperature; pH: 6.5–7.5 (5% solids); Biodegradable
Composition: 50% active ingred.
Reaction: Acid (slight)
Containers: Lacquered drums

Rewocid U 185. Dutton & Reinisch Ltd.
Chemical Description: Monoethanolamide of undecylenic acid
Category: Fungicide, detergent, nonionic
Applications: Improves foam quality and stability, superfatting, and increases viscosity; fully compatible with all anionic, cationic, amphoteric, and other nonionic surface-active agents
Form: Flakes; Color: Yellow; Odor: Typical (strong)
Properties: Solubility: Sol. in alcohol, glycol; disp. in warm water; M.P.: ≈ 50 C
Composition: 100% active ingred.
Reaction: Alkaline

Rewo-Derm S 1333. Dutton & Reinisch Ltd.
Chemical Description: Disodium sulfosuccinate of a specially selected vegetable fatty acid
Category: Detergent, anionic
Applications: Used for washing-up liquids; liquid and powder detergents; all types of shampoos; foam bath; skin creams; face lotions; make-up foundations; all types of

cleaners; drilling- and cutting-oils
Form: Liquid; Color: Amber
Properties: pH: 6.5–7.5 (5% solids)
Composition: 40% active ingred.

Rewoderm SB/LAN 5. Dutton & Reinisch Ltd.
Chemical Description: Wool wax alcohol sulfosuccinate
Category: Detergent, anionic
Applications: Moisturizing skin protective agent for creams, lotions, and cleansing formulations
Form: Liquid; Color: Brown
Properties: Solubility: Slightly sol. in alcohols, glycols, and other solvents; pH: 6.5–7.5 (5% solids); Biodegradable
Composition: 50% active ingred.

Rewo-Lan 5. Dutton & Reinisch Ltd.
See Rewoderm SB/LAN 5

Rewomat B 2003. Dutton & Reinisch Ltd.
Chemical Description: Tetrasodium (1,2-dicarboxyethyl)-N-alkyl sulfosuccinamide
Category: Detergent, foaming agent and stabilizer, solubilizer, dispersant, emulsifier, anionic
Applications: Emulsion polymerization additive; industrial and metal cleaners; dishwashing liquids, household, and all-purpose cleaners; solubilizer for other detergent raw materials; carpet and upholstery cleaners; soldering aid; reduction of gel formation; dispersion of pigments; emulsifier for wax and oil; cosmetics, e.g., cold creams, skin and hand lotions; esp. good in presence of inorganic salts; high calcium tolerance
Form: Clear liquid; Color: Amber
Properties: Solubility: Gives a clear sol'n. in water, alkali, and electrolytes; Stability: Good in presence of inorganic salts; pH: 7–9 (10% solids)
Composition: ≈ 35% solids
References: Folder: section 6/1277/0602

Rewomat TMS. Dutton & Reinisch Ltd.
Chemical Description: Disodium alkyl sulfosuccinamide
Category: Foaming agent and stabilizer, emulsifier, wetting agent, anionic
Applications: Foam additive for natural and synthetic lattices, esp. in foam backings for carpets; emulsion polymerization; wetting agent for latex impregnation
Form: Soft creamy paste or liquid
Properties: Stability: Stable in alkaline and acid media; pH: 9–10 (10% solids); Biodegradable
Composition: ≈ 35% solids
References: Folder: section 6/1277/0604

Rewomid 203/S. Dutton & Reinisch Ltd.
Chemical Description: Lauric acid diethanolamide
Category: Detergent, nonionic
Applications: Foam stabilizer and viscosity modifier for shampoos and detergents
Form: Solid; Color: Off-white;

Odor: Typical (slight)
Properties: M.P.: ≈ 40 C
Composition: 100% active ingred.
Reaction: Alkaline
Containers: Mild steel drums

Rewomid AC 28. Dutton & Reinisch Ltd.
Chemical Description: Groundnut fatty acid polydialkanolamide
Category: Detergent, emulsifier, nonionic
Applications: Used as an anti-corrosion additive for emulsifiable and nonemulsifiable mineral-oil-based metal-working lubricants, including drilling-, cutting-, and grinding-oils
Form: Liquid; Color: Brown
Properties: Density: 0.98 g/cm^3 @ 20 C; Sp. gr.: ≈ 1.0; Visc.: Medium; Acid No.: 14–16.5
Composition: 100% active ingred.
Reaction: Alkaline
Containers: Mild steel drums

Rewomid C 212. Dutton & Reinisch Ltd.
Chemical Description: Coconut fatty acid monoethanolamide
Category: Detergent, nonionic
Applications: Can be incorporated in all types of detergent products to provide: improvement of foam quantity, stability, and texture; increased viscosity; superfatting effect/skin protection; improved detergency; stabilizer of emulsions
Form: Flakes; Color: Off-white; Odor: Amidic (slight)
Properties: M.P.: ≈ 73 C
Composition: 100% active ingred.
Reaction: Alkaline
Containers: Paper sacks

Rewomid DC 212/S. Dutton & Reinisch Ltd.
Chemical Description: Coconut fatty acid diethanolamide
Category: Detergent, nonionic
Applications: Foam stabilizer and viscosity modifier used for shampoos and detergents
Form: Liquid; Color: Light yellow; Odor: Typical (slight)
Properties: Density: 0.99 g/cm^3 @ 20 C; Visc.: Medium
Composition: 100% active ingred.
Reaction: Alkaline
Containers: Mild steel drums

Rewomid DC 220/SE. Dutton & Reinisch Ltd.
Chemical Description: Coconut fatty acid diethanolamide
Category: Detergent, nonionic
Applications: Thickening agent, foam stabilizer, superfatting agent
Form: Soft paste
Composition: 100% active ingred.

Rewomid DL 203. Dutton & Reinisch Ltd.
Chemical Description: Lauric acid polydialkanolamide
Category: Detergent, nonionic
Applications: Increases foam quality, quantity, and stability; increases viscosity; good superfatting; improves washing power; stabilization of emulsions; fixation of perfumes

Form: Wax; Color: White
Properties: Density: 0.97 g/cm^3 @ 20 C; Sp. gr.: ≈ 1.0; Visc.: High
Composition: 100% active ingred.
Reaction: Alkaline

Rewomid DL 240. Dutton & Reinisch Ltd.
Chemical Description: Coconut fatty acid polydiethanolamide
Category: Detergent, nonionic
Applications: Can be added to liquid and paste cosmetic preparations and all types of cleaning- and washing-agents to produce the following improvements: increased viscosity; improved foam quality, quantity, and stability; stabilization of emulsions; superfatting; improved washing power
Form: Liquid; Color: Yellow; Odor: Typical (faint)
Properties: Visc.: Medium
Composition: 100% active ingred.
Reaction: Alkaline
Containers: Mild steel drums

Rewomid DO 280. Dutton & Reinisch Ltd.
Chemical Description: Oleic acid polydiethanolamide
Category: Detergent, nonionic
Applications: Wide range of liquid and paste products in the cosmetic, cleaning-agent, and detergent industries
Form: Liquid; Color: Amber; Odor: Typical (slight)
Properties: Sp. gr.: ≈ 1.0; Visc.: High
Composition: 100% active ingred.
Reaction: Alkaline

Rewomid DO 280/SE. Dutton & Reinisch Ltd.
Chemical Description: Oleic acid diethanolamide
Category: Detergent, nonionic
Applications: Recommended for cosmetic products for which improved skin-compatibility, foam quality, and viscosity characteristics are required
Form: Liquid; Color: Honey
Properties: Density: 0.96 g/cm^3 @ 20 C
Composition: ≈ 100% active ingred.

Rewomid IPE 280. Dutton & Reinisch Ltd.
Chemical Description: Oleic monoisopropanolamide
Category: Detergent, nonionic
Applications: Skin-protecting additive for liquid and paste products; in addition to its excellent superfatting properties, it helps to produce a fine-structured and creamy foam
Form: Semisolid; Color: Light yellow; Odor: Typical (slight)
Composition: 100% active ingred.
Reaction: Alkaline
Containers: Lacquered drums

Rewomid IPL 203. Dutton & Reinisch Ltd.
Chemical Description: Lauric monoisopropanolamide
Category: Detergent, nonionic
Applications: Additive for solid and paste end-products, whose properties are as follows: increased foam quality and stability; in-

creased viscosity; good superfatting; improved washing power; stabilization of emulsions
Form: Flakes; Color: White; Odor: Typical (slight)
Properties: M.P.: ≈ 55 C
Composition: 100% active ingred.
Reaction: Alkaline
Containers: Paper sacks

Rewomid IPP 240. Dutton & Reinisch Ltd.
Chemical Description: Coconut fatty acid monoisopropanolamide
Category: Detergent, nonionic
Applications: See Rewomid IPL 203
Form: Flakes; Color: White; Odor: Typical (slight)
Properties: M.P.: ≈ 50 C
Composition: 100% active ingred.
Reaction: Alkaline
Containers: Mild steel drums

Rewomid L203. Dutton & Reinisch Ltd.
Chemical Description: Lauric monoethanolamide
Category: Detergent, nonionic
Applications: Suitable for all types of solid, paste, and powder detergent preparations; improvement of foam texture, volume, and stability; increased viscosity; superfatting effect/skin protection; fixation of perfumes; stabilization of emulsions; improvement of detergency
Form: Flakes; Color: White; Odor: Typical (slight)
Properties: M.P.: ≈ 84 C
Composition: 100% active ingred.
Reaction: Alkaline
Containers: Paper sacks

Rewomid M 365. Dutton & Reinisch Ltd.
Chemical Description: Rolling- and hydraulic-oil additive
Category: Raw material, high-pressure additive, lubricant, corrosion inhibitor, nonionic
Applications: Used as an additive for water-soluble metal working liquids; for rolling- and hydraulic-oil bathes on polyglycolether derivatives; raw material for metal-working preparations, such as screw-thread cutting lubricants, ball-bearing polishes; widely used as a component in many rolling-oil formulations; compatible with anionic, cationic, amphoteric, and other nonionic products over a wide pH range
Form: Viscous liquid; Color: Amber
Properties: Solubility: Sol. in water; Cloud Pt.: 37–43 C (2% solids in dist. water); pH: 7–9 (1% solids); Biodegradable
Composition: ≈ 100% active ingred.

Rewomid RE. Dutton & Reinisch Ltd.
Chemical Description: Unadulterated combination of nonionic fatty acid blends
Category: Anticorrosion additive, emulsifier, nonionic
Applications: Additive for emulsi-

fiable metal-working lubricants such as drilling-, cutting-, and grinding oils; inhibits corrosion; good emulsifying power; mild to the skin; increases lubricity; low toxicity
Form: Liquid; Color: Dark brown
Properties: Density: 0.98 g/cm^3 @ 20 C; Biodegradable
Composition: 100% active ingred.

Rewomid S 280. Dutton & Reinisch Ltd.
Chemical Description: Stearic acid monoethanolamide
Category: Detergent, nonionic
Applications: Excellent superfatting effect; increased mildness to the skin; improved foam quality; increased viscosity; improved washing power; stabilization of emulsions
Form: Flakes; Color: White; Odor: Typical (slight)
Properties: M.P..: ≈ 92 C
Composition: 100% active ingred.
Reaction: Alkaline
Containers: Paper sacks

Rewomine IM-BT. Dutton & Reinisch Ltd.
Chemical Description: Imidazoline derivative
Category: Adhesive additive, corrosion inhibitor, emulsifier, cationic
Applications: Adhesive additive for bitumen; adhesive agent and emulsifier in normal acidic bitumen emulsions; rust inhibitor in acidic media
Form: Liquid; Color: Green-brown, 13 max. (Gardner); Odor: Amine
Properties: Density: 0.94 g/cm^3 @ 20 C
Composition: 100% active ingred.
Reaction: Alkaline
Toxicity: Avoid prolonged contact with skin
Containers: Iron containers

Rewomine IM-CA. Dutton & Reinisch Ltd.
Chemical Description: 1-Hydroxyethyl-2-heptadecenyl imidazoline
Category: Corrosion inhibitor, emulsifier, penetrant, wetting agent, additive, cationic
Applications: Leather industry—used in pickling baths as a corrosion inhibitor; metal-working industry—used as an additive for corrosion inhibiting oils, in cooling-fluids, and in cutting oils; small quantities will improve in the penetration and binding power of paint or dyes in the priming of wood and metal surfaces; wetting agent for carbonization baths; additive for the preparation of glass fiber
Form: Wax; Color: Yellow
Properties: m.w.: 285; Solubility: Sol. in most polar and nonpolar solvents; disp. in water; Stability: Stable in neutral and acidic media
Composition: 100% active ingred.
Reaction: Alkaline
Toxicity: Avoid prolonged contact with skin
Containers: Iron containers
Warnings and Caution: Should be

stored in air-tight containers as it is a tertiary amine and can react with carbon dioxide

Rewomine IM-OA. Dutton & Reinisch Ltd.
Chemical Description: 1-Hydroxyethyl-2-heptadecenyl imidazoline
Category: Rust inhibitor, adhesive agent, coupling agent, emulsifier, detergent, antistat, raw material, cationic
Applications: Can be used whenever inhibition of corrosion and/or an improved adhesive effect is required; in heating oils, fatty acid combination provides good rust prevention; used as a coupling agent where rust prevention is required; in pickling baths, used as an inhibitor to reduce attack on the steel or iron surface and to reduce hydrogen brittleness; excellent cationic emulsifier for the emulsification of oils and bitumens; acidic solutions can be prepared which are suitable for the formulations of metal cleaners, toilet bowl cleaners, rust removers; antistatic effect when dissolved in aqueous and alcoholic solutions; raw material for quaternization reactions; suitable for the flotation of phosphates, potassium, feldspar, iron, and other ores; used in the paint industry for pigment flushing and to improve coupling during priming of surfaces
Form: Liquid; Color: Amber
Properties: m.w.: 345; Solubility: Sol. in most polar and nonpolar solvents; disp. in water; Density: 0.94 g/cm^3 @ 20 C; Visc.: 800 cps; Stability: High thermal stability; pH: 10–11 (1% aq. sol'n.)
Composition: 100% active ingred.
Reaction: Alkaline
Toxicity: Avoid prolonged contact with skin
Containers: Iron containers
Warnings and Caution: This product will tend to crystallize at a rate dependent on length and temperature of storage

Rewominoxid B 204. Dutton & Reinisch Ltd.
Chemical Description: Alkylamido-propyldimethylamine oxide
Category: Foam booster, emulsifier, dispersant, softener, nonionic
Applications: Foam booster in wash-active formulations; calcium soap dispersant; used in shampoos; improves dermatological properties of other anionics
Form: Clear liquid; Color: Pale yellow; Odor: Almost odorless
Properties: pH: 5–7 (10% solids); Biodegradable
Composition: 35% min. solids
References: Folder: section 14/1277/1401

Rewopal C 6. Dutton & Reinisch Ltd.
Chemical Description: Coconut amide polyglycolether
Category: Dispersant, wetting agent, foaming agent, nonionic
Applications: Excellent calcium soap dispersing and wetting properties; used advantageously as a solvent for perfumes in shampoo

and foam-bath formulations; improves foam quality when used in combination with other surface-active raw materials; can be easily combined with all anionic, amphoteric, cationic, and nonionic materials; can be used for the formulation of spray- and immersion bath-cleaners
Form: Liquid; Color: Amber
Properties: Cloud Pt.: 80–90 C (2% in dist. water); Stability: Stable in dilute acid and alkaline sol'ns.; pH: 8–10 (1% solids); Biodegradable
Composition: 100% active ingred.

Rewopal HV 5. Dutton & Reinisch Ltd.
Chemical Description: Nonylphenol polyglycolether
Category: Emulsifier, raw material, wetting agent, degreasing agent, nonionic
Applications: Used as an emulsifier for preparations of all sorts of emulsions and insecticides; raw material for rinsing, washing, and cleaning agents of all types; wetting agent in the paper and cellulose industry; raw material for textile treatment; degreasing agent in pickling and alkaline immersion-bath cleaners
Form: Liquid; Color: 2 max. (Gardner)
Properties: Solubility: Sol. in mineral oil, alcohol, ketone, toluene, xylene, chlorinated hydrocarbons; Cloud Pt.: 59–62 C (10% in 25% aq. butyl diglycol); Stability: Stable in presence of dilute acids and alkalis, in hard water, and oxidizing agents; pH: 5–7 (1% solids)
Composition: 100% active ingred.

Rewopal HV 8. Dutton & Reinisch Ltd.
Chemical Description: Nonylphenol polyglycolether
Category: Emulsifier, raw material, wetting agent, degreasing agent, nonionic
Applications: See Rewopal HV 5
Form: Liquid; Color: 2 max. (Gardner)
Properties: Solubility: Sol. in mineral oil, alcohol, ketone, toluene, xylene, chlorinated hydrocarbons; Cloud Pt.: 41–45 C (10% in 25% aq. butyl diglycol); Stability: Stable in the presence of dilute acids and alkalis, in hard water, and oxidizing agents; pH: 5–7 (1% solids)
Composition: 100% active ingred.

Rewopal HV 9. Dutton & Reinisch Ltd.
Chemical Description: Nonylphenol polyglycolether
Category: Emulsifier, raw material, wetting agent, degreasing agent, nonionic
Applications: See Rewopal HV 5
Form: Liquid; Color: 2 max. (Gardner)
Properties: Solubility: Sol. in mineral oil, alcohol, ketone, toluene, xylene, chlorinated hydrocarbons; Cloud Pt.: 59–62 C (2% in dist. water); Stability: Stable in the presence of dilute acids and alkalis,

in hard water, and oxidizing agents; pH: 5–7 (1% solids)
Composition: 100% active ingred.

Rewopal HV 10. Dutton & Reinisch Ltd.
Chemical Description: Nonylphenol polyglycolether
Category: Emulsifier, raw material, wetting agent, degreasing agent, nonionic
Applications: See Rewopal HV 5
Form: Liquid; Color: 2 max. (Gardner)
Properties: Solubility: Sol. in mineral oil, alcohol, ketone, toluene, xylene, chlorinated hydrocarbons; Cloud Pt.: 70–73 C (2% in dist. water); Stability: Stable in the presence of dilute acids and alkalis, in hard water, and oxidizing agents; pH: 5–7 (1% solids)
Composition: 100% active ingred.

Rewopal HV 14. Dutton & Reinisch Ltd.
Chemical Description: Nonylphenol polyglycolether
Category: Emulsifier, raw material, wetting agent, degreasing agent, nonionic
Applications: See Rewopal HV 5
Form: Soft wax; Color: White
Properties: Solubility: Sol. in mineral oil, alcohol, ketone, toluene, xylene, chlorinated hydrocarbons; Cloud Pt.: 60–63 C (2% in 10% salt sol'n.); Stability: Stable in the presence of dilute acids and alkalis, in hard water, and oxidizing agents; pH: 5–7 (1% solids)
Composition: 100% active ingred.

Rewopal HV 25. Dutton & Reinisch Ltd.
Chemical Description: Nonylphenol polyglycolether
Category: Emulsifier, raw material, wetting agent, degreasing agent, nonionic
Applications: See Rewopal HV 5
Form: Wax; Color: White
Properties: Solubility: Sol. in mineral oil, alcohol, ketone, toluene, xylene, chlorinated hydrocarbons; Cloud Pt.: 73–76 C (2% in 10% salt sol'n.); Stability: Stable in the presence of dilute acids and alkalis, in hard water, and oxidizing agents; pH: 5–7 (1% solids)
Composition: 100% active ingred.

Rewopal LA 3. Dutton & Reinisch Ltd.
Chemical Description: Fatty alcohol polyglycolether
Category: Emulsifier, coupling agent, raw material, nonionic
Applications: Because of its low level of ethoxylation, this product has a definite hydrophobic character; ideal raw material for the production of various ether sulfates with excellent foaming properties; can be easily incorporated into sulfosuccinate-based shampoos to increase the viscosity and to provide a mild superfatting effect
Form: Liquid; Color: Clear
Properties: Solubility: Sol. in mineral oil and many organic solvents; Hydroxyl No.: 165–180; pH: 5–7 (1% solids); Biodegradable

Rewopal O 8. Dutton & Reinisch Ltd.
Chemical Description: Oleic acid amide polyglycolether
Category: Wetting and dispersing agent, emulsifier, detergent, nonionic
Applications: Excellent wetting and calcium soap dispersing agent; good oil-in-water emulsifier; can be combined with all anionic, cationic, amphoteric, and nonionic materials; suitable for machine washing formulations
Form: Liquid; Color: Brown
Properties: Solubility: Sol. in water; fairly sol. in alcohol, ketones, esters, chlorinated hydrocarbons, benzene, fatty oils; sparingly sol. in toluene, xylene; Cloud Pt.: 75–85 C; Stability: Stable in dilute acid and alkaline sol'ns.; pH: 8–10 (1% solids); Biodegradable
Composition: 100% active ingred.

Rewopal TA 11. Dutton & Reinisch Ltd.
Chemical Description: Fatty alcohol polyglycolether
Category: Wash-active base, nonionic
Applications: Has excellent wetting-, dispersing-, emulsifying-, cleaning-power
Form: Soft wax; Color: White
Properties: Solubility: Sol. in water and many organic solvents; Cloud Pt.: 69–73 C (2% in dist. water); Stability: Stable in dilute acids, alkalines, hard water sol'ns.; pH: 5–7 (1% solids); Biodegradable

Rewopal TA 25. Dutton & Reinisch Ltd.
Chemical Description: Fatty alcohol polyglycolether
Category: Wash-active base, nonionic
Applications: See Rewopal TA 11
Form: Powder; Color: White
Properties: Solubility: Sol. in water and many organic solvents; Cloud Pt.: 75–90 C (2% in 10% salt sol'n.); Stability: Stable in dilute acids, alkalines, hard water sol'ns.; pH: 5–7 (1% solids); Biodegradable

Rewopal TA 50. Dutton & Reinisch Ltd.
Chemical Description: Fatty alcohol polyglycolether
Form: Powder; Color: White
Properties: Cloud Pt.: 72–76 C (2% in 10% salt sol'n.); pH: 5–7 (1% solids); Biodegradable

Rewophat E 1027. Dutton & Reinisch Ltd.
Chemical Description: Alkylphenol polyglycolether phosphate
Category: Corrosion inhibitor, emulsifier, dispersant, wetting agent, anionic
Applications: Used in pure oil for cutting, rolling, and extruding metals; reduces friction, prevents wear, cleans, and has antistatic properties
Form: Liquid; Color: Yellow
Properties: Acid No.: 100 ± 5; Stability: Stable in neutral and alkaline solutions
Composition: 100% active ingred.

Rewophat EAK 8190. Dutton & Reinisch Ltd.
Chemical Description: Fatty alcohol polyglycolether phosphate
Category and Applications: See Rewophat E 1027
Form: Liquid; Color: Yellow
Properties: Solubility: Disp. in water; Acid No.: 150 ± 10; Stability: Stable in neutral and alkaline sol'ns.; Biodegradable
Composition: ≈ 100% active ingred.

Rewopol-Emulsifier BW. Dutton & Reinisch Ltd.
Chemical Description: Modified nonionic fatty acid derivative
Category: Emulsifier, nonionic
Applications: Forms an opalescent cleansing liquid with water; formulated with organic solvents to produce hand cleaners, cleaners for oily or greasy surfaces; mild to the skin
Form: Clear, viscous liquid; Color: Yellow
Properties: Solubility: Oil-sol.; Acid Value: 1 max.
Composition: 100% active ingred.
References: Folder: section 4/ 1277/0401

Rewopol-Emulsifier BWA. Dutton & Reinisch Ltd.
Chemical Description: Modified nonionic fatty acid derivative
Category: Emulsifier, nonionic
Applications: Metal cleaning formulations, e.g., engine cleaners, solvent degreasers; emulsifier for paraffin wax and for cold cleaners used to dewax cars; mild to the skin
Form: Clear, viscous liquid; Color: Yellow
Properties: Acid Value: 1 max.
Composition: 100% active ingred.
References: Folder: section 4/ 1277/0402

Rewopol FBR. Dutton & Reinisch Ltd.
Chemical Description: Detergent combination
Category: Detergent, raw material, corrosion inhibitor, anionic
Applications: Raw material for weakly alkaline household and general-purpose cleaners; concentrated cleaners; products for the removal of latex-based floor waxes and polishes; corrosion inhibitor; preparation of oil or solvent emulsifiers with water
Form: Clear, viscous liquid; Color: 3–6 (Gardner)
Properties: Density: ≈ 0.98 g/cm^3 (20 C)
Composition: 100% active ingred.
References: Folder: section 3/ 1277/0312

Rewopol FLM. Dutton & Reinisch Ltd.
Chemical Description: Blend of foaming agents
Category: Foam fire-extinguisher base, anionic
Applications: Low-density foam fire-extinguisher base; produces stable foam blanket
Form: Clear liquid; Color: Light yellow

Properties: Density: ≈ 1.0–1.1 g/cm^3 (20 C); Clear Pt.: −12 C; pH: 6.5–7.5 (1% solids)
Composition: 25% active ingred.
References: Folder: section 3/1277/0313

Rewopol M 5047. Dutton & Reinisch Ltd.
Chemical Description: Detergent combination
Category: Detergent, anionic
Applications: Preconcentrate for surface cleaners
Form: Clear liquid (may become turbid at low temps. or with long storage); Color: Colorless
Properties: pH: 9–10 (1% solids)
Composition: 45% min. solids

Rewopol NL, 3, 28%. Dutton & Reinisch Ltd.
Chemical Description: Sodium lauryl ether sulfate
Category: Detergent, foaming agent, anionic
Applications: Cleaning formulations; high quality shampoos and foam bath preparations
Form: Clear liquid; Color: Almost colorless; Odor: Low
Properties: Visc.: 50–150 cps; Stability: Very good temperature stability; pH: ≈ 7–8 (5% solids)
Composition: 28% active ingred.
References: Folder: section 2/1277/0201

Rewopol NLS 28. Dutton & Reinisch Ltd.
Chemical Description: Sodium lauryl sulfate
Category: Detergent, emulsifier, anionic
Applications: Raw materials for washing and cleaning agents in liquid, paste, or powder form; cosmetics, e.g., foam bath, shampoos; emulsifier in emulsion polymerization systems
Form: Liquid (may become turbid at low temps. and on long storage); Color: Light yellow
Properties: m.w.: 300; pH: 7–8 (1% solids)
Composition: 28% active ingred.
References: Folder: section 2/1277/0202

Rewopol NLS 90-Powder. Dutton & Reinisch Ltd.
Chemical Description: Sodium lauryl sulfate
Category and Applications: See Rewopol NLS 28
Form: Powder; Color: White to light yellow
Properties: m.w.: 300; pH: 7–9 (1% solids)
Composition: 85% min. active
References: Folder: section 2/1277/0202

Rewopol OFR. Dutton & Reinisch Ltd.
Category: Detergent
Applications: Surface cleaner, e.g., cleaning of synthetic car seats, furniture coverings, white-walled tires, baths, sinks, enameled surfaces, table tops, floors, mirrors, etc.
Form: Clear liquid; Color: Colorless
Properties: pH: 9–10 (1% solids)

Composition: 16% min. solids
References: Folder: section 3/ 1277/0316

Rewopol S 1590. Dutton & Reinisch Ltd.
Chemical Description: Detergent combination
Category: Detergent, anionic
Applications: Preconcentrate for preparation of dishwashing liquids
Form: Clear, viscous liquid; Color: Yellow
Properties: Stability: Temperature stable down to 3 C; pH: 6.0–7.0 (1% solids)
Composition: 44% solids
References: Folder: section 3/ 1277/0310

Rewopol S 1605. Dutton & Reinisch Ltd.
Chemical Description: Detergent combination
Category: Detergent, wetting agent, emulsifier, anionic
Applications: Raw material for liquid washing, rinsing, and cleaning agents in the household and in industry; has good wetting, emulsifying properties, foam stability, and mildness to skin
Form: Clear, viscous liquid; Color: Yellow
Properties: pH: 7–8 (1% solids)
Composition: 48% solids
References: Folder: section 3/ 1277/0311

Rewopol SB-DO 70. Dutton & Reinisch Ltd.
Chemical Description: Sodium diisooctyl sulfosuccinate
Category: Wetting agent, solubilizer, anionic
Applications: Cosmetic preparations, household cleaners, e.g., cleaning lotions, nail varnish removers, shampoos, foam baths, dishwashing agents, window and glass cleaners; in metal cleaning; as antismear agent for car windscreens; in paper industry; in textile industry as wetter for hydrophobic fibers; in paint industry; in pickling agents; in dye industry; in dry cleaning
Form: Liquid; Color: Almost colorless
Properties: Stability: Stable @ pH 3–9; Surface Tension: $<$30 dynes/ cm (0.1% sol'n. @ 20 C); pH: 6.5–7.5 (5% solids); Biodegradable
Composition: 64% min. active
References: Folder: section 1/ 1277/0108

Rewopol SBF 12. Dutton & Reinisch Ltd.
Chemical Description: Disodium lauryl alcohol sulfosuccinate
Category: Detergent, foaming aid, anionic
Applications: Mild detergent raw material with foam-improving properties; for cosmetics, carpet and upholstery shampoos (hardens the residue for easy removal), synthetic soap bars, etc.
Form: Paste; Color: White
Properties: pH: 6.5–7.5 (5% solids); Biodegradable
Composition: 39% solids
References: Folder: section 1/ 1277/0104

Rewopol SBR 12-Powder. Dutton & Reinisch Ltd.
Chemical Description: Disodium lauryl alcohol sulfosuccinate
Category: Detergent, anionic
Applications: Mild detergent for synthetic soaps, normal soap bars, dry shampoos, enzyme washing powders, moisture sensitive technical products; improves spray-drying properties of softer materials
Form: Powder; Color: White or light yellow
Properties: Solubility: Water-sol.; Density: ≈ 250 g/l (packing, drum-dried); Stability: Stable in weakly acid or weakly alkaline media; pH: 6.5–7.5 (5% solids); Biodegradable
Composition: 95% solids
References: Folder: section 1/1277/0105

Rewopol SBFA 30, 40%. Dutton & Reinisch Ltd.
Chemical Description: Disodium lauryl alcohol polyglycol-ether sulfosuccinate
Category: Detergent, anionic
Applications: Mild detergent raw material with good foaming for shampoos, foam bath products, intimate hygiene and baby preparations, washing-up liquids, handcleaners
Form: Clear liquid (may become turbid at low temps. or on long storage); Color: Almost colorless
Properties: Visc.: 100–200 cps; pH: 6.5–7.5 (5% solids); Biodegradable
Composition: 39% min. solids
References: Folder: section 1/1277/0101

Rewopol SBL 203. Dutton & Reinisch Ltd.
Chemical Description: Disodium lauric acid ethanolamide sulfosuccinate
Category: Detergent, anionic
Applications: Mild detergent for cosmetics; improves foam quality and produces milder effect on skin; aids spray-drying by increasing hardness of particles; used in carpet and upholstery shampoos; synthetic soaps
Form: Turbid liquid (tends to separate at room temp.); Color: Yellow
Properties: pH: 6.5–7.5 (5% solids); Biodegradable
Composition: 39% min. solids
References: Folder: section 1/1277/0102

Rewopol SBV. Dutton & Reinisch Ltd.
Chemical Description: Modified sulfosuccinate
Category: Detergent, anionic
Applications: Mild detergent with good foaming for preparation of mild, nonirritant shampoos and foam baths, rinsing and cleaning agents with skin-protecting properties
Form: Clear liquid; Color: Light yellow
Properties: Visc.: ≈ 2500 cps; pH: 6.5–7.5 (1% solids); Biodegradable

Composition: 28% min. solids
References: Folder: section 3/ 1277/0306

Rewopol SCK 2040. Dutton & Reinisch Ltd.
Chemical Description: Detergent combination
Category: Detergent, anionic
Applications: Raw material for liquid cleaning agents, shampoos, etc.
Form: Clear, viscous liquid; Color: Yellow
Properties: Visc.: ≈ 600 cps (room temp.); pH: 7–8.5 (1% solids); Biodegradable
Composition: 39% min. solids
References: Folder: section 3/ 1277/0314

Rewopol TS 25. Dutton & Reinisch Ltd.
Chemical Description: Sulfosuccinate-base detergent combination
Category: Detergent, foaming agent, anionic
Applications: Foam-cleaner concentrate for cleaning of coarse-fibered textiles, e.g., carpets, upholstery, car seat covers
Form: Liquid (may become turbid at low temps. or on long storage); Color: Light yellow
Properties: Clear Pt.: 8 C max.; pH: 6.5–7.5 (1% solids); Biodegradable
Composition: 25% solids
References: Folder: section 3/ 1277/0301

Rewopon AM-2C. Dutton & Reinisch Ltd.
Chemical Description: Coconut based ampholyte
Category: Detergent, dispersant, foam booster, wetting agent, amphoteric
Applications: Stabilizes the foam, and acts as a calcium soap dispersant when used in hard water; used in toiletries, mainly in mild shampoo and foam-bath preparations, and esp. in baby products; compatible with all anionic, cationic, and nonionic wash active substances
Form: Liquid; Color: Golden-yellow
Properties: Biodegradable
Composition: 49% active ingred.

Rewopon AM-2L. Dutton & Reinisch Ltd.
Chemical Description: Imidazoline-based ampholyte
Category: Detergent, amphoteric
Applications: Mild foam baths, shampoos, personal hygiene toiletries
Form: Liquid
Composition: 50% active ingred.

Rewopon AM-B 13. Dutton & Reinisch Ltd.
Chemical Description: Alkyl amido betaine
Category: Detergent, foam booster, amphoteric
Applications: Suitable for the formulation of shampoos, foam baths, and other toiletries; can be combined with anionic wash active

agents to give clear solutions; has a strong foam stabilizing effect even in the presence of oil and soap; compatible with all anionic, cationic, nonionic, and other amphoteric wash active substances
Form: Liquid; Color: Yellow
Properties: pH: 4–6 (10% solids); Biodegradable
Composition: 30% active ingred.

Rewopon AM-CA. Dutton & Reinisch Ltd.
Chemical Description: Coconut based ampholyte
Category: Detergent, foaming agent, amphoteric
Applications: Excellent foaming properties which are almost unaffected by the presence of salts and calcium soaps; highly suitable for the formulation of shampoos and for improving the soap compatibility of foam baths
Form: Clear liquid; Color: Yellow
Properties: pH: 8–8.5 (10% solids)
Composition: 30% active ingred.

Rewopon AM-V. Dutton & Reinisch Ltd.
Chemical Description: Short-chain fatty acid-based ampholyte
Category: Detergent, wetting agent, amphoteric
Applications: Exhibits excellent wetting properties in acid and in alkaline media; high tolerance towards the presence of large amounts of electrolytes; suitable for the preparation of clear stable cleaning agents with a high salt content; used as a wetting agent for all types of acidic and alkaline cleaners; reduces surface tension enabling more rapid and complete penetration of the electrolyte
Form: Liquid; Color: Dark amber
Properties: pH: 10–12 (10% solids)
Composition: 28% active ingred.

Reworyl-Sulfonic Acid K. Dutton & Reinisch Ltd.
Chemical Description: Dodecylbenzene sulfonic acid
Category: Detergent base, anionic
Applications: Can be neutralized with various bases for dishwashing, washing, and cleaning formulations
Form: Viscous liquid; Color: Brown
Properties: Acid Value: 180 ± 5
Composition: 97% min. active
References: Folder: section 5/1277/0501

Reworyl T 65. Dutton & Reinisch Ltd.
Chemical Description: p-Toluene sulfonic acid, 65%
Category: Catalyst, electrolyte, intermediate, anionic
Applications: Catalyst for the preparation of esters, for acid hydrolysis, polymerization, dehydration, and alkylation reactions; as an electrolyte in several electrolytic processes; intermediate for the preparation of cresol; leather tanning agent; in highly toxic crop-dusting agents; the production of synthetic fibers; in the vulcanization of buna-rubber; so-

lidification and hardening of phenol resins for foundry industry
Form: Clear liquid; Color: Pale; Odor: Characteristic
Properties: Acid Value: 224 min.
Composition: 65% active ingred.
Toxicity: Strong acid–skin irritant
Warnings and Caution: Wear protective goggles and gloves

Reworyl TKS 90/F. Dutton & Reinisch Ltd.
Chemical Description: Triammonium dodecylbenzenesulfonate
Category: Detergent, anionic
Applications: Detergent raw material for highly conc. cleansing agents, esp. for stream-jet and high pressure cleaners, defatting agents, car shampoos; also for chain greasing, and for cleaning steel ball bearings
Form: Viscous liquid (may become turbid at low temps. or on long storage); Color: Dark amber
Properties: pH: 7.0–7.5 (1% solids); Biodegradable
Composition: 85% min. active
References: Folder: section 5/1277/0504

Rexfoam 30 U. Graden Chem. Co.
Category: Defoamer
Applications: Defoamer in the manufacture of adhesives from synthetic latex systems; recommended for PVA, acrylic, and styrene butadiene emulsions
Form: Slightly hazy liquid
Properties: Solubility: Easily disp. in water; Density: 7.3 lb/gal
Composition: 100% active ingred.
Containers: 400 lb net drums

Rexfoam 40-U. Graden Chem. Co.
Category: Defoamer
Applications: Defoamer for latex paint systems, e.g., acrylic, butadiene-styrene, polyvinyl acetate
Form: Slightly hazy liquid
Properties: Density: 7.3 lb/gal
Composition: 100% active ingred.
Containers: 400 lb net drums
Warnings and Caution: Protect from freezing

Rexfoam 70-D. Graden Chem. Co.
Category: Defoamer
Applications: Emulsifiable defoamer and internal release aid for latex coatings
Form: Opaque liquid; Color: Cream
Properties: Density: 7.9 lb/gal (80 F); Visc.: 600 ± 100 cps (80 F); Flash Pt.: > 300 F; Pour Pt.: < 10 F; pH: 6.7 (5% disp.)
Containers: 400 lb net (55 gal) bung-type drums

Rexfoam 90-P. Graden Chem. Co.
Category: Defoamer
Applications: Defoamer for latex paints, esp. butadiene-styrene, acrylic, and polyvinyl acetate types; latex adhesives, and most aq. systems to reduce or eliminate foaming or frothing
Form: Opaque liquid; Color: Off-white
Properties: Solubility: Easily disp. in all latex paints and adhesive

systems; Density: 7.5 lb/gal
Containers: 400 lb net (55 gal) steel drums

Rexfoam 100 G. Graden Chem. Co.
Category: Defoamer
Applications: Defoamer in the manufacture of paints and adhesives from synthetic latex systems, esp. PVA, acrylic, and vinyl acrylic emulsions
Form: Opaque liquid; Color: Off-white
Properties: Density: 7.5 lb/gal; Pour Pt.: 0 F
Containers: 400 lb net (55 gal) steel drums

Rexfoam 2211. Graden Chem. Co.
Category: Defoamer
Applications: High efficiency foam control agent for aq. paint systems, esp. for polyvinyl acetate and acrylics
Form: Liquid; Color: Off-white
Properties: Solubility: Insol. in water; Density: 7.3 lb/gal; Pour Pt.: 0 F
Containers: 400 lb net (55 gal) steel drums

Rexfoam A. Graden Chem. Co.
Chemical Description: Fortified silicone defoamer
Category: Defoamer
Applications: Defoamer for specialized use; eliminates severe and difficult foam problems where ordinary defoamers are ineffective
Form: Creamy, translucent liquid
Properties: Solubility: Can be diluted with kerosene, toluene, Stoddard solvent, 2 ethyl hexanol, naphtha, and similar solvents; Density: 8.5 lb/gal (80 F); Visc.: 1330 cps; Flash Pt.: $>$ 300 F
Containers: 5 gal containers and 55 gal bung-type drums (425 lb net)

Rexol 25/4. Hart Chem. Ltd.
Chemical Description: Alkyl phenol ethoxylate
Category: Emulsifier, nonionic
Applications: Intermediate for anionic sulfonates
Form: Liquid; Color: Clear
Properties: Ref. Index: 1.5000; Solubility: Oil-sol.; Sp. gr.: 1.02; Visc.: 200 cps; HLB: 8.7
Composition: 100% active ingred.
Reaction: Neutral
Toxicity: Skin and eye irritant
Containers: Drums, T/T
Warnings and Caution: Avoid contact with skin and eyes

Rexol 25/6. Hart Chem. Ltd.
Chemical Description: Alkyl phenol ethoxylate
Category: Emulsifier, intermediate, dispersant, nonionic
Applications: Oil-sol. dispersant for petroleum oils; intermediate for anionic sulfates; antistatic plasticizer for plastics
Form: Liquid; Color: Clear
Properties: Ref. Index: 1.4909; Solubility: Oil-sol.; Sp. gr.: 1.04; Visc.: 250 cps; HLB: 10.8
Composition: 100% active ingred.
Reaction: Neutral

Toxicity: Skin and eye irritant
Containers: Drums, T/T
Warnings and Caution: Avoid contact with skin and eyes

Rexol 25/7. Hart Chem. Ltd.
Chemical Description: Alkyl phenol ethoxylate
Category: Emulsifier, nonionic
Form: Liquid; Color: Clear
Properties: HLB: 11.6
Composition: 100% active ingred.
Reaction: Neutral
Toxicity: Skin and eye irritant
Containers: Drums, T/T
Warnings and Caution: Avoid contact with skin and eyes

Rexol 25/8. Hart Chem. Ltd.
Chemical Description: Alkyl phenol ethoxylate
Category: Detergent, wetting agent, emulsifier, nonionic
Applications: Base surfactant for household and industrial detergents; low foaming; used for leather, paint, textile, pulp and paper industries
Form: Liquid; Color: Clear
Properties: Solubility: Clear sol. in water
Composition: 100% active ingred.
Reaction: Neutral
Toxicity: Skin and eye irritant
Containers: Drums, T/T
Warnings and Caution: Avoid contact with skin and eyes

Rexol 25/11. Hart Chem. Ltd.
Chemical Description: Alkyl phenol ethoxylate
Category: Detergent, wetting agent, nonionic
Applications: Scouring agents for natural and synthetic textile fibers
Form: Liquid; Color: Clear
Properties: Ref. Index: 1.4893; Solubility: Clear sol.; Sp. gr.: 1.06; Visc.: 314 cps; HLB: 13.8
Composition: 100% active ingred.
Reaction: Neutral
Toxicity: Skin and eye irritant
Containers: Drums, T/T
Warnings and Caution: Avoid contact with skin and eyes

Rexol 25/14. Hart Chem. Ltd.
Chemical Description: Alkyl phenol ethoxylate
Category: Detergent, wetting agent, nonionic
Applications: Detergent and wetting agent effective at elevated temperatures and in electrolyte systems; used for metal cleaning
Form: Liquid; Color: Clear
Properties: Ref. Index: 1.4848; Solubility: Clear sol.; Sp. gr.: 1.055; Visc.: 250 cps
Composition: 100% active ingred.
Reaction: Neutral
Toxicity: Skin and eye irritant
Containers: Drums, T/T
Warnings and Caution: Avoid contact with skin and eyes

Rexol 25/15. Hart Chem. Ltd.
Chemical Description: Alkyl phenol ethoxylate
Category: Detergent, wetting agent, nonionic
Applications: Hard surface detergency
Form: Paste; Color: Clear

Properties: Ref. Index: 1.4870; Solubility: Clear sol.; Sp. gr.: 1.06; Visc.: 400 cps (35 C); HLB: 14.8
Composition: 100% active ingred.
Reaction: Neutral
Toxicity: Skin and eye irritant
Containers: Drums, T/T
Warnings and Caution: Avoid contact with skin and eyes

Rexol 25/407. Hart Chem. Ltd.
Chemical Description: Alkyl phenol ethoxylate
Category: Detergent, wetting agent, emulsifier, nonionic
Applications: Stabilizer for synthetic latices; emulsifier for vinyl acetate and acrylate polymerization; demulsification of petroleum oils; pressurized textile scouring
Form: Liquid; Color: Yellow
Properties: Solubility: Highly water-sol.; Sp. gr.: 1.10
Composition: 70% active ingred.
Reaction: Neutral
Toxicity: Skin and eye irritant
Containers: Drums, T/T
Warnings and Caution: Avoid contact with skin and eyes

Rexol 25J. Hart Chem. Ltd.
Chemical Description: Alkyl phenol ethoxylate
Category: Detergent, wetting agent, emulsifier, nonionic
Applications: Detergent and wetting agent for textile, pulp and paper, leather, paint, and pesticide industries
Form: Liquid; Color: Clear
Properties: Ref. Index: 1.4823; Solubility: Clear sol.; Sp. gr.: 1.06; Visc.: 250 cps; HLB: 13.0
Composition: 100% active ingred.
Reaction: Neutral
Toxicity: Skin and eye irritant
Containers: Drums, T/T
Warnings and Caution: Avoid contact with skin and eyes

Rexol 25JM1. Hart Chem. Ltd.
Chemical Description: Alkyl phenol ethoxylate
Category: Detergent, wetting agent, emulsifier, nonionic
Applications: Kraft and sulfite digesters, screen rooms, bleach plants, and paper machines
Form: Liquid; Color: Yellow
Properties: Solubility: Clear sol.; Sp. gr.: 1.05; Visc.: 290 cps
Composition: 100% active ingred.
Reaction: Neutral
Toxicity: Skin and eye irritant
Containers: Drums, T/T
Warnings and Caution: Avoid contact with skin and eyes

Rexol 25JWC. Hart Chem. Ltd.
Chemical Description: Alkyl phenol ethoxylate
Category: Detergent, wetting agent, emulsifier, nonionic
Applications: Efficient for scouring of raw wool; foaming agent for Kuster dyeing
Form: Liquid; Color: Clear
Properties: Solubility: Clear sol.; HLB: 3.4
Composition: 100% active ingred.
Reaction: Neutral
Toxicity: Skin and eye irritant

Containers: Drums, T/T
Warnings and Caution: Avoid contact with skin and eyes

Rexol 35/100. Hart Chem. Ltd.
Chemical Description: Ethoxylated fatty alcohol
Category: Detergent, wetting agent, emulsifier, dispersant, nonionic
Applications: Wetting agent in the scouring of paper machine felts; preparation of industrial detergents
Form: Paste; Color: Clear
Properties: Solubility: Clear sol.; Sp. gr.: 1.003; Visc.: 100 cps
Composition: 100% active ingred.
Reaction: Neutral
Toxicity: Skin and eye irritant
Containers: Drums, T/L
Warnings and Caution: Avoid contact with skin and eyes

Rexol 130. Hart Chem. Ltd.
Chemical Description: Ethoxylated alcohol
Category: Emulsifier, nonionic
Applications: Dyeing assistant for vat, naphthal, direct, and basic dyestuffs or anionic/cationic dyestuff combination
Form: Liquid; Color: Clear
Properties: Ref. Index: 1.3850; Solubility: Clear sol.; Sp. gr.: 1.02; Visc.: 93 cps
Composition: 35% active ingred.
Reaction: Neutral
Toxicity: Skin and eye irritant
Containers: Drums, T/L
Warnings and Caution: Avoid contact with skin and eyes

Rexol 2000 HWM. Hart Chem. Ltd.
Chemical Description: Modified ethoxylated amide
Category: Emulsifier, nonionic
Applications: Antistatic emulsifier for manufacture of economical mineral oil-based textile lubricants used on natural and synthetic fibers
Form: Liquid; Color: Amber
Properties: Ref. Index: 1.4618; Sp. gr.: 0.982; Visc.: 300 cps
Composition: 77% active ingred.
Reaction: Alkaline
Toxicity: Skin and eye irritant
Containers: Drums, T/L
Warnings and Caution: Avoid contact with skin and eyes

Rexol AE-3. Hart Chem. Ltd.
Chemical Description: Linear alcohol ethoxylate
Category: Detergent, emulsifier, nonionic
Applications: Manufacture of ether sulfates; emulsifier for hydrocarbons
Form: Liquid; Color: Clear
Properties: Ref. Index: 1.4494; Solubility: Dispersible; Sp. gr.: 0.901; Visc.: 24 cps; HLB: 8.2
Composition: 100% active ingred.
Toxicity: Skin and eye irritant
Containers: Drums, T/T
Warnings and Caution: Avoid contact with skin and eyes

Rexol CCN. Hart Chem. Ltd.
Chemical Description: Castor oil ethoxylate
Category: Emulsifier, lubricant, nonionic

Applications: Synthetic fiber manufacturing
Form: Liquid; Color: Clear
Properties: Solubility: Clear sol.; Sp. gr.: 0.98; Visc.: 610 cps
Composition: 90% active ingred.
Reaction: Neutral
Toxicity: Skin and eye irritant
Containers: Drums
Warnings and Caution: Avoid contact with skin and eyes

Rexonic 6X. Hart Chem. Ltd.
Chemical Description: Linear alcohol ethoxylate
Category: Detergent, wetting agent, emulsifier, nonionic
Applications: Biodegradable wetting agent and detergent for the cleaning of paper machine wet felts
Form: Liquid; Color: Clear
Properties: Ref. Index: 1.3695; Solubility: Clear sol.; Sp. gr.: 1.01; Visc.: 400 cps
Composition: 30% active ingred.
Reaction: Neutral
Toxicity: Skin and eye irritant
Containers: Drums, T/T
Warnings and Caution: Avoid contact with skin and eyes

Rexonic 1012-6. Hart Chem. Ltd.
Chemical Description: Linear alcohol ethoxylate
Category: Detergent, wetting agent, emulsifier, nonionic
Applications: Biodegradable detergent and wetting agent for scouring of natural and synthetic textile fibers
Form: Liquid; Color: Clear
Properties: Ref. Index: 1.4532; Solubility: Clear sol.; Sp. gr.: 0.986; Visc.: 34 cps; HLB: 12; Biodegradable
Composition: 100% active ingred.
Reaction: Neutral
Toxicity: Skin and eye irritant
Containers: Drums, T/T
Warnings and Caution: Avoid contact with skin and eyes

Rexonic 1218-6. Hart Chem. Ltd.
Chemical Description: Linear alcohol ethoxylate
Category: Detergent, wetting agent, nonionic
Applications: Biodegradable detergent and wetting agent for synthetic and cellulosic textile fibers; detergent for soaping off dyed or printed goods
Form: Liquid; Color: Clear
Properties: Ref. Index: 1.4492; Solubility: Clear sol.; Sp. gr.: 1.002; Visc.: 140 cps; Biodegradable
Composition: 90% active ingred.
Reaction: Neutral
Toxicity: Skin and eye irritant
Containers: Drums, T/T
Warnings and Caution: Avoid contact with skin and eyes

Rexonic N-4. Hart Chem. Ltd.
Chemical Description: Linear alcohol ethoxylate
Category: Intermediate, nonionic
Applications: Preparation of sulfates that have particular value in high foaming liquid detergents; used for wetting, back wetting, sizing, emulsification, and dispers-

ing of petroleum oils
Form: Liquid; Color: Clear
Properties: Solubility: Oil-sol.; Sp. gr.: 0.92; Visc.: 50 cps; HLB: 8.0; Biodegradable
Composition: 100% active ingred.
Reaction: Alkaline
Toxicity: Skin and eye irritant
Containers: Drums, T/T
Warnings and Caution: Avoid contact with skin and eyes

Rexonic N6.6. Hart Chem. Ltd.
Chemical Description: Linear alcohol ethoxylate
Category: Detergent, wetting agent, emulsifier, nonionic
Applications: Formulation of high foaming liquid detergents; foam stabilizer
Form: Paste; Color: Clear
Properties: Solubility: Clear sol. in water; Sp. gr.: 0.98; HLB: 11.9; Biodegradable
Composition: 100% active ingred.
Reaction: Neutral
Toxicity: Skin and eye irritant
Containers: Drums, T/T
Warnings and Caution: Avoid contact with skin and eyes

Rexonic N7. Hart Chem. Ltd.
Chemical Description: Linear alcohol ethoxylate
Category: Detergent, nonionic
Applications: Heavy duty liquid detergents; retains high performance over wide range of temperature and water hardness
Form: Solid; Color: Clear
Properties: Solubility: Clear sol.; Sp. gr.: 1.001; HLB: 14.0
Composition: 100% active ingred.
Reaction: Neutral
Toxicity: Skin and eye irritant
Containers: Drums, T/T
Warnings and Caution: Avoid contact with skin and eyes

Rexonic N7.2. Hart Chem. Ltd.
Chemical Description: Linear alcohol ethoxylate
Category: Detergent, wetting agent, emulsifier, nonionic
Applications: Excellent hard surface detergent
Form: Solid; Color: Clear
Properties: Solubility: Clear sol.; Sp. gr.: 1.0
Composition: 100% active ingred.
Reaction: Neutral
Toxicity: Skin and eye irritant
Containers: Drums, T/T
Warnings and Caution: Avoid contact with skin and eyes

Rexonic N7.6. Hart Chem. Ltd.
Chemical Description: Linear alcohol ethoxylate
Category: Detergent, wetting agent, emulsifier, nonionic
Applications: Base surfactants for household and industrial detergents
Form: Solid; Color: Clear
Properties: Solubility: Clear sol.; HLB: 15.2
Composition: 100% active ingred.
Reaction: Neutral
Toxicity: Skin and eye irritant
Containers: Drums, T/T
Warnings and Caution: Avoid contact with skin and eyes

Rexonic N23-3. Hart Chem. Ltd.
Chemical Description: Linear alcohol ethoxylate
Category: Detergent, nonionic
Applications: Oil-sol. surfactant and chemical intermediate used in solvent emulsion cleaners and dry cleaning detergents
Form: Liquid; Color: Clear
Properties: Solubility: Oil-sol.; Sp. gr.: 0.93; Visc.: 19 cps
Composition: 100% active ingred.
Reaction: Neutral
Toxicity: Skin and eye irritant
Containers: Drums, T/T
Warnings and Caution: Avoid contact with skin and eyes

Rexonic N23-6.5. Hart Chem. Ltd.
Chemical Description: Linear alcohol ethoxylate
Category: Detergent, wetting agent, emulsifier, nonionic
Applications: Industrial maintenance, metal cleaning, paper, paint, and leather processing
Form: Liquid; Color: Clear
Properties: Solubility: Clear sol. in water, Sp. gr.: 0.96; Visc.: 30 cps; HLB: 11.2
Composition: 100% active ingred.
Reaction: Neutral
Toxicity: Skin and eye irritant
Containers: Drums, T/T
Warnings and Caution: Avoid contact with skin and eyes

Rexonic N25-7. Hart Chem. Ltd.
Chemical Description: Linear alcohol ethoxylate
Category: Detergent, wetting agent, emulsifier, nonionic
Applications: Formulation of high foaming detergents; good wetting agent
Form: Paste; Color: Clear
Properties: Solubility: Clear sol. in water; Sp. gr.: 0.958
Composition: 100% active ingred.
Reaction: Neutral
Toxicity: Skin and eye irritant
Containers: Drums, T/L
Warnings and Caution: Avoid contact with skin and eyes

Rexonic P-1. Hart Chem. Ltd.
Chemical Description: Ethoxylated, propoxylated linear alcohol
Category: Detergent, wetting agent, emulsifier, dispersant, nonionic
Applications: Where low or nonfoaming characteristics are required on both natural and synthetic fibers
Form: Liquid; Color: Clear
Properties: Ref. Index: 1.455; Solubility: Clear sol. in water; Sp. gr.: 0.97; Visc.: 90 cps; Biodegradable
Composition: 100% active ingred.
Reaction: Neutral
Toxicity: Skin and eye irritant
Containers: Drums, T/T
Warnings and Caution: Avoid contact with skin and eyes

Rexopal 3928. Hart Chem. Ltd.
Chemical Description: Modified linear alcohol ethoxylate
Category: Detergent, wetting agent, nonionic
Applications: Low foaming biodegradable scouring and wetting a-

gent for natural and synthetic fibers
Form: Liquid; Color: Yellow
Properties: Ref. Index: 1.4465; Solubility: Clear sol. in water; Sp. gr.: 1.0; Visc.: 45 cps
Composition: 85% active ingred.

Rexopal 3960. Hart Chem. Ltd.
Chemical Description: Modified linear alcohol ethoxylate
Category: Detergent, nonionic
Applications: Optimum detergency with minimum foam for prescouring of polyester yarns, knit, and cellulosics
Form: Liquid; Color: Yellow
Properties: Ref. Index: 1.4480; Solubility: Clear sol.; Sp. gr.: 0.977
Composition: 85% active ingred.
Reaction: Neutral
Toxicity: Skin and eye irritant
Containers: Drums, T/T
Warnings and Caution: Avoid contact with skin and eyes

Rexopal 4356. Hart Chem. Ltd.
Chemical Description: Modified linear alcohol ethoxylate
Category: Detergent, wetting agent, nonionic
Applications: Low foaming scouring and wetting agent; caustic compatible
Form: Liquid; Color: Clear
Properties: Solubility: Good dispersion, 37% caustic
Composition: 78% active ingred.
Reaction: Neutral
Toxicity: Skin and eye irritant
Containers: Drums, T/T
Warnings and Caution: Avoid contact with skin and eyes

Rexopal C52. Hart Chem. Ltd.
Chemical Description: Blend of sulfonates and linear alcohol ethoxylate
Category: Detergent, anionic/nonionic
Applications: Unique size removal properties obtained when prescouring woven fabrics
Form: Liquid; Color: Yellow
Properties: Ref. Index: 1.4058; Solubility: Cloudy; Sp. gr.: 1.0; Visc.: 32.5 cps
Composition: 40% active ingred.
Reaction: Neutral
Toxicity: Skin and eye irritant
Containers: Drums, T/T
Warnings and Caution: Avoid contact with skin and eyes

Rexopal EP1. Hart Chem. Ltd.
Chemical Description: Polyoxyalkylene
Category: Detergent, nonionic
Applications: Detergent for the prescouring of polyester and blends where minimum foam is required
Form: Liquid; Color: Clear
Properties: Solubility: Clear sol. in water; Sp. gr.: 0.990; Visc.: 112 cps
Composition: 50% active ingred.
Reaction: Neutral
Containers: Drums, T/T
Warnings and Caution: Avoid contact with skin and eyes

Rexopal NB7. Hart Chem. Ltd.
Chemical Description: Modified linear alcohol ethoxylate
Category: Detergent, nonionic
Applications: Nonfoaming detergent; increased efficiency obtained in one and two stage scouring and dyeing of polyester knits
Form: Liquid; Color: Clear
Properties: Solubility: Clear sol.; Sp. gr.: 0.978
Composition: 40% active ingred.
Reaction: Neutral
Toxicity: Skin and eye irritant
Containers: Drums, T/T
Warnings and Caution: Avoid contact with skin and eyes

Rexopal SN3. Hart Chem. Ltd.
Category: Detergent, emulsifier, anionic
Applications: Synergistic blend of solvent and surface active agents; exceptional oil emulsification properties offer high efficiency in prescouring synthetics
Form: Liquid; Color: White
Properties: Solubility: Clear; Sp. gr.: 1.08; Visc.: 2000 cps
Composition: 40% active ingred.
Reaction: Alkaline
Toxicity: Skin and eye irritant
Containers: Drums, T/T
Warnings and Caution: Avoid contact with skin and eyes

Ritachol. R.I.T.A. Corp.
Chemical Description: Derivative of cholesterol and pure lanolin-related sterols
Category: Emulsifier, moisturizer, emollient, lubricant, stabilizer, nonionic
Applications: Cosmetics and toiletries, e.g., cold cream, cleansing creams and lotions, antiperspirants, deodorants, make-up, moisturizing preparations, lip-related products, nail and cuticle formulations; stabilizes emulsions, dispersions, and suspensions; primary or auxiliary emulsifier; epidermal moisturizer, lubricant, and emollient
Form: Liquid; Color: Pale straw; Odor: Faint
Properties: Solubility: Miscible with mineral oil, isopropyl myristate and palmitate, natural and synthetic glycerides; Sp. gr.: 0.84–0.86; Sapon. No.: 2 max.; Acid No.: 1.0 max.; Iodine No.: 12 max.
Composition: 100% active ingred.

Ritaderm. R.I.T.A. Corp.
Category: Emollient, moisturizer, lubricant, nonionic
Applications: Cosmetics; dual-action natural moisturizer, lubricant, skin and hair conditioner in oil phase creams, gels, and lotions; substantive to skin
Form: Cream; Color: Yellow
Properties: M.P.: 42–48 C
Composition: 100% active ingred.

Ritalan C. R.I.T.A. Corp.
Chemical Description: Enriched essence of lanolin, the esterified lanolin sterol esters
Category: Emollient, solubilizer, nonionic
Applications: Blending agent,

emollient, and solubilizer for certain waxes and other oil-sol. or -disp. materials; shampoos, bubble baths, creams, lotions
Form: Liquid; Color: 2 max. (ASTM); Odor: Bland
Properties: Solubility: Sol. in mineral oil
Composition: 100% active ingred.

Robane. Robeco Chem. Inc.
Chemical Description: Special purified form of hexamethyltetracosane (squalane)
Category: Moisturizer, emollient, lubricant, humectant
Applications: Aids the spread of topical agents over the skin, increases skin respiration, prevents insensible water loss, imparts a suppleness to skin without a greasy feel; cosmetics and pharmaceuticals
Form: Liquid oil; Color: Colorless; Odor: Odorless; Taste: Tasteless

Robeyl. Robeco Chem. Inc.
Chemical Description: Squalene and hydrogenated shark liver oil
Category: Emollient
Applications: A natural polyunsaturated emollient; for high color gloss
Form: Clear liquid; Color: Almost colorless, very faint yellow; Taste: Bland, mild
Properties: Ref. Index: 1.4750–1.4850 (20 C); Sp. gr.: 0.8375–0.8495 (20 C); Visc.: 15–20 cps (20 C); Iodine Value: 235–260

Robuoy. Robeco Chem. Inc.
Chemical Description: Purified branched isoprenoid saturated paraffinic hydrocarbon of animal origin
Category: Lubricant, buoyant
Applications: An adjunct to oceanographic research and arctic lubrication
Form: Liquid oil; Color: Colorless; Odor: Faint
Properties: m.w.: 268.53; Ref. Index: 1.435–1.440 (20 C); Solubility: Miscible with vegetable and mineral oils, organic solvents, and lipophilic substances; Sp. gr.: 0.775–0.795 (20 C); Visc.: 6.4 cs; F.P.: ≈ –60 C; B.P.: ≈ 290 C; Flash Pt.: ≈ 175 C; Sapon. No.: 0–5; Acid No.: 0–5; Iodine No.: 0–7.5
Composition: 90% min. purity

Roccal 50% Technical. Hilton-Davis
Chemical Description: Alkyl dimethyl benzyl ammonium chloride
Category: Disinfectant, sanitizer, cationic
Applications: For general disinfection of floors, walls, showers, toilet seats; for sanitizing utensils and equipment; for general room sanitizing; for use in food or beverage processing plants, food service establishments, industrial plants, and farm use (not for use in hospitals)
Form: Liquid (may crystallize after prolonged storage at low temp.); Color: Light yellow; Odor: Mild
Properties: Solubility: Freely sol.

in water, lower alcohols, ketones, and glycols; Density: 8.2 lb/gal; Sp. gr.: 0.98; Stability: Excellent tolerance to hard water; pH: 6.8–8.0 (10% sol'n. in dist. water)
Composition: 50% active ingred.
Toxicity: Corrosive–severe eye and skin irritant
Warnings and Caution: Wear protective goggles and gloves; harmful or fatal if swallowed; avoid breathing spray or mist; avoid contamination of food
References: Tech. bulletin A-5-7/76

Roccal II 50%. Hilton-Davis
Chemical Description: Benzalkonium chloride USP
Category: Antimicrobial agent, disinfectant, sanitizer, cationic
Applications: See Roccal 50% Technical; Roccal II is noted for its high phenol coefficient and hard water tolerance; also as an antistatic agent for fabrics
Form: Liquid (may crystallize on prolonged storage at low temps.); Color: Light yellow; Odor: Mild, pleasant
Properties: m.w.: 359 (avg. of active); Solubility: Freely sol. in water, lower alcohols, ketones, and glycols; Density: 7.8 lb/gal; Sp. gr.: 0.94; pH: 6.8–8.0 (10% sol'n. in dist. water)
Composition: 50% active ingred.
Toxicity: Corrosive–severe eye and skin irritant
Warnings and Caution: Wear protective gloves and goggles; harmful or fatal if swallowed; avoid breathing spray or mist; avoid contamination of food
References: Tech. bulletin A-4-7/76

Roccal MC-14. Hilton-Davis
Chemical Description: Tetradecyl benzyl dimethyl ammonium chloride dihydrate
Category: Sanitizer, disinfectant, cationic
Applications: Formulated in liquid and solid chemical specialty products, e.g., detergent-sanitizers, germicidal laundry products, swimming pool algicides and slimicides, medicated cosmetics, and toiletries
Form: Free-flowing powder; Color: White
Properties: m.w.: 404; Solubility: 25% w/v in water; 75% w/v in 95% ethyl alcohol; also very sol. in lower ketones, alcohols, and glycols; M.P.: 60 C min.; pH: 7.5 ± 0.5 (10% sol'n. in dist. water)
Composition: 90% active ingred.
Warnings and Caution: Do not mix with strong oxidizing agents–potentially explosive
References: A-6-12/77

Romie 802. Tokaiseiyu Ind. Co., Ltd.
Chemical Description: Polyquaternary ammonium salt
Category: Softener, cationic
Applications: Softener for acrylic and other synthetic fibers
Form: Paste
Composition: 15% conc.

S

Sanac C. Armstrong Chem. Co.
Chemical Description: 2-(2-carboxyethoxy) ethyl 2-(2-(2-hydroxy) ethoxyethyl methyldodecylammonium methyl sulfate, potassium salt
Category: Detergent, wetting and foaming agent, solubilizer, amphoteric
Applications: Heavy duty surface cleaners, metal treatment
Form: Semiviscous liquid; Color: Straw to amber
Properties: Solubility: Completely sol. in water in all dilutions and in polar solvents; Density: 8.7 lb/gal; Sp. gr.: 1.05 (80 F); Stability: Aq. sol'ns. are stable to acids and alkalis over a pH range of 1–14; unaffected by hard water; pH: 7.5–8.5; Biodegradable
Composition: 35% active ingred.
Reaction: Compatible with acids and alkalis

Sanac S. Armstrong Chem. Co.
Chemical Description: [2-(2-carboxyethoxy) ethyl] [2-[2-(hydroxy) ethoxyethyl] methyloctadecenylammonium methyl sulfate, potassium salt
Category and Applications: See Sanac C
Form: Semiviscous liquid; Color: Straw to amber; Odor: Mild and typical
Properties: Solubility: Completely sol. in water in all dilutions; Density: 8.7 lb/gal; Sp. gr.: 1.05 (80 F); Stability: Unaffected by hard water; aq. sol'ns. are stable to acids and alkalis over a pH range of 1–14; pH: 7.5–8.5
Composition: 35% active ingred.
Reaction: Compatible with acids and alkalis

Sancowad AN Liquid. Sandoz Colors & Chem.
Category: Detergent, wetting agent, emulsifier, foaming agent
Applications: Foam producing agent for Sancowad dyeing process; disperse dyes on polyester; acid dyes on nylon; direct or reactive dyes on cotton; polyester/acrylic blends; wool dyeing with acid or reactive dyes

Form: Liquid; Color: Colorless
Properties: Solubility: Miscible with water; Sp. gr.: 1.02
Composition: 20–21% active
Reaction: Neutral

Sancowad CN Paste. Sandoz Colors & Chem.
Category: Detergent, wetting agent, emulsifier, foaming agent
Applications: Foam producing agent for Sancowad dyeing process; dyeing medium to deep shades on nylon with acid dyes and blends of nylon with Lycra and/or spandex; dyeing wool/nylon blends
Form: Paste; Color: Colorless
Properties: Solubility: Miscible with water; Sp. gr.: 1.06
Composition: 73% active ingred.
Reaction: Neutral

Sancowad NI Liquid. Sandoz Colors & Chem.
Category: Detergent, wetting agent, emulsifier, foaming agent, antiprecipitant
Applications: Foam producing agent for Sancowad dyeing process; dyeing acrylics with basic dyes; dyeing acrylic/nylon with basic/acid dyes where it acts as an antiprecipitant
Form: Liquid; Color: Colorless
Properties: Sp. gr.: 1.06
Composition: 30% active ingred.
Reaction: Neutral

Sandoclean MC Liquid. Sandoz Colors & Chem.
Category: Detergent, wetting agent, emulsifier, dispersing agent
Applications: Heavy duty cleaner for removal of tars and scale from all dyeing equipment; used in conjunction with caustic soda and hydrosulfite for removal of trimers, oils, dyestuff tars, etc. from all dyeing equipment
Form: Liquid; Color: Colorless
Properties: Solubility: Miscible with water; Visc.: 12 cps; F.P.: 14 F
Composition: 23% active ingred.

Sandogen C-AR Liquid. Sandoz Colors & Chem.
Category: Dyeing assistant
Applications: Dyeing nylon with acid dyes; leveling agent esp. for deep acid-dyeable nylon fibers to ensure a uniform dyeing
Form: Liquid

Sandogen C-CM Liquid. Sandoz Colors & Chem.
Category: Dye leveling agent, antiprecipitant
Applications: Antiprecipitant for dyeing multicolored shades on nylon carpet; prevents coprecipitation of acid and basic dyes; produces level dyeings; maintains maximum contrast on differential dyeing nylon
Form: Liquid; Color: Yellow
Properties: Solubility: Miscible with water; Sp. gr.: 1.04
Composition: 31% active ingred.
Reaction: Alkaline

Sandogen C-CS Liquid. Sandoz Colors & Chem.
Category: Dyeing assistant
Applications: Dyeing solid shades

on nylon carpet; level dyeing of solid shades on nylon carpet to assure homogeneous distribution of dyes from top to bottom (continuous and exhaust applications)
Form: Liquid; Color: Yellowish
Properties: Solubility: Miscible with water; Sp. gr.: 1.03
Composition: 17.5% active ingred.
Reaction: Alkaline

Sandogen CN Liquid. Sandoz Colors & Chem.
Category: Dyeing assistant
Applications: Low foam leveling and barré covering agent for nylon dyeing in Gaston County jets, beams, and packages; dyeing tufted carpets of regular or differential dyeing nylon fibers
Form: Liquid

Sandogen NH Liquid. Sandoz Colors & Chem.
Category: Dyeing assistant
Applications: Dyeing acid/premetallized dyes on nylon; leveling, migrating, and retarding acid dyes on nylon; used with Lyogen P for one step dyeing at the boil
Form: Liquid; Color: Yellow
Properties: Solubility: Miscible with water; Sp. gr.: 1.0
Composition: 45% active ingred.
Reaction: Neutral

Sandogen PY Liquid. Sandoz Colors & Chem.
Category: Dyeing assistant
Applications: Dyeing polyester fibers with disperse dyes; promotes leveling and migration of disperse dyes on polyester, allows shorter dye cycles, and reduces trimer deposits in jets, beams, becks, and package dyeing
Form: Liquid; Color: Yellow
Properties: Solubility: Miscible with water; Sp. gr.: 1.07
Composition: 26% active ingred.
Reaction: Neutral

Sandopan CBH Paste. Sandoz Colors & Chem.
Chemical Description: Sulfonate/modified ethylene oxide adduct
Category: Detergent, emulsifier, wetting agent, anionic
Applications: Continuous pretreatment of cotton and cotton blends where high electrolyte stability is required; concentrated reinforcing liquors in HT-bleaching ranges (Kleinewefers) in single stage (boil off and bleach in on stage) and two stage procedures where high caustic and silicate stability is needed
Form: Paste; Color: Yellow; Odor: Mild
Properties: Solubility: Miscible with water; F.P.: –4 F
Composition: 51% active ingred.
Reaction: Alkaline

Sandopan CPE Liquid. Sandoz Colors & Chem.
Category: Detergent, wetting agent
Applications: Preparation of polyester and polyester/cotton; desizing, neutral, and mildly alkaline scouring, caustic soda saturator (up to 2%); desizing and boiling

off cottons
Form: Liquid; Color: Yellowish; Odor: Mild
Properties: Solubility: Miscible with water; Sp. gr.: 1.05; Flash Pt.: 110 F; pH: 6.5–7.5
Composition: 62% active ingred.
Reaction: Neutral

Sandopan DTC. Sandoz Colors & Chem.
Chemical Description: Ethoxylated complex
Category: Detergent, wetting agent, anionic
Applications: Cosmetics, household, and specialty cleaner; dishwashing detergent, industrial, and institutional cleaners
Form: Gel; Color: Clear
Properties: Solubility: Sol. in water in all proportions; Sp. gr.: 1.06 (20 C); Visc.: 5000 cps; F.P.: 0 C; B.P.: 220 F; Stability: Very stable; Emulsifiability: Good
Composition: 68% active ingred.
Containers: 499 lb drums
References: Tech. bulletin 7-166

Sandopan DTC-100. Sandoz Colors & Chem.
Chemical Description: Ethoxylated complex
Category: Detergent, anionic
Applications: Household and specialty cleaner; textile, cement, industrial, leather applications
Form: Liquid; Color: Yellow
Properties: Solubility: Sol. in water in all proportions; Sp. gr.: 1.05 (20 C); Visc.: 128 cps; Stability: Stable to acids, alkali, reducing and oxidizing agents
Composition: 70% active ingred; other components: water and isopropanol
Reaction: Alkaline, neutral
References: Tech. bulletin 3-441/75

Sandopan DTC-100 Liquid. Sandoz Colors & Chem.
Chemical Description: Ethoxylated complex
Category: Detergent, wetting agent, emulsifier, anionic
Applications: Preparation of all fibers, esp. cotton and blends; stable to caustic; enzyme desizing; solvent emulsion scouring; alkaline boil-off to 8% caustic; scouring and bleaching of fibers on all open equipment; soaping-off dyeings
Form: Liquid; Color: Yellowish
Properties: Solubility: Miscible with Varsol, xylene, kerosene, water; Sp. gr.: 1.05; Flash Pt.: 110 F; pH: 7.0 ± 0.25
Composition: 69% active ingred.
Reaction: Neutral

Sandopan DTC-Acid. Sandoz Colors & Chem.
Chemical Description: Ethoxylated complex, free acid form
Category: Detergent
Applications: Household, specialty, and industrial cleaner; cosmetics products
Form: Liquid; Color: Light yellow
Properties: Solubility: Sol. in isopropanol, ethanol, acetone, Varsol, mineral spirits, petroleum

ether; Visc.: 225 cps
Composition: 90% active ingred.
Reaction: Acid

Sandopan DTC Gel. Sandoz Colors & Chem.
Chemical Description: Ethoxylated complex
Category: Detergent, wetting agent, emulsifier, anionic
Applications: Preparation of all fibers, esp. cotton and blends; stable to caustic; enzyme desizing; solvent emulsion scouring; alkaline boil-off up to 8% caustic; scouring and bleaching of fibers on all open equipment; soaping-off dyeings (vats, sulfurs)
Form: Paste; Color: Yellowish; Odor: Mild
Properties: Solubility: Miscible with Varsol, xylene, kerosene; miscible with water @ 100 F; Sp. gr.: 1.06
Composition: 73% active ingred.
Reaction: Neutral

Sandopan DTC Linear Gel. Sandoz Colors & Chem.
Chemical Description: Ethoxylated complex
Category: Detergent, wetting agent, emulsifier, anionic
Applications: Preparation of all fibers, esp. cotton and cotton blends; enzyme desizing; solvent emulsion scouring; alkaline boil-off (up to 8% caustic); scouring and bleaching of fibers on all open equipment; soaping-off of dyeings
Form: Gel; Color: Light yellow; Odor: Mild
Properties: Solubility: Clearly miscible with water and Varsol, xylene, and kerosene; Sp. gr.: 1.05 (20 C); Visc.: > 500 cps; F.P.: 0 C; B.P.: 220 F; Stability: Stable; pH: 7.0 ± 0.25; Biodegradable
Composition: 70% active ingred.
Reaction: Alkaline
Containers: 400 lb fiber drums

Sandopan DTC Linear Liquid. Sandoz Colors & Chem.
Chemical Description: Ethoxylated complex
Category: Detergent, wetting agent, emulsifier, anionic
Applications: Preparation of all fibers, esp. cotton and cotton blends; enzyme desizing; solvent emulsion scouring; alkaline boil-off (up to 8% caustic); scouring and bleaching of fibers in all open equipment; soaping-off dyeings
Form: Liquid; Odor: Mild
Properties: Solubility: Clearly miscible with water; Sp. gr.: 1.05; Biodegradable
Composition: 58% active ingred.

Sandopan DTC Linear Special. Sandoz Colors & Chem.
Chemical Description: Ethoxylated complex
Category: Detergent, wetting agent, emulsifier, anionic
Applications: See Sandopan DTC Linear Liquid
Form: Liquid; Color: Yellowish; Odor: Mild
Properties: Solubility: Miscible

with water; Sp. gr.: 1.05; Visc.: 60 cps; Biodegradable
Composition: 35% active ingred.
Reaction: Alkaline

Sandopan KD (U) Liquid. Sandoz Colors & Chem.
Chemical Description: Modified sodium alkyl sulfate
Category: Detergent, anionic
Applications: Light duty applications, e.g., prescouring and postscouring of various fibers; prescouring cotton and cotton/polyester knits; prescouring man-made fibers; prescouring and fulling of wool; postscouring various dyeings
Form: Liquid; Color: Clear; Odor: Mild
Properties: Solubility: Miscible with water; Sp. gr.: 1.02; Biodegradable
Composition: 21% active ingred.
Reaction: Neutral

Sandopan LF Liquid. Sandoz Colors & Chem.
Chemical Description: Ethylene oxide adduct
Category: Detergent, emulsifier, nonionic
Applications: Low foaming surfactant for use with Dilatin LB for one-bath scouring/dyeing of polyester knits under jet, pressure beck, or atmospheric; prescouring of polyester knits by batch or continuously
Form: Liquid; Color: Clear; Odor: Mild
Properties: Solubility: Miscible with water; Sp. gr.: 1.02; pH: 6.0
Composition: 25% active ingred.
Reaction: Slightly acid

Sandopan SD Liquid. Sandoz Colors & Chem.
Category: Detergent, wetting agent, emulsifier
Applications: Continuous preparation of cotton and cotton blends in saturator/J box, esp. where good Ca/Ng silicate dispersibility is needed; open width caustic boiloff; saturator/kier bleaching; desizing cottons
Form: Liquid; Color: Yellowish
Properties: Solubility: Miscible with water; F.P.: 14 F
Composition: 46–49% active
Reaction: Neutral

Sandopan TFL Conc. Sandoz Colors & Chem.
Chemical Description: Sulfonated fatty amide
Category: Detergent, amphoteric
Applications: Cosmetic and toiletries products, e.g., shampoos and bubble bath; also in leather industry
Form: Gel; Color: Amber; Odor: Mild fatty
Properties: Solubility: Sol. in water in all proportions; Sp. gr.: 1.14 (20 C); F.P.: > 0 C; B.P.: 220 F; Stability: Good
Composition: 48% conc.
Reaction: Alkaline
Containers: 375 lb fiber drums
References: Bulletin No. 7-212/74

Sandopan TFL Double Strength Liquid. Sandoz Colors & Chem.
Chemical Description: Fatty acid derivative
Category: Detergent, wetting agent, emulsifier, amphoteric
Applications: Dyeing nylon carpets with disperse dyes; dyeing acetate/triacetate with disperse dyes; lubricant and dispersant in boil-off and dyeing of synthetic fabrics
Form: Liquid; Color: Yellow; Odor: Mild
Properties: Solubility: Miscible with water
Reaction: Alkaline

Sandopan TFL Liquid. Sandoz Colors & Chem.
Chemical Description: Fatty acid derivative
Category: Detergent, wetting agent, emulsifier, amphoteric
Applications: Dyeing nylon carpets with disperse dyes; dyeing acetate/triacetate with disperse dyes; lubricant and dispersant in boil-off and dyeing of synthetic fabrics
Form: Liquid; Color: Yellow; Odor: Mild
Properties: Solubility: Miscible with water; Sp. gr.: 1.03; pH: 9.0
Composition: 11.5% active ingred.
Reaction: Alkaline

Sandopure BNA Liquid. Sandoz Colors & Chem.
Category: Detergent
Applications: Print washing for basic dye prints on acetate/nylon blends; removal of unfixed dye and print paste chemicals with minimum staining of white background for all basic dyed acetate/nylon
Form: Liquid; Color: Brownish
Properties: Solubility: Miscible with water; Sp. gr.: 1.10
Composition: 28% active ingred.
Reaction: Alkaline

Sandopure DK Liquid. Sandoz Colors & Chem.
Category: Detergent, wetting agent
Applications: Washing acid dye prints; leveling agent recommended for dyeing cellulose fibers with direct, vat, and sulfur dyes
Form: Liquid; Color: Amber
Properties: Solubility: Miscible with water; Sp. gr.: 1.02
Composition: 16.6% active ingred.
Reaction: Alkaline

Sandopure DK Paste. Sandoz Colors & Chem.
Category and Applications: See Sandopure DK Liquid
Form: Liquid; Color: Brownish
Properties: Solubility: Sol. in water; Sp. gr.: 1.06
Composition: 49.5% active ingred.
Reaction: Alkaline

Sandopure EC Liquid. Sandoz Colors & Chem.
Category: Detergent
Applications: Print washing of disperse dye or acid dye prints; washing disperse dye prints on acetate, triacetate, polyester, nylon, and blends; washing acid dye prints on

nylon; scouring of all synthetic fibers, esp. when heavily soiled or with high oil content
Form: Liquid; Color: Yellowish; Odor: Mild
Properties: Solubility: Miscible with water; Sp. gr.: 1.02
Reaction: Alkaline

Sandopure PW Liquid. Sandoz Colors & Chem.
Category: Detergent
Applications: Print washing of disperse dye and acid dye prints and synthetic fiber detergent; washing disperse dye prints on acetate, triacetate, polyester, nylon, and blends; washing acid dye prints on nylon; scouring of all synthetics, esp. when heavily soiled or with high oil content
Form: Liquid; Color: Amber; Odor: Mild
Properties: Solubility: Sol. in water; Sp. gr.: 1.04
Composition: 82% active ingred.
Reaction: Neutral

Sandopure SO Liquid. Sandoz Colors & Chem.
Category: Detergent
Applications: Soaping off prints, esp. vats, Rapidogens, and aniline black; after soaping sulfur dyeings
Form: Liquid; Color: Colorless
Properties: Solubility: Sol. in water in all proportions; Sp. gr.: 1.04
Composition: 18% active ingred.
Reaction: Alkaline

Sandozin C-PK Liquid. Sandoz Colors & Chem.
Category: Wetting agent
Applications: Used in Kusters or TAK units on nylon and polyester carpeting; foaming agent in continuous dyeing of polyester carpets in the Kusters with disperse dyes; used in carpet printing formulations
Form: Liquid; Color: Colorless; Odor: Mild
Properties: Solubility: Miscible with water; Sp. gr.: 1.02
Composition: 20–21% active
Reaction: Neutral

Sandozin C-WF Liquid. Sandoz Colors & Chem.
Category: Detergent, wetting and foaming agent
Applications: Continuous carpet dyeing; removal of spinning oil and yarn sizes; in dyeing differential dyeing nylon with no loss in substantivity
Form: Liquid; Color: Colorless
Properties: Sp. gr.: 1.06
Composition: 100% active ingred.
Reaction: Neutral

Sandozin D-100 Liquid. Sandoz Colors & Chem.
Category: Detergent, wetting agent
Applications: Detergent and wetting agent for acid, neutral, and alkaline medium; continuous scouring and scouring/brightening procedures for nylon and polyester; scouring synthetics and blends with cotton; wetting/scouring/fulling of wool
Form: Liquid; Color: Yellowish; Odor: Mild
Properties: Solubility: Miscible

with water; Sp. gr.: 1.06
Composition: 100% active ingred.
Reaction: Neutral

Sandozol KB. Sandoz Ltd.
Chemical Description: Sulfonated oil
Category: Anionic surfactant
Applications: Finishing and sanforizing oil
Form: Liquid

Sandozin N High Conc. Liquid. Sandoz Colors & Chem.
Category: Detergent, wetting agent, emulsifier, dispersant, lubricant
Applications: Wetting, dispersing, and lubricating agent for wetting out textiles prior to dyeing, esp. at low temps.; wetting agent prior to dyeing (Beck, jig, wetting of skeins, packages, and raw stock); emulsifier for trichlorobenzene carriers
Form: Liquid; Color: Brownish
Properties: Solubility: Miscible with water and trichlorobenzene; Sp. gr.: 1.07
Composition: 62% active ingred.
Reaction: Neutral

Sani Det. Coastal Industries
Category: Detergent
Applications: Cleaning, disinfecting, and deodorizing garbage storage areas, lavatories, operating rooms; cleaners of street, buildings, hospitals
Form: Liquid; Color: Water white; Odor: Mild
Properties: Solubility: Sol. in water in all proportions at ambient temp.; Density: 8.19 lb/gal; Sp. gr.: 0.981
Composition: 15% active ingred.
Reaction: Neutral
Toxicity: Eye irritant; drying to skin
Warnings and Caution: Avoid contact with eyes and prolonged contact with skin

Sanimal 250 C, 275 N, K 51. Nippon Nyukazai Co., Ltd.
Category: Emulsifier, nonionic/anionic
Applications: Used as emulsifier for sanitary pesticides

Santelle-EOM K. Durkee Industrial Foods Group
Chemical Description: Ethoxylated mono- and diglycerides
Category: Emulsifier, nonionic
Applications: Whipped toppings, frozen desserts, coffee whiteners; improves volume, stabilizes emulsion, aids dispersion
Form: Paste
Properties: M.P.: 80–85 F; HLB: 13.1
Composition: 100% active ingred.

Santone 3-1-S. Durkee Industrial Foods Group
Chemical Description: Triglycerol monostearate
Category: Emulsifier
Applications: Aeration of lipid systems; solubilization; color/flavor dispersion in water
Form: Bead; Color: Cream
Properties: M.P.: 128–130 F; HLB:

7.2; Sapon. No.: 122–139; Hydroxyl No.: 333–368; Iodine No.: 3.0 max.
Composition: 27–34% total mono- and diglycerides

Santone 10-10-O. Durkee Industrial Foods Group
Chemical Description: Polyglycerol esters of fatty acids
Category: Emulsifier
Applications: Coffee whiteners, frozen desserts; solubilization; emulsion stabilizer; aids dispersion; controlled agglomeration; flavor oil dispersions
Form: Liquid
Properties: HLB: < 1.0; Sapon. No.: 165–180; Hydroxyl No.: 25–45; Iodine No.: 65–80

Sarkosyl L. Ciba-Geigy
Chemical Description: Lauroyl sarcosine
Category: Detergent, corrosion inhibitor, foam booster and stabilizer, wetting agent, lubricant, emulsifier, anionic
Applications: Dentifrices, shampoos, aerosol products, antidandruff preparations, deodorants, pharmaceuticals (e.g., nasal and throat sprays, mouth washes, douches, bath oils, lotions, balms), detergents (e.g., alkali detergents, rug/upholstery shampoos, window cleaners, dishwashing, fine fabric, liquid hand cleaners, toilet soaps, scouring pads), metal processing and finishing (e.g., plating, solvent degreasing, metalworking and cutting oils)
Form: Powder
Properties: m.w.: 264–285; Solubility: Sol. in most organic solvents, incl. glycerin, glycols, silicones, phosphate esters, and aliphatic hydrocarbons; insol. in water; Sp. gr.: 0.969; M.P.: 35–37 C; Stability: Good at elevated temps.
Composition: 94% min. purity
Toxicity: Skin irritant on prolonged contact with conc. acid

Sarkosyl LC. Ciba-Geigy
Chemical Description: Cocoyl sarcosine
Category and Applications: See Sarkosyl L
Form: Powder
Properties: m.w.: 285–300; Solubility: Sol. in most organic solvents, incl. glycerin, glycols, silicones, phosphate esters, and aliphatic hydrocarbons; insol. in water; Sp. gr.: 0.969; M.P.: 23–28 C
Composition: 94% min. purity
Toxicity: Skin irritant on prolonged contact with conc. acid

Sarkosyl NL-30. Ciba-Geigy
Chemical Description: Sodium lauroyl sarcosinate
Category: See Sarkosyl L
Applications: Dentifrices, shampoos, rug and window cleaners, and fine fabric detergents
Form: Liquid; Color: Colorless
Properties: pH: 7.5–8.5 (1% sol'n.)
Composition: 30% active ingred.

Sarkosyl O. Ciba-Geigy
Chemical Description: Oleoyl sarcosine
Category and Applications: See Sarkosyl L
Form: Powder
Properties: m.w.: 340–360; Solubility: Sol. in most organic solvents, incl. glycerin, glycols, silicones, phosphate esters, and aliphatic hydrocarbons; insol. in water; Sp. gr.: 0.948
Composition: 94% min. purity
Toxicity: Skin irritant on prolonged contact with conc. acid

Schercamox C-AA. Scher Chem. Inc.
Chemical Description: Coco amido propyl dimethyl amine oxide
Category: Detergent, wetting agent, foaming agent, nonionic (in sol'ns. with pH < 7.0, cationic in nature)
Applications: Skin conditioner in shampoos, rinses, and shaving creams; provides antistatic properties; improves foam stability in dishwashing liquids
Form: Clear to slightly hazy liquid; Color: Water white
Properties: m.w.: 320 (avg.); Solubility: Sol. in water and most hydrophilic solvents incl. alcohols, glycols, triols, polyols, and glycol ethers; Sp. gr.: 0.986; pH: 7.0 ± 0.5; Biodegradable
Composition: 30% active ingred.
Reaction: Neutral
Containers: 55 gal poly-lined drums
References: Bulletin No. 501

Schercamox CMA. Scher Chem. Inc.
Chemical Description: Dihydroxyethyl cocoamine oxide
Category: Detergent, foam stabilizer, nonionic (in sol'ns. with pH < 7.0, cationic in nature)
Applications: Hair conditioner and foam stabilizer in shampoos; imparts emolliency, lubricity, viscosity, and increased foam to hair and body shampoos, bubble baths, shave creams, and soaps
Form: Clear liquid; Color: Water white; Odor: Mild
Properties: m.w.: 301 (avg.); Solubility: Sol. in water and most hydrophilic solvents, e.g., alcohols, glycols, triols, polyols, and glycol ethers; Sp. gr.: 0.99 ± 0.05; Stability: Good; pH: 7.0 ± 0.5 (1% sol'n.)
Composition: 40% active ingred.
Reaction: Neutral
Containers: 55 gal poly-lined drums
References: Bulletin No. 500

Schercamox DMA. Scher Chem. Inc.
Chemical Description: Alkyl dimethyl amine oxide
Category: Detergent, wetting agent, foam booster, nonionic (in sol'ns. with pH < 7.0, cationic in nature)
Applications: Wetting and foam booster in bubble baths; secondary surfactant in shampoos, bubble baths, and liquid dishwashing products to increase foaming and mildness

Form: Clear liquid; Color: Water white; Odor: Mild
Properties: m.w.: 256 (avg.); Solubility: Sol. in water and most hydrophilic solvents, e.g., alcohols, glycols, triols, polyols, and glycol ethers; Stability: Good; pH: 7.0 ± 1.0
Composition: 30% active ingred.
Containers: 55 gal poly-lined drums
References: Bulletin No. 506

Schercamox DML. Scher Chem. Inc.
Chemical Description: Dimethyl lauryl amine oxide
Category: Wetting agent, emulsifier, foaming agent, nonionic (in sol'ns. with pH < 7.0, cationic in nature)
Applications: Wetting agent and foamer in shampoos and bath preparations; shampoo surfactant to give foam boosting, viscosity, and substantivity to skin and hair for conditioning
Form: Clear liquid; Color: Water white; Odor: Mild
Properties: m.w.: 235 (avg.); Solubility: Sol. in water and most hydrophilic solvents, e.g., alcohols, glycols, triols, polyols, and glycol ethers; Sp. gr.: 0.99 ± 0.01; pH: 7.0 ± 1.0 (1% sol'n.); Stability: Good
Composition: 30% active ingred.
Reaction: Neutral
Containers: 55 gal poly-lined drums
References: Bulletin No. 502

Schercamox DMM. Scher Chem. Inc.
Chemical Description: Dimethyl myristyl amine oxide
Category: Detergent, wetting agent, emulsifier, antistatic agent, foaming agent, stabilizer, nonionic (in sol'ns. with pH < 7.0, cationic in nature)
Applications: Wetting and foaming agent in shampoos and bubble baths, liquid dishwashing products; mildly substantive to hair; also as viscosity builder in shampoos
Form: Clear liquid; Color: Water white; Odor: Mild
Properties: m.w.: 263 (avg.); Solubility: Sol. in water and most hydrophilic solvents, e.g., alcohols, glycols, triols, polyols, and glycol ethers; Sp. gr.: 0.98 ± 0.05; pH: 7.0 ± 1.0; Stability: Good
Composition: 30% active ingred.
Reaction: Neutral
Containers: 55 gal poly-lined drums
References: Bulletin No. 503

Schercamox DMS. Scher Chem. Inc.
Chemical Description: Stearic dimethyl amine oxide
Category: Detergent, softener, viscosity builder, foam stabilizer, skin emollient, hair conditioner, nonionic (in sol'ns. with pH < 7.0, cationic in nature)
Applications: Hair rinses, shampoos, lotions, cream rinses, shaving creams
Form: Paste; Color: White to off-

white; Odor: Typical
Properties: m.w.: 326 (avg.); Solubility: Sol. in most alcohols, glycols, triols, polyols, and glycol ethers; slightly sol. in water; Sp. gr.: 0.90; Stability: Good; pH: 7.0 ± 1.0 (1% sol'n.)
Composition: 25% active ingred.
Reaction: Neutral
Containers: 55 gal poly-lined drums
References: Bulletin No. 504

Schercomid CDO-Extra. Scher Chem. Inc.
Chemical Description: Cocoamide DEA
Category: Detergent, wetting agent, foam stabilizer, viscosity builder, nonionic/anionic
Applications: Shampoos, household cleaners, industrial cleaners with builders, liquid dishwashing compounds, rug and floor cleaners, personal care formulations
Form: Viscous liquid; Color: Yellow; Odor: Slight
Properties: Solubility: Sol. in water, alcohols, diols, triols, glycol ethers, polyols, aromatic and chlorinated hydrocarbons; partly sol. in mineral oil and natural fats; Sp. gr.: 1.00 ± 0.01; Visc.: 950 cps min.; Flash Pt.: > 170 C (OC); Acid Value: 4–5; pH: 10 ± 0.5 (10% sol'n.)
Composition: 100% active ingred.
Reaction: Alkaline
Containers: 55 gal poly-lined drums
References: Bulletin No. 317

Schercomid EAC. Scher Chem. Inc.
Chemical Description: Modified fatty amide
Category: Detergent, wetting agent, emulsifier, nonionic
Applications: Powerful detergent for scouring all textile fibers; most effective for scouring and acid fulling of wool; compatible with all scour bath additives, incl. alkalis and solvents
Form: Liquid; Color: Amber; Odor: Mildly ammoniacal
Properties: Solubility: Sol. in water over a wide temp. range; Density: 8.16 lb/gal (50 C); Sp. gr.: 0.98; pH: 9.8 ± 1.0 (1% sol'n.); Biodegradable
Composition: 100% active ingred.
Reaction: Alkaline
Containers: 55 gal steel drums
References: Bulletin No. 334

Schercomid LD. Scher Chem. Inc.
Chemical Description: Lauramide DEA
Category: Detergent, emulsifier, foam stabilizer and booster, viscosity builder, nonionic/anionic
Applications: Detergent formulations, hard surface cleaning formulations, shampoos, bubble baths, cosmetics; foam booster and stabilizer for all types of sulfate and sulfonate detergents
Form: Viscous liquid; Color: Yellow; Odor: Slight, typical
Properties: Solubility: Sol. in water and most organic solvents, e.g., alcohols, glycols, glycol ethers, polyols, aromatic and chlorinated hydrocarbons; disp. in aliphatic

hydrocarbons, mineral oil, and natural fats; Density: 8.4 lb/gal; Sp. gr.: 1.01 ± 0.01; Visc.: 1000 cps min.; Flash Pt.: > 170 C (OC); Acid Value: 20–26; Stability: Good; pH: 9.7 ± 0.5 (10% sol'n.)
Composition: 100% active ingred.
Reaction: Alkaline
Containers: 55 gal steel drums
References: Bulletin No. 320

Schercomid ODA. Scher Chem. Inc.
Chemical Description: Oleamide DEA
Category: Detergent, emulsifier, viscosity builder, dispersant, foam booster and stabilizer, nonionic/anionic
Applications: Detergent in gel-type industrial pine cleaners; W/O emulsifier for all types of aromatic and aliphatic hydrocarbon solvents, esp. those of high molecular weight; dispersant for mineral clays or pigments; shampoo formulations; substantive hair conditioning agent
Form: Liquid; Color: Amber; Odor: Mild, characteristic
Properties: Solubility: Sol. in most alcohols, glycols, glycol ethers, aliphatic and chlorinated hydrocarbons; partly sol. in aromatic hydrocarbons; disp. in water; Density: 7.9 lb/gal; Sp. gr.: 0.950 ± 0.01; Visc.: 1200 cps min.; F.P.: 10 C max.; Flash Pt.: > 170 C (OC); Acid Value: 12–16; pH: 9.9 ± 0.5 (10% disp.)
Composition: 100% active ingred.
Reaction: Alkaline
Containers: 55 gal steel drums
References: Bulletin No. 331

Schercomid SLE. Scher Chem. Inc.
Chemical Description: Linoleamide DEA
Category: Emulsifier, viscosity builder, stabilizer, nonionic
Applications: Good viscosity builder for shampoos; low temp. cloud pt. shampoos; conditioning effect on hair; W/O emulsifier; emulsion stabilizer
Form: Clear liquid; Color: Amber; Odor: Mild fruity
Properties: Solubility: Sol. in most organic solvents, e.g., alcohols, glycols, glycol ethers, glycerol, aliphatic, aromatic, and chlorinated hydrocarbons, mineral and natural fats; disp. in water; Density: 8.0 lb/gal; Sp. gr.: 0.960; Flash Pt.: > 180 C (OC); Acid Value: 1.0; pH: 10.2 ± 0.5 (10% disp.)
Composition: 100% active ingred.
Reaction: Alkaline
Containers: 55 gal steel drums
References: Bulletin No. 303

Schercomid SL-Extra. Scher Chem. Inc.
Chemical Description: Lauramide DEA
Category: Detergent, wetting agent, foam stabilizer and booster, thickener, emulsifier, nonionic
Applications: Gel, paste, and cream shampoos, shampoo concentrates, bubble bath formulations, manual dishwashing formulations, liquid heavy duty laundry detergents, liquid soaps, all-purpose cleaning products; emulsifier for aliphatic, aromatic hydrocarbons and oils for O/W emulsions
Form: Crystalline solid; Color:

White; Odor: Mild fruity
Properties: Solubility: Sol. in most alcohols, glycols, glycol ethers, polyols, aliphatic (lower members), aromatic, and chlorinated hydrocarbons, and natural fats; disp. in water and mineral oils; Density: 8.1 lb/gal (45 C); Sp. gr.: 0.97 ± 0.01 (45 C); Visc.: 1800 cps min. (10% aq. disp.); M.P.: 42 ± 3 C; Acid Value: 1.0 max.; pH: 10.2 ± 0.5 (10% aq. disp.)
Composition: 100% active ingred.
Reaction: Alkaline
Containers: 55 gal steel drums
References: Bulletin No. 301

Schercomid SO-A. Scher Chem. Inc.
Chemical Description: 1:1 Oleic acid diethanolamide
Category: Emulsifier, nonionic/anionic
Applications: W/O emulsifier; couples water into solvents to reduce flammability
Form: Clear liquid; Color: Light amber; Odor: Slight, typical oleic
Properties: Solubility: Sol. in most organic solvents, e.g., alcohols, glycols, glycol ethers, glycerol, aliphatic, aromatic, and chlorinated hydrocarbons, mineral oil, and natural fats; disp. in water; Density: 7.9 lb/gal; Sp. gr.: 0.95 ± 0.01; Visc.: 450 cps min.; Acid Value: 8.0 max.; pH: 9.8 ± 0.5 (10% disp.)
Composition: 100% active ingred.
Reaction: Alkaline
Containers: 55 gal bung head drums
References: Bulletin No. 307

Schercopol CMIS-Na. Scher Chem. Inc.
Chemical Description: Disodium monolauramido MIPA-sulfosuccinate
Category: Detergent, foaming agent, nonionic
Applications: Low irritating shampoo ingredient; mild detergent for shampoos, bubble baths, rug and upholstery cleaners, light duty detergents
Form: Clear liquid; Color: Yellow; Odor: Mild, characteristic
Properties: m.w.: 491 (avg.); Solubility: Complete in water; Density: 9.4 lb/gal; Sp. gr.: 1.12 ± 0.01; Visc.: 400 cps max.; Cloud Pt.: 1.0 C max.; pH: 5–7
Composition: 40% active ingred.
Reaction: Neutral
Containers: 55 gal poly-lined drums
References: Bulletin No. 701

Schercopol CMS-Na. Scher Chem. Inc.
Chemical Description: Disodium monococoamido sulfosuccinate
Category: Detergent, foaming agent, softener, anionic
Applications: Primary surfactant in hair and skin care products, e.g., shampoos, bath oils, bubble baths, skin cleansers
Form: Clear liquid; Color: Yellow; Odor: Mild, characteristic

Properties: m.w.: 471 (avg.); Solubility: Complete in water; Density: 9.3 lb/gal; Sp. gr.: 1.12 ± 0.01; Visc.: 100 cps max.; Cloud Pt.: 5.0 C; pH: 5–7
Composition: 30% active ingred.
Reaction: Neutral
Containers: 55 gal poly-lined drums
References: Bulletin No. 702

Schercopol OMIS-Na. Scher Chem. Inc.
Chemical Description: Disodium monooleamido MIPA sulfosuccinate
Category: Detergent, foaming agent, anionic
Applications: Primary surfactant in nonsting baby shampoo formulations, nonirritating baby and adult shampoos; leaves soft, elegant feel on hair and skin
Form: Clear liquid; Color: Yellow; Odor: Mild, characteristic
Properties: Solubility: Complete in water; Density: 9.16 lb/gal; Sp. gr.: 1.1 ± 0.05; Visc.: 500 cps max.; Cloud Pt.: 5.0 C max.; pH: 6.0 ± 1.0
Composition: 35% active ingred.
Reaction: Neutral
Containers: 55 gal poly-lined drums
References: Bulletin No. 703

Schercopol OMS-Na. Scher Chem. Inc.
Chemical Description: Disodium monooleamido MEA sulfosuccinate
Category: Detergent, foaming agent, anionic
Applications: Nonirritating detergent in bath oils, body soaps, shampoos, dishwashing detergents; imparts conditioning effect on hair, soft emollient feel on skin
Form: Clear liquid; Color: Yellow; Odor: Mild, characteristic
Properties: Solubility: Complete in water; Density: 9.16 lb/gal; Sp. gr.: 1.10 ± 0.05; Visc.: 500 cps max.; Cloud Pt.: 5.0 C max.; pH: 6.0 ± 1.0
Composition: 35% active ingred.
Reaction: Neutral
Containers: 55 gal poly-lined drums
References: Bulletin No. 700

Schercotaine CAB. Scher Chem. Inc.
Chemical Description: Cocamidopropyl betaine
Category: Detergent, wetting agent, foaming agent, antistat, amphoteric
Applications: Detergent in mild shampoos, skin cleansing compositions, bath products; softening effect on skin and hair; antistat in shampoos
Form: Clear liquid; Color: Pale yellow; Odor: Slight, characteristic
Properties: m.w.: 386 (avg.); Solubility: Complete in water; Density: 8.75 lb/gal; Sp. gr.: 1.05 ± 0.01; Visc.: 100 cps max.; Cloud Pt.: –2.0 C max.; Stability: Stable over wide pH range; compatible in high electrolyte formulations; pH: 5.5 ± 1.0; Biodegradable

Composition: 30% active ingred.
Reaction: Neutral
Containers: 55 gal poly-lined drums
References: Bulletin No. 901

Schercotaine SCAB. Scher Chem. Inc.
Chemical Description: Cocamidopropyl hydroxy sultane
Category: Detergent, wetting agent, foaming agent, cloud pt. depressant, amphoteric
Applications: Counter-irritant in mild shampoos; very low temp. cloud pt. and mild detergent for use in shampoos, bubble baths, and skin cleansers; good foaming in hard or soft water
Form: Clear liquid; Color: Light lemon; Odor: Slight, typical
Properties: m.w.: 500 (avg.); Solubility: Complete in water; Density: 8.9 lb/gal; Sp. gr.: 1.07 ± 0.01; Visc.: 100 cps max.; Cloud Pt.: –7 C; Stability: Stable over wide pH range; effective in acid or alkaline media; pH: 6.0 ± 1.0; Biodegradable
Composition: 35% active ingred.
Reaction: Neutral
Containers: 55 gal poly-lined drums
References: Bulletin No. 900

Schercoterge 140. Scher Chem. Inc.
Chemical Description: Modified, ethoxylated fatty amide
Category: Detergent, wetting agent, emulsifier, nonionic
Applications: General textile scouring agent; wool fulling and scouring agent; dyeing assistant; post dye scouring agent; for use in any textile wet scouring operation on jig, dye beck, dye box, and continuous equipment; dye bath stabilizer
Form: Liquid; Color: Lemon; Odor: Mild ammoniacal
Properties: Solubility: Sol. in hot or cold water; Density: 8.3 lb/gal; Sp. gr.: 1.0; Stability: Stable in acid and alkaline media; pH: 9.5 ± 0.5 (1% sol'n.); Biodegradable
Composition: 98% active ingred.
Reaction: Alkaline
Containers: 55 gal lined steel drum or tank wagon
References: Bulletin No. 1111

Schercoteric MS-SF-2 (38%). Scher Chem. Inc.
Chemical Description: Salt free coconut dicarboxylated imidazolinium amphoteric
Category: Detergent, foaming agent, amphoteric
Applications: Low irritation shampoo base; reduces irritantcy of other ingredients
Form: Clear liquid; Color: Yellow; Odor: Slight, typical
Properties: Sp. gr.: 1.07; Cloud Pt.: –8 C; pH: 9.5 ± 0.5 (50%)
Composition: 38% active ingred.
References: Bulletin No. 100

Schercoteric MS-SF-2 (Conc.-70%). Scher Chem. Inc.
Chemical Description: Salt free coconut dicarboxylated imidazolinium amphoteric
Category and Applications: See

Schercoteric MS-SF-2 (38%)
Form: Pourable gel; Color: Light amber; Odor: Slight, typical
Properties: Solubility: Sol. in water; Sp. gr.: 1.10; Stability: Good; pH: 9.5 ± 0.5 (50%)
Composition: 70% active ingred.
Reaction: Neutral
Containers: 55 gal steel drums
References: Bulletin No. 100

Schercoteric MS-SF 38%. Scher Chem. Inc.
Chemical Description: Coco imidazolinium derivative monocarboxylate
Category: Detergent, amphoteric
Applications: Salt free, low irritating shampoo base; mild dishwashing and industrial cleaners
Form: Clear liquid; Color: Yellow; Odor: Mild, typical
Properties: m.w.: 406 (theoretical); Density: 8.7 lb/gal; Sp. gr.: 1.05; pH: 8–10 (as is)
Composition: 38% active ingred.
References: Bulletin No. 109

Schercoteric MS-SF Conc.-70. Scher Chem. Inc.
Chemical Description: Coco imidazolinium derivative monocarboxylate
Category: Detergent, amphoteric
Applications: Salt free, low irritating shampoo base; mild dishwashing and industrial cleaners
Form: Soft gel; Color: Light amber; Odor: Mild, typical
Properties: m.w.: 406 (theoretical); Solubility: Complete in water; Density: 9.1 lb/gal; Sp. gr.: 1.1; Stability: Stable and effective over wide pH range; pH: 8–10 (as is)
Composition: 70% active ingred.
Reaction: Neutral
Containers: 55 gal steel drums
References: Bulletin No. 109

Scherpol LSB. Scher Chem. Inc.
Category: Detergent, dyeing assistant, nonionic
Applications: Nonfoaming scouring and dyeing assistant for polyester; removes excess dye carrier, oils, and loose color from dyed polyester fabric thus reducing smoke formation during drying cycle
Form: Liquid; Color: Water white; Odor: Odorless
Properties: Solubility: Water-sol.; Sp. gr.: 0.9; Stability: Good
Composition: 70% active ingred.
Reaction: Neutral

SCMS. Hilton-Davis
Chemical Description: Sodium coconut monoglyceride sulfonate
Category: Wetting agent, dispersant, emulsifier, detergent, foaming agent, anionic
Applications: Water-base systems; shampoos, toothpaste, etc.
Form: Free-flowing powder; Color: White; Odor: Mild, characteristic
Properties: m.w.: 366 (nominal); Solubility: Very low water solubility: HLB: ≈ 40; pH: 6–8 (aq. slurry); Biodegradable
Composition: 93% min. active

Scourtex 1063. Hart Chem. Ltd.
Chemical Description: Fatty alkanolamide
Category: Scouring agent, anionic
Applications: Alkaline fulling agent, rinses freely at lower temps., and gives fabrics softness and lubricity
Form: Liquid; Color: Amber
Properties: Ref. Index: 1.4794; Solubility: Disp. in water; Sp. gr.: 1.012; Visc.: 700 cps; Biodegradable
Composition: 92% active ingred.
Reaction: Alkaline
Toxicity: Skin and eye irritant
Containers: Drums, T/T
Warnings and Caution: Avoid contact with skin and eyes

Scripset 520. Monsanto
Chemical Description: Copolymer of styrene:maleic anhydride
Category: Dispersant, emulsifier, binder, sizing agent, viscosity modifier, stabilizer
Applications: Starch modifier, pigment dispersant, protective colloid, sizing, coating, water-paint calsomines, adhesives, printing, preparation of emulsified paints
Form: Free-flowing fine powder; Color: Off-white; Odor: Faint, aromatic
Properties: m.w.: ≈ 50,000; Solubility: Sol. in alkaline aq. systems, incl. ammonium hydroxide, sodium hydroxide, soda ash, sodium borate, amines, and in acetone; Stability: Good sol'n. stability under suitable conditions
Containers: 50 lb bags
Warnings and Caution: Avoid prolonged contact with skin; avoid breathing in dust
References: Product information sheet SC-2

Scripset 540. Monsanto
Chemical Description: Copolymer of styrene:maleic anhydride
Category and Applications: See Scripset 520
Form: Free-flowing fine powder; Color: Off-white; Odor: Faint, aromatic
Properties: m.w.: ≈ 20,000; Solubility: Sol. in alkaline aq. systems, and in many organic solvents incl. alcohols, ketones, ethyl acetate, and cosolvents (xylene/butanol); Stability: Good sol'n. stability under suitable conditions
Containers: 50 lb bags
Warnings and Caution: Avoid prolonged contact with skin; avoid breathing in dust
References: Product information sheet SC-3

Scripset 550. Monsanto
Chemical Description: Copolymer of styrene:maleic anhydride
Category and Applications: See Scripset 520
Form: Free-flowing fine powder; Color: Off-white; Odor: Faint, aromatic
Properties: m.w.: ≈ 10,000; Solubility: Sol. in alkaline aq. systems, and many organic solvents incl. alcohols, ketones, ethyl acetate, and cosolvents (xylene/butanol); Stability: Good sol'n. stabili-

ty under suitable conditions
Containers: 50 lb bags
Warnings and Caution: Avoid prolonged contact with skin; avoid breathing in dust
References: Product information sheet SC-4

Scripset 700. Monsanto
Chemical Description: 30% aq. sol'n. of a styrene:maleic anhydride copolymer
Category and Applications: See Scripset 520
Form: Liquid; Odor: Faint, aromatic
Properties: m.w.: ≈ 50,000; Solubility: Easily diluted with water; Density: 9.5 lb/gal; Stability: Good under suitable conditions (maintain 7.5 pH)
Composition: 30% active ingred.
Containers: 55 gal drums and bulk
Warnings and Caution: Avoid prolonged contact with skin
References: Product information sheet SC-5

Scripset 720. Monsanto
Chemical Description: 25% aq. sol'n. of a styrene:maleic anhydride copolymer
Category and Applications: See Scripset 520
Form: Liquid; Odor: Faint, aromatic
Properties: m.w.: ≈ 50,000; Solubility: Easily diluted with water; Stability: Good sol'n. stability under suitable conditions (maintain 8.0 pH)
Composition: 25% active ingred.
Containers: 55 gal drums and bulk
Warnings and Caution: Avoid prolonged contact with skin
References: Product information sheet SC-6

Sellogen 641. Diamond Shamrock/Process Chem. Div.
Chemical Description: Alkyloxyalkyl sodium sulfate
Category: Detergent, dispersant, emulsifier, anionic
Applications: Textile detergent; emulsifies natural hide grease resulting in much cleaner stock entering the liming operation; also used during pickling; efficient in hard water
Form: Clear liquid; Color: Yellow
Properties: Solubility: Clear 10% aq. sol'n.; Cloud Pt.: 5 C; Flash Pt.: 43 C (PMCC); Stability: Compatible with acids and alkalis; pH: 8.5 (2% sol'n.); Biodegradable
Composition: 30% active ingred.
Containers: 55 gal fiber drums with plastic liners
Warnings and Caution: Store at room temp.
References: Prod. bulletin LE-58 rev.

Sellogen HR. Diamond Shamrock/Process Chem. Div.
Chemical Description: Sodium alkyl naphthalene sulfonate
Category: Wetting agent, dispersant, anionic
Applications: Heavy duty household and industrial cleaners, e.g., wall and floor cleaners, floor wax removers, scouring powders, metal

polishes, glass and bottle washing compounds, brick and tile cleaners, automobile radiator cleaners, pickling, industrial metal cleaners, soaps and detergents, fire control, paint strippers, agricultural chemical formulations, emulsion and suspension polymerization aids, latex paints
Form: Powder; Color: Off-white
Properties: Solubility: Readily sol. in water; Density: 0.40 g/cc; Stability: Stable in high electrolyte concs.; effective in acid and alkaline media; Surface Tension: 44.1 dynes/cm (0.25% conc.); pH: 9.0 (5% sol'n.)
Composition: 75.5% active ingred.
Containers: 55 gal polylined Leverpak fiber drums
References: Prod. bulletin GEN-22 Rev.

Sellogen W. Diamond Shamrock/ Process Chem. Div.
Chemical Description: Sodium alkyl naphthalene sulfonate
Category and Applications: See Sellogen HR
Form: Powder; Color: Off-white
Properties: Solubility: Readily sol. in water; Density: 0.38 g/cc; Stability: Stable in high electrolyte concs.; effective in alkaline or acid media; Surface Tension: 42.1 dynes/cm (0.25% conc.); pH: 9.0 (5% sol'n.)
Composition: 66.5% active ingred.
Containers: 40 lb multiwall paper bags and bulk
References: Prod. bulletin GEN-22 Rev.

Sellogen WL. Diamond Shamrock/ Process Chem. Div.
Chemical Description: Sodium alkyl naphthalene sulfonate
Category and Applications: See Sellogen HR
Form: Liquid; Color: Dark amber
Properties: Solubility: Readily sol. in water; Density: 9.5 lb/gal; Sp. gr.: 1.140; Stability: Stable in high electrolyte concs.; effective in acid or alkaline media; Surface Tension: 48.9 dynes/cm (0.25% conc.); pH: 9.5 (10% sol'n.)
Composition: 34% active ingred.
References: Prod. bulletin GEN-22 Rev.

Separol AF 27. BASF AG
Chemical Description: High-molecular adducts of EO and PO
Category: Demulsifier, nonionic
Applications: Separates saline formation water from crude oil emulsions; removes residual traces of salt from crude oils in refineries; used in all types of dehydrating equipment
Form: Clear liquid; Color: Yellow
Properties: Solubility: Sol. in alcohols and their blends with aromatic solvents; Density: 0.96 g/cc (20 C); Visc.: ≈ 30 cs (20 C); Flash Pt.: > 23 C; Pour Pt.: –30 C
Containers: 200 kg drums
Warnings and Caution: Use with adequate ventilation

Separol WF 22. BASF AG
Chemical Description: High-molecular adduct of EO and PO
Category and Applications: See

Separol AF 27
Form: Clear liquid; Color: Yellow
Properties: Solubility: Sol. in alcohols and water; Density: ≈ 1.02 g/cc (20 C); Visc.: ≈ 200 cs (20 C); Flash Pt.: > 23 C; Pour Pt.: –50 C
Containers: 200 kg drums
Warnings and Caution: Use with adequate ventilation

Separol WF 34. BASF AG
Chemical Description: High-molecular adduct of EO and PO
Category and Applications: See Separol AF 27
Form: Clear liquid; Color: Yellow
Properties: Solubility: Sol. in alcohols and their blends with aromatic solvents; Density: ≈ 0.95 g/cc (20 C); Visc.: ≈ 200 cs (20 C); Flash Pt.: 16 C; Pour Pt.: –50 C
Containers: 190 kg drums
Warnings and Caution: Use with adequate ventilation

Separol WF 41. BASF AG
Chemical Description: High-molecular adduct of EO and PO
Category and Applications: See Separol AF 27
Form: Clear liquid; Color: Yellow
Properties: Solubility: Sol. in alcohols and their blends with aromatic solvents; Density: ≈ 0.95 g/cc (20 C); Visc.: ≈ 50 cs (20 C); Flash Pt.: 17 C; Pour Pt.: < –50 C
Containers: 190 kg drums
Warnings and Caution: Use with adequate ventilation

Separol WF 221. BASF AG
Chemical Description: High-molecular adduct of EO and PO
Category and Applications: See Separol AF 27
Form: Clear liquid; Color: Yellow
Properties: Solubility: Sol. in alcohols and water; Density: ≈ 1.02 g/cc (20 C); Visc.: ≈ 200 cs (20 C); Flash Pt.: > 23 C; Pour Pt.: –50 C
Containers: 200 kg drums
Warnings and Caution: Use with adequate ventilation

Separol WK 25. BASF AG
Chemical Description: High-molecular adduct of EO and PO
Category and Applications: See Separol AF 27
Form: Clear liquid; Color: Yellow
Properties: Solubility: Sol. in alcohols and water; Density: ≈ 1 g/cc (20 C); Visc.: ≈ 300 cs (20 C); Flash Pt.: 24 C; Pour Pt.: –50 C
Containers: 200 kg drums
Warnings and Caution: Use with adequate ventilation

Seyco G-88. AZS Chem. Co.
Category: Detergent, wetting agent
Applications: Dyeing, textile finishing
Form: Liquid; Color: Yellow; Odor: None
Properties: Sp. gr.: 0.95; F.P.: < 0 C; B.P.: 60 C; Stability: Good
Composition: 51% active ingred.
Toxicity: Eye irritant
Containers: Drums
Warnings and Caution: Do not breathe in vapors

Seycol E-66. AZS Chem. Co.
Category: Detergent, wetting agent
Applications: Textile finishing
Form: Hazy liquid
Properties: Sp. gr.: 0.95; F.P.: 0 C; B.P.: 100 C; Stability: Good
Composition: 55% active ingred.
Toxicity: Eye irritant
Containers: Drums
Warnings and Caution: Flammable material; do not breathe in vapors

Seycopon A-95 LCP. AZS Chem. Co.
Category: Detergent, foam stabilizer
Applications: Foam stabilizer for latex
Form: Liquid; Color: Light yellow
Properties: Sp. gr.: 1.02; F.P.: 0 C; B.P.: 100 C; Stability: Good
Composition: 30% active ingred.
Toxicity: Eye irritant
Containers: Drums

Seycopen NI Conc. AZS Chem. Co.
Category: Detergent, wetting agent, emulsifier, penetrant
Applications: Textile finishing, dyeing, desizing and boiling off cottons
Form: Liquid; Color: Colorless; Odor: None
Properties: Solubility: Sol. in water; Sp. gr.: 1.02; F.P.: 0 C; Stability: Good
Composition: 100% active ingred.
Toxicity: Eye irritant
Containers: Drums

Seycopen SL. AZS Chem. Co.
Category: Dyeing assistant
Applications: Dyeing and textile finishing
Form: Liquid; Color: Pale yellow
Properties: Sp. gr.: 1.0; F.P.: 0 C; B.P.: 100 C; Stability: Good
Composition: 40% active ingred.
Reaction: Neutral
Toxicity: Eye irritant
Containers: Drums
Warnings and Caution: Do not breathe in vapors

Seycosolv B-5. AZS Chem. Co.
Category: Detergent, wetting agent, emulsifier
Applications: Textile finishing, dyeing, desizing and boiling off cottons
Form: Liquid; Color: Pale straw
Properties: Sp. gr.: 0.855; F.P.: $<$ 0 C; B.P.: $>$ 40 C; Stability: Good
Composition: 90% active ingred.
Reaction: Neutral
Toxicity: Eye irritant
Containers: Drums
Warnings and Caution: Do not breathe in vapors

SF 18. General Electric
Chemical Description: Polydimethylsiloxane
Category: Antifoam, lubricant, mold release
Applications: Rubber and plastic lubricant; base fluid for grease; mold release for rubber and plastic and food applictions; antifoam in food applications and aq. defoaming formulations
Form: Clear, oily liquid; Color:

Water-white
Properties: Ref. Index: 1.4030; Sp. gr.: 0.973; Visc.: 350 cs; Flash Pt.: 500 F (CC); Pour Pt.: –58 F; Stability: Good thermal, oxidative, and shear stability

SF 69. General Electric
Chemical Description: Polydimethylsiloxane polymer fluid
Category: Paint additive, flow control agent
Applications: Used as an antifloat, flow-control, and anticratering agent in protective coatings such as varnishes, paints, enamels; also reduces "organe peel" and produces "wetter" spray coatings
Form: Liquid
Properties: Ref. Index: 1.4020; Sp. gr.: 0.965; Visc.: 20 cs; Flash Pt.: 125 F (CC); Pour Pt.: –40 F

SF 81. General Electric
Chemical Description: Polydimethylsiloxane
Applications: Useful as low temp. damping media, heat transfer media dielectric coolant, and is a base fluid for low temp. silicone compounds
Form: Liquid
Properties: Ref. Index: 1.4030; Sp. gr.: 0.972; Visc.: 50 cs; Flash Pt.: 460 F (CC); Pour Pt.: –120 F; Stability: Good thermal, oxidative, and shear stability

SF 96. General Electric
Chemical Description: Polydimethylsiloxane
Category: Lubricant, defoamer, chemical specialties, nonionic
Applications: Plastic and rubber lubrication; as dampening or heat transfer fluids; oil defoamer; in chemical specialties, antifoams, paint additives; mold release for tires, rubber, plastics; textile lubricant and softener/modifier
Form: Liquid (available in various viscosity grades from 5–1000 cs); Color: Water-white
Properties: Ref. Index: 1.3970–1.4035; Solubility (5 cs grade): Sol. in aliphatic, aromatic, and chlorinated hydrocarbons, higher alcohols and ketones, lower alcohols, some higher hydrocarbons; Sp. gr.: 0.916–0.974; Visc.: 5–1000 cs; Flash Pt.: 280–500 F (CC); Pour Pt.: –120 to –58 F; Stability: Good thermal, oxidative, and shear stability; Surface Tension: 19.7–21.1 dynes/cm
Composition: 100% active ingred.

SF 97. General Electric
Chemical Description: Polydimethylsiloxane
Category: Silicone fluid
Applications: Electrical grade silicone fluids similar to SF 96 series; for use in capacitors, pulse transformers, specialty transformers, air-borne and landbased radar equipment, television circuit components, and in systems where heat transfer properties are required as well as dielectric behavior
Form: Clear, oily liquid; Color: Water-white (available in 50 and 100 cs visc. grades)
Properties (50 and 100 cs grades

respectively): Ref. Index: 1.4020, 1.4030; Sp. gr.: 0.963, 0.968; Visc.: 50 cs, 100 cs; Flash Pt.: 460 F, 461 F (CC); Pour Pt.: –67 F, –67 F; Stability: Good thermal, oxidative, and shear stability

SF 99. General Electric
Chemical Description: Polydimethylsilocane
Category: Dielectric fluid
Applications: Water-repellent surface coatings and films on glass or other ceramics to obtain reliable operation of electronic equipment under varying conditions of temp. and humidity; coatings for paper, fabrics, wood, and metals; treatment of magnesium oxide in heating elements or appliances (protects against leakage currents)
Form: Clear, low viscosity liquid
Properties: Ref. Index: 1.4020; Sp. gr.: 0.970; Visc.: 10 cs; Flash Pt.: 145 F (CC); Pour Pt.: –40 F; Stability: Good thermal, oxidative, and shear stability

SF 1023. General Electric
Chemical Description: Phenyl containing silicone fluid
Category: Flow control agent, dispersion aid
Applications: Flow control agent in polyester coatings; aids pigment dispersion and improves gloss and gloss retention in organic finish systems
Properties: Sp. gr.: 1.030; Visc.: 55 cs; Flash Pt.: 180 F (CC)
Composition: 100% solids

SF 1029. General Electric
Chemical Description: Chlorophenylmethyl silicone fluid
Category: Lubricant
Applications: API thread lube formulations; hydraulic fluids; base fluid for grease; sliding metal on metal lubrication; excellent high temp. stability, resists gelation for long periods at extreme temps., when in contact with air
Form: Liquid
Properties: Ref. Index: 1.4290; Sp. gr.: 1.030; Visc.: 75 cs; Flash Pt.: 425 F (CC); Pour Pt.: –100 F; Stability: Excellent thermal stability

SF 1066. General Electric
Chemical Description: Polydimethylsiloxane/ethylene oxide-polypropylene oxide copolymer
Category: Lubricant, mold release, chemical specialties, nonionic
Applications: Cosmetics and toiletries; rubber and plastic mold release; paintable release; textile softeners/modifiers; thread and fiber lubes
Form: Liquid
Properties: Ref. Index: 1.4470; Solubility: Sol. in aromatic and chlorinated hydrocarbons, alcohols, higher ketones, water, some higher hydrocarbons; Sp. gr.: 1.04; Visc.: 800 cs; Flash Pt.: 200 F (CC); Pour Pt.: –13 F; Stability: Stable in water, alcohol, and some hydrocarbons; good thermal, oxidative, and shear stability; Surface Tension: 25.5 dynes/cm
Composition: 100% active ingred.

SF 1080. General Electric
Chemical Description: Methyl alkyl polysiloxane silicone fluid
Category: Mold release agent, lubricant, nonionic
Applications: Lubricating, die cast, and release applications in rubber, plastic, and metal industries; useful in applications requiring paintability and in areas where prevention of secondary paint contamination should be avoided; as internal mold release agent in vinyl slush molding; as aluminum die cast mold release agent
Form: Liquid; Color: Light yellow
Properties: Ref. Index: 1.4930; Solubility: Sol. in aliphatic, aromatic, and chlorinated hydrocarbons, higher alcohols, and higher ketones; Sp. gr.: 1.035; Visc.: 1500 cs; Flash Pt.: 400 F (CC); Pour Pt.: –50 F; Stability: Excellent thermal stability; Surface Tension: 27.0 dynes/cm
Composition: 100% active ingred.

SF 1091. General Electric
Chemical Description: Methyl alkyl polysiloxane silicone fluid
Category: Mold release agent, lubricant, paint additive
Applications: Mold release fluid; high temp. lubricant for soft metals, e.g., aluminum, zinc, and copper, and as a rubber and plastic lubricant, esp. rubber on steel or aluminum and rubber on rubber; thread and fiber lubes; paint additives for flow control, mar resistance, and gloss
Form: Liquid
Properties: Ref. Index: 1.4480; Sp. gr.: 0.906; Visc.: 500 cs; Flash Pt.: 400 F (CC); Pour Pt.: –72 F; Stability: Good thermal, oxidative, and shear stability

SF 1093. General Electric
Chemical Description: Polydimethylsiloxane
Category: High temp. silicone fluid, mold release agent
Applications: High temp. heat transfer applications, laboratory constant temp. baths, aerosol oven sprays, high temp. fluid clutches, military applications requiring extreme temp. operation, mechanical damping, thermostats, and as a mold release
Form: Liquid; Color: Amber; Odor: Essentially odorless (available in 50 and 100 cs visc. grades)
Properties (50 and 100 cs grades respectively): Ref. Index: 1.4040, 1.4040; Sp. gr.: 0.963, 0.968; Visc.: 50 cs, 100 cs; Flash Pt.: 425 F, 425 F (CC); Pour Pt.: –100 F, –90 F; Stability: Good thermal, oxidative, and shear stability

SF 1134. General Electric
Chemical Description: Methyl alkyl polysiloxane
Category: Lubricant
Applications: Formulated thread and fiber lubricants; sliding metal on metal lubrication; cosmetics and toiletries; compatible with organic materials, e.g., paraffinic

waxes
Form: Wax
Properties: Sp. gr.: 0.93; Pour Pt.: 97 F; Flash Pt.: 399 F (CC)

SF 1147. General Electric
Chemical Description: Methyl alkyl polysiloxane
Category: Lubricant
Applications: High temp. hydraulic and lubricating fluid with excellent metal-on-metal lubricating properties; used for impregnating sintered metal bearings for fans and motors, fluid for machine operations with aluminum–drilling, tapping, hobbing, broaching, turning, spinning, drawing, extruding; instrument lubrication; vacuum pumps
Form: Liquid
Properties: Ref. Index: 1.4433; Solubility: Sol. in aliphatic, aromatic, and chlorinated hydrocarbons, higher alcohols, higher ketones; Sp. gr.: 0.890; Visc.: 50 cs; Flash Pt.: 400 F (CC); Pour Pt.: –60 F; Surface Tension: 26.0 dynes/cm

SF 1153. General Electric
Chemical Description: Polydimethyldiphenyl siloxane
Category: Lubricant, dielectric coolant
Applications: Low temp. damping applications such as aircraft instruments and electronic equipment; as heat transfer media, dielectric coolant, and low temp. lubricant involving light to moderate bearing loads; base fluid in silicone greases for ball bearing lubrication from –100 to 400 F; good low temp. properties with high temp. performance
Form: Liquid
Properties: Ref. Index: 1.4219; Sp. gr.: 0.980; Visc.: 100 cs; Flash Pt.: 435 F (CC); Pour Pt.: –100 F; Stability: Good thermal, oxidative, and shear stability

SF 1154. General Electric
Chemical Description: Polydimethyldiphenyl siloxane
Category: Lubricant, dielectric coolant, coupler
Applications: High temp. heat transfer applications; base fluid in high temp. greases; dielectric coolant; high temp. ultrasonic coupler; high temp. bath and oxide protector for solder baths; outstanding heat resistance; useful at –40 to 500 F
Form: Clear liquid; Color: Light straw
Properties: Ref. Index: 1.4980; Sp. gr.: 1.050; Visc.: 190 cs; Flash Pt.: 440 F (CC); Pour Pt.: –40 F; Stability: Excellent heat stability

SF 1173. General Electric
Chemical Description: Polydimethylsiloxane
Applications: Cosmetic ingred. in both aerosol and nonaerosol hair care products, aerosol deodorant and antiperspirant, glass cleaner additive, panel cleaner
Form: Liquid
Properties: Ref. Index: 1.3935; Solubility: Miscible with lower

alcohols, and typical aliphatic and aromatic solvents, incl. the halogenated solvents; Sp. gr.: 0.960; Visc.: 2.4 cs; Flash Pt.: 131 F (CC); Pour Pt.: 63 F

SF 1221. General Electric
Chemical Description: Copolymer of a polydimethylsiloxane and a polyoxyalkylene ether
Category: Lubricant, mold release agent
Applications: Fiber and thread lube, hot system defoaming, paintable mold release agent, flow control paint additive
Form: Liquid
Properties: Ref. Index: 1.4470; Solubility: Sol. in water, lower alcohols, and some hydrocarbon solvents; Sp. gr.: 1.040; Visc.: 800 cs; Flash Pt.: 200 F (CC); Pour Pt.: −13 F; Stability: Stable in water, lower alcohols, and some hydrocarbon solvents

SF 1250. General Electric
Chemical Description: Chlorophenyl methyl siloxane
Category: Lubricant, mold release agent
Applications: Hydraulic systems; fluid transmission; servomotors and mechanisms; instruments, clocks, and timers; machine tool components; antifriction, rolling, and sliding mechanisms; shock absorbers and damping devices
Form: Liquid
Properties: Ref. Index: 1.4280; Sp. gr.: 1.050; Visc.: 70 cs; Flash Pt.: 410 F (CC); Pour Pt.: −100 F

Sipex 7WC. Alcolac Inc.
Chemical Description: Specialty surfactant
Category: Foaming agent, wetting agent, anionic
Applications: Produces dense, stable foam for bubble baths, rug shampoos, etc.; also in textile scouring of dyed or printed fabrics where its rapid wetting and good stability toward pH and temp. variations is advantageous
Form: Viscous liquid; Color: Pale yellow; Odor: Mild
Properties: Visc.: 2500 cps; Cloud Pt.: 16 C; Stability: Stable to pH and temp. variations; pH: 7.2 (10% sol'n.)
Composition: 31% active ingred.

Sipex 206-C. Alcolac Inc.
Category: Detergent, foaming agent
Applications: Fabric shampoo concentrate for carpets and upholstery; good detergency and soil suspension; produces copious, stable "dry" foam; for use on cotton, wool, and synthetics; dries to a powder which can be vacuumed
Properties: Visc.: 500 cps; Cloud Pt.: 14 C; pH: 7
Composition: 31% active ingred.

Sipex 206C-SP. Alcolac Inc.
Chemical Description: Modified alkyl phenol ethoxylate
Category and Applications: See Sipex 206-C
Form: Clear liquid
Properties: Visc.: 80 cps; Cloud Pt.: < 10 C; pH: 7
Composition: 29% active ingred.

Sipex BOS. Alcolac Inc.
Chemical Description: Sodium 2-ethylhexyl sulfate
Category: Wetting agent, emulsifier, detergent, stabilizer, anionic
Applications: Moderate foamer; rinse aid for automatic dishwashers; post-stabilizer in latex paints; metal treatment (nickel brightener); mercerization of gray cotton goods; fruit and vegetable washing (removes field soil, microorganisms, pesticide residues) and peeling (aids wetting/penetration of caustic bath); viscosity control in plywood manufacture; hard surface cleaners
Form: Clear liquid
Properties: Visc.: 50 cps; HLB: 42; Cloud Pt.: < 10 C; Stability: Stable to high pH and temp., and in relatively high concs. of caustic soda or other electrolytes; pH: 9.5–10.5 (10% sol'n.); Biodegradable
Composition: 39–40% active
Warnings and Caution: Avoid contact with eyes or prolonged skin contact

Sipex CAV. Alcolac Inc.
Chemical Description: Sodium isodecyl sulfate
Category: Wetting agent, emulsifier, stabilizer, anionic
Applications: Moderate foamer; rinse aid for automatic dishwashers; post-stabilizer in latex paints; nickel brightening baths; rapid mercerization of gray cotton goods; aid for emulsion polymerization of vinyl, acrylic, SBR, and PVC systems; hard surface cleaners
Form: Clear liquid
Properties: Density: 8.95 lb/gal; Visc.: 150 cps; Cloud Pt.: < 10 C; Stability: Stable to high pH and temp., and relatively high concs. of caustic soda or other electrolytes; pH: 9.5–10.5 (10% sol'n.)
Composition: 39–40% active
Warnings and Caution: Avoid contact with eyes or prolonged skin contact

Sipex EC-111. Alcolac Inc.
Chemical Description: Sodium cetyl sulfate
Category: Lubricant, softener, emulsifier, detergent, flotation agent, anionic
Applications: Collector in ore flotation; softener/lubricant for cotton or rayon fibers; softener for desized cotton piece goods; shampoos, hand creams, lotions, detergent bars, facial cleansers, and other cosmetics
Form: Thick paste; Color: White
Properties: pH: 8 (10% sol'n.)
Composition: 25% active ingred.

Sipex EST-30. Alcolac Inc.
Chemical Description: Sodium tridecyl ether sulfate
Category: Wetting agent, emulsifier, foaming agent, detergent base, anionic
Applications: Sole emulsifier in emulsion polymerization of styrene, styrene-acrylic, and styrene-butadiene systems, and esp. for monomers that are only sparingly sol. in water; post-stabilizer for

emulsion polymerization; household and industrial cleaners, liquid dishwashing compounds; heavy duty scouring of rayon and synthetic fibers
Form: Liquid; Color: 2 max. (Gardner)
Properties: Cloud Pt.: 14 C max.; pH: 7.5–8.5 (10% sol'n.)
Composition: 29–30% active
Warnings and Caution: Avoid contact with eyes and prolonged skin contact

Sipex EST-60. Alcolac Inc.
Chemical Description: Sodium tridecyl ether sulfate
Category: Wetting agent, detergent
Applications: Household cleaning products for walls, windows, woodwork, floors, porcelain, enamel tiles; can be compounded with alkaline ingred. for oven cleaners and industrial metal cleaning processes
Form: Clear liquid
Properties: Cloud Pt.: None; Stability: Stable over wide range of pH and temp.; pH: 9 (10% sol'n.)
Composition: 57% active ingred.

Sipex EST-75. Alcolac Inc.
Chemical Description: Sodium tridecyl ether sulfate
Category: Emulsifier, wetting agent, anionic
Applications: Shampoos and other cosmetics; can be compounded with strongly alkaline ingred. for heavy duty cleaners, wax strippers, aerosol cleaners, etc.
Form: Liquid
Properties: Stability: Stable to alkalis; pH: 9.0–10.5 (10% sol'n.)
Composition: 70% active ingred.
Warnings and Caution: Avoid contact with eyes or prolonged skin contact

Sipex FA-38. Alcolac Inc.
Category: Foaming agent, anionic
Applications: Production of foam in emulsion polymer systems; produces high flash foam which resists collapse at curing temps. as high as 250 F, will not discolor, and is unaffected by pigments or electrolytes; the foamed latex is used in furniture and textile fields as backing for draperies, carpets, and apparel
Form: Thixotropic paste; Color: White
Properties: M.P.: ≈ 35 C; pH: 8
Composition: 32% active ingred.

Sipex JGT. Alcolac Inc.
Category: Detergent, anionic/nonionic
Applications: Textile scouring–removal of oils and greases from natural and synthetic fibers; minimizes agglomeration of the dye; aids rapid and thorough penetration
Form: Clear liquid; Color: Amber
Properties: Cloud Pt.: 68 F; Stability: Stable in acids, alkalis, or metal salts, and up to 10% caustic soda sol'ns.; Surface Tension: 34 dynes/cm (1 g/100 ml water); pH:

7–8 (1% sol'n.)
Composition: 28% active ingred.
Containers: 460 lb net drums

Sipex JKB. Alcolac Inc.
Category: Anionic surfactant
Applications: Emulsion polymerization of highly carboxylated acrylic monomers used for latex thickeners, alkali-sol. resins, metalized floor finishes, textile sizing agents, hand-modifiers
Form: Clear liquid
Properties: pH: 6.5 (10% sol'n.)
Composition: 30% active ingred.

Sipex LCP. Alcolac Inc.
Chemical Description: Sodium lauryl sulfate
Applications: Low cloud pt. surfactant which imparts good stability and uniform particle size to vinyl chloride, vinylidene chloride, styrene, butadiene, and acrylic ester emulsion polymers
Properties: Visc.: 50 cps; Cloud Pt.: 1 C; pH: 8.0 (10% sol'n.)
Composition: 30% active ingred.

Sipex OLS. Alcolac Inc.
Chemical Description: Sodium octyl sulfate
Category: Wetting agent, dispersant, anionic
Applications: Low foaming rinse aid for automatic dishwashers; mercerizing agent for grey cotton goods (stable to 20–25% caustic sol'ns.); surfactant in electrolyte baths for metal cleaning; hard surface cleaning; Neoprene dispersant
Form: Clear liquid
Properties: Visc.: 100 cps; HLB: 42; Stability: Stable under varying conditions of pH and temp.; pH: 8 (10% sol'n.)
Composition: 33% active ingred.

Sipex OS. Alcolac Inc.
Chemical Description: Sodium oleyl sulfate
Category: Emulsifier, emollient, anionic
Applications: Emulsion polymerization, esp. for monomers relatively insol. in water; contributes emolliency and mildness to hair, scalp, skin formulations and other cosmetics; FDA-approved as component of food packaging cellophane
Form: Clear liquid
Properties: HLB: 42; pH: 8 (10% sol'n.)
Composition: 26% active ingred.

Sipex SB. Alcolac Inc.
Chemical Description: Sodium lauryl sulfate
Category: Emulsifier, anionic
Applications: Emulsion polymerization; vegetable and fruit washing; food packaging applications; metal cleaning; textile scouring; commercial car washing
Form: Clear liquid
Properties: Visc.: 150 cps; Cloud Pt.: 20 C; Stability: Good; pH: 8 (10% sol'n.)
Composition: 29% active ingred.

Sipex SD. Alcolac Inc.
Chemical Description: Sodium lauryl sulfate

Category: Anionic surfactant
Applications: High active, spray dried grade for use in metal processing, electroplating, textile manufacturing, leather treatment, rug and upholstery shampoos
Form: Powder; Color: White
Properties: HLB: 40; pH: 7.5 (3% sol'n.)
Composition: 95% active ingred.

Sipex TDS. Alcolac Inc.
Chemical Description: Sodium tridecyl sulfate
Category: Emulsifier, wetting agent, anionic
Applications: Emulsion polymerization of vinyl chloride, styrene, and styrene/acrylic monomers; detergent formulations
Form: Liquid; Color: Colorless
Properties: Visc.: 100 cps; Cloud Pt.: 10 C; pH: 8.5 (10% sol'n.)
Composition: 25% active ingred.

Sipex TEA. Alcolac Inc.
Chemical Description: Triethanolamine lauryl sulfate
Category: Detergent, emulsifier, dispersant, stabilizer
Applications: Pigment dispersant and stabilizer; wool scouring; ore flotation; rug shampoos; latex paint formulations; dishwashing detergents
Form: Clear liquid
Properties: Visc.: 30 cps; Cloud Pt.: 2 C; pH: 7 (10% sol'n.)
Composition: 40% active ingred.

Sipex UB. Alcolac Inc.
Chemical Description: Sodium lauryl sulfate
Category: Emulsifier, anionic
Applications: Low salt and low unsulfated alcohol surfactant for vinyl chloride, vinylidene chloride, styrene, butadiene, and acrylic ester emulsion polymers
Form: Liquid
Properties: Visc.: 50 cps; HLB: 40; Cloud Pt.: 18 C; Stability: Stable to alkalis; pH: 8 (10% sol'n.)
Composition: 30% active ingred.

Sipon EA. Alcolac Inc.
Chemical Description: Ammonium lauryl ether sulfate
Category: Detergent, foaming and wetting agent, solubilizer, penetrant, anionic
Applications: Generates stable, dense foam with small bubbles for "luxurious" feel in cosmetics and toiletries; clear liquid shampoos, rich lather bubble baths, antiseptic creams, contraceptive foams, diaper rash preventives, deodorant preparations, mild liquid detergents for woolens and synthetics, pet shampoos
Form: Liquid; Odor: Mild, pleasant
Properties: Visc.: 300 cps; Cloud Pt.: –5 C; pH: 6.7 (10% sol'n.)
Composition: 27% active ingred.
Warnings and Caution: Releases ammonia at pH $>$ 7

Sipon EAY. Alcolac Inc.
Chemical Description: Ammonium lauryl ether sulfate
Category and Applications: See Sipon EA
Form: Thixotropic product

Properties: Cloud Pt.: 0 C; pH: 6.7 (10% sol'n.)
Composition: 26% active ingred.
Warnings and Caution: Release ammonia at pH > 7

Sipon L-22. Alcolac Inc.
Chemical Description: Ammonium lauryl sulfate
Category: Detergent, foaming agent, anionic
Applications: High foaming, mild detergent for shampoos, antiseptic soaps, vaginal preparations, shaving creams, pet shampoos; also as a wool scouring agent, assistant for pigment dispersion; emulsion polymerization aid; fire fighting foams
Form: Clear liquid
Properties: Visc.: 1000 cps; HLB: 31; Cloud Pt.: 14 C pH: 6.8 (10% sol'n.)
Composition: 28% active ingred.
Warnings and Caution: Product slowly degrades above pH 7

Sipon LM. Alcolac Inc.
Chemical Description: Magnesium lauryl sulfate
Category: Detergent, anionic
Applications: Low cloud pt. surfactant for clear liquid shampoos, bubble baths, rug and upholstery shampoos (dries to a powder for easy removal), food packaging applications; reduces interfacial tension
Form: Clear liquid
Properties: Visc.: 50 cps; Cloud Pt.: 2 C; pH: 6.5 (10% sol'n.)
Composition: 27% active ingred.

Sipon LS. Alcolac Inc.
Chemical Description: Sodium lauryl sulfate
Category: Detergent, anionic
Applications: Surfactant with optimum balance of opacity and viscosity for use in formulating cream or paste shampoos, ointments, toothpastes, shaving creams, and other cosmetics
Form: Viscous liquid
Properties: Visc.: 25,000 cps; HLB: 40; Cloud Pt.: 20 C; pH: 8 (10% sol'n.)
Composition: 28% active ingred.

Sipon LSB. Alcolac Inc.
Chemical Description: Sodium lauryl sulfate
Category: Detergent, anionic
Applications: Low salt base ingred. for shampoos in liquid, lotion, or cream form, bubble baths, shaving creams, medicinal ointments, and other cosmetics; produces luxuriant, dense white foam even under adverse conditions of greasy soil load or water hardness
Form: Clear liquid; Color: Pale
Properties: Visc.: 150 cps; HLB: 40, Cloud Pt.: 20 C; pH: 8 (10% sol'n.)
Composition: 29% active ingred.

Sipon LT-6. Alcolac Inc.
Chemical Description: Triethanolamine lauryl sulfate
Category: Shampoo ingred., anionic
Applications: Shampoo ingred. with mildness and good foaming properties

Form: Clear liquid; Color: Pale yellow
Properties: Visc.: 50 cps; HLB: 34.0; Cloud Pt.: 2 C; pH: 7 (10% sol'n.)
Composition: 40% active ingred.

Sipon MLS. Alcolac Inc.
Chemical Description: Pharmaceutical grade magnesium lauryl sulfate
Category: Lubricant
Applications: Soluble lubricant in drug tableting operations; does not have waterproofing liabilities of magnesium stearates
Form: Spray-dried
Properties: pH: 6.5 (3% sol'n.)
Composition: 88% active ingred.

Sipon WD. Alcolac Inc.
Chemical Description: High purity sodium lauryl sulfate
Category: Emulsifier, foaming agent, anionic
Applications: Food processing emulsifier and as a film former on fresh citrus; pharmaceuticals, esp. foaming toothpastes; household formulations, incl. rug shampoos; bath products, e.g., shampoos, bubble baths, and other foaming toiletries; industrial applications
Form: Crystal
Properties: Density: 5 lb/gal (bulk); HLB: 40; pH: 8.6 (1% sol'n.)
Composition: 98% active ingred.

Siponate 330. Alcolac Inc.
Chemical Description: Anionic surfactant
Category: Emulsifier, solubilizer, dispersing aid, solvent, penetrant, anionic
Applications: Emulsifier and dispersing aid in agricultural sprays; solvent for organics in herbicides and pesticides; degreasers and wax removers; spot removal aid in dry cleaning; latex paints; pigment dispersion aid; metal cleaning processes; automotive specialties; cutting oils; acid aluminum cleaners; oil slick emulsifiers
Form: Liquid; Color: Amber
Properties: Sp. gr.: 1.03; Visc.: 6500 cps; HLB: 11.7; pH: 3–6 (5% sol'n.)
Composition: 90% active ingred.

Siponate DS-4. Alcolac Inc.
Chemical Description: Sodium dodecyl benzene sulfonate
Category: Emulsifier, anionic
Applications: Emulsifier for styrene-butadiene, vinyl chloride, or acrylic polymers, producing latices of uniform particle size with good heat stability; food packaging applications; surfactant for washing fruits and vegetables
Form: Liquid; Color: Dark brown
Composition: 23% active ingred.

Siponate DS-10. Alcolac Inc.
Chemical Description: Sodium dodecyl benzene sulfonate
Category: Emulsifier, anionic
Applications: Surfactant in styrene-butadiene, vinyl chloride, vinyl acetate, styrene, and acrylic emulsion polymerization systems, producing latices with good uni-

formity of particle size and mechanical stability; films have good heat stability; heavy-duty household or institutional cleaning compounds; secondary oil recovery operations; washing fruits and vegetables; food packaging materials
Form: Flake
Properties: pH: 7.5 (10% sol'n.)
Composition: 98% active ingred.

Siponic 218. Alcolac Inc.
Chemical Description: Dodecyl mercaptan ethoxylate
Category: Emulsifier, wetting agent, detergent, nonionic
Applications: Metal cleaning; inhibitors during acid pickling of steel; textile treatment; may be compounded with alkaline salts to aid detergency; readily compatible with quaternary germicidal agents; agricultural formulations; reduces irritation of cationics in hair products; wood pulp and paper industry; emulsifier for petroleum oils, chlorinated solvents, silicones, metallic soaps, and emulsion polymers
Form: Liquid; Color: 6 max. (Gardner)
Properties: Solubility: Most water-sol. of the Siponic thioethers; good sol. in hard water and in presence of alkaline detergent builders; Density: 8.6 lb/gal; Sp. gr.: 1.05 (20/20 C); Cloud Pt.: 52 C (1% sol'n.); Stability: Stable in wide range of pH and temp., except strongly acidic media; Surface Tension: 28 dynes/cm (0.05% sol'n.)
Composition: 95% conc.
Warnings and Caution: Avoid contact with skin and eyes; unsuitable for use with oxidizing agents

Siponic 260. Alcolac Inc.
Chemical Description: Dodecyl mercaptan ethoxylate
Category: Emulsifier, wetting agent, detergent, nonionic
Applications: Heavy industrial metal cleaning (removes grease and oils in organic solvent systems); inhibitor during acid pickling of steel; agricultural formulations; wood pulp and paper industries; emulsifier for petroleum oils, chlorinated solvents, silicones, metallic soaps, and emulsion polymers
Form: Liquid; Color: 6 max. (Gardner)
Properties: Solubility: Very sol. in common organic solvents; Density: 8.4 lb/gal; Sp. gr.: 1.01 (20/20 C); Surface Tension: 29 dynes/cm (0.05% sol'n.)
Composition: 98% active ingred.
Warnings and Caution: Avoid contact with skin and eyes; unsuitable for use with oxidizing agents

Siponic E5. Alcolac Inc.
Chemical Description: Stearyl-cetyl alcohol ethoxylate, 10 moles EO
Category: Emulsifier, lubricant, dye assistant, nonionic
Applications: Conditioning agents for shampoos providing lubricity and emolliency; dyeing assistants,

leveling agents, and fulling assistants in acid media softeners for textiles; emulsion polymerization; self-emulsifying floor waxes; paper finishes; rubber emulsifiers; pharmaceuticals
Form: Wax
Properties: HLB: 12.4; Cloud Pt.: 73 C (1% NaCl); pH: 6.5 (1% sol'n.)
Composition: 100% active ingred.

Siponic E7. Alcolac Inc.
Chemical Description: Stearyl-cetyl alcohol ethoxylate
Category and Applications: See Siponic E5; E7 is esp. suitable in textile lubricants and finishes
Form: Wax
Properties: HLB: 14.1; Cloud Pt.: 93 C (1% NaCl); pH: 6.5 (1% sol'n.)
Composition: 100% active ingred.

Siponic E15. Alcolac Inc.
Chemical Description: Stearyl-cetyl alcohol ethoxylate
Category and Applications: See Siponic E5
Form: Wax
Properties: HLB: 16.7; Cloud Pt.: > 95 C (1% NaCl); pH: 6.5 (1% sol'n.)
Composition: 100% active ingred.

Siponic L1. Alcolac Inc.
Chemical Description: Lauryl alcohol ethoxylate
Category: Emulsifier, thickener, stabilizer, emulsifier, nonionic
Applications: Thickener in shampoos and bubble baths; stabilizer for emulsion polymers; supplemental emulsifiers in specific monomer systems; self-emulsifying floor waxes; paper finishes; rubber emulsifiers; emollients for pharmaceuticals
Form: Liquid
Properties: HLB: 3.7; pH: 6.5 (1% sol'n.)
Composition: 100% active ingred.

Siponic L4. Alcolac Inc.
Chemical Description: Lauryl alcohol ethoxylate
Category: Emulsifier, thickener, stabilizer, nonionic
Applications: See Siponic L1; also outstanding emulsifier for cosmetic creams and lotions
Form: Liquid
Properties: HLB: 9.4; pH: 6.5 (1% sol'n.)
Composition: 100% active ingred.

Siponic L7. Alcolac Inc.
Chemical Description: Lauryl alcohol ethoxylate
Category: Emulsifier, stabilizer, nonionic
Applications: Emulsifier for cosmetic creams and lotions; stabilizer for emulsion polymers; supplemental emulsifiers in specific monomer systems; self-emulsifying floor waxes; paper finishes; rubber emulsifiers; emollients for pharmaceuticals
Form: Wax (also available as Siponic L7-90, a 90% aq. sol'n.)
Properties: HLB: 12.5; Cloud Pt.: 56 C (1% NaCl); pH: 6.5 (1% sol'n.)
Composition: 100% active ingred.

Siponic L12. Alcolac Inc.
Chemical Description: Lauryl alcohol ethoxylate
Category and Applications: See Siponic L7
Form: Wax
Properties: HLB: 14.6; Cloud Pt.: 89 C (1% NaCl); pH: 6.5 (1% sol'n.)
Composition: 100% active ingred.

Siponic L16. Alcolac Inc.
Chemical Description: Lauryl alcohol ethoxylate
Category and Applications: See Siponic L7
Form: Wax
Properties: HLB: 15.8; Cloud Pt.: > 95 C (1% NaCl); pH: 6.5 (1% sol'n.)
Composition: 100% active ingred.

Siponic L25. Alcolac Inc.
Chemical Description: Lauryl alcohol ethoxylate
Category and Applications: See Siponic L7
Form: Wax
Properties: HLB: 16.9; Cloud Pt.: > 95 C (1% NaCl); pH: 6.5 (1% sol'n.)
Composition: 100% active ingred.

Siponic SK. Alcolac Inc.
Chemical Description: Dodecyl mercaptan ethoxylate
Category: Emulsifier, wetting agent, detergent, nonionic
Applications: Metal cleaning; car washing formulations; heavy-duty detergents for grease and grime; engine degreasing (cuts through carbonaceous soils); inhibitor during acid pickling of steel; low temp. scouring of synthetic blends, greasy yarns, and complex knitted fabrics; agricultural formulations; antiskinning agent for oil-based paints; reduces irritation of cationics in hair products; wood pulp and paper industry; emulsifier for petroleum oils, chlorinated solvents, silicones, metallic soaps, and emulsion polymers
Form: Liquid; Color: 6 max. (Gardner)
Properties: Solubility: Good sol. in hard water and in the presence of alkaline builders; Density: 8.5 lb/gal; Sp. gr.: 1.03 (20/20 C); Cloud Pt.: 28 C (1% sol'n.); Stability: Stable in wide range of pH and temp., except strongly acidic media; Surface Tension: 31 dynes/cm (0.05% sol'n.)
Composition: 98% active ingred.
Toxicity: Mild skin irritant
Warnings and Caution: Avoid contact with skin and eyes; unsuitable for use with oxidizing agents

Siponic TD-6. Alcolac Inc.
Category: Emulsifier, solubilizer, nonionic
Applications: Solubilizes water in dry cleaning solvents, kerosene, petroleum naphtha, etc.; oil-sol. emulsifier; agricultural sprays; engine degreasing; metal parts cleaning; similar applications where detergency in solvent-based systems is required
Form: Clear, oily liquid
Properties: Solubility: Oil-sol.;

Density: 8.2 lb/gal; Sp. gr.: 0.98; Visc.: ≈ 80 cps; Flash Pt.: ≈ 355 F; Acid No.: > 2; pH: 6 (1% sol'n.)
Composition: 99% active ingred.
Containers: 55 gal drums, tank-trailers, truckloads

Siponic Y500-70. Alcolac Inc.
Chemical Description: Oleyl alcohol ethoxylate (25 moles EO)
Category: Emulsifier, wetting agent, stabilizer, dispersant, foaming agent, nonionic
Applications: Emulsifier for polymerization; stabilizer for natural and synthetic latices; dispersant in polyurethane foams; foaming agent for styrene-butadiene and natural latex; acid degreasing baths for wool; dyeing aid for wool-synthetic fiber blends; self-emulsifying floor waxes; pharmaceutical emulsions; acid metal cleaning baths; fat liquoring for leather; food packaging applications
Form: Liquid; Odor: Mild
Properties: Visc.: 1650 cps; HLB: 16.2; Cloud Pt.: 69 C (1% sol'n. in NaCl); Stability: Stable to pH variations and in the presence of multivalent cations; pH: 7.5 (1% sol'n.)
Composition: 70% active ingred.
Warnings and Caution: Store at room temp. to prevent gel formation

Siponic Y501. Alcolac Inc.
Chemical Description: Oleyl alcohol ethoxylate (20 moles EO)
Category: Emulsifier, wetting agent, stabilizer, dispersant, nonionic
Applications: Emulsifier for polymerization; stabilizer for natural and synthetic latices; dispersant in polyurethane foams; acid degreasing baths for wool; dyeing aid for wool-synthetic fiber blends; self-emulsifying floor waxes; pharmaceutical emulsions; acid metal cleaning baths; fat liquoring for leather; food packaging applications
Form: Solid; Odor: Mild
Properties: Solubility: Sol. in water at room temp., and in polar organic solvents such as methanol and acetone; disp. in petroleum ether; M.P.: ≈ 35 C; HLB: 15.4; Cloud Pt.: 65 C (1% sol'n. in NaCl); Stability: Stable to pH variations and in the presence of multivalent cations; pH: 7.5 (1% sol'n.)
Composition: 100% active ingred.

Slik No. 5. NL Baroid/NL Industries, Inc.
Chemical Description: Alkanolamide and an oxyethylated alkylphenol
Category: Degreaser, nonionic
Applications: Solvent cleaner or water-sol. detergent for use in shops for washing engine parts, cleaning carburetors, machinery, tools, and shop floors; general cleaning around rigs; in building maintenance to remove floor wax; in service stations to wash white sidewall tires, clean windshields, tools, grease racks, driveways; as

an oil slick dispersant; as a cuttings wash where oil muds are used
Form: Liquid
Properties: Density: 8.38 lb/gal; Sp. gr.: 1.004 (68 F); Visc.: 25 SUS (70 F); F.P.: 32 F; Flash Pt.: None (TOC); Pour Pt.: 32 F
Composition: 18% conc.
Containers: Specially lined 5 gal cans and 55 gal drums
Warnings and Caution: Store to prevent freezing where possible; slightly corrosive to mild steel

SM 2033. General Electric
Chemical Description: Silicone O/W emulsion of 100 cs dimethyl fluid
Category: Silicone emulsion for chemical specialty, nonionic
Applications: Auto polishes, vinyl polishes, furniture polishes; provides a quick-breaking emulsion, ease of rub-out, and more uniform deposition in spray-and-wipe
Form: Liquid
Properties: Density: 8.2–8.3 lb/gal; Visc.: < 20 cs
Composition: 35% silicone

SM 2035. General Electric
Chemical Description: Silicone O/W emulsion of 1000 cs dimethyl fluid
Category and Applications: See SM 2033
Form: Liquid
Properties: Density: 8.2–8.3 lb/gal; Visc.: < 20 cs
Composition: 35% silicone

SM 2040. General Electric
Chemical Description: Silicone O/W emulsion of a 10,000 cs Viscasil dimethyl fluid
Category: Slip agent, nonionic
Applications: Slip agent in aerosol starches and other chemical specialties where slip or release are important; furniture polish
Form: Liquid
Properties: Density: 8.2–8.3 lb/gal; Visc.: < 20 cs
Composition: 35% silicone

SM 2061. General Electric
Chemical Description: Silicone O/W emulsion of a 60,000 cs Viscasil dimethyl fluid
Category: Slip agent, nonionic
Applications: Vinyl polishes; optimum slip in aerosol spray starches
Form: Liquid; Color: White; Odor: Slight sorbic acid
Properties: Solubility: Appreciable sol. in water; Density: 8.2–8.3 lb/gal; Sp. gr.: 1.01; Visc.: < 20 cs; B.P.: 212 F; Flash Pt.: > 575 F; Stability: Stable
Composition: 35% silicone
Toxicity: May cause minor eye irritation

SM 2068. General Electric
Chemical Description: Silicone anionic O/W emulsion of a dimethyl fluid of a visc. > 100,000 cs
Category: Slip agent, anionic
Applications: Slip agent in aerosol starches or fabric sizes
Form: Liquid; Color: White; Odor: Slight
Properties: Solubility: Appreciable

sol. in water; Density: 8.2–8.3 lb/gal; Sp. gr.: 1.01; Visc.: 20 cs; B.P.: 212 F; Stability: Good dilution and shear stability
Composition: 35% silicone
Toxicity: Mild transitory eye irritation
Warnings and Caution: Safety glasses recommended

SM 2140. General Electric
Chemical Description: 10,000 cs dimethyl polysiloxane resin (emulsion in water)
Category: Mold release agent, nonionic
Applications: For use in rubber and plastic production operations, in foundry release; good lubricity
Form: Liquid; Color: White; Odor: Essentially odorless
Properties: Solubility: Appreciable sol. in water; Density: 8.25 lb/gal; Sp. gr.: 1.01; Visc.: 1500 cps; B.P.: 212 F; Stability: High temp. stability; generally stable in hard water
Composition: 50% silicone
Toxicity: May cause slight transitory eye irritation
Warnings and Caution: Safety glasses recommended; keep from freezing

SM 2155. General Electric
Chemical Description: Silicone emulsion of dimethyl fluid [SF 96 (1000 cs)]
Category and Applications: See SM 2140
Form: Liquid; Color: White
Properties: Density: 8.25 lb/gal; Visc.: 1500 cps max.; Stability: High temp. stability; generally stable in hard water
Composition: 50% silicone
Toxicity: Transitory eye irritant
Warnings and Caution: Avoid freezing of the product

SM 2162. General Electric
Chemical Description: Silicone emulsion of a dimethyl fluid [SF 96 (350 cs)]
Category and Applications: See SM 2140
Form: Liquid; Color: White; Odor: Mild
Properties: Solubility: Appreciable sol. in water; Density: 8.25 lb/gal; Sp. gr.: 1.04; Visc.: 1000 cps max.; B.P.: 212 F; Flash Pt.: > 575 F; Stability: High temp. stability; generally stable in hard water
Composition: 50% silicone
Toxicity: Transitory eye irritant
Warnings and Caution: Avoid freezing of the product

S-Maz 20. Mazer Chem. Inc.
Chemical Description: Sorbitan monolaurate
Category: Nonionic surfactant
Applications: Preparing W/O emulsions; as antistats, textile softeners, lubricants, process defoamers, opacifiers, coemulsifiers; used with T-Maz series as O/W emulsifiers for cosmetics, food formulations, industrial oils, and household products
Form: Liquid; Color: Amber
Properties: Solubility: Sol. in etha-

nol, naphtha; disp. hot in water; miscible in certain proportions with acetone, toluol, mineral oil, vegetable oil; Sp. gr.: 1.0; Visc.: 4500 cps; HLB: 8.6; Sapon. No.: 158–170; Acid No.: 7 max.; Hydroxyl No.: 330–358
Composition: 100% conc.

S-Maz 40. Mazer Chem. Inc.
Chemical Description: Sorbitan monopalmitate
Category and Applications: See S-Maz 20
Form: Flakes; Color: Tan
Properties: Solubility: Partly sol. hot in ethanol, acetone, toluol, mineral and vegetable oils; miscible hot in certain proportions with naphtha; M.P.: 45–46 C; HLB: 6.7; Sapon. No.: 140–150; Acid No.: 7.5 max.; Hydroxyl No.: 275–305
Composition: 100% conc.

S-Maz 60. Mazer Chem. Inc.
Chemical Description: Sorbitan monostearate
Category and Applications: See S-Maz 20
Form: Flakes; Color: Cream
Properties: Solubility: Partly sol. hot in ethanol, toluol, mineral and vegetable oils; M.P.: 50–53 C; HLB: 4.7; Sapon. No.: 147–157; Acid No.: 10 max.; Hydroxyl No.: 235–260
Composition: 100% conc.

S-Maz 65. Mazer Chem. Inc.
Chemical Description: Sorbitan tristearate
Category and Applications: See S-Maz 20
Form: Flakes; Color: Cream
Properties: Solubility: Partly sol. in toluol; disp. in acetone, naphtha, mineral and vegetable oils; Sp. gr.: 1.0; M.P.: 53–55 C; HLB: 2.1; Sapon. No.: 176–188; Acid No.: 15 max.; Hydroxyl No.: 66–80
Composition: 100% conc.

S-Maz 67. Mazer Chem. Inc.
Chemical Description: Sorbitan monoisostearate
Category and Applications: See S-Maz 20
Form: Liquid; Color: Light yellow
Properties: Solubility: Sol. in mineral and vegetable oils; miscible in certain proportions with ethanol, toluol, naphtha; disp. in water; Sp. gr.: 1.0; Visc.: 1000 cps; HLB: 4.7; Sapon. No.: 146–157; Acid No.: 10 max.; Hydroxyl No.: 230–255
Composition: 100% conc.

S-Maz 80. Mazer Chem. Inc.
Chemical Description: Sorbitan monooleate
Category and Applications: See S-Maz 20
Form: Liquid; Color: Amber
Properties: Solubility: Sol. in mineral and vegetable oils; miscible in certain proportions with ethanol, toluol, naphtha; disp. in water; Sp. gr.: 1.0; Visc.: 1000 cps; HLB: 4.3; Sapon. No.: 149–160; Acid No.: 7.5 max.; Hydroxyl No.: 193–209
Composition: 100% conc.

S-Maz 85. Mazer Chem. Inc.
Chemical Description: Sorbitan trioleate
Category and Applications: See S-Maz 20
Form: Liquid; Color: Amber
Properties: Solubility: Sol. in toluol, naphtha, mineral and vegetable oils; miscible hot in certain proportions with ethanol, acetone; disp. in water; Sp. gr.: 1.0; Visc.: 200 cps; HLB: 1.8; Sapon. No.: 172–186; Acid No.: 14 max.; Hydroxyl No.: 56–68
Composition: 100% conc.

S-Maz 90. Mazer Chem. Inc.
Chemical Description: Sorbitan monotallate
Category and Applications: See S-Maz 20
Form: Liquid; Color: Amber
Properties: Solubility: Sol. in mineral and vegetable oils; miscible in certain proportions with ethanol, toluol, naphtha; disp. in water; Sp. gr.: 1.0; Visc.: 800 cps; HLB: 4.1; Sapon. No.: 145–160; Acid No.: 10 max.; Hydroxyl No.: 180–210
Composition: 100% conc.

Sodasil P-598. Witco Chem./Organics Div.
Chemical Description: Sodium silicate/sodium carbonate-based builder
Category: Detergent
Applications: Booster beads for high-, medium-, or low-foaming household and industrial cleaners
Form: Powder
Composition: 85% conc.

Sodium Hydrosulfite, Conc. Cities Service Co.
Category: Reducing agent
Applications: Powerful reducing agent; used in vat color reduction, bleaching soaps, paper stock, sugar, clay, etc.; sodium formate and zinc process materials
Form: Crystalline powder; Color: White
Containers: 250-lb steel or fiber drums

Sodium Lactate, 60%. Patco Cosmetic Prod./C.J. Patterson Co.
Chemical Description: Sodium lactate
Category: Humectant
Applications: For use as a moisturizer
Form: Clear liquid; Color: 30 max. (APHA); Odor: Essentially odorless
Properties: Sp. gr.: 1.31–1.34; pH: 8–9 (as is)
Containers: 600 lb net open head steel drum with polyethylene liner, 50 lb net polyethylene container-in-a-carton
References: Bulletin No. 119

Sofnon TK-8. Toho Chem. Ind. Co.
Chemical Description: Blend of nonionics, anionics, wool fat, and mineral oil
Category: Wetting agent
Applications: Spinning oil for

rough wool
Form: Liquid
Composition: 100% active ingred.

Softigen 701. Dynamit Nobel Chem.
Chemical Description: Partial glyceride mixture of an unsaturated fatty acid rich in hydroxyl groups
Category: Refatting and skin protecting agent, emulsifier, nonionic
Applications: Pharmaceutical and cosmetic ointments, creams, suppositories, soaps, bath products; W/O emulsions; adhesive power to and compatibility with skin
Form: Clear, oily liquid; Color: Yellowish-white; Odor: Characteristic
Properties: Solubility: Sol. in ether, benzene, toluene, xylene, chloroform, and dichlorethylene; Density: 0.979–0.981 (30 C); Visc.: 640–660 cps (30 C); Sapon. No.: 155–170; Acid No.: 3 max.; Iodine No.: 72–78
Composition: 100% conc.
Containers: 25 kg net tinned casks, 200 kg net barrels

Softigen 767. Dynamit Nobel Chem.
Chemical Description: Partial glyceride blend of natural, saturated vegetable fatty acids with medium chain length
Category: Refatting agent, solubilizer, wetting agent, emollient, nonionic
Applications: Fat producing and restoring agent in cosmetics and pharmaceuticals, e.g., foam bath preparations, low-degree alcoholic preparations such as after-shaves; solubilizer for perfumes, essential oils, and volatile crystalline substances; wetting properties for making cloudy detergents clear; treatment of psoriatic diseases
Form: Viscous, oily liquid; Color: Yellowish; Odor: Characteristic
Properties: Solubility: Sol. in water, acetone, ethyl acetate, butyl acetate; miscible in any proportion with isopropanol, 1,2-propylene glycol, glycerin (30%), castor oil; Density: ≈ 1.068 g/ml; Visc.: 110–130 cps; HLB: 19; Sapon. No.: 90–100; Acid No.: 1 max.; Iodine No.: 1 max.; Stability: Stable in aq. acid sol'ns. of pH > 2; Surface Tension: 29.7 dynes/cm (1% sol'n. @ 20 C)
Composition: 100% conc.

Softisan 100. Dynamit Nobel Chem.
Chemical Description: Triglyceride of a mixture of saturated vegetable fatty acids having a chain length C_{10}–C_{18}
Category: Ointment base, nonionic
Applications: Oxidation-resistant solid fat for use in cosmetic and pharmaceutical industry; improves consistency, melting properties, and gloss in ointments, creams, lipsticks, make-up, etc.
Form: Solid; Color: White; Odor and Taste: Neutral
Properties: Ref. Index: 1.4490–1.4510 (40 C); Solubility: Sol. in ether, benzene, toluene, xylene,

chloroform, carbon tetrachloride, dichlorethylene, dioxane; Sp. gr.: 0.950–0.980 (20 C); Visc.: ≈ 30 cps (40 C); M.P.: 33–36 C; Solid. Pt.: 29–33 C; Sapon. No.: 230–240; Acid No.: 0.5 max.; Hydroxyl No.: 15 max.; Iodine No.: 3 max.
Containers: 20 kg net cartons, in pellet form

Softisan 378. Dynamit Nobel Chem.
Chemical Description; Caprylic/capric/stearic triglyceride
Category: Ointment base, stabilizer, nonionic
Applications: Neutral ointment base with good skin compatibility and resorption characteristics, for the preparation of nonaqueous ointments and creams
Form: Waxy
Properties: Sapon. No.: 240–255; Acid No.: < 1; Hydroxyl No.: < 20
Composition: 100% active ingred.

Softisan 601. Dynamit Nobel Chem.
Chemical Description: Glyceryl cocoate and hydrogenated coconut oil and cetereth-25
Category: Ointment base, nonionic
Applications: Cosmetic and pharmaceutical ointment and cream base for O/W-type products
Form: Solid; Color: White-yellow
Properties: Sp. gr.: 0.97 (60 C); Sapon No.: 120–140; Acid No.: 1 max.; Hydroxyl No.: 240–270; Iodine No.: 5 max.
Composition: 100% conc.

Solar 25. Swift Chem.
Chemical Description: Nonionic coconut oil fatty acid amine condensate and anionic amine sulfonate
Category: Wetting agent, detergent, nonionic/anionic
Applications: Air entraining agent in gypsum products, cement, and concrete; cleaning compounds, incl. hand dishwashing products, car wash, or other surface cleaning where high sudsing is desired
Form: Viscous liquid; Color: Straw
Properties: Sp. gr.: 1.05; Pour Pt.: 0 C; Surface Tension: 34.5 dynes/cm (0.1% aq.); pH: 8.5 (1% sol'n.)
Composition: 100% active ingred.
Containers: Pails, 450 lb drums

Solar 40 BD Flakes & Granules. Swift Chem.
Chemical Description: Alkyl aryl sulfonate
Category: Detergent, wetting agent, anionic
Applications: Sudsing detergent base for built compounds; produces uniform and nondusty product, active in hard or soft, hot or cold water, and in acid or alkaline media; used in metal cleaning, dairy cleaning, dishwashing compounds, building maintenance, agricultural chemicals, automotive specialties
Form: Drum dried flakes or granules; Color: Creamy white; Odor:

Very slight
Properties: Solubility: 6% in water; Density: 37.4 lb/ft^3 (bulk, granules); Surface Tension: 35.8 dynes/cm (0.1%); pH: 9.05 (1% sol'n.); Biodegradable
Composition: 40% active ingred.
Containers: Bags
References: Tech. bulletin #204

Solar 80 Flakes & Granules. Swift Chem.
Chemical Description: Alkyl aryl sulfonate
Category: Detergent, wetting agent, anionic
Applications: Manufacture of detergents, bubble baths, dishwashing products, industrial cleaners, auto wash products; rewetting agent in paper or wettable powders; paint and cement industries
Form: Flakes and granules; Color: Creamy white; Odor: Very slight, pleasant
Properties: Solubility: 16% in water; Density: 21.8 lb/ft^3 (bulk, flakes), 33 lb/ft^3 (bulk, granules); Surface Tension: 35.8 dynes/cm (0.1%); pH: 7.7 (1% sol'n.); Biodegradable
Composition: 85% active ingred.
Containers: Bags
References: Tech. bulletin #205

Solar Air. Swift Chem.
Chemical Description: Coconut fatty acid amine condensate
Category: Detergent, nonionic
Applications: Air-entraining agent for concrete blocks, pipes, precast structures, prestressed beams, etc.; advantages are: greater "green" strength, elimination of hairline cracks, lighter and more uniform color, less wear and tear on equipment
Form: Liquid; Color: Yellow
Properties: Ref. Index: 1.3427; Sp. gr.: 1.0015 (30 C); Visc.: 5 cps (30 C)
Composition: 25% active ingred.
Containers: 55 gal drums

Solar Air-2. Swift Chem.
Category and Applications: See Solar Air
Form: Liquid; Color: Light
Properties: Ref. Index: 1.3520; Sp. gr.: 1.0040 (30 C); Visc.: 10 cps (30 C)
Composition: Double strength Solar Air

Solar Air-4. Swift Chem.
Category and Applications: See Solar Air
Form: Liquid; Color: Light
Properties: Ref. Index: 1.3695; Sp. gr.: 1.0090 (30 C); Visc.: 22 cps (30 C)
Composition: Quadruple strength Solar Air

Solar CI-387. Swift Chem.
Chemical Description: Imidazoline hydrochloride
Category: Wax-rinse base, cationic
Applications: Car wax compound and cleaner; imparts a wax-like water repellent film to automotive bodies in the rinse operation; retards rusting; nonsmearing
Form: Liquid; Color: Amber;

Odor: Mild
Properties: Sp. gr.: 0.951; Visc.: 40 cps; pH: 5–7
Composition: 52% active ingred.
Containers: 40 lb pails, 400 lb drums

Solar CO. Swift Chem.
Chemical Description: Coconut oil fatty acid amine condensate
Category: Detergent, intermediate, nonionic
Applications: Cosmetic and detergent formulations incl. liquid dishwashing and other high sudsing liquids where suds stabilization is desired; intermediate in manufacture of modified amides; dedusting and suds stabilization of dry detergents
Form: Viscous liquid; Color: Light straw; Odor: Slight
Properties: Sp. gr.: 1.02; Flash Pt.: 166 C; Pour Pt.: 23 C; pH: 9.5 (1% sol'n.)
Composition: 100% active ingred.
Containers: 450 lb drums, pails
References: Tech. bulletin #209

Solar DC. Swift Chem.
Category: Detergent
Applications: Industrial dry cleaning detergent
Form: Liquid; Color: Amber
Properties: pH: 9.2
Composition: 100% active ingred.
Reaction: Alkaline
Containers: Pails, drums

Solar DW. Swift Chem.
Chemical Description: Proprietary blend of petrochemicals and mild fatty chemicals
Category: Detergent, anionic
Applications: Detergent base for hard surface cleaners, liquid hand dishwashing compounds; produces copious suds even in presence of grease, oil, and hard water
Form: Liquid; Color: Amber; Odor: Slight
Properties: pH: 10.5
Composition: 100% active ingred.
Reaction: Alkaline
Containers: Pails, drums
References: Tech. bulletin # 235

Solar F221. Swift Chem.
Chemical Description: Modified amide
Category: Emulsifier, nonionic
Applications: Produces heat stable W/O emulsions; making of cutting and soluble oils
Form: Liquid; Color: Dark amber; Odor: Practically odorless
Properties: Solubility: Slightly sol. in diesel oil, kerosene, paraffin oil, isopropyl and ethyl alcohol, methylene chloride, and chloroform; Density: 8.5 lb/gal; Pour Pt.: 10 F
Composition: 100% active ingred.
Containers: Pails, drums
References: Tech. bulletin #227

Solar F342. Swift Chem.
Chemical Description: Modified amide
Category: Emulsifier, nonionic
Applications: O/W emulsions; laundering; making of cutting and soluble oils
Form: Liquid; Color: Dark amber; Odor: Practically odorless

Properties: Solubility: Sol. in ethyl alcohol, isopropyl alcohol, diesel oil, kerosene, and chloroform; disp. in water; Density: ≈ 8 lb/gal; Flash Pt.: 210 F (CC); Pour Pt.: –6 F
Composition: 100% active ingred.
Containers: Pails, 450 lb drums
References: Tech. bulletin #228

Solar HD. Swift Chem.
Chemical Description: Blend of detergents, water softeners, and conditioners
Category: Detergent, anionic
Applications: Laundry product of moderate alkalinity for use by commercial laundries, hospitals, rug cleaners, households; produces moderate volume of stable suds in hard or soft water
Form: Free-flowing powder; Color: White; Odor: Pleasant
Properties: Solubility: Sol. in hot or cold water; Density: 55 lb/ft^3 (bulk); pH: 11.0 (1%)
Composition: 18% active ingred.
Reaction: Alkaline
Containers: 50 lb bags
References: No. 206

Solar LLC-361. Swift Chem.
Chemical Description: Phosphate-free formulated compound
Category: Detergent
Applications: Concentrated laundry compound with brighteners and suspending agents for commercial and institutional use
Form: Liquid; Color: Amber
Properties: Density: 8.52 lb/gal; Visc.: 4.2 poises; Cloud Pt.: > 212 F; Pour Pt.: 50 F
Reaction: Alkaline
Containers: Pails, 450 lb drums

Solar NP. Swift Chem.
Chemical Description: Nonyl phenoxy polyoxyethylene ethanol
Category: Detergent, emulsifier, wetting agent, dispersant, penetrant, nonionic
Applications: Base for cleaning compounds (floor cleaners, hand soaps, shampoos, sanitizers, acid type bowl cleaners, etc.); cosmetics (deodorants, synthetic bars); dairy cleaning; laundry and dry cleaning; leather wetting, defatting, softening; metal cleaning, pickling, plating; paint, plastics, and rubber dispersing; textile scouring, dyeing, desizing, bleaching aid; agricultural wetting and emulsification; brewing industry (bottle washing); used in acid or alkaline sol'ns. and in hard water or sol'ns. of metal salts
Form: Viscous, clear, liquid; Odor: Nil
Properties: Solubility: Sol. in water, ethylene glycol, ethanol carbon tetrachloride, xylene; Sp. gr.: 1.06; Cloud Pt.: 125–133 F (1%); Flash Pt.: 535–555 F; Pour Pt.: 37 F; Solid. Pt.: 26 F; Surface Tension: 30 dynes/cm (0.01% sol'n.); pH: 7.0 (1% sol'n.)
Composition: 100% active ingred.
Reaction: Neutral
Containers: Pails, drums

Solar Regular. Swift Chem.
Chemical Description: Coconut oil fatty acid amine condensate

modified with diethanolamine oleate
Category: Detergent, wetting agent, corrosion inhibitor, nonionic
Applications: Base for cleaning compounds; thickening and average sudsing properties; floor cleaners, metal cleaning, burnishing, general degreasing, rust retardation, textiles, laundering, cleaners of streets, buildings, hospitals, making of cutting and soluble oils
Form: Viscous liquid; Color: Amber; Odor: Sweet
Properties: Solubility: Complete in water; Sp. gr.: 1.07; Flash Pt.: 179 C; Pour Pt.: –4 C; pH: 9.5 (1% sol'n.)
Composition: 100% active ingred.
Reaction: Alkaline
Containers: 5 gal pails, 45 gal drums

Solar TE. Swift Chem.
Chemical Description: Specially modified amide produced primarily from coconut oil
Category: Detergent, lubricant, coupling agent, nonionic
Applications: Hard surface cleaners, built formulas, liquid floor scrubbing compounds, liquid steam cleaners; textile fulling and scouring of woolens and raw stock; textile lubricant; couples most alkalis
Form: Liquid; Color: Amber
Properties: Flash Pt.: 360 F; Fire Pt.: 370 F; Pour Pt.: 20 F; Surface Tension: 32.0 dynes/cm (0.1%); pH: 9.15 (1% sol'n.)
Containers: Pails, drums
References: Tech. bulletin #224

Solricin 135. N.L. Ind./Ind. Chem. Div.
Chemical Description: Potassium ricinoleate
Category: Emulsifier, anionic
Applications: Mild germicide, glycerized rubber lubricant, emulsifier and foam stabilizer in foamed rubber; making of cutting and sol. oils; household and cosmetic products
Form: Liquid; Color: 2 (Gardner)
Properties: Density: 8.6 lb/gal; Sp. gr.: 1.034; Visc.: 0.9 stokes
Composition: 32% active ingred.
Containers: 55 gal drums
Warnings and Caution: Keep from contact with oxidizing materials

Solulan 5. Amerchol
Chemical Description: Ethoxylated lanolin alcohol (5 moles EO)
Category: Emulsifier, stabilizer, solubilizer, wetting and dispersing agent, conditioner, emollient, nonionic
Applications: Creams and lotions; soaps, shampoos, and detergent systems (superfatting, foam stabilizer); hair care products (plasticizer for hair spray and styling resins, conditioner); aftershaves, astringents, and friction lotions (solubilizer for perfumes and emollient oils); cold waves, hair dyes, and bleaches; dermatological vehicles
Form: Semisold; Color: Amber; Odor: Faint, pleasant
Properties: Solubility @ 5%: Sol. in anhydrous ethanol; disp. cloudy) in water, castor oil, iso-

propyl myristate, mineral oil; Sapon. Value: 14 max.; Acid Value: 3 max.; Hydroxyl Value: 120–135
Composition: 100% conc.

Solulan 16. Amerchol
Chemical Description: Ethoxylated lanolin alcohol (16 moles EO)
Category and Applications: See Solulan 5; most effect foam stabilizer for detergent systems
Form: Waxy solid; Color: Light tan; Odor: Faint, pleasant
Properties: Solubility @ 5%: Sol. in water, 30% ethanol; forms cloudy gel in castor oil, isopropyl myristate; HLB: 15; Cloud Pt.: 64–70 C; Sapon. Value: 8 max.; Acid Value: 1.5 max.; Hydroxyl Value: 60–70; pH: 4.5–7.5 (10% aq.)
Composition: 100% conc.

Solulan 25. Amerchol
Chemical Description: Ethoxylated lanolin alcohol (25 moles EO)
Category and Applications: See Solulan 5
Form: Waxy solid; Color: Yellow; Odor: Faint, pleasant
Properties: Solubility @ 5%: Sol. in water, 30% ethanol; disp. (cloudy) in isopropyl myristate; HLB: 16; Cloud Pt.: 85–92 C; Sapon. Value: 8 max.; Acid Value: 1 max.; Hydroxyl Value: 40–50; pH: 4.5–7.5 (10% aq.)
Composition: 100% conc.

Solulan 75. Amerchol
Chemical Description: Ethoxylated lanolin (75 moles EO)
Category and Applications: See Solulan 5
Form: Waxy solid; Color: Light yellow-amber; Odor: Faint, pleasant
Properties: Solubility @ 5%; Sol. in water, 30% ethanol; forms cloudy gel in glycerin, castor oil; HLB: 15; Cloud Pt.: 80–87 C; Sapon. Value: 10–20; Acid Value: 1.5 max.; Hydroxyl Value: 40–50; pH: 4.5–7.5 (10% aq.)
Composition: 100% conc.

Solulan 97. Amerchol
Chemical Description: Acetylated, ethoxylated lanolin (9 moles EO)
Category and Applications: See Solulan 5
Form: Viscous liquid; Color: Light amber; Odor: Faint, pleasant
Properties: Solubility @ 5%: Sol. in water, ethanol (30% to anhydrous), castor oil; HLB: 15; Sapon. Value: 110–130; Acid Value: 3 max.; Hydroxyl Value: 10 max.; pH: 4.5–7.5 (10% aq.)
Composition: 100% conc.

Solulan 98. Amerchol
Chemical Description: Partially acetylated lanolin-derived complex (10 moles EO)
Category and Applications: See Solulan 5; also aids pearling of stearic acid emulsions
Form: Viscous liquid; Color: Light amber; Odor: Faint, pleasant
Properties: Solubility @ 5%: Sol. in water, ethanol (30% to anhy-

drous), castor oil; HLB: 13; Sapon. Value: 65–75; Acid Value: 2 max.; Hydroxyl Value: 57–67; pH: 4.5–7.5 (10% aq.)
Composition: 100% conc.

Solulan C-24. Amerchol
Chemical Description: Polyoxyethylene cholesterol (24 moles EO)
Category and Applications: See Solulan 5; most effective in preventing gelation of soap-based emulsions
Form: Waxy solid; Color: Off-white to pale yellow; Odor: Faint, pleasant
Properties: Solubility @ 5%: Sol. in water; sol. @ 1% in 30% ethanol, propylene glycol; forms cloudy gel in castor oil, isopropyl myristate; HLB: 14; Cloud Pt.: 88–95 C; Sapon. Value: 3 max.; Acid Value: 1.5 max.; Hydroxyl Value: 35–45
Composition: 100% conc.

Solulan L-575. Amerchol
Chemical Description: Ethoxylated lanolin (75 moles EO)
Category and Applications: See Solulan 5
Form: Clear, viscous liquid; Color: Light yellow-amber; Odor: Faint, pleasant
Properties: Solubility @ 5%: Sol. in water, 30% ethanol; forms cloudy gel in glycerin; HLB: 15; Cloud Pt.: 80–87 C; Sapon. Value: 10 max.; Acid Value: 1 max.; Hydroxyl Value: 20–30
Composition: 50% conc.

Solusol 75%. American Cyanamid Co.
Chemical Description: Dioctyl ester of sodium sulfosuccinic acid
Category: Wetting agent, anionic
Applications: Cosmetic formulations in aq. systems
Form: Liquid; Color: Nearly colorless
Properties: Sp. gr.: 1.09
Composition: 75% active ingred.; other components: lower alcohol and water
Toxicity: Eye irritant
Containers: Steel or Liquipak drums
Warnings and Caution: Avoid eye contact

Solusol 84%. American Cyanamid Co.
Chemical Description: Dioctyl ester of sodium sulfosuccinic acid
Category: Wetting agent, anionic
Applications: Cosmetic formulations
Form: Liquid; Color: Nearly colorless
Properties: Sp. gr.: 1.10; B.P.: 350 F
Composition: 84% active ingred.; other components: 16% propylene glycol
Toxicity: Eye irritant
Containers: Liquipak drums
Warnings and Caution: Avoid eye contact

Solusol 85%. American Cyanamid Co.
Chemical Description: Dioctyl ester of sodium sulfosuccinic acid

Category: Wetting agent, anionic
Applications: Cosmetic formulations in aq. systems
Form: Solid; Color: White
Properties: Sp. gr.: 1.1; M.P.: < 300 C
Composition: 85% active ingred.; other components: 15% sodium benzoate
Toxicity: Eye irritant
Containers: Leverpak drums
Warnings and Caution: Avoid eye contact

Solusol 100%. American Cyanamid Co.
Chemical Description: Dioctyl ester of sodium sulfosuccinic acid
Category: Wetting agent, anionic
Applications: Cosmetic formulations, esp. in nonaqueous systems
Form: Solid; Color: White
Properties: Solubility: 1.5 g in 100 ml water @ 25 C; Sp. gr.: 1.1; M.P.: 153–157 C; Surface Tension: 28.7 dynes/cm (0.1% solids)
Composition: 100% active ingred.
Toxicity: Eye irritant
Containers: Fiber and leverpak drums
Warnings and Caution: Avoid eye contact

Solvent Scour 2527. Hart Chem. Ltd.
Chemical Description: Blend of alkanolamides and alkyl phenol ethoxylate
Category: Scouring agent, nonionic
Applications: Removing persistent stains and greasy products which are usually difficult to scour out
Form: Liquid; Color: Yellow
Properties: Solubility: Forms stable emulsion; Density: 0.899
Composition: 100% active ingred.
Reaction: Acid
Toxicity: Skin and eye irritant
Containers: Drums, T/T
Warnings and Caution: Avoid contact with skin and eyes

Solwax C-24. Van Schuppen Chemie
Chemical Description: Polyethoxylated cholesterol
Category: Emulsifier, stabilizer, so solubilizer, nonionic
Applications: O/W emulsifier; stabilizer for W/O emulsions; used in alcoholic and hydrophilic cosmetics, e.g., hair sprays, hair lotions, hair gels, shampoos, shaving lotions; solubilizer for steroids in aq. systems
Form: Firm waxy solid; Color: Light yellow; Odor: Faint, pleasant
Properties: Solubility: Sol. in water and alcohol; M.P.: 45–48 C; Cloud Pt.: 85–95 C (1% sol'n. in 5% NaCl aq.); Sapon. Value: 5.0 max.; Acid Value: 1.0 max.; Hydroxyl Value: 45–60
Composition: 100% active ingred.
References: Prod. information no. 22

Solwax L 20. Van Schuppen Chemie
Chemical Description: Alkoxylated derivative of pharmaceutical lanolin

Category: Emulsifier, foam stabilizer, softener, nonionic
Applications: O/W emulsifier; used in hydrophilic cosmetics, e.g., creams, hand lotions, milks, deodorants, brilliantines; in shampoos and shaving compositions, to enhance foam stability; in liquid detergent compositions for softening and skin protection
Form: Firm waxy solid; Color: Light yellow; Odor: Faint, pleasant
Properties: Solubility: Dissolves in water and alcohol; Softening Pt.: 45–55 C; Cloud Pt.: 75–80 C (1% sol'n. in 5% NaCl aq.); Sapon. Value: 10–30; Acid Value: 3.0 max.; Hydroxyl Value: 15–45
Composition: 100% active ingred.
References: Prod. information no. 21 A

Solwax LG 35. Van Schuppen Chemie
Chemical Description: Alkoxylated derivative of lanolin alcohols
Category: Emulsifier, emollient, solubilizer, nonionic
Applications: In alcoholic and hydrophilic cosmetics, e.g., hair sprays, hair lotions, hair gels, shampoos, shaving lotions; emollient, conditioning, solubilizing, and gelling properties; emulsifier for O/W emulsions
Form: Firm waxy solid; Color: Yellow-brownish; Odor: Faint, pleasant
Properties: Solubility: Sol. in water and alcohol; Softening Pt.: 40–50 C; Cloud Pt.: 75–85 C (1% sol'n. in 5% NaCl aq.); Sapon. Value: 5–15; Acid Value: 5.0 max.; Hydroxyl Value: 45–65
Composition: 100% active ingred.
References: Prod. bulletin no. 23

Sontex 19. Marathon Morco Co.
Chemical Description: White mineral oil, USP
Category: Lubricant, spreading agent, release agent, emollient, binder, plasticizer
Applications: USP grade suitable for internal and medicinal use; drugs/pharmaceutical, personal health care; cosmetics; food processing and packaging; animal feed, medication, and care; paper; plastic; rubber processing; soap; textiles; baking, yeast, beet sugar, candy
Form: Oil; Color: Water-white; Odor and Taste: None
Properties: m.w.: 391; Density: 7.28 lb/gal; Sp. gr.: 0.874 (60/60 F); Visc.: 39.7 cs (100 F); Flash Pt.: 374 F (COC); Pour Pt.: 0 F
Reaction: Neutral

Sontex 21. Marathon Morco Co.
Chemical Description: White mineral oil, USP
Category and Applications: See Sontex 19
Form: Oil; Color: Water-white; Odor and Taste: None
Properties: m.w.: 396; Density: 7.29 lb/gal; Sp. gr.: 0.875 (60/60 F); Visc.: 44.1 cs (100 F); Flash Pt.: 385 F (COC); Pour Pt.: 0 F
Reaction: Neutral

Sontex 35. Marathon Morco Co.
Chemical Description: White mineral oil, USP
Category and Applications: See Sontex 19
Form: Oil; Color: Water-white; Odor and Taste: None
Properties: m.w.: 430; Density: 7.34 lb/gal; Sp. gr.: 0.881 (60/60 F); Visc.: 75.5 cs (100 F); Flash Pt.: 425 F (COC); Pour Pt.: 0 F
Reaction: Neutral

Sontex 55. Marathon Morco Co.
Chemical Description: White mineral oil, NF
Category: Lubricant, spreading agent, release agent, emollient, binder, plasticizer
Applications: NF grade suitable for use in cosmetic, food processing, and many medicinal formulations; also animal feed, medication, and care; paper; plastic; rubber processing; soap; textiles
Form: Oil; Color: Water-white; Odor and Taste: None
Properties: m.w.: 283; Density: 6.99 lb/gal; Sp. gr.: 0.8398 (60/60 F); Visc.: 8.8 cs (100 F); Flash Pt.: 325 F (COC); Pour Pt.: 10 F
Reaction: Neutral

Sontex 70. Marathon Morco Co.
Chemical Description: White mineral oil, NF
Category and Applications: See Sontex 55
Form: Oil; Color: Water-white; Odor and Taste: None
Properties: m.w.: 298; Density: 7.19 lb/gal; Sp. gr.: 0.8639 (60/60 F); Visc.: 13.8 cs (100 F); Flash Pt.: 330 F (COC); Pour Pt.: −40 F
Reaction: Neutral

Sontex 75. Marathon Morco Co.
Chemical Description: White mineral oil, NF
Category and Applications: See Sontex 55
Form: Oil; Color: Water-white; Odor and Taste: None
Properties: m.w.: 340; Density: 7.00 lb/gal; Sp. gr.: 0.8408 (60/60 F); Visc.: 14.4 cs (100 F); Flash Pt.: 370 F (COC); Pour Pt.: 10 F
Reaction: Neutral

Sontex 75 T. Marathon Morco Co.
Chemical Description: White mineral oil, technical grade
Category: Lubricant, spreading agent, release agent, emollient, binder, plasticizer
Applications: Cosmetics, pharmaceuticals, plastics, textiles, rubber processing, paper, etc.; technical grades have optimum chemical stability and purity, but do not meet one or more of the USP or NF requirements
Form: Oil; Color: Water-white; Odor and Taste: None
Properties: m.w.: 319; Density: 7.05 lb/gal; Sp. gr.: 0.8473 (60/60 F); Visc.: 14.4 cs (100 F); Flash Pt.: 350 F (COC); Pour Pt.: 10 F

Sontex 85. Marathon Morco Co.
Chemical Description: White mineral oil, NF
Category and Applications: See

Sontex 55
Form: Oil; Color: Water-white; Odor and Taste: None
Properties: m.w.: 335; Density: 7.11 lb/gal; Sp. gr.: 0.8545 (60/60 F); Visc.: 16.9 cs (100 F); Flash Pt.: 365 F (COC); Pour Pt.: 10 F
Reaction: Neutral

Sontex 85 T. Marathon Morco Co.
Chemical Description: White mineral oil, technical grade
Category and Applications: See Sontex 75 T
Form: Oil; Color: Water-white; Odor and Taste: None
Properties: m.w.: 335; Density: 7.10 lb/gal; Sp. gr.: 0.8524 (60/60 F); Visc.: 16.9 cs (100 F); Flash Pt.: 355 F (COC); Pour Pt.: 10 F

Sontex 95 T. Marathon Morco Co.
Chemical Description: White mineral oil, technical grade
Category and Applications: See Sontex 75 T
Form: Oil; Color: Water-white; Odor and Taste: None
Properties: m.w.: 364; Density: 7.12 lb/gal; Sp. gr.: 0.8550 (60/60 F); Visc.: 19.3 cs (100 F); Flash Pt.: 360 F (COC); Pour Pt.: 10 F

Sontex 100. Marathon Morco Co.
Chemical Description: White mineral oil, NF
Category and Applications: See Sontex 55
Form: Oil, Color: Water-white; Odor and Taste: None
Properties: m.w.: 364; Density: 7.08 lb/gal; Sp. gr.: 0.8500 (60/60 F); Visc.: 20.5 cs (100 F); Flash Pt.: 370 F (COC); Pour Pt.: 10 F
Reaction: Neutral

Sontex 150. Marathon Morco Co.
Chemical Description: White mineral oil, NF
Category and Applications: See Sontex 55
Form: Oil; Color: Water-white; Odor and Taste: None
Properties: m.w.: 365; Density: 7.26 lb/gal; Sp. gr.: 0.8720 (60/60 F); Visc.: 32.0 cs (100 F); Flash Pt.: 365 F (COC); Pour Pt.: 0 F
Reaction: Neutral

Sorbit P. Ciba-Geigy Corp.
Chemical Description: Alkyl naphthalene sulfonate
Category: Wetting agent, foaming agent, emulsifier, detergent, anionic
Applications: Germicides, textile leveling
Form: Powder
Properties: Stability: Stable to hydrolysis in acid or alkali
Composition: 65% conc.

Sorbitol. Quad Chem. Corp.
Chemical Description: Sorbitol, 70% USP sol'n.
Form: Liquid; Color: Water-white; Odor: Odorless; Taste: Sweet
Properties: Ref. Index: 1.458; Sp. gr.: 1.30; Visc.: 110 cps; pH: 7
Composition: 70% active ingred.

Soromine AL. GAF Corp.
Chemical Description: Complex fatty amido amphoteric surfactant

Category: Softener, lubricant, amphoteric
Applications: Imparts an excellent hand, good body and draping qualities, and lubricity to synthetic, cellulosic, and animal fibers; finishing agents for all types of processing equipment (pads, jigs, package machines, dye becks, etc.); on leather in fat-liquoring; resists yellowing
Form: Smooth paste; Color: White; Odor: Essentially odorless
Properties: Solubility: Disp. in warm water; Stability: Stable to acid and alkali in concs. normally found in textile industry; stable in thermosetting baths; pH: 6.0–7.5 (10% disp.)
Composition: 26% min. active
Reaction: Function as substantive (cationic) softeners in acid liquors and as nonsubstantive (nonionic and/or anionic) softeners in neutral or alkaline liquors; 1% sol'n. is essentially neutral
References: Tech. bulletin 7543-176

Soromine AT. GAF Corp.
Chemical Description: Complex fatty amido amphoteric surfactant
Category and Applications: See Soromine AL
Form: Smooth paste; Color: White; Odor: Essentially odorless
Properties: Solubility: Disp. in warm water; Stability: Stable to acid and alkali in concs. normally found in textile industry; stable in thermosetting baths; pH: 6.0–7.5 (10% disp.)
Composition: 20% min. active
Reaction: See Soromine AL
References: Tech. bulletin 7543-176

Sorpol Series. Toho Chem. Ind. Co.
Chemical Description: Blend of polyoxyethylene alkylaryl ether, polyoxyethylene sorbitan alkylate, and special anionic sulfonate
Category: Emulsifier, nonionic/anionic
Applications: Emulsifier for agricultural pesticides
Form: Liquid
Composition: 100% active ingred.

Sotex 3CW. Morton Chem. Co.
Chemical Description: Long chain fatty acid ester containing multiple ether linkages
Category: Dispersant, wetting agent, nonionic
Applications: Dispersions of pigments in long oil alkyds, epoxy esters, and oleoresinous vehicles for paints and coatings; dispersions in lithographic ink vehicles; wetting agent when dry blending dyes with polystyrene, polypropylene, and other thermoplastic resins ; in combination with Sotex COS # 2 for dispersions of high oil absorption pigments and phospho-tungstated pigments
Form: Viscous liquid
Properties: Solubility: Sol. in aliphatic and aromatic hydrocarbons; very slightly sol. in polar solvents; Sp. gr.: 0.930–0.960; B.P.: > 200 C

Composition: 100% active ingred.
Containers: 440 lb (55 gal) steel drums

Sotex COS # 2. Morton Chem. Co.
Chemical Description: Long chain fatty acid ester of alkyl amino linkage
Category: Dispersant, cationic
Applications: Dispersions of high oil absorption pigments in oils, oleoresinous vehicles, plasticizers, etc. for paints and coatings; surface treating pigments; yields treated pigments with soft texture, reduced oil absorption, stability, high-color development in shorter milling times
Form: Liquid
Properties: Solubility: Oil-sol.; Sp. gr.: 0.900–0.940; B.P.: > 200 C
Composition: 100% active ingred.
Containers: 400 lb (55 gal) steel drums

Sotex CW. Morton Chem. Co.
Chemical Description: Long chain fatty acid ester containing multiple ether linkages
Category: Viscosity depressant and stabilizer, dispersant, cosolvent, nonionic
Applications: Viscosity depressant and stabilizer in vinyl plastisols; dispersions in amine resin-modified short oil alkyds for paints and coatings; with Sotex COS # 2 for dispersions of high oil absorption pigments in plasticizers; as polar-nonpolar cosolvent
Form: Viscous liquid
Properties: Solubility: Sol. in systems of hydrocarbons mixed with alcohols (urea or melamine modified alkyd resins), or with esters (nitrocellulose formulations), or with ketones (vinyl systems), and with plasticizers; Sp. gr.: 0.970–1.010; B.P.: > 200 C
Composition: 100% active ingred.
Containers: 440 lb (55 gal) steel drums

Sotex N. Morton Chem. Co.
Chemical Description: Long chain fatty acid esters containing multiple ether linkage
Category: Dispersant, nonionic
Applications: Dispersion of pigments in flexographic ink, nitrocellulose, and vinyl sol'ns.; in amine-modified alkyd formulas to improve gloss and eliminate craters caused by dispersed moisture; dispersions in all polar solvents
Form: Viscous liquid
Properties: Solubility: Sol. in all polar solvents (water, alcohol, esters, ketones, etc.); Sp. gr.: 1.02–1.06; B.P.: > 200 C
Composition: 100% active ingred.
Containers: 450 lb (55 gal) steel drums

Sovatex AW/2. Standard Chem. Co.
Category: Detergent, anionic
Applications: High temp. solvent detergent for textile scouring and spotting agents
Properties: Solubility: Water-disp.; Stability: Stable in bleach baths and with desizing enzymes

Sovatex BX/O. Standard Chem. Co.
Category: Detergent, anionic
Applications: High temp. solvent detergent for textile scouring and spotting agents
Properties: Solubility: Water-disp.

Sovatex C.130. Standard Chem. Co.
Category: Detergent, wetting agent, anionic
Applications: Textile treatment
Form: Liquid

Sovatex DS/C2. Standard Chem. Co.
Category: Detergent, wetting agent, anionic/nonionic
Applications: Textile treatment
Form: Liquid

Sovatex NIP/2. Standard Chem. Co.
Category: Detergent, nonionic
Applications: Medium/high temp. nonflammable solvent detergent for textile scouring and spotting agents
Properties: Solubility: Water-disp.

Sovatex PN/O. Standard Chem. Co.
Category: Dyebath auxiliary
Applications: Detergent penetrant and dyebath assistant

Sovatex SA/8. Standard Chem. Co.
Category: Detergent, anionic
Applications: Low temp. solvent detergent for textile scouring and spotting agents; esp. safe on delicate fabrics, rubber, etc.
Properties: Solubility: Water-sol.

Sovatex T/L. Standard Chem. Co.
Category: Detergent, anionic
Applications: Low temp. nonflammable solvent detergent for textile spotting agents and scouring
Properties: Solubility: Water-disp.

Sovatex WA/60. Standard Chem. Co.
Category: Wetting agent, anionic
Applications: Dyebath auxiliary particularly for acidic, neutral, and mild alkali conditions

Span 20. I.C.I. United States Inc.
Chemical Description: Sorbitan monolaurate
Category: Emulsifier, stabilizer, thickener, lubricant, softener, antistatic agent, nonionic
Applications: Foods, pharmaceuticals, cosmetics, cleaning compounds, textiles (lubricants, softeners, antistats, finishing aids, spinning assistants)
Form: Liquid; Color: Amber
Properties: Solubility @ 1%: Sol. in isopropyl alcohol, perchloroethylene, xylene, cottonseed oil, mineral oil; sol. (hazy) in propylene glycol; Visc.: 4250 cs; HLB: 8.6
Composition: 100% active ingred.

Span 40. I.C.I. United States Inc.
Chemical Description: Sorbitan monopalmitate
Category and Applications: See Span 20
Form: Solid; Color: Tan
Properties: Solubility @ 1%: Sol. in isopropyl alcohol, xylene; sol.

(hazy) in perchloroethylene; HLB: 6.7; Pour Pt.: 48 C
Composition: 100% active ingred.

Span 60. I.C.I. United States Inc.
Chemical Description: Sorbitan monostearate
Category and Applications: See Span 20
Form: Solid; Color: Tan
Properties: Solubility @ 1%: Sol. in isopropyl alcohol; sol. (hazy) in perchloroethylene, xylene; HLB: 4.7; Pour Pt.: 53 C
Composition: 100% active ingred.

Span 65. I.C.I. United States Inc.
Chemical Description: Sorbitan tristearate
Category and Applications: See Span 20
Form: Solid; Color: Cream
Properties: Solubility @ 1%: Sol. in isopropyl alcohol, perchloroethylene, xylene; HLB: 2.1; Pour Pt.: 53 C
Composition: 100% active ingred.

Span 80. I.C.I. United States Inc.
Chemical Description: Sorbitan monooleate
Category and Applications: See Span 20
Form: Liquid; Color: Amber
Properties: Solubility @ 1%: Sol. in isopropyl alcohol, perchloroethylene, xylene, cottonseed and mineral oils; Visc.: 1000 cs; HLB: 4.3
Composition: 100% active ingred.

Span 85. I.C.I. United States Inc.
Chemical Description: Sorbitan trioleate
Category and Applications: See Span 20
Form: Liquid; Color: Amber
Properties: Solubility @ 1%: Sol. in isopropyl alcohol, perchloroethylene, xylene, cottonseed and mineral oils; Visc.: 210 cs; HLB: 1.8
Composition: 100% active ingred.

Sponto 101. Witco Chem./Organics Div.
Chemical Description: Blend of oil-sol. sulfonates and polyoxyethylene ethers
Category: Emulsifier, anionic/nonionic
Applications: Pesticide formulations, esp. as lipophilic member of a matched pair (with Sponto 203)
Form: Liquid
Composition: 100% conc.

Sponto 140T. Witco Chem./Organics Div.
Chemical Description: Blended emulsifiers
Category: Emulsifier, anionic/nonionic
Applications: Emulsifier for high-conc. organic phosphate pesticide formulations; auxiliary emulsifiers for Sponto 232/234, 300T/500T, and 101/203 paired systems

Sponto 150T. Witco Chem./Organics Div.
Chemical Description: Blended

emulsifier
Category and Applications: See Sponto 140T

Sponto 168-D. Witco Chem./Organics Div.
Chemical Description: Mixture of mono and diesters of phosphoric acid and an alkylphenoxy polyethoxyethanol
Category: Emulsifier, anionic
Applications: Emulsifier and compatibility agent for pesticide/liquid fertilizer combinations
Form: Liquid
Composition: 100% conc.

Sponto 200. Witco Chem./Organics Div.
Chemical Description: Blended emulsifier
Category: Emulsifier, nonionic
Applications: Citrus spray oils and other crop oils

Sponto 203. Witco Chem./Organics Div.
Chemical Description: Blend of oil-sol. sulfonates and polyoxyethylene ethers
Category: Emulsifier, anionic/nonionic
Applications: Single emulsifier system for 4 lb/gal methyl parathion; broad range emulsifier for organic phosphate toxicants when used with Sponto 101
Form: Liquid
Composition: 100% conc.

Sponto 207. Witco Chem./Organics Div.
Chemical Description: Blended emulsifier
Category: Emulsifier, anionic
Applications: Emulsifier for pentachlorophenol in agricultural formulations

Sponto 232. Witco Chem./Organics Div.
Chemical Description: Blend of oil-sol. sulfonates and polyoxyethylene ethers
Category: Emulsifier, anionic/nonionic
Applications: Lipophilic member of matched emulsifier pair for a wide range of pesticide formulations; used with Sponto 234
Form: Liquid
Composition: 100% conc.

Sponto 234. Witco Chem./Organics Div.
Chemical Description: Blend of oil-sol. sulfonates and polyoxyethylene ethers
Category: Emulsifier, anionic/nonionic
Applications: Hydrophilic member of matched emulsifier pair for a wide range of pesticide formulations; used with Sponto 232
Form: Liquid
Composition: 100% conc.

Sponto 300T. Witco Chem./Organics Div.
Chemical Description: Blend of oil-sol. sulfonates and polyoxyethylene ethers
Category: Emulsifier, anionic/nonionic
Applications: Lipophilic member

of matched emulsifier pair for a wide range of pesticide formulations; used with Sponto 500T
Form: Liquid
Composition: 100% conc.

Sponto 305. Witco Chem./Organics Div.
Chemical Description: Blended emulsifiers
Category: Emulsifier, anionic/nonionic
Applications: See Sponto 140T

Sponto 500T. Witco Chem./Organics Div.
Chemical Description: Blended emulsifier
Category: Emulsifier, anionic/nonionic
Applications: Hydrophilic member of a matched emulsifier pair for a wide range of pesticide formulations; used with Sponto 300T

Sponto 639. Witco Chem./Organics Div.
Chemical Description: Blended emulsifiers
Category and Applications: See Sponto 140T

Sponto 2029. Witco Chem./Organics Div.
Chemical Description: Blended emulsifiers
Category: Emulsifier, anionic/nonionic
Applications: Emulsifier for aromatic solvents used in aquatic weed control

Sponto 2125. Witco Chem./Organics Div.
Chemical Description: Blended emulsifiers
Category: Emulsifier, anionic/nonionic
Applications: Emulsifier for 3 lb/gal propanil in agricultural formulations

Sponto 2210. Witco Chem./Organics Div.
Chemical Description: Blended emulsifiers
Category: Emulsifier, stabilizer, nonionic
Applications: Emulsifier and adjuvant for spray-oil systems; emulsion stabilizer and compatibility agent for a wide range of herbicides

Sponto 2229. Witco Chem./Organics Div.
Chemical Description: Blended emulsifiers
Category: Emulsifier, anionic/nonionic
Applications: Emulsifier for flowable agricultural formulations

Sponto 2236. Witco Chem./Organics Div.
Chemical Description: Blended emulsifiers
Category and Applications: See Sponto 2229

Sponto AD4-10. Witco Chem./Organics Div.
Chemical Description: Blended

emulsifiers
Category and Applications: See Sponto 2029

Sponto AL26-43. Witco Chem./Organics Div.
Chemical Description: Blended emulsifiers
Category and Applications: See Sponto 2125

Sponto AL69-49. Witco Chem./Organics Div.
Chemical Description: Blended emulsifier
Category: Emulsifier, nonionic
Applications: Emulsifier for dimethoate agricultural formulations

Sponto AM2-07. Witco Chem./Organics Div.
Chemical Description: Blended emulsifier
Category: Emulsifier, anionic/nonionic
Applications: Emulsifier for 3 or 4 lb/gal propanil for agricultural formulations

Sponto N-139B. Witco Chem./Organics Div.
Chemical Description: Blended emulsifiers
Category and Applications: See Sponto 140T

Sponto N-710, N-712, N-714. Witco Chem./Organics Div.
Chemical Description: Blend of oil-sol. sulfonates with polyoxyethylene ethers
Category: Emulsifier, anionic/nonionic
Applications: Emulsifier for phenoxy-type herbicide formulations
Form: Liquid

Spreading Agent ET0672. Croda Inc.
Chemical Description: Complex alkyl polyglycol ether
Category: Emulsifier, spreading agent, dispersant
Applications: O/W emulsifier; bath oil surface spreading agent and dispersing agent
Form: Clear liquid; Color: Very pale straw

Squalane. Croda Inc.
Chemical Description: Hexamethyl tetracosane
Category: Emollient, penetrant
Applications: Skin conditioning creams and lotions, make-up, hair and pharmaceutical preparations
Form: Clear liquid; Color: Water-white; Odor: Odorless

SR 131. General Electric
Chemical Description: Silicone resin
Applications: Emulsion car polishes for improved film durability; furniture polishes; easily incorporated into both solvent and emulsion-based formulations
Form: Resin; Color: Transparent
Properties: Solubility: Sol. in mineral spirits; Sp. gr.: 1.06–1.09; Visc.: 20–70 cps; Flash Pt.: 120 F (CC)
Composition: 70% active ingred.; solvent: Solvesso 150

SS 4098. General Electric
Chemical Description: Solvent sol'n. of silicone polymers and other film-forming components
Category: Water repellent
Applications: Useful for pressurized packaging to impart water repellency to textiles and leather; may be applied by dipping, swabbing, or mechanical spraying; does not discolor most fabrics; cures without heat by reacting with the moisture in the air
Form: Color: Colorless to light straw
Properties: Solubility: Dilutable in common aliphatic and aromatic hydrocarbons; Sp. gr.: 0.90; Visc.: < 10 cps; Flash Pt.: 70 F (CC)
Composition: 50% active ingred.; solvent: mineral spirits
Warnings and Caution: Water sensitive–should be diluted with water-free solvents and not exposed to moist air for long periods

SS 4129. General Electric
Chemical Description: Solvent sol'n. of silicone polymers and other film-forming components
Category: Water repellent
Applications: Imparts water repellency to leather goods; cures without heat by reacting with moisture in the air
Form: Liquid; Color: Lt. yellow
Properties: Solubility: Dilutable in common aliphatic and aromatic hydrocarbons; Sp. gr.: 1.01; Visc.: 750 cps; Flash Pt.: 150 F min. (CC)
Composition: 90% min. active; solvent: mineral spirits
Warnings and Caution: Water sensitive–should be diluted with water-free solvents and not exposed to moist air for long periods

Stabilizer AWN. Sandoz Ltd.
Chemical Description: Aq. sol'n. of an inorganic salt containing a wetting agent
Category: Stabilizer, weakly anionic
Applications: Stabilizing agent for hydrogen peroxide bleaching baths
Form: Liquid

Stafoam. Nippon Oil & Fats Co., Ltd.
Chemical Description: Fatty acid diethanol amide of a special composition (1:2 type)
Category: Detergent, rust inhibitor
Applications: Liquid detergents for laundry or drycleaning use; textile softening auxiliaries, dyeing auxiliaries; metal cleaning agents; liquid soaps; dirt dispersers; good viscosity-increasing action
Form: Liquid; Color: Light yellowish
Properties: Solubility: Readily sol. in water; sol. > 20% in ethanol, isopropyl alcohol, propylene glycol, diethylene glycol, tetrachloroethylene, benzene, xylene; Sp. gr.: 0.996 (30 C); Visc.: 810 cps (30 C); Surface Tension: 8 dynes/cm (0.1%); pH: 9.5–10.5 (1% aq.)
Containers: 17 kg C/N, 200 kg D/M, or bulk

Stafoam DF-1. Nippon Oil & Fats Co., Ltd.
Chemical Description: Coconut-oil fatty acid diethanolamide (1:1 type) meeting Cosmetics Materials standards
Category: Detergent, foam booster, softener, antistat
Applications: Superior cleaning power and viscosity- and foam-increasing effects; used in shampoo, dishwashing detergents, neutral liquid laundry detergents, drycleaning, textile softening auxiliaries, antistatic, metal cleaning; more effective when mixed with Stafoam DL
Form: Liquid (solidifies in winter); Color: Light yellowish
Properties: Solubility: Sol. > 20% in ethanol, isopropyl alcohol, propylene glycol, diethylene glycol, petroleum ether, tetrachloroethylene, benzene, xylene; Sp. gr.: 0.981 (30 C); Visc.: 625 cps (30 C); Surface Tension: 14 dynes/cm (0.1%); pH: 9.6–10.6 (1% aq.)
Containers: 17 kg C/N, 190 kg D/M, or bulk

Stafoam DF-4. Nippon Oil & Fats Co., Ltd.
Chemical Description: Coconut-oil fatty acid diethanolamide (1:1 type) meeting the Cosmetics Materials standards
Category and Applications: See Stafoam DF-1
Form: Liquid (solidifies in winter); Color: Light yellowish
Properties: Solubility: Sol. > 20% in ethanol, isopropyl alcohol, propylene glycol, diethylene glycol, petroleum ether, tetrachloroethylene, benzene, xylene; Sp. gr.: 0.974 (30 C); Visc.: 675 cps (30 C); M.P.: 15 C max.; Surface Tension: 16 dynes/cm (0.1%); pH: 9.6–10.6 (1% aq.)
Containers: 17 kg C/N, 190 kg D/M, or bulk

Stafoam DL. Nippon Oil & Fats Co., Ltd.
Chemical Description: Lauric acid diethanolamide (1:1 type) meeting the Cosmetics Materials standards
Category: Detergent, foam stabilizer and booster, viscosity modifier, softener
Applications: Shampoos, dishwashing detergents, neutral liquid laundry detergents, drycleaning, cosmetics components, textile softening auxiliaries, dyeing auxiliaries, anticlouding agents
Form: Solid; Color: White
Properties: Solubility: Sol. > 2% in ethanol, isopropyl alcohol, propylene glycol, diethylene glycol, tetrachloroethylene, benzene, xylene; Sp. gr.: 0.969 (50 C); Visc.: 179 cps (50 C); M.P.: 42–47 C; Surface Tension: 5 dynes/cm (0.1%); pH: 9.6–10.6 (1% aq.)
Containers: 17 kg C/N or bulk

Stafoam F. Nippon Oil & Fats Co., Ltd.
Chemical Description: Coconut-oil fatty acid diethanolamide (1:2 type)
Category: Detergent, rust inhibi-

tor, foaming agent, penetrant
Applications: Heavy-duty liquid detergents, metal cleaning, dirty-dispersal auxiliaries, liquid laundry and drycleaning detergents, textile softening auxiliaries, dyeing auxiliaries, liquid soaps
Form: Liquid; Color: Light yellowish
Properties: Solubility: Readily sol. in water; sol. > 20% in ethanol, isopropyl alcohol, propylene glycol, diethylene glycol, tetrachloroethylene, benzene, xylene; Sp. gr.: 0.995 (30 C); Visc.: 755 cps (30 C); Surface Tension: 7 dynes/cm (0.1%); pH: 9.5–10.5 (1% aq.)
Containers: 17 kg C/N, 200 kg D/M, or bulk

Standamid 100. Henkel Argentina
Chemical Description: Coconut fatty acid monoethanolamide
Category: Foam stabilizer, superfatting and thickening agent
Applications: Superfatting and thickening agent for medicated and cosmetic shampoos, bubble baths, and other detergents; consistency-giving factor for cosmetic preparations in stick form; skin protecting component for industrial ointments; additive to hair treatment preparations
Form: Solid
Composition: ≈ 90% amide content

Standamid CD. Henkel Inc.
Chemical Description: Short chain coconut oil diethanolamide (1:2 fatty acid:diethanolamide)
Category: Detergent, foam stabilizer and booster, wetting agent, superfatting, nonionic
Applications: Formulating cosmetic, household, and industrial detergents, and other chemical specialties; produces flash foam even in extremely high concs. of electrolytes
Form: Viscous liquid; Color: Amber
Properties: Stability: Good in acid, alkaline, or neutral salts in moderate conc.; pH: 9.0–10.0 (10% sol'n.)
Composition: 96–98% active

Standamid CM. Henkel Inc.
Chemical Description: Cocamide MEA
Category: Foaming agent and stabilizer, nonionic
Applications: Powdered or beaded bubble baths
Form: Granules
Properties: M.P.: 140–148 F; Acid No.: 0–1.0; Alkali No.: 8.0–12.0; Stability: Chemically stable at pH 5–14; pH: 9.5–10.5 (10% sol'n.)
Composition: 100% conc.

Standamid CMG. Henkel Inc.
Chemical Description: Cocamide MEA; super amide (1:1) derived from whole coconut oil
Category: Foam booster, stabilizer, thickener
Applications: Shampoos, bubble baths, dishwashing detergents; contains no amide ester
Form: Solid

Properties: Stability: Excellent stability characteristics
Composition: 100% conc.

Standamid KD. Henkel Inc.
Chemical Description: Coconut fatty acid diethanolamide
Category: Thickening agent, superfatting agent, foam stabilizer, emulsifier, nonionic
Applications: Liquid shampoos and bubble baths; emulsifier and dissolving intermediate
Form: Paste; Color: Amber
Properties: M.P.: 100 F; Acid No.: 1.0 max.; Alkali No.: 27.0 max.; Stability: Chemically stable at pH 5–12; pH: 9.0–9.5 (10% sol'n.)
Composition: 85–90% amide content

Standamid KDM. Henkel Inc.
Chemical Description: Coconut fatty acid alkanolamine condensate (1:2 fatty acid:diethanolamine)
Category: Foam builder and stabilizer, thickener, wetting agent, dispersant, nonionic
Applications: Shampoos, bubble baths, household detergents, liquid soaps; good lime-soap dispersing and anti-redeposition properties
Form: Viscous liquid; Color: Amber
Properties: Solubility: Readily sol. in water, alcohol, acetone, and glycols; Acid No.: 0–5; pH: 9.5–10.0 (1% sol'n.)
Composition: 100% conc.

Standamid LD. Henkel Inc.
Chemical Description: Lauric acid diethanolamide
Category: Foam stabilizer, superfatting agent, thickening agent, nonionic
Applications: Shampoos, bath products; fortifier for perfumes in soaps
Form: Paste
Properties: M.P.: 105 F; Acid No.: 0–1.0; Alkali No.: 30–40; Stability: Chemically stable at pH 5–12; pH: 9.5–10.5 (10% sol'n.)
Composition: $>$ 90% amide content

Standamid LD 80/20. Henkel Inc.
Chemical Description: Modified lauric/myristic acid diethanolamide
Category: Foam booster and stabilizer, detergency builder, emulsifier, thickener, conditioner, nonionic
Applications: Shampoos, bubble baths, dishwashing detergents, creams, and lotions
Form: Clear liquid; Color: Amber
Properties: Cloud Pt.: $\approx$ 0 C; Acid Value: 0.5 max.; Alkali Value: 25.0–35.0; pH: 9.5–10.5 (10% sol'n.)
Composition: 80% conc.

Standamid LDM. Henkel Inc.
Chemical Description: Modified lauric diethanolamide
Category: Foam stabilizer, viscosity builder, nonionic
Applications: Shampoos and bubble baths

Form: Clear, slightly viscous liquid
Properties: Cloud Pt.: 0 C; Acid Value: 3.0%
Composition: 100% conc.

Standamid LP. Henkel Argentina
Chemical Description: Lauric acid isopropanolamide
Category: Foam stabilizer, superfatting and thickening agent, nonionic
Applications: Shampoos and other detergents; fortifier for perfumes in soaps
Form: Solid
Properties: Acid Value: < 2%
Composition: > 90% amide content

Standamid PD. Henkel Inc.
Chemical Description: Cocamide DEA, Kritchevsky-type amide (2:1)
Category: Foam builder, stabilizer, thickener
Applications: Shampoos, bubble baths with anionics, nonionics, and amphoterics
Form: Viscous liquid; Color: Amber
Composition: 100% conc.

Standamid SD. Henkel Inc.
Chemical Description: Cocamide DEA
Category: Foam stabilizer, nonionic
Applications: Bath gels, liquid bubble baths, shampoos, dishwashing and laundry detergents
Form: Liquid
Properties: M.P.: 25 F; Acid No.: 0–1.0; Alkali No.: 30.0–40.0; Stability: Chemically stable at pH 5–12; pH: 9.5–10.5 (10% sol'n.)
Composition: 100% conc.

Standamox 01. Henkel Inc.
Chemical Description: Oleamine oxide
Category: Lubricant, softener
Applications: Thixotropic: 1% in water yields viscous liquid; 3–5% yields gel; plate out effect from aq. systems gives beading, lubricity, and softening
Form: Clear liquid; Color: Amber
Composition: 55% conc.

Standamox CAW. Henkel Inc.
Chemical Description: Cocamidopropylamine oxide
Category: Foam builder, stabilizer, thickener, emollient
Applications: Low irritation; ideal for baby shampoos, bubble baths, skin care preparations
Form: Clear liquid
Composition: 29.5% conc.

Standamox C 30. Henkel Argentina
Chemical Description: Coco amido amine oxide
Category: Wetting agent, foam stabilizer, viscosity builder
Applications: Shampoos, bubble baths, bath oils, dishwash concentrates, and other detergents
Form: Liquid
Properties: pH: 6–7 (as is)
Composition: 28–30% active

Standamul 302. Henkel Inc.
Chemical Description: Propylene glycol/dicaprylate/dicaprate

Category: Base, emollient
Applications: Carrier for lipid soluble substances in cosmetic and pharmaceutical topical preparations, pigmented cosmetic systems where oxidative stability is critical; quality creams, lotions, makeup bases, skin cleansers, bath oils, and antiperspirants
Form: Clear, low viscosity oily liquid; Odor: Faint, characteristic
Properties: Solubility: Sol. in mineral oil, isopropyl myristate, oleyl alcohol, castor oil, anhydrous ethanol, and silicone fluid; Sp. gr.: ≈ 0.921 (20 C); Gel Pt.: –30 C max.; HLB: 9; Sapon. No.: 315–335; Acid No.: 0.5 max.; Iodine No.: 2.0 max.; Stability: Excellent low temp. stability
Composition: 100% conc.
Containers: 385 lb net steel drums (closed head), or bulk

Standamul 318. Henkel Inc.
Chemical Description: Caprylic/capric triglyceride
Category: Base
Applications: Carrier for lipid soluble substances for cosmetic, pharmaceutical topical and oral preparations, and food products; quality bath oils, body oils, and antiperspirants; extender for flavors, esp. where alcohol solubility is desired
Form: Clear, low viscosity oily liquid; Color: Almost colorless; Odor: Odorless
Properties: Solubility: Sol. in mineral oil, isopropyl myristate, oleyl alcohol, castor oil, and anhydrous ethanol; Sp. gr.: ≈ 0.950 (20 C); Visc.: ≈ 25 cps; Gel Pt.: –10 C max.; HLB: 12; Cloud Pt.: –5 C max.; Sapon. No.: 340–350; Acid No.: 0.5 max.; Hydroxyl No.: 5.0 max.; Iodine No.: 0.5 max.; Stability: Excellent low temp. stability
Composition: 100% conc.
Containers: 374 lb net steel drums (closed head), or bulk

Standamul 1414-E. Henkel Inc.
Chemical Description: Myreth-3 myristate
Category: Carrier, self-emulsifying base, emulsifier, coupling agent
Applications: Carrier for lipid soluble substances; self-emulsifying base for cosmetic and pharmaceutical topical preparations; emollient and moisturizer for quality creams, lotions, skin cleansers, deodorants, dispersible-type bath oils, body oils, shampoos, and hydroalcoholic preparations
Form: Clear, low viscosity oily liquid
Properties: Solubility: Sol. in mineral oil, isopropyl myristate, oleyl alcohol, castor oil, anhydrous ethanol; sparingly sol. in water; HLB: 12; Cloud Pt.: 25 C max.; Sapon. No.: 90–100; Hydroxyl No.: 10–20
Composition: 100% conc.
Containers: 450 lb net closed head steel drums or bulk

Standamul 1616. Henkel Inc.
Chemical Description: Cetyl palmitate

Category: Spermaceti replacement, consistency giving agent
Applications: Cosmetic and pharmaceutical topical applications; quality creams, lotions, ointments, hydroalcoholic stick preparations, pigmented anhydrous stick cosmetics
Form: Waxy solid (flake); Color: White; Odor: Faint characteristic
Properties: Solubility: Sol. in mineral oil, isopropyl myristate, oleyl alcohol, castor oil; disp. in hot ethanol and polyols; M.P.: 45–55 C; HLB: 9; Sapon. No.: 115–130; Acid No.: 1.0 max.; Hydroxyl No.: 5.0 max.; Iodine No.: 3.0 max.
Composition: 100% conc.
Containers: 110 lb net paper bags and fiber drums

Standamul 7061. Henkel Inc.
Chemical Description: Isocetyl stearate
Category: Base, emollient
Applications: Carrier for lipid soluble substances for cosmetic and pharmaceutical topical preparations; nongreasy emolliency and low cloud pt.; used for bath oils, body oils, skin enrichment lotions and creams, and cleansing lotions
Form: Clear, low viscosity oil; Color: Colorless; Odor: Faint characteristic
Properties: Solubility: Sol. in mineral oil, isopropyl myristate, oleyl alcohol, castor oil; Sp. gr.: ≈ 0.865; Visc.: ≈ 25 cps; HLB: 8; Cloud Pt.: 0 C max.; Sapon. No.: 105–115; Acid No.: 0.5 max.; Hydroxyl No.: 5.5 max.; Iodine No.: 5.0 max.
Composition: 100% conc.
Containers: 400 lb net closed head steel drums or bulk

Standamul 7063. Henkel Inc.
Chemical Description: Octyl dodecyl stearate
Category and Applications: See Standamul 7061
Form: Clear, low viscosity oil; Color: Colorless; Odor: Faint characteristic
Properties: Solubility: Sol. in mineral oil, isopropyl myristate, oleyl alcohol, castor oil; Sp. gr.: ≈ 0.865; Visc.: ≈ 100 cps; HLB: 8; Cloud Pt.: 0 C max.; Sapon. No.: 90–105; Acid No.: 0.5 max.; Hydroxyl No.: 5.0 max.; Iodine No.: 5.0 max.
Composition: 100% conc.
Containers: 400 lb net closed head steel drums or bulk

Standamul 7105. Henkel Inc.
Chemical Description: Caprylic/capric triglyceride
Category: Base
Applications: Carrier for lipid soluble substances for cosmetic and pharmaceutical preparations; pigmented cosmetic systems where oxidative stability is critical; used for quality creams, lotions, skin cleansers, perfume oils, bath oils, body oils, and antiperspirants
Form: Clear, low viscosity oil; Color: Almost colorless; Odor: Odorless

Properties: Solubility: Sol. in mineral oil, isopropyl myristate, oleyl alcohol, castor oil, anhydrous ethanol; Sp. gr.: ≈ 1.064; Visc.: 100 cps max.; HLB: 12; Cloud Pt.: –5 C max.; Sapon. No.: 340–350; Acid No.: 0.5 max.; Hydroxyl No.: 5.0 max.; Iodine No.: 0.5 max.
Composition: 100% conc.
Containers: 374 lb net closed head steel drums or bulk

Standamul Conc. 1002. Henkel Inc.
Chemical Description: Cetearyl alcohol, PEG-40 hydrogenated castor oil, and stearalkonium chloride
Category: Emulsifier, conditioner
Applications: Proprietary concentrate for cream rinses and hair conditioners; incorporates a nonionic emulsifier and a cationic conditioner; compatible with proteins, herbal extracts, etc.; nonirritating to skin and eyes
Form: Waxy flakes; Color: Ivory
Composition: 99% conc.

Standamul CTA. Henkel Inc.
Chemical Description: Hexyl laurate
Category: Spreading agent, base, dispersant
Applications: Spreading qualities on skin, deep skin penetration, and pigment dispersion; carrier for lipid soluble substances in cosmetics and pharmaceutical topical preparations; used for quality creams, lotions, makeup bases and removers, bath oils, body oils
Form: Clear, low viscosity oil; Color: Colorless; Odor: Practically odorless
Properties: Solubility: Sol. in mineral oil, isopropyl myristate, oleyl alcohol, castor oil, anhydrous ethanol, silicone fluid; Sp. gr.: 0.85–0.87 (20 C); Gel Pt.: –5 C max.; HLB: 9; Cloud Pt.: 0 C max.; Sapon. No.: 190–205; Acid No.: 0.5 max.; Hydroxyl No.: 2.0 max.; Iodine No.: 1.0 max.
Composition: 100% conc.
Containers: 385 lb net closed head steel drums or bulk

Standamul CTV. Henkel Inc.
Chemical Description: Decyl oleate
Category and Applications: See Standamul CTA
Form: Clear, low viscosity liquid; Odor: Faint characteristic
Properties: Solubility: Sol. in mineral oil, isopropyl myristate, oleyl alcohol, castor oil, anhydrous ethanol; Sp. gr.: 0.86–0.87 (20 C); Gel Pt.: 5 C max.; HLB: 9; Cloud Pt.: 10 C max.; Sapon. No.: 130–150; Acid No.: 0.5 max.; Hydroxyl No.: 5.0 max.; Iodine No.: 55–65
Composition: 100% conc.
Containers: 385 lb net closed head steel drums or bulk
Warnings and Caution: Subject to oxidation

Standamul G. Henkel Inc.
Chemical Description: 2-octyl dodecanol
Category: Carrier, dispersant, coupling agent
Applications: Carrier for oil-sol. active ingred.; dissolves dyes; pig-

ment dispersant; coupling agent for waxes and other fatty materials in cosmetics and toiletries; gentle baby oils
Form: Crystal clear liquid; Color: Water-white; Odor: Odorless
Properties: Cloud Pt.: ≈ −50 C; Sapon. No.: 5.0 max.; Acid No.: 0.5 max.; Hydroxyl No.: 170–185; Iodine Value: 8.0 max.
Composition: 100% conc.

Standamul G-16. Henkel Inc.
Chemical Description: 2-hexyldecanol
Category: Extender, solvent, binder, emollient
Applications: Extender and solvent for pigments and fragrance oils; imparts softness and suppleness to skin; binder in pressed powdered eye and face makeup
Form: Clear liquid; Color: Water-white
Properties: Visc.: 100 cps; Cloud Pt.: ≈ −50 C; Sapon. No.: 5.0 max.; Acid No.: 0.5 max.; Hydroxyl No.: 210–230; Iodine Value: 8.0 max.; Stability: Stable to acids and alkalis and on long storage
Composition: 100% conc.

Standamul G-3236 Stearate. Henkel Inc.
Chemical Description: Myristyl eicosyl stearate
Category: Stabilizer, base, lubricant, opacifier, emollient
Applications: Cosmetic and pharmaceutical topical preparations; provides structure, stability, and lubricity to cosmetic stick products, e.g., blushes, lipsticks, eyeshadows, deodorants, fragrance sticks; adds slip and nontackiness to antiperspirant sticks; opacifier and emollient in specialty bath oils
Form: Soft wax; Color: Light; Odor: Faint characteristic
Properties: Solubility: Sol. in mineral oil, isopropyl myristate, oleyl alcohol, castor oil; disp. in polyols and hydroalcoholic sol'ns.; M.P.: 33–40 C; HLB: 7; Sapon. No.: 65–75; Acid No.: 1.0 max.; Stability: Good hydrolytic, oxidative, and color stability
Containers: 385 lb net open head steel drums

Standamul HE. Henkel Inc.
Chemical Description: PEG-7 glyceryl cocoate
Category: Emollient, refatting agent, solubilizer, coupling agent, emulsifier, nonionic
Applications: Self-emulsifying oil for cosmetics and pharmaceutical topical preparations; recommended where emolliency is required in aq. formulations, e.g., shampoos, bubble baths, hydroalcoholics, skin cleansers, creams, lotions; refatting agent; solubilizer for essential oils and lipid soluble active ingredients
Form: Clear, low viscosity oil; Odor: Faint characteristic
Properties: Solubility: Sol. in oleyl alcohol, castor oil, anhydrous ethanol, water; sparingly sol. in isopropyl myristate and most

polyols; Sp. gr.: ≈ 1.050; Visc.: ≈ 200 cps; HLB: 16; Cloud Pt.: ≈ 0 C; Sapon. No.: 90–100; Acid No.: 5.0 max.; Hydroxyl No.: 180–195; Iodine No.: 5.0 max.
Composition: 100% conc.
Containers: 385 lb net closed head steel drums or bulk

Standamul O-5. Henkel Inc.
Chemical Description: Oleth-5
Category: Emulsifier, emollient, nonionic
Applications: Series can be blended to obtain special emulsification effects (O/W and W/O emulsions); emolliency, body, and conditioning properties; mild to the skin
Form: Viscous liquid
Properties: Solubility: Sol. in alcohols, hydrocarbons, and most organic solvents; Stability: Stable to many alkalis and acids under extreme pH conditions
Composition: 100% conc.

Standamul O-10. Henkel Inc.
Chemical Description: Oleth-10
Category and Applications: See Standamul O-5
Form: Soft waxy solid
Properties: Solubility: Sol. in alcohols, hydrocarbons, and most organic solvents; Stability: Stable to many alkalis, acids under extreme pH conditions
Composition: 100% conc.

Standamul O-20. Henkel Inc.
Chemical Description: Oleth-20
Category and Applications: See Standamul O-5
Form: Waxy solid
Properties: Solubility: Sol. in alcohols, hydrocarbons, and most organic solvents; solubility or dispersibility in water increases as POE increases in series; Stability: Stable to many alkalis and acids under extreme pH conditions
Composition: 100% conc.

Standamul OXL. Henkel Inc.
Chemical Description: PPG-10 Ceteareth-20
Category: Emollient, superfatting agent, nonionic
Applications: Emollient with both hydrophilic and lipophilic properties; superfatting agent for bubble baths, bath oils, and shampoos; completely nonirritating to skin and eyes
Form: Liquid; Color: Water-white
Properties: Solubility: Sol. in cold water; miscible in vegetable, mineral, and perfume oils
Composition: 98% conc.

Standamul STC-25. Henkel Inc.
Chemical Description: Stearalkonium chloride
Category: Antistatic agent, cationic
Applications: Cationic surfactant that imparts antistatic properties to hair, synthetic and natural fibers; effective in cream rinse hair preparations
Form: Slurry; Color: White
Composition: 25% conc.

Standapol 125-E Conc. Henkel Inc.
Chemical Description: Sodium Laureth-12 sulfate
Category: Detergent, foaming agent
Applications: Low irritation shampoos, bath products, cleansing cosmetic preparations, baby shampoo, skin care products
Form: Clear liquid; Color: Straw-yellow
Properties: Solubility: Sol. in water and a variety of hydrophilic solvents; Cloud Pt.: 10 C max.; Stability: Stable at pH 5–12 under normal temp. conditions; compatible with anionics, nonionics, amphoterics, soap, and alkaline builders; pH: 7.5–9.0 (10% aq.)
Composition: 58–60% active
Warnings and Caution: Contains ethanol–use proper precautionary methods for storage and handling

Standapol 130E. Henkel Inc.
Chemical Description: Sodium Laureth-12 sulfate
Category and Applications: See Standapol 125-E Conc.
Form: Clear liquid; Color: Pale-yellow
Properties: Solubility: Readily sol. in water and a variety of hydrophilic solvents; Gel Pt.: 10 C max.; Stability: Stable at pH 5–12 under normal temp. conditions; pH: 7.5–9.0 (10% aq.)
Composition: 28–30% active

Standapol 230E Conc. Henkel Inc.
Chemical Description: Ammonium Laureth-12 sulfate
Category and Applications: See Standapol 125-E Conc.; 230E is ideal for "acid pH shampoos" when stabilized at pH 4.5–6.5
Form: Clear liquid; Color: Straw-yellow
Properties: Solubility: Readily sol. in water and a variety of hydrophilic solvents; Gel Pt.: 10 C max.; Cloud Pt.: ≈ 0 C; Stability: Stable at pH 2–7 under normal temp. conditions; pH: 6.5–7.0 (10% aq.)
Composition: 28–30% active

Standapol A. Henkel Inc.
Chemical Description: Ammonium lauryl sulfate
Category: Detergent, foamer, anionic
Applications: Water-white or pastel-colored liquid and gel shampoos, bubble baths, cleansing cosmetic preparations, "acid-pH" shampoos
Form: Viscous liquid; Color: Water-white
Properties: Visc.: 1500–3000 cps; Stability: Compatible with natural and synthetic gums and resins; pH: 6.5–7.0 (10% aq.)
Composition: 27.5–28.5% active

Standapol AB-45. Henkel Inc.
Chemical Description: Coco betaine
Category: Foaming, conditioning, antistatic, and gelling agent, amphoteric
Applications: Substantive to hair; in combination with alkyl sulfates and alkyl ether sulfates, and in the presence of soap, acts as foam booster and viscosity builder, and

reduces irritation potential of these surfactants
Form: Liquid; Color: Pale yellow
Properties: Solubility: Sol. in alcohols, glycols, alkanolamides, and water; Cloud Pt.: < 0 C; Stability: Stable in strong acid and alkaline sol'ns.; pH: 6.5–7.5 (10% sol'n.)
Composition: 36.5% min. active

Standapol AMS-100. Henkel Inc.
Chemical Description: Ammonium lauryl sulfate
Category: Detergent, foaming agent
Applications: Water-white or pastel-colored liquid and gel shampoos, bubble baths, cleansing cosmetic preparations, "acid-pH" shampoos
Form: Viscous liquid; Color: Water-white
Properties: Visc.: 1500–3000 cps; Cloud Pt.: <10 C; Stability: Compatible with natural and synthetic gums and resins; pH: 6.5–7.0 (10% aq.)
Composition: 27.5–28.5% active

Standapol BAW. Henkel Inc.
Chemical Description: Cocamidopropyl betaine
Category: Conditioning agent, antistat, foaming agent, amphoteric
Applications: Mild surfactant and conditioner for hair, bath, and skin care products; pronounced substantivity, antistatic, and foaming properties
Form: Liquid; Color: Light amber
Properties: Stability: Stable in strong acid and alkaline sol'ns.
Composition: 34% conc.

Standapol BC-35. Henkel Inc.
Chemical Description: Cocamidopropyl betaine
Category: Conditioning agent, emollient, viscosity builder, gelling agent, amphoteric
Applications: Mild conditioning surfactant for shampoos, bath, and skin care preparations
Form: Liquid; Color: Light amber
Composition: 35% conc.

Standapol CIM-40. Henkel Inc.
Chemical Description: Acyl (coco fatty) amido glycine betaine
Category: Amphoteric surfactant
Applications: Extremely mild surfactant for nonirritating shampoo and bubble bath preparations; remains clear at low pH
Form: Clear liquid; Color: Light amber; Odor: Pleasant, mild
Properties: Cloud Pt.: ≈ 0 C; pH: 8.0–9.0 (10% sol'n.)
Composition: 40–42% active

Standapol Conc. 7021. Henkel Inc.
Chemical Description: DEA-lauryl sulfate and TEA-lauryl sulfate
Category: Detergent, foaming agent, anionic
Applications: Proprietary blend; conc. shampoo/bubble baths; high foaming anhydrous detergent for gelatin capsules
Form: Anhydrous slurry
Composition: 100% conc.

Standapol Conc. 7023. Henkel Inc.
Chemical Description: Cocamide DEA and DEA-myreth sulfate
Category: Detergent, foaming agent, solubilizer, anionic
Applications: Concentrated shampoos and bath products where high foaming is required; gelatin capsule bath beads; solubilizer for fatty alcohols, fatty acid esters, and mineral oil
Form: Clear, viscous liquid; Color: Straw-yellow
Properties: Visc.: 250 cps max. (50 C); Cloud Pt.: 10 C max.; pH: 8.5–9.5 (1% aq.)
Composition: 100% conc.

Standapol CS-50. Henkel Inc.
Chemical Description: Sodium lauryl sulfate, sodium cetyl sulfate, and Laureth-3
Category: Detergent, foamer
Applications: Shampoo concentrate for pearlescent paste and lotion shampoos and bubble baths; rich, creamy, stable lather with excellent feel
Form: Viscous liquid; Color: Pearly white; Odor: Mild, characteristic
Properties: pH: 6.5–7.5 (10% aq.)
Composition: 28.5–30.0% active

Standapol CS Paste. Henkel Inc.
Chemical Description: Sodium lauryl sulfate, sodium cetyl sulfate, and Laureth-3
Category and Applications: See Standapol CS-50
Form: Creamy paste; Color: Pearly white; Odor: Mild, characteristic
Properties: pH: 6.5–7.5 (10% aq.)
Composition: 57–60% active

Standapol DEA. Henkel Inc.
Chemical Description: DEA-lauryl sulfate
Category: Base, detergent, foamer, anionic
Applications: Base for clear liquid shampoos, bubble baths; creamy, stable lather with soapy feel
Form: Clear liquid; Color: Light yellow
Composition: 36% conc.

Standapol EA-1. Henkel Inc.
Chemical Description: Ammonium laureth sulfate
Category: Base, detergent, foaming agent, anionic
Applications: Clear liquid and gel shampoos, bubble baths, foaming bath oils, cleansing cosmetic preparations, "acid-pH" shampoos; good flash foam in soft and hard water, and in the presence of greasy soil
Form: Viscous liquid; Color: Water-white
Properties: Visc.: ≈ 2500 cps; Cloud Pt.: 5 C max.; pH: 6.5–7.0 (10% aq.)
Composition: 25.5–26.5% active

Standapol EA-2. Henkel Inc.
Chemical Description: Ammonium laureth sulfate
Category and Applications: See Standapol EA-1
Form: Viscous liquid; Color: Water-white
Properties: Visc.: 2500 cps max.; Cloud Pt.: 0 C max.; pH: 6.5–7.0 (10% aq.)
Composition: 24.5–25.5% active

Standapol EA-40 Conc. Henkel Inc.
Chemical Description: Ammonium myreth sulfate
Category: Detergent, foaming agent, anionic
Applications: Liquid, semiliquid, paste, gel, and cream shampoos, bubble baths, cleansing cosmetic preparations, "acid-pH" shampoos, professional shampoo concentrates; good flash foam in hard or soft water and in presence of greasy soil
Form: Clear liquid; Color: Pale-yellow
Properties: Cloud Pt.: 10 C max.; Stability: Stable over wide pH range; pH: 6.0–7.0 (10% aq.)
Composition: 58–60% active
Warnings and Caution: Contains ethanol–use proper precautionary methods for storage and handling

Standapol ES-1. Henkel Inc.
Chemical Description: Sodium laureth sulfate
Category: Detergent, foaming agent, anionic
Applications: Base for liquid and semiliquid shampoos, bubble baths, and cleansing cosmetic preparations; good flash foam in hard and soft water and in the presence of greasy soil
Form: Viscous liquid; Color: Water-white
Properties: Visc.: 500–1200 cps; Cloud Pt.: 5 C max.; pH: 7.5–8.5 (10% aq.)
Composition: 24.5–25.5% active

Standapol ES-2. Henkel Inc.
Chemical Description: Sodium laureth sulfate
Category and Applications: See Standapol ES-1
Form: Viscous liquid; Color: Water-white
Properties: Visc.: 4000–5000 cps; Cloud Pt.: 0 C max.; pH: 6.5–7.5 (10% aq.)
Composition: 25–26% active

Standapol ES-3. Henkel Inc.
Chemical Description: Sodium laureth sulfate
Category and Applications: See Standapol ES-1
Form: Liquid; Color: Water-white
Properties: Visc.: 500 cps max.; Cloud Pt.: $\approx$ 0 C; pH: 6.5–8.0 (10% aq.)
Composition: 28–29% active

Standapol ES-40 Conc. Henkel Inc.
Chemical Description: Sodium myreth sulfate
Category: Detergent, foaming agent, anionic
Applications: Base for liquid, semiliquid, paste, gel, and cream shampoos, bubble baths, cleansing cosmetic preparations; good flash foam in hard or soft water and in the presence of greasy soil; professional shampoo concentrates
Form: Clear liquid; Color: Pale-yellow
Properties: Cloud Pt.: 10 C max.; Stability: Stable over wide pH range; pH: 7.5–9.0 (10% aq.)

Composition: 58–60% active
Warnings and Caution: Contains ethanol–use proper precautionary methods for storage and handling

Standapol ES-7099. Henkel Inc.
Chemical Description: Sodium laureth sulfate
Category and Applications: See Standapol ES-1
Form: Viscous liquid; Color: Water-white
Properties: Visc.: 4000–5000 cps; Cloud Pt.: 0 C max.; pH: 6.5–7.5 (10% aq.)
Composition: 25–26% active

Standapol MG. Henkel Inc.
Chemical Description: Magnesium lauryl sulfate
Category: Detergent, foaming agent, anionic
Applications: Clear and colored liquid and gel shampoos, bubble baths, and other cosmetic cleansing preparations; produces copious, dense stable foam and desirable feel
Form: Liquid; Color: Straw yellow
Properties: Cloud Pt.: 3 C max.; pH: 6–7 (10% aq.)
Composition: 28–30% active

Standapol MLS. Henkel Inc.
Chemical Description: MEA-lauryl sulfate
Category: Detergent, foaming agent
Applications: Base for clear and colored liquid shampoos, bubble baths, cosmetic cleansing preparations; creamy, stable lather with soapy feel
Form: Clear, viscous liquid; Color: Light yellow
Properties: Visc.: 3000–5000 cps; Cloud Pt.: ≈ 0 C; pH: 7–8 (10% aq.)
Composition: 32–34% active

Standapol OLB-30. Henkel Inc.
Chemical Description: Oleyl betaine
Category: Detergent, conditioning agent, foaming agent, viscosity builder, amphoteric
Applications: Hair and skin care; aids in processing; compatible with anionics, nonionics, cationics, and other amphoterics
Form: Viscous liquid; Color: Amber
Composition: 30% conc.

Standapol OLB-50. Henkel Inc.
Chemical Description: Oleyl betaine
Category: Detergent, conditioning agent, foaming agent, viscosity builder, amphoteric
Applications: Hair and skin care products; compatible with anionics, nonionics, cationics, and other amphoterics
Form: Gel; Color: Light amber
Composition: 50% conc.

Standapol S. Henkel Inc.
Chemical Description: Sodium lauryl sulfate, sodium laureth sulfate, lauramide MIPA, cocamide MEA, glycol stearate, and coceth-8

Category: Detergent, foaming agent, anionic
Applications: Formulated shampoo concentrate; produces rich, creamy, stable lather in shampoos and bubble bath formulations
Form: Viscous liquid; Color: Pearly-white
Properties: Visc.: 1200–1700 cps; pH: 6.6–7.4 (10% aq.)
Composition: 36.8–37.3% active

Standapol SH-100. Henkel Inc.
Chemical Description: Disodium monooleamido PEG-2 sulfosuccinate
Category: Detergent, anionic
Applications: Nonirritating shampoo base for adults and babies, and gel shampoo concentrates; synergistic foaming effect with alcohol sulfates, ether sulfates; acts as anti-irritant with alkyl sulfates and sulfonates, alkylolamides, and amine oxides
Form: Liquid; Color: Light amber; Odor: Mild, pleasant
Properties: Cloud Pt.: $<$ 0 C; pH: 6.0–7.0 (5% sol'n.)
Composition: 29–31% active

Standapol SH-135 Special. Henkel Inc.
Chemical Description: Disodium monooleamido PEG-2 sulfosuccinate
Category and Applications: See Standapol SH-100
Form: Liquid; Color: Light amber; Odor: Mild, pleasant
Properties: Cloud Pt.: 10 C max.; pH: 6–7 (5% sol'n.)
Composition: 34–36% active

Standapol SH-200. Henkel Inc.
Chemical Description: Ammonium monooleamido PEG-2 sulfosuccinate
Category and Applications: See Standapol SH-100
Form: Liquid; Color: Light amber; Odor: Mild, pleasant
Properties: Cloud Pt.: $<$ 0 C; pH: 6–7 (5% sol'n.)
Composition: 29–31% active

Standapol SH-300. Henkel Inc.
Chemical Description: TEA monooleamido PEG-2 sulfosuccinate
Category and Applications: See Standapol SH-100
Form: Liquid; Color: Light amber; Odor: Characteristic
Properties: Cloud Pt.: $<$ 0 C; pH: 6–7 (5% sol'n.)
Composition: 39–41% active

Standapol SHC-101. Henkel Inc.
Chemical Description: Disodium monooleamido PEG-2 sulfosuccinate and sodium lauryl sulfate
Category: Detergent, foaming agent, anionic
Applications: Mild liquid, clear, and lotion shampoos or bubble baths for adults or babies; high foaming with dense foam structure; substantive to hair; nonirritating
Form: Liquid; Color: Light amber; Odor: Mild, pleasant
Properties: Visc.: 1000 cps max.; Cloud Pt.: 0 C max.; pH: 6.5–7.5 (5% sol'n.)
Composition: 30% min. solids

Standapol SHC-301. Henkel Inc.
Chemical Description: TEA monooleamido PEG-2 sulfosuccinate and TEA lauryl sulfate
Category and Applications: See Standapol SHC-101
Form: Liquid; Color: Light amber; Odor: Mild, pleasant
Properties: Visc.: 1200 cps max.; Cloud Pt.: 0 C max.; pH: 6–7 (5% sol'n.)
Composition: 39–41% solids

Standapol T. Henkel Inc.
Chemical Description: TEA-lauryl sulfate
Category: Detergent, foaming agent, anionic
Applications: Base for liquid and gel shampoos, bubble baths; superior color
Form: Liquid; Color: Water-white
Composition: 39% conc.

Standapol TS-100. Henkel Inc.
Chemical Description: TEA-lauryl sulfate
Category and Applications: See Standapol T
Form: Liquid; Color: Water-white
Composition: 39% conc.

Standapol WAAC. Henkel Inc.
Chemical Description: Sodium lauryl sulfate
Category: Detergent, foaming agent, anionic
Applications: Base for liquid, semiliquid, and cream shampoos, bubble baths, and cleansing cosmetic preparations; rich, creamy, stable lather; pearlescent products
Form: Paste; Color: Pearly white
Properties: Visc.: 20,000 cps min.; Cloud Pt.: 25 C max.; pH: 7.5–8.5 (10% aq.)
Composition: 28–30% conc.

Standapol WAQ-115. Henkel Inc.
Chemical Description: Sodium lauryl sulfate
Category: Detergent, foaming agent
Applications: Liquid, semiliquid, paste, and cream shampoos, bubble baths, cleansing cosmetic preparations; fine, dense foam; low salt
Form: Viscous liquid; Color: Water-white
Properties: Visc.: 1000 cps max.; Cloud Pt.: 12 C max.; pH: 7.5–8.5 (10% aq.)
Composition: 29–31% conc.

Standapol WAQ-LC. Henkel Inc.
Chemical Description: Sodium lauryl sulfate
Category: Detergent, foaming agent
Applications: Liquid and semiliquid shampoos, bubble baths, and cleansing cosmetic preparations; lowest salt FAS
Form: Liquid; Color: Water-white
Properties: Visc.: 100 cps max.; Cloud Pt.: 8 C max.; pH: 7–9 (10% aq.)
Composition: 28.5–31.5% conc.

Standapol WAQ Special. Henkel Inc.
Chemical Description: Sodium lauryl sulfate

Category and Applications: See Standapol WAQ-115
Form: Viscous liquid; Color: Water-white
Properties: Visc.: 1500 cps max.; Cloud Pt.: 15 C max.; pH: 7.5–8.5 (10% aq.)
Composition: 28–30% conc.

Standapol WAS-100. Henkel Inc.
Chemical Description: Sodium lauryl sulfate
Category and Applications: See Standapol WAQ-115
Form: Viscous liquid; Color: Water-white
Properties: Visc.: 1500 cps max.; Cloud Pt.: 15 C max.; pH: 7.5–8.5 (10% aq.)
Composition: 28–31% conc.

Starfol D. Ashland Chem. Co.
Chemical Description: Glyceryl monostearate, ethoxylated
Category: Emulsifier, nonionic
Applications: Emulsifier in bakery products
Form: Solid; Color: 250 max. (APHA)
Properties: M.P.: ≈ 26 C; HLB: 13.6; Sapon. Value: 65–75; Acid Value: 2 max.
Composition: 100% conc.

Starfol IPM. Ashland Chem. Co.
Chemical Description: Isopropyl myristate
Category: Emollient, nonionic
Form: Liquid; Color: 30 max. (APHA)
Properties: Sapon. Value: 202–212; Acid Value: 1 max.; Iodine Value: 1 max.
Composition: 100% conc.

Starfol IPP. Ashland Chem. Co.
Chemical Description: Isopropyl palmitate
Category: Emollient, nonionic
Form: Liquid; Color: 30 max. (APHA)
Properties: Sapon. Value: 182–192; Acid Value: 1 max.; Iodine Value: 1 max.
Composition: 100% conc.

Starfol Wax CG. Ashland Chem. Co.
Chemical Description: Difatty ester
Category: Nonionic
Applications: Synthetic spermaceti wax
Form: Solid; Color: 100 max. (APHA)
Properties: M.P.: 46–49 C; Sapon. Value: 109–117; Acid Value: 2 max.; Iodine Value: 1.0 max.
Composition: 100% conc.

Steinamat B 2003. Dutton & Reinisch Ltd.
Chemical Description: Tetrasodium (1,2-dicarboxyethyl)-N-alkyl sulfosuccinamide
Category: Detergent, foaming agent and stabilizer, solubilizer, dispersant, emulsifier, anionic
Applications: Emulsion polymerization additive; industrial and metal cleaners; dishwashing liquids, household, and all-purpose cleaners; solubilizer for other de-

tergent raw materials; carpet and upholstery cleaners; soldering aid; reduction of gel formation; dispersion of pigments; emulsifier for wax and oil; cosmetics, e.g., cold creams, skin and hand lotions; esp. good in presence of inorganic salts; high calcium tolerance
Form: Clear liquid; Color: Amber
Properties: Solubility: Gives a clear sol'n. in water, alkali, and electrolytes; Stability: Good in presence of inorganic salts; pH: 7–9 (10% solids)
Composition: ≈ 35% solids
References: Folder: section 6/1277/0602

Steinamat TMS. Dutton & Reinisch Ltd.
Chemical Description: Disodium alkyl sulfosuccinamide
Category: Foaming agent and stabilizer, emulsifier, wetting agent, anionic
Applications: Foam additive for natural and synthetic lattices, esp. in foam backings for carpets; emulsion polymerization; wetting agent for latex impregnation
Form: Soft creamy paste or liquid
Properties: Stability: Stable in alkaline and acid media; pH: 9–10 (10% solids); Biodegradable
Composition: ≈ 35% solids
References: Folder: section 6/1277/0604

Steinamid 203/S. Dutton & Reinisch Ltd.
Chemical Description: Lauric acid diethanolamide
Category: Detergent, nonionic
Applications: Foam stabilizer and viscosity modifier for shampoos and detergents
Form: Solid; Color: Off-white; Odor: Typical (slight)
Properties: M.P.: ≈ 40 C
Composition: 100% active ingred.
Reaction: Alkaline
Containers: Mild steel drums

Steinamid AC 28. Dutton & Reinisch Ltd.
Chemical Description: Groundnut fatty acid polydialkanolamide
Category: Detergent, emulsifier, nonionic
Applications: Used as an anticorrosion additive for emulsifiable and nonemulsifiable mineral-oil-based metal-working lubricants, including drilling-, cutting-, and grinding-oils
Form: Liquid; Color: Brown
Properties: Density: 0.98 g/cm^3 @ 20 C; Sp. gr.: ≈ 1.0; Visc.: Medium; Acid No.: 14–16.5
Composition: 100% active ingred.
Reaction: Alkaline
Containers: Mild steel drums

Steinamid C 212. Dutton & Reinisch Ltd.
Chemical Description: Coconut fatty acid monoethanolamide
Category: Detergent, nonionic
Applications: Can be incorporated in all types of detergent products to provide: improvement of foam quantity, stability, and texture; increased viscosity; superfatting effect/skin protection; improved

detergency; stabilizer of emulsions
Form: Flakes; Color: Off-white; Odor: Amidic (slight)
Properties: M.P.: D 73 C
Properties: M.P.: ≈ 73 C
Composition: 100% active ingred.
Reaction: Alkaline
Containers: Paper sacks

Steinamid DC 212/S. Dutton & Reinisch Ltd.
Chemical Description: Coconut fatty acid diethanolamide
Category: Detergent, nonionic
Applications: Foam stabilizer and viscosity modifier used for shampoos and detergents
Form: Liquid; Color: Light yellow; Odor: Typical (slight)
Properties: Density: 0.99 g/cm^3 @ 20 C; Visc.: Medium
Composition: 100% active ingred.
Reaction: Alkaline
Containers: Mild steel drums

Steinamid DC 220/SE. Dutton & Reinisch Ltd.
Chemical Description: Coconut fatty acid diethanolamide
Category: Detergent, nonionic
Applications: Thickening agent, foam stabilizer, superfatting agent
Form: Soft paste
Composition: 100% active ingred.

Steinamid DL 240. Dutton & Reinisch Ltd.
Chemical Description: Coconut fatty acid polydiethanolamide
Category: Detergent, nonionic
Applications: Can be added to liquid and paste cosmetic preparations and all types of cleaning- and washing-agents to produce the following improvements: increased viscosity; improved foam quality, quantity, and stability; stabilization of emulsions; superfatting; improved washing power
Form: Liquid; Color: Yellow; Odor: Typical (faint)
Properties: Visc.: Medium
Composition: 100% active ingred.
Reaction: Alkaline
Containers: Mild steel drums

Steinamid DO 280. Dutton & Reinisch Ltd.
Chemical Description: Oleic acid polydiethanolamide
Category: Detergent, nonionic
Applications: Wide range of liquid and paste products in the cosmetic, cleaning-agent, and detergent industries
Form: Liquid; Color: Amber; Odor: Typical (slight)
Properties: Sp. gr.: ≈ 1.0; Visc.: Medium
Composition: 100% active ingred.
Reaction: Alkaline

Steinamid DO 280/SE. Dutton & Reinisch Ltd.
Chemical Description: Oleic acid diethanolamide
Category: Detergent, nonionic
Applications: Recommended for cosmetic products for which improved skin-compatibility, foam quality, and viscosity characteristics are required
Form: Liquid; Color: Honey

Properties: Density: 0.96 g/cm^3 @ 20 C
Composition: ≈ 100% active ingred.

Steinamid IPE 280. Dutton & Reinisch Ltd.
Chemical Description: Oleic monoisopropanolamide
Category: Detergent, nonionic
Applications: Skin-protecting additive for liquid and paste products; in addition to its excellent superfatting properties, it helps to produce a fine-structured and creamy foam
Form: Semisolid; Color: Light yellow; Odor: Typical (slight)
Composition: 100% active ingred.
Reaction: Alkaline
Containers: Lacquered drums

Steinamid IPL 203. Dutton & Reinisch Ltd.
Chemical Description: Lauric monoisopropanolamide
Category: Detergent, nonionic
Applications: Additive for solid and paste end-products, whose properties are as follows: increased foam quality and stability; increased viscosity; good superfatting; improved washing power; stabilization of emulsions
Form: Flakes; Color: White; Odor: Typical (slight)
Properties: M.P.: ≈ 55 C
Composition: 100% active ingred.
Reaction: Alkaline
Containers: Paper sacks

Steinamid IPP 240. Dutton & Reinisch Ltd.
Chemical Description: Coconut fatty acid monoisopropanolamide
Category: Detergent, nonionic
Applications: See Rewomid IPL 203
Form: Flakes; Color: White; Odor: Typical (slight)
Properties: M.P.: ≈ 50 C
Composition: 100% active ingred.
Reaction: Alkaline
Containers: Mild steel drums

Steinamid L203. Dutton & Reinisch Ltd.
Chemical Description: Lauric monoethanolamide
Category: Detergent, nonionic
Applications: Suitable for all types of solid, paste, and powder detergent preparations; improvement of foam texture, volume, and stability; increased viscosity; superfatting effect/skin protection; fixation of perfumes; stabilization of emulsions; improvement of detergency
Form: Flakes; Color: White; Odor: Typical (slight)
Properties: M.P.: ≈ 84 C
Composition: 100% active ingred.
Reaction: Alkaline
Containers: Paper sacks

Steinamid RE. Dutton & Reinisch Ltd.
Chemical Description: Unadulterated combination of nonionic fatty acid blends
Category: Anticorrosion additive,

emulsifier, nonionic
Applications: Additive for emulsifiable metal-working lubricants such as drilling-, cutting-, and grinding oils; inhibits corrosion; good emulsifying power; mild to the skin; increases lubricity; low toxicity
Form: Liquid; Color: Dark brown
Properties: Density: 0.98 g/cm^3 @ 20 C; Biodegradable
Composition: 100% active ingred.

Steinamid S 280. Dutton & Reinisch Ltd.
Chemical Description: Stearic acid monoethanolamide
Category: Detergent, nonionic
Applications: Excellent superfatting effect; increased mildness to the skin; improved foam quality; increased viscosity; improved washing power; stabilization of emulsions
Form: Flakes; Color: White; Odor: Typical (slight)
Properties: M.P..: ≈ 92 C
Composition: 100% active ingred.
Reaction: Alkaline
Containers: Paper sacks

Steinamin IM-BT. Dutton & Reinisch Ltd.
Chemical Description: Imidazoline derivative
Category: Adhesive additive, corrosion inhibitor, emulsifier, cationic
Applications: Adhesive additive for bitumen; adhesive agent and emulsifier in normal acidic bitumen emulsions; rust inhibitor in acidic media
Form: Liquid; Color: Green-brown, 13 max. (Gardner); Odor: Amine
Properties: Density: 0.94 g/cm^3 @ 20 C
Composition: 100% active ingred.
Reaction: Alkaline
Toxicity: Avoid prolonged contact with skin
Containers: Iron containers

Steinamin IM-CA. Dutton & Reinisch Ltd.
Chemical Description: 1-Hydroxyethyl-2-heptadecenyl imidazoline
Category: Corrosion inhibitor, emulsifier, penetrant, wetting agent, additive, cationic
Applications: Leather industry—used in pickling baths as a corrosion inhibitor; metal-working industry—used as an additive for corrosion inhibiting oils, in cooling-fluids, and in cutting oils; small quantities will improve in the penetration and binding power of paint or dyes in the priming of wood and metal surfaces; wetting agent for carbonization baths; additive for the preparation of glass fiber
Form: Wax; Color: Yellow
Properties: m.w.: 285; Solubility: Sol. in most polar and nonpolar solvents; disp. in water; Stability: Stable in neutral and acidic media
Composition: 100% active ingred.
Reaction: Alkaline
Toxicity: Avoid prolonged contact with skin

Containers: Iron containers
Warnings and Caution: Should be stored in air-tight containers as it is a tertiary amine and can react with carbon dioxide

Steinamin IM-OA. Dutton & Reinisch Ltd.
Chemical Description: 1-Hydroxyethyl-2-heptadecenyl imidazoline
Category: Rust inhibitor, adhesive agent, coupling agent, emulsifier, detergent, antistat, raw material, cationic
Applications: Can be used whenever inhibition of corrosion and/or an improved adhesive effect is required; in heating oils, fatty acid combination provides good rust prevention; used as a coupling agent where rust prevention is required; in pickling baths, used as an inhibitor to reduce attack on the steel or iron surface and to reduce hydrogen brittleness; excellent cationic emulsifier for the emulsification of oils and bitumens; acidic solutions can be prepared which are suitable for the formulations of metal cleaners, toilet bowl cleaners, rust removers; antistatic effect when dissolved in aqueous and alcoholic solutions; raw material for quaternization reactions; suitable for the flotation of phosphates, potassium, feldspar, iron, and other ores; used in the paint industry for pigment flushing and to improve coupling during priming of surfaces
Form: Liquid; Color: Amber
Properties: m.w.: 345; Solubility: Sol. in most polar and nonpolar solvents; disp. in water; Density: 0.94 g/cm^3 @ 20 C; Visc.: 800 cps; Stability: High thermal stability; pH: 10–11 (1% aq. sol'n.)
Composition: 100% active ingred.
Reaction: Alkaline
Toxicity: Avoid prolonged contact with skin
Containers: Iron containers
Warnings and Caution: This product will tend to crystallize at a rate dependent on length and temperature of storage

Steinaminoxid B 204. Dutton & Reinisch Ltd.
Chemical Description: Alkylamido-propyldimethylamine oxide
Category: Foam booster, emulsifier, dispersant, softener, nonionic
Applications: Foam booster in wash-active formulations; calcium soap dispersant; used in shampoos; improves dermatological properties of other anionics
Form: Clear liquid; Color: Pale yellow; Odor: Almost odorless
Properties: pH: 5–7 (10% solids); Biodegradable
Composition: 35% min. solids
References: Folder: section 14/1277/1401

Steinapal C 6. Dutton & Reinisch Ltd.
Chemical Description: Coconut amide polyglycolether
Category: Dispersant, wetting agent, foaming agent, nonionic
Applications: Excellent calcium soap dispersing and wetting prop-

erties; used advantageously as a solvent for perfumes in shampoo and foam-bath formulations; improves foam quality when used in combination with other surface-active raw materials; can be easily combined with all anionic, amphoteric, cationic, and nonionic materials; can be used for the formulation of spray- and immersion bath-cleaners
Form: Liquid; Color: Amber
Properties: Cloud Pt.: 80–90 C (2% in dist. water); Stability: Stable in dilute acid and alkaline sol'ns.; pH: 8–10 (1% solids); Biodegradable
Composition: 100% active ingred.

Steinapal HV 5. Dutton & Reinisch Ltd.
Chemical Description: Nonylphenol polyglycolether
Category: Emulsifier, raw material, wetting agent, degreasing agent, nonionic
Applications: Used as an emulsifier for preparations of all sorts of emulsions and insecticides; raw material for rinsing, washing, and cleaning agents of all types; wetting agent in the paper and cellulose industry; raw material for textile treatment; degreasing agent in pickling and alkaline immersion-bath cleaners
Form: Liquid; Color: 2 max. (Gardner)
Properties: Solubility: Sol. in mineral oil, alcohol, ketone, toluene, xylene, chlorinated hydrocarbons; Cloud Pt.: 59–62 C (10% in 25% aq. butyl diglycol); Stability: Stable in presence of dilute acids and alkalis, in hard water, and oxidizing agents; pH: 5–7 (1% solids)
Composition: 100% active ingred.

Steinapal HV 8. Dutton & Reinisch Ltd.
Chemical Description: Nonylphenol polyglycolether
Category: Emulsifier, raw material, wetting agent, degreasing agent, nonionic
Applications: See Steinapal HV 5
Form: Liquid; Color: 2 max. (Gardner)
Properties: Solubility: Sol. in mineral oil, alcohol, ketone, toluene, xylene, chlorinated hydrocarbons; Cloud Pt.: 41–45 C (10% in 25% aq. butyl diglycol); Stability: Stable in the presence of dilute acids and alkalis, in hard water, and oxidizing agents; pH: 5–7 (1% solids)
Composition: 100% active ingred.

Steinapal HV 9. Dutton & Reinisch Ltd.
Chemical Description: Nonylphenol polyglycolether
Category: Emulsifier, raw material, wetting agent, degreasing agent, nonionic
Applications: See Steinapal HV 5
Form: Liquid; Color: 2 max. (Gardner)
Properties: Solubility: Sol. in mineral oil, alcohol, ketone, toluene, xylene, chlorinated hydrocarbons; Cloud Pt.: 59–62 C (2% in dist. water); Stability: Stable in the

presence of dilute acids and alkalis, in hard water, and oxidizing agents; pH: 5–7 (1% solids)
Composition: 100% active ingred.

Steinapal HV 10. Dutton & Reinisch Ltd.
Chemical Description: Nonylphenol polyglycolether
Category: Emulsifier, raw material, wetting agent, degreasing agent, nonionic
Applications: See Steinapal HV 5
Form: Liquid; Color: 2 max. (Gardner)
Properties: Solubility: Sol. in mineral oil, alcohol, ketone, toluene, xylene, chlorinated hydrocarbons; Cloud Pt.: 70–73 C (2% in dist. water); Stability: Stable in the presence of dilute acids and alkalis, in hard water, and oxidizing agents; pH: 5–7 (1% solids)
Composition: 100% active ingred.

Steinapal HV 14. Dutton & Reinisch Ltd.
Chemical Description: Nonylphenol polyglycolether
Category: Emulsifier, raw material, wetting agent, degreasing agent, nonionic
Applications: See Steinapal HV 5
Form: Soft wax; Color: White
Properties: Solubility: Sol. in mineral oil, alcohol, ketone, toluene, xylene, chlorinated hydrocarbons; Cloud Pt.: 60–63 C (2% in 10% salt sol'n.); Stability: Stable in the presence of dilute acids and alkalis, in hard water, and oxidizing agents; pH: 5–7 (1% solids)
Composition: 100% active ingred.

Steinapal HV 25. Dutton & Reinisch Ltd.
Chemical Description: Nonylphenol polyglycolether
Category: Emulsifier, raw material, wetting agent, degreasing agent, nonionic
Applications: See Steinapal HV 5
Form: Wax; Color: White
Properties: Solubility: Sol. in mineral oil, alcohol, ketone, toluene, xylene, chlorinated hydrocarbons; Cloud Pt.: 73–76 C (2% in 10% salt sol'n.); Stability: Stable in the presence of dilute acids and alkalis, in hard water, and oxidizing agents; pH: 5–7 (1% solids)
Composition: 100% active ingred.

Steinapal LA 3. Dutton & Reinisch Ltd.
Chemical Description: Fatty alcohol polyglycolether
Category: Emulsifier, coupling agent, raw material, nonionic
Applications: Because of its low level of ethoxylation, this product has a definite hydrophobic character; ideal raw material for the production of various ether sulfates with excellent foaming properties; can be easily incorporated into sulfosuccinate-based shampoos to increase the viscosity and to provide a mild superfatting effect
Form: Liquid; Color: Clear
Properties: Solubility: Sol. in min-

eral oil and many organic solvents; Hydroxyl No.: 165–180; pH: 5–7 (1% solids); Biodegradable

Steinapal M 365. Dutton & Reinisch Ltd.
Chemical Description: Rolling- and hydraulic-oil additive
Category: Raw material, high-pressure additive, lubricant, corrosion inhibitor, nonionic
Applications: Used as an additive for water-soluble metal working liquids; for rolling- and hydraulic-oil bathes on polyglycolether derivatives; raw material for metal-working preparations, such as screw-thread cutting lubricants, ball-bearing polishes; widely used as a component in many rolling-oil formulations; compatible with anionic, cationic, amphoteric, and other nonionic products over a wide pH range
Form: Viscous liquid; Color: Amber
Properties: Solubility: Sol. in water; Cloud Pt.: 37–43 C (2% solids in dist. water); pH: 7–9 (1% solids); Biodegradable
Composition: ≈ 100% active ingred.

Steinapal O 8. Dutton & Reinisch Ltd.
Chemical Description: Oleic acid amide polyglycolether
Category: Wetting and dispersing agent, emulsifier, detergent, nonionic
Applications: Excellent wetting and calcium soap dispersing agent; good oil-in-water emulsifier; can be combined with all anionic, cationic, amphoteric, and nonionic materials; suitable for machine washing formulations
Form: Liquid; Color: Brown
Properties: Solubility: Sol. in water; fairly sol. in alcohol, ketones, esters, chlorinated hydrocarbons, benzene, fatty oils; sparingly sol. in toluene, xylene; Cloud Pt.: 75–85 C; Stability: Stable in dilute acid and alkaline sol'ns.; pH: 8–10 (1% solids); Biodegradable
Composition: 100% active ingred.

Steinapal TA 11. Dutton & Reinisch Ltd.
Chemical Description: Fatty alcohol polyglycolether
Category: Wash-active base, nonionic
Applications: Has excellent wetting-, dispersing-, emulsifying-, cleaning-power
Form: Soft wax; Color: White
Properties: Solubility: Sol. in water and many organic solvents; Cloud Pt.: 69–73 C (2% in dist. water); Stability: Stable in dilute acids, alkalines, hard water sol'ns.; pH: 5–7 (1% solids); Biodegradable

Steinapal TA 25. Dutton & Reinisch Ltd.
Chemical Description: Fatty alcohol polyglycolether
Category: Wash-active base, nonionic
Applications: See Steinapal TA 11
Form: Powder; Color: White
Properties: Solubility: Sol. in wa-

ter and many organic solvents; Cloud Pt.: 75–90 C (2% in 10% salt sol'n.); Stability: Stable in dilute acids, alkalines, hard water sol'ns.; pH: 5–7 (1% solids); Biodegradable

Steinapal TA 50. Dutton & Reinisch Ltd.
Chemical Description: Fatty alcohol polyglycolether
Form: Powder; Color: White
Properties: Cloud Pt.: 72–76 C (2% in 10% salt sol'n.); pH: 5–7 (1% solids); Biodegradable

Steinaphat E 1027. Dutton & Reinisch Ltd.
Chemical Description: Alkylphenol polyglycolether phosphate
Category: Corrosion inhibitor, emulsifier, dispersant, wetting agent, anionic
Applications: Used in pure oil for cutting, rolling, and extruding metals; reduces friction, prevents wear, cleans, and has antistatic properties
Form: Liquid; Color: Yellow
Properties: Acid No.: 100 ± 5; Stability: Stable in neutral and alkaline solutions
Composition: 100% active ingred.

Steinaphat EAK 8190. Dutton & Reinisch Ltd.
Chemical Description: Fatty alcohol polyglycolether phosphate
Category and Applications: See Steinaphat E 1027
Form: Liquid; Color: Yellow
Properties: Solubility: Disp. in water; Acid No.: 150 ± 10; Stability: Stable in neutral and alkaline sol'ns.; Biodegradable
Composition: ≈ 100% active ingred.

Steinapol-Emulsifier BW. Dutton & Reinisch Ltd.
Chemical Description: Modified nonionic fatty acid derivative
Category: Emulsifier, nonionic
Applications: Forms an opalescent cleansing liquid with water; formulated with organic solvents to produce hand cleaners, cleaners for oily or greasy surfaces; mild to the skin
Form: Clear, viscous liquid; Color: Yellow
Properties: Solubility: Oil-sol.; Acid Value: 1 max.
Composition: 100% active ingred.
References: Folder: section 4/1277/0401

Steinapol-Emulsifier BWA. Dutton & Reinisch Ltd.
Chemical Description: Modified nonionic fatty acid derivative
Category: Emulsifier, nonionic
Applications: Metal cleaning formulations, e.g., engine cleaners, solvent degreasers; emulsifier for paraffin wax and for cold cleaners used to dewax cars; mild to the skin
Form: Clear, viscous liquid; Color: Yellow
Properties: Acid Value: 1 max.
Composition: 100% active ingred.
References: Folder: section 4/1277/0402

Steinapol FBR. Dutton & Reinisch Ltd.
Chemical Description: Detergent combination
Category: Detergent, raw material, corrosion inhibitor, anionic
Applications: Raw material for weakly alkaline household and general-purpose cleaners; concentrated cleaners; products for the removal of latex-based floor waxes and polishes; corrosion inhibitor; preparation of oil or solvent emulsifiers with water
Form: Clear, viscous liquid; Color: 3–6 (Gardner)
Properties: Density: ≈ 0.98 g/cm^3 (20 C)
Composition: 100% active ingred.
References: Folder: section 3/ 1277/0312

Steinapol FLM. Dutton & Reinisch Ltd.
Chemical Description: Blend of foaming agents
Category: Foam fire-extinguisher base, anionic
Applications: Low-density foam fire-extinguisher base; produces stable foam blanket
Form: Clear liquid; Color: Light yellow
Properties: Density: ≈ 1.0–1.1 g/cm^3 (20 C); Clear Pt.: –12 C; pH: 6.5–7.5 (1% solids)
Composition: 25% active ingred.
References: Folder: section 3/ 1277/0313

Steinapol M 5047. Dutton & Reinisch Ltd.
Chemical Description: Detergent combination
Category: Detergent, anionic
Applications: Preconcentrate for surface cleaners
Form: Clear liquid (may become turbid at low temps. or with long storage); Color: Colorless
Properties: pH: 9–10 (1% solids)
Composition: 45% min. solids

Steinapol NL, 3, 28%. Dutton & Reinisch Ltd.
Chemical Description: Sodium lauryl ether sulfate
Category: Detergent, foaming agent, anionic
Applications: Cleaning formulations; high quality shampoos and foam bath preparations
Form: Clear liquid; Color: Almost colorless; Odor: Low
Properties: Visc.: 50–150 cps; Stability: Very good temperature stability; pH: ≈ 7–8 (5% solids)
Composition: 28% active ingred.
References: Folder: section 2/ 1277/0201

Steinapol NLS 28. Dutton & Reinisch Ltd.
Chemical Description: Sodium lauryl sulfate
Category: Detergent, emulsifier, anionic
Applications: Raw materials for washing and cleaning agents in liquid, paste, or powder form; cosmetics, e.g., foam bath, shampoos; emulsifier in emulsion polymerization systems
Form: Liquid (may become turbid at low temps. and on long storage); Color: Light yellow

Properties: m.w.: 300; pH: 7–8 (1% solids)
Composition: 28% active ingred.
References: Folder: section 2/ 1277/0202

Steinapol NLS 90-Powder. Dutton & Reinisch Ltd.
Chemical Description: Sodium lauryl sulfate
Category and Applications: See Steinapol NLS 28
Form: Powder; Color: White to light yellow
Properties: m.w.: 300; pH: 7–9 (1% solids)
Composition: 85% min. active
References: Folder: section 2/ 1277/0202

Steinapol OFR. Dutton & Reinisch Ltd.
Category: Detergent
Applications: Surface cleaner, e.g., cleaning of synthetic car seats, furniture coverings, white-walled tires, baths, sinks, enameled surfaces, table tops, floors, mirrors, etc.
Form: Clear liquid; Color: Colorless
Properties: pH: 9–10 (1% solids)
Composition: 16% min. solids
References: Folder: section 3/ 1277/0316

Steinapol S 1590. Dutton & Reinisch Ltd.
Chemical Description: Detergent combination
Category: Detergent, anionic
Applications: Preconcentrate for preparation of dishwashing liquids
Form: Clear, viscous liquid; Color: Yellow
Properties: Stability: Temperature stable down to 3 C; pH: 6.0–7.0 (1% solids)
Composition: 44% solids
References: Folder: section 3/ 1277/0310

Steinapol S 1605. Dutton & Reinisch Ltd.
Chemical Description: Detergent combination
Category: Detergent, wetting agent, emulsifier, anionic
Applications: Raw material for liquid washing, rinsing, and cleaning agents in the household and in industry; has good wetting, emulsifying properties, foam stability, and mildness to skin
Form: Clear, viscous liquid; Color: Yellow
Properties: pH: 7–8 (1% solids)
Composition: 48% solids
References: Folder: section 3/ 1277/0311

Steinapol SB-DO 70. Dutton & Reinisch Ltd.
Chemical Description: Sodium diisooctyl sulfosuccinate
Category: Wetting agent, solubilizer, anionic
Applications: Cosmetic preparations, household cleaners, e.g., cleaning lotions, nail varnish removers, shampoos, foam baths, dishwashing agents, window and glass cleaners; in metal cleaning;

as antismear agent for car windscreens; in paper industry; in textile industry as wetter for hydrophobic fibers; in paint industry; in pickling agents; in dye industry; in dry cleaning
Form: Liquid; Color: Almost colorless
Properties: Stability: Stable @ pH 3–9; Surface Tension: <30 dynes/cm (0.1% sol'n. @ 20 C); pH: 6.5–7.5 (5% solids); Biodegradable
Composition: 64% min. active
References: Folder: section 1/1277/0108

Steinapol SBF 12. Dutton & Reinisch Ltd.
Chemical Description: Disodium lauryl alcohol sulfosuccinate
Category: Detergent, foaming aid, anionic
Applications: Mild detergent raw material with foam-improving properties; for cosmetics, carpet and upholstery shampoos (hardens the residue for easy removal), synthetic soap bars, etc.
Form: Paste; Color: White
Properties: pH: 6.5–7.5 (5% solids); Biodegradable
Composition: 39% solids
References: Folder: section 1/1277/0104

Steinapol SBFA 30, 40%. Dutton & Reinisch Ltd.
Chemical Description: Disodium lauryl alcohol polyglycol-ether sulfosuccinate
Category: Detergent, anionic
Applications: Mild detergent raw material with good foaming for shampoos, foam bath products, intimate hygiene and baby preparations, washing-up liquids, handcleaners
Form: Clear liquid (may become turbid at low temps. or on long storage); Color: Almost colorless
Properties: Visc.: 100–200 cps; pH: 6.5–7.5 (5% solids); Biodegradable
Composition: 39% min. solids
References: Folder: section 1/1277/0101

Steinapol SBL 203. Dutton & Reinisch Ltd.
Chemical Description: Disodium lauric acid ethanolamide sulfosuccinate
Category: Detergent, anionic
Applications: Mild detergent for cosmetics; improves foam quality and produces milder effect on skin; aids spray-drying by increasing hardness of particles; used in carpet and upholstery shampoos; synthetic soaps
Form: Turbid liquid (tends to separate at room temp.); Color: Yellow
Properties: pH: 6.5–7.5 (5% solids); Biodegradable
Composition: 39% min. solids
References: Folder: section 1/1277/0102

Steinapol SBR 12-Powder. Dutton & Reinisch Ltd.
Chemical Description: Disodium lauryl alcohol sulfosuccinate
Category: Detergent, anionic

Applications: Mild detergent for synthetic soaps, normal soap bars, dry shampoos, enzyme washing powders, moisture sensitive technical products; improves spray-drying properties of softer materials
Form: Powder; Color: White or light yellow
Properties: Solubility: Water-sol.; Density: ≈ 250 g/l (packing, drum-dried); Stability: Stable in weakly acid or weakly alkaline media; pH: 6.5–7.5 (5% solids); Biodegradable
Composition: 95% solids
References: Folder: section 1/ 1277/0105

Steinapol SBV. Dutton & Reinisch Ltd.
Chemical Description: Modified sulfosuccinate
Category: Detergent, anionic
Applications: Mild detergent with good foaming for preparation of mild, nonirritant shampoos and foam baths, rinsing and cleaning agents with skin-protecting properties
Form: Clear liquid; Color: Light yellow
Properties: Visc.: ≈ 2500 cps; pH: 6.5–7.5 (1% solids); Biodegradable
Composition: 28% min. solids
References: Folder: section 3/ 1277/0306

Steinapol SCK 2040. Dutton & Reinisch Ltd.
Chemical Description: Detergent combination
Category: Detergent, anionic
Applications: Raw material for liquid cleaning agents, shampoos, etc.
Form: Clear, viscous liquid; Color: Yellow
Properties: Visc.: ≈ 600 cps (room temp.); pH: 7–8.5 (1% solids); Biodegradable
Composition: 39% min. solids
References: Folder: section 3/ 1277/0314

Steinapol TS 25. Dutton & Reinisch Ltd.
Chemical Description: Sulfosuccinate-base detergent combination
Category: Detergent, foaming agent, anionic
Applications: Foam-cleaner concentrate for cleaning of coarse-fibered textiles, e.g., carpets, upholstery, car seat covers
Form: Liquid (may become turbid at low temps. or on long storage); Color: Light yellow
Properties: Clear Pt.: 8 C max.; pH: 6.5–7.5 (1% solids); Biodegradable
Composition: 25% solids
References: Folder: section 3/ 1277/0301

Steinapon AM-2C. Dutton & Reinisch Ltd.
Chemical Description: Coconut based ampholyte
Category: Detergent, dispersant, foam booster, wetting agent, amphoteric
Applications: Stabilizes the foam, and acts as a calcium soap disper-

sant when used in hard water; used in toiletries, mainly in mild shampoo and foam-bath preparations, and esp. in baby products; compatible with all anionic, cationic, and nonionic wash active substances
Form: Liquid; Color: Golden-yellow
Properties: Biodegradable
Composition: 49% active ingred.

Steinapon AM-2L. Dutton & Reinisch Ltd.
Chemical Description: Imidazoline-based ampholyte
Category: Detergent, amphoteric
Applications: Mild foam baths, shampoos, personal hygiene toiletries
Form: Liquid
Composition: 50% active ingred.

Steinapon AM-B 13. Dutton & Reinisch Ltd.
Chemical Description: Alkyl amido betaine
Category: Detergent, foam booster, amphoteric
Applications: Suitable for the formulation of shampoos, foam baths, and other toiletries; can be combined with anionic wash active agents to give clear solutions; has a strong foam stabilizing effect even in the presence of oil and soap; compatible with all anionic, cationic, nonionic, and other amphoteric wash active substances
Form: Liquid; Color: Yellow
Properties: pH: 4–6 (10% solids); Biodegradable
Composition: 30% active ingred.

Steinapon AM-CA. Dutton & Reinisch Ltd.
Chemical Description: Coconut based ampholyte
Category: Detergent, foaming agent, amphoteric
Applications: Excellent foaming properties which are almost unaffected by the presence of salts and calcium soaps; highly suitable for the formulation of shampoos and for improving the soap compatibility of foam baths
Form: Clear liquid; Color: Yellow
Properties: pH: 8–8.5 (10% solids)
Composition: 30% active ingred.

Steinapon AM-V. Dutton & Reinisch Ltd.
Chemical Description: Short-chain fatty acid-based ampholyte
Category: Detergent, wetting agent, amphoteric
Applications: Exhibits excellent wetting properties in acid and in alkaline media; high tolerance towards the presence of large amounts of electrolytes; suitable for the preparation of clear stable cleaning agents with a high salt content; used as a wetting agent for all types of acidic and alkaline cleaners; reduces surface tension enabling more rapid and complete penetration of the electrolyte
Form: Liquid; Color: Dark amber
Properties: pH: 10–12 (10% solids)
Composition: 28% active ingred.

Steinaryl-Sulfonic Acid K. Dutton & Reinisch Ltd.
Chemical Description: Dodecylbenzene sulfonic acid

Category: Detergent base, anionic
Applications: Can be neutralized with various bases for dishwashing, washing, and cleaning formulations
Form: Viscous liquid; Color: Brown
Properties: Acid Value: 180 ± 5
Composition: 97% min. active
References: Folder: section 5/1277/0501

Steinaryl T 65. Dutton & Reinisch Ltd.
Chemical Description: p-Toluene sulfonic acid, 65%
Category: Catalyst, electrolyte intermediate, anionic
Applications: Catalyst for the preparation of esters, for acid hydrolysis, polymerization, dehydration, and alkylation reactions; as an electrolyte in several electrolytic processes; intermediate for the preparation of cresol; leather tanning agent; in highly toxic crop-dusting agents; the production of synthetic fibers; in the vulcanization of buna-rubber; solidification and hardening of phenol resins for foundry industry
Form: Clear liquid; Color: Pale; Odor: Characteristic
Properties: Acid Value: 224 min.
Composition: 65% active ingred.
Toxicity: Strong acid–skin irritant
Warnings and Caution: Wear protective goggles and gloves

Steinaryl TKS 90/F. Dutton & Reinisch Ltd.
Chemical Description: Triammonium dodecylbenzenesulfonate
Category: Detergent, anionic
Applications: Detergent raw material for highly conc. cleansing agents, esp. for stream-jet and high pressure cleaners, defatting agents, car shampoos; also for chain greasing, and for cleaning steel ball bearings
Form: Viscous liquid (may become turbid at low temps. or on long storage); Color: Dark amber
Properties: pH: 7.0–7.5 (1% solids); Biodegradable
Composition: 85% min. active
References: Folder: section 5/1277/0504

Steinazid DU 185. Dutton & Reinisch Ltd.
Chemical Description: Undecylenic acid polyethanolamide
Category: Detergent, emulsifier, nonionic
Applications: Used as bacteriocide, thickening, foam stabilizing agent in shampoos
Form: Viscous liquid; Color: Yellow
Properties: Solubility: Sol. in water, oil, alcohols, glycols; Sp. gr.: ≈ 1.0; Visc.: High
Composition: 100% active ingred.
Reaction: Acid
Containers: Lacquered drums

Steinazid SBU 185. Dutton & Reinisch Ltd.
Chemical Description: Sodium salt of the monosulfosuccinate of undecylenic acid monoethanolamide
Category: Fungicide, detergent,

foaming agent, anionic
Applications: Exhibits exceptional antiseborrheoic and antiallergic properties; can be combined with all anionic, nonionic, and amphoteric products without any reduction in its acitivity; used in foambaths, shampoos, skin-cleaners, foot baths, washing agents
Form: Powder; Color: Pale yellow; Odor: Typical (slight)
Properties: Solubility: Sol. in alcohol, glycol; Sp. gr.: ≈ 1.0; Stability: Stable at room temperature; pH: 6.5–7.5 (5% solids); Biodegradable
Composition: 50% active ingred.
Reaction: Acid (slight)
Containers: Lacquered drums

Steinazid U 185. Dutton & Reinisch Ltd.
Chemical Description: Monoethanolamide of undecylenic acid
Category: Fungicide, detergent, nonionic
Applications: Improves foam quality and stability, superfatting, and increases viscosity; fully compatible with all anionic, cationic, amphoteric, and other nonionic surface-active agents
Form: Flakes; Color: Yellow; Odor: Typical (strong)
Properties: Solubility: Sol. in alcohol, glycol; disp. in warm water; M.P.: ≈ 50 C
Composition: 100% active ingred.
Reaction: Alkaline

Steol 4N. Stepan Chem. Co.
Chemical Description: Lauryl ether sulfate, sodium salt
Category: Detergent, emulsifier, foaming agent, anionic
Applications: Shampoos, liquid detergents, foaming agents, heavy duty alkaline cleaners, textile mill applications (kier boiling, bleach assistants, neutral and alkaline scouring), emulsion polymerization; esp. suited for bubble baths and shampoos
Form: Liquid; Color: Water-white
Properties: Solubility: Water-sol.; Sp. gr.: 1.045; Visc.: 31 cps; Cloud Pt.: –4 C; Stability: Stable in hard water, neutral and alkaline media; pH: 8.0 (1% in dist. water); Biodegradable
Composition: 28% active ingred.

Steol 7N. Stepan Chem. Co.
Chemical Description: Linear fatty alcohol ether sulfate, sodium salt
Category: Detergent, wetting and foaming agent, anionic
Applications: Very mild surfactant for shampoo and bubble baths
Form: Liquid; Color: Pale yellow
Properties: Solubility: Water-sol.; Sp. gr.: 1.054; Visc.: 100 cps; Cloud Pt.: –2 C; Surface Tension: 37.7 dynes/cm (0.1%); pH: 8.0 (10%); Biodegradable
Composition: 28% active ingred.

Steol 7T. Stepan Chem. Co.
Chemical Description: Linear fatty alcohol ether sulfate, triethanolamine salt
Category and Applications: See Steol 7N
Form: Liquid; Color: Pale yellow

Properties: Solubility: Water-sol.; Sp. gr.: 1.046; Visc.: 100 cps; Cloud Pt.: −1 C; Surface Tension: 38.4 dynes/cm (0.1%); pH: 7.3 (10%); Biodegradable
Composition: 28% active ingred.

Steol CA-460. Stepan Chem. Co.
Chemical Description: Lauryl fatty alcohol sulfate, ammonium salt
Category: Detergent, emulsifier, foaming agent, anionic
Applications: Shampoos, liquid detergents, foaming agents, heavy duty alkaline cleaners, textile mill applications (kier boiling, bleach assistants, neutral and alkaline scouring), emulsion polymerization; esp. suited for high foaming detergents, e.g., dishwashing detergents and industrial cleaners
Form: Liquid; Color: Pale yellow
Properties: Solubility: Water-sol.; Sp. gr.: 1.016; Visc.: 67 cps; Cloud Pt.: 19 C; Stability: Stable in hard water; pH: 7.0 (1% in dist. water); Biodegradable
Composition: 59% active ingred.
Warnings and Caution: Product will liberate ammonia above pH 7

Steol CS-460. Stepan Chem. Co.
Chemical Description: Lauryl fatty alcohol sulfate, sodium salt
Category and Applications: See Steol CA-460
Form: Liquid; Color: Pale yellow
Properties: Solubility: Water-sol.; Sp. gr.: 1.037; Visc.: 69 cps; Cloud Pt.: 17 C; Stability: Stable in hard water, neutral and alkaline media; pH: 8.0 (1% in dist. water); Biodegradable
Composition: 59.5% active ingred.

Steol CS-760. Stepan Chem. Co.
Chemical Description: Linear fatty alcohol ether sulfate, sodium salt
Category and Applications: See Steol 7N
Form: Liquid; Color: Pale yellow
Properties: Solubility: Water-sol.; Sp. gr.: 1.059; Visc.: 100 cps; Cloud Pt.: 1 C; Surface Tension: 46.2 dynes/cm (0.1%); pH: 8.0 (10%); Biodegradable
Composition: 59% active ingred.; contains 12–15% ethanol
Warnings and Caution: Contains ethanol–use proper precautionary methods for storage and handling

Steol KA-460. Stepan Chem. Co.
Chemical Description: Linear fatty alcohol ether sulfate, ammonium salt
Category: Detergent, foaming and wetting agent, anionic
Applications: Shampoos, bubble baths, dish detergents, degreasers; designed to eliminate the preliminary gelling stage frequently encountered on dilution
Form: Liquid; Color: Straw
Properties: Solubility: Water-sol.; Density: 8.65 lb/gal; Cloud Pt.: 0 C; Stability: Effective in hard and soft water; pH: 7.0 (10% in dist. water); Biodegradable
Composition: 60% active ingred.
Warnings and Caution: Product will liberate ammonia above pH 7.5

Steol KS-460. Stepan Chem. Co.
Chemical Description: Linear fatty alcohol ether sulfate, sodium salt
Category and Applications: See Steol KA-460
Form: Liquid; Color: Straw
Properties: Solubility: Water-sol.; Density: 8.94 lb/gal; Cloud Pt.: 0 C; Stability: Effective in hard or soft water; compatible with other anionics, nonionics, and alkaline builders; pH: 8.0 (10% in dist. water); Biodegradable
Composition: 60% active ingred.

Stepan D-50. Stepan Chem. Co.
Chemical Description: Isopropyl myristate
Category: Emollient
Applications: Bubbling bath oils, bath oils, aerosol antiperspirants and deodorants, electric shave preparations, cosmetic lotions and creams
Form: Clear oily liquid; Color: Water-white; Odor: Bland, very slightly fatty odor
Properties: Ref. Index: 1.4325; Solubility: Oil-sol.; Acid Value: 0.22; Iodine Value: 0.02
Composition: 100% active ingred.

Stepan D-70. Stepan Chem. Co.
Chemical Description: Isopropyl palmitate
Category and Applications: See Stepan D-50
Form: Clear liquid; Color: Water-white; Odor: Essentially odorless
Properties: Ref. Index: 1.437; Solubility: Oil-sol.; Sapon. Value: 185–195; Iodine Value: 1.0 max.
Composition: 100% active ingred.

Stepan HDA-7. Stepan Chem. Co.
Chemical Description: Fatty acid alkylolamide, sodium cation
Category: Detergent, foaming agent, anionic/nonionic
Applications: Limited foam base for all purpose and floor cleaners
Form: Liquid; Color: Amber
Properties: Solubility: Water-sol.
Composition: 90% active ingred.

Stepan P-621. Stepan Chem. Co.
Chemical Description: Lauric-myristic alkylolamide
Category: Foam stabilizer, nonionic
Applications: Liquid detergent foam stabilizer
Form: Solid; Color: Light
Properties: Solubility: Water-disp.
Composition: High active

Stepanate AM. Stepan Chem. Co.
Chemical Description: Ammonium xylene sulfonate
Category: Hydrotrope, coupling agent, solubilizer, anionic
Applications: Coupling agent in detergent field for surfactants in phosphate and silicate sol'ns.; heavy duty cleaners, wax strippers, complex dishwash detergents; solvent or fluidizer; freeze-thaw stabilizer; additive for high electrolyte or brine systems
Form: Liquid; Color: Colorless
Properties: Solubility: Water-sol.;

Density: 9.05 lb/gal; pH: 6.5–7.2 (1% aq.)
Composition: 40% active ingred.

Stepanate T. Stepan Chem. Co.
Chemical Description: Sodium toluene sulfonate
Category and Applications: See Stepanate AM
Properties: Density: 10.1 lb/gal; pH: 7–9 (1% aq.)
Composition: 40% active ingred.

Stepanate X. Stepan Chem. Co.
Chemical Description: Sodium xylene sulfonate
Category and Applications: See Stepanate AM; also as cloud point depressant
Form: Liquid; Color: Colorless
Properties: Solubility: Water-sol.; Density: 9.83 lb/gal; pH: 7–9 (1% aq.)
Composition: 40% active ingred.

Stepanhold R-1. Stepan Chem. Co.
Chemical Description: PVP/ethyl methacrylate/methacrylic acid terpolyer
Category: Terpolymer resin
Applications: Hair spray and wave set formulations; good holding power, high luster, conditioning, extreme hard film, corrosion resistance
Properties: Solubility: Sol. in alcohols, water; Density: 8.00 lb/gal (70 F); Visc.: 10,000 cps; Acid No.: 60–65
Composition: 50% solids; solvent: SDA-40 anhydrous alcohol

Stepanol 317. Stepan Chem. Co.
Chemical Description: Anionic blend, amine cation
Category: Detergent, foaming agent, anionic
Applications: Completely formulated dilutable base for shampoos and bubble baths; viscosity increases on dilution; useful in hard and soft waters
Form: Viscous, clear liquid; Color: Yellow
Properties: Solubility: Water-sol.; Sp. gr.: 1.029; Visc.: 750 cps; 1000 cps @ 1:7 dilution; pH: 7.7 (5% aq.); Biodegradable
Composition: 92.5% active ingred.

Stepanol AM. Stepan Chem. Co.
Chemical Description: Ammonium lauryl sulfate
Category: Detergent, foaming agent, anionic
Applications: Cosmetics (liquid, cream, aerosol, and dry shampoos, hand lotions, cleansing creams, shaving creams, bubble bath), rug and upholstery shampoos, railway car cleaners, car shampoos, industrial cleaners, metal cleaners, fruit washing, dairy cleaners, household liquid detergents, emulsion polymerization, insecticides, clay dispersions, textile and leather processing, pharmaceuticals
Form: Viscous liquid; Color: Pale yellow
Properties: Solubility: Water-sol.; pH: 6–7 (10% sol'n.)
Composition: 28–30% active

Stepanol AM-V. Stepan Chem. Co.
Chemical Description: Ammonium lauryl sulfate
Category and Applications: See Stepanol AM
Form: Viscous liquid to light gel
Properties: Solubility: Water-sol.; pH: 6–7 (10% sol'n.)
Composition: 27–29% active

Stepanol DEA. Stepan Chem. Co.
Chemical Description: DEA lauryl sulfate
Category and Applications: See Stepanol AM
Form: Clear liquid; Color: Pale yellow
Properties: Solubility: Water-sol.; pH: 7.3–7.7 (10% sol'n.)
Composition: 33–35% active

Stepanol ME Dry. Stepan Chem. Co.
Chemical Description: Sodium lauryl sulfate
Category and Applications: See Stepanol AM
Form: Powder; Color: White
Properties: Solubility: Water-sol.; pH: 7.5–10 (10% sol'n.)
Composition: 93% min. active

Stepanol MG. Stepan Chem. Co.
Chemical Description: Magnesium lauryl sulfate
Category and Applications: See Stepanol AM
Form: Clear, viscous liquid; Color: Pale yellow
Properties: Solubility: Water-sol.; pH: 6.5–7.5 (10% sol'n.)
Composition: 28–30% active

Stepanol WA-100. Stepan Chem. Co.
Chemical Description: Sodium lauryl sulfate, high purity dentifrice grade
Category: Detergent, foaming, wetting, and suspending agent, anionic
Applications: Dentifrice formulations; pharmaceuticals; emulsion polymerization; wetting and suspending agent for clay dispersions
Form: Powder; Color: White
Properties: Solubility: Water-sol.; pH: 7.5–10 (10% sol'n.)
Composition: 98.5% min. active

Stepanol WAC. Stepan Chem. Co.
Chemical Description: Sodium lauryl sulfate
Category and Applications: See Stepanol AM
Form: Clear liquid; Color: Pale yellow
Properties: Solubility: Water-sol.; pH: 7.5–8.5 (10% sol'n.)
Composition: 28–30% active

Stepanol WA Extra. Stepan Chem. Co.
Chemical Description: Sodium lauryl sulfate
Category and Applications: See Stepanol AM
Form: Clear liquid
Properties: Solubility: Water-sol.; pH: 7.5–8.5 (10% sol'n.)
Composition: 28–30% active

Stepanol WA Paste. Stepan Chem. Co.
Chemical Description: Sodium lauryl sulfate

Category and Applications: See Stepanol AM
Form: Clear paste (30 C); Color: Nearly water-white
Properties: Solubility: Water-sol.; pH: 7.5–8.5 (10% sol'n.)
Composition: 28–30% active

Stepanol WAQ. Stepan Chem. Co.
Chemical Description: Sodium lauryl sulfate
Category and Applications: See Stepanol AM
Form: Clear, viscous liquid (30 C); Color: Nearly water-white
Properties: Solubility: Water-sol.; pH: 7.5–8.5 (10% sol'n.)
Composition: 28–30% active

Stepanol WA Special. Stepan Chem. Co.
Chemical Description: Sodium lauryl sulfate
Category and Applications: See Stepanol AM
Form: Clear liquid; Color: Nearly water-white
Properties: Solubility: Water-sol.; pH: 7.5–8.5 (10% sol'n.)
Composition: 28–30% active

Stepanol WAT. Stepan Chem. Co.
Chemical Description: TEA lauryl sulfate
Category and Applications: See Stepanol AM
Form: Clear liquid; Color: Nearly water-white
Properties: Solubility: Water-sol.; pH: 7.0–8.5 (10% sol'n.)
Composition: 39–41.5% active

Stepantan A. Stepan Chem. Co.
Chemical Description: Sodium sulfonate of a naphthalene-formaldehyde condensation product
Category: Dispersant, tanning agent, anionic
Applications: Synthetic tanning assistant, pitch control agent in papermaking, dye mordant, pigment dispersant, etc.; effective dispersant in acid or alkaline media
Form: Powder; Color: Cream; yields pale straw-colored sol'ns. in water
Properties: Solubility: Very sol. in water; Density: 0.37–0.42 g/cc (tamped); pH: 7.4 (2% aq.)

Stepantex DA-52. Stepan Chem. Co.
Chemical Description: Alkyamine alkylaryl sulfonate
Category: Detergent, emulsifier, dispersant, demulsifier, coupling agent
Applications: Promotes water coupling with solvents to give clear blends, e.g., for treating mops and dust cloths for better dirt pick-up; emulsion degreasers; hand cleaners; emulsifier for dye carriers; dispersant for dyes and pigments; drycleaning detergent; demulsifier in oil recovery
Properties: Solubility: Sol. in kerosene, perchlorethylene, Stoddard solvent, xylene, isopropanol; disp. in water, propylene glycol, mineral oil

Stepantex Q-90B. Stepan Chem. Co.
Chemical Description: Fatty quaternary sulfate
Category: Softener, cationic
Applications: Fabric softener concentrate for home and commercial laundry use; nonyellowing and rewetting properties
Form: Clear liquid; Color: Pale amber (100 F)
Properties: Solubility: Readily disp. in water @ 120–125 F; Density: 8.03 lb/gal; Sp. gr.: 0.96; Pour Pt.: 77 F
Composition: 90% active ingred.
Containers: 425 lb net open head steel drums

Sterling 2XS. Canada Packers Ltd.
Chemical Description: Sodium xylene sulfonate
Category: Hydrotrope, coupling agent, cloud point depressant
Applications: Used for liquid detergent or general purpose applications
Form: Liquid
Composition: 40% active ingred.

Sterling 66/33. Canada Packers Ltd.
Chemical Description: Cocamide/lauramide DEA
Category: Foam stabilizer, auxiliary detergent, viscosity builder, nonionic
Form: Liquid or pasty
Composition: 100% active ingred.

Sterling AB-40. Canada Packers Ltd.
Chemical Description: Sodium linear dodecyl benzene sulfonate
Category: Detergent, anionic
Applications: Powdered detergents, carwash, and general cleaners
Form: Powder
Composition: 40% active ingred.

Sterling AB-80. Canada Packers Ltd.
Chemical Description: Sodium linear dodecyl benzene sulfonate
Category and Applications: See Sterling AB-40
Form: Flake
Composition: 80% active ingred.

Sterling AM. Canada Packers Ltd.
Chemical Description: Ammonium lauryl sulfate
Category: Detergent, anionic
Applications: Shampoos, bubble baths, detergents; used for high viscosity and clear products; contains some propylene glycol
Form: Viscous liquid
Composition: 27.7% active ingred.

Sterling AM-HV. Canada Packers Ltd.
Chemical Description: Ammonium lauryl sulfate
Category and Applications: Same as Sterling AM but contains no propylene glycol
Form: Very viscous liquid
Composition: 28.0% active ingred.

Sterling Amide 374. Canada Packers Ltd.
Chemical Description: Cocamide DEA

Category: Foamer, viscosity control agent, nonionic
Applications: Base for increasing viscosity for low solids economical detergent formulations; good oil solubilizer
Form: Liquid
Composition: 100% active ingred.

Sterling Amide 460. Canada Packers Ltd.
Chemical Description: Lauramide/myristamide DEA
Category: Foam stabilizer, auxiliary detergent, viscosity builder
Form: Liquid
Composition: 76% active ingred.; contains some propylene glycol

Sterling AOS. Canada Packers Ltd.
Chemical Description: Sodium $C_{14}-C_{16}$ olefin sulfonate
Category: Detergent, foaming agent and stabilizer, anionic
Applications: Shampoo, bubble baths, rug shampoos, general purpose cleaners, liquid detergents; excellent foaming and foam stability characteristics
Form: Liquid
Composition: 39% active ingred.

Sterling DEA. Canada Packers Ltd.
Chemical Description: Cocamide DEA
Category: Foam stabilizer, auxiliary detergent, viscosity builder, nonionic
Form: Liquid
Composition: 100% active ingred.

Sterling Emulsifier #3. Canada Packers Ltd.
Chemical Description: Pareth-25-3
Category: Emulsifier, nonionic
Applications: Emulsifier and detergent additive
Form: Liquid
Properties: Solubility: Oil-sol.
Composition: 100% active ingred.

Sterling Emulsifier #15. Canada Packers Ltd.
Chemical Description: Pareth-25-15
Category: Emulsifier, solubilizer, nonionic
Applications: Emulsifier and detergent additive; solubilizer and water-sol. emulsifier
Form: Paste
Properties: Solubility: Water-sol.
Composition: 100% active ingred.

Sterling ES 600. Canada Packers Ltd.
Chemical Description: Ammonium pareth-25 sulfate
Category: Detergent, anionic
Applications: Shampoos, bubble baths, detergents, degreasers, liquid detergents, powdered cleaners
Form: Liquid
Composition: 57.5% active ingred.; contains 14% ethanol

Sterling Granulated Wax. Canada Packers Ltd.
Chemical Description: Cocamide MEA
Category: Detergent, viscosity

builder, foam stabilizer
Form: Flake
Composition: 100% active ingred.

Sterling LA Acid-C. Canada Packers Ltd.
Chemical Description: Linear dodecyl benzene sulfonic acid
Category: Intermediate, detergent, anionic
Applications: Intermediate for detergent preparations, liquid detergents, industrial cleaners; liquid and powder; produces lower viscosity and cloud pt. liquid products than LA Acid Reg.
Form: Liquid
Composition: 96.3% active ingred.

Sterling LA Acid Reg. Canada Packers Ltd.
Chemical Description: Linear dodecyl benzene sulfonic acid
Category: Intermediate, detergent, anionic
Applications: Intermediate for detergent preparations, liquid detergents, industrial cleaners (liquids and powders)
Form: Liquid
Composition: 96.3% active ingred.

Sterling LA Oil. Canada Packers Ltd.
Chemical Description: TEA-dodecyl benzene sulfonate
Category: Detergent, anionic
Applications: Liquid detergents and general purpose cleaners
Form: Liquid; Color: Amber
Composition: 57% active ingred.

Sterling LA Paste 55%. Canada Packers Ltd.
Chemical Description: Sodium linear dodecyl benzene sulfonate
Category: Intermediate, detergent, anionic
Applications: Intermediate for detergent preparations; liquid detergents and liquid general purpose and industrial cleaning formulations
Form: Paste
Composition: 55% active ingred.

Sterling LA Paste 60%. Canada Packers Ltd.
Chemical Description: Sodium linear dodecyl benzene sulfonate
Category and Applications: See Sterling LA Paste 55%; LA Paste 60% contains 3% sodium xylene sulfonate
Form: Paste
Composition: 57% active ingred.

Sterling LDEA-90. Canada Packers Ltd.
Chemical Description: Lauramide DEA
Category: Foam stabilizer, auxiliary detergent, viscosity builder, nonionic
Applications: Used in cosmetics
Form: Solid
Composition: 100% active ingred.

Sterling LM-70. Canada Packers Ltd.
Chemical Description: Lauramide/myristamide DEA
Category and Applications: See Sterling LDEA-90; LM-70 has

greater viscosity building properties
Form: Solid
Composition: 100% active ingred.

Sterling NPX. Canada Packers Ltd.
Chemical Description: Nonoxynol-9
Category: Intermediate, detergent, emulsifier, nonionic
Applications: Detergent intermediate; household and industrial cleaner formulations; emulsifier
Form: Liquid
Composition: 100% active ingred.

Sterling Snow White 750. Canada Packers Ltd.
Chemical Description: Ditallow imidazolinium quaternary
Category: Softener
Applications: Substantive high mole weight quaternary for softening fabrics in home and industrial applications
Form: Liquid
Composition: 75% active ingred.

Sterling Super 1E. Canada Packers Ltd.
Chemical Description: Sodium laureth sulfate
Category: Detergent, anionic
Applications: Shampoos, bubble baths, detergents; yields high viscosity, low cloud point products
Form: Liquid
Composition: 25% active ingred.

Sterling Super 2E-25%. Canada Packers Ltd.
Chemical Description: Sodium laureth-2 sulfate
Category and Applications: See Sterling Super 1E
Form: Liquid
Composition: 25% active ingred.

Sterling Super 2E-60%. Canada Packers Ltd.
Chemical Description: Sodium laureth-2 sulfate
Category and Applications: See Sterling Super 1E; 2E-60% contains 14.5% ethanol
Form: Liquid
Composition: 58% active ingred.

Sterling Super 3.5E. Canada Packers Ltd.
Chemical Description: Sodium laureth-3.5 sulfate
Category: Detergent, emulsifier, anionic
Applications: Shampoos, bubble baths, detergents; polymerization emulsifier
Form: Liquid
Composition: 28% active ingred.

Sterling Super A. Canada Packers Ltd.
Chemical Description: Sodium pareth-25 sulfate
Category: Detergent, anionic
Applications: Shampoos, bubble baths, detergents, degreasers, liquid detergents, powdered cleaners; contains 11% ethanol
Form: Liquid
Composition: 57.5% active ingred.

Sterling WADE. Canada Packers Ltd.
Chemical Description: DEA lauryl

sulfate
Category: Detergent, anionic
Applications: Shampoos, bubble baths, detergents; used in clear concentrate cleansing preparations
Form: Liquid
Composition: 35% active ingred.

Sterling WA Paste. Canada Packers Ltd.
Chemical Description: Sodium lauryl sulfate
Category: Detergent, anionic
Applications: Shampoos, bubble baths, cosmetic base material
Form: Viscous liquid
Composition: 28.2% active ingred.

Sterling WAQ-CH. Canada Packers Ltd.
Chemical Description: Sodium lauryl sulfate
Category: Detergent, emulsifier, anionic
Applications: Shampoos, bubble baths, detergents, polymerization emulsifier
Form: Liquid
Composition: 29.3% active ingred.

Sterling WAQ-Cosmetic. Canada Packers Ltd.
Chemical Description: Sodium lauryl sulfate
Category: Detergent, anionic
Applications: Shampoo, bubble baths, detergent; higher free fatty alcohol content
Form: Viscous liquid
Composition: 29.5% active ingred.

Sterling WAQ-LO. Canada Packers Ltd.
Chemical Description: Sodium lauryl sulfate
Category: Detergent, emulsifier
Applications: Shampoos, bubble baths, detergents, polymerization emulsifier; very low cloud point
Form: Liquid
Composition: 29.5% active ingred.

Sterling WAQ-P. Canada Packers Ltd.
Chemical Description: Pearlescent sodium lauryl sulfate
Category: Detergent
Applications: Shampoo, bubble bath, cosmetic base material; higher inorganic salt content; for pearlescent shampoo formulations
Form: Viscous liquid
Composition: 29% active ingred.

Sterling WAT. Canada Packers Ltd.
Chemical Description: TEA-lauryl sulfate
Category: Detergent, anionic
Applications: Shampoo, bubble baths, detergents
Form: Liquid
Composition: 40% active ingred.

Sterling XE. Canada Packers Ltd.
Chemical Description: Pareth-25-10
Category: Detergent, wetting agent, emulsifier, nonionic
Applications: Detergent and wetting agent for use in industrial and household formulations; textile scouring; emulsifier
Form: Liquid
Composition: 85% active ingred.

Sterophos 1001. Witco Chem./ Israel
Category: Detergent, disinfectant
Applications: Disinfecting and cleaning of milking equipment and dairies; water softening, disinfecting, deodorizing, noncorrosive

Sterophos 1002. Witco Chem./ Israel
Category and Applications: See Sterophos 1001
Form: Liquid

Sterox AJ. Monsanto Ind. Chem.
Chemical Description: Tridecyl alcohol 8.5 mole ethoxylate
Category: Detergent, wetting agent, emulsifier, nonionic
Applications: In general purpose liquid and dry products; rewetting; deinking; agricultural emulsifiers; industrial cleaners; textile processing; laundering; transport industry
Form: Liquid; Color: Clear yellow; Odor: Mild
Properties: Ref. Index: 1.4469; Solubility: Water-sol.; Sp. gr.: 1.002; Visc.: 116 cps; HLB: 9.7; Cloud Pt.: 39–45 C; Pour Pt.: –12 F; Flash Pt.: 380 F; Stability: Good; Surface Tension: 26.2 dynes/cm
Composition: 85% active ingred.
Reaction: Neutral
Toxicity: Eye and skin irritant
Containers: Steel drums and bulk
References: Tech. bulletin IC/DP-305

Sterox AJ-100. Monsanto Ind. Chem.
Chemical Description: Tridecyl alcohol 8.5 mole ethoxylate
Category and Applications: See Sterox AJ
Form: Paste; Color: White; Odor: Mild
Properties: Ref. Index: 1.4600; Solubility: Water-sol.; Sp. gr.: 1.003; HLB: 9.7; Cloud Pt.: 39–45 C; Flash Pt.: 385 F; Pour Pt.: 82 F; Stability: Good; Surface Tension: 26.4 dynes/cm
Composition: 99% active ingred.
Reaction: Neutral
Toxicity: Skin and eye irritant
Containers: Steel drums and bulk
References: Tech. bulletin IC/DP-305

Sterox CD. Monsanto Ind. Chem.
Chemical Description: Tall oil 10 mole ethoxylate
Category: Detergent, wetting agent, emulsifier, nonionic
Applications: In general purpose liquid and dry products, dedusting, rewetting paper, industrial hard surface cleaners, textiles, laundering, transport
Form: Liquid; Color: Clear brown; Odor: Mild
Properties: Ref. Index: 1.4830; Solubility: Miscible in water; Sp. gr.: 1.06–1.07; HLB: 12.3; Flash Pt.: 555 F; Pour Pt.: 46 F; Stability: Good; Surface Tension; 38.6 dynes/cm
Composition: 97.8% active ingred.
Reaction: Neutral
Toxicity: Slight skin and eye irri-

tant
Containers: Steel drums and bulk
References: Tech. bulletin IC/DP-305

Sterox DF. Monsanto Ind. Chem.
Chemical Description: Dodecyl phenol 6 mole ethoxylate
Category: Wetting agent, emulsifier, nonionic
Applications: Oil-sol. emulsifier, paper rewetting, intermediate for anionic surfactant mfg., in metal cleaning compounds, O/W emulsifier, agricultural emulsifier, dedusting agents, felt washing, latex emulsion stabilizer, preparation of wettable powders, paints, cutting and soluble oils
Form: Viscous liquid; Color: Clear yellow; Odor: Mild
Properties: m.w.: 535; Ref. Index: 1.4922; Solubility: Miscible with water; sol. in nonpolar solvents; Sp. gr.: 1.0277; Visc.: 343 cps; HLB: 10.0; Flash Pt.: 495 F; Pour Pt.: 0 F; Hydroxyl No.: 102–111; Surface Tension: 29.9 dynes/cm; pH: 6.0–7.5 (1%)
Composition: 99.5% active ingred.
Reaction: Neutral
Toxicity: Skin and eye irritant
Containers: Steel drums and bulk
References: Tech. bulletin IC/DP-305

Sterox DJ. Monsanto Ind. Chem.
Chemical Description: Dodecyl phenol 10.5 mole ethoxylate
Category: Detergent, wetting agent, emulsifier, nonionic
Applications: In general purpose dry and liquid products, household and industrial cleaners, textiles, laundering, transport, paper and leather processing, metal cleaning, emulsifiable toxicant concentrates, latex formulations
Form: Viscous liquid; Color: Clear yellow; Odor: Mild
Properties: m.w.: 715; Ref. Index: 1.4868; Solubility: Water-sol.; Sp. gr.: 1.0469; Visc.: 1310 cps; HLB: 12.6; Cloud Pt.: 38–42 C; Flash Pt.: 555 F; Pour Pt.: 40 F; Stability: Good; Surface Tension: 29.7 dynes/cm; pH: 6.0–7.5 (1%)
Composition: 99.1% active ingred.
Reaction: Neutral
Toxicity: Causes skin irritation, eye burns
Containers: Steel drums and bulk
References: Tech. bulletin IC/DP-305

Sterox ND. Monsanto Ind. Chem.
Chemical Description: Nonyl phenol 4 mole ethoxylate
Category: Detergent, wetting agent, emulsifier, nonionic
Applications: Intermediate for anionic surfactant mfg.; in solvent cleaners; as additive in petroleum products; dedusting; making of cutting and soluble oils; polyvinyl acetate plasticizer; latex emulsion stabilizer; agricultural emulsions; animal skin cleaners
Form: Viscous liquid; Color: Clear yellow; Odor: Mild
Properties: m.w.: 396; Ref. Index: 1.4962; Solubility: Miscible with water; sol. in nonpolar solvents; Sp. gr.: 1.0216; Visc.: 252 cps;

HLB: 8.9; Flash Pt.: 435 F; Pour Pt.: –24 F; Hydroxyl No.: 133–150; Stability: Good; Surface Tension: 29.7 dynes/cm; pH: 6.0–7.5 (1%)
Composition: 99.5% active ingred.
Reaction: Neutral
Toxicity: Causes eye burn, skin irritation
Containers: Steel drums and bulk
References: Tech. bulletin IC/DP-305

Sterox NE. Monsanto Ind. Chem.
Chemical Description: Nonyl phenol 5 mole ethoxylate
Category: Detergent, wetting agent, emulsifier, nonionic
Applications: Intermediate for anionic surfactant mfg.; in solvent cleaners; additives in petroleum products; abatement of dust; making of cutting and soluble oils
Form: Liquid; Color: Clear yellow; Odor: Mild
Properties: Ref. Index: 1.4955; Solubility: Miscible with water; sol. in nonpolar solvents; Sp. gr.: 1.0322; Visc.: 236 cps; HLB: 10.0; Flash Pt.: 445 F; Pour Pt.: –30 F; Stability: Good; Surface Tension: 28.7 dynes/cm
Composition: 99.5% active ingred.
Reaction: Neutral
Toxicity: Skin and eye irritant
Containers: Steel drums and bulk
References: Tech. bulletin IC/DP-305

Sterox NF. Monsanto Ind. Chem.
Chemical Description: Nonyl phenol 6 mole ethoxylate
Category and Applications: See Sterox NE; also in preparation of emulsified paints
Form: Liquid; Color: Clear yellow; Odor: Mild
Properties: Ref. Index: 1.4940; Solubility: Miscible with water; sol. in nonpolar solvents; Sp. gr.: 1.0385; Visc.: 207 cps; Flash Pt.: 480 F; Pour Pt.: –20 F; Stability: Good; Surface Tension: 28.8 dynes/cm
Composition: 99.5% active ingred.
Reaction: Neutral
Toxicity: Skin and eye irritant
Containers: Steel drums and bulk
References: Tech. bulletin IC/DP-305

Sterox NG. Monsanto Ind. Chem.
Chemical Description: Nonyl phenol 6.5 mole ethoxylate
Category and Applications: See Sterox NE; also as additive in solvent-base paints
Form: Liquid; Color: Clear yellow; Odor: Mild
Properties: Ref. Index: 1.4933; Solubility: Miscible with water; sol. in nonpolar solvents; Sp. gr.: 1.0426; Visc.: 229 cps; HLB: 11.3; Flash Pt.: 485 F; Pour Pt.: 0 F; Stability: Good; Surface Tension: 29.1 dynes/cm
Composition: 99.5% active ingred.
Reaction: Neutral
Toxicity: Skin and eye irritant
Containers: Steel drums and bulk
References: Tech. bulletin IC/DP-305

Sterox NJ. Monsanto Ind. Chem.
Chemical Description: Nonyl phenol 9.5 mole ethoxylate
Category: Detergent, wetting agent, emulsifier, nonionic
Applications: General purpose surfactant; dry and liquid detergent products; textile applications; metal and hard surface cleaning; paper processing; latex paint wetting; agricultural emulsifiers; vegetable washing; dairy cleaning; laundering
Form: Viscous liquid; Color: Clear yellow; Odor: Mild
Properties: m.w.: 625; Ref. Index: 1.4884; Solubility: Water-sol.; Sp. gr.: 1.0538; Visc.: 241 cps; HLB: 13.0; Cloud Pt.: 51–56 C; Flash Pt.: 540 F; Pour Pt.: 26 F; Stability: Good; Surface Tension: 29.8 dynes/cm; pH: 6.0–7.5 (1%)
Composition: 99.5% active ingred.
Reaction: Neutral
Toxicity: Causes skin irritation, eye burns
Containers: Steel drums and bulk
Warnings and Caution: May turn cloudy below 50 F–mix before using
References: Tech. bulletin IC/DP-305

Sterox NK. Monsanto Ind. Chem.
Chemical Description: Nonyl phenol 10.5 mole ethoxylate
Category: Detergent, wetting agent, emulsifier, nonionic
Applications: In general purpose dry and liquid products; household and industrial cleaners; paper processing; metal cleaning; latex paint wetting; agricultural emulsions; textile processing; vegetable washing; dairy cleaning; laundering
Form: Viscous liquid; Color: Clear yellow; Odor: Mild
Properties: m.w.: 673; Ref. Index: 1.4879; Solubility: Water-sol.; Sp. gr.: 1.062; Visc.: 304 cps; HLB: 13.5; Cloud Pt.: 66–70 C; Flash Pt.: 560 F; Pour Pt.: 36 F; Stability: Good; Surface Tension: 30.9 dynes/cm; pH: 6.0–7.5 (1%)
Composition: 99.5% active ingred.
Reaction: Neutral
Toxicity: Causes skin irritation, eye burns
Containers: Steel drums and bulk
Warnings and Caution: May turn cloudy below 55 F–mix before using
References: Tech. bulletin IC/DP-305

Sterox NL. Monsanto Ind. Chem.
Chemical Description: Nonyl phenol 11 mole ethoxylate
Category: Detergent, wetting agent, emulsifier, nonionic
Applications: In general purpose dry and liquid products; household and industrial cleaners; textiles; laundering; transport
Form: Liquid; Color: Clear yellow; Odor: Mild
Properties: Ref. Index: 1.4867; Solubility: Sol. in water; Sp. gr.: 1.065; Visc.: 283 cps; Cloud Pt.: 74–79 C; Flash Pt.: 535 F; Pour Pt.: 34 F; Stability: Good; Surface Tension: 32.3 dynes/cm
Composition: 99.5% active ingred.
Reaction: Neutral

Toxicity: Skin and eye irritant
Containers: Steel drums and bulk
References: Tech. bulletin IC/DP-305

Sterox NM. Monsanto Ind. Chem.
Chemical Description: Nonyl phenol 12.5 mole ethoxylate
Category: Detergent, wetting agent, emulsifier, nonionic
Applications: In general purpose dry and liquid products; household and industrial cleaners; paper processing; latex paint wetting; agricultural emulsions; vegetable washing; metal cleaning; dairy cleaning; laundering; transport
Form: Viscous liquid; Color: Clear yellow; Odor: Mild
Properties: m.w.: 770; Ref. Index: 1.4860; Solubility: Water-sol.; Sp. gr.: 1.0674; Visc.: 275 cps; HLB: 14.3; Cloud Pt.: 88–94 C; Flash Pt.: 535 F; Pour Pt.: 52 F; Stability: Good; Surface Tension: 33.1 dynes/cm; pH: 6.0–7.5 (1%)
Composition: 99.5% active ingred.
Reaction: Neutral
Toxicity: Causes eye burns, may cause skin irritation
Containers: Steel drums and bulk
References: Tech. bulletin IC/DP-305

Strodex MO-100. Dexter Chem. Corp.
Chemical Description: 2-ethylhexyl polyphosphoric ester acid anhydride
Category: Emulsifier, dispersant, anionic
Applications: Used in oil base paints and leather coating specialties as a pigment dispersant
Form: Liquid
Properties: Solubility: Sol. in most polar and nonpolar solvents; insol. in water
Composition: 100% conc.

Strodex MOK-70. Dexter Chem. Corp.
Chemical Description: Potassium salt of Strodex MO-100
Category: Dispersant, wetting and rewetting agent, anionic
Applications: Dispersant for carbon black pigments in water systems, and for zinc pigments and lead silicate in exterior latex paints; good wetting and rewetting properties
Form: Paste
Composition: 70% conc.

Strodex MR-100. Dexter Chem. Corp.
Chemical Description: Complex organic polyphosphoric ester acid anhydride
Category: Emulsifier, dispersant, anionic
Applications: Dispersant for organic pigments in polar and nonpolar solvents
Form: Liquid
Composition: 100% conc.

Strodex MRK-98. Dexter Chem. Corp.
Chemical Description: Potassium salt of Strodex MR-100
Category: Dispersant, emulsifier,

anionic
Applications: Recommended for grinding and dispersing organic pigments
Form: Liquid
Composition: 98% conc.

Strodex P-100. Dexter Chem. Corp.
Chemical Description: Complex organic polyphosphoric ester acid anhydride
Category: Emulsifier, dispersant, coupling agent, anionic
Applications: Dispersant for many organic and inorganic pigments in most polar and nonpolar solvents; coupling agent for conventional nonionic emulsifiers in both aq. and nonaq. systems; used in metal and tile cleaners; as pigment grinding aid
Form: Liquid
Composition: 100% conc.

Strodex PK-90. Dexter Chem. Corp.
Chemical Description: Potassium salt of Strodex P-100
Category: Dispersant, wetting agent, emulsifier, anionic
Applications: Dispersant for titanium dioxide and extender pigments in latex paints; oxidation-corrosion inhibitor; does not hydrolyze in contact with flake caustic; sol. in high electrolyte conc.; powerful wetting agent which drops surface tension to 25 dyne range; used in heavy-duty alkaline cleaners; dispersant for barium sulfate and low oil absorbency iron oxides in aq. sol'ns.
Form: Liquid
Composition: 90% conc.

Strodex PSK-28. Dexter Chem. Corp.
Chemical Description: Potassium salt of complex organic phosphate esters
Category: Emulsifier, dispersant, wetting agent, anionic
Applications: Pigment dispersant
Form: Liquid
Properties: Solubility: Water-sol.
Composition: 58% conc.

Strodex SE-100. Dexter Chem. Corp.
Chemical Description: Complex organic polyphosphoric ester acid anhydride
Category: Emulsifier, dispersant, anionic
Applications: Used as an emulsifier and coemulsifier in aq. and nonaq. systems
Form: Liquid
Properties: Solubility: Oil-sol.
Composition: 100% conc.

Strodex SEK-50. Dexter Chem. Corp.
Chemical Description: Potassium salt of Strodex SE-100
Category: Stabilizer, emulsifier, dispersant, anionic
Applications: Freeze-thaw and heat stabilizer for latex emulsions and latex paints; used where a calcium-insensitive emulsifier is required; as a stabilizer for titanium dioxide and extender pig-

ments
Form: Liquid
Composition: 50% conc.

Strodex Super V-8. Dexter Chem. Corp.
Chemical Description: Potassium salt of complex organic phosphate esters
Category: Emulsifier, detergent, dispersant, wetting agent, anionic
Applications: Preparation of cotton and polyester/cotton blends (desizing, kier boiling, peroxide and chlorine bleaching, open width boil-off, jig or beck preparation), preparation of synthetics (continuous, beck or jig, beam, package), dyeing, afterscouring, printing
Form: Clear liquid
Properties: Solubility: Readily sol. in hot or cold water, caustic sol'ns. up to 8%, and in most aliphatic, aromatic, and chlorinated solvents; compatible with sodium perborate, sodium hypochlorite, hydrogen peroxide, most optical bleaches, many thermosetting and thermoplastic resins, starches, and gums; Sp. gr.: 1.09–1.11; Stability: Stable in alkaline and mildly acidic sol'ns., and at extremes of temp. and pressure; pH: 7–8 (1%)
Composition: 75–80% active
Patent: U.S. Patent 3,380,927

STXS. Continental Oil
Chemical Description: Sodium toluene xylene sulfonate
Category: Hydrotrope
Applications: Hydrotrope for detergent formulations
Form: Clear liquid; Color: Nearly colorless; Odor: Mild
Properties: Density: 9.94; Sp. gr.: 1.18; M.P.: 12 F; B.P.: 212 F
Composition: 42% active ingred.

Sufatol LX/C. Standard Chem. Co.
Chemical Description: Sulfated fatty alcohol
Category: Detergent
Applications: Textile scouring
Form: Paste

Sulfamin. Berol Kemi AB
Chemical Description: Triethanolamine alkyl aryl sulfonate
Category: Detergent, foaming and wetting agent, anionic
Applications: Liquid washing and cleaning agents, esp. with large quantities of builders; textile wetting and penetrating agent
Form: Clear liquid; Color: 5 (Gardner)
Properties: Solubility @ 5%: Sol. in water, ethanol, propylene glycol; disp. in acetone, trichloroethylene; Density: 1.080 g/cc (20 C); Visc.: 300 cps (20 C); Clear Pt.: $<$ 5 C; Flash Pt.: $\geqslant$ 100 C; Stability: Stable in the presence of weak acids and bases; Surface Tension: 34 dynes/cm; pH: 6.5–7.5 (1% aq.); Biodegradable
Composition: 50% active ingred.
Warnings and Caution: Store and handle above clear pt. of 5 C

Sulfatol 33 MO. Aarhus Oliefabrik
Chemical Description: Monoethanolamine lauryl sulfate

Category: Detergent, foaming agent
Applications: Shampoos, foaming baths, toilet applications, light duty detergents, dishwashing detergents, carpet shampoos, various industrial applications
Form: Liquid
Composition: 30% active ingred.

Sulfatol 33 Pasta. Aarhus Oliefabrik
Chemical Description: Sodium lauryl sulfate
Category and Applications: See Sulfatol 33 MO
Form: Paste
Composition: ≈38% active ingred.

Sulfatol 33 Pasta 13. Aarhus Oliefabrik
Chemical Description: Sodium lauryl sulfate
Category and Applications: See Sulfatol 33 MO
Form: Liquid/paste
Composition: 25% active ingred.

Sulfatol 33 TR. Aarhus Oliefabrik
Chemical Description: Triethanolamine lauryl sulfate
Category and Applications: See Sulfatol 33 MO
Form: Liquid
Composition: 40% active ingred.

Sulfatol B 6. Aarhus Oliefabrik
Chemical Description: Ammonium/triethanolamine lauryl sulfate
Category and Applications: See Sulfatol 33 MO
Form: Liquid
Composition: 30% active ingred.

Sulfatol B 7. Aarhus Oliefabrik
Chemical Description: Mixture based on lauryl ether sulfate
Category and Applications: See Sulfatol 33 MO
Form: Liquid
Composition: 30% active ingred.

Sulfatol B 10. Aarhus Oliefabrik
Chemical Description: Mixture based on lauryl ether sulfate
Category and Applications: See Sulfatol 33 MO
Form: Pearly liquid
Composition: 35% active ingred.

Sulfatol E 3. Aarhus Oliefabrik
Chemical Description: Sodium lauryl ether sulfate
Category and Applications: See Sulfatol 33 MO
Form: Liquid
Composition: 28% active ingred.

Sulfatol KE 3. Aarhus Oliefabrik
Chemical Description: Sodium lauryl ether sulfate
Category and Applications: See Sulfatol 33 MO
Form: Liquid/paste
Composition: 56% active ingred.

Sul-fon-ate AA-9. Tennessee Corp./ Subsidiary of Cities Service Co.
Chemical Description: Sodium dodecyl benzene sulfonate
Category: Wetting agent, anionic
Applications: Air pollution control; cement industry; cosmetics; fertilizers, insecticides; leather processing (degreasing, fat liquoring); metal cleaning, electroplat-

ing, etching, pickling; mining (dust control, ore flotation); street cleaning; fire fighting; paper processing; petroleum refining (emulsion breaking, oil well acidulizing and flooding); rubber processing; surface coatings; sanitary supplies (building, car, dairy, general household cleaners, commercial laundries, fruit and vegetable processing, scouring powders); soaps; textile processing (scouring, carbonizing, fulling, desizing, bleaching, dyeing, stripping)
Form: Crisp flake; Color: White; Odor: Odorless
Properties: Stability: Stable in hard water, mild acids or alkalis, and sol'ns. of electrolytes; Surface Tension: 30.6 dynes/cm (86 F); pH: 7–8 (1% sol'n.)
Composition: 90% active ingred.
Containers: 51-gal LeverPak fiber drums

Sul-fon-ate AA-10. Tennessee Corp./ Subsidiary of Cities Service Co.
Chemical Description: Sodium dodecyl benzene sulfonate
Category and Applications: See Sul-fon-ate AA-9; AA-10 has esp. low inorganic salt content; esp. useful in specialty applications, e.g., organic solvents, combinations with soap or nonionics, and other systems where inorganics are not compatible
Form: Crisp flake; Color: White
Properties: Stability: Stable in hard water, mild acids or alkalis, and sol'ns. of electrolytes; pH: 7–8 (1% sol'n.)
Composition: 96% active ingred.
Containers: 51-gal LeverPak fiber drums

Sul-fon-ate OA-5. Cities Service Co.
Chemical Description: Sulfonated oleic acid, sodium salt
Category: Wetting agent, anionic
Applications: Anionic wetting agent efficient at extremes of pH; low foamer
Form: Liquid; Color: Light brown
Properties: Stability: Good; Biodegradable
Composition: 41% active ingred.
Containers: 500 lb net heresite-lined steel drums, tank cars

Sul-fon-ate OE-500. Cities Service Co.
Chemical Description: Amyl-ester of the sodium salt of sulfonated oleic acid
Category: Solubilizer, emulsifier
Applications: Solubilizer and emulsifier for dyes and other additives used in polyesters
Containers: 55 gal steel drums

Sulfonate Slurry C-550. Continental Oil
Category: Detergent, wetting agent, emulsifier
Applications: Liquid detergent formulations, emulsion polymerization
Form: Liquid; Color: 60 (Klett, 5% sol'n.); Odor: Nil
Properties: Density: 8.5; Sp. gr.: 1.05; Visc.: 40,000; M.P.: 30
Composition: 50% active ingred.

Sulfonic Acid LS. Hart Chem. Ltd.
Chemical Description: Linear alkylbenzene sulfonic acid
Category: Detergent, intermediate, anionic
Applications: High active intermediate for liquid and powdered detergent formulations
Form: Liquid; Color: Amber
Properties: Density: 1.055; Visc.: 1000 cps
Composition: 96% active ingred.
Reaction: Acid
Toxicity: Skin and eye irritant
Containers: Drums, T/T
Warnings and Caution: Avoid contact with skin and eyes

Sulfopon WA 30. Henkel Argentina
Chemical Description: Sodium lauryl sulfate
Category: Raw material
Applications: Raw material for the manufacture of cream shampoos, medicated shampoos, and bubble baths
Form: Paste
Properties: pH: 6.5–8.0 (5% sol'n.)
Composition: 29–31% active

Sulfosil P-491. Witco Chem./Organics Div.
Chemical Description: Sodium silicate/sodium sulfate-based builder
Category: Detergent
Applications: Booster beads for high-, medium-, or low-foaming household and industrial cleaners
Form: Powder
Composition: 85% conc.

Sulfosoft. Berol Kemi AB
Chemical Description: Linear alkyl benzene sulfonic acid
Category: Intermediate, anionic
Applications: Intermediate for the production of detergents; in neutralized form in high-foaming household and industrial cleaning and washing compounds; oil-sol. salts in emulsifiers for solvents and oils; sodium salt in emulsion polymerization of vinyl chloride, vinyl acetate, various acrylates, etc.
Form: Viscous liquid; Color: ⩽ 150 (Hazen)
Properties: m.w.: 320; Solubility @ 5%: Sol. in water, ethanol, benzene, white spirit, light fuel oil; Density: 1.04 g/cc; Visc.: 3500 cps (20 C) to 500 cps (60 C); Pour Pt.: ≈ 6 C; Stability: Stable in relatively high concs. of strong acids and bases; Surface Tension: 33 dynes/cm; pH: 2.2 (1% aq.); Biodegradable
Composition: 97 ± 1.0% active
Toxicity: Acid–irritating to skin, eyes, respiratory system
Warnings and Caution: Wear protective goggles and gloves

Sulframin 40. Witco Chem./Organics Div.
Chemical Description: Linear alkylaryl sodium sulfonate
Category: Detergent base, wetting agent, dispersant, defoamer, anionic
Applications: Detergent base and wetting agent for household and industrial specialty compounds;

metal cleaning prior to treatment such as coating and plating; textile industry as lubricants, emulsifiers, dye dispersants, penetrants, scouring agents, antistats; granular form used in petroleum industry as defoamer for water-based drilling fluids, detergent, wetter, and dispersant
Form: Flakes or granular
Composition: 40% conc.

Sulframin 40DA. Witco Chem./Organics Div.
Chemical Description: Linear alkylaryl sodium sulfonate
Category: Detergent, wetting agent, lubricant, emulsifier, antistat, anionic
Applications: General-purpose spray-dried high-foaming detergent for car washing, household, and industrial cleaning; metal cleaning prior to treatment; textile surfactant
Form: Beads
Composition: 40% anionic
Reaction: Neutral

Sulframin 40RA. Witco Chem./Organics Div.
Chemical Description: Linear alkylaryl sodium sulfonates
Category and Applications: See Sulframin DA; also base for shampoos, bubble baths, and foaming body cleansers
Form: Beads
Composition: 40% conc.

Sulframin 40T. Witco Chem./Organics Div.
Chemical Description: Linear alkylaryl triethanolamine sulfonate
Category: Detergent, emulsifier, wetting agent, lubricant, antistat, anionic
Applications: Detergent base and wetting agent; emulsifier for household and industrial specialty compounds; metal cleaning; textile surfactant
Form: Liquid
Composition: 40% conc.

Sulframin 45. Witco Chem./Organics Div.
Chemical Description: Linear alkylaryl sodium sulfonate
Category: Detergent, wetting agent, anionic
Applications: High-foaming base for liquid detergents, household, and industrial specialty compounds; metal cleaning; textile surfactant; contains hydrotrope
Form: Liquid
Properties: Cloud Pt.: 14 C; Clear Pt.: 15 C
Composition: 45% conc.

Sulframin 45LX. Witco Chem./Organics Div.
Chemical Description: Linear alkylaryl sodium sulfonate
Category and Applications: See Sulframin 45
Form: Liquid
Composition: 44% conc.

Sulframin 60T. Witco Chem./Organics Div.
Chemical Description: Linear alkylaryl triethanolamine sulfonate
Category and Applications: See

Sulframin 40T; also for shampoos, bubble baths, and foaming body cleansers
Form: Liquid
Composition: 60% conc.

Sulframin 85. Witco Chem./Organics Div.
Chemical Description: Linear alkylaryl sodium sulfonate
Category: Detergent, wetting agent, lubricant, emulsifier, antistat, anionic
Applications: Highly conc., low-salt detergent base and wetter for household and industrial specialty compounds; metal cleaning; textile surfactant
Form: Flakes or powder
Composition: 85% conc.

Sulframin 90. Witco Chem./Organics Div.
Chemical Description: Linear alkylaryl sodium sulfonate
Category and Applications: See Sulframin 85; also used as surfactant for emulsion polymerization and latex stabilization
Form: Flakes
Composition: 91% conc.

Sulframin 1240. Witco Chem./Organics Div.
Chemical Description: Linear alkylaryl sodium sulfonate
Category: Detergent, wetting agent, lubricant, emulsifier, antistat, anionic
Applications: Detergent base and wetting agent for household and industrial specialty compounds; metal cleaning; textile surfactant; contains hydrotrope
Form: Slurry
Composition: 40% conc.

Sulframin 1245. Witco Chem./Organics Div.
Chemical Description: Linear alkylaryl sodium sulfonate
Category and Applications: See Sulframin 1240
Form: Slurry
Composition: 45% conc.

Sulframin 1250. Witco Chem./Organics Div.
Chemical Description: Linear alkylaryl sodium sulfonate
Category: Detergent, wetting agent, lubricant, emulsifier, antistat, anionic
Applications: Detergent base and wetting agent for household and industrial specialty compounds; metal cleaning; textile surfactant; contains no hydrotrope; base for shampoos, bubble baths, and foaming body cleansers
Form: Slurry
Composition: 51% conc.

Sulframin 1260. Witco Chem./Organics Div.
Chemical Description: Linear alkylaryl sodium sulfonate
Category and Applications: See Sulframin 1240; also base for shampoos, bubble baths, and foaming body cleansers
Form: Slurry
Composition: 60% conc.

Sulframin 1288. Witco Chem./Organics Div.
Chemical Description: Linear alkylaryl sulfonic acid
Category: Intermediate, wetting agent, penetrant, lubricant, emulsifier, dispersant, detergent, antistat, anionic
Applications: Intermediate for wetting agents, emulsifiers, and detergents for household and industrial specialties; textile surfactant
Form: Liquid
Composition: 88% conc.

Sulframin 1298. Witco Chem./Organics Div.
Chemical Description: Linear alkylaryl sulfonic acid
Category: Intermediate, wetting agent, penetrant, lubricant, emulsifier, dispersant, detergent, antistat, stabilizer, anionic
Applications: Intermediate for wetting agents, emulsifiers, and detergents in household and industrial specialties; intermediate for bases for shampoos, bubble baths, and foaming body cleansers; textile surfactant; surfactant for emulsion polymerization and latex stabilization
Form: Liquid
Composition: 98% conc.

Sulframin 1388. Witco Chem./Organics Div.
Chemical Description: Linear alkylaryl sulfonic acid
Category and Applications: See Sulframin 1288
Form: Liquid
Composition: 88% conc.

Sulframin 4010D, 4010R. Witco Chem./Organics Div.
Chemical Description: Linear alkylaryl sodium sulfonate-based detergent
Category: Detergent
Applications: Spray-dried, high-foaming detergent for car washing, household, and industrial cleaning
Form: Beads

Sulframin AOS. Witco Chem./Organics Div.
Chemical Description: Alpha olefin sodium sulfonate
Category: Detergent
Applications: High-foaming detergent base for liquid detergent, shampoo, and bubble bath formulations; surfactant for emulsion polymerization and latex stabilization
Form: Liquid

Sulframin AOS 90. Witco Chem./Organics Div.
Chemical Description: Alpha olefin sodium sulfonate
Category: Detergent
Applications: High-foaming detergent base for liquid detergent, shampoo, and bubble bath formulations
Form: Flakes

Sulframin CSA. Witco Chem./Organics Div.
Chemical Description: Cumene

sulfonic acid
Category: Hydrotrope, coupling agent, solubilizer, catalyst
Applications: Hydrotrope; coupling agent; solubilizer for liquid detergents; catalyst for organic reactions

Sulframin HD. Witco Chem./Organics Div.
Chemical Description: Linear alkylaryl sodium sulfonate-based detergent
Category: Detergent, anionic
Applications: Completely formulated, spray-dried, high-foaming laundry and all-purpose detergent
Form: Beads

Sulframin Low-Foam HD. Witco Chem./Organics Div.
Chemical Description: Linear alkylaryl sodium sulfonate-based detergent
Category: Detergent, anionic
Applications: Completely formulated, spray-dried, low-foaming laundry and all-purpose detergent
Form: Beads

Sulframin LX. Witco Chem./Organics Div.
Chemical Description: Modified linear alkylaryl sodium sulfonate
Category: Detergent, wetting agent, anionic
Applications: Detergent base and wetting agent for household and industrial specialty products; high active content; fast solution rate; liquid alkali compatibility; contains hydrotrope; metal cleaning; textile surfactant
Form: Flakes
Composition: 85% conc.

Sulframin Phos-Free HD. Witco Chem./Organics Div.
Chemical Description: Linear alkylaryl sodium sulfonate-based detergent
Category: Detergent, anionic
Applications: Completely formulated, spray-dried, moderate-foaming laundry and all-purpose detergent containing no phosphates or nitrolotriacetate (NTA); a low foam version of this product is also manufactured
Form: Beads

Sulframin TX. Witco Chem./Organics Div.
Chemical Description: Modified toluene sulfonic acid
Category: Hydrotrope, coupling agent, solubilizer, catalyst
Applications: Hydrotrope; coupling agent; solubilizer for liquid detergents;

Sulfur Dioxide. Cities Service Co.
Applications: Chemical manufacturing, cellulose pulp, clay, glue, petroleum refining, food processing, glass making, soya protein, metallurgical processing, chromium waste treatment, starch refining, sugar refining, textiles and water treatment
Form: Compressed gas
Composition: 99.9% active ingred.
Containers: 150-lb and 1-ton cylinders, tank cars and trucks

Sulfur Trioxide. Cities Service Co.
Category: Sulfonating agent
Applications: Sulfonating agent used in production of synthetic detergents, wetting agents, and other chemicals
Form: Clear liquid
Containers: Tank cars

Super Amide GR. Onyx Chem. Co.
Chemical Description: Cocoamide DEA (1:1)
Category: Foam stabilizer, emulsifier, thickener, nonionic
Applications: Household, industrial, and cosmetic formulations; maximum cost/performance
Form: Liquid
Properties: Sp. gr.: 0.99 (25/20 C); Flash Pt.: 146 F; Acid No.: 10 max.; Alkali No.: 32 max.
Composition: 80% conc.

Super Amide L9. Onyx Chem. Co.
Chemical Description: Lauramide DEA (1:1)
Category: Thickening agent, foam stabilizer, nonionic
Applications: Cosmetic, industrial, and household formulations, e.g., shampoos, bubble baths, light duty dishwash detergents, industrial cleaners
Form: Paste
Properties: Sp. gr.: 0.98 (25/20 C); Flash Pt.: 178 F; Acid No.: 9 max.; Alkali No.: 45 max.
Composition: 90% conc.

Super Amide L9C. Onyx Chem. Co.
Chemical Description: Lauramide DEA (1:1)
Category and Applications: See Super Amide L9
Form: Paste
Properties: Sp. gr.: 0.96 (25/20 C); Flash Pt.: 148 F; Acid No.: 2 max.; Alkali No.: 16 max.
Composition: 97% conc.

Super Amide LM. Onyx Chem. Co.
Chemical Description: Lauric myristic alkanolamide (1:1)
Category and Applications: See Super Amide L9
Form: Paste
Properties: Sp. gr.: 0.98 (25/20 C); Flash Pt.: 178 F; Acid No.: 10 max.; Alkali No.: 27 max.

Super-Pro #5A. Stepan Chem. Co.
Chemical Description: Collagen protein hydrolysate condensed with higher saturated fatty acids and partially cross-linked to higher molecular weight polyols; neutralized with blend of potassium hydroxide and triethanolamine
Category: Detergent, conditioner, emulsifier, moisturizer, foaming agent, anionic
Applications: Shampoos, facial cleansers, bath gels, cosmetic creams and lotions; very mild; skin texturizer; good foaming and foam stability
Form: Liquid; Color: 10 max. (Gardner)
Properties: Sp. gr.: 1.15 min.; Visc.: 400 cps min.; pH: 7.0–7.5
Composition: 68% active ingred.

Super Pro 100. Stepan Chem. Co.
Chemical Description: Blend of protein-derived surfactants
Applications: Substantive to hair; maximum absorption in both damaged and undamaged hair; used in shampoos, rinses, grooming aids; skin care products, e.g., lotions, creams, and bar soaps
Form: Clear liquid; Color: Amber
Properties: Solubility: Water-sol.; Sp. gr.: 1.20; Visc.: 400 cps; pH: 6.75
Composition: 64% solids

Super-Sat. R.I.T.A. Corp.
Chemical Description: Hydrogenated lanolin
Category: Plasticizer, emollient, nonionic
Applications: Cosmetics and pharmaceutical preparations, e.g., make ups, night creams, shaving creams
Form: Solid; Color: White; Odor: Trace
Properties: Sp. gr.: 0.85–0.86; M.P.: 48–52 C
Composition: 100% active ingred.

Super-Sat AWS Series. R.I.T.A. Corp.
Chemical Description: Ethoxylated hydrogenated lanolin, 6 different EO concs.
Category: Emollient, plasticizer, emulsifier, nonionic
Applications: Wide range of cosmetic preparations
Form: Solid; Color: 2 max. (ASTM)
Composition: 100% active ingred.

Supercol U. Henkel Inc.
Chemical Description: Guar gum
Category: Viscosity builder, texture and slip modifier
Applications: Creams and lotions
Form: Powder; Color: White
Composition: 100% conc.

Supratol VF. Hart Chem. Ltd.
Chemical Description: Highly sulfonated castor oil
Category: Dispersant, lubricant, penetrant, anionic
Applications: Penetrating, lubricating, and conditioning agent in leather processing; leveling and dispersing agent for acid and disperse dyestuffs
Form: Liquid; Color: Amber
Properties: Ref. Index: 1.4022; Solubility: Clear sol.; Density: 1.112; Visc.: 90 cps
Composition: 30% active ingred.
Reaction: Acid
Toxicity: Skin and eye irritant
Containers: Drums, T/T
Warnings and Caution: Avoid contact with skin and eyes

Surco 55S. Onyx Chem. Co.
Chemical Description: Sodium alkyl aryl sulfonate
Category: Detergent, anionic
Applications: Neutralized sodium salt of Surco DDBSA; used as the detergent base for a variety of liquid cleaning systems; free of hydrotrope
Form: Slurry
Properties: Sp. gr.: 1.02 (25/20 C); Flash Pt.: $>$ 200 F; pH: 7–8
Composition: 49–52% active

Surco 56. Onyx Chem. Co.
Chemical Description: Formulated anionic blend
Category: Detergent, anionic
Applications: A completely formulated synthetic floor cleaning and general purpose concentrate; contains organic builders
Form: Liquid
Properties: Sp. gr.: 1.09; Flash Pt.: > 200 F; pH: 10–11
Composition: 56% min. active

Surco 58B. Onyx Chem. Co.
Chemical Description: Formulated anionic blend
Category: Detergent, anionic
Applications: Blended ammonium salt detergent concentrate specifically designed for use as a light duty cleaner or dishwash; low cloud point
Form: Liquid
Properties: Sp. gr.: 1.09; Flash Pt.: > 200 F; pH: 6–7
Composition: 60% min. active

Surco 60S. Onyx Chem. Co.
Chemical Description: Sodium alkyl aryl sulfonate
Category: Detergent, anionic
Applications: Neutralized sodium salt of Surco DDBSA; used as the detergent base for a variety of liquid cleaning systems; contains 2–4% sodium xylene sulfonate
Form: Slurry
Properties: Sp. gr.: 1.02 (25/20 C); Flash Pt.: > 200 F; pH: 7–8
Composition: 58–61% active

Surco 60T. Onyx Chem. Co.
Chemical Description: Triethanolamine alkyl aryl sulfonate
Category: Detergent, anionic
Applications: A high active triethanolamine salt of Surco DDBSA that remains homogeneous at ambient temp.; handles easier than the nonhomogeneous slurries
Form: Liquid
Properties: Sp. gr.: 1.08 (25/20 C); Flash Pt.: > 200 F; pH: 6–7
Composition: 56–60% active

Surco 128T. Onyx Chem. Co.
Chemical Description: Cocoamide DEA
Category: Foam stabilizer, thickener, emulsifier, nonionic
Applications: Used in a variety of household, industrial, and cosmetic formulations; provide maximum cost/performance
Form: Liquid
Properties: Sp.gr.: 0.99 (25/20 C); Flash Pt.: 146 F; Acid No.: 2 max.; Alkali No.: 32 max.
Composition: 100% conc.

Surco 168B. Onyx Chem. Co.
Chemical Description: Formulated anionic blend
Category: Detergent, foaming agent, anionic
Applications: Blended detergent concentrate for dishwash and car wash applications; high foaming, good cleaning; can easily be built to excellent viscosities in a variety of systems; low cloud pt.
Form: Liquid

Properties: Sp. gr.: 1.04; Flash Pt.: > 200 F; pH: 7–8
Composition: 59% min. active

Surco AA 10 MM. Onyx Chem. Co.
Chemical Description: Modified mixed fatty acid diethanolamide
Category: Detergent, nonionic
Applications: Designed for a variety of hard surface cleaners; develops excellent viscosities at low concs.; self-coupling; high tolerance for inorganics
Form: Liquid
Properties: Sp. gr.: 0.98 (25/20 C); Flash Pt.: > 200 F; Acid No.: 65–72; Alkali No.: 85–112
Composition: 100% conc.

Surco CMEA. Onyx Chem. Co.
Chemical Description: Coconut monoethanolamide
Category: Foam stabilizer, thickener, opacifier, nonionic
Applications: Industrial, household, and cosmetic formulations
Form: Wax
Properties: Sp. gr.: 0.92 (25/20 C); Flash Pt.: 146 F; Acid No.: 2 max.; Alkali No.: 19 max.
Composition: 100% conc.

Surco DC-47. Onyx Chem. Co.
Chemical Description: Formulated anionic blend
Category: Detergent, foam stabilizer, anionic
Applications: Blended sodium salt detergent concentrate that exhibits excellent detergency and foam stability; provides excellent viscosities throughout a wide range of dilutions; high active to solid ratio generates maximum performance
Form: Liquid
Properties: Sp. gr.: 1.07; Flash Pt.: > 200 F; pH: 7–8
Composition: 47% min. active

Surco DC-60. Onyx Chem. Co.
Chemical Description: Formulated anionic blend, sodium salt
Category: Detergent, anionic
Applications: Blended detergent concentrate with higher solids content than Surco DC-47; provides excellent performance, foam stability, good viscosities throughout a wide range of use dilutions
Form: Liquid
Properties: Sp. gr.: 1.09; Flash Pt.: > 200 F; pH: 7.0–7.7
Composition: 60% min. active

Surco DDBSA. Onyx Chem. Co.
Chemical Description: Linear dodecyl benzene sulfonic acid
Category: Detergent intermediate, anionic
Applications: High purity detergent intermediate used in the preparation of alkyl aryl sulfonates; the neutralized salts provide the base for both liquid and powdered detergent systems
Form: Liquid
Properties: Sp. gr.: 1.05 (25/20 C); Flash Pt.: > 200 F
Composition: 96% min. active

Surco EA58M. Onyx Chem. Co.
Chemical Description: Ammonium lauryl ether sulfate/alkanola-

mide/urea blend
Category: Detergent, foaming agent, anionic
Applications: A low cost, high active blend used in both concentrated and regular types of shampoos and bubble baths
Form: Liquid
Properties: Sp. gr.: 1.03; pH: 6–7
Composition: 58% min. active

Surco HDL Base. Onyx Chem. Co.
Chemical Description: Coconut fatty acid/mixed fatty acid-DDBSA
Category: Detergent, foaming agent, anionic
Applications: Designed for high alkali and high inorganic hard surface cleaner systems; anionic portion enhances foaming
Form: Liquid
Properties: Sp. gr.: 1.02 (25/20 C); Flash Pt.: > 200 F; Acid No.: 46–65; Alkali No.: 50–70
Composition: 98% conc.

Surco LL-50. Onyx Chem. Co.
Chemical Description: Formulated anionic blend
Category: Detergent, nonionic/anionic
Applications: Formulated heavy duty liquid laundry detergent complete with builders and optical brighteners
Form: Liquid
Properties: Sp. gr.: 1.07; Flash Pt.: > 200 F; pH: 9.0–9.8
Composition: 50% min. active

Surco MA. Onyx Chem. Co.
Chemical Description: Coconut fatty acid/mixed fatty acid DDBSA
Category: Detergent, foaming agent, nonionic
Applications: Provides a unique balance of high foaming, self-coupling, viscosity building, and degreasing properties in a variety of hard surface cleaners
Form: Liquid
Properties: Sp. gr.: 1.00 (25/20 C); Flash Pt.: > 200 F; Acid No.: 59–80; Alkali No.: 96–118

Surco MG-LS. Onyx Chem. Co.
Chemical Description: Magnesium lauryl sulfate
Applications: Widely used in rug and upholstery shampoos; low cloud pt.
Form: Liquid
Properties: Sp. gr.: 1.02; Flash Pt.: > 200 F

Surco MGS-35. Onyx Chem. Co.
Chemical Description: Magnesium lauryl sulfate/cocomonoethanolamide blend
Category: Detergent, foaming agent, anionic
Applications: Used as a rug and upholstery shampoo; exhibits excellent viscosity on dilution
Form: Liquid
Properties: Sp. gr.: 1.02; Flash Pt.: > 200 F; pH: 6–7
Composition: 33% min. active

Surco ODC-55. Onyx Chem. Co.
Chemical Description: Formulated anionic blend
Category: Detergent, emulsifier, anionic
Applications: A premium light duty dishwash concentrate which is an excellent degreaser and emulsifier; low cloud point; combination of sodium and ammonium salts
Form: Liquid
Properties: Sp. gr.: 1.04; Flash Pt.: > 200 F; pH: 6.8–7.5
Composition: 54% min. active

Surco SF42B. Onyx Chem. Co.
Chemical Description: Sodium alkyl aryl sulfonate
Category: Detergent, anionic
Applications: Exhibits excellent rinsing properties; low salt; contains ethanol
Form: Liquid
Properties: Sp. gr.: 1.02 (25/20 C); Flash Pt.: 81 F; pH: 6.5–7.5
Composition: 40–44% active

Surco SR-100. Onyx Chem. Co.
Chemical Description: Modified coconut/oleic fatty acid diethanolamide
Category: Detergent, emulsifier, wetting agent, nonionic
Applications: Low-foaming component for a variety of hard surface cleaners; develops high viscosities and is an effective emulsifier
Form: Liquid
Properties: Sp. gr.: 0.98 (25/20 C); Flash Pt.: > 200 F; Acid No.: 42–53; Alkali No.: 148–174
Composition: 100% conc.

Surco SR-200. Onyx Chem. Co.
Chemical Description: Coconut fatty acid/DDBSA diethanolamide
Category: Detergent, foaming agent, viscosity builder, anionic
Applications: Hard surface cleaning systems
Form: Liquid
Properties: Sp. gr.: 1.03 (25/20 C); Flash Pt.: > 200 F; Acid No.: 50–75; Alkali No.: 105–150
Composition: 98% conc.

Surco SXS. Onyx Chem. Co.
Chemical Description: Sodium xylene sulfonate
Category: Hydrotrope, coupling and solubilizing agent, anionic
Form: Liquid
Properties: Sp. gr.: 1.17 (25/20 C); Flash Pt.: > 200 F; pH: 7–10
Composition: 39–42% active

Surco TY. Onyx Chem. Co.
Chemical Description: Coconut/mixed fatty acid diethanolamide
Category: Nonionic
Applications: Used as a component in conveyor lubricants as well as a wide variety of hard surface cleaners
Form: Liquid
Properties: Sp. gr.: 0.99 (25/20 C); Flash Pt.: > 200 F; Acid No.: 18–26; Alkali No.: 95–120

Surco WC Conc. Onyx Chem. Co.
Chemical Description: Cocoamide DEA
Category: Detergent, foam booster and stabilizer, thickening agent, nonionic
Applications: High purity amide for shampoos, bubble baths, and dishwash formulations; semiliquid state allows ease of handling
Form: Liquid
Properties: Sp. gr.: 0.99 (25/20 C); Flash Pt.: 178 F; Acid No.: 2 max.; Alkali No.: 21–46
Composition: 100% conc.

Surco WHC. Onyx Chem. Co.
Chemical Description: Coconut/oleic fatty acid diethanolamide
Category: Emulsifier, nonionic
Applications: Emulsifier for waterless hand cleaners
Form: Liquid
Properties: Sp. gr.: 1.02 (25/20 C); Flash Pt.: > 200 F; Acid No.: 13 max.; Alkali No.: 40 max.
Composition: 100% conc.

Surfac 18-EHP. Mars Chem. Corp.
Chemical Description: Primary amine phosphate salt
Category; Corrosion inhibitor, cationic
Applications: Rust or corrosion inhibition in oil-based fluids, metal working oils (esp. in soluble oil formulations when compounded to emulsify in water); improved EP and wear properties of lubricants
Form: Viscous liquid; Color: Amber
Properties: Solubility: Oil-sol.; Density: 7.9 lb/gal; Sp. gr.: 0.95; Acid No.: 130–140
Composition: 100% active ingred.
Containers: 5 gal pails, 55 gal steel drums, bulk tanktrucks or tankcars

Surfac P14B. Mars Chem. Corp.
Chemical Description: Primary ether amine acetate
Category: Cationic surfactant
Properties: m.w.: 268 (of amine); Solubility: Readily disp. in water; Density: 7.60 lb/gal
Composition: 100% conc.
Reaction: Mildly alkaline
Containers: 5 gal pails, 55 gal drums, bulk tanktruck and tankcar
Warnings and Caution: Avoid contact with skin or eyes

Surfac P24M. Mars Chem. Corp.
Chemical Description: Primary ether amine acetate
Category: Cationic surfactant
Properties: m.w.: 261 (of amine); Solubility: Readily disp. in water; Density: 7.61 lb/gal
Composition: 100% conc.
Reaction: Mildly alkaline
Containers: 5 gal pails, 55 gal drums, bulk tanktruck or tankcar
Warnings and Caution: Avoid contact with skin and eyes

Surfactol-13. NL Ind./Indus. Chem. Div.
Chemical Description: Chemically modified castor oil
Category: Wetting agent, wax plasticizer, mold release agent, anti-

foam agent, nonionic
Applications: Textiles; leather; paints; household; cosmetics; dyeing; tanning; finishing; sizing; making of cutting and soluble oils; dispersing waxes, pigments, resins, dry powders
Form: Liquid; Color: 4 (Gardner); Odor: Faint
Properties: Solubility: 80% sol. in water; Density: 8.36 lb/gal; Sp. gr.: 1.005; Visc.: 17 stokes
Composition: 100% conc.
Containers: 55 gal drums
Warnings and Caution: Keep containers closed

Surfactol-318. NL Ind./Indus. Chem. Div.
Chemical Description: 5 mole ethoxylate of castor oil
Category: Emulsifier, solubilizer, nonionic
Applications: Emulsifier for oils, waxes; solubilizer for fragrances; emollient; used in textiles, paints, household, cosmetics, dyeing, rewetting dried skins, tanning, finishing, sizing, insecticides, herbicides, fungicides, kier boiling, making of cutting and soluble oils, dispersing waxes, pigments, resins; preparation of emulsified paints
Form: Liquid; Color: 3 (Gardner); Odor: Faint
Properties: Solubility: Sol. in toluene, butyl acetate, MEK; Sp. gr.: 0.984; Visc.: 690 cps; HLB: 3.6
Composition: 100% conc.
Containers: 55 gal drums
Warnings and Caution: Keep containers closed

Surfactol-365. NL Ind./Indus. Chem. Div.
Chemical Description: 20 mole ethoxylate of castor oil
Category: Emulsifier, solubilizer, nonionic
Applications: Emulsifier for oils, waxes; solubilizer for fragrances; emollient; used in textiles, leather, dyeing, paints, household, cosmetics, tanning, finishing, rewetting dried skins, sizing, insecticides, herbicides, fungicides, kier boiling, making of cutting and soluble oils, dispersing waxes, pigments, resins; preparation of emulsified paints
Form: Liquid; Color: 3 (Gardner)
Properties: Solubility: Sol. in toluene, butyl acetate, MEK, ethanol, water; Sp. gr.: 1.054; Visc.: 500 cps; HLB: 13
Composition: 100% conc.
Containers: 55 gal drums
Warnings and Caution: Keep from contact with oxidizing materials; keep containers closed

Surfactol 575. NL Ind./Indus. Chem. Div.
Chemical Description: PEG-66 trihydroxystearin
Category: Emulsifier, solubilizer
Applications: Emulsifier for oils, fats, waxes; solubilizer for fragrances, dyes, and essential oils; used in cosmetic creams and lotions, perfume and fragrance systems, antiperspirant sticks, lubricants, furniture polish, textiles, leather, dyeing, household products, tanning, finishing, rewetting

dried skins, sizing, insecticides, herbicides, fungicides, making of cutting and soluble oils, dispersing waxes, pigments, resins, dry powders, preparation of emulsified paints
Form: Solid; Color: White
Properties: Solubility: Sol. in water, alcohol; Sp. gr.: 1.08; M.P.: 39 C; HLB: 15
Containers: 55 gal drums
Warnings and Caution: Keep from contact with oxidizing materials; keep containers closed

Surfactol 590. NL Ind./Indus. Chem. Div.
Chemical Description: PEG-200 trihydroxystearin
Category: Emulsifier, solubilizer
Applications: Emulsifier for oils, fats, waxes; solubilizer for fragrances, dyes, and essential oils; used in cosmetic creams and lotions, perfume and fragrance systems, antiperspirant sticks, lubricants, furniture polish, textiles, leather, dyeing, household products, coating, insecticides, herbicides, fungicides, kier boiling, making of cutting and soluble oils, dispersing waxes, pigments, resins, preparation of emulsified paints
Form: Solid; Color: White
Properties: Solubility: Sol. in water, alcohol; Sp. gr.: 1.18; M.P.: 53 C; HLB: 18
Containers: 55 gal drums
Warnings and Caution: Keep from contact with oxidizing materials; keep containers closed

Surfam D89DA. Mars Chem. Corp.
Chemical Description: Alkyl ether diamine
Category: Corrosion inhibitor, flotation agent
Applications: Flotation of phosphate and iron ores; corrosion inhibitor formulations; combines with acids to form water-disp. salts
Form: Liquid; Color: Light; Odor: Mild ammonia
Properties: Density: 7.3 lb/gal; Pour Pt.: 13 C
Reaction: Alkaline
Warnings and Caution: Avoid contact with skin and eyes

Surfam P5 Distilled. Mars Chem. Corp.
Chemical Description: Methoxypropylamine
Category: Intermediate
Applications: Used in boiler feed waters to neutralize acid forming compounds in the feed or condensate return lines; amine soaps of Surfam 5 used as emulsifiers in wax emulsions for floor care products, natural and synthetic resins for textile finishing and water-based paints, and metal working oils; also used where a mild organic base is needed, e.g., as dye solvents, insecticide emulsions
Form: Clear, brilliant liquid; Color: 2 max. (Gardner)
Properties: m.w.: 90 (avg.); Solubility: Completely sol. in water and most common organic solvents; Density: 7.29 lb/gal; F.P.: –74 C; B.P.: 115–123 C; Flash

Pt.: 102 F
Composition: 98% min. active
Toxicity: Severely irritating to eyes, skin, and mucous membranes
Warnings and Caution: Wear eye protection

Surfam P5 Technical. Mars Chem. Corp.
Chemical Description: Methoxypropylamine
Category and Applications: See Surfam P5 Distilled
Form: Clear and brilliant liquid; Color: 7 max. (Gardner)
Properties: m.w.: 91 (avg.); Solubility: Completely sol. in water and most common organic solvents; Density: 7.30 lb/gal
Composition: 90% min. active
Toxicity: Severely irritating to eyes, skin, and mucous membranes
Warnings and Caution: Wear eye protection

Surfam P10. Mars Chem. Corp.
Chemical Description: Primary ether amine
Category: Intermediate, corrosion inhibitor, cationic
Applications: Primary ether amines and their salts can be used in corrosion inhibitor formulations; when ethoxylated, used as cationic emulsifiers and lubricants in fiber and fabric processing; raw material to produce sulfosuccinimates, amphoterics, quaternary ammonium compounds, and as intermediates in the formulation of anticaking agents, specialty surfactants, gasoline and fuel additives, etc.
Form: Clear liquid; Color: 4 max. (Gardner)
Properties: m.w.: 169 (avg.); Density: 7.09 lb/gal
Composition: 100% conc.
Reaction: Mildly alkaline
Containers: 55 gal drums and bulk
Warnings and Caution: Avoid contact with skin and eyes

Surfam P12B. Mars Chem. Corp.
Chemical Description: Primary ether amine
Category and Applications: See Surfam P10
Form: Clear liquid; Color: 4 max. (Gardner)
Properties: m.w.: 200 (avg.); Density: 7.11 lb/gal
Composition: 100% conc.
Reaction: Mildly alkaline
Containers: 55 gal drums and bulk
Warnings and Caution: Avoid contact with skin and eyes

Surfam P14B. Mars Chem. Corp.
Chemical Description: Primary ether amine
Category and Applications: See Surfam P10
Form: Clear liquid; Color: 4 max. (Gardner)
Properties: m.w.: 229 (avg.); Density: 7.11 lb/gal
Composition: 100% conc.
Reaction: Mildly alkaline
Containers: 55 gal drums and bulk
Warnings and Caution: Avoid contact with skin and eyes

Surfam P17B. Mars Chem. Corp.
Chemical Description: Primary ether amine
Category and Applications: See Surfam P10
Form: Clear liquid; Color: 4 max. (Gardner)
Properties: m.w.: 279 (avg.); Density: 7.30 lb/gal
Composition: 100% conc.
Reaction: Mildly alkaline
Containers: 55 gal drums and bulk
Warnings and Caution: Avoid contact with skin and eyes

Surfam P24M. Mars Chem. Corp.
Chemical Description: Primary ether amine
Category and Applications: See Surfam P10
Form: Clear liquid; Color: 4 max. (Gardner)
Properties: m.w.: 212 (avg.); Density: 7.10 lb/gal
Composition: 100% conc.
Reaction: Mildly alkaline
Containers: 55 gal drums and bulk
Warnings and Caution: Avoid contact with skin and eyes

Surfam P86M. Mars Chem. Corp.
Chemical Description: Primary ether amine
Category and Applications: See Surfam P10
Form: Clear liquid; Color: 4 max. (Gardner)
Properties: m.w.: 280 (avg.); Density: 7.05 lb/gal
Composition: 100% conc.
Reaction: Mildly alkaline
Containers: 55 gal drums and bulk
Warnings and Caution: Avoid contact with skin and eyes

Surfam P89MB. Mars Chem. Corp.
Chemical Description: Primary ether amine
Category and Applications: See Surfam P10
Form: Clear liquid; Color: 4 max. (Gardner)
Properties: m.w.: 305 (avg.); Density: 7.10 lb/gal
Composition: 100% conc.
Reaction: Mildly alkaline
Containers: 55 gal drums and bulk
Warnings and Caution: Avoid contact with skin and eyes

Surfam P-MEPA. Mars Chem. Corp.
Chemical Description: Methoxyethoxypropylamine
Category: Specialty chemical, intermediate
Applications: Used in textile dye chemicals, emulsifier intermediate in floor waxes, and agricultural products
Form: Liquid; Color: 1 (Gardner)
Properties: Density: 7.82 lb/gal (20 C); Sp. gr.: 0.93–0.94; F.P.: 0 C; B.P.: 210–230 C; Flash Pt.: < 212 F (COC)
Composition: 97% min. active
Toxicity: Severely irritating to eyes, skin, and mucous membranes
Warnings and Caution: Wear eye protection

Surflo-OW-1. NL Ind./Baroid Div.
Chemical Description: Oxyethylated glycerol ester of a fatty acid
Category: Emulsifier, nonionic
Applications: O/W emulsifier for elimination of oil slicks caused by spills or oily effluent waters; reduces fire hazards, limits contamination, accelerates natural bacterial decomposition of oil slicks, and prevents adhesion to beaches, piers, marine life, etc.
Form: Liquid
Properties: Solubility: Sol. in isopropyl alcohol and fresh, brackish or sea water; disp. in aliphatic and aromatic hydrocarbons; Density: 8.56 lb/gal; Sp. gr.: 1.028 (65 F); Visc.: 160 SUS (70 F); F.P.: 0 F; Flash Pt.: > 175 F (TOC); Pour Pt.: 5 F; pH: 7.5; Biodegradable
Composition: 58% conc.
Containers: Specially lined 5 gal cans or 55 gal drums
Warnings and Caution: Avoid contact with skin, eyes, and clothing

Surflo-S24. NL Ind./Baroid Div.
Chemical Description: Oxyethylated alkylphenol and an oxyethylated alcohol
Category: Wetting agent, dispersant, detergent, surface and interfacial tension reducing agent, nonionic
Applications: For use in producing wells, water injection wells, gas storage wells, cooling towers, surface lines, vessels, and other equipment, and in flotation units
Form: Liquid
Properties: Sol. in water at chloride concs. from zero to saturation and at temps. up to 200 F; Density: 8.30 lb/gal; Sp. gr.: 0.996 (68 F); Visc.: 125 SUS (70 F); Flash Pt.: 82 F (PMCC); Pour Pt.: < −10 F
Composition: 30% conc.
Containers: 55 gal drums
Warnings and Caution: Take care in handling around open flames and sparks; avoid contact with skin, eyes, and clothing; slightly corrosive to mild steel

Surflo-S30. NL Ind./Baroid Div.
Chemical Description: Ammonium salt of a sulfated alcohol ethoxylate
Category: Detergent, foaming agent, anionic
Applications: Used as a surfactant or at higher concs. as a foaming agent in producing wells, water injection wells, gas storage wells, cooling towers, surface lines, vessels, and other production equipment, and in air drilling applications
Form: Liquid
Properties: Solubility: Highly water-sol.; Density: 8.16 lb/gal; Sp. gr.: 0.979 (68 F); Visc.: 52 SUS (70 F); Flash Pt.: 81 F (TOC); Pour Pt.: −3 F; Biodegradable
Composition: 40% conc.
Containers: 5 gal cans and 55 gal drums

Surflo-S32. NL Ind./Baroid Div.
Chemical Description: Sodium alkyl aryl sulfonate
Category: Detergent, dispersant,

emulsifier, anionic
Applications: Used as a surfactant in oil industry producing wells, water injection wells, gas storage wells, cooling towers, surface lines, vessels, and other equipment
Form: Liquid
Properties: Solubility: Highly water-sol.; Density: 8.16 lb/gal; Sp. gr.: 0.980 (68 F); Visc.: 44 SUS (70 F); Flash Pt.: 80 F (TOC); Pour Pt.: < −10 F
Composition: 20% conc.
Containers: Specially lined 5 gal cans and 55 gal drums
Warnings and Caution: Take care in handling around open flames or sparks; avoid contact with skin, eyes, and clothing

Surflo-S40. NL Ind./Baroid Div.
Chemical Description: Polyglycol and an ammonium salt of a sulfated alcohol ethoxylate
Category: General surfactant, demulsifier, antifoulant, anionic
Applications: In oil industry as a surfactant in salt water disposal and injection systems, cooling tower cleaning, well stimulation, and gas storage wells; excellent characteristics for oil displacement from sand and other solids
Form: Liquid
Properties: Solubility: Water-sol.; Density: 8.22 lb/gal; Sp. gr.: 0.9873 (68 F); Visc.: 43 SUS (70 F); Flash Pt.: 90 F (TCC); Pour Pt.: −10 F; pH: 7.3 (1% sol'n.)
Composition: 27% conc.
Containers: 55 gal drums
Warnings and Caution: Take care in handling around open flames or sparks; avoid contact with skin, eyes, and clothing

Surflo-S41. NL Ind./Baroid Div.
Chemical Description: Polyglycol
Category: General surfactant, demulsifier, foam stabilizer, antifoulant, nonionic
Applications: In oil industry as a general surfactant in salt water disposal and injection systems, open salt water pits, cooling tower cleaning, well stimulation, gas storage wells; excellent characteristics for oil displacement from sand and other solids
Form: Liquid
Properties: Solubility: Water-sol.; Density: 8.04 lb/gal; Sp. gr.: 0.9662 (68 F); Visc.: 47 SUS (70 F); Flash Pt.: 120 F (TOC); Pour Pt.: −5 F; pH: 5.5 (1% sol'n.)
Composition: 20% conc.
Containers: 55 gal drums
Warnings and Caution: Take care in handling around open flames or sparks; avoid contact with skin, eyes, and clothing

Surflo-S362. NL Ind./Baroid Div.
Chemical Description: Ammonium salt of a sulfated alcohol ethoxylate
Category: Foaming agent, anionic
Applications: Foaming agent in air drilling opeations; effective in fresh waters and heavy brines
Form: Liquid
Properties: Solubility: Highly water-sol.; Density: 8.65 lb/gal; Sp. gr.: 1.038 (68 F); Visc.: 95 SUS

(70 F); Flash Pt.: 79 F (TCC); Pour Pt.: −10 F; pH: 8.8 (5% sol'n.); Biodegradable
Composition: 100% conc.
Containers: 55 gal drums
Warnings and Caution: Take care in handling aroudn open flames or sparks; avoid contact with skin, eyes, or clothing

Surflo-S375. NL Ind./Baroid Div.
Chemical Description: Ammonium salt of sulfated alcohol ethoxylates
Category: Foaming agent, anionic
Applications: Foaming agent in air drilling operations; effective in fresh and hard waters, as well as in heavy brines
Form: Liquid
Properties: Density: 8.50 lb/gal; Sp. gr.: 1.020 (68 F); Visc.: 192 SUS (70 F); Flash Pt.: 74 F (PMCC); Pour Pt.: −10 F; pH: 6.4 (1% sol'n.); Biodegradable
Composition: 53% conc.
Containers: 55 gal drums
Warnings and Caution: Take care in handling around open flames or sparks; avoid contact with skin, eyes, and clothing

Surflo-S390. NL Ind./Baroid Div.
Chemical Description: Ammonium salt of sulfate alcohol ethoxylates
Category and Applications: See Surflo-S375
Form: Liquid
Properties: Density: 8.46 lb/gal; Sp. gr.: 1.016; Visc.: 195 SUS (70 F); Flash Pt.: 82 F (PMCC); Pour Pt.: −10 F; pH: 6.4 (1% sol'n.); Biodegradable
Composition: 38% conc.
Containers: 55 gal drums
Warnings and Caution: Take care in handling around open flames or sparks; avoid contact with skin, eyes, and clothing

Surfonic J-4. Jefferson Chem. Co.
Chemical Description: Alkyl polyoxyalkalene ether
Category: Detergent, wetting agent, nonionic
Applications: Industrial cleaning compounds, incl. sanitizers, metal cleaning and germicidal detergents; household detergents incl. floor, oven, tile, hard surface cleaners, laundry, dishwash; paint brush cleaners; wetting agents for clays in brick and ceramics mfg.; concrete additive; dust control compositions; textile and paper wetting; wallpaper removal; decreased caking of fertilizers
Form: Clear liquid
Properties: Solubility: Water-sol.; Density: 8.4 lb/gal; Sp. gr.: 1.0072 (20 C); Visc.: 267 SUS (68 F); F.P.: 21.1 F; Cloud Pt.: 61 C (1% aq.); Flash Pt.: 375 F (COC); Pour Pt.: 23 F; Hydroxyl No.: 92; Surface Tension: 29.24 dynes/cm (0.1%); pH: 5–7 (1% aq.); Biodegradable
Composition: 100% active ingred.
Containers: 55 gal unlined, closed-head drums or bulk
Warnings and Caution: Avoid prolonged contact with conc. form

Surfonic LF-6. Jefferson Chem. Co.
Chemical Description: Modified aliphatic alcohol-ethylene oxide adducts
Category: Low-foaming nonionic surfactant
Applications: Mechanical dishwashing, laundry detergents, pulp and paper, metal cleaning, and other uses requiring low foam
Form: Clear liquid; Color: Pale yellow
Properties: Ref. Index: 1.4559 (20 C); Solubility: Sol. in water, acetone, benzene, toluene, perchloroethylene, carbon tetrachloride, kerosene, and at higher concs. is clear sol. in petroleum ether and naphtha; Sp. gr.: 1.0023 (20/20 C); Visc.: 60 cs (100 F); Cloud Pt.: 35.5 C (1% aq.); Surface Tension: 29.6 dynes/cm (0.1% conc.); pH: 6–7 (1% aq.)
Composition: 100% conc.
Toxicity: Slight eye irritant; skin irritant on prolonged or repeated contact
Containers: Drums and bulk

Surfonic LF-7. Jefferson Chem. Co.
Chemical Description: Modified aliphatic alcohol-ethylene oxide adduct
Category and Applications: See Surfonic LF-6
Form: Clear liquid; Color: Pale yellow
Properties: Ref. Index: 1.4565 (20 C); Solubility: Sol. in water, acetone, benzene, toluene, perchloroethylene, carbon tetrachloride, kerosene, petroleum ether, and naphtha; Sp. gr.: 1.0068; Visc.: 67 cs (100 F); HLB: 12.0; Cloud Pt.: 30.5 (1% aq.); Surface Tension: 30.0 dynes/cm (0.1% conc.); pH: 6–7 (1% aq.)
Composition: 100% conc.
Toxicity: Slight eye irritant; skin irritant on prolonged or repeated contact only
Containers: Drums and bulk

Surfonic LF-17. Jefferson Chem. Co.
Chemical Description: Modified primary alcohol-ethylene oxide adduct
Category: Low-foaming nonionic surfactant
Applications: Mechanical dishwashing, laundry detergents, pulp and paper processing, metal cleaning, and other uses where foam control and biodegradability are important; also useful in controlling the gelling tendency which nonionics exhibit when mixed with water
Form: Clear liquid
Properties: HLB: 12.2; Cloud Pt.: 34.8 C (1% aq.); Surface Tension: 33.0 dynes/cm (0.1%); Biodegradable
Containers: 460 lb net drums or bulk

Surfonic N Series. Jefferson Chem. Co.
Chemical Description: Adduct of nonylphenol and ethylene oxide
Category: Emulsifier, wetting agent, detergent, penetrant, solubilizer, dispersant, nonionic

Applications: Agricultural chemicals (esp. N-95), industrial cleaning compounds (N-40, N-60, N-95), lime soap dispersion (esp. N-95), dry cleaning detergent (N-60), ceramics and concrete (N-95), dust control (N-95), paper (N-95), wallpaper removal (N-95, N-120), photographic film developing (N-95), heavy duty detergents (N-85), fire fighting, emulsion polymerization (N-300, N-400), indirect food additives (N-10, N-40, N-60, N-85, N-95, N-100, N-120, N-150, N-200, N-300, N-400); other uses include components of cutting oil emulsifiers, stabilizers for rubber latex, drilling-mud additives, emulsifier for styrene-butadiene latices (N-60), degreaser for leather industry, O/W and W/O emulsions for cosmetics industry, antifoaming agents in certain systems (N-10, N-31.5)
Form: Liquid to solid
Reaction: Essentially neutral
Toxicity: May cause skin and eye irritation
Containers: 55 gal drums and bulk
Warnings and Caution: Avoid contact with eye or prolonged exposure to skin

Surfonic N-10. Jefferson Chem. Co.
Chemical Description: Adduct of nonylphenol and ethylene oxide
Category and Applications: See Surfonic N Series
Form: Liquid
Properties: m.w.: 264; Ref. Index: 1.5090 (20 C); Solubility: Sol. in acetone, methanol, xylene, carbon tetrachloride, mineral oil; Density: 8.1 lb/gal (20 C); Sp. gr.: 0.98 (20/20 C); Visc.: 675 SUS (100 F); HLB: 3.4; Flash Pt.: 355 F (OC); Stability: Stable to acids, bases, and salts
Composition: 100% active ingred.

Surfonic N-31.5. Jefferson Chem. Co.
Chemical Description: Adduct of nonylphenol and ethylene oxide
Category and Applications: See Surfonic N Series
Form: Liquid
Properties: m.w.: 358; Ref. Index: 1.4950 (20 C); Solubility: Sol. in acetone, methanol, xylene, carbon tetrachloride, mineral oil; Density: 8.4 lb/gal (20 C); Sp. gr.: 1.01 (20/20 C); Visc.: 480 SUS (100 F); HLB: 7.7; Flash Pt.: 410 F (OC); Stability: Stable to acids, bases, salts
Composition: 100% active ingred.

Surfonic N-40. Jefferson Chem. Co.
Chemical Description: Adduct of nonylphenol and ethylene oxide
Category and Applications: See Surfonic N Series
Form: Liquid
Properties: m.w.: 396; Ref. Index: 1.4979 (20 C); Solubility: Sol. in acetone, methanol, xylene, carbon tetrachloride, Stoddard solvent; Density: 8.5 lb/gal (20 C); Sp. gr.: 1.026 (20/20 C); Visc.: 445 SUS (100 F); HLB: 8.9; Flash Pt.: 435 F (OC); Stability: Stable to acids,

bases, salts; Surface Tension: 27.5 dynes/cm (0.1%)
Composition: 100% active ingred.

Surfonic N-60. Jefferson Chem. Co.
Chemical Description: Adduct of nonylphenol and ethylene oxide
Category and Applications: See Surfonic N Series
Form: Liquid
Properties: m.w.: 484; Ref. Index: 1.4938 (20 C); Solubility: Sol. in acetone, methanol, xylene, carbon tetrachloride, Stoddard solvent, disp. in water; Density: 8.7 lb/gal (20 C); Sp. gr.: 1.041 (20/20 C); Visc.: 440 SUS (100 F); F.P.: < 0 C; HLB: 10.9; Flash Pt.: 475 F (OC); Stability: Stable to acids, bases, salts; Surface Tension: 28.7 dynes/cm (0.1%); Biodegradable > 90%
Composition: 100% active ingred.

Surfonic N-85. Jefferson Chem. Co.
Chemical Description: Adduct of nonylphenol and ethylene oxide
Category and Applications: See Surfonic N Series
Form: Liquid
Properties: m.w.: 596; Ref. Index: 1.4923 (20 C); Solubility: Sol. in acetone, methanol, xylene, carbon tetrachloride, water; Density: 8.8 lb/gal (20 C); Sp. gr.: 1.056 (20/20 C); Visc.: 485 SUS (100 F); F.P.: < 0 C; HLB: 12.6; Cloud Pt.: 44 C (1% aq.); Flash Pt.: 500 F (OC); Stability: Stable to acids, bases, salts; Surface Tension: 30.5 dynes/cm (0.1%); Biodegradable > 90%
Composition: 100% active ingred.

Surfonic N-95. Jefferson Chem. Co.
Chemical Description: Adduct of nonylphenol and ethylene oxide
Category and Applications: See Surfonic N Series
Form: Liquid
Properties: m.w.: 632; Ref. Index: 1.4893 (20 C); Solubility: Sol. in acetone, methanol, xylene, carbon tetrachloride, water; Density: 8.8 lb/gal (20 C); Sp. gr.: 1.061 (20/20 C); Visc.: 510 SUS (100 F); F.P.: 5 C; HLB: 12.9; Cloud Pt.: 54.2 C (1% aq.); Flash Pt.: 500 F (OC); Stability: Stable to acids, bases, salts; Surface Tension: 30.8 dynes/cm (0.1%); Biodegradable > 90%
Composition: 100% active ingred.

Surfonic N-100. Jefferson Chem. Co.
Chemical Description: Adduct of nonylphenol and ethylene oxide
Category and Applications: See Surfonic N Series
Form: Liquid
Properties: m.w.: 660; Ref. Index: 1.4888 (20 C); Solubility: Sol. in acetone, methanol, xylene, carbon tetrachloride, water; Density: 8.8 lb/gal (20 C); Sp. gr.: 1.064 (20/20 C); Visc.: 69 SUS (210 F); F.P.: 8 C; HLB: 13.2; Cloud Pt.: 65 C (1% aq.); Flash Pt.: 510 F (OC); Stability: Stable to acids, bases, salts; Surface Tension: 31.0 dynes/cm (0.1%); Biodegradable > 90%
Composition: 100% active ingred.

Surfonic N-102. Jefferson Chem. Co.
Chemical Description: Adduct of nonylphenol and ethylene oxide
Category and Applications: See Surfonic N Series
Form: Liquid
Properties: m.w.: 668; Ref. Index: 1.4884 (20 C); Solubility: Sol. in acetone, methanol, xylene, carbon tetrachloride, water; Density: 8.8 lb/gal (20 C); Sp. gr.: 1.065 (20/20 C); F.P.: 9 C; HLB: 13.4; Cloud Pt.: 81 C (1% aq.); Flash Pt.: 510 F (OC); Stability: Stable to acids, bases, salts; Surface Tension: 31.2 dynes/cm (0.1%); Biodegradable > 90%
Composition: 100% active ingred.

Surfonic N-120. Jefferson Chem. Co.
Chemical Description: Adduct of nonylphenol and ethylene oxide
Category and Applications: See Surfonic N Series
Form: Liquid
Properties: m.w.: 748; Ref. Index: 1.4869 (20 C); Solubility: Sol. in acetone, methanol, xylene, carbon tetrachloride, water; Density: 8.9 lb/gal (20 C); Sp. gr.: 1.070 (20/20 C); Visc.: 560 SUS (100 F); F.P.: 14 C; HLB: 14.1; Cloud Pt.: 81 C (1% aq.); Flash Pt.: 525 F (OC); Stability: Stable to acids, bases, salts; Surface Tension: 32.3 dynes/cm (0.1%); Biodegradable > 90%
Composition: 100% active ingred.

Surfonic N-150. Jefferson Chem. Co.
Chemical Description: Adduct of nonylphenol and ethylene oxide
Category and Applications: See Surfonic N Series
Form: Semisolid; Color: White
Properties: m.w.: 880; Ref. Index: 1.4815 (30 C); Solubility: Sol. in acetone, methanol, xylene, carbon tetrachloride, water; Density: 8.95 lb/gal (20 C); Sp. gr.: 1.065 (30/4 C); Visc.: 89 SUS (210 F); F.P.: 23 C; HLB: 15.0; Cloud Pt.: 94 C (1% aq.); Flash Pt.: > 500 F (OC); Stability: Stable to acids, bases, salts; Surface Tension: 34.2 dynes/cm (0.1%)
Composition: 100% active ingred.

Surfonic N-200. Jefferson Chem. Co.
Chemical Description: Adduct of nonylphenol and ethylene oxide
Category and Applications: See Surfonic N Series
Form: Semisolid; Color: White
Properties: m.w.: 1100; Ref. Index: 1.4720 (50 C); Solubility: Sol. in acetone, methanol, xylene, carbon tetrachloride, water; Density: 9.0 lb/gal (20 C); Visc.: 105 SUS (210 F); F.P.: 34 C; HLB: 15.8; Cloud Pt.: > 100 C (1% aq.); Flash Pt.: > 500 F (OC); Stability: Stable to acids, bases, salts
Composition: 100% active ingred.

Surfonic N-300. Jefferson Chem. Co.
Chemical Description: Adduct of

nonylphenol and ethylene oxide
Category and Applications: See Surfonic N Series
Form: Waxy solid; Color: White
Properties: m.w.: 1540; Ref. Index: 1.4690 (50 C); Solubility: Sol. in acetone, methanol, xylene, carbon tetrachloride, water; Density: 9.1 lb/gal (20 C); Visc.: 160 SUS (210 F); F.P.: 44 C; HLB: 17.1; Cloud Pt.: > 100 C (1% aq.); Flash Pt.: > 500 F (OC); Stability: Stable to acids, bases, salts
Composition: 100% active ingred.

Surfonic N-400. Jefferson Chem. Co.
Chemical Description: Adduct of nonylphenol and ethylene oxide
Category and Applications: See Surfonic N Series
Form: Waxy solid; Color: White
Properties: m.w.: 1980; Solubility: Sol. in acetone, methanol, xylene, carbon tetrachloride, water; Density: 9.1 lb/gal (20 C); Visc. 210 SUS (210 F); F.P.: 45 C; HLB: 17.8; Cloud Pt.: > 100 C (1% aq.); Flash Pt.: > 500 F (OC); Stability: Stable to acids, bases, salts
Composition: 100% active ingred.

Surfonic NB-5. Jefferson Chem. Co.
Chemical Description: 70% aq. sol'n. of Surfonic N-300
Category and Applications: See Surfonic N Series
Form: Clear liquid
Composition: 70% active ingred.

Surfonic NB-14. Jefferson Chem. Co.
Chemical Description: 70% aq. sol'n. of Surfonic N-400
Category and Applications: See Surfonic N Series
Form: Clear liquid
Composition: 70% active ingred.

Surfynol 61. Air Products & Chem.
Chemical Description: Dimethyl hexynol
Category: Wetting agent, nonionic
Applications: Volatile wetting agent for paper coatings, floor polishes, and glass cleaning formulations
Form: Clear liquid; Color: Colorless
Properties: Sp. gr.: 0.855 (20 C); B.P.: 151 C; Surface Tension: 50.2 (0.1% aq.)
Composition: 100% conc.

Surfynol 82. Air Products & Chem.
Chemical Description: Dimethyl octynediol
Category: Defoamer, wetting agent, nonionic
Applications: Defoamer/wetting agent in pesticide concentrates; cosmetic ingredient; low foam wetting agent in developer compounds; defoamer in electroplating baths
Form: Free-flowing crystalline solid; Color: White
Properties: Sp. gr.: 0.933 (20 C); M.P.: 51 C; B.P.: 222 C; Surface Tension: 55.3 (0.1% aq.)
Composition: 100% conc.

Surfynol 104. Air Products & Chem.
Chemical Description: Tetramethyl decynediol
Category: Wetting agent, defoamer, emulsifier, viscosity reducer, nonionic
Applications: Increased wetting and low foam with other surfactants; defoamer and dye dispersant in paint and ink formulations; low foaming surfactant for use in rinse aids; substrate pigment wetting agent for industrial coatings and adhesives; defoamer in dyestuff production; wetter, defoamer, and emulsifier for pesticide formulations; wetting agent for industrial cleaners; viscosity reducer for vinyl dispersions
Form: Waxy solid; Color: White
Properties: Sp. gr.: 0.893 (20 C); M.P.: 37 C; B.P.: 260 C; Surface Tension: 31.6 (0.1% aq.)
Composition: 100% conc.

Surfynol 104A. Air Products & Chem.
Chemical Description: Sol'n. of Surfynol 104 in 2-ethyl hexanol
Category and Applications: See Surfynol 104
Form: Clear liquid; Color: Light yellow
Properties: Sp. gr.: 0.869; Surface Tension: 33.0 (0.1% aq.)
Composition: 50% conc.

Surfynol 104E. Air Products & Chem.
Chemical Description: Sol'n. of Surfynol 104 in ethylene glycol
Category and Applications: See Surfynol 104
Form: Clear liquid; Color: Light yellow
Properties: Sp. gr.: 1.001; Surface Tension: 36.2 (0.1% aq.)
Composition: 50% conc.

Surfynol 104H. Air Products & Chem.
Chemical Description: Sol'n. of Surfynol 104 in ethylene glycol
Category and Applications: See Surfynol 104
Form: Clear liquid; Color: Light yellow
Properties: Sp. gr.: 0.951; Surface Tension: 33.8 (0.1% aq.)
Composition: 75% conc.

Surfynol 440. Air Products & Chem.
Chemical Description: Ethoxylated Surfynol 104 (3.5 moles)
Category: Defoamer, rewetting agent, leveling agent, nonionic
Applications: Low-foaming nonionic for water-based industrial finishes; defoamer, rewetting and leveling agent for paperboard coatings and paper glazes; metal cleaning and plating bath additive
Form: Clear liquid; Color: Straw
Properties: Sp. gr.: 0.982; Surface Tension: 31.0 (0.1% aq.)
Composition: 100% conc.

Surfynol 465. Air Products & Chem.
Chemical Description: Ethoxylated Surfynol 104 (10 moles)
Category: Nonionic surfactant

Applications: Emulsion polymerization surfactant when blended with Surfynol 485; electroplating additive
Form: Clear liquid; Color: Straw
Properties: Sp. gr.: 1.038; Surface Tension: 33.2 (0.1% aq.)
Composition: 100% conc.

Surfynol 485. Air Products & Chem.
Chemical Description: Ethoxylated Surfynol 104 (30 moles)
Category: Nonionic surfactant
Applications: Emulsion polymerization surfactant when blended with Surfynol 465; electroplating additive
Form: Clear liquid; Color: Straw
Properties: Sp. gr.: 1.080 (30 C); Surface Tension: 40.1 (0.1% aq.)
Composition: 100% conc.

Surfynol PC. Air Products & Chem.
Chemical Description: Acetylenic glycol blend
Category: Defoamer, nonionic
Applications: Defoamer for paper coating latices and systems where the foaming influence is a water-sol. polymer; antishock agent for paper coatings
Form: Clear liquid; Color: Light yellow
Properties: Sp. gr.: 0.9585 (20 C)
Composition: 100% conc.

Surfynol TG. Air Products & Chem.
Chemical Description: Acetylenic glycol blend in ethylene glycol
Category: Wetting agent, nonionic
Applications: Pigment and substrate wetting agent which demonstrates low foaming in latex and water-reducible paints, adhesives, paper coatings, and other pigmented aq. systems
Form: Clear liquid; Color: Light yellow
Properties: Sp. gr.: 1.000; Surface Tension: 27.6 (0.1% aq.)
Composition: 83% conc.

Swanic 6L. Swastik
Chemical Description: Ethoxylated nonylphenol
Category: Emulsifier ingred., detergent, wetting agent, nonionic
Applications: Emulsifier and detergent used in wool scouring, paint formulation, and textile processing
Form: Clear liquid
Properties: pH: 6–7 (1% in dist. water)
Composition: 100% active ingred.

Swanic 7L. Swastik
Chemical Description: Ethoxylated nonylphenol
Category and Applications: See Swanic 6L
Form: Clear liquid
Properties: pH: 6–7 (1% in dist. water)
Composition: 25% active ingred.

Swanic 51. Swastik
Chemical Description: Coconut monoethanolamide (oil-based)
Category: Foam booster, foam stabilizer
Applications: Used as a foam booster/stabilizer in detergent

powders
Form: Solid paste; Color: White
Reaction: Slightly alkaline

Swanic 52L. Swastik
Chemical Description: Coconut alkylolamide
Category: Foam booster, viscosity controller, anionic
Applications: Used as a foam stabilizer and viscosity builder in shampoos and liquid detergent formulations
Form: Liquid; Color: Yellow
Composition: 100% conc.
Reaction: Very mildly alkaline

Swanic 110. Swastik
Chemical Description: Diluted nonionic
Category: Dye retarding/leveling agent, nonionic
Applications: Textile dyeing
Form: Liquid
Reaction: Practically neutral

Swanic D-16, D-24, D-52, D-60, D-80, D-120. Swastik
Chemical Description: Polyoxyethylene ether
Category: Emulsifier, detergent, wetting agent, nonionic
Applications: Detergent, emulsifier ingred. in the formulation of emulsifiers for pesticides
Form: Liquid, paste, solid
Reaction: Practically neutral

Swanic J. Swastik
Chemical Description: Blend of nonionics
Category: Emulsifier, nonionic
Applications: Jute batching oils at very low concentrations
Form: Liquid
Composition: 100% conc.

Swaryl 9-P. Swastik
Chemical Description: Blended anionic detergent with alkaline builders
Category: Detergent, degreaser, emulsifier
Applications: Alkaline degreasing agent in metal industry
Form: Physically mixed powder; Color: White
Properties: Solubility: Water-sol.
Reaction: Alkaline

Swaryl 10-C. Swastik
Chemical Description: Formulated product
Category: Detergent, wetting agent, anionic
Applications: For use in wool processing industry as an acid milling agent
Form: Paste; Color: Brownish
Composition: 60% active ingred.
Reaction: Highly acidic when dissolved in water

Swascofix CD-38. Swastik
Chemical Description: Formulated blended product
Category: Emulsifier
Applications: Emulsifier for Chlordane EC
Form: Liquid
Reaction: Slightly acidic

Swascofix DT-50. Swastik
Chemical Description: Formulated blend

Category: Emulsifier
Applications: Emulsifier for DDT EC
Form: Liquid
Reaction: Slightly acidic

Swascofix DU-39. Swastik
Chemical Description: Formulated blend
Category: Emulsifier
Applications: Emulsifier for Dursban EC
Form: Liquid
Reaction: Slightly acidic

Swascofix E-45. Swastik
Chemical Description: Blended emulsifier
Category: Emulsifier
Applications: Emulsifier ingred. for Endosulfan
Form: Liquid
Reaction: Slightly acidic

Swascofix ES-36. Swastik
Chemical Description: Formulated product
Category: Emulsifier
Applications: Emulsifier in the formulation of EC's based on Endosulfan
Form: Liquid
Reaction: Slightly acidic

Swascofix F-28. Swastik
Chemical Description: Formulated blend
Category: Emulsifier
Applications: Emulsifier for Phenthoate EC
Form: Liquid
Reaction: Slightly acidic

Swascofix FN-33, FN-34. Swastik
Chemical Description: Formulated blend
Category: Emulsifier
Applications: Emulsifier for Fenitrothion EC
Form: Liquid
Reaction: Slightly acidic

Swascofix L-20. Swastik
Chemical Description: Formulated blend
Category: Emulsifier
Applications: Emulsifier for Lindane EC
Form: Liquid
Reaction: Slightly acidic

Swascofix M-21. Swastik
Chemical Description: Formulated blend
Category: Emulsifier
Applications: Emulsifier for Malathion EC
Form: Liquid
Reaction: Slightly acidic

Swascofix ME-25. Swastik
Chemical Description: Blended product
Category: Emulsifier
Applications: Used as emulsifier in the formulation of Metasystox EC
Form: Liquid
Reaction: Slightly acidic

Swascofix ON-12. Swastik
Chemical Description: Formulated blend
Category: Emulsifier
Applications: Emulsifier for sili-

cone EC
Form: Liquid
Reaction: Neutral

Swascofix OP-50. Swastik
Chemical Description: Formulated product
Category: Emulsifier
Applications: Emulsifier for mineral oils; used in the mfg. of soluble cutting oils
Form: Liquid
Reaction: Slightly alkaline

Swascofix PH-25. Swastik
Chemical Description: Formulated blend
Category: Emulsifier
Applications: Emulsifier for Phosalone EC
Form: Liquid
Reaction: Slightly acidic

Swascofix R-14. Swastik
Chemical Description: Formulated blend
Category: Emulsifier
Applications: Emulsifier for Dimethoate EC
Form: Liquid
Reaction: Slightly acidic

Swascofix TD-33, TD-43. Swastik
Chemical Description: Formulated blend
Category: Emulsifier
Applications: Emulsifier for Tetradifon EC
Form: Liquid
Reaction: Slightly acidic

Swascol 1PC. Swastik
Chemical Description: Sodium lauryl sulfate (C_{12}:C_{14} 70:30)
Category: Detergent, foaming agent, anionic
Applications: High foaming detergent for use in the manufacture of shampoos, toothpaste, liquid detergents
Form: Crystalline or amorphous powder
Composition: 96% active ingred.
Reaction: Slightly alkaline

Swascol 3L. Swastik
Chemical Description: Sodium lauryl sulfate
Category: Wetting agent, detergent, foaming agent, anionic
Applications: Used in the formulation of shampoos, cosmetics, liquid detergents, etc.
Form: Liquid/paste
Composition: 30% active ingred.
Reaction: Slightly alkaline

Swascol 4L. Swastik
Chemical Description: TEA salt of a sulfated lauryl alcohol
Category: Foaming agent, detergent, anionic
Applications: Base material for shampoos and liquid detergents
Form: Liquid
Composition: 40% active ingred.
Reaction: Slightly alkaline

Swascol L-327. Swastik
Chemical Description: Sodium salt of sulfated ethoxylated fatty alcohol
Category: Foaming agent, deter-

gent, anionic
Applications: Based material in the formulation of shampoos
Form: Liquid
Properties: Cloud Pt.: 0 C
Composition: 27% active ingred.
Reaction: Slightly alkaline

Swascol RWT. Swastik
Chemical Description: Sulfonated ester
Category: Wetting/rewetting agent, anionic
Applications: Recommended for use as an instant wetting/rewetting agent in textile industry
Form: Liquid
Composition: 50% conc.
Reaction: Practically neutral

Swascol T-51. Swastik
Chemical Description: Formulated product
Category: Wetting agent, anionic
Applications: Powerful wetting agent in alkaline conditions; used in mercerizing bath in textile industry
Form: Liquid
Reaction: Alkaline

Swascoclin M-22. Swastik
Chemical Description: Formulated product
Category: Degreaser
Applications: Neutral metal cleaning compound for nonferrous metals (based on use of organic solvents); degreasing agent for aluminum and other soft metalware
Form: Liquid
Properties: Solubility: Emulsifiable in water; degreases in cold as well as at slightly elevated temps.
Reaction: Practically neutral

Swastik Detergent Powder. Swastik
Chemical Description: Alkyl aryl sulfonate sodium salt built with alkaline builders, STPP, and other chemical additives
Category: Detergent, anionic
Applications: Detergent powder for heavily soiled clothes, cottons
Form: Spray-dried powder; Color: Blue
Reaction: Alkaline

Sway. Swastik
Chemical Description: Alkyl aryl sulfonate sodium salt built with STPP, alkaline builders, and other chemical additives
Category: Detergent
Applications: Heavy duty synthetic detergent powder; recommended for cottons
Form: Spray-dried powder; Color: Blue
Reaction: Alkaline

SXS. Continental Oil
Chemical Description: Sodium xylene sulfonate
Category: Hydrotrope
Applications: Hydrotrope for detergent formulations
Form: Liquid; Color: Nearly colorless; Odor: Mild
Properties: Density: 9.80; Sp. gr.: 1.17; Visc.: 6.0; M.P.: 31 F; B.P.: 212 F
Composition: 41% active ingred.

Syncrowax BB4, BB5. Croda Inc.
Chemical Description: Synthetic waxes
Category: Beeswax substitute
Applications: Creams and sticks
Form: Solid; Color: Pale cream

Syncrowax ERLC. Croda Inc.
Chemical Description: Ethylene glycol ester of fatty wax acids
Category: Reducer of bleeding and sweating, gloss improver
Applications: Cosmetic sticks and creams
Form: Solid; Color: Light tan

Syncrowax HGLC. Croda Inc.
Chemical Description: Glycerol ester of fatty wax acids
Category: Strength improver, stabilizer, gloss improver
Applications: Confers strength and rigidity to stick compositions, e.g., lip and eye make-up sticks
Form: Solid; Color: Light tan

Syncrowax HRC. Croda Inc.
Chemical Description: Modified triglyceride of natural vegetable fatty acids
Category: Gloss improver
Applications: Lipstick and make-up compositions, nonalcoholic antipersipirant sticks, anhydrous make-up and eyebrow pencils, hair and skin creams
Form: Solid; Color: Off-white

Syncrowax PRLC. Croda Inc.
Chemical Description: Propylene glycol ester of fatty wax acids
Category and Applications: See Syncrowax ERLC
Form: Solid; Color: Light tan

Synotol 119 N. PVO International
Chemical Description: Cocamide DEA
Category: Detergent, wetting agent, dispersant, thickener, foam booster and stabilizer, lime soap dispersant, nonionic
Applications: Foam builders, viscosity improvers, detergents, shampoos, bubble baths, shaving creams, cleansers, clear gels, germicidal skin cleansers
Form: Liquid
Composition: 50% amide

Synotol CN 60. PVO International
Chemical Description: Cocamide DEA
Category and Applications: See Synotol 119 N
Form: Liquid
Composition: 50% amide

Synotol CN 80. PVO International
Chemical Description: Cocamide DEA
Category and Applications: See Synotol 119 N
Form: Liquid
Composition: 75% amide

Synotol CN 90. PVO International
Chemical Description: Cocamide DEA
Category: Foam booster, wetting agent, thickener, foam stabilizer, superfatting agent

Applications: Household and cosmetic preparations
Form: Liquid
Composition: 85% conc.

Synotol Detergent E. PVO International
Chemical Description: Cocamide DEA
Category and Applications: See Synotol 119 N
Form: Liquid
Composition: 50% amide

Synotol L 60. PVO International
Chemical Description: Lauramide DEA
Category and Applications: See Synotol 119 N
Form: Liquid
Composition: 50% amide

Synotol L 90. Drew Produtos
Chemical Description: Lauric acid diethanolamide
Category and Applications: See Synotol CN 90
Form: Solid
Composition: 88% conc.

Synotol LM 60. PVO International
Chemical Description: Lauramide DEA
Category and Applications: See Synotol 119 N
Form: Liquid
Composition: 50% amide

Synotol LM 90. Drew Productos
Chemical Description: Lauric myristic acid diethanolamide
Category and Applications: See Synotol CN 90
Form: Paste
Composition: 88% conc.

Synperonic 3S25. ICI Petrochemicals Div.
Chemical Description: Sodium salt of Synperonic A3 sulfate
Applications: Widely used in cosmetic formulations such as shampoos, bubble baths, shower gels, etc.
Form: Mobile liquid; Color: Colorless to pale yellow
Properties: Density: 1.040 g/ml (20 C); Surface Tension: 40.2 dynes/cm (0.1%); pH: 7.0–8.5 (5% aq.)
Composition: 24.5–25.5% active

Synperonic 3S60A. ICI Petrochemicals Div.
Chemical Description: Ammonium salt of Synperonic A3 sulfate
Category: Detergent, foaming agent, anionic
Applications: High foaming with good color and low odor for use in household formulations, e.g., dishwash liquids; also used to enhance the solubility of less soluble nonionics and anionics
Form: Clear to hazy mobile liquid; Color: Yellow
Properties: Density: 1.020 g/ml (20 C); Surface Tension: 39.4 dynes/cm (0.1%); pH: 6.5–8.0 (5% aq.)
Composition: 59–61% active

Synperonic 3S60S. ICI Petrochemicals Div.
Chemical Description: Sodium

salt of Synperonic A3 sulfate
Category and Applications: See Synperonic 3S60A
Form: Mobile liquid; Color: Colorless to pale yellow
Properties: Density: 1.046 g/ml (20 C); Surface Tension: 40.5 dynes/cm (0.1%); pH: 7.0–8.5 (5% aq.)
Composition: 59–61% active

Synperonic A2. ICI Petrochemicals Div.
Chemical Description: Synthetic primary alcohol ethoxylate
Category: Detergent, emulsifier, intermediate, nonionic
Applications: Oil-sol. detergents, emulsifiers, and coemulsifiers capable of emulsifying a wide range of oils, waxes, and solvents; bases for toxicant emulsifiers, textile antistatics, scouring agents, wetting agents, and specialty cleaners; intermediate for sulfation, the resultant high foaming anionics used in liquid detergents, shampoos, and bubble baths
Form: Liquid; Color: 50 max. (Hazen)
Properties: Solubility: Sol. in alcohol, glycol ethers, kerosene, mineral oil; Density: 0.897 g/ml; Visc.: 25 cps; F.P.: 7 C; HLB: 5.9; Pour Pt.: 2 C; Hydroxyl Value: 188; pH: 6–8 (1% aq.); Biodegradable
Composition: 99% min. active

Synperonic A3. ICI Petrochemicals Div.
Chemical Description: Synthetic primary alcohol ethoxylate
Category and Applications: See Synperonic A2
Form: Liquid; Color: 50 max. (Hazen)
Properties: Solubility: Sol. in alcohol, glycol ethers, kerosene, mineral oil; Density: 0.918 g/ml; Visc.: 29 cps; F.P.: 12 C; HLB: 7.8; Pour Pt.: 5 C; Hydroxyl Value: 161; pH: 6–8 (1% aq.); Biodegradable
Composition: 99% min. active

Synperonic A4. ICI Petrochemicals Div.
Chemical Description: Synthetic primary alcohol ethoxylate
Category and Applications: See Synperonic A2
Form: Liquid; Color: 50 max. (Hazen)
Properties: Solubility: Sol. in alcohol, glycol ethers, kerosene; sol./disp. in mineral oil; Density: 0.937 g/ml; Visc.: 34 cps; F.P.: 17 C; HLB: 9.1; Pour Pt.: 9 C; Hydroxyl Value: 145; pH: 6–8 (1% aq.); Biodegradable
Composition: 99% min. active

Synperonic A6. ICI Petrochemicals Div.
Chemical Description: Synthetic primary alcohol ethoxylate
Category and Applications: See Synperonic A2
Form: Liquid; Color: 50 max. (Hazen)
Properties: Solubility: Sol. in alcohol, glycol ethers; partly sol. in water and kerosene; disp. in

mineral oil; Density: 0.966 g/ml; Visc.: 46 cps; F.P.: 23 C; HLB: 11.2; Pour Pt.: 17 C; Hydroxyl Value: 118; Surface Tension: 29.8 dynes/cm (0.1%); pH: 6–8 (1% aq.); Biodegradable
Composition: 99% min. active

Synperonic A7. ICI Petrochemicals Div.
Chemical Description: Synthetic primary alcohol ethoxylate
Category: Wetting agent, detergent, leveling agent, emulsifier, nonionic
Applications: Water-sol. wetting agents, general purpose detergents, scouring and dye leveling agents; as emulsifiers or coemulsifiers they cover a wide range of medium polarity oils, solvents, and waxes
Form: Viscous liquid; Color: 50 max. (Hazen)
Properties: Solubility: Sol. in water, alcohol, glycol ethers; disp. in kerosene; Density: 0.977 g/ml; Visc.: 60 cps; F.P.: 26 C; HLB: 12.2; Cloud Pt.: 44–48 C (1% aq.); Pour Pt.: 21 C; Hydroxyl Value: 104; Surface Tension: 30.2 dynes/cm (0.1%); pH: 6–8 (1% aq.); Biodegradable
Composition: 99% min. active

Synperonic A9. ICI Petrochemicals Div.
Chemical Description: Synthetic primary alcohol ethoxylate
Category and Applications: See Synperonic A7
Form: Paste; Color: White
Properties: Solubility: Sol. in water, alcohol, glycol ethers; Density: 0.984 g/ml (40 C); Visc.: 35 cps (40 C); F.P.: 29 C; HLB: 12.5; Cloud Pt.: 62–68 C (1% aq.); Pour Pt.: 24 C; Hydroxyl Value: 93; Surface Tension: 31.4 dynes/cm (0.1%); pH: 6–8 (1% aq.); Biodegradable
Composition: 99% min. active

Synperonic A11. ICI Petrochemicals Div.
Chemical Description: Synthetic primary alcohol ethoxylate
Category and Applications: See Synperonic A7; also as low foam and controlled foam domestic powder detergents
Form: Paste; Color: White
Properties: Solubility: Sol. in water, alcohol, glycol ethers; Density: 0.998 g/ml (40 C); Visc.: 42 cps (40 C); F.P.: 31 C; HLB: 13.9; Cloud Pt.: 85–89 C (1% aq.); Pour Pt.: 27 C; Hydroxyl Value: 81; Surface Tension: 33.2 dynes/cm (0.1%); pH: 6–8 (1% aq.); Biodegradable
Composition: 99% min. active

Synperonic A14. ICI Petrochemicals Div.
Chemical Description: Synthetic primary alcohol ethoxylate
Category: Detergent, solubilizer, dispersant, stabilizer, nonionic
Applications: Water-sol. detergent of low solubility in nonpolar solvents; main uses as detergents, solubilizers, dispersants, and stabilizers, esp. in alkaline formulations

and at high temps.
Form: Wax; Color: White
Properties: Solubility: Sol. in water, alcohol, glycol ethers; Density: 1.014 g/ml (40 C); Visc.: 55 cps (40 C); F.P.: 35 C; HLB: 14.9; Cloud Pt.: > 100 C (1% aq.); Pour Pt.: 30 C; Hydroxyl Value: 68; Surface Tension: 36.5 dynes/cm (0.1%); pH: 6–8 (1% aq.); Biodegradable
Composition: 99% min. active

Synperonic A20. ICI Petrochemicals Div.
Chemical Description: Synthetic primary alcohol ethoxylate
Category and Applications: See Synperonic A14
Form: Hard wax; Color: White
Properties: Solubility: Sol. in water, alcohol, glycol ethers; Density: 1.028 g/ml (50 C); Visc.: 58 cps (50 C); F.P.: 41 C; HLB: 16.2; Cloud Pt.: > 100 C (1% aq.); Pour Pt.: 37 C; Hydroxyl Value: 50; Surface Tension: 40.1 dynes/cm (0.1%); pH: 6–8 (1% aq.); Biodegradable
Composition: 99% min. active

Synperonic BD100. ICI Petrochemicals Div.
Chemical Description: Synthetic primary alcohol ethoxylate with a small amount of propylene oxide in the EO chain
Category: Detergent, emulsifier, nonionic
Applications: Low-foaming, water-sol. detergent liquid at ambient temps. with novel emulsifying properties
Form: Liquid; Color: 150 max. (Hazen)
Properties: Solubility: Sol. in water, alcohol, glycol ethers, kerosene; Density: 0.985 (20 C); Visc.: 92 cps (20 C); F.P.: 11 C; HLB: 12.0; Pour Pt.: 5 C; Surface Tension: 29.7 dynes/cm (0.1%); pH: 6–8 (1% aq.)
Composition: 95% min. active

Synperonic DM Series. ICI Petrochemicals Div.
Applications: Used to promote the rapid separation of water in crude oil emulsions; generally used at the well-head and for refinery desalting

Synperonic K87. ICI Petrochemicals Div.
Chemical Description: Synthetic primary alcohol ethoxylate with small amount of propylene oxide in the EO chain
Category: Detergent, nonionic
Applications: Low pour point, water-sol., low-foaming detergent esp. formulated for textile and raw wool scouring
Form: Liquid; Color: 150 max. (Hazen)
Properties: Solubility: Sol. in water, alcohol, glycol ethers; disp. in kerosene; Density: 0.980 g/ml (20 C); Visc.: 72 cps (20 C); F.P.: 18 C; HLB: 12.0; Cloud Pt.: 44–49 C (1% in 5% w/v isopropanol/water); Pour Pt.: 11 C; Hydroxyl Value: 106; Surface Tension: 29.5 dynes/

cm (0.1%); pH: 6–8 (1% aq.); Biodegradable
Composition: 99% min. active

Synperonic NP4. ICI Petrochemicals Div.
Chemical Description: Nonyl phenol ethoxylate
Category: Detergent, emulsifier, intermediate, nonionic
Applications: Oil-sol. detergent and emulsifier; intermediate for sulfation, the resultant products being functional as anionic detergents, lubricants, and antistatic agents; emulsifying agents (and coemulsifiers compatible with all other surfactants) capable of emulsifying a wide range of oils, waxes, and solvents
Form: Liquid; Color: 150 max. (Hazen)
Properties: Solubility: Sol. in alcohol, glycol ethers, kerosene, mineral oil; Density: 1.022 g/ml (20 C); Visc.: 400 cps (20 C); F.P.: < 0 C; HLB: 8.9; Pour Pt.: < 0 C; Hydroxyl Value: 141; pH: 6–8 (1% aq.)
Composition: 99% min. active
Toxicity: Eye irritant; skin irritant on prolonged contact with concentrates

Synperonic NP5. ICI Petrochemicals Div.
Chemical Description: Nonyl phenol ethoxylate
Category and Applications: See Synperonic NP4
Form: Liquid; Color: 150 max. (Hazen)
Properties: Solubility: Sol. in alcohol, glycol ethers, kerosene, mineral oil; insol./disp. in water; Density: 1.035 g/ml (20 C); Visc.: 350 cps (20 C); F.P.: < 0 C; HLB: 10.5; Pour Pt.: < 0 C; Hydroxyl Value: 130; pH: 6–8 (1% aq.)
Composition: 99% min. active
Toxicity: Eye irritant; skin irritant on prolonged contact with conc. form

Synperonic NP6. ICI Petrochemicals Div.
Chemical Description: Nonyl phenol ethoxylate
Category and Applications: See Synperonic NP4
Form: Liquid; Color: 150 max. (Hazen)
Properties: Solubility: Sol. in alcohol, glycol ethers, mineral oil; sol./disp. in kerosene; disp. in water; Density: 1.041 g/ml (20 C); Visc.: 355 cps (20 C); F.P.: < 0 C; HLB: 10.9; Pour Pt.: < 0 C; Hydroxyl Value: 113; pH: 6–8 (1% aq.)
Composition: 99% min. active
Toxicity: Eye irritant; skin irritant on prolonged contact with conc. form

Synperonic NP8. ICI Petrochemicals Div.
Chemical Description: Nonyl phenol ethoxylate
Category: Detergent, wetting agent, emulsifier, nonionic
Applications: Water-sol. general purpose detergents (esp. high performance detergents for textile

scouring), wetting agents, and emulsifiers (covering a wide range of medium polarity oils and solvents)
Form: Liquid; Color: 150 max. (Hazen)
Properties: Solubility: Sol. in water, alcohol, glycol ethers; disp./sol. in kerosene; Density: 1.053 g/ml (20 C); Visc.: 355 cps (20 C); F.P.: < 0 C; HLB: 12.3; Cloud Pt.: 30–34 C (1% aq.); Pour Pt.: < 0 C; Hydroxyl Value: 95; Surface Tension: 29.4 dynes/cm (0.1%, 20 C); pH: 6–8 (1% aq.)
Composition: 99% min. active
Toxicity: Eye irritant; skin irritant on prolonged contact with conc. form

Synperonic NP9. ICI Petrochemicals Div.
Chemical Description: Nonyl phenol ethoxylate
Category and Applications: See Synperonic NP8
Form: Liquid; Color: 150 max. (Hazen)
Properties: Solubility: Sol. in water, alcohol, glycol ethers; disp. in kerosene; Density: 1.058 g/ml (20 C); Visc.: 340 cps (20 C); F.P.: < 0 C; HLB: 12.8; Cloud Pt.: 51–56 C (1% aq.); Pour Pt.: 0 C; Hydroxyl Value: 89; Surface Tension: 30.6 dynes/cm (0.1%, 20 C); pH: 6–8 (1% aq.)
Composition: 99% min. active
Toxicity: Eye irritant; skin irritant on prolonged contact with conc. form

Synperonic NP10. ICI Petrochemicals Div.
Chemical Description: Nonyl phenol ethoxylate
Category: Detergent, solubilizer, dispersant, stabilizer, nonionic
Applications: Sol. in water and polar solvents but show little or no solubility in nonpolar solvents and oils; used as detergents, detergent additives, solubilizers, dispersants, stabilizers
Form: Liquid; Color: 150 max. (Hazen)
Properties: Solubility: Sol. in water, alcohol, glycol ethers; Density: 1.061 g/ml (20 C); Visc.: 360 cps (20 C); F.P.: 5 C; HLB: 13.3; Cloud Pt.: 62–67 C (1% aq.); Pour Pt.: 5 C; Hydroxyl Value: 85; Surface Tension: 30.6 dynes/cm (0.1%, 20 C); pH: 6–8 (1% aq.)
Composition: 99% min. active
Toxicity: Eye irritant; skin irritant on prolonged contact with conc. form

Synperonic NP12. ICI Petrochemicals Div.
Chemical Description: Nonyl phenol ethoxylate
Category and Applications: See Synperonic NP10
Form: Liquid; Color: 150 max. (Hazen)
Properties: Solubility: Sol. in water, alcohol, glycol ethers; Density: 1.062 g/ml; Visc.: 265 cps; F.P.: 11 C; HLB: 13.9; Cloud Pt.: 79–84 C (1% aq.); Pour Pt.: 14 C; Hydroxyl Value: 76; Surface Ten-

sion: 35.2 dynes/cm (0.1%, 20 C); pH: 6–8 (1% aq.)
Composition: 99% min. active
Toxicity: Eye irritant; skin irritant on prolonged contact with conc. form

Synperonic NP13. ICI Petrochemicals Div.
Chemical Description: Nonyl phenol ethoxylate
Category: Emulsifier, nonionic
Applications: Used in conjunction with an oil-sol. anionic surfactant as emulsifier for a wide range of solvents and agrochemical pesticides and herbicides
Form: Paste; Color: 150 max. (Hazen)
Properties: Solubility: Sol. in water, alcohol, glycol ethers; Density: 1.068; Visc.: 280 cps; F.P.: 14 C; HLB: 14.4; Cloud Pt.: 87–92 C (1% aq.); Pour Pt.: 17 C; Hydroxyl Value: 71; Surface Tension: 34.8 dynes/cm (0.1%, 20 C); pH: 6–8 (1% aq.)
Composition: 99% min. active
Toxicity: Eye irritant; skin irritant on prolonged contact with conc. form

Synperonic NP15. ICI Petrochemicals Div.
Chemical Description: Nonyl phenol ethoxylate
Category and Applications: See Synperonic NP13
Form: Paste; Color: 150 max. (Hazen)
Properties: Solubility: Sol. in water, alcohol, glycol ethers; Density: 1.058 g/ml (40 C); Visc.: 125 cps (40 C); F.P.: 21 C; HLB: 15.0; Cloud Pt.: 95–99 C (1% aq.); Pour Pt.: 21 C; Hydroxyl Value: 64; Surface Tension: 33.4 dynes/cm (0.1%, 20 C); pH: 6–8 (1% aq.)
Composition: 99% min. active
Toxicity: Eye irritant; skin irritant on prolonged contact with conc. form

Synperonic NP20. ICI Petrochemicals Div.
Chemical Description: Nonyl phenol ethoxylate
Category: Solubilizing agent, emulsifier, nonionic
Applications: Solubilizer and emulsifier for highly polar substrates
Form: Solid; Color: 150 max. (Hazen)
Properties: Solubility: Sol. in water, alcohol, glycol ethers; Density: 1.073 g/ml (40 C); Visc.: 168 cps (40 C); F.P.: 30 C; HLB: 16.0; Cloud Pt.: > 100 C (1% aq.); Pour Pt.: 30 C; Hydroxyl Value: 49; Surface Tension: 41.7 dynes/cm (0.1%, 20 C); pH: 6–8 (1% aq.)
Composition: 99% min. active
Toxicity: Eye irritant; skin irritant on prolonged contact with conc. form

Synperonic NP30. ICI Petrochemicals Div.
Chemical Description: Nonyl phenol ethoxylate
Category and Applications: See Synperonic NP20
Form: Solid; Color: 150 max.

(Hazen)
Properties: Solubility: Sol. in water, alcohol, glycol ethers; Density: 1.074 g/ml (50 C); Visc.: 150 cps (50 C); F.P.: 40 C; HLB: 17.1; Cloud Pt.: > 100 C (1% aq.); Pour Pt.: 40 C; Hydroxyl Value: 37; Surface Tension: 42.8 dynes/cm (0.1%, 20 C); pH: 6–8 (1% aq.)
Composition: 99% min. active
Toxicity: Eye irritant; skin irritant on prolonged contact with conc. form

Synperonic OP10. ICI Petrochemicals Div.
Chemical Description: Octyl phenol ethoxylate
Category: Detergent, wetting agent, emulsifier, nonionic
Applications: Water-sol. general purpose detergents, wetting agents, and emulsifiers with good sol'n. properties in the presence of alkalis and at higher temps.
Form: Liquid; Color: 150 max. (Hazen)
Properties: Solubility: Sol. in water, alcohol, glycol ethers; Density: 1.062 g/ml (20 C); Visc.: 393 cps (20 C); F.P.: 7 C; HLB: 13.6; Cloud Pt.: 63–67 C (1% aq.); Pour Pt.: 7 C; Hydroxyl Value: 90; Surface Tension: 31.8 dynes/cm (0.1%, 20 C); pH: 6–8 (1% aq.)
Composition: 99% min. active
Toxicity: Eye irritant; skin irritant on prolonged contact with conc. form

Synperonic OP11. ICI Petrochemicals Div.
Chemical Description: Octyl phenol ethoxylate
Category and Applications: See Synperonic OP10
Form: Liquid; Color: 150 max. (Hazen)
Properties: Solubility: Sol. in water, alcohol, glycol ethers; Density: 1.067 g/ml (20 C); Visc.: 371 cps (20 C); F.P.: 8 C; HLB: 14.0; Cloud Pt.: 79–82 C (1% aq.); Pour Pt.: 8 C; Hydroxyl Value: 87; Surface Tension: 31.0 dynes/cm (0.1%, 20 C); pH: 6–8 (1% aq.)
Composition: 99% min. active
Toxicity: Eye irritant; skin irritant on prolonged contact with conc. form

Synperonic OSD 20. ICI Petrochemicals Div.
Category: Dispersant
Applications: High efficiency, concentrated, low toxicity oil slick dispersant

Synperonic PE30/10. ICI Petrochemicals Div.
Chemical Description: Block copolymer of propylene oxide and ethylene oxide
Category: Defoamer, demulsifier, wetting agent, nonionic
Form: Liquid
Properties: m.w.: ≈ 1900; Sp. gr.: 1.017 (20 C); Visc.: 325 cs; M.P.: < 0 C; HLB: 3; Cloud Pt.: 24 C (1% aq.); pH: 6–8 (2.5%)
Composition: 99% min. active
Containers: 210 liter drum or bulk
Warnings and Caution: Avoid contact with eye and prolonged contact with skin

Synperonic PE30/20. ICI Petrochemicals Div.
Chemical Description: Block copolymer of propylene oxide and ethylene oxide
Category: Wetting and dispersing agent, demulsifier, nonionic
Applications: Wetting, dispersing, demulsifier, and as a rinse aid additive for machine dishwashing
Form: Liquid
Properties: m.w.: ≈ 2500; Sp. gr.: 1.032 (20 C); Visc.: 420 cs; M.P.: < 0 C; HLB: 7; Cloud Pt.: 32 C (1% aq.); Surface Tension: 42 dynes/cm (0.1%, 20 C); pH: 6–8 (2.5%)
Composition: 99% min. active
Containers: 210 liter drums or bulk
Warnings and Caution: Avoid contact with eye and prolonged contact with skin

Synperonic PE30/40. ICI Petrochemicals Div.
Chemical Description: Block copolymer of propylene oxide and ethylene oxide
Category: Detergent, wetting agent, dispersant, nonionic
Form: Viscous liquid
Properties: m.w.: ≈ 2900; Sp. gr.: 1.05 (20 C); Visc.: 600 cs; M.P.: < 0 C; HLB: 15; Cloud Pt.: 58 C (1% aq.); Surface Tension: 41.1 dynes/cm (0.1%, 20 C); pH: 6–8 (2.5%)
Composition: 99% active ingred.
Containers: 210 liter drums or bulk
Warnings and Caution: Avoid contact with eyes and prolonged contact with skin

Synperonic PE30/80. ICI Petrochemicals Div.
Chemical Description: Block copolymer of propylene oxide and ethylene oxide
Category: Emulsifier, dispersant, stabilizer, nonionic
Applications: Emulsifier/stabilizer in emulsion polymerization processes; dispersant/stabilizer for pigments in emulsion paints, inks, etc.
Form: Flake
Properties: m.w.: ≈ 8500; M.P.: ≈ 50 C; HLB: 29; Cloud Pt.: > 100 C (1% aq.); Surface Tension: 48 dynes/cm (0.1%, 20 C); pH: 6–8 (2.5%)
Composition: 99% min. active
Containers: 100 kg net fiberboard kegs

Synperonic PE39/70. ICI Petrochemicals Div.
Chemical Description: Block copolymer of propylene oxide and ethylene oxide
Category: Emulsifier, dispersant, nonionic
Form: Flake; Color: White
Properties: m.w.: ≈ 7500; M.P.: ≈ 49 C; HLB: 24; Cloud Pt.: > 100 C (1% aq.); Surface Tension: 45.2 dynes/cm (0.1%, 20 C); pH: 6–8 (2.5%)
Composition: 99% min. active
Containers: 50 kg net fiberboard kegs

Synperonic SC Series. ICI Petrochemicals Div.
Chemical Description: Blend of nonionics
Applications: Facilitates the rapid emulsification into water of heavy wax deposits in crude and fuel oil storage tanks

Synperonic TC 26. ICI Petrochemicals Div.
Chemical Description: Blend of nonionics in solvent
Category: Detergent, degreaser, nonionic
Applications: Solvent-based water-emulsifiable tank cleaner and degreaser which will give fast separation of emulsified oil from water under static conditions
Form: Liquid
Composition: 90% conc.

Synperonic TC 30C. ICI Petrochemicals Div.
Chemical Description: Concentrated blend of nonionics in an oxygenated aliphatic coupling solvent
Category: Degreaser, nonionic
Applications: Used with other solvents (e.g., kerosene, fresh water, salt water) for the removal and emulsification of hydrophobic (oily) soils; tank and drum cleaning, metal and engine degreasing, concrete degreasing; ideal for marine, refinery, or off-shore oil rig cleaning
Form: Liquid
Composition: 100% conc.

Synprol Alcohol. ICI Petrochemicals Div.
Chemical Description: Fully saturated synthetic primary alcohol (67% C_{13} and 33% C_{15})
Category: Chemical intermediate
Applications: Intermediate for the production of ethoxylates, sulfates, and ether sulfates
Properties: Ref. Index: 1.4457; Sp. gr.: 0.837 (20/20 C); Visc.: 29.1 cps (20 C); F.P.: $\approx$ 14 C; Flash Pt.: 290 F (PMCC); Acid Value: $<$ 1; Hydroxyl Value: 268
Composition: 99% active ingred.

Synprol Sulphate. ICI Petrochemicals Div.
Applications: Usually used as the sodium salt; used in cosmetic field for bubble baths, shampoos, etc.
Reaction: Neutral

Synprolam 35. ICI Petrochemicals Div.
Chemical Description: Synthetic primary amine
Category: Anticaking agent, flotation agent, corrosion inhibitor, dispersant, emulsifier, intermediate
Applications: Fertilizer anticaking, mineral flotation, corrosion inhibitors, metal cleaning formulations, pigment dispersions, rubber processing auxiliaries, bitumen emulsifiers; intermediate for the production of amine salts, amine ethoxylates, and sulfosuccinamates
Form: Clear liquid; Color: 50 (Hazen)

Properties: Sp. gr.: 0.8 (20/20 C); Visc.: 4.5 cps (20 C); F.P.: ≈ 12 C; Flash Pt.: 127 C (PMCC); Iodine No.: 2
Composition: 99% primary amine
Reaction: Strongly alkaline
Warnings and Caution: Avoid contact with skin and eyes; must not be ingested

Synprolam 35DM. ICI Petrochemicals Div.
Chemical Description: Synthetic tertiary fatty amine
Category: Chemical intermediate
Applications: Intermediate for the production of quaternary salts with high biocidal activity, and amine oxides (used in cosmetics production); also used as catalysts for polyurethane foam and as corrosion inhibitors
Form: Clear liquid; Color: 50 (Hazen)
Properties: Sp. gr.: 0.79 (20/20 C); Visc.: 3.9 cps (20 C); F.P.: –21 C; Flash Pt.: 143 C (PMCC); Iodine No.: 2
Composition: 97% tertiary amine
Reaction: Strongly alkaline
Warnings and Caution: Avoid contact with skin and eyes; must not be ingested

Synprolam 35DMO. ICI Petrochemicals Div.
Chemical Description: 30% aq. sol'n. of alkyl dimethyl amine oxide based on Synprolam 35DM
Category: Foam booster and stabilizer, thickening agent, mildly cationic
Applications: Foam booster/stabilizer and thickening aid in shampoos and bubble baths; substantive to hair; considerable surface activity (sol'ns. in water have lower surface and interfacial tensions with liquid paraffin than the coconut type amine oxides)
Form: Clear liquid; Color: 50 (Hazen)
Properties: Sp. gr.: 0.97 (20/20 C); F.P.: 0 C; Surface Tension: 29 dynes/cm (0.1% aq.)
Composition: 30% active ingred.
Toxicity: Moderately severe skin irritant and very severe eye irritant
Containers: Store in stainless steel, glass-lined, reinforced polyester, or polypropylene containers
Warnings and Caution: Wear protective gloves, goggles, and clothing

Synprolam 35X Series. ICI Petrochemicals Div.
Chemical Description: Amine ethoxylates
Category: Emulsifier, stabilizer
Applications: Emulsifying agents and emulsion stabilizers

Syntergent DIC Series. Diamond Shamrock Europe/Process Chem. Div.
Chemical Description: Various nonionic and nonionic/penetrant blends
Category: Deinking agent, mainly nonionic
Applications: Deinking agents for waste paper
Form: Liquids
Composition: 100% conc.

Syntergent TER-1. Diamond Shamrock Europe/Process Chem. Div.
Chemical Description: Phosphate ester
Category: Wetting agent, dispersant, anionic
Applications: Low-foam wetting agent, general purpose textile surfactant, calcium sulfate dispersant
Form: Liquid
Composition: 82% conc.

Synthrapol KB. ICI United States
Chemical Description: Ethylene oxide condensate of aliphatic alcohol
Category: Wetting agent, rewetting agent, detergent, nonionic
Applications: Rapid wetting at low concs. for textile processing, esp. where complete penetration of fabric or high speed operation is important; used in bleaching, enzyme desizing, dyeing, finishing, rewetting treated fabrics, removal of soil, aftersoaping of dyed and printed fabrics, and in processing of natural fabrics; also increases stability of application baths of fluorochemical oil and water repellents
Form: Clear liquid; Color: Nearly coloress; Odor: Alcohol
Properties: Solubility: Very sol. in cold water; Sp. gr.: ≈ 1.0; Cloud Pt.: ≈ 40 C; Stability: Unaffected by hard water and metal salts such as calcium and magnesium; pH: 5.5–7.0 (1% sol'n.)
Composition: 96 ± 2% active
Reaction: Neutral
References: Tech. bulletin 269R

Synthrapol N. ICI United States
Chemical Description: Aq. sol'n. of an alcohol ethoxylate
Category: Detergent, wetting agent, emulsifier, penetrant/dyeing assistant, nonionic
Applications: Raw wool scouring and scouring of fabrics to remove dirt, lubricants, and sizes; desizing; kier boiling; graphite removal; penetrant/dyeing assistant for packaged and skin-dyed acrylics; good cold water agent
Form: Liquid; Color: Nearly colorless; Odor: Mild, pleasant
Properties: Sp. gr.: ≈ 1.0; F.P.: ≈ 0 C; B.P.: ≈ 100 C; pH: 5.5–7.0
Composition: 26% conc.
Reaction: Neutral
Toxicity: May cause eye irritation
References: Tech. bulletin 298R

Synthrapol SP. ICI United States
Chemical Description: Blend of nonionic and anionic detergents
Category: Detergent, emulsifier, nonionic/anionic
Applications: Prescouring of textiles to remove oils, waxes, and lubricants; postscouring for removal of unfixed dyestuff; used for synthetic and natural fibers; compatible with electrolytes, acids, hard water
Form: Liquid; Color: Pale yellow; Odor: Faint
Properties: Solubility: Readily sol. in hot or cold, hard or soft water; Sp. gr.: ≈ 1.0; F.P.: ≈ 0 C; B.P.: ≈ 100 C; pH: ≈ 6.5 (2%)
Composition: ≈ 35% active ingred.
Toxicity: In conc. form, may cause

eye irritation and skin burns
Warnings and Caution: Wear protective goggles and gloves
References: Tech. bulletin 256R

T

Tairygent CA-1. Formosa Chem. & Fibre Corp.
Chemical Description: Linear dodecyl benzene sulfonic acid
Form: Paste; Color: Brown
Composition: 96% min. active

Tairygent CA-2. Formosa Chem. & Fibre Corp.
Chemical Description: Branched dodecyl benzene sulfonic acid
Form: Paste; Color: Brown
Composition: 96% min. active

Tairygent CB-1. Formosa Chem. & Fibre Corp.
Chemical Description: Linear sodium dodecyl benzene sulfonate
Form: Paste; Color: White
Properties: pH: 7 ± 1 (1% sol'n.)
Composition: 60 ± 2% active

Tairygent CB-2. Formosa Chem. & Fibre Corp.
Chemical Description: Branched sodium dodecyl benzene sulfonate
Form: Paste; Color: White
Properties: pH: 7 ± 1 (1% sol'n.)
Composition: 60 ± 2% active

Tamol 731-25%. Rohm and Haas Co.
Chemical Description: Sodium salt of carboxylated polyelectrolyte
Category: Dispersant, anionic
Applications: Dispersant for dyes and pigments
Form: Liquid; Color: Light yellow
Properties: Density: 9.2 lb/gal; Visc.: 200 cps; F.P.: 28 F
Composition: 25% active ingred.

Tamol 731-SD. Rohm and Haas Co.
Chemical Description: Sodium salt of carboxylated polyelectrolyte
Category and Applications: See Tamol 731-25%
Form: Powder; Color: Tan
Properties: Density: 24 lb/ft^3 (bulk); Surface Tension: 36 dynes/cm (1%)
Composition: 93% active ingred.

Tamol 850. Rohm and Haas Co.
Chemical Description: Sodium salt of carboxylated polyelectrolyte

Category: Dispersant, anionic
Applications: Low-foaming dispersant for inorganic pigments
Form: Liquid; Color: Light yellow
Properties: Density: 10.1 lb/gal; Visc.: 345 cps; F.P.: 23 F; Surface Tension: 71 dynes/cm (1%)
Composition: 30% active ingred.

Tamol 901. Rohm and Haas Co.
Chemical Description: Ammonium salt of carboxylated polyelectrolyte
Category: Dispersant, anionic
Form: Liquid; Color: Colorless
Properties: Density: 10 lb/gal; Visc.: 100–400 cps; F.P.: 23 F
Composition: 30% active ingred.

Tamol 960. Rohm and Haas Co.
Chemical Description: Sodium salt of carboxylated polyelectrolyte
Category: Dispersant, anionic
Form: Liquid; Color: 125 (APHA)
Properties: Density: 10.6 lb/gal; Visc.: 550 cps
Composition: 40% active ingred.

Tamol L. Conc. Rohm and Haas Co.
Chemical Description: Sodium salt of condensed naphthalenesulfonic acid
Category: Dispersant, anionic
Applications: FDA cleared for food contact uses
Form: Liquid; Color: Amber
Properties: Density: 10.4 lb/gal; Visc.: 20 cps; F.P.: 23 F; Surface Tension: 71 dynes/cm (1%)
Composition: 47.5% active ingred.

Tamol NN. BASF AG
Chemical Description: 30% aq. sol'n. of Tamol NNOK
Category: Stabilizer, dispersant, precipitant, grinding aid, anionic
Applications: Stabilizer for aq. dispersions and emulsions; grind-aids; dispersant for pigments; precipitant for basic dyes and cationic compounds; dispersing and plastifying agents for concrete; auxiliaries in metal finishing
Form: Clear liquid; Color: Pale brown
Properties: Stability: Highly resistant to acids, alkalis, and salts; pH: 9.5–10.5 (10% aq.)
Composition: 33% active ingred.
Toxicity: Mucous membrane irritant
Warnings and Caution: Protect against frost
References: Tech. leaflet M 1750 e

Tamol NNO. BASF AG
Chemical Description: Sodium salt of a naphthalenesulfonic acid condensation product
Category and Applications: See Tamol NN
Form: Powder; Color: Pale brown
Properties: Stability: Highly resistant to acids, alkalis, and salts; pH: 6.5–7.5 (10% aq.)
Composition: 95% active ingred.
Toxicity: Mucous membrane irritant
References: Tech. leaflet M 1750 e

Tamol NNOK. BASF AG

Chemical Description: Sodium salt of a naphthalenesulfonic acid condensation product
Category and Applications: See Tamol NN
Form: Powder; Color: Pale brown
Properties: Stability: Highly resistant to alkalis, acids, and salts; pH: 9.5–10.5 (10% aq.)
Composition: 95% active ingred.
Toxicity: Mucous membrane irritant
References: Tech. leaflet M 1750 e

Tamol PA Liquid. BASF AG

Chemical Description: Sodium salt of a polycarboxylic acid
Category: Dispersant, solubilizer, protective colloid, anionic
Applications: Chemical and chemical downstream industries; used in water treatment and as builders for detergents that act as soil dispersers, dispersing agents, and complexing agents for calcium and magnesium salts; also in car shampoos; floor, plastics, and metal cleaners; detergents and clear rinses for glass and ceramics
Form: Clear liquid; Color: Slightly yellowish
Properties: Solubility: Readily miscible at low and high temps. with soft and hard water and aq. sol'ns. of acids, alkalis, and salts; Density: 1.20–1.25 g/cc (20 C); Visc.: 80–250 cps; pH: 6–8 (undiluted)
Composition: 35 ± 1% solids
References: TI/P 2528 e

Tamol PA Powder. BASF AG

Chemical Description: Sodium salt of a polycarboxylic acid
Category and Applications: See Tamol PA Liquid
Form: Fine powder; Color: White to yellowish
Properties: Solubility: Readily miscible at low and high temp. with soft and hard water and aq. sol'ns. of acids, alkalis, and salts; sparingly sol. in organic solvents; Density: ≈ 470 g/l (bulk); pH: 6.5–8.0 (20% aq.)
Composition: > 91% conc.
References: TI/P 2528 e

Tamol S. BASF AG

Chemical Description: Sodium salt of a polycarboxylic acid
Category: Dispersant, anionic
Applications: Dispersant, esp. for water treatment
Form: Liquid, powder
Composition: 45% conc. (liquid); 91% conc. (powder)

Tamol SG-1. Rohm and Haas Co.

Chemical Description: Ammonium salt of carboxylated polyelectrolyte
Category: Dispersant, anionic
Form: Liquid; Color: Colorless
Properties: Density: 9.6 lb/gal; Visc.: 260 cps; F.P.: 32 F
Composition: 35% active ingred.

Tamol SN. Rohm and Haas Co.
Chemical Description: Sodium salt of condensed naphthalene-sulfonic acid
Category: Dispersant, anionic
Applications: Dispersant for pigments, carbon black, clay, dyestuffs, etc.; cleared for food contact and agricultural uses
Form: Powder; Color: Tan
Properties: Density: 36 lb/ft^3 (bulk); Flash Pt.: 300 F (TOC); Surface Tension: 71 dynes/cm (1%)
Composition: 93% active ingred.

Tanapon INF. Tanatex Chem. Co.
Category: Detergent, wetting agent, dispersant, emulsifier, nonionic
Applications: Nonfoaming detergent for prescouring textiles for removal of dirt and oils on the Jet dyeing machine, package machines, beams, and continuous washers
Form: Viscous emulsion; Color: White; Odor: Mild
Properties: Density: 8.6 lb/gal; Sp. gr.: 1.03
Reaction: Slightly alkaline
Warnings and Caution: Avoid contact with skin and eyes; avoid ingestion and prolonged inhalation of vapors
References: TB: 061501-2

Tanapon LTS. Tanatex Chem. Co.
Category: Detergent, emulsifier, wetting agent, penetrant, modified nonionic
Applications: Textile scouring, effective at low temps.; emulsifies oils and waxes; rapid wetting penetration aids removal of dirt and graphite; used in desizing, prescouring, and afterscouring in batch or continuous processes
Form: Liquid; Color: Yellow; Odor: Mild
Properties: Solubility: Dissolves readily in cold or warm water; Density: 8.6 lb/gal; Sp. gr.: 1.03; Stability: Stable in alkaline sol'ns. containing up to 7.5 g/l caustic soda; compatible with enzymes, weak acids, sodium hydrosulfite; Biodegradable
Reaction: Essentially neutral
Warnings and Caution: Avoid contact with skin and eyes; avoid ingestion and prolonged inhalation of vapors
References: TB: A060704-1

Tanapon X-70. Tanatex Chem. Co.
Category: Detergent, wetting agent, penetrant, modified nonionic
Applications: Textile prescouring, afterscouring, and desizing in batch or continuous processes; emulsifies oils and waxes; rapid wetting and penetration aids removal of dirt and graphite; high cloud point
Form: Liquid; Color: Amber; Odor: Mild
Properties: Solubility: Dissolves readily in cold or warm water; Density: 8.3 lb/gal; Sp. gr.: 1.0; Cloud Pt.: $>$ 212 F; Stability: Stable in alkaline sol'ns. containing up to 7.5 g/l caustic soda; compat-

ible with weak acids and sodium hydrosulfite; Biodegradable
Reaction: Slightly alkaline
Warnings and Caution: Avoid contact with skin and eyes; avoid ingestion and prolonged inhalation of vapors
References: TB: 060700-5

Tanaterge 627. Tanatex Chem. Co.
Category: Leveling, scouring, and wetting agent, anionic
Applications: Prescouring and afterscouring agent; leveling agent for disperse dyes; helps remove synthetic sizing material; resin stripping before repairs; continuous and batch operations
Form: Liquid; Color: Pale yellow; Odor: Mild
Properties: Solubility: May be added directly to cold or warm bath; Density: 8.6 lb/gal; Sp. gr.: 1.03; Stability: Stable @ pH 2–12
Reaction: Very slightly acidic
Warnings and Caution: Avoid contact with skin and eyes; avoid ingestion and prolonged inhalation of vapors
References: TB: 060171-2

Tanaterge D-6. Tanatex Chem. Co.
Category: Wetting agent, emulsifier, dyebath auxiliary, nonionic
Applications: Compatibilizer for anionic and cationic dyes; prevents scumming in cationic dyeing of acrylics, modacrylics, and cationic dyeable polyester; emulsifies residual yarn lubricants; thorough wetting and penetration
Form: Liquid; Color: Light yellow; Odor: Mild
Properties: Solubility: Readily sol. in water; Density: 8.3 lb/gal; Sp. gr.: 1.0; Stability: Incompatible with strong acids and alkalis
Reaction: Essentially neutral
Warnings and Caution: Avoid contact with skin and eyes; avoid ingestion and prolonged inhalation of vapors
References: TB: 110300-3

Tanaterge ESS. Tanatex Chem. Co.
Category: Dispersant, detergent, anionic
Applications: Removes WD size prior to dyeing, and processing oils from the goods so that a separate prescour for oil removal is eliminated
Form: Liquid; Color: Yellow
Properties: Density: 8.7 lb/gal; Sp. gr.: 1.04
Reaction: Essentially neutral
References: PB: 620701-2

Tanaterge FTD. Tanatex Chem. Co.
Chemical Description: Phosphated derivative
Category: Afterwashing surfactant, modified anionic
Applications: For continuous afterwashing of dyed or printed polyesters and their blends; clears disperse dye stain from cellulosic fibers without hydrosulfite treatment; removes and suspends unfixed dyes
Form: Liquid; Color: Straw; Odor: Mild
Properties: Solubility: Dissolves readily in warm water; Density:

9.1 lb/gal; Sp. gr.: 1.1; Stability: Compatible with alkalis and most other chemicals used in textile wet processing
Composition: 67% conc.
Reaction: Essentially neutral
Warnings and Caution: Avoid contact with skin and eyes; avoid ingestion and prolonged inhalation of vapors
References: TB: 066700-2

Tanaterge LV Conc. Tanatex Chem. Co.
Category: Wetting agent, penetrant, scouring agent, nonionic
Applications: Wetting, penetrating, and scouring agent for 100% cotton and polyester/cotton; rapidly wets out goods for improved bleaching efficiency; afterscour to improve crockfastness; used in all package machines, jigs, and continuous equipment
Form: Liquid; Color: Water-white; Odor: Mild
Properties: Density: 8.3 lb/gal; Sp. gr.: 1.0; Stability: Stable to strong acids, alkalis, peroxide, and 5% caustic sol'n.
Composition: 50% active ingred.
Reaction: Neutral
Warnings and Caution: Avoid contact with skin and eyes; avoid ingestion and prolonged inhalation of vapors
References: TB: 060709-2

Tanaterge RK. Tanatex Chem. Co.
Category: Wetting agent, nonionic/anionic
Applications: Wetting agent for bleaching/scouring systems, esp. designed for continuous operations
Form: Liquid; Color: Water-white; Odor: Mild
Properties: Solubility: Dissolves in cold or warm water; Density: 8.4 lb/gal; Sp. gr.: 1.0; Stability: Stable in 5% caustic sol'n.
Reaction: Neutral
Warnings and Caution: Avoid contact with skin and eyes; avoid ingestion and prolonged inhalation of vapors
References: PTB: 060745-1

Tanaterge SFW. Tanatex Chem. Co.
Category: Wetting and scouring agent, nonionic
Applications: Rapid and uniform wetting for removing oil and dirt and prevents their redeposition; low-foaming; esp. effective in the wet-out of nylon carpet in continuous operations; also for scouring cellulosics, polyester, and nylon in batch operations
Form: Liquid; Color: Water-white; Odor: Mild
Properties: Density: 8.1 lb/gal; Sp. gr.: 0.97
Reaction: Neutral
Warnings and Caution: Avoid contact with skin and eyes; avoid ingestion and prolonged inhalation of vapors
References: TB: 083406-3

Tandem 5K. ICI United States
Chemical Description: Mono- and diglycerides with 0.02% BHA and 0.01% citric acid (60%) and polysorbate 60 (40%), kosher grade

Category: Emulsifier, conditioner, softener, nonionic
Applications: Food emulsifier; conditioner/softener for yeast-raised baked goods
Form: Plastic solid; Color: Creamy white
Properties: Solubility: Sol. above its melting pt. at low levels in vegetable oils; disp. in water; M.P.: ≈ 128 F; HLB: 8.1; Flash Pt.: > 300 F; Fire Pt.: > 300 F; Iodine Value: ≈ 30
Composition: 100% conc.

Tandem 8. ICI United States
Chemical Description: Mono- and diglycerides with 0.02% BHA and 0.01% citric acid (60%) and polysorbate 60 (40%)
Category and Applications: See Tandem 5K
Form: Plastic solid; Color: Creamy white
Properties: Solubility: Sol. above its melting pt. at low levels in vegetable oils; disp. in water; M.P.: 128 F; HLB: 8.1; Flash Pt.: > 300 F; Fire Pt.: > 300 F; Iodine Value: ≈ 23
Composition: 100% conc.

Tandem 9. ICI United States
Chemical Description: Mono- and diglycerides and polysorbate 60 (25%)
Category and Applications: See Tandem 5K
Form: Flakes; Color: Creamy white; Odor and Taste: Bland
Properties: Solubility: Disp. above its melting pt. in water and vegetable oils; M.P.: ≈ 135 F; HLB: 6.0; Flash Pt.: > 300 F; Fire Pt.: > 300 F; Iodine Value: ≈ 1
Composition: 100% conc.

Tandem 11H. ICI United States
Chemical Description: Hydrated mono- and diglycerides and polysorbate 60 (12.5%) with sodium propionate and lactic acid as preservatives
Category and Applications: See Tandem 5K
Form: Soft plastic; Color: Creamy white
Properties: Solubility: Disp. in water and vegetable oils; HLB: 6.3; Flash Pt.: > 300 F; Fire Pt.: > 300 F; Iodine Value: ≈ 2; pH: ≈ 5
Composition: 50% conc.

Tandem 552. ICI United States
Chemical Description: Mono- and diglycerides with 0.02% BHA and 0.01% citric acid with polysorbate 60 (36%), and water
Category and Applications: See Tandem 5K
Form: Clear liquid; Color: Golden
Properties: Solubility: Sol. at low levels in vegetable oils; disp. in water; M.P.: ≈ 45 F; HLB: 8.1; Flash Pt.: > 300 F; Fire Pt.: > 300 F; Iodine Value: ≈ 39
Composition: 100% conc.
Patent: U.S. 3,785,993

TC 9754. Pro-Chem, Inc.
Category: Detergent, rust inhibitor
Applications: Used to simultane-

ously clean steel and inhibit rust; used as a replacement for solvents; also for household, street, buildings, hospital cleaners, making of cutting and soluble oils
Form: Liquid; Color: Water-white; Odor: Nil
Properties: Solubility: Sol. in water @ 0–95 C; Density: 8.7 lb/gal; Sp. gr.: 1.04; Visc.: 2 cps (room temp.); F.P.: –1 C; B.P.: 220 C; Stability: Very good; Surface Tension: ≈ 33 dynes/cm (0.1%); pH: 9.5
Composition: 20% active ingred.
Reaction: Alkaline
Toxicity: May cause eye sting; drying to skin
Containers: Steel drums

T-Det A-026. Thompson-Hayward Chem. Co.
Chemical Description: Reaction product of ethylene oxide and a blend of straight-chain primary alcohols (C_{10}–C_{12})
Category: Wetting agent, detergent, nonionic
Applications: Textile processing; low-foaming surfactant for formulated detergents for laundry use and for home and industrial general purpose cleaners
Form: Clear liquid; Color: 100 max. (APHA)
Properties: Solubility: Sol. in water, ethanol, glycol ether, HAN, isopropanol, xylene; slightly sol. in kerosene and trichlorethylene; Density: 8.2 lb/gal (60/60 F); Sp. gr.: 0.98 (60/60 F); Visc.: 21 cps (100 F); F.P.: ≈ 41 F; HLB: 13.5; Cloud Pt.: ≈ 36 C (1%); Flash Pt.: 250 F (OC); Stability: Stable at normal use concs. with most oxidizing or bleaching agents used in the textile industry; compatible with acids, alkalis, and other builders; pH: 6–8 (1%); Biodegradable
Composition: 100% conc.
Warnings and Caution: Do not use with high concs. of strong oxidizing agents—potentially explosive

T-Det C-40. Thompson-Hayward Chem. Co.
Chemical Description: Reaction product of castor oil with 40 moles of ethylene oxide
Category: Emulsifier, solubilizer, degreaser, lubricant, dispersant, leveling agent, defoamer, stabilizer, penetrant, nonionic
Applications: Used in fat liquoring, dyeing, and degreasing of leather; metal processing, soluble maching oils; paints for dispersing pigment slurries, leveling, defoaming, stabilizing; paper manufacture; wax and polish preparations; textile processing (antistat, dyeing assistant, kier boiling, softeners); mimeograph inks; rubber lubricants
Form: Liquid; Color: Yellow; Odor: Mild, oily
Properties: Solubility: Sol. in acetone, butyl Cellosolve, carbon tetrachloride, ethanol, ether, ethylene glycol, methanol, vegetable oil, water, xylene; Density: ≈ 8.7 lb/gal; Sp. gr.: ≈ 1.05; HLB: 14.2; Flash Pt.: > 200 F (TOC); Pour

Pt.: ≈ 60 F; Stability: Moderately stable to acids, bases, and salts; pH: ≈ 6.0 (1%)
Composition: 99.5% min. active
Containers: 470 lb net steel closed head drum

T-Det DD-5. Thompson-Hayward Chem. Co.
Chemical Description: Reaction product of dodecyl phenol with 5 moles of ethylene oxide
Category: Emulsifier, nonionic
Applications: Oil-sol. surfactant with excellent emulsifying properties for aromatic and aliphatic hydrocarbon solvents; effective dry cleaning soap additive
Form: Liquid; Color: Pale yellow; Odor: Mild, aromatic
Properties: Solubility: Oil-sol.; Density: ≈ 8.42 lb/gal (20 C); Sp. gr.: ≈ 1.01 (20 C); HLB: 9.6; Flash Pt.: > 300 F; Pour Pt.: < 0 F; Stability: Stable to acids, bases, and salts; pH: ≈ 7 (1% sol'n.)
Composition: 99.5% min. active
Containers: 460 lb net drums and bulk
Warnings and Caution: Do not mix with conc. oxidizing or reducing agents—potentially explosive

T-Det DD-7. Thompson-Hayward Chem. Co.
Chemical Description: Reaction product of dodecyl phenol with 7 moles of ethylene oxide
Category: Emulsifier, nonionic
Applications: Water-disp. surfactant with excellent emulsifying properties for certain solvents; solvent and emulsion cleaners
Form: Liquid; Color: Pale yellow; Odor: Mild, aromatic
Properties: Solubility: Disp. in water; Density: ≈ 8.66 lb/gal (20 C); Sp. gr.: ≈ 1.04 (20 C); HLB: 11.1; Flash Pt.: > 300 F; Pour Pt.: < 0 F; Stability: Stable to acids, bases, and salts; pH: ≈ 7 (1% sol'n.)
Composition: 99.5% min. active
Containers: 460 lb net drums and bulk
Warnings and Caution: Do not mix with conc. oxidizing or reducing agents—potentially explosive

T-Det DD-9. Thompson-Hayward Chem. Co.
Chemical Description: Reaction product of dodecyl phenol with 9 moles of ethylene oxide
Category: Detergent, emulsifier, wetting agent, nonionic
Applications: Low-foaming surfactant
Form: Liquid; Color: Pale yellow; Odor: Mild, aromatic
Properties: Solubility: Disp. in water; sol. in many aromatic hydrocarbons; Density: ≈ 8.7 lb/gal (20 C); Sp. gr.: ≈ 1.05 (20 C); HLB: 12.0; Flash Pt.: > 300 F; Pour Pt.: ≈ 42 F; Stability: Stable to acids, bases, and salts; pH: ≈ 7 (1% sol'n.)
Composition: 99.5% min. active
Containers: 460 lb net drums and bulk
Warnings and Caution: Do not mix with conc. oxidizing or reducing agents—potentially explosive

T-Det DD-10. Thompson-Hayward Chem. Co.
Chemical Description: Reaction product of dodecyl phenol with 10 moles of ethylene oxide
Category: Emulsifier, nonionic
Applications: Water-disp. surfactant with excellent emulsifying properties for certain solvents
Form: Liquid; Color: Pale yellow; Odor: Mild, aromatic
Properties: Solubility: Water-disp.; Density: ≈ 8.75 lb/gal (20 C); Sp. gr.: ≈ 1.05 (20 C); Cloud Pt.: ≈ 112 F (1% sol'n.); Flash Pt.: > 300 F; Pour Pt.: ≈ 48 F; Stability: Stable to acids, bases, and salts; pH: ≈ 7 (1% sol'n.)
Composition: 99.5% min. active
Containers: 460 lb net drums and bulk
Warnings and Caution: Do not mix with conc. oxidizing or reducing agents—potentially explosive

T-Det EPO-61. Thompson-Hayward Chem. Co.
Chemical Description: Reaction product of ethylene oxide and propylene oxide
Category: Wetting agent, emulsifier, defoaming and foam control agent, nonionic
Applications: Crude oil demulsifier; foam control in paper manufacture; water treating compounds; emulsifier and wetting agent in agricultural toxicant formulations
Form: Essentially clear liquid
Properties: Solubility: Sol. in ethanol, toluene, xylene, perchlorethylene; Density: 8.34 lb/gal; Sp. gr.: ≈ 1.01; Visc.: ≈ 271 cps; Cloud Pt.: ≈ 23 C (1%); Flash Pt.: > 200 C; Pour Pt.: –28 C; Stability: Stable to acids, bases, and salts
Composition: 99.5% min. active
Containers: 450 lb net closed head steel drums
Warnings and Caution: Do not mix with conc. oxidizing or reducing agents—potentially explosive

T-Det N-4. Thompson-Hayward Chem. Co.
Chemical Description: Reaction product of nonylphenol with 4 moles of ethylene oxide
Category: Detergent, emulsifier, wetting agent, intermediate, stabilizer, nonionic
Applications: Agricultural toxicant formulations; intermediate for conversion to the sulfate ester used in household detergents; dry cleaning "soap" additive for chlorinated and petroleum-based solvents; leather processing; solvent cleaners for metal processing; color development aid, stabilizer, and emulsifier for latex paints; gasoline additive; sludge dispersant additive in fuel oils
Form: Oily liquid; Color: Pale yellow; Odor: Mild, aromatic
Properties: Solubility: Sol. in butyl Cellosolve, carbon tetrachloride, corn oil, ethanol, kerosene, mineral oil, mineral spirits, perchloroethylene, Stoddard solvent, toluene; Density: 8.5 lb/gal; Sp. gr.: ≈ 1.02; Visc.: ≈ 52 SUS (210 F); HLB: 8.9; Flash Pt.: > 400 F (OC); Pour Pt.: < 0 F; Stability:

Stable to acids, bases, salts; pH: ≈ 7
Composition: 99.5% min. active
Containers: 460 lb net closed head steel drums
Warnings and Caution: Do not mix with conc. oxidizing or reducing agents–potentially explosive

T-Det N-6. Thompson-Hayward Chem. Co.
Chemical Description: Reaction product of nonylphenol with 6 moles of ethylene oxide
Category: Emulsifier, wetting agent, detergent, dispersant, nonionic
Applications: Agricultural toxicant formulations; dry cleaning "soap" additive for chlorinated and petroleum-based solvents; emulsifier for silicones and household waxes and polishes; fat liquoring in leather processing; metal drawing and forming compounds; solvent cleaners; deinking agent for paper; detergent and dispersant for petroleum oils and fuel oils
Form: Liquid; Color: Pale yellow; Odor: Mild, aromatic
Properties: Solubility: Sol. in butyl Cellosolve, carbon tetrachloride, corn oil, diesel fuel, ethanol, kerosene, Stoddard solvent, xylene; Density: 8.7 lb/gal; Sp. gr.: ≈ 1.04; Visc.: ≈ 58 SUS (210 F); HLB: 10.9; Flash Pt.: > 400 F (OC); Pour Pt.: < 0 F; Stability: Stable to acids, bases, and salts; pH: ≈ 7
Composition: 99.5% min. active
Containers: 460 lb net closed head steel drums
Warnings and Caution: Do not mix with conc. oxidizing or reducing agents–potentially explosive

T-Det N-8. Thompson-Hayward Chem. Co.
Chemical Description: Reaction product of nonylphenol with 8 moles of ethylene oxide
Category: Detergent, wetting agent, dispersant, emulsifier, solubilizer, nonionic
Applications: Textile processing; agricultural toxicant formulations; household and industrial cleaning formulations; degreasing and fat liquoring in leather industry; emulsion paint systems; pigment dispersing; rewetting, deinking, felt washing, leveling in paper manufacture; secondary oil recovery operations
Form: Essentially clear, viscous liquid; Odor: Mild, aromatic
Properties: Density: 8.75 lb/gal; Sp. gr.: ≈ 1.05; Visc.: ≈ 63 SUS (210 F); F.P.: ≈ 30 F; HLB: 12.4; Cloud Pt.: 78 F (1%); Flash Pt.: 500 F (OC); Stability: Stable to acids, bases, and salts; pH: ≈ 6 (1%)
Composition: 99.5% min. active
Toxicity: Eye irritant
Containers: 460 lb net closed head steel drums
Warnings and Caution: Do not mix with conc. oxidizing or reducing agents–potentially explosive

T-Det N-9.5. Thompson-Hayward Chem. Co.
Chemical Description: Reaction product of nonylphenol with 9.5 moles of ethylene oxide
Category: Detergent, wetting agent, dispersant, emulsifier, solubilizer, foaming agent, nonionic
Applications: See T-DET N-8
Form: Essentially clear, viscous liquid; Odor: Mild, aromatic
Properties: Solubility: Sol. in butyl Cellosolve, carbon tetrachloride, ethanol, ethylene glycol, HAN, toluene, water, xylene; Density: 8.8 lb/gal; Sp. gr.: ≈ 1.06; Visc.: ≈ 68 SUS (210 F); F.P.: ≈ 41 F; HLB: 13.1; Cloud Pt.: ≈ 135 F (1%); Flash Pt.: 500 F (OC); Stability: Stable to acids, bases, and salts; pH: ≈ 7 (1%)
Composition: 99.5% min. active
Toxicity: Eye irritant
Containers: 460 lb net closed head steel drums
Warnings and Caution: Do not mix with conc. oxidizing or reducing agents–potentially explosive

T-Det N-10.5. Thompson-Hayward Chem. Co.
Chemical Description: Reaction product of nonylphenol with 10.5 moles of ethylene oxide
Category: Detergent, wetting agent, dispersant, emulsifier, solubilizer, nonionic
Applications: Textile processing; household and industrial cleaning products, e.g., all-purpose liquid detergents, controlled-foam heavy duty laundry detergents, copper, floor, porcelain, and window cleaners, waterless hand cleaners, acid cleaners, alkaline soak tank cleaners, and electrolytic cleaners
Form: Essentially clear, viscous liquid; Odor: Mild, aromatic
Properties: Solubility: Sol. in butyl Cellosolve, ethanol, ethylene glycol, HAN, toluene, water, xylene; Density: 8.8 lb/gal; Sp. gr.: ≈ 1.06; F.P.: ≈ 49 F; HLB: 10.4; Cloud Pt.: ≈ 155 F; Flash Pt.: 500 F (OC); Stability: Stable to acids, bases, and salts; pH: ≈ 7 (1%)
Composition: 99.5% min. active
Toxicity: Eye irritant
Containers: 460 lb net closed head steel drums
Warnings and Caution: Do not mix with conc. oxidizing or reducing agents–potentially explosive

T-Det N-12. Thompson-Hayward Chem. Co.
Chemical Description: Reaction product of nonylphenol with 12 moles of ethylene oxide
Category: Detergent, wetting agent, nonionic
Applications: Detergent base for liquid detergents and certain sanitizer formulations; wetting agent for use in soak tank metal cleaners; excellent nonionic for blending with powdered detergent formulations
Form: Opaque liquid; Odor: Mild, aromatic
Properties: Density: 8.9 lb/gal; Sp. gr.: ≈ 1.07; HLB: 12; Cloud Pt.: ≈ 180 F (1%); Flash Pt.: >

400 F (OC); Pour Pt.: ≈ 57 F; Stability: Stable to acids, bases, and salts; pH: ≈ 7 (1%)
Composition: 99.5% min. active
Containers: 460 lb net closed head steel drums
Warnings and Caution: Do not mix with conc. oxidizing or reducing agents—potentially explosive

T-Det N-14. Thompson-Hayward Chem. Co.
Chemical Description: Reaction product of nonylphenol with 14 moles of ethylene oxide
Category: Detergent, emulsifier, wetting agent, penetrating agent, nonionic
Applications: High temp. detergent; emulsifier for oils, fats, and waxes; wetting agent for high temp. soak tank metal cleaners; wetter and penetrant in caustic sol'ns. and corrosion inhibitors
Form: Semisolid; Color: Off-white; Odor: Mild, aromatic
Properties: Solubility: Sol. in butyl Cellosolve, carbon tetrachloride, 2% corn oil, ethanol, ethylene glycol, water, xylene; Density: 8.9 lb/gal (68 F); Sp. gr.: ≈ 1.07 (68 F); Visc.: ≈ 89 SUS (210 F); HLB: 14.7; Cloud Pt.: ≈ 203 F (1%); Flash Pt.: > 400 F (OC). Pour Pt.: ≈ 67 F; Stability: Stable to acids, bases, and salts; pH: ≈ 6 (1%)
Composition: 99.5% min. active
Toxicity: Eye irritant
Containers: 460 lb net closed head steel drums
Warnings and Caution: Do not mix with conc. oxidizing or reducing agents—potentially explosive

T-Det N-30. Thompson-Hayward Chem. Co.
Chemical Description: Reaction product of nonylphenol with 30 moles of ethylene oxide
Category: Detergent, emulsifier, wetting agent, nonionic
Applications: High temp. detergent; pressure scouring textile operations; emulsifier for oils, fats, and waxes; wetting agent in applications involving high concs. of electrolytes
Form: Waxy solid; Color: Pale yellow; Odor: Mild, aromatic
Properties: Solubility: Sol. in butyl Cellosolve, ethanol, ethylene glycol, water, xylene; Sp. gr.: ≈ 1.07 (132 F); HLB: 17.0; Cloud Pt.: Clear @ 212 F (1%); Flash Pt.: > 500 F; Pour Pt.: ≈ 110 F; Stability: Stable to acids, bases, and salts; pH: 6.5–8.0 (1%)
Composition: 99.5% min. active
Containers: 480 lb net steel removable head drums
Warnings and Caution: Do not mix with conc. oxidizing or reducing agents—potentially explosive

T-Det N-40. Thompson-Hayward Chem. Co.
Chemical Description: Reaction product of nonylphenol with 40 moles of ethylene oxide
Category and Applications: See T-DET N-30
Form: Waxy solid; Color: Pale

yellow; Odor: Mild, aromatic
Properties: Sp. gr.: ≈ 1.07 (132 F); HLB: 17.7; Cloud Pt.: Clear @ 212 F (1% sol'n.); Flash Pt.: > 500 F; Pour Pt.: ≈ 120 F; Stability: Stable to acids, bases, and salts; pH: 6.5–8.0 (1%)
Composition: 99.5% min. active
Containers: 480 lb net removable head steel drums
Warnings and Caution: Do not mix with conc. oxidizing or reducing agents–potentially explosive

T-Det N-50. Thompson-Hayward Chem. Co.
Chemical Description: Reaction product of nonylphenol with 50 moles of ethylene oxide
Category and Applications: See T-DET N-30
Form: Waxy solid; Color: Pale yellow; Odor: Mild, aromatic
Properties: Sp. gr.: ≈ 1.07 (132 F); HLB: 18.0; Cloud Pt.: Clear @ 212 F (1% sol'n.); Flash Pt.: > 500 F; Pour Pt.: ≈ 120 F; Stability: Stable to acids, bases, and salts; pH: 6.5–8.0 (1%)
Composition: 99.5% min. active
Containers: 480 lb net removable head steel drums
Warnings and Caution: Do not mix with conc. oxidizing or reducing agents–potentially explosive

T-Det N-70. Thompson-Hayward Chem. Co.
Chemical Description: Reaction product of nonylphenol with 70 moles of ethylene oxide
Category: Detergent, emulsifier, stabilizer, nonionic
Applications: High-temp. detergent; pressure scouring textile operations; emulsifier for oils, fats, and waxes; emulsifier/stabilizer in latex and alkyd emulsions; useful in emulsion polymerization
Form: Waxy solid; Color: Pale yellow; Odor: Mild, aromatic
Properties: Density: 9.0 lb/gal (135 F); Sp. gr.: ≈ 1.08 (135 F); Cloud Pt.: Clear @ 212 F (1% sol'n.); Flash Pt.: > 500 F; Pour Pt.: ≈ 125 F; Stability: Stable to acids, bases, and salts; pH: 6.0–7.0 (1%)
Composition: 99.5% min. active
Warnings and Caution: Do not mix with conc. oxidizing or reducing agents–potentially explosive

T-Det N-100. Thompson-Hayward Chem. Co.
Chemical Description: Reaction product of nonylphenol with 100 moles of ethylene oxide
Category and Applications: See T-DET N-30; also used as emulsifier for asphalt
Form: Waxy solid; Color: Pale yellow; Odor: Mild, aromatic
Properties: Solubility: Sol. in ethylene dichloride, ethanol, and water; Sp. gr.: ≈ 1.08 (135 F); Cloud Pt.: Clear @ 212 F (1% sol'n.); Flash Pt.: > 500 F; Pour Pt.: ≈ 127 F; Stability: Stable to acids, bases, and salts; pH: 6–7 (1%)
Composition: 99.5% min. active
Containers: 480 lb net removable head steel drums

Warnings and Caution: Do not mix with conc. oxidizing or reducing agents—potentially explosive

T-Det N-307. Thompson-Hayward Chem. Co.
Chemical Description: 70% sol'n. of T-DET N-30 (nonylphenol-ethylene oxide adduct)
Category and Applications: See T-DET N-30
Form: Liquid; Color: 100 (APHA)
Properties: Density: 9.1 lb/gal; Visc.: 806 (Brookfield); Cloud Pt.: > 100 C (1%); Flash Pt.: > 212 F (TOC); Pour Pt.: ≈ 30 F
Composition: 70 ± 0.5% active
Containers: 480 lb net closed head lined steel drums
Warnings and Caution: Do not mix with conc. oxidizing or reducing agents—potentially explosive

T-Det N-407. Thompson-Hayward Chem. Co.
Chemical Description: 70% sol'n. of T-DET N-40 (nonylphenol-ethylene oxide adduct)
Category and Applications: See T-DET N-30; also as emulsifier in paint systems
Form: Liquid; Color: 100 (APHA)
Properties: Density: 9.0 lb/gal; Sp. gr.: 1.08; Visc.: 900 cps (Brookfield); Cloud Pt.: > 212 F (1%); Flash Pt.: > 212 F (TOC); Pour Pt.: ≈ 30 F; Stability: Stable to acids, bases, and salts
Composition: 70 ± 0.5% active
Containers: 480 lb net closed head lined steel drums
Warnings and Caution: Do not mix with conc. oxidizing or reducing agents—potentially explosive

T-Det N-507. Thompson-Hayward Chem. Co.
Chemical Description: 70% sol'n. of T-DET N-50 (nonylphenol-ethylene oxide adduct)
Category and Applications: See T-DET N-30; also as emulsifier in various paint systems
Form: Liquid; Color: 100 (APHA)
Properties: Density: 9.1 lb/gal; Sp. gr.: 1.09; Visc.: 760 cps (Brookfield); Cloud Pt.: > 100 C (1%); Flash Pt.: > 212 F (TOC); Pour Pt.: ≈ 40 F; Stability: Stable to acids, bases, and salts
Composition: 70 ± 0.5% active
Containers: 480 lb net closed head lined steel drums
Warnings and Caution: Do not mix with conc. oxidizing or reducing agents—potentially explosive

T-Det N-705. Thompson-Hayward Chem. Co.
Chemical Description: 50% sol'n. of T-DET N-70 (nonylphenol-ethylene oxide adduct)
Category and Applications: See T-DET N-70
Form: Clear to slightly hazy liquid; Odor: Mild, aromatic
Properties: Density: 9.0 lb/gal (68 F); Sp. gr.: ≈ 1.08 (68 F); Cloud Pt.: Clear @ 212 F (1%); Flash Pt.: > 500 F; Pour Pt.: 40 F; Stability: Stable to acids, bases, and salts; pH: 5.0–6.5 (1% sol'n.)
Composition: 50% min. active

Containers: Stainless steel or plastic-lined tanks
Warnings and Caution: Do not mix with conc. oxidizing or reducing agents—potentially explosive

T-Det N-1007. Thompson-Hayward Chem. Co.
Chemical Description: 70% sol'n. of T-DET N-100 (nonylphenol-ethylene oxide adduct)
Category and Applications: See T-DET N-100
Form: Clear to opaque, viscous liquid; Odor: Mild, aromatic
Properties: Sp. gr.: 1.1; Cloud Pt.: Clear @ 212 F (1%); Flash Pt.: > 500 F; Pour Pt.: ≈ 62 F; pH: 6–7 (1% sol'n.)
Composition: 70 ± 0.5% active

T-Det O-407. Thompson-Hayward Chem. Co.
Chemical Description: Reaction product of octyl phenol with 40 moles of ethylene oxide
Category: Detergent, emulsifier, stabilizer, nonionic
Applications: Highly water-sol., high-temp. detergent and emulsifier/stabilizer in various applications
Form: Liquid; Color: 100 max. (APHA)
Properties: Solubility: Highly water-sol.; Density: ≈ 9.1 lb/gal; Visc.: 600 cps; Cloud Pt.: > 100 C (1%); Flash Pt.: > 100 C (TOC); Pour Pt.: ≈ 30 F; Stability: Stable to acids, bases, and salts; pH: ≈ 7 (1% sol'n.)
Composition: 70 ± 0.5% active

Containers: 480 lb net closed head lined steel drums
Warnings and Caution: Do not mix with conc. oxidizing or reducing agents—potentially explosive

T-Det TDA-65. Thompson-Hayward Chem. Co.
Chemical Description: Reaction product of tridecyl alcohol with ≈ 65% ethylene oxide
Category: Detergent, wetting agent, dispersant, emulsifier, solubilizer, nonionic
Applications: Textile processing; hard surface detergent; useful in dispersing and suspending solids such as carbon black, clay, color pigments, and soil; wetting and rewetting agent for dust settling, corrosion inhibitors, metal cleaning and pickling
Form: Hazy liquid; Odor: Mild, aliphatic
Properties: Solubility: Water-sol.; Density: 8.20 lb/gal; Cloud Pt.: ≈ 130 F (1%); Flash Pt.: > 290 F (PMCC); Pour Pt.: ≈ 62 F; Stability: Stable to acids, bases, and salts; pH: 5–7
Composition: 99.5% min. active
Warnings and Caution: Do not mix with conc. oxidizing or reducing agents—potentially explosive

Teepol CM44. Shell Chem.
Chemical Description: Aq. sol'n. of the sodium salt of C_9–C_{13} alkylbenzene sulfonate
Category: Detergent, emulsifier, anionic

Applications: Neutralized anionic detergent with good surface active and emulsification properties; used in dishwashing and general cleaning liquids where it enables cold stability characteristics to be obtained
Form: Liquid; Color: Pale amber
Properties: Solubility: Miscible with water in all proportions; Density: 1.05 kg/l (20 C); Visc.: 100–500 cs (20 C); Clear Pt.: > 16 C; pH: 8.0–9.5 (5% aq.); Biodegradable > 90%
Composition: > 25% active ingred.
Toxicity: Eye irritant; skin irritant on prolonged contact
Containers: 223 kg net lacquer-lined mild steel drums
References: Data sheet 2A 175

Teepol GC 56P. Shell Chem.
Chemical Description: Blend of alkylbenzene sulfonate and primary alcohol ethoxysulfate
Category: Detergent
Applications: Light duty liquid detergents for dishwashing and general cleaning applications; esp. suitable for repacking operations
Form: Clear liquid
Properties: Solubility: Miscible with water in all proportions; Density: 1.06 kg/l (20 C); Visc.: 1900 cs (33% conc., 20 C); Cloud Pt.: 1 C; Clear Pt.: 3 C; pH: 6.5–8.0 (5% sol'n.); Biodegradable > 90%
Composition: > 33% active ingred.
Toxicity: Eye irritant; skin irritant on prolonged contact
Containers: 225 kg net lacquer-lined mild steel drums
References: Data sheet 2A 172

Teepol GD7P. Shell Chem.
Chemical Description: Blend of alkylbenzene sulfonate, primary alcohol ethoxysulfate, and primary alcohol ethoxylate in aq. sol'n.
Category: Detergent, anionic/nonionic
Applications: Light duty liquid detergent for dishwashing and general cleaning; can be diluted before repacking while maintaining a substantial viscosity (down to 24% active)
Form: Clear liquid
Properties: Solubility: Miscible with water in all proportions; Density: 1,07 kg/l (20 C); Visc.: 620 cs (40% conc., 20 C); Cloud Pt.: 2 C; Clear Pt.: 6 C; pH: 6.5–8.0 (5% sol'n.)
Composition: > 40% active ingred.
Toxicity: Eye irritant; skin irritant on prolonged contact
Containers: 223 kg net lacquer-lined mild steel drums
References: Data Sheet 2A 170

Teepol GD53. Shell Chem.
Chemical Description: Blend of alkylbenzene sulfonate, primary alcohol ethoxysulfate, and primary alcohol ethoxylate in aq. sol'n.
Category: Detergent, anionic/nonionic
Applications: Light duty liquid detergent for dishwashing and general cleaning; suitable for repacking; can be diluted down to

17.5% active while maintaining a substantial viscosity
Form: Clear liquid
Properties: Solubility: Readily miscible with water in all proportions; Density: 1.07 kg/l (20 C); Visc.: 24–300 cs (30% conc., 20 C); Cloud Pt.: 1 C; Clear Pt.: 3 C; pH: 6.5–8.0 (5% sol'n.)
Composition: > 30% active ingred.
Toxicity: Eye irritant; skin irritant on prolonged contact
Containers: 223 kg net lacquer-lined mild steel drums
References: Data sheet 2A 171

Teepol HB6. Shell Chem.
Chemical Description: Aq. sol'n. of sodium salts of sulfated broad-cut primary alcohol
Category: Solubilizer, emulsifier
Applications: Components of hard surface cleaners and germicidal cleaners
Form: Liquid; Color: Pale amber
Properties: Solubility: Sol. in water at all dilutions; Density: 1.06 kg/l (20 C); Visc.: 20 cs (20 C); Clear Pt.: 10 C; pH: 8.0 (5% aq.); Biodegradable > 95%
Composition: 34.2% active
Toxicity: Eye irritant; skin irritant on prolonged contact
Containers: 215 kg net lined steel drums
References: Data sheet 2A 173

Teepol PB. Shell Chem.
Chemical Description: Aq. sol'n. of primary alcohol ethoxysulfate (sodium cation)
Category: Foaming agent
Applications: For use in plasterboard (gypsum wallboard) mfg.
Form: Clear liquid
Properties: Solubility: Miscible with water in all proportions; Density: 1.06 kg/l (20 C); Visc.: 50 cs (20 C); Clear Pt.: 0 C; pH: 6.5–8.0 (5% sol'n.); Biodegradable > 95%
Composition: > 32% active ingred.
Toxicity: Eye irritant; skin irritant on prolonged contact
Containers: 215 kg net lacquer-lined mild steel drums
Warnings and Caution: Storage temp. should not exceed 40 C to minimize hydrolysis
References: Data sheet 2A 174

Tek Wet 752. Hamblet & Hayes Co.
Category: Detergent, wetting agent, emulsifier
Applications: Skin, hide, leather degreasing; cleaners of street, buildings, hospitals; textile processing; tanning; wetting pulp in beater; fat liquoring
Form: Liquid; Color: Colorless; Odor: Slight
Properties: Solubility: Sol. in water and various organic solvents; Sp. gr.: 0.975; Visc.: 63 cs (20 C); F.P.: –2 C; Flash Pt.: 430 F; Surface Tension: 28 dynes/cm; Biodegradable
Composition: 100% active ingred.
Reaction: Neutral
Toxicity: Slight skin irritant; defats tissues; serious eye irritant
Containers: 55 gal drums and bulk
Warnings and Caution: Protect eyes; avoid prolonged skin contact

Tek Wet 951. Hamblet & Hayes Co.
Category and Applications: See Tek Wet 752
Form: Liquid; Color: Colorless; Odor: Mild
Properties: Solubility: Sol. in water and various organic solvents; Sp. gr.: 1.006; Visc.: 86 cs (20 C); F.P.: 13 C; Flash Pt.: 470 F; Surface Tension: 29 dynes/cm; Biodegradable
Composition: 100% active ingred.
Reaction: Neutral
Toxicity: Slight skin irritant; defats tissues; serious eye irritant
Containers: 55 gal drums and bulk
Warnings and Caution: Protect eyes; avoid prolonged skin contact

Tenlo 70. Diamond Shamrock/Process Chem. Div.
Category: Wetting agent, dispersant, nonionic
Applications: Pigment grinding aid for oil-solvent and latex paint systems; disperses pigments, clays, and other paint additives, increases solids loading, lowers viscosity, decreases sedimentation rates; wetting agent for iron oxide pigments in vinyl sol'ns. used in mfg. of magnetic inks and magnetic tapes
Form: Liquid; Color: Amber
Properties: Solubility @ 5%: Sol. in mineral spirits, aromatic hydrocarbons, petroleum oil, linseed oil, isopropyl acetate; forms hazy sol'n. in hot and cold water, and slightly hazy in ketones; Density: 8.2 lb/gal; pH: 9.5 (2% sol'n.)
Composition: 97% active ingred.
Containers: 55 gal tighthead steel drums
Warnings and Caution: Protect from freezing
References: PAD-34 Rev.

Terg-A-Zyme. Alconox Inc.
Chemical Description: Blend of alkyl aryl sulfonates, lauryl alcohol sulfates, phosphates, and carbonates, with the addition of a protease enzyme
Category: Detergent, wetting agent, sequestering and synergistic agents, anionic
Applications: Enzyme-activated detergent for use in hospitals, laboratories, dairies; for presoaking hard-to-remove stains; cleaning agent in reverse osmosis equipment in dairy and pollution processing
Form: Powder
Properties: pH: 9.0–9.5
Containers: 2 lb box, 25 and 50 lb carton, 100 and 300 lb drums
References: Bulletin #5

Tergenol 1122. Hart Chem. Ltd.
Chemical Description: Blended product
Category: Detergent, anionic
Applications: Heavy duty detergent for woven and knitted fabrics
Form: Liquid; Color: Amber
Properties: Ref. Index: 1.4462; Solubility: Clear soluble; Density: 1.03; Visc.: 170 cps; Stability: Stable to acids, alkalis, and oxidizing agents
Composition: 80% active ingred.

Reaction: Neutral
Toxicity: Skin and eye irritant
Containers: Drums, T/T

Tergenol G. Hart Products Corp.
Chemical Description: Sodium n-methyl-n-oleoyl taurate
Category: Detergent, wetting agent, emulsifier
Applications: Used in textiles, skins, leather, laundering, dyeing, household and industrial cleaners, cosmetics products, desizing and boiling off cottons, bleaching, rewetting dried skins, spot removal, dispersing waxes, pigments, resins, and dry powders, and in printing; also as corrosion inhibitor in chlorite bleaching
Form: Gel; Color: Light
Properties: Solubility: Water-sol.
Composition: 15% active ingred.
Containers: Drums

Tergenol S Liquid. Hart Products Corp.
Chemical Description: Sodium-n-methyl-n-oleoyltaurate
Category and Applications: See Tergenol G; anionic surfactant
Form: Liquid; Color: Amber
Composition: 38% active ingred.
Reaction: Neutral
Containers: Drums

Tergenol Slurry. Hart Products Corp.
Chemical Description: Sodium-n-methyl-n-oleoyltaurate
Category and Applications: See Tergenol G; anionic
Form: Paste; Color: White
Properties: Solubility: Water-sol.
Composition: 60% active ingred.
Reaction: Neutral
Containers: Drums

Tergitol 15-S-3. Union Carbide Corp.
Chemical Description: Polyethylene glycol ether of linear alcohol (3 moles EO)
Category: Detergent, emulsifier, wetting agent, nonionic
Applications: Textile surfactant, used in cotton scouring, alkaline scouring of polyester/cotton, wool scouring, bleaching, polyethylene and dye carrier emulsification, dyeing and printing assistants, soaping off, textile lubricants
Form: Liquid; Color: 75 max. (Pt-Co)
Properties: m.w.: 332 (avg.); Solubility @ 10%: Sol. in perchloroethylene, isopropanol, butyl Cellosolve, xylene, Isopar L, Aromatic 150, Varsol solvent, white mineral oil; Density: 7.68 lb/gal (20 C); Sp. gr.: 0.924 (20/20 C); Visc.: 37 cs (20 C); HLB: 8.0; Cloud Pt.: < 0 C (1% aq.); Flash Pt.: 370 F (COC); Pour Pt.: –32 C; Hydroxyl No.: 169; Biodegradable
Composition: 100% conc.
Toxicity: Eye irritant; defatting to skin
Containers: 1, 5 gal containers, 55 gal drums

Tergitol 15-S-5. Union Carbide Corp.
Chemical Description: Polyethyl-

ene glycol ether of linear alcohol (5 moles EO)
Category and Applications: See Tergitol 15-S-3
Form: Liquid; Color: 75 max. (Pt-Co)
Properties: m.w.: 420 (avg.); Solubility @ 10%: Sol. in perchloroethylene, isopropanol, butyl Cellosolve, xylene, Isopar L, Aromatic 150, Varsol solvent, white mineral oil; Density: 8.03 lb/gal (20 C); Sp. gr.: 0.965 (20/20 C); Visc.: 51 cs (20 C); HLB: 10.5; Cloud Pt.: $<$ 0 C (1% aq.); Flash Pt.: 410 F (COC); Pour Pt.: –15 C; Hydroxyl No.: 134; pH: 6–8 (1%); Biodegradable
Composition: 100% conc.
Toxicity: Causes eye burns and skin irritation
Containers: 1 and 5 gal containers, 55 gal drums

Tergitol 15-S-7. Union Carbide Corp.
Chemical Description: Polyethylene glycol ether of linear alcohol (7 moles EO)
Category and Applications: See Tergitol 15-S-3; esp. for desizing and soaping off
Form: Liquid; Color: 75 max. (Pt-Co)
Properties: m.w.: 508 (avg.); Solubility @ 10%: Sol. in water, perchloroethylene, isopropanol, butyl Cellosolve, Aromatic 150, Varsol solvent; Density: 8.26 lb/gal (20 C); Sp. gr.: 0.994 (20/20 C); Visc.: 67 cs (20 C); HLB: 12.1; Cloud Pt.: 37 C (1% aq.); Flash Pt.: 440 F (COC); Pour Pt.: 7 C; Hydroxyl No.: 110; Surface Tension: 29.1 dynes/cm (0.1%); pH: 6–8 (1%); Biodegradable
Composition: 100% conc.
Toxicity: Causes eye burns and skin irritation
Containers: 1 and 5 gal containers, 55 gal drums

Tergitol 15-S-9. Union Carbide Corp.
Chemical Description: Polyethylene glycol ether of linear alcohol (9 moles EO)
Category and Applications: See Tergitol 15-S-3; esp. for desizing and soaping off
Form: Liquid; Color: 50 max. (pt-Co)
Properties: m.w.: 596 (avg.); Solubility @ 10%: Sol. in water, perchloroethylene, isopropanol, butyl Cellosolve, xylene, Aromatic 150, Varsol solvent; Density: 8.37 lb/gal (20 C); Sp. gr.: 1.006 (20/20 C); Visc.: 78 cs (20 C); HLB: 13.3; Cloud Pt.: 60 C (1% aq.); Flash Pt.: 470 F (COC); Pour Pt.: 13 C; Hydroxyl No.: 94; Surface Tension: 30.3 dynes/cm (0.1%); pH: 6–8 (1%); Biodegradable
Composition: 100% conc.
Toxicity: Causes eye burns
Containers: 1 and 5 gal containers, 55 gal drums

Tergitol 15-S-12. Union Carbide Corp.
Chemical Description: Polyethylene glycol ether of linear alcohol (12 moles EO)

Category and Applications: See Tergitol 15-S-3; also for desizing
Form: Semisolid; Color: 100 max. (Pt-Co)
Properties: m.w.: 728 (avg.); Solubility @ 10%: Sol. in water, perchloroethylene, isopropanol, butyl Cellosolve, xylene, Aromatic 150; Density: 8.49 lb/gal (30 C); Sp. gr.: 1.020 (30/20 C); Visc.: 120 cs (20 C); HLB: 14.5; Cloud Pt.: 90 C (1% aq.); Flash Pt.: 460 F (COC); Pour Pt.: 17 C; Hydroxyl No.: 77; Surface Tension: 31.5 dynes/cm (0.1%); pH: 6–8 (1%); Biodegradable
Composition: 100% conc.
Toxicity: Causes eye and skin irritation
Containers: 1 and 5 gal containers, 55 gal drums

Tergitol 15-S-15. Union Carbide Corp.
Chemical Description: Polyethylene glycol ether of linear alcohol (15 moles EO)
Category and Applications: See Tergitol 15-S-3
Form: Solid; Color: 100 max. (Pt-Co)
Properties: m.w.: 860 (avg.); Density: 8.40 (55 C); Sp. gr.: 1.025 (40/20 C); Visc.: 63 cs (40 C); HLB: 15.4; Cloud Pt.: > 100 C (1% aq.); Flash Pt.: 490 F (COC), Pour Pt.: 23 C; Hydroxyl No.: 65; pH: 6–8 (1% aq.); Biodegradable
Composition: 100% conc.
Toxicity: Defatting to skin
Containers: 1 and 5 gal containers, 55 gal drums

Tergitol 15-S-20. Union Carbide Corp.
Chemical Description: Polyethylene glycol ether of linear alcohol (20 moles EO)
Category and Applications: See Tergitol 15-S-3
Form: Solid; Color: 100 max. (Pt-Co)
Properties: m.w.: 1080 (avg.); Density: 8.66 lb/gal (40 C); Sp. gr.: 1.040 (40/20 C); Visc.: 102 cs (40 C); HLB: 16.3; Cloud Pt.: > 100 C (1% aq.); Flash Pt.: 480 F (COC); Pour Pt.: 34 C; Hydroxyl No.: 52; pH: 6–8 (1%); Biodegradable
Composition: 100% conc.
Toxicity: Defatting to skin
Containers: 1 and 5 gal containers, 55 gal drums

Tergitol 15-S-30. Union Carbide Corp.
Chemical Description: Polyethylene glycol ether of linear alcohol (30 moles EO)
Category and Applications: See Tergitol 15-S-3
Form: Solid; Color: 100 max. (Pt-Co)
Properties: m.w.: 1520 (avg.); Density: 8.78 lb/gal (55 C); Sp. gr.: 1.063 (45/20 C); Visc.: 102 cs (40 C); HLB: 17.4; Cloud Pt.: > 100 C (1% aq.); Flash Pt.: 505 F (COC); Pour Pt.: 40 C; Hydroxyl No.: 37; pH: 6–8 (1%); Biodegradable
Composition: 100% conc.

Toxicity: Defatting to skin
Containers: 1 and 5 gal containers, 55 gal drums

Tergitol 15-S-40. Union Carbide Corp.
Chemical Description: Polyethylene glycol ether of linear alcohol (40 moles EO)
Category and Applications: See Tergitol 15-S-3
Form: Solid; Color: 100 max (Pt-Co)
Properties: m.w.: 1960 (avg.); Density: 8.83 lb/gal (55 C); Sp. gr.: 1.066 (50/20 C); HLB: 18.0; Cloud Pt.: > 100 C (1% aq.); Flash Pt.: 515 F (COC); Pour Pt.: 45 C; Hydroxyl No.: 29; pH: 6–8 (1%); Biodegradable
Composition: 100% conc.
Toxicity: Defatting to skin
Containers: 1 and 5 gal containers, 55 gal drums

Tergitol 25-L-3. Union Carbide Corp.
Chemical Description: Polyethylene glycol ether of a primary alcohol (3 moles EO)
Category: Detergent, wetting agent, emulsifier, nonionic
Applications: Textile surfactant; used in cotton scouring, alkaline scouring of polyester/cotton, wool scouring, bleaching, polyethylene and dye carrier emulsification, dyeing and printing assistants, textile lubricants
Form: Liquid; Color: 50 max. (Pt-Co)
Properties: m.w.: 341 (avg.); Solubility @ 10%: Sol. in perchloroethylene, isopropanol, butyl Cellosolve, xylene, Aromatic 150; disp. in white mineral oil; Density: 7.73 lb/gal (20 C); Sp. gr.: 0.929 (20/20 C); Visc.: 36 cs (20 C); HLB: 7.7; Cloud Pt.: < 0 C (1% aq.); Flash Pt.: 260 F (COC); Pour Pt.: 8 C; Hydroxyl No.: 165; pH: 5–7 (1% sol'n. in 1:1 isopropanol/water); Biodegradable
Composition: 100% conc.
Toxicity: Defatting to skin
Containers: 1 and 5 gal containers, 55 gal drums

Tergitol 25-L-5. Union Carbide Corp.
Chemical Description: Polyethylene glycol ether of a primary alcohol (5 moles EO)
Category and Applications: See Tergitol 25-L-3
Form: Liquid; Color: 50 max. (Pt-Co)
Properties: m.w.: 440 (avg.); Solubility @ 10%: Sol. in perchloroethylene, isopropanol, butyl Cellosolve, xylene, Aromatic 150; disp. in water and white mineral oil; Density: 7.94 lb/gal (30 C); Sp. gr.: 0.955 (30/20 C); Visc.: 95 cs (20 C); HLB: 10.4; Cloud Pt.: < 0 C (1% aq.); Flash Pt.: 325 F (COC); Pour Pt.: 16 C; Hydroxyl No.: 127; pH: 5–7 (1%); Biodegradable
Composition: 100% conc.
Toxicity: Eye and skin irritant
Containers: 1 and 5 gal containers, 55 gal drums

Tergitol 25-L-7. Union Carbide Corp.
Chemical Description: Polyethylene glycol ether of a primary alcohol (7 moles EO)
Category and Applications: See Tergitol 25-L-3; also for desizing
Form: Liquid; Color: 50 max. (Pt-Co)
Properties: m.w.: 550 (avg.); Solubility @ 10%: Sol. in water, perchloroethylene, isopropanol, butyl Cellosolve, xylene, Aromatic 150; Density: 8.19 lb/gal (30 C); Sp. gr.: 0.985 (30/20 C); Visc.: 51 cps (38 C); HLB: 12.4; Cloud Pt.: 50 C (1% aq.); Flash Pt.: 340 F (COC); Pour Pt.: 23 C; Hydroxyl No.: 102; Surface Tension: 28.7 dynes/cm (0.1%); pH: 5–7 (1%); Biodegradable
Composition: 100% conc.
Toxicity: Causes eye burns and skin irritation
Containers: 1 and 5 gal containers, 55 gal drums

Tergitol 25-L-9. Union Carbide Corp.
Chemical Description: Polyethylene glycol ether of a primary alcohol (9 moles EO)
Category and Applications: See Tergitol 25-L-3; also in desizing
Form: Semisolid; Color: 50 max. (Pt-Co)
Properties: m.w.: 585 (avg.); Solubility @ 10%: Sol. in water, perchloroethylene, isopropanol, butyl Cellosolve, xylene, Aromatic 150; Density: 8.18 lb/gal (40 C); Sp. gr.: 0.991 (30/20 C); Visc.: 33 cps (40 C); HLB: 12.8; Cloud Pt.: 60 C (1% aq.); Flash Pt.: 285 F (COC); Pour Pt.: 24 C; Hydroxyl No.: 96; Surface Tension: 29.0 dynes/cm (0.1%); pH: 5–7 (1%), Biodegradable
Composition: 100% conc.
Toxicity: Skin and eye irritant
Containers: 1 and 5 gal containers, 55 gal drums

Tergitol 25-L-12. Union Carbide Corp.
Chemical Description: Polyethylene glycol ether of a primary alcohol (12 moles EO)
Category and Applications: See Tergitol 25-L-3; also for desizing
Form: Solid; Color: 100 max. (Pt-Co)
Properties: m.w.: 730 (avg.); Solubility @ 10%: Sol. in water, perchloroethylene, isopropanol, butyl Cellosolve, xylene, Aromatic 150; Density: 8.37 lb/gal (40 C); Sp. gr.: 1.013 (30/20 C); Visc.: 51 cs (38 C); HLB: 14.2; Cloud Pt.: 90 C (1% aq.); Flash Pt.: 410 F (COC); Pour Pt.: 30 C; Hydroxyl No.: 77; Surface Tension: 32.0 dynes/cm (0.1%); pH: 5–7 (1%); Biodegradable
Composition: 100% conc.
Toxicity: Eye irritant
Containers: 1 and 5 gal containers, 55 gal drums

Tergitol LN-60. Union Carbide Corp.
Chemical Description: Primary alcohol-based nonionic
Category: Detergent, penetrant,

emulsifier, wetting agent, chemical intermediate, nonionic
Applications: Textile surfactant; used for cotton scouring, alkaline scouring of polyester/cotton, bleaching, polyethylene and dye carrier emulsification, dyeing and printing assistant; intermediate in phosphating, sulfating, etc.
Form: Liquid; Color: 50 max. (Pt-Co)
Properties: m.w.: 770 (avg.); Solubility @ 10%: Sol. in water, perchloroethylene, isopropanol, butyl Cellosolve, xylene, Aromatic 150; Density: 8.43 lb/gal (20 C); Sp. gr.: 1.014 (20/20 C); Visc.: 104 cs (20 C); HLB: 12.8; Cloud Pt.: 60 C (1% aq.); Flash Pt.: 480 F (COC); Pour Pt.: 27 C; Hydroxyl No.: 73; pH: 5–8 (1%); Biodegradable
Composition: 100% conc.
Toxicity: Skin and eye irritant
Containers: 1 and 5 gal containers, 55 gal drums

Tergitol Min-Foam 1X. Union Carbide Corp.
Chemical Description: Modified nonionics based on secondary alcohol
Category: Nonionic surfactant
Applications: Low-foaming textile surfactant for jet dyeing
Form: Liquid; Color: 150 max. (Pt-Co)
Properties: Density: 8.28 lb/gal (20 C); Sp. gr.: 0.995 (20/20 C); Visc.: 57 cs (20 C); Cloud Pt.: 40 C (1% aq.); Flash Pt.: 455 F (COC); Pour Pt.: –42 C; pH: 5–8 (1%)
Composition: 100% conc.
Toxicity: Causes eye burns
Containers: 1 and 5 gal containers, 55 gal drums

Tergitol Min-Foam 2X. Union Carbide Corp.
Chemical Description: Modified nonionic based on secondary alcohol
Category and Applications: See Tergitol Min-Foam 1X
Form: Liquid; Color: 150 max. (Pt-Co)
Properties: Density: 8.14 lb/gal (20 C); Sp. gr.: 0.978 (20/20 C); Visc.: 66 cs (20 C); Cloud Pt.: 21 C (1% aq.); Flash Pt.: 450 F (COC); Pour Pt.: –47 C; pH: 5–8 (1%)
Composition: 100% conc.
Toxicity: Defatting to skin
Containers: 1 and 5 gal containers, 55 gal drums

Tergitol NP-14. Union Carbide Corp.
Chemical Description: Nonylphenol polyethylene glycol ether (4 moles EO)
Category: Detergent, wetting agent, emulsifier, intermediate, nonionic
Applications: Cotton scouring, alkaline scouring of polyester/cotton blends, bleaching, polyethylene and dye carrier emulsification, dyeing and printing assistants, intermediate for phosphating, sulfating, etc.
Form: Liquid; Color: 100 max.

(Pt-Co)
Properties: m.w.: 396 (avg.); Density: 8.57 lb/gal (20 C); Sp. gr.: 1.031 (20/20 C); Visc.: 448 cs (20 C); HLB: 8.9; Cloud Pt.: $<$ 0 C (1% aq.); Flash Pt.: 480 F (COC); Solid. Pt.: –40 C; Hydroxyl No.: 142; pH: 5–8 (10%); Fair biodegradability
Composition: 100% conc.
Toxicity: Eye irritant
Containers: 1 and 5 gal containers, 55 gal drums

Tergitol NP-27. Union Carbide Corp.
Chemical Description: Nonylphenol polyethylene glycol ether (7 moles EO)
Category and Applications: See Tergitol NP-14
Form: Liquid; Color: 75 max. (Pt-Co)
Properties: m.w.: 528 (avg.); Density: 8.71 lb/gal (20 C); Sp. gr.: 1.038 (30/20 C); Visc.: 338 cs (20 C); HLB: 11.7; Cloud Pt.: 22 C (1% aq.); Flash Pt.: 525 F (COC); Solid. Pt.: –6 C; Hydroxyl No.: 106; Surface Tension: 30.4 dynes/cm (0.1%); pH: 5–8 (10%); Fair biodegradability
Composition: 100% conc.
Toxicity: Eye irritant
Containers: 1 and 5 gal containers, 55 gal drums

Tergitol NP-44. Union Carbide Corp.
Chemical Description: Nonylphenol polyethylene glycol ether (40 moles EO)
Category and Applications: See Tergitol NP-14
Form: Solid; Color: 5 max. (Gardner)
Properties: m.w.: 1980 (avg.); Solubility: Highly water-sol.; Density: 8.97 lb/gal (55 C); Sp. gr.: 1.082 (58/20 C); HLB: 17.8; Cloud Pt.: $>$ 100 C (1% aq.); Flash Pt.: 525 F (COC); Solid. Pt.: 48 C; Hydroxyl No.: 28; pH: 4–8 (10%); Fair biodegradability
Composition: 100% conc.
Toxicity: Eye irritant
Containers: 1 and 5 gal containers, 55 gal drums

Tergitol NPX. Union Carbide Corp.
Chemical Description: Nonylphenol ethoxylate (10.5 moles EO)
Category and Applications: See Tergitol NP-14
Form: Liquid; Color: 50 max. (Pt-Co)
Properties: m.w.: 682 (avg.); Solubility @ 10%: Sol. in water, perchloroethylene, isopropanol, butyl Cellosolve, xylene, Aromatic 150; Density: 8.84 lb/gal (20 C); Sp. gr.: 1.062 (20/20 C); Visc.: 318 cs (20 C); HLB: 13.6; Cloud Pt.: 63 C (0.5% aq.); Flash Pt.: 500 F (COC); Solid. Pt.: 7 C; Hydroxyl No.: 83; pH: 5–8 (10%); Fair biodegradability
Composition: 100% conc.
Toxicity: Eye irritant
Containers: 1 and 5 gal containers, 55 gal drums

Tergitol TMN-3. Union Carbide Corp.
Chemical Description: Trimethyl-

nonanol PEG ether (3 moles EO)
Category: Wetting agent, nonionic
Applications: Textile processing
Form: Liquid
Properties: m.w.: 318 (avg.); Density: 7.74 lb/gal (20 C); Sp. gr.: 0.936 (20/20 C); HLB: 8.3; Cloud Pt.: $<$ 0 C (1% aq.); Flash Pt.: 320 F (COC); Solid. Pt.: –40 C; Hydroxyl No.: 176; pH: 5–8 (10%)
Composition: 100% conc.
Toxicity: Eye irritant
Containers: 1 and 5 gal containers, 55 gal drums

Tergitol TMN-6. Union Carbide Corp.
Chemical Description: Trimethylnonanol PEG ether (6 moles EO)
Category and Applications: See Tergitol TMN-3
Form: Liquid; Color: 150 max. (Pt-Co)
Properties: m.w.: 450 (avg.); Density: 8.40 lb/gal (20 C); Sp. gr.: 1.024 (20/20 C); Visc.: 105 cs (20 C); HLB: 11.7; Cloud Pt.: 36 C (0.5% aq.); Flash Pt.: None; Solid. Pt.: –22 C; Hydroxyl No.: 125; pH: 5–8 (10%)
Composition: 90% conc.
Toxicity: Causes eye burns
Containers: 1 and 5 gal containers, 55 gal drums

Tergitol TMN-10. Union Carbide Corp.
Chemical Description: Trimethylnonanol PEG ether (10 moles EO)
Category and Applications: See Tergitol TMN-3
Form: Liquid; Color: 200 max. (Pt-Co)
Properties: m.w.: 626 (avg.); Density: 8.69 lb/gal (20 C); Sp. gr.: 1.044; Visc.: 153 cs (20 C); HLB: 14.1; Cloud Pt.: 76 C (1% aq.); Flash Pt.: None; Solid. Pt.: –7.5 C; Hydroxyl No.: 90; pH: 5-8 (10%)
Composition: 90% conc.
Toxicity: Causes eye burns
Containers: 1 and 5 gal containers, 55 gal drums

Tergitol TP-9. Union Carbide Corp.
Chemical Description: Nonylphenol ethoxylate (9 moles EO)
Category and Applications: See Tergitol NP-14
Form: Liquid; Color: 50 max. (Pt-Co)
Properties: m.w.: 616 (avg.); Density: 8.80 lb/gal (20 C); Sp. gr.: 1.057 (20/20 C); Visc.: 318 cs (20 C); HLB: 12.9; Cloud Pt.: 54 C (0.5% aq.); Flash Pt.: 540 F (COC); Solid. Pt.: 0 C; Hydroxyl No.: 91; Surface Tension: 31.2 dynes/cm (0.1%); pH: 5-8 (10%); Fair biodegradability
Composition: 100% conc.
Toxicity: Eye irritant
Containers: 1 and 5 gal containers, 55 gal drums

Tergitol XD. Union Carbide Corp.
Chemical Description: Polyalkylene glycol ether
Category: Dispersant, emulsifier, nonionic
Applications: Improve the dispersion of dye pigments in textile baths; emulsifier for latex systems

Form: Solid; Color: 100 (Pt-Co) of melt
Properties: m.w.: 2500 (avg.); Density: 8.66 lb/gal (40 C); Sp. gr.: 1.041 (40/20 C); Visc.: 428 cs (40 C); Cloud Pt.: 75 C (0.5% aq.); Flash Pt.: 480 F (COC); Solid. Pt.: 33 C; Hydroxyl No.: 22; pH: 4–7 (10%)
Composition: 100% conc.
Toxicity: Eye irritant
Containers: 1 and 5 gal containers, 55 gal drums

Tergitol XH. Union Carbide Corp.
Chemical Description: Polyalkylene glycol ether
Category and Applications: See Tergitol XD
Form: Solid; Color: 100 (Pt-Co) of melt
Properties: m.w.: 3500 (avg.); Density: 8.72 lb/gal (50 C); Sp. gr.: 1.048 (50/20 C); Visc.: 325 cs (40 C); Cloud Pt.: 95 C (0.5% aq.); Flash Pt.: None; Solid. Pt.: 41 C; Hydroxyl No.: 16; pH: 4–7 (10%)
Composition: 100% conc.
Toxicity: Eye irritant
Containers: 1 and 5 gal containers, 55 gal drums

Tergitol XJ. Union Carbide Corp.
Chemical Description: Polyalkylene glycol ether
Category and Applications: See Tergitol XD
Form: Solid; Color: 2 (Gardner) of melt
Properties: m.w.: 2300 (avg.); Density: 8.51 lb/gal (45 C); Sp. gr.: 1.027 (40/20 C); Cloud Pt.: 50 C (1% aq.); Flash Pt.: 495 F (COC); Solid. Pt.: 26 C; Hydroxyl No.: 24; pH: 5.5–8.0 (1%)
Composition: 100% conc.
Toxicity: Eye irritant
Containers: 1 and 5 gal containers, 55 gal drums

Teric 12A3. ICI Australia Ltd.
Chemical Description: Ethylene oxide condensate based on straight chain synthetic alcohols, premium grade (2.6 moles EO)
Category: Wetting and dispersing agent, coemulsifier, detergent, intermediate, nonionic
Applications: Wetting and dispersing agent in oil and solvent-based systems, e.g., winter and summer spray oils for horticultural use and for the dispersion of pigments in solvents and mineral oils; coemulsifier for W/O-type emulsions, esp. with vegetable oils, aliphatic mineral oils, white oils, and kerosene; domestic and industrial cleaners; intermediate for manufacture of sulfate and phosphate surfactants
Form: Liquid; Color: 100 (Hazen)
Properties: Solubility: Sol. in benzene, ethyl acetate, ethyl Icinol, perchlorethylene, ethanol, kerosene, mineral oil, vegetable oil, paraffin oil, olein; Sp. gr.: 0.912 (20 C); Visc.: 28 cps (20 C); M.P.: < 0 C; HLB: 7.1; Stability: Good in hard or saline waters and in reasonable conc. of acids and alkalis; pH: 6–8 (1% aq.); Biodegradable
Composition: 100% active ingred.

Toxicity: May cause skin or eye irritation
Warnings and Caution: Spillages are slippery

Teric 12A6. ICI Australia Ltd.
Chemical Description: Ethylene oxide condensate based on straight chain synthetic alcohols, premium grade (6 moles EO)
Category: Wetting and dispersing agent, detergent, intermediate, nonionic
Applications: Specialty formulations for textile processing, hard surface abrasives, metal descaling, liquid and powdered detergents for domestic and industrial use in laundries, metal cleaning, sanitizing, dairy detergents; chemical intermediate
Form: Liquid to paste; Color: 100 (Hazen)
Properties: Solubility: Sol. in benzene, ethyl acetate, ethyl Icinol, perchlorethylene, ethanol, kerosene, mineral and vegetable oil, paraffin oil, olein; disp. in water; Sp. gr.: 0.955 (50 C); Visc.: 24 cps (50 C); M.P.: 10 ± 2 C; HLB: 11.2; Cloud Pt.: 40 ± 2 C (1% in hard water); Stability: Good in hard or saline waters and in reasonable concs. of acids and alkalis; Surface Tension: 27.8 dynes/cm (0.1%, 20 C); pH: 6–8 (1% aq.); Biodegradable
Composition: 100% active ingred.
Toxicity: May cause eye or skin irritation
Warnings and Caution: Spillages are slippery

Teric 12A9. ICI Australia Ltd.
Chemical Description: Ethylene oxide condensate based on straight chain synthetic alcohols, premium grade (9 moles EO)
Category and Applications: See Teric 12A6
Form: Paste; Color: 100 (Hazen)
Properties: Solubility: Sol. in water, benzene, ethyl acetate, ethyl Icinol, perchlorethylene, ethanol, olein; disp. in vegetable oil; Sp. gr.: 0.984 (50 C); Visc.: 32 cps (50 C); M.P.: 20 ± 2 C; HLB: 13.2; Cloud Pt.: 82 ± 2 C (1% in hard water); Stability: Good in hard or saline waters and in reasonable concs. of acids and alkalis; Surface Tension: 30.2 dynes/cm (0.1%, 20 C); pH: 6–8 (1% aq.); Biodegradable
Composition: 100% active ingred.
Toxicity: May cause skin and eye irritation
Warnings and Caution: Spillages are slippery

Teric 12A12. ICI Australia Ltd.
Chemical Description: Ethylene oxide condensate based on straight chain synthetic alcohols, premium grade (11.5 moles EO)
Category and Applications: See Teric 12A6
Form: Solid; Color: 100 (Hazen)
Properties: Solubility: Sol. in benzene, ethyl acetate, ethyl Icinol, perchlorethylene, ethanol, olein; disp. in vegetable oil; Sp. gr.: 0.993 (50 C); Visc.: 37 cps (50 C); M.P.: 24 ± 2 C; HLB: 14.4; Cloud Pt.: 92 ± 2 C (1% in hard

water); Stability: Good in hard or saline waters and in reasonable concs. of acids and alkalis; Surface Tension: 35.6 dynes/cm (0.1%, 20 C); pH: 6–8 (1%-aq.); Biodegradable
Composition: 100% active ingred.
Toxicity: May cause skin and eye irritation
Warnings and Caution: Spillages are slippery

Teric 12A16. ICI Australia Ltd.
Chemical Description: Ethylene oxide condensate based on straight chain synthetic alcohols, premium grade (16 moles EO)
Category: Wetting and dispersing agent, emulsifier, detergent, nonionic
Applications: Wetting agent esp. in conditions of high alkalinity or salinity; emulsifier of fats and oily residues in degreasing compounds and industrial and domestic hard surface cleaners; used in alkaline cleaners, in the solubilization of perfumes for detergent products, and in specialty preparations such as deodorant sanitizer blocks
Form: Solid; Color: 100 (Hazen)
Properties: Solubility: Sol. in water, benzene, ethyl acetate, ethyl Icinol, perchlorethylene, ethanol, olein; Sp. gr.: 1.009 (50 C); Visc.: 38 cps (50 C); M.P.: 27 ± 2 C; HLB: 15.3; Cloud Pt.: 98 ± 2 C (1% in hard water); Stability: Stable to highly alkaline or saline sol'ns.; Surface Tension: 33.7 dynes/cm (0.1%, 20 C); pH: 6–8 (1% aq.); Biodegradable
Composition: 100% active ingred.
Toxicity: May cause skin or eye irritation
Warnings and Caution: Spillages are slippery

Teric 12A23. ICI Australia Ltd.
Chemical Description: Ethylene oxide condensate based on straight chain synthetic alcohols, premium grade (23 moles EO)
Category and Applications: See Teric 12A16
Form: Solid; Color: 100 (Hazen)
Properties: Solubility: Sol. in water, benzene, ethyl acetate, ethyl Icinol, perchlorethylene, ethanol, olein; Sp. gr.: 1.032 (50 C); Visc.: 75 cps (50 C); M.P.: 35 ± 2 C; HLB: 16.6; Cloud Pt.: > 100 C (1% in hard water); Stability: Stable to highly alkaline and saline sol'ns.; Surface Tension: 35.7 dynes/cm (0.1%, 20 C); pH: 6–8 (1% aq.); Biodegradable
Composiiton: 100% active ingred.
Toxicity: May cause eye and skin irritation
Warnings and Caution: Spillages are slippery

Teric 12M2. ICI Australia Ltd.
Chemical Description: Cocoamine + 2 moles EO
Category: Wetting and dispersing agent, emulsifier, stabilizer, nonionic
Applications: Wetting of hydrophobic surfaces, e.g., alloy steel, stone, and fibers; dewatering of washed metal pieces prior to plating; dewatering of stone aggregate

prior to being incorporated into bituminous road mixes; textile processing; formulation of lubricants and dye bath auxiliaries; emulsion stabilization; fat liquoring compounds; preparation of wax emulsions for paper impregnation
Form: Liquid
Properties: Solubility: Sol. in benzene, ethyl acetate, ethyl Icinol, perchlorethylene, ethanol, kerosene, mineral and vegetable oil, paraffin oil, olein; Sp. gr.: 0.913; Visc.: 166 cps (20 C); M.P.: –1 ± 2 C; HLB: 11.4; Stability: Good in hard or saline waters and in reasonable concs. of acids and alkalis; Surface Tension: 28.2 dynes/cm (0.01%, 20 C); pH: 8–10 (1% aq.)
Composition: 100% active ingred.
Toxicity: May cause skin and eye irritation
Warnings and Caution: Spillages are slippery

Teric 12M5. ICI Australia Ltd.
Chemical Description: Cocoamine + 5 moles EO
Category and Applications: See Teric 12M2
Form: Liquid
Properties: Solubility: Sol. in benzene, ethyl acetate, ethyl Icinol, perchlorethylene, ethanol, mineral and vegetable oil, paraffin oil, olein; disp. in water and kerosene; Sp. gr.: 0.971; Visc.: 311 cps (20 C); M.P.: –9 ± 2 C; HLB: 12.4; Stability: Good in hard or saline waters and in reasonable concs. of acids and alkalis; Surface Tension: 29.2 dynes/cm (0.01%, 20 C); pH: 8–10 (1% aq.)
Composition: 100% active ingred.
Toxicity: May cause skin and eye irritation
Warnings and Caution: Spillages are slippery

Teric 12M15. ICI Australia Ltd.
Chemical Description: Cocoamide + 15 moles EO
Category and Applications: See Teric 12M2
Form: Liquid
Properties: Solubility: Sol. in water, benzene, ethyl acetate, ethyl Icinol, perchlorethylene, ethanol, vegetable oil, olein; partly sol. in kerosene, mineral and paraffin oil; Sp. gr.: 1.038; Visc.: 343 cps (20 C); M.P.: –11 ± 2 C; HLB: 15.7; Cloud Pt.: > 100 C (1% in hard water); Stability: Good in hard or saline waters and in reasonable concs. of acids and alkalis; Surface Tension: 39.6 dynes/cm (0.01%, 20 C); pH: 8–10 (1% aq.)
Composition: 100% active ingred.
Toxicity: May cause skin and eye irritation
Warnings and Caution: Spillages are slippery

Teric 13A9. ICI Australia Ltd.
Chemical Description: Tridecyl alcohol condensed with ≈9 moles EO per mole of alcohol
Category: Wetting and dispersing agent, emulsifier, detergent, nonionic
Applications: Pigment dispersions and wettable powders for agricul-

tural and horticultural applications; formulation of self-emulsifiable O/W systems; solvent emulsion cleaners; nonbiodegradable
Form: Liquid; Color: 150 (Hazen)
Properties: Solubility: Sol. in water, ethyl acetate, ethyl Icinol, ethanol, olein; Sp. gr.: 1.018 (20 C); Visc.: 142 cps (20 C); M.P.: < 0 C; HLB: 13.3; Cloud Pt.: 63 ± 2 C (1% in hard water); Surface Tension: 27.9 dynes/cm (0.1%, 20 C); pH: 6–8 (1% aq.)
Composition: 90% active ingred.
Toxicity: May cause skin and eye irritation
Warnings and Caution: Spillages are slippery

Teric 16A16. ICI Australia Ltd.
Chemical Description: Cetyl-stearyl alcohol + 16 moles EO
Category: Wetting and dispersing agent, emulsifier, detergent, solubilizer, nonionic
Applications: Dispersant in difficult hydrophobic conditions, e.g., the milling of carbon black into rubber stock; dye and pigment carriers at high temps.; textile processing aids, e.g., dye-bath additives; manufacture of wax emulsions and polishes for industrial and domestic use; self-scouring textile lubricants; antistatic properties; barrier creams and waterless hand cleaners; heavy duty cleaners
Form: Flakes; Color: 100 (Hazen)
Properties: Solubility: Sol. in water, benzene, ethyl acetate, ethyl Icinol, perchlorethylene, ethanol, olein; Sp. gr.: 1.01 (50 C); Visc.: 65 cps (50 C); M.P.: 38 ± 2 C; HLB: 14.9; Cloud Pt.: > 100 C (1% in hard water); Stability: Stable to acids, alkalis, and salts; Surface Tension: 37.1 dynes/cm (0.1%, 20 C); pH: 6–8 (1% aq.); Biodegradable
Composition: 100% active ingred.
Toxicity: May cause skin and eye irritation
Warnings and Caution: Spillages are slippery

Teric 16A22. ICI Australia Ltd.
Chemical Description: Cetyl-stearyl alcohol + 22 moles EO
Category and Applications: See Teric 16A16
Form: Flakes; Color: 100 (Hazen)
Properties: Solubility: Sol. in water, benzene, ethyl acetate, ethyl Icinol, perchlorethylene, ethanol, olein; Sp. gr.: 1.01 (50 C); Visc.: 92 cps (50 C); M.P.: 40 ± 2 C; HLB: 15.8; Cloud Pt.: > 100 C (1% in hard water); Stability: Stable to acids, alkalis, and salts; Surface Tension: 41.3 dynes/cm (0.1%, 20 C); pH: 6–8 (1% aq.); Biodegradable
Composition: 100% active ingred.
Toxicity: May cause skin and eye irritation
Warnings and Caution: Spillages are slippery

Teric 16A29. ICI Australia Ltd.
Chemical Description: Cetyl-stearyl alcohol + 29 moles EO
Category: Wetting and dispersing agent, emulsifier, detergent, sol-

ubilizer, nonionic
Applications: Dispersant in difficult hydrophobic conditions, e.g., the milling of carbon black into rubber stock; dye and pigment carriers at high temps.; textile processing aids, e.g., dye-bath additives; manufacture of wax emulsions and polishes for industrial and domestic use; self-scouring textile lubricants; antistatic properties; heavy duty cleaners
Form: Flakes; Color: 100 (Hazen)
Properties: Solubility: Sol. in water, benzene, ethyl acetate, ethyl Icinol, perchlorethylene, ethanol, olein; Sp. gr.: 1.05 (50 C); Visc.: 105 cps (50 C); M.P.: 45 ± 2 C; HLB: 17.7; Cloud Pt.: > 100 C (1% in hard water); Stability: Stable to acids, alkalis, and salts; Surface Tension: 46.0 dynes/cm (0.1%, 20 C); pH: 6–8 (1% aq.); Biodegradable
Composition: 100% active ingred.
Toxicity: May cause skin and eye irritation
Warnings and Caution: Spillages are slippery

Teric 16M2. ICI Australia Ltd.
Chemical Description: Soya amine + 2 moles EO
Category: Wetting and dispersing agent, emulsifier, nonionic
Applications: Emulsifier for waxes and fats; formulation of leather dressing compounds, fiber lubricants, and wax emulsions for the waterproofing of fiber board packaging and of particle board for the building industry
Form: Liquid
Properties: Solubility: Sol. in benzene, ethyl acetate, ethyl Icinol, perchlorethylene, ethanol, kerosene, mineral and vegetable oil, paraffin oil, olein; Sp. gr.: 0.906; Visc.: 191 cps (20 C); M.P.: –15 ± 2 C; HLB: 9.5; Stability: Good in hard or saline waters and in reasonable concs. of acids and alkalis; Surface Tension: 26.6 dynes/cm (0.01%, 20 C); pH: 8–10 (1% aq.)
Composition: 100% active ingred.
Toxicity: May cause skin and eye irritation
Warnings and Caution: Spillages are slippery

Teric 16M5. ICI Australia Ltd.
Chemical Description: Soya amine + 5 moles EO
Category: Wetting and dispersing agent, emulsifier, nonionic
Applications: Wetting adjuvant in cleaning compounds, agricultural sprays, and dye leveling compounds; formulations of metal cleaning compounds; emulsifier for fats and waxes; formulation of leather dressing compounds, fiber lubricants, and wax emulsions for the waterproofing of fiber board packaging and of particle board for the building industry
Form: Liquid
Properties: Solubility: Sol. in benzene, ethyl acetate, ethyl Icinol, perchlorethylene, ethanol, kerosene, mineral and vegetable oil, paraffin oil, olein; partly sol. in water; Sp. gr.: 0.954; Visc.: 182 cps (20 C); M.P.: –16 ± 2 C; HLB:

10.5; Stability: Good in hard or saline waters and in reasonable concs. of acids and alkalis; Surface Tension: 29.7 dynes/cm (0.01%, 20 C); pH: 8–10 (1% aq.)
Composition: 100% active ingred.
Toxicity: May cause skin and eye irritation
Warnings and Caution: Spillages are slippery

Teric 16M10. ICI Australia Ltd.
Chemical Description: Soya amine + 10 moles EO
Category: Wetting and dispersing agent, emulsifier, nonionic
Applications: Emulsifier for fats and waxes; used in formulations of leather dressing compounds, fiber lubricants, and wax emulsions for the waterproofing of fiber board packaging and of particle board for the building industry
Form: Liquid
Properties: Solubility: Sol. in water, benzene, ethyl acetate, ethyl Icinol, perchlorethylene, ethanol, kerosene, mineral and vegetable oil, paraffin oil, olein; Sp. gr.: 0.994; Visc.: 219 cps (20 C); M.P.: −22 ± 2 C; HLB: 12.1; Cloud Pt.: 89 ± 2 C (1% in hard water); Stability: Good in hard or saline waters and in reasonable concs. of acids and alkalis; Surface Tension: 36.1 dynes/cm (0.01%, 20 C); pH: 8–10 (1% aq.)
Composition: 100% active ingred.
Toxicity: May cause skin and eye irritation
Warnings and Caution: Spillages are slippery

Teric 16M15. ICI Australia Ltd.
Chemical Description: Soya amine + 15 moles EO
Category: Wetting and dispersing agent, emulsifier, nonionic
Applications: Wetting adjuvant in cleaning compounds, agricultural sprays, and dye leveling compounds; wetting and dispersing agent in dye-baths and in highly acid or highly saline conditions; emulsifier for fats and waxes; formulations of leather dressing compounds, fiber lubricants, and wax emulsions for the waterproofing of fiber board packaging and of particle board for the building industry
Form: Liquid
Properties: Solubility: Sol. in water, benzene, ethyl acetate, ethyl Icinol, perchlorethylene, ethanol; partly sol. in kerosene, vegetable and mineral oil, paraffin oil, olein; Sp. gr.: 1.021; Visc.: 266 cps (20 C); M.P.: −18 ± 2 C; HLB: 13.8; Cloud Pt.: > 100 C (1% in hard water); Stability: Good in hard or saline waters and in reasonable concs. of acids and alkalis; Surface Tension: 37.4 dynes/cm (0.01%, 20 C); pH: 8–10 (1% aq.)
Composition: 100% active ingred.
Toxicity: May cause skin and eye irritation
Warnings and Caution: Spillages are slippery

Teric 17A2. ICI Australia Ltd.
Chemical Description: Cetyl-oleyl alcohol + 2 moles EO
Category: Emulsifier, coemulsifier,

detergent, intermediate, nonionic
Applications: Manufacture of self-emulsifiable and self-scouring textile lubricants, solvent cleaners, waterless hand cleaners; coemulsifiers; detergent additives in petroleum oils, esp. crude fuel oils (aids sludge prevention and minimizes deposition and blockages in fuel lines); intermediate for the production of phosphated and sulfated anionic surfactants
Form: Liquid; Color: 150 (Hazen)
Properties: Solubility: Sol. in benzene, ethyl acetate, ethyl Icinol, perchlorethylene, ethanol, kerosene, vegetable and mineral oil, paraffin oil, olein; disp. in water; Sp. gr.: 0.902 (20 C); Visc.: 37 cps (20 C); M.P.: 11 ± 2 C; HLB: 6.0; Stability: Good in hard or saline waters and in reasonable concs. of acids and alkalis; pH: 6–8 (1% aq.); Biodegradable
Composition: 100% active ingred.
Toxicity: May cause skin and eye irritation
Warnings and Caution: Spillages are slippery

Teric 17A3. ICI Australia Ltd.
Chemical Description: Cetyl-oleyl alcohol + 3 moles EO
Category and Applications: See Teric 17A2
Form: Liquid; Color: 150 (Hazen)
Properties: Solubility: Sol. in benzene, ethyl acetate, ethyl Icinol, perchlorethylene, ethanol, kerosene, mineral and vegetable oil, paraffin oil, olein; disp. in water; Sp. gr.: 0.918 (20 C); Visc.: 41 cps (20 C); M.P.: 13 ± 2 C; HLB: 7.5; Stability: Good in hard or saline waters and in reasonable concs. of acids and alkalis; pH: 6–8 (1% aq.); Biodegradable
Composition: 100% active ingred.
Toxicity: May cause skin and eye irritation
Warnings and Caution: Spillages are slippery

Teric 17A6. ICI Australia Ltd.
Chemical Description: Cetyl-oleyl alcohol + 6 moles EO
Category and Applications: See Teric 17A2
Form: Liquid; Color: 150 (Hazen)
Properties: Solubility: Sol. in benzene, ethyl acetate, ethyl Icinol, perchlorethylene, ethanol, vegetable oil, olein; disp. in water; Sp. gr.: 0.957 (20 C); Visc.: 60 cps (20 C); M.P.: 15 ± 2 C; HLB: 10.2; Stability: Good in hard or saline waters and in reasonable concs. of acids and alkalis; pH: 6–8 (1% aq.); Biodegradable
Composition: 100% active ingred.
Toxicity: May cause skin and eye irritation
Warnings and Caution: Spillages are slippery

Teric 17A8. ICI Australia Ltd.
Chemical Description: Cetyl-oleyl alcohol + 8 moles EO
Category: Wetting and dispersing agent, emulsifier, nonionic
Applications: Wetting agent in antistatic and desizing baths for yarns and fabrics; promotes uniform dyeing of fiber blends such

as wool/nylon; emulsifier in winter and summer spray oils for horticultural use and salt-water stable dispersants for removal of oil slicks
Form: Paste; Color: 150 (Hazen)
Properties: Solubility: Sol. in water, benzene, ethyl acetate, ethyl Icinol, perchlorethylene, ethanol, olein; Sp. gr.: 0.963 (50 C); Visc.: 35 cps (50 C); M.P.: 17 ± 2 C; HLB: 11.6; Cloud Pt.: 41 ± 2 C (1% in hard water); Stability: Good in hard or saline waters and in reasonable concs. of acids and alkalis; Surface Tension: 30.8 dynes/cm (0.1%, 20 C); pH: 6–8 (1% aq.); Biodegradable
Composition: 100% active ingred.
Toxicity: May cause skin and eye irritation
Warnings and Caution: Spillages are slippery

Teric 17A10. ICI Australia Ltd.
Chemical Description: Cetyl-oleyl alcohol + 10 moles EO
Category and Applications: See Teric 17A8; also as solubilizer for essential oils and perfumes in sanitation chemicals
Form: Solid; Color: 150 (Hazen)
Properties: Solubility: Sol. in water, benzene, ethyl acetate, ethyl Icinol, perchlorethylene, ethanol, olein; Sp. gr.: 0.973 (50 C); Visc.: 43 cps (50 C); M.P.: 24 ± 2 C; HLB: 12.7; Cloud Pt.: 72 ± 2 C (1% in hard water); Stability: Good in hard or saline waters and in reasonable concs. of acids and alkalis; Surface Tension: 32.9 dynes/cm (0.1%, 20 C); pH: 6–8 (1% aq.); Biodegradable
Composition: 100% active ingred.
Toxicity: May cause skin and eye irritation
Warnings and Caution: Spillages are slippery

Teric 17A13. ICI Australia Ltd.
Chemical Description: Cetyl-oleyl alcohol + 13 moles EO
Category and Applications: See Teric 17A10
Form: Solid; Color: 150 (Hazen)
Properties: Solubility: Sol. in water, benzene, ethyl acetate, ethyl Icinol, perchlorethylene, ethanol, olein; Sp. gr.: 0.986 (50 C); Visc.: 45 cps (50 C); M.P.: 25 ± 2 C; HLB: 13.8; Cloud Pt.: 84 ± 2 C (1% in hard water); Stability: Good in hard or saline waters and in reasonable concs. of acids and alkalis; Surface Tension: 33.0 dynes/cm (0.1%, 20 C); pH: 6–8 (1% aq.); Biodegradable
Toxicity: May cause skin and eye irritation
Warnings and Caution: Spillages are slippery

Teric 17A25. ICI Australia Ltd.
Chemical Description: Cetyl-oleyl alcohol + 25 moles EO
Category and Applications: See Teric 17A10
Form: Solid; Color: 150 (Hazen)
Properties: Solubility: Sol. in water, benzene, ethyl acetate, ethyl Icinol, perchlorethylene, ethanol; Sp. gr.: 1.030 (50 C); Visc.: 92 cps (50 C); M.P.: 37 ± 2 C; HLB:

16.2; Cloud Pt.: > 100 C (1% in hard water); Stability: Good in saline or hard waters and in reasonable concs. of acids and alkalis; Surface Tension: 42.0 dynes/cm (0.1%, 20 C); Biodegradable
Composition: 100% active ingred.
Toxicity: May cause skin or eye irritation
Warnings and Caution: Spillages are slippery

Teric 17M2. ICI Australia Ltd.
Chemical Description: Tallow amine + 2 moles EO
Category: Wetting and dispersing agent, emulsifier, corrosion inhibitor, nonionic
Applications: Dewatering agent displacing water from surfaces such as printed electronic circuits, electrical components, and roadmaking aggregate; emulsifier for fats, waxes, and mineral oils used in cutting oils and metal working lubricants; corrosion inhibitor in protective oils and metal working compounds
Form: Paste
Properties: Solubility: Sol. in benzene, ethyl acetate, ethyl Icinol, perchlorethylene, ethanol, kerosene, mineral and vegetable oil, paraffin oil, olein; Sp. gr.: 0.896; Visc.: 56 cps (40 C); M.P.: 25 ± 2 C; HLB: 9; Stability: Good in hard or saline waters and in reasonable concs. of acids and alkalis; Surface Tension: 26.5 dynes/cm (0.01%, 20 C); pH: 8–10 (1% aq.)
Composition: 100% active ingred.
Toxicity: May cause skin and eye irritation
Warnings and Caution: Spillages are slippery

Teric 17M5. ICI Australia Ltd.
Chemical Description: Tallow amine + 5 moles EO
Category and Applications: See Teric 17M2
Form: Semiliquid
Properties: Solubility: Sol. in benzene, ethyl acetate, ethyl Icinol, perchlorethylene, ethanol, kerosene, mineral and vegetable oil, paraffin oil, olein; partly sol. in water; Sp. gr.: 0.953; Visc.: 177 cps (20 C); M.P.: 11 ± 2 C; HLB: 10; Stability: Good in hard or saline waters and in reasonable concs. of acids and alkalis; Surface Tension: 31.5 dynes/cm (0.01%, 20 C); pH: 8–10 (1% aq.)
Composition: 100% active ingred.
Toxicity: May cause skin and eye irritation
Warnings and Caution: Spillages are slippery

Teric 17M15. ICI Australia Ltd.
Chemical Description: Tallow amine + 15 moles EO
Category and Applications: See Teric 17M2
Form: Liquid
Properties: Solubility: Sol. in water, benzene, ethyl acetate, ethyl Icinol, perchlorethylene, ethanol, vegetable oil, olein; partly sol. in kerosene, mineral oil, paraffin oil; Sp. gr.: 1.028; Visc.: 428 cps (20

C); M.P.: −19 ± 2 C; HLB: 13.3; Cloud Pt.: > 100 C (1% in hard water); Stability: Good in hard or saline waters and in reasonable concs. of acids or alkalis; Surface Tension: 40.0 dynes/cm (0.01%, 20 C); pH: 8–10 (1% aq.)
Composition: 100% active ingred.
Toxicity: May cause skin and eye irritation
Warnings and Caution: Spillages are slippery

Teric 18M2. ICI Australia Ltd.
Chemical Description: Octadecyl amine + 2 moles EO
Category: Wetting and dispersing agent, emulsifier, corrosion inhibitor, nonionic
Applications: Emulsion stabilizer; emulsifier used for agricultural toxicants, and in the processing of textiles, paper, leather, and building board; corrosion inhibitor in specialty lubricants and greases
Form: Solid
Properties: Solubility: Sol. in benzene, ethyl acetate, ethyl Icinol, perchlorethylene, ethanol, kerosene, mineral and vegetable oil, paraffin oil, olein; Sp. gr.: 0.887 (50 C); Visc.: 75 cps (50 C); M.P.: 50 ± 2 C; HLB: 8.5; Stability: Good in hard or saline waters and in reasonable concs. of acids and alkalis; pH: 8–10 (1% aq.)
Composition: 100% active ingred.
Toxicity: May cause skin and eye irritation
Warnings and Caution: Spillages are slippery

Teric 18M5. ICI Australia Ltd.
Chemical Description: Octadecyl amine + 5 moles EO
Category and Applications: See Teric 18M2; also for wetting, softening, and antistatic aids in the solvent cleaning of textiles and the compounding of plastics
Form: Solid
Properties: Solubility: Sol. in benzene, ethyl acetate, ethyl Icinol, perchlorethylene, ethanol, kerosene, mineral oil, vegetable oil, paraffin oil, olein; disp. in water; Sp. gr.: 0.935 (50 C); Visc.: 46 cps (50 C); M.P.: 35 ± 2 C; HLB: 9.5; Stability: Good in hard or saline waters and in reasonable concs. of acids and alkalis; pH: 8–10 (1% aq.)
Composition: 100% active ingred.
Toxicity: May cause skin and eye irritation
Warnings and Caution: Spillages are slippery

Teric 18M10. ICI Australia Ltd.
Chemical Description: Octadecyl amine + 10 moles EO
Category and Applications: See Teric 18M5
Form: Liquid
Properties: Solubility: Sol. in water, benzene, ethyl acetate, ethyl Icinol, perchlorethylene, ethanol, kerosene, vegetable oil, olein; partly sol. in mineral and paraffin oil; Sp. gr.: 0.997 (20 C); Visc.: 255 cps (20 C); M.P.: 18 ± 2 C; HLB: 11.2; Cloud Pt.: > 100 C (1% in hard water); Stability: Good in hard or saline waters and

in reasonable concs. of acids and alkalis; Surface Tension: 34.0 dynes/cm (0.1%, 20 C); pH: 8–10 (1% aq.)
Composition: 100% active ingred.
Toxicity: May cause skin and eye irritation
Warnings and Caution: Spillages are slippery

Teric 18M20. ICI Australia Ltd.
Chemical Description: Octadecyl amine + 20 moles EO
Category and Applications: See Teric 18M2; also as wetting and dispersing agents to ensure the level application of dyes to yarns and fabrics; esp. effective in acid sol'ns.
Form: Paste
Properties: Solubility: Sol. in water, benzene, ethyl acetate, ethyl Icinol, perchlorethylene, ethanol, olein; partly sol. in vegetable oil; Sp. gr.: 1.025 (50 C); Visc.: 66 cps (50 C); M.P.: 28 ± 2 C; HLB: 14.5; Cloud Pt.: > 100 C (1% in hard water); Stability: Good in hard and saline waters and in reasonable concs. of acids and alkalis; Surface Tension: 40.3 dynes/cm (0.1%, 20 C); pH: 8–10 (1% aq.)
Composition: 100% active ingred.
Toxicity: May cause skin and eye irritation
Warnings and Caution: Spillages are slippery

Teric 18M30. ICI Australia Ltd.
Chemical Description: Octadecyl amine + 30 moles EO
Category and Applications: See Teric 18M20
Form: Solid
Properties: Solubility: Sol. in water, benzene, ethyl acetate, ethyl Icinol, perchlorethylene, ethanol; Sp. gr.: 1.038 (50 C); Visc.: 116 cps (50 C); M.P.: 28 ± 2 C; HLB: 17.8; Cloud Pt.: > 100 C (1% in hard water); Stability: Good in hard or saline waters and in reasonable concs. of acids and alkalis; Surface Tension: 43.0 dynes/cm (0.1%, 20 C); pH: 8–10 (1% aq.)
Composition: 100% active ingred.
Toxicity: May cause skin and eye irritation
Warnings and Caution: Spillages are slippery

Teric 121. ICI Australia Ltd.
Chemical Description: Fatty acid EO condensate
Category: Antifoam agent, emulsifier, nonionic
Applications: Coemulsifier of vegetable and mineral oils, paraffinic waxes, and common solvents; self-scouring lubricant for textile processing; foam control agent in paper manufacture, industrial fermentation and distillation, textile dyeing, the formulation of adhesives, and in waste water treatment
Form: Liquid; Color: > 250 (Hazen)
Properties: Solubility: Sol. in benzene, ethyl acetate, ethyl Icinol, perchloroethylene, ethanol, vegetable oil, olein; Sp. gr.: 0.987 (20 C); Visc.: 353 cps (20 C); M.P.: < 0 C; HLB: 7.4; Stability: Good

sol'n. stability in hard or saline waters and in reasonable concs. of acids and alkalis; Self-emulsifying; pH: 8 (1% aq.); Biodegradable
Composition: 100% active ingred.
Toxicity: May cause eye and skin irritation
Warnings and Caution: Spillages are slippery

Teric 124. ICI Australia Ltd.
Chemical Description: Fatty acid EO condensate
Category: Antifoam agent, emulsifier, nonionic
Applications: Coemulsifier of vegetable and mineral oils, paraffinic waxes, and common solvents; emulsifier for mineral oil lubricants for the processing of yarns and cloth; foam control agent in paper mfg., industrial fermentation and distillation, textile dyeing, the formulation of adhesives, and in waste water treatment
Form: Liquid; Color: > 250 (Hazen)
Properties: Solubility: Sol. in benzene, ethyl acetate, ethyl Icinol, perchloroethylene, ethanol, mineral and vegetable oils, olein; disp. in paraffin oil; Sp. gr.: 0.966 (20 C); Visc.: 118 cps (20 C); M.P.: < 0 C; HLB: 4.5; Stability: Good in hard or saline waters and in reasonable amounts of acids and alkalis; Self-emulsifying; pH: 8 (1% aq.); Biodegradable
Composition: 100% active ingred.
Toxicity: May cause eye and skin irritation
Warnings and Caution: Spillages are slippery

Teric 127. ICI Australia Ltd.
Chemical Description: Fatty acid EO condensate
Category and Applications: See Teric 124
Form: Liquid; Color: > 250 (Hazen)
Properties: Solubility: Sol. in benzene, ethyl acetate, ethyl Icinol, perchloroethylene, ethanol, mineral and vegetable oils, olein; disp. in paraffin oil; Sp. gr.: 0.978 (20 C); Visc.: 87 cps (20 C); M.P.: < 0 C; HLB: 7.7; Stability: Good in hard or saline waters and in reasonable amounts of acids and alkalis; Self-emulsifying; pH: 8 (1% aq.); Biodegradable
Composition: 100% active ingred.
Toxicity: May cause eye and skin irritation
Warnings and Caution: Spillages are slippery

Teric 151. ICI Australia Ltd.
Chemical Description: Fatty acid EO condensate
Category: Emulsifier, antistatic agent, lubricant, nonionic
Applications: Textile processing aids, antistats, and lubricants for synthetic fibers; forms O/W emulsions with paraffinic mineral oils, vegetable oils, animal fats, and common solvents, e.g., kerosene and turpentine
Form: Solid; Color: > 250 (Hazen)

Properties: Solubility: Sol. in benzene, ethyl acetate, ethyl Icinol, perchlorethylene, ethanol, kerosene, vegetable oil, olein; disp. in water; Sp. gr.: 0.994 (50 C); Visc.: 45 cps (50 C); M.P.: 30 ± 2 C; HLB: 11.6; Stability: Good in hard and saline waters and in reasonable amounts of acids and alkalis; pH: 7–9 (1% aq.); Biodegradable
Composition: 100% active ingred.
Toxicity: May cause eye and skin irritation
Warnings and Caution: Spillages are slippery

Teric 152. ICI Australia Ltd.
Chemical Description: Nonionic blend
Category: Emulsifier, nonionic
Applications: Formulation of textile processing aids; emulsifier for paraffin oil and for a variety of petroleum fractions; useful component of solvent degreasers and cleaners used in metal cleaning and cleaning of workshops and garages
Form: Liquid; Color: > 250 (Hazen)
Properties: Solubility: Sol. in benzene, ethyl acetate, ethyl Icinol, perchlorethylene, ethanol, kerosene, mineral and vegetable oils, olein; disp. in paraffin oil, water; Sp. gr.: 0.998 (20 C); Visc.: 250 cps (20 C); M.P.: < 0 C; HLB: 6.7; Stability: Good in hard and saline waters and in reasonable amounts of acids and alkalis; pH: 5–7 (1%); Biodegradable
Composition: 100% active ingred.
Toxicity: May cause eye and skin irritation
Warnings and Caution: Spillages are slippery

Teric 154. ICI Australia Ltd.
Chemical Description: Fatty acid EO condensate
Category: Emulsifier, antistatic agent, lubricant, nonionic
Applications: Formulation of textile processing aids; antistat and lubricant for natural fibers; emulsifier of vegetable oils and light petroleum oils; self-emulsifying
Form: Liquid; Color: > 250 (Hazen)
Properties: Solubility: Sol. in water, benzene, ethyl acetate, ethyl Icinol, perchlorethylene, ethanol, kerosene, vegetable oil, olein; Sp. gr.: 1.031 (20 C); Visc.: 89 cps (20 C); M.P.: 7 ± 2 C; HLB: 13.3; Stability: Good in hard and saline waters and in reasonable amounts of acids and alkalis; Self-emulsifying; Surface Tension: 30.0 dynes/cm (0.1%, 20 C); pH: 6–8 (1% aq.); Biodegradable
Composition: 100% active ingred.
Toxicity: May cause eye and skin irritation
Warnings and Caution: Spillages are slippery

Teric 157. ICI Australia Ltd.
Chemical Description: Nonionic blend
Category: Wetting and dispersing agent, nonionic
Applications: Formulation of dis-

persible powders for agricultural, surface coatings, and ceramic applications
Form: Fine powder; Color: >250 (Hazen)
Properties: HLB: 12.8; Stability: Good in hard and saline waters and in reasonable amounts of acids and alkalis; pH: 7–9 (1% aq.)
Toxicity: May cause eye and skin irritation
Warnings and Caution: Spillages are slippery

Teric 158. ICI Australia Ltd.
Chemical Description: Blended alkyl phenol ethoxylate
Category and Applications: See Teric 157
Form: Liquid; Color: > 250 (Hazen)
Properties: Solubility: Sol. in benzene, ethyl acetate, ethyl Icinol, perchlorethylene, ethanol, olein; disp. in water and vegetable oil; Sp. gr.: 1.046 (20 C); Visc.: 322 cps (20 C); M.P.: < 0 C; HLB: 11.6; Stability: Good in hard and saline waters and in reasonable amounts of acids and alkalis; pH: 6–8 (1% aq.)
Composition: 100% active ingred.
Toxicity: May cause eye and skin irritation
Warnings and Caution: Spillages are slippery

Teric 160. ICI Australia Ltd.
Chemical Description: Alkylaryl polyoxyalkylene ether
Category: Wetting and dispersing agent, detergent, nonionic
Applications: Manufacture of surface coatings, e.g., paints and latex emulsions, esp. for dispersing of sulfurs and clays; low foam detergent powders and detergent sanitizers
Form: Liquid; Color: 250 (Hazen)
Properties: Solubility: Sol. in water, benzene, ethyl acetate, ethyl Icinol, perchlorethylene, ethanol, vegetable oil, olein; Sp. gr.: 1.032 (20 C); Visc.: 312 cps (20 C); M.P.: < 0 C; HLB: 11.3; Cloud Pt.: 29 ± 2 C (1% in hard water); Stability: Good in hard or saline waters and in reasonable concs. of acids and alkalis; Surface Tension: 30.4 dynes/cm (0.1%, 20 C); pH: 6–8 (1% aq.)
Composition: 100% active ingred.
Toxicity: May cause eye and skin irritation
Warnings and Caution: Spillages are slippery

Teric 161. ICI Australia Ltd.
Chemical Description: Alkylaryl polyoxyalkylene ether
Category and Applications: See Teric 160
Form: Liquid; Color: 250 (Hazen)
Properties: Solubility: Sol. in water, benzene, ethyl acetate, ethyl Icinol, perchlorethylene, ethanol, vegetable oil, and olein; Sp. gr.: 1.034 (20 C); Visc.: 340 cps (20 C); M.P.: < 0 C; HLB: 12.3; Cloud Pt.: 35 ± 2 C (1% in hard water); Stability: Good in hard or saline waters and in reasonable concs. of acids and alkalis; Surface Tension: 33.4 dynes/cm (0.1%, 20 C); pH:

6–8 (1% aq.)
Composition: 100% active ingred.
Toxicity: May cause eye and skin irritation
Warnings and Caution: Spillages are slippery

Teric 162. ICI Australia Ltd.
Chemical Description: Nonionic EO derivative
Category: Wetting and dispersing agent, detergent, nonionic
Applications: Manufacture of surface coatings, e.g., paints and latex emulsions, esp. for wetting pigment such as TiO_2; low-foam detergent powders for machine dishwashing, detergent sanitizers, and iodophor manufacture
Form: Liquid; Color: 250 (Hazen)
Properties: Solubility: Sol. in water, benzene, ethyl acetate, ethyl Icinol, perchlorethylene, ethanol, olein; Sp. gr.: 1.055 (20 C); Visc.: 680 cps (20 C); M.P.: 10 ± 2 C; HLB: 10.5; Cloud Pt.: 39 ± 2 C (1% in hard water); Stability: Good in hard or saline waters and in reasonable concs. of acids and alkalis; Surface Tension: 33.1 dynes/cm (0.1%, 20 C); pH: 6–8 (1% aq.)
Composition: 100% active ingred.
Toxicity: May cause eye and skin irritation
Warnings and Caution: Spillages are slippery

Teric 163. ICI Australia Ltd.
Chemical Description: Alkylaryl polyoxyalkylene ether
Category: Demulsifier, detergent, nonionic
Applications: Low-foam detergent powders for machine dishwashing, in-situ cleaning of milking plants, butter, vats, etc., detergent sanitizers, and iodophor manufacture
Form: Liquid; Color: 250 (Hazen)
Properties: Solubility: Sol. in water, benzene, ethyl acetate, ethyl Icinol, perchlorethylene, ethanol, vegetable oil, olein; Sp. gr.: 1.110 (20 C); Visc.: 1240 cps (20 C); M.P.: $<$ 0 C; HLB: 14.9; Cloud Pt.: 97 ± 2 C (1% in hard water); Stability: Good in hard and saline waters and in reasonable concs. of acids and alkalis; Surface Tension: 50.4 dynes/cm (0.1%, 20 C); pH: 4–6 (1% aq.)
Composition: 100% active ingred.
Toxicity: May cause eye and skin irritation
Warnings and Caution: Spillages are slippery

Teric 164. ICI Australia Ltd.
Chemical Description: Modified alcohol ethoxylate
Category: Detergent, nonionic
Applications: Manufacture of low-foam detergents for machine dishwashing, in-situ cleaning of milking plants, butter vats, etc., detergent sanitizers, iodophor manufacture
Form: Liquid; Color: 250 (Hazen)
Properties: Solubility: Sol. in water, benzene, ethyl acetate, ethyl Icinol, perchlorethylene, ethanol, kerosene, vegetable oil, olein; Sp. gr.: 0.937 (20 C); Visc.: 65 cps (20 C); M.P.: 2 ± 2 C; HLB: 10.8;

Cloud Pt.: 40 ± 2 C (1% in hard water); Stability: Good in hard or saline waters and in reasonable concs. of acids and alkalis; Surface Tension: 30.5 dynes/cm (0.1%, 20 C); pH: 6–8 (1% aq.); Biodegradable
Composition: 100% active ingred.
Toxicity: May cause eye and skin irritation
Warnings and Caution: Spillages are slippery

Teric 165. ICI Australia Ltd.
Chemical Description: Modified alcohol ethoxylate
Category: Wetting agent, dedusting agent, nonionic
Applications: Low-foaming additive for strongly alkaline cleaning powders, metal cleaning, rinse aids, bottle washing, and mechanical dishwashing; dedusting agent for powdered caustic and powdered detergents
Form: Liquid; Color: Dark brown
Properties: Solubility: Sol. in water, ethyl Icinol, ethanol, olein; Sp. gr.: 1.035 (20 C); Visc.: 68 cps (20 C); HLB: 12.5; Cloud Pt.: 30–35 C (1% aq.); Pour Pt.: 2–3 C; Acid No.: 3 max.; Stability: Good in strong alkalis; Surface Tension: 28.2 dynes/cm (0.1%); pH: 5; Biodegradable
Composition: 89% min. active
Toxicity: May cause eye and skin irritation
Warnings and Caution: Spillages are slippery

Teric 200. ICI Australia Ltd.
Chemical Description: Alkylene oxide derivative
Category: Emulsifier, nonionic
Applications: Manufacture of self-emulsifiable concentrates of insecticides, pesticides, and herbicides in a variety of solvents; esp. for phosphate ester-types, e.g., parathion, malathion, and highly conc. hormone ester herbicides
Form: Paste; Color: 150 (Hazen)
Properties: Solubility: Sol. in benzene, water, ethyl acetate, ethyl Icinol, perchlorethylene, ethanol, olein; Sp. gr.: 1.034 (50 C); Visc.: 123 cps (50 C); M.P.: 30 ± 2 C; HLB: 16.1; Cloud Pt.: 90 ± 2 C (1% in hard water); Stability: Good in hard or saline waters and in reasonable concs. of acids and alkalis; Surface Tension: 36.0 dynes/cm (0.1%, 20 C); pH: 6–8 (1% aq.)
Composition: 100% active ingred.
Toxicity: May cause eye and skin irritation
Warnings and Caution: Spillages are slippery

Teric 210. ICI Australia Ltd.
Category: Emulsifier, nonionic
Applications: Manufacture of self-emulsifiable concentrates of pesticides, insecticides, and herbicides in a variety of solvents; esp. for hormone-ester weed killers
Form: Paste; Color: 150 (Hazen)
Properties: Solubility: Sol. in water, benzene, ethyl acetate, ethyl Icinol, perchlorethylene, ethanol,

olein; Sp. gr.: 1.046 (50 C); Visc.: 75 cps (50 C); M.P.: 17 ± 2 C; HLB: 13.3; Cloud Pt.: 62 ± 2 C (1% in hard water); Stability: Good in hard or saline waters and in reasonable concs. of acids and alkalis; Surface Tension: 29.2 dynes/cm (0.1%, 20 C); pH: 6–8 (1% aq.)

Teric 300. ICI Australia Ltd.
Category: Detergent, wetting agent, lubricant, intermediate, anionic
Applications: Textile industry as self-scouring lubricant, yarn scouring detergent, and wetting agent; intermediate in the manufacture of superphosphates
Form: Liquid; Color: > 250 (Hazen)
Properties: Solubility: Sol. in water, benzene, ethyl acetate, ethyl Icinol, perchlorethylene, ethanol; Sp. gr.: 1.048 (20 C); Visc.: 264 cps (20 C); M.P.: 5 ± 2 C; HLB: 15.1; Surface Tension: 30.0 dynes/cm (0.1%, 20 C); pH: 8–10

Teric 303. ICI Australia Ltd.
Category: Stabilizer, corrosion inhibitor, anionic
Applications: Stabilization of iodophor concentrates; corrosion and scale formation inhibitor in feed water systems
Form: Paste; Color: > 250 (Hazen)
Properties: Solubility: Sol. in water, benzene, ethyl acetate, ethyl Icinol, perchlorethylene, ethanol, olein; partly sol. in vegetable oil; Sp. gr.: 1.015 (50 C); Visc.: 68 cps (50 C); M.P.: 28 ± 2 C; HLB: 14.5; Cloud Pt.: > 100 C; Surface Tension: 44.8 dynes/cm (0.1%, 20 C); pH: 8–10

Teric 304. ICI Australia Ltd.
Category: Demulsifier, anticaking agent, stabilizer, corrosion inhibitor, anionic
Applications: Separation of oil from O/W emulsions in crude oil recovery; anticaking agent for salts and compounded fertilizers; inhibition of corrosion and scale formation in feed water systems
Form: Liquid; Color: > 250 (Hazen)
Properties: Solubility: Sol. in water, benzene, ethyl acetate, ethyl Icinol, perchlorethylene, ethanol, vegetable oil, olein; Sp. gr.: 1.044 (20 C); Visc.: 412 cps (20 C); M.P.: < 0 C; Cloud Pt.: > 100 C; Surface Tension: 44.2 dynes/cm (0.1%, 20 C); pH: 5–7

Teric 305. ICI Australia Ltd.
Chemical Description: Phosphate ester of a nonionic detergent
Category: Wetting and dispersing agent, emulsifier, detergent, hydrotrope, lubricant, corrosion inhibitor, anionic
Applications: Wetting agent, detergent, and dispersant in acid pickling and cleaning baths, alkaline degreasers, and heavy duty detergent powders; in emulsion

polymerization and emulsifier systems, e.g., cutting oils; hydrotrope for stabilizing aq. sol'ns. of salts, esp. in heavy duty liquid detergents; dedusting agent in powdered detergents; lubricant and corrosion inhibitor in metal working and wire drawing lubricants
Form: Paste; Color: > 250 (Hazen)
Properties: Solubility: Sol. in water, benzene, ethyl acetate, ethyl Icinol, perchlorethylene, ethanol, vegetable oil, olein; Sp. gr.: 1.140 (50 C); Visc.: 2500 cps (50 C); M.P.: 25 ± 2 C; Cloud Pt.: > 100 C; Surface Tension: 34.5 dynes/cm (0.1%, 20 C); pH: 1–3

Teric 306. ICI Australia Ltd.
Chemical Description: Acid form of phosphated nonionic
Category and Applications: See Teric 305
Form: Liquid; Color: > 250 (Hazen)
Properties: Solubility: Sol. in water, ethyl Icinol, ethanol, olein; disp. in ethyl acetate; Sp. gr.: 1.400 (20 C); Visc.: 5000 cps (50 C); M.P.: 10 ± 2 C; Cloud Pt.: > 100 C; Surface Tension: 45.6 dynes/cm (0.1%, 20 C); pH: 1–3

Teric 350. ICI Australia Ltd.
Chemical Description: Fatty acid condensate
Category: Emulsifier, lubricant, corrosion inhibitor, nonionic
Applications: Emulsifier and coemulsifier for a variety of solvents and mineral oils, for formulations of self-emulsifiable systems of solvents for use as bases for solvent cleaners, degreasing and decarbonizing preparations; lubricant and corrosion inhibitor in oil-based lubricant systems esp. soluble oils for metal working industry
Form: Liquid; Color: > 250 (Hazen)
Properties: Solubility: Sol. in benzene, ethyl Icinol, perchlorethylene, ethanol, olein; disp. in water, ethyl acetate; Sp. gr.: 1.010 (20 C); Visc.: 600 cps (20 C); HLB: 16.6; pH: 8–10 (1% aq.)
Composition: 100% conc.

Teric 351B. ICI Australia Ltd.
Category and Applications: See Teric 350
Form: Liquid; Color: > 250 (Hazen)
Properties: Solubility: Sol. in perchlorethylene, olein; disp. in water; Sp. gr.: 1.017 (20 C); Visc.: 750 cps (20 C); HLB: 8.2; pH: 8–10 (1% aq.)

Teric BL8. ICI Australia Ltd.
Chemical Description: Polyalkylene oxide derivative of a synthetic alcohol
Category: Wetting and dispersing agent, detergent, emulsifier
Applications: Wetting and dispersing agent in printing inks, textiles, woolscouring, dust suppression, surface cleaning; domestic and industrial surface cleaners, polishes, liquid and powdered formulations; emulsifier of common solvents,

greases, and oils, emulsification of wool grease in woolscouring
Form: Liquid; Color: 250 (Hazen)
Properties: Solubility: Sol. in water, benzene, ethyl acetate, ethyl Icinol, perchlorethylene, ethanol, kerosene, vegetable oil, olein; Sp. gr.: 1.003 (20 C); Visc.: 63 cps (20 C); M.P.: < 0 C; HLB: 13.7; Cloud Pt.: 59 ± 2 C (1% in hard water); Stability: Stable in mild acids, mild alkalis, and in presence of electrolytes; Surface Tension: 30.3 dynes/cm (0.1%, 20 C); pH: 6–8 (1% aq.); Biodegradable
Composition: 100% active ingred.
Toxicity: May cause skin and eye irritation
Warnings and Caution: Spillages are slippery

Teric BL9. ICI Australia Ltd.
Chemical Description: Polyalkylene oxide derivative of a synthetic alcohol
Category and Applications: See Teric BL8
Form: Liquid; Color: 250 (Hazen)
Properties: Solubility: Sol. in water, benzene, ethyl acetate, ethyl Icinol, perchlorethylene, ethanol, vegetable oil, olein; Sp. gr.: 1.018 (20 C); Visc.: 100 cps (20 C); M.P.: < 0 C; HLB: 14.5; Cloud Pt.: 67 ± 2 C (1% in hard water); Stability: Stable in mild acids, mild alkalis, and in presence of electrolytes; Surface Tension: 30.0 dynes/cm (0.1%, 20 C); pH: 6–8 (1% aq.); Biodegradable
Composition: 98% active ingred.
Toxicity: May cause skin and eye irritation
Warnings and Caution: Spillages are slippery

Teric C12. ICI Australia Ltd.
Chemical Description: Castor oil + 12 moles EO
Category: Emulsifier, detergent, lubricant, nonionic
Applications: Emulsifier and coemulsifier in the formulation of soluble oils, solvent cleaners, and temporary protective coatings; corrosion resistant; component in industrial and institutional cleaning preparations; die lubricant in metal forming operations; mold release agent in plastics; fiber lubricant in textile processing
Form: Liquid; Color: > 250 (Hazen)
Properties: Solubility: Sol. in water, benzene, ethyl acetate, ethyl Icinol, ethanol, vegetable oil, olein; Sp. gr.: 1.048 (20 C); Visc.: 5000 cps (20 C); M.P.: 3 ± 2 C; HLB: 12.7; Cloud Pt.: 67–69 C (1% in hard water); Stability: Good in hard or saline waters and in reasonable concs. of acids and alkalis; Surface Tension: 42.0 dynes/cm (0.1%, 20 C); pH: 6–8 (1% aq.)
Composition: 100% active ingred.
Toxicity: May cause skin and eye irritation
Warnings and Caution: Spillages are slippery

Teric CDE. ICI Australia Ltd.
Chemical Description: Coconut fatty acid diethanolamide
Category: Emulsifier, detergent,

foam booster, nonionic
Applications: Emulsifier of perchlorethylene, power kerosene, etc. in water; foam booster, thickener, and detergent in domestic dishwashing formulations
Form: Liquid; Color: > 250 (Hazen)
Properties: Solubility: Sol. in benzene, ethyl Icinol, perchlorethylene, ethanol, olein; disp. in ethyl acetate; disp. in water up to 1%; Sp. gr.: 1.010 (20 C); Visc.: 600 cps (20 C); M.P.: < 0 C; HLB: 16.6; Stability: Stable to mild acids and alkalis but tends to hydrolyze under extreme conditons; pH: 8–10; Biodegradable
Composition: 100% active ingred.
Toxicity: May cause skin and eye irritation
Warnings and Caution: Spillages are slippery

Teric CME3. ICI Australia Ltd.
Chemical Description: Coconut monoethanolamide + 3 moles EO
Category: Wetting agent, dispersant, coemulsifier, detergent, foam booster, nonionic
Applications: Wetting agent for textile applications; coemulsifier for paraffinic, naphthenic, and various vegetable oils; emulsion stabilizer; increases hard water tolerance; foam booster in liquid and powder detergent blends
Form: Liquid to paste; Color: > 250 (Hazen; Odor: Low
Properties: Solubility: Sol. in benzene, ethyl acetate, ethyl Icinol, perchlorethylene, ethanol, olein; disp. in water; Sp. gr.: 0.972 (50 C); Visc.: 56 cps (50 C); M.P.: 18 ± 2 C; HLB: 13.8; Stability: Good in hard or saline waters and in reasonable concs. of acids and alkalis; pH: 8–10 (1% aq.); Biodegradable
Composition: 100% active ingred.
Toxicity: May cause skin and eye irritation
Warnings and Caution: Spillages are slippery

Teric CME7. ICI Australia Ltd.
Chemical Description: Coconut monoethanolamide + 7 moles EO
Category: Wetting agent, dispersant, detergent, foam stabilizer, nonionic
Applications: Wetting agent for textile applications; ingred. in liquid and powder cleaning compounds
Form: Liquid; Color: > 250 (Hazen); Odor: Low
Properties: Solubility: Sol. in water, benzene, ethyl acetate, ethyl Icinol, perchlorethylene, ethanol; Sp. gr.: 1.048 (20 C); Visc.: 264 cps (20 C); M.P.: 5 ± 2 C; HLB: 15.1; Stability: Good in hard or saline waters and in reasonable concs. of acids and alkalis; Surface Tension: 30.0 dynes/cm (0.1%, 20 C); pH: 8–10 (1% aq.); Biodegradable
Composition: 100% active ingred.
Toxicity: May cause skin and eye irritation
Warnings and Caution: Spillages are slippery

Teric DD5. ICI Australia Ltd.
Chemical Description: Dodecyl phenol + 5 moles EO
Category: Emulsifier, nonionic
Applications: Coemulsifier in solvent emulsion cleaners and decarbonizing preparations; formulation of self-emulsifiable concentrates for use as metal lubricants, cutting, milling, and grinding aids, textile processing aids, and mineral drilling lubricants
Form: Liquid; Color: 150 (Hazen)
Properties: Solubility: Sol. in benzene, ethyl acetate, ethyl Icinol, perchlorethylene, ethanol, kerosene, mineral and vegetable oil, olein; Sp. gr.: 1.014 (20 C); Visc.: 550 cps (20 C); M.P.: <0 C; HLB: 8.6; Stability: Good in hard or saline waters and in reasonable concs. of acids and alkalis; Surface Tension: 29.5 dynes/cm (0.1%, 20 C); pH: 6–8 (1% aq.); Biodegradability: 80%
Composition: 100% active ingred.
Toxicity: May cause skin and eye irritation
Warnings and Caution: Spillages are slippery

Teric DD9. ICI Australia Ltd.
Chemical Description: Dodecyl phenol + 9 moles EO
Category and Applications: See Teric DD5
Form: Liquid; Color: 150 (Hazen)
Properties: Solubility: Sol. in benzene, ethyl Icinol, ethanol, kerosene, olein; disp. in ethyl acetate, perchlorethylene, vegetable oil; Sp. gr.: 1.040 (20 C); Visc.: 453 cps (20 C); M.P.: < 0 C; HLB: 12.1; Stability: Good in hard or saline waters and in reasonable concs. of acids and alkalis; Surface Tension: 28.2 dynes/cm (0.1%, 20 C); pH: 6–8 (1% aq.)
Composition: 100% active ingred.
Toxicity: May cause skin and eye irritation
Warnings and Caution: Spillages are slippery

Teric G9A5. ICI Australia Ltd.
Chemical Description: Synthetic C_9–C_{11} alcohol + 5 moles EO
Category: Emulsifier, nonionic
Applications: Emulsifier for solvents; formulations of hard surface cleaners, degreasers, and dispersants in industrial and domestic applications
Form: Liquid; Color: 250 (Hazen)
Properties: Solubility: Sol. in water, vegetable oil, olein; Sp. gr.: 0.976 (20 C); Visc.: 52 cps (20 C); M.P.: < 0 C; HLB: 11.6; Cloud Pt.: 36 ± 2 C (1% in hard water); Stability: Good in hard or saline waters and in reasonable concs. of acids and alkalis; Surface Tension: 20.1 dynes/cm (0.1%, 20 C); pH: 5–7 (1% aq.)
Composition: 100% active ingred.
Toxicity: May cause skin and eye irritation
Warnings and Caution: Spillages are slippery

Teric G9A6. ICI Australia Ltd.
Chemical Description: Synthetic C_9–C_{11} alcohol + 6 moles EO
Category and Applications: See

Teric G9A5
Form: Liquid; Color: 250 (Hazen)
Properties: Solubility: Sol. in water, benzene, ethyl acetate, ethyl Icinol, perchlorethylene, ethanol, vegetable oil, olein; Sp. gr.: 0.987 (20 C); Visc.: 38 cps (20 C); M.P.: < 0 C; HLB: 12.4; Cloud Pt.: 53 ± 2 C (1% in hard water); Stability: Good in hard or saline waters and in reasonable concs. of acids and alkalis; Surface Tension: 24.0 dynes/cm (0.1%, 20 C); pH: 5–7 (1% aq.)
Composition: 100% active ingred.
Toxicity: May cause skin and eye irritation
Warnings and Caution: Spillages are slippery

Teric G9A8. ICI Australia Ltd.
Chemical Description: Synthetic C_9–C_{11} alcohol + 8 moles EO
Category: Emulsifier, dispersant, wetting agent, nonionic
Applications: Used in liquid and powdered detergent formulations for domestic use, as degreasers in textile applications such as scouring of raw wools, etc.
Form: Paste; Color: 250 (Hazen)
Properties: Solubility: Sol. in water, benzene, ethyl acetate, ethyl Icinol, perchlorethylene, ethanol, vegetable oil, olein; Sp. gr.: 0.988 (50 C); Visc.: 21 cps (50 C); M.P.: 13 ± 2 C; HLB: 13.7; Cloud Pt.: 78 ± 2 C (1% in hard water); Stability: Good in hard or saline waters and in reasonable concs. of acids and alkalis; Surface Tension: 27.0 dynes/cm (0.1%, 20 C); pH: 5–7 (1% aq.)
Composition: 100% active ingred.
Toxicity: May cause skin and eye irritation
Warnings and Caution: Spillages are slippery

Teric G9A12. ICI Australia Ltd.
Chemical Description: Synthetic C_9–C_{11} alcohol + 12 moles EO
Category and Applications: See Teric G9A8
Form: Solid; Color: 250 (Hazen)
Properties: Solubility: Sol. in water, benzene, ethyl acetate, ethyl Icinol, perchlorethylene, ethanol, vegetable oil, olein; disp. in paraffin oil; Sp. gr.: 1.023 (20 C); Visc.: 40 cps (50 C); M.P.: 24 ± 2 C; HLB: 15.4; Cloud Pt.: > 100 C (1% in hard water); Stability: Good in hard or saline waters and in reasonable concs. of acids and alkalis; Surface Tension: 34.5 dynes/cm (0.1%, 20 C); pH: 5–7 (1% aq.)
Composition: 100% active ingred.
Toxicity: May cause skin and eye irritation
Warnings and Caution: Spillages are slippery

Teric G12A4. ICI Australia Ltd.
Chemical Description: Ethylene oxide condensate based on straight chain synthetic alcohols, technical grade (4.5 moles EO)
Category: Wetting and dispersing agent, coemulsifier, detergent, intermediate, nonionic
Applications: Dust suppression;

aq. pigment dispersions; coemulsifier for W/O-type emulsion concentrates; liquid and powdered detergents for industrial and domestic laundry, dishwashing, metal cleaning, sanitizing, textile processing aids, abrasive cleaners; intermedaite in the manufacture of sulfate and phosphate surfactants
Form: Liquid; Color: 250 (Hazen)
Properties: Solubility: Sol. in benzene, ethyl acetate, ethyl Icinol, perchlorethylene, ethanol, kerosene, mineral and vegetable oils, paraffin oil, olein; disp. in water; Sp. gr.: 0.950 (20 C); Visc.: 40 cps (20 C); M.P.: 3 ± 2 C; HLB: 9.8; Stability: Good in hard or saline waters and in reasonable concs. of acids and alkalis; pH: 6–8 (1% aq.); Biodegradable
Composition: 100% active ingred.
Toxicity: May cause skin or eye irritation
Warnings and Caution: Spillages are slippery

Teric G12A6. ICI Australia Ltd.
Chemical Description: Ethylene oxide condensate based on straight chain synthetic alcohols, technical grade (6 moles EO)
Category: Wetting and dispersing agent, detergent, intermediate, nonionic
Applications: Specialty formulations for textile processing, hard surface abrasives, metal descaling, liquid and powdered detergents for domestic and industrial use in laundries, metal cleaning, sanitizing, dairy detergents; intermediate for highly alkaline metal cleaning formulations
Form: Liquid to paste; Color: 250 (Hazen)
Properties: Sp. gr.: 0.955 (20 C); Visc.: 24 cps (50 C); M.P.: 10 ± 2 C; HLB: 11.2; Cloud Pt.: 40 ± 2 C (1% in hard water); Stability: Good in hard or saline waters and in reasonable concs. of acids and alkalis; Surface Tension: 27.8 dynes/cm (0.1%, 20 C); pH: 6–8 (1% aq.); Biodegradable
Composition: 100% active ingred.
Toxicity: May cause eye or skin irritation
Warnings and Caution: Spillages are slippery

Teric G12A8. ICI Australia Ltd.
Chemical Description: Ethylene oxide condensate based on straight chain synthetic alcohols, technical grade (7.8 moles EO)
Category and Applications: See Teric G12A6
Form: Paste; Color: 250 (Hazen)
Properties: Solubility: Sol. in benzene, ethyl acetate, ethyl Icinol, perchlorethylene, ethanol, kerosene, mineral and vegetable oil, paraffin oil, olein; disp. in water; Sp. gr.: 0.971 (50 C); Visc.: 31 cps (50 C); M.P.: 16 ± 2 C; HLB: 12.6; Cloud Pt.: 58 ± 2 C (1% in hard water); Stability: Good in hard or saline waters and in reasonable concs. of acids and alkalis; Surface Tension: 27.2 dynes/cm (0.1%, 20 C); pH: 6–8 (1% aq.); Biodegradable
Composition: 100% active ingred.

Toxicity: May cause eye or skin irritation
Warnings and Caution: Spillages are slippery

Teric G12A12. ICI Australia Ltd.
Chemical Description: Ethylene oxide condensate based on straight chain synthetic alcohols, technical grade (11.5 moles EO)
Category and Applications: See Teric 12A6
Form: Solid; Color: 250 (Hazen)
Properties: Sp. gr.: 0.993 (20 C); Visc.: 37 cps (20 C); M.P.: 24 ± 2 C; HLB: 14.4; Cloud Pt.: 92 ± 2 C (1% in hard water); Stability: Good in hard or saline waters and in reasonable concs. of acids and alkalis; Surface Tension: 35.6 dynes/cm (0.1%, 20 C); pH: 6–8 (1% aq.); Biodegradable
Composition: 100% active ingred.
Toxicity: May cause skin and eye irritation
Warnings and Caution: Spillages are slippery

Teric GN5, GN8, GN9, GN10, GN12, GN13, GN15. ICI Australia Ltd.
See Teric N5, N8, N9, N10, N12, N13, N15 respectively

Teric LA4. ICI Australia Ltd.
Chemical Description: Liquid form of Teric G12A4
Category and Applications: See Teric G12A4
Form: Liquid
Properties: Solubility: Sol. in water, ethyl acetate, ethyl Icinol, ethanol, kerosene, mineral oil, olein; disp. in vegetable oil; Sp. gr.: 0.955 (20 C); Visc.: 45 cps (20 C); M.P.: 0 ± 2 C; HLB: 9.8; Stability: Good in hard or saline waters and in reasonable concs. of acids and alkalis; pH: 6–8 (1% aq.); Biodegradable
Composition: 88% active ingred.
Toxicity: May cause eye or skin irritation
Warnings and Caution: Spillages are slippery

Teric LA8. ICI Australia Ltd.
Chemical Description: Liquid form of Teric G12A8
Category and Applications: See Teric G12A8
Form: Liquid; Color: Light amber
Properties: Solubility: Sol. in water, benzene, ethyl acetate, ethyl Icinol, perchlorethylene, ethanol; Sp. gr.: 1.002 (20 C); Visc.: 100 cps (20 C); M.P.: 10 ± 2 C; HLB: 12.6; Cloud Pt.: 60 ± 2 C (1% in hard water); Stability: Good in hard and saline waters and in reasonable concs. of acids and alkalis; Surface Tension: 27.2 dynes/cm (0.1%, 20 C); pH: 6–8 (1% aq.); Biodegradable
Composition: 88% active ingred.
Toxicity: May cause skin and eye irritation
Warnings and Caution: Spillages are slippery

Teric LAN70. ICI Australia Ltd.
Chemical Description: High grade lanolin + 70 moles EO

Category: Lubricant, emulsifier, dispersant, nonionic
Applications: Textile lubricant
Form: Soft waxy solid; Color: Yellowish-amber
Properties: Solubility: Sol. in water, benzene, ethyl Icinol, perchlorethylene, ethanol, olein; disp. in ethyl acetate; Sp. gr.: 1.13 (20 C); Visc.: 246 cps (50 C); M.P.: 42 ± 2 C; HLB: 16.3; Cloud Pt.: > 100 C (1% in hard water); Stability: Good in hard or saline waters and in reasonable concs. of acids and alkalis; Surface Tension: 48.1 dynes/cm (0.1%, 20 C); pH: 4.5–7.0 (1% aq.)
Composition: 100% active ingred.
Toxicity: May cause skin and eye irritation
Warnings and Caution: Spillages are slippery

Teric N2. ICI Australia Ltd.
Chemical Description: Nonyl phenol + 2 moles EO
Category: Wetting and dispersing agent, emulsifier, defoamer, intermediate, nonionic
Applications: Wetting and dispersing in solvent or oil-based systems; coemulsifier for O/W emulsions; emulsifier W/O emulsions; used in agricultural toxicant products, industrial solvent-cleaning systems, surface coating preparations; intermediate for the production of sulfated and phosphated anionics
Form: Liquid; Color: 100 (Hazen)
Properties: Solubility: Sol. in benzene, ethyl acetate, ethyl Icinol, perchlorethylene, ethanol, kerosene, mineral and vegetable oil, paraffin oil, olein; disp. in water; Sp. gr.: 1.001 (20 C); Visc.: 620 cps (20 C); M.P.: < 0 C; HLB: 5.7; Stability: Good in hard or saline waters and in reasonable concs. of acids and alkalis; pH: 6–8 (1% aq.)
Composition: 100% active ingred.
Toxicity: May cause skin and eye irritation
Warnings and Caution: Spillages are slippery

Teric N3. ICI Australia Ltd.
Chemical Description: Nonyl phenol + 3.5 moles EO
Category and Applications: See Teric N2
Form: Liquid; Color: 100 (Hazen)
Properties: Solubility: Sol. in benzene, ethyl acetate, ethyl Icinol, perchlorethylene, ethanol, kerosene, mineral and vegetable oil, olein; disp. in water and paraffin oil; Sp. gr.: 1.017 (20 C); Visc.: 395 cps (20 C); M.P.: < 0 C; HLB: 8.2; Stability: Good in hard or saline waters and in reasonable concs. of acids and alkalis; pH: 6–8 (1% aq.)
Composition: 100% active ingred.
Toxicity: May cause skin and eye irritation
Warnings and Caution: Spillages are slippery

Teric N4. ICI Australia Ltd.
Chemical Description: Nonyl phenol + 4 moles EO
Category and Applications: See Teric N2
Form: Liquid; Color: 100 (Hazen)

Properties: Solubility: Sol. in benzene, ethyl acetate, ethyl Icinol, perchlorethylene, ethanol, kerosene, mineral and vegetable oil, olein; disp. in water; Sp. gr.: 1.023 (20 C); Visc.: 370 cps (20 C); M.P.: < 0 C; HLB: 8.9; Stability: Good in hard or saline waters and in reasonable concs. of acids and alkalis; pH: 6–8 (1% aq.)
Composition: 100% active ingred.
Toxicity: May cause skin and eye irritation
Warnings and Caution: Spillages are slippery

Teric N5. ICI Australia Ltd.
Chemical Description: Nonyl phenol + 5.5 moles EO
Category and Applications: See Teric N2
Form: Liquid; Color: 100 (Hazen); also avail. in tech. grade, Teric GN5 with Hazen color of 250
Properties: Solubility: Sol. in benzene, ethyl acetate, ethyl Icinol, perchlorethylene, ethanol, kerosene, vegetable oil, olein; disp. in water; Sp. gr.: 1.035 (20 C); Visc.: 355 cps (20 C); M.P.: < 0 C; HLB: 10.5; Stability: Good in hard or saline waters and in reasonable concs. of acids and alkalis; pH: 6–8 (1% aq.)
Composition: 100% active ingred.
Toxicity: May cause skin and eye irritation
Warnings and Caution: Spillages are slippery

Teric N8. ICI Australia Ltd.
Chemical Description: Nonyl phenol + 8.5 moles EO
Category: Wetting and dispersing agent, emulsifier, solubilizer, detergent, nonionic
Applications: Wettable powders, pigments, and ceramics for concrete mfg., in agricultural sprays, in solvent cleaners, in paints (esp. based on acrylics, PVA, or alkyd resins); heavy duty detergents where biodegradability is not important
Form: Liquid; Color: 100 (Hazen); also avail. in tech. grade, Teric GN8 with Hazen color of 250
Properties: Solubility: Sol. in water, benzene, ethyl acetate, ethyl Icinol, perchlorethylene, ethanol, vegetable oil, olein; Sp. gr.: 1.056 (20 C); Visc.: 350 cps (20 C); M.P.: 0 ± 2 C; HLB: 12.3; Cloud Pt.: 32 ± 2 C (1% in hard water); Stability: Good in hard or saline waters and in reasonable concs. of acids and alkalis; Surface Tension: 29.4 dynes/cm (0.1%, 20 C); pH: 6–8 (1% aq.)
Composition: 100% active ingred.
Toxicity: May cause skin and eye irritation
Warnings and Caution: Spillages are slippery

Teric N9. ICI Australia Ltd.
Chemical Description: Nonyl phenol + 9 moles EO
Category and Applications: See Teric N8
Form: Liquid; Color: 100 (Hazen); also avail. in tech. grade, Teric GN9 with Hazen color of 250
Properties: Solubility: Sol. in water, benzene, ethyl acetate, ethyl

Icinol, perchlorethylene, vegetable oil, olein; Sp. gr.: 1.060 (20 C); Visc.: 330 cps (20 C); M.P.: 0 ± 2 C; HLB: 12.8; Cloud Pt.: 52 ± 2 C (1% in hard water); Stability: Good in hard or saline waters and in reasonable concs. of acids and alkalis; Surface Tension: 30.6 dynes/cm (0.1%, 20 C); pH: 6–8 (1% aq.)
Composition: 100% active ingred.
Toxicity: May cause skin and eye irritation
Warnings and Caution: Spillages are slippery

Teric N10. ICI Australia Ltd.
Chemical Description: Nonyl phenol + 10 moles EO
Category and Applications: See Teric N8
Form: Liquid; Color: 100 (Hazen); also avail. in tech. grade, Teric GN10 with Hazen color of 250
Properties: Solubility: Sol. in water, benzene, ethyl acetate, ethyl Icinol, perchlorethylene, vegetable oil, olein; Sp. gr.: 1.063 (20 C); Visc.: 360 cps (20 C); M.P.: 5 ± 2 C; HLB: 13.3; Cloud Pt.: 67 ± 2 C (1% in hard water); Stability: Good in hard or saline waters and in reasonable concs. of acids and alkalis; Surface Tension: 30.6 dynes/cm (0.1%, 20 C); pH: 6–8 (1% aq.)
Composition: 100% active ingred.
Toxicity: May cause skin and eye irritation
Warnings and Caution: Spillages are slippery

Teric N11. ICI Australia Ltd.
Chemical Description: Nonyl phenol + 11 moles EO
Category and Applications: See Teric N8
Form: Liquid; Color: 100 (Hazen)
Properties: Solubility: Sol. in water, benzene, ethyl acetate, ethyl Icinol, perchlorethylene, ethanol, olein; Sp. gr.: 1.069 (20 C); Visc.: 410 cps (20 C); M.P.: 7 ± 2 C; HLB: 13.7; Cloud Pt.: 74 ± 2 C (1% in hard water); Stability: Good in hard or saline waters and in reasonable concs. of acids and alkalis; Surface Tension: 34.8 dynes/cm (0.1%, 20 C); pH: 6–8 (1% aq.)
Composition: 100% active ingred.
Toxicity: May cause skin and eye irritation
Warnings and Caution: Spillages are slippery

Teric N12. ICI Australia Ltd.
Chemical Description: Nonyl phenol + 12 moles EO
Category and Applications: See Teric N8
Form: Liquid; Color: 100 (Hazen); also avail. in tech. grade, Teric GN12 with Hazen color of 250
Properties: Solubility: Sol. in water, benzene, ethyl acetate, ethyl Icinol, perchlorethylene, ethanol, olein; Sp. gr.: 1.045 (50 C); Visc.: 67 cps (50 C); M.P.: 11 ± 2 C; HLB: 13.9; Cloud Pt.: 82 ± 2 C (1% in hard water); Stability: Good in hard or saline waters and in reasonable concs. of acids and alkalis; Surface Tension: 35.2

dynes/cm (0.1%, 20 C); pH: 6–8 (1% aq.)
Composition: 100% active ingred.
Toxicity: May cause skin and eye irritation
Warnings and Caution: Spillages are slippery

Teric N13. ICI Australia Ltd.
Chemical Description: Nonyl phenol + 13 moles EO
Category and Applications: See Teric N8
Form: Liquid; Color: 100 (Hazen); also avail. in tech. grade, Teric GN13 with a Hazen color of 250
Properties: Solubility: Sol. in water, benzene, ethyl acetate, ethyl Icinol, perchlorethylene, ethanol, olein; Sp. gr.: 1.049 (50 C); Visc.: 75 cps (50 C); M.P.: 14 ± 2 C; HLB: 14.4; Cloud Pt.: 89 ± 2 C (1% in hard water); Stability: Good in hard or saline waters and in reasonable concs. of acids and alkalis; Surface Tension: 34.9 dynes/cm (0.1%, 20 C); pH: 6–8 (1% aq.)
Composition: 100% active ingred.
Toxicity: May cause skin and eye irritation
Warnings and Caution: Spillages are slippery

Teric N15. ICI Australia Ltd.
Chemical Description: Nonyl phenol + 15 moles EO
Category: Wetting agent, coemulsifier, nonionic
Applications: Wetting in the presence of high electrolyte concs. and in powder or flake formulations; coemulsifier in relatively complex systems such as emulsion paints; solubilizes perfumes; low-foaming detergents and sanitizers
Form: Paste; Color: 100 (Hazen); also avail. in tech. grade, Teric GN15 with a Hazen color of 250
Properties: Solubility: Sol. in water, benzene, ethyl acetate, ethyl Icinol, perchlorethylene, ethanol, olein; Sp. gr.: 1.051 (50 C); Visc.: 82 cps (50 C); M.P.: 21 ± 2 C; HLB: 15.0; Cloud Pt.: 95 ± 2 C (1% in hard water); Stability: Good in hard or saline waters and in reasonable concs. of acids and alkalis; Surface Tension: 33.4 dynes/cm (0.1%, 20 C); pH: 6–8 (1% aq.)
Composition: 100% active ingred.
Toxicity: May cause skin and eye irritation
Warnings and Caution: Spillages are slippery

Teric N20. ICI Australia Ltd.
Chemical Description: Nonyl phenol + 20 moles EO
Category and Applications: See Teric N15
Form: Solid; Color: 100 (Hazen)
Properties: Solubility: Sol. in water, benzene, ethyl acetate, ethyl Icinol, perchlorethylene, ethanol, olein; Sp. gr.: 1.061 (50 C); Visc.: 100 cps (50 C); M.P.: 30 ± 2 C; HLB: 16.0; Cloud Pt.: > 100 C (1% in hard water); Stability: Good in hard or saline waters and in reasonable concs. of acids and alkalis; Surface Tension: 41.7 dynes/cm (0.1%, 20 C); pH: 6–8

(1% aq.)
Composition: 100% active ingred.
Toxicity: May cause skin and eye irritation
Warnings and Caution: Spillages are slippery

Teric N30. ICI Australia Ltd.
Chemical Description: Nonyl phenol + 30 moles EO
Category and Applications: See Teric N15
Form: Flakes; Color: 100 (Hazen)
Properties: Solubility: Sol. in water, benzene, ethyl acetate, ethyl Icinol, perchlorethylene, ethanol; Sp. gr.: 1.066 (50 C); Visc.: 150 cps (50 C); M.P.: 40 ± 2 C; HLB: 17.2; Cloud Pt.: > 100 C (1% in hard water); Stability: Good in hard or saline waters and in reasonable concs. of acids and alkalis; Surface Tension: 42.8 dynes/cm (0.1%, 20 C); pH: 6–8 (1% aq.)
Composition: 100% active ingred.
Toxicity: May cause skin and eye irritation
Warnings and Caution: Spillages are slippery

Teric N40. ICI Australia Ltd.
Chemical Description: Nonyl phenol + 40 moles EO
Category and Applications: See Teric N15
Form: Flakes; Color: 100 (Hazen)
Properties: Solubility: Sol. in water, benzene, ethyl acetate, ethyl Icinol, perchlorethylene, ethanol; Sp. gr.: 1.080 (50 C); Visc.: 250 cps (50 C); M.P.: 45 ± 2 C; HLB: 18.0; Cloud Pt.: > 100 C (1% in hard water); Stability: Good in hard or saline waters and in reasonable concs. of acids and alkalis; Surface Tension: 41.0 dynes/cm (0.1%, 20 C); pH: 6–8 (1% aq.)
Composition: 100% active ingred.
Toxicity: May cause skin and eye irritation
Warnings and Caution: Spillages are slippery

Teric N100. ICI Australia Ltd.
Chemical Description: Nonyl phenol + 100 moles EO
Category and Applications: See Teric N15
Form: Flakes; Color: 100 (Hazen)
Properties: Solubility: Sol. in water, benzene, ethyl acetate, ethyl Icinol, perchlorethylene, ethanol; Sp. gr.: 1.113 (60 C); M.P.: 52 ± 2 C; HLB: 19.1; Cloud Pt.: > 100 C (1% in hard water); Stability: Good in hard or saline waters and in reasonable concs. of acids and alkalis; Surface Tension: 46.0 dynes/cm (0.1%, 20 C); pH: 6–8 (1% aq.)
Composition: 100% active ingred.
Toxicity: May cause skin and eye irritation
Warnings and Caution: Spillages are slippery

Teric OF4. ICI Australia Ltd.
Chemical Description: Oleic acid + ethylene oxide
Category: Emulsifier, antistatic agent, nonionic
Applications: Used in O/W emulsions of vegetable and mineral oils, fats, waxes, and solvents such as

kerosene and turpentine; formulation of cutting oils, as metal lubricants, in metal cleaning formulations, and in solvent degreasers; textile and paper finishing applications; antistat in textile processing and the manufacture of synthetic fibers
Form: Liquid; Color: > 250 (Hazen)
Properties: Solubility: Sol. in benzene, ethyl acetate, ethyl Icinol, perchlorethylene, ethanol, kerosene, vegetable oil, olein; disp. in water; Sp. gr.: 0.978 (20 C); Visc.: 87 cps (20 C); M.P.: < 0 C; HLB: 7.7; Stability: Good in hard or saline waters and in reasonable concs. of acids and alkalis; pH: 8 (1% aq.); Biodegradable
Composition: 100% active ingred.
Toxicity: May cause skin and eye irritation
Warnings and Caution: Spillages are slippery

Teric OF6. ICI Australia Ltd.
Chemical Description: Oleic acid + 6 moles EO
Category and Applications: See Teric OF4
Form: Liquid; Color: > 250 (Hazen)
Properties: Solubility: Sol. in benzene, ethyl acetate, ethyl Icinol, perchlorethylene, ethanol, kerosene, vegetable oil, olein; disp. in water; Sp. gr.: 0.992 (20 C); Visc.: 82 cps (20 C); M.P.: < 0 C; HLB: 9.7; Stability: Good in hard or saline waters and in reasonable concs. of acids and alkalis; Surface Tension: 33.9 dynes/cm (0.1%, 20 C); pH: 7–9 (1% aq.); Biodegradable
Composition: 100% active ingred.
Toxicity: May cause skin and eye irritation
Warnings and Caution: Spillages are slippery

Teric OF8. ICI Australia Ltd.
Chemical Description: Oleic acid + 8 moles EO
Category: Emulsifier, antistatic agent, nonionic
Applications: Auxiliary product in textile and paper finishing applications; antistatic agent in textile processing and the manufacture of synthetic fibers
Form: Liquid; Color: > 250 (Hazen)
Properties: Solubility: Sol. in benzene, ethyl acetate, ethyl Icinol, perchlorethylene, ethanol, kerosene, vegetable oil, olein; disp. in water; Sp. gr.: 1.012 (20 C); Visc.: 120 cps (20 C); M.P.: < 0 C; HLB: 11.1; Stability: Good in hard or saline waters and in reasonable concs. of acids and alkalis; pH: 8-10 (1% aq.); Biodegradable
Composition: 100% active ingred.
Toxicity: May cause skin and eye irritation
Warnings and Caution: Spillages are slippery

Teric PE61. ICI Australia Ltd.
Chemical Description: Polyoxypropylene + 5.7 moles EO
Category: Wetting and dispersing agent, emulsifier, detergent, de-

mulsifier, defoamer, intermediate, nonionic
Applications: Low-foaming wetting and dispersing agent for pigments in latex paints; emulsifier-stabilizer in latex paints, esp. in butadiene, vinyl acetate, and acrylic latices; detergent sanitizers and low-foam alkaline cleaners where biodegradability is not important; dewatering aids for treatment of crude oil emulsions at the well-head; defoamer in paper pulp liquors, starch sizing, glues, and boiler water systems; component of urethane foams
Form: Liquid; Color: 100 (Hazen)
Properties: Solubility: Sol. in benzene, ethyl acetate, ethyl Icinol, perchlorethylene, vegetable oil, olein; Sp. gr.: 1.017 (20 C); Visc.: 388 cps (20 C); M.P.: $<$ 0 C; HLB: 3; Cloud Pt.: 24 ± 2 C (1% in hard water); Stability: Good in hard and saline waters and in reasonable concs. of acids and alkalis; Surface Tension: 39.0 dynes/cm (0.1%, 20 C); pH: 6 (1% aq.)
Composition: 100% active ingred.
Toxicity: May cause skin and eye irritation
Warnings and Caution: Spillages are slippery

Teric PE62. ICI Australia Ltd.
Chemical Description: Polyoxypropylene + 17 moles EO
Category and Applications: See Teric PE61; also used for bloat control in cattle
Form: Liquid; Color: 100 (Hazen)
Properties: Solubility: Sol. in water, benzene, ethyl acetate, ethyl Icinol, perchlorethylene, ethanol, vegetable oil, olein; Sp. gr.: 1.032 (20 C); Visc.: 440 cps (20 C); M.P.: $<$ 0 C; HLB: 7; Cloud Pt.: 32 ± 2 C (1% in hard water); Stability: Good in hard or saline waters and in reasonable concs. of acids and alkalis; Surface Tension: 42.1 dynes/cm (0.1%, 20 C); pH: 6 (1% aq.)
Composition: 100% active ingred.
Toxicity: May cause skin and eye irritation
Warnings and Caution: Spillages are slippery

Teric PE64. ICI Australia Ltd.
Chemical Description: Polyoxypropylene + 25.5 moles EO
Category: Wetting and dispersing agent, emulsifier, detergent, intermediate, demulsifier, nonionic
Applications: Low-foaming wetting and dispersing agent for pigments in latex paints; emulsifier-stabilizer in latex paints; iodine-phosphoric acid dairy cleaners and sanitizers; bloat control in cattle; oil-well acidizing and water-flooding techniques
Form: Liquid; Color: 100 (Hazen)
Properties: Solubility: Sol. in water, benzene, ethyl acetate, ethyl Icinol, perchlorethylene, ethanol, olein; Sp. gr.: 1.051 (20 C); Visc.: 1332 cps (20 C); M.P.: 8 ± 2 C; HLB: 15; Cloud Pt.: 58 ± 2 C (1% in hard water); Stability: Good in hard or saline waters and in reasonable concs. of acids and alkalis; Surface Tension: 43.2 dynes/cm

(0.1%, 20 C); pH: 6 (1% aq.)
Composition: 100% active ingred.
Toxicity: May cause skin and eye irritation
Warnings and Caution: Spillages are slippery

Teric PE68. ICI Australia
Chemical Description: Polyoxypropylene + 150 moles EO
Category: Wetting and dispersing agent, emulsifier, detergent, intermediate, nonionic
Applications: Dispersant-stabilizer in the manufacture of latex paints; talc dispersant in paper coating; iodine-phosphoric acid dairy cleaners and sanitizers; plasticizer for polyurethane and phenolic aminoresins
Form: Flakes; Color: 100 (Hazen)
Properties: Solubility: Sol. in water, benzene, ethyl acetate, ethyl Icinol, perchlorethylene, ethanol, olein; M.P.: 50 ± 2 C; HLB: 29; Cloud Pt.: > 100 C (1% in hard water); Stability: Good in hard or saline waters and in reasonable concs. of acids and alkalis; Surface Tension: 49.0 dynes/cm (0.1%, 20 C); pH: 6 (1% aq.)
Composition: 100% active ingred.
Toxicity: May cause skin and eye irritation
Warnings and Caution: Spillages are slippery

Teric SF9. ICI Australia Ltd.
Chemical Description: Stearic acid + 9 moles EO
Category: Coemulsifier, nonionic
Applications: Preparation of self-emulsifiable waxes for waterproofing of fiber board, chip board building sheets, and cardboard packaging; component of floor wax emulsions and wax strippers
Form: Solid; Color: 250 (Hazen)
Properties: Solubility: Sol. in benzene, ethyl acetate, ethyl Icinol, perchlorethylene, ethanol, kerosene, vegetable oil, olein; Sp. gr.: 0.994 (50 C); Visc.: 45 cps (50 C); M.P.: 30 ± 2 C; HLB: 10.8; Stability: Good in hard or saline waters and in reasonable concs. of acids and alkalis; pH: 8 (1% aq.); Biodegradable
Composition: 100% active ingred.
Toxicity: May cause skin and eye irritation
Warnings and Caution: Spillages are slippery

Teric SF15. ICI Australia Ltd.
Chemical Description: Stearic acid + 15 moles EO
Category: Coemulsifier, nonionic
Applications: Component in textile and leather dressing oils and in dye bath additives to promote leveling
Form: Solid; Color: 250 (Hazen)
Properties: Solubility: Sol. in water, benzene, ethyl acetate, ethyl Icinol, perchlorethylene; Sp. gr.: 1.014 (50 C); Visc.: 55 cps (50 C); M.P.: 32 ± 2 C; HLB: 14.0; Stability: Good in hard or saline waters and in reasonable concs. of acids and alkalis; Surface Tension: 40.7 dynes/cm (0.1%, 20 C); pH: 8 (1% aq.); Biodegradable
Composition: 100% active ingred.

Toxicity: May cause skin and eye irritation
Warnings and Caution: Spillages are slippery

Teric T2. ICI Australia Ltd.
Chemical Description: Tall oil ethoxylate
Category: Wetting and dispersing agent, emulsifier, detergent, nonionic
Applications: Wetting and dispersing for oil-based systems; grinding aids for the wet milling of pigments and resins; manufacture of self-emulsifiable, water-washable lubricants for the metal and textile industries, solvent cleaners, and powder detergent formulations
Form: Liquid; Color: > 250 (Hazen)
Properties: Solubility: Sol. in benzene, ethyl acetate, ethyl Icinol, perchlorethylene, ethanol, mineral and vegetable oil, olein; disp. in water; Sp. gr.: 0.966 (20 C); Visc.: 118 cps (20 C); M.P.: < 0 C; HLB: 4.5; Stability: Good in hard or saline waters and in reasonable concs. of acids and alkalis; pH: 6–8 (1% aq.); Biodegradable
Composition: 100% active ingred.
Toxicity: May cause skin and eye irritation
Warnings and Caution: Spillages are slippery

Teric T5. ICI Australia Ltd.
Chemical Description: Tall oil + 5 moles EO
Category and Applications: See Teric T2
Form: Liquid; Color: > 250 (Hazen)
Properties: Solubility: Sol. in benzene, ethyl acetate, ethyl Icinol, perchlorethylene, ethanol, mineral and vegetable oil, olein; disp. in water; Sp. gr.: 1.018 (20 C); Visc.: 196 cps (20 C); M.P.: < 0 C; HLB: 8.2; Stability: Good in hard or saline waters and in reasonable concs. of acids and alkalis; pH: 8–10 (1% aq.); Biodegradable
Composition: 100% active ingred.
Toxicity: May cause skin and eye irritation
Warnings and Caution: Spillages are slippery

Teric T7. ICI Australia Ltd.
Chemical Description: Tall oil + 7 moles EO
Category and Applications: See Teric T2
Form: Liquid; Color: > 250 (Hazen)
Properties: Solubility: Sol. in benzene, ethyl acetate, ethyl Icinol, perchlorethylene, ethanol, vegetable oil, olein; Sp. gr.: 1.031 (20 C); Visc.: 151 cps (20 C); M.P.: < 0 C; HLB: 10.0; Stability: Good in hard or saline waters and in reasonable concs. of acids and alkalis; pH: 8–10 (1% aq.); Biodegradable
Composition: 100% active ingred.
Toxicity: May cause skin and eye irritation
Warnings and Caution: Spillages are slippery

Teric T10. ICI Australia Ltd.
Chemical Description: Tall oil + 10 moles EO
Category and Applications: See Teric T2; also in heavy duty washing powders where low foam is important
Form: Liquid; Color: > 250 (Hazen)
Properties: Solubility: Sol. in benzene, ethyl acetate, ethyl Icinol, perchlorethylene, ethanol, vegetable oil, olein; disp. in water; Sp. gr.: 1.042 (20 C); Visc.: 253 cps (20 C); M.P.: 5 ± 2 C; HLB: 11.7; Stability: Good in hard or saline waters and in reasonable concs. of acids and alkalis; pH: 8–10 (1% aq.); Biodegradable
Composition: 100% active ingred.
Toxicity: May cause skin and eye irritation
Warnings and Caution: Spillages are slippery

Teric X5. ICI Australia Ltd.
Chemical Description: Octyl phenol + 5 moles EO
Category: Wetting and dispersing agent, emulsifier, detergent, nonionic
Applications: Wettable powders, metal cleaning compounds, alkaline cleaners, sol'ns. for acid treatment of metals, dispersible agricultural powders, solvent-toxicant emulsions, pigment dispersions, textile auxiliaries; as emulsifiers in aromatic and aliphatic solvents, mineral and vegetable spray oil emulsions, wax dispersions, solvent emulsion cleaners (polishes, paints, and floor wax removers); detergents where biodegradability is not important, i.e., for emulsifiable solvent cleaners, minimum foam nonionics, disinfectant detergents and sanitizers, detergent products containing cationic softeners, floor polishes, etc.
Form: Liquid; Color: 100 (Hazen)
Properties: Solubility: Sol. in benzene, ethyl acetate, ethyl Icinol, perchlorethylene, ethanol, kerosene, vegetable oil, paraffin oil, olein; partly sol. in mineral oil; disp. in water; Sp. gr.: 1.046 (20 C); Visc.: 465 cps (20 C); M.P.: < 0 ± 2 C; HLB: 10.4; Stability: Good in hard or saline waters and in reasonable concs. of acids and alkalis; pH: 6–8 (1% aq.)
Composition: 100% active ingred.
Toxicity: May cause skin and eye irritation
Warnings and Caution: Spillages are slippery

Teric X7. ICI Australia Ltd.
Chemical Description: Octyl phenol + 7.5 moles EO
Category and Applications: See Teric X5
Form: Liquid; Color: 100 (Hazen)
Properties: Solubility: Sol. in water, benzene, ethyl acetate, ethyl Icinol, perchlorethylene, ethanol, vegetable oil, olein; Sp. gr.: 1.059 (20 C); Visc.: 420 cps (20 C); M.P.: < 0 ± 2 C; HLB: 12.3; Cloud Pt.: 28 ± 2 C (1% in hard water); Stability: Good in hard or saline waters and in reasonable concs. of acids and alkalis; Surface Tension:

29.5 dynes/cm (0.1%, 20 C); pH: 6–8 (1% aq.)
Composition: 100% active ingred.
Toxicity: May cause skin and eye irritation
Warnings and Caution: Spillages are slippery

Teric X8. ICI Australia Ltd.
Chemical Description: Octyl phenol + 8.5 moles EO
Category and Applications: See Teric X5
Form: Liquid; Color: 100 (Hazen)
Properties: Solubility: Sol. in water, benzene, ethyl acetate, ethyl Icinol, perchlorethylene, ethanol, olein; partly sol. in vegetable oil; Sp. gr.: 1.060 (20 C); Visc.: 402 cps (20 C); M.P.: $< 0 \pm 2$ C; HLB: 12.6; Cloud Pt.: 47 ± 2 C (1% in hard water); Stability: Good in hard or saline waters and in reasonable concs. of acids and alkalis; Surface Tension: 30.2 dynes/cm (0.1%, 20 C); pH: 6--8 (1% aq.)
Composition: 100% active ingred.
Toxicity: May cause skin and eye irritation
Warnings and Caution: Spillages are slippery

Teric X10. ICI Australia Ltd.
Chemical Description: Octyl phenol + 10 moles EO
Category and Applications: See Teric X5
Form: Liquid; Color: 100 (Hazen)
Properties: Solubility: Sol. in water, benzene, ethyl acetate, ethyl Icinol, perchlorethylene, ethanol, olein; partly sol. in vegetable oil; Sp. gr.: 1.062 (20 C); Visc.: 393 cps (20 C); M.P.: 7 ± 2 C; HLB: 13.6; Cloud Pt.: 65 ± 2 C (1% in hard water); Stability: Good in hard or saline waters and in reasonable concs. of acids and alkalis; Surface Tension: 31.8 dynes/cm (0.1%, 20 C); pH: 6–8 (1% aq.)
Composition: 100% active ingred.
Toxicity: May cause skin and eye irritation
Warnings and Caution: Spillages are slippery

Teric X11. ICI Australia Ltd.
Chemical Description: Octyl phenol + 11 moles EO
Category and Applications: See Teric X5
Form: Liquid; Color: 100 (Hazen)
Properties: Solubility: Sol. in water, benzene, ethyl acetate, ethyl Icinol, perchlorethylene, ethanol, olein; partly sol. in vegetable oil; Sp. gr.: 1.067 (20 C); Visc.: 371 cps (20 C); M.P.: 8 ± 2 C; HLB: 14.0; Cloud Pt.: 80 ± 2 C (1% in hard water); Stability: Good in hard or saline waters and in reasonable concs. of acids and alkalis; Surface Tension: 31.0 dynes/cm (0.1%, 20 C); pH: 6–8 (1% aq.)
Composition: 100% active ingred.
Toxicity: May cause skin and eye irritation
Warnings and Caution: Spillages are slippery

Teric X13. ICI Australia Ltd.
Chemical Description: Octyl phenol + 13 moles EO
Category and Applications: See

Teric X5
Form: Liquid; Color: 100 (Hazen)
Properties: Solubility: Sol. in water, benzene, ethyl acetate, ethyl Icinol, perchlorethylene, ethanol, olein; partly sol. in vegetable oil; Sp. gr.: 1.052 (50 C); Visc.: 70 cps (50 C); M.P.: 15 ± 2 C; HLB: 14.7; Cloud Pt.: 90 ± 2 C (1% in hard water); Stability: Good in hard or saline waters and in reasonable concs. of acids and alkalis; Surface Tension: 32.0 dynes/cm (0.1%, 20 C); pH: 6–8 (1% aq.)
Composition: 100% active ingred.
Toxicity: May cause skin and eye irritation
Warnings and Caution: Spillages are slippery

Teric X16. ICI Australia Ltd.
Chemical Description: Octyl phenol + 16 moles EO
Category and Applications: See Teric X5
Form: Semisolid; Color: 100 (Hazen)
Properties: Solubility: Sol. in water, benzene, ethyl acetate, ethyl Icinol, perchlorethylene, ethanol, olein; Sp. gr.: 1.059 (50 C); Visc.: 95 cps (50 C); M.P.: 24 ± 2 C; HLB: 15.2; Cloud Pt.: > 100 C (1% in hard water); Stability: Good in hard or saline waters and in reasonable concs. of acids and alkalis; Surface Tension: 35.6 dynes/cm (0.1%, 20 C); pH: 6–8 (1% aq.)
Composition: 100% active ingred.
Toxicity: May cause skin and eye irritation
Warnings and Caution: Spillages are slippery

Teric X40. ICI Australia Ltd.
Chemical Description: Octyl phenol + 40 moles EO
Category and Applications: See Terix X5
Form: Solid; Color: 100 (Hazen)
Properties: Solubility: Sol. in water, benzene, ethyl acetate, ethyl Icinol, perchlorethylene, ethanol; Sp. gr.: 1.082 (50 C); Visc.: 245 cps (50 C); M.P.: 44 ± 2 C; HLB: 17.6; Cloud Pt.: > 100 C (1% in hard water); Stability: Good in hard or saline waters and in reasonable concs. of acids and alkalis; Surface Tension: 36.4 dynes/cm (0.1%, 20 C); pH: 6–8 (1% aq.)
Composition: 100% active ingred.
Toxicity: May cause skin and eye irritation
Warnings and Caution: Spillages are slippery

Tetronic 50R1. BASF Wyandotte/Ind. Chem. Group
Chemical Description: Compound based on ethylenediamine to which is added ethylene oxide and propylene oxide, sequentially
Category: Demulsifier, defoaming agent, wetting and dispersing agent, penetrant, plasticizer, antistat
Applications: Demulsification of crude oil, industrial waste oil; cutting and grinding fluids; machine dishwashing; solid rinse aids; wet-

ting and dispersing agent for cement and dyes; wetting agent for inks; pigment dispersant and viscosity regulator for paint; tablet coating; plasticizer; antistatic agent for plastics; rubber activator
Form: Liquid
Properties: Solubility: Sol. 10 g/100 ml in toluene, xylene, propylene glycol, isopropyl myristate; sol. (cloudy) in ethanol (95%); Sp. gr.: 1.02; Visc.: 672 cps; Cloud Pt.: 29 C (1%); Pour Pt.: –18 C; Surface Tension: 40.4 dynes/cm (0.1%)
Composition: 100% active ingred.

Tetronic 50R4. BASF Wyandotte/Ind. Chem. Group
Chemical Description: See Tetronic 50R1
Category and Applications: See Tetronic 50R1
Form: Liquid
Properties: Solubility: Sol. 10 g/100 ml in ethanol (95%, 100%), toluene, xylene, carbon tetrachloride, propylene glycol; Sp. gr.: 1.04; Visc.: 1080 cps; Cloud Pt.: 53 C (1%); Pour Pt.: –4 C; Surface Tension: 46.3 dynes/cm (0.1%)
Composition: 100% active ingred.

Tetronic 50R8. BASF Wyandotte/Ind. Chem. Group
Chemical Description: See Tetronic 50R1
Category and Applications: See Tetronic 50R1
Form: Solid
Properties: Solubility: Sol. 10 g/100 ml in ethanol (100%), toluene; sol. (cloudy) in ethanol (95%) and xylene; M.P.: 38 C; Cloud Pt.: 88 C (1%); Surface Tension: 52.7 dynes/cm (0.1%)
Composition: 100% active ingred.

Tetronic 70R1. BASF Wyandotte/Ind. Chem. Group
Chemical Description: See Tetronic 50R1
Category and Applications: See Tetronic 50R1
Form: Liquid
Properties: Solubility: Sol. 10 g/100 ml in ethanol (95%, 100%), toluene, xylene, carbon tetrachloride, isopropyl myristate; Sp. gr.: 1.02; Visc.: 800 cps; Cloud Pt.: 25 C (1%); Pour Pt.: –17 C; Surface Tension: 38.7 dynes/cm (0.1%)
Composition: 100% active ingred.

Tetronic 70R2. BASF Wyandotte/Ind. Chem. Group
Chemical Description: See Tetronic 50R1
Category and Applications: See Tetronic 50R1
Form: Liquid
Properties: Solubility: Sol. 10 g/100 ml in ethanol (95%, 100%), toluene, xylene, carbon tetrachloride, isopropyl myristate; Sp. gr.: 1.03; Visc.: 884 cps; Cloud Pt.: 31 C (1%); Pour Pt.: –22 C; Surface Tension: 39.8 dynes/cm (0.1%)
Composition: 100% active ingred.

Tetronic 70R4. BASF Wyandotte/ Ind. Chem. Group
Chemical Description: See Tetronic 50R1
Category and Applications: See Tetronic 50R1
Form: Liquid
Properties: Solubility: Sol. 10 g/ 100 ml in ethanol (95%, 100%), toluene, xylene, propylene glycol; Sp. gr.: 1.05; Visc.: 2160 cps; Cloud Pt.: 42 C (1%); Pour Pt.: 6 C; Surface Tension: 42.9 dynes/ cm (0.1%)
Composition: 100% active ingred.

Tetronic 90R1. BASF Wyandotte/ Ind. Chem. Group
Chemical Description: See Tetronic 50R1
Category and Applications: See Tetronic 50R1
Form: Liquid
Properties: Solubility: Sol. 10 g/ 100 ml in toluene, xylene, carbon tetrachloride, isopropyl myristate; sol. (cloudy) in ethanol (100%); Sp. gr.: 1.02; Visc.: 945 cps; Cloud Pt.: 21 C (1%); Pour Pt.: –17 C
Composition: 100% active ingred.

Tetronic 90R4. BASF Wyandotte/ Ind. Chem. Group
Chemical Description: See Tetronic 50R1
Category and Applications: See Tetronic 50R1
Form: Liquid
Properties: Solubility: Sol. 10 g/ 100 ml in ethanol (95%), toluene, xylene, propylene glycol; Sp. gr.: 1.05; Visc.: 3867 cps; Cloud Pt.: 38 C (1%); Pour Pt.: 12 C; Surface Tension: 42.7 dynes/cm (0.1%)
Composition: 100% active ingred.

Tetronic 90R8. BASF Wyandotte/ Ind. Chem. Group
Chemical Description: See Tetronic 50R1
Category and Applications: See Tetronic 50R1
Form: Solid
Properties: Solubility: Sol. 10 g/ 100 ml in ethanol (95%); M.P.: 47 C; Cloud Pt.: 81 C (1%); Surface Tension: 50.4 dynes/cm (0.1%)
Composition: 100% active ingred.

Tetronic 110R1. BASF Wyandotte/ Ind. Chem. Group
Chemical Description: See Tetronic 50R1
Category and Applications: See Tetronic 50R1
Form: Liquid
Properties: Solubility: Sol. 10 g/ 100 ml in ethanol (100%), toluene, xylene, carbon tetrachloride, isopropyl myristate; Sp. gr.: 1.02; Visc.: 1000 cps; Cloud Pt.: 21 C (1%); Pour Pt.: –18 C
Composition: 100% active ingred.

Tetronic110R2. BASF Wyandotte/ Ind. Chem. Group
Chemical Description: See Tetronic 50R1
Category and Applications: See Tetronic 50R1
Form: Liquid
Properties: Solubility: Sol. 10 g/ 100 ml in ethanol (100%), toluene,

xylene, carbon tetrachloride, isopropyl myristate; Sp. gr.: 1.03; Visc.: 1320 cps; Cloud Pt.: 27 C (1%); Pour Pt.: –20 C; Surface Tension: 37.5 dynes/cm (0.1%)
Composition: 100% active ingred.

Tetronic 110R7. BASF Wyandotte/ Ind. Chem. Group
Chemical Description: See Tetronic 50R1
Category and Applications: See Tetronic 50R1
Form: Solid
Properties: Solubility: Sol. 10 g/ 100 ml in carbon tetrachloride; sol. (cloudy) in ethanol (100%) and xylene; M.P.: 47 C; Cloud Pt.: 52 C (1%); Surface Tension: 46.5 dynes/cm (0.1%)
Composition: 100% active ingred.

Tetronic 130R1. BASF Wyandotte/ Ind. Chem. Group
Chemical Description: See Tetronic 50R1
Category and Applications: See Tetronic 50R1
Form: Liquid
Properties: Solubility: Sol. 10 g/ 100 ml in ethanol (95%), toluene, carbon tetrachloride; sol. (cloudy) in xylene and isopropyl myristate; Sp. gr.: 1.02; Visc.: 1240 cps; Cloud Pt.: 20 C (1%); Pour Pt.: –18 C
Composition: 100% active ingred.

Tetronic 130R2. BASF Wyandotte/ Ind. Chem. Group
Chemical Description: See Tetronic 50R1
Category and Applications: See Tetronic 50R1
Form: Liquid
Properties: Solubility: Sol. 10 g/ 100 ml in toluene, carbon tetrachloride, isopropyl myristate; sol. (cloudy) in ethanol (95%, 100%), xylene; Sp. gr.: 1.03; Visc.: 1880 cps; Cloud Pt.: 25 C (1%); Pour Pt.: –3 C; Surface Tension: 35.0 dynes/cm (0.1%)
Composition: 100% active ingred.

Tetronic 150R1. BASF Wyandotte/ Ind. Chem. Group
Chemical Description: See Tetronic 50R1
Category and Applications: See Tetronic 50R1
Form: Liquid
Properties: Solubility: Sol. 10 g/ 100 ml in toluene, xylene, carbon tetrachloride, isopropyl myristate; Sp. gr.: 1.01; Visc.: 1840 cps; Cloud Pt.: 20 C (1%); Pour Pt.: –17 C
Composition: 100% active ingred.

Tetronic 150R4. BASF Wyandotte/ Ind. Chem. Group
Chemical Description: See Tetronic 50R1
Category and Applications: See Tetronic 50R1
Form: Paste
Properties: Solubility: Sol. 10 g/ 100 ml in toluene, xylene; M.P.: 34 C; Cloud Pt.: 25 C (1%); Surface Tension: 38.9 dynes/cm (0.1%)
Composition: 100% active ingred.

Tetronic 150R8. BASF Wyandotte/ Ind. Chem. Group
Chemical Description: See Tetronic 50R1
Category and Applications: See Tetronic 50R1
Form: Solid
Properties: Solubility: Sol. 10 g/ 100 ml in ethanol (95%); M.P.: 53 C; Cloud Pt.: 38 C (1%); Surface Tension: 44.4 dynes/cm (0.1%)
Composition: 100% active ingred.

Tetronic 304. BASF Wyandotte/ Ind. Chem. Group
Chemical Description: Compound formed by the addition of propylene oxide to ethylene diamine, followed by the addition of EO
Category: Emulsifier, thickener, wetting agent, dispersant, solubilizer, demulsifier, detergent, antistat, stabilizer, nonionic
Applications: Cosmetics and pharmaceuticals, e.g., cream sachet, hair creams and gels, insect repellents, sunscreen cream; demulsifier in petroleum industry; detergent ingredient; antistat for polyethylene and some synthetic resin molding powders; metal treatment; emulsion polymerization; used in latex-based paints, aq.-based synthetic cutting fluids, in the vulcanization of rubber, etc.
Form: Liquid; Color: Colorless
Properties: m.w.: 1650; Ref. Index: 1.4649; Solubility: Water-sol.; sol. 10 g/100 ml in ethanol (95%), toluene, xylene, benzene, propylene glycol, carbon tetrachloride, perchlorethylene; Sp. gr.: 1.06; Visc.: 450 cps; HLB: 16; Cloud Pt.: 94 C (1%); Pour Pt.: –11 C; Surface Tension: 53.0 dynes/cm (0.1%)
Composition: 100% active ingred.

Tetronic 504. BASF Wyandotte/ Ind. Chem. Group
Chemical Description: See Tetronic 304
Category and Applications: See Tetronic 304
Form: Liquid; Color: Colorless
Properties: m.w.: 3400; Ref. Index: 1.4612; Solubility: Water-sol.; sol. 10 g/100 ml in ethanol (95%), toluene, xylene, benzene, propylene glycol, carbon tetrachloride, perchlorethylene; Sp. gr.: 1.04; Visc.: 800 cps; HLB: 15.5; Cloud Pt.: 68 C (1%); Pour Pt.: 7 C; Surface Tension: 44.2 dynes/cm (0.1%)
Composition: 100% active ingred.

Tetronic 701. BASF Wayndotte/ Ind. Chem. Group
Chemical Description: See Tetronic 304
Category and Applications: See Tetronic 304
Form: Liquid; Color: Colorless
Properties: m.w.: 3400; Ref. Index: 1.4553; Solubility: Sol. 10 g/ 100 ml in ethanol (95%), toluene, xylene, benzene, propylene glycol, carbon tetrachloride, perchlorethylene; Sp. gr.: 1.02; Visc.: 575 cps; HLB: 3; Cloud Pt.: 22 C (1%); Pour Pt.: –21 C; Surface Tension: 36.1 dynes/cm (not completely sol. @ 0.1%)

Composition: 100% active ingred.
Toxicity: Moderate skin and eye irritant

Tetronic 702. BASF Wyandotte/ Ind. Chem. Group
Chemical Description: See Tetronic 304
Category and Applications: See Tetronic 304
Form: Liquid; Color: Colorless
Properties: m.w.: 4000; Ref. Index: 1.4572; Solubility: Water-sol.; sol 10 g/100 ml in ethanol (95%), toluene, xylene, benzene, propylene glycol, carbon tetrachloride, perchlorethylene; Sp. gr.: 1.03; Visc.: 770 cps; HLB: 7; Cloud Pt.: 27 C (1%); Pour Pt.: –7 C; Surface Tension: 36.8 dynes/cm (0.1%)
Composition: 100% active ingred.

Tetronic 704. BASF Wyandotte/ Ind. Chem. Group
Chemical Description: See Tetronic 304
Category and Applications: See Tetronic 304
Form: Liquid; Color: Colorless
Properties: m.w.: 5500; Ref. Index: 1.4613; Solubility: Water-sol.; sol. 10 g/100 ml in ethanol (95%), toluene, xylene, benzene, propylene glycol, carbon tetrachloride, perchlorethylene; Sp. gr.: 1.04; Visc.: 850 cps; HLB: 15; Cloud Pt.: 65 C (1%); Pour Pt.: 18 C; Surface Tension: 40.3 dynes/cm (0.1%)
Composition: 100% active ingred.

Tetronic 707. BASF Wyandotte/ Ind. Chem. Group
Chemical Description: See Tetronic 304
Category and Applications: See Tetronic 304
Form: Flakes
Properties: m.w.: 12,000; Solubility: Water-sol.; sol. 10 g/100 ml in ethanol (95%), toluene, benzene, carbon tetrachloride, perchlorethylene; M.P.: 49 C; HLB: 27; Cloud Pt.: > 100 C (1%); Surface Tension: 47.6 dynes/cm (0.1%)
Composition: 100% active ingred.
Toxicity: Mild skin and eye irritant

Tetronic 901. BASF Wyandotte/ Ind. Chem. Group
Chemical Description: See Tetronic 304
Category and Applications: See Tetronic 304
Form: Liquid; Color: Colorless
Properties: m.w.: 4750; Ref. Index: 1.4545; Solubility: Sol. 10 g/100 ml in ethanol (95%), toluene, xylene, benzene, carbon tetrachloride, perchlorethylene; Sp. gr.: 1.02; Visc.: 700 cps; HLB: 2.5; Cloud Pt.: 20 C (1%); Pour Pt.: –23 C
Composition: 100% active ingred.

Tetronic 904. BASF Wyandotte/ Ind. Chem. Group
Chemical Description: See Tetronic 304
Category and Applications: See Tetronic 304
Form: Liquid; Color: Colorless

Properties: m.w.: 7500; Ref. Index: 1.4604; Solubility: Water-sol.; sol. 10 g/100 ml in ethanol (95%), toluene, xylene, benzene, propylene glycol, carbon tetrachloride, perchlorethylene; Sp. gr.: 1.04; Visc.: 6000 cps; HLB: 14.5; Cloud Pt.: 64 C (1%); Pour Pt.: 29 C; Surface Tension: 35.4 dynes/cm (0.1%)
Composition: 100% active ingred.

Tetronic 908. BASF Wyandotte/ Ind. Chem. Group
Chemical Description: See Tetronic 304
Category and Applications: See Tetronic 304
Form: Flakes
Properties: m.w.: 27,000; Solubility: Water-sol.; sol. 10 g/100 ml in ethanol (95%), benzene; M.P.: 58 C; HLB: 30.5; Cloud Pt.: > 100 C (1%); Surface Tension: 45.7 dynes/cm (0.1%)
Composition: 100% active ingred.
Toxicity: Mild skin and eye irritant

Tetronic 1101. BASF Wyandotte/ Ind. Chem. Group
Chemical Description: See Tetronic 304
Category and Applications: See Tetronic 304
Form: Liquid; Color: Colorless
Properties: m.w.: 5600; Ref. Index: 1.4540; Solubility: Sol. 10 g/100 ml in ethanol (95%), toluene, xylene, benzene, carbon tetrachloride, perchlorethylene; Sp. gr.: 1.02; Visc.: 700 cps; HLB: 2; Cloud Pt.: 17 C (1%); Pour Pt.: −15 C; Surface Tension: 34.0 dynes/cm (not completely sol. @ 0.1%)
Composition: 100% active ingred.
Toxicity: Mild skin irritant, minimal eye irritant

Tetronic 1102. BASF Wyandotte/ Ind. Chem. Group
Chemical Description: See Tetronic 304
Category and Applications: See Tetronic 304
Form: Liquid; Color: Colorless
Properties: m.w.: 6300; Ref. Index: 1.4557; Solubility: Water-sol.; sol. 10 g/100 ml in ethanol (95%), toluene, xylene, benzene, carbon tetrachloride, perchlorethylene; Sp. gr.: 1.03; Visc.: 820 cps; HLB: 6; Cloud Pt.: 31 C (1%); Pour Pt.: 7 C; Surface Tension: 33.7 dynes/cm (0.1%)
Composition: 100% active ingred.

Tetronic 1104. BASF Wyandotte/ Ind. Chem. Group
Chemical Description: See Tetronic 304
Category and Applications: See Tetronic 304
Form: Paste
Properties: m.w.: 8300; Solubility: Water-sol.; sol. 10 g/100 ml in ethanol (95%), toluene, xylene, benzene, propylene glycol, carbon tetrachloride, perchlorethylene; M.P.: 34 C; HLB: 14; Cloud Pt.: 72 C (1%); Surface Tension: 36.3 dynes/cm (0.1%)
Composition: 100% active ingred.

Tetronic 1107. BASF Wyandotte/ Ind. Chem. Group
Chemical Description: See Tetronic 304
Category and Applications: See Tetronic 304
Form: Solid
Properties: m.w.: 14,500; Solubility: Water-sol.; sol. 10 g/100 ml in ethanol (95%), toluene, benzene, carbon tetrachloride, perchlorethylene; M.P.: 51 C; HLB: 24; Cloud Pt.: > 100 C; Surface Tension: 42.9 dynes/cm (0.1%)
Composition: 100% active ingred.
Toxicity: Minimal skin and eye irritant

Tetronic 1301. BASF Wyandotte/ Ind. Chem. Group
Chemical Description: See Tetronic 304
Category and Applications: See Tetronic 304
Form: Liquid; Color: Colorless
Properties: m.w.: 6800; Ref. Index: 1.4545; Solubility: Sol. 10 g/ 100 ml in ethanol (95%), toluene, xylene, benzene, carbon tetrachloride, perchlorethylene; Sp. gr.: 1.02; Visc.: 1000 cps; HLB: 1.5; Cloud Pt.: 16 C (1%); Pour Pt.: −9 C; Surface Tension: 33.4 dynes/cm (not completely sol. @ 0.1%)
Composition: 100% active ingred.

Tetronic 1302. BASF Wyandotte/ Ind. Chem. Group
Chemical Description: See Tetronic 304
Category and Applications: See Tetronic 304
Form: Liquid; Color: Colorless
Properties: m.w.: 7800; Ref. Index: 1.4562; Solubility: Water-sol.; sol. 10 g/100 ml in ethanol (95%), toluene, xylene, benzene, carbon tetrachloride, perchlorethylene; Sp. gr.: 1.03; Visc.: 1300 cps; HLB: 5.5; Cloud Pt.: 20 C (1%); Pour Pt.: 13 C; Surface Tension: 34.1 dynes/cm (0.1%)
Composition: 100% active ingred.
Toxicity: Minimal skin and eye irritant

Tetronic 1304. BASF Wyandotte/ Ind. Chem. Group
Chemical Description: See Tetronic 304
Category and Applications: See Tetronic 304
Form: Paste
Properties: m.w.: 10,500; Solubility: Water-sol.; sol. 10 g/100 ml in ethanol (95%), toluene, xylene, benzene, propylene glycol, carbon tetrachloride, perchlorethylene; M.P.: 36 C; HLB: 13.5; Cloud Pt.: 78 C (1%); Surface Tension: 35.5 dynes/cm (0.1%)
Composition: 100% active ingred.

Tetronic 1307. BASF Wyandotte/ Ind. Chem. Group
Chemical Description: See Tetronic 304
Category and Applications: See Tetronic 304
Form: Solid
Properties: m.w.: 18,600; Solubility: Water-sol.; sol. 10 g/100 ml in ethanol (95%), toluene, benzene,

carbon tetrachloride, perchlorethylene; M.P.: 54 C; HLB: 23.5; Cloud Pt.: > 100 C (1%); Surface Tension: 43.8 dynes/cm (0.1%)
Composition: 100% active ingred.
Toxicity: Minimal eye and skin irritant

Tetronic 1501. BASF Wyandotte/ Ind. Chem. Group
Chemical Description: See Tetronic 304
Category and Applications: See Tetronic 304
Form: Liquid; Color: Colorless
Properties: m.w.: 7900; Ref. Index: 1.4537; Solubility: Sol. 10 g/ 100 ml in ethanol (95%), toluene, xylene, benzene, carbon tetrachloride, perchlorethylene; Sp. gr.: 1.02; Visc.: 1170 cps; HLB: 1.0; Cloud Pt.: 15 C (1%); Pour Pt.: –4 C; Surface Tension: 33.9 dynes/cm (not completely sol. @ 0.1%)
Composition: 100% active ingred.

Tetronic 1502. BASF Wyandotte/ Ind. Chem. Group
Chemical Description: See Tetronic 304
Category and Applications: See Tetronic 304
Form: Liquid; Color: Colorless
Properties: m.w.: 9000; Ref. Index: 1.4560; Solubility: Water-sol.; sol. 10 g/100 ml in ethanol (95%), toluene, xylene, benzene, carbon tetrachloride, perchlorethylene; Sp. gr.: 1.03; Visc.: 1570 cps; HLB: 5.0; Cloud Pt.: 70 C (1%); Pour Pt.: 18 C; Surface Tension: 34.1 dynes/cm (0.1%)
Composition: 100% active ingred.

Tetronic 1504. BASF Wyandotte/ Ind. Chem. Group
Chemical Description: See Tetronic 304
Category and Applications: See Tetronic 304
Form: Paste
Properties: m.w.: 12,500; Solubility: Water-sol.; sol. 10 g/100 ml in ethanol (95%), toluene, xylene, benzene, propylene glycol, carbon tetrachloride, perchlorethylene; M.P.: 41 C; HLB: 13.0; Cloud Pt.: 90 C (1%); Surface Tension: 37.3 dynes/cm (0.1%)
Composition: 100% active ingred.
Toxicity: Minimal skin and eye irritant

Tetronic 1508. BASF Wyandotte/ Ind. Chem. Group
Chemical Description: See Tetronic 304
Category and Applications: See Tetronic 304
Form: Solid
Properties: m.w.: 27,000; Solubility: Water-sol.; sol. 10 g/100 ml in ethanol (95%), benzene; M.P.: 60 C; HLB: 27; Cloud Pt.: > 100 C (1%); Surface Tension: 43.8 dynes/ cm (0.1%)
Composition: 100% active ingred.
Toxicity: Minimal skin irritant; slight eye irritant

Texapon K-12. Henkel Inc.
Chemical Description: Sodium lauryl sulfate

Category: Wetting, dispersing, and foaming agent, anionic
Applications: Matches USP grade purity; used in dentifrice manufacture, bubble baths, cosmetic creams, and emulsions
Form: Fine powder; Color: White
Properties: pH: 7.5–9.0 (1% aq.)
Composition: 90% min. active

Texapon L-100. Henkel Inc.
Chemical Description: Sodium lauryl sulfate
Category: Wetting, dispersing, foaming, and cleansing agent
Applications: Dentifrice manufacture, mouthwash formulations; low salt and no unsulfated matter
Form: Fine powder; Color: White
Properties: pH: ≈ 7.0 (1% aq.)
Composition: ≈ 99% active ingred.

Texapon N 40. Henkel Argentina
Chemical Description: Sodium lauryl ether sulfate
Category: Raw material, anionic
Applications: Raw material for the manufacture of liquid shampoos and bubble baths
Form: Liquid
Properties: pH: 6–7.5 (5% sol'n.)
Composition: 25–26% active

Texapon N 42. Henkel Argentina
Chemical Description: Triethanolamine lauryl sulfate
Category and Applications: See Texapon N 40
Form: Liquid
Properties: pH: 6–7 (5% sol'n.)
Composition: 39–41% active

Texapon NA. Henkel Argentina
Chemical Description: Ammonium lauryl ether sulfate
Category and Applications: See Texapon N 40
Form: Liquid
Properties: pH: 6.5–7.0 (5% sol'n.)
Composition: 23–24% active

Texapon SBN. Henkel Argentina
Chemical Description: Mixture of special washing active substances, based on fatty alcohol ether sulfates with sulfosuccinic acid esters
Category: Detergent, anionic
Applications: Detergent base in manufacture of baby and children preparations
Form: Liquid
Composition: 30% active ingred.

Texapon V Highly Conc. Needles. Henkel Inc.
Chemical Description: Sodium lauryl sulfate
Category: Detergent, wetting and foaming agent, emulsifier, anionic
Applications: Detergent and cosmetic base; emulsifier for cosmetic creams and O/W emulsions; produces fine, stable foam
Form: Fine needles; Color: White
Properties: Stability: Unaffected by water hardness; pH: 7–8 (1% aq.)
Composition: ≈ 90% active

Texapon Z Highly Conc. Needles. Henkel Inc.
Chemical Description: Sodium lauryl sulfate
Category: Detergent, wetting and

foaming agent, anionic
Applications: Detergent and cosmetic base; produces fine, stable foam
Form: Fine needles; Color: White
Properties: Stability: Unaffected by water hardness; pH: 6.8–7.5 (1% aq.)
Composition: 85–89% active

Texapon Z Highly Conc. Powder. Henkel Inc.
Chemical Description: Sodium lauryl sulfate
Category and Applications: See Texapon V Highly Conc. Needles
Form: Fine powder; Color: White
Properties: Stability: Unaffected by water hardness; pH: 7.5–8.5 (0.25% aq.)
Composition: 90% min. active

Texnol IL. Nippon Nyukazai Co., Ltd.
Chemical Description: Alkyl imidazoline laurate
Category: Antistatic agent, nonionic
Form: Solid
Composition: 100% active ingred.

Texnol R 5. Nippon Nyukazai Co., Ltd.
Chemical Description: Alkyl benzyl ammonium salt
Category: Emulsifier, fungicide, softener, anionic
Form: Liquid
Composition: 100% active ingred.

Texnol SPT. Nippon Nyukazai Co., Ltd.
Chemical Description: Special blended nonionics
Category: Antistatic agent, nonionic
Form: Solid
Composition: 100% active ingred.

Tex-Wet 1001. Intex Products, Inc.
Chemical Description: Sodium dioctyl sulfosuccinate
Category: Wetting and rewetting agent, anionic
Applications: Used in textile wet processing where alkalis are not encountered; provides fabric which can be run at high speeds in subsequent processes, such as sanforizing, finishing, etc.
Form: Clear liquid
Properties: Density: 8.8 lb/gal; Stability: Stable in hard water, mild acid, and mild alkaline liquors; efficient in hot and cold sol'ns.; pH: 6 ± 0.5 (1% sol'n.); Biodegradable
Composition: 60 ± 1% solids

Tex-Wet 1002. Intex Products, Inc.
Chemical Description: Phosphated alcohol
Category: Wetting agent, dispersant, penetrant, sequestrant, anionic
Applications: Low foaming wetting and penetrating agent for textile wet processing; dyeing assistant for woolen goods, and with sulfur colors; sequestering properties makes it effective in hard water
Form: Liquid; Color: Light amber
Properties: Solubility: Readily sol. in water; Density: 8.75 lb/gal; Stability: Stable in mild concs. of ac-

ids and alkalis; forms a hazy sol'n. in 3–4% caustic concs. when cold, but remains clear while hot; pH: 7.0 ± 0.5
Composition: 55% conc.

Tex-Wet 1004. Intex Products, Inc.
Chemical Description: Blend of alkyl aryl sulfonate and sulfated nonionic
Category: Detergent, wetting and rewetting agent, anionic
Applications: Wet processing, preparation and/or desizing assistant, after soaping operations, dyeing assistant
Properties: Density: 8.2 lb/gal; pH: 7–8; Biodegradable
Composition: 35 ± 0.5% solids

Tex-Wet 1143. Intex Products, Inc.
Chemical Description: Acid form of complex organic phosphate ester
Category: Detergent, emulsifier, antistatic agent, wetting agent, anionic
Applications: Continuous caustic boil-off assistant; kier boil assistant; bleaching assistant; detergent intermediate; emulsifier and antistat
Form: Clear, viscous liquid
Properties: pH: 1.5–2.0
Composition: 100% active ingred.

Tex-Wet 1158. Intex Products, Inc.
Chemical Description: Sodium alcohol ether sulfate
Category: Wetting agent, dispersant, emulsifier, detergent, anionic
Applications: Dyeing assistant; fabric scouring agent; cotton processing (improves penetration and wetting of bleaching sol'ns., aids removal of surface-adhering dyes after dyeing); stabilizes dye bath; dyeing of synthetics
Form: Liquid; Color: Pale amber
Properties: Density: 8.7 lb/gal; Stability: Stable to high concs. of alkalis and over a wide pH range; insensitive to hard water; Biodegradable
Composition: 57% min. active
Containers: Stainless steel or plastic-lined tanks
Warnings and Caution: Avoid contact with eyes and prolonged contact with skin

Tex-Wet 1197. Intex Products, Inc.
Chemical Description: Dodecylbenzene sulfonic acid
Category: Detergent intermediate, emulsifier, anionic
Applications: Most effective in the salt form when neutralized with an alkali; amine salts used as solvent emulsifier or coemulsifier, i.e., solvent scours or dye carriers; when neutralized with caustic soda used in textile dyeing and scouring; used in blended detergents, and in dish detergent formulations and general cleaners
Form: Liquid; Color: Reddish-brown
Properties: Density: 8.8 lb/gal; Acid No.: 180–190; Biodegradable
Composition: 97% active ingred.
Containers: Type 316 stainless steel, alloy, or clad; glass-lined

steel; lead-lined; fiber glass-reinforced polyester; heresite-lined steel

T-H Crystal Inhibitor No. 5. Thompson-Hayward Chem. Co.
Category: Sequestering agent
Applications: Prevents the precipitation of 2,4-D or MCP in 4 lb/gal amine salt formulations in hard water containing calcium and magnesium; prevents sludge formation and avoids clogging of spray nozzles during field applications
Form: Liquid; Color: Yellow
Properties: Density: 8.75 lb/gal; Sp. gr.: ≈ 1.05; Flash Pt.: > 200 F (TOC); Pour Pt.: 60 F

Tispray. Swastik
Chemical Description: Self-emulsifiable mineral oil
Category: Self-emulsifying mineral oil in water
Applications: Agricultural spray oil esp. for use in orchards and tea gardens
Form: Liquid
Reaction: Slightly alkaline

TL 724. Berol Kemi AB
Chemical Description: Acid form of phosphate ester
Category: Anionic surfactant
Applications: Makes it possible to use nonionics in many electrolyte sol'ns. where they would normally be salted out
Form: Clear, viscous liquid; Color: 8 (Gardner)
Properties: Solubility @ 5%: Sol. in ethanol, acetone, xylene, trichloroethylene, propylene glycol, white spirit; disp. in water; Density: 1.23 g/cc (30 C); Visc.: 23,000 cps (30 C); Clear Pt.: –10 C max. (upon thawing); Flash Pt.: 100 C min.; Flow Pt.: 12 C (upon thawing); pH: 1.7 (1% aq. disp.)
Composition: ≈ 100% active

T-Maz 20. Mazer Chem., Inc.
Chemical Description: POE 20 sorbitan monolaurate
Category: Emulsifier, solubilizer, wetting agent, antistat, stabilizer, dispersant, viscosity modifier, nonionic
Applications: Used in the food, cosmetic, drug, textile, and metalworking industries
Form: Liquid; Color: Yellow
Properties: Solubility: Sol. in water, ethanol, acetone; miscible in certain proportions with toluol, veg. oil; Sp. gr.: 1.1; Visc.: 400 cps; HLB: 16.7; Sapon. No.: 40–50; Acid No.: 2 max.; Hydroxyl No.: 96–108
Composition: 97% min. active

T-Maz 40. Mazer Chem., Inc.
Chemical Description: POE 20 sorbitan monopalmitate
Category and Applications: See T-Maz 20
Form: Liquid; Color: Yellow
Properties: Solubility: Sol. in water, ethanol, acetone; miscible in certain proportions with toluol, naphtha, veg. oil; Sp. gr.: 1.0; Visc.: 600 cps; HLB: 15.6; Sapon. No.: 43–49; Acid No.: 2 max.; Hydroxyl No.: 89–105
Composition: 97% min. active

T-Maz 60. Mazer Chem., Inc.
Chemical Description: POE 20 sorbitan monostearate Polysorbate 60
Category and Applications: See T-Maz 20
Form: Liquid; Color: Yellow
Properties: Solubility: Sol. in water, toluol; miscible in certain proportions with ethanol, acetone, naphtha; Sp. gr.: 1.1; Visc.: 550 cps; HLB: 14.9; Pour Pt.: 23–25 C; Sapon. No.: 45–55; Acid No.: 2 max.; Hydroxyl No.: 81–96
Composition: 97% min. active

T-Maz 65. Mazer Chem., Inc.
Chemical Description: POE 20 sorbitan tristearate Polysorbate 65
Category and Applications: See T-Maz 20
Form: Solid; Color: Tan
Properties: Solubility: Sol. in ethanol, acetone, naphtha; sol. hot in mineral and veg. oil; disp. in water and toluol; Sp. gr.: 1.1; M.P.: 30–32 C; HLB: 10.5; Sapon. No.: 88–98; Acid No.: 2 max.; Hydroxyl No.: 44–60
Composition: 97% min. active

T-Maz 67. Mazer Chem. Inc.
Chemical Description: POE 20 sorbitan monoisostearate
Category and Applications: See T-Maz 20
Form: Liquid; Color: Yellow
Properties: Solubility: Sol. in water, toluol; miscible in certain proportions with ethanol, acetone, naphtha; Sp. gr.: 1.0; Visc.: 400 cps; HLB: 14.8; Sapon. No.: 45–55; Acid No.: 2 max.; Hydroxyl No.: 80–95
Composition: 97% min. active

T-Maz 80. Mazer Chem., Inc.
Chemical Description: POE 20 sorbitan monooleate Polysorbate 80
Category and Applications: See T-Maz 20
Form: Liquid; Color: Yellow
Properties: Solubility: Sol. in water, ethanol, veg. oil, toluol; poorly sol. in acetone; Sp. gr.: 1.0; Visc.: 400 cps; HLB: 15.0; Sapon. No.: 45–55; Acid No.: 2 max.; Hydroxyl No.: 65–80
Composition: 97% min. active

T-Maz 85. Mazer Chem., Inc.
Chemical Description: POE 20 sorbitan trioleate
Category and Applications: See T-Maz 20
Form: Liquid; Color: Yellow
Properties: Solubility: Sol. in ethanol; miscible hot in certain proportions with acetone, toluol, naphtha, veg. oil; disp. in water; Sp. gr.: 1.0; Visc.: 300 cps; HLB: 11.0; Sapon. No.: 83–93; Acid No.: 2 max.; Hydroxyl No.: 39–52
Composition: 95% min. active

T-Maz 90. Mazer Chem., Inc.
Chemical Description: POE 20 sorbitan monotallate
Category and Applications: See T-Maz 20
Form: Liquid; Color: Amber
Properties: Solubility: Sol. in wa-

ter, ethanol, toluol, veg. oil; poorly sol. in acetone; Sp. gr.: 1.0; Visc.: 500 cps; HLB: 15.0; Sapon. No.: 45–55; Acid No.: 2 max.; Hydroxyl No.: 65–80
Composition: 97% min. active

T-Mulz 63. Thompson-Hayward Chem. Co.
Chemical Description: Blend of anionic and nonionic components
Category: Emulsifier, anionic
Applications: Specially developed emulsifier for the 6 lb/gal ethyl parathion + 3 lb/gal methyl parathion formulation; also used in other highly conc. formulations
Form: Opaque liquid; Color: Amber
Properties: Density: 8.4 lb/gal; Sp. gr.: 1.01 (20 C); Visc.: 650 cps; Flash Pt.: 101 F (TCC); Pour Pt.: 30 F; pH: 3 (5% disp.)

T-Mulz 339. Thompson-Hayward Chem. Co.
Chemical Description: Nonionic blend
Category: Emulsifier, nonionic
Applications: Developed for dimethoate formulations containing cyclohexanone and xylene as the solvent system
Form: Liquid; Color: Lt. yellow
Properties: Density: 8.27 lb/gal; Sp. gr.: ≈ 0.992 (20 C); Flash Pt.: 98 F (TOC); Pour Pt.: –5 C
Containers: 450 lb net steel drums

T-Mulz 391. Thompson-Hayward Chem. Co.
Category: Emulsifier, coemulsifier, nonionic
Applications: Emulsifier or coemulsifier for use in oil-based flowable pesticide formulations
Form: Clear liquid; Color: Pale yellow
Properties: Density: 8.5 lb/gal (20 C); Sp. gr.: ≈ 1.02 (20 C); HLB: 9.6; pH: 5–7 (1% IPA-water sol'n.)

T-Mulz 392. Thompson-Hayward Chem. Co.
Category and Applications: See T-Mulz 391
Form: Clear liquid; Color: Pale yellow
Properties: Density: 8.7 lb/gal (20 C); Sp. gr.: ≈ 1.04 (20 C); HLB: 10.8; pH: 5–7 (1% IPA-water sol'n.)

T-Mulz 565. Thompson-Hayward Chem. Co.
Chemical Description: Complex organic phosphate ester, free acid
Category: Emulsifier, detergent, hydrotrope, anionic
Applications: Emulsifier for a wide range of solvents, esp. when hard waters are a problem; emulsifier systems for agricultural and industrial formulations; hard surface detergent; hydrotrope to solubilize nonionics in alkaline built liquid detergent formulations
Form: Clear liquid; Color: Light straw
Properties: Solubility: Good sol. in alkaline sol'ns.; Density: 9.2 lb/gal; Sp. gr.: 1.1 (20 C); Flash Pt.: > 200 F (TOC); Pour Pt.: 36 F; pH: 2–3 (1% aq.)
Composition: 99.5% active ingred.

Toxicity: Acidic material
Containers: 55 gal drums and bulk
Warnings and Caution: Do not mix with conc. oxidizing or reducing agents—potentially explosive

T-Mulz 596. Thompson-Hayward Chem. Co.
Chemical Description: Complex organic phosphate ester, free acid
Category: Emulsifier, anionic
Applications: Emulsion polymerization
Form: Waxy solid; Color: Light straw
Properties: Density: 9.80 lb/gal (20 C); Sp. gr.: 1.18 (20 C); Flash Pt.: > 100 C (CC); Pour Pt.: 43 C; pH: 2–3 (1% aq.)
Toxicity: Acidic material
Containers: 55 gal drums and bulk
Warnings and Caution: Do not mix with conc. oxidizing or reducing agents—potentially explosive

T-Mulz 598. Thompson-Hayward Chem. Co.
Chemical Description: Complex organic phosphate ester, free acid
Category and Applications: See T-Mulz 565; also in the manufacture of mechanical dishwashing formulations
Form: Clear liquid; Color: Light straw
Properties: Solubility: Good solubility in alkaline systems; Density: 9.2 lb/gal; Sp. gr.: 1.1 (20 C); Flash Pt.: > 200 F (TOC); Pour Pt.: 23 F; pH: 2–3 (1% aq.)
Composition: 99.5% active ingred.
Toxicity: Acidic material
Containers: 55 gal drums and bulk
Warnings and Caution: Do not mix with conc. oxidizing or reducing agents—potentially explosive

T-Mulz 619. Thompson-Hayward Chem. Co.
Chemical Description: Anionic/nonionic blend
Category: Emulsifier
Applications: Specially developed for use at 20% level in making orthodichlorobenzene emulsifiable concentrates
Form: Slightly hazy liquid; Color: Amber
Properties: Density: 8.27 lb/gal; Visc.: ≈ 25 cps (80 F); Flash Pt.: 138 F (TOC); Pour Pt.: ≈ 14 F; pH: ≈ 6 (5%)
Composition: 99.5% active ingred.

T-Mulz 699. Thompson-Hayward Chem. Co.
Chemical Description: Anionic/nonionic blend
Category: Emulsifier
Applications: For use in 3 and 4 lb/gal Propanil emulsifiable concentrates
Form: Liquid; Color: Amber
Properties: Density: 8.56 lb/gal; Sp. gr.: 1.03 (20 C); Visc.: ≈ 200 cps; Flash Pt.: 90 F; Pour Pt.: < 5 F
Composition: 99.5% active ingred.

T-Mulz 808 A. Thompson-Hayward Chem. Co.
Chemical Description: Blended polyoxyethylene ether
Category: Emulsifier, nonionic

Applications: Concentrated agricultural crop oil formulations; enhances performance of herbicides such as atrazine wettable powders and water-disp. suspensions in combination with a suitable crop oil
Form: Clear liquid; Color: Straw
Properties: Density: 8.33 lb/gal (68 F); Sp. gr.: 1.00 (68 F); Flash Pt.: > 200 F (TCC); Pour Pt.: < 10 F; pH: 4–6 (1% aq. disp.)

T-Mulz 979. Thompson-Hayward Chem. Co.
Chemical Description: Blend of anionic-nonionic components
Category: Emulsifier, anionic/nonionic
Applications: Emulsifier for insecticides and herbicides; effective at levels of 2.5–3% for low volume aerial spraying and 4–6% for conventional ground spraying
Form: Opaque liquid
Properties: Density: 8.66 lb/gal; Sp. gr.: 1.040 (68 F); Flash Pt.: 108 F (CPMCC); Pour Pt.: 70 F; pH: 5.5–6.5 (1% sol'n.)
Composition: 99.5% active ingred.

T-Mulz A02. Thompson-Hayward Chem. Co.
Chemical Description: Blend of nonionic components
Category: Emulsifier, nonionic
Applications: Agricultural spray oils which have a broad range of applications from wettable powders to orchard spray oils
Form: Clear liquid
Properties: Solubility: Oil-sol.; Density: 8.34 lb/gal; Sp. gr.: ≈ 1.00 (20 C); Visc.: 1700 cps (20 C); HLB: 11.3; Flash Pt.: >500 F (PMCC); Pour Pt.: 20 F

T-Mulz FCO. Thompson-Hayward Chem. Co.
Category: Emulsifier, nonionic
Applications: Emulsifier for Florida citrus oils
Form: Liquid; Color: Pale yellow
Properties: Density: 8.6 lb/gal; Sp. gr.: ≈ 1.03 (20 C); Visc.: ≈ 57 SUS (210 F); Flash Pt.: >200 F (TOC); Pour Pt.: <0 F
Composition: 99.5% active ingred.

T-Mulz Mal-1S. Thompson-Hayward Chem. Co.
Chemical Description: Blend of anionic and nonionic components
Category: Emulsifier, anionic
Applications: Emulsifier for 5 lb/gal Malathion formulations
Form: Translucent liquid; Color: Straw
Properties: Density: 8.2 lb/gal; Sp. gr.: 0.99 (20 C); Visc.: 100 cps; HLB: 13.5; Flash Pt.: 55 F (TOC); Pour Pt.: 10 F; pH: 6.5 (5% disp.)
Composition: 99.5% active ingred.

T-Mulz O. Thomson-Hayward Chem. Co.
Chemical Description: Blend of anionic and nonionic compounds
Category: Emulsifier, anionic
Applications: Lipophilic emulsifier for good soft water performance; used with T-Mulz W for a wide range of pesticide formu-

lations
Form: Translucent liquid; Color: Straw
Properties: Solubility: Oil-loving; Density: 8.7 lb/gal; Sp. gr.: 1.04 (20 C); HLB: 10.7; Flash Pt.: 108 F; Pour Pt.: 55 F; pH: 6.5 (5% disp.)
Composition: 99.6% active ingred.

T-Mulz PB. Thompson-Hayward Chem. Co.
Chemical Description: Blend of anionic and nonionic components
Category: Emulsifier
Applications: Pyrethrin formulations with deodorized kerosene as the solvent; wide range of specialty Pyrethrin formulations
Form: Clear liquid; Color: Pale yellow
Properties: Density: 8.7 lb/gal (20 C); Sp. gr.: ≈ 1.04 (20 C); Flash Pt.: > 300 F; Pour Pt.: < 0 F; pH: 5–7 (1% IPA-water sol'n.)
Containers: 460 lb net drums

T-Mulz W. Thompson-Hayward Chem. Co.
Chemical Description: Blend of anionic and nonionic components
Category: Emulsifier, nonionic
Applications: Hydrophilic emulsifier for hard water performance; used with T-Mulz O for a wide range of pesticide formulations
Form: Translucent liquid; Color: Light straw
Properties: Solubility: Water-loving; Density: 8.7 lb/gal; Sp. gr.: 1.04 (20 C); HLB: 14.4; Flash Pt.: 132 F; Pour Pt.: 64 F; pH: 6.5 (5% disp.)
Composition: 99.6% active ingred.

Toluenesulfonic Acid. Ruetgers-Nease Chem. Co.
Chemical Description: Toluene sulfonic acid
Category: Hydrotrope, solvent, intermediate, catalyst, anionic
Applications: Solubilizer for organic compounds; as solvents for phthalocyanine pigments, in pickling baths for magnesium, and other metal cleaning; tanning agent; intermediate; crosslinking, condensation, esterification, curing, and other catalytic uses, esp. suited as an esterification catalyst in the manufacture of phthalates and other plasticizers
Form: Semicrystalline solid; Color: Amber to brown; Odor: Slight odor of toluene
Properties: m.w.: 172.21; Solubility: Sol. in acetone, ethanol, methanol; slightly sol. in benzene, toluene, xylene, and ether; sol. 270 g/100 ml in water; Density: 10.8 lb/gal (bulk); Sp. gr.: 1.30 (50 C); Crystallization Range: 35–50 C; Stability: Contact with water will cause hydrolysis
Composition: 95% active ingred.
Toxicity: Strong acid
Containers: 100 lb net and 575 lb net steel drums
Warnings and Caution: Wear protective goggles and gloves; corrosive to steel; store in glass or type 304 stainless steel

Toluenesulfonic Acid 70%: 30% Methanol Blend. Ruetgers-Nease Chem. Co.
Category and Applications: See Toluenesulfonic Acid; the blends are generally used to catalyze or set furfuryl alcohol, foundry resins where dilution is necessary
Properties: Crystallization Range: < –25 C
Composition: 65% min. conc.
Toxicity: Corrosive
Containers: Truckload drums and LTL drums
Warnings and Caution: Wear protective goggles and gloves; corrosive to steel–store in glass or type 316 stainless steel

Toluenesulfonic Acid 70%: 30% Water Blend. Ruetgers-Nease Chem. Co.
Category and Applications: See Toluenesulfonic Acid 70%: 30% Methanol Blend
Properties: Crystallization Range: < –5 C
Composition: 65% min. conc.
Toxicity: Corrosive
Containers: Truckload drums and LTL drums
Warnings and Caution: Wear protective gloves and goggles; corrosive to steel–store in glass or type 316 stainless steel

Toximul 360-A. Stepan Chem. Co.
Chemical Description: Blend of anionic sulfonates and nonionic polyether surfactants
Category: Emulsifier, anionic/nonionic
Applications: Matched pair emulsifier with Toximul 360-B for premium grade formulations of chlordane and heptachlor
Form: Liquid; Color: Amber
Properties: Density: 8.56 lb/gal; Sp. gr.: 1.029; Visc.: 1300 cps; HLB: 10.5; Flash Pt.: 116 F (Setaflash closed tester); Pour Pt.: –3 C; pH: 5.0–5.5 (5% in 50% aq. isopropanol)
Composition: 99.5% conc.
Containers: 450 lb net lined closed-head drums and bulk
Warnings and Caution: Combustible liquid

Toximul 360-B. Stepan Chem. Co.
Chemical Description: Blend of alkylbenzene sulfonate, nonionic surfactant, and fluidizing solvent
Category: Emulsifier, anionic/nonionic
Applications: Matched pair emulsifier with Toximul 360-A for premium grade formulations of chlordane and heptachlor
Form: Liquid; Color: Amber
Properties: Density: 8.52 lb/gal; Sp. gr.: 1.024; Visc.: 350 cps; HLB: 13.5; Flash Pt.: 77 F (Setaflash closed tester); Pour Pt.: 15 C; pH: 5.0–5.5 (5% in 50% aq. isopropanol)
Composition: 99.5% conc.
Containers: 450 lb net lined closed-head drums or bulk
Warnings and Caution: Flammable liquid

Toximul 360-B-HF. Stepan Chem. Co.
Chemical Description: Higher

flash version of Toximul 360-B
Category and Applications: See Toximul 360-B
Form: Liquid; Color: Amber
Properties: Density: 8.37 lb/gal; Sp. gr.: 1.006; Visc.: 135 cps; HLB: 13.5; Flash Pt.: 110 F (Setaflash closed tester); Pour Pt.: 15 C; pH: 5.0–5.5 (5% in 50% aq. isopropanol)
Composition: 99.5% min. conc.
Containers: 460 lb net lined closed-head drums or bulk
Warnings and Caution: Combustible liquid

Toximul 425. Stepan Chem. Co.
Chemical Description: Blend of anionic and nonionic surfactants
Category: Emulsifier
Applications: Matched emulsifier pair with Toximul 475 for 5 lb/gal Malathion formulations; useful in broad ranges of temp. and water hardness; 425 is the more hydrophobic of the pair
Form: Liquid
Properties: Density: 8.25 lb/gal; Sp. gr.: 0.99; Flash Pt.: 125 F (TOC); pH: 6.2 (5% in 50% aq. isopropanol)
Composition: 99.7% conc.

Toximul 475. Stepan Chem. Co.
Chemical Description: Blend of anionic and nonionic surfactants
Category: Emulsifier
Applications: Matched emulsifier pair with Toximul 425 for 5 lb/gal Malathion formulations; 475 is the more hydrophilic of the pair
Form: Liquid
Properties: Density: 8.18 lb/gal; Sp. gr.: 0.98; Flash Pt.: 125 F (TOC); pH: 6.2 (5% in 50% aq. isopropanol)
Composition: 99.7% conc.

Toximul 500, 600. Stepan Chem. Co.
Chemical Description: Blend of calcium alkylbenzene sulfonates, nonionic surfactants, and fluidizing solvents
Category: Emulsifier
Applications: Pesticide emulsifiable concentrate formulations
Form: Liquid; Color: Amber
Properties: Density: 8.55 lb/gal; Sp. gr.: 1.027; Visc.: 1600 cps; HLB: 10.5; Flash Pt.: 117 F (Setaflash closed tester); Pour Pt.: –3 C; pH: 5.5–6.5 (5% in 50% aq. isopropanol)
Composition: 99.4% conc.
Containers: 450 lb net lined closed-head drums and bulk
Warnings and Caution: Combustible liquid

Toximul 555. Stepan Chem. Co.
Chemical Description: Blend of calcium alkylbenzene sulfonates, nonionic surfactants, and fluidizing solvents
Category: Emulsifier
Applications: Pesticide emulsifiable concentrate formulations
Form: Liquid; Color: Amber
Properties: Density: 8.35 lb/gal; Sp. gr.: 1.001; Visc.: 540 cps; HLB: 10.0; Flash Pt.: 121 F (Setaflash closed tester); Pour Pt.: 3 C; pH: 5.5–6.5 (5% in 50% aq. iso-

propanol)
Composition: 99.5% conc.
Containers: 450 lb net lined closed-head drums and bulk
Warnings and Caution: Combustible liquid

Toximul 603. Stepan Chem. Co.
Chemical Description: Blend of calcium alkylbenzene sulfonates, nonionic surfactants, and fluidizing solvents
Category: Emulsifier, anionic/nonionic
Applications: Pesticide emulsifiable concentrate formulations
Form: Paste
Properties: Density: 9.10 lb/gal; Sp. gr.: 1.093; HLB: 14.0; Flash Pt.: 163 F (Setaflash closed tester); Pour Pt.: 36 C; pH: 5–6 (5% in 50% aq. isopropanol)
Composition: 99.5% conc.
Containers: 450 lb net lined closed-head drums and bulk
Warnings and Caution: Combustible liquid

Toximul AP. Stepan Chem. Co.
Category: Emulsifier
Applications: For chlorinated and phosphate insecticides

Toximul D. Stepan Chem. Co.
Chemical Description: Blend of calcium alkylbenzene sulfonates, nonionic surfactants, and fluidizing solvents
Category: Emulsifier, anionic/nonionic
Applications: Matched emulsifier pair with Toximul H for chlorinated, phosphate, and multiple-toxicant pesticide formulations; useful over broad range of temp. and water hardness
Form: Liquid; Color: Dark amber
Properties: Density: 8.58 lb/gal; Sp. gr.: 1.031; Visc.: 1900 cps; HLB: 10.5; Flash Pt.: 119 F (Setaflash closed tester); Pour Pt.: 3 C; pH: 5–6 (5% in 50% aq. isopropanol)
Composition: 99.5% conc.
Containers: 450 lb net lined closed-head drums or bulk
Warnings and Caution: Combustible liquid

Toximul H. Stepan Chem. Co.
Chemical Description: Blend of calcium alkylbenzene sulfonates, nonionic surfactants, and fluidizing solvents
Category: Emulsifier, anionic/nonionic
Applications: Matched emulsifier pair with Toximul D for chlorinated, phosphate, and multiple-toxicant pesticide formulations; useful over broad range of temp. and water hardness
Form: Liquid; Color: Dark amber
Properties: Density: 8.45 lb/gal; Sp. gr.: 1.016; Visc.: 300 cps; HLB: 13.5; Flash Pt.: 73 F (Setaflash closed tester); Pour Pt.: 9 C; pH: 5–6 (5% in 50% aq. isopropanol)
Composition: 99.4% conc.
Containers: 450 lb net lined closed-head drums and bulk
Warnings and Caution: Flammable liquid

Toximul H-HF. Stepan Chem. Co.
Chemical Description: Higher flash version of Toximul H
Category and Applications: See Toximul H
Form: Liquid; Color: Amber
Properties: Density: 8.30 lb/gal; Sp. gr.: 0.997; Visc.: 375 cps; HLB: 13.5; Flash Pt.: 117 F (Setaflash closed tester); Pour Pt.: 15 C; pH: 5–6 (5% in 50% aq. isopropanol)
Composition: 99.4% conc.
Containers: 450 lb net lined closed-head drums and bulk
Warnings and Caution: Combustible liquid

Toximul MP. Stepan Chem. Co.
Chemical Description: Blend of calcium alkylbenzene sulfonates, nonionic surfactants, and fluidizing solvents
Category: Emulsifier, anionic/nonionic
Applications: Pesticide E.C. formulations
Form: Liquid; Color: Amber
Properties: Density: 8.50 lb/gal; Sp. gr.: 1.021; Visc.: 700 cps; HLB: 11.0; Flash Pt.: 98 F (Setaflash closed tester); Pour Pt.: 3 C; pH: 5.5–6.5 (5% in 50% aq. isopropanol)
Composition: 99.4% conc.
Containers: 450 lb net lined closed-head drums and bulk
Warnings and Caution: Flammable liquid

Toximul MP-8. Stepan Chem. Co.
Chemical Description: Blend of calcium alkylbenzene sulfonates, nonionic surfactants, and fluidizing solvents
Category: Emulsifier, anionic/nonionic
Applications: Pesticide E.C. formulations
Form: Liquid; Color: Amber
Properties: Density: 8.52 lb/gal; Sp. gr.: 1.024; Visc.: 400 cps; HLB: 12.5; Flash Pt.: 94 F (Setaflash closed tester); Pour Pt.: 15 C; pH: 5.5–6.5 (5% in 50% aq. isopropanol)
Composition: 99.5% conc.
Containers: 450 lb net lined closed-head drums and bulk
Warnings and Caution: Flammable liquid

Toximul MP-8B. Stepan Chem. Co.
Category: Emulsifier, anionic/nonionic
Applications: High concentration Cythion and Malathion emulsifiable liquids
Form: Liquid; Color: Amber
Properties: Density: 8.15 lb/gal; Sp. gr.: 0.98; Flash Pt.: 165 F (TOC); pH: 4.7 (5% in 50% aq. isopropanol)
Composition: 99.5% conc.

Toximul MP-8-HF. Stepan Chem. Co.
Chemical Description: Higher flash version of Toximul MP-8
Category and Applications: See Toximul MP-8
Form: Liquid; Color: Amber
Properties: Density: 8.27 lb/gal; Sp. gr.: 0.994; Visc.: 400 cps; HLB: 12.5; Flash Pt.: 114 F (Seta-

flash closed tester); Pour Pt.: 15 C; pH: 5.5–6.5 (5% in 50% aq. isopropanol)
Composition: 99.5% conc.
Containers: 450 lb net lined closed-head drums and bulk
Warnings and Caution: Combustible liquid

Toximul MP-10. Stepan Chem. Co.
Chemical Description: Blend of calcium alkylbenzene sulfonates, nonionic surfactants, and fluidizing solvents
Category: Emulsifier, anionic/nonionic
Applications: High-performance emulsifier for the 6 lb ethyl parathion/3 lb methyl parathion formulation
Form: Liquid; Color: Amber
Properties: Density: 8.42 lb/gal; Sp. gr.: 1.012; Visc.: 700 cps; HLB: 12.0; Flash Pt.: 89 F (Setaflash closed tester); Pour Pt.: 6 C; pH: 5–6 (5% in 50% aq. isopropanol)
Composition: 99.4% conc.
Containers: 450 lb net lined closed-head drums and bulk
Warnings and Caution: Flammable liquid

Toximul MP-10-HF. Stepan Chem. Co.
Chemical Description: Higher flash version of Toximul MP-10
Category and Applications: See Toximul MP-10
Form: Liquid; Color: Amber
Properties: Density: 8.41 lb/gal; Sp. gr.: 1.011; Visc.: 840 cps; HLB: 12.0; Flash Pt.: 125 F (Setaflash closed tester); Pour Pt.: 12 C; pH: 5–6 (5% in 50% aq. isopropanol)
Composition: 99.4% conc.
Containers: 450 lb net lined closed-head drums and bulk
Warnings and Caution: Combustible liquid

Toximul MP-HF. Stepan Chem. Co.
Chemical Description: Higher flash version of Toximul MP
Category and Applications: See Toximul MP
Form: Liquid; Color: Amber
Properties: Density: 8.42 lb/gal; Sp. gr.: 1.012; Visc.: 540 cps; HLB: 11.0; Flash Pt.: 119 F (Setaflash closed tester); Pour Pt.: 6 C; pH: 5.5–6.5 (5% in 50% aq. isopropanol)
Composition: 99.4% conc.
Containers: 450 lb net lined closed-head drums and bulk
Warnings and Caution: Combustible liquid

Toximul R. Stepan Chem. Co.
Chemical Description: Blend of calcium alkylbenzene sulfonates, nonionic surfactants, and fluidizing solvents
Category: Emulsifier, anionic/nonionic
Applications: Matched emulsifier pair with Toximul S for all toxicant systems, incl. chlorinated, brominated, phosphated insecticides, herbicides, nematocides; useful in a broader range of temp. and water hardness; products have

excellent storage stability
Form: Liquid; Color: Amber
Properties: Solubility: Sol. in toxicant systems; Density: 8.51 lb/gal; Sp. gr.: 1.023; Visc.: 750 cps; HLB: 10.5; Flash Pt.: 92 F (Setaflash closed tester); Pour Pt.: 0 C; pH: 5.5–6.5 (5% in 50% aq. isopropanol)
Composition: 99.5% conc.
Containers: 450 lb net lined closed-head drums and bulk
Warnings and Caution: Flammable liquid

Toximul R-HF. Stepan Chem. Co.
Chemical Description: Higher flash version of Toximul R
Category and Applications: See Toximul R
Form: Liquid; Color: Amber
Properties: Density: 8.30 lb/gal; Sp. gr.: 0.997; Visc.: 770 cps; HLB: 10.5; Flash Pt.: 121 F (Setaflach closed tester); Pour Pt.: 6 C; pH: 5.5–6.5 (5% in 50% aq. isopropanol)
Composition: 99.5% conc.
Containers: 450 lb net lined closed-head drums and bulk
Warnings and Caution: Combustible liquid

Toximul S. Stepan Chem. Co.
Chemical Description: Blend of calcium alkylbenzene sulfonates, nonionic surfactants, and fluidizing solvents
Category: Emulsifier, anionic/nonionic
Applications: Matched emulsifier pair with Toximul R for all toxicant systems, incl. chlorinated, brominated, phosphated insecticides, herbicides, nematocides; useful in a broader range of temp. and water hardness; products have excellent storage stability
Form: Liquid; Color: Amber
Properties: Solubility: Sol. in toxicant systems; Density: 8.58 lb/gal; Sp. gr.: 1.031; Visc.: 400 cps; HLB: 13.0; Flash Pt.: 129 F (Setaflash closed tester); Pour Pt.: 9 C; pH: 5–6 (5% in 50% aq. isopropanol)
Composition: 99.5% conc.
Containers: 450 lb net lined closed-head drums and bulk
Warnings and Caution: Combustible liquid

Tri-Basic Copper Sulfate. Cities Service Co.
Category: Fungicide
Applications: Neutral, compatible fungicide applied to foliage as spray or dust for control of disease on vegetables, fruits, and ornamentals
Form: Fine powder; Color: Green
Properties: Solubility: Water-insol.
Composition: 53% copper
Containers: 50-lb multiwall bags, 4, 6, and 10 lb bags

Triton 770 Conc. Rohm and Haas Co.
Chemical Description: Sodium alkylaryl polyether sulfate, 30% aq. sol'n. containing 23% 2-propanol
Category: Detergent, wetting agent, penetrant, anionic

Applications: Used for detergents, waxes, cleaners, rag cooking, textile processing, degreasing of skins prior to tanning, etc.
Form: Liquid; Color: Amber
Properties: Density: 8.3 lb/gal; Visc.: 15 cps; Flash Pt.: 78 F (PM-CC); Pour Pt.: −20 F; Surface Tension: 28 dynes/cm (1%)
Composition: 30% active ingred.

Triton B-1956. Rohm and Haas Co.
Chemical Description: Modified phthalic glycerol alkyd resin
Category: Emulsifier, detergent, nonionic
Applications: Insecticide formulations; insecticide and fungicide sprays for fruits and vegetables
Form: Liquid; Color: Amber
Properties: Density: 8.8 lb/gal; Visc.: 100 cps; Flash Pt.: 74 F (Setaflash CC); Pour Pt.: 25 F; Surface Tension: 32 dynes/cm (0.01%)
Composition: 77% active ingred.; solvent: ethylene dichloride

Triton BG-10. Rohm and Haas Co.
Category: Nonionic surfactant
Applications: Used in bottle washing and agricultural formulations
Form: Liquid; Color: Dark brown
Properties: Solubility: Sol. in 50% sodium hydroxide sol'n.; Density: 9.6 lb/gal; Visc.: 1900 cps; Cloud Pt.: > 100 C (1%); Flash Pt.: > 300 F (TOC); Pour Pt.: 10 F; Surface Tension: 34 dynes/cm (0.01%); Biodegradable
Composition: 70% active ingred.

Triton CF-10. Rohm and Haas Co.
Chemical Description: Alkylaryl polyether
Category: Detergent, dyeing assistant, nonionic
Applications: Low-foaming detergent for mechanical dishwashing, rinse additives, automatic laundering, metal cleaning, dairy cleaning; nylon dyeing assistant
Form: Liquid; Color: 4 (VCS)
Properties: Density: 8.9 lb/gal; Visc.: 250 cps; HLB: 14.0; Cloud Pt.: 28 C (1%); Flash Pt.: > 500 F (TOC); Pour Pt.: 60 F; Surface Tension: 32 dynes/cm (1%)
Composition: 100% active ingred.

Triton CF-21. Rohm and Haas Co.
Chemical Description: Alkylaryl polyether
Category: Detergent, nonionic
Applications: Low-foaming detergent for mechanical dishwashing, final rinse additive, automatic laundering, metal cleaning, dairy cleaning
Form: Liquid; Color: 125 (APHA)
Properties: Density: 8.7 lb/gal; Visc.: 250 cps; Cloud Pt.: 40 C (1%); Flash Pt.: > 480 F (TOC); Pour Pt.: −30 F; Surface Tension: 31 dynes/cm (1%)
Composition: 100% active ingred.

Triton CF-32. Rohm and Haas Co.
Chemical Description: Amine polyglycol condensate
Category: Detergent, wetting agent, antifoaming agent, nonionic
Applications: Very low-foaming detergent for mechanical dishwash-

ing, metal cleaning, dairy cleaning
Form: Liquid; Color: 5 (VCS)
Properties: Density: 8.6 lb/gal; Visc.: 550 cps; Cloud Pt.: 25 C (1%); Flash Pt.: > 300 F (TOC); Pour Pt.: 15 F; Surface Tension: 32 dynes/cm (1%)
Composition: 95% active ingred. in water

Triton CF-54. Rohm and Haas Co.
Chemical Description: Modified polyethoxy adduct
Category: Detergent, nonionic
Applications: Low-foaming detergent for mechanical dishwashing, rinse additive, automatic laundering, metal cleaning, dairy cleaning
Form: Liquid; Color: 2 (VCS)
Properties: Density: 8.7 lb/gal; Visc.: 180 cps; HLB: 13.6; Cloud Pt.: 38 C (1%); Flash Pt.: > 200 F (PMCC); Pour Pt.: 35 F; Surface Tension: 29 dynes/cm (1%)
Composition: 100% active ingred.

Triton CF-76. Rohm and Haas Co.
Chemical Description: Modified polyethoxy adduct
Category: Detergent, nonionic
Applications: Low-foaming detergent; dishwashing formulations
Form: Liquid; Color: Lt. yellow
Properties: Density: 8.7 lb/gal; Visc.: 295 cps; Cloud Pt.: 31 C (1%); Flash Pt.: > 300 F (TOC); Pour Pt.: 40 F; Surface Tension: 29 dynes/cm (0.01%)
Composition: 100% active ingred.

Triton CF-87. Rohm and Haas Co.
Chemical Description: Modified terminated alkylaryl ether
Category: Nonionic surfactant
Applications: Low-foaming nonionic for use in rinse aids in mechanical dishwashing
Form: Liquid; Color: Straw
Properties: Density: 9.0 lb/gal; Visc.: 240 cps; Cloud Pt.: 32 C (1%); Flash Pt.: > 215 F (TOC); Pour Pt.: 30 F; Surface Tension: 32 dynes/cm (0.01%)
Composition: 90% active ingred.

Triton CG-400. Rohm and Haas Co.
Chemical Description: Stearyl dimethyl benzylammonium chloride (aq. disp. containing 5% 2-propanol)
Category: Cationic surfactant
Form: Liquid; Color: White
Properties: Density: 8.0 lb/gal (50 C); Flash Pt.: 111 F (Setaflash CC); Pour Pt.: 110 F
Composition: 25% active ingred.

Triton CG-500. Rohm and Haas Co.
Chemical Description: Stearyl dimethyl benzylammonium chloride (aq. disp. containing 11% 2-propanol)
Category: Cationic surfactant
Form: Liquid; Color: White
Properties: Density: 7.7 lb/gal (50 C); Flash Pt.: 90 F (Setaflash CC); Pour Pt.: 110 F
Composition: 50% active ingred.

Triton CS-7. Rohm and Haas Co.
Chemical Description: Blend of alkylaryl polyether alcohols and organic sulfonates (aq. sol'n. containing 8% 2-propanol)
Category: Spreading-binding a-

gent, anionic
Applications: Used in agricultural concentrates, aerial and low-volume applications with insecticides, fungicides, and miticides
Form: Liquid; Color: Yellow
Properties: Density: 8.5 lb/gal; Flash Pt.: 98 F (Setaflash CC); Pour Pt.: –35 F
Composition: 60% active ingred.

Triton DF-12. Rohm and Haas Co.
Chemical Description: Modified polyethoxylated alcohol
Category: Detergent, nonionic
Applications: Low-foaming detergent for mechanical dishwashing, rinse additives, automatic laundering, metal cleaning, automatic floor scurbbing, dairy cleaning
Form: Liquid; Color: Lt. straw
Properties: Density: 8.6 lb/gal; Visc.: 60 cps; Cloud Pt.: 17 C (1%); Flash Pt.: 430 F (COC); Pour Pt.: 30 F; Surface Tension: 33 dynes/cm (0.01%)
Composition: 100% active ingred.

Triton DF-16. Rohm and Haas Co.
Chemical Description: Terminated ethoxylated linear alcohol
Category: Detergent, nonionic
Applications: Low-foaming detergent esp. suited for rinse additives
Form: Liquid; Color: < 200 (APHA)
Properties: Density: 8.2 lb/gal; Visc.: 35 cps; Cloud Pt.: 36 C (1%); Flash Pt.: > 200 F (PMCC); Pour Pt.: 20 F; Stability: Excellent acid stability: Surface Tension: 28 dynes/cm (1%)
Composition: 100% active ingred.

Triton DF-18. Rohm and Haas Co.
Chemical Description: Modified polyethoxylated alcohol
Category: Detergent, nonionic
Applications: Low-foaming detergent for dishwashing, cleaners for food processing equipment, spray metal cleaners, etc.
Form: Liquid; Color: Yellow
Properties: Density: 8.5 lb/gal; Visc.: 140 cps; Cloud Pt.: < 0 C (1%); Flash Pt.: > 220 F (TOC); Pour Pt.: 65 F; Surface Tension: 32 dynes/cm (0.01%); Biodegradable
Composition: 90% active ingred.

Triton DF-20. Rohm and Haas Co.
Chemical Description: Modified ethoxylate, acid form
Category: Detergent, anionic
Applications: Low-foaming detergent for use in highly alkaline built liquid concentrate cleaners, metal cleaners, etc.
Form: Liquid; Color: Clear straw
Properties: Density: 8.5 lb/gal; Visc.: 630 cps; Flash Pt.: > 300 F (TOC); Pour Pt.: –15 F; Surface Tension: 28 dynes/cm (0.1% @ pH 11)
Composition: 100% active ingred.

Triton DN-14. Rohm and Haas Co.
Chemical Description: Alkylpolyether alcohol
Category: Detergent, emulsifier, leveling agent, nonionic
Applications: Dye leveling, scouring, and emulsifying
Form: Liquid; Color: Yellow
Properties: Density: 8.8 lb/gal; Visc.: 400 cps; Cloud Pt.: 90 C

(1%); Flash Pt.: > 300 F (TOC); Pour Pt.: 55 F; Surface Tension: 39 dynes/cm (1%); Biodegradable
Composition: 90% active ingred. in water

Triton GR-5M. Rohm and Haas Co.
Chemical Description: Dioctyl sodium sulfosuccinate (aq. sol'n. containing 20% 2-propanol)
Category: Wetting agent, emulsifier, anionic
Applications: High-speed wetting, rewetting, and emulsifying
Form: Liquid; Color: 200 (APHA)
Properties: Density: 8.4 lb/gal; Visc.: 40 cps; Flash Pt.: 75 F (Setaflash CC); Pour Pt.: –60 F; Surface Tension: 23 dynes/cm (1%)
Composition: 60% active ingred.

Triton GR-7M. Rohm and Haas Co.
Chemical Description: Dioctyl sodium sulfosuccinate
Category: Emulsifier, anionic
Applications: Drycleaning detergent, agricultural emulsions
Form: Liquid; Color: 2 (VCS)
Properties: Density: 8.4 lb/gal; Visc: 110 cps; Flash Pt.: 130 F (TOC); Pour Pt.: –70 F
Composition: 64% active ingred.; solvent: light petroleum distillate

Triton H-55. Rohm and Haas Co.
Chemical Description: Phosphate ester, potassium salt
Category: Hydrotrope, anionic
Applications: Used for conventional nonionic and anionic surfactants in alkaline liquid concentrates
Form: Liquid; Color: 3 (VCS)
Properties: Density: 11.2 lb/gal; Visc.: 40 cps; Flash Pt.: 154 F (PMCC)
Composition: 50% active ingred. in water

Triton H-66. Rohm and Haas Co.
Chemical Description: Phosphate ester, potassium salt
Category and Applications: See Triton H-55
Form: Liquid; Color: Amber
Properties: Density: 10.5 lb/gal; Visc.: 120 cps; Flash Pt.: > 200 F (PMCC); Pour Pt.: –5 F
Composition: 50% active ingred. in water

Triton N-57. Rohm and Haas Co.
Chemical Description: Nonylphenoxy polyethoxy ethanol (5 EO)
Category: Detergent, emulsifier, intermediate, nonionic
Form: Liquid; Color: 100 (APHA)
Properties: m.w.: 440 (avg.); Solubility: Oil-sol; water-disp.; Density: 8.5 lb/gal; Visc.: 240 cps; HLB: 10.0; Cloud Pt.: < 0 C (1%); Flash Pt.: > 515 F (COC); Pour Pt.: –25 F; Surface Tension: 29 dynes/cm (1%)
Composition: 100% active ingred.

Triton N-60. Rohm and Haas Co.
Chemical Description: Nonylphenoxy polyethoxy ethanol (6 EO)
Category: See Triton N-57
Form: Liquid; Color: 125 (APHA)
Properties: m.w.: 484 (avg.); Density: 8.7 lb/gal; Visc.: 300 cps;

HLB: 10.9; Cloud Pt.: 53 C (1% in 12.5% aq. isopropyl alcohol); Pour Pt.: –25 F; Surface Tension: 28 dynes/cm (1%)
Composition: 100% active ingred.

Triton N-101. Rohm and Haas Co.
Chemical Description: Nonylphenoxy polyethoxy ethanol (9–10 EO)
Category: Wetting agent, detergent, dispersant, emulsifier, nonionic
Applications: Household and industrial cleaners, textile processing; emulsifier for pesticides
Form: Liquid; Color: 100 (APHA)
Properties: m.w.: 642 (avg.); Density: 8.7 lb/gal; Visc.: 240 cps; HLB: 13.4; Cloud Pt.: 54 C (1%); Flash Pt.: > 300 F (TOC); Pour Pt.: 40 F; Surface Tension: 29 dynes/cm (1%)
Composition: 100% active ingred.

Triton N-111. Rohm and Haas Co.
Chemical Description: Nonylphenoxy polyethoxy ethanol (11 EO)
Category: Wetting agent, detergent, emulsifier, nonionic
Applications: Textile processing
Form: Liquid; Color: 100 (APHA)
Properties: m.w.: 704 (avg.); Solubility: Water-sol.; Density: 8.8 lb/gal; Visc.: 310 cps; HLB: 13.8; Cloud Pt.: 72 C (1%); Flash Pt.: > 300 F (TOC); Pour Pt.: 55 F
Composition: 100% active ingred.

Triton N-150. Rohm and Haas Co.
Chemical Description: Nonylphenoxy polyethoxy ethanol (15 EO)
Category: Detergent, emulsifier, nonionc
Form: Liquid; Color: 160 (APHA)
Properties: m.w.: 880 (avg.); Density: 9.0 lb/gal; Visc.: 4350 cps; HLB: 15.0; Cloud Pt.: 95 C (1%); Pour Pt.: 65 F; Surface Tension: 33 dynes/cm (1%)
Composition: 100% active ingred.

Triton N-401. Rohm and Haas Co.
Chemical Description: Nonylphenoxy polyethoxy ethanol (40 EO)
Category: Emulsifier, nonionic
Applications: Emulsion polymerization
Form: Liquid; Color: 250 (APHA)
Properties: m.w.: 1980 (avg.); Density: 9.2 lb/gal; Visc.: 850 cps; HLB: 17.8; Cloud Pt.: > 100 C (1%); Flash Pt.: > 230 F (Setaflash CC); Pour Pt.: 18 F
Composition: 70% active ingred.

Triton QS-15. Rohm and Haas Co.
Chemical Description: Sodium salt of amphoteric surfactant
Category: Detergent, anionic
Applications: Used for metal cleaning and latex stability
Form: Liquid; Color: Amber
Properties: Solubility: Good sol. in aq. sol'ns. of strong alkalis; Density: 9.3 lb/gal; Visc.: 8500 cps; Flash Pt.: > 300 F (TOC); Pour Pt.: 25 F; Stability: Good stability in aq. sol'ns. of strong alkalis incl. caustic; Surface Tension: 40 dynes/cm (0.1%)
Composition: 100% active ingred.

Triton QS-30. Rohm and Haas Co.
Chemical Description: Phosphate ester, acid form

Category: Detergent, anionic
Applications: Metal cleaning
Form: Liquid; Color: 12 (VCS)
Properties: Solubility: Good sol. in strong alkalis; Density: 9.8 lb/gal; Visc.: 10,000 cps; Flash Pt.: > 200 F (PMCC); Pour Pt.: 30 F; Stability: Good in strong alkalis; Surface Tension: 33 dynes/cm (0.1% @ pH 7.4)
Composition: 90% active ingred. in water

Triton QS-44. Rohm and Haas Co.
Chemical Description: Phosphate ester, acid form
Category: Detergent, hydrotrope, anionic
Applications: Metal cleaning; cleared for food contact and agricultural uses
Form: Liquid; Color: 7 (VCS)
Properties: Solubility: Good sol. in strong alkalis; Density: 9.8 lb/gal; Visc.: 8000 cps; Flash Pt.: > 200 F (PMCC); Pour Pt.: 35 F; Stability: Good in strong alkalis; Surface Tension: 33 dynes/cm (0.1%)
Composition: 80% active ingred. in water

Triton W-30 Conc. Rohm and Haas Co.
Chemical Description: Sodium alkylaryl ether sulfate (aq. sol'n. containing 27% 2-propanol)
Category: Wetting agent, emulsifier, penetrant, anionic
Applications: Used for sizing, dyeing, desizing; wetting agent for textile processing; emulsifier in emulsion polymerization
Form: Liquid; Color: Amber
Properties: Density: 8.2 lb/gal; Visc.: 9 cps; Flash Pt.: 74 F (PMCC); Pour Pt.: −15 F; Surface Tension: 29 dynes/cm (1%)
Composition: 27% active ingred.

Triton X-15. Rohm and Haas Co.
Chemical Description: Octylphenoxy ethoxy ethanol (1 EO)
Category: Emulsifier, coupling agent, nonionic
Form: Liquid; Color: 250 (APHA)
Properties: m.w.: 250 (avg.); Density: 8.2 lb/gal; Visc.: 790 cps; HLB: 3.6; Flash Pt.: > 300 F (TOC); Pour Pt.: 15 F
Composition: 100% active ingred.

Triton X-35. Rohm and Haas Co.
Chemical Description: Octylphenoxy polyethoxy ethanol (3 EO)
Category: See Triton X-15
Form: Liquid; Color: 125 (APHA)
Properties: m.w.: 338 (avg.); Density: 8.5 lb/gal; Visc.: 370 cps; HLB: 7.8; Flash Pt.: > 300 F (TOC); Pour Pt.: −10 F; Surface Tension: 29 dynes/cm (0.01%)
Composition: 100% active ingred.

Triton X-45. Rohm and Haas Co.
Chemical Description: Octylphenoxy polyethoxy ethanol (5 EO)
Category: Detergent, dispersant, emulsifier, wetting agent, nonionic
Applications: Drycleaning detergent; insecticide emulsifier; dispersing and wetting agent in organic systems

Form: Liquid; Color: 100 (APHA)
Properties: m.w.: 426 (avg.); Density: 8.7 lb/gal; Visc.: 290 cps; HLB: 10.4; Cloud Pt.: <0 C (1%); Flash Pt.: > 300 F (TOC); Pour P.t: –15 F; Surface Tension: 28 dynes/cm (1%)
Composition: 100% active ingred.

Triton X-67. Rohm and Haas Co.
Chemical Description: Alkylpolyether alcohol
Category: Nonionic surfactant
Form: Solid; Color: White
Properties: HLB: 16.8; Cloud Pt.: > 100 C (1%); Flash Pt.: > 300 F (TOC); Surface Tension: 43 dynes/cm (1%)
Composition: 100% active ingred.

Triton X-100. Rohm and Haas Co.
Chemical Description: Octylphenoxy polyethoxy ethanol (9–10 EO)
Category: Wetting agent, dispersant, detergent, emulsifier, nonionic
Applications: Household and industrial cleaners; textile processing, wool scouring; emulsifier for insecticides and herbicides
Form: Liquid; Color: 100 (APHA)
Properties: m.w.: 628 (avg.); Density: 8.9 lb/gal; Visc.: 240 cps; HLB: 13.5; Cloud Pt.: 65 C (1%); Flash Pt.: > 300 F (TOC); Pour Pt.: 45 F; Surface Tension: 30 dynes/cm (1%)
Composition: 100% active ingred.

Triton X-102. Rohm and Haas Co.
Chemical Description: Octylphenoxy polyethoxy ethanol (12–13 EO)
Category: Detergent, wetting agent, nonionic
Applications: Metal cleaning, industrial and household liquid detergents and cleaners
Form: Liquid; Color: 100 (APHA)
Properties: m.w.: 756 (avg.); Density: 8.9 lb/gal; Visc.: 330 cps; HLB: 14.6; Cloud Pt.: 88 C (1%); Flash Pt.: > 300 F (TOC); Pour Pt.: 60 F; Surface Tension: 32 dynes/cm (1%)
Composition: 100% active ingred.

Triton X-114. Rohm and Haas Co.
Chemical Description: Octylphenoxy polyethoxy ethanol (7–8 EO)
Category: Detergent, wetting agent, nonionic
Applications: Controlled-foam detergent for household and industrial laundry uses
Form: Liquid; Color: 100 (APHA)
Properties: m.w.: 536 (avg.); Density: 8.8 lb/gal; Visc.: 260 cps; HLB: 12.4; Cloud Pt.: 22 C (1%); Flash Pt.: > 300 F (TOC); Pour Pt.: 15 F; Surface Tension: 29 dynes/cm (1%)
Composition: 100% active ingred.

Triton X-114SB. Rohm and Haas Co.
Chemical Description: 80% sol'n. of Triton X-114 in isopropyl alcohol
Category and Applications: See Triton X-114
Form: Liquid; Color: 100 (APHA)
Properties: m.w.: 536 (avg.); Density: 8.2 lb/gal; HLB: 12.4; Flash

Pt.: 61 F (Setaflash CC); Pour Pt.: –5 F
Composition: 80% active ingred.

Triton X-120. Rohm and Haas Co.
Chemical Description: Octylphenoxy polyethoxy ethanol (9–10 EO)
Category: Wetting and dispersing agent, nonionic
Applications: Agricultural wettable powders
Form: Solid; Color: White
Properties: m.w.: 628 (avg.)
Composition: 40% active ingred.

Triton X-151. Rohm and Haas Co.
Chemical Description: Blend of alkylaryl polyether alcohols and organic sulfonates
Category: Emulsifier, anionic
Applications: FDA exempt emulsifier for E.C. for insecticides
Form: Liquid
Properties: Density: 8.7 lb/gal; Visc.: 80 cps; Flash Pt.: 56 F (Setaflash CC); Pour Pt.: 10 F; Surface Tension: 37 dynes/cm (0.01%)
Composition: 100% active ingred.

Triton X-152. Rohm and Haas Co.
Chemical Description: Blend of alkylaryl polyether alcohols and organic sulfonates
Category: Anionic surfactant
Form: Liquid; Color: 11 (VCS)
Properties: Density: 8.5 lb/gal; Visc.: 155 cps; Flash Pt.: 56 F (Setaflash CC); Pour Pt.: 35 F
Composition: 100% active ingred.

Triton X-161. Rohm and Haas Co.
Chemical Description: Blend of alkylaryl polyether alcohols and organic sulfonates
Category: Anionic surfactant
Form: Liquid; Color: 8–10 (VCS)
Properties: Density: 8.7 lb/gal; Visc.: 175 cps; Flash Pt.: 185 F (TOC); Pour Pt.: –45 F
Composition: 100% active ingred.

Triton X-165. Rohm and Haas Co.
Chemical Description: Octylphenoxy polyethoxy ethanol (16 EO)
Category: Detergent, emulsifier, wetting agent, nonionic
Applications: Surfactant for use where higher water solubility and performance at elevated water temps. are needed
Form: Liquid; Color: 125 (APHA)
Properties: m.w.: 910 (avg.); Density: 9.0 lb/gal; Visc.: 540 cps; HLB: 15.8; Cloud Pt.: > 100 C (1%); Flash Pt.: > 300 F (TOC); Pour Pt.: 55 F; Surface Tension: 34 dynes/cm (1%)
Composition: 70% active ingred. in water

Triton X-171. Rohm and Haas Co.
Chemical Description: Blend of alkylaryl polyether alcohols and organic sulfonates
Category and Applications: See Triton X-151
Form: Liquid
Properties: Density: 8.7 lb/gal; Visc.: 270 cps; Flash Pt.: 44 F (Setaflash CC); Pour Pt.: –20 F
Composition: 100% active ingred.

Triton X-172. Rohm and Haas Co.
Chemical Description: Blend of alkylaryl polyether alcohols and organic sulfonates
Category: Anionic surfactant
Form: Liquid; Color: 14 (VCS)
Properties: Density: 8.7 lb/gal; Visc.: 6640 cps; Flash Pt.: 71 F (Setaflash CC); Pour Pt.: 5 F
Composition: 100% active ingred.

Triton X-180. Rohm and Haas Co.
Chemical Description: Blend of alkylaryl polyether alcohols and organic sulfonates
Category: Emulsifier, anionic
Applications: FDA-exempt emulsifier for a wide range of pesticide formulations
Form: Liquid; Color: Hazy amber
Properties: Density: 8.6 lb/gal; Visc.: 130 cps; Flash Pt.: 54 F (Setaflash CC); Pour Pt.: 40 F
Composition: 100% active ingred.

Triton X-185. Rohm and Haas Co.
Chemical Description: Blend of alkylaryl polyether alcohols and organic sulfonates
Category and Applications: See Triton X-180
Form: Liquid; Color: Clear amber
Properties: Density: 8.7 lb/gal; Visc.: 465 cps; Flash Pt.: 34 F (Setaflash CC); Pour Pt.: 40 F
Composition: 100% active ingred.

Triton X-185A. Rohm and Haas Co.
Chemical Description: Blend of alkylaryl polyether alcohols and organic sulfonates
Category: Anionic surfactant
Form: Solid; Color: Lt. brown
Properties: Density: 8.8 lb/gal; Flash Pt.: 88 F (Setaflash CC); Pour Pt.: 45 F
Composition: 100% active ingred.

Triton X-190. Rohm and Haas Co.
Chemical Description: Blend of alkylaryl polyether alcohols and organic sulfonates
Category and Applications: See Triton X-180
Form: Liquid; Color: Clear amber
Properties: Density: 8.8 lb/gal; Visc.: 715 cps; Flash Pt.: 46 F (Setaflash CC); Pour Pt.: –15 F
Composition: 100% active ingred.

Triton X-193. Rohm and Haas Co.
Chemical Description: Blend of alkylaryl polyether alcohols and organic sulfonates
Category: Emulsifier, anionic
Applications: FDA-exempt emulsifier for highly active phosphate pesticide formulations
Form: Liquid; Color: Hazy amber
Properties: Density: 8.7 lb/gal; Visc.: 740 cps; Flash Pt.: 74 F (Setaflash CC); Pour Pt.: 60 F
Composition: 100% active ingred.

Triton X-200. Rohm and Haas Co.
Chemical Description: Sodium alkylaryl polyether sulfonate
Category: Detergent, wetting agent, emulsifier, leveling agent, stabilizer, anionic
Applications: Metal cleaning, pickling and plating baths; household cleaners; polymerization

emulsifier and latex post-stabilizer; wetting agent in strong alkaline baths; dispersant for fulling in lime soaps; dye leveling agent for acid dyestuffs
Form: Liquid; Color: White
Properties: Density: 8.9 lb/gal; Visc.: 7000 cps; Flash Pt.: > 300 F (TOC); Pour Pt.: 25 F; Surface Tension: 29 dynes/cm (1%)
Composition: 28% active ingred. in an aq. dispersion

Triton X-202. Rohm and Haas Co.
Chemical Description: Sodium alkylaryl polyether sulfonate
Category: Detergent, wetting agent, emulsifier, stabilizer, anionic
Applications: Metal cleaning, pickling and plating baths; household cleaners; latex post-stabilization
Form: Liquid; Color: White
Properties: Density: 8.9 lb/gal; Visc.: 4800 cps; Flash Pt.: > 212 F (TOC); Pour Pt.: 45 F; Surface Tension: 30 dynes/cm (1%)
Composition: 30% active ingred. in an aq. dispersion

Triton X-207. Rohm and Haas Co.
Chemical Description: Alkylaryl polyether alcohol
Category: Emulsifier, nonionic
Applications: Oil emulsifier for use in certain agricultural sprays and in industrial oils
Form: Liquid; Color: 3 (VCS)
Properties: Density: 8.2 lb/gal; Visc.: 600 cps; HLB: 10.7; Flash Pt.: > 200 F (PMCC); Pour Pt.: −10 F
Composition: 100% active ingred.

Triton X-301. Rohm and Haas Co.
Chemical Description: Sodium alkylaryl polyether sulfate
Category: Detergent, wetting agent, penetrant, anionic
Applications: High-foaming detergent for use in household and industrial cleaners
Form: Liquid; Color: Cream
Properties: Density: 8.8 lb/gal; Visc.: 4200 cps; Flash Pt.: > 300 F (TOC); Pour Pt.: 30 F; Surface Tension: 28 dynes/cm (1%)
Composition: 20% active ingred. in an aq. dispersion

Triton X-305. Rohm and Haas Co.
Chemical Description: Octylphenoxy polyethoxy ethanol (30 EO)
Category: Nonionic surfactant
Form: Liquid; Color: 150 (APHA)
Properties: m.w.: 1526 (avg.); Density: 9.1 lb/gal; Visc.: 470 cps; HLB: 17.3; Cloud Pt.: > 100 C (1%); Flash Pt.: > 300 F (TOC); Pour Pt.: 35 F; Surface Tension: 41 dynes/cm (1%)
Composition: 70% active ingred. in water

Triton X-363M. Rohm and Haas Co.
Chemical Description: Alkylaryl polyether alcohol
Category: Emulsifier, nonionic
Applications: FDA-exempt emulsifier for agricultural spray oils, industrial oils, etc.; also cleared for food contact uses
Form: Liquid; Color: Lt. straw
Properties: Density: 8.2 lb/gal; Visc.: 100 cps; HLB: 9.1; Flash

Pt.: > 198 F (Setaflash CC); Pour Pt.: –25 F
Composition: 100% active ingred.

Triton X-400. Rohm and Haas Co.
Chemical Description: Stearyl dimethyl benzylammonium chloride (aq. disp. containing 5% 2-propanol)
Category: Cationic surfactant
Form: Liquid; Color: Creamy white
Properties: Density: 8.0 lb/gal; Flash Pt.: > 212 F (TOC); Surface Tension: 34 dynes/cm (1%)
Composition: 25% active ingred.

Triton X-405. Rohm and Haas Co.
Chemical Description: Octylphenoxy polyethoxy ethanol (40 EO)
Category: Nonionic surfactant
Form: Liquid; Color: 250 (APHA)
Properties: m.w.: 1966 (avg.); Density: 9.2 lb/gal; Visc.: 490 cps; HLB: 17.9; Cloud Pt.: > 100 C (1%); Flash Pt.: > 212 F (TOC); Pour Pt.: 25 F; Surface Tension: 44 dynes/cm (1%)
Composition: 70% active ingred. in water

Triton X-700. Rohm and Haas Co.
Chemical Description: Blend of alkylaryl polyether alcohols and organic sulfonates
Category: Anionic surfactant
Form: Liquid; Color: 8–10 (VCS)
Properties: Density: 8.5 lb/gal; Visc.: 175 cps; Flash Pt.: 58 F (Setaflash CC); Pour Pt.: 45 F
Composition: 100% active ingred.

Triton X-705. Rohm and Haas Co.
Chemical Description: Octylphenoxy polyethoxy ethanol (70 EO)
Category: Nonionic surfactant
Form: Solid; Color: White
Properties: m.w.: 3286 (avg.); Density: 9.4 lb/gal; HLB: 18.7; Cloud Pt.: > 100 C (1%); Flash Pt.: > 500 F (COC); Surface Tension: 41 dynes/cm (1%)
Composition: 100% active ingred.

Triton X-705-50%. Rohm and Haas Co.
Chemical Description: 50% aq. sol'n. of Triton X-705
Category: Nonionic surfactant
Form: Liquid; Color: 200 (APHA)
Properties: m.w.: 3286 (avg.); Density: 9.0 lb/gal; Visc.: 625 cps; HLB: 18.7; Cloud Pt.: > 100 C (1%); Flash Pt.: > 212 F (TOC); Pour Pt.: 5 F; Surface Tension: 41 dynes/cm (1%)
Composition: 50% active ingred.

Triton X-800. Rohm and Haas Co.
Chemical Description: Blend of alkylaryl polyether alcohols and organic sulfonates
Category: Anionic surfactant
Form: Liquid; Color: 8–11 (VCS)
Properties: Density: 8.8 lb/gal; Visc.: 1010 cps; Flash Pt.: 56 F (Setaflash CC); Pour Pt.: –20 F
Composition: 100% active ingred.

Trycol 5951. Emery Industries, Inc.
Chemical Description: Alkyl polyoxyethylene glycol
Category: Wetting agent, emulsi-

fier, penetrant, nonionic
Applications: Rapid wetting agent and emulsifier for use in solvent-based heavy duty liquid cleaners; low to moderate foaming properties
Form: Liquid; Color: 1 (Gardner)
Properties: Solubility @ 5%: Sol. in water, butyl stearate, glycerol trioleate, Stod. solvent; disp. in mineral oil; Density: 8.2 lb/gal; Visc.: 23 cs; HLB: 11.6; Cloud Pt.: < 25 C
Composition: 100% active ingred.

Trycol DA-4. Emery Industries, Inc.
Chemical Description: POE (4) decyl alcohol
Category: Wetting agent, penetrant, intermediate, nonionic
Applications: Wetting agent with moderately low foaming properties; penetrant for yarn and other fabrics in atmospheric and pressure dyeing systems, and in resin pad-bath applications; intermediate in the production of anionics
Form: Liquid; Color: < 1 (Gardner)
Properties: Solubility: Sol. in glycerol trioleate; disp. in water, mineral oil, butyl stearate, perchloroethylene, Stod. solvent; Density: 7.9 lb/gal; Visc.: 30 cs; HLB: 10.5; Cloud Pt.: < 25 C; Stability: Stable to hydrolysis by acids and alkalis
Composition: 100% active ingred.
Warnings and Caution: Conc. mineral acids will react chemically with product

Trycol DA-6. Emery Industries, Inc.
Chemical Description: POE (6) decyl alcohol
Category: Wetting agent, emulsifier, nonionic
Applications: High-speed wetting agent improving permeability of clay soils; wetter for fire fighting; pad-bath penetrant in water and grease repellent applications; emulsifier for polyethylene emulsions for water repellent applications
Form: Liquid; Color: 1 (Gardner)
Properties: Solubility @ 5%: Sol. in water, butyl stearate, glycerol trioleate, Stod. solvent; disp. in mineral oil, perchloroethylene; Density: 8.3 lb/gal; Visc.: 15 cs; HLB: 12.5; Cloud Pt.: 40 C; Stability: Stable to hydrolysis by acids and alkalis
Composition: 100% active ingred.
Warnings and Caution: Conc. mineral acids will react chemically with product

Trycol DNP-8. Emery Industries, Inc.
Chemical Description: POE (8) dinonyl phenol
Category: Emulsifier, foam control agent, spreading agent, stabilizer, intermediate
Applications: Emulsifier for polar and nonpolar solvents, and in textile jet dye carrier applications; spreading agent in pigment printing; post-stabilizer in emulsion polymerization; intermediate in mfg. of anionics; surfactant component in acidic cleaners, aerosols,

insecticides, and wax emulsions
Form: Liquid; Color: 2 (Gardner)
Properties: Solubility @ 5%: Sol. in glycerol trioleate, perchloroethylene; disp. in water, mineral oil, butyl stearate, Stod. solvent; Density: 8.4 lb/gal; Visc.: 410 cs; HLB: 10.4; Cloud Pt.: < 25 C; Stability: Stable to hydrolysis by acids and alkalis
Composition: 100% active ingred.

Trycol DNP-150. Emery Industries, Inc.
Chemical Description: POE (150) dinonyl phenol
Category: Emulsifier, dispersant, wetting agent
Applications: Controlled-suds component in built, heavy duty detergents, hard surface cleaners, alkaline bottle washing compounds, dairy detergents; dispersant and wetting agent for pesticides; in textile industry as emulsifier for flaked biphenyl dye carriers, detergent, leveling agent for acid, cationic, and premetalized dyeings of wool and acrylics
Form: Solid/flakes; Color: 1 (Gardner)
Properties: Solubility @ 5%: Sol. in water, perchloroethylene; M.P.: 55 C; HLB: 19.0; Cloud Pt.: > 100 C; Stability: Stable to hydrolysis by acids and alkalis
Composition: 100% active ingred.

Trycol DNP-150/50. Emery Industries, Inc.
Chemical Description: POE (150) dinonyl phenol, 50%
Category and Applications: See Trycol DNP-150
Form: Liquid; Color: < 1 (Gardner)
Properties: Solubility @ 5%: Sol. in water; Density: 9.0 lb/gal; HLB: 19.0; Cloud Pt.: > 100 C; Stability: Stable to hydrolysis by acids and alkalis
Composition: 50% active ingred.

Trycol LAL-4. Emery Industries, Inc.
Chemical Description: POE (4) lauryl alcohol
Category: Emulsifier, intermediate, nonionic
Applications: Emulsifier for oils and fats in industrial lubricants, textile coning oils; coemulsifier for silicone in cleaner polishes and mold release agents; when sulfated used as a high-foaming anionic surfactant
Form: Liquid; Color: 1 (Gardner)
Properties: Solubility @ 5%: Sol. in butyl stearate, glycerol trioleate, perchloroethylene; disp. in water, mineral oil, Stod. solvent; Density: 7.7 lb/gal; Visc.: 35 cs; HLB: 9.2; Cloud Pt.: < 25 C; Stability: Stable to hydrolysis by acids and alkalis; Biodegradable
Composition: 100% active ingred.

Trycol LAL-8. Emery Industries, Inc.
Chemical Description: POE (8) lauryl alcohol
Category: Detergent, emulsifier, wetting agent, nonionic
Form: Liquid; Color: 1 (Gardner)

Properties: Solubility @ 5%: Sol. in water, glycerol trioleate, perchloroethylene; disp. in butyl stearate, Stod. solvent; Density: 8.2 lb/gal; HLB: 12.6; Cloud Pt.: 62 C; Stability: Stable to hydrolysis by acids and alkalis; Biodegradable
Composition: 100% active ingred.

Trycol LAL-12. Emery Industries, Inc.
Chemical Description: POE (12) lauryl alcohol
Category: Detergent, wetting agent, emulsifier, nonionic
Form: Solid; Color: 1 (Gardner)
Properties: Solubility @ 5%: Sol. in water; disp. in butyl stearate, glycerol trioleate, Stod. solvent; M.P.: 32 C; HLB: 14.0; Cloud Pt.: 90 C; Stability: Stable to hydrolysis by acids and alkalis; Biodegradable
Composition: 100% active ingred.
Warnings and Caution: Conc. mineral acids will react chemically with product

Trycol LAL-23. Emery Industries, Inc.
Chemical Description: POE (23) lauryl alcohol
Category: Emulsifier, solubilizer, nonionic
Applications: Coemulsifier for silicone in cleaner polishes and mold release agents; solvent emulsifier for textile dye carriers
Form: Solid; Color: < 1 (Gardner)
Properties: Solubility @ 5%: Sol. in water; disp. in perchloroethylene; M.P.: 46 C; HLB: 16.9; Cloud Pt.: > 100 C; Stability: Stable to hydrolysis by acids and alkalis
Composition: 100% active ingred.
Warnings and Caution: Conc. mineral acids will react chemically with product

Trycol LF-1. Emery Industries, Inc.
Chemical Description: Nonionic ether
Category: Foam control agent, penetrant, nonionic
Applications: Controls foam and provides penetration in textile wet processing; also used in metal processing rinses
Form: Liquid; Color: 1 (Gardner)
Properties: Solubility @ 5%: Sol. in water, glycerol trioleate, Stod. solvent; disp. in butyl stearate, perchloroethylene; Density: 8.5 lb/gal; Visc.: 65 cs; Cloud Pt.: 25 C
Composition: 100% active ingred.

Trycol MC-24. Emery Industries, Inc.
Chemical Description: Nonionic ether
Category: Emulsifier, leveling agent, nonionic
Applications: Textile dye leveling agent; emulsifier in wet processing products
Form: Solid
Properties: Solubility @ 5%: Sol. in water, perchloroethylene; disp. in glycerol trioleate, Stod. solvent;

M.P.: 47 C; HLB: 16.3; Cloud Pt.: > 100 C
Composition: 100% active ingred.

Trycol NP-1. Emery Industries, Inc.
Chemical Description: POE (1) nonyl phenol
Category: Coemulsifier, defoaming agent, nonionic
Applications: Oil-sol. surfactant and coemulsifier; in combination with water-sol. surfactants as a defoaming agent
Form: Liquid; Color: 1 (Gardner)
Properties: Solubility @ 5%: Sol. in mineral oil, perchloroethylene, Stod. solvent; Density: 8.3 lb/gal; Visc.: 460 cs; HLB: 4.6; Cloud Pt.: < 25 C; Stability: Stable to hydrolysis by acids and alkalis
Composition: 100% active ingred.

Trycol NP-4. Emery Industries, Inc.
Chemical Description: POE (4) nonyl phenol
Category: Coemulsifier, corrosion inhibitor, nonionic
Applications: Oil-sol. coemulsifier often used as a corrosion inhibitor in two cycle engine oils
Form: Liquid; Color: 1 (Gardner)
Properties: Solubility @ 5%: Sol. in perchloroethylene, Stod. solvent; sol./disp. in mineral oil; Density: 8.5 lb/gal; Visc.: 300 cs; HLB: 8.9; Cloud Pt.: < 25 C; Stability: Stable to hydrolysis by acids and alkalis
Composition: 100% active ingred.

Trycol NP-7. Emery Industries, Inc.
Chemical Description: POE (7) nonyl phenol
Category: Wetting agent, nonionic
Applications: Wetting agent in solvent emulsions, cleaners, acid cleaners, and detergents; good sol. in aromatic solvents
Form: Liquid; Color: 1 (Gardner)
Properties: Solubility @ 5%: Sol. in perchloroethylene, Stod. solvent; sol./disp. in mineral oil; disp. in water; Density: 8.6 lb/gal; Visc.: 325 cs; HLB: 12.3; Cloud Pt.: 20 C; Stability: Stable to hydrolysis by acids and alkalis
Composition: 100% active ingred.

Trycol NP-9. Emery Industries, Inc.
Chemical Description: POE (9) nonyl phenol
Category: Detergent, wetting agent, emulsifier, nonionic
Form: Liquid; Color: 1 (Gardner)
Properties: Solubility @ 5%: Sol. in water, perchloroethylene, Stod. solvent; Density: 8.8 lb/gal; Visc.: 325 cs; HLB: 12.9; Cloud Pt.: 55 C; Stability: Stable to hydrolysis by acids and alkalis
Composition: 100% active ingred.

Trycol NP-30. Emery Industries, Inc.
Chemical Description: POE (30) nonyl phenol
Category: Detergent, wetting agent, emulsifier, nonionic
Applications: Emulsifier for fats, oils, and waxes and in emulsion polymerization; water-sol. detergent and wetting agent
Form: Solid; Color: 1 (Gardner)
Properties: Solubility @ 5%: Sol. in water; M.P.: 40 C; HLB: 17.1;

Cloud Pt.: > 100 C; Stability: Stable to hydrolysis by acids and alkalis
Composition: 100% active ingred.

Trycol OAL-23. Emery Industries, Inc.
Chemical Description: POE (23) oleyl alcohol
Category: Emulsifier, dispersant, solubilizer, detergent, stabilizer, lubricant, nonionic
Applications: Stabilizer and anticoagulant for natural and synthetic latices and dye pastes; dyeing assistant for wool/acrylic blends; detergent and lubricant for fiber/fabric scouring; emulsifier for waxes used in coating citrus fruits
Form: Solid; Color: 2 (Gardner)
Properties: Solubility @ 5%: Sol. in water, mineral oil, glycerol trioleate, perchloroethylene; disp. in Stod. solvent; M.P.: 40 C; HLB: 15.9; Cloud Pt.: > 100 C; Stability: Stable to hydrolysis by acids and alkalis
Composition: 100% active ingred.
Warnings and Caution: Conc. mineral acids will react chemically with products

Trycol TDA-3. Emery Industries, Inc.
Chemical Description: POE (3) tridecyl alcohol
Category: Emulsifier, intermediate, nonionic
Applications: Emulsifier and antifoam in textile formulations; intermediate for the production of anionics by sulfation or phosphation
Form: Liquid; Color: < 1 (Gardner)
Properties: Solubility @ 5%: Sol. in butyl stearate, glycerol trioleate, Stod. solvent; disp. in water, mineral oil, perchloroethylene; Density: 7.8 lb/gal; Visc.: 30 cs; HLB: 8.0; Cloud Pt.: < 25 C; Stability: Stable to hydrolysis by acids and alkalis
Composition: 100% active ingred.
Warnings and Caution: Conc. mineral acids will react chemically with product

Trycol TDA-6. Emery Industries, Inc.
Chemical Description: POE (6) tridecyl alcohol
Category: Emusifier, dispersant, wetting agent, nonionic
Applications: Emulsifier and detergent in degreasers and cutting oils; low temp. wool scouring; moderately low foaming; intermediate for production of surfactants by sulfation or phosphation
Form: Liquid; Color: < 1 (Gardner)
Properties: Solubility @ 5%: Sol. in butyl stearate, glycerol trioleate, perchloroethylene, Stod. solvent; disp. in water and mineral oil; Density: 8.2 lb/gal; Visc.: 50 cs; HLB: 11.4; Cloud Pt.: < 25 C; Stability: Stable to hydrolysis by acids and alkalis
Composition: 100% active ingred.
Warnings and Caution: Conc. mineral acids will react chemically with product

Trycol TDA-8. Emery Industries, Inc.
Chemical Description: POE (8) tridecyl alcohol
Category: Foam builder, solubilizer, coemulsifier, nonionic
Applications: Foam builder and solubilizer for alkylaryl sulfonates, essential oils, aromatic solvents, waxes, and fats; also as coemulsifier
Form: Liquid; Color: < 1 (Gardner)
Properties: Solubility @ 5%: Sol. in water, glycerol trioleate, Stod. solvent; disp. in mineral oil, butyl stearate, perchloroethylene; Density: 8.3 lb/gal; Visc.: 65 cs; HLB: 12.5; Cloud Pt.: 44 °C; Stability: Stable to hydrolysis by acids and alkalis
Composition: 100% active ingred.
Warnings and Caution: Conc. mineral acids will react chemically with product

Trycol TDA-9. Emery Industries, Inc.
Chemical Description: POE (9) tridecyl alcohol
Category: Wetting and rewetting agent, detergent, coemulsifier, nonionic
Applications: General purpose surfactant used with alkylaryl sulfonates in light and heavy duty high-foaming cleaners; rewetter for paper towels; raw wool detergent; coemulsifier for aromatic and aliphatic solvents
Form: Liquid; Color: < 1 (Gardner)
Properties: Solubility @ 5%: Sol. in water, glycerol trioleate, Stod. solvent; disp. in mineral oil, butyl stearate, perchloroethylene; Density: 8.3 lb/gal; Visc.: 75 cs; HLB: 13.2; Cloud Pt.: 55 C; Stability: Stable to hydrolysis by acids and alkalis
Composition: 100% active ingred.
Warnings and Caution: Conc. mineral acids will react chemically with product

Trycol TDA-11. Emery Industries, Inc.
Chemical Description: POE (11) tridecyl alcohol
Category: Detergent, nonionic
Applications: Light and heavy duty cleaning formulations
Form: Liquid; Color: < 1 (Gardner)
Properties: Solubility @ 5%: Sol. in water, glycerol trioleate, Stod. solvent; disp. in mineral oil, butyl stearate, perchloroethylene; Density: 8.4 lb/gal; HLB: 13.7; Cloud Pt.: 74 C; Stability: Stable to hydrolysis by acids and alkalis
Composition: 100% active ingred.
Warnings and Caution: Conc. mineral acids will react chemically with product

Trycol TDA-18. Emery Industries, Inc.
Chemical Description: POE (18) tridecyl alcohol
Category: Emulsifier, dispersant, solubilizer, detergent, intermediate, leveling agent, nonionic
Applications: Emulsifier in clean-

ers; intermediate in the mfg. of anionic surfactants; textile acid dye leveling agent and wet processing detergent with excellent foam stabilizing properties
Form: Solid; Color: < 1 (Gardner)
Properties: Solubility @ 5%: Sol. in water; disp. in butyl stearate, glycerol trioleate, perchloroethylene, Stod. solvent; M.P.: 39 C; HLB: 16.0; Cloud Pt.: > 100 C; Stability: Stable to hydrolysis by acids and alkalis
Composition: 100% active ingred.
Warnings and Caution: Conc. mineral acids will react chemically with product

Trycol TP-2. Emery Industries, Inc.
Chemical Description: POE (2) tridecyl phenol
Category: Coemulsifier, foam reducing agent, nonionic
Form: Liquid; Color: 6 (Gardner)
Properties: Solubility @ 5%: Sol. in mineral oil, perchloroethylene, Stod. solvent; Density: 8.1 lb/gal; Visc.: 600 cs; HLB: 4.6; Cloud Pt.: < 25 C; Stability: Stable to hydrolysis by acids and alkalis
Composition: 100% active ingred.

Trycol TP-6. Emery Industries, Inc.
Chemical Description: POE (6) tridecyl phenol
Category: Emulsifier, dispersant, detergent, nonionic
Applications: Emulsifier in metal cleaning, cutting oil formulations, emulsifiable solvent cleaners, and in drycleaning formulations
Form: Liquid; Color: 6 (Gardner)
Properties: Solubility @ 5%: Sol. in perchloroethylene, Stod. solvent; sol./disp. in mineral oil; disp. in water; Density: 8.5 lb/gal; Visc.: 300 cs; HLB: 9.9; Cloud Pt.: < 25 C; Stability: Stable to hydrolysis by acids and alkalis
Composition: 100% active ingred.

Trydet 20. Emery Industries, Inc.
Chemical Description: Ethoxylated mixed esters of rosin and fatty acids
Category: Detergent, emulsifier, nonionic
Applications: Low-foaming detergent and emulsifier, esp. in conjunction with alkylaryl sulfonates and soaps; textile leveling agent for package and skein dyeing and in soaping off vat and naphthol dyeings; coemulsifier for xylene, kerosene, trichlorobenzene, o-dichlorobenzene, and similar solvents when used with Trylon 3300 or Tryfac 610-K
Form: Liquid; Color: 10 (Gardner)
Properties: Solubility @ 5%: Sol. in water, mineral oil; disp. in glycerol trioleate; Density: 9.0 lb/gal; Visc.: 480 cs; HLB: 13.4; Cloud Pt.: > 100 C; Stability: Stable in neutral and moderately acidic or alkaline systems
Composition: 96% active ingred.

Trydet ISA-4. Emery Industries, Inc.
Chemical Description: POE (4) isostearic acid
Category: Lubricant, nonionic
Applications: Exhibits unique sur-

factant and lubricant properties resulting from the liquid, long-chain, saturated fatty acid from which it is made; used in fiber lubricants, textile fiber processing, conc. liquid fabric softeners
Form: Liquid; Color: 2 (Gardner)
Properties: Solubility @ 5%: Disp. in water, mineral oil, butyl stearate, glycerol trioleate, perchlorethylene; Density: 8.2 lb/gal; Visc.: 100 cs; HLB: 7.7; Cloud Pt.: < 25 C; Stability: Stable in neutral or moderately acidic or alkaline systems
Composition: 100% active ingred.

Trydet ISA-9. Emery Industries, Inc.
Chemical Description: POE (9) isostearic acid
Category and Applications: See Trydet ISA-4
Form: Liquid; Color: 1 (Gardner)
Properties: Solubility @ 5%: Disp. in water, mineral oil, butyl stearate, glycerol trioleate, perchloroethylene; Density: 8.5 lb/gal; Visc.: 125 cs; HLB: 11.7; Cloud Pt.: < 25 C; Stability: Stable in neutral or moderately acid or alkaline systems
Composition: 100% active ingred.

Trydet LA-7. Emery Industries, Inc.
Chemical Description: POE (7) lauric acid
Category: Emulsifier, lubricant, viscosity control agent, nonionic
Applications: Lubricant component and scrooping agent for textile fibers and yarn; applied either from aq. or emulsifiable mineral oil systems; also as viscosity control agent for plastisols
Form: Liquid; Color: 2 (Gardner)
Properties: Solubility @ 5%: Disp. in water, mineral oil, butyl stearate, glycerol trioleate, perchloroethylene, Stod. solvent; Density: 8.4 lb/gal; Visc.: 55 cs; HLB: 11.8; Cloud Pt.: < 25 C; Stability: Stable in neutral and moderately acidic or alkaline systems
Composition: 100% active ingred.

Trydet OA-5. Emery Industries, Inc.
Chemical Description: POE (5) oleic acid
Category: Emulsifier, lubricant, softener, nonionic
Applications: Emulsifier for mineral oils, fatty oils, and solvents; used as cutting oil emulsifier, solvent emulsifier in metal cleaners and degreasers, a W/O emulsifier for consumer pesticide aerosols, and as lubricant component in textile processing; softens and lubricates leather during tanning
Form: Liquid; Color: 3 (Gardner)
Properties: Solubility @ 5%: Sol. in glycerol trioleate; disp. in water, mineral oil, butyl stearate, Stod. solvent; Density: 8.1 lb/gal; Visc.: 55 cs; HLB: 8.2; Cloud Pt.: < 25 C; Stability: Stable in neutral and moderately acidic or alkaline systems
Composition: 100% active ingred.

Trydet OA-7. Emery Industries, Inc.
Chemical Description: POE (7) oleic acid
Category: Emulsifier, lubricant, nonionic
Applications: See Trydet OA-5; also used as a self-emulsifying component in formulating textile softeners
Form: Liquid; Color: 3 (Gardner)
Properties: Solubility @ 5%: Disp. in water, mineral oil, butyl stearate, glycerol trioleate, perchloroethylene, Stod. solvent; Density: 8.3 lb/gal; Visc.: 160 cs; HLB: 10.2; Cloud Pt.: < 25 C; Stability: Stable in neutral and moderately acidic or alkaline systems
Composition: 100% active ingred.

Trydet OA-10. Emery Industries, Inc.
Chemical Description: POE (10) oleic acid
Category: Emulsifier, lubricant, rewetting agent, nonionic
Applications: Emulsifier for solvents in pesticide carriers, metal cleaners, and for neatsfoot oil in leather fat liquoring; textile specialty detergents and dyeing assistants; rewetting agent for paper
Form: Liquid; Color: 2 (Gardner)
Properties: Solubility @ 5%: Disp. in water, mineral oil, butyl stearate, glycerol trioleate, perchloroethylene, Stod. solvent; Density: 8.5 lb/gal; Visc.: 110 cs; HLB: 12.1; Cloud Pt.: < 25 C; Stability: Stable in neutral or moderately acidic or alkaline systems
Composition: 100% active ingred.

Trydet SA-5. Emery Industries, Inc.
Chemical Description: POE (5) stearic acid
Category: Emulsifier, lubricant, softener, nonionic
Applications: Emulsifier for mineral oils and fats; lubricant in leather industry; napping softener in textile industry, providing scroop and fiber-to-metal lubricity; used in mineral oil emulsions for polishes and lard oil emulsions for metal buffing compounds
Form: Solid; Color: 1 (Gardner)
Properties: Solubility @ 5%: Disp. in water, mineral oil, glycerol trioleate, perchloroethylene, Stod. solvent; M.P.: 35 C; HLB: 8.7; Cloud Pt.: < 25 C; Stability: Stable in neutral and moderately acidic or alkaline systems
Composition: 100% active ingred.

Trydet SA-7. Emery Industries, Inc.
Chemical Description: POE (7) stearic acid
Category: Emulsifier, lubricant, softener, nonionic
Applications: Waxy emulsifier for oils and fats in industrial lubricants; imparts softening and lubricating properties to textiles and leather
Form: Solid; Color: < 1 (Gardner)
Properties: Solubility @ 5%: Disp. in water, mineral oil, butyl stearate, glycerol trioleate, perchloroethylene, Stod. solvent; M.P.: 30 C; HLB: 11.1; Cloud Pt.: < 25 C; Stability: Stable in neutral and

moderately acidic or alkaline systems
Composition: 100% active ingred.

Trydet SA-8. Emery Industries, Inc.
Chemical Description: POE (8) stearic acid
Category: Lubricant, stabilizer, nonionic
Applications: Self-emulsifying lubricant, softener, and scrooping agent for synthetic fibers for textile and other industrial applications; stabilizer at low levels in starch sol'ns.
Form: Solid; Color: < 1 (Gardner)
Properties: Solubility @ 5%: Disp. in water, mineral oil, butyl stearate, glycerol trioleate, perchloroethylene, Stod. solvent; M.P.: 30 C; HLB: 11.1; Cloud Pt.: < 25 C; Stability: Stable in neutral and moderately acidic or alkaline systems
Composition: 100% active ingred.

Trydet SA-23. Emery Industries, Inc.
Chemical Description: POE (23) stearic acid
Category: Emulsifier, thickener, antigellant, nonionic
Applications: Emulsifier for glycerol monostearate in nonionic textile lubricants and softeners; emulsifier and thickener in cosmetic creams, lotions, and pharmaceutical ointments; antigellant in starch sol'ns.
Form: Solid; Color: 1 (Gardner)
Properties: Solubility @ 5%: Sol. in water, mineral oil, glycerol trioleate, perchloroethylene; M.P.: 45 C; HLB: 15.6; Cloud Pt.: > 100 C; Stability: Stable in neutral and moderately acidic or alkaline systems
Composition: 100% active ingred.

Trydet SA-40. Emery Industries, Inc.
Chemical Description: POE (40) stearic acid
Category: Emulsifier, stabilizer, antigellant, lubricant, nonionic
Applications: Emulsifier for glycerol monostearate and other waxy esters in the the production of concentrated, pourable textile lubricants and softeners; stabilizer and antigellant for starch sol'ns.; emulsifier for cosmetics and pharmaceuticals; as a nonabrasive coating for glass bottles
Form: Solid/flakes; Color: < 1 (Gardner)
Properties: Solubility @ 5%: Sol. in water, mineral oil; disp. in glycerol trioleate, perchloroethylene; M.P.: 50 C; HLB: 17.3; Cloud Pt.: < 25 C; Stability: Stable in neutral and moderately acidic or alkaline systems
Composition: 100% active ingred.

Tryfac 610-A. Emery Industries, Inc.
Chemical Description: Free acid form of phosphate ester
Category: Intermediate, anionic
Applications: Neutralized salts are used as emulsifiers for aliphatic and aromatic solvents

Form: Liquid; Color: 1 (Gardner)
Properties: Solubility @ 5%: Sol. in water, butyl stearate, perchloroethylene, Stod. solvent; disp. in glycerol trioleate; Density: 8.8 lb/gal; Visc.: 290 cs
Composition: 100% active ingred.

Tryfac 610-K. Emery Industries, Inc.
Chemical Description: Potassium salt of phosphate ester
Category: Emulsifier, wetting agent, detergent, antistat, corrosion inhibitor, anionic
Applications: Emulsifier for chlorinated hydrocarbons in heavy duty cleaners and alkali-tolerant detergents; component in metalworking compounds; textile emulsifier and detergent in solvent scouring, emulsifier for dye carriers, scouring agent for cotton goods, antistat in processing oils for natural and synthetic fibers; corrosion inhibitor and dispersant
Form: Liquid; Color: < 1 (Gardner)
Properties: Solubility @ 5%: Sol. in water, butyl stearate, glycerol trioleate, perchloroethylene, Stod. solvent; Density: 9.2 lb/gal; Visc.: 700 cs
Composition: 88% active ingred.

Tryfac 910-K. Emery Industries, Inc.
Chemical Description: Potassium salt of phosphate ester
Category: Stabilizer, detergent, wetting agent, anionic
Applications: Polymer emulsification stabilizer; detergent and wetting agent in the conversion of starches in textile desizing without interfering with enzymatic action
Form: Liquid; Color: 1 (Gardner)
Properties: Solubility @ 5%: Sol. in water, perchloroethylene, Stod. solvent; disp. in butyl stearate and glycerol trioleate; Density: 9.4 lb/gal; Visc.: 480 cs
Composition: 90% active ingred.

Tryfac 5555. Emery Industries, Inc.
Chemical Description: Free acid form of a complex phosphate ester
Category: Emulsifier, detergent, anionic
Applications: Textile emulsifier in low-foaming dye carrier systems and in solvent scouring systems; emulsifier for various flame retardants; its stability allows its use at high temps.
Form: Liquid; Color: 5 (Gardner)
Properties: Solubility @ 5%: Sol. in mineral oil, butyl stearate, glycerol trioleate, perchloroethylene, Stod. solvent; disp. in water; Density: 8.8 lb/gal; Visc.: 4000 cs
Composition: 100% active ingred.

Trylon 3300. Emery Industries, Inc.
Category: Emulsifier, detergent, anionic
Applications: Solvent emulsifier in degreasing cleaners of metal parts and engine blocks; textile emulsifier for dye carriers in high-temp. dye applications, and for solvent scouring; emulsifier for a wide range of aliphatic and aro-

matic solvents incl. Stod. solvent, kerosene, xylene, and chlorinated solvents
Form: Liquid; Color: 6 (Gardner)
Properties: Solubility @ 5%: Sol. in butyl stearate, perchloroethylene, Stod. solvent; disp. in water, mineral oil, glycerol trioleate; Density: 8.5 lb/gal; Visc.: 2100 cs; Cloud Pt.: > 100 C
Composition: 100% active ingred.

Trylon 6733. Emery Industries, Inc.
Chemical Description: Long-chain, liquid, saturated fatty ester
Category: Lubricant, nonionic
Applications: Textile and industrial processing lubricants
Form: Liquid; Color: 1 (Gardner)
Properties: Solubility @ 5%: Sol. in butyl stearate, glycerol trioleate, perchloroethylene, Stod. solvent; disp. in water and mineral oil; Density: 8.0 lb/gal; Visc.: 65 cs; Stability: Good heat stability
Composition: 100% active ingred.

Trylon 6735. Emery Industries, Inc.
Category: Wetting agent, emulsifier, penetrant, coemulsifier, nonionic
Applications: Low-foaming surfactant for use in industrial, institutional, and consumer detergents; textile scouring, wetting, and penetrating operations where low foam is required; coemulsifier for solvents in solvent and dye carriers
Form: Liquid; Color: 1 (Gardner)
Properties: Solubility @ 5%: Sol. in water, butyl stearate, glycerol trioleate, perchloroethylene, Stod. solvent; disp. in mineral oil; Density: 8.2 lb/gal; Visc.: 40 cs; Cloud Pt.: 37 C
Composition: 100% active ingred.

Trylon EW. Emery Industries, Inc.
Chemical Description: Nonionic ether
Category: Emulsifier, penetrant, stabilizer, nonionic
Applications: Emulsifier for solvents, vegetable oils, and waxes; emulsifier for leather tanning chemicals; textile resin bath penetrant, polymer stabilizer, solvent scour emulsifier, cold water scour for felted fabrics, and enzyme bath penetrant
Form: Liquid; Color: 1 (Gardner)
Properties: Solubility @ 5%: Sol. in butyl stearate, glycerol trioleate, perchloroethylene; disp. in water and Stod. solvent; Density: 8.6 lb/gal; Visc.: 270 cs; HLB: 11.5; Cloud Pt.: 25 C
Composition: 100% active ingred.

Trylox 1086. Emery Industries, Inc.
Chemical Description: Ethoxylated sorbitol hexaoleate
Category: Emulsifier, nonionic
Applications: O/W balanced emulsifier for petroleum oils, vegetable oils, and solvents; used for oil-based metal lubricants, textile processing aids, and as emulsifier component for aliphatic and aromatic solvents in low-foaming textile dye carriers

Form: Liquid; Color: 6 (Gardner)
Properties: Solubility @ 5%: Sol. in butyl stearate, glycerol trioleate, perchloroethylene, Stod. solvent; disp. in water and mineral oil; Density: 8.4 lb/gal; Visc.: 235 cs; HLB: 10.2; Cloud Pt.: < 25 C
Composition: 100% active ingred.

Trylox 1186. Emery Industries, Inc.
Chemical Description: Ethoxylated sorbitol hexatallate
Category and Applications: Similar to Trylox 1086
Form: Liquid; Color: 9 (Gardner)
Properties: Solubility @ 5%: Sol. in butyl stearate, glycerol trioleate, Stod. solvent; disp. in water, mineral oil, perchloroethylene; Density: 8.5 lb/gal; Visc.: 330 cs; HLB: 10.9; Cloud Pt.: < 25 C
Composition: 100% active ingred.

Trylox 6747. Emery Industries, Inc.
Chemical Description: Ethoxylated sorbitol hexaoleate
Category: Emulsifier, nonionic
Applications: Low-foaming emulsifier for organic solvents; can also be used for paraffin oils in industrial lubricants
Form: Liquid; Color: 4 (Gardner)
Properties: Solubility @ 5%: Sol. in mineral oil, butyl stearate, glycerol trioleate; disp. in water, perchloroethylene, Stod. solvent; Density: 8.4 lb/gal; Visc.: 300 cs; HLB: 11.4; Cloud Pt.: < 25 C
Composition: 100% active ingred.

Trylox CO-5. Emery Industries, Inc.
Chemical Description: POE (5) castor oil
Category: Emulsifier, dispersant, foam control agent, nonionic
Applications: Coemulsifier with foam control properties for chlorinated and aromatic solvents; aids dispersion of pigment slurries and improves gloss and freeze-thaw resistance in water-based paints; clay dispersant; carrier for paper coatings; textile foam control agent and emulsifier in low-foaming dye carriers
Form: Liquid; Color: 4 (Gardner)
Properties: Solubility @ 5%: Sol. in glycerol trioleate; disp. in water, butyl stearate, perchloroethylene, Stod. solvent; Density: 8.2 lb/gal; Visc.: 800 cs; HLB: 3.8; Cloud Pt.: < 25 C
Composition: 100% active ingred.

Trylox CO-16. Emery Industries, Inc.
Chemical Description: POE (16) castor oil
Category: Emulsifier, lubricant, nonionic
Applications: Coemulsifier for metalworking oils and hydraulic fluids; component in textile leveling and dispersing agents for vat and naphthol dyes, and as coemulsifier for rayon delustrants and fiber lubricants
Form: Liquid; Color: 3 (Gardner)
Properties: Solubility @ 5%: Disp. in water, butyl stearate, glycerol trioleate, perchloroethylene, Stod.

solvent; Density: 8.5 lb/gal; Visc.: 600 cs; HLB: 8.6; Cloud Pt.: < 25 C
Composition: 100% active ingred.

Trylox CO-30. Emery Industries, Inc.
Chemical Description: POE (30) castor oil
Category: Emulsifier, dispersant, degreaser, lubricant, stabilizer, nonionic
Applications: Emulsifier for oils, solvents, and waxes; pigment dispersant; degreaser, emulsifier, and lubricant in fat liquoring; viscosity and emulsion stabilizer for polyvinyl acetate and water-based paints; emulsifier and dispersant for urethane foams and polyester resins; coemulsifier in fabric softener and dye carrier systems; dyeing assistant; emulsifier in synthetic fiber lubricants; aids softness and absorbency of wet-strength papers
Form: Liquid; Color: 2 (Gardner)
Properties: Solubility @ 5%: Sol. in water; disp. in butyl stearate, glycerol trioleate, perchloroethylene, Stod. solvent; Density: 8.8 lb/gal; Visc.: 720 cs; HLB: 11.7; Cloud Pt.: > 100 C
Composition: 100% active ingred.

Trylox CO-36. Emery Industries, Inc.
Chemical Description: POE (36) castor oil
Category: Emulsifier, lubricant, softener, nonionic
Applications: Emulsifier for solvents and oils; lubricant and softener for textiles and leather; coemulsifier with anionic emulsifiers in formulating textile dye carriers based on aromatic and aliphatic solvents
Form: Liquid; Color: 2 (Gardner)
Properties: Solubility @ 5%: Sol. in water; disp. in butyl stearate, glycerol trioleate, perchloroethylene, Stod. solvent; Density: 8.8 lb/gal; Visc.: 700 cs; HLB: 12.6; Cloud Pt.: > 100 C
Composition: 100% active ingred.

Trylox CO-40. Emery Industries, Inc.
Chemical Description: POE (40) castor oil
Category and Applications: See Trylox CO-36
Form: Liquid; Color: 2 (Gardner)
Properties: Solubility @ 5%: Sol. in water; disp. in butyl stearate, glycerol trioleate, perchloroethylene, Stod. solvent; Density: 8.8 lb/gal; Visc.: 650 cs; HLB: 12.9; Cloud Pt.: > 100 C
Composition: 100% active ingred.

Trylox CO-200. Emery Industries, Inc.
Chemical Description: POE (200) castor oil
Category: Emulsifier, lubricant, antistat, nonionic
Applications: Used in textile fiber processing as an antistatic humectant and scrooping agent, and as an emulsifier for hydrophobic glycerides and sorbitan esters in blended lubricants for fibers and

yarns
Form: Solid; Color: 1 (Gardner)
Properties: Solubility @ 5%: Sol. in water; M.P.: 40 C; HLB: 18.1; Cloud Pt.: > 100 C
Composition: 100% active ingred.

Trylox CO-200/50. Emery Industries, Inc.
Chemical Description: 50% aq. sol'n. of Trylox CO-200
Category and Applications: See Trylox CO-200
Form: Liquid; Color: < 1 (Gardner)
Properties: Solubility @ 5%: Sol. in water; Density: 8.6 lb/gal; HLB: 18.1; Cloud Pt.: > 100 C
Composition: 50% active ingred.

Trylox HCO-5. Emery Industries, Inc.
Chemical Description: POE (5) hydrogenated castor oil
Category: Emulsifier, lubricant, softener, nonionic
Applications: Coemulsifier for synthetic esters used in the textile industry; fiber lubricant and softener
Form: Liquid; Color: 1 (Gardner)
Properties: Solubility @ 5%: Sol. in mineral oil, glycerol trioleate, Stod. solvent; disp. in water, butyl stearate, perchloroethylene; Density: 7.9 lb/gal; Visc.: 1200 cs; HLB: 3.8; Cloud Pt.: < 25 C
Composition: 100% active ingred.

Trylox HCO-16. Emery Industries, Inc.
Chemical Description: POE (16) hydrogenated castor oil
Category: Emulsifier, lubricant, softener, nonionic
Applications: Emulsifier for castor oil and provides lubricity and softness in fabric softeners and aerosol fabric sprays
Form: Liquid; Color: 1 (Gardner); Odor: Very low
Properties: Solubility @ 5%: Sol. in mineral oil, glycerol trioleate; disp. in water, butyl stearate, perchloroethylene, Stod. solvent; Density: 8.4 lb/gal; Visc.: 2100 cs; HLB: 8.2; Cloud Pt.: < 25 C; Stability: Excellent heat stability
Composition: 100% active ingred.

Trylox HCO-25. Emery Industries, Inc.
Chemical Description: POE (25) hydrogenated castor oil
Category: Emulsifier, nonionic
Applications: Emulsifier for textile resin finishing and softener-lubricant systems; coemulsifier for lanolin and in dye carrier solvent systems
Form: Liquid; Color: 1 (Gardner)
Properties: Solubility @ 5%: Sol. in mineral oil, glycerol trioleate; disp. in water, butyl stearate, perchloroethylene, Stod. solvent; Density: 8.6 lb/gal; Visc.: 1100 cs; HLB: 10.8; Cloud Pt.: 25 C
Composition: 100% active ingred.

Trylox HCO-200/50. Emery Industries, Inc.
Chemical Description: POE (200) hydrogenated castor oil, 50% aq. sol'n.

Category: Emulsifier, lubricant, nonionic
Form: Clear liquid; Color: < 1 (Gardner)
Properties: Solubility @ 5%: Sol. in water; Density: 8.8 lb/gal; HLB: 18.1; Cloud Pt.: > 100 C
Composition: 50% active ingred.

Trylox SS-20. Emery Industries, Inc.
Chemical Description: POE (20) sorbitol
Category: Humectant, plasticizer, intermediate, nonionic
Applications: Humectant, plasticizer; intermediate in the synthesis of fatty acid esters and improves the pourability and clarity of high solid "clear concentrate" surfactant dilutions
Form: Liquid; Color: 1 (Gardner)
Properties: Solubility @ 5%: Sol. in water; Density: 9.7 lb/gal; Visc.: 400 cs; HLB: 16.1; Cloud Pt.: > 100 C
Composition: 100% active ingred.

Trymeen CAM-10. Emery Industries, Inc.
Chemical Description: POE (10) coco amine
Category: Coemulsifier and emulsifier, antistat, dispersant, cationic
Applications: Coemulsifier and antistat for textile processing oils; dispersant for inorganic salts in viscose spinning; emulsifier for fats and oils in industrial lubricants
Form: Liquid; Color: 8 (Gardner)
Properties: Solubility @ 5%: Sol. in water, glycerol trioleate, perchloroethylene; disp. in butyl stearate and Stod. solvent; Density: 8.4 lb/gal; Visc.: 260 cs; HLB: 13.8; Cloud Pt.: > 100 C
Composition: 100% active ingred.
Reaction: Yield slightly alkaline sol'ns. when dissolved in water

Trymeen CAM-15. Emery Industries, Inc.
Chemical Description: POE (15) coco amine
Category: Emulsifier, antistat, softener, lubricant, cationic
Applications: Substantive to clay, metals, minerals, and textiles; additive to improve clarity of alkali cellulose in the mfg. of rayon fiber; prevents sludge build-up on spinnerets in spin baths; textile processing; industrial lubricants
Form: Liquid; Color: 6 (Gardner)
Properties: Solubility @ 5%: Sol. in water, glycerol trioleate, perchloroethylene; Density: 8.6 lb/gal; Visc.: 120 cs; HLB: 15.4; Cloud Pt.: > 100 C
Composition: 100% active ingred.
Reaction: Yields slightly alkaline sol'ns. when dissolved in water

Trymeen OAM-30/80. Emery Industries, Inc.
Chemical Description: POE (30) oleyl amine
Category: Emulsifier, dyeing assistant, cationic
Applications: Antiprecipitant in cross dyeing; mild stripping agent and dye leveler for acid dyes
Form: Liquid; Color: 6 (Gardner)
Properties: Solubility @ 5%: Sol.

in water; Density: 8.9 lb/gal; HLB: 15.3; Cloud Pt.: > 100 C
Composition: 80% active ingred.
Reaction: Yields slightly alkaline sol'ns. when dissolved in water

Trymeen SAM-50. Emery Industries, Inc.
Chemical Description: POE (50) stearyl amine
Category: Emulsifier, antistat, lubricant, leveling agent, cationic
Applications: Emulsifier in metal buffing compounds, for stearic acid and to prevent premature coagulation in mild acid or salt baths in latex rubber compounding; lubricant for fiberglass; leveling agent for acid, cationic, and disperse dyeings
Form: Solid; Color: 4 (Gardner)
Properties: Solubility @ 5%: Sol. in water and perchloroethylene; M.P.: 50 C; HLB: 17.8; Cloud Pt.: > 100 C
Composition: 100% active ingred.
Reaction: Yields slightly alkaline sol'ns. when dissolved in water

Trymeen TAM-8. Emery Industries, Inc.
Chemical Description: POE (8) tallow amine
Category: Emulsifier, cationic
Applications: Moderately low-foaming emulsifier for mineral oil, waxes, solvents, and vegetable oils; contributes to the wetting, lubricating, and softening properties of the end product; solvent emulsifier for low-foaming dye carriers; dyeing assistant; emulsifier for polyethylene textile softeners, and for fats and oils in industrial lubricants
Form: Liquid; Color: 8 (Gardner)
Properties: Solubility @ 5%: Sol. in water, butyl stearate, glycerol trioleate, perchloroethylene, Stod. solvent; Density: 8.2 lb/gal; Visc.: 200 cs; HLB: 11.5; Cloud Pt.: 80 C
Composition: 100% active ingred.
Reaction: Yields slightly alkaline sol'ns. when dissolved in water

Trymeen TAM-15. Emery Industries, Inc.
Chemical Description: POE (15) tallow amine
Category: Emulsifier, leveling agent, intermediate, antistat, cationic
Applications: Antiprecipitant for mixed dye baths; leveling agent for acid dyes; migrating agent for dispersed dyes; intermediate for quaternary ammonium compounds; antistat for processing synthetic fibers
Form: Liquid; Color: 8 (Gardner)
Properties: Solubility @ 5%: Sol. in water, glycerol trioleate, perchloroethylene; disp. in butyl stearate and Stod. solvent; Density: 8.5 lb/gal; Visc.: 175 cs; HLB: 14.2; Cloud Pt.: > 100 C
Composition: 100% active ingred.
Reaction: Yields slightly alkaline sol'ns. when dissolved in water

Trymeen TAM-20. Emery Industries, Inc.
Chemical Description: POE (20)

tallow amine
Category: Antistat, lubricant, coemulsifier, leveling agent, cationic
Applications: Textile antistat and lubricant for wool and synthetic fiber processing; coemulsifier and antistat in synthetic fiber spin finishes; antiprecipitant, leveling, and migrating agent in dyeing; antistat in carpet shampoos
Form: Liquid; Color: 6 (Gardner)
Properties: Solubility @ 5%: Sol. in water, perchloroethylene; disp. in glycerol trioleate; Density: 8.7 lb/gal; Visc.: 240 cs; HLB: 15.4; Cloud Pt.: > 100 C
Composition: 100% active ingred.
Reaction: Yields slightly alkaline sol'ns. when dissolved in water

Trymeen TAM-25. Emery Industries, Inc.
Chemical Description: POE (25) tallow amine
Category and Applications: See Trymeen TAM-20
Form: Liquid; Color: 6 (Gardner)
Properties: Solubility @ 5%: Sol. in water, perchloroethylene; Density: 8.8 lb/gal; Visc.: 170 cs; HLB: 16.1; Cloud Pt.: > 100 C
Composition: 100% active ingred.
Reaction: Yields slightly alkaline sol'ns. when dissolved in water

Trymeen TAM-40. Emery Industries, Inc.
Chemical Description: POE (40) tallow amine
Category: Emulsifier, antistat, dyeing assistant, stabilizer, cationic
Applications: Antistat additive for commercial carpet maintenance; dyeing assistant and stabilizer for natural and synthetic latices to prevent coagulation by acids
Form: Solid; Color: 6 (Gardner)
Properties: Solubility @ 5%: Sol. in water, perchloroethylene; M.P.: 40 C; HLB: 17.4; Cloud Pt.: > 100 C
Composition: 100% active ingred.
Reaction: Yields slightly alkaline sol'ns. when dissolved in water

Tween 20. ICI United States
Chemical Description: Polyoxyethylene 20 sorbitan monolaurate (Polysorbate 20)
Category: Emulsifier, solubilizer, antistat, lubricant, nonionic
Applications: O/W emulsifier, esp. for use in conjunction with corresponding Span products; also used in the textile industry as antistatic agent and fiber lubricants
Form: Liquid; Color: Pale yellow
Properties: Solubility: Sol. in water, methanol, ethanol, isopropanol, propylene glycol, ethylene glycol, cottonseed oil; Sp. gr.: ≈ 1.1; Visc.: ≈ 400 cps; HLB: 16.7; Flash Pt.: > 300 F; Fire Pt.: > 300 F; Sapon. No.: 40–50; Acid No. 2 max.; Hydroxyl No.: 96–108
Composition: 100% active ingred.

Tween 20 SD. Atlas Chemical Industries N.V.
Chemical Description: Polyoxyethylene sorbitan monolaurate (specially deodorized)

Category and Applications: See Tween 20
Form: Liquid
Composition: 100% active ingred.

Tween 21. ICI United States
Chemical Description: POE (4) sorbitan monolaurate
Category and Applications: See Tween 20
Form: Oily liquid; Color: Yellow
Properties: Solubility: Sol. in corn oil, dioxane, Cellosolve, carbon tetrachloride, methanol, ethanol, aniline; sol. at high concs. in mineral and cottonseed oils, diethyl ether, ethyl acetate, ethylene glycol; sol. (hazy) in mineral spirits and toluol; disp. in water (gels at high concs.); Sp. gr.: ≈ 1.1; Visc.: ≈ 500 cps; HLB: 13.3; Flash Pt.: > 300 F; Fire Pt.: > 300 F; Sapon. No.: 100–115; Acid No.: 3 max.; Hydroxyl No.: 225–255
Composition: 100% active ingred.

Tween 40. ICI United States
Chemical Description: POE (20) sorbitan monopalmitate
Category and Applications: See Tween 20
Form: Liquid (may gel on standing); Color: Pale yellow
Properties: Solubility: Sol. in water, methanol, ethanol, isopropanol, ethylene glycol, cottonseed oil; Sp. gr.: ≈ 1.08; HLB: 15.6; Flash Pt.: > 300 F; Fire Pt.: > 300 F; Sapon. No.: 41–52; Acid No.: 2 max.; Hydroxyl No.: 90–105
Composition: 100% active ingred.

Tween 60. ICI United States
Chemical Description: POE (20) sorbitan monostearate
Category and Applications: See Tween 20
Form: Liquid (may gel on standing); Color: Pale yellow
Properties: Solubility: Sol. in water, isopropyl alcohol, ethyl alcohol; Sp. gr.: ≈ 1.1; Visc.: ≈ 600 cps; HLB: 14.9; Flash Pt.: > 300 F; Fire Pt.: > 300 F; Sapon. No.: 45–55; Acid No.: 2 max.; Hydroxyl No.: 81–96
Composition: 100% active ingred.

Tween 61. ICI United States
Chemical Description: POE (4) sorbitan monostearate
Category and Applications: See Tween 20
Form: Waxy solid; Color: Tan
Properties: Solubility: Sol. in methanol, ethanol; disp. in distilled water and hard water (200 ppm); Sp. gr.: ≈ 1.06; HLB: 9.6; Flash Pt.: > 300 F; Fire Pt.: > 300 F; Pour Pt.: ≈ 100 F; Sapon. No.: 95–115; Acid No.: 2 max.; Hydroxyl No.: 170–200
Composition: 100% active ingred.

Tween 65. ICI United States
Chemical Description: Polysorbate 65
Category and Applications: See Tween 20
Form: Waxy solid; Color: Tan
Properties: Solubility: Sol. in isopropyl alcohol, ethyl alcohol, vegetable and mineral oil; Sp. gr.: ≈ 1.05; HLB: 10.5; Flash Pt.: >

300 F; Fire Pt.: > 300 F; Pour Pt.: ≈ 92 F; Sapon. No.: 88–98; Acid No.: 2 max.; Hydroxyl No.: 44–60
Composition: 100% active ingred.

Tween 80. ICI United States
Chemical Description: POE (20) sorbitan monooleate (Polysorbate 80 USP)
Category and Applications: See Tween 20
Form: Liquid; Color: Pale yellow
Properties: Solubility: Sol. in water, isopropyl alcohol, ethanol; Sp. gr.: ≈ 1.08; Visc.: ≈ 425 cps; HLB: 15.0; Flash Pt.: > 300 F; Fire Pt.: > 300 F; Sapon. No.: 45–55; Acid No.: 2 max.; Hydroxyl No.: 65–80
Composition: 100% active ingred.

Tween 81. ICI United States
Chemical Description: POE (5) sorbitan monooleate
Category and Applications: See Tween 20
Form: Oily liquid (may gel on standing); Color: Amber
Properties: Solubility: Sol. in mineral oil, corn oil, dioxane, Cellosolve, methanol, ethanol, ethyl acetate, aniline; sol. at certain concs. in toluol, mineral spirits, cottonseed oil, acetone, carbon tetrachloride; disp. in water, ether, ethylene glycol, and 5% sol'ns. of sulfuric acid, sodium hydroxide, sodium sulfate, and aluminum chloride; Sp. gr.: ≈ 1; Visc.: ≈ 450 cps; HLB: 10.0; Flash Pt.: > 300 F; Fire Pt.: > 300 F; Sapon. No.: 96–104; Acid No.: 2 max.; Hydroxyl No.: 134–150
Composition: 100% active ingred.

Tween 85. ICI United States
Chemical Description: POE (20) sorbitan trioleate (Polysorbate 85)
Category and Applications: See Tween 20
Form: Liquid (may gel); Color: Amber
Properties: Solubility: Sol. in most vegetable oils, Cellosolve, lower alcohols, aromatic solvents, ethyl acetate, most mineral oils, mineral spirits, acetone, dioxane, carbon tetrachloride, and ethylene glycol; disp. in water; Sp. gr.: ≈ 1.0; Visc.: ≈ 300 cps; HLB: 11.0; Flash Pt.: > 300 F; Fire Pt.: > 300 F; Sapon. No.: 80–95; Acid No.: 2 max.; Hydroxyl No.: 39–52
Composition: 100% active ingred.

Tween-Mos 100 K Special. ICI United States
Chemical Description: Mono- and diglycerides (80%) and Polysorbate 80 (20%)
Category: Emulsifier, nonionic
Applications: Food emulsifier; for emulsifier-stabilizer mixes in ice cream and frozen dessert manufacture; certified kosher by the Union of Orthodox Jewish Congregations of America Ⓤ
Form: Beads; Color: Ivory white
Properties: Solubility: Sol. above its melting pt. in isopropyl alcohol, vegetable and mineral oils;

M.P.: ≈ 135 F; HLB: 5.2; Flash Pt.: > 300 F; Fire Pt.: > 300 F; Iodine Value: ≈ 6
Composition: 100% active ingred.

TX-112. Theobald Industries
Chemical Description: Nonphosphate built detergent
Category: Detergent, anionic
Applications: Heavy duty, all purpose detergent; nonphosphate laundry detergent; hard surface cleaning
Form: Beads; Color: White/blue
Properties: Solubility: Sol. in water; Density: 0.32 g/cc; Biodegradable
Composition: 14% active ingred.; other components: sodium carbonate, sodium silicate, various fillers and cleansing aids
Reaction: Acid
Containers: Cartons, bags, drums
Warnings and Caution: Avoid contact with eyes, mucous membranes, and prolonged skin contact

TX-228. Theobald Industries
Chemical Description: Spray-dried oxygen bleach, light density
Category: Bleach
Applications: Oxygen-type laundry bleach
Form: Beads; Color: White or light blue
Properties: Solubility: Sol. in water; Density: 0.25 g/cc; Stability: Evolution of O_2 with water
Composition: Sodium perborate, sodium silicate plus various fillers and cleansing agents
Reaction: Acid
Toxicity: Eye irritant
Containers: Cartons, bags, drums
Warnings and Caution: Avoid contact with eyes, mucous membranes, and prolonged skin contact

TX-318. Theobald Industries
Chemical Description: Built detergent with phosphate
Category: Detergent, anionic
Applications: Low phosphate laundry detergent
Form: Beads; Color: White/blue
Properties: Solubility: Sol. in water; Density: 0.32 g/cc
Composition: 16% active ingred.; other components: sodium polyphosphate, sodium silicate, plus various fillers and cleansing aids
Reaction: Acid
Containers: Cartons, bags, drums
Warnings and Caution: Avoid contact with eyes, mucous membranes, and prolonged skin contact

TX-338. Theobald Industries
Chemical Description: Built cold water detergent, nonphosphate
Category: Detergent, anionic
Applications: Nonphosphate cold water laundry detergent
Form: Beads; Color: Blue
Properties: Solubility: Sol. in water; Density: 0.32 g/cc
Composition: 17.5% active ingred.; other components: sodium carbonate, sodium silicate, plus various fillers and cleansing aids
Reaction: Acid
Containers: Cartons, bags, drums

Warnings and Caution: Avoid contact with eyes, mucous membranes, and prolonged skin contact

TX-358. Theobald Industries
Chemical Description: Built cold water detergent with phosphate
Category: Detergent, mixed active system
Applications: Low phosphate, cold water laundry detergent
Form: Beads; Color: White
Properties: Solubility: Sol. in water; Density: 0.32 g/cc
Composition: 16% active ingred.; other components: sodium polyphosphate, sodium silicate plus various fillers and cleansing aids
Reaction: Acid
Containers: Cartons, bags, drums
Warnings and Caution: Avoid contact with eyes, mucous membranes, and prolonged skin contact

U

Ultra Blend 60. Witco Chemical/ Organics Div.
Chemical Description: Blends of linear alkylaryl sulfonate, alkanolamide, and ethoxylated adduct
Category: Detergent, anionic
Applications: High-foaming liquid detergent and car washing concentrate
Form: Liquid
Properties: Biodegradable
Composition: 60% conc.

Ultra Blend 100. Witco Chemical/ Organics Div.
Chemical Description: See Ultra Blend 60
Category and Applications: See Ultra Blend 60
Form: Liquid
Properties: Biodegradable
Composition: 100% conc.

Ultra Blend P-280. Witco Chemical/ Organics Div.
Chemical Description: Blend of linear alkylaryl sulfonate, alcohol ether sulfate, and ethoxylated adduct
Category: Detergent, anionic
Applications: High-foaming liquid detergent concentrate
Form: Liquid
Properties: Biodegradable

Ultra NCS Liquid. Witco Chemical/ Organics Div.
Chemical Description: Ammonium cumene sulfonate
Category: Hydrotrope, coupling agent, solubilizer, anionic
Applications: Hydrotrope; coupling agent; solubilizer; cloud point depressant for liquid detergents; antiblocking agent for powdered detergents
Form: Liquid
Composition: 40% conc.

Ultra NXS Liquid. Witco Chemical/ Organics Div.
Chemical Description: Ammonium xylene sulfonate
Category and Applications: See Ultra NCS Liquid
Form: Liquid
Composition: 45% conc.

Ultra SCS Liquid. Witco Chemical/Organics Div.
Chemical Description: Sodium cumene sulfonate
Category and Applications: See Ultra NCS Liquid
Form: Liquid; also available in powder form
Composition: 40% conc.

Ultra Sulfate AE-3. Witco Chemical/Organics Div.
Chemical Description: Ammonium alcohol ether sulfate
Category: Detergent, wetting agent, penetrant, lubricant, emulsifier, dye dispersant, scouring aid, antistat, anionic
Applications: Base for shampoos, bubble baths, and foaming body cleansers; high-foaming detergent for specialty household and industrial cleaners; textile surfactant
Form: Liquid
Composition: 58% conc.

Ultra Sulfate SE-5. Witco Chemical/Organics Div.
Chemical Description: Sodium alcohol ether sulfate
Category and Applications: See Ultra Sulfate AE-3
Form: Liquid
Composition: 58% conc.

Ultra Sulfate SL-1. Witco Chemical/Organics Div.
Chemical Description: Sodium lauryl alcohol sulfate
Category and Applications: See Ultra Sulfate AE-3; also used as surfactant for emulsion polymerization and latex stabilization
Form: Liquid
Composition: 30% conc.

Ultra SXS Liquid. Witco Chemical/Organics Div.
Chemical Description: Sodium xylene sulfonate
Category and Applications: See Ultra NCS Liquid
Form: Liquid
Composition: 40% conc.

Ultra SXS Powder. Witco Chemical/Organics Div.
Chemical Description: Sodium xylene sulfonate
Category and Applications: See Ultra NCS Liquid
Form: Powder
Composition: 90% conc.

Unamide C-2. Lonza, Inc.
Chemical Description: Ethoxylated coco alkanolamide
Category: Foam booster/stabilizer, viscosity builder, emulsifier, nonionic
Applications: Personal care, industrial, and household products, incl. shampoos, bubble baths, light duty liquids, hard surface cleaners
Form: Solid; Color: White
Properties: Sapon. No.: 15 max.; Hydroxyl No.: 190–220; pH: 9–10 (5% sol'n.)
Composition: 100% active ingred.

Unamide C-5. Lonza, Inc.
Chemical Description: Ethoxylated coco alkanolamide

Category and Applications: See Unamide C-2
Form: Liquid; Color: Light yellow
Properties: Sapon. No.: 12 max.; Hydroxyl No.: 145–165; pH: 9–10 (5% sol'n.)
Composition: 100% active ingred.

Unamide C-72-3. Lonza, Inc.
Chemical Description: 2:1 Coco diethanolamide
Category: Viscosity builder, foam stabilizer, nonionic
Applications: Used for light duty liquids, industrial/household hard surface cleaners, surfactants, emulsifiers, corrosion inhibitors, lubricants
Form: Liquid; Color: Amber
Properties: Acid No.: 0–10; Amine Value: 128–138; pH: 9–10.5 (5%)

Unamide CDX. Lonza, Inc.
Chemical Description: 1:1 Coco diethanolamide
Category: Foam booster/stabilizer, viscosity builder, nonionic
Applications: Cosmetic and toiletries, esp. high performance shampoos, bubble bath, etc.
Form: Liquid; Color: Straw
Properties: Solubility: Sol. (hazy) in water (< 10%); Density: 8.2 lb/gal; Sp. gr.: 0.99; Acid No.: 0–0–3; Amine Value: 27–35; pH: 9–10.5 (5% sol'n.); Biodegradability: 100%
Composition: 100% active ingred.

Unamide CMX. Lonza, Inc.
Chemical Description: 1:1 Coco monoethanolamide
Category and Applications: See Unamide CDX
Form: Solid; Color: Light
Properties: Acid No.: 0–3; Amine Value: 0–18; pH: 9–10.5 (5%)
Composition: 100% active ingred.

Unamide D-10. Lonza, Inc.
Chemical Description: 2:1 Modified coco diethanolamide
Category and Applications: See Unamide C-72-3
Form: Liquid; Color: Dark amber
Properties: Acid No.: 55–75; Amine Value: 110–117; pH: 9–10.5 (5% sol'n.)
Composition: 100% active ingred.

Unamide GG-75. Lonza, Inc.
Chemical Description: 2:1 Modified coco diethanolamide
Category and Applications: See Unamide C-72-3
Form: Liquid; Color: Yellow
Properties: Acid No.: 80–100; Amine Value: 184–192; pH: 9–10.5 (5% sol'n.)
Composition: 100% active ingred.

Unamide J-56. Lonza, Inc.
Chemical Description: 2:1 Lauric diethanolamide
Category and Applications: See Unamide C-72-3
Form: Liquid; Color: Amber
Properties: Acid No.: 18–25; Amine Value: 155–170; pH: 9–10.5 (5% sol'n.)
Composition: 100% active ingred.

Unamide JJ-35. Lonza, Inc.
Chemical Description: 1:1 Coco oil diethanolamide

Category and Applications: See Unamide CDX
Form: Liquid; Color: Lt. straw
Properties: Acid No.; 0–10; Amine Value: 20–35; pH: 9–10.5 (5% sol'n.)
Composition: 100% active ingred.

Unamide L-2. Lonza, Inc.
Chemical Description: Ethoxylated lauric alkanolamide
Category and Applications: See Unamide C-2
Form: Solid; Color: White
Properties: Sapon. No.: 15 max.; Hydroxyl No.: 190–220; pH: 9–10 (5% sol'n.)
Composition: 100% active ingred.

Unamide L-5. Lonza, Inc.
Chemical Description: Ethoxylated lauric alkanolamide
Category and Applications: See Unamide C-2
Form: Soft wax; Color: White
Properties: Sapon. No.: 12 max.; Hydroxyl No.: 145–165; pH: 9–10 (5% sol'n.)
Composition: 100% active ingred.

Unamide LDL. Lonza, Inc.
Chemical Description: 1:1 Coco oil diethanolamide
Category and Applications: See Unamide C-72-3
Form: Liquid; Color: Lt. straw
Properties: Density: 8.2 lb/gal; Sp. gr.: 0.98; Acid No.: 0–10; Amine Value: 20–35; pH: 9–10.5 (5% sol'n.); Biodegradability: 100%
Composition: 100% active ingred.

Unamide LDX. Lonza, Inc.
Chemical Description: 1:1 Lauric diethanolamide
Category and Applications: See Unamide CDX; also used in premium light duty detergents
Form: Solid; Color: White
Properties: Solubility: Sol. (hazy) in water (< 10%); Density: 8.2 lb/gal; Sp. gr.: 0.99; Acid No.: 0–3; Amine Value: 0–16; pH: 9–10.5 (5% sol'n.); Biodegradability: 100%
Composition: 100% active ingred.

Unamide LMDX. Lonza, Inc.
Chemical Description: 1:1 Lauric/myristic diethanolamide
Category and Applications: See Unamide CDX; also for use in premium light duty detergents, and where optimum thickening and foam stabilization are required
Form: Solid; Color: White
Properties: Solubility: Sol. (hazy) in water (< 10%); Density: 8.2 lb/gal; Sp. gr.: 0.99; Acid No.: 0–0–3; Amine Value: 32–40; pH: 9–10.5 (5% sol'n.); Biodegradability: 100%
Composition: 100% active ingred.

Unamide N-72-3. Lonza, Inc.
Chemical Description: 2:1 Coco diethanolamide
Category and Applications: See Unamide C-72-3
Form: Liquid; Color: Yellow
Properties: Acid No.: 0–10; Amine Value: 128–138; pH: 9–10.5 (5% sol'n.)
Composition: 100% active ingred.

Unamide S. Lonza, Inc.
Chemical Description: 1:1 Stearic diethanolamide
Category: Viscosity builder, opacifying agent, lubricant, nonionic
Applications: Thickener and opacifier in personal care products; viscosity builder, gelling agent, lubricant, opacifier in industrial and household products; textile lubricant/finishing agent
Form: Waxy solid; Color: White
Properties: Acid No.: 11–15; Amine Value: 44–52
Composition: 100% active ingred.

Unamide SI. Lonza, Inc.
Chemical Description: 1:1 Stearic (ethyl amino hydroxyethyl) amide
Category: Viscosity builder, gelling agent, lubricant, opacifier, nonionic
Applications: Used in household and industrial products; textile lubricant/finishing agent
Form: Waxy solid; Color: White
Properties: Acid No.: 0–5; Amine No.: 54–58

Unamide W. Lonza, Inc.
Chemical Description: 1:1 Stearic diethanolamide
Category and Applications: See Unamide SI
Form: Liquid; Color: Amber
Properties: Acid No.: 9–12; Amine Value: 44–52

Uniperol A. BASF AG
Chemical Description: Mixture of nonionic and anionic products
Category: Leveling agent, nonionic/anionic
Applications: Leveling agent for polyamide with affinity for the fiber; used to control the rate of uptake of acid and metal complex dyes; prevents stripiness in dyeing
Form: Viscous liquid; Color: Brown
Properties: Stability: Stable in neutral, acid, and weakly alkaline liquors; hard water causes no precipitations
Reaction: Aq. sol'ns. have slightly acid reaction
Toxicity: May cause irritation on prolonged contact with the skin and mucous membranes
References: Tech. leaflet M 5536 e

Uniperol AC. BASF AG
Chemical Description: Oxyethylated fatty amine
Category: Dyeing auxiliary, leveling agent, slightly cationic
Applications: Dyeing auxiliary for wool and synthetic fibers; leveling agent with affinity for the fiber; recommended for dyeing of wool and polyamides with acid and metal complex dyes; prewashing and afterwashing of synthetic fibers in dyeing and printing; partial and complete stripping of dyeings and prints
Form: Clear, viscous liquid; Color: Light brown
Properties: Solubility: Readily sol. in cold water; Cloud Pt.: None at acid pH; Stability: Resistant to hard water, acids, alkalis, and metal salts
Reaction: Slightly alkaline to neu-

tral reaction at concs. commonly used
Toxicity: In conc. form, causes severe irritation of the skin and mucous membranes
References: Tech. leaflet M 5636 e

Uniperol AC Highly Conc. BASF AG
Chemical Description: Oxyethylated fatty amine
Category and Applications: See Uniperol AC (100 parts Uniperol AC = 40 parts Uniperol AC Highly Conc.)
Form: Highly viscous oil; Color: Brown
Properties: Solubility: Dissolves readily in water yielding clear sol'ns.
Toxicity: In conc. form, causes severe irritation of the skin and mucous membranes
References: Tech. bulletin M 5636

Uniperol EL. BASF AG
Chemical Description: Oxyethylation product of a vegetable oil
Category: Dispersant, emulsifier, leveling agent, nonionic
Applications: Used in the dyeing of synthetic fibers with disperse dyes, in the afterwashing of textiles that have been dyed with disperse dyes, and as an emulsifying agent for O/W emulsions
Form: Oily liquid (pasty or solid at low temps., but liquefies on heating); Color: Yellow
Properties: Solubility: Readily sol. in water, liquid or molten fatty acids, fats, greases, waxes, and many organic solvents; Stability: Unaffected by hard water; saponified by strong alkalis, but stable to weak alkalis and acids, esp. at low temps.
Composition: Almost 100% active ingred.
Reaction: Aq. sol'ns. have a neutral reaction
References: Tech. leaflet M 1900 e

Uniperol KA. BASF AG
Chemical Description: Mixture of nonionic surfactants
Category: Dyeing auxiliary, nonionic
Applications: Auxiliary for dyeing blends of acrylic fibers with wool, polyamide, and cellulosic fibers; permits one-bath dyeing of these blends with selected cationic and anionic dyes
Form: Viscous liquid; Color: Yellow-brown
Properties: Solubility: Readily sol. in cold or warm water; Stability: Stable in neutral, acid, and weakly alkaline sol'ns.; no precipitations with hard water
Toxicity: May cause irritation and inflammation on prolonged contact with the skin and mucous membranes
References: Tech. leaflet M 5573 e

Uniperol O. BASF AG
Chemical Description: Fatty alcohol-polyglycol ether
Category: Dyeing auxiliary, leveling agent, wetting agent, dispersant, detergent, nonionic
Applications: Leveling agent in

wool dyeing with 1:1 metal complex dyes; in wool-union dyeing; degreasing and cleansing textiles in acid dyebaths; auxiliary for one-bath bulking and precleansing of high-bulk acrylic fiber yarns; leveling/penetrating agent in dyeing with direct and basic dyes; leveling/retarding agent in dyeing cotton with vat dyes; degumming silk

Form: Clear, low-viscosity liquid (becomes pasty at < 0 C); Color: Almost colorless

Properties: Solubility: Dissolves in water in all proportions; Stability: Resistant to hard water, acids, alkalis, and metal salts

Composition: 20% active ingred.

Reaction: Aq. sol'ns. have an almost neutral reaction

Toxicity: May cause irritation and inflammation on prolonged contact with the skin and mucous membranes

References: Tech. leaflet M 1917 e

Uniperol O Micropearl. BASF AG

Chemical Description: Fatty alcohol-polyglycol ether

Category and Applications: See Uniperol O (1 part Uniperol O Micropearl is equivalent to 3 parts Uniperol O)

Form: Pourable, dust-free powder; Color: Off-white

Properties: Solubility: Dissolves readily in water @ 35–40 C

Toxicity: May cause irritation and inflammation on prolonged contact with skin and mucous membranes

References: Tech. leaflet M 1917 e

Uniperol SE. BASF AG

Chemical Description: Oxyethylation product with affinity for the dye

Category: Leveling agent

Applications: Leveling agent for milling dyes and for 1:1 and 1:2 metal complex dyes

Form: Viscous, pourable liquid; Color: Pale brown

Properties: Solubility: Dissolves readily in water forming clear sol'ns.; Stability: Stable to hard water, acids, alkalis, and metal salts

Reaction: Aq. sol'ns. are practically neutral

Toxicity: Has a caustic effect on skin and mucous membranes in undiluted form

References: Tech. leaflet M 1912 e

Uniperol W. BASF AG

Chemical Description: Aliphatic ethoxylation product

Category: Dyeing auxiliary, dispersant, leveling agent, anionic

Applications: Dyeing auxiliary for wool and synthetic fibers; leveling agent for dyeing wool and its blends with metal complex, acid, and chroming dyes; dispersant in dyeing synthetics with disperse dyes even under high-temp. conditions; promotes leveling and penetration of dyed shades on hard fibers

Form: Almost clear, highly viscous liquid; Color: Yellow brown

Properties: Solubility: Readily sol. in water producing clear sol'ns.; Stability: Stable to acids, alkalis, hard water, salts of heavy metals

and electrolytes
Composition: 40% active ingred.
Reaction: Aq. sol'ns. have a neutral to weakly alkaline reaction
References: Tech. leaflet M 1947 e

Uniperol W Flakes. BASF AG
Chemical Description: Aliphatic ethoxylation product (water-free concentrate)
Category and Applications: See Uniperol W (2 parts Uniperol W is equivalent to 1 part Uniperol W Flakes)
Form: Flakes; Color: Pale yellow to light brown
Properties: Solubility: Readily sol. in water producing clear sol'ns.
Reaction: Aq. sol'ns. have a neutral to weakly alkaline reaction
Composition: 40% active ingred.
References: Tech. leaflet M 1947 e

Uniquat CB-50. Lonza, Inc.
Chemical Description: 1-Hydroxyethyl-1-benzyl-2-alkyl imidazolinium chloride (50% in isopropyl alcohol)
Category: Microbiocide, cationic
Applications: Controls bacteria (incl. slime formers), fungi, yeast, mold, and algae in recirculating cooling tower waters; helps clean and loosen slime debris from cooling and flooding system surfaces
Form: Clear to turbid liquid; Color: Dark amber
Properties: Density: 7.76 lb/gal; Sp. gr.: 0.932; Visc.: 56 cps (20 C); F.P.: 25 C; Flash Pt.: 89 F (COC)
Composition: 50% active ingred.
Toxicity: Corrosive; causes severe eye and skin damage; harmful if absorbed through the skin; harmful or fatal if swallowed; toxic to fish
Warnings and Caution: Treated effluent should not be discharged where it will drain into lakes, streams, ponds, or public water; flammable—do not use or store near heat or open flame

V

Valdet 561. Valchem Chem. Div.
Chemical Description: Ethylene oxide condensate of a nonyl phenol
Category: Wetting agent, dispersant, detergent, penetrant, emulsifier, nonionic
Applications: Desizing cotton and rayon; wetting and dispersing agent in dyeing; scouring wool, cotton, synthetics; soaping of fast color dyed cotton and rayon; as a resin finish penetrant; emulsifier for wool lubricants; kier boiling of cotton
Form: Clear to slightly hazy liquid; Color: 150 max. (APHA); Odor: Faintly aromatic
Properties: Solubility: 100% sol. in water; Density: 8.7 lb/gal; Sp. gr.: 1.055–1.065; Cloud Pt.: 158–165 F (1%); Flash Pt.: 427–445 F; Stability: Good; pH: 7.0 ± 1.0 (10% sol'n.)
Composition: 99% min. solids
Containers: 450 lb net steel drums
Warnings and Caution: Avoid strong oxidizing agents; avoid contact with eyes

Valdet 4016. Valchem Chem. Div.
Chemical Description: Straight chain alcohol ethoxylate
Category: Wetting agent, stabilizer, nonionic
Applications: Nonrewetting surfactant and emulsion stabilizer used in fluorocarbon finishes imparting water and oil repellency; solvent and solvent emulsion scouring operations; emulsifier for Stod. solvent
Form: Clear to slightly hazy liquid; Color: 60 max. (APHA)
Properties: Solubility: Water-sol.; Sp. gr.: 1.011; Cloud Pt.: 39–45 C (1%); Acid Value: 1 max.; pH: 7.0 ± 0.5 (10% sol'n.); Biodegradable
Composition: 90% active ingred.
Containers: 450 lb net steel drums
Warnings and Caution: Avoid eye contact with conc. form

Valdet 5020. Valchem Chem. Div.
Category: Stripping agent, dye retarder, cationic
Applications: Stripping agent and dye retarder for azoic dyes; also in

the application of cationic dyes to acrylic fibers
Form: Clear liquid; Color: 100 max. (APHA)
Properties: Density: 8.2 lb/gal; Sp. gr.: 0.975 ± 0.01; pH: 7.5 ± 0.5
Composition: 15 ± 1% solids
Containers: 425 lb net steel drums

Valdet 5022. Valchem Chem. Div.
Chemical Description: Surfactant blend
Category: Lubricant, leveling agent, wetting and rewetting agent, anionic
Applications: Lubricating and leveling agent for the prevention of crack and chafe marks on cellulosics, synthetics, and their blends; used in prescour or dye bath
Form: Clear liquid; Color: 9 max. (Gardner)
Properties: Solubility: Dissolves easily in warm water; Density: 8.5 lb/gal; Sp. gr.: 1.019; pH: 6.3 ± 0.5
Composition: 39% active ingred.
Containers: 450 lb net liquipaks

Valdet AC-40. Valchem Chem. Div.
Chemical Description: Amine condensate
Category: Detergent, anionic
Applications: Detergent for textile wet processing incl. scouring and soaping operations, fabric printing and dyeing, boil-off of synthetics; also as a general cleaning agent for machinery, floors, and fixtures
Form: Clear liquid; Color: Amber
Properties: Solubility: Miscible with water; Density: 8.3 lb/gal; Sp. gr.: 1.010–1.030; pH: 11.0 ± 1
Composition: 40% active ingred.
Containers: 475 lb net steel drums

Valdet CC. Valchem Chem. Div.
Chemical Description: Amine condensate
Category and Applications: Similar uses to Valdet AC-40 but amphoteric in nature
Form: Clear liquid; Color: Dark amber; Odor: Coconut oil
Properties: Solubility: Miscible with water; Density: 8.3 lb/gal; Sp. gr.: 0.990–1.010; Flash Pt.: 194 F (TCC); Stability: Good; pH: 9.0 ± 1.0
Composition: 100% active ingred.
Toxicity: Irritant to skin and mucous membranes on extended contact
Containers: 450 lb net steel drums

Valdet MT-36. Valchem Chem. Div.
Chemical Description: Concentrated amide sulfonate detergent
Category: Detergent, wetting agent, dyeing and bleaching assistant, finishing agent, anionic
Applications: Textile wetting and scouring, kier assistant, open boil-off assistant, bleaching assistant, dyeing assistant; also for acid dyeing and acid milling of wool; good wetting and detergency at elevated temps.
Form: Clear liquid
Properties: Solubility: Dissolves readily in water to form a clear

sol'n.; Density: 8.3 lb/gal; Stability: Stable to acids, salts, alkalis, bleaching agents, and anionic or nonionic dyeing auxiliaries; pH: 6.5–8.5
Composition: 40 ± 1% solids
Containers: 450 lb net steel drums

Vannox AC 2, ACH, ACL. Nippon Nyukazai Co., Ltd.
Category: Emulsifier, nonionic/anionic
Applications: Emulsifier for organic chlorinated insecticides
Form: Liquid

Vannox PW. Nippon Nyukazai Co., Ltd.
Category: Wetting agent, anionic
Applications: Wetting agent for wettable powder
Form: Powder

Vantoc CL. ICI Australia Ltd.
Chemical Description: Quaternary ammonium
Category: Cationic surfactant and biocide

Varamide A-2. Sherex
Chemical Description: Refined coconut diethanolamide
Category: Detergent, nonionic
Composition: 100% active ingred.

Varamide A-7. Sherex
Chemical Description: Oleic diethanolamide
Category: Detergent, nonionic
Composition: 100% active ingred.

Varamide A-10. Sherex
Chemical Description: Modified coco diethanolamide
Category: Detergent, nonionic
Applications: Detergent for viscous floor cleaners
Form: Liquid
Composition: 100% active ingred.

Varamide A-12. Sherex
Chemical Description: Modified coco diethanolamide
Category: Detergent, nonionic
Applications: Detergent for high-phosphate cleaners
Form: Liquid
Composition: 100% active ingred.

Varamide A-83. Sherex
Chemical Description: Modified coco diethanolamide
Category: Detergent, nonionic
Applications: Detergent in general purpose cleaners
Form: Liquid
Composition: 100% active ingred.

Varamide L-1. Sherex
Chemical Description: Lauric diethanolamide
Category: Detergent, foam stabilizer, nonionic
Applications: Thickener and foam booster/stabilizer for shampoos
Form: Paste
Composition: 100% active ingred.

Varamide MA-1. Sherex
Chemical Description: Coco superamide
Category: Detergent, foam stabilizer, thickener, foam booster, nonionic
Applications: Used in dishwashing

compounds and shampoos
Form: Liquid
Composition: 100% active ingred.

Varamide ML-1. Sherex
Chemical Description: Lauric superamide
Category and Applications: See Varamide MA-1
Form: Wax
Composition: 100% active ingred.

Varamide ML-4. Sherex
Chemical Description: Lauric-myristic superamide
Category and Applications: See Varamide MA-1
Form: Paste
Composition: 100% active ingred.

Varamide SL-9. Sherex
Chemical Description: Lauric superamide
Category: Detergent, foam stabilizer, thickener, nonionic
Applications: Used for shampoos, light and heavy duty detergents, with improved handling properties
Form: Solid
Composition: 100% active ingred.

Varamide SL-910. Sherex
Chemical Description: Lauric-palmitic superamide
Category and Applications: See Varamide SL-9
Form: Liquid
Composition: 100% active ingred.

Varifoam MS-80. Sherex
Chemical Description: Formulated shampoo concentrate
Category: Detergent, anionic
Applications: Baby shampoo concentrate
Composition: 80% solids

Varifoam YM Special. Sherex
Chemical Description: Formulated shampoo concentrate
Category: Anionic surfactant
Applications: Shampoos and bubble baths

Varifos 2039, 2058, 2611. Sherex
Chemical Description: Free acid of complex organic phosphate ester
Category: Detergent, emulsifier, wetting agent, anionic
Applications: Used for highly alkaline, heavy duty cleaners
Form: Liquid
Composition: 100% active ingred.

Varine C. Sherex
Chemical Description: Coco hydroxyethyl imidazoline
Category: Emulsifier, shampoos base, cationic
Form: Liquid
Composition: 100% active ingred.

Varine O. Sherex
Chemical Description: Oleic hydroxyethyl imidazoline
Category: Emulsifier, corrosion inhibitor, cationic
cationic
Form: Liquid
Composition: 100% active ingred.

Varion CADG. Sherex
Chemical Description: Coco amido betaine

Category: Detergent, foaming agent, conditioner, solubilizer, amphoteric
Applications: Shampoo detergent, foaming, and conditioning agent; solubilizer for lauryl sulfate
Form: Liquid
Composition: 32% active ingred.

Varion CAS. Sherex
Chemical Description: Coco sulfobetaine
Category: Detergent, amphoteric

Varion CDG. Sherex
Chemical Description: Coco betaine
Category: Foaming agent, detergent, solubilizer, amphoteric
Applications: Detergent, foaming, and conditioning agent in shampoos; solubilizer for lauryl sulfate in conc. shampoos
Form: Liquid
Composition: 31% active ingred.

Variquat 50MC. Sherex
Chemical Description: Alkyl dimethylbenzyl ammonium chloride
Category: Germicide, algicide, disinfectant, deodorant, cationic
Applications: EPA-registered pesticides; concentrate for the manufacture of disinfectants, sanitizers; used in food processing, dairy, restaurant, and household products
Form: Liquid
Composition: 50% active ingred.

Variquat 60LC. Sherex
Chemical Description: Alkyl dimethylbenzyl ammonium chloride
Category: Algicide, cationic
Applications: Algicide for use in swimming pools and in industrial water treatment
Form: Liquid
Composition: 50% active ingred.

Variquat 80MC. Sherex
Chemical Description: Alkyl dimethylbenzyl ammonium chloride
Category: Germicide, disinfectant, deodorant, cationic
Applications: EPA-registered for pesticides; concentrate for the manufacture of disinfectants and sanitizers
Form: Liquid
Composition: 80% active ingred.

Variquat 638. Sherex
Chemical Description: Ethoxylated quaternary
Category: Antistat, emulsifier, cationic
Applications: Antistat, plating bath foam blanket, emulsifier
Form: Liquid
Composition: 75% active ingred.

Variquat A200. Sherex
Chemical Description: Allyl trimethyl ammonium chloride
Category: Emulsifier, cationic
Form: Liquid
Composition: 50% active ingred.

Variquat B200. Sherex
Chemical Description: Benzyltrimethyl ammonium chloride
Category: Dispersant, dye leveler, emulsifier, cationic
Form: Liquid
Composition: 60% active ingred.

Variquat E228. Sherex
Chemical Description: Cetyl trimethyl ammonium chloride
Category: Softener, conditioner, cationic
Applications: Hair softener and conditioner
Form: Liquid

Variquat LC80. Sherex
Chemical Description: Alkyl dimethylbenzyl ammonium chloride
Category: Algicide, cationic
Applications: Algicide for use in swimming pools and in industrial water treatment
Form: Liquid
Composition: 80% active ingred.

Varisoft 110. Sherex
Chemical Description: Complex quaternary sulfate
Category: Softener, cationic
Applications: Fabric softener concentrate for use in formulating softeners used in home and commercial laundries and textile processing
Form: Paste
Composition: 75% solids

Varisoft 135. Sherex
Chemical Description: Formulated dryer-type concentrate
Category and Applications: See Varisoft 110
Form: Paste
Composition: 92% solids

Varisoft 137. Sherex
Chemical Description: Dimethyl dihydrogenated tallow ammonium methosulfate
Category and Applications: See Varisoft 110
Form: Paste
Composition: 90% solids

Varisoft 190-100P. Sherex
Chemical Description: Dimethyl distearyl ammonium methyl sulfate
Category and Applications: See Varisoft 110
Form: Powder
Composition: 97% solids

Varisoft 222, 238. Sherex
Chemical Description: Complex difatty quaternary sulfate
Category and Applications: See Varisoft 110
Form: Paste
Composition: 75–90% solids

Varisoft 445. Sherex
Chemical Description: Methyl (1) hydrogenated tallowamidoethyl (2) hyd. tallow imidazolinium methosulfate
Category and Applications: See Varisoft 110
Form: Liquid
Composition: 77% solids

Varisoft 475. Sherex
Chemical Description: Methyl (1) tallowamidoethyl (2) tallow imidazolinium methosulfate
Category and Applications: See Varisoft 110
Form: Liquid
Composition: 77% solids

Varisoft 3262. Sherex
Chemical Description: Fatty quaternary sulfate
Category: Cream rinse concentrate, cationic
Form: Solid
Composition: 75% conc.

Varisoft 3690. Sherex
Chemical Description: Methyl (1) oleylamidoethyl (2) oleyl imidazolinium methosulfate
Category and Applications: See Varisoft 110
Form: Liquid
Composition: 77% solids

Varisoft 6112. Sherex
Chemical Description: Tallow-based bis-imidazolinium methosulfate (diquat)
Category and Applications: See Varisoft 110
Form: Liquid/solids
Composition: 77% solids

Varisoft SDC. Sherex
Chemical Description: Stearyl dimethylbenzyl ammonium chloride
Category: Softener, cationic
Applications: Softener hair conditioning cream rinse
Form: Paste
Composition: 25% conc.

Varonic 320. Sherex
Chemical Description: Polyoxyethylene oleyl alcohol
Category: Emulsifier, lubricant, nonionic

Varonic 400DL. Sherex
Chemical Description: PEG 400 dilaurate
Category: Emulsifier, nonionic
Applications: Emulsifier for cosmetics
Form: Liquid
Composition: 100% active ingred.

Varonic 400MS. Sherex
Chemical Description: PEG 400 monostearate
Category: Emulsifier, thickener, nonionic
Applications: For emollients, creams, and lotions
Form: Solid
Composition: 100% active ingred.

Varonic 1000MS. Sherex
Chemical Description: PEG 1000 monostearate
Category: Solubilizer, humectant, nonionic surfactant
Form: Solid

Varonic 1800MS. Sherex
Chemical Description: PEG 1800 monostearate
Category: Solubilizer, humectant, nonionic surfactant
Form: Solid

Varonic K 202, K 205, K 205LC, K 210, K 210LC, K 215, K 215LC. Sherex
Chemical Description: Ethoxylated coco amines
Category: Antistat, emulsifier, dispersant, nonionic
Applications: Dispersant for oils,

pigments; antistatic agent, emulsifier
Form: Liquid
Composition: 100% active ingred.

Varonic L 202, L 205, L 205LC, L 230. Sherex
Chemical Description: Ethoxylated soya amines
Category: Emulsifier, antistat, dispersant, nonionic
Applications: Emulsifier for latex, dyes, oils
Form: Liquid
Composition: 100% active ingred.

Varonic L 230 (60%), L 230 (80%). Sherex
Chemical Description: Ethoxylated soya amines
Category and Applications: See Varonic L 202, etc.
Composition: 60 and 80% active respectively

Varonic LI-63, LI-67. Sherex
Chemical Description: Ethoxylated monoglycerides
Category: Modifier, nonionic
Applications: Modifier for sodium lauryl ether sulfate to produce low-irritation shampoos and other personal care products

Varonic T 202, T 205, T 205LC, T 210, T 215, T 215LC. Sherex
Chemical Description: Ethoxylated tallow amines
Category: Emulsifier, antistat, dispersant, nonionic
Applications: Emulsifier for latex, dyes, oils
Form: Liquid
Composition: 100% active ingred.

Varonic U 202, U 205, U 205LC, U 215, U 250. Sherex
Chemical Description: Ethoxylated hydrogenated tallow amines
Category: Antistat, emulsifier, dispersant, dye intermediate, nonionic
Form: Semiliquid to solids
Composition: 100% active ingred.

Varox 185E, 185 PH. Sherex
Chemical Description: Alkyl ether amine oxide
Category: Foam booster, detergent, nonionic

Varox 365. Sherex
Chemical Description: Dimethyl lauryl amine oxide
Category: Foam booster, detergent, nonionic

Varox 1770. Sherex
Chemical Description: Coco amido propyl dimethyl amine oxide
Category: Foam booster, detergent, nonionic
Form: Liquid
Composition: 50% active ingred.

Varstat 55. Sherex
Chemical Description: Quaternary
Category: Antistat, cationic
Applications: Antistatic agent for treating fibers and textiles
Properties: Hydrophilic

Varstat 66. Sherex
Chemical Description: Ethyl bis (poly hydroxyethyl) alkyl ammonium ethyl sulfate
Category: Antistat, cationic
Applications: Antistatic agent for treating fibers and textiles
Form: Liquid
Composition: 68–76% conc.

Varstat K22, T22. Sherex
Chemical Description: Ethoxylated fatty amine
Category: Antistat, cationic
Applications: Tertiary amine-type antistatic agent for internal use in polymeric resins
Form: Liquid
Composition: 100% conc.

Varsulf 60L. Sherex
Chemical Description: Triethanolamine alkylaryl sulfonate
Category: Detergent, foaming agent, anionic
Applications: General purpose detergent and foamer
Form: Liquid
Composition: 60% conc.

Veegum HS. R.T. Vanderbilt Co.
Chemical Description: Magnesium aluminum silicate
Category: Thickening and suspending agent
Applications: Developed for improved electrolyte compatibility when used as a thickening and suspending agent
Form: Small flakes; Color: White
Properties: Solubility: Insol. in water and alcohol; swells many times original volume in water; Visc.: 150 cps ± 50% (5.5% aq. disp.); pH: ≈ 9 (5% aq. disp.)
Composition: 92% min. conc.
Containers: Fiber drums
Warnings and Caution: Store under dry conditions

Versilan MX 112. Diamond Shamrock Europe/Process Chem. Div.
Chemical Description: Modified alcohol ethoxylate
Category: Solvent, nonionic/anionic
Applications: Used in solvent and water-based degreasers, affording fast phase separation, reduced cleaning times, and cleaner effluent
Form: Liquid

Versilan MX 123. Diamond Shamrock Europe/Process Chem. Div.
Chemical Description: Modified alcohol ethoxylate
Category: Nonionic surfactant
Applications: Electrolyte-sol. surfactant for use in highly built liquids
Form: Liquid

Versilan MX 134. Diamond Shamrock Europe/Process Chem. Div.
Chemical Description: Modified ethylene oxide propylene oxide copolymer
Category: and Applications: See Versilan MX 123
Form: Liquid

Versilan MX 145. Diamond Shamrock Europe/Process Chem. Div.
Chemical Description: Modified

quaternary ammonium compound
Category: Cationic surfactant
Applications: Concentrate for vehicle cleaning formulations, esp. removal of traffic film
Form: Liquid

Versilan MX 167. Diamond Shamrock Europe/Process Chem. Div.
Chemical Description: Alcohol ether sulfate/nonionic blend
Category: Anionic/nonionic surfactant
Applications: Shampoo and bubble bath concentrate
Form: Liquid

Victawet 12. Stauffer Chem. Co.
Chemical Description: Ethylene oxide reaction product with 2-ethyl hexanol and P_2O_5
Category: Wetting agent, penetrant, dispersant, stabilizer, nonionic
Applications: Low foaming surfactant for package dyeing of nylon, acid-type cleaners, emulsion polymerization, starch coatings
Form: Liquid; Color: 500 max. (APHA)
Properties: Solubility: Sol. in alcohol, acetone, toluene; disp. in water (forms milky sol'n.); Density: 9.34 lb/gal; Sp. gr.: 1.121; Visc.: 91.2 cs (100 F); B.P.: 325 F; Flash Pt.: 250 F (COC); Pour Pt.: –48 F; Acid No.: 0.1 max.; Stability: Effective in both acid and alkaline sol'ns.; disp. in 5% sulfuric acid and sodium hydroxide; Surface Tension: 29.4 dynes/cm^2 (0.2% conc.); pH: 7.0–7.4 (0.5% sol'n.)
Composition: 100% active ingred.
Toxicity: Irritating to skin, eyes, and mucous membranes
Containers: Phenolic-lined 5 gal pail or 55 gal drum

Victawet 35B. Stauffer Chem. Co.
Chemical Description: 2-Ethyl hexanol P_2O_5, NaOH reaction product
Category: Wetting agent, penetrant, dispersant, stabilizer, anionic
Applications: Low-foaming surfactant
Form: Soft paste; Color: Light tan to white
Properties: Solubility: Sol. in dist. water, red oil, ethylene glycol; disp. in hard water, 5% NaOH, toluene, corn oil, cottonseed oil, and alcohol; Density: 9.7 lb/gal; Sp. gr.: 1.17; Visc.: > 100,000 cps; B.P.: ≈ 200 F; Flash Pt.: 200 F (COC); Stability: Slowly hydrolyzes (in hours) with strong aq. acid; Surface Tension: 26.2 dynes/cm^2 (0.2% sol'n. @ 27 C); pH: 6.9–7.4 (0.5% sol'n.)
Composition: 68–72% active
Toxicity: May cause local irritation on contact with skin, eyes, and mucous membranes
Containers: Polyethylene-lined fiber cartons

Victawet 58B. Stauffer Chem. Co.
Chemical Description: Capryl alcohol, P_2O_5, NaOH reaction product

Category: Wetting agent, penetrant, dispersant, stabilizer, anionic
Applications: Low-foaming surfactant
Form: Soft paste; Color: Light tan to white
Properties: Solubility: Sol. in dist. water, red oil, ethylene glycol, 5% Na_2SO_4; disp. in hard water, 5% NaOH, toluene, cottonseed oil, alcohol; Density: 9.77 lb/gal; Sp. gr.: 1.121; Visc.: > 100,000 cps; Flash Pt.: 180 F (COC); Stability: Slowly hydrolyzes (in hours) with strong aq. acid; Surface Tension: 29.0 dynes/cm^2 (0.2% sol'n. @ 27 C); pH: 6.8–7.3 (5% sol'n.)
Composition: 68–72% active
Toxicity: May cause local irritation on contact with skin, eyes, and mucous membranes
Containers: Polyethylene-lined fiber cartons

Vifcoll CCN-40. Nikko Chem. Co., Ltd.
Chemical Description: 30% aq. sol'n. of Vifcoll CCN-40 powder
Category: Detergent, emulsifier
Applications: Preparation of cosmetics, pharmaceuticals, home-use products incl. shampoos, kitchen utensil cleansing agents, face and hand cleansers, cleaner for fruits or vegetables, emulsified cosmetic, pharmaceutical, or household products
Form: Liquid; Color: Pale yellow
Properties: m.w.: ≈ 600; Stability: Senstive to water hardness; pH: 5.5–7.5 (10% sol'n.)
Composition: 30% active ingred.

Vifcoll CCN-40 Powder. Nikko Chem. Co., Ltd.
Chemical Description: Sodium salt of N-cocoyl collagen peptide
Category and Applications: See Vifcoll CCN-40
Form: Powder; Color: White or pale yellow
Properties: m.w.: ≈ 600; Stability: Aq. sol'ns. are sensitive to water hardness; pH: 5.5–7.5 (3% aq. sol'n.)
Warnings and Caution: Hygroscopic—avoid moisture during handling and storage

Viscasil 5M. General Electric
Chemical Description: Silicone dimethyl polysiloxane
Category: Nonionic
Applications: Useful for polishes, window cleaners, lube sprays, mold release, hair care products
Form: Clear liquid; Color: Colorless; Odor: Essentially odorless
Properties: Solubility: Negligible water solubility; Sp. gr.: 0.98; Flash Pt.: >500 F (TOC); Stability: Generally stable; thermal decomposition at temp. > 500 F will produce CO_2 and SiO_2
Composition: 100% active ingred.
Toxicity: May cause minor eye irritation on eye contact

Viscasil 10M. General Electric
Chemical Description: Silicone dimethyl polysiloxane
Category and Applications: See Viscasil 5M
Form: Clear liquid; Color: Colorless; Odor: Essentially odorless

Properties: Solubility: Negligible water-sol.; Sp. gr.: 0.98; Flash Pt.: > 500 F (TOC); Stability: Generally stable; combustion will produce CO, CO_2, and SiO_2

Viscasil 60M. General Electric
Chemical Description: Silicone dimethyl polysiloxane
Category and Applications: See Viscasil 5M
Form: Clear liquid; Color: Colorless; Odor: Essentially odorless
Properties: Solubility: Negligible water-sol.; Sp. gr.: 0.98; Flash Pt.: > 600 F (TOC); Stability: Generally stable; thermal decomposition at temp. > 600 F will produce CO_2 and SiO_2
Composition: 100% active ingred.
Toxicity: May cause minor eye irritation on eye contact

Volpo Series. Croda Chem. Ltd.
Chemical Description: Alkylpolyglycolethers
Category: Wetting, scouring, leveling agents, dispersants, solubilizers, emulsifiers, antistats, plasticizers, nonionic
Applications: Series of biodegradable surfactants varying in alcohol base and degree of ethoxylation; used as emulsifiers in cosmetics and pharmaceuticals (esp. CS); solubilizers for perfumes and essential oils; wetting agent and detergent in hard surface cleaners (esp. T and 25D); emulsifier and dispersants for solvent cleaners and degreasers (O5, 25D5); liquid detergents, car washing compounds, metal, dairy, and glass cleaners; intermediates for the production of iodophors, alkyl ether phosphates, and sulfated detergents; in metal processing compounds, rolling and cutting oils (25D and O with 3-5 EO), metal pickling (T10, T15); emulsifiers in textile finishes, lubricants, antistats, scouring, wetting, dye-leveling, lime soap dispersants; wetting and detergent properties in textile processing (T), bleaching and carbonizing of wool and scouring, desizing, and Kier boiling (T10); emulsifiers in textile oils, fiber lubricants (O, C, 25D); herbicides and pesticides; wetting and dust control (T10, 25D10); oil slick dispersants and marine tank cleaners; paper processing (T10); emulsion polymerization and latex stabilization (CS, O, and 25D20 for vinyl resins and rubber latex); polishes (CS, O, 15, and 20); adhesives (T10, 25D10)
Form and Properties: See individual listings
Composition: 97% min. active
Toxicity: May cause eye irritation
Containers: Store in resin-lined or stainless steel containers

Volpo 25D3. Croda Chem. Ltd.
Chemical Description: POE 3 $C_{12}-C_{15}$ ether
Form: Liquid; Color: Water-white
Properties: Solubility: Sol. in ethanol, mineral oil, kerosene, trichloroethylene, oleic acid, arachis oil, oleyl alcohol; partly sol. in butyl stearate; HLB: 7.8; Acid Value:

1.0 max.; Hydroxyl Value: 165–170; Iodine Value: 2.0 max.; Stability: Stable to strong mineral acids and alkalis; pH: 6.0–7.5 (3% sol'n.); Biodegradable

Volpo 25D5. Croda Chem. Ltd.
Chemical Description: POE 5 C_{12}–C_{15} ether
Form: Opaque liquid; Color: Off-white
Properties: Solubility: Sol. in ethanol, trichloroethylene, oleic acid; partly sol. in water, mineral oil, butyl stearate, arachis oil; HLB: 10.3; Acid Value: 1.0 max.; Hydroxyl Value: 130–140; Iodine Value: 1.0 max.; Stability: Stable to strong mineral acids and alkalis; Surface Tension: 29.5 dynes/cm (0.1%); pH: 6.0–7.5 (3%); Biodegradable

Volpo 25D10. Croda Chem. Ltd.
Chemical Description: POE 10 C_{12}–C_{15} ether
Form: Soft paste; Color: Off-white
Properties: Solubility: Sol. in water, ethanol, trichloroethylene; partly sol. in mineral oil, kerosene, butyl stearate, oleic acid, arachis oil, oleyl alcohol; HLB: 13.6; Cloud Pt.: 83 C (1% aq.); Acid Value: 1.0 max.; Hydroxyl Value: 85–95; Iodine Value: 1.0 max.; Stability: Stable to strong mineral acids and alkalis; Surface Tension: 33.5 dynes/cm (0.1% aq.); pH: 6.0–7.5 (3%); Biodegradable

Volpo 25D15. Croda Chem. Ltd.
Chemical Description: POE 15 C_{12}–C_{15} ether
Form: Soft solid; Color: White
Properties: Solubility: Sol. in water, ethanol, trichloroethylene; partly sol. in kerosene, butyl stearate, oleic acid, oleyl alcohol; HLB: 15.3; Cloud Pt.: 73 C (brine); Acid Value: 1.0 max.; Hydroxyl Value: 65–75; Iodine Value: 1.0 max.; Stability: Stable to alkalis and strong mineral acids; Surface Tension: 36.5 dynes/cm (0.1% aq.); pH: 6.0–7.5 (3%); Biodegradable

Volpo 25D20. Croda Chem. Ltd.
Chemical Description: POE 25 C_{12}–C_{15} ether
Form: Soft solid; Color: White
Properties: Solubility: Sol. in water, ethanol, trichloroethylene; partly sol. in kerosene, butyl stearate, oleic acid, oleyl alcohol; HLB: 16.1; Cloud Pt.: 80 C (brine); Acid Value: 1.0 max.; Hydroxyl Value: 50–60; Iodine Value: 1.0 max.; Stability: Stable to alkalis and strong mineral acids; Surface Tension: 38.5 dynes/cm (0.1% aq.); pH: 6.0–7.5 (3%); Biodegradable

Volpo CS3. Croda Chem. Ltd.
Chemical Description: POE 3 ceto stearyl ether
Form: Soft waxy solid; Color: Off-white
Properties: Solubility: Sol. in ethanol, kerosene, trichloroethyl-

ene, oleic acid, oleyl alcohol; partly sol. in butyl stearate, arachis oil; almost sol. in mineral oil; HLB: 7.3; Acid Value: 1.0 max.; Hydroxyl Value: 145–155; Iodine Value: 2.0 max.; Stability: Stable to alkalis and strong mineral acids; pH: 6.0–7.5 (3%); Biodegradable

Volpo CS5. Croda Chem. Ltd.
Chemical Description: POE 5 ceto stearyl ether
Form: Soft waxy solid; Color: Off-white
Properties: Solubility: Sol. in ethanol, trichloroethylene, oleic acid; partly sol. in kerosene, butyl stearate, arachis oil; almost sol. in mineral oil, oleyl alcohol; HLB: 9.5; Acid Value: 1.0 max.; Hydroxyl Value: 120–130; Iodine Value: 2.0 max.; Stability: Stable to alkalis and strong mineral acids; Surface Tension: 33.0 dynes/cm (0.1% aq.); pH: 6.0–7.5 (3%); Biodegradable

Volpo CS10. Croda Chem. Ltd.
Chemical Description: POE 10 ceto stearyl ether
Form: Soft waxy solid; Color: Off-white
Properties: Solubility: Sol. in ethanol, trichloroethylene, oleic acid; partly sol. in water, mineral oil, kerosene, butyl stearate, arachis oil; almost sol. in oleyl alcohol; HLB: 12.9; Cloud Pt.: 68 C (1% aq.); Acid Value: 1.0 max.; Hydroxyl Value: 85–95; Iodine Value: 2.0 max.; Stability: Stable to alkalis and strong mineral acids; Surface Tension: 34.5 dynes/cm (0.1% aq.); pH: 6.0–7.5 (3%); Biodegradable

Volpo CS15. Croda Chem. Ltd.
Chemical Description: POE 15 ceto stearyl ether
Form: Waxy solid; Color: Off-white
Properties: Solubility: Sol. in ethanol, trichloroethylene, oleic acid; partly sol. in arachis oil; almost sol. in water, oleyl alcohol; HLB: 14.6; Cloud Pt.: 94 C (1% aq.); Acid Value: 1.0 max.; Hydroxyl Value: 65–75; Iodine Value: 2.0 max.; Stability: Stable to alkalis and strong mineral acids; Surface Tension: 35.5 dynes/cm (0.1% aq.); pH: 6.0–7.5 (3%); Biodegradable

Volpo CS20. Croda Chem. Ltd.
Chemical Description: POE 20 ceto stearyl ether
Form: Hard waxy solid; Color: Off-white
Properties: Solubility: Sol. in water, ethanol, trichloroethylene, oleic acid; partly sol. in arachis oil, oleyl alcohol; HLB: 15.6; Cloud Pt.: 78 C (brine); Acid Value: 1.0 max.; Hydroxyl Value: 50–60; Iodine Value: 2.0 max.; Stability: Stable to alkalis and strong mineral acids; Surface Tension: 41.5 dynes/cm (0.1% aq.); pH: 6.0–7.5 (3%); Biodegradable

Volpo O3. Croda Chem. Ltd.
Chemical Description: POE 3 oleyl ether

Form: Liquid; Color: Pale straw
Properties: Solubility: Sol. in ethanol, mineral oil, kerosene, trichloroethylene, oleic acid, arachis oil, oleyl alcohol; partly sol. in butyl stearate; HLB: 6.7; Acid Value: 1.0 max.; Hydroxyl Value: 140–150; Iodine Value: 50–52; Stability: Stable to alkalis and strong mineral acids; pH: 6.0–7.5 (3%); Biodegradable

Volpo O5. Croda Chem. Ltd.
Chemical Description: POE 5 oleyl ether
Form: Liquid; Color: Pale straw
Properties: Solubility: Sol. in ethanol, trichloroethylene, oleic acid; partly sol. in water, mineral oil, kerosene, butyl stearate, oleyl alcohol; almost sol. in arachis oil; HLB: 9.0; Acid Value: 1.0 max.; Hydroxyl Value: 115–125; Iodine Value: 42–48; Stability: Stable to alkalis and strong mineral acids; Surface Tension: 31.0 dynes/cm (0.1% aq.); pH: 6.0–7.5 (3%); Biodegradable

Volpo O10. Croda Chem. Ltd.
Chemical Description: POE 10 oleyl ether
Form: Paste; Color: Pale straw
Properties: Solubility: Sol. in water, ethanol, trichloroethylene; partly sol. in mineral oil, kerosene, butyl stearate, oleic acid, arachis oil, oleyl alcohol; HLB: 12.4; Cloud Pt.: 57 C (1% aq.); Acid Value: 1.0 max.; Hydroxyl Value: 80–90; Iodine Value: 28–33; Stability: Stable to alkalis and strong mineral acids; Surface Tension: 33.0 dynes/cm (0.1% aq.); pH: 6.0–7.5 (3%); Biodegradable

Volpo O15. Croda Chem. Ltd.
Chemical Description: POE 15 oleyl ether
Form: Paste; Color: Pale straw
Properties: Solubility: Sol. in water, ethanol, trichloroethylene; partly sol. in mineral oil, kerosene, butyl stearate, oleic acid, arachis oil, oleyl alcohol; HLB: 14.2; Cloud Pt.: 91 C (1% aq.); Acid Value: 1.0 max.; Hydroxyl Value: 60–70; Iodine Value: 20–25; Stability: Stable to alkalis and strong mineral acids; Surface Tension: 34.5 dynes/cm (0.1% aq.); pH: 6.0–7.5 (3%); Biodegradable

Volpo O20. Croda Chem. Ltd.
Chemical Description: POE 20 oleyl ether
Form: Soft solid; Color: Pale straw
Properties: Solubility: Sol. in water, ethanol, trichloroethylene; partly sol. in butyl stearate, oleic acid, arachis oil; HLB: 15.5; Cloud Pt.: 78 C (brine); Acid Value: 1.0 max.; Hydroxyl Value: 50–60; Iodine Value: 15-20; Stability: Stable to alkalis and strong mineral acids; Surface Tension: 41.0 dynes/cm (0.1% aq.); pH: 6.0–7.5 (3%); Biodegradable

Volpo T3. Croda Chem. Ltd.
Chemical Description: POE 3 tridecyl ether
Form: Liquid; Color: Water-white
Properties: Solubility: Sol. in

ethanol, mineral oil, kerosene, trichloroethylene, oleic acid, arachis oil, oleyl alcohol; partly sol. in butyl stearate; HLB: 8.0; Acid Value: 1.0 max.; Hydroxyl Value: 170–180; Iodine Value: 1.0 max.; Stability: Stable to alkalis and strong mineral acids; pH: 6.0–7.5 (3%); Biodegradable

Volpo T5. Croda Chem. Ltd.
Chemical Description: POE 5 tridecyl ether
Form: Opaque liquid; Color: Water-white
Properties: Solubility: Sol. in ethanol, kerosene, trichloroethylene, oleic acid; partly sol. in water, mineral oil, butyl stearate, arachis oil; almost sol. in oleyl alcohol; HLB: 10.4; Acid Value: 1.0 max.; Hydroxyl Value: 135–145; Iodine Value: 1.0 max.; Stability: Stable to alkalis and strong mineral acids; Surface Tension: 29.0 dynes/cm (0.1% aq.); pH: 6.0–7.5 (3%); Biodegradable

Volpo T10. Croda Chem. Ltd.
Chemical Description: POE 10 tridecyl ether
Form: Soft paste; Color: Off-white
Properties: Solubility: Sol. in water, ethanol, trichloroethylene; partly sol. in kerosene, butyl stearate, oleic acid, arachis oil, oleyl alcohol; HLB: 13.7; Cloud Pt.: 61 C (1% aq.); Acid Value: 1.0; Hydroxyl Value: 90–100; Iodine Value: 1.0 max.; Stability: Stable to alkalis and strong mineral acids; Surface Tension: 29.5 dynes/cm (0.1% aq.); pH: 6.0–7.5 (3%); Biodegradable

Volpo T15. Croda Chem. Ltd.
Chemical Description: POE 15 tridecyl ether
Form: Soft paste; Color: Off-white
Properties: Solubility: Sol. in water, ethanol, trichloroethylene; partly sol. in butyl stearate, oleic acid, arachis oil, oleyl alcohol; HLB: 15.5; Cloud Pt.: 94 C (1% aq.); ; Acid Value: 1.0 max.; Hydroxyl Value: 65–75; Iodine Value: 1.0 max.; Stability: Stable to alkalis and strong mineral acids; Surface Tension: 32.0 dynes/cm (0.1% aq.); pH: 6.0–7.5 (3%); Biodegradable

Volpo T20. Croda Chem. Ltd.
Chemical Description: POE 20 tridecyl ether
Form: Soft solid; Color: Off-white
Properties: Solubility: Sol. in water, ethanol, trichloroethylene; partly sol. in butyl stearate, oleic acid, oleyl alcohol; HLB: 16.3; Cloud Pt.: 75 C (brine); Acid Value: 1.0 max.; Hydroxyl Value: 50–55; Iodine Value: 1.0 max.; Stability: Stable to alkalis and strong mineral acids; Surface Tension: 36.5 dynes/cm (0.1% aq.); pH: 6.0–7.5 (3%); Biodegradable

W

Wecobee FS. PVO International
Chemical Description: Triglyceride derived from lauric-based fats
Applications: Used in cosmetics and pharmaceuticals for cocoa butter replacement, lipsticks and glossers, pomades, solid fragrances, antiperspirant sticks, suppositories, emollient creams and lotions; used in food industry in chocolate-type coatings, vegetable dairy products, e.g., coffee whiteners, whipped toppings, sour dressings, dips, etc.; good mold release characteristics
Form: Solid; Color: 30/3.0 (Lovibond); Odor and Taste: Bland
Properties: Visc.: 13.5 cps (150 F); M.P.: 107–109 F (Wiley); Sapon. Value: 230–250; Iodine Value: 4; Stability: Good

Wecobee FW. PVO International
Chemical Description: Triglyceride derived from lauric-based fats
Applications: See Wecobee FS
Form: Solid; Color: 30/3.0 (Lovibond); Odor and Taste: Bland
Properties: Visc.: 14.0 cps (150 F); M.P.: 101–103 F (Wiley); Sapon. Value: 230–250; Iodine Value: 4; Stability: Good

Wecobee M. PVO International
Chemical Description: Triglyceride derived from lauric-based fats
Applications: See Wecobee FS
Form: Solid; Color: 30/3.0 (Lovibond); Odor and Taste: Bland
Properties: Visc.: 13.5 cps (150 F); M.P.: 96–98 F (Wiley); Sapon. Value: 238–250; Iodine Value: 4; Stability: Good

Wecobee R. PVO International
Chemical Description: Triglyceride derived from lauric-based fats
Applications: See Wecobee FS
Form: Solid; Color: 30/3.0 (Lovibond); Odor and Taste: Bland
Properties: Visc.: 14.0 cps (150 F); M.P.: 101–103 F (Wiley); Sapon. Value: 230–250; Iodine Value: 4; Stability: Good

Wecobee S. PVO International
Chemical Description: Triglyceride derived from lauric-based fats

Applications: See Wecobee FS
Form: Solid; Color: 30/3.0 (Lovibond); Odor and Taste: Bland
Properties: Visc.: 14.0 cps (150 F); M.P.: 113–115 F (Wiley); Sapon. Value: 230–250; Iodine Value: 4; Stability: Good

Wecobee SS. PVO International
Chemical Description: Triglyceride derived from lauric-based fats
Applications: See Wecobee FS
Form: Solid; Color: 30/3.0 (Lovibond); Odor and Taste: Bland
Properties: Visc.: 13.5 cps (150 F); M.P.: 116–118 F (Wiley); Sapon. Value: 230–250; Iodine Value: 4; Stability: Good

Wecobee W. PVO International
Chemical Description: Triglyceride derived from lauric-based fats
Applications: See Wecobee FS
Form: Solid; Color: 30/3.0 (Lovibond); Odor and Taste: Bland
Properties: Visc.: 13.5 cps (150 F); M.P.: 94–96 F (Wiley); Sapon. Value: 240–254; Iodine Value: 4; Stability: Good

Wecotop 1013. PVO International
Chemical Description: Spray-beaded vegetable fat derived from soya and palm oils
Category: Food emulsifier
Applications: Used in dry blends such as pie crust mixes, cake mixes, doughnut mixes, gravy mixes, breading mixes; improves stability in icings; improves standup properties in caramels; pan release agent
Form: Solid; Color: 20/2.0 (Lovibond); Taste: Bland
Properties: M.P.: 120–125 F; Iodine Value: 45; Stability: Stable towards oxidative and hydrolytic breakdown

White Swan. Croda Chem. Ltd.
Chemical Description: Anhydrous lanolin BP
Category: Emollient, emulsifier, conditioner, moisturizer
Applications: W/O emulsifier and emollient for cosmetics; esp. useful in skin and hair care conditioning products and make up compositions; superfatting agent in soap; aids dispersion of pigments into anhydrous systems
Form: Unctuous mass; Color: Yellow
Properties: Lipophilic

Witbreak DGE-60, DGE-75, DGE-85, DGE-128A, DGE-156A, DGE-159A, DGE-169, DGE-182. Witco Chemical/Organics Div.
Chemical Description: Glycol esters
Category: Demulsifier
Applications: Demulsifier for the petroleum industry and for waste disposal of processing oils in metal processing

Witbreak DPG-15, DPG-25, DPG-40, DPG-482, DTG-62. Witco Chemical/Organics Div.
Chemical Description: Polyoxyalkylene glycols
Category: Demulsifier
Applications: Demulsifier for the

petroleum industry and for waste disposal of processing oils in metal processing

Witbreak DRA-20, DRA-21, DRA-22, DRA-50, DRB-11, DRB-127, DRB-271, DRB-272, DRC-163, DRC-164, DRC-165, DRC-167, DRC-168, DRC-232. Witco Chemical/Organics Div.
Chemical Description: Oxyalkylated phenolic resins
Category: Demulsifier
Applications: Demulsifiers for the petroleum industry and for waste disposal of processing oils in metal processing industry

Witbreak GBG Series. Witco Chemical/Organics Div.
Chemical Description: Polyoxyalkylated modified resins
Category: Demulsifier
Applications: Demulsifier for the petroleum industry and for waste disposal of processing oils in metal processing

Witbreak RTC-323. Witco Chemical/Organics Div.
Chemical Description: Polymeric amine
Category: Demulsifier, anionic
Applications: Reverse demulsifier concentrate; O/W demulsifier; used in petroleum industry and for waste disposal of processing oils in metal processing
Form: Liquid
Properties: Solubility: Water-sol.

Witbreak RTC-1315, RTC-1316. Witco Chemical/Organics Div.
Chemical Description: Polymeric amine salts
Category: Demulsifier, anionic
Applications: Reverse demulsifier concentrates; O/W demulsifier; used in petroleum industry and for waste disposal of processing oils in metal processing
Form: Liquid
Properties: Solubility: Water-sol.

Witbreak RTC-6010, RTC-6020, RTC-6030, RTC-6040. Witco Chemical/Organics Div.
Chemical Description: Polymeric amines
Category: Demulsifier
Applications: Reverse demulsifier in petroleum industry

Witcamide 61. Witco Chemical/Organics Div.
Chemical Description: Oleic isopropanolamide
Category: Conditioner, emulsifier, lubricant, nonionic
Applications: Cosmetics and toiletries

Witcamide 70. Witco Chemical/Organics Div.
Chemical Description: Fatty amide
Category: Opacifier, conditioner, lubricant, gelling agent, mold release agent, binder, nonionic
Applications: Cosmetics and toiletries
Form: Wax

Witcamide 82. Witco Chemical/Organics Div.
Chemical Description: Cocodiethanolamide
Category: Foam stabilizer, viscosity modifier, lubricant, conditioner, emulsifier, wetting agent, penetrant, dye dispersant, scouring aid, antistat, nonionic
Applications: Cosmetics and toiletries, base for scrub soap; general thickening agent; industrial surfactant; metal processing; textile surfactant
Form: Liquid
Composition: 100% conc.

Witcamide 272. Witco Chemical/Organics Div.
Chemical Description: Modified alkanolamide
Category: Detergent, emulsifier, lubricant, wetting agent, penetrant, dye dispersant, scouring aid, antistat, nonionic
Applications: Industrial surfactant; metal processing; drilling fluid surfactant; textile surfactant
Form: Liquid
Properties: Solubility: Water-sol.; sol. in high concs. of electrolytes; Stability: Stable to strong acids and alkalis
Composition: 90% conc.

Witcamide 511. Witco Chemical/Organics Div.
Chemical Description: Alkanolamide
Category: Emulsifier, dispersant, lubricant, wetting agent, penetrant, dye dispersant, scouring aid, antistat, nonionic
Applications: W/O emulsifier for aerosol and industrial applications; dispersant for oil or solvent systems; drilling mud surfactant for petroleum industry; metal processing surfactant; textile surfactant; functions over wide pH range
Form: Liquid and paste

Witcamide 512. Witco Chemical/Organics Div.
Chemical Description: Blend of a polyfunctional anionic and a modified alkanolamide
Category: Detergent, emulsifier, lubricant, wetting agent, penetrant, dye dispersant, scouring aid, antistat, anionic
Applications: Industrial surfactant; metal processing; drilling mud surfactant in petroleum industry; textile surfactant
Form: Liquid
Properties: Solubility: Water-sol.; sol. in high concs. of electrolytes; Stability: Stable to strong acids and alkalis

Witcamide 823/10. Witco Chem. Ltd., Israel
Chemical Description: Isostearic alkanolamide
Category: Conditioner, lubricant, nonionic
Applications: Hair conditioner and lubricant esp. for spray application; improves slip and combout properties in hair care products; adds lubricity to creams and lotions
Form: Liquid; Color: Light amber

Properties: Solubility @ 10%: Sol. in aromatic solvents, chlorinated hydrocarbons, mineral spirits, kerosene, mineral oil, vegetable oil, ethyl alcohol; disp. in water (gel); Cloud Pt.: −5 C; Acid No.: 10 max.; Alkali No.: 60 max.; pH: 9.5 ± 0.5 (10% disp.)
Composition: 100% active ingred.

Witcamide 1017. Witco Chemical/ Organics Div.
Chemical Description: Modified alkanolamide
Category: Drilling mud surfactant, nonionic
Form: Liquid
Composition: 100% conc.

Witcamide 5130. Witco Chemical/ Organics Div.
Chemical Description: Modified alkanolamide
Category: Emulsifier, viscosity modifier, lubricant, dispersant, suspending agent, wetting agent, penetrant, dye dispersant, scouring aid, antistat, foam stabilizer, anionic/nonionic
Applications: Aerosol formulations, textile surfactant; metal processing; industrial surfactant
Form: Liquid
Properties: Solubility: Water-sol.
Composition: 98% conc.

Witcamide 5133. Witco Chemical/ Organics Div.
Chemical Description: Alkanolamide
Category: Foam stabilizer, viscosity modifier, lubricant, conditioner, emulsifier, wetting agent, penetrant, dye dispersant, scouring aid, antistat, nonionic
Applications: Cosmetics and toiletries, base for scrub soap; industrial surfactant; metal processing; textile surfactant
Form: Liquid
Composition: 100% conc.

Witcamide 5138. Witco Chemical/ Organics Div.
Chemical Description: Alkanolamide
Category: Emulsifier, dispersant, detergent, lubricant, wetting agent, penetrant, dye dispersant, scouring aid, antistat, nonionic
Applications: Industrial surfactant; dry cleaning agent; wetting agent in oil-based systems; sludge dispersant in fuel oils; metal processing; textile surfactant
Form: Liquid
Properties: Solubility: Water-disp., oil-sol.

Witcamide 5140. Witco Chemical/ Organics Div.
Chemical Description: Modified alkanolamide
Category: Detergent, foam stabilizer, air-entraining agent, wetting agent, dispersant, nonionic
Applications: Industrial cleaning agent; foam stabilizer for drilling mud surfactants and in aq. media; air-entraining agent
Form: Liquid

Witcamide 5145M. Witco Chemical/ Organics Div.
Chemical Description: Modified

alkanolamide
Category: Detergent, anionic/nonionic
Applications: Detergent for built liquid detergent products; dry cleaning detergent
Form: Liquid
Composition: 100% conc.

Witcamide 5168. Witco Chemical/Organics Div.
Chemical Description: Modified alkanolamide
Category: Cationic surfactant
Applications: Film-forming alkanolamide containing a tertiary amine group for cosmetics, textiles, detergents, industrial uses
Form: Liquid
Composition: 100% conc.

Witcamide 5195. Witco Chemical/Organics Div.
Chemical Description: Alkanolamide
Category: Foam stabilizer, viscosity modifier, lubricant, conditioner, foaming agent, lubricant, emulsifier, wetting agent, penetrant, dye dispersant, scouring aid, antistat, thickening agent, nonionic
Applications: Cosmetics and toiletries; industrial foaming agent and foam stabilizer; metal processing; textile surfactant; aerosol formulations

Witcamide AL69-58. Witco Chemical/Organics Div.
Chemical Description: Alkanolamide
Category: Emulsifier, lubricant, wetting agent, penetrant, dye dispersant, scouring aid, antistat, cationic/nonionic
Applications: W/O emulsifier in industrial applications; metal processing; textile surfactant
Form: Liquid
Composition: 100% conc.

Witcamide MAS, MM. Witco Chemical/Organics Div.
Chemical Description: Fatty amides
Category: Opacifier, conditioner, lubricant, gelling agent, mold release agent, binder, nonionic
Applications: Cosmetics
Form: Wax

Witcamine 209. Witco Chemical/Organics Div.
Chemical Description: 1-(2-Aminoethyl)-2-n-alkyl-2-imidazoline
Category: Corrosion inhibitor intermediate, emulsifier, lubricity agent, wetting agent, penetrant, dye dispersant, scouring aid, antistat, cationic
Applications: Corrosion inhibitor intermediate in petroleum industry and metal processing; textile surfactant
Form: Liquid

Witcamine 210. Witco Chemical/Orgnaics Div.
Chemical Description: Alkyl amidoamine
Category: Corrosion inhibitor intermediate, emulsifier, lubricity agent, cationic

Applications: Corrosion inhibitor intermediate in petroleum industry and for metal processing

Witcamine 211. Witco Chemical/Organics Div.
Chemical Description: Mixed 1-(2-aminoethyl)-2-imidazoline
Category and Applications: See Witcamine 209
Form: Liquid

Witcamine 235. Witco Chemical/Organics Div.
Chemical Description: 1-Polyaminoethyl-2-n-alkyl-2-imidazoline
Category and Applications: See Witcamine 209

Witcamine 3164. Witco Chemical/Organics Div.
Chemical Descrription: Alkylamidoamine
Category and Applications: See Witcamine 210

Witcamine AL42-12. Witco Chemical/Organics Div.
Chemical Description: Imidazoline
Category: Emulsifier, beading agent, antistat, corrosion inhibitor intermediate, lubricant, wetting agent, penetrant, dye dispersant, scouring aid, cationic
Applications: Industrial, textile, and metal processing surfactant
Form: Liquid

Witcamine PA-78B. Witco Chemical/Organics Div.
Chemical Description: Fatty imidazoline salt
Catetory: Dispersant, corrosion inhibitor, wetting agent, cationic
Applications: Dispersant for inorganic solids in organic media; corrosion inhibitor for halogenated solvents; pigment dispersant and wetting agent in oil-based systems; aerosol formulations
Form: Liquid
Properties: Solubility: Oil-sol.
Composition: 40% conc.

Witcamine RAD-0500, RAD-0515, RAD-1100, RAD-1110. Witco Chemical/Organics Div.
Chemical Description: Polyoxyethylated amines
Category: Corrosion inhibitor intermediate, emulsifier, scouring agent, softener, dye assistants, cationic/nonionic
Applications: Inhibitors for HCl in acid pickling, hydrogen sulfide, carbon dioxide, and organic acids; emulsifiers for acid-stable emulsions
Form: Liquid
Composition: 100% conc.

Witco 912. Witco Chemical/Organics Div.
Chemical Description: Blend of polyoxethylene carboxylic acid esters and sulfonates
Category: Dispersant, solubilizer
Applications: Pigment dispersant for aq. systems; solubilizer for methyl cellulose; paint surfactant

Witco 915. Witco Chemical/Organics Div.
chemical Description: Imidazoline

blend
Category: Dispersant, suspending agent, wetting and grinding aid, emulsifier, cationic
Applications: Pigment dispersant and suspending agent, wetting and grinding aid for oil-based paint systems; emulsifier for latex paints

Witco 916. Witco Chemical/Organics Div.
Chemical Description: Fatty acid ester
Category: Spreading agent
Applications: Spreading agent for aq. paint systems

Witco 918. Witco Chemical/Organics Div.
Chemical Description: Alkylaryl sulfonate
Category: Wetting and grinding aid, dispersant
Applications: Pigment wetting and grinding aid for paints; pigment dispersant for aq. systems; surfactant for universal tube colors
Properties: Solubility: Oil-sol.

Witco 934. Witco Chemical/Organics Div.
Chemical Description: Alkanolamide
Category: Dispersant, suspending agent, wetting and grinding aid
Applications: Pigment dispersant and suspending agent, wetting and grinding aid for oil-based paint systems; surfactant for latex paints and universal tube colors

Witco 936. Witco Chemical/Organics Div.
Chemical Description: Alkylphenol/ethylene oxide adduct
Category: Grinding and wetting aid
Applications: Pigment wetting and grinding aid in oil-based paint systems

Witco 942. Witco Chemical/Organics Div.
Chemical Description: Glycerol monooleate
Category: Emulsifier
Applications: Emulsifier for latex paints

Witco 960. Witco Chemical/Organics Div.
Chemical Description: Alkylphenol/ethylene oxide adduct
Category: Wetting and grinding aid, emulsifier
Applications: Pigment wetting and grinding aid for aq. paint systems; emulsifier for polymerization

Witco DTA-150, DTA-180. Witco Chemical/Organics Div.
Chemical Description: Fatty acid polymers
Category: Corrosion inhibitor intermediates
Applications: Petroleum industry

Witco Sulphonic Acid LABS. Witco Chem. Ltd., Israel
Category: Chemical intermediate, anionic

Applications: Intermediate for the production of detergents in the form of liquids, powders, and solids
Form: Low-viscosity liquid; Color: Light brown
Properties: m.w.: 325 ± 2 (avg.); Acid Value: 235 ± 5

Witcodet 2006. Witco Chemical/Organics Div.
Chemical Description: Formulated, biodegradable concentrate
Category: Detergent, anionic/nonionic
Applications: Detergent concentrate for dishwashing and carwashing formulations, and shampoos
Form: Liquid
Properties: Biodegradable
Composition: 70% conc.

Witcolate 58. Witco Chemical/Organics Div.
Chemical Description: Ammonium alcohol ether sulfate
Category: Detergent, wetting agent, penetrant, lubricant, emulsifier, dye dispersant, scouring aid, antistat
Applications: High-foaming liquid detergent base; detergent for metal cleaning prior to treatment such as coating and plating; textile surfactant

Witcolate 1276. Witco Chemical/Organics Div.
Chemical Description: Alcohol ether sulfate
Category: Foaming agent, detergent, wetting agent, penetrant, lubricant, emulsifier, dye dispersant, scouring aid, antistat, anionic
Applications: Foaming agent for wallboard manufacture and for the petroleum industry; detergent for metal cleaning prior to treatment such as coating and plating; textile surfactant; emulsion polymerization and latex stabilization
Form: Liquid

Witcolate D-510. Witco Chemical/Organics Div.
Chemical Description: Sodium 2-ethylhexyl sulfate
Category: Wetting agent, penetrant, anionic
Applications: Wetting agent and penetrant for industrial use and polymerization reactions
Form: Liquid
Properties: Solubility: Water-sol.

Witcomul 78. Witco Chemical/Organics Div.
Chemical Description: Sorbitan monotallate
Category: Emulsifier, thickening agent, emulsion stabilizer
Applications: W/O emulsifier; thickening agent and emulsion stabilizer for industrial applications; drilling fluid additive

Witcomul 1202, 1287, 1305, 1317. Witco Chemical/Organics Div.
Chemical Description: Blended surfactants
Applications: Dye carrier and solvent emulsifiers in textile industry

Witcomul 1557. Witco Chemical/Organics Div.
Chemical Description: Polyoxyethylene sorbitan monotallate
Applications: Drilling fluid additive in petroleum industry

Witcomul 3020. Witco Chemical/Organics Div.
Chemical Description: Complex polyfunctional surfactant
Applications: Drilling fluid additive for the petroleum industry

Witcomul 3107. Witco Chemical/Organics Div.
Chemical Description: Complex oxyethylene sulfate
Applications: Drilling fluid additive for the petroleum industry

Witcomul 3126. Witco Chemical/Organics Div.
Chemical Description: Amine ester
Applications: Drilling fluid additive for the petroleum industry

Witcomul 4016. Witco Chemical/Organics Div.
Chemical Description: Complex alkylate
Applications: Drilling fluid additive for the petroleum industry

Witcomul AM2-10C, CC37-45E, CC86-32G. Witco Chemical/Organics Div.
Chemical Description: Blended surfactants
Category: Emulsifier
Applications: Emulsifier for flame retardants in the textile industry

Witcomul CC38-10M. Witco Chemical/Organics Div.
Chemical Description: Blended emulsifier
Category: Emulsifier
Applications: Emulsifier for orthodichlorobenzene in industrial applications

Witcomul H-50A. Witco Chemical/Organics Div.
Chemical Description: Blended emulsifier
Category: Emulsifier, dispersant
Applications: Oil-soluble emulsifier for emulsion cleaning; sludge dispersant for fuel oils
Properties: Solubility: Oil-sol.

Witconate 79S. Witco Chemical/Organics Div.
Chemical Description: Amine sulfonate
Category: Emulsifier, wetting agent
Applications: Industrial surfactant
Properties: Solubility: Water-sol.

Witconate 93S. Witco Chemical/Organics Div.
Chemical Description: Amine sulfonate
Category: Emulsifier, wetting agent
Applications: Industrial surfactant
Properties: Solubility: Oil-sol.

Witconate 605T. Witco Chemical/Organics Div.
Chemical Description: Alkylaryl sulfonate
Category: Emulsifier, dispersant,

wetting agent, penetrant, lubricant, dye dispersant, scouring aid, antistat
Applications: O/W emulsifier, dispersant, oil wetting agent for industrial uses; additive for lube oils; textile surfactant
Properties: Solubility: Oil-sol.

Witconate 702. Witco Chemical/Organics Div.
Chemical Description: Ammonium salt of high molecular weight alkylaryl sulfonic acid
Category: Demulsifier
Applications: Demulsifier for the petroleum industry

Witconate 1840X. Witco Chemical/Organics Div.
Chemical Description: Fatty acid sulfonate
Category: Detergent, coupling agent, emulsifier, wetting agent, penetrant, lubricant, dye dispersant, scouring aid, antistat
Applications: Industrial surfactant for alkaline systems; used for metal processing; textile surfactant

Witconate 3009-15. Witco Chemical/Organics Div.
Chemical Description: Sodium alkylaryl ether sulfate
Category: Emulsifier, stabilizer
Applications: Latex emulsifier and stabilizer for polymerization reactions

Witconate CHB. Witco Chemical/Organics Div.
Chemical Description: High molecular weight alkylaryl sulfonic acid
Category: Demulsifier
Applications: Used in petroleum industry

Witconate P10-59. Witco Chemical/Organics Div.
Chemical Description: Amine salt of dodecyl benzene sulfonic acid
Category: Detergent, wetting agent, dispersant, emulsifier, solubilizer, anionic
Applications: Detergent and wetting agent for oil-based systems; dispersant in oil- and water-based systems; detergent and water solubilizer in dry-cleaning surfactants
Form: Liquid
Properties: Solubility: Oil-sol.; water-disp.

Witconate P-1073F. Witco Chemical/Organics Div.
Chemical Description: Amine salt of alkylaryl sulfonic acid
Category: Drilling fluid additive
Applications: Petroleum industry

Witconol 14, 18L. Witco Chemical/Organics Div.
Chemical Description: Polyglycerol fatty acid esters
Category: Emulsifier, lubricant, antifoaming agent, nonionic
Applications: W/O emulsifier, lubricant, and antifoaming agent for industrial use; also used in aerosol formulations; 18L used as solubilizer
Form: Liquid
Composition: 100% conc.

Witconol 171, 172. Witco Chemical/Organics Div.
Chemical Description: Polyalkylene glycol ethers
Category: Defoaming agent
Applications: Defoaming agents for soap and detergent foam control; low-foaming surfactants

Witconol 1206. Witco Chemical/Organics Div.
Chemical Description: Alkyl polyoxyethylene glycol ethers
Category: Wetting agent, viscosity and flow control agents
Applications: Low-foaming industrial wetting agent; latex viscosity and flow control agent for polymerization reactions

Witconol 1207. Witco Chemical/Organics Div.
Chemical Description: Alkyl polyoxyethylene glycol ether
Category: Wetting agent
Applications: Low-foaming industial wetting agent

Witconol APEM, APM, APS. Witco Chemical/Organics Div.
Chemical Description: Alkoxylated alcohols
Category: Synthetic oil, lubricant, emulsifier, wetting agent, penetrant, dye dispersant, scouring aid, antistat, nonionic
Applications: Synthetic oils for creams, lotions, and cosmetic gels; metal processing; textile surfactant
Form: Liquid

Witconol CA. Witco Chemical/Organics Div.
Chemical Description: Glycerol mono- and distearate, self-emulsifying
Category: Lubricant, bodying agent, emulsifier, opacifier, nonionic
Applications: Used in industrial, cosmetic, and aerosol formulations
Form: Wax
Composition: 100% conc.

Witconol CAD. Witco Chemical/Organics Div.
Chemical Description: Diethylene glycol stearate
Category: Lubricant, bodying agent, nonionic
Applications: Self-emulsifying lubricant and opacifying agent for cosmetics and toiletries
Form: Wax
Composition: 100% conc.

Witconol CD-17. Witco Chemical/Organics Div.
Chemical Description: Polypropoxylated polyol
Category: Synthetic oil, emulsifier, lubricant, antistat, foam modifier, nonionic
Applications: Synthetic oils for bath oil and hair preparations; cosmetic emulsifier; synthetic lubricant oil and foam modifier for aerosol formulations

Witconol CD-18. Witco Chemical/Organics Div.
Chemical Description: Polypro-

poxylated polyol
Category: Synthetic oil, emulsifier, lubricant, antistat, nonionic
Applications: Synthetic oils for bath oil and hair preparations; cosmetic emulsifier

Witconol DOS. Witco Chemical/Organics Div.
Chemical Description: Diethylene glycol oleate
Category: Emulsifier, nonionic
Applications: Self-emulsifying O/W emulsifier for cosmetics and industrial formulations
Form: Liquid
Composition: 100% conc.

Witconol F26-46. Witco Chemical/Organics Div.
Chemical Description: Polypropylene glycol fatty acid ester
Category: Spreading and coupling agent, dispersant, anticaking agent, flow control agent, nonionic
Applications: Cosmetic oil, spreading and coupling agent for oil systems; dispersant for industrial use; spreading and anticaking agent for aerosol antiperspirant sprays;latex viscosity and flow control agent
Form: Liquid

Witconol GOT. Witco Chemical/Organics Div.
Chemical Description: Glycerol mono- and dioleate
Category: Emulsifier, nonionic
Applications: Surfactant and emulsifier for industrial use
Form: Liquid
Properties: Solubility: Oil-sol.
Composition: 100% conc.

Witconol H-31A. Witco Chemical/Organics Div.
Chemical Description: Polyethylene glycol 400 monooleate
Category: Emulsifier, spreading agent, nonionic
Applications: Emulsifier and spreading agent for cosmetic creams, lotions, and bath oils, and in aerosol and industrial formulations
Form: Liquid
Composition: 100% conc.

Witconol H-35A. Witco Chemical/Organics Div.
Chemical Description: Polyethylene glycol 400 monostearate
Category: Emulsifier, nonionic
Applications: O/W emulsifier for cosmetic and industrial uses; bubble-strength controller for quick-breaking foams in aerosol formulations
Form: Wax

Witconol MST. Witco Chemical/Organics Div.
Chemical Description: Glycerol mono- and distearate
Category: Emulsifier, lubricant, plasticizer, flow control agent, nonionic
Applications: Cosmetic, pharmaceutical, aerosol formulations; internal lubricant, plasticizer, and emulsifier in industrial applica-

tions; flow control agent for polymerization reactions
Form: Solid
Composition: 100% conc.

Witconol NP-100. Witco Chemical/Organics Div.
Chemical Description: Alkyl polyether
Category: Dispersant, stabilizer, leveling agent
Applications: Pigment dispersant, latex stabilizer and leveling agent for polymerization reactions

Witconol O. Witco Chemical/Organics Div.
Chemical Description: Glycerol mono- and dioleate
Category: Emulsifier
Applications: Industrial surfactant and emulsifier
Properties: Solubility: Oil-sol.

Witconol RDC-D. Witco Chemical/Organics Div.
Chemical Description: Diglycol coconate
Category: Surfactant, emulsifier, nonionic
Applications: Cosmetic and industrial use
Form: Liquid
Properties: Solubility: Oil-sol.

Witconol RHP. Witco Chemical/Organics Div.
Chemical Description: Propylene glycol fatty acid ester
Category: Emollient, stabilizer, nonionic
Applications: Cosmetics, toiletries, and aerosol formulations
Form: Paste
Composition: 100% conc.

Witconol RHT. Witco Chemical/Organics Div.
Chemical Description: Glycerol mono- and distearate, self-emulsifying
Category: Lubricant, plasticizer, emulsifier, nonionic
Applications: Internal lubricant, plasticizer, and emulsifier for industrial use
Form: Paste
Composition: 100% conc.

Witcor CI-1. Witco Chemical/Organics Div.
Chemical Description: Cationic surfactant containing free amines and solubilizers
Category: Corrosion inhibitor intermediate, cationic
Applications: Petroleum industry

Witcor CI-6. Witco Chemical/Organics Div.
Chemical Description: Complex cationic compound
Category and Applications: See Witcor CI-1

Witcor CI-3117. Witco Chemical/Organics Div.
Chemical Description: Amide
Category and Applications: See Witcor CI-1

Witcor SI-3065. Witco Chemical/Organics Div.
Chemical Description: Complex

phosphate ester
Category: Scale inhibitor, anionic
Applications: Petroleum industry
Form: Liquid

Witcosperse 101. Witco Chemical/Organics Div.
Chemical Description: Oil-soluble metal sulfonate
Category: Dispersant, anionic
Applications: Free-flowing powder form dispersing agent for wettable-powder pesticide formulations
Form: Powder
Properties: Solubility: Oil-sol.
Composition: 50% conc.

Witcosperse 201. Witco Chemical/Organics Div.
Chemical Description: Alkylaryl polyethoxyethanol
Category: Wetting agent, nonionic
Applications: Free-flowing powder-form wetting agent for wettable-powder pesticide formulations
Form: Powder
Composition: 50% conc.

Witcosperse 205. Witco Chemical/Organics Div.
Chemical Description: Polyoxyalkylated glycol
Category and Applications: See Witcosperse 201
Form: Powder
Composition: 50% conc.

Witcosperse 206. Witco Chemical/Organics Div.
Chemical Description: Ethoxylated sorbitan ester
Category and Applications: See Witcosperse 201
Form: Powder
Composition: 50% conc.

Witcosperse 207. Witco Chemical/Organics Div.
Chemical Description: Alkyl polyoxyalkylene ether
Category and Applications: See Witcosperse 201
Form: Powder
Composition: 50% conc.

Witcosperse 214. Witco Chemical/Organics Div.
Chemical Description: Surfactant blend
Category and Applications: See Witcosperse 201
Form: Powder

Witcosperse 305. Witco Chemical/Organics Div.
Chemical Description: Surfactant blend
Category: Surfactant, anionic/nonionic
Applications: Surface-active agent for wettable-powder pesticide formulations
Form: Powder

Witcosperse 401, 402, 403. Witco Chemical/Organics Div.
Chemical Description: Surfactant blend
Category: Wetting agent, dispersant, defoaming agent, anticaking agent, anionic/nonionic

Applications: Wettable-powder pesticide formulations
Form: Free-flowing powder

WT-707 Stabilizer. PVO International
Chemical Description: Blend of microcrystalline cellulose, dextrose, sodium carboxymethyl cellulose, and hydroxypropyl methylcellulose
Category: Stabilizer
Applications: For use in whipped toppings and nondairy fillings

XYZ

Xylenesulfonic Acid, Refined. Ruetgers-Nease Chem. Co.
Category: Catalyst, hydrotrope, solvent, intermediate
Applications: Solubilizer of organic compounds in strongly acid systems; metal cleaning solutions; catalytic applications
Form: Liquid
Properties: Crystallization Range: Very viscous at 10–15 C
Composition: 90% min. conc.

Xylenesulfonic Acid, Technical. Ruetgers-Nease Chem. Co.
Category and Applications: See Xylenesulfonic Acid, Refined; tech. grade used in metal cleaners and in formulations which can tolerate a higher percentage of free sulfuric acid
Form: Liquid
Properties: Crystallization Range: Very viscous at –15 C
Composition: 65% min. conc.
Toxicity: Corrosive
Containers: LTL drums
Warnings and Caution: Wear protective goggles and gloves; corrosive to steel–store in glass or type 316 stainless steel

Yeoman. Croda Chem. Ltd.
Chemical Description: Anhydrous lanolin BP
Category: Superfatting agent, dispersion aid
Applications: Used in all cosmetics; esp. in skin and hair care conditioning products and make up compositions; superfatting agent in soap; aids dispersion of pigments into anhydrous systems
Form: Unctuous mass; Color: Yellow
Properties: Lipophilic

Zelec NE. DuPont
Chemical Description: Fatty alcohol phosphate
Category: Antistat, anionic
Applications: Antistatic agent for

textiles, plastics, and films; suitable for use in aq. systems
Form: Viscous paste; Color: Yellow; Odor: Mild, fatty alcohol
Properties: Solubility: Very sol. in water; sol. to at least 1% in ethanol, isopropanol, ethylene glycol, benzene, toluene, or Freon TF solvent; Density: 9.3 lb/gal; Sp. gr.: 1.10; Flash Pt.: > 210 F (PM-CC); Stability: Stable to alkalis; good thermal stability; pH: 7.0–7.5 (40% sol'n.)
Composition: 100% active ingred.

Zelec NK. DuPont
Chemical Description: Fatty alcohol phosphate
Category and Applications: See Zelec NE; NK is preferred for use in solvent systems
Form: Viscous paste; Color: Light yellow; Odor: Mild, fatty alcohol
Properties: Solubility: Disp. in water; sol. to at least 20% in ethanol, isopropanol, ethylene glycol, benzene, toluene, or Freon TF solvent; Density: 8.7 lb/gal; Sp. gr.: 1.05; Flash Pt.: > 210 F (PM-CC); Stability: Stable to alkalis; good thermal stability; pH: 7.0–7.5 (10% disp.)
Composition: 100% active ingred.

Zelec TY. DuPont
Chemical Description: Alcohol phosphate composition
Category: Antistat, anionic
Applications: Antistatic agent for use on polyolefin fibers, plastics, and films; does not reduce water repellency of textiles
Form: Liquid; Color: Light amber; Odor: Mild
Properties: Solubility: Sol. in water in all proportions; Density: 9.8 lb/gal; Flash Pt.: Nonflammable (TCC); Stability: Stable in alkaline sol'ns.; separates below pH 4; pH: 6.7–7.3 (10% sol'n.)

Zelec UN. DuPont
Chemical Description: Unneutralized fatty alcohol phosphate
Category: Lubricant, antistat, anionic
Applications: Additive for oils to improve lubricity and provide corrosion protection; blended with oils and solvents in the formulation of textile lubricants and finishes; release agent in plastic molding operations
Form: Liquid; Color: Pale yellow
Properties: Solubility: Sol. in perchloroethylene, kerosene, Stod. solvent, o-dichlorobenzene, toluene, glycerin, Tetralin solvent; disp. in water; Density: 8.22 lb/gal; Sp. gr.: 0.99; pH: 2 (5% disp.)
Composition: 100% active ingred.

Zonyl A. DuPont
Chemical Description: Complex ethoxylate; ethylene oxide-ester condensate
Category: Wetting agent, emulsifier, lubricant, nonionic
Applications: Textile wetting agent for fibers and fabrics during dyeing; in paper industry as leveling agent and antifoam in pigment coatings; provides low surface ten-

sion in metal processing, manufacture of pigment finishes
Form: Liquid; Color: Clear to pale yellow; Odor: Mild
Properties: Solubility: Sol. in water, polar and nonpolar solvents; Sp. gr.: 1.07; HLB: 6.7; Stability: Stable in acids, bases; stable to cold, heat, and storage; Surface Tension: 26 dynes/cm (0.1% aq.)
Composition: 100% active ingred.

Zonyl FSA. DuPont
Chemical Description: Fluorochemical surfactant
Category: Wetting agent, emulsifier, dispersant, corrosion inhibitor, anionic
Applications: Used in adhesives, agriculture (insecticides, bacteriocides, etc.), building construction, leveling agent for floor polishes, emulsifier for polymerization, esp. of fluoromonomers, pigment grinding, metal corrosion inhibitor, additive for alkaline and solvent cleaners, leveling/antifogging agent for films and coatings, additive for finishes and paints, wetting agent for fire fighting, paper industry, inks, oil well treatments, drilling muds, plastics industry, textile finishes, lubricants, dyeing and bleaching aids, etc.
Form: Liquid
Properties: Solubility: Sol. > 2% in water, methyl alcohol; slightly sol. in ethyl acetate, toluene; Density: 9.8 lb/gal; Flash Pt.: 69 F (PMCC); Stability: Stable to heat, alkalis, oxidizing agents; Surface Tension: 18 dynes/cm (0.1%)
Composition: 50% solids; diluent: isopropanol/water
Toxicity: Minimal eye irritant

Zonyl FSB. DuPont
Chemical Description: Fluorochemical surfactant
Category: Wetting agent, foaming agent, emulsifier, dispersant, corrosion inhibitor, amphoteric
Applications: See Zonyl FSA; also foaming agent for solvent cleaners, elastomers, fire fighting foams; froth flotation agent
Form: Liquid
Properties: Solubility: Sol. > 2% in water, isopropyl alcohol, 1:1 water/isopropyl alcohol, methyl alcohol; slightly sol. in ethyl acetate, toluene; Density: 8.8 lb/gal; Flash Pt.: 67 F (PMCC); Stability: Stable to heat, acids, alkalis, oxidizing agents; effective in hard water; Surface Tension: 17 dynes/cm (0.1%)
Composition: 40% solids; diluent: isopropanol/water
Toxicity: Severe eye irritant

Zonyl FSC. DuPont
Chemical Description: Flurochemical surfactant
Category: Wetting agent, foaming agent, emulsifier, dispersant, corrosion inhibitor, cationic
Applications: See Zonyl FSB
Form: Liquid
Properties: Solubility: Sol. > 2% in water, isopropyl alcohol, 1:1 water/isopropyl alcohol, methyl alcohol, acetone; slightly sol. in tetrahydrofuran; Density: 9.7

lb/gal; Flash Pt.: 70 F (PMCC); Stability: Stable to heat, acids, alkalis, oxidizing agents; Surface Tension: 18 dynes/cm (0.1%)
Composition: 50% solids; diluent: isopropanol/water
Toxicity: Severe eye irritant; mild skin sensitizer

Zonyl FSJ. DuPont
Chemical Description: Fluorochemical surfactant
Category: Wetting agent, anionic
Applications: Wetting agent for adhesives, agriculture, biochemicals, automotive cleaners and waxes, fire fighting, plywood veneer mfg., inks, oil well treatments, drilling muds, photographic coatings, resin molds, textile finishes, dyeing, etc.
Form: Liquid
Properties: Solubility: Sol. > 2% in water, 1:1 water/isopropyl alcohol, methyl alcohol; slightly sol. in acetone, ethyl acetate; Density: 9.3 lb/gal; Flash Pt.: 83 F (PMCC); Stability: Stable to heat, alkalis; Surface Tension: 20 dynes/cm (0.1%)
Composition: 40% solids; diluent: isopropanol/water
Toxicity: Minimal eye irritant; skin irritant in undiluted form

Zonyl FSN. DuPont
Chemical Description: Fluorochemical surfactant
Category: Wetting agent, emulsifier, dispersant, corrosion inhibitor, nonionic
Applications: See Zonyl FSA
Form: Liquid
Properties: Solubility: Sol. > 2% in water, isopropyl alcohol, 1:1 water/isopropyl alcohol, methyl alcohol, acetone, ethyl acetate, tetrahydrofuran; Density: 8.8 lb/gal; Flash Pt.: 72 F (PMCC); Stability: Stable to heat, acids, alkalis, oxidizing agents; Surface Tension: 23 dynes/cm (0.1%)
Composition: 40% solids; diluent: isopropanol/water
Toxicity: Minimal eye irritant

Zonyl FSP. DuPont
Chemical Description: Fluorochemical surfactant
Category and Applications: See Zonyl FSA
Form: Liquid
Properties: Solubility: Sol. > 2% in water, 1:1 water/isopropyl alcohol, methyl alcohol; slightly sol. in ethyl acetate; Density: 9.6 lb/gal; Flash Pt.: 75 F (PMCC); Stability: Stable to heat, alkalis, oxidizing agents; Surface Tension: 19 dynes/cm (0.1%)
Composition: 35% solids; diluent: isopropanol/water
Toxicity: Minimal eye irritant

Zonyl RP. DuPont
Chemical Description: Fluorochemical surfactant
Category: Fluoridizer, wetting agent, anionic
Applications: Paper fluoridizer; imparts excellent oil, grease, and solvent repellency to paper and paperboard

Form: Liquid; Color: Amber
Properties: Solubility: Sol. in water in all proportions; Density: 9.5 lb/gal; Visc.: 10 cps; Flash Pt.: $>$ 28 C; pH: 7–8
Composition: 34% conc.